Microbial Life in Antarctica
The Global Impact of Microbiology

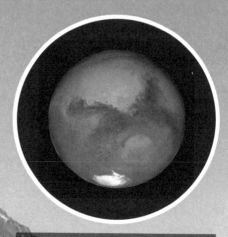

Antarctic microbes offer a model for possible life on Mars.
Page 5

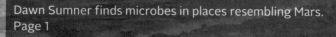

Dawn Sumner finds microbes in places resembling Mars.
Page 1

NSF-sponsored scientists drill a hole through five meters of ice to reach lake microbes.
Page 832

Cyanobacterial mat emerges from ice, dispersing a microbial ecosystem.
Page 2

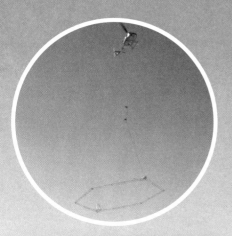

Jill Mikucki's airborne magnetic sensor reveals brine underground, full of iron-metabolizing microbes.
Page 872

Colonial choanoflagellates may resemble Earth's earliest multicellular organism. Cultured from an Antarctic lake by Rachael Morgan-Kiss.
Page 773

Wei Li filters Antarctic lake phototrophs in a field lab, to assess their contributions to global carbon fixation.
Page 773

Throughout this book, examples of Antarctic microbiology are indicated with this icon to make them easier to find.

FOURTH EDITION

Microbiology
An Evolving Science

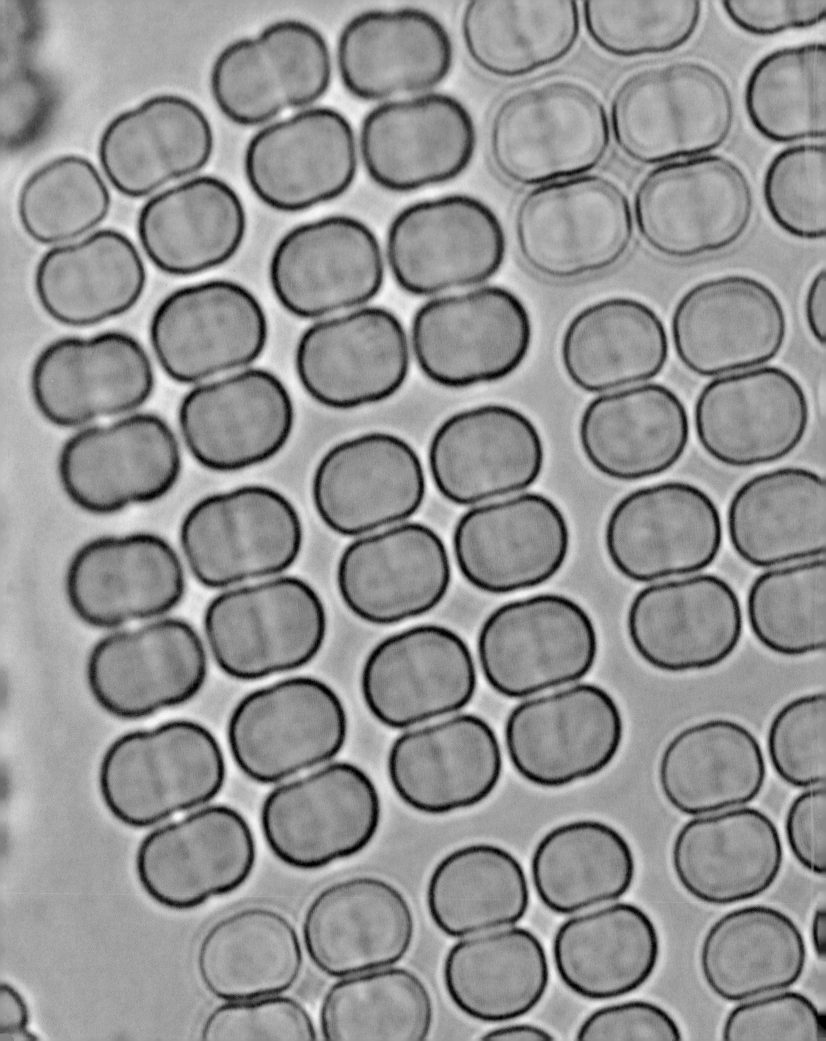

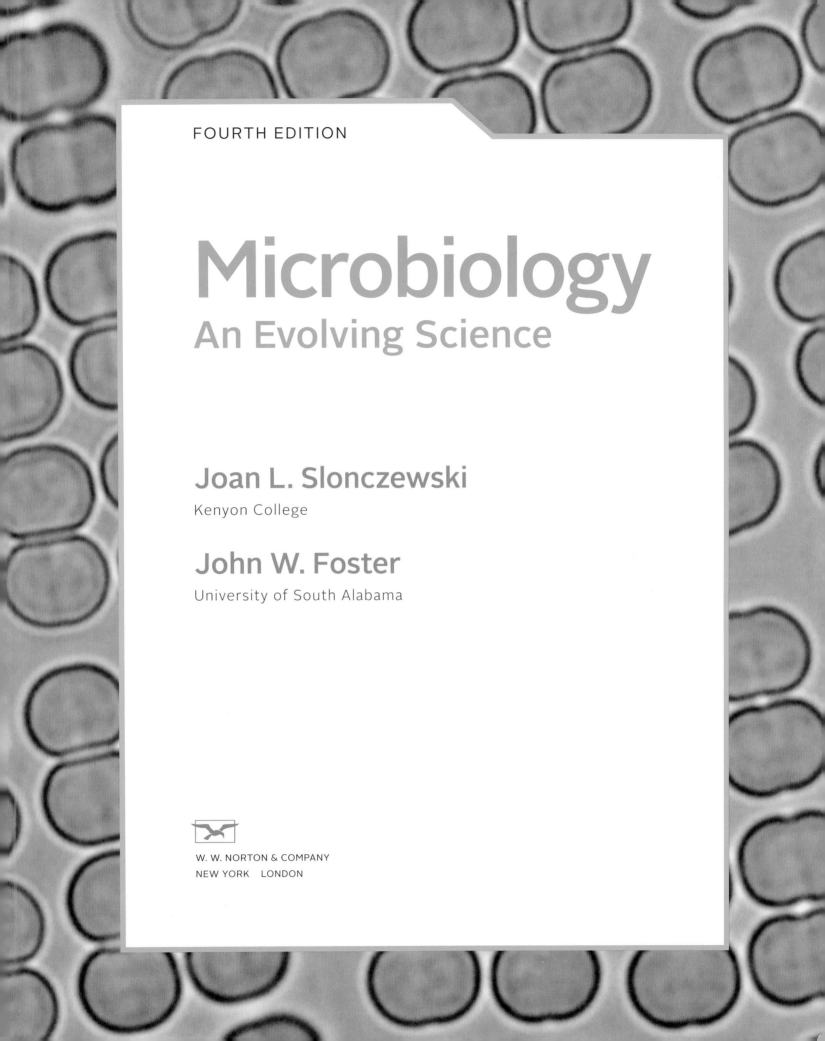

FOURTH EDITION

Microbiology
An Evolving Science

Joan L. Slonczewski
Kenyon College

John W. Foster
University of South Alabama

W. W. NORTON & COMPANY

NEW YORK   LONDON

W. W. Norton & Company has been independent since its founding in 1923, when William Warder Norton and Mary D. Herter Norton first published lectures delivered at the People's Institute, the adult education division of New York City's Cooper Union. The firm soon expanded its program beyond the Institute, publishing books by celebrated academics from America and abroad. By midcentury, the two major pillars of Norton's publishing program—trade books and college texts— were firmly established. In the 1950s, the Norton family transferred control of the company to its employees, and today—with a staff of four hundred and a comparable number of trade, college, and professional titles published each year—W. W. Norton & Company stands as the largest and oldest publishing house owned wholly by its employees.

Editor: Betsy Twitchell
Developmental editor: Michael Zierler
Senior project editor: Thomas Foley
Manuscript editor: Stephanie Hiebert
Associate Director of Production, College: Benjamin Reynolds
Editorial assistant: Taylere Peterson
Art director: Rubina Yeh
Designer: Lissi Sigillo
Managing editor, college: Marian Johnson
Managing Editor, College Digital Media: Kim Yi
Media editor: Kate Brayton
Associate media editor: Cailin Barrett-Bressack
Media project editor: Jesse Newkirk
Media assistant editor: Victoria Reuter
Media editorial assistant: Gina Forsythe
Marketing manager: Todd Pearson
Photo manager: Trish Marx
Permissions Manager: Megan Jackson Schindel
Permissions specialist: Bethany Salminen
Illustrations for Editions 1–3: Precision Graphics
Illustrations for the 4th Edition: Dragonfly Media Group
Composition by MPS North America LLC
Project manager at MPS: Jackie Strohl
Manufacturing by Transcontinental

Library of Congress Cataloging-in-Publication Data

Names: Slonczewski, Joan, author. | Foster, John Wade, author.
Title: Microbiology : an evolving science / Joan L. Slonczewski, Kenyon
    College, John W. Foster, University of South Alabama.
Description: Fourth edition. | New York : W. W. Norton & Company, 2016. |
    Includes bibliographical references and index.
Identifiers: LCCN 2016051604 | ISBN 9780393602340 (hardcover)
Subjects: LCSH: Microbiology—Textbooks.
Classification: LCC QR41.2 .S585 2016 | DDC 579—dc23 LC record available at
https://lccn.loc.gov/2016051604

W. W. Norton & Company, Inc., 500 Fifth Avenue, New York, N.Y. 10110
wwnorton.com

W. W. Norton & Company Ltd., 15 Carlisle Street, London W1D 3BS

2 3 4 5 6 7 8 9

Dedication

We dedicate this Fourth Edition to the memory of our colleagues, three brilliant scientists and educators: Fred Neidhardt (1931–2016), extraordinary physiologist and founder of the field of microbial stress response; Bob Kadner (1942–2005), renowned microbial physiologist whose discoveries of biosynthesis and transport shaped the field of metabolism; and Katrina Edwards (1968–2014), pioneering geomicrobiologist who founded the Center for Dark Energy Biosphere Investigations (C-DEBI). We, the authors, are grateful for their contributions and deeply moved by their passing.

BRIEF CONTENTS

CONTENTS

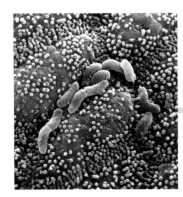

PART 1
The Microbial Cell

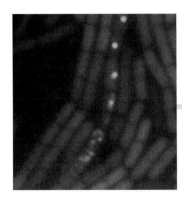

PART 2
Genes and Genomes

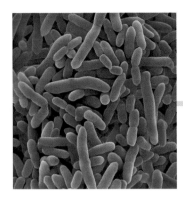

PART 3
Metabolism and Biochemistry

PART 4
Microbial Diversity and Ecology

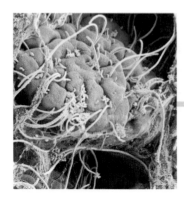

PART 5
Medicine and Immunology

APPENDIX 2
Taxonomy A-11

eTOPIC CONTENTS

Access to the eTopics is available through both the ebook and the Norton Coursepack.

Our first three editions established *Microbiology: An Evolving Science* as the defining core text of our generation—the book that inspires undergraduate science majors to embrace the microbial world. This Fourth Edition continues our commitment to the fundamentals, but also highlights two current and breathtaking themes of discovery: Antarctic microbiology and our intestinal microbiome. Antarctic microbes offer models for life on Mars, exotic ecosystems, and opportunities for biotechnology. This new material includes data and images from Joan Slonczewski's own field work in the McMurdo Dry Valleys of Antarctica. Closer to home, our intestinal microbiome reveals extraordinary connections to human health and behavior, as now promoted by the National Microbiome Initiative (NMI). The microbiome story is vividly told by John Foster, a leading investigator of gut bacteria.

In this Fourth Edition, we have maintained our signature balance between cutting-edge ecology and medicine, including the use of case histories in the medical section. Our balanced depiction of women and minority scientists, including young researchers, continues to draw enthusiastic responses from our adopters. Our focus on evolution, and our modern organization reflecting changes in the field, proved so successful that other textbooks have since adjusted their chapter sequence to parallel *Microbiology: An Evolving Science*. We have kept this chapter organization to facilitate year-to-year course transitions for instructors.

In many chapters, we relate topics to current events, to keep students interested in and informed on the role of microbiology in the world today. One example is synthetic biology, the construction of microbes with genetic circuits engineered for commercial use (Chapter 12, Biotechniques and Synthetic Biology). Another example is the use of viral replication cycles to develop lentiviral treatments for cancer and inherited disorders, including the first possible "cure" for pediatric leukemia (presented in Chapter 11, Viral Molecular Biology).

The Fourth Edition still holds to the idea that this text is a community project, drawing not only on the authors' experience as researchers and educators, but also on the input of hundreds of colleagues from around the world to create a comprehensive microbiology book for the twenty-first century. We present the full story of molecular microbiology and microbial ecology from its classical history of Koch, Pasteur, and Winogradsky, right up to the research of twenty-first-century researchers Rita Colwell and Bonnie Bassler. We have included countless contributions recommended by colleagues from around the globe, at institutions such as Washington University, University of California–Davis, University of Wisconsin–Madison, Cornell University, Florida State University, University of Toronto, University of Edinburgh, University of Antwerp, Seoul National University, Chinese University of Hong Kong, and many more. We are grateful to you all.

While we have expanded and developed new topics, we also recognized the need to keep the length and "core" of the book to a size reasonable enough for the undergraduate student. The content in virtually all of the chapters in this book has been limited to a maximum of six numbered sections, helping to keep the coverage from straying too far from the fundamentals. In addition, several chapters underwent major revision in this Fourth Edition, including Chapter 3, Cell Structure and Function, with a tightened opener and a new section on cell aging; and Chapter 21, Microbial Ecology, which opens with a new section on metagenomics and the culturing of "unculturables." Along with these new additions, we have also tightened the content overall, actually decreasing the size of the book from the Third Edition.

In order to contain length while adding new material, we continue to transfer certain topics online as "eTopics." The eTopics are called out in the text, hyperlinked to the ebook, and their key terms are fully indexed in the printed book. Therefore, returning adopters can be confident of keeping access to all of the material they taught from the Second and Third Editions, but now they also have new topics on *Mycobacterium tuberculosis* cell aging and drug resistance (Chapter 3) and on bacteria that convert phage genes into toxin secretion systems (Chapter 25), and much more.

Major Features

Our book targets the science major in biology, microbiology, or biochemistry. Several important features make our book the best text available for undergraduates today:

NEW **Themes of discovery: Antarctic microbiology and our intestinal microbiome.** The Fourth Edition features new content in every chapter on two exciting and relevant new themes. Marginal icons highlight examples of each theme, such as:

- In Chapter 5, research by Asim Bej (University of Alabama Birmingham) and others on psychrophiles reveals the composition and metabolic capabilities of the South Pole microbiome. Novel compounds discovered by members of the polar microbiome have anticancer and antimicrobial potential.

- In Chapter 7, Ruth Lay (Cornell University) used metagenomics to discover that the abundance of some members of the intestinal microbiome are influenced by host genetics. One such organism, *Christensenella minuta*, also influenced weight gain when orally "transplanted" into mouse intestines.

A new "mini-interview" opens each chapter, offering a total of twenty-eight new perspectives from cutting-edge researchers. Examples include:

- A Chapter 1 interview with Dawn Sumner, geomicrobiologist, explaining how cyanobacterial mats colonize Antarctic lakes.

- A Chapter 2 interview with Grant Jensen, whose 3D cryotomography offers an extraordinary view of chlamydia parasitizing a cell.

Research on contemporary themes such as evolution, genomics, metagenomics, molecular genetics, and biotechnology enrich students' understanding of foundational topics and highlight the current state of the field. Every chapter presents numerous current research examples within the up-to-date framework of molecular biology. Examples of current research include tools to explore evolution in aging bacterial colonies; determining the "pangenome," the overall set of genes available to a species; simultaneously profiling gene expression patterns in host and microbe during an infection; and the spectroscopic measurement of carbon flux from microbial communities.

An updated art program with engaging figures that are also dynamic learning tools. Process diagrams have been rendered more accessible by reducing the length of supporting captions and expanding the use of in-figure bubble captions. In addition, scientists pursuing research today are presented alongside the traditional icons of the field. For example, Chapter 1 introduces historical figures such as Koch and Pasteur alongside marine microbiologist Heide Schulz-Vogt and undergraduate students currently conducting experimental evolution of *E. coli*.

Core concepts are presented in a student-friendly way that motivates learning. Ample Thought Questions throughout every chapter challenge students to think critically about core concepts, the way a scientist would.

An innovative media package, including a new Smartwork5 online homework course, provides powerful tools of visualization and assessment. Smartwork5 includes review, critical thinking, visual, and animation questions for every chapter. Each activity builds on the text and pedagogy to help students master key concepts, think critically, and apply what they've learned.

Additional features of the Fourth Edition include:

- **Genetics and genomics are presented as the foundation of microbiology.** Molecular genetics and genomics are thoroughly integrated with core topics throughout the book. This approach gives students an understanding of how genomes reveal potential metabolic pathways in diverse organisms, and how genomics and metagenomics reveal the character of microbial communities.

- **Microbial ecology and medical microbiology receive equal emphasis,** with particular attention paid to the merging of these fields. Throughout the book, phenomena are presented with examples from both ecology and medicine; for example, when discussing horizontal transfer of "genomic islands" we present symbiosis islands associated with nitrogen fixation, as well as pathogenicity islands associated with disease (Chapter 9).

- **Unlike most microbiology textbooks,** our text provides size scale information for nearly every micrograph.

- **Viruses are presented in molecular detail and in ecological perspective.** For example, in marine ecosystems, viruses play key roles in limiting algal populations while selecting for species diversity (Chapter 6). Similarly, a constellation of bacteriophages influences enteric flora.

- **Microbial diversity that students can grasp.** We present microbial diversity in a manageable framework that enables students to grasp the essentials of the most commonly presented taxa, the continual discovery of organisms ranging from anammox bacteria to emerging pathogenic *Escherichia* strains.

- **Appendices for review and further study.** Our book assumes a sophomore-level understanding of introductory biology and chemistry, with on-line eAppendices for those in need of review.

Organization

The topics in this book are arranged so that students can progressively develop an understanding of microbiology from key concepts and research tools. The chapters of Part 1 present key foundational topics: history, visualization, the bacterial cell, microbial growth and control, and virology.

The six chapters in Part 1 present many topics that are then developed in further detail throughout Parts 2 through 5. Part 2 presents modern genetics and genomics. Part 3 presents cell metabolism and biochemistry, although the chapters in Part 3 are written in such a way that they can be presented before the genetics material if so desired. Part 4 explores microbial ecology and diversity and discusses the roles of microbial communities in local ecosystems and global cycling. And then the chapters of Part 5 (Chapters 23–28) present medical and disease microbiology from an investigative perspective, founded on the principles of genetics, metabolism, and microbial ecology.

What's New in the Fourth Edition?

Throughout the Fourth Edition of *Microbiology: An Emerging Science*, research examples have been updated to highlight the newest experimental techniques and important topics of interest in microbiology today, including current examples of the two new themes—Antarctic microbiology and the intestinal microbiome. The content in each chapter has been focused around fewer numbered chapter sections to help students master the fundamentals. The art has been updated and numerous Thought Questions and Special Topic boxes have been updated. Every chapter opens with a new research interview that features the work of established scientists, postdocs, and graduate students from around the world. A review of of these changes by chapter are featured in the following list.

CHAPTER 1: Microbial Life: Origin and Discovery. The chapter opener describes research on cyanobacterial mats that grow at the bottom of Antarctic lakes. Other new and unusual microbes are presented, including *Pyrodictium abyssi*, which lives off of the sulfides spewed from oceanic thermal vents, and the giant marine bacterium, *Thiomargarita namibiensis*.

CHAPTER 2: Observing the Microbial Cell. In the chapter opener, pathogenic chlamydias are seen in a whole new light—using 3D cryotomographic microscopy. Exciting imaging techniques that continue to push forward our understanding of cell behavior are presented. Super resolution imaging enables single molecules to be tracked within living cells, and a new special topic on NanoSIMS (nanoscale secondary ion mass spectrometry) describes how this chemical imaging method is being used to probe intestinal microbiomes.

CHAPTER 3: Cell Structure and Function. The molecular processes that coordinate DNA replication and cell division are the subject of the chapter opener. Several microbial cell biology topics are expanded and updated, including cell fractionation, a discussion of the cell envelope, and polar aging.

CHAPTER 4: Bacterial Culture, Growth, and Development. The chapter opener reveals the microbial diversity and evolution occurring in old bacterial colonies. The explanation of generation time and the description of continuous culture is revised for clarity. Recent research alters our view of what's going on during stationary phase within a liquid culture. A new special topic follows antibiotic hunters as they search for new medicines among "unculturable" microorganisms. New data on biofilm-busting peptides are presented.

CHAPTER 5: Environmental Influences and Control of Microbial Growth. Interspecies cell signaling is discussed in the chapter opener. The impacts of human activities, including global climate change, on microbial ecosystems are updated with some of the latest research. A new special topic looks at bacteriophage therapy to treat biofilm infections.

CHAPTER 6: Viruses. This first of two chapters on viruses opens with viral ecology, to highlight the critical roles that viruses play in ecosystems. Gut Bacteriophages in the gut microbiome is the topic of Section 6.4. The modern use of genome sequencing to classify viruses is presented. A special topic presents research on the role of a recently identified virus that confers thermotolerance to both its host fungus and the fungi's symbiotic plant host.

CHAPTER 7: Genomes and Chromosomes. The chapter opener describes research using molecular biology and fluorescence cell imaging to follow the fate of plasmids and the chromosome in *E. coli*. Section 7.5 is expanded to include the latest information about archaeal chromosomes. New data is included about intestinal metagenomics and single-cell genomics. A new special topic describes the molecular mechanisms that keep sister chromosomes from being severed by the growing septum during bacterial cell division.

CHAPTER 8: Transcription, Translation, and Bioinformatics. The chapter opener presents current research on the structural biology of coupled transcription and translation in bacteria. Much of the chapter is revised. Discussion of bioinformatics is revised extensively, including new material on how bioinformatics is revealing the complex interactions within the human gut microbiome. The processes of transcription, translation, and transertion are described using new research examples and art that makes translation more realistic and easier to understand. A new special topic looks at new discoveries on the structural biology of ribosome translocation during protein synthesis.

CHAPTER 9: Gene Transfer, Mutations, and Genome Evolution. The selection pressure of antibiotics on *Streptococcus pneumoniae* evolution is discussed in the chapter opener. Information on CRISPR is updated with new examples to emphasize its growing utility in molecular biology research. The discussion of horizontal gene transfer includes new data from studies on Antarctic microbes and on movement of genes between bacteria and eukaryotes. A new special topic on studies of DNA repair proteins containing [4Fe-4S] clusters hypothesizes that these proteins use electrons to locate damaged DNA.

CHAPTER 10: Molecular Regulation. A new chapter opener looks at how enterohemorrhagic *E. coli* virulence genes are regulated. The roles of RNA thermometers in the heat-shock response and of riboswitches in regulating gene expression are introduced. Discussion of the class of regulatory RNA molecules known as small RNA (sRNA) is updated and expanded with numerous examples, including sRNAs in Archaea that have counterparts in both Eukarya and Bacteria.

CHAPTER 11: Viral Molecular Biology. This chapter on viruses is extensively revised. It opens with current research on the presence of endogenous retroviral particles that are expressed in human embryos. The section on bacteriophage uses lambda phage as the model organism, and includes examples from genome analysis and synthetic biology to highlight contemporary approaches to understanding its role in the human gut microbiome. New images using cryo-electron tomography illustrate in great detail the organization of RNA within an influenza virion. The exciting antitumor tool, T-VEC, an engineered HSV-1 virus, is described.

CHAPTER 12: Biotechniques and Synthetic Biology. The chapter opener introduces Ribo-T, an engineered ribosome, which highlights some of the possibilities of biomolecular engineering and synthetic biology. The use of the CRISPR/Cas9 system as an editing tool and method of regulating gene expression in experiments is described. The idea of using synthetic auxotrophy as a biocontainment method for engineered *E. coli* (and potentially other microbes) is described.

CHAPTER 13: Energetics and Catabolism. The gut microbiome is featured in a new chapter opener, which describes how *Bacteroides* species in the human intestines secrete catabolic enzymes, thus providing essential catabolites for many microbial species in the gut. New research from Antarctica is presented on psychotrophs that metabolism phenanthrene.

CHAPTER 14: Electron Flow in Organotrophy, Lithotrophy, and Phototrophy. A new chapter opener describes extracellular electron transfer among *Geobacter* species—a form of microbial electricity. In the gut, *Salmonella enterica* subvert neutrophils by using tetrathionate to assist in pathogenesis. A new special topic surveys our attempts to harness microbial electricity to power our electrical devices.

CHAPTER 15: Biosynthesis. The chapter opener highlights a simple assay for screening soil-dwelling actinomycetes that produce novel glycopeptide antibiotics. The discussion of ways in which microbes control the energetic costs of biosynthesis is expanded to include recently discovered examples of resource sharing within Antarctic marine phototrophs and among *E. coli* that form nanotubes to exchange amino acids.

CHAPTER 16: Food and Industrial Microbiology. A novel lipase isolated from an Antarctic psychrophile, *Candida antarctica*, is the subject of the new chapter opener. The discovery of microbial biologicals and their commercialization is featured. For example, bioprospecting identified a biological fungicide, whose active ingredient is *Streptomyces lydicus*, which suppresses fungi that attack plant roots and leaves.

CHAPTER 17: Origins and Evolution. A new chapter opener presents evidence of competitive and cooperative evolution in *E. coli*. Some of the latest data from Richard Lenski's long-term evolution experiment is included. The chapter presents recent clues to the nature of early life that are found in the Dry Valleys of Antarctica, where cyanobacteria form thick microbial mats. The criteria for defining a species is updated to include current information from genome and rRNA analysis along with ecotype sharing.

CHAPTER 18: Bacterial Diversity. The chapter opener presents fascinating data about the biofilms found growing on colorectal tumors. Information has been updated about the mosaicism of deep-branching thermophile genomes from Aquificae and Thermotogae. A new special topic introduces bizarre looking filamentous bacteria that form symbiotic relationships with cells in the mammalian gut.

CHAPTER 19: Archaeal Diversity. Our understanding of the Archaeal world continues to change rapidly. The chapter opener describes *Altiarchaeum hamiconexum*, a marsh-dwelling archaeon that uses its grappling hook appendages to link together into biofilms. The latest updates on archaeal phylogeny are included, like reclassification of the Miscellaneous Crenarchaeota Group to the Bathyarchaeota. Examples of ammonia-oxidizing archaea are discussed, including the deep ocean psychrophile, *Cenarchaeum symbiosum*, which is an endosymbiont of a marine sponge. A new special topic looks at current research on methanogens living in our intestines and what factors cause them to colonize some hosts, but not others.

CHAPTER 20: Eukaryotic Diversity. A new chapter opener discusses choanoflagellates from Antarctica that cycle between single-cell and colonial forms. Cellular and genetic traits of these protists make them an excellent organism for the study of metazoan origins. Recent data is described about inducing multicellularity in the green algae, *Chlamydomonas reinhardtii*. The lineup of parasitic protozoa discussed in the chapter is expanded to include a number of intestinal parasites, like *Cryptosporidium parvum*, *Balantidium coli*, and *Encephalitozoon intestinalis*.

CHAPTER 21: Microbial Ecology. The chapter opener gets "crabby." It describes the symbiotic relationship between chemosynthetic bacteria and the yeti crab at hydrothermal vents located 2.5 miles down beneath the surface of the Antarctic Southern Ocean. This chapter is extensively reorganized and updated with new material on metagenome sequencing, a new discussion on the human colonic microbiome's roles in host digestion, brain health and immunity, and recent research findings on oceanic microbes, like *Prochlorococcus*. A new special topic presents results of studying the ecology and migration of cyanobacterial mats found in many Antarctic lakes.

CHAPTER 22: Microbes in Global Elemental Cycles. The discovery of *Nitrospira* species that perform both ammonia and nitrite oxidation is presented in the chapter opener. The special topic presents evidence that a million-year-old underground river carrying iron and sulfur bacteria courses 500 meters below the Taylor Valley in Antarctica. The chapter presents recent scientific modeling data, which cautions that global warming is increasing microbial activity in the once frozen permafrost. Release of carbon stores from the permafrost due to microbial metabolism could accelerate global warming.

CHAPTER 23: Human Microbiota and Innate Immunity. The new chapter opener presents cryo-electron microscopy data illustrating the assembly of the multiprotein inflammasome complex. This chapter is thoroughly reorganized with updated presentations of innate immunity and a new emphasis on the human body as an ecosystem. In particular, Section 23.2 focuses on the gut microbiota, presenting numerous examples of beneficial gut microbes and introducing the concept of dysbiosis, the accidental penetration of organisms beyond a site of colonization or an imbalance in microbiome composition.

CHAPTER 24: The Adaptive Immune Response. The chapter opener shows how bystander B cells in lymph nodes are essential for T cell migration, an essential early step in B cell maturation into a plasma cell. This chapter is extensively reorganized. Sections on antibody structure and production are merged and streamlined. There is a new section on gut mucosal immunity. A discussion about vaccinations has been moved into this chapter from Chapter 26. A new special topic provides evidence that endogenous retroviruses may help B cells respond to the presence of T cell-independent antigens.

CHAPTER 25: Microbial Pathogenesis. A new chapter opener looks at recent experimental results that reveal how *Yersinia pestis* co-opts host proteins to aid in its pathogenesis. The molecular mechanisms of other pathogenic microbes are presented, such as, the role of the adhesin molecule MAM7 in initiating contact between a Gram-negative pathogen and its host cell. New information is presented on how pathogens control virulence factor gene expression based on their environment and how they thwart antigen presentation by the host immune system. The section on experimental tools used to study pathogenesis is completely rewritten. It now focuses on the current methods in genomic sequencing and bioinformatics analysis, transcriptomics, and imaging and fluorescent probe techniques to address cell biological questions.

CHAPTER 26: Microbial Diseases. The chapter opener summarizes current facts about Zika virus, reminding us that emerging infectious diseases are still a serious threat to human health. Information about numerous examples of microbial disease have been updated, including a new discussion of osteomyelitis, updates on the virulence factors and molecular mechanisms of infection used by *Helicobacter pylori*, and a new discussion of diarrhea and its impacts on the gut microbiome.

CHAPTER 27: Antimicrobial Therapy. Antimicrobial hunters and their treasures are the subject of the new chapter opener. In this example, an Antarctic sponge is the source of darwinolide, a novel diterpene that is effective against MRSA. New technologies are presented that could be used for rapid identification of pathogens in the clinical laboratory, such as multiplex PCR and miniaturized magnetic resonance machines. A new special topic looks at the use of monoclonal antibodies as antimicrobials.

CHAPTER 28: Clinical Microbiology and Epidemiology. The new chapter opener looks at novel genome-based tests to profile the gut microbiome in patients and correlate the profile with various gastrointestinal diseases. Updates on rapid and automated clinical detection methods are described, including the use of next-gen sequencing and programmable RNA sensors to identify pathogens in patient samples. The section on detecting emerging infectious diseases is completely rewritten. It now includes discussions of Zika virus and the role of climate change on the emergence and spread of microbial diseases.

Resources

SMARTWORK5 ONLINE HOMEWORK. Norton's powerful and accessible online homework platform features answer specific feedback, a variety of engaging question types, and the integration of the stunning art from the book and process animations to help students master microbiology concepts. Smartwork5 integrates with campus LMS's such as Blackboard and Canvas and features a simple, intuitive interface making it the easiest-to-use online homework system for instructors and students.

PRESENTATION TOOLS. Every figure and photograph in the textbook is available in JPEG and PowerPoint format for use in lecture. In order to provide stunning, high-quality visuals, every image has been hand-examined to make sure colors will not fade when projected and to optimize font size and composition for clear, legible viewing even in the back row. Labeled and unlabeled versions are available. In addition, Lecture PowerPoint decks including key figures form the text, links to animations, and clicker questions, are available for download at wwnorton.com/instructors.

MICROGRAPH DATABASE. The Micrograph Database includes searchable access to most of the micrographs in the textbook, tagged by characteristics such as taxonomy, shape, and habitat. The Micrograph Database can be accessed at wwnorton.com/instructors.

PROCESS ANIMATIONS. Sixty process animations depicting key processes of microbiology are offered in multiple formats and and embedded in PowerPoint files. These animations are all based on the art found in the textbook and were developed under the careful supervision of the textbook authors. Student access to the animations is available in the ebook, Smartwork5 online homework course or via the Coursepack. Instructor access to the process animations is available at wwnorton.com/instructors.

Animation Topics Include:

Microscopy

Replisome Movement in a Dividing Cell

Chemotaxis

Phosphotransferase System (PTS) Transport

Dilution Streaking Technique

Biofilm Formation

Endospore Formation

Lysis and Lysogeny

Supercoiling and Topoisomerases

DNA Replication

Rolling Circle Mechanism of Plasmid Replication

PCR

Protein Synthesis

Protein Export

SecA-Dependent General Secretion Pathway

ABC Transporters

Bacterial Conjugation

Recombination

DNA Repair Mechanisms: Methyl Mismatch Repair

DNA Repair Mechanisms: Nucleotide Excision Repair

DNA Repair Mechanisms: Base Excision Repair

Transposition

The *lac* Operon

Transcriptional Attenuation

Chemotaxis: Molecular Events

Quorum Sensing

Influenza Virus Entry into a Cell

Influenza Virus Replication

HIV Replication

Herpes Virus Replication

Construction of a Gene Therapy Vector

Tagging Proteins for Easy Purification

Real-Time PCR

A Bacterial Electron Transport System

ATP Synthase Mechanism

Oxygenic Photosynthesis

Agrobacterium: A Plant Gene Transfer Vector

Phylogenetic Trees

DNA Shuffling

Listeria Infection

Light-Driven Pumps and Sensors

Malaria: A Cycle of Transmission between Mosquito and Human

The Basic Inflammatory Response

Phagocytosis

The Activation of the Humoral and Cell-Mediated Pathways

Cholera Toxin Mode of Action

Process of Type III Secretion

Retrograde Movement of Tetanus Toxin to an Inhibitory Neuron

DNA Sequencing

TEST BANK. Thoroughly revised for the Fourth Edition and using the Norton Assessment Guidelines, each chapter of the Test Bank consists of five question types classified according to the first five levels of Bloom's taxonomy of knowledge types: Remembering, Understanding, Applying, Analyzing, and Evaluating. Questions are further classified by section and difficulty, making it easy to construct tests and quizzes that are meaningful and diagnostic according to instructors' needs. Questions are multiple-choice and short-answer. The Test Bank is available in *ExamView Assessment Suite*, Word RTF, and PDF formats, downloadable from wwnorton.com/instructors.

COURSEPACKS. At no cost to professors or students, Norton Coursepacks are available in a variety of formats, including all versions of Blackboard and WebCT. With just a simple download, an adopter can bring high-quality Norton digital media into a new or existing online course (no extra student passwords required), and it's theirs to keep. Content includes chapter-based assignments, quizzes, animation activities and more. Coursepacks can be downloaded at wwnorton.com/instructors.

ENHANCED EBOOK. An affordable and convenient alternative to the print book, Norton Ebooks retain the content and design of the print book and allow students to highlight and take notes, print chapters as needed, and search the text with ease. The enhanced ebook includes:

- **Process animations** based on the text art and developed under the watchful eyes of the textbook authors.

- **Links to eTopics** written by Joan Slonczewski and John Foster, which supplement and enrich concepts covered in the text.

- **Flashcards** of all the key terms in the book and their definitions.

Acknowledgments

We are very grateful for the help of many people in developing and completing this book, including Norton editors John Byram, Vanessa Drake-Johnson, Mike Wright, and especially Betsy Twitchell, whose heroic efforts assured completion of the Fourth Edition. Our developmental editor, Michael Zierler, contributed greatly to the clarity of presentation. Trish Marx did an amazing job of tracking down all kinds of images from sources all over the world. Kate Brayton's coordination of electronic media development has resulted in a superb suite of resources for students and instructors alike. We thank associate media editor Cailin Barrett-Bressack, media assistant editor Victoria Reuter, and assistant media editor Gina Forsythe for producing the IM and the Test Bank, as well as contributing in many other ways to the development of the digital resources. Without senior project editor Thom Foley's incredible attention to detail, the innumerable moving parts of this project would never have become a finished book. Marian Johnson, Norton's managing editor in the college department, helped coordinate the complex process involved in shaping the manuscript over the years. Ben Reynolds ably and calmly managed the manufacturing of this book. Editorial assistant Taylere Peterson coordinated the transfer of many drafts among many people. We thank marketing manager Todd Pearson for ensuring microbiology instructors know about our exciting Fourth Edition. Finally, we thank Roby Harrington, Drake McFeely, and Julia Reidhead for their support of this book over its nearly decade in print.

For the quality of our new illustrations in the Fourth Edition, we thank the many artists at Dragonfly Media Group, who developed attractive and accurate representations and showed immense patience in getting the details right.

We thank the numerous colleagues over the years who encouraged us in our project, especially the many attendees at the Microbial Stress Gordon Conferences. We greatly appreciate the insightful reviews and discussions of the manuscript provided by our colleagues, and the many researchers who contributed their micrographs and personal photos. We thank the American Society for Microbiology journals for providing many valuable resources. Reviewers Erik Zinser, Lynn Thomason, and Robert Barrington offered particularly insightful comments on the metabolism and genetics sections, and Richard Lenski and Zachary Blount provided particularly insightful comments on experimental evolution. We would also like to thank the following reviewers:

Fourth Edition Reviewers

Emma Allen-Vercoe, University of Guelph

Alexandra Armstrong, University of Arizona and Pima Community College

Daniel Aruscavage, Kutztown University

Dennis Arvidson, Michigan State University

Nazir A. Barekzi, Old Dominion University

Miriam Barlow, University of California, Merced

Suzanne S. Barth, University of Texas at Austin

Hazel Barton, University of Akron

Yan Boucher, Univeristy of Alberta

Linda Bruslind, Oregon State University

Kathleen L. Campbell, Emory University

John Carmen, Northern Kentucky University

Carlton Rodney Cooper, University of Delaware

Vaughn Cooper, University of Pittsburgh, School of Medicine

John Dennehy, Queens College

Kathleen A. Feldman, University of Connecticut

Kelly A. Flanagan, Mount Holyoke College

Clifton Franklund, Ferris State University

Heather Fullerton, Western Washington University

Bethany Henderson-Dean, The University of Findlay

Karen Huffman, Genesee Community College

Edward Ishiguro, University of Victoria

Mack Ivey, University of Arkansas

Ece Karatan, Appalachian State University

Robert J. Kearns, University of Dayton

Alexandra M. Kurtz, Georgia Gwinnett College

Manuel Llano, The University of Texas at El Paso

Shawn Massoni, Mount Holyoke College

Ann G. Matthysse, University of North Carolina at Chapel Hill

William R. McCleary, Brigham Young University

James A. Nienow, Valdosta State University

C. O. Patterson, Texas A&M University

Ronald D. Porter, The Pennsylvania State University

Veronica Riha, Madonna University

Benjamin G. Rohe, University of Delaware

Joseph Romeo, San Fransciso State University

Pratibha Saxena, University of Texas at Austin

Richard Seyler, Virginia Tech

Alastair Simpson, Dalhousie University

Marek Sliwinski, University of Northern Iowa

Amy Springer, UMass Amherst

Vincent J. Starai, University of Georgia

Nikhil Thomas, Dalhousie University

Mitch Walkowicz, UMass Amherst

Susan Wang, Washington State University

Cheryl Whistler, University of New Hampshire

Adam C. Wilson, Georgia State University

Erik Zinser, University of Tennessee

Pre-Revision Survey Reviewers

Eric Allen, University of California, San Diego

Emma Allen-Vercoe, University of Guelph

Jason Andrus, Meredith College

Catalina Arango Pinedo, St. Joseph's University

Alexandra Armstrong, University of Arizona, Pima Community College

Nazir Barekzi, Old Dominion University

Miriam Barlow, University of California, Merced

Prakash H. Bhuta, Eastern Washington University

Cheryl Boice, Florida Gateway College

Blaise Boles, University of Iowa

Suzanna Bräuer, Appalachian State University

Alison Buchan, University of Tennessee

Robert Carey, Lebanon Valley College

Christian Chauret, Indiana University Kokomo

Cindy Cisar, Northeastern State University

Jeff Copeland, Eastern Mennonite University

Bela Dadhich, Delaware County Community College

Jaiyanth Daniel, Indiana University-Purdue University Fort Wayne

Diane Davis, Rutgers University

Sandra G. Devenny, Delaware County Community College

Eugene Dunkley, Greenville College

Kathleen A. Feldman, University of Connecticut

Pat M. Fidopiastis, California Polytechnic State University

David Fulford, Edinboro University of Pennsylvania

Heather Fullerton, Western Washington University

Michelle Furlong, Clayton State University

Eileen Gregory, Rollins College

Julianne Grose, Brigham Young University

Haidong Gu, Wayne State University

Julie Harless, Lone Star College Montgomery

Geoffrey Holm, Colgate University

Edward Ishiguro, University of Victoria

Mark Kainz, Ripon College

Dubear Kroening, University of Wisconsin-Fox Valley

Douglas F. Lake, Arizona State University

Maia Larios-Sanz, University of St. Thomas

Craig Laufer, Hood College

Maureen Leonard, Mount Mary University

Alex Lowrey, University of North Georgia-Gainesville

Aaron Lynne, Sam Houston State University

Ann Matthysse, University of North Carolina at Chapel Hill

Brendan Mattingly, University of Kansas Edwards

William R. McCleary, Brigham Young University

Robert McLean, Texas State University

Aaron Mitchell, Carnegie Mellon University

Naomi Morrissette, University of California, Irvine

Annika Mosier, University of Colorado, Denver

Jacalyn Newman, formerly of University of Pittsburgh
Tanya Noel, University of Windsor
Florence Okafor, Alabama A&M University
Lorraine Olendzenski, St. Lawrence University
Samantha Oliphant, Nevada State College, Marian University
    Indianapolis
Hyun-Woo Park, California Baptist University
Todd Primm, Sam Houston State University
Veronica Riha, Madonna University
Joseph Romeo, San Francisco State University
Silvia Rossbach, Western Michigan University
Natividad Ruiz, The Ohio State University
Robert Rychert, Boise State University
Pratibha Saxena, University of Texas at Austin
Matthew M. Schmidt, Stony Brook University (SUNY)
Adam Silver, University of Hartford
David Singleton, York College of Pennsylvania
Marek Sliwinski, University of Northern Iowa
Amy Springer, University of Massachusetts Amherst
Vincent Starai, University of Georgia
Sang-Jin Suh, Auburn University
James R. Walker, University of Texas at Austin
Dara L. Wegman-Geedey, Augustana College
Elizabeth Wenske-Mullinax, University of Kansas
Gordon Wolfe, California State University, Chico
Marie Yeung, California Polytechnic State University, San Luis
    Obispo
Virginia Young, Mercer University
Noha Youssef, Oklahoma State University
Fanxiu Zhu, Florida State University

Third Edition Reviewers
Emma Allen-Vercoe, University of Guelph
Gregory Anderson, Indiana University–Purdue University
    Indianapolis
Lisa Antoniacci, Marywood University
Bruce M. Applegate, Purdue University
Dennis Arvidson, Michigan State University
Vicki Auerbuch Stone, University of California, Santa Cruz
Tom Beatty, University of British Columbia
Melody Bell, Vernon College
Prakash Bhuta, Eastern Washington University
Blaise Boles, University of Michigan
Suzanna Bräuer, Appalachian State University
Ginger Brininstool, Lousisiana State University–Baton Rouge
    Campus
Kathleen L. Campbell, Emory University
Jeff Cardon, Cornell College
Rob Carey, Lebanon Valley College
Maria Castillo, New Mexico State University
Todd Ciche, Michigan State University
Sharron Crane, Rutgers University
Nicola Davies, University of Texas Austin
Angus Dawe, New Mexico State University
Janet Donaldson, Mississippi State

Erastus Dudley, Huntingdon College
Kathleen Dunn, Boston College
Valerie Edwards-Jones, Manchester Metropolitan University
Lehman Ellis, Our Lady of Holy Cross College
David Esteban, Vassar College
Xin Fan, West Chester University
Babu Fathepure, Oklahoma State University
Michael Gadsden, York University
Veronica Godoy-Carter, Northeastern University
Stjepko Golubic, Boston University
Vladislav Gulis, Coastal Carolina University
Ernest Hannig, University of Texas–Dallas
Julian Hurdle, The University of Texas at Arlington
Edward Ishiguro, University of Victoria
Choong-Min Kang, Wayne State University
Bessie Kebaara, Baylor University
John Lee, The City College of The City University of New York
Manuel Llano, University of Texas–El Paso Campus
Aaron Lynne, Sam Houston State University
Ghislaine Mayer, Virginia Commonwealth University
Bob McLean, Texas State University
Sladjana Malic, Manchester Metropolitan University
Gregory Marczynski, McGill University
Naomi Morrissette, University of California, Irvine
Kenneth Murray, Florida International University
Kari Naylor, University of Central Arkansas
Tracy O'Connor, Mount Royal University
Rebecca Parales, University of California, Davis
Roger Pickup, University of Lancaster
Robert Poole, The University of Sheffield
Geert Potters, Antwerp Maritime Academy
Ines Rauschenbach, Rutgers University
Veronica Riha, Madonna University
Marie-Claire Rioux, John Abbott College
Jason A. Rosenzweig, Texas Southern University
Ronald Russell, University of Dublin
Matt Schrenk, East Carolina University
Gary Schultz, Marshall University
Chola Shamputa, Mount Saint Vincent University
Nilesh Sharma, Western Kentucky University
Donald Sheppard, McGill University
Garriet Smith, The University of South Carolina Aiken
Vincent J. Starai, University of Georgia
Lisa Stein, University of Alberta
Karen Sullivan, Louisiana State University
Kapil Tahlan, Memorial University of Newfoundland and Labrador
Liang Tang, The University of Kansas
Tzuen-Rong Jeremy Tzeng, Clemson University
Claire Vieille, Michigan State University
James R. Walker, University of Texas Austin
Susan C. Wang, Washington State University
Chris Weingart, Dennison College
John Zamora, Middle Tennessee State University
Stephanie Zamule, Nazareth College
Fanxiu Zhu, The Florida State University

Second Edition Reviewers

Michael Allen, University of North Texas
Gladys Alexandre, University of Tennessee Knoxville
Hazel Barton, Northern Kentucky University
Suzanne S. Barth, University of Texas at Austin
Barry Beutler, The College of Eastern Utah
Michael J. Bidochka, Brock University
Dwayne Boucaud, Quinnipiac University
Derrick Brazill, Hunter College
Graciela Brelles-Mariño, California State Polytechnic University, Pomona
Jay Brewster, Pepperdine University
Linda Bruslind, Oregon State University
Marion Brodhagen, Western Washington University
Alison Buchan, University of Tennessee Knoxville
Jeffrey Byrd, St. Mary's College of Maryland
Silvia T. Cardona, University of Manitoba
Andrea Castillo, Eastern Washington University
Miguel Cervantes-Cervantes, Rutgers University
Tin-Chun Chu, Seton Hall University
Paul Cobine, Auburn University
Tyrrell Conway, University of Oklahoma
Scott Dawson, University of California–Davis
Jose de Ondarza, SUNY Plattsburgh
Donald W. Deters, Bowling Green State University
Clarissa Dirks, The Evergreen State College
William T. Doerrler, Louisiana State University
Janet R. Donaldson, Mississippi State University
Xin Fan, West Chester University
Babu Z. Fathepure, Oklahoma State University
Clifton Franklund, Ferris State University
Gregory D. Frederick, University of Mary Hardin-Baylor
Christopher French, University of Edinburgh
Jason M. Fritzler, Stephen F. Austin State University
Katrina Forest, University of Wisconsin–Madison
Kimberley Gilbride, Ryerson University
Stjepko Golubic, Boston University
Enid T. Gonzalez, California State University, Sacramento
John E. Gustafson, New Mexico State University
Lynn E. Hancock, Kansas State University
Martina Hausner, Ryerson University
J. D. Hendrix, Kennesaw State University
Michael C. Hudson, University of North Carolina–Charlotte
Jane E. Huffman, East Stroudsburg University
Michael Ibba, Ohio State University
Gilbert H. John, Oklahoma State University
John A. Johnson, University of New Brunswick St. John
Mark C. Johnson, Georgetown College
Carol Ann Jones, University of California Riverside
Ece Karatan, Appalachian State University
Daniel B. Kearns, Indiana University Bloomington
Robert J. Kearns, University of Dayton
Susan Koval, University of Western Ontario
Deborah Kuzmanovic, University of Michigan
Peter Kennedy, Lewis & Clark College

Greg Kleinheinz, University of Wisconsin Oshkos
Jesse J. Kwiek, Ohio State University
Andrew Lang, Memorial University of Newfoundland
Margaret Liu, University of Michigan
Thomas W. De Lany, Kilgore College
Maia Larios-Sanz, University of St. Thomas
Beth Lazazzera, University of California, Los Angeles
Dr. Lee H. Lee, Montclair State University
Mark Liles, Auburn University
Jun Liu, University of Toronto
Manuel Llano, University of Texas at El Paso
Zhongjing Lu, Kennesaw State University
Aaron Lynne, Sam Houston State University
John C. Makemson, Florida International University
Donna L. Marykwas, California State University Long Beach
Ann G. Matthysse, University of North Carolina at Chapel Hill
Ghislaine Mayer, Virginia Commonwealth University
Robert Maxwell, Georgia State University
William R. McCleary, Brigham Young University
Nancy L. McQueen, California State University, Los Angeles
Scott A. Minnich, University of Idaho
Philip F. Mixter, Washington State University
Christian D. Mohr, University of Minnesota
Craig Moyer, Western Washington University
Scott Mulrooney, Michigan State University
Kari Murad, The College of Saint Rose
William Wiley Navarre, University of Toronto
Ivan J. Oresnik, University of Manitoba
Cleber Costa Ouverney, San Jose State University
Deborah Polayes, George Mason University
Pablo J. Pomposiello, University of Massachusetts Amherst
Joan Press, Brandeis University
Todd P. Primm, Sam Houston State University
Sharon R. Roberts, Auburn University
Michelle Rondon, University of Wisconsin–Madison
Silvia Rossbach, Western Michigan University
Ben Rowley, University of Central Arkansas
Chad R. Sethman, Waynesburg University
Matthew O. Schrenk, East Carolina State University
Anthony Siame, Trinity Western University
Lyle Simmons, University of Michigan
Daniel R. Smith, Seattle University
Garriet W. Smith, University of South Carolina Aiken
Geoffrey B. Smith, New Mexico State University
Ruth Sporer, Rutgers University Camden
Anand Sukhan, Northeastern State University
Karen Sullivan, Louisiana State University
Virginia Stroeher, Bishop's University
Dorothea K. Thompson, Purdue University
Wendy C. Trzyna, Marshall University
Bernard Turcotte, McGill University
Dave Westenberg, Missouri University of Science and Technology
Ann Williams, University of Tampa
Charles F. Wimpee, University of Wisconsin–Milwaukee
Jianping Xu, McMaster University

First Edition Reviewers

Laurie A. Achenbach, Southern Illinois University, Carbondale

Stephen B. Aley, University of Texas, El Paso

Mary E. Allen, Hartwick College

Shivanthi Anandan, Drexel University

Brandi Baros, Allegheny College

Gail Begley, Northeastern University

Robert A. Bender, University of Michigan

Michael J. Benedik, Texas A&M University

George Bennett, Rice University

Kathleen Bobbitt, Wagner College

James Botsford, New Mexico State University

Nancy Boury, Iowa State University of Science and Technology

Jay Brewster, Pepperdine University

James W. Brown, North Carolina State University

Whitney Brown, Kenyon College undergraduate

Alyssa Bumbaugh, Pennsylvania State University, Altoona

Kathleen Campbell, Emory University

Alana Synhoff Canupp, Paxon School for Advanced Studies, Jacksonville, FL

Jeffrey Cardon, Cornell College

Tyrrell Conway, University of Oklahoma

Vaughn Cooper, University of New Hampshire

Marcia L. Cordts, University of Iowa

James B. Courtright, Marquette University

James F. Curran, Wake Forest University

Paul Dunlap, University of Michigan

David Faguy, University of New Mexico

Bentley A. Fane, University of Arizona

Bruce B. Farnham, Metropolitan State College of Denver

Noah Fierer, University of Colorado, Boulder

Linda E. Fisher, late of the University of Michigan, Dearborn

Robert Gennis, University of Illinois, Urbana-Champaign

Charles Hagedorn, Virginia Polytechnic Institute and State University

Caroline Harwood, University of Washington

Chris Heffelfinger, Yale University graduate student

Joan M. Henson, Montana State University

Michael Ibba, Ohio State University

Nicholas J. Jacobs, Dartmouth College

Douglas I. Johnson, University of Vermont

Robert J. Kadner, late of the University of Virginia

Judith Kandel, California State University, Fullerton

Robert J. Kearns, University of Dayton

Madhukar Khetmalas, University of Central Oklahoma

Dennis J. Kitz, Southern Illinois University, Edwardsville

Janice E. Knepper, Villanova University

Jill Kreiling, Brown University

Donald LeBlanc, Pfizer Global Research and Development (retired)

Robert Lausch, University of South Alabama

Petra Levin, Washington University in St. Louis

Elizabeth A. Machunis-Masuoka, University of Virginia

Stanley Maloy, San Diego State University

John Makemson, Florida International University

Scott B. Mulrooney, Michigan State University

Spencer Nyholm, Harvard University

John E. Oakes, University of South Alabama

Oladele Ogunseitan, University of California, Irvine

Anna R. Oller, University of Central Missouri

Rob U. Onyenwoke, Kenyon College

Michael A. Pfaller, University of Iowa

Joseph Pogliano, University of California, San Diego

Martin Polz, Massachusetts Institute of Technology

Robert K. Poole, University of Sheffield

Edith Porter, California State University, Los Angeles

S. N. Rajagopal, University of Wisconsin, La Crosse

James W. Rohrer, University of South Alabama

Michelle Rondon, University of Wisconsin–Madison

Donna Russo, Drexel University

Pratibha Saxena, University of Texas, Austin

Herb E. Schellhorn, McMaster University

Kurt Schesser, University of Miami

Dennis Schneider, University of Texas, Austin

Margaret Ann Scuderi, Kenyon College

Ann C. Smith Stein, University of Maryland, College Park

John F. Stolz, Duquesne University

Marc E. Tischler, University of Arizona

Monica Tischler, Benedictine University

Beth Traxler, University of Washington

Luc Van Kaer, Vanderbilt University

Lorraine Grace Van Waasbergen, The University of Texas, Arlington

Costantino Vetriani, Rutgers University

Amy Cheng Vollmer, Swarthmore College

Andre Walther, Cedar Crest College

Robert Weldon, University of Nebraska, Lincoln

Christine White-Ziegler, Smith College

Jianping Xu, McMaster University

Finally, we offer special thanks to our families for their support. Joan's husband Michael Barich offered unfailing support. John's wife Zarrintaj ("Zari") Aliabadi contributed to the text development, especially the sections on medical microbiology and public health.

To the Reader: Thanks!

We greatly appreciate your selection of this book as your introduction to the science of microbiology. As our textbook continues to evolve, it benefits greatly from the input of its many readers, students as well as professors. We truly welcome your comments, especially if you find text or figures that are in error or unclear. Feel free to contact us at the addresses listed below.

Joan L. Slonczewski
slonczewski@kenyon.edu
John W. Foster
jwfoster@southalabama.edu

JOAN L. SLONCZEWSKI received her B.A. from Bryn Mawr College and her Ph.D. in Molecular Biophysics and Biochemistry from Yale University, where she studied bacterial motility with Robert M. Macnab. After postdoctoral work at the University of Pennsylvania, she has since taught undergraduate microbiology in the Department of Biology at Kenyon College, where she earned a Silver Medal in the National Professor of the Year program of the Council for the Advancement and Support of Education. She has published numerous research articles with undergraduate coauthors on bacterial pH regulation, and has published five science fiction novels including *The Highest Frontier* and *A Door into Ocean*, both of which earned the John W. Campbell Memorial Award. She conducted field work on microbial ecosystems in Antarctica, sponsored by the National Science Foundation. She has served as At-large Member representing Divisions on the Council Policy Committee of the American Society for Microbiology, and as a member of the Editorial Board of the journal *Applied and Environmental Microbiology*.

JOHN W. FOSTER received his B.S. from the Philadelphia College of Pharmacy and Science (now the University of the Sciences in Philadelphia), and his Ph.D. from Hahnemann University (now Drexel University School of Medicine), also in Philadelphia, where he worked with Albert G. Moat. After postdoctoral work at Georgetown University, he joined the Marshall University School of Medicine in West Virginia; he is currently teaching in the Department of Microbiology and Immunology at the University of South Alabama College of Medicine in Mobile, Alabama. Dr. Foster has coauthored three editions of the textbook *Microbial Physiology* and has published over 100 journal articles describing the physiology and genetics of microbial stress responses. He has served as Chair of the Microbial Physiology and Metabolism division of the American Society for Microbiology and as a member of the editorial advisory board of the journal *Molecular Microbiology*.

CHAPTER 1
Microbial Life: Origin and Discovery

Microbes grow in frozen Antarctica, and everywhere else. A human body contains many more microbes than it does human cells, including 100 trillion bacteria in the digestive tract. Throughout history, humans had a hidden partnership with microbes ranging from food production to mining minerals. Microscopes revealed the tiny organisms at work in our bodies and in our environment. In the twentieth century, microbial genetics led to recombinant DNA and sequenced genomes. Today, microbes lead discoveries in medicine and global ecology.

CURRENT RESEARCH highlight

Cyanobacterial mats under ice. At the bottom of an Antarctic lake, beneath 3 meters of ice, cyanobacteria capture enough light for photosynthesis. The cyanobacteria grow flame-shaped pinnacles that produce oxygen gas. Protists and other microscopic consumers flourish in a complex ecosystem, at temperatures just above freezing. They survive dark winters while surrounding environments are frozen to −80°C. Their existence suggests the possibility of nearly frozen life on other planets, such as Mars.

Source: Ian Hawes et al. 2013. *Biology* **2**:151.

DAWN SUMNER AND DALE ANDERSEN

AN INTERVIEW WITH

DAWN SUMNER, GEOMICROBIOLOGIST, UC DAVIS

COURTESY OF DAWN Y. SUMNER

How does your study of microbes extend the work of historic Antarctic explorers?

Our work in Antarctic lakes expands our knowledge of the most biologically productive ecosystems in the Antarctic Dry Valleys. Early explorers recognized microbial growth in shallow ponds and streams but didn't have access to the luxuriant mats that coat the lake floors until diving started in the 1980s. We now use modern molecular and computational approaches to better understand how these communities have adapted to the extreme seasonal changes of the lake environments.

What might Antarctic microbes suggest about potential life on Mars?

The widespread distribution of cyanobacterial mat organisms shows that they can be transported by wind and survive freeze-drying. These mats provide an excellent model for life on Earth prior to the evolution of macroscopic grazers and burrowers, as well as for possible ecologies on other planets, such as Mars. Mars transitioned from a planet with flowing liquid water to one dominated by ice, and during this transition, ice-covered lakes were likely common.

Life began early in the history of planet Earth, with microscopic organisms, or "microbes." Over the eons, those microbes evolved to shape our atmosphere, our geology, and the energy cycles of all ecosystems. For the first 2 billion years, all life was microbial. What did it look like? To imagine it, we can look at the parts of the world that still support only microscopic life, such as the McMurdo Dry Valleys of Antarctica (**Fig. 1.1**). In summer, the temperature rarely rises above freezing, while in winter it plunges below −60°C. The only native life-forms are bacteria, lichens and protists, and tiny invertebrates such as nematodes. Cyanobacteria form thick mats below the ice of frozen lakes, where they support microscopic ecosystems (see the Current Research Highlight). Remarkably, some cyanobacterial mats can survive for years trapped in ice, to emerge when dry winds sublime the surface (**Fig. 1.1** inset). The emerging microbes blow away and colonize new habitats.

Could Antarctic microbiology help us find life on other planets? The Dry Valleys are the closest model we have to what life might look like on Mars (see **Special Topic 1.1**). As of this writing, the existence of microbial life on Mars is still unknown, but here on Earth, many terrestrial microbes remain as mysterious as Mars. Barely 0.1% of the microbes in our biosphere can be cultured in the laboratory; even the digestive tract of a newborn infant contains species of bacteria unknown to science.

We find microbes throughout our biosphere, from the superheated black smoker vents at the ocean floor to the interiors of our own bodies. Bacteria such as *Escherichia coli* are among the 100 trillion inhabitants of our intestines, where they help digest our food. Alternatively, *E. coli* may colonize the plants we eat (**Fig. 1.2A**). Microbial eukaryotes (cells with nuclei) such as the voracious *Stentor* engulf aquatic prey (**Fig. 1.2B**). Archaea are a lifeform distinct from both bacteria and eukaryotes. Some archaea grow in extreme environments, such as concentrated salt (**Fig. 1.2C**). And all kinds of life host viruses. For example, herpes simplex virus infects human cells (**Fig. 1.2D**).

Yet before we devised microscopes in the seventeenth century, we humans were unaware of the unseen living organisms that surround us, that float in our air and water, and that inhabit our own bodies. Microbes generate the very air we breathe, including nitrogen gas and much of the oxygen and carbon dioxide. They fix nitrogen for plants, and they make vitamins, such as vitamin B_{12}. In the ocean, microbes produce biomass for the food web that feeds the fish we eat, and microbes consume toxic wastes such as the oil from the *Deepwater Horizon* spill in the Gulf of Mexico in 2010. At the same time, virulent pathogens take our lives—and researchers risk their lives to study them. Working with pathogens such as Ebola virus requires sealed suits and respiratory equipment (**Fig. 1.3**). Despite all

FIGURE 1.1 ■ Antarctic "Dry Valleys," where all native life is microbial. Surface temperatures range from 0°C (summer) to −40°C (winter). **Inset:** A cyanobacterial mat emerges from the ice.

A. Bacteria: *E. coli*

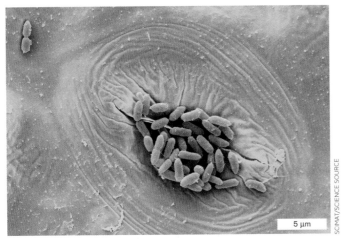

5 µm

SCIMAT/SCIENCE SOURCE

B. Eukaryote: *Stentor*

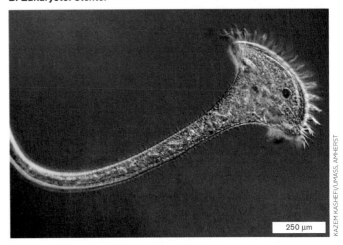

250 µm

KAZEM KASHEFI/UMASS, AMHERST

C. Archaea: Halophiles

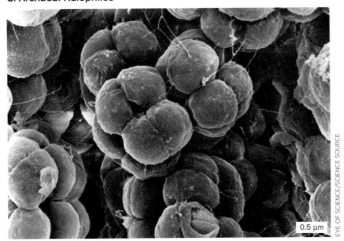

0.5 µm

EYE OF SCIENCE/SCIENCE SOURCE

D. Virus: Herpes simplex virus

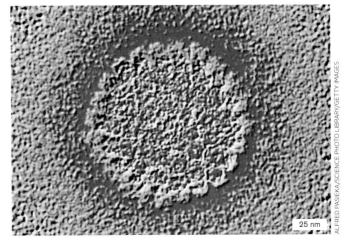

25 nm

ALFRED PASIEKA/SCIENCE PHOTO LIBRARY/GETTY IMAGES

FIGURE 1.2 ■ Representative kinds of microbes. A. Bacteria: *E. coli* on leaf stomate. **B.** Eukaryote: *Stentor*. **C.** Archaea: Halococcus. **D.** Virus: Herpes simplex virus.

LAURENT CIPRIANI / ASSOCIATED PRESS

FIGURE 1.3 ■ Researching deadly pathogens. This laboratory in Lyon, France, studies Ebola virus and influenza virus.

our advances in medicine and public health, humans continue to die of microbial diseases. Each year, millions of children succumb to waterborne pathogens and respiratory infections.

Today we discover surprising new kinds of microbes in places they were previously thought absent, such as 3 kilometers down in a South African gold mine, or within human breast milk. Microbes shape our biosphere and provide new tools that impact human society. For example, the use of heat-stable bacterial DNA polymerase (a DNA-replicating enzyme) in a technique called the **polymerase chain reaction** (**PCR**) allows us to detect minute amounts of DNA in traces of blood or fossil bone. Microbial technologies led us from the discovery of the double helix to the sequence of the human genome, the total genetic information that defines our species.

In Chapter 1 we introduce the concept of a microbe, and we survey the history of human discovery. We explain how to show which pathogen causes a disease.

SPECIAL TOPIC 1.1   How Did Life Originate?

If all life on Earth shares descent from a microbial ancestor, how did the first microbe arise? The earliest fossil evidence of cells in the geological record appears in sedimentary rock that formed as long ago as 3.8 billion years. Although the nature of the earliest reported fossils remains controversial, it is generally accepted that "microfossils" from over 2 billion years ago were formed by living cells. Moreover, the living cells that formed these microfossils looked remarkably similar to bacterial cells today, forming chains of simple rods or spheres (**Fig. 1**).

The exact composition of the first environment for life is controversial. The components of the first living cells may have formed from spontaneous reactions sparked by ultraviolet absorption or electrical discharge. American chemists Stanley Miller (1930–2007) and Harold C. Urey (1893–1981) argued that the environment of early Earth contained mainly reduced compounds—compounds that have a strong tendency to donate electrons, such as ferrous iron, methane, and ammonia. More recent evidence has modified this view, but it is agreed that the strong electron acceptor oxygen gas (O_2) was absent until the first photosynthetic microbes produced it. Today, all our cells are composed of highly reduced molecules that are readily oxidized (lose electrons to O_2). This seemingly hazardous composition may reflect our cellular origin in the chemically reduced environment of early Earth.

In 1953, Miller attempted to simulate the highly reduced conditions of early Earth to test whether ultraviolet absorption or electrical discharge could cause reactions producing the fundamental components of life (**Fig. 2A**). He boiled a solution of water containing hydrogen gas, methane, and ammonia and applied an electrical discharge (comparable to a lightning strike). The electrical discharge excites electrons in the molecules and causes them to react. Astonishingly, the reaction produced a number of amino acids, including glycine, alanine, and aspartic acid. A similar experiment in 1961 by Spanish-American researcher Juan Oró (1923–2004) (**Fig. 2B**) combined hydrogen cyanide and ammonia under electrical discharge to obtain adenine, a fundamental component of DNA and of the energy carrier adenosine triphosphate (ATP).

Molecules that spontaneously formed in Miller's and Oró's experiments are also found in meteorites and comets. This observation led Oró to propose that the first chemicals of life could have come from outer space, perhaps carried by comets. Furthermore, at the time life arose on Earth, Earth's geochemistry resembled that of other planets, such as Mars. Could Mars, too, have originated life?

In 2012, to seek evidence for life on Mars, NASA landed the *Mars Science Laboratory*, or *Curiosity* rover, near the base of a mountain on the planet Mars. The car-sized rover had a laser to drill into rock, X-ray and fluorescence analyzers, and camera microscopes. It began its mission to test the Martian soil

FIGURE 1 ■ Microfossils of ancient cyanobacteria. These fossils, from the Bitter Springs Formation, Australia, are about 850 million years old.

FIGURE 2 ■ Simulating early Earth's chemistry. A. Stanley Miller with the apparatus of his early-Earth simulation experiment. **B.** Biochemist Juan Oró demonstrated the formation of adenine and other biochemicals from reaction conditions found in comets.

for water, organic compounds, and other potential evidence of microbial life.

Among the most fascinating bits of evidence to come back from Mars were the photographs of rock formations dating to 3.7 billion years ago, near the time we believe life began on Earth (**Fig. 3**). Geologist Nora Noffke, at Old Dominion University, examined these photographs using the same criteria geologists use to define fossils of early life on Earth. She observed many features of Martian rock that resemble ancient terrestrial fossils of microbial mats—possibly bacteria similar to those of Antarctic lakes. For example, flat stretches of sediment show layers that appear folded or even rolled up, as if a flexible mat of living cells had torn and rolled back onto itself. Noffke compiled such a large group of features that ultimately, she argues, if such forms indicate fossil life on Earth, they must also indicate fossil life on Mars.

If microbial life once existed on Mars, could some forms live on, perhaps under the rock strata where pockets of water remain? As of this writing, *Curiosity's* quest for data rolls on.

RESEARCH QUESTION

How can we be sure that a fossil formation arose biogenically—that is, from a life-form? Could you define quantitative criteria for microbial fossils?

Noffke, Nora. 2015. Ancient sedimentary structures in the <3.7 Ga Gillespie Lake Member, Mars, that resemble macroscopic morphology, spatial associations, and temporal succession in terrestrial microbialites. *Astrobiology* **15**:169–192.

B.

A.

FIGURE 3 ■ **Mars fossil life.** **A.** Sedimentary formations on Mars resemble those of fossil cyanobacterial mats on Earth. **Inset:** *Curiosity* rover (shown in this artist's depiction) took the photograph. **B.** Nora Noffke argues that the Mars formations are microbially induced sedimentary structures, comparable to those found on Earth.

Finally, we address the exciting century of molecular microbiology, in which microbial genetics, genomics, and evolution have transformed the practice of medicine and our understanding of the natural world.

1.1 From Germ to Genome: What Is a Microbe?

From early childhood, we hear that we are surrounded by microscopic organisms, or "germs," which we cannot see. What are these microbes? Our modern concept of a microbe has deepened through the use of two major research tools: advanced microscopy and the sequencing of genomic DNA. Microscopy is covered in Chapter 2, and genome sequencing is presented in Chapter 7. Exciting tools of "synthetic biology," to engineer new kinds of microbes, are described in Chapter 12.

A Microbe Is a Microscopic Organism

A **microbe** is commonly defined as a living organism that requires a microscope to be seen. Microbial cells range in size from millimeters (mm) down to 0.2 micrometer (μm), and viruses may be tenfold smaller (**Table 1.1**). Some microbes consist of a single cell, the smallest unit of life, a membrane-enclosed compartment of water solution containing molecules that carry out metabolism. Each microbe contains a genome used to reproduce its own kind. Microbial cells acquire food, gain energy to build themselves, and respond to environmental change. Microbes evolve at rapid rates—often fast enough to observe in the laboratory (discussed in Chapter 17).

Our simple definition of a microbe, however, leaves us with contradictions:

■ **Super-size microbial cells.** Most single-celled organisms require a microscope to render them visible, and thus they fit the definition of a microbe. Nevertheless, some species of protists, such as giant amebas, grow to sizes large enough to see with the unaided eye. The marine sulfur bacterium *Thiomargarita namibiensis*, called the "sulfur pearl of Namibia," grows to 0.7 mm, larger than the eye of a fruit fly (**Fig. 1.4**). Even more surprising, a single cell of the "killer alga" *Caulerpa taxifolia* covers acres beneath the coastal waters of California. The alga expands and forms leaflike extensions without cell division.

■ **Microbial communities.** Many microbes form complex multicellular assemblages, such as mushrooms, kelps, and biofilms. In these structures, cells are differentiated into distinct types that complement each other's functions, as in multicellular organisms. And yet, some multicellular worms and arthropods require a microscope for us to see but are <u>not</u> considered microbes.

■ **Viruses.** A **virus** is a noncellular particle containing genetic material that takes over the metabolism of a cell to generate more virus particles. Some viruses consist of only a short chromosome packed in protein. Other kinds of viruses, such as pandoraviruses that infect amebas, show the size and complexity of a cell. Although viruses are not fully functional cells, some viral genomes may have evolved from cells.

Note: Each section of text contains Thought Questions that may have various answers. Possible responses are provided at the back of the book.

Thought Questions

1.1 The minimum size of known microbial cells is about 0.2 μm. Could even smaller cells be discovered? What factors may determine the minimum size of a cell?

1.2 If viruses are not functional cells, are they "alive"?

TABLE 1.1	Sizes of Some Microbes	
Microbe	**Description**	**Approximate size**
Varicella-zoster virus 1	Virus that causes chickenpox and shingles	100 nanometers (nm) = 10^{-7} meter (m)
Prochlorococcus	Photosynthetic marine bacteria	500 nm = 5×10^{-7} m
Escherichia coli	Bacteria growing within human intestine	1 micrometer (μm) = 10^{-6} m
Spirogyra	Aquatic algae that form long filaments of cells	40 μm = 4×10^{-5} m (cell width)
Pelomyxa	Ameba (a protist) that consumes bacteria in soil or water	5 millimeters (mm)

A.

B.

FIGURE 1.4 ▪ **A giant microbial cell:** *Thiomargarita namibiensis.* **A.** *T. namibiensis,* a marine sulfur bacterium (light microscopy). **B.** Heide Schulz-Vogt, Max Planck Institute for Marine Microbiology (holding the flask), examines a culture of the sulfur bacteria with a colleague.

In practice, our definition of a microbe derives from tradition as well as genetic considerations. In this book we consider microbes to include **prokaryotes** (cells lacking a nucleus, including bacteria and archaea) as well as certain classes of **eukaryotes** (cells with a nucleus), such as algae, fungi, and protists (discussed in Chapter 20). The bacteria, archaea, and eukaryotes—known as the three "domains"—evolved from a common ancestral cell (see Section 1.5 and Chapter 21). Viruses and even smaller infectious particles are discussed in Chapters 6 and 11.

Note: The formal names of the three domains are **Bacteria**, **Archaea**, and **Eukarya**. Members of these domains are called **bacteria** (singular, **bacterium**), **archaea** (singular, **archaeon**), and **eukaryotes** (singular, **eukaryote**), respectively. The microbiology literature includes alternative spellings for some of these terms, such as "archaean" and "eucaryote."

Microbial Genomes Are Sequenced

How have we learned how microbes work? A key tool is the study of microbial genomes. A **genome** is the total genetic information contained in an organism's chromosomal DNA. The genes in a microbe's genome and the sequence of DNA tell us a lot about how that microbe grows and associates with other species. For example, if a microbe's genome includes genes for nitrogenase, a nitrogen-fixing enzyme, that microbe probably can fix nitrogen from the atmosphere into proteins—its own proteins and those of associated plants. And by comparing DNA sequences of different microbes, we can figure out how closely related they are and how they evolved.

The first method of DNA sequencing that was fast enough to sequence large genomes was developed by Fred Sanger (1918–2013) at the University of Cambridge (**Fig. 1.5**). This achievement—which jump-started the study of molecular biology—earned Sanger the 1980 Nobel Prize in Chemistry, together with Walter Gilbert and Paul Berg. Sanger and colleagues used the new method to sequence DNA containing tens of thousands of base pairs, such as the DNA of the human mitochondrion. In **Figure 1.5**, Sanger inspects a "DNA ladder," an X-ray exposure of radiolabeled DNA fragments in a classic sequencing reaction.

But most genomes of cells contain millions, or even billions, of base pairs. In 1995, scientists completed the first genome sequence of a cellular microbe, the bacterium

FIGURE 1.5 ▪ **Fred Sanger devised the method of DNA sequence analysis used to sequence the first genomes.**

Genome of *Haemophilus influenzae*

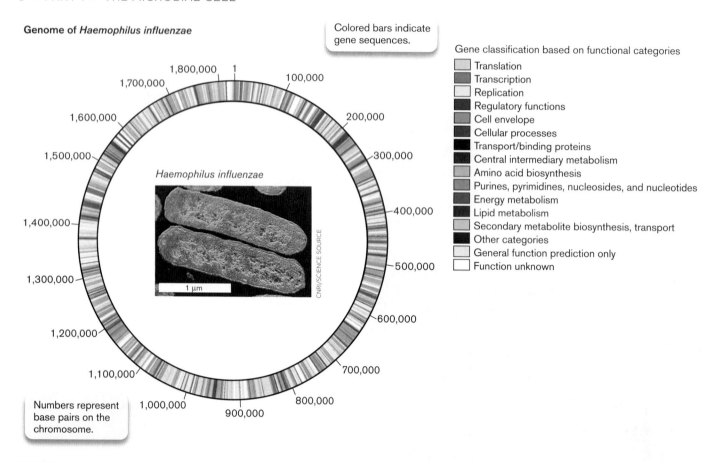

Colored bars indicate gene sequences.

Gene classification based on functional categories
- Translation
- Transcription
- Replication
- Regulatory functions
- Cell envelope
- Cellular processes
- Transport/binding proteins
- Central intermediary metabolism
- Amino acid biosynthesis
- Purines, pyrimidines, nucleosides, and nucleotides
- Energy metabolism
- Lipid metabolism
- Secondary metabolite biosynthesis, transport
- Other categories
- General function prediction only
- Function unknown

Haemophilus influenzae

1 µm

CNRI/SCIENCE SOURCE

Numbers represent base pairs on the chromosome.

FIGURE 1.6 ■ **The first sequenced genome.** The genome of *Haemophilus influenzae*, a bacterium that causes ear infections and meningitis, was the first DNA sequence completed for a cellular organism. **Inset:** Colorized electron micrograph.

THE INSTITUTE FOR GENOMIC RESEARCH

FIGURE 1.7 ■ **Claire Fraser, past president of The Institute for Genomic Research (TIGR), sequenced numerous microbial genomes.**

Haemophilus influenzae (**Fig. 1.6**). *H. influenzae* causes meningitis in children, a disease now prevented by the Hib vaccine. The *H. influenzae* genome has nearly 2 million base pairs, which specify about 1,700 genes. The sequence was determined by a large team of scientists led by Craig Venter, Hamilton Smith, and Claire Fraser (**Fig. 1.7**) at The Institute for Genomic Research (TIGR). The TIGR team devised a special computational strategy for assembling large amounts of sequence data—a strategy later used to sequence the human genome.

Today we sequence new bacterial genomes daily. In addition to sequencing individual genomes, computational strategies are used to sequence thousands of genomes of microbes sampled from a natural environment, such as the acid drainage from an iron mine. The collection of sequences taken directly from the environment is called a **metagenome**. The first metagenome of an acid mine was sequenced by Jill Banfield and co-workers at UC Berkeley in 2004. Now, metagenomes are sequenced for microbial communities of medical interest, such as that of the human colon. Human gut microbes contain 100 times more genes in their metagenomes than the human genome contains—and many of these microbial genes contribute to our health!

Comparing genomes has revealed a set of core genes shared by all organisms. These core genes add further evidence that all life on Earth, including humans, shares a common ancestry. Genomes are discussed further in Chapter 7, and the evolution of genomes and metagenomes is discussed in Chapters 17 and 21.

To Summarize

- **A microbe is a living organism that requires a microscope to be seen.** Some organisms exist in both microbial and macroscopic forms.

- **Microbes grow in communities.** A community may include both microbial and macroscopic species.

- **Microbial capabilities are defined by their genome sequences.**

1.2 Microbes Shape Human History

Today our knowledge of microbes is enormous, and it keeps growing. Yet throughout most of human history we were unaware of how microbes shaped our culture. Yeasts and bacteria made foods such as bread and cheese (**Fig. 1.8A**), as well as alcoholic beverages (discussed in Chapter 16). "Rock-eating" bacteria, known as "lithotrophs," leached copper and other metals from ores exposed by mining, enabling ancient human miners to obtain these metals. The lithotrophic oxidation of minerals for energy generates strong acids, which accelerate breakdown of the ore. Today about 20% of the world's copper, as well as some uranium and zinc, is produced by bacterial leaching. Unfortunately, microbial acidification also consumes the stone of ancient monuments (**Fig. 1.8B**)—a process intensified by airborne acidic pollution.

How did people find out about microbes? (**Table 1.2**, pages 10–11). Microscopists in the seventeenth and eighteenth centuries formulated key concepts about microbes and their existence, including their means of reproduction and death. In the nineteenth century, the "golden age" of microbiology, scientists established the fundamental principles of disease pathology and microbial ecology that are still in use today. This period laid the foundation for modern biology, in which genetics and molecular biology provide powerful tools for scientists to manipulate microorganisms for medicine, research, and industry.

Microbial Disease Devastates Human Populations

Microbial diseases such as bubonic plague and AIDS have profoundly affected human history (**Fig. 1.9**). The plague,

FIGURE 1.9 ■ Microbial disease in history and culture.
A. Medieval church procession to ward off the Black Death (bubonic plague). **B.** The AIDS Memorial Quilt spread before the Washington Monument for display in 1992. Each panel of the quilt memorializes an individual who died of AIDS.

A.

CHRONICLE OF AEGIDIUS LI MUISIS/PRIVATE COLLECTION/ BRIDGEMAN IMAGES

A.

BOYER/ROGER VIOLLET/GETTY IMAGES

B.

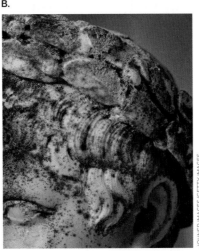

FIGURE 1.8 ■ Production and destruction by microbes. **A.** Roquefort cheeses ripening in France. **B.** Statue at the cathedral of Cologne, Germany, decaying from the action of lithotrophic microbes. The process is accelerated by acid rain.

JOHNER IMAGES/GETTY IMAGES

B.

VANESSA VICK/GETTY IMAGES

TABLE 1.2 Microbes and Human History

Date	Microbial discovery	Discoverer(s)
Microbes impact human culture without detection		
10,000 BCE	Food and drink are produced by microbial fermentation.	Egyptians, Chinese, and others
1500 BCE	Tuberculosis, polio, leprosy, and smallpox are evident in mummies and tomb art.	Egyptians
50 BCE	Copper is recovered from mine water acidified by sulfur-oxidizing bacteria.	Roman metal workers under Julius Caesar
1546 CE	Syphilis and other diseases are observed to be contagious.	Girolamo Fracastoro (Padua)
Early microscopy and the origin of microbes		
1676	Microbes are observed under a microscope.	Antonie van Leeuwenhoek (Netherlands)
1688	Spontaneous generation is disproved for maggots.	Francesco Redi (Italy)
1717	Smallpox is prevented by inoculation of pox material, a rudimentary form of immunization.	Turkish women taught Lady Mary Montagu, who brought the practice to England
1765	Microbe growth in organic material is prevented by boiling in a sealed flask.	Lazzaro Spallanzani (Padua)
1798	Cowpox vaccination prevents smallpox.	Edward Jenner (England)
1835	Fungus causes disease in silkworms (first pathogen to be demonstrated in animals).	Agostino Bassi de Lodi (Italy)
1847	Chlorine as antiseptic wash for doctor's hands decreases pathogens.	Ignaz Semmelweis (Hungary)
1881	Bacterial spores survive boiling but are killed by cyclic boiling and cooling.	John Tyndall (Ireland)
"Golden age" of microbiology: principles and methods established		
1855	Sanitation shows statistical correlation with mortality (Crimean War).	Florence Nightingale (England)
1857	Microbial fermentation produces lactic acid or alcohol.	Louis Pasteur (France)
1864	Microbes fail to appear spontaneously, even in the presence of oxygen.	Louis Pasteur (France)
1866	Microbes are defined as a class distinct from animals and plants.	Ernst Haeckel (Germany)
1867	Antisepsis during surgery prevents patient death.	Joseph Lister (England)
1877	Bacteria are a causative agent of anthrax.	Robert Koch (Germany)
1881	First artificial vaccine is developed (against anthrax).	Louis Pasteur (France)
1882	First pure culture of colonies is grown on solid medium, *Mycobacterium tuberculosis*.	Robert Koch (Germany)
1884	Koch's postulates are published, based on anthrax and tuberculosis.	Robert Koch (Germany)
1884	Gram stain is devised to distinguish bacteria from human cells.	Hans Christian Gram (Denmark)
1886	Intestinal bacteria include *Escherichia coli*, the future model organism.	Theodor Escherich (Austria)
1889	Bacteria oxidize iron and sulfur (lithotrophy).	Sergei Winogradsky (Russia)
1889	Bacteria isolated from root nodules are proposed to fix nitrogen.	Martinus Beijerinck (Netherlands)
1892, 1899	The concept of a virus is proposed to explain tobacco mosaic disease.	Dmitri Ivanovsky (Russia) and Martinus Beijerinck (Netherlands)
Cell biology, biochemistry, and genetics		
1908	Antibiotic chemicals are synthesized and identified (chemotherapy).	Paul Ehrlich (USA)
1911	Viruses are found to be a cause of cancer in chickens.	Peyton Rous (USA)
1917	Bacteriophages are recognized as viruses that infect bacteria.	Frederick Twort (England) and Félix d'Herelle (France)
1924	The ultracentrifuge is invented and used to measure the size of proteins.	Theodor Svedberg (Sweden)
1928	*Streptococcus pneumoniae* bacteria are transformed by material from dead cells.	Frederick Griffith (England)
1929	Penicillin, the first widely successful antibiotic, is made by a fungus. The molecule is isolated in 1941.	Alexander Fleming (Scotland), Howard Florey (Australia), and Ernst Chain (England)
1933–1945	The transmission electron microscope is invented and used to observe cells.	Ernst Ruska and Max Knoll (Germany)
1937	The tricarboxylic acid cycle is discovered.	Hans Krebs (Germany)
1938	The microbial "kingdom" is subdivided into prokaryotes (Monera) and eukaryotes.	Herbert Copeland (USA)
1938	*Bacillus thuringiensis* spray is produced as the first bacterial insecticide.	Insecticide manufacturers (France)
1941	One gene encodes one enzyme in *Neurospora*.	George Beadle and Edward Tatum (USA)
1941	Poliovirus is grown in human tissue culture.	John Enders, Thomas Weller, and Frederick Robbins (USA)
1944	DNA is the genetic material that transforms *S. pneumoniae*.	Oswald Avery, Colin MacLeod, and Maclyn McCarty (USA)
1945	The bacteriophage replication mechanism is elucidated.	Salvador Luria (Italy) and Max Delbrück (Germany), working in the USA
1946	Bacteria transfer DNA by conjugation.	Edward Tatum and Joshua Lederberg (USA)
1946–1956	X-ray diffraction crystal structures are obtained for the first complex biological molecules: penicillin and vitamin B_{12}.	Dorothy Hodgkin, John Bernal, and co-workers (England)
1950	Anaerobic culture technique is devised to study anaerobes of the bovine rumen.	Robert Hungate (USA)
1950	The *E. coli* K-12 genome carries a latent bacteriophage lambda.	Esther Lederberg (USA) and André Lwoff (France)
1951	Transposable elements in DNA are discovered in maize and later shown in bacteria.	Barbara McClintock (USA)
1952	DNA is injected into a cell by a bacteriophage.	Martha Chase and Alfred Hershey (USA)

TABLE 1.2 | Microbes and Human History (*continued*)

Date	Microbial discovery	Discoverer(s)
Molecular biology and recombinant DNA		
1953	Overall structure of DNA is identified by X-ray diffraction analysis as a double helix.	Rosalind Franklin and Maurice Wilkins (England)
1953	Double-helical DNA consists of antiparallel chains connected by the hydrogen bonding of AT and GC base pairs.	James Watson (USA) and Francis Crick (England)
1959	Expression of the messenger RNA for the *E. coli lac* operon is regulated by a repressor protein.	Arthur Pardee (England); François Jacob and Jacques Monod (France)
1960	Radioimmunoassay for detection of biomolecules is developed.	Rosalyn Yalow and Solomon Bernson (USA)
1961	The chemiosmotic theory, which states that biochemical energy is stored in a transmembrane proton gradient, is proposed and tested.	Peter Mitchell and Jennifer Moyle (England)
1966	The genetic code by which DNA information specifies protein sequences is deciphered.	Marshall Nirenberg, H. Gobind Khorana, and others (USA)
1967	Bacteria can grow at temperatures above 80°C in hot springs at Yellowstone National Park.	Thomas Brock (USA)
1968	Serial endosymbiosis is proposed to explain the evolution of mitochondria and chloroplasts.	Lynn Margulis (USA)
1969	Retroviruses contain reverse transcriptase, which copies RNA to make DNA.	Howard Temin, David Baltimore, and Renato Dulbecco (USA)
1972	Inner and outer membranes of Gram-negative bacteria (*Salmonella*) are separated by ultracentrifugation.	Mary Osborn (USA)
1973	A recombinant DNA molecule is made in vitro (in a test tube).	Stanley Cohen, Annie Chang, Robert Helling, and Herbert Boyer (USA)
1974	A rotary motor drives the bacterial flagellum.	Howard Berg, Michael Silverman, and Melvin Simon (USA)
1975	mRNA-rRNA base pairing initiates protein synthesis in *E. coli*.	Joan Steitz and Karen Jakes (USA); Lynn Dalgarno and John Shine (Australia)
1975	The dangers of recombinant DNA are assessed at the Asilomar Conference.	Paul Berg, Maxine Singer, and others (USA)
1975	Monoclonal antibodies are produced indefinitely in tissue culture by hybridomas, antibody-producing cells fused to cancer cells.	George Köhler (Germany) and Cesar Milstein (UK)
1977, 1980	A DNA sequencing method is invented and used to sequence the first genome of a virus.	Fred Sanger, Walter Gilbert, and Allan Maxam (England and USA)
1977	Archaea are identified as a third domain of life, the others being eukaryotes and bacteria.	Carl Woese (USA)
1978	The first protein catalog, based on 2D gels, is compiled for *E. coli*.	Fred Neidhardt, Peter O'Farrell, and colleagues (USA)
1978	Biofilms are a major form of existence of microbes.	William Costerton and others (Canada)
1979	Smallpox is declared eliminated—a global triumph of immunology and public health.	World Health Organization
Genomics, structural biology, and molecular ecology		
1981	Invention of the polymerase chain reaction (PCR) makes available large quantities of DNA.	Kary Mullis (USA)
1981–1986	Self-splicing and self-replicating RNA is discovered in the protist *Tetrahymena*.	Thomas Cech, Sidney Altman, Jennifer Doudna, and Jack Szostak (USA)
1982	Archaea are discovered with optimal growth above 100°C.	Karl Stetter (Germany)
1982	Viable but noncultured bacteria contribute to ecology and pathology.	Rita Colwell and Norman Pace (USA)
1982	Prions, infectious agents consisting solely of protein, are characterized.	Stanley Prusiner (USA)
1983	Human immunodeficiency virus (HIV) is discovered as the cause of AIDS.	Françoise Barré-Sinoussi and Luc Montagnier (France); Robert Gallo (USA)
1983	Genes are introduced into plants by use of *Agrobacterium tumefaciens* plasmid vectors.	Eugene Nester, Mary-Dell Chilton, and colleagues (USA)
1984	Acid-resistant *Helicobacter pylori* grow in the stomach, where they cause gastritis.	Barry Marshall and J. Robin Warren (Australia)
1987	*Geobacter* bacteria that can generate electricity are discovered.	Derek Lovley and colleagues (USA)
1988	*Prochlorococcus* is identified as Earth's most abundant marine phototroph.	Sallie Chisholm and colleagues (USA)
1993	A giant bacterium (*Epulopiscium*)—large enough to see—is identified.	Esther Angert and Norman Pace (USA)
1995	First genome is sequenced for a cellular organism, *Haemophilus influenzae*.	Craig Venter, Hamilton Smith, Claire Fraser, and others (USA)
2001	Ribosome structure is obtained at near-atomic level by X-ray diffraction.	Marat Yusupov, Harry Noller, and colleagues (USA)
2004	Mimivirus genome shows that large DNA viruses evolved from cells.	Didier Raoult and colleagues (France)
2006	First metagenomes are sequenced, from Iron Mountain acid mine drainage and from the Sargasso Sea.	Jill Banfield, Craig Venter, and others (USA)
2006	Vaccine prevents genital human papillomavirus (HPV), the most common sexually transmitted infection.	Patented by Georgetown University and other institutions (USA and Australia)
2013	For the first time, a person is cured of cancer by a genetically modified form of HIV virus.	Michael Kalos, Stephan Grupp, Carl June, and colleagues (USA)
1988–2016	*Escherichia coli* long-term evolution experiment reaches 50,000 generations and continues.	Richard Lenski, Zachary Blount, and colleagues (USA)

which wiped out a third of Europe's population in the four-teenth century, was caused by *Yersinia pestis*, a bacterium spread by rat fleas. Ironically, the plague-induced population decline enabled the social transformation that led to the Renaissance, a period of unprecedented cultural advancement. In the nineteenth century, the bacterium *Mycobacterium tuberculosis* stalked overcrowded cities, and tuberculosis was so common that the pallid appearance of tubercular patients became a symbol of tragic youth in European arts, such as Puccini's opera *La Bohème*. Today, societies throughout the world are devastated by the epidemic of acquired immunodeficiency syndrome (AIDS), caused by the human immunodeficiency virus (HIV). The United Nations estimates that 35 million people are living with HIV infection today, and this year 1.2 million will die of AIDS.

Historians traditionally emphasize the role of warfare in shaping human destiny, and the brilliance of leaders, or the advantage of new technology, in determining which civilizations rise or fall. Yet the fate of human societies is often determined by microbes. For example, much of the native population of North America was exterminated by smallpox introduced by European invaders. Throughout history, more soldiers have died of microbial infections than of wounds in battle.

The significance of disease in warfare was first recognized by the British nurse and statistician Florence Nightingale (1820–1910) (**Fig. 1.10A**). Better known as the founder of professional nursing, Nightingale also founded the science of medical statistics. She used methods invented by French statisticians to demonstrate the high mortality rate due to disease among British soldiers during the Crimean War. To show the deaths of soldiers due to various causes, she devised the "polar area chart" (**Fig. 1.10B**). In this chart, blue wedges represent deaths due to infectious disease, red wedges represent deaths due to wounds, and black wedges represent all other causes of death. Infectious disease accounts for more than half of all mortality.

Before Nightingale, no one understood the impact of disease on armies, or on other crowded populations, such as in cities. Nightingale's statistics convinced the British government to improve army living conditions and to upgrade the standards of army hospitals. In modern epidemiology, statistical analysis continues to be a crucial tool in determining the causes of disease.

Microscopes Reveal the Microbial World

The seventeenth century was a time of growing inquiry and excitement about the "natural magic" of science and patterns of our world, such as the laws of gravitation and motion formulated by Isaac Newton (1642–1727). Robert Boyle (1627–1691) performed the first controlled experiments on the chemical conversion of matter. Physicians attempted new treatments for disease involving the application of "stone and minerals" (that is, chemicals)—what today we would call "chemotherapy." Minds were open to

A.

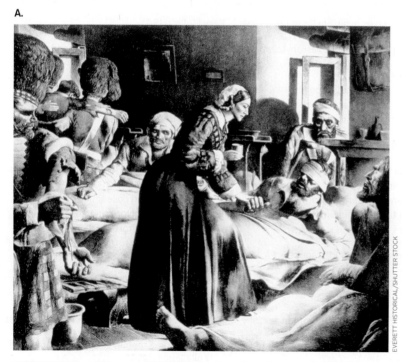

EVERETT HISTORICAL/SHUTTER STOCK

B.

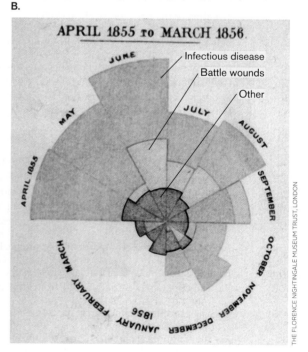

THE FLORENCE NIGHTINGALE MUSEUM TRUST, LONDON

FIGURE 1.10 ■ Florence Nightingale, founder of medical statistics. **A.** Florence Nightingale was the first to use medical statistics to demonstrate the significance of mortality due to disease. **B.** Nightingale's polar area chart of mortality data during the Crimean War.

consider the astounding possibility that our surroundings, indeed our very bodies, were inhabited by tiny living beings.

Robert Hooke observes the microscopic world.

The first microscopist to publish a systematic study of the world as seen under a microscope was Robert Hooke (1635–1703). As curator of experiments for the Royal Society of London, Hooke built the first compound microscope—a magnifying instrument containing two or more lenses that multiply their magnification in series. With his microscope, Hooke observed biological materials such as nematode "vinegar eels," mites, and mold filaments. Hooke published drawings of these microbes in *Micrographia* (1665), the first publication of objects observed under a microscope (**Fig. 1.11**).

Hooke was the first to observe distinct units of living material, which he called "cells." Hooke first named the units cells because the shape of hollow cell walls in a slice of cork reminded him of the shape of monks' cells in a monastery. But his crude lenses achieved at best 30-fold power (30×), and he never observed single-celled bacteria.

Antonie van Leeuwenhoek observes bacteria with a single lens.

Hooke's *Micrographia* inspired other microscopists, including Antonie van Leeuwenhoek (1632–1723),

FIGURE 1.11 ■ **Robert Hooke's *Micrographia*.** Mold sporangia, drawn by Hooke in 1665 from his observations of objects using a compound microscope.

who became the first individual to observe single-celled microbes (**Fig. 1.12A**). As a young man, Leeuwenhoek lived in the Dutch city of Delft, where he worked as a cloth draper, a profession that introduced him to magnifying glasses. The magnifying glasses were used to inspect the quality of the cloth, enabling the worker to count the number of threads. Later in life, Leeuwenhoek took up the hobby of grinding ever-stronger lenses to see into the world of the unseen.

Leeuwenhoek ground lenses stronger than Hooke's, which he used to build single-lens magnifiers, complete with sample holder and focus adjustment (**Fig. 1.12B**). First he observed insects, including lice and fleas; then the relatively large single cells of protists and algae; then, ultimately, bacteria. One day he applied his microscope to observe matter extracted from between his teeth. He wrote, "To my great surprise [I] perceived that the aforesaid matter contained very many small living Animals, which moved themselves very extravagantly."

Over the rest of his life, Leeuwenhoek recorded page after page on the movement of microbes, reporting their size and shape so accurately that in many cases we can determine the species he observed (**Fig. 1.12C**). He performed experiments, comparing, for example, the appearance of "small animals" from his teeth before and after drinking hot coffee. The disappearance of microbes from

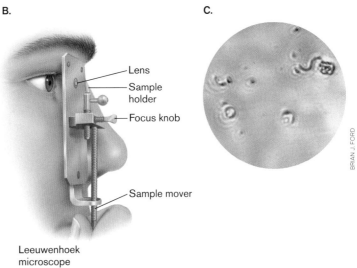

Lens
Sample holder
Focus knob
Sample mover

Leeuwenhoek microscope (circa late 1600s)

FIGURE 1.12 ■ **Antonie van Leeuwenhoek. A.** A portrait of Leeuwenhoek, the first person to observe individual microbes. **B.** "Microscope" (magnifying glass) used by Leeuwenhoek. **C.** Spiral bacteria viewed through a replica of Leeuwenhoek's instrument.

his teeth after drinking a hot beverage suggested that heat killed microbes—a profoundly important principle for the study and control of microbes ever since.

Leeuwenhoek is believed to have died, ironically, of a disease contracted from sheep whose bacteria he had observed. Historians have often wondered why it took so many centuries for Leeuwenhoek and his successors to determine the link between microbes and disease. Although observers such as Agostino Bassi de Lodi (1773–1856) noted cases of microbes associated with pathology (see **Table 1.2**), the very ubiquity of microbes—most of them actually harmless—may have obscured their more deadly roles. In addition, it was hard to distinguish between microbes and the single-celled components of the human body, such as blood cells and sperm. It was not until the nineteenth century that human tissues could be distinguished from microbial cells by the application of differential chemical stains (discussed in Chapter 2).

Thought Question

1.3 Why do you think it took so long for humans to connect microbes with infectious disease?

Spontaneous Generation: Do Microbes Have Parents?

The observation of microscopic organisms led priests and philosophers to wonder where these tiny beings came from. In the eighteenth century, scientists and church leaders intensely debated the question of **spontaneous generation**. Spontaneous generation is the concept that living creatures such as maggots could arise spontaneously, without parental organisms. Chemists of the day tended to support spontaneous generation, as it appeared similar to the way chemicals changed during reaction. Christian church leaders, however, supported the biblical view that all organisms have "parents" going back to the first week of creation.

The Italian priest Francesco Redi (1626–1697) showed that maggots in decaying meat were the offspring of flies. Meat kept in a sealed container, excluding flies, did not produce maggots. Thus, Redi's experiment argued against spontaneous generation for macroscopic organisms. The meat still putrefied, however, producing microbes that seemed to arise "without parents."

To disprove spontaneous generation of microbes, another Italian priest, Lazzaro Spallanzani (1729–1799), showed that a sealed flask of meat broth sterilized by boiling failed to grow microbes. Spallanzani also noticed that microbes often appeared in pairs. Were these two parental microbes coupling to produce offspring, or did one microbe become two? Through long and tenacious observation, Spallanzani

watched a single microbe grow in size until it split in two. Thus he demonstrated cell fission, the process by which cells arise by the splitting of preexisting cells.

Even Spallanzani's experiments, however, did not put the matter to rest. Proponents of spontaneous generation argued that the microbes in the priest's flask lacked access to oxygen and therefore could not grow. The pursuit of this question was left to future microbiologists, including the famous French microbiologist Louis Pasteur (1822–1895) (**Fig. 1.13A**). In addressing spontaneous generation and related questions, Pasteur and his contemporaries laid the foundations for modern microbiology.

Louis Pasteur reveals the biochemical basis of microbial growth. Pasteur began his scientific career as a chemist and wrote his doctoral thesis on the structure of organic crystals. He discovered the fundamental chemical property of chirality, the fact that some organic molecules exist in two forms that differ only by mirror symmetry. In other words, the two structures are mirror images of one another, like the right and left hands. Pasteur found that when microbes were cultured on a nutrient substance containing both mirror forms, only one mirror form was consumed.

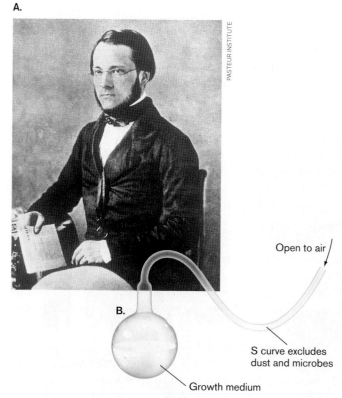

A.

PASTEUR INSTITUTE

Open to air

B.

S curve excludes dust and microbes

Growth medium

FIGURE 1.13 ▪ Louis Pasteur, founder of medical microbiology and immunology. A. Pasteur's contributions to the science of microbiology and immunology earned him lasting fame. **B.** Swan-necked flask. Pasteur showed that, after boiling, the contents in such a flask remain free of microbial growth, despite access to air.

He concluded that the metabolic preference for one mirror form was a fundamental property of life. Subsequent research has confirmed that most molecules of organisms, such as DNA and proteins, are found in only one of their mirror forms.

As a chemist, Pasteur was asked to help with a widespread problem encountered by French manufacturers of wine and beer. The alcohol in beverages comes from **fermentation**, a process by which microbes gain energy by converting sugars into alcohol. In the time of Pasteur, however, the conversion of grapes or grain to alcohol was believed to be a spontaneous chemical process. No one could explain why some fermentation mixtures produced vinegar (acetic acid) instead of alcohol. Pasteur discovered that fermentation is actually caused by living yeast, a single-celled fungus. In the absence of oxygen, yeast produces alcohol as a terminal waste product. But when the yeast culture is contaminated with bacteria, the bacteria outgrow the yeast and produce acetic acid instead of alcohol. (Fermentative metabolism is discussed in Chapter 13.)

Pasteur's work on fermentation led him to test a key claim made by proponents of spontaneous generation. The proponents claimed that Spallanzani's failure to find spontaneous appearance of microbes was due to lack of oxygen. From his studies of yeast fermentation, Pasteur knew that some microbial species do not require oxygen for growth. So he devised an unsealed flask with a long, bent "swan neck" that admitted air but kept the boiled contents free of dust that carried microbes (**Fig. 1.13B**). The famous swan-necked flasks remained free of microbial growth for many years, but when a flask was tilted to enable contact of broth with dust, microbes grew immediately. Thus, Pasteur disproved that lack of oxygen was the reason for the failure of spontaneous generation in Spallanzani's flasks.

But even Pasteur's work did not prove that microbial growth requires preexisting microbes. The Irish scientist John Tyndall (1820–1893) attempted the same experiment as Pasteur but sometimes found the opposite result. Tyndall found that the broth sometimes gave rise to microbes no matter how long it was sterilized by boiling. The microbes appear because some kinds of organic matter, particularly hay infusion, are contaminated with a heat-resistant form of bacteria called "endospores" (or "spores"). The spore form can be eliminated only by repeated cycles of boiling and resting, in which the spores germinate to the growing, vegetative form that is killed at 100°C.

It was later discovered that endospores could be killed by boiling under pressure, as in a pressure cooker, which generates higher temperatures than can be obtained at atmospheric pressure. The steam pressure device called the **autoclave** became the standard way to sterilize materials for the controlled study of microbes. (Microbial control and antisepsis are discussed further in Chapter 5.)

Although spontaneous generation was discredited as a continual source of microbes, at some point in the past the first living organisms must have originated from nonliving materials. How did the first microbes arise? Experiments that aim to replicate the conditions of early Earth yield simple amino acids and bases of DNA and RNA (see **Special Topic 1.1**). The earliest fossil evidence of cells appears in sedimentary rock that formed more than 2 billion years ago, and amazingly similar formations are found on Mars. **The experimental basis for the origin of life and evolution is discussed in detail in Chapter 17.**

To Summarize

- **Microbes affected human civilization** for centuries before humans guessed at their existence through their contributions to our environment, food and drink production, and infectious diseases.

- **Florence Nightingale** statistically quantified the impact of infectious disease on human populations.

- **Robert Hooke** and **Antonie van Leeuwenhoek** were the first to record observations of microbes through simple microscopes.

- **Spontaneous generation** is the theory that microbes arise spontaneously, without parental organisms. **Lazzaro Spallanzani** showed that microbes arise from preexisting microbes and demonstrated that heat sterilization can prevent microbial growth.

- **Louis Pasteur** discovered the microbial basis of fermentation. He also showed that providing oxygen does not enable spontaneous generation.

- **John Tyndall** showed that repeated cycles of heat were necessary to eliminate spores formed by certain kinds of bacteria.

1.3 Medical Microbiology

Over the centuries, thoughtful observers such as Fracastoro and Bassi (see **Table 1.2**) noted a connection between microbes and disease. Ultimately, researchers developed the **germ theory of disease**, the theory that many diseases are caused by microbes. Research today pursues the secrets of many microbial diseases, such as cholera, the focus of Rita Colwell, first microbiologist to direct the National Science Foundation (**eTopic 1.1**).

The first scientific basis for determining that a specific microbe causes a disease was devised by the

German physician Robert Koch (1843–1910) (**Fig. 1.14**). As a college student, Koch conducted biochemical experiments on his own digestive system. Koch's curiosity about the natural world led him to develop principles and methods crucial to microbial investigation, including the pure-culture technique and the famous Koch's postulates for identifying the causative agent of a disease. He applied his methods to numerous lethal diseases around the world, including anthrax and tuberculosis in Europe, bubonic plague in India, and malaria in New Guinea (**Fig. 1.15**).

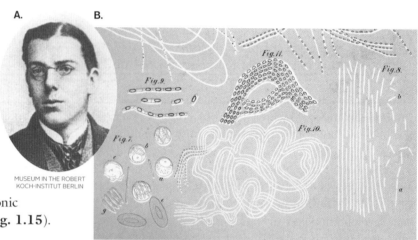

MUSEUM IN THE ROBERT KOCH-INSTITUT BERLIN

FIGURE 1.14 ■ Robert Koch, founder of the scientific method of microbiology. **A.** Koch as a university student. **B.** Koch's sketch of anthrax bacilli in mouse blood.

Growth of Microbes in Pure Culture

Unlike Pasteur, who was a university professor, Koch took up a medical practice in a small Polish-German town. To make space in his home for a laboratory to study anthrax and other deadly diseases, his wife curtained off part of his patient examination room.

Anthrax interested Koch because its epidemics in sheep and cattle caused economic hardship among local farmers. Today, anthrax is no longer a major problem for agriculture, because its transmission is prevented by effective environmental controls and vaccination. It has, however, gained notoriety as a bioterror agent because anthrax bacteria can survive for long periods in the dormant desiccated form of an endospore. In 2001, anthrax spores sent through the mail contaminated post offices as well as an office building of the U.S. Senate, causing several deaths.

To investigate whether anthrax was a transmissible disease, Koch used blood from an anthrax-infected cow carcass to inoculate a rabbit. When the rabbit died, he used its blood to inoculate a second rabbit, which then died in turn. The blood of the unfortunate animal had turned black with long, rod-shaped bacilli. Upon introduction of these bacilli into healthy animals, the animals became ill with anthrax. Thus, Koch demonstrated an important principle of epidemiology: the **chain of infection**, or transmission of a disease. In retrospect, his choice of anthrax was fortunate, because anthrax microbes generate disease very quickly, multiply in the blood to high numbers, and remain infective outside the body for long periods.

Koch and his colleagues then applied their experimental logic and culture methods to a more challenging disease: tuberculosis. In Koch's day, tuberculosis caused one-seventh of all reported deaths in Europe; today, tuberculosis bacteria continue to infect millions of people worldwide. Koch's approach to anthrax, however, was less applicable to tuberculosis, a disease that develops slowly after many years of dormancy. Furthermore, the causative bacterium, *Mycobacterium tuberculosis*, is small and difficult to distinguish

MUSEUM IN THE ROBERT KOCH INSTITUTE, BERLIN

FIGURE 1.15 ■ Robert Koch visits New Guinea. Koch (second from left) investigating malaria in 1899.

from human tissue or from different bacteria of similar appearance associated with the human body. How could Koch prove that a particular bacterium caused a particular disease?

What was needed was to isolate a **pure culture** of microorganisms, a culture grown from a single "parental" cell. Previous researchers had achieved pure cultures by a laborious process of serially diluting suspended bacteria until a culture tube contained only a single cell. Alternatively, inoculating a solid surface such as a sliced potato could produce isolated **colonies**—distinct populations of bacteria, each grown from a single cell. For *M. tuberculosis*, Koch inoculated serum, which then formed a solid gel after heating. Later he refined the solid-substrate technique by adding gelatin to a defined liquid medium, which could then be chilled to form a solid medium in a glass dish. A covered

A.

B.

FIGURE 1.16 ■ **Angelina and Walther Hesse.** **A.** Portrait of the Hesses, who first used agar to make solid-substrate media for bacterial growth. **B.** Colonies from a streaked agar plate.

version called the **petri dish** (or "petri plate") was invented by a colleague, Julius Richard Petri (1852–1921). The petri dish is a round dish with vertical walls covered by an inverted dish of slightly larger diameter. Today the petri dish, generally made of disposable plastic, remains an indispensable part of the microbiological laboratory.

Another improvement in solid-substrate culture was the replacement of gelatin with materials that remain solid at higher temperatures, such as the gelling agent **agar** (a polymer of the sugar galactose). The use of agar was recommended by Angelina Hesse (1850–1934), a microscopist and illustrator, to her husband, Walther Hesse (1846–1911), a young medical colleague of Koch (**Fig. 1.16**). Agar comes from red algae (seaweed), which is used by East Indian birds to build nests; it is the main ingredient in the delicacy "bird's nest soup." Dutch colonists used agar to make jellies and preserves, and a Dutch colonist from Java introduced it to Angelina Hesse. The Hesses used agar to develop the first effective growth medium for tuberculosis bacteria. Pure culture and growth conditions are discussed further in Chapters 4 and 5.

Note that some kinds of microbes cannot be grown in pure culture—that is, without other organisms. For example, the marine cyanobacterium *Prochlorococcus*, a major source of Earth's oxygen, requires heterotrophic bacteria to remove toxic oxygen radicals. And all viruses can be cultured only within their host cells (see Chapter 6). The discovery of viruses is explored at the end of this section.

Koch's Postulates

For his successful determination of the bacterium that causes tuberculosis, *Mycobacterium tuberculosis*, Koch was awarded the 1905 Nobel Prize in Physiology or Medicine. Koch formulated his famous set of criteria for establishing a causative link between an infectious agent and a disease (**Fig. 1.17**). These four criteria are known as **Koch's postulates**:

1. The microbe is found in all cases of the disease but is absent from healthy individuals.
2. The microbe is isolated from the diseased host and grown in pure culture.
3. When the microbe is introduced into a healthy, susceptible host (or animal model), the host shows the same disease.
4. The same strain of microbe is obtained from the newly diseased host. When cultured, the strain shows the same characteristics as before.

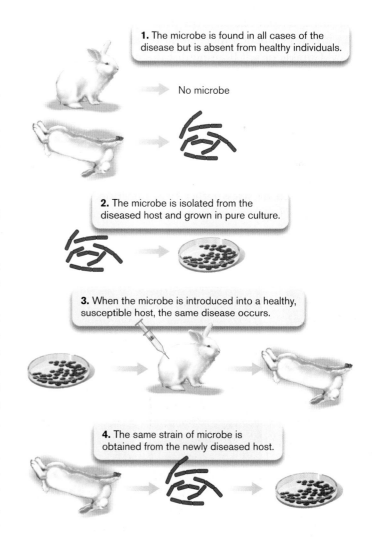

FIGURE 1.17 ■ **Koch's postulates defining the causative agent of a disease.**

Koch's postulates continue to be used to determine whether a given strain of microbe causes a disease. An example is Lyme disease (borreliosis), a tick-borne infection first described in New England, and shown in 1981 to be caused by the bacterium *Borrelia burgdorferi*. Nevertheless, the postulates remain only a guide; individual diseases and pathogens may confound one or more of the criteria. For example, tuberculosis bacteria are now known to cause symptoms in only 10% of the people infected. If Koch had been able to detect these silent bacilli, they would not have fulfilled his first criterion. In the case of AIDS, the concentration of HIV virus is so low that initially no virus could be detected in patients with fully active symptoms. It took the invention of the polymerase chain reaction (PCR), a method of producing any number of copies of DNA or RNA sequences, to detect the presence of HIV. Modern research on microbial disease is presented in Chapter 25.

Another difficulty with AIDS and many other human diseases is the absence of an animal host that exhibits the same disease. For AIDS, even chimpanzees (our closest relatives) are not susceptible, although they acquire a similar disease from related retroviruses. For diseases without a cure, experimental inoculation of humans is banned by law. Curable or self-limiting diseases may be tested on volunteers in clinical trials. In rare cases, researchers have voluntarily exposed themselves to a proposed pathogen. For example, Australian researcher Barry Marshall ingested *Helicobacter pylori* to convince skeptical colleagues that this organism could colonize the extremely acidic stomach. *H. pylori* turned out to be the causative agent of gastritis and stomach ulcers, conditions that had long been thought to be caused by stress rather than infection. For the discovery of *H. pylori* and its role in gastritis, Marshall and

colleague J. Robin Warren won the 2005 Nobel Prize in Physiology or Medicine.

Thought Question

1.4 How could you use Koch's postulates (**Fig. 1.17**) to demonstrate the causative agent of influenza? What problems not encountered with anthrax would you need to overcome?

Immunization Prevents Disease

Identifying the cause of a disease is, of course, only the first step in developing an effective therapy and preventing further transmission. Early microbiologists achieved some remarkable insights on how to control pathogens (see **Table 1.2**).

The first clue of how to protect an individual from a deadly disease came from the dreaded smallpox. In the eighteenth century, smallpox infected a large fraction of the European population, killing or disfiguring many people. In countries of Asia and Africa, however, the incidence of smallpox was decreased by the practice of deliberately inoculating children with material from smallpox pustules. Inoculated children usually developed a mild case of the disease and were protected from smallpox thereafter.

The practice of smallpox inoculation was introduced from Turkey to Europe in 1717 by Lady Mary Montagu, a smallpox survivor (**Fig. 1.18A**). While traveling in Turkey, Lady Montagu learned that many elderly women there had perfected the art of inoculation: "The old woman comes with a nut-shell full of the matter of the best sort of small-pox, and asks what vein you please to have opened." During a

A.

B.

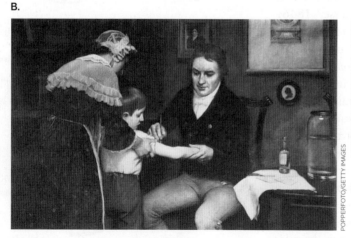

C.

FIGURE 1.18 ▪ **Smallpox vaccination. A.** Lady Mary Wortley Montagu, shown in Turkish dress. The artist avoided showing Montagu's facial disfigurement from smallpox. **B.** Dr. Edward Jenner, depicted vaccinating 8-year-old James Phipps with cowpox matter from the hand of milkmaid Sarah Nelmes, who had caught the disease from a cow. **C.** Eighteenth-century newspaper cartoon depicting public reaction to cowpox vaccination.

period outside the host, the virus becomes "attenuated"—that is, loses some of its molecular structure required for infection. The attenuated virus stimulates the immune system with much lower mortality than does the fully virulent virus. Lady Montagu arranged for the procedure on her own son and then brought the practice back to England. A similar practice of smallpox inoculation was introduced to the American colonies by a slave, Onesimus, from the Coromantee people of Africa. Onesimus convinced his master, Reverend Dr. Cotton Mather, to promote smallpox inoculation as a defense against an epidemic that was devastating Boston.

Preventive inoculation with smallpox was dangerous, however, because some infected individuals still contracted serious disease and were contagious. Thus, doctors continued to seek a better method of prevention. In England, milkmaids claimed that they were protected from smallpox after they contracted cowpox (caused by vaccinia virus), a related but much milder disease. English physician Edward Jenner (1749–1823) confirmed this claim by deliberately infecting patients with matter from cowpox lesions (**Fig. 1.18B**). The practice of cowpox inoculation was called **vaccination**, after the Latin word *vacca*, meaning "cow." At the time, the practice was highly controversial, as people feared they would somehow turn into cows (**Fig. 1.18C**). Today, unfortunately, modern immunizations still raise irrational concerns. Failure to accept immunization leads to outbreaks of preventable disease, such as the Disneyland measles outbreak in 2015.

Pasteur was aware of vaccination as he studied the course of various diseases in experimental animals. In the spring of 1879, he was studying fowl cholera, a transmissible disease of chickens with a high death rate. He had isolated and cultured the bacteria that had killed the chickens, but he left his work during the summer for a long vacation. No refrigeration was available to preserve cultures, and when he returned to work, the aged bacteria failed to cause disease in his chickens. Pasteur then obtained fresh bacteria from an outbreak of disease elsewhere, as well as some new chickens. But the fresh bacteria failed to make the original chickens sick (those that had been exposed to the aged bacteria). All of the new chickens, exposed only to the fresh bacteria, contracted the disease. Grasping the clue from his mistake, Pasteur had the insight to recognize that an attenuated strain of microbe, altered somehow to eliminate its potency to cause disease, could still confer immunity to the virulent disease-causing form.

Pasteur was the first to recognize the significance of attenuation and extend the principle to other pathogens. We now know that the molecular components of pathogens generate **immunity**, the resistance to a specific disease, by stimulating the **immune system**, an organism's exceedingly complex cellular mechanisms of defense (see Chapters 23 and 24). Understanding the immune system awaited the techniques of molecular biology a century later, but nineteenth-century

physicians developed several effective examples of **immunization**, the stimulation of an immune response by deliberate inoculation with an attenuated pathogen.

The way to attenuate a strain depends on the pathogen. Heat treatment or aging for various periods often turns out to be the most effective approach. The original success of prophylactic smallpox inoculation was due to natural attenuation of the virus during the time between acquisition of smallpox matter from a diseased individual and inoculation of the healthy patient. A far more elaborate treatment was required to combat the most famous disease for which Pasteur devised a vaccine: rabies.

The rabid dog loomed large in folklore, and rabies was dreaded for its particularly horrible and inevitable course of death. Pasteur's vaccine for rabies required a highly complex series of heat treatments and repeated inoculations. Its success led to his instant fame (**Fig. 1.19**). Grateful survivors of rabies founded the Pasteur Institute for medical research, one of the world's greatest medical research institutions, whose scientists in the twentieth century discovered the virus HIV, which causes AIDS.

Antiseptics and Antibiotics

Before the work of Koch and Pasteur, many patients died of infections transmitted unwittingly by their own doctors.

FIGURE 1.19 ■ Pasteur cures rabies. This cartoon in a French newspaper depicts Louis Pasteur protecting children from rabid dogs.

In 1847, Hungarian physician Ignaz Semmelweis (1818–1865) noticed that the death rate of women in childbirth due to puerperal fever was much higher in his own hospital than in a birthing center run by midwives. He guessed that the doctors in his hospital were transmitting pathogens from cadavers that they had dissected. So he ordered the doctors to wash their hands in chlorine, an **antiseptic** agent (a chemical that kills microbes). The mortality rate fell, but this revelation displeased other doctors, who refused to accept Semmelweis's findings.

In 1865, the British surgeon Joseph Lister (1827–1912) noted that half of his amputee patients died of sepsis. Lister knew from Pasteur that microbial contamination might be the cause. So he began experiments to develop the use of antiseptic agents, most successfully carbolic acid, to treat wounds and surgical instruments. After initial resistance, Lister's work, with the support of Pasteur and Koch, drew widespread recognition. In the twentieth century, surgeons developed fully **aseptic** environments for surgery—that is, environments completely free of microbes.

The problem with most antiseptic chemicals that killed microbes was that if taken internally, they would also kill the patients. Researchers sought a "magic bullet," an **antibiotic** molecule that would kill only microbes, leaving their host unharmed.

An important step in the search for antibiotics was the realization that microbes themselves produce antibiotic compounds. This conclusion followed from the famous accidental discovery of penicillin by the Scottish medical researcher Alexander Fleming (1881–1955) (**Fig. 1.20A**). In 1929, Fleming was culturing *Staphylococcus*, which infects wounds. He found that one of his plates of *Staphylococcus* was contaminated with a mold, *Penicillium notatum*, which he noticed was surrounded by a clear region free of *Staphylococcus* colonies (**Fig. 1.20B**). Following up on this observation, Fleming showed that the mold produced a substance that killed bacteria. We now know this substance as penicillin.

In 1941, biochemists Howard Florey (1898–1968) and Ernst Chain (1906–1979) purified the penicillin molecule, which we now know inhibits formation of the bacterial cell wall. Penicillin saved the lives of many Allied troops during World War II, the first war in which an antibiotic became available to soldiers.

The second half of the twentieth century saw the discovery of many new and powerful antibiotics. Most of the new antibiotics, however, were made by little-known bacteria and fungi from endangered ecosystems—a circumstance that focused attention on wilderness preservation. Furthermore, the widespread and often indiscriminate use of antibiotics selects for pathogens to evolve resistance to antibiotics. As a result, antibiotics have lost their effectiveness against certain strains of major pathogens. For example, multidrug-resistant *Mycobacterium tuberculosis* and

A.

B.

Penicillium mold

FIGURE 1.20 ▪ Alexander Fleming, discoverer of penicillin. **A.** Fleming in his laboratory. **B.** Fleming's original plate of bacteria with *Penicillium* mold inhibiting the growth of bacterial colonies.

methicillin-resistant *Staphylococcus aureus* (MRSA) are now serious threats to public health. To combat evolving drug resistance, we continually need to research and develop new antibiotics. Microbial biosynthesis of antibiotics is discussed in Chapter 15, and the medical use of antibiotics is discussed in Chapter 27.

> **Thought Questions**
>
> **1.5** Why do you think some pathogens generate immunity readily, whereas others evade the immune system?
>
> **1.6** How do you think microbes protect themselves from the antibiotics they produce?

The Discovery of Viruses

Viruses are much smaller than the host cells they infect; most are too small to be seen by a light microscope. So how were they discovered? In 1892, the Russian botanist Dmitri Ivanovsky (1864–1920) studied tobacco mosaic disease, a condition in which the leaves become mottled and the crop yield is decreased or destroyed altogether. Ivanovsky knew that some kind of microbe from the diseased plants transmitted the disease, and he wondered how small it was. He was surprised to find that the agent of transmission could pass through a porcelain filter having a pore size (0.1 μm) that blocked known microbes. Later, the Dutch plant microbiologist Martinus Beijerinck (1851–1931) conducted similar filtration experiments. Beijerinck concluded that because the agent of disease passed through a filter that retained bacteria, it could not be a bacterial cell.

The "filterable agent" of disease was ultimately purified by the American scientist Wendell Stanley (1904–1971), who processed 4,000 kilograms (kg) of infected tobacco leaves. Stanley obtained a sample of infective virus particles pure enough to crystallize, in a 3D array comparable to crystals composed of inert chemicals. The crystal was analyzed by X-ray crystallography (discussed in Chapter 2) to reveal the molecular structure of tobacco mosaic virus (**Fig. 1.21A**)—a feat that earned Stanley the 1946 Nobel Prize in Chemistry. The fact that an object capable of biological reproduction could be stable enough to be crystallized amazed scientists, ultimately leading to a new, more mechanical view of living organisms. Today, we consider viruses "subcellular organisms." Even smaller subcellular entities that replicate within cells include plasmids and prions (see Chapter 6).

The individual particle of tobacco mosaic virus consists of a helical tube of protein subunits containing its genetic material coiled within (**Fig. 1.21B**). Stanley thought the virus was a catalytic protein, but colleagues later determined that it contained RNA as its genetic material. The structure of the coiled RNA was solved through X-ray crystallography by the British scientist Rosalind Franklin (1920–1958). Other viruses that have RNA genomes include influenza and HIV (AIDS); viruses with DNA genomes include polio and herpes viruses. We now know that all kinds of animals, plants, and microbial cells can be infected by viruses—and carry endogenous viruses that may benefit their hosts. Viral function and disease are discussed in Chapters 6, 11, and 26.

To Summarize

- **Robert Koch** devised techniques of pure culture to study a single species of microbe in isolation. A key technique is culture on solid medium using agar, as developed by Angelina and Walther Hesse, in a double-dish container devised by Julius Petri.

- **Koch's postulates** provide a set of criteria to establish a causative link between an infectious agent and a disease.

- **Edward Jenner** established the practice of vaccination, inoculation with cowpox to prevent smallpox. Jenner's discovery was based on earlier observations by Lady Mary Montagu and others that a mild case of smallpox could prevent future cases.

- **Louis Pasteur** developed the first vaccines based on attenuated strains, such as the rabies vaccine.

- **Ignaz Semmelweis** and **Joseph Lister** showed that antiseptics could prevent transmission of pathogens from doctor to patient.

- **Alexander Fleming** discovered that the *Penicillium* mold generates a substance that kills bacteria.

- **Howard Florey** and **Ernst Chain** purified the substance penicillin, the first commercial antibiotic to save human lives.

- **Dmitri Ivanovsky** and **Martinus Beijerinck** discovered viruses as filterable infective particles. Viruses, which may be harmful or beneficial, infect all kinds of cells.

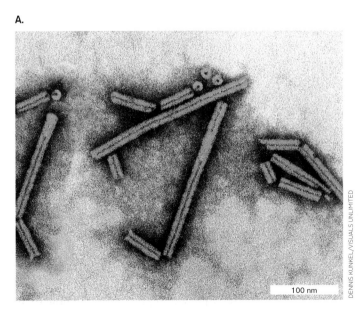

A.

100 nm

DENNIS KUNKEL/VISUALS UNLIMITED

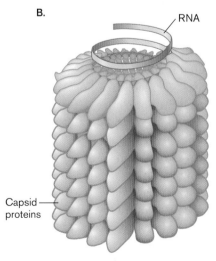

B.

RNA

Capsid proteins

FIGURE 1.21 ■ Tobacco mosaic virus (TMV).
A. Particles of tobacco mosaic virus (colorized transmission electron micrograph). **B.** In TMV, a protein capsid surrounds an RNA chromosome.

1.4 Microbial Ecology

Koch's growth of microbes in pure culture was a major advance, enabling the systematic study of microbial physiology and biochemistry. But how does pure culture relate to "natural" environments, such as a forest or a human intestine, where countless kinds of microbes interact?

In hindsight, the invention of pure culture eclipsed the equally important study of microbial ecology (discussed in Chapters 21 and 22). Microbes cycle the many minerals essential for all life, including all atmospheric nitrogen gas and much of the oxygen. Yet less than 0.1% of all microbial species can be cultured in the laboratory. Indeed, unculturable microbes make up the majority of Earth's entire biosphere. Only the outer skin of Earth supports complex multicellular life. The depths of Earth's crust, to at least 3 kilometers (km) down, as well as the atmosphere 15 km out into the stratosphere, remain the domain of microbes. So, to a first approximation, Earth's ecology <u>is</u> microbial ecology.

Microbes Support Natural Ecosystems

The first microbiologists to culture microbes in the laboratory selected the kinds of nutrients that feed humans, such as beef broth or potatoes. Some of Koch's contemporaries, however, suspected that other kinds of microbes living in soil or wetlands consume more exotic fare. Soil samples were known to oxidize hydrogen gas, and this activity was eliminated by treatment with heat or acid, suggesting microbial origin. Ammonia in sewage was oxidized by donating electrons to oxygen, forming nitrate. Nitrate formation was eliminated by antibacterial treatment. These findings suggested the existence of microbes that "eat" hydrogen gas or ammonia instead of beef or potatoes, but no one could isolate these microbes in culture.

Among the first to study microbes in natural habitats was the Russian scientist Sergei Winogradsky (1856–1953). Winogradsky waded through marshes to discover microbes with metabolisms quite alien from human digestion. For example, he discovered that species of the bacterium *Beggiatoa* oxidize hydrogen sulfide (H_2S) to sulfuric acid (H_2SO_4). *Beggiatoa* fixes carbon dioxide into biomass without consuming any organic food. Organisms that feed solely on inorganic minerals are

known as chemolithotrophs, or lithotrophs (discussed further in Chapters 4 and 14).

The lithotrophs studied by Winogradsky could not be grown on Koch's plate media containing agar or gelatin. The bacteria that Winogradsky isolated can grow only on inorganic minerals; in fact, some species are actually poisoned by organic food. For example, nitrifiers convert ammonia to nitrate, forming a crucial part of the nitrogen cycle in natural ecosystems. Winogradsky cultured nitrifiers on a totally inorganic solution containing ammonia and silica gel, which supported no other kind of organism. This experiment was an early example of **enrichment culture**, the use of selective growth media that support certain classes of microbial metabolism while excluding others.

Instead of isolating pure colonies, Winogradsky built a model wetland ecosystem containing regions of enrichment for microbes of diverse metabolism. This model is called the **Winogradsky column** (**Fig. 1.22**). The model consists of a glass tube containing mud (a source of wetland bacteria) mixed with shredded newsprint (an organic carbon source) and calcium salts of sulfate and carbonate (an inorganic carbon source for autotrophs). After exposure to light for several weeks, several zones of color develop, full of mineral-metabolizing bacteria. At the top, cyanobacteria conduct **photosynthesis**, using light energy to split water and produce molecular oxygen. Below, purple sulfur bacteria use photosynthesis to split hydrogen sulfide, producing sulfur. At the bottom, with O_2 exhausted, bacteria reduce (donate electrons to) alternative electron acceptors such as sulfate. Sulfate-reducing bacteria produce hydrogen sulfide and precipitate iron.

Like a battery cell, the gradient from oxygen-rich conditions at the surface to highly reduced conditions below generates a voltage potential. We now know that the entire Earth's surface acts as a battery—for humans, a potential source of renewable energy. A fuel cell to generate electricity is described in Chapter 14.

Winogradsky and later microbial ecologists showed that bacteria perform unique roles in **geochemical cycling**, the global interconversion of inorganic and organic forms of nitrogen, sulfur, phosphorus, and other minerals. Without these essential conversions (nutrient cycles), no plants or animals could live. Bacteria and archaea fix nitrogen (N_2) by reducing it to ammonia (NH_3), the form of nitrogen assimilated by plants. This process is called **nitrogen fixation** (**Fig. 1.23**). Other bacterial species oxidize ammonium ions (NH_4^+) in several stages back to nitrogen gas. Nitrogen fixation and geochemical cycling are discussed further in Chapters 21 and 22.

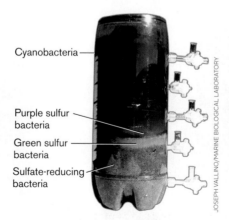

Cyanobacteria

Purple sulfur bacteria

Green sulfur bacteria

Sulfate-reducing bacteria

JOSEPH VALLINO/MARINE BIOLOGICAL LABORATORY

FIGURE 1.22 ▪ Winogradsky column. A wetland model ecosystem designed by Sergei Winogradsky.

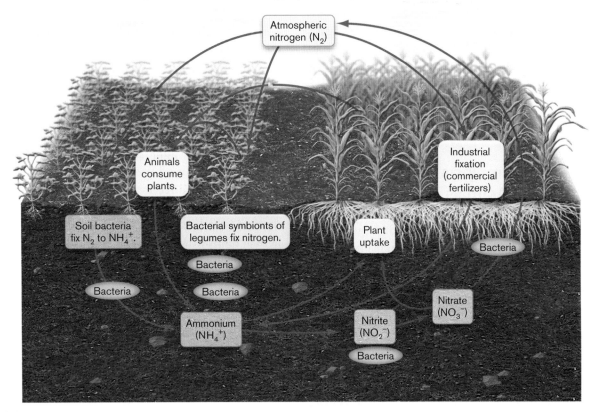

FIGURE 1.23 ▪ **The global nitrogen cycle.** All life depends on these oxidative and reductive conversions of nitrogen—most of which are performed only by microbes.

Thought Question

1.7 Why don't all living organisms fix their own nitrogen?

Today, microbes with unusual properties, such as the ability to digest toxic wastes or withstand extreme temperatures, have valuable applications in industry and bioremediation. For this reason, microbial ecology is a priority for funding by the National Science Foundation (NSF). Microbial ecologist Rita Colwell (**Fig. 1.24A**) directed the NSF from 1998 to 2004 and founded the Biocomplexity Initiative to study complex interactions between microbes and other life in the environment. Such research includes discovery of **extremophiles**, microbes from environments with extreme heat, salinity, acidity, or other factors. One such extremophile microbe is the archaeon *Geogemma*, which reduces rust (iron oxide, Fe_2O_3) to the magnetic

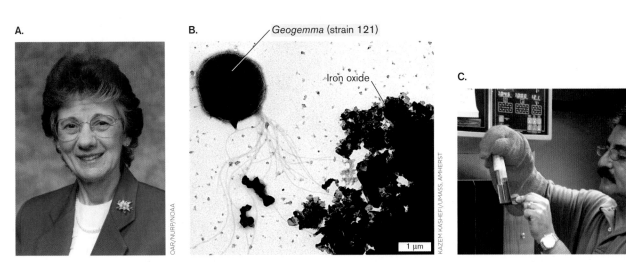

FIGURE 1.24 ▪ **An extreme thermophile reduces iron oxide to magnetite.** **A.** Rita Colwell, director of the National Science Foundation from 1998 to 2004, promoted the study of environmental microbiology. See **eTopic 1.1** for an interview with her. **B.** *Geogemma* is a round archaeon with a tuft of flagella (transmission electron micrograph). **C.** Kazem Kashefi, now at Michigan State University, pulls a live culture of "strain 121" (*Geogemma*) out of an autoclave generally used to kill all living organisms at 121°C (250°F).

mineral magnetite (Fe_3O_4) while growing in an auto-clave at 121°C, a temperature high enough to kill all other known organisms (**Fig. 1.24B**). Microbiologist Kazem Kashefi uses a magnet to show that *Geogemma* converts nonmagnetic iron oxide (rust, Fe_2O_3) to the magnetic mineral magnetite (Fe_3O_4) (**Fig. 1.24C**).

Bacterial Endosymbiosis with Plants and Animals

Within plant cells, certain bacteria fix nitrogen as **endosymbionts**, organisms living symbiotically inside larger organisms. Endosymbiotic bacteria known as rhizobia induce the roots of legumes to form special nodules to facilitate bacterial nitrogen fixation. Rhizobial endosymbiosis was first observed by Martinus Beijerinck. Microbial endosymbiosis, in a variety of diverse forms, is widespread in all ecosystems. Many interesting cases involve animal or human hosts. Endosymbiotic microbes make essential nutritional contributions to host animals. Ruminant animals such as cattle, as well as insects such as termites, require digestive bacteria to break down cellulose and other plant polymers. Even humans obtain about 15% of their nutrition from colonic bacteria (**Fig. 1.25**). Intestinal bacteria such as *E. coli* and *Bacteroides* species grow as **biofilms**, organized multispecies communities adhering to a surface. Biofilms play major roles in all ecosystems and within parts of the human body (discussed in Chapters 4 and 21). The biofilm shown in Figure 1.25B is attached to the surface of a digested food particle. Most kinds of plant fibers that we consume actually require microbial enzymes to digest (see Chapter 21).

Many invertebrates, such as hydras and corals, harbor endosymbiotic phototrophs that provide products of photosynthesis in return for protection and nutrients. Other kinds of endosymbiosis involve more than nutrition. In the light organs of squid, luminescent bacteria, such as *Aliivibrio fischeri*, produce light (bioluminescence) that helps the host evade nocturnal predators. The host squid controls the amount and direction of light that leaves the organ so that it matches the illumination from the moon, rendering the squid nearly invisible. Bacteria that normally inhabit the human intestine and skin protect our bodies from infection by pathogens. Gut bacteria regulate development of our immune system, and even send signals to the brain.

The human **microbiota** or **microbiome** is the collection of all microbes associated with the human body. Physicians increasingly consider the human microbiome to be a part of the body, as essential as a limb or an organ. In 2016, the U.S. government announced the National Microbiome Initiative to advance understanding of how microbiomes contribute to our health and the environment. Microbiomes are discussed further in Chapter 21.

Note: The terms **microbiota** and **microbiome** have similar meaning. "Microbiota" refers to the ecological community of microbes living within or upon the human body. "Microbiome" refers to the community of microbes associated with a human organ, or with a different defined habitat such as soil or plants. Another term, **metagenome**, refers to the collective genetic sequences found in all the microbes of a microbiome.

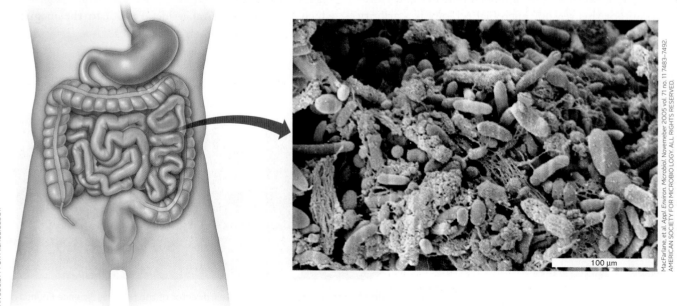

FIGURE 1.25 ■ Intestines contain biofilms. Bacteria growing on the surface of a residual food particle aid human digestion.

To Summarize

- **Sergei Winogradsky** developed the first system of enrichment culture, called the Winogradsky column, to grow microbes from natural environments.

- **Chemolithotrophs (or lithotrophs)** metabolize inorganic minerals, such as ammonia, instead of the organic nutrients used by the microbes isolated by Koch.

- **Geochemical cycling** depends on bacteria and archaea that cycle nitrogen, phosphorus, and other minerals throughout the biosphere.

- **Endosymbionts** are microbes that live within host organisms and provide essential functions for their hosts, such as nitrogen fixation for legume plants, or digestion of complex food molecules for humans and other animals.

- **Martinus Beijerinck** was the first to demonstrate that nitrogen-fixing rhizobia grow as endosymbionts within leguminous plants.

1.5 The Microbial Family Tree

The bewildering diversity of microbial life-forms presented nineteenth-century microbiologists with a seemingly impossible task of classification. So little was known about life under the lens that natural scientists despaired of ever learning how to distinguish microbial species. The famous classifier of species, Swedish botanist Carl von Linné (Carolus Linnaeus, 1707–1778), called the microbial world "chaos."

Microbes Are a Challenge to Classify

Early taxonomists faced two challenges as they attempted to classify microbes. First, the resolution of the light microscope visualized little more than the outward shape of microbial cells, and vastly different kinds of microbes looked more or less alike (discussed in Chapter 2). This challenge was overcome as advances in biochemistry and microscopy made it possible to distinguish microbes by metabolism and cell structure, and ultimately by DNA sequence.

Second, microbes do not readily fit the classic definition of a species—that is, a group of organisms that interbreed. Unlike multicellular eukaryotes, microbes generally reproduce asexually. When they do exchange genes, they may do so with related strains or with distantly related species (discussed in Chapter 9). Nevertheless, microbiologists have devised working definitions of microbial species that enable us to usefully describe populations while being flexible enough to accommodate continual revision and change (discussed in Chapter 17). The most useful definitions are based on genetic similarity. For example, two distinct species generally share no more than 95% similarity of DNA sequence.

Note: The names of microbial species are occasionally changed to reflect new understanding of genetic relationships. For example, the causative agent of bubonic plague was formerly called *Bacterium pestis* (1896), *Bacillus pestis* (1900), and *Pasteurella pestis* (1923), but it is now called *Yersinia pestis* (1944). The older names, however, still appear in the literature—a point to remember when carrying out research. Names of bacteria and archaea are compiled in the List of Prokaryotic Names with Standing in Nomenclature (LPSN).

Microbes Include Eukaryotes and Prokaryotes

In the nineteenth century, taxonomists had no DNA information. As they tried to incorporate microbes into the tree of life, they faced a conceptual dilemma because microbes could not be categorized as either animals or plants, which since ancient times had been considered the two "kingdoms" or major categories of life. Taxonomists attempted to apply these categories to microbes—for example, by including algae and fungi with plants. But German naturalist Ernst Haeckel (1834–1919) recognized that microbes differed from both plants and animals in fundamental aspects of their lifestyle, cellular structure, and biochemistry. Haeckel proposed that microscopic organisms constituted a third kind of life—neither animal nor plant—which he called Monera.

In the twentieth century, biochemical studies revealed profound distinctions even within the Monera. In particular, microbes such as protists and algae contain a nucleus enclosed by a nuclear membrane, whereas bacteria do not. Herbert Copeland (1902–1968) proposed a system of classification that divided Monera into two groups: the eukaryotic protists (protozoa and algae) and the prokaryotic bacteria. Copeland's four-kingdom classification (plants, animals, eukaryotic protists, and prokaryotic bacteria) was later modified by Robert Whittaker (1920–1980) to include fungi as another kingdom of eukaryotic microbes. Whittaker's system thus generated five kingdoms: bacteria, protists, fungi, and the multicellular plants and animals.

Eukaryotes Evolved through Endosymbiosis

The five-kingdom system was modified dramatically by Lynn Margulis (1938–2011) at the University of Massachusetts

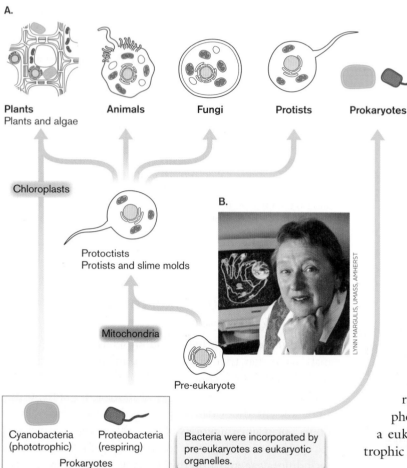

A.

Plants
Plants and algae

Animals

Fungi

Protists

Prokaryotes

Chloroplasts

Protoctists
Protists and slime molds

B.

Mitochondria

Pre-eukaryote

LYNN MARGULIS, UMASS, AMHERST

Cyanobacteria
(phototrophic)

Proteobacteria
(respiring)

Prokaryotes

Bacteria were incorporated by
pre-eukaryotes as eukaryotic
organelles.

FIGURE 1.26 ■ **Lynn Margulis and the serial endosymbiosis theory.** **A.** Five-kingdom scheme, modified by the endosymbiosis theory. **B.** Margulis proposed that organelles evolve through endosymbiosis.

(**Fig. 1.26**). Margulis tried to explain how it is that eukaryotic cells contain mitochondria and chloroplasts, membranous organelles that possess their own chromosomes. She proposed that eukaryotes evolved by merging with bacteria to form composite cells by intracellular endosymbiosis, in which one cell internalizes another that grows within it. The endosymbiosis may ultimately generate a single organism whose formerly independent members are now incapable of independent existence.

Margulis proposed that early in the history of life, respiring bacteria similar to *E. coli* were engulfed by pre-eukaryotic cells, where they evolved into mitochondria, the eukaryote's respiratory organelles. Similarly, she proposed that a phototroph related to cyanobacteria was taken up by a eukaryote, giving rise to the chloroplasts of phototrophic algae and plants.

A.

100°C

OAR/NURP/NOAA

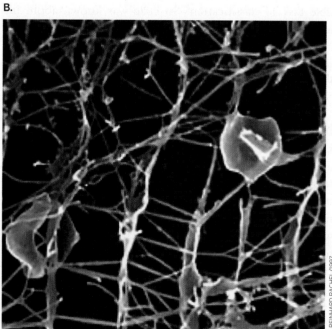

B.

© REINHARD RACHEL/1997

FIGURE 1.27 ■ **Archaea include extremophiles such as** ***Pyrodictium abyssi.*** **A.** Thermal vents of superheated water and sulfides rise from the ocean floor. **B.** The vent sulfides feed *P. abyssi* networks of interconnected cells, at temperatures above 100°C (scanning electron micrograph).

The endosymbiosis theory was highly controversial because it implied a **polyphyletic**, or multiple, ancestry of living species, inconsistent with the long-held assumption that species evolve only by divergence from a common ancestor (**monophyletic** ancestry). Ultimately, DNA sequence analysis produced compelling evidence of the bacterial origin of mitochondria and chloroplasts. Both of these classes of organelles contain circular molecules of DNA, whose sequences show unmistakable homology (similarity) to those of bacteria. DNA sequences and other evidence established the common ancestry between mitochondria and respiring bacteria, and between chloroplasts and cyanobacteria. The symbiotic origins and evolution of mitochondria and chloroplasts are discussed further in Chapter 17.

Archaea Differ from Bacteria and Eukaryotes

Are there cellular microbes that differ from both bacteria and eukaryotes? Gene sequence analysis led to another startling advance in our understanding of how cells evolved. In 1977, Carl Woese (1928–2012), at the University of Illinois, was studying a group of recently discovered prokaryotes that live in seemingly hostile environments, such as the boiling sulfur springs of Yellowstone, or that conduct unusual kinds of metabolism, such as production of methane (methanogenesis). Woese used the sequence of the gene for 16S ribosomal RNA (16S rRNA) as a "molecular clock," a gene whose sequence differences can be used to measure the time since the divergence of two species (discussed in Chapter 17). The divergence of rRNA genes showed that the newly discovered prokaryotes were a distinct form of life: **archaea** (**Fig. 1.27**).

The archaea resemble bacteria in their relatively simple cell structure, in their lack of a nucleus, and in their ability to grow in a wide range of environments. In fact, many archaea grow along with bacteria within the human colon, or in common soil or water. But certain types of archaea, such as the autoclave-cultured archaeon *Geogemma*, grow in environments more extreme than any that support bacteria. The genetic sequences of archaea differ as much from those of bacteria as from those of eukaryotes; in fact, their gene expression machinery is more similar to that of eukaryotes.

Woese's discovery replaced the classification scheme of five kingdoms with three equally distinct groups, now called the three "domains": Bacteria, Archaea, and Eukarya (**Fig. 1.28**). In the three-domain model, the bacterial ancestor of mitochondria derives from ancient proteobacteria (shaded pink in the figure), whereas chloroplasts derive from ancient cyanobacteria (shaded green).

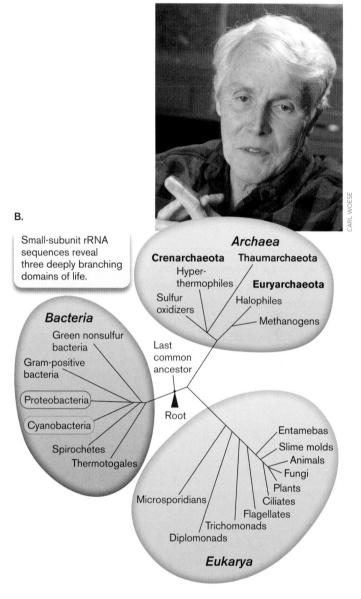

FIGURE 1.28 ■ Carl Woese and the three domains of life.
A. Woese proposed that archaea constitute a third domain of life. **B.** Three domains form a monophyletic tree based on small-subunit rRNA sequences. The length of each branch approximates the time of divergence from the last common ancestor. For a more detailed tree, see Chapter 17.

The three-domain classification is largely supported by the sequences of microbial genomes, although some genes are transferred both within and between the domains (discussed in Chapter 17).

Thought Question

1.8 What arguments support the classification of Archaea as a third domain of life? What arguments support the classification of archaea and bacteria together, as prokaryotes, distinct from eukaryotes?

To Summarize

- **Classifying microbes** was a challenge historically because of the difficulties in observing distinguishing characteristics of different categories.

- **Ernst Haeckel** recognized that microbes constitute a form of life distinct from animals and plants.

- **Herbert Copeland** and **Robert Whittaker** classified prokaryotes as a form of microbial life distinct from eukaryotic microbes such as protists.

- **Lynn Margulis** proposed that eukaryotic organelles such as mitochondria and chloroplasts evolved by endosymbiosis from prokaryotic cells engulfed by pre-eukaryotes.

- **Carl Woese** discovered a domain of prokaryotes, Archaea, whose genetic sequences diverge equally from those of bacteria and those of eukaryotes. Archaea grow in a wide range of environments; some species grow under conditions that exclude bacteria and eukaryotes.

1.6 Cell Biology and the DNA Revolution

During the twentieth century, amid world wars and societal transformations, the field of microbiology exploded with new knowledge (see **Table 1.2**). More than 99% of what we know about microbes today was discovered after 1900 by scientists too numerous to cite in this book. Advances in biochemistry and microscopy revealed the fundamental structure and function of cell membranes and proteins. The revelation of the structures of DNA and RNA led to the discovery of the genetic programs of model bacteria, such as *E. coli*, and the lambda bacteriophage. Beyond microbiology, these advances produced the technology of "recombinant DNA," or genetic engineering, the construction of molecules that combine DNA sequences from unrelated species. These microbial tools offered unprecedented applications for human medicine and industry (discussed in Chapters 7–12).

Cell Membranes and Macromolecules

In 1900, the study of cell structure was still limited by the limited resolution of the light microscope and by the absence of tools that could take apart cells to isolate their components. Both of these limitations were overcome by the invention of powerful instruments. Just as society was being transformed by machines ranging from jet airplanes to vacuum cleaners, the study of microbiology was also being transformed by machines. Two instruments had exceptional impact: The **electron microscope** revealed the internal structure of cells (Chapter 2), and the **ultracentrifuge** enabled isolation of subcellular parts (Chapter 3).

The electron microscope. In the 1920s, at the Technical University in Berlin, student Ernst Ruska (1906–1988) was invited to develop an instrument for focusing rays of electrons. Ruska recalled, from his childhood, that his father's microscope could magnify fascinating specimens of plants and animals, but that its resolution was limited by the wavelength of light. He was eager to devise lenses that could focus beams of electrons, with wavelengths far smaller than that of light, to reveal living details never seen before. Ultimately, Ruska built lenses to focus electrons using specially designed electromagnets. Magnetic lenses were used to complete the first electron microscope in 1933 (**Fig. 1.29**). Early transmission electron microscopes achieved about tenfold greater magnification than the light microscope, revealing details such as the ridged shell of a diatom. Further development steadily increased magnification, to as high as a millionfold.

For the first time, cells were seen to be composed of a cytoplasm containing macromolecules and bounded by a phospholipid membrane. For example, the electron micrograph in **Figure 1.30** shows a "thin section" of the photosynthetic bacterium *Chlorobium*, including its nucleoid

Column contains magnetic lenses.

MARINE BIOLOGICAL LABORATORY/WHOI

FIGURE 1.29 ■ **An early transmission electron microscope.**

(DNA) and its light-harvesting chlorosomes. Electron microscopy is discussed further in Chapter 2.

Subcellular structures, however, raised many questions about cell function that visualization alone could not answer. Biochemists showed that cell function involves numerous chemical transformations mediated by enzymes. A milestone in the study of metabolism was the elucidation by German biochemist Hans Krebs (1900–1981) of the tricarboxylic acid cycle (TCA cycle, or Krebs cycle), by which the products of sugar digestion are converted to carbon dioxide. The TCA cycle provides energy for many bacteria and for the mitochondria of eukaryotes. But even Krebs understood little of how metabolism is organized within a cell; he and his contemporaries considered the cell a "bag of enzymes." The full understanding of cell structure required experiments on isolated parts of cells.

The ultracentrifuge. Centrifugation can separate whole cells from the fluid in which they are suspended. The first centrifuges spun samples in a rotor with centrifugal force of a few thousand times that of gravity. In the nineteenth century, biochemists proposed that even greater centrifugal forces could separate components of lysed cells, even macromolecules such as proteins. The Swedish chemist Theodor Svedberg (1884–1971), at the University of Uppsala, built such a machine: the ultracentrifuge. By the twentieth century, ultracentrifuges achieved rates so high that they required a vacuum to avoid burning up like a space reentry vehicle. Ultracentrifuges isolated protein complexes such as ribosomes, and DNA molecules such as plasmids.

Experiments combining electron microscopy and ultracentrifugation revealed how membranes govern energy transduction within bacteria and within organelles such as mitochondria and chloroplasts. In the 1960s, English biochemists Peter Mitchell (1920–1992) and Jennifer Moyle proposed and tested a revolutionary idea called the **chemiosmotic theory**. The chemiosmotic theory states that the reduction-oxidation (redox) reactions of the electron transport system store energy in the form of a gradient of protons (hydrogen ions) across a membrane, such as the bacterial cell membrane or the inner membrane of the mitochondrion. The energy stored in the proton gradient, in turn, drives the synthesis of ATP (discussed in Chapter 14).

Microbial Genetics Leads the DNA Revolution

As the form and function of living cells emerged in the early twentieth century, a largely separate line of research

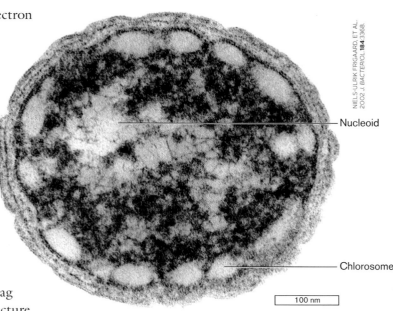

NIELS-ULRIK FRIGAARD, ET AL. 2002 J. BACTERIOL. **184** 3368.

Nucleoid

Chlorosome

100 nm

FIGURE 1.30 ■ Electron micrograph of *Chlorobium* species, a photosynthetic bacterium. The thin section reveals the nucleoid (containing DNA), the light-harvesting chlorosomes, and envelope membranes.

revealed patterns of heredity of cell traits. In eukaryotes, the Mendelian rules of inheritance were rediscovered and connected to the behavior of subcellular structures called chromosomes. Frederick Griffith (1879–1941) showed in 1928 that an unknown substance from dead bacteria could carry genetic information into living cells, transforming harmless bacteria into a strain capable of killing mice—a process called **transformation**. Some kind of "genetic material" must be inherited to direct the expression of inherited traits, but no one knew what that material was or how its information was expressed.

Then, in 1944, Oswald Avery (1877–1955) and colleagues showed that the genetic material for transformation is deoxyribonucleic acid, or DNA. An obscure acidic polymer, DNA had been previously thought too uniform in structure to carry information; its precise structure was unknown. As World War II raged among nations, scientists embarked on an epic struggle: the quest for the structure of DNA.

The double helix. The tool of choice to discover the structure of molecules was X-ray crystallography, a method developed by British physicists in the early 1900s. The field of X-ray analysis included an unusual number of women, including Dorothy Hodgkin (1910–1994), who later won a Nobel Prize for the structures of penicillin and vitamin B_{12}. In 1953, crystallographer Rosalind Franklin joined a laboratory at King's College London to study the structure

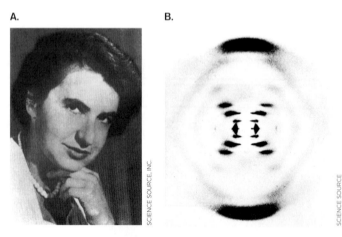

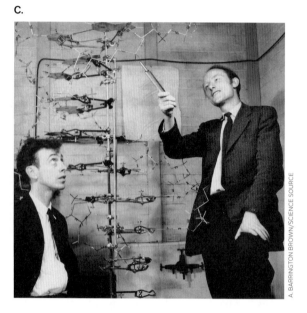

FIGURE 1.31 ■ **The DNA double helix.** **A.** Rosalind Franklin discovered that DNA forms a double helix. **B.** X-ray diffraction pattern of DNA, obtained by Franklin. **C.** James Watson (left) and Francis Crick discovered the complementary pairing between bases of DNA and the antiparallel form of the double helix.

of DNA (**Fig. 1.31A**). As a woman and as a Jew who supported relief work in Palestine, Franklin felt socially isolated at the male-dominated Protestant university; her work was disparagingly called "witchcraft." Nevertheless, her exceptional X-ray micrographs (**Fig. 1.31B**) revealed for the first time that the standard B-form DNA was a double helix.

Without Franklin's knowledge, her colleague Maurice Wilkins showed her data to a competitor, James Watson at the University of Cambridge. The pattern led Watson and Francis Crick (1916–2004) to guess that the four bases of the DNA "alphabet" were paired in the interior of Franklin's double helix (**Fig. 1.31C**). They published their model in the journal *Nature*, while denying that they had used Franklin's data. The discovery of the double helix earned Watson, Crick, and Wilkins the 1962 Nobel Prize in Physiology or Medicine. Franklin died of ovarian cancer before the prize was awarded. Before her death, however, she had turned her efforts to the structure of ribonucleic acid (RNA). She determined the helical form of the RNA chromosome within tobacco mosaic virus, the first viral RNA to be characterized.

Modern X-ray crystallography (discussed in Chapter 2) reveals with atomic precision the structure of DNA, including its complementary base pairs (**Fig. 1.32A**). The complementary pairing of DNA bases led to the development of techniques for **DNA sequencing**, the reading of a sequence of DNA base pairs. **Figure 1.32B** shows a portion of the

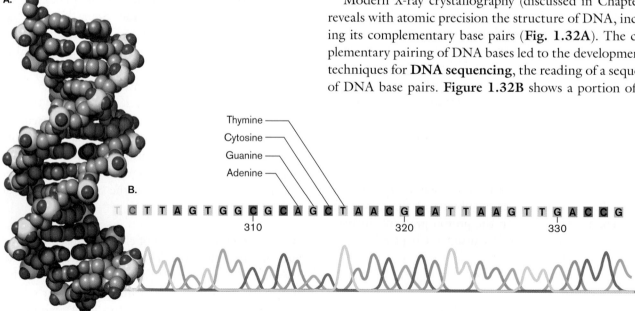

FIGURE 1.32 ■ **DNA.** **A.** The structure of DNA, based on modern X-ray crystallography. **B.** A DNA sequence fluorogram obtained from bacterial genomic DNA. Each colored trace represents the fluorescence of one of the four bases terminating a chain of DNA. Units represent number of DNA bases.

DNA sequence from bacterial DNA isolated by an undergraduate student. (The sequencing process is described in Chapter 7.) In the data, each color represents a fluorescent signal from one of the four bases (adenine, guanine, cytosine, or thymine). Each peak represents a DNA fragment terminating in that particular base. The order of fragment lengths yields the sequence of bases in one strand. Reading the DNA sequence enabled microbiologists to determine the beginning and endpoint of microbial genes, and ultimately entire genomes, as discussed in Section 1.1.

Reading the genomes enabled microbiologists to see the history of microbial evolution, reaching back to a time even before the advent of DNA—to a pre-DNA world when the cell's chromosomes were actually composed of ribonucleic acid, RNA. This hypothetical world without DNA is called the **RNA world**. How did life function in the RNA world? We hypothesize that cells used RNA for all the functions of DNA and protein, including information storage and replication, and biochemical catalysis. RNA molecules capable of catalysis, called ribozymes, were discovered in 1982 by Thomas Cech at the University of Colorado, and Sidney Altman at Harvard University, who earned the Nobel Prize in Chemistry in 1989. That same year, Jennifer Doudna and Jack Szostak at Harvard showed how an RNA molecule, a self-splicing intron from *Tetrahymena*, could catalyze its own replication (**Fig. 1.33**). In 2009, Gerald Joyce and Tracey Lincoln at the Scripps Research Institute constructed the first self-replicating ribozyme, an RNA that can catalyze reactions and copy itself indefinitely. These achievements support the theory that early organisms were composed primarily of RNA.

How do DNA and RNA sequences convey information in the cell? To read the DNA language required deciphering the genetic code—how triplets of DNA "letters" specify the amino acid units of proteins. This story is discussed in Chapter 8.

The DNA revolution began with bacteria. What amazed the world about DNA was that such a simple substance, composed of only four types of subunits, is the genetic material that determines all the different organisms on Earth. The promise of this insight was first fulfilled in bacteria and bacteriophages, whose small genomes and short generation times made key experiments possible (see Chapters 7–9). Bacterial tools were later extended to animals and plants; for example:

- **Bacteria readily recombine DNA from unrelated organisms.** The mechanisms of bacterial recombination led to construction of artificially recombinant DNA, or "gene cloning." Recombinant DNA ultimately enabled us to transfer genes between the genomes of virtually all types of organisms.

- **Bacterial DNA polymerases are used for polymerase chain reaction (PCR) amplification of DNA.** A hot spring in Yellowstone National Park yielded the bacterium *Thermus aquaticus*, whose DNA polymerase could survive many rounds of cycling to near-boiling temperature.

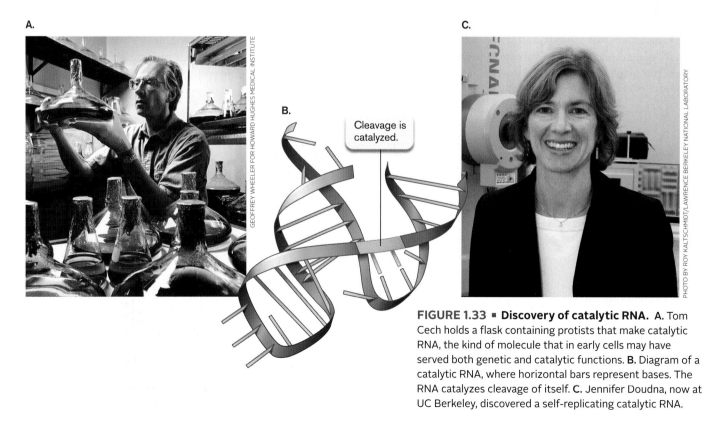

A. GEOFFREY WHEELER FOR HOWARD HUGHES MEDICAL INSTITUTE

B. Cleavage is catalyzed.

C. PHOTO BY ROY KALTSCHMIDT/LAWRENCE BERKELEY NATIONAL LABORATORY

FIGURE 1.33 ■ Discovery of catalytic RNA. A. Tom Cech holds a flask containing protists that make catalytic RNA, the kind of molecule that in early cells may have served both genetic and catalytic functions. **B.** Diagram of a catalytic RNA, where horizontal bars represent bases. The RNA catalyzes cleavage of itself. **C.** Jennifer Doudna, now at UC Berkeley, discovered a self-replicating catalytic RNA.

The Taq polymerase formed the basis of a multibillion-dollar industry of PCR amplification of DNA, with applications ranging from genome sequencing to forensic identification.

- **Gene regulation discovered in bacteria provided models for animals and plants.** The first key discoveries of gene expression were made in bacteria and bacteriophages. Regulatory DNA-binding proteins were discovered in bacteria and then subsequently found in all classes of living organisms.

In the 1970s, when the DNA revolution began, its implications drew public concern. The use of recombinant DNA to make hybrid organisms—organisms combining DNA from more than one species—seemed "unnatural." We now know that in natural environments, genes frequently move between species. Furthermore, recombinant DNA technology raised the specter of placing deadly genes that produce toxins such as botulin into innocuous human-associated bacteria such as *E. coli*.

The unknown consequences of recombinant DNA so concerned molecular biologists that in 1975 they held a conference to assess the dangers and restrict experimentation on recombinant DNA. The conference, led by Paul Berg and Maxine Singer at Asilomar (Pacific Grove, California) was possibly the first time in history that a group of scientists organized and agreed to regulate and restrict their own field.

On the positive side, the emerging world of molecular biology excited the imagination of young scientists and entrepreneurs. The pioneering biotechnology company Genentech was founded in 1976 by Robert Swanson and bacterial geneticist Robert Boyer, from UC San Francisco. Growing numbers of students entered the field

of molecular biology, seeking to invent medical cures—or even to clone dinosaurs, as in Michael Crichton's novel and film *Jurassic Park* (1993). While the idea of cloning a dinosaur remains science fiction, the tools of microbial genetics opened a window into the past by letting us read the DNA of long-dead organisms preserved in museums.

Thought Question

1.9 Do you think engineered strains of bacteria should be patentable? What about sequenced genes or genomes?

Microbial Discoveries Transform Medicine and Industry

Twentieth-century microbiology transformed the practice of medicine and generated entire new industries of biotechnology and bioremediation. Following the discovery of penicillin, Americans poured millions of dollars of private and public funds into medical research. The March of Dimes campaign for private donations to prevent polio led to the successful development of a vaccine that has nearly eliminated the disease. Since then, research on microbes and other aspects of biology has grown with support from U.S. government agencies, such as the National Institutes of Health and the National Science Foundation, as well as from governments of other countries, particularly the European nations and Japan. Further support comes from private foundations, such as the Pasteur Institute, the Wellcome Trust, and the Howard Hughes Medical Institute. Medical research generates astonishing advances, such as the use of human immunodeficiency virus (HIV, the cause

TABLE 1.3	Fields of Research in Microbiology
Field	**Subject of study**
Experimental microbiology	Fundamental questions about microbial form and function, genetics, and ecology
Medical microbiology	The mechanism, diagnosis, and treatment of microbial disease
Epidemiology	Distribution and causes of disease in humans, animals, and plants
Immunology	The immune system and other host defenses against infectious disease
Food microbiology	Fermented foods and food preservation
Industrial microbiology	Production of drugs, cloned gene products, and biofuels
Environmental microbiology	Microbial diversity and microbial processes in natural and artificial environments
Bioremediation	The use of microbial metabolism to remediate human wastes and industrial pollutants
Forensic microbiology	Analysis of microbial strains as evidence in criminal investigations
Astrobiology	The origin of life in the universe and the possibility of life outside Earth

FIGURE 1.34 ■ **Microbiologists at work.** Students at Kenyon College conduct research on bacterial gene expression.

of AIDS) to devise gene therapy agents that cure cancer (discussed in Chapter 11).

Research in microbiology includes fields as diverse as medicine and space science (**Table 1.3**). (A scientist whose career combines medicine and space science is profiled in **eTopic 1.2**). These fields all recruit microbiologists (**Fig. 1.34**). Industrial and applied biology (see Chapter 16) use bacteria to clone and produce therapeutic proteins, such as insulin for diabetics. Recombinant viruses make safer vaccines. At the frontiers of science, we use microbes for "synthetic biology," the construction of novel organisms with useful functions (see Chapter 12). For example, synthetic biology may design bacteria with an on/off switch to report the presence of arsenic in environmental samples. On a global level, the management of our planet's biosphere, with the challenges of pollution and global warming, increasingly depends on our understanding of microbial populations (see Chapter 22).

To Summarize

- **Genetics of bacteria, bacteriophages, and fungi** in the early twentieth century revealed fundamental insights about gene transmission that apply to all organisms.

- **Structure and function of the genetic material, DNA,** emerged from a series of experiments in the twentieth century.

- **Molecular microbiology generated key advances,** such as the cloning of the first recombinant molecules and the invention of DNA sequencing technology.

- **Genome sequence determination and bioinformatic analysis** became the tools that shape the study of biology in the twenty-first century.

- **Microbial discoveries transformed medicine and industry.** Biotechnology produces new kinds of pharmaceuticals and industrial products. Synthetic biology engineers new kinds of organisms with useful functions.

Concluding Thoughts

Advances in microbial science raise important questions for society. How can medical research control emerging diseases? How do microbes contribute to global cycles of carbon and nitrogen? How can microbial metabolism clean up polluted environments, such as the vast oil spill from the explosion of the *Deepwater Horizon* wellhead in 2010? How can we engineer viruses to treat human diseases?

This book explores the current explosion of knowledge about microbial cells, genetics, and ecology. We introduce the applications of microbial science to human affairs, from medical microbiology to environmental science. Most important, we discuss research methods—how scientists make the discoveries that will shape tomorrow's view of microbiology. Chapter 2 presents the imaging tools that make possible our increasingly detailed view of the structures of cells (Chapter 3) and viruses (Chapter 6). Chapters 4 and 5 introduce microbial nutrition and growth in diverse habitats, including what we discover from microbial genomes. Throughout, we invite readers to share with us the excitement of discovery in microbiology.

CHAPTER REVIEW

Review Questions

1. Explain the apparent contradictions in defining microbiology as the study of microscopic organisms or as the study of single-celled organisms.
2. What is the genome of an organism? How do genomes of viruses differ from those of cellular microbes?
3. Under what conditions might microbial life have originated? What evidence supports current views of microbial origin?
4. List the ways in which microbes have affected human life throughout history.
5. Summarize the key experiments and insights that shaped the controversy over spontaneous generation. What questions were raised, and how were they answered?
6. Explain how microbes are cultured in liquid and on solid media. Compare and contrast the culture methods of Koch and Winogradsky. How did their different approaches to microbial culture address different questions in microbiology?
7. Explain how a series of observations of disease transmission led to the development of immunization to prevent disease.
8. Summarize key historical developments in our view of microbial taxonomy. What attributes of microbes have made them challenging to classify?
9. Explain how various discoveries in "natural" bacterial genetics were used to develop recombinant DNA technology.

Thought Questions

1. How do Earth's microbes contribute to human health? Include examples of environmental microbes outside the human body, as well as microbes associated with the human body.
2. When space scientists seek evidence for life on Mars, why do you think they expect to find microbes rather than creatures like the "alien monsters" often depicted in science fiction?
3. Why do you think so many environmental microbes cannot be cultured in laboratory broth or agar media?
4. Outline the different contributions to medical microbiology and immunology of Louis Pasteur, Robert Koch, and Florence Nightingale. What methods and assumptions did they have in common, and how did they differ?
5. Outline the different contributions to environmental microbiology of Sergei Winogradsky and Martinus Beijerinck. Why did it take longer for the significance of environmental microbiology to be recognized, as compared with pure-culture microbiology?
6. What kinds of evidence support the common ancestry of life from cells with RNA chromosomes? Could cells with RNA chromosomes exist today? Why or why not?

Key Terms

agar (17)
antibiotic (20)
antiseptic (20)
Archaea (7, 27)
archaeon (7)
aseptic (20)
autoclave (15)
Bacteria (7)
bacterium (7)
biofilm (24)

chain of infection (16)
chemiosmotic theory (29)
colony (16)
DNA sequencing (30)
electron microscope (28)
endosymbiont (24)
enrichment culture (22)
Eukarya (7)
eukaryote (7)
extremophile (23)

fermentation (15)
genome (7)
geochemical cycling (22)
germ theory of disease (15)
immune system (19)
immunity (19)
immunization (19)
Koch's postulates (17)
metagenome (8, 24)
microbe (6)

microbiome (24)
microbiota (24)
monophyletic (27)
nitrogen fixation (22)
petri dish (17)
photosynthesis (22)

polymerase chain reaction (PCR) (3)
polyphyletic (27)
prokaryote (7)
pure culture (16)
RNA world (31)
spontaneous generation (14)

transformation (29)
ultracentrifuge (28)
vaccination (19)
virus (6)
Winogradsky column (22)

Recommended Reading

Albers, Sonja-Verena, Patrick Forterre, David Prangishvili, and Christa Schleper. 2013. The legacy of Carl Woese and Wolfram Zillig: From phylogeny to landmark discoveries. *Nature Reviews. Microbiology* **11**:713–719.

Blaser, Martin, Peer Bork, Claire Fraser, Rob Knight, and Jun Wang. 2013. The microbiome explored: Recent insights and future challenges. *Nature Reviews. Microbiology* **11**:213–217.

Blount, Zachary D., Christina Z. Borland, and Richard E. Lenski. 2008. Historical contingency and the evolution of a key innovation in an experimental population of *Escherichia coli. Proceedings of the National Academy of Sciences USA* **105**:7899–7906.

Breitbart, Mya, Luke R. Thompson, Curtis A. Suttle, and Matthew B. Sullivan. 2007. Exploring the vast diversity of marine viruses. *Oceanography* **20**:135–139.

Brock, Thomas D. 1999. *Robert Koch: A Life in Medicine and Bacteriology.* ASM Press, Washington, DC.

Dinc, Gulten, and Yesim I. Ulman. 2007. The introduction of variolation "A La Turca" to the West by Lady Mary Montagu and Turkey's contribution to this. *Vaccine* **25**:4261–4265.

Dubos, Rene. 1998. *Pasteur and Modern Science.* Translated by Thomas Brock. ASM Press, Washington, DC.

Fleishmann, Robert D., Mark D. Adams, Owen White, Rebecca A. Clayton, Ewen F. Kirkness, et al. 1995. Whole-genome random sequencing and assembly of *Haemophilus influenzae* Rd. *Science* **269**:496–512.

Gann, Alexander, and Jan Witkowski. 2010. The lost correspondence of Francis Crick. *Nature* **467**:519–524.

Hesse, Wolfgang. 1992. Walther and Angelina Hesse—early contributors to bacteriology. *ASM News* **58**:425–428.

Luef, Birgit, Kyle R. Frischkorn, Kelly C. Wrighton, Hoi-Ying N. Holman, Giovanni Birarda, et al. 2015. Diverse uncultivated ultra-small bacterial cells in groundwater. *Nature Communications* **6**:6372.

Maddox, Brenda. 2002. *The Dark Lady of DNA.* HarperCollins, New York.

Margulis, Lynn. 1968. Evolutionary criteria in Thallophytes: A radical alternative. *Science* **161**:1020–1022.

Raoult, Didier. 2005. The journey from *Rickettsia* to Mimivirus. *ASM News* **71**:278–284.

Sherman, Irwin W. 2006. *The Power of Plagues.* ASM Press, Washington, DC.

Thomas, Gavin. 2005. Microbes in the air: John Tyndall and the spontaneous generation debate. *Microbiology Today* (November 5): 164–167.

Ward, Naomi, and Claire Fraser. 2005. How genomics has affected the concept of microbiology. *Current Opinion in Microbiology* **8**:564–571.

CHAPTER 2
Observing the Microbial Cell

Microscopy reveals the vast realm of micro-organisms invisible to the unaided eye. The **microscope** enables us to count the number of microbes in the human bloodstream or in dilute natural environments such as the ocean. It shows us how microbes swim and respond to signals such as a new food source. Fluorescence microscopy captures single molecules within a living cell. Electron microscopy explores the cell's interior and models viruses, even catching a virus in the act of infection.

CURRENT RESEARCH highlight

3D cryotomography of a chlamydia. Chlamydias are obligate intracellular pathogens, some of which cause trachoma and sexually transmitted disease. The environmental chlamydia *Simkania* propagates within an ameba, a model for human infection. Grant Jensen at the California Institute of Technology visualized a *Simkania* cell within its host cell, using electron microscopy at supercold temperatures (cryo-electron microscopy). Cryo-electron microscopy reveals how the intracellular chlamydia recruits host membranes to surround itself. Thus, *Simkania* induces its host cell to remodel a compartment where the chlamydia can grow.

Source: Martin Pilhofer et al. 2014. *Environ. Microbiol.* **16**:417.

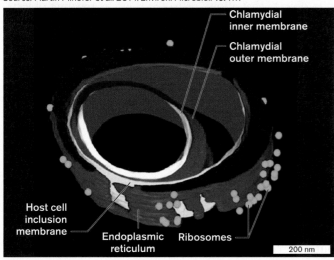

AN INTERVIEW WITH

GRANT J. JENSEN, MICROSCOPIST, CALIFORNIA INSTITUTE OF TECHNOLOGY

How has your 3D cryotomography changed our view of the bacterial cell?

We've pioneered the application of 3D electron cryotomography to intact bacterial cells. Because the cells are frozen-hydrated (rather than chemically fixed and dehydrated as in traditional EM), authentic molecular details are preserved. This allows us to visualize all the large macromolecular complexes, such as cell walls, cytoskeletal filaments, DNA aggregates, and flagellar motors, with unprecedented clarity. We have seen how bacterial cells are not simply "bags of enzymes," but rather highly organized and finely tuned machines.

What future developments in microscopy do you foresee?

The biggest opportunity I see today is correlating light and electron microscopy images of the same specimen. Correlation would allow us to tag an object of interest inside a cell with a fluorophore, watch it move in real time during some dynamic cellular event, and then, in a moment of interest, quick-freeze the cell, transfer the specimen into an electron microscope, and record a 3D image of the same cell at molecular resolution.

How did people first see microbes? As we saw in Chapter 1, the microscope of Antonie van Leeuwenhoek first revealed the tiny life-forms on his teeth; his superior lenses were key to his success. Since the time of Leeuwenhoek, microscopists have devised ever-more-powerful instruments to search for microbes in familiar and unexpected habitats.

An example of such a habitat is the interior of a human cell exploited by *Chlamydia trachomatis*, an obligate intracellular pathogen (**Fig. 2.1**). *C. trachomatis* causes sexually transmitted infections that may escape the patient's notice until they cause infertility. Intracellular chlamydias also escape the immune response—but electron microscopy catches them in the act. The scanning electron micrograph of **Figure 2.1** images an infected cell sliced open. There, the chlamydias multiply within a membrane vesicle that takes up much of the human host cell. Protected from cell defenses, chlamydias fission and form "elementary bodies" that are inactive but carry infection to a new cell. The electron micrograph shows numerous progeny chlamydias protected by the vesicle.

What does this protective compartment look like, and how does it form? In the Current Research Highlight, we saw how Grant Jensen's advanced technique of cryo-EM tomography reveals the contours of an environmental chlamydia (*Simkania* species) infecting an ameba. In tomography, a number of digital "slices" through the sample are assembled by computer to form a 3D model. The ameba provides clues to the process of human chlamydial infection. The computer-generated model shows how *Simkania*'s own envelope (inner and outer membranes) is enclosed by host cell membranes. The pathogen remodels the host intracellular membranes to protect itself from host defenses. To investigate the function of environmental and medical microbes, we present scanning electron microscopy (SEM) and cryo-EM tomography in Section 2.6.

First in Chapter 2, we present light microscopy, and the theory and practice of observing microbes, which are essential for every student and professional in the field or clinic. We then explore exciting advanced tools for research. Fluorescence and super-resolution imaging, electron microscopy, and scanning probe microscopy push ever farther the frontiers of the unseen. And we continue to invent new kinds of microscopy, such as chemical imaging microscopy, to reveal microbial metabolism at work.

2.1 Observing Microbes

Most microbes are too small to be seen; that is, they are microscopic, requiring the use of a microscope to be seen. But why can't we see microbes without magnification? The answer is surprisingly complex. In fact, our definition of "microscopic" is based on the properties of our eyes. We define what is visible and what is microscopic in terms of the human eye.

Resolution of Objects by Our Eyes

What determines the smallest object we can see? The size at which objects become visible depends on the eye's ability to resolve detail. **Resolution** is the smallest distance between two objects that allows us to see them as separate objects. The eyes of humans and other animals observe an object by focusing its image on a retina packed with light-absorbing photoreceptor cells (**Fig. 2.2** ▶). The image appears sharp, in **focus**, if the eye's lens and cornea bend all the light rays from each point of the object to converge at one point on the retina. Nearby points are then resolved as separate.

In the human eye, the finest resolution of two separate points is perceived by the fovea, the portion of the retina where the photoreceptors are packed at the highest density. The foveal photoreceptors are cone cells, which detect primary colors (red, green, or blue) and finely resolved detail. A group of cones with their linked neurons forms one unit of detection, comparable to a pixel on a computer screen. The distance between two foveal "pixels"

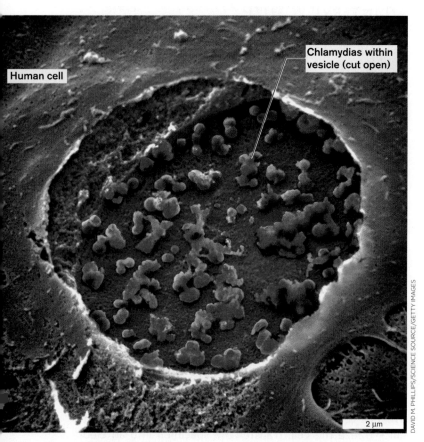

Human cell

Chlamydias within vesicle (cut open)

2 μm

DAVID M. PHILLIPS/SCIENCE SOURCE/GETTY IMAGES

FIGURE 2.1 ▪ *Chlamydia trachomatis* **multiplying within a human cell.** Colorized scanning electron microscopy (SEM).

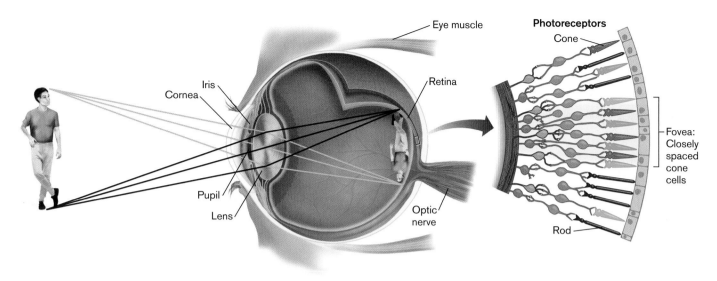

FIGURE 2.2 ■ **Defining the microscopic.** Within the human eye, the lens focuses an image on the retina. ▶

(groups of cones with neurons) limits our resolution to 100–200 micrometers (μm)—that is, one- or two-tenths of a millimeter. So, a tenth of a millimeter is about the smallest object that most of us can see (resolve distinctly) without a magnifier.

What if our eyes were formed differently? The retinas of eagles have cones packed more closely than ours, so an eagle can resolve objects eight times as small (or eight times as far away) as a human can; hence, the phrase "eagle-eyed" means "sharp-sighted." On the other hand, insect compound eyes have photoreceptors farther apart than ours, so insect eyes have poorer resolution. The best they can do is resolve objects 100-fold larger than those we can resolve. If a science-fictional giant ameba had eyes with photoreceptors 2 meters apart, it would perceive humans as "microscopic."

> **Thought Question**
>
> **2.1** As shown in **Figure 2.2**, the image passing through your cornea and lens is inverted on your retina. Why, therefore, does the world appear right side up?

Note: In this book, we use standard metric units for size:

1 millimeter (mm) = one-thousandth of a meter (m) = **10^{-3} m**

1 micrometer (μm) = one-thousandth of a millimeter = **10^{-6} m**

1 nanometer (nm) = one-thousandth of a micrometer = **10^{-9} m**

1 picometer (pm) = one-thousandth of a nanometer = **10^{-12} m**

Some authors still use the traditional unit angstrom (Å), which equals a tenth of a nanometer, or 10^{-10} meter.

Resolution Differs from Detection

Can we <u>detect</u> the presence of objects whose size we cannot resolve? Yes, we can detect their presence as a group. For example, our eyes can detect a large population of microbes, such as a spot of mold on a piece of bread (about a million cells) or a cloudy tube of bacteria in liquid culture (a million cells per milliliter) (**Fig. 2.3A**). **Detection**, the ability to determine the presence of an object, differs from resolution. When the unaided eye detects the presence of mold or bacteria, it cannot resolve distinct cells.

To resolve most kinds of microbial cells, our eyes need assistance—that is, **magnification**. Magnification reveals the shapes of individual bacteria such as the wetland phototroph *Rhodospirillum rubrum* (**Fig. 2.3B**). Magnifying an object means increasing the object's apparent dimensions. As the distance increases between points of detail, our eyes can now resolve the object's shape as a magnified image.

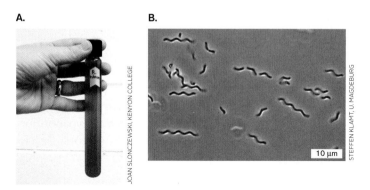

A.

B.

FIGURE 2.3 ■ **Detecting and resolving bacteria. A.** A tube of bacterial culture, *Rhodospirillum rubrum*. The presence of bacteria is detected, though individual cells are not resolved. **B.** Individual cells of *R. rubrum* are resolved by light microscopy.

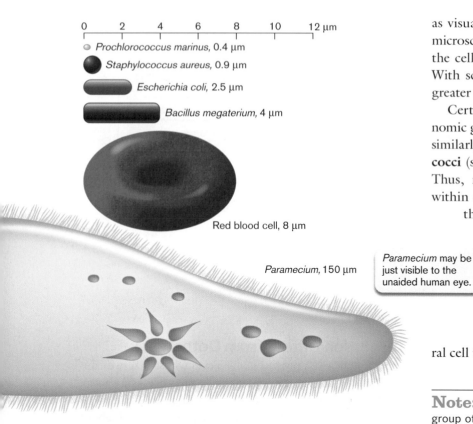

FIGURE 2.4 ■ **Relative sizes of different cells.**

as visualized by light microscopy or by scanning electron microscopy. Note that with bright-field light microscopy, the cell shape is just discernible under the highest power. With scanning electron microscopy, cell shapes appear in greater detail (higher resolution).

Certain shapes of bacteria are common to many taxonomic groups. For example, both bacteria and archaea form similarly shaped rods, or **bacilli** (singular, **bacillus**), and **cocci** (spheres; singular, **coccus**), as shown in **Figure 2.6**. Thus, rods and spherical shapes evolved independently within different taxa. In contrast, a unique bacterial shape that evolved in only one taxon is the **spirochete**, a tightly coiled spiral. Species of spirochetes cause diseases such as syphilis and Lyme disease. The spiral form of the cell is maintained by internal axial filaments and flagella, as well as an outer sheath. (For more on spirochetes, see Section 18.6.) A different, unrelated spiral form is the "spirillum" (plural, spirilla), a wide, rigid spiral cell that is similar to a rod-shaped bacillus.

Note: The genus name *Bacillus* refers to a specific taxonomic group of bacteria, but the term "bacillus" (plural, bacilli) refers to any rod-shaped bacterium or archaeon.

Microbial Size and Shape

Different kinds of microbes differ in size, over a range of several orders of magnitude, or powers of ten (**Fig. 2.4**). Eukaryotic microbes are found across the full range of cell size, from photosynthetic picoeukaryotes abundant in the oceans (0.2–2.0 µm) to giant amebas that reach nearly a centimeter (discussed in Chapter 20). Within a eukaryotic cell larger than 5 µm, a typical student's light microscope may resolve intracellular compartments such as the nucleus and vacuoles containing digested food (**Fig. 2.5**). Protists show complex shapes and appendages. For example, an ameba from an aquatic ecosystem shows a large nucleus and pseudopods to engulf prey (**Fig. 2.5A**). Pseudopods can be seen moving by the streaming of their cytoplasm. Another protist readily observed by light microscopy is *Trypanosoma brucei*, an insect-borne blood parasite that causes African sleeping sickness (**Fig. 2.5B**). In the trypanosome, we observe a nucleus and a flagellum. Eukaryotic flagella propel the cell by a whiplike action. For more on microbial eukaryotes, see Chapter 20.

The most commonly studied prokaryotes (bacteria and archaea) are smaller than 10 µm. Their overall shape can be seen, but most of their internal structures (discussed in Chapter 3) are too small to resolve by light microscopy. **Figure 2.6** shows some common cell shapes of bacteria,

Microscopy for Different Size Scales

To resolve microbes and microbial structures of different sizes requires different kinds of microscopes. **Figure 2.7** shows the different techniques used to resolve microbes and structures of various sizes. For example, a paramecium

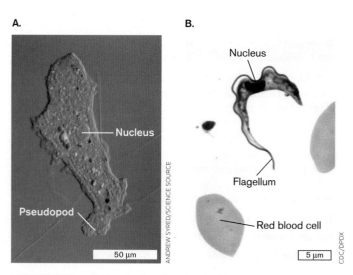

FIGURE 2.5 ■ **Eukaryotic microbial cells.** Eukaryotic microbes are large enough that details of internal and external organelles can be seen under a light microscope. **A.** *Amoeba proteus.* **B.** *Trypanosoma brucei* (cause of sleeping sickness) among blood cells.

A. Filamentous rods (bacilli).
Lactobacillus lactis, Gram-positive bacteria (LM).

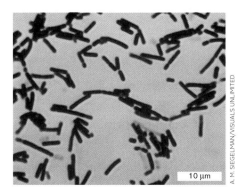

10 µm

A. M. SIEGELMAN/VISUALS UNLIMITED

C. Spirochetes.
Borrelia burgdorferi, cause of Lyme disease, among human blood cells (LM).

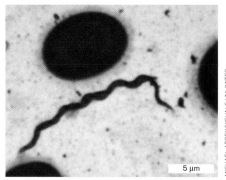

5 µm

MICHAEL ABBEY/VISUALS UNLIMITED

E. Cocci in pairs (diplococci).
Streptococcus pneumoniae, a cause of pneumonia. Methylene blue stain (LM).

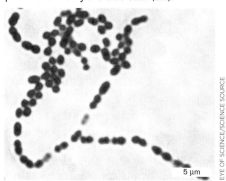

5 µm

EYE OF SCIENCE/SCIENCE SOURCE

B. Rods (bacilli).
Lactobacillus acidophilus, Gram-positive bacteria (SEM).

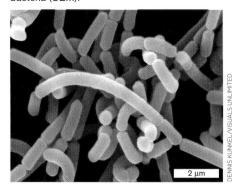

2 µm

DENNIS KUNKEL/VISUALS UNLIMITED

D. Spirochetes.
Leptospira interrogans, cause of leptospirosis in animals and humans (SEM).

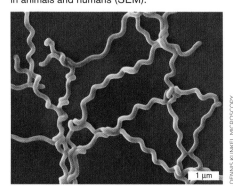

1 µm

DENNIS KUNKEL MICROSCOPY

F. Cocci in chains.
Anabaena sp., filaments of cyanobacteria. Producers for the marine food chain (SEM).

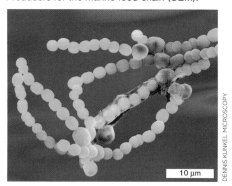

10 µm

DENNIS KUNKEL MICROSCOPY

FIGURE 2.6 ■ Common shapes of bacteria. A, C, E. The shapes of most bacterial cells can be discerned with light microscopy (LM), but their subcellular structures and surface details cannot be seen. **B, D, F.** Surface detail is revealed by scanning electron microscopy (SEM). These SEM images are colorized to enhance clarity.

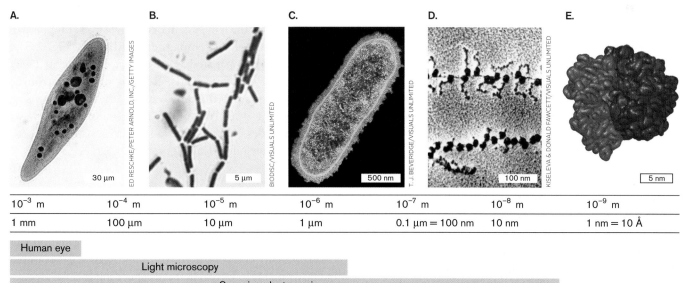

| 10⁻³ m | 10⁻⁴ m | 10⁻⁵ m | 10⁻⁶ m | 10⁻⁷ m | 10⁻⁸ m | 10⁻⁹ m |

10^{-3} m 10^{-4} m 10^{-5} m 10^{-6} m 10^{-7} m 10^{-8} m 10^{-9} m

1 mm   100 µm   10 µm   1 µm   0.1 µm = 100 nm   10 nm   1 nm = 10 Å

Human eye

Light microscopy

Scanning electron microscopy

Transmission electron microscopy

Atomic force microscopy

X-ray crystallography

FIGURE 2.7 ■ Microscopy and X-ray crystallography, range of resolution.
A. *Paramecium* (stained light microscopy, LM).
B. *Pseudomonas* (stained LM). **C.** *Escherichia coli* (transmission electron microscopy, TEM).
D. Ribosomes on messenger RNA (TEM). **E.** Ribosome model (X-ray crystallography).

can be resolved under a light microscope, but an individual ribosome (20 nm in diameter) requires electron microscopy.

■ **Light microscopy** (**LM**) resolves images of individual bacteria by their absorption of light. The specimen is commonly viewed as a dark object against a light-filled field, or background; this is called **bright-field microscopy** (seen in **Fig. 2.7A** and **B**). Advanced techniques, based on special properties of light, include fluorescence, dark-field, and phase-contrast microscopy.

■ **Electron microscopy** (**EM**) uses beams of electrons to resolve details several orders of magnitude smaller than those seen under light microscopy. In **scanning electron microscopy** (**SEM**), the electron beam is scattered from the metal-coated surface of an object, generating an appearance of 3D depth. In **transmission electron microscopy** (**TEM**) (**Fig. 2.7C** and **D**), the electron beam travels through the object, where the electrons are absorbed by an electron-dense metal stain.

■ **Atomic force microscopy** (**AFM**) uses intermolecular forces between a probe and an object to map the 3D topography of a cell, or of cell parts.

■ **X-ray crystallography** detects the interference pattern of X-rays entering the crystal lattice of a molecule. From the interference pattern, researchers build a computational model of the structure of the individual molecule, such as a protein or a nucleic acid, or even a molecular complex such as a ribosome (**Fig. 2.7E**).

Thought Question

2.2 (refer to **Fig. 2.7**) You have discovered a new kind of microbe, never observed before. What kinds of questions about this microbe might be answered by light microscopy? What questions would be better addressed by electron microscopy?

Chemical Imaging Microscopy

A new class of imaging techniques known as **chemical imaging microscopy** goes beyond size and shape to reveal the chemical composition of the microbial object. These techniques apply mass spectroscopy to a microscopic object. The object, such as a microbe, is bombarded with a beam of ions that vaporizes organic molecules from the object's surface. The molecular fragments that come off are identified through mass analysis. The microbe may be cultured with heavy isotopes of an element such as carbon or nitrogen, whose incorporation into biomass alters the mass spectra. These imaging techniques show where in the cell different nutrients are fixed into biomass. **Special Topic 2.1** presents

an example of chemical imaging of intestinal bacteria. Another example of chemical imaging, showing nitrogen-fixing cyanobacteria, is presented in **eTopic 2.1**.

To Summarize

■ **Detection** is the ability to determine the presence of an object.

■ **Resolution** is the smallest distance by which two objects can be separated and still be distinguished as separate.

■ **Magnification** is an increase in the apparent size of an image.

■ **Eukaryotic microbes may be large enough to resolve subcellular structures** under a light microscope. Other eukaryotic cells are as small as bacteria.

■ **Bacteria and archaea are generally too small for subcellular resolution by a light microscope.** Their shapes include characteristic forms such as rods and cocci.

■ **Different kinds of microscopy** resolve cells and subcellular structures of different sizes. Chemical imaging microscopy reveals the chemical composition of a cell.

2.2 Optics and Properties of Light

How do light rays magnify an image? Light microscopy directly extends the lens system of our own eyes. Light is part of the spectrum of **electromagnetic radiation** (**Fig. 2.8**), a form of energy propagated as waves that are associated with electrical and magnetic fields. Regions of the electromagnetic spectrum are defined by wavelength, which for visible light is about 400–750 nm. Radiation of longer wavelengths includes infrared and radio waves, whereas shorter wavelengths include ultraviolet rays and X-rays.

Light Carries Information

All forms of electromagnetic radiation carry information from the objects they interact with. The information carried by radiation can be used to detect objects; for example, radar (using radio waves) detects a speeding car. All electromagnetic radiation travels through a vacuum at the same speed: about 3×10^8 meters per second (m/s), the speed of light. The speed of light (c) is equal to the wavelength (λ) of the radiation multiplied by its frequency (v), the number of wave cycles per unit time:

$$c = \lambda v$$

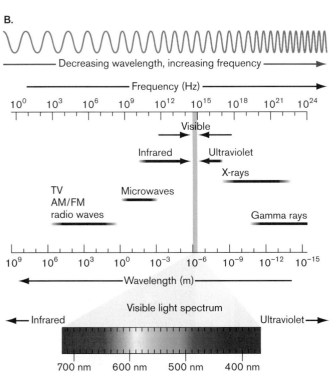

FIGURE 2.8 ■ **Electromagnetic energy. A.** Electromagnetic radiation is composed of electrical and magnetic waves perpendicular to each other. **B.** The electromagnetic spectrum includes the visible range of light.

Because c is constant, the longer the wavelength λ is, the lower the frequency v is. Frequency is usually measured in hertz (Hz), reciprocal seconds (λ/s).

The wavelength λ limits the size of objects that can be resolved as separate from neighboring objects. Resolution requires:

■ **Contrast between the object and its surroundings.** **Contrast** is the difference in light and dark. If an object and its surroundings absorb or reflect radiation equally, then the object will be undetectable. It is hard to observe a cell of transparent cytoplasm floating in water, because the aqueous cytoplasm and the extracellular water tend to transmit light similarly, producing little contrast.

■ **Wavelength smaller than the object.** For an object to be resolved, the wavelength of the radiation must be equal to or smaller than the size of the object. If the wavelength of the radiation is larger than the object, then most of the wave's energy will simply pass through the object, like an ocean wave passing around a dock post. Thus, radar, with a wavelength of 1–100 centimeters (cm), cannot resolve microbes, though it easily resolves cars and people.

■ **Magnification.** The human retina absorbs radiation within a range of wavelengths, 400–750 nm (0.40–0.75 μm), which we define as visible light. But the smallest distance our retina can resolve is 150 μm, about 300 times the wavelength of light. Thus, we are unable to access all of the information contained in the light that enters our eyes. To use more of the information carried within the light, we must spread the light rays apart far enough for our retina to perceive the resolved image.

Light Interacts with an Object

The physical behavior of light resembles in some ways a beam of particles and in other ways a waveform. The particles of light are called "photons." Each photon has an associated wavelength that determines how the photon will interact with a given object. The combined properties of particle and wave enable light to interact with an object in several different ways:

■ **Absorption** means that the absorbing object gains the photon's energy (**Fig. 2.9A** ▶). The energy is converted

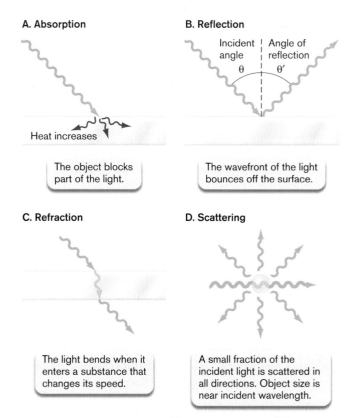

FIGURE 2.9 ■ **Interaction of light with matter.** ▶

SPECIAL TOPIC 2.1 | Catch Your Bacteria Snacking

Light rays and electron beams offer glimpses into the astonishing world within a cell. But what are all those intriguing knobs and tubules made of? Chemical imaging microscopy shows more than the outline of bacteria; it shows what they've been eating.

A high-resolution method for chemical imaging is called nanoscale secondary ion mass spectrometry (NanoSIMS).

The imaging method starts with an ionizing probe, a source of energy that breaks up the large organic molecules of a sample (**Fig. 1**). The molecular fragments, called "secondary ions," fly off from the source and are captured by a mass spectrometer. This instrument measures fragment masses of the secondary ions, generating a mass spectrum. Mass spectra are taken from thousands of locations, scanned across a bacterial cell.

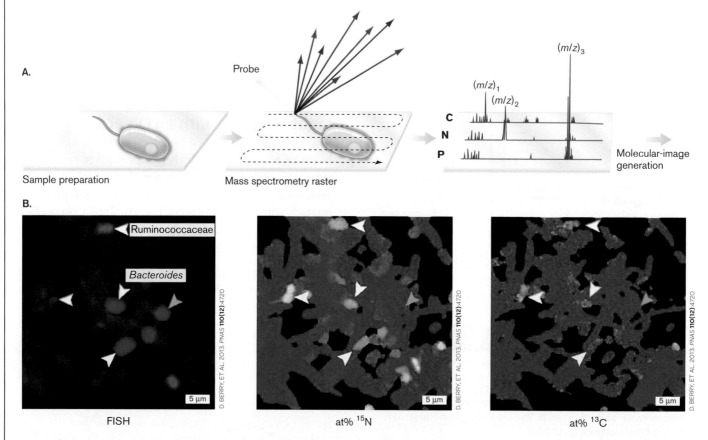

FIGURE 1 ■ Imaging mass spectrometry. Mass spectra are obtained from thousands of locations throughout the sample surface. **A.** Molecular fragments are selected for analysis of mass-to-charge ratio (m/z). Selected isotopes may label specific atoms—for example, C, N, or P. The relative intensities of individual compounds are visualized using false-color gradients. **B.** Fluorescence in situ hybridization (FISH, left) and NanoSIMS (middle and right) of bacteria feeding on secretions from the intestinal epithelium, showing atomic percentages (at%) of intracellular nitrogen (^{15}N) and carbon (^{13}C).

to a different form, usually heat. (That is why a live specimen eventually "cooks" on the slide if observed for too long.) When a microbial specimen absorbs light, it can be observed as a dark spot against a bright field, as in bright-field microscopy. Some molecules that absorb light of a specific wavelength reemit energy as light with a longer wavelength; this is called **fluorescence**. Fluorescence microscopy is discussed in Section 2.4.

■ **Reflection** means that the wavefront redirects from the surface of an object at an angle equal to its incident angle

(**Fig. 2.9B**). Reflection of light waves is analogous to the reflection of water waves. Reflection from a silvered mirror or a glass surface is used in the optics of microscopy.

■ **Refraction** means that light bends as it enters a substance that slows its speed (**Fig. 2.9C**). Such a substance is said to be refractive and, by definition, has a higher **refractive index** than air. Refraction is the key property that enables a lens to magnify an image.

What information does NanoSIMS yield? Different kinds of isotope labeling can focus on specific chemical questions. For example, the cell's uptake of nitrogen-rich protein can be detected by incorporation of the heavy isotope ^{15}N. The increased weight of ^{15}N compared to the normally predominant ^{14}N (as indicated by the mass ratio $^{15}N/^{14}N$) is detected quantitatively in the molecular fragment masses. Alternatively, we can detect isotopes of carbon ($^{13}C/^{12}C$).

At the University of Vienna, Michael Wagner and colleagues (**Fig. 2**) used NanoSIMS to address the question: Which of our intestinal bacteria feed on our own secretions, instead of the food we consume? Our intestinal epithelium secretes a mucus layer, which helps prevent colonization by pathogens. But the mucus also "farms" bacteria, whose feeding helps moderate the buildup of secretions.

Wagner's group used different types of bacteria to colonize gnotobiotic (germ-free) mice. The mice were injected intravenously with a nutrient (the amino acid threonine) containing the heavy-atom isotopes ^{13}C and ^{15}N. The threonine is readily incorporated into mouse proteins and secretions, but the bacteria inoculated into the mouse's intestinal lumen have no direct access to the threonine, only to mouse secretions containing it. Thus, NanoSIMS of the intestinal bacteria reveals which ones are snacking on mucus.

In **Figure 1B**, various bacteria on the mouse epithelium are identified by fluorescence in situ hybridization (FISH, discussed in Chapter 21). The FISH technique uses short species-specific nucleic acid probes to hybridize to the ribosomal RNA of different bacteria. In the FISH panel shown, the fluorescent probes identify single cells of *Bacteroides acidifaciens* (blue) and Ruminococcaceae OTU_5807 (red). The parallel panels show the atomic percent (at%) of the given isotopes ^{15}N and ^{13}C, imaged on a color scale quantifying the amount of isotope enrichment. The panels show that most of the bacterial cells corresponding to *B. acidifaciens* and Ruminococcaceae acquire nitrogen from the host (gold or yellow color, white arrowhead). The third panel shows that some but not all of the bacteria show carbon uptake. Overall, out of the mixed gut microbes, the *B. acidifaciens*

FREDERIK SCHULZ, UNIVERSITÄT WIEN

FIGURE 2 ■ David Berry, Alexander Loy, and Michael Wagner use NanoSIMS to image bacteria foraging on the gut epithelium.

and Ruminococcaceae strains were shown to be the main snackers on host mucus. These observations will help medical researchers characterize the composition and dynamics of a healthy gut microbiome.

RESEARCH QUESTION

The number of *Bacteroides* foraging on host secretions changes when other species are introduced. Why might this happen? Can you propose an experiment using NanoSIMS to test your hypothesis?

Berry, D., B. Stecher, A. Schintlmeister, J. Reichert, S. Brugiroux, et al. 2013. Host-compound foraging by intestinal microbiota revealed by single-cell stable isotope probing. *Proceedings of the National Academy of Sciences USA* **110**:4720.

■ **Scattering** means that a portion of the wavefront is converted to a spherical wave originating from the object (**Fig. 2.9D**). If a large number of particles simultaneously scatter light, we see a haze—for example, the haze of bacteria suspended in a culture tube. Special optical arrangements (such as dark-field microscopy, discussed in Section 2.5) can use scattered light to detect (but not resolve) microbial shapes smaller than the wavelength of light.

Magnification by a Lens

Magnification requires the bending of light rays, as in refraction. As a wavefront of light enters a refractive material, such as glass, the region of the wave that first reaches the material is slowed, while the rest of the wave continues at its original speed until it also passes into the refractive material (**Fig. 2.10A** ▶). As the entire wavefront continues through the refractive material, its path continues, bent at an angle from its original direction.

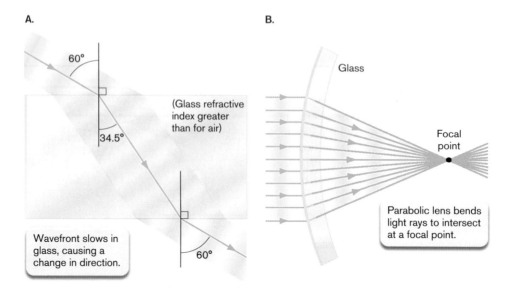

FIGURE 2.10 ■ Refraction of light waves. **A.** Wavefronts of light shift direction as they enter a substance of higher refractive index, such as glass. **B.** Glass with parabolic curvature (a lens) bends light rays to intersect at a focal point. ▶

Thought Question

2.3 Explain what happens to the refracted light wave as it emerges from a piece of glass of even thickness. How do its new speed and direction compare with its original (incident) speed and direction?

How does refraction accomplish magnification? Refraction magnifies an image when light passes through a refractive material shaped so as to spread its rays. One shape that spreads light rays is a parabolic curve. When light rays enter a **lens** of refractive material with a parabolic surface (**Fig. 2.10B**), parallel rays each bend at an angle such that all of the rays meet at a certain point, called the **focal point**. From the focal point behind the lens, the light rays continue, spreading out with an expanding wavefront. This expansion magnifies the image carried by the wave. The distance from the lens to the focal point (called the focal distance) is determined by the degree of curvature of the lens, and by the refractive index of its material.

In **Figure 2.11** ▶, the object under observation is placed near the focal point (F) in front of a lens. The light rays trace a path opposite to that of **Figure 2.10B**. The rays expanding from point F are bent by the lens into a

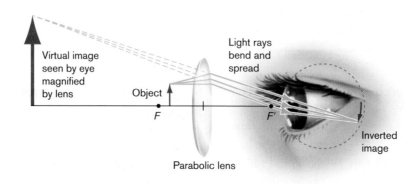

FIGURE 2.11 ■ A lens magnifies an image. The object is placed near the focal point (F) in front of the lens. The lens bends and spreads the light rays, inverting the magnified image. ▶

nearly parallel path entering the eye. The eye perceives the expanded light rays as a virtual image—that is, an image that appears to represent a much larger object farther away. The expansion of light rays, or magnification, increases the distances between points of the image. The details of the magnified image become larger than the spacing of photoreceptor units in the retina. Thus, our eye can perceive details that the unaided eye cannot see.

Resolution of Detail

What limits the effect of magnification? The spreading of light rays does not in itself increase resolution. For example, an image composed of dots does not gain detail when enlarged on a photocopier, nor does an image composed of pixels gain detail when enlarged on a computer screen. In these cases, magnification fails to show details because the individual details of the image expand in proportion to the expansion of the overall image. Magnification without increasing detail is called **empty magnification**.

The resolution of detail in microscopy is limited by the wave nature of light. In theory, a perfect lens that focuses all of the light from an object should form a perfect image as its rays converge through the focal point. But light rays actually form wavefronts of infinite extent. Since the width of the lens is finite, only part of the wavefront enters, causing **interference**. The converging edges of the wave interfere with each other to form alternating regions of light and dark (**Fig. 2.12** ▶). Thus, a point source of light (such as a point of detail in a specimen) forms an image of a bright central peak surrounded by interference rings of light and dark. Even a well-focused bright object appears as a bright disk surrounded by faint rings.

Suppose an object consists of a collection of point sources of light. Each point source generates a central peak of intensity. The width of this central peak will define the resolution, or separation distance, between any two points of the object (**Fig. 2.12**). This resolution determines the degree of detail that can be observed. In practice, any object, such as a stained microbe against a bright field, can be considered a large collection of points of light that act as partly resolved peaks of intensity.

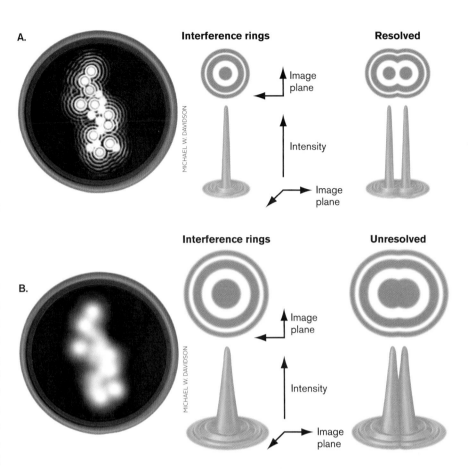

FIGURE 2.12 ■ **Interference of light waves at the focal point generates concentric rings surrounding the peak intensity.** **A.** Broad wavefronts generate narrow interference rings with peaks well resolved. **B.** Narrow wavefronts generate wide interference rings that are unresolved. ▶

What factors limit resolution of an image? The wavelength of light limits the sharpness of the peak intensity of a point of detail. The finite width of the wavefront captured by the lens leads to interference and widens the peak intensity. Thus, bright-field light microscopy resolves only details that are greater than half the wavelength of light, about 200 nm (0.2 μm). Nevertheless, in advanced optical methods such as fluorescence microscopy (see Section 2.4), computation can extract positional detail from light rays. Such methods, called **super-resolution imaging**, enable us to track cellular molecules at a precision of 20–40 nm (discussed in Section 2.4).

To Summarize

■ **Electromagnetic radiation** interacts with an object and acquires information we can use to detect the object. **Contrast** between object and background makes it possible to detect the object and resolve its component parts.

- **The wavelength of the radiation** must be equal to or smaller than the size of the object for a microscope to resolve the object's shape.

- **Absorption** means that the energy from light (or other electromagnetic radiation) is acquired by the object. **Reflection** means that the wavefront bounces off the surface of a particle at an angle equal to its incident angle. **Scattering** means that a wavefront interacts with an object of smaller dimension than the wavelength. Light scattering enables the detection of objects whose detail cannot be resolved.

- **Refraction** is the bending of light as it enters a substance that slows its speed. Refraction through a curved lens **magnifies** an image, enlarging its details beyond the spacing between our eye's photoreceptors.

- **Interference** between wavefronts converts a point source of light to a peak of intensity surrounded by rings. The width of the peak limits the **resolution** of details of an image.

2.3 Bright-Field Microscopy

The most common kind of light microscopy is called **bright-field microscopy**, in which an object such as a bacterial cell is perceived as a dark silhouette blocking the passage of light (for examples, see **Fig. 2.6A**, **C**, and **E**). Details of the object are defined by the points of light surrounding its edge. Here we explain how a typical student's microscope works, and how to use it to image microbes.

Magnification

How do the optics of a bright-field microscope maximize the observation of detail? We consider the following factors:

- **Wavelength and resolution.** Our eyes can resolve a distance as small as 100–200 μm, while the resolution limit from the wavelength of light is 200 nm (0.2 μm)—that is, 500- to 1,000-fold smaller. Thus, the greatest magnification that can improve our perception of detail is about 1,000×. Any greater magnification expands the image size, but the width of the image peaks expands without resolution (see **Fig. 2.12**). As we noted in the previous section, this expansion without increasing resolution is called empty magnification.

- **Light and contrast.** For any given lens system, a balanced amount of light yields the highest contrast between the dark specimen and the light background. High contrast is needed to perceive the full resolution at a given magnification.

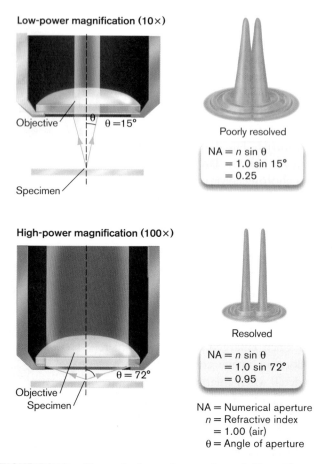

Low-power magnification (10×)

Objective θ θ = 15°

Specimen

Poorly resolved

$$NA = n \sin \theta$$
$$= 1.0 \sin 15°$$
$$= 0.25$$

High-power magnification (100×)

θ = 72°

Objective
Specimen

Resolved

$$NA = n \sin \theta$$
$$= 1.0 \sin 72°$$
$$= 0.95$$

NA = Numerical aperture
n = Refractive index
  = 1.00 (air)
θ = Angle of aperture

FIGURE 2.13 ■ Numerical aperture and resolution. The numerical aperture (NA) equals the refractive index (n) of the medium containing the light cone, multiplied by the sine of the angle of the light cone (θ). Higher NA allows greater resolution.

- **Lens quality.** All lenses contain inherent **aberrations** that detract from perfect curvature. Optical properties limit the perfection of a single lens, but manufacturers construct microscopes with a series of lenses that multiply each other's magnification and correct for aberrations.

Let's first consider magnification of an image by a single lens. **Figure 2.13** shows an **objective lens**, a lens situated directly above an object or specimen that we wish to observe at high resolution. How can we maximize the resolution of details?

An object at the focal point of a lens sits at the tip of a cone of light formed by rays from the lens converging at the object. The angle of the light cone is determined by the curvature and refractive index of the lens. The lens fills an aperture, or hole, for the passage of light; and for a given lens the light cone is defined by an angle θ (theta) projecting from the midline, known as the **angle of aperture**. As θ increases and the horizontal width of the light cone ($\sin \theta$) increases, a wider cone of light passes through the specimen. The wider the cone of light rays, the less the interference between wavefronts—and the narrower the peak intensities in the image. Thus, a wider light cone allows us to resolve smaller details. The greater the angle of aperture of the lens (sin θ), the better the resolution.

Resolution also depends on the refractive index of the medium that contains the light cone, which is usually air. The refractive index (n) is the ratio of the speed of light in a vacuum to its speed in another medium. For air, n is extremely close to 1. For water, n = 1.33; for lens material, n ranges from 1.4 to 1.6. As light passing through air or water enters a lens of higher refractive index, the light bends, at angles up to a maximum (θ). The product of the refractive index (n) of the medium multiplied by sin θ is the **numerical aperture** (NA):

$$NA = n \sin \theta$$

In **Figure 2.13** we see the calculation of NA for an objective lens of magnification 10×, and for a lens of magnification 100×. As NA increases, the peak intensities of an image narrow and the distance between two objects that can be resolved decreases. The minimum resolution distance R varies inversely with NA:

$$R = \frac{\lambda}{2NA}$$

where λ represents the wavelength of incident light. Notice that this equation limits resolution to approximately half the wavelength of light (λ/2).

As the lens strength increases and the light cone widens, the lens must come nearer the object. Defects in lens curvature become more of a problem, and focusing becomes more challenging. As θ becomes very wide, too much of the light from the object is lost from refraction at the glass-to-air interface. To collect and focus more light, we need to increase the refractive index of the medium

between the object and the objective lens (air, as shown so far). For the highest-power objective lens, generally 100×, a zone of constant refractive index is maintained by replacing air with **immersion oil** between the object and the lens. Immersion oil has a refractive index comparable to that of the lens (n = 1.5) (**Fig. 2.14**). Immersion oil minimizes the loss of light rays by refraction and makes it possible to reach 100× magnification with minimal distortion. In a two-lens system, 100× objective is multiplied by 10× ocular magnification to yield 1,000× total magnification.

Thought Question

2.4 (refer to **Fig. 2.13**) For a single lens, what angle θ might offer magnification even greater than 100X? What practical problem would you have in designing such a lens to generate this light cone?

The Compound Microscope

A **compound microscope** is a system of multiple lenses designed to correct or compensate for lens **aberrations** (deviations from perfect curvature). Why do we use a compound microscope instead of a single perfect lens? The manufacture of high-power lenses is difficult because as the glass curvature increases, the effects of aberration increase faster than the magnification. Instead of one thick lens, a series of lower-power lenses can multiply their magnification with minimal aberration. **Figure 2.15** ▶ shows a typical arrangement of a compound microscope: the light source is placed at the bottom, shining upward through a series of lenses, including the condenser, objective, and ocular lenses.

Between the light source and the condenser sits a diaphragm, a device to cut the diameter of the light column. Lower-power lenses require lower light levels because the excess light makes it impossible to observe the darkening effect of specimen absorbance. This difference between the dark (absorbing) specimen and the bright (transparent) field is called **contrast**. Higher-power lenses spread the light rays farther and thus require an open diaphragm to collect sufficient light for contrast.

Above the diaphragm, the **condenser** consists of one or more lenses that collect a beam of rays from the light source onto a small area of the slide, where light may be absorbed by the object or specimen. Condenser lenses increase light available for contrast but do <u>not</u> participate in magnification. The **objective lens** is the first to form a magnified image (I) of the object (**Fig. 2.15A**). As the image forms, each light ray traces a path toward a position opposite its point of origin; thus, the image is

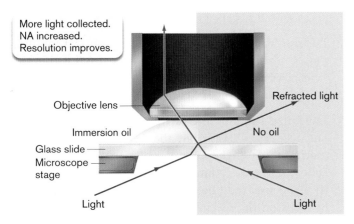

FIGURE 2.14 ■ Use of immersion oil in microscopy. Immersion oil with a refractive index comparable to that of glass (n = 1.5) prevents light rays from bending away from the objective lens.

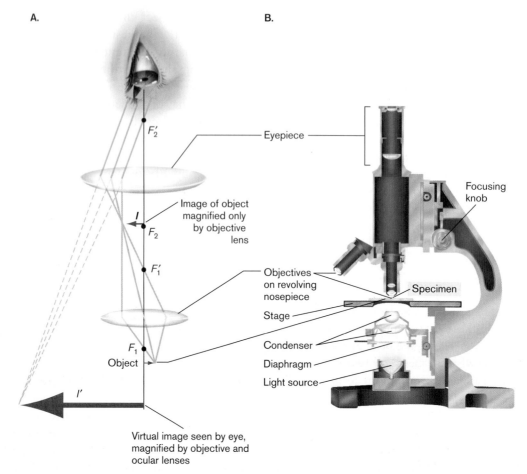

A.

F'_2

Eyepiece

I

F_2

Image of object magnified only by objective lens

F'_1

Objectives on revolving nosepiece

F_1

Object

Stage

I'

Virtual image seen by eye, magnified by objective and ocular lenses

B.

Focusing knob

Specimen

Condenser

Diaphragm

Light source

FIGURE 2.15 ▪ **Anatomy of a compound microscope.** **A.** Light path through the microscope. **B.** Cutaway view. ▶

mirror-reversed. Keep this mirror reversal in mind when exploring a field of cells.

The first image of the object (I) is then amplified by a secondary magnification step through the **ocular lens** within the eyepiece. The final image (I') is comparable to the virtual image of **Figure 2.11**, but I' includes the total magnification of the object by objective times ocular lenses. The magnification factor of the ocular lens is multiplied by the magnification factor of the objective lens to generate the **total magnification** (power). For example, a 10× ocular multiplied by a 40× objective generates 400× total magnification.

The nosepiece of a compound microscope typically holds three or four objective lenses of different magnifying power, such as 4×, 10×, 40×, and 100× (requiring immersion oil). These lenses are arranged so as to rotate in turn into the optical column. In a high-quality instrument the lenses are set at different heights from the slide so as to be **parfocal**. In a parfocal system, when an object is focused with one lens, it remains in focus, or nearly so, when another lens is rotated to replace the first.

Note: Objective lenses can be obtained in several different grades of quality, manufactured with different kinds of correction for aberrations. Lenses should feature at minimum the following corrections: "plan" correction for field curvature, to generate a field that appears flat; and "apochromat" correction for spherical and chromatic aberrations.

Observing a specimen under a compound microscope requires several steps:

▪ **Position the specimen centrally in the optical column.** Only a small area of a slide can be visualized within the field of view of a given lens. The higher the magnification, the smaller the field of view that will be seen.

▪ **Optimize the amount of light.** At lower power, too much light will wash out the light absorption of the specimen. At higher power, more light needs to be collected, lest everything appear dark. To optimize light, the condenser must be set at the correct vertical position to focus on the specimen, and the diaphragm must be adjusted to transmit the amount of light that produces the best contrast.

■ **Focus the objective lens.** The focusing knob permits adjustment of the focal distance between the objective lens and the specimen on the slide so as to bring the specimen into the focal plane, the plane that contains the focal points for light entering the lens from all directions. Typically, we focus first using a low-power objective, which generates a greater **depth of field**—that is, a range of planes in which the object appears in or near focus. After focusing under low power, we can rotate a higher-power lens into view and then fine-tune the adjustment.

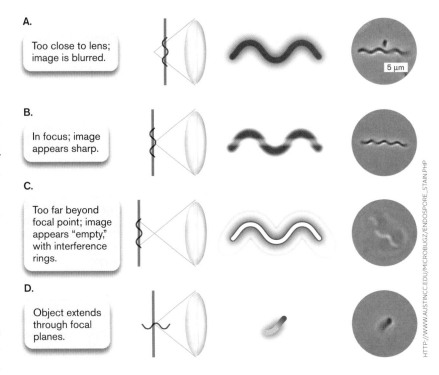

FIGURE 2.16 ■ **Bacteria observed at different levels of focus.**

Preparing a Specimen for Microscopy

A simple way to observe microbes is to place them in a drop of water on a slide with a coverslip. This is called a **wet mount** preparation. The advantage of the wet mount is that the organism is viewed in as natural a state as possible, without artifacts resulting from chemical treatment; and we can observe live behavior such as swimming (**Figs. 2.3B** and **2.16**). The disadvantage of the wet mount is that most living cells are transparent and therefore show little contrast with the external medium. With limited contrast, the cells can barely be distinguished from background, and both detection and resolution are minimal.

Another disadvantage of the wet mount is that the sample rapidly converts absorbed light to heat, thus tending to overheat and dry out. To avoid overheating, researchers use a temperature-controlled flow cell, in which fresh medium passes through the specimen (**Fig. 2.17**). The microbe to be observed must adhere to a specially coated slide within the flow cell. The adherent cells may grow and multiply as a biofilm, nourished continually by fresh medium.

Focusing the Object

An object appears in focus (that is, it is situated within the focal plane of the lens) when its edge appears sharp and distinct from the background. The shape of the dark object is actually defined by the points of light surrounding its edge. At higher power, as we reach the resolution limit, these points of light are only partly resolved. The partial resolution of these points of light generates interference effects, such as extra rings of light surrounding an object.

In **Figure 2.16** we observe *Rhodospirillum rubrum*, a photosynthetic bacterium that enriches wetlands. As *R. rubrum* cells swim in and out of the focal plane, their appearance changes through optical effects. When a bacterium swims out of the focal plane too close to the lens, resolution declines and the image blurs (**Fig. 2.16A**). When the bacterium swims within the focal plane, its image

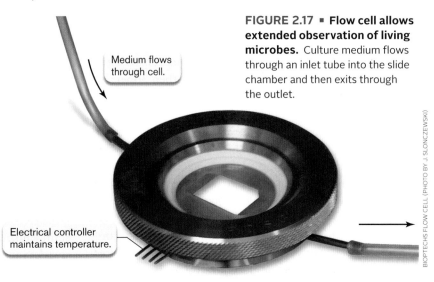

FIGURE 2.17 ■ **Flow cell allows extended observation of living microbes.** Culture medium flows through an inlet tube into the slide chamber and then exits through the outlet.

appears sharp, with a bright line along its edge. A helical cell such as *R. rubrum* shows well-focused segments alternating with hazier segments that are too close to the lens (**Fig. 2.16B**). When the cell swims too far past the focal plane, the bright interference lines collapse into the object's silhouette, which now appears bright or "hollow," or surrounded by rings (**Fig. 2.16C**). In fact, the bacterium is not hollow at all; only its image has changed.

When the cell extends across several focal planes, different portions appear out of focus (either too near or too far from the lens). In addition, when the end of a cell points toward the observer, light travels through the length of the cell before reaching the observer, so the cell absorbs more light and appears dark (**Fig. 2.16D**). Motile bacteria swimming in and out of the focal plane present a challenge even to experienced microscopists. The higher the magnification, the narrower the depth of the focal plane; thus, observing swimming organisms requires a trade-off between magnification and depth of field.

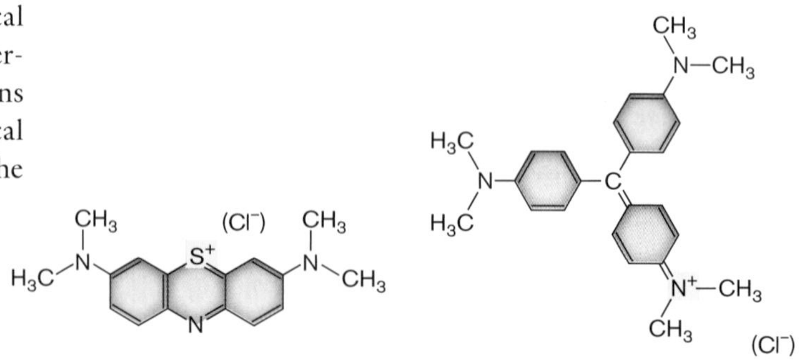

Thought Question

2.5 Under starvation, a bacterium such as *Bacillus thuringiensis*, the biological insecticide, packages its cytoplasm into a spore, leaving behind an empty cell wall. Suppose, under a microscope, you observe what appears to be a hollow cell. How can you tell whether the cell is indeed hollow or is simply out of focus?

Fixation and Staining Improve Resolution and Contrast

Detection and resolution of cells under a microscope are enhanced by **fixation** and **staining**, procedures that usually kill the cell. Fixation is a process by which cells are made to adhere to a slide in a fixed position. We can fix cells with methanol or by heat treatment to denature the cell's proteins, whose exposed side chains then adhere to the glass. A stain absorbs much of the incident light, usually over a wavelength range that results in a distinctive color. The use of chemical stains was developed in the nineteenth century, when German chemists used organic synthesis to invent new coloring agents for clothing. Clothing was made of natural fibers such as cotton or wool, so a substance that dyed clothing would be likely to react with biological specimens.

How do stains work? Most stain molecules contain conjugated double bonds or aromatic rings that absorb visible light (**Fig. 2.18**) and one or more positive charges, such as

FIGURE 2.18 ▪ **Chemical structure of stains.** Methylene blue and crystal violet are cationic (positively charged) dyes. Chloride (Cl^-) is the counter-ion.

phosphoryl groups on membrane phospholipids (discussed in Section 3.2). The positively charged groups react with negatively charged groups in the bacterial cell envelope. Different stains vary with respect to the strength of their binding and the degree of binding to different parts of a cell.

Simple stains. A **simple stain** adds dark color specifically to cells, but not to the external medium or surrounding tissue (in the case of pathological samples). The most commonly used simple stain is methylene blue, originally used by Robert Koch to stain bacteria (see **Figs. 2.6E** and **2.7B**). A typical procedure for fixation and staining is shown in **Figure 2.19**. First we fix a drop of culture on a slide by treating with methanol or by heating on a slide warmer. Either treatment denatures cell proteins, exposing side chains that bind to the glass. We then flood the slide with methylene blue solution. The positively charged molecule binds to the negatively charged cell envelope of fixed bacteria. After excess stain is washed off and the slide has been dried, we observe it under high-power magnification using immersion oil.

Differential stains. A **differential stain** stains one kind of cell but not another. The most famous differential stain is the **Gram stain**, devised in 1884 by the Danish physician Hans Christian Gram (1853–1938). Gram first used the Gram stain to distinguish pneumococcal (*Streptococcus pneumoniae*) bacteria from human lung tissue. In **Figure 2.20A**, Gram-stained *S. pneumoniae* bacteria appear dark purple among human white blood cells. Other species of bacteria, such as *Proteus mirabilis* (a cause of urinary infections), fail to retain the purple stain (**Fig. 2.20B**). Different bacterial species are classified as Gram-positive or Gram-negative, depending on whether they retain the stain.

A.

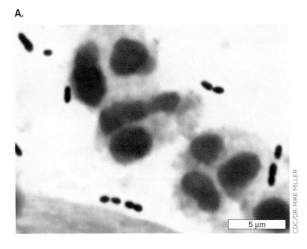

CDC/DR. MIKE MILLER

B.

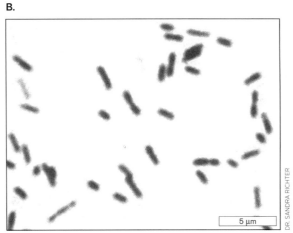

DR. SANDRA RICHTER

FIGURE 2.20 ■ **Gram staining of bacteria. A.** Gram stain of a sputum specimen from a patient with pneumonia, containing Gram-positive *Streptococcus pneumoniae* (purple diplococci) among white blood cells in pus. The white blood cell nuclei stain pink (counterstain). **B.** Gram-negative *Proteus mirabilis* (pink rods).

In the Gram stain procedure (**Fig. 2.21**), a dye such as crystal violet binds to the bacteria; it also binds to the surface of human cells, but less strongly. After the excess stain is washed off, a **mordant**, or binding agent, is applied. The mordant used is iodine solution, which contains iodide ions (I^-). The iodide complexes with the positively charged crystal violet molecules trapped inside the cells. The crystal violet–iodide complex is now held more strongly within the cell wall. The thicker the cell wall, the more crystal violet–iodide molecules are held.

Next, a decolorizer, ethanol, is added for a precise time interval (typically 20 seconds). The decolorizer removes loosely bound crystal violet–iodide, but Gram-positive cells retain the stain tightly. The **Gram-positive** cells that retain the stain appear dark purple, while the **Gram-negative** cells are colorless. Timing the decolorizer step is critical because if it lasts too long, the Gram-positive cells, too, will release their crystal violet stain. In the final step, a **counterstain**,

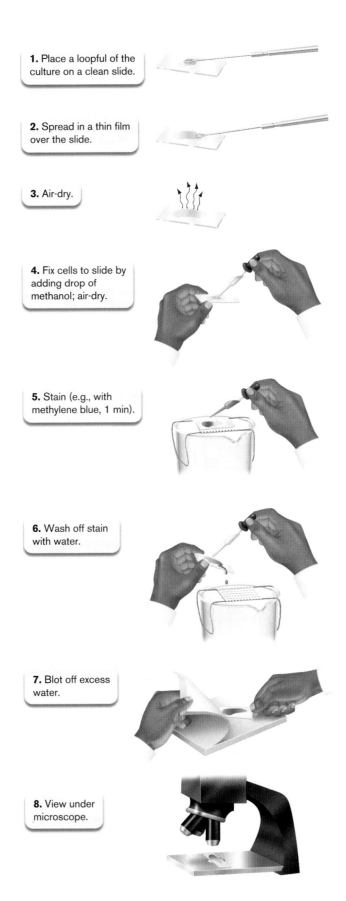

1. Place a loopful of the culture on a clean slide.

2. Spread in a thin film over the slide.

3. Air-dry.

4. Fix cells to slide by adding drop of methanol; air-dry.

5. Stain (e.g., with methylene blue, 1 min).

6. Wash off stain with water.

7. Blot off excess water.

8. View under microscope.

FIGURE 2.19 ■ **Procedure for simple staining with methylene blue.**

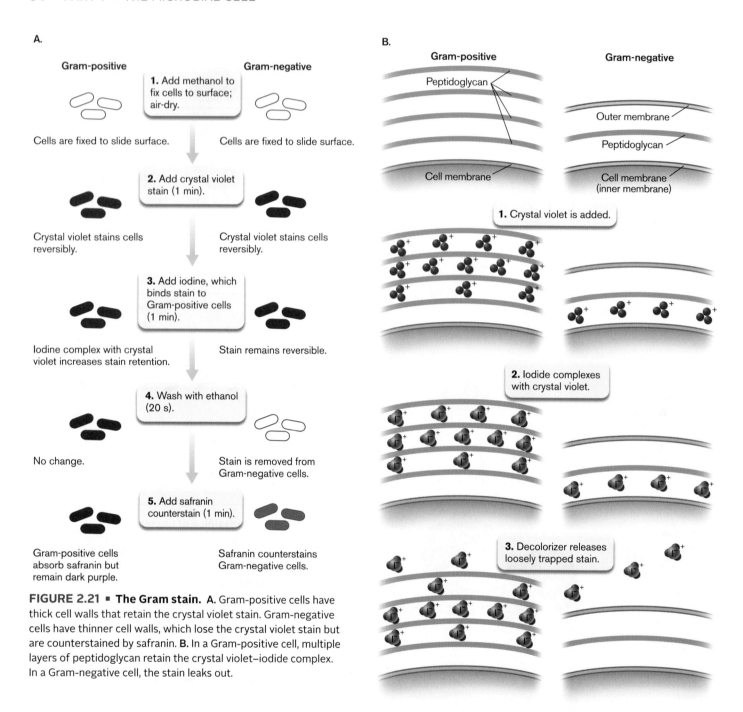

FIGURE 2.21 ■ The Gram stain. A. Gram-positive cells have thick cell walls that retain the crystal violet stain. Gram-negative cells have thinner cell walls, which lose the crystal violet stain but are counterstained by safranin. **B.** In a Gram-positive cell, multiple layers of peptidoglycan retain the crystal violet–iodide complex. In a Gram-negative cell, the stain leaks out.

safranin, is applied. This process allows the visualization of Gram-negative material, which is stained pale pink by the safranin. Gram-positive cells also retain safranin; thus, if the cells are decolorized too long, both Gram-positive and Gram-negative cells will appear pink due to safranin.

How does the Gram stain distinguish different cell types? Most Gram-negative species of bacteria possess a cell wall that is thinner and more porous than that of Gram-positive species (discussed in Chapter 3). A Gram-negative cell wall has only one to three layers of peptidoglycan (sugar chains cross-linked by peptides), whereas a Gram-positive cell has five or more layers. The multiple

layers of peptidoglycan retain enough stain complex that the cell appears purple.

The Gram stain became a key tool for identifying pathogens in the clinical laboratory. As we'll see in Chapter 3, the Gram stain effectively distinguishes Proteobacteria (a clade of bacteria with a thin cell wall and an outer membrane) from Firmicutes (bacteria with a thick cell wall and no outer membrane.) Proteobacteria include *Escherichia coli* and many related intestinal bacteria. Another important Gram-negative group is the Bacteroidetes, whose species work with Proteobacteria to digest our food (discussed in Chapters 13 and 21). Our colon also contains Gram-positive

A.

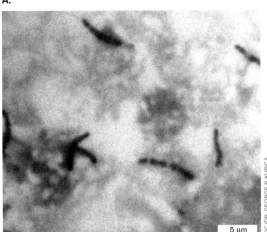

B.

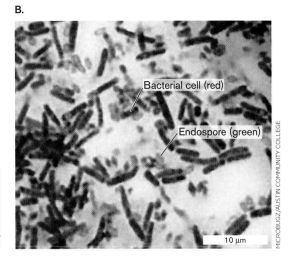

Bacterial cell (red)

Endospore (green)

5 µm

CDC/DR. GEORGE P. KUBICA

10 µm

MICROBUGZ/AUSTIN COMMUNITY COLLEGE

FIGURE 2.22 ■
Differential stains.
A. Acid-fast stain of *Mycobacterium tuberculosis* (red) in sputum.
B. Endospore stain of *Clostridium tetani*.

Firmicutes such as species of *Clostridium* and *Enterococcus*. The digestive contributions of Firmicutes are less clear. Most intestinal bacteria are mutualists; that is, they share positive contributions with their host (the human body). However, the gut community may be invaded by deadly pathogens, such as *Escherichia coli* O157:H7, *Enterococcus faecalis*, and *Clostridium difficile*.

Still other groups of bacteria and archaea have different kinds of cell walls that may stain Gram-positive, Gram-negative, or variable (discussed in Chapters 18 and 19). Moreover, even Firmicutes such as *Bacillus* species show variable stain results depending on their growth state and environmental conditions.

Other differential stains reveal components specific to certain classes of bacteria:

- **Acid-fast stain** (Ziehl-Neelsen). Carbolfuchsin specifically stains mycolic acids of *Mycobacterium tuberculosis* and *M. leprae*, the causative agents of tuberculosis and leprosy, respectively (**Fig. 2.22A**).

- **Spore stain.** When samples are boiled with malachite green, the stain binds specifically to the endospore coat (**Fig. 2.22B**). It detects spores of *Bacillus* species such as *B. thuringiensis* (the insecticide) and *B. anthracis* (the cause of anthrax), as well as spores of *Clostridium botulinum*, which produces botulinum toxin.

- **Negative stain.** A negative stain is a suspension of opaque particles such as India ink added to darken the surrounding medium and reveal transparent components such as the outer capsule of a pathogen (presented in Chapter 3). Other kinds of negative stains are used for electron microscopy (see Section 2.6).

- **Antibody tags.** Stains linked to antibodies can identify precise strains of bacteria or even specific molecular components of cells. The antibody (which binds a specific cell protein) is linked to a reactive enzyme for

detection, or to a fluorophore (fluorescent molecule) for immunofluorescence microscopy (discussed next, in Section 2.4).

To Summarize

- **In bright-field microscopy**, image quality depends on the wavelength of light; the magnifying power of a lens; and the position of the focal plane, the region where the specimen is in focus (that is, where the sharpest image is obtained).

- **A compound microscope** achieves magnification and resolution through the objective and ocular lenses.

- **A wet mount** specimen contains living microbes.

- **Fixing and staining** a specimen kills it but improves contrast and resolution.

- **The Gram stain** differentiates between two major bacterial taxa, which stain either Gram-positive or Gram-negative. Eukaryotes stain negative.

- **Acid-fast, spore, negative, and antibody stains** are other kinds of differential stains.

2.4 Fluorescence Microscopy and Super-Resolution Imaging

Fluorescence microscopy (also called epifluorescence microscopy) is a powerful tool for identifying specific kinds of microbes, such as pathogens or members of environmental communities. Fluorescence also reveals specific cell parts at work, such as division proteins in the act of accomplishing cell fission. The profound importance of fluorescence for microbial discovery was acknowledged by

A.

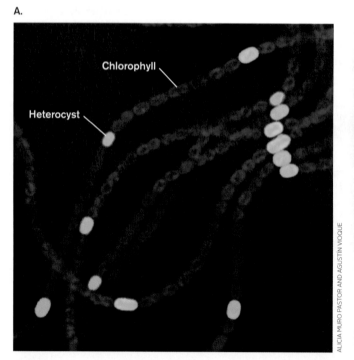

B.

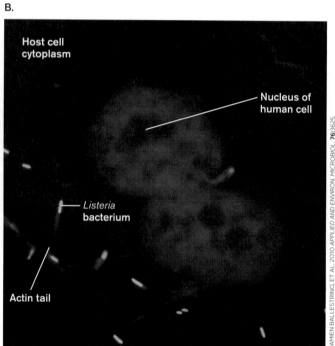

FIGURE 2.23 ■ Fluorescence microscopy. A. Cyanobacteria show chlorophyll-based autofluorescence (red) and fluorescence from heterocysts (green) expressing a nitrogen-stress gene fused to GFP. **B.** *Listeria* bacteria (expressing GFP) invade human cells, propelled by polymerizing "tails" of actin. Human cell nuclei fluoresce red (DAPI stain for DNA).

the awarding of the 2008 Nobel Prize in Chemistry for the discovery and development of green fluorescent protein, GFP (to Osamu Shimomura, Martin Chalfie, and Roger Tsien) and the 2014 Nobel in Chemistry for the development of super-resolved fluorescence microscopy, or super-resolution imaging (to Eric Betzig, Stefan Hell, and William Moerner).

In fluorescence microscopy, the specimen absorbs light of a defined wavelength and then emits light of lower energy, thus longer wavelength; that is, the specimen "fluoresces." Some microbes, such as cyanobacteria and algae, fluoresce on their own (autofluorescence), owing to endogenous fluorescent molecules such as chlorophyll. For other aims, specific parts of the cell are labeled with a **fluorophore**, a fluorescent dye or protein. In **Figure 2.23A**, cells of the cyanobacterium *Anabaena* show autofluorescence (red) arising from their chlorophyll. Every tenth cell or so, however, develops as a nitrogen-fixing heterocyst that lacks chlorophyll. In **Figure 2.23A**, the heterocysts specifically express a nitrogen-stress gene fused to a gene encoding GFP.

Fluorescence microscopy is also used by marine ecologists to reveal tiny bacteria and plankton growing in seawater, a highly dilute natural environment (discussed in Chapter 21). Such observations support the study of microbial responses to climate change. Microbes, including viruses, bacteria, and protists, are detected by fluorescence of DNA-specific stains such as DAPI (4′, 6-diamidino-2-phenylindole). The

advantage of DAPI fluorescent stain is that it detects only live cells whose DNA is intact, distinguishing them from environmental debris.

For medicine, fluorescence enables detection of microbial pathogens that colonize the relatively large cells and organ systems of the human body. **Figure 2.23B** shows infection by *Listeria monocytogenes*, the cause of listeriosis, a dangerous food-borne disease. In this study, the *Listeria* strain was engineered with a gene expressing GFP (green fluorescence). The bacteria (green) propel themselves through a human cell by polymerizing "tails" of the cell's actin (violet, stained by an actin-specific fluorophore, phalloidin). The human cell nuclei are labeled red with DAPI. Thus, fluorescence can visualize three different components within one microscopic field.

Excitation and Emission

How and when does a molecule fluoresce? Fluorescence occurs when a molecule absorbs light of a specific wavelength (the **excitation wavelength**) that has just the right energy needed to raise an electron to a higher-energy orbital (**Fig. 2.24**). Because this higher-energy electron state is unstable, the electron decays to an orbital of slightly lower energy, while losing some energy as heat. The electron then falls to its original level by emitting a photon of less energy and longer wavelength (the **emission wavelength**). The

A.

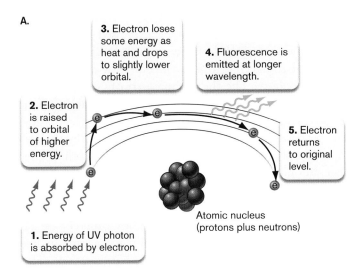

3. Electron loses some energy as heat and drops to slightly lower orbital.

4. Fluorescence is emitted at longer wavelength.

2. Electron is raised to orbital of higher energy.

5. Electron returns to original level.

Atomic nucleus (protons plus neutrons)

1. Energy of UV photon is absorbed by electron.

B.

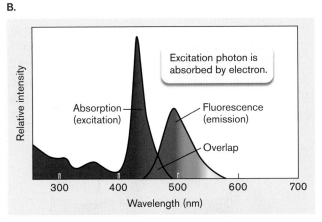

Excitation photon is absorbed by electron.

Absorption (excitation)

Fluorescence (emission)

Overlap

Relative intensity

300   400   500   600   700

Wavelength (nm)

FIGURE 2.24 ■ **Fluorescence.** Energy gained from UV absorption is released as heat and as a photon of longer wavelength in the visible region. **A.** Fluorescence on the molecular level. **B.** Comparison of absorption and emission spectra for a fluorophore.

The reflected blue light enters the objective lens, which focuses it onto the specimen, where it excites fluorophores to fluoresce green. The fluorescence emanates in all directions from the specimen (like a point source). Since the light rays point in all directions, a small portion of the fluorescent light (green) returns through the objective lens to reach the dichroic mirror. The green light now has a longer wavelength, above the penetration limit of the mirror, so it continues through to the ocular lens. The ocular lens focuses the fluorescent light onto the photodetectors of a highly sensitive digital camera.

Fluorescence can be observed in live organisms. The fluorescent organisms are commonly held in place on the slide—for example, by a pad of agarose gel.

Fluorophores for Labeling

What determines the properties of a fluorophore? The molecular structure of each fluorophore determines its peak wavelengths of excitation and emission, as well as its binding properties. For example, the aromatic rings of DAPI mimic a base pair, enabling intercalation between base pairs of DNA (**Fig. 2.26A**). DAPI absorbs in the UV and emits in the blue range. Note, however, that the computer driving the optical system can convert the fluorescence signal to any color chosen by the microscopist; for example, in **Figure 2.23B** the DAPI stain is shown red.

emitted photon has a longer wavelength (less energy) because part of the electron's energy of absorption was lost as heat.

The optical system for fluorescence microscopy uses filters to limit incident light to the wavelength of excitation and emitted light to the wavelength of emission (**Fig. 2.25**). The wavelengths of excitation and emission are determined by the choice of fluorophore; in this case we show excitation light as blue, and emission as green. Because only a small portion of the spectrum is used, fluorescence requires a high-intensity light source such as a tungsten arc lamp. The light passes through a filter that screens out all but the peak wavelengths of excitation. The excitation light (blue) is then reflected by a dichroic mirror (dichroic filter), a material that reflects light below a certain wavelength but transmits light above that wavelength.

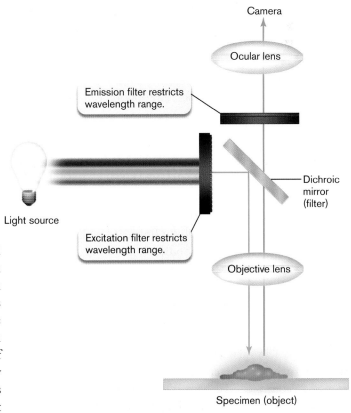

Camera

Ocular lens

Emission filter restricts wavelength range.

Dichroic mirror (filter)

Light source

Excitation filter restricts wavelength range.

Objective lens

Specimen (object)

FIGURE 2.25 ■ **Fluorescence microscopy.**

A.

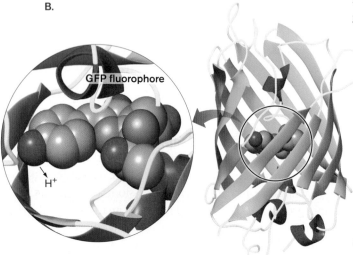

4′,6-Diamidino-2-phenylindole (DAPI)

B.

GFP fluorophore

H⁺

FIGURE 2.26 ■ Fluorescent molecules (fluorophores) commonly used in microscopy. The conjugated double bonds in these molecules provide closely spaced molecular orbitals that give rise to fluorescence. **A.** DAPI specifically labels DNA. **B.** Green fluorescent protein (GFP) is expressed endogenously by the cell. **Inset:** Three GFP amino acid residues (serine, tyrosine, and glycine) condense to form the fluorophore.

The cell specificity of the fluorophore can be determined by:

■ **Chemical affinity.** Certain fluorophores have chemical affinity for certain classes of biological molecules. For example, the fluorophore FM4-64 (green excitation, red emission) specifically binds membranes.

■ **Labeled antibodies.** Antibodies that specifically bind a cell component are chemically linked to a fluorophore molecule. The use of antibodies linked to fluorophores is known as immunofluorescence.

■ **DNA hybridization.** A short sequence of DNA attached to a fluorophore will hybridize to a specific sequence in the genome. Thus we can label one position in the chromosome.

■ **Gene fusion reporter.** Cells can be engineered with a gene fusion, a fused gene that expresses one of their own bacterial proteins combined with a gene encoding GFP or one of many variants expressing different colors.

Originally isolated from a jellyfish, *Aequorea victoria*, the Nobel-winning GFP can be expressed from a gene spliced into the DNA of any organism; even monkeys have been engineered to glow green. How does GFP act as a fluorophore? The fluorophore part of GFP consists of three amino acid residues that fuse to form an aromatic ring structure, embedded within a beta barrel protein tube. The properties of the fluorophore are modified by the surrounding protein, so mutation of the gene encoding GFP generates numerous variants with different spectral ranges.

Another interesting property of GFP is that its fluorophore can ionize (labeled in **Fig. 2.26B**). The ionized and protonated forms have slightly different excitation ranges, thus affording a way to measure hydronium ion concentration (pH) within a cell. This property enables GFP to report on environmental stress responses of bacteria.

The wide range of colors and environmental sensitivities of GFP variants provides an extraordinary set of probes for cell structure, as we will see in Chapter 3. For example, different fluorophores label specific molecules during DNA replication within a growing cell of *Bacillus subtilis* (**Fig. 2.27B**). The red

A.

COURTESY OF MELANIE BERKMEN, MIT

B.

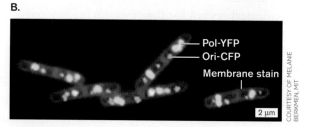

Pol-YFP
Ori-CFP
Membrane stain

2 μm

COURTESY OF MELANIE BERKMEN, MIT

FIGURE 2.27 ■ The replisome and the DNA origin of replication. A. Melanie Berkmen, working in the laboratory of Alan Grossman, obtains the fluorescence micrograph shown. **B.** Fluorescence microscopy reveals the DNA origin, labeled blue by a protein fused to cyan fluorescent protein, binding at a sequence near the origin (Ori-CFP). Replisomes are labeled yellow by fusion of a DNA polymerase subunit to yellow fluorescent protein (Pol-YFP) in dividing cells of *Bacillus subtilis*.

color arises from the membrane-specific fluorophore FM4-64. The DNA origin of replication is labeled blue by cyan fluorescent protein (CFP), a color variant of GFP. The gene encoding CFP is fused to a gene encoding a DNA-binding protein specific to the *B. subtilis* origin of replication. The replisomes (DNA polymerases) are labeled yellow, owing to a fused gene encoding yellow fluorescent protein (YFP). The replisomes usually locate together near the center of the cell, but sometimes they separate, visible as two yellow spots.

> **Thought Question**
>
> **2.6** What experiment could you devise to determine the actual order of events in *Bacillus subtilis* DNA replication?

A concern with the use of GFP fluorescence is that proteins fused to GFP may behave differently from the original nonfused protein. In some cases, the GFP portion causes fusion proteins to form complexes at the cell poles that are absent in non-GFP cells. Thus, in cellular biology, it is always important to confirm data with the results of a different kind of technique—for example, localization of the target protein with a labeled antibody.

Note also in **Figure 2.27B** that the labeled membranes and DNA origin appear diffuse—that is, unresolved.

The emitted light travels in all directions from the point source of the object, and its resolution is limited by the wavelength. Thus, fluorescence cannot resolve the detailed shape of a protein, or distinguish two proteins that are close together. But we can detect the location of a DNA-binding protein within a cell and resolve it as distinct from another fluorescent object located elsewhere. Furthermore, computational techniques called **super-resolution imaging** enable us to pinpoint the protein's location with a precision tenfold greater than the resolution of ordinary optical microscopy.

Super-Resolution Imaging

Cell function requires the interaction of key single molecules, such as the chromosomal DNA with a protein binding its origin. William Moerner at Stanford University was the first to demonstrate the possibility of single-molecule tracking of fluorescent proteins in bacteria (**Fig. 2.28**).

How can we track a single molecule, which is far smaller than the resolution limit of light ($\lambda/2 = 200$ nm)? Recall the shape of the magnified image of a point source of light (**Fig. 2.12**). Upon magnification, each image of a point source appears as a peak intensity surrounded by rings of much lower intensity. The sharpness of the main peak is limited by the wavelength of light. But the precision with which we know the peak's central position is much narrower (**Fig. 2.28A**). In other words, the uncertainty of

A. Single-molecule localization

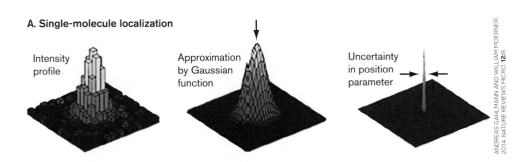

Intensity profile

Approximation by Gaussian function

Uncertainty in position parameter

ANDREAS GAHLMANN AND WILLIAM MOERNER. 2014. *NATURE REVIEWS MICRO* **12**:9.

B. Single-molecule tracking: monitoring protein motion

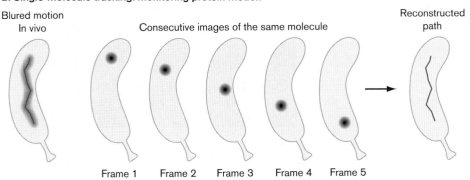

Blured motion
In vivo

Consecutive images of the same molecule

Reconstructed path

Frame 1 Frame 2 Frame 3 Frame 4 Frame 5

C.

L.A. CICERO/STANDFORD NEWS

FIGURE 2.28 ▪ Single-molecule localization by computation: a form of super-resolution imaging. **A.** The uncertainty of the central peak position. **B.** Tracking a single molecule in a cell. **C.** William Moerner was the first to demonstrate the possibility of single-molecule tracking of fluorescent proteins in bacteria.

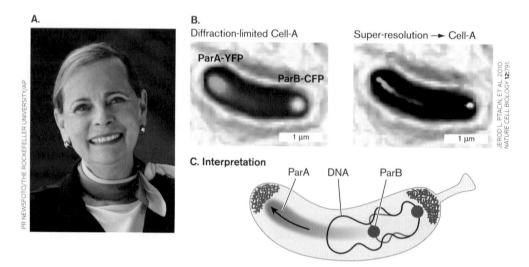

FIGURE 2.29 ■ **Super-resolution imaging reveals movement of origin-binding proteins across a cell.** **A.** Lucy Shapiro, winner of the National Medal of Science in 2013. **B.** Super-resolution imaging reveals the migration of the ParA cell fission protein across the cell of *Caulobacter crescentus*. The gene encoding ParA is fused to YFP, and the gene encoding ParB is fused to CFP. **C.** ParA migrates clear across the cell, generating a track to guide ParB to pull the newly replicated DNA origin across from one pole to the opposite pole. The arrow indicates movement.

the central peak position is about a tenth the width of the intensity profile. Computation based on the intensity profile can reveal the peak positions with high precision. The peak positions show how individual proteins move within a living cell (**Fig. 2.28B**).

In an early application of super-resolution imaging, Moerner worked with Lucy Shapiro at Stanford University to track the movement of DNA-binding proteins during cell fission of *Caulobacter crescentus* (**Fig. 2.29**). Shapiro received the National Medal of Science from President Obama in 2013 for her studies of the intriguing developmental cycle of this bacterium, which involves a transition between stalked and flagellar cells (presented in Chapter 3). Moerner, Shapiro, and their students used super-resolution imaging to track the migration of the ParA cell fission protein (**Fig. 2.29B** and **C**). In this experiment, a gene encoding ParA is fused to a gene for yellow fluorescent protein (YFP), whereas ParB is fused to cyan fluorescent protein (CFP). ParA migrates clear across the cell, generating a spindle-like track to guide ParB pulling the newly replicated DNA origin from one pole to the opposite pole. This dramatic feat is one of many intracellular mechanisms we explore in the next chapter.

Some advanced forms of fluorescence microscopy use laser beams to resolve subcellular details, and even to visualize cells in 3D. One such method is **confocal laser scanning microscopy** (or confocal microscopy). In confocal microscopy, a microscopic laser light source scans across the specimen. **Figure 2.30** shows a biofilm composed of the pathogenic bacterium *Pseudomonas*

aeruginosa, treated with the antibiotic tobramycin. The biofilm is treated with fluorophores that reveal live cells (green) beneath the dead cells, killed by tobramycin (red). The hidden live cells cause problems for medical therapy. Confocal microscopy enables us to visualize the 3D structure of the biofilm. The method of confocal microscopy is explained in **eTopic 2.2**.

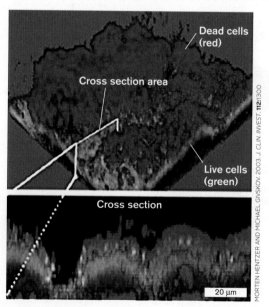

FIGURE 2.30 ■ **Biofilm with live/dead fluorophore, observed by confocal laser scanning microscopy.** *Pseudomonas aeruginosa* cells growing in a biofilm, treated with the antibiotic tobramycin.

Chemical Imaging Microscopy

Fluorescence microscopy can now be coupled to new tools of chemical imaging microscopy. Chemical imaging uses mass spectrometry (analysis of molecular fragments by mass) to visualize the distribution of chemicals in a biological sample, such as a microbial colony. The chemical distribution can actually be observed within a cell, at 100-nm resolution, using nanoscale secondary ion mass spectrometry (NanoSIMS). Chemical imaging offers extraordinary opportunities to map the structure and function of cells (see **Special Topic 2.1**). In natural communities, such as soil or the intestinal microbiome, the distribution of chemicals can be mapped onto specific microbes or subcellular compartments defined by fluorescence microscopy.

To Summarize

- **Fluorescence microscopy** uses fluorescence by a fluorophore to reveal specific cells or cell parts.

- **The specimen absorbs light at one wavelength and then emits light at a longer wavelength.** Color filters allow only light in the excitation range to reach the specimen, and only emitted light to reach the photodetector.

- **A fluorophore can label a cell part by** chemical affinity for a component such as a membrane, attachment to an antibody stain, or attachment to a short nucleic acid that hybridizes to a DNA sequence.

- **Fluorescent proteins such as GFP can be fused to a specific protein expressed by the cell.** Endogenous GFP-type proteins can track intracellular movement of cell parts and can report environmental stress responses.

- **Super-resolution imaging** can define the position of a fluorescent protein with a precision of 20–40 nm, tenfold better than the resolution limit of light magnification.

- **Chemical imaging microscopy** maps the distribution of chemicals within a cell.

2.5 Dark-Field and Phase-Contrast Microscopy

Advanced optical techniques enable us to visualize structures that are difficult or impossible to detect under a bright-field microscope, either because their size is below the limit of resolution of light or because their cytoplasm is transparent. These techniques take advantage of special properties of light waves, such as light scattering (dark field) and phase contrast.

Dark-Field Microscopy Detects Unresolved Objects

Dark-field microscopy enables microbes to be visualized as halos of bright light against the darkness, just as stars are observed against the night sky. A tiny object whose size is well below the wavelength of light, such as a virus particle, can be detected by light scattering. An important clinical application of dark-field microscopy is the detection of *Treponema pallidum*, the cause of syphilis. *T. pallidum* cells are so narrow (0.1 μm) that their shape cannot be resolved by bright-field microscopy. Nevertheless, *T. pallidum* can be detected by dark-field microscopy (**Fig. 2.31**). The spiral cells brightly scatter light, although the width of the narrow cell is not resolved.

The optics of dark-field microscopy involve light scattering. The wavefront of scattered light is spherical, like a wave emitted by a point source (see **Fig. 2.9D**). The scattered wave has a much smaller amplitude than that of the incident (incoming) wave. Therefore, with ordinary bright-field optics, scattered light is washed out. Detection of scattered light requires a modified condenser arrangement that excludes all light that is transmitted directly (**Fig. 2.32**). The condenser contains a "spider light stop," an opaque disk held by three "spider legs" across an open ring. The

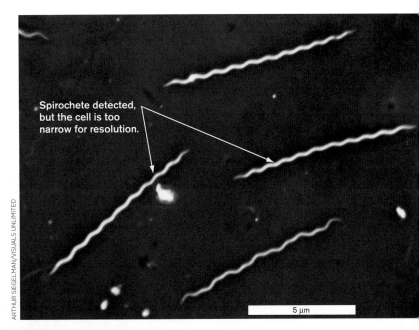

Spirochete detected, but the cell is too narrow for resolution.

ARTHUR SIEGELMAN/VISUALS UNLIMITED

5 μm

FIGURE 2.31 ■ Dark-field observation of bacteria. *Treponema pallidum* specimen from a patient with syphilis.

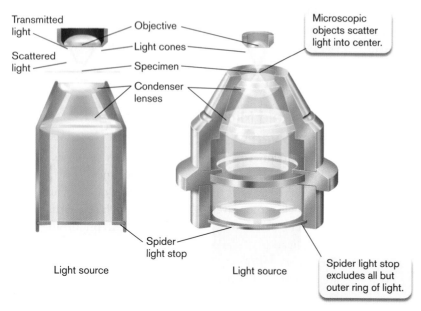

FIGURE 2.32 ■ A dark-field condenser system with a spider light stop. Below the condenser, the spider light stop excludes all but an annular ring of light from the light source. The annular ring converges as a hollow cone of light focused on the specimen. Only light scattered by the specimen enters the objective lens; the field background appears dark.

bacterial diseases such as urethritis, in which the pathogen needs to swim up through the urethra. The bacterial swimming apparatus consists of helical filaments called **flagella** (singular, **flagellum**), which are rotated by a motor device embedded in the bacterial cell wall (for flagellar structure, see Section 3.7). The "swimming strokes" of bacteria were first elucidated by Howard Berg and Robert Macnab (1940–2003), using dark-field optics to view the helical flagella. The flagella are too narrow to resolve, but dark-field optics can detect them, as seen in the dark-field micrograph of *Salmonella enterica* (**Fig. 2.33**). Detecting flagella requires such high light intensity the bacterial cell itself appears "over-exposed"; its shape is unresolved. Another disadvantage of dark-field microscopy is that any tiny particle, including specks of dust (see **Fig. 2.31**), can scatter light and interfere with visualization of the specimen. Unless the medium is extremely clear, it can be difficult to distinguish microbes of interest from particulates.

ring permits only a hollow cone of light to focus on the object. The incident light converges at the object and then generates an inverted hollow cone radiating outward.

For dark field, the objective lens is positioned in the central region, where it completely misses the directly transmitted light. For this reason, the field appears dark. However, light scattered by the object radiates outward in a spherical wave. A sector of this spherical wave enters the objective lens and is detected as a halo of light.

An intriguing application of dark-field optics is the study of bacterial motility. Motility is important in

Thought Questions

2.7 Some early observers claimed that the rotary motions observed in bacterial flagella could not be distinguished from whiplike patterns, comparable to the motion of eukaryotic flagella. Can you imagine an experiment to distinguish the two and prove that the flagella rotate? *Hint:* Bacterial flagella can get "stuck" to the microscope slide or coverslip.

2.8 Compare and contrast fluorescence microscopy with dark-field microscopy. What similar advantage do they provide, and how do they differ?

FIGURE 2.33 ■ Motile bacteria observed under dark-field microscopy. A. Flagellated *Salmonella enterica* observed with low light, which limits scattering. Only cell bodies are detected; no flagella. **B.** The light intensity is increased. Flagella are detected, although their fine structure is not resolved, because their width is below the threshold of resolution by light.

A. Low light intensity

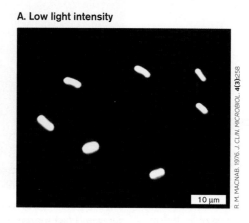

B. High light intensity

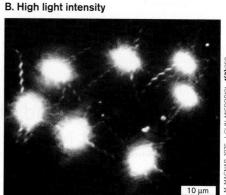

R. M. MACNAB. 1976. J. CLIN. MICROBIOL. **4**(3):258

R. M. MACNAB. 1976. J. CLIN. MICROBIOL. **4**(3):258

Phase-Contrast Microscopy

Phase-contrast microscopy (**PCM**) exploits differences in refractive index between the cytoplasm and the surrounding medium or between different organelles. The example of phase-contrast microscopy in **Figure 2.34A** shows a community of microbes cultured from a cyanobacterial liftoff mat in the ice of an Antarctic lake. This mat sample survived a month stored at −80°C and then was placed under light in dilute salts at 10°C. Filamentous cyanobacteria (*Microcoleus antarcticus*) soon appeared, bubbling oxygen. The filaments hosted a protist, a stalked cell with cilia

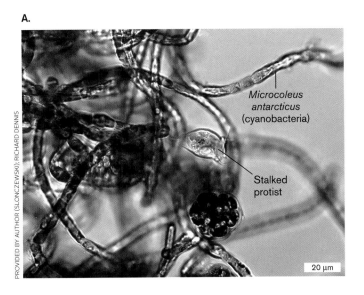

A.

Microcoleus antarcticus (cyanobacteria)

Stalked protist

20 μm

PROVIDED BY AUTHOR (SLONCZEWSKI); RICHARD DENNIS

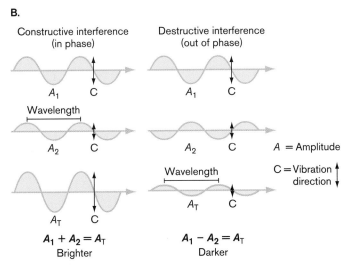

B.

Constructive interference (in phase)

Destructive interference (out of phase)

A_1 C

A_1 C

Wavelength

A_2 C

A_2 C A = Amplitude

C = Vibration direction

Wavelength

A_T C

A_T C

$A_1 + A_2 = A_T$
Brighter

$A_1 - A_2 = A_T$
Darker

FIGURE 2.34 ■ **Phase-contrast microscopy (PCM). A.** PCM of cyanobacteria with stalked protist. Cultured from Antarctica, Lake Fryxell cyanobacterial liftoff mat. **B.** Interference is the principle behind PCM. In constructive interference (left), the peaks of the two wave trains rise together; their amplitudes are additive ($A_1 + A_2$), forming a wave of greater total amplitude (A_T). In destructive interference (right), the peaks of the waves are opposite one another, so their amplitudes cancel ($A_1 - A_2$), forming a wave of lesser amplitude.

beating to engulf bacteria. Under phase contrast, the cell outlines appear dark because light passes entirely through the cell envelope, whose refractive index differs from that of cytoplasm.

Phase-contrast optics depend on the principle of interference (introduced in Section 2.2). In interference, two wavefronts (or two portions of a wavefront) interact with each other by addition (amplitudes in phase) or subtraction (amplitudes out of phase) (**Fig. 2.34B**). The result of interference between two waves is a pattern of alternating zones of constructive and destructive interference (brightness and darkness).

The optical system for phase contrast was invented in the 1930s by the Dutch microscopist Frits Zernike (1888–1966), for which he earned the Nobel Prize in Physics. In this system, slight differences in the refractive index of the various cell components are transformed into differences in the intensity of transmitted light. Zernike's scheme makes use of the fact that living cells have relatively high contrast because of their high concentration of solutes. Given the size and refractive index of commonly observed cells, light is retarded by approximately one-quarter of a wavelength when it passes through the cell. In other words, after passing through a cell, light exits the cell about one-quarter of a wavelength behind the phase of light transmitted directly through the medium.

The Zernike optical system is designed to retard the refracted light by an additional quarter of a wavelength, so that the light refracted through the cell is slowed by a total of half a wavelength compared with the light transmitted through the medium. When two waves are out of phase by half a wavelength, they produce destructive interference, canceling each other's amplitude (**Fig. 2.34B**, right). The result is a region of darkness in the image of the specimen.

As in dark-field microscopy, the light transmitted through the medium in phase-contrast microscopy needs to be separated from the light interacting with the object—in this case, light waves slowed by refraction. This separation is performed by a ring-shaped slit, called an "annular ring," similar in function to the spider light stop. The annular ring stops light from passing directly through the center of the lens system, where the specimen is located, and generates a hollow cone of light, which is focused through the specimen and generates an inverted cone above it (**Fig. 2.35**). Light passing through the specimen, however, is not only retarded; it is also refracted and thus bent into the central region within the inverted cone.

Both the refracted light from the specimen and the outer cone of transmitted light enter the phase plate. The phase plate consists of refractive material that is thinner in the region met by the outer (transmitted) light cone. The refracted light passing through the center of the phase plate

FIGURE 2.35 ▪ Phase-contrast optics. A. The specimen retards light by approximately one-quarter of a wavelength. The phase plate contains a central disk of refractive material that retards light from the specimen by another quarter wavelength, increasing the phase difference to half a wavelength. The light from the specimen and the transmitted light are now fully out of phase; they cancel, making the specimen appear dark. **B.** In the phase-contrast microscope, the annular ring forms a hollow cone of light that passes through the refractive material of the specimen. When the transmitted and refracted light cones re-join at the focal point, they are out of phase; their amplitudes cancel each other, and that region of the image appears dark against a bright background.

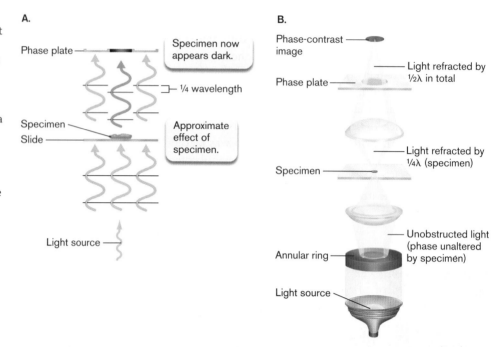

is retarded by an additional one-quarter wavelength compared with the transmitted light passing through the thinner region on the outside; the overall difference approximates half a wavelength. When the light from the inner and outer regions focuses at the ocular lens, the amplitudes of the waves cancel and produce a region of darkness. In this system, small differences in refractive index can produce dramatic differences in contrast between the offset phases of light.

Figure 2.36 shows a eukaryotic microbe, the baker's yeast *Saccharomyces cerevisiae*, imaged by two different optical systems involving phase contrast. Both systems distinguish intracellular compartments such as nucleus and vacuoles. The classical phase contrast with Zernike optics (**Fig. 2.36A**) shows a dark line of phase contrast at the cortex surrounding each yeast cell. **Figure 2.36B** shows a different kind of optical system, called **differential interference contrast microscopy** (**DIC**), also known as Nomarski microscopy. DIC optics require a polarized light source (with light rays restricted to one plane). The optical system enhances contrast by superimposing an image of the specimen onto a second beam

of light that generates interference fringes. These interference patterns are highly sensitive to slight differences in the refractive index of the specimen. They produce an illusion of shadowing across the specimen. The "false 3D" effect can be seen across the yeast cells and their intracellular vacuoles.

A. Phase contrast (Zernike optics)

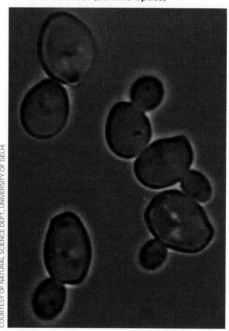

B. DIC (Nomarski optics)

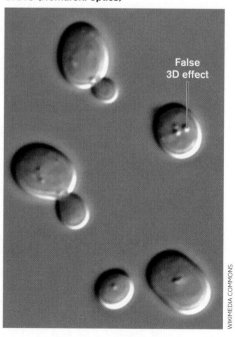

FIGURE 2.36 ▪ Phase contrast (Zernike optics) compared to differential interference contrast (DIC, Nomarski optics). A. *Saccharomyces cerevisiae* (baker's yeast) observed by phase contrast. The dense cell wall, highly refractive, appears as a dark rim. **B.** Yeast observed by DIC. The gradient of contrast across the cell generates a false 3D effect.

To Summarize

- **Dark-field microscopy** uses scattered light to detect objects too small to be resolved by light rays. Extremely small microbes and thin structures can be detected. The shapes of objects are not resolved.

- **Phase-contrast microscopy** with Zernike optics superimposes refracted light and transmitted light shifted out of phase so as to reveal differences in refractive index as patterns of light and dark. Live cells with transparent cytoplasm, and the organelles of eukaryotes, can be observed with high contrast.

- **Differential interference contrast microscopy (DIC)** with Nomarski optics superimposes interference bands on an image, accentuating small differences in refractive index.

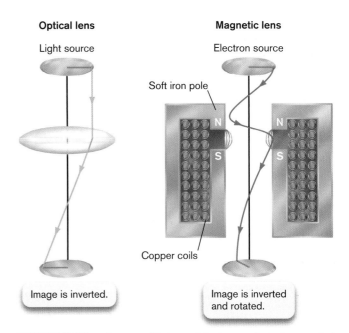

FIGURE 2.37 ■ A magnetic lens. The beam of electrons spirals around the magnetic field lines. The U-shaped magnet acts as a lens, focusing the spiraling electrons much as a refractive lens focuses light rays.

2.6 Electron Microscopy, Scanning Probe Microscopy, and X-Ray Crystallography

All cells are built of macromolecular structures. The foremost tool for observing the shapes of these structures is **electron microscopy** (**EM**). In electron microscopy, magnetic lenses focus beams of electrons to image cell membranes, chromosomes, and ribosomes with a resolution a thousand times that of light microscopy. Other kinds of microscopy are emerging, such as scanning probe microscopy, which images the contours of live bacteria. For atom-level detail of a macromolecule, the tool of choice remains X-ray crystallography.

Electron Microscopy

How does an electron microscope work? Electrons are ejected from a metal subjected to a voltage potential. Like photons, the electrons travel in a straight line, interact with matter, and carry information about their interaction. And also like photons, electrons can exhibit the properties of waves. The wavelength associated with an electron is 100,000 times smaller than that of a photon; for example, an electron accelerated over a voltage of 100 kilovolts (kV) has a wavelength of 0.0037 nm, compared with 400–750 nm for visible light. However, the actual resolution of electrons in microscopy is limited not by the wavelength, but by the aberrations of the lensing systems used to focus electrons. The magnetic lenses used to focus electrons never achieve the precision required to utilize the full potential resolution of the electron beam.

Electrons are focused by means of a magnetic field directed along the line of travel of the beam (**Fig. 2.37**). As a beam of electrons enters the field, it spirals around the magnetic field lines. The shape of the magnet can be designed to generate field lines that will focus the beam of electrons in a manner analogous to the focusing of photons by a refractive lens. The electron beam, however, forms a spiral because electrons travel around magnetic field lines. Magnetic lenses generate large aberrations; thus, we need a series of corrective magnetic lenses to obtain a resolution of about 0.2 nm. This resolution is a thousand times greater than the 200-nm resolution of light microscopy.

Thought Question

2.9 Like a light microscope, an electron microscope can be focused at successive powers of magnification. At each level, the image rotates at an angle of several degrees. Given the geometry of the electron beam, as shown in **Figure 2.37**, why do you think the image rotates?

Transmission EM and scanning EM. Two major types of electron microscopy are **transmission electron microscopy** (**TEM**) and **scanning electron microscopy** (**SEM**). In TEM, electrons are transmitted through the specimen as in light microscopy to reveal internal structure. In SEM, the electron beams scan across the surface of the specimen and are reflected to reveal the contours of its 3D surface.

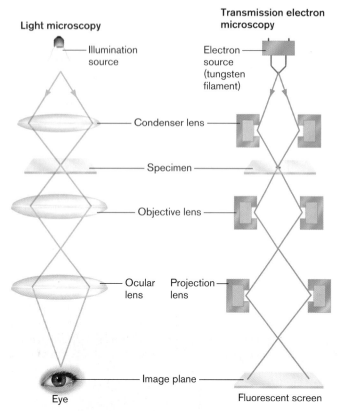

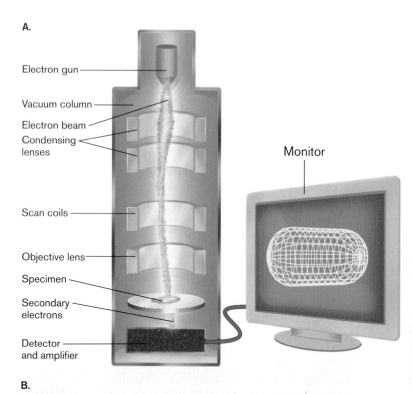

FIGURE 2.38 ▪ Transmission electron microscopy (TEM).
The light source is replaced by an electron source consisting of a high-voltage current applied to a tungsten filament, which gives off electrons when heated. Each magnetic lens shown (condenser, objective, projection) actually represents a series of lenses.

The transmission electron microscope closely parallels the design of a bright-field microscope, including a source of electrons (instead of light), a magnetic condenser lens, a specimen, and an objective lens (**Fig. 2.38**). The light source is replaced by an electron source consisting of a high-voltage current applied to a tungsten filament, which gives off electrons when heated. The electron beam is focused onto the specimen by a magnetic condenser lens. The specimen image is then magnified by a magnetic objective lens. The projection lens, analogous to the ocular lens of a light microscope, focuses the image on a fluorescent screen.

The scanning electron microscope is arranged somewhat differently from the TEM, in that a series of condenser lenses focuses the electron beam onto the surface of the specimen. Reflected electrons are then picked up by a detector (**Fig. 2.39**).

Sample preparation for EM. Standard electron microscopy of biological specimens at room temperature poses special problems. The entire optical column must be maintained under vacuum to prevent the electrons from colliding with the gas molecules in air. The requirement for a vacuum precludes the viewing of live specimens, which in any case would be quickly destroyed by the electron beam. Moreover, the

FIGURE 2.39 ▪ Scanning electron microscopy (SEM). A. In the scanning electron microscope, the electron beam is scanned across a specimen coated in gold, which acts as a source of secondary electrons. The incident electron beam ejects secondary electrons toward a detector, generating an image of the surface of the specimen. **B.** Loading a specimen into the vacuum column.

structure of most specimens lacks sufficient electron density (ability to scatter electrons) to provide contrast. Thus, the specimen requires an electron-dense **negative stain** using salts of heavy-metal atoms such as tungsten or uranium. The heavy atoms collect outside the surfaces of cell structures such as membranes, where their electron scatter reveals the outline of the structure. Staining, however, can be avoided for **cryo-electron microscopy** (see the next section).

We can prepare a specimen by embedding it in a polymer for thin sections. A special knife called a microtome cuts slices through the specimen, each slice a fraction of a micrometer thick. Alternatively, a specimen consisting of, for example, virus particles or isolated organelles can be sprayed onto a copper grid. In either case, the electron beam penetrates the object as if it were transparent. The electrons are actually absorbed by the heavy-atom stain, which collects at the edge of biological structures.

Figure 2.40 shows examples of transmission electron microscopy. The TEM of *Bacillus anthracis* in **Figure 2.40A** shows a thin section through a bacillus, including a cell wall, membranes, and glycoprotein filaments. The section is stained with uranyl acetate (a salt of uranium ion). The image includes electron density throughout the depth of the section. In **Figure 2.40B**, flagellar motors have been isolated from a bacterial cell envelope and spread on a grid

for TEM. The motor protein complexes are visualized by an electron-dense negative stain (phosphotungstate) deposited in the crevices around the complexes on a copper grid. The micrograph reveals details of each motor, including the axle and individual rings. The details are well resolved by the beam of electrons—an achievement far beyond that possible with a light microscope.

Scanning electron microscopy can show whole cells in 3D view, with much greater resolution than light microscopy can accomplish. SEM is particularly effective for visualizing cells within complex communities such as a biofilm. **Figure 2.41A** shows a biofilm of hyperthermophilic archaea from a deep-sea thermal vent, cultured at 95°C. The bulbous cells of *Pyrococcus furiosus* extend flagella that attach

A. TEM of *Bacillus anthracis* showing envelope and cytoplasm

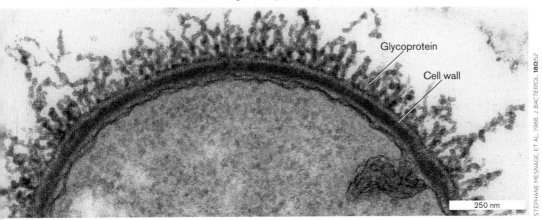

B. TEM of flagellar motors from *Salmonella enterica*

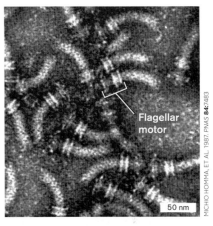

FIGURE 2.40 ■ **Transmission electron micrographs.** **A.** *Bacillus anthracis* thin section showing envelope and cytoplasm. Stained with uranyl acetate. **B.** Flagellar motors from *Salmonella enterica*, shadowed with colloidal gold.

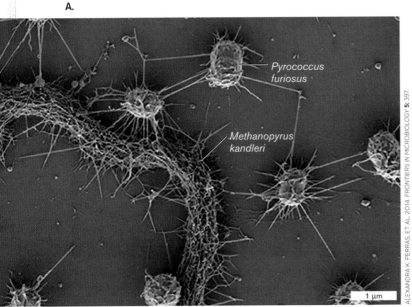

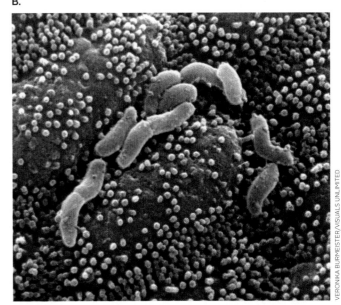

FIGURE 2.41 ■ **Scanning electron micrographs.** **A.** *Pyrococcus furiosus* extend flagella that attach the rod-shaped *Methanopyrus kandleri*. **B.** *Helicobacter pylori* adheres to the villi (small bulges) of the gastric epithelium. Bacteria are colorized green.

the rod-shaped *Methanopyrus kandleri*. The shapes of these archaea show up clearly in relation to one another via SEM.

In a clinical example, **Figure 2.41B** shows the pathogen *Helicobacter pylori* colonizing the gastric epithelium (stomach lining). *H. pylori* bacteria are helical rods (colorized green). Note that the colorizing consists of interpretation by a photo artist; no actual colors are observed by electron microscopy, since colors are defined not by electrons but by visible light. The bacterium *Helicobacter pylori* was first reported by Australian scientist Barry Marshall, but it proved difficult to isolate and culture. Ultimately, electron microscopy confirmed the existence of *H. pylori* in the stomach and helped document its role in gastritis and stomach ulcers.

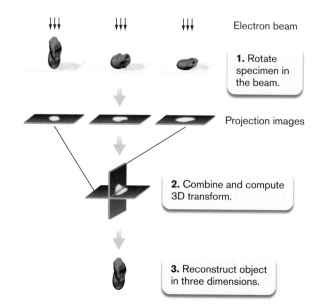

FIGURE 2.42 ■ **3D image construction in cryo-electron tomography.** Cryo-EM images are obtained in multiple focal planes throughout an object. The images are combined through a mathematical transformation to model the entire object in 3D.

Thought Question

2.10 What kinds of research questions could you investigate using SEM? What questions could you answer using TEM?

An important limitation of traditional electron microscopy, whether TEM or SEM, is that the fixatives and heavy-atom stains can introduce artifacts into the image, especially at finer details of resolution. In some cases, different preparation procedures have led to substantially different interpretations of subcellular structure. For example, an oval that appears hollow might be interpreted as a cell when in fact it represents a deposit of staining material. A microscopic structure that is interpreted incorrectly is termed an **artifact**. Avoiding artifacts is an important concern in microscopy.

Cryo-Electron Microscopy and Tomography

How can electron microscopy achieve finer resolution and avoid artifacts due to staining? High-strength electron beams now permit low-temperature **cryo-electron microscopy** (**cryo-EM**), also known as **electron cryomicroscopy**. In cryo-EM, the specimen does not require staining, because the high-intensity electron beams can detect smaller signals (contrast in the specimen) than earlier instruments could. The specimen must, however, be flash-frozen—that is, suspended in water and frozen rapidly in a refrigerant of high heat capacity (ability to absorb heat). The rapid freezing avoids ice crystallization, leaving the water solvent in a glass-like amorphous phase. The specimen retains water content and thus closely resembles its living form, although it is still ultimately destroyed by electron bombardment.

Another innovation made possible by cryo-EM is **tomography**, the acquisition of projected images from different angles of a transparent specimen (see the Current Research Highlight). **Cryo-electron tomography**, or

electron cryotomography, avoids the need to physically slice the sample for thin-section TEM. The images from tomography are combined digitally to visualize the entire object in 3D. Repeated scans can be summed computationally to obtain an image at high resolution (**Fig. 2.42**). The scans are taken either at different angles or within different focal planes. Each different scan images slightly different parts of the object. The scans are summed computationally to generate a 3D model.

One use of cryo-electron tomography is to generate high-resolution models of virus particles. For a highly symmetrical virus, images of multiple particles can be averaged together. The digitally combined images can achieve high resolution, nearly comparable to that of X-ray crystallography. Wah Chiu at Baylor College of Medicine (**Fig. 2.43A**) pioneered the use of cryo-EM to visualize virus particles at high resolution. An example is rice dwarf virus (**Fig. 2.43B**), one of the world's most economically destructive agricultural pathogens. Other viruses modeled recently include herpes virus and human immunodeficiency virus (HIV). Cryo-EM is especially useful for particles that cannot be crystallized for X-ray diffraction analysis, the most common means of molecular visualization (see pages 71–73).

Cryo-EM models of a virus are impressive, but can we build a 3D model of an entire cell? Grant Jensen and colleagues at the California Institute of Technology use cryo-electron tomography to visualize an entire flash-frozen bacterium. Such a model includes all the cell's parts and their cytoplasmic connections—and reveals new structures never seen before.

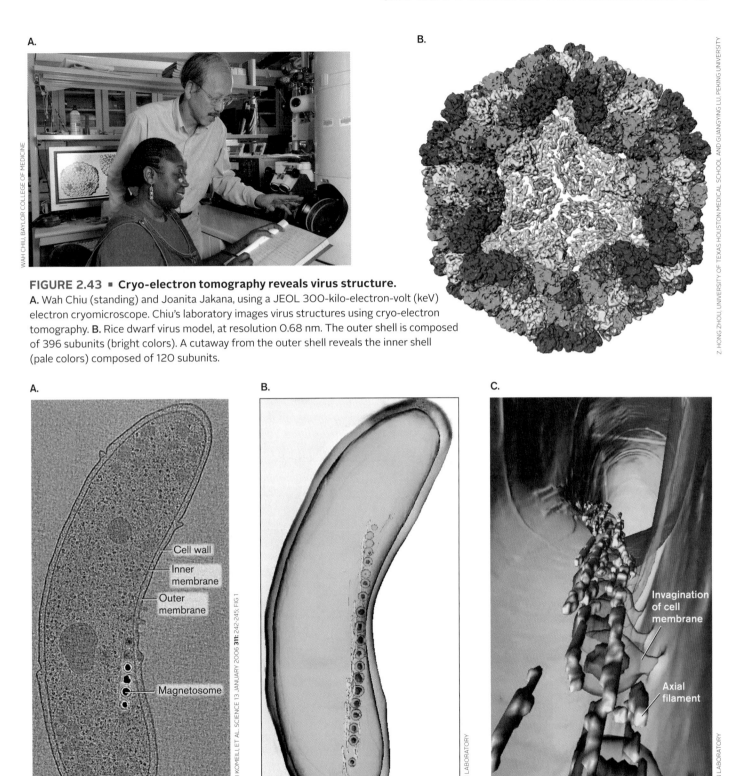

FIGURE 2.43 ■ **Cryo-electron tomography reveals virus structure.**
A. Wah Chiu (standing) and Joanita Jakana, using a JEOL 300-kilo-electron-volt (keV) electron cryomicroscope. Chiu's laboratory images virus structures using cryo-electron tomography. **B.** Rice dwarf virus model, at resolution 0.68 nm. The outer shell is composed of 396 subunits (bright colors). A cutaway from the outer shell reveals the inner shell (pale colors) composed of 120 subunits.

FIGURE 2.44 ■ **Magnetotactic cell visualized by cryo-electron tomography.** **A.** A single cryo-EM scan lengthwise through *Magnetospirillum magneticum*. **B.** 3D model of *M. magneticum* based on multiple scans. **C.** Expanded view of the cell interior.

The cell modeled in **Figure 2.44** is *Magnetospirillum magneticum*, a bacterium that can swim along magnetic field lines because the cell contains a string of magnetic particles composed of the mineral magnetite (iron oxide, Fe_3O_4). A cryo-EM section through the bacterium (**Fig. 2.44A**) shows fine details, including the inner membrane (equivalent to the cell membrane), peptidoglycan cell wall, and outer membrane, an outer covering found in Gram-negative bacteria. Four dark magnetosomes (particles of magnetite) appear in a chain, each surrounded by a vesicle of membrane.

Figure 2.44B models the magnetosomes, reconstructed using multiple cryo-EM scans across the volume of the cell. The magnetosomes are colorized red, each surrounded by a membrane vesicle (green). The vesicles are organized within the cell by a series of protein axial filaments, colorized yellow. **Figure 2.44C** shows an expanded image of the magnetosomes viewed from the cell interior. This expansion reveals that the magnetosome vesicles consist of invaginations from the cell membrane. Thus, the 3D model shows how the magnetite particles are fixed in position by invaginated membranes and held in a line by axial filaments.

Chapter 3 presents additional structures visualized by cryo-EM tomography, such as the flagellar motors of spirochete bacteria, and the carbon dioxide–fixing structures of marine cyanobacteria. Most surprisingly, cryo-EM reveals bacterial microtubules—structures never seen before, which offer a promising new target for antibiotics.

Scanning Probe Microscopy

Scanning probe microscopy (SPM) enables nanoscale observation of cell surfaces. Unlike electron microscopy, some forms of SPM can be used to observe live bacteria in water or exposed to air.

SPM techniques measure a physical interaction, such as the "atomic force" between the sample and a sharp tip. **Atomic force microscopy (AFM)** measures the van der Waals forces between the electron shells of adjacent atoms of the cell surface and the sharp tip. In AFM, an instrument probes the surface of a sample with a sharp tip a couple of micrometers long and often less than 10 nm in diameter (**Fig. 2.45A**). The tip is located at the free end of a lever that

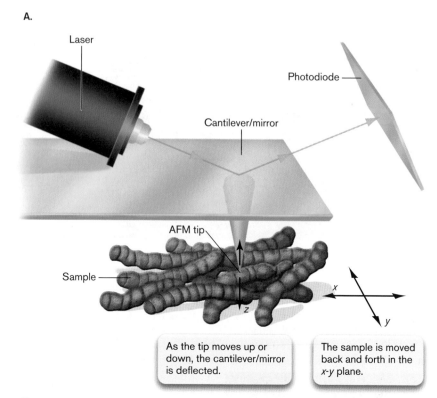

As the tip moves up or down, the cantilever/mirror is deflected.

The sample is moved back and forth in the x-y plane.

NICOLE HANSMEIER, ET AL. 2006. *MICROBIOLOGY* **152**:923–935

FIGURE 2.45 ▪ **Atomic force microscopy enables visualization of untreated cells. A.** The atomic force microscope (AFM) has a fine-pointed tip attached to a cantilever that moves over a sample. The tip interacts with the sample surface through atomic force. As the tip is pushed away, or pulled into a depression, the cantilever is deflected. The deflection is measured by a laser light beam focused onto the cantilever and reflected into a photodiode detector. **B.** This AFM image shows live bacteria collected on a filter, from seawater off the coast of California. Two round bacteria and a helical bacterium can be seen (raised regions, green-white).

is 100–200 μm long. The lever is deflected by the force between the tip and the sample surface. Deflection of the lever is measured by a laser beam reflected off a cantilever attached to the tip as the sample scans across. The measured deflections allow a computer to map the topography of cells in liquid medium with a resolution below 1 nm.

In **Figure 2.45B**, AFM was used to observe live bacteria collected on a filter, from seawater off the coast of California. Two round bacteria and a helical bacterium can be seen (raised regions, green-white). The cells were observed in water suspension, without stain. Thus, AFM can help assess the ecological contributions of marine bacteria that cannot be cultured.

X-Ray Diffraction Analysis

To know a cell, we need to isolate the cell's individual molecules. The major tool used at present to visualize a molecule is **X-ray diffraction analysis**, or **X-ray crystallography**. Much of our knowledge of microbial genetics (Chapters 7–12) and metabolism (Chapters 13–16) comes from crystal structures of key macromolecules.

Unlike microscopy, X-ray diffraction does not present a direct view of a sample, but generates computational models. Dramatic as the models are, they can only represent particular aspects of electron clouds and electron density

that are fundamentally "unseeable." That is why we represent molecular structures in different ways that depend on the context—by electron density maps, as models defined by van der Waals radii, or as stick models. Proteins are frequently presented in a cartoon form that shows alpha helix and beta sheet secondary structures.

For a molecule that can be crystallized, X-ray diffraction makes it possible to fix the position of each individual atom in the molecule. Atomic resolution is possible because the wavelengths of X-rays are much shorter than the wavelengths of visible light and are comparable to the size of atoms. X-ray diffraction is based on the principle of wave interference (see **Fig. 2.34B**). The interference pattern is generated when a crystal containing many copies of an isolated molecule is bombarded by a beam of X-rays (**Fig. 2.46A**). The wavefronts associated with the X-rays are diffracted as they pass through the crystal, causing interference patterns. In the crystal, the diffraction pattern is generated by a symmetrical array of many sample molecules (**Fig. 2.46B**). The more copies of the molecule in the array, the narrower the interference pattern and the greater the resolution of atoms within the molecule. Diffraction patterns obtained from the passage of X-rays through a crystal (**Fig. 2.46C**) can be analyzed by computation to develop a precise structural model for the molecule, detailing the position of every atom in the structure.

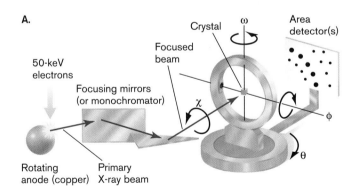

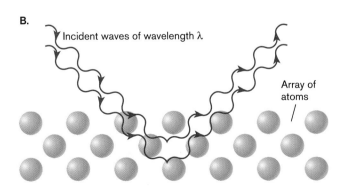

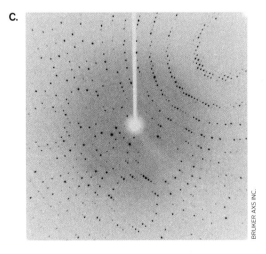

FIGURE 2.46 ■ Visualizing molecules by X-ray crystallography. A. Apparatus for X-ray crystallography. The X-ray beam is focused onto a crystal, which is rotated over all angles to obtain diffraction patterns. The intensity of the diffracted X-rays is recorded on film or with an electronic detector. **B.** X-rays are diffracted by rows of identical molecules in a crystal. The diffraction pattern is analyzed to generate a model of the individual molecules. **C.** Diffraction pattern from a crystal.

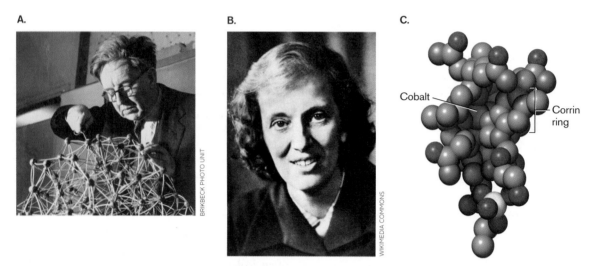

FIGURE 2.47 ▪ **Pioneering X-ray crystallography. A.** John Bernal developed X-ray crystallography to solve the structure of complex biological molecules. **B.** Dorothy Hodgkin was awarded the 1964 Nobel Prize in Chemistry for her work in X-ray crystallography. **C.** Vitamin B_{12}, whose structure was originally solved by Hodgkin. The corrin ring structure is built around an atom of cobalt (pink). Carbon atoms here are gray; oxygen, red; nitrogen, blue; phosphorus, yellow. Hydrogen atoms are omitted for clarity.

The application of X-ray crystallography to complex biological molecules was pioneered by the Irish crystallographer John Bernal (1901–1971) (**Fig. 2.47A**). Bernal was particularly supportive of women students and colleagues, including Rosalind Franklin (1920–1958), who made important discoveries about DNA and RNA, and Nobel laureate Dorothy Crowfoot Hodgkin (1910–1994). Hodgkin (**Fig. 2.47B**) won the 1964 Nobel Prize in Chemistry for solving the crystal structures of penicillin and vitamin B_{12} (**Fig. 2.47C**). She later solved one of the first protein structures—that of the hormone insulin.

Today, X-ray data undergo digital analysis to generate sophisticated molecular models, such as the one seen in **Figure 2.48** of anthrax lethal factor, a toxin produced by *Bacillus anthracis* that kills the infected host cells. The model for anthrax lethal factor was encoded in a Protein Data Bank (PDB) text file that specifies coordinates for all atoms of the structure. The Protein Data Bank is a worldwide database of solved X-ray structures, freely available on the Internet. Visualization software is used to present the structure as a "ribbon" of amino acid residues, color-coded for secondary structure. In **Figure 2.48**, the red coils represent alpha helix structures, whereas the blue arrows represent beta sheets (for a review of protein structures, see Appendix 1).

Note: Molecular and cellular biology increasingly rely on visualization in 3D. Many of the molecules illustrated in this book are based on structural models deposited in the Protein Data Bank, as indicated by the PDB file code. You may view these structures in 3D by downloading the PDB file and viewing it with a free plug-in such as Jmol.

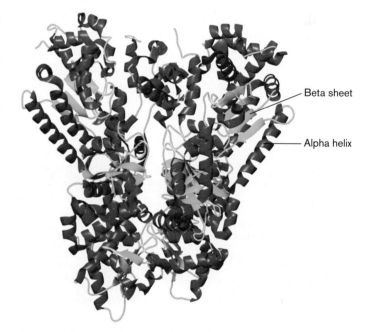

FIGURE 2.48 ▪ **X-ray crystallography of a protein complex, anthrax lethal factor.** The toxin consists of a butterfly-shaped dimer of two peptide chains. This cartoon model is based on X-ray-crystallographic data, showing alpha helix (red coils) and beta sheet (blue arrows). (PDB code: 1J7N)

A limitation of X-ray analysis is the unavoidable deterioration of the specimen under bombardment by X-rays. The earliest X-ray diffraction models of molecular complexes such as the ribosome relied heavily on components from thermophilic bacteria and archaea that grow at high temperatures. Because thermophiles have evolved to grow under higher thermal stress, their macromolecular complexes form more stable crystals than do their homologs in organisms growing at moderate temperatures.

X-ray diffraction analysis of crystals from a wide range of sources was made possible by cryocrystallography. In cryocrystallography, as in cryo-EM, crystals are frozen rapidly to liquid-nitrogen temperature. The frozen crystals have greatly decreased thermal vibrations and diffusion, thus lessening the radiation damage to the molecules. Models based on cryocrystallography can present multisubunit structures such as the bacterial ribosome complexed with transfer RNAs and messenger RNA.

To Summarize

- **Electron microscopy (EM)** focuses beams of electrons on an object stained with a heavy-metal salt that scatters electrons. Much higher resolution can be obtained than with light microscopy.

- **Transmission electron microscopy (TEM)** transmits electron beams through a thin section.

- **Scanning electron microscopy (SEM)** involves scanning of a 3D surface with an electron beam.

- **Cryo-electron microscopy (cryo-EM)** involves the observation of samples flash-frozen in water solution. **Tomography** combines multiple images by computation to achieve high resolution.

- **Atomic force microscopy (AFM)**, a form of scanning probe microscopy (SPM), uses intermolecular force measurement to observe cells in water solution.

- **X-ray diffraction analysis**, or **X-ray crystallography**, uses X-ray diffraction (interference patterns) from crystallized macromolecules to model the form of a molecule at atomic resolution.

- **Cryocrystallography** uses frozen crystals with greatly decreased thermal vibrations and diffusion, enabling the determination of structures of large macromolecular complexes, such as the ribosome.

Concluding Thoughts

The tools of microscopy and molecular visualization described in this chapter have shaped our current understanding of microbial cells—how they grow and divide, organize their DNA and cytoplasm, and interact with other cells. Our current models of cell structure and function are explored in Chapter 3. In Chapter 4 we discuss how cells use their structures to obtain energy, reproduce, and develop dormant forms that can remain viable for thousands of years.

CHAPTER REVIEW

Review Questions

1. What principle defines an object as microscopic?
2. Explain the difference between detection and resolution.
3. How do eukaryotic and prokaryotic cells differ in appearance under the light microscope?
4. Explain how electromagnetic radiation carries information and why different kinds of radiation can resolve different kinds of objects.
5. Describe how light interacts with an object through absorption, reflection, refraction, and scattering.
6. Explain how refraction enables magnification of an image.
7. Explain how magnification increases resolution and why empty magnification fails to increase resolution.
8. Explain how angle of aperture and resolution change with increasing lens magnification.
9. Summarize the optical arrangement of a compound microscope.
10. Explain how to focus an object and how to tell when the object is in or out of focus.
11. Explain the relative advantages and limitations of wet mount and stained preparations for observing microbes.
12. Explain the significance (and limitations) of the Gram stain for bacterial taxonomy.
13. Explain the basis of dark-field, phase-contrast, and fluorescence microscopy. Give examples of applications of these advanced techniques.
14. Explain how super-resolution imaging enables tracking of intracellular molecules.
15. Explain the difference between transmission and scanning electron microscopy, and the different applications of each.

Thought Questions

1. Explain which features of bacteria you can study by (a) light microscopy; (b) fluorescence microscopy; (c) scanning EM; (d) transmission EM.

2. Explain how resolution is increased by magnification. Why can't the details be resolved by your unaided eye? Explain why magnification reaches a limit. Why can it not go on resolving greater detail?

3. Explain why artifacts appear in microscopic images, even with the best lenses. Explain how you can tell the difference between an optical artifact and an actual feature of an image.

4. How can "detection without resolution" be useful in microscopy? Explain with specific examples.

Key Terms

aberration (48, 49)
absorption (43)
acid-fast stain (55)
angle of aperture (48)
antibody tags (55)
artifact (68)
atomic force microscopy (AFM) (42, 70)
bacillus (40)
bright-field microscopy (42, 48)
chemical imaging microscopy (42)
coccus (40)
compound microscope (49)
condenser (49)
confocal laser scanning microscopy (60)
contrast (43, 47)
counterstain (53)
cryo-electron microscopy (cryo-EM) (electron cryomicroscopy) (66, 68)
cryo-electron tomography (electron cryotomography) (68)
dark-field microscopy (61)
depth of field (51)
detection (39)

differential interference contrast microscopy (DIC) (64)
differential stain (52)
electromagnetic radiation (42)
electron cryomicroscopy (68)
electron microscopy (EM) (42, 65)
emission wavelength (56)
empty magnification (47)
excitation wavelength (56)
fixation (52)
flagellum (62)
fluorescence (44)
fluorophore (56)
focal point (46)
focus (38)
Gram-negative (53)
Gram-positive (53)
Gram stain (52)
immersion oil (49)
interference (47)
lens (46)
light microscopy (LM) (42)
magnification (39)
microscope (37)
mordant (53)
negative stain (55, 66)
numerical aperture (49)

objective lens (48, 49)
ocular lens (50)
parfocal (50)
phase-contrast microscopy (PCM) (63)
reflection (44)
refraction (44)
refractive index (44)
resolution (38)
scanning electron microscopy (SEM) (42, 65)
scanning probe microscopy (SPM) (70)
scattering (45)
simple stain (52)
spirochete (40)
spore stain (55)
staining (52)
super-resolution imaging (47, 59)
tomography (68)
total magnification (50)
transmission electron microscopy (TEM) (42, 65)
wet mount (51)
X-ray crystallography (X-ray diffraction analysis) (42, 71)

Recommended Reading

Altindal, Tuba, Suddhashil Chattopadhyay, and Xiao-Lun Wu. 2011. Bacterial chemotaxis in an optical trap. *PLoS One* 6:e18231.

Chiu, W., M. L. Baker, W. Jiang, and Z. H. Zhou. 2002. Deriving folds of macromolecular complexes through electron cryomicroscopy and bioinformatics approaches. *Current Opinion in Structural Biology* 12:263–269.

Gahlmann, Andreas, and William E. Moerner. 2014. Exploring bacterial cell biology with single-molecule tracking and super-resolution imaging. *Nature Reviews. Microbiology* 12:9–22.

Graumann, Peter L., and Richard Losick. 2001. Coupling of asymmetric division to polar placement of replication origin regions in *Bacillus subtilis*. *Journal of Bacteriology* **183**:4052–4060.

Jiang, W., J. Chang, J. Jakana, P. Weigele, J. King, et al. 2006. Structure of epsilon15 bacteriophage reveals genome organization and DNA packaging-injection apparatus. *Nature* **439**:612–616.

Komeili, A., Z. Li, D. K. Newman, and G. J. Jensen. 2006. Magnetosomes are cell membrane invaginations organized by the actin-like protein MamK. *Science* **311**:242–245.

Lucic, Vladan, Friedrich Förster, and Wolfgang Baumeister. 2005. Structural studies by electron tomography: From cells to molecules. *Annual Review of Biochemistry* **74**:833–865.

Matias, Valério R. F., Ashruf Al-Amoudi, Jacques Dubochet, and Terry J. Beveridge. 2003. Cryo-transmission electron microscopy of frozen-hydrated sections of *Escherichia coli* and *Pseudomonas aeruginosa*. *Journal of Bacteriology* **185**:6112–6118.

Murphy, Douglas B. 2001. *Fundamentals of Light Microscopy and Electronic Imaging*. Wiley-Liss, Hoboken, NJ.

Popescu, Aurel, and R. J. Doyle. 1996. The Gram stain after more than a century. *Biotechniques in Histochemistry* **71**:145–151.

Ptacin, Jerod L., Steven F. Lee, Ethan C. Garner, Esteban Toro, Michael Eckart, et al. 2010. A spindle-like apparatus guides bacterial chromosome segregation. *Nature Cell Biology* **12**:791–798.

Tocheva, E., Z. Li, and G. Jensen. 2010. Electron cryotomography, p. 213–232. *In* Lucy Shapiro and Richard M. Losick (eds.), *Cell Biology of Bacteria: A Subject Collection from Cold Spring Harbor Perspectives in Biology*. Cold Spring Harbor Laboratory Press, Cold Spring Harbor, NY.

CHAPTER 3
Cell Structure and Function

With just a few thousand genes in its genome, the bacterial cell continually builds new parts while timing its DNA replication with fission. A cell expands its membrane and builds new ribosomes to make proteins. Flagella propel bacteria and archaea to new habitats. And the microbes secrete signals to contact fellow cells. Cell-surface proteins give us targets for new antibiotics or vaccines. Other features, such as flagellar motors, inspire us to build microscopic machines.

CURRENT RESEARCH highlight

Growing bacterium expands its nucleoid ahead of cell division. *Escherichia coli* DNA is coiled into a region called a nucleoid (blue fluorophore). Completing cell division requires the Z-ring formed by polymerization of the protein FtsZ (red fluorophore). With rich nutrients available, DNA replicates well ahead of cell division, forming two to four nucleoids. Norbert Hill in Petra Levin's lab identified a nutrient-dependent regulator that delays FtsZ polymerization. By delaying Z-ring formation, the regulator allows the bacterial cell to grow larger before it divides.

Source: Norbert S. Hill et al. 2013. *PLoS Genetics* **9**:e1003663.

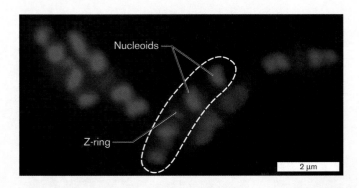

AN INTERVIEW WITH

PETRA LEVIN, MICROBIOLOGIST, WASHINGTON UNIVERSITY IN ST. LOUIS

COURTESY OF DOROTHY GREGG

What excites you about the "cell division odyssey"?

Determining how cells coordinate division with cell growth, DNA replication, and chromosome segregation is akin to solving an extremely challenging brainteaser. Each time we identify a component of the signal transduction pathway coordinating cell division with cell growth or an elusive part of the cell division machinery, I experience not only the thrill of discovery but also the thrill of knowing we have opened the door to yet another mystery!

How do you think our view of bacterial cell division will be transformed in the future?

The new microfluidic devices, and the high-throughput image capture technology that facilitates their use, enable us to follow single cells from birth to division across tens of generations. Another recent advance is super-resolution microscopy. In combination, these advances will help us answer long-standing questions about the forces behind bacterial cytokinesis and the mechanisms that coordinate cell division with other cell cycle processes. I can't wait!

To grow and multiply, a microbial cell must gain nutrients faster than its competitors do. For example, when your intestinal Gram-negative bacteria have plenty of food, they delay cell division at the Z-ring (see the Current Research Highlight) so that they can absorb nutrients and grow larger. At the same time, each cell must exclude toxins such as your intestinal bile salts. Such toxins are screened by the cell's outer membrane. Microbial cells have evolved an amazing range of molecular parts, from rotary machines that make ATP to enzyme factories that build antibiotics. Chapter 3 explores the key parts of bacteria and archaea, including specialized structures such as magnetosomes.

Most bacteria share these traits:

▪ **Thick, complex outer envelope.** The envelope protects the cell from environmental stress and mediates exchange with the environment.

▪ **Compact genome.** Prokaryotic genomes are compact, with relatively little noncoding DNA. Small genomes maximize the production of cells from limited resources.

▪ **Tightly coordinated functions.** The cell's parts work together in a highly coordinated mechanism, which may enable a high rate of reproduction.

Beyond the fundamentals, bacteria evolve amazing diversity, which we will explore in Chapter 18. We also discuss diversity of archaea in Chapter 19, and of microbial eukaryotes such as fungi and protists in Chapter 20. Overall, the three domains show these key traits:

▪ **Archaea, like bacteria, are prokaryotes (cells that lack a nucleus).** Archaea have unique membrane and envelope structures that enable survival in extreme environments.

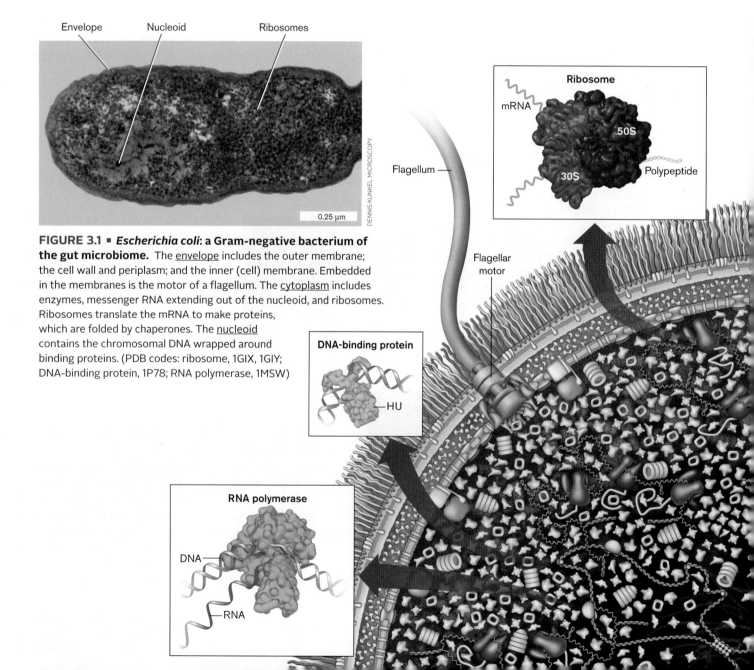

FIGURE 3.1 ▪ ***Escherichia coli*: a Gram-negative bacterium of the gut microbiome.** The envelope includes the outer membrane; the cell wall and periplasm; and the inner (cell) membrane. Embedded in the membranes is the motor of a flagellum. The cytoplasm includes enzymes, messenger RNA extending out of the nucleoid, and ribosomes. Ribosomes translate the mRNA to make proteins, which are folded by chaperones. The nucleoid contains the chromosomal DNA wrapped around binding proteins. (PDB codes: ribosome, 1GIX, 1GIY; DNA-binding protein, 1P78; RNA polymerase, 1MSW)

■ **Eukaryotic cells have extensive membranous organelles.** Organelles such as endoplasmic reticulum and Golgi complex are reviewed in eAppendix 2. The mitochondria and chloroplasts of eukaryotic cells evolved by endosymbiosis with engulfed bacteria (see Chapter 17). Diverse microbial eukaryotes, such as fungi and protists, are explored in Chapter 20.

Now we embark on a tour of a typical Gram-negative bacterial cell, a common resident of your gut, *Escherichia coli* (**Fig. 3.1**). Along the way, we learn how our understanding of this model cell has emerged from microscopy, cell fractionation, and genetic analysis.

3.1 The Bacterial Cell: An Overview

A cell is more than a "soup" full of ribosomes and enzymes. In fact, the cell's parts fit together in a structure that is ordered, though flexible. Our model of the bacterial cell (**Fig. 3.1**) offers an interpretation of how the major components of one cell fit together. The model represents

Escherichia coli. Its general features apply to many kinds of bacteria, particularly the Gram-negative inhabits of your colon such as Proteobacteria and Bacteroidetes. Remember that we cannot literally "see" the molecules within a cell, but microscopy and subcellular analysis provide the basis for our model.

Model of a Bacterial Cell

Within a cell, the cytoplasm consists of a gel-like network composed of proteins and other macromolecules. The cytoplasm is contained by a **cell membrane**, or **plasma membrane**. For *E. coli*, a Gram-negative bacterium (discussed in Section 3.3), the cell membrane is called the **inner membrane**, in order to distinguish it from the

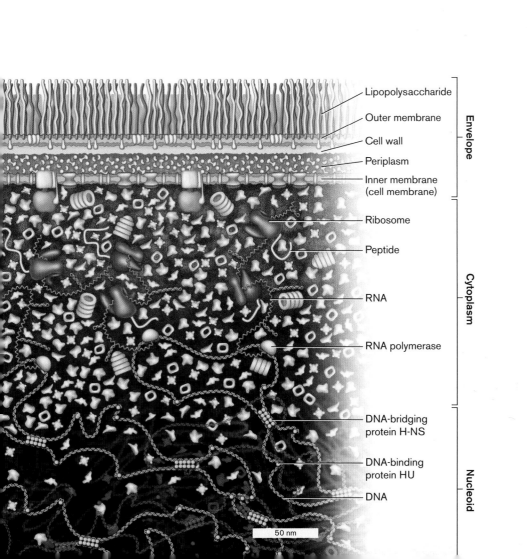

Lipopolysaccharide
Outer membrane
Cell wall
Periplasm
Inner membrane (cell membrane)
Envelope

Ribosome
Peptide
RNA
RNA polymerase
Cytoplasm

DNA-bridging protein H-NS
DNA-binding protein HU
DNA
Nucleoid

50 nm

Bacterial Cell Components (Examples)

Outer membrane proteins:

Sugar porin (10 nm)

Braun lipoprotein (8 nm)

Inner membrane proteins:

Transporter

Secretory complex (Sec)

ATP synthase (20 nm diameter in inner membrane; 32 nm total height)

Periplasmic proteins:

Arabinose-binding protein (3 × 3 × 6 nm)

Acid resistance chaperone (HdeA) (3 × 3 × 6 nm)

Disulfide bond protein (DsbA) (3 × 3 × 6 nm)

Cytoplasmic proteins:

Pyruvate kinase (5 × 10 × 10 nm)

Phosphofructokinase (4 × 7 × 7 nm)

Proteasome (12 × 12 × 15 nm)

Chaperone GroEL (18 × 14 nm)

Other proteins

Transcription and translation complexes:

RNA polymerase (10 × 10 × 16 nm)

Ribosome (21 × 21 × 21 nm)

Nucleoid components:

DNA (2.4 nm wide × 3.4 nm/10 bp)

DNA-binding protein (3 × 3 × 5 nm)

DNA-bridging protein (3 × 3 × 5 nm)

additional "outer membrane." The inner membrane is composed of phospholipids, transporter proteins, and other molecules. This membrane prevents cytoplasmic proteins from leaking out and maintains gradients of ions and nutrients. Between the inner and outer membranes lies the **cell wall**, a fortress-like structure composed of sugar chains linked covalently by peptides (peptidoglycan). The cell wall forms a single molecule that surrounds the cell. The wall material is flexible, but it limits expansion of the cytoplasm, keeping the cell membrane intact when water flows in. Like a balloon, the resulting turgor pressure makes the cell rigid.

In a Gram-negative bacterium, the cell wall lies within the **periplasm**, a water-filled space containing nutrient-binding proteins and secretion machines. Outside the cell wall lies the **outer membrane** of phospholipids and **lipopolysaccharides** (**LPS**), a class of lipids attached to long polysaccharides (sugar chains). The LPS layer may be surrounded by a thick capsule. The capsule polysaccharides form a slippery layer that inhibits phagocytosis by macrophages (presented in Chapter 26). The inner membrane, cell wall, and outer membrane constitute the Gram-negative cell **envelope**.

The bacterial envelope includes cell-surface proteins that enable the bacterium to interact with specific host organisms. For example, *E. coli* cell-surface proteins help the bacterium colonize the human intestinal epithelium. The nitrogen-fixing symbiont *Sinorhizobium* has cell-surface proteins that help the bacteria to colonize legume plants for nitrogen fixation. Another common external structure is the **flagellum** (plural, **flagella**), a helical protein filament whose rotary motor propels the cell in search of a more favorable environment. Flagella appear on many Proteobacteria, such as *E. coli*, but are absent in the Bacteroidetes such as *Bacteroides* and *Prevotella* species.

Within the cell, the cell membrane and envelope provide an attachment point for one or more chromosomes. The chromosome is organized within the cytoplasm as a system of looped coils called the **nucleoid** (visualized by DAPI fluorescence in the Current Research Highlight). Unlike the round, compact nucleus of eukaryotic cells, the bacterial nucleoid is <u>not</u> enclosed by a membrane. In fact, loops of DNA extend throughout the cytoplasm. The DNA is transcribed by RNA polymerase to form messenger RNA (mRNA), as well as transfer RNA (tRNA) and ribosomal RNA (rRNA). As the mRNA transcripts grow, they bind ribosomes to start synthesizing polypeptide chains. As the polypeptides grow, protein complexes called chaperones help them fold into their functional conformations. The concept of information flow from DNA to RNA to protein is presented in Chapters 7–10.

How can we know how all the cell parts shown in **Figure 3.1** interact and work together? Fluorescence microscopy can show us how the nucleoid DNA expands as the cell grows. As we learned in Chapter 2, fluorescence can pinpoint a molecule or complex within a cell, such as the Z-ring where the cell divides (see the Current Research Highlight). But to visualize the cell's interior as a whole requires higher resolution, as in electron microscopy.

Biochemical Composition of Bacteria

The bacterial cell model in **Figure 3.1** represents the shape and size of cell parts but tells us little about the chemistry of the cell, or of the cell's environment. Chemistry explains, for example, why wiping a surface with ethanol kills microbes, whereas water has little effect. Water is a universal constituent of cytoplasm but is excluded by cell membranes. Ethanol, however, dissolves both polar and nonpolar substances; thus, ethanol disintegrates membranes and destroys the folded structure of proteins. For a review of elementary chemistry, see eAppendix 1.

All cells share common chemical components:

■ **Water**, the fundamental solvent of life

■ **Essential ions**, such as potassium, magnesium, and chloride ions

■ **Small organic molecules**, such as lipids and sugars, that are incorporated into cell structures and that provide nutrition by catabolism

■ **Macromolecules**, such as nucleic acids and proteins, that contain information, catalyze reactions, and mediate transport, among many other functions

The details of bacterial chemistry emerged in the 1950s through pioneering studies by Fred Neidhardt at the University of Michigan, and many colleagues. Cell composition varies with species, growth phase, and environmental conditions (as we will see in later chapters). **Table 3.1** summarizes the chemical components of a cell for the model bacterium *Escherichia coli* during exponential growth.

Small molecules and ions. The *E. coli* cell consists of about 70% water, the essential solvent required to carry out fundamental metabolic reactions and to stabilize proteins. The water solution contains inorganic ions, predominantly potassium, magnesium, and phosphate. Inorganic ions store energy in the form of transmembrane gradients, and they serve essential roles in enzymes. For example, a magnesium ion is required at the active site of RNA polymerase to help catalyze the linking of ribonucleotides into RNA.

The cell also contains many kinds of small charged organic molecules, such as phospholipids and enzyme cofactors. A major class of organic cations is the polyamines, molecules with multiple amine groups that are

TABLE 3.1	Molecules of a Bacterial Cell, *Escherichia coli*, During Balanced Exponential Growth[a]		
Component	Percentage of total weight[b]	Approximate number of molecules/cell	Number of different kinds
Water	70	20,000,000,000	1
Proteins	16	2,400,000	4,000[c]
RNA:			
rRNA, tRNA, and other small RNA (sRNA) molecules	6	250,000	200
mRNA	0.7	4,000	2,000[c]
Lipids:			
Phospholipids (membrane)	3	25,000,000	50
Lipopolysaccharides (outer membrane)	1	1,400,000	1
DNA	1	2[d]	1
Metabolites and biosynthetic precursors	1.3	50,000,000	1,000
Peptidoglycan (murein sacculus)	0.8	1	1
Inorganic ions	0.1	250,000,000	20
Polyamines (mainly putrescine and spermidine)	0.1	6,700,000	2

[a]Values shown are for a hypothetical "average" cell cultured with aeration in glucose medium with minimal salts at 37°C.

[b]The total weight of the cell (including water) is about 10^{-12} gram (g), or 1 picogram (pg).

[c]The number of different kinds is difficult to estimate for proteins and mRNA because some genes are transcribed at extremely low levels and because proteins and RNA include kinds that are rapidly degraded.

[d]In rapidly growing cells, cell fission typically lags approximately one generation behind DNA replication—hence, two identical DNA copies per cell.

Source: Modified from F. Neidhardt and H. E. Umbarger. 1996. Chemical composition of *Escherichia coli*, p. 14. In F. C. Neidhardt (ed.), *Escherichia coli and Salmonella: Cellular and Molecular Biology*, 2nd ed. ASM Press, Washington, DC.

positively charged when the pH is near neutral. Polyamines balance the negative charges of the cell's DNA and stabilize ribosomes during translation.

Macromolecules. Many cells have similar content of water and small molecules, but their specific character is defined by their macromolecules, especially their nucleic acids (DNA and RNA) and their proteins. DNA and RNA molecules can be isolated by size using agarose gel **electrophoresis**, in which the negatively charged molecules migrate in an electrical field (see eAppendix 3). The nucleic acid content of bacteria is relatively high, nearly 8% for *E. coli*—much higher than in multicellular eukaryotes. For microbes, the high nucleic acid content allows the cell to maximize reproduction of its chromosome while minimizing cell resources.

The high proportion of nucleic acids actually makes bacterial cells inedible for humans, who lack the enzymes to digest the uric acid waste product of digested nucleotides. That is why we cannot eat most kinds of bacteria as a major part of our diet. Nevertheless, we do consume bacteria within vegetables, as all plants have bacteria growing within their transport tissues.

The cell's genomic DNA directs expression of its proteins (discussed in Chapters 7–10). A given cell uses different genes to make different proteins, depending on environmental conditions such as temperature, nutrient levels, and entry into a host organism. Individual proteins are made in very different amounts, from 10 per cell to 10,000 per cell. The proteins expressed by a cell under given conditions are known collectively as a proteome. (We discuss the proteome in Chapter 8, as a function of the DNA genome and RNA transcriptome.) Other kinds of macromolecules are found in the cell wall and outer membrane. The bacterial cell wall consists of **peptidoglycan**, an organic polymer of peptide-linked sugars that constitutes nearly 1% of the cell mass, approximately the same mass as that of DNA. Peptidoglycan limits the volume of the enclosed cell, so as water rushes in, it generates turgor pressure. This investment of biomass in the cell wall shows the importance (for most species) of maintaining turgor pressure in dilute environments, where water would otherwise enter by osmosis, causing osmotic shock (see Appendix 1 and Section 3.2).

Thought Question

3.1 Which chemicals do we find in the greatest number in a bacterial cell? The smallest number? Why does a cell contain 100 times as many lipid molecules as strands of RNA?

Cell Fractionation

Cell fractionation (Fig. 3.2) is how we separate cellular components such as membranes, ribosomes, and flagella. We can study these isolated parts in detail, though we lose information about interactions with other parts of the cell.

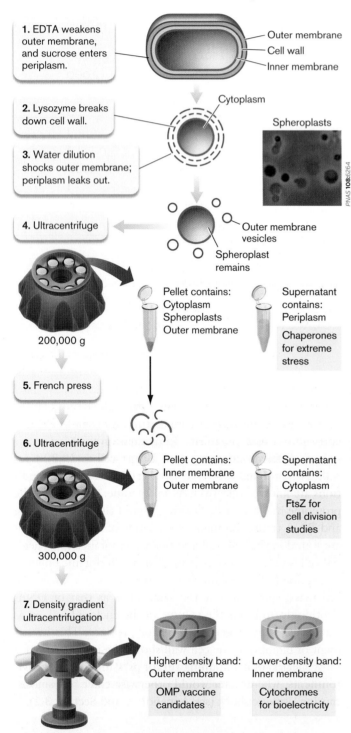

1. EDTA weakens outer membrane, and sucrose enters periplasm.

— Outer membrane
— Cell wall
— Inner membrane

2. Lysozyme breaks down cell wall.

Cytoplasm

Spheroplasts

PNAS **108**:6264

3. Water dilution shocks outer membrane; periplasm leaks out.

4. Ultracentrifuge

Outer membrane vesicles

Spheroplast remains

200,000 g

Pellet contains:
Cytoplasm
Spheroplasts
Outer membrane

Supernatant contains:
Periplasm

Chaperones for extreme stress

5. French press

6. Ultracentrifuge

Pellet contains:
Inner membrane
Outer membrane

Supernatant contains:
Cytoplasm

FtsZ for cell division studies

300,000 g

7. Density gradient ultracentrifugation

Higher-density band:
Outer membrane

Lower-density band:
Inner membrane

OMP vaccine candidates

Cytochromes for bioelectricity

FIGURE 3.2 ■ Fractionation of Gram-negative cells. Cell periplasm fills with sucrose, and lysozyme breaks down the cell wall. Dilution in water causes osmotic shock to the outer membrane, and periplasmic proteins leak out. Subsequent centrifugation steps separate the proteins of the periplasm, cytoplasm, and inner and outer membranes. *Photo Source:* Lars D. Rennera and Douglas B. Weibel. *PNAS* **108(15):**6264.

Cell fractionation also provides purified proteins that act as antigens for candidate vaccines. For example, a vaccine against *Neisseria meningitidis* type B (meningococcus) contains a highly immunogenic outer membrane protein.

How can we disassemble a cell to isolate its parts? Early-twentieth-century microbiologists wondered how to separate cell parts, without all the molecules mixing together. The answer was ultracentrifugation, the separation of molecules subject to high *g* forces (explained in eAppendix 3). Mary Jane Osborn at the University of Connecticut Health Center discovered that the inner and outer membranes of Gram-negative bacteria have different densities, and thus can be separated by density gradient ultracentrifugation.

Cell fractionation requires techniques that **lyse** (break open) the cell. The lysis method must generate enough force to separate the membrane lipids (held together by hydrophobic force) but not enough to disintegrate complexes of protein and RNA. For a Gram-negative cell, the method requires further specificity to separate "compartments" containing different sets of proteins: the inner and outer membranes, and the aqueous cytoplasm and periplasm.

Cell wall lysis and spheroplast formation. First, we suspend the cells in a sucrose solution containing ethylenediaminetetraacetic acid (EDTA), a molecule whose charged groups disrupt the outer membrane (**Fig. 3.2**, step 1). The disrupted membrane allows sucrose to cross. Sucrose fills the periplasm, maintaining an osmotically stable solution. Next, lysozyme cleaves peptidoglycan and thus breaks down the cell wall (step 2). Lacking the turgid cell wall, the cell swells into a sphere called a **spheroplast**.

The periplasm of the spheroplasts then undergoes osmotic shock, by transfer to distilled water (step 3). Water rushes in through the EDTA-weakened outer membrane, and the periplasm leaks out into the extracellular medium.

Note that other means of cell disruption may work better for other kinds of cells. Mild-detergent lysis can dissolve membranes without denaturing proteins. Alternatively, sonication is a way to lyse a cell by intense ultrasonic vibrations that are above the range of human hearing. For especially tough cells, such as cyanobacteria, a "bead beater" with microscopic glass beads can tear the cells open.

Ultracentrifugation. A key tool of cell fractionation is the **ultracentrifuge**, a device in which tubes containing solutions of cell components are spun at very high speed. The high rotation rate generates centrifugal forces strong enough to separate subcellular particles (see eAppendix 3). The particles are collected in fractions of sample from the tube, and then the fractions are observed by electron microscopy.

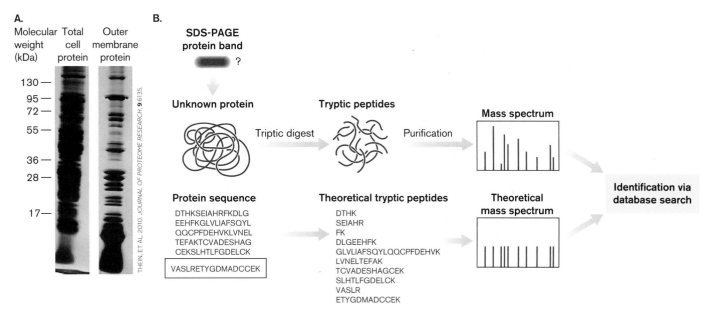

FIGURE 3.3 ■ Protein analysis. A. Gel electrophoresis of total cell proteins compared to outer membrane proteins from cell fractionation. **B.** Outer membrane proteins are identified by tryptic digest and mass spectrum analysis. The resulting peptide sequence is compared with those predicted from genome data.

Following osmotic shock, ultracentrifugation of the spheroplasts (**Fig. 3.2**, step 4) separates the periplasmic contents from the other three types of cell compartments (cytoplasm, inner membrane, and outer membrane fragments, which form vesicles). The periplasm is a valuable source of interesting proteins such as sugar transporters and chaperones (proteins that help other proteins fold under stress).

Spheroplast lysis. The pellet can now be further processed by a French press (**Fig. 3.2**, step 5), a device that squeezes cell contents through a narrow tube to break open the membranes. The broken membranes coalesce into tiny vesicles. A second step of ultracentrifugation now pellets the inner and outer membrane vesicles, while removing the cytoplasm in the supernatant (step 6). The cytoplasm provides many types of proteins for study, such as FtsZ, as well as complexes (multiprotein structures) such as the ribosomes and DNA polymerases.

Finally, the membrane fraction contains a mixture of very different components of the inner membrane (such as electron transport proteins) and the outer membrane (such as cell-surface proteins that we could use to make vaccines). We can separate the inner and outer membranes by density gradient ultracentrifugation (step 7). The gradient of solution density is generated by forming a gradient of sucrose concentration. In the gradient, the inner and outer membrane vesicles separate because of differences in their density, not their particle size. The lower-density fractions contain inner membrane vesicles, whereas the higher-density fractions contain outer membrane vesicles.

The membrane vesicle proteins are analyzed on electrophoretic gels (**Fig. 3.3**). We can identify the protein bands on the gel by enzyme digest and mass spectroscopy. The enzyme trypsin cleaves proteins only at aminoacyl residues lysine or arginine, thus generating peptides of defined composition. The peptide mass sizes are used to identify the cleaved protein, based on comparison with the protein sequence predicted by the organism's genome.

> **Thought Question**
>
> **3.2** Suppose we wish to isolate flagellar motors, which are protein complexes that span the envelope from inner membrane to outer membrane (**Fig. 3.1**). How might we modify the cell fractionation procedure?

Genetic Analysis

A limitation of cell fractionation is that it provides little information about processes that require an intact cell, such as cell division. How can we remove or alter a part of a cell without breaking it open?

An approach that is complementary to cell fractionation is genetic analysis (discussed in detail in Chapters 7–12). In genetic analysis, we can mutate a strain so as to lose or alter a gene; then we select mutant strains for loss of a given function. The phenotype of the mutant cell may yield clues about the function of the altered part.

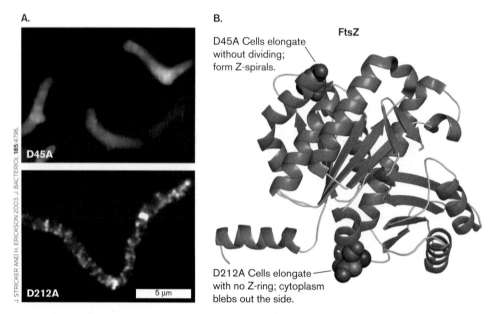

A.

B.

FtsZ

D45A Cells elongate without dividing; form Z-spirals.

D212A Cells elongate with no Z-ring; cytoplasm blebs out the side.

5 μm

J. STRICKER AND H. ERICKSON 2003 *J. BACTERIOL.* **185**:4796.

FIGURE 3.4 ■ **Genetic analysis of FtsZ. A.** *E. coli* with aspartate (D) at position 45 replaced by alanine (A) (D45A) elongate abnormally, forming blebs from the side, with no Z-rings. Cells with aspartate replaced by alanine at position 212 (D212A) elongate to form extended nondividing cells that contain spiral FtsZ complexes. FtsZ was visualized by immunofluorescence. **B.** Model of FtsZ protein monomer based on X-ray crystallography shows the position of the mutant residues, D212A and D45A.

For example, **Figure 3.4A** shows what happens to *E. coli* cells containing mutations in the gene encoding FtsZ. In certain strains, the *ftsZ* gene has one base pair altered by site-directed mutagenesis (see Chapter 12) so as to replace aspartate (D) at position 212 with alanine (A), designated D212A. The effect is to remove a negative charge at a defined position on FtsZ (**Fig. 3.4B**). As a result, cells extend into long filaments while failing to divide, and the FtsZ protein forms abnormal spiral complexes. A different mutant has aspartate replaced by alanine at position 45. The position-45 mutant cells elongate while forming abnormal blebs of cytoplasm off the side of the cell where fission fails. These experiments reveal functional parts of the FtsZ structure and clues to the formation of the Z-ring.

Different kinds of experimental methods may complement each other, to confirm or extend the conclusions from each. Another example is the analysis of the ribosome by cell fractionation, crystallography, and genetic analysis. Ribosome analysis is presented in **eTopic 3.1**.

To Summarize

- **Bacterial cells are protected by a thick cell envelope.** The envelope includes a cell membrane and a peptidoglycan cell wall. A Gram-negative cell includes an outer membrane, and the cell membrane is called the inner membrane.

- **Bacteria are composed of nucleic acids, proteins, phospholipids, and other organic and inorganic chemicals.** Proteins in the cell vary, depending on the species and environmental conditions.

- **The bacterial cytoplasm is highly structured.** DNA replication, RNA transcription, and protein synthesis occur coordinately within the cytoplasm.

- **Microscopy reveals cell structure.** Transmission electron microscopy shows how cell parts fit within the cell as a whole. Fluorescence microscopy reveals the location and dynamics of individual components.

- **Cell fractionation isolates cell parts for structural and biochemical analysis.** The compartments of a Gram-negative cell can be separated by cell lysis and ultracentrifugation.

- **Genetic analysis of mutants reveals functions of cell parts.** Genetic analysis complements cell fractionation.

3.2 The Cell Membrane and Transport

How does a cell distinguish "itself" from what is outside? The structure that defines the existence of a cell is the cell membrane (**Fig. 3.5**). Overall, the membrane contains the cytoplasm within the external medium, mediating exchange between the two. The cell membrane consists of a phospholipid bilayer containing lipid-soluble proteins. It behaves as a two-dimensional fluid, within which proteins and lipids can diffuse. The proteins form about half the mass of the membrane and provide specific functions, such as nutrient transport. For review of elementary cell structure, see eAppendix 2.

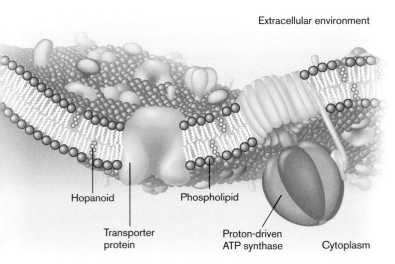

Hopanoid

Phospholipid

Transporter
protein

Proton-driven
ATP synthase

Cytoplasm

Extracellular environment

FIGURE 3.5 ■ Bacterial cell membrane. The cell membrane consists of a phospholipid bilayer, with hydrophobic fatty acid chains directed inward, away from water. The bilayer contains stiffening agents such as hopanoids. Half the membrane volume consists of proteins.

Membrane Lipids

Most membrane lipids are **phospholipids**. A phospholipid possesses a charged phosphoryl "head" that contacts the water interface, as well as a hydrophobic "tail" of fatty acids packed within the bilayer. Lipid biosynthesis is a key process that is vulnerable to antibiotics. For example, the bacterial enzyme enoyl reductase, which synthesizes fatty acids (discussed in Chapter 15) is the target of triclosan, a common antibacterial additive in detergents and cosmetics.

A typical phospholipid consists of glycerol with ester links to two fatty acids and a phosphoryl polar head group, which at neutral pH is deprotonated (negatively charged) (**Fig. 3.6**). This kind of phospholipid is called a phosphatidate. The negatively charged head group of the phosphatidate can contain various organic groups, such as glycerol to form phosphatidylglycerol (**Fig. 3.6A**). In other lipids, the polar head group has a side chain with positive charge. The positive charge commonly resides on an amine group, such as ethanolamine in phosphatidylethanolamine (**Fig. 3.6B**). Phospholipids with positive charge or with mixed charges are concentrated in portions of the membrane that interact with DNA, which has negative charge.

In the bilayer, all phospholipids face each other tail to tail, keeping their hydrophobic side chains away from the water inside and outside the cell. The two layers of phospholipids in the bilayer are called **leaflets**. One leaflet of phospholipids faces the cell interior; the other faces the exterior. As a whole, the phospholipid bilayer imparts fluidity and gives the membrane a consistent thickness (about 8 nm).

Membrane Proteins

Membrane proteins serve functions such as transport, communication with the environment, and structural support.

■ **Structural support.** Some membrane proteins anchor together different layers of the cell envelope (discussed in Section 3.3). Other proteins attach the membrane to the cytoskeleton, or form the base of structures extending out from the cell, such as flagella.

A.

B.

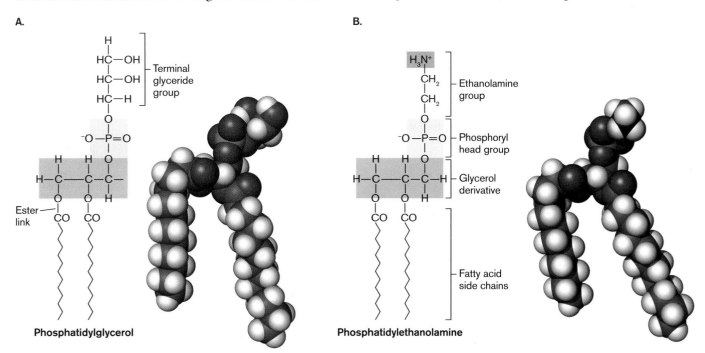

Phosphatidylglycerol

Phosphatidylethanolamine

FIGURE 3.6 ■ Phospholipids. A. Phosphatidylglycerol consists of glycerol with ester links to two fatty acids, and a phosphoryl group linked to a terminal glycerol. **B.** Phosphatidylethanolamine contains a glycerol linked to two fatty acids, and a phosphoryl group with a terminal ethanolamine. The ethanolamine carries a positive charge.

- **Detection of environmental signals.** In *Vibrio cholerae*, the causative agent of cholera, the membrane protein ToxR detects acidity and elevated temperature—signs that the bacteria is in the host's digestive tract. The ToxR domain facing the cytoplasm then binds to a DNA sequence, activating expression of cholera toxin.

- **Secretion of virulence factors and communication signals.** Membrane protein complexes export toxins and cell signals across the envelope. For example, symbiotic nitrogen-fixing rhizobia require membrane proteins NodI and NodJ to transport nodulation signals out to the host plant roots, inducing the plant to form root nodules containing the bacteria.

- **Ion transport and energy storage.** Transport proteins manage ion flux between the cell and the exterior. Ion transport generates gradients that store energy.

An example of a bacterial membrane protein is the leucine transporter LeuT (**Fig. 3.7**). LeuT drives uptake of leucine, coupled to a gradient of sodium ions. The protein complex was purified for X-ray diffraction from *Aquifex aeolicus*, a thermophile whose heat-stable proteins form durable crystals. Remarkably, LeuT is homologous (shares common ancestry) with a human neuron protein that transports neurotransmitters. Thus, this bacterial protein serves as a model for study of neuron function.

Proteins embedded in a membrane require a hydrophobic portion that is soluble within the membrane. Typically, several hydrophobic alpha helices thread back and forth through the membrane. Other peptide regions extend outside the membrane, containing charged and polar amino acids that interact favorably with water. **Figure 3.7** shows the LeuT charge distribution. Hydrophobic amino acid residues (white) make the protein soluble in the membrane, while portions with negative charge (red) and positive charge (blue) lock the protein in.

Transport across the Cell Membrane

The cell membrane acts as a barrier to keep water-soluble proteins and other cell components within the cytoplasm. But how do nutrients from outside get into the cell—and how do secreted products such as toxins get out? Specific membrane proteins transport molecules across the membrane between the cytoplasm and the outside. Selective transport is essential for cell survival; it means the ability to acquire scarce nutrients, exclude waste, and transmit signals to neighbor cells.

Passive diffusion. Small uncharged molecules, such as O_2, CO_2, and water, easily permeate the membrane. Some molecules, such as ethanol, also disrupt the membrane—an action that can make such molecules toxic to cells. By contrast, large, strongly polar molecules such as sugars, and charged molecules such as amino acids, generally cannot penetrate the hydrophobic interior of the membrane, and thus require transport by specific proteins. Water molecules permeate the membrane, but their rate of passage is increased by protein channels called aquaporins.

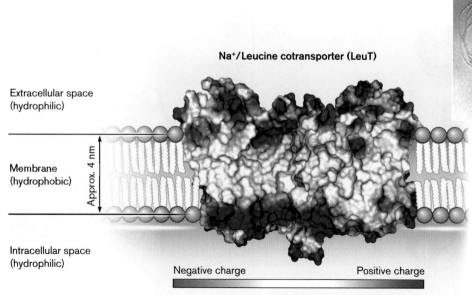

Na+/Leucine cotransporter (LeuT)

Extracellular space (hydrophilic)

Membrane (hydrophobic) — Approx. 4 nm

Intracellular space (hydrophilic)

Negative charge — Positive charge

FIGURE 3.7 ■ **A cell membrane–embedded transport protein: the LeuT sodium/leucine cotransporter of *Aquifex* bacteria.** The protein complex carries leucine across the cell membrane into the cytoplasm, coupled to sodium ion influx. (PDB code: 3F3E) **Inset:** *Aquifex aeolicus* grows at 96°C in hot springs. *Source:* Karl Stetter and Reinhardt Rachel, U. Regensburg, Germany.

Osmosis. Most cells maintain a concentration of total solutes (molecules in solution) that is higher inside the cell than outside. As a result, the internal concentration of water is lower than the concentration outside the cell. Because water can cross the membrane but charged solutes cannot, water tends to diffuse across the membrane into the cell, causing the expansion of cell volume, in a process called osmosis. The resulting pressure on the cell membrane is called **osmotic pressure** (see figure in Appendix 1). Osmotic pressure will cause a cell to burst, or lyse, in the absence of a countering pressure such as that provided by the cell wall. That is how bacteria are killed by penicillin, which disrupts cell wall synthesis.

Membrane-permeant weak acids and bases. A special case of movement across cell membranes is that of **membrane-permeant weak acids** and **weak bases** (**Fig. 3.8**), which exist in equilibrium between charged and uncharged forms:

$$\text{Weak acid: HA} \rightleftharpoons \text{H}^+ + \text{A}^-$$

$$\text{Weak base: B} + \text{H}_2\text{O} \rightleftharpoons \text{BH}^+ + \text{OH}^-$$

Membrane-permeant weak acids and weak bases cross the membrane in their uncharged form: HA (weak acid) or B (weak base). On the other side, entering the aqueous cytoplasm, the acid dissociates (HA to A⁻ and H⁺) or the base reassociates with H⁺ (B to BH⁺). In effect, membrane-permeant acids conduct acid (H⁺) across the membrane, causing acid stress; similarly, membrane-permeant bases conduct OH⁻ across the membrane, causing alkali stress. If the H⁺ concentration (acidity) outside the cell is greater than inside, it will drive weak acids into the cell.

Many key substances in cellular metabolism are membrane-permeant weak acids and bases, such as acetic acid. Most pharmaceutical drugs—therapeutic agents delivered to our tissues via the bloodstream—are weak acids or bases whose uncharged forms exist at sufficiently low concentration to cross the membrane without disrupting it. Examples of weak acids that deprotonate (acquiring negative charge) at neutral pH include aspirin (acetylsalicylic acid) and penicillin (**Fig. 3.8A**). Examples of weak bases that protonate (acquiring positive charge) at neutral pH include Prozac (fluoxetine) and tetracycline (**Fig. 3.8B**).

> **Thought Question**
>
> **3.3** Amino acids have acidic and basic groups that can dissociate. Why are they <u>not</u> membrane-permeant weak acids or weak bases? Why do they fail to cross the phospholipid bilayer?

Transmembrane ion gradients. Molecules that carry a fixed charge, such as hydrogen and sodium ions (H⁺ and Na⁺), cannot cross the phospholipid bilayer. Such ions

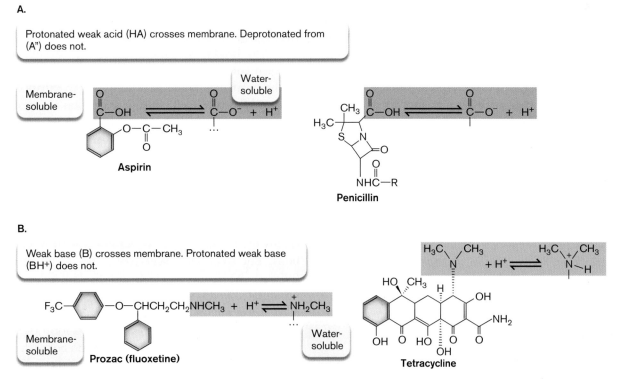

FIGURE 3.8 ■ Common drugs are membrane-permeant weak acids and bases. In its charged form (A⁻ or BH⁺), each drug is soluble in the bloodstream. The uncharged form (HA or B) is hydrophobic and penetrates the cell membrane.

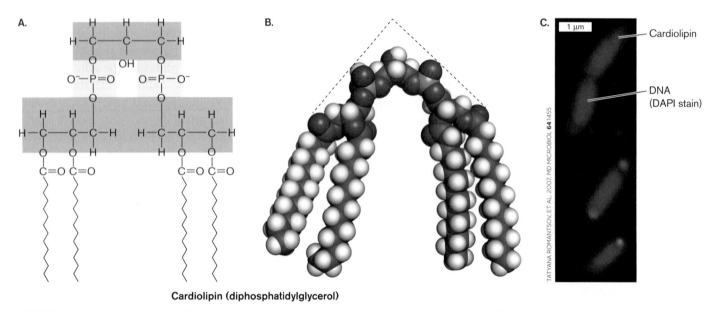

Cardiolipin (diphosphatidylglycerol)

FIGURE 3.9 ▪ Cardiolipin localizes to the poles. **A.** Cardiolipin is a double phospholipid joined by a third glycerol. **B.** A space-filling model of cardiolipin shows its triangular shape. **C.** Cardiolipin localizes to the bacterial cell poles, as shown by microscopy with a cardiolipin-specific fluorophore.

usually exist in very different concentrations inside and outside the cell. An **ion gradient** (ratio of ion concentrations) across the cell membrane can store energy for nutrition or to drive the transport of other molecules. Inorganic ions require transport through specific **transport proteins**, or **transporters**. So, too, do organic molecules that carry a charge at cytoplasmic pH, such as amino acids and vitamins. Transport may be passive or active. In **passive transport**, molecules accumulate or dissipate along their concentration gradient. **Active transport**—that is, transport from lower to higher concentration—requires cells to spend energy. A transporter protein obtains energy for active transport by cotransport of another substance down its gradient from higher to lower concentration, or by coupling transport to a chemical reaction (discussed in Chapter 4).

Membrane Lipid Diversity

Membranes require a uniform thickness and stability to maintain structural integrity and function. So why do individual membrane lipids differ in structure? Different environments favor different forms of membrane lipids. For example, lipid structure helps determine whether an organism can grow in a hot spring, or whether it can colonize human lungs.

Environmental stress. Starvation stress increases bacterial production of lipids with an unusual type of phosphoryl head group. **Cardiolipin**, or diphosphatidylglycerol,

is actually a double phospholipid linked by a glycerol (**Fig. 3.9A**). Cardiolipin concentration increases in bacteria grown to starvation or stationary phase (discussed in Chapter 4). Within a cell, cardiolipin does not diffuse at random; it concentrates in patches called "domains" near the cell poles. The polar localization of cardiolipin was demonstrated by fluorescence microscopy, in which a cardiolipin-specific fluorophore localized to the poles of *E. coli* (**Fig. 3.9B**). The "wedge" shape of cardiolipin, with its narrow head group and wide fatty acid group, is thought to form concave domains of lipid that stabilize the curve of the polar membrane. Cardiolipin may enhance the formation of smaller cells during starvation. How does this benefit the cell? At the cell pole (**Fig. 3.9C**), cardiolipin binds certain environmental stress proteins, such as a protein that transports osmoprotectants when the cell is under osmotic stress. Thus, a phospholipid can have specific functions associated with specific membrane proteins.

The fatty acid component of phospholipids also varies. The most common bacterial fatty acids are hydrogenated chains of varying length, typically between 6 and 22 carbons. But some fatty acid chains are partly unsaturated (possess one or more carbon-carbon double bonds). Most unsaturated bonds in membranes are *cis*, meaning that both alkyl chains are on the same side of the bond, so the unsaturated chain has a "kink," as in the *cis* form of oleic acid (**Fig. 3.10**). Because the kinked chains do not pack as closely as the straight hydrocarbon chains do, the membrane is more "fluid." This is why, at room temperature, unsaturated vegetable oils are fluid, whereas highly saturated butterfat

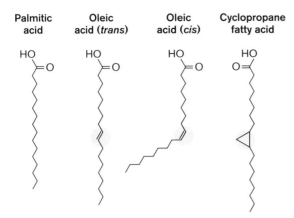

FIGURE 3.10 ■ Phospholipid side chains.

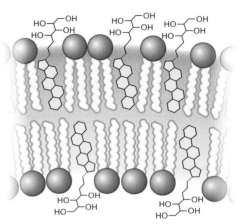

FIGURE 3.11 ■ Hopanoids add strength to membranes. Hopanoids limit the motion of phospholipid tails, thus stiffening the membrane.

is solid. The enhanced fluidity of a kinked phospholipid improves the function of the membrane at low temperature; hence, bacteria can respond to cold and heat by increasing or decreasing their synthesis of unsaturated phospholipids.

Another interesting structural variation is cyclization of part of the chain to form a stiff planar ring with decreased fluidity. The double bond of unsaturated fatty acids can incorporate a carbon from *S*-adenosyl-L-methionine to form a three-membered ring, generating a cyclopropane fatty acid (**Fig. 3.10**). Bacteria convert unsaturated fatty acids to cyclopropane during starvation and acid stress, conditions under which membranes require stiffening. Cyclopropane conversion is an important factor in the pathogenesis of *Mycobacterium tuberculosis* and in the acid resistance of food-borne toxigenic *E. coli*.

In addition to phospholipids, membranes include planar molecules that fill gaps between hydrocarbon chains. These stiff, planar molecules reinforce the membrane, much as steel rods reinforce concrete. In eukaryotic membranes, the reinforcing agents are sterols, such as **cholesterol**. In some bacteria, the same function is filled by pentacyclic (five-ring) hydrocarbon derivatives called **hopanoids**, or hopanes (**Fig. 3.11**). Like cholesterol, hopanoids fit between the fatty acid side chains of membranes and limit their motion, thus stiffening the membrane. Hopanoids appear in geological sediments, where they indicate ancient bacterial decomposition; they provide useful data for petroleum exploration.

Archaea have unique membrane lipids. The membrane lipids of archaea differ fundamentally from those of bacteria and of eukaryotes. All archaeal phospholipids replace the ester link between glycerol and fatty acid with an ether

link, C–O–C (**Fig. 3.12**). Ethers are much more stable than esters, which hydrolyze easily in water. This is one reason why some archaea can grow at higher temperatures than all other forms of life. Another modification is that archaeal hydrocarbon chains are branched **terpenoids**, polymeric structures derived from isoprene, in which every fourth carbon extends a methyl branch. The branches strengthen the membrane by limiting movement of the hydrocarbon chains.

The most extreme hyperthermophiles, which live beneath the ocean at 110°C, have terpenoid chains linked at the tails, forming a tetraether monolayer. In some species, the terpenoids cyclize to form cyclopentane rings. These planar rings stiffen the membrane under stress to an even greater extent than the cyclopropyl chains of bacteria. For more on archaeal cells, see Chapter 19.

FIGURE 3.12 ■ Terpene-derived lipids of archaea. In archaea, the hydrocarbon chains are ether-linked to glycerol, and every fourth carbon has a methyl branch. In some archaea, the tails of the two facing lipids of the bilayer are fused, forming tetraethers; thus, the entire membrane consists of a monolayer.

Interestingly, both hopanoids and cholesterol are synthesized from the same precursor molecules as the unique lipids of archaea (**Fig. 3.13**). It may be that hopanoids and cholesterol persist in bacteria and eukaryotes as derivatives of lipids that were once possessed by a common ancestor of all three domains.

To Summarize

- **The cell membrane** consists of a phospholipid bilayer containing hydrophobic membrane proteins. Membrane proteins serve diverse functions, including transport, cell defense, and cell communication.

- **Small uncharged molecules**, such as oxygen, can penetrate the cell membrane by diffusion.

- **Membrane-permeant weak acids and weak bases** exist partly in an uncharged form that can diffuse across the membrane and increase or decrease, respectively, the H^+ concentration within the cell.

- **Polar molecules and charged molecules** require membrane **proteins to mediate** transport. Such facilitated transport can be active or passive.

- **Ion gradients** generated by membrane pumps store energy for cell functions. Active transport (up the concentration gradient) requires input of energy, whereas passive transport (down a gradient) does not.

- **Diverse fatty acids** are found in different microbial species and in microbes adapted to different environments.

- **Archaeal membranes have ether-linked terpenoids**, which confer increased stability at high temperature and extreme acidity. Some archaea have diglycerol tetraethers, which form a monolayer.

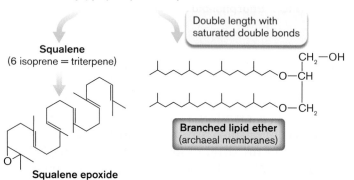
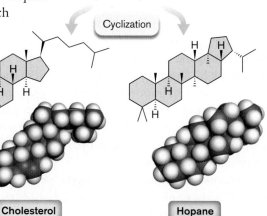

FIGURE 3.13 ■ **Synthesis of terpene-derived lipids.** Archaea synthesize their membrane lipids from isoprene chains, to form terpenoids such as squalene. In bacteria and eukaryotes, squalene is converted to cholesterol; and in some bacteria, squalene cyclizes to form hopanes (hopanoids).

3.3 The Cell Wall and Outer Layers

How do bacteria and archaea protect their cell membrane? For most species, the cell envelope includes at least one structural supporting layer, like an external skeleton, located outside the cell membrane. The most common structural support is the cell wall (see **Fig. 3.1**). Many species possess additional coverings, such as an outer membrane or an S-layer. Nevertheless, a few prokaryotes, such as the mycoplasmas, have a cell membrane with no outer layers, depending on host fluids for osmotic balance. Some archaea with only a cell membrane grow in extreme acid (pH zero); how they survive is unknown.

The Cell Wall Is a Single Molecule

The bacterial cell wall, also known as the **sacculus**, consists of a single interlinked molecule that envelops the cell. The sacculus has been isolated from *E. coli* and visualized by TEM; in **Figure 3.14A**, the isolated sacculus appears flattened on the sample grid like a deflated balloon. Its

A.

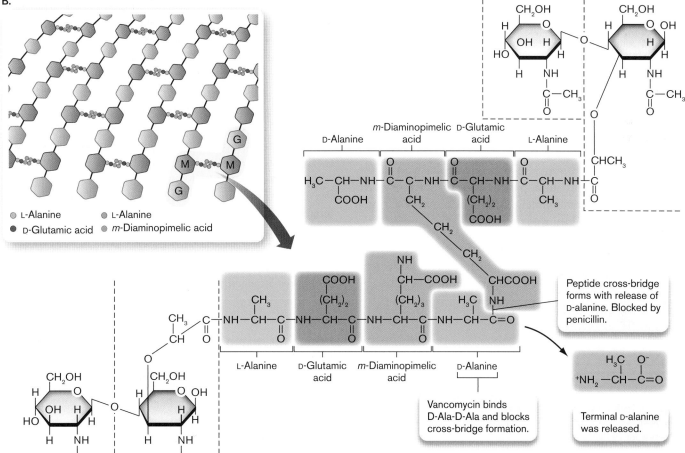

FEMS MICROBIOLOGY REVIEWS/OXFORD JOURNALS. **32(2)** 149, FIG 3A

FIGURE 3.14 ■ The peptidoglycan sacculus and peptidoglycan cross-bridge formation. **A.** Isolated sacculus from *Escherichia coli* (TEM). **B.** A disaccharide unit of glycan has an attached peptide of four to six amino acids.

geometrical structure encloses maximal volume with minimal surface area. The sacculus—unlike the membrane—is a single-molecule cage-like structure, highly porous to ions and organic molecules. The mesh grows by strand insertion and elongation in arcs around the cell. The cage-like form is not rigid; it is more like a flexible mesh bag with unbreakable joints. Turgor pressure within the enclosed cytoplasm fills out the cell's shape, such as a rod or a spherical coccus.

Peptidoglycan structure. Most bacterial cell walls are composed of peptidoglycan, a polymer of peptide-linked chains of amino sugars. Peptidoglycan is synonymous with **murein** ("wall molecule"). The molecule consists of parallel polymers of disaccharides called **glycan** chains cross-linked with peptides of four amino acids (**Fig. 3.14B**). Peptidoglycan is unique to bacteria, although some archaea build analogous structures whose overall physical nature is similar. (Archaeal cell walls are presented in Chapter 19.)

The long chains of peptidoglycan consist of repeating units of a disaccharide composed of *N*-acetylglucosamine (an amino sugar derivative) and *N*-acetylmuramic acid (glucosamine plus a lactic acid group) (**Fig. 3.14B**). The lactate group of muramic acid forms an amide link with the amino terminus of a short peptide containing four to six amino acid residues. The peptide extension can form **cross-bridges** connecting parallel strands of glycan.

The peptide contains two amino acids in the unusual D mirror form: D-glutamic acid and D-alanine. The third amino acid, *m*-diaminopimelic acid, has an extra amine group, which forms an amide link to a cross-bridged peptide. The amide link forms with the fourth amino acid of the adjacent peptide, D-alanine (1) (**Fig. 3.14B**). The cross-bridge forms by removal of a second D-alanine (2) at the end of the chain. The cross-linked peptides of neighboring glycan strands form the cage of the sacculus.

Note: Amino acids have two forms that are mirror opposites, D and L, of which only the L form is incorporated by ribosomes into protein. The D-form amino acids, however, are used by microbes for many nonprotein structural molecules.

The details of peptidoglycan structure vary among bacterial species. Some Gram-positive species, such as *Staphylococcus aureus* (a cause of toxic shock syndrome), have peptides linked by bridges of pentaglycine instead of the D-alanine link to *m*-diaminopimelic acid. In Gram-negative species, the *m*-diaminopimelic acid is linked to the outer membrane, as discussed shortly.

Peptidoglycan synthesis as a target for antibiotics. Synthesis of peptidoglycan requires many genes encoding enzymes to make the special sugars, build the peptides, and seal the cross-bridges. Many of these enzymes bind the antibiotic penicillin and are thus known as **penicillin-binding proteins**. Because peptidoglycan is unique to bacteria, enzymes of peptidoglycan biosynthesis make excellent targets for new antibiotics (see **Fig. 3.14B**). For example, the transpeptidase that cross-links the peptides is the target of penicillin. Vancomycin, a major defense against *Clostridium difficile* and drug-resistant staphylococci, prevents cross-bridge formation by binding the terminal D-Ala-D-Ala dipeptide. Vancomycin binding prevents release of the terminal D-alanine.

Unfortunately, the widespread use of such antibiotics selects for evolution of resistant strains. One of the most common agents of resistance is the enzyme beta-lactamase, which cleaves the lactam ring of penicillin, rendering it ineffective as an inhibitor of transpeptidase. In a different mechanism, strains resistant to vancomycin contain an altered enzyme that adds lactic acid to the end of the branch peptides in place of the terminal D-alanine. The altered peptide is no longer blocked by vancomycin. As new forms of drug resistance emerge, researchers continue to seek new antibiotics that target cell wall formation (discussed in Chapter 27).

How does peptidoglycan grow, overall, throughout the elongating cell? Interestingly, different kinds of bacteria have evolved to organize their cell wall growth differently. Yves Brun and his student Erkin Kuru, at Indiana University, devised an ingenious way to reveal the growth pattern (**Fig. 3.15**). Kuru designed D-amino acids with fluorophores attached that the growing cell wall incorporates.

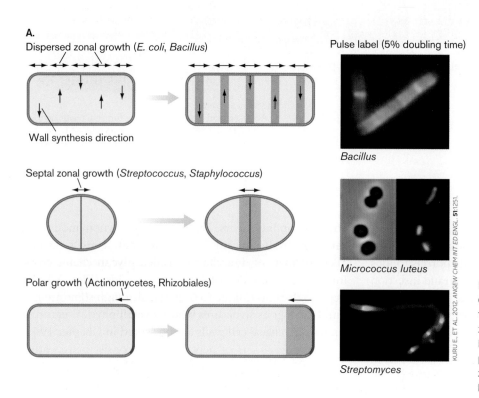

A.
Dispersed zonal growth (*E. coli*, *Bacillus*)
Wall synthesis direction

Septal zonal growth (*Streptococcus*, *Staphylococcus*)

Polar growth (Actinomycetes, Rhizobiales)

Pulse label (5% doubling time)
Bacillus
Micrococcus luteus
Streptomyces

KURU E., ET AL. 2012. ANGEW CHEM INT ED ENGL. **51**:1251.

© JEAN-FRANÇOIS GOUT

B.

FIGURE 3.15 ■ Peptidoglycan growth in different species. A. Different species of bacteria synthesize new peptidoglycan in dispersed zones, at the septum only, or at the poles. Fluorescent D-amino acids are added for short periods (pulse labeling) to reveal the growth zones. **B.** Erkin Kuru, in the laboratory of Yves Brun at Indiana University, devised the fluorescent D-amino acids to probe cell wall growth.

Because these are D-amino acids and not L-amino acids, ribosome-directed translation does not use them. Thus, the fluorophores label only cell wall, not proteins. Kuru found that bacteria such as *E. coli* and *Bacillus subtilis* synthesize cell wall in zones dispersed throughout the cell. Gram-positive cocci, however, such as *Streptococcus* and *Staphylococcus*, synthesize cell wall only at the midpoint (or septum), where the cell is about to fission (discussed in the next section). Still other bacteria, such as the actinomycete *Streptomyces*, form new cell wall only at the poles.

Thought Question

3.4 What other ways can you imagine that bacteria might mutate to become resistant to vancomycin?

Cell Envelope of Bacteria

Most bacteria have additional envelope layers that provide structural support and protection from predators and host defenses (**Fig. 3.16**). Additional molecules are attached to the cell wall and cell membrane, and some thread through the layers. Here we present the envelope composition of three major kinds of bacteria, two of which (Firmicutes and Proteobacteria) are distinguished by the Gram stain. The third, Mycobacteria, is distinguished by the acid-fast stain (discussed in Chapter 2).

- **Firmicutes (Gram-positive)** have a thick cell wall with 3–20 layers of peptidoglycan, interpenetrated by teichoic acids. The phylum Firmicutes consists of Gram-positive species such as *Bacillus thuringiensis* and *Streptococcus pyogenes*, the cause of strep throat.

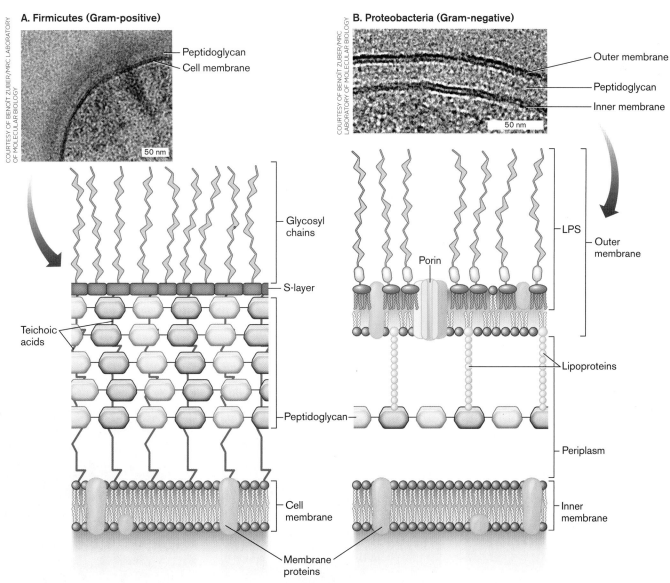

FIGURE 3.16 ▪ Cell envelope: Gram-positive (Firmicutes) and Gram-negative (Proteobacteria). A. Firmicutes (Gram-positive) cells have a thick cell wall with multiple layers of peptidoglycan, threaded by teichoic acids. **A inset:** Gram-positive envelope of *Bacillus subtilis* (TEM). **B.** Proteobacteria (Gram-negative) cells have a single layer of peptidoglycan covered by an outer membrane; the cell membrane is called the inner membrane. **B inset:** Gram-negative envelope of *Pseudomonas aeruginosa* (TEM).

▪ **Proteobacteria (Gram-negative)** have a thin cell wall with one or two layers of peptidoglycan, enclosed by an outer membrane. The phylum Proteobacteria consists of Gram-negative species such as *Escherichia coli* and nitrogen-fixing *Sinorhizobium meliloti*.

▪ **Mycobacteria** of the phylum Actinomycetes have a complex multilayered envelope that includes defensive structures such as mycolic acids. Examples include *Mycobacterium tuberculosis*, the cause of tuberculosis, and *M. leprae*, the cause of leprosy.

Note that other important kinds of bacteria, such as cyanobacteria and spirochetes, have very different envelopes. These different envelope structures may stain Gram-positive, Gram-negative, or variable (discussed in Chapter 18). Archaea have yet other diverse kinds of envelopes that cannot be distinguished by Gram stain (see Chapter 19).

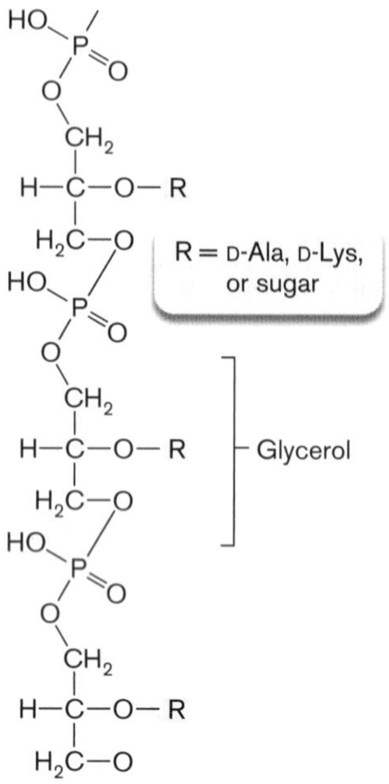

Thought Question

3.5 Figure 3.16 highlights the similarities and differences between the cell envelopes of Gram-negative and Gram-positive bacteria. What do you think are the advantages and limitations of a cell's having one layer of peptidoglycan (Gram-negative) versus several layers (Gram-positive)?

Firmicute Cell Envelope—Gram-Positive

A section of a Firmicute cell envelope (Gram-positive) is shown in **Figure 3.16**. The multiple layers of peptidoglycan are reinforced by **teichoic acids** threaded through its multiple layers (**Fig. 3.17**). Teichoic acids are chains of phosphodiester-linked glycerol or ribitol, with sugars or amino acids linked to the middle –OH groups. The

FIGURE 3.17 ▪ Teichoic acids. Teichoic acids in the Gram-positive cell wall consist of glycerol or ribitol phosphodiester chains.

R = D-Ala, D-Lys, or sugar

Glycerol

negatively charged cross-threads of teichoic acids, as well as the overall thickness of the Gram-positive cell wall, help retain the Gram stain.

How does the cell wall attach extracellular structures? Gram-positive bacteria have a type of enzyme called "sortase" that forms a peptide bond from a cell wall cross-bridge to a protein extending from the cell. Proteins attached by sortases can help the cell acquire nutrients, or help the cell adhere to a substrate. Sortases are now used in the protein engineering industry.

S-layer. Free-living bacteria and archaea often possess a tough surface layer called the **S-layer**. **Figure 3.18** shows the S-layer of *Lysinibacillus sphaericus*, a Gram-positive

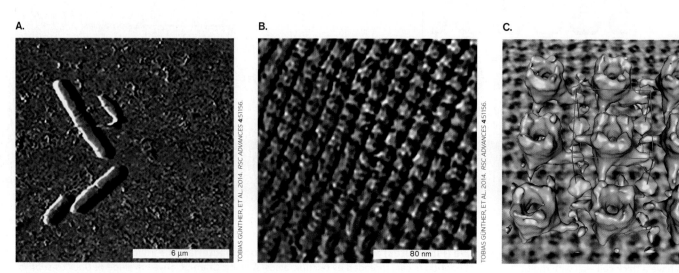

A.　6 µm
B.　80 nm
C.

TOBIAS GUNTHER, ET AL. 2014. *RSC ADVANCES* 4:51156.
LUIS COMOLLI

FIGURE 3.18 ▪ S-layer of *Lysinibacillus* bacteria. A. *Lysinibacillus* imaged by atomic force microscopy (AFM). **B.** S-layer surface imaged by AFM. **C.** Computational model of S-layer based on cryo-EM.

A. Gram-negative envelope

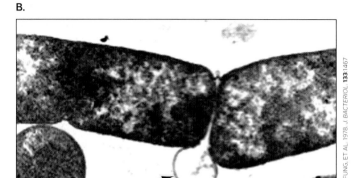

FIGURE 3.19 ■ Gram-negative cell envelope.
A. Murein lipoprotein has an N-terminal cysteine triglyceride inserted in the inward-facing leaflet of the outer membrane. The C-terminal lysine forms a peptide bond with the *m*-diaminopimelic acid of the peptidoglycan (murein) cell wall. **B.** Lack of murein lipoprotein in mutant *Salmonella* causes the outer membrane to balloon out (arrow), when the cell tries to divide (TEM).

JOAN FUNG, ET AL. 1978. *J. BACTERIOL.* **133**:1467

environments, such as the human lung, where they are protected from osmotic shock.

Capsule. Another common extracellular structure is the **capsule**, a slippery layer of loosely bound polysaccharides. The capsule of pathogens such as *Staphylococcus aureus* can prevent phagocytosis by white blood cells, thereby enabling the pathogen to persist in the blood.

Thought Question

3.6 Why would laboratory culture conditions select for evolution of cells lacking an S-layer?

Proteobacterial Cell Envelope— Gram-Negative

A cell envelope of Proteobacteria (Gram-negative) is shown in **Figures 3.16B** and **3.19**. The envelope includes one or two layers of peptidoglycan covered by an outer membrane. The Gram-negative outer membrane confers defensive abilities and toxigenic properties on many pathogens, such as *Salmonella* species and enterohemorrhagic *E. coli* (strains that cause hemorrhaging of the colon). Between the outer and inner (cell) membranes, the aqueous compartment (containing the cell wall) is called the periplasm.

Lipoprotein and lipopolysaccharide (LPS). The inward-facing leaflet of the outer membrane has a phospholipid composition similar to that of the cell membrane (which

bacterium found on the surface of beets and carrots. The S-layer is a crystalline sheet of thick subunits consisting of protein or glycoprotein (proteins with attached sugars). Each subunit contains a pore large enough to admit a wide range of molecules (**Fig. 3.18**). As modeled by cryo-EM tomography (see Chapter 2), the S-layer proteins are arranged in a highly ordered array, either hexagonally or tetragonally. The S-layer is rigid, but it also flexes and allows substances to pass through it in either direction. S-layers help pathogens such as *Bacillus anthracis* bind and attack host cells. An S-layer contributes to biofilm formation (the periodontal bacterium *Tannerella forsythia*) and swimming (the aquatic cyanobacterium *Synechococcus* species).

The functions of the S-layer are hard to study in the laboratory because the S-layer is often lost by bacteria after repeated subculturing. Traits commonly diminish in the absence of selective pressure for genes encoding them— a process called reductive evolution (discussed in Chapter 17). For example, the mycoplasmas are close relatives of Gram-positive bacteria that have permanently lost their cell walls, as well as the S-layer. Mycoplasmas have no need for cell walls, because they are parasites living in host

in Gram-negative species is called the **inner membrane** or inner cell membrane). The outer membrane's inward-facing leaflet includes lipoproteins that connect the outer membrane to the peptide bridges of the cell wall. The major lipoprotein is called **murein lipoprotein**, also known as Braun lipoprotein (**Fig. 3.19A**). Murein lipoprotein consists of a protein with an N-terminal cysteine attached to three fatty acid side chains. The side chains are inserted in the inward-facing leaflet of the outer membrane. The C-terminal lysine forms a peptide bond with the *m*-diaminopimelic acid of peptidoglycan (murein). What happens to a mutant cell that fails to make murein lipoprotein? As the cell grows and divides, it fails to attach its outer membrane to the growing cell wall, causing the outer membrane to balloon out in the region where the daughter cells separate (**Fig. 3.19B**).

The outward-facing leaflet of the outer membrane has very different lipids from the inner leaflet. The main outward-facing phospholipids are called **lipopolysaccharide** (**LPS**) (**Fig. 3.20**). LPS is of crucial medical importance because it acts as an **endotoxin**. An endotoxin is a cell component that is harmless as long as the pathogen remains intact; but when released by a lysed cell, endotoxin overstimulates host defenses, inducing potentially lethal endotoxic shock. Thus, antibiotic treatment of an LPS-containing pathogen can kill the cells but can also lead to death of the patient.

The membrane-embedded anchor of LPS is **lipid A**, a molecule shaped like a six-legged giraffe (**Fig. 3.20A**). The lipid A moiety is the endotoxic part of LPS. The molecule's six fatty acid "legs" have shorter chains than those of the inner cell membrane, and two pairs are branched. The fatty acids have ester or amide links to the "body," a dimer of glucosamine (an amino sugar also found in peptidoglycan). Analogous to the glycerol of glyceride phospholipids, each glucosamine has a phosphoryl group whose negative charge interacts with water. One glucosamine extends the long "neck" of the core polysaccharide, a sugar chain that reaches far outside the cell (**Fig. 3.20B**). The core polysaccharide consists of five to ten sugars with side chains such as phosphoethanolamine. It extends to an **O antigen** or O polysaccharide, a chain of as many as 200 sugars. The O polysaccharide may be longer than the cell itself. These chains of sugars form a layer that helps bacteria resist phagocytosis by white blood cells. The combination of sugar units in the O antigen varies greatly among different strains and species of Gram-negative bacteria.

Outer membrane proteins. The outer membrane contains unique proteins not found in the inner membrane. Outer membranes contain a class of transporters called **porins** that permit the entry of nutrients such as sugars and peptides (**Fig. 3.21**). Outer membrane porins such as OmpF have a distinctive cylinder of beta sheet conformation (reviewed in eAppendix 1), also known as a beta barrel. A typical outer membrane porin exists as a trimer of beta barrels, each of which acts as a pore for nutrients.

Outer membrane porins have limited specificity, allowing passive uptake of various molecules—including antibiotics such as ampicillin. Ampicillin is a form of penicillin, which must get through the outer membrane to access the cell wall in order to block the formation of peptide cross-bridges. Ampicillin contains two charged groups and is thus unlikely to diffuse

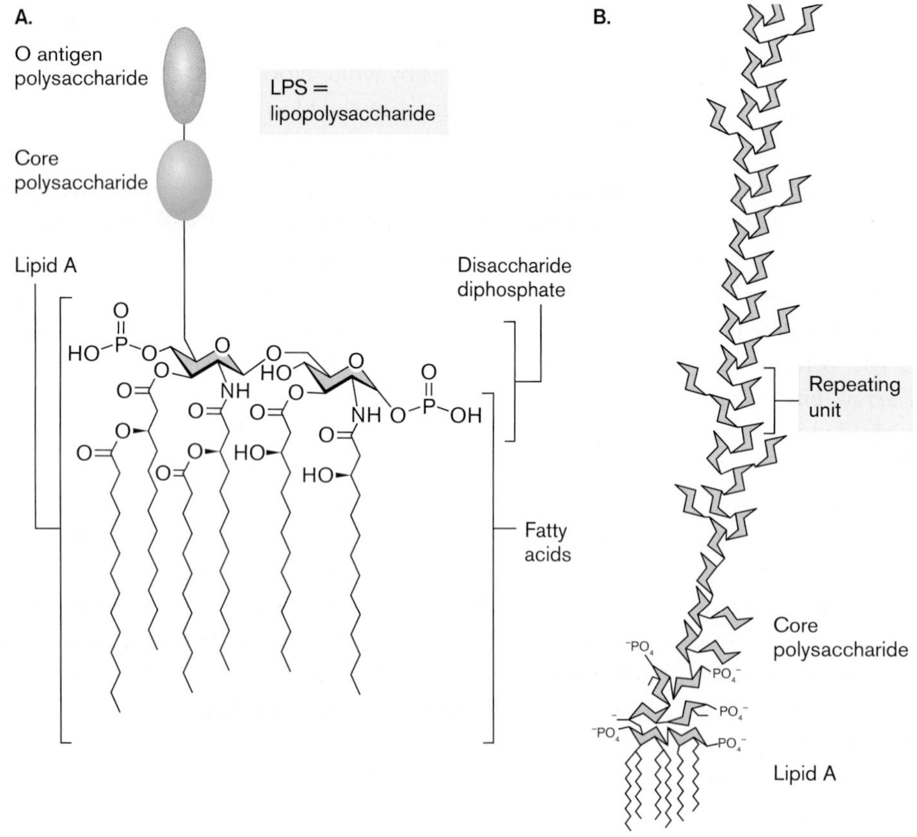

FIGURE 3.20 ■ Lipopolysaccharide (LPS). **A.** Lipopolysaccharide (LPS) consists of core polysaccharide and O antigen linked to a lipid A. Lipid A consists of a dimer of phosphoglucosamine esterified or amidated to six fatty acids. **B.** Repeating polysaccharide units of O antigen extend from lipid A.

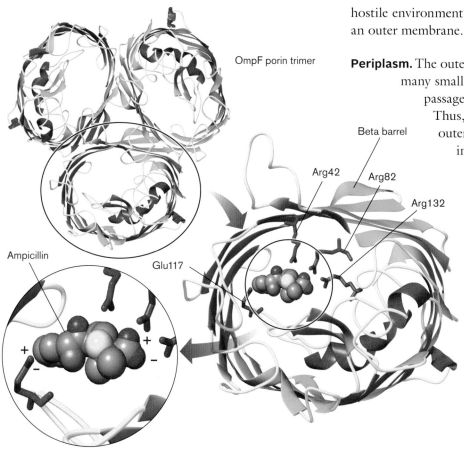

OmpF porin trimer

Beta barrel

Arg42

Arg82

Arg132

Glu117

Ampicillin

FIGURE 3.21 ■ OmpF porin transports ampicillin. This model of the OmpF trimer is based on X-ray crystallography. (PDB code: 2OMF) Within each tubular porin monomer, charged amino acid residues contact ampicillin. *Source: Modified from Ekaterina M. Nestorovich et al. 2002. PNAS* **99**:9789–9794.

hostile environment for Gram-positive bacteria, which lack an outer membrane.

Periplasm. The outer membrane is porous to most ions and many small organic molecules, but it prevents the passage of proteins and other macromolecules. Thus, the region between the inner and outer membranes of Gram-negative cells, including the cell wall, defines a separate membrane-enclosed compartment of the cell known as the periplasm (see **Fig. 3.19**). The periplasm contains specific enzymes and nutrient transporters not found within the cytoplasm, such as periplasmic transporters for sugars, amino acids, or other nutrients. Periplasmic proteins are subjected to fluctuations in pH and salt concentration, because the outer membrane is porous to ions. Some periplasmic proteins help refold proteins unfolded by oxidizing agents or by acidification.

Overall, the outer membrane, periplasm, inner membrane, and cytoplasm define four different cellular compartments within a Gram-negative cell: two membrane-soluble compartments (outer and inner membrane), and two aqueous compartments (periplasm and cytoplasm). Each type of cellular protein is typically found in only one of these locations. For example, the proton-translocating ATP synthase is found only in the inner membrane fractions, whereas sugar-accepting porins are only in the outer membrane.

through a lipid bilayer. But the molecule crosses the *E. coli* outer membrane by passing through OmpF, where its charged groups are attracted to charged amino acid residues extending inside the porin (**Fig. 3.21**). Ampicillin's positively charged amine group is attracted to the carboxylate of glutamate-117, and its negatively charged carboxylate is attracted to the arginine amines. Experiments demonstrating ampicillin passage through OmpF are described in **eTopic 3.2**.

If porins can admit dangerous molecules as well as nutrients, should a cell make porins or not? In fact, cells express different outer membrane porins under different environmental conditions. In a dilute environment, cells express porins of large pore size, maximizing the uptake of nutrients. In a rich environment—for example, within a host—cells down-regulate the expression of large porins and express porins of smaller pore size, selecting only smaller nutrients and avoiding the uptake of toxins. For example, the porin regulation system of Gram-negative bacteria enables them to grow in the colon containing bile salts—a

S-layer and capsule. Some Gram-negative bacteria also form an S-layer and/or a capsule exterior to the outer membrane. For example, an S-layer is found in *Caulobacter*, an aquatic proteobacterium; and in species of the genus *Campylobacter*, which cause diarrheal disease and systemic infections of immunocompromised patients. A capsule is found in virulent strains of *Haemophilus influenzae*, which was the leading cause of childhood meningitis before development of the Hib vaccine (see Chapter 24).

Thought Question

3.7 Why would proteins be confined to specific cell locations? Why would a protein not be able to function everywhere in the cell?

Mycobacterial Cell Envelope

Exceptionally complex cell envelopes are found in actinomycetes, a large and diverse family of soil bacteria that produce antibiotics and other industrially useful products (discussed in Chapter 18). The most complex envelopes known are those of actinomycete-related bacteria, the mycobacteria. Mycobacteria include the famous pathogens *Mycobacterium tuberculosis* (the cause of tuberculosis) and *Mycobacterium leprae* (the cause of leprosy). The mycobacterial envelope includes features of both Gram-positive and Gram-negative cells, as well as structures unique to mycobacteria (**Fig. 3.22**).

FIGURE 3.22 ■ Mycobacterial envelope structure. A complex cell wall includes a peptidoglycan layer linked to a chain of galactose polymer (galactan) and arabinose polymer (arabinan). Arabinan forms ester links to mycolic acids, which form an outer bilayer with phenolic glycolipids.

The thick mycobacterial cell wall offers physical protection comparable to the cell wall of Gram-positive bacteria. In mycobacteria, the peptidoglycan is linked to chains of galactose, called galactans. The galactans are attached to arabinans, polymers of the five-carbon sugar arabinose. The arabinan-galactan polymers are known as arabinogalactans. Arabinogalactan biosynthesis is inhibited by two major classes of anti-tuberculosis drugs: ethambutol and the benzothiazinones.

The ends of the arabinan chains form ester links to mycolic acids (uncharged mycolates). Mycolic acids provide the basis for acid-fast staining, in which cells retain the dye carbolfuchsin, an important diagnostic test for mycobacteria and actinomycetes (described in Chapter 28). Mycolic acids contain a hydroxy acid backbone with two hydrocarbon chains—one comparable in length to typical membrane lipids (about 20 carbons), the other about threefold longer. The long chain includes ketones, methoxyl groups, and cyclopropane rings. Hundreds of different forms are known. The mycolic acids form a bilayer interleaved with sugar mycolates—a kind of outer membrane, analogous to the Gram-negative outer membrane. This mycolate layer even contains porins homologous to Gram-negative beta barrel porins such as OmpA. Other mycolate-embedded proteins include virulence factors such as fibronectin-binding protein (Fbp). Fbp enhances the ability of *M. tuberculosis* to invade macrophages.

The outer ends of the sugar mycolates are interleaved with phenolic glycolipids, which include phenol groups linked to sugar chains. The extreme hydrophobicity of the phenol derivatives generates a waxy surface that prevents phagocytosis by macrophages. Overall, the thick, waxy envelope excludes many antibiotics and offers exceptional protection from host defenses, enabling the pathogens of tuberculosis and leprosy to colonize their hosts over long periods. However, the thick envelope also retards uptake of nutrients. As a result, *M. tuberculosis* and *M. leprae* grow extremely slowly and are a challenge to culture in the laboratory.

Bacterial Cytoskeleton

In eukaryotes, cell shape has long been known to be maintained by a cytoskeleton of protein microtubules and filaments (reviewed in eAppendix 2). But what determines the shape of bacteria? We saw earlier that bacterial shape is in part maintained by the cell wall and the resulting turgor pressure. But research over the past decade shows that bacteria also possess protein cytoskeletal components—and remarkably, some of them are homologous to eukaryotic cytoskeletal proteins. For example, the cell division protein FtsZ is a homolog of the eukaryotic protein tubulin (the subunit of eukaryotic microtubules).

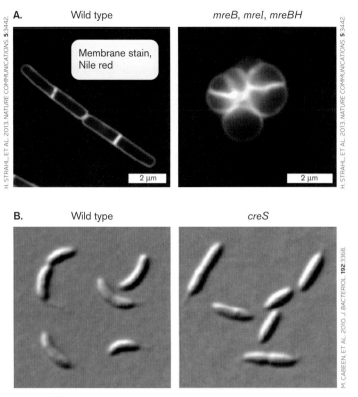

FIGURE 3.23 ■ Cytoskeletal mutants. Wild-type cells compared with mutants in *Bacillus subtilis* (**A**; fluorescence microscopy) and *Caulobacter crescentus* (**B**; DIC microscopy).

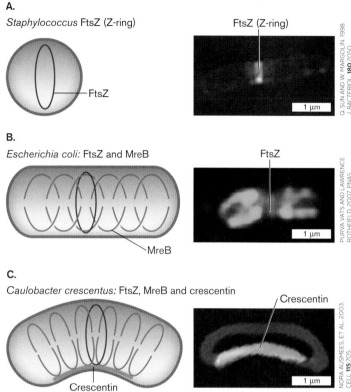

FIGURE 3.24 ■ Shape-determining proteins. **A.** Cell diameter is maintained by FtsZ polymerization to form the Z-ring. **B.** Elongation of a rod-shaped cell requires Mre proteins. MreB polymerizes around an *E. coli* cell (MreB-YFP fluorescence) along with a Z-ring of FtsZ (fluorescent anti-FtsZ antibody). **C.** Crescent-shaped cells possess a third shape-determining protein, CreS (crescentin), which polymerizes along the inner curve of the crescent. Crescentin protein fused to green fluorescent protein (CreS-GFP) localizes to the inner curve of *Caulobacter crescentus*. Membrane-specific stain FM4-64 (red fluorescence) localizes to the membrane around the cell.

Another bacterial cytoskeletal protein, MreB, is a homolog of the eukaryotic microfilament protein actin.

The bacterial cytoskeletal proteins are revealed by gene defects that drastically alter the cell shape. For example, **Figure 3.23A** compares wild-type cells of *Bacillus subtilis* with cells containing mutations in three *mreB* homologs (*mreB*, *mreI*, and *mreBH*). The wild-type cells have a defined rod shape, whereas the mutant shape is round and undefined. The mutant lacks the MreB complex that regulates peptidoglycan synthesis and thereby defines the rod-shaped cell. Another example of a shape-altering mutation affects the comma-shaped cell of *Caulobacter crescentus* (**Fig. 3.23B**). A mutation in gene *creS* results in cells that are straight instead of curved. The *creS* gene expresses the cytoskeletal protein CreS (crescentin).

How do the various cytoskeletal proteins work together to generate the overall shape of a bacterial cell? The functions of cytoskeletal proteins are probed by fluorescent protein fusions (**Fig. 3.24**). In both spherical bacteria (cocci) and rod-shaped bacilli, the FtsZ complex Z-ring determines the cell diameter. Elongation of a rod-shaped cell requires polymerization of a second cytoskeletal protein, MreB (**Fig. 3.24B**). MreB localizes in arc-shaped patches, just beneath the cell membrane of the rod-shaped cell. Some models propose a helical coil of MreB around the entire cell, but more recent data indicate arcs of MreB driven by

elongating strands of cell wall. If the rod shape is curved (called a vibrio, or crescent shape), the third cytoskeletal protein, crescentin, polymerizes along the inner curve of the crescent (**Fig. 3.24C**). The cell's outer curve is visualized by a membrane-specific fluorophore. These cytoskeletal proteins, and their variants that have evolved in other species, work together within cells to generate the shapes of bacteria.

To Summarize

- **The cell wall maintains turgor pressure.** The cell wall is porous, but its network of covalent bonds generates turgor pressure that protects the cell from osmotic shock.
- **The Gram-positive cell envelope** has multiple layers of peptidoglycan, threaded by teichoic acids.
- **The S-layer of proteins** is highly porous but can prevent phagocytosis and protect cells in extreme environments. In archaea, the S-layer serves the structural function of a cell wall.

- **The capsule**, composed of polysaccharide and glycoprotein filaments, protects cells from phagocytosis. Either Gram-positive or Gram-negative cells may possess a capsule.

- **The Gram-negative outer membrane** regulates nutrient uptake and excludes toxins. The outer membrane contains LPS and protein porins of varying selectivity.

- **The mycobacterial cell wall includes features of both Gram-positive and Gram-negative cells.** The arabinogalactan layer adds thickness to the cell wall. The mycolate outer membrane and phenolic glycolipids limit uptake of nutrients and antibiotics.

- **The bacterial cytoskeleton** includes proteins that regulate cell size, play a role in determining the rod shape of bacilli, and generate curvature in vibrios.

3.4 The Nucleoid and Cell Division

How are DNA and its expression machinery organized within the cell? Bacteria organize their DNA very differently from eukaryotes. For example, **Figure 3.25** shows enteropathogenic *E. coli* cells growing upon a cultured human cell. Enteropathogenic *E. coli* (EPEC) are diarrheal pathogens that cause the host cell to form actin "pedestals" where the bacteria inject toxins (discussed in Chapter 25). In this thin-section TEM, each bacterium contains

a filamentous nucleoid region that extends through the cytoplasm. In contrast, the nucleus of the eukaryotic cell, only a fraction of which is visible in the figure, is many times larger than the entire bacterial cell, and the chromosomes it contains are separated from the cytoplasm by the nuclear membrane.

DNA Is Organized in the Nucleoid

All living cells on Earth possess chromosomes consisting of DNA. The genetic functions of microbial DNA are discussed in detail in Chapters 7–12. Here we focus on the physical organization of DNA within the nucleoid of bacterial and archaeal cells (**Fig. 3.26A**).

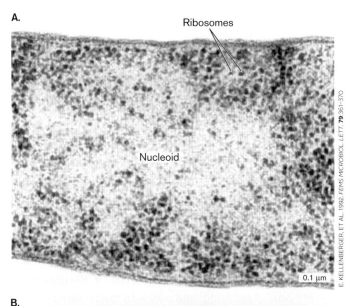

A.

Ribosomes

Nucleoid

0.1 μm

E. KELLENBERGER, ET AL. 1992. *FEMS MICROBIOL. LETT.* **79**:361–370

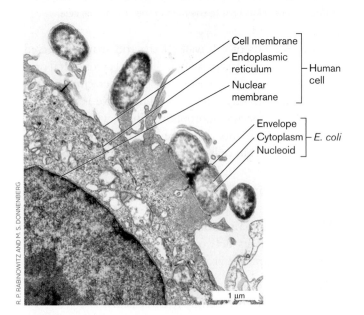

Cell membrane
Endoplasmic reticulum — Human cell
Nuclear membrane

Envelope
Cytoplasm — *E. coli*
Nucleoid

1 μm

R. P. RABINOWITZ AND M. S. DONNENBERG

FIGURE 3.25 ▪ **Bacterial nucleoid and eukaryotic nucleus.** Enteropathogenic *Escherichia coli* bacteria attached to the surface of a tissue-cultured human cell (TEM). The human cell has a well-defined nucleus delimited by a nuclear membrane, whereas the *E. coli* nucleoid DNA extends throughout the cell.

B.

DNA   DNA origin

DNA-binding protein   DNA domain

FIGURE 3.26 ▪ **The nucleoid is organized in domains.**
A. The *E. coli* nucleoid appears as clear regions that exclude the dark-staining ribosomes and contain DNA strands (cryo-TEM).
B. The nucleoid forms chromosome loops called domains, which radiate from the center.

Note: In bacteria and archaea, the genome typically consists of a single circular chromosome, but some species have a linear chromosome or multiple chromosomes. In this chapter we focus on the simple case of a single circular chromosome.

In a bacterial cell the DNA is organized in loops called **domains**. The DNA domains extend throughout the cytoplasm. The midpoint on the DNA is the origin of replication, which is attached to the cell envelope at a point on the cell's equator, halfway between the two poles (**Fig. 3.26B**). At the origin, the DNA double helix is melted open by binding proteins, and then DNA polymerase synthesizes new strands in both directions (bidirectionally). The origin and other aspects of DNA replication are covered in detail in Chapter 7.

How does all of the cell's DNA fit neatly into the nucleoid? In some bacteria, the domains loop back to the center of the cell, near the origin of replication. Within the domains, the DNA is compacted by supercoils. Supercoils (or superhelical turns) are extra twists in the chromosome, beyond those inherent in the structure of the DNA double helix (discussed in Chapter 7). The supercoiling causes portions of DNA to double back and twist upon itself, resulting in compaction of the chromosome. Supercoils are generated by enzymes such as gyrase, a major target for antibiotics such as quinolones.

DNA is also compacted by **DNA-binding proteins** (see **Fig. 3.26B**). Some binding proteins, such as H-NS and HU, also function as regulators of gene expression. Binding proteins can respond to the state of the cell; for example, under starvation conditions, when most RNA synthesis ceases, the binding protein Dps is used to organize the DNA into a protected crystalline structure. Such "biocrystallization" by Dps and related proteins may be a key to the extraordinary ability of microbes to remain viable for long periods in stationary phase or as endospores.

Note: In biology, the word "domain" is used in several different ways, each referring to a defined portion of a larger entity.

■ **DNA domains** of the nucleoid are distinct loops of DNA that extend from the origin.

■ **Protein domains** are distinct functional or structural regions of a protein.

■ **Lipid domains** are clusters of one type of lipid within a membrane.

■ **Taxonomic domains** are genetically distinct classes of organisms, such as Bacteria, Archaea, and Eukarya.

Transcription and Translation

The information encoded in DNA is "read" by the processes of transcription and translation to yield gene products. The initial product of a gene is a single strand of ribonucleic acid (RNA), produced when DNA is transcribed by RNA polymerase (**Fig. 3.27**). In some cases, the newly made RNA has a function of its own, such as being one of the RNA components of a ribosome, the cell's protein-making machine. For most genes, however, the RNA is a messenger RNA (mRNA) that immediately binds to a ribosome for translation. A growing bacterial cell typically invests 40% of its energy in the translation of mRNA by ribosomes. The ribosome, with its large number of protein and RNA components, is probably the most complex subcellular machine to be discovered. Because of its complexity, the ribosome is targeted by many kinds of antibiotics, such as tetracyclines. The process of translation is discussed further in Chapter 8.

Coupling of transcription and translation. In bacteria and archaea, translation is tightly coupled to transcription; the ribosomes bind to mRNA and begin translation even before the mRNA strand is complete. Thus, a growing bacterial cell is full of mRNA strands dotted with ribosomes

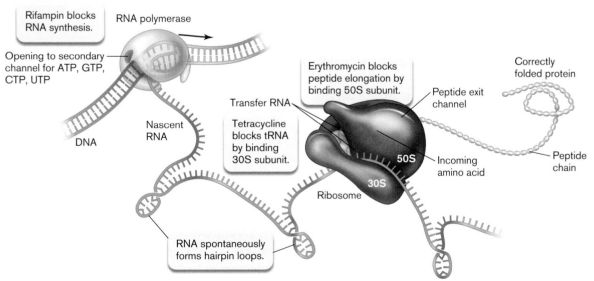

FIGURE 3.27 ■ RNA synthesis is coupled to protein synthesis. As the DNA sequence is transcribed into a strand of messenger RNA, ribosomes already start translating the RNA to synthesize proteins.

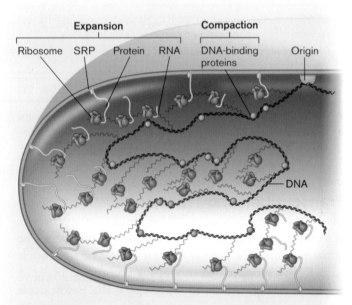

FIGURE 3.28 ■ Coupling protein synthesis and secretion.
Bacterial transcription of DNA to RNA is coordinated with translation of RNA to make proteins. Growing peptides destined for the membrane bind the signal recognition particle (SRP) for insertion into the membrane.

(**Fig. 3.28**). In rapidly growing bacteria, the DNA is transcribed and the messenger RNA translated to proteins, while the DNA itself is being replicated. This remarkable coordination of replication, transcription, and translation explains why some bacterial cells can divide in as little as 10 minutes.

By contrast, how do eukaryotes organize DNA replication and gene expression? Eukaryotes replicate their DNA during S phase of the cell cycle, while expressing most genes and proteins during G phase of interphase (reviewed in eAppendix 2). Within the nucleus, nascent (elongating) mRNA is translated by ribosomes, which check the transcript for errors and target faulty transcripts for destruction (see Reid and Nicchitta, 2012, *J. Cell Biology* 197:7). However, the majority of eukaryotic mRNA is translated outside the nucleus, in the cytoplasm. Whereas bacteria coordinate their DNA replication and gene expression, eukaryotes separate these processes both spatially (nucleus versus cytoplasm) and temporally (different growth phases).

Inserting proteins into the membrane. Some of the newly translated proteins are destined for the membrane or for secretion outside the cell. Proteins destined for membrane insertion must be hydrophobic and hence are poorly soluble in the aqueous cytoplasm where their mRNA is translated. How can proteins that are insoluble in water be folded correctly in the cytoplasm? In prokaryotes, membrane proteins and secreted proteins are synthesized in association with the cell membrane (**Fig. 3.28**). This coupling of transcription and translation to membrane insertion has the effect of expanding the nucleoid into distal parts of the cell, partly counteracting the condensation of DNA by DNA-binding

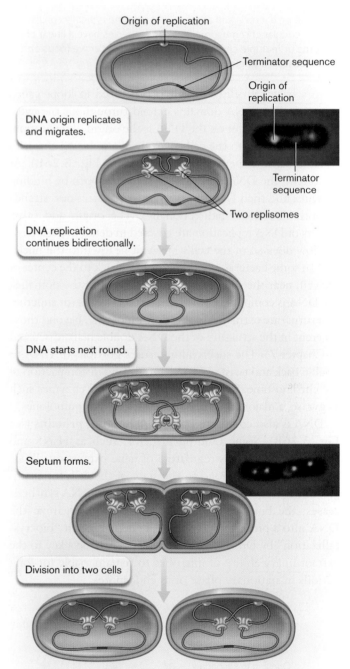

FIGURE 3.29 ■ Replisome movement within a dividing cell.
The DNA origin-of-replication sites (green) move apart in the expanding cell as the two replisomes (yellow) stay near the middle, where they replicate around the entire chromosome, completing the terminator sequence last (red). *Source:* Ivy Lau et al. 2003. *Mol. Microbiol.* **49**:731. ▶

proteins. Membrane protein maturation and secretion are discussed in Chapter 8.

Cell Division

How does a growing bacterial cell divide, or fission, into daughter cells? Bacterial cell fission requires highly coordinated growth and expansion of all the cell's parts. Unlike eukaryotes, prokaryotes synthesize RNA and proteins continually while the cell's DNA undergoes replication.

Bacterial DNA replication is coordinated with the expansion of the cell wall and ultimately the separation of the cell into two daughter cells. **Bacteria do <u>not</u> undergo mitosis or meiosis** (these eukaryotic processes are reviewed in eAppendix 2). Bacterial DNA replication is outlined here as it relates to cell division; the molecular details of the process are discussed in Chapter 7.

Replication begins at the origin of replication (**Fig. 3.29** ▶), a unique DNA sequence in the chromosome. At the origin sequence, the DNA double helix begins to unzip, forming two replication forks. At each replication fork, DNA is synthesized by DNA polymerase. The complex of DNA polymerase with its accessory components is called a **replisome**. The replisome actually includes two DNA polymerase enzymes: one to replicate the "leading strand," the other for the "lagging strand." The lag time is short compared with the overall time of replication; thus, as the replisome travels along the DNA, it converts one helix into two progeny helices almost simultaneously.

Note: Each replisome contains two DNA polymerases (for leading and lagging strands). Each dividing nucleoid requires two replisomes (for bidirectional replication), and thus four DNA polymerases overall.

Within the cell, replication proceeds outward in both directions around the genome. Thus, bidirectional replication requires two replisomes, one for each replicating fork. Fluorescent probes show that two replisomes are located near the middle of the growing cell (**Fig. 3.29**). Each replisome forms a replicating fork that directs two daughter strands of DNA toward opposite poles. The two copies of the DNA origin of replication (green in the figure), attached to the cell envelope, move apart as the cell expands. The termination site (red) remains in the middle of the cell, where the two replisomes continue replication at both forks. Finally, as the termination site replicates, the two replisomes separate

from the DNA. At each new origin site, however, two pairs of new replisomes have formed. In a fast-growing cell, the new origin sites begin a second round of replication, even before termination of the previous round.

Note that the contents of the cytoplasm and envelope must expand coordinately with DNA replication for the cell to generate progeny equivalent to the parent. In a rod-shaped cell, the cell envelope and cell wall must elongate in a process coordinated by cytoskeletal proteins so as to maintain progeny of even girth and length. As discussed earlier, the girth and length of the elongating cell depend on shape-determining proteins such as FtsZ and MreB.

Septation Completes Cell Division

For the cell to divide, DNA replication must be complete. Replication of the DNA termination site triggers growth of the dividing partition of the envelope, called the **septum** (plural, **septa**). The septum grows inward from the sides of the cell, at last constricting and sealing off the two daughter cells. This process is called **septation**. Septation and envelope extension require rapid biosynthesis of all envelope components, including membranes and cell wall. Recall how the lack of murein lipoprotein connecting the outer membrane to the cell wall causes deformation at septation (see **Fig. 3.19B**). The biosynthetic enzymes required for cell division are all of great interest as antibiotic targets. Cell wall biosynthesis poses an interesting theoretical problem: How is it possible to expand the covalent network of the sacculus without breaking links to insert new material, thus weakening the wall? The answer remains unclear.

Septation of spherical cells. In spherical cells (cocci), such as *Staphylococcus aureus*, the process of septation generates most of the new cell envelope to enclose the expanding cytoplasm (**Fig. 3.30**). Furrows form in the cell envelope,

A.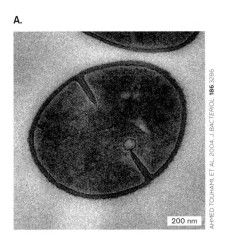

B.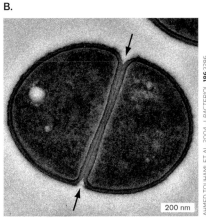

C.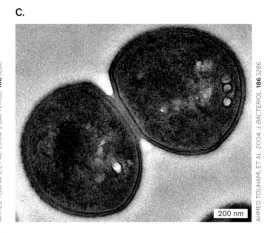

FIGURE 3.30 ▪ Septation in *Staphylococcus aureus*. A. Furrows appear in the cell envelope, all around the cell equator, as new cell wall grows inward (TEM). **B.** Two new envelope partitions are complete. **C.** The two daughter cells peel apart. The facing halves of each cell contain entirely new cell wall.

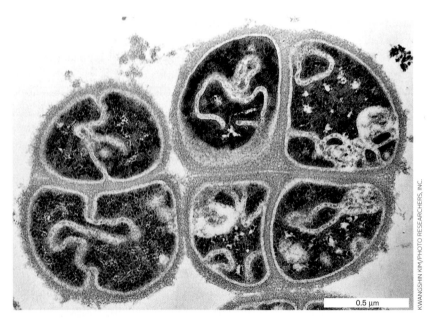

KWANGSHIN KIM/PHOTO RESEARCHERS, INC.

0.5 μm

FIGURE 3.31 ■ **Septation orientation determines the arrangement of progeny cells.** For *Micrococcus tetragenus,* septation in two planes generates a tetrad (TEM with negative stain).

in a ring all around the cell equator, as new cell wall grows inward. The wall material must consist of two separable partitions. When the partitions are complete, the two progeny cells peel apart. The facing halves of each cell consist of entirely new cell wall.

The spatial orientation of septation has a key role in determining the shape and arrangement of cocci. If the cell always septates in parallel planes, as in *Streptococcus* species, cells form chains (**Fig. 3.30**). If, however, the cell septates in random orientations, or if cells reassociate loosely after septation, they form compact hexagonal arrays similar to the grape clusters portrayed in classical paintings—hence the Greek-derived term **staphylococci** (*staphyle* means "bunch of grapes"). Such clusters are found in colonies of *Staphylococcus aureus.* If subsequent septation occurs at right angles to the previous division, the cells may form tetrads and even cubical octads called "sarcinae" (singular, sarcina). Tetrads are formed by *Micrococcus tetragenus,* a cause of pulmonary infections (**Fig. 3.31**).

Septation of rod-shaped cells. In rod-shaped cells, unlike cocci, cell division requires the envelope to elongate before septation, followed by the formation of a new polar envelope for each progeny cell. The process of septation involves an intricate series of molecular signals that is at the frontier of current research. Mutations in a gene such as *ftsZ* cause *E. coli* to form long filaments instead of dividing normally. This filamentation results from a failure to form a septum between cells. The mutant genes causing this behavior

were called *fts* for "filamentation temperature sensitive" because their mutation is temperature sensitive. The temperature-sensitive cells divide normally at a "permissive" temperature but fail to septate at the "restrictive" temperature, forming long filaments. Temperature-sensitive mutants offer the researcher a way to grow and study bacteria that have severe defects in cell growth.

To Summarize

■ **DNA is organized in the nucleoid.** The bacterial DNA is generally attached to the envelope at the origin of replication, on the cell's equator. Loops of DNA called domains are supercoiled and bound to DNA-binding proteins.

■ **During transcription, the ribosome translates RNA to make proteins.** Proteins are folded by chaperones and in some cases secreted at the cell membrane.

■ **Membrane-inserted proteins are translated at the membrane.** The coupling of transcription, translation, and membrane insertion expands the nucleoid into distal parts of the cell.

■ **DNA is replicated bidirectionally by the replisome.** During bacterial DNA replication, genes continue transcription and translation.

■ **Cell expansion and septation are coordinated with DNA replication.** Septation may occur in one plane (forming a chain of cells) or at right angles to the previous septation (forming a tetrad).

3.5 Cell Polarity and Aging

Are bacterial cells symmetrical? Some bacteria have different structures at either pole, and their cell division generates two different cell types. But even superficially symmetrical bacilli such as *E. coli* show underlying chemical and physical asymmetry. The two cell poles differ in their origin and age—a phenomenon called **polar aging**. Polar aging has surprising consequences for the interactions of bacteria with their environment and for antibiotic resistance.

Bacterial Cell Differentiation

Bacteria whose poles have different structures generate two different forms of progeny. The wetland bacterium

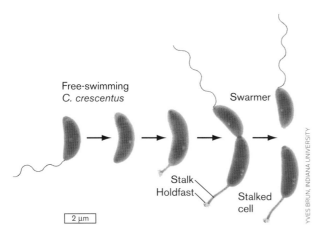

Free-swimming
C. crescentus

Swarmer

Stalk
Holdfast

Stalked
cell

2 μm

YVES BRUN, INDIANA UNIVERSITY

FIGURE 3.32 ▪ Asymmetrical cell division: a model for development. A swarmer cell of *Caulobacter crescentus* loses its flagellum and grows a stalk. The stalked cell divides to produce a swarmer cell (TEM).

Caulobacter crescentus has one plain pole and one pole with either a flagellum or a cytoplasmic extension called a stalk (**Fig. 3.32**). A flagellated cell swims about freely in an aqueous habitat, such as a pond or a sewage bed. After swimming for about half an hour, if the bacterium finds a place with enough nutrients, the cell sheds its flagellum and replaces it with a stalk. The stalked cell

attaches to sediment and then immediately starts to replicate its DNA and divides, producing a flagellated daughter cell, as well as a daughter cell containing the original stalk.

How does *C. crescentus* organize itself to produce two different cell types, each with a different organelle at one pole? The process is a rudimentary form of cell differentiation, comparable to the differentiation processes that animal cells undergo in the embryo. The process has been studied through genetic analysis (introduced in Section 3.1).

The *C. crescentus* life cycle is governed by regulator proteins such as TipN, studied by students of Christine Jacobs-Wagner at Yale University (see **eTopic 3.3**). Mutants lacking TipN make serious mistakes in development. Instead of making a single flagellum at the correct cell pole, the cell makes multiple flagella at various locations, even on the stalk (**Fig. 3.33A**). Jacobs-Wagner proposed that TipN is a landmark protein that correctly marks the site of a new cell pole, and the polar placement of flagella. **Figure 3.33B** shows cells expressing TipN fused to GFP, which is then detected by fluorescence. The fluorescent TipN-GFP localizes to the cell pole opposite the stalk. As the cell prepares to divide, TipN leaves the pole, delocalizing around the cell; but it relocalizes at the septum, where the new poles appear.

A. *Caulobacter* mutants lack gene for TipN

2 μm

HUBERT LAM, ET AL. 2006. CELL
124:I011

B. Cells express TipN-GFP

Time

DIC

TipN-GFP

2 μm

HUBERT LAM AND CHRISTINA JACOBS-WAGNER.

● TipN-GFP

New

Old

Old

New

Old

C.

© JASON VARNEY/ VARNEYPHOTO.COM

FIGURE 3.33 ▪ A landmark protein for the cell pole.
A. *Caulobacter* mutants lacking TipN protein make mistakes: flagella grow out of stalks, or at the stalked pole (fluorescence microscopy). **B.** The protein TipN appears at the pole of a *Caulobacter* stalk cell, visualized as TipN-GFP [differential interference contrast microscopy (DIC) and fluorescence microscopy]. As the cell grows, TipN delocalizes and then localizes again at the septum. Septation yields two daughter cells with TipN at the pole of each. **C.** Christine Jacobs-Wagner, winner of the American Society for Microbiology's Eli Lilly Award for her studies of *Caulobacter* development.

FIGURE 3.34 ■ **Cell cycle of Caulobacter.** A swarmer cell loses its flagellum and grows a stalk. PodJ protein (purple) is at the flagellar pole, while DivJ protein (red) is at the stalk. TipN (yellow) is found at "new" poles (newly septated). TipN delocalizes and then localizes at the cell equator, midway between poles. The pole with PodJ grows a flagellum. The cell septates, forming two new poles, each containing TipN. The stalked cell still has DivJ at the stalk, and the new swarmer cell has PodJ at the flagellum. *Source:* Modified from Melanie Lawler and Yves Brun. 2006. *Cell* **124**:891.

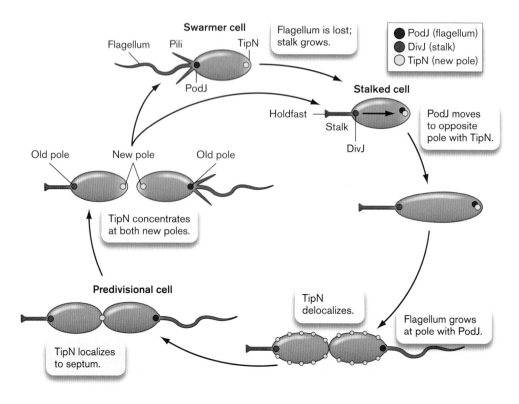

Cell development involves many such proteins working together. **Figure 3.34** shows how TipN interacts with two other polar proteins: the flagellar marker PodJ, and the stalk marker DivJ. Each young cell (swarmer cell at top of cycle) has a new pole containing TipN. PodJ migrates to the pole with TipN. The stalk marker DivJ is produced at the pole where PodJ was previously. These redistributions of proteins cause the flagellated swarmer cell to lose its flagellum, replacing it with a stalk as the cell settles in a favorable environment. As the stalked cell grows, TipN delocalizes around the cell; then it localizes again at the middle, where the cell septates and divides. Once division is complete, TipN concentrates at both new poles. As PodJ moves to the new pole, that pole grows a flagellum. Overall, throughout the cycle a series of polar proteins localize and delocalize to define the polar functions.

Thought Question

3.8 **Figure 3.33** presents data from an experiment that allows the function of the TipN protein of *Caulobacter* to be visualized by microscopy. Can you propose an experiment with mutant strains of *Caulobacter* to test the hypothesis that one of the proteins shown in **Figure 3.34** is required for one of the cell changes shown?

Polar Aging

Does an apparently symmetrical cell such as *E. coli* actually show polar differences? In fact, the actual process of cell division generates daughter cells with chemically different poles (**Fig. 3.35**). Each cell starts out with one "old" pole (red in the figure) and one "new" pole (blue) where the parental cell septated. As the next cell divides, two daughter cells form, each with another "new" pole. But meanwhile, the "old" poles continue to age. With each generation, the polar cell wall material degrades slightly, increasing the chance of cell lysis. In a population of *E. coli* under environmental stress, at each cell division some members of the population die—of polar old age.

The cause of polar aging in stressed *E. coli* is the preferential accumulation of protein aggregates, which are nonfunctional and cannot be unfolded or degraded. For unknown reasons, proteins aggregate more frequently under a stress condition, such as low pH, or the presence of an antibiotic. Proteins damaged by a stress condition are packed away in the cell's older pole, allowing the new-pole cells to remain intact and grow faster.

Why does polar aging matter? One consequence of polar aging is that cells of different polar ages may differ in their resistance to antibiotics. This phenomenon could cause problems for antibiotic therapy, as discussed in **Special Topic 3.1**. *Mycobacterium tuberculosis* shows an extreme

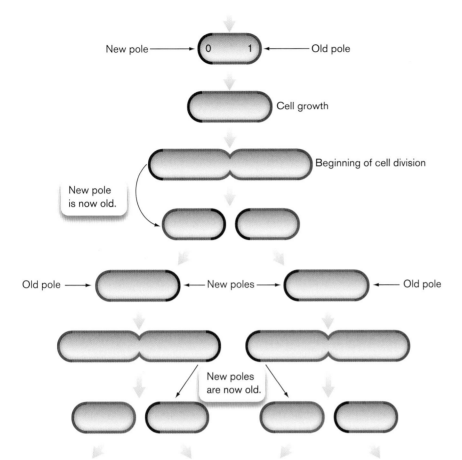

FIGURE 3.35 ■ Bacterial cell division generates cells with an old pole and a new pole. Succeeding generations have cells with diverse combinations of new poles (blue), old poles (red), and very old poles (two or more generations, also red). *Source: Eric Stewart, et al. 2005. PLoS Biol.* **3:e45**

To Summarize

- **The poles of a bacterial cell may differ in form and function.** *Caulobacter crescentus* has one plain pole, and one pole that has either a flagellum or a stalk. A stalked cell fissions to produce one stalked cell and one flagellar cell.

- **The two bacterial poles differ in age.** One pole arises from the septum of the parental cell, whereas the other pole arises from a parental pole.

- **In *E. coli*, successive cell divisions yield progeny with a mixture of polar ages.** Cells with a very old pole may cease replication and die.

- **Polar aging is increased by stress.** Environmental stress, such as an antibiotic or low pH, causes protein aggregates to collect at the cell's older pole.

- **In actinomycetes and mycobacteria, cells extend at the poles, which alternate between fast and slow extension.** Mycobacterial cells can thus reverse polar aging.

version of alternate polar aging that generates variable resistance to antibiotics. The result may give tuberculosis bacteria the opportunity to "try out" resistance to various antibiotics applied in chemotherapy.

Gram-positive bacteria show other adaptations of polar asymmetry. Spore formers such as *Bacillus* species can undergo an asymmetrical cell division to form an endospore at one end. In this process, one daughter nucleoid forms an inert endospore capable of remaining dormant but viable for many years (discussed in Chapter 4). Other species expand their cells by extending one pole only. The actinomycete *Corynebacterium glutamicum*, a soil bacterium useful for industrial production, has its replisome at one cell pole. In *Corynebacterium*, as DNA replication begins, a second replisome moves to the opposite pole, while new cell wall forms at the poles. Unequal or unipolar cell extension is common among actinomycetes and mycobacteria.

3.6 Specialized Structures

We have introduced the major structures that all cells need to contain and organize their contents, maintain their DNA, and synthesize new parts. Besides these fundamental structures, different species have evolved specialized devices adapted to diverse metabolic strategies and environments.

Thylakoids, Carboxysomes, and Storage Granules

Photosynthetic bacteria, also called phototrophs, produce food and oxygen for marine and aquatic ecosystems. In the water, phototrophs must collect as much light as possible to drive photosynthesis (see Section 14.6). To maximize the collecting area of their photosynthetic membranes, phototrophs have evolved specialized systems of extensively folded intracellular membrane called **thylakoids**

SPECIAL TOPIC 3.1 — Senior Cells Make Drug-Resistant Tuberculosis

Tuberculosis, caused by *Mycobacterium tuberculosis*, is a global health problem—amplified by the specter of resistance to nearly every known antibiotic. Effective treatment of tuberculosis can require months or years of therapy with multiple antibiotics, and when treatment lapses, resistant strains appear. How do cells become resistant so fast? The secret may lie in the diversity of growth phenotypes that arise from the unusual course of cell division in mycobacteria. The diverse growth states cause different cells to resist different antibiotics.

At the Harvard School of Public Health, microbiologist Sarah Fortune (**Fig. 1A**) and her students designed an experiment to test the hypothesis that different growth states of mycobacteria have different antibiotic sensitivities. First, Fortune established a way to grow the bacteria in a microfluidic chamber. The chamber contains ridges to keep the growing cells in place during many hours of perfusion with fresh medium. The chamber was used to observe growth starting with a single isolated cell of *M. tuberculosis*, or of *M. smegmatis*, a nonpathogenic model organism with very similar growth traits.

In **Figure 1B**, a single septating cell is labeled with a green fluorophore, while new cell growth appears red. If this cell were *E. coli*, what would happen? The newly septated *E. coli*

cells would each expand throughout, like a balloon, and septate again. But the mycobacterial cell does no such thing. Instead, each cell extends only at its old pole (2 h). Once growth from the old poles is complete, the cell divides (5 h) and the "new poles" start to elongate. In other words, each cell alternates between elongating from an old pole and then elongating from the opposite pole.

The process of alternating pole extension yields classes of cells that differ in their growth traits (**Fig. 2A**). The daughter cell arising from the rapidly extended pole is called an "accelerator." The daughter cell left behind at the alternate pole (composed mainly of old material) is called an "alternator." The alternator cell delays a generation and then extends its older pole. The alternator grows more slowly than the accelerator. Differing in both growth rate and size, the alternator and accelerator cells were expected to show distinctly different patterns of metabolism. These metabolic states are assigned age classes: age 1 and age 2, respectively.

In the next generation, division of the alternator cell again yields cells of the classes age 1 and age 2. But division of the accelerator cell yields a superaccelerator that extends even faster than the parent; this metabolic class is called age 3. **Figure 2B** shows some of the data that demonstrate the statistical differences in growth rate among cells observed in

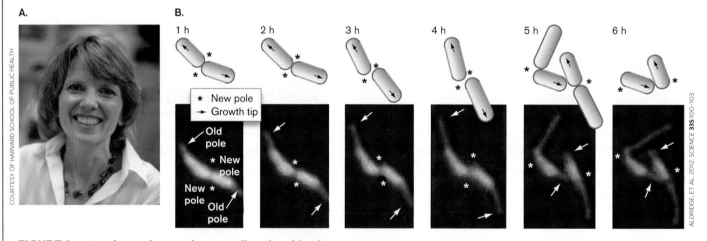

FIGURE 1 ▪ Mycobacteria grow by extending the old pole. A. Sarah Fortune studies how mycobacteria show age-related differences in antibiotic resistance. B. *Mycobacterium smegmatis* cells extend the old poles (1 h to 4 h), and then start extending the new poles.

(**Fig. 3.36A**). Thylakoids consist of layers of folded sheets (lamellae) or tubes of membranes packed with chlorophylls and electron carriers. Cyanobacteria containing thylakoids structurally resemble eukaryotic chloroplasts, which evolved from a common ancestor of modern cyanobacteria.

The thylakoids conduct only the "light reactions" of photon absorption and energy storage. The energy obtained is

rapidly spent to fix carbon dioxide—a process that occurs within **carboxysomes**. Carboxysomes are polyhedral, protein-covered bodies packed with the enzyme Rubisco for CO_2 fixation.

How do phototrophs keep themselves at the top of the water column? **Gas vesicles** increase buoyancy and keep the cell afloat. **Figure 3.36B** shows a cross section

A.

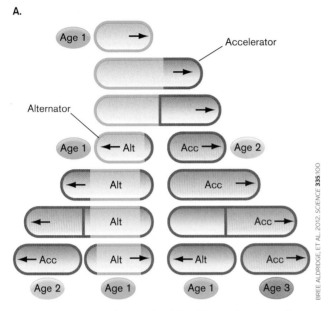

B.

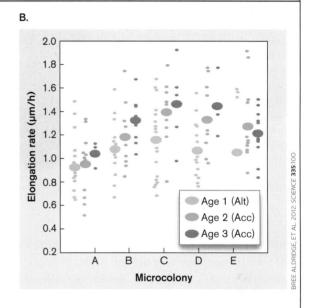

BREE ALDRIDGE, ET AL. 2012. *SCIENCE* **335**:100

BREE ALDRIDGE, ET AL. 2012. *SCIENCE* **335**:100

FIGURE 2 ■ **Mycobacteria with older poles grow faster. A.** Mycobacterial cells extend first the older pole (accelerator cell, Acc) and then the newer pole (alternator cell, Alt). **B.** Cells with older poles (age 2, age 3) extend faster than cells with newer poles (age 1). Extension rates also vary greatly within each age class.

the three age classes. Subsequent cell divisions of age 3 cells yield mixed data, Fortune reports; at some point, an unknown mechanism resets the growth rate to that of the ancestral cell. Thus, mycobacterial cells (unlike *E. coli*) can reverse polar aging.

How do these different cell types respond to antibiotics? Fortune compared the survival rates of alternator and accelerator cells. Alternator cells showed greater resistance to the antibiotics cycloserine and meropenem, which inhibit peptidoglycan synthesis. This makes sense because alternator cells form relatively small amounts of cell wall, compared to the rapidly extending accelerator cells. On the other hand, accelerator cells were more resistant to rifampicin, which inhibits RNA polymerase. The rifampicin sensitivity of alternator cells might be explained by the rapid RNA turnover required to initiate extension of the alternate pole.

Overall, we see that the "senior cells" from mycobacterial cell division play distinctive roles in generating antibiotic resis-

tance. And the data scatter in **Figure 2** suggests that other factors besides polar age remain to be discovered. No doubt, the mycobacterial cell cycle has further surprises in store for medical researchers developing therapies to cure and curb the spread of tuberculosis.

RESEARCH QUESTION

Which age class of *M. tuberculosis* (age 1 or age 3) do you think would be more resistant to quinolones (which block gyrase that supercoils DNA)? Or to TMC207 (which inhibits ATP synthase)? Propose experiments to test your hypotheses.

Aldridge, Bree B., Marta Fernandez-Suarez, Danielle Heller, Vijay Ambravaneswaran, Daniel Irimia, et al. 2012. Asymmetry and aging of mycobacterial cells lead to variable growth and antibiotic susceptibility. *Science* **335**:100–103.

of *Microcystis*, a cyanobacterium that forms toxic algal blooms in lakes polluted by agricultural runoff. *Microcystis* shows typical gas vesicles, which are protein vacuoles that trap and collect gases such as hydrogen or carbon dioxide produced by the cell's metabolism. Each vesicle is a tube with two conical ends. The tubes pack in hexagonal arrays.

When light is scarce, phototrophs may digest their thylakoids for energy and as a source of nitrogen. Alternatively, the cell may digest energy-rich materials from storage granules composed of glycogen or other polymers, such as polyhydroxybutyrate (PHB) and poly-3-hydroxyalkanoate (PHA). PHB and PHA polymers are of interest as biodegradable plastics, and bacteria have been engineered to

FIGURE 3.36 ▪ Organelles of phototrophs. A. The marine phototroph *Prochlorococcus marinus* (TEM). Beneath the envelope lie the photosynthetic double membranes called thylakoids. Carboxysomes are polyhedral, protein-covered bodies packed with the Rubisco enzyme for CO_2 fixation. **B.** *Microcystis* gas vesicles in hexagonal arrays provide buoyancy, enabling the phototroph to remain at the surface of the water, exposed to light.

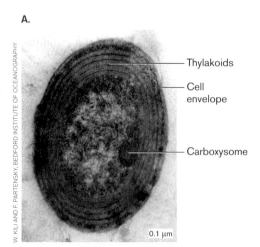

A.

Thylakoids

Cell envelope

Carboxysome

0.1 μm

W. KILI AND F. PARTENSKY, BEDFORD INSTITUTE OF OCEANOGRAPHY

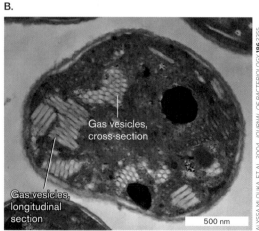

B.

Gas vesicles, cross-section

Gas vesicles, longitudinal section

500 nm

ALYSSA MLOUKA, ET AL. 2004. *JOURNAL OF BACTERIOLOGY* **186**:2355.

produce them industrially. Similar storage granules are also produced by nonphototrophic soil bacteria.

Another type of storage device is sulfur—granules of elemental sulfur produced by purple and green phototrophs through photolysis of hydrogen sulfide (H_2S). Instead of disposing of the sulfur, the bacteria store it in granules, either within the cytoplasm (purple phototrophs) or as "globules" attached to the outside of the cell. Sulfur-reducing bacteria also make extracellular sulfur globules (**Fig. 3.37**). The sulfur may be usable as an oxidant when reduced substrates become available. And the presence of potentially toxic sulfur granules may help cells avoid predation.

Pili, Stalks, and Nanotubes

In a favorable habitat, such as a running stream full of fresh nutrients or the epithelial surface of a host, it is advantageous for a cell to adhere to a substrate. Adherence, the ability to attach to a substrate, requires specific structures.

A common adherence structure is the **pilus** (plural, **pili**), also called fimbria (plural, fimbriae), which is constructed of straight filaments of protein monomers called pilin. For example, the sexually transmitted pathogen *Neisseria gonorrhoeae* uses pili to attach to the mucous membranes of the reproductive tract (**Fig. 3.38**). In Gram-negative enteric bacteria, pili of a different kind, also called the sex pili, attach a "male" donor cell to a "female" recipient cell for transfer of DNA. This process of DNA transfer is called conjugation. The genetic consequences of conjugation are discussed in Chapter 9.

COURTESY OF JUERGEN WIEGEL AND MANFRED ROHDE

1 μm

FIGURE 3.37 ▪ External sulfur particles. Sulfur globules dot the surface of *Thermoanaerobacter sulfurigignens*, an anaerobic thermophilic bacterium that gains energy by reducing thiosulfate ($S_2O_3^{2-}$) to elemental sulfur (S^0).

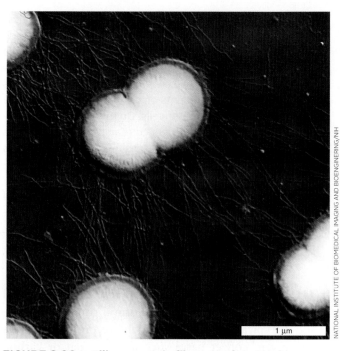

NATIONAL INSTITUTE OF BIOMEDICAL IMAGING AND BIOENGINEERING/NIH

1 μm

FIGURE 3.38 ▪ Pili are protein filaments for attachment. *Neisseria gonorrhoeae*, cause of the sexually transmitted infection gonorrhea, use pili to attach to the host mucous membrane (SEM).

Another kind of attachment organelle is a membrane-embedded extension of the cytoplasm called a **stalk**, seen earlier in the stalked cell of *Caulobacter* (**Fig. 3.32**). The tip of the stalk secretes adhesion factors called holdfasts, which firmly attach the bacterium in an environment that has proved favorable. A stalk and holdfast enable iron-oxidizing bacteria to form large orange biofilms. An example is *Gallionella ferruginea*, an iron-oxidizing species that grows long, twisted stalks (**Fig. 3.39**). The stalks of adherent *Gallionella* cells become coated by orange iron hydroxides, tinting aquatic streams contaminated by iron drainage.

Some kinds of bacterial structures, such as pili and extracellular polysaccharides, connect cells to form

A.

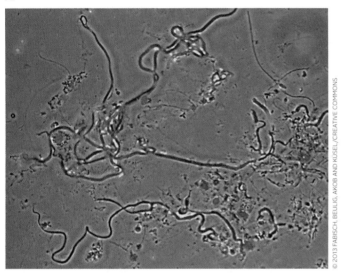

B.

FIGURE 3.39 ▪ *Gallionella* **sp. are iron-oxidizing, stalked bacteria. A.** The twisted filaments of *Gallionella*, taken from a slow-moving freshwater stream at Audley in New South Wales, Australia. **B.** *Gallionella* bacteria oxidize iron, coloring streams like this one—Gessenbach Creek in Germany—orange because of metal pollution.

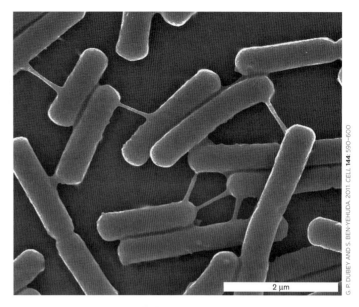

FIGURE 3.40 ▪ **Intercellular nanotubes.** *Bacillus subtilis* bacteria with intercellular connections called nanotubes, which pass material from one cell to the next.

multicellular communities called biofilms (discussed in Chapters 4 and 10). Structures that connect the cells' cytoplasm include intercellular nanotubes. Nanotubes are extensions of cell envelopes that connect the cytoplasm or periplasm between two different cells. **Figure 3.40** shows how nanotubes interconnect cells of *Bacillus subtilis*. Nanotubes can transmit materials from one cell to another—even between cells of different species. Unfortunately, viruses can generate nanotubes in their host cells in order to transmit progeny viruses to a new cell without exposure to the immune system.

Rotary Flagella

What happens when the cell's environment runs out of nutrients, or becomes filled with waste? In rapidly changing environments, cell survival requires **motility**, the ability to move and relocate. Many bacteria and archaea can swim by means of rotary **flagella** (singular, **flagellum**). Flagellar motility benefits the cell by causing the population to disperse, decreasing competition. Motility also enables cells to swim toward a favorable habitat (chemotaxis, discussed shortly.)

Flagellar motility. Flagella are helical propellers that drive the cell forward like the motor of a boat. Howard Berg at the California Institute of Technology originally described the bacterial flagellar motor, which was the first rotary device to be discovered in a living organism. Different bacterial species have different numbers and arrangements of flagella. Peritrichous cells, such as *E. coli* and *Salmonella* species, have flagella randomly distributed

A.

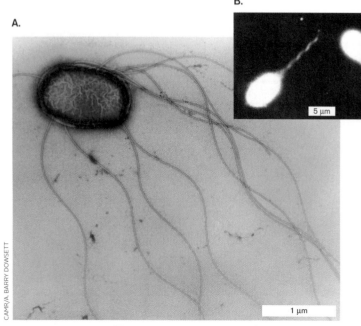

CAMR/A. BARRY DOWSETT

1 µm

B.

ROBERT MACNAB 1976.
J CLIN MICROBIOL 4:258

5 µm

FIGURE 3.41 ▪ Flagellated *Salmonella* bacteria. A. *Salmonella enterica* has multiple flagella (colorized TEM). **B.** The flagella collect in a bundle behind a swimming cell. Under dark-field microscopy, the cell body appears overexposed, about five times as large as the actual cell.

around the cell (**Fig. 3.41A**). The flagella rotate together in a bundle behind the swimming cell (**Fig. 3.41B**). Lophotrichous cells, such as *Rhodospirillum rubrum*, have flagella attached at one or both ends. In monotrichous (polar) species, such as the *Caulobacter* swarmer cell (see **Fig. 3.32**), the cell has a single flagellum at one end.

How does a rotary flagellum work? Each flagellum has a spiral filament of protein monomers called flagellin (protein FliC) (**Fig. 3.42**). The filament actually rotates by means of

a motor driven by the cell's transmembrane proton current—the same proton potential that drives the membrane-embedded ATP synthase (presented in Chapter 14). The flagellar motor is embedded in the layers of the cell envelope. The motor possesses an axle and rotary parts, all composed of specific proteins. For example, protein MotB forms part of the ion channel whose flux of hydrogen ions powers rotation, similarly to the proton flux that drives ATP synthase. Another protein, FliG, forms part of the device that generates torque (rotary force). Much of the motor's structure and function was elucidated by Scottish microbiologist Robert Macnab (1940–2003) at Yale University.

What kinds of experiments reveal the motor components? An example of an experiment dissecting the flagellar motor is shown in **Figure 3.43**. Japanese microbiologist Tohru Minamino and colleagues constructed strains of *Salmonella enterica* in which the gene encoding fluorescent GFP is fused to a gene encoding a flagellar motor protein, MotB or FliG. Fluorescence microscopy reveals the concentration of GFP fluorescence at one or two positions within the cell (green dots). A second fluorophore, Alexa, is conjugated to an anti-flagellin antibody. The Alexa fluorescence (red) reveals the flagellar filament. When the green and red fluorescence images are merged, the red flagellar filaments

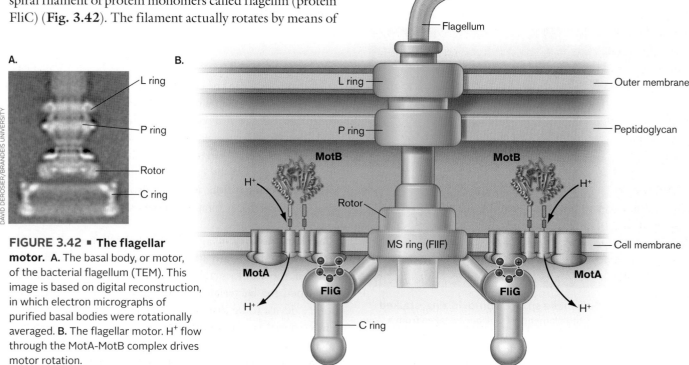

A.

DAVID DEROSIER/BRANDEIS UNIVERSITY

- L ring
- P ring
- Rotor
- C ring

FIGURE 3.42 ▪ The flagellar motor. A. The basal body, or motor, of the bacterial flagellum (TEM). This image is based on digital reconstruction, in which electron micrographs of purified basal bodies were rotationally averaged. **B.** The flagellar motor. H⁺ flow through the MotA-MotB complex drives motor rotation.

B.

FliC

Flagellum

L ring — Outer membrane

P ring — Peptidoglycan

MotB   MotB

H⁺   H⁺

Rotor

MS ring (FliF)

MotA   MotA

FliG   FliG

H⁺   H⁺

— Cell membrane

C ring

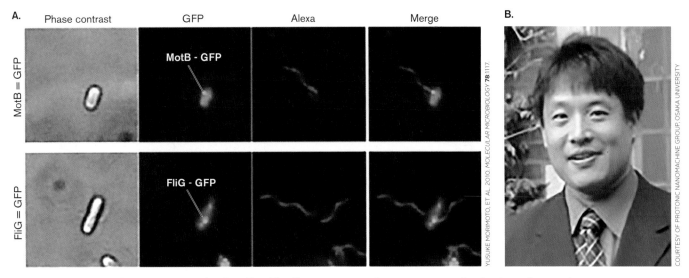

FIGURE 3.43 ■ **Flagellar motor proteins localized by fluorescence microscopy.** **A.** Cells of *Salmonella enterica* express a GFP fused to flagellar motor protein MotB or to FliG. Bright green dots (white arrows) indicate GFP-protein complexed with a flagellar motor. The flagellar filament is visualized via the red fluorophore Alexa, conjugated to an anti-flagellin antibody. The merged image shows how each flagellar filament (red) extends from a motor containing either MotB or FliG (green, protein fused to GFP). **B.** Tohru Minamino, Osaka University, investigates the structure and function of the flagellar motor.

appear to extend from the motor positions that contain either MotB or FliG. Further experiments dissect the roles of key amino acid residues in the function of these proteins.

Note: Bacterial flagella differ completely from the whiplike flagella and cilia of eukaryotes. Eukaryotic flagella are much larger structures containing multiple microtubules enclosed by a membrane (shown in Chapter 20). They move with a whiplike motion, powered by ATP hydrolysis all along the flagellum.

Chemotaxis. How do cells decide where to swim? Most flagellated cells have an elaborate sensory system for taxis, the ability to swim toward favorable environments (attractant signals, such as nutrients) and away from inferior environments (repellent signals, such as waste products). Taxis to specific chemicals is called **chemotaxis**. For chemotaxis, the attractants and repellents are detected by a polar array of chemoreceptors. Chemoreceptors act like a "nose," telling the bacterium when it is swimming toward a source of attractant such as a sugar or an amino acid.

Another form of chemotaxis is **magnetotaxis**, the ability to sense and respond to magnetism. Found in pond water, magnetotactic bacteria orient themselves along the Earth's lines of magnetic field. Magnetotactic bacteria can be collected from the environment by placing a magnet in a jar of pond water; bacteria orienting by the field lines collect nearby. The bacteria sense the magnetic field by means of **magnetosomes**, microscopic membrane-embedded crystals of the magnetic mineral magnetite, Fe_3O_4 (shown in the previous chapter, Fig. 2.44). The magnetosomes orient bacterial swimming toward the bottom of the pond.

Magnetotactic bacteria are anaerobes or microaerophiles (require low oxygen), which prefer the lower part of the water column, where oxygen concentration is lowest. In the northern latitudes, where Earth's magnetic field lines point downward, bacteria that are magnetotactic swim "downward" toward magnetic north.

Thought Question

3.9 How would a magnetotactic species have to behave if it was in the Southern Hemisphere instead of the Northern Hemisphere?

Chemotaxis requires a mechanism for the rotary flagella to propel the cell toward attractants or away from repellents. This movement is accomplished by flagellar rotation either clockwise or counterclockwise relative to the cell (**Fig. 3.44A**). When a cell is swimming toward an attractant chemical, the flagella rotate counterclockwise (CCW), enabling the cell to swim smoothly for a long stretch. When the cell veers away from the attractant, receptors send a signal that allows one or more flagella to switch rotation clockwise (CW), against the twist of the helix. This switch in the direction of rotation disrupts the bundle of flagella, causing the cell to tumble briefly and end up pointed in a random direction. The cell then swims off in the new direction. The resulting pattern of movement generates a "biased random walk" in which the cell tends to migrate toward the attractant (**Fig. 3.44B**). How do bacteria connect their flagellar switching mechanism to the signal reception of their chemoreceptors? The signal transduction mechanism is explained in Chapter 10.

A.

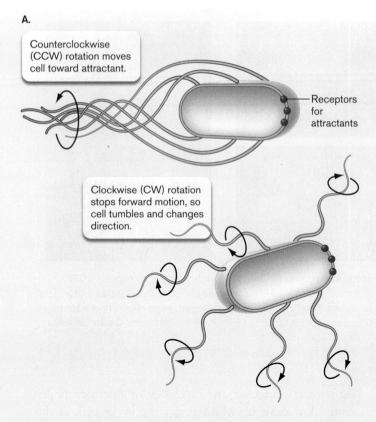

Counterclockwise (CCW) rotation moves cell toward attractant.

Receptors for attractants

Clockwise (CW) rotation stops forward motion, so cell tumbles and changes direction.

B.

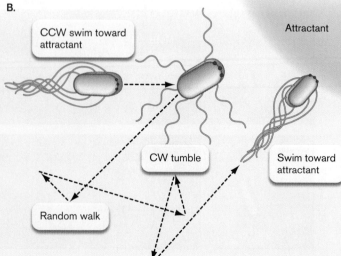

CCW swim toward attractant

Attractant

CW tumble

Swim toward attractant

Random walk

FIGURE 3.44 ■ Chemotaxis. A. Flagella are oriented in a bundle extending behind one pole. When the cell veers away from the attractant, the receptors send a signal that allows one or more flagella to switch rotation from counterclockwise (CCW) to clockwise (CW). This switched rotation disrupts the bundle of flagella, causing the cell to tumble briefly before it swims off in a new direction. **B.** The pattern of movement is a "biased random walk" in which the cell sometimes moves randomly but overall tends to migrate toward the attractant.

Besides motility and chemotaxis, surprisingly, flagella have also evolved an alternate function: adherence of cells to a substrate to begin forming a biofilm (discussed in Chapter 4). Thus, an organism can evolve a structure that serves one function but later evolves to serve another function.

Thought Question

3.10 Most laboratory strains of *E. coli* and *Salmonella* commonly used for genetic research lack flagella. Why do you think this is the case? How can a researcher maintain a motile strain?

In addition to flagellar rotation, other forms of bacterial motility are just beginning to be understood. For example, "twitching motility" is a process by which bacteria such as *Pseudomonas aeruginosa* use pili to drag themselves across a surface. Twitching motility is involved in biofilm formation (discussed in Section 4.5). Another kind of motility, called "gliding," is observed in cyanobacteria and in myxobacteria.

To Summarize

- **Phototrophs** possess thylakoid membrane organelles packed with photosynthetic apparatus and carboxysomes for carbon dioxide fixation. Gas vesicles provide buoyancy in the water column.

- **Storage granules** store sulfur or carbon polymers for energy.

- **Adherence structures** enable prokaryotes to remain in an environment with favorable environmental factors. Major adherence structures include pili or fimbriae (protein filaments), and the holdfast (a cell extension).

- **Flagellar motility** involves rotary motion of helical flagella.

- **Chemoreceptors and magnetosomes** provide information that directs flagellar motility. Chemotaxis is a mechanism by which the cell migrates up a gradient of an attractant substance or down a gradient of repellents.

Concluding Thoughts

The form and function of cells continue to amaze us with each new discovery, such as the polar aging of bacteria and the intricate mechanism of their flagellar motors.

The elaborate cell forms evolved by microbes challenge the inventors of antibiotics, as well as designers of molecular machines. As a journalist observed in *Science*, "When it comes to nanotechnology, physicists, chemists, and materials scientists can't hold a candle to the simplest bacteria."

CHAPTER REVIEW

Review Questions

1. What are the major features of a bacterial cell, and how do they fit together for cell function as a whole?
2. What fundamental traits do most prokaryotes have in common with eukaryotic microbes? What traits are different?
3. Explain how cell fractionation allows us to separate proteins of the Gram-negative outer membrane, periplasm, inner membrane, and cytoplasm.
4. Outline the structure of the peptidoglycan sacculus, and explain how it expands during growth. Cite two different kinds of experimental data that support our current views of the sacculus.
5. Compare and contrast the structure of Gram-positive and Gram-negative cell envelopes. Explain the strengths and weaknesses of each kind of envelope.
6. Explain how DNA transcription to RNA is integrated with translation and with protein processing and secretion.
7. Explain how the process of DNA replication is coordinated with cell wall septation.
8. Explain how the asymmetry of a bacterial cell generates daughter cells with different structure and function. Explain the significance for environmental adaptation and for antibiotic therapy.
9. What kinds of subcellular structures are found in certain cells with different functions, such as photosynthesis or magnetotaxis?
10. Compare and contrast bacterial structures for attachment and motility. Explain the molecular basis of chemotaxis.

Thought Questions

1. The aquatic bacterium *Caulobacter crescentus* alternates between two cell forms: a cell with a flagellum that swims, and a stalked cell that adheres to particulate matter. The flagellar cell can discard its flagellum to grow a stalk and adhere, and then the stalked cell divides to give one stalked cell and one flagellated cell. What would be the adaptive advantage of this alternating morphology?
2. Suppose that one cell out of a million has a mutant gene blocking S-layer synthesis, and suppose that the mutant strain can grow twice as fast as the S-layered parent. How many generations would it take for the mutant strain to constitute 90% of the population?
3. Explain two different ways that an aquatic phototroph might use its subcellular structures to maximize its access to light. Explain how an aerobe (an organism requiring molecular oxygen for growth) might remain close to the surface, with access to air.
4. How do pathogenic bacteria avoid being engulfed by phagocytes of the human bloodstream? How do you think various aspects of the cell structure can prevent phagocytosis?

Key Terms

active transport (88)
capsule (95)
carboxysome (108)
cardiolipin (88)
cell fractionation (82)

cell membrane (79)
cell wall (80)
chemotaxis (113)
cholesterol (89)
cross-bridge (92)

DNA-binding protein (101)
domain (of proteins) (101)
electrophoresis (81)
endotoxin (96)
envelope (80)

flagellum (80, 111)
gas vesicle (108)
glycan (91)
hopanoid (hopane) (89)
inner membrane (79, 96)
ion gradient (88)
leaflet (85)
lipid A (96)
lipopolysaccharide (LPS) (80, 96)
lysis (lyse) (82)
magnetosome (113)
magnetotaxis (113)
membrane-permeant weak acid (87)
membrane-permeant weak base (87)
motility (111)

murein (91)
murein lipoprotein (96)
nucleoid (80)
O antigen (96)
osmotic pressure (87)
outer membrane (80)
passive transport (88)
penicillin-binding protein (92)
peptidoglycan (81)
periplasm (80)
phospholipid (85)
pilus (110)
plasma membrane (79)
polar aging (104)
porin (96)

replisome (103)
S-layer (94)
sacculus (90)
septation (103)
septum (103)
spheroplast (82)
stalk (111)
staphylococcus (104)
teichoic acid (94)
terpenoid (89)
thylakoid (107)
transport protein
    (transporter) (88)
ultracentrifuge (82)

Recommended Reading

Aldridge, Bree B., Marta Fernandez-Suarez, Danielle Heller, Vijay Ambravaneswaran, Daniel Irimia, et al. 2012. Asymmetry and aging of mycobacterial cells lead to variable growth and antibiotic susceptibility. *Science* **335**:100–103.

Celler, Katherine, Roman I. Koning, Abraham J. Koster, and Gilles P. van Wezel. 2013. Multidimensional view of the bacterial cytoskeleton. *Journal of Bacteriology* **195**:1627–1636.

Clark, Michelle W., Anna M. Yie, Elizabeth K. Eder, Richard G. Dennis, Preston J. Basting, et al. 2015. Periplasmic acid stress increases cell division asymmetry (polar aging) of *Escherichia coli. PLoS ONE* **10**:e0144650.

Feucht, Andrea, and Jeff Errington. 2005. *ftsZ* mutations affecting cell division frequency, placement and morphology in *Bacillus subtilis. Microbiology* **151**:2053–2064.

Lam, Hubert, Whitman B. Schofield, and Christine Jacobs-Wagner. 2006. A landmark protein essential for establishing and perpetuating the polarity of a bacterial cell. *Cell* **124**:1011–1023.

Lele, Uttara N., Ulfat I. Baig, and Milind G. Watve. 2011. Phenotypic plasticity and effects of selection on cell division symmetry in *Escherichia coli. PLoS One* **6**:e14516.

Lenz, Peter, and Lotte Søgaard-Andersen. 2011. Temporal and spatial oscillations in bacteria. *Nature Reviews. Microbiology* **9**:565–577.

Libby, Elizabeth A., Manuela Roggiani, and Mark Gouliana. 2012. Membrane protein expression triggers chromosomal locus repositioning in bacteria. *Proceedings of the National Academy of Sciences USA.* **109**(19):7445–7450.

Nilsen, Trine, Arthur W. Yan, Gregory Gale, and Marcia B. Goldberg. 2005. Presence of multiple sites containing polar material in spherical *Escherichia coli* cells that lack MreB. *Journal of Bacteriology* **187**:6187–6196.

Partridge, Jonathan D., Vincent Nieto, and Rasika M. Harshey. 2015. A new player at the flagellar motor: FliL controls both motor output and bias. *mBio* **6**:e02367.

Renner, Lars D., and Douglas B. Weibel. 2011. Cardiolipin microdomains localize to negatively curved regions of *Escherichia coli* membranes. *Proceedings of the National Academy of Sciences USA* **108**:6264–6269.

Ruiz, Natividad, Daniel Kahne, and Thomas J. Silhavy. 2006. Advances in understanding bacterial outer-membrane biogenesis. *Nature Reviews. Microbiology* **4**:57–66.

Saier, Milton H., Jr. 2008. Structure and evolution of prokaryotic cell envelopes. *Microbe* **3**:323–328.

Schwechheimer, Carmen, and Meta J. Kuehn. 2015. Outer-membrane vesicles from Gram-negative bacteria: biogenesis and functions. *Nature Reviews Microbiology* **13**:605–619.

Stewart, Eric J., Richard Madden, Gregory Paul, and François Taddei. 2005. Aging and death in an organism that reproduces by morphologically symmetric division. *PLoS Biology* **3**:e45.

CHAPTER 4
Bacterial Culture, Growth, and Development

Where do bacteria find food, and how do they grow? Bacteria struggle to survive in natural habitats because they constantly compete for food. Over eons, however, evolution has enabled bacteria to find new nutritional niches that require the microbes to consume strange foods (mothballs, for instance), harness energy from light, or survive boiling temperatures around deep-sea thermal vents. In the process, single-celled organisms developed ways to chemically "talk" to each other and form intricate multicellular communities called biofilms. The one process that links all of these activities is growth.

CURRENT RESEARCH highlight

Evolution in aging colonies. Bacteria in old colonies struggle to survive because nutrients dwindle and toxic products rise. Ivan Matic and Claude Saint-Ruf found aging colonies to be breeding grounds for diversity. This 10-day-old *E. coli* colony has numerous mutant islands expressing different patterns of two fluorescent proteins (green and red). **Inset:** An island with dead cells (blue) in the center surrounded by growing cells (green) and dormant cells (red). The patterns suggest that each island evolves independently and that death of the first mutant cells provides nutrients for their descendants.

Source: Saint-Ruf et al. 2014. *J. Bacteriol.* **196**:3059–3073.

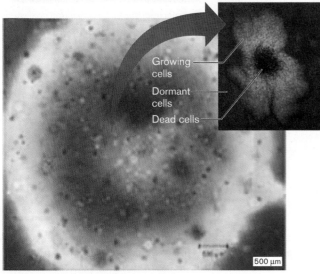

Growing cells
Dormant cells
Dead cells

500 μm

AN INTERVIEW WITH

IVAN MATIC, MICROBIOLOGIST, UNIVERSITÉ PARIS DESCARTES

© SEBASTIEN FLEURIER

How have studies on aging colonies altered our view of bacterial diversity?
The capacity of microbes to colonize practically all environmental niches on Earth fascinates us. Our work demonstrates that microhabitats within structured environments are important for shaping this diversity. For instance, environmental parameters of microhabitats in a bacterial colony continuously change as a consequence of bacterial activity. The disparate selective pressures that result favor the emergence and coexistence of diverse variants that could evolve to inhabit new environmental niches. In a homogeneous environment, by contrast, natural selection reduces diversity by eliminating all but the fittest variants.

What application(s) might stem from your work?
Our study illustrates how novel metabolic capabilities can emerge in bacteria, transforming them, for example, into more successful host invaders. Our work may also explain how heritable antibiotic tolerance develops even in the absence of antimicrobial agents. Learning how drug tolerance evolves in structured environments may influence how we use antibiotics to treat bacterial infections.

Like all living things, microorganisms need sources of carbon and energy to grow. As those resources are depleted, organisms can do one of the following: they can die (most do), evolve to better use what little resource remains, or cannibalize other, less fortunate members of the community. All three of these possible outcomes were observed within the single colony described in the Current Research Highlight.

What is the evidence that cells taken from the growth islands in aging colonies have actually evolved? Claude Saint-Ruf (**Fig 4.1A**), working with Ivan Matic, isolated numerous variants from islands in aging colonies. She then tested how well these variant cells competed with their parental strains for growth in reconstructed aging colonies. Equal numbers of parental and variant cells were mixed, spotted onto solid growth media (agar), and incubated for 7 days. In **Figure 4.1B**, you can see in the control (top panel) that two parents labeled with different fluorescent proteins grew equally well on agar when mixed. However, the lower panel clearly shows that a variant strain (green) has outcompeted its parent (red) in the reconstructed aging colony. The results show that evolution in aging colonies will generate mutants that better utilize dwindling nutrients or the end products of metabolism.

Understanding how bacteria use food to increase cell mass and, ultimately, cell number enables us to control their growth and even manipulate them to make useful products. Yeast, for example, consume glucose and break it down to ethanol and carbon dioxide gas. These end products are mere waste to the yeast, but extremely important to humans who enjoy beer.

Chapter 4 provides a broad perspective of microbial growth, introducing topics that will be expanded on in later chapters. We start by discussing the nutrients bacteria need to grow and the ways nutrients are used. For example, how do different bacteria obtain carbon and nitrogen, where do they get their energy, and what mechanisms do they use to gather the nutrients they need? (The details of metabolism emerge in Chapters 13, 14, and 15.) Next we explain how scientists use our knowledge of microbial nutrition to culture bacteria in the laboratory and measure their growth. And we ponder, why do most bacterial species growing in the world fail to grow in the lab? We end by describing the

communities that bacterial cells form while they grow and how some species differentiate into unique forms able to survive starvation or explore for new sources of food—all so they can grow again.

4.1 Microbial Nutrition

Bacterial cells look simple but are remarkably complex and efficient replication machines. One cell of *Escherichia coli*, for example, can divide to form two cells every 20–30 minutes. At a rate of 30 minutes per division, one cell could potentially multiply to over 1×10^{14} cells in 24 hours—that is, 100 trillion organisms! Although these 100 trillion cells would weigh only about 1 gram altogether, the mass of cells would explode to 10^{14} grams (that is, 10^7 <u>tons</u>) after another 24 hours (two days total) of replicating every 30 minutes. Why, then, are we not buried under mountains of *E. coli*?

Nutrient Supplies Limit Microbial Growth

One factor limiting growth is the finite supply of nutrients. Microbes, in fact, often live where nutrients are scarce.

A.

© SEBASTIEN FLEURIER

B.

Parental (red)/Parental (green) mixed colony

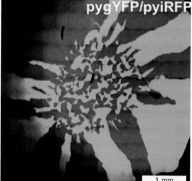

pygYFP/pyiRFP

1 mm

Parental (red)/variant (green) mixed colony

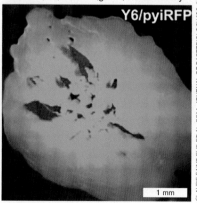

Y6/pyiRFP

1 mm

CLAUDE SAINT-RUF, ET AL. 2014. *JOURNAL OF BACTERIOLOGY* **196:** 3059-3073

FIGURE 4.1 Survival of the fittest in aging colonies. A. Claude Saint-Ruf examined growth competition between parental *E. coli* and evolved variants found within an aging colony (Current Research Highlight). **B.** 100 cells of a variant and 100 cells of its parent were mixed and spotted onto growth agar. Two parental strains labeled with different fluorescent proteins can be seen in the top panel; the bottom panel shows a parental strain (red) mixed with a variant strain (green).

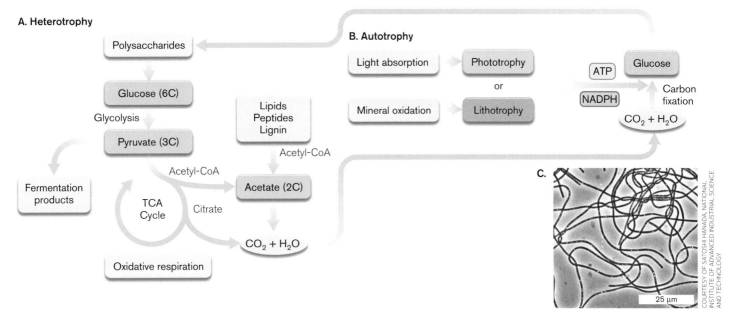

FIGURE 4.2 ■ The carbon cycle. The carbon cycle requires both autotrophs and heterotrophs. **A.** Heterotrophs gain energy from degrading complex organic compounds (such as polysaccharides) to smaller compounds (such as glucose and pyruvate). The carbon from pyruvate moves through the tricarboxylic acid (TCA) cycle and is released as CO_2. In the absence of a TCA cycle, the carbon can end up as fermentation products, such as ethanol or acetic acid. **B.** Autotrophs use light energy or energy derived from the oxidation of minerals to capture CO_2 and convert it to complex organic molecules. **C.** *Chloroflexus aggregans*, originally isolated from hot springs in Japan, possesses extraordinary metabolic versatility. It grows anaerobically (without oxygen) as a photoheterotroph and aerobically (with oxygen) as a chemoheterotroph.

Essential nutrients are those compounds that a microbe <u>must</u> have but cannot make. The organism needs to find and import these nutrients from the immediate environment. When an essential nutrient becomes depleted, microbes cannot grow. How organisms cope with these periods of starvation until nutrients are restored will be discussed later.

All microorganisms require a minimum set of **macronutrients**, nutrients needed in large quantities (as discussed in Chapter 3). Six macronutrients—carbon, nitrogen, phosphorus, hydrogen, oxygen, and sulfur—make up the carbohydrates, lipids, nucleic acids, and proteins of the cell. Four other macronutrients are cations that serve as **cofactors** for specific enzymes (Mg^{2+}, Fe^{2+}, and K^+) or act as regulatory signaling molecules (Ca^{2+}). All cells also require very small amounts of certain trace elements, called **micronutrients**. These include cobalt, copper, manganese, molybdenum, nickel, and zinc, which are ubiquitous trace contaminants on glassware and in water. As a result, these elements are <u>not</u> added to laboratory media unless heroic measures have been taken to first remove the elements from the medium. Micronutrients are required by cells as essential components of enzymes or cofactors. Cobalt, for example, is part of the cofactor vitamin B_{12}.

Some organisms, such as the common laboratory bacterium *E. coli*, make all their proteins, nucleic acids, and cell wall and membrane components from this very simple blend of chemical elements and compounds. For many other microbes, this basic set of nutrients is insufficient. *Borrelia burgdorferi*, for example, the cause of Lyme disease, requires an extensive mixture of complex organic supplements to grow.

Microbes Build Biomass through Autotrophy or Heterotrophy

Maintaining life on this planet is an amazing process. All of Earth's life-forms are based on carbon, which they acquire through delicately choreographed processes that recycle key nutrients. The carbon cycle, a critical part of this process, involves two counterbalancing kinds of metabolism: **heterotrophy** and **autotrophy**. The metabolic details of these pathways are discussed later, in Chapters 13 and 14.

Heterotrophs (such as *E. coli*) rely on other organisms to form the organic compounds (such as glucose) that they use as carbon sources. Most heterotrophs are organotrophs. **Organotrophy** is a form of metabolism in which organic carbon sources are disassembled in ways that generate energy and then reassembled to make cell constituents such as proteins and carbohydrates (**Fig. 4.2A**). This process converts a large amount of the organic carbon source to CO_2, which is then released to the atmosphere. Thus, left on their own, heterotrophs would deplete the world of organic carbon sources (converting them to unusable CO_2) and then starve to death. For life to continue, CO_2 must be recycled.

Autotrophs (such as cyanobacteria) assimilate CO_2 as a carbon source, reducing it (adding hydrogen atoms) to generate complex cell constituents made up of C, H, and O (for example, carbohydrates, which have the general formula CH$_2$O) (**Fig. 4.2B**). When autotrophs later die or are eaten, these organic compounds can be used as carbon sources by heterotrophs. Autotrophs are classified as photoautotrophs or chemolithoautotrophs by how they obtain energy. **Photoautotrophs** use light energy to fix CO$_2$ into biomass, whereas **chemolithoautotrophs** fix CO$_2$ using chemical reactions without light. Most chemolithoautotrophs gain energy by oxidizing inorganic substances such as iron or ammonia (described next). Both photoautotrophs and chemolithoautotrophs carry out the autotrophic process of CO$_2$ fixation, which forms part of the carbon cycle between autotrophs and heterotrophs (**Fig. 4.2**). In addition, many microorganisms (for example, phototrophic soil bacteria) can use both heterotrophy and autotrophy to gain carbon.

> **Thought Question**
>
> **4.1** In a mixed ecosystem of autotrophs and heterotrophs, what happens if the autotroph begins to outgrow the heterotroph and makes excess organic carbon?

Microbes Obtain Energy through Phototrophy or Chemotrophy

Although the macronutrients mentioned earlier (C, N, P, H, O, and S) provide the essential building blocks to make proteins and other cell structures, all synthetic processes require an energy source. Depending on the organism, energy can be obtained from chemical reactions triggered by the absorption of light (**phototrophy**, or photosynthesis) or from oxidation-reduction reactions that transfer electrons from high-energy compounds to make products of lower energy (**chemotrophy**), capturing the energy difference to do work. Chemotrophic organisms fall into two classes that use different sources of electron donors: **lithotrophs** (also called chemolithotrophs) and **organotrophs** (chemoorganotrophs). Lithotrophs remove electrons (oxidize) inorganic chemicals (for example, H$_2$, H$_2$S, NH$_4^+$, NO$_2^-$, and Fe^{2+}) for energy, whereas organotrophs oxidize organic compounds (for example, sugars). Oxidation and reduction is reviewed in eAppendix 1.

What if a microbe conducts more than one type of metabolism? Many free-living soil and aquatic bacteria can obtain energy from heterotrophy, lithotrophy, and phototrophy, all in one cell. Such microbes are called mixotrophs or **photoheterotrophs**. A photoheterotroph has multiple gene systems that are expressed under different conditions and yield products that carry out different functions. For example, *Rhodospirillum rubrum* grows by photoheterotrophy when light is available and oxygen is

absent, but by respiration, without absorbing light, when O$_2$ is available.

Note: The following prefixes for "-trophy" terms help distinguish different forms of biomass-building and energy-yielding metabolism.

Carbon source for biomass:

Auto-: CO$_2$ is fixed and assembled into organic molecules.
Hetero-: Preformed organic molecules are acquired from outside, broken down for carbon, and the carbon reassembled to make biomass.

Energy source:

Photo-: Light absorption captures energy.
Chemo-: Chemical electron donors are oxidized.

Electron source:

Litho-: Inorganic molecules donate electrons.
Organo-: Organic molecules donate electrons.

In chemotrophy, the amount of energy harvested from oxidizing a compound depends on the compound's reduction state. The more reduced the compound is, the more electrons it has to give up and the higher is the potential energy yield. A reduced compound, such as glucose, can donate electrons to a less reduced (more oxidized) compound, such as nicotinamide adenine dinucleotide (NAD), releasing energy (in the form of donated electrons) and becoming oxidized in the process. NAD is a cell molecule critical to energy metabolism and is discussed, along with oxidation-reduction reactions, in Chapter 13.

In short, microbes are classified on the basis of their carbon and energy acquisition as follows:

■ **Autotroph (autotrophy).** Autotrophs build biomass by fixing CO$_2$ into complex organic molecules. Autotrophs gain energy through one of two general metabolic routes that either use or ignore light.

Photoautotroph (photoautotrophy). Photoautotrophs generate energy through light absorption by the photolysis (light-activated breakdown) of H$_2$O or H$_2$S. The energy is used to fix CO$_2$ into biomass.

Chemolithoautotroph (chemolithoautotrophy). Chemolithoautotrophs produce energy from oxidizing inorganic molecules such as iron, sulfur, or nitrogen. This energy is used to fix CO$_2$ into biomass.

■ **Heterotroph (heterotrophy).** Heterotrophs break down organic compounds from other organisms to gain energy and to harvest carbon for building their own biomass. Heterotrophic metabolism can be divided into two classes, also based on whether light is involved.

Photoheterotroph (photoheterotrophy). Photoheterotrophs obtain energy from the catabolism of organic compounds and through light absorption. Organic compounds are broken down and used to build biomass.

FIGURE 4.3 ■ **Bacterial metabolism in an Antarctic desert.** Mars Oasis, Alexander Island (average winter temperature: –40°C; average summer temperature: 3°C), where scientists warm the soil to test the ecological effects of global climate change. **Inset:** Endolithic cyanobacteria, and the endospore-forming heterotroph *Paenibacillus wynnii.*

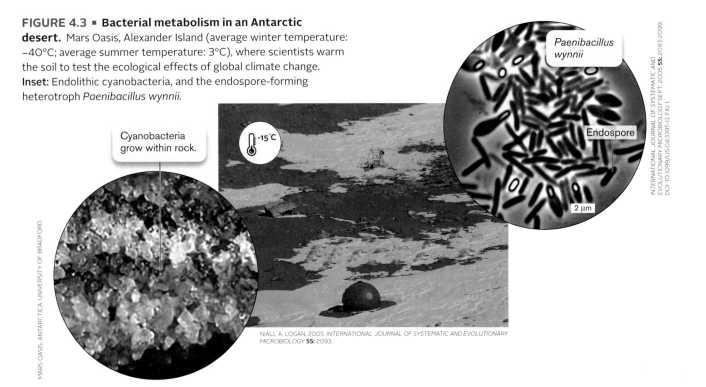

Cyanobacteria grow within rock.

Paenibacillus wynnii

Endospore

2 μm

−15°C

INTERNATIONAL JOURNAL OF SYSTEMATIC AND EVOLUTIONARY MICROBIOLOGY SEPT. 2005 **55:** 2093–2099, DOI 10.1099/IJS.0.63395–0, FIG 1.

NIALL A. LOGAN, 2005. *INTERNATIONAL JOURNAL OF SYSTEMATIC AND EVOLUTIONARY MICROBIOLOGY* **55:** 2093.

Chemoheterotroph (chemoheterotrophy or organotrophy). Chemoheterotrophs obtain energy and carbon for biomass solely from organic compounds. Chemoheterotrophy is also called chemoorganotrophy or, commonly, just heterotrophy.

The survival and metabolism of any one group of organisms depend on the survival and metabolism of other groups of organisms. For example, the cyanobacteria, a type of photosynthetic microorganism that originated 2.5–3.5 billion years ago, produce most of the oxygen that we breathe. Cyanobacteria also depend on heterotrophic bacteria to consume the molecular oxygen that the cyanobacteria produce, since molecular oxygen is toxic to cyanobacteria.

A present-day cyanobacteria-containing ecosystem that may resemble that of ancient Earth is the polar desert Mars Oasis, on Alexander Island, Antarctica (**Fig. 4.3**). In this harsh habitat, cyanobacteria grow in soil and even as "endoliths" within rock. In the soil we find heterotrophs such as *Paenibacillus wynnii. P. wynnii* consumes oxygen to catabolize many carbohydrates to CO_2, and also fixes its own nitrogen from the atmosphere. The bacteria can make endospores (see Section 4.6), which survive during long periods of dormancy, such as the Antarctic winter. Researchers set up experiments to test the effect of heating the soil, as a simulation of climate change, on the ecosystem.

Today, cyanobacteria form the base of Earth's marine food chain. The autotrophic cyanobacteria fix carbon in the ocean and are eaten by heterotrophic protists. The protists are then devoured by fish, and the fish produce the CO_2 fixed by the cyanobacteria. And eventually, we eat the fish.

Energy Is Stored for Later Use

Whatever the source, energy, once obtained, must be converted to a form useful to the cell. This form can be chemical energy, such as that contained in the high-energy phosphate bonds in adenosine triphosphate (ATP), or it can be electrochemical energy, which is stored in the form of an electrical potential generated between compartments separated by a membrane (see Chapter 14). Energy stored by an electrical potential across the membrane is known as the **membrane potential**. A membrane potential is generated when chemical (or light) energy is used to pump protons outside of the cell, making the proton concentration greater outside the cell than inside. For example, membrane proteins such as cytochrome oxidases use energy from respiration to pump hydrogen ions across the cell membrane, generating a hydrogen ion gradient. This movement of positively charged ions produces an electrical gradient across the cell membrane, making the inside of the cell more negatively charged than the outside. The hydrogen ion gradient plus the charge difference (voltage potential) across the membrane form an **electrochemical potential**. Because this electrochemical potential includes the hydrogen ion gradient, it is called the **proton potential**, or **proton motive force**. The energy stored in the proton motive force can be used by specific transport proteins to move nutrients into the cell (see Section 3.3), to directly drive motors that rotate flagella, and to drive the synthesis of ATP by a membrane-embedded **ATP synthase** (**Fig. 4.4A**).

The membrane-embedded F-ATP synthase, also called F_1F_o ATP synthase, provides most of the ATP for aerobic respiring cells such as *E. coli*. Essentially the same complex

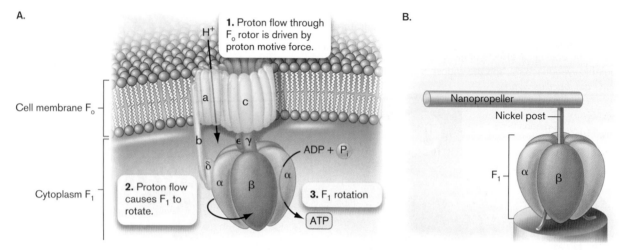

FIGURE 4.4 ■ Bacterial membrane ATP synthase. A. The F_o portion of the F_1F_o complex of ATP synthase is embedded in the cell membrane. **B.** An artificial "biomolecular motor" was built from an ATP synthase F_1 unit attached to a nickel post and a nanopropeller.

mediates ATP generation in our own mitochondria. ATP synthase is a complex of many different proteins. The enzyme includes a channel (F_o) that allows H^+ to move across the membrane and drive rotation of the ATP synthase complex (F_1). Rotation of F_1 mediates the formation of ATP. The role of the proton potential in metabolism is discussed in detail in Chapters 13 and 14.

The idea that a living organism could contain rotating parts was controversial when such parts were first discovered in bacterial flagella (discussed in Section 3.6). The discovery of rotary biomolecules has inspired advances in nanotechnology, the engineering of microscopic devices. For example, a "biomolecular motor" was devised using an ATP synthase F_1 complex to drive a metal submicroscopic propeller (**Fig. 4.4B**). In the future, such biomolecular design may be used to build microscopic robots that enter the bloodstream to perform microsurgery.

The Nitrogen Cycle

Nitrogen is an essential component of proteins, nucleic acids, and other cellular constituents, and as such, is required in large amounts by living organisms. So how do bacteria get nitrogen? Nitrogen gas (N_2) makes up nearly 79% of Earth's atmosphere, but most organisms are unable to use nitrogen gas, because the triple bond between the two nitrogen atoms is highly stable and requires considerable energy to be broken. For nitrogen to be used for growth, it must first be "fixed," or converted to ammonium ions (NH_4^+). Nitrogen gas is converted to NH_4^+ by **nitrogen-fixing bacteria**. The ammonium is then used by all microbes to make amino acids and other nitrogenous compounds needed for growth.

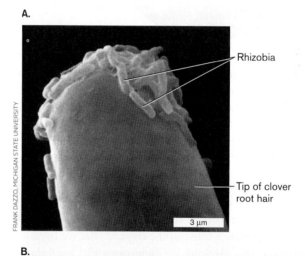

FIGURE 4.5 ■ *Rhizobium* and a legume. A. Symbiotic *Rhizobium* cells clustered on a clover root tip (SEM). Although the rhizobia (about 0.9 μm × 3 μm) are shown clustered on the surface of the root, they will soon invade the root and begin a symbiotic partnership that will benefit both organisms. **B.** Root nodules. After the rhizobia invade the plant root, symbiosis between plant and microbe produces nodules.

Nitrogen-fixing bacteria may be free-living in soil or water, or they may form symbiotic associations with plants or other organisms. A **symbiont** is an organism that lives in intimate association with a second organism. For example, *Rhizobium*, *Sinorhizobium*, and *Bradyrhizobium* species are nitrogen-fixing symbionts of leguminous plants such as soybeans, chickpeas, and clover (**Fig. 4.5**). Although symbionts are the most widely known nitrogen-fixing bacteria, the majority of nitrogen in soil and marine environments is fixed by free-living bacteria and archaea.

Once fixed, how does nitrogen get back into the atmosphere? As with the carbon cycle, various groups of organisms collaborate to recycle ammonium ions and nitrate ions (NO_3^-) into nitrogen gas in what is called the nitrogen cycle, and they collect energy in the process (**Fig. 4.6**). One group of bacteria gains energy by converting, or oxidizing, ammonia to form nitrate in a process called **nitrification** (*Nitrosomonas*, *Alcaligenes*, and *Nitrobacter*). Nitrification is a form of lithotrophy. Other heterotrophic microbes (*Paracoccus*) can <u>reduce</u> nitrate to N_2 via **denitrification**, a process that uses nitrate and related inorganic forms of nitrogen as terminal electron acceptors for certain electron transport chains. Denitrifying bacteria send an amount of nitrogen into the atmosphere that roughly balances the amount removed by nitrogen fixation. Like the carbon cycle, the nitrogen cycle illustrates how nature manages to replenish planet Earth. For the environmental significance of nitrogen metabolism, see Chapter 22.

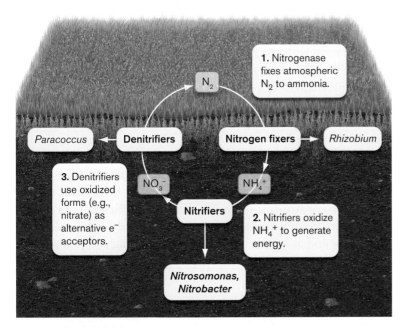

FIGURE 4.6 ■ **The nitrogen cycle.** Different sets of bacteria fix nitrogen from atmospheric nitrogen gas (N_2) to form ammonia (NH_4^+), convert NH_4^+ to nitrate (NO_3^+), and return N_2 to the atmosphere.

To Summarize

- **Microorganisms** require certain essential macro- and micronutrients to grow.

- **Autotrophs** use CO_2 as a carbon source, either through photosynthesis or through lithotrophy, and make organic compounds as biomass.

- **Heterotrophs** consume the organic compounds made by autotrophs to gain carbon.

- **Energy gained by autotrophy or chemotrophy** is stored either as proton motive force or chemical energy (ATP).

- **Nitrogen fixers (only bacteria and archaea)** incorporate nitrogen into biomass and contribute organic nitrogen to the rest of the ecosystem.

- **Chemotrophic nitrifying bacteria gain energy** by converting NH_4^+ (made by nitrogen-fixing bacteria) into nitrate and nitrite.

- **Heterotrophic denitrifying organisms** use nitrate and nitrite as electron acceptors to make nitrogen gas.

4.2 Nutrient Uptake

How do bacteria gather nutrients? Whether a microbe swims by flagella toward a favorable habitat or, lacking motility, drifts through its environment, the organism must be able to find nutrients and move them across the membrane into the cytoplasm. The membrane, however, presents a daunting obstacle. Membranes are designed to separate what is outside the cell from what is inside. So, for a cell to gain sustenance from the environment, the membrane must be selectively permeable to nutrients the cell can use. A few compounds, such as oxygen and carbon dioxide, can passively diffuse across the membrane, but most cannot. Selective permeability is achieved in three ways:

- Substrate-specific carrier proteins (**permeases**) in the membrane

- Nutrient-binding proteins that patrol the periplasmic space

- Membrane-spanning protein channels, or pores, that discriminate between substrates

Microbes must also overcome the problem of low nutrient concentrations in the natural environment (for example, lakes or streams). If the intracellular concentration of nutrients were no greater than the extracellular

A.

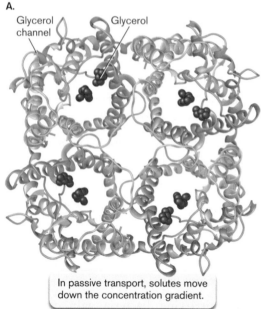

Glycerol channel

Glycerol

In passive transport, solutes move down the concentration gradient.

B.

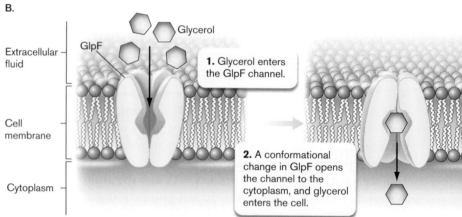

Extracellular fluid

Cell membrane

Cytoplasm

Glycerol

GlpF

1. Glycerol enters the GlpF channel.

2. A conformational change in GlpF opens the channel to the cytoplasm, and glycerol enters the cell.

FIGURE 4.7 ■ Facilitated diffusion. A. The glycerol transporter of *E. coli*, viewed from the external side of the membrane, consists of four channels that span the cell membrane. Each channel (blue and yellow) can transport two glycerol molecules (magenta) at a time. (PDB code: 1FX8) **B.** Facilitated diffusion of glycerol through GlpF. The protein facilitates the movement of the compound from outside the cell (where the concentration of glycerol is high) to inside the cell (where the concentration of glycerol is low).

concentration, the cell would remain starved of most nutrients. To solve this dilemma, most organisms have evolved efficient transport systems that concentrate nutrients inside the cell relative to outside. However, moving molecules against a concentration gradient requires some form of energy.

In contrast to environments where nutrients are available but exist at low concentrations (for example, aqueous environments), certain habitats have plenty of nutrients but those nutrients are locked in a form that cannot be transported into the cell. Starch, a large, complex carbohydrate, is but one example. Many microbes unlock these nutrient "vaults" by secreting digestive enzymes that break down complex carbohydrates or other molecules into smaller compounds that are easier to transport. The amazing mechanisms that cells use to extrude these large digestive proteins through the membrane and into their surrounding environment are discussed in Chapter 8.

Facilitated Diffusion

Although most transport systems use cellular energy to bring compounds into the cell, a few do not. **Facilitated diffusion** uses the concentration gradient of a compound to move that compound across the membrane from a compartment of higher concentration to a compartment of lower concentration. The best these passive systems can do is to equalize the internal and external concentrations of a solute; facilitated transport cannot move a molecule against its gradient. Facilitated diffusion systems are used for compounds that are either too large or too polar to diffuse on their own.

The most important facilitated diffusion transporters are those of the aquaporin family that transport water and small polar molecules such as glycerol (which is used for energy and for building phospholipids). Glycerol transport is performed by an integral membrane protein in *E. coli* called GlpF. The structure of a glycerol channel is shown in **Figure 4.7A**, where the complex is viewed from the outer face of the membrane. The complex is a tetramer of four channels, each of which transports a glycerol.

In the membrane, GlpF reversibly (and randomly) assumes two conformations. One form exposes the glycerol-binding site to the external environment, whereas the second form exposes this site to the cytoplasm. When the concentration of glycerol is greater outside than inside the cell, the form with the binding site exposed to the exterior is more likely to find and bind glycerol. After binding glycerol, GlpF changes shape, closing itself to the exterior and opening to the interior (**Fig. 4.7B**). Bound glycerol is then released (diffuses) into the cytoplasm (influx). Of course, this form of GlpF could also bind a glycerol molecule in the cytoplasm and release it outside the cell. Thus, once the cytoplasmic concentration of glycerol equals the concentration outside the cell, bound glycerol can be released to either compartment. However, facilitated diffusion normally promotes glycerol influx, because the cell consumes the compound as it enters the cytoplasm, keeping cytoplasmic concentrations of glycerol low.

Active Transport Requires Energy

Most forms of transport expend energy to take up molecules from outside the cell and concentrate them inside. The ability to import nutrients against their natural

A. Symport

1. Energy is released as one substrate (red) moves down its concentration gradient.

2. This energy moves a second substrate (blue) against its gradient and into the cell.

Outside

Inside

B. Antiport

1. Antiporter binds substrate A (red) on the cytoplasmic side of the membrane.

2. Antiporter opens to the outside, where the concentration of A is less.

3. Substrate A leaves its binding site, and substrate B (blue) then binds to its site.

Outside

Cell membrane

Inside

4. Antiporter opens to the inside of the cell. Substrate B is released in exchange for substrate A.

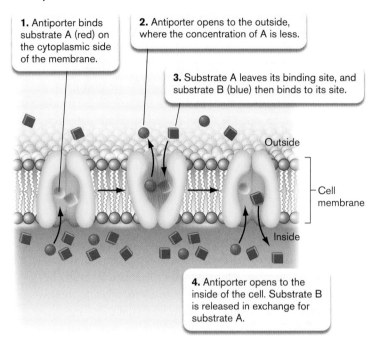

C.

COURTESY OF RONALD KABACK

FIGURE 4.8 ■ **Coupled transport.** In both symport **(A)** and antiport **(B)**, substrate B (blue) is taken up against its gradient because of the energy released by substrate A (red) traveling down its gradient. **C.** Ronald Kaback (left) of UCLA elucidated the mechanism of transport by the proton-driven lactose symporter LacY.

concentration gradients is critical in aquatic habitats, where nutrient concentrations are low, and in soil habitats, where competition for nutrients is high.

The simplest way to use energy to move molecules across a membrane is to exchange the energy of one chemical gradient for that of another. The most common chemical gradients used are those of ions, particularly the positively charged ions Na^+ and K^+. These ions are kept at different concentrations on either side of the cell membrane. When an ion moves down its concentration gradient (from high to low), energy is released. Some transport proteins harness that free energy and use it to drive transport of a second molecule up, or against, its concentration gradient in the process called **coupled transport**.

The two types of coupled transport systems are **symport**, in which the two molecules travel in the same direction (**Fig. 4.8A**), and **antiport**, in which the actively transported molecule moves in the direction opposite that of the driving ion (**Fig. 4.8B**). An example of a symporter is the lactose permease LacY of *E. coli*, one of the first transport proteins to have its function elucidated. This work was carried out by the pioneering membrane biochemist H. Ronald Kaback (**Fig. 4.8C**). LacY moves lactose inward, powered by a proton that is also moving inward (symport). LacY proton-driven transport is said to

be <u>electrogenic</u> because an unequal distribution of charge results (for example, symport of a neutral lactose molecule with H^+ results in net movement of positive charge).

An example of <u>electroneutral</u> coupled transport, in which there is no net transfer of charge, is that of the Na^+/H^+ antiporter. The Na^+/H^+ antiporter couples the export of Na^+ with the import of a proton (antiport). Because molecules of like charges are merely exchanged, there is no net movement of charge. Sodium exchange is important for all organisms and is particularly critical for organisms living in high-salt habitats (see Section 5.4).

Thought Questions

4.2 In what situation would antiport and symport be passive rather than active transport?

4.3 What kind of transporter, other than an antiporter, could produce electroneutral coupled transport?

Symport and antiport transporter proteins function by alternately opening one end or the other of a channel that spans the cell membrane. The channel contains solute-binding sites (**Fig. 4.8A** and **B**). When the channel is open to the high-concentration side of the membrane, the driving ion (solute) attaches to the binding sites. The transport protein then changes shape to open that site to the low-concentration side of the membrane, and the ion leaves. When and where the second (cotransported) solute binds depends on whether the transport protein is an antiporter or a symporter.

With all this ion traffic across the membrane going on, a careful accounting must be kept of how many ions are inside the cell relative to outside. The cell must recirculate ions back and forth across the membrane to maintain certain gradients if the organism is to survive. Because key ATP-producing systems require an electrochemical gradient across the membrane, it is especially important to keep the interior of the cell negatively charged relative to the exterior. However, because the movement of many compounds is coupled to the import of positive ions, the electrochemical gradient will eventually dissipate, or depolarize, unless positive ions are also exported. Depolarization must be avoided, because a depolarized cell loses membrane integrity and cannot carry out the transport functions needed to sustain growth. A healthy cell maintains a proper charge balance by using the electron transport chain to move protons out of the cell and by using antiporters to exchange negatively and positively charged ions as needed.

While symport and antiport systems directly link the transport of different ions, the movement of one ion can also be linked indirectly to movement of another molecule. For example, proton concentrations are typically greater outside the cell than inside. The inwardly directed proton gradient is called proton motive force. Proton motive force can impel the exit of Na^+ through the Na^+/H^+ antiporter. The resulting Na^+ gradient can then drive the symport of amino acids into the cell. In this case, Na^+ moves back into the cell down its gradient, and the energy released is tied to the import of an amino acid against its gradient.

ABC Transporters Are Powered by ATP

As we pointed out in Section 3.2, a major function of proton transport is to form the proton motive force that powers ATP synthesis (for details on the proton motive force, see Chapter 14). The energy stored in ATP can then drive membrane transport of nutrients.

The largest family of energy-driven transport systems is the ATP-binding cassette superfamily, also known as **ABC transporters**. These transporters are found in bacteria, archaea, and eukaryotes. All of them appear to have arisen from a common ancestral porter, so they share a considerable amount of amino acid sequence homology. Many different ABC transporters mediate the transport of a wide variety of substrates.

It is impressive that nearly 5% of the *E. coli* genome is dedicated to producing the components of 70 different varieties of uptake and efflux ABC transporters. The uptake ABC transporters are critical for transporting nutrients such as maltose, histidine, arabinose, and galactose. The efflux ABC transporters are generally used as multidrug efflux pumps that allow microbes to survive exposures to hazardous chemicals. *Lactococcus*, for example, can use a pump called LmrP to export a broad range of antibiotics, including tetracyclines, streptogramins, quinolones, macrolides, and aminoglycosides, thus conferring resistance to those drugs (see Section 27.3).

An ABC transporter typically consists of two hydrophobic proteins that form a membrane channel and two cytoplasmic proteins that contain a highly conserved amino acid motif, called the ATP-binding cassette, that binds ATP. The ABC transporter superfamily contains both uptake and efflux transport systems. The uptake systems (but not the efflux systems) possess an additional, extracytoplasmic protein, called a substrate-binding protein, that initially binds the substrate (also called solute). In Gram-negative bacteria, these substrate-binding proteins float in the periplasmic space between the inner and outer membranes. In Gram-positive bacteria, which lack an outer membrane, the proteins must be tethered to the cell surface.

ABC transport (**Fig. 4.9**) starts with the substrate-binding protein snagging the appropriate solute, either as it floats by a Gram-positive microbe or as the molecule enters the periplasm of a Gram-negative cell. Most substrates

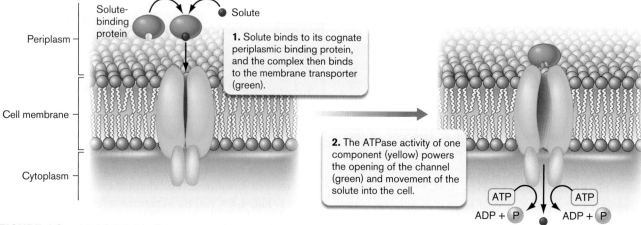

FIGURE 4.9 ■ **ABC (ATP-binding cassette) transporter in a Gram-negative organism.**

A.

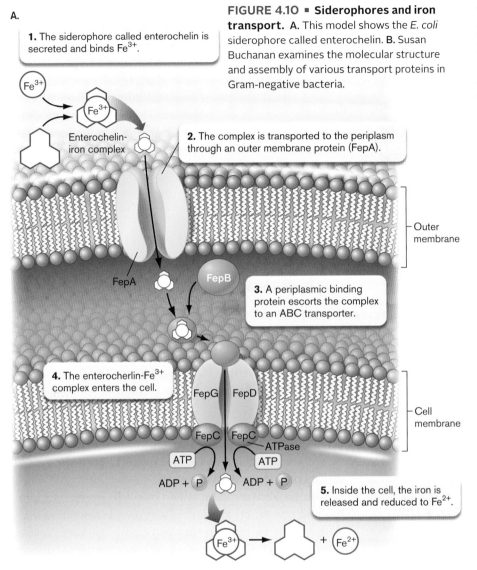

1. The siderophore called enterochelin is secreted and binds Fe^{3+}.

Enterochelin-iron complex

2. The complex is transported to the periplasm through an outer membrane protein (FepA).

Outer membrane

FepA

FepB

3. A periplasmic binding protein escorts the complex to an ABC transporter.

4. The enterocherlin-Fe^{3+} complex enters the cell.

FepG   FepD

Cell membrane

FepC   FepC

ATPase

ATP   ATP

ADP + P   ADP + P

5. Inside the cell, the iron is released and reduced to Fe^{2+}.

Fe^{3+} → + Fe^{2+}

FIGURE 4.10 ■ Siderophores and iron transport. A. This model shows the *E. coli* siderophore called enterochelin. **B.** Susan Buchanan examines the molecular structure and assembly of various transport proteins in Gram-negative bacteria.

B.

PHOTOGRAPH BY RHODA BAER

to move the substrate into the periplasm. Because substrate-binding proteins have a high affinity for their cognate (matched) solutes, their use increases the efficiency of transport when concentrations of solute are low.

Once united with its solute, the binding protein binds to the periplasmic face of the channel protein and releases the solute, which now moves to a site on the channel protein. This interaction triggers a structural (or conformational) change in the channel protein (the green "membrane transporter" in **Fig. 4.9**) that is telegraphed to the nucleotide-binding proteins on the cytoplasmic side (yellow). On receiving this signal, the nucleotide-binding proteins start hydrolyzing ATP and send a return conformational change through the channel, signaling the channel to open its cytoplasmic side and allow the solute to enter the cell.

Siderophores Are Secreted to Scavenge Iron

Iron, an essential nutrient of most cells, is mostly locked up in nature as $Fe(OH)_3$, which is insoluble and unavailable for transport. Many bacteria and fungi have solved this transport dilemma by synthesizing and secreting specialized molecules called **siderophores** (Greek for "iron bearer"), which have a very high affinity for the small amounts of soluble ferric iron available in the environment. These iron scavenger molecules are produced and secreted by cells when the intracellular iron concentration is low (**Fig. 4.10**). In most Gram-negative organisms, the siderophore (for example, enterochelin in *E. coli*) binds iron in the environment, and the siderophore-iron complex then attaches to specific receptors in the outer membrane. At this point, either the iron is released directly and is passed to other transport proteins or the complex is transported across the cytoplasmic membrane by a dedicated ABC transporter. The iron is released intracellularly and reduced from Fe^{3+} to Fe^{2+} for biosynthetic use. Other Gram-negative microorganisms, such as *Neisseria gonorrhoeae* (the causative agent of gonorrhea), do not use siderophores at all but employ receptors on their surface that bind human iron complexes (for example, transferrin or lactoferrin) and wrest the iron from them.

nonspecifically enter the periplasm of Gram-negative organisms through the outer membrane pores, although some high-molecular-weight substrates, like vitamin B_{12}, require the assistance of a specific outer membrane protein

Because microbial iron-transporting proteins are critical to the pathogenesis of many microbes, solving the molecular

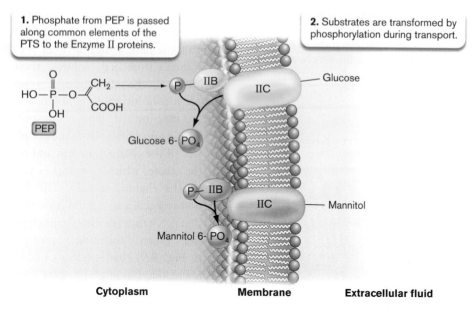

1. Phosphate from PEP is passed along common elements of the PTS to the Enzyme II proteins.

2. Substrates are transformed by phosphorylation during transport.

Cytoplasm   **Membrane**   **Extracellular fluid**

FIGURE 4.11 ■ Group translocation: the phosphotransferase system (PTS) of *E. coli*. The phosphate group from phosphoenolpyruvate (PEP) is ultimately passed along proteins common to all PTS sugar transport systems and to Enzyme II components of the PTS that are specific to individual substrates (such as glucose or mannitol). The substrate is then phosphorylated during transport, making it different from the external sugar. As a result, the sugar is always traveling down its concentration gradient into the cell. A more detailed look at the cellular location of different PTS proteins is found in **eTopic 4.1**. ▶

structures of these proteins is of great interest. Learning their structural and functional vulnerabilities could lead to the development of new antibiotics that stop iron transport. Susan Buchanan from the National Institutes of Health is working to solve the structures of several iron and heme transport systems in pathogenic microbes (**Fig. 4.10B**).

Group Translocation Avoids "Uphill" Battles

The uptake transporters that we have just considered increase concentrations of solute inside the cell relative to the outside. They move nutrients "uphill" against a concentration gradient. An entirely different system, known as **group translocation**, cleverly accomplishes the same result but without really moving a substance uphill. Group translocation alters the substrate during transport by attaching a new group (for example, phosphate) to it. Because the modified nutrient inside the cell is chemically different from the related compound outside, the parent solute entering the cell is always moving down its concentration gradient, regardless of how much solute has already been transported. Note that this process uses energy to chemically alter the solute. ABC transporters and group translocation systems both involve active transport, but group translocation systems are not ABC transporters.

The **phosphotransferase system** (**PTS**) is a well-characterized group translocation system present in many bacteria. It uses energy from phosphoenolpyruvate (PEP),

an intermediate in glycolysis, to attach a phosphate to specific sugars during their transport into the cell (**Fig. 4.11** ▶). Glucose, for example, is converted during transport to glucose 6-phosphate. The system has a modular design that accommodates different substrates. Protein elements that initiate phosphotransfer from PEP are used for all sugars transported by the PTS (step 1). Later PTS proteins, however, are unique to a given carbohydrate (step 2). For example, different enzyme IIB and membrane-embedded IIC proteins transfer phosphate to glucose and mannitol during transport. Because the sugar (glucose, for instance) is phosphorylated during transport, glucose itself never accumulates in the cytoplasm. As a result, the sugar transported via PTS always travels <u>down</u> its concentration gradient into the cell. More details about the PTS are given in **eTopic 4.1**. In Chapter 9 we will see how this physiological system impacts the genetic control of many other systems.

Like prokaryotic cells, eukaryotic cells possess antiporters and symporters and use ABC transport systems as multidrug efflux pumps, but they also employ another process, called endocytosis, which often precedes nutrient transport across membranes. Endocytosis is discussed in **eTopic 4.2** and eAppendix 2.

To Summarize

- **Transport systems** move nutrients across phospholipid bilayer membranes.

- **Facilitated diffusion** helps solutes move across a membrane from a region of high concentration to one of lower concentration.

- **Antiport** and **symport** are coupled transport systems in which energy released by moving a driving ion (H^+ or Na^+) from a region of high concentration to one of low concentration is harnessed and used to move a solute against its concentration gradient.

- **ABC transporters** use the energy from ATP hydrolysis to move solutes "uphill," against their concentration gradients.

- **Siderophores** are secreted to bind ferric iron (Fe^{3+}) and transport it into the cell, where it is reduced to the more useful ferrous form (Fe^{2+}). Siderophore-iron complexes enter cells with the help of ABC transporters.

■ **Group translocation systems** use energy to chemically modify the solute during transport.

4.3 Culturing and Counting Bacteria

How do we capture bacteria and study them? Microbes in nature usually exist in complex, intertwined multispecies communities. For detailed studies of a single species, however, cells of the species are usually grown in pure culture. This section will describe how that is done. But after 120 years of trying to grow microbes in the laboratory, we have succeeded in culturing less than 1% of the microorganisms around us. The vast majority of the microbial world has yet to be tamed. At the end of this section we will discuss some new, innovative methods that enable us to culture at least some of these "unculturable" microbes.

Bacteria Are Grown in Culture Media

For those organisms that can be cultured, we have access to a variety of culturing techniques that can be used for different purposes. Bacterial culture media may be either liquid or solid. A liquid, or broth, medium, in which organisms can move about freely, is useful for studying the growth characteristics of a single strain of a single species (that is, a **pure culture**). Liquid media are also convenient for examining growth kinetics and microbial biochemistry at different phases of growth. Solid media, usually gelled with agar, are useful for trying to separate mixtures of different organisms as they are found in the natural environment or in clinical specimens.

Dilution Streaking and Spread Plates

Solid media are basically liquid media to which a solidifying agent has been added. The most versatile and widely used solidifying agent is agar (for the development of agar medium, see Section 1.3). Derived from seaweed, agar forms an unusual gel that liquefies at 100°C but does not solidify again until cooled to about 40°C. Liquefied agar medium poured into shallow, covered petri dishes cools and hardens to provide a large, flat surface on which a mixture of microorganisms can be streaked to separate individual cells. Each cell will divide and grow to form a distinct, visible colony of cells (**Fig. 4.12**).

As shown in **Figure 4.13** ▶, a drop of liquid culture is collected with an inoculating loop and streaked across the agar plate surface in a pattern called **dilution streaking**. Organisms fall off the loop as it moves along the agar surface.

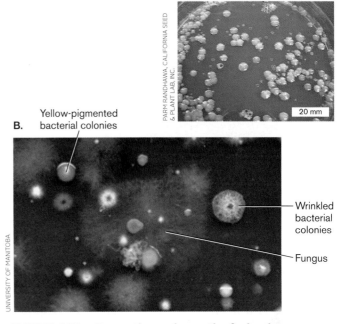

FIGURE 4.12 ■ **Separation and growth of microbes on an agar surface.** **A.** Colonies (diameter 1–5 mm) of *Acidovorax citrulli* separated on an agar plate. *A. citrulli* is a plant pathogen that causes watermelon fruit blotch. **B.** A mixture of yellow-pigmented bacterial colonies, wrinkled bacterial colonies, and fungus separated by dilution on an agar plate. As time passed, the fungal colony overgrew adjacent bacterial colonies.

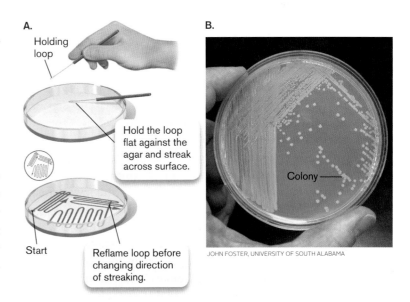

FIGURE 4.13 ■ **Dilution streaking technique.** **A.** A liquid culture is sampled with a sterile inoculating loop and streaked across the plate in three or four areas, with the loop flamed between areas to kill bacteria still clinging to it. Dragging the loop across the agar diminishes the number of organisms clinging to the loop until only single cells are deposited at a given location. **B.** *Salmonella enterica* culture obtained by dilution streaking. ▶

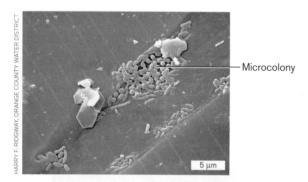

FIGURE 4.14 ▪ **Bacterial colonies.** A 3-day-old microcolony (biofilm) on the surface of a membrane used to treat municipal wastewater in Fountain Valley, California (SEM).

Toward the end of the streak, few bacteria remain on the loop, so individual cells will land and stick to different places on the agar surface. If the medium, whether artificial (for example, laboratory medium) or natural (for example, crab carapace) contains the proper nutrients and growth factors, a single cell will multiply into many millions of offspring, forming a microcolony. At first visible only under a microscope, the microcolony (**Fig. 4.14**) grows into a visible droplet called a **colony** (**Fig. 4.13B**). A pure culture of the species or one strain of a species can be obtained by touching a single colony with a sterile inoculating loop and inserting that loop into fresh liquid medium.

It is important to note that the "one cell equals one colony" paradigm does not hold for all bacteria. Organisms such as *Streptococcus* and *Staphylococcus* usually do not exist as single cells, but grow as chains or clusters of several cells. Thus, a cluster of ten *Staphylococcus* cells will form only one colony on an agar medium and so is called a colony-forming unit (CFU).

Another way to isolate pure colonies is the **spread plate** technique. Starting from a liquid culture of bacteria, a series of tenfold dilutions is made, and a small amount of each dilution is placed directly on the surface of separate agar plates (**Fig. 4.15**). The sample is spread over the surface of the plate with an alcohol-sterilized, bent glass rod. The early dilutions, those containing the most bacteria, will produce **confluent** growth that covers the entire agar surface. Later dilutions, containing fewer and fewer organisms, yield individual colonies. As we will see later, spread plates not only enable us to isolate pure cultures but also can be used to enumerate the number of **viable** bacteria in the original growth tube. A viable organism is one that successfully replicates to form a colony. **Thus, each colony on an agar plate represents one viable organism (or CFU) present in the original liquid culture.**

Complex versus Synthetic Media

We often culture bacteria in a type of medium, called **complex medium** (or "rich medium") that is nutrient-rich but has poorly defined components. Alternatively, we can select

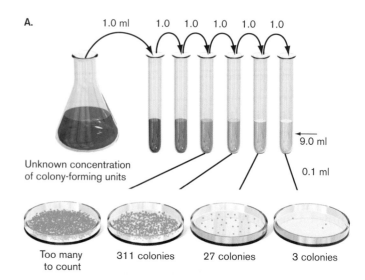

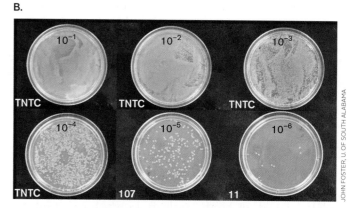

FIGURE 4.15 ▪ **Tenfold dilutions, plating, and viable counts.**
A. A culture containing an unknown concentration of cells is serially diluted. One milliliter (ml) of culture is added to 9.0 ml of diluent broth and mixed, and then 1 ml of this 1/10 dilution is added to another 9.0 ml of diluent (10^{-2} dilution). These steps are repeated for further dilution, each of which lowers the cell number tenfold. After dilution, 0.1 ml of each dilution is spread onto an agar plate.
B. Plates prepared as in (A) are incubated at 37°C to yield colonies. By multiplying the number of countable colonies (107 colonies on the 10^{-5} plate) by 10, you get the number of cells in 1.0 ml of the 10^{-5} dilution. Multiplying that number by the reciprocal of the dilution factor, you can calculate the number of cells (colony-forming units, or CFUs) per milliliter in the original broth tube ($10^7 \times 10^1 \times 10^5 = 1.1 \times 10^8$ CFUs/ml). TNTC = too numerous to count.

a defined or synthetic medium in which all chemical components are known. Bacteria grow more slowly on defined media because the organisms must synthesize more of their own components. A **minimal defined medium** contains only those nutrients that are essential for growth of a given microbe (**Table 4.1**). However, not every microbe that can be cultured on a complex medium can be grown on a defined medium. Recipes for complex media usually contain several ingredients, such as yeast extract or beef extract, whose exact composition is unknown. These additives include a rich variety of amino acids, peptides, nucleosides, vitamins, and some sugars. Some organisms are particularly fastidious, requiring that components of blood

TABLE 4.1	Composition of Commonly Used Media		
Medium	**Ingredients per liter**		**Organisms cultured**
Luria Bertani (complex)	Bacto tryptone[a]	10 g	Many Gram-negative and Gram-positive organisms (such as *Escherichia coli* and *Staphylococcus aureus*, respectively)
	Bacto yeast extract	5 g	
	NaCl	10 g	
	Adjust to pH 7		
M9 medium (defined)	Glucose	2.0 g	Gram-negative organisms such as *E. coli*
	Na_2HPO_4	6.0 g (42 mM)	
	KH_2PO_4	3.0 g (22 mM)	
	NH_4Cl	1.0 g (19 mM)	
	NaCl	0.5 g (9 mM)	
	$MgSO_4$	2.0 mM	
	$CaCl_2$	0.1 mM	
	Adjust to pH 7		
Sulfur oxidizers (defined)	NH_4Cl	0.52 g	*Acidithiobacillus thiooxidans*
	KH_2PO_4	0.28 g	
	$MgSO_4 \cdot 7H_2O$	0.25 g	
	$CaCl_2$	0.07 g	
	Elemental sulfur	1.56 g	
	CO_2	5%	
	Adjust to pH 3		

[a]Bacto tryptone is a pancreatic digest of casein (bovine milk protein).

be added to a basic complex medium. With these additions, the complex medium is called an **enriched medium**.

Complex media provide many of the chemical building blocks that a cell would otherwise have to synthesize on its own. For example, instead of making proteins that synthesize tryptophan, all the cell needs is a membrane transport system to harvest prefabricated tryptophan from the medium. Likewise, fastidious organisms that require blood in their media may reclaim the heme released from red blood cells as their own, using it as an "enzyme prosthetic group," a group critical to enzyme function (for example, the heme group in cytochromes). All of this saves the scavenging cell a tremendous amount of energy, and as a result, bacteria tend to grow fastest in complex media.

The metabolism of a microbe growing in a complex medium is hard to characterize. How would you know whether *E. coli* possesses the ability to make tryptophan if the bacterium grew only in complex media? Questions like this can be asked of organisms able to grow in fully defined synthetic media. In preparing a synthetic medium, we start with water and then add various salts, carbon, nitrogen, and energy sources in precise amounts. For self-reliant organisms like *E. coli* or *Bacillus subtilis*, that is all that's needed. Other organisms, such as *Shigella* species or mutant strains of *E. coli* or *B. subtilis*, require additional ingredients to satisfy requirements imposed by the absence of specific metabolic pathways (growth factors are discussed shortly). However, newer molecular tools, such as whole-genome sequencing, can reveal missing biosynthetic pathways and help predict which nutrients are needed to grow a particular species in a synthetic medium.

> **Thought Question**
>
> **4.4** What would be the phenotype (growth characteristic) of a cell that lacks the *trp* genes (genes required for the synthesis of tryptophan)? What would be the phenotype of a cell missing the *lac* genes (genes whose products catabolize the carbohydrate lactose)?

Selective and Differential Media

Microorganisms are remarkably diverse with respect to their metabolic capabilities and resistance to certain toxic agents. These differences are exploited in **selective media**, which favor the growth of one organism over another, and in **differential media**, which expose biochemical differences between two species that grow equally well. For example, Gram-negative bacteria, with their outer membrane, are much more resistant than Gram-positive bacteria to detergents like bile salts and certain dyes, such as crystal violet. A solid medium containing bile salts and crystal violet is considered selective because it favors the growth of Gram-negative over Gram-positive bacteria.

On the other hand, a differential medium is needed to distinguish between organisms that differ not in their ability to grow, but in a particular biochemical function they possess. For example, *E. coli* and *Salmonella enterica*, a major cause of diarrhea, are both Gram-negative, but only *E. coli* can ferment lactose. Both organisms will grow on solid media containing a nonfermentable carbon source like peptone, a dye called "neutral red," and lactose. Unlike *S. enterica*, *E. coli* ferments the lactose and produces acidic end products. These products lower the pH surrounding the colony, the neutral-red dye enters the cells, and the colony of *E. coli* turns red. *S. enterica* grows well on the nonfermentable peptides but cannot ferment lactose. Consequently, acidic end products are not produced, and the colonies of *S. enterica* remain white, their natural color. In this example, growth in differential media easily distinguishes colonies of lactose fermenters from nonfermenters.

Several media used in clinical microbiology are both selective and differential. MacConkey medium, for example, contains bile salts and crystal violet to prevent the growth of Gram-positive bacteria and allow the growth of Gram-negative bacteria (**Fig. 4.16**). The medium also includes lactose, neutral red, and peptones to differentiate lactose fermenters (Lac$^+$, red colonies) from nonfermenters (Lac$^-$, uncolored colonies). This medium is often used to identify bacteria that cause diarrheal disease because most normal microbiota that grow on this medium are lactose fermenters, whereas two important pathogens, *Salmonella* and *Shigella*, are lactose nonfermenters.

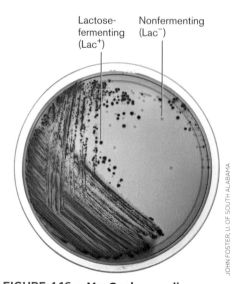

Lactose-fermenting (Lac$^+$)   Nonfermenting (Lac$^-$)

JOHN FOSTER, U. OF SOUTH ALABAMA

FIGURE 4.16 ■ **MacConkey medium, a culture medium both selective and differential.** Only Gram-negative bacteria grow on lactose MacConkey (selective). Only a species capable of fermenting lactose produces pink colonies (differential), because only fermenters can take up the neutral red and peptones that are also in the medium. Gram-negative nonfermenters appear as uncolored colonies.

of the relevant metabolic pathway through random mutations. Once that happens, a species will require **growth factors** (**Table 4.2**) to grow in laboratory media. Growth factors are specific nutrients not required by other species. Why, for example, should *Streptococcus pyogenes* make glutamic acid or alanine if both are readily available in its normal environment, the human oral cavity? Because it never needs to make glutamic acid or alanine in its natural habitat, *S. pyogenes* has lost the genes needed to synthesize these amino acids. For *S. pyogenes* to grow in the laboratory, the culture medium must contain alanine and glutamic acid along with the macro- and micronutrients mentioned previously.

Some species have adapted so well to their natural habitats that we still do not know how to grow them in the

---

Thought Questions

4.5 If lactose was left out of MacConkey medium (**Fig. 4.16**), would lactose-fermenting *E. coli* bacteria grow, and if so, what color would their colonies be?

4.6 The addition of sheep's blood to agar produces a very rich medium called blood agar. Do you think blood agar is a selective medium? A differential medium? *Hint:* Some bacteria can lyse red blood cells.

Growth Factors, Unculturable Microbes, and Obligate Intracellular Bacteria

Why do some bacterial species fail to grow in minimal medium, or fail to grow at all in the laboratory? Such failures are a consequence of evolution and the organism's natural growth environment. Long-term growth in an ecological niche that continually provides a compound an organism would otherwise have to make can lead to loss

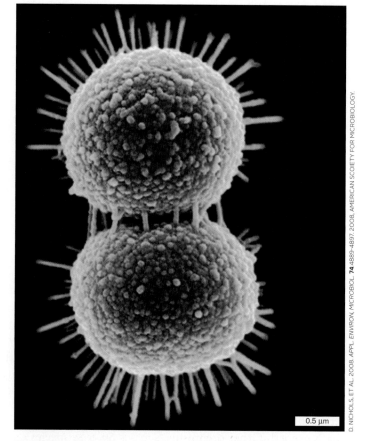

0.5 μm

D. NICHOLS, ET AL. 2008. *APPL. ENVIRON. MICROBIOL.* **74**:4889–4897. 2008, AMERICAN SOCIETY FOR MICROBIOLOGY.

FIGURE 4.17 ■ **"Unculturable" marine organism MSC33.** To grow in natural environments, many bacterial species rely on factors produced by other species within their niche. The microbe shown, MSC33, will not grow in laboratory media unless a peptide growth factor from another species is included.

TABLE 4.2	Growth Factors and Natural Habitats of Organisms Associated with Disease		
Organism	**Diseases**	**Natural habitats**	**Growth factors**
Shigella	Bloody diarrhea	Humans	Nicotinamide (NAD)[a]
Haemophilus	Meningitis, chancroid	Humans and other animal species, upper respiratory tract	Hemin, NAD
Staphylococcus	Boils, osteomyelitis	Widespread	Complex requirement
Abiotrophia	Osteomyelitis	Humans and other animal species	Vitamin K, cysteine
Legionella	Legionnaires' disease	Soil, refrigeration cooling towers	Cysteine
Bordetella	Whooping cough	Humans and other animal species	Glutamate, proline, cysteine
Francisella	Tularemia	Wild deer, rabbits	Complex, cysteine
Mycobacterium	Tuberculosis, leprosy	Humans	Nicotinic acid (NAD),[a] alanine (*M. leprae* is unculturable)
Streptococcus pyogenes	Pharyngitis, rheumatic fever	Humans	Glutamate, alanine

[a]Both nicotinamide and nicotinic acid are derived from NAD, nicotinamide adenine dinucleotide.

laboratory. As mentioned earlier, over 99% of bacterial species that are present in water or soil, or that grow in or on animals, will not form colonies on an agar plate. Some of these "unculturable" organisms depend on growth factors, such as siderophores, provided by other species that cohabit their natural environment. Some of these factors even appear to act like hormones, somehow stimulating replication of the unculturable organism. For example, **Figure 4.17** shows a marine organism (MSC33) that will not grow in the laboratory unless you include a peptide made by a companion microbe from the same natural environment. The peptide is not a nutrient, but instead has a signaling function that induces cell division. Today, new methods of culturing these "unculturable" species are being used to discover new antibiotic-producing microbes (**Special Topic 4.1**).

How do we know that unculturable microbes exist if we cannot grow them? All known microorganisms have a set of genes encoding RNA molecules present in ribosomes. The ribosomal RNA molecules are highly conserved across the phylogenetic tree. A DNA-amplifying procedure called the polymerase chain reaction (PCR, described in Section 7.6) can screen for the presence of these genes in soil and water samples. Comparing the DNA sequences of the PCR products from environmental samples with the DNA sequences of similar genes from known, culturable organisms reveals

that nature harbors many undiscovered microbes. Even though we don't know the growth and nutritional requirements of these phantom microbes, modern genomic techniques can expose their existence and can even provide remarkable insight into their physiologies (see Section 8.6).

Another class of bacteria, known as obligate intracellular bacteria, could also be called unculturable because they, too, will not grow on laboratory media. These species first evolved to penetrate and then grow only within a eukaryotic cell. For example, *Rickettsia prowazekii*, the cause of epidemic typhus fever, has adapted to grow within the cytoplasm of eukaryotic cells and nowhere else (**Fig. 4.18**). As it evolved, this obligate intracellular bacterium lost key pathways needed for independent growth because the host cell supplied them.

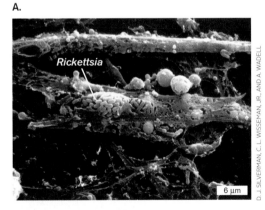

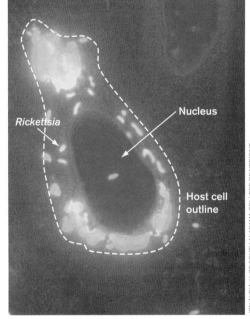

FIGURE 4.18 ■ *Rickettsia prowazekii growing within eukaryotic cells.* A. *R. prowazekii* growing within the cytoplasm of a chicken embryo fibroblast (SEM). B. Fluorescent stain of *R. prowazekii* (approx. 0.5 μm long), growing within a cultured human cell (outline marked by dotted line). The rickettsias are green (FITC-labeled antibody, arrow), the host cell nucleus is blue (Hoechst stain), and the mitochondria are red (Texas Red MitoTracker). The bacterium grows only in the cytoplasm, not in the nucleus.

| SPECIAL TOPIC 4.1 | Antibiotic Hunters Culture the "Unculturable" |

There is a quiet panic among infectious disease doctors, who worry that we will soon run out of effective antibiotics to treat deadly infections. Since antibiotics were introduced over 70 years ago, bacterial pathogens have continually evolved to resist their effects. Scientists, for their part, have struggled to find new antibiotics to replace the old, ineffective ones. Chapter 27 will discuss this "arms race" in more detail. The classic approach for finding new antibiotics is to screen bacteria and fungi snatched from exotic environments such as the Amazon for secreted antimicrobial compounds. This method worked for decades, but over the past 20 years it has yielded very few new antibiotics. A new way of thinking was needed if we humans are to maintain dominance over microbes.

Fewer than 1% of the microbes in the world whose DNA we can detect can be cultured in the laboratory. This means we have missed capturing the antibiotic potential of over 99% of living microbes. These "unculturable" organisms are present in soil, water, and even the human body. Why do the microbes grow there but not in the lab? They grow in their natural environments because of interspecies cooperation. In this environmental "buddy system," one species provides a growth factor to another species that cannot make its own. Can we find a way to tame such uncooperative organisms and screen them for new antibiotics?

Kim Lewis (Northeastern University, Boston) and colleagues have devised an elegant method to do just that. To harvest previously unculturable soil microbes, samples prepared from soil were diluted such that one bacterial cell was delivered into each channel of a multichannel iChip (**Fig. 1A**). The channels contained agar plugs in which a microbe could grow. Both sides of the device were covered with semipermeable membranes and placed back into the soil (**Fig. 1B**). Nutrients and growth factors made by organisms in the soil diffused into the chamber and allowed many of the unculturable microbes to form colonies. Once a colony formed, the previously uncultured organism became "domesticated," able to grow on agar medium without further assistance. The mechanism of domestication is not yet known.

The scientists made extracts from 10,000 iChip isolates and tested each one for antibiotic activity against *Staphylococcus aureus*, an important pathogen that can cause infections ranging from boils to sepsis. One extract was highly effective. The Gram-negative organism that produced this antibiotic was sequenced and named *Eleftheria terrae* (**Fig. 2**). The structure of the antibiotic, teixobactin, is shown in **Figure 2B**. It is a nonribosomal peptide synthesized by enzymes instead of a ribosome (nonribosomal peptide synthesis is described in Special Topic 15.1).

The cellular target of teixobactin is peptidoglycan synthesis in Gram-positive bacteria (the peptide cannot penetrate the Gram-negative outer membrane). The new antibiotic kills bacteria by binding to the lipid carrier molecule, bactoprenol, that ferries the peptidoglycan subunits *N*-acetylglucosamine and *N*-acetylmuramic acid across the bacterial membrane (described in Chapter 3). As a result, growing cells lyse because they cannot make new cell wall. Chapter 27 describes the details of peptidoglycan synthesis and the antibiotics that target steps in this process.

A.

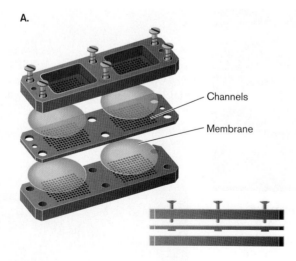

Channels

Membrane

B.

SLAVA EPSTEIN/NORTHEASTERN UNIVERSITY.

FIGURE 1 ■ **The multichannel iChip.** The iChip was used to find the antibiotic teixobactin.

Teixobactin is unusual among antibiotics because attempts to generate antibiotic resistance to it have failed. Most antibiotics bind to bacterial proteins whose sequence can easily change by mutation to become antibiotic resistant. Teixobactin, however, binds to a highly conserved, <u>nonprotein</u> component of the bacterial cell (bactoprenol). Altering bactoprenol to become teixobactin resistant would require the bacterium to make new kinds of enzymes. Thus, the likelihood that pathogens can develop resistance is low. This feature makes this drug highly attractive as a potential therapeutic agent. Whether or not teixobactin proves useful in treating human infections, the new in situ culturing strategies used to discover this new drug have reenergized the field of antibiotic discovery and should help scientists solve the mysteries of the "unculturable."

RESEARCH QUESTION

Propose a mechanism for the domestication of unculturable microbes. Why would "domestication" not happen in the natural environment?

Ling, Losee L., Tanja Schneider, Aaron J. Peoples, Amy L. Spoering, Ina Engels, et al. 2015. A new antibiotic kills pathogens without detectable resistance. *Nature* **517**:455–459.

A.

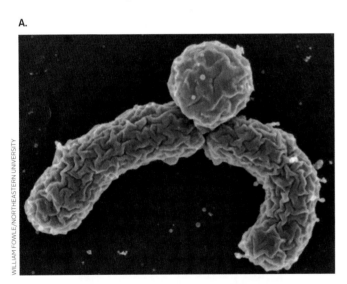

B.

C.

FIGURE 2 ■ ***Eleftheria terrae* and teixobactin.** A previously uncultured Gram-negative bacterium, *Eleftheria terrae* (**A**), makes teixobactin (**B**), a new antibiotic. Color highlights mark each residue. End = enduracididine, a nonprotein amino acid; modified from alanine. **C.** Kim Lewis (right), with post-doctoral researcher Brian Conlon.

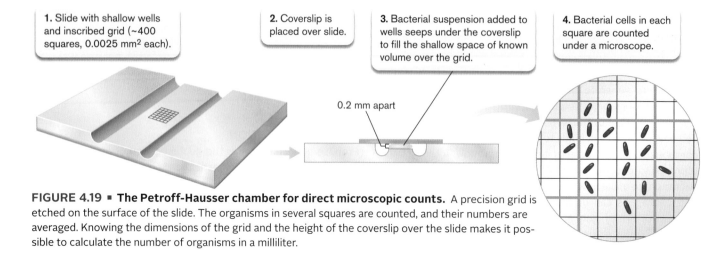

1. Slide with shallow wells and inscribed grid (~400 squares, 0.0025 mm² each).

2. Coverslip is placed over slide.

3. Bacterial suspension added to wells seeps under the coverslip to fill the shallow space of known volume over the grid.

4. Bacterial cells in each square are counted under a microscope.

0.2 mm apart

FIGURE 4.19 ▪ The Petroff-Hausser chamber for direct microscopic counts. A precision grid is etched on the surface of the slide. The organisms in several squares are counted, and their numbers are averaged. Knowing the dimensions of the grid and the height of the coverslip over the slide makes it possible to calculate the number of organisms in a milliliter.

However, we still do not know what those factors are. We can grow *Rickettsia* using animal cell tissue culture or in chicken eggs (the bacteria grow inside the endothelial cells of blood vessels formed in fertilized eggs). But despite extensive efforts to grow them outside of a host cell (called axenic growth), *R. prowazekii* has proved uncooperative.

Techniques for Counting Bacteria

Growing culturable bacteria in laboratory media is important for studying their physiology and genetics, but knowing how many bacteria are present in the medium is critical for interpreting results. You might also ask how we determine whether a lake is contaminated with fecal bacteria or our peanut butter contains *Salmonella*. In addition to finding organisms in these cases, we have to count how many there are to determine the extent of contamination. Counting or quantifying microorganisms is surprisingly difficult because each of the available techniques measures a different physical or biochemical aspect of growth. Thus, a cell density value (given as cells per milliliter) derived from one technique will not necessarily agree with the value obtained by a different method. Here are some commonly used methods for counting microbes.

Direct counting of living and dead cells. Microorganisms can be counted directly using a microscope. A dilution of a bacterial culture is placed on a special microscope slide called a hemocytometer (or, more specifically for bacteria, a Petroff-Hausser counting chamber) (**Fig. 4.19**). Etched on the surface of the slide is a grid of precise dimensions, and placing a coverslip over the grid forms a space of precise volume. The number of organisms counted within that volume is used to calculate the concentration of cells in the original culture.

However, "seeing" an organism under the microscope does not mean that the organism is alive, because living and dead cells are indistinguishable by this basic approach. Living cells may be distinguished from dead cells by fluorescence microscopy using fluorescent chemical dyes, as discussed in Chapter 2. For example, propidium iodide, a red dye, intercalates between DNA bases but cannot freely penetrate the energized membranes of living cells. Thus, only dead cells stain red under a fluorescence scope. Another dye, Syto-9, enters both living and dead cells, staining them both green. By combining Syto-9 with propidium iodide, living and dead cells can be distinguished: Living cells stain green, whereas dead cells appear orange or yellow because

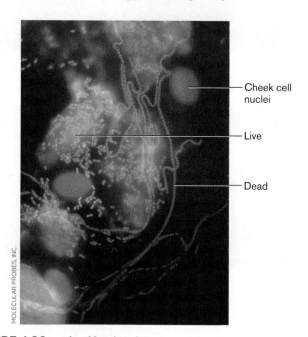

Cheek cell nuclei

Live

Dead

MOLECULAR PROBES, INC.

FIGURE 4.20 ▪ Live/dead stain. Live and dead bacteria visualized on freshly isolated human cheek epithelial cells using the LIVE/DEAD *Bac*Light Bacterial Viability Kit. Dead bacterial cells fluoresce orange or yellow because propidium (red) can enter the cells and intercalate the base pairs of DNA. Live cells fluoresce green because Syto-9 (green) enters the cell. The faint green smears are the outlines of cheek cells.

Thought Question

4.7 Use the information in **Figure 4.19** to determine the concentration (in cells per milliliter) of bacteria shown.

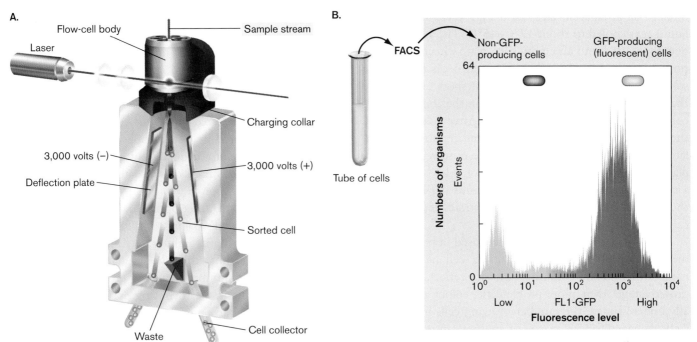

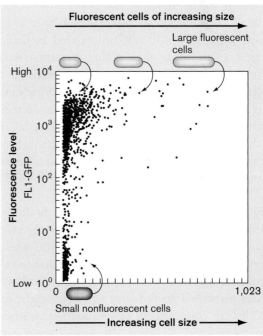

FIGURE 4.21 ■ Fluorescence-activated cell sorter (FACS).
A. Schematic of a FACS apparatus (bidirectional sorting). **B.** Separation of GFP-producing *E. coli* from non-GFP-producing *E. coli*. The low-level fluorescence in the cells on the left is baseline fluorescence (autofluorescence). The scatterplot displays the same FACS data, showing the size distribution of cells (*x*-axis) with respect to the level of fluorescence (*y*-axis). The larger cells may be cells that are about to divide.

both dyes enter and Syto-9 (green) plus propidium (red) gives a yellow appearance (**Fig. 4.20**).

Direct counting without microscopy can be achieved using an electronic technique that not only counts but also separates populations of bacterial cells according to their distinguishing properties. The instrument is called a **fluorescence-activated cell sorter** (**FACS**) or flow cytometer. In the FACS technique, bacterial cells that synthesize a fluorescent protein (such as cyan fluorescent protein; see Section 2.4), or that have been labeled with a fluorescent antibody or chemical, are passed single file through a small orifice and then through a laser beam (**Fig. 4.21A**). Detectors measure light scatter in the forward direction (an indicator of particle size) and to the side (which indicates shape or granularity). In addition, the laser activates the fluorophore in the fluorescent antibody, and a detector measures fluorescence intensity.

Within a single culture, one subpopulation of cells may fluoresce more than or less than another because of the presence or absence of a targeted protein. Thus the FACS technique enables us to use cell size and the level of fluorescence to identify and count different populations of cells. For example, FACS analysis can determine the expression of a gene in different subpopulations of cells in culture. This can be done by placing the green fluorescent protein gene (*gfp*) under the control of a specific bacterial gene (making a gene fusion). Researchers can then use the flow cytometer to count cells expressing that gene, and even sort high-expressing cells from low-expressing cells. FACS analysis makes it possible to determine which conditions allow expression of the gene and whether all cells in the population express that gene at the same time and to the same extent (**Fig. 4.21B**).

Viable counts. Viable cells, as noted previously, are those that can replicate and form colonies on a plate. To obtain a viable cell count, dilutions of a liquid culture can be plated directly on an agar surface (as in **Fig. 4.15**) or added to liquid agar cooled to about 42°C–45°C. The agar is subsequently poured into an empty petri plate (this is called the **pour plate technique**), where the agar cools further and solidifies. Because many bacteria resist short exposures to that temperature, individual cells retain the ability to form colonies on, and in, the pour plate. After colonies form, they are counted, and the original cell number is calculated. For example, if 100 colonies are observed in a pour plate made with 100 microliters (μl) of a 10^{-3} dilution of a culture, then there were 10^6 organisms per milliliter in the original culture. **Figure 4.15** illustrates these types of calculations.

Although viable counts are widely used in research, this method is problematic for measuring cell number. One issue is that colony counting does not reflect cell size or growth stage. Even more problematic, colony counts usually underestimate the number of <u>living</u> cells in a culture. Cells damaged for one reason or another can remain metabolically active and alive but may be too compromised to divide. Because these cells will not form colonies on an agar plate, they will not be counted as living. Comparing a viable count with a direct count obtained from a live/dead stain can expose the presence of these damaged cells. Another challenge is any organism that grows in chains, such as *Streptococcus*, because each colony originates from a group of cells, not a single cell. Consequently, counting colonies underestimates actual cell number. For this reason, viable counts are reported as colony-forming units (CFUs) rather than cells.

Biochemical assays. In contrast to methods that visualize individual cells, assays of cell biomass, protein content, or metabolic rate measure the overall size of a population of cells. The most straightforward but time-consuming biochemical approach to monitoring population growth is to measure the dry weight of a culture. Cells are collected by centrifugation, washed, dried in an oven, and weighed. Because bacterial cells weigh very little, a large volume of culture must be harvested to obtain measurements, making this technique quite insensitive. A more accurate alternative is to measure increases in protein levels, which correlate with increases in cell number. Protein levels are more easily measured with relatively sensitive assays.

Optical density measures growth in real time. The phenomenon of light scattering was introduced in Chapter 2. Recall that the presence of bacteria in a tube of medium can be detected by how cloudy the medium appears as the cells scatter light. The decrease in intensity of a light beam due to the scattering of light by a suspension of particles is measured as **optical density**.

The optical density of light scattered by bacteria is a very useful tool for estimating population size. The method is quick and easy, but because light scattering is a complex function of cell number and cell volume, optical density provides only an approximate result that must be corrected using a standard curve. Typically, a standard curve is obtained that plots viable counts for cultures versus optical densities as measured by a spectrophotometer. Thereafter, the optical density of a growing culture is measured, and the standard curve is used to estimate cell number at any given point during growth.

One problem with using optical density to estimate cell numbers is that a cell's volume can vary depending on its growth stage, altering its light-scattering properties. Thus, the cell number estimated by the standard curve may deviate from the true number. Another problem is that dead cells also scatter light. Clearly, using optical density to estimate viable count can be misleading, especially when measuring populations in stationary phase.

To Summarize

■ **Microbes in nature** usually exist in complex, multispecies communities, but for detailed studies they must be grown separately in pure culture.

■ **Bacteria can be cultured** on solid or liquid media.

■ **Defined or synthetic media** contain only chemical components that are known. **Minimal defined media** contain only the defined nutrients essential for growth of a given organism.

■ **Complex**, or **rich**, **media** contain many nutrients. Other media exploit physiological differences between organisms and can be defined as **selective**, **differential**, or both (for example, **MacConkey medium**).

■ **Microbes can evolve to become unculturable or to require specific growth factors** depending on the nutrient richness of their natural ecological niche.

■ **Obligate intracellular bacteria** lose metabolic pathways provided by their hosts and develop requirements for growth factors supplied by their hosts.

■ **Microorganisms in culture may be counted directly** under a microscope, with or without staining, or by use of a fluorescence-activated cell sorter (FACS).

■ **Microorganisms can be counted indirectly** by viable counts, measurement of dry weight, protein levels, or optical density.

■ **A viable bacterial organism** is defined as being capable of replicating.

4.4 The Growth Cycle

How do microbes grow? What determines their rate of growth? And when does growth restart in a nongrowing population? In nature, the answers to these questions are difficult to determine because most microbes exist in complex, mixed communities fixed to solid surfaces. Yet these same multicellular communities can also send off **planktonic cells**, free-living organisms that grow and multiply on their own. Growth of these planktonic cells is easier to measure.

All species at one time or another exhibit rapid growth, nongrowth, and many phases in between. For clarity, we present here the principles of rapid growth, while bearing in mind the diversity of growth situations in nature. We care about growth because how rapidly a microbe grows influences how fast a pathogen causes disease, or how quickly contaminated food spoils. Similarly, a bacterial species that consumes oil is useful for cleaning up oil spills only if the organism can grow rapidly.

Survival of any species ultimately depends on its ability to make more of its own kind. A typical bacterium that can be cultured in the laboratory grows by increasing in length and mass, which facilitates expansion of its nucleoid as its DNA replicates (see Section 3.4). As DNA replication nears completion, the cell, in response to complex genetic signals, begins to synthesize an equatorial septum that ultimately separates the two daughter cells. In this overall process, called **binary fission**, one parental cell splits into two equal daughter cells (**Fig. 4.22A**).

Although a majority of culturable bacteria divide symmetrically in two equal halves, some species divide asymmetrically. For example, the bacterium *Caulobacter* forms a stalked cell that remains fixed to a solid surface but reproduces by budding from one end to produce small, unstalked motile cells. The marine organism *Hyphomicrobium* also replicates asymmetrically by budding, releasing a smaller cell from a stalked parent (**Fig. 4.22B**).

Eukaryotic microbes divide by a special form of cell fission involving mitosis, the segregation of pairs of chromosomes within the nucleus (see Section 20.2 and eAppendix 2). Some eukaryotes also undergo more complex life cycles involving budding and diverse morphological forms.

Exponential Growth

The process of reproduction has implications for growth not only of the individual, but also of populations. If we assume that growth is unbounded, what happens to the population? The unlimited growth of any population obeys a simple law: The **growth rate**, or rate of increase in cell numbers or biomass, is proportional to the population size at a given time. Such a growth rate is called "exponential" because it generates an exponential curve, a curve whose slope increases continually.

How does binary fission of cells generate an exponential curve? If each cell produces two cells per generation, then the population size at any given time is proportional to 2^n, where the exponent n represents the number of generations (that is, cell divisions in which offspring replace parents) that have taken place between two time points. Thus, cell number rises exponentially. Many microbes, however, have replication cycles based on numbers other than 2. For example, some cyanobacteria form cell aggregates that divide by multiple fissions, releasing dozens of daughter cells. The cyanobacterium enlarges without dividing, and then suddenly divides many times without separating. The cell mass breaks open to release hundreds of progeny cells.

> ### Thought Question
>
> **4.8** A virus such as influenza virus might produce 800 progeny virus particles from one host cell infected by one virus. How would you mathematically represent the exponential growth of the virus? What practical factors might limit such growth?

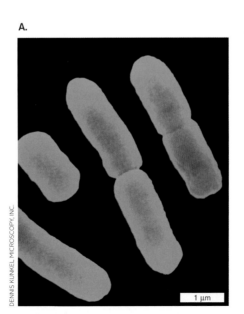

A.

DENNIS KUNKEL MICROSCOPY, INC.

1 µm

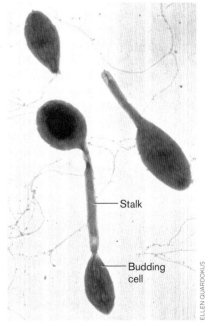

B.

Stalk

Budding cell

ELLEN QUARDOKUS

FIGURE 4.22 ■ Symmetrical and asymmetrical cell division. A. Symmetrical cell division, or binary fission, in *Lactobacillus* sp. (SEM). **B.** Asymmetrical cell division via budding in the marine bacterium *Hyphomicrobium* (approx. 4 µm long).

Generation Time

A variable not accounted for in bacterial growth curves is the length of time from one generation to the next. In an environment with unlimited resources, bacteria divide at a constant interval called the **generation time**. The length of that interval varies with respect to many parameters, including the bacterial species, type of medium, temperature, and pH. The generation time for cells in culture is also known as the **doubling time**, because the population of cells doubles over one generation. For example, one cell of E. coli placed into a complex medium will divide every 20 minutes. After 1 hour of growth (three generations), that one cell will have become eight (1 to 2, 2 to 4, 4 to 8). Because cell number (N) doubles with each division, the increase in cell number over time is exponential, not linear. A linear increase would occur if cell number rose by a fixed amount after every generation (for example, 1 to 2, 2 to 3, 3 to 4).

Why do we care about generation time? As noted already, generation time is important when we're trying to understand how rapidly a pathogen can cause disease symptoms. In biotechnology, how fast a producing organism grows will affect how quickly a commercially useful by-product can be made. Let's review how we calculate generation time.

Starting with any number of organisms (N_0), the number of organisms after n generations will be $N_0 \times 2^n$. For example, a single cell after three generations ($n = 3$) will produce

$$1 \text{ cell} \times 2^3 = 8 \text{ cells}$$

The number of generations that an exponential culture undergoes in a given time period can be calculated if the number of cells at the start of the period (N_0) and the number of cells at the end of the period (N_t) are known.

Methods such as viable counts are used to make those determinations. Thus,

$$N_t = N_0 \times 2^n$$

can be expressed as

$$\log_2 N_t = \log_2 N_0 + n \log_2 2$$
$$= n + \log_2 N_0$$

Solving for n:

$$n = \log_2 N_t - \log_2 N_0$$
$$= \log_2 (N_t / N_0)$$

Once the number of generations (n) over a given time is known, generation time (g) is calculated as

$$g = t/n \text{ [if 2 hours (120 min) yielded}$$
$$6 \text{ generations, then } g = 20 \text{ min]}$$

The rate of exponential growth is expressed as the mean **growth rate constant** (k), which is the number of generations (n) per unit time (usually generations per hour). This is written as

$$k = n/t, \text{ where } t \text{ is 1 hour; or } k = 1/g$$

Thus, if the generation time is 20 minutes (0.33 hour) for a given bacterial species in a given medium, the growth rate constant $k = 1/0.33$ hour = 3 generations per hour. If the generation time is 2 hours, the growth rate constant is 0.5 generation per hour.

In practice, exponential growth lasts for only a short period when all nutrients are in full supply and the concentration of waste products has not become a limiting factor.

The growth rate constant can also be calculated from the slope of $\log_2 N$ over time, where N is a relative measure of culture density, such as the optical density measured in a spectrophotometer (**Fig. 4.23A**). The units of N do not

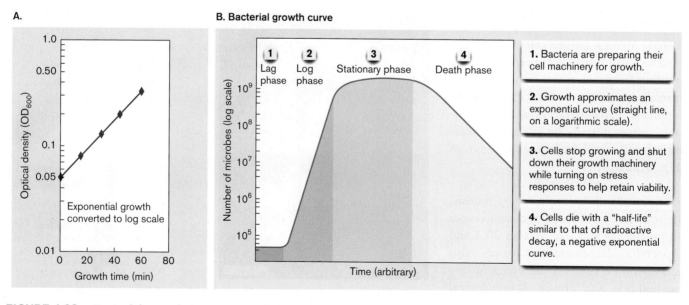

FIGURE 4.23 ■ Bacterial growth curves. A. Theoretical growth curve of a bacterial suspension measured by optical density (OD) at a wavelength of 600 nm. **B.** Phases of bacterial growth in a typical batch culture.

matter, because we are always looking at ratios of cell numbers relative to an earlier level (N_1/N_0). For example, we can use the following series of optical density (OD) measurements to determine N:

Time (min)	OD$_{600}$	log$_2$ OD$_{600}$
0	0.05	−4.32
15	0.08	−3.65
30	0.13	−2.94
45	0.20	−2.32
60	0.33	−1.59

If we plot $\log_2$ OD$_{600}$ versus time, we obtain a line with a slope of 0.0452 per minute. The growth rate constant, in doublings per hour, becomes

$$k = (0.0452/\text{min})(60 \text{ min/h})$$
$$= 2.7 \text{ generations per hour}$$

The steeper the slope, the faster the organisms are dividing.

Note: An alternative formula uses logarithms to base 10, requiring conversion to base 2:

$$\log_{10} N_t = \log_{10} N_0 + n \log_{10} 2$$
$$= \log_{10} N_0 + n 0.301$$
$$n = 3.3 \log_{10} (N_t/N_0)$$

Thought Questions

4.9 Suppose you ingest 20 cells of *Salmonella enterica* in a peanut butter cookie. They all survive the stomach and enter the intestine. Suppose further that the sum total of subsequent bacterial replication and death (caused by the host) produces an average generation time of 2 hours and you will feel sick when there are 1,000,000 bacteria. How much time will elapse before you feel sick?

4.10 Suppose one cell of the nitrogen fixer *Sinorhizobium meliloti* colonizes a plant root. After 5 days (120 hours), there are 10,000 bacteria fixing N$_2$ within the plant cells. What is the bacterial doubling time?

The mathematics of exponential growth is relatively straightforward, but remember that microbes grow differently in pure culture (very rare in nature) than they do in mixed communities, where neighboring cells produce all kinds of substances that may feed or poison other microbes. In mixed communities, the microbes may grow planktonically (floating in liquid), as in the open ocean, or as a biofilm on solid matter suspended in that ocean. In each instance, the mathematics of exponential growth applies, at least until the community reaches a density at which different species begin to compete or nutrients become scarce.

Thought Question

4.11 It takes 40 minutes for a typical *E. coli* cell to completely replicate its chromosome and about 20 minutes to prepare for another round of replication. Yet the organism enjoys a 20-minute generation time growing at 37°C in complex medium. How is this possible? *Hint:* How might the cell overlap the two processes?

Stages of Growth in Batch Culture

Exponential growth never lasts indefinitely, because nutrient consumption and toxic by-products eventually slow the growth rate until it halts altogether. The simplest way to model the effects of changing conditions is to culture bacteria in liquid medium within a closed system, such as a flask or test tube. This is called **batch culture**. In batch culture, no fresh medium is added during incubation; thus, nutrient concentrations decline and waste products accumulate during growth.

The changing conditions of a batch culture profoundly affect bacterial physiology and growth, and they illustrate the remarkable ability of bacteria to adapt to their environment. As medium conditions deteriorate, alterations occur in membrane composition, cell size, and metabolic pathways, all of which impact generation time. Microbes possess intricate, self-preserving genetic and metabolic mechanisms that slow growth before their cells lose viability. Because many bacteria replicate by binary fission, plotting culture growth (as represented by the logarithm of the cell number) versus incubation time makes it possible to see the effect of changing conditions on generation time and reveals the stages of growth shown in **Figure 4.23B**—namely, lag phase, log phase, stationary phase, and death phase.

Lag phase. Cells transferred from an old culture to fresh growth media need time to detect their environment, express specific genes, and synthesize components needed to grow rapidly. As a result, bacteria inoculated into fresh media typically experience a lag period, or **lag phase**, during which cells do not divide. Several factors influence lag phases. Cells taken from an aged culture may be damaged and require time for repair. Carbon, nitrogen, or energy sources different from those originally used by the seed culture must be sensed, and the appropriate enzyme systems must be synthesized. The length of the lag phase also varies with changes in temperature, pH, and salt concentration, as well as nutrient richness. For example, transferring cells from one complex medium to a fresh complex medium results in a very short lag phase, whereas cells grown in a complex medium and then plunged into a minimal defined medium experience a protracted lag phase. During lag phase in the latter case, cells must synthesize all the amino acids, nucleotides, and other metabolites previously supplied by the complex medium.

Early log, or exponential, phase. Once cells have retooled their physiology to accommodate the new environment, they begin to grow exponentially and enter what is called **exponential**, or **logarithmic** (**log**), **phase.** Exponential growth is balanced growth, in which all cell components are synthesized at constant rates relative to each other—the assumption behind the generation time calculation given in the previous section. At this stage, represented by the linear part of the growth curve, cells are growing and dividing at the maximum rate possible in the medium and growth conditions provided (such as temperature, pH, and osmolarity). Cells are largest at this stage of growth. If cell division were synchronized and all cells divided at the same time, the growth curve during this period would appear as a series of steps with cell numbers doubling instantly after every generation time. But batch cultures are <u>not</u> synchronous. Every cell has an equal generation time, but each cell divides at a slightly different moment, making the cell number rise smoothly.

Cells enjoying balanced, exponential growth are temporarily thrown into metabolic chaos (unbalanced growth) when their medium is abruptly changed. Nutritional downshift (moving cells from a good carbon source such as glucose to a poorer carbon source such as succinate) or nutritional upshift (moving cells to a better carbon source) casts cells into unbalanced growth. Downshifting to a carbon source with a lower energy yield means not only that a different set of enzymes must be made and employed to use the carbon source, but also that the previous high rate of macromolecular synthesis (such as ribosome synthesis) used to support a fast generation time is now too rapid relative to the lower energy yield. Failure to adjust will lead to increased mistakes in RNA, protein, and DNA synthesis, depletion of key energy stores, and ultimately death.

How do microbes adjust to nutrient shifts? During downshifts, energy levels (ATP) drop faster than macromolecular synthesis systems can stop making macromolecules. Molecular fail-safe systems quickly slow rates of

macromolecular synthesis until DNA, RNA, ribosome, and cell wall synthesis come into balance with the rest of metabolism. In contrast, nutritional upshift means that the cell will start making more energy than it needs at its current growth rate, in which case cell metabolism will again be out of balance with cell division. The same fail-safe mechanisms used during downshift will, during upshift, kick macromolecular synthesis into a higher gear, once again establishing balanced growth. Nutritional upshift causes cells to reenter log phase, but with a shorter generation time.

Late log phase. As cell density (number of cells per milliliter) rises during log phase, the rate of doubling eventually slows, and a new set of growth phase–dependent genes is expressed. At this point, some species can also begin to detect the presence of others by sending and receiving chemical signals in a process known as quorum sensing (discussed in Chapter 10).

Stationary phase. Eventually, cell numbers stop rising, owing to the lack of a key nutrient or the buildup of waste products. At this point the growth curve levels off and the culture is said to be in **stationary phase** (**Fig. 4.23B**). It was thought that in stationary phase, the rate of cell division equaled the rate of cell death. That is, cells were either dead or they were alive. That model has recently been challenged by Nathalie Balaban and colleagues from the Hebrew University of Jerusalem. They elegantly show that the vast majority of cells thought to have died in stationary phase are actually alive but growth-arrested—and can still make protein (**Fig. 4.24**). The nature of growth arrest is unclear but may involve the molecule guanosine tetraphosphate, which is associated with the stringent response (discussed in Chapter 10).

If they did not change their physiology, microbes would be very vulnerable upon entering stationary phase. Cells in stationary phase are not as metabolically nimble as cells in exponential phase, so damage from oxygen radicals and the toxic by-products of metabolism readily kill them. As an avoidance strategy, some bacteria differentiate into very resistant spores in response to nutrient depletion (see Section 4.6), while other bacteria undergo less dramatic but very effective molecular reprogramming. The microbial model organism *E. coli*, for example, adjusts to stationary phase by decreasing its size, thus minimizing the volume of its cytoplasm compared to the volume of its nucleoid. Fewer nutrients are then required to sustain the smaller cell. New stress resistance enzymes are also synthesized to handle oxygen radicals, protect DNA and proteins, and increase cell wall strength through increased peptidoglycan crosslinking. As a result, *E. coli* cells in stationary phase

A.

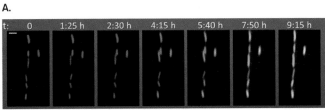

O. GEFEN, ET AL., 2014. *PROC. NATL. ACAD. SCI* **111**:556-561.

B.

COURTESY OF HEBREW UNIVERSITY

FIGURE 4.24 ■ **Observing dormant *E. coli* making protein in stationary phase. A.** *E. coli* cells trapped in a microfluidic channel express a red fluorescent protein in exponential phase. After 15 hours in stationary phase, a chemical inducer molecule was added to observe what percentage of cells could still synthesize a new, green fluorescent protein (time 0). At least 90% of cells produced the new protein, meaning these cells are alive and dormant, not dead. **B.** Nathalie Balaban (center) with student Eitam Rotem and lab manager Irene Ronin, viewing fluorescent bacteria growing and dying.

become more resistant to heat, osmotic pressure, pH changes, and other stresses that they might encounter while waiting for a new supply of nutrients.

Thought Question

4.12 The bacterium *Acidithiobacillus thiooxidans* is an extremophile that grows using sulfur as an energy source. (a) Draw the approximate growth curves that you would expect to see, extending from log phase to stationary phase, if four cultures with different starting numbers of bacteria were grown in the same concentration of sulfur. Use **Figure 4.23B** as the model, and 4×10^5, 4×10^6, 4×10^7, and 4×10^8 as the starting cell densities. Maximum growth yield is 10^9 cells per milliliter. (b) Draw a second graph showing how the curves would change if the initial population density was constant but the concentration of sulfur varied.

Death phase. Without reprieve in the form of new nutrients, cells in stationary phase will eventually succumb to

toxic chemicals present in the environment. Cells begin to die, and the culture enters the death phase. Like the growth rate, the **death rate**—the rate at which cells die—is logarithmic. In **death phase**, the number of cells that die in a given time period is proportional to the number that existed at the beginning of the time period. Thus, the death rate is a negative exponential function. The death rate can be expressed as a half-life, the time over which a population declines by half.

Determining microbial death rates is critical to the study of food preservation and to the development of antibiotics (further discussed in Chapters 5, 16, and 27). Although death curves are basically logarithmic, exact death rates are difficult to define because mutations arise that promote survival, and some cells grow by cannibalizing others. Consequently, the death phase is extremely prolonged. In fact, a portion of the cells will often survive for months, years, or even decades.

Note that the nice, smooth exponential growth phase shown in **Figure 4.23B** does not always hold for microbes growing in natural environments or in complex laboratory media containing multiple carbon and energy sources. Some bacterial species produce odd-looking growth curves in complex media as the population depletes one carbon source and must switch physiology to use another. Even *E. coli* doesn't really experience a uniform log-phase metabolism growing in complex medium; instead, it smoothly transitions through a series of metabolic states.

Thought Questions

4.13 What can happen to the growth curve when a culture medium contains two carbon sources, if one is a preferred carbon source of growth-limiting concentration and the second is a nonpreferred source?

4.14 How would you modify the equations describing microbial growth rate to describe the rate of death?

4.15 Why are cells in log phase larger than cells in stationary phase?

Continuous Culture

In the classic growth curve that develops in closed systems, the exponential phase spans only a few generations. In open systems, however, where fresh medium is continuously added to a culture and an equal amount of culture is constantly siphoned off, bacterial populations can be maintained in exponential phase at a constant cell mass for extended periods of time. In this type of growth pattern, known as **continuous culture**, all cells in a population achieve a steady state, which permits a detailed analysis of

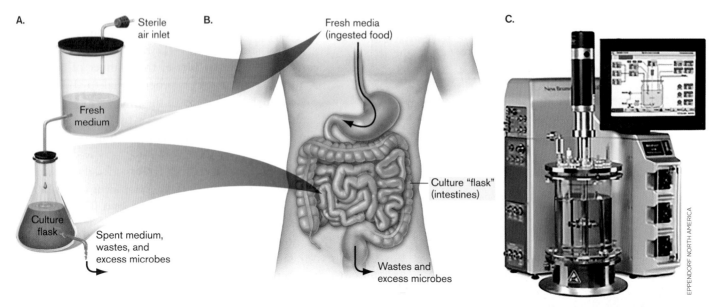

FIGURE 4.25 ▪ **Chemostats and continuous culture. A.** The basic chemostat ensures logarithmic growth by constantly adding and removing equal amounts of culture media. **B.** The human gastrointestinal tract is engineered much like a chemostat, in that new nutrients are always arriving from the throat while equal amounts of bacterial culture exit in fecal waste. **C.** A modern chemostat.

microbial physiology at different growth rates. The **chemostat** is a continuous culture system in which the diluting medium contains a limiting amount of one essential nutrient (**Fig. 4.25**). You could think of your gastrointestinal tract as a kind of crude chemostat. Nutrient enters through your mouth and passes through your intestine, where it feeds your microbiome, and your microbiome exits in fecal waste. In both cases the numbers of microbes in the chamber (or gut) remain relatively constant. The GI tract is different from a chemostat, of course, in that nutrient intake and fecal exit are not continuous and water is absorbed.

The complex relationships among dilution rate, cell mass, and generation time in a chemostat are illustrated in **Figure 4.26**. The curves in this figure represent a typical experimental result. At very low dilution (flow) rates, the nutrient is so limiting that cells will divide very slowly and cell mass will remain low. Any increase in flow rate, however, will increase the availability of the limiting nutrient such that cells grow faster and cell mass increases. The rate of cell division and the cell mass are kept constant when the flow rate is kept constant, because the amount of culture (and cells) removed from the vessel exactly compensates for the increased rate of cell division. Constant cell mass, or density, can be maintained only over a certain range of flow rates. At faster and faster flow rates, cells are eventually removed more quickly than they can be replenished by division, so cell density (cell mass) decreases in the vessel— a phenomenon called "washout."

Continuous cultures are used to study large numbers of cells at a constant growth rate and cell mass for both research and industrial applications. Its advantage over batch culture is that the physiology of cells in continuous culture is homogeneous. Consequently, continuous

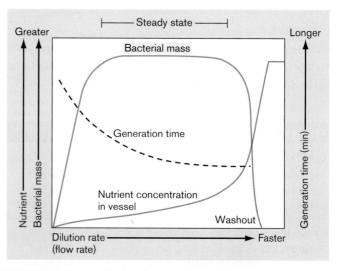

FIGURE 4.26 ▪ **Relationships among chemostat dilution rate, cell mass, and generation time.** As the dilution rate (*x*-axis) increases, the generation time decreases and the mass of the culture increases. When the rate of dilution exceeds the division rate, cells are washed from the vessel faster than they can be replaced by division, and the cell mass decreases. The *y*-axis varies depending on the curve, as labeled.

cultures are used in industry to optimize the production of antibiotics, beer, and other microbial products. In research, continuous culture is used to examine what happens to metabolic flux (essentially the rate at which molecules move through metabolic pathways) before and after a biochemical step is altered, and to conduct long-term studies of bacterial evolution.

One disadvantage of the continuous culture apparatus shown in **Figure 4.25C** is the large volume of media required to maintain a 500-ml culture chamber over a period of days or months. To solve this problem, microfluidic principles have been used to miniaturize the process onto a 5-cm computer-controlled lab chip. One milliliter of culture moves along three chambers that monitor oxygen, optical density, and pH. Culture conditions are maintained or changed by a computer that controls delivery of media and chemicals from on-chip reservoirs. The heightened ability to control culture parameters in small volumes can be used to characterize genetic switches, environmental shock responses, directed evolution, co-metabolism, and the dynamics of mixed-species cultures.

To Summarize

- The **growth cycle** of organisms grown in liquid batch culture consists of lag phase, log (exponential) phase, stationary phase, and death phase.

- **The physiology of a bacterial population** changes with growth phase.

- **Generation time** is the length of time it takes for a population of cells to double in number.

- **The generation time for a single species will vary** as culture conditions change. **During exponential growth**, generation time remains constant.

- **Continuous culture** can be used to sustain a population of bacteria at a specified growth rate and cell density.

4.5 Biofilms

Can bacteria collaborate? Bacteria are typically thought of as unicellular, but many, if not most, bacteria in nature form specialized, surface-attached, collaborative communities called **biofilms**. Indeed, within aquatic environments bacteria are found mainly associated with surfaces—a fact that underscores the importance of biofilms in nature. Soil biofilms help filter groundwater through the soil, where the bacteria break down organic waste. This process is especially important in wetlands, which provide an important "ecosystem service" for human communities (discussed in Chapter 22). Other biofilms play critical roles in microbial pathogenesis and environmental degradation, which costs billions of dollars each year in equipment damage, product contamination, and medical infections. For example, pseudomonad or staphylococcal biofilms can damage ventilators used to assist respiration. Biofilms can also form inside indwelling catheter tubes that deliver fluids and medications to patients. In both of these examples, the biofilms serve as direct sources of infection, so developing ingenious ways to prevent biofilm formation on medical instrumentation is a major goal of biomedical research. Making catheter surfaces mimic the architecture of sharkskin is one intriguing approach (**eTopic 4.3**).

Biofilms can be constructed by a single species or by multiple, collaborating species, as in the mixture of facultative and anaerobic bacteria that form dental plaque. Biofilms can form on a range of organic and inorganic surfaces (**Fig. 4.27** ▶). The Gram-negative bacterium *Pseudomonas aeruginosa*, for example, can form a single-species biofilm on the lungs of cystic fibrosis patients or on medical implants. Distinct stages in biofilm development (the biofilm "life cycle") include initiation, maturation, maintenance, and dissolution (or dispersal). Bacterial biofilms form when nutrients are plentiful. The goal of biofilms in nature is to stay where food is abundant. Why should a microbe travel off to hunt for food when it is already available? Once nutrients become scarce, however, individuals detach from the community to forage for new sources of nutrients.

Biofilms in nature can take many different forms and serve different

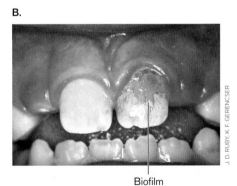

A.

B.

Biofilm

FIGURE 4.27 ■ Biofilms. A. A greenish-brown slime biofilm found on cobbles of the streambed in High Ore Creek, Montana. B. The biofilm that forms on teeth is called plaque. ▶

functions for different species. Confocal images of biofilms are shown in **Figure 4.28** (inset) and Figure 2.30. The formation of biofilms can be cued by different environmental signals in different species. These signals include pH, iron concentration, temperature, oxygen availability, and the presence of certain amino acids. Nevertheless, a common pattern emerges in the formation of many kinds of biofilms (**Fig. 4.28** ▶).

First, the specific environmental signal induces a genetic program in planktonic cells. The planktonic cells then start to attach to nearby inanimate surfaces by means of flagella, pili, lipopolysaccharides, or other cell-surface appendages, and they begin to coat that surface with an organic monolayer of polysaccharides or glycoproteins to which more planktonic cells can attach. At this point, cells may move along surfaces using a **twitching motility** that involves the extension and retraction of a specific type of pilus. Ultimately, they stop moving and firmly attach to the surface. As more and more cells bind to the surface, they can begin to communicate with each other by sending and receiving chemical signals in a process called **quorum sensing**. These chemical signaling molecules are continually made and secreted by individual cells. Once the population reaches a certain number (analogous to an organizational "quorum"), the chemical signal achieves a specific concentration that the cells can sense. This concentration triggers genetically regulated changes that cause cells to bind tenaciously to the substrate and to each other. Quorum sensing serves

the biofilm in many ways. Among other functions, quorum sensing triggers the increased resistance of biofilms to antibiotics and to phagocytosis by white blood cells—subjects we explore in **eTopic 4.4** and Special Topic 27.1.

Once established, the cells in a microcolony form a thick extracellular matrix of polysaccharide polymers and entrapped organic (DNA and proteins) and inorganic materials (**Fig. 4.28**). These **exopolysaccharides** (**EPSs**), such as alginate produced by *P. aeruginosa* and colanic acid produced by *E. coli*, increase the antibiotic resistance of residents within the biofilm. As the biofilm matures, the amalgam of adherent bacteria and matrix takes on complex 3D shapes such as columns and streamers, forming channels through which nutrients flow. For many bacteria, sessile (nonmoving) cells in a biofilm chemically "talk" to each other in order to build microcolonies and keep water channels open. *Bacillus subtilis* also spins out a fibril-like amyloid protein called TasA, which tethers cells and strengthens the biofilm. The effect of TasA on a floating biofilm formed on liquid medium is evident in **Figure 4.29**.

Bacteria growing in biofilms also exhibit a type of cell differentiation brought about by different physiological

FIGURE 4.28 ■ Biofilm development. The stages of biofilm development in *Pseudomonas*, which generally apply to the formation of many kinds of biofilms. **Inset:** A mucoid environmental strain of *P. aeruginosa* produces uneven, lumpy biofilms in an experimental flow cell. Cells in the biofilm were stained green with the fluorescent DNA-binding dye Syto-9 (3D confocal laser scanning microscopy). *Source:* H. C. Flemming and J. Wingender. 2010. *Nat. Rev. Microbiol.* **8:**623–633. ▶

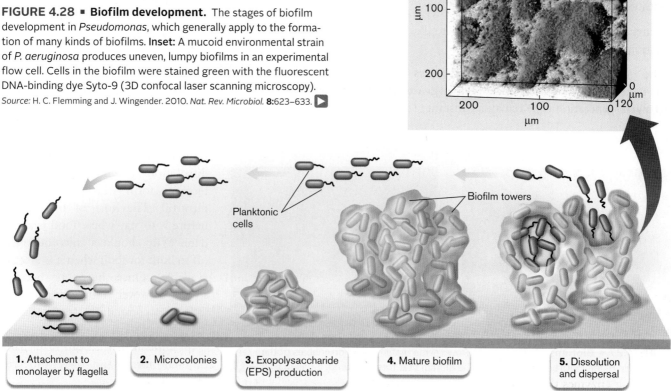

1. Attachment to monolayer by flagella
2. Microcolonies
3. Exopolysaccharide (EPS) production
4. Mature biofilm
5. Dissolution and dispersal

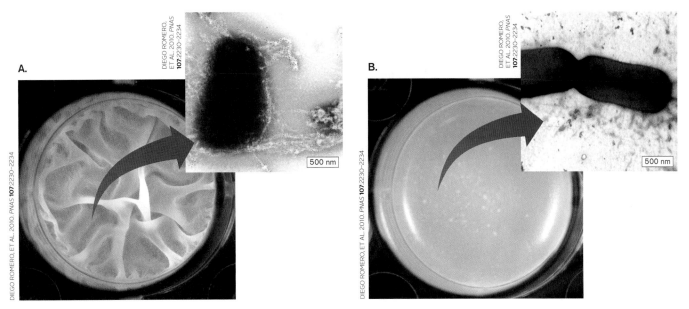

FIGURE 4.29 ▪ Floating biofilm (pellicle) formation of *Bacillus subtilis*. Cells were grown in a broth for 48 hours without agitation at 30°C. The pellicles formed by wild-type and *tasA* mutant *B. subtilis* are strikingly different. Wild-type pellicles are extremely wrinkly **(A)**, whereas *tasA* mutant pellicles are flat and fragile **(B)**. **Insets:** Electron micrographs of wild-type (A) and *tasA* mutant (B) cells.

conditions in different layers of the biofilm. For example, oxygen does not penetrate deep into biofilms. So in a colony growing on agar, cells at the colony surface will be exposed to oxygen, whereas cells near the agar will not. Colonies of microbes like *E. coli* that do not need oxygen to grow will have an actively growing zone at the agar-colony interface because cells there can ferment nutrients diffusing up from the agar. Cells at the colony-air interface will be in stationary phase because the nutrients are consumed before reaching them (**Fig. 4.30**). The reverse pattern happens for oxygen-requiring bacteria such as *Pseudomonas*. Cells deep in the colony at the agar surface are in stationary phase because they can't get oxygen. Nutrients from the agar are not consumed and diffuse toward the oxygen-rich colony surface, where they feed a growth zone.

How do cells escape from a biofilm? When a sessile biofilm (or a part of it) begins to starve or experiences oxygen depletion, some cells start making enzymes that dissolve

the matrix. *P. aeruginosa*, for instance, produces an alginate lyase that can strip away the EPS. Cells of *B. subtilis* sever their links to TasA fibers in some as yet unknown way. The

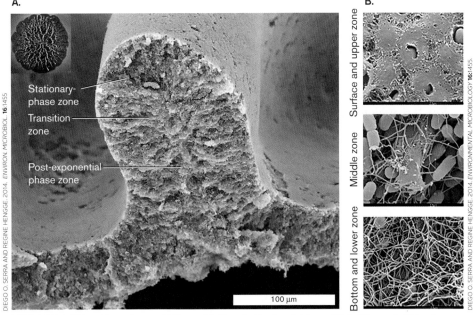

FIGURE 4.30 ▪ Two-layer differentiation in *Escherichia coli* biofilms. **A.** Side view of a cross section through a ridge of a macrocolony (see inset) grown on salt-free LB agar plates for 5 days (SEM). Areas false-colored in red and blue represent zones of cells exhibiting stationary-phase and post-exponential-phase physiologies, respectively. The narrow purple area represents the physiological transition zone between the lower and upper layers. **B.** SEM images showing (top) stationary phase cells on the macrocolony surface covered with secreted cellulose, (middle) transition zone cells covered with pili and cellulose next to "naked" cells, and (bottom) mesh-entangled flagella of post-exponential-growth cells.

A. No peptide

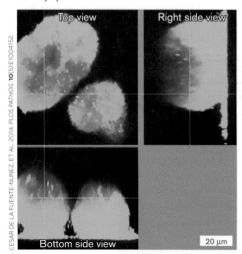

B. + 1018 peptide (0.8 µg/ml)

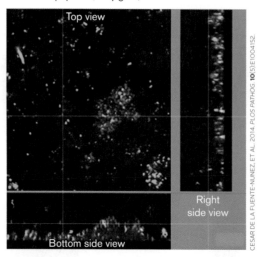

C.

cyanobacterial biofilms are common in thermal springs. Suspended particles called "marine snow" are found in oceans and appear to be floating biofilms comprising many unidentified organisms (discussed in Chapter 21). The particles seem to be capable of methanogenesis, nitrogen fixation, and sulfide production, indicating that the architecture of biofilms enables anaerobic metabolism to occur in an otherwise aerobic environment.

FIGURE 4.31 ▪ Effect of an antibiofilm peptide on a *Pseudomonas biofilm.* Confocal images of 2-day-old *Pseudomonas* biofilms (top and two side views) after 23 hours of no peptide (**A**) or with antibiofilm peptide treatment (**B**). The peptide promotes degradation of the intracellular signaling molecule (p)ppGpp. Cells were stained with live/dead stain in which live cells fluoresce green and dead cells stain red. Most cells from the treated biofilm were dead or had dispersed. **C.** Robert Hancock (right) and master's student Pat Taylor.

To Summarize

- **Biofilms** are complex, multicellular, surface-attached microbial communities.

- **Chemical signals** enable bacteria to communicate (**quorum sensing**) and in some cases to form biofilms.

- **Biofilm development** involves the adherence of cells to a substrate, the formation of microcolonies, and, ultimately, the formation of complex channeled communities that generate new planktonic cells.

sessile biofilm then sends out "scouts" called dispersal cells to initiate new biofilms. For scouts to form, genes whose products produce EPSs must be turned off, and for bacteria capable of motility, genes needed for flagella must be activated.

Recall that biofilms are important for chronic infections, so preventing or reversing their formation could prove therapeutic. Two intracellular signaling molecules critical to biofilm development and dispersal in many bacteria are the unusual nucleotides cyclic dimeric guanosine monophosphate (cdiGMP) and guanosine tetra- and pentaphosphate [(p)ppGpp]. High concentrations of either molecule promote biofilm formation, whereas low levels promote dispersal. Robert Hancock and colleagues from the University of British Columbia may have found a way to undo a biofilm's architecture by targeting one of these signal nucleotides. They recently discovered a synthetic peptide that binds the signaling molecule (p)ppGpp and promotes its degradation in various pathogens. When used in vitro, this peptide caused the death and dispersal of biofilm cells and may do the same in an infection (**Fig. 4.31**).

Organisms adapted to life in extreme environments also form biofilms. Archaea form biofilms in acid mine drainage (pH 0), where they contribute to the recycling of sulfur, and

4.6 Cell Differentiation

Can bacteria change shape? Many bacteria faced with environmental stress undergo complex molecular reprogramming that includes changes in cell structure. Some species, like *E. coli*, experience relatively simple changes in cell structure, such as the formation of smaller cells or thicker cell surfaces. However, select species, such as *Caulobacter crescentus*, undergo elaborate cell differentiation processes. *Caulobacter* cells convert from the swimming form to the holdfast form before cell division (see Section 3.5). Each cell cycle then produces one sessile cell attached to its substrate by a holdfast, while its sister cell swims off in search of another habitat.

Eukaryotic microbes also have highly complex life cycles. For example, *Dictyostelium discoideum* is a seemingly unremarkable ameba that grows as separate, independent cells.

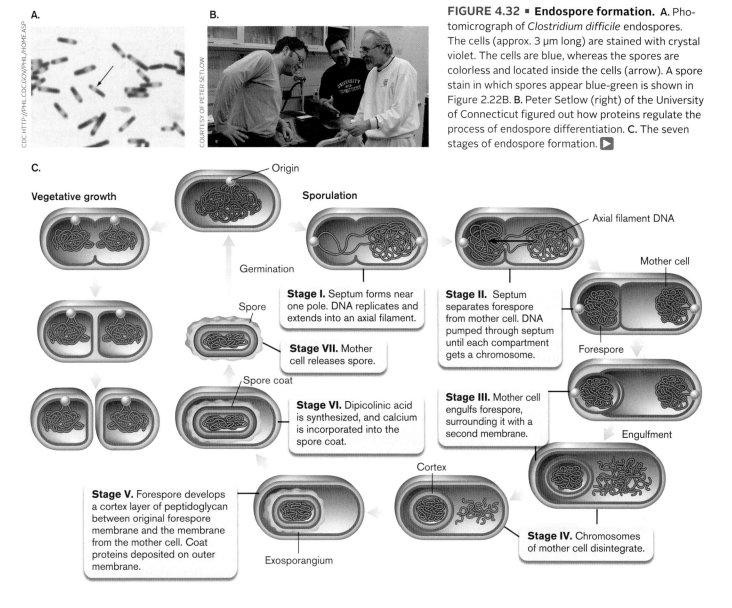

FIGURE 4.32 ■ Endospore formation. A. Photomicrograph of *Clostridium difficile* endospores. The cells (approx. 3 μm long) are stained with crystal violet. The cells are blue, whereas the spores are colorless and located inside the cells (arrow). A spore stain in which spores appear blue-green is shown in Figure 2.22B. **B.** Peter Setlow (right) of the University of Connecticut figured out how proteins regulate the process of endospore differentiation. **C.** The seven stages of endospore formation. ▶

CDC HTTP://PHIL.CDC.GOV/PHIL/HOME.ASP

COURTESY OF PETER SETLOW

C.

Origin

Vegetative growth

Sporulation

Axial filament DNA

Germination

Mother cell

Spore

Stage I. Septum forms near one pole. DNA replicates and extends into an axial filament.

Stage II. Septum separates forespore from mother cell. DNA pumped through septum until each compartment gets a chromosome.

Forespore

Stage VII. Mother cell releases spore.

Spore coat

Stage VI. Dipicolinic acid is synthesized, and calcium is incorporated into the spore coat.

Stage III. Mother cell engulfs forespore, surrounding it with a second membrane.

Engulfment

Cortex

Stage V. Forespore develops a cortex layer of peptidoglycan between original forespore membrane and the membrane from the mother cell. Coat proteins deposited on outer membrane.

Stage IV. Chromosomes of mother cell disintegrate.

Exosporangium

When challenged by adverse conditions such as starvation, however, *D. discoideum* secretes chemical signaling molecules that choreograph a massive interaction of individuals to form elaborate multicellular structures. The developmental cycle of this organism may be compared to that of other eukaryotic microbes that are human parasites (discussed in Chapter 20).

Endospores Are Bacteria in Suspended Animation

Certain Gram-positive genera, including important pathogens such as *Clostridium tetani* (tetanus), *Clostridium botulinum* (botulism), and *Bacillus anthracis* (anthrax), have the remarkable ability to develop dormant spores that are heat and desiccation resistant. Spores are particularly hearty because they do not grow and do not need nutrients

until they germinate. Resistance to heat and desiccation (and its lethal toxin) makes *B. anthracis* spores a potential bioweapon.

Most of what we know about bacterial sporulation comes from the Gram-positive soil bacterium *Bacillus subtilis*. When growing in rich media, this microbe undergoes normal vegetative growth and can replicate every 30–60 minutes. However, starvation initiates an elaborate 8-hour genetic program that directs an asymmetrical cell division process and ultimately yields a spore (see Section 10.3).

As shown in **Figure 4.32** ▶, sporulation can be divided into seven discrete stages based primarily on cell morphology. Stage 0 (not shown) represents the point at which the vegetative cell "decides" to use one of two potential polar division sites to begin septum formation instead of the central division site used for vegetative growth. In stage I, the DNA is replicated and stretched into a long axial filament that spans the length of the cell. There are two chromosome copies at this point. Ultimately, one of the polar division sites wins out and forms a septum. In stage II, the septum divides the cell into two unequal compartments: the smaller **forespore**, which will ultimately become the spore, and the larger **mother cell**, from which the forespore is derived. Each compartment eventually contains one of the replicated chromosomes. (Most of the chromosome in the forespore has to be pumped in <u>after</u> septum formation.)

In stage III of sporulation, the mother cell membrane engulfs the forespore. Next, the mother cell chromosome is destroyed (stage IV) and a thick peptidoglycan layer (cortex) is placed between the two membranes surrounding the forespore protoplast (stage V). Layers of coat proteins are then deposited on the outer membrane, also in stage V. Stage VI completes the development of spore resistance to heat and chemical insults. This last process includes the synthesis of dipicolinic acid (which stabilizes and protects spore DNA) and the uptake of calcium into the coat of the spore. Finally, the mother cell, now called a sporangium, releases the mature spore (stage VII).

Spores resist many environmental stresses that would kill vegetative cells. Spores owe this resistance, in part, to their desiccation (they have only 10%–30% of a vegetative cell's water content). But, as discovered by Peter Setlow and colleagues, spores are also packed with small acid-soluble proteins (SASPs) that bind to and protect DNA. The SASP coat protects the spore's DNA from damage by ultraviolet light and various toxic chemicals.

A fully mature spore can exist in soil for at least 50–100 years, and spores have been known to last thousands of years. Once proper nutrient conditions arise, another genetic program, called **germination**, is triggered to wake the dormant cell, dissolve the spore coat, and release a viable vegetative cell.

While sporulation is an effective survival strategy, bacteria that sporulate actually go out of their way <u>not</u> to sporulate—even to the point of cannibalism. During nutrient limitation, these bacteria will secrete proteins that can kill their siblings and then use the released nutrients to prevent starvation and, thus, sporulation. When (and only when) that strategy fails, they sporulate.

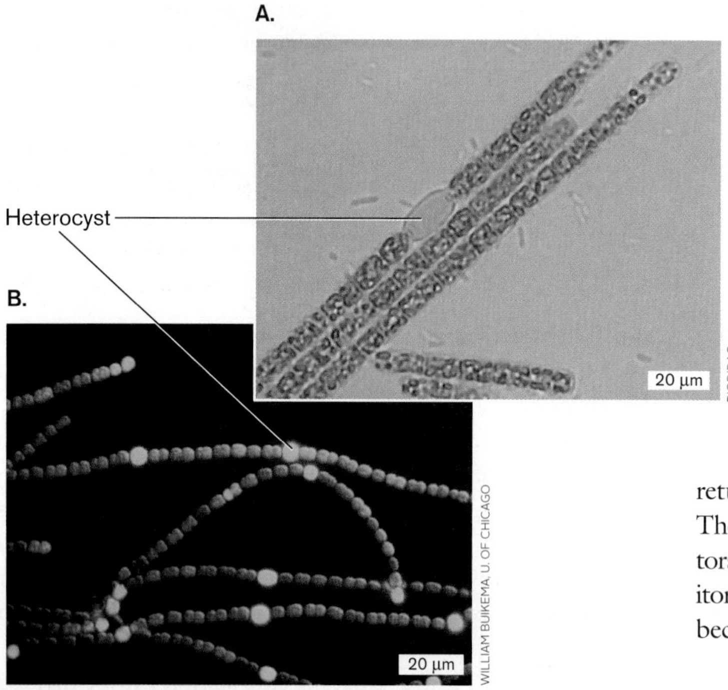

Cyanobacteria Differentiate into Nitrogen-Fixing Heterocysts

Some autotrophic cyanobacteria, such as *Anabaena*, not only carry out photosynthesis but also "fix" atmospheric nitrogen to make ammonia. This is surprising because nitrogenase, the enzyme required to fix nitrogen, is very sensitive to oxygen present in air and produced as a by-product of photosynthesis. So one might expect that photosynthesis and nitrogen fixation would be two mutually exclusive physiological activities. *Anabaena* solved this dilemma by developing specialized cells, called heterocysts, that can fix nitrogen (**Fig. 4.33**). A tightly regulated genetic program converts every tenth photosynthetic, vegetative cell to a heterocyst. As part of the differentiation process, heterocysts make nitrogenase, produce three additional cell walls, and form a specialized envelope that provides a barrier to atmospheric O_2 and degrades the photosynthesis machinery that produces O_2. The heterocyst then supplies nitrogen compounds to the adjacent vegetative cells, which, in return, send carbon sugars to the heterocyst (see **eTopic 2.1**). The precise spacing of heterocysts relies on the ratio of inhibitors relative to activators in different cells in the chain. Inhibitors predominate in vegetative cells, whereas cells fated to become heterocysts contain higher levels of an activator.

FIGURE 4.33 ■ Cyanobacteria and heterocyst formation.
A. Light-microscope image of the cyanobacterium *Phanizomenon*. B. The cyanobacterium *Anabaena*. The expression of genes in heterocysts is different from their expression in other cells. All cells in the figure contain a cyanobacterial gene to which the gene for green fluorescent protein (GFP) has been spliced. Only cells that have formed heterocysts are expressing the fused gene, which makes the cell fluoresce bright green.

Starvation Induces Differentiation into Fruiting Bodies

Certain species of bacteria, in the microbial equivalent of a barn raising, produce architectural marvels called fruiting bodies. The Gram-negative species *Myxococcus xanthus* uses a **gliding motility** (involving a type of pilus,

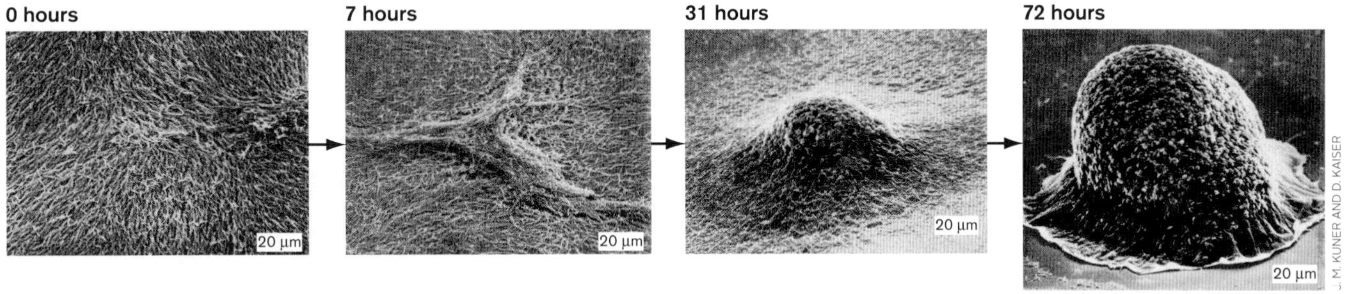

0 hours   7 hours   31 hours   72 hours

M. KUNER AND D. KAISER

FIGURE 4.34 ■ *Myxococcus* **swarm, erecting a fruiting body.** Approximately 100,000 cells begin to aggregate, and over the course of 72 hours they erect a fruiting body.

not a flagellum) to travel on surfaces as individuals or to move together as a mob (**Fig. 4.34**). Starvation triggers a developmental cycle in which 100,000 or more individuals aggregate, rising into a mound called a fruiting body. Myxococci within the interior of the fruiting body differentiate into thick-walled, spherical spores that are released into the surroundings. The random dispersal of spores is an attempt to find new sources of nutrients. This differentiation process requires many cell-cell interactions and a complex genetic program that we still do not fully understand.

Some Bacteria Differentiate to Form Filamentous Structures

The actinomycetes (see Chapter 18), such as *Streptomyces*, are bacteria that form mycelia and sporangia analogous to the filamentous structures of eukaryotic fungi (**Fig. 4.35**). Several developmental programs tied to nutrient availability are at work in this process (**Fig. 4.36**). Under favorable nutrient conditions, a germ tube emerges from a germinating spore, grows from its tip (tip extension), and forms

branches that grow along, and within, the surface of its food source (**Fig. 4.36**, step 1). This type of growth produces an intertwined network of long multinucleate filaments (**hyphae**; singular, **hypha**) collectively called substrate **mycelia** (singular, **mycelium**). After a few days, a signaling molecule made by the organisms accumulates to a level that activates a new set of genes, including one encoding a surfactant, that allow hyphae to grow into the atmosphere, rising above the surface to form aerial mycelia (**Fig. 4.36**, steps 2a and 2b). Compartments at the tips of these aerial hyphae contain 20–30 copies of the genome. Aerial hyphae stop growing as nutrients decline, triggering a developmental program that synthesizes antibiotics. Meanwhile, the older ends of the filaments senesce, and their decomposing cytoplasm attracts scavenger microbes—which are killed by the antibiotics. The younger streptomycete cells then feast on the dead scavengers.

The aerial hyphae ultimately produce spores (arthrospores) that are fundamentally different from bacterial endospores. This program lays down multiple septa that subdivide the compartment into single-genome prespores (**Fig. 4.36**, step 3). The shape of the prespore then

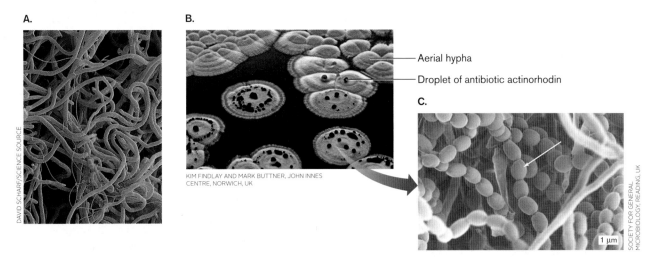

A.

DAVID SCHARF/SCIENCE SOURCE

B.

KIM FINDLAY AND MARK BUTTNER, JOHN INNES CENTRE, NORWICH, UK

— Aerial hypha
— Droplet of antibiotic actinorhodin

C.

SOCIETY FOR GENERAL MICROBIOLOGY, READING, UK

FIGURE 4.35 ■ **Mycelia.** **A.** *Streptomyces* substrate mycelia. **B.** Filamentous colonies of *Streptomyces coelicolor*, an actinomycete known for producing antibiotics (blue pigment in water droplets). **C.** *Streptomyces* aerial hyphae. The arrow points to a hyphal spore (approx. 1 μm each).

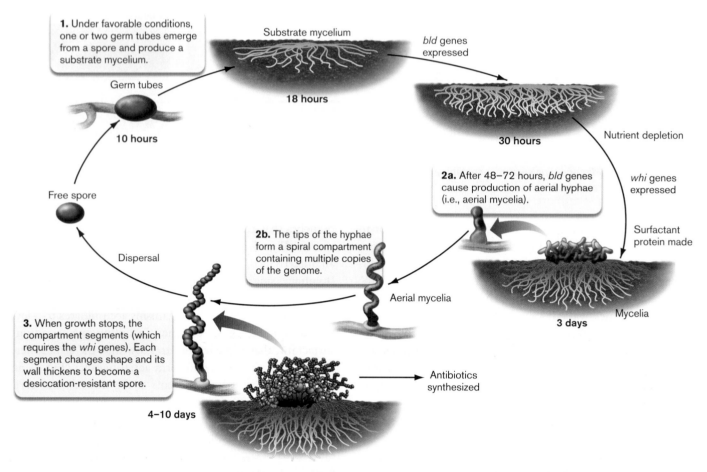

1. Under favorable conditions, one or two germ tubes emerge from a spore and produce a substrate mycelium.

Germ tubes

10 hours

Substrate mycelium

18 hours

bld genes expressed

30 hours

Nutrient depletion

whi genes expressed

Surfactant protein made

2a. After 48–72 hours, *bld* genes cause production of aerial hyphae (i.e., aerial mycelia).

Free spore

Dispersal

2b. The tips of the hyphae form a spiral compartment containing multiple copies of the genome.

Aerial mycelia

Mycelia

3 days

3. When growth stops, the compartment segments (which requires the *whi* genes). Each segment changes shape and its wall thickens to become a desiccation-resistant spore.

Antibiotics synthesized

4–10 days

FIGURE 4.36 ■ **Developmental cycle of *Streptomyces coelicolor*.**

changes, its cell wall thickens, and deposits are made in the spore that increase resistance to desiccation. These organisms are of tremendous interest, both for their ability to make antibiotics and for their fascinating developmental programs.

Thought Question

4.16 How might *Streptomyces* and *Actinomyces* species avoid "committing suicide" when they make their antibiotics?

To Summarize

- **Microbial development** involves complex changes in cell forms.

- **Endospore development** by *Bacillus* and *Clostridium* species is a multistage process that includes asymmetrical cell division to make a forespore and a mother cell, forespore engulfment by the mother cell, deposition of coat proteins around the forespore, and steps that increase chemical and heat resistance of the endospore.

- **Heterocyst development** enables cyanobacteria to fix nitrogen anaerobically while maintaining oxygenic photosynthesis.

- **Multicellular fruiting bodies** in *Myxococcus* and **mycelia** in actinomycetes develop in response to starvation, dispersing dormant cells to new environments.

Concluding Thoughts

Bacteria, as simple as they seem, perform incredibly complex and highly orchestrated processes to achieve growth. They coordinate the gathering of food with the synthesis of biomass and then adjust those processes as nutrients change in the local environment, or when they are challenged by stress. Most, if not all, bacteria also undergo elegant developmental processes ranging from biofilms and swarming, to sporulation or heterocyst formation. As with metabolic adaptations, developmental programs are launched in response to environmental pressures such as starvation, desiccation, changes in temperature, and even crowding. The environmental influence over these microbial processes impacts the composition and interspecies collaboration among water and soil ecosystems, as well as within the microbial communities that inhabit the human body. In Chapter 5 we build on these concepts and explore the remarkable ways that microbes respond to environmental stress and change.

CHAPTER REVIEW

Review Questions

1. What nutrients do microbes need to grow?
2. Explain how autotrophy, heterotrophy, phototrophy, and chemotrophy differ.
3. Explain the basics of the carbon and nitrogen cycles.
4. Describe the various mechanisms of transporting nutrients in prokaryotes and eukaryotes. What are facilitated diffusion, coupled transport, ABC transporters, group translocation, and endocytosis?

5. Why is it important to grow bacteria in pure culture?
6. Under what circumstances would you use a selective medium? A differential medium?
7. What factors define the growth phases of bacteria grown in batch culture?
8. Describe the important features of biofilms.
9. Name three kinds of bacteria that differentiate, and give highlights of the differentiation processes.

Thought Questions

1. Bile salts are used in certain selective media. What are bile salts, and why might they be more harmful to Gram-positive organisms than to Gram-negatives?
2. Why is *Rickettsia prowazekii*, which can grow only in the cytoplasm of a eukaryotic cell, considered a living organism but viruses are not?
3. Suppose 1,000 bacteria are inoculated in a tube containing a minimal salts medium, where they double once an hour, and 10 bacteria are inoculated into rich medium, where they double in 20 minutes. Which tube will have more bacteria after 2 hours? After 4 hours?
4. An exponentially growing culture has an optical density at 600 nm (OD_{600}) of 0.2 after 30 minutes and an OD_{600} of 0.8 after 80 minutes. What is the doubling time?
5. What are the generation times for (a) *Clostridium perfringens*, the cause of gas gangrene; (b) *Mycobacterium leprae*, the cause of leprosy; (c) *Thermus aquaticus*, a hot-springs bacterium; and (d) *Psychromonas antarcticus*, a cold-loving organism that grows at high pressure?
6. Mercuric ions (Hg^{2+}) and methylmercury [$(CH_3Hg)^+$] are major, human-generated, toxic contaminants of water and soil. Some species of bacteria can bioremediate these compounds by transporting them into the cell and reducing them to elemental mercury (Hg^0). One of the transport proteins is called MerC. Use an Inter-

net search engine to determine how many organisms have a MerC homolog.

7. Environmental bacteria were isolated from the water reservoirs of insect-digesting pitcher plants. The isolated strains were cultured in tryptone–yeast extract broth, which contains many different peptide and carbohydrate nutrients. The different growth curves obtained are shown in the graph. Explain how these curves differ from the "standard" growth curve and propose hypotheses as to why they differ.

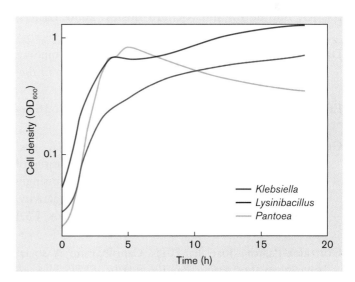

Key Terms

ABC transporter (126)
antiport (125)
ATP synthase (121)

autotroph (119)
autotrophy (119, 120)
batch culture (141)

binary fission (139)
biofilm (145)
chemoheterotrophy (121)

chemolithoautotrophy (120)
chemostat (144)
chemotrophy (120)
cofactor (119)
colony (130)
complex medium (130)
confluent (130)
continuous culture (143)
coupled transport (125)
death phase (143)
death rate (143)
denitrification (123)
differential medium (131)
dilution streaking (129)
doubling time (140)
electrochemical potential (121)
enriched medium (131)
essential nutrient (119)
exopolysaccharide (EPS) (146)
exponential phase (142)
facilitated diffusion (124)
fluorescence-activated cell sorter (FACS) (137)

forespore (150)
generation time (140)
germination (150)
gliding motility (150)
group translocation (128)
growth factor (132)
growth rate (139)
growth rate constant (140)
heterotrophy (119, 120)
hypha (151)
lag phase (141)
lithotrophy (120)
logarithmic (log) phase (142)
macronutrient (119)
membrane potential (121)
micronutrient (119)
minimal defined medium (130)
mother cell (150)
mycelium (151)
nitrification (123)
nitrogen-fixing bacterium (122)
optical density (138)
organotrophy (119, 121)

permease (123)
phosphotransferase system (PTS) (128)
photoautotrophy (120)
photoheterotrophy (120)
phototrophy (120)
planktonic cell (139)
pour plate technique (138)
proton potential (proton motive force) (121)
pure culture (129)
quorum sensing (146)
selective medium (131)
siderophore (127)
spread plate (130)
stationary phase (142)
symbiont (123)
symport (125)
twitching motility (146)
viable (130)

Recommended Reading

Bassler, Bonnie L., and Richard Losick. 2006. Bacterially speaking. *Cell* **125**:237–246.

Davidson, Amy L., and Jue Chen. 2004. ATP-binding cassette transporters in bacteria. *Annual Reviews in Biochemistry* **73**:241–268.

Errington, John. 2010. From spores to antibiotics. *Microbiology* **156**:1–13.

Gefena, Orit, Ofer Fridmana, Irine Ronina, and Nathalie Q. Balabana. 2014. Direct observation of single stationary-phase bacteria reveals a surprisingly long period of constant protein production activity. *Proceedings of the National Academy of Sciences USA* **111**:556–561.

Gonzalez-Pastor, Jose E. 2011. Cannibalism: A social behavior in sporulating *Bacillus subtilis*. *FEMS Microbiology Reviews* **35**:415–424.

Laverty, Garry, Sean P. Gorman, and Brendan F. Gilmore. 2014. Biomolecular mechanisms of *Pseudomonas aeruginosa* and *Escherichia coli* biofilm formation. *Pathogens* **3**:596–632.

Lopian, Livnat, Yair Elisha, Anat Nussbaum-Shochat, and Orna Amster-Choder. 2010. Spatial and temporal organization of the *E. coli* PTS components. *EMBO Journal* **29**:3630–3645.

Omsland, Anders, Ted Hackstadt, and Robert A. Heizen. 2013. Bringing culture to the uncultured: *Coxiella burnetii* and lessons for obligate intracellular bacterial pathogens. *PLoS Pathogens* **9**:e1003540.

Romero, Diego, Hera Vlamakis, Richard Losick, and Roberto Kolter. 2011. An accessory protein required for anchoring and assembly of amyloid fibres in *B. subtilis* biofilms. *Molecular Microbiology* **80**:1155–1168.

Saier, Milton H., Jr., Bin Wang, Wei Hao Zheng, Eric I. Sun, and Ming Ren Yen. 2010. Mosaic energy-coupled transporters. *Microbe* **5**:105–109.

Serra, Diego O., and Regina Hengge. 2014. Stress responses go three dimensional—The spatial order of physiological differentiation in bacterial macrocolony biofilms. *Environmental Microbiology* **16**:1455–1471.

Setlow, Peter. 2014. Germination of spores of *Bacillus* species: What we know and do not know. *Journal of Bacteriology* **196**:1297–1305.

Skaar, Eric P. 2010. The battle for iron between bacterial pathogens and their vertebrate hosts. *PLoS Pathogens* **6**:e1000949.

Skerker, Jeffrey M., and Michael T. Laub. 2004. Cell-cycle progression and the generation of asymmetry in *Caulobacter crescentus*. *Nature Reviews. Microbiology* **2**:325–337.

Teschler, Jennifer K., David Zamorano-Sánchez, Andrew S. Utada, Christopher J. A. Warner, Gerard C. L. Wong, et al. 2015. Living in the matrix: Assembly and control of *Vibrio cholerae* biofilms. *Nature Reviews. Microbiology* **13**:255–268.

Wood, Thomas K. 2014. Biofilm dispersal: Deciding when it is better to travel. *Molecular Microbiology* **94**:747–750.

CHAPTER 5

Environmental Influences and Control of Microbial Growth

Microbes have both the fastest and the slowest growth rates of any known organisms. Some hot-springs bacteria can double in as little as 10 minutes, whereas deep-sea-sediment microbes may take as long as 100 years. What determines these differences in growth rate? Nutrition is one factor, but niche-specific physical parameters such as temperature, pH, and osmolarity are equally important. In Chapter 5 we explore the limits of microbial growth and show how this knowledge helps us control the microbial world.

CURRENT RESEARCH highlight

Manipulating your neighbor. Can one bacterial species manipulate differentiation of another? Elizabeth Shank and student Matthew Powers explored this question by searching for soil bacteria that could inhibit biofilm gene expression in *Bacillus subtilis*. One species, *Pseudomonas protegens*, uses 2,4-diacetylphloroglucinol (DAPG) for that purpose. A DAPG-secreting *P. protegens* colony (arrow) alters the wrinkle morphology of a nearby *B. subtilis* colony (below). Subinhibitory concentrations of DAPG also decreased *Bacillus* spore formation tenfold, proving that some microbiome species secrete signaling molecules that manipulate the development of "niche-mates." Microbiome composition could be radically changed by such interactions.

Source: Matthew J. Powers et al. 2015. *J. Bacteriol.* **197**:2129–2138.

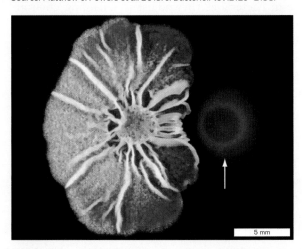

5 mm

AN INTERVIEW WITH

ELIZABETH SHANK, MICROBIOLOGIST, UNIVERSITY OF NORTH CAROLINA

COURTESY OF UNIVERSITY OF NORTH CAROLINA

How might tinkering with sporulation affect a microbial ecosystem?

Cross-species interactions between *Bacillus subtilis* and *Pseudomonas protegens* could allow them to co-colonize plant roots by partitioning growing cells of each species in different regions of the root. Alternatively, *P. protegens* might have a competitive growth advantage if *B. subtilis* cannot sporulate to survive harsh conditions. Altering growth or survival of one member of a microbial community can also have ripple effects on other members of the ecosystem.

Will the signal molecule DAPG affect sporulation of other bacteria, and if so, are there potential applications?

Whether DAPG affects other species depends on how it impacts sporulation. DAPG probably acts early in the SpoOA signaling pathway, increasing the chances that DAPG will affect other bacteria with conserved SpoOA pathways. If DAPG does affect other bacteria, it might be used to help plants resist particular pathogens or, if DAPG prevents sporulation of the biopesticide *Bacillus thuringiensis*, it could extend the efficacy of this biopesticide. Regardless, DAPG is a useful chemical tool to probe how *P. protegens* and *B. subtilis* may chemically manipulate each other in nature.

157

Microbes encounter numerous environmental stresses during their existence. Some are physical, some are chemical, and some are imposed by their competitors, as illustrated in the Current Research Highlight. Although a species' physiology is geared to work within only a narrow range of physical parameters, in nature the environment can change quickly and dramatically. Many marine microbes, for instance, move within seconds from deep-sea cold to the searing heat of a thermal vent. How do these organisms survive? Most species have stopgap measures called stress responses that temporarily protect the organism from brief forays into potentially lethal environments. But some organisms evolve to thrive, not just survive, in extreme environments. How do these so-called extremophiles grow under conditions that would kill most living things? In addition to environmental threats, microbes in nature wage wars among themselves by secreting antimicrobial compounds. Why doesn't the toxin-producing microbe kill itself? Is it somehow immune to its own toxin? Knowledge about how microbes cope with their environments, and each other, is exploited by humans to prevent the growth of bacteria in foods, destroy disease-causing microbes, and develop new microbe-based industries.

We begin this chapter by describing how physical and chemical changes in the environment impact the growth of different groups of microbes. We also explore how microorganisms adapt to different environments in ways both transient (involving temporary expression of genes) and permanent (modifications of the gene pool). The permanent genetic changes have led to biological diversity. Finally, we examine the different ways humans try to limit the growth of microorganisms to protect plants, animals, and ourselves.

As you proceed through this chapter, you will encounter two recurring themes: that different groups of microbes live in vastly different environments, and that microbes can respond in diverse ways when confronted by conditions outside their niche, or comfort zone. Chapter 21 builds on this information to discuss how communities of microbes interact with their environment, and presents current research methods that are used to study microbial ecology.

5.1 Environmental Limits on Growth

What is normal? With our human frame of reference, we tend to think that "normal" growth conditions are those found at sea level with a temperature between 20°C and 40°C, a near-neutral pH, a salt concentration of 0.9%, and ample nutrients. Any ecological niches outside this window are labeled "extreme," and the organisms inhabiting them are called **extremophiles**. The term "extremophile" was first used by NASA biochemist Robert MacElroy, who sought evidence for kinds of life that might inhabit other planets.

Extremophiles are microbes (bacteria, archaea, and some eukaryotes) that are able to grow in conditions extremely different from those optimal for humans. For example, one group of organisms can grow at temperatures above the boiling point of water (100°C), while another group requires a strongly acidic (pH 2) environment to grow. According to our definition of what is "normal," conditions on Earth when life began were certainly extreme. The earliest microbes, then, likely grew in these extreme environments. Organisms that grow under conditions that seem normal to humans likely evolved from an ancient extremophile that gradually adapted as the environment changed to that of our present-day Earth.

A single environment can simultaneously encompass multiple extremes. In Yellowstone National Park, for instance, an acid pool can be found next to an alkali pool, both at extremely high temperatures. Thus, extremophiles typically evolve to survive multiple extreme environments.

Extremophiles may provide insight into the workings of extraterrestrial microbes we may one day encounter, since outer space certainly qualifies as an extreme environment. Our experiences with extremophiles should alert us to the dangers of underestimating the precautions necessary in handling extraterrestrial samples. For example, we should not assume that irradiation will sterilize samples from future planetary or interstellar missions. Such treatments do not even kill *Deinococcus radiodurans*, an extremophile found on Earth.

How do we even begin to study organisms that grow in boiling water or sulfuric acid solutions or organisms that we cannot culture in the laboratory? Genome sequences present new opportunities for investigating these questions. Bioinformatic analysis, which uses the DNA sequence of a gene to predict the function of its protein product, allows us to study the biology of organisms that we cannot culture. Transcriptomic and proteomic techniques can examine all the gene and protein expression, respectively, that take place in bacteria as they adjust to changing environments such as temperature or pH. These techniques are discussed further in Chapters 7, 8, and 12.

We have already mentioned the fundamental physical conditions (temperature, pH, osmolarity) that define an environment and favor (select for) the growth of specific groups of organisms. Within a given microbial community, each species is further localized to a specific niche defined by a narrower range of environmental factors (see the Current Research Highlight). Species find their niche, in part, because every protein and macromolecular structure within a cell is affected by changes in environmental conditions. For example, a single enzyme works best under a unique set of temperature, pH, and salt conditions because those conditions allow it to fold into its optimum shape, or conformation. Deviations

TABLE 5.1	Basic Environmental Classification of Microorganisms			
Environmental parameter	**Classification**			
Temperature	Hyperthermophile* (growth above 80°C)	Thermophile* (growth between 50°C and 80°C)	Mesophile (growth between 15°C and 45°C)	Psychrophile* (growth below 15°C)
pH	Alkaliphile* (growth above pH 9)	Neutralophile (growth between pH 5 and pH 8)	Acidophile* (growth below pH 3)	
Osmolarity	Halophile* (growth in high salt, >2 M NaCl)			
Oxygen	Strict aerobe (growth only in O_2)	Facultative microbe (growth with or without O_2)	Microaerophile (growth only in small amounts of O_2)	Strict anaerobe (growth only without O_2)
Pressure	Barophile* (growth at high pressure, greater than 380 atm)		Barotolerant (growth between 10 and 500 atm)	

*Considered extremophiles.

from these optimal conditions cause the protein to fold a little differently and become less active. While not all enzymes within a cell boast the same physical optima, these optima must at least be similar and matched to the organism's environment for the organism to function effectively.

As you might guess from the preceding discussion, microbes are commonly classified by their environmental niche. **Table 5.1** summarizes these environmental classes.

To Summarize

- **Different species** exhibit different optimal growth values of temperature, pH, and osmolarity.

- **Extremophiles** inhabit fringe environments with conditions that do not support human life.

- **The environmental habitat** (such as high salt or acidic pH) inhabited by a particular species is defined by the tolerance of that organism's proteins and other macromolecular structures to the physical conditions within that niche.

- **Global approaches** used to study gene expression allow us to view how organisms respond to changes in their environment.

5.2 Temperature and Pressure

How do microbes react to hot and cold, or to high pressure? Let's start with temperature. Unlike humans (and mammals in general), microbes cannot control their temperature; thus, bacterial cell temperature matches that of the immediate environment. Because temperature affects the average rate of molecular motion, changes in temperature impact every aspect of microbial physiology, including membrane fluidity, nutrient transport, DNA stability, RNA stability, and enzyme structure and function. Every organism has an "optimum" temperature at which it grows most quickly, as well as minimum and maximum temperatures that define the limits of growth.

Growth temperature limits are imposed, in part, by the thousands of proteins in a cell, all of which must function within the same temperature range. A species grows most quickly at temperatures where all of the cell's proteins work most efficiently as a group to produce energy and synthesize cell components. Growth stops when rising temperatures cause critical enzymes or cell structures (such as the cell membrane) to fail. At cold temperatures, growth ceases because enzymatic processes become too sluggish and the cell membrane becomes too rigid. The membrane needs to remain fluid so that it can expand as cells grow larger and so that proteins needed for solute transport can be inserted into the membrane.

The different branches of life reflect narrowing tolerance to heat. Different archaeal species, for example, can grow in extremely hot or extremely cold temperatures, and some can grow in the middle range. Bacteria, for the most part, tolerate temperatures between the archaeal extremes. Eukaryotes are even less temperature tolerant than bacteria, with individual species capable of growth between 10°C and 65°C.

Growth Rate and Temperature

In general, microbes that grow at higher temperatures can achieve higher rates of growth (**Fig. 5.1A**). Remarkably, for any one species the relationship between growth temperature and the growth rate constant k (the number of generations per hour; see Section 4.4) obeys the Arrhenius equation for simple chemical reactions (**eTopic 5.1**).

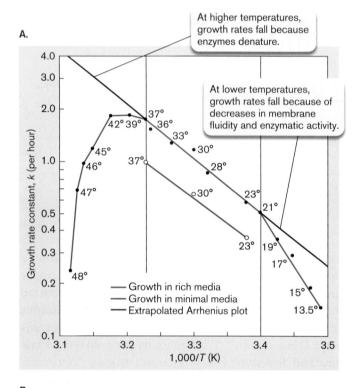

A.

At higher temperatures, growth rates fall because enzymes denature.

At lower temperatures, growth rates fall because of decreases in membrane fluidity and enzymatic activity.

— Growth in rich media
— Growth in minimal media
— Extrapolated Arrhenius plot

$1,000/T$ (K)

B.

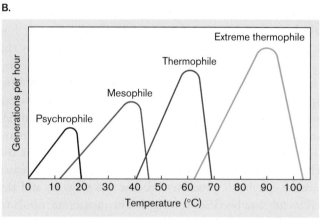

Extreme thermophile

Thermophile

Mesophile

Psychrophile

Temperature (°C)

FIGURE 5.1 ■ Relationship between temperature and growth rate. A. The growth rate constant (k) of the enteric organism *Escherichia coli* is plotted against the inverse of the growth temperature on the Kelvin scale ($1,000/T$ is used to give a convenient scale on the x-axis). This is a more detailed view of a mesophilic growth temperature curve. As temperature rises above or falls below the optimum range, growth rate decreases faster than is predicted by the Arrhenius equation. **B.** The relationship between temperature and growth rates of different groups of microbes. Note that the peak growth rate increases linearly with temperature and obeys the Arrhenius equation. *Source:* Part A from Sherrie L. Herendeen et al. 1979. *J. Bacteriol.* **139**:185.

The general result of the Arrhenius equation is that growth rate roughly doubles for every 10°C rise in temperature (**Fig. 5.1A**). This same relationship is observed for most chemical reactions.

At the upper and lower limits of the growth range, however, the Arrhenius effect breaks down. Critical proteins denature at high temperatures, whereas lower temperatures decrease membrane fluidity and limit the conformational mobility of enzymes, thereby lowering their activity. As a result, growth stops at temperature extremes. The typical temperature growth range for most bacteria spans the organism's optimal growth temperature by 30°C–40°C, but some organisms have a much narrower tolerance. Even within a species, we can find mutants that are more sensitive to one extreme or the other (heat sensitive or cold sensitive). These mutations often define key molecular components of stress responses, such as the heat-shock proteins (discussed in Chapter 10).

Thermodynamic principles limit a cell's growth to a narrow temperature range. For example, heat increases molecular movement within proteins. Too much or too little movement will interfere with enzymatic reactions. A great diversity exists among microbes because different groups have evolved to grow within very different thermal ranges. A species grows within a specific thermal range because its proteins have evolved to tolerate that range. Outside that range, proteins will denature or function too slowly for growth. The upper limit for protists is about 50°C, while some fungi can grow at temperatures as high as 60°C. Prokaryotes, however, have been found to grow at temperatures ranging from below 0°C to above 100°C. Temperatures over 100°C are usually found near thermal vents deep in the ocean. Vent water temperature can rise to 350°C, but the pressure is sufficiently high to keep the water in a liquid state.

Thought Question

5.1 Why haven't cells evolved so that all their enzymes have the same temperature optimum? If they did, wouldn't they grow even more rapidly?

Microorganisms Are Classified by Growth Temperature

Using range of growth temperature, microorganisms can be classified as mesophiles, psychrophiles, or thermophiles (**Fig. 5.1B**).

Mesophiles include the typical "lab rat" microbes, such as *Escherichia coli* and *Bacillus subtilis*. Their growth optima range between 20°C and 40°C, with a minimum of 15°C and a maximum of 45°C. Because they are easy to

A.

ASIM BEJ, UNIVERSITY OF ALABAMA AT BIRMINGHAM

DR. ALFONSO DAVILA/SETI INSTITUTE

B.

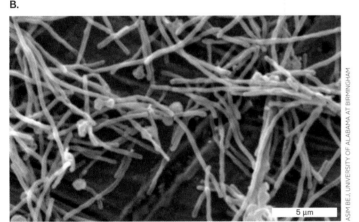

5 μm

ASIM BEJ, UNIVERSITY OF ALABAMA AT BIRMINGHAM

FIGURE 5.2 ■ Psychrophilic environments and microbes.
A. The continent of Antarctica is an extreme environment populated by many species of psychrophilic microorganisms, most of them unknown. In addition to being brutally cold, this extreme environment is nutrient-poor and subject to high levels of solar UV irradiation. **Inset:** Asim Bej, University of Alabama at Birmingham, collects samples from the ecosystem at Schirmacher Oasis (location as marked). Genome sequences from the captured microbes, like those shown in (B), reveal the composition and metabolic capabilities of the South Pole microbiome. **B.** Psychrotolerant *Flavobacterium* (grows between 0°C and 22°C) from a South Pole lake made from glacial meltwater in summer (high temperature = 0.9°C) (SEM). Novel compounds made by members of the polar microbiome are screened for anticancer and antimicrobial potential.

grow and because most human pathogens are mesophiles, much of what we know about protein, membrane, and DNA structure came from studying this group of organisms. However, detailed 3D views of protein structures are frequently based on studies of two other classes of organisms whose optimum growth temperature ranges flank that

of the mesophiles—namely, psychrophiles (on the low-temperature side) and thermophiles (on the high-temperature side). Because proteins from thermophiles are more stable structures than proteins from mesophiles, it's easier to crystallize proteins from extremophiles and determine their structure by X-ray crystallography (see Section 2.6).

Psychrophiles are microbes that grow at temperatures as low as −10°C, but their optimum growth temperature is usually around 15°C. Psychrophiles are prominent members of microbial communities beneath icebergs in the Arctic and Antarctic (**Fig. 5.2**). In addition to true psychrophiles, there are cold-resistant mesophiles (psychrotolerant bacteria or psychrotrophs) that grow between 0°C and 35°C. Both psychrophiles and psychrotrophs can be isolated from Antarctic lakes, beneath several meters of ice (**Fig. 5.2B**). Entire microscopic ecosystems flourish there, capable of surviving −40°C all winter, and then growing at near zero during the sunlit summer. Polar microorganisms may be screened for novel compounds with anticancer and antimicrobial potential. Closer to home, in human environments, psychrotolerant bacteria cause milk to spoil in the refrigerator. Even some pathogens, such as *Listeria monocytogenes* (one cause of food poisoning and septic abortions), can grow at refrigeration temperatures.

Why do these organisms grow so well in the cold? One reason is that the proteins of psychrophiles are more flexible than those of mesophiles and require less energy (heat) to function. Of course, the downside to the increased flexibility of psychrophilic proteins is that they denature at lower temperatures than their mesophilic counterparts. As a result, psychrophiles grow poorly, if at all, when temperatures rise above 20°C. Another reason psychrophiles favor cold is that their membranes are more fluid at low temperatures (because they contain a high proportion of unsaturated fatty acids); at higher temperatures their membranes are <u>too</u> flexible and fail to maintain cell integrity. Finally, bacteria and archaea that grow at 0°C in glaciers also contain antifreeze proteins and other cryoprotectants (such as trehalose) that can depress the freezing point by 2°C. So, although these organisms can grow in ice, they will not freeze. Interestingly, some psychrophilic and psychrotolerant bacteria actually stimulate ice formation in their surrounding environment (**eTopic 5.2**).

Psychrophilic enzymes are of commercial interest because their ability to carry out reactions at low temperature is useful in food processing and bioremediation. Enzymes help brew beer more quickly, break down lactose

A.

B.

C.

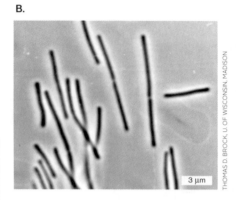

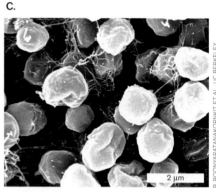

FIGURE 5.3 ■ **Thermophilic environments and thermophiles. A.** Yellowstone National Park hot spring. **B.** *Thermus aquaticus*, a hyperthermophile first isolated at Yellowstone by Thomas Brock. Cell length varies from 3 to 10 μm. **C.** Thermophile *Methanocaldococcus jannaschii*, grown at 78°C and 30 psi.

in milk, and can remove cholesterol from various foods. The production of foods at lower temperatures is beneficial too, because the lower processing temperatures minimize the growth of typical mesophiles that degrade and spoil food. Genetically engineered psychrophilic organisms can safely degrade toxic organic contaminants (for example, petroleum) in the cold. Arctic environments are particularly sensitive to pollution because contaminants are slow to degrade in the freezing temperatures. Consequently, the ability to seed Arctic oil spills with psychrophilic organisms armed with petroleum-degrading enzymes could more rapidly restore contaminated environments.

Thermophiles (**Fig. 5.3**) are species adapted to growth at high temperatures (typically 55°C and higher). **Hyperthermophiles**, which grow at temperatures as high as 121°C, are found near thermal vents that penetrate Earth's crust on the ocean floor and on land (for example, hot springs). The thermophile *Thermus aquaticus* was the first source of a high-temperature DNA polymerase used for PCR amplification of DNA. *T. aquaticus* was discovered in a hot spring at Yellowstone National Park by microbiologist Thomas Brock, a pioneer in the study of thermophilic organisms. Its application to the polymerase chain reaction revolutionized molecular biology (discussed in Chapter 7).

Extreme thermophiles often have specially adapted membranes and protein sequences. The thermal limits of these structures determine the specific high-temperature ranges in which various species can grow. Because enzymes in thermophiles (thermozymes) do not unfold as easily as mesophilic enzymes, they more easily hold their shape at higher temperatures. Thermophilic enzymes are stable, in part, because they contain relatively low amounts of glycine, a small amino acid that contributes to an enzyme's flexibility (glycines do not contain

side chains, so they cannot form stabilizing intramolecular bonds). In addition, the amino termini of proteins in these organisms often are "tied down" by hydrogen bonding to other parts of the protein, making them harder to denature.

Like all microbes, thermophiles have chaperone proteins that help refold other proteins as they undergo thermal denaturation. Thermophile genomes are packed with numerous DNA-binding proteins that stabilize DNA. In addition, these organisms possess special enzymes that tightly coil DNA in a way that makes it more thermostable and less likely to denature (think of a coiled phone cord that has twisted and bunched up on itself).

Special membranes also help give cells additional stability at high temperatures. Unlike the typical lipid bilayers of mesophiles, the membranes of thermophiles manage to "glue" together parts of the two hydrocarbon layers that point toward each other, making them more stable. They do this by incorporating more saturated linear lipids into their membranes. Saturated lipids form straight hydrocarbon tails that align well with neighboring lipids and form a highly organized structure that is stable to heat. The membranes of mesophiles are composed mostly of unsaturated lipids that bend against each other and align poorly. Consequently, the membranes of mesophiles are more fluid at lower temperatures.

The membranes of hyperthermophilic archaea impart an amazing level of heat resilience by being lipid monolayers, not bilayers (see Fig. 3.12 and eAppendix Fig. A2.3B). Lipid bilayers peel apart under withering heat. Monolayers, built for extremophile living, do not. Monolayer membranes are heat stable because long hydrocarbon chains directly tether glycerophosphates on opposite sides of the membrane. The chains (40 carbons long) do not contain fatty acids, but are made of isoprene units bonded by ether

FIGURE 5.4 ▪ Bioreactor used to grow thermophilic microorganisms. Dr. Robert Kelly and students stand next to a 20-liter bioreactor that they use for the engineering analysis of biofuel-producing microbes. Left to right: Aaron Hawkins, Andrew Loder, Hong Lian, Kelly, and Yejun Han.

PHOTO PROVIDED BY ROBERT KELLY

linkages to glycerol phosphate. More on thermophiles can be found in Chapter 19.

How can we "see" the genome sequences of organisms that will not grow at usual laboratory temperatures? And could we use those sequences to invent biofuels? The DNA of <u>uncultured</u> hyperthermophiles can be amplified by the polymerase chain reaction (PCR), a technique that multiplies a small sample of DNA (see Section 7.6). PCR can amplify DNA directly from the natural environment in which the organisms are found. Genomic comparison with known models such as *E. coli* quickly reveals whether an organism under study may possess specific metabolic pathways and regulatory responses. The ability to peek at the genome of hyperthermophiles led Robert Kelly and colleagues at North Carolina State University to mix genes and enzymes from different species (**Fig. 5.4**). Their lab attempts to engineer a new organism able to convert abundantly available CO_2 and H_2 directly into high-energy liquid fuels.

The Heat-Shock Response

As insurance against extinction, most microorganisms possess elegant genetic programs that remodel their physiology into one that can temporarily survive inhospitable conditions. Rapid temperature changes experienced during growth activate batches of stress response genes, resulting in the **heat-shock response** (discussed in Chapter 10). The protein products of these heat-activated genes include chaperones that maintain protein

shape and enzymes that change membrane lipid composition. The heat-shock response, first identified in *E. coli* by Tetsuo Yamamori and Takashi Yura in 1982, has since been documented in all living organisms examined thus far.

Thought Question

5.2 If microbes lack a nervous system, how can they sense a temperature change?

Adaptation to Pressure

Living creatures at Earth's surface (sea level) are subjected to a pressure of 1 atmosphere (atm), which is equal to 0.101 megapascal (MPa) or 14 pounds per square inch (psi). At the bottom of the ocean, however—thousands of meters deep—hydrostatic pressure averages a crushing 400 atm and can reach as high as 1,000 atm (101 MPa, or 14,600 psi) in ocean trenches (**Fig. 5.5**). Organisms

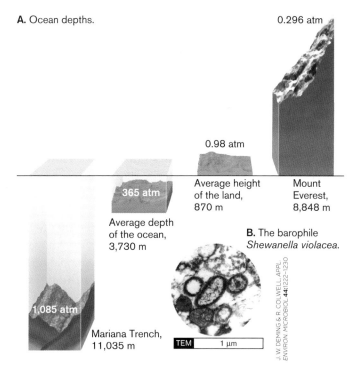

A. Ocean depths.

0.296 atm

0.98 atm

365 atm

Average height of the land, 870 m

Mount Everest, 8,848 m

Average depth of the ocean, 3,730 m

B. The barophile *Shewanella violacea.*

1,085 atm

Mariana Trench, 11,035 m

TEM   1 μm

J. W. DEMING & R. COLWELL, APPL. ENVIRON. MICROBIOL. **44**:1222–1230

FIGURE 5.5 ▪ Barophilic environments and piezophiles. **A.** Ocean depths. The deepest part of the ocean is at the bottom of the Mariana Trench, a depression in the floor of the western Pacific Ocean, just east of the Mariana Islands. The Mariana Trench is 2,500 km (1,554 miles) long and 70 km (44 miles) wide. Near its southwestern extremity, about 340 km (210 miles) southwest of Guam, lies the deepest point on Earth. This point, referred to as the Challenger Deep, plunges to a depth of 11,035 meters (nearly 7 miles). The pressure there (110 MPa) is over 1,000 times higher than what we experience on land (0.1 MPa). **B.** The barophile *Shewanella violacea.*

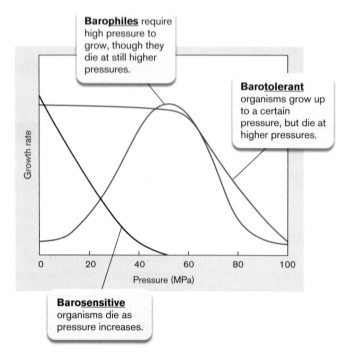

Barophiles require high pressure to grow, though they die at still higher pressures.

Barotolerant organisms grow up to a certain pressure, but die at higher pressures.

Barosensitive organisms die as pressure increases.

FIGURE 5.6 ■ **Relationship between growth rate and pressure.**

adapted to grow at these overwhelmingly high pressures are called **barophiles** or **piezophiles**. From the curves in **Figure 5.6**, notice that barophiles actually require elevated pressure to grow, while barotolerant organisms grow well in the range of 1–50 MPa, but their growth falls off thereafter.

Many barophiles are also psychrophilic, because the average temperature at the ocean's floor is 2°C. However, barophilic hyperthermophiles form the basis of thermal vent communities that support symbiotic worms and giant clams (see Chapter 21).

How bacteria survive pressures of 80–100 MPa (11,600–14,500 psi) is still a mystery. It is known, though, that increased hydrostatic pressure and cold temperatures reduce membrane fluidity. Because fluidity of the cell membrane is critical to survival, the phospholipids of deep-sea bacteria commonly have high levels of polyunsaturated fatty acids to increase membrane fluidity. It is thought that in addition to these membrane changes, internal structures must be pressure adapted. For example, ribosomes in the barosensitive organism *E. coli* (maximum growth pressure 50 MPa) dissociate at pressures above 60 MPa, so barophiles must contain uniquely designed ribosome structures that can withstand even higher pressures.

Clues to the physiological changes needed for growth at high pressure may come from *E. coli* adaptive-evolution studies. Over a period of 500 generations (126 days),

Douglass Bartlett and colleagues at the Scripps Institution of Oceanography gradually increased the pressure under which *E. coli* was grown to 62 MPa. One *E. coli* mutant evolved to successfully grow at this pressure. The evolved strain developed a mutation in a key fatty acid synthesis protein, which changed the ratio of its membrane fatty acids. Further studies could uncover mutations in additional genes and improve our understanding of the evolutionary steps necessary for microbial growth under extreme pressure.

To Summarize

- **The Arrhenius equation** applies to the growth of microorganisms: within a specific growth temperature range, the growth rate doubles for every 10°C rise in temperature.

- **Membrane fluidity** varies with the composition of lipids in a membrane, which in turn dictates the temperature and pressure at which an organism can grow.

- **Mesophiles, psychrophiles, and thermophiles** are groups of organisms that grow at moderate, low, and high temperatures, respectively.

- **The heat-shock response** produces a series of protective proteins in organisms exposed to temperatures near the upper edge of their growth range.

- **Barophiles (piezophiles)** can grow at pressures up to 1,000 atm but fail to grow at low pressures. Growth at high pressure requires specially designed membranes and protein structures.

Thought Question

5.3 What could be a relatively simple way to grow barophiles in the laboratory?

5.3 Osmolarity

Water is critical to life, but environments differ in the amount of water actually available to growing organisms. Microbes, for instance, can use only water that is not bound at any given instant to ions or other solutes in solution. Water availability is measured as **water activity** (a_w), a quantity approximated by concentration. Because interactions with solutes lower water activity, the more solutes there are in a solution, the less water there is available for microbes to use for growth. Water activity is typically measured as the ratio of the solution's vapor pressure relative to that of pure water. A solution is placed in a sealed chamber, and the amount of water vapor is determined at equilibrium. If the air above the sample is 97% saturated relative to the moisture present over pure water, the relative humidity is 97% and the water activity is 0.97. Most bacteria growing on land or in freshwater habitats require water activity to be greater than 0.91 (the water activity of seawater). Fungi can tolerate water activity levels as low as 0.86.

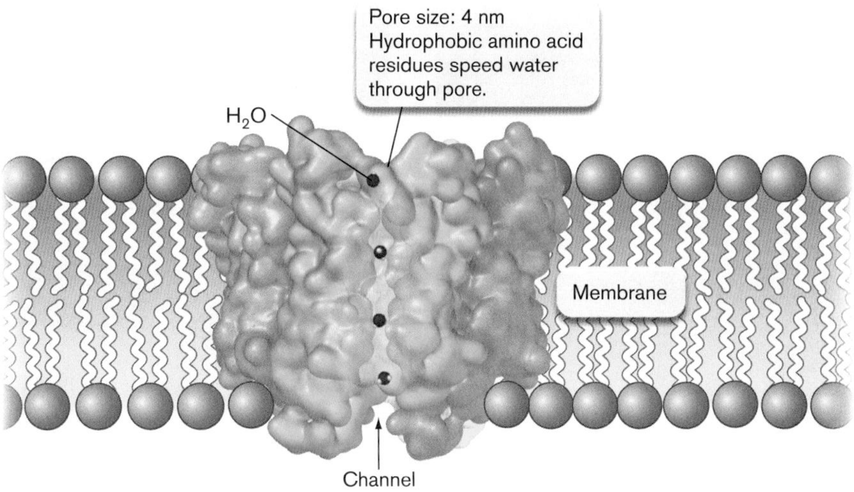

FIGURE 5.7 ■ Aquaporin. Transverse view of the channel through which water molecules move. Complementary halves of the channel are formed by adjacent protein monomers. (PDB code: 1J4N)

Osmotic Stress

Osmolarity is a measure of the number of solute molecules in a solution and is inversely related to a_w. The more particles there are in a solution, the greater the osmolarity and the lower the water activity. Osmolarity is also important for a cell because of the cell's semipermeable plasma membrane. This membrane allows the osmolarity inside the cell to be different from the osmolarity outside. The principles of physical chemistry dictate that a solute present at different concentrations in two chambers separated by a semipermeable membrane will tend to equilibrate. But if the semipermeable membrane does not allow solutes to move through the membrane, then osmolarity will equilibrate by water moving between the chambers. For a cell in a hypertonic medium, where the external osmolarity is higher than the internal, water will leave the cell in an attempt to equalize osmolarity across the membrane. In contrast, suspending a cell in a hypotonic medium (one of lower osmolarity than the cell) will cause an influx of water (see eAppendix 2).

Water does not move across cell membranes primarily by simple diffusion. Instead, special membrane water channels formed by proteins called aquaporins enable water to traverse the membrane much faster than by unmediated diffusion. Rapid movement of water helps protect cells against osmotic stress (**Fig. 5.7**). However, too much water moving in or out of a cell is detrimental. Cells may ultimately explode or implode, depending on the direction the water moves. Even bacteria with a rigid cell wall suffer. They may not explode like a human cell, but the forces placed on the cell wall are great.

Protection against Osmotic Stress

In addition to moving water, microbes have at least two other mechanisms to minimize osmotic stress across membranes. When stranded in a hypertonic medium (higher osmolarity than the cell), bacteria try to protect their internal water from leaving the cell by synthesizing or importing compatible solutes that increase intracellular osmolarity. Compatible solutes are small molecules that do not disrupt normal cell metabolism even at high intracellular concentrations. Increasing intracellular levels of these compounds (such as proline, glutamic acid, potassium, or betaine) elevates cytoplasmic osmolarity without any detrimental effects, making it unnecessary for water to leave the cell. In contrast, ions such as Na^+ are not compatible solutes and will disturb metabolism at high intracellular concentrations.

Cells also contain pressure-sensitive (mechanosensitive) channels that can be used to leak solutes out of the cell. It is believed that these channels are activated by rising internal pressures in cells immersed in a hypotonic medium (lower osmolarity than the cell). When activated, the channels allow solutes to escape, thereby lowering internal osmolarity and preventing too much water from entering the cell.

Outside a certain range of external osmolarity, these housekeeping strategies become ineffective at controlling internal osmolarity. To adapt, microbes launch a global response in which cellular physiology is transformed to tolerate brief encounters with potentially lethal salt (or other solute) concentrations. Some changes are similar to those provoked by heat shock, such as the increased synthesis of chaperones that protect critical cell proteins from denaturation. Other changes include alterations in outer membrane pore composition (for Gram-negative organisms).

A.

B.

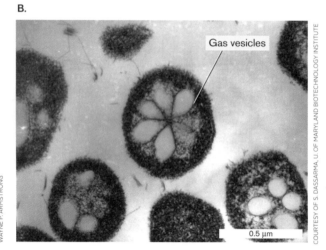

Gas vesicles

0.5 μm

C.

WAYNE P. ARMSTRONG

COURTESY OF S. DASSARMA, U. OF MARYLAND BIOTECHNOLOGY INSTITUTE

COURTESY OF S. DASSARMA, U. OF MARYLAND BIOTECHNOLOGY INSTITUTE

FIGURE 5.8 ■ **Halophilic salt flats and halophilic bacteria.** **A.** The halophilic salt flats along Highway 50 east of Fallon, Nevada, are colored pinkish red by astronomical numbers of halophilic bacteria. **B.** Cross section of the archaeon *Halobacterium* sp. (TEM). Gas vesicles allow the organism to float in liquid and acquire more oxygen. **C.** Shiladitya DasSarma and colleagues at the University of Maryland completed the genome sequence of *Halobacterium* species NRC-1.

> **Thought Question**
>
> **5.4** How might the concept of water availability be used by the food industry to control spoilage?

Halophiles Require High Salt

Some species of archaea and bacteria have evolved to require high salt (NaCl) concentration in order to grow. These microbes are called **halophiles** (**Fig. 5.8**). In striking contrast to most bacteria, which require salt concentrations from 0.05 to 1 M (0.2%–5% NaCl), the extremely halophilic archaea can grow at an a_w of 0.75 and actually require NaCl at concentrations of 2–4 M (10%–20%) to grow. For comparison, seawater is about 3.5% NaCl. All cells, even halophiles, prefer to keep a relatively low intracellular Na^+ concentration, because some solutes are moved into the cell by symport with Na^+. To achieve a low internal Na^+ concentration, halophilic microbes use special ion pumps to excrete sodium and replace it with other cations, such as potassium, which is a compatible solute. In fact, the proteins and cell components (for example, ribosomes) of halophiles require remarkably high intracellular potassium levels to maintain their structure. Halophilic archaea are presented in Chapter 19.

To Summarize

- **Water activity** (a_w) is a measure of how much water in a solution is available for a microbe to use.

- **Osmolarity** is a measure of the number of solute molecules in a solution and is inversely related to a_w.

- **Aquaporins** are membrane channel proteins that allow water to move quickly across membranes to equalize internal and external pressures.

- **Compatible solutes** are used to minimize pressure differences across the cell membrane.

- **Mechanosensitive channels** can leak solutes out of the cell when internal pressure rises.

- **Halophilic organisms** require high salt concentrations to grow.

5.4 Hydronium (pH) and Hydroxide Ion Concentrations

As with salt and temperature, the concentration of hydrogen ions (H^+)—actually, hydronium ions (H_3O^+)—also has a direct effect on the cell's macromolecular structures. Extreme concentrations of either hydronium or hydroxide ions (OH^-) in a solution will limit growth. In other words, too much acid or base is harmful to cells. Despite this sensitivity to pH extremes, living cells tolerate a greater range in environmental concentration of H^+ than of virtually any other chemical substance. *E. coli*, for example, tolerates a pH range from 2 to 10, a 10-million-fold difference (but grows only between pH 4.5 and 9). For a brief review of pH, refer to eAppendix 1.

pH Optima, Minima, and Maxima

The charges on various amino or carboxyl groups within a protein help forge the intramolecular bonds that dictate

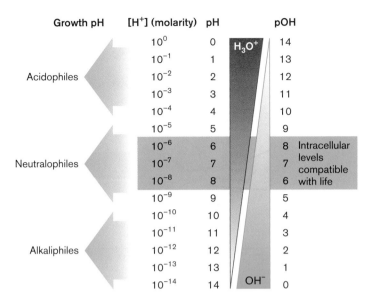

FIGURE 5.9 ■ Classification of organisms according to their optimum growth pH. pOH is the $\log_{10}$ of the reciprocal of the hydroxide ion (OH^-) concentration; that is, $pOH = -\log[OH^-]$.

protein shape and thus protein activity. Because H^+ concentration, $[H^+]$, affects the protonation of these ionizable groups, changing the pH can alter the charges on these groups, as well as protein structure and activity. The result is that all enzyme activities exhibit optima, minima, and maxima with regard to pH, much as they do for temperature. As we saw with temperature, groups of microbes have evolved to inhabit diverse niches, for which pH values can range from 0 to 11.5 (**Fig. 5.9**). However, species differences in optimum growth pH are not dictated by the pH limits at which critical cell proteins function.

Generally speaking, the majority of enzymes, regardless of the pH at which their source organism thrives, tend to operate best between pH 5 and 8.5 (which, if you think about it, is still a 3,000-fold range in hydrogen ion concentration). Yet many microbes grow in even more acidic or alkaline environments.

Unlike its temperature, the intracellular pH of a microbe, as well as its osmolarity, is not necessarily the same as that of its environment. Biological membranes are relatively impermeable to protons—a fact that allows the cell to maintain an internal pH compatible with protein function when growing in extremely acidic or alkaline environments. When the difference between the intracellular and extracellular pH (ΔpH) is very high, protons can leak through either directly or via proteins that thread the membrane. Excessive influx or efflux of protons can cause problems by altering internal pH.

Membrane-permeant organic acids, also called weak acids (discussed in Chapter 3), can accelerate the leakage of protons into a cell. Unlike H^+, the uncharged form

of an organic acid (HA) can freely permeate cell membranes and dissociate intracellularly, releasing a proton that then acidifies the internal pH (**eTopic 5.3** describes how). An example of organic acid stress involves the lactic acid produced and secreted by lactobacilli during the formation of yogurt. The buildup of lactic acid limits its bacterial growth, leaving yogurt with plenty of food value. The food industry has taken advantage of this phenomenon by preemptively adding citric acid or sorbic acid to certain foods. This practice controls microbial growth under pH conditions that do not destroy the flavor or quality of the food. **Figure 5.10** provides the pH values of various everyday items. Food microbiology is discussed further in Chapter 16.

Neutralophiles, Acidophiles, and Alkaliphiles Grow in Different pH Ranges

Cells have evolved to live under different pH conditions not by drastically changing the pH optima of their enzymes, but by using novel pH homeostasis strategies that maintain intracellular pH between pH 5 and pH 8, even when the cell is immersed in pH environments well above or below that range.

Hydrogen ion molarity		Example of solutions at this pH
1×10^0	pH = 0	Battery acid (strong), hydrofluoric acid
1×10^{-1}	pH = 1	Hydrochloric acid secreted by stomach lining
1×10^{-2}	pH = 2	Lemon juice, gastric acid, vinegar
1×10^{-3}	pH = 3	Grapefruit, orange juice, soda
1×10^{-4}	pH = 4	Acid rain, tomato juice
1×10^{-5}	pH = 5	Soft drinking water, black coffee
1×10^{-6}	pH = 6	Urine, saliva
1×10^{-7}	pH = 7	"Pure" water
1×10^{-8}	pH = 8	Seawater
1×10^{-9}	pH = 9	Baking soda solution
1×10^{-10}	pH = 10	Great Salt Lake, milk of magnesia
1×10^{-11}	pH = 11	Ammonia solution
1×10^{-12}	pH = 12	Soapy water
1×10^{-13}	pH = 13	Bleaches, oven cleaner
1×10^{-14}	pH = 14	Liquid drain cleaner

FIGURE 5.10 ■ pH values of common substances.

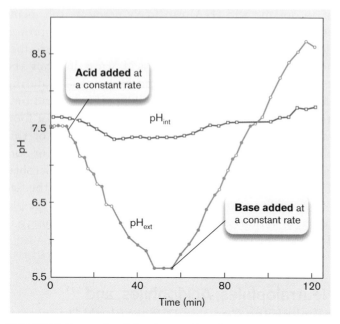

FIGURE 5.11 ■ **Maintaining internal pH (pH homeostasis) over a wide range of external pH.** Internal pH (pH_{int}) of the neutralophile *Escherichia coli* measured following the addition of acid to change external pH (pH_{ext}) and the subsequent addition of base. Internal pH was determined using nuclear magnetic resonance to measure changes in methyl phosphate. The two phosphate species titrate over different pH ranges. *Source:* Joan L. Slonczewski et al. 1981. *PNAS* **78**:6271.

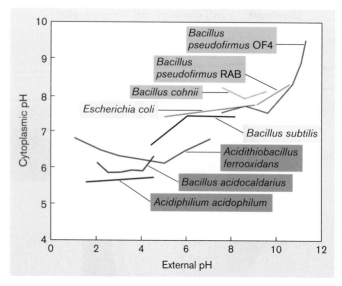

FIGURE 5.12 ■ **Cytoplasmic pH as a function of the external pH among acidophiles, neutralophiles, and alkaliphiles.** *Source:* Joan L. Slonczewski et al. 2009. *Adv. Microb. Physiol.* **55**:1–79.

Three classes of organisms are differentiated by the pH of their growth range: neutralophiles, acidophiles, and alkaliphiles (see **Fig. 5.9**).

Neutralophiles, which generally grow between pH 5 and pH 8, include most human pathogens. Many neutralophiles, including *E. coli* and *Salmonella enterica*, adjust their metabolism to maintain an internal pH slightly above neutrality, which is where their enzymes work best. They maintain this pH even in the presence of moderately acidic or basic external environments (**Fig. 5.11**). Other neutralophiles allow their internal pH to fluctuate with external pH but usually maintain a pH difference (ΔpH) of about 0.5 pH unit across the membrane at the upper and lower limits of growth pH. The ΔpH value is an important component of the transmembrane proton potential, a source of energy for the cell (see Chapter 14).

Note: The older term "neutrophile" used for this group of organisms is similar to the descriptor for a specific type of white blood cell ("neutrophil," discussed in Chapter 23). To avoid confusion, the term "neutrophil" should be reserved for the white blood cell and the term "neutralophile" used to designate microbes with growth optima near neutral pH (pH 7).

Acidophiles are bacteria and archaea that live in acidic environments. They are often lithotrophs (chemolithoautotrophs) that oxidize reduced metals and generate strong acids, such as sulfuric acid. Consequently, they grow between pH 0 and pH 5. Acidophiles generally maintain an internal pH that is considerably more acidic than that of neutralophiles but still less acidic than their growth environment (**Fig. 5.12**). The ability to grow at this pH is due partly to altered membrane lipid profiles (high levels of tetraether lipids) that decrease proton permeability, as well as to ill-defined proton extrusion mechanisms. Often, an organism that is an extremophile with respect to one environmental factor is an extremophile with respect to others as well. *Sulfolobus acidocaldarius*, for example, is a thermophile and an acidophile (**Fig. 5.13**). It uses sulfur as an energy source and grows in acidic hot springs rich in sulfur.

Alkaliphiles occupy the opposite end of the pH spectrum, growing best at values ranging from pH 9 to pH 11. They are commonly found in saline soda lakes, which have high salt concentrations and pH values as high as pH 11. Soda lakes, like Lake Magadi in Kenya's Great Rift Valley (**Fig. 5.14A**), are steeped in carbonates, which explains their extraordinarily alkaline pH. An alkaliphilic organism first identified in Lake Magadi is *Halobacterium salinarum* (also known as *Natronobacterium gregoryi*), a halophilic archaeon (**Fig. 5.14B**).

The cyanobacterium *Spirulina* is another alkaliphile that grows in soda lakes. Its high concentration of carotene gives the organism a distinctive pink color (note the color of the lake in **Fig. 5.14A**). *Spirulina* is also a major

A.

B.

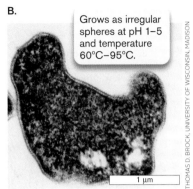

Grows as irregular spheres at pH 1–5 and temperature 60°C–95°C.

1 μm

FIGURE 5.13 ■ Sulfur Caldron acid spring and *Sulfolobus acidocaldarius.* A. Sulfur Caldron, in the Mud Volcano area of Yellowstone National Park, is one of the most acidic springs in the park. It is rich in sulfur and in *Sulfolobus*, an archaeon that thrives in hot, acidic waters with temperatures from 60°C to 95°C and a pH of 1–5. **B.** Thin-section electron micrograph of *S. acidocaldarius.*

food for the famous pink flamingos indigenous to these African lakes and is, in fact, the reason pink flamingos are pink. After the birds ingest these organisms, digestive processes release the carotene pigment to the circulation, which then deposits it in the birds' feathers, turning them pink (**Fig. 5.14C**). Humans also consume *Spirulina*, as a health food supplement, but they do not turn pink, because the cyanobacteria are only a small component of their diet.

The internal enzymes of alkaliphiles, like those of acidophiles, exhibit rather ordinary pH optima (around pH 8). The key to the survival of alkaliphiles is the cell-surface barrier that sequesters fragile cytoplasmic enzymes away from harsh extracellular pH. Key structural features of the cell wall, such as the presence of acidic polymers and an excess of hexosamines in the peptidoglycan, appear to be essential. The reason is unclear. At the membrane, some alkaliphiles also possess a high level of diether lipids (more stable than ester-linked phospholipids), which prevent protons from leaking out of the cell (see Section 3.2).

Because external protons are in such short supply at alkaline pH, most alkaliphiles use a sodium motive force in addition to a proton motive force to do much of the work of the cell (see Section 14.2). They also rely heavily on

A.

B.

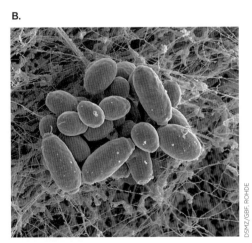

C.

FIGURE 5.14 ■ A soda lake ecosystem. A. Lake Magadi in Kenya. Its pink color is due to the cyanobacterium *Spirulina*. **B.** Alkaliphile *Halobacterium salinarum* (aka *Natronobacterium gregoryi*). Cell size, approx. 1 μm × 3 μm. **C.** Pink flamingos turn pink because they ingest large quantities of *Spirulina*.

FIGURE 5.15 ■ Na⁺ circulation in alkaliphiles. Cells of alkaliphiles use Na⁺ in place of H⁺ to do some of the work of the cell. They require an inwardly directed sodium gradient to rotate flagella and transport nutrient solutes. The Na⁺/H⁺ antiporter is also used to keep internal pH lower than external pH.

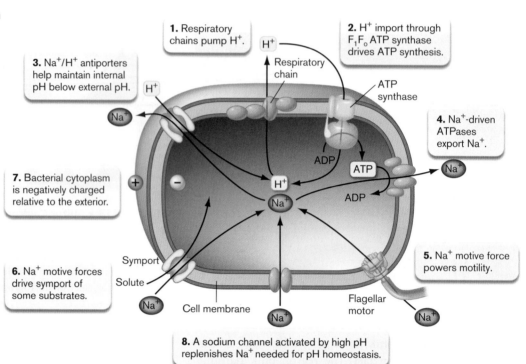

1. Respiratory chains pump H⁺.

2. H⁺ import through F₁Fₒ ATP synthase drives ATP synthesis.

3. Na⁺/H⁺ antiporters help maintain internal pH below external pH.

4. Na⁺-driven ATPases export Na⁺.

7. Bacterial cytoplasm is negatively charged relative to the exterior.

5. Na⁺ motive force powers motility.

6. Na⁺ motive forces drive symport of some substrates.

8. A sodium channel activated by high pH replenishes Na⁺ needed for pH homeostasis.

Na⁺/H⁺ antiporters (see Section 4.2) to bring protons into the cell. This H⁺ influx keeps the internal pH well below the extremely alkaline external pH. The Na⁺/H⁺ antiporters partly explain why many alkaliphiles are resistant to high salt (NaCl) concentrations: Sodium ions are expelled while protons are sucked in. Important aspects of sodium circulation in alkaliphiles are depicted in **Figure 5.15**.

In contrast to proteins <u>within</u> the cytoplasm, enzymes <u>secreted</u> from alkaliphiles are able to work in very alkaline environments. The inclusion of base-resistant enzymes like proteases, lipases, and cellulases in laundry detergents helps get our "whites whiter and our brights brighter." Other commercially useful alkaliphilic enzymes include cyclodextrin glucanotransferase, which produces cyclodextrins from starch (discussed in **eTopic 5.4**).

Thought Question

5.5 Recall from Section 4.2 that an antiporter couples movement of one ion down its concentration gradient with movement of another molecule uphill, against its gradient. If this is true, how could a Na⁺/H⁺ antiporter work to bring protons into a haloalkaliphile growing in high salt at pH 10? Since the Na⁺ concentration is lower inside the cell than outside and the H⁺ concentration is higher inside than outside (see **Fig. 5.15**), both ions are moving against their gradients.

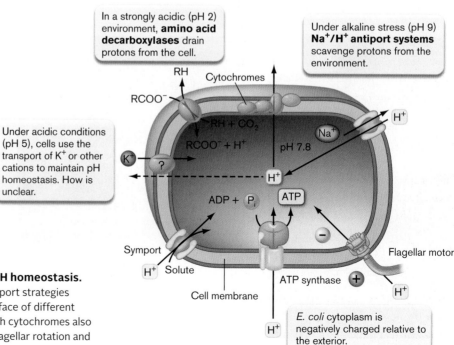

In a strongly acidic (pH 2) environment, **amino acid decarboxylases** drain protons from the cell.

Under alkaline stress (pH 9) **Na⁺/H⁺ antiport systems** scavenge protons from the environment.

Under acidic conditions (pH 5), cells use the transport of K⁺ or other cations to maintain pH homeostasis. How is unclear.

FIGURE 5.16 ■ Proton circulation and pH homeostasis. A typical *E. coli* cell uses various proton transport strategies to maintain an internal pH near pH 7.8 in the face of different external pH stresses. Proton pumping through cytochromes also establishes a proton gradient, which drives flagellar rotation and solute transport.

E. coli cytoplasm is negatively charged relative to the exterior.

pH Homeostasis and Acid Resistance

When cells are placed in pH conditions below their optimum, protons can enter the cell and lower internal pH to lethal levels. Microbes can prevent the unwanted influx of protons in a variety of ways (**Fig. 5.16**). *E. coli*, for example, can counter proton influx by transporting a variety of cations, such as K^+ or Na^+. How cation transport accomplishes H^+ efflux is unclear. Some evidence suggests a link to the role of K^+ in osmoprotection. At the other extreme, under extremely alkaline conditions, the cells can use the Na^+/H^+ antiporters mentioned previously (and in Section 4.2) to recruit protons into the cell in exchange for expelling Na^+. Some organisms can also change the pH of the medium using various amino acid decarboxylases and deaminases. For instance, *E. coli* consumes organic acids when growing at low pH, but produces these acids while trying to grow under alkaline conditions. *Helicobacter pylori*, the causative agent of gastric ulcers, employs an exquisitely potent urease to generate massive amounts of ammonia, which neutralizes the acid pH environment. These acid stress and alkaline stress protection systems are usually not made or at least do not become active until the cell encounters an extreme pH.

Many, if not all, microbes also possess an emergency global response system referred to as acid tolerance or acid resistance. In a process analogous to the heat-shock response, bacterial physiology undergoes a major molecular reprogramming in response to hydrogen ion stress. The levels of a large number of proteins increase, while the levels of others decrease. Many of the genes and proteins involved in the acid stress response overlap with other stress response systems, including the heat-shock response. These physiological responses include modifications in membrane lipid composition, enhanced pH homeostasis, and numerous other changes with unclear purpose. Some pathogens, such as *Salmonella*, sense a change in external pH as part of the signal indicating that the bacterium has entered a host cell environment (see **eTopic 5.5**).

To Summarize

- **Hydrogen ion concentration** affects protein structure and function. Thus, enzymes have pH optima, minima, and maxima.

- **Microbes use pH homeostasis mechanisms** to keep their internal pH near neutral when in acidic or alkaline media.

- **Adding weak acids** to certain foods undermines bacterial pH homeostasis mechanisms, thereby preventing food spoilage and killing potential pathogens.

- **Neutralophiles, acidophiles, and alkaliphiles** prefer growth under neutral, low, and high pH conditions, respectively.

- **Acid and alkaline stress responses** result when a given species is placed under pH conditions that slow its growth. The cell increases the levels of proteins designed to mediate pH homeostasis and protect cell constituents.

5.5 Oxygen

Many microorganisms can grow in the presence of molecular oxygen (O_2). Some even use oxygen as a terminal electron acceptor in the **electron transport system (ETS)**—also known as the electron transport chain (ETC)—a series of membrane proteins (for instance, cytochromes) that can help convert energy trapped in nutrients to a biologically useful form. The use of O_2 as the terminal electron acceptor is called **aerobic respiration** (see Chapter 14).

Oxygen Has Benefits and Risks

Electrons pulled from various energy sources (for example, glucose) possess intrinsic energy that electron transport systems incrementally extract. Once the cell has drained as much energy as possible from an electron, that electron must be passed to a final (terminal) electron acceptor molecule, such as oxygen gas (O_2), that diffuses away in the medium. This clears the way for another electron to be passed down the chain (**Fig. 5.17**). The electron

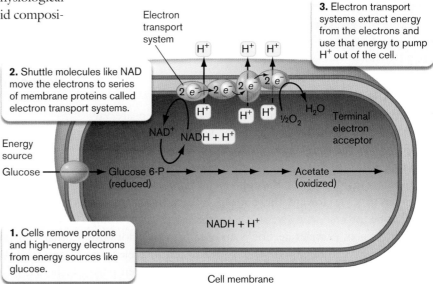

FIGURE 5.17 ■ The role of oxygen as a terminal electron acceptor in respiration. Pumping protons out of the cell by electron transport systems produces more positive charges outside the cell than inside, resulting in an electrochemical gradient (also called proton motive force). At the end of the ETS, the electron must be passed to a final (terminal) electron acceptor (for example, O_2), thus clearing the path for the next electron. This net process is called respiration.

TABLE 5.2	Examples of Aerobes and Anaerobes		
Aerobic microbes	**Facultative microbes**	**Microaerophilic microbes**	**Anaerobic microbes**
***Neisseria* spp.** Causative organisms of meningitis, gonorrhea	***Escherichia coli*** Normal gut biota; additional pathogenic strains	***Helicobacter pylori*** Cause of gastric ulcers	***Azoarcus tolulyticus*** Degrades toluene
Pseudomonas fluorescens Found in soil; degrades TNT and aromatic hydrocarbons	***Saccharomyces cerevisiae*** Yeast; used in baking	***Lactobacillus* spp.** Ferment milk to form yogurt	***Bacteroides* spp.** Normal gut biota
***Azotobacter* spp.** Soil microorganisms; fix atmospheric nitrogen	***Bacillus anthracis*** Cause of anthrax	***Campylobacter* spp.** One cause of gastroenteritis	***Clostridium* spp.** Soil microorganisms; causative agents of tetanus and botulism
***Rhizobium* spp.** Soil microorganisms; plant symbionts	***Vibrio cholerae*** Cause of cholera	***Treponema pallidum*** Cause of syphilis	***Actinomyces* spp.** Soil microorganisms; synthesize antibiotics
	***Staphylococcus* spp.** Found on skin; cause boils		***Desulfovibrio* spp.** Reduce sulfate

transport system pumps protons (H^+) out of the cell. The resulting unequal distribution of H^+ across the membrane produces a transmembrane electrochemical gradient, a kind of "biobattery" called the proton motive force (details are discussed in Section 14.2). The overall process of electron transport and proton motive force generation is called respiration if the electron donor is from an organic source and lithotrophy if the source is inorganic (discussed in Chapters 13 and 14).

Despite its importance for aerobic respiration, oxygen and its breakdown products are dangerously reactive—a serious problem for all cells. As a result, different species have evolved to either tolerate or avoid oxygen altogether. **Table 5.2** gives examples of microbes that grow at different levels of oxygen.

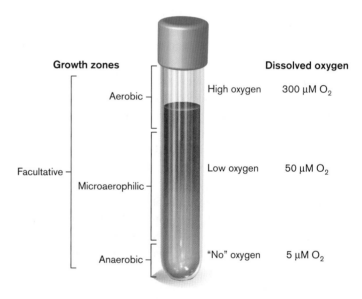

FIGURE 5.18 ∎ **Oxygen-related growth zones in a standing test tube.**

Aerobes versus Anaerobes

The relationships between microbes and oxygen are varied. **Figure 5.18** shows a test tube with growth medium. The top of the tube, closest to air, is oxygenated; the lower part of the tube has much lower levels of oxygen. Some microbes grow only at the top of the tube, while others prefer to grow toward the bottom. Where in the tube a microbe grows depends on that organism's relationship with oxygen. A **strict aerobe** is an organism that not only exists in oxygen but also uses oxygen as a terminal electron acceptor. The strict aerobe grows <u>only</u> with oxygen present and consumes oxygen during metabolism (aerobic respiration). An aerobe will grow only at the top of the tube shown in **Figure 5.18**. In contrast, a **strict anaerobe** dies in the least bit of oxygen (>5 μM dissolved O_2).

Strict anaerobes do not use oxygen as an electron acceptor, but this is not why they die in air. Some anaerobes die

because they are vulnerable to reactive oxygen molecules (also called reactive oxygen species, or ROS) produced by their own metabolism. Other anaerobes have enzymes that can protect them from ROS, but the dissolved oxygen raises the redox potential to a point that interferes with the use of alternative (non-oxygen) electron acceptors that the organism needs to make energy. Anaerobes will grow at the bottom of the tube shown in **Figure 5.18**. As we will discuss later, some bacteria previously considered strict anaerobes are not so strict.

What makes reactive oxygen species? Any organism that possesses NADH dehydrogenase 2—aerobe or anaerobe—will, in the presence of oxygen, inadvertently autooxidize the FAD (flavin adenine dinucleotide) cofactor within the enzyme and produce dangerous amounts of superoxide radicals ($^{\bullet}O_2^{-}$) (**Fig. 5.19**). Superoxide will degrade to hydrogen peroxide (H_2O_2), another reactive molecule. Iron, present as a cofactor in several enzymes, can then catalyze a reaction with hydrogen peroxide to produce the highly toxic hydroxyl radical ($^{\bullet}OH$). All of these molecules seriously damage DNA, RNA, proteins, and lipids. Consequently, oxygen is actually an extreme environment in which survival requires special talents. Aerobes destroy reactive oxygen species with an ample supply of enzymes such as superoxide dismutase (to remove superoxide) and peroxidase and catalase (to remove hydrogen peroxide). Aerobes also have resourceful enzyme systems that detect and repair macromolecules damaged by oxidation.

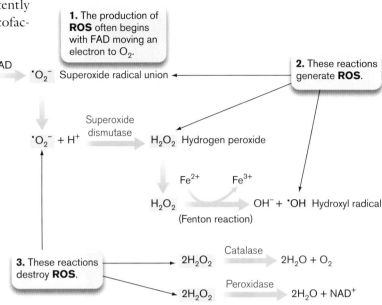

FIGURE 5.19 ■ **Generation and destruction of reactive oxygen species (ROS).** ROS species are marked yellow. The autooxidation of flavin adenine dinucleotide (FAD) and the Fenton reaction occur spontaneously to produce superoxide and hydroxyl radicals, respectively. The other reactions require enzymes. FAD is a cofactor for a number of enzymes (for example, NADH dehydrogenase 2). Catalase and peroxidase detoxify hydrogen peroxide.

Anaerobes versus Facultative Microbes

Anaerobic microbes fall into several categories. Some anaerobes actually do respire by means of electron transport systems, but instead of using oxygen, they rely on alternative terminal electron acceptors like nitrate (NO_3^{-}) to conduct **anaerobic respiration** and produce energy. Anaerobes of another ilk do not possess cytochromes, cannot respire, and so must rely on carbohydrate **fermentation** for energy (that is, they conduct **fermentative metabolism**). In fermentation, ATP energy is produced through substrate-level phosphorylation in a process that does not involve cytochromes. In either case, tolerance for ROS is low.

Facultative organisms are microbes that can live with or without oxygen and grow throughout the tube shown in **Figure 5.18**. **Facultative anaerobes** (such as *E. coli*) possess enzymes that destroy toxic oxygen by-products, but they have both fermentative <u>and</u> respiratory potential. Whether a member of this group uses aerobic respiration, anaerobic respiration, or fermentation depends on the availability of oxygen and the amount of carbohydrate present. **Aerotolerant anaerobes** use only fermentation to provide energy but contain superoxide dismutase and catalase (or peroxidase) to protect them from ROS. These enzymes allow aerotolerant anaerobes to grow in air (containing oxygen) while retaining a fermentation-based (anaerobic) metabolism. Aerotolerant anaerobes will grow throughout the tube in **Figure 5.18**. Microorganisms that possess <u>decreased</u> levels of superoxide dismutase and/or catalase will be **microaerophilic**, meaning they will grow only at low oxygen concentrations.

Some organisms previously considered anaerobes (for example, *Bacteroides fragilis*) are really transiently aerotolerant. These "anaerobes" tolerate oxygen because they possess low levels of ROS-protective enzymes and can even use very low levels of oxygen as terminal electron acceptors. *B. fragilis*, part of the normal gastrointestinal microbiota, may even help lower O_2 levels in the intestine.

The fundamental composition of all cells reflects their evolutionary origin as anaerobes. Lipids, nucleic acids, and amino acids are all highly reduced—which is why our bodies are combustible. We never would have evolved that way if molecular oxygen had been present from the beginning. Even today, the majority of all microbes are anaerobic, growing buried in the soil, within our anaerobic digestive tract, or within biofilms on our teeth.

Thought Questions

5.6 If anaerobes cannot live in oxygen, how do they incorporate oxygen into their cellular components?

5.7 How can anaerobes grow in the human mouth, where there is so much oxygen?

Culturing Anaerobes in the Laboratory

Many anaerobic bacteria cause horrific human diseases, such as tetanus, botulism, and gangrene. Some of

A.

Catalyst in lid mediates reaction. $H_2 + \frac{1}{2}O_2 \rightarrow H_2O$

GasPak envelope generates H_2 and CO_2.

JACK BOSTRACK/VISUALS UNLIMITED

B.

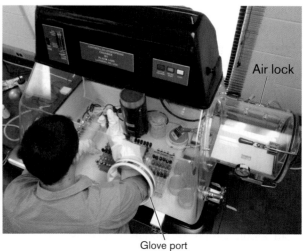

Air lock

Glove port

FIGURE 5.20 ▪ Anaerobic growth technology. A. An anaerobe jar. **B.** Student researcher using an anaerobic chamber with glove ports.

JOAN SLONCZEWSKI

these organisms or their secreted toxins are even potential weapons of terror (for example, *Clostridium botulinum*). Because of their ability to wreak havoc on humans, culturing these microorganisms was an early goal of microbiologists. Despite the difficulties involved, conditions were eventually contrived in which all, or at least most, of the oxygen could be removed from a culture environment.

Three oxygen-removing techniques are used today. Special reducing agents (for example, thioglycolate) or enzyme systems (such as Oxyrase) that eliminate dissolved oxygen can be added to ordinary liquid media. Anaerobes can then grow beneath the culture surface. A second, very popular way to culture anaerobes, especially on agar plates, is to use an anaerobe jar (**Fig. 5.20A**). Agar plates streaked with the organism are placed into a sealed jar with a foil packet that releases H_2 and CO_2 gases. A palladium packet hanging from the jar lid catalyzes a reaction between the H_2 and O_2 in the jar to form H_2O, and effectively removes O_2 from the chamber. The CO_2 released is required by some reactions to produce key metabolic intermediates. Some microaerophilic microbes, like the pathogens *Helicobacter pylori* (the major cause of stomach ulcers) and *Campylobacter jejuni* (a major cause of diarrhea), require low levels of O_2 but elevated amounts of CO_2. These conditions are obtained by using similar gas-generating packets.

For strict anaerobes exquisitely sensitive to oxygen, even more heroic efforts are required to establish an oxygen-free environment. A special anaerobic glove box must be used in which the atmosphere is removed by vacuum and replaced with a precise mixture of N_2 and CO_2 gases (**Fig. 5.20B**).

Thought Question

5.8 What evidence led people to think about looking for anaerobes? *Hint:* Look up "Spallanzani," "Pasteur," and "spontaneous generation" on the Internet.

To Summarize

- **Oxygen is a benefit to aerobes**, organisms that can use it as a terminal electron acceptor to extract energy from nutrients.

- **Oxygen is toxic** to all cells—for example, anaerobes—that do not have enzymes capable of efficiently destroying reactive oxygen species (ROS).

- **Anaerobic metabolism** can be either **fermentative** or **respiratory**. Anaerobic respiration requires the organism to possess cytochromes that can transfer electrons to terminal electron acceptors other than oxygen. Fermentative metabolism uses substrate-level phosphorylation to generate ATP in a process that does not involve cytochromes.

- **Aerotolerant anaerobes** grow in either the presence or the absence of oxygen, but use fermentation as their primary, if not only, means of gathering energy. These microbes also have enzymes that destroy ROS, allowing them to grow in oxygen.

- **Facultative anaerobes** grow with or without oxygen and have enzymes that destroy ROS. In addition, they possess both the ability for fermentative metabolism and respiration (anaerobic and aerobic). They can use oxygen as a terminal electron acceptor.

5.6 Nutrient Deprivation and Starvation

It is intuitively obvious that limiting the availability of a carbon source or other essential nutrient will limit growth. Less obvious are the dramatic molecular events that cascade through a starving cell. Optimizing growth rate when nutrient levels are suboptimal is an important aim of free-living bacteria, given that intestinal, soil, and marine environments rarely offer excess nutrients.

Starvation Activates Survival Genes

Numerous gene systems are affected when nutrients decline (see Sections 10.2 and 10.3). Growth rate slows, and daughter cells become smaller and begin to experience what is called a "starvation" response, in which the microbe senses a dire situation developing but still strives to find new nourishment. The resulting metabolic slowdown generates increased concentrations of critically important small signaling molecules, such as cyclic adenosine monophosphate (cyclic AMP or cAMP) and guanosine tetraphosphate (ppGpp), which globally transform gene expression. The highly soluble nature of these small molecules means they can quickly diffuse throughout the cell, promoting a fast response. During this metabolic retooling, transport systems for potential nutrients are produced even if the matching substrates are unavailable. Cells begin to make and store glycogen, presumably as an internal emergency store in case no other nutrient is found. Some organisms growing on nutrient-limited agar plates can even form colonies with intricate geometrical shapes that help the population cope, in some unknown way, with nutrient stress (**Fig. 5.21**).

As nutrient conditions worsen, the organism prepares for famine by activating many different stress survival genes.

The products of these genes afford protection against stressors such as reactive oxygen radicals or temperature and pH extremes. No cell can predict the precise stresses it might encounter while incapacitated, so it is advantageous to be prepared for as many as possible. As described in Section 4.6, some species undergo elaborate developmental processes that ultimately produce dormant spores.

When severely stressed by starvation, some members of a bacterial population appear to sacrifice themselves to save others by undergoing what is termed **programmed cell death**. The dying cells release nutrients that neighboring cells use to survive. One of the mechanisms for programmed cell death involves so-called toxin-antitoxin systems. For each TA pair, the toxin protein will stop growth or kill the cell, but the antitoxin (sometimes a protein, sometimes a small RNA molecule) can inactivate the toxin.

An important toxin-antitoxin system in *E. coli* is the MazE (antitoxin)–MazF (toxin) module (**Fig. 5.22A**). Because toxin and antitoxin are simultaneously made, healthy cells live. However, MazE is unstable (degraded by the ClpAP protease) and must continually be replenished by synthesis to inactivate MazF. If cells are starved, they stop making MazE and F. MazE antitoxin is degraded, leaving the more stable MazF free to cleave many cellular mRNA molecules. As a result, the cell first enters stasis, from which it can recover if more MazE is made. But if MazE is not forthcoming, the cell dies and releases nutrients. The dying cell will also signal nearby cells to undergo programmed cell death. A peptide cleaved from glucose-6-phosphate dehydrogenase is released from the dying cell and enters nearby cells. The peptide binds to MazE and prevents it from neutralizing MazF. The MazF toxin, now active, will eventually kill the bystander cell. Combined, enough nutrients are released to rescue a subset of the population. Nancy Woychik at Rutgers University studies how toxin-antitoxin systems contribute to latency of the pathogen *Mycobacterium tuberculosis* (**Fig. 5.22B**).

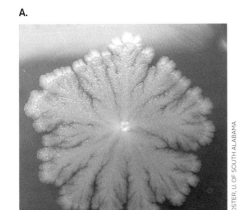

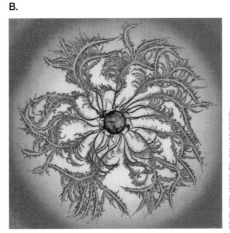

FIGURE 5.21 ■ Effects of starvation on colony morphology. A. Starving *E. coli* colony (diameter, 6 cm). **B.** *Paenibacillus dendritiformis* C morphotype grown on hard agar (1.75%) under starvation conditions. The colony consists of branches with chiral twists (colored green), all with the same handedness.

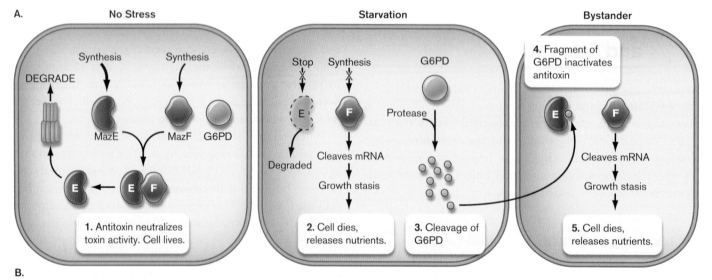

A.

No Stress

Synthesis   Synthesis

DEGRADE

MazE   MazF   G6PD

E   E F

1. Antitoxin neutralizes toxin activity. Cell lives.

Starvation

Stop   Synthesis   G6PD

E   F

Protease

Degraded   Cleaves mRNA

Growth stasis

2. Cell dies, releases nutrients.   **3.** Cleavage of G6PD

Bystander

4. Fragment of G6PD inactivates antitoxin

E   F

Cleaves mRNA

Growth stasis

5. Cell dies, releases nutrients.

B.

PHOTO COURTESY OF B.N. SINGH

FIGURE 5.22 ■ **Programmed cell death in response to starvation.**
A. The *E. coli* MazEF toxin-antitoxin system is thought to play an important role in bacterial survival during stress. MazE antitoxin is continually degraded by ClpAP and must be replenished to neutralize MazF toxin. Other toxin-antitoxin systems are found in other bacterial species. Stresses such as starvation, oxidative stress, or antibiotics can activate this system. G6PD = glucose 6-phosphate dehydrogenase. **B.** Nancy Woychik studies homologs of MazEF toxin-antitoxin systems in *Mycobacterium tuberculosis*, the cause of tuberculosis. Her lab explores how these toxin-antitoxin systems contribute to the latency of this pathogen.

Microbes Encounter Multiple Stresses in Real Life

Bacterial stress responses have traditionally been studied by exposing organisms to individual stresses. *Escherichia coli*, for example, synthesizes a specific set of proteins when exposed to high temperature and a different set of proteins when exposed to high salt. Some proteins, however, may be highly expressed under both conditions, but each stress response also includes proteins unique to that stress.

In the world outside of the laboratory, by contrast, environmental situations can be quite complex, involving multiple, not just single, stresses. An organism could simultaneously undergo carbon starvation in a high-salt, low-pH environment. A classic study by Kelly Abshire and Fred Neidhardt examined this situation using the pathogen *Salmonella enterica*, a cause of diarrhea. *S. enterica*

invades human macrophage cells and survives in phagocytic vacuoles, where numerous stresses, such as low pH, oxidative stress, and nutrient limitations, are simultaneously imposed on the bacteria. Comparing the proteins synthesized by *Salmonella* growing in this compartment with the proteins synthesized under single stresses in the laboratory revealed an unexpected response pattern. Although many stress-related proteins were induced in the intracellular environment, no one set of stress-induced proteins was induced in its entirety. Furthermore, several bacterial proteins were induced by growth only within the macrophage phagolysosome, suggesting the presence of unknown intracellular stresses. Thus, caution is advised when trying to predict cell responses to real-world situations based solely on controlled laboratory studies that alter only single parameters.

Humans Influence Microbial Ecosystems

Human activities have striking impacts on microbial ecosystems. One example comes from the mining of coal and minerals from the earth or, rather, from what happens after those mines are abandoned. Mining usually takes place below the water table, so water is continually pumped out to prevent flooding. Once a mine is abandoned, pumping stops and the mine floods. Acid mine drainage develops from the oxidation of pyrite (FeS_2) unearthed by the mining operations. The exposed pyrite oxidizes in air to form sulfuric acid that, along with soluble Fe^{2+}, can drain from the mine and destroy natural ecosystems. Acidophiles such as *Acidithiobacillus ferrooxidans* are key contributors to pyrite oxidation.

Mining is not the only human activity affecting Earth's microbiota. Another example is eutrophication. Natural ecosystems are typically low in nutrients (oligotrophic; **eTopic 5.6**) but teem with diversity, so that numerous species compete for the same limiting nutrients. Maximum diversity in a given ecosystem is maintained, in part, by the different nutrient-gathering profiles of competing microbes. However, the sudden infusion of large quantities of a formerly limiting nutrient, a process called **eutrophication**, can lead to a "bloom" of microbes, typically autotrophic cyanobacteria (formerly called blue-green algae; see Section 18.2). One species initially held in check by the limiting nutrient now exhibits unrestricted growth, consuming other nutrients as it grows to a degree that threatens the existence of competing species.

Humans cause eutrophication in several ways. Fertilizer runoff from agricultural fields, urban lawns, and golf courses is one source. Untreated or partially treated domestic sewage is another. Spilling large amounts of phosphates or nitrogen into lakes—Lake Erie, for example (**Fig. 5.23**)—powerfully stimulates cyanobacterial growth. The resulting bacterial "blooms" (wrongly called "algal blooms")

can deplete the oxygen in the water and lead to fish kills. Native fish species can disappear, to be replaced by species more tolerant of the new conditions. The concept of limiting nutrients in ecosystems is covered further in Chapter 21.

Climate change caused by human activity is another process that will gradually alter microbial ecosystems. Put simply, the spewing of heat-trapping CO_2 into the atmosphere by burning hydrocarbons is speeding Earth's warming. Since 1980, the average atmospheric temperature has risen 0.6°C (1.2°F). Konstantinos Konstantinidis from the Georgia Institute of Technology, and Jizhong Zhou from the University of Oklahoma recently completed a ten-year study examining how a mere 2°C difference in soil temperature can affect a community of soil microorganisms. Using an infrared light, the team warmed a patch of Oklahoma prairie soil 2°C above that of an adjacent control patch of soil. They then used DNA-based techniques to catalog the microbial genera present (**Fig. 5.24**). Some taxonomic groups of organisms became more dominant (Actinobacteria), while others became less abundant (Proteobacteria and Acidobacteria). What future effects these kinds of changes will have on the carbon and nitrogen cycles, as well as on farm productivity, remain to be seen.

FIGURE 5.23 ■ **Eutrophication in Lake Erie.** Cyanobacterial bloom (bright blue-green color) in Lake Erie caused by excessive phosphorous eutrophication.

NASA IMAGE BY JEFF SCHMALTZ, LANCE/ELSDIS MODIS RAPID RESPONSE, VISIBLE EARTH MARCH 21, 2012.

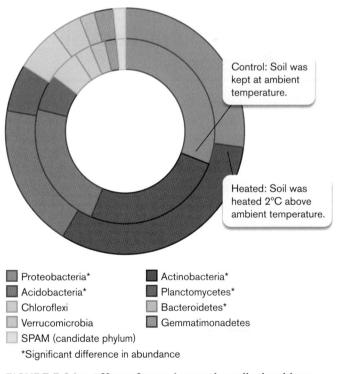

Control: Soil was kept at ambient temperature.

Heated: Soil was heated 2°C above ambient temperature.

■ Proteobacteria*   ■ Actinobacteria*
■ Acidobacteria*   ■ Planctomycetes*
■ Chloroflexi   ■ Bacteroidetes*
■ Verrucomicrobia   ■ Gemmatimonadetes
■ SPAM (candidate phylum)
*Significant difference in abundance

FIGURE 5.24 ■ **Effect of warming on the soil microbiome.** Two patches of Oklahoma soil were monitored for microbial taxa over ten years. The two rings illustrate the average abundances of phyla. SPAM = candidate division, microbes discovered in <u>sp</u>ring in <u>a</u>lpine and <u>me</u>adow soils.

To Summarize

■ **Starvation** is a stress that can elicit a molecular response in many microbes. Enzymes are produced to increase the efficiency of nutrient gathering and to protect cell macromolecules from damage.

■ **The starvation response** is usually triggered by the accumulation of small signaling molecules, such as cyclic AMP or guanosine tetraphosphate.

■ **Human activities can cause eutrophication**, which damages delicately balanced ecosystems by introducing nutrients that can allow one member of the ecosystem to flourish at the expense of other species.

5.7 Physical, Chemical, and Biological Control of Microbes

We have seen how microbes live; now, how do they die? A primary goal of our health care system is to control or kill microbes that can potentially harm us. Within the recent past, infectious disease was an imminent and constant threat to most of the human population. The average family in the United States prior to 1900 had four or five children, but parents could expect half of them to succumb to deadly infectious diseases. What today would be a simple infected cut, in years past held a serious risk of death, and a trip to the surgeon was tantamount to playing Russian roulette with an unsterilized scalpel. Improvements in sanitation procedures and antiseptics and the advent of antibiotics have, to a large degree, curtailed the incidence and lethal effects of many infectious diseases. Success in this endeavor has played a major role in extending life expectancy and in contributing to the population explosion.

A variety of terms are used to describe antimicrobial control measures. The terms convey subtle, yet vitally important, differences in control strategies and outcomes.

■ **Sterilization** is the process by which <u>all</u> living cells, spores, and viruses are destroyed on an object.

■ **Disinfection** is the killing, or removal, of <u>disease-producing</u> organisms from inanimate surfaces; it does not necessarily result in sterilization. Pathogens are killed, but other microbes may survive.

■ **Antisepsis** is similar to disinfection, but it applies to removing pathogens from the surface of living tissues, like the skin. Antiseptic chemicals are usually not as toxic as disinfectants, which frequently damage living tissues.

■ **Sanitation** is closely related to disinfection. It consists of reducing the microbial population to safe levels and usually involves both cleaning and disinfecting an object.

Antimicrobials can also be classified on the basis of the specific groups of microbes destroyed, leading to the terms "microbicide," "bactericide," "algicide," "fungicide," and "virucide." These agents can be classified further as either "-static" (inhibiting growth) or "-cidal" (killing cells). For example, antibacterial agents may be **bacteriostatic** or **bactericidal**. Chemical substances are **germicidal** if they kill pathogens (and many nonpathogens), but germicidal agents do not necessarily kill spores.

Although these descriptions emphasize the killing of pathogens, it is important to note that antimicrobial agents can also kill or prevent the growth of nonpathogens. Many public health standards are based on total numbers of microorganisms on an object, regardless of pathogenic potential. For example, to gain public health certification, the restaurants we frequent must demonstrate low numbers of bacteria (pathogenic or not) in their food preparation areas.

Cells Treated with Antimicrobials Die at a Logarithmic Rate

Exposing microbes to lethal chemicals or conditions does not instantly kill all microorganisms. Microbes die according to a negative exponential curve, where cell numbers are reduced in equal fractions at constant intervals. The efficacy of a given lethal agent or condition is measured as **decimal reduction time** (**D-value**), which is the length of time it takes that agent (or condition) to kill 90% of the population (a drop of one log unit, or a drop to 10% of the original value). **Figure 5.25** illustrates the exponential death profile of a bacterial culture heated to 100°C. The D-value is

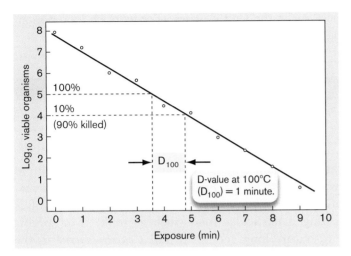

FIGURE 5.25 ■ The death curve and the determination of D-values. Bacteria were exposed to a temperature of 100°C, and survivors were measured by viable count. The D-value is the time required to kill 90% of cells (that is, the time it takes for the viable cell count to drop by one $\log_{10}$ unit).

a little over 1 minute. The food industry uses D-value and several other parameters to evaluate the efficiency of killing (see Section 16.5).

Thought Question

5.11 If a disinfectant is added to a culture containing 1×10^6 CFUs per milliliter and the D-value of the disinfectant is 2 minutes, how many viable cells are left after 4 minutes of exposure?

Several factors influence the ability of an antimicrobial agent to kill microbes. These include the initial population size (the larger the population, the longer it takes to reduce it to a specific number), the population composition (are spores involved?), the concentration of the antimicrobial agent, and the duration of exposure. Although the effect of concentration seems intuitively obvious, an increase in concentration is matched by an increase in death rate only over a narrow concentration range. Increases above a certain level might not accelerate killing at all. For example, 70% ethanol is actually better than pure ethanol at killing organisms, because some water is needed to help ethanol penetrate cells. The ethanol then dehydrates cell proteins.

Why, then, is death an exponential function? Why don't all cells in a population die instantly when treated with lethal heat or chemicals? The reason is based, in part, on the random probability that an agent will cause a lethal "hit" in a given cell. Cells contain thousands of different

proteins and thousands of molecules of each one. Not all proteins and not all genes in a chromosome are damaged by an agent at the same time. Damage accumulates. Only when enough molecules of an <u>essential</u> protein or a gene encoding that protein are damaged will the cell die. Cells that die first are those that accumulate lethal hits early. Members of the population that die later have, by random chance, absorbed more hits on nonessential proteins or genes, sparing the essential ones.

Why, if 90% of a population is killed in 1 minute, isn't the remaining 10% killed in the next minute? It seems logical that all should have perished. Yet after the second minute, 1% of the original population remains alive. This phenomenon can also be explained by the random-hit concept. Although fewer viable cells remain after 1 minute, each has the same random chance of having a lethal hit as when the treatment began. Thus, death rate is an exponential function, much like radioactive decay is an exponential function.

A final consideration is the overall fitness of individual cells. It is a mistake to assume that all cells in a population are identical. At any given time, for instance, one cell may express a protein that another cell has just stopped expressing (for example, superoxide dismutase). In that instant, the first cell might contain a bit more of that protein. If the protein is essential or confers a level of stress protection (such as against superoxide), the cell with more of that protein can absorb more punishment before it dies. The presence of lucky individuals expressing the right repertoire of proteins might also explain why death curves commonly level off after a certain point.

Physical Agents That Kill Microbes

Physical agents are often used to kill microbes or control their growth. Commonly used physical control measures include temperature extremes, pressure (usually combined with temperature), filtration, and irradiation.

High temperature and pressure. Even though microbes were discovered less than 400 years ago, thermal treatment of food products to render them safe has been practiced for over 5,000 years. Moist heat is a much more effective killer than dry heat, thanks to the ability of water to penetrate cells. Many bacteria, for instance, easily withstand 100°C dry heat but not 100°C boiling water. We humans are not so different, finding it easier to endure a temperature of 32°C (90°F) in dry Arizona than in humid Louisiana.

While boiling water (100°C) can kill most vegetative (actively growing) organisms, spores are built to withstand this abuse, and thermophiles prefer it. Killing spores and

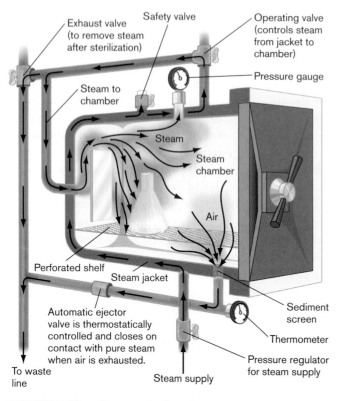

FIGURE 5.26 ■ **Steam autoclave.**

thermophiles usually requires combining high pressure and temperature. At high pressure, the boiling point of water rises to a temperature rarely experienced by microbes living at sea level. Even endospores quickly die under these conditions. This combination of pressure and temperature is the principle behind sterilization using the steam autoclave (**Fig. 5.26**). Standard conditions for steam sterilization are 121°C (250°F) at 15 psi for 20 minutes—a set of conditions that experience has taught us will kill all spores (except those of some thermophiles, which do not affect food or human health). These are also the conditions produced in pressure cookers used for home canning of vegetables.

Failure to adhere to these heat and pressure parameters can have deadly consequences, even in your own home. For instance, *Clostridium botulinum* is a spore-forming soil microbe that commonly contaminates fruits and vegetables used in home canning. The improper use of a pressure cooker while canning these goods will allow spores of this pathogen to survive. Once the can or jar is cool, the spores will germinate and begin producing their deadly toxin. All of this happens while the canned goods sit on a shelf waiting to be opened and consumed. Once ingested, the toxin makes its way to the nervous system and paralyzes the victim. Several incidents of this disease, called botulism, occur

each year in the United States. (For more on food poisoning, see Chapter 16.)

Pasteurization. Originally devised by Louis Pasteur to save products of the French wine industry from devastating bacterial spoilage, **pasteurization** today involves heating a particular food (such as milk) to a specific temperature long enough to kill *Coxiella burnetii*, the causative agent of Q fever, the most heat-resistant non-spore-forming pathogen known. In the process, pasteurization also kills other disease-causing microbes. Three U.S. government–approved time and temperature combinations can be used for pasteurization of milk. The LTLT (low-temperature/long-time) process involves bringing milk to a temperature of 63°C (145°F) for 30 minutes. In contrast, the HTST (high-temperature/short-time) method (also called "flash pasteurization") brings the milk to a temperature of 72°C (161°F) for only 15 seconds. Both processes accomplish the same thing—the destruction of *C. burnetii* and other bacteria—but they do not sterilize milk. A third process, known as ultra-high-temperature (UHT) pasteurization—134°C (273°F) for 1–2 seconds—reduces bacterial content even more than the LTLT or HTST methods. UHT pasteurization produces nearly sterile milk with an unrefrigerated shelf life of up to 6 months. This is important, especially in developing countries, where refrigeration is not always available.

Cold. Low temperatures have two basic purposes in microbiology: to temper growth and to preserve strains. Bacteria not only grow more slowly in cold, but also die more slowly. Refrigeration temperatures (4–8°C, or 39–43°F) are used for food preservation because most pathogens are mesophilic and grow poorly, if at all, at those temperatures. One exception is the Gram-positive bacillus *Listeria monocytogenes*, which can grow reasonably well in the cold and causes disease when ingested.

Long-term storage of bacteria usually requires placing solutions in glycerol at very low temperatures (–70°C). Glycerol prevents the production of razor-sharp ice crystals that can pierce cells from without or within. This deep-freezing suspends growth altogether and keeps cells from dying. Another technique, called **lyophilization**, freeze-dries microbial cultures for long-term storage. In this technique, cultures are quickly frozen at very low temperatures (quick-freezing also limits ice crystal

Syringe filter

Syringe with fluid attached here.

Sterile membrane filter

Sterile solution collected in a sterile container.

Bottle top filter

Filtration device

Unsterile media

Sterile membrane filter

Connect to vacuum.

Sterile solution collected.

FIGURE 5.27 ■ Membrane filtration devices.

formation) and placed under vacuum, where the resulting sublimation removes all water from the media and cells, leaving just the cells in the form of a powder. These freeze-dried organisms remain viable for years. Finally, viruses and mammalian cells must be kept at extremely low temperatures (–196°C), submerged in liquid nitrogen.

Liquid nitrogen freezes cells so quickly that ice crystals do not have time to form.

Filtration. Filtration through micropore filters with pore sizes of 0.2 µm can remove microbial cells, but not viruses, from solutions. Samples from 1 milliliter to several liters can be drawn through a membrane filter by vacuum or can be forced through it using a syringe (**Fig. 5.27**). Filter sterilization avoids the use of heat, which can damage certain materials. Strictly speaking, though, the solutions are not really sterile, because these filters do not trap viruses.

Air can also be sterilized by filtration. This process forms the basis of several personal protective devices. A surgical mask is a crude example, while **laminar flow biological safety cabinets** are more elaborate (and more effective). These cabinets force air through high-efficiency particulate air (HEPA) filters and remove over 99.9% of airborne particulate material 0.3 µm in size or larger. Biosafety cabinets are critical to protect individuals working with highly pathogenic material (**Fig. 5.28A** and **B**). Newer technologies have been developed that embed antimicrobial agents or enzymes directly into the fibers of the filter (**Fig. 5.28C**). Organisms entangled in these fibers are not just trapped; they are attacked by the antimicrobials and lysed.

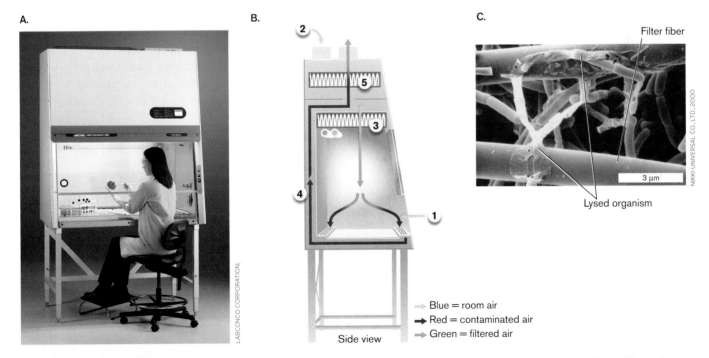

A.

LABCONCO CORPORATION;

B.

Side view

C.

Filter fiber

Lysed organism

NIKKI-UNIVERSAL CO., LTD, 2000

3 µm

→ Blue = room air
→ Red = contaminated air
→ Green = filtered air

FIGURE 5.28 ■ Biological safety cabinet. **A.** A scientist examines a sample under the hood. **B.** Schematic of the safety cabinet. Air from the room enters the cabinet through the cabinet opening (1), or is pumped in (2) through a HEPA filter (3). It then passes behind the negative-pressure exhaust plenum (4) and is passed from the cabinet through another HEPA filter (5). **C.** Immobilized enzyme filter. The primary function of this enzyme filter is to kill airborne microorganisms caught on the surface of the filter, thus protecting against secondary contamination by microorganisms in air filtration systems. The photo shows lysed bacteria (*Bacillus subtilis*). The cell walls have been hydrolyzed by enzymatic action, and cell membranes are broken as a result of osmotic pressure pushing outward against the membrane.

Irradiation. Public health authorities worldwide are increasingly concerned about food contaminated with pathogenic microorganisms such as *Salmonella* species, *E. coli* O157:H7, *Listeria monocytogenes*, and *Yersinia enterocolitica*. Irradiation, the bombardment of foods with high-energy electromagnetic radiation, has long been a potent, if politically sensitive, strategy for sterilizing food after harvesting. The food consumed by NASA astronauts, for example, has for some time been sterilized by irradiation as a safeguard against food-borne illness in space, but public pressure severely limited the mainstream use of this technology over concerns about product safety and exposure of workers. Resistance to irradiation largely disappeared, however, after 2001, when anthrax spores were mailed across the eastern United States and irradiation was used to sterilize our mail. Numerous studies have proved that foods do not become radioactive when irradiated, nor are long-lived reactive molecules produced that are dangerous to humans.

Aside from ultraviolet light, which, owing to its poor penetrating ability, is useful only for surface sterilization, there are three other sources of irradiation: gamma rays, electron beams, and X-rays. Radiation dosage is usually measured in a unit called the gray (Gy), which is the amount of energy transferred to the food, microbe, or other substance being irradiated. A single chest X-ray delivers roughly half a milligray (1 mGy = 0.001 Gy). To kill *Salmonella*, freshly slaughtered chicken can be irradiated at up to 4.5 kilograys (kGy)—about 7 million times the energy of a single chest X-ray. The Food and Drug Administration has also approved the use of irradiation (4 kGy) on beef, pork, fruits, vegetables, oysters, seeds, shell eggs, and spices.

When microbes present in food are irradiated, water and other intracellular molecules absorb the energy and form very short-lived reactive chemicals that damage DNA. Unless the organism repairs this damage, it will die while trying to replicate. Microbes differ greatly in their sensitivity to irradiation, depending on the size of their genome, the rate at which they can repair damaged DNA, and other factors. Whether the food to be irradiated is frozen or fresh also matters, as it takes a higher dose of radiation to kill microbes in frozen foods.

The size of the DNA "target" is a major factor in radiation efficacy. Parasites and insect pests, which have large amounts of DNA, are rapidly killed by extremely low doses of radiation, typically with D-values of less than 0.1 kGy (in this instance, the D-value is the <u>dose</u> of radiation needed to kill 90% of the organisms). It takes more radiation to kill bacteria (D-values in the range of 0.3–0.7 kGy) because they have less DNA per cell unit (less target per cell). It takes even more radiation to kill a bacterial spore (D-values on the order of 2.8 kGy) because they contain little water, the source of most ionizing damage to DNA. Viral pathogens have the smallest amount of nucleic acid, making them resistant to irradiation doses approved for foods (viruses have D-values of 10 kGy or higher). Infectious agents that do not contain nucleic acids are an even bigger problem. Prions, for example, are misfolded brain proteins that "self-replicate" and cause neurodegenerative diseases (see Section 26.7). Because prions do not contain nucleic acids, the agent can be inactivated by irradiation only at extremely high doses. Thus, irradiation of food is effective in eliminating parasites and bacteria, but is woefully inadequate for eliminating viruses or prions.

Note: Electromagnetic radiation emitted by microwave ovens does not directly kill bacteria. However, the heat generated when electromagnetic radiation excites water molecules in an organism will kill the organism if the temperature attained is high enough.

Some Bacteria Are Highly Resistant to Physical Control Measures

Deinococcus radiodurans could be nicknamed "Conan the Bacterium" and serve as the poster microbe for extremophiles (**Fig. 5.29**). It was discovered in 1956 in a can of meat that had spoiled despite having been sterilized by radiation. The microbe has the greatest ability to survive radiation of any known organism and could probably survive an atomic blast. The bacterium's ability to withstand radiation may have evolved as a side effect of developing

A.

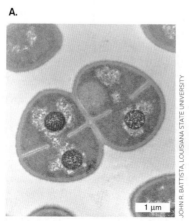

B.

FIGURE 5.29 ■ *Deinococcus radiodurans.* A. The amount of radiation *D. radiodurans* can survive is equivalent to that of an atomic blast. The nature of the dark inclusion bodies in three of the four cells in this quartet is currently not known. **B.** John Battista of Louisiana State University showed that *D. radiodurans* has exceptional capabilities for repairing radiation-damaged DNA.

resistance to extreme drought, since dehydration and radiation produce similar types of DNA damage.

What properties account for this microbe's amazing resistance to radiation? One is that *D. radiodurans* possesses an unusual capacity for repairing damaged DNA. Many mechanisms are involved, including a highly efficient double-strand-break DNA repair system involving homologous recombination (see Section 9.5). Each cell of this organism contains four to six copies of its two chromosomes and two plasmids. So even when its DNA is irradiated and broken into thousands of fragments, overlapping intact fragments can be found and spliced together in proper order. A second property that contributes to radiation resistance is the ability of *D. radiodurans* to aggressively protect its proteins, which are even more susceptible to radiation than is DNA. This mechanism, discovered by Michael Daly at the Uniformed Services University of the Health Sciences in Maryland, involves the intracellular accumulation of large amounts of manganese that can remove highly damaging free radicals generated by radiation.

On the basis of this research, *D. radiodurans* was genetically engineered to treat radioactive mercury–contaminated waste from nuclear reactors—a process called **bioremediation** (discussed in Chapter 22). The genes for mercury conversion were spliced from a strain of *E. coli* that is resistant to particularly toxic forms of mercury and inserted into *D. radiodurans*. The genetically altered superbug was able to withstand the ionizing radiation and convert toxic waste into forms that could be removed safely. Fortunately, there is little need to worry about its becoming a superpathogen, because the organism does not cause disease and is susceptible to antibiotics.

Chemical Agents

Disinfection by physical agents is very effective, but in numerous situations their use is impractical (kitchen countertops) or plainly impossible (skin). In these instances, chemical agents are the best approach. A number of factors influence the efficacy of a given chemical agent. These include:

- **The presence of organic matter.** A chemical placed on a dirty surface will bind to any inert organic material present, lowering the agent's effectiveness against microbes. It is not always possible to clean a surface prior to disinfection (as in a blood spill), but the presence of organic material must be factored into estimates of how long to disinfect a surface or object.

- **The kinds of organisms present.** Ideally, the agent should be effective against a broad range of pathogens.

- **Corrosiveness.** The disinfectant should not corrode the surface or, in the case of an antiseptic, damage skin.

- **Stability, odor, and surface tension.** The chemical should be stable during storage, possess a neutral or pleasant odor, and have a low surface tension so that it can penetrate cracks and crevices.

The phenol coefficient. Phenol, first introduced by Joseph Lister in 1867 to reduce the incidence of surgical infections, is no longer used as a disinfectant, because of its toxicity, but its derivatives, such as cresols and orthophenylphenol, are still in use. The household product Lysol is a mixture of phenolics. Phenolics are useful disinfectants because they denature proteins, are effective in the presence of organic material, and remain active on surfaces long after application.

Although phenol is no longer used as a disinfectant, its potency makes it the benchmark against which other disinfectants are measured. The **phenol coefficient test** consists of inoculating a fixed number of bacteria—for example, *Salmonella enterica* or *Staphylococcus aureus*—into dilutions of the test agent. At timed intervals, samples are withdrawn from each dilution and inoculated into fresh broth (which contains no disinfectant). The phenol coefficient is based on the highest dilution (lowest concentration) of a disinfectant that will kill all the bacteria in a test after 10 minutes of exposure but leaves survivors after only 5 minutes of exposure. This concentration is known as the maximum effective dilution. Dividing the reciprocal of the maximum effective dilution for the test agent (for example, ethanol) by the reciprocal of the maximum effective dilution for phenol gives the phenol coefficient (**Table 5.3**). For example, if the maximum effective dilution for agent X

TABLE 5.3	Phenol Coefficients for Various Disinfectants	
Chemical agent	*Staphylococcus aureus*	*Salmonella enterica*
Phenol	1.0	1.0
Chloramine	133.0	100.0
Cresols	2.3	2.3
Ethanol	6.3	6.3
Formalin	0.3	0.7
Hydrogen peroxide	—	0.001
Lysol	5.0	3.2
Mercury chloride	100.0	143.0
Tincture of iodine	6.3	5.8

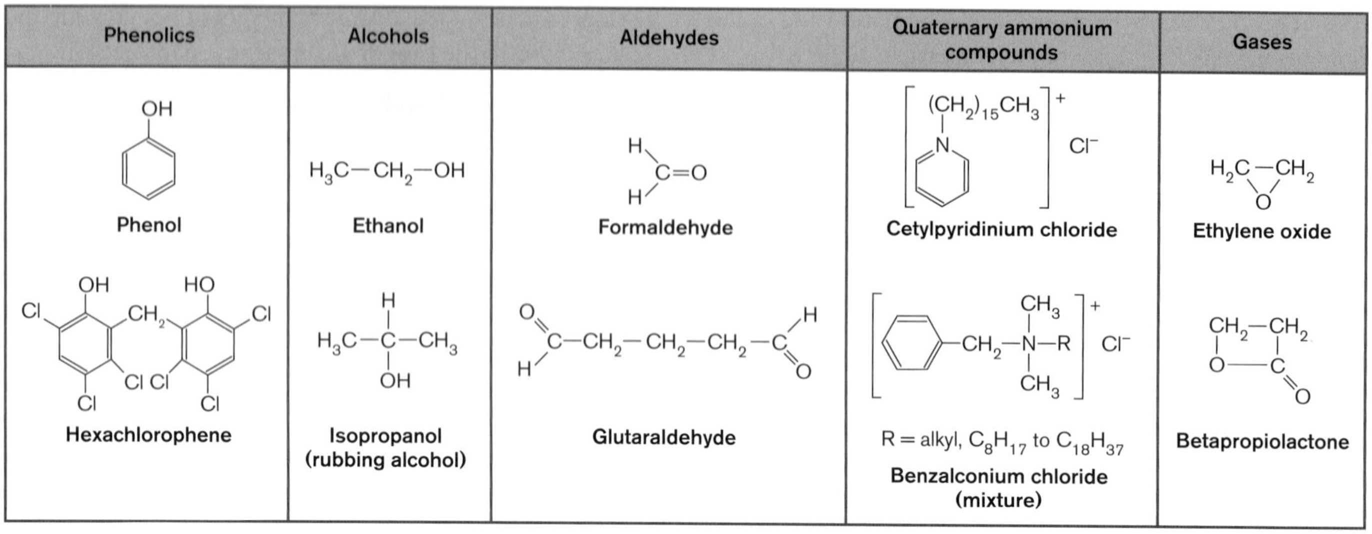

Phenolics	Alcohols	Aldehydes	Quaternary ammonium compounds	Gases
Phenol	Ethanol	Formaldehyde	Cetylpyridinium chloride	Ethylene oxide
Hexachlorophene	Isopropanol (rubbing alcohol)	Glutaraldehyde	R = alkyl, C_8H_{17} to $C_{18}H_{37}$ Benzalconium chloride (mixture)	Betapropiolactone

FIGURE 5.30 ■ Structures of some common disinfectants and antiseptics.

is 1/900 and that of phenol is 1/90, then the phenol coefficient of X is 900/90 = 10; the higher the coefficient, the higher the efficacy of the disinfectant.

Commercial Disinfectants

Ethanol, iodine, chlorine, and surfactants (for example, detergents) are all used to reduce or eliminate microbial content from commercial products (**Fig. 5.30**). The first three are compounds that damage proteins, lipids, and DNA. Highly reactive iodine complexed with an organic carrier forms an iodophor, a compound that is water-soluble, stable, nonstaining, and capable of releasing iodine slowly to avoid skin irritation. Wescodyne and Betadine (trade names) are iodophors used, respectively, for the surgical preparation of skin and for wounds. Chlorine is another highly reactive disinfectant with universal application. It is recommended for general laboratory and hospital disinfection and kills the HIV virus.

Detergents can also be antimicrobial agents. The hydrophobic and hydrophilic ends of detergent molecules (the coexistence of which makes the molecules amphipathic) will emulsify fat into water. Cationic (positively charged) but not anionic (negatively charged) detergents are useful as disinfectants because the positive charges can gain access to the negatively charged bacterial cell and disrupt membranes. Anionic detergents are not antimicrobial but do help in the mechanical removal of bacteria from surfaces.

Low-molecular-weight aldehydes such as formaldehyde (HCHO) are highly reactive, combining with and inactivating proteins and nucleic acids. This characteristic makes them useful disinfectants.

Disposable plasticware like petri dishes, syringes, sutures, and catheters are not amenable to heat sterilization or liquid disinfection. These materials are best sterilized using antimicrobial gases. Ethylene oxide gas (EtO) is a very effective sterilizing agent; it destroys cell proteins, is microbicidal and sporicidal, and rapidly penetrates packing materials, including plastic wraps. Using an instrument resembling an autoclave, EtO at 700 milligrams per liter (mg/l) will sterilize an object after 8 hours at 38°C or 4 hours at 54°C if the relative humidity is kept at 50%. Unfortunately, EtO is explosive. A less hazardous gas sterilant is betapropiolactone. It does not penetrate as well as EtO, but it decomposes after a few hours, which makes it easier to dispose of than EtO.

A new procedure, known as gas discharge plasma sterilization, may replace EtO because it is less harmful to operators. Gas discharge plasma is made by passing certain gases through a radio frequency electrical field to produce highly reactive chemical species that can damage membranes, DNA, and protein. It is not yet widely used.

Antimicrobial touch surfaces. Despite widespread use of disinfectants and antibiotics by medical personnel, hospital-acquired infections remain a major concern. A promising antimicrobial technology that can help reduce infections in hospitals and elsewhere involves embedding antimicrobial compounds such as copper (Cu) in the surfaces that people touch. For example, a bed rail made with copper can kill pathogens deposited by one person before the organism can be transmitted to a second person touching the same surface. Upon contact with bacteria, metallic copper releases toxic Cu^+ ions that trigger the lysis of bacterial membranes within minutes, although the mechanism involved is unclear (neither DNA damage nor reactive oxygen species are involved). Companies are incorporating metallic copper into objects such as handrails, door releases, and hospital bed rails.

Bacteria Can Develop Resistance to Disinfectants

It is widely known that bacteria can develop resistance to antibiotics used to treat infections. This is a serious concern in the medical community. So, one might wonder whether bacteria can also develop resistance against chemical disinfectants used to <u>prevent</u> infections. The answer is yes—and no. It is difficult for a bacterium to develop resistance to chemical agents that have multiple targets and can easily diffuse into a cell. Iodine, for example, has both of these characteristics. However, disinfectants that have multiple targets at <u>high</u> concentrations may have only a single target at lower concentrations—a situation that can foster the development of resistance. For instance, triclosan (a halogenated bisphenol compound used in many soaps, deodorants, and toothpastes) targets several cell constituents, making it nicely bactericidal at high concentrations. However, at low concentrations triclosan only inhibits fatty acid synthesis and is merely bacteriostatic. Organisms have developed resistance to triclosan at low concentrations by altering the fatty acid synthesis protein normally targeted by triclosan.

Low-level resistance also can be achieved through membrane-spanning, multidrug efflux pumps (described in Section 4.2). For instance, the MexCD-OprJ efflux system of *Pseudomonas aeruginosa*, a Gram-negative bacterium that causes infections in burn and cystic fibrosis patients, can pump several different biocides, detergents, and organic solvents out of the cell, thereby reducing their efficacy. This finding and other reports of *Pseudomonas* gaining resistance to disinfectants have led many clinicians to advocate caution in the widespread use of certain chemical disinfectants.

Biofilm formation is another ingenious way that bacteria survive exposures to disinfectants. A biofilm is a 3D community of bacterial cells attached to a solid surface (see Section 4.5). A biofilm can protect cells in several ways. For example, the extracellular matrix proteins and polysaccharides that hold biofilms together also bind disinfectants, slowing their penetration into the deeper recesses of the structure. Slower penetration means that cells deep in the biofilm have time to activate protective stress response systems before destructive levels of disinfectant reach them. Biofilms also exhibit stratified growth patterns based in part on nutrient access (see Fig. 4.30). Cells at the periphery have ample access to oxygen and nutrients, while cells within do not. Nutrient starvation worsens the farther a

A. BRIDIER ET AL. 2011. *BIOFOULING* **27**:1017–1032

FIGURE 5.31 ■ Mixed biofilm. 3D projection of a mixed 24-hour biofilm of *E. coli* expressing mCherry fluorescent protein (red) and *Pseudomonas aeruginosa* expressing green fluorescent protein (GFP; green). Mixed-population biofilms can have properties distinct from monospecies biofilms.

cell is from the surface. Because nutrient starvation activates stress response systems, each biofilm has stratified layers of increasingly stress-resistant cells that can better tolerate chemical insults.

Finally, biofilms, which contain multiple species in nature, are opportunities for protective, interspecies collaborations (**Fig. 5.31**). Protective enzymes from one species could protect a nonproducing species from a chemical insult, much as a big brother protects a little brother from a bully. Multispecies biofilms can also be more massive than monospecies biofilms. The food pathogen *Escherichia coli* O157:H7 forms a biofilm with 400 times more volume when grown with *Acinetobacter calcoaceticus*, an organism found in meatpacking plants. Increased volume alone will slow the penetration of the disinfectant and protect the collective.

Antibiotics Selectively Control Bacterial Growth

Antibiotics as made in nature are chemical compounds synthesized by one microbe to either kill or stop the growth of other competing microbial species. Antimicrobial compounds that kill other cells are called bactericidal, whereas compounds that merely stop growth are called bacteriostatic. Chapter 27 describes the various antibiotic modes of action. When purified and administered to patients suffering from an infectious disease, antibiotics can produce seemingly miraculous recoveries.

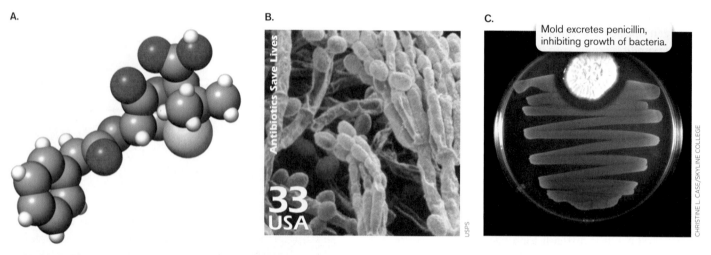

FIGURE 5.32 ■ **Penicillin.** **A.** Space-filling model of the penicillin G molecule produced by the *Penicillium* mold. **B.** U.S. postage stamp from 1999 showing *Penicillium notatum*. **C.** *P. notatum* excretes penicillin, which inhibits the growth of *Staphylococcus aureus*.

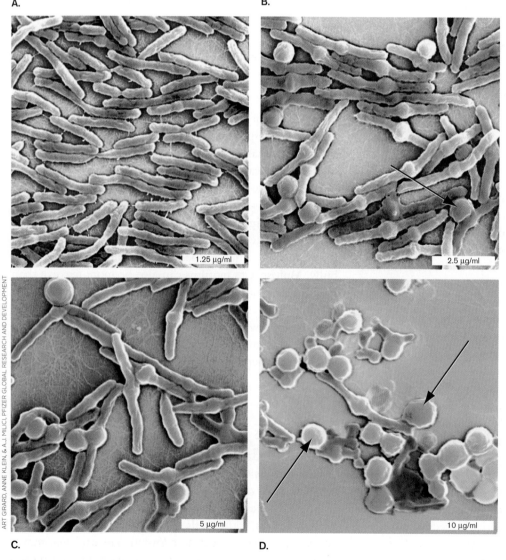

FIGURE 5.33 ■ **Effect of ampicillin (a penicillin derivative) on *E. coli*.** Cells were incubated for 1 hour at the antibiotic concentrations shown. Swollen areas of cells in panels (B)–(D) reflect weakening cell walls (arrows). Cells shown are approx. 2 μm long.

As we saw in Chapter 1, penicillin (**Fig. 5.32**), produced by *Penicillium notatum*, was discovered serendipitously in 1929 by Alexander Fleming. **Figure 5.32A** shows this molecule, which mimics a part of the cell wall. Because of this mimicry, penicillin binds to biosynthetic proteins involved in peptidoglycan synthesis and prevents cell wall formation. The drug is bactericidal because <u>actively growing</u> cells lyse without the support of the cell wall (**Fig. 5.33**). Other antibiotics target protein synthesis, DNA replication, cell membranes, and various enzyme reactions. These interactions are described throughout Parts 1–3 of this text.

So how do antibiotic-producing microbes avoid suicide? In some instances, the producing organism lacks the target molecule. *Penicillium* mold, for instance, lacks peptidoglycan and is immune to penicillin by default. Some bacteria produce antimicrobial compounds that target other members of the same species. In this case, the producing organism can modify its own receptors to no longer recognize the compound (as with some bacterial colicins).

Another strategy is to modify the antibiotic if it reenters the cell. This is the case with streptomycin produced by *Streptomyces griseus*. Streptomycin inhibits bacterial protein synthesis and does not discriminate between the protein synthesis machinery of *Streptomyces* and that of others. However, while making enzymes that synthesize and secrete streptomycin, *S. griseus* simultaneously makes the enzyme streptomycin 6-kinase, which remains locked in the cell. If any secreted streptomycin reenters the cell, this enzyme renders the drug inactive by attaching a phosphate group to it.

Because many microorganisms have become resistant to commonly used antibiotics, pharmaceutical companies continually search for new antibiotics using a variety of approaches (see Special Topic 4.1). Traditional procedures include scouring soil and ocean samples collected from all over the world for new antibiotic-producing organisms and chemically redesigning existing antibiotics so that they can bypass microbial resistance strategies. Newer techniques allow scientists to "mine the genomes" of microbes for potential drug targets and use computer-based methods to predict the structure and function of potential new antibiotics.

Biological Control of Micro bes

Pitting microbe against microbe is an effective way to prevent disease in humans and animals. One of the hallmarks of a healthy ecosystem is the presence of a diversity of organisms. This is true not only for tropical rain forests and coral reefs, but also for the complex ecosystems of human skin and the intestinal tract. In these environments, the presence of harmless microbial communities can retard the growth of undesired pathogens. The pathogenic fungus *Phytophthora cinnamomi*, for example, causes root rot in plants but is biologically controlled by *Myrothecium* fungi. *Staphylococcus* species normally present on human skin produce short-chain fatty acids that retard the growth of pathogenic strains. Another illustration is the human intestine, which is populated by as many as 500 microbial species. Most of these species are nonpathogenic organisms that exist in symbiosis with their human host. Vigorous competition with members of the normal intestinal microbiota and the production of permeant weak acids by fermentation helps control the growth of numerous pathogens. A prominent example of this phenomenon is the pathogen *Clostridium difficile*, whose growth is normally kept in check by gut microbiota.

Microbial competition has been widely exploited for agricultural purposes and to improve human health

through the intake of **probiotics**. In general, a probiotic is a food or supplement that contains live microorganisms and improves intestinal microbial balance. Newborn baby chicks, for instance, are fed a microbial cocktail of normal gut microbes designed to quickly colonize the intestinal tract and prevent colonization by *Salmonella*, a frequent contaminant of factory-farmed chicken. In another example, *Lactobacillus* and *Bifidobacterium* have been used to prevent and treat diarrhea in children.

Russian biologist Ilya Mechnikov, winner of the 1908 Nobel Prize in Physiology or Medicine, was the first to suggest that a high concentration of lactobacilli in the gut microbiome is important for health and longevity in humans. Yogurt is a probiotic that contains *Lactobacillus acidophilus* and a number of other lactobacilli. It is often recommended as a way to restore a normal balance to gut microbiota (for example, after the microbiota has been disturbed by antibiotic treatment), and it appears useful in the treatment of some forms of inflammatory bowel disease.

Phage therapy, another **biocontrol** method, was first described in 1907 by Félix d'Herelle at France's Pasteur Institute, long before antibiotics were discovered. Bacteriophages are viruses that prey on bacteria (discussed in Section 6.1). Each bacterial species is susceptible to a limited number of specific phages. Because a phage infection often causes bacterial lysis, it was considered feasible to treat infectious diseases with a phage targeted to the pathogen. At one time, doctors used phages as medical treatment for illnesses ranging from cholera to typhoid fever. In some cases, a liquid containing the phage was poured into an open wound. In other cases, phages were given orally, introduced via aerosol, or injected. Sometimes the treatments worked; sometimes they did not. When antibiotics came into the mainstream, phage therapy largely faded. Now that strains of bacteria that are resistant to standard antibiotics are on the rise, the idea of phage therapy has enjoyed renewed interest from the worldwide medical community (**Special Topic 5.1**).

Commercial phage products are now available to target the food-borne pathogens *E. coli* O157:H7, *Salmonella enterica*, and *Listeria monocytogenes*. *E. coli* O157:H7, for instance, is a pathogen that can contaminate hamburger and cause bloody diarrhea. The phage product contains several different phages and is sprayed onto the hides of cattle 1–4 hours before slaughter. Cattle carcasses are steam-pasteurized and acid-washed to diminish bacterial contamination. *E. coli* cells that manage to survive these treatments will be infected and killed by the phage, reducing the consumer's risk of disease. Using multiple phages in each preparation nearly eliminates the risk that phage resistance will develop in the pathogen.

SPECIAL TOPIC 5.1   Phage "Smart Bombs" Target Biofilms

Biofilms play an important role in many infectious disease processes. Examples include *Pseudomonas aeruginosa* lung infections that plague cystic fibrosis patients and life-threatening heart valve infections caused by *Enterococcus faecalis*. It is difficult to cure these types of infections, because many of the pathogens responsible are resistant to antibiotics and because biofilm architecture may hinder antibiotic and immune cell access to bacteria nestled in the deep recesses of the structure. What's more, biofilms provide a reservoir of bacteria for chronic infections throughout the body. With antibiotics failing, is there an alternative? A recent study by Ronen Hazan, Leron Khalifa, and Nurit Beyth from Hebrew University (**Fig. 1**) suggests that phage therapy can be used to destroy biofilm infections without antibiotics.

FIGURE 1 ■ **Nurit Beyth, Leron Khalifa, and Ronen Hazan.**

The model they used was tooth infection following endodontic dental procedures. Endodontic procedures remove the nerve of a tooth because the pulp around it is infected. Drilling to remove infected pulp and the inflamed nerve will inadvertently allow entry of oral microbiota, such as *Enterococcus faecalis*, into dentinal tubules. Dentinal tubules are thin, branching tubes that extend radially from the pulp (the center of the tooth), through the dentine and then stop at the enamel. Once it is within the tubules and the canal is sealed, *E. faecalis* can form a biofilm infection of the dentine.

The authors used an *Enterococcus*-specific bacteriophage called EFDG1 (**Fig. 2**) to first determine whether the phage could destroy a 2-week-old biofilm of *E. faecalis* grown in a plastic dish. The confocal fluorescent images in **Figure 3** show a biofilm without phage treatment and a similar biofilm 7 days after phage treatment. The results suggest that phages are promising candidates for killing well-established *E. faecalis* biofilms.

The scientists next tested whether phage treatment would prevent root canal infections. Extracted teeth with exposed nerve canals were autoclaved and contaminated with a suspension of *E. faecalis*. Root canal procedures were performed on the teeth, and the canals were sealed. After 2 days, cross sections were made and stained with live/

dead stain to show bacteria in a tooth prepared without any contamination (**Fig. 4A**), in a tooth prepared with *E. faecalis* contamination (**Fig. 4B**), and in a contaminated tooth simultaneously treated with phage (**Fig. 4C**). Live bacteria (green) are seen within the dentin tubules in **Figure 4B** but not in

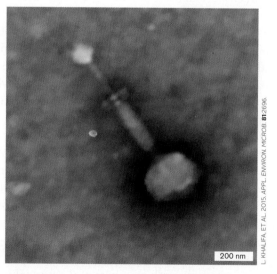

FIGURE 2 ■ *Enterococcus* **phage EFDG1.**

the tubules treated with phage (**Fig. 4C**). Note that the root canal itself stained nonspecifically, even in the absence of bacteria (**Fig. 4A**). The results show that EFDG1 is an efficient killer of *E. faecalis* biofilms in this ex vivo model of root canal infection and may be a complementary strategy to antibiotic treatment, especially for antibiotic resistant microbes.

RESEARCH QUESTION

How would you test whether *Enterococcus faecalis* can develop resistance to the EFDG1 phage? List other potential drawbacks to using phage as a treatment for human infections.

Leron, Khalifa, Yair Brosh, Daniel Gelman, Shunit Coppenhagen-Glazer, Shaul Beyth, et al. 2015. Targeting *Enterococcus faecalis* biofilms with phage therapy. *Applied and Environmental Microbiology* 81:2696–2705.

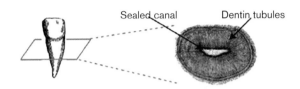

A. No bacteria

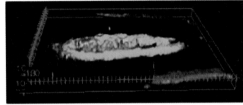

B. *E. faecalis*

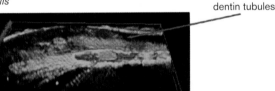

C. *E. faecalis* + EFDG1

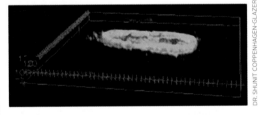

FIGURE 4 ■ Horizontal root sections of teeth subjected to endodontic treatment. Confocal fluorescence microscopy. **A.** Tooth without bacteria or phage. **B.** Tooth was irrigated with bacteria only. **C.** Tooth was irrigated with bacteria and phage EFDG1. Stained bacteria (green from live/dead viability stain) are depicted in the dentinal tubules surrounding the root canal (red arrow). Note that the root canal is stained nonspecifically, even in the absence of bacteria.

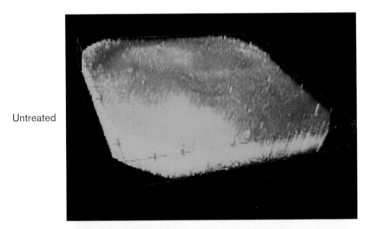

Untreated

+ EFDG1

FIGURE 3 ■ Two-week-old biofilm of *Enterococcus faecalis* was either untreated (top) or treated with phage EFDG1 for 7 days (bottom). Confocal fluorescence microscopy.

L. KHALIFA, ET AL. 2015. *APPL. ENVIRON. MICROB.* **81**:2696.

DR. SHUNIT COPPENHAGEN-GLAZER

To Summarize

- **Physical and chemical agents** kill microbes by denaturing proteins or DNA, or by disrupting lipid bilayers. **Biocontrol** is the use of one microbe to control the growth of another.

- **Antisepsis** is the removal of potential pathogens from the surfaces of living tissues, while **disinfection** kills pathogens on inanimate objects. **Sterilization** kills all living organisms.

- **Antimicrobial** compounds can be **bacteriostatic** or **bactericidal**.

- **The D-value** is the time (or dose, for irradiation) needed to reduce viable cells to 10% of the original value.

- **An autoclave** uses high pressure to achieve temperatures that will sterilize objects.

- **Food can be preserved** by pasteurization, refrigeration, filtration, and irradiation.

- **The phenol coefficient** is used to compare one disinfectant to another.

- **Antibiotics** are compounds produced by one living microorganism that kill other microorganisms.

- **Probiotics** contain certain microbes that, when ingested, aim to restore balance to the intestinal microbiome.

- **Phage therapy** offers a possible alternative to antibiotics in the face of rising antibiotic resistance.

Concluding Thoughts

Microbiology as a science was founded on the need to understand and control microbial growth. The initial impetus was to control the diseases of humans, as well as the diseases of plants and animals. But, as we will see in later chapters, microbiology has developed into a science that has helped us understand the molecular processes of life. Concepts such as biological diversity, food microbiology, microbial disease, and antibiotics will be revisited in later chapters.

CHAPTER REVIEW

Review Questions

1. Explain the nature of extremophiles and discuss why these organisms are important.
2. What parameters define any growth environment?
3. List and define the classifications used to describe microbes that grow in different physical growth conditions.
4. What do thermophiles have to do with PCR technology?
5. Why is water activity important to microbial growth? What changes water activity?
6. How do cells protect themselves from osmotic stress?
7. Why do changes in H^+ concentration affect cell growth?
8. How do acidophiles and alkaliphiles manage to grow at the extremes of pH?
9. If an organism can live in an oxygenated environment, does that mean the organism uses oxygen to grow? If an organism can live in an anaerobic environment, does that mean it cannot use oxygen as an electron acceptor? Why or why not?
10. What happens when a cell exhausts its available nutrients?
11. List and briefly explain the various means by which humans control microbial growth. What is a D-value? What is a phenol coefficient?
12. How do microbes prevent the growth of other microbes?

Thought Questions

1. Given a natural lake environment with 100 species of bacteria, why does the species with the fastest generation time not overwhelm the others? Or does it?

2. *Escherichia coli* is a facultative species, able to grow with or without oxygen. What would it take to make this organism an anaerobe?

3. Two spore formers, *Bacillus stearothermophilus* and *Bacillus coagulans*, have D-values at 121°C of 5 minutes and 0.07 minute, respectively. How could the spores from these organisms have such different D-values? *Hint:* Find the optimum growth temperatures for these organisms.

4. Phage therapy is touted by some as a solution to antibiotic resistance. Explain why phage therapy may not be able to solve this problem.

5. With respect to bacteria, is the physiological effect of HCl at pH 4 the same as that of an organic acid at pH 4?

Key Terms

acidophile (168)
aerobic respiration (171)
aerotolerant anaerobe (173)
alkaliphile (168)
anaerobic respiration (173)
antibiotic (185)
antisepsis (178)
bactericidal (178)
bacteriostatic (178)
barophile (164)
biocontrol (187)
bioremediation (183)
decimal reduction time (D-value) (178)
disinfection (178)
electron transport system (ETS) (171)

eutrophication (177)
extremophile (158)
facultative (173)
facultative anaerobe (173)
fermentation (fermentative metabolism) (173)
germicidal (178)
halophile (166)
heat-shock response (163)
hyperthermophile (162)
laminar flow biological safety cabinet (181)
lyophilization (180)
mesophile (160)
microaerophilic (173)

neutralophile (168)
osmolarity (165)
pasteurization (180)
phenol coefficient test (183)
piezophile (164)
probiotic (187)
programmed cell death (175)
psychrophile (161)
sanitation (178)
sterilization (178)
strict aerobe (172)
strict anaerobe (172)
thermophile (162)
water activity (164)

Recommended Reading

Atomi, Haruyuki. 2005. Recent progress towards the application of hyperthermophiles and their enzymes. *Current Opinion in Chemical Biology* 9:163–173.

Blasius, Melanie, Ulrich Hubscher, and Suzanne Sommer. 2008. *Deinococcus radiodurans*: What belongs to the survival kit? *Critical Reviews in Biochemistry and Molecular Biology* 43:221–238.

Daly, Michael, Elena Gaidamakova, Vera Matrosova, Alexander Vasilenko, Min Zhai, et al. 2007. Protein oxidation implicated as the primary determinant of bacterial radioresistance. *PLoS Biology* 5:769–779.

D'Amico, Salvino, Tony Collins, Jean-Claude Marx, Georges Feller, and Charles Gerday. 2006. Psychrophilic microorganisms: Challenges for life. *EMBO Reports* 7:385–389.

Edgar, Rotem, Nir Friedman, Shahar Molshanski-Mor, and Udi Qimron. 2012. Reversing bacterial resistance to antibiotics by phage-mediated delivery of dominant sensitive genes. *Applied and Environmental Microbiology* 78:744–751.

Hoehler, Tori M., and Bo Barker Jørgensen. 2013. Microbial life under extreme energy limitation. *Nature Reviews. Microbiology* 11:83–94.

Kota, Swathi, and Hari S. Misra. 2008. Identification of a DNA processing complex from *Deino-coccus radiodurans. Biochemistry and Cell Biology* 86:448–458.

Krulwich, Terry A., George Sachs, and Etana Padan. 2011. Molecular aspects of bacterial pH sensing and homeostasis. *Nature Reviews. Microbiology* 9:330–343.

Luo, Chengwei, Luis M. Rodriguez-R, Eric R. Johnston, Liyou Wu, Lei Cheng, et al. 2014. Soil microbial community responses to a decade of warming as revealed by comparative metagenomics. *Applied and Environmental Microbiology* 80:1777–1786.

Marietou, Angeliki, Alice T. Nguyen, Eric E. Allen, and Douglas H. Bartlett. 2015. Adaptive laboratory evolution of *Escherichia coli* K-12 MG1655 for growth at high hydrostatic pressure. *Frontiers in Microbiology* 5:749.

Nobrega, Franklin L., Ana Rita Costa, Leon D. Kluskens, and Joana Azeredo. 2015. Revisiting phage therapy: New applications for old resources. *Trends in Microbiology* 23:185–191.

Slonczewski, Joan, James A. Coker, and Shiladitya DasSarma. 2010. Microbial growth with multiple stressors. *Microbe* 5:110–116.

van Gestel, Jordi, Hera Vlamakis, and Roberto Kolter. 2015. From cell differentiation to cell collectives: *Bacillus subtilis* uses division of labor to migrate. *PLoS Biology* 13(4): e1002141.

CHAPTER 6

Viruses

All kinds of cells, including bacteria, eukaryotes, and archaea, can be infected by viruses. Viruses infect a host cell and use the cell's machinery to form progeny virions. In ecosystems, viruses cycle nutrients, control host populations, and promote host diversity. Viruses may kill their host cells, or they may copy themselves into their host genome. In humans, endogenous viral DNA evolved into many portions of our genome. And now, researchers engineer viruses to attack tumors, and to conduct gene therapy.

CURRENT RESEARCH highlight

Chlorovirus infects chlorella algae. This giant marine virus has genes for metabolic enzymes and tRNA—surprising for a virus. In Michael Rossmann's laboratory, Xinzheng Zhang determined the chlorovirus 3D structure using cryo-electron microscopy. Cryo-EM reveals the inner coil of DNA surrounded by the protein capsid and envelope. The capsid spike contacts the algal surface and then opens, allowing viral DNA to penetrate the host cell. Within the cell, the viral DNA forms virus factories that make about a thousand progeny virions.

Source: Adapted from Zhang et al. 2011. *PNAS* **108**:14837.

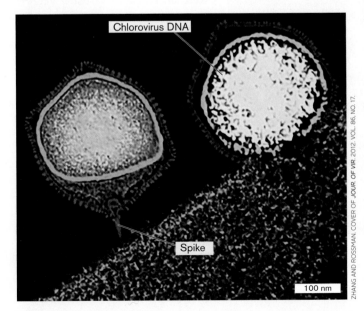

AN INTERVIEW WITH

XINZHENG ZHANG, VIROLOGIST AND ELECTRON MICROSCOPIST, INSTITUTE OF BIOPHYSICS, CHINESE ACADEMY OF SCIENCES

What exciting discovery has your microscopy revealed about this virus of algae?

Using cryo-electron microscopy, we were able to capture several essential steps of how the chlorovirus enters the host cells. The infection starts when the spike on the unique vertex of the virus recognizes the cellular receptors. The virus then digests the cell wall by its encoded enzyme. Meanwhile, the virus undergoes uncoating around the unique vertex. Without the cell wall and the virus capsid shell, the viral membrane can fuse with the cellular membrane to release the viral genome into the host cell. It is amazing that the virus can accomplish such a comprehensive job.

What is the importance of algal viruses in the natural world?

Phytoplankton (marine algae) play an important role in Earth's carbon cycle. In natural environments, viruses such as algal viruses help control the population of phytoplankton. Algal viruses are important in ecosystems for controlling algal blooms.

Viruses are everywhere. In the oceans, viruses are the most numerous and genetically diverse forms of life. The number of viruses infecting bacteria and algae can reach 10^7 (10 million) per milliliter. Viruses act as a dominant consumer of marine microbes. **Figure 6.1** shows an example of marine viruses infecting the algae *Emiliana huxleyi* (**Fig. 6.1A**) and *Pavlova* (**Fig. 6.1B**). When marine algae overgrow, they can generate a bloom that covers thousands of square kilometers (**Fig. 6.1C**). The pale clouds in the water are the reflected light from billions of calcite plates, or "coccoliths," that coat each algal cell. Yet within a few days, this gigantic bloom is dissipated by viruses.

Throughout the ocean, viruses play a decisive role in controlling algal blooms. Consumer organisms apparently cannot grow fast enough to control such blooms, but viruses spread rapidly through the population. By lysing the algae as they grow, viruses return algal carbon and minerals to the surface water before the algae starve to death and their bodies sink. When biomass sinks, its minerals become unavailable for phototrophs and their consumers. Thus, viruses are major players in the cycling of atmospheric CO_2.

All cellular organisms are infected by viruses. In humans, most of our own viruses go unnoticed, and some actually contribute to our health. But the viruses we hear about are those that cause epidemics, such as seasonal influenza and the AIDS pandemic. In research, viruses provided both tools and model systems for our discovery of the fundamental principles of molecular biology. The first genes mapped, the first regulatory switches defined, and the first genomes sequenced were all those of viruses. Vectors for gene cloning and gene therapy continue to be derived from viruses (see Chapter 12). Our understanding of viruses, particularly bacterial viruses, called bacteriophages, provides a background for the molecular biology we will encounter in Part 2 of this book (Chapters 7–12). And remarkably, we now engineer human viruses to deliver gene therapy and kill cancers (discussed in Chapter 11).

In Chapter 6 we introduce the major themes of virus structure and function, and the ways that viruses manipulate host cells for

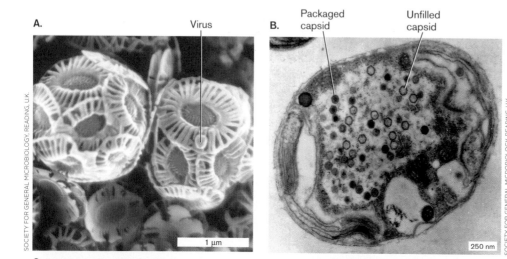

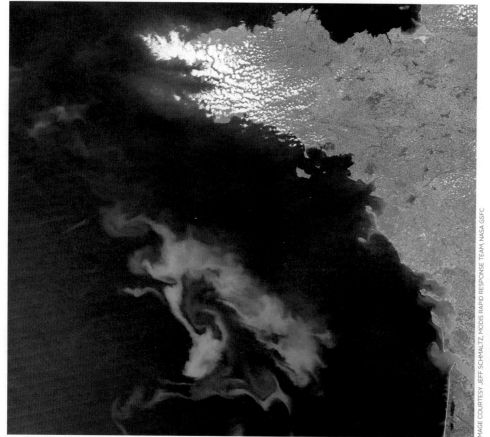

FIGURE 6.1 ■ Viruses control an algal bloom. A. A virus attaches to the surface of a marine alga, *Emiliana huxleyi* (SEM). **B.** Progeny virions assemble within an alga (*Pavlova* sp.). **C.** Bloom of *E. huxleyi* off Plymouth, England, detected by satellite remote sensing. Marine algal blooms are controlled by viruses.

their own reproduction. The molecular biology of viral infection and replication is explored further in Chapter 11. Viral disease pathology and epidemiology are discussed in Chapters 25–27.

6.1 Viruses in Ecosystems

What is a **virus**? A virus may be defined as a noncellular particle, or **virion**, that infects a host cell, where it reproduces (**Fig. 6.2**). The virion includes a genome composed of DNA or RNA, enclosed by a **capsid** of protein. In more

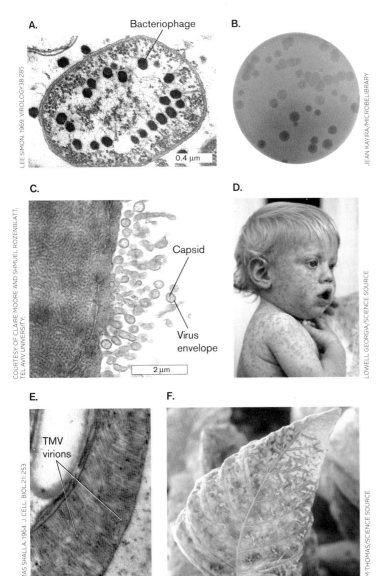

FIGURE 6.2 ■ **Virus infections and disease. A.** Bacteriophage T2 particles pack in a regular array within an *E. coli* cell (TEM). **B.** Bacteriophage infection forms plaques of lysed cells on a lawn of bacteria. **C.** Measles virions bud out of human cells in tissue culture (TEM). **D.** Child infected with measles shows a rash of red spots. **E.** Tobacco leaf section is packed with tobacco mosaic virus particles. **F.** Tomato leaf infected by tobacco mosaic virus shows mottled appearance.

complex viruses, the protective protein may be called a head coat or a core particle (discussed in Sections 6.3–6.5.)

Viruses Infect Cells

Different types of viruses infect different specific host cells. A virus that infects bacteria is called a **bacteriophage** (**Fig. 6.2A**). Bacteriophage T2 infects the host bacterium, *E. coli*. The T2 and T4 phages are "tailed" phages, whose capsid has a tail that inserts the viral genome into the host cell. Within the host, the phage genes direct production and assembly of progeny virions. Virions are released when the host cell lyses. As cells lyse, their disappearance can be observed as a **plaque**, a clear spot within a lawn of bacterial cells (**Fig. 6.2B**). Each plaque arises from a single virion, or phage particle, that lyses a host cell and spreads progeny to infect adjacent cells. The plaque count represents the number of individual infective virions from the phage suspension that was spread on the plate.

A virus that infects humans is the measles virus (**Fig. 6.2C**). The measles virus has an envelope that is derived from the host cell plasma membrane as the virus exits the host cell. As the virus infects a new host cell, the envelope fuses with the host cell plasma membrane, releasing the viral contents into the cytoplasm of the cell. After replicating within the infected cell, newly formed measles virions become enveloped by host cell membrane as they bud out of the host cell. The spreading virus generates a rash of red spots on the skin of infected patients (**Fig. 6.2D**) and can be fatal (one in 500 cases).

Plants are infected by viruses such as tobacco mosaic virus (TMV). Within the plant cell, virions accumulate to high numbers (**Fig. 6.2E**) and travel through interconnections to neighboring cells. Infection by tobacco mosaic virus results in mottled leaves and stunted growth (**Fig. 6.2F**). Plant viruses cause major economic losses in agriculture worldwide.

Historically, viruses were defined as nonliving particles, because at first the virion was the only form to be visualized and understood. The Russian botanist Dmitri Ivanovsky (1864–1920) and the Dutch microbiologist Martinus Beijerinck (1851–1931) first proposed the existence of viruses as infectious agents that passed through a filter too small for cells to pass. Certain kinds of these infectious agents were actually crystallized from solution. Wendell Stanley (1904–1971) earned the 1946 Nobel Prize in Chemistry for the first crystallization of a virus, tobacco mosaic virus (TMV). He later crystallized poliovirus. The crystallization of viruses reinforced their definition as nonliving.

But what happens after a virus infects a cell? The first replication cycles discovered were those of viruses that infect bacteria, known as **bacteriophages** or **phages**. Bacteriophage infections were first studied by English

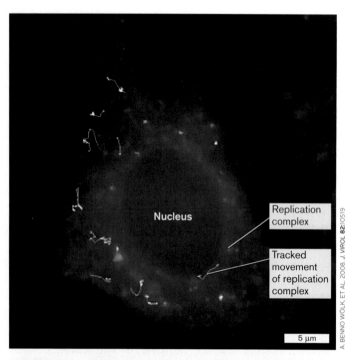

A. BENNO WOLK, ET AL. 2008. J. VIROL 82:10519

FIGURE 6.3 ■ Replication complexes of hepatitis C virus. Membranous replication complexes surround the nucleus of an infected liver cell in tissue culture. The virus expresses replication proteins fused to a fluorescent protein (GFP), shown pink.

bacteriologist Frederick William Twort (1877–1950) and by French microbiologist Félix d'Herelle (1873–1949). To actually reveal what went on in the host cell, Ernst Ruska (1906–1988) used the newly invented electron microscope (discussed in Chapter 2).

The electron micrographs in **Figure 6.2** show startling views of progeny virions infecting cells. However, they fail to show the dynamic nature of virus replication. A remarkable view of viral replication emerges from fluorescence microscopy (**Fig. 6.3**). For an RNA virus such as poliovirus or hepatitis C virus, virions are assembled within "virus factories," virus-induced cell compartments called a **replication complex**. The micrograph in **Figure 6.3** shows a cell from rat liver tissue culture infected by hepatitis C virus. The liver cell nucleus is surrounded by replication complexes composed of viral proteins and host-derived membranes (pink-orange fluorescence). Within each replication complex, fluorescence arises from a viral gene fused to a gene encoding a fluorescent protein (GFP). The replication complexes fluoresce against the dark background. The complexes move around within the cell, as tracked by video recording. It can be argued that the viral replication complexes themselves behave like intracellular parasites.

Integrated Viral Genomes

Some viruses do more than replicate within a cell: they integrate their own genomes into the host genome. In effect, such viruses become a part of the host organism. A virus that integrates its genome into the DNA of a bacterial genome is called a **prophage**. Within a human cell, an integrated viral genome is called a **provirus**. A permanently integrated provirus transmitted via the germ line is called an **endogenous virus**.

How were integrated viral genomes discovered? In the 1950s, genetic analysis of bacteriophage lambda in *E. coli* showed evidence of phage gene expression from within host genomes. **Bacteriophage lambda is discussed at length in Chapter 11.** Early in the twentieth century, American cancer researcher Peyton Rous (1879–1970) discovered that some avian viruses cause tumors. The tumor generation results from integration of a viral genome. For this discovery, Rous won the 1966 Nobel Prize in Physiology or Medicine.

But the biggest surprise emerged from decades of study of the genomes of host organisms. Many host cell traits turn out to be expressed by integrated viral genomes, or prophages, that actually confer benefits on their host. For example, *Nostoc* cyanobacteria possess a prophage that encodes nitrogenase, an enzyme for nitrogen fixation. Many human pathogens, such as *Staphylococcus aureus*, contain prophages encoding toxins. Human genomes show evidence of much genetic material arising from viral genomes that integrated into host DNA, only to mutate and get "stuck" in the host. As much as half of the human genome may have arisen from viruses or from virus-like elements of DNA called transposons (discussed in Chapter 9). For example, some endogenous viral genomes express placental proteins that are essential for early development of human embryos.

Dynamic Nature of Viruses

Emerging evidence supports a dynamic view of viral existence, and leads us to revisit the question: What is a virus? Is a virus a living organism? We now know that a virus may interconvert among three very different forms (**Fig. 6.4**):

■ **Virion, or virus particle.** The **virion** is an inert particle consisting of nucleic acid enclosed by a protein **capsid**. Some viruses package enzymes and possess a lipid envelope. A virion does not carry out any metabolism or energy conversion.

■ **Intracellular replication complex.** Within a host cell, the viral gene products direct the cell's enzymes to assemble progeny virions at "virus factories" called replication complexes. Assembly requires host ribosomes, as well as intricate collaboration between host and viral proteins.

■ **Viral genome integrated within host DNA.** Some types of viral genomes may integrate within a host chromosome and replicate as part of the host. Integration may be permanent; alternatively, the viral genome may be reactivated to start assembling virions.

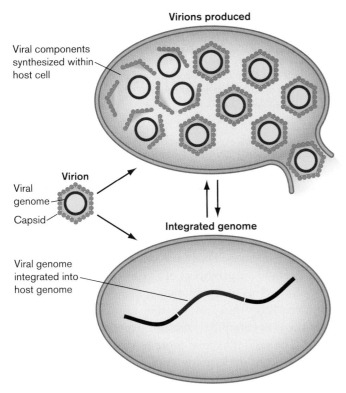

FIGURE 6.4 ■ Virus as a subcellular organism. The virion, or virus particle, consists of a nucleic acid genome contained by a protein capsid. A virion may infect a host cell and cause the cell's enzymes to synthesize progeny virions. Alternatively, infection may lead to integration of the viral genome within the host cell genome. Integration may last indefinitely, or else may lead to production of virions.

The inert nature of the virion (particle), which lacks metabolism, argues that viruses are nonliving. However, we know that cellular life-forms such as bacteria can convert into inert forms such as endospores that remain viable for thousands of years. The virion assembly process argues that viruses are living organisms. Assembly includes metabolism and production of progeny, processes we see for obligate intracellular bacteria such as chlamydias. Furthermore, the genomes of large viruses such as chloroviruses show evidence of reductive evolution (evolutionary loss of genes) from a cell.

On the other hand, genome integration argues for a view of viruses as host cell components. Many bacteria regulate phage-encoded genes along with their own genes (discussed in Chapter 10). For example, *Vibrio cholerae*, the cause of cholera, uses bacterial genes to regulate expression of the cholera toxin, which is encoded by an integrated prophage. For the *V. cholerae* bacterium, the viral prophage is a functional part of the cell.

How and where did viruses originate? Did they form within host cells, or did they arise as independent entities? Giant viral genomes encode numerous metabolic enzymes, including aminoacyl-tRNA synthetases (enzymes that attach an amino acid to the transfer RNA for translation).

The size and complexity of giant viruses argue that these viruses evolved by reductive evolution of an intracellular parasitic bacterium. Similar arguments are made for large viruses such as the smallpox and herpes viruses. On the other hand, small viruses with small RNA genomes, such as influenza virus and HIV, look more like something that arose from parts of a cell. For example, the key retroviral enzyme reverse transcriptase (which copies RNA to DNA) shows homology to telomerase, a host cell enzyme that maintains chromosome ends. Viral origins are discussed further in **eTopic 6.1**.

Viral Ecology

While viruses are the tiniest of biological entities, they play starring roles in ecosystems. Acute viruses (which rapidly kill their hosts) act as predators or parasites to limit host population density. They also recycle nutrients from their host bodies. An increase in host population density increases the rate of transmission of viral pathogens. As the host population declines, viruses are less likely to find a new host before they lose infectivity, while many of the remaining hosts have undergone selection for resistance. Thus, viruses can limit host density without causing extinction of the host.

In a community of multiple host species, the overall effect of viruses is to increase host diversity. Each viral species has a limited host range and requires a critical population density to sustain the chain of infection. In marine phytoplankton, a virus limits its host species to a population density far lower than what is sustainable by the available resources, such as light for photosynthesis. The resources then support other species resistant to the given virus (but susceptible to other ones). Thus, overall, marine viruses prevent the dominance of any one species and foster the evolution of many distinct host species. The continual threat of marine viral epidemics explains the great diversity of phytoplankton ranging from silica-shelled diatoms to toxin-producing dinoflagellates.

Persistent viruses (viruses that don't immediately kill the host) may benefit the host population overall. Most important, viruses transfer genes across species, conferring traits such as the ability to metabolize new food sources. Humans harbor many obscure herpes and cold-type viruses that stimulate normal development of the immune system. In natural mammal populations, a persistent virus can act as a bioweapon that kills off competitor populations susceptible to the virus. An example is the koala retrovirus, KoRV, which has become endogenous (persistent) in many koala populations of Australia. When KoRV is transmitted to koalas that lack the retrovirus, it kills most of those newly infected—except for the few in which the virus persists. In these infected animals, the immune system controls the virus but cannot eradicate it.

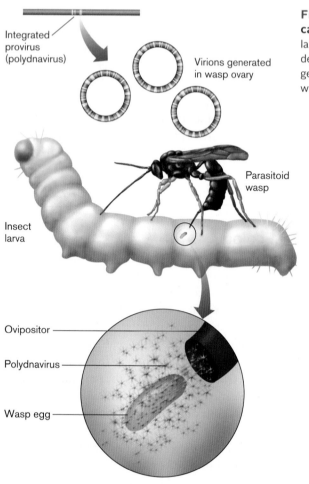

FIGURE 6.5 ■ The relationship of polydnaviruses, wasps, and caterpillars. Parasitoid wasps lay their eggs inside a living insect caterpillar. When a female wasp deposits her eggs inside the caterpillar, she also deposits her symbiogenic polydnavirus virions. The virions express wasp genes in the caterpillar, where they prevent the encapsulation process that would otherwise wall off the wasp egg and kill it.

Integrated viral genomes can evolve complex symbiotic interactions with multiple hosts, combining mutualism and parasitism. An example is the relationship of polydnaviruses, wasps, and caterpillars (**Fig. 6.5**). Parasitoid wasps lay their eggs inside a living insect caterpillar, where the larvae must hatch and feed without stimulating the caterpillar's defense. The wasp genome contains an integrated genome (or **provirus**) of a virus called polydnavirus. The polydnavirus forms its own genomes and virions only within the wasp ovary. When the wasp deposits her eggs inside the host caterpillar, she also deposits her mutualistic virions. The virions express proteins that prevent the caterpillar cells from undergoing encapsulation, a process that would otherwise wall off the wasp eggs and kill them.

In marine ecosystems, viruses play a significant role in the cycling of food molecules (**Fig. 6.6**). In ocean food webs, phytoplankton is consumed by grazers, and grazers are eaten by higher trophic consumers (discussed in Chapter 21). But at each level, some of the carbon fixed as biomass sinks to the ocean floor. There, benthic microbes metabolize so slowly that most carbon is effectively removed from the carbon cycle. At the same time, viral infection and lysis convert the bodies of phytoplankton, grazers, and carnivores into detritus consisting of small organic particles and soluble molecules. The viral detritus gets taken up by heterotrophs, and minerals are absorbed by autotrophs. This uptake of viral detritus is called the **viral shunt**. The viral shunt returns some organic matter to microbial consumers in the upper region of the ocean (discussed in Chapter 21). Some carbon is thus diverted from the ocean sink and instead plays a significant part in the CO_2 cycle. The viral shunt also plays a major part in cycling phosphorus in the oceans.

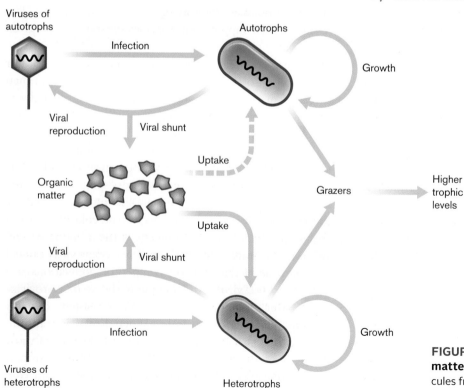

FIGURE 6.6 ■ Marine viruses recycle organic matter. Viruses divert the flow of organic molecules from marine phytoplankton and bacteria back to metabolism, avoiding loss by sedimentation.

Viral Disease

When viral infection causes harm to an animal or plant, the host has a disease. Each species of virus infects a particular group of host species, known as its **host range**. Some viruses can infect only a single species; for example, HIV infects only humans. Close relatives of humans, such as the chimpanzee, are not infected by HIV, although they are susceptible to a highly related virus, simian immunodeficiency virus (SIV). By contrast, West Nile virus, transmitted by mosquitoes, infects many species of birds and mammals. West Nile virus has a much broader host range than HIV and SIV.

For humans, viruses cause many forms of illness. The influence of viral diseases on human history and culture would be hard to overstate. More people died of influenza in the global epidemic of 1918 than in the battles of World War I. Poliovirus is famous for causing the outbreaks of poliomyelitis that swept the United States during the first half of the twentieth century. President Franklin Roosevelt, himself a victim of the disease, established a national foundation called the March of Dimes to develop a vaccine. The spectacular success of the March of Dimes set the pattern for future public support of research on cancer and AIDS.

Chronic viral infections, however, are more common than acute disease, and are an everyday part of our lives. The most frequent infections of college students are due to respiratory pathogens such as rhinovirus (the common cold) and Epstein-Barr virus (infectious mononucleosis), as well as sexually transmitted viruses such as herpes simplex virus (HSV) and papillomavirus (genital warts). Some viruses, such as rhinoviruses, are eliminated rapidly by our immune system, whereas others, such as herpes, establish a lifelong latent infection. Viruses also impact human industry; for example, bacteriophages (literally, "bacteria-eaters") infect cultures of *Lactococcus* during the production of yogurt and cheese. Plant pathogens such as cauliflower mosaic virus and rice dwarf virus cause substantial losses in agriculture.

In contrast to our vast arsenal of antibiotics (effective against bacteria), the number of antiviral drugs remains small. Because the machinery of viral growth is largely that of the host cell, viruses present relatively few targets that can be attacked by antiviral drugs without harming the host. An exception is human immunodeficiency virus (HIV), the cause of AIDS. Molecular studies of HIV have yielded several major classes of antiviral drugs, such as AZT and protease inhibitors. The molecular biology behind anti-HIV drugs is discussed in Chapters 11 and 27.

To Summarize

- **Viruses** can exist as a virion (inert virus particle), as the intracellular assembly of virion components, or as a viral genome integrated within host cell DNA.

- **All classes of organisms are infected by viruses.** Usually the hosts are limited to a particular host range of closely related strains or species.

- **Acute virus infection limits host population density.** Virus-associated mortality may increase the genetic diversity of host species.

- **Persistent viruses remain in hosts**, where they may evolve traits that confer positive benefits in a virus-host mutualism.

- **Marine viruses** infect most phytoplankton, releasing their minerals in the upper water, where they are available for other phototrophs. Viral activity substantially impacts the global carbon balance.

- **Viruses transfer genes between host genomes.** Viral gene transfer is a major source of cell genome evolution.

6.2 Virus Structure

The structure of a virion keeps the viral genome intact, and it enables infection of the appropriate host cell. First, the stable capsid protects the viral genome from degradation and enables it to be transmitted outside the host. Second, in order for the viral genome to reproduce, the virion must either insert its genome into the host cell or disassemble within the host. In the process, the original particle loses its stable structure and its own identity as such. But if viral reproduction succeeds, then it yields numerous progeny virions.

A virion possesses a genome of either DNA or RNA, contained by proteins that compose the capsid. The form of the virion depends on the species of virus. The virion may be symmetrical or asymmetrical, or combine aspects

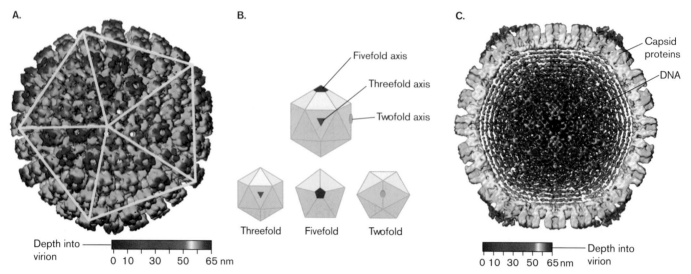

FIGURE 6.7 ▪ Herpes: icosahedral capsid symmetry. A. Icosahedral capsid of herpes simplex 1 (HSV-1), with envelope removed. Imaging of the capsid structure is based on computational analysis of cryo-TEM. Images of 146 virus particles were combined digitally to obtain this model of the capsid at 2-nm resolution. **B.** Icosahedral symmetry includes fivefold, threefold, and twofold axes of rotation. **C.** The icosahedral capsid contains spooled DNA. *Source:* A, C. Z. Hong Zhou et al. 1999. *J. Virol.* **73**:3210.

of both. The papillomavirus virion, for example, consists solely of a capsid containing a genome. Some viruses, such as the tailed bacteriophages, have an elaborate delivery device to transfer the viral genome into the host cell. Other viruses possess a membrane envelope. The protein capsid of an enveloped virus is called a core particle.

Understanding virus structure is crucial for devising vaccines and drug therapies. For example, the Gardasil vaccine for human papillomavirus, HPV (recommended for all children before adolescence), is composed of capsid proteins from four different HPV strains. The molecular basis of viral structure and drugs is presented in Chapter 11.

> **Thought Question**
>
> **6.3** What will happen if a virus particle remains intact within a host cell and fails to release its genome?

Symmetrical Virions

A symmetrical virus may have a capsid of one of two types: icosahedral or filamentous (helical). Some have an intermediate form, such as the HIV core. The advantage of geometrical symmetry is that it provides a way to form a package out of repeating protein units generated by a small number of genes and encoded by a short chromosomal sequence. The smaller the viral genome, the more genome copies can be synthesized from the host cell's limited supply of nucleotides. Nevertheless, some symmetrical viruses, such as herpes virus and mimivirus, have much larger genomes. Large genomes offer a greater range of functions for viral components.

Icosahedral viruses. Many viruses package their genome in an **icosahedral** (20-sided) **capsid**. An icosahedral capsid takes the form of a polyhedron with 20 identical triangular faces. In the capsid, each triangle can be composed of three identical but asymmetrical protein units. An icosahedral capsid is found in the herpes simplex virus (**Fig. 6.7A**). Each triangular face of the capsid is determined by the same genes encoding the same protein subunits. The actual form of viral subunits can vary greatly, generating very different complex shapes for different viral species. But no matter what the pattern of subunits in the triangular unit, the structure overall shows the rotational symmetry of an icosahedron (**Fig. 6.7B**): threefold symmetry around the axis through two opposed triangular faces, fivefold symmetry around an axis through opposite points, and twofold symmetry around an axis through opposite edges.

> **Thought Question**
>
> **6.4** For a viral capsid, what is the advantage of an icosahedron, as shown in **Figure 6.7**, instead of some other polyhedron?

Virus particles can be observed by standard transmission electron microscopy (TEM), but the details of capsid structure as in **Figure 6.7A** require digital reconstruction of cryo-EM images (discussed in Section 2.6). Recall from Chapter 2 that in cryo-EM, the viral samples for TEM are flash-frozen, preventing the formation of ice crystals. Flash-freezing enables observation without

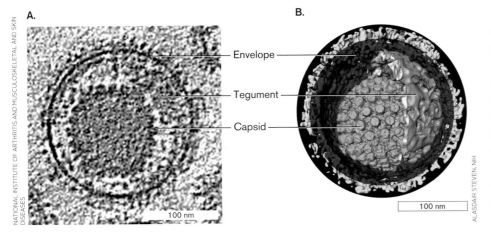

A.

B.

Envelope

Tegument

Capsid

100 nm

100 nm

FIGURE 6.8 ■ **Envelope and tegument surround the herpes capsid.** **A.** Section showing envelope and tegument proteins surrounding capsid (cryo-EM). **B.** Cutaway reconstruction of the herpes virion (cryo-EM tomography).

The space between the capsid and the envelope contains "tegument" proteins, such as enzymes that interact with the host cell to depress its defense responses. Tegument proteins are expressed during infection of a host cell, and then get packaged in the virion. Both viral and host proteins may be packaged as tegument. The mature envelope bristles with glycoprotein **spike proteins** that attach it to the capsid. The spike proteins enable the virus to attach and infect the next host cell. Further details of herpes structure and function are presented in Section 11.5.

stain. The electron beams penetrate the object; thus, images of individual capsids actually provide a glimpse of the virus's internal contents. By digitally combining and processing cryo-TEM images from a number of capsids, a 3D reconstruction is built for the entire virus particle. Within the icosahedral capsid, the herpes genome is spooled and tightly packed. **Figure 6.7C** shows that the double-stranded DNA of the herpes genome is packed under high pressure. A molecular motor powered by ATP drives viral DNA into the host nucleus (discussed in Chapter 11).

In some icosahedral viruses, such as HIV, the capsid (called a core particle) is enclosed in an **envelope** membrane. The envelope is composed of host-derived membrane from the host cell in which the virion formed; it also contains proteins specified by the viral genome. The HIV envelope derives from host plasma membrane, whereas other viruses, such as herpes virus, derive their envelope from intracellular membranes such as the nuclear membrane or endoplasmic reticulum. The envelope and capsid contents of herpes virus are shown in **Figure 6.8**.

Note: Distinguish the viral envelope (phospholipid membrane derived from a host cell membrane) from the bacterial cell envelope (protective layers outside the bacterial cell membrane). The bacterial envelope is discussed in Chapter 3.

Filamentous viruses. A second major category of virus structure is that of **filamentous viruses**. A famous example is Ebola virus (**Fig. 6.9A**), which causes a fatal disease of humans and related primates; the virus caused a major epidemic in 2014 (discussed in Chapter 28). A filamentous bacteriophage that infects *E. coli* is phage M13 (**Fig. 6.9B**). Filamentous bacteriophages cause problems for human medicine and industry. One infects the Gram-positive species *Propionibacterium freudenreichii*, a key fermenting agent for Swiss cheese. Another filamentous phage, CTXφ, integrates its sequence into the genome of *Vibrio cholerae*, where it carries the deadly toxin genes required for cholera. On the other hand, filamentous phages have been used in nanotechnology to nucleate the growth of crystalline "nanowires" for electronic devices.

The filamentous bacteriophage M13 (**Fig. 6.9B**) consists of a relatively simple capsid of protein monomers.

A.

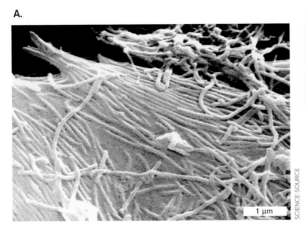

1 μm

B.

Tail fibers

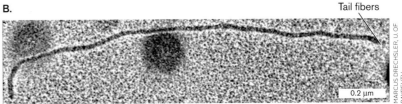

0.2 μm

FIGURE 6.9 ■ **Filamentous viruses.** **A.** Ebola virus filaments (SEM). **B.** The filamentous bacteriophage M13 has a relatively simple helical capsid that surrounds the genome coiled within (TEM).

A.

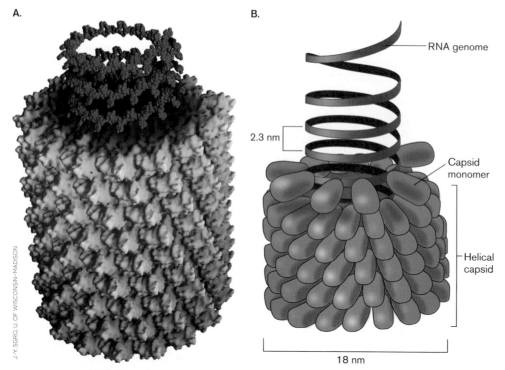

B.

RNA genome

2.3 nm

Capsid monomer

Helical capsid

18 nm

FIGURE 6.10 ■ Tobacco mosaic virus: helical symmetry. A. The helical filament of tobacco mosaic virus (TMV) contains a single-stranded RNA genome coiled inside. **B.** Components of the TMV virion.

The monomers are stacked around a coiled genome consisting of a circle of single-stranded DNA. At one end of the filament, short tail fibers mediate specific attachment to the host. The filament attaches to the F pilus of an *E. coli* bacterium containing an F plasmid that encodes pili.

Filamentous viruses show helical symmetry. The pattern of capsid monomers forms a helical tube around the genome, which usually winds helically within the tube. In a helical capsid, the genome is a single-stranded DNA (as in phage M13) or RNA (as in tobacco mosaic virus). **Figure 6.10** shows how the RNA strand of tobacco mosaic virus winds in a spiral within a tube of capsid monomers

laid down in a spiral array. Such a tube can be imagined as a planar array of subunits that coils around such that each row connects to the row above, generating a spiral. The length of the helical capsid may extend up to 50 times its width, generating a flexible filament.

Unlike an icosahedral capsid, which usually has a fixed size, a helical capsid can vary in length to accommodate different lengths of nucleic acid. Furthermore, some viruses package several genome segments into separate helical capsids. For example, influenza virus packages several different genome segments into separate helical packages of different sizes, contained together within a membrane envelope. Separate chromosome packaging enables influenza virus to pack different numbers of RNA segments into different virions. Such virions are thus defective, but the process enables rapid evolution of new strains (discussed further in Chapter 11).

Tailed Viruses

Some bacteriophages supplement the icosahedral capsid or head coat with an elaborate delivery device. For example, bacteriophage T4 (**Fig. 6.11A**) has an icosahedral "head" containing the pressure-packed DNA, attached to a helical "neck" that channels the nucleic acid into the host cell.

A.

Genome

Head: 100 nm

Collar

Neck

Tail fibers

Tailsheath

Baseplate

B.

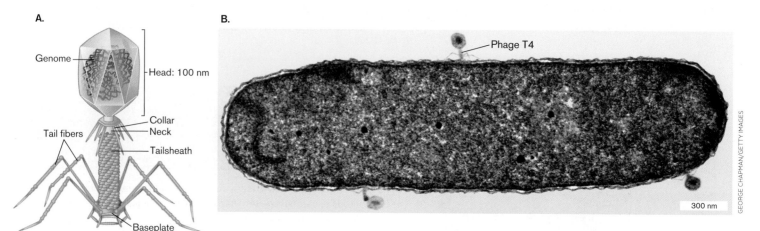

Phage T4

300 nm

FIGURE 6.11 ■ Bacteriophage T4 capsid. A. Phage T4 particle with protein capsid containing packaged double-stranded DNA genome. The capsid has a sheath with tail fibers that facilitate attachment to the surface of the host cell. After attachment, the sheath contracts and the core penetrates the cell surface, injecting the phage genome. **B.** *E. coli* infected by phage T4 (colorized blue, TEM).

The tail baseplate has six jointed tail fibers that stabilize the structure on the host cell surface (**Fig. 6.11A** and **B**). After phage infection, the production of virions within a cell requires a factorylike assembly line of phage tail parts (described for phage T4 in **eTopic 11.1**).

Asymmetrical Virions

Many viruses have a capsid or core particle that is asymmetrical. Influenza viruses are RNA viruses that lack capsid symmetry. Instead, the RNA segments are coated with nucleocapsid proteins (discussed in detail in Chapter 11).

The poxviruses, such as vaccinia (cowpox, the source of smallpox vaccine) have a double-stranded DNA genome stabilized by covalent connection of its two strands at each end (**Fig. 6.12A** and **B**). The core envelope encloses the nucleocapsid-coated DNA, as well as a large number of accessory proteins. The accessory proteins are needed early in viral infection, such as initiation proteins for the transcription of viral genes and RNA-processing enzymes that modify viral mRNA molecules. The proteins may be found either inside the capsid or in the tegument between the core envelope and the outer membrane, which is coated with surface tubules of protein. Large asymmetrical viruses contain so many enzymes that they appear to have evolved from degenerate cells.

Vaccinia virus offers opportunities for new applications, such as conversion to vaccines for other viruses, and engineering to combat tumors. For example, students in Paulo Verardi's laboratory at the University of Connecticut (**Fig. 6.12C**) recently engineered an on/off switch for vaccinia that will make the virus safer for immunocompromised patients. The switch involves a recombinant tetracycline repressor gene, whose product confers resistance to tetracycline. Repressor controls and engineering are discussed in Chapters 10 and 12.

Viroids and Prions

Are there infectious agents even simpler than small viruses? For some infectious agents, the nucleic acid genome is itself the entire infectious particle; there is no protective capsid. Such agents are called **viroids**, which are infectious agents

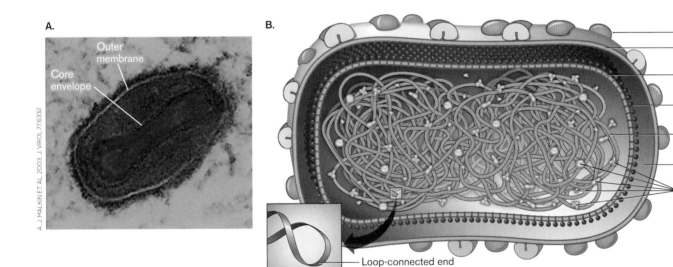

FIGURE 6.12 ■ Vaccinia poxvirus. A. Vaccinia virion observed in aqueous medium by atomic force microscopy (AFM). **B.** A pox virion includes an outer membrane and a core envelope membrane containing envelope proteins enclosing the double-stranded DNA genome and accessory proteins. The DNA is stabilized by a hairpin loop at each end. **C.** Allison Titong and Paulo Verardi examine vaccinia virus in tissue culture, to develop advanced vaccines and cancer treatments.

Potato spindle tuber viroid—circular ssRNA

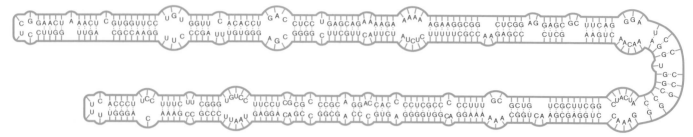

FIGURE 6.13 ■ Viroids: infective RNA. Potato spindle tuber viroid consists of a circular single-stranded RNA (ssRNA) that hybridizes internally.

of plants. A viroid encodes no genes, but it hijacks the plant cell's RNA polymerase to replicate itself. Viroids infect many kinds of fruits and vegetables, entering the plant cell through a damaged cell wall. For example, citrus viroids cause economic losses in the citrus industry.

The potato spindle tuber viroid (**Fig. 6.13**) consists of a circular, single-stranded molecule of RNA that doubles back on itself to form base pairs interrupted by short, unpaired loops. This unusual circularized form avoids breakdown by host RNase enzymes. The RNA folds up into a globular structure that interacts with host cell proteins. The RNA is replicated by host RNA polymerase. The host RNA polymerase normally requires a DNA template, but the replication process is modified during viroid infection. During infection, the RNA polymerase thus replicates progeny copies of the viroid, which encodes no products other than itself. Viroids can cause as much host destruction as "true viruses," and some authors, particularly in plant pathology, classify them as "viruses without capsids."

Some viroids have catalytic ability, comparable to enzymes made of protein. An RNA molecule capable of catalyzing a reaction is called a ribozyme (discussed in Chapter 8). Viral ribozymes may be able to cleave themselves or other specific RNA molecules. Their ability to cleave very specific RNA sequences has applications in medical research, such as cleaving human mRNA involved in cancer.

Can an infectious agent propagate without a genome of its own? A remarkable class of infectious agents is believed to consist solely of protein. These agents, known as **prions**, are thought to be aberrant proteins arising from the host cell. Prions gained notoriety when they were implicated in brain infections such as Creutzfeldt-Jakob disease, a variant form of which is known as "mad cow" disease because it may be transmitted through defective proteins in beef from diseased cattle. Other diseases believed to be caused by prion transmission include scrapie, a disease of sheep; and kuru, a degenerative brain disease that was found in a tribe of people who customarily consumed the brains of deceased relatives.

In prion-associated diseases, the infective agent is unaffected by treatments that destroy RNA or DNA, such as nucleases or UV irradiation. A prion is an aberrant form of a normal cell protein that assumes an abnormal conformation or tertiary structure (**Fig. 6.14A**). The prion form of the protein acts by binding to normally folded proteins of the same class and altering their conformation to that of the prion. The multiplying prion then alters the conformation of other normal subunits, forming harmful aggregates in the cell and ultimately leading to cell death. In the brain, prion-induced cell death leads to tissue deterioration and dementia

A.

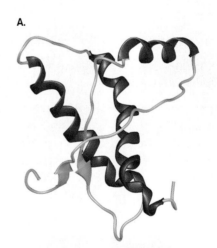

Normal conformation   Abnormal conformation

B.

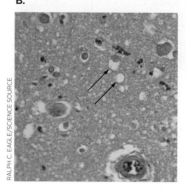

RALPH C. EAGLE/SCIENCE SOURCE

FIGURE 6.14 ■ Prion disease.
A. The normal conformation of a prion, compared to the abnormal conformation. The abnormal form "recruits" normally folded proteins and changes their conformation into the abnormal form. (PDB code: 1AG2)
B. Section of a human brain showing "spongiform" holes typical of Creutzfeldt-Jakob disease.

(**Fig. 6.14B**). Prion diseases are unique in that they can be transmitted by an infective protein instead of by DNA or RNA, and they propagate the conformational change of existing molecules without synthesizing entirely new infective molecules.

A prion disease can be initiated by infection with an aberrant protein. More rarely, the cascade of protein misfolding can start with the spontaneous misfolding of an endogenous host protein. The chance of spontaneous misfolding is greatly increased in individuals who inherit certain alleles encoding the protein; thus, Creutzfeld-Jakob disease can be inherited genetically from one person with one mutant allele.

To Summarize

- A **virion** consists of a capsid or a core particle enclosing a nucleic acid genome. For some viruses, the capsid is further enclosed by an envelope of membrane with embedded proteins.

- A **viral capsid** is composed of repeated protein subunits—a structure that maximizes the structural capacity while minimizing the number of genes needed for construction.

- **The capsid or core particle packages the viral genome** and delivers it into the host cell.

- **Icosahedral capsids** have regular, icosahedral symmetry.

- **Filamentous (helical) capsids** have uniform width, generating a flexible filamentous virion.

- **Enveloped virions** consist of a genome (RNA or DNA) packaged by nucleocapsid proteins. The packaged genome (or core particle) and tegument or accessory proteins are enclosed within phospholipid membrane derived from the host cell. The envelope includes virus-specific spike proteins.

- **Viroids** that infect plants consist of RNA hairpins with no capsid.

- **Prions** are infectious proteins that induce a cell's native proteins to fold incorrectly and impair cell function.

6.3 Viral Genomes and Classification

Viral genomes are structurally more diverse than those of cells. A given type of virus may have a genome that consists of either RNA or DNA, single- or double-stranded, linear or circular. The form of the genome has key consequences for the mode of infection, and for the course of a viral disease. Viral genomes are used as the basis of virus classification.

Viral Genomes: Small or Large

Small viruses commonly have a small genome, encoding fewer than ten genes. For example, cauliflower mosaic virus (diameter 50 nm) has a genome encoding only seven genes (**Fig. 6.15A**), which actually overlap each other in sequence. This overlap in sequence is made possible by the use of different reading frames—start positions to define the first base of the codon for translation to an amino acid sequence (discussed in Chapter 8).

Many small viral genomes consist of RNA. RNA viruses include some of today's most important human pathogens, such as influenza virus, hepatitis C virus, and human immunodeficiency virus (HIV). The RNA genome of avian leukosis virus (**Fig. 6.15B**) has protein-encoding genes grouped by functional categories of core capsid, replication enzymes, and envelope proteins (proteins embedded in the envelope phospholipid bilayer).

At the opposite end of the scale, the "giant viruses" have genomes of double-stranded DNA comprising 500–2,500

A. Cauliflower mosaic virus genome (DNA)

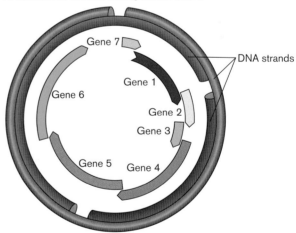

B. Avian leukosis virus genome (RNA)

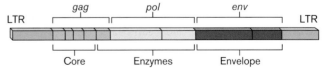

FIGURE 6.15 ▪ Simple viral genomes. A. Cauliflower mosaic virus has a circular genome of double-stranded DNA, whose strands are interrupted by nicks. The genome encodes seven overlapping genes. **B.** Avian leukosis virus, a single-stranded RNA retrovirus resembling eukaryotic mRNA. Three genes (*gag, pol,* and *env*) encode polypeptides that are eventually cleaved to form a total of nine functional products. LTR = long terminal repeat.

A.

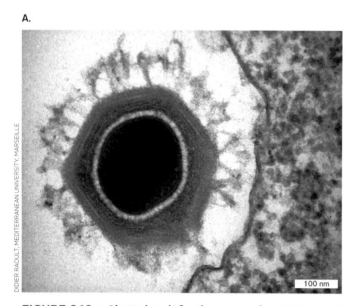

B.

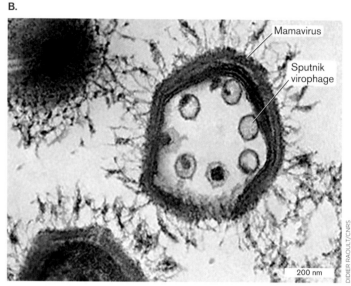

Mamavirus

Sputnik virophage

100 nm

200 nm

FIGURE 6.16 ▪ **Giant virus infecting an ameba.** **A.** Mimivirus is larger than some bacteria (TEM). **B.** Sputnik virophage infects Mamavirus, a relative of Mimivirus (TEM).

genes. A giant virus called mimivirus (diameter 300 nm), which infects amebas and may cause human pneumonia, is itself as large as some bacteria (**Fig. 6.16A**) and has a bacterium-sized genome of 1.2 million base pairs, encoding more than 1,000 genes. Discovered in 2003 by French virologist Didier Raoult and colleagues, mimivirus has a genome that encodes numerous cellular functions, including DNA repair and protein folding by chaperones. The virus gains entry to its ameba host by phagocytosis, because its large particle size makes the virus particle resemble a bacterium that the ameba could engulf for food. Once taken up, the mimivirus capsid opens to release viral enzymes and DNA. Within the ameba's cytoplasm, the viral enzymes generate a replication complex that produces progeny mimiviruses. Mimiviruses are so large that they can actually become infected themselves by smaller viruses, such as the Sputnik virophage, or "virus-eater" (**Fig. 6.16B**).

A surprising source of giant viruses is the frozen polar environments of the Arctic and Antarctic regions. Antarctic lakes reveal giant viruses related to Mimivirus and the Sputnik virophage. The Arctic tundra of Siberia reveals even more remarkable viruses, such as Pithovirus (**Fig. 6.17**). Pithovirus is an unusual lozenge-shaped virus, nearly the size of bacteria such as *E. coli*. Jean-Marie Claverie and colleagues from Aix-Marseille University isolated Pithovirus from Siberian tundra frozen for 30,000 years. Remarkably, a virion so old was successfully inoculated into an ameba host, where it generated progeny virions. In 2015, yet another Siberian giant virus, Mollivirus sibericum, was revived and found to package ribosomal proteins from its host ameba—the first virus ever shown to contain parts of a ribosome. The revival of ancient viruses leads to concern

A.

B.

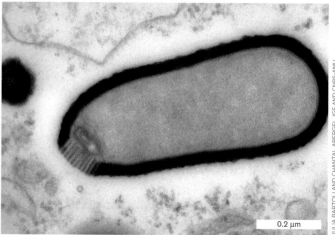

0.2 μm

FIGURE 6.17 ▪ **Giant virus from Siberia.** **A.** The Siberian tundra is melting. **B.** Pithovirus, a virus as large as *E. coli*, was isolated from tundra frozen for 30,000 years (TEM).

that as tundra melts during inevitable climate change, the melting permafrost will release human pathogens from long-dead hosts such as buried victims of smallpox.

Giant viruses are intriguing because their genomes specify so many enzymes with house-keeping cell functions, such as nutrient transport, mRNA translation, and even cell motility (**Fig. 6.18**). Such large cell-like genomes suggest the likelihood that a virus evolved from a parasitic cell.

The International Committee on Taxonomy of Viruses

How do we classify the bewildering diversity of viruses? Today we classify organisms by the relatedness of their gene sequences. The definition of a virus species, however, is problematic, given the small size and high mutability of viral genomes and the ability of different viruses to recombine their genome segments within an infected host cell. Furthermore, not all viruses are monophyletic—that is, descended from a common ancestor. In fact, different classes of viruses appear to have evolved from different sources—for example, from parasitic cells or from host cell components such as DNA replication enzymes. Viruses are classified by genome composition, virion structure, and host range.

For purposes of study and communication, a working classification system has been devised by the International Committee on Taxonomy of Viruses (ICTV). The ICTV classification system is based on several criteria:

- **Genome composition.** The nucleic acid of the viral genome can vary remarkably with respect to physical structure: It may consist of DNA or RNA, it may be single- or double-stranded, it may be linear or circular, and it may be whole or **segmented** (that is, divided into separate "chromosomes"). Genomes are classified by the Baltimore method (discussed next).

- **Capsid symmetry.** The protein capsid or core particle may be helical or icosahedral, with various levels of symmetry.

- **Envelope.** The presence of a host-derived envelope, and the envelope structure if present, are characteristic of related viruses.

- **Size of the virus particle.** Related viruses generally share the same size range; for example, enteroviruses such as poliovirus are only 30 nm across (about the size of a ribosome), whereas poxviruses are 200–400 nm, as large as a small bacterium.

- **Host range.** Closely related viruses usually infect the same or related hosts. However, viruses with extremely

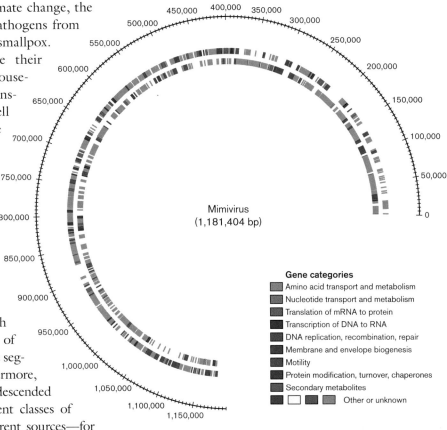

FIGURE 6.18 ■ Genome of Mimivirus. The genome of this giant virus specifies numerous enzymes with cell functions.

different hosts can show surprising similarities in genetics and structure. For example, both rabies virus and potato yellow dwarf virus are enveloped, bullet-shaped viruses of the rhabdovirus family.

Note: In nomenclature, families of viruses are designated by Latin names with the suffix "-viridae": for example, *Papillomaviridae*. Nevertheless, the common forms of such family names are also used—for example, "papillomaviruses." Within a family, a virus species is simply capitalized, as in "Papillomavirus."

The Baltimore Virus Classification

Given the wide variety of viral structures, how do we determine their relatedness? In general, viruses of the same genome class (such as double-stranded DNA) show greater evidence of shared ancestry with each other than with viruses of a different class of genome (such as RNA). In 1971, David Baltimore proposed that the primary distinctions among classes of viruses be the genome composition (RNA or DNA) and the route used to express messenger RNA (mRNA). Baltimore, together with Renato Dulbecco (1914–2012) and Howard Temin (1943–1994), was awarded the 1975 Nobel Prize in Physiology or Medicine for discovering how tumor viruses cause cancer.

A.

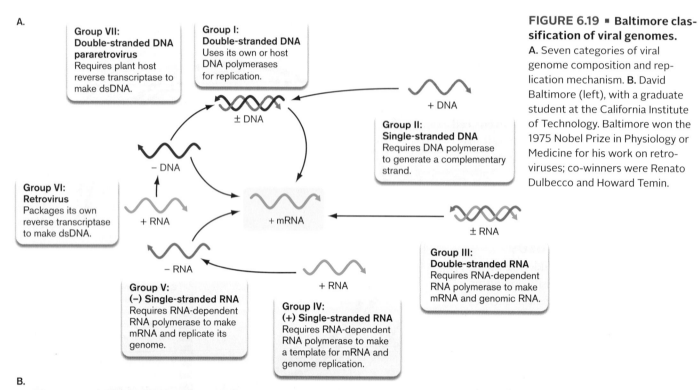

Group VII:
Double-stranded DNA
pararetrovirus
Requires plant host
reverse transcriptase to
make dsDNA.

Group I:
Double-stranded DNA
Uses its own or host
DNA polymerases
for replication.

± DNA

+ DNA

Group II:
Single-stranded DNA
Requires DNA polymerase
to generate a complementary
strand.

− DNA

Group VI:
Retrovirus
Packages its own
reverse transcriptase
to make dsDNA.

+ RNA

+ mRNA

± RNA

Group III:
Double-stranded RNA
Requires RNA-dependent
RNA polymerase to make
mRNA and genomic RNA.

− RNA

+ RNA

Group V:
(−) Single-stranded RNA
Requires RNA-dependent
RNA polymerase to make
mRNA and replicate its
genome.

Group IV:
(+) Single-stranded RNA
Requires RNA-dependent
RNA polymerase to make
a template for mRNA and
genome replication.

FIGURE 6.19 ■ Baltimore classification of viral genomes.
A. Seven categories of viral genome composition and replication mechanism. **B.** David Baltimore (left), with a graduate student at the California Institute of Technology. Baltimore won the 1975 Nobel Prize in Physiology or Medicine for his work on retroviruses; co-winners were Renato Dulbecco and Howard Temin.

B.

All cells and viruses need to make messenger RNA to produce their fundamental protein components. The production of mRNA from the viral genome is central to a virus's ability to propagate its kind. Cellular genomes always make mRNA by copying double-stranded DNA. For viruses, however, different kinds of genomes require fundamentally different mechanisms to produce mRNA. The different means of mRNA production generate distinct groups of viruses with shared ancestry.

So far, the genome composition and mechanisms of replication and mRNA expression define seven fundamental groups of viral species (**Fig. 6.19**). Examples of these seven fundamental groups are shown in **Table 6.1**. A more extensive taxonomic survey of viral species is presented in Appendix 2.

Group I. Double-stranded DNA viruses such as the herpes and smallpox viruses make their own DNA poly-

merase or use that of the host for genome replication. Their genes can be transcribed directly by a standard RNA polymerase, in the same way that a cellular chromosome would be transcribed. The RNA polymerase used can be that of the host cell, or it can be encoded by the viral genome.

Group II. Single-stranded DNA viruses such as canine parvovirus require the host DNA polymerase to generate the complementary DNA strand. The double-stranded DNA can then be transcribed by host RNA polymerase.

Group III. Double-stranded RNA viruses require a viral **RNA-dependent RNA polymerase** to generate messenger RNA by transcribing directly from the RNA genome. Since the RNA polymerase is required immediately upon infection, such viruses package a viral RNA polymerase with their genome before exiting the host cell. A major class of double-stranded RNA viruses is the reoviruses, including rotavirus, a cause of diarrhea in children. Another reovirus has been engineered to destroy human tumors without infecting normal cells. This "oncolytic" reovirus, called Reolysin, is now in clinical trials for cancer therapy.

Group IV. (+) sense single-stranded RNA viruses consist of a positive-sense (+) strand (the coding strand) that can serve directly as mRNA to be translated to viral proteins. Replication of the RNA genome, however, requires synthesis of the template (−) strand [complementary to the (+) strand] by a viral RNA-dependent RNA polymerase to form a double-stranded RNA intermediate.

TABLE 6.1 **Groups of Viruses—Baltimore Classification (Expanded Table Appears in Appendix 2)**

Virus example	Taxonomic group with examples
Phage lambda	**Group I. Double-stranded DNA viruses** ± DNA **Bacteriophage lambda** infects *Escherichia coli*. **Chloroviruses** infect algae, controlling algal blooms. **Herpes viruses** cause chickenpox, genital infections, and birth defects. **Human papillomavirus** strains cause warts and tumors.
Geminivirus	**Group II. Single-stranded DNA viruses** + DNA **Anelloviruses** are found in human blood plasma; they cause no known harm. **Bacteriophage M13** infects *E. coli*. **Geminiviruses** infect tomatoes and other plants. **Parvoviruses** cause disease in cats, dogs, and other animals.
Rotavirus	**Group III. Double-stranded RNA viruses** ± RNA **Birnaviruses** infect fish. **Cystiviruses** infect bacteria. **Reoviruses** such as rotavirus cause severe diarrhea in infants. Other reoviruses are in clinical trials to fight tumors (oncolysis).
Rhinovirus	**Group IV. (+) sense single-stranded RNA viruses** + RNA **Coronaviruses** such as SARS cause severe respiratory disease. **Flaviviruses** cause hepatitis C, West Nile disease, yellow fever, and dengue fever. **Poliovirus** infects human intestinal epithelium and nerves. **Tobacco mosaic virus** infects plants.
Rabies virus	**Group V. (−) sense single-stranded RNA viruses** − RNA **Filoviruses** such as Ebola virus cause severe hemorrhagic disease. **Orthomyxoviruses** cause influenza. **Paramyxoviruses** cause measles and mumps. **Rhabdovirus** causes rabies.
Human immuno-deficiency virus	**Group VI. Retroviruses (RNA reverse-transcribing viruses)** + RNA → DNA **Feline leukemia virus (FeLV)**, **Rous sarcoma virus (RSV)**, and **avian leukosis virus (ALV)** cause cancer. **Lentiviruses** include **human immunodeficiency virus (HIV)**, the cause of AIDS. Engineered "lentivectors" are used for gene therapy.
Caulimovirus	**Group VII. Pararetroviruses (DNA reverse-transcribing viruses)** DNA + RNA **Caulimoviruses** (such as cauliflower mosaic virus, or CaMV) infect many kinds of vegetables. CaMV provides the best vector tools for plant biotechnology. **Hepadnaviruses** such as hepatitis B virus infect the human liver.

Positive-sense (+) RNA viruses include enteroviruses such as poliovirus, the coronaviruses, and the flaviviruses that cause West Nile encephalitis and hepatitis C.

Group V. (−) sense single-stranded RNA viruses such as influenza virus have genomes that consist of template, or "negative-sense," RNA. Thus, they need to package a viral RNA-dependent RNA polymerase for transcribing (−) RNA to (+) mRNA. The (−) strand RNA viral genomes are often segmented; that is, they consist of multiple separate linear chromosomes—a key factor in the evolution of killer strains of influenza (see Section 11.2).

Group VI. Retroviruses, or RNA reverse-transcribing viruses, such as HIV and feline leukemia virus, have genomes that consist of (+) strand RNA. Instead of RNA polymerase, they package a **reverse transcriptase**, which transcribes the RNA into a double-stranded DNA (for details, see Section 11.3). The double-stranded DNA is then integrated into the host genome, where it directs the expression of the viral genes.

Group VII. Pararetroviruses, or DNA reverse-transcribing viruses, have a replication cycle that requires reverse transcriptase. For example, hepatitis B virus (a hepadnavirus) first copies its double-stranded DNA genomes into RNA, and then reverse-transcribes the RNA to progeny DNA using a reverse transcriptase packaged in the original virion. In contrast, plant pararetroviruses, such as cauliflower mosaic virus (CaMV, a caulimovirus), generate an RNA intermediate that replicates using a reverse transcriptase made by the host cell. Many plant genomes include a gene for reverse transcriptase. Cauliflower mosaic virus is of enormous agricultural significance for its use as a vector to construct pesticide-resistant food crops.

Molecular Evolution of Viruses

The phylogeny, or genetic relatedness, of viruses can be determined within families. For example, the herpes family includes double-stranded DNA viruses that cause several human and animal diseases, such as chickenpox, oral and genital herpes infection, and respiratory and genital infections in horses. Herpes genomes consist of double-stranded DNA, 120–220 kilobases (kb) encoding about 70–200 genes; an example is that of varicella-zoster virus, the causative agent of chickenpox (**Fig. 6.20A**). The genome includes two "unique" segments of genes, one long and one short (U_L and U_S), joined by two inverted repeats (IRs). Other herpes genomes share similar structure, though they differ in gene order and IR position.

The relatedness of different herpes viruses that evolved from a common ancestor can be measured by comparing

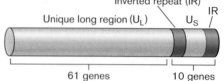

A. Varicella-zoster virus (VZV) genome (120 kb)

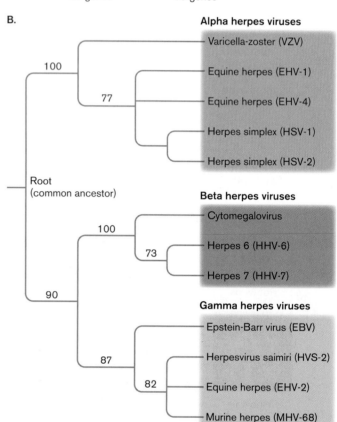

FIGURE 6.20 ■ Phylogeny of herpes viral genomes.
A. Genome structure of human varicella-zoster virus (VZV), the causative agent of chickenpox. **B.** Phylogeny of human and animal herpes viruses, based on whole-genome sequence analysis comparing clusters of orthologous groups of genes. Numbers measure percentage of DNA sequence shared by two divergent genomes.

their genome sequences. Comparison is based on orthologous genes, or **orthologs**. Orthologs are genes of common ancestry in two genomes that share the same function (a topic discussed in Chapter 8). For cellular organisms, we often use the ribosomal RNA gene sequences to measure relatedness. Viruses have no ribosomal RNA, but closely related viruses share other orthologous genes. In pairs of orthologs, the amount of difference in sequence correlates approximately with the time following divergence from a common ancestor (a topic discussed in Chapter 17). A tree of genomic divergence (or phylogeny) was devised for herpes viruses (**Fig. 6.20B**). The comparison of all gene pairs places herpes strains into three related classes designated alpha, beta, and gamma (**Fig. 6.20B**). The alpha class includes human varicella-zoster virus and the oral and genital herpes viruses (HSV-1 and HSV-2), as well as equine herpes virus. The beta class includes cytomegalovirus, a common cause of congenital infections (present at birth), as well as two lesser-known viruses. The gamma class includes Epstein-Barr virus (the cause of infectious mononucleosis), as well as several viruses of animals. Divergence of phylogenetic trees is discussed further in Chapter 17.

Gene comparison generates a tree for closely related viruses such as herpes. But how can we assess phylogeny of more distantly related viruses that share no genes—and even have genomes of different kinds of nucleic acids? Unlike cells, viruses do not possess genes common to all species. Furthermore, viral genomes are highly mosaic; that is, they evolved from multiple sources. Mosaic genomes result from recombination or reassortment of chromosomes from different viruses coinfecting a host.

In some cases, phylogeny is inconsistent with the fundamental chemical composition of the genome. For example,

some DNA bacteriophages actually share closer ancestry with RNA bacteriophages than they do with DNA animal viruses. This is because two or more viruses can coinfect a cell and exchange genetic components. Thus, the genomic content of a virus can be influenced by its host range.

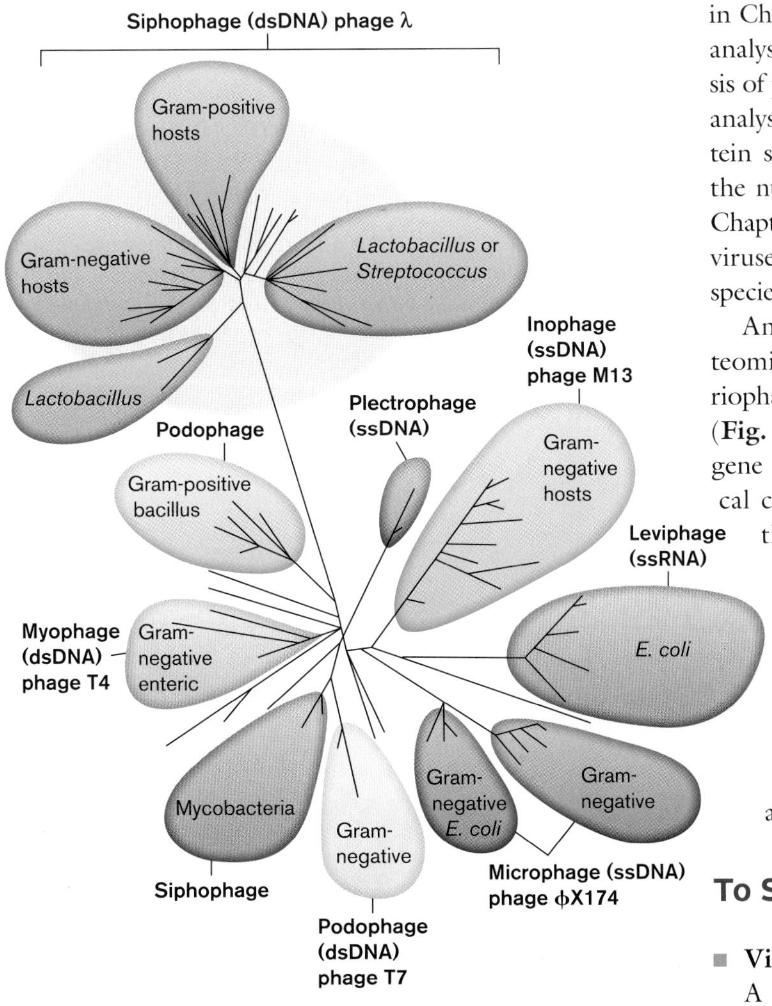

<hr>

Thought Question

6.5 How could viruses with different kinds of genomes (RNA versus DNA) combine and share genetic content in their progeny?

<hr>

As more viral genomes are sequenced, classification methods have been devised to take advantage of sequence information without requiring gene products common to all species. One promising approach is that of proteomics, analysis of the **proteome**, the proteins encoded by genomes (discussed in Chapter 10). Proteins are identified through biochemical analysis of virus particles and through bioinformatic analysis of protein sequences encoded in the genomes. Proteomic analysis is useful for distantly related viruses because protein sequences often show relatedness that is obscured in the nucleic acid sequence by silent mutations (discussed in Chapter 9). Even if only a few proteins are shared by any two viruses, statistical comparison of all proteins in a set of viral species reveals underlying degrees of relatedness.

An example of virus classification based on proteomic analysis is that of the "proteomic tree" of bacteriophages proposed by Forest Rohwer and Rob Edwards (**Fig. 6.21**). Unlike earlier trees based on a single common gene sequence, the proteomic tree is based on the statistical comparison of phage protein sequences predicted by the genomic DNA of many different species of phages. The proteomic analysis predicts major evolutionary categories of phage species that share common host bacteria. For example, phages that infect Gram-negative hosts will show more genetic commonality with each other than with phages that infect Gram-positive hosts. Shared hosts have a significant impact on phage evolution because coinfecting a host enables phages to exchange genes.

FIGURE 6.21 ■ The bacteriophage proteomic tree.
Comparison of all proteins encoded by each genome predicts distinct groups of bacteriophages. Within each group, there are subgroups of phages with shared hosts, since sharing of hosts facilitates genetic recombination and horizontal transfer of genes between different phages. *Source:* Based on Forest Rohwer and Rob A. Edwards. 2002. *J. Bacteriol.* **184**:4529.

To Summarize

- **Viruses contain infective genomes of RNA or DNA.** A viral genome may be single- or double-stranded, linear or circular.

- **Giant viruses** have 200–2,000 genes and may have evolved from intracellular parasitic cells.

- **Smaller viral genomes** comprise fewer than ten genes. Such viruses might have evolved from cell parts.

- **Classification of viruses** includes genome composition, virion structure, and host range.

- **The Baltimore virus classification** emphasizes the form of the genome (DNA or RNA, single- or double-stranded) and the route to generate messenger RNA.

- **Phylogeny of closely related viruses** can be calculated by comparing all related pairs of genes from the viral genomes.

- **Proteomic classification of distantly related viruses** includes information from all viral proteins. Statistical analysis reveals common descent of viruses infecting a common host.

6.4 Bacteriophages: The Gut Virome

Viruses display a remarkable diversity of ways to replicate within a host cell. Here, we discuss the key modes of bacteriophage replication, and their consequences for host cells. We focus on the best-known bacteriophages, those of the mammalian intestinal community. Gut bacteriophages, or "coliphages," are part of a microbial community that modulates human digestion, the immune system, and mental health.

Historically, bacteriophages have provided some of the most fundamental insights in molecular biology. In 1952, Alfred Hershey and Martha Chase showed that the transmission of DNA by a bacteriophage to a host cell led to the production of progeny bacteriophages, thus confirming that DNA is the hereditary material. In 1950, André Lwoff and Antoinette Gutman showed that a phage genome could integrate itself within a bacterial genome—the first recognition that genes could enter and leave a cell's genome. Other fundamental concepts of the genetic unit and the basis of gene transcription came from experiments on bacteriophages, as discussed in Chapters 7–9.

Bacteriophages Infect a Host Cell

To commence an infection cycle, bacteriophages need to contact and attach to the surface of an appropriate host cell. Contact and attachment are mediated by **cell-surface receptors**, proteins on the host cell surface that are specific to the host species and that bind to a specific viral component. A cell-surface receptor for a virus is actually a protein with an important function for the host cell, but the virus has evolved to take advantage of the protein. For example, phages that infect *Salmonella enterica* can use outer membrane proteins such as OmpF or TolC, which is part of a drug efflux complex (**Fig. 6.22**). Alternatively, a phage might bind to LPS (lipopolysaccharide; see Chapter 3). The phage-receptor binding is usually highly specific; a bacterium can evolve resistance through single-amino-acid mutations in its protein. The lambda phage receptor protein (maltose porin) of *E. coli* is described in Chapter 11.

Most bacteriophages (phages) inject only their genome into a cell through the cell envelope, thus avoiding the need for the capsid to penetrate the molecular barrier of the cell wall. For example, the phage T4 virion has a sheath that contracts, bringing the head near the cell surface to inject its DNA (**Fig. 6.23A** ▶). The pressure of the spooled DNA—as high as 50 atmospheres (atm)—is released,

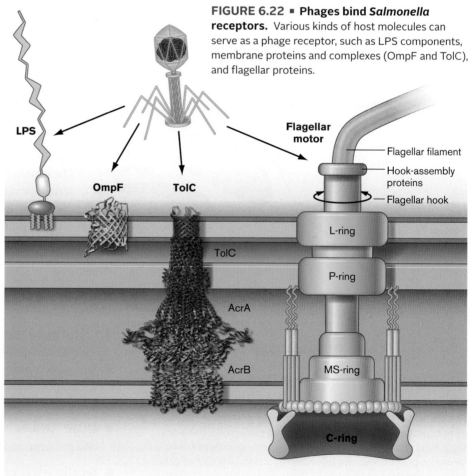

FIGURE 6.22 ■ Phages bind *Salmonella* receptors. Various kinds of host molecules can serve as a phage receptor, such as LPS components, membrane proteins and complexes (OmpF and TolC), and flagellar proteins.

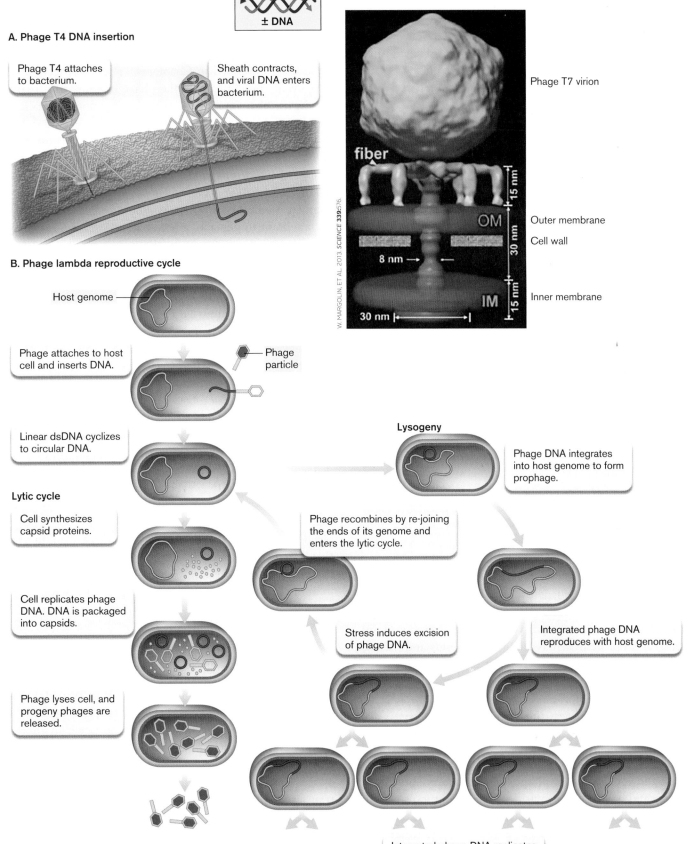

± DNA

A. Phage T4 DNA insertion

Phage T4 attaches to bacterium.

Sheath contracts, and viral DNA enters bacterium.

Phage T7 virion

fiber

15 nm

OM — Outer membrane

Cell wall

30 nm

8 nm

IM — Inner membrane

15 nm

30 nm

W. MARGOLIN, ET AL. 2013. SCIENCE **339**:576.

B. Phage lambda reproductive cycle

Host genome

Phage attaches to host cell and inserts DNA.

Phage particle

Linear dsDNA cyclizes to circular DNA.

Lytic cycle

Cell synthesizes capsid proteins.

Cell replicates phage DNA. DNA is packaged into capsids.

Phage lyses cell, and progeny phages are released.

Lysogeny

Phage DNA integrates into host genome to form prophage.

Phage recombines by re-joining the ends of its genome and enters the lytic cycle.

Stress induces excision of phage DNA.

Integrated phage DNA reproduces with host genome.

Integrated phage DNA replicates with host genome.

FIGURE 6.23 ▪ Bacteriophage reproduction: lysis and lysogeny. A. Phage T4 attaches to the cell surface by its tail fibers and then contracts to inject its DNA. The empty capsid remains outside as a "ghost." **Inset:** Cryo-EM model of phage T7 injecting DNA. **B.** Lysis occurs when the phage genome reproduces progeny phage particles, as many as possible, and then lyses the cell to release them. In phage lambda, lysogeny can occur when the phage genome integrates itself into that of the host. The phage genome is replicated along with that of the host cell. The phage DNA, however, can direct its own excision by expressing a site-specific DNA recombinase. This excised phage chromosome then initiates a lytic cycle. ▶

expelling the DNA into the cell. After the genome has been inserted, the phage capsid remains outside, attached to the cell surface. The empty capsid is termed a "ghost" because of its pale appearance in an electron micrograph.

The lytic cycle. A lytic cycle of replication generates a large number of progeny phages and then lyses the defunct cell. The lytic replication cycle requires these steps:

- **Host recognition and attachment.** A phage particle must contact a receptor molecule and adhere to a host cell.

- **Genome entry.** The phage genome must enter the host cell and gain access to the cell's machinery for gene expression.

- **Assembly of phages.** Phage components must be expressed and assembled. Components usually "self-assemble"; that is, the joining of their parts is favored thermodynamically.

- **Exit and transmission.** Progeny phages must exit the host cell, and then reach new host cells to infect.

In a lytic cycle, when a phage particle injects its genome into a cell, it immediately reproduces as many progeny phage particles as possible (**Fig. 6.23B**). The process of reproduction involves replicating the phage genome, as well as expressing phage mRNA to make enzymes and capsid proteins. Some phages, such as T4, digest the host DNA to increase the efficiency of phage production. Phage particles assemble, and the host cell lyses, releasing progeny phages.

The phage adsorbs to its host receptor and inserts its double-stranded DNA into the host cytoplasm. The phage genes are then expressed by the host cell RNA polymerase and ribosomes. "Early genes" are expressed early in the lytic cycle. Other phage-expressed proteins then work together with the cellular enzymes and ribosomes to replicate the phage genome and produce phage capsid proteins. The capsid proteins self-assemble into capsids and package the phage genomes—a process that takes place in defined stages, like a factory assembly line. At last, a "late gene" from the phage genome expresses an enzyme that lyses the host cell wall, releasing the mature virions. **Lysis** is also referred to as a burst, and the number of virus particles released is called the **burst size**.

Lysogeny. A **temperate phage**, such as phage lambda, can infect and lyse cells like a virulent phage, but it also has an alternative pathway: to integrate its genome as a prophage (see **Fig. 6.23B**). The phage is said to "lysogenize" the host, leading to a state called **lysogeny**. Phage lambda has a linear genome of double-stranded DNA, which circularizes upon entry into the cell. The circularized genome then recombines into that of the host by **site-specific recombination** of DNA. In site-specific recombination, a recombinase enzyme aligns the phage genome with the host DNA and exchanges the DNA backbone linkages with those of the host genome. (This process of DNA backbone recombination is explained in Chapter 9.) The DNA recombination event thus integrates the phage genome into that of the host. Now integrated, the phage genome is called a prophage. The presence of the prophage prevents further infection (superinfection) by other virions of the same type.

In lysogeny, the prophage DNA is replicated along with that of the host cell as the host reproduces (**Fig. 6.23B**). The host gains the benefit of resistance to superinfection. Implicit in the term "lysogeny," however, is the ability of such a strain to generate a lytic burst of phage. For a lysogen to enter lysis, the prophage directs its own excision from the host genome by an intramolecular process of site-specific recombination. The two ends of the phage genome exchange their DNA linkages so as to come apart from the host DNA, which now closes its circle with the prophage removed. As the phage DNA exits the host genome, it circularizes and initiates a lytic cycle, destroying the host cell and releasing phage particles.

How does a lysogen "decide" to reactivate and begin a lytic cycle? The decision between lysogeny and lysis is determined by proteins that bind DNA and repress the transcription of genes for virus replication (discussed in Chapter 11). Exit from lysogeny into lysis can occur at random, or it can be triggered by environmental stress such as UV light, which damages the cell's DNA. The regulatory switch of lysogeny responds to environmental cues indicating the likelihood that the host cell will survive and continue to propagate the phage genome. If a cell's growth is strong, it is more likely that the phage DNA will remain inactive, whereas events that threaten host survival will trigger a lytic burst. Similarly, in animal viral infections such as herpes, an environmental stress triggers reactivation of a virus that was dormant within cells (a latent infection). Reactivation of a latent herpes infection results in painful outbreaks of skin lesions.

During the exit from lysogeny, the virus can acquire host genes and pass them on to other host cells. The process of transferring host genes is known as **transduction**. Sometimes a transducing bacteriophage picks up a bit of host genome and transfers it to a new host cell. In another kind of transduction, the entire phage genome is replaced by host DNA packaged in the phage capsid, resulting in a virus particle that transfers only host DNA. Host DNA transferred by viruses can become permanently incorporated into the infected host genome, providing genes that express products useful to the host (**Table 6.2**).

The mechanisms of phage-mediated transduction are discussed in Chapter 9. In natural environments, phage transduction mediates much of the recombination of bacterial genomes. In the laboratory, the ability of phages to transfer genes provided some of the first vectors for recombinant DNA technology (see Chapter 12).

TABLE 6.2	Integrated Viral Genomes that Provide Host Traits		
Prophage (host infected)	**Bacterial product from prophage gene**	**Human endogenous retrovirus (HERV)**	**Human protein or regulator provided by HERV**
Phage C1 (*Clostridium botulinum*)	Botulnum toxin (*c1*)	HERV-W	Syncytin-1 (retroviral Env protein; placental fusion)
Beta phage (*Corynebacterium diphtheriae*)	Diphtheria toxin (*tox*)	HERV-FRD	Syncytin-2 (retroviral Env protein; placental fusion)
Lambda (*E. coli* O157:H7)	Cell envelope protein (*bor*)	HERV	INSL4 (insulin-like protein 4; placental development)
Phage 933 (*E. coli* O157:H7)	Shiga toxin (*stx*)	HERV-E	MID1 (midline development; prevents Opitz syndrome)
Epsilon 34 (*Salmonella enterica*)	LPS synthesis (*rfb*)	HERV-E	Apolipoprotein C1 (liver function)
Epsilon 34 (*Salmonella enterica*)	Type III secreted toxin (*sopE*)	HERV-E	Endothelin type-B receptor (placenta function)
TSST-1 (*Staphylococcus aureus*)	Toxic shock syndrome (*speA*)	HERV-L	Beta-1,3-galactosyltransferase (colon and mammary gland function)
CTXphi (*Vibrio cholerae*)	Cholera toxin (*ctxAB*)	LINE-1	ATRN (soluble attractin; modulates inflammation)

The slow-release cycle. A slow-release replication cycle differs from lysis and lysogeny in that phage particles reproduce without destroying the host cell (**Fig. 6.24**). Slow release is performed by filamentous phages such as phage M13. In slow-release replication, the single-stranded circular DNA of M13 serves as a template to synthesize a double-stranded intermediate. The double-stranded intermediate slowly generates single-stranded progeny genomes, which are packaged by supercoiling and coated with capsid proteins. The phage particles then extrude through the cell envelope without lysing the cell. The host cell continues to reproduce, though more slowly than uninfected cells do, because many of its resources are diverted to virus production.

Thought Question

6.6 What are the relative advantages and disadvantages (to a virus) of the slow-release strategy, compared with the strategy of a temperate phage, which alternates between lysis and lysogeny?

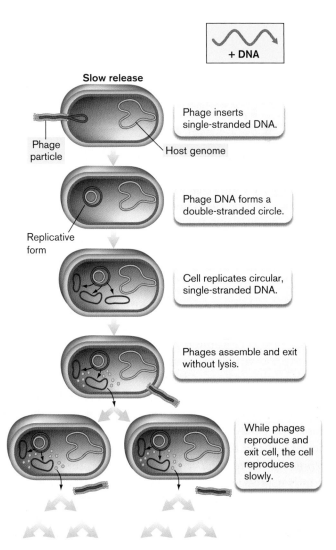

FIGURE 6.24 ■ Bacteriophage replication cycle: slow release. In the slow-release replication cycle, a filamentous phage produces phage particles without lysing the cell. The host continues to reproduce itself, but more slowly than uninfected cells do because many of its resources are being used to make phages.

| SPECIAL TOPIC 6.1 | Virus to the Rescue |

When your environment heats up, what can help you survive? Viruses can, according to Marilyn Roossinck, at the Samuel Roberts Noble Foundation in Oklahoma, and her student Regina Redman. Redman and Roossinck identified a mysterious virus in a fungus, *Curvularia protuberata*, which grows associated with a panic grass, *Dichanthelium lanuginosum*, in Yellowstone National Park. The fungus, *C. protuberata*, grows as an "endophyte" within the vasculature of the grass—a highly common form of symbiosis (discussed in Chapter 21). What is remarkable about this fungus-plant pair is that each partner confers thermal tolerance (heat resistance) on the other (**Fig. 1**). Together, the plant and fungus grow at temperatures as high as 65°C—which commonly occur in the vicinity of Yellowstone's thermal springs. Separately, the fungus and plant each grow at temperatures no higher than 38°C, typical of their species.

Roossinck knew that fungi harbor many diverse viruses, and that viruses often mediate fungus-plant interactions. She hypothesized that a fungal virus might be involved in the thermotolerance. To hunt for a virus, she used northern blot (probing RNA transcripts with labeled DNA) to analyze the fungal RNA for small double-stranded segments (dsRNA) typical of fungal viruses (double-stranded RNA; **Table 6.1**, Group III). The northern blot revealed two viral dsRNAs that form the genome of a previously unknown virus. Roossinck ultimately named the virus Curvularia thermal tolerance virus (CThTV).

But does thermal tolerance require the virus? To answer this question, Roossinck had to isolate CThTV virions from the virus's fungal host (**Fig. 2A**). She then conducted experiments comparing plants colonized by fungi with and without the virus. She found that only the fungi infected with CThTV could enable the plant *Dichanthelium lanuginosum* to grow

A.

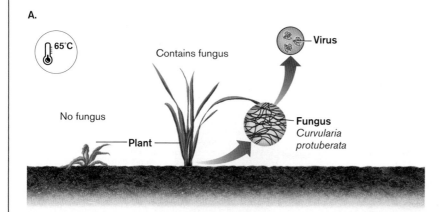

B.

FIGURE 1 ▪ **Plants colonized by virus-infected fungus survive heat shock. A.** A Yellowstone grass colonized by fungi harboring a virus grows taller than grass lacking fungi. **B.** Regina Redman, now at Symbiogenics, discovered the role of the fungal virus in thermal resistance. *Source:* Modified from Marilyn Roossinck. 2011. *Nat. Rev. Microbiol.* **9**:99.

Integrated Viruses as Host Cell Parts

Many prophages and endogenous viruses function as part of their host cell. **Table 6.2** shows examples of prophages that express virulence genes for pathogenic bacteria—a benefit to the pathogen, enabling it to better colonize the host animal. For example, phage C1 in *Clostridium botulinum* expresses the peptide botulinum toxin. This toxin causes the paralysis associated with botulism (and is also the basis of the Botox cosmetic treatment.) In *Corynebacterium diphtheriae*, the cause of diphtheria, the diphtheria toxin is expressed by beta phage. Other virulence factors (proteins that enhance disease) expressed by prophages include envelope proteins that help defend the bacterium from the immune system.

Analogous to bacterial prophages, many human genes and gene control elements have evolved from human viruses, particularly human endogenous retroviruses (HERVs). Retroviruses are a category of virus that includes HIV, the cause of AIDS (discussed in Section 6.5, and in greater detail in Chapter 11). HERVs are reverse-transcribing RNA viruses whose DNA copies became permanently fixed in our chromosomes. For example, placental proteins called "syncytin" mediate cell fusion during an early stage of placental development, allowing the fused syncytium to implant in the uterine lining and access the maternal blood supply. Two genes for syncytin evolved from HERV genes encoding retroviral envelope proteins. The details of the remarkable story of endogenous retroviruses—and the

at high temperature. Remarkably, the virus-infected fungus could also confer thermal tolerance on tomato plants (**Fig. 2B**). Thus, the virus-fungus mutualism may confer a specific thermotolerance property on various plant species—including plants important for agriculture.

Redman has now joined the Symbiogenics agricultural research firm, to extend her work using endophytic fungi to protect commercial rice plants from heat and salt stress. The viral components of these fungus-plant mutualisms are as yet unknown, but they offer exciting prospects for sustaining agriculture during climate change.

RESEARCH QUESTION

Researchers hypothesize that the mechanism by which the CThTV-infected fungus protects the plant is to inactivate the reactive oxygen ions produced by the plant's stress response to high temperature. Can you propose an experiment to test this hypothesis?

Márquez, Luis M., Regina S. Redman, Russell J. Rodriguez, and Marilyn J. Roossinck. 2007. A virus in a fungus in a plant: Three-way symbiosis required for thermal tolerance. *Science* **315**:513–515.

Redman, Regina S., Yong Ok Kim, Claire J. D. A. Woodward, Chris Greer, Luis Espino, et al. 2011. Increased fitness of rice plants to abiotic stress via habitat adapted symbiosis: A strategy for mitigating impacts of climate change. *PLoS One* **6**:e14823.

A.

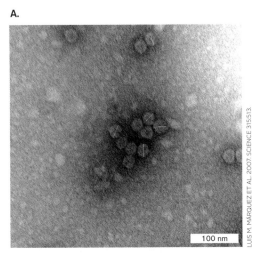

B.

Tomato plants with fungus: 65°C shock

No virus   Fungus has virus

100 nm

FIGURE 2 ■ **Virus confers thermotolerance on tomato plants. A.** Virus (CThTV) isolated from a fungus that confers thermotolerance. **B.** Tomato plants with fungus lacking virus (left) die after exposure at 65°C; with virus present (right), plants survive.

related story of retroviral gene therapy—are discussed in Chapter 11.

The virus-host relationship often evolves in response to environmental stress. The role of virus-host mutualism in stressed plants offers exciting possibilities for agriculture (see **Special Topic 6.1**).

Bacterial Defenses

In natural environments, viruses commonly outnumber cellular microbes by tenfold or more. So how have host cells evolved to defend themselves? Several remarkable resistance mechanisms have evolved. Their molecular basis is explained in greater detail in Chapters 7 and 9.

Genetic resistance. All bacteria acquire random mutations in their genomes as they reproduce (see Chapters 7 and 9). When attacked by bacteriophages, bacterial populations undergo natural selection; mutants that happen to be harder to infect will survive. Bacteria resist phage infection by expressing a gene that encodes an altered host receptor protein, which fails to bind the viral coat protein. Alternatively, a different cellular protein evolves to block phage binding to the receptor. An evolutionary "arms race" ensues, in which phages may evolve enzymes that cleave the host defense molecules.

Restriction endonucleases. Bacteria modify their DNA by adding methyl groups to bases within certain sequences. The bacteria then express restriction endonucleases, enzymes that cleave DNA lacking the methylated patterns—which includes potential virus DNA (see Chapter 7). However, phage genomes composed of RNA or of modified DNA (such as phage T4, discussed in Chapter 11) escape cleavage by these enzymes.

CRISPR: a bacterial immune system. Amazingly, bacteria possess an adaptive defense against viruses that is analogous to an immune system. (Adaptive immunity of humans is presented in Chapter 24.) The bacterial adaptive defense involves short DNA sequences homologous to DNA of phages that could infect the cell. The sequences are called <u>c</u>lustered <u>r</u>egularly <u>i</u>nterspaced <u>s</u>hort <u>p</u>alindromic <u>r</u>epeats (**CRISPR**). When a phage attacks a bacterium, if bacterial enzymes succeed in destroying the phage DNA, they may copy a tiny piece of it as a CRISPR segment, inserted as a spacer at the head of a long line of about 30 CRISPR sequences (**Fig. 6.25**).

> **Thought Question**
>
> **6.7** How might a phage evolve resistance to the CRISPR host defense (outlined in **Fig. 6.25**)?

Now the adapted host cell "remembers" infection by the specific phage—along with many other previous phages from previous infections that had inserted other spacers. The next time the adapted host cell is attacked by the same phage, all of its genomic CRISPR sequences are expressed as RNA. The CRISPR RNA is cleaved into small sequences (crRNA) containing one spacer from the original bacterial DNA. The crRNA binds to the Cascade protein complex (or Cas complex), which now detects phage DNA homologous to its virus-derived crRNA. The Cas-crRNA complex proceeds to cleave the phage DNA, preventing phage replication. For more details of this exciting molecular defense system, see Chapter 9.

The Gut Bacteriophage Community

How do bacteriophages impact microbial communities? The best-understood phage community is that of the gut **virome**. A virome is a community of viruses within a host ecosystem (for example, the human or other animal intestinal tract). We present some of the major phage-host interactions within the human intestinal tract in **Figure 6.26**. Note that this simplified diagram omits the viruses that

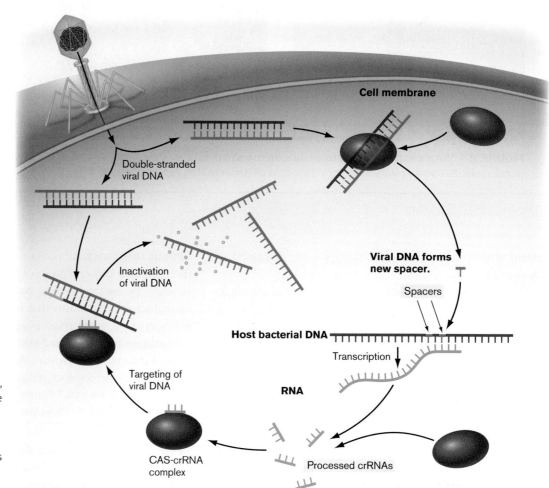

FIGURE 6.25 ■ CRISPR defense of bacterial cell. A piece of phage DNA gets copied as a "spacer" into the host genome. If the bacterium survives infection, later reinfection by the same kind of phage causes transcription of the spacers into CRISPR RNA. A processed spacer (crRNA) joins the Cas complex to recognize and cleave the phage DNA.

Cell membrane

Double-stranded viral DNA

Inactivation of viral DNA

Viral DNA forms new spacer.

Spacers

Host bacterial DNA

Transcription

RNA

Targeting of viral DNA

CAS-crRNA complex

Processed crRNAs

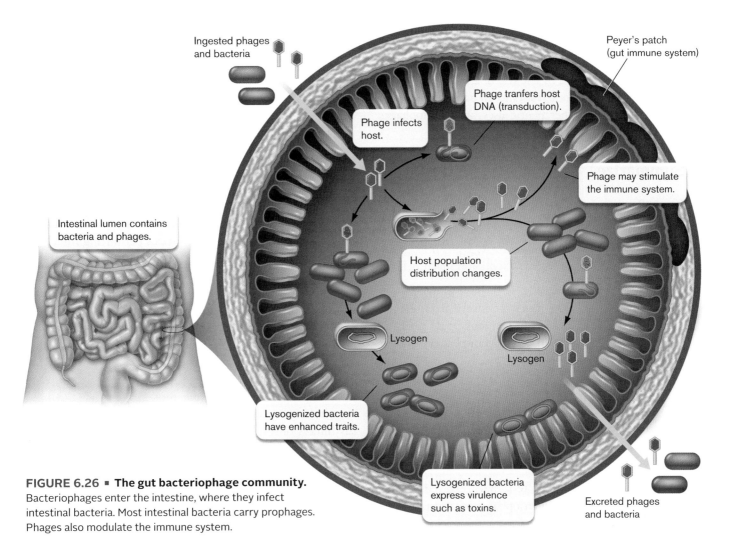

FIGURE 6.26 ■ The gut bacteriophage community.
Bacteriophages enter the intestine, where they infect
intestinal bacteria. Most intestinal bacteria carry prophages.
Phages also modulate the immune system.

infect human body cells; we focus here on the phages that
infect gut bacteria.

The human intestines contain a remarkably dense microbial community. Bacterial and archaeal cells are estimated
at 10^{11}–10^{13} cells per gram, whereas phage particles are
estimated at 10^9 per gram. The phage count is likely under-
estimated, as it is based on detection by electron microscopy and plate counts of viability (discussed in Section 6.6).
In addition, more than half the genomes sequenced from
gut bacteria show lysogeny, commonly with several different types of prophage in a given genome. The prophages
may protect the gut bacteria from superinfection by some
of the many types of phages present in the gut.

The gut community receives continual influx of new
bacteria and phages, while shedding present members
(**Fig. 6.26**). When a new type of phage enters, what happens? If the phage encounters a susceptible host bacterium, it may undergo a lytic cycle, lysogeny, or slow release,
depending on the phage's genetic program. Cell lysis may
have the effect of depopulating a dominant population of
the bacterial community. What is the result? One possible
result is intestinal "dysbiosis," deterioration of health due

to loss of health-enhancing bacteria. Another bacterial species, less healthy for the host, may increase in population.
But now, a different phage species (DNA shown red) may
depopulate the newly risen species, restoring equilibrium
in the community. Still other phages may transfer useful
genes from newcomer bacteria, such as genes encoding
enzymes to metabolize new kinds of food molecules.

Some prophages of pathogenic bacteria express virulence
factors, such as the Shiga toxin of *Shigella* and of *E. coli*
O157:H7 (**Table 6.2**). At the same time, research yields
intriguing evidence for phage effects that are positive:

■ **Phages may limit the bacterial numbers to levels that
the human immune system can tolerate.** Lysogenized
bacteria may use quorum sensing to detect host cell populations and "decide" whether to start a lytic cycle.

■ **Phage particles may modulate immune system activity** by suppressing T-cell activation and tumor formation
(T-cells are discussed in Chapter 24).

■ **Phages may attack biofilms.** Biofilms of pathogens such as
Pseudomonas aeruginosa may be eroded by phage infection.

The positive potential of bacteriophages has led researchers to investigate the engineering of phages for phage therapy. An idea dating back to the early twentieth century, phage therapy was eclipsed by the rise of antibiotics. Today, as we face growing antibiotic resistance in pathogens, we are taking another look at therapeutic uses of bacteriophages (Chapters 5 and 11).

To Summarize

- **Host cell-surface receptors** mediate the attachment of bacteriophages to a cell and confer host specificity.

- **Lytic cycle.** A bacteriophage injects its DNA into a host cell, where it utilizes host gene expression machinery to produce progeny virions.

- **Lysogeny.** Some bacteriophages can insert their genome into that of the host cell, which then replicates the phage genome along with its own. A lysogenic bacterium can initiate a lytic cycle.

- **Gene transfer.** Genes are transferred by phage processes of transduction and lysogeny.

- **Slow release.** Some bacteriophages use the host machinery to make progeny that bud from the cell slowly, slowing growth of the host without lysis.

- **Bacterial host defense.** Bacteria have evolved several forms of defense against bacteriophage infection, such as altered receptor proteins, restriction endonucleases, and CRISPR integration of phage DNA sequences.

- **The gut bacteriophage community** includes phages that infect, lyse, or lysogenize bacterial hosts. Gut phages impact bacterial community structure, transfer genes among bacteria, and modulate the human gut immune system.

6.5 Animal and Plant Viruses

Animal and plant viruses solve problems similar to those faced by bacteriophages: host attachment, genome entry and gene expression, virion assembly, and virion release. The more complex structure of eukaryotic cells, however, requires viral replication cycles that are more complex. Viral reproduction may involve intracellular compartments such as the nucleus or secretory system and may depend on tissue and organ development in multicellular organisms. Studies of animal virus replication reveal potential targets for antiviral drugs, such as protease inhibitors for HIV (discussed in Chapter 11).

Note: The virus replication cycles in this chapter are simplified. For greater molecular detail of selected viruses, see Chapter 11.

Animal Viruses Show Tissue Tropism

Like bacteriophages, animal viruses evolve by the fitness advantage of binding specific receptor proteins on their host cell. An example of a human virus-receptor interaction is that of rhinovirus (see **Table 6.1**, Group IV), which causes the common cold. Rhinovirus attaches to ICAM-1, a human glycoprotein (protein with sugar chains) needed for intercellular adhesion (**Fig. 6.27A**). The rhinovirus binds to a domain of ICAM-1 essential for ICAM-1 to bind a lymphocyte protein called integrin.

The host receptors play a key role in determining the host range, the group of host species permitting infection. Within a host, receptor molecules can also determine the viral **tropism**, or ability to infect a particular tissue type within a host. Some viruses, such as Ebola virus, exhibit broad tropism, infecting many kinds of host tissues, whereas others, such as papillomavirus, show tropism for only one type (in the case of papillomavirus, the epithelial tissues). Tropism may depend on the virus's ability to interact with the cytoplasm, or it may require the presence of an appropriate host cell receptor protein that can bind the viral surface attachment protein. For example, poliovirus infects only a specific class of human cells that display the immunoglobulin-like receptor protein PVR. Mice lack the PVR protein on their cell surfaces, and they are not normally infected by polio, but when transgenic mice were engineered to express PVR on their cells, the mice could then be infected.

Tissue-specific receptors determine the host tropism of avian influenza strain H5N1. The H5N1 strain infects birds by binding to a glycoprotein receptor on cell surfaces of the avian respiratory tract. The H5N1 strain requires a receptor protein with a sialic acid sugar chain terminating in galactose linked at the C-3 position (alpha-2,3), common in the avian respiratory tract. In humans, however, most nasal upper respiratory cells have receptors with galactose linked at the C-6 position (alpha-2,6). Human cells with the C-3 linkage are more common in the lower respiratory tract. That is why avian influenza H5N1 infection of humans has been relatively rare. However, only a small mutation in the H5N1 envelope protein could enable it to bind to the alpha-2,6 receptor more effectively, allowing rapid transmission between humans. The molecular basis of influenza infection is discussed further in Chapter 11.

Thought Question

6.8 How could humans undergo natural selection for resistance to rhinovirus infection? Is such evolution likely? Why or why not?

A. Host receptor binding

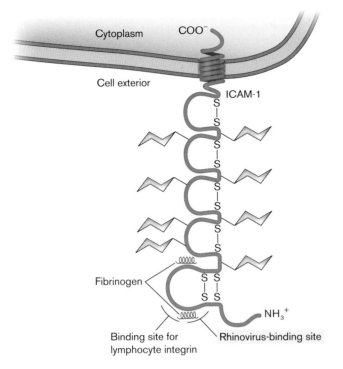

B. Coated RNA genome enters cytoplasm

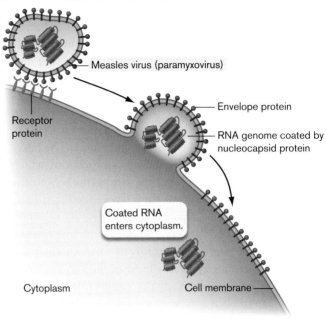

C. Uncoating within endosomes

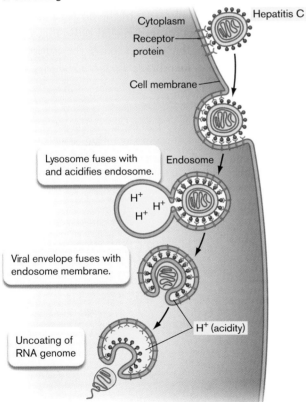

D. Uncoating at the nuclear membrane

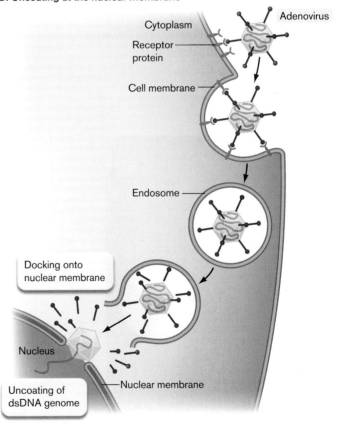

FIGURE 6.27 ▪ Receptor binding and genome uncoating. **A.** Rhinovirus attaches to the intercellular adhesion molecule (ICAM-1), a glycoprotein required by the host cell to bind a lymphocyte integrin, a cell-surface matrix protein required for cell-cell adhesion. After binding a specific receptor on the host cell membrane, an animal virus enters the cell, where its genome is uncoated. **B.** Measles virus: The coated RNA genome enters the cytoplasm. **C.** Hepatitis C: The genome is uncoated within an endosome. **D.** Adenovirus: The genome is uncoated at the nuclear membrane. *Source:* Part A based on J. Bella et al. 1998. *PNAS* **95**:4140.

Genome entry and uncoating. Most animal viruses, unlike bacteriophages, enter the host cell as virions. The contents of the virion (genome and matrix proteins) may then interact with the cell in several different ways. For example, measles virus, a paramyxovirus, enters the cell by binding host receptor proteins, thereby causing the viral envelope to fuse with the host cell membrane (**Fig. 6.27B**). The measles RNA genome coated by nucleocapsid proteins is released directly into the cytoplasm. For other types of viruses, the entire virion is internalized within an endosome. The internalized capsid undergoes **uncoating**, a process in which the capsid comes apart, releasing the viral genome into the cytoplasm. Uncoating is shown for a flavivirus, hepatitis C virus (**Fig. 6.27C**). The hepatitis C virion is first taken up by **endocytosis**. In endocytosis, the cell membrane forms a vesicle around the virion and engulfs it, forming an endocytic vesicle. The endocytic vesicle fuses with a lysosome, whose acidity activates entry of the capsid into the cytoplasm. The capsid then comes apart, and the viral genome is uncoated.

Yet other kinds of viruses, such as adenovirus, enter the cell by endocytosis but require transport to the nucleus (**Fig. 6.27D**). Following endocytosis and lysosome fusion, the adenoviral genome loses some of its capsid proteins (partial uncoating). A capsid protein then disrupts the endocytic membrane, allowing the remaining capsid to exit. The capsid then docks at a nuclear pore and injects its DNA genome into the nucleus. The adenoviral DNA is replicated by its own adenoviral DNA polymerase (carried in the virion), but it uses host nuclear histones for DNA packing, as well as host transcription factors available in the nucleus.

Animal Virus Replication Cycles

How does a virus replicate within an animal cell? The replication cycle of a given virus depends on the form of the viral genome. A DNA genome can use some or all of the host replication enzymes. An RNA genome, however, must encode either an RNA-dependent RNA polymerase to generate an RNA template or a reverse transcriptase to generate a DNA template, in the case of retroviruses.

DNA virus replication. An example of a double-stranded DNA virus is human papillomavirus (HPV) (see **Table 6.1**, Group I), the cause of genital warts (**Fig. 6.28A**). HPV is the most common sexually transmitted disease in the United States and one of the most common worldwide. Certain strains infect the skin, whereas others infect the mucous membranes through genital or anal contact (sexual transmission). Like phage lambda, papillomavirus has an active reproduction cycle and a dormant cycle in which the viral genome integrates into that of the host.

HPV initially infects basal epithelial cells (**Fig. 6.28B**). In the basal cells, papilloma virions enter the cytoplasm by receptor binding and membrane fusion. The virion then

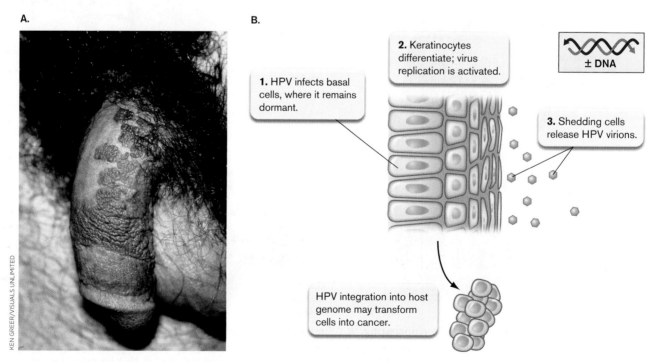

A.

B.

1. HPV infects basal cells, where it remains dormant.

2. Keratinocytes differentiate; virus replication is activated.

± DNA

3. Shedding cells release HPV virions.

HPV integration into host genome may transform cells into cancer.

KEN GREER/VISUALS UNLIMITED

FIGURE 6.28 ■ Human papillomavirus. A. Certain strains of human papillomavirus (HPV) cause warts on the genitals or anus. **B.** HPV infects basal epithelial cells, where the DNA uncoats but remains dormant. As cells differentiate, new virions are synthesized and released by shedding cells. Some HPV proteins transform host cells into cancer cells.

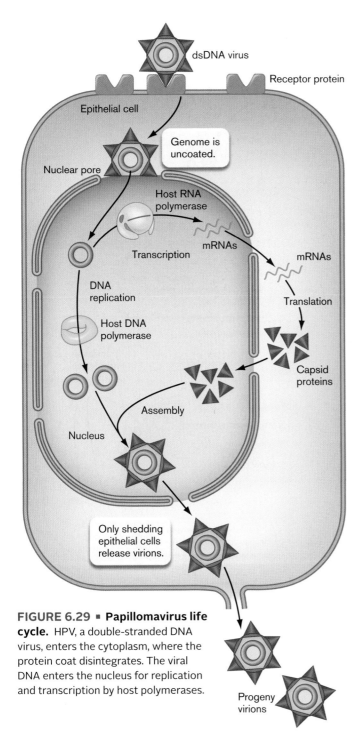

FIGURE 6.29 ■ Papillomavirus life cycle. HPV, a double-stranded DNA virus, enters the cytoplasm, where the protein coat disintegrates. The viral DNA enters the nucleus for replication and transcription by host polymerases.

Viral replication is largely inhibited until the basal cells start to differentiate into keratinocytes. Host cell differentiation induces the viral DNA to replicate and undergo transcription by host polymerases. The mRNA transcripts then exit the nuclear pores, as do host mRNAs, for translation in the cytoplasm. The translated capsid proteins, however, return to the nucleus for assembly of the virion. Nuclear virion assembly is typical of DNA viruses (with the exception of poxviruses, which replicate entirely in the cytoplasm).

How do HPV virions disseminate, and how may they cause cancer? As the keratinocytes containing HPV complete differentiation, the cells start to come apart and are shed from the epithelial surface. Cells release HPV virions during this shedding process. But in the basal cells, HPV has an alternative pathway of integrating its genome into that of the host (analogous to phage lysogeny). The integrated genome can transform host cells into cancer cells. Cells are transformed to cancer through increased expression of viral oncogenes (cancer-causing genes; see **eTopic 26.1**). The oncogenes inhibit the expression of host tumor suppressor genes. Certain HPV strains are more likely to cause cancer than others. The most common strains are preventable by the Gardasil vaccine.

RNA virus replication. The picornaviruses include poliovirus, the cause of paralytic poliomyelitis; and rhinoviruses, which cause the common cold (see **Table 6.1**, Group IV). Picornavirus genomes contain (+) strand RNA, allowing the virus to replicate entirely in the cytoplasm, without any DNA intermediates.

A picornavirus binds to a surface receptor, such as ICAM-1 for rhinovirus or the PVR receptor for poliovirus. The (+) strand RNA is uncoated by insertion through the cell membrane into the cytoplasm—much as a bacteriophage inserts its genome into a cell (**Fig. 6.30**). The role of endocytosis is debated; poliovirus requires no endocytosis, but the rhinovirus genome may require endocytosis and low-pH induction.

After uncoating, a gene in the picornavirus RNA is translated by host ribosomes to make RNA-dependent RNA polymerase. The polymerase uses the viral RNA template to make (−) strand RNA. The (−) strand RNA then serves as a template for other viral mRNAs, as well as for progeny genomic (+) RNA. These RNA molecules are synthesized within virus-induced vesicles or replication complexes formed from the endoplasmic reticulum. Capsid proteins are synthesized by host ribosomes, and the capsids self-assemble in the cytoplasm. Virions then assemble at the cell membrane and are released by subverting lysosomes, which attempt to digest them. For greater detail of the poliovirus replication cycle, see **eTopic 11.3**.

undergoes uncoating via disintegration of the protein capsid or coat, releasing its circular, double-stranded genome (**Fig. 6.29**). The uncoated viral genome enters the nucleus, where it is replicated by the host DNA polymerase and transcribed by the host RNA polymerase. The viral mRNA molecules then return to the cytoplasm for translation of capsid proteins, which return to the nucleus for assembly of virions.

The process of HPV reproduction is complicated by the developmental progression of basal cells into keratinocytes (mature epithelial cells), and ultimately cells to be shed or sloughed off from the surface (see **Fig. 6.28B**).

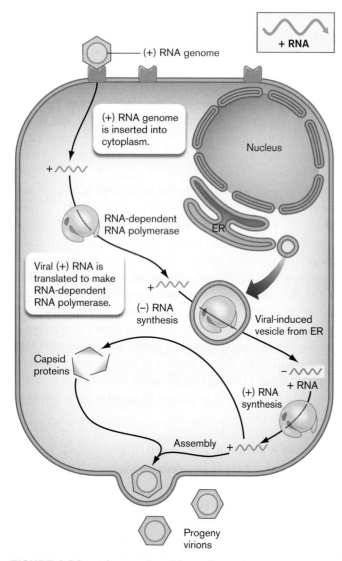

FIGURE 6.30 ■ Picornavirus life cycle. A picornavirus inserts its (+) strand RNA into the cell. Reproduction occurs entirely in the cytoplasm. A key step is the early translation of a viral gene to make RNA-dependent RNA polymerase.

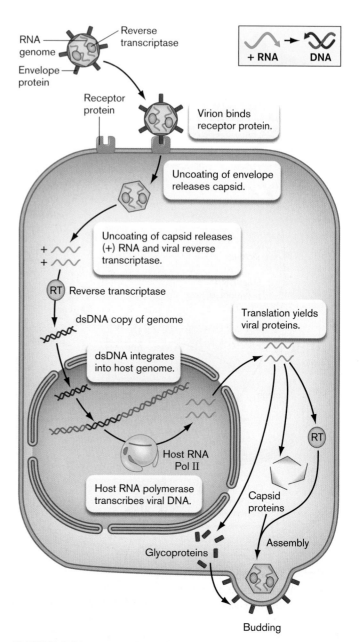

FIGURE 6.31 ■ Retrovirus life cycle. A retrovirus such as human immunodeficiency virus (HIV) uses reverse transcriptase to copy its RNA into double-stranded DNA. The DNA then enters the nucleus to recombine in the host genome, where a host RNA polymerase generates viral mRNA and viral genomic RNA.

Note that other kinds of RNA viruses, such as influenza virus, encapsidate a (−) strand genome. In this case, the (−) strand must serve as the template to generate mRNA, as well as a (+) secondary template for (−) strand progeny genomes. Influenza virus replication includes other interesting molecular complications, discussed in Chapter 11.

RNA retroviruses. Retroviruses include human immunodeficiency virus (HIV), the causative agent of AIDS; and feline leukemia virus (FeLV), the cause of feline leukemia, a disease that commonly afflicts domestic cats (see **Table 6.1**, Group VI). A retrovirus such as HIV uses a reverse transcriptase to make a DNA copy of its RNA genome (**Fig. 6.31**). Instead of being translated from an early gene, the reverse transcriptase is actually carried

within the virion, bound to the RNA genome with a primer in place. The virion contains two (+) strand RNA copies of the HIV genome, each carrying its own reverse transcriptase. Virions infect helper T cells and other lymphocytes of the active immune system (discussed in Chapter 24). After uncoating in the cytoplasm, the viral RNA is copied into double-stranded DNA.

The DNA copy then enters the nucleus, where a viral integrase enzyme integrates the DNA into the host genome. To generate virions, a host RNA polymerase transcribes the viral genome into viral mRNA and viral genomic RNA. The reverse transcriptase has a high error rate, generating slightly different versions of the virus—some of which can evade host defenses and antiviral drugs. The viral mRNA reenters the cytoplasm for translation to produce coat proteins, reverse transcriptase, and envelope proteins. The coat proteins are transported by the endoplasmic reticulum (ER) to the cell membrane, where virions self-assemble and bud out. Alternatively, a retrovirus may transmit from cell to cell through cytoplasmic nanotubes (introduced in Chapter 3), which avoid viral destruction by the immune system.

At a certain point, infected cells can suddenly begin to generate large numbers of virions, destroying the immune system. The cause of accelerated reproduction is poorly understood, although it involves cytokines released under stress conditions such as poor health or pregnancy. HIV replication is activated via protein regulators encoded by the virus. The full process of HIV reproduction is described in Chapter 11.

Thought Question

6.9 From the standpoint of a virus, what are the advantages and disadvantages of replication by the host polymerase, compared with using a polymerase encoded by its own genome?

Oncogenic viruses. As many as 20% of human cancers are caused by **oncogenic viruses**, such as Epstein-Barr virus (which causes lymphomas) and hepatitis C virus (which causes liver cancer). (Hepatitis C replication is discussed in **eTopic 11.4**.) When these viruses infect a cell, instead of destroying it the virus may **transform** the cell to divide and grow out of control. For a virus, the advantage of cancer transformation is that it expands the population of infected cells that produce virus particles, or that replicate a viral genome hidden within a host chromosome.

How do oncogenic viruses cause cancer? Several different mechanisms enable different types of viruses to transform normal host cells to proliferate abnormally and form tumors.

■ **Oncogenes.** A retrovirus such as feline leukemia virus (FeLV) may carry an oncogene, which can transform the host cells. Usually an oncogene encodes an abnormal form of a host protein that controls cell proliferation.

■ **Genome integration.** Certain DNA viruses, such as human papillomavirus, can integrate their genome into a host chromosome. The integrated viral genome expresses proteins that stimulate host cell division and may ultimately lead to growth of tumors.

■ **Cell cycle control.** Oncogenic viruses such as papillomaviruses express oncogenes for proteins that interact with cell cycle controls and can stimulate uncontrolled growth.

Viral capacities for gene transfer and host genome control may be manipulated artificially and used for gene therapy. In fact, some of the most dangerous viruses, such as HIV, are being engineered to make nonvirulent gene delivery devices (discussed in Section 11.4).

Chronic viral infections. Some human viruses appear in our bodies indefinitely, causing either mild disease or none. For example, anelloviruses commonly appear in our blood plasma, without known disease. Children frequently shed enteroviruses (viruses of the intestinal tract, related to poliovirus) with no known effects. Some evidence supports the idea that low-level chronic viral infections might actually enhance the function of our immune system.

Plant Virus Replication Cycles

All kinds of plants are subject to viral infection. Plant viruses pose enormous challenges to agriculture, especially where the concentrated growth of a single strain of food crop (monoculture) provides ideal conditions for a virus to spread.

Plant virus entry to host cells. In contrast to animal viruses and bacteriophages, plant viruses infect cells by mechanisms that do <u>not</u> involve specific membrane receptors. The reason may be that plant cell membranes are covered by thick cell walls impenetrable to virion uptake or genome insertion. Thus, the entry of plant viruses usually requires **mechanical transmission**, nonspecific access through physical damage to tissues, such as abrasions of the leaf surface caused by a feeding insect. Mechanical transmission of plant viruses is limited by the cell wall. Most plant viruses gain entry to cells by one of three routes:

■ **Contact with damaged tissues.** Viruses such as tobacco mosaic virus appear to require nonspecific entry into broken cells.

■ **Transmission by an animal vector.** Insects and nematodes transmit many kinds of plant viruses. For example, the geminiviruses are inoculated into cells by plant-eating insects such as aphids, beetles, and grasshoppers.

■ **Transmission through seed.** Some plant viruses enter the seed and infect the next generation.

An economically important plant virus is the potyvirus called plum pox virus, a major pathogen of plums,

FIGURE 6.32 ▪ Plum pox is caused by potyvirus. A. Potyvirus, a filamentous (+) strand RNA virus, approximately 800 nm in length (TEM). **B.** Potyvirus is transmitted by aphids, which suck the plant sap and release the virus into the damaged tissues. **C.** Streaking of flowers caused by potyvirus infection. **D.** Ring-shaped pockmarks appear on the infected fruit.

peaches, and other stone fruits. Plum pox virus, a Group IV (+) strand RNA virus, is transmitted by aphids (**Fig. 6.32**). Following infection, the spread of the virus generates streaked leaves and flowers, as well as ring-shaped pockmarks on the surfaces of the fruit and of the stone within.

Plant virus transmission through plasmodesmata. Within a plant, the thick cell walls prevent a lytic burst or budding out of virions. Instead, plant virions spread to uninfected cells by traveling through **plasmodesmata** (singular, **plasmodesma**). Plasmodesmata are membrane channels that connect adjacent plant cells (**Fig. 6.33**). The outer channel connects the cell membranes of the two cells; the inner channel connects the endoplasmic reticulum.

Passage through the plasmodesmata requires action by movement proteins whose expression is directed by the viral genome. In some cases, the movement proteins transmit the entire plant virion; in other cases, only the nucleic acid itself is small enough to pass through. The infection strategies of plant viral genomes may have features in common with those of viroids, which lack capsids altogether.

DNA pararetroviruses. Pararetroviruses possess a DNA genome that requires transcription to RNA in the cytoplasm, followed by reverse transcription to form DNA genomes for progeny virions. Some pararetroviruses, such as hepadnavirus, infect humans, but the best-known pararetrovirus is cauliflower mosaic virus (CaMV), a caulimovirus

(see **Table 6.1**, Group VII). CaMV is an important tool for biotechnology because it has a highly efficient promoter for gene transcription, allowing high-level expression of cloned genes. Vectors derived from caulimoviruses are used to construct transgenic plants.

CaMV is transmitted by secretions from an insect whose bite damages plant tissues, providing access to the

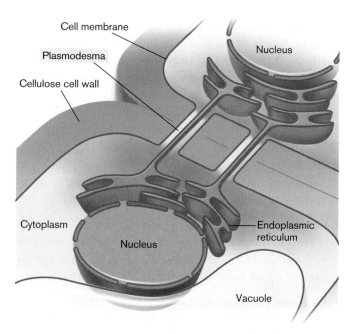

FIGURE 6.33 ▪ Plant cells connected by plasmodesmata. Plasmodesmata offer a route for plant viruses to reach uninfected cells.

cytoplasm (**Fig. 6.34**). The CaMV genome moves from the cytoplasm to the cell nucleus through a nuclear pore. Within the nucleus, two promoters on its DNA genome direct transcription to RNA. The two RNA transcripts exit the nucleus for translation by host ribosomes to make viral proteins. A host reverse transcriptase, present in plant cells, copies the RNA into DNA viral genomes. After virions are assembled in the cytoplasm, virus-encoded proteins, called movement proteins, help transfer the virions through plasmodesmata into an adjacent cell.

A CaMV promoter sequence is commonly used in gene transfer vectors for plant biotechnology because transcription of the gene (such as one that confers pesticide resistance on the host) linked to the viral promoter is very efficient. In the field, 10% of cruciferous vegetables are typically infected with cauliflower mosaic virus. Some critics of gene technology fear that the prevalence of the CaMV promoter in transgenic crops may lead to the evolution of new pararetroviruses.

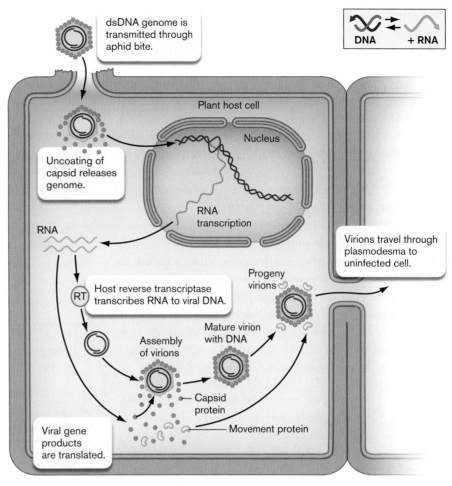

FIGURE 6.34 ■ Caulimovirus life cycle. The cauliflower mosaic virus (CaMV), a DNA pararetrovirus, uses host RNA polymerase to copy its DNA into RNA and uses host reverse transcriptase (RT) to make DNA copies.

Animal and Plant Host Defenses

How do animals and plants defend themselves from virus infection? Since viruses are ubiquitous, a wide range of defense mechanisms have evolved. Defenses important for humans are part of our immune system, presented in Chapters 23 and 24.

Genetic resistance. As we saw for bacteria, animal and plant hosts continually experience mutations, some of which lead to strains that resist viral infection by halting adsorption or some other key step of the virus's replication cycle. When a virus becomes widespread, natural selection favors resistant strains. But commercial livestock and crops are typically a monoculture in which no resistant variants are available. Thus, when an outbreak arises, an entire crop may be destroyed. To save a crop, it may be interbred with a wild strain that possesses genes conferring resistance.

An example of genetic resistance in humans is resistance to HIV/AIDS. In populations where HIV prevalence is high, rare individuals emerge whose cells are not infected, or are infected at such low rates that the virus is eventually cleared. The basis of this resistance is a defective allele encoding a T-lymphocyte cell-surface protein that acts as a coreceptor, which is required for binding of virus HIV-1. The role of coreceptors in HIV infection and resistance is discussed further in Chapter 11.

Immune system. The immune system of humans and other animals possesses extensive cellular machinery to thwart viral infection. A component of our innate immunity is the class of proteins called interferons, which recognize general signs of viral infections such as the presence of double-stranded RNA (see Chapter 23). For adaptive immunity, viral proteins expressed in the cell membrane of an infected cell are recognized by specific antibodies that stimulate immune cells to destroy the infected cell and halt its viral production. The antibodies recognize a specific virus strain, such as a strain of influenza virus to which the individual has been exposed previously. For example, during the 2009 flu epidemic, many individuals over the age of 50 had some protection arising from exposure to a similar strain in an earlier epidemic (see Chapter 24).

RNA interference. RNA interference, or RNAi, is a mechanism by which mRNA molecules expressed by a viral genome are recognized by a host protein-RNA complex that shuts down further expression. RNA interference was first discovered in plants, where the system is most extensive, but it is now known to be widespread among all eukaryotes and archaea (discussed in Chapter 9). The mechanisms of RNA interference are now being engineered for use in gene therapy to halt gene expression in cancer and in inherited diseases.

Emergence of Viral Pathogens

Where does a "new virus" come from? Most human pathogens come from other humans, or from related animals called vectors. Certain human-infecting viruses persist in the wild, such as rabies virus and West Nile virus. Their persistence requires broad host ranges; rabies infects many different mammals, whereas West Nile virus infects birds as well as humans and horses. Understanding the epidemiology of rabies or of West Nile encephalitis requires understanding the behavior and seasonal migration patterns of wild organisms. In 2002, the SARS respiratory disease was caused by a coronavirus previously unknown to science. The coronavirus was eventually traced genetically to viruses found in bats and in civet cats marketed for food, in the area of Guangdong, China, where the human outbreak began.

Other emerging viruses arise as variants of endemic milder pathogens. Viruses long associated with a host, such as the common-cold viruses (rhinoviruses), tend to have evolved a moderate disease state that provides ample opportunities for host transmission. A virus that "jumps" from an animal host, however, may cause a more acute syndrome with higher mortality. The best-known cases are the exceptionally virulent emerging strains of influenza, which generally result from intracellular recombination of human strains with strains from pigs or ducks (discussed in Chapter 11). For example, in 2013 the avian influenza strain H7N9 emerged from poultry in China, where it killed several people before it was contained. Changing distribution patterns of insect vectors and animal hosts can generate new epidemics of a pathogen in regions where the virus could not spread before. Such changes in distribution can be brought about by many factors, including global climate change (**eTopic 6.2** and Chapter 28).

To Summarize

- **Host cell-surface receptors** mediate animal virus attachment to a cell and confer host specificity and tropism.

- **Animal DNA viruses** either inject their genome or enter the host cell by endocytosis. The viral genome requires uncoating for gene expression.

- **RNA viruses** use an RNA-dependent RNA polymerase to transcribe their messenger RNA.

- **Retroviruses** use a reverse transcriptase to copy their genomic sequence into DNA for insertion in the host chromosome.

- **Oncogenic viruses** transform the host cell to become cancerous. Mechanisms of oncogenesis by different types of viruses include insertion of an oncogene into the host genome, integration of the entire viral genome, and expression of viral proteins that interfere with host cell cycle regulation.

- **Plant viruses** enter host cells by transmission through a wounded cell surface or an animal vector. Plant viruses travel to adjacent cells through plasmodesmata.

- **Pararetroviruses** contain DNA genomes but generate an RNA intermediate that requires reverse transcription to DNA for progeny virions.

- **Emerging viral pathogens** increase during environmental change.

6.6 Culturing Viruses

To learn how microbes grow, we culture them in the laboratory. So how do we culture a virus? A complication of virus culture is the need to grow the virus within a host cell. Therefore, any virus culture system must be a double culture of host cells plus viruses. Culturing viruses of multicellular animals and plants involves additional complications, as viruses show tropism for particular tissues or organs. Viruses may replicate in tissue culture, but the tissue culture does not show all the properties of an organ within a living organism. Therefore, a virus propagated in tissue culture will evolve to lose some of the virulence factors needed to infect an animal.

Batch Culture

Batch culture, or culture in an enclosed vessel of liquid medium, enables growth of a large population of viruses for study. Bacteriophages can be inoculated into a growing culture of bacteria, usually in a culture tube or a flask. The culture fluid is then sampled over time and assayed for phage particles. The growth pattern usually takes the form of a step curve (**Fig. 6.35**).

To observe one cycle of phage reproduction, phages are added to host cells at a **multiplicity of infection** (**MOI**, ratio of phage to cells) such that every host cell is infected. The phage particles immediately adsorb to surface receptors of host cells and inject their DNA. As a result, intact virions

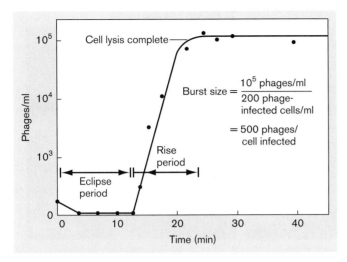

FIGURE 6.35 ■ One-step growth curve for a bacteriophage. After initial infection of a liquid culture of host cells, the titer of virus drops near zero as all virions attach to the host. During the eclipse period, progeny phages are being assembled within the cell. As cells lyse (the rise period), virions are released until they reach the final plateau.

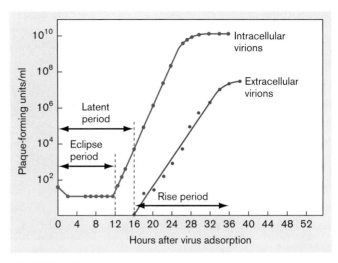

FIGURE 6.36 ■ One-step growth curve for a virus. The titer of extracellular virus drops to near zero during the latent period, as all virions adsorb to the host. Then progeny virions begin to emerge by budding out from the infected cell. For an animal virus, the growth curve may take hours to level off; the "burst" event is not defined as clearly as for phages.

are virtually undetectable in the growth medium. This short period after infection is called the **eclipse period**. For some species, it is possible to distinguish between the eclipse period and a **latent period**, the time during which the inserted phage genome directs production of progeny virions. The virions accumulate within the cell but have not yet emerged in the medium. In animal viruses, the latent period is less distinct because large numbers of virions usually generate progeny through budding out of the host cell (**Fig. 6.36**).

Note: The "latent period" of a lytic virus is the period between initial phage-host contact and the first appearance of progeny phage. This must be distinguished from the "latent infection" of a virus that maintains its genome within a host cell without reproducing virions.

As cells begin to lyse and liberate progeny viruses, the culture enters the **rise period**, during which virus particles appear in the growth medium. The rise period ends when all the progeny viruses have been liberated from their host cells. If the number of viruses that go on to inoculate additional host cells is small, then the virus concentration at the end point divided by the original concentration of inoculated phage approximates the **burst size**—that is, the number of viruses produced per infected host cell. To estimate the burst size, we can divide the concentration of progeny virions by the concentration of inoculated virions, assuming that all the original virions infect a cell.

The burst size, together with the cell density prior to lysis, determines the concentration of the resultant

suspension of virus particles, called a **lysate**. In the case of bacteriophages, a lysate of phage particles can be extremely stable, remaining infective at room temperature for many years. Eukaryotic viruses, however, tend to be less stable and need to be maintained in culture or deep freeze.

Thought Question

6.10 Why does bacteriophage reproduction give a step curve, whereas cellular reproduction generates an exponential growth curve? Could you design an experiment in which viruses generate an exponential curve? Under what conditions does the growth of cellular microbes give rise to a step curve?

Tissue Culture of Animal Viruses

In the case of animal and plant viruses, the multicellular nature of the host is an important factor in the pathology and transmission of the pathogen (discussed in Chapters 25 and 26). Animal viruses can be cultured within whole animals by serial inoculation, where virus is transferred from an infected animal to an uninfected one. Culture within animals ensures that the virus strain maintains its original **virulence** (ability to cause disease). But the process is expensive and laborious, involving the large-scale use of animals.

A historic event in 1949 was the first successful growth of a virus in tissue culture. Poliovirus, the causative agent

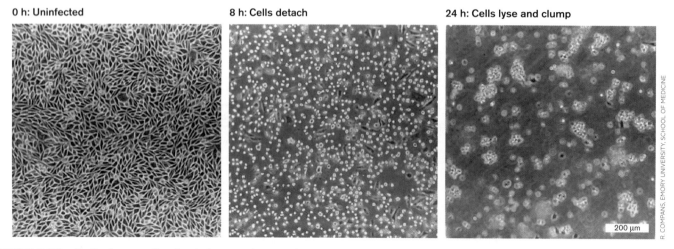

FIGURE 6.37 ■ **Poliovirus replication in human tissue culture.** Before infection (0 hours), the cultured cells grow in a smooth layer. At 8 hours, infected cells have detached from the culture dish. By 24 hours, cells have lysed or in some cases clumped with other cells.

of the devastating childhood disease poliomyelitis, was grown in human cell tissue culture (**Fig. 6.37**) by John F. Enders, Thomas J. Weller, and Frederick Robbins at Children's Hospital in Boston. As heralded that year in *Scientific American*, "It means the end of the 'monkey era' in poliomyelitis research.... Tissue-culture methods have provided virologists with a simple in vitro method for testing a multitude of chemical and antibiotic agents." Since then, tissue culture has remained the most effective way to study the molecular biology of animal and plant viruses and to develop vaccines and antiviral agents.

Some viruses can be grown in a tissue culture of cells growing confluently on a surface. The cells must be immortalized—that is, genetically altered—to continue cell division indefinitely. The fluid bathing the tissue layer is sampled for virus concentration. As in the case of bacteriophage batch culture, we can define an eclipse period, a latent period before appearance of the first progeny virions in the culture fluid, and a rise period. In tissue culture, the time course of animal virus replication is usually much longer (hours or days) than that of bacteriophages (typically less than an hour under optimal conditions). The burst size of animal viruses, however, is typically several orders of magnitude larger than that of phages. The reason for the larger burst size is probably that the volume of a host cell is much larger than that of a bacterial host, thus providing a larger supply of materials to build virions.

Thought Question

6.11 What kinds of questions about viruses can be addressed in tissue culture, and what questions require infection of an animal model?

Plaque Isolation and Assay of Bacteriophages

For the investigation of cellular microbes, an important tool is the culturing of individual colonies on a solid substrate that prevents dispersal throughout the medium, as described in Chapters 1 and 4. Plate culture of colonies enables us to isolate a population of microbes descended from a common progenitor. But viruses cannot be isolated as "colonies." The reason is that although viruses can be obtained at incredibly high concentrations, they disperse in suspension. Even on a solid medium, viruses never form a solid visible mass comparable to the mass of cells that constitutes a cellular colony.

In viral plate culture, viruses from a single progenitor lyse their surrounding host cells, forming a clear area called a **plaque**. Each plaque arises from a single infected bacterium that bursts, its phage particles diffusing to infect neighboring cells. To perform a plaque assay of bacteriophages, a diluted suspension of phages is mixed with bacterial cells in soft agar, and the mixture is then poured over a nutrient agar plate (**Fig. 6.38**). Where no bacteriophages are present, the bacteria grow homogeneously as a "lawn," an opaque sheet over the surface (confluent growth). Where there is a bacteriophage, it infects a cell, replicates, and spreads progeny phages to adjacent cells, killing them as well (**Fig. 6.39**). The loss of cells results in a round, clear area seemingly cut out of the bacterial lawn. Plaques can be counted and used to calculate the concentration of phage particles, or **plaque-forming units** (**PFUs**), in a given suspension of liquid culture. The liquid culture can be analyzed by serial dilution in the same way one would analyze a suspension of bacteria.

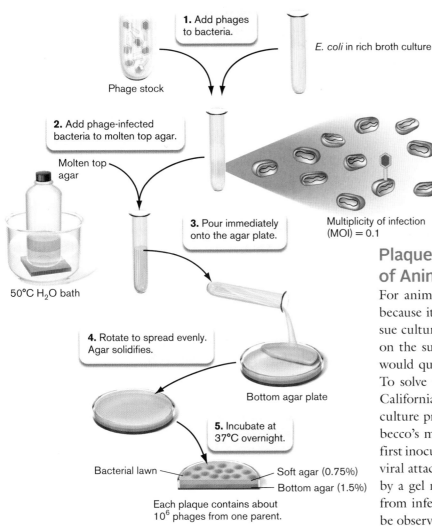

1. Add phages to bacteria.

E. coli in rich broth culture

Phage stock

2. Add phage-infected bacteria to molten top agar.

Molten top agar

3. Pour immediately onto the agar plate.

Multiplicity of infection (MOI) = 0.1

50°C H₂O bath

4. Rotate to spread evenly. Agar solidifies.

Bottom agar plate

5. Incubate at 37°C overnight.

Bacterial lawn

Soft agar (0.75%)
Bottom agar (1.5%)

Each plaque contains about 10⁶ phages from one parent.

FIGURE 6.38 ■ Plating a phage suspension to count isolated plaques. A suspension of bacteria in rich broth culture is inoculated with a low proportion of phage particles (multiplicity of infection is approximately 0.1). Each plaque arises from a single infected bacterium that bursts, its phage particles diffusing to infect neighboring cells.

Plaques offer a convenient way to isolate a recombinant DNA molecule contained in a bacteriophage vector (discussed in Chapter 12). In **Figure 6.39B**, the blue plaques result from a phage vector carrying the gene encoding the enzyme beta-galactosidase. This enzyme converts a colorless compound into a blue dye. When the indicator gene is interrupted by an inserted recombinant gene, the phage produces white plaques, which indicate the successful production of recombinant DNA phages.

Plaque Isolation and Assay of Animal Viruses

For animal viruses, the plaque assay has to be modified because it requires infection of cells in tissue culture. Tissue culture usually involves growth of cells in a monolayer on the surface of a dish containing fluid medium, which would quickly disperse any viruses released by lysed cells. To solve this problem, in 1952 Renato Dulbecco, at the California Institute of Technology, modified the tissue culture procedure for plaque assays (**Fig. 6.40A**). In Dulbecco's method, the tissue culture with liquid medium is first inoculated with virus. After sufficient time to allow for viral attachment to cells, the fluid is removed and replaced by a gel medium. The gel retards the dispersal of viruses from infected cells, and as the host cells die, plaques can be observed. **Figure 6.40B** shows a plate culture of human coronavirus infection of colon carcinoma cells.

Animal viruses that do not kill their host cells require a different kind of assay, based on identification of a "focus" (plural, foci), a group of cells infected by the virus. A focus may be identified using a fluorescent antibody, called a fluorescent-focus assay. Another type of focus assay can be used to isolate oncogenic viruses, which transform their host cells into cancer cells. The cancer cells lose contact inhibition; they grow up in a pile

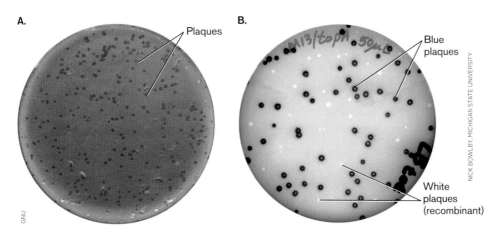

A. Plaques

B. Blue plaques

White plaques (recombinant)

NICK BOWLBY, MICHIGAN STATE UNIVERSITY

FIGURE 6.39 ■ Phage plaques on a lawn of bacteria. **A.** Phage lambda plaques on a lawn of *Escherichia coli* K-12. **B.** Plaques of recombinant phage M13 on *E. coli*. The original phage expresses beta-galactosidase, an enzyme that makes a blue product (blue plaques). White plaques are produced by phage particles whose genome is recombinant (contains a cloned gene interrupting the gene for beta-galactosidase).

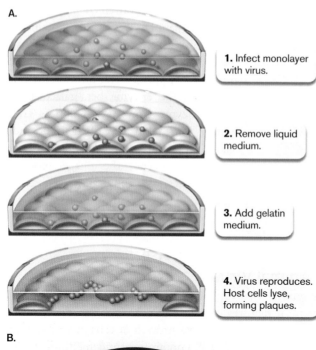

1. Infect monolayer with virus.

2. Remove liquid medium.

3. Add gelatin medium.

4. Virus reproduces. Host cells lyse, forming plaques.

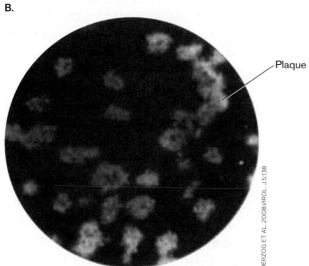

Plaque

HERZOG ET AL. 2008/VIROL J.5:138

FIGURE 6.40 ▪ Plate culture of animal viruses. A. Modified plaque assay for animal viruses. The gelled medium retards the dispersal of progeny virions from infected cells, restricting new infections to neighboring cells. The result is a visible clearing of cells (a plaque) in the monolayer. **B.** Plaque assay in which human coronavirus suspension was plated on a monolayer of colon carcinoma cells in tissue culture.

instead of remaining in the normal monolayer. These piles of transformed cells, or transformed foci, can easily be visualized and counted. This procedure is known as the transformed-focus assay.

To Summarize

- **Culturing viruses** requires growth in host cells.

- **Batch culture** of viruses generates a step curve.

- **Plate culture** involves replication of phages on a bacterial lawn, or animal viruses in tissue culture.

- **Plaque assay** is a method of culturing viruses in which single phages or virions each generate an isolated clearing of host cells. The plaques can be counted to enumerate the infectious virions in a suspension, called plaque-forming units.

Concluding Thoughts

In this chapter we have covered broadly the structure and function of all kinds of viruses. As devastating as viruses can be to their hosts, they also provide engines of genomic change (see Chapter 9) and ecological balance, including mutualistic relationships with a host (see Chapter 21). Understanding virus function prepares us to discuss microbial genetics in Chapters 7–10. Chapter 11 then explores in depth the molecular biology of selected viruses that cause human disease. Viral molecular biology is a field of growing importance for medical and agricultural research, for genetic engineering and gene therapy, and for understanding natural ecosystems.

CHAPTER REVIEW

Review Questions

1. Compare and contrast the form of icosahedral and filamentous (helical) viruses, citing specific examples.
2. How do viral genomes gain entry into cells in bacteria, plants, and animals?

3. Explain the key structural features that define the seven Baltimore groups of viral genomes. Explain the consequences of each structure for viral replication.
4. How do viral genomes interact with host genomes, and what are the consequences for host evolution?

5. Compare and contrast the lytic, lysogenic, and slow-release replication cycles of bacteriophages. What are the strengths and limitations of each?

6. Compare and contrast the replication cycles of RNA viruses and DNA viruses in animal hosts. What are the strengths and limitations of each?

7. Explain the plate count procedure for enumerating viable bacteriophages. How must this procedure be modified to measure the concentration of animal viruses? Oncogenic viruses?

8. Explain how a pure isolate of a virus can be obtained. How do the procedures differ from those used for isolating bacteria?

9. Explain the generation of the step curve of virus proliferation. Why is virus proliferation generally observed as a single step, or generation, in contrast to the life cycles of cellular microbes, outlined in Chapter 4?

10. Explain the key contributions of viruses to natural ecosystems. What may happen in an ecosystem where viruses are absent or fail to cause significant infection?

Thought Questions

1. Discuss the functions of different structural proteins of a virion, such as capsid, nucleocapsid, tegument, and envelope proteins. How do these functions compare and contrast with functions of cellular proteins?

2. What are the relative advantages of the virulent phage replication cycle of phage T4, the lysis-lysogeny options of phage lambda, and the slow-release replication of phage M13? Under what conditions might each strategy be favored over the others?

3. Given the basis of viral tropism, how might an animal evolve traits that confer resistance to a virus infection?

4. If viruses return a substantial fraction of marine CO_2 to the atmosphere, can you imagine any ways to modulate virus proliferation so as to divert the carbon into sedimenting biomass?

Key Terms

bacteriophage (195)
batch culture (228)
burst size (214, 229)
capsid (195, 196)
cell-surface receptor (212)
CRISPR (218)
DNA reverse-transcribing virus (210)
eclipse period (229)
endocytosis (222)
endogenous virus (196)
envelope (201)
filamentous virus (201)
host range (199)
icosahedral capsid (200)
latent period (229)
lysate (229)
lysis (214)

lysogeny (214)
mechanical transmission (225)
multiplicity of infection (MOI) (228)
oncogenic virus (225)
ortholog (orthologous gene) (210)
pararetrovirus (210)
phage (195)
plaque (195, 230)
plaque-forming unit (PFU) (230)
plasmodesma (226)
prion (204)
prophage (196)
proteome (211)
provirus (196, 198)
replication complex (196)
retrovirus (210)
reverse transcriptase (210)

rise period (229)
RNA-dependent RNA polymerase (208)
RNA reverse-transcribing virus (210)
segmented genome (207)
site-specific recombination (214)
spike protein (201)
temperate phage (214)
transduction (214)
transform (225)
tropism (220)
uncoating (222)
viral shunt (198)
virion (195, 196)
viroid (203)
virome (218)
virulence (229)
virus (195)

Recommended Reading

Abergel, Chantal, Matthieu Legendre, and Jean-Michel Claverie. 2015. The rapidly expanding universe of giant viruses: Mimivirus, Pandoravirus, Pithovirus and Mollivirus. *FEMS Microbiology Reviews* **39**:779–796.

Chaturongakul, Soraya, and Puey Ounjai. 2014. Phage–host interplay: Examples from tailed phages and Gram-negative bacterial pathogens. *Frontiers in Microbiology* **5**:442.

Chaudhry, Rabia M., J. E. Drewes, and Kara Nelson. 2015. Mechanisms of pathogenic virus removal in a full-scale membrane bioreactor. *Environmental Science & Technology* **49**:2815–2822.

Denner, Joachim, and Paul R. Young. 2013. Koala retroviruses: Characterization and impact on the life of koalas. *Retrovirology* **10**:108.

Grigg, Patricia, Allison Titong, Leslie A. Jones, Tilahun D. Yilma, and Paulo H. Verardi. 2013. Safety mechanism assisted by the repressor of tetracycline (SMART) vaccinia virus vectors for vaccines and therapeutics. *Proceedings of the National Academy of Sciences USA* **110**:15407–15412.

Grünewald, Kay, Prashant Desai, Dennis C. Winkler, J. Bernard Heymann, David M. Belnap, et al. 2003. Three-dimensional structure of herpes simplex virus from cryo-electron tomography. *Science* **302**:1396–1398.

Harris, Audray, Giovanni Cardone, Dennis C. Winkler, J. Bernard Heymann, Matthew Brecher, et al. 2006. Influenza virus pleiomorphy characterized by cryoelectron tomography. *Proceedings of the National Academy of Sciences USA* **103**:19123–19127.

Horvath, Phillippe, and Randolphe Barrangou. 2010. CRISPR/Cas, the immune system of bacteria and archaea. *Science* **327**:167–170.

Legendrea, Matthieu, Julia Bartoli, Lyubov Shmakova, Sandra Jeudy, Karine Labadie, et al. 2013. Thirty-thousand-year-old distant relative of giant icosahedral DNA viruses with a pandoravirus morphology. *Proceedings of the National Academy of Sciences USA* **111**:4274–4279.

Maggie, Fabricio, and Mauro Bendinelli. 2009. Immunobiology of the Torque teno viruses and other anelloviruses. *Current Topics in Microbiology and Immunology* **331**:65–90.

Mills, Susan, Fergus Shanahan, Catherine Stanton, Colin Hill, Aidan Coffey, et al. 2013. Movers and shakers. *Gut Microbes* **4**:4–16.

Raoult, Didier, Stéphane Audic, Catherine Robert, Chantel Abergel, and Patricia Renesto. 2004. The 1.2-megabase genome sequence of Mimivirus. *Science* **306**:1344–1350.

Rohwer, Forest, and Rebecca Vega Thurber. 2009. Viruses manipulate the marine environment. *Nature* **459**:207–212.

Roossinck, Marilyn J. 2011. The good viruses: Viral mutualistic symbioses. *Nature Reviews. Microbiology* **9**:99.

Srinivasiah, Sharath, Jaysheel Bhavsar, Kanika Thapar, Mark Liles, Tom Schoenfeld, et al. 2008. Phages across the biosphere: Contrasts of viruses in soil and aquatic environments. *Research in Microbiology* **159**:349–357.

Suttle, Curtis A. 2007. Marine viruses—major players in the global ecosystem. *Nature Reviews. Microbiology* **5**:801–812.

Trask, Shane D., Sarah M. McDonald, and John T. Patton. 2012. Structural insights into the coupling of virion assembly and rotavirus replication. *Nature Reviews. Microbiology* **10**:165–177.

CHAPTER 7
Genomes and Chromosomes

A genome is all of the genetic information that defines an organism. For a bacterial species, this can include its chromosome, plasmids, and viruses. Today, information stored in almost any genome can be revealed and modified through new technologies that rapidly sequence DNA and manipulate its content. In Chapter 7 we discuss how bacterial genomes replicate, organize, and migrate through the dividing bacterial cell.

CURRENT RESEARCH highlight

Do bacterial plasmids and chromosomes mix? Plasmids are small DNA elements that replicate separately from the chromosome. But do chromosomes and plasmids intermingle or remain separated in the cell? Rodrigo Reyes-Lamothe and colleagues fluorescently tagged *E. coli* plasmid DNA (green) and chromosomal DNA (nucleoid, red) with different GFP proteins and examined their locations. Plasmids consistently localized in nucleoid-free regions at the cell poles (shown), indicating that the two forms of DNA do not readily mingle. How the nucleoid excludes plasmids is unclear.

Source: Reyes-Lamothe et al. 2014. *Nucleic Acids Res.* **42**:1042–1051.

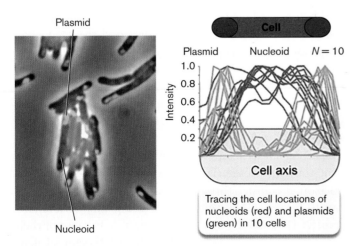

Plasmid

Nucleoid

Tracing the cell locations of nucleoids (red) and plasmids (green) in 10 cells

AN INTERVIEW WITH

RODRIGO REYES-LAMOTHE, MICROBIOLOGIST, MCGILL UNIVERSITY

How has your work on DNA segregation altered our view of replication?

One noteworthy discovery was when we found that the DNA disentangling activity of topoisomerases follows very soon after the DNA replication fork moves along the chromosome. Another revelation came when we discovered that the preferred localization of plasmids at the cell poles does not overlap with that of the chromosome. This localization makes plasmids a good tool, in combination with fluorescence microscopy, for dissecting the timing of DNA replication events.

Are there potential antimicrobial application(s) for your work?

Many components of DNA replication have great potential as drug targets. DNA replication is an underexploited target of current antibiotics, despite the fact that bacterial DNA replication is an essential cellular process whose machinery is evolutionarily distinct from ours.

How does a bacterium organize and replicate its genome? As shown in the Current Research Highlight, Dr. Rodrigo Reyes-Lamothe and his collaborator, professor Marcelo Tolmasky (**Fig. 7.1**), found that plasmids and chromosomes, by and large, occupy different spaces in the cell. This segregation makes sense if the organism is to replicate and make RNA efficiently without entangling components of its genome. Compartmentalizing bacterial chromosomes and plasmids is one form of genome organization. Are there others?

The question of bacterial chromosome organization was first raised in the mid-twentieth century after one of our intestinal bacteria became the focus of efforts to understand genes and genetics. In one remarkable experiment, a cell of *E. coli* was lysed, releasing its chromosome for electron microscopy. What spewed out of this single cell was a strand of DNA 1,500 times longer than *E. coli* itself (**Fig. 7.2** ▶). Scientists wondered how this enormous molecule could fit into a single cell, and how an enzyme could duplicate that much DNA (more than 4.6 million base pairs) without the whole DNA molecule getting tangled up. And how did it manage to do this in under 20 minutes, the doubling time of the organism? Over half a century later, those questions continue to intrigue us.

In this chapter we discuss the structure and organization of prokaryotic genomes, as well as the mechanisms for their replication. We examine how microbial enzymes that manipulate DNA were used to develop techniques such as DNA sequencing and the polymerase chain reaction (PCR). We also introduce the concept of **information flow**—namely, how information stored in DNA gets passed to new generations and becomes protein. **DNA replication** copies the information held in a DNA sequence and makes a new DNA molecule passed to offspring. **Transcription** converts DNA information into a "readable" form called RNA. **Translation** then "reads" the information in RNA to make protein. Transcription and translation are discussed in Chapter 8. The remaining chapters of Part 2 discuss gene recombination and transmission (Chapter 9), the regulation of gene expression (Chapter 10), the specialized

FIGURE 7.1 ■ **Marcelo Tolmasky of McGill University collaborated with Rodrigo Reyes-Lamothe in discovering that plasmids and chromosomes occupy separate spaces in a bacterial cell.**

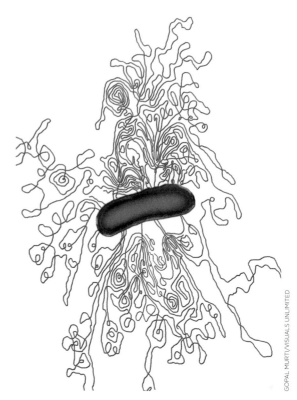

GOPAL MURTI/VISUALS UNLIMITED

FIGURE 7.2 ■ **Osmotically disrupted bacterial cell with its DNA released.** Length of the bacterium is approximately 2 micrometers. ▶

genetic mechanisms of viruses (Chapter 11), and the various molecular techniques and technologies that spawned the field of synthetic biology (Chapter 12).

7.1 DNA: The Genetic Material

Where do genetic traits come from, and how are they passed from one generation to the next? The ability of plants and animals to transfer genetic traits was known well before we knew that DNA was the carrier of genetic information. Early researchers understood the Mendelian model of gene inheritance and how it dictated the passage of traits from parent to offspring in eukaryotes—a gene transmission mechanism known as **vertical gene transfer**. Bacteria, however, appeared to operate differently, because in addition to vertical gene transfer, bacteria showed **horizontal gene transfer**, the movement of genetic information from one cell into another. The study of horizontal gene transfer in bacteria ultimately led to the discovery of horizontal transfer throughout animals and plants. Relative to bacteria, however, animals and plants horizontally transfer DNA on a much slower timescale because of their longer generation times. Mechanisms of horizontal gene transfer are discussed in Chapter 9.

Our path to understanding DNA began in 1928 when Frederick Griffith (1879–1941) discovered that he could kill mice with live but seemingly harmless (avirulent) *Streptococcus pneumoniae*, but only if he coinjected the mice with <u>dead</u> cells from a virulent strain of the bacteria. Something from the dead bacteria transformed the innocuous live bacteria into killers. This form of horizontal gene transfer, called **transformation**, led to the discovery in 1944 that genetic information is embedded in the base sequence of deoxyribonucleic acid (DNA) (see Section 1.6). Another example of transformation is the toxic shock gene that can be moved horizontally among strains of *Staphylococcus aureus*. As a result of these and many other experiments, we now appreciate that chromosomes are made of contiguous packets of information, called **genes**.

Genes are units of information composed of a sequence of DNA nucleotides of four different types: adenine (A), guanine (G), thymine (T), and cytosine (C). (For a review at the level of introductory biology, see eAppendix 1.) A **structural gene** is a string of nucleotides that can be decoded by an enzyme to produce a functional RNA molecule. A structural gene usually produces an RNA molecule that in turn encodes a protein. A **DNA control sequence**, on the other hand, regulates the <u>expression</u> of a structural gene. DNA control sequences do not encode RNA or protein (so they are not considered genes), but they do regulate RNA production from an adjacent structural gene. Control sequences include promoters that launch RNA synthesis from a structural gene, and binding sites for regulatory proteins that can activate or inactivate that promoter.

As noted previously, the entire genetic complement of DNA in a cell that defines it as an organism is called its **genome**. Our primary goal in Chapter 7 is to convey how genomes are maintained and replicated. Mining the informational content of DNA sequences and genes is described more fully in Chapters 8 and 10.

Bacterial Genomes

In the early twentieth century, the chromosomes of bacteria, unlike those of eukaryotes, could not be observed by light microscopy. The reason is that bacteria, unlike eukaryotes, do not undergo mitosis, a process in which chromosomes condense and thicken about a thousandfold, making them visible with light microscopy. Important clues to bacterial chromosome structure were gleaned, however, from painstaking genetic studies. In the 1950s it was discovered that **conjugation** (see Section 9.1), a horizontal gene transfer mechanism requiring cell-to-cell contact, could transfer large segments of some bacterial chromosomes—not all at once, as in the established Mendelian model of plants and animals, but sequentially over a period of time (it takes 100 minutes to move the entire *E. coli* chromosome from one cell to another).

Thus, even though the bacterial chromosome could not be seen, conjugation allowed genes to be mapped relative to each other according to time of transfer. For example, a donor strain that can synthesize the amino acids alanine and proline can directly transfer the encoding genes to a recipient cell defective in those genes. The transfer process is nonspecific, so any gene can be transferred in this way. Completion of transfer also requires **recombination**, in which the donor DNA fragment replaces the recipient DNA fragment. Successful transfer of the genes for amino acid synthesis enables the formerly defective recipient to form colonies on minimal media lacking either amino acid. Because chromosome transfer takes time and starts from a fixed point, not all genes are transferred simultaneously. It might take one gene 10 minutes of cell contact to be transferred, while the second gene, farther away from the starting point of DNA transfer, takes an additional 20 minutes.

Because eukaryotic chromosomes are linear, scientists initially expected that bacterial chromosomes would be linear too. However, the early genetic maps drawn from conjugation experiments just would not fit together in a manner consistent with a linear model—for the simple reason that the bacterial chromosome in *E. coli*, the organism under study at the time, is circular. We now know that most bacteria and archaea have circular chromosomes. Some species, however, have linear chromosomes, or even a mix of linear and circular chromosomes. Examples include the Lyme disease agent *Borrelia burgdorferi*, and the plant tumor agent *Agrobacterium tumefaciens* (**Table 7.1**).

The size range of genomes across the phylogenetic tree, from viruses to humans, is enormous. Generally speaking, the simpler the organism, the smaller its genome. At some point, as DNA content is trimmed, the organism loses independence and lives only if it has parasitized another organism. For examples, see the evolution of mitochondria and chloroplasts (Section 17.6) and obligate intracellular pathogens (Chapter 23).

To Summarize

- **A genome** is all of the genetic information that defines an organism.

- **Genomes** of bacteria and archaea are made up of chromosomes and plasmids consisting of DNA.

- **Chromosomes** of bacteria and archaea can be circular or linear, as can plasmids.

- **Functional units** of DNA sequences include structural genes and regulatory sequences.

TABLE 7.1	**Genomes of Representative Bacteria and Archaea**		
Species (strain)	**Chromosome(s)* (kilobase pairs, kb)** **Circular and linear**	**Plasmid(s)* (kb)** **Circular**	**Total (kb)**
Bacteria			
Mycobacterium tuberculosis Tuberculosis	4,400		4,400
Mycoplasma genitalium Normal flora, human skin	580		580
Burkholderia cepacia Respiratory infections in immunocompromised patients	3,870 + 3,217 + 876	93	8,056
***Escherichia coli* K-12 (W3110)** Model strain for *E. coli* research	4,600		4,600
***Anabaena* species (PCC 7120)** Cyanobacteria: major photosynthetic producer of carbon source for aquatic ecosystems	6,370	110 + 190 + 410	7,080
Borrelia burgdorferi Lyme disease	911	21; sizes 9 to 58	>1,250
Agrobacterium tumefaciens Tumors in plants; genetic engineering vector	2,840 + 2,070	214 + 542	5,666
Archaea			
Methanocaldococcus jannaschii Methanogen from thermal vent	1,660	16 + 58	1,734
Haloarcula marismortui Halophile from volcanic vent	3,130 + 288	33 + 33 + 39 + 50 + 155 + 132 + 410	4,270
	1,000 kb	500 kb	

*Purple circles and lines indicate relative sizes of genomic elements and whether these are circular or linear. Size bars are provided under each column.

7.2 Genome Organization

Do microbial genomes differ in structure or functional organization? New techniques for constructing physical maps of genomes and determining the sequences of whole genomes have revealed tremendous diversity in the size and organization of prokaryotic genomes.

Genomes Vary in Size

Bacterial and archaeal chromosomes range in size from approximately 130 to 14,000 kilobase pairs (kb). For comparison, eukaryotic chromosomes range from 2,900 kb (Microsporidia) to over 100,000,000 kb (flowering plants). The human genome is over 3,000,000 kb.

Note: The designation "kb" can refer to the length of a double-stranded or single-stranded DNA molecule. A bacterial genome is, by definition, double-stranded. Some viral genomes can be single-stranded.

One of the smallest cellular genomes sequenced thus far is that of *Mycoplasma genitalium*. These pathogens rely on their host environment for many products but can still grow outside a host cell. The complete genome of *M. genitalium* consists of only 580 kb and encodes 480 proteins.

It lacks the genes required for many biosynthetic functions, including the synthesis of amino acids, the construction of cell walls, and a functional tricarboxylic acid (TCA) cycle. In contrast, free-living bacteria that can grow in soil have larger genomes and dedicate many genes to the synthesis or acquisition of amino acids or TCA cycle intermediates. Even different strains of one species, such as *Salmonella enterica*, may vary considerably in gene distribution. **Figure 7.3** illustrates genes of *Salmonella enterica* serovar Typhimurium (abbreviated S. Typhimurium) transcribed in the clockwise direction (outer circle) and the counterclockwise direction (inner circle), which means that they are transcribed off of opposite strands of DNA. The color coding of different genes indicates presence in S. Typhimurium versus other *Salmonella* species and seven additional members of the family Enterobacteriaceae. For example, genes marked orange are present in all eight species, while those marked blue are present only in S. Typhimurium.

Another feature that distinguishes bacterial and archaeal genomes from those of eukaryotes is the amount of so-called noncoding DNA (DNA that does not encode proteins). Many, but not all, eukaryotes contain huge amounts of noncoding DNA scattered between genes. In some species (such as humans), over 90% of the total DNA is noncoding. Some noncoding regions include **enhancer** sequences needed to drive transcription of eukaryotic promoters and DNA expanses that separate enhancers. Enhancer sequences can function at large distances from the gene they regulate. A **promoter** is the DNA sequence immediately in front of, and sometimes within, a gene needed to activate the gene's expression. Most of the noncoding spacers appear to be remnants of genes lost over the course of evolution and pieces of defunct viral genomes. Noncoding regions, however, may provide raw material for future evolution.

In contrast to many eukaryotes, bacteria and archaea tend to have very little noncoding DNA (typically less than 15% of the genome). Archaeal genomes do, however, contain a few genes with internal noncoding DNA sequences that resemble the introns of eukaryotes.

Functional Units of Genes

In the simplest case, a gene can operate independently of other genes. The RNA produced from a single gene is said to be "monocistronic," which means it codes for one protein. Alternatively, a gene may exist in tandem with other genes in a unit called an **operon** (**Fig. 7.4A**). All genes in an operon are situated head to tail on the chromosome and

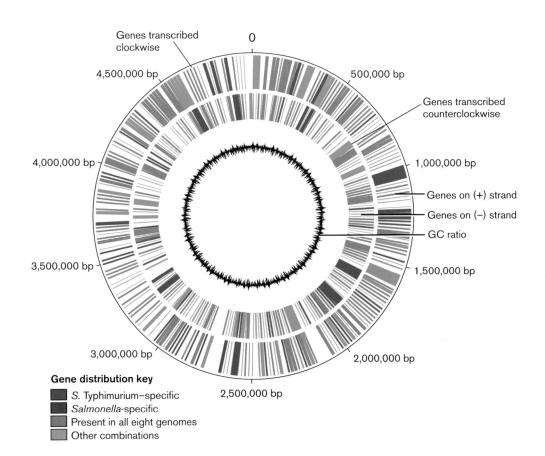

Gene distribution key
- S. Typhimurium–specific
- *Salmonella*-specific
- Present in all eight genomes
- Other combinations

FIGURE 7.3 ■ **The circular genome of *Salmonella enterica* serovar Typhimurium LT2.** The figure illustrates, by color coding, whether a gene is specific to *Salmonella* or is present in *Salmonella* and seven other species in the family Enterobacteriaceae. The black, inner circle is the GC content of the LT2 DNA (peaks pointing outward and inward indicate GC-rich and AT-rich areas, respectively).

A. Single gene vs. operon

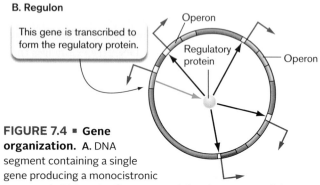

Promoter   Promoter

Gene 1 | Gene 2 | Gene 3

Monocistronic RNA   Polycistronic RNA

Transcription start site

B. Regulon

This gene is transcribed to form the regulatory protein.

Operon

Regulatory protein

Operon

FIGURE 7.4 ■ Gene organization. A. DNA segment containing a single gene producing a monocistronic message (which codes for one protein) and an operon of three genes producing a polycistronic message (from which three different proteins are made). **B.** Diagram of a circular bacterial genome, containing genes and operons coordinately controlled by a single regulatory protein.

are controlled by a single regulatory sequence located in front of the first gene. The single RNA molecule produced from the operon contains all the information from all the genes in that operon and is called "polycistronic."

On a functional level, a collection of genes and operons at different positions on the chromosome form a **regulon** when they have a unified biochemical purpose (such as amino acid biosynthesis) and are regulated by the same regulatory protein (**Fig. 7.4B**). The various mechanisms regulating expression of these functional genetic units are discussed in Chapter 10.

Note: In bacteria, the names of genes are given as a three-letter abbreviation of the encoded enzyme's name (for example, the gene *dam* encodes deoxyadenosine methylase) or the function of related genes (for example, genes designated *proA*, *proB*, and *proC* encode enzymes involved in proline biosynthesis). Bacterial gene names are written in Italics with lowercase letters. For example, the genes involved in catabolizing lactose are the *lac* genes. If several genes are involved in the pathway, a fourth letter, capitalized, is used. Thus, the three genes *lacZ*, *lacY*, and *lacA* are all associated with lactose catabolism. When speaking of a gene product, a nonItalic (roman) font is used, and the first letter is capitalized. Thus, *lacZ* is the gene and LacZ is the protein product of that gene.

A. Chemical structure of DNA

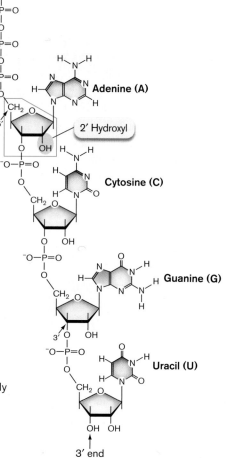

Thymine (T)   3′ end

Adenine (A)

5′ end

2′-Deoxyribose

The 2′-deoxy position that distinguishes DNA from RNA

Phosphodiester bond

Numbered carbons of deoxyribose

Cytosine (C)

Guanine (G)

3′ end

B. Chemical structure of RNA

5′ end

Adenine (A)

Ribose   2′ Hydroxyl

Cytosine (C)

Guanine (G)

Uracil (U)

5′ end

3′ end

FIGURE 7.5 ■ Structures of DNA and RNA. A. In the cell, DNA bases are added only to a preexisting 3′ OH of a nucleoside monophosphate, so the 5′ ends in this figure are drawn as nucleoside monophosphates. (Dotted lines indicate hydrogen bonds between bases.) **B.** Cellular RNA molecules, however, begin with a 5′ triphosphate.

DNA Function Depends on Its Chemical Structure

DNA is composed of four different nucleotides linked by a phosphodiester backbone (**Fig. 7.5A**). Each nucleotide consists of a **nucleobase** (also called a **nitrogenous base**) attached through a ring nitrogen to carbon 1 of 2-deoxyribose in the phosphodiester backbone. The 2-deoxy position that distinguishes DNA from RNA (**Fig. 7.5B**) is highlighted in the figure. A **phosphodiester bond** (also marked in **Fig. 7.5A**) joins adjacent deoxyribose molecules in DNA to form the phosphodiester backbone. Phosphodiester bonds link the 3′ carbon of one ribose to the 5′ carbon of the next ribose. The two backbones are **antiparallel** so that at either end of the DNA molecule, one DNA strand ends with a 3′ hydroxyl group and the complementary strand ends with a 5′ phosphate. This antiparallel arrangement is necessary so that complementary bases protruding from the two strands can pair properly via hydrogen bonding. Base pairing is not possible if DNA strands are modeled in a parallel arrangement.

The nucleobases in DNA are planar heteroaromatic structures arranged perpendicular to the phosphodiester backbone and parallel to each other like a stack of coins. **Purines** (bicyclic nucleobases: adenine or guanine) pair with **pyrimidines** (monocyclic nucleobases: thymine or cytosine). Under physiological conditions of salt (about 0.85% NaCl) and pH (pH 7.8), the hydrogen bonding of the bases permits adenine to pair only with thymine (via two hydrogen bonds) and likewise guanine with cytosine (via three hydrogen bonds). These complementary base interactions enable the two phosphodiester backbones to wrap around each other to form the classic double helix, or duplex.

The thousands of H-bonds that form between purines and pyrimidines along the interior of a DNA duplex (**Fig. 7.5**) make the bonding of the two complementary strands of DNA highly specific, so that a duplex forms only between complementary strands. Although H-bonds govern the specificity of strand pairing, the thermal stability of the helix is due predominantly to the stacking of the hydrophobic base pairs. Stacking of base pairs allows water and ions to interact with the hydrophilic, negatively charged phosphate backbone of DNA while avoiding the hydrophobic interior of the helix.

The stacking of DNA bases also appears to permit DNA to act as a "wire" capable of conducting electrons over long distances. The conduction of electrons along DNA may be sensed by certain proteins bound to DNA (discussed in Chapter 9).

Thought Question

7.1 What do you think happens to two single-stranded DNA molecules isolated from <u>different</u> genes when they are mixed together at very high concentrations of salt? *Hint:* High salt concentrations favor bonding between hydrophobic groups.

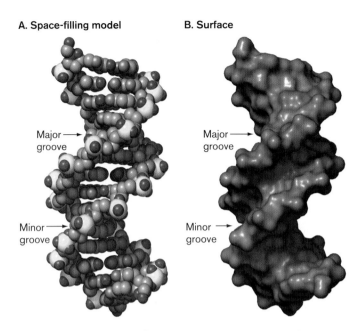

A. Space-filling model   **B. Surface**

Major groove

Major groove

Minor groove

Minor groove

FIGURE 7.6 ■ Models of DNA. **A.** Space-filling model of DNA. **B.** DNA surface, modeled using nuclear magnetic resonance. (PDB code: 1K8J)

In the space-filling model of DNA in **Figure 7.6A** and the contour map in **Figure 7.6B,** notice that the DNA double helix has grooves: a wide major groove and a narrower minor groove. The two grooves are generated by the angles at which the paired bases meet each other. These grooves provide DNA-binding proteins access to base sequences buried in the center of the molecule, so that the proteins can interact with the bases without the strands being separated. **Figure 7.7** shows an example of an important DNA-binding protein, DtxR of *Corynebacterium*

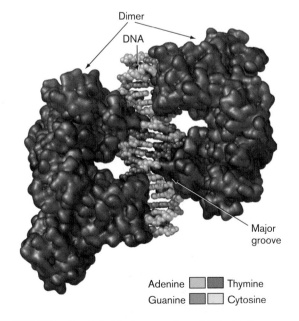

Dimer

DNA

Major groove

Adenine ▢▢ Thymine
Guanine ▢▢ Cytosine

FIGURE 7.7 ■ Protein that recognizes DNA. DtxR repressor protein dimer binds in the major groove of DNA. (PDB code: 1COW)

diphtheriae, the cause of diphtheria. The DtxR dimer (pair of subunits) binds the major groove of DNA at the promoter for the diphtheria toxin gene (*dtxA*). Toxin expression is repressed when iron concentrations are high, which signals to the bacteria that they are in an environment outside a human host. Gene regulation is discussed further in Chapter 10.

At high temperatures (50°C–90°C), the hydrogen bonds in DNA break and the duplex falls apart, or **denatures**, into two single strands. The temperature required to denature a DNA molecule depends on the GC/AT ratio of a sequence. More energy is required to break the three H-bonds of a GC base pair than the two H-bonds of an AT base pair. Thus, DNA with a high GC content requires a higher denaturing temperature than does similar-sized DNA with a lower GC content. The black center ring in **Figure 7.3** illustrates how the GC content of bases can change around a single chromosome.

When DNA has been heated to the point of strand separation, lowering the temperature permits the two single strands to find each other and reanneal into a stable double helix. The kinetics of DNA renaturation is much <u>slower</u> than that of denaturation, because renaturation is a random, hit-or-miss process of complementary sequences finding each other. This melting/reannealing property of DNA is exploited in a number of molecular techniques (see the discussion of the polymerase chain reaction in Section 7.6). Note, however, that bacteria and archaea growing at extreme pH or temperature protect their DNA from denaturation through the use of remarkable DNA-binding proteins such as the archaeal histone proteins Hmf and Htz. The role of DNA-binding proteins in microbial survival is the subject of considerable research.

Thought Question

7.2 How do the kinetics of denaturation and renaturation depend on DNA concentration?

RNA Differs Slightly from DNA

Why do cells make both RNA and DNA? As we learned in Chapter 3, DNA and RNA have different missions. The cell uses DNA to stably archive the information needed to make a functional cell. The growing cell continually accesses this information by making <u>temporary</u> copies of its genes in the form of RNA (ribonucleic acid) molecules that direct the synthesis of proteins. The cell also makes small RNA molecules that adjust information flow by modulating DNA gene expression. To keep their roles separate, DNA and RNA must have slightly different structures.

DNA in a cell is usually double-stranded, whereas RNA is usually single-stranded. DNA and RNA are chemically similar, except that in RNA the sugar ribose replaces deoxyribose and the pyrimidine base uracil replaces thymine (see **Fig. 7.5B**). Functionally, these two differences prevent enzymes meant to work on DNA, such as DNA polymerases, from acting on RNA. They also prevent RNA nucleases (RNases) from degrading DNA. However, uracil can still base-pair with adenine, which means that hybrid RNA-DNA double-stranded molecules can form (hybridize) when base sequences are complementary. In fact, this **hybridization** is a necessary step in the decoding of genes to make proteins.

Although RNA molecules are commonly thought of as single-stranded, all RNA molecules in nature have RNA-RNA double-stranded regions called "hairpins." Hairpin structures form when complementary nucleotide sequences within the primary RNA sequence bend back and hybridize. These double-stranded hairpins have a variety of biological functions.

Bacterial Chromosomes Are Compacted into a Nucleoid

The chromosome of *E. coli* has over 4.6 million bases in one strand, or over 9 million, counting both strands. This is a huge molecule. At the normal pH of the cell (7.8), all the phosphates in the backbone (all 9 million of them) are unprotonated and negatively charged, so this one molecule contributes greatly to the overall negative charge of the cytoplasm.

Figure 7.2 shows DNA spewing out of a damaged bacterial cell. Laid out, the chromosome is 1,500 times longer than the cell. It is obvious from this photomicrograph that an intact, healthy cell must compact a huge bundle of DNA into a very small volume. DNA is the second-largest molecule in the cell (only peptidoglycan is larger) and comprises a large portion of a bacterial cell's dry mass, about 3%–4%. Although packaging 3% of a cell's dry weight may not seem like a challenge, realize that DNA is further confined only to ribosome-free areas of the cell, so the chromosome-packing density reaches about 15 mg/ml. In a test tube, DNA at 15 mg/ml is almost a gel, so how can anything move inside a cell? And how does all this DNA keep from getting hopelessly tangled?

As introduced in Chapter 3, cells pack their DNA into a manageable form that still allows ready access to DNA-binding proteins. Although bacteria lack a nuclear membrane, they pack their DNA into a series of protein-bound domains collectively called the **nucleoid** (see Section 3.4). Unlike the compact nucleus of eukaryotes, the bacterial nucleoid is distributed throughout the cytoplasm.

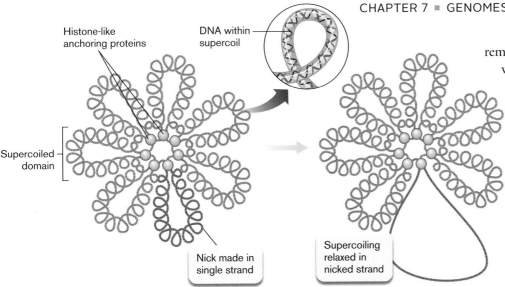

FIGURE 7.8 ■ **Bacterial nucleoid.** A nucleoid showing domain loops after gentle release from cells. The single-strand nick unwinds (relaxes) only one loop. ▶

DNA Supercoiling Compacts the Chromosome

A nucleoid gently released from *E. coli* appears as 30–100 tightly wound loops (**Fig. 7.8** ▶). The boundaries of each loop are defined by anchoring proteins called histone-like proteins for their similarity to histones, the DNA-binding proteins of eukaryotes. The double helix within each domain is itself helical, or supercoiled. The easiest way to envision supercoiling is to picture a coiled telephone cord. After much use, a phone cord twists, or supercoils, upon itself. Supercoiled phone cords are quite compact, taking up less space than a relaxed cord. Circular DNA works the same way—a property used by the cell to pack its chromosome.

Note that DNA cannot form supercoils unless its ends are tethered. In a circular chromosome, the DNA ends are tethered to each other. Introducing an extra twist by breaking one or both strands, twisting one end, and then resealing the strands means that the increased, or decreased, torsional (twisting) stress is trapped in the final circular molecule. It cannot then spontaneously unwind.

Remarkably, the nucleoid with its 30–100 loops (or "domains") can maintain different loops at different superhelical densities. The independence of supercoiled domains was demonstrated by introducing one single-strand nick in the phosphodiester backbone of one domain (see **Fig. 7.8**). You can do this by adding very small amounts of a nuclease (an enzyme that cleaves a nucleic acid). The ends of the broken strand, driven by the energy inherent in the supercoil, rotate about the unbroken complementary strand of the duplex and relax the supercoil. A single nick in a genome, however, removes supercoils from only one domain. How is this possible if the chromosome is one circular molecule? The unaffected chromosomal domains

remained supercoiled because they were constrained at their bases by anchoring proteins, such as HU and H-NS (histone-like proteins), that prevent rotation (for nucleoid organization, see Fig. 3.26).

How does DNA achieve the supercoiled state? The bacterial cell produces enzymes that can twist DNA into supercoils and other enzymes that relieve supercoils. A single twist introduced into a small (300-bp) circular DNA molecule forms a single supercoil as shown in **Figure 7.9** ▶. To introduce the DNA twist, a

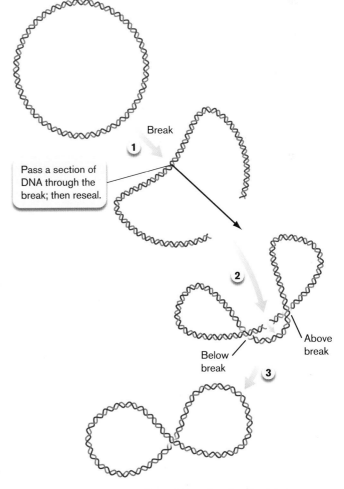

FIGURE 7.9 ■ **Supercoiling of 300-bp circular DNA.** A supercoil can be introduced into a double-stranded, circular DNA molecule by (1) cleaving both strands at one site in the molecule, (2) passing an intact part of the molecule <u>between</u> ends of the cut site, and (3) reconnecting the free ends. ▶

supercoiling enzyme makes a double-strand break at one point in the circle, passes another part of the DNA through the break, and reseals it. The result is the same as if one end of the broken circle were twisted one full turn.

To put this in context, most DNA in nature is right-handed. Right-handed, helical DNA turns clockwise when you look down the length of the double strand. Now think of a right-handed helical phone cord. Twist the cord so as to <u>decrease</u> the number of helix twists (<u>underwinding</u> the cord). As you keep underwinding, the torsional stress of removing clockwise helical twists is relieved when the phone cord (or DNA) flips around itself (supercoils) in a clockwise rotation. Because the DNA is underwound, the compensatory supercoil is called a <u>negative</u> supercoil. <u>Overwinding</u> a clockwise helix (like DNA) introduces additional clockwise helical twists; the stress of which is relieved by the DNA supercoiling in a counterclockwise direction. Because the DNA helix is overwound, the compensatory supercoil is called a <u>positive</u> supercoil.

The nucleoids of bacteria and most archaea, as well as the nuclear DNA of eukaryotes, are kept <u>negatively</u> supercoiled. Because the DNA is underwound, the two strands of negatively supercoiled DNA are easier to separate than those of positively supercoiled DNA. This is important for transcription enzymes, such as RNA polymerase, that must separate strands of DNA to make RNA.

Note, however, that some archaeal species living in acid at high temperature have nucleoids that are <u>positively</u> supercoiled to keep DNA double-stranded in these inhospitable environments (discussed next). Positively supercoiled DNA is harder to denature, because it takes excess energy to separate overwound DNA.

Topoisomerases Supercoil DNA

Supercoiling changes the topology of DNA. Topology is a description of how spatial features of an object are connected to each other. Thus, enzymes that change DNA supercoiling are called **topoisomerases**. To maintain proper DNA supercoiling levels, a cell must delicately balance the activities of two types of topoisomerases. Type I topoisomerases are typically single proteins that cleave only one strand of a double helix, while type II enzymes have multiple subunits that cleave both strands of a DNA molecule. Type I enzymes relieve or unwind supercoils. As shown in **Figure 7.10** ▶, topoisomerase I cleaves one strand of a negatively supercoiled double helix and holds on to both ends of the break. The release of intrinsic energy allows the enzyme to pass the intact strand through the break and re-ligate the strand, which reintroduces a helical turn. The molecule is released with one less negative supercoil.

The action of DNA gyrase is more involved (**Fig. 7.11** ▶). DNA gyrase is an example of a type II topoisomerase whose function is to introduce negative supercoils in DNA (see **Fig. 7.11A**). The active gyrase complex is a tetramer composed of two GyrA and two GyrB proteins. The gyrase B subunits first grab a section of the double helix. Then GyrA, in an ATP-dependent process catalyzed by GyrB, introduces a double-strand break, passes a

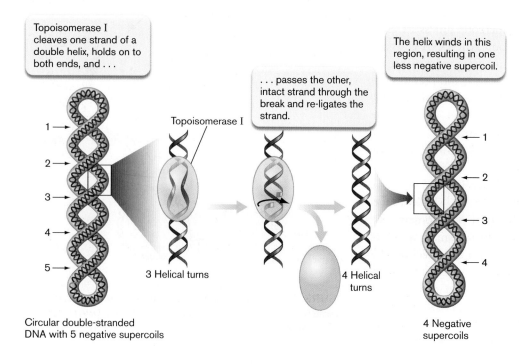

Topoisomerase I cleaves one strand of a double helix, holds on to both ends, and . . .

Topoisomerase I

3 Helical turns

. . . passes the other, intact strand through the break and re-ligates the strand.

4 Helical turns

The helix winds in this region, resulting in one less negative supercoil.

Circular double-stranded DNA with 5 negative supercoils

4 Negative supercoils

FIGURE 7.10 ▪ Mechanism of action for type I topoisomerases (topo I of *E. coli*). Topoisomerase I relaxes a negatively supercoiled DNA molecule by introducing a single-strand nick. ▶

A.

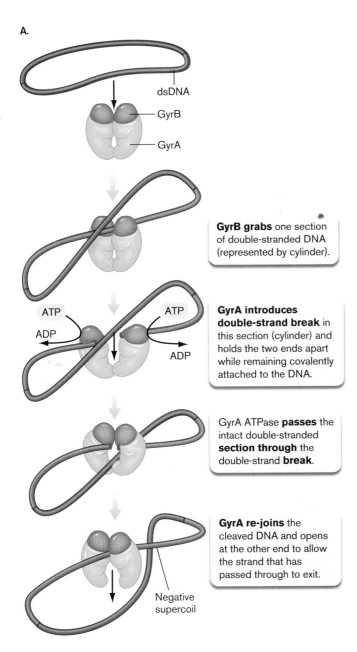

GyrB grabs one section of double-stranded DNA (represented by cylinder).

GyrA introduces double-strand break in this section (cylinder) and holds the two ends apart while remaining covalently attached to the DNA.

GyrA ATPase **passes** the intact double-stranded **section through** the double-strand **break**.

GyrA re-joins the cleaved DNA and opens at the other end to allow the strand that has passed through to exit.

B.

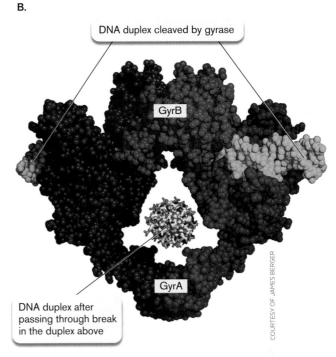

DNA duplex cleaved by gyrase

GyrB

GyrA

DNA duplex after passing through break in the duplex above

COURTESY OF JAMES BERGER

FIGURE 7.11 ■ **Mechanism of action for type II topoisomerases. A.** Mode of action of DNA gyrase from *E. coli*. **B.** Three-dimensional representation of DNA gyrase. The gyrase complex grips a broken DNA duplex (shown in green) and has transported a second duplex (the multicolored rosette) through the break. ▶

different part of the double helix through the break, and reseals the break. The other end of GyrA then opens to release the DNA, now with one more negative supercoil. **Figure 7.11B** shows a 3D representation of DNA gyrase in the midst of generating a supercoil.

Enzymes that make or manage bacterial DNA, RNA, and proteins are common targets for antibiotics. For instance, the **quinolone** antibiotics specifically target bacterial type II topoisomerases. They do not affect eukaryotic topoisomerases. A modern quinolone, ciprofloxacin, was the treatment of choice for anthrax pneumonia during the 2001 anthrax attacks. The progenitors of this drug family, nalidixic and oxolinic acids, were used to map *gyrA* and *gyrB*, the first drug resistance genes identified in *E. coli*. The modern successors of these drugs, the

fluoroquinolones, are among the most widely used antimicrobials in the world. These drugs do not block topoisomerase action but stabilize the complex in which DNA gyrase is covalently attached to DNA (see **Fig. 7.11**). The stuck complex forms a physical barrier in front of the DNA replication complex, and the bacterial cell dies.

Extreme thermophiles (hyperthermophilic archaea) possess an unusual gyrase called reverse DNA gyrase. In contrast to the DNA gyrase from mesophiles, reverse gyrase introduces <u>positive</u> supercoils into the chromosome. It is proposed that tightening the coil helps protect the chromosome against thermal denaturation. Because the DNA has <u>extra</u> turns, it takes more energy (heat) to separate the strands.

Thought Questions

7.3 DNA gyrase is essential to cell viability. Why, then, are nalidixic acid–resistant cells that contain mutations in *gyrA* still viable?

7.4 Bacterial cells contain many enzymes that can degrade linear DNA. How, then, do linear chromosomes in organisms like *Borrelia burgdorferi* (the causative agent in Lyme disease) avoid degradation?

To Summarize

- **Noncoding DNA** can constitute a large amount of a eukaryotic genome, while prokaryotes have very little noncoding DNA.

- **DNA is composed of two antiparallel chains** of purine and pyrimidine nucleotides in which phosphate links the 5′ carbon of one nucleotide with the 3′ carbon of the next in the chain. The result is a double helix containing a deep major groove and a more shallow minor groove.

- **Hydrogen bonding and interactions between the stacked bases** hold together complementary strands of DNA.

- **Supercoiling** by topoisomerases compacts DNA into an organized nucleoid.

- Bacteria, eukaryotes, and most archaea possess **negatively supercoiled DNA**. Archaea living in extreme environments have **positively supercoiled genomes**.

- **Type I topoisomerases** cleave one strand of a DNA molecule and <u>relieve</u> supercoiling; **type II topoisomerases** cleave both strands of DNA and use ATP to <u>introduce</u> supercoils.

7.3 DNA Replication

How do bacteria replicate their DNA quickly and minimize errors? Quick and accurate replication of DNA can help a microorganism grow and compete with other species. Replication efficiency is one reason why bacterial pathogens such as *Salmonella* can cause disease so quickly after ingestion. In this respect, bacteria differ from multicellular organisms, which need to regulate cell division carefully within their tissues because unregulated growth within tissues leads to cancer. The process of bacterial replication involves over 20 proteins coming together in a complex machine. Operation of the replication complex is all the more remarkable, considering that some bacteria, such as thermophilic *Bacillus* species that live in hot springs, can double their population in less than 15 minutes.

The molecular details of bacterial DNA replication are important to understand because they provide targets for new antibiotics and tools for biotechnology, such as the polymerase chain reaction (PCR) (see Section 7.6). In addition, the bacterial proteins of DNA repair have homologs in the human genome, defects in which produce inherited human diseases, such as xeroderma pigmentosum, that predispose the carrier to certain cancers.

Overview of Bacterial DNA Replication

To replicate a molecule containing millions of base pairs poses formidable challenges. How does replication begin and end? How is accuracy checked and maintained?

Semiconservative replication. Replication of cellular DNA is **semiconservative**, meaning that each daughter cell receives one parental strand and one newly synthesized strand (**Fig. 7.12**). At the **replication fork**, the advancing DNA synthesis machine separates the parental strands while extending the new, growing strands. The semiconservative mechanism provides a means for each daughter duplex to be checked for accuracy against its parental strand.

Enzymes that synthesize DNA or RNA can connect nucleotides only in a 5′-to-3′ direction. That is, every newly made strand begins with a 5′ triphosphate and ends with a 3′ hydroxyl group (see **Fig. 7.5**). A polymerase (a chain-lengthening enzyme complex) fastens the 5′ alpha-phosphate of an incoming nucleoside triphosphate to the 3′ hydroxyl end of the growing chain, thus forming a phosphodiester bond. (The alpha-phosphate is the phosphate closest to the sugar.) This 5′-to-3′ enzymatic constraint produces an interesting mechanistic puzzle: If polymerases can synthesize DNA only in a 5′-to-3′ direction and the

FIGURE 7.12 ■ Semiconservative replication. A replication bubble with two replication forks. Replication is called "semiconservative" because one parental strand is conserved and inherited by each daughter cell genome. It is called "bidirectional" because it begins at a fixed origin and progresses in opposite directions.

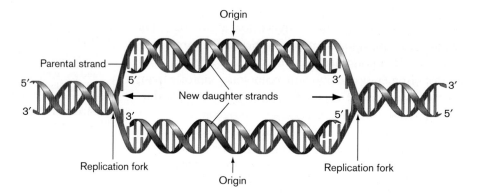

two phosphodiester backbones of the double helix are anti-parallel, then how are both strands of a moving replication fork synthesized simultaneously? One strand presents no problem, because it is synthesized in a 5'-to-3' direction toward the fork, but synthesizing the other new strand in a 5'-to-3' direction would seem to dictate that it move <u>away</u> from the fork (**Fig. 7.13**). How is this possible?

Note: The term "nucleoside triphosphate" (nucleobase-ribose condensed with three phosphoryl groups) is also commonly written "nucleotide triphosphate."

Thought Question

7.5 Suppose you can label DNA in a bacterium by growing cells in medium containing either ^{14}N nitrogen or the heavier isotope, ^{15}N; you can isolate pure DNA from the organism; and you can subject DNA to centrifugation in a cesium chloride solution, a solution that forms a density gradient when subjected to centrifugal force, thereby separating the light (^{14}N) and heavy (^{15}N) forms of DNA to different locations in the test tube. Given these capabilities, how might you prove that DNA replication is semiconservative?

The process of DNA replication is divided into three phases: (1) initiation, which is the melting (unwinding) of the helix and the loading of the DNA polymerase enzyme complex; (2) elongation, which is the sequential addition of deoxyribonucleotides to a growing DNA chain, followed by proofreading; and finally (3) termination, in which the DNA duplex is completely duplicated, the negative supercoils are restored, and key sequences of new DNA are methylated.

Replication from a single origin. Replication in bacteria begins at a single defined DNA sequence called the **origin (oriC)**. Following initiation, a circular bacterial chromosome replicates <u>bidirectionally</u> (in both directions away from the origin; see **Fig. 7.13**, step 1) until it terminates at defined **termination (ter) sites** located on the opposite side of the molecule. Once the process has begun, the cell is committed to completing a full round of DNA synthesis. As a result, the decision of when to start copying the genome is critical. If it starts too soon, the cell accumulates unneeded chromosome copies; if it starts too late, the dividing cell's septum "guillotines" the chromosome, killing both daughter cells. Consequently, elaborate fail-safe mechanisms link the initiation of DNA replication with cell mass, generation time, and cellular health, making the timing of initiation remarkably precise.

Note: Some single-celled microbes have chromosomes with more than one origin. The archaeon *Sulfolobus acidocaldarius,* for instance, has a single chromosome with <u>three</u> replication origins.

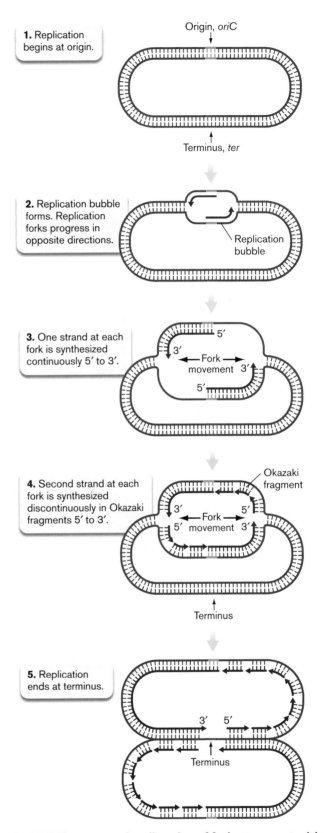

FIGURE 7.13 ■ **Comparing direction of fork movement with direction of DNA synthesis.**

Initiation of replication

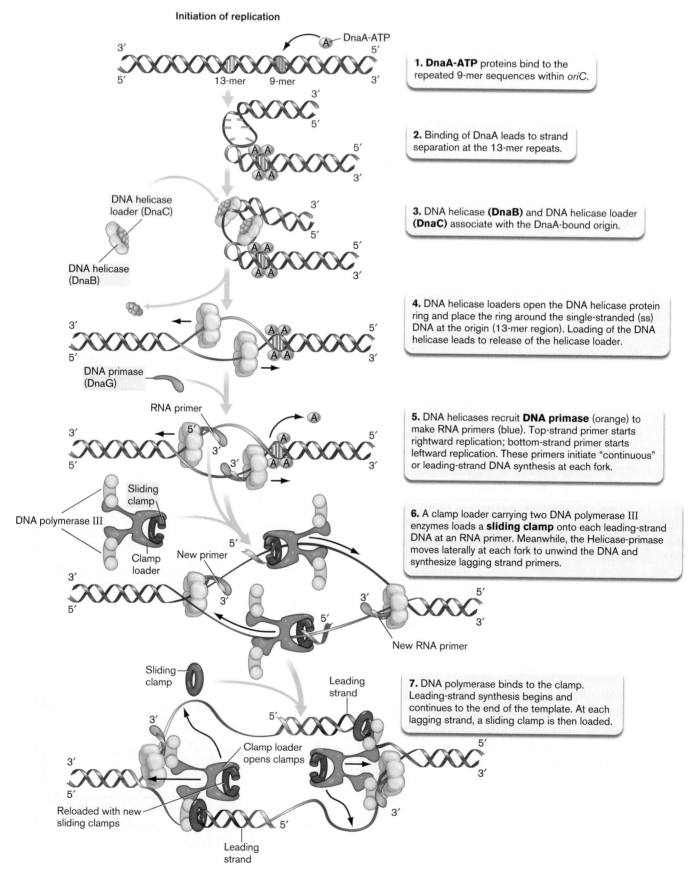

1. DnaA-ATP proteins bind to the repeated 9-mer sequences within *oriC*.

2. Binding of DnaA leads to strand separation at the 13-mer repeats.

3. DNA helicase **(DnaB)** and DNA helicase loader **(DnaC)** associate with the DnaA-bound origin.

4. DNA helicase loaders open the DNA helicase protein ring and place the ring around the single-stranded (ss) DNA at the origin (13-mer region). Loading of the DNA helicase leads to release of the helicase loader.

5. DNA helicases recruit **DNA primase** (orange) to make RNA primers (blue). Top-strand primer starts rightward replication; bottom-strand primer starts leftward replication. These primers initiate "continuous" or leading-strand DNA synthesis at each fork.

6. A clamp loader carrying two DNA polymerase III enzymes loads a **sliding clamp** onto each leading-strand DNA at an RNA primer. Meanwhile, the Helicase-primase moves laterally at each fork to unwind the DNA and synthesize lagging strand primers.

7. DNA polymerase binds to the clamp. Leading-strand synthesis begins and continues to the end of the template. At each lagging strand, a sliding clamp is then loaded.

FIGURE 7.14 ■ Initiation of DNA replication. ▶

Fundamentals of DNA replication. The basic process of chromosome replication is outlined in **Figure 7.13**. After initiation of replication, a replication bubble forms at the origin. The bubble contains two replication forks that move in opposite directions around the chromosome (**Fig. 7.13**, step 2). DNA polymerases synthesize DNA in a 5′-to-3′ direction. At each fork, therefore, one new DNA strand can be synthesized continuously until the terminus region (step 3). However, because the two DNA strands are antiparallel and the DNA polymerases synthesize only 5′ to 3′, the other daughter strand has to be synthesized discontinuously, in stages—seemingly backward relative to the moving fork (step 4). The fragments of DNA formed on this discontinuously synthesized strand are called **Okazaki fragments**, after the Japanese scientists, Reiji and Tsuneko Okazaki, who discovered them. As we will discuss later, the Okazaki fragments are progressively stitched together to make a continuous, unbroken strand. Ultimately, the two replicating forks meet at the terminal sequence (step 5), and the two daughter chromosomes separate.

Overall, copying the entire chromosome in *E. coli* takes about 40 minutes. Chromosome-partitioning processes then move each chromosome to different ends of the cell so that a cell wall can form at midcell (the cell "equator"). Once the cell wall is complete (the amount of time varies but is generally about 20 minutes), the two daughter cells, with their new chromosomes, can separate. So you might imagine the whole process from the start of replication to cell separation would take about 60 minutes for *E. coli*. But when nutrients are plentiful, *E. coli* can divide in 20 minutes. How is this possible? As we saw in Chapter 3, the answer is that in fast-growing cells, a partially replicated chromosome can begin new rounds of replication even before the first round is complete.

Now let's examine each step in molecular detail to answer some important questions about this mechanism.

Initiating Replication

What determines when replication begins? Initiation is controlled by DNA methylation, and by the binding of a specific initiator protein to the origin sequence. Further molecular events load the elaborate DNA polymerase complex and generate the first RNA primer for the new DNA strand. **Figure 7.14** ▶ presents an overview of the initiation process.

DNA methylation controls timing. The *E. coli* origin of replication (*oriC*) is a 245-base-pair sequence. Initiation of replication at *oriC* is activated by one protein, DnaA, and inhibited by another, SeqA. Immediately after a cell has divided, the level of active DnaA (DnaA bound to ATP) is low, and the inhibitor protein SeqA binds to *oriC* to prevent ill-timed initiations (before the cell has grown enough to divide again).

How does SeqA know to bind just after the origin has replicated? The key is DNA methylation. *E. coli* uses the enzyme deoxyadenosine methylase (Dam) to attach a methyl group to the N-6 position of adenine in the sequence GATC (in **Fig. 7.5A**, see the N of the NH_2 group attached to the six-membered ring). GATC sequences are scattered along the chromosome on both strands. Just after the origin has replicated, there is a short lag before the newly synthesized strand is methylated by Dam methylase. As a result, the origin is temporarily hemimethylated—a situation in which only one of the two complementary strands is methylated. Because SeqA has a high affinity for hemimethylated origins, this inhibitor will bind most tightly immediately after the origin has been replicated. Thus bound, SeqA will prevent another initiation event. Eventually, the Dam methylase will methylate the new strand and decrease SeqA binding.

The replication initiator protein, DnaA. Timing of initiation is determined by the concentration of the replication initiator protein DnaA complexed with ATP (DnaA-ATP). DnaA-ATP recognizes specific 9-bp repeats at *oriC*. As the cell grows, the level of active DnaA-ATP rises until it is sufficient to bind to these repeats (**Fig. 7.14**, step 1). Binding of DnaA to the origin facilitates melting of DNA and initiates the assembly of a membrane-attached replication hyperstructure (the replisome), a complex of numerous proteins that come together and bind at *oriC*.

After it is replicated, the origin cannot trigger another round of replication for two reasons: inhibition by SeqA and decreasing levels of unbound DnaA-ATP. Another round of replication can begin only after (1) the origin becomes fully methylated, (2) SeqA dissociates, and (3) the DnaA-ATP concentration rises.

What happens to new replication origins after replication begins? **Figure 7.15** reveals that, even though the origin

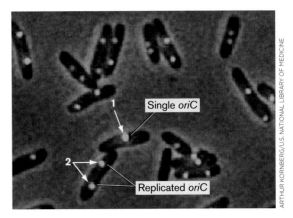

FIGURE 7.15 ■ Movement of newly replicated origins. The *oriC* loci on chromosomes were visualized through a GFP protein that binds only at *oriC*.

starts in the center of the cell, the newly replicated origins (green fluorescence) move toward opposite cell poles. Moving origins toward the cell poles is part of the partitioning mechanism that gets chromosomes out of harm's way before the division septum forms at midcell (described below).

Initiation requires RNA polymerases. An unexpected feature of DNA replication is that its initiation actually requires two RNA polymerases. The first is the housekeeping RNA polymerase used to make most of the RNA in the cell (discussed in Chapter 8). The second is the DNA primase (discussed shortly).

The housekeeping RNA polymerase transcribes DNA at *oriC*, which helps separate the two DNA strands (**Fig. 7.14**, step 2; this RNA polymerase is not shown). Strand separation at *oriC* allows a special DNA helicase (DnaB), in association with a DNA helicase loader (DnaC), to bind the two replication forks formed during initiation (step 3). DnaC facilitates proper placement of DNA helicase at the fork, and then it disengages and leaves.

The DnaB helicase uses energy from ATP hydrolysis to unwind the DNA helix <u>before</u> DNA moves into the DNA polymerase replicating complex. The ringlike DnaB is assembled around one DNA strand at each replication fork. After loading DnaB at the origin, DnaC is released (**Fig. 7.14**, step 4). As the DNA unwinds, small single-stranded DNA-binding proteins (SSBs, seen in **Fig. 7.17**) coat the exposed single-stranded DNA, protecting it from nucleases patrolling the cell and preventing re-formation of double-stranded DNA. The origin is almost ready to receive DNA polymerase.

DNA-dependent DNA polymerases possess the unique ability to "read" the nucleotide sequence of a DNA template and synthesize a complementary DNA strand. The discovery of this activity earned Arthur Kornberg (1918–2007) (**Fig. 7.16**) the 1959 Nobel Prize in Physiology or Medicine. However, as remarkable as these enzymes are, no DNA polymerase can start synthesizing DNA unless there is a pre-existing DNA or RNA fragment to extend—that is, a primer fragment. The primer fragment possesses a 3′ OH end that receives incoming deoxyribonucleotides. Consequently, once the helicase DnaB is bound to DNA, the next step is to make RNA primers at each fork (**Fig. 7.14**, step 5). In contrast to DNA polymerases, RNA polymerases <u>can</u> synthesize RNA without a primer. The RNA polymerase required for DNA replication is called DNA **primase** (DnaG). Primase synthesizes short (10–12 nucleotides) RNA primers at the origin that can launch DNA replication. One primase is loaded at each of the two replication forks.

Why do DNA polymerases require RNA primers? RNA primers probably remain from when all life was RNA based—an exciting model of molecular evolution, discussed in Chapter 17.

FIGURE 7.16 ▪ **Arthur Kornberg and Sylvy Kornberg, circa 1960.** A biochemist in her own right, Sylvy Kornberg contributed in the effort to purify DNA polymerase I.

A sliding clamp tethers DNA polymerase to DNA. At this point, the DNA is almost ready for DNA polymerase. But first a **sliding clamp** protein (the beta subunit of DNA polymerase III) must be loaded to keep the DNA polymerase affixed to the DNA (**Fig. 7.14**, step 6). Without this clamp, DNA polymerase would frequently "fall off" the DNA molecule (see **eTopic 7.1**). A multisubunit complex (called the clamp-loading complex) places the beta clamp, along with an attached pair of DNA polymerase molecules, onto DNA. DNA polymerase (specifically DNA Pol III, discussed later) then binds to the 3′ OH terminus of the primer RNA molecule and begins to synthesize new DNA (step 7). The molecular structure of the beta clamp loaded onto DNA is shown in **eTopic 7.1**.

Elongation of Replicating DNA

Escherichia coli contains five different DNA polymerase proteins, designated Pol I through Pol V. All DNA polymerases catalyze the synthesis of DNA in the 5′-to-3′ direction. However, only Pol III and Pol I participate directly in chromosome replication. The other polymerases conduct operations to rescue stalled replication forks and repair DNA damage.

DNA polymerase III. The main replication polymerase, Pol III, is a complex, multicomponent enzyme. Pol III was another discovery by Arthur Kornberg. The DNA synthesis activity of Pol III is held in the alpha subunit of the complex, while other subunits are used for improving fidelity (accuracy of replication) and processivity (a measure of how long the polymerase remains attached to, and replicates, a template). The Pol III epsilon subunit (DnaQ), for example, contains a **proofreading** activity that corrects mistakes and improves fidelity.

Proofreading activities within DNA polymerases scan for mispaired bases that have been mistakenly added to a growing chain. A mispaired base is more mobile than the correct base in a DNA molecule because a mispaired base does not properly hydrogen-bond to the template base. This motion halts DNA elongation by Pol III because the base is not properly positioned at the enzyme's active site. Stalling of Pol III activity triggers an intrinsic 3′-to-5′ **exonuclease** activity in the epsilon subunit. Exonucleases degrade DNA starting from either the 5′ end or the 3′ end, depending on the enzyme. The exonuclease activity of Pol III cleaves the phosphodiester bond, releasing the improperly paired base from the growing chain. Once the wayward base has been excised, Pol III can resume elongation.

Both DNA strands are elongated simultaneously. After initiation, each replication fork contains one elongating 5′-to-3′ strand, called the "leading strand" (look back at **Fig. 7.14**, step 7). But how is the opposite strand at each fork replicated? There are no known DNA polymerases capable of synthesizing DNA in the 3′-to-5′ direction, which would seem to be needed if both strands are to be synthesized simultaneously. DNA synthesis of one strand continuing all the way back to the origin is not a solution, because it would leave the unreplicated strand at each fork exposed to possible degradation for too long and would double the time needed to complete DNA replication.

The cell has solved this dilemma by coordinating the activity of two DNA Pol III enzymes in one complex—one for each strand. The two associated Pol III complexes, together with DNA primase and helicase, form the **replisome**. As the dsDNA unwinds at the fork, the problem strand loops out and primase (DnaG) synthesizes a primer. The second Pol III enzyme binds to the primed section of the loop and synthesizes DNA in the 5′-to-3′ direction (imagine the lower template strand in **Figure 7.17** threading from left to right, through the polymerase ring). All the while, the second polymerase moves with the first polymerase (on the leading strand) toward the fork (**Fig. 7.17**, step 1). Recent evidence indicates that the two *E. coli* replisomes move along DNA toward opposite poles of the cell, but generally stay within the middle third of the cell. They eventually meet again at the terminator located at midcell.

Note that simultaneous extension of the two strands at a single fork requires that synthesis of the looped strand must lag behind synthesis of the leading strand and that new RNA primers must be synthesized by DNA primase (DnaG) every 1,000 bases or so. Thus, the lagging strand is synthesized discontinuously, in pieces called Okazaki fragments, while the leading strand can be synthesized continuously. As the leading strand moves forward, advancing the fork, there remains a long stretch of lagging strand complementary to the already replicated leading strand. This lagging strand is single-stranded but protected by single-stranded DNA-binding proteins (SSBs) (**Fig. 7.17**, step 2). After about 1,000 bases, DNA primase reenters and synthesizes a new RNA primer in anticipation of lagging-strand DNA synthesis (step 3). At some point, the lagging-strand polymerase bumps into the 5′ end of the previously synthesized fragment. This interaction causes DNA polymerase to disengage from that strand (step 4), and the clamp loader loads a new clamp near the new RNA primer (step 5). The DNA polymerase binds to that clamp and begins synthesizing another Okazaki fragment (step 6). This process repeats every 1,000 bases or so around the chromosome.

The model just presented assumes that the replisome contains two DNA polymerase III molecules. However, the replisome actually consists of three polymerases—one on the leading strand and two on the lagging strand. The second polymerase on the lagging strand comes into play only when a large gap of unreplicated DNA remains on the lagging strand. For simplicity, the third polymerase is not included in the model shown in **Figure 7.17**.

DNA polymerase I. Discontinuous DNA synthesis results in a daughter strand containing long stretches of DNA punctuated by tiny patches of RNA primers. This RNA must be replaced with DNA to maintain chromosome integrity. To remove the RNA, cells typically use the 5′-to-3′ exonuclease activity of Pol I or an RNase enzyme specific for RNA-DNA hybrid molecules (called RNase H). A DNA Pol I enzyme then synthesizes a DNA patch using the 3′ OH end of the preexisting DNA fragment as a priming site (**Fig. 7.18**). When DNA Pol I reaches the next fragment, the enzyme removes the 5′ nucleotide and resynthesizes it. This process of replicating DNA increases accuracy and decreases mutations.

Once DNA Pol I stops synthesizing, it cannot join the 3′ OH of the last added nucleotide with the 5′ phosphate of the abutting fragment. The resulting nick in the phosphodiester backbone is repaired by **DNA ligase**, which in *E. coli* and many other bacteria uses energy gained by cleaving nicotinamide adenine dinucleotide (NAD) to form the phosphodiester bond (see **Fig. 7.18**). NAD is not used in its usual way, as a reductant that oxidizes substrates. Energy inherent in the diphosphate bond of NAD is captured upon cleavage by DNA ligase and used to re-join the 3′ OH and 5′ phosphate ends present at the nick. DNA ligase from eukaryotes and some other microbes use ATP rather than NAD in this capacity.

Elongation of DNA synthesis

1. The **leading-strand DNA Pol III** enzyme replicates the leading strand. **SSBs** cover and protect the unreplicated single strand. The DNA helicase remains on the lagging strand, unwinding the dsDNA moving into the replisome complex.

2. Lagging-strand DNA polymerase synthesizes the lagging strand, which loops out after passing through the polymerase.

3. After DNA helicase has moved approximately 1,000 bases, another **RNA primer** is synthesized on each lagging strand.

4. When the lagging-strand polymerase bumps into the 5′ end of a previously synthesized fragment, the **DNA polymerase is released** and the clamp is disengaged.

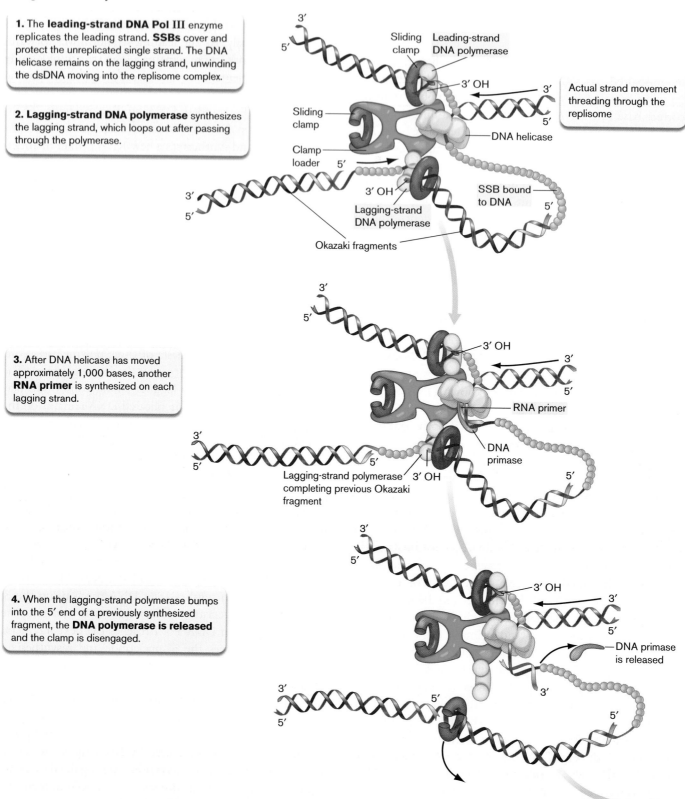

FIGURE 7.17 ■ The DNA polymerase dimer acting at a replication fork. The leading and lagging strands are synthesized simultaneously in the 5′-to-3′ direction. For clarity, the beta clamp on the lagging strand is shown on the opposite side of Pol III as compared to its position on the leading strand.

7.6 How fast does *E. coli* DNA polymerase synthesize DNA (in nucleotides per second) if the genome is 4.6 million base pairs, replication is bidirectional, and the chromosome completes a round of replication in 40 minutes?

DNA replication generates supercoils. As template DNA is threaded through the replisome, the helicase continually pulls apart the two strands of the DNA helix. As a result, the DNA ahead of the fork twists, introducing positive supercoils. (Try this yourself: Twist two pieces of string together, staple one end of the duplex to a piece of cardboard, and then pull the two strands apart from the free end. Notice the supercoiling that takes place beyond, or downstream of, the moving fork.) The increasing torsional stress in the chromosome could stop replication by making strand separation more and more difficult. What prevents the buildup of torsional stress is the DNA gyrase (see **Fig. 7.11**) that is located ahead of the fork, removing the positive supercoils as they form.

Terminating Replication and Segregating Sister Chromosomes

Bidirectional replication of a circular bacterial chromosome results in the two replication forks trying to replicate through the same DNA sequences 180° from the origin—that is, halfway around the chromosome. What tells the polymerases to stop? The *E. coli* chromosome has as many as 10 terminator sequences (*ter*) that polymerases enter

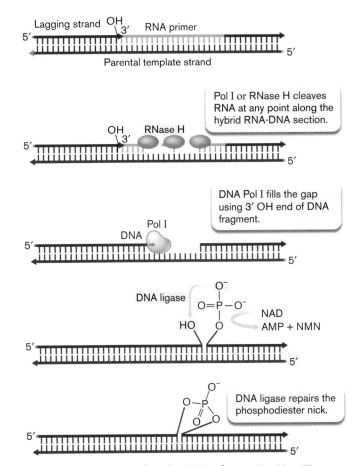

FIGURE 7.18 ■ Removing the RNA primer. The 3′-to-5′ exonuclease activity of RNase H or the 5′-to-3′ exonuclease of Pol I cleaves the RNA primer (blue). In either case, DNA polymerase I uses the preexisting 3′ OH end of the DNA fragment to fill the gap. Finally, DNA ligase repairs the phosphodiester nick using energy derived from the cleavage of NAD. AMP = adenosine monophosphate; NMN = nicotinamide monophosphate.

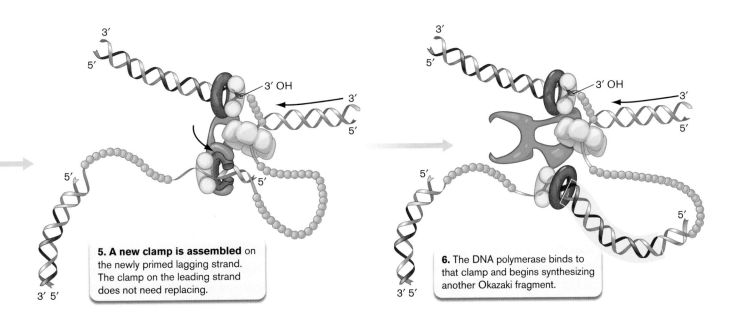

5. A new clamp is assembled on the newly primed lagging strand. The clamp on the leading strand does not need replacing.

6. The DNA polymerase binds to that clamp and begins synthesizing another Okazaki fragment.

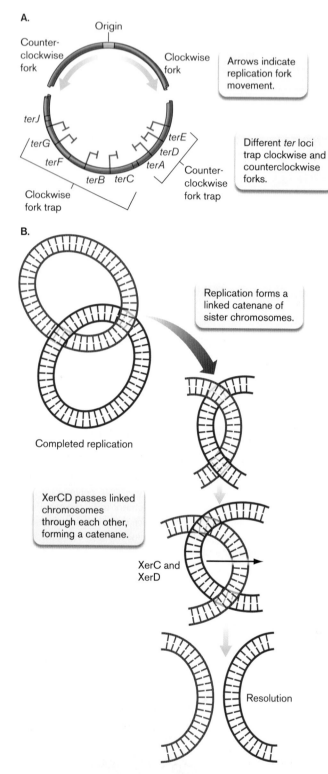

A.

Origin

Counter-clockwise fork

Clockwise fork

Arrows indicate replication fork movement.

terJ

terG

terF

terB   terC

Clockwise fork trap

terE

terD

terA

Counter-clockwise fork trap

Different *ter* loci trap clockwise and counterclockwise forks.

B.

Completed replication

Replication forms a linked catenane of sister chromosomes.

XerCD passes linked chromosomes through each other, forming a catenane.

XerC and XerD

Resolution

FIGURE 7.19 ■ Terminating replication of the chromosome.
A. Terminator regions for DNA replication on the *E. coli* chromosome. **B.** Resolution of DNA replication catenanes by XerC and XerD.

called Tus (terminus utilization substance) binds to these sequences. Tus stops DnaB helicase activity and, thus, polymerization. Multiple terminator sites ensure that the polymerase complex does not escape and continue replicating DNA. Which terminator site is used depends in part on whether replication of one fork has lagged behind replication of the other.

Thought Question

7.7 Reexamine **Figure 7.15**. If you GFP-tagged a protein bound to the *ter* region, where would you expect fluorescence to appear in the cell with two origins?

During and immediately after replication, the cell is faced with what could be called "knotty" problems. The first develops as soon as replication begins. Replicated DNA molecules occasionally form knots between homologous genes on sister chromosomes, holding them together. These knots must be removed before sister chromosomes can segregate to opposite cell poles. Resolving these knots requires the enzyme topoisomerase IV, a type II topoisomerase similar to DNA gyrase. Topo IV activity is stimulated by SeqA, the protein that binds to newly replicated, hemimethylated DNA. Recall that SeqA also prevents reinitiation at newly replicated origins. This one protein, therefore, prevents the start of a premature round of replication and removes topological constraints to chromosome segregation.

The second "knotty" problem is encountered when a chromosome finishes replicating. Because of the topology of the chromosome, the two daughter molecules will appear as a **catenane**, a pair of linked rings. The rings must be unlinked so that sister chromosomes can segregate at termination (**Fig. 7.19B**). The bacterial cell uses proteins called XerC and XerD to unlink these rings. XerCD recognizes a specific site (called *dif*) on both DNA molecules and catalyzes a series of cutting and re-joining steps that essentially pass one molecule through the other. Once the two daughter chromosomes have resolved and move toward the poles, the cell can begin to divide, forming the cell septum. A mutant bacterium defective in XerC or D will have difficulty dividing (**Fig. 7.20**).

Once replicated, how do sister chromosomes move to opposite poles of a bacterial cell? In some species, such as *Bacillus subtilis*, a mechanical segregation system consisting of the proteins ParA and ParB and the DNA-binding site *parS* pulls the chromosomes toward different ends of the cell. *E. coli*, however, does not possess a *parABS* system for segregating sister chromosomes. It seems that the process of replication itself, which takes place mostly

but rarely, if ever, leave (**Fig. 7.19A**). One set of terminators deals solely with the clockwise-replicating polymerase, while the other set halts DNA polymerases replicating counterclockwise relative to the origin. A protein

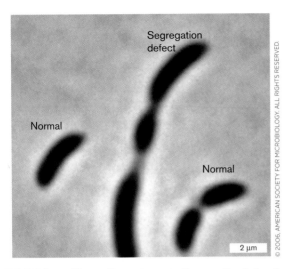

FIGURE 7.20 ■ **Effect of a *xerD* mutation on *Caulobacter* cell division.** *Caulobacter* replicates by producing a motile cell at one end of a nonmotile, fixed, stalked cell. The photomicrograph shows cells of a *xerD* mutant. The middle cell is elongated with multiple constrictions because of the *xerD* defect in chromosomal segregation. Even though Topo IV is present in the *xerD* mutant, both mechanisms are needed to ensure that linked replicating chromosomes always clear the division plane. *Source:* R. B. Jensen. 2006. *J. Bacteriol.* **188**:6016–6019.

in the middle third of the cell (see Fig. 3.29), propels *E. coli* chromosomes toward opposite cell poles. The terminator region remains in the center of the cell until after it is replicated. The replicating chromosome, then, takes on the appearance of a butterfly with its wings outstretched toward either cell pole. This configuration places the genome in a vulnerable position. A bacterial cell must avoid making a cell division septum too early or risk slicing through its still-replicating chromosomes. How bacteria prevent this disaster is discussed in **Special Topic 7.1**.

Thought Questions

7.8 Individual cells in a population of *E. coli* typically initiate replication at different times (asynchronous replication). However, depriving the population of a required amino acid can synchronize reproduction of the population. Ongoing rounds of DNA synthesis finish, but new rounds do not begin. Replication stops until the amino acid is once again added to the medium—an action that triggers simultaneous initiation in all cells. Why does this synchronization of reproduction happen?

7.9 The antibiotic rifampin inhibits transcription by RNA polymerase, but not by primase (DnaG). What happens to DNA synthesis if rifampin is added to a synchronous culture?

To Summarize

- **Replication is semiconservative**, with newly synthesized strands lengthening in a 5′-to-3′ direction. Replication consists of initiation, elongation, and termination steps.

- **Replication is initiated** from a fixed DNA origin attached to the cell membrane. Initiation depends on the mass and size of the growing cell. It is controlled by the accumulation of initiator and repressor proteins and by methylation at the origin.

- **Elongation** requires that primase (DnaG) must lay down an RNA primer, DNA polymerase III must act as a dimer at each replication fork, and a sliding clamp must keep DNA Pol III attached to the template DNA molecule.

- **The 3′-to-5′ proofreading** activity of Pol III corrects accidental errors during polymerization.

- **DNA ligase** joins Okazaki fragments in the lagging strand.

- **Two replisomes, each containing a pair of DNA Pol III complexes**, move along the DNA in *E. coli*.

- **Termination** involves stopping replication forks halfway around the chromosome at *ter* sites that inhibit helicase (DnaB) activity.

- **Ringed catenanes** formed at the completion of replication are separated by the proteins XerC and XerD.

7.4 Plasmids

Two kinds of extragenomic DNA molecules can interact with bacterial genomes: horizontally transferred **plasmids** and the genomes of bacteriophages (viruses that infect bacterial cells). Plasmid-encoded functions can contribute to the physiology of the cell (for example, antibiotic resistance). In some cases, the plasmid or phage DNA will integrate into the bacterial genome (see Section 9.2). Eukaryotic genomes and the genomes of eukaryote-specific plasmids and viruses also share proteins and undergo chromosomal interactions, such as insertions of viral DNA into the host chromosome. Viruses of eukaryotes and their applications in genetic engineering are discussed in greater detail in Chapters 11 and 12. In this section we discuss aspects of plasmid replication. Discussion of bacteriophage replication can be found in Chapter 6 and **eTopic 7.2**.

SPECIAL TOPIC 7.1 Nucleoid Occlusion Factors and the Septal "Guillotine"

During bacterial cell division, it is crucial that sister chromosomes dodge the septum forming at midcell. Segregation proteins, such as topoisomerase IV (ParC) in *Escherichia*, work to untangle replicated sister chromosomes so that they can move away from midcell toward the safety of opposite cell poles. But because replication and segregation take time, it is equally important for a cell to <u>prevent</u> septal formation as long as replicating chromosomes occupy midcell. If cell division starts too early, the growing septum could slice through ("guillotine") a replicating chromosome, killing the cell. Bacteria such as *E. coli* and *Bacillus subtilis* use nucleoid exclusion proteins to prevent septal guillotining of chromosomes. The protein in *E. coli* is called SlmA.

How does SlmA work? Recent evidence by Shishen Du and Joe Lutkenhaus at the University of Kansas Medical Center sheds new light on the mechanism. Recall from Chapter 3 that before a septum can form at midcell, a protein called FtsZ must polymerize to form a ring around the circumference of the membrane. However, SlmA can bind to chromosomal DNA and prevent FtsZ polymerization and, thus, septum formation. **Figure 1** shows in vitro evidence for this inhibition. FtsZ in solution will form long, rod-like polymers, but not in the presence of SlmA molecules bound to <u>S</u>lmA

A. FtsZ

B. FtsZ + SlmA + SBS

FtsZ filament

©2014 DU LUTKEN HAUS. OPEN ACCESS LICENSE UNDER THE CREATIVE COMMONS ATTRIBUTION LICENSE.

FIGURE 1 ▪ **Nucleoid exclusion protein SlmA prevents FtsZ polymerization.** **A.** In vitro polymerization of FtsZ. **B.** When SlmA and SBS sequences are present, FtsZ fails to polymerize in vitro.

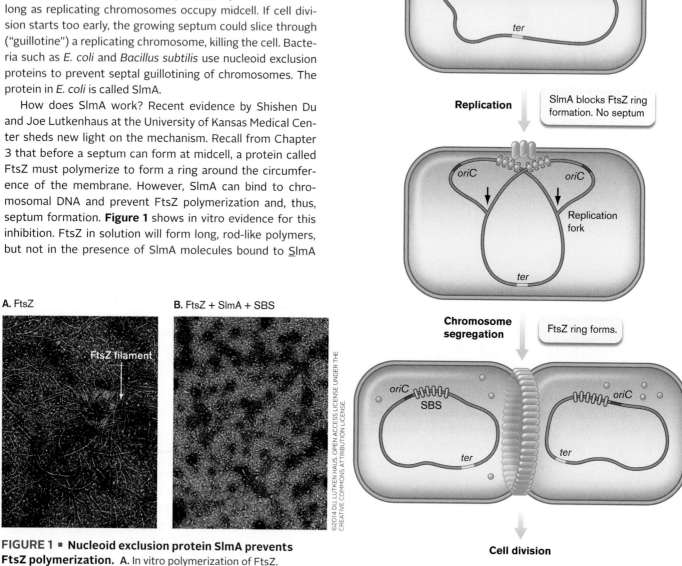

Replication

SlmA blocks FtsZ ring formation. No septum

Chromosome segregation

FtsZ ring forms.

Cell division

FIGURE 2 ▪ **Model for SlmA function.**

Plasmids Replicate Autonomously

Plasmids are much smaller than chromosomes (**Fig. 7.21**) and are found in archaea, bacteria, and eukaryotic microbes. Plasmids are usually circular, and circular plasmids, like circular chromosomes, are typically negatively supercoiled. Replication of many of these extrachromosomal elements is not tied to chromosome replication. Each plasmid contains its own origin sequence for DNA

DNA-binding sites (SBS). In the chromosome, SBS sequences cluster mostly near the origin of replication. Consequently, the nucleoid exclusion protein SlmA prevents septum formation by binding to SBS sequences near the origin and will also bind to part of FtsZ, as long as the nucleoid remains near midcell (**Fig. 2**).

Part of the evidence for this model is shown in **Figure 3**. Du and Lutkenhaus examined cells whose nucleoids were stuck near midcell because of a *parC* mutation and estimated the frequency of septum formation sites. Comparisons were made between cells expressing native SlmA (**Fig. 3A**), cells lacking SlmA (**Fig. 3B**), and cells that express native SlmA but have an altered FtsZ that will polymerize even when bound to SlmA. Wild-type SlmA$^+$ cells rarely formed inappropriate septa over their nucleoids (**Fig. 3A**), while *slmA* mutants frequently formed inappropriate septa (**Fig. 3B**). Cells with the mutant form of FtsZ that was resistant to SlmA also exhibited higher numbers of inappropriate septa (**Fig. 3C**). These data support the proposed model of SlmA in preventing septal formation.

So how does a septum ever form? Because SlmA binds to the SBS sequences near the origin of replication, SlmA will be stripped from FtsZ as the sister origins and chromosomes segregate toward opposite cell poles following replication. With the nucleoids nestled at either end of the dividing cell and SlmA removed from FtsZ, septum formation can safely proceed and the cell can divide.

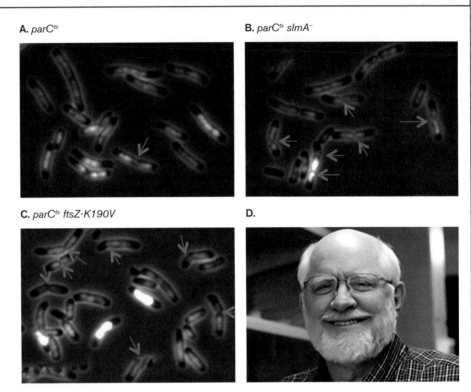

A. *parC^{ts}*

B. *parC^{ts} slmA$^-$*

C. *parC^{ts} ftsZ-K190V*

D.

FIGURE 3 ■ Formation of inappropriate septa in cells with stalled nucleoids.
A–C. A series of temperature-sensitive *parC* mutants (*parC^{ts}*) were shifted from 30°C, a growth temperature at which *ParC^{ts}* will work, to 42°C for one hour, a temperature at which *ParC^{ts}* will not help sister nucleoids move toward cell poles. Nucleoids were stained with the fluorescent dye DAPI (white), and constrictions representing the beginning of septum formation were observed. DAPI was added 20 minutes before imaging. Red arrows indicate apparent constrictions. **D.** Joe Lutkenhaus led research into how SlmA works.
Source: A–C: S. Du and J. Lutkenhaus. 2014. *PloS Genetics* **10**:e1004460. D: Provided by J. Lutkenhaus. Courtesy of University of Kansas Medical Center.

RESEARCH QUESTION
Design an experiment in which fluorescently tagged proteins could be used in vivo to confirm SlmA-FtsZ interaction.

Du, Shishen, and Joe Lutkenhaus. 2014. SlmA antagonism of FtsZ assembly employs a two-pronged mechanism like MinCD. *PLoS Genetics* **10**:e1004460.

replication, but only a few of the genes needed for replication. Thus, even when the timing of plasmid replication is not linked to that of the chromosome, many of the proteins used for plasmid replication are actually host enzymes. Which host proteins are used depends on the plasmid.

Plasmids can replicate in two different ways. Bidirectional replication starts at a single origin and moves in

A.

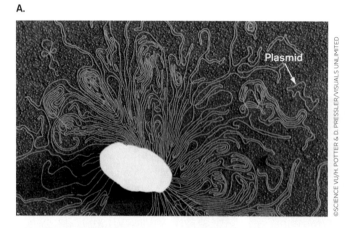

©SCIENCE VU/H. POTTER & D. PRESSLER/VISUALS UNLIMITED

B.

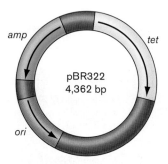

FIGURE 7.21 ■ **Plasmid map. A.** Note the huge difference in size between a circular plasmid DNA molecule (arrow) and chromosomal DNA after both are gently released from a cell (approx. 1 μm). **B.** Map of plasmid pBR322. This plasmid contains an origin of replication (*ori*) and genes encoding resistance to ampicillin (*amp*) and tetracycline (*tet*).

two directions simultaneously. Rolling-circle replication (**Fig. 7.22** ▶) is unidirectional. In rolling-circle replication, a replication initiator (RepA, encoded by a plasmid gene) binds to the origin of replication and nicks one strand. RepA holds on to one end (5′ PO$_4$) of the nicked strand, while the other end (3′ OH) serves as a primer for host DNA polymerase to replicate the intact, complementary strand. The RepA initiator protein recruits a helicase that unwinds DNA, which becomes coated by single-stranded DNA-binding proteins. As replication proceeds, the nicked strand progressively peels off without replicating, until the strand is completely displaced. Then the two ends of the displaced but nicked single strand are re-joined by the RepA protein and released. The single-stranded circular molecule is protected by host single-stranded DNA-binding proteins until host enzymes replicate a complementary strand and regenerate a double-stranded molecule. Although most known plasmids use

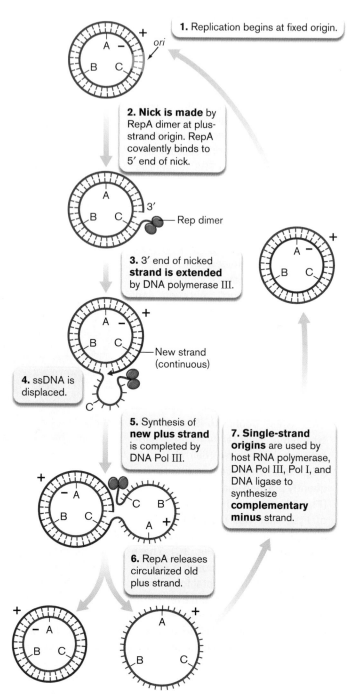

1. Replication begins at fixed origin.

2. Nick is made by RepA dimer at plus-strand origin. RepA covalently binds to 5′ end of nick.

Rep dimer

3. 3′ end of nicked **strand is extended** by DNA polymerase III.

New strand (continuous)

4. ssDNA is displaced.

5. Synthesis of **new plus strand** is completed by DNA Pol III.

7. Single-strand origins are used by host RNA polymerase, DNA Pol III, Pol I, and DNA ligase to synthesize **complementary minus** strand.

6. RepA releases circularized old plus strand.

FIGURE 7.22 ■ **Plasmid replication: rolling-circle model.** ▶

only one of the two replication strategies (bidirectional or unidirectional), a few can use either one, depending on the circumstance of the cell.

Plasmid "Tricks" Ensure Inheritance

If plasmids replicate autonomously, can cells easily lose their plasmids? When a plasmid-containing host cell divides, is it just chance that determines whether both

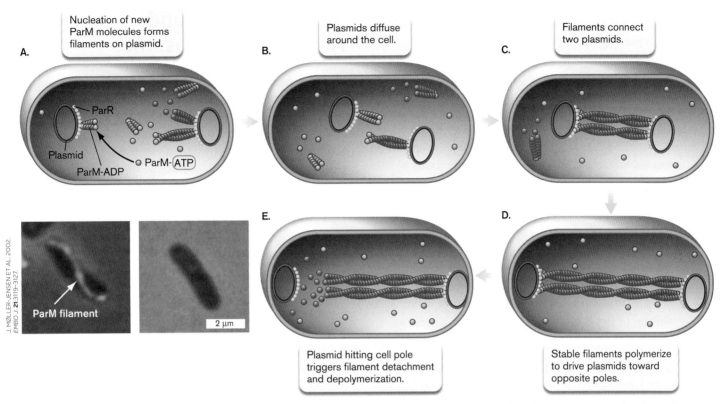

FIGURE 7.23 ■ Molecular model of plasmid segregation. The two images at far right show a cell with a *parM* mutant plasmid unable to make a ParM filament, and a cell with a *parM*⁺ plasmid making a ParM actin-like filament (green) to partition daughter plasmids (equivalent to panel D). ParM filament was seen by combined phase-contrast and immunofluorescence microscopy using rabbit anti-ParM antibodies and fluorescent Alexa 488–conjugated goat antirabbit IgG antibodies. *Source:* Adapted from Christopher S. Campbell and R. Dyche Mullins. 2007. *J. Cell Biol.* **179**:1059–1066, fig. 4.

daughter cells inherit the plasmid? In some instances, the answer is yes (see the Current Research Highlight). Some plasmids make many copies of themselves before the bacterial cell divides. These are called high-copy-number plasmids and can attain 50–700 copies per cell, depending on the plasmid. As a result, there is a high probability that both daughter cells will, by random chance, contain at least one plasmid.

Not all plasmids are high-copy-number, however. Low-copy-number plasmids limit how many copies they make to avoid draining the cell of energy. Cells forced to waste energy making many plasmid copies could be at a growth disadvantage when competing with plasmid-less cells in a natural environment. Low-copy-number plasmids evolved clever partitioning systems that ensure both daughter cells will contain copies of the plasmid. **Figure 7.23** illustrates a model for how one type of partitioning works for plasmid R1, a *Salmonella* plasmid that imparts multidrug antibiotic resistance to host bacteria. The process employs three known plasmid-encoded genes: *parC*, *parM*, and *parR*. The *parC* DNA sequence

is analogous to the centromere of eukaryotic chromosomes. The DNA-binding protein ParR binds to the *parC* sequence and forms a ParR-*parC* plasmid complex. ParM protein is an actin-like molecule that forms long filaments as it hydrolyzes ATP. The ParM filaments are dynamically unstable, constantly elongating and shortening. However, each end of a ParM actin-like filament can bind to a ParR-*parC* plasmid complex. When both ends of a ParM filament contact ParR-*parC* plasmid complexes, the filament stabilizes and elongates until the two plasmids hit the opposite poles of the cell. The plasmid is thus dislodged from the filament, and the filament quickly dissociates.

Some plasmids ensure their inheritance by carrying genes whose functions benefit the host bacterium under certain conditions. For instance, bacterial plasmids can carry genes that confer antibiotic resistance (discussed in Figure 7.21B and Chapter 27). As long as the antibiotic is present in the environment, any cell that loses the plasmid will be killed or stop growing. Antibiotic resistance plasmids benefit bacteria, but they are a major problem for modern hospitals, where plasmids carrying multiple drug resistance genes are transmitted from harmless bacteria into pathogens. On the other hand, plasmids (such as pBR322) containing drug resistance genes are the workhorses of genetic technology and have benefited society tremendously.

Other kinds of host survival genes carried by plasmids include genes providing resistance to toxic metals, genes encoding toxins that aid pathogenesis, and genes encoding proteins that enable symbiosis. These are discussed in later chapters. Most of the genes involved in the nitrogen-fixing symbiosis of *Rhizobium*, for example, are plasmid-borne. Far from being freeloaders, plasmids often contribute significantly to the physiology of an organism.

Some plasmids, however, use what could be called "strong-arm measures" to ensure their maintenance. These plasmids come equipped with self-preservation genes, called "addiction modules," that force the host cell to keep the plasmid or die (**eTopic 7.3**).

Plasmids Are Transmitted between Cells

Plasmids have played a pivotal role in evolution because of how easily they can be passed back and forth between bacterial species. This is particularly evident today in the spread of antibiotic resistance, whose genetic basis is often a plasmid-encoded gene.

Some plasmids are self-transferable via conjugation, a process that requires cell-to-cell contact to move the plasmid from a donor cell to a recipient (as discussed in Section 9.1). Other plasmids are incapable of conjugation (nontransmissible). A third group can be transferred only if a self-transferable plasmid resides in the same cell. In this case, the conjugation mechanism produced by one plasmid will act on the other plasmid. Any plasmid released from dead cells can also be taken up intact by some bacteria in a process called transformation (see Section 9.1). Finally, plasmids can be transmitted in nature by accidentally being packaged into bacteriophage head coats—in other words, by bacteriophage transduction (also covered in Section 9.1).

To Summarize

- **Plasmids are autonomously replicating** circular or linear DNA molecules that are part of a cell's genome.

- **Plasmids replicate** by rolling-circle and/or bidirectional mechanisms.

- **Plasmids can be transferred** between cells.

7.5 Eukaryotic and Archaeal Chromosomes

The chromosomes of eukaryotic microbes such as protists and algae have much in common with those of prokaryotes. Both consist of double-stranded DNA, for example, and both usually replicate bidirectionally. Nevertheless, important differences exist as well, particularly in genome structure. Eukaryotic chromosomes are linear and contained within a nucleus, and their replication involves mitosis (reviewed in eAppendix 2). Archaeal chromosomes, similar to bacterial chromosomes, are circular and lack a nuclear membrane, but archaeal proteins involved in transcription and replication appear more eukaryotic than bacterial.

Eukaryotic Genomes Are Large and Linear

Overall, the genomes of eukaryotic nuclei are larger than those of bacteria, sometimes by several orders of magnitude. For example, *Saccharomyces cerevisiae* and *Mycoplasma pneumoniae* genomes are 12,052 kb and 800 kb, respectively. Eukaryotes typically have duplicate copies of multiple chromosomes (diploid number ranges between 10 and 100). All eukaryotic chromosomes are linear, whereas many, if not most, bacterial chromosomes are circular. Eukaryotic chromosomes use mitosis to replicate and segregate to daughter cells (see Appendix 1). Each eukaryotic chromosome has numerous origins of replication that collectively generate hundreds of replication bubbles, although not all of them "fire" during each replication cycle. Termination zones essential to bacterial chromosomes are not found in eukaryotes.

The ends of eukaryotic chromosomes, called **telomeres**, have a special problem replicating. Because their chromosomes are linear, eukaryotes cannot replicate their 3′ ends using DNA-dependent DNA polymerase. The result is that the 5′ ends of a chromosome (the result of lagging-strand synthesis) are always shorter than 3′ ends. Left like this, chromosomes would become progressively shorter and genetic information would be lost. The solution to this problem is an enzyme complex called **telomerase**. Telomerase is actually a reverse transcriptase that reads RNA as a template to synthesize DNA. Telomerase uses an intrinsic RNA (an RNA that is part of the enzyme) as a template to add numerous DNA repeat sequences to the 3′ ends of chromosomes. Extension of the 3′ end by telomerase will permit synthesis of the RNA primer necessary to replicate the 5′ end of the chromosome. The process prevents the loss of genetic information during division.

Telomerase may have evolved from the ancient progenitor cells that contained RNA rather than DNA genomes (see Section 17.2). The reverse transcriptase may also be the evolutionary source of retroviruses (see **eTopic 6.1**).

Note: Bacteria with linear chromosomes do not have telomerases. How do they replicate their 3′ ends? Some cap the ends of their chromosomes with covalently bound terminal proteins that prime DNA replication (*Streptomyces*). Others form covalently closed hairpin ends (essentially making one long, circular, single-stranded molecule). The ends appear to recombine after being replicated (*Borrelia*).

Unlike prokaryotic cells, eukaryotic cells pack their DNA within the confines of a nucleus where a series of proteins called **histones** compact the DNA. Histones are rich in arginine and lysine, so they are positively charged, basic proteins that easily bind to the negatively charged DNA. The DNA becomes wrapped around the histones to form units called nucleosomes. Histones also play a regulatory role through methylation and acetylation. Bacteria, too, have DNA-packaging proteins, but they are more diverse and less essential for function (see Sections 3.4 and 7.2).

The detailed structures of the genomes of eukaryotes and prokaryotes reveal surprising differences (**Fig. 7.24**). The genome of the bacterium *E. coli* is packed with genes (green in the figure) encoding proteins or RNA molecules. These genes are separated by very little unused or noncoding DNA (purple), and by only an occasional mobile element, such as an insertion sequence, that can move from one DNA molecule to another. Most prokaryotic genes are organized in coordinately regulated, multiple-gene operons. In contrast, 98% of the human genome consists of noncoding sequences (sequences that do not encode proteins). Noncoding sequences include sequences that regulate protein-coding genes; genes that encode untranslated, small regulatory RNAs; and the fossil genomes of ancient viruses. Coding genes are separated by large stretches of noncoding sequences and usually are not clustered together in operons. Moreover, coding genes in the human genome are interrupted by **introns** (DNA within a gene that is not part of the coding sequence for a protein, shown as yellow in **Fig. 7.24**) and ancient gene duplications that have decayed into nonfunctional, vestigial **pseudogenes**. Bacteria also have pseudogenes, but they account for much less of the genome than in eukaryotes. The more rapid replication of bacteria causes pseudogenes to be lost more quickly.

Note: Pseudogenes differ from noncoding DNA in that at least part of a pseudogene's sequence is similar to that of a gene with a known function. Also note that some pseudogenes express mRNA, but the proteins produced have no obvious function.

By themselves, promoters of eukaryotic genes have very low activity and require enhancer DNA sequences to drive transcriptional activity. Enhancer sequences act at long distances from promoters, and scientists suspect that once enhancers became important, it was necessary to place enough DNA between them to reduce the activation of other, unrelated but adjacent promoters.

Archaeal Genomes Combine Features of Bacteria and Eukaryotes

All characterized archaea have a single circular chromosome and polygenic operons; and, like bacteria, their reproduction is predominantly asexual. Archaea are true prokaryotes

A. A genome section of the bacterium *Escherichia coli*. Sections include coding nenes (green), noncoding sequences (purple), and mobile elements such as insertion sequences (IS).

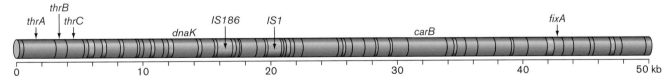

B. Noncoding DNA sequences (purple) comprise 98% of the human genome; protein-coding genes in eukaryotes are interrupted by introns (yellow).

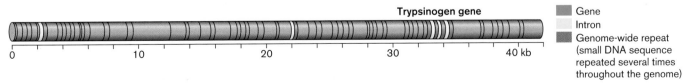

FIGURE 7.24 ■ **Genome structure in a prokaryote and a eukaryote.**

because their cells lack a nuclear membrane, but the structures of the DNA-packing proteins, RNA polymerase, and ribosomal components more closely resemble those of eukaryotes. **Figure 7.25** compares the core RNA polymerase (RNAP) subunits of the archaeon *Sulfolobus* to those of the eukaryotic RNA polymerase II (RNAPII). The archaeal TATA-binding protein (TBP, green) and transcription factor B (TFB, pink) also show homology to eukaryotic counterparts that help position and activate RNAPII at some eukaryotic promoters. One consequence of eukaryotic-like transcription and translation in archaea is that archaea are resistant to antibacterial antibiotics that target those processes.

Oddly, these organisms have two different types of DNA polymerase: one similar to those of eukaryotes and the other unique to archaea. The archaeal origin of the replication sequence also shows greater similarity to those of eukaryotes. Finally, it should be noted that archaeal genomes encode certain unique components, such as the metabolic pathway of methanogenesis. (For more on archaea, see Chapter 19.)

What experimental data allow us to make such comparisons? Overwhelmingly, we rely on new data from the growing number of microbial genomes sequenced (over 30,000 as of 2015). Comparison of genomes has revolutionized our understanding of evolutionary relationships among microbes.

We will now examine the tools of DNA sequence analysis that have made these studies possible. The most important of these tools—restriction mapping, DNA sequencing, and amplification by the polymerase chain reaction—actually harness the molecular apparatus used by bacteria to replicate or protect their own chromosomes.

To Summarize

- **Eukaryotic chromosomes** are always linear, double-stranded DNA molecules that replicate by mitosis.

- **A reverse transcriptase** called telomerase is needed to finish replicating the ends of a eukaryotic chromosome.

- **Histones** (eukaryotic DNA-packing proteins) play a critical role in forming chromosomes.

- **Introns and pseudogenes** are noncoding DNA sequences that make up a large portion of eukaryotic chromosomes.

- **Archaeal chromosomes** resemble those of bacteria in size and shape, but archaeal DNA polymerases are more closely related to eukaryotic enzymes.

7.6 DNA Sequence Analysis

How do we study a genome? We have just described the core concepts of DNA structure, packaging, and replication. This knowledge is crucial to understanding genomics and the fundamentals of genomic analysis. But what about the basic techniques used to manipulate DNA? These include isolating genomic DNA from cells, snipping out DNA fragments with surgical precision, splicing them into plasmid vehicles, and reading their nucleotide sequences. These are the techniques that drive the genomic revolution.

DNA Isolation and Purification

The chemical uniformity of DNA means that simple and reliable purification methods can be used to isolate it. A variety of techniques are used to extract DNA from bacterial cells. The cells may be lysed using lysozyme to degrade the peptidoglycan of the cell wall, followed by treatment with detergents to dissolve the cell membranes. Lysozyme is an enzyme that cleaves the bond linking residues of peptidoglycan. Once the cell contents are released, most proteins are removed by being precipitated in a high-salt solution. The precipitated proteins are then removed by centrifugation, and the cleared lysate containing DNA is passed through a column containing a silica resin that specifically binds DNA. The remaining proteins are washed out of the column because they do not stick to the resin, and the DNA is eluted with water. The DNA is then concentrated via alcohol

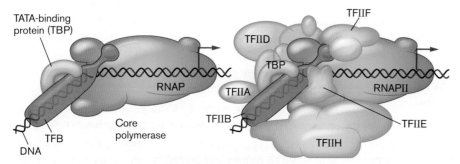

A. Archaeal RNA polymerase

TATA-binding protein (TBP)

RNAP

TFB

DNA

Core polymerase

B. Eukaryotic RNA polymerase II

TFIIF

TFIID

TBP

TFIIA

TFIIB

RNAPII

TFIIE

TFIIH

FIGURE 7.25 ■ RNA polymerases from Archaea and Eukarya exhibit homology. The core RNA polymerase (RNAP) subunits of the archaeon *Sulfolobus* **(A)** show homology to those of the eukaryotic RNA polymerase II (RNAPII). **(B)**. The TATA-binding protein (TBP, green) and transcription factor B (TFB, pink) of *Sulfolobus* also show homology to eukaryotic counterparts.

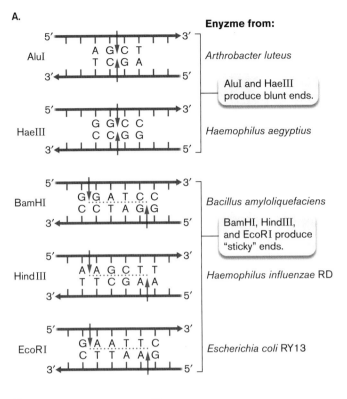

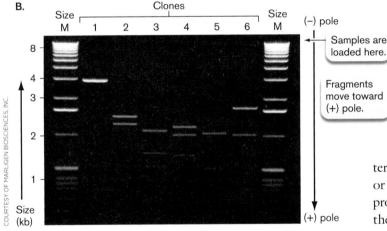

FIGURE 7.26 ■ Bacterial DNA restriction endonucleases. A. Target sequences of sample enzymes. The names of these enzymes reflect the genus and species of the source organism. **B.** Agarose-gel size analysis of EcoRI-restricted DNA fragments. Smaller fragments move toward the bottom faster than large fragments. DNA bands are viewed using chemical stains that bind DNA and fluoresce under UV light.

precipitation (ethanol or isopropanol). DNA may be precipitated from aqueous solution by the addition of ethanol and salt because ethanol removes water associated with DNA's phosphoryl groups (negatively charged), which then bind the Na^+ ions from the salt. The DNA, now neutralized, precipitates out of the polar water-ethanol solution. The precipitated DNA is then dissolved in water or

a weak buffer. At this point, the extracted DNA can be examined with a variety of analytical tools.

There are other methods for isolating plasmid DNA, such as equilibrium density gradient centrifugation. (See **eTopic 7.4** for a discussion of equilibrium density gradient centrifugation and its use in the classic Meselson-Stahl experiment.)

Restriction Endonucleases and the Birth of Recombinant DNA

What seemed to be an esoteric line of research, exploring how bacteria cleave the "foreign" DNA of invading plasmids and viruses, ultimately led to a tool that revolutionized biology. One of the ways bacteria rid themselves of foreign DNA is through proteins called **restriction endonucleases** that cleave unfamiliar DNA molecules at specific sequences called **restriction sites** (usually 4–8 bp long). Different species of bacteria use restriction endonucleases that recognize different DNA sequences. A few examples of restriction enzymes and target sequences are shown in **Figure 7.26A**. The most useful restriction endonucleases for molecular biology recognize restriction sites that are four to six bases in length. Notice that the sequence of each of these restriction sites is a **palindrome** in which the top and bottom strands read the same in the 5′-to-3′ direction. A palindromic DNA restriction site enables a restriction endonuclease to attack both strands of a duplex using the same substrate recognition site.

Restriction endonucleases cleave the phosphodiester backbones of opposite strands at locations either near or within the center of the restriction site. The cut may produce blunt ends or staggered ends. In staggered ends, the top strand is cut at one end of the site, and the bottom strand is cut at the other end (**Fig. 7.26A**). Staggered ends are also called "cohesive ends" because the protruding strand at one end can base-pair with a complementary protruding strand from any DNA fragment cut with the same restriction endonuclease, regardless of the source organism. The ability of cohesive ends from different organisms to base-pair initially made recombinant DNA technology possible.

But how do bacteria making these scissor-like DNA enzymes cut foreign DNA and not destroy their own? They protect their chromosome with a methylating enzyme (a modification enzyme) that places methyl groups on both strands of restriction sites recognized by a specific restriction endonuclease. A restriction endonuclease cannot cut a site if either one of the strands is methylated. Thus, newly replicated double-stranded molecules are protected, since one strand, the template, remains methylated at all times.

Foreign DNA, such as phage DNA, that originates from a cell with one type of restriction modification system will <u>not</u> be protected upon entering another cell with a different restriction modification system. The foreign DNA, even though it is methylated, will not be protected, because it will not be methylated in the <u>right</u> places. As a result, foreign DNA is more likely to be destroyed than to undergo protective methylation.

Agarose gels can be used to analyze DNA fragments generated by restriction endonuclease digestion (agarose-gel electrophoresis is explained in eAppendix 3). Each lane of the gel shown in **Figure 7.26B** represents a different population of DNA molecules cut with the same restriction endonuclease. Because DNA is negatively charged, all DNA molecules travel to the positive pole during electrophoresis. Pore sizes in the agarose are such that they allow small molecules to speed through the gel, while larger molecules take longer to move. Thus, DNA fragments in this gel separate on the basis of size—the smaller the fragment, the farther it travels down the lane. The DNA fragments are of different sizes, so they travel different distances in the gel. The sum of the sizes of the fragments in each lane yields the size of the original, intact, uncut molecule.

1. Plasmid and foreign DNA are cut with a restriction endonuclease (EcoRI) to produce identical cohesive ends.

2. Cut vector and foreign DNA fragments are mixed. Cohesive ends anneal.

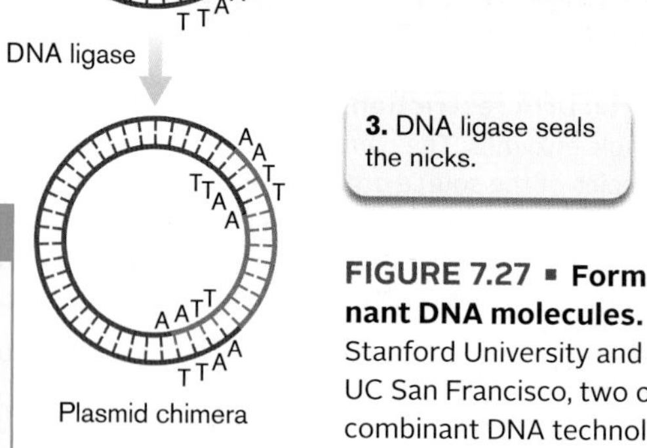

3. DNA ligase seals the nicks.

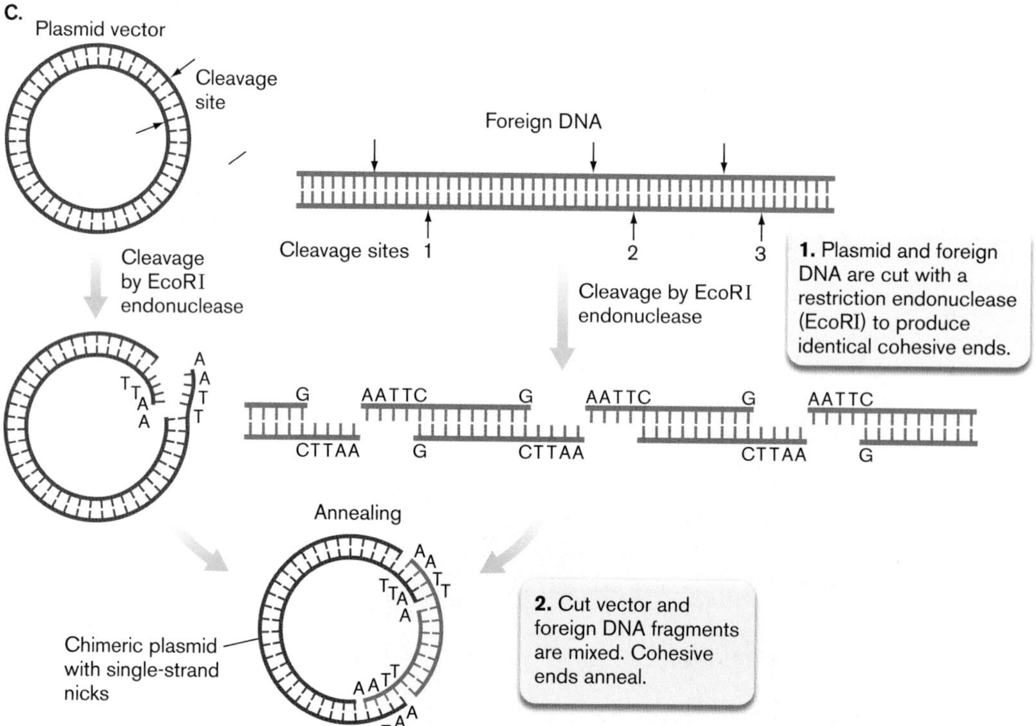

FIGURE 7.27 ■ **Formation of recombinant DNA molecules.** Stanley Cohen **(A)** of Stanford University and Herbert Boyer **(B)** of UC San Francisco, two of the founders of recombinant DNA technology. **C.** Cloning a gene.

Thought Question

7.10 **Figure 7.26B** illustrates how agarose gels separate DNA molecules. If you ran a highly supercoiled plasmid in one lane of the gel and the same plasmid with one phosphodiester bond cut in another lane, where would the two molecules end up relative to each other in this type of gel?

Cloning. How did scientists ever conceive of cloning a gene from one organism into the DNA of an unrelated organism? The beginning of the recombinant DNA revolution can be traced to 1972. Scientists knew that many restriction endonucleases generate cohesive ends and that some plasmids contain a single site for certain restriction endonucleases. However, the significance of these facts went unrecognized until Stanley Cohen (**Fig. 7.27A**), Herb Boyer (**Fig. 7.27B**), and Stanley Falkow were relaxing together after a scientific meeting in 1972. They suddenly realized that a piece of DNA cut from one organism's chromosome could be grafted to a plasmid cut with the same restriction endonuclease (**Fig. 7.27C**). They also knew that DNA ligase (see **Fig. 7.18**) would seal the fragment to the plasmid and form a new artificial DNA molecule. They hypothesized that the recombinant plasmid could be introduced into *E. coli* and replicated. Three months later, the strategy, called **cloning**, worked. This seminal work was published by Boyer and Cohen along with students Annie Chang and Robert Helling. Over the next decade, the technique became known as "gene cloning," an immensely powerful technology that thrust biology into the genomic era.

For many years, scientists used the restriction endonuclease strategy discovered by Boyer, Cohen, Chang, and Helling to randomly generate genome libraries containing all the genes in an organism's chromosome. Today, new approaches can more efficiently probe the microbial genome. A microbe's entire genome can now be sequenced within a few days—a process that once took years. Once obtained, the sequence can be analyzed using **bioinformatics**, a computer-assisted process that predicts the beginnings and ends of genes, determines the amino acid sequences of gene-encoded proteins, and forecasts probable functions for those proteins (Chapter 8 describes bioinformatics in more detail). Interesting genes can then be amplified using PCR (described later) and cloned into plasmids using restriction endonucleases. The cloned gene can be reintroduced into mutant or wild-type organisms using various techniques and the effects on physiology or pathogenicity determined.

PCR Amplifies Specific Genes from Complex Genomes

Imagine having the ability to amplify one copy of a gene into millions more within one or two hours. The technique, called the **polymerase chain reaction** (**PCR**), has revolutionized biological research, medicine, and forensic analysis.

The PCR technique, outlined in **Figure 7.28**, requires the design of specific oligonucleotide primers (usually between 20 and 30 bp) that anneal to known DNA sequences flanking the DNA you want to amplify. A thermostable DNA polymerase uses these primers to replicate the target DNA. Heat-stable DNA polymerases are used because the PCR reaction mixture must undergo repeated cycles of heating to 95°C (to separate DNA strands so that they are available for primer annealing), cooling to 55°C (to allow primer annealing), and heating to 72°C (the optimum reaction temperature for the polymerase). The polymerase Taq, from the thermophile *Thermus aquaticus*, was originally used for this purpose, but polymerases from other thermophiles are also used. Because PCR reactions require 25–30 heating and cooling cycles, a machine called a thermocycler is used to reproducibly and rapidly deliver these cycles. It is important to realize that PCR technology emerged from studies of a seemingly obscure class of microorganisms: extremophiles.

This basic PCR technique has been modified to serve many purposes. Primers can be engineered to contain specific restriction sites that simplify subsequent cloning, as described already. If the primers used for PCR are highly specific for a gene that is present in only one microorganism, PCR can be used to detect the presence of that organism in a complex environment, such as the presence of the pathogen *E. coli* O157:H7 in hamburger. Multiplex PCR, involving several primer sets, can be used to detect multiple pathogens in a single reaction (see eAppendix 3).

PCR has also profoundly changed our judicial system by making it possible to amplify the tiniest amounts of DNA

Thought Question

7.11 Given that *E. coli* possesses restriction endonucleases that cleave DNA from other species, how is it possible to clone a gene from one organism to another without the cloned gene sequence being degraded?

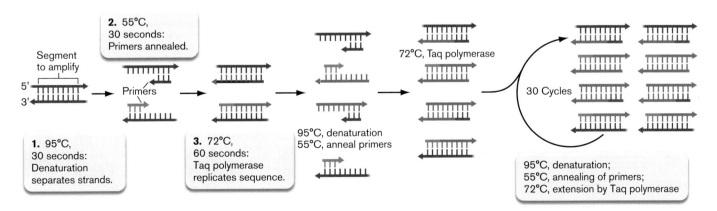

FIGURE 7.28 ■ The polymerase chain reaction. The cyclic PCR reaction makes a large number of copies of a small piece of DNA. Potentially 10^{30} copies of a fragment can be made from a single DNA molecule.

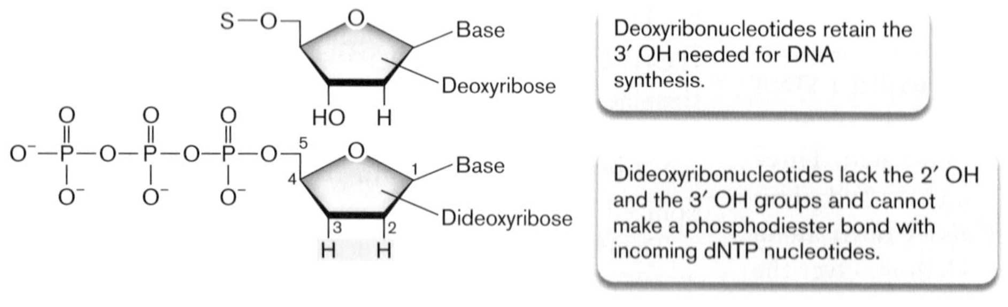

FIGURE 7.29 ■ **Dideoxyribonucleotide.** Review Figure 7.5 to see how missing the 3′ OH group would terminate DNA synthesis. dNTP = deoxyribonucleoside triphosphate, or deoxyribonucleotide.

Deoxyribonucleotides retain the 3′ OH needed for DNA synthesis.

Dideoxyribonucleotides lack the 2′ OH and the 3′ OH groups and cannot make a phosphodiester bond with incoming dNTP nucleotides.

contaminating a crime scene. The technique provides conclusive evidence in court cases where no other evidence exists. Increasingly, individual human genomes are being sequenced as a standard medical test—with profound ethical and societal implications. The hopes and fears raised by the invention of PCR-based genome sequencing inspired the science fiction film *Gattaca*, directed by Andrew Niccol in 1997, depicting an imaginary future in which everyone's destiny is determined by his or her DNA sequence. The film imagined that knowledge of DNA sequence could impact which jobs are available to someone, whether insurance coverage can be withheld, and even whether two individuals can marry. Today, the Genetic Information Nondiscrimination Act passed by the U.S. Congress in 2008 protects individuals from "genetic discrimination" based on their DNA. In a more compassionate application, human sequence information is beginning to be used to predict which drugs will best treat individual patients—for example, patients with specific types of tumors.

> **Thought Question**
>
> **7.12** PCR is a powerful technique, but a sample can be easily contaminated and the wrong DNA amplified—perhaps sending an innocent person to jail. What might you do to minimize this possibility?

DNA Sequencing

Even with the development of gene cloning, our transition into the genomic age would not have been possible without the ability to rapidly sequence entire genomes. Previous sequencing methods used an ingeniously simple strategy but were not rapid. The dideoxy chain termination method developed by Fred Sanger in the 1970s relied on the fact that the 3′ hydroxyl group on 2′-deoxyribonucleotides (shown in **Fig. 7.5A**) is absolutely <u>required</u> for a DNA chain to grow. Thus, incorporation of a 2′,3′-dideoxyribonucleotide (also known as dideoxynucleotide; **Fig. 7.29**) into a growing chain prevents further elongation, because without the 3′ OH, extension of the phosphodiester backbone is impossible.

Details of the Sanger dideoxy sequencing method are provided in eAppendix 3. But briefly, the method begins by cloning a gene to be sequenced into a plasmid vector. Next, a short (20- to 30-bp) oligonucleotide DNA primer molecule is designed to anneal at a site immediately adjacent to the insert you want to sequence. The mixture is heated to 95°C to separate the DNA strands, and then cooled, which allows the primer to bind. DNA polymerase can then use the 3′ OH end of this primer to synthesize DNA from the cloned-insert template. A small amount of <u>di</u>deoxyribo<u>nu</u>cleoside <u>tri</u>phosphate (ddNTP—for example, dideoxy ATP) is included in the DNA synthesis reaction, which already contains all four normal <u>de</u>oxyribonucleoside <u>tri</u>phosphates (dNTPs), also called deoxyribonucleotides. Because normal 2′-deoxyadenosine is present at high concentration, chain elongation usually proceeds normally. However, the sequencing reaction contains just enough dideoxy ATP to make DNA polymerase occasionally substitute the dead-end

Cluster generation

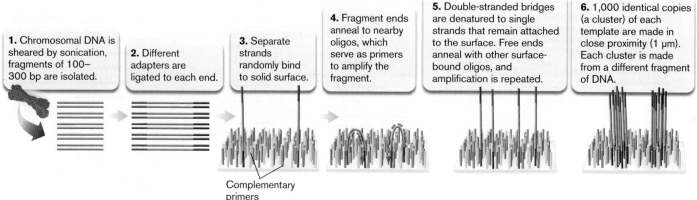

1. Chromosomal DNA is sheared by sonication, fragments of 100–300 bp are isolated.

2. Different adapters are ligated to each end.

3. Separate strands randomly bind to solid surface.

4. Fragment ends anneal to nearby oligos, which serve as primers to amplify the fragment.

5. Double-stranded bridges are denatured to single strands that remain attached to the surface. Free ends anneal with other surface-bound oligos, and amplification is repeated.

6. 1,000 identical copies (a cluster) of each template are made in close proximity (1 μm). Each cluster is made from a different fragment of DNA.

Complementary primers

FIGURE 7.30 ■ **DNA sequencing; sequencing by synthesis.** **Steps 1–6:** Generation of clusters by bridge amplification. **Steps 7–10:** Sequencing by synthesis using reversible fluorescent termination.

base for the natural one, at which point the chain stops growing. The result is a population of DNA strands of varying length, each one truncated at a different adenine position. How the tagged fragments are used to determine sequence is described in eAppendix 3. The Sanger method has been replaced by even more rapid techniques, all of which include an element of the Sanger chain termination approach.

Next-Generation Sequencing

The old way of sequencing genomes was to randomly clone DNA fragments into plasmid or phage vectors and then separately sequence each fragment. It took months, if not years, to sequence an entire genome. Today, new sequencing technologies, called "next-generation sequencing," have combined the power of robotics, computers, and fluidics such that an entire bacterial genome can be sequenced within days.

One of these techniques, pyrosequencing, is described in **eTopic 7.5.** The most used technique today is called sequencing by synthesis (a technology developed by Solexa, a company now part of Illumina, Inc.); the Illumina process is outlined in **Figure 7.30.** Basically, the genome to be sequenced is sonically fragmented into 100- to 300-bp segments, and different linker oligonucleotides are ligated to each end (**Fig. 7.30,** steps 1 and 2). Small fragments are used because the technique can sequence only 100–300 bp from any one fragment. Strands of each fragment are then separated and the mixture added to an optical flow cell. The fragments, millions of them, are randomly fixed to the solid surface (step 3). Each individual fragment sticks to a different area of the cell. The glass surface also has a dense lawn of oligonucleotides fixed at their 5′ ends to the glass surface. These oligonucleotides are complementary to the fragment linker ends but do not hybridize to them at this point.

Next, each single-stranded DNA (ssDNA) fragment is converted into a tight cluster of identical fragments through a series of so-called bridge amplifications (**Fig. 7.30,** steps 4 and 5). Ends of each fragment anneal to nearby matched oligos (fixed to the glass surface), which then serve as primers to amplify the fragments further. Multiple amplifications result in millions of different ssDNA fragment clusters dotting each lane of each slide (step 6).

All clusters are simultaneously sequenced in a repetitive series of single-step, reversible chain termination reactions that attach fluorescently labeled nucleotides one at a time to the clusters (**Fig. 7.30,** steps 7 and 8). Because each individual cluster is a tuft of identical DNA molecules, all of the molecules in that cluster will have the same tagged base added and will fluoresce the same color. Note that the sequence of either strand can be determined by adding one or the other primer. After a base is added, a snapshot captures the colors, which reveal the bases added to each cluster (step 9). Next, the fluorescent marker (fluor) is removed from the growing chain, which also reverses chain termination, and the slide is again flooded with tagged bases (step 10). Snapshots taken after each sequencing round sequentially capture the fluorescent colors of each cluster as each base is added.

The old Sanger dideoxy method could sequence 1,000 bp per template but required considerable time, effort, and space to do so. In contrast, each flow-cell lane in the sequencing-by-synthesis technique produces 10–120 million DNA fragments, thus generating up to 600 gigabases (Gb) of DNA sequence per run (1 Gb = 1 billion bases). Each read is short (about 100 bases per template), but because millions of templates are read, the massively parallel sequencing yields hundreds of millions of bases. Computer programs then find the overlaps among the millions of sequence results and assemble them into the sequence of the entire chromosome. With this technology, you can have your own genome sequenced for less than $1,000.

Sequencing by synthesis

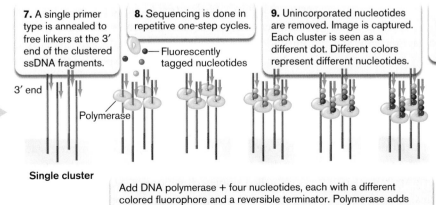

7. A single primer type is annealed to free linkers at the 3′ end of the clustered ssDNA fragments.

8. Sequencing is done in repetitive one-step cycles.

Fluorescently tagged nucleotides

3′ end

Polymerase

Single cluster

Add DNA polymerase + four nucleotides, each with a different colored fluorophore and a reversible terminator. Polymerase adds one nucleotide, which will be the same for all members of the cluster.

9. Unincorporated nucleotides are removed. Image is captured. Each cluster is seen as a different dot. Different colors represent different nucleotides.

10. Fluor and terminator are removed, and the flow cell is again flooded with the sequencing reactants that will add the next base in the sequence. As before, the color is captured by laser snapshot.

Sequence snapshot

How the information from sequences is mined is discussed further in Chapter 8. Many important discoveries have resulted from this technology. A particularly intriguing discovery was finding that the entire genome of the bacterium *Wolbachia* (which infects 20% of the world's insect population) is embedded within the genome of *Drosophila*, the common fruit fly. The evolutionary implications of this discovery are under intense investigation.

Microbiomes and Metagenomics

Microorganisms in nature do not generally exist as pure cultures. They grow instead as complex consortia containing numerous species (microbiomes), most of which have never been grown in a laboratory. With newer and ever-faster DNA sequencing methodologies and PCR techniques, we can now "peer" deeper into the natural world. Like the explorers of old, scientists are finding new species never before seen and are beginning to understand the intricate ways bacterial species interact to form what could be called biological support groups.

Metagenomics, a term coined in 1998 by microbiologist Jo Handelsman (Yale University), is the use of modern genomic techniques to study microbial communities directly in their natural environments, bypassing the need for isolating and cultivating individual species in the laboratory. The field began with the culture-independent retrieval of 16S rRNA genes from the environment by amplifying DNA directly from environmental samples using PCR protocols. Since genes encoding 16S rRNA from small ribosomal subunits (also called small-subunit rRNA) are highly conserved, it is possible to determine whether a given sequence of 16S rRNA represented a new species and to which known species it was most closely related. Since then, metagenomics has revolutionized microbiology by shifting the focus even further away from cultivatable organisms toward the estimated 99% of microbial species that cannot be cultivated. Since 2003, hundreds of metagenomic projects have been carried out or are ongoing. Examples of the metagenomic analysis of ecosystems are presented in Chapter 21.

Identifying membership of a microbiome can be very informative but says little about the genetic and metabolic potential of the community. Randomly sequencing DNA from a microbiome, however, will identify most, if not all, of the genes present in all the organisms that the consortium comprises. All of these genes make up what is known as the **metagenome**—that is, the combined yet unsorted genomic potential of all the organisms in the sample. Thus, metagenomics not only identifies the principal partners of a microbial consortium, but also reveals many of their metabolic capabilities. Communities of microorganisms can actually form new metabolic

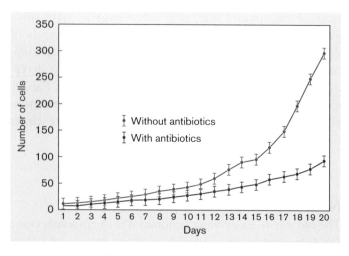

FIGURE 7.31 ■ Effect of microbiota on the growth of the ciliate *Euplotes focardii* at 4°C. Protozoans were either untreated or treated with streptomycin and penicillin G. Antibiotic treatment lasted for 15 days after which the protozoans were transferred to antibiotic-free media. All organisms were fed the green alga *Dunaliella tertiolecta*. *Source:* Pucciarelli, et al. 2015. Microbial Consortium Associated with the Antarctic Marine Ciliate *Euplotes focardii*: An Investigation from Genomic Sequences. *Microbial Ecology* **70**:484-497, fig 6B.

pathways in which a useless by-product of one member can be used by another member in a reaction whose product can benefit a third member.

Metagenomic analysis has also been used to explore how a microbiome might expand the ecological niche of some eukaryotic microbes. One intriguing study recently identified the bacterial microbiome of the Antarctic marine protozoan *Euplotes focardii*, a strictly psychrophilic ciliate isolated from Terra Nova Bay. This organism will not survive above 10°C. Sandra Pucciarelli and colleagues from the University of Camerino, Italy, discovered that *E. focardii* "cured" of their bacterial microbiota by antibiotics had difficulty growing at 4°C (**Fig. 7.31**). Using Illumina DNA sequencing, the scientists analyzed the 16S rRNA sequences present in *E. focardii* DNA extracts. The data revealed that most bacterial species were either from **Gammaproteobacteria**, **Alphaproteobacteria**, or Bacteroidetes. When protein-encoding sequences were examined, many were homologous to ice-binding or antifreeze proteins found in other cold-adapted bacterial species (Section 5.2). The researchers conclude that one or more members of the bacterial consortium have an important role in facilitating growth of this Antarctic protozoan at low temperatures.

The science of metagenomics has even progressed to a point where the genome of a single cell can be sequenced—a process called single-cell genomics (SCG). SCG requires special protocols, equipment, and reagents to isolate a single cell from an environment and, without ever growing it, extract

A.

ROBERT BARKER/CORNELL UNIVERSITY

B.

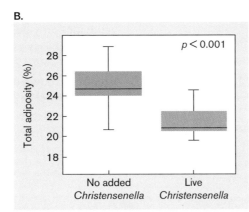

FIGURE 7.32 ■ Heritable member of the human microbiome. A. Ruth Ley led a team that used metagenomics to discover that the abundance of some members of the intestinal microbiome are influenced by host genetics. **B.** One such organism, *Christensenella minuta*, also influenced weight gain and adiposity when orally "transplanted" into mouse intestines. The graph shows that adiposity 21 days after a fecal transplant was significantly lower in germ-free mice transplanted with *C. minuta*–containing fecal preparations (blue) than with preparations lacking this organism (orange). *Source:* Goodrich et al. 2014. *Cell* **159**:789–799, fig. 7B.

and amplify its DNA. Once amplified, the genomic DNA is readily sequenced using the next-generation sequencing tools described earlier. SCG can identify the "microbial dark matter" of undiscovered taxa in a microbiome, and then tell us which genes in a metagenome belong to which organism. Using SCG, Tonja Woyke at the U.S. Department of Energy's Joint Genome Institute and her colleagues sequenced single cells from 20 major uncultured archaeal and bacterial lineages. They discovered 20,000 new hypothetical protein families, found genes encoding bacterial sigma factors in archaea, revealed an archaeal purine synthesis pathway in bacteria, and identified several new candidate phyla of Bacteria and Archaea. Single-cell genomic studies promise to yield many more surprises.

The power of metagenomics has also been directed toward solving the human microbiome. One hundred trillion microbes (up to 10 microbial cells for every one nucleated human cell) live in the human body, but we did not know the identity of most of them, because we could not grow them in the laboratory. How we interact with these organisms and they with us is a major question in modern microbiology—one that has led to the formation of the Human Microbiome Project, which is examining gut, skin, and oral microbiomes. We already know that the composition of intestinal microbiota can influence blood chemistry and may contribute to irritable bowel syndrome (inflammation of the gastrointestinal tract), but microbiota are also suspected of influencing an individual's susceptibility to diabetes and obesity.

One recent paper reveals that human genes can influence the makeup of our microbiome. Ruth E. Ley (Cornell University; **Fig. 7.32A**) and collaborators cataloged the intestinal microbiomes of identical and fraternal human twins to determine whether any part of the microbiome might be heritable. They used PCR to amplify 16S rRNA sequences from human fecal samples and sequenced the products. The 16S rRNA sequences were matched to a public database to identify bacterial species. They found that *Christensenella minuta*, a recently discovered firmicute, was the most highly heritable among identical twins. Even more striking was that this microbe was found more often in lean than in obese subjects, and it prevented weight gain when transplanted into germ-free mice (**Fig. 7.32B**). Knowing the identities of these microbes and their collective metabolic potential will undoubtedly lead to new advances in the treatment and prevention of human diseases. More on metagenomics and microbial ecology is found in Section 21.1.

To Summarize

- **Restriction endonucleases** used for DNA analysis cleave DNA at specific recognition sequences, which are usually 4–6 bp in length and produce either a blunt or a staggered cut.

- **Agarose-gel electrophoresis** will separate DNA molecules on the basis of their size.

- **Restriction endonuclease–digested DNA** molecules were first cloned into plasmids in the early 1970s.

- **The polymerase chain reaction (PCR)** will amplify one or two copies of a specific gene a millionfold.

- **The most commonly used DNA sequencing** technique (the Sanger technique) uses dideoxyribonucleotides—nucleotides that lack both the 2′ OH and 3′ OH groups—to terminate chain elongation.

- **Entire genomes** are sequenced by fragmenting chromosomal DNA, amplifying the fragments in clusters, and sequencing each cluster. Overlapping DNA sequences from different clusters are matched by computer analysis until the entire genome is reconstructed.

- **Metagenomics** uses rapid DNA sequencing and other genomic techniques to study consortia of microbes directly in their natural environment.

Concluding Thoughts

We used to think of bacteria as simple, single-celled organisms. In fact, they are far from simple. Complexity is evident in the elegant mechanisms they use to organize and replicate their genomes. The next three chapters will expand on this idea and discuss the way bacteria make proteins, regulate genes, exchange DNA, and, in the process, evolve. Knowing how a bacterial cell replicates and genetically controls the repertoire of available proteins and enzymes provides a unique perspective from which to study the physiology of microbial growth, explored in Part 3 of this textbook. Furthermore, understanding the bacterial genome allows us to investigate the physiology of pathogenic, as well as nonculturable, organisms and gain greater insight into the evolutionary diversity of species that we will discuss in Part 4.

CHAPTER REVIEW

Review Questions

1. What is the difference between vertical and horizontal gene transfer?
2. Explain the structural types of bacterial genomes. What is a structural gene?
3. Describe the three functional levels of gene organization.
4. What are the differences between DNA and RNA?
5. Explain DNA supercoiling. Why is it important to microbial genomes?
6. Discuss the mechanisms of topoisomerases. What drug targets a topoisomerase?
7. What are the basic mechanisms involved in DNA replication?
8. How does the bacterial cell regulate the initiation of chromosome replication?
9. What is the clamp loader? Primase? DNA helicase (DnaB)? Helicase loader (DnaC)? DNA proofreading?
10. How is the problem of replicating both strands at a replication fork solved?
11. What is a catenane? What does it have to do with DNA replication?
12. Explain the polymerase chain reaction (PCR).
13. What is rolling-circle replication?
14. What is DNA restriction and modification? Why is it important to bacteria? Why is it important to forensic scientists?
15. Explain next-generation DNA sequencing.

Thought Questions

1. When you sequence a genome, the DNA used is in fragments. How do you know where the base pairs in the genome are located?
2. During rapid growth, why would a bacterial cell die if an antibiotic drug formed, as the text says, "a physical barrier in front of the DNA replication complex"?
3. Why would bacteria, which replicate more quickly than eukaryotic organisms, lose pseudogenes more rapidly than eukaryotes do?

Key Terms

Alphaproteobacteria (268)
antiparallel (241)
bioinformatics (265)
catenane (254)
cloning (265)
conjugation (237)
denature (242)

DNA control sequence (237)
DNA ligase (251)
DNA replication (236)
enhancer (239)
exonuclease (251)
Gammaproteobacteria (268)
gene (237)

genome (237)
histone (261)
horizontal gene transfer (236)
hybridization (242)
information flow (236)
intron (261)
metagenome (268)

metagenomics (268)
nitrogenous base (241)
nucleobase (241)
nucleoid (242)
Okazaki fragments (249)
operon (239)
origin (*oriC*) (247)
palindrome (263)
phosphodiester bond (241)
plasmid (255)
polymerase chain reaction (PCR) (265)
primase (250)

promoter (239)
proofreading (251)
pseudogene (261)
purine (241)
pyrimidine (241)
quinolone (245)
recombination (237)
regulon (240)
replication fork (246)
replisome (251)
restriction endonuclease (263)
restriction site (263)

semiconservative (246)
sliding clamp (250)
structural gene (237)
telomerase (260)
telomere (260)
termination (*ter*) site (247)
topoisomerase (244)
transcription (236)
transformation (237)
translation (236)
vertical gene transfer (236)

Recommended Reading

Bouet, Jean-Yves, Mathiu Stouf, Elise Lebailly, and Francois Cornet. 2014. Mechanisms for chromosome segregation. *Current Opinion in Microbiology* **22**:60–65.

Campbell, Christopher S., and R. Dyche Mullins. 2007. In vivo visualization of type II plasmid segregation: Bacterial actin filaments pushing plasmids. *Journal of Cell Biology* **179**:1059–1066.

Costa, Allesandro, Iris V. Hood, and James M. Berger. 2013. Mechanisms for initiating cellular DNA replication. *Annual Reviews of Biochemistry* **82**:25–54.

Dame, Remus T., Olga J. Kalmykowa, and David C. Grainger. 2011. Chromosomal macrodomains and associated proteins: Implications for DNA organization and replication in Gram-negative bacteria. *PLoS Genetics* **7**:e1002123.

Duggin, Ian G., R. Gerry Wake, Stephen D. Bell, and Thomas M. Hill. 2008. The replication fork trap and termination of chromosome replication. *Molecular Microbiology* **70**:1323–1333.

Falkow, Stanley. 2001. I'll have chopped liver please, or how I learned to love the clone. *ASM News* **67**:555.

Georgescu, Roxana E., Isabel Kurth, and Mike E. O'Donnell. 2011. Single-molecule studies reveal the function of a third polymerase in the replisome. *Nature Structural & Molecular Biology* **19**:113–116.

Goodrich, J. K., J. L. Waters, A. C. Poole, J. L. Sutter, O. Koren, et al. 2014. Human genetics shape the gut microbiome. *Cell* **159**:789–799.

Hedlund, Brian P., Jeremy A. Dodsworth, Senthil K. Murugapiran, Christian Rinke, and Tanja Woyke. 2014. Impact of single-cell genomics and metagenomics on the emerging view of extremophile "microbial dark matter." *Extremophiles* **18**:865–875.

Kuzminov, Andrei. 2014. The precarious prokaryotic chromosome. *Journal of Bacteriology* **196**:1793–1806.

Le, Tung Bk, and Michael T. Laub. 2014. New approaches to understanding the spatial organization of bacterial genomes. *Current Opinion in Microbiology* **22**:15–21.

Lindas, A. C., and R. Bernander. 2013. The cell cycle of archaea. *Nature Reviews. Microbiology* **11**:627–638.

McHenry, Charles S. 2011. Bacterial replicases and related polymerases. *Current Opinion in Chemical Biology* **15**:587–594.

Reyes-Lamothe Rodrigo, Nicolas Emilien, and David J. Sherratt. 2012. Chromosome replication and segregation in bacteria. *Annual Review of Genetics* **46**:121–143.

Robinson, Andrew, and Antoine M. van Oijen. 2013. Bacterial replication, transcription and translation: Mechanistic insights from single-molecule biochemical studies. *Nature Reviews. Microbiology* **11**:303–315.

Rodrigo, Reyes-Lamothe, Tung Tran, Diane Meas, Laura Lee, Alice M. Li, et al. 2014. High-copy bacterial plasmids diffuse in the nucleoid-free space, replicate stochastically and are randomly partitioned at cell division. *Nucleic Acids Research* **42**:1042–1051.

Sanchez-Romero, Maria A., Ignatio Cota, and Josep Casadesus. 2015. DNA methylation in bacteria: From the methyl group to the methylome. *Current Opinion in Microbiology* **25**:9–16.

Song, Dan, and Joseph J. Loparo. 2015. Building bridges within the bacterial chromosome. *Trends in Genetics* **31**:164–173.

Wolanski, Marcin, Rafal Donczew, Anna Zawilak-Pawlik, and Jolanta Zakrzewska-Czerwinska. 2014. *oriC*-encoded instructions for the initiation of bacterial chromosome replication. *Frontiers in Microbiology* **5**:735.

CHAPTER 8
Transcription, Translation, and Bioinformatics

The cell accesses the vast store of data in its genome using tiny molecular machines. One machine (RNA polymerase) reads the DNA template to make RNA (transcription). Another machine (the ribosome) decodes the RNA to assemble a protein (translation). Once made, each polypeptide must fold properly and somehow find its correct cellular or extracellular location. Proteins that have outlived their usefulness must be destroyed and their amino acids recycled. In Chapter 8 we explain how the cell makes proteins and guides their fate. What emerges is a picture of remarkable biomolecular integration, controlled to maintain balanced growth and ensure survival.

CURRENT RESEARCH highlight

Molecular coupling. Transcription by RNA polymerase (RNAP), and translation by ribosomes make protein. Bacteria couple these processes so that ribosomes can make protein even while RNAP is transcribing the RNA. For some genes coupling requires a protein to link the ribosome to a transcribing RNAP. RfaH, for instance, simultaneously binds RNAP and ribosomal protein S10. Irina Artsimovitch and Paul Rösch found that once RfaH binds to RNA polymerase, a major domain of RfaH dramatically refolds to expose an S10 binding site. RfaH then couples the ribosome to RNAP and prevents premature termination of transcription.

Source: Tomar et al. 2013. *Nucleic Acids Res.* **41**:10077–10085.

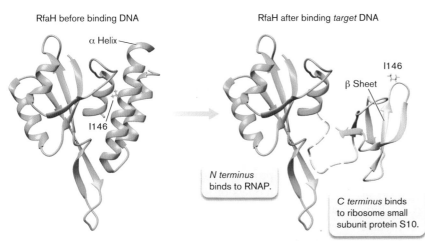

RfaH before binding DNA

α Helix

I146

RfaH after binding *target* DNA

I146

β Sheet

N terminus binds to RNAP.

C terminus binds to ribosome small subunit protein S10.

AN INTERVIEW WITH

IRINA ARTSIMOVITCH, MICROBIOLOGIST, THE OHIO STATE UNIVERSITY

DR. SUSHIL TOMAR

How have your structural biology studies altered our view of transcription and protein folding?

Transcription includes a series of transient intermediates that serve as targets for regulation. Structural analysis allows us to visualize these elusive states. We showed that factor RfaH bound to a transcribing RNA polymerase undergoes a dramatic structural change in which an entire domain refolds from an alpha helix into a beta barrel. This finding overturns the "one sequence–one fold" paradigm for proteins and suggests the existence of other "transformer" proteins.

What future applications do you see for your team's discovery?

Dramatic alpha-to-beta refolding of prion proteins underlies pathological processes that cause some neurodegenerative diseases. The RfaH model of fold conversion could help guide the development of therapies for these devastating diseases.

Today we know a great deal about transcription and translation. The Current Research Highlight, for instance, illustrates how Irina Artsimovitch, Paul Rösch, and Monali NandyMazumdar (**Fig. 8.1**) exposed a mechanism by which these processes are coupled in bacteria. But how did the journey begin? By 1960 we knew that the DNA in a cell holds the genetic code, but the code itself and how the code produces protein remained mysterious. RNA was a suspected, but unproven, intermediate until Marshall Nirenberg and his postdoctoral student Heinrich Matthaei at the National Institutes of Health (**Fig. 8.2A**) designed a cell-free system (a cell lysate of *E. coli*) to test the RNA hypothesis. They used RNA molecules, synthesized by Maxine Singer (**Fig. 8.2B**), that contained simple, known, repeated sequences, such as poly-A (consisting only of adenylic acid), poly-U (polyuridylic acid), poly-AAU, and poly-ACAC. These RNA molecules were tested to see whether the repetitive sequence might direct incorporation of a specific amino acid into a protein. Each synthetic polynucleotide was tested in the presence of a radiolabeled amino acid. If the radioactive amino acid was incorporated into a polypeptide, the polypeptide would also be radioactive. On the morning of May 27, 1960, the results of experiment 27Q showed that the poly-U RNA specified the assembly of radioactive polyphenylalanine. It was the first break in the genetic code. For his work, Nirenberg, along with Har Gobind Khorana (who developed methods for making synthetic nucleic acids) and Robert Holley (who solved the structure of yeast transfer RNA), won the 1968 Nobel Prize in Physiology or Medicine.

In Chapter 8 we explore the way microbes, primarily bacteria, interpret the information held within a nucleotide sequence of DNA and convert that information into a string of amino acids—a protein. From there we look at what the cell does with those proteins once they are made. For instance, specific proteins are moved into the periplasm, while other proteins must be inserted into membranes. We also show how damaged proteins are selectively degraded. We conclude with bioinformatics, now an essential tool of microbiology. Bioinformatic computer programs can compare

FIGURE 8.2 ■ **Many scientists have contributed to our understanding of the genetic code. A.** Heinrich Matthaei (left) and Marshall Nirenberg (right) were the first to crack the genetic code. An early hand-drawn model of what would eventually be known as translation is behind them. **B.** Maxine Singer, a key contributor to the genetic code experiments, also helped develop guidelines for recombinant DNA research.

a new gene sequence with hundreds of thousands of known genes. These comparisons help predict for any given microbe what genes it has, what proteins it makes, and even what food it consumes. Patterns in DNA sequences across species also allow us to pose new questions about microbial life, disease, and evolution.

8.1 RNA Polymerases and Sigma Factors

To survive and reproduce, every cell needs to access information encoded within DNA. How does this happen? Chromosomal DNA is large and cumbersome, so the first step in the process is to make multiple copies of the information in snippets of RNA that can move around the cell and, like disposable photocopies, can be destroyed once the encoded protein is no longer needed. This copying of DNA to RNA is called **transcription**.

FIGURE 8.1 ■ **Monali Nandy-Mazumdar.** NandyMazumdar worked with Irina Artsimovitch and Paul Rösch to elucidate how bacteria use RfaH to couple transcription and translation.

RNA Polymerase Transcribes DNA to RNA

An enzyme complex called **RNA polymerase**, also known as **DNA-dependent RNA polymerase**, carries out transcription, making RNA copies (called **transcripts**) of a DNA template. The DNA **template strand** specifies the base sequence of the new complementary strand of RNA.

RNA polymerase in bacteria consists of a core polymerase and a sigma factor. Core polymerase contains the proteins required to elongate an RNA chain. **Sigma factor** is a protein needed only for initiation of RNA synthesis, not for its elongation. Together, core polymerase plus sigma factor are called the holoenzyme.

A single core RNA polymerase is a complex of four different subunits: two alpha (α) subunits, one beta (β) subunit, and one beta-prime (β′) subunit (**Fig. 8.3**). A fifth subunit, omega (ω), is shown in the figure but is not required for transcription. The beta-prime subunit houses the Mg^{2+}-containing catalytic site for RNA synthesis, as well as sites for the rNTP (ribonucleoside triphosphate, or ribonucleotide) substrates, the DNA substrates, and the RNA products. The 3D structure of RNA polymerase shows that DNA fits into a cleft formed by the beta and beta-prime subunits (**Fig. 8.3**). The alpha subunit assembles the other two subunits (β and β′) into a functional complex. It also communicates through physical "touch" with various regulatory proteins that can bind DNA. These protein-protein interactions inform RNA polymerase what to do after binding DNA.

The composition of RNA polymerase and the sequences of its subunits are well conserved among the Bacteria. However, differences in function are notable when examining this enzyme from organisms living in extreme environments. *Pseudomonas syringae* strain Lz4W, for example, is a Gram negative, Antarctic soil species that grows at 0°C. RNA polymerase from *E. coli* does not function at that temperature. M. K. Ray and colleagues from the Centre for Cellular and Molecular Biology in India extracted RNA polymerase from *P. syringae* and found that it did function in vitro at 0°C. The optimum temperature was 37°C but the enzyme retained 10%–15% activity at 0°C, enough so it could live and grow.

Sigma factors. Where does a gene begin and end? The helical structure of a DNA sequence appears largely homogeneous, making it hard to distinguish one area from another. In fact, core RNA polymerase randomly binds and transcribes DNA, "blind" to where a gene starts. Some random transcript generation may be useful (see Chapter 17), but totally random transcription initiation would be extremely wasteful. Consequently, there must be a mechanism to tell RNA polymerase where a real gene starts.

Sigma (σ) factors are the proteins that guide RNA polymerase to the beginnings of genes (**Fig. 8.3**). Similar to how a visually impaired person uses touch to identify the beginning of a sentence written in braille, sigma factors use "touch" to identify the beginning of a gene. A sigma factor first binds to RNA polymerase through the beta and beta-prime subunits

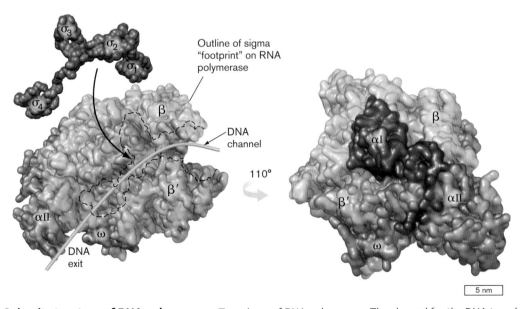

FIGURE 8.3 ■ **Subunit structure of RNA polymerase.** Two views of RNA polymerase. The channel for the DNA template is shown by the yellow line. Subunits (αI, αII, β, β′, and ω) are color-coded dark green, medium green, light green, cyan, and gold, respectively. The function of the omega (ω) subunit is currently unclear. Recent evidence suggests it may have a role in sigma factor competition for core polymerase. Sigma factor (red), which recognizes promoters on DNA, is shown separate from core polymerase in the left-hand panel. Different functional areas of sigma are labeled sigma 1 through sigma 4 (σ_1–σ_4). Sigma factor interacts with the alpha (α), beta (β), and beta-prime (β′) subunits. The molecule on the left is rotated 110° to give the image on the right. To view stereo images of RNA polymerase, locate code 1L9Z in the RCSB Protein Data Bank on the Internet. *Source:* Robert D. Finn et al. 2000. *EMBO J.* **19**:6833–6844.

A. Strong E. coli promoters

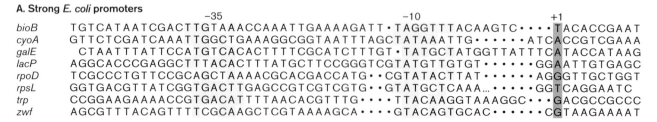

```
                                    -35                                                    -10                          +1
bioB   TGTCATAATCGACTTGTAAACCAAATTGAAAGATT·TAGGTTTACAAGTC····TACACCGAAT
cyoA   GTTCTCGATCAAATTGGCTGAAAGGCGGTAATTTAGCTATAAATTG······ATCACCGTCGAAA
galE    CTAATTTATTCCATGTCACACTTTTCGCATCTTTGT·TATGCTATGGTTATTTCATACCATAAG
lacP   AGGCACCCGAGGCTTTACACTTTATGCTTCCGGGTCGTATGTTGTGT······GGAATTGTGAGC
rpoD   TCGCCCTGTTCCGCAGCTAAAACGCACGACCATG··CGTATACTTAT······AGGGTTGCTGGT
rpsL   GGTGACGTTATCGGTGACTTGAGCCGTCGTCGTG··GTATGCTCAAA...·····GGTCAGGAATC
trp    CCGGAAGAAAACCGTGACATTTTAACACGTTTG····TTACAAGGTAAAGGC···GACGCCGCCC
zwf    AGCGTTTACAGTTTTCGCAAGCTCGTAAAAGCA····GTACAGTGCAC······CGTAAGAAAAT
```

B. Consensus sequences of σ⁷⁰ promoters

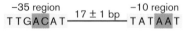

-35 region   -10 region
TTGACAT —— 17 ± 1 bp —— TATAAT

C. Sequence of lacP promoter

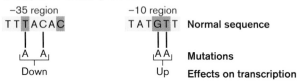

-35 region   -10 region
TTTACAC   TATGTT   Normal sequence
   │ │   │ │
   A A   A A   Mutations
  Down   Up   Effects on transcription

D.

Sigma factor	Promoter recognized	Promoter consensus sequence	
		−35 region	−10 region
RpoD σ^{70}	Most genes	TTGACAT	TATAAT
RpoH σ^{32}	Heat shock–induced genes	TCTCNCCCTTGAA	CCCCATNTA
RpoF σ^{28}	Genes for motility and chemotaxis	CTAAA	CCGATAT
RpoS σ^{38}	Stationary-phase and stress response genes	TTGACA	TCTATACTT
		−24 region	−12 region
RpoN σ^{54}	Genes for nitrogen metabolism and other functions	CTGGNA	TTGCA

FIGURE 8.4 ■ −10 and −35 sequences of E. coli promoters. A. Alignment of sigma-70 (σ⁷⁰)-dependent promoters from different genes. Dots were added to help visualize alignments: biotin synthesis (*bioB*); cytochrome o (*cyoA*); galactose utilization (*galE*); lactose utilization (*lacP*); RNA polymerase (*rpoD*); small-subunit ribosome protein (*rpsL*); tryptophan synthesis (*trp*); glucose 6-phosphate dehydrogenase (*zwf*, zwischenferment). Yellow indicates conserved nucleotides; brown denotes transcript start sites (+1). **B.** The alignment in (A) generates a consensus sequence of σ⁷⁰-dependent promoters (red-screened letters indicate nucleotide positions where different promoters show a high degree of variability). "N" indicates that any of the four standard nucleotides can occupy the position. **C.** Mutations in the *lacP* promoter that affect promoter strength (*lac* genes encode proteins that are used to metabolize the carbohydrate lactose). Some mutations can cause decreased transcription (called "down mutations"), while others cause increased transcription ("up mutations"). **D.** Some *E. coli* promoter sequences recognized by different sigma factors.

(see the dotted black outline in **Fig. 8.3**). The bound sigma factor helps the core enzyme detect a specific DNA sequence, called the **promoter**, marking the beginning of a gene. A single bacterial species can make several different sigma factors. Each sigma factor helps core RNA polymerase find the start of a different subset of genes. However, a single core polymerase complex can bind only one sigma factor at a time. Hundreds of sigma factor genes have been sequenced from numerous species of bacteria and archaea. Although they all encode somewhat different amino acid sequences, there are enough sequence and structural similarities to make them recognizable as sigma factors.

Promoters. Every cell has a "housekeeping" sigma factor that keeps essential genes and pathways operating. In the case of *E. coli* and other rod-shaped, Gram-negative bacteria, that factor is sigma-70, so named because it is a 70-kilodalton (kDa) protein (its gene designation is *rpoD*). Genes recognized by sigma-70 all contain similar promoter sequences that consist of two parts. The DNA base

corresponding to the start of the RNA transcript is called nucleotide +1 (+1 nt). Relative to this landmark, promoter sequences are characteristically centered at −10 and −35 nt before the start of transcription (see **Fig. 8.4**). Other sigma factors typically recognize different consensus sequences at one or both of these positions (or at nearby locations in some cases), or they shape the overall polymerase complex to bind the promoter.

The DNA promoter sequence recognized by a given sigma factor can be determined by comparing known promoter sequences of different genes whose expression requires the same sigma. Similarities among these different promoter DNA sequences define a **consensus sequence** likely recognized by the sigma factor (**Fig. 8.4A** and **B**). A consensus sequence consists of the most likely base (or bases) at each position of the predicted promoter. Although promoters are double-stranded DNA (dsDNA) sequences, convention is to present the promoter as the single-stranded DNA (ssDNA) sequence of the sense strand (nontemplate strand), which has the same sequence as the RNA product.

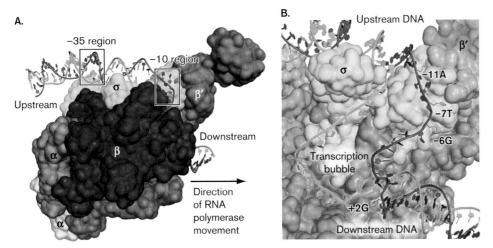

FIGURE 8.5 ■ RNA polymerase holoenzyme bound to a promoter. A. The initial open complex forms when holoenzyme binds to a promoter. DNA −10 and −35 contacts with sigma factor are shown. Nontemplate strand is color-coded magenta; template strand, green. **B.** Blowup of (A), with the beta subunit removed to view the transcription bubble. Some bases in the nontemplate strand are flipped outward (yellow) to interact with sigma factor or the beta subunit after the transcription bubble is formed. The base at position +1 is the first base transcribed.

Some positions in a consensus sequence are highly conserved, meaning that the same base is found in that position in every promoter. Other, less conserved positions can be occupied by different bases. Few promoters actually have the most common base at every position. Even highly efficient promoters usually differ from the consensus at one or two positions.

Sigma factor recognition of promoters. How do sigma factors, or any other DNA-binding proteins for that matter, recognize specific DNA sequences when the DNA is a double helix? The phosphodiester backbone is quite uniform, and the interior of paired bases appears inaccessible. However, proteins can still recognize DNA sequences because the alpha-helical parts of proteins that bind DNA recognize (via noncovalent bonding) side groups that protrude from the bases into the major and minor grooves of DNA (see **Fig. 7.6**). Portions of sigma-70 from *E. coli* wrap around DNA, allowing certain parts of the protein to fit into DNA grooves.

Figure 8.5A shows the holoenzyme RNA polymerase complex positioned at a promoter and the points (−10 and −35 nt) where sigma factor contacts DNA. Sigma factors generally contain four highly conserved amino acid sequences, called regions (see **Fig. 8.3**, σ_1–σ_4). Part of region 2 of the sigma-70 family recognizes −10 sequences, whereas region 4 recognizes the −35 sites. Part of region 1 helps separate the DNA strands to make a transcription "bubble" in preparation for RNA synthesis (**Fig. 8.5B**).

Sigma factor control of complex physiological responses. A single cell will wander into and out of many stressful environments. Each time this happens, the cell must readjust its physiology, sometimes very quickly. Large readjustments can be made when a single sigma factor coordinately (simultaneously) activates a required set of genes. Gene sets controlled by different sigma factors include those dealing with nitrogen metabolism, flagellar synthesis, heat stress, starvation, sporulation, and many other physiological responses.

How does a sigma factor control gene expression? In response to a given environment, microbes will increase synthesis or decrease destruction of the appropriate sigma factor (described in Chapter 10). The resulting high concentration of the specialty sigma factor will dislodge other sigma factors from core polymerase. In this way, the cell redirects RNA polymerase to the promoters of genes best suited for growth or survival in the new environment.

Thought Questions

8.1 If each sigma factor recognizes a different promoter, how does the cell manage to transcribe genes that respond to multiple stresses, each involving a different sigma factor?

8.2 Imagine two different sigma factors with different promoter recognition sequences. What would happen to the overall gene expression profile in the cell if one sigma factor were artificially overexpressed? Could there be a detrimental effect on growth?

8.3 Why might some genes contain multiple promoters, each one specific for a different sigma factor?

To Summarize

- **RNA polymerase holoenzyme**, consisting of core RNA polymerase and a sigma factor, initiates transcription of a DNA template strand.

- **Sigma factors** help core RNA polymerase locate consensus promoter sequences near the beginning of a gene. The sequences identified and bound by *E. coli* sigma-70 are located at −10 and −35 bp upstream of the transcription start site.

- **Dynamic changes in gene expression follow** changes in the relative levels of different sigma factors.

8.2 Transcription of DNA to RNA

Like DNA replication, transcription of DNA to RNA occurs in three stages:

1. **Initiation**, in which RNA polymerase binds to the beginning of the gene, melts open the DNA helix, and catalyzes placement of the first RNA nucleotide.
2. **Elongation**, the sequential addition of ribonucleotides to the 3′ OH end of a growing RNA chain.
3. **Termination**, whereby sequences at the end of the gene trigger release of the polymerase and the completed RNA molecule.

The newly released RNA polymerase can then engage another sigma factor to seek a new promoter.

Transcription Initiation

RNA polymerase constantly scans DNA for promoter sequences (**Fig. 8.6**, step 1). It binds DNA loosely and comes off repeatedly. Once bound to the promoter, RNA polymerase holoenzyme forms a loosely bound, closed complex with DNA, which remains annealed and double-stranded—that is, unmelted (step 2). To successfully transcribe a gene, this closed complex must become an open complex through the unwinding of one helical turn, which causes DNA to become unpaired in this area (step 3) (see also **Fig. 8.5B**). After promoter unwinding, RNA polymerase in the open complex becomes tightly bound to DNA. Region 1 of the sigma factor is important for DNA unwinding, as is a rudder complex in the beta-prime subunit that separates the two strands.

The open-complex form of RNA polymerase begins transcription. The first ribonucleoside triphosphate (rNTP) of the new RNA chain is usually a purine (A or G). The purine base-pairs to the position designated +1 on the DNA template, which marks the start of the gene. As the enzyme complex moves along the template, subsequent rNTPs diffuse through a channel in the polymerase and into position at the DNA template. After the first base is in place, each subsequent rNTP transfers a ribonucleoside monophosphate (rNMP) to the growing chain while releasing a pyrophosphate (PP_i):

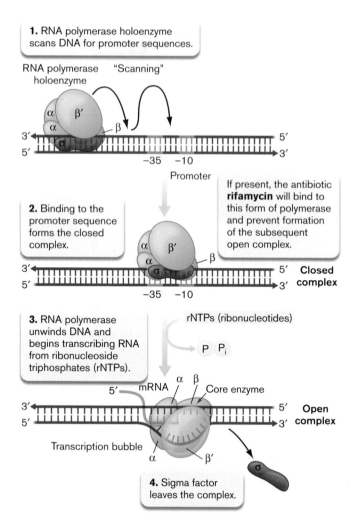

FIGURE 8.6 ■ The initiation of transcription. Sigma factor helps RNA polymerase find promoters but is discarded after the first few RNA bases are polymerized. (Omega is not shown.)

The "energy released" (actually the free energy change) following cleavage of the rNTP triphosphate groups is used to form the phosphodiester link to the growing polynucleotide chain. (Free energy change, ΔG, is presented in Chapter 13.)

As seen in **Figure 8.5B**, the beta-prime subunit forms part of the pocket, or cleft, in which the DNA template is read. Another part of the beta-prime cleft is involved in hydrolyzing rNTPs. Deep at the base of the cleft is the active site of polymerization, defined by three evolutionarily conserved aspartate residues that chelate (bond with) two magnesium ions (Mg^{2+}). As we saw for DNA polymerases, these metal ions play a key role in catalyzing the polymerization reaction.

Transcription Elongation

A transcribing complex retains the sigma factor until about nine bases have been joined, and then it dissociates

(**Fig. 8.6**, step 4). The newly liberated sigma factor can then recycle onto an unbound core RNA polymerase to direct another round of promoter binding. Meanwhile, the original RNA polymerase continues to move along the template, synthesizing RNA at approximately 45 bases per second. As the DNA helix unwinds, a 17-bp transcription bubble forms and moves with the polymerase complex. DNA unwinding produces positive DNA supercoils ahead of the advancing bubble. The supercoils are removed by DNA topoisomerase enzymes (see Section 7.2).

Transcription Termination

How does RNA polymerase know when to stop? Again, the secret is in the sequence. All bacterial genes use one of two known transcription termination signals, called either Rho-dependent or Rho-independent. Rho-dependent termination relies on a protein called Rho and an ill-defined sequence at the 3′ end of the gene that appears to be a strong pause site. Pause sites are areas of DNA that slow or stall the movement of RNA polymerase. Some pause sites occur within the gene's **open reading frame** (**ORF**), the part of a gene that actually encodes a protein, to allow time for the ribosome to catch up to a transcribing RNA polymerase (discussed in Section 8.3). A protein called NusG binds to RNA polymerase to prevent premature termination at internal pause sites. The transcription termination pause site, however, is located after the ORF, beyond the translation stop codon. A stop codon (described later) is a

three-base sequence in mRNA that stops ribosome movement. If transcription stopped before the ribosome reached the translation stop codon, an incomplete protein would be made.

Rho factor binds to an exposed region of RNA after the ORF at C-rich sequences that lack obvious secondary structure. This is the transcription terminator pause site. Rho monomers assemble as a hexamer around the RNA (**Fig. 8.7A**). Then, like a person pulling a raft to shore by a rope tied to a tree, Rho pulls itself to the paused RNA polymerase by threading downstream RNA through the ring via an intrinsic ATPase activity. Once Rho touches the polymerase, an RNA-DNA helicase activity built into Rho appears to unwind the RNA-DNA heteroduplex, which releases the completed RNA molecule and frees the RNA polymerase.

The second type of termination event, called Rho-independent termination, or intrinsic termination, occurs in the absence of Rho. Rho-independent termination requires a GC-rich region of RNA roughly 20 bp upstream from the 3′ terminus, as well as four to eight consecutive uridine residues at the terminus (**Fig. 8.7B**). The GC-rich sequence

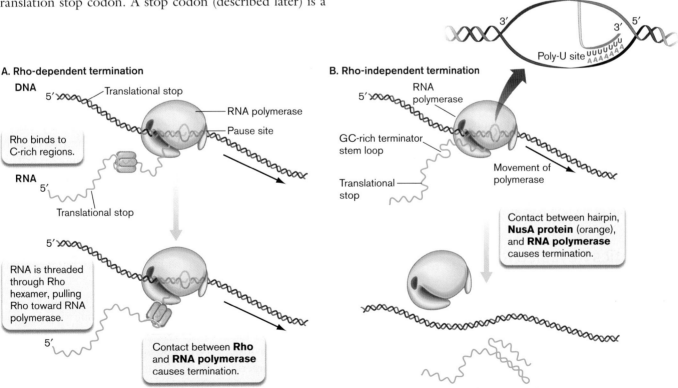

FIGURE 8.7 ■ The termination of transcription. A. Rho-dependent termination requires Rho factor but not NusA. **B.** Rho-independent termination requires NusA but not Rho.

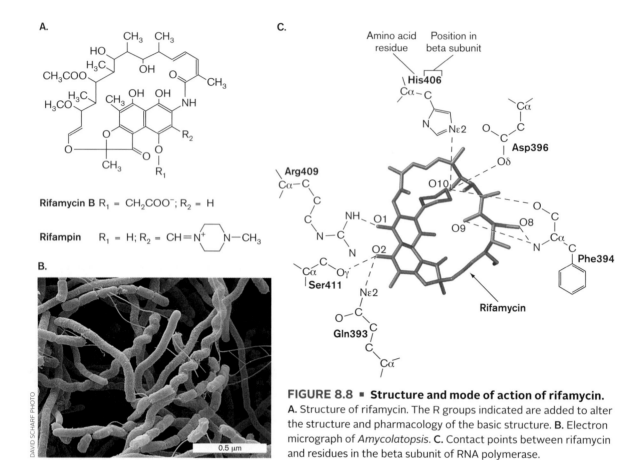

FIGURE 8.8 ■ Structure and mode of action of rifamycin.
A. Structure of rifamycin. The R groups indicated are added to alter the structure and pharmacology of the basic structure. **B.** Electron micrograph of *Amycolatopsis*. **C.** Contact points between rifamycin and residues in the beta subunit of RNA polymerase.

contains complementary bases and forms an RNA stem, or stem loop, a structure that contacts RNA polymerase. Contact halts nucleotide addition, causing RNA polymerase to pause. A protein called NusA stimulates transcription pause at these sites. While the polymerase is paused, the DNA-RNA duplex is weakened because the poly-U/poly-A base pairs at the 3′ terminus contain only two hydrogen bonds per pair, so they are easier to melt. Melting the hybrid molecule releases the transcript and halts transcription. The pause in polymerase movement, and thus transcription, is important to prevent the formation of tighter base-pairing downstream of the UA region.

Antibiotics Revealed RNA Synthesis Machines

To be useful in medicine, all antibiotics must meet two fundamental criteria: They must kill or retard the growth of a pathogen, and they must not harm the host. Antibiotics used against bacteria, for instance, must selectively attack features of bacterial targets not shared with eukaryotes. Many antibiotics possess highly specific modes of action, binding to and altering the activity of one specific "machine" of one type of cell.

One example is the antibiotic rifamycin B (**Fig. 8.8A**), produced by the actinomycete *Amycolatopsis mediterranei* (**Fig. 8.8B**). Rifamycin B selectively targets bacterial RNA polymerase and binds to the polymerase's beta subunit near the Mg^{2+} active site, thus blocking the RNA exit channel (**Fig. 8.8C**). RNA polymerase can carry out two or three polymerization steps, but then it stops because the nascent RNA cannot exit. Rifamycin does not prevent RNA polymerase from binding to promoters, and because an mRNA molecule already passing through the exit channel masks the rifamycin-binding site, the drug cannot bind to RNA polymerases already transcribing DNA. The semisynthetic derivative of this drug, rifampin, is used for treating tuberculosis and leprosy and is given to contacts (family members or roommates) of individuals with bacterial meningitis caused by *Neisseria meningitidis*.

Actinomycin D (**Fig. 8.9A**) is another antibiotic produced by an actinomycete. Its phenoxazone ring is a planar structure that "squeezes," or intercalates, between GC base pairs within DNA. The side chains then extend in opposite directions along the minor groove (**Fig. 8.9B**). Because it mimics a DNA base, actinomycin D blocks the elongation phase of transcription; but because it binds any DNA, it is not selective for bacteria. It can be used to treat

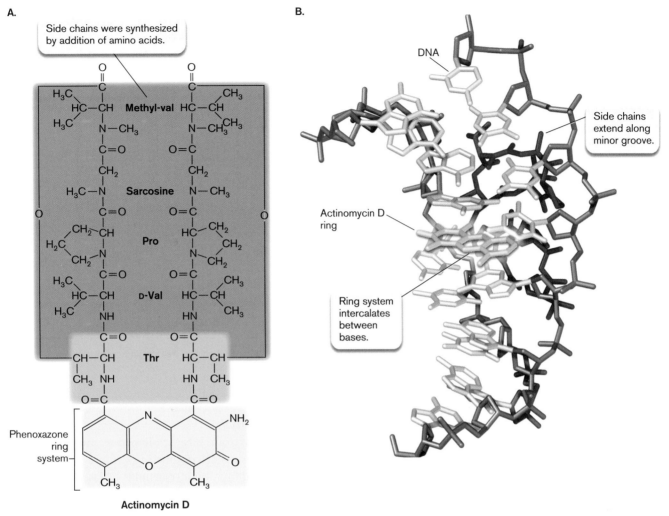

FIGURE 8.9 ■ **Structure and mode of action of actinomycin D.** This antibiotic inserts its ring structure **(A)** between parallel DNA bases and wraps its side chains along the minor groove **(B)**. (PDB code for B: 1DSC)

human cancers because cancer cells replicate rapidly, but it has severe side effects because it inhibits DNA synthesis in normal cells too. (Antibiotics are discussed further in Chapter 27.)

How did scientists determine the details of RNA and protein synthesis? First, purified RNA polymerase and ribosomes were used to demonstrate transcription and translation in cell-free systems (in test tubes). Then, protein subunits of RNA polymerase extracted from antibiotic-resistant and sensitive strains of *E. coli* were mixed together to reconstruct chimeric molecules. To determine which RNA polymerase subunit was targeted by the antibiotic rifamycin, for instance, RNA polymerases from rifamycin-sensitive and rifamycin-resistant strains were purified from cell extracts. The component parts of the polymerases were separated and a chimeric RNA polymerase was reassembled (like a 3D jigsaw puzzle) by mixing subunits from antibiotic-sensitive and antibiotic-resistant polymerases. The reconstituted polymerase preparation was tested to see whether it could make RNA in the presence of rifamycin. Resistance will happen only when the subunit conveying resistance has been added—in this case, the beta subunit. Investigators used this information in subsequent in vitro assays and X-ray crystallography to learn more about the enzymatic mechanism of the beta subunit.

Different Classes of RNA Have Different Functions

It is important to note that not all RNA molecules are translated to protein. There are several classes of RNA, each designed for a different purpose (**Table 8.1**). The class of RNA molecules that encode proteins is called **messenger RNA (mRNA)**. Molecules of mRNA average 1,000–1,500 bases in length but can be much longer or shorter, depending on the size of the protein they encode. Another class of RNA, called **ribosomal RNA (rRNA)**, forms the scaffolding on which ribosomes are built. As will be

TABLE 8.1	Classes of RNA in *E. coli*[a]					

RNA class	Function	Number of types	Average size	Approximate half-life	Unusual bases
mRNA (messenger RNA)	Encodes protein	Thousands	1,500 nt	3–5 minutes	No
rRNA (ribosomal RNA)	Synthesizes protein as part of ribosome	3	5S, 120 nt; 16S, 1,542 nt; 23S, 2,905 nt	Hours	Yes
tRNA (transfer RNA)	Shuttles amino acids	41 (86 genes)	80 nt	Hours	Yes
sRNA[b] (small RNAs, or regulatory RNAs)	Controls transcription, translation, or RNA stability	20–30	<100 nt	Variable	No
tmRNA (properties of transfer and messenger RNA)	Frees ribosomes stuck on damaged mRNA	Roughly one per species	300–400 nt	3–5 minutes	No
Catalytic RNA	Carries out enzymatic reactions (e.g., RNase P)	?	Varies	3–5 minutes	No

[a] Six major classes of RNA exist in all bacteria. They differ in their function, quantity, average sizes, and half-lives, and whether they contain modified bases.

[b] Small RNAs include antisense RNA and micro-RNA.

discussed later, rRNA also forms the catalytic center of the ribosome. **Transfer RNA (tRNA)** molecules ferry amino acids to the ribosome. A unique property of rRNA and tRNA is the presence of unusual modified bases not found in other types of RNA.

The fourth important class of RNA is called **small RNA (sRNA)**. Molecules of sRNA do not encode proteins but are used to regulate the stability or translation of specific mRNAs into proteins (for example, RNA thermometers are discussed in Chapter 10). As will be discussed later in more detail, all mRNA molecules contain untranslated leader sequences that precede the actual coding region. Complementary sequences within some of these RNA leader regions snap back on themselves to form double-stranded stems and stem loop structures that can obstruct ribosome access and limit translation. Regulatory sRNAs can base-pair with these regions in mRNA and either disrupt or stabilize intrastrand stem structures.

The final two classes of RNA are **tmRNA**, which has properties of both tRNA and mRNA (discussed later), and **catalytic RNA**. Catalytic RNA molecules, also referred to as **ribozymes**, are usually found associated with proteins in which the enzymatic (catalytic) activity actually resides in the RNA portion of the complex rather than in the protein. All of these functional RNAs may represent remnants of the ancient "RNA world," where the earliest ancestral cells were built of RNA parts whose functions were later assumed by proteins. The RNA-world model is discussed in Chapter 17.

RNA Stability

Once released from RNA polymerase, most prokaryotic mRNA transcripts are doomed to a short existence owing to their degradation by intracellular RNases. Why would a cell tolerate this seemingly wasteful practice? Remember, bacteria face extremely rapid changes in their environment, including changes in temperature, salinity, or nutrients, to name a few. To survive, cells must be prepared to react quickly and halt synthesis of superfluous or even detrimental proteins. An effective way to do this is to rapidly destroy the mRNAs that encode those proteins. Unfortunately, rapid mRNA degradation makes it necessary to continually transcribe those genes as long as their products are needed.

RNA stability is measured in terms of half-life, which is the length of time the cell needs to degrade half the molecules of a given mRNA species. The average half-life for mRNA is 1–3 minutes but can be as little as 15 seconds. Other RNAs can be fairly stable, with half-lives of 10–20 minutes or longer. However, the stabilities of the different kinds of RNAs differ drastically. The mRNAs are usually unstable compared to rRNAs and tRNAs, which contain modified bases less susceptible to RNase digestion. These more stable RNA molecules have half-lives on the order of hours.

The major RNA-degrading machine of *E. coli*, the degradosome, is a four-protein complex including an RNase, an RNA helicase, and two metabolic enzymes. The degradosome complex is organized in a helical cytoskeletal structure associated with the cell membrane. These findings suggest that the RNA degradosome is compartmentalized within the cell.

Compartmentalization may regulate RNA degradation by limiting access of the RNase to cytoplasmic RNA.

To Summarize

- **Initial binding of RNA polymerase** to promoter DNA forms the **closed complex**.

- **The DNA strands separate** (melt) and form a "bubble" of DNA around the polymerase called the **open complex**.

- **Antibiotics that inhibit bacterial transcription** include **rifamycin B** (which binds to RNA polymerase to inhibit transcription initiation) and **actinomycin D** (which binds DNA to nonselectively inhibit transcription elongation).

- **Rho-dependent and Rho-independent** mechanisms mediate transcription termination.

- **Messenger RNA** (mRNA) molecules encode proteins.

- **Noncoding RNAs** include **ribosomal RNAs** (rRNAs), **transfer RNAs** (tRNAs), and **small RNAs**. Small RNAs can regulate gene expression (sRNA), have

catalytic activity (catalytic RNA), or function as a combination of tRNA and mRNA (tmRNA).

- **Half-lives of RNA** species vary within the cell.

8.3 Translation of RNA to Protein

Once a gene has been copied into mRNA, the next stage is **translation**, the decoding of the RNA message to synthesize protein. An mRNA molecule can be thought of as a sentence in which triplets of nucleotides, called **codons**, represent individual words, or amino acids (**Fig. 8.10**). Ribosomes are the machines that read the language of mRNA and convert, or translate, it into protein. Translation involves numerous steps, the first of which is the search by ribosomes for the beginning of an mRNA protein-coding region. Because the code consists of triplet codons, a ribosome must start translating at precisely the right base (in the right frame), or the product will be gibberish. Before we discuss how the ribosome finds the right reading frame, let's review the code itself and the major players in translation.

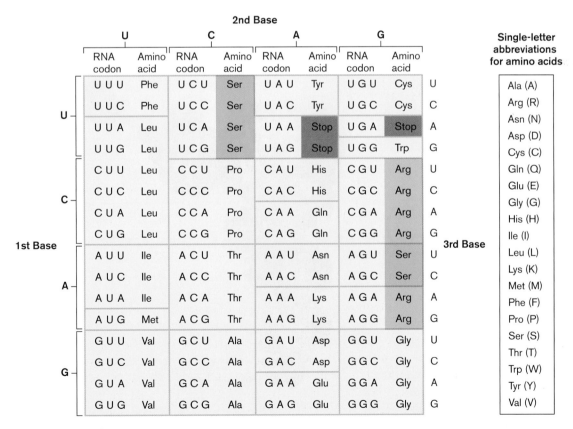

FIGURE 8.10 ■ The genetic code. Codons within a single box encode the same amino acid. Color-highlighted amino acids are encoded by codons in two boxes. Stop codons are highlighted red. Often, single-letter abbreviations for amino acids are used to convey protein sequences (see legend above).

The Genetic Code and tRNA Molecules

When learning a new language, we need a dictionary that converts words from one language into the other. In the case of gene expression, how do we convert from the 4-base language of RNA to the protein language of 20 amino acids? Through painstaking effort, Marshall Nirenberg, Har Gobind Khorana, and colleagues cracked the molecular code and found that each codon (a triplet of nucleotides) represents an individual amino acid. Remarkably, and with few exceptions, the code operates universally across species. (Exceptions include UGA, which is normally read as a stop codon but encodes tryptophan in vertebrate mitochondria, *Saccharomyces*, and mycoplasmas; and CUG, which encodes serine instead of leucine in *Candida albicans*.) In the code presented in **Figure 8.10**, we can see that most amino acids have multiple codon synonyms. Glycine (Gly), for example, is translated from four different codons, and leucine (Leu) from six, but methionine (Met) and tryptophan (Trp) each have only one codon. Because more than one codon can encode the same amino acid, the code is said to be degenerate or redundant. Notice that in almost every case, synonymous codons differ only in the <u>last</u> base. How the cell handles this degeneracy (redundancy) in the code will be explained later.

Only 61 out of a possible 64 codons specify amino acids. Four of these act as codons that can mark the beginning of the protein "sentence." In order of preference, they are AUG (90%), GUG (8.9%), UUG (1%), and CUG (0.1%). Because of its inefficiency in starting translation, the use of a rare start codon, such as CUG, limits the translation of any ORF in which it is found. Although these **start codons** are used to begin all proteins, they are not restricted to that role; they also encode amino acids found in the middle of coding sequences. Consequently, there must be something else about mRNA that signifies whether an AUG codon, for example, marks the start of a protein or resides internally, where it codes for an amino acid (as will be discussed shortly).

The remaining three triplets (UAA, UAG, and UGA) are equally important, for they tell the ribosome when to stop reading a gene sentence. Called **stop codons**, they trigger a series of events (to be described later) that dismantle ribosomes from mRNA and release a completed protein.

Note: There are actually 21 or 22 amino acids found in the proteins of some microbes. The extra amino acids, however, are modifications of one of the basic 20. Selenocysteine, for instance, is synthesized by modifying a serine bound to a tRNA molecule. The selenocysteinyl tRNA recognizes UGA, normally a stop codon. There is no pool of free selenocysteine in the cell.

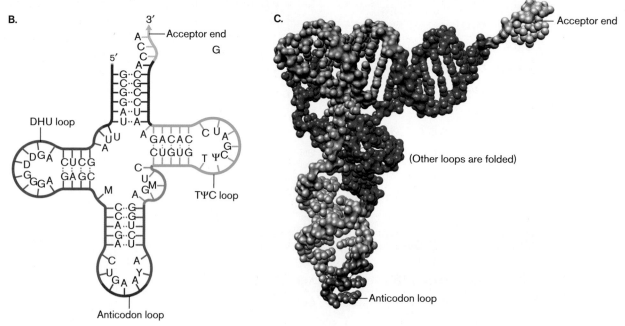

A.
5′ GCGGAUUUAGCUCAGDDGGGAGAGCMCCAG CUGAAYAUCUGGAGMUCCUGUGTΨCGAUCCACAAUUCGCACCA 3′

FIGURE 8.11 ■ Transfer RNA. A. Primary sequence. The letters D, M, Y, T, and Ψ stand for modified bases found in tRNA. **B.** Cloverleaf structure. DHU (or D) is dihydrouracil, which occurs only in this loop; TΨC consists of thymine, pseudouracil, and cytosine bases that occur as a triplet in this loop. The DHU and TΨC loops are named for the modified nucleotides that are characteristically found there. **C.** Three-dimensional structures. The anticodon loop binds to the codon, while the acceptor end binds to the amino acid. (PDB code: 1GIX)

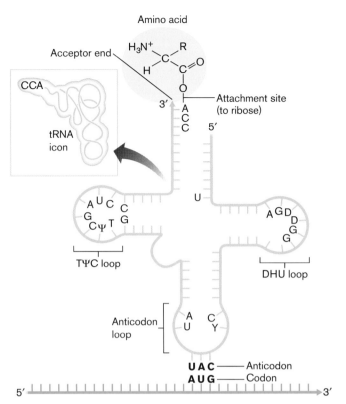

FIGURE 8.12 ■ Codon-anticodon pairing. The tRNA anticodon consists of three nucleotides at the base of the anticodon loop. The anticodon hydrogen-bonds with the mRNA codon in an antiparallel fashion. This tRNA is "charged" with an amino acid covalently attached to the 3′ end.

Thought Question

8.4 How might the redundancy of the genetic code be used to establish evolutionary relationships between different species? *Hints:* 1. Genomes of different species have different overall GC content. 2. Within a given genome one can find segments of DNA sequence with a GC content distinctly different from that found in the rest of the genome.

The decoder (or adapter) molecules that convert the language of RNA (codons) to the language of proteins (amino acids) are the tRNAs. Their job is to travel through the cell and return to the ribosome with amino acids in tow. Approximately 80 bases in length, a tRNA laid flat looks like a clover leaf with three loops (**Fig. 8.11A** and **B**). In nature, however, the molecule folds into the "boomerang-like" 3D structure shown in **Figure 8.11C**. The very bottom, or middle, loop of all tRNA molecules (as depicted in **Fig. 8.11B** and **C**) harbors an **anticodon** triplet that base-pairs with codons in mRNA (**Fig. 8.12**). As a result, this loop is called the anticodon loop. Notice that codon and anticodon pairings are aligned in an antiparallel manner. Most tRNA molecules begin with a 5′ G, and all end with a 3′ CCA, to which an amino acid attaches (see **Figs. 8.11** and **8.12**). Because the 3′ end of the tRNA accepts the amino acid, it is called the acceptor end.

Transfer RNA molecules contain a large number of unusual, modified bases, which accounts for the strange letter codes seen in **Figure 8.11A** (D, M, Y, T, and Ψ). Structures of some of the odd bases are shown in **Figure 8.13**. Wybutosine (yW), for example, has three rings instead of the two found in a normal purine. How do these unusual bases end up in tRNA? During transcription of the tRNA genes, normal, unmodified bases are incorporated into the transcript. Some of these are modified later by specific enzymes to make inosine and other odd bases. The remarkable stability of tRNA molecules is explained, in part, by these unusual bases, because they are poor substrates for RNases.

In the cloverleaf structure of tRNA, two of the loops are named after modifications that are invariantly present in those loops. One is called the TΨC loop because this loop in every tRNA has the nucleotide triplet consisting of thymidine, pseudouridine (Ψ), and cytidine. The other

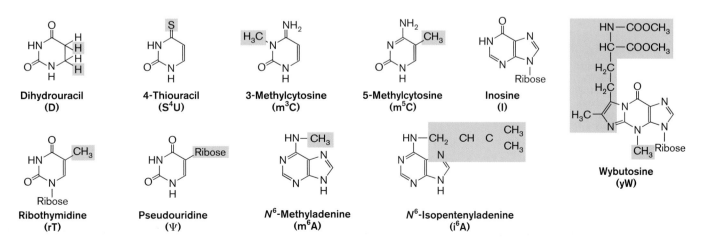

FIGURE 8.13 ■ Modified bases present in tRNA and rRNA molecules. Modifications are highlighted in green.

loop always contains dihydrouridine and is called the DHU loop (see **Figs. 8.11** and **8.12**). These loops are recognized by the enzymes that match tRNA to the proper amino acid.

As noted previously, the genetic code is redundant in that many amino acids have codon synonyms. Redundancy is found primarily in the third position of the codon (for example, UU<u>U</u> and UU<u>C</u> both encode phenylalanine). A single tRNA can recognize both codons because of "wobble" in the first position of the anticodon, which corresponds to the third position of the codon (remember, as with DNA, base-pairing between RNA strands is antiparallel). The wobble is due in part to the curvature of the anticodon loop and to the use of an unusual base (inosine) at this position in some tRNA molecules. The wobble structure allows one anticodon to pair with several codons differing only in the third position.

Aminoacyl-tRNA Synthetases Attach Amino Acids to tRNA

When a tRNA anticodon (for instance, GCG) pairs with its complementary codon (CGC in this case), the ribosome has no way of checking that the tRNA is attached (that is, charged) to the "correct" amino acid (here, arginine). Consequently, each tRNA must be charged with the proper amino acid <u>before</u> it encounters the ribosome. How are amino acids correctly matched to tRNA molecules and affixed to their 3′ ends? The charging of tRNAs is carried out by a set of enzymes called **aminoacyl-tRNA synthetases**. Each cell has generally 20 of these "match and attach" proteins—one for each amino acid. Some bacteria, however, have only 18, choosing instead to modify already-charged glutamyl-tRNA and aspartyl-tRNA by adding an amine to make glutamine and asparagine derivatives. Each aminoacyl-tRNA synthetase recognizes all of the tRNA molecules that transport the same amino acid.

Every aminoacyl-tRNA synthetase enzyme has a specific binding site for its cognate (matched) amino acid. Each enzyme also has a site that recognizes the target tRNA and an active site that joins the carboxyl group of the amino acid to the 3′ OH (class II synthetases) or 2′ OH (class I synthetases) of the tRNA by forming an ester (**Fig. 8.14**). The tRNA synthetase disengages after the amino acid has been attached to the tRNA. Amino acids initially attached to the 2′ OH are then moved to the 3′ OH position.

Each aminoacyl-tRNA synthetase must recognize its own tRNA but <u>not</u> bind to any other tRNA. Specificity is based on recognizing unique features of the tRNA anticodon loop, the TΨC loop, and the DHU loop. Thus, each tRNA has its own set of interaction sites that match only the proper aminoacyl-tRNA synthetase.

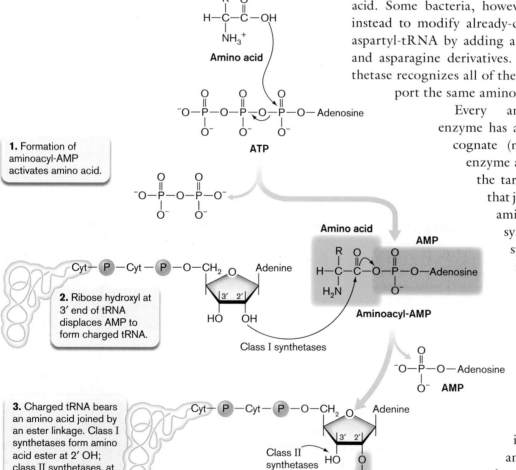

FIGURE 8.14 ■ Charging of tRNA molecules by aminoacyl-tRNA synthetases. At the end of this process, each amino acid is attached to the 3′ end of CCA on a specific tRNA molecule. Curved arrows indicate nucleophilic attack by electrons.

A. Small subunit (30S)   **B. Large subunit (50S)**   **C. Ribosome (70S)**

FIGURE 8.15 ■ **Bacterial ribosome structure.** As this schematic illustrates, note that a section of the 30S subunit **(A)** fits into the valley of the 50S subunit **(B)** when forming the 70S ribosome **(C)**.

The Ribosome, a Translation Machine

The decoding process of the ribosome may be compared to a language translation machine that converts one language into another. The ribosome can be viewed as translating the language of the mRNA code into the amino acid sequences of proteins that conduct the activities of the cell. Ribosomes are composed of two complex subunits, each of which includes rRNA and protein components. In prokaryotes, the subunits are named 30S and 50S for their "size" in Svedberg units. A Svedberg unit reflects the rate at which a molecule sediments under the centrifugal force of a centrifuge (discussed in eAppendix 3). Within the living cell, the two subunits exist separately but come together on mRNA to form the translating 70S ribosome (**Fig. 8.15**). (Note that Svedberg units are not directly additive, because they represent a rate of sedimentation, not a mass.)

The smaller, 30S subunit (900,000 Da) contains 21 ribosomal proteins (named S1–S21; S = "small") assembled around one 16S rRNA molecule. The 50S subunit (1.6 million Da) consists of 31 proteins [designated L1–L34 (three numbers were not used); L = "large"] formed around two rRNA molecules (5S and 23S). **Figure 8.16** presents the 3D spatial arrangement of rRNA and protein in the 50S subunit. Note that the majority of the ribosome is RNA.

How does such a complex molecular machine get built? As discussed in greater detail in **eTopic 8.1**, assembly begins with the transcription of ribosomal RNA genes (DNA), sometimes referred to as rDNA. The 16S, 23S, and 5S rRNAs are initially transcribed as one RNA molecule and posttranscriptionally processed into separate rRNA molecules. During rRNA transcription, ribosomal proteins begin to assemble within the secondary rRNA structures. Thus, the ribosome is built by precise, timed molecular RNA-RNA, RNA-protein, and protein-protein interactions.

Thought Question

8.5 The synthesis of ribosomal RNAs and ribosomal proteins represents a major energy drain on the cell. How might the cell regulate the synthesis of these molecules when the growth rate slows because amino acids are limiting? *Hint:* Predict what happens to any translating ribosome when an amino acid is limiting. Look up RelA protein on the Internet.

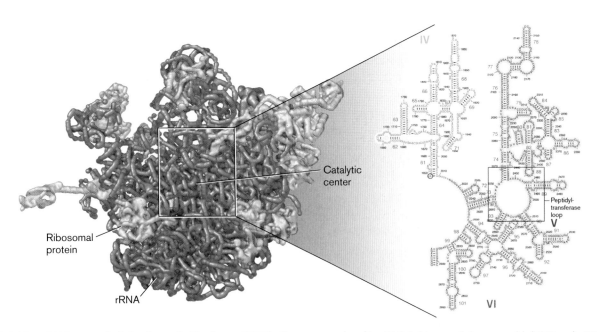

FIGURE 8.16 ■ **RNA-protein interfaces in the large (50S) ribosome subunit.** rRNA is blue; proteins are gold. (PDB code: 1GIY) **Blowup:** Partial secondary structure of 23S rRNA, which includes the peptidyltransferase catalytic site. Note the many double-stranded hairpin structures that fold into domains (Roman numerals). Nucleotides are numbered in black; stem structures, in blue.

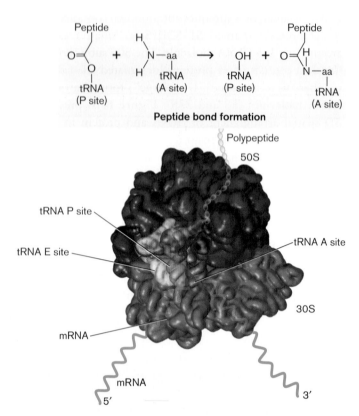

FIGURE 8.17 ▪ **Binding of tRNA.** X-ray-crystallographic model of *Thermus thermophilus* ribosome with associated tRNAs. 5OS is red, 3OS is magenta, and tRNAs in the A, P, and E sites are blue, green, and yellow, respectively.) **Inset:** The formation of a peptide bond between the peptidyl-tRNA in the P site and aminoacyl-tRNA in the A site. The mRNA (light blue) travels along the 3OS subunit, and the growing peptide (yellow) exits from a channel formed in the 5OS subunit. (PDB codes: 1GIX and 1GIY) ▶

The 70S ribosome harbors three binding sites for tRNA (**Fig. 8.17** ▶). During translation, each tRNA molecule moves through the sites in an assembly-line fashion, first entering at one site, and then moving progressively through the others before being jettisoned from the ribosome. The first position is the aminoacyl-tRNA **acceptor site** (**A site**), which binds an incoming aminoacyl-tRNA. The growing peptide is attached to tRNA berthed in the **peptidyl-tRNA site** (**P site**). Finally, a tRNA recently stripped of the polypeptide is located in the **exit site** (**E site**). **Figure 8.17** also illustrates the path mRNA takes into the ribosome and the nascent (that is, growing) peptide emerging from it.

The Ribosome Is a "Ribozyme"

The ribosome makes the peptide bonds that stitch amino acids together, and it does so by using a remarkable enzymatic activity called **peptidyltransferase**, present in the 50S subunit. Contrary to expectation, the peptidyltransferase activity resides not in ribosomal proteins, but in the rRNA. Peptidyltransferase is actually a **ribozyme** (an RNA molecule that carries out catalytic activity), and it is a part of 23S rRNA that is highly conserved across all species (**Fig. 8.16** inset, loop V). Proteins surrounding this active center offer structural assistance to ensure that the RNA is folded properly and to interact with tRNA substrates.

The sequences of ribosomal RNAs from all microbes are very similar, in large part because rRNA plays an important catalytic role in the activities of ribosomes. But there are differences in rRNA sequences that increase in relation to the evolutionary distance between species. As a result, rRNA serves as a molecular clock that measures the approximate time since two species diverged. More information on molecular clocks can be found in Section 17.3.

How Do Ribosomes Find the Right Reading Frame?

Every mRNA has three potential reading frames, depending on which base the ribosome happens to start with. A ribosome that chooses the wrong frame will produce proteins with totally different amino acid sequences. In addition, a shifted reading frame often generates an inadvertent stop codon that prematurely terminates the peptide. So how does the ribosome find the right frame? The key is a section of untranslated RNA sequence positioned upstream of the protein-coding segment of mRNA. The upstream leader RNA contains a purine-rich consensus sequence (5′-AGGAGGU-3′) located four to eight bases upstream of the start codon. This upstream sequence is called the **ribosome-binding site** or **Shine-Dalgarno sequence**, after Lynn Dalgarno and her student John Shine at Australian National University, who discovered it in 1974.

The Shine-Dalgarno site is complementary to a sequence at the 3′ end of 16S rRNA (5′-ACCUCCU-3′), found in the 30S ribosome subunit. Binding of mRNA to this site positions the start codon (such as AUG or GUG) precisely in the ribosome P site, ready to pair with an *N*-formyl-methionyl-tRNA. Elegant proof that the Shine-Dalgarno sequence actually binds to the 3′ end of 16S rRNA is described in **eTopic 8.2**.

Recall from Section 7.2 that many genes in the bacterial chromosome are arranged in tandem (an operon) and are transcribed together to produce a polygenic (formerly polycistronic) mRNA. Because the beginning of each gene in a polygenic mRNA carries its own ribosome-binding site, different ribosomes can bind simultaneously to the start of each gene. Mutating the first gene in a three-gene operon can halt translation of the first protein, but the second and third proteins can still be translated because the RNA sequence for each has its own ribosome-binding site.

Note: Eukaryotic ribosomes have a different mechanism for finding the start codon. They generally start translating at the first AUG after the 5′ cap. The 5′ cap is a 7-methylguanosine added post-transcriptionally to the 5′ end of eukaryotic mRNA molecules. The normal role of the cap is thought to be as a signal to transport RNA out of the nucleus and/or to stabilize mRNA.

Polysomes, Coupled Transcription/Translation, and Transertion

Polysomes. Once a ribosome begins translating mRNA and moves off of the ribosome-binding site, another ribosome can immediately jump onto that site. The result is an RNA molecule with multiple ribosomes moving along its length at the same time. The multiribosome structure is known as a **polysome** (**Fig. 8.18A**). Ribosomes in a

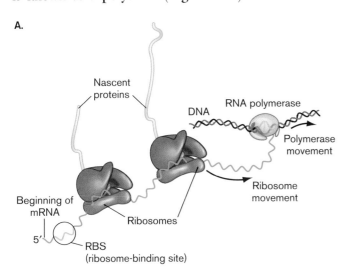

A.

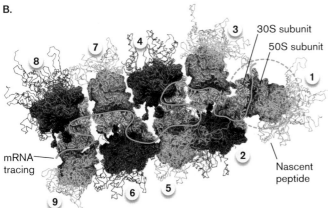

B.

FIGURE 8.18 ■ Coupled transcription and translation in bacteria. A. During coupled transcription and translation in prokaryotes, ribosomes attach at mRNA ribosome-binding sites and start synthesizing protein before transcription of the gene is complete. **B.** Model of *E. coli* polysome showing the nascent polypeptides (numbered) exiting from each ribosome. A representative ribosome is shown in the dashed circle. Note the helical arrangement of ribosomes along the chain, which are held together by mRNA (blue tracing). The closer the ribosome is to the 3′ end of the mRNA, the longer the synthesized protein molecules grow.

polysome are closely packed and arranged helically along the mRNA (**Fig. 8.18B**). Polysomes help protect the message from degrading RNases and enable the speedy production of protein from just a single mRNA molecule.

Coupled transcription and translation. Because bacteria and archaea lack nuclear membranes, ribosome subunits floating through the cytoplasm have an opportunity to bind to the 5′ end of mRNA and begin making protein even before RNA polymerase has finished making the mRNA molecule. The simultaneous building of both mRNA and proteins is called coupled transcription and translation (**Fig. 8.18A**). For certain genes, transcription-translation coupling requires a protein to physically link the ribosome to RNA polymerase. One such protein is RfaH, described in the Current Research Highlight.

The coupling of transcription and translation in bacteria makes it possible for the cell to use translation as a means of regulating transcription. One process, attenuation, is explained further in Chapter 10. Note, however, that much, if not most, translation—at least in *E. coli* and *Bacillus subtilis*—seems to occur at the polar ends of cells, which are polyribosome-rich and have little to no DNA (**Fig. 8.19**). Thus, although transcription-translation coupling happens when transcripts are first made, most translation occurs on free mRNA transcripts that have diffused away from the nucleoid and into polyribosome-rich areas.

The coupling of transcription and translation presents a potential problem. Ribosomes generally travel along mRNA more slowly than mRNA is generated by RNA polymerase, so RNA polymerase can potentially scoot ahead of the ribosome and leave large tracts of RNA unprotected and susceptible to nucleases. The cell handles this problem by modulating transcription speed. The rate of RNA synthesis averages about 45 nt per second, which roughly equals the average rate of translation (16 amino acids per second, or

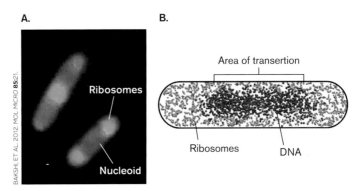

A.

B.

FIGURE 8.19 ■ Most translation in *E. coli* is not coupled to transcription. A. Most ribosomes exist in nucleoid-free areas of the cell, primarily at the cell poles. DNA was visualized using a red fluorescent dye. Ribosomes contained a protein (S2) fused to yellow fluorescent protein, which here appears green. **B.** Computerized model showing the distribution of polysomes in the cell.

48 nt/s). However, these speeds vary. For example, RNA polymerase pauses momentarily at sites rich in GC content (GC base pairs, with three hydrogen bonds, are harder to melt than AT pairs, which have only two hydrogen bonds). Once sigma factor exits the transcription complex, proteins called NusA and NusG enter the complex and can actually lengthen the pause in RNA polymerase activity to allow time for the trailing protein-synthesizing ribosome to catch up to the polymerase. Pausing allows the ribosome to follow RNA polymerase closely and protect the RNA.

Transertion. Transcription-translation coupling has another benefit for the bacterial cell: expanding, or decompacting, the nucleoid. Many membrane proteins are inserted into membranes while they are still being translated (Section 8.5 discusses protein secretion mechanisms). The cotranslational insertion of membrane proteins sets up a situation whereby transcription, translation, and membrane insertion can all be coupled—a process called **transertion**. Transertion links the nucleoid to the cell membrane along the cylindrical portion of the cell and expands the width of the nucleoid. This expansion may help, among other things, to counteract nucleoid contraction produced by transcription-generated DNA supercoils. Decompaction will facilitate diffusion of ribosome subunits, RNA polymerase, and regulatory proteins into the nucleoid.

Transcription in eukaryotic microbes is not coupled to translation. In contrast to bacteria and archaea, eukaryotic microbes use separate cellular compartments to carry out most of their transcription and translation. They transcribe genes in the nucleus where internal, noncoding parts of the mRNA (introns) are removed by a process known as RNA splicing. The processed transcripts are then transported to the cytoplasm and translated. Most eukaryotic DNA viruses do the same. However, a small amount of translation is performed in the eukaryotic nucleus, where it can also be coupled to transcription.

Defining a Gene

Before we describe the mechanics of translation, it will help to illustrate the alignments between the DNA sequence of a structural gene (a gene encoding a protein), and the mRNA transcript containing translation signals and the protein-coding sequences. **Figure 8.20** shows the "sense" and "template" DNA strands of a two-gene operon. The sequence of the sense strand matches that of the mRNA transcript, but with T's substituting for U's. The template strand is the strand actually "read" by RNA polymerase. Note that +1 marks the DNA base where the mRNA transcript starts and that a single transcript includes both operon genes (polygenic). In the mRNA transcript, an untranslated "leader" sequence precedes the gene A protein-coding region, located between translation start and stop signals (discussed later). Downstream of the translation stop signal for gene B lies an untranslated "trailer." The leader and trailer sequences help regulate gene expression. As you proceed in this chapter, refer back to **Figure 8.20** to clarify important relationships connecting a DNA operon, its composite genes, the polygenic transcript, and the resulting proteins.

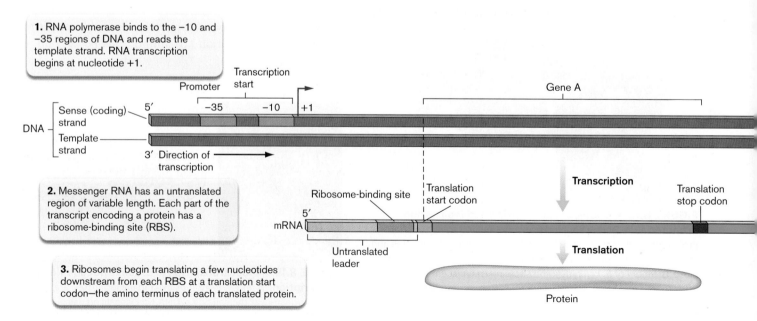

1. RNA polymerase binds to the −10 and −35 regions of DNA and reads the template strand. RNA transcription begins at nucleotide +1.

2. Messenger RNA has an untranslated region of variable length. Each part of the transcript encoding a protein has a ribosome-binding site (RBS).

3. Ribosomes begin translating a few nucleotides downstream from each RBS at a translation start codon—the amino terminus of each translated protein.

FIGURE 8.20 ■ Alignment of structural genes in a bacterial operon, the mRNA transcript, and protein products. In this figure, the term "gene" refers to the region of DNA corresponding to the entire mRNA transcript, including upstream and downstream untranslated areas.

Thought Question

8.6 Figure 8.20 illustrates an operon and its relationship to transcripts and protein products. Imagine that a mutation that stops translation (TAA, for example) was substituted for a normal amino acid about midway through the DNA sequence encoding gene *A*. What happens to the expression of the gene *A* and gene *B* proteins?

The Three Stages of Protein Synthesis

Once the ribosome has been properly positioned on the message, peptide synthesis can begin. Like transcription, peptide synthesis has three stages: initiation, which brings the two ribosomal subunits together, placing the first amino acid in position; elongation, which sequentially adds amino acids as directed by the mRNA transcript; and termination. Several steps in this process can be inhibited by certain antibiotics (described later in this section).

Initiation of translation. In bacteria such as *E. coli*, initiation of protein synthesis requires three small proteins called initiation factors (IF1, IF2, and IF3).

Initially, IF3 binds to the 30S subunit to separate the 30S and 50S subunits. Separating the subunits allows other initiation factors and mRNA access to the 30S subunit (**Fig. 8.21** ▶, step 1). IF1 and mRNA join the 30S subunit. IF1 blocks the tRNA A site, and the mRNA sequence for the ribosome-binding site aligns with its complementary sequence on 16S rRNA (step 2). Next, IF2 complexed to GTP escorts the initiator *N*-formylmethionyl-tRNA (fMet-tRNA) to the start codon located at what will be the

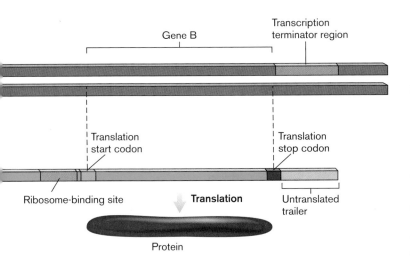

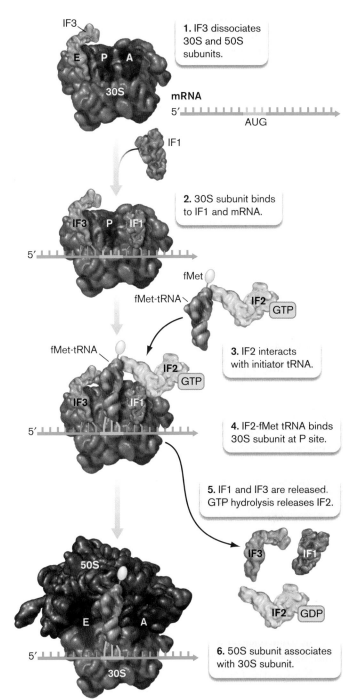

1. IF3 dissociates 30S and 50S subunits.

2. 30S subunit binds to IF1 and mRNA.

3. IF2 interacts with initiator tRNA.

4. IF2-fMet tRNA binds 30S subunit at P site.

5. IF1 and IF3 are released. GTP hydrolysis releases IF2.

6. 50S subunit associates with 30S subunit.

FIGURE 8.21 ■ Translation initiation. The end result of translation initiation is assembly of the 50S-30S-mRNA complex with the initiator tRNA-fMet set in the P site. See text for details. ▶

P site (steps 3 and 4). *N*-formylmethionyl-tRNA binds to all start codons and is the only aminoacyl-tRNA to bind directly to the P site. Once the initiator tRNA is in place, GTP is hydrolyzed, and IF1 and IF3 are released (step 5). The 50S subunit then docks to the 30S subunit (step 6). The ribosome is now "locked and loaded," ready for elongation. Note that initiation is much more complex in the Archaea, involving as many as six different IF proteins.

Elongation of the peptide. In elongation, three basic steps are repeated: (1) An aminoacyl-tRNA binds to the A (acceptor) site; (2) a peptide bond forms between the new amino acid and the growing peptide pre-positioned in the P site; (3) the message must move by one codon (**Fig. 8.22** ▶). First an elongation factor (EF-Tu) associates with GTP to form a complex (EF-Tu-GTP) that binds to most charged aminoacyl-tRNAs (except for the initiator tRNA). Sequences within the 50S rRNA recognize features of the EF-Tu-GTP–aminoacyl-tRNA complex, guiding it into the A site (**Fig. 8.22**, step 1). Correct selection of a tRNA is guided in large part by codon-anticodon pairing, but conformational changes sensed by various ribosomal proteins customize the fit. If the fit is not perfect, the tRNA is rejected.

Once an aminoacyl-tRNA is in the A site, a peptide bond is formed between the amino acid moored there and the terminal amino acid linked to tRNA in the P site. Simultaneously, a GTP is hydrolyzed and the resulting EF-Tu-GDP is expelled. Peptide bond formation effectively transfers the peptide from the tRNA in the P site to tRNA in the A site (**Fig. 8.22**, steps 2 and 3).

For protein synthesis to continue, the ribosome must advance by one codon, which moves the peptidyl-tRNA from the A site into the P site, leaving the A site vacant. The process, called **translocation**, involves another elongation factor, EF-G, associated with GTP. EF-G-GTP binds to the ribosome, GTP is hydrolyzed, and the 30S subunit rotates clockwise to ratchet the 50S subunit ahead on the message by one codon (step 4). The 30S subunit then rotates back (step 5). How EF-G uses a fingerlike motion to complete translocation is explained in **Special Topic 8.1**. The translocation maneuver opens up the A site, moves the peptidyl-tRNA into the P site, and slides uncharged tRNA into the E (exit) site. The next aminoacyl-tRNA that enters the A site stimulates a conformational change in the ribosome that telegraphs through to the E site and ejects the uncharged tRNA.

Notice that EF-Tu and EF-G recycle sequentially on and off the ribosome by binding to the same area of the ribosome. Because EF-G-GTP and the EF-Tu-GTP–aminoacyl-tRNA complex are structurally similar (an example of molecular mimicry) and because both bind to the same site, the two factors must cycle on and off the ribosome sequentially. Although this process seems incredibly complex, the ribosomes of *E. coli* manage to link together 16 amino acids per second. Scientists have now visualized the staggered movements of a single ribosome as it translates an mRNA molecule (**eTopic 8.3**).

Termination of translation. Eventually, the ribosome arrives at the end of the coding region, but not the end of the RNA. As noted previously, the end of the coding

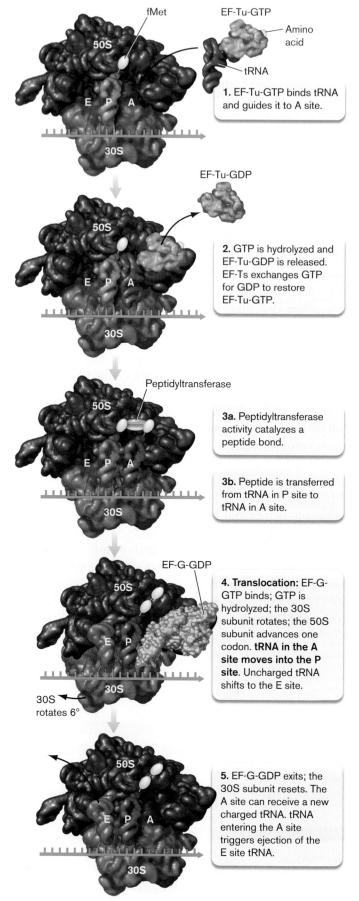

1. EF-Tu-GTP binds tRNA and guides it to A site.

2. GTP is hydrolyzed and EF-Tu-GDP is released. EF-Ts exchanges GTP for GDP to restore EF-Tu-GTP.

3a. Peptidyltransferase activity catalyzes a peptide bond.

3b. Peptide is transferred from tRNA in P site to tRNA in A site.

4. Translocation: EF-G-GTP binds; GTP is hydrolyzed; the 30S subunit rotates; the 50S subunit advances one codon. **tRNA in the A site moves into the P site.** Uncharged tRNA shifts to the E site.

30S rotates 6°

5. EF-G-GDP exits; the 30S subunit resets. The A site can receive a new charged tRNA. tRNA entering the A site triggers ejection of the E site tRNA.

FIGURE 8.22 ∎ **Elongation of the peptide.** At the end of each elongation cycle, the ribosome has moved forward by one codon, which positions the tRNA with nascent peptide in the P site and the tRNA from which the peptide was passed in the E site, ready for ejection. ▶

SPECIAL TOPIC 8.1	Translocation: EF-G Gets Physical

Translocation of the ribosome down an mRNA molecule is a critical part of translation. Ribosomes must move one codon along an mRNA molecule while tRNAs in the ribosome A (acceptor) site and P (peptidyl-tRNA) site shift to the P site and E (exit) site, respectively. Elongation factor G (EF-G) plays a major role in this process. When EF-G-GTP binds to a <u>pre</u>translocation ribosome (see **Fig. 8.22**), it triggers the ratchet-like movement of the 30S subunit relative to the 50S subunit. During subunit ratcheting, the CCA acceptor ends of tRNAs move from the A and P sites to the P and E sites, leaving the anticodon ends in the original sites. Hydrolysis of GTP bound to EF-G then moves the anticodon ends of these tRNAs, along with mRNA, to the new sites. The result is the <u>post</u>translocation ribosome in which mRNA has moved one codon and tRNAs in the A and P sites have moved to the P and E sites, respectively (see **Fig. 8.22**).

One question, however, has troubled structural biologists for some time. Cryo-electron microscopy and X-ray-crystallographic examinations of EF-G bound to posttranslocation ribosomes show an <u>elongated</u> form of EF-G with its domain IV projecting into the decoding center of the ribosome—a place where tRNA in the A site binds to its codon. The question is, How can EF-G bind to a pretranslocation ribosome without colliding with tRNA already in the A site? A team led by Thomas Steitz at Yale University (**Fig. 1**) discovered the answer using a new way to trap EF-G bound to <u>pre</u>translocation ribosomes from *Thermus thermophilus*.

Part of their strategy involved placing a nonhydrolyzable peptidyl-tRNA in the P site to prevent peptide transfer to tRNA in the A site. The result was a ribosome stalled in the pretranslocation state. When they added EF-G, they were able to crystallize the pretranslocation complex and perform X-ray crystallography to determine EF-G structure. The results, shown in **Figure 2A**, revealed that domain IV of EF-G was <u>not</u> extended <u>before</u> translocation, but was tucked away in a manner that did not impinge on tRNA in the A site. The authors propose that once EF-G-GTP binds to the ribosome, the GTP-binding site becomes ordered, triggering a conformational change in EF-G to an extended fingerlike form (**Fig. 2B**). The extended form reaches into the A site to complete translocation of the anticodon ends of tRNAs and moves the associated mRNA down the ribosome. The model is similar to a finger flicking a paper clip along a table.

FIGURE 1 ▪ **Thomas Steitz (right) and associate research scientists Matthieu Gagnon (left) and Jinzhong Lin examine the EF-G–binding site on a model of the 70S ribosome.**

The study by Steitz's team shows that EF-G is far more flexible than previously thought and suggests that controlling flexibility by GTP binding and stabilization is key to completing translocation. GTP hydrolysis, then, is required to dissociate EF-G from the posttranslocation ribosome.

RESEARCH QUESTION

Propose three ways that an antibiotic might inhibit protein synthesis by interfering with EF-G.

Lin, Jinzhong, Matthieu G. Gagnon, David Bulkley, and Thomas A. Steitz. 2015. Conformational changes of elongation factor G on the ribosome during tRNA translocation. *Cell* **160**:219–227.

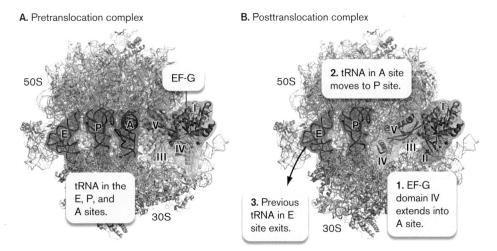

A. Pretranslocation complex

50S

EF-G

E P A V IV III

tRNA in the E, P, and A sites.

30S

B. Posttranslocation complex

50S

2. tRNA in A site moves to P site.

E P V IV III II I

3. Previous tRNA in E site exits.

1. EF-G domain IV extends into A site.

30S

FIGURE 2 ▪ **Structures of EF-G bound to the pretranslocation (A) and posttranslocation (B) ribosome.**

region is marked by one of three stop codons. Translocation after forming the last peptide bond brings the mRNA stop codon into the A site (**Fig. 8.23** ▶, step 1). No tRNA binds, but one of two **release factors** (RF1 or RF2) will enter (also step 1). RF binding leads to ejection of tRNA in the E site and activates the peptidyltransferase, thereby cutting the bond that tethers the completed peptide to tRNA in the P site (step 2).

With the protein released, the ribosome must disassemble. RF3 causes RF1 or RF2 to depart the ribosome (**Fig. 8.23**, step 3). Then, ribosome recycling factor (RRF), along with EF-G, binds at the A site, and an accompanying GTP hydrolysis undocks the two ribosomal subunits (step 4). IF3 then reenters the 30S subunit to eject the remaining uncharged tRNA and mRNA (step 5), thereby preventing the 30S and 50S subunits from redocking. The liberated ribosomal subunits are now free to diffuse through the cell, ready to bind yet another mRNA and begin the translation sequence anew.

> **Thought Question**
>
> **8.7** While working as a member of a pharmaceutical company's drug discovery team, you find that a soil microbe snatched from the jungles of South America produces an antibiotic that will kill even the most deadly, drug-resistant form of *Enterococcus faecalis*, which is an important cause of heart valve vegetations in bacterial endocarditis. Your experiments indicate that the compound stops protein synthesis. How could you more precisely determine the antibiotic's mode of action? *Hint:* Can you use mutants resistant to the antibiotic?

Antibiotics That Affect Translation

Streptomycin (**Fig. 8.24A**), a well-known member of the aminoglycoside family of antibiotics, is produced by a species of *Streptomyces*. The drug targets bacterial small ribosomal subunits by binding to a region of 16S rRNA in the 30S subunit that forms part of the decoding A site and to protein S12, a protein critical for maintaining the specificity of codon-anticodon binding. Streptomycin bound to 16S rRNA makes decoding of mRNA at the A site "sloppy" by permitting illicit codon-anticodon matchups that result in a mistranslated protein sequence.

Bacteria can become resistant to streptomycin via spontaneous mutations in the S12 gene (*rpsL*) or 16S

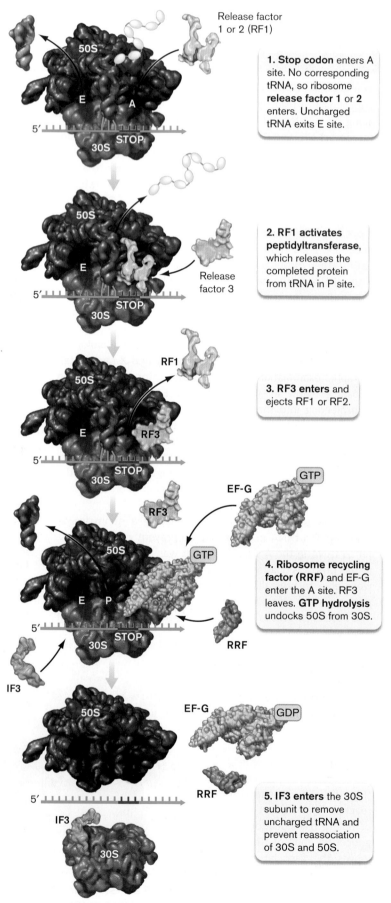

Release factor 1 or 2 (RF1)

1. Stop codon enters A site. No corresponding tRNA, so ribosome **release factor 1** or **2** enters. Uncharged tRNA exits E site.

Release factor 3

2. RF1 activates peptidyltransferase, which releases the completed protein from tRNA in P site.

RF1

3. RF3 enters and ejects RF1 or RF2.

RF3

EF-G GTP

GTP

RF3

GTP

4. Ribosome recycling factor (RRF) and EF-G enter the A site. RF3 leaves. **GTP hydrolysis** undocks 50S from 30S.

RRF

IF3

EF-G GDP

RRF

IF3

5. IF3 enters the 30S subunit to remove uncharged tRNA and prevent reassociation of 30S and 50S.

FIGURE 8.23 ■ Termination of translation. The completed protein is released, and the ribosome subunits are recycled. ▶

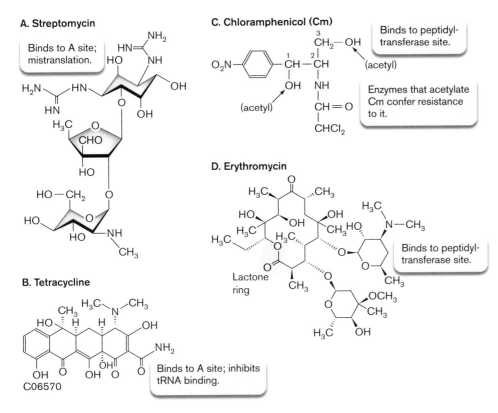

A. Streptomycin

Binds to A site; mistranslation.

C. Chloramphenicol (Cm)

Binds to peptidyltransferase site.

(acetyl)

Enzymes that acetylate Cm confer resistance to it.

(acetyl)

D. Erythromycin

Binds to peptidyltransferase site.

Lactone ring

B. Tetracycline

C06570

Binds to A site; inhibits tRNA binding.

FIGURE 8.24 ■ **Antibiotics that inhibit protein synthesis in bacteria.** Streptomycin (**A**) and tetracycline (**B**) bind to the A site. Streptomycin causes mistranslation, tetracycline inhibits tRNA binding. Chloramphenicol (**C**) and erythromycin (**D**) bind to the peptidyltransferase site, thus inhibiting peptide bond formation.

rRNA. The altered ribosomal protein or rRNA remains functional but does not bind streptomycin. Some bacteria gain resistance by acquiring an aminoglycoside phosphotransferase that modifies streptomycin so that it cannot bind to its target. Additional mechanisms are discussed in Chapter 27. Other therapeutically important aminoglycosides include gentamicin, tobramycin, and amikacin.

Tetracycline and derivatives such as doxycycline (**Fig. 8.24B**) also target the 30S ribosomal subunit, where they bind to 16S rRNA near the A site. But instead of causing mistranslation, the tetracyclines prevent aminoacyl-tRNA from binding to the A site. Resistance to tetracycline can be conferred by an efflux transport system that effectively removes the antibiotic from the bacterial cell. Genes encoding this drug resistance transport system are usually carried on mobile genetic elements like plasmids that can be passed from cell to cell (see Section 9.1). Other resistance mechanisms are described in Chapter 27.

Another species of *Streptomyces* produces chloramphenicol (**Fig. 8.24C**), which attacks the 50S subunit. It binds to 23S rRNA at the peptidyltransferase active site and inhibits peptide bond formation. Resistance to this drug comes from an ability to synthesize an enzyme, chloramphenicol acetyltransferase,

that modifies chloramphenicol in a way that destroys its activity.

Erythromycin, made by *Streptomyces erythraeus*, is one of a large group of related antibiotics called macrolides, whose hallmark is a large lactone ring of 12–22 carbon atoms (**Fig. 8.24D**). They all attack the 50S subunit by binding to 23S rRNA in the nascent peptide exit tunnel near the peptidyltransferase center. Binding alters peptidyltransferase structure and interferes with peptide bond formation. Resistance to macrolides usually involves an efflux pump or methylation of the relevant area of 23S rRNA by an enzyme produced from a mobile genetic element.

Other translation-targeting antibiotics interfere with mRNA binding to the ribosome (kasugamycin), prevent translocation by targeting EF-G (fusidic acid), or use structural similarity to tRNA (molecular mimicry) to trick peptidyltransferase into action without having a bona fide tRNA in the A site (puromycin). We chronicle the discovery and use of antibiotics more completely in Chapter 27.

Thought Questions

8.8 How might one gene code for two proteins with different amino acid sequences?

8.9 Why involve RNA in protein synthesis? Why not translate directly from DNA?

8.10 Codon 45 of a 90-codon gene was changed into a translation stop codon, producing a shortened (truncated) protein. What kind of mutant cell could produce a full-length protein from the gene <u>without</u> removing the stop codon? *Hint:* What molecule recognizes a codon?

Unsticking Stuck Ribosomes: tmRNA and Protein Tagging

Premature transcription termination or cleavage of mRNA near its 3′ end leaves an mRNA without a translational stop codon. What happens when a ribosome translates these damaged mRNA molecules? Without a stop codon, there is nothing to trigger ribosome release when the ribosome reaches the end of the message. So the ribosome is stuck at the end of the mRNA with a peptidyl-tRNA in the ribosome P site. How can the cell fix this potentially toxic situation?

The answer is a translation rescue molecule, tmRNA, with properties of tRNA and mRNA. The tRNA end has an attached amino acid (alanine). The mRNA end encodes a peptide tag (**Fig. 8.25A**). When tmRNA finds a stalled ribosome, the aminoacyl-tRNA end enters the unoccupied ribosome A site (**Fig. 8.25B**, steps 1 and 2). A peptide bond forms between the stalled polypeptide and the alanine on the tmRNA (step 3). The tmRNA is now attached to the nascent peptide and translocates to the P site (step 4). The mRNA section of tmRNA then displaces the old mRNA and is translated (step 5). Translation adds a 10-amino-acid peptide tag to the carboxyl end of the incomplete protein, marking the protein for destruction. A stop codon in tmRNA triggers protein release and ribosome disassembly (step 6). Finally, a helper protein (SspB) recognizes the proteolysis tag and brings the useless aberrant polypeptide to the ClpXP protease (discussed in Section 8.4) for degradation. Mutant cells lacking this rescue system often struggle to grow.

> **Thought Question**
>
> **8.11** What would happen if the tyrosine residue UAC in the mRNA-like domain of tmRNA was altered by mutation to UAA?

To Summarize

- **Triplet nucleotide codons** in mRNA encode specific amino acids. **Transfer RNA molecules** interpret the genetic code and bring specific amino acids to the A (acceptor) site in the ribosome.

- **Specific codons** mark the beginning and end of a gene. The Shine-Dalgarno sequence in mRNA located before the start codon helps the ribosome find the correct reading frame in the mRNA.

- **Transcription and translation** are coupled in bacteria and archaea.

- **Initiation of protein synthesis** in bacteria requires three initiation factors that bring the ribosomal subunits together on an mRNA molecule.

- **Peptidyltransferase** activity of the ribosome is carried out by ribosomal RNA, not protein. The peptide elongates by one amino acid when the ribosome ratchets one codon length along the mRNA.

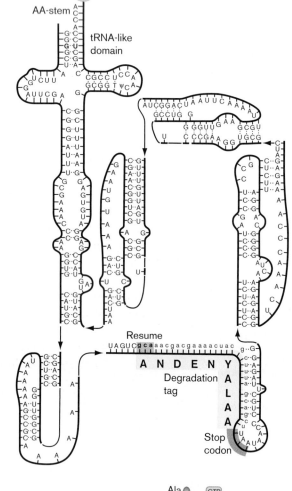

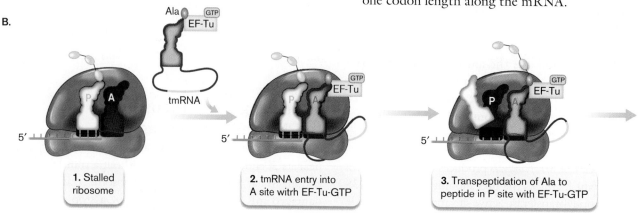

FIGURE 8.25 ■ tmRNA and protein tagging. **A.** Structure of SsrA, a type of tmRNA in *E. coli*. The tRNA-like domain is at the top left, and the mRNA-like portion encoding the proteolysis tag is at the bottom. The stop codon is highlighted orange. **B.** Mechanism of tmRNA tagging in *E. coli*.

■ **Translation terminates** upon reaching a stop codon. A release factor enters the A site and triggers peptidyltransferase activity, thus freeing the completed protein from tRNA in the P site.

■ **Ribosome release factor** and EF-G bind to the A site to dissociate the two ribosomal subunits from the mRNA.

■ **Antibiotics that affect translation** can cause ribosomes to misread mRNA (streptomycin; binds 30S subunit), inhibit aminoacyl-tRNA binding to the A site (tetracycline; binds 30S subunit), interfere with peptidyltransferase (chloramphenicol; binds 50S subunit), trigger peptidyltransferase prematurely (puromycin; binds A site, 50S subunit), cause translocation to abort (erythromycin; binds 50S subunit), or prevent translocation (fusidic acid; binds EFG).

■ **tmRNA** rescues ribosomes stuck on damaged mRNA that lacks a stop codon.

8.4 Protein Modification, Folding, and Degradation

Once a protein is made, is it functional? For many proteins, translation is not the last step in producing a functional molecule. Often a protein must be modified <u>after</u> translation either to achieve an appropriate 3D structure or to regulate its activity. Primary, secondary, and tertiary structures of proteins can be modified after the primary protein sequence has been assembled by the ribosomes. And what happens when a protein is damaged or is no longer needed? A healthy cell "cleans house" by degrading damaged or unneeded proteins. The precious amino acids are then recycled into making new proteins.

Protein Processing after Translation

Completed proteins released from the ribosome contain *N*-formylmethionine (fMet) at the N terminus (as previously described). With some proteins, the *N*-formyl group is "sliced" off the N terminus by methionine deformylase, leaving methionine. For other amino acids, methionine aminopeptidase removes the entire amino acid. *N*-formylmethionine (fMet) is important during the course of an infection because fMet peptides are produced only by bacteria and mitochondria, not by archaea or by the cytoplasmic ribosomes of eukaryotes. Our white blood cells can detect low concentrations of fMet peptides (around 10^{-12} M) as a sign of invading bacteria or of necrotic (dying) host cells releasing mitochondria.

Other types of protein processing, such as the addition of acetyl groups or AMP, can change protein function. Proteolytic cleavages can either activate or inactivate a protein. Examples of modified bacterial enzymes include glutamine synthetase (adenylylation), isocitrate dehydrogenase (phosphorylation), and a variety of ribosomal proteins (acetylation). These modifications directly regulate enzyme activity (glutamine synthetase and isocitrate dehydrogenase) or alter tertiary structure (ribosomal proteins).

Protein Folding: Assume the Position

As a new protein emerges from the ribosome, how does it fold into exactly the correct shape to do its job? Christian Anfinsen (1916–1995) won the 1972 Nobel Prize in Chemistry for demonstrating that, for some proteins, folding is governed solely by the protein itself. In other words, the optimal 3D structure of a protein is determined solely by the linear sequence of amino acid residues. But three decades later, other scientists discovered that the folding of many proteins requires assistance from other proteins. These helper proteins are called **chaperones** (or chaperonins). Chaperones associate with target proteins during some phase of the folding process and then dissociate, usually after folding of the target protein is completed. Although chaperones exhibit some specificity, a given chaperone can help fold many different types of proteins.

The major chaperone family in most species includes GroEL (60 kDa), GroES (10 kDa), DnaK (70 kDa), DnaJ (40 kDa), and trigger factor (48 kDa). Because their levels

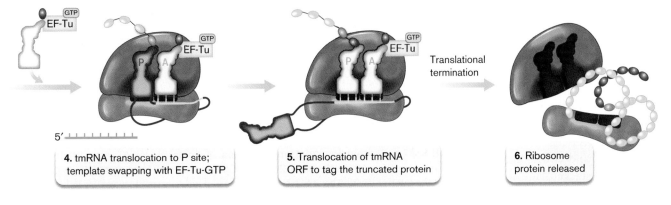

5. **4.** tmRNA translocation to P site; template swapping with EF-Tu-GTP

5. Translocation of tmRNA ORF to tag the truncated protein

6. Ribosome protein released

A. GroEL chaperone

ATP causes structural change in molecule.

GroES controls access to GroEL.

Chaperoned protein fits inside the GroEL chamber.

GroEL-ATP   GroEL-GroES-ATP   GroEL-GroES (top view)

B. DnaK chaperone

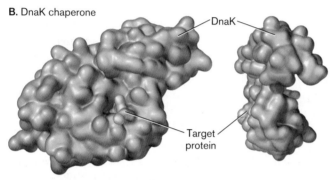

DnaK

Target protein

FIGURE 8.26 ■ *E. coli* GroEL-GroES and DnaK (HSP70) structures. **A.** Three-dimensional reconstructions of GroEL-ATP, GroEL-GroES-ATP, and GroEL-GroES from cryo-EM. The first two panels are side views; the third panel is a top view. GroES is red. (PDB codes: 2C7E, 1PCQ) **B.** DnaK clamping down on a peptide (yellow). (PDB code: 1DKX)

in *E. coli* increase in response to high-temperature stress, these chaperones were originally named **heat-shock proteins** (**HSPs**) and are, in fact, more resistant to heat denaturation than is the average protein. Representatives of these chaperones are found in all species. Because their molecular weights are similar, DnaK examples throughout nature are called HSP70s, while homologs of GroEL and DnaJ are called, respectively, HSP60s and HSP40s. Chaperones increase in response to heat stress because they are needed to help refold heat-damaged proteins.

The GroEL and GroES chaperones form a stacked ring with a hollow center like a barrel (**Fig. 8.26**). The chaperoned

protein fits inside. The small capping protein GroES controls entrance to the chamber, as shown in **Figure 8.26A** and **eTopic 8.4**. Cycles of ATP binding and hydrolysis cause conformational changes within the chamber that can reconfigure target proteins. DnaK (HSP70) chaperones have a very different structure (**Fig. 8.26B**). They do not form rings like the GroEL and GroES chaperones, but can clamp down on a peptide to assist folding. Proteins emerging from a bacterial ribosome enter a folding pathway that involves a hierarchy of these chaperones.

Protein Degradation: Cleaning House

What happens when a cell no longer needs a specific protein or when a cell synthesizes a protein with incorrect amino acids? Because the cell's needs constantly change, the presence of useless proteins can adversely affect the cell and must be destroyed. This is particularly true of regulatory proteins, whose concentrations must change with time or in response to alterations in the cellular condition.

Many normal proteins contain degradation signals called degrons that dictate the stability of a protein. The **N-terminal rule** describes one type of degron. The rule states that the N-terminal amino acid of a protein correlates with its stability. For example, proteins beginning with leucine, phenylalanine, tryptophan, or tyrosine experience a short half-life (2 minutes or less), whereas proteins with aspartic acid, glutamic acid, or cysteine in the lead position have a longer half-life. A protein called ClpS facilitates degradation of these short-lived proteins. ClpS recognizes the

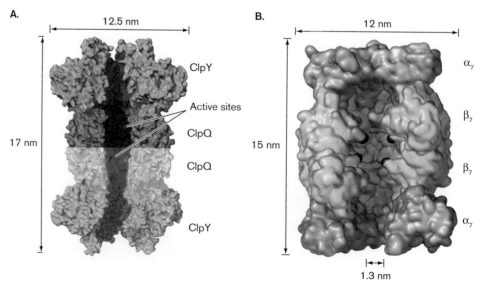

A.

12.5 nm

17 nm

ClpY

Active sites

ClpQ

ClpQ

ClpY

B.

12 nm

15 nm

1.3 nm

α_7

β_7

β_7

α_7

FIGURE 8.27 ■ Protein degradation machines. **A.** Bacterial ClpY ATPase and ClpQ protease (*Haemophilus influenzae*). (PDB code 1G3I) Two of the six subunits from each ring were removed to reveal the interior cavity. The active sites involved in peptide bond cleavage are indicated in pink. **B.** The 20S proteasome from the methanoarchaeon *Methanosarcina thermophila*. (PDB code: 1GOU)

destabilizing N-terminal amino acids and then presents the protein to the bacterial ClpAP protease (described below).

Abnormally folded proteins are recognized by proteases in part because hydrophobic regions that are normally buried within the protein's 3D structure become exposed. The protein is progressively degraded into smaller and smaller pieces by a series of these proteases. Initial cuts, usually involving ATP-dependent endoproteases like Lon protein or ClpP, are followed by digestion with tripeptidases and dipeptidases. Endopeptidases cleave proteins somewhere within the sequence but not from the ends of the sequence. Many peptidases use ATP hydrolysis to help unfold the target protein prior to digestion. Unfolding is necessary for the protein to slide into a barrel-shaped protease such as ClpAP, ClpXP, or ClpYQ in bacteria (**Fig. 8.27**).

Bacterial Clp proteases have a proteolytic core made of two homoheptameric protein rings of either ClpP or ClpY. The ClpP protease has interchangeable homohexameric ATPase caps made of ClpX, ClpA, ClpB, or ClpC, each of which recognizes different substrates. ClpY plays a similar capping role for ClpQ. The accessory proteins recognize

and present different substrate proteins to the ClpP protease, thereby regulating which proteins are degraded. Protein-degrading enzymes are classified as serine, cysteine, or threonine proteases, depending on the key residue in their active sites.

Bacterial Clp proteases are structurally similar to eukaryotic proteasomes, which are even more complex protein-degrading machines. Proteasomes are found primarily in eukaryotes and archaea, although a few bacteria, such as the lung pathogen *Mycobacterium tuberculosis*, have them. Proteasomes are described in **eTopic 8.5**. Eukaryotic proteasomes recognize and then degrade proteins tagged by ubiquitin, a 76-amino-acid peptide. Some bacterial and viral pathogens exploit ubiquitination to reroute host metabolism.

What happens to proteins damaged by stress? Are they always degraded? Microbes are constantly exposed to environmental insults such as high temperature or pH extremes, which damage proteins and cause them to misfold. As an energy-saving device and to prevent interruption of protein function, injured proteins go through a kind of triage process that evaluates whether they are salvageable or must be destroyed before they can endanger the cell. Chaperones constantly hunt for misfolded (or otherwise damaged) proteins and attempt to refold them. But if the protein is released from a chaperone and remains misfolded, it can, by chance, either reengage the chaperone or bind a protease that destroys it (**Fig. 8.28**). This fold-or-destroy triage system is essential if a microbe is to survive environmental stress.

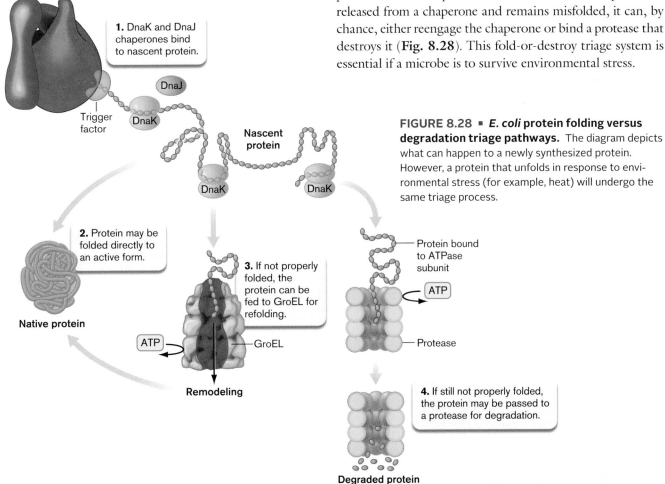

FIGURE 8.28 ▪ *E. coli* protein folding versus degradation triage pathways. The diagram depicts what can happen to a newly synthesized protein. However, a protein that unfolds in response to environmental stress (for example, heat) will undergo the same triage process.

1. DnaK and DnaJ chaperones bind to nascent protein.

Trigger factor

DnaJ

DnaK

Nascent protein

DnaK

DnaK

2. Protein may be folded directly to an active form.

Native protein

3. If not properly folded, the protein can be fed to GroEL for refolding.

ATP

GroEL

Remodeling

Protein bound to ATPase subunit

ATP

Protease

4. If still not properly folded, the protein may be passed to a protease for degradation.

Degraded protein

To Summarize

- **Protein modifications** are made after translation.

- **The N-terminal amino acid** (*N*-formylmethionine, or fMet) can be removed by methionine aminopeptidase, or just the formyl group can be removed by methionine deformylase.

- **An inactive precursor protein** can be cleaved into a smaller active protein, or other groups can be added to the protein (for example, phosphate or AMP).

- **Chaperone proteins** help translated proteins fold properly.

- **All proteins in all cells are eventually degraded** by specific devices such as proteases or proteasomes.

- **The N-terminal rule** describes one type of degradation signal (degron) that marks the half-life of a protein (that is, how long it takes 50% of the protein to degrade).

- **ATP-dependent proteases** such as Lon or ClpP usually initiate the degradation of a large protein.

- **Damaged proteins** randomly enter chaperone-based refolding pathways or degradation pathways until the protein is repaired or destroyed.

8.5 Secretion: Protein Traffic Control

Microorganisms, especially Gram-negative bacteria, face a challenge in delivering proteins to different target locations in the cell. Recall that Gram-negative microbes are surrounded by two layers of membrane (the inner membrane, or cell membrane, and an outer membrane), between which lies a periplasmic space (see Section 3.3). Many proteins are specifically destined for one or another of these cell compartments. Other proteins are secreted completely out of the cell into the surrounding environment (for example, hemolysins that lyse red blood cells). But how do these diverse proteins know where to go? Protein traffic out of the cell is directed by an elaborate set of protein secretion systems. Each system selectively delivers a set of proteins originally made in the cytoplasm to various extracytoplasmic locations.

The term "secretion" is used to describe movement of a protein <u>out</u> of the cytoplasm. There are protein secretion systems that move proteins out of the cytoplasm into the cytoplasmic membrane and across the membrane to the periplasm, others that move proteins to the outer membrane, and still others that deliver proteins across both of the membranes and into the surrounding environment.

An added complication of protein export is that periplasmic proteins are usually delivered unfolded into the periplasm and require another set of chaperones to fold properly in this cell compartment.

Protein Export Out of the Cytoplasm

Proteins destined for the bacterial cell membrane (such as membrane transport proteins) or envelope regions—including periplasm (binding proteins), outer membrane (porins), or extracellular spaces (proteases)—require special export systems. These systems manage to move hydrophilic proteins through one or more hydrophobic membrane barriers. Proteins meant for the inner membrane (for example, cytochromes) contain very hydrophobic N-terminal **signal sequences** of 15–30 amino acids. Signal sequences tether nascent proteins to the membrane and confer conformations that allow the proteins to melt into the fabric of the membrane. Inner membrane proteins also contain hydrophobic transmembrane-spanning regions (20–25 amino acids)

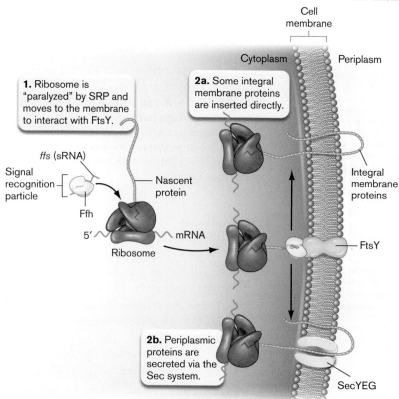

1. Ribosome is "paralyzed" by SRP and moves to the membrane to interact with FtsY.

2a. Some integral membrane proteins are inserted directly.

2b. Periplasmic proteins are secreted via the Sec system.

Cell membrane

Cytoplasm   Periplasm

ffs (sRNA)

Signal recognition particle

Nascent protein

Ffh

Ribosome

5′   mRNA

Integral membrane proteins

FtsY

SecYEG

FIGURE 8.29 ■ SRP and cotranslational export in *E. coli*. A ribosome "paralyzed" by an SRP does not resume translating protein until encountering FtsY in the membrane. Translation can then recommence. Some proteins designated for integral membrane location are inserted directly (top). Other integral membrane proteins and proteins destined for the periplasm are inserted or secreted via the Sec system (bottom). ▶

that aid in this insertion process. These hydrophobic regions are important because they are compatible with the hydrophobicity of the membrane itself. A nutrient transport protein often has 12 such membrane-spanning regions, which weave back and forth across the membrane.

One special export system begins with a complex called the **signal recognition particle** (**SRP**), which targets proteins for inner membrane insertion. A second export mechanism uses a protein called trigger factor that assists proteins destined for the periplasm. (Trigger factor was mentioned earlier as a chaperone.) These two protein traffic pathways converge on a general secretion complex composed of three proteins, collectively called the SecYEG translocon, embedded in the cell (inner) membrane. Depending on the exported protein, SecYEG will assist export to the periplasm or insertion into the membrane.

Protein Export to the Cell Membrane

The pathway leading proteins to the inner (cell) membrane begins with an SRP. In *E. coli*, the SRP consists of a 54-kDa protein (Ffh) complexed with a small RNA molecule (*ffs*). SRP binds to the signal sequences of integral cell membrane proteins as they are being translated (**Fig. 8.29** ▶) and halts further translation in the cytoplasm. The nascent protein with its paralyzed, nontranslating ribosome is delivered to the membrane-embedded protein FtsY, where translation resumes. The partially translated protein is now subject to one of two fates: It may be cotranslationally inserted directly into the cell membrane, meaning that the protein is inserted even as it is still being translated; or it may be completely synthesized, after which it is delivered to the SecYEG translocon for insertion. The route to membrane insertion depends on the protein to be inserted.

Protein Export to the Periplasm: The Sec-Dependent General Secretion Pathway

The periplasm contains important proteins that bind nutrients for transport into the cell and other proteins

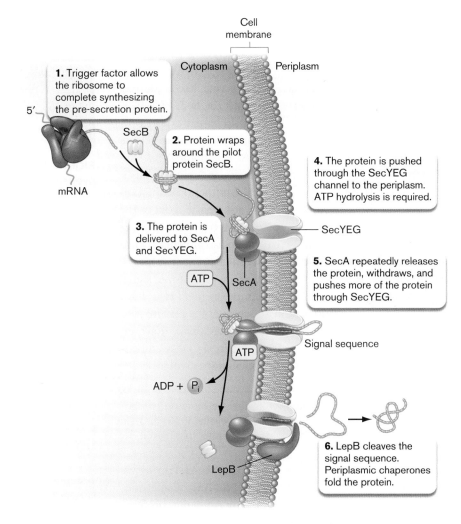

1. Trigger factor allows the ribosome to complete synthesizing the pre-secretion protein.

2. Protein wraps around the pilot protein SecB.

3. The protein is delivered to SecA and SecYEG.

4. The protein is pushed through the SecYEG channel to the periplasm. ATP hydrolysis is required.

5. SecA repeatedly releases the protein, withdraws, and pushes more of the protein through SecYEG.

6. LepB cleaves the signal sequence. Periplasmic chaperones fold the protein.

Cell membrane

Cytoplasm   Periplasm

5′

SecB

mRNA

SecYEG

ATP   SecA

ATP

ADP + Pi

Signal sequence

LepB

FIGURE 8.30 ▪ The SecA-dependent general secretion pathway. This pathway exports many proteins across the cell membranes of Gram-negative and Gram-positive bacteria. ▶

that carry out enzymatic reactions. For example, one form of superoxide dismutase (SOD), an enzyme that degrades superoxide, is a periplasmic protein in *Salmonella enterica* and other Gram-negative bacteria. Many periplasmic proteins, such as SOD and maltose-binding protein (which imports the sugar maltose), are delivered to the periplasm by a common pathway called the general secretion pathway.

The general secretion pathway has several steps. First, the peptide is completely translated in the cytoplasm (**Fig. 8.30** ▶, step 1). Trigger factor interacts with newly synthesized protein as the protein exits the ribosome and keeps pre-secreted proteins in a loosely folded conformation, awaiting interaction with the next component of the secretion machinery. The completed pre-secretion protein is then captured by a piloting protein called SecB (step 2), which unfolds the pre-secretion protein and delivers it to SecA, a protein peripherally associated with the membrane-spanning SecYEG translocon (step 3). Keeping a pre-secretion protein unfolded in the cytoplasm assists the secretion process because sliding an unfolded protein through a membrane is far easier than trying to deliver a folded one.

One model has the SecA ATPase acting like a plunger (**Fig. 8.30**, step 4). It inserts deep into the SecYEG channel, shoving about 20 amino acids of the target export protein into the channel. ATP hydrolysis causes SecA to release the protein and withdraw (step 5). At this point, SecA can bind fresh ATP, rebind the target protein, and reinsert, pushing another 20 amino acids through. Another protein complex (not shown) may pull the translocating protein out of the channel. Proteins needed in the periplasm have cleavable signal sequences at their amino-terminal ends. Immediately following translocation of the amino-terminal sequence into the periplasm, the sequence is snipped off by periplasmic signal peptidases (LepB is one of several examples in *E. coli*). This cleaving completes translocation and releases the mature protein into the periplasm (step 6). Signal peptidases, however, will not cleave signals from proteins destined to stay embedded within the membrane (integral membrane proteins).

Note: "Translocation" can refer to the movement of a ribosome along mRNA, or it can describe the movement of a protein from one cell compartment (cytoplasm) to another (periplasm).

Periplasmic proteins delivered by the Sec system arrive unfolded and inactive. Because the folding chaperones mentioned earlier are cytoplasmic, periplasmic proteins need another set of dedicated periplasmic chaperones to guide their tertiary folding. Another problem with periplasmic proteins is that the oxidizing environment of aerobic cells can oxidize cysteines within a protein and produce inappropriate cysteine disulfide bonds that destroy enzyme function. Special periplasmic disulfide reductases are required to reduce these S–S bonds back to two SH groups. Many periplasmic proteins, however, need certain disulfide bonds to be active, so the periplasm also contains a disulfide bond catalyst (DsbA) to make those bonds.

Gram-positive bacteria must also export proteins across the cell membrane and then fold and process them once they are secreted. However, Gram-positive bacteria lack a periplasmic space that could facilitate interactions between newly secreted proteins and the accessory processing proteins. Many streptococci solve this problem by clustering their secretion systems and accessory factors at the cytoplasmic membrane in an anionic phospholipid microdomain called the ExPortal. The ExPortal is located near the cell septum and appears linked to peptidoglycan synthesis (**Fig. 8.31**). Proteins

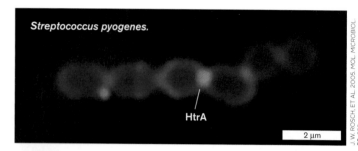

Streptococcus pyogenes.

HtrA

2 μm

FIGURE 8.31 ■ Location of the ExPortal of *Streptococcus pyogenes*. HtrA was identified using immunofluorescence. Note that HtrA is located at the septum.

in the ExPortal include HtrA (assists in pili formation and covalent attachment of proteins to the cell wall) and sortase (aids maturation of secreted proteins), as well as Sec system components and chaperones. Some proteins that pass through the ExPortal are truly secreted; others are not. The latter proteins have a transmembrane domain (missing in Gram-negative homologs) that anchors the proteins to the Gram-positive cytoplasmic membrane. The anchored protein is extracellular but will not float away. Eukaryotic microbes such as the yeast *Saccharomyces cerevisiae* also possess secretion systems that move proteins to the membrane and beyond. However, the eukaryotic Sec systems are more complex and are evolutionarily distinct from bacterial Sec systems. Archaeal secretion systems are actually more similar to those of eukaryotes than to those of bacteria.

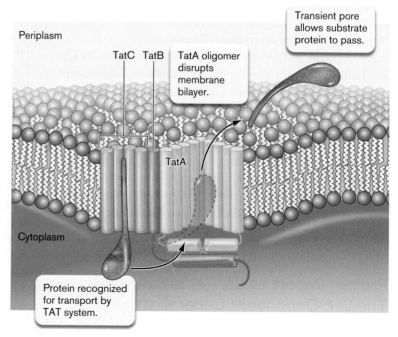

Periplasm

TatC TatB

TatA oligomer disrupts membrane bilayer.

Transient pore allows substrate protein to pass.

TatA

Cytoplasm

Protein recognized for transport by TAT system.

FIGURE 8.32 ■ The twin arginine translocase (TAT). Model for the Tat protein translocase, which includes proteins TatA, TatB, and TatC.

Export of Prefolded Proteins to the Periplasm

In a dramatic departure from Sec-dependent transport systems, some proteins, like TorA, a component of an anaerobic respiratory chain, can be transported fully folded across the membrane to their periplasmic destination. These proteins contain the amino acid motif RRXFXK within their N-terminal signal sequence (where R = arginine, F = phenylalanine, K = lysine, and X = any amino acid). This sequence, called the "twin arginine motif" targets the protein to the membrane-embedded twin arginine translocase (TAT), a transport complex that ships fully folded proteins across the cell membrane to the periplasm (**Fig. 8.32**). Whereas Sec-dependent transport is ATP driven, the TAT system is powered by the proton motive force (see Chapters 4 and 14). TatA oligomers transiently rupture the membrane bilayer to form temporary pores that allow protein substrates to pass.

Journeys to the Outer Membrane

Outer membrane proteins (OMPs) are made in the cytoplasm and exported to the periplasm by the SecA-dependent secretion system. Some OMPs have hydrophobic C-terminal signal sequences that facilitate insertion into the outer membrane, but all have a beta barrel structure that ultimately suits them for outer membrane placement (for instance, see TolC, **Fig. 8.33** ▶). Periplasmic chaperones prevent aggregation

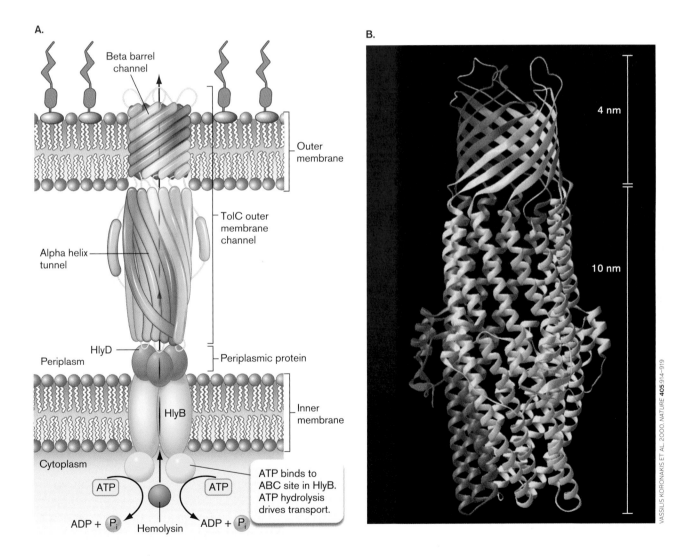

VASSILIS KORONAKIS ET AL. 2000, *NATURE* **405**:914–919

FIGURE 8.33 ■ **Type I secretion: the HlyABC transporter.** **A.** Hemolysin (HlyA) is transported directly from the cytoplasm into the extracellular medium through a multicomponent ABC transport system. The HlyB and D proteins are dedicated to HlyA transport. TolC is shared with other transport systems. Not drawn to scale. **B.** Molecular model of TolC. The beta barrel channel spans the outer membrane, and the alpha helix tunnel extends into the periplasm. Three monomers (red, yellow, and blue) make up the channel. *Source:* Part A modified from Moat et al. 2002. *Microbial Physiology*, 4th ed. Wiley-Liss. ▶

of OMPs as they traverse the periplasm and deliver the proteins to a multisubunit, outer membrane machine called the BAM (beta barrel assembly machine) complex that facilitates OMP assembly in the outer membrane. Although the players seem to be known, the mechanism by which beta barrels are folded and inserted into the outer membrane bilayer remains unclear. BAM has also been implicated recently in the transport of so-called autotransporter proteins such as the attachment protein pertactin of *Bordetella pertussis*. Autotransporters, which also possess a beta barrel domain, were previously thought to mediate their own transport across the outer membrane unassisted by other transporters.

Journeys through the Outer Membrane

For many reasons, Gram-negative bacteria need to export proteins completely out of the cell and into their surrounding environment. Some exported proteins digest extracellular peptides for carbon and nitrogen sources; others act as free-floating toxins that bind and kill host cells. Still others are injected directly into eukaryotic cells by pathogenic or symbiotic microbes to commandeer host metabolic processes. Seven secretion systems, identified as type I–type VII, have evolved to ship these proteins out of the cell (**Table 8.2**). A few start with the Sec system just to get the protein into the periplasm, where dedicated outer membrane systems take over and complete export. Other systems provide nonstop service, delivering the protein directly from the cytoplasm to the extracellular space.

The diversity of system design is impressive. It is the result, in some instances, of selective evolutionary pressures appropriating established cell processes (for example, pilus assembly). New systems evolve through the accidental duplication of one set of genes followed by random mutations that innovatively redesign the duplicated set for a new purpose. We know this because the footprints of genetic divergence have been left behind in the DNA sequence. Type I secretion is described here. Other systems will be covered during the discussion of pathogenesis in Chapter 25.

Type I Protein Secretion

Chapter 4 describes the family of ATP-binding cassette (ABC) influx transporters, whose signature is an amino acid motif that binds ATP. (The term "cassette" refers to a sequence of amino acids that is conserved in many proteins with similar functions.) In addition to ABC influx transporters, similar ABC transporters function in the opposite direction to export various toxins, proteases, and lipases, as well as antimicrobial drugs (multidrug efflux transporters). These ABC transporters are the simplest of

---

TABLE 8.2 **Comparison of Mechanisms that Secrete Proteins across the Outer Membrane**

Type[a]	Mechanism	Structure[b]	Location of protein substrate[c]	Location of secretory signals[d]	Number of components
I	Coupled to TolC	ABC type	Cytoplasm	N terminus, not cleaved	3
II	Extending and contracting	Pilus-like structure	Periplasm	Cleaved N terminus for Sec-dependent transport	12–16
III	Proteins injected directly into host cell cytoplasm	Syringe, related to flagella biogenesis	Cytoplasm	None	20
IV	Conjugation-like	Multicomponent	Some are cytoplasmic; others are periplasmic	None	8 or 9
V	Autotransport	Self-transporting channel in outer membrane formed by C-terminal domain	Periplasm	N terminus, cleaved	1
VI	Proteins injected directly into host cell cytoplasm	Contractile piston, phage tail origin	Cytoplasm	None	12–20
VII	Unknown	Unknown	Cytoplasm	Unknown	At least 4

[a] The seven classes are grouped on the basis of their structure.

[b] Types II, III, and IV are evolutionarily related to mechanisms that assemble pili (type II) or flagella (type III) or that carry out conjugation (type IV). Type VI systems are related to phage tail proteins. Type VII is unrelated to the rest.

[c] Some systems pick up protein substrates from the cytoplasm and transport them across both membranes. Substrates for other systems are collected in the periplasm.

[d] Substrates differ as to the presence of N-terminal signal sequences.

the protein secretion systems and make up what is called type I protein secretion (for example, the Hly system that secretes hemolysin; **Fig. 8.33**). Type I systems all have three protein components, one of which contains an ATP-binding cassette. One component is an outer membrane channel, the second is an ABC protein at the inner membrane, and the third is a periplasmic protein lashed to the inner membrane.

Proteins secreted through type I systems never contact the periplasm, because they pass through a continuous channel that extends from the cytoplasm to the outer membrane. The inner membrane and periplasmic subunits are generally substrate specific, but numerous ABC export systems share the channel protein TolC. TolC is an intriguing protein composed of a beta barrel channel embedded in the outer membrane and an alpha helix tunnel spanning the periplasm (**Fig. 8.33B**). The type I transport shown in **Figure 8.33A** exports a hemolysin (HlyA) from *E. coli* that lyses red blood cell membranes. HlyB and HlyD are the ABC and periplasmic components, respectively.

Six other protein secretion systems are briefly summarized in **Table 8.2**. Some move proteins directly from the cytoplasm to the outside, similar to the type I system, whereas others pick up proteins deposited in the periplasm by the Sec system. They all play important roles in microbial pathogenesis and are more fully discussed in Chapter 25.

To Summarize

■ **Special protein export mechanisms** are used to move proteins to the inner membrane, the periplasm, the outer membrane, and the extracellular surroundings.

■ **N-terminal amino acid signal sequences** help target membrane proteins to the membrane.

■ **The general secretory system** involving the SecYEG translocon can move unfolded proteins to the inner membrane or periplasm.

■ **The signal recognition particle (SRP)** pauses the translation of a subset of proteins that will be placed into the membrane.

■ **SecB protein** binds to certain unfolded proteins that will eventually end up in the periplasm, and pilots them to the SecYEG translocon.

■ **The twin arginine translocase (TAT)** can move a subset of already folded proteins across the inner membrane and into the periplasm.

■ **Type I secretion systems** are ATP-binding cassette (ABC) mechanisms that move certain secreted proteins directly from the cytoplasm to the extracellular environment.

8.6 Bioinformatics: Mining the Genomes

Discovering how biological information flows from DNA to RNA to protein was a major accomplishment of the twentieth century. So, too, was the ability to sequence DNA and analyze protein structure and function. We can now call on the vast store of information gathered over the last half century to make predictions about the genetics and physiology of microbes even when we cannot grow them in the laboratory. These predictions are made using **bioinformatics**, an interdisciplinary field that combines computer science, mathematics, and statistics. The following sections reveal basic principles of this field.

Annotating a Genome Sequence

Bioinformatics has forever changed how the science of microbiology is conducted. Since 1998, the complete genomes of almost 3,500 microbial species have been published, and many others have been partially sequenced. Bioinformatic scientists compare genes from different species to deduce an organism's physiology and evolutionary development, even if the organism cannot be grown in the laboratory. Recall that less than 1% of microorganisms can be cultured. Bioinformatics, then, is critical to understanding the complexity of the uncultured microbial world.

Chapter 7 explained how we sequence a genome and introduced you to metagenomics. But how do we learn what genes are encoded in a newly sequenced genome? The first step is **annotation**. Annotation is analogous to identifying separate sentences and words in an unknown language. Annotating a DNA sequence is done with computers using the known rules of transcription and translation to identify the start and stop sites of potential genes. DNA sequences are then translated *in silico* to determine amino acid sequences. When a sequence appears to encode a protein, the DNA sequence is called an open reading frame (ORF).

Seeking ORFs. Computer programs such as ORF finders use the universal genetic code to deduce protein sequences potentially formed in all reading frames on RNA molecules transcribed from either direction on the chromosome

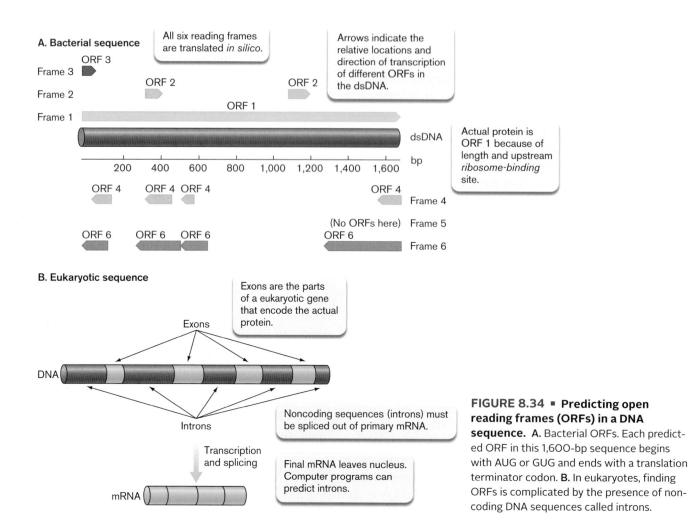

A. Bacterial sequence

All six reading frames are translated *in silico*.

Arrows indicate the relative locations and direction of transcription of different ORFs in the dsDNA.

Frame 3 — ORF 3
Frame 2 — ORF 2
Frame 1 — ORF 1

dsDNA

Actual protein is ORF 1 because of length and upstream *ribosome-binding* site.

bp

200 400 600 800 1,000 1,200 1,400 1,600

ORF 4 ORF 4 ORF 4   ORF 4   Frame 4

(No ORFs here) Frame 5
ORF 6 ORF 6 ORF 6   ORF 6   Frame 6

B. Eukaryotic sequence

Exons are the parts of a eukaryotic gene that encode the actual protein.

Exons

DNA

Introns

Noncoding sequences (introns) must be spliced out of primary mRNA.

Transcription and splicing

Final mRNA leaves nucleus. Computer programs can predict introns.

mRNA

FIGURE 8.34 ■ **Predicting open reading frames (ORFs) in a DNA sequence. A.** Bacterial ORFs. Each predicted ORF in this 1,600-bp sequence begins with AUG or GUG and ends with a translation terminator codon. **B.** In eukaryotes, finding ORFs is complicated by the presence of noncoding DNA sequences called introns.

(**Fig. 8.34A**). An ORF, the equivalent of a sentence in our analogy, is defined as a DNA sequence that can potentially encode a string of amino acids of minimum length—say, 50 residues. The 50 residues of the ORF could encode a 5,500-Da (5.5-kDa) protein, since the average weight of an amino acid is 110 Da. Each ORF begins with a translation start codon (usually ATG, or more rarely, GTG or TTG). A translation start codon marks where ribosomes start to read a messenger RNA molecule. An ORF ends with a translation termination codon (in the DNA, these are TAA, TAG, or TGA). In addition, the computer can identify an ORF by looking for potential ribosome-binding sites upstream of the start codon, but these ribosome-binding sites may differ between species, so finding one is not essential to declaring the presence of an ORF.

While the preceding methods work for prokaryotes, identifying ORFs in eukaryotes is more difficult. In eukaryotes, most genes contain **introns** (long, noncoding sequences that occur in the middle of genes), as do the initial mRNA products made from those genes (**Fig. 8.34B**). The vast majority of known bacterial and archaeal genes do not contain introns. Introns serve several regulatory

functions in eukaryotes that determine whether a protein product is made. However, the eukaryotic cell must use special splicing mechanisms to remove the intron sequences and re-join the protein-encoding sequences, called **exons**, of the mRNA prior to translation. (In eukaryotic organisms, the RNA transcribed directly from DNA is called the **preliminary mRNA transcript** or **pre-mRNA**; the "final" mRNA transcript is produced when the introns have been spliced out.) Because there are few identifying sequence characteristics that mark an intron, extremely sophisticated computer analysis is needed to determine where to remove introns in order to derive the ORF coding sequence of a eukaryotic gene.

Assigning function. We know from considerable experimental precedent that proteins with the same function, regardless of species, usually have critical amino acid sequences in common because they evolved from the same ancestral sequence. Sequences common to two different protein or DNA molecules are called homologous sequences. During the annotation process, ORFs are compared to hundreds of thousands of known proteins to look for sequence homology. ORFs with sufficient homology to known proteins can then be assigned putative function. Powerful computer programs are available to carry out these analyses quickly (for example, visit the Joint Genome Institute website).

If a query protein sequence does not show high levels of similarity with known proteins, "annotation by function" can be assigned when a protein domain exhibits characteristics or motifs similar to proteins of known function. For example, periodic hydrophobic regions may indicate a membrane protein, or an amino acid sequence (or motif) matching those of known ATPases may suggest a similar function for the query protein. Annotation by function provides an extremely valuable starting point toward unraveling a protein's function, but biological validation is required for confirmation.

Genes encoding transfer RNAs and ribosomal RNAs do not encode proteins but can be identified because of the conservation of these sequences across vast phylogenetic distances. The results of annotation can be represented in many different forms. **Figure 8.35** shows a commonly used method of presenting the results—namely, a circular display map (in this case chromosome I of the Gram-negative pathogen *Vibrio vulnificus*). Genes are color-coded by their predicted function. *V. vulnificus* causes lethal blood or wound infections associated with eating, or catching, raw shellfish. Knowing the sequence of *V. vulnificus* helps identify which genes are associated with disease.

Homologs, Orthologs, and Paralogs

Homologies found between genes or proteins suggest an evolutionary relatedness. Genes or proteins that are homologous (**homologs**) probably evolved from a common ancestral gene. Homologous genes or proteins can be classified as either orthologous or paralogous. Orthologous genes (**orthologs**) have functions essentially the same but occur in two or more different species. For example, the gene for glutamine synthase (*gltB*) in *E. coli* is orthologous to *gltB* in *Vibrio cholerae* and to *gltB* in the Gram-positive bacterium *Bacillus subtilis*. Paralogous genes (**paralogs**) arise by duplication within the same species (or progenitor) but evolve to carry out different

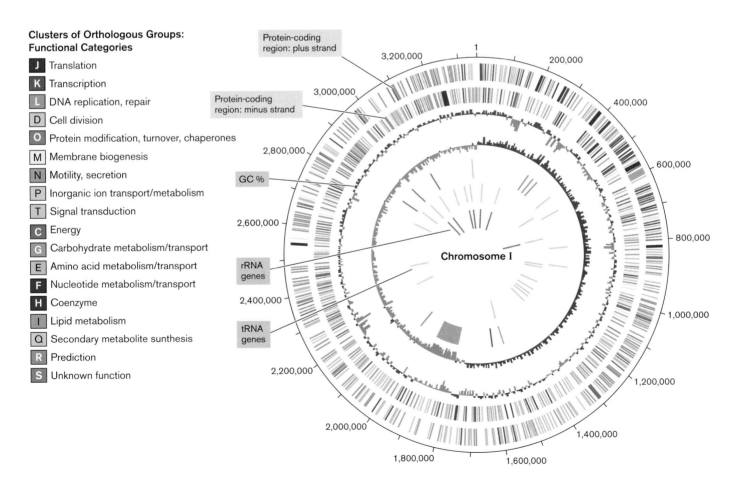

Clusters of Orthologous Groups: Functional Categories

J Translation
K Transcription
L DNA replication, repair
D Cell division
O Protein modification, turnover, chaperones
M Membrane biogenesis
N Motility, secretion
P Inorganic ion transport/metabolism
T Signal transduction
C Energy
G Carbohydrate metabolism/transport
E Amino acid metabolism/transport
F Nucleotide metabolism/transport
H Coenzyme
I Lipid metabolism
Q Secondary metabolite sunthesis
R Prediction
S Unknown function

FIGURE 8.35 ■ **Circular display map of the *Vibrio vulnificus* genome (chromosome I; 3,377 kb).** The color code indicates gene clustering by function. The two outer rings represent genes transcribed from different DNA strands. The innermost circle uses peaks to show GC content levels above (red) or below (green) the genome average.

functions (**Fig. 8.36**). For example, genes encoding type III protein secretion systems, which export virulence proteins, have evolved from paralogous genes encoding flagellar biogenesis. The two sets of genes share a common ancestor but have since evolved completely different functions (in this case, secretion and motility). Paralogous genes are maintained in a microbial genome because their distinct functions contribute to the organism's adaptive potential. Duplicate genes of identical function usually result in loss of one copy or the other by degenerative evolution (see Chapter 9).

Many computer programs and resources used to analyze DNA and protein sequences are freely available on the Web. Here are some useful programs. You can find the program by simply querying for it in any search engine.

- **BLAST** (<u>B</u>asic <u>L</u>ocal <u>A</u>lignment <u>S</u>earch <u>T</u>ool), by the National Center for Biotechnology Information (NCBI), compares a sequence of interest with all other DNA or protein sequences deposited in sequence databases.

- **Multiple Sequence Alignment**, using various tools of the European Bioinformatics Institute (EBI), aligns sequences of genes identified as homologous by BLAST analysis.

- **KEGG** (<u>K</u>yoto <u>E</u>ncyclopedia of <u>G</u>enes and <u>G</u>enomes) outlines biochemical pathways in many sequenced organisms. Areas within this site graphically display reference biochemical pathways (for example, glycolysis) and then indicate which proteins in that pathway are predicted to occur in any organism with a sequenced genome.

- **Motif Search** searches DNA or proteins for sequence signatures such as ATP-binding sites.

- **ExPASy** (<u>Ex</u>pert <u>P</u>rotein <u>A</u>nalysis <u>Sy</u>stem) contains many molecular tools, including Swiss-Prot, the definitive index of known proteins.

- **Joint Genome Institute Genome Portal** has compiled known genome sequences of microbes and eukaryotes.

Some of these programs are designed to align DNA sequences or deduced protein sequences from different species. The goal can be to identify genetic relatedness between species or assign function of a gene or protein. The relatedness between species is best determined by aligning DNA sequences (discussed in Section 17.3). To assign function to a newly sequenced gene, it is best to compare protein products rather than the DNA sequence. Recall that one amino acid can be encoded by more than one DNA codon. For instance, a DNA strand (nontemplate) encoding the peptide Ala-Leu-Ser could be 5′ GCT CCT TCC or 5′ GCA CCG TCA. It is easier to see the homology using the deduced peptide. On the other hand, not all genes encode proteins. Prime examples are the genes encoding stable ribosomal RNA molecules. The DNA sequences for these rRNAs are highly conserved across species, and their similarities reveal homology even among the three ancestral domains (Bacteria, Archaea, and Eukarya).

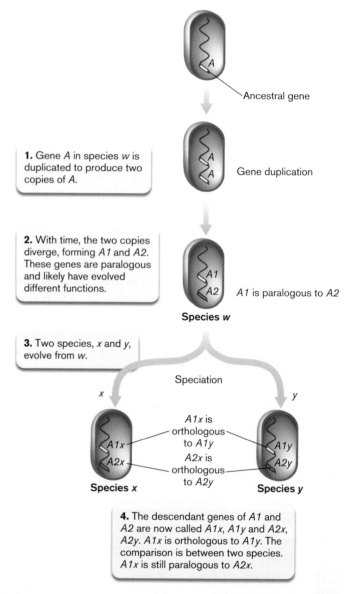

1. Gene *A* in species *w* is duplicated to produce two copies of *A*.

2. With time, the two copies diverge, forming *A1* and *A2*. These genes are paralogous and likely have evolved different functions.

3. Two species, *x* and *y*, evolve from *w*.

4. The descendant genes of *A1* and *A2* are now called *A1x*, *A1y* and *A2x*, *A2y*. *A1x* is orthologous to *A1y*. The comparison is between two species. *A1x* is still paralogous to *A2x*.

Ancestral gene

Gene duplication

A1 is paralogous to *A2*

Species *w*

Speciation

A1x is orthologous to *A1y*

A2x is orthologous to *A2y*

Species *x*

Species *y*

FIGURE 8.36 ■ **Paralogous versus orthologous genes.** An ancestral gene can undergo a duplication to evolve an orthologous or paralogous gene.

Note: As first mentioned in Section 8.3, do not confuse template and nontemplate strands of DNA for a given gene. The template strand, read 3′ to 5′ by RNA polymerase, is transcribed to make the 5′-to-3′ RNA transcript. The nontemplate DNA strand, also called the sense strand, has the <u>same</u> sequence as the RNA transcript, with the substitution of T's for U's.

A word of caution: It is enticing to make definitive proclamations about gene or protein function based on the computer analysis of a DNA sequence. As good as a prediction may seem, it is only a prediction—a well-educated guess. Biochemical confirmation of function must be made, if possible. Nevertheless, the predictions we can make are powerful. In one recent example, a metabolic model was constructed for the Gram-positive anaerobe *Faecalibacterium prausnitzii*, a major member of the human intestinal microbiome (**Fig. 8.37**). This organism has anti-inflammatory properties and produces a major energy source, butyrate, used by epithelial cells lining the large intestine. The bioinformatically

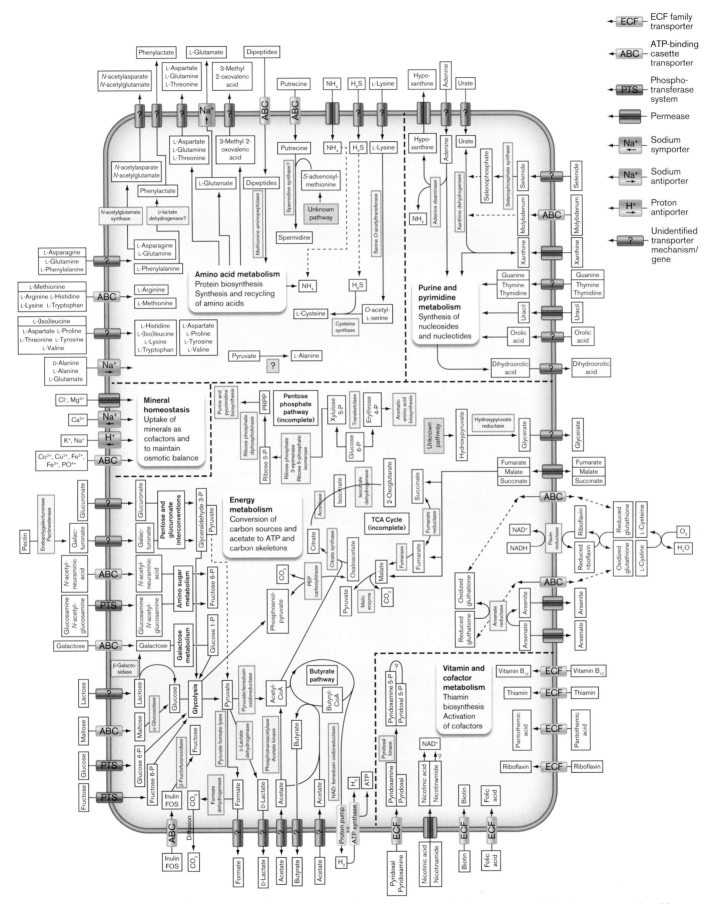

FIGURE 8.37 ▪ **Overview of proposed pathways in *Faecalibacterium prausnitzii*, a member of the human gut microbiome.** The pathways are predicted from bioinformatic genome annotation, the exometabolome, and growth experiments in defined minimal media. Putative pathways and transport systems are indicated with a question mark. Transporter icons are shown in the key. ECF = extracytoplasmic function; FOS = fructooligosaccharide; PRPP = phosphoribosyl pyrophosphate.

generated model was used to design a defined minimal medium that would grow the organism. The defined medium had several amino acid and vitamin growth factors added because genes encoding their synthetic pathways were missing from the bacterium. The organism can now be grown in defined media. This is an example of what has been called **functional genomics**. Functional genomics is an integrative process in which bioinformatic approaches enable scientists to make predictions about function for a set of genes and then test those predictions experimentally.

> **Thought Question**
>
> **8.12** An ORF 1,200 bp in length could encode a protein of what size and molecular weight? *Hint:* Find the molecular weight of an average amino acid.

Bioinformatics and the Human Gut Microbiome

Bioinformatic analysis has contributed much to our understanding of the human microbiome, including its formation. One interesting example comes from a study of carbohydrate-active enzymes (CAZymes)—enzymes that humans need to digest complex polysaccharides. Oddly, the genes encoding these enzymes are not found in the human genome but are present in the Gram-negative anaerobe *Bacteroides thetaiotaomicron*, a prominent member of the human intestinal microbiota. Recently, a new CAZyme (porphyranase) was identified by Mirjam Czjzek, Gurvan Michel, and PhD student Jan-Hendrik Hehemann at Université Pierre et Marie Curie (**Fig. 8.38**) from a marine bacterium called *Zobellia galactanivorans*. This enzyme specifically degrades a sulfonated polysaccharide, porphyran, present in edible marine red algae (seaweed). The team performed a bioinformatic BLAST analysis looking for this enzyme in other species and discovered that all orthologs of porphyranase, except one, were encoded by marine microbes. This makes sense because the biological source of the substrate is found in the sea.

The one exception was *Bacteroides plebeius*, a human gut microbe found only in Japanese individuals. Metagenomic studies found no examples of a porphyranase gene in gut microbiota from North American or French individuals, or from any terrestrial source. Why the Japanese? Curiously, red seaweed, known as nori, is a common food in Japan, and the only source of the porphyran substrate for this CAZyme. Sequence analysis of the *B. plebeius* genome showed that this human intestinal microbe received an unusual set of genes from a marine bacterium, probably by horizontal gene transfer. One of these was the porphyranase CAZyme gene. Horizontal gene transfer is discussed in Sections 7.1 and 9.6. **Figure 8.38** shows the present-day sequence homologies between genes transferred to the ancestor of *B. plebeius* from ancestors of the marine bacteria *Z. galactanivorans* and *Microscilla* species. These data support the idea of horizontal transfer.

How might Japanese gut bacteria have acquired the marine microbe's porphyranase gene? In the eighth century, the Japanese government accepted seaweed as a form of payment, suggesting that seaweed was very important in Japanese culture. The authors propose that consumption of red seaweed with its attendant marine bacteria containing porphyranase is the likely route by which the gene was transferred to the gut microbe. The gene, now part of the Japanese intestinal microbiome, enables humans to digest the porphyran polysaccharide in red seaweed.

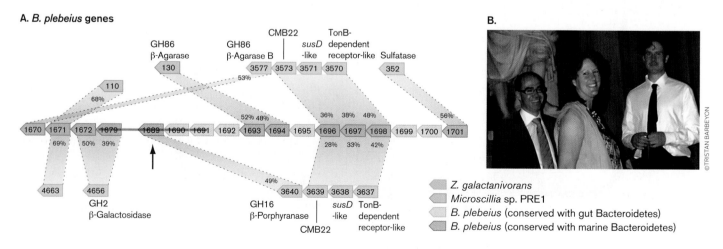

FIGURE 8.38 ▪ **Origin of *Bacteroides plebeius* CAZyme gene.** **A.** Percent homology between *Bacteroides plebeius* gene sequences and genes horizontally transferred from *Zobellia galactanivorans* and *Microscilla*. The *B. plebeius* porphyranase CAZyme is gene 1689. **B.** Gurvan Michel, Mirjam Czjzek, and Jan-Hendrik Hehemann. CMB = unannotated gene; GH = glycoside hydrolase family. *Source:* Hehemann, et al. 2010. *Nature* **464:**908–912

To Summarize

- **Annotation** requires computers that look for patterns in DNA sequences. Annotation predicts regulators, ORFs, rDNA, and tRNA genes. Similarities in protein sequence (deduced from the DNA sequence) are used to predict protein structure and function.

- An **open reading frame (ORF)** is a sequence of DNA predicted by various sequence cues to encode an actual protein.

- **Eukaryotic genes** contain introns and exons, making computer predictions of an ORF more difficult than it is for bacterial or archaeal genes.

- **DNA alignments** of similar genes or proteins can reveal evolutionary relationships.

- **Paralogs and orthologs** arise from gene duplications and speciation events, respectively. Paralogous genes coexist in the same genome but have different functions. Orthologous genes occur in the genomes of different species but produce proteins with similar functions.

Concluding Thoughts

The cellular transcription, translation, and secretion pathways described in this chapter are essential for life because they efficiently assemble biochemical pathways without wasting energy. They also enable pathogens to deliver toxins that can subdue a host and help microbes in mixed communities make antibiotics to eliminate competitors. Efficiency in these processes is maintained through elaborate control mechanisms that sense the organism's physiological state and environment and then trigger changes in replication, transcription, translation, and/or protein processing. How bacteria regulate gene expression in response to environmental stimuli, including threats to survival, will be discussed in Chapter 10.

But first, in Chapter 9, we explore how natural selection randomly redesigns genomes to adapt to ecological niches. Microbes use a variety of DNA exchange mechanisms, gene duplications, and alterations to evolve into forms better adapted to their environments and, in the process, may produce entirely new species.

CHAPTER REVIEW

Review Questions

1. What are some characteristics of an ORF?
2. What is a DNA sequence alignment, and what can it tell you?
3. Describe the differences between a pair of orthologous genes and a pair of paralogous genes.
4. How can bioinformatics predict a metabolic pathway for an organism that cannot be grown in the laboratory?
5. What defines a promoter?
6. What are sigma factors, and what role do they play in gene expression?
7. Describe the three stages of transcription.
8. Explain the degeneracy of the genetic code. What is the wobble in codon-anticodon recognition?
9. Describe the stages of protein synthesis. Why is the ribosome called a ribozyme?
10. Discuss some antibiotics that affect transcription or translation.
11. What is coupled transcription and translation? Does it occur in eukaryotic cells?
12. How do bacterial cells release ribosomes that are stuck onto damaged mRNA molecules lacking termination codons?
13. What can happen to misfolded proteins?
14. Why are only certain proteins secreted from the bacterial cell? What are some secretion mechanisms?
15. In what major way do proteins transported by the twin arginine translocase (TAT) differ from other exported proteins?
16. Compare protein degradation in eukaryotes and bacteria.
17. What is annotation? How does it apply to bioinformatics?

Thought Questions

1. The process of transcription generates positive super-coils in front of the polymerase as it moves along a DNA template. Why doesn't the DNA in front of the polymerase become so knotted that the polymerase can no longer separate the DNA strands?

2. Why do cells secrete some proteins into their environments?

3. Type I protein secretion systems transport certain proteins from the cytoplasm of Gram-negative bacteria directly to the outside of the cell, across two membranes. How might the system "know" which proteins to transport?

Key Terms

acceptor site (A site) (288)
aminoacyl-tRNA synthetase (286)
annotation (305)
anticodon (285)
bioinformatics (305)
catalytic RNA (282)
chaperone (297)
codon (283)
consensus sequence (276)
DNA-dependent RNA
   polymerase (275)
exit site (E site) (288)
exon (306)
functional genomics (310)
heat-shock protein (HSP) (298)
homolog (307)
intron (306)

messenger RNA (mRNA) (281)
N-terminal rule (298)
open reading frame (ORF) (279)
ortholog (307)
paralog (307)
peptidyltransferase (288)
peptidyl-tRNA site (P site) (288)
polysome (289)
preliminary mRNA transcript
   (pre-mRNA) (306)
promoter (276)
release factor (294)
Rho factor (279)
ribosome-binding site (288)
ribosomal RNA (rRNA) (281)
ribozyme (282, 288)
RNA polymerase (275)

Shine-Dalgarno sequence (288)
sigma factor (275)
signal recognition particle (SRP) (301)
signal sequence (300)
small RNA (sRNA) (282)
start codon (284)
stop codon (284)
template strand (275)
tmRNA (282)
transcript (275)
transcription (274)
transertion (290)
transfer RNA (tRNA) (282)
translation (283)
translocation (292)

Recommended Reading

Bakshi, Somenath, Heejun Choi, Jagannath Mondal, and James C. Weisshaar. 2014. Time-dependent effects of transcription- and translation-halting drugs on the spatial distributions of the *Escherichia coli* chromosome and ribosomes. *Molecular Microbiology* **94**:871–887.

Bakshi, Somenath, A. Siryaporn, Mark Goulian, and J. C. Weisshaar. 2012. Superresolution imaging of ribosomes and RNA polymerase in live *Escherichia coli* cells. *Molecular Microbiology* **85**:21–38.

Brandt, Florian, Stephanie A. Etchells, Julio O. Ortiz, Adrian H. Elcock, F. Ulrich Hartl, et al. 2009. The native 3D organization of bacterial polysomes. *Cell* **136**:261–271.

Burger, Adelle, Chris Whiteley, and Aileen Boshoff. 2011. Current perspectives of the *Escherichia coli* RNA degradosome. *Biotechnology Letters* **33**:2337–2350.

Costa, Tiago R., Catarina Felisberto-Rodrigues, Amit Meir, Marie S. Prevost, Adam Redzej, et al. 2015. Secretion systems in Gram-negative bacteria: Structural and mechanistic insights. *Nature Reviews. Microbiology* **13**:343–359.

Edwards, David J., and Kathryn E. Holt. 2013. Beginner's guide to comparative bacterial genome analysis using next-generation sequence data. *Microbial Informatics and Experimentation* **3**:2.

Hui, Monica P., Patricia L. Foley, and Joel G. Belasco. 2014. Messenger RNA degradation in bacterial cells. *Annual Reviews of Genetics* **48**:537–559.

Keiler, Kenneth C. 2015. Mechanisms of ribosome rescue in bacteria. *Nature Reviews. Microbiology* **13**:285–297.

Murakami, Katsuhiko S., Shoko Masuda, Elizabeth A. Campbell, Oriana Muzzin, and Seth Darst.

2002. Structural basis of transcription initiation: RNA polymerase holoenzyme-DNA complex. *Science* **296**:1285–1290.

Petrov, Anton S., Chad R. Bernier, Chiaolong Hsiao, Ashlyn M. Norris, Nicholas A. Kovacs, et al. 2014. Evolution of the ribosome at atomic resolution. *Proceedings of the National Academy of Sciences USA* **111**:10251–10256.

Preissler, Steffen, and Elke Deuerling. 2012. Ribosome-associated chaperones as key players in proteostasis. *Trends in Biochemical Sciences* **37**:274–283.

Proshkin, Sergey, A. Rachid Rahmouni, Alexander Mironov, and Evgeny Nudler. 2010. Cooperation between translating ribosomes and RNA polymerase in transcription elongation. *Science* **328**:504–508.

Schulze, Ryan J., Joanna Komar, Mathieu Botte, William J. Allen, Sarah Whitehouse, et al. 2014. Membrane protein insertion and proton-motive-force-dependent secretion through the bacterial holo-translocon SecYEG-SecDF-YajC-YidC. *Proceedings of the National Academy of Sciences USA* **111**:4844–4849.

Vega, Luis A., Gary C. Port, and Michael G. Caparon. 2013. An association between peptidoglycan synthesis and organization of the *Streptococcus pyogenes* ExPortal. *MBio* **4**:e00485–13.

Washburn, Robert S., and Max E. Gottesman. 2015. Regulation of transcription elongation and termination. *Biomolecules* **5**:1063–1078.

Wojtas, Magdelena, Bibiana Peralta, Marina Ondiviela, Maria Mogni, Stephen D. Bell, et al. 2011. Archaeal RNA polymerase: The influence of the protruding stalk in crystal packing and preliminary biophysical analysis of the Rpo13 subunit. *Biochemical Society Transactions* **39**:25–30.

Yakhnin, Alexander V., and Paul Babitzke. 2014. NusG/Spt5: Are there common functions of this ubiquitous transcription elongation factor? *Current Opinions in Microbiology* **18**:68–71.

CHAPTER 9

Gene Transfer, Mutations, and Genome Evolution

9.1 Mosaic Genomes and Gene Transfer

9.2 Recombination

9.3 Mutations

9.4 DNA Repair

9.5 Mobile Genetic Elements

9.6 Genome Evolution

DNA is a dynamic molecule. Genome sequences change over generations through mutations, DNA rearrangements, and gene transfers between species. Bacterial and archaeal genomes sometimes shuttle large clusters of genes between members of different taxonomic domains. All this interspecies and intraspecies DNA traffic has led ecologists to expand the definition of the "microbial genome" to include DNA in a cell plus DNA from other organisms to which the microbe has potential access.

CURRENT RESEARCH highlight

Antibiotics and evolution. *Streptococcus pneumoniae* produces invasive disease such as pneumonia but also causes recurrent ear infections. The most recent ear infection strain will contain genes incorporated from earlier strains. Jan-Willem Veening found that antibiotics may stimulate this evolution. *S. pneumoniae* takes up DNA from dead cells via transformation, a process initiated by competence genes located near the replication origin. Antimicrobials that stall replication forks but allow initiation can increase copy number (gene dosage) and expression of origin-proximal genes. Veening showed that a sublethal dose of 6-(*p*-hydroxyphenylazo)-uracil (HPUra), which reversibly binds DNA polymerase III, will stall replication forks and trigger expression of competence-regulated gene *ssbB-gfp* (green cells). Thus, some antibiotics may stimulate in vivo pneumococcal evolution by activating transformation.

Source: Slager et al. 2014. Cell **157**:395–406.

AN INTERVIEW WITH

JAN-WILLEM VEENING, MICROBIOLOGIST, UNIVERSITY OF GRONINGEN AND UNIVERSITY OF LAUSANNE

© ANNE DE JONG

How might gene dosage–induced competence enhance evolution during *Streptococcus pneumoniae* infections?

Many types of stress, including some antibiotics, cause gene dosage–induced competence in *S. pneumoniae*. Because humans are often co-colonized by multiple pneumococcal strains, competence and DNA uptake can help one strain acquire resistance and virulence factors from others. Circumstantial evidence strongly suggests that gene dosage–induced competence plays a very important role for successful pneumococcal carriage. It will be interesting to see whether other key regulators located close to the origin of replication in the pneumococcus, or in other bacteria, are affected by gene dosage and how the cell's physiology changes in response.

Are there potential applications for your findings?

Antibiotics that stall replication elongation include the commonly used fluoroquinolones. Our work suggests that physicians should, whenever possible, treat pneumococcal infections with alternative antibiotics that do not activate competence, like beta-lactam antibiotics. In addition, we are trying to identify novel compounds that can inhibit competence development. Such compounds might be used together with existing antibiotics to reduce the spread of antibiotic resistance.

Control (uninduced)   + Competence factor   + HPUra

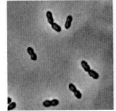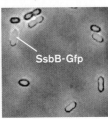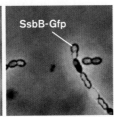

SsbB-Gfp

SsbB-Gfp

SsbB-Gfp

We often think of meaningful evolution as having to take place over thousands if not millions of years. But that is not always so. One striking example involves the pathogen *Streptococcus pneumoniae*, which causes serious lung infections in any age group, and also recurrent ear infections in infants. A team of researchers headed by Fen Hu and Garth Erhlich (Drexel University) found convincing evidence of evolution in a strain of *S. pneumoniae* isolated from an 8-month-old child stricken with recurrent ear infections over 7 months. Genome sequences of strains isolated during separate bouts revealed 16 distinct recombination events leading from the earliest to the latest isolate. During her short life, the child had been colonized by multiple strains of pneumococcus that readily exchanged DNA via transformation. As the Current Research Highlight suggests, the high level of horizontal gene transfer was likely influenced by antibiotic treatments. The resultant gene transfers allowed the species to quickly evolve and cause recurrent disease.

Two other scientists, John Sullivan and Clive Ronson from New Zealand, witnessed microbial evolution taking place in soil within a mere decade. These researchers were studying the microbe *Mesorhizobium loti*, a symbiotic bacterial species that forms nitrogen-fixing nodules on plant roots. Genes encoding symbiosis are located as a group on the mesorhizobial chromosome. In addition to these symbiotic rhizobia, there are many nonsymbiotic species that are incapable of forming root nodules. In a remarkable experiment, Sullivan and Ronson inoculated a single strain of *M. loti* into an area of land devoid of natural nodulating rhizobia. Seven years later, they discovered that the area contained many genetically diverse symbiotic mesorhizobia now able to nodulate the flowering plant *Lotus corniculatus*. These microbes did not exist before the experiment. The 500-kb genome segment encoding symbiosis had somehow made its way from *M. loti* into these other bacteria—essentially generating new species. How did this happen so quickly?

In Chapter 9 we address the mechanisms that mediate transient, as well as heritable, DNA movements between species. Next we describe the competing processes of mutagenesis and mutation repair required for self-preservation and evolutionary change. We close the chapter by explaining how all these processes collaborate over millennia to remodel genomes and build biological diversity. The consequences of gene flow during evolution will be discussed in Chapter 17.

9.1 Mosaic Genomes and Gene Transfer

Genomic analysis shows that over millennia, microbes undergo extensive gene loss and gain. Archaea, for example, arose from a common eukaryotic-archaeal phylogenetic branch and, as a result, possess many traits in common with eukaryotes, such as the structure and function of their DNA and RNA polymerases. However, many archaeal genes whose products are involved with intermediary metabolism look purely bacterial. In fact, 37% of the proteins found in the archaeon *Methanocaldococcus jannaschii* are found in all three domains—Archaea, Eukarya, and Bacteria. Another 26% are otherwise found only among bacteria, while a mere 5% are confined to archaea and eukaryotes. Archaea, then, enjoy a mixed heritage.

Another surprise arising from bioinformatic studies is the mosaic nature of the *E. coli* genome and, indeed, of all microbial genomes. Though we have intensively studied *E. coli* for over 100 years, we now find that the organism's DNA is rife with pathogenicity islands, fitness islands, inversions, deletions (when compared to similar species), paralogous genes, and orthologous genes. How did all this genomic blending happen? The answer appears to involve heavy gene traffic between species (horizontal gene transfer), recombination events occurring within a species, and a variety of mutagenic and DNA repair strategies. All of these processes accelerated natural selection, where trial and error shaped genome content.

The Discovery of Horizontal Gene Transfer

In 1928, a perceptive English medical officer, Frederick Griffith (1879–1941), found that he could kill mice by injecting them with dead cells of a virulent pneumococcus (*Streptococcus pneumoniae*, a cause of pneumonia), together with live cells of a nonvirulent mutant. Even more extraordinary was that he recovered live, virulent bacteria from the dead mice. Were the dead bacteria brought back to life? Unfortunately, Griffith was killed by a German bomb during an air raid on London in 1941 and never learned the answer.

In a landmark series of experiments published in 1944, Oswald Avery (1877–1955), Colin MacLeod (1909–1972), and Maclyn McCarty (1911–2005) proved that Griffith's experiment was not a case of resurrecting dead cells. Instead, DNA released from dead cells of the virulent strain entered into the harmless living strain of *S. pneumoniae*—an event that transformed the live strain into a killer. The process of importing free DNA into bacterial cells is now known as **transformation**.

Transformation provided the first clue that gene exchange can occur in microorganisms. Why do bacteria carry out this and other forms of gene transfer? Genome comparisons show that the fundamental purpose of bacterial gene transfer is to acquire genes that might be useful as the environment changes. The natural movement of genes between species or genera is called **horizontal gene transfer** or lateral gene transfer. Horizontal gene transfer between cells differs from vertical gene transfer, which is the generational passing of genes from parent to offspring (as in cell division). If horizontal acquisition of a gene system improves the competitiveness of the cell, the new genes will be retained and the descendants will have evolved into a new kind of organism.

Before we can discuss how bacterial genomes evolve, we need to understand the various gene exchange mechanisms available to them. Following the explanations of gene exchange, we will discuss the processes that incorporate newly acquired DNA into genomes (for example, recombination).

Transformation of Naked DNA

How do live bacteria import DNA from dead bacteria? Many bacteria can import DNA fragments and plasmids released from nearby dead cells via the process of transformation. Natural transformation requires specific protein complexes called **transformasomes**. Organisms in which transformation is a natural part of the growth cycle include Gram-positive bacteria like *Streptococcus* and *Bacillus*, as well as Gram-negative species such as *Haemophilus* and *Neisseria*. At least 82 species have been shown to be naturally competent.

Other bacteria, such as *E. coli* and *Salmonella*, do not possess the equipment needed to import DNA naturally. They require artificial manipulations to drive DNA into their cells. These laboratory techniques include perturbing the membrane by chemical ($CaCl_2$) and electrical (**electroporation**) methods. $CaCl_2$ alters the membrane, making these cells chemically competent so that DNA can pass. Electroporation, on the other hand, uses a brief electrical pulse to "shoot" DNA across the membrane.

Why do species such as *Neisseria* undergo natural transformation? First, species that indiscriminately import DNA may use the transformed DNA as food. Second, the more finicky species that transform only compatible DNA sequences may use DNA released from dead compatriots to repair their own damaged genomes. Note that the transformation process is random. A cell does not <u>choose</u> which genes to import or keep. Finally, transformation can influence evolution, enabling species to adjust to new environments by acquiring new genes from other species via horizontal gene transfer. Transformation may have enabled pathogens such as *Neisseria gonorrhoeae*, the cause of gonorrhea, to acquire genes whose protein products now help the organism evade the host immune system.

Gram-positive organisms, competence, and quorum sensing. Natural transformation in Gram-positive organisms typically involves the growth phase–dependent assembly of a transformasome complex across the cell membrane (**Fig. 9.1**). The transformasome is composed of a binding

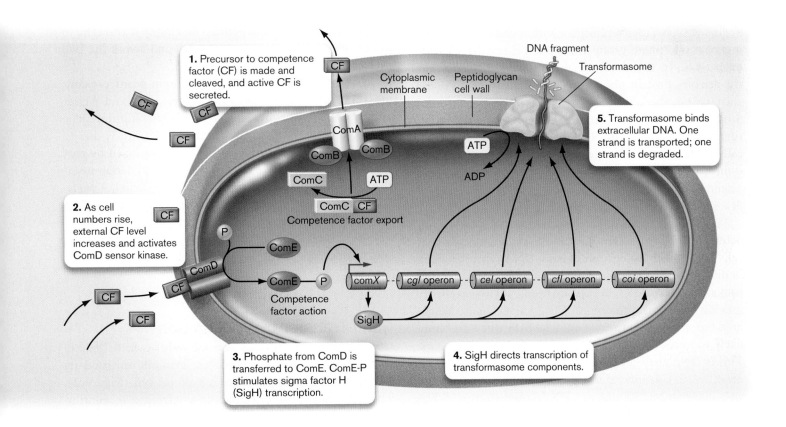

FIGURE 9.1 ■ Transformation in *Streptococcus*. The process of transformation in this organism begins with the synthesis of a signaling molecule (competence factor, CF) and concludes with the import of a single-stranded DNA strand through a transformasome complex.

protein that captures extracellular DNA floating in the environment, plus proteins that form a transmembrane pore. A nuclease degrades one strand of a double-stranded DNA molecule while pulling the other strand intact through the pore and into the cell. Once inside, the strand can be incorporated into the chromosome by recombination, a process we will discuss in Section 9.2.

Once the transformasome is assembled, the cell is **competent**, meaning that it can import free DNA fragments and incorporate them into its genome by recombination. What triggers growth phase–dependent competence? For some Gram-positive bacteria, competence for transformation is generated by a chemical conversation (quorum sensing; see Sections 4.5 and 10.5) that takes place between members of the culture. Every individual in a growing population produces and secretes a small, 15- to 20-amino-acid peptide, generically called competence factor (CF), that progressively accumulates in the medium until it induces a genetic program that makes the population competent (**Fig. 9.1**, step 1). The sequence of the CF peptide is unique to each species, as are the specifics of the induction process. For *Streptococcus pneumoniae*, the level of CF in the medium increases (step 2) as the population increases—that is, as the cell density increases.

Above a certain concentration threshold, competence factor is able to bind to a sensory protein built into the cell membrane (ComD for *S. pneumoniae*). This binding begins what is called a phosphorylation cascade (the passing of a phosphate group from one protein to another; eTopic 4.1). In the competence phosphorylation cascade, the sensory protein phosphorylates itself using ATP and then passes the phosphate to a cytoplasmic regulatory protein, ComE, which stimulates expression of a novel sigma factor, SigH (**Fig. 9.1**, step 3). Sigma H is specifically used to transcribe genes encoding the transformasome (step 4). The protein products of these genes are assembled at the membrane, and the cell becomes competent (step 5).

Why would organisms regulate transformation competence by quorum sensing? One hypothesis holds that cells are unlikely to encounter stray DNA when growing in dilute natural environments such as ponds, where other bacteria are scarce. So, in this situation, why waste energy making the transformasome? When these same cells are growing at high density, as in a biofilm, they are more likely to encounter DNA released from dying neighbors. This is DNA they could use to repair their own damaged genomes, to consume as food, or to sample for a new survival mechanism, should the DNA come from a different species present in a biofilm consortium. Regulation by quorum sensing would ensure that the transformasome would not form until there was a good chance that free foreign DNA was available.

Thought Question

9.1 Figure 9.1; illustrates the process of transformation in *Streptococcus pneumoniae*. Will a mutant of *Streptococcus* lacking ComD be able to transform DNA?

Gram-negative species transform DNA without competence factors. Gram-negative species capable of natural transformation do not appear to make competence factors. Either they are always competent, like *Neisseria*, or they become competent when starved (for example, *Haemophilus*). Competence development does not depend on cell-cell communication. Gram-negative organisms have a cytoplasmic membrane transformasome similar to that employed by Gram-positive microbes, but Gram-negative bacteria also must overcome an outer membrane barrier to DNA importation. *Neisseria* species, for example, appear to import DNA through a system paralogous to type IV pilus assembly (discussed in Section 8.5). Type IV pili reversibly assemble and disassemble at the cell surface, expanding and contracting as a result. This happens constantly during the growth of a culture and can help cells move across solid surfaces. During transformation in species of *Neisseria*, disassembly of a pilus is thought to drag transforming DNA into the cell and across the two membranes.

In another departure from the Gram-positive example, transformation in species of *Haemophilus* and *Neisseria* is species and sequence specific, thereby limiting gene exchange between different genera. Specificity is due to particular sequences in DNA that are recognized by part of the uptake apparatus. However, not all Gram-negative competence systems display such specificity; *Acinetobacter calcoaceticus*, a lung pathogen, is able to take up DNA from any source at very high frequency. This is possible because the DNA-binding part of its transformation system does not need to recognize a specific nucleotide sequence.

Gene Transfer by Conjugation

Conjugation, sometimes erroneously called "bacterial sex," requires cell-cell contact typically initiated by a special pilus protruding from a donor cell (**Fig. 9.2**). The pilus can also dramatically contribute to biofilm formation (eTopic 9.1). Conjugation occurs in many species of bacteria and archaea, even in hyperthermophiles

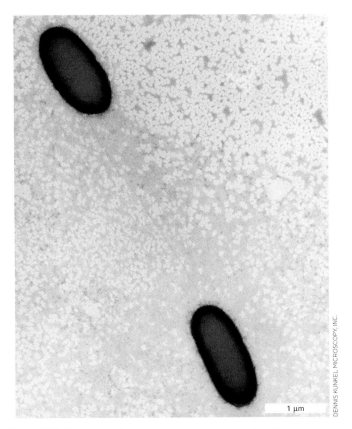

FIGURE 9.2 ■ Sex pilus connecting two *E. coli* cells.
Pseudocolor added.

such as *Sulfolobus* species. In nature, mixed-species biofilms are thought to enable DNA transfer between different species, albeit at a low frequency. As for the mechanism of DNA transfer by conjugation, most species use type IV–like protein secretion systems because a protein is transferred along with DNA (as described next and in Section 8.5).

In the case of the Gram-negative *E. coli*, the tip of the specialized plasmid-encoded sex pilus attaches to a receptor on the recipient cell and then contracts, drawing the two cells closer. The two cell envelopes fuse and generate a conjugation complex, through which single-stranded DNA passes from donor to recipient. The conjugation complex is similar to the transformation complex.

Bacterial conjugation requires the presence of special transferable plasmids that usually contain all the genes needed for pilus formation and DNA export. In some cases, plasmids can also contain genes encoding antibiotic resistance. A well-studied, transferable plasmid in *E. coli* is called **fertility factor**, or **F factor** (**Fig. 9.3** ▶). F factor can transfer itself to another recipient cell, but if it integrates into the chromosome of its host, it can also transfer host genes. To carry out plasmid replication and DNA

transfer, F factor uses two replication origins, *oriV* and *oriT*, located at different positions on the plasmid. The origin *oriV* is used to replicate and maintain the plasmid in nonconjugating cells, whereas *oriT* is used only to replicate DNA during DNA transfer. F factor also contains several *tra* genes, whose protein products carry out the DNA transfer process.

Conjugation begins with cell-cell contact between a donor **F⁺ cell** that carries the plasmid and a recipient **F⁻ cell** (**Fig. 9.3**, step 1). Formation of a fused membrane conjugation conduit (step 2) triggers synthesis of an F factor–encoded helicase/endonuclease (also called TraI or relaxase) that nicks the phosphodiester backbone at the *nic* site in *oriT* (step 3). TraI forms a relaxosome complex with other F factor–encoded proteins. The relaxosome unwinds donor double-stranded DNA and begins the transfer process. The blowup at step 3 in **Figure 9.3** shows a model of the multifactor relaxosome complex called a conjugation bridge. The translocated TraI relaxase remains bound to the 5′ end of the nicked strand, and the 5′ DNA-protein complex is transferred through a pore protein in the recipient. The translocated relaxase and the 5′ end of the strand remain in the membrane, while the rest of the strand is transferred through the pore. The intact circular strand remains in the donor. DNA polymerase III (Pol III) is then recruited to *oriT* in the donor, where replication begins (step 4). In contrast to bidirectional replication, which is used to duplicate the chromosome, replication for conjugative transfer occurs unidirectionally, by rolling-circle replication (step 5) (see Section 7.4).

DNA polymerase III synthesis of the replacement strand in the donor cell drives movement of the transferred strand through the pore. The polymerase uses the untransferred intact circular strand as a template (see **Fig. 9.3**, steps 4 and 5).

Once the transfer is complete, the relaxase re-ligates the 5′ end it was holding to the 3′ tail of the transferred strand in the recipient. Thus, the last portion of F factor moved to the recipient is *oriT*. Once circularized, the plasmid replicates and the F⁻ recipient cell becomes a new F⁺ donor cell (**Fig. 9.3**, step 6). Ultimately, the conjugation complex spontaneously comes apart, and the membranes seal. This transfer process is very quick, taking less than five minutes to transfer the entire 110-kb F factor.

Is it possible for two donors to exchange DNA with each other? The answer is generally no; donors have mechanisms in place to prevent such exchange. For example, a membrane protein encoded by the F factor will inhibit formation of a conjugation complex with another donor that possesses the same protein. This mechanism prevents the pointless transfer of a plasmid to a cell that already has that plasmid.

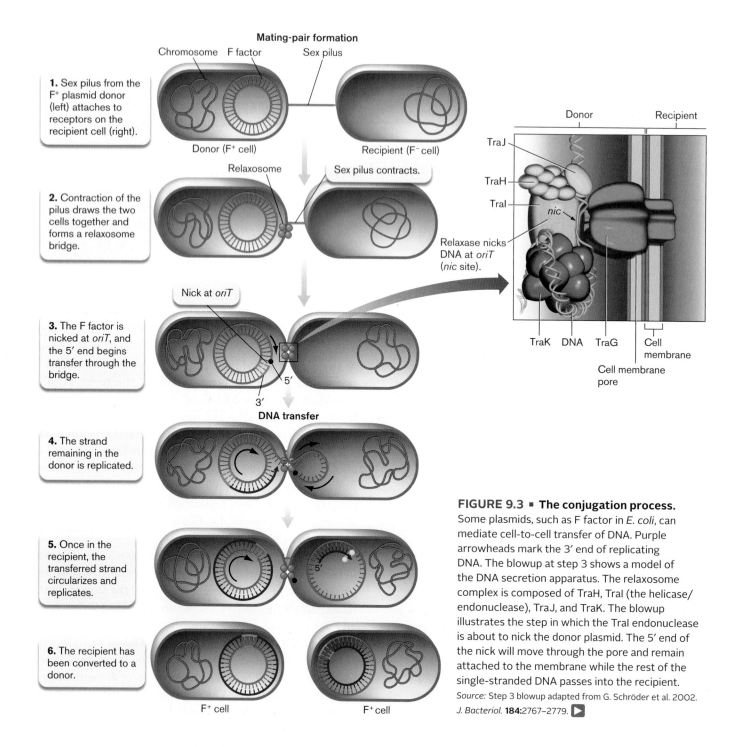

Mating-pair formation

1. Sex pilus from the F⁺ plasmid donor (left) attaches to receptors on the recipient cell (right).

Chromosome F factor Sex pilus

Donor (F⁺ cell) Recipient (F⁻ cell)

2. Contraction of the pilus draws the two cells together and forms a relaxosome bridge.

Relaxosome Sex pilus contracts.

3. The F factor is nicked at *oriT*, and the 5′ end begins transfer through the bridge.

Nick at *oriT*

5′

3′

DNA transfer

4. The strand remaining in the donor is replicated.

5. Once in the recipient, the transferred strand circularizes and replicates.

5′

6. The recipient has been converted to a donor.

F⁺ cell F⁺ cell

Donor Recipient

TraJ
TraH
TraI
nic

Relaxase nicks DNA at *oriT* (*nic* site).

TraK DNA TraG Cell membrane

Cell membrane pore

FIGURE 9.3 ■ The conjugation process.
Some plasmids, such as F factor in *E. coli*, can mediate cell-to-cell transfer of DNA. Purple arrowheads mark the 3′ end of replicating DNA. The blowup at step 3 shows a model of the DNA secretion apparatus. The relaxosome complex is composed of TraH, TraI (the helicase/endonuclease), TraJ, and TraK. The blowup illustrates the step in which the TraI endonuclease is about to nick the donor plasmid. The 5′ end of the nick will move through the pore and remain attached to the membrane while the rest of the single-stranded DNA passes into the recipient.
Source: Step 3 blowup adapted from G. Schröder et al. 2002. *J. Bacteriol.* **184:**2767–2779. ▶

The F-factor plasmid can integrate into the chromosome. An F-factor plasmid can remain a plasmid or can be integrated into the chromosome of the recipient cell (**Fig. 9.4**). Integration of the plasmid into the genome involves recombination between two circular DNA duplexes—a process explained in Section 9.2. Because the plasmid can exist in extrachromosomal and integrated forms, it is sometimes called an episome (*epi*, Greek for "over"; and *some* from "chromosome").

A strain with the F factor integrated into the chromosome is called an **Hfr** (<u>h</u>igh-<u>f</u>requency <u>r</u>ecombination)

strain. The strain is called "high-frequency" because every cell is capable of transferring <u>chromosomal</u> DNA to an F⁻ cell. Very few cells with an integrated F factor are present in an F⁺ culture. A cell with F factor inserted into the chromosome is capable of transferring all or part of the chromosome into a recipient cell. Essentially, the entire chromosome becomes the F factor. However, the integrated F factor is actually the last bit of DNA transferred during Hfr conjugation. Because it takes nearly 100 minutes to transfer the entire *E. coli* chromosome

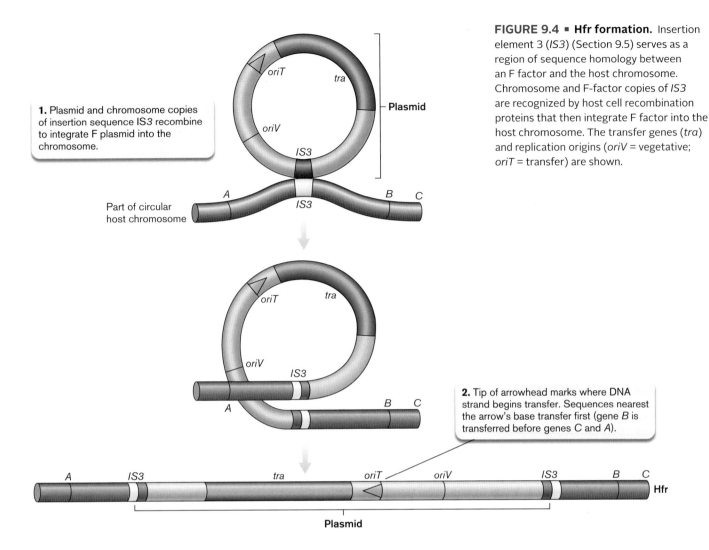

1. Plasmid and chromosome copies of insertion sequence IS3 recombine to integrate F plasmid into the chromosome.

Part of circular host chromosome

FIGURE 9.4 ■ Hfr formation. Insertion element 3 (*IS3*) (Section 9.5) serves as a region of sequence homology between an F factor and the host chromosome. Chromosome and F-factor copies of *IS3* are recognized by host cell recombination proteins that then integrate F factor into the host chromosome. The transfer genes (*tra*) and replication origins (*oriV* = vegetative; *oriT* = transfer) are shown.

2. Tip of arrowhead marks where DNA strand begins transfer. Sequences nearest the arrow's base transfer first (gene *B* is transferred before genes *C* and *A*).

(as opposed to only 5 minutes for the free F plasmid), the integrated F factor is rarely transferred before the conjugation bridge breaks, so the recipient almost never becomes an F⁺ or Hfr cell. How Hfr conjugation was used in the past to map *E. coli* genes is described in **eTopic 9.2**. Today, rapid DNA sequencing is used to map genes in any organism. Nevertheless, conjugation remains important for evolution and as a tool for scientists to move genes between species.

An integrated F factor can excise from the chromosome. Another type of gene shuffling can occur when an integrated F factor reverses course and excises from the chromosome via host recombination mechanisms. The F factor is usually excised completely and restored to its original form. Occasionally, however, it is excised along with some neighboring DNA from the host chromosome, yielding a product that remains an F-factor plasmid but also contains some chromosomal DNA. The derivative F plasmid containing host DNA is called an F-prime (F′) plasmid or **F-prime (F′) factor (Fig. 9.5)**.

In contrast to the Hfr transfer of chromosomal genes, genes hitchhiking on an F′ plasmid do not have to recombine into the recipient chromosome to be maintained. The extra genes can be expressed as part of the F′ plasmid. The resultant strain is called a "partial diploid" because it contains two copies of those few genes—one set on the chromosome, the other on the F′ factor. Partial diploids are useful to the cell because the second copy of the gene can be the raw material that evolves into a new gene. Partial diploids are useful to geneticists because they can reveal whether a mutant allele of a gene is dominant over a normal (wild-type) allele. Partial diploids can also help geneticists understand gene organization.

Many different types of plasmids can be found in the microbial world. Most are not transferable by themselves. However, one group of plasmids that cannot transfer themselves can be <u>mobilized</u> if a transferable plasmid such as F factor is also present in the same cell. Mobilizable plasmids usually contain an *oriT*-like DNA replication origin recognized by the conjugation apparatus of the transferable plasmid. As a result, when the transferable plasmid begins

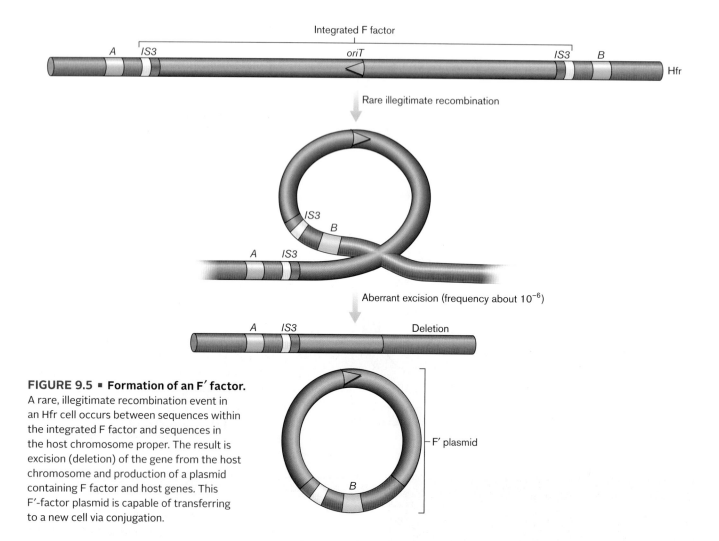

FIGURE 9.5 ■ Formation of an F′ factor.
A rare, illegitimate recombination event in an Hfr cell occurs between sequences within the integrated F factor and sequences in the host chromosome proper. The result is excision (deletion) of the gene from the host chromosome and production of a plasmid containing F factor and host genes. This F′-factor plasmid is capable of transferring to a new cell via conjugation.

conjugating, so does the mobilizable plasmid. This is one way in which antibiotic resistance genes on plasmids called R factors can be spread throughout a microbial population. Note that not all mobilizable plasmids contain antibiotic resistance genes.

Thought Question

9.2 Transfer of an F factor from an F$^+$ cell to an F$^-$ cell converts the recipient to F$^+$. Why doesn't transfer of an Hfr do the same?

***Agrobacterium* practices interdomain conjugation.** You may be surprised to learn that some bacteria can actually transfer genes across biological domains. One striking example is *Agrobacterium tumefaciens*, which causes crown gall disease in plants (**Fig. 9.6A**). This Gram-negative plant pathogen contains a tumor-inducing (Ti) plasmid that can be transferred via conjugation to plant cells. The bacteria detect and swim toward phenolic compounds released from the wound of a damaged plant where they attach

(**Fig. 9.6B**). Subsequent transfer, integration, and expression of the Ti plasmid in the plant cell genome trigger the release of plant hormones that stimulate tumorous growth of the plant. Plant cells within the tumor release amino acid derivatives, called "opines," that the microbe can then use as a source of carbon and nitrogen. The unique mode of action of *A. tumefaciens* has made this bacterium an indispensable tool for plant breeding and enables entirely new (nonplant) genes to be engineered into crops (discussed in Chapter 16).

Note: Conjugation in bacteria and archaea is mechanistically different from conjugation among eukaryotic microbes. Some eukaryotic microbes exchange nuclei through a structure, called a conjugation bridge, very different from that of bacteria (see Chapter 20).

Gene Transfer by Phage Transduction

Bacteriophages harbor their own genomes, distinct from those of their host cells, but could phages also pluck host

A.

ROBERT CALENTINE/VISUALS UNLIMITED

B.

Agrobacterium cell

Plant cell

1 μm

MARTHA HAWES, UNIVERSITY OF ARIZONA

FIGURE 9.6 ■ An example of gene transfer between bacteria and plants. A. Crown gall disease tumor caused by the bacterium *Agrobacterium tumefaciens*. **B.** Electron micrograph of *A. tumefaciens* attached to plant cells.

genes from one cell and move them to another? Norton Zinder (1928–2012) first suspected that bacteriophages could taxi chromosomal DNA between cells. To test this idea, he grew *Salmonella* in two tubes separated by a fine filter through which viruses could pass, but not bacteria. The experiments, reported in 1966, showed that a filterable agent, or virus, could carry genetic material between bacterial strains; direct contact between bacteria was

unnecessary. Thus, bacteriophages can accidentally move bacterial genes between cells as an offshoot of the phage life cycle (see Section 6.4 and **eTopic 7.2**). The process in which bacteriophages carry payloads of host DNA from one cell to another is known as **transduction**. There are two basic types of transduction: generalized and specialized. **Generalized transduction** can take any gene from a donor cell and transfer it to a recipient cell, whereas **specialized transduction** (also known as restricted transduction) can transfer only a few closely linked genes between cells.

How does transduction happen? Bacteriophages capable of generalized transduction have trouble distinguishing their own DNA from that of the host when attempting to package DNA into their capsids, so pieces of bacterial host DNA accidentally become packaged in the phage capsid instead of phage DNA. In the case of *Salmonella* P22 phage, the packaging system recognizes a certain DNA sequence on P22 DNA called a *pac* site. During rolling-circle replication, P22 DNA forms long concatemers containing many P22 genomes arranged in tandem. The *pac* site defines the ends of the phage genome, marking where the packaging system cuts the P22 DNA and starts packaging DNA into the next empty phage head. However, certain DNA sequences on the *Salmonella* chromosome also "look" like *pac* sites. As a result, the packaging system sometimes mistakenly packages host DNA instead of phage DNA.

The result of this mistake is that 1% of the phage particles in any population of P22 contain no phage DNA but carry host DNA plucked from around the chromosome (**Fig. 9.7**). The phages that carry host DNA are called transducing particles. Any single transducing particle will contain only one segment of host DNA, but different particles in the phage population will contain different segments of host DNA. When a transducing particle injects its DNA into a cell, no new phages are made, but the hijacked host DNA can recombine, or exchange, with sequences in the host chromosome of the newly infected cell, thus changing the genetic makeup of the recipient.

Thought Question

9.3 A prophage genome embedded in the chromosome of a lysogenic cell will prevent the replication of any new phage DNA (from the same phage group) that has infected the cell. By preventing the replication of superinfecting phage DNA, the prophage genome protects itself from destruction by lysis. Is a lysogen also protected from generalized transduction carried out by a similar phage? For example, can a P22 lysogen of *Salmonella* be transduced by P22 grown on a different strain of *Salmonella*? Explain why or why not.

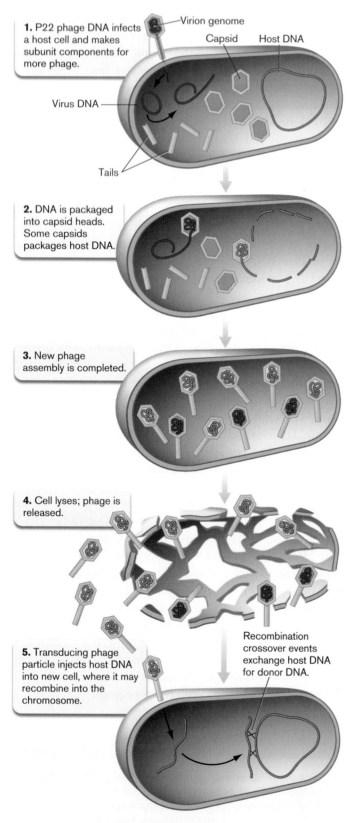

1. P22 phage DNA infects a host cell and makes subunit components for more phage.

Virion genome

Capsid   Host DNA

Virus DNA

Tails

2. DNA is packaged into capsid heads. Some capsids packages host DNA.

3. New phage assembly is completed.

4. Cell lyses; phage is released.

5. Transducing phage particle injects host DNA into new cell, where it may recombine into the chromosome.

Recombination crossover events exchange host DNA for donor DNA.

FIGURE 9.7 ■ Generalized transduction. Generalized transduction by phage vectors can move any segment of donor chromosome to a recipient cell. The number of genes transferred in any one phage capsid is limited, however, to what can fit in the phage head.

Specialized (restricted) transduction is a phage-mediated gene transfer mechanism similar to the F′ factors described previously. In contrast to generalized transduction, specialized transduction can move only a limited number of host genes. *E. coli* phage lambda (discussed in Chapter 11) provides the classic example of specialized transduction. Lambda phage DNA is linear when it first enters the cell; then it circularizes at cohesive ends called *cos* sites. Next, a small DNA sequence in lambda called *attP* can recombine with a similar host DNA sequence (called *attB*) located between the *gal* (galactose catabolism) and *bio* (biotin synthesis) genes on the *E. coli* chromosome (**Fig. 9.8**, step 1). This process, carried out by the phage integrase protein, produces a chromosome with an integrated phage genome, referred to as a prophage, flanked by chimeric *att* sites called *attL* and *attR*. These sites are called "chimeric" because each one is made half from the bacterial and half from the phage *att* sites. Prophage DNA remains latent until something happens to the cell to activate it.

Specialized transduction begins with the reactivation of this prophage DNA, often by DNA damage. Usually, the phage enzymes that excise the lambda DNA do so precisely. The chromosome and viral DNAs are restored to their native states. The viral DNA will then replicate and make more phage particles containing normal phage DNA. On rare occasions, however, improper excision (mediated by host recombination enzymes) can take place between host DNA sequences that lie adjacent to the phage insertion site (*attB* in **Fig. 9.8**, step 2) and similar DNA sequences within the prophage. Improper excision yields a virus that will lack a few viral genes (tail genes missing in **Fig. 9.8**, step 3) but will include host genes lying adjacent to the phage attachment site (the galactose utilization gene *gal* in step 3). With some phages, the specialized transducing particles can replicate unaided. However, in **Figure 9.8** the result is a defective, specialized transducing phage DNA (lambda d*gal*, or λd*gal*). Specialized transducing phage particles that are defective cannot replicate by themselves, but require the presence of a helper phage to supply missing gene products.

Once formed, specialized transducing phages can deliver the hybrid DNA molecule to a new recipient cell. This is the transduction process. Once in that cell, the phage DNA can integrate into the host *attB* site, carrying the donor host gene(s) with it (**Fig. 9.8**, steps 4 and 5). The result will be another partial diploid situation in which the new recipient contains two copies of a host gene: one originally present on its chromosome, and one brought in by the transducing DNA.

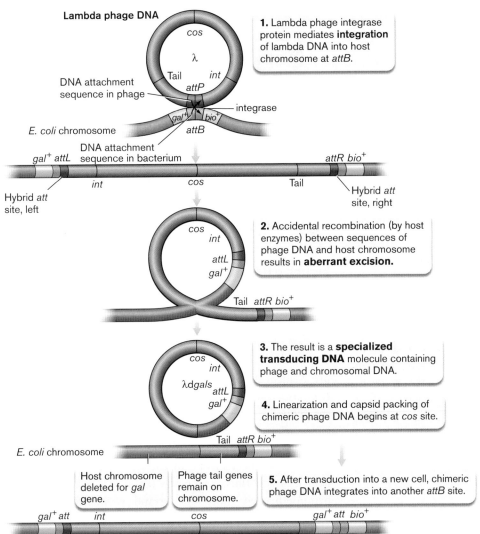

Lambda phage DNA

1. Lambda phage integrase protein mediates **integration** of lambda DNA into host chromosome at *attB*.

2. Accidental recombination (by host enzymes) between sequences of phage DNA and host chromosome results in **aberrant excision.**

3. The result is a **specialized transducing DNA** molecule containing phage and chromosomal DNA.

4. Linearization and capsid packing of chimeric phage DNA begins at *cos* site.

5. After transduction into a new cell, chimeric phage DNA integrates into another *attB* site.

Host chromosome deleted for *gal* gene.

Phage tail genes remain on chromosome.

FIGURE 9.8 ■ **Specialized transduction is restricted to moving host genes flanking the phage attachment site.** The number of genes transferred is limited by the size of the phage head. The resulting recipient, or transductant, chromosome becomes partially diploid for the transferred gene (in this case, *gal*).

DNA Restriction and Modification

There are dangers to the cell associated with the indiscriminate transfer of DNA between bacteria. The most obvious risk involves bacteriophages whose goal is to replicate at the expense of target cells. A bacterium that can digest invading phage DNA while protecting its own chromosome has a far better chance of surviving in nature than do cells unable to make this distinction. As a result, bacteria have developed a kind of "Halt! Who goes there?" approach to gene exchange. It is an imperfect approach, however, that still leaves room for beneficial genetic exchanges.

This protection system, called "restriction and modification," involves the enzymatic cleavage (restriction) of alien DNA and the protective methylation (modification) of self DNA (**Fig. 9.9**). Most bacteria produce DNA **restriction endonucleases** (also known as restriction enzymes), enzymes that recognize specific short DNA sequences

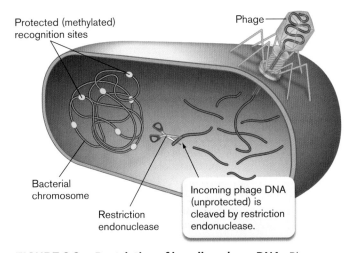

FIGURE 9.9 ■ **Restriction of invading phage DNA.** Phage DNA is injected into a host, where restriction endonucleases can digest it. Host DNA is protected because specific methylations of its own DNA prevent the enzymes from cutting it.

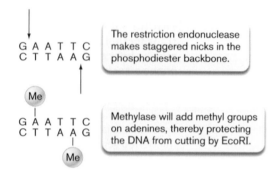

FIGURE 9.10 ■ EcoRI restriction site. EcoRI is an example of a class II restriction-modification system. Shown here are cleavage (top) and methyl modifications (bottom). The DNA sequence shown is specifically recognized by the endonuclease and methylase enzymes.

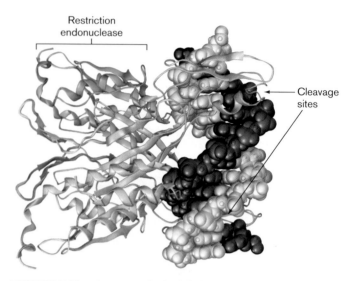

FIGURE 9.11 ■ DNA restriction endonuclease attached to a DNA restriction sequence. The part of the protein shown in gold in this side view is about to cut the dark gray DNA backbone at the top. The light blue part of the protein will cut the light gray DNA at the bottom. (PDB code: 1QRI)

(known as recognition sites) and cleave DNA at or near those sequences (**Fig. 9.10**).

There are three main types of restriction endonucleases (**Table 9.1**), called types I, II, and III. Types I and III restriction endonucleases have their restriction and modification activities combined in one multifunctional protein and cleave DNA some distance away from the recognition site. Type II restriction endonucleases (which are used most often for cloning) possess only endonuclease activity; a separate type II modification protein methylates the same restriction site. Type II restriction endonucleases generally recognize palindromic DNA sequences and cleave at those sites. **Figure 9.11** illustrates how the class II restriction endonuclease EcoRI holds the DNA as it makes a staggered cut at the cleavage (restriction) site.

Curiously, the chromosome of an organism producing a restriction endonuclease also contains target sequences for that enzyme. So how do bacteria avoid "committing suicide" with their own restriction endonucleases? They protect themselves with specific modification enzymes that use *S*-adenosylmethionine to attach methyl groups to the restriction site sequences (see **Fig. 9.10**). Methylation makes the sequence invisible to the cognate (matched) restriction endonuclease (because the modified "A" no longer looks like adenine to the restriction endonuclease). Only one strand of the sequence needs to be methylated to protect the duplex from cleavage; thus, even newly replicated, and consequently hemimethylated (only one strand is methylated), DNA sequences are invisible to the restriction endonuclease.

Note: Restriction endonucleases are named according to the species from which the enzyme was isolated. Thus, EcoRI is an enzyme from *E. coli*. Previously written as *Eco*RI, these enzymes no longer have the first three letters italicized.

TABLE 9.1	Main Types of Restriction-Modification Systems				
Restriction-modification system	**Restriction and modification activities**	**Number of subunits**	**Recognition site characteristics**	**Cleavage and modification sites**	**Examples**
Type I	Present in one multifunctional protein	Three different subunits	5–7 bp, asymmetrical	Located 100 bp or more from recognition site	EcoK in *E. coli*; StyLTIII in *Salmonella enterica*
Type II	Separate methylase and restriction endonuclease enzymes	One or two identical subunits per activity	4–6 bp, palindromic	At or near recognition site	EcoRI in *E. coli*; HindIII in *Haemophilus influenzae*
Type III	Present in one multifunctional protein	Two different subunits	5–7 bp, asymmetrical	Located 24–26 bp from recognition site	Eco571 in *E. coli*; BceSI in *Bacillus cereus*

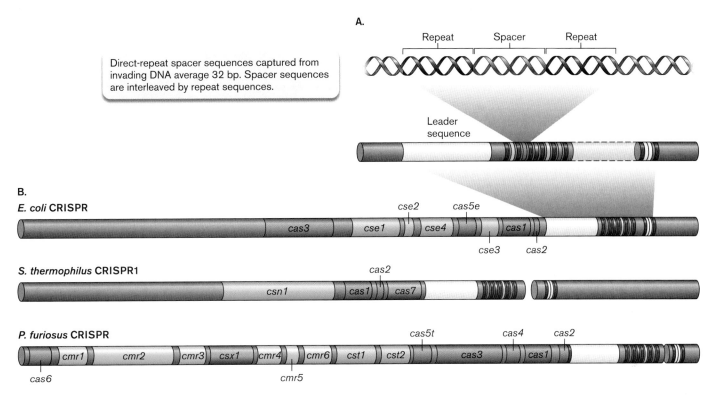

FIGURE 9.12 ▪ **Anatomy of a CRISPR locus.** **A.** Spacer regions derived from invader DNA. The number of repeat-spacer units varies greatly. A conserved leader sequence (gray) of several hundred base pairs is located on one side of the cluster. **B.** CRISPR-associated (*cas*) genes surround the CRISPR locus. Three examples of well-studied CRISPR loci are shown (*Escherichia coli, Streptococcus thermophilus,* and *Pyrococcus furiosus*). Core cas genes are depicted in red, subtype-specific genes in blue, and the RAMP module in green. Unclassified genes are shown in dark gray.

Thought Question

9.4 How do you think phage DNA containing restriction sites evades the restriction-modification screening systems of its host?

CRISPR Interference: Adaptive Immunity of Bacteria and Archaea

One of the most formidable wars on Earth is waged by phages and viruses against Bacteria and Archaea, respectively. Phage predators outnumber their prokaryotic prey 10:1 and destroy between 4% and 50% of prokaryotic life-forms. In an attempt to save themselves, Bacteria and Archaea have developed intriguing defense systems. One system masks receptors for phage attachment. A second uses enzymes to digest invading phage DNA while strategically protecting their own (see the preceding discussion). The third is an adaptive system, first introduced in Chapter 6, called **CRISPR** (clustered regularly interspaced short palindromic repeats), in which an organism that manages to survive a phage attack captures a piece of the invader's genome and wields it as defense against future attack. Nearly 50% of Bacteria and 85% of Archaea possess one or more of these CRISPR loci.

CRISPR anatomy. A CRISPR locus on a bacterial chromosome (**Fig. 9.12**) is composed of short direct-repeat sequences (averaging 32 bp) separated by spacers of uniform length (20–72 bp, depending on the species). Although the sequences of direct repeats are, by definition, nearly identical, the sequences of the spacers are different from one another. Note that repeats and spacers do <u>not</u> encode proteins. Near these sequence clusters lie CRISPR-associated gene families (*cas*) that <u>do</u> encode proteins. A single species can have one or more of these *cas* genes, as well as *cas* subtype genes (called *cse* for *E. coli*). Some CRISPR loci also contain a set of genes called the RAMP (repeat-associated mysterious proteins) module, which includes the gene encoding a putative polymerase.

CRISPR function. Clues about the function of the variable spacer regions were uncovered using bioinformatics, which revealed that some spacers bear sequence homology to bacteriophage or plasmid genes. It turned out that cells harboring these spacers were immune to the corresponding invaders, but related species lacking these spacers were susceptible. Thus, CRISPR is perceived as a primitive microbial immune system.

How does a CRISPR locus work? Bacteria with a CRISPR locus acquire new spacers by incorporating a piece of an

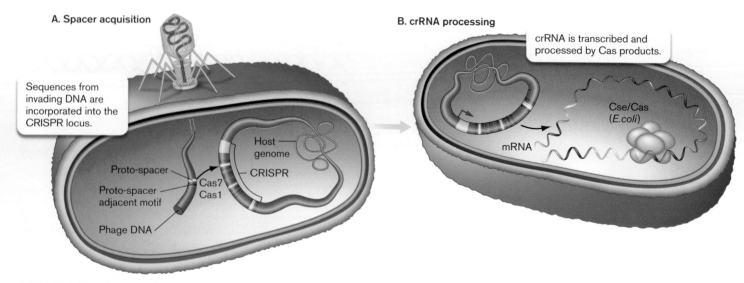

FIGURE 9.13 ■ An overall model of CRISPR/Cas activity. *Source:* Adapted from Karginov, Fedor V., and Gregory J. Hannon. 2010. The CRISPR System: Small RNA-guided defense in bacteria and archaea. *Mol. Cell* **37**:7–19 (Fig. 4).

invader's DNA (**Fig. 9.13A**; spacer acquisition). One or more CRISPR Cas proteins may cleave part of an invading phage genome and integrate it as the lead spacer in the CRISPR region. It is then thought that the CRISPR locus is transcribed starting from an upstream leader sequence. The RNA transcript is then cleaved and trimmed (processed) by some of the Cas products into small RNAs composed of a single spacer sequence (crRNA, also called guide RNA) (**Fig. 9.13B**; crRNA processing stage). Each guide RNA associates with Cas proteins to form a riboprotein complex that binds to a homologous sequence from an infecting phage or plasmid and directs cleavage of the foreign DNA. The process may be aided by RAMP module proteins (**Fig. 9.13C**; effector stage). As a result, the infected cell is spared destruction.

The CRISPR system has also been modified for in vivo genetic engineering purposes (see Chapter 12). Guide RNA molecules can be designed to direct Cas nucleases to cleave almost any in vivo DNA sequence. CRISPR-Cas technology has gained widespread use for modifying eukaryotic genes. Emmanuelle Charpentier, currently at the Helmholtz Centre for Infection Research in Germany, and Jennifer Doudna at UC Berkeley (**Fig. 9.14**) played integral roles in deciphering the CRISPR mechanism.

Novel CRISPR functions. A number of studies have now shown that CRISPR loci contribute to cell function beyond providing immunity from phage infections. One novel function of CRISPR loci, discovered from work in George O'Toole's Dartmouth laboratory, may be to banish a lysogenic cell from a biofilm, so that phage produced by the lysogen cannot kill the rest of the biofilm community. **Figure 9.15** demonstrates this phenomenon with *Pseudomonas aeruginosa*.

A.

B.

FIGURE 9.14 ■ Emmanuelle Charpentier (A) and Jennifer Doudna (B), who performed some of the seminal work on the CRISPR mechanism.

C. Effector stage

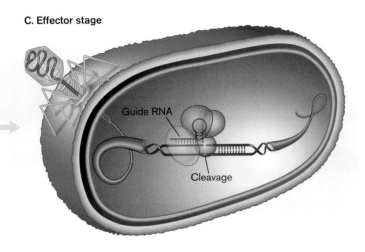

Guide RNA

Cleavage

Figure 9.15A shows that a lysogenized (infected) cell of *P. aeruginosa* does <u>not</u> form a biofilm. If this lysogen were a single cell among many biofilm-producing nonlysogens, then it would not associate with the biofilm. This self-exile saves the biofilm from infection, should the phage become active. The CRISPR locus is essential for this loss, since disruption of the CRISPR or several *cas* genes restores a lysogen's ability to form biofilms (**Fig. 9.15B**). Introducing a wild-type copy of the gene (**Fig. 9.15C**) into the mutant cell once again prevents biofilm formation. Thus, a mechanism in which an infected cell would impose self-exile would save the population. The mechanism itself starts with Cas3 interacting with a spacer derived from the

prophage. The interaction stimulates the emergency SOS DNA repair system (described later), which triggers the expression of some prophage genes that kill the cell.

Another novel function for a CRISPR-Cas system was discovered in the lung pathogen *Legionella pneumophila*, a Gram-negative bacillus that causes a pneumonia known as Legionnaire's disease. In 2015, twelve people in New York City died during an outbreak of Legionnaires' disease after inhaling *L. pneumophila*–contaminated aerosols emitted from a hotel rooftop air-conditioning system. The pathogen persists in these systems by infecting amebas that inhabit the water-cooling towers. Nicholas Cianciotto and Felizza Gunderson at Northwestern University discovered that intracellular *L. pneumophila* highly expresses several *cas* genes when infecting amebas. The scientists also found that by knocking out one of the *cas* genes (*cas2*), they could significantly limit the ability of *Legionella* to infect amebas. Other components of the CRISPR system did not have this effect. The authors hypothesize that Cas2, which is an RNase, could modify the level or structure of another RNA that influences or encodes factors needed for infection. Thus, preventing *cas2* expression could curtail the infection of amebas and disrupt the pathogenesis of *L. pneumophila*.

To Summarize

- **Transformation** is the uptake by living cells of free-floating DNA from dead, lysed cells.

- **Competence** for transformation in some organisms is triggered by a genetically programmed physiological change.

- **Conjugation** is a DNA transfer process mediated by a transferable plasmid that requires cell-cell contact and formation of a protein complex between mating cells.

- **Some bacteria can transfer DNA across phylogenetic domains.** For example, *Agrobacterium tumefaciens* conjugates with plant cells.

- **Transduction** is the process whereby bacteriophages transfer fragments of bacterial DNA from one bacterium to another. In **generalized transduction**, a phage preparation can move any gene in a bacterial genome to another bacterium. In **specialized transduction**, a phage can move only a limited number of bacterial genes.

- **Restriction endonucleases** protect bacteria from invasion by foreign DNA. Restriction-modification enzymes methylate restriction target sites in the host DNA to prevent self-digestion.

- **The CRISPR interference system** is a small interfering RNA system in which an organism captures a piece of an invader's DNA and uses it to fend off future attacks.

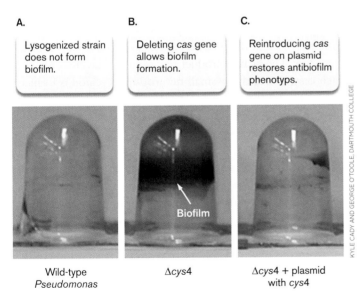

A.	B.	C.
Lysogenized strain does not form biofilm.	Deleting *cas* gene allows biofilm formation.	Reintroducing *cas* gene on plasmid restores antibiofilm phenotyps.

Biofilm

Wild-type *Pseudomonas*   Δ*cys*4   Δ*cys*4 + plasmid with *cys*4

FIGURE 9.15 ▪ The effect of CRISPR on *Pseudomonas aeruginosa* biofilm formation. Shown are upside-down tubes in which *P. aeruginosa* lysogenized with DMS3 phage were grown. Cells that form a biofilm stick to the side of the tube and are not dislodged by washing. The biofilm is revealed after staining with crystal violet. *cys4* is a Cas-encoding gene.

9.2 Recombination

Once a new piece of DNA enters a cell through one of the gene exchange systems, what happens to it? The answer depends on the nature of the acquired DNA. If the DNA is a plasmid, capable of autonomous replication, it can coexist in the cell separate from the host chromosome. If the DNA is incapable of autonomous replication, however, then it may be incorporated into the chromosome through recombination or, as is usually the case, it may be degraded by nucleases.

Two DNA molecules in a single cell can recombine by one of two main mechanisms: generalized or site-specific recombination. **Generalized recombination** requires that the two recombining molecules have a considerable stretch of homologous DNA sequence. **Site-specific recombination**, on the other hand, requires very little sequence homology between the recombining DNA molecules, but it does require a short sequence (10–20 bp) recognized by the recombination enzyme.

Generalized Recombination and RecA

Enzymes participating in generalized recombination are able to align homologous stretches of DNA in two different DNA molecules and catalyze an exchange of strands. The result can be a **cointegrate** molecule in which a single recombination event (that is, a single crossover) joins the two participating DNA molecules. For instance, when two circular DNA molecules are joined by a single recombination event, the two molecules are added together and form one large circular molecule (for an example, see **Fig. 9.4**). A second crossover separated by some distance from the first, however, will lead to an equal exchange of DNA in which neither molecule increases in length.

Why is recombination advantageous? There are three probable functions for generalized recombination in the microbial cell:

- Recombination probably first evolved as an internal method of DNA repair, useful to fix mutations or restart stalled replication forks. This role does not involve foreign DNA.

- Cells with damaged chromosomes use DNA donated by others of the same species to repair their damaged genes.

- Recombination is part of a "self-improvement" program that samples genes from other organisms for an ability to enhance the competitive fitness of the cell.

The central, but by no means only, player in generalized recombination is a protein called RecA. RecA molecules are able to scan DNA molecules for homology and align the homologous regions, forming a triplex DNA molecule, or synapse. Homologs of RecA are found in many other species.

The clearest model of RecA-mediated recombination comes from *E. coli* (**Fig. 9.16** ▶). Before RecA can find homology between two DNA molecules, the donor double-stranded DNA molecule is converted to a single strand by the RecBCD enzyme. The RecBCD complex enters at the end of a DNA fragment and begins to unwind it (**Fig. 9.16**, steps 1 and 2). The complex changes activity when it encounters 8-bp sequences called Chi (<u>c</u>rossover <u>h</u>ot-spot <u>i</u>nstigator) that are scattered throughout most DNA molecules. RecBCD nicks DNA at a Chi site and continues unwinding the strand. RecA then loads onto the single strand as a filament (step 2).

When RecA finds homology between donor and recipient DNA (the minimum required is about 50 bp), the donor single-stranded DNA invades the homologous region in the double-stranded DNA recipient molecule and displaces its like strand. The process is called strand invasion (**Fig. 9.16**, step 3). At this point, a single-strand crossover has been made. The single-strand crossover produces a molecule with four double-stranded ends.

Other recombination proteins, RuvA and RuvB, assemble at the crossover point and extend the invasion in a process called "strand assimilation" or branch migration (**Fig. 9.16**, step 4 and blowup). RuvAB proteins pull matched hybrid strands in opposite directions, thereby extending the base-pairing between homologous donor and recipient strands.

Ultimately, the end of the displaced recipient strand is cleaved (**Fig. 9.16**, step 5) and ligated to the donor strand (step 6). The resulting structure is known as a **Holliday junction**, after the scientist (Robin Holliday) who first proposed it (step 7; also seen in the blowup). The final step is resolution of the crossover Holliday junction. To visualize this, we mentally rotate the right half of the Holliday structure as shown in **Figure 9.16**, step 7. RuvC protein cleaves across the junction (step 8), and the products are ligated to form a complete crossover. The ligated product represents a true crossover. In each case, notice that a small **heteroduplex** region is formed in which one strand comes from the donor and the complementary region comes from the recipient.

If we were viewing the end result of transduction, where the recipient is a circular chromosome and the donor is a double-stranded linear fragment, the recombinant molecule would appear as shown in **Figure 9.16**, step 9. A second crossover farther along the donor strand is then required to maintain circular integrity of the recipient (step 10). This double crossover means that any genes residing between the first and second crossovers are simultaneously exchanged and are said to be genetically linked.

Thought Question

9.5 In a transductional cross between an $A^+B^+C^+$ genotype donor and an $A^-B^-C^-$ genotype recipient, 100 A^+ recombinants were selected. Of those 100, 15% were also B^+, while 75% were C^+. Is gene *B* or gene *C* closer to gene *A*?

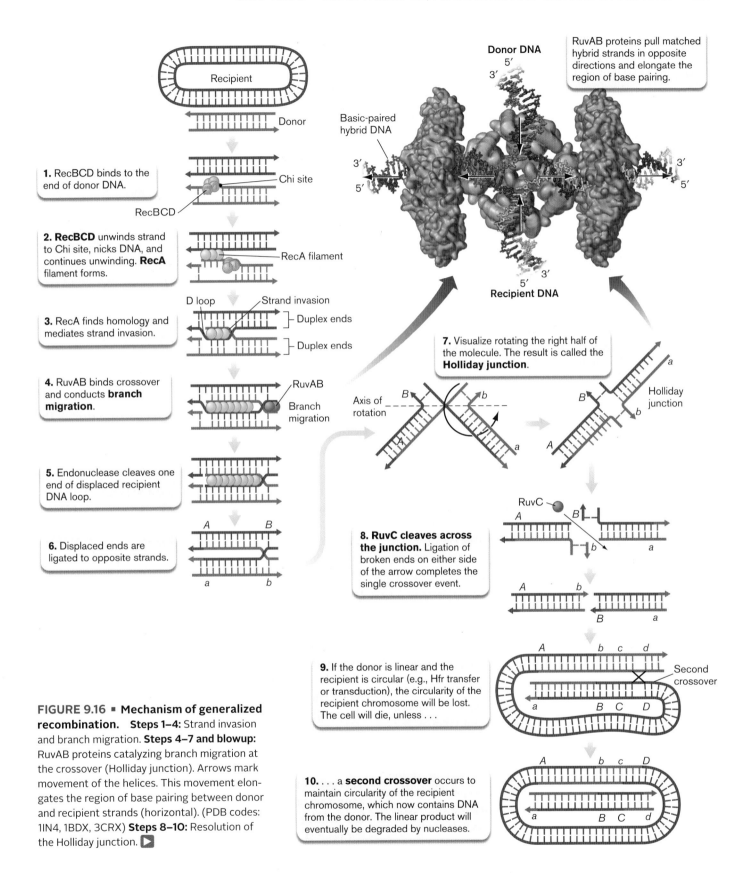

1. RecBCD binds to the end of donor DNA.

2. RecBCD unwinds strand to Chi site, nicks DNA, and continues unwinding. **RecA** filament forms.

3. RecA finds homology and mediates strand invasion.

4. RuvAB binds crossover and conducts **branch migration**.

5. Endonuclease cleaves one end of displaced recipient DNA loop.

6. Displaced ends are ligated to opposite strands.

RuvAB proteins pull matched hybrid strands in opposite directions and elongate the region of base pairing.

7. Visualize rotating the right half of the molecule. The result is called the **Holliday junction**.

8. RuvC cleaves across the junction. Ligation of broken ends on either side of the arrow completes the single crossover event.

9. If the donor is linear and the recipient is circular (e.g., Hfr transfer or transduction), the circularity of the recipient chromosome will be lost. The cell will die, unless . . .

10. . . . a **second crossover** occurs to maintain circularity of the recipient chromosome, which now contains DNA from the donor. The linear product will eventually be degraded by nucleases.

FIGURE 9.16 ■ Mechanism of generalized recombination. Steps 1–4: Strand invasion and branch migration. **Steps 4–7 and blowup:** RuvAB proteins catalyzing branch migration at the crossover (Holliday junction). Arrows mark movement of the helices. This movement elongates the region of base pairing between donor and recipient strands (horizontal). (PDB codes: 1IN4, 1BDX, 3CRX) **Steps 8–10:** Resolution of the Holliday junction. ▶

When trying to understand recombination (and gene exchange overall), it is important to remember that the processes are generally random. The lucky bacterium that receives and recombines the right piece of DNA will benefit. This is why a genetic experiment usually requires screening hundreds of millions of cells to find a few recombinants that grow into visible colonies.

Not all recombination pathways follow the model shown, in which RecA protein forms a filament first on ssDNA and then searches dsDNA for homology. The extremely radiation-resistant species *Deinococcus radiodurans*, for example, does the exact opposite. Its RecA first binds dsDNA and then searches ssDNA for homology. This inverse DNA strand exchange accounts, in part, for the remarkable radiation resistance of this microbe (read more about it in **eTopic 9.3**).

Site-Specific Recombination Is RecA Independent

In contrast to generalized recombination mechanisms that require RecA protein and can recombine any region of the chromosome, site-specific recombination does not utilize RecA and moves only a limited number of genes. This form of recombination involves very short regions of homology between donor and target DNA molecules. Dedicated enzyme systems specifically recognize those sequences and catalyze a crossover between them to produce a cointegrate molecule. The integration of phage lambda, described earlier, is one example of site-specific recombination involving a 15-bp *att* site and the integrase enzyme (see **Fig. 9.8**).

Other examples of site-specific recombination are flagellar **phase variation** in the pathogen *Salmonella enterica* and phase variation in the expression of type I pili in *E. coli*. In these systems, a segment of DNA is inverted (flipped) at a low frequency, resulting in the on-or-off regulation of adjacent gene expression. Phase variation is frequently employed by pathogens to evade the host immune system by changing the expression of cell-surface proteins. For more on flagellar phase variation, see Section 10.1.

To Summarize

- **Recombination** is the process by which DNA sequences can be exchanged between DNA molecules.

- **General recombination** involves large regions of sequence homology between recombining DNA molecules.

- **RecA protein** mediates generalized recombination.

- **The extremely radiation-resistant *Deinococcus radiodurans*** is thought to use RecA protein to patch together homologous ends of fragmented DNA in a way that reconstructs the chromosome after extreme radiation damage.

- **Site-specific recombination** requires little homology between donor and recipient DNA molecules. Site-specific recombination enables phage DNA to integrate into bacterial chromosomes and is a process that can turn on or turn off certain genes, as in flagellar phase variation in *Salmonella*.

9.3 Mutations

What are mutations, and how do they affect evolution? Any permanent, heritable alteration in a DNA sequence, whether harmful, beneficial, or neutral, is called a **mutation**. Two basic requirements are needed to produce a heritable mutation: Namely, there must be a change in the base sequence, and the cell must fail to repair the change before the next round of replication. Repair mechanisms will be discussed in Section 9.4. Although it is tempting to think that all mutations destroy a protein's activity, this is not the case. A given mutation may or may not affect the informational content or phenotype of the organism.

Mutations come in several different physical and structural forms:

- A **point mutation** is a change in a single nucleotide (**Fig. 9.17A** and **B**). Changing a purine to a different purine or a pyrimidine to a different pyrimidine is called a **transition**. Swapping a purine for a pyrimidine (or vice versa) is a **transversion**.

- **Insertions** and **deletions** involve, respectively, the addition or subtraction of one or more nucleotides (**Fig. 9.17C** and **D**), making the sequence either longer or shorter than it was originally.

- An **inversion** results when a fragment of DNA is flipped in orientation relative to DNA on either side (**Fig. 9.17E**).

- A **reversion** restores a sequence altered by mutation to its <u>original</u> sequence.

Mutations can also be categorized into several informational classes (refer to Figure 8.10 for the genetic code). Mutations that do not change the amino acid sequence of a translated open reading frame (ORF) are called **silent mutations**. For example, a point mutation changing TTT to TTC in the sense DNA strand (corresponding to a UUU-to-UUC codon change in mRNA) still codes for phenylalanine. Thus, even though the DNA sequence has changed, the protein sequence remains the same (a synonymous substitution). However, if the UUU codon were changed to UUA (a U-to-A transversion), then the protein would have a leucine where a phenylalanine had been (see **Fig. 9.17A**). This type of mutation is a **missense mutation** because the amino acid sequence of the protein has changed. (Remember: RNA polymerase makes mRNA by reading the DNA template strand, while the complementary, or sense, DNA strand has the same sequence as the mRNA.)

The amino acid substitution resulting from a missense mutation may or may not alter protein function. The outcome depends on the structural importance of the original amino acid and how close in structure the replacement amino acid is to the original. Missense mutations result in either conservative amino acid replacements, in which the new amino acid

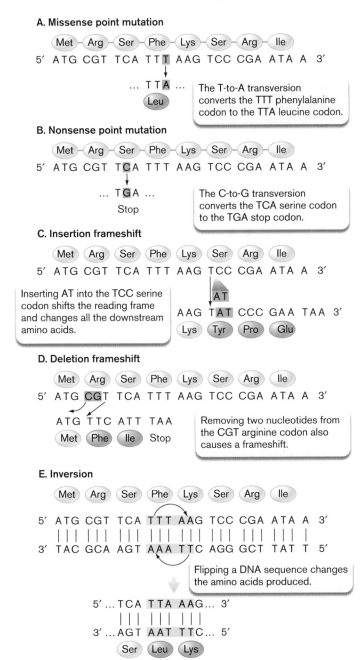

FIGURE 9.17 ■ Changes in a DNA sequence that result in different classes of mutations.

is structurally similar to the original (for example, leucine is substituted for isoleucine), or nonconservative replacements, in which a very different amino acid is substituted (for example, tyrosine for alanine). A missense change may decrease or eliminate the activity of the protein (a **loss-of-function mutation**), or it may make the protein more active. It could even gain a new activity, such as an expanded substrate specificity or a completely different substrate specificity (these are called **gain-of-function mutations**).

A mutation that <u>eliminates</u> function is known as a **knockout mutation** (see **Fig. 9.17B**). Knockout mutations can include multiple-base insertions and deletions, as well as nonsense mutations. A **nonsense mutation** is a point mutation that changes an amino acid codon into a

translation termination codon—for example, UCA (serine) to UAA. The result will be a truncated protein most likely lacking any function—another example of a knockout mutation. Typically, these defective, truncated proteins are degraded by cellular proteases (see Section 8.4).

Insertions and deletions can alter the reading frame of the DNA sequence (see **Fig. 9.17C** and **D**). Remarkable as it is, the ribosome simply reads RNA sequences one codon at a time, stringing amino acids together in the process. It translates each codon "word" but cannot understand the overall protein "sentence." It does not recognize when bases have been added or removed through mutation; instead, it keeps reading the sequence in triplets. If the number of bases inserted or deleted is not a multiple of three, the ribosome will read the wrong triplets. The result is a **frameshift mutation**, which produces a garbled protein "sentence." Frameshift mutations often cause the ribosome to encounter a premature stop codon, originally in a different reading frame. If the insertion or deletion involves multiples of three bases, the reading frame is not changed, but one or more amino acids are added or removed.

An inversion mutation flips a DNA sequence (see **Fig. 9.17E**). Imagine the sequence highlighted in the figure rotating 180° while the adjacent sequences (in black) remain right where they are. The rotation would retain the 5'-to-3' polarity in the new molecule. But if the inversion occurred within a gene, it would likely change the codons in the area and alter the resulting protein. What **Figure 9.17E** shows is a small inversion; however, inversions often involve large tracts of DNA encompassing several genes. If an entire gene with its promoter inverts, the gene will likely remain functional, and its encoded protein may very well still be made. Inversions occur within a genome as a result of recombination events between similar DNA sequences or as a consequence of mobile genetic elements jumping between different areas of a genome.

Mutations can affect both the genotype and the phenotype of an organism. The genotype of an organism reflects its genomic sequences. Regardless of whether a mutation causes a change in a biochemical trait (phenotype), <u>every</u> mutation causes a change in the genotype. In contrast to genotype, phenotype comprises only observable characteristics, such as biochemical, morphological, or growth traits. For instance, the product of *argA* allows an organism to make enough arginine to support growth in media devoid of arginine. Any mutation in *argA* is a genotypic change, but if the result is a synonymous amino acid substitution, then there is no phenotypic change; the organism can still make its own arginine. However, if the mutation destroys the activity of the *argA* product, then the microbe can no longer make arginine, and the amino acid must be supplied in the medium. This is a phenotypic change, too. A mutant that has lost the ability to synthesize a substance required for growth is called an **auxotroph**.

TABLE 9.2	Mutagenic Agents and Their Effects

Mutagenic agent	Effects
Chemical agent	
Base analog *Examples:* caffeine, 5-bromouracil	Substitutes "look-alike" molecule for normal nitrogenous base during DNA replication: point mutation
Alkylating agent *Example:* nitrosoguanidine	Adds alkyl group, such as methyl group ($-CH_3$), to nitrogenous base, resulting in incorrect pairing: point mutation
Deaminating agent *Examples:* nitrous acid, nitrates, nitrites	Removes amino group ($-NH_2$) from nitrogenous base: point mutation
Acridine derivative *Examples:* acridine dyes, quinacrine	Inserts (intercalates) into DNA ladder between backbones to form a new rung, distorting the helix: can cause frameshift mutations
Electromagnetic radiation	
Ultraviolet rays	Link adjacent pyrimidines to each other, as in thymine dimer formation, thereby impairing replication; lethal if not repaired
X-rays and gamma rays	Ionize and break molecules in cells to form free radicals, which in turn break DNA; lethal if not repaired

It is also important to realize that small mutations (such as a single base substitution) can have large effects on phenotype, whereas large mutations (such as the insertion of a 5-kb transposon between two genes) may have little or no effect. To illustrate, consider that a single point mutation in the *hpr* gene of the phosphotransferase sugar transport system (discussed in Section 4.2 and **eTopic 4.1**) will render a bacterium incapable of growing on many sugars, but inserting 5 kb of DNA just past the *hpr* stop codon will have no effect on cell growth. For mutations, it's often not the size that counts; it's the location.

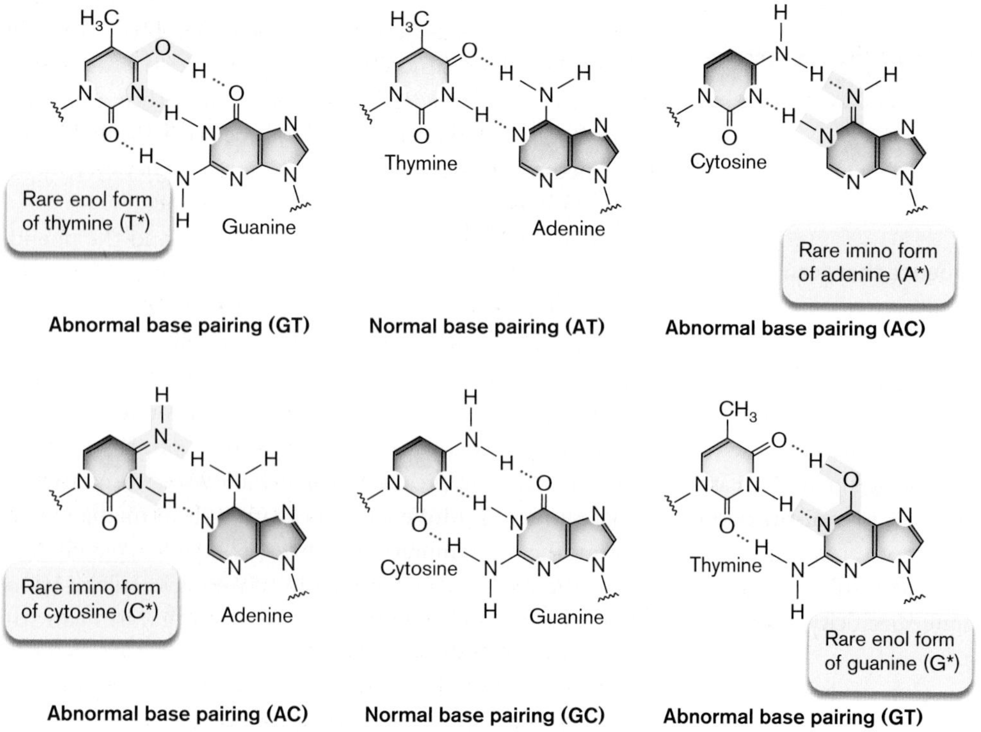

Mutations Arise by Diverse Mechanisms

DNA is susceptible to damage inflicted by a variety of physical and chemical agents. For example, irradiation by X-rays can cause a massive number of DNA strand breaks. When this happens, the integrity of the chromosome is lost and the cell dies. Other agents can directly modify the bases in DNA while leaving its overall structure intact. In this case, the modified bases have altered hydrogen bond base-pairing properties that result in the incorporation of an inappropriate base during replication. When that happens, a mutation results.

Mutations can be caused by **mutagens**, chemical agents that can damage DNA (**Table 9.2**). Even in the absence of a mutagen, though, mutations arise spontaneously. Because DNA proofreading and repair pathways are so efficient, spontaneous mutations are rare, having a frequency of occurrence ranging from 10^{-6} to 10^{-8} per cell division in a given gene (*E. coli*).

Spontaneous mutations in a genome arise for many reasons—for example, tautomeric shifts in the chemical structure of the bases (**Fig. 9.18**). Tautomeric shifts involve a change in the bonding properties of amino ($-NH_2$) and keto ($C=O$) groups. Normally, the amino and keto forms predominate (over 85%), but when an

FIGURE 9.18 ■ **Rare tautomeric forms of bases have altered base-pairing properties.** Tautomeric transitions can lead to permanent mutations.

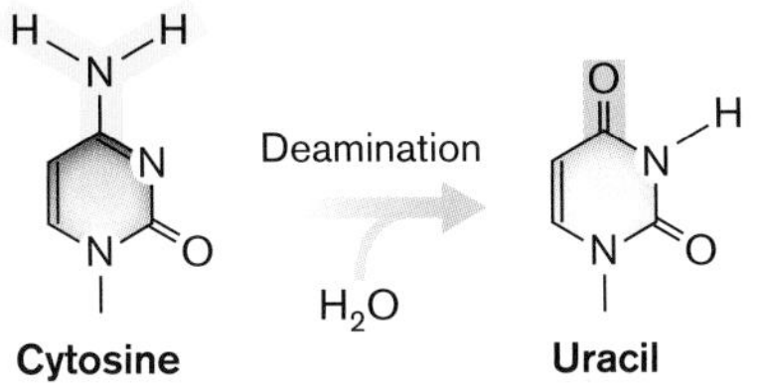

FIGURE 9.19 ■ **Spontaneous deamination of cytosine.** Oxidative deamination changes cytosine to uracil, which will base-pair with adenine. The result is a transition mutation.

amino group shifts to an imino ($=NH$) group, for example, then base-pairing changes. A cytosine that normally base-pairs with guanine will, in its rare imino form, base-pair with adenine. Tautomeric shifts that occur during DNA replication will increase the number of mutational events. Even though the replication apparatus is very accurate with various proof-reading and repair functions, such as the 3′-to-5′ exonuclease activity of DNA polymerase III (see the discussion of DnaQ in Section 7.3), mistakes do occur, albeit at a very low rate.

Besides misincorporation mistakes, naturally occurring intracellular chemical reactions with water can damage DNA. These endogenous reactions are an important source of spontaneous mutations. For example, cytosine spontaneously deaminates to yield uracil (**Fig. 9.19**). The result is a GC-to-AT transition. In addition, purines are particularly susceptible to spontaneous loss from DNA via breakage of the glycosidic bond connecting the base to the sugar backbone (**Fig. 9.20**). The result of this loss is the formation of an **apurinic site** (one missing a purine base) in the DNA. Lack of a purine would obviously hinder transcription and replication.

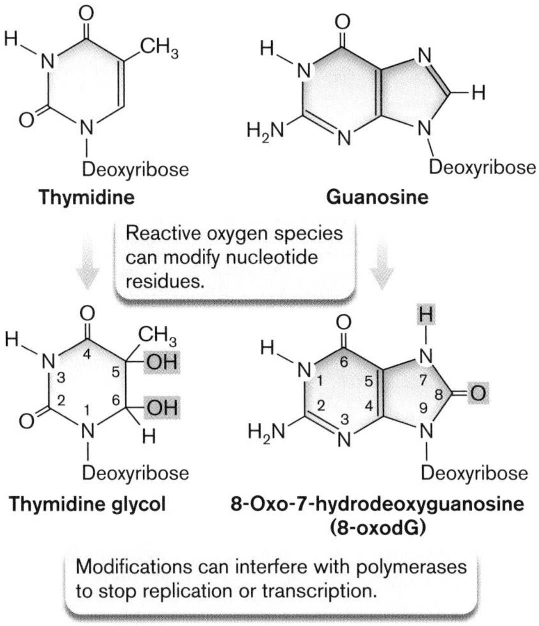

FIGURE 9.21 ■ **Examples of damage caused by reactive oxygen species.** The modifications can interfere with polymerase function and stop replication or interfere with the transcription of affected genes. The blue highlighting identifies modifications to thymidine and guanosine residues.

DNA can also be damaged by metabolic activities of the cell that produce reactive oxygen species, such as hydrogen peroxide (H_2O_2), superoxide radicals ($^{\bullet}O_2^{-}$), and hydroxyl radicals ($^{\bullet}OH$). Even though bacteria have biochemical mechanisms to detoxify reactive oxygen species, the systems can be overwhelmed. Oxidative damage causes the production of thymidine glycol or 8-oxo-7-hydrodeoxyguanosine in DNA (**Fig. 9.21**).

Naturally occurring intracellular methylation agents (for example, S-adenosylmethionine) can spontaneously methylate DNA to produce a variety of altered bases. The spontaneous methylation of the N-7 position of guanine, for example, weakens the glycosidic bond and spontaneously releases the base (forming an apurinic site) or opens the imidazole ring (forming a methylformamide pyrimidine). In addition to mispairing, some of these spontaneous events can lead to major chromosomal rearrangements, such as duplications, inversions, and deletions.

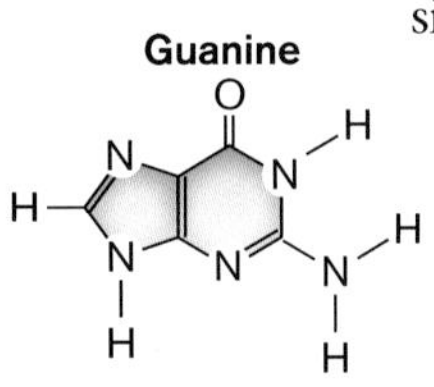

Mutagens tend to increase the mutation rate by increasing the number of mistakes in a DNA molecule, as well as by inducing repair pathways that themselves introduce mutations (error-prone DNA polymerases such as Pol IV and Pol V discussed in Section 9.4).

Ultraviolet (UV) light will produce striking structural alterations in DNA molecules. Pyrimidines (more than purines) are highly susceptible to UV radiation. The energy

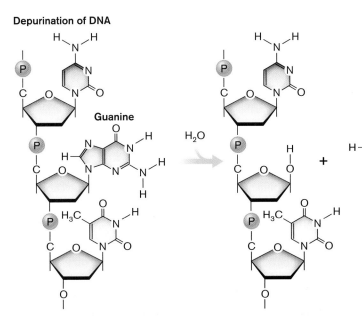

FIGURE 9.20 ■ **Spontaneous formation of an apurinic site.**

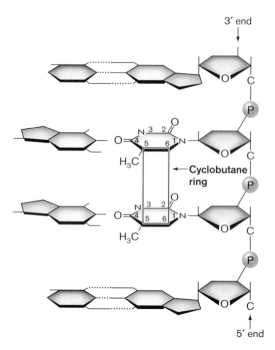

FIGURE 9.22 ■ Production of a pyrimidine dimer. The energy from UV irradiation can be absorbed by pyrimidine molecules. The excited electrons of carbons 5 and 6 on adjacent pyrimidines can then be shared to form a four-membered cyclobutane ring between adjacent pyrimidines. The pyrimidine dimer blocks replication and transcription.

absorbed by a pyrimidine hit with UV light boosts the energy of its electrons to the point where the molecule is unstable. If two pyrimidines are neighbors on a single DNA strand, their energized electrons can react to form a four-membered cyclobutane ring. The result is a pyrimidine dimer that will block replication and transcription (**Fig. 9.22**).

Although DNA can be damaged in numerous ways, the cell can repair that damage before it becomes fixed as a mutation. But the repair mechanisms are not perfect. Repair errors contribute heavily to the formation of heritable mutations and, thus, to evolution. Although most mutations decrease the fitness of a species, some can improve fitness by, among other things, enabling an organism to more efficiently use or compete for available food sources.

Mutation Rates and Frequencies

The **mutation rate** for a given gene is often defined as the number of mutations formed per cell division. Since the number of cells in a typical culture starts out very low and grows to a very high density, the number of cell divisions is approximately equal to the number of cells in the culture. For example, to go from one cell to eight cells requires seven cell divisions (not generations). Mutation rate can be estimated according to the formula $a = m$ per number of cell divisions, where a is the mutation rate and m is the number of mutations that occur as the number

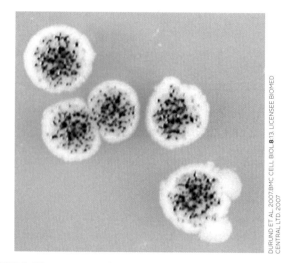

FIGURE 9.23 ■ Sectored colonies of bacteria. This *E. coli* mutator strain is a lactose-negative mutant that does not ferment lactose and thus does not turn the indicator blue. But because the strain has a mutation in one of the mutator genes, there is a high rate of reversion to the lactose fermenter strain, which shows up as blue papillae on the white colony.

of cells increases from N_0 to N. Thus, a culture growing from 10^3 (1,000) cells to 10^7 (10,000,000) cells requires approximately 10^7 cell divisions (10,000,000 − 1,000 = 9,999,000 cell divisions). If 10 mutants developed in a specific gene, then the mutation rate would be approximately $10/10^7 = 1 \times 10^{-6}$. More on mutation rate can be found in **eTopic 9.4**.

Note: Distinguish between the number of cell divisions, the number of generations, and the number of cell generations. For instance, the replication of 1,000 cells to give 2,000 cells requires 1,000 cell divisions but takes place over only one generation (meaning one doubling). You will sometimes see the phrase mutation rate "per <u>cell</u> generation," which is essentially equivalent to "per cell division."

A less complicated calculation is **mutation frequency**. Mutation frequency tells us how many mutant cells are present in a population. It is calculated simply as the ratio of mutants per total cells in the population.

The colonies in **Figure 9.23** graphically illustrate the result of a high mutation rate. The example shows an *E. coli* mutator strain that has a point mutation in the lactose operon. It will not ferment lactose. The medium contains lactose and peptides as carbon sources, and an indicator that turns blue if the organism can ferment lactose. This medium helps scientists follow the development of mutations in the gene needed to ferment lactose. Because the original cell in this case was Lac⁻ (due to a mutation in that gene), most of the colony is white. However, the strain is a mutator strain defective in a DNA repair gene (see Section 9.4). This defect causes a high rate of reversion to Lac⁺. Each cell in the growing colony that reverted to become a lactose fermenter produced offspring that can turn the indicator blue. These mutants appear as blue papillae (nipple-like outgrowths) on the surface of the white colony. Each papilla represents a separate mutational event.

Thought Question

9.6 You have been asked to calculate the mutation rate for a gene in *Salmonella enterica*, so you dilute an 18-hour broth culture to 1×10^4 cells/ml and let it grow to 1×10^9 cells/ml. At that point you dilute and plate cells on the appropriate agar medium and estimate that there were 10,000,000 mutants in the culture. What you don't know is that the original dilution of 1×10^4 cells/ml already had 10 mutants defective in that gene. Does the presence of those 10 mutants affect your estimate of the mutation rate? If so, then how?

Identifying Mutagens Using Bacterial "Guinea Pigs"

In a world where we are continually exposed to new chemicals, it is important to determine which chemicals are potential mutagens. Bruce Ames and his colleagues invented a simple alternative that uses bacteria as a rapid initial screen, for which Ames received the National Medal of Science in 1998. The method, called the Ames test, relies on a mutant of *Salmonella enterica* that is defective in the *hisG* gene, whose product is involved in histidine biosynthesis (**Fig. 9.24**). The *hisG* mutant cannot grow on a minimal defined medium lacking histidine. However, if a reversion mutation occurs in the *hisG* gene and <u>restores</u> the gene to its original functional state (phenotypic reversion),

the new mutant cell will form a colony even in the absence of histidine. This method is called a reversion test.

Ames used this *his* reversion test to screen compounds for potential mutagenicity. If a chemical can mutate one gene (*hisG*), it can potentially affect any gene. As shown in **Figure 9.24**, a mutagen-containing disk is placed in the middle of an agar plate spread with the original *his* mutant. As the mutagen diffuses into the medium and causes reversion mutations, colonies start to appear, forming a ring around the disk. The longer the plate is incubated, the more revertants are produced.

The plate reversion test goes only so far, however. The technique will not expose potential mutagens that require processing by mammalian enzymes to become mutagenic. In humans, many nonmutagenic chemicals can be transformed into potent mutagens by the liver. The liver is the chief organ for detoxifying the body—a task that liver enzymes accomplish by chemically modifying foreign substances. To expose the mutagenic potential of a chemical, the chemical is treated with a rat liver extract before the mixture is applied to bacteria for the *his* reversion test (**Fig. 9.25**). If the enzymes of

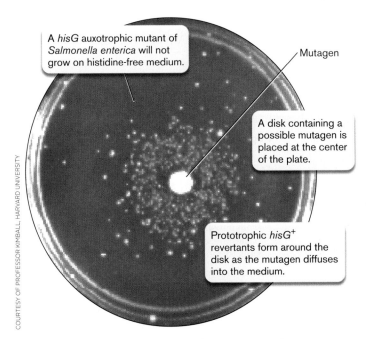

FIGURE 9.24 ■ Basic Ames test for mutagenesis. *Salmonella enterica hisG* mutant was spread uniformly over the surface of a histidine-free medium.

A *hisG* auxotrophic mutant of *Salmonella enterica* will not grow on histidine-free medium.

Mutagen

A disk containing a possible mutagen is placed at the center of the plate.

Prototrophic *hisG*+ revertants form around the disk as the mutagen diffuses into the medium.

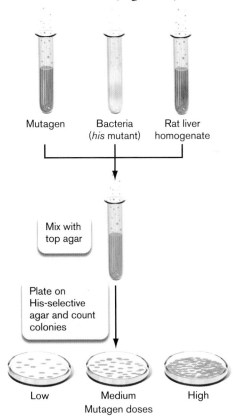

Mutagen

Bacteria (*his* mutant)

Rat liver homogenate

Mix with top agar

Plate on His-selective agar and count colonies

Low

Medium

High

Mutagen doses

FIGURE 9.25 ■ The modified Ames assay to test for the mutagenic properties of chemicals processed through the liver. The potential mutagen, *his*-mutant bacteria, and liver homogenate are combined and mixed with agar. The combination is poured into a petri plate. If the liver extract enzymes act on the test compound and the metabolites produced are mutagenic, then increasing numbers of His+ revertants will be observed with increasing doses of mutagen. If the compound is <u>not</u> mutagenic, few relevant colonies will be seen on any plate.

the liver convert the compound to a mutagen (they do this accidentally, not on purpose), then His⁺ revertant colonies will be seen. This method accelerates the process of drug discovery by providing an inexpensive preliminary screen for weeding out mutagenic chemicals before more expensive animal testing is undertaken.

More involved assays for mutagenicity involve transgenic mice. Mice are engineered to have the bacterial *lacZ* gene, for instance, inserted into their chromosomes. The *lacZ* gene encodes an enzyme that is easily assayed, so it is useful as a reporter for mutagenic activity. This reporter gene will be distributed throughout the mouse in all organs. A potential mutagen is administered to the mouse, and after some time the mouse is sacrificed, its organs are removed, and the DNA is extracted and examined for mutations arising in the *lacZ* gene. This approach allows scientists to track where the mutagen is distributed in the mouse and to determine whether certain organs convert harmless precursor chemicals into dangerous mutagenic compounds.

To Summarize

■ **A mutation** is any heritable change in DNA sequence, regardless of whether a change in gene function results.

■ **Genotype** reflects the genetic makeup of an organism, whereas **phenotype** reflects its physical traits.

■ **Classes of mutations** include silent mutations, missense mutations, nonsense mutations, point mutations, insertions, deletions, and frameshift mutations.

■ **Spontaneous mutations** reflect tautomeric shifts in DNA nucleotides during replication, accidental incorporation of noncomplementary nucleotides during

replication, or "natural" levels of chemical or physical (irradiation) mutagens in the environment.

■ **Chemical mutagens** can alter purine and pyrimidine structure and change base-pairing properties.

■ **The mutagenicity** of a chemical can be assessed by its effect on bacterial cultures.

9.4 DNA Repair

Microorganisms are equipped with a variety of molecular tools that repair DNA damage before the damage becomes a heritable mutation (see **Table 9.3**). The type of repair mechanism used (and when it is used) depends on two things: the type of mutation needing repair and the extent of damage. Some repair mechanisms are error proof and do not introduce mutations; others are error prone and require "emergency" DNA polymerases expressed under dire circumstances. These polymerases sacrifice replication accuracy to rescue the damaged genome. Whether damage is introduced by mutagens or by inaccurate DNA synthesis, microbial survival depends on the ability to repair DNA.

We will first discuss the **error-proof repair** pathways that prevent mutations. These include methyl mismatch repair, photoreactivation, nucleotide excision repair, base excision repair, and recombinational repair. Then we will turn our attention to **error-prone repair** pathways. These pathways risk introducing mutations and operate only when damage is so severe that the cell has no other choice but to die. Realize that many repair enzymes are in short supply. **Special Topic 9.1** describes a new DNA wire model for how some important DNA repair enzymes locate DNA damage.

TABLE 9.3	**Types of DNA Repair**			
System	**Genes**	**Mutations recognized**	**Repair mechanism**	**Result**
Photoreactivation	*phrB*	Pyrimidine dimers	Cyclobutane ring cleaved	Accurate
Nucleotide excision	*uvrABCD*	Helical destabilization (e.g., pyrimidine dimers)	Patch of nucleotides excised	Accurate
Base excision	*fpg, ung, tag, mutY, nfo*	Various modified bases	Glycosylases remove base from phosphodiester backbone; apurinic (AP) sites formed	Accurate
Methyl mismatch	*mutHSL, dam*	Transitions, transversions	Nick on nonmethylated strand; excision of nucleotides	Accurate
Recombination	*recA*	Single-strand gaps	Recombination	Accurate
Translesion bypass synthesis	*umuDC*	Gaps	Part of SOS system	Error prone (generates mutations)

Error-Proof Repair Pathways

Methyl mismatch repair. What happens if DNA polymerase simply makes a mistake and incorporates a normal but incorrect base? The inherent error rate of DNA polymerase III is approximately one mistake per 10^8 bases synthesized, after proofreading. However, the mutation rate in a live cell is actually only 10^{-10} per base pair replicated. But how can a cell repair a mutation after it has already been introduced during replication? One method is **methyl mismatch repair**, which is based on repair enzymes recognizing the methylation pattern in DNA bases. As discussed in Section 7.3, many bacteria tag their parental DNA by methylating it at specific sites. In *E. coli*, for example, deoxyadenosine methylase (Dam) methylates the palindromic sequence GATC to produce GA^{ME}TC. The Dam methylase does this soon, but not immediately, after replication of a DNA sequence.

Misincorporation of a base during replication produces a mismatch between the incorrect base in the newly synthesized but <u>unmethylated</u> strand and the correct base residing in the parental, methylated strand. Methyl-directed mismatch repair enzymes (MutS, MutL, and MutH) bind to the mismatch. MutS first binds to the mismatch and recruits MutL and MutH (**Fig. 9.26** ▶, steps 1 and 2). MutL recognizes the methylated strand (GA^{ME}TC) and brings it in a loop to meet MutS and MutH (steps 3 and 4). Then, MutH cleaves the unmethylated strand containing the mutation, near the GATC sequence (step 5). A DNA helicase called UvrD then unwinds the cleaved strand, exposing it to a variety of exonucleases (step 6). The result is a gap that is filled in by DNA polymerase I (Pol I) and sealed by DNA ligase.

The methyl-directed mismatch repair proteins (and genes) are called Mut (and *mut*) because a high mutation rate results in strains that are defective in one of these proteins. A bacterial strain with a high mutation rate is called a **mutator strain**.

Note: Not all DNA methylations signal methyl mismatch repair. DNA methylation by restriction-modification systems, for example, prevents cleavage by DNA restriction endonucleases, but is not used to designate parental DNA after replication.

Thought Question

9.7 During transformation in *Streptococcus pneumoniae*, a single strand of donor DNA is transferred into the recipient. This single strand of DNA can be recombined into the recipient's genome at the homologous area. However, if the segment of recipient DNA had a mutation in it relative to the donor DNA strand, then the result would be a mismatch after recombination. Explain how, in the face of mismatch repair, the recipient cell ever retains the wild-type sequence from the donor.

Photoreactivation and nucleotide excision. Other DNA repair mechanisms do not distinguish between parental and newly synthesized strands. Many microbes in the environment commonly encounter ultraviolet light. The pyrimidine dimers that form as a result of ultraviolet irradiation

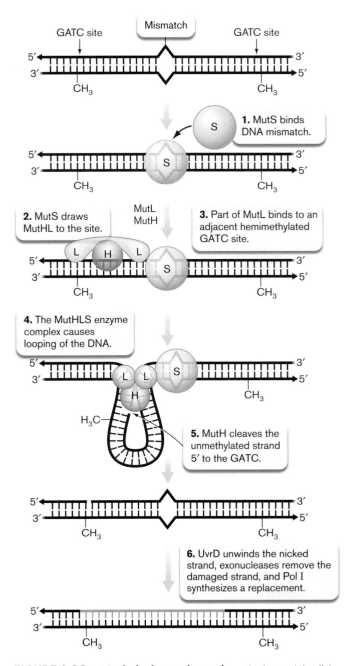

FIGURE 9.26 ■ Methyl mismatch repair. The bacterial cell, in this case *E. coli*, can use specific methylations on DNA to recognize parental DNA strands for preferential DNA repair. Newly replicated strands are not immediately methylated. So, when a mismatch is found, the mismatch repair system views the newly synthesized strand as suspect and replaces the section of unmethylated DNA encompassing the mismatch. The steps following cleavage are similar to what is shown in Figure 9.27 (steps 7 and 8). ▶

| SPECIAL TOPIC 9.1 | DNA as a Live Wire: Using Electrons to Find DNA Damage |

How do DNA repair proteins find DNA damage? Do they randomly float in the cytoplasm until they happen to bump into an abasic site or a mispaired base? Or do they bind DNA and slide along its length until running into something? Jacqueline Barton from the California Institute of Technology (**Fig. 1**) thinks that neither option is likely for repair proteins containing [4Fe-4S] groups. The [4Fe-4S] group is a tetrahedral cluster (**Fig. 2**, inset) with important roles in oxidation-reduction chemistry involving electron transfer. The position of [4Fe-4S] clusters in proteins is coordinated by four cysteine residues that bind to the Fe molecules.

How might a redox center help repair proteins locate DNA damage? Many DNA repair enzymes are in very short supply. The DNA glycosylase MutY in *E. coli*, for instance, is present at only about 30 proteins per cell. It would take a long time for those few molecules to scan that much DNA. It would be nice if different repair proteins could collaborate in some way to scan large stretches of DNA for defects. Barton and her collaborators think that is exactly what happens. Their

FIGURE 1 ■ **Jacqueline Barton (center).**

research indicates that repair enzymes with [4Fe-4S] clusters communicate with each other through long-distance oxidation-reduction reactions using DNA as a "wire." Barton was awarded the National Medal of Science in 2011 for her discoveries.

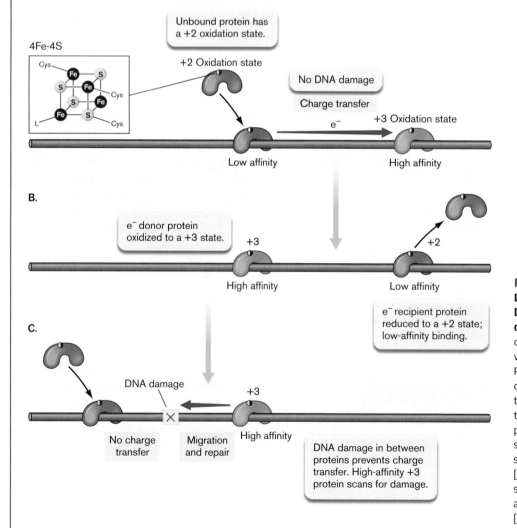

FIGURE 2 ■ **Model of how DNA charge transport helps DNA repair proteins find DNA damage. A, B.** Charge transport occurs between repair proteins when there is no DNA damage. Repair proteins cycle on and off undamaged DNA. **C.** Charge transport fails when there is damage to intervening DNA between repair proteins. High-affinity (+3 oxidation state) proteins accumulate near the site of damage. The redox-active [4Fe-4S] clusters in the proteins are signified by clusters of two orange and two yellow spheres. **Inset:** [4Fe-4S] cluster.

The stacked core of aromatic bases that extends down the helical axis of DNA allows DNA to conduct electrons, but only if the DNA is intact. DNA damage destroys the integrity of the wire. Repair proteins with [4Fe-4S] clusters that bind DNA can pass an electron along an undamaged DNA "wire" (a process called DNA charge transport, or CT) to another protein with a [4Fe-4S] cluster. Here is how the model works (**Fig. 2**):

- The [4Fe-4S] cluster in unbound repair proteins has a +2 oxidation state (purple).
- When bound to DNA, the +2 cluster is oxidized to the +3 state (turquoise), but <u>only</u> if an electron is transferred along DNA to another [4Fe-4S] DNA-bound repair protein. The connecting DNA must be free of damage or the electron will not transfer.
- The recipient protein then converts (is reduced) from +3 back to a +2 oxidation state.
- The +2 recipient protein now has a lower DNA-binding affinity and dissociates, only to bind DNA again elsewhere. The process of scanning by charge transport then repeats from the new location.
- If the repair protein binds near damaged DNA (a "broken wire"), it cannot transfer an electron to a downstream +3 recipient protein. The bound, oxidized (+3) repair protein migrates two-dimensionally to find and repair the damage.

The [4Fe-4S] repair proteins use charge transport, or rather the lack of it, to accumulate in the vicinity of DNA damage.

Figure 3 provides evidence that the system works in cells. The experiment starts with a mutant strain of *E. coli* (called InvA) that requires DinG protein to grow. DinG is a DNA helicase that has a [4Fe-4S] cluster. This protein unwinds DNA-RNA hybrids called R loops that accumulate at stalled replication forks. An InvA mutant that lacks the *dinG* gene has a severe growth defect because of these R loops (not shown). The question Barton's group asked was whether another [4Fe-4S]–containing repair enzyme, Endo III, would help DinG find R-loop damage. In the experiment, growth (turbidity) means that DinG successfully found and removed R loops. Conversely, no growth means that DinG did not efficiently find R loops. If the hypothesis is correct and Endo III charge transport function helps DinG find R loops, then an InvA strain carrying an Endo III protein that retains CT function would be expected to grow (tubes 1, 5, and 6). Any InvA strain without Endo III or that carried Endo III lacking CT function would not grow (tubes 2, 3, and 4). The endonuclease catalytic activity of Endo III, however, would not be important (tube 5). The results in **Figure 3** match the prediction and indicate that Endo III charge transport helped DinG find R loops.

DNA charge transport seems to be an efficient way to redistribute repair proteins to sites of DNA lesions for repair

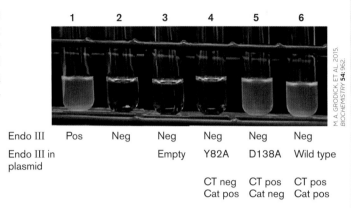

	1	2	3	4	5	6
Endo III	Pos	Neg	Neg	Neg	Neg	Neg
Endo III in plasmid			Empty	Y82A	D138A	Wild type
				CT neg	CT pos	CT pos
				Cat pos	Cat neg	Cat pos

M. A. GRODICK, ET AL. 2015. *BIOCHEMISTRY* **54**: 962.

FIGURE 3 ▪ **Charge transport (CT) works in vivo.** All tubes contain an InvA mutant dependent on DinG for growth. Growth indicates that CT-proficient Endo III helps guide DinG to R-loop damage at stalled replication forks. Some strains (tubes 2–6) lack the chromosomal gene for Endo III, but certain Endo III mutants contain plasmids that encode replacement Endo III proteins. CT neg or CT pos = mutant Endo III that is, respectively, deficient or proficient in charge transport; Cat neg or Cat pos = mutant Endo III that is, respectively, defective or functional in catalytic activity.

and could provide a general way for any DNA-bound [4Fe-4S] proteins to signal across the genome.

RESEARCH QUESTION

Atomic force microscopy can identify proteins bound to DNA molecules. Describe what you would observe, given the following: a mixture of DNA molecules of two different lengths. The larger DNA molecules are undamaged, while the smaller DNA molecules have a base pair mismatch in the middle. Immediately after adding Endo III to the mix, the protein is seen bound to all molecules. What would you observe after an incubation period?

Grodick, M. A., N. B. Muren, and J. K. Barton. 2015. DNA charge transport within the cell. *Biochemistry* **54**:962–973.

can be repaired by a light-activated mechanism called **photoreactivation**. In photoreactivation, the enzyme photolyase binds to the dimer and cleaves the cyclobutane ring linking the two adjacent, damaged nucleotides. The damage is repaired without any bases being excised.

While photolyase is specific for pyrimidine dimer repair and requires light, a system called **nucleotide excision repair** (**NER**) operates in the dark (or light) and is used to excise other kinds of damaged DNA, as well as pyrimidine dimers. In this system, a three-subunit endonuclease (UvrABC) excises a patch of 12–13 nucleotides that includes the dimer (**Fig. 9.27** ▶). The basic mechanism involves UvrA, associated with UvrB, recognizing the damaged base (**Fig. 9.27**, steps 1–3). UvrA is ejected (step 4), and UvrB binds UvrC (step 5), which actually cleaves the damaged strand at two sites flanking the damaged base (step 6). The small fragment is removed, and the gap is repaired by DNA polymerase I (steps 7 and 8). Besides pyrimidine dimers, NER can repair damage such as O^6-methylguanine, N^2-guanine, N^3-adenine, and other modified bases that significantly alter base pairing and helix conformation. Nucleotide excision repair is also used to excise other kinds of damaged DNA besides pyrimidine dimers.

Nucleotide excision repair can be enhanced by ongoing RNA transcription. Bacteria and eukaryotic cells preferentially repair genes that are being transcribed. The preferential repair of transcriptionally active genes makes sense, since a cell unable to transcribe such a gene because of recent DNA damage would be at a survival disadvantage. An RNA polymerase that encounters a UV dimer or other unrecognizable base on a template DNA strand will stall during transcription. The stalled RNA polymerase is then recognized by a protein that mediates a process called transcription-coupled repair (TCR). One TCR pathway dislodges RNA polymerase (Mfd dependent); the other pushes RNA polymerase back from the lesion (UvrD and NusA dependent). The transcription-coupled repair protein then snares a nearby UvrAB to begin nucleotide excision repair.

Base excision repair snips damaged bases from DNA. Although nucleotide excision repair can recognize and repair several types of damage, it does not work in all instances. Another error-proof process, known as **base excision repair** (**BER**), employs a battery of glycosylase enzymes that can recognize and clip certain damaged bases from the phosphodiester backbone. Uracil DNA glycosylase, hypoxanthine DNA glycosylase, and 3-methyladenine glycosylase recognize, respectively, uracil, hypoxanthine, and 3-methyladenine when these bases are present in DNA. Uracil can be found in DNA either as a spontaneous deamination product of cytosine (see **Fig. 9.19**) or as a result of inappropriate incorporation during replication. Spontaneous deamination of adenine residues produces hypoxanthine. Uracil

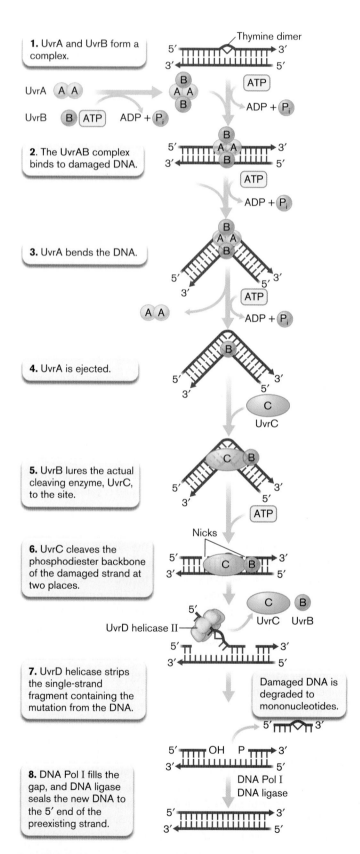

FIGURE 9.27 ▪ Nucleotide excision repair. This repair mechanism cuts out a single-stranded segment of DNA containing the mutation and uses DNA replication to repair the gap. ▶

and hypoxanthine have base-pairing properties very different from those of the original bases, so their formation can lead to mutations. Mutagens such as methyl methanesulfonate can alkylate adenine to produce the adduct 3-methyladenine. A 3-methyladenine residue will totally block DNA replication, making this a lethal form of damage. Consequently, repairing these mutations is vital for survival.

The glycosylases just noted cleave the bond connecting the damaged base to deoxyribose in the phosphodiester backbone (**Fig. 9.28** ▶, step 1). The result is an intact phosphodiester backbone missing a base (an abasic site). The site is called an **AP site** because it is missing either a purine (apurinic) or a pyrimidine (apyrimidinic). The next step in base excision repair involves AP endonucleases that specifically cleave the phosphodiester backbone at AP sites (step 2). The 5′-to-3′ exonuclease activity of the gap-filling DNA polymerase I will degrade the cleaved strand downstream of the AP site and at the same time synthesize in its stead a replacement strand containing the proper base (step 3). DNA ligase seals the remaining nick, and the repair process is complete (step 4).

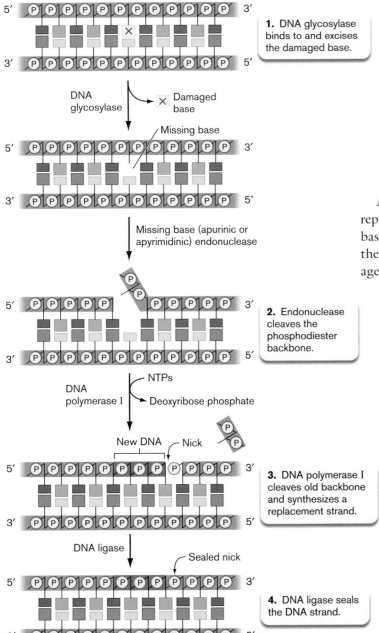

1. DNA glycosylase binds to and excises the damaged base.

2. Endonuclease cleaves the phosphodiester backbone.

3. DNA polymerase I cleaves old backbone and synthesizes a replacement strand.

4. DNA ligase seals the DNA strand.

FIGURE 9.28 ■ Base excision repair. Specialized enzymes remove damaged bases from DNA without breaking the phosphodiester backbone. NTPs = nucleoside triphosphates. ▶

Thought Question

9.8 It has been reported that hypermutable bacterial strains are overrepresented in clinical isolates. Out of 500 isolates of *Haemophilus influenzae*, for example, 2%–3% were mutator strains having mutation rates 100–1,000 times higher than lab reference strains. Why might mutator strains be beneficial to pathogens?

All of the repair processes described thus far (mismatch repair, photoreactivation, nucleotide excision repair, and base excision repair) are error-proof pathways that rely on the presence of a good template strand opposite the damaged strand. Replication of the template strand is used to replace the damaged bases accurately. But what happens when both strands are damaged or when there is only one strand?

Recombinational repair. Figure 9.29 illustrates one repair mechanism that engages when DNA replication takes place before a UV-induced dimer can be excised by nucleotide excision repair. Because DNA polymerase III (the main polymerase for chromosome replication) cannot decode a dimer, it skips over the damaged area and restarts with a new RNA primer, leaving a gap opposite the dimer. At this point, the Uvr complex cannot excise the UV dimer, because chromosome integrity would be lost. However, the RecA protein involved with recombination can bind to the gap and initiate a genetic exchange in which a piece of the undamaged, properly replicated strand is spliced into the gap. This is called **recombinational repair** and happens mostly near replication forks. Once the gap is filled, the Uvr complex can remove the dimer, and DNA polymerase I will fill the gap in the other strand.

The recombinational repair system is also an error-proof repair pathway. Note that the gap formed in the donor

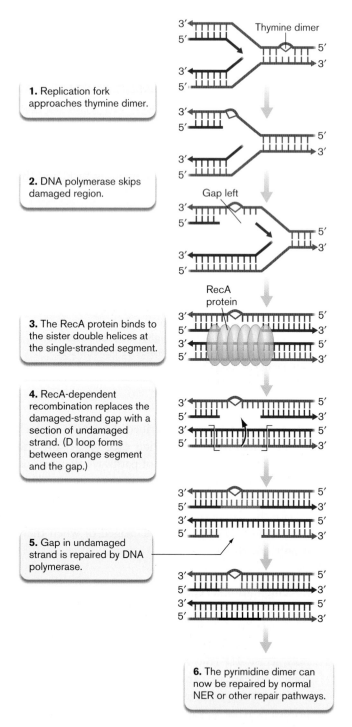

1. Replication fork approaches thymine dimer.

2. DNA polymerase skips damaged region.

3. The RecA protein binds to the sister double helices at the single-stranded segment.

4. RecA-dependent recombination replaces the damaged-strand gap with a section of undamaged strand. (D loop forms between orange segment and the gap.)

5. Gap in undamaged strand is repaired by DNA polymerase.

6. The pyrimidine dimer can now be repaired by normal NER or other repair pathways.

FIGURE 9.29 ■ Recombinational repair. Recombination can be used to repair DNA damage of replicating DNA when one daughter strand is undamaged. A single-stranded segment of the undamaged daughter strand can be used to replace a gap in the damaged daughter strand.

strand is easily replaced by the gap-filling DNA polymerase I. Recombinational repair is not limited to pyrimidine dimers. It will work on any damage that causes gaps during replication.

Error-Prone DNA Repair

SOS ("Save Our Ship") repair. When DNA damage is extensive, repair strategies that excise pieces of DNA or that require recombination will destroy the circularity of the chromosome and kill the cell. To save the chromosome, and itself, the cell must take more drastic measures and induce the **SOS response**, a system that introduces mutations into severely damaged DNA. The cell relaxes replication fidelity to maintain a circular chromosome even if incorrect bases are introduced. In the SOS system, the RecA protein, normally located at the cell poles, senses the extent of DNA damage by monitoring the level of single-stranded DNA produced. For example, excessive ultraviolet irradiation produces numerous ssDNA gaps because DNA polymerase cannot replicate through pyrimidine dimers. RecA interacts with the ssDNA and disengages from the poles.

Formation of RecA filaments (see **Fig. 9.16**, step 2) on ssDNA activates a second function of RecA, called coprotease activity, that stimulates autodigestion of the LexA repressor, a protein that normally prevents SOS activation. LexA protein binds to the promoters of DNA repair genes and prevents their transcription (**Fig. 9.30A** and **B**). Cleavage of LexA then unleashes the production of DNA repair enzymes. Among these enzymes are two "sloppy" DNA polymerases that lack proofreading activity: DinB (also called DNA polymerase IV, or Pol IV) and UmuDC (also called DNA polymerase V, or Pol V) (**Fig. 9.30C**). The UmuDC enzyme is perfect for replicating through damaged bases (a process called translesion bypass replication) because it sacrifices accuracy for continuity. When the enzyme encounters an undecipherable damaged base, it will insert whatever nucleotide is available. The result, of course, will be numerous permanent mutations, but the benefit is that the cell has a chance to live if it can tolerate the mutations. The choice really is "mutate or die," because housekeeping DNA polymerases like Pol III cannot move through damaged DNA. (Note that the term "housekeeping" is applied to proteins or enzymes that keep the cell running at all times.)

When the cell is severely compromised by mutation, it temporarily stops dividing. Like a pit stop during an auto race, this pause in cell division allows time for repair enzymes to fix the damage. The product of another SOS-regulated gene, called *sulA*, causes this pause by binding to the FtsZ cell division protein, keeping it from initiating cell division (see Sections 3.1 and 3.4). Once the damage has been repaired, RecA coprotease is inactivated and LexA repressor accumulates, turning off all the SOS genes, including *sulA*. The lingering SulA protein is then degraded by a protease called Lon. The protein is called

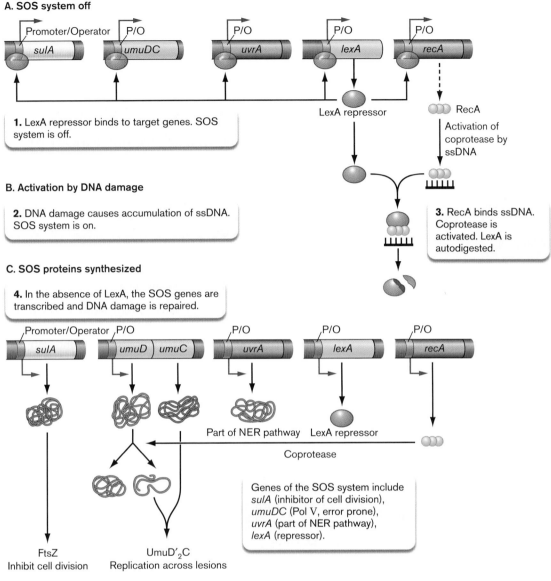

A. SOS system off

1. LexA repressor binds to target genes. SOS system is off.

B. Activation by DNA damage

2. DNA damage causes accumulation of ssDNA. SOS system is on.

C. SOS proteins synthesized

4. In the absence of LexA, the SOS genes are transcribed and DNA damage is repaired.

RecA

Activation of coprotease by ssDNA

3. RecA binds ssDNA. Coprotease is activated. LexA is autodigested.

Part of NER pathway   LexA repressor

Coprotease

Genes of the SOS system include *sulA* (inhibitor of cell division), *umuDC* (Pol V, error prone), *uvrA* (part of NER pathway), *lexA* (repressor).

FtsZ
Inhibit cell division

UmuD′$_2$C
Replication across lesions

5. Products of the target genes inhibit cell division, mediate replication across lesions (Pol V), and participate in nucleotide excision repair.

FIGURE 9.30 ■ Regulation of the SOS response system. The emergency DNA repair system known as the SOS response is induced when there is extensive DNA damage. The system is not a single repair mechanism, but a set of different mechanisms that collaborate to rescue the cell.

Lon because when it is missing, SulA inappropriately accumulates, inhibits cell division, and produces <u>long</u> filamentous cells.

The coprotease activity of RecA can also activate prophages by stimulating autocleavage of proteins that prevent phage replication. This phenomenon is used by some organisms to displace competitors in an environmental niche. For instance, *Streptococcus pneumoniae*, a cause of pneumonia, displaces *Staphylococcus aureus* in the human nasopharynx by producing hydrogen peroxide. Hydrogen peroxide damages *S. aureus* DNA, which activates the SOS response and triggers replication of resident bacteriophages that lyse the cell. The loss of *Staphylococcus aureus* means more room for *Streptococcus pneumoniae*.

Nonhomologous end joining. Double-strand breaks in DNA are particularly dangerous to a cell because the loss of chromosome integrity is immediate. In rapidly growing bacteria such as *E. coli*, double-strand breaks can be repaired by homologous recombination as long as another copy of the chromosome is present to guide the repair. But in slow-growing bacteria such as *Mycobacterium*

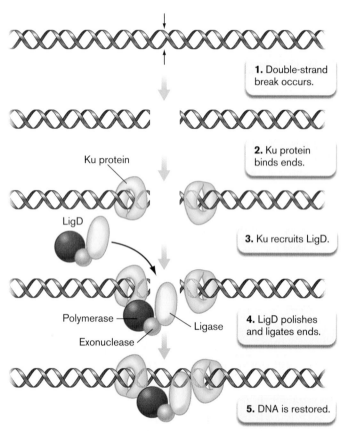

FIGURE 9.31 ▪ Nonhomologous end joining. NHEJ, a common repair mechanism in eukaryotes, has been documented in some bacteria, such as *Mycobacterium tuberculosis*. These pathogens can survive within macrophages that produce antibacterial compounds (nitric oxide and hydrogen peroxide) that generate double-strand DNA breaks. NHEJ mechanisms can repair those dsDNA breaks.

1. Double-strand break occurs.

2. Ku protein binds ends.

Ku protein

LigD

3. Ku recruits LigD.

Polymerase — Ligase
Exonuclease

4. LigD polishes and ligates ends.

5. DNA is restored.

tuberculosis, a second chromosome copy is usually not available. These bacteria can use an intriguing repair mechanism called **nonhomologous end joining** (**NHEJ**), which is also used by mammals (**Fig. 9.31**).

Two bacterial proteins, Ku and LigD, carry out NHEJ repair. Ku protein binds to the ends of a double-strand break and recruits LigD, a protein with polymerase and 3′ exonuclease activities that fill in or remove single-strand overhangs, and a ligase activity that joins two double-strand breaks. Because the system does not require homology, it can be error prone, causing the loss or addition of a few nucleotides at the break site or even the joining of two previously unlinked DNA molecules. NHEJ has been reported in *Mycobacterium* and *Bacillus* species.

Human homologs of bacterial repair genes. About 30% of *E. coli* genes have human homologs. The functions of the human genes may be similar to those of *E. coli*, or the human genes may have newly acquired functions. These genes often turn out to play key roles in human genetic

diseases. An example is *mutS*, of the methyl mismatch DNA repair system. In humans, the ancestor of *mutS* evolved to have multiple repair functions, as well as to participate in postmeiotic segregation. Defects in this repair mechanism have been linked to increased risk of certain colon cancers. Numerous other human genetic diseases are caused by mutations in homologs of bacterial repair genes. For example, defective excision repair genes in humans cause xeroderma pigmentosum, a disease that that leads to blindness and skin cancers. A deficiency in transcription-coupled repair causes Cockayne syndrome in humans, a devastating neurodegenerative disease.

To Summarize

- **DNA repair pathways** in microorganisms include error-proof and error-prone mechanisms.

- **Methyl mismatch repair** uses methylation of the parental DNA strand to distinguish it from newly replicated DNA. The premise is that the parental strand will contain the proper DNA sequence.

- **Photoreactivation** cleaves the cyclobutane rings of pyrimidine dimers.

- **Nucleotide excision repair (NER)** clips out a patch of single-stranded DNA containing certain types of damaged bases.

- **Base excision repair (BER)** excises structurally altered bases without cleaving the phosphodiester backbone. The resulting AP (apurinic or apyrimidinic) site is targeted by AP nucleases.

- **Recombinational repair** takes place at a replication fork. A "good" strand of DNA is used to replace a homologous damaged strand.

- **Extensive DNA damage leads to induction of the SOS response**, producing increased levels of the error-proof repair systems, as well as error-prone translesion bypass DNA polymerases that introduce mutations.

- **Nonhomologous end joining (NHEJ) mechanisms** repair double-strand DNA breaks in some, but not all, bacteria.

9.5 Mobile Genetic Elements

In 1948, Barbara McClintock (1902–1992) noticed that certain genetic traits of corn defied the laws of Mendelian inheritance. The genes encoding these traits, sometimes called "jumping genes," seemed to hop from one

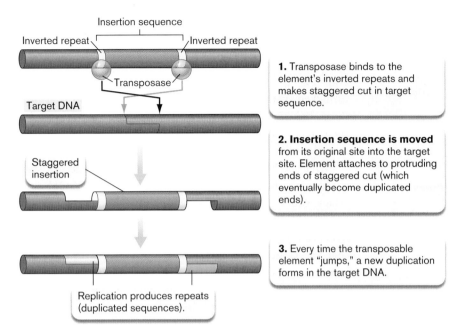

1. Transposase binds to the element's inverted repeats and makes staggered cut in target sequence.

2. **Insertion sequence is moved** from its original site into the target site. Element attaches to protruding ends of staggered cut (which eventually become duplicated ends).

3. Every time the transposable element "jumps," a new duplication forms in the target DNA.

FIGURE 9.32 ■ Basic transposition and the origin of target site duplication. Enzymes that catalyze transposition generate duplications in the target site by ligating the ends of the insertion element to the long ends of a staggered cut at the target DNA site. ▶

Note: An inverted repeat is a DNA sequence identical to another downstream sequence. The repeats have reversed sequences and are separated from each other by intervening sequences:

5′-AATCGAT . . . . ATCGATT-3′
3′-TTAGCTA . . . . TAGCTAA-5′

Recall that you must consider both strands of the DNA to see the inversion. Compare each strand in the 5′-to-3′ direction. View the top strand left to right and bottom strand right to left. When no nucleotides intervene between the inverted sequences, the whole structure is called a palindrome.

Transposition is the process of moving a transposable element <u>within</u> or <u>between</u> DNA molecules. During transposition, a short target DNA sequence on the destination DNA molecule is duplicated so that one copy of the sequence will flank each end of the element (**Fig. 9.32**). The transposase randomly selects one of many possible target sequences where it will move the insertion element.

Transposable elements transpose by one of two mechanisms—namely, nonreplicative or replicative transposition (**Fig. 9.33** ▶). In nonreplicative transposition, the insertion sequence excises itself out of one host DNA while integrating into the destination DNA. In replicative transposition, the sequence copies itself into the new host DNA while a copy remains within the original host.

chromosome to another. Although McClintock's theories were provocative at the time, we now know that these types of genes, referred to as **transposable elements**, exist in virtually all life-forms and can move both within and between chromosomes. These mobile DNA elements have contributed greatly to genome rearrangements during the evolution of all species.

Transposable elements are not autonomous. Unlike plasmids, these mobile elements are incapable of existing outside of a larger DNA molecule. They exist only as hitchhikers integrated into some other DNA molecule (**Fig. 9.32** ▶). All transposable elements include a gene encoding a **transposase**, an enzyme that catalyzes the transfer or copying of the element from one DNA molecule into another. There are two types of transposable elements in Bacteria: insertion sequences and transposons. An **insertion sequence** (**IS**) is a simple transposable element (typically 700–1,500 bp) consisting of a transposase gene flanked by short inverted-repeat sequences that are targets of the transposase (see **Fig. 9.32**). **Transposons** are more complex than simple insertion elements, because they carry other genes in addition to those required for transposition.

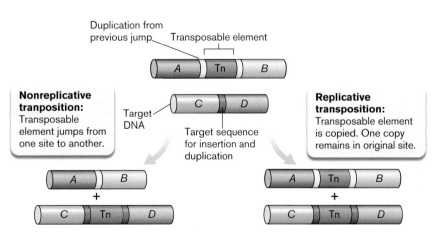

FIGURE 9.33 ■ Products of nonreplicative and replicative transposition. Nonreplicative transposition moves an insertion element from one DNA site to another without leaving a copy of the element at the original site. Replicative transposition leaves the element at the original site and moves a replicated copy to the new site. ▶

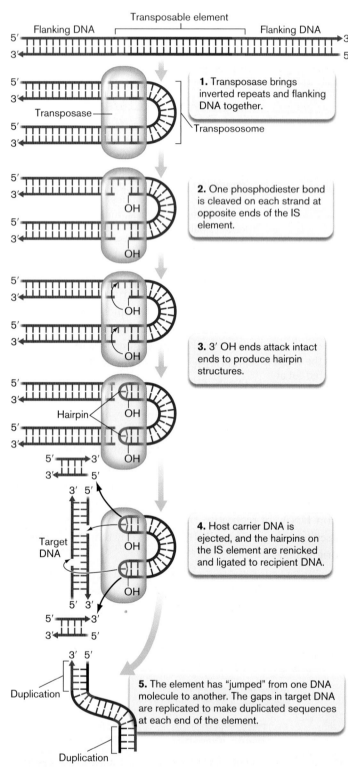

1. Transposase brings inverted repeats and flanking DNA together.

2. One phosphodiester bond is cleaved on each strand at opposite ends of the IS element.

3. 3′ OH ends attack intact ends to produce hairpin structures.

4. Host carrier DNA is ejected, and the hairpins on the IS element are renicked and ligated to recipient DNA.

5. The element has "jumped" from one DNA molecule to another. The gaps in target DNA are replicated to make duplicated sequences at each end of the element.

FIGURE 9.34 ■ Nonreplicative transposition. The transposome complex includes the transposase binding to the ends of the transposable element and the target DNA. The black DNA segments marked "Duplication" in the last panel correspond to the duplications represented in Figure 9.32. ▶

The nonreplicative model of transposition is shown in **Figure 9.34**. The transposase protein binds to the inverted-repeat ends of the transposable element and

to the target DNA, forming a transpososome complex (**Fig. 9.34** ▶, step 1). The transposase cuts the phosphodiester backbone (step 2), severing one strand at one end of the insertion sequence and the other strand at the other end. The 3′ OH ends of the IS element then attack the unnicked strands in a transesterification reaction that produces hairpin structures (step 3). In this instance, joining the ends of the double-stranded molecule produces a single strand with a hairpin.

After the host carrier DNA is ejected from the transpososome, the hairpin ends are renicked and the 3′ OH ends attack the target DNA molecule in a staggered manner (**Fig. 9.34**, step 4). The element has successfully "jumped" from one molecule to another without replicating. Note that the target DNA segment replicates as a result of the insertion process, thereby producing a copy of the target sequence at each end of the IS element (step 5).

Nonreplicative transposition is carried out by insertion sequence elements and by composite transposons. A composite transposon typically consists of <u>two</u> insertion sequences that flank an antibiotic resistance gene, or one or more catabolic genes (for example, genes for benzene catabolism). Often the interior inverted repeats of the IS elements have degenerated, so the transposase acts on primarily the two outermost inverted repeats, causing the whole transposon to move as one unit. Tn*10* is one such example (**Fig. 9.35**).

In contrast to composite transposons, replicative transposition is typically observed with <u>complex</u> transposons, such as the Tn*3* family of transposons (**Fig. 9.36A**). Tn*3* not only possesses transposase and antibiotic resistance genes, but also includes a gene whose product is called resolvase. The transposase acts on the inverted-repeat ends of the element, causing simultaneous replication and insertion of the transposon into target DNA. Replicative transposition produces a cointegrate between two different DNA molecules, one of which was the source of the transposon (**Fig. 9.36B**). As a result of transposition and replication, the cointegrate molecule will contain two copies of the element flanking the integrated plasmid. Resolvase is a site-specific recombinase that mediates recombination between the duplicated transposons at the internal Tn*3 res* sites. The result of this recombination is restoration of the two original circular replicons, each one now containing a copy of the transposon.

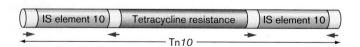

FIGURE 9.35 ■ Tn*10*, a composite transposon. Yellow fragments represent inverted repeats, and the red arrows indicate their orientation. In Tn*10*, the internal repeats contain mutations that prevent the individual IS elements from jumping off on their own.

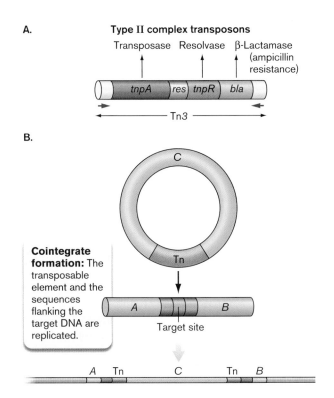

A.

Type II complex transposons

Transposase   Resolvase   β-Lactamase
(ampicillin
resistance)

tnpA *res* *tnpR* *bla*

Tn*3*

B.

C

Tn

Cointegrate formation: The transposable element and the sequences flanking the target DNA are replicated.

A   B

Target site

A   Tn   C   Tn   B

FIGURE 9.36 ▪ Transposition-mediated cointegrate formation between two DNA elements. During replicative transposition, the insertion element is copied and simultaneously ligated to the target site. The result is a transient cointegrate molecule in which donor DNA and target DNA become one molecule.

Thought Question

9.9 Diagram how a composite transposon like Tn*10* can generate inversions or deletions in target DNA during transposition. *Hint:* This happens when transposition within the chromosome occurs using the inverted repeats closest to the tetracycline resistance gene (see **Fig. 9.35**). If you draw it right, you will see, in the end, that the tetracycline resistance gene is lost.

All of the transposition events mentioned so far involve transfer between DNA molecules within the same cell. Some transposons, called **conjugative transposons**, are able to transfer from one cell to another by conjugation. One example is the transposon Tn*916* of *Enterococcus*, which encodes tetracycline resistance. These conjugative transposons excise from donor DNA and form a circular intermediate, like a plasmid, just prior to conjugation, but they do not replicate autonomously. They must be part of a larger self-replicating DNA entity (for example, plasmid, chromosome, or phage). Once transferred into a new host, the conjugative transposon integrates into the recipient's chromosome. Tn*916* and some other conjugative transposons evolved to transfer when a population is facing danger. For example, in the

absence of tetracycline, only a few cells in a population may have Tn*916*. However, the presence of just a small amount of tetracycline triggers conjugative transfer of the element, spreading it throughout the population.

The study of transposons may provide insight into the workings of the human immunodeficiency virus (HIV), which causes AIDS. The structure of transposase from transposon Tn*5* bears a striking similarity to the HIV integrase needed to embed the reverse-transcribed HIV viral DNA into the human genome (see Section 11.3). Another type of mobile element, called an integron, has evolved in bacteria to capture DNA cassettes that encode antibiotic resistance genes (see **eTopic 9.5**).

To Summarize

- **Transposable elements and insertion sequences** ("jumping genes") move from one DNA molecule to another, usually without replicating separately (that is, they are not plasmids).

- **Transposase** is an enzyme that forms a transpososome complex with the transposable element and target DNAs.

- **Insertion sequences** are simple DNA transposable elements containing a transposase gene flanked by short inverted-repeat sequences.

- **Transposable elements move** by nonreplicative or replicative mechanisms.

- **Transposons** are complex transposable elements carrying additional genes (encoding, for example, drug resistance).

- **Composite transposons** have two duplicate insertion sequence elements that flank additional genes. Often the interior inverted repeats of these elements contain mutations.

- **Transposons can carry a variety of genes**, including antibiotic resistance genes.

9.6 Genome Evolution

Can bacterial genes evolve to become part of eukaryotic genomes—and vice versa? A sense of evolution, of what went before, pervades efforts to understand microbial genomes. Genes can be removed (deletion), added (insertion), rearranged (recombination), or divided to the point where the genome bears only a slight resemblance to what it once was. The evolution of a genome is a random process driven by natural selection. Chapter 17 describes the many features of microbial evolution, but here in Chapter 9 we discuss the molecular mechanisms that drive that evolution.

The result of evolution can be seen in the genomes of various strains of the gut microbe *Escherichia coli*. Much of this evolution presumably took place in the intestines of human and other animal hosts, the normal habitat of this organism. The K-12 strain (MG1655), for example, contains 4,641,652 base pairs (compared with the 3 billion base pairs of the human genome). Of this genome, 87.8% encodes proteins, 0.8% encodes tRNA and rRNA, and another 0.7% is DNA with no known function. Approximately 11% of the chromosome is involved with various forms of regulation. Even though *E. coli* is the best-studied organism on the planet, about 20% of the 4,489 genes still remain a mystery to us, having no known function other than vague annotations such as "oxidoreductase." Close examination of the sequences reveals how the current organism we call *E. coli* came about through the evolution of an ancestral genome. Two basic processes are thought to contribute to genome restructuring: horizontal gene transfers and duplications followed by functional divergence through mutation.

Horizontal Gene Transfer

Microbial genomes evolve by randomly assembling an eclectic array of genes from many sources. For example, it has been estimated that nearly 20% of the *E. coli* genome may have originated in other microbes. *E. coli* strain O157:H7, the culprit in several fatal outbreaks of food-borne disease throughout the world, contains 1,387 genes that are not in strain K-12. These additional genes represent about 25% of the O157:H7 genome and encode virulence factors, metabolic pathways, and prophages (phage genes integrated into host chromosomes), all of which were acquired from other species. The impact of horizontal gene transfer on genome evolution is estimated to be 100 times greater than the impact of mutations.

Evidence for horizontal transfer. Evidence of gene shuffling comes from comparing the proportion of GC base pairs along the chromosome (also known as GC content). On average, the *E. coli* genome consists of 50.8% GC pairs. But 15% of the K-12 genome and 26% of the O157:H7 genome show GC proportions that differ significantly from the rest of the genome and show a different codon usage, suggesting that these genes came from other bacterial species and were acquired by *E. coli* more recently. These horizontal gene transfers, occurring via conjugation, transduction, or transformation, are estimated to happen at the rate of about 16 kb, or 0.4% of the *E. coli* genome, every million years. Noticing a sudden change in the GC content in a chromosome DNA sequence is the equivalent of an evolutionary "footprint" marking a horizontal gene transfer. The region where such a change is observed is called a **genomic island**. Genomic islands are large, transferred genetic elements

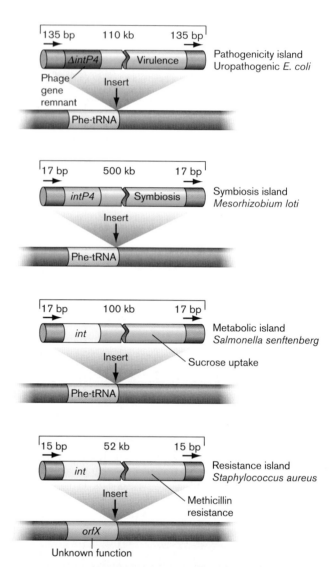

FIGURE 9.37 ■ Genomic islands from different microorganisms. Genomic islands are often found inserted adjacent to tRNA genes and have direct repeats at their ends (arrows). Functions for each island are identified.

containing numerous genes that serve a common function, such as virulence (**pathogenicity islands**), symbiosis (**symbiosis islands**), or survival (**fitness islands**).

Genomic islands are often flanked by boundary regions such as direct repeats or insertion elements, usually found near tRNA genes (**Fig. 9.37**). The association of genomic islands with tRNA genes may reflect the fact that some tRNA genes are dispensable because often multiple copies of a given tRNA gene are present within a genome. Genomic islands as a whole are often genetically unstable, subject to further rearrangements or deletions, and very often contain functional or remnant mobility genes encoding transposases, integrases, or other enzymes associated with recombination.

Pathogenicity and symbiosis islands can encode remarkable protein secretion systems (called type III and type IV secretion systems) capable of injecting effector proteins directly from the bacterium straight into a eukaryotic

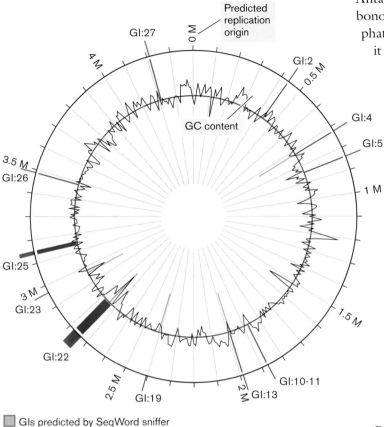

FIGURE 9.38 ▪ **Positions of genomic islands in the chromosome of the Antarctic microbe *Oleispira antarctica*.** Genomic island (GIs) were predicted by various computer programs (color-coded) and numbered GI:1 through GI:27. GC content of the genome is shown along the inner circle. Values inside the circle represent GC content below the genome average; those outside the circle reflect GC content above the genome average. The genome-wide GC content is 42%.

Antarctic coastal marine environments. It is a hydrocarbonoclastic organism whose metabolism is restricted to aliphatic, saturated and unsaturated hydrocarbons. Because it voraciously eats oil, *O. antarctica* is considered well suited for remediating oil spills in polar environments. Gene sequencing and functional genome analysis have revealed 27 genomic islands in this species, covering 12% of the entire genome (**Fig. 9.38**). Systems associated with these islands include antibiotic resistance genes, chaperones, heavy-metal efflux pumps, and several other efflux pumps. All of these genes may be important for the survival of *O. antarctica* in the hostile environments of natural or industrial oil leaks.

Horizontal gene transfers between Bacteria and Eukarya. In eukaryotes, the sexual exchange of genes is usually only within a single species. Microbes, on the other hand, are more promiscuous. Matings between microbial genera are common and perhaps even desirable from an evolutionary viewpoint. Sharing genes between species allows each one to sample genes from the other and keep genes (through natural selection) that increase fitness. But what about transfers between Bacteria and Eukarya? Do they happen? If so, are they useful? Work from Gos Micklem and colleagues at the University of Cambridge suggest that interdomain transfer of DNA over the course of evolution may be more common than previously realized. They report that DNA sequences from many eukaryotes, including humans, appear to include a variety of genes horizontally transferred from bacteria (**Fig. 9.39**). How the genes jumped domains is unclear.

In a bizarre example, Joseph Mougous (University of Washington) and colleagues described a eukaryotic gene called *dae* (domesticated amidase effector) that was

cell. Once inside the target cell, these effector proteins alter the function of eukaryotic cell proteins, making the host organism, in the case of symbiosis, more accepting of the symbiont. Type III and IV secretion systems are discussed further in Chapter 25.

Genomic islands in Antarctic microbes. Horizontal gene transfers have also been seen in bacteria isolated from the frigid waters of the Antarctic (about −2°C). *Oleispira antarctica*, for example, is a psychrophilic Gram-negative marine organism isolated from

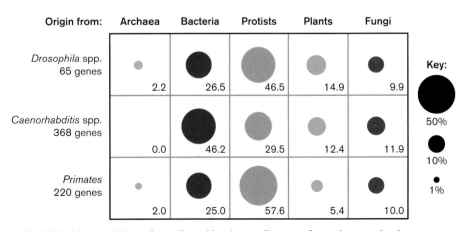

FIGURE 9.39 ▪ **Origins of predicted horizontally transferred genes in the genomes of *Drosophila*, *Caenorhabditis*, and primates.** Numbers show percent contribution.

Origin from:	Archaea	Bacteria	Protists	Plants	Fungi
Drosophila spp. 65 genes	2.2	26.5	46.5	14.9	9.9
Caenorhabditis spp. 368 genes	0.0	46.2	29.5	12.4	11.9
Primates 220 genes	2.0	25.0	57.6	5.4	10.0

Key:
50%
10%
1%

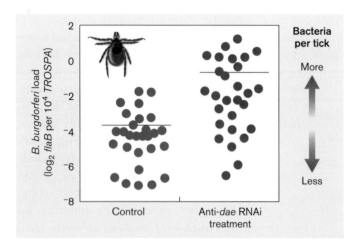

FIGURE 9.40 ■ The *dae* product helps the deer tick control the levels of *Borrelia burgdorferi*. Deer tick nymphs (inset) fed on spirochete-infected mice, and two weeks later *B. burgdorferi* levels were quantified using PCR to identify the number of *B. burgdorferi flaB* genes. One *flaB* gene equals one organism. To estimate relative numbers of bacteria per tick (*B. burgdorferi* load), the ratio of *B. burgdorferi*–specific *flaB* gene copies relative to copies of the tick-specific gene *TROSPA* were determined. Control nymphs received no treatment. Treated nymphs were administered an interfering RNA (RNAi) that binds to *dae* mRNA and reduces *dae* expression. Each data point represents three nymphs. Horizontal lines indicate mean values.

transferred from Bacteria to eukaryotic ticks and mites several times over millions of years. The bacterial version of the protein, called Tae, degrades peptidoglycan and is delivered as the tip of a spear from one bacterium to another via type VI secretion (discussed in Special Topic 25.1). Once injected into the "enemy" bacterium, Tae kills the victim by disassembling the bacterial cell wall. Among the eukaryotic recipients of the *dae* gene is the deer tick *Ixodes scapularis*, which serves as the reservoir for the Lyme disease spirochete *Borrelia burgdorferi* (Lyme disease is described in Chapter 26). After their horizontal transfer, the *dae* genes became expressed and eventually acquired eukaryotic protein secretion signals. **Figure 9.40** illustrates that the *I. scapularis dae* gene helps the insect control the levels of *B. burgdorferi* colonizing the tick's midgut. The bacterial gene, therefore, became part of the *Ixodes* innate immune system against *B. burgdorferi* and probably other bacteria. Innate immunity is described in Chapter 23. Evidence of more massive DNA transfers from some endosymbiotic bacteria to their eukaryotic hosts is discussed in **eTopic 9.6**.

Can interdomain DNA transfers take place in the opposite direction—from humans to bacteria, for instance? Mark Anderson and Han Seifert from Northwestern University found that up to 11% of *Neisseria gonorrhoeae* isolates (the cause of gonorrhea) contain human-derived DNA sequences known as L1, or long interspersed nuclear elements (LINEs). LINEs are human DNA sequences that copy and insert themselves from one gene into another. Since the L1 sequence in *N. gonorrhoeae* was not found in the closely related species *N. meningitidis*, the transfer from L1 to *N. gonorrhoeae* probably occurred relatively recently in evolutionary history (after *N. gonorrhoeae* and *N. meningitidis* split from their common ancestor). The function of *Neisseria* L1 is currently unknown.

The Intestine: Cauldron of HGT

The human gastrointestinal tract is ideal for exchanging genetic information between widely diverse species (Section 23.1). The gut contains a high bacterial cell density and mixed-species biofilms that enable close and frequent contact between organisms, both living (a requisite for conjugation) and dead (a source of transforming DNA). The intestine is also rife with phages that can mediate transduction. Section 8.6 described a spectacular example of horizontal transfer that likely took place within the human gut. A gene enabling digestion of a complex carbohydrate found only in red seaweed was transferred from a marine bacterium to an ancestor of the human intestinal bacterium *Bacteroides plebeius*. Human digestion, helped by this new bacterium, could now harvest carbon and energy from this unique carbohydrate.

Perhaps the most intriguing and worrisome evidence of horizontal gene transfer (HGT) in the gut involves antibiotic resistance. The gut microbiome can be considered a repository of many antibiotic resistance genes, usually within anaerobic members of the microbiome. Studies that examined the gut metagenomes of hundreds of individuals have identified upwards of 1,000 antibiotic resistance genes "lurking" among the microbial population. Their presence is not always evident until someone is treated with an antibiotic. For example, a plasmid encoding high-level resistance to carbapenem antibiotics was horizontally transferred from *Klebsiella pneumoniae* to *Escherichia coli* in the intestine of a 91-year-old patient. The resistance enzyme, carbapenemase, confers resistance to all cephalosporins, monobactams, and carbapenems used as first-line drugs for hospitalized patients (Chapter 27). The man was being treated for sepsis (a severe blood infection) with ertapenem. When the man's stool was tested after treatment, it contained carbapenem-resistant *K. pneumoniae*, but not *E. coli*. One month later, carbapenem-resistant *E. coli* containing the same plasmid was recovered from his stool.

Antibiotic exposure may even induce HGT among microbiome members. Human volunteers whose stools contained erythromycin-susceptible *Bacteroides* strains were given erythromycin. Within 7 days, erythromycin-resistant *Bacteroides* strains were found. Conjugation of erythromycin resistance plasmids from unknown members of the gut

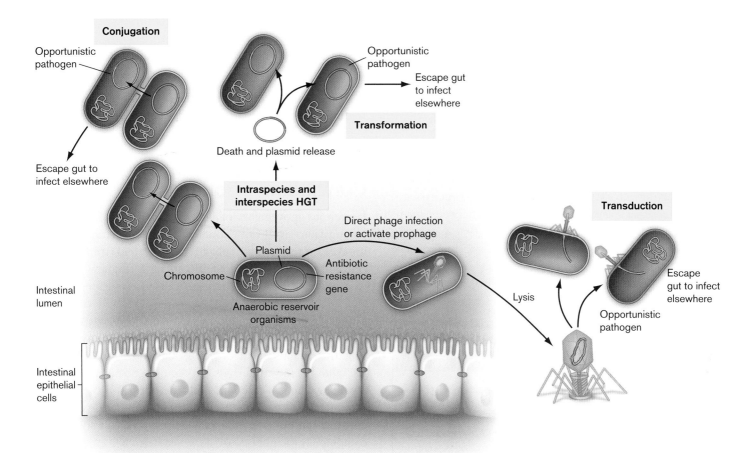

FIGURE 9.41 ■ Model for horizontal gene transfer (HGT) between gut microbiota. High cell density and biofilm formation in the intestine makes the horizontal transfer of antibiotic resistance (or other genes) easier. Many anaerobic members of the human microbiome (such as *Bacteroides*) serve as reservoirs for antibiotic resistance genes. Various gene transfer mechanisms will randomly transfer the genes, often on plasmids, to other members of the microbiota, some of which are opportunistic pathogens (*E. coli*, for instance). Opportunistic pathogens can escape the gut to cause disease in other body locations. *Source:* Derived from Shaick, 2015. The gut resistome. *Philosophical Transactions of the Royal Society B* **270**:20140087.

microbiome were implicated. Antibiotics can also induce the SOS response in bacteria. The SOS response will activate prophage replication (Section 9.4), and the resulting phage can mediate transduction of genes encoding antibiotic resistance, or other functions, to gut coinhabitants. **Figure 9.41** illustrates the mechanisms by which antibiotic resistance genes can be transferred between reservoir gut microbes to opportunistic pathogens that can escape the intestine, either by defecation or intestinal lesion, to infect elsewhere.

Genome Reduction

It is important to note that evolution involves gene loss as well as gene acquisition. The large-scale loss of genes through evolution is known as genome reduction. For example, pathogenic shigellae (the cause of bacillary dysentery) exhibit chromosomal "black holes," regions lacking genes that occur in the closely related *E. coli*. Absence of these genes is required for a fully virulent phenotype (why is not clear). Genome reduction is discussed in Chapter 17.

Sometimes evidence of genome reduction in the making can be observed. Many genomes contain pseudogenes, genes that by homology appear to encode an enzyme but are nonfunctional because a portion is missing through deletion. These appear to be the remnants of genes that were useful to an ancestral species but were made superfluous when one of its descendants adapted to a new environmental niche. When evolutionary pressure to retain a functional form of the gene no longer exists, the imperative to repair mutations within that gene also evaporates. The pseudogene, therefore, is in the process of being eliminated. One pathogen caught in the act of losing its genes because of its lifestyle is *Mycobacterium leprae*, the cause of Hansen's disease (leprosy). Over half of the *M. leprae* genome is made of pseudogenes, presumably formed from genes it no longer needs after becoming an obligate intracellular pathogen.

Duplications and Divergence

Gene duplication is the most important mechanism for generating new genes and biochemical processes. Duplication frees a gene from its previous functional constraints and allows divergent evolution through mutation. Duplications can arise in several ways, including the transposition of a transposable element and recombination after replication between multiple direct or inverted-repeat sequences located in the chromosome. A description of how recombination between direct repeats at a replication fork can lead to a duplication is found in **eTopic 9.7**. Superfamilies of proteins arising from divergent evolution share structural and functional features but may catalyze different reactions. One example is the superfamily of ABC (ATP-binding cassette) transporters (see Section 4.2). Homology between members of the ABC family may be as little as 10%, suggesting that once an enzyme adopts a new function, the sequence diverges rapidly. Thus, gene duplications and mutation-driven divergence are processes that directly generate orthologous and paralogous genes, discussed in Chapter 8.

Thought Question

9.10 Gene homologs of *dnaK* encoding the heat-shock chaperone HSP70 exist in all three domains of life. All bacteria contain HSP70, but only some species of archaea encode a *dnaK* homolog. The archaeal homologs are closely related to those of bacteria. Knowing this information, how do you suppose *dnaK* genes arose in archaea?

To Summarize

- **Horizontal gene transfers** occur between species by conjugation, transformation, and transduction.

- A DNA sequence with a **GC content** different from that of flanking chromosomal DNA is one sign of horizontal gene transfer.

- **Genomic islands** are the result of horizontal gene transfers that expand the competitiveness of a recipient by contributing to its pathogenicity, symbiosis, or fitness in a hostile environment.

- **Superfamilies of functional proteins** result from gene duplications and mutations that cause divergent evolution of function.

Concluding Thoughts

Fortuitous movements of genes between species through conjugation (cell-cell contact), transformation (DNA uptake), transduction (phage vector), and transposition (jumping genes) have contributed in a fundamental way to evolution. These mechanisms caused evolutionary leaps and enabled ancestral bacteria to develop (and trade) the symbiotic and pathogenic mechanisms needed to inhabit new ecological niches. DNA repair mechanisms that prevent the establishment of deleterious mutations nevertheless allow some mutations to occur. An imperfect DNA repair mechanism offers the cell the opportunity to throw the genetic dice. The vast majority of mutations will be detrimental, but a key few may help the variant strain outcompete its neighbors in its ecosystem. Given our new awareness of genetic mobility, how do we now define a species? For instance, numerous strains of *Helicobacter pylori* show large differences in gene sequence and organization. Should each be called a different species? Currently, we use an arbitrary setting of ribosomal RNA gene divergence to assign species, but some wonder whether this is still a valid benchmark. For further discussion of microbial evolution, see Chapter 17.

CHAPTER REVIEW

Review Questions

1. What are the basic ways microorganisms exchange DNA?
2. Discuss horizontal versus vertical gene transfer.
3. Describe competence and how it comes about in a population.
4. What is an F factor, and how does it (and other factors like it) contribute to gene exchange?
5. What is a partial diploid? Explain how a cell can become a partial diploid.
6. What does microbial gene exchange have to do with the plant disease called crown gall disease?
7. Compare specialized versus generalized transduction.
8. Discuss how bacteria protect themselves against invading bacteriophages.
9. What is the value of recombination to a species?
10. List the major proteins that contribute to recombination, and specify their roles in the process.

11. Define "Holliday junction," "strand assimilation," and "cointegrate."
12. List and explain the different types of mutations.
13. How are genotype and phenotype different?
14. Describe six different DNA repair mechanisms. Which ones contribute to mutations?
15. Explain the basic process of transposition. Why are insertion sequences always flanked by direct repeats of host DNA? How are transposons different from plasmids?
16. What are pathogenicity islands and fitness islands? What are their characteristics?

Thought Questions

1. Calculate the mutation rate for a gene in which the number of mutations increases from 0 to 500 as the cell number increases from 10^1 to 10^8.
2. Why does the competence factor for initiating synthesis and assembly of the transformasome have to be exported out of the cell to bind ComD? Why doesn't the molecule bind internally?
3. Rapidly growing bacteria can initiate a second round of replication before a previous round is completed. Would such rapid growth introduce more mutations (due to replication errors) than slower growth would by the same species?
4. *Agrobacterium tumefaciens* Ti plasmid has been used as a tool to genetically modify plants. Why would a plant biologist use a tumor-causing plasmid to breed new plants? Wouldn't the genetically altered plant develop a tumor?
5. Though we can identify sets of genes that have been horizontally transferred from one species of microbe to another, rarely can we identify the source species. What might account for this failure?
6. You have just isolated a new temperate bacteriophage for *Salmonella enterica*. How can you determine whether this phage mediates generalized or specialized transduction?

Key Terms

AP site (343)
apurinic site (335)
auxotroph (333)
base excision repair (BER) (342)
cointegrate (330)
competent (318)
conjugation (318)
conjugative transposon (349)
CRISPR (327)
deletion (332)
electroporation (317)
error-prone repair (338)
error-proof repair (338)
F^- cell (319)
F^+ cell (319)
F-prime (F′) factor (321)
fertility (F) factor (319)
fitness island (350)
frameshift mutation (333)
gain-of-function mutation (333)
generalized recombination (330)
generalized transduction (323)

genomic island (350)
heteroduplex (330)
Hfr strain (320)
Holliday junction (330)
horizontal gene transfer (316)
insertion (332)
insertion sequence (IS) (347)
inversion (332)
knockout mutation (333)
loss-of-function mutation (333)
methyl mismatch repair (339)
missense mutation (332)
mutagen (334)
mutation (332)
mutation frequency (336)
mutation rate (336)
mutator strain (339)
nonhomologous end joining (NHEJ) (346)
nonsense mutation (333)
nucleotide excision repair (NER) (342)

pathogenicity island (350)
phase variation (332)
photoreactivation (342)
point mutation (332)
recombinational repair (343)
restriction endonuclease (325)
reversion (332)
silent mutation (332)
site-specific recombination (330)
SOS response (344)
specialized transduction (323)
symbiosis island (350)
transduction (323)
transformasome (317)
transformation (316)
transition (332)
transposable element (347)
transposase (347)
transposition (347)
transposon (347)
transversion (332)

Recommended Reading

Bondy-Denomy, Joseph, and Alan R. Davidson. 2014. To acquire or resist: The complex biological effects of CRISPR–Cas systems. *Trends in Microbiology* **22**:218–225.

Cabezón, Elena, Jorge Ripoll-Rozada, Alejandro Pena, Fernando de la Cruz, and Ignacio Arechaga. 2015. Towards an integrated model of bacterial conjugation. *FEMS Microbiology Reviews* **39**:81–95.

Chou, Seemay, Matthew D. Daugherty, S. Brook Peterson, Jacob Biboy, Youyun Yang, et al. 2015. Transferred interbacterial antagonism genes augment eukaryotic innate immune function. *Nature* **518**:98–101.

Dini-Andreote, Francisco, Fernando D. Andreote, Welington L. Araujo, Jack T. Trevors, and Jan D. van Elsas. 2012. Bacterial genomes: Habitat specificity and uncharted organisms. *Microbial Ecology* **64**:1–7. doi:10.1007/s00248-012-0017-y.

Doyle, Marie, Maria Fookes, Al Ivens, Michael W. Mangan, John Wain, et al. 2007. An H-NS-like stealth protein aids horizontal DNA transmission in bacteria. *Science* 5:654–666.

Dunning Hotopp, Julie C., Michael E. Clark, Deodoro C. Oliveira, Jeremy M. Foster, Peter Fischer, et al. 2007. Widespread lateral gene transfer from intracellular bacteria to multicellular eukaryotes. *Science* **317**:1753–1756.

Frost, Laura S., Raphael Leplae, Anne O. Summers, and Ariane Toussaint. 2005. Mobile genetic elements: The agents of open source evolution. *Nature Reviews. Microbiology* 3:722–732.

Fuchs, Robert P., and Shingo Fujii. 2013. Translesion DNA synthesis and mutagenesis in prokaryotes. *Cold Spring Harbor Perspectives in Biology* 5:a012682.

Johnston, Calum, Bernard Martin, Gwennaele Fichant, Patrice Polard, and Jean-Pierre Claverys. 2014. Bacterial transformation: Distribution, shared mechanisms and divergent control. *Nature Reviews. Microbiology* **12**:181–196.

Louwen, Rogier, Raymond H. J. Staals, Hubert P. Endtz, Peter van Baarlen, and John van der Oost. 2014. The role of CRISPR-Cas systems in virulence of pathogenic bacteria. *Microbiology and Molecular Biology Reviews* **78**:74–88.

Million-Weaver, Samuel, Ariana Nakta Samadpour, and Houra Merrikha. 2015. Replication restart after replication-transcription conflicts requires RecA in *Bacillus subtilis. Journal of Bacteriology* **197**:2374–2382.

Polz, Martin F., Eric J. Alm, and Willian P. Hanage. 2013. Horizontal gene transfer and the evolution of bacterial and archaeal population structure. *Trends in Genetics* **29**:170–175.

Selva, Laura, David Viana, Gill Regev-Yochay, Krzysztof Trzcinski, Juan M. Corpa, et al. 2009. Killing niche competitors by remote-control bacteriophage induction. *Proceedings of the National Academy of Sciences USA* **106**:1234–1238.

van der Veen, Stijn, and Christoph M. Tang. 2015. The BER necessities: The repair of DNA damage in human-adapted bacterial pathogens. *Nature Reviews. Microbiology* **13**:83–94.

Van Houten, Bennett, and Neil Kad. 2014. Investigation of bacterial nucleotide excision repair using single-molecule techniques. *DNA Repair* (Amsterdam) **20**:41–48.

CHAPTER 10
Molecular Regulation

A bacterial genome encodes thousands of different proteins needed to handle many different environmental contingencies. But to compete in any mixed-species environment, a microbe cannot waste energy making unneeded proteins. Cells achieve molecular efficiency by using elegant control systems that selectively increase or decrease gene transcription, mRNA translation, or mRNA degradation, as well as by degrading or sequestering regulatory proteins. Chapter 10 describes how a cell "knows" when it needs to alter its physiology and explains the regulatory mechanisms used to effect that change.

CURRENT RESEARCH highlight

Controlling uninvited genes. Enterohemorrhagic *E. coli* (EHEC) expresses many virulence genes from horizontally transferred pathogenicity islands (PAIs). Certain PAI proteins injected into intestinal cells compel pedestals to form that cup the bacteria (panel A). Vanessa Sperandio and Charley Gruber discovered that this PAI process is controlled by a non-PAI regulatory system (GlmZ) that normally controls cell wall synthesis. Low GlmZ levels (panel B) <u>increase</u> pedestal formation compared to wild type (panel A), whereas high levels of GlmZ (panel C) <u>decrease</u> pedestal formation. Thus, the core chromosome-encoded Glm system was co-opted to control a horizontally transferred pathogenicity island.

*Source: Gruber and Sperandio. 2015. Infect. Immun. **83**:1286–1295.*

A. Wild type — Pedestals / Actin is stained green.

B. Low GlmZ — ΔglmY

C. High GlmZ — ΔglmY ΔrapZ

AN INTERVIEW WITH

VANESSA SPERANDIO, MOLECULAR MICROBIOLOGIST,
UNIVERSITY OF TEXAS, SOUTHWESTERN MEDICAL CENTER

Why do PAI genes co-opt non-PAI regulators to control their expression?

The most likely reason horizontally acquired islands appropriate recipient regulatory systems is that those systems already sense the cell's environment. The Glm system, for example, controls synthesis of amino sugars used for cell wall biosynthesis. It was "smart" for EHEC PAI genes involved in type III secretion (T3SS) to co-opt Glm control because assembly of this secretion system requires cell wall rearrangement.

You also discovered the EHEC epinephrine-sensing system (QseEF) that controls virulence. Does QseEF affect the Glm system?

Yes. QseEF is a two-component signal transduction system that binds epinephrine to sense entry into a host. QseEF initiates pedestal formation by activating several PAI genes. Boris Gorke's group found that QseF also regulates synthesis of the sRNA *glmY* that stabilizes GlmZ. So, during infection, QseE senses epinephrine and phosphorylates QseF. QseF-P activates the transcription of *glmY*, which then stabilizes GlmZ. Finally, elevated GlmZ limits pedestal formation.

The Current Research Highlight illustrates how a horizontally acquired gene becomes integrated into the regulatory network of a bacterial cell. Integration ensures that the new gene is expressed only when it is beneficial to the cell. But what constitutes a regulatory network? Microbes use numerous mechanisms to sense their internal and external environments. The information collected directs synthesis of waves of proteins whose concentrations change with changing environments. The cell's surface, for example, contains an array of sensing proteins that monitor osmolarity, pH, temperature, and the chemical content of the surroundings. Quorum sensing, in which bacteria secrete and sense chemical signaling molecules, allows members of microbial communities to communicate, cooperate, and even promote symbiotic relationships.

Take the Hawaiian bobtailed squid, *Euprymna scolopes*. During the day, this tiny squid remains buried in the sand of shallow reef flats around Hawaii. After sunset, the animal emerges from its hiding place and begins its search for food. As it swims in the moonlit night, its light organ projects light downward in an apparent attempt to camouflage the squid from predatory fish swimming below. Looking up, the fish see only light (called counterillumination), not a squid's shadow moving against the surface-filtered light of the moon. The light, however, is not made by the squid. Inside the squid's light organ are luminescent bacteria called *Aliivibrio fischeri* (formerly *Vibrio fischeri*). Bacteria, not the squid, produce the light. However, these microbes do not <u>constantly</u> glow. They light up only when their cell number and the concentration of a secreted signaling molecule rise above a threshold level. The critical density of bacteria is attained by nightfall, and the genes needed to make light are "turned on." Symbiosis is achieved because the bacteria obtain nutrients growing in the light organ, and the squid, for its part, survives another night.

Chapter 10 explores the fundamental principles of gene regulation in microbes and discusses how these individual systems are woven into global regulatory networks that interconnect many processes throughout the cell.

10.1 Gene Expression: Levels of Control

How many ways can a gene's expression be controlled? The expression of genes and their products (mRNA and protein) can be controlled at various levels, with each level of control offering different advantages to the cell. The major levels of control can be categorized as DNA sequence controls, mRNA stability controls, transcriptional controls, translational controls, and posttranslational modifications.

In general, DNA sequence–level control is the most drastic and the least reversible, whereas control at the protein (translational/posttranslational) level is the most rapid and most reversible. We summarize these control levels here and then discuss examples.

- **Alteration of DNA sequence.** Some microbes use random or programmed changes of DNA sequence to activate or disable a particular gene. One example is phase variation, in which reversible flipping of a DNA segment enables a pathogen to turn on or turn off expression of cell-surface proteins. The switch in protein structure helps the microbe evade a host's immune system.

- **Control of transcription.** Many types of gene regulation occur at the level of transcription. The most common mechanisms of transcriptional control in prokaryotes involve operons. Recall that an operon is a string of two or more genes in a chromosome that are expressed from a common promoter located in front of the first gene in the operon. Genes in an operon are coordinately regulated by protein repressors, activators, and sigma factors, as well as by **small RNAs** (**sRNAs**). Coordinate regulation means that the expression levels of all genes in the operon increase or decrease simultaneously.

- **Control of mRNA stability.** Levels of specific mRNA molecules are regulated by RNase activity, which degrades some mRNA molecules as fast as they are transcribed. In some cases, sRNA molecules bind RNA transcripts and help or hinder degradation.

- **Translational control.** Translation by ribosomes can be regulated by sequestering mRNA ribosome-binding sites or by concealing mRNA sequences that recognize specific translational repressor proteins. Translational control mechanisms are often coupled to transcriptional mechanisms, offering "fine-tuning" of operon control. Attenuation, for instance, is a mechanism that uses <u>translation</u> to sense the level of an amino acid in the cell and then increase or decrease <u>transcription</u> of the genes encoding the enzymes that synthesize the amino acid.

- **Posttranslational control.** Once proteins are made, their activity can be controlled by modifying protein structure—for example, by protein cleavage, phosphorylation, methylation, or acetylation. These modifications can activate, deactivate, or even lead to the destruction of the protein.

All of these control measures can be used in various combinations by the cell to build integrated control circuits that coordinately regulate multiple systems throughout the cell—a key to homeostasis and balanced growth (see nitrogen regulation, discussed in Section 15.5).

We begin our discussion of gene expression by describing how DNA rearrangements can flip genes on and off. Next we discuss how bacterial cells monitor conditions inside and outside of the cell to make "informed" decisions about altering gene expression. In Section 10.2 we will then explore several paradigms of operon control.

DNA Rearrangements That Alter Gene Expression

Most regulatory mechanisms that alter gene expression use interactions between proteins and DNA, proteins and RNA, or RNA and RNA. These control mechanisms are easily reversible. A more drastic means of control, however, involves altering the DNA sequence itself. A classic example of this strategy is phase variation. Phase variation helps microbial pathogens avoid the immune system.

Any infection will trigger the production of antibodies specific to the invading microbe's component parts, such as pili, flagella, and lipopolysaccharides (discussed in Chapter 24). Antibodies that bind to these microbial surface structures are useful for clearing an infection. However, some microbes use gene regulation to periodically change their immunological appearance, like a chameleon changing its color, by changing the amino acid composition of a particular surface protein. This "shape-shifting" by the microbe, called **phase variation**, renders useless those antibodies specific for the old structure. The embattled immune system must start all over again making new antibodies, thus prolonging the course of infection. Two types of DNA rearrangement can be used to generate phase variations: gene inversions and slipped-strand mispairing.

Gene inversion: an on/off switch. Flagellar phase variation in the Gram-negative bacterium *Salmonella enterica* involves a DNA recombination event known as gene inversion that flips the orientation of a gene or DNA segment in the chromosome. *S. enterica* has two genes, widely separated on the chromosome, that encode different forms of flagellin, the protein from which flagella are made. A reversible DNA inversion turns off one gene while turning on the other. The invertible switch is a 993-bp DNA fragment (or cassette), called the H region, that contains an outwardly directed promoter and a gene called *hin*, whose product, Hin recombinase (also called Hin invertase), mediates the recombination (**Fig. 10.1**). In antigenic parlance, the term "H antigen" refers to flagella, so the acronym "Hin" stands for H inversion. The *hin* DNA cassette is flanked by short (26-bp) inverted repeats called *hixL* (left) and *hixR* (right).

Note: An **inverted repeat** is a sequence found in identical (but inverted) forms at two sites on the same double helix (for example, 5'-ATCGATCGnnnnnnCGATCGAT-3'). A **direct repeat** is a sequence found in identical form at two sites on the same double helix (for example, 5'-ATCGATCGnnnnnnATCGATCG-3'). A **tandem repeat** is a direct repeat without any intervening DNA sequence (for example, ATCGATCGATCGATCGATCGATCG).

Hin recombinase collaborates with other less specific DNA remodeling proteins, such as Fis, to link the 26-bp left (*hixL*) and right (*hixR*) ends of the invertible DNA element. The two ends, each bound to a Hin monomer, are brought together by Hin-Hin protein interactions. DNA within the cassette then forms a loop. Hin cuts within the center of each *hix* site, producing staggered ends. An exchange of Hin subunits leads to strand inversion, so that the orientation of the DNA cassette is reversed relative to the flanking DNA on either side.

In one orientation, the outwardly directed promoter of the H region directs expression of H2 flagellin (encoded by *fljB*) and a repressor (FljA) that prevents transcription of the other flagellin gene, *fliC* (see **Fig. 10.1A**). After the inversion, however, the promoter points in the wrong direction, so there is no production of H2 flagellin or FljA, the repressor of *fliC* (see **Fig. 10.1D**). Lacking this repressor, the *fliC* flagellin gene is expressed. Thus, H1 flagellin (present in phase 1 cells) is synthesized instead of H2 flagellin (present in phase 2 cells). The amino acid sequences, and thus the antigenicity, of the two flagellar proteins are different. Inversion of the H region allows *Salmonella* to change how it appears to a host immune system. In each generation, the rate of the reversible switch varies from about one cell in 10^3 to one in 10^5. Note, however, that this switch would not accomplish much if the infecting population of bacteria started out mixed—that is, producing both types of flagella. The initial infection must be of one phenotype.

Slipped-strand mispairing. A different type of phase variation relies on multiple, short sequence repeats within a gene. The repeats "confuse" DNA polymerase as it replicates, causing it to slip occasionally during replication. Slippage either adds a repeat to, or deletes a repeat from, the gene and alters its translational reading frame. If the mRNA produced during transcription is out of frame, the protein is not made. This random process can alternately turn a gene off and then back on again in subsequent generations. A detailed example is discussed in **eTopic 10.1** using the *opa* genes of *Neisseria gonorrhoeae*, the causative agent of gonorrhea. Eukaryotic microbes, especially pathogenic sporozoa, possess elaborate mechanisms of phase variation. The trypanosome that causes "sleeping sickness" undergoes extensive genetic shuffling and mutation of its coat proteins over successive generations, essentially overwhelming the host immune system by presenting every possible form of antigen.

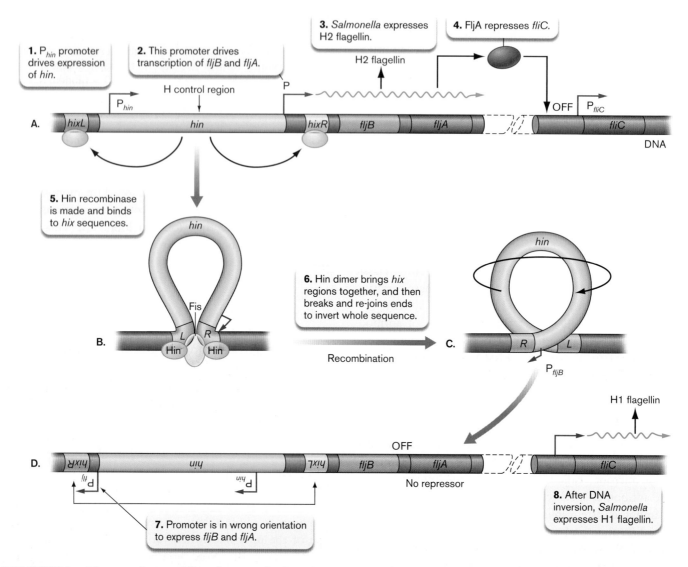

1. P*hin* promoter drives expression of *hin*.

2. This promoter drives transcription of *fljB* and *fljA*.

3. *Salmonella* expresses H2 flagellin.

4. FljA represses *fliC*.

H2 flagellin

H control region

P*hin*

P

OFF P*fliC*

A. *hixL* *hin* *hixR* *fljB* *fljA* *fliC*

DNA

5. Hin recombinase is made and binds to *hix* sequences.

hin

Fis

6. Hin dimer brings *hix* regions together, and then breaks and re-joins ends to invert whole sequence.

hin

B. L R Hin Hin

C. R L

Recombination

P*fljB*

H1 flagellin

OFF

D. *hixR* *hin* *hixL* *fljB* *fljA* *fliC*

P*flj* P*hin*

No repressor

8. After DNA inversion, *Salmonella* expresses H1 flagellin.

7. Promoter is in wrong orientation to express *fljB* and *fljA*.

FIGURE 10.1 ▪ **Phase variation of flagellar proteins in *Salmonella enterica*.** An invertible region containing a promoter controls the expression of two unlinked flagellar protein genes. In one orientation (**A**), the *fljB* promoter drives synthesis of H2 flagellin (*fljB*) and a repressor (FljA) of the H1 flagellin gene (*fliC*). Action by Hin recombinase causes the segment to invert (**B, C**), thereby reorienting the promoter. Because the repressor FljA is not formed, the gene for H1 can be expressed (**D**).

Thought Question

10.1 While viewing **Figure 10.1**, imagine the phenotype of a cell in which *fljB* has been deleted but *fljA* is still expressed. Would cells be motile? What type of flagella would be produced? Would the cells undergo phase variation? What would happen if *fliC* alone was deleted?

General Concepts of Transcriptional Control

DNA rearrangements that alter gene expression are random, controlled only by the efficiency of the recombinase catalyzing an inversion or the fidelity by which a DNA polymerase

replicates sequence repeats. However, transcriptional and translational controls are not random. For those mechanisms, a cell must monitor two compartments—the cytoplasm within the cell and the environment outside—to know when to alter gene expression and adjust its physiology.

Intracellularly, the concentrations of vitamins, amino acids, and nucleotides must be sufficient and balanced to supply the biosynthetic and energetic needs of the cell. To achieve balance, the cell must control de novo (new) synthesis of these compounds and sense the carbon and energy sources that are present in order to assemble the proper catabolic pathway.

The microbe also needs to detect hazardous conditions outside the cell and to distinguish whether it is floating

A.

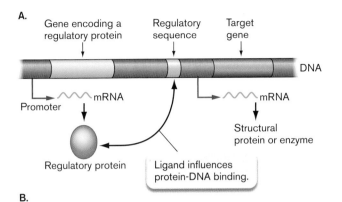

B.

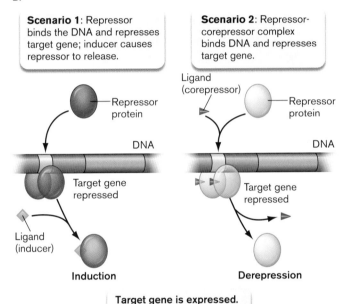

C.

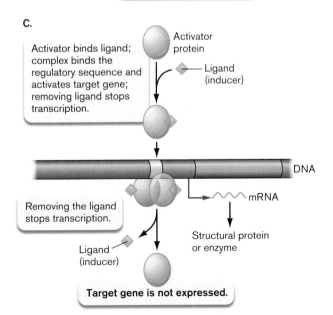

FIGURE 10.2 ▪ **General aspects of transcriptional regulation by repressor and activator proteins. A.** Schematic drawing of a regulatory system. The product of the regulatory gene (the regulatory protein) binds to DNA sequences near the promoter of the target gene and controls whether transcription occurs. Specific ligands influence binding. **B.** Repressor proteins bind to DNA sequences and prevent transcription. **C.** Activator proteins generally bind to specific chemical ligands in the cytoplasm before the protein can bind to DNA sequences near target genes. Activator proteins stimulate transcription.

in a pond of water, a gastrointestinal tract, or a mammalian host cell. Once the environment is sensed, the cell can change the repertoire of genes it expresses to meet its needs or to protect itself.

Cells use different mechanisms to sense and respond to conditions within the cell and outside the cell membrane. Sensing conditions within the cell is relatively straightforward. **Regulatory proteins** bind specific low-molecular-weight compounds called ligands (**Fig. 10.2A**). Different regulatory proteins bind different ligands. For example, one regulator binds a carbohydrate ligand and alerts the cell that a new carbon source is available, while a different regulator senses whether enough of the amino acid tryptophan is present in the cytoplasm to carry out protein synthesis. The ligand, once bound, then alters the ability of the regulatory protein to latch onto specific DNA regulatory sequences located near the promoters of target genes. The regulatory gene may be located near to or quite far away from the target gene on the chromosome.

Genes encoding regulatory proteins are usually, but not always, transcribed separately from the target gene (**Fig. 10.2A**). Regulatory proteins come in two forms: **repressors** and **activators** (note that activator proteins differ from activator <u>sequences of DNA</u>). Repressor proteins bind to regulator sequences and prevent the transcription of target genes—an event known as **repression**. Repression happens in one of two ways (scenarios 1 and 2 in **Fig. 10.2B**), depending on the repressor. In scenario 1, the repressor binds a specific DNA sequence and prevents transcription of a target gene. Relief from repression requires that a specific ligand, called an **inducer**, bind to the repressor protein, causing it to release from the DNA sequence. Because a small inducer molecule is required, the increased expression of the target gene is called **induction**. The lactose operon, discussed in Section 10.2, is one example of an inducible system.

Note: DNA regulatory sequences are sometimes called "operator sequences" if binding decreases expression of the target genes, or "activator sequences" if binding increases expression.

Other repressor proteins (scenario 2) bind poorly to DNA regulatory sequences unless they first bind a small ligand called a **corepressor**. As the ligand disappears from the cell, it is no longer available to bind to the repressor protein. When this happens, the repressor releases from the DNA and the target gene is expressed. This process is called **derepression** rather than induction. The tryptophan operon, also discussed later, is an example of a repression/derepression system.

Activator proteins also bind DNA, but they stimulate transcription by touching an RNA polymerase stuck at a nearby promoter, spurring it to initiate transcription (**Fig. 10.2C**). Most activator proteins bind poorly to DNA sequences, unless an inducer is present. When the intracellular concentration of inducer falls, the activator protein (without inducer) either leaves the DNA or moves to a nearby site from which it can no longer contact RNA polymerase. As a result, transcription of that target gene stops.

Figure 10.3 illustrates how one repressor protein, CI from lambda phage, binds to DNA. The protein

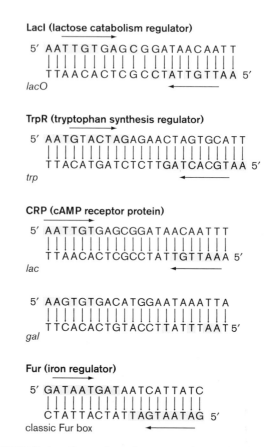

LacI (lactose catabolism regulator)

5′ AATTGTGAGCGGATAACAATT
    |||||||||||||||||||||
    TTAACACTCGCCTATTGTTAA 5′
lacO

TrpR (tryptophan synthesis regulator)

5′ AATGTACTAGAGAACTAGTGCATT
    ||||||||||||||||||||||||
    TTACATGATCTCTTGATCACGTAA 5′
trp

CRP (cAMP receptor protein)

5′ AATTGTGAGCGGATAACAATTT
    |||||||||||||||||||||
    TTAACACTCGCCTATTGTTAAA 5′
lac

5′ AAGTGTGACATGGAATAAATTA
    |||||||||||||||||||||
    TTCACACTGTACCTTATTTAAT 5′
gal

Fur (iron regulator)

5′ GATAATGATAATCATTATC
    |||||||||||||||||||
    CTATTACTATTAGTAATAG 5′
classic Fur box

FIGURE 10.4 ■ Examples of DNA regulatory sequences. The sequences shown are located upstream of the genes noted in italics. Inverted repeats are shown in yellow. Note that in some inverted repeats, occasionally a base is not repeated (only those bases that repeat are highlighted in the diagram). Arrows indicate the direction of symmetry.

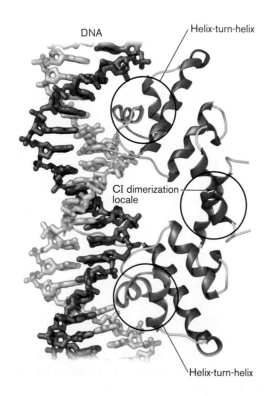

FIGURE 10.3 ■ Binding of a repressor protein to DNA. The dimer of lambda CI repressor binding to DNA. Note the helix-turn-helix motif located in two successive major grooves. Helices from this motif are brown and purple, whereas the turns are green. (PDB code: 1LMB)

forms a dimer, and one part of each molecule, called the DNA-binding domain, interacts with DNA in the major groove. DNA sites that bind regulatory proteins typically exhibit a sequence symmetry that involves an inverted repeat (**Fig. 10.4**). Dimers of a regulatory protein bind to the DNA, with each member of the duo binding half of the symmetrical DNA sequence. Protein-DNA binding can be demonstrated in several ways, as described in Section 12.3.

It is important to realize that all DNA-binding proteins can actually bind at a low level to any DNA sequence. However, each individual binding protein will bind more tightly to a specific sequence near a particular promoter. Thus, DNA-binding proteins exhibit a relative specificity for DNA target sequences, not absolute specificity. These proteins essentially "scan" DNA for high-affinity binding sites. Relative specificity helps explain how a promoter, through evolution, "recruits" the service of a regulatory protein when a nearby weak binding-site sequence changes by mutation into a tighter binding site.

Sensing the Extracellular Environment

Sensing what goes on outside the cell is more challenging than sensing intracellular conditions because intracellular regulatory proteins cannot reach through the membrane and touch what is outside. A common mechanism used by Gram-positive and Gram-negative organisms to collect information from outside of the cell relies on a series of two-member protein phosphorylation relay systems called **two-component signal transduction systems**. Each two-component system will regulate a different set of genes. The first protein in each relay, the **sensor kinase**, spans the membrane (**Fig. 10.5**). A kinase transfers a phosphoryl group from ATP to a protein. The sensory domain of most sensor kinase proteins contacts the outside environment (or periplasm), while the other end (the kinase domain) protrudes into the cytoplasm.

Each sensor protein of a two-component system recognizes a different molecule or condition (for example, PhoQ in *Salmonella* senses magnesium). Once activated, the external sensory domain triggers a conformational change in the kinase domain that activates a self-phosphorylation reaction. Phosphate from ATP is attached to a specific histidine residue located in a different part of the protein. Then, like two relay runners passing a baton, the phosphorylated sensor kinase protein passes the phosphate to a cognate (matched) cytoplasmic protein called a **response regulator**. This transfer, called transphosphorylation, occurs at a specific aspartate residue within the response regulator. The phosphorylated response regulator commonly binds to regulatory DNA sequences in front of one or more specific genes and activates or represses expression.

By linking different forms of regulation, the cell produces overlapping controls and complex integrated circuits that coordinate many aspects of cell physiology.

To Summarize

- **Expression or function of a gene product can be controlled at several levels:** DNA sequence, transcription, mRNA stability, translation, or posttranslation (by modifying the protein).

- **Small regulatory RNAs** also play key roles in affecting the transcription, mRNA turnover, and translation of specific genes or gene products.

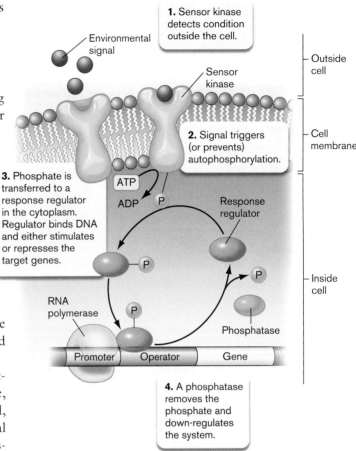

1. Sensor kinase detects condition outside the cell.

Environmental signal

Sensor kinase

Outside cell

Cell membrane

2. Signal triggers (or prevents) autophosphorylation.

3. Phosphate is transferred to a response regulator in the cytoplasm. Regulator binds DNA and either stimulates or represses the target genes.

ATP

ADP P

Response regulator

Inside cell

RNA polymerase

P

P

Phosphatase

Promoter   Operator   Gene

4. A phosphatase removes the phosphate and down-regulates the system.

FIGURE 10.5 ■ **Two-component signal transduction systems sense the external environment.** A transmembrane sensor kinase protein senses an environmental condition outside the cell (as in Gram-positive bacteria) or in the periplasm (as in Gram-negative bacteria).

- **Gene rearrangement controls** include invertible promoter switches or repetitive DNA sequences within a coding region that cause DNA polymerase to "slip" during DNA synthesis.

- **Regulatory proteins** help a cell sense changes in its internal environment and alter gene expression to match.

- **Repressor and activator proteins** bind to operator and activator DNA sequences, respectively, in front of target genes. Repressors prevent transcription; activators stimulate it.

- **Two-component signal transduction systems** help the cell sense and respond to its environment, both inside and outside.

- **Integrated control circuits** connect many individual regulatory systems to coordinate regulation throughout the cell.

A.

B.

FIGURE 10.6 ■ Discoverers of gene regulation. A. François Jacob (foreground) and Jacques Monod in 1971. **(B)** André Lwoff. This trio of scientists worked at the Pasteur Institute and won the 1965 Nobel Prize in Physiology or Medicine for their groundbreaking work on induction and gene regulation.

10.2 Operon Control

In 1961, French scientists Jacques Monod (1910–1976) and François Jacob (1920–2013) (**Fig. 10.6A**) proposed the revolutionary idea that the expression of genes could be regulated. They, among others, noticed that the enzyme used by *Escherichia coli* to consume the carbohydrate lactose was produced only when lactose was added to growth media. The term "induction" was coined to describe this phenomenon. The enzymes required to metabolize glucose, however, behaved differently. They were always

present (that is, constitutive) in the cell. The groundbreaking work of Jacob and Monod, and of André Lwoff (1902–1994) (**Fig. 10.6B**) for his study of phage lysogeny, won them a Nobel Prize and launched the field of gene regulation, a scientific realm where discoveries continue to surprise us.

Paradigm of the Lactose Operon

It took many years after Monod and Jacob's discovery to learn exactly how the lactose-degrading enzyme beta-galactosidase is induced. Many of the concepts presented here

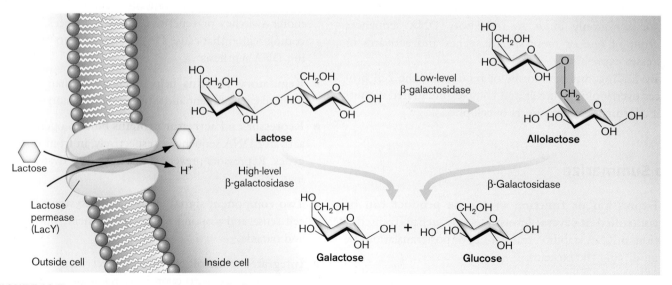

FIGURE 10.7 ■ Lactose transport and catabolism. A dedicated lactose permease uses proton motive force to move lactose (and a proton) into the cell. Once there, the enzyme beta-galactosidase (LacZ) can cleave the disaccharide into its component parts (galactose and glucose) or alter the linkage between the monosaccharides to produce allolactose, an important chemical needed to induce the genes that encode this pathway.

apply to numerous other bacterial gene systems, including some required for *Erwinia* infection of plants and *Streptococcus pneumoniae* infection of humans.

Lactose catabolism. Lactose is a disaccharide sugar made of glucose and galactose that can be used as a carbon and energy source (**Fig. 10.7**). The *E. coli* integral membrane protein that imports (transports) lactose from the extracellular environment is called LacY. The enzyme that subsequently cleaves lactose into its component parts is called

beta-galactosidase (LacZ). The products, glucose and galactose, are subsequently degraded by the enzymes of glycolysis to harness energy and capture carbon (see Section 13.5). Without LacY and LacZ, the catabolic energy of lactose is unavailable to the cell.

Lactose induces the *lac* operon. Figure 10.8A ▶ shows the genes in *E. coli* that encode the simple regulatory circuit for lactose catabolism. The genes *lacZ*, *lacY*, and *lacA* form an **operon**, which is a group of genes cotranscribed from a common promoter. The role of *lacA*, which encodes thiogalactoside transacetylase (LacA), is unclear. It is not needed to ferment lactose but may detoxify a harmful by-product of lactose metabolism.

> **Note:** "Lactose operon," "*lac* operon," and "*lacZYA* operon" all refer to the same system.

In the absence of lactose, the *lac* operon is transcribed at extremely low levels (fewer than ten molecules of LacZ per cell). The reason for low expression is that transcription of *lacZYA* is repressed by the protein product of the regulator gene *lacI* (**Fig. 10.8B**). The *lacI* gene is situated immediately upstream of *lacZYA* and is transcribed from a different promoter. A tetramer of LacI repressor protein forms in the cell and binds to two operator regions of DNA. One operator sequence is called *lacO*, which partially overlaps the *lacZYA* promoter (P_{lacZYA} or *lacP*). The second operator site is found within *lacI* and is called $lacO_I$ (**Fig. 10.8A**). The *lac* operators control whether the three structural genes are transcribed. The LacI tetramer

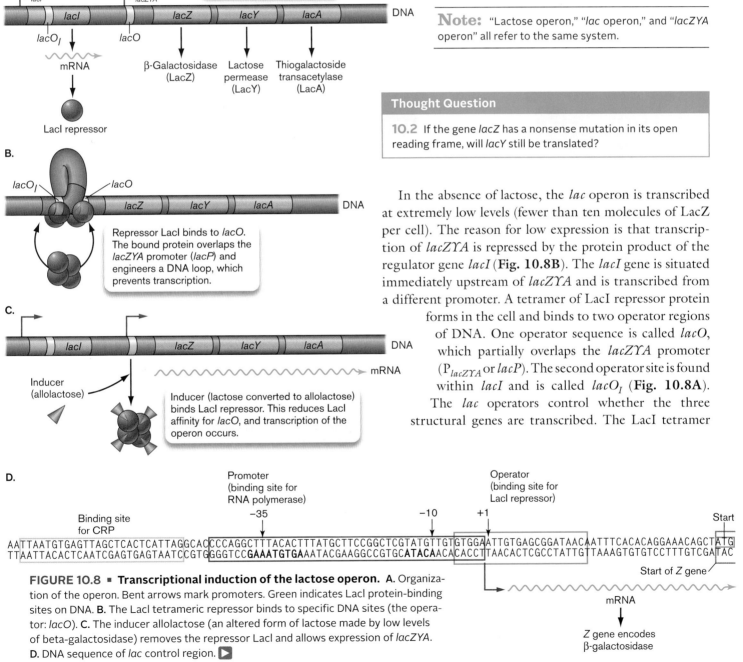

FIGURE 10.8 ■ Transcriptional induction of the lactose operon. A. Organization of the operon. Bent arrows mark promoters. Green indicates LacI protein-binding sites on DNA. **B.** The LacI tetrameric repressor binds to specific DNA sites (the operator: *lacO*). **C.** The inducer allolactose (an altered form of lactose made by low levels of beta-galactosidase) removes the repressor LacI and allows expression of *lacZYA*. **D.** DNA sequence of *lac* control region. ▶

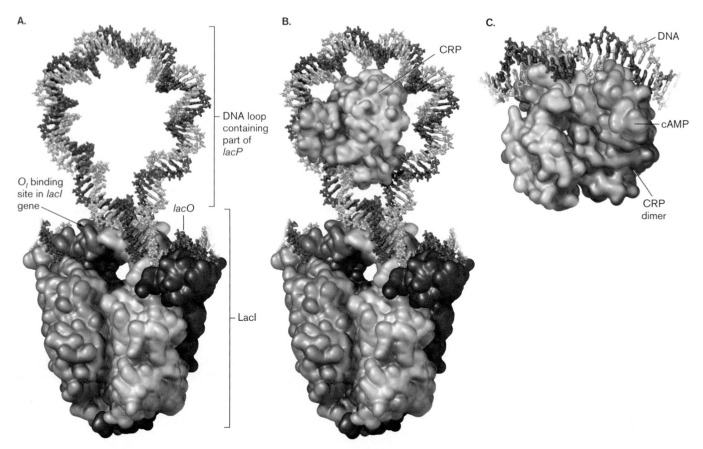

FIGURE 10.9 ■ **Regulatory protein interactions with DNA at the *lacZYA* control region.** **A.** LacI repressor binds to two operator regions, *lacO* and *lacO_I*, so that the DNA forms a loop. (PDB codes: 1ZO4, 1K8J) **B.** If no lactose is present, LacI will prevent cAMP-CRP from activating transcription. (PDB codes: 1ZO4, 1K8J, 2CGP) **C.** If only lactose is present (no glucose), cAMP-CRP binds to the binding site near the *lac* promoter, bends DNA, and interacts with RNA polymerase.

simultaneously binds the *lacO* and *lacO_I* operators (a dimer at each site). As a result, the intervening DNA loops out (**Figs. 10.8B** and **10.9A**) and prevents RNA polymerase from continuing transcription into the structural *lacZYA* genes. (A third binding site for LacI exists but is not important to our discussion here.)

Once lactose is added to the medium, the *lac* operon is expressed at 100-fold higher levels. How does lactose get into the cell to induce the operon if the operon encoding its transport protein is repressed? As noted already, even when repressed, the *lacZYA* operon is transcribed at a low level. (Most genes are expressed at low constitutive levels even when uninduced.) This means that a small amount of lactose can be transported into the cell because a few molecules of the lactose transporter LacY are made. The tiny amount of beta-galactosidase (LacZ) expressed in uninduced cells is also important. At this low level, the enzyme does not completely cleave the glycosidic bond of lactose, but rearranges it to form allolactose, the form of lactose that activates the operon (whose structure is shown in **Fig. 10.7**). Allolactose binds the

repressor and "unlocks" the protein (by altering the conformation of LacI) so that it is released from the operator (**Fig. 10.8C**). Once this happens, RNA polymerase guided by sigma factor (not shown in **Fig. 10.8C**) can find the *lac* promoter sequences and initiate transcription of the *lacZYA* structural genes.

cAMP and cAMP receptor protein stimulate transcription. Another important mechanism that governs the level of *lacZYA* transcription in *E. coli* involves a small molecule called cyclic AMP (cAMP), which accumulates when a cell is starved for carbon. Cyclic AMP is a derivative of AMP (adenosine monophosphate) in which the 5′ phosphate is linked to the 3′ OH group of the ribose, making a cyclic structure. How cellular levels of cAMP change in response to carbon starvation has been extensively studied, but questions still remain. What is clear is that intracellular cAMP levels fluctuate with the activities of enzymes that can make (adenylyl cyclase), degrade (phosphodiesterases), or export cAMP. Various internal and external signals alter the balance of these processes (see **eTopic 10.2**).

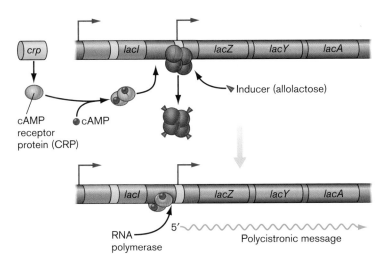

FIGURE 10.10 ■ Activation of the *lacZYA* operon by cAMP and CRP. Although the inducer allolactose removes the repressor LacI and allows expression of *lacZYA*, maximum expression requires the presence of cAMP and cAMP receptor protein (CRP), which bind to a separate site at the *lacZYA* promoter. Bound CRP can interact with RNA polymerase and increase the rate of transcription <u>initiation</u>.

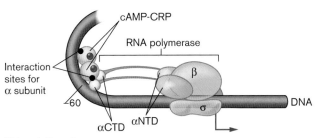

Abbreviations key

CRP: cAMP receptor protein
αCTD: C-terminal domain of RNA polymerase alpha subunit
αNTD: N-terminal domain of RNA polymerase alpha subunit

FIGURE 10.11 ■ CRP interactions with RNA polymerase. Promoters like the one at the *lacZYA* operon possess a CRP-binding site positioned about −60 bp from the transcriptional start. CRP on these promoters interacts with the alpha subunit C-terminal domain (αCTD) of RNA polymerase. *Source:* Adapted from Moat et al. 2002. *Microbial Physiology*, 4th ed. Wiley-Liss.

Note: Recall that a gene name written in roman (not italic) type and with a capital first letter is really the name of the protein product of that gene. Thus, *lacZ* is the gene, but LacZ is the protein. In presenting a genotype, gene names written with a superscript + are considered wild-type genes. Gene names written <u>without</u> a superscript + are considered mutant genes.

As it accumulates, cyclic AMP controls the expression of many genes by combining with a dimeric regulatory protein called cAMP receptor protein (CRP). The cAMP-CRP complex binds to specific DNA sequences located near many bacterial genes and modifies their transcription, usually acting as an activator. This is the case for the *lacZYA* operon (**Figs. 10.10** and **10.8D**).

Note: The original name for CRP was "catabolite activator protein" (CAP), given to reflect its role in catabolite repression. Though some investigators still use the "CAP" designation, the term "CRP" is generally preferred today. Thus, the gene encoding the protein is called *crp*.

How does the cAMP-CRP complex ultimately activate the expression of *lacZYA*? RNA polymerase cannot easily form an open complex (see Section 8.2) and is essentially stuck at *lacP*, even in the absence of a LacI repressor. This problem is overcome when the cAMP-CRP complex binds to a 22-bp DNA sequence located −60 bp from the start of transcription (just upstream of the *lacZYA* operator), causing the DNA to bend (see **Figs. 10.8D** and **10.9C**). CRP can then directly interact with the alpha subunit of RNA polymerase bound at the *lac* promoter and activate transcription (**Fig. 10.11** and **eTopic 10.3**). Notice in **Figure 10.9B** that when neither glucose nor lactose is present, cAMP-CRP cannot by itself activate *lacZYA* transcription, because it binds within the DNA loop formed by the LacI (repressor) tetramer, leaving no room for RNA polymerase to bind to DNA.

Thought Questions

10.3 Predict the effects of the following null mutations on the induction of beta-galactosidase by lactose, and predict whether the *lacZ* gene is expressed at high or low levels in each case: The inactivated, mutant genes to consider are *lacI*, *lacO*, *lacP*, *crp*, and *cya* (the gene encoding adenylyl cyclase). What effects will those mutations have on catabolite repression?

10.4 Predict what will happen to the expression of *lacZ* when a second copy of the *lac* operon region containing various mutations is present on a plasmid. The genotypes of these partial diploid strains are presented as chromosomal genes/plasmid gene. (a) *lacI⁻ lacO⁺P⁺Z⁺Y⁺A⁺*/plasmid *lacI⁺*; (b) *lacO⁻ lacI⁺P⁺Z⁺Y⁺A⁺*/plasmid *lacO⁺*; (c) *crp⁻ lacI⁺O⁺P⁺Z⁺Y⁺A⁺*/plasmid *crp⁺*.

Glucose represses the *lac* operon. What happens if, in addition to lactose, the medium contains an alternative carbon source, such as glucose? Enzymes for glucose catabolism (glycolysis) are always produced at high levels because glucose is the favored catabolite (carbon source), providing the quickest source of energy. Many carbohydrates, including lactose, must first be converted to glucose to be catabolized. So, in the interest of greater efficiency, *E. coli* forestalls inducing the *lac* operon while glucose is present.

A.

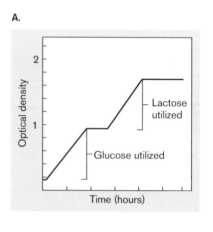

B.

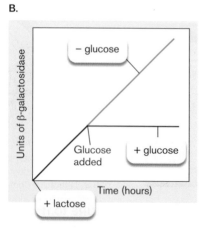

FIGURE 10.12 ▪ Catabolite repression of the *lacZYA* operon. A. The diauxic growth curve of *E. coli* growing on a mixture of glucose and lactose. **B.** Glucose repression of LacZ (beta-galactosidase) production. Lactose was added at the beginning of the experiment to parallel cultures, and beta-galactosidase activity was measured. At the point indicated, glucose was added to one culture. From that point on, synthesis of beta-galactosidase continued to increase in the culture that lacked glucose but stopped in the culture that had glucose.

The phenomenon is known as **catabolite repression**, in which the presence of a more favorable catabolite (commonly glucose) prevents expression of operons enabling catabolism of a second carbohydrate.

When glucose and lactose are both present in the medium, cells initially grow by breaking down glucose until the glucose is depleted. Growth temporarily stops while lactose present in the medium induces the *lacZYA* operon. Once induced, cells can consume lactose and resume growth. The resultant biphasic growth curve is often called **diauxic growth** (**Fig. 10.12A**). But if lactose was present from the start, why was *lacZYA* turned off? It was turned off because glucose indirectly prevents the induction of *lacZYA*. **Figure 10.12B** illustrates that even when *lacZYA* is already induced, adding glucose stops (or represses) induction.

Catabolite repression via inducer exclusion. Failure of lactose to induce *lacZYA* during growth on glucose (**Fig. 10.12B**, red horizontal line) is due mainly to the fact that growth on glucose keeps lactose out of the cell. This phenomenon is known as **inducer exclusion**. If lactose cannot enter the cell, the *lacZYA* operon cannot be induced. The key to inducer exclusion is that a component of the glucose transport system (phosphotransferase system, or PTS; see **eTopic 4.1**), while transporting glucose, will bind to and inhibit LacY permease (**Fig. 10.13**).

The PTS transfers a phosphate from phosphoenolpyruvate (PEP) along a series of proteins to glucose during transport. When glucose <u>is</u> present, the glucose transport proteins continually transfer phosphate to glucose and, so, are usually left without phosphate (**Fig. 10.13A**). Unphosphorylated enzyme IIA interacts with LacY in the membrane and inhibits LacY activity. So, lactose cannot enter the cell, and the *lac* operon remains uninduced. When glucose is <u>not</u> present, the PTS proteins remain phosphorylated; enzyme IIA-P does not interact with LacY, so LacY transports lactose into the cell; and the *lac* operon is induced (**Fig. 10.13B**).

A. Glucose present

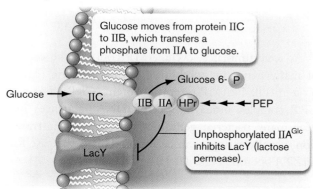

B. Glucose absent

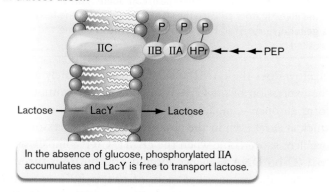

FIGURE 10.13 ▪ Glucose transport via the phosphotransferase system inhibits LacY (lactose permease). A. Inducer exclusion. Phosphoenolpyruvate (PEP) "feeds" phosphate into the PTS, which relays the phosphate to glucose during transport. The level of unphosphorylated IIA^Glc is high because glucose continually siphons off the phosphate. Unphosphorylated IIA^Glc inhibits LacY (lactose permease) activity to keep lactose from entering the cell. **B.** In the absence of glucose, the phosphorylated forms of glucose-specific IIA^Glc and IIBC^Glc accumulate and cannot inhibit LacY. LacY transports lactose, and the *lac* operon is induced.

TABLE 10.1	Examples of The AraC/XylS Family of Transcriptional Regulators	
Regulator	**Organism**	**Function**
ExsA	*Pseudomonas aeruginosa* (human pathogen)	Controls type III secretion system in response to Ca^{2+}
ToxT	*Vibrio cholerae* (human pathogen)	Controls several virulence genes, including one encoding cholera toxin
TxtR	*Streptomyces scabies* (taproot pathogen)	Controls synthesis of thaxtomin, a plant toxin
RipA	*Corynebacterium glutamicum* (soil bacterium)	Represses aconitase and other iron-containing proteins
YbtA	*Yersinia pestis* (human pathogen)	Controls iron uptake transporters
NitR	*Rhodococcus rhodochrous* (plant pathogen)	Regulates synthesis of indole-3-acetic acid

Transport of some sugars through the PTS also affects adenylyl cyclase activity, but this mechanism is not a major cause of glucose/lactose diauxic growth (see **eTopic 10.2**).

> **Thought Question**
>
> **10.5** Researchers often use isopropyl-β-D-thiogalactopyranoside (IPTG) rather than lactose to induce the *lacZYA* operon. IPTG resembles lactose, which is why it can interact with the LacI repressor, but it is not degraded by beta-galactosidase. Why do you think the use of IPTG is preferred in these studies?

Other Systems of Operon Control

Besides the *lac* operon model, bacteria have a surprising number of different ways to regulate operons. In this section we discuss proteins that have dual regulatory functions—performing as both activators (positive control) and repressors (negative control)—and we'll see how an idling ribosome sends a chemical signal that affects the expression of many genes and operons.

The AraC/XylS Family of Transcriptional Regulators

What could be better than having a regulatory protein that either represses (like LacI) or activates (like CRP) expression of an operon? How about a regulator that can repress and activate gene expression, depending on whether a carbohydrate substrate is available? A regulator this versatile will provide very tight control over operon expression. One such regulator, called AraC, regulates the genes encoding arabinose catabolism. When arabinose is absent, AraC represses expression of the genes that break down arabinose; but when arabinose is present, AraC activates these same genes. The products of these genes ultimately convert the five-carbon sugar L-arabinose to D-xylulose 5-phosphate, an intermediate in the pentose phosphate

shunt, a pathway that provides reducing energy for biosynthesis (see Section 13.5).

Bioinformatic computer analysis of microbial genomes reveals a family of over 10,000 regulators with homology to AraC and the closely related XylS activator (xylose catabolism). For instance, YbtA is an AraC-type regulator of the *Yersinia pestis* pesticin/yersiniabactin receptor for iron uptake. The signature sequence of these proteins is found at the C-terminal end and contains DNA-binding motifs. The AraC/XylS family members are a diverse group of intracellular sensors that collectively regulate a variety of different cell functions, including carbon metabolism and virulence, as well as many responses to environmental conditions (**Table 10.1**).

The AraC/XylS advantage. One advantage of AraC/XylS family regulators, based on the AraC prototype, is that they remain affixed to their target genes. A drawback of the LacI repressor strategy is that the repressors in this family must fully dissociate from operator DNA during induction and disperse throughout the cell. Reestablishing repression requires slow, random diffusion of bulky LacI proteins back to the *lacO* DNA sequence. This takes some time. The AraC/XylS family strategy keeps the regulatory protein bound near target operons. The regulators simply shuttle back and forth between repressor and activator DNA-binding sites, shortening the delay between induction and repression. The known small chemical inducer molecules for these regulators quickly diffuse through the cytoplasm to find their cognate regulatory proteins already camped at target promoters.

The AraC strategy. The model for regulation by AraC/XylS regulators is based primarily on AraC itself, the most intensely studied family member. AraC is a 33-kDa protein with a DNA-binding C-terminal domain attached by a flexible linker peptide to the N-terminal dimerization domain (**Fig. 10.14A**). AraC forms a dimer in vivo that can assume one of two conformations, depending on whether arabinose is available. When arabinose is absent, the dimer exists

FIGURE 10.14 ▪ Regulation of the *araBAD* operon by the AraC regulator. A. View of AraC showing dimerization and DNA-binding domains. **B.** Alternative conformations of AraC dimers. The different conformations change the location of where the dimer can bind DNA. The region shown is about 400 bp. Note that O_2 is an operator, and I_1 and I_2 are other DNA-binding sites that, when occupied by AraC, induce expression.

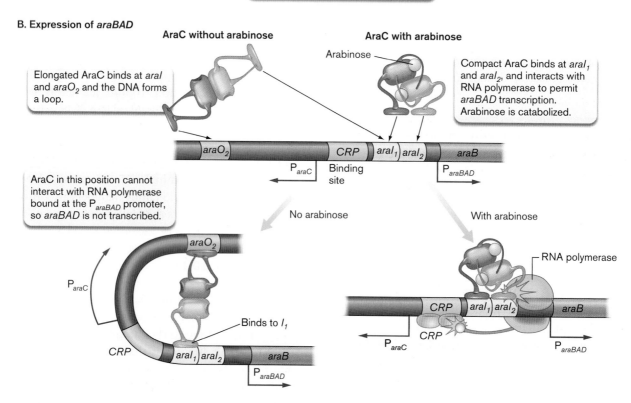

A. Structure of AraC

C-terminal DNA-binding domain

Flexible linker

N-terminal arm

N-terminal dimerization domain

Arabinose-binding site

AraC forms a dimer. Without arabinose, the N-terminal arm binds its own C-terminal DNA-binding domain. When bound to arabinose, the N-terminal arm binds to the dimerization domain of the other monomer.

B. Expression of *araBAD*

AraC without arabinose

Elongated AraC binds at *araI* and *araO₂* and the DNA forms a loop.

AraC with arabinose

Arabinose

Compact AraC binds at *araI₁* and *araI₂*, and interacts with RNA polymerase to permit *araBAD* transcription. Arabinose is catabolized.

araO₂ | CRP Binding site | araI₁ | araI₂ | araB

P$_{araC}$ P$_{araBAD}$

AraC in this position cannot interact with RNA polymerase bound at the P$_{araBAD}$ promoter, so *araBAD* is not transcribed.

No arabinose

With arabinose

araO₂

P$_{araC}$

Binds to I_1

CRP | araI₁ | araI₂ | araB

P$_{araBAD}$

RNA polymerase

CRP | araI₁ | araI₂ | araB

CRP

P$_{araC}$ P$_{araBAD}$

---

in a rigid, elongated form because the N-terminal arm of each monomer binds to its own C-terminal domain. This form represses expression of the *araBAD* operon for arabinose degradation. When arabinose is present, however, the dimer assumes a more compact form because the N-terminal arm of one monomer binds to the N-terminal domain of the other monomer. This compact form activates expression of *araBAD* (**Fig. 10.14B**).

How does AraC act as both repressor and activator? AraC is able to bind to three important DNA-binding sites. Two of them, O_2 and I_1, are widely separated and flank a binding site for cAMP-CRP that, as discussed in Section 10.2, is a global activator of many bacterial genes, including *araBAD*. The elongated AraC dimer (no arabinose) can bind these two sites. The result is a looped DNA structure that blocks the CRP-binding site and limits activation by cAMP-CRP. The *araBAD* operon is not expressed. However, the I_1 site sits next to another site, I_2, near the promoter. The compact AraC dimer (with

arabinose) can bind only to these sites. Binding to $I_1 I_2$ prevents DNA loop formation, allows cAMP-CRP access to the CRP DNA-binding site, and brings AraC close enough to physically contact an RNA polymerase already bound (but stuck) at the *araBAD* promoter. Transcription of the *araBAD* operon begins.

According to this model, the DNA region is rarely free of AraC. Arabinose moves the dimer from one site (repression) to the other (induction). AraC-like proteins show strong family resemblance at their DNA-binding domains, but homology usually disappears at the other end of the protein, the dimerization domain. The dimerization domain is where these proteins appear to bind or respond to a particular ligand (for example, arabinose). For most of the AraC/XylS family members, the identity of the ligand remains a mystery that, when solved, can yield interesting discoveries. For example, the AraC-like regulator MarR binds salicylate (an aspirin metabolite) to activate multiple genes conferring antibiotic resistance.

10.6 When the *lacI* gene of *E. coli* is missing because of mutation, the *lacZYA* operon is highly expressed regardless of whether lactose is present in the medium. Based on **Figure 10.14** and its illustration of arabinose operon expression, what would happen to *araBAD* expression if AraC was missing? Why?

Repression of Anabolic (Biosynthetic) Pathways

Repressing biosynthetic pathways is fundamentally different from repressing catabolic (degradative) systems. Repressor proteins that control catabolic pathways, such as lactose degradation, typically bind the initial substrate or a closely related product (for example, allolactose in the case of the *lac* operon). Binding the substrate decreases repressor protein affinity for operator DNA. Thus, increased concentration of the substrate or inducer actually removes the repressor from the operator and <u>derepresses</u> expression of the operon. This derepression makes sense because the cell "wants" to make the enzymes that use the substrate as a carbon and energy source.

In contrast, genes encoding biosynthetic enzymes are regulated by repressors (called inactive aporepressors) that must bind the end product of the pathway (for example, tryptophan for the *trp* operon) to become active repressors. The pathway product that binds the aporepressor is called a corepressor. Binding of the corepressor (end product) to the repressor increases the repressor's affinity for the operator sequence upstream of the target gene or operon. As the concentration of end product, such as an amino acid or nucleotide, increases in the cell beyond what is needed to support growth, the cell will shut the biosynthetic pathway down and not waste energy making a superfluous pathway or compound.

The Tryptophan Operon: Repression and Attenuation

As just noted, many amino acid biosynthetic pathways are controlled by transcriptional repression, in which a repressor protein binds to a DNA operator sequence to prevent transcription. For instance, when internal tryptophan levels exceed cellular needs, the excess tryptophan (acting as a corepressor) will bind to an inactive repressor protein, TrpR, converting it to an active repressor (**Fig. 10.15**). TrpR repressor then binds to an operator DNA sequence positioned upstream of the tryptophan (*trp*) operon,

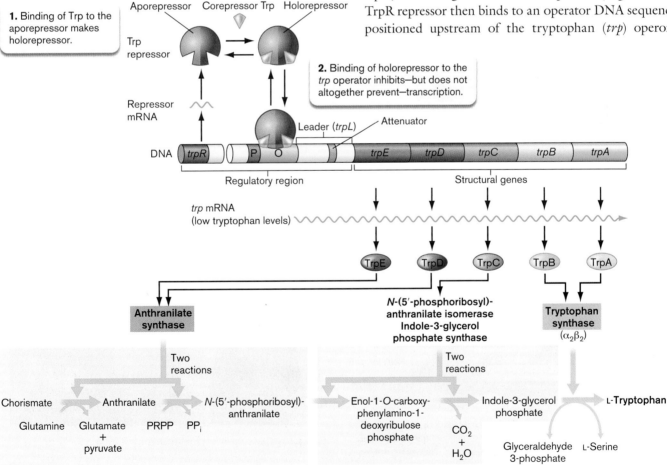

FIGURE 10.15 ■ **The tryptophan biosynthetic pathway in *E. coli* and repression of the tryptophan operon.** The tryptophan biosynthetic enzymes and their encoding genes are shown. TrpR aporepressor (inactive repressor) binds excess tryptophan when intracellular concentration exceeds need. The holorepressor (active repressor) then binds to the *trp* operator and greatly reduces transcription. Note the long polycistronic message in blue. Repression lowers expression about 100-fold. PRPP = phosphoribosyl pyrophosphate.

which encodes the enzymes required for tryptophan biosynthesis. Repressor bound to the *trp* operator represses expression by blocking RNA polymerase.

Repression, however, is not the whole story in regulating the *trp* operon. Many amino acid biosynthetic operons, including the *trp* operon, have adopted a second strategy for down-regulating tryptophan synthesis, which can be used alone or in conjunction with repression. This second mechanism is called **transcriptional attenuation**. Attenuation uses the ribosome as a sensor of amino acid levels. It is a transcriptional control mechanism in which the ability to translate part of an mRNA determines whether the transcribing RNA polymerase can continue transcription. Because attenuation halts transcription in progress, it affords an even quicker response to changing amino acid levels than does simple repression.

Transcriptional attenuation was discovered by Charles Yanofsky (**Fig. 10.16A** ▶) and his colleagues at Stanford University. While examining the beginning of the *trp* operon in *E. coli*, they discovered an odd DNA region, called the leader sequence, located between the *trp* operator and *trpE*, the first structural gene of the operon. The leader sequence encodes a short peptide, but the peptide has no enzymatic function. Its main, if not only, role is "to be or not to be."

Key features of the leader sequence include a pair of tryptophan codons (UGGUGG), a translational stop codon, and

B. Stem loop structures in attenuator region

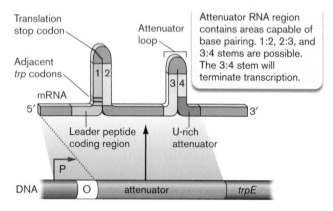

C. High tryptophan levels

1. Ribosome translates through *trp* codons and encounters translation stop codon.

2. Ribosome stops, covering mRNA regions 1 and 2. Polymerase continues to transcribe regions 3 and 4. The 3:4 termination loop forms.

3. The 3:4 loop binds RNA polymerase and causes its release before reaching *trpE*.

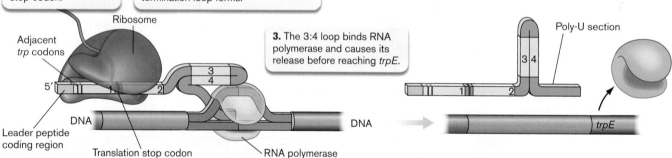

D. Low tryptophan levels

1. Ribosome translates leader.

2. Scarce tRNA^Trp makes ribosome stall at *trp* codons. Polymerase continues through attenuator.

3. Stalled ribosome covers region 1, allowing 2:3 stem loop to form. The less energetically favorable 3:4 transcription terminator loop cannot form.

4. Polymerase transcribes *trpE*.

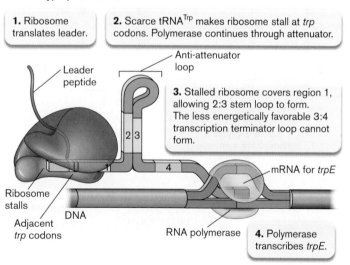

A.

FIGURE 10.16 ■ **The transcriptional attenuation mechanism at the *trp* operon.** **A.** Charles Yanofsky was instrumental in discovering attenuation and several other gene regulatory mechanisms while studying tryptophan metabolism. Here he is seen receiving the National Medal of Science from President George W. Bush in 2003. **B.** Relationship between the mRNA attenuator region and encoding DNA. **C.** Attenuation when *E. coli* is growing in high tryptophan concentrations. **D.** Transcriptional read-through when *E. coli* is growing in low tryptophan concentrations. tRNA^Trp = tryptophanyl-tRNA. ▶

four complementary nucleotide stretches within the leader mRNA. These regions, numbered 1–4, can base-pair to form competing stem loop structures (**Fig. 10.16B**). Two of the stem loop structures are critical to the mechanism. These are the **anti-attenuator stem loop** formed by regions 2 and 3 and the **attenuator stem loop** (or terminator stem loop) formed by regions 3 and 4. If the 3:4 attenuator stem loop forms, then RNA polymerase is ejected and transcription stops (**Fig. 10.16C**). Formation of the 2:3 anti-attenuator stem loop, however, prevents formation of the 3:4 stem loop because the 2:3 stem is longer and more thermodynamically stable than the 3:4 stem. The anti-attenuator stem, if it forms, allows RNA polymerase to transcribe into *trpE* (**Fig. 10.16D**). But what controls which stem loop forms?

High tryptophan levels. Because the ribosome is very large, it can barrel through RNA stem loop structures. When the cell is replete with charged tryptophanyl-tRNA and needs no more (**Fig. 10.16C**), the ribosome quickly translates through the key tryptophan codons in the leader sequence but runs into a translation stop codon between regions 1 and 2. The ribosome <u>stops</u> in this position, enveloping region 2 and preventing formation of the 2:3 stem. As a result, once RNA polymerase transcribes through region 4, the 3:4 attenuator stem snaps together. The attenuator stem then interacts with the RNA polymerase ahead of it and halts transcription. As you would expect, the ribosome dissociates after reaching the region 2 stop codon, but because the 3:4 stem loop is already in place, the anti-attenuator 2:3 stem loop does not form. Subsequent ribosome release leads to formation of a 1:2 stem structure, precluding all possibility of regions 2 and 3 annealing.

Low tryptophan levels. However, if the level of charged tryptophanyl-tRNA is low (**Fig. 10.16D**), then the ribosome following behind RNA polymerase stalls over the tryptophan codons. Because these codons occur right at the beginning of region 1, the ribosome does not cover region 2. So, as soon as RNA polymerase transcribes region 3, the 2:3 anti-attenuator stem loop forms and stops formation of the 3:4 attenuator stem loop. The result is that RNA polymerase can continue into the structural genes, and ultimately more tryptophan is made. (A new ribosome binds to a ribosome-binding site at the *trpE* message.)

Realize that for the *trp* operon, attenuation is a fine-tuning mechanism. The repressor provides the majority of control. However, transcriptional attenuation is a common regulatory strategy used to control many operons that code for amino acid biosynthesis. Note that even though translation is part of the attenuation control mechanism, attenuation is <u>not</u> considered translational control. The reason is that RNA polymerase, rather than the ribosome, is the target of the control. An example of translational control is discussed shortly.

The Stringent Response

During transitions from nutrient-rich to nutrient-poor conditions, microbes must contend with dramatic fluctuations in growth rate. This variation presents a problem. When a cell is growing rapidly, its molecular machinery is geared for peak performance. The pace of synthesizing new ribosomes is frenetic, trying to keep up with rapid cell division. The more ribosomes a cell contains, the faster that cell can make new proteins and the faster it can grow. But what happens when the party's over—when poor carbon and energy sources cannot supply enough energy to maintain rapid cell division? Without a way of curbing ribosome construction, cells would soon fill with idle ribosomes. Under these conditions, bacteria undergo a process called the **stringent response**. The stringent response causes a decrease in the number of rRNA transcripts made for ribosome assembly and alters the expression of numerous other genes.

In the stringent-response strategy, idling ribosomes produce a small signaling molecule called guanosine tetraphosphate (ppGpp), which interacts with RNA polymerase and lowers the enzyme's ability to transcribe genes encoding ribosomal RNA (**Fig. 10.17**). How is ppGpp made? When an uncharged tRNA binds at the ribosome A site, which can happen during amino acid starvation, a ribosome-associated protein called RelA transfers phosphate from ATP to GTP to form ppGpp. This signal nucleotide interacts with the beta subunit of RNA polymerase and diminishes its recognition of promoters for operons producing rRNA and tRNA. The result is the down-regulation of rRNA and tRNA synthesis. The less rRNA there is available for building ribosomes, the fewer ribosomes will be produced.

This raises another question. Even though the synthesis of ribosomal RNA has been curtailed, won't the cell

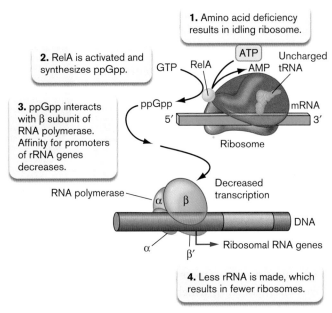

1. Amino acid deficiency results in idling ribosome.

2. RelA is activated and synthesizes ppGpp.

GTP   RelA   ATP   Uncharged
AMP   tRNA

3. ppGpp interacts with β subunit of RNA polymerase. Affinity for promoters of rRNA genes decreases.

ppGpp

mRNA

5′   3′

Ribosome

RNA polymerase

α   β

Decreased transcription

DNA

α   β′   Ribosomal RNA genes

4. Less rRNA is made, which results in fewer ribosomes.

FIGURE 10.17 ■ **Ribosome-dependent synthesis of guanosine tetraphosphate and the stringent response.**

continue to waste resources on the synthesis of ribosomal proteins? It turns out that some ribosomal proteins can bind to the mRNA that encodes them and inhibit translation. So, when rRNA levels in the cell are low, free ribosomal proteins accumulate in the cytoplasm, unassociated with ribosomes. These excess ribosomal proteins begin to bind to their own mRNA molecules and inhibit the translation of their own coding regions, as well as the coding regions of other ribosomal proteins residing on the same polycistronic mRNA. This process is called **translational control** because regulation affects the translation of an mRNA by ribosomes rather than transcription by RNA polymerase.

To Summarize

- **DNA-binding proteins** often recognize symmetrical DNA sequences.

- **The *lacZYA* operon of *E. coli* is controlled by repression and activation.** The **tetrameric LacI repressor** binds operator DNA sequences to prevent RNA polymerase from accessing the promoter. **Allolactose** (rearranged lactose) binds LacI, reduces repressor affinity for the operator, and allows induction of the operon.

- **The cAMP-CRP complex** activates *lacZYA* transcription by interacting with the C-terminal domain of the RNA polymerase alpha subunit. cAMP-CRP regulates many types of operons.

- **In catabolite repression**, a preferred carbon source (for example, glucose) prevents the induction of an operon (for example, *lac*) that enables catabolism of a different carbon source (lactose). Glucose transport through the phosphotransferase system causes catabolite repression by inhibiting LacY permease activity (inducer exclusion) and lowers cAMP levels.

- **Many anabolic pathway genes** (for example, for tryptophan biosynthesis) are repressed by the end product of the pathway (for example, the amino acid), which binds to a corepressor that inhibits its transcription.

- **AraC-like proteins** are a large family of regulators, present in many bacterial species, that can activate and repress a target operon by assuming different conformations.

- **Attenuation** is a transcriptional regulatory mechanism in which translation of a leader peptide affects mRNA structure to influence transcription of a downstream structural gene.

- **The stringent response** is triggered during nutrient limitation when low cellular amino acid levels cause ribosomes to idle and synthesize the signaling molecule ppGpp. Binding of ppGpp to RNA polymerase decreases synthesis of rRNA, which slows the rate of new ribosome synthesis. The overall rate of translation, then, will match growth rate.

10.3 Sigma Factors and Regulatory RNAs

We have seen how repressor and activator proteins can sense the substrates or end products of biochemical pathways and influence the transcription of individual genes and operons. Might the cell also control the expression of large numbers of genes by modulating the levels or activities of sigma factors? And might RNA molecules themselves sense changes in cell physiology and posttranscriptionally regulate the translation or stability of other RNA molecules? The answer to both of these questions is yes.

Sigma Factor Activity Can Be Regulated

Many bacteria employ alternative sigma factors to direct transcription of distinct sets of genes (see Section 8.2). For example, many Gram-negative bacteria, such as *E. coli* or *Salmonella enterica*, use the sigma factor called sigma S (also called sigma-38, RpoS, σ^S, or σ^{38}) to initiate the

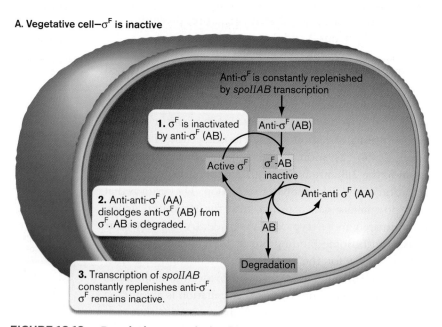

A. Vegetative cell—σ^F is inactive

Anti-σ^F is constantly replenished by *spoIIAB* transcription

1. σ^F is inactivated by anti-σ^F (AB).

Anti-σ^F (AB)

Active σ^F

σ^F-AB inactive

Anti-anti σ^F (AA)

2. Anti-anti-σ^F (AA) dislodges anti-σ^F (AB) from σ^F. AB is degraded.

AB

Degradation

3. Transcription of *spoIIAB* constantly replenishes anti-σ^F. σ^F remains inactive.

FIGURE 10.18 ▪ Regulating sporulation by genetic asymmetry. A. Summary of mechanisms that control sigma F activity in growing cells of *Bacillus subtilis*. The interactions of sigma F, anti-sigma F (AB), and anti-anti-sigma F (AA) collectively keep sigma F inactive. **B.** Sigma F is activated in the forespore until the forespore chromosome is pumped into the forespore. Activation of sigma F leads to the synthesis of sigma G in the forespore (not shown), which directs the next stage of spore development.

transcription of a variety of genes that promote survival in stationary phase. In large part, *Salmonella* orchestrates the stationary-phase accumulation of sigma S by modulating proteolysis of the factor (protein degradation is discussed in Section 8.4). ClpXP protease degrades sigma S rapidly during exponential growth. When cells enter stationary phase or experience an environmental stress that slows growth, degradation of sigma S stops. Sigma S levels increase in these situations and trigger the expression of stress survival genes.

Other sigma factors are controlled by **anti-sigma factor** proteins that inhibit sigma factor activity. Anti-sigma factor proteins target specific sigma factors and block access to core RNA polymerase. The anti-sigma strategy prevents expression of target genes until they are needed. For example, *Salmonella* uses an anti-sigma factor (FlgM) to time events involved in constructing flagella. The transcription factor sigma F is required to synthesize proteins used in the final stages of flagellar biosynthesis. The anti-sigma factor FlgM, however, keeps sigma F function at bay until membrane assembly of the flagellar basal bodies is complete. Once complete, the basal body selectively secretes the anti-sigma factor from the cell. Secreting FlgM from the cell, then, frees intracellular sigma F to direct transcription of the final set of flagellar assembly genes.

Anti-sigma factors can themselves be neutralized by **anti-anti-sigma factors** that bind the anti-sigma factor more tightly than sigma factor does. The anti-anti-sigma factor acts as a decoy to release anti-sigma factor from the actual sigma factor. Freed sigma factor can then join core RNA polymerase and direct transcription of target genes.

Many mechanisms can liberate sigma factors from anti-sigma factors. The next section describes one linked to a cell differentiation event.

In Sporulation, Different Sigma Factors Are Activated in the Mother Cell and Forespore

Bacillus and *Clostridium* species such as *B. anthracis* (anthrax) and *C. tetani* (tetanus) are soil organisms that produce spores when nutrients become scarce. Sporulation requires an asymmetrical cell division in which one of the compartments, called the forespore, ultimately becomes the spore (see Section 4.6). The regulation of this system is quite complex, involving programs of transcription in the mother cell that are separate from those in the forespore compartment. How is this possible? Part of the answer is that at the time of septum formation, when the forespore is first produced, only 30% of the chromosome (the part near the replication origin) is actually <u>inside</u> the forespore. The remainder of the chromosome is slowly pumped in over a 15-minute period. The sporulating cell takes advantage of this delay to selectively express different genes in different compartments.

For example, sigma F is present in <u>both</u> forespore and mother cell compartments, but it directs the transcription of select genes only in the forespore. Compartment-dependent activation starts with an intricate timing mechanism that involves anti-sigma and anti-anti-sigma factors (**Fig. 10.18**). Like all sporulation sigma factors, sigma F is inactive when first synthesized. Sigma F is inactive because it binds to an anti-sigma F protein dubbed SpoIIAB that is cosynthesized with sigma F in the predivisional cell (**Fig. 10.18A**, step 1). After division, both proteins become equally partitioned between the two compartments. However, in addition to an anti-sigma, there is an anti-anti-sigma F protein, called SpoIIAA

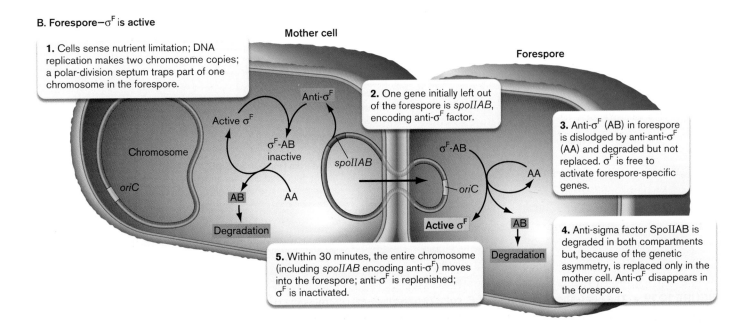

B. Forespore—σ^F is active

Mother cell

1. Cells sense nutrient limitation; DNA replication makes two chromosome copies; a polar-division septum traps part of one chromosome in the forespore.

Active σ^F

Anti-σ^F

σ^F-AB inactive

Chromosome

spoIIAB

oriC

AB   AA

Degradation

Forespore

2. One gene initially left out of the forespore is *spoIIAB*, encoding anti-σ^F factor.

σ^F-AB

oriC

AA

Active σ^F   AB

Degradation

3. Anti-σ^F (AB) in forespore is dislodged by anti-anti-σ^F (AA) and degraded but not replaced. σ^F is free to activate forespore-specific genes.

4. Anti-sigma factor SpoIIAB is degraded in both compartments but, because of the genetic asymmetry, is replaced only in the mother cell. Anti-σ^F disappears in the forespore.

5. Within 30 minutes, the entire chromosome (including *spoIIAB* encoding anti-σ^F) moves into the forespore; anti-σ^F is replenished; σ^F is inactivated.

(AA for "anti-anti"), which is also equally distributed (steps 2 and 3). Anti-anti sigma factor frees sigma F from anti-sigma F, which is degraded by the ClpPC protease. This is not a problem in the mother cell, because that compartment just makes more AB (anti-sigma), so sigma F is never active.

These events occur in both compartments (forespore and mother cell), so why is sigma F activation limited to the forespore? Recall that at the time of septum formation, one chromosome is <u>slowly</u> pumped into the forespore. The gene encoding the AB anti-sigma factor is located far from the origin and, as such, is not in the forespore immediately after septum formation (**Fig. 10.18B**, step 2). Consequently, no new AB (anti-sigma) is made in the forespore during this time, and what was there is dislodged from σ^F by anti-anti factor and degraded by ClpPC (steps 3 and 4). Depletion of anti-sigma in the forespore frees sigma F to act in the forespore compartment for a limited time. Liberated sigma F transcribes the genes needed for subsequent sporulation steps. The gene targets for sigma F do reside in the part of the genome trapped early on in the forespore.

Eventually, the *spoIIAB* gene enters the forespore and replenishes anti-σ^F (step 5), and the next stage of sporulation begins.

Elegant proof for this model was obtained in the laboratory of Richard Losick, a leading scientist in the field of microbial development. In Losick's experiment, the gene for SpoIIAB (anti-sigma) was moved from its normal location far from the origin to a position near the origin. As a result, the anti-sigma F gene entered the forespore early, continually replenished anti-sigma F lost by degradation, and stopped sporulation.

The simple, unequal distribution of the chromosome during septal formation is heavily exploited during sporulation to trigger differential gene expression in two different cell compartments.

Thought Question

10.7 Predict the phenotype of a *spoIIAA* mutant that completely lacks SpoIIAA anti-anti-sigma factor (see **Fig. 10.18**).

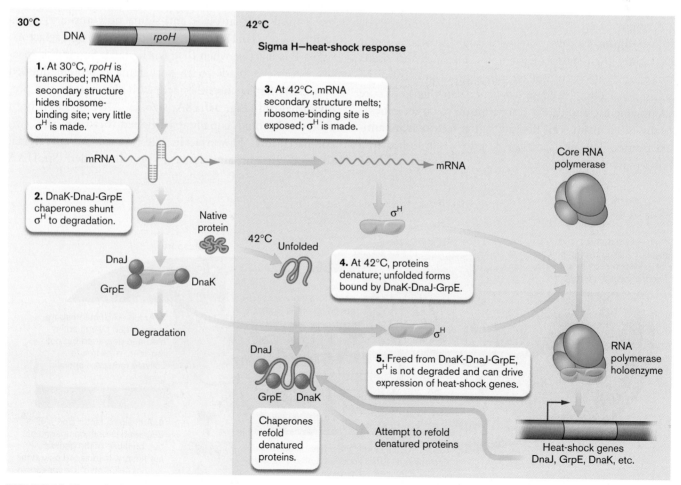

FIGURE 10.19 ■ The heat-shock response of *E. coli*. Two mechanisms control sigma H levels. The small amount of sigma H that can be made at 30°C is met by the DnaK-DnaJ-GrpE chaperone system and shuttled toward degradation. At 42°C, however, misfolded cytoplasmic proteins siphon off the chaperone trio and release sigma H to direct transcription of the heat-shock genes.

Sigma Factor Control by RNA Thermometers and Proteolysis

Regulation of the heat-shock sigma factor sigma H (also called sigma-32, RpoH, σ^H, or σ^{32}) in *E. coli* illustrates how the synthesis of a sigma factor can be controlled at the level of translation and degradation (**Fig. 10.19**). Excessive heat, above 42°C for *E. coli*, causes proteins to denature and membrane structure to deteriorate. All cells subjected to heat above their comfort zone (optimal growth range) will express a set of proteins called heat-shock proteins. These proteins include chaperones that refold damaged proteins, and a variety of other proteins that affect DNA and membrane integrity. The transcription of many *E. coli* heat-shock genes requires sigma H. So, one of the first consequences of growth at elevated temperature is an increase in the amount of sigma H present in the cell.

The level of sigma H is regulated by two temperature-dependent mechanisms; one controls the speed (rate) of sigma H synthesis, while the other determines its rate of degradation. The gene encoding sigma H is *rpoH*. At 30°C, *rpoH* mRNA adopts a secondary structure at the 5′ end that buries a ribosome-binding site, so *rpoH* mRNA is poorly translated. The 5′ region of *rpoH* mRNA is called the ROSE element for <u>r</u>epression <u>o</u>f heat <u>s</u>hock gene <u>e</u>xpression. A sudden rise in temperature, however, melts this secondary structure and exposes the ribosome-binding site, allowing translation to occur more readily. Thus, heat shock increases sigma H synthesis, which in turn increases transcription of the heat-shock genes whose products include chaperones and proteases.

Proteolysis also controls sigma H accumulation. At 30°C, the *rpoH* message is poorly translated, owing to secondary structure, as previously described, but some sigma H protein is made. Inappropriate expression of heat-shock genes at 30°C is prevented by the DnaK-DnaJ-GrpE chaperone system, which interacts with sigma H and shuttles it to various proteases for digestion (see **Fig. 10.19**). At 42°C, however, proteolysis of sigma H decreases, and sigma H is allowed to accumulate. Degradation decreases because at the higher temperature, the chaperones are siphoned away from sigma H by the many other heat-denatured proteins formed. The new goal of the chaperones is to refold and rescue those damaged proteins. Chaperone redeployment frees sigma H to transcribe the heat-shock genes, which include the chaperone genes *dnaK*, *dnaJ*, and *grpE*. These genes also have promoters that depend on other sigma factors to drive basal expression. Thus, as the temperature rises, the amount of sigma H is increased by two temperature-dependent mechanisms: One increases translation by exposing the ribosome-binding site (a so-called RNA thermometer), while the second redeploys the chaperones that direct its proteolysis.

Many other examples of RNA thermometers exist, some of which are involved in pathogenesis. Bacteria that infect mammals have RNA thermometers set to body temperature, 37°C—a set point that helps the microbe sense entry into the mammalian host. The Gram-positive bacillus *Listeria monocytogenes*, for instance, causes mild gastroenteritis or serious meningitis in humans. When the organism is ingested, it synthesizes a regulatory protein, PrfA, that activates transcription of a number of virulence genes. The *prfA* mRNA is transcribed regardless of temperature, but its 5′ end contains a ROSE element that prevents translation until the temperature rises—after ingestion.

Regulatory RNAs

Are the regions of DNA between protein-encoding genes merely spacers, or are they somehow important for regulating traditional genes? And what about the nontemplate strands of genes? Do they have a function other than simply maintaining complementarity to the template strand? A surprising fraction of a bacterial chromosome does not encode mRNA, rRNA, or tRNA, but instead makes untranslated RNA with regulatory functions. The laboratories of Susan Gottesman and Gisela Storz at the National Institutes of Health (**Fig. 10.20**) were instrumental in

A.

PHOTOGRAPH BY RHODA BAER

B.

COURTESY OF GIESLA STORZ

FIGURE 10.20 ■ **Susan Gottesman and Gisela Storz.** Gottesman (**A**, center) and Storz (**B**) played instrumental roles in establishing the importance of sRNA molecules in bacteria.

discovering that regions between genes (intergenic regions) can encode **small RNA (sRNA)** molecules (100–200 nt) that affect the expression of many other genes.

In addition to intergenic sRNAs, there are so-called *cis*-antisense RNA (asRNA) molecules. A *cis*-antisense RNA is transcribed from the nontemplate DNA strand that lay opposite an mRNA-encoding template strand (see Fig. 8.20). *Cis*-antisense regulatory transcripts can base-pair with cognate sense mRNAs and control their expression. Regulatory RNAs help control a variety of processes, such as plasmid replication, transposition, phage development, viral replication, bacterial virulence, environmental stress responses, and developmental control in lower eukaryotes.

Mechanisms of sRNA function. Regulatory sRNA molecules typically affect gene expression <u>after</u> transcription (posttranscriptional control) by binding to complementary sequences located within target mRNA transcripts. These hybridizations can either stimulate or prevent translation, or affect degradation. Many but not all sRNAs require an RNA chaperone protein called Hfq to stabilize the sRNA and promote its regulatory effects on target mRNAs. Hfq is a hexameric ring protein with sRNA- and mRNA-binding faces. **Table 10.2** groups sRNA molecules by mechanism.

Most known sRNAs <u>inhibit translation</u> by base-pairing to a region of mRNA that overlaps the ribosome-binding site (RBS). Binding, therefore, prevents the ribosome from accessing the RBS (**Fig. 10.21A**). Other sRNA molecules can <u>enhance translation</u> by binding to part of a long 5′ untranslated mRNA sequence located upstream of an RBS (**Fig 10.21B**). Without the sRNA, the untranslated mRNA sequence folds in a way that occludes the RBS and prevents translation. However, when sRNA binds the target mRNA sequence, the untranslated region refolds to expose the RBS (an example is DsrA). Ribosomes can now bind the mRNA and translate the message.

Some sRNA molecules <u>promote degradation</u> of mRNA by altering mRNA folding to expose an RNase cleavage site (**Fig. 10.21C**)—for instance, RNAIII, discussed later. The second mechanism in this group <u>prevents degradation</u> of the message by binding to and occluding an RNase E– binding site (the sRNA SgrS) (**Fig. 10.21D**).

Another class of sRNA affects <u>mRNA processing</u>. Assisted by sRNA, a long, unstable, polycistronic transcript can be processed to make shorter, more stable monocistronic mRNAs. For instance, an sRNA such as GadY of *E. coli*, can bind between two coding regions of a polycistronic message and expose an RNase cleavage site required for processing (**Fig. 10.21E**). Members of the last sRNA regulatory group can bind to select regulator proteins and interfere with their function (**Fig. 10.21F**). The sRNA— CsrB, for example—can prevent the regulator from binding to its DNA target sequence.

sRNA molecules expand the reach of regulatory proteins. Large regulatory proteins (for example, LacI or TrpR) typically control only a few genes, but their synthesis is

---

TABLE 10.2   Classes of sRNA Molecules

Mechanism	sRNA	Function
1. Inhibits translation by blocking ribosome-binding site (RBS).	OxyS	Regulates oxidative stress gene expression (*E. coli*).
	CyaR	Represses porin OmpX (*E. coli*).
	ChiX	Prevents transport of chitosugars by preventing the synthesis of ChiP porin (*E. coli*).
	SprD	Regulates Sbi (*Staphylococcus aureus* <u>b</u>inder of <u>I</u>gG) immune evasion molecule.
	Qrr3	Controls quorum sensing by sequestering *luxO* (*Vibrio cholerae*).
	sRNA$_{162}$	Inhibits translation of regulator MM2241; affects methyltransferases (*Methanosarcina mazei*).
2. Permits translation by exposing RBS.	DsrA	Increases translation of RpoS mRNA (*E. coli*).
3. Promotes degradation of mRNA.	RNAIII	Regulates global regulator of *agr*-controlled virulence genes; also encodes a delta hemolysin (*S. aureus*).
	RyhB	Expands regulation by Fur repressor (*E. coli*).
	Qrr3	Controls quorum sensing via *luxR, luxM* (*V. cholerae*).
4. Inhibits degradation of mRNA.	SgrS	Controls sugar transport by sequestering an RNase E site (*E. coli*).
5. Stimulates processing of mRNA to make more stable transcripts.	GadY	Regulates acid resistance (*E. coli*).
6. Titrates regulatory proteins away from target mRNAs.	CsrB	Regulates carbon storage (*E. coli*).

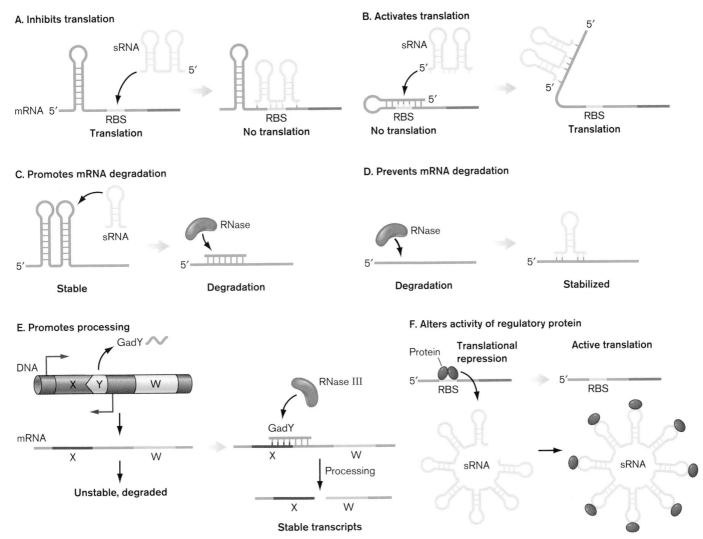

FIGURE 10.21 ■ **Mechanisms of regulatory sRNA function.** Some sRNA molecules will **(A)** inhibit or **(B)** activate translation by blocking or exposing a ribosome-binding site (RBS). Other groups of sRNA molecules can **(C)** promote or **(D)** prevent the degradation of target mRNA; **(E)** mediate processing of long, unstable multigene mRNAs into more stable, shorter molecules; or **(F)** interfere with regulator protein activity.

energetically expensive. Small RNAs can expand a protein's regulatory reach and, thus, maximize return on investment. The benefits of sRNAs are that they do not require protein synthesis to control gene expression, they diffuse rapidly, and they typically act on preexisting messages. Small RNAs represent one of the most economical ways to inhibit gene expression.

One example of an sRNA molecule in *E. coli* is RyhB, which regulates iron storage by expanding the reach of Fur, the ferric uptake regulator. Iron is hugely important for pathogens growing in the human body and for the quality of soil, where it determines which communities of bacteria and plants can grow. However, too much iron increases oxidative stress and damages the cell. As a result, iron content must be tightly controlled. In many bacteria, including intestinal *E. coli*, iron uptake is regulated by Fur, which senses iron and regulates several genes whose products either scavenge iron from the environment or store iron

in the cell (see Section 4.2). When intracellular iron levels are high, Fur represses expression of scavenging genes but induces production of iron storage proteins (**Fig. 10.22A**). Fur represses gene expression by directly binding DNA operator sequences in front of target genes. But how does Fur activate iron storage genes? One mechanism is indirect and involves Fur repression of the sRNA RyhB (90 nt) (**Fig. 10.22B**). RyhB sRNA is made when iron levels are low (as within the human body) because the Fur repressor is not active. RyhB then helps destroy mRNAs of several iron-storing and iron-using proteins (such as succinate dehydrogenase made from the *sucCDAB* operon). As a result, dwindling iron reserves can be put to more productive use. When iron is plentiful, however, Fur will directly repress *ryhB*. The lack of RyhB sRNA stabilizes the expression of the iron storage genes. Thus, succinate dehydrogenase is made, binds intracellular iron, and enables the use of succinate as a carbon and energy source.

FIGURE 10.22 ■ **Activity of a small regulatory RNA molecule.** **A.** When iron levels are high, Fur repressor protein binds to the *ent* and *ryhB* Fur box DNA sequence (a short, specific DNA sequence in front of the genes regulated by Fur) and represses their expression. Enterochelin is no longer made, but the *sucCDAB* message encoding succinate dehydrogenase can be translated. **B.** Under low iron conditions, RyhB sRNA is expressed. RyhB sRNA binds to the *sucCDAB* message and renders it susceptible to an RNase.

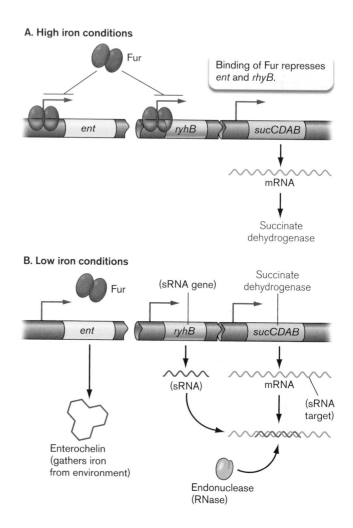

A. High iron conditions

Fur

Binding of Fur represses *ent* and *rhyB.*

ent *ryhB* *sucCDAB*

mRNA

Succinate dehydrogenase

B. Low iron conditions

Fur (sRNA gene) Succinate dehydrogenase

ent *ryhB* *sucCDAB*

(sRNA) mRNA (sRNA target)

Enterochelin (gathers iron from environment)

Endonuclease (RNase)

> **Thought Question**
>
> **10.8** The relationship between the small RNA RyhB, the iron regulatory protein Fur, and succinate dehydrogenase is shown in **Figure 10.22**. Based on this regulatory circuit, will a *fur* mutant grow on succinate?

Figure 10.23 illustrates the mechanism of another sRNA. The Gram-positive pathogen *Staphylococcus aureus* produces an sRNA, called RNAIII (514 nt long), that controls a large number of virulence genes. The 3′ end of RNAIII forms hairpin loops that can interact with hairpin loops at the 5′ ends of target mRNA molecules (for instance, coagulase), one of which contains a ribosome-binding site. Sections of the loops hybridize, block the mRNA ribosome-binding site, and inhibit translation. The RNA-RNA hybrid is also an alluring target for a ribonuclease that will degrade the message. RNAIII is expressed after the bacterial population reaches a certain density during infection

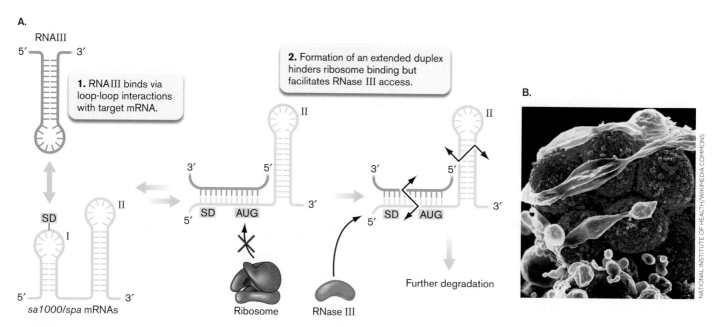

A.

RNAIII

5′ 3′

1. RNAIII binds via loop-loop interactions with target mRNA.

2. Formation of an extended duplex hinders ribosome binding but facilitates RNase III access.

II

SD I II

5′ 3′

sa1000/spa mRNAs

SD AUG

Ribosome

RNase III

SD AUG

Further degradation

B.

FIGURE 10.23 ■ **Model for RNA III small RNA function.** **A.** RNA III in *Staphylococcus aureus* binds to its target mRNAs. SD (the Shine-Dalgarno ribosome-binding site) and AUG are marked green, RNA III is blue, and the mRNA target (*sa1000/spa*) is light blue. Bent arrows indicate degradation. **B.** Methicillin-resistant *S. aureus*, or MRSA (red), killing a human neutrophil (SEM). *Source:* Part A modified from S. Boisset et al. 2007. *Genes Dev.* **21**:1353–1366; part B from Adam D. Kennedy et al. 2008. *PNAS* cover image **105**(4). © 2008 by The National Academy of Sciences of the U.S.

(see the discussion of quorum sensing in Section 10.5). It then inhibits the translation of, and helps degrade, mRNAs no longer required for the infection (for example, those for some exotoxins and surface attachment proteins).

sRNAs in Archaea. Archaeal species also use sRNA molecules to fine-tune their physiology. One class of sRNA in Archaea, called small nucleolar RNAs (snoRNAs), is also prominent in eukaryotes. The snoRNA designation was retained for Archaea even though Archaea lack a nucleolus, or nucleus, for that matter. Some archaeal snoRNAs help guide the activity of enzymes. For example, one snoRNA guides a methylation enzyme to methylate rRNA at specific sites (*Pyrococcus* and *Sulfolobus*). Another snoRNA guides the conversion of uridine to pseudouridine in rRNAs. A second type of archaeal sRNA, called tRF, comes from tRNA degradation. One example is a tRF produced by the halophile *Haloferax volcanii* during alkaline stress. This tRF binds to *Haloferax* ribosomes and inhibits translation, thereby helping the organism survive high pH.

Archaea also produce sRNAs similar to those of bacteria, but their functions are largely unknown. One archaeal sRNA (sRNA$_{162}$) whose function is known inhibits the translation of an mRNA that encodes methyltransferase in *Methanosarcina mazei* (a methanogen). Numerous other archaeal sRNAs have been identified, but more must be learned about their functions if we are to understand their place in archaeal biology.

***Cis*-antisense RNA.** We are increasingly finding that the sense DNA strands from many protein-encoding genes are transcribed to make so-called ***cis*-antisense RNAs** (**asRNAs**) with regulatory functions. Typically, asRNAs are 700–3,000 nt long, originate within a protein-coding gene, and affect <u>only</u> that gene. These features distinguish asRNA from sRNA. Deep sequencing of the *E. coli* transcriptome revealed the presence of about 1,000 different *cis*-antisense RNA genes out of 4,290 protein-encoding genes. In fact, every microbe examined produces asRNA molecules.

Different asRNAs have different effects on their target genes. When bound to their sense mRNA counterparts, asRNAs can engineer attenuator loops that stop transcription, prevent mRNA translation, or trigger mRNA degradation. Even the simple act of asRNA transcription can produce collisions between converging RNA polymerase complexes that prematurely terminate transcription. An example of the collision mechanism was found for *Clostridium acetobutylicum*. The *ubiG-mccBA* operon encodes proteins needed to convert methionine to cysteine (**Fig. 10.24**), but it is expressed only when methionine

A. The *ubiG-mccB-mccA* operon

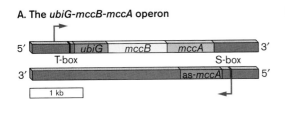

S-box (specific for *S*-adenosylmethionine, or SAM) and T-box (specific for cysteine) sequences are riboswitches. SAM binding of the S riboswitch terminates transcription of antisense message.

B. Collision (low level of *S*-adenosylmethionine)

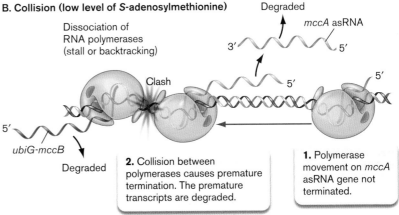

1. Polymerase movement on *mccA* asRNA gene not terminated.

2. Collision between polymerases causes premature termination. The premature transcripts are degraded.

FIGURE 10.24 ■ A *cis*-antisense RNA gene produces colliding RNA polymerases. The MccA protein from the *Clostridium acetobutylicum ubiG-mccB-mccA* operon helps convert methionine to cysteine when cellular methionine levels are high. **A.** Gene arrangement for *ubiG-mccB-mccA* operon. Red arrows indicate promoters and transcription start points for the *ubiG* operon and the antisense gene. **B.** Antisense RNA forms when methionine levels are low. (Black and red horizontal arrows indicate direction of polymerase movement.) The MccA transcript is not made, and the premature transcript is degraded. **C.** Antisense message does not form when methionine concentration is high. Transcription of the *ubiG-mccBA* operon is completed, and cysteine can be made.

C. No collision (high level of *S*-adenosylmethionine)

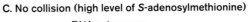
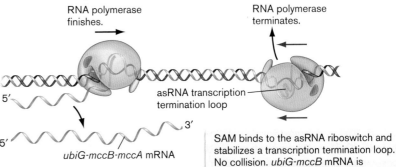

SAM binds to the asRNA riboswitch and stabilizes a transcription termination loop. No collision. *ubiG-mccB* mRNA is completed, and cysteine is made.

levels are high. When cytoplasmic methionine concentration is low, the RNA polymerase that transcribes the antisense *mccA* gene will collide with, and stop, the RNA polymerase transcribing the *ubiG-mccBA* operon (**Fig. 10.24B**). The sense MccA transcript is not made, and the premature transcript is degraded. However, when methionine concentration is high, *S*-adenosylmethionine (made from methionine) binds to the asRNA at the S-box (a riboswitch, described next) and stabilizes a transcription termination loop (**Fig. 10.24C**). The *mccA* antisense RNA is not produced, and because the RNA polymerase making asRNA disengages, there is no polymerase collision. The *ubiG-mccBA* operon mRNA is completed, MccA is made, and cysteine is synthesized. Evidence suggests that asRNAs in bacteria have a broad impact on microbial processes.

Riboswitches Sense Cytoplasmic Molecules

Proteins are not the only molecules capable of recognizing and binding to low-molecular-weight metabolites. RNA molecules can do the same thing by folding into three-dimensional structures. These RNA 3D structures bind specific target metabolites much as a protein binds to its substrate. Because of this property, some RNA molecules, called **ribozymes** (or catalytic RNA), catalyze enzymatic reactions. We already discussed how 23S rRNA forms the peptidyltransferase activity of ribosomes (see Chapter 8). Other RNA molecules, called **riboswitches**, resemble activator or repressor proteins in that they bind to a cell metabolite to control gene expression.

A riboswitch is typically found at the 5′ untranslated end of an mRNA molecule, upstream of a coding sequence. The riboswitch can assume two alternative stem loop structures that switch back and forth in response to a target metabolite (**Fig. 10.25**). Translation of the message or transcription of the gene can be affected. For a riboswitch that affects translation (**Fig. 10.25A**), excess ligand (an amino acid or vitamin) will bind to the riboswitch and stabilize a 3D structure that sequesters a ribosome-binding site (RBS). The buried RBS <u>prevents</u> translation of the coding region. Without ligand, an alternative structure forms that exposes the RBS, and translation ensues. For riboswitches that affect transcription (**Fig. 10.25B**), binding of ligand <u>prevents</u> transcription termination. The coding region is expressed.

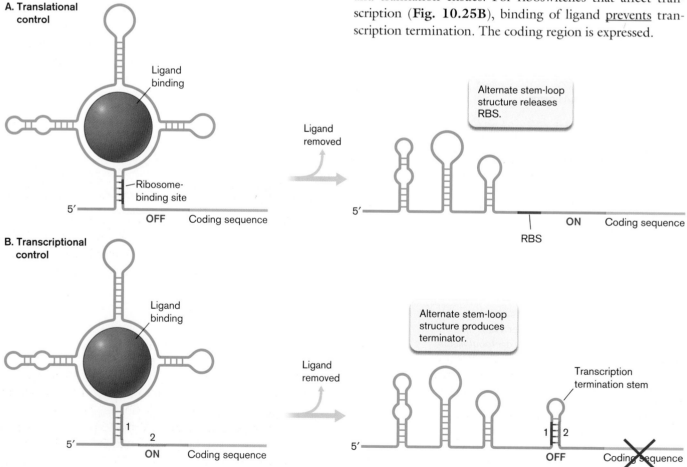

FIGURE 10.25 ▪ **Riboswitch regulation. A.** Translation control. Ligand binds to a riboswitch and stabilizes a structure that sequesters a ribosome-binding site. The coding region is not translated. Removing the ligand results in an alternate structure that releases RBS and enables translation. **B.** Transcriptional control. Ligand stabilizes a secondary structure that prevents formation of a transcription termination stem. Coding region is transcribed. Removing the ligand results in an alternate structure that prevents transcription.

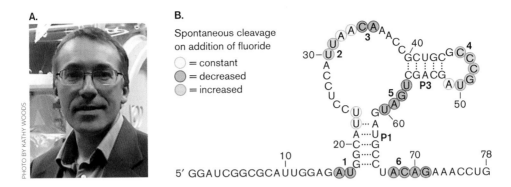

FIGURE 10.26 ■ A fluoride riboswitch. Ron Breaker **(A)** discovered the fluoride riboswitch called WT 78 Psy **(B)**. Colored circles mark bases that become susceptible to cleavage after the riboswitch binds fluoride—an indication that the structure of the RNA molecule changes. A pseudoknot contains at least two stem loop structures that form a knot-shaped three-dimensional conformation.

Riboswitches can recognize a diverse set of ligands, including amino acids, vitamins, *S*-adenosylmethionine (see above), thiamine pyrophosphate, and magnesium. Ron Breaker's laboratory at Yale University (**Fig. 10.26**) discovered an unusual riboswitch in bacteria and archaea that specifically senses fluoride, not a cell metabolite. Fluoride has been used for decades to inhibit dental caries, but it is also abundant in Earth's crust. The fluoride riboswitch was found attached to genes such as enolase that are inhibited by this halide. Fluoride buildup in bacteria will be detected by the fluoride riboswitch, which triggers increased translation of the enzyme being inhibited. The result is increased fluoride resistance of the microbe. Recent work has shown that riboswitches can also regulate expression of noncoding RNAs and control access to transcription termination factor Rho or RNase E.

To Summarize

- **Changing the synthesis or activity of a sigma factor** will coordinately regulate a set of related genes. Sigma factors can be controlled by altered transcription, translation, proteolysis, and anti-sigma factors.

- **Sporulation** relies on the hierarchical activation of a series of different sigma factors. Control mechanisms include **temporary asymmetrical chromosome distribution** between the forespore and the mother cell, and **anti-sigma** and **anti-anti-sigma factors**.

- **Small regulatory RNAs** found within bacterial intergenic regions regulate the transcription or stability of specific mRNA molecules and broaden the reach of protein regulators.

- **A *cis*-antisense RNA** is an RNA produced from the sense DNA strand of a protein-encoding gene and affects expression of only that gene. These RNAs bind to complementary target mRNA and either stabilize the mRNA or make it susceptible to degradation.

- **RNA thermometers and riboswitches** are secondary structures at the 5′ end of specific mRNA molecules that can obscure access to ribosome-binding sites or transcriptional terminator stems. Conditions that stabilize the secondary structures (cold for RNA thermometers, and small molecules for riboswitches) will alter translation of an mRNA or prematurely stop transcription.

10.4 Integrated Control Circuits

The regulatory circuits outlined to this point have involved mostly single-switch mechanisms. But microbes build on these mechanisms by combining multiple switches with redundant feedback controls that can form a kind of integrated gene circuit. Integrated circuits can send a virus down alternate lifestyle paths (lysis or lysogeny), couple the genetic and biochemical control of a metabolic pathway (ammonia assimilation), or time the events of a developmental cycle such as sporulation.

The deeper we delve into the regulatory workings that govern the expression of a cell's genome, the more clearly we see the hierarchy of control. Groups of simple control circuits are collectively regulated by other, more complex circuits. One illustration of this hierarchy was already presented—namely, the role of cAMP and CRP in regulating lactose and arabinose catabolism. The *lacZYA* and *araBAD* operons are each controlled separately by dedicated regulatory switches—LacI and AraC, respectively. At a higher level, however, both operons are coregulated by cAMP and

CRP in response to energy status (whether or not glucose is present and utilized). The systems we are about to discuss represent complex examples of nature's inventiveness and need for control.

The first example we will look at here actually has nothing to do with transcriptional/translational or post-translational control of protein production. It does illustrate the concept of two-component signal transduction, whereby one protein (the sensor kinase) senses a change in the environment and transmits a signal (via phosphorylation) to the second component in the system (the response regulator). Two-component signal transduction systems can be used to control transcription, but they can also be used to affect the physiology and behavior of a cell without changing transcription. The example we will use is chemotaxis, the ability of a bacterium to move toward environments favorable to growth and away from harsh environments.

Chemotaxis: How Bacteria Know Where They Are Going

Bacteria that use flagella to move do not simply dash about "hoping" they will swim in a favorable direction (presented in Chapter 3). They have a sense about where they want to go and where not to go. Bacteria such as *E. coli* purposefully move toward certain amino acids and carbohydrates present in media. They swim toward these chemicals by sensing their gradients. **Chemotaxis** is the ability of an organism to sense chemical gradients and modify its motility in response.

Before we discuss how these chemical gradients are sensed, let's first talk about how bacteria can change direction. Recall that the flagellar motor can rotate clockwise or counterclockwise (see Section 3.6, Fig. 3.44) and that, on any one bacterium, all motors coordinate their rotations. Thus, all flagellar motors on a single cell will rotate clockwise (or counterclockwise), and then suddenly switch to the opposite rotation. In the counterclockwise mode, all flagella sweep behind the cell, forming a rotating bundle that propels the organism forward in what is called smooth swimming or a "run." When flagellar rotors suddenly switch to a clockwise rotation, the bundle is disrupted and the bacterium "tumbles" in a random fashion. Once the flagellar rotors switch back to counterclockwise rotation, the bacterium will have reoriented itself in a different position, changing its swimming direction.

The key to chemotaxis is a mechanism that suppresses the number of tumbles an organism makes when it moves from a lower to higher concentration of an attractant chemical. For instance, an organism moving toward an attractant

may tumble only twice in 5 seconds. In contrast, an organism moving in the wrong direction (that is, toward a lower concentration of attractant) may tumble eight times in 5 seconds. If, all of a sudden, the organism finds itself going in the right direction, the sensory transduction system will suppress the number of tumbles that occur, and the cell will continue moving in the right direction.

This is the same strategy you use to find a hamburger that you can smell cooking on a grill in your neighborhood but cannot see. You randomly walk around until the smell gets stronger, and then you keep moving in that direction until someone hands you a burger.

How does the bacterium suppress tumble frequency? Let's start at the end of the system and work backward. The natural bias of the flagellar rotor is counterclockwise rotation (in other words, smooth swimming). In order to tumble, the cell phosphorylates a protein called CheY. CheY-P interacts with a rotor protein and flips the rotor in reverse to turn clockwise (**Fig. 10.27** ▶, step 1). CheA protein is the kinase that phosphorylates CheY and is the "key" to chemotaxis. The more CheA kinase activity there is, the more CheY-P is made, resulting in a higher frequency of tumbling (reorienting) events. Another protein, CheZ, continually dephosphorylates CheY. When *E. coli* senses an attractant chemical like an amino acid at the cell surface, the result is decreased CheA kinase activity, which leads to lower CheY-P levels. The presence of fewer CheY-P molecules allows longer periods of counterclockwise rotation and, consequently, extended smooth swimming toward the attractant.

All this makes sense, but how does a cell "know" that it has moved into an area with attractant? The answer begins with clusters of special membrane-spanning proteins called **methyl-accepting chemotaxis proteins** (**MCPs**, or chemoreceptors) located at the poles of *E. coli* and other bacteria (**Fig. 10.27**, step 1). The periplasmic domains of different MCPs bind to different attractant molecules that flow into the periplasm as the cell moves. The cytoplasmic domain of each MCP binds to the CheA kinase noted earlier (via an intermediary protein, CheW) and controls CheA activity. When a chemoattractant chemical, such as serine, binds to the periplasmic side of the MCP, the conformation of the cytoplasmic domain changes and inhibits CheA kinase activity (step 2). The CheY-P level then decreases (because CheZ continues to dephosphorylate CheY-P), and smooth swimming ensues.

This, too, makes sense when an organism first encounters a chemoattractant, but how does the cell know to keep moving into even higher concentrations? As it happens, the conformational change in the MCP that inactivates CheA kinase activity also subjects the cytoplasmic side of the MCP to methylation (via *S*-adenosylmethionine) by CheR

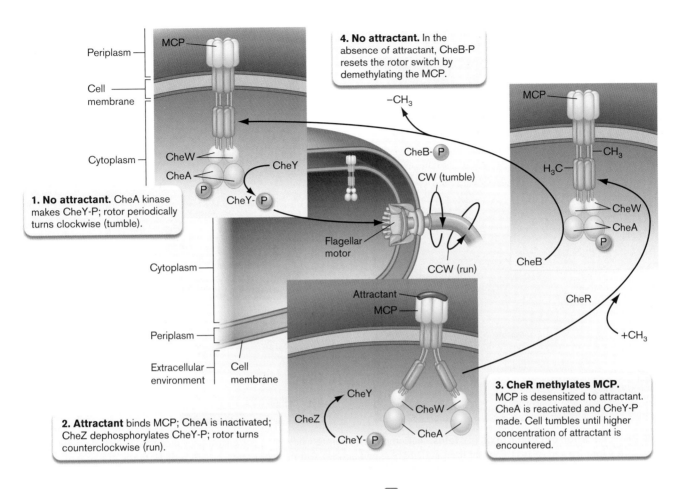

1. No attractant. CheA kinase makes CheY-P; rotor periodically turns clockwise (tumble).

2. Attractant binds MCP; CheA is inactivated; CheZ dephosphorylates CheY-P; rotor turns counterclockwise (run).

4. No attractant. In the absence of attractant, CheB-P resets the rotor switch by demethylating the MCP.

3. CheR methylates MCP. MCP is desensitized to attractant. CheA is reactivated and CheY-P made. Cell tumbles until higher concentration of attractant is encountered.

FIGURE 10.27 ■ Chemotaxis. Chemotaxis signaling pathway in *E. coli.* ▶

methylase (**Fig. 10.27**, step 3). Methylation of glutamate residues in the cytoplasmic struts of the MCP reactivates the CheA kinase but desensitizes the MCP so that an even higher concentration of attractant must be present in order to bind the MCP and inhibit CheA kinase. So, if the organism is moving into a higher concentration of attractant, CheA is again inactivated, tumbling is suppressed, and the cell prolongs the run. If it is not moving into higher concentrations of attractant, CheY-P is made again and the organism will tumble. If the bacterium moves to a lower concentration of attractant, the cell will keep tumbling at a high frequency until it moves back into a higher concentration, at which point tumbling is suppressed to allow a smooth run.

When the cell moves away from attractant, the methylation switch is reset by another protein, CheB-P, which removes methyl groups so that the system is resensitized to attractant (**Fig. 10.27**, step 4). The system is elegantly fine-tuned in that CheA kinase is also the protein that

phosphorylates CheB. So, as the cell moves away from attractant, CheA kinase phosphorylates CheY to produce tumble, and it phosphorylates CheB to reset the sensitization switch.

MCP sensory proteins are not uniformly distributed around the cell surface, but assemble as clusters at both poles of *E. coli*. This clustering seems logical because the bacterium points with its pole in the direction it swims. Karen Lipkow (Cambridge Systems Biology Centre, University of Cambridge) and Peijun Zhang (University of Pittsburgh) have used computer modeling and cryo-EM tomography, respectively, to image the dynamics of chemotactic signaling and chemosensory arrays (**Fig. 10.28**). **Figure 10.28A** depicts one such MCP cluster (yellow) at one pole of the bacterium and illustrates the diffusion of CheY-P toward the opposite pole. CheY-P will interact with flagellar rotors all along the cell surface, causing tumble. The physical structure of a chemoreceptor array is shown in **Figure 10.28B**.

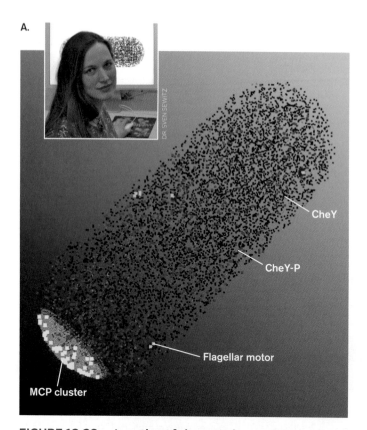

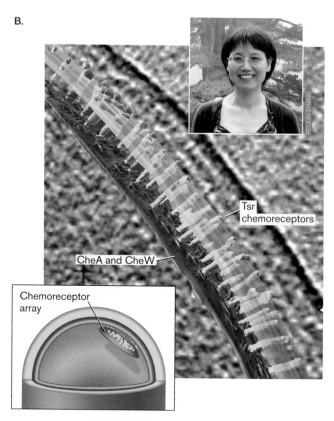

FIGURE 10.28 ▪ Location of chemotaxis proteins. A. Computer-generated view of *E. coli* chemotaxis proteins. Depicted are the MCP cluster of receptors (orange and yellow), flagellar motors (light blue and purple), CheY (black), and CheY-P (red). Note the gradient of CheY-P. **A inset:** Karen Lipkow. **B.** Chemoreceptor array structure at a pole of *E. coli*. Hexagonal arrays of Tsr serine-detecting chemoreceptors (yellow) are held together by signal transduction proteins CheA and CheW (blue) (cryo-EM tomography). **B inset:** Peijun Zhang. Cryo-EM tomography. *Source:* Part A from Karen Lipkow et al. 2005. *J. Bacteriol.* **187**:45–53 (cover photo); part B from Peijun Zhang et al. 2007. *PNAS* **104**:3777–3781.

It is interesting to note that while the basic chemotactic mechanism is evolutionarily conserved in bacteria, different genera use it differently. In the Gram-positive *Bacillus subtilis*, for example, ligand binding to an MCP stimulates CheA kinase, and CheY-P stimulates counterclockwise flagellar rotation and, thus, extended runs. This mechanism is the exact opposite of that used in *E. coli*.

Chemotaxis is important for some obvious and perhaps not so obvious reasons. It clearly provides a useful survival strategy in nature, keeping bacteria moving toward nutrient and away from trouble (for example, toxic compounds). For natural commensal bacteria or pathogens, however, chemotaxis can also be used to move the organism toward a cell surface to which it can attach. For example, the intestinal lining can exude chemical attractants that, like a beacon, will lead the bacteria in the intestine toward the cell surface where they can attach. In sum, chemotactic sensory perception plays a major role in structuring microbial communities, in affecting microbial activities, and in influencing various microbial interactions with the surroundings. Other important regulatory mechanisms operate without affecting gene expression. One intriguing system involving toxin-antitoxin modules is described in **eTopic 10.4**.

Thought Question

10.9 Using antibody to flagella, we can tether a cell of *E. coli* to a glass slide via a single flagellum. Looking through a microscope, you will then see the bacterium rotate in opposite directions as the flagellar rotor switches from clockwise to counterclockwise rotation and back again (see **Fig. 10.27**). Which way will the bacillus rotate when an attractant is added? What would the interval between clockwise and counterclockwise switching be if you tethered the mutants *cheY*, *cheA*, *cheZ*, and *cheR* to the slide and then added attractant?

Biofilms and Second Messengers: Cyclic Di-GMP

To change cell physiology, some stress-sensing proteins produce **second messenger** molecules, which bind to regulatory (or effector) proteins and change their activities. We have already discussed two second messenger molecules: cAMP (involved in carbon/energy metabolism; see Section 10.2) and ppGpp (ribosome synthesis; see Section 10.2). These molecules freely diffuse through

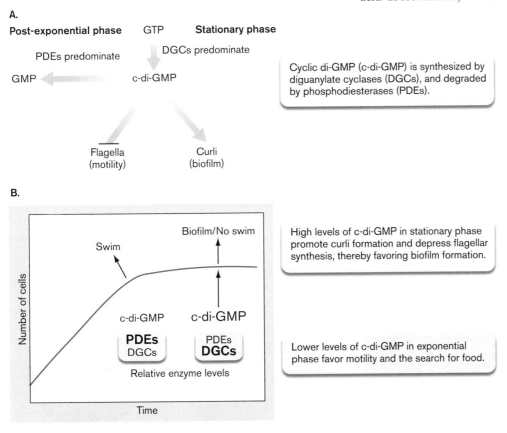

FIGURE 10.29 ■ Cyclic di-GMP [bis-(3'-5')-cyclic dimeric guanosine monophosphate], or cdiGMP.

the bacterial cell in search of effector proteins. Another important second messenger is cyclic di-GMP (c-di-GMP) (**Fig. 10.29**), which is one key to timing biofilm formation. *E. coli* cells transition between a motile, single-cell state (planktonic) and an adhesive multicellular biofilm. The transition can be seen clearly in batch cultures. The highly motile state appears during post-exponential growth when nutrient limitation forces *E. coli* to "forage" for food. When resources diminish further (stationary phase), the organism changes strategy: growth slows, motility decreases (a huge energy savings), and the synthesis of adhesins such as curli pili increases. After adhering to a surface, the cells start synthesizing exopolysaccharide matrix (see Chapter 4). The molecule c-di-GMP coordinates the transition by repressing flagellar synthesis genes and activating biofilm-promoting genes such as those encoding curli pili.

How is coordination achieved? Synthesis of c-di-GMP from GTP is carried out by many diguanylate cyclases (DGCs) in the cell, all of which contain the amino acid motif GGDEF (see Figure 8.10 for explanation of the single-letter amino acid abbreviations). However, dedicated phosphodiesterases (PDEs) with an EAL motif initiate degradation of c-di-GMP to GMP (**Fig. 10.30A**). *Salmonella* and *Pseudomonas* contain 19 and 41 different GGDEF/EAL proteins, respectively. Each DGC and PDE protein becomes activated by a different signal, thus increasing or decreasing c-di-GMP levels under different conditions. As shown in **Figure 10.30B**, PDEs predominate in post-exponential phase cells, which keeps c-di-GMP level low. Low c-di-GMP concentration favors motility and scavenging. In contrast, DGCs predominate in stationary phase and increase c-di-GMP level. High c-di-GMP inhibits motility and scavenging. Thus, highly motile cells are made in post-exponential phase while sessile, adherent cells able to produce biofilms are produced in stationary phase. **Figure 10.30C** shows that c-di-GMP is required for *Salmonella* to make biofilms. Deleting all known GGDEF motif proteins eliminates c-di-GMP synthesis and halts biofilm formation.

Cyclic di-GMP is a ubiquitous second messenger in bacteria and is not found in

FIGURE 10.30 ■ c-di-GMP coordinates the switch from planktonic growth to biofilm formation. A, B. The relative levels of PDEs and DGCs at different growth phases. C. Surface biofilm production by *Salmonella* requires c-di-GMP.

eukaryotes or archaea. But why does one cell possess so many GGDEF and EAL proteins? One hypothesis is that cognate pairs of these proteins may operate at separate locations within a single cell and generate localized changes in c-di-GMP concentrations. Multiple, focused c-di-GMP gradients could produce different outputs in different parts of a single cell.

Other examples of integrated control circuits are found throughout the microbial world. Regulation of nitrogen assimilation, for example, integrates physiological and genetic controls (discussed in Chapter 15). A classic example of integrated control involves the "lifestyle" decisions made by bacteriophage lambda, which are described in Chapter 11.

To Summarize

- **Bacterial genes are regulated by a hierarchy of regulators** that form integrated gene circuits.

- **Gene switches** can regulate "decision-making" processes in microbes.

- **Chemotaxis** is a behavior in which motile microbes swim toward favorable environments (chemoattractants) or away from unfavorable environments (chemorepellents).

- **The direction of flagellar motor rotation** determines the type of movement. Counterclockwise rotation results in smooth swimming; clockwise rotation results in tumbling.

- **Random movement toward an attractant** causes a drop in CheY-P levels, which enables counterclockwise rotation and smooth swimming.

- **Methyl-accepting chemotaxis proteins (MCPs)** clustered at cell poles bind chemoattractants and initiate a series of events lowering CheY-P levels. Reversible methylation or demethylation of MCPs desensitizes or sensitizes MCPs, respectively.

- **The second messenger cyclic di-GMP (c-di-GMP)** is made by numerous proteins containing a GGDEF amino acid motif. Many cellular functions are influenced by c-di-GMP, including biofilm formation and motility.

10.5 Quorum Sensing: Chemical Conversations

A discovery that fundamentally changed the way we think about microbes was made during studies of *Aliivibrio fischeri*, a peculiar marine microorganism that colonizes the light organ of the Hawaiian bobtailed squid (*Euprymna scolopes*) (**Fig. 10.31**). As discussed at the beginning of this chapter, *A. fischeri* is bioluminescent but glows only at high cell densities—a situation achieved naturally in the squid's colonized light organ or artificially in vitro (**Fig. 10.31C and D**). What accounts for the dependence of gene expression on cell density? How do cells "know" they are crowded?

The phenomenon was originally dubbed **quorum sensing** because it seemed akin to parliamentary rules of order that require a minimum number of members (a quorum) to be present at a meeting in order to conduct business. In the microbial world, however, gene regulation is only loosely associated

FIGURE 10.31 ■ **Visual demonstration of quorum sensing. A.** The luminescent bacterium *Aliivibrio fischeri* (formerly *Vibrio fischeri*) colonizes the light organ of the Hawaiian bobtailed squid (*Euprymna scolopes*). The light organ is deep inside each squid and therefore cannot be seen. **B.** Edward Ruby (University of Hawaii) has studied various aspects of the symbiotic relationship between *A. fischeri* and its squid host. **C.** When colonies of *A. fischeri* are observed in a well-lit place, the light emitted by the bacteria is not visible. **D.** If the same colonies are viewed in darkness, the intensity of luminescence is remarkable.

CHRIS FRAZEE AND MARGARET MCFALL-NGAI/ WIKIMEDIA COMMONS

COURTESY OF EDWARD RUBY

COURTESY OF PROFESSOR MADDEN

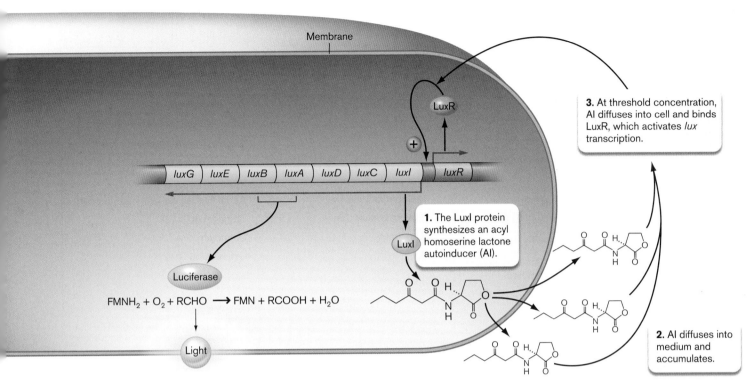

FIGURE 10.32 ■ **Microbial communication through quorum sensing.** The *lux* system of *Aliivibrio fischeri* mediates that organism's bioluminescence. Synthesis and accumulation of an autoinducer (AI) trigger expression of the *lux* operon. The greater the cell number and the smaller the container, the faster AI will accumulate. The resulting luciferase enzymes catalyze bioluminescence. The luciferase reaction, catalyzed by LuxA and LuxB, uses oxygen and reduced flavin mononucleotide (FMN) to oxidize a long-chain aldehyde (RCHO) and, in the process, produces blue-green light. Other *lux* gene products are involved in synthesis of the aldehyde. ▶

with actual cell numbers. Induction of a quorum-sensing gene system really requires the accumulation of a secreted small molecule called an **autoinducer** (usually a homoserine lactone, although Gram-positive organisms are known to use short peptides). The more cells there are in a discrete space, the faster the critical level of autoinducer is reached.

At a certain extracellular concentration, the secreted autoinducer reenters cells and binds to a regulatory molecule. In the case of *A. fischeri*, the regulatory molecule is LuxR (**Fig. 10.32** ▶). The LuxR-autoinducer complex activates transcription of the luciferase target genes that confer bioluminescence. Luciferase is the enzyme that produces light in *A. fischeri*. The apparent cell density requirement can be bypassed by the simple addition of purified autoinducer to a low-density cell culture. Many other microbes use these chemical languages to coordinate behavior of the population (**Table 10.3**). For instance, quorum sensing is used by pathogens to time the production of virulence factors for optimum effect on the host.

Quorum Sensing and Pathogenesis

Pseudomonas aeruginosa is a human pathogen that commonly infects patients with cystic fibrosis, a genetic disease

of the lung. The organism forms a biofilm in the lung and secretes virulence factors (such as proteases and other degradative enzymes) that destroy lung tissues (and thereby severely compromise lung function). These virulence proteins, however, are not made until cell density is fairly high—that is, at a point where the organism might have a chance of overwhelming its host. Made too early, virulence proteins would alert the host to launch an immune response. The induction mechanism involves two interconnected quorum-sensing systems, called Las and Rhl, both composed of regulatory proteins homologous to LuxR and LuxI of *Aliivibrio fischeri*. Many pathogens besides *Pseudomonas* appear to use chemical signaling to control virulence genes. These include *Salmonella*, *Escherichia coli*, *Vibrio cholerae*, the plant symbiont *Rhizobium*, and many others.

Interspecies Communication

Some microbial species not only chemically talk among themselves but can communicate with other species. *Vibrio harveyi*, for example, uses two different, but converging, quorum-sensing systems to coordinate control of its luciferase. Both sensing pathways are very different from the *Aliivibrio fischeri* system. One utilizes an acyl homoserine

TABLE 10.3 Examples of Microbial Quorum-Sensing Systems

System	Organism	Autoinducer family	Function
LuxR/LuxI	*Aliivibrio fischeri*	Homoserine lactone	Bioluminescence
LuxQ/LuxS	*Vibrio harveyi*	Furanosyl borate diester	Bioluminescence
LuxN/LuxLM	*Vibrio harveyi*	Homoserine lactone	Bioluminescence
LasR/LasI	*Pseudomonas aeruginosa*	Homoserine lactone	Exoenzyme production
Rhl	*Pseudomonas aeruginosa*	Homoserine lactone	Exoenzyme production
Agr	*Staphylococcus*	Peptide	Exotoxin production
StrR	*Streptomyces griseus*	γ-Butyrolactone	Aerial hyphae; antibiotic production
TraR/TraI	*Agrobacterium tumefaciens*	Homoserine lactone	Conjugation
YpeR/YpeI	*Yersinia pestis*	Unknown	Unknown
SdiA	*Escherichia coli*	Unknown	Cell division

Structures of different autoinducer families:

Homoserine lactone family   **γ-Butyrolactone**   **Furanosyl borate diester**

lactone (AHL) as an autoinducer (AI-1) to communicate with other *V. harveyi* cells. The second system produces a different autoinducer (AI-2), which contains borate. Because many species appear to produce this second signaling molecule, it is thought that mixed populations of microbes use it to "talk" to each other. Recently, Karina Xavier (Universidade Nova de Lisboa, Portugal) demonstrated that *E. coli* engineered to overproduce AI-2 could manipulate the composition of gut microbiota in mice.

In the case of *V. harveyi*, specific membrane sensor kinase proteins are used to sense each autoinducer (**Fig. 10.33**). At low cell densities (no autoinducer), both sensor kinases initiate phosphorylation cascades that converge on a shared response regulator, LuxO, to produce phosphorylated LuxO. LuxO-P activates expression of small RNAs called Qrr1, 2, 3, 4, and 5 that promote degradation of mRNA encoding the regulator LuxR, which is needed to activate expression of the *lux* genes. Thus, at low

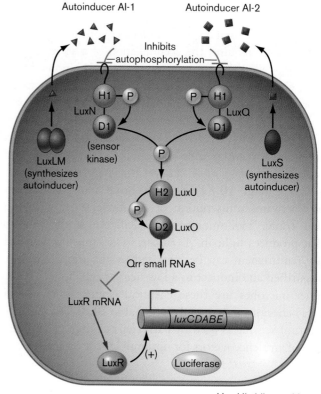

FIGURE 10.33 ■ Two quorum-sensing systems of *Vibrio harveyi*. In the absence of autoinducers (AI-1 and AI-2), both sensor kinases trigger converging phosphorylation cascades that end with the phosphorylation of LuxO. Phosphorylated LuxO (LuxO-P) activates the expression of Qrr small RNAs that promote degradation of LuxR mRNA. Luciferase is not made. As autoinducer concentrations increase, they inhibit autophosphorylation of the sensor kinases and the phosphorylation cascade. As a result, Qrr levels decrease, allowing synthesis of the LuxR regulator, which activates the *lux* operon.

H = Histidine residue
D = Aspartate residue

FIGURE 10.34 ■ **Bonnie Bassler.** Bassler was instrumental in characterizing interspecies communication among bacteria.

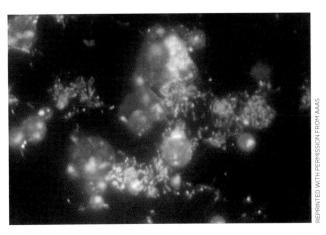

FIGURE 10.36 ■ *Enteromorpha zoospores.* Zoospores of the alga *Enteromorpha* (red) attach to biofilm-producing bacteria (blue) in response to lactones produced by the bacteria.

cell densities the culture does <u>not</u> display bioluminescence. At high cell densities, the autoinducers prevent signal transmission by inhibiting phosphorylation. The cell stops making Qrr sRNA and starts making LuxR (<u>not</u> a homolog of the *A. fischeri* LuxR). LuxR activates the *lux* operon and the "lights" are turned on. Bonnie Bassler at Princeton University (**Fig. 10.34**) and Peter Greenberg at the University of Washington (**Fig. 10.35**) are two of the leading scientists whose studies revealed the complex elegance of quorum sensing in *Vibrio* and *Pseudomonas* species. Other organisms, such as *Salmonella*, have been shown to activate the AI-2 pathway of *V. harveyi*, dramatically supporting the concept of cross-species communication.

A report by Ian Joint and his colleagues showed that bacteria can even communicate across the prokaryotic-eukaryotic boundary. The green seaweed *Enteromorpha* (a eukaryote) produces motile zoospores that explore and attach to *Vibrio anguillarum* bacterial cells in biofilms (**Fig. 10.36**). They attach and remain there because the bacterial cells produce AHL molecules that the zoospores sense. Part of the evidence for this interdomain communication involved showing that the zoospores would even attach to biofilms of *E. coli* carrying the *Vibrio* genes for the synthesis of AHL. The implications of possible interdomain conversations are staggering. Does our microbiome "speak" to us? Do we "speak" back? For further discussion of molecular communication between prokaryotes and eukaryotes, see Chapter 21.

This knowledge begs the question, Why would a squid want to harbor this bioluminescent microbe? Buried in the sand by day, the squid emerges at night from its safe hiding place to hunt for food. In moonlight, the squid would appear as a dark silhouette from below, marking it as easy prey for predators. It is thought that the squid camouflages itself by projecting light downward from its light organ, giving the appearance of surface-filtered light to predators below.

Quorum sensing is a wonderful way to coordinate group behaviors, but what about more private communication between individual cells? Is that possible? **Special Topic 10.1** shows that it is.

FIGURE 10.35 ■ **Peter Greenberg, one of the pioneers of cell-cell communication research.** Greenberg has studied quorum sensing in *Vibrio* species and various other pathogenic bacteria, such as *Pseudomonas*.

Thought Questions

10.10 Genes encoding luciferase can be used as "reporters." What would happen if the promoter for an SOS response gene was fused to the luciferase open reading frame?

10.11 What would happen if a culture was coinoculated with an *Aliivibrio fischeri luxI* mutant and a *luxA* mutant, neither of which produces light?

| SPECIAL TOPIC 10.1 | Networking with Nanotubes |

Microbial communication by secreted molecules (quorum sensing) is now a well-established phenomenon known to synchronize group behaviors. But quorum sensing is a system more like a loudspeaker than a phone. Everyone in the neighborhood, friend or foe, can potentially "hear" the signal. Another drawback to quorum sensing is that only small molecules can be sent and received, and these molecules are subject to degradation by extracellular factors. But what if microbes had a one-on-one communication system (like texting) in which not only signaling molecules, but also enzymes, DNA, or RNA, could be passed privately from one bacterium to another without ever facing the extracellular environment—temporarily changing the phenotype of the receiving cell?

Such a system was recently discovered by Gyanendra P. Dubey and Sigal Ben-Yehuda from the Hebrew University of Jerusalem (**Fig. 1A**). Using *Bacillus subtilis* as a model, these scientists witnessed adjacent bacilli exchanging cellular materials such as green fluorescent protein (GFP), while cells separated by a very small distance could not. Electron microscopy revealed how this was possible. Unexpectedly, the adjacent cells were connected by small tubes called nanotubes (**Fig. 1B**). These tubes were much larger (about 100 nm) than the bore of an average pilus (about 5 nm). Not only were proteins transferred, but nanotubes also served as portals through which DNA, RNA, and other molecules could be transferred. Unlike quorum sensing, nanotube communications were private, between one cell and another, although small, multicellular networks connected by nanotubes were also observed.

Dubey and Ben-Yehuda found that nanotubes could also form between <u>different</u> species of bacteria. Nanotube connections developed between two different Gram-positive species (*B. subtilis* and *Staphylococcus aureus*), and even between Gram-positive (*B. subtilis*) and Gram-negative (*E. coli*) species

(**Fig. 2A**). **Figure 2B** shows the transfer of GFP from a *gfp⁺ B. subtilis* to *E. coli*. The faint green speckled cells in **Figure 2B** are *E. coli* that received GFP via nanotube transfer from an adjacent *B. subtilis*.

Nanotube transfer of proteins could prove very beneficial to members of mixed-species biofilms. Imagine that one cell type in a biofilm is resistant to the antibiotic lincomycin, while another cell type is sensitive to this antibiotic but resistant to chloramphenicol. Could nanotubes help cells exchange resistance proteins, thereby providing each with a transient antibiotic resistance phenotype? **Figure 3** illustrates that transient nonhereditary phenotypes can be acquired by nanotube exchanges. **Figure 3A** diagrams how two cells—one expressing lincomycin resistance, the other expressing chloramphenicol resistance—can exchange resistance proteins and possibly mRNA through the nanotubes. **Figure 3B** shows the result of an actual experiment. *B. subtilis* Cm^R cells (P1) and Lin^R cells (P2) were grown on LB complex agar as individual or mixed patches (column 1). Patches were transferred to fresh media containing the drugs indicated. Notice that only the P1 + P2 mixed cells grew on the medium containing both Cm and Lin. To see whether the resistance phenotype was heritable, material from the patch was streaked for single colony isolation on LB (no drug), and individual colonies (arising from single cells) were tested for drug resistance. Amazingly, each colony exhibited resistance to one or the other drug only, indicating that double drug resistance was a transient, nonheritable phenotype (not the result of DNA transfer).

This mechanism of directly sharing resistance proteins could be extremely valuable to a biofilm under assault by multiple agents or even microbes under nutrient limitation. A report in 2015 from Christian Kost's laboratory at the Max Planck Institute for Chemical Ecology shows that nanotubes

A.

G.P. DUBEY, S. BEN-YEHUDA, *CELL*, FEB 18; **144**(4):590–600, FIG 3C

B.

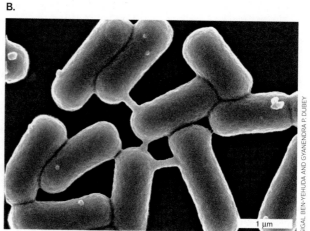

1 μm

SIGAL BEN-YEHUDA AND GYANENDRA P. DUBEY

FIGURE 1 ■ Nanotubes form between neighboring *Bacillus subtilis* cells. A. The research team of Professor Sigal Ben-Yehuda (left) and Gyanendra P. Dubey discovered nanotubes. **B.** Cells were grown on agar for 6 hours and visualized by high-resolution SEM. Notice that several cells are connected.

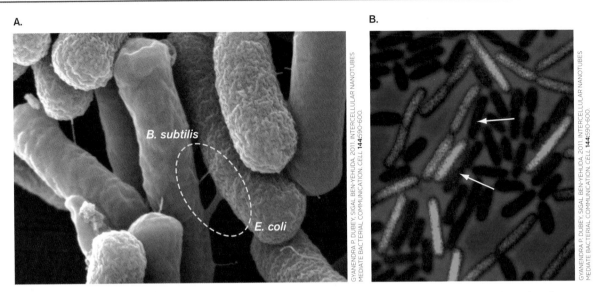

FIGURE 2 ■ Interspecies nanotubes. A. SEM showing nanotubes forming between *Bacillus subtilis* and *Escherichia coli*. **B.** *B. subtilis* containing GFP and *E. coli* without GFP were plated together on agarose. The fluorescent image was taken 30 minutes after plating. The bright green organisms are *B. subtilis*, but notice that *E. coli* cells adjacent to *B. subtilis* have started to turn green (arrows) because of the transfer of GFP protein.

can mediate nutrient transfers between auxotrophic organisms under starvation conditions. Communicating by nanotubes, therefore, gives bacteria a means of straightforward immediate transfer of information and materials across inherent species barriers. Moreover, molecules channeled by nanotubes are protected from degrading enzymes and harsh environmental conditions. Though untested, nanotubes might also be an efficient defensive strategy used to kill competitors by directly delivering toxic molecules.

RESEARCH QUESTION

Knowing that nanotubes can temporarily transfer antibiotic resistance from one cell to a second cell, how might you select for a mutant defective in nanotube formation? Discuss one potential pitfall with your approach.

Dubey, Gyanendra P., and Sigal Ben-Yehuda. 2011. Intercellular nanotubes mediate bacterial communication. *Cell* **144**:590–600.

FIGURE 3 ■ Transient nonhereditary phenotypes acquired by nanotubes. A. Chloramphenicol-resistant (Cm^R) and lincomycin-resistant (Lin^R) cells forming nanotube connections that exchange resistance proteins and possibly mRNA. **B.** *Bacillus subtilis* wild-type cells (WT), Cm^R cells (P1), and Lin^R cells (P2) were grown for 4 hours on LB complex agar individually or mixed (column 1). Patches were transferred to fresh LB medium with the drugs indicated (columns 2–4) or with no drugs added (column 5). Growth on the Cm + Lin plate was achieved only when the two strains were mixed in a single patch.

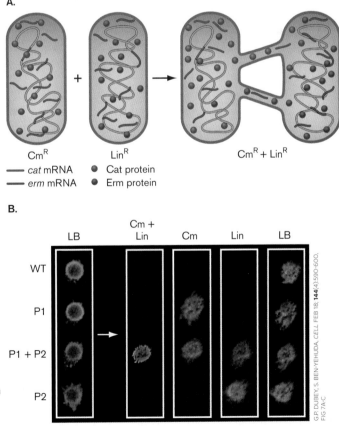

To Summarize

■ **Quorum sensing** involves the synthesis, secretion, and extracellular accumulation of small autoinducer signaling molecules. Cells within a population sense a threshold concentration of autoinducer and simultaneously respond by expressing a subset of genes.

■ Quorum sensing **enables communication between cells** of a single species or between multiple species.

■ **Pathogens use quorum sensing** to time the expression of virulence genes during growth within a host.

10.6 Transcriptomics and Proteomics

The studies exposing many of the regulatory systems discussed in this chapter were accomplished by examining only a small number of genes at any one time. Now, however, new technologies enable scientists to examine thousands of genes or proteins in a single experiment.

Transcriptomics

A pebble tossed into a pond makes a wave that spreads and ultimately disturbs the entire pond. Similarly, altering one biochemical pathway in a cell produces a physiological "wave" that impacts, to a greater or lesser degree, all other pathways. What typically initiates and propagates a physiological "wave" is a transcriptional change in gene expression. Scientists now have the ability to see these events unfold by monitoring the expression of every gene in a cell—simultaneously. The result is known as the **transcriptome**.

This global view of gene expression is possible using next-generation, high-throughput DNA sequencing technology (see Section 7.6). However, mRNA is easily degraded and has to be converted, molecule by molecule, into more stable DNA in order to be counted. Basically, all the RNA is extracted from a culture of cells, and then stable RNAs (tRNA and rRNA) comprising the bulk of RNA are removed by hybridization. The remaining RNA molecules are fragmented by sonic disruption, and the fragments converted to complementary DNA (cDNA) with

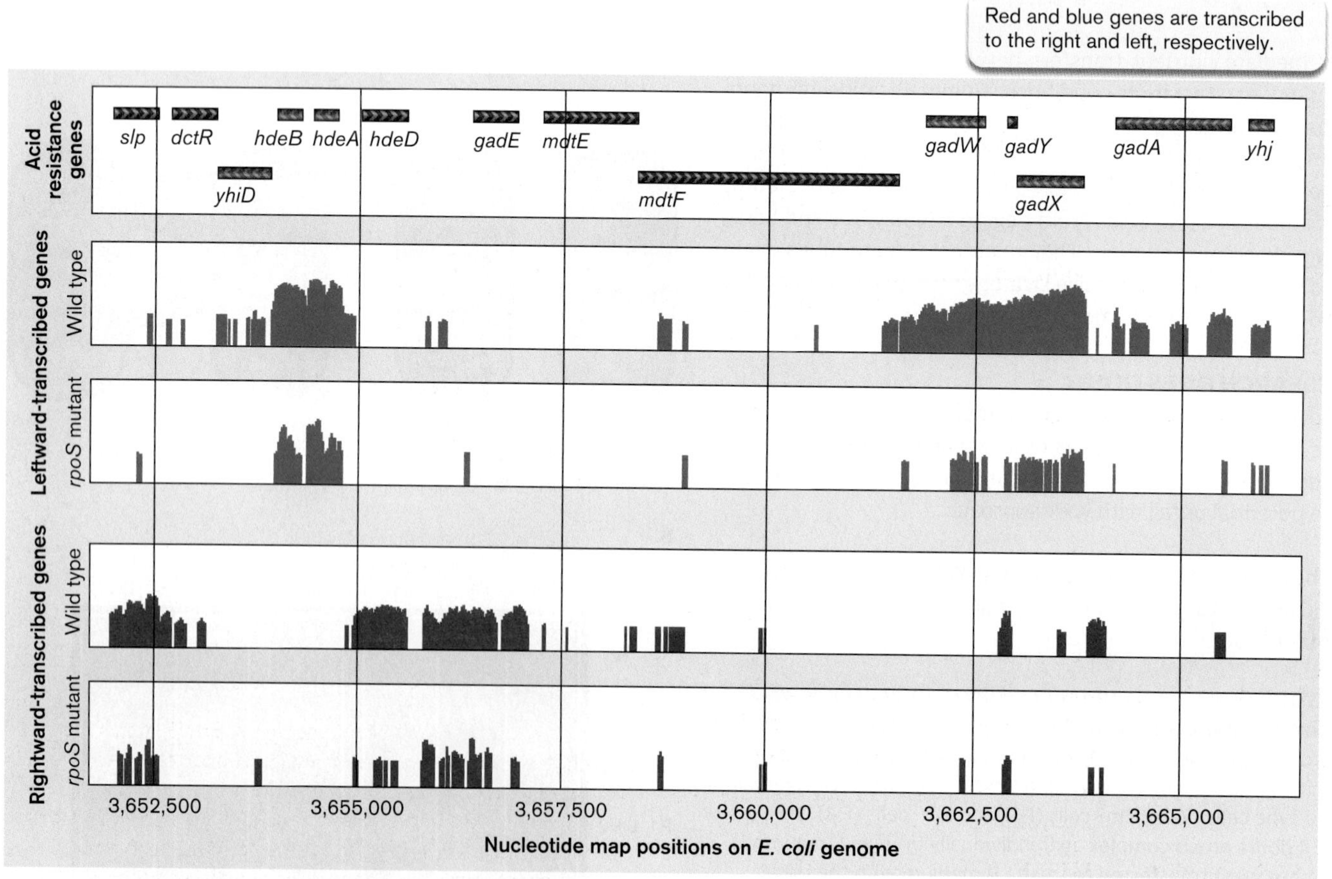

FIGURE 10.37 ■ **Transcriptomics of *Escherichia coli* acid resistance genes.** Deep sequencing was used to quantify the relative numbers of transcripts made from each gene in wild-type and *rpoS* mutant strains. Each vertical bar in the bottom four panels represents a portion of a transcript. The experiments actually captured data from all genes expressed from the genome, but only some of these data are shown.
Source: Modified from University of Oklahoma Gene Expression Database.

reverse transcriptase. The number of cDNA fragments made corresponds to the number of transcripts present in the cell extract. DNA adapters are ligated to each end of the cDNA molecules. Each molecule, with or without amplification, is then sequenced in a high-throughput manner (for example, Illumina's sequencing by synthesis) to obtain short sequences from one end or both ends of each fragment (corresponding to transcript copy). The reads, millions of them, are typically 30–400 bp. A computer program aligns each sequence with its genome position (gene) and then tabulates the number of fragments sequenced at any one location. The number of RNA sequences (RNA-seq data) found for a particular gene is proportional to how much that gene was expressed.

Comparing results between cells grown under different conditions, or between a mutant and a wild-type strain, will expose how the expression of every gene is influenced by the change. **Figure 10.37**, for instance, illustrates the effect of sigma S on the expression of acid resistance genes in *E. coli*. *E. coli* can survive pH 2 exposures for long periods of time—a property known as acid resistance. This capability requires numerous genes, as shown in the top panel of the figure. Tyrrell Conway and colleagues at the University of Oklahoma used deep sequencing to quantify the transcript levels of these genes in wild-type and *rpoS* mutants of *E. coli* lacking sigma S. *E. coli rpoS* mutants are highly sensitive to acidic environments like stomach contents. **Figure 10.37** compares the expression of these genes in wild-type (panels 2 and 4) and RpoS-deficient strains (panels 3 and 5). Note that transcripts of all of these genes are decreased in the acid-sensitive *rpoS* mutant.

Another powerful technique, called DNA microarray, can also simultaneously examine the expression of every gene in a cell. For an organism like *Salmonella* Typhi, the causative agent of typhoid fever, this requires seeing more than 4,600 genes. All of this can be done on a slide the size of a postage stamp. The microarray technique is described in eAppendix 3.

Proteomics

If all the expressed mRNAs in a microbe make up its transcriptome, then all the expressed proteins form its **proteome**. Unlike the DNA genome, which is fixed, the transcriptome and proteome continually change in response to a changing environment. RNAseq and gene array technologies can capture what happens transcriptionally, but gene arrays do not necessarily reveal which proteins are actually expressed, because stress can change the translation of a message without changing its transcription. Thus, the level of a protein can increase significantly without any change in the amount of mRNA. In addition, proteins undergo posttranslational modifications (for example, acetylation, phosphorylation)

in response to different environments. The powerful techniques of 2D gel electrophoresis and mass spectrometry reveal fluctuations in the proteome.

Two-dimensional gel electrophoresis. The science of 2D gels is described in eAppendix 3. Briefly, the technique uses polyacrylamide gel electrophoresis to separate proteins from cell extracts in one dimension by their charge and in a second dimension by their molecular weight. When combined, isoelectric focusing and SDS-PAGE display a majority of the cell's proteins in a 2D array, as shown in **Figure 10.38** (some proteins do not appear, because their quantities or solubilities are too low). The proteins are visualized by staining with fluorescent dyes after separation. Subsequent computer analysis of the proteome patterns obtained from cells grown under two different conditions will reveal proteins whose levels increase or decrease in response to the changing environment. **Figure 10.38** shows a dual-channel image analysis that compares *Bacillus subtilis* proteomes from cultures grown in a minimal glucose medium with or without amino acid supplementation. The different colors indicate whether

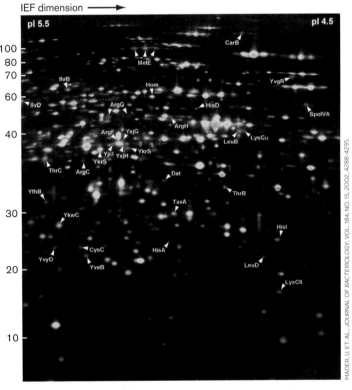

MADER, U. ET. AL., JOURNAL OF BACTERIOLOGY, VOL. 184, NO. 15, 2002, 4288-4295.

FIGURE 10.38 ■ **Proteomic profile of *Bacillus subtilis* cells grown in minimal media with and without mixed amino acid supplementation (casamino acids).** The isoelectric focusing (IEF) gradient used in the first dimension was pH 4.5–5.5; this represents only a part of the entire proteome. The image is the result of dual-channel analysis of silver-stained gels. A computer assigns the color red to proteins expressed in minimal media and green to proteins expressed in amino acid–supplemented media. If the proteins are expressed under both conditions, the red and green colors combine to form yellow or orange.

the level of a protein is higher, lower, or the same in the two cultures. Proteins of interest can be plucked from the gel and their identities determined though mass spectrometry techniques, discussed next (see also eAppendix 3).

Mass spectrometry. Knowing the sequence of an organism's genome gives us a wealth of useful information and new ways to probe its physiology. One of those ways involves mass spectrometry, which experimentally determines the exact mass of an unknown protein or peptide fragment and uses that mass to identify the protein. The process is easy for a single protein picked from a gel, but new techniques make it possible to identify and quantify proteins from whole-cell extracts.

To begin the analysis, proteins present in cell extracts are digested into distinct peptide fragments with a site-specific protease (trypsin) (**Fig. 10.39**). The fragment mixture is passed through an analytical column that separates peptides according to differences in hydrophobicity or some other parameter (**Fig. 10.39**, steps 1–3). The column effluent is then directly fed through tandem mass spectrometry (MS-MS) instrumentation (step 4). In MS-MS, each proteolytic fragment is subfragmented by ionization to produce progressively smaller secondary fragments missing one or more amino acids. Because the weight of each amino acid is distinct, MS-MS analysis can determine the amino acid sequence of the initial proteolytic fragment (step 5). Sophisticated computer programs identify the proteins by comparing the amino acid sequences of each protein fragment with the predicted sequences of proteolytic fragments from all ORFs in a genome (steps 6 and 7).

Can we also use mass spectrometry to quantify a given protein in two different samples? Determining relative quantities of a protein between experimental samples typically involves labeling peptides with isobaric tags before MS-MS analysis (**Fig. 10.40**). Isobaric tags have two parts: a reporter group and a balance group. The masses of reporter groups from different tags differ slightly. However, the masses of the balance groups also differ, such that all the intact isobaric tags have the same initial mass. Cell proteins from different experimental samples are digested by trypsin, and the peptide fragments from each sample are collectively labeled with different isobaric tags. That is, one sample's peptides are all modified with an isobaric tag whose reporter group's mass differs slightly from the reporter group of an isobaric tag used to mark peptides from a different sample. But because each reporter group is linked to a balance group, all tags for all samples have the same initial mass (that is, they are isobaric). The two samples are then mixed and run through MS-MS analysis. Each peptide peak will contain representatives from both samples because their masses are the same. Subsequent ionization of a single peptide will release the different mass

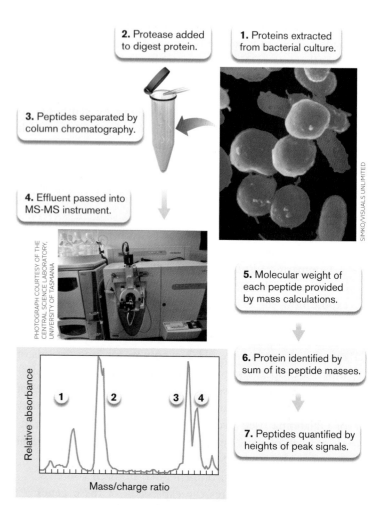

FIGURE 10.39 ■ Identifying proteins directly from whole-cell extracts by mass spectrometry. Proteins extracted from a bacterial culture are digested into peptides with trypsin. The peptides are separated by column chromatography and analyzed by mass spectrometry. In tandem mass spectrometry (MS-MS), the mass of each peptide is determined first (peaks 1–4 in the graph), and then selected peptides are subjected to additional fragmentation by ion spray (not shown). Each resulting peptide fragment will differ in size by one or more amino acids. Knowing the mass of each amino acid and the masses of the different peptide fragments enables extrapolation of the original peptide's sequence.

reporter groups and cleave the peptide backbone to determine the peptide's sequence. Relative peptide quantities in samples are determined by comparing the intensities of the released reporter ion signals present in the MS-MS scan. This technology has been used, for instance, to identify and evaluate changes in the *Streptomyces coelicolor* proteome during mycelial development. About two-thirds of industrial antibiotics are synthesized by *S. coelicolor*.

Thought Question

10.12 Can tandem mass spectrometry (MS-MS) identify where modifications such as phosphorylation or acetylation happen on proteins?

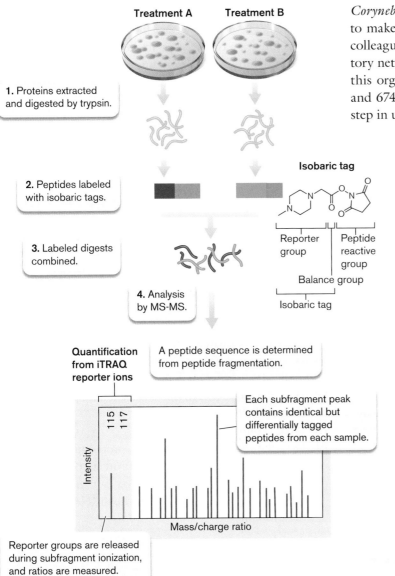

FIGURE 10.40 ■ Isobaric labeling and quantitation of peptides by mass spectrometry. Isobaric tags (iTRAQ, the isobaric tags for relative and absolute quantitation technique) contain a reporter group and a balance group so that different tags still have the same masses. The peptide reactive group is used to link the tag to peptide amines.

Transcriptomic and proteomic technologies have been used for many microbial applications. One fascinating example centers on the transcriptional events that occur during the developmental cycle of *Caulobacter crescentus*. This organism progresses from a nonreplicating swimming form (swarmer cell) to a sessile stalked form that can replicate to make more swarming offspring. DNA microarray and proteomic analysis have revealed several potential regulators of this complex developmental process. Bioinformatic approaches have also been applied to human body odor.

Corynebacterium jeikeium strain K411 degrades skin lipids to make volatile odorous products. Helena Barzantny and colleagues reconstructed a putative transcriptional regulatory network from the complete DNA sequence analysis of this organism. The result was a network of 48 regulators and 674 gene regulatory interactions that provide an early step in understanding how body odor develops.

To Summarize

- **The transcriptome and proteome** constitute all of a cell's mRNA molecules and proteins, respectively. Transcriptomes and proteomes change as environmental conditions change.

- **DNA microarrays** and **RNAseq next-generation sequencing** are used to monitor the levels of thousands of individual mRNA molecules made during growth.

- **Two-dimensional gels** separate proteins by isoelectric point and molecular weight. They offer a snapshot of the proteome at any given point of growth or under any given growth condition.

- **Mass spectrometry** can determine the exact molecular weight of a protein (or the sequence of a component peptide) in a cell extract or from a spot on a 2D gel. The exact molecular weight or peptide sequence is compared against the database of ORFs deduced from the genomic sequence of the organism. A successful match determines the protein's identity.

Concluding Thoughts

Bacteria employ a vast array of regulatory mechanisms, both simple and complex, to make survival decisions. But do they possess cognitive ability? This is a controversial idea, but one worth pondering. Cognition in its broadest sense is the process of acquiring, processing, storing, and acting on information from the environment. Memory, for instance, is critical to cognition. Chemotactic bacteria use arrays of methyl-accepting chemotaxis proteins to remember where it is in a chemical gradient. The information acquired is then used to guide decisions about cell movement. Two-component signal transduction systems, prominent in bacteria, contribute to cognition if the systems cross-talk. Bacterial signal transduction systems cross-talk when a histidine kinase phosphorylates a noncognate response regulator. Small RNA molecules also contribute to system cross talk by influencing disparate regulatory systems. Higher-order manifestations of cognition include communication and sociability. Bacteria communicate using autoinducer molecules and exhibit social behaviors by building biofilms or by swarming. These examples, and many others, support the idea that bacteria have a type of cognitive ability. They lack self-awareness, certainly, but they can solve problems, communicate, and cooperate with each other. The next great task is to explore how individual bacteria integrate the massive amounts of information they collect from hundreds of signaling pathways to form coherent, adaptive responses.

CHAPTER REVIEW

Review Questions

1. List regulatory mechanisms discussed in this chapter that work at the DNA level, transcriptional level, translational level, and posttranslational level.
2. How does lactose induce the *lacZYA* operon?
3. If *lacY* is induced only when lactose is present, how does external lactose induce the system?
4. How does tryptophan repress the tryptophan operon?
5. Describe a two-component signal transduction system.
6. Discuss how glucose impacts the utilization of lactose as a carbon source.
7. Name a regulatory protein that can activate and repress an operon's expression. How does it do that?
8. What is the regulatory mechanism that uses translation to control transcription? How does it work?

9. When growth of *E. coli* slows, how does the cell slow its production of ribosomes?
10. Describe four ways that sigma factor production/activity can be regulated.
11. Discuss the different forms of small regulatory RNA molecules.
12. Describe how *E. coli* senses a chemical gradient and changes its behavior.
13. What is quorum sensing?
14. Describe transcriptomics and proteomics. Can you think of a situation in which a protein could be shown to increase in response to a stress but that same stress would not increase synthesis of the mRNA that encodes the protein?

Thought Questions

1. What will happen to the expression of the tryptophan operon if you replace the key tryptophan codons in the attenuator region with tyrosine codons?
2. Adding tryptophan to *E. coli* will cause repression of the *trp* operon genes. Mutations in the *trpR* repressor gene and the *trp* operator will have the same phenotype; that is, adding tryptophan will no longer repress expression of the *trp* genes. What will happen to the phenotype if you transform each mutant with a plasmid carrying the wild-type *trpR* gene or the wild-type *trp* operator region?
3. The mosquito that transmits yellow fever, *Aedes aegypti*, requires water-filled, human-made containers for egg laying. Normally, gravid females deposit their eggs in multiple containers but are particularly stimulated to

deposit on fermenting leaves in water. Scientists tested the relative egg deposits made by gravid mosquitoes given a choice between a container with sterile water and a second container inoculated with 14 bacterial species isolated from a bamboo infusion (which was previously shown to be a favorite deposit medium). Ninety percent of the eggs were deposited on the surface of the bacterial infusion, and only 10% were deposited on the sterile water. How did the presence of the bacteria influence the choice?
4. This chapter has outlined the many ways bacteria sense their environment, their neighbors, and their location (rock, intestine, ocean). Can we then say that bacteria are conscious?

Key Terms

activator (361)
anti-anti-sigma factor (375)
anti-attenuator stem loop (373)
anti-sigma factor (375)
attenuator stem loop (373)
autoinducer (389)
catabolite repression (368)
chemotaxis (384)
cis-antisense RNA (asRNA) (378, 381)
corepressor (362)
derepression (362)
diauxic growth (368)
direct repeat (359)

inducer (361)
inducer exclusion (368)
induction (361)
inverted repeat (359)
methyl-accepting chemotaxis protein
  (MCP) (384)
operon (365)
phase variation (359)
proteome (395)
quorum sensing (388)
regulatory protein (361)
repression (361)
repressor (361)

response regulator (363)
riboswitch (382)
ribozyme (382)
second messenger (386)
sensor kinase (363)
small RNA (sRNA) (358, 378)
stringent response (373)
tandem repeat (359)
transcriptional attenuation (372)
transcriptome (394)
translational control (374)
two-component signal transduction
  system (363)

Recommended Reading

Babski, Julia, Lisa-Katharina Maier, Ruth Heyer, Katharina Jaschinski, Daniela Prasse, et al. 2014. Small regulatory RNAs in Archaea. *RNA Biology* 11:484–493.

Bandara, H. M., O. L. Lam, L. J. Jin, and L. Samaranayake. 2012. Microbial chemical signaling: A current perspective. *Critical Reviews in Microbiology* 38:217–249.

Battesti, Aurelia, Nadim Majdalani, and Susan Gottesman. 2011. The RpoS-mediated general stress response in *Escherichia coli*. *Annual Review of Microbiology* 65:189–213.

Commichau, Fabian M., Achim Dickmanns, Jan Gundlach, Ralf Ficner, and Jörg Stülke. 2015. A jack of all trades: The multiple roles of the unique essential second messenger cyclic di-AMP. *Molecular Microbiology* 97:189–204.

Dalebroux, Zachary D., and Michelle S. Swanson. 2012. ppGpp: Magic beyond RNA polymerase. *Nature Reviews. Microbiology* 10:203–212.

Fozo, Elizabeth M., Matthew R. Hemm, and Gisela Storz. 2008. Small toxic proteins and the antisense RNAs that repress them. *Microbiology and Molecular Biology Reviews* 72:579–589.

Henke, Jennifer M., and Bonnie L. Bassler. 2004. Bacterial social engagements. *Trends in Cell Biology* 14:648–656.

Higgins, Douglas, and Jonathan Dworkin. 2012. Recent progress in *Bacillus subtilis* sporulation. *FEMS Microbiology Reviews* 36:131–148.

Jones, Christopher W., and Judith P. Armitage. 2015. Positioning of bacterial chemoreceptors. *Trends in Microbiology* 23:247–256.

Lewis, Mitchell. 2005. The *lac* repressor. *Critical Reviews in Biology* 328:521–548.

Lyon, Patricia. 2015. The cognitive cell: Bacterial behavior reconsidered. *Frontiers in Microbiology* 6:264.

Mellin, J. R., and Pascale Cossart. 2015. Unexpected versatility in bacterial riboswitches. *Trends in Genetics* 31:150–156.

Merino, Enrique, and Charles Yanofsky. 2005. Transcription attenuation: A highly conserved regulatory strategy used by bacteria. *Trends in Genetics* 21:260–264.

Papenfort, Kai, and Carin K. Vanderpool. 2015. Target activation by regulatory RNAs in bacteria. *FEMS Microbiology Reviews* 39:362–378.

Parker, Christopher T., and Vanessa Sperandio. 2009. Cell-to-cell signaling during pathogenesis. *Cellular Microbiology* 11:363–369.

Ponnusamy, Longanathan, Ning Xu, Satoshi Nojima, Dawn M. Wesson, Coby Schal, et al. 2008. Identification of bacteria and bacteria-associated chemical cues that mediate oviposition site preferences by *Aedes aegypti*. *Proceedings of the National Academy of Sciences USA* 105:9262–9267.

Schleif, Robert. 2010. AraC protein, regulation of the L-arabinose operon in *Escherichia coli*, and the light switch mechanism of AraC action. *FEMS Microbiology Reviews* 34:779–796.

Sesto, Nina, Omri Wurtzel, Cristel Archambaud, Rotem Sorek, and Pascale Cossart. 2013. The excludon: A new concept in bacterial antisense RNA-mediated gene regulation. *Nature Reviews. Microbiology* **11**:75–82.

Staron, Anna, and Thorsten Mascher. 2010. Extracytoplasmic function sigma factors come of age. *Microbe* **5**:164–170.

Vink, Cornelis, Gloria Rudenko, and H. Steven Seifert. 2012. Microbial antigenic variation mediated by homologous DNA recombination. *FEMS Microbiology Reviews* **36**:917–948.

Williams, Paul. 2007. Quorum sensing, communication and cross-kingdom signaling in the bacterial world. *Microbiology* **153**:3923–3938.

Yang, Ji, Marija Tauschek, and Roy M. Robins-Browne. 2011. Control of bacterial virulence by AraC-like regulators that respond to chemical signals. *Trends in Microbiology* **19**:128–135.

CHAPTER 11
Viral Molecular Biology

Viruses such as influenza virus, hepatitis virus, and HIV sicken hundreds of millions of people worldwide. Yet many other viruses coexist with us without harm. What makes one virus deadly and another helpful? The answer lies in the molecules. Endogenous viral genomes contribute molecular parts for our bodies, such as a protein for placental fusion. Even the viruses most deadly to humans can be converted to vectors that cure a human disease.

CURRENT RESEARCH highlight

Human embryo makes retroviral particles. Human genomes contain many sequences of endogenous retroviruses (HERVs) that infected human ancestors millions of years ago. In some cases, the retroviral genes evolved into human genes that express essential proteins. Human embryos at the blastocyst stage express proteins of the endogenous retrovirus HERV-K, including virus-like particles labeled by HERV-K capsid antibody stain (red). The blastocyst nuclei are labeled with DAPI fluorophore (blue). HERV-K virus-like particles arise when cell nuclei express an embryonic regulator protein (labeled green). Activation of viral proteins may protect the embryo from infection by exogenous viruses.

Source: Edward Grow and Joanna Wysocka et al. 2015. *Nature* **522**:221.

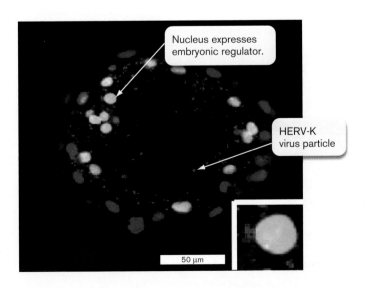

Nucleus expresses embryonic regulator.

HERV-K virus particle

50 μm

AN INTERVIEW WITH

JOANNA WYSOCKA, STEM CELL BIOLOGIST, STANFORD UNIVERSITY

CHRISTOPHER VAUGHAN/STANFORD UNIVERSITY COMMUNICATION OFFICE

How does an endogenous retrovirus contribute to the human genome?

Endogenous retroviruses (ERVs) are remnants of ancient retroviral infections and comprise about 8% of the human genome. Major evolutionary innovations, such as placentation, have been associated with the co-option of ERV sequences for regulatory functions. One human ERV (HERV) subtype, HERV-K, is of particular interest, as it is the most recently acquired HERV in the human genome from which multiple insertions have retained protein-coding potential. Remarkably, HERV-K virus-like particles and Gag proteins are readily detected in human blastocysts, indicating that early human development proceeds in the presence of retroviral products.

What do you find most exciting about the implications for virology and human biology?

From the human development perspective, these results argue that complex interactions between HERV-encoded retroviral products and host factors fine-tune regulatory properties of early human development. Moreover, HERVs can rewire developmental networks via their functions as promoters or enhancers for human genes. Furthermore, HERVs are also interesting from a human evolution perspective, as they can be exapted (evolve a new function) in a lineage-specific manner as human-specific enhancers.

Buried in the human genome are copies of an amazing time traveler, the endogenous retrovirus HERV-K. Retroviruses are defined by reverse transcriptase, an enzyme that copies viral RNA genomes into DNA (discussed in Section 11.3). Sometimes, after the molecular process of reverse transcription, the retroviral genome gets "stuck" in the host genome—and even enters the host germ line, where it passes on to future generations. Over the generations, perhaps tens of millions of years, this endogenous viral genome mutates (see Section 11.4). Most of the mutating viral genes lose function—but sometimes the genes evolve into useful parts of the host. Joanna Wysocka's research shows that HERV-K may actually generate virus particles within the human blastocyst, an early stage of the human embryo (see the Current Research Highlight). To produce the virus-like particles, cells must express a major embryonic regulator protein. How could these virus particles help the embryo develop? Wysocka shows that HERV-K particles may stimulate the embryonic cells' innate immune defenses—and thus protect the cells from dangerous viral infections.

What makes a virus good or bad? Molecular structure makes the difference. For instance, the seasonal strains of influenza that kill up to half a million people each year use specific envelope proteins to infect our respiratory tissues (see Section 11.2). The structure of these proteins determines whether a type of influenza virus can bind the glycoproteins on the surface of a human cell. Once inside a host cell, the viral genome and packaged molecules must collaborate with host enzymes to build virions, and to evade defense molecules of the immune system. Alternatively, a viral regulator may cause the virus to remain dormant within the cell.

In Chapter 10 we discussed how molecular mechanisms such as repressors and DNA inversions regulate the function of bacteria. Now, Chapter 11 presents the molecular mechanisms of viral infection. For background, we assume a fundamental understanding of the nature of viruses, presented in Chapter 6. Chapter 11 explores in depth four important viruses. First we consider the bacteriophage lambda, a type of phage found in your digestive tract and used by scientists for synthetic biology. We then examine in detail three viruses that infect humans: influenza (a negative-strand RNA virus), HIV (a retrovirus), and herpes simplex (a large DNA virus). We will see that all viral infection processes share common themes but also use molecular mechanisms unique to each virus.

Section 11.4 presents our emerging awareness of endogenous retroviruses such as HERV-K—and how we build retroviral vectors for lifesaving gene therapy.

Note: Online eTopics cover four additional viruses in depth:
- **eTopic 11.1** Phage T4: The Classic Molecular Model
- **eTopic 11.2** The Filamentous Phage M13: Vaccines and Nanowires
- **eTopic 11.3** Poliovirus: (+) Strand RNA Virus
- **eTopic 11.4** Hepatitis C: (+) Strand RNA Virus

11.1 Phage Lambda: Enteric Bacteriophage

Bacteriophage lambda infects *Escherichia coli* within the human gut, among the trillions of phages in our intestinal community (**Fig. 11.1**). Historically, phage lambda's interaction with *E. coli* was the first living system simple enough to dissect at the molecular level. The phage yielded fundamental discoveries in gene regulation, most notably the control of lysogeny (introduced in Chapter 6). In microbial communities, lysogeny provides a way for phages to transfer genes between bacterial genomes. When lambda lysogeny was first discovered, we had no idea that human genomes similarly carry endogenous retroviruses such as HERV-K.

The "lambda switch" between lysis and lysogeny provided clues to the bacterial regulons presented in Chapter 10, as well as molecular mechanisms of animals and plants. Today, the well-studied phage lambda provides tools for synthetic biosensors and even DNA chip devices.

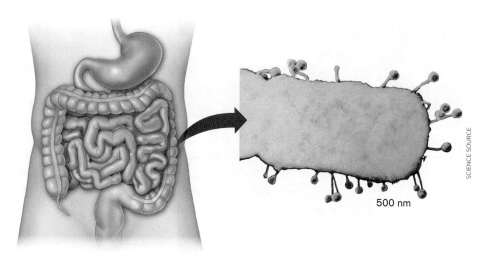

FIGURE 11.1 ■ Within the human intestine, *Escherichia coli* hosts phage lambda. Colorized SEM.

FIGURE 11.2 ▪ **Esther Lederberg discovered phage lambda.** Lederberg in her laboratory at Stanford, holding phage stocks in cotton-stoppered tubes.

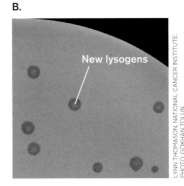

FIGURE 11.3 ▪ **A lysogen reveals phage lambda. A.** The lysogenic *E. coli* strain (vertical streak) releases phages that cause plaques in a susceptible *E. coli* strain (horizontal streak). **B.** Lambda plaques show a cloudy center where newly formed lysogens appear. **C.** Undergraduate Kathleen Morrill holds a plate of lambda *lacZ* reporter strains on MacConkey agar; with Lynn Thomason, at the National Cancer Institute.

Discovery of Phage Lambda

Bacteriophage lambda was discovered in 1950 by Esther Lederberg (1922–2006), pioneering bacterial geneticist at the University of Wisconsin–Madison, and later Stanford University (**Fig. 11.2**). At the time, it was known that some kinds of bacteria spontaneously lyse and release phages, but the mechanism, even the source of phage from within the cells, was a mystery. The strain *Escherichia coli* K-12 had been isolated in 1922 from a healthy patient's colon and was used for decades thereafter in teaching and research. In 1950, Lederberg was performing a genetic cross between two strains of *E. coli* K-12: a standard laboratory stock, and a strain that she had mutagenized with ultraviolet light. After the two strains were mixed, the mutant strain formed colonies that were "nibbled and plaqued" (**Fig. 11.3A**). The plaques resulted from phage particles lysing the mutant strain. The mutant had lost its lambda prophage and was therefore susceptible to infection. The infecting phage particles came from the lysogenic *E. coli* K-12, which Lederberg had streaked across.

Further experiments confirmed that at that time, all known stocks of K-12, the most widely used strain of *E. coli*, were lysogenic for phage lambda. **Lysogeny** means that the genome of a phage is incorporated into the genome of the host cell (discussed in Chapter 6). As the host cell then reproduces, phage particles no longer exist as such, but the phage DNA replicates indefinitely within the host, as an integrated **prophage**. The presence of the prophage confers resistance to infection by the same type of phage. Phage lambda is thus called a "temperate phage," rather than a "virulent phage" such as T4, which always kills its host. But occasionally (about once in a million cells), a molecular signal tells the prophage genes to make progeny phages and lyse the cell. Thus, unknown to researchers at the time, a given tube of growing *E. coli* lysogen typically carried about a million lambda phages per milliliter. The phages released can form plaques on a strain that is not a lysogen (has no lambda prophage) (**Fig. 11.3B**). The plaques have a cloudy center because some of the infected cells become lysogens, which start growing up where the infection began.

Lederberg and others learned to "cure" the K-12 lysogen to eliminate the phage from stock cultures, including most strains used today. Other notable discoverers of the nature of lysogeny include French microbiologists André Lwoff (1902–1994) and François Jacob (1920–2013). The molecular basis of the lysis/lysogeny "switch" was elucidated by Mark Ptashne and colleagues (discussed shortly).

A.

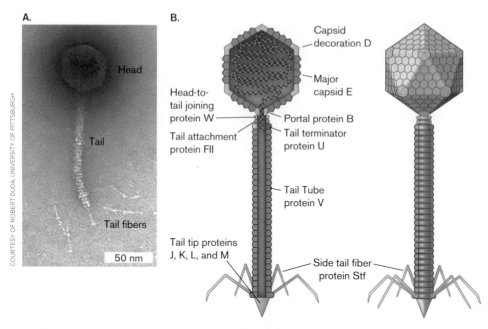

COURTESY OF ROBERT DUDA, UNIVERSITY OF PITTSBURGH

Head

Tail

Tail fibers

50 nm

B.

Capsid decoration D

Head-to-tail joining protein W

Major capsid E

Portal protein B

Tail attachment protein FII

Tail terminator protein U

Tail Tube protein V

Tail tip proteins J, K, L, and M

Side tail fiber protein Stf

FIGURE 11.4 ■ The lambda virion. A. Phage particle visualized by heavy-atom negative stain (TEM). **B.** Diagram of components, colored to match coding genes in the genome (see Fig. 11.5).

Full regulation of the switch includes numerous viral and host proteins, whose details continue to be discovered by researchers such as Lynn Thomason and her student Kathleen Morrill at the National Cancer Institute (**Fig. 11.3C**).

Thought Question

11.1 Plaques from phage lambda quickly fill with resistant lysogens. Could there be a different way for the host cells to become resistant, without forming lysogens?

Phage Structure and Genome Replication

The phage lambda particle (virion) is typical of tailed phages of the siphophage family (**Fig. 11.4**). Various siphophages infect Gram-negative and Gram-positive bacteria. Siphophages commonly appear in soil and water, as well as in enteric communities.

Head contains DNA. The part of a tailed phage that contains its genome is called the "head." Usually the head consists of an icosahedral protein complex, equivalent to the capsid of a tail-less virion. The head of phage lambda contains coiled double-stranded DNA. Thus, according to genome classification, phage lambda falls under Baltimore Group I (see Chapter 6). The protein coat is composed of two types of subunits: the face subunit E and the "decoration" subunit D. The D subunits form a pentamer at each vertex of the icosahedron, except for the tail connector. The tail connector comprises four proteins encoded by different genes (B, FII, U, W). Each connector subunit is found in multiple copies that form a ring around the tail connector.

Tail tube. The tail itself consists of a long tube of 32 hexamer rings of subunit V. Remarkably, the exact length of the tail (the number of hexamers) depends on a "tape

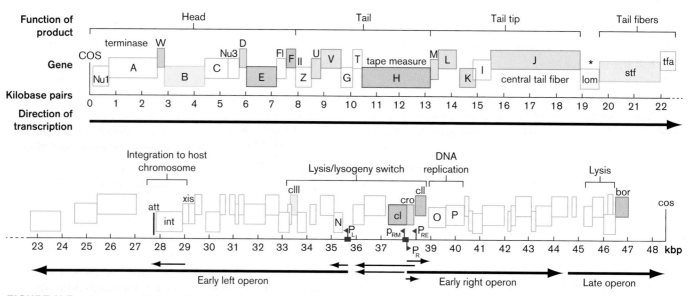

FIGURE 11.5 ■ Genome of phage lambda. For each gene, vertical offset indicates the reading frame.

measure protein" (protein H). If we delete part of the gene encoding the tape measure protein, the phage will assemble a shorter tail. On the other hand, if the gene is lengthened by addition of nucleotides, the phage will make a longer tail. How the "tape measure" works is unclear.

With respect to function, the tail of all siphophages is noncontractile (unlike the contractile tail of myophages such as phage T4; **eTopic 11.1**). The tail tube has a tip and tail fibers, both of which help the phage attach to its host cell. Upon host attachment, the lambda DNA must uncoil and pass through the entire length of the tail to reach the host cytoplasm.

Genome. Phage lambda has a genome of 48.5 kilobase pairs (kb) that includes about 70 genes. In **Figure 11.5**, the left side of the genome encodes mainly components of the virion, including head, tail, and tail fibers. Most of these genes are color-coded to match their products, shown in **Figure 11.4**. The right side of the genome includes enzymes for key functions such as host integration, DNA replication and lysis, and the famous lysis/lysogeny switch (discussed shortly).

Within a phage head, the packaged genome is linear. The left and right ends each possess *cos*, a sequence of 200 bp that was cleaved by a terminase enzyme when the phage DNA was packaged. The cleavage generated a staggered cut or "sticky ends" (**Fig. 11.6**). As the phage DNA now enters the new host cell, the *cos* sticky ends anneal together, and their backbones are sealed by DNA ligase (a replication enzyme discussed in Chapter 7). The genome is now circular, as the sealed ends form a long operon transcribed "rightward" from P_R, all the way through the virion structural components (**Fig. 11.5**). Another operon is transcribed "leftward," from P_L. Note that both the P_R and P_L operons use all three reading frames, and that some coding genes overlap. Long mRNA transcripts with gene overlap in three reading frames are a common feature of viral genomes.

Besides the P_R and P_L promoters, which function during the lytic cycle, other promoters are used for key events of the lysis/lysogeny switch.

During the lytic cycle, the circularized DNA molecule undergoes several rounds of bidirectional replication via host DNA polymerase (**Fig. 11.6**). The DNA is sometimes called a "theta form" because of the shape of the circle while replication is in progress. The completed circles are then nicked (cleaved on one strand), generating a 3′ OH end. The 3′ OH end of DNA serves as a primer for **rolling-circle replication**. Rolling-circle replication is commonly used by plasmids (see Chapter 7), as well as by circularized viral genomes such as that of herpes viruses (see Section 11.5). In the rolling-circle process, one strand of DNA extends its 3′ end continually around the circular template, while the 5′ end "rolls away." The 5′ extension

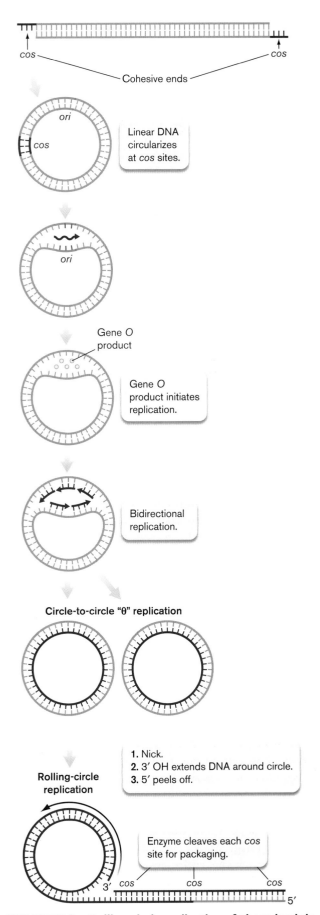

FIGURE 11.6 ■ Rolling-circle replication of phage lambda. Initially, the linear genome circularizes and replicates bidirectionally. Then it replicates multiple genomes end to tail, forming a concatemer.

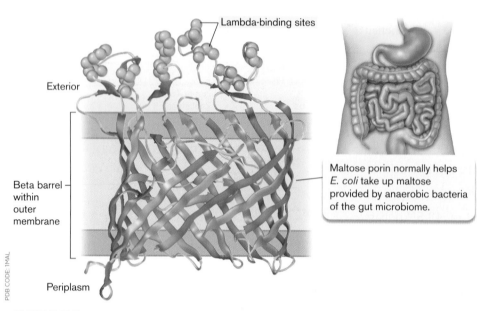

Lambda-binding sites

Exterior

Beta barrel within outer membrane

Periplasm

PDB CODE: 1MAL

Maltose porin normally helps *E. coli* take up maltose provided by anaerobic bacteria of the gut microbiome.

FIGURE 11.7 ▪ **Host receptor: maltose porin.** The maltose porin (PDB code: 1MAL) is embedded in the outer membrane of *E. coli*.

generates a long line of tandemly repeated genomes called a **concatemer**. The complementary strand of DNA fills in later. Once the long double-stranded molecule is complete, the terminase enzyme cleaves the concatemer into pieces that fit into phage head coats. For phage lambda, the genomes are cleaved at the *cos* site, which acts as a signal to package the progeny genome into the protein head coat. Cleavage restores the two *cos* ends to form linear genomes coiled into the phage head.

Thought Question

11.2 What advantages does rolling-circle replication offer a phage?

Phage Attachment and Infection

To infect a host cell, a virus needs to attach to the host surface and insert its genome into the cytoplasm (discussed in Chapter 6). Most types of bacteriophages extrude their genome from the head or capsid and thread it across the bacterial envelope while leaving the capsid outside. An exception is phage M13 (**eTopic 11.2**), whose filamentous capsid penetrates the entire envelope to replicate slowly within the cytoplasm. Phage lambda, however, uses the more common mechanism of binding to a specific cell-surface receptor. The phage adsorbs (attaches) to the receptor by contact with its tail fibers.

The receptor for phage lambda is the *E. coli* uptake complex for maltose (glucose dimer) and other short-chain sugars. In the colon, *E. coli* obtain these short glucose chains from anaerobes such as *Bacteroides* species that break down large complex polysaccharides from plant material (discussed in Chapters 13 and 21). Sugar chain uptake is crucial for *E. coli*, so the bacteria are unlikely to lose the complex by evolution, even though phage lambda takes advantage of it for infection.

The phage binds specifically to maltose porin, the outer membrane pore that transports maltose into the cell (**Fig. 11.7**). Porins are a large family of proteins that share a distinctive "beta barrel" structure (presented in Chapter 3). The beta barrel of maltose porin (blue in **Fig. 11.7**) is buried in the outer membrane. The phage-binding sites (green) were identified by amino acid substitution mutations that prevent phage binding and confer host resistance to lambda. The maltose porin was first called "lambda receptor," LamB, because it was discovered by its function as the lambda receptor. Of course, the protein actually evolved in the host as a way to obtain nutrients.

Attaching to the cell surface is just the first challenge the phage faces to establish infection (**Fig. 11.8**, step 1). How will the phage get its DNA all the way across the periplasmic space and the inner membrane? Unlike phage T4 (whose tail contracts to expel DNA under pressure), phage lambda uses an extrusion mechanism that is not fully understood. Somehow the DNA threads across the periplasm and through the maltose transport complex of the inner membrane, to reach the cytoplasm (step 2). Within the cytoplasm, the staggered ends of the *cos* sites anneal and are ligated, circularizing the genome (step 3). Now, the host RNA polymerase can begin to transcribe the phage operons.

At first, though, the newly introduced phage genome supports transcription of just a few key proteins, most notably the "control proteins" Cro and CII (**Fig. 11.8**, step 4):

- **Cro protein leads to lysis.** Small amounts of Cro activate the lytic cycle.

- **CII protein leads to lysogeny.** Small amounts of CII block expression of Cro and lytic proteins, and CII induces expression of CI (known as "lambda repressor").

So, which protein wins—Cro or CII? The answer depends on numerous factors in the gut environment,

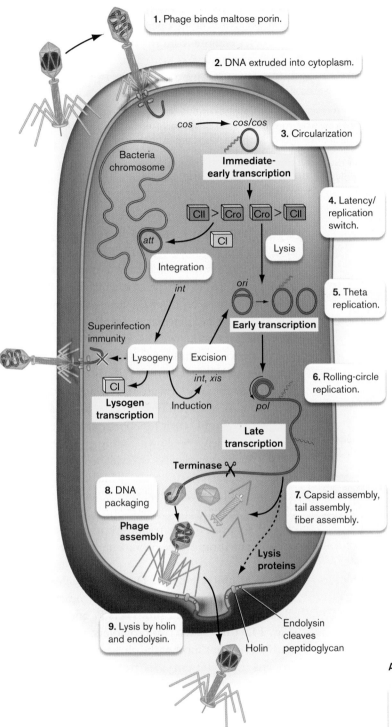

FIGURE 11.8 ■ Phage lambda infection of *E. coli* K-12.

suppose multiple phage particles coinfect the cell at the same time. This event implies that the host population is outnumbered by phages, and thus progeny phages will have poor opportunities to find a host. For this reason the phage has evolved a mechanism to avoid lytic reproduction: The multiple phages express a higher level of CII, some of which now evades the host protease and blocks Cro. When CII blocks Cro, the fortunate host cell becomes a lysogen instead of lysing. A phage genome integrates in the host genome as a prophage, at the *att* site. The phage DNA integrates with *att* by site-specific recombination—that is, recombination between the backbones of two DNAs that share a short sequence in common (discussed in Chapter 9).

The integrated prophage now expresses only a few proteins, including CI (lambda repressor). CI repressor prevents the lytic cycle by blocking transcription of lytic promoters. The CI repressor also prevents superinfection by other lambda phages. This is what happens to most of your intestinal bacteria between meals: the bacteria stay in stationary phase with their lysogenic phages repressed. The phage lambda CI repressor is famous, because its binding to DNA was the first protein-DNA binding event to be described (**Fig. 11.9**). Mark Ptashne at Memorial Sloan Kettering Cancer Center (**Fig. 11.9B**) showed how CI protein forms a dimer that binds a specific DNA sequence. Each CI subunit binds DNA through an interaction between an alpha helix of the protein and a major-groove sequence of DNA (**Fig. 11.9A**). Similar alpha helix binding to the major groove mediates the function of many genetic regulators of animals and plants, as well as bacteria.

When phage lambda infects *E. coli*, what if the Cro control wins, instead of CI? Cro protein represses CI expression, thus promoting the lytic cycle. The lytic cycle leads to **lysis** (host cell destruction, and release of progeny phages). The circularized phage DNA replicates, first bidirectionally (**Fig. 11.8**, step 5, "theta replication"), and then by

whose signals combine to lead to one decision or the other. One such signal is a surge in nutrients, such as when you eat a meal, sending rich organic substrates down to your intestines. High nutrient concentrations cause the *E. coli* protease HflB to cleave the phage protein CII, leaving Cro to induce the lytic cycle. Low nutrient concentrations inhibit HflB, leaving CII available to block Cro. Alternatively,

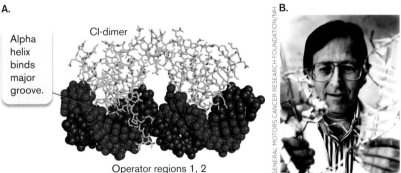

FIGURE 11.9 ■ CI repressor binds DNA. **A.** The CI repressor dimer binds the operator sequence by fitting the DNA sequence at the major groove. (PDB: 1LMB) **B.** Mark Ptashne worked out the binding of CI repressor to DNA, and the mechanism of the lysis/lysogeny switch.

rolling-circle replication (step 6). Structural proteins are synthesized and assembled to form the empty capsid (or head coat), tail tube, and tail fibers (step 7). The *cos* end of the DNA concatemer gets stuffed into a progeny head coat (step 8). The stuffed portion of the concatemer is then cleaved at the next *cos* site by a phage enzyme called terminase. Terminase also helps attach the filled head to the tail tube. When most of the progeny phages have assembled, a phage-encoded protein called holin punctures the inner cell membrane, providing a channel through which

endolysin reaches the cytoplasm. Endolysin cleaves peptidoglycan (step 9, lysis). Now holes open across the envelope, and phage particles emerge from the destroyed cell.

Thought Questions

11.3 A researcher adds phage lambda to an *E. coli* population whose cells fail to express maltose porin. After several days, the *E. coli* are now lysed by phage. What could be the explanation?

11.4 Suppose a mutant lambda phage lacks holin and endolysin. What will happen when this mutant infects *E. coli*?

A. Lysis pathway

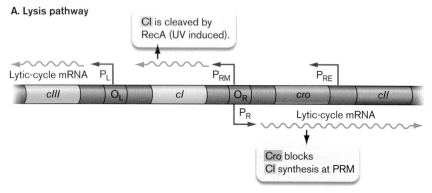

B. Lysogeny pathway: Make CII and CI

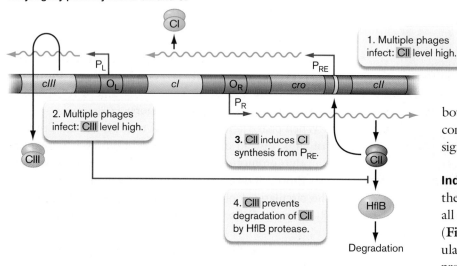

C. Lysogeny pathway: Block lytic cycle

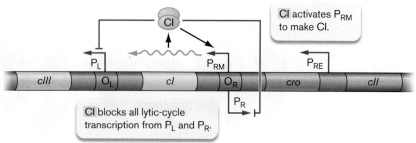

FIGURE 11.10 ■ **Lysogeny: to lyse or not? A.** The CI repressor maintaining lysogeny is cleaved by RecA (induced by UV exposure). Cro blocks CI synthesis, allowing expression of the lytic promoters P_L and P_R. **B.** If multiple phages infect simultaneously, the genome expresses enough CII protein to bind P_{RE}, activating expression of CI. **C.** The CI protein blocks promoters P_L and P_R (lytic cycle) and activates P_{RM} (to make more CI).

Lysogeny: To Lyse or Not?

As a lysogen, the host *E. coli* replicates just as it did without any prophage integrated. The prophage expresses CI repressor, which blocks nearly all expression of phage genes dangerous to the cell and prevents superinfection by other phage lambda particles. Yet at any time, a spontaneous event may override the CI repressor and trigger a lytic cycle. The lysis/lysogeny decision centers around CI and Cro proteins, with multiple other regulators that modify their function, only some of which are shown in **Fig. 11.10**. Such complexity is evidence of a lengthy evolution, in which multiple "adjustments" had time to occur in both phage and host genomes. The result is a control network highly responsive to diverse signal inputs from the environment.

Induction of lytic cycle. During lysogeny, the CI repressor blocks expression of nearly all other genes from the promoters P_L and P_R (**Fig. 11.10B**). These two promoters, particularly P_R, express most of the phage structural proteins and lytic enzymes from the circularized lambda genome (see **Fig. 11.5**). But the number of repressor molecules present in a cell is small enough for fluctuation to lead to rare events. Occasionally, perhaps one in 10^7 cells, the CI concentration is lowered to a level at which P_R becomes exposed, allowing expression of Cro. CI is further decreased by host RecA, which cleaves this protein along with many host repressors (**Fig. 11.10A**). RecA is activated by DNA damage during UV light exposure, so UV light can induce lysis in 100% of a lysogenic population.

When Cro is expressed, this protein binds to both operators (O_L and O_R) for the

P_L and P_R operons, respectively. Cro regulates these operons during the lytic cycle and, at the same time, blocks expression of CI from promoter P_{RM}. Thus, no further repressor can be made to block phage production, and the cell is committed to lysis.

Maintaining lysogeny. To avoid lysis requires continual expression of CI repressor (**Fig. 11.10B**). Early in lysogeny, CI is expressed from promoter P_{RE}. CI expression requires P_{RE} to bind CII protein. But CII protein is vulnerable to cleavage by host HflB (in a high-nutrient medium). HflB cleavage can be prevented by CIII, a protein expressed early from promoter P_L.

Once CI protein attains sufficient concentration, it activates its own expression from a different promoter, P_{RM}. P_{RM} activation requires CI dimers binding to O_R. Besides activating P_{RM}, the CI binding also blocks expression of P_R, the major lytic operon, as well as the leftward-transcribing operon from P_L. From then on, CI maintains its own expression while preventing induction of lysis. Lysis occurs only when stress such as DNA damage activates RecA to cleave CI.

The full repression of lysis requires eight molecules of CI in all: a pair of dimers at O_R, and another pair of dimers at O_L (**Fig. 11.10C**). The two pairs of dimers actually bind each other as an octamer, with DNA looped between them (not shown). The multiple molecules working together show "cooperativity," a property whereby the binding of two regulators is stronger than the sum of the individual regulators binding DNA. Cooperativity increases the on/off character of a molecular switch, lessening the occurrence of partial states in between. Other examples of cooperative molecular regulators are presented in Chapter 10.

Synthetic Biology

Within natural microbiomes such as that of our intestine, lysogenic phages carry genes that offer useful functions to their host bacterium. An example is the *bor* gene (found at the right-hand end of the lambda genome; see **Fig. 11.5**). The *bor* gene is not needed for phage function; it encodes Bor lipoprotein, which resides in the outer membrane of the host bacterium. The presence of Bor protects a lysogen from destruction by the serum complement cascade (described in Chapter 23). The location of the *bor* gene near the end of the phage genome suggests that it could have been picked up from an ancestral host, by a process of specialized transduction (see Chapter 9). In specialized transduction, a prophage initiates lysis and picks up a small adjacent piece of host DNA while copying its genome out of the integration site. The host DNA is then copied into progeny phage and is transferred to the next host infected.

Today, we use genetic elements of the lambda switch for synthetic biology—in effect, imitating natural gene transfer

mechanisms to construct bacteria with functions useful to us. For example, Pamela Silver's students used a CI/Cro switch to build a bacterial recorder that detects an environmental signal within the intestine. This kind of bacterial device offers promise for future development of diagnostic and therapeutic tools.

Silver's bacterial recorder possesses a stripped-down version of the CI/Cro switch (**Fig. 11.11A**). The *cro* gene is fused to *lacZ*, whose product generates blue colonies when expressed on indicator plates. (Gene fusion techniques are presented in Chapter 12.) But *cro* expression from P_R is blocked by CI repressor, which in turn is expressed from P_{RM}. Cro would block P_{RM}—if it were expressed. The net result is colonies that are white.

The same bacteria contain another fragment of the lambda switch, the "trigger element," in which Cro is expressed from a gene under control of the antibiotic-inducible promoter *tetP*. Now, suppose we add the inducing antibiotic (ATC) to the bacteria. The *tetP-cro* trigger element now expresses Cro, which represses expression of CI by binding P_{RM}. With CI expression turned off, the *cro-lacZ* fusion indefinitely expresses its protein, causing blue colonies (**Fig. 11.11B**). The blue phenotype recurs for many generations.

What happens when we put this bacterial recorder into the microbiome of a mouse? First, to colonize the mouse intestine, the bacterium needs to include a selective gene for competitive advantage within the microbial community—a gene conferring resistance to streptomycin. The mice are treated with streptomycin (**Fig. 11.11C**) until the bacterial recorders are established. Then the researcher adds the test antibiotic ATC. After ATC is removed, bacteria are sampled from mice fecal pellets. For up to 7 days following ATC removal, blue colonies appear, showing that the bacteria record ATC exposure and report it long after the signal has gone.

Other applications of phage lambda sound more like science fiction. In 2015, Ido Yosef and colleagues at Tel Aviv University constructed a lambda phage that specifically kills antibiotic-resistant bacteria, while sparing bacteria that lack antibiotic resistance. The phage actually delivers a CRISPR-Cas defense to its host, protecting the host from phage lysis. In a different kind of experiment, physicists at the Weizmann Institute of Science in Rehovot, Israel, used a Cro switch to build an "artificial cell" on a DNA chip. This partly biological device is used to study biochemical networks relevant to medicine. There is no end in sight to future applications of the familiar phage lambda.

To Summarize

■ **Bacteriophage lambda was discovered in an *E. coli* lysogen from a human colon.** An *E. coli* lysogen released phage that infected a sensitive strain. The human gut microbiome is full of phage lysogens.

A. Memory element with reporter

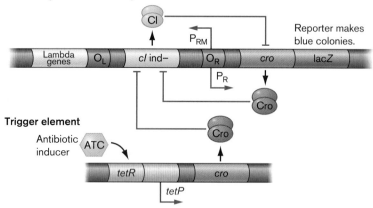

B. Antibiotic triggers Cro; makes colonies blue

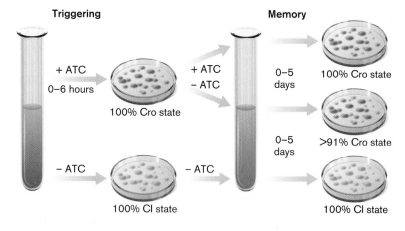

C. Within a mouse, bacteria record antibiotic exposure

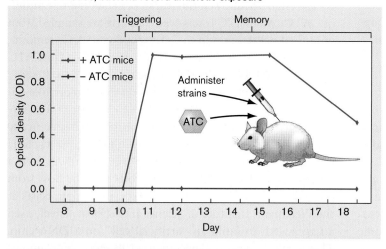

FIGURE 11.11 ■ **Antibiotic reporter bacterium uses a CI/Cro switch.**
A. The CI/Cro switch was engineered to record exposure to an antibiotic (ATC). The Cro "memory" is triggered by an inducer that turns off CI expression over many cell generations. **B.** The ATC signal turns colonies blue. Blue colonies persist for 5 days. **C.** The bacterial detector works when administered to mice. Seven days after addition and removal of ATC, the engineered bacteria still make blue colonies (Cro state). Strep = streptomycin.
Source: Jonathan Kotula et al. 2014. *PNAS* **111**:4838.

■ **The phage lambda virion** consists of a head containing its DNA genome and accessory proteins, a tail composed of an internal tube, and tail fibers.

■ **Phage lambda binds to the host maltose porin.** Phage DNA is inserted into the cytoplasm, where early phage genes are expressed.

■ **Control proteins bind to DNA, leading to lysis or lysogeny.** Single-phage infection (Cro protein) leads to lysis, whereas multiple-phage infection (CII protein) more likely leads to lysogeny.

■ **In a lytic cycle, rolling-circle replication generates progeny genomes.** The progeny genomes are packaged into head coats and then cleaved from the concatemer; the filled heads are attached to tails. Late-expressed proteins lead to lysis.

■ **CII up-regulates expression of CI repressor, which maintains lysogeny.** CI then induces its own expression, while repressing expression of Cro and proteins of the lytic cycle. Human gut bacteria frequently switch between lysis and lysogeny.

■ **The CI/Cro switch is used for synthetic biology.** The molecular switch is built into biomedical devices.

11.2 Influenza Virus: (−) Strand RNA Virus

We now turn to viruses that infect humans, focusing on three whose biology is well studied: influenza virus, a negative-strand RNA virus; human immunodeficiency virus (HIV), the retrovirus that causes AIDS; and herpes simplex, a double-stranded DNA virus causing oral and genital herpes. Viruses of humans show mechanisms that resemble those of bacteriophages. For example, retroviruses integrate the DNA copy of their genomes into the host cell genome, analogous to lysogeny by phage lambda. Other viruses, such as influenza, have no known latent form. The influenza virus is highly virulent and lyses infected cells with little or no latent state.

Influenza virus is an important human pathogen, causing up to half a million deaths per year worldwide. Besides the seasonal strains, influenza shows a cyclic appearance of strains that cause pandemic mortality, such as the famous pandemic of 1918, which infected 20% of the world's population and killed more people than World War I did. The 1918 strain arose as a mutant form of an influenza strain infecting birds. In 2009, a highly transmissible strain related to swine

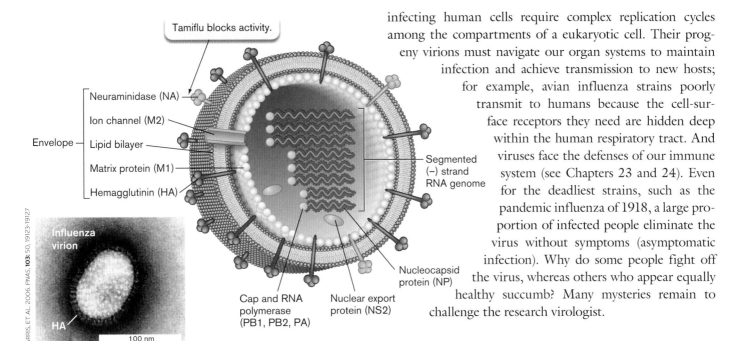

FIGURE 11.12 ■ Structure of influenza A. Diagram of virion structure, showing envelope (colored tan), envelope proteins, matrix protein (yellow), RNA segments (blue) with attached polymerase, and the nuclear packing protein NS2. **Inset:** Influenza A virion (TEM). The brush-like border coating the envelope consists of glycoproteins, hemagglutinin (HA), and neuraminidase (NA).

influenzas ("swine flu") spread rapidly around the world but caused relatively mild illness. A future strain might emerge combining the high transmission seen in swine flu with the high human mortality seen in the avian strain.

Research focuses on the mechanisms of influenza infection as targets for new antiviral agents. But human viruses are challenging to study—more so than bacteriophages. Viruses

infecting human cells require complex replication cycles among the compartments of a eukaryotic cell. Their progeny virions must navigate our organ systems to maintain infection and achieve transmission to new hosts; for example, avian influenza strains poorly transmit to humans because the cell-surface receptors they need are hidden deep within the human respiratory tract. And viruses face the defenses of our immune system (see Chapters 23 and 24). Even for the deadliest strains, such as the pandemic influenza of 1918, a large proportion of infected people eliminate the virus without symptoms (asymptomatic infection). Why do some people fight off the virus, whereas others who appear equally healthy succumb? Many mysteries remain to challenge the research virologist.

Virion Structure and Genome

The influenza virion has an asymmetrical structure (**Fig. 11.12**). Instead of an icosahedral capsid, the (−) strand RNA genome segments are individually coated by **nucleocapsid proteins** (**NPs**). The term "nucleocapsid" refers generally to proteins coating a viral genome and packaged within or as part of the virion. The NP-coated RNA segments are loosely contained by a shell of **matrix proteins** (M1). The matrix layer is further enclosed by the envelope. The envelope derives from the phospholipid membrane of the host cell, which incorporates viral proteins such as hemagglutinin (HA) and neuraminidase (NA). These viral envelope proteins join the membrane to the enclosed matrix, maintaining its structure. When a newly formed virion exits its host cell, neuraminidase acts as an enzyme to cleave a cell membrane glycoprotein, thus releasing the virion outside the cell. Neuraminidase can be blocked by the antiviral agent Tamiflu (oseltamivir), one of the main drugs available to treat influenza.

Negative-strand RNA. The replication cycle for a (−) strand RNA genome is complex because it needs to synthesize a complementary (+) strand RNA before it can express any genes in the host. Thus, within the virion, each NP-coated RNA segment possesses its own RNA-dependent RNA polymerase complex (proteins PB1, PB2, PA) (**Figs. 11.12** and **11.13A**). These proteins were

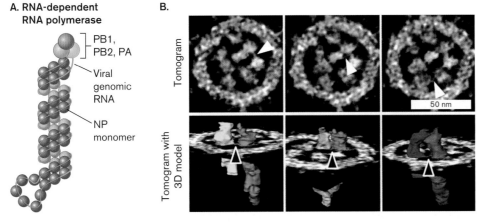

FIGURE 11.13 ■ The eight ribonucleoprotein (RNP) complexes are linked and packed in the virion. **A.** Structure of an RNP complex, including the RNA chromosome segment helically wrapped around NP monomers and complexed with the RNA-dependent RNA polymerase (PB1, PB2, PA). **B.** Influenza RNP complexes packed within a virion (colorized cryo-EM). Three tomography sections reveal links between specific RNPs (arrowheads). *Source:* Part A from Amie Eisfeld et al. 2015. *Nat. Rev. Microbiol.* **13**:28; part B from Takeshi Noda et al. 2012. *Nat. Commun.* **3**:639.

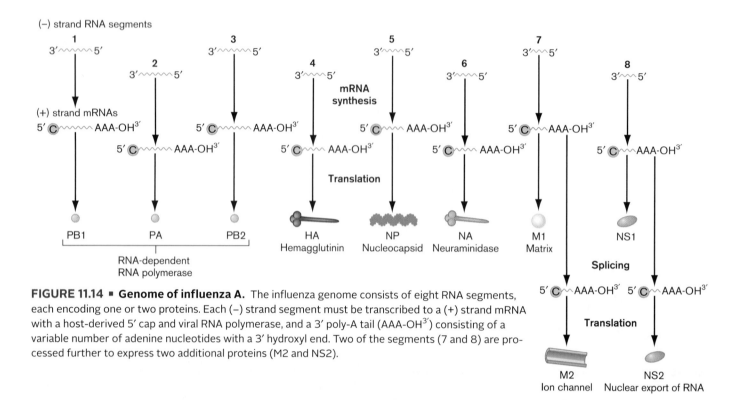

FIGURE 11.14 ■ **Genome of influenza A.** The influenza genome consists of eight RNA segments, each encoding one or two proteins. Each (–) strand segment must be transcribed to a (+) strand mRNA with a host-derived 5' cap and viral RNA polymerase, and a 3' poly-A tail (AAA-OH$^{3'}$) consisting of a variable number of adenine nucleotides with a 3' hydroxyl end. Two of the segments (7 and 8) are processed further to express two additional proteins (M2 and NS2).

expressed in the previous host, before the virions formed and the cell lysed. Within the virion, each RNA segment forms a loop, complexed with NP subunits that condense the RNA in a helix. (The helix does not involve base pairs, and differs from the standard helical forms of RNA.) The two ends of RNA are complexed with the RNA-dependent RNA polymerase, poised for RNA synthesis early in infection.

Segmented genome. Influenza virus has a **segmented genome** consisting of multiple separate nucleic acids, like the multiple chromosomes of a eukaryotic cell. The influenza genome includes eight segments, each a separate linear (–) strand of RNA encoding one or two products (**Figs. 11.12** and **11.14**). Within the infected cell, each (–) strand segment must be transcribed to a (+) strand mRNA with a host-derived 5' cap and viral RNA polymerase, and a 3' poly-A tail (AAA-OH$^{3'}$) typical of eukaryotic mRNA. If two different strains of influenza virus infect a host simultaneously, their segments can reassort to generate a novel hybrid strain. Because influenza genomes are capable of **reassortment**, they can rapidly generate a new strain that our immune system fails to recognize, such as the pandemic H1N1 strain of 2009 (discussed shortly). At the same time, like the other viruses, influenza viruses continually acquire small mutations that can lead to new phenotypes with respect to drug resistance and host range. The double threat of drastic change and subtle smaller changes in its genome explains why influenza presents a huge global challenge for public health.

Note: Distinguish between **reassortment** (two different viruses contribute separate genome segments to a reassortant genome) and **recombination** (two different viruses contribute genetic material to a recombinant molecule.)

Thought Question

11.5 Could the influenza genome change by recombination of segments, rather than reassortment? What about the lambda phage genome?

Given that viral infection requires all eight segments, how does the assembly mechanism package exactly eight segments, one of each? Until recently, EM imaging of influenza virions suggested that the RNA genome segments are packaged at random, avoiding the energetic expense of an accurate packaging mechanism. Random packaging would result in a vast majority of defective particles. But the relatively large eukaryotic host cell can produce 10,000 progeny virions, more than enough infective particles to propagate the virus.

More advanced imaging, however, shows that, in fact, the eight segments package precisely within the virion. The segments appear to link to each other in order as they arrange themselves. **Figure 11.13B** shows the evidence from cryo-electron tomography (a high-resolution EM technique described in Chapter 2). Cryo-EM combined images from frozen unstained virus particles, whose structure appeared highly consistent (unlike those imaged by earlier kinds of EM). The computed images show sections across the influenza virion, in which all eight RNA segments stand side by side, like a bundle of sticks. Sections taken from different depths through the particle reveal tiny molecular connections between adjacent segments. Further experiments with genetic constructs and fluorescence microscopy confirm that all eight unique segments link together in a defined, reproducible pattern.

Reassortment of virulent strains. The key advantage of a segmented genome is that it enables reassortment between two strains coinfecting the same cell. Reassortment leads to "antigenic shift," in which a very different strain evades the host immune system. Even strains that infect animals such as ducks or swine may reassort their segments with those of a coinfecting human virus. Influx of genes from a distantly related strain can sharply increase virulence and mortality. For example, the 1968 Hong Kong flu strain, which killed over 33,000 people in the United States, derived three segments from avian strains. Major epidemics of exceptionally virulent influenza arise as a result of reassortment with genome segments from strains that evolved within ducks or swine, agricultural animals that live in close proximity to humans. In each genome, the "H" and "N" numbers designate alleles of the genes encoding envelope proteins hemagglutinin and neuraminidase, respectively. For example, the Hong Kong flu strain had alleles H3 and N2 (which have since become prevalent in "seasonal" human flu strains).

A.

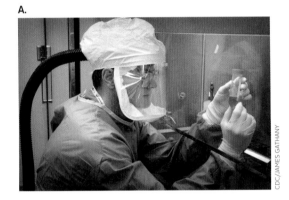

CDC/JAMES GATHANY

FIGURE 11.15 ■ Reassortment between human, avian, and swine strains generates exceptionally virulent strains of influenza A. A. The reconstructed strain of the 1918 pandemic influenza virus is studied by Terrence Tumpey, a microbiologist at the Centers for Disease Control and Prevention, Atlanta. **B.** The 2009 strain of H1N1 influenza A arose from a series of reassortments of the eight RNA segments from avian, swine, and human influenza strains. The numbers following "H" and "N" refer to different alleles of the genes encoding hemagglutinin and neuraminidase, respectively. Source: Part B modified from Gavin J. D. Smith et al. 2009. *Nature* **459**:1122.

B.

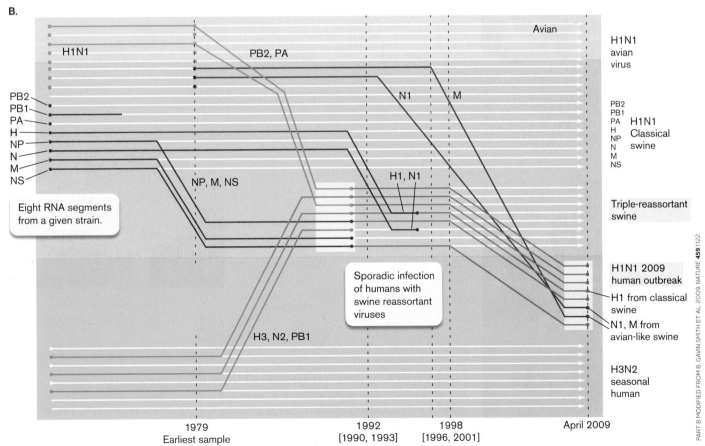

PART B MODIFIED FROM B. GAVIN SMITH ET AL. 2009. *NATURE* **459**:1122.

In 1979, in Europe, an avian flu strain was found to have "jumped" into swine (the "avian-like" swine strain). In 1992, a triple-reassortant strain was identified that included segments PB2 and PA (encoding RNA-dependent RNA polymerase) from an avian virus; PB1 (polymerase subunit), NP (nucleocapsid), and M (matrix protein) from a swine virus; and PB, H, and N from a human seasonal strain of influenza A. Since the H and N envelope proteins came from a human strain, the triple reassortant could be transmitted readily between humans. Today, molecular surveillance reveals many emerging strains of human influenza A that combine avian and swine alleles. The "swine" strain in 2009 had alleles H1 and N1, similar to the pandemic 1918 strain (**Fig. 11.15B**). In 2009, prompt public health measures such as quarantine helped keep the disease rate low. Yet another reassortant avian strain, H7N9, emerged in China in 2013. So far, these avian strains have been contained—but the next time we may be less fortunate.

Attachment and Host Cell Entry

What determines which animals can be infected by a given flu strain? One factor is the requirement for a host cellular protease to cleave the hemagglutinin protein on the virion envelope. Cleavage of hemagglutinin enables a small peptide called the **fusion peptide** to mediate viral entry into the host cell (discussed shortly; see **Fig. 11.17**). The presence of the protease is one **host factor** that determines which kind of host may be infected, and which tissues within the host support viral replication. Host factors of many kinds mediate viral infections.

Influenza receptor is a sialic acid glycoprotein. Another important host factor for influenza is cell-surface glycoproteins that contain a terminal sialic acid (**Fig. 11.16**). The sialic acid polysaccharide of the glycoprotein binds hemagglutinin, attaching the virion and enabling endocytosis. The precise structure of the sialic acid host receptor may determine whether a strain such as avian influenza H5N1 will spread directly between humans. For example, the sialic acid connection in the receptor polysaccharide can involve different OH groups of the sugar galactose: a linkage to the OH-3 (alpha-2,3) or to the OH-6 (alpha-2,6) bond (**Fig. 11.16**). The influenza strain H5N1 recognizes mainly the alpha-2,3-linked protein, found in birds. In humans, the upper respiratory tract contains mainly alpha-2,6-linked receptors; alpha-2,3-linked receptors are found only deeper within the lungs. But swine carry receptors

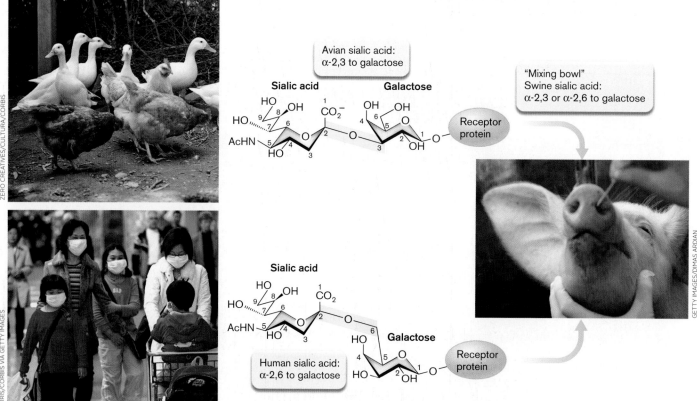

FIGURE 11.16 ■ Influenza receptors in different hosts. The avian host receptor polysaccharide contains sialic acid with an alpha-2,3 bond to galactose, whereas the human receptor in the upper respiratory tract has an alpha-2,6 bond. Swine receptors include both forms of sialic acid; thus, swine can be infected by both avian and human strains and may act as a "mixing bowl" for reassortment.

of both types. For this reason, swine are believed to act as a "mixing bowl" for strains from birds and humans, as well as swine. Thus, swine incubated the avian strain in 1979, and then allowed later reassortment with human-flu genome segments, leading eventually to the 2009 strain—which sickened both pigs and humans. Today, large swine facilities are monitored for appearance of novel reassortant strains.

The avian influenza strain H5N1 causes exceptionally high mortality in humans, but it is rarely transmitted from one person to another. More rapid transmission could arise from antigenic drift, a small mutation in the gene encoding avian hemagglutinin. Rapid person-to-person transmission of H5N1 might cause an influenza pandemic with high mortality. That is why public health organizations were so concerned when, in 2011, researchers announced that they had identified mutations conferring H5N1 transmission in the ferret model system, which closely resembles the human system.

Endocytosis and membrane fusion. Endocytosis of the influenza virion involves a key step of acid-mediated membrane fusion, which offers a target for antiviral agents (**Fig. 11.17** ▶). As the influenza virion binds its sialic acid receptor (**Fig. 11.17**, step 1), a host protease cleaves each HA, forming a fusion peptide (step 2). The hemagglutinin trimer now contains three N-terminal fusion peptides. A fusion peptide is a portion of an envelope protein (cleaved from hemagglutinin in the case of influenza virus) that changes conformation so as to facilitate envelope fusion with the host cell membrane.

When the virion is taken up by endocytosis, the endocytic vesicle fuses with a lysosome and its interior acidifies (**Fig. 11.17**, step 3). The lowered pH (increased H⁺ concentration) drives H⁺ ions into the virion through the M2 ion channel in the matrix layer (see **Fig. 11.12** for virion structure). The influx of acid causes the matrix proteins to dissociate from the NP-coated RNA. M2 ion channel is the target of amantadine, one of the first anti-influenza drugs; unfortunately, most strains today have evolved resistance to the drug. Low pH also induces a conformational change shifting the C-terminal ends back and the N-terminal fusion peptides outward to face the vesicle membrane.

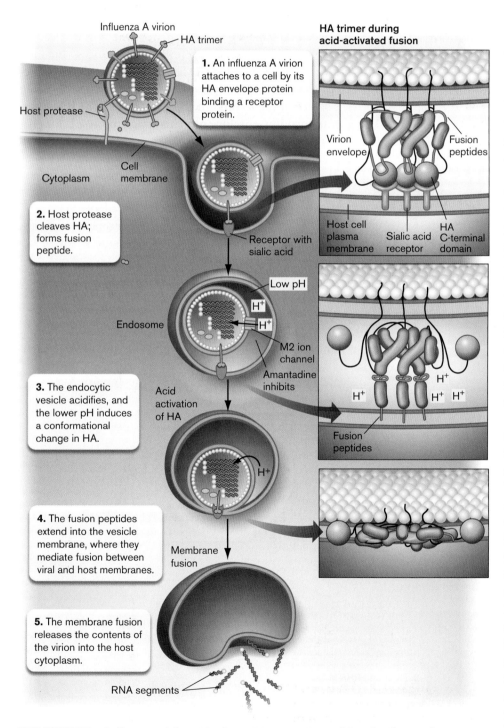

FIGURE 11.17 ■ **Influenza virion attachment to receptor, acid activation, and release of genome in the cytoplasm.** ▶

The peptides extend into the membrane (step 4), where they mediate fusion between viral and host membranes. The fusion process expels the contents of the virion into the host cytoplasm (step 5).

> **Thought Question**
>
> **11.6** How could a swine "mixing bowl" enable avian influenza strain H7N9 to become a deadly pandemic strain?

Replication Cycle of Influenza A

The replication of influenza virus requires that the viral components travel in and out of the nucleus (**Fig. 11.18 ▶**). Several viral enzymes and structural proteins travel along with its genetic material. As progeny envelope proteins are made, they require transport through the ER and Golgi to the cell membrane. The overall replication cycle is highly complex—even in the simplified diagram shown here.

Synthesis of (+) strand mRNA. After the influenza virion is endocytosed (**Fig. 11.18**, step 1), all the viral (−) RNA segments are released in the cytoplasm (step 2). Each RNA retains its coat of nucleocapsid proteins, as well as a prepackaged RNA-dependent RNA polymerase. The NP-coated RNA segments individually pass through a nuclear pore into the nucleus (step 3). Within the nucleus, each genomic (−) RNA segment with its prepackaged polymerase synthesizes (+) strand RNA for mRNA (step 4). Each mRNA synthesis initiates with a 7-methylguanine-"capped" RNA fragment (labeled "C" in the figure). The influenza polymerase obtains the cap fragments from the host by cleaving them from host nuclear pre-mRNA—a process quaintly known as "cap snatching." The (+) strand mRNA molecules return to the cytoplasm for translation (step 5), using the snatched cap to bind the host ribosome. The RNA segments encoding envelope proteins attach to the ER for protein synthesis and transport to the host cell membrane (step 6). The newly synthesized nucleocapsid proteins (NPs), as well as RNA-dependent RNA polymerase components,

FIGURE 11.18 ■ Replication of influenza virus. ▶

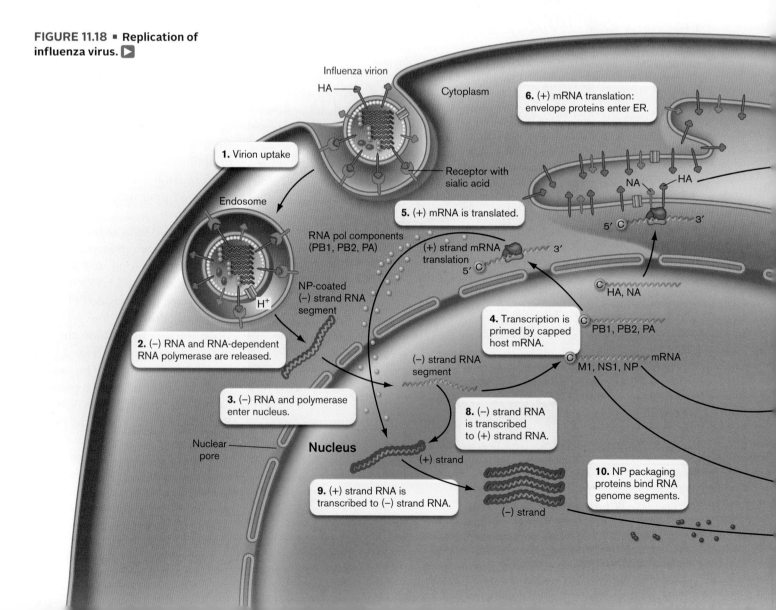

subsequently return to the nucleus (step 7). Other genome-packaging proteins (M1 and NS2) also return to the nucleus.

Synthesis of (+) strand and (−) strand genomic RNA. Back in the nucleus, the original (−) strand RNA segments also serve as templates for RNA synthesis, <u>without</u> cap snatching (**Fig. 11.18**, step 8). The uncapped (+) strand RNA then becomes coated with the newly made NP subunits imported from the cytoplasm. The NP-coated (+) strand serves as template to synthesize (−) strand RNA (step 9), which also becomes coated with NP (step 10).

The NP-coated (−) RNA associates with a newly made polymerase for a future cycle of viral replication. The RNA is then complexed with matrix protein (M1) and nuclear export protein (NS2)—proteins that were imported from the cytoplasm earlier. At last, the fully packaged (−) RNA segments exit the nucleus to the cytoplasm (**Fig. 11.18**, step 11), where they approach the cell membrane for packaging into progeny virions (step 12).

Envelope synthesis and assembly. The envelope proteins synthesized at the ER include hemagglutinin (HA) and neuraminidase (NA). Within the ER lumen, these proteins are glycosylated by host enzymes and then transferred to the Golgi (**Fig. 11.18**, step 13) for export to the cell membrane (step 14). Within the cell membrane, the envelope proteins assemble around a group of (−) RNA segments complexed with their matrix and packaging proteins, completing the virion particle (step 15).

To exit the cell, the virion buds out (**Fig. 11.18**, step 16). Viral exit requires a final step of host release: Neuraminidase (the envelope protein N) cuts the sialic acid link of the host glycoproteins (step 17), releasing the virion out into the bloodstream. This host release activity of neuraminidase is inhibited by oseltamivir (Tamiflu), the main antiviral agent currently useful against influenza. Tamiflu remains useful today, but resistant strains of the 2009 H1N1 virus have emerged, so we urgently need new antivirals ahead of the next influenza pandemic.

Experimental evidence. How do researchers figure out all the steps of replication shown in **Figure 11.18?** A key technique is fluorescence microscopy, as described in Chapters 2 and 3. Fluorophores such as GFP, conjugated antibodies, and labeled nucleic acid probes reveal the presence of viral components within a cell, even tracking their motion over

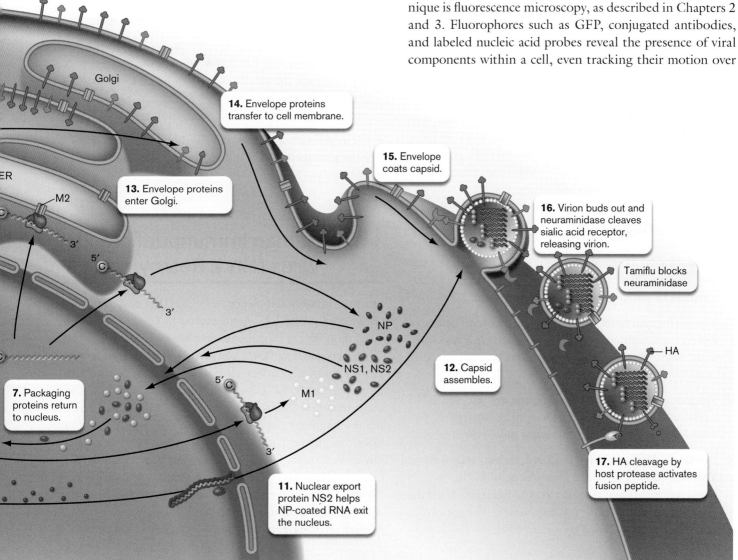

Golgi

14. Envelope proteins transfer to cell membrane.

15. Envelope coats capsid.

ER

M2

13. Envelope proteins enter Golgi.

3′

5′

3′

16. Virion buds out and neuraminidase cleaves sialic acid receptor, releasing virion.

Tamiflu blocks neuraminidase

NP

HA

NS1, NS2

12. Capsid assembles.

5′

M1

7. Packaging proteins return to nucleus.

3′

17. HA cleavage by host protease activates fusion peptide.

11. Nuclear export protein NS2 helps NP-coated RNA exit the nucleus.

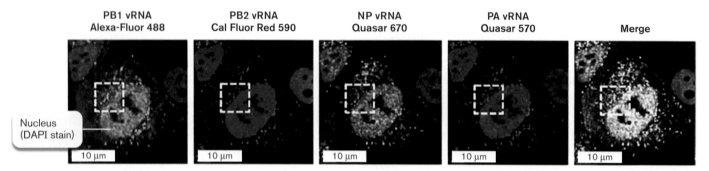

FIGURE 11.19 ■ **Localization of RNA segments returning to the cytoplasm.** Four different viral RNA segments (vRNAs) of the IAV genome were labeled by FISH (hybridization to fluorophore-labeled probes). The "merge" panel shows that most RNA segments enter the cytoplasm grouped with two or three other segments (and thus appear white). Arrowheads mark examples of grouped segments. *Source:* Seema Lakdawala et al. 2014. *PLoS Pathog.* **10**:e1003971.

time. An example of fluorescence data is shown in **Figure 11.19**, from Seema Lakdawala and colleagues at the National Institutes of Health. In this experiment, influenza A infection was observed using MDCK cells, a standard laboratory culture derived from canine kidney tissue. The infected cells were simultaneously probed with RNA molecules complementary to four different influenza segments (PB1, PB2, NP, PA). The complementary RNA probes were linked to four different fluorophores with different excitation-emission properties. Thus, the positions of progeny viral RNA segments (vRNAs) were detected and assigned four different colors. The "merge" panel shows how many of the different fluorophore positions overlap (white dots). The overlap of colors from different types of segments suggests that vRNAs are already linking together as they exit the nucleus (**Fig. 11.18**, step 11) and enter the cytoplasm for assembly into virions. Such ingenious experiments continually require us to rewrite the textbooks.

To Summarize

- **Influenza virus** causes periodic pandemics of respiratory disease. New virulent strains arise through reassortment of human, avian, and swine strains.

- **The influenza virus consists of segmented (–) strand RNA.** Each segment is packaged with nucleocapsid proteins. Segments from different strains reassort through coinfection.

- **Nucleocapsid and matrix proteins** enclose the RNA segments of the influenza virus. The matrix is surrounded by an envelope containing spike proteins.

- **Envelope HA proteins mediate virion attachment.** The HA protein includes a fusion peptide that undergoes conformational change to cause fusion between the viral envelope and host cell membrane. For influenza, the virion is internalized by endocytosis.

- **Lysosome fusion with endosomes** triggers viral envelope fusion with the endosome membrane. The viral genome and proteins are then released into the cytoplasm. Viral (–) strand RNA segments attached to RNA-dependent RNA polymerase enter the nucleus.

- **Influenza mRNA synthesis initiates with a capped RNA fragment** cleaved from host mRNA. The capped viral mRNAs return to the cytoplasm for translation.

- **Genomic RNA synthesis generates a (+) strand RNA as a template for (–) RNA segments.** Progeny RNA segments are then packaged in newly made nucleocapsid protein and exported to the cytoplasm, for coating with matrix, host cell membrane, and viral envelope proteins.

- **Neuraminidase cleaves the sialic acid connection to host glycoproteins.** This key step of virion release can be blocked by Tamiflu.

11.3 Human Immunodeficiency Virus (HIV): Retrovirus

Human immunodeficiency virus (HIV) is a **retrovirus** (Baltimore group VI). As discussed in Chapter 6, a retrovirus requires the enzyme **reverse transcriptase** to copy its RNA genome into DNA. Retroviruses are a large family of viruses known to infect all types of vertebrate and invertebrate animals; for examples, see **Table 11.1**. The "simple retroviruses" have genomes of just four genes, such as feline leukemia virus (FeLV), the number one killer of outdoor cats in the United States. Simple retroviruses generally cause cancer. The **lentiviruses**, or "slow viruses," cause diseases that progress slowly over many years. Lentiviruses possess additional regulator genes that modulate host interactions. In addition, most animal genomes show evidence of **endogenous retroviruses**, sequences from an ancient

TABLE 11.1	Retroviruses of Animals (Examples)		
Genus	**Virus**	**Disease(s)**	**Hosts**
Simple retroviruses			
Alpharetrovirus	Avian leukosis virus (ALV)	Leukemia	Birds
	Rous sarcoma virus (RSV)	Sarcoma (tumor)	Birds
Betaretrovirus	Mouse mammary tumor virus (MMTV)	Mammary tumor	Mice
Gammaretrovirus	Feline leukemia virus (FeLV)	Lymphoma, immunodeficiency	Cats
	Moloney murine leukemia virus (MMLV)	Leukemia	Mice
Deltaretrovirus	Bovine leukemia virus (BLV)	Leukemia	Cattle
	Primate T-lymphotrophic virus (PTLV-1) [formerly human T-cell leukemia virus (HTLV)]	Leukemia	Humans
Epsilonretrovirus	Walleye dermal sarcoma virus (WDSV)	Sarcoma	Fish
Lentiviruses	Human immunodeficiency virus (HIV-1, HIV-2)	AIDS	Humans
	Simian immunodeficiency virus (SIV)	Simian AIDS	Monkeys
	Equine infectious anemia virus (EIAV)	Anemia	Horses
	Maedi-Visna virus (MV)	Neurological disease	Sheep

retrovirus whose genome integrated and became "fixed" by mutation, such as HERV-K (discussed in the Current Research Highlight). Surprisingly, endogenous retroviruses evolve into essential parts of host genomes (see Section 11.4).

The most famous lentivirus is **human immunodeficiency virus** (**HIV**), the cause of **acquired immunodeficiency syndrome** (**AIDS**). The virus HIV and its causative role in AIDS were discovered by French virologist Luc Montagnier, building on Robert Gallo's studies of retroviruses. In 2013, according to the United Nations, 35 million people globally were living with HIV, and AIDS claimed 1.5 million lives. Yet three decades of research have yielded drugs that can prevent AIDS and eliminate HIV from populations. And the virus is engineered to make "lentivectors," our most successful agents of gene therapy (see Section 11.4).

History of HIV and AIDS

HIV is a lentivirus that evolved from viruses infecting African monkeys. Two major types are recognized: HIV-1, the cause of most infections at present; and HIV-2, another type that appears to have evolved independently from a different strain infecting monkeys. The virus is transmitted through blood and through genital or oral-genital contact. HIV can hide in the host cell for many years, with only gradual buildup of virus particles, most of which are eliminated by the host. Eventually, however, the virus destroys the body's T lymphocytes, leaving the host defenseless against many organisms that normally would be harmless.

The first U.S. cases of AIDS were reported in 1981. A historical view of AIDS in the United States emerges in the book *And the Band Played On* by Randy Shilts, adapted as an award-winning film in 1993. The book and film show how American society failed for many years to grasp the significance of AIDS because the syndrome first appeared in societal groups considered marginal (homosexual men and certain ethnic immigrants), although it spread to all social classes. In addition, the virus proved extremely difficult to detect and grow in culture. The discovery of HIV sparked controversy because Gallo failed to acknowledge his use of a virus-producing cell line from Montagnier, the first to isolate HIV-1. Since that time, the two scientists and many others have collaborated to develop a test for HIV-1 infection and to search for a vaccine (**Fig. 11.20A**). In 2008, the Nobel Prize in Physiology or Medicine was awarded to Montagnier and Françoise Barré-Sinoussi (**Fig. 11.20B**) for their discovery of HIV and its role in AIDS.

Thirty years later, HIV infects one in every 100 adults worldwide, equally among women and men. In developed countries, treatment effectively prolongs the life of people with AIDS. Treatment involves mixtures of antiretroviral drugs—all based on the molecular mechanisms of HIV infection, described in this chapter. HIV treatment with multiple drugs is called "highly active antiretroviral therapy," or HAART. A surprising benefit of HAART was discovered through humanitarian treatment programs in Africa, such as the President's Emergency Plan for AIDS Relief (PEPFAR), initiated in 2003 by President George W. Bush and continued under President Barack Obama. The PEPFAR program showed that, in communities that

A. Luc Montagnier and Robert Gallo

B. Françoise Barré-Sinoussi

C. HIV virions

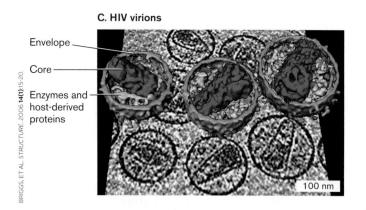

Envelope

Core

Enzymes and host-derived proteins

100 nm

BRIGGS, ET AL. *STRUCTURE.* 2006 **14(1)**:15-20.

D. HIV core

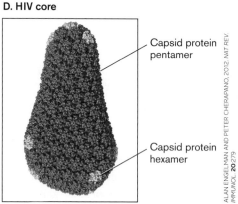

Capsid protein pentamer

Capsid protein hexamer

ALAN ENGELMAN AND PETER CHERAPANO, 2012. *NAT. REV. IMMUNOL.* **20**:279

FIGURE 11.20 ■ **HIV discovery. A.** Luc Montagnier (left) and Robert Gallo agree to collaborate on development of an AIDS vaccine, 2002. **B.** Françoise Barré-Sinoussi, at the Pasteur Institute, worked with Montagnier to discover the virus that causes AIDS. **C.** HIV virions (cryo-EM tomography). **D.** HIV core, model from cryo-EM and X-ray crystallography.

deliver HAART to all members, regardless of infection status, the virus production decreases so much that transmission declines—and HIV can be eliminated from the population. Thus, testing and treating an entire population could actually wipe out AIDS.

In the United States, since 2012 the CDC has recommended routine HIV testing for all people 13 years of age and older. Individuals at risk for HIV can take one daily pill to greatly decrease their chance of infection. But still, infected individuals must take expensive drugs with side effects, and the HIV eventually acquires resistance.

We need a vaccine, but decades of research have failed. Why? The answers to this question are complex.

■ **High mutation rate.** The mutation rate of HIV is among the highest known for any virus. Within one patient, the virus evolves into a quasispecies whose different strains attack different organs and predominate at different stages of the disease.

■ **Complex regulation of replication.** The lentiviruses express a greater number of regulator proteins than do the

"simple" retroviruses. These complex regulatory options of a lentivirus enable HIV to hide itself within host cells.

HIV Structure and Genome

The structure of HIV as visualized by TEM consists of an electron-dense **core particle** (or capsid) surrounded by a phospholipid envelope (**Fig. 11.20C**). The conical core is composed of capsid (CA) subunits whose arrangement is partly icosahedral (**Fig. 11.20D**). The membrane around the core contains **spike proteins**, which join the membrane to the matrix, as in influenza virus. The envelope forms around the core from host cell membrane, when a progeny virion is budding out. In HIV, budding out does not rapidly lyse the cell but has other devastating consequences for cell function (discussed shortly).

HIV core. The core contains two distinct single-stranded copies of the RNA genome (**Fig. 11.21A**). Unlike influenza segments, each of the two RNAs contains a complete "map" of HIV genes. However, the two RNAs can

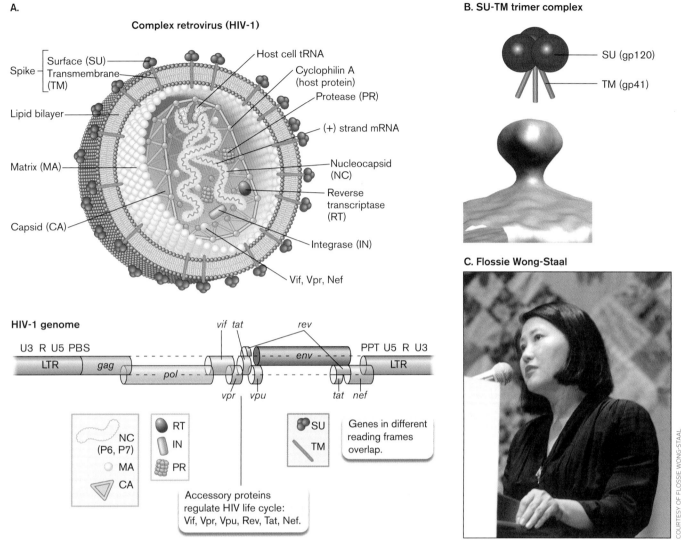

FIGURE 11.21 ■ **HIV-1 structure and genome.** **A.** Internal structure of the HIV-1 virion, color-coded to match the genome. In the genome sequence, the staggered levels indicate three different reading frames. **B.** Envelope spike complex (model based on cryo-EM tomography). **C.** Flossie Wong-Staal, pioneering AIDS researcher at UC San Diego, was the first to clone the HIV genome. Wong-Staal now pursues gene therapy approaches to AIDS prevention and develops lentiviral gene vectors. *Source:* Part A modified from Briggs et al. 2006. *Structure* **14**:15–20; part B modified from P. Zhu. 2008. *PLoS Pathog.* **11**:e1000203.

have slightly different alleles arising from distinct replication events. Thus, the HIV virion is genetically "diploid." A nonfunctional mutant gene on one genome may be complemented by a functional allele on the other.

Each RNA genome is coated with nucleocapsid (NC) proteins similar in function to the NP proteins of influenza virus. Unlike influenza virus, each RNA of HIV requires a primer for DNA synthesis: a tRNA derived from the previously infected host cell. The host-derived tRNA is packaged in place on the RNA template, ready to go. The primed and packaged RNA is contained within the core composed of CA subunits. The core also contains about 50 copies of reverse transcriptase (RT) and protease (PR), as well as a DNA integration factor (IN). Unique to type

HIV-1, subunits of a host chaperone named cyclophilin A are incorporated into the structure—about one for every ten core subunits. An HIV-1 mutant that fails to incorporate cyclophilin A can attach to a host cell and insert its capsid, but the core fails to come apart, and infection is halted.

The core is surrounded by a matrix (MA subunits), which reinforces the host-derived phospholipid membrane. The membrane is pegged to the matrix by spike proteins composed of the envelope subunits TM and SU (**Fig. 11.21B**). As in influenza virus, the spike proteins play crucial roles in host attachment and entry.

HIV genome. The genome of HIV (**Fig. 11.21A**) was first cloned for molecular study by Flossie Wong-Staal, now at

TABLE 11.2	Accessory Proteins of HIV-1	

Protein	Function	Effect of mutation
Vif	Virion component: • Protects reverse transcriptase from error-inducing host protein APOBEC3G.	Virions produced are noninfective
Vpr	Virion component: • Transcription factor; activates HIV transcription during G_2 phase of cell cycle; arrests T-cell growth. • Imports DNA across nuclear membrane; avoids need to infect rapidly dividing cells in which mitosis dissolves the nucleus.	Lower production of virions
Nef	Virion component: • Internalizes and degrades CD4 receptors, to avoid superinfection by more HIV virions, and to lessen immune response to the infected cell. • Decreases expression of major histocompatibility complex (MHC) proteins that stimulate cytotoxic T cells.	Slower progression to AIDS
Vpu	Membrane protein: • Degrades CD4, releasing bound spike proteins. • Promotes virion assembly and release from cell-surface tetherins.	Early death of host cell; lower production of virions
Rev	Nuclear phosphoprotein, combines with host cell proteins: • Stabilizes certain mRNAs in nucleus. • Exports mRNA out of nucleus into cytoplasm, inducing shift from latent phase to virion-producing phase.	Failure of infection
Tat	Transcription factor: • Binds TAR site on nascent RNA to activate transcription. • Associates with histone acetylases and kinases to activate transcription of integrated viral DNA.	Failure of chromosome replication

UC San Diego (**Fig. 11.21C**). Born in China in 1947, Wong-Staal immigrated to the United States and then worked with Gallo on the early discoveries of HIV. Wong-Staal now pursues gene therapy approaches to combat HIV infection and is developing retroviral vectors for human gene therapy (discussed in Section 11.4).

The HIV genome includes three main open reading frames that are found in all retroviruses: *gag*, *pol*, and *env*. The *gag* sequence encodes capsid, nucleocapsid, and matrix proteins; *pol* encodes reverse transcriptase (RT), integrase, and protease; and *env* encodes envelope proteins. The *gag* and *pol* sequences overlap, but because they are translated in different reading frames (different ways to align the triplet code), the ribosome expresses each independently of the other. During infection, each reading frame is transcribed and translated as a polyprotein; then, at subsequent stages, each polyprotein is cleaved by proteases to form the mature products.

In HIV-1, the *gag* and *pol* sequences overlap, and *env* overlaps with genes encoding **accessory proteins**, proteins that modify and regulate retroviral infection. Accessory proteins are unique to lentiviruses, mediating host response and maintaining the long-term slow progression of lentiviral disease. The accessory proteins are expressed within the infected host cell and regulate the replication cycle (**Table 11.2**). For example, Tat protein activates transcription of the viral genome. The HIV-1 genome encodes

at least six accessory proteins—a greater number than in any other retrovirus. They are major targets for research and drug discovery aimed at preventing HIV proliferation.

Origin and evolution of HIV. Where did HIV come from? The origin of HIV has been traced back to the early twentieth century, based on genome sequence comparison with related viruses infecting other primates, called simian immunodeficiency viruses (SIV) (**Fig. 11.22A**). Sequence comparison of different strains of HIV and SIV reveals that an immunodeficiency virus actually entered the human population more than once, from SIV strains derived from related primates in Africa. It is thought that human consumption of primates for meat may have introduced SIV strains that then adapted to human infection. Today, the vast majority of HIV-infected patients show the HIV-1 strain M; but some people have been infected by HIV-2, which derived independently from another SIV strain.

HIV is the most rapidly evolving pathogen known; its replication generates about one mutation per progeny virion. For this reason, physicians always prescribe a combination of antiretroviral drugs, with different molecular targets. The hope is that if any one mutation confers resistance to one drug, the mutant virus will still be blocked by another. This strategy of drug combination and continual testing for resistance enables many treated HIV carriers to

A. Origin of HIV

B. Evolution of drug-resistant HIV

FIGURE 11.22 ■ Origin and evolution of HIV. A. Strains of HIV and SIV (simian immunodeficiency virus) arose independently multiple times over several decades, from a common origin in monkeys. *P.t.s. = Pan troglodytes schweinfurthii*; *P.t.t. = Pan troglodytes troglodytes*. **B.** When HIV infects a patient, different drugs may select strains with different resistance mutations. The mutant strains may then recombine to generate a double-resistant strain.

remain free of AIDS for decades. But in some cases, recombination of different mutants can generate strains resistant to multiple antiviral agents (**Fig. 11.22B**).

Within a single infected patient, the high mutation rates generate multiple virus strains with differing properties of replication, tissue tropism, and resistance to antibiotics. This dynamic population of diverse mutant strains is called a **quasispecies** (**Fig. 11.23**). The quasispecies forms when an infective virion commences replication with rapid mutation and trait diversification. Many of the progeny virions have sequences and tropisms so different from their ancestor that, in isolation, they would be classified as different species. Different clonal variants of the virus colonize different tissues and organs. Virus types within a quasispecies may interact cooperatively on a functional level, by

serving complementary roles in the disease state, and thus collectively define the traits of the viral population. But what happens when an infected "donor" introduces HIV to a recipient? Within the recipient, only one HIV type proliferates—close to the original type that generated the quasispecies. After this acute infection, the HIV population again diversifies into the quasispecies. It's as if a relatively narrow range of genotype carries the HIV "germ line," whereas the mutant types sustain the infected state and maintain immunosuppression.

HIV Attachment and Host Cell Entry

Like other viruses, HIV needs to recognize specific receptor molecules on the surface of its target cells. The primary

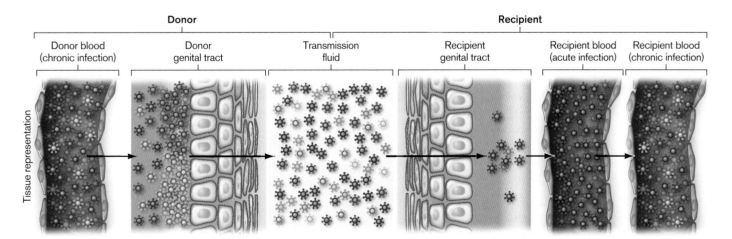

FIGURE 11.23 ■ Quasispecies development. HIV rapidly generates mutant progeny (different colors) in a chronically infected "donor." The variants may colonize different tissues and organs. Only one of these types (purple) is optimized to infect the next "recipient." Within the recipient, the new virions replicate and regenerate the quasispecies. *Source:* Modified from Sarah Joseph, 2015. *Nat. Rev. Microbiol.* **13**:414.

receptor for HIV is the CD4 surface protein on CD4 T lymphocytes (T cells). The normal function of CD4 surface proteins is to connect the T cell with an antigen-presenting cell, which activates the T cell to turn on B-cell production of antibodies (discussed in Section 24.2). Disruption of this antibody production is the main cause of the AIDS-related susceptibility to opportunistic infections. Note, however, that CD4 proteins appear on many other cell types, such as microglia (macrophage-like cells in the central nervous system) and Langerhans cells (immune cells of the epidermis). Their presence may make other cells susceptible to infection by HIV.

Spike proteins mediate membrane fusion. The binding of HIV to CD4 receptors involves the envelope spike protein SU (**Fig. 11.24**). Spike proteins are the main external proteins accessible to the host immune system.

HIV attachment to the cell membrane requires a fusion peptide rearrangement similar to that of influenza virus, except that it takes place at the cell surface (**Fig. 11.24**). When SU (gp120) binds to CD4, the spike transmembrane component TM (gp41) unfolds and extends its fusion peptide into the host cell membrane. In addition, SU binds to secondary receptors in the membrane called **chemokine receptors** (**CCRs**), such as the macrophage receptor CCR5. Chemokines are signaling molecules for the immune system, but their receptor proteins can bind viruses that evolve to take advantage of them. After the spike protein SU binds

receptors and the TM fusion peptide inserts, the HIV-1 envelope fuses with the plasma membrane. A CCR such as CCR5 is also called a **coreceptor**, a protein acting with CD4 to bind the HIV spike proteins.

The requirement for CCR attachment varies among different types of HIV. Some chemokine receptors are found on neurons, and their involvement in HIV infection may mediate the neurological disorders seen in AIDS. Furthermore, the predominance of different viral envelope types with different receptor preferences varies over the course of infection. As HIV evolves a quasispecies, the virions target the CCR5 receptor early in infection, whereas later-evolved virions target a different host surface protein, called CXCR4. These "X4" virions can infect early-stage T cells that do not yet carry CCR5; thus the AIDS disease accelerates. The X4 virions are less infective when transmitted; thus, most new infections start with a CCR5 strain.

An exciting discovery was that individuals who lack the CCR5 protein because of a genetic defect show a high degree of resistance to HIV infection. This finding prompted the Pfizer company to develop an antiviral blocker of CCR5, maraviroc (see **Fig. 11.24**), which was approved for therapy in 2007 (see **eTopic 11.5**).

After HIV binds to membrane receptors, how does its genome enter the cell? The HIV envelope fuses with the cell membrane, enabling the HIV core to enter the cytoplasm directly. This HIV entry mechanism differs from

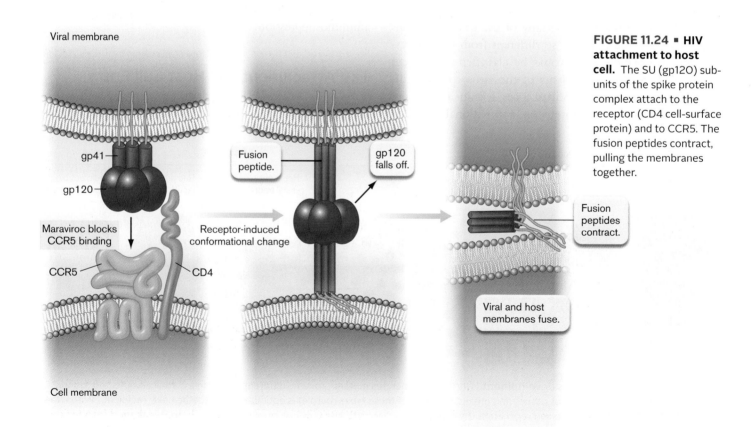

FIGURE 11.24 ▪ HIV attachment to host cell. The SU (gp120) subunits of the spike protein complex attach to the receptor (CD4 cell-surface protein) and to CCR5. The fusion peptides contract, pulling the membranes together.

that of influenza virus, in which endocytosis and lysosome fusion are required to open the capsid and release the genome into the cell. The HIV core (composed of CA and the host-derived cyclophilin A) dissolves, releasing the two RNA genomes, along with associated viral enzymes, into the cytoplasm.

> **Thought Question**
>
> **11.7** How do attachment and entry of HIV resemble attachment and entry of influenza virus? How do attachment and entry differ between these two viruses?

The two RNA genomes each possess a 5′ "cap" and a 3′ poly-A "tail" that enable them to mimic host nuclear mRNA. Each RNA is hybridized to a tRNA primer for DNA synthesis. The primer is a lysine-specific tRNA from the previous infected cell. The primer might be expected to hybridize at the 3′ end of the template, where its 3′ OH "points" toward the opposite end, positioned to synthesize all the way down. Surprisingly, however, the 3′ OH end of the tRNA actually binds near the 5′ end, where initially it can generate only a brief sequence. These early sequences bind key regulatory factors for transcription and for DNA insertion into the host genome (discussed next).

Reverse Transcriptase Copies RNA to DNA

A retrovirus, unlike other RNA viruses, must integrate its entire genome into the host genome in order to replicate viral genomes and produce new progeny virions. Thus, the RNA genome needs to serve as a template to synthesize a DNA complement; but then the original RNA template must be degraded and replaced by a DNA strand, for host integration. All these processes are accomplished by **reverse transcriptase (RT)**, the defining enzyme of a retrovirus. Reverse transcriptase is the source of the high error rate of retroviral replication—on average, one or two errors per copy of HIV. This high error rate generates the quasi-species of different strains within an HIV-infected person.

Reverse transcriptase is the target of the first clinically useful drug to treat HIV infection, the nucleotide analog **azidothymidine (AZT)**. AZT is incorporated into the growing DNA chain in place of a thymidine, but because its 3′ OH is replaced by an azido group ($-N_3$), no further nucleotides can be added.

The reverse transcriptase complex actually possesses three different activities:

■ **DNA synthesis from the RNA template.** Synthesis of DNA is first primed by the host tRNA, which was hybridized to the chromosome within the virion.

■ **RNA degradation.** After DNA synthesis, the template RNA is gradually removed through an RNase H activity of the RT complex. Removal of RNA enables replacement of the entire original RNA template by DNA.

■ **DNA-dependent DNA synthesis.** To make the DNA complementary strand replacing the RNA, the RT needs to use the newly made DNA as its template. Thus, RT has the rare ability to use either DNA or RNA as a template.

As shown in **Figure 11.25A**, the RNA template with its short RNA primer is threaded through the RT complex between the "thumb" and "fingers"—a configuration typical of other RNA polymerases (discussed in Chapter 8). The RT complex adds successive deoxynucleotides from deoxyribonucleoside triphosphates (dNTPs) starting at the 3′ OH end of the RNA primer. As DNA elongates, however, the RNA template is cleaved from behind by the RT complex. Thus, the new DNA actually replaces preexisting RNA sequence. This "destructive replication" is unique to retroviruses. The details are important, as they suggest possible targets for new antiviral drugs.

Reverse transcription. Reverse transcription of the HIV genome involves unexpected complications (**Fig. 11.25B**). First, the host-derived tRNA primer initiates synthesis (by reverse transcriptase, RT) of a DNA strand complementary to the RNA chromosome. DNA is elongated toward the 5′ end of the HIV chromosome, generating a short segment (**Fig. 11.25B**, step 1). The original RNA template for this short segment is then degraded by the RNase H activity of RT, leaving only the DNA extension of the tRNA primer (step 2).

The new DNA primes the second template. The original RNA template had repeated ends (labeled "r," lowercase, for RNA in **Fig. 11.25B**), and the exposed DNA copy of the 5′ end has a complementary sequence ("R," uppercase, for DNA). The "R" DNA from the second tRNA extension hybridizes to the 3′ end of the original RNA (**Fig. 11.25B**, step 3). The hybridized DNA elongates along the rest of the chromosome (step 4), up to the primer-binding site (pbs) for mRNA transcription. DNA completion is followed by degradation of the remaining RNA template, except for occasional short fragments to serve as primers, such as the polypurine tract (ppt). A complementary DNA strand is then synthesized through the PPT primer, leaving a nick at U3 (step 5).

Interestingly, human host cells have evolved a protein, APOBEC3G, that deaminates viral cytosines and thus increases the error rate of reverse transcriptase. APOBEC3G can be packaged into progeny virions, decreasing production of infective virions in the next host. However, HIV has evolved an accessory protein, Vif (see **Table 11.2**), that binds APOBEC3G to prevent it from being packaged into virions.

A. Reverse transcriptase

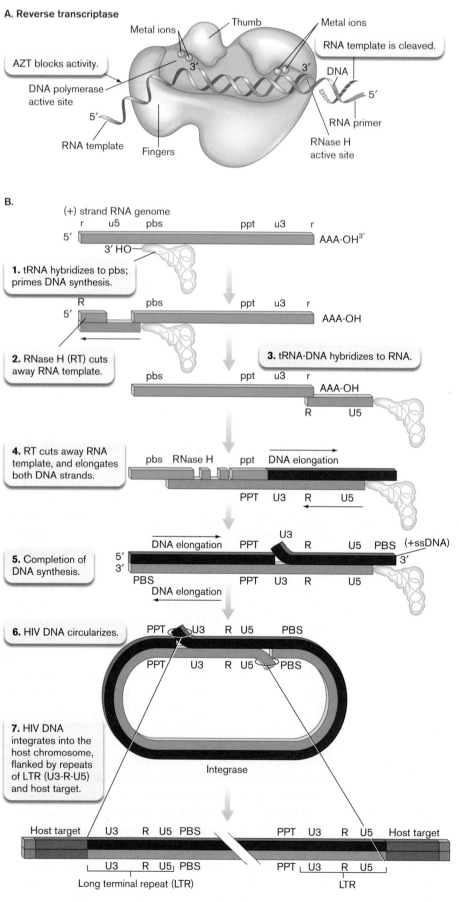

B.

FIGURE 11.25 ▪ **Reverse transcription of the HIV genome and integration into host DNA.** **A.** The RNA template with its short RNA primer is threaded through the RT complex between the "thumb" and "fingers"—a configuration typical of other RNA polymerases. The RT complex adds successive dNTPs, as in regular DNA synthesis. As DNA elongates, the RNA template is cleaved from behind by the RNase H site of reverse transcriptase (RT). **B.** The tRNA primer initiates a short sequence of DNA complementary to the 5′ end of the HIV chromosome. The corresponding template is then degraded, and the DNA-primer complex is transferred to the opposite end, where it can complete synthesis of the genome. Remaining RNA is cleaved by RT and replaced by DNA. The double-stranded DNA circularizes at the U3-R-U5 sequence and then integrates somewhere in the host genome.

Integration into the host genome. The final step of genome processing requires integration into host DNA. The double-stranded DNA copy of the HIV genome (plus a short repeat of the 5′ end) circularizes (**Fig. 11.25B**, step 6). The circular molecule then integrates into the host chromosome by site-specific recombination at a host target sequence (step 7), mediated by the HIV integrase protein (IN). This step forms an integrated viral genome, or **provirus**. Integration generates two copies of the provirus 5′ end (sequence U3-R-U5), which is called a **long terminal repeat** (**LTR**). The provirus and LTR ends are also flanked by two copies of the host target sequence. Proviral sequences can now be expressed, directing production of progeny virions.

An alternative to virion production is that the integrated HIV genome lies dormant, like the lambda prophage in *E. coli*. The integrated HIV genome is replicated passively within the genome of its host cell, hiding for many years with only infrequent production of virions. The few virions shed by the patient, however, can infect an unsuspecting individual who has sexual contact with or is exposed to the blood of the patient.

Replication Cycle of HIV

The steps of HIV replication are outlined in **Figure 11.26** ▶. The main points of viral entry and replication are typical of retroviruses. HIV, however, has an exceptionally large number of accessory proteins that govern the level of viral production and the duration of the quiescent phase, when the integrated chromosome replicates with the host cell.

Synthesis of HIV mRNA and progeny genomic RNA. After the HIV virion attaches to the host receptors, its envelope fuses with the host membrane (**Fig. 11.26**, step 1). Unlike influenza virus, the HIV core enters the cytoplasm directly, without endocytosis (step 2). The core partly uncoats, while the RNA chromosomes within are reverse-transcribed to make double-stranded DNA (step 3). The double-stranded DNA enters the nucleus through a nuclear pore (step 4)—a key step facilitated by Vpr accessory protein. Vpr enables infection of nondividing cells, which only lentiviruses can do; other retroviruses, such as those causing lymphoma, must infect dividing cells, in which the nuclear membrane dissolves during mitosis.

Upon entering the nucleus, the DNA copy of the HIV genome integrates its sequence as a provirus at a random position in a host chromosome (step 5). Integration is catalyzed by integrase (the IN protein; see **Fig. 11.21**). Integrase inhibitors such as raltegravir are an important class of anti-HIV drugs. Within the nucleus, full-length RNA transcripts are made by host RNA polymerase II, including a 5′ cap and a 3′ poly-A tail (step 6). Some of the RNAs exit the nucleus (step 7) to serve as mRNA for translation of polyproteins. Polyproteins are translated in alternative versions, such as Gag-Pol. Other full-length RNA transcripts exit the nucleus to form RNA dimers for progeny virions (step 8). Still other RNA transcripts within the nucleus are cut and spliced to complete the *env* gene sequence for translation of Env (envelope) proteins (step 9).

The Env proteins are made within the endoplasmic reticulum (ER) (**Fig. 11.26**, step 10). They pass through the Golgi for glycosylation and packaging (step 11) and are exported to the cell membrane (step 12). At the membrane, Env proteins plug into the core particle as it forms from the RNA dimers plus Gag-Pol peptides (step 13).

Virion assembly and exit. The core particles are packaged with envelope derived from host cell membrane containing Env spike proteins (**Fig. 11.26**, step 14). To escape the host cell, emerging virions require accessory protein Vpu to bind a "tetherin," a host adhesion protein induced by interferon to cause reuptake and digestion of virions. Vpu causes proteosomal degradation of the tetherin. Emerging virions, some still tethered to the cell, are seen in **Figure 11.27A**.

As the virion buds off, the protease (PR) cleaves the Gag-Pol peptide to complete maturation of the core structure containing Gag subunits as well as reverse transcriptase (RT) (**Fig. 11.26**, step 15). The Gag subunits now form the conical core structure. Proteases that cleave Gag-Pol offer important drug targets, which have led to the development of anti-HIV drugs known as **protease inhibitors**.

However, HIV has alternative means of cell-to-cell transmission that avoid exposing virions to the immune system. One alternative is cell fusion, mediated by binding of Env in the membrane to CD4 receptors on a neighboring cell (see **Fig. 11.26**). The two cells then fuse, and HIV core particles can enter the new cell through their fused cytoplasm. The fusion of many cells can form a giant multinucleate cell called a syncytium. Cell fusion with formation of syncytia enables HIV to infect neighboring cells without ever exiting a cell. Another means of cell-to-cell transmission is to travel through a "nanotube" connection between two T cells (**Fig. 11.27B**).

The intricate scheme in **Figure 11.26** actually omits many functions of HIV accessory proteins that enhance the virulence of HIV infection (see **Table 11.2**). Mutation of genes for accessory proteins often decreases virulence; thus, these proteins are potential targets for chemotherapy. Surprisingly, even a modest decrease of HIV infectivity can have major benefits for the patient, suggesting that HIV is so crippled by its high mutation rate that the slightest interference greatly decreases production of infective virions. Thus, numerous effective antiviral agents are now known—but all have side effects, and all select for resistant strains.

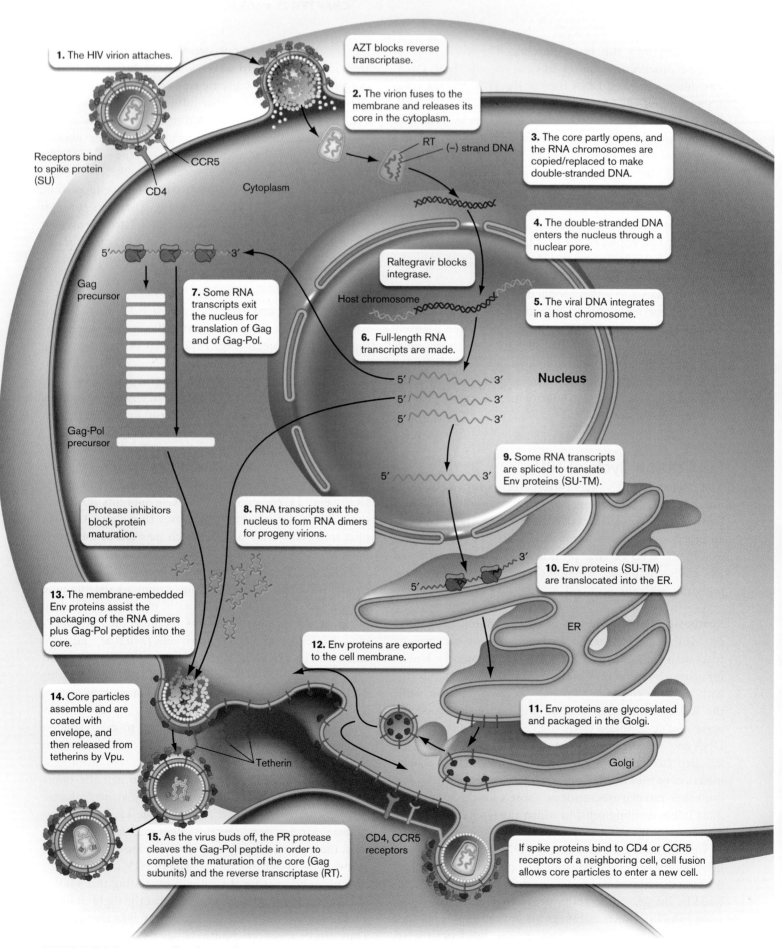

1. The HIV virion attaches.

AZT blocks reverse transcriptase.

2. The virion fuses to the membrane and releases its core in the cytoplasm.

Receptors bind to spike protein (SU)

CCR5

CD4

Cytoplasm

RT   (−) strand DNA

3. The core partly opens, and the RNA chromosomes are copied/replaced to make double-stranded DNA.

4. The double-stranded DNA enters the nucleus through a nuclear pore.

5′   3′

Gag precursor

7. Some RNA transcripts exit the nucleus for translation of Gag and of Gag-Pol.

Raltegravir blocks integrase.

Host chromosome

5. The viral DNA integrates in a host chromosome.

6. Full-length RNA transcripts are made.

5′ 3′
5′ 3′
5′ 3′

Nucleus

Gag-Pol precursor

9. Some RNA transcripts are spliced to translate Env proteins (SU-TM).

5′ 3′

Protease inhibitors block protein maturation.

8. RNA transcripts exit the nucleus to form RNA dimers for progeny virions.

10. Env proteins (SU-TM) are translocated into the ER.

5′   3′

ER

13. The membrane-embedded Env proteins assist the packaging of the RNA dimers plus Gag-Pol peptides into the core.

12. Env proteins are exported to the cell membrane.

11. Env proteins are glycosylated and packaged in the Golgi.

14. Core particles assemble and are coated with envelope, and then released from tetherins by Vpu.

Tetherin

Golgi

15. As the virus buds off, the PR protease cleaves the Gag-Pol peptide in order to complete the maturation of the core (Gag subunits) and the reverse transcriptase (RT).

CD4, CCR5 receptors

If spike proteins bind to CD4 or CCR5 receptors of a neighboring cell, cell fusion allows core particles to enter a new cell.

FIGURE 11.26 ▪ HIV replication cycle. The HIV virion attaches its receptor and fuses with the host cell membrane, releasing its contents in the cytoplasm to undergo a replication cycle. ▶

A.

NEIL STUART ET AL. 2008. NATURE 451:425

B.

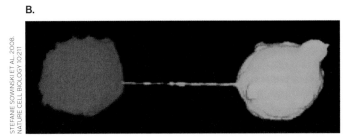

STEFANIE SOWINSKI ET AL. 2008. NATURE CELL BIOLOGY 10:211

FIGURE 11.27 ■ **HIV exits from an infected cell.** **A.** As virions emerge, they remain tethered to the cell surface by host tetherins, requiring a release step mediated by accessory protein Vpu (TEM). **B.** A T cell infected with HIV can transfer virions to an uninfected cell through a nanotubular connection. The two T cells are tagged here with different fluorescent labels (red versus green).

To Summarize

- **Human immunodeficiency virus (HIV)** causes an ongoing pandemic of acquired immunodeficiency syndrome (AIDS). Molecular biology has led to drugs that control the infection.

- **HIV is a retrovirus** whose RNA genome is reverse-transcribed into double-stranded DNA, which integrates into the DNA of the host cell.

- **The HIV core** contains two different copies of its RNA genome, each bound to a primer (host tRNA) and reverse transcriptase (RT). The core is surrounded by an envelope containing spike protein trimers.

- **HIV binds the CD4 receptor** of T lymphocytes together with the chemokine receptor CCR5. Following virion-receptor binding and envelope-membrane fusion, HIV core particle is released into the cytoplasm, where it partly uncoats.

- **Reverse transcriptase synthesizes DNA from the HIV RNA template**, primed by the tRNA. **RNA degradation** allows formation of a double-stranded DNA. Entering the nucleus, the retroviral **DNA integrates** into the host genome.

- **Retroviral mRNAs are exported to the cytoplasm** for translation. Envelope proteins are translated at the ER and exported to the cell membrane.

- **Retroviruses are assembled at the cell membrane**, where virions are released slowly, without lysis. Alternative routes of cell-to-cell transmission involve cell fusion (forming syncytia) or travel through an intercellular nanotube.

- **Accessory proteins regulate virion formation** and the latent phase, in which double-stranded DNA persists without reproduction of progeny virions.

11.4 Endogenous Retroviruses and Gene Therapy

Suppose an integrated HIV genome mutated and lost the ability to produce progeny. What would happen to its genome? The integrated genome would be "trapped" within a cell, an endogenous retrovirus. If the cell entered the host germ line, over many host generations in its host the retroviral sequence would inevitably accumulate more mutations. It could even provide the material for evolution of a new trait. Such endogenous retroviruses inspired the idea that researchers could intentionally manipulate a retrovirus to impart a useful trait and use it for gene therapy.

Retroelements in the Human Genome

The human genome is riddled with remains of retroviral genomes, in various states of decay. These decaying genomes are collectively known as retroelements (**Fig. 11.28**). Endogenous retroviruses (in humans, HERVs) are retroelements that retain all the genomic elements of a retrovirus, including *gag*, *env*, and *pol* genes. An example is HERV-K, an endogenous retrovirus that activates during normal development of human embryos (see Current Research Highlight).

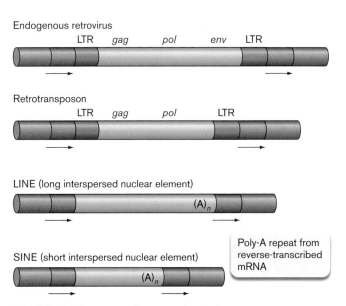

FIGURE 11.28 ■ **Retroelements in the human genome.** Endogenous retroviruses and other retroelements in the human genome may arise from progressive degeneration of ancestral retroviruses, or they may be progenitors of new retroviruses.

Other elements, known as **retrotransposons**, retain only partial retroviral elements but may maintain a reverse transcriptase to copy themselves into other genomic locations. An example of a retrotransposon is the well-known Alu sequence, a short sequence found in about a million copies in the human genome. In some cases, a retrotransposon such as Alu can interrupt a key human gene, leading to a genetic defect such as a defective lipoprotein receptor associated with abnormally high cholesterol level and heart failure. Still other retroelements, known as LINEs (long interspersed nuclear elements) and SINEs (short interspersed nuclear elements), show more vestigial remnants of retroviral genomes. Amazingly, retroelements and transposons appear to have generated about half the sequence of the human genome.

Many HERVs and other mammalian ERVs are now known to express viral components, such as reverse transcriptase, as part of normal host function. An example in mice is an ERV participating in the B-cell immune response to carbohydrate antigens (**Fig. 11.29**). In mice, normal function of the B-cell response involves B-cell antigen receptors, antibody-like molecules that reside in the plasma membrane (discussed in Chapter 24). Like antibodies, B-cell receptors specifically recognize certain types of foreign molecules—in this case, carbohydrates that might come from a pathogen. The B-cell receptor signal requires a cascade of intracellular signals culminating in B-cell proliferation and antibody production. Part of the cascade requires induction of an ERV to express retroviral RNA and reverse transcriptase. Both the RNA and the reverse-transcribed DNA activate genes for antibody production. The requirement is demonstrated by treating mice with inhibitors of reverse transcriptase, which consequently inhibit antibody production.

Gene Therapy with Lentivectors

The exceptional ability of lentiviruses to deliver their own DNA to human cells makes them our most effective **gene transfer vectors**. A gene transfer vector is a DNA sequence that can express a recombinant gene within an animal or plant cell, either from a plasmid or from a sequence integrated into the host genome. Vectors derived from HIV are particularly useful for their ability to transfer genes into nonmitotic cells.

The viruses used as gene transfer vectors are those whose replication cycles establish a viral genome within the host nucleus. Some of the first viral vectors were made from double-stranded DNA viruses such as adenoviruses, which cause mild respiratory illness. An adenoviral genome enters the cell nucleus, where it circularizes and replicates separately from the host chromosomes, in a cycle similar to that of herpes viruses. Thus, adenoviral vectors avoid the long-term risks of inserting DNA permanently into the genome of the host cell. But the disadvantage is that the adenoviral genes are eventually lost, and thus the treatment must be repeated. Repeated exposure to the vector eventually stimulates an immune response that destroys it.

An even more exciting class of gene transfer vectors derives from the lentivirus HIV, called lentiviral vector, or **lentivector**. Lentivectors integrate genes into a host chromosome, providing longer-lasting therapy (**Fig. 11.30** ▶). The lentivector is engineered to remove viral genes that cause disease, and to express an altered envelope protein from a different virus, such as vesicular stomatitis virus (VSV). The VSV envelope protein increases the viral host range and tropism, allowing treatment of a wide range of tissues. The vector

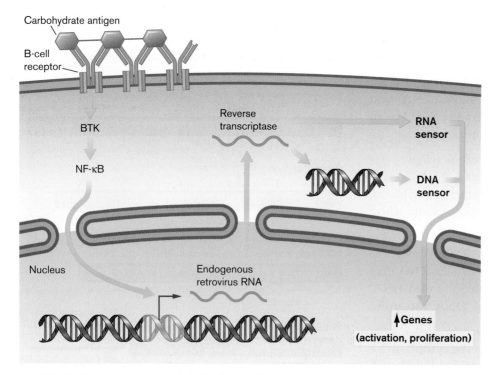

FIGURE 11.29 ■ Endogenous retrovirus expression mediates B-cell response to carbohydrate antigens. Transcribed retroviral RNA and reverse-transcribed DNA mediate the signal of carbohydrate antigen binding to the B-cell receptor. BTK = Bruton's tyrosine kinase; NF-κB = nuclear factor kappa beta.

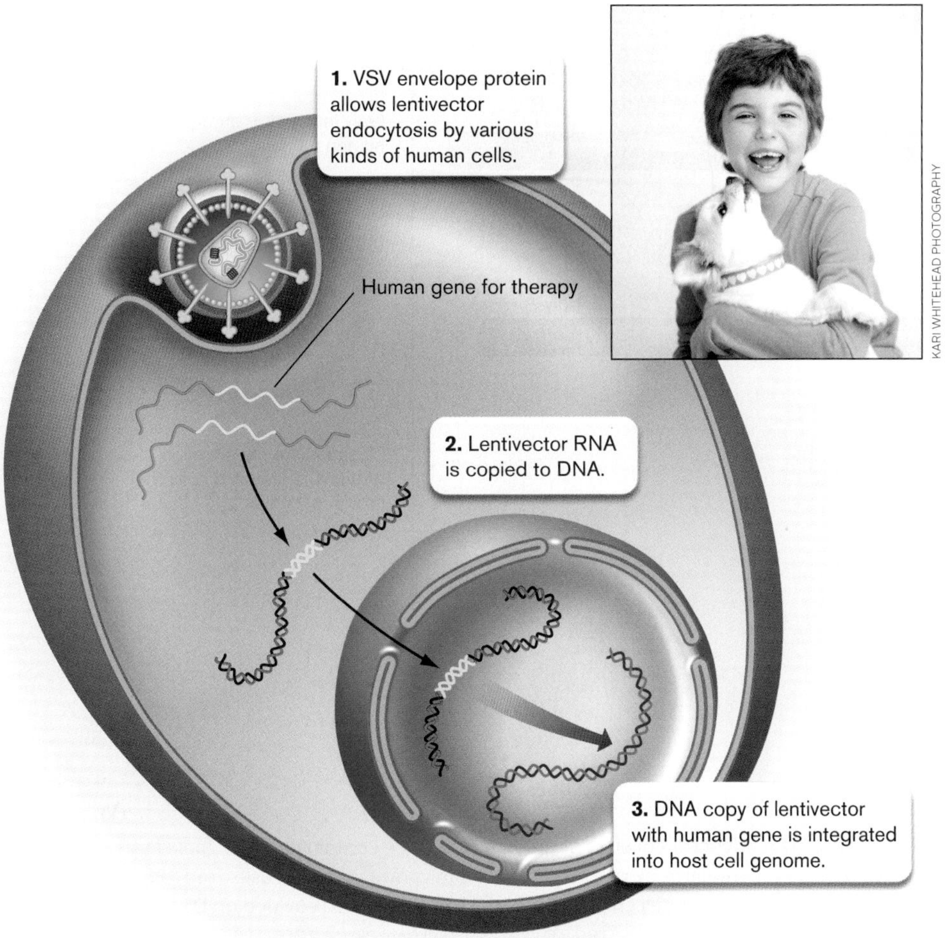

1. VSV envelope protein allows lentivector endocytosis by various kinds of human cells.

Human gene for therapy

2. Lentivector RNA is copied to DNA.

3. DNA copy of lentivector with human gene is integrated into host cell genome.

KARI WHITEHEAD PHOTOGRAPHY

FIGURE 11.30 ■ **Lentiviral gene therapy.** A lentiviral vector (lentivector) derived from HIV consists of virions coated with a VSV envelope protein that allows uptake by various kinds of cells. The engineered viral genome lacks disease-causing genes but possesses a transgene needed by the patient. The lentivector RNA with the transgene becomes copied into DNA and integrated into a host chromosome. **Inset:** Emily Whitehead was the first child to be considered cured of an illness (B-cell leukemia) by a lentiviral vector. ▶

contains a human transgene that becomes integrated into the host genome.

Could lentiviral integration be dangerous for the patient? Lentiviral integration into host DNA poses the danger of activating an adjacent proto-oncogene—that is, a human gene that causes cancer when expressed at the wrong time. To avoid this problem, lentivectors are modified so that their promoters can activate only the gene of choice, not an adjacent cancer gene. Furthermore, lentiviral vectors offer a way to integrate DNA into nondividing cells of differentiated tissues such as brain neurons. For example, in 2009 a lentivector derived from HIV halted progression of a fatal brain disease, adrenoleukodystrophy (ALD), in two young boys. The lentivector inserted a gene replacing a defective gene in each boy's blood stem cells.

In a different case, in 2012, the first leukemia patients were successfully treated by lentiviral gene therapy. The HIV-derived vector reprogrammed the patients' own T cells to attack cancerous B cells. A 7-year-old girl, Emily Whitehead (**Fig. 11.30** inset), was the first child to be considered fully "cured" of disease by a lentiviral vector.

How a Lentivector Works

To construct a safe and effective vector, the HIV genome is modified extensively. In the example shown in **Figure 11.31** ▶, accessory genes *vpr*, *vpu*, *nef*, and *vif*, which encode HIV virulence factors for disease, were removed. Other protein-encoding genes necessary for virion production were put into DNA helper plasmids, to be provided only in tissue culture for vector production. These genes provide the capsid monomer (*gag*), reverse transcriptase (*pol*), envelope glycoprotein (*env*), and a regulator of mRNA export from the nucleus (*rev*). The HIV *env* gene is replaced by an *env* gene from another virus, vesicular stomatitis virus (VSV). The VSV envelope protein has a broad tropism (host range) and thus enables the lentivector to infect a broad range of host cell types. Each helper plasmid drives its gene expression from a well-studied promoter of another virus, such as cytomegalovirus (CMV) or respiratory syncytial virus (RSV). The original HIV vector genome retains only the LTR end sequences (R-U5) required for genome integration, a packaging signal from the start of *gag*, and the infectivity-enhancing polypurine tract (PPT).

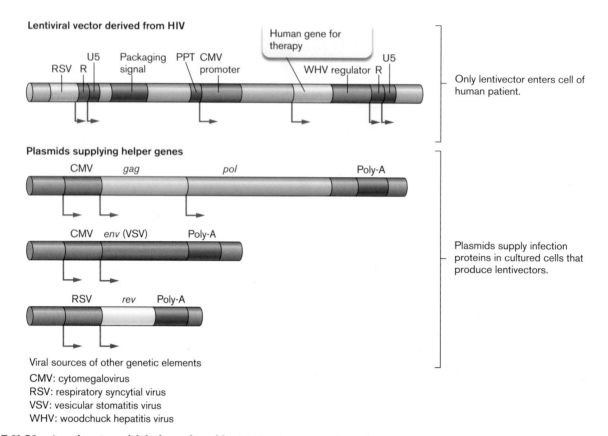

FIGURE 11.31 ■ Lentivector with helper plasmids. The lentivector consists of an RNA sequence containing HIV signal elements required for genomic integration (dark blue), with promoter and regulator elements derived from various other viruses (red) and the therapeutic human transgene (yellow). In order to produce the virions in cell culture, essential HIV genes are provided on DNA helper plasmids. Their expression is driven by regulatory elements from other viruses (red). *Source:* Modified from Blesch. 2003. *Methods* **33**:164. ▶

To express the **transgene** (the gene of interest for expression in the host), a CMV promoter was inserted in the HIV-derived RNA vector. To further enhance transgene expression, a genetic enhancer sequence was added from woodchuck hepatitis virus (WHV). Note that a "lentiviral" gene transfer system, in fact, includes genetic elements from a diverse set of human and animal viruses, all found in previous studies to contribute specific properties to infection. Although these genes originate from different viruses, they nevertheless function together like parts of a machine.

To produce infective virions, the HIV-derived RNA vector plus the three helper plasmids are introduced into a special tissue culture line. The vector and plasmids enter the cells by a process, called **transfection**, in which calcium phosphate treatment promotes the uptake of nucleic acids across the cell membrane. The host tissue culture cells are derived from human embryonic kidney 293T cells containing a gene for a protein from simian virus 40 (SV40) that enables the replication of DNA plasmids containing an SV40 replication site. The 293T cells allow efficient expression of the viral genes on the helper plasmids, as well as a full virion production cycle in which the vector RNA is replicated and packaged into virions.

Safety of Lentivectors

The use of disease-causing viruses for human therapy raises important concerns about safety. The viruses might express toxins, induce cancer, or trigger a damaging immune response. These concerns need to be weighed against the risks of the conditions they are used to treat, such as severe combined immunodeficiency (SCID), cystic fibrosis, and cancer. In general, gene therapy is approved only for conditions that are life-threatening and for which alternative therapies are inadequate.

Viruses used for gene therapy are engineered extensively to decrease risks, using techniques of genetic engineering presented in Chapter 12. Safety features of viral vectors include:

- **Deleting virulence genes.** Viral genes that promote disease and virion proliferation, but are not required for establishment of the viral DNA in the nucleus, are deleted from the vector genome. To produce the vector in tissue culture, the viral proliferation genes are provided on helper plasmids.

- **Avoiding genome insertion next to oncogenes.** Vectors engineered from adenoviruses are usually designed to avoid host chromosome integration altogether. The disadvantage of avoiding integration is that the separate viral DNA is soon lost from host tissues, and the therapy requires frequent repetition. In lentivectors, molecular modification avoids activation of proto-oncogenes.

■ **Altering tissue specificity.** The tropism, or tissue specificity, can be altered by replacing the gene for the viral envelope glycoprotein (spike protein) with the envelope gene from a different virus. For example, a rabies virus glycoprotein can be used to target the vector to brain cells. The alteration of viral tissue specificity by envelope gene replacement is called "pseudotyping." Pseudotyping can be used either to narrow the host range or to broaden it, depending on the needs of the vector.

■ **Avoiding germ-line infection.** Current medical standards prohibit alteration of the germ line, the egg and sperm cells that transmit genes to the next generation. Because the long-term risks of gene therapy are unknown, only somatic gene therapy (gene insertion into somatic, or body, cells) is permitted.

Thought Question

11.8 What do you think are the arguments for or against lentivectors editing the germ line?

To Summarize

■ **Ancient retroviral sequences persist within animal genomes**, including the human genome. Endogenous retroviruses (ERVs) retain the four main retroviral genes. Other sequences decay by mutation and are called retroelements.

■ **Gene transfer** vectors are made from viruses.

■ **Virulence genes are deleted from a lentivector (lentiviral vector).** Efficient promoter sequences from other viruses are inserted.

■ **Adenoviral vectors circularize and replicate separately from the host genome.**

■ **Lentivectors integrate within the host genome.**

11.5 Herpes Simplex Virus: DNA Virus

Many important viruses of humans and other animals contain genomes of double-stranded DNA (**Table 11.3**). DNA viruses include the causative agents of well-known diseases such as smallpox, chickenpox, and infectious mononucleosis (mono). Most DNA viruses are considerably larger than

---

TABLE 11.3	**DNA Viruses of Animals (Examples)**		
Virus	**DNA replication**	**Disease(s)**	**Host(s)**
Adenoviruses (many strains)	Viral DNA polymerase, single-strand binding protein, and protein primer	Enteritis or respiratory diseases	Humans, other mammals, birds
Papovavirus (simian virus 40, SV40)	Cellular DNA polymerases	Asymptomatic	Monkeys
Herpes viruses			
Herpes simplex virus 1 and 2	All viral components (DNA polymerase, primase, etc.)	Epithelial and genital lesions, latency in neurons	Humans
Varicella-zoster virus	Viral components	Chickenpox, shingles	Humans
Epstein-Barr virus	All cellular components (DNA polymerase, etc.)	Infectious mononucleosis, Hodgkin's lymphoma	Humans
Other strains		Epithelial lesions, cancer	Monkeys, cattle, horses
Papillomaviruses	Viral DNA helicase; cellular polymerase		
Human papillomaviruses (many strains)		Genital warts, cervical and penile cancer, skin warts	Humans
Other papillomaviruses		Warts, cancer	Rabbits, cattle, sheep
Poxviruses	All viral components		
Variola major virus		Smallpox	Humans
Vaccinia virus		Cowpox	Cattle, humans
Other poxviruses		Monkeypox	Monkeys, camels, birds, humans

FIGURE 11.32 ■ Genital herpes infection. Lesions on the elbow of an 11-year-old, female patient caused by HSV-2 infection.

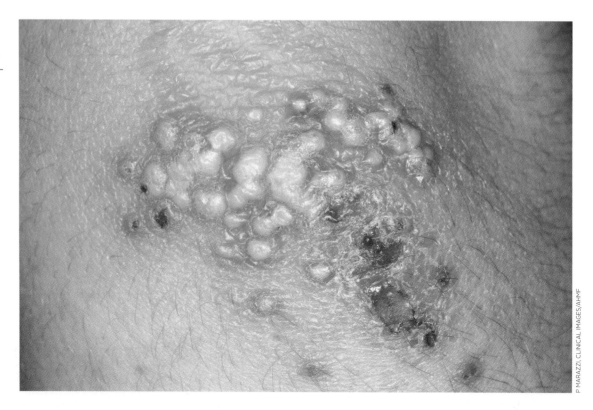

RNA viruses and encode a wider range of viral enzymes; for example, the vaccinia genome encodes nearly 200 different proteins. The complexity of viruses such as vaccinia and herpes approaches that of small cells.

Herpes viruses have been associated with humans and our ape ancestors for hundreds of millions of years (see Chapter 6). Different herpes viruses cause diseases ranging from chickenpox (varicella-zoster virus) to birth defects (cytomegalovirus). Today, in a remarkable twist, we have engineered a "tumor-eating herpes" to treat metastatic tumors (discussed shortly).

Herpes viral DNA replicates by mechanisms similar to those used in the replication of prokaryotic and phage genomes, either bidirectionally from an origin of replication (as in bacteria) or by the rolling-circle method (as in phages such as T4). Like cellular genomes, herpes viral DNA replication requires more than a polymerase; enzymes such as helicase, primase, and single-strand binding proteins are also needed.

Herpes Simplex Virus Infects the Oral or Genital Mucosa

An important example of a DNA virus is herpes simplex virus (HSV). Strains HSV-1 and HSV-2 cause one of the most common infections in the United States. Approximately 60% of Americans acquire herpes simplex, usually HSV-1, in epithelial lesions commonly known as cold sores. About 30%–60% acquire genital herpes, usually HSV-2, through sexual contact (oral or vaginal). Genital herpes

causes recurrent eruptions of infection in the reproductive tract (**Fig. 11.32**). Many of those infected are unaware of symptoms, but they can still transmit the disease to others.

Herpes simplex virus typically infects cells of the oral or genital mucosa, causing ulcerated sores. The primary infection is epithelial, followed by latent infection within neurons of the ganglia. A common site of infection is the trigeminal ganglion, which processes nerve impulses between the face and eyes and the brain stem.

The latent infection of the ganglia later leads to new outbreaks of virus, often triggered by stress such as menstruation, sunlight exposure, or depression of the immune system. Progeny virions travel back down the dendrites to the epithelia, causing lytic infection. In the trigeminal ganglion, herpes reactivation can lead to eye disease or lethal brain infection. In most cases, herpes symptoms can be controlled by antiviral agents such as acyclovir (discussed in Chapter 27). No cure or means of preventing future outbreaks exists. In pregnant women, HSV can be transmitted to the fetus, with serious complications for the child.

Herpes simplex virus is closely related to varicella-zoster virus, the cause of chickenpox, also an epithelial infection. Varicella, too, can hide in ganglial neurons, emerging decades later to cause painful skin lesions called shingles.

Herpes Simplex Virus Structure

The herpes virion comprises a double-stranded DNA chromosome packed within an icosahedral capsid

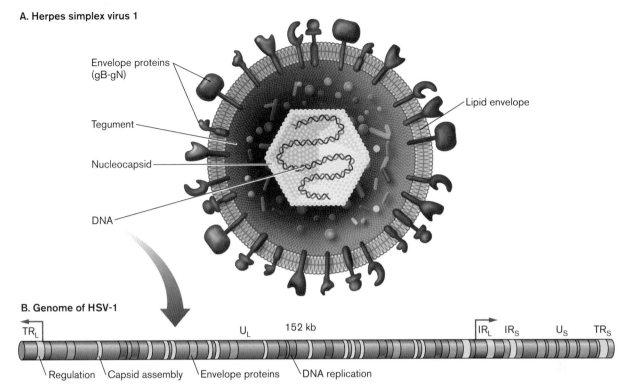

A. Herpes simplex virus 1

Envelope proteins (gB-gN)

Tegument

Nucleocapsid

DNA

Lipid envelope

B. Genome of HSV-1

TR_L

U_L 152 kb

IR_L IR_S U_S TR_S

Regulation   Capsid assembly   Envelope proteins   DNA replication

FIGURE 11.33 ▪ Herpes simplex virus 1: virion and genome. **A.** The HSV-1 virion consists of a double-stranded DNA chromosome packaged within an icosahedral capsid. The capsid is surrounded by tegument, a collection of virus-encoded and host-derived proteins. The tegument is contained within a host-derived membrane envelope, including several kinds of envelope proteins. **B.** The genome of HSV-1 spans 152,000 base pairs, encoding more than 70 gene products. The HSV sequence consists of two segments: long (U_L) and short (U_S). Each segment contains a unique region (U_L or U_S) flanked by two inverted-repeat regions—terminal (TR_L or TR_S) and internal (IR_L or IR_S)—where the two segments meet. ▶

(**Fig. 11.33A** ▶). The capsid is surrounded by **tegument**, a collection of about 15 different kinds of virus-encoded proteins, as well as proteins from the previous host. The tegument is contained within a host-derived membrane envelope with several kinds of spike proteins. The HSV-1 genome spans 152 kb, encoding more than 70 gene products (**Fig. 11.33B**). The sequence includes two unique segments, long (U_L) and short (U_S), each flanked by a terminal repeat (TR_L or TR_S) and an internal repeat (IR_L or IR_S). Within the host, the genome circularizes, so the genetic linkage map appears circular.

Herpes genes and gene regulatory elements resemble those of eukaryotic genes. Each gene has a promoter with eukaryotic control sequences such as the TATA box. Genes are transcribed and translated individually; there is no polyprotein. Genes fall into temporal classes within the viral replication cycle: immediate, early, and late expressed genes. Each class has specific regulatory elements. Thus, the virus expresses only the gene products needed for a given phase of infection. An additional class, consisting of the LAT genes, is expressed for latent infection (discussed next).

Herpes Simplex Attachment and Host Cell Entry

Unlike HIV, the relatively large herpes virion has several envelope proteins that can bind to several alternative receptor molecules on the host cell surface, such as a homolog of tumor necrosis factor receptor called HveA or intercellular adhesion molecules called nectins (**Fig. 11.34** ▶, step 1). As in the case of HIV, the entire herpes capsid enters the cytoplasm (step 2); but unlike HIV, whose core particle partly uncoats, the intact herpes capsid travels down a scaffold of microtubules (step 3) to the nuclear membrane. During this stage, the virion host shutoff factor (Vhs) degrades host mRNA, thus shutting off host protein synthesis. At a nuclear pore complex, the herpes capsid injects its DNA (step 4). The DNA is forced out of the capsid and into the nucleus by the high pressure of double-stranded DNA packing in the capsid, similar to the high-pressure injection of phage T4 DNA into a bacterial cell. The DNA then circularizes (step 5) to form a plasmid-like intermediate.

The herpes genome now takes one of two alternative directions: expression of mRNA for proteins of the infection cycle,

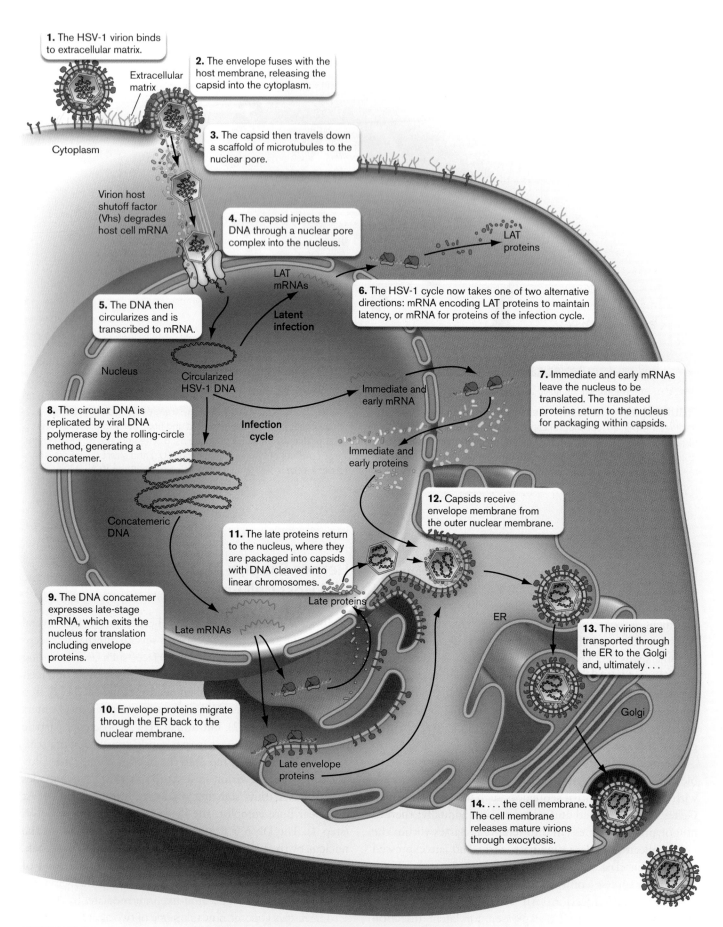

1. The HSV-1 virion binds to extracellular matrix.

Extracellular matrix

Cytoplasm

Virion host shutoff factor (Vhs) degrades host cell mRNA

2. The envelope fuses with the host membrane, releasing the capsid into the cytoplasm.

3. The capsid then travels down a scaffold of microtubules to the nuclear pore.

4. The capsid injects the DNA through a nuclear pore complex into the nucleus.

LAT proteins

LAT mRNAs

Latent infection

5. The DNA then circularizes and is transcribed to mRNA.

Nucleus

Circularized HSV-1 DNA

6. The HSV-1 cycle now takes one of two alternative directions: mRNA encoding LAT proteins to maintain latency, or mRNA for proteins of the infection cycle.

Immediate and early mRNA

7. Immediate and early mRNAs leave the nucleus to be translated. The translated proteins return to the nucleus for packaging within capsids.

8. The circular DNA is replicated by viral DNA polymerase by the rolling-circle method, generating a concatemer.

Infection cycle

Immediate and early proteins

Concatemeric DNA

12. Capsids receive envelope membrane from the outer nuclear membrane.

11. The late proteins return to the nucleus, where they are packaged into capsids with DNA cleaved into linear chromosomes.

9. The DNA concatemer expresses late-stage mRNA, which exits the nucleus for translation including envelope proteins.

Late proteins

ER

13. The virions are transported through the ER to the Golgi and, ultimately . . .

Late mRNAs

10. Envelope proteins migrate through the ER back to the nuclear membrane.

Late envelope proteins

Golgi

14. . . . the cell membrane. The cell membrane releases mature virions through exocytosis.

FIGURE 11.34 ■ Replication cycle of HSV-1. The HSV-1 virion binds to receptors on the host cell membrane and releases its capsid in the cytoplasm. The DNA chromosome is transferred into the host nucleus to conduct the replication cycle. ▶

or expression of mRNA encoding LAT proteins to maintain latency (**Fig. 11.34**, step 6). If the latent course is taken, the DNA circle can persist within the cell for decades before switching to lytic infection. Latent infection is seen most often in nerve cells, such as those of the trigeminal ganglion.

Replication of Herpes Simplex Virus

In the nucleus, herpes DNA is transcribed to mRNA by host RNA polymerase II (see **Fig. 11.34**, step 5). If mRNA for lytic infection is produced, it exits the nucleus to be translated by ribosomes. Many different mRNAs are produced and exported, including those required for "immediate" and "early" stages of infection (**Fig. 11.34**, step 7). The translated proteins return to the nucleus for packaging within capsids.

To generate progeny genomes, the circular DNA is replicated by viral enzymes, including DNA polymerase, single-strand binding protein, and a proofreading endonuclease (**Fig. 11.34**, step 8). Additional enzymes are provided by the host cell. DNA is replicated by the rolling-circle method, generating a concatemer similar to that of phage T4. Unlike T4, however, the herpes DNA is eventually cut into segments defined by the terminal repeat sequences.

The newly synthesized DNA expresses late-stage mRNA (**Fig. 11.34**, step 9), which exits the nucleus for translation. Translated envelope proteins are inserted into the ER membrane, through which they migrate to the nuclear membrane (step 10). Other late proteins reenter the nucleus for assembly into capsids containing DNA genomes (step 11). The envelope forms from the outer nuclear membrane (step 12). The virions are then transported through the ER, where they undergo secondary envelopment (step 13). The secondary-enveloped virions move to the Golgi, and ultimately to the cell membrane (step 14). The secondary envelope fuses with the cell membrane, releasing mature virions through exocytosis (step 14). Rapid release of virions destroys cells, causing the characteristic sores of herpes infection.

> ### Thought Question
>
> **11.9** Compare and contrast the fate of the HSV chromosome with that of the HIV chromosome.

Persistent Viral Infections

Herpes viruses can infect humans and other animals indefinitely, following initiation of latency by the viral LAT proteins (see **Fig. 11.34**, step 6). Most humans are infected by several latent herpes viruses, such as human herpes viruses 6 and 7, Epstein-Barr virus, and cytomegalovirus (see **Special Topic 11.1**). We may be infected all our lives without knowing it; in fact, our bodies may actually benefit from the presence of these viruses. For example, in 2007, Herbert Virgin and colleagues at the Washington University School of Medicine in St. Louis showed that mice infected with a gamma herpes virus similar to Epstein-Barr virus resist infection by the bacterial pathogens *Listeria monocytogenes* (the cause of listeriosis) and *Yersinia pestis* (the cause of bubonic plague). The mechanism of antibacterial resistance may involve stimulation of the immune response. Thus, some of our silent herpes viral "partners" may have coevolved with humans in a mutually beneficial relationship, or mutualism (discussed in Chapter 21).

The maintenance of persistent or latent viral infection involves several kinds of processes that are surprisingly similar in DNA viruses such as herpes and in retroviruses such as HIV. These processes include:

- **Infection of cell types suitable for long-term persistence.** After infecting skin cells for rapid viral replication, some HSV virions infect neurons. Neurons are long-lived cells that provide the virus with an everlasting home in the host. Similarly, Epstein-Barr virus persists within long-lived memory-B-cell lymphocytes.

- **Regulation of viral gene expression.** The LAT proteins suppress expression of viral genes for lytic replication, thus preventing host cell destruction where the latent viral DNA resides. Suppression is the result of assembly of "heterochromatin," chromosome-associated host proteins that inactivate all viral genes except those encoding LAT proteins.

- **Viral subversion of cellular apoptosis.** To maintain latent infection, a viral DNA must prevent host cell apoptosis, a form of programmed self-destruction in response to viral infection. For example, cytomegalovirus prevents apoptosis by mimicking one host apoptosis protein and inhibiting another. Similarly, the HIV retroviral accessory protein Tat prevents apoptosis by inhibiting protein p53, which suppresses tumors through apoptosis.

- **Evasion of immune responses.** Both herpes viruses and retroviruses express proteins that inhibit signaling molecules of the immune system, or that mimic immunosuppressive signals.

Herpes Viruses Engineered to Kill Tumors

The long-standing adaptation of herpes viruses to the human body confers properties useful for virotherapy. Some kinds of viruses have a preference for infection of tumor cells. Such a virus is called "oncolytic." Oncolytic viruses are being developed to treat cancer. An example of an oncolytic virus for tumor therapy is the engineered form of HSV-1 virus called T-VEC, developed by a company acquired by Amgen. In 2015, the United States approved the use of T-VEC to treat melanoma in patients with inoperable tumors.

Cytomegalovirus

We often think of herpes viruses as something to avoid "catching," like chickenpox or herpes simplex. But some members of this ancient and diverse virus family are hard to avoid. Cytomegalovirus (CMV) infects between 50% and 80% of us before age 40. The virus is transmitted by person-to-person contact, usually causing no symptoms. But CMV may start to replicate in immunocompromised people—for example, after an organ transplant. And during pregnancy, new CMV exposure or reactivation of a latent infection can expose the fetus. CMV infection is the cause of most of the prenatal birth defects in the United States. About one of every 100 infants is born infected, and a quarter of these will have severe birth defects or developmental problems later. Birth defects may include microcephaly, neurological problems, and loss of hearing or vision.

How does CMV infection cause birth defects? Lee Fortunato at the University of Idaho, Moscow, is studying this question (**Fig. 1A**). Fortunato investigated the effect of CMV infection

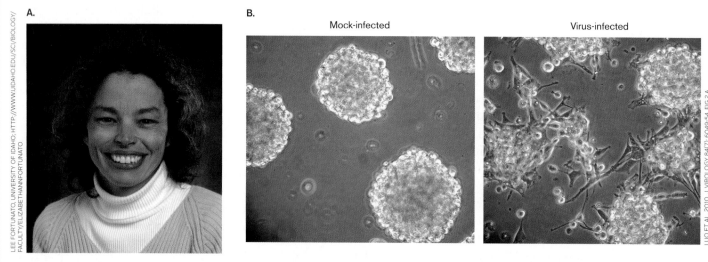

A. Mock-infected Virus-infected

FIGURE 1 ■ Cytomegalovirus (CMV) alters neural development. A. Lee Fortunato studies cytomegalovirus. B. Mock-infected embryonic neural progenitor cells (NPCs, left) grow as clumps of round neurospheres. NPCs infected with virus (right) form long projections and commence migration—a sign of premature differentiation.

T-VEC infects tumor cells and replicates, but it cannot replicate within normal cells. When T-VEC infects tumor cells, it also releases tumor cell fragments that attract cells of the immune system to recognize the tumors for destruction. Because T-VEC activates the immune system, it helps destroy tumors throughout the body. This antitumor effect of a virus is particularly promising for metastatic tumors that are hard to treat by other means.

To Summarize

- **Herpes simplex virus** causes recurring eruptions of sores in the oral or genital mucosa. Initial transmission is by oral or genital contact, followed by eruptions from reactivated virus latent in ganglial neurons.

- **The herpes virion** contains a double-stranded DNA genome, packed in an icosahedral capsid. The capsid is surrounded by numerous matrix proteins and by an envelope.

- **HSV attachment** may involve several alternative receptors. A microtubular scaffold transports the herpes virions to the nucleus, where the DNA genome is inserted. The DNA circularizes for transcription.

- **LAT protein expression** leads to latent infection, usually in nerve cells, where the DNA persists silently for months or years.

- **DNA genomes of HSV** are synthesized by the rolling-circle method, using viral DNA polymerase supplemented by viral and host-generated components.

- **Infectious mRNA expression** leads to production of capsid, matrix, and envelope proteins for assembly of HSV.

- **HSV assembly** takes place at the nuclear membrane or other membranes. The virions are released from the cell by exocytosis. Rapid release leads to mucosal pathology.

- **HSV can be engineered to specifically kill tumors.**

on human neural progenitor cells (NPCs), the cells that develop into neurons. Neural progenitor cells were obtained from the brain of a premature infant that died of natural causes. When cultured in a dish, the neural cells grow as round cells in a clump, remaining in position. The cells can be induced to differentiate (become more advanced neural cells). But what happens when the cells are infected by CMV? Fortunato found that infected NPCs begin to differentiate prematurely, at the wrong stage of development, thus causing a defect. In **Figure 1B**, the CMV-infected cells (right-hand photo) are differentiating, extending their cytoplasm, and starting to move. Thus, Fortunato shows how cytomegalovirus could alter the differentiation of embryonic cells during neonatal development.

To uncover the details of CMV infection, we need an immortalized cell line in which the virus replicates its genome indefinitely. Fortunato developed such a line, using T98G glioblastoma cells (**Fig. 2**). Glioblastoma is a cancer of the glia, accessory cells to neurons. The cancer cells have lost their growth inhibition and replicate continually in tissue culture. **Figure 2** shows infected cells with the nuclei stained blue (Hoechst stain) and the CMV viral genomes stained pink (FISH, fluorescent DNA probe). We hope that improved understanding of this viral process can yield insights that will help prevent future birth defects.

RESEARCH QUESTION

How could you use the glioblastoma cell line to determine which viral proteins initiate or prevent the CMV replication cycle?

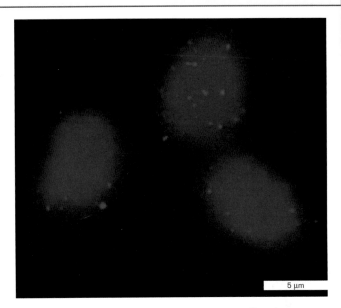

FIGURE 2 ■ Cytomegalovirus genomes replicating within glioblastoma cells. Cell nuclei are stained blue by the DNA-specific Hoechst stain. CMV genomes are labeled pink by DNA probes bound to fluorescent antibodies (FISH). *Source:* Ying-Liang Duan et al. 2014. *J. Virol.* **88**:3861.

Duan, Ying-Liang, Han-Qing Ye, Anamaria G. Zavala, Cui-Qing Yang, Ling-Feng Miao, et al. 2014. Maintenance of large numbers of virus genomes in human cytomegalovirus-infected T98G glioblastoma cells. *Journal of Virology* **88**:3861–3873.

Concluding Thoughts

In considering the mechanisms of viral infection, a picture emerges of interesting commonalities, as well as intriguing diversity:

■ **Viral infection requires host membrane receptors.** Different viruses evolve to take advantage of different cell membrane proteins in order to recognize and gain entry into host cells. Mutation of host receptors confers resistance to the virus.

■ **Viruses recruit host cells to replicate their genomes.** Viral genomes are remarkably diverse, including RNA or DNA, single- or double-stranded, linear or circular. Within the host cell, replication proceeds by a variety of methods, with varying dependence on viral and host enzymes.

■ **Virus particles (and their genomes) range in size from a few components to assemblages approaching the complexity of cells.** The advantage of simplicity is the minimal requirement for resources. The advantage of complexity is the fine-tuning of infection mechanisms for evading host defenses.

The study of viral molecular biology raises intriguing questions about the nature of viral replication and about viral origins (see **eTopic 6.1**). Research in virology offers hope for new drugs and cures for humanity's worst plagues, as well as devastating diseases of agricultural plants and animals. At the same time, it is sobering to note that despite the enormous volumes we now know about viruses such as HIV and influenza, the AIDS pandemic continues to grow, and we face the likely emergence of a new deadly flu strain. Molecular research can succeed only in partnership with epidemiology and public health (discussed in Chapter 28).

CHAPTER REVIEW

Review Questions

1. How does phage lambda attach to its correct host cell and insert its genome for replication?
2. How do the CI repressor and Cro protein modulate the switch between lysogeny and lysis?
3. How do influenza virions gain access to the host cytoplasm? Explain the role of fusion peptides.
4. How does influenza virus manage the replication and packaging of its segmented genome? What is the consequence of genome segmentation for virus evolution?
5. How does HIV provide ready-made components for replication of its genome? How does reverse transcriptase convert the single-stranded RNA genome to double-stranded DNA?
6. What is the role of protease in HIV replication? What is the significance of protease for AIDS therapy?
7. How did ancient retroviruses participate in evolution of the human genome?
8. For gene therapy, how can we construct a lentivector so as to avoid replication of the virus?
9. How does herpes virus compartmentalize the expression and replication of its DNA genome?
10. Compare and contrast the needs of DNA genome replication with those of RNA genome replication.

Thought Questions

1. RNA viruses and DNA viruses represent fundamentally different reproductive strategies. How do their different strategies impact the immune response and the development of antiviral agents?
2. Discuss the roles of host-modulating viral proteins in HIV infection, and in herpes virus infection. What various kinds of functions do these proteins serve for the virus, and what are their effects on the host cell?
3. Hemophilia is a life-threatening genetic blood disorder (the absence of a clotting factor) for which current therapies involve providing blood products and recombinant proteins. Discuss what might be the advantages and relative risks of using viral vectors to treat hemophilia.

Key Terms

accessory protein (422)
acquired immunodeficiency syndrome (AIDS) (419)
azidothymidine (AZT) (425)
chemokine receptor (CCR) (424)
concatemer (406)
core particle (420)
coreceptor (424)
endogenous retroviruses (418)
fusion peptide (414)
gene transfer vector (430)
host factor (414)

human immunodeficiency virus (HIV) (419)
lentivector (430)
lentivirus (418)
long terminal repeat (LTR) (427)
lysis (407)
lysogeny (403)
matrix protein (411)
nucleocapsid protein (NP) (411)
prophage (403)
protease inhibitor (427)
provirus (427)
quasispecies (423)

reassortment (412)
recombination (412)
retrotransposon (430)
retrovirus (418)
reverse transcriptase (RT) (418, 425)
rolling-circle replication (405)
segmented genome (412)
spike protein (420)
tegument (435)
transfection (432)
transgene (432)

Recommended Reading

Barton, Erik S., Douglas W. White, Jason S. Cathelyn, Kelly A. Brett-McClellan, Michael Engle, et al. 2007. Herpesvirus latency confers symbiotic protection from bacterial infection. *Nature* **447**:326–329.

Cartier, Nathalie, Salima Hacein-Bey-Abina, Cynthia C. Bartholomae, Gabor Veres, Manfred Schmidt, et al. 2009. Hematopoietic stem cell gene therapy with a lentiviral vector in X-linked adrenoleukodystrophy. *Science* **326**:818–823.

Cockrell, Adam S., and Tal Kafri. 2007. Gene delivery by lentivirus vectors. *Molecular Biotechnology* **36**:184–204.

Engelman, Alan, and Peter Cherepanov. 2012. The structural biology of HIV-1: Mechanistic and therapeutic insights. *Nature Reviews. Microbiology* **10**:279–290.

Göke, Jonathan, Xinyi Lu, Yun-Shen Chan, Huck-Hui Ng, Lam-Ha Ly, Friedrich Sachs, and Iwona Szczerbinska. 2015. Dynamic transcription of distinct classes of endogenous retroviral elements marks specific populations of early human embryonic cells. *Cell Stem Cell*, **16**: 135–141.

Grow, Edward J., Ryan A. Flynn, Shawn L. Chavez, Nicholas L. Bayless, Mark Wossidlo, et al. 2015. Intrinsic retroviral reactivation in human preimplantation embryos and pluripotent cells. *Nature* **522**:221–225.

Heeney, Jonathan L., Angus G. Dalgleish, and Robin A. Weiss. 2006. Origins of HIV and the evolution of resistance to AIDS. *Science* **313**:462–466.

Joseph, Sarah B., Ronald Swanstrom, Angela D. M. Kashuba, and Myron S. Cohen. 2015. Bottlenecks in HIV-1 transmission: Insights from the study of founder viruses. *Nature Reviews. Microbiology* **13**:414–425.

Kane, Melissa, and Tatyana Golovkina. 2010. Common threads in persistent viral infections. *Journal of Virology* **84**:4116–4123.

Karzbrun, Eyal, Alexandra M. Tayar, Vincent Noireaux, and Roy H. Bar-Ziv. 2014. Programmable on-chip DNA compartments as artificial cells. *Science* **345**:829–832.

Kotula, Jonathan W., S. Jordan Kerns, Lev A. Shaket, Layla Siraj, James J. Collins, et al. 2014. Programmable bacteria detect and record an environmental signal in the mammalian gut. *Proceedings of the National Academy of Sciences USA* **111**:4838–4843.

Lakdawala, Seema S., Yicong Wu, Peter Wawrzusin, Juraj Kabat, Andrew J. Broadbent, et al. 2014. Influenza A virus assembly intermediates fuse in the cytoplasm. *PLoS Pathogens* **10**:e1003971.

Smith, Gavin J. D., Dhanasekaran Vijaykrishna, Justin Bahl, Samantha J. Lycett, Michael Worobey, et al. 2009. Origins and evolutionary genomics of the 2009 swine-origin H1N1 influenza A epidemic. *Nature* **459**:1122–1126.

Stevens, James, Ola Blixt, Terrence M. Tumpey, Jeffrey K. Taubenberger, James C. Paulson, et al. 2006. Structure and receptor specificity of the hemagglutinin from an H5N1 influenza virus. *Science* **312**:404–410.

Yosef, Ido, Miriam Manor, Ruth Kiro, and Udi Qimron. 2015. Temperate and lytic bacteriophages programmed to sensitize and kill antibiotic-resistant bacteria. *Proceedings of the National Academy of Sciences USA* **112**:7267–7272.

Zeng, Ming, Zeping Hu, Xiaolei Shi, Xiaohong Li, Xiaoming Zhan, et al. 2014. MAVS, cGAS, and endogenous retroviruses in T-independent B cell responses. *Science* **346**:1486–1492.

CHAPTER 12
Biotechniques and Synthetic Biology

The science of biotechnology uses living organisms or their products to improve human health or to perform specific industrial or manufacturing processes. Achieving these goals, however, requires the constant invention of new technologies. Chapter 12 describes these technologies and explores how they are used.

CURRENT RESEARCH highlight

"Shadow" genetics. Can a "shadow" protein synthesis system exist in a bacterial cell? Special ribosomes would be needed whose subunits would not mix with native ribosomes but would recognize unnatural mRNA ribosome-binding sites (RBSs). Michael C. Jewett and Alexander S. Mankin engineered such a system by linking 16S rRNA in the 30S subunit to 23S rRNA of the 50S subunit. The tethered ribosome, Ribo-T, is shown with proteins removed. Ribo-T also recognizes an alternative RBS sequence. Ribo-T could even be engineered to polymerize unnatural amino acids to make proteins with novel functions.

Source: From Cédric Orelle et al. 2015. *Nature* **524**:119–124.

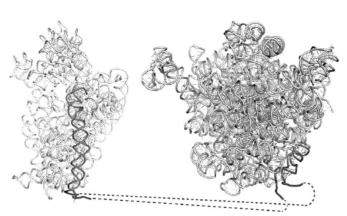

Tethered sections of 16S (orange) and 23S (blue) rRNA molecules

AN INTERVIEW WITH

MICHAEL JEWETT, BIOLOGICAL ENGINEER, NORTHWESTERN UNIVERSITY

NORTHWESTERN UNIVERSITY

What potential uses do you envision for this technology?

We can now ask previously unanswerable questions about ribosome function. For example, we assumed that the two ribosomal subunits had to separate for successful protein synthesis. Ribo-T proves this assumption is inaccurate. We can also use Ribo-T to create the first ever fully independent ribosome mRNA system in cells, where mRNA decoding, catalysis of polypeptide synthesis, and protein excretion can be optimized for new substrates and functions. This opens a transformational new direction in biomolecular engineering and synthetic biology.

How did the two of you first decide to collaborate?

I believe we decided to collaborate over lunch. After moving from Boston to start my lab at Northwestern, I connected with Shura [Alexander Mankin], as he is one of the world's experts on the ribosome. We share a fascination with this molecular machine, and only through interdisciplinary science at the boundaries of traditional disciplines could we pursue the adventure that led to Ribo-T.

The engineering of new ribosomes by Michael Jewett and Alexander Mankin (**Fig. 12.1**) is an amazing achievement in biotechnology. Although we think of biotechnology as a new field, the earliest biotechnologists actually lived about 10,000 years ago, when our ancestors unwittingly figured out how to use yeast to make alcohol, bread, and dairy products (discussed in Chapter 16). The field changed little over the millennia, until 1928, when the Scotsman Alexander Fleming discovered that microbes produce antibiotics. That watershed event, along with unraveling the structure of DNA and deciphering the genetic code, ushered in the high-tech era of biotechnology. We have since learned a great deal about the molecular biology of the microbes our ancestors used. Now, through genetic engineering and synthetic biology, we can make bacteria produce hormones, engineer vaccines in plants, devise previously unimagined biochemical pathways, and design bacteria that detect explosives and flash a warning.

Biotechnology can also improve human existence. One example is the engineering of potatoes that resist devastation by a parasitic fungus. Today's potato farmers must spray huge amounts of antifungal agents on their crops to combat potato blight disease, the cause of the potato famine that wiped out 30% of the Irish population in the nineteenth century. Spraying plants as many as 25 times a season with chemicals, the typical U.S. potato farmer spends about $250 per acre to fight the disease—a huge incentive to find less expensive means of prevention. Biotechnologists have discovered a wild potato from Argentina that is highly resistant to *Phytophthora infestans*, the cause of potato blight disease. Although the Argentine potato has no commercial value (it is inedible), the gene conveying resistance (*Rpi-vnt1*) was genetically inserted by recombination into commercial potato breeds, making them resistant to the fungus. This discovery will eventually improve the health of potato crops, reduce our exposure to currently used antifungal chemicals, and contribute to our understanding of disease resistance in plants. Because of an abundance of caution and opposition to genetically modified crops, however, efforts to approve the use of these modified potatoes are still ongoing in the European Union.

We begin our discussion of biotechnology by explaining the techniques used to probe the inner workings of microbes. Many of these techniques, such as DNA sequencing and proteomics, are described throughout the book (**Table 12.1**). We will repeatedly refer you to these sections as we describe more cutting-edge techniques in this chapter. These new techniques include CRISPR-Cas gene engineering, as well as molecular techniques such as differential RNA sequencing and chromatin immunoprecipitation. Finally, we describe some applications of microbial biotechnology and discuss the exciting new field of synthetic biology.

12.1 Genetic Analyses

The identification and manipulation of genes has been a mainstay of microbial genetics research for many decades. Mutating genes and observing the effects of those mutations on cell behaviors has solved many mysteries of biology. Mysteries like how microbes make antibiotics without killing themselves, what happens when cells divide, and why some species sporulate. We start this section with a research case history to illustrate how many of these techniques can be applied to a single scientific question.

Acid Survival: A Research Case Study

The human stomach not only digests food; it also guards the rest of the gastrointestinal tract against ingested pathogens. The stomach's gastric acidity (about pH 2) quickly kills the vast majority of microbes that we ingest. However, strains of the intestinal bacterium *E. coli*, such as enterohemorrhagic *E. coli* O157:H7, can survive pH 2 conditions for hours. This means that a person needs to ingest only a few cells of this *E. coli* strain to become colonized or infected. That organism is said to have a low infectious dose. What makes *E. coli* so special? The molecular tools that we describe here helped reveal the novel mechanisms of acid resistance used by this microbe and the regulatory networks that control them. Our story begins with the discovery of acid-sensitive *E. coli* mutants that <u>failed</u> to survive at pH 2.

FIGURE 12.1 ■ Alexander Mankin. Mankin (University of Illinois in Chicago) is a world expert on the ribosome and was instrumental in conceiving and engineering Ribo-T with Michael Jewett.

TABLE 12.1	Techniques Described Elsewhere in This Book	
Molecular technique	**Designed use**	**Location in book**
Electrophoresis	To separate DNA, RNA, protein, or other molecules according to size or charge	eAppendix 3
Fluorescence in situ hybridization (FISH)	To identify specific genes or RNA targets in a permeabilized cell	Chapter 18; eAppendix 3
Polymerase chain reaction (PCR)	To amplify specific segments of the genome sequence	Chapter 7
Quantitative (real-time) PCR amplification	To quantify specific nucleic acids (DNA or RNA)	Chapter 12
Restriction digestion	To excise sections of DNA that can be cloned into plasmids or other vectors	Chapter 7
Gene cloning (recombinant DNA)	To make a recombinant DNA molecule (by splicing DNA from one organism into DNA vectors) that can be introduced into a different organism	Chapter 7
DNA sequencing	To read the order of deoxynucleotides in a DNA molecule	Chapter 7
Metagenomics	To use DNA sequencing to discover, identify, and characterize members of a mixed population of microbes in their natural environment	Chapter 7
Transcriptomics	To quantify the environmentally induced expression of RNA molecules from all genes in a genome	Chapter 10
Microarrays	To quantify changes in gene expression using hybridization technology	Chapter 10; eAppendix 3
Deep sequencing	To repeatedly and rapidly sequence short segments of DNA or RNA to reconstruct a microbial genome or to quantify transcripts	Chapter 7 (DNAseq); Chapter 10 (RNAseq)
Proteomics	To quantify the environmentally induced levels of all proteins in a cell	Chapter 10
SDS polyacrylamide gels	To separate proteins according to their molecular size	Chapter 3
Isoelectric focusing	To separate proteins according to charge	Chapter 10; eAppendix 3
Two-dimensional gels	To separate proteins according to charge and size	Chapter 10; eAppendix 3
Mass spectrometry	To rapidly determine a protein's mass and (by MS-MS) to determine its amino acid sequence	Chapter 10; eAppendix 3
Molecular databases	To search known DNA or protein sequences for similarities to newly sequenced molecules	Chapter 8
Site-directed mutagenesis	To target specific amino acids in a protein for mutagenesis	eTopic 12.7

Transposons Inactivate and Mark Target Genes

How were *E. coli* acid survival genes found and their function discovered? An early step taken to gain insight into *E. coli* acid resistance involved isolating mutants defective in that process. The initial goal was to inactivate relevant genes and tag them with an antibiotic resistance marker. To do this, a transposon (see Section 9.5) containing a tetracycline resistance gene was randomly inserted into the chromosome by transposition. When a transposon inserts into a target gene sequence, the target gene is inactivated and the mutant cell becomes antibiotic resistant. In our example, each tetracycline-resistant cell resulting from transposition grew into a colony; the colony was picked with a sterile toothpick and transferred into a well of a microtiter plate containing growth medium (**Fig. 12.2A**).

After about 10,000 colonies were transferred to microtiter dishes and grown, a multipronged replicator device (a square with 48 metal prongs arranged in six rows of eight) was used to transfer small amounts of each culture to new microtiter plate wells that contained a pH 2 medium. After several hours of incubation, the multipronged replicator was again used to sterily transfer samples of the pH 2 culture from each well to an agar plate. A few of the tetracycline-resistant mutants failed to grow on the agar, meaning they had been killed by the pH 2 environment. The transposon in these acid-sensitive mutants had inserted into a gene required for acid resistance, rendering it inactive. The acid-sensitive mutants were retrieved from the original growth medium microtiter plates and subjected to further analysis.

Thought Questions

12.1 How would you modify the technique illustrated in **Figure 12.2A** to identify mutants defective in the synthesis of an amino acid, such as glutamic acid? *Hint:* You can use a minimal agar medium.

12.2 In this postgenomic era, how might you generally design a strategy to construct a strain of *E. coli* that cannot synthesize histidine?

A.

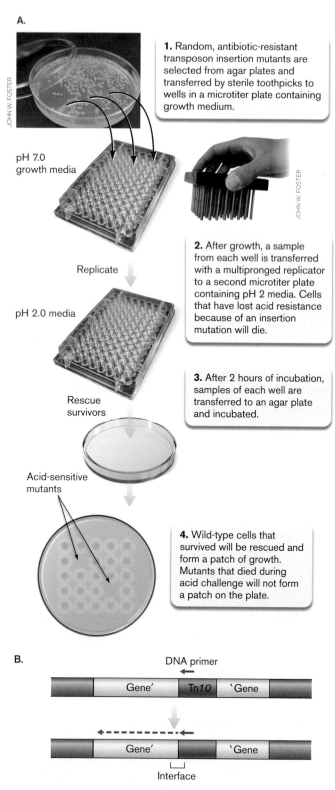

1. Random, antibiotic-resistant transposon insertion mutants are selected from agar plates and transferred by sterile toothpicks to wells in a microtiter plate containing growth medium.

pH 7.0 growth media

Replicate

pH 2.0 media

2. After growth, a sample from each well is transferred with a multipronged replicator to a second microtiter plate containing pH 2 media. Cells that have lost acid resistance because of an insertion mutation will die.

Rescue survivors

3. After 2 hours of incubation, samples of each well are transferred to an agar plate and incubated.

Acid-sensitive mutants

4. Wild-type cells that survived will be rescued and form a patch of growth. Mutants that died during acid challenge will not form a patch on the plate.

B.

DNA primer

| Gene′ | Tn*10* | ′Gene |

| Gene′ | | ′Gene |

Interface

FIGURE 12.2 ■ Selection for acid-sensitive mutants.
A. Mutagenesis with transposons led to the discovery of several *E. coli* genes involved in acid resistance. The technique is applicable to many other biochemical systems. **B.** Strategy to identify the gene target of a transposon insertion. A primer that anneals to a sequence in the transposon and a second primer that can, at a low annealing temperature, bind random sequences around the chromosome (including downstream of the transposon) are used to amplify the junction between the transposon and the gene into which it has inserted. Subsequent sequencing of the product using only the transposon primer identifies the gene and the insertion point.

DNA Sequencing of Insertion Sites and Computer Analysis

Once mutants in a project are identified, it is important to know which genes the insertions occurred in. In our example, an oligonucleotide primer (a short, single-stranded DNA molecule) that anneals to one end of the transposon was used to sequence across the insertion joint and into the adjacent *E. coli* DNA (**Fig. 12.2B**). The DNA sequences obtained were subjected to computer-based homology searches with the sequenced genome of *E. coli*. The resulting matches indicated that insertions had occurred in four genes. Two of these genes encode isoforms of glutamate decarboxylase (GadA and GadB) that differ by only a few amino acids. A third gene produces a glutamate/GABA antiporter (GadC), and a fourth gene encodes a putative DNA-binding protein with no known function, although it was annotated as a possible regulator.

The nature of the genes' functions suggests how the system might protect cells submerged in acid (**Fig. 12.3**). When *E. coli* is placed in a pH 2 environment, the cytoplasmic pH falls to dangerously low levels. Glutamate decarboxylase removes a carboxyl group from glutamate and replaces

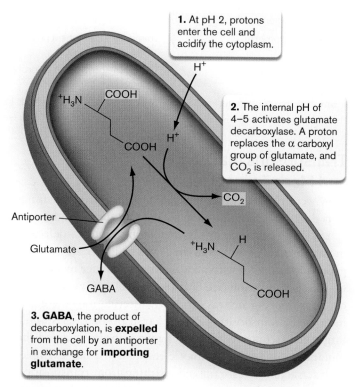

1. At pH 2, protons enter the cell and acidify the cytoplasm.

2. The internal pH of 4–5 activates glutamate decarboxylase. A proton replaces the α carboxyl group of glutamate, and CO_2 is released.

Antiporter

Glutamate

GABA

3. GABA, the product of decarboxylation, is **expelled** from the cell by an antiporter in exchange for **importing glutamate**.

FIGURE 12.3 ■ Model of acid resistance in *E. coli*. DNA sequence data identified the location of the transposon insertions leading to acid sensitivity. The genes identified included two isoforms of glutamate decarboxylase and a putative antiporter that brings glutamate into the cell in exchange for the decarboxylation product gamma-aminobutyric acid (GABA), which is transported out of the cell. The combined action of the decarboxylase and antiporter removes H^+ from the cell.

it with a proton from the cytoplasm, thereby decreasing acidity and elevating cytoplasmic pH. The decarboxylation product of the reaction is gamma-aminobutyric acid (γ-aminobutyric acid), also known as GABA. But to continue draining protons from the cytoplasm, *E. coli* must bring in more glutamate. The glutamate/GABA antiporter does this by importing glutamate while expelling GABA.

Alternative strategies have also been used to identify genes potentially related to acid resistance. Microarrays and RNA deep sequencing techniques have been used to identify all cellular transcripts regulated by growth of *E. coli* under milder acid environments. Bioinformatic strategies outlined in Chapter 8 were then used to hypothesize function of the products (regulator, transporter, and so on). Some of the acid pH-regulated genes were then selectively deleted using in vivo molecular tools that precisely delete the gene of interest, replacing it with an antibiotic drug marker. Once these genes were deleted, researchers could determine the effects of the missing gene on acid resistance or on the expression of other acid-regulated genes.

Exploring Gene Regulation: Transcriptomics versus Reporter Fusions

Another important question in our research example was whether acid resistance genes are regulated in response to changes in the environment. Are they always expressed, making the cell ever-ready to encounter acid stress? Or does the cell sense when it might encounter life-threatening low pH and only then express acid resistance genes? Recall that the cell can control production of a protein by regulating transcription and/or translation, or by altering mRNA or protein stability (processes described in Chapters 8 and 10). How do microbiologists differentiate among these possibilities or determine whether there is any regulation in the first place?

As will be described in Section 12.2, quantitative (real-time) PCR can be used to measure changes in the amount of a specific RNA transcript when cells are subjected to environmental or genetic perturbations. RNA deep sequencing (RNAseq), described in Section 10.6, can measure the levels of all transcripts in a cell. However, those techniques have drawbacks if you want to continually monitor gene expression in a live cell or if you want to screen for regulatory genes that control the expression of the target gene.

One technique that allows continuous monitoring of a gene's expression involves linking (or fusing) the promoter of the gene of interest to what is called a promoter-less **reporter gene**, such as *lacZ* (which encodes beta-galactosidase) or *gfp* (GFP, green fluorescent protein; see Chapter 2). Reporter genes encode proteins whose activities are easily assayed. In the case of *lacZ*, beta-galactosidase (β-galactosidase) converts the substrate *o*-nitrophenyl galactoside (ONPG) to a

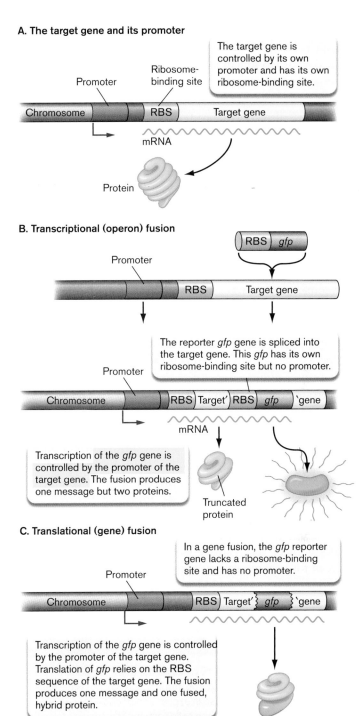

FIGURE 12.4 ■ **Transcriptional (operon) and translational (gene) fusions.** Aspects of a gene's regulation can be monitored by fusing a second gene, one that produces an easily assayed protein, to the gene under study. **A.** The gene targeted for construction of a reporter fusion. **B.** Construction of a transcriptional fusion. **C.** Construction of a translational fusion. The *gfp* open reading frame must be fused in-frame to the amino-terminal end of the target gene.

yellow product that is easily measured with a spectrophotometer. Fusions to GFP are detected by a fluorometer, which detects fluorescence.

A promoter-less reporter gene fused to the 3′ end of a target gene produces an artificial operon (**Fig. 12.4A and B**).

Because there is only one promoter, only one mRNA transcript is made. The 5′ end of the newly engineered polycistronic mRNA will contain the target gene transcript followed at the 3′ end by the reporter gene transcript. Now, any factor that controls expression of the target gene's promoter will also control production of the reporter. Because the reporter retains its own ribosome-binding site, it will not respond to translational controls over the target transcript. The fusion just described is known as a **transcriptional fusion**, which can monitor transcriptional control of the target gene.

Another type of reporter lacks both its promoter and its RBS. This reporter produces gene fusions (also known as protein fusions or **translational fusions**). For the reporter protein to be synthesized, the reporter gene sequence must be joined to the target gene sequence in the proper codon reading frame (**Fig. 12.4C**). Now, not only will the mRNA transcripts of the target and reporter genes be fused, but the peptides of the truncated target gene and the reporter gene will also be linked. In this case, anything controlling the transcription or translation of the target gene will also control reporter protein levels.

Each type of fusion can be constructed in vitro using plasmids and then transferred into recipient cells in a way that promotes the exchange of the reporter fusion for the resident gene. Once constructed, these fusions allow the researcher to monitor how different growth conditions, or whether suspected regulators, influence the transcription or translation of the gene. These fusions provide convenient phenotypes (for example, lactose fermentation or fluorescence) that can be exploited in mutant hunts to screen for new regulators. eAppendix 3 describes how *lacZ* fusions can be used to find regulatory genes.

dips. Now imagine looking down at a neighborhood of houses. You see that the heater in one house turns on as another house's heater turns off. The result appears to be a stochastic (random) process of heaters being turned on and off, but the population as a whole stays warm.

Do bacterial cells respond to stress in the same way? What technique could even address this question in single cells? LacZ fusions are inadequate because LacZ enzyme activity is measured as an average of all cells in the population. However, green fluorescent protein (GFP), or a derivative such as yellow fluorescent protein (YFP), fused to a gene does allow scientists to view (in real time) what happens inside single cells within a population. Realizing this, Michael Elowitz and colleagues from the California Institute of Technology used a YFP fusion to elegantly reveal stochastic gene expression in *Bacillus subtilis* subjected to energy stress.

Elowitz and co-workers placed the promoter and ribosome-binding site of the gene encoding sigma B, a major stress sigma factor from *B. subtilis*, in front of *yfp*. Cells in which sigma B was expressed would fluoresce green. A sigma B–YFP cell was then treated with mycophenolic acid to induce energy stress. What happened next was observed using quantitative time-lapse microscopy. Surprisingly, the constant energy stress imposed on the population triggered unsynchronized pulses of sigma B activation in individual cells (**Fig. 12.5**). As the stress increased, the pulses became more frequent. Thus, sigma B controls its target genes through sustained on/off pulsing (like the neighborhood house heaters) rather than by continuous activation.

As this example illustrates, GFP fusions are more versatile than LacZ fusions in many ways. GFP fusions can be

---

Thought Question

12.3 You have monitored expression of a gene fusion in which *gfp* is fused to *gadA*, the glutamate decarboxylase gene. You find a 50-fold increase in expression of *gadA* (based on fluorescence level) when the cells carrying this fusion are grown in media at pH 5.5 as compared to pH 8. What additional experiments must you do to determine whether the control is transcriptional or translational?

Gene expression in a single cell versus the population. When a population of bacteria experiences a stress, do all cells simultaneously activate their stress response genes? And once activated, do the genes stay on until the stress dissipates? Think of your own house as a single cell. When it becomes cold outside (and then inside), the thermostat activates your heater. At some point, however, the heater turns off and then turns on again whenever the inside temperature

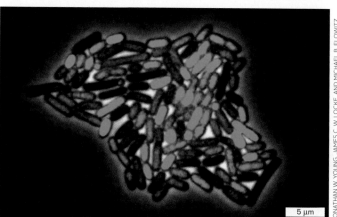

FIGURE 12.5 ∎ **Flashing bacteria.** YFP fusion (manipulated to appear green) revealed the stochastic expression of the sigma B gene among single cells of *Bacillus subtilis* subjected to energy stress. Individual cells flash on and off much like a thermostat-controlled house heating system. A movie of the process can be seen at: http://www.sciencemag.org/content/suppl/2011/10/05/science.1208144.DC1/1208144s1.mov.

used to track gene expression, identify the location of proteins inside cells, and follow the movement of pathogens within host animals.

Targeted Gene Editing and Regulation with CRISPR-Cas9 Technology

Chapter 9 described a form of bacterial and archaeal "immune system" called CRISPR (clustered regularly interspaced short palindromic repeats). Recall that when a bacterial cell survives a viral attack, a short DNA sequence from the virus is captured and inserted into the CRISPR locus as a kind of memory of the attack (see Fig. 9.13). The CRISPR locus includes sequences captured from many viruses, forming a kind of rogues' gallery of viral suspects. The virus-derived sequences are separated from each other by short (32 bp) palindromic repeat sequences.

Once captured, transcripts of the virus memory sequences are periodically incorporated into a "search and destroy" endonuclease called Cas (Fig. 9.13). If the same virus reinvades the cell at a later date, the CRISPR transcript (called csRNA) guides Cas to the target viral DNA sequence and cleaves it, thereby destroying the virus. This is how the system works in nature. One CRISPR-Cas system has now been repurposed into a powerful molecular tool that can genetically edit any gene in a eukaryotic cell. The technology has enabled precision gene editing in species where this has never been possible.

The CRISPR-Cas9 editing tool. The system most often used is the CRISPR-Cas9 from *Streptococcus pyogenes*. The key to this tool is in programming a synthetic csRNA (now called a guide RNA, or gRNA) to direct Cas9 endonuclease to a specific eukaryotic gene. The programmed gRNA is fused to what is called the tracer RNA, a molecule that binds to and activates Cas9 (**Fig. 12.6**). How are these genes introduced into eukaryotic cells? The gene that encodes Cas9 and the gene for the programmed guide RNA are inserted into plasmids that are introduced into eukaryotic cells via transformation, and the genes are transiently expressed in the nucleus. The guide RNA binds to the Cas9 endonuclease and helps it recognize the target eukaryotic sequence. The endonuclease

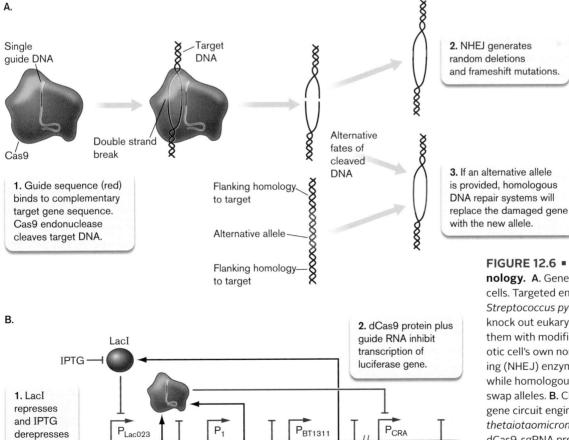

FIGURE 12.6 ■ CRISPR-Cas9 technology. **A.** Gene editing in eukaryotic cells. Targeted endonuclease Cas9 from *Streptococcus pyogenes* can efficiently knock out eukaryotic genes or replace them with modified forms. The eukaryotic cell's own nonhomologous end joining (NHEJ) enzymes achieve knockouts while homologous DNA repair systems swap alleles. **B.** CRISPR interference gene circuit engineered in *Bacteroides thetaiotaomicron*. When IPTG is added, dCas9-sgRNA prevents transcription of the luciferase gene *nanoluc*.

then introduces a double-strand break in the target DNA—a critical feature of this gene-editing tool. To survive, the eukaryotic cell will try to re-join the blunt ends using nonhomologous end joining (NHEJ) repair (**Fig. 12.6A**; described in Section 9.4). However, the process often deletes a few bases, which may cripple the gene. Once mutated, the effect of the altered gene on cell physiology can be assessed.

The targeted gene can also be changed or replaced using CRISPR-Cas9 technology. In this case, a third plasmid is introduced that expresses a modified version (or allele) of the gene. Both ends of the replacement gene are homologous to DNA flanking the target gene, so homologous recombination systems will replace the broken gene with the modified allele. Any type of insertion, deletion, or change in sequence can be introduced.

A striking example of this technology's potential was demonstrated when it was used to disrupt latent HIV provirus in infected cells. As described in Chapter 11, HIV DNA made by reverse transcriptase integrates into the host genome. Current drug therapies can stop HIV replication and transmission, but will not expunge the provirus from infected cells. Scientists from Kyoto University led by Hirotaka Ebina used the CRISPR-Cas9 system to disrupt HIV provirus, offering hope that this system may provide a tool for curing HIV infection.

Manipulating microbial gene expression in the intestine. The CRISPR-Cas9 system has also been modified to serve as a regulator of gene expression. CRISPR interference (CRISPRi), for example, uses an inactive version of Cas9 (dCas9) that no longer works as an endonuclease but can still be guided to target DNA sequences. Once bound to its target, dCas9 blocks transcription. Timothy Lu's laboratory at MIT recently used CRISPRi to control gene expression in *Bacteroides thetaiotaomicron* growing in a mouse intestine. *B. thetaiotaomicron* is an important bacterial member of mammalian gut microbiomes (discussed in Sections 13.4 and 23.1). The scientists designed an sgRNA (single guide RNA) to target a luciferase gene that was part of the organism's genome. The sgRNA and the gene for dCas9 were also integrated into the microbe's genome (**Fig. 12.6B**). The gene for dCas9 also included the *lacO* operator, which means that dCas9 can be induced by the addition of IPTG. Without IPTG, the circuit allows luciferase to be made, and cells glow. Adding IPTG, however, induces dCas9 production. The sgRNA guides dCas9 to the luciferase gene to turn it off. The authors demonstrated that the system worked when the programmed organism resided in the mouse intestine and IPTG was added to drinking water. The ability to precisely modulate gene expression in commensal organisms should enable functional studies of the microbiome, noninvasive monitoring of in vivo environments, and long-term targeted therapeutics.

To Summarize

- **Transposons** can be used to mutate genes and mark the defective gene with antibiotic resistance.

- **The location of a transposon insertion** can be determined by sequencing from the end of the transposon across the insertion junction. The junction sequence is examined by online comparative BLAST analysis to identify the gene in the genome.

- **Annotations of the gene** may provide clues about the function of the protein.

- **Regulation of a gene** can be determined by fusing that gene to an easily assayed reporter gene.

- **Transcriptional fusions** to reporter genes reflect transcriptional control of the subject gene.

- **Translational fusions** to reporter genes (gene fusions) reflect transcriptional and translational control of the subject gene.

- **The CRISPR-Cas9 microbial "immune system"** has been adapted as a tool to quickly and efficiently edit eukaryotic genes.

12.2 Molecular Techniques

In many cases, making gene fusions is not possible. For instance, some fusion proteins can aggregate and block membranes. In these situations, deep sequencing and mass spectrometry techniques can be used for whole-cell transcriptomic and proteomic purposes, respectively (see Chapter 10). But sometimes, examining only a specific gene or protein is desired. Classic molecular techniques can then be used, such as the northern blot (for RNA), the Southern blot (for DNA), and the western blot (for protein). All rely on electrophoresis to separate relevant macromolecules (see Section 3.2). Details of the northern and Southern blot techniques are found in eAppendix 3. Here we briefly review the techniques and show how they helped reveal details of *E. coli* acid resistance.

Northern and Southern Blots

Northern blots are used to analyze the presence, size, and processing of a specific RNA molecule in a cell extract. In the northern blot technique, RNA is extracted from the cell, and individual molecules are separated by size, using electrophoresis through an agarose-formaldehyde gel (see eAppendix 3). The separated fragments are transferred by simple capillary action onto a nylon or nitrocellulose membrane, forming the blot. After the RNA has been transferred, the membrane is

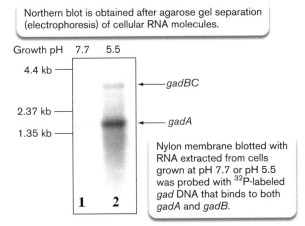

Northern blot is obtained after agarose gel separation (electrophoresis) of cellular RNA molecules.

Growth pH   7.7   5.5

4.4 kb —

2.37 kb —

1.35 kb —

← gadBC

← gadA

Nylon membrane blotted with RNA extracted from cells grown at pH 7.7 or pH 5.5 was probed with ^{32}P-labeled *gad* DNA that binds to both *gadA* and *gadB*.

1   **2**

FIGURE 12.7 ■ **Northern blot to view mRNA levels.** Northern blots can be used to monitor the quantity and breakdown of RNA. The blot shown illustrates that *gadB* and *gadC* are transcribed as an operon (note the size markers on the left). The *gadA* gene is located elsewhere in the genome and is transcribed separately. The data also illustrate that *gadA* and *gadBC* mRNAs accumulate after growth at pH 5.5. The accumulation is due to increased transcription. A second blot using a *gadC* probe was used to show that the larger transcript also contains *gadC* (not shown). *Source:* Modified from Zhuo Ma et al. 2003. *Mol. Microbiol.* **49**:1309–1320.

hybridized (or "probed") with a small, labeled DNA fragment (usually made via PCR) that will anneal to a specific mRNA species on the blot. A similar probing strategy, called the **Southern blot** technique, is used to detect the presence of specific DNA bands (genes) among electrophoretically separated DNA fragments (see eAppendix 3).

In our acid resistance example, northern blots were used to ask whether the *E. coli* acid resistance genes are regulated. The northern blot in **Figure 12.7** reveals that when cells are grown under acidic conditions (low pH), the *gadA* gene and the operon *gadBC* are both induced. The DNA probe used can bind to either the *gadA* or *gadB* mRNA. Because *gadB* and *gadC* form an operon, a longer mRNA molecule appears on the blot. The RNA bands are also more intense in the lane from acid-grown cultures (pH 5.5) than in the lane from cultures grown at high pH (pH 7.7), suggesting that the transcription of these genes is induced by acid. An alternative procedure for quantifying levels (but not sizes) of specific RNA molecules is quantitative (real-time) PCR, which is discussed later in this section.

Western Blots

Another way to examine gene regulation is to detect the protein products themselves using a technique called the **western blot**. The western blot is frequently used in today's laboratories. To begin with, the protein extract is subjected to SDS polyacrylamide gel electrophoresis (SDS-PAGE) (see Section 10.6 and eAppendix 3). The protein bands are electrophoretically transferred from the gel to a membrane made of polyvinylidene fluoride (PVDF) (**Fig. 12.8A**).

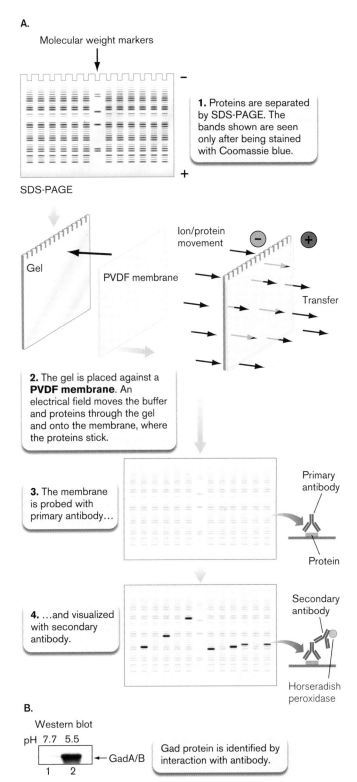

A.

Molecular weight markers

1. Proteins are separated by SDS-PAGE. The bands shown are seen only after being stained with Coomassie blue.

SDS-PAGE

Gel

PVDF membrane

Ion/protein movement

Transfer

2. The gel is placed against a **PVDF membrane**. An electrical field moves the buffer and proteins through the gel and onto the membrane, where the proteins stick.

3. The membrane is probed with primary antibody...

Primary antibody

Protein

4. ...and visualized with secondary antibody.

Secondary antibody

Horseradish peroxidase

B.

Western blot

pH   7.7   5.5

← GadA/B

1   **2**

Gad protein is identified by interaction with antibody.

FIGURE 12.8 ■ **Western blot analysis.** Specific proteins within a mixture of many proteins can be identified by western blot. **A.** The western blot procedure. **B.** Western blot analysis of the glutamate decarboxylase content of *E. coli* grown at pH 7.7 versus pH 5.5. The production of the protein is induced by growth under acidic conditions. *Source:* Modified from Zhuo Ma et al. 2003. *Mol. Microbiol.* **49**:1309.

A.

In a polyacrylamide gel, DNA bound to a DNA-binding protein travels more slowly than does free DNA.

Direction of electrophoresis

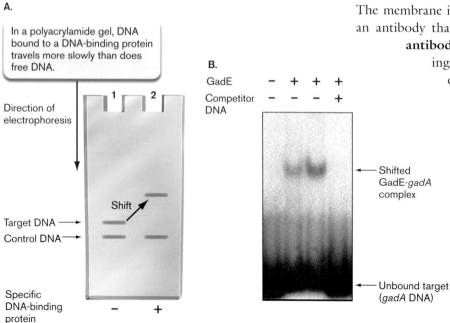

Target DNA →

Control DNA →

Shift

Specific DNA-binding protein   −   +

B.

GadE   −   +   +   +
Competitor   −   −   −   +
DNA

← Shifted GadE-*gadA* complex

← Unbound target (*gadA* DNA)

FIGURE 12.9 ■ Electrophoretic mobility shift assay (EMSA). Protein-DNA interactions can be monitored by EMSA. **A.** Diagrammatic representation of EMSA, showing slowed mobility of a hypothetical DNA-protein complex in a polyacrylamide gel. **B.** Gel shift experiment showing that the GadE regulatory protein binds to a radiolabeled promoter fragment of *gadA*. Lane 1 contains radiolabeled target DNA only. Lanes 2 and 3 contain radiolabeled target DNA and GadE protein. Lane 4 is the same as lane 3 but includes excess unlabeled promoter fragment. The unlabeled fragments effectively outcompete labeled fragments for the protein, so no shift in the radiolabeled fragment is seen. The treatment in lane 4 provides a check on the specificity of the binding. *Source:* Modified from Zhuo Ma et al. 2003. *Mol. Microbiol.* **49**:1309.

The membrane is then washed with a solution containing an antibody that binds to the specific protein (**primary antibody**). Primary antibodies are made by injecting purified protein into animals and then extracting the serum after a few months.

Depending on the source of the primary antibody (mouse, rabbit, or other), the membrane is then probed with a **secondary antibody** that will bind to the primary antibody. If the primary antibody was made in mice, the secondary antibody will be an anti-mouse IgG antibody that specifically binds to an amino acid sequence common to all mouse primary antibodies. The secondary antibody may be tagged with horseradish peroxidase to help visualize where the antibody binds to the blot. The result is an antibody "sandwich" in which the primary antibody binds to the target protein and the secondary antibody binds to the primary antibody. Adding hydrogen peroxide (which is reduced by horseradish peroxidase) and luminol (a chemical that emits light when oxidized by horse radish peroxidase) sets off a luminescent reaction wherever an antibody sandwich has assembled. The light emitted is detected by **autoradiography**, which

FIGURE 12.10 ■ Tagging proteins with a His$_6$ tag for easy purification. **A.** His$_6$ tag vector. **B.** Purification of the fusion protein on a nickel column. **C.** SDS polyacrylamide gel electrophoresis (SDS-PAGE) of fractions from the purification of His$_6$-GadB. Lane 1: Protein molecular weight markers. Lane 2: Uninduced cells. Lane 3: Induced cells. Lanes 4–9: His$_6$-GadB eluted from nickel column with imidazole. *Source:* From Zhuo Ma et al. 2003. *Mol. Microbiol.* **49**:1309. ▶

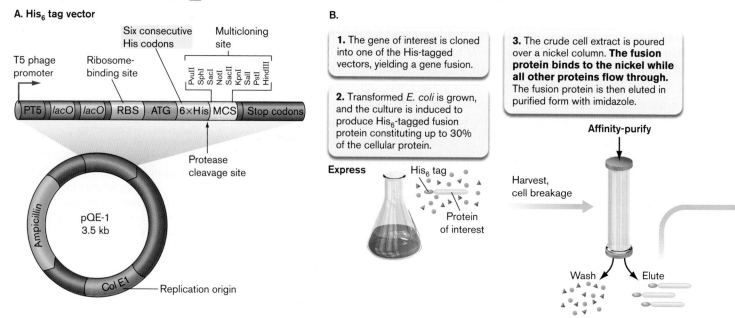

A. His$_6$ tag vector

Six consecutive His codons

Multicloning site

T5 phage promoter

Ribosome-binding site

PvuII
SphI
SacI
NotI
SacII
KpnI
SalI
PstI
HindIII

PT5 | *lacO* | *lacO* | RBS | ATG | 6×His | MCS | Stop codons

Protease cleavage site

Ampicillin

pQE-1
3.5 kb

Col E1 — Replication origin

B.

1. The gene of interest is cloned into one of the His-tagged vectors, yielding a gene fusion.

2. Transformed *E. coli* is grown, and the culture is induced to produce His$_6$-tagged fusion protein constituting up to 30% of the cellular protein.

3. The crude cell extract is poured over a nickel column. **The fusion protein binds to the nickel while all other proteins flow through.** The fusion protein is then eluted in purified form with imidazole.

Express

His$_6$ tag

Protein of interest

Harvest, cell breakage

Affinity-purify

Wash   Elute

uses X-ray film to capture the light, or phosphorimaging (a phosphorimager is a machine that records the energy emitted as light or radioactivity from gels or membranes). Only protein bands to which the primary antibody has bound will be detected. The technique is used to determine whether, and at what levels, specific proteins are present in a cell extract; it can also reveal whether a protein has been cleaved (processed) from a larger to a smaller form.

In the acid resistance example (**Fig. 12.8B**), western blot analysis was performed to detect glutamate decarboxylase protein. The blot shows that the level of this protein is induced by growth under acidic conditions. Thus, the acid-induced increase in mRNA from transcription (as shown by northern blot) correlates with an increase in translated protein (as shown by western blot).

> **Thought Question**
>
> **12.5** What would you conclude if the northern blot of the *gad* genes showed no difference in mRNA levels in cells grown at pH 5.5 and pH 7.7, but the western blot showed more protein at pH 5.5 than at pH 7.7?

DNA Mobility Shifts

Once a gene encoding a putative transcriptional regulatory protein has been identified by sequence annotation, proof of function requires a demonstration that the protein will bind to a DNA sequence present in the target promoter area. One approach is to add purified protein to the putative target DNA fragment and then see whether the protein causes a shift in the electrophoretic mobility of that fragment in a gel. An example of this method, as applied to our acid resistance problem, is illustrated in **Figure 12.9A**. Linear DNA molecules travel in agarose or polyacrylamide gels at an inverse rate relative to their size; that is, larger molecules travel more slowly than smaller ones. If a protein binds to a DNA molecule, the complex is larger than the DNA alone and will travel even more slowly in the gel. This method of analyzing protein-DNA binding is called the **electrophoretic mobility shift assay** (**EMSA**) (see Section 10.1).

In our research example, one of the genes that affected acid resistance (*yhiE*) was annotated in databases as a potential transcriptional regulator. The encoded protein was purified and tested by EMSA for its ability to bind to the promoter regions of the two potential targets, *gadA* and *gadBC*. As shown in **Figure 12.9B**, the purified YhiE protein, renamed GadE because of its role in regulating glutamic acid decarboxylase expression, did bind the *gadA* sequence and retard its mobility. Thus, GadE is a true regulator of the acid resistance genes.

Purifying Proteins by Affinity Tag

How do researchers purify proteins? Today, a protein with a known sequence can be purified in about a week. The trick is to fuse the gene encoding the protein to a DNA sequence encoding a peptide tag that strongly binds a particular small ligand molecule. The target ligand is attached to beads, and the cell extract containing the tagged protein is passed over the beads. The tagged protein binds to the beads, while other proteins do not. This technique, called **affinity chromatography**, allows researchers to essentially "fish" the tagged protein from a complex mixture of proteins present in a cell extract.

There are several commonly used peptide tags. One tag is a series of six histidines, called a His_6 tag, which tightly binds to nickel. A commercially available plasmid that is used to make His-tagged proteins includes an inducible promoter constructed by linking the promoter from T5 phage to two copies of *lacO* (**Fig. 12.10A** ▶). Adding

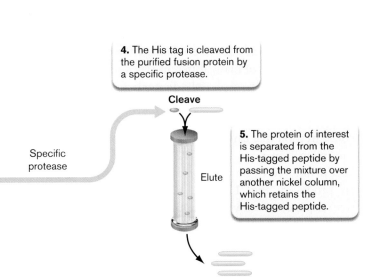

4. The His tag is cleaved from the purified fusion protein by a specific protease.

Cleave

Specific protease

Elute

5. The protein of interest is separated from the His-tagged peptide by passing the mixture over another nickel column, which retains the His-tagged peptide.

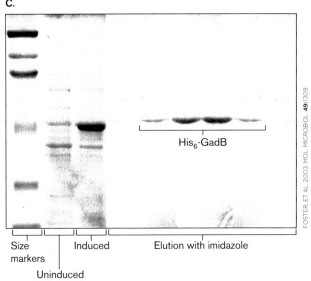

C.

His_6-GadB

Size markers

Uninduced

Induced

Elution with imidazole

FOSTER, ET AL. 2003. MOL. MICROBIOL. **49**:1309

the lactose analog IPTG will induce transcription (see Chapter 10) and control when the fusion protein is produced. To make the fusion, the **vector** (the DNA carrier molecule) sequentially contains a ribosome-binding site (RBS), a start ATG codon, a sequence encoding six histidine residues, and a multicloning site (MCS) to insert the desired open reading frame (ORF). The translational reading frame of the His codons and the codons of the ORF must be the same to yield a protein fusion.

To purify the resulting protein, nickel-coated beads are loaded into a column, and then a crude extract containing the tagged protein is added to the top (**Fig. 12.10B**). Proteins that do not bind to the nickel (that is, those without the His_6 tag) are eluted with buffer. The tagged protein molecules that did bind are then eluted from the column with an elution buffer containing imidazole. Imidazole has a stronger affinity for the nickel, so it displaces the His_6-tagged protein from the beads. Analysis of the progressive purification of GadB by the His_6 tag fusion method is shown in **Figure 12.10C**.

Mapping Transcriptional Start Sites

Recall from Section 8.1 that promoters have consensus sequences located at −10 and −35 bp from the transcription start site. To begin searching for upstream promoter sequences, we need to define where each transcript begins. One method to determine transcript length is called **primer extension** (details are provided in eAppendix 3). In brief, a single DNA primer is designed to anneal near the 5′ end of an mRNA. Reverse transcriptase, which synthesizes DNA from an RNA template, is then used to extend the primer to the 5′ start of the message. The result is complementary DNA (cDNA) of a precise length. When electrophoresed next to a DNA sequencing ladder of the region (made using the same primer), the cDNA product aligns with the base corresponding to the transcription start site. eAppendix 3 describes how primer extension was used to identify the *gadA* promoter.

Primer extension will identify transcripts for single genes, but how can transcript start sites for an entire genome be determined all at once? This is possible using modern deep sequencing technologies (see Section 10.6). Briefly, whole-cell RNA transcripts are extracted from cells and made into small fragments that are converted to cDNA. The cDNA is sequenced using next-generation sequencing platforms. Computer programs then assemble the fragments and align them with the genomic DNA sequence. An example of this strategy is shown in **Figure 12.11A**. The 5′ end of each assembled, nonoverlapping transcript is presumed to be the transcriptional start site. So the segment in the figure contains two transcripts—or does it?

RNA molecules can undergo a type of processing in which a long transcript starting from a bona fide promoter is cleaved into smaller fragments. So, is transcript B in **Figure 12.11A** made from its own promoter, or is it a product made from a larger transcript encompassing A and B?

Recall that the 5′ nucleotide in all primary RNA transcripts is a triphosphate, whereas the first nucleotide of a cleaved (processed) transcript has to be a monophosphate (**Fig. 12.11B**). This distinction led to a novel **differential RNA sequencing (dRNAseq)** approach that was developed to discover primary transcriptional start sites on a genome-wide scale. It uses the 5′-monophosphate-dependent terminator exonuclease (TEX) that specifically degrades 5′-monophosphorylated RNA species, such as processed RNA (including mature rRNA and tRNA). Any 5′-triphosphorylated RNA species (primary transcripts) remain intact. This approach makes it possible to identify primary-transcript start sites by comparing TEX-treated transcripts with untreated libraries. The dRNAseq procedure has been used to identify primary transcripts in numerous species, including small RNAs (sRNAs) in the plant pathogen *Xanthomonas campestris*.

DNA Protection Analysis Identifies DNA-Protein Contacts

Although EMSA identifies DNA fragments bound by a transcriptional regulatory protein, the technique does not reveal which nucleotides are bound. There are several techniques that can do this. One method asks which nucleotides are protected from nuclease digestion when a regulatory protein binds to a DNA fragment. The protected bases represent what is called the "footprint" of the binding protein. Identifying specific nucleotide-protein interactions between transcriptional regulators and their target sites (for instance, an activator of a pathogenicity operon) provides insight into how these interactions promote or inhibit transcription.

Because a regulatory protein covers DNA at its binding-site sequence, an enzyme that might attack the sequence cannot get close enough to do damage. Deoxyribonuclease I (DNase I), one of many different types of DNases, is one such enzyme. It cleaves phosphodiester bonds between bases. **Figure 12.12** presents the DNase I footprinting technique. In this method, DNase I is added to a DNA fragment at a concentration that can nick each molecule only once during the incubation period (**Fig. 12.12A**). When the DNA is later denatured by heat, arrays of single-stranded fragments are produced (**Fig. 12.12B**). Each fragment is a different size, depending on where the nick occurred. Because only one end of one strand of the DNA is labeled (by radiation or fluorescence), only the fragments containing that label will be seen following polyacrylamide gel electrophoresis and autoradiography.

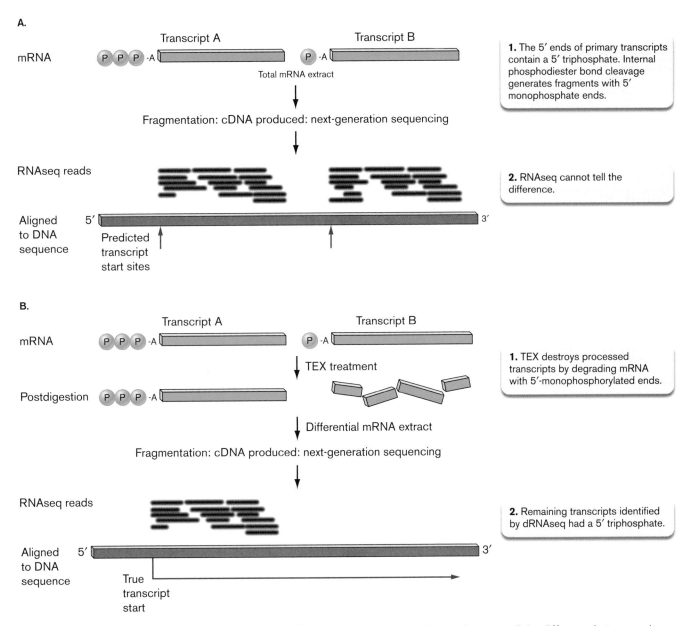

A.

mRNA — Transcript A / Transcript B

Total mRNA extract

Fragmentation: cDNA produced: next-generation sequencing

RNAseq reads

Aligned to DNA sequence — 5′ ... 3′

Predicted transcript start sites

1. The 5′ ends of primary transcripts contain a 5′ triphosphate. Internal phosphodiester bond cleavage generates fragments with 5′ monophosphate ends.

2. RNAseq cannot tell the difference.

B.

mRNA — Transcript A / Transcript B

TEX treatment

Postdigestion

1. TEX destroys processed transcripts by degrading mRNA with 5′-monophosphorylated ends.

Differential mRNA extract

Fragmentation: cDNA produced: next-generation sequencing

RNAseq reads

2. Remaining transcripts identified by dRNAseq had a 5′ triphosphate.

Aligned to DNA sequence — 5′ ... 3′

True transcript start

FIGURE 12.11 ▪ **RNAseq analysis of transcript start sites.** **A.** RNA sequencing (RNAseq) cannot tell the difference between primary transcripts and cleavage products. Black lines above the DNA sense sequence strand represent independent RNAseq reads made from RNA. **B.** RNAseq results after treating mRNA transcripts with the 5′-monophosphate-dependent terminator exonuclease (TEX). Comparing the RNAseq results of the whole transcript with TEX-treated transcripts using differential RNAseq (dRNAseq) will reveal true primary-transcript start sites.

Figure 12.12C presents the actual footprint of GadX, another regulator of *E. coli* acid resistance. The regulatory protein GadX covers a region around the promoter of *gadA* and *gadB* (only *gadA* is shown). The region is immediately upstream of the promoter. It is thought that both regulatory proteins GadE (previously described) and GadX communicate with RNA polymerase at the promoter, allowing it to transcribe the acid resistance genes.

DNA protection assays do not <u>definitively</u> show contact points between protein and DNA. Areas of the protein only have to be close enough to the DNA to hinder nuclease access. Protein-DNA interaction sites can be better defined

by other, more difficult methods that cause a cross-link between amino acids of the protein and the specific bases they contact in the DNA. Nevertheless, the DNA protection assay remains a useful tool.

Thought Question

12.6 Another regulator of glutamate decarboxylase (Gad) production does not affect the production of *gad* mRNA, but is required to accumulate Gad protein. What two regulatory mechanisms might account for this phenotype?

A. Random nicking by DNase I

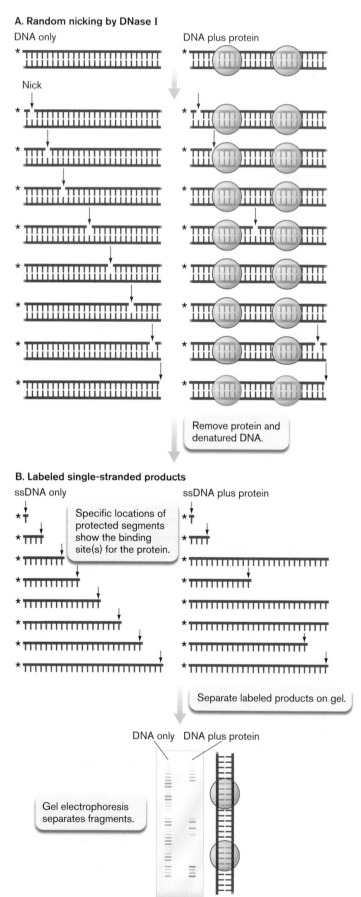

Remove protein and denatured DNA.

B. Labeled single-stranded products

Specific locations of protected segments show the binding site(s) for the protein.

Separate labeled products on gel.

DNA only   DNA plus protein

Gel electrophoresis separates fragments.

Quantitative (real-time) PCR

Another modification of the PCR technique is called **quantitative (real-time) PCR** (**RT PCR**). Real-time PCR uses fluorescence to monitor the progress of PCR as it occurs (that is, in real time). Data are collected throughout the PCR process rather than just at the end of the reaction. Quantitative PCR can be used to quantify the level of DNA or RNA in a sample. DNA is quantified by how long it takes to <u>first</u> detect an amplified product while the polymerase chain reaction is still running. The higher the starting copy number of the nucleic acid target, the sooner a significant increase in fluorescence is observed. Thus, the time at which fluorescence increases is a reflection of the amount of nucleic acid in the original sample.

How does quantitative PCR work? Two techniques are commonly used. In one, a compound called SYBR green is

FIGURE 12.12 ■ DNA protection assay using DNase I.
A. DNase I randomly nicks unprotected double-stranded DNA. Conditions are designed so that there will be only one nick per molecule of DNA. Small arrows indicate nicks. **Left:** Unprotected DNA. **Right:** DNA-binding proteins block DNase I activity. **B.** Labeled single-stranded products of the reactions in (A) after heat denaturation. Fragments are separated on polyacrylamide gels. **C.** Gel autoradiograph showing DNase I footprint resulting from a regulator of *E. coli* acid resistance, GadX, binding to the *gadA* promoter region. The actual protein used was an MBP (maltose-binding protein)–GadX fusion protein. The first two lanes represent the fragment's DNA sequence (G and A lanes only). Orange boxes marked with Roman numerals on the right indicate areas protected by GadX. The pink box marks the sequence bound by GadE, the other regulator.
Source: Modified from Angela Tramonti et al. 2002. *J. Bacteriol.* **184**:2603.

C. *E. coli* GadX footprint

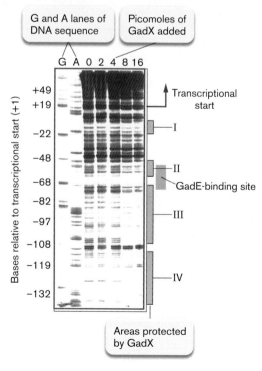

A.

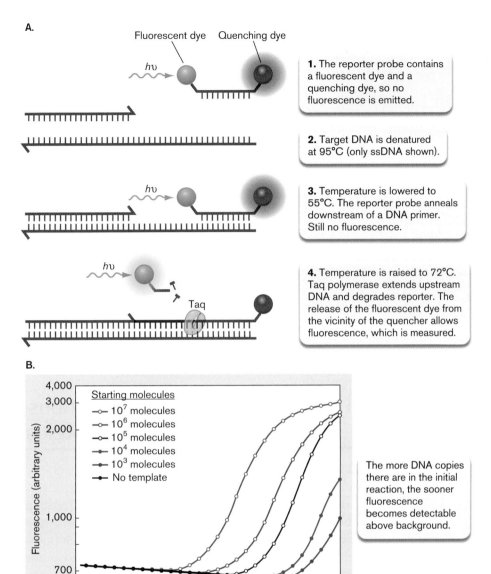

Fluorescent dye   Quenching dye

1. The reporter probe contains a fluorescent dye and a quenching dye, so no fluorescence is emitted.

2. Target DNA is denatured at 95°C (only ssDNA shown).

3. Temperature is lowered to 55°C. The reporter probe anneals downstream of a DNA primer. Still no fluorescence.

4. Temperature is raised to 72°C. Taq polymerase extends upstream DNA and degrades reporter. The release of the fluorescent dye from the vicinity of the quencher allows fluorescence, which is measured.

B.

Starting molecules
- 10^7 molecules
- 10^6 molecules
- 10^5 molecules
- 10^4 molecules
- 10^3 molecules
- No template

The more DNA copies there are in the initial reaction, the sooner fluorescence becomes detectable above background.

FIGURE 12.13 ■ Quantitative (real-time) PCR. A. The Taq DNA polymerase, extending an upstream primer, reaches the downstream reporter probe and degrades the probe, releasing the fluorescent dye from the vicinity of the quencher. **B.** Amplification plot of *Rhodococcus* with primer BPH4. Different numbers of DNA copies were used for each reaction mixture. ▶

to synthesize DNA from the upstream primer (**Fig. 12.13**). However, Taq polymerase also has 5′-to-3′ exonuclease activity. So, Taq will run into and cleave the reporter probe, separating the fluorescent dye from the quencher dye. Fluorescence is emitted. Meanwhile, primer extension by Taq polymerase continues to the end of the template, and the template is amplified. After each annealing cycle, more reporter probe binds to the newly made templates and is cleaved by Taq polymerase during each round of polymerization. As a result, fluorescence continues to increase as the amount of amplified template increases. The more DNA copies there are in the initial reaction, the sooner fluorescence becomes detectable above background.

A modification of qPCR, called reverse transcription quantitative PCR (RT-qPCR), measures the level of specific mRNA transcripts in the cell. In this procedure, (discussed in Chapter 28) cellular mRNA is converted into cDNA with reverse transcriptase and then quantitative PCR quantifies the cDNA. RT-qPCR was recently used to identify which of 39 RNA helicase genes are induced by freezing in the Antarctic psychrophilic alga *Chlamydomonas* species ICE-L. *Chlamydomonas* are algae that grow in freezing temperature and give Antarctic snow an eerie red or green color (**Fig. 12.14**). RNA helicases are essential for unwinding double-stranded RNA molecules, especially in cold environments. The

colder the environment, the more important these enzymes become, but the harder it is for them to function. Chenlin Liu and Xiaohang Huang from the First Institute of Oceanography in China discovered that three RNA helicase genes in this organism were significantly induced after 5 hours of freezing treatment. The authors suggest a prominent role for these enzymes in the acclimation of *Chlamydomonas* to freezing temperatures.

added to the reaction mix. This dye binds to double-stranded DNA and fluoresces. The fluorescence emitted increases with the amount of double-stranded PCR product produced.

The second quantitative PCR procedure, sometimes called **TaqMan** (**Fig. 12.13** ▶) uses a reporter oligonucleotide probe containing a fluorescent dye on its 5′ end and a quencher dye on its 3′ end. As long as the probe remains intact, the quencher absorbs the energy emitted by the fluorescent dye—a process called **fluorescence resonance energy transfer** (**FRET**). The reporter probe does not itself prime DNA synthesis, but anneals to the target downstream of a priming oligonucleotide.

The priming oligonucleotide and the reporter probe both anneal to the target DNA sequence. Taq polymerase begins

Note: Be careful not to confuse "reverse transcription" and "real-time," both of which may be found in the literature abbreviated as "RT". In this book, "RT" never means "real-time", but stands instead for either "reverse transcription" (as in "RT-qPCR") or "reverse transcriptase".

FIGURE 12.14 ■ Antarctic red snow algae. Psychrophilic *Chlamydomonas* algae can grow at freezing temperatures. Their pigmented chlorophyll will color snow red or green, depending on the species.

SETH RESNICK/GETTY IMAGES

Thought Question

12.7 How would you have to modify standard quantitative PCR to quantify the level of a specific mRNA?

To Summarize

- **Northern blot technology** uses labeled DNA from a specific gene to probe the levels and sizes of RNA made from that gene. **Southern blot technology** uses labeled DNA from a specific gene to probe genomes for the presence of that gene.

- **Western blot technology** uses antibodies to detect the quantity and size of specific proteins in cell extracts.

- **Electrophoretic mobility shift assay (EMSA)** examines protein interactions with DNA. **DNA protection assays** reveal the bases in a DNA sequence protected by a DNA-binding protein.

- **Affinity chromatography** purifies fusion proteins for biotechnology purposes.

- **Primer extension analysis** uses labeled DNA probes and reverse transcriptase to identify the 5′ end of a specific transcript (mapping the transcriptional start site). **Differential RNA sequencing (dRNAseq)** reveals transcript start sites from an entire genome.

- **Quantitative (real-time) PCR** is used to quantify specific DNA or RNA molecules present in cell extracts.

12.3 Visualizing the Interactions and Movements of Proteins

When a cell encounters a change in its environment, the impact on the cell radiates along and between pathways affecting many different genes and proteins. Tracking these global regulatory influences can show how a species adjusts its physiology to better tolerate, or even flourish in, a new environment. Transcriptomic and proteomic techniques described in Chapters 7, 8, and 10 reveal the global changes in mRNA and protein levels that cells undergo when entering new environments. Here we will show how to reveal interaction networks between proteins and genes, and between proteins and proteins. We will also see how to track protein movements through cells.

Whole-Genome DNA-Binding Analysis: ChIP-seq

The techniques discussed in Section 12.2 are great for identifying a protein-DNA binding site where you already suspect that the protein binds. But is there a way to blindly determine all the sites in a genome to which a given protein binds? There is, and the basic process, called **ChIP sequencing (ChIP-seq)** is outlined in **Figure 12.15**. In the cell, the DNA-binding protein of interest (a transcription factor, TF) binds to all of its DNA target sites in the genome. The cells are treated with formaldehyde to covalently cross-link proteins to DNA. The DNA and its bound proteins are isolated and sheared into small fragments. Next, the transcription factor–DNA complexes are "fished" from the extract using tiny beads with

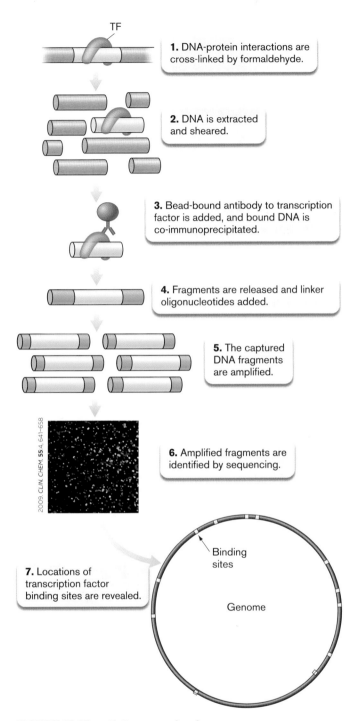

1. DNA-protein interactions are cross-linked by formaldehyde.

2. DNA is extracted and sheared.

3. Bead-bound antibody to transcription factor is added, and bound DNA is co-immunoprecipitated.

4. Fragments are released and linker oligonucleotides added.

5. The captured DNA fragments are amplified.

6. Amplified fragments are identified by sequencing.

7. Locations of transcription factor binding sites are revealed.

2009. CLIN. CHEM. 55:4, 641–658

Binding sites

Genome

FIGURE 12.15 ■ **ChIP-seq technology.**

are then sequenced by massively parallel DNA sequencing (next-generation sequencing, discussed in Section 7.6). The sequence identifies precisely where the transcription factor bound to the genome and which genes it likely controls.

Mapping the Interactome: Protein-Protein Interactions

Many cellular proteins interact with and influence the function of other proteins in the cell. For example, anti-sigma factors help control the timing of sporulation in *Bacillus* by interacting with sigma proteins. Many proteins, such as RNA polymerase and the ribosome, function as components of multisubunit complexes. However, there are other intracellular situations in which protein complexes can simplify physiological processes. A complex containing fatty acid biosynthesis enzymes, for instance, allows intermediates to pass quickly from one subunit to another without having to diffuse through the cell. Thus, all the protein-protein interactions of a living cell, which are collectively referred to as the **interactome**, are essential to the normal workings of the cell. Unraveling these interactions is an invaluable way of understanding protein function. How, then, does one map an interactome?

Yeast two-hybrid analysis. It is important to probe protein-protein interactions in vivo, where the proteins actually work. An ingenious tool to measure in vivo protein-protein interaction is the two-hybrid system. Two-hybrid analysis can be used to mine for unknown "prey" proteins that interact with a known "bait" protein.

While there are many types of two-hybrid techniques, a classic example is the **yeast two-hybrid system** shown in **Figure 12.16**. In yeast two-hybrid analysis, genes encoding two potentially interacting proteins are fused to separated parts of the yeast GAL4 transcription factor, which is normally one protein (**Fig. 12.16A**). One gene is fused to the GAL4 activation domain, while the other gene is fused to the GAL4 DNA-binding domain. If the two proteins of interest interact with each other, then the two parts of GAL4 are brought together. The DNA-binding domain binds to the yeast *GAL1* gene promoter region, and the activation domain (towed behind by the interacting proteins) is positioned to bind RNA polymerase and stimulate *GAL1* transcription.

The target reporter gene is typically a chromosomal *GAL1-lacZ* transcriptional fusion, although other reporter genes can be used. When the two hybrid proteins interact, the complex activates transcription of *GAL1-lacZ*, which is visualized on agar plates containing X-Gal. X-Gal is a colorless substrate of beta-galactosidase that, when cleaved, produces a blue product.

The power of this technique is that it can also be used to find unknown proteins that interact with a known protein. A known protein fused to one of the GAL4 domains

antibodies attached that bind only to that particular TF protein. The beads are pelleted by centrifugation, which "pulls down" antibody bound to the TF protein–DNA complexes. Unbound DNA fragments are washed away. This process is called **chromatin immunoprecipitation**, or **ChIP**.

Next, the protein-DNA cross-links are removed (usually by heat), and the released DNA fragments are amplified by PCR. (To amplify the unknown fragments, the fragments are ligated at both ends to linker oligonucleotides that can be amplified by known primers containing fluorescent dye.) The fluorescently tagged, amplified fragments

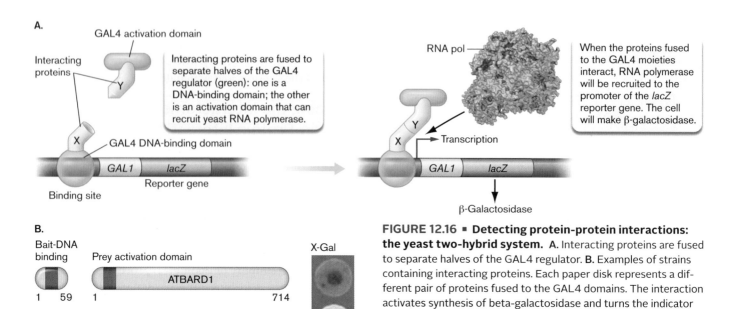

A.

B.

FIGURE 12.16 ■ **Detecting protein-protein interactions: the yeast two-hybrid system.** **A.** Interacting proteins are fused to separate halves of the GAL4 regulator. **B.** Examples of strains containing interacting proteins. Each paper disk represents a different pair of proteins fused to the GAL4 domains. The interaction activates synthesis of beta-galactosidase and turns the indicator blue. Proteins in the first pair interacted, but removing 128 amino acids from the prey construct destroyed the interaction. *Source:* Modified from W. Reidt et al. 2006. *EMBO J.* **25**:4326–4337.

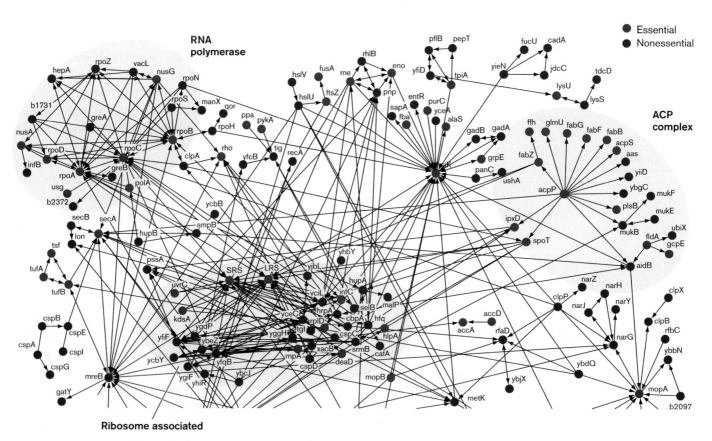

FIGURE 12.17 ■ **Protein interaction network of *E. coli.*** Partial predicted network. Each small, filled circle represents a gene or protein. Red and blue nodes indicate essential and nonessential proteins, respectively. Interconnecting lines indicate interaction between two proteins. Large, yellow circles reflect interactions taking place with RNA polymerase and acyl carrier protein, or ACP (fatty acid biosynthesis). *Source:* From Gareth Butland et al. 2005. *Nature* **433**:531–537.

can be used as "bait" to find the "prey" proteins expressed by randomly cloned "prey" genes fused to the other GAL4 domain. Plasmids containing the bait and prey fusions are transformed into yeast cells, and the transformants are plated onto a medium containing X-Gal. Colonies that express beta-galactosidase are the result of protein-protein interactions between the bait and prey fusion proteins. Sequencing the DNA insertion can then identify the gene encoding the prey protein. An example of two-hybrid analysis is shown in **Figure 12.16B**.

Another approach to mapping interactomes uses individual bait proteins to pluck <u>all</u> of a particular protein's interacting partners out of the cell. A small-peptide affinity tag added to one end of the bait protein will allow the bait protein and its interacting proteins to be purified by affinity chromatography. The process is described in **eTopic 12.1**. A graphical representation of part of an interactome is shown in **Figure 12.17**.

> ### Thought Question
>
> **12.8** You suspect that proteins A and C can simultaneously bind to protein B, and that this interaction then allows A to interact with C in the complex. How could you use two-hybrid analysis to determine whether these <u>three</u> proteins can interact?

Tracking Cells with Light

The use of green fluorescent protein (GFP) to track the location of plasmids and nucleoids in living bacteria is described at the start of Chapter 7. GFP technology can also be used to target how microbial proteins move through eukaryotic cells. For instance, the potato virus X (or potexvirus) is a positive-strand RNA virus (a positive-strand RNA is equivalent to mRNA). One viral protein, called TBGp3, is required for the virus to move between plant cells. Investigators fused GFP to the TBGp3 protein and asked where the protein ended up in the plant leaves. They delivered the genes to plant cells by bombarding leaves with the fusion plasmids. As shown in **Figure 12.18**, GFP alone

was diffusely located throughout the leaf. However, the GFP-TBGp3 fusion protein inserted only into the reticulated network of veins. Altering specific residues within the protein via mutation prevented this localization, causing the fusion protein to remain diffused throughout the plant cells. These results suggest that TBGp3 protein helps guide the virus through the plant's reticulated network.

This type of gene fusion technology has been used extensively to trace the movement and final location of numerous proteins delivered into host cells by pathogenic bacteria. GFP tagging has also been used to track the intermingling of different bacteria forming a biofilm (**eTopic 12.2**).

To Summarize

- **ChIP-seq technology** can identify all of the sequences in a genome to which a given protein will bind.

- **Two-hybrid analysis** uses hybrid proteins to detect protein-protein interactions, and can be used to build protein interaction maps.

- **A protein's intracellular location or movement** can be visualized by fusing it to green fluorescent protein (GFP).

12.4 Applied Biotechnology

It is remarkable how many useful products can be imagined and are being developed using the tools of biotechnology (see Chapter 16). You may be surprised to learn that plants have been engineered to produce microbial vaccines, and that vaccines can be delivered not only by injecting a protein, but by simply injecting DNA or RNA that encodes the antigenic protein (DNA vaccines are discussed in **eTopic 12.3**). We even have technology that uses a genetically engineered virus to repair a human genetic defect (**eTopic 12.4**). These are all remarkable additions to our scientific toolbox and are already helping to improve human life.

Bacterial Genes Save Crops

Protecting crops such as corn and cotton from hungry insects is an age-old problem. Insects such as moths or beetles can cause widespread damage to a crop in a short period of time. In the nineteenth and twentieth centuries, chemical pesticides were widely used to kill these pests. While generally effective, chemical pesticides can contaminate soil and groundwater and can impact human health. Through bioengineering, we now have a better, safer way.

A.
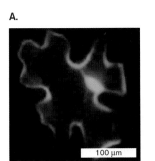

B.
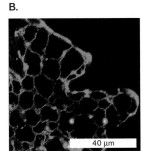

C.
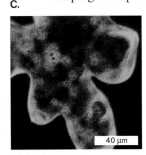

FIGURE 12.18 ■ **Using GFP protein to track the potato virus protein TBGp3 in a plant leaf. A.** Plasmid containing GFP only. Fluorescence is distributed throughout the leaf. **B.** Plasmid containing GFP-TBGp3 fusion. Fluorescence is restricted to the reticulated network of the leaf. **C.** Plasmid containing mutant GFP-TBGp3 fusion gene. The mutant protein is not properly targeted to the reticulated network. *Source:* From K. Krishnamurthy et al. 2003. *Virology* **309**:135.

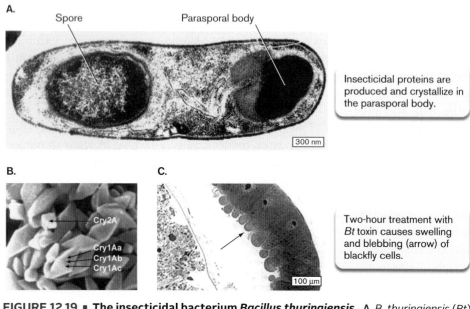

A.

Spore   Parasporal body

Insecticidal proteins are produced and crystallize in the parasporal body.

300 nm

B.   **C.**

Cry2A
Cry1Aa
Cry1Ab
Cry1Ac

Two-hour treatment with *Bt* toxin causes swelling and blebbing (arrow) of blackfly cells.

100 μm

FIGURE 12.19 ■ **The insecticidal bacterium *Bacillus thuringiensis*. A.** *B. thuringiensis* (*Bt*) in the process of sporulation. **B.** Parasporal crystals (Cry1Aa, Cry1Ab, Cry1Ac, and Cry2A) from *B. thuringiensis* subsp. *kurstaki* are used to control caterpillar pests. **C.** Effect of *Bt* toxin on midgut columnar cells of blackfly larvae. *Source:* Parts A and B courtesy of Brian Federici, American Academy of Microbiology; part C from C. F. Cavados et al. 2004. *Mem. Inst. Oswaldo Cruz* **99**:493–498.

other sensitive sites where chemical pesticides can cause adverse effects.

Although spreading *B. thuringiensis* on crops is helpful, an even more effective means of delivering the toxin has been developed. The genes encoding insecticidal proteins have been spliced right into the genomes of agriculturally important plants such as cotton and corn. The gene is placed under the control of a plant promoter, so that the transgenic plant makes its own insecticide. As of 2013, over 75% of the cotton and corn planted in the United States contains *B. thuringiensis* genes. Rigorous scientific studies have shown that foods from these transgenic plants are safe.

Bacillus thuringiensis (*Bt*), a close relative of *Bacillus anthracis*, sporulates on the surfaces of plants. In addition to the spore, the sporulating cell makes a separate parasporal body that cradles the spore (**Fig. 12.19**). The parasporal body contains crystallized proteins that are toxic to insects feeding on the plant. A single subspecies of *B. thuringiensis* can produce multiple insecticidal proteins. After the insect or insect larva ingests the insecticidal protein crystals, the alkaline environment in the insect midgut dissolves the crystals, and insect proteases inadvertently activate the proteins. Activated insecticidal proteins insert into the membrane of the midgut cells and form pores that lead to a loss of membrane potential, cell lysis, and death of the insect through starvation. It is also proposed that death is due to septicemia by enteric bacteria that escape the damaged midgut (antibiotic treatment reduces death). Because of its effectiveness, farmers have taken to spreading *B. thuringiensis* directly on their crops, where even dead cells are effective.

B. thuringiensis is considered a highly beneficial pesticide with few downsides. Unlike most insecticides, *B. thuringiensis* insecticides do not have a broad spectrum of activity, so they do not kill animals or beneficial insects—including the natural enemies of harmful insects (predators and parasites) and beneficial pollinators (such as honeybees). Therefore, *B. thuringiensis* integrates well with other natural crop controls. Perhaps the major advantage is that *B. thuringiensis* is nontoxic to people, pets, and wildlife, so it is safe to use on food crops or in

Thought Question

12.9 Would insect resistance to an insecticidal protein be a concern when developing a transgenic plant? Why or why not? How would you design a transgenic plant to limit the possibility of insects developing resistance?

Vaccine Proteins Produced in Plants

Starting with the war against smallpox in the late 1700s, vaccines have been used to inoculate humans against many terrible diseases (see Chapters 25 and 26). However, the cost of producing them is enormous, and the price, of course, is passed on to those who are vaccinated or to governments that conduct vaccination programs. Because the cost is problematic for impoverished nations, 20% of the world's infants are not vaccinated properly, resulting in over 2 million preventable deaths annually.

One potential solution to this problem was to engineer plants to express vaccine proteins. The gene encoding a vaccine protein is inserted into the genome of a plant so that the plant makes the vaccine protein as part of itself. The vaccine gene must be placed under the control of a well-expressed plant promoter (recently, plastid genes have been used). Transgenic plants expressing bacterial or viral virulence proteins can be used in two ways: They can be direct vaccine delivery systems, as through edible fruits and vegetables (such as potatoes, tomatoes, and corn);

or, as with transgenic tobacco crops, they can be a cheap way to grow large amounts of the vaccine protein that can be extracted later and used in more conventional ways. Producing vaccine in edible fruits was an attractive idea because the fruit could be grown locally and distributed to residents of poor countries even without refrigeration. But practical and ethical problems have, so far, prevented implementation. The use of engineered plants to mass-produce and purify vaccine proteins, however, especially for use in animals, is happening. Plants used to make vaccines are listed in **Table 12.2**.

Phage Display Technology

Phage display is a technique in which DNA sequences encoding nonphage peptides are cloned into a phage capsid gene. These peptides are synthesized as part of the capsid protein and then "displayed" on the surface of the phage (**Fig. 12.20A**). The DNA sequences can code for random

TABLE 12.2	Potential Vaccine Antigens Expressed in Plants	
Source of protein	**Vaccine protein or peptide**	**Plant**
Enterotoxigenic *E. coli*	Heat-labile enterotoxin B subunit (LT-B)	Tobacco, potato, tobacco chloroplast, maize kernels
Vibrio cholerae	Cholera toxin B subunit (CT-B)	Potato, tobacco chloroplast
Hepatitis B virus	Hepatitis B surface antigen	Tobacco, potato
Norwalk virus	Norwalk virus capsid protein (NVCP)	Tobacco, potato
Rabies virus	Glycoprotein	Tomato
Foot-and-mouth disease virus	Viral protein I	Alfalfa
Clostridium tetani	TetC	Tobacco chloroplast

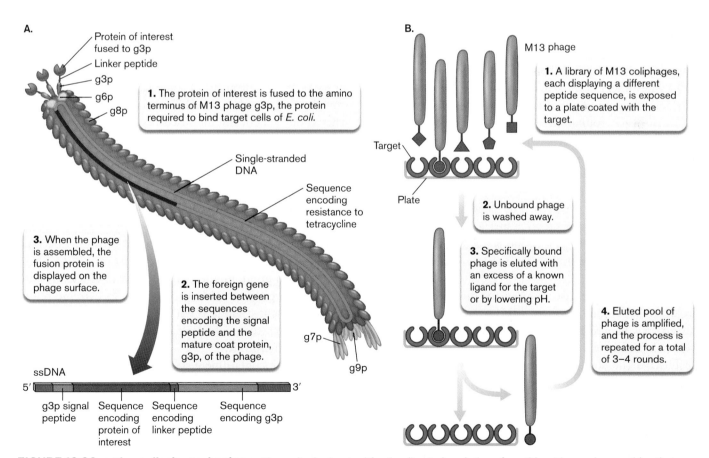

FIGURE 12.20 ■ Phage display technology. Phage display is a tool for the directed evolution of peptides. Discovering peptides that bind to specific target proteins is simplified by phage display techniques. **A.** Construction of the phage display protein. The product of gene 8 (g8p) is a small 5.2-kDa protein that forms the cylinder of the capsid. The other coat proteins (g3p, g6p, g7p, and g9p) cap the ends of the cylinder. The g3p protein is an attachment protein. **B.** Biopanning a library of M13 display phage for a peptide that specifically binds to a target molecule.

peptides, antibodies cloned from natural sources, or libraries (collections) of mutant enzymes that one wishes to study. Phage display enables scientists to screen for peptide variants with altered properties such as increased affinity for ligand.

The phage vectors used most often are the filamentous bacteriophages fd and M13 (discussed in **eTopic 11.2**). These phages infect *E. coli* strains containing the F plasmid (the phages attach to the sex pili) and replicate without killing the host cell. The phage particle consists of a single-stranded DNA molecule encapsulated in a long, cylindrical capsid made up of five proteins. The product of gene 3 (g3p)—present in three to five copies—confers phage infectivity. Most phage display libraries have been produced by cloning into gene 3.

Peptides displayed on filamentous phages have been used to select peptides, including antibodies, with high affinities for receptors or antigens (see Section 24.2). Recombinant phages that display high-affinity binding peptides are isolated from libraries in a process called biopanning (**Fig. 12.20B**). In biopanning, the target ligand—for example, a human virus attachment protein—is attached to a plastic plate or other support, and phages displaying random peptides are added. The phages that do not stick to the human virus protein are washed away. Those that do bind are subsequently released by adding free ligand (for example, human virus protein). This process is analogous to the way miners would pan rivers for gold. Several rounds of biopanning are required to obtain high-affinity clones. Following each round, the selected phages are enriched by repropagation in *E. coli*. Phage display was successfully used to identify high-affinity peptides that bind to attachment proteins of some eukaryotic viruses. These virus-binding peptides effectively block viral infection. More recently, phage display was used to identify specific single-chain antibodies that bind to a surface protein uniquely present on cancer cells. The antibody disrupts the function of this protein and significantly decreases viability of the cancer cell.

Phage display can also be used for directed evolution projects that retool enzymes (directed evolution is discussed in Chapter 17). Natural enzymes, because they were selected by evolution to sustain life, do not necessarily have the stability or catalytic activity that would make them suitable for more harsh biotechnological applications. Phage display, with its ability to screen millions of mutants, can quickly find the variant with the desired trait. An example of how phage display was used to direct the evolution of a more effective beta-lactamase enzyme that cleaves penicillin is described in **eTopic 12.5**. Other in vitro evolution techniques—DNA shuffling and site-directed mutagenesis—are described in **eTopics 12.6 and 12.7**.

Thought Question

12.10 You have just employed phage display technology to select for a protein that tightly binds and blocks the eukaryotic cell receptor targeted by anthrax toxin. When this receptor is blocked, the toxin cannot get into the target cell. You suspect that the protein may be a useful treatment for anthrax. How would you recover the gene following phage display and then express and purify the protein product?

In addition to what we describe here, microbes have been genetically engineered for many other applications that we will discuss in later chapters. These applications include the bioremediation of organic wastes (Chapter 13) and inorganic wastes (Chapter 14); mine leaching (Chapter 14); engineering of producer microbes (Chapter 16); and oil-spill cleanup, mercury removal, and wastewater treatment (**eTopic 22.2**).

Biocontainment: "Who Let the Bugs Out?"

Since the advent of genetic engineering in the 1970s, scientists and ethicists have struggled with how to prevent the accidental escape of a genetically engineered microbial "Frankenstein" from the lab. Physical containment such as air locks at the entrances of some labs is necessary to prevent the escape of lethal pathogens (see Section 28.6), but what about genetically modified organisms (GMOs)? The potential consequences of releasing GMOs into an environment are largely unknown. A GMO designed for laboratory use but accidentally released into the natural environment might alter ecological balances through competition or by sharing modified genes with other organisms in the environment. A highly improbable, but not inconceivable, concern is that a GMO microbe could evolve to become pathogenic to plants, animals, or humans. With these unknowns in mind, might there be a way to efficiently contain these organisms without resorting to extreme physical containment strategies? For instance, could we engineer the organisms to quickly self-destruct if they escape the laboratory environment?

Existing biocontainment strategies employ natural auxotrophies that are not easily met in nature, or conditional suicide switches introduced into the organism's genome (described in Section 12.5). But many of these mechanisms can be compromised through metabolic cross-feeding or genetic mutation. However, a promising new approach involving a bioengineered *synthetic* auxotrophy has emerged from the Yale University laboratory of Farren Isaacs. This team of scientists altered the genetic code of an *E. coli* strain to require an unnatural amino acid found only in laboratories. To recode the bacterium, they replaced all UAG stop codons with UAA stop codons. They then converted UAG into a sense codon by introducing an orthologous aminoacyl-tRNA synthetase–tRNA pair derived from the archaean *Methanocaldococcus jannaschii*. The tRNA recognizes the UAG codon, and the amino acid binding pocket of the tRNA synthetase was modified to accept the unnatural amino acid *p*-azido-L-phenylalanine (pAzF). The new tRNA synthetase will charge the UAG tRNA with pAzF.

Finally, UAG codons were introduced into three essential genes encoding DnaA (initiation of replication), MurG (peptidoglycan synthesis), and SerS (serine aminoacyl tRNA synthetase). For this strain to grow, the unnatural amino acid pAzF must be incorporated by translation into those three essential proteins. Because pAzF is not found in the environment, any recoded microbe released from the lab will die and not pose a threat. Synthetic auxotrophies like the one just described will allow us to control when and where genetically engineered microbes of unknown risk can grow.

To Summarize

- **Genes encoding insecticidal proteins** from *Bacillus thuringiensis* have been engineered into plant genomes to protect crops from insect devastation.

- **Proteins from pathogenic bacteria and viruses** cloned into plants may provide a cost-effective means to vaccinate large populations.

- **Genetically engineered viruses** are being tested as gene therapy vehicles to deliver DNA to correct inherited diseases in humans.

- **Phage display** is a form of in vitro evolution in which DNA sequences encoding random peptides are cloned into phage capsid protein genes. Phage particles displaying peptides with a desired binding property are pulled from the general phage population by biopanning.

- **Biocontainment** strategies are necessary to counteract an accidental release of a genetically engineered microbe into the natural environment.

12.5 Synthetic Biology: Biology by Design

What if we could, with all our knowledge of cells and genes and gene circuits, make bacteria do things they wouldn't ordinarily do, such as sense explosives in mines and then tell us about it? Or detect pathogens or keep time? Or

maybe even remember things—much like a computer? These are all goals of **synthetic biology**. Synthetic biology is the design and construction of new biological parts, devices, and systems for a desired purpose, similar to building electronic circuits.

How do we make these new parts or components? All living organisms already contain an instruction set encoded in DNA that determines what the creature looks like and what it does. Humans have been altering the genetic codes of plants and animals for millennia by selectively breeding individual plants and animals with desirable features. Today, scientists easily take pieces of genetic information from one organism and insert them directly into another, bypassing the need to breed. This is the basis of genetic engineering.

Recent technological advances now allow scientists to synthesize and manipulate DNA in ways never before possible. By applying engineering principles to these genetic manipulations, researchers can take components of genes from several different organisms, link them together like Lego blocks, and design new organisms that do new things. For instance, Jay Keasling's laboratory (UC Berkeley) used genes from several species to engineer a new pathway for artemisinin, an important antimalarial drug. By exploiting *E. coli*, Keasling's team eliminated the effort required to chemically synthesize a structurally complex molecule. In another example, Chang Li and colleagues (Harvard Medical School) engineered an *E. coli* that prevented cancer in mice. The new strain contained the invasin gene from *Yersinia pseudotuberculosis* and the hemolysin gene from *Listeria monocytogenes*. The invasin gene allowed the modified *E. coli* to invade mouse cells. The hemolysin enabled delivery of an inhibitory small RNA molecule that reduced expression of a tumor-initiating gene. Amazingly, this novel *E. coli* prevented cancer in this mouse model.

Principles of Synthetic Biology

The key to synthetic biology is engineering. Among the more visually intriguing ways to apply engineering principles to biological systems involves the use of GFP as a visual on-or-off output; that is, cells light up or go dark. If you build the biological circuit correctly, you can make a population of bacteria alert you to the presence of a toxic compound by flashing on and off (**Special Topic 12.1**).

What engineering principles are employed by synthetic-biology scientists? Many of them are borrowed from the field of electrical engineering and involve "logic gates." An electronic logic gate (as in semiconductors) receives a tiny current as an input and produces voltage as an output. A genetic logic gate is simply a promoter and a gene. An input signal such as a regulatory protein affects the

SPECIAL TOPIC 12.1   Bacteria "Learn" to Keep Time and Signal Danger

We know that bacteria "talk" to each other using signaling molecules, but with synthetic biology can we also train them to keep time? The answer appears to be yes. Jeff Hasty's laboratory at UC San Diego constructed a network circuit that relies on quorum sensing to propagate a signal through a population of cells (**Fig. 1**). The network design is shown in **Figure 2**. The organism, *E. coli*, was built to contain the *luxL* gene from *Vibrio fischeri* and the *aiiA* gene from *Bacillus thuringiensis*. LuxL synthase produces an acyl homoserine lactone (AHL) autoinducer (see Chapter 10); AiiA is a secreted enzyme that degrades extracellular AHL. Both *luxL* and *aiiA* have a *luxL* promoter that can be activated by LuxR bound to AHL. LuxR was also engineered into this synthetic organism, but it is constitutively expressed. The output signal is a form of GFP, so the cells glow when the circuit is on.

How does it work? **Figure 3** shows what happens when these cells are grown as a streak on an agar plate. The system produces a flash of emission that propagates bidirectionally along the streak and then shuts off. The flash starts at a point in the colony where AHL stochastically accumulates at a low level. The secreted AHL reenters nearby cells, where AHL binds LuxR and activates the expression of all three circuits in the cell cluster. Those cells then glow (see the 100-minute panel in **Fig. 3**). The new burst of AHL diffuses and activates more nearby cells, which also start to glow (the 106- and 118-minute panels). Because the AHL signal diffuses away from the producing cells, the light wave propagates through the streak. However, the AiiA enzyme that degrades AHL is also made and secreted by the glowing cells. AiiA degrades AHL and shuts the system off beginning where the light wave started (the 138-minute panel). The remaining GFP is degraded by cell proteases, and cells stop glowing. But now, without AHL, AiiA is no longer made. Consequently, small amounts of AHL can again accumulate at some point in the streak and the migrating flash begins again (the 170- and 180-minute panels).

FIGURE 1 ■ Synchronous-clock scientists. Jim Hasty (right) and Arthur Prindle examine a culture used for their synchronous-clock experiments.

The system was then modified to be a sensor for arsenic. For this purpose, *luxR* was placed under the control of a promoter repressed by ArsR. Arsenic inactivates the ArsR repressor. In the absence of arsenic, flashing was not seen, because LuxR is needed to activate the *luxP-gfp* fusion. When even small

promoter, which drives the output signal (mRNA and protein). The output signal for one gate can be an input signal for another part of the logic circuit. The final output signal of the circuit does something useful or eye-catching (think GFP fluorescence).

Many of the symbols used in building an electronic circuit are also used when building a complex genetic circuit. **Figure 12.21** shows three common logic gates used in synthetic biology and the symbols that represent them. The first is a "buffer gate" that amplifies signals (**Fig. 12.21A**). For a simple gene, the protein input activates a promoter;

a message is then transcribed and a protein is made. With a "NOT gate," a protein input <u>represses</u> a promoter and the output protein is <u>not</u> made (**Fig. 12.21B**). Finally, an "OR gate" involves several genes. For instance, two alternative gene output proteins can activate a third gene. So, in **Figure 12.21C**, the alternative input proteins I_1 and I_2 activate gene 1 or gene 2, respectively, to make product 1 or product 2 (Pr_1 or Pr_2). Either of those output signals can activate gene 3 to make product 3, Pr_3 (such as GFP). Once you understand how these gates work, you can make any kind of circuit.

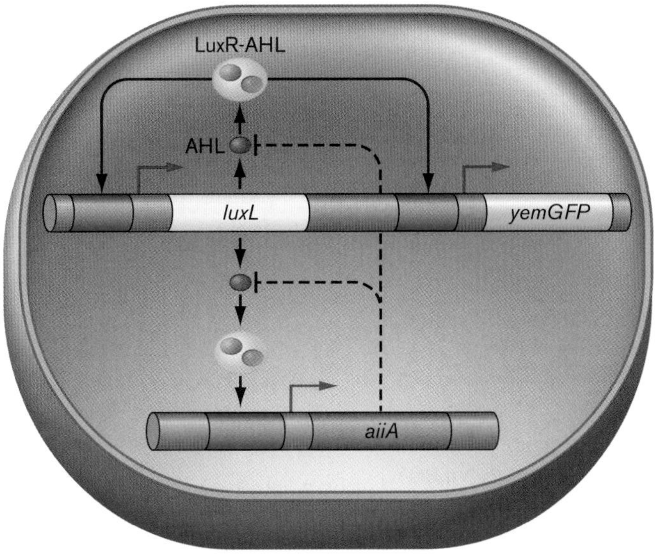

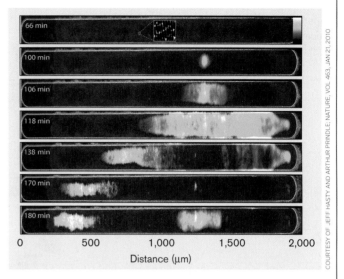

FIGURE 2 ■ **Network diagram for a synchronized oscillating clock.** Arrows indicate activation; "T" lines mark inhibition. Although it was not drawn this way, realize that AiiA protein is secreted and degrades only extracellular acyl homoserine lactone (AHL). This degradation stops AHL from accumulating extracellularly. The AHL concentration becomes too low to diffuse back into the cell, so *luxL*, *aiiA*, and *yemGFP* promoters are no longer activated.

FIGURE 3 ■ **Oscillatory flashing from engineered *E. coli*.**

COURTESY OF JEFF HASTY AND ARTHUR PRINDLE; NATURE, VOL. 463, JAN 21, 2010

amounts of arsenic were encountered, however, colonies trapped in multiple chambers of a microfluidic device begin to flash. (For a movie of this flashing, go to http://www.nature.com/nature/journal/v481/n7379/extref/nature10722-s3.mov.)

The authors successfully built a liquid crystal display (LCD)–like macroscopic clock that can sense arsenic and alert us to its presence. Given the vast sensing capabilities of bacteria, it now seems possible to build low-cost genetic bio-sensors able to detect heavy metals and even pathogens in the field.

RESEARCH QUESTION

Can you think of a potential pitfall with using the "arsenic alarm" in real life? How might you overcome this problem? *Hint:* Look up the effect of arsenic on bacterial cells. What does the ArsR protein regulate?

Prindle, A., P. Samayoa, I. Razinkov, T. Danino, L. S. Tsimring, et al. 2012. A sensing array of radically coupled genetic "biopixels." *Nature* **481**:39–44.

Toggle Switches

Electrical systems rely heavily on toggle switches to control whether a system is turned on or off. Similarly, synthetic-biology circuits depend on biological toggle switches. **Figure 12.22** illustrates a basic, genetically engineered toggle switch designed to control whether a *gfp* gene is turned on or off. The circuit involves two repressor genes (called NOT gates) whose products can repress each other's transcription. The switch depends on whether one of two different inducer signals is present. Inducer 1 inactivates repressor 1, which means repressor 2 is produced. Repressor 2, in turn, stops transcription of the repressor 1 gene and the reporter gene. So, when inducer 1 is added, GFP is <u>not</u> made and the system is stably toggled off. Alternatively, inducer 2 inactivates repressor 2, which means that the genes for repressor 1 and GFP are transcribed and the cell lights up. The system is stably toggled on because repressor 1 halts transcription of repressor 2. So, a genetic engineer can control the on/off switch of these cells by adding one or the other inducer. Of course, for this system to toggle, at least a small amount of both repressor proteins must be made at all times—just enough to bind to inducer molecules and have an effect on gene expression.

FIGURE 12.21 ■ Examples of gate symbols used in electronics, and their synthetic-biology equivalents. In (A), for example, when inducer is added, the transcript for Pr_1 is made. Without inducer, Pr_1 is not made.

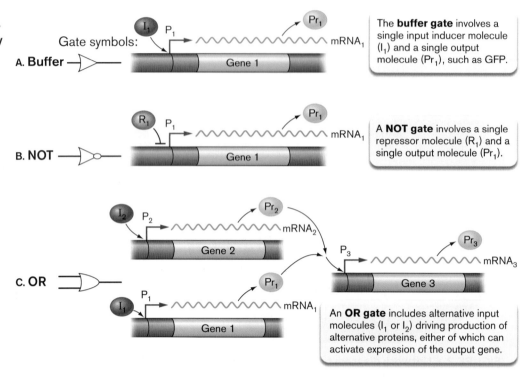

Gate symbols:

A. Buffer

The **buffer gate** involves a single input inducer molecule (I_1) and a single output molecule (Pr_1), such as GFP.

B. NOT

A **NOT gate** involves a single repressor molecule (R_1) and a single output molecule (Pr_1).

C. OR

An **OR gate** includes alternative input molecules (I_1 or I_2) driving production of alternative proteins, either of which can activate expression of the output gene.

Thought Question

12.11 What would happen with a toggle switch if the genetic engineer could set repressor 1 and repressor 2 protein levels to be <u>exactly</u> equal? Imagine that this is done without either inducer present.

A Bacterial Oscillator Switch

An oscillator is another important component of many electronic systems. In an electronic circuit, the oscillator produces a repetitive electronic signal viewed as a wave. Can synthetic-biology scientists also make an oscillating genetic circuit? **Figure 12.23A** shows a basic genetic oscillator switch designed by Jim Hasty's laboratory at UC San Diego. The components come from some of the systems discussed in Chapter 10. One component is the gene for the AraC activator protein, which was linked by Hasty's group to a hybrid operator region that included an *ara* activator sequence and a LacI repressor control sequence. The next component is the gene encoding the repressor LacI, which was also spliced to the hybrid *ara* activator/LacI repressor control sequences. The resulting circuit contains negative and positive feedback loops. The *lac* and *ara* operons are discussed in Section 10.2.

FIGURE 12.22 ■ Genetic toggle switch. Notice that repressor 1 and the reporter *gfp* are transcribed colinearly from promoter 2. Two outputs are possible, depending on which inducer is added. Adding inducer 1 stops fluorescence, whereas adding inducer 2 triggers fluorescence. What happens if you add them both?

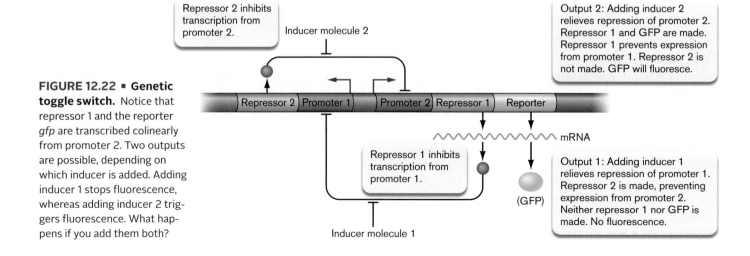

Repressor 2 inhibits transcription from promoter 2.

Inducer molecule 2

Output 2: Adding inducer 2 relieves repression of promoter 2. Repressor 1 and GFP are made. Repressor 1 prevents expression from promoter 1. Repressor 2 is not made. GFP will fluoresce.

| Repressor 2 | Promoter 1 | Promoter 2 | Repressor 1 | Reporter |

mRNA

Repressor 1 inhibits transcription from promoter 1.

Output 1: Adding inducer 1 relieves repression of promoter 1. Repressor 2 is made, preventing expression from promoter 2. Neither repressor 1 nor GFP is made. No fluorescence.

(GFP)

Inducer molecule 1

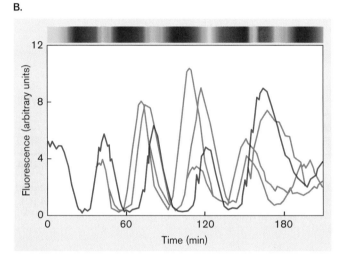

FIGURE 12.23 ■ A dual-feedback oscillator constructed in *Escherichia coli.* **A.** Network diagram. A hybrid promoter, P*lac/ara* (small pink and green boxes), drives the transcription of *araC* and *lacI*, forming positive and negative feedback loops. **B.** Single-cell fluorescence oscillations induced with 0.7% arabinose plus 2-mM IPTG (red line in graph) or 1-mM IPTG (gray lines). The points represent experimental fluorescence values. The colored bar across the top represents the intensity of the fluorescence (red = highest; blue = lowest). *Source:* Modified from J. Stricker et al. 2008. *Nature* **456**:516–519.

To see how this oscillator works, IPTG (the chemical that inactivates LacI) and L-arabinose (the sugar that activates AraC) are added at the same time. The increase in AraC production drives expression of GFP but also increases LacI repressor, which eventually represses *araC*. However, the IPTG <u>inactivates</u> LacI protein, which allows renewed *araC* and *lacI* expression. As shown in **Figure 12.23B**, the differential activity of the two feedback loops drives oscillation back and forth between fluorescence (on) and no fluorescence (off). Changing the concentrations of arabinose and IPTG will modulate the oscillation frequency, making the circuit tunable. **Special Topic 12.1** presents an example of an oscillatory genetic circuit that keeps time and senses arsenic. Reporter organisms equipped with this circuit could announce the presence of dangerous toxins such as arsenic in water, soil, or a variety of food materials.

System Noise

Before we discuss more complex circuits, we should first talk about "noise." **Noise** (variance from a mean) is a problem in any electrical circuit but also arises in biological circuits, because of fluctuations of gene expression among single cells in a population (see Section 12.1). The concept of noise in a biological or electrical system might be best explained by the following analogy. Imagine you are at a concert and your friend is trying to talk to you. You can't understand what she is saying, because the sound of her voice is not rising above the music and all of the background noise. The band would have to go quiet in order for you to hear her. The same is true for anything you want to measure. The less noise there is, the easier it is to measure what rises above it. In bacteria, translation inefficiency is a major contributor to noise, although many factors can influence the expression of a gene (**Fig. 12.24**). To be useful, synthetic biological circuits must operate with transcriptional and translational controls balanced in a way that minimizes noise. Choosing the right ribosome-binding sequence and promoter, as well as using a small RNA to control message stability or translation, can help to decrease noise.

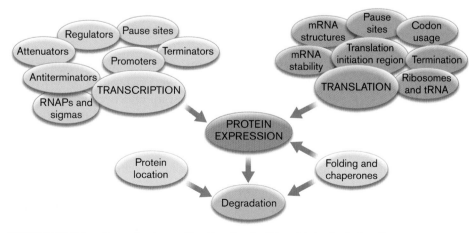

FIGURE 12.24 ■ Some factors affecting "noise" in a biological circuit.

Engineered Riboswitches and Switchboards

Chapter 10 described how the cell can use riboswitches to sense cell metabolites and control the translation of mRNA molecules. Synthetic-biology scientists seized upon this concept and have learned to tailor small RNAs, called synthetic riboswitches, to control the translation of nearly any gene they want. **Figure 12.25** reveals that a basic

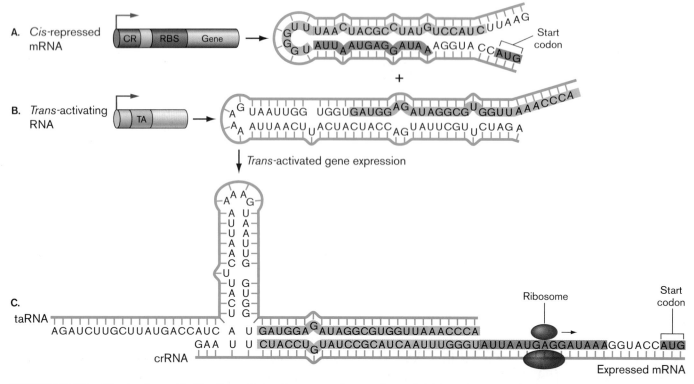

FIGURE 12.25 ■ **The basic synthetic riboswitch.** *Cis*-repressed mRNA (crRNA) folds to obscure a ribosome-binding site (RBS) needed to translate a downstream output gene **(A)**. *Trans*-activating RNA (taRNA), driven by a different promoter **(B)**, can base-pair with crRNA **(C)**, thus opening up the hairpin to expose the RBS. Whatever activates the taRNA will activate translation of the output genes' mRNA.

riboswitch consists of two parts: a *cis*-repressed mRNA (crRNA) and a *trans*-activating RNA (taRNA). The crRNA sequence is linked to an output gene and, by forming a hairpin, hides a ribosome-binding site to prevent translation. The taRNA, once it is made, will promote translation of the output gene by base-pairing with the crRNA. The base pairing releases the RBS for translation. Transcription and translation of the output gene can now be manipulated by whatever activates the promoters used to express the crRNA and taRNA genes. This process can reduce noise in a system.

Members of James Collins's laboratory (Harvard University) designed a series of matched crRNA/taRNA riboswitches and linked them to various promoters and output genes. The design and response of these circuits to various environmental signals is shown in **Figure 12.26**. Notice that each riboswitch responded to only one signal.

The Collins group then linked these riboswitches to genes encoding carbon metabolism enzymes and placed them all in the same cell, essentially making a metabolic switchboard that could channel carbon flow through alternative metabolic pathways depending on which riboswitch was activated. Metabolic switchboards would have important industrial applications. For instance, the device could simultaneously sense a variety of metabolic states in a large

batch fermentation system and maximize the efficiency of an industrially attractive pathway.

Kill Switches

Most of what we have described about synthetic biology has little chance of dangerous unintended consequences, but what about the long term? Biologists would like to engineer new strains that do new tasks, such as gobble up toxins in the environment. Remediating toxin contamination in nature, however, would require the release of genetically modified organisms outside the laboratory. This raises concerns about potential, unknown havoc these organisms might cause.

To allay these concerns, scientists must engineer failsafe mechanisms that kill the organism at a predetermined point. Hence the quest for effective "kill switches." A genetically modified organism equipped with a kill switch can be made to commit suicide once the bacterium's job is done. One proof-of-principle kill switch engineered by synthetic-biology techniques is shown in **Figure 12.27**.

The gene for CcdB, a potent DNA-damaging toxin, was linked to the promoter P_{LtetO}, which is repressed by the TetR protein. Repression is relieved (and CcdB is made) only when an analog of tetracycline is present. Tetracycline binds and inactivates TetR, which then releases from the *ccd* promoter. As a result, the kill toxin gene *ccdB* is expressed and can kill the cell. The engineers, however, added a fail-safe mechanism to prevent the organism from killing itself too early. They made sure that the translation of CcdB mRNA is blocked by a crRNA. A second component of the system, activated by arabinose, encodes the compensatory taRNA that can expose the RBS buried within the crRNA-CcdB message. So, both tetracycline and arabinose must be added for the cell to effectively produce CcdB and kill itself.

Although this system, as designed, is impractical for real-world use, its success shows that kill switches can be made. How could this kill switch be modified so that it could be used in the real world? What if repression of CcdB was tied to the presence of a toxic product found in the environment? Once the organism destroyed the toxin, the kill switch would be "thrown," and the bacterium would kill itself.

FIGURE 12.26 ■ Riboswitchboard.
A. Four separate circuits were designed. Each circuit is controlled by a different promoter that senses a different environmental parameter: P_{LuxI} (acyl homoserine lactone), P_{LlexO} (mitomycin C), P_{LfurO} (iron), or P_{MgrB} (magnesium). Different riboswitch pairs (labeled 42, 10, 12, and 12y) were used to control the translation of the output reporters for each circuit.
B–E. These graphs show that each circuit responded to only a single environmental input signal. Inducers for GFP, mCherry, and LacZ (B, C, and D) were added at time 0. The inoculum for the luciferase circuit (E) was grown in low Mg^{2+} and was active even at time 0.

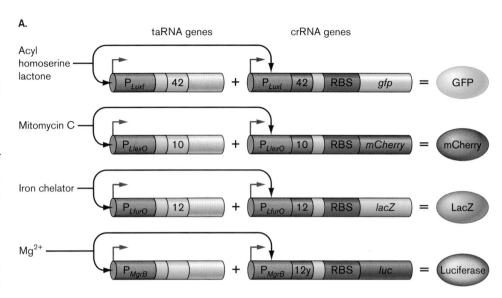

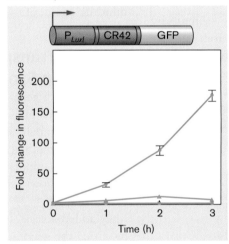

B. GFP expression

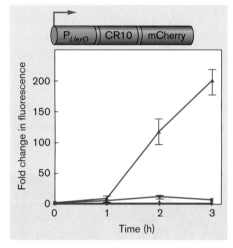

C. mCherry expression

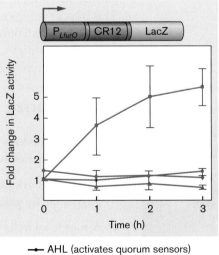

D. LacZ expression

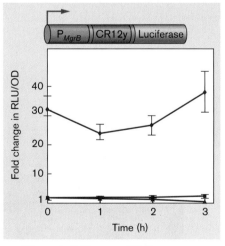

E. Luciferase expression

- AHL (activates quorum sensors)
- Iron chelator (promotes iron starvation)
- Mitomycin C (damages DNA)
- No $MgCl_2$ (promotes Mg^{2+} limitation)

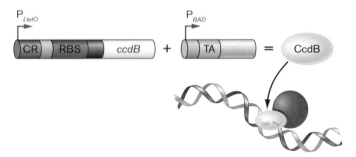

FIGURE 12.27 ■ A synthetic-biology "kill switch," or suicide module. The addition of tetracycline relieves TetR repression of the *ccdB* gene, but the gene is equipped with a *cis*-repressed RNA (CR) that prevents translation. The addition of L-arabinose enables AraC to activate the complementary *trans*-activating RNA gene (TA). The *ccdB* mRNA is translated, and CcdB causes DNA damage, which kills the cell.

BioBricks and Do-It-Yourself Synthetic Biology

The science of synthetic biology makes new logic circuits by linking promoters from one system to genes from another system. The art of synthetic biology is to combine these new genetic logic circuits in ways that produce new functional systems. But where do all the genetic "Lego-like" blocks come from? Laboratories around the world construct and deposit their "building materials" into a central registry at the Massachusetts Institute of Technology. An

FIGURE 12.28 ■ Winners of the 2015 iGEM competition Grand Prize trophy. The College of William & Mary team (mentored by Professor Margaret Saha) won Grand Prize at the 2015 competition for their project characterizing promoter-driven transcriptional noise in *E. coli*. **Top:** The BioBricks trophy.

annual weekend-long synthetic-biology showdown called the International Genetically Engineered Machine (iGEM) competition is also held at MIT. Competitors from universities over the years have assembled thousands of connectible pieces of DNA that they call BioBricks and have deposited them in the BioBricks Foundation registry at MIT (the iGEM competition trophy is a large aluminum Lego; **Fig. 12.28**). BioBricks range from those that kill cells to one that makes cells smell like bananas.

Despite its potential for good, one concern about synthetic biology is that anyone, theoretically, can do it. In fact, the lure of constructing a new organism, combined with the ever-lowering cost of equipment needed to carry out these experiments, has spawned a community of do-it-yourself (DIY) genetic engineers, including high school students and so-called garage scientists. Most of these DIY efforts are well-intentioned and could yield useful products as a result of outside-the-box thinking, but we should be mindful, again, of unintended consequences.

Genome Transplants: The Ultimate in Synthetic Biology

Tinkering with gene circuits is one thing, but carrying out a complete genome transplant is another. We have been able to manipulate small pieces of bacterial genomes for many years via mutation, cloning, and recombination techniques. But manipulating an entire genome has been out of reach—until now. Scientists have taken the genome of one bacterial species, *Mycoplasma mycoides*, moved it to yeast (they added a yeast centromere so that it would replicate), modified the genome using yeast genetics, and then transplanted the modified genome into a different species, *Mycoplasma capricolum*, replacing the *M. capricolum* genome. This biotechnological advance is especially exciting because genetic manipulation of *M. mycoides* was previously impossible. Using this technique, we might redesign bacterial systems and even engineer new species.

To Summarize

- **Synthetic biology** applies engineering principles to design and construct new biological parts for a desired purpose.

- **Synthetic-biology circuits use genetic logic gates**, such as buffer gates, NOT gates, and OR gates.

- **A toggle switch**, built from two NOT gates, can turn a gene on or off by sensing two different chemical signals.

- **An oscillating switch** will repeatedly turn on and off in response to a signal.

- **System noise** is the fluctuation of gene expression among single cells of a population. System noise can dull the clarity of a response circuit. **Riboswitches** engineered into a genetic circuit can minimize noise.

■ **Genetic kill switches** can eliminate a genetically modified organism when its job is done.

■ **Complete genome transplants** will enable the genetic manipulation of organisms in which traditional methods have failed.

Concluding Thoughts

In this chapter we have discussed only some of the available molecular techniques and developing technologies (such as synthetic biology) used to probe biological processes and manipulate them to our benefit. Keep them in mind as we progress deeper into the physiology of microbial growth and the mechanisms of microbial pathogenesis. Many of the approaches described here and in eAppendix 3 were used to elucidate the concepts presented in later chapters. Realize, too, that the power of biotechnology brings with it a sobering responsibility. If nature has taught us anything, it's that tampering with it is risky. For instance, what would happen if an *Escherichia coli* that produced human growth hormone—or worse yet, one that had been engineered to make botulism toxin—managed to colonize the human gut? While these possibilities may seem unlikely, future advances in biotechnology must be guided by serious considerations of bioethics and an eye toward unintended consequences.

CHAPTER REVIEW

Review Questions

1. What are the important considerations when designing a mutant selection strategy?
2. Once the DNA sequence of a gene is known, what specific methods can be used to gain clues as to the possible function of the gene product?
3. What are the key features of reporter genes used to measure transcriptional control? Translational control?
4. What is the difference between a northern blot and a Southern blot?
5. Explain the principle of a western blot.
6. What techniques are used to determine whether a protein binds to a DNA sequence?
7. Explain the important points for constructing a plasmid that makes His$_6$-tagged protein.
8. Explain how one can determine the start site of a transcript.
9. How can protein-protein interactions be determined in vitro?
10. Describe two-hybrid analysis. What are bait and prey proteins?
11. How can the cellular location of a protein be tracked in vivo?
12. Explain how PCR can be used in a single reaction to identify which of two species of pathogenic bacteria are present in a food sample.
13. Discuss how phage display might be used to produce a more toxic toxin. What are the ramifications of technologies such as this?
14. Discuss some ways biotechnology has helped the agricultural industry. How has the field impacted vaccine delivery or our approaches to the treatment of metabolic diseases in humans?
15. Describe how BioBricks and logic gates are used in synthetic biology.

Thought Questions

1. You have identified two regulatory proteins, RegA and RegB, which affect the expression of gene *yxx*, a gene involved in quorum sensing. Mutants lacking RegA or RegB fail to express *yxx*. Describe two alternative models that would explain this. Next, describe how results from ChIP-seq analysis could help distinguish between those possibilities.

2. The Gram-negative bacillus *Shigella dysenteriae*, the agent causing bacillary dysentery, will live and grow inside eukaryotic cells. Despite lacking flagella, the organism moves through the cytoplasm of the host cell by polymerizing host actin at one pole of the cell. This pushes the organism through the cytoplasm with such force that the bacterium will poke into an adjacent cell.

You have identified a *Shigella* protein that you think helps mediate actin tail formation and suspect that the protein might localize to one pole of the cell. What biotechnique would you use to test your hypothesis?

3. In Chapter 10 we discussed the *lac* operon and the *Salmonella* phase variation system in which DNA recombination controls a switch between two different flagellin proteins. You would like to understand more about the mechanism of regulation, so you have genetically engineered a strain of *E. coli* to contain the following genes: (1) the *hin* gene controlled by the *lacP* promoter, (2) a transcriptional activator gene flanked by *hix* sites, and (3) a green fluorescent protein gene (*gfp*) controlled by the transcriptional activator. Briefly describe what happens if (a) the strain is grown in lactose, (b) the strain is grown without lactose, (c) you delete the *lacI* gene, and (d) you eliminate the *lac* operator.

Key Terms

affinity chromatography (453)	interactome (459)	TaqMan (457)
autoradiography (452)	noise (469)	transcriptional fusion (448)
ChIP sequencing (ChIP-seq) (458)	northern blot (450)	translational fusion (448)
chromatin immunoprecipitation (ChIP) (459)	phage display (463)	vector (454)
differential RNA sequencing (dRNAseq) (454)	primary antibody (452)	western blot (451)
	primer extension (454)	yeast two-hybrid system (459)
electrophoretic mobility shift assay (EMSA) (453)	quantitative PCR (qPCR) (456)	
	reporter gene (447)	
fluorescence resonance energy transfer (FRET) (457)	secondary antibody (452)	
	Southern blot (451)	
	synthetic biology (465)	

Recommended Reading

Ahmad, Parvaiz, Muhammad Ashraf, M. Muhammad Younis, Xiangyang Hu, Ashwani Kumar, et al. 2012. Role of transgenic plants in agriculture and biopharming. *Biotechnology Advances* **30**:524–540.

Berens, Christian, Florian Groher, and Beatrix Suess. 2015. RNA aptamers as genetic control devices: The potential of riboswitches as synthetic elements for regulating gene expression. *Biotechnology Journal* **10**:246–257.

Bouveret, Emmanuelle, and Christine Brun. 2012. Bacterial interactomes: From interactions to networks. *Methods in Molecular Biology* **804**:15–33.

Ebina, Hirotaka, Naoko Misawa, Yuka Kanemura, and Yoshio Koyanagi. 2013. Harnessing the CRISPR/Cas9 system to disrupt latent HIV-1 provirus. *Scientific Reports* **3**(art. 2510). doi:10.1038/srep02510.

Huang, Johnny X., Sharon L. Bishop-Hurley, and Matthew A. Cooper. 2012. Development of anti-infectives using phage display: Biological agents against bacteria, viruses and parasites. *Antimicrobial Agents and Chemotherapy* **56**:4569–4582.

Locke, James C., Jonathan W. Young, Michelle Fontes, Maria J. Hernandez Jimenez, and Michael B. Elowitz. 2011. Stochastic pulse regulation in bacterial stress response. *Science* **334**:366–369.

Martínez-Garcia, Esteban, and Victor de Lorenzo. 2012. Transposon-based and plasmid-based genetic tools for editing genomes of Gram-negative bacteria. *Methods in Molecular Biology* **813**:267–283.

Mimee, Mark, Alex C. Tucker, Christopher A. Voigt, and Timothy K. Lu. 2015. Programming a human commensal bacterium, *Bacteroides thetaiotaomicron*, to sense and respond to stimuli in the murine gut microbiota. *Cell Systems* **1**:62–71.

Palmieri, Dario, Timothy Richmond, Claudia Piovan, Tyler Sheetz, Nicola Zanesi, et al. 2015. Human anti-nucleolin recombinant immunoagent for cancer therapy. *Proceedings of the National Academy of Sciences USA* **112**:9418–9423.

Qi, Lei S., and Adam P. Arkin. 2014. A versatile framework for microbial engineering using synthetic noncoding RNAs. *Nature Reviews. Microbiology* **12**:341–354.

Rajagopala, Seesandra V., Patricia Sikorski, Ashwani Kumar, Roberto Mosca, James Vlasblom, et al. 2014. The binary protein-protein interaction landscape of *Escherichia coli*. *Nature Biotechnology* **32**:285–290.

Rovner, Alex J., Adrian D. Haimovich, Spencer R. Katz, Zhe Li, Michael W. Grome, et al. 2015. Recoded organisms engineered to depend on synthetic amino acids. *Nature* **518**:89–93.

Ruder, Warren C., Ting Lu, and James J. Collins. 2011. Synthetic biology moving into the clinic. *Science* **333**:1248–1252.

Saade, Fadi, and Nikolai Petrovsky. 2012. Technologies for enhanced efficacy of DNA vaccines. *Expert Review of Vaccines* **11**:189–209.

Selle, Kurt, and Rodolphe Barrangou. 2015. Harnessing CRISPR–Cas systems for bacterial genome editing. *Trends in Microbiology* **23**:225–232.

Trimble, Cornelia L., Shiwen Peng, Ferdynand Kos, Patti Gravitt, Raphael Viscidi, et al. 2009. A phase I trial of a human papillomavirus DNA vaccine for HPV16+ cervical intraepithelial neoplasia 2/3. *Clinical Cancer Research* **15**:361–367.

Uhlig, Christiane, Fabian Kilpert, Stephan Frickenhaus, Jessica U. Kegel, Andreas Krell, et al. 2015. In situ expression of eukaryotic ice-binding proteins in microbial communities of Arctic and Antarctic sea ice. *ISME Journal* **9**:2537–2540.

Zhang, Weiwen, and David R. Nielsen. 2014. Synthetic biology applications in industrial microbiology. *Frontiers in Microbiology* **5**(art. 451).

CHAPTER 13
Energetics and Catabolism

To grow and multiply, all living cells need energy. Energy-yielding reactions such as photolysis and catabolism enable cells to build biomass and grow. Catabolism is the step-by-step process of breaking down complex molecules into smaller ones. Microbes catabolize food molecules within our own digestive tract and in the soil and water all around us. Our uses of microbial catabolism range from producing alcohol to remediating hazardous wastes, from our local wastewater plant to remote stations of Antarctica.

CURRENT RESEARCH highlight

Gut bacteria share their catabolic enzymes. *Bacteroides fragilis* are prominent Gram-negative bacteria of our colonic microbiome. Colonic bacteria compete intensely for the carbon sources ingested by the host. But *Bacteroides* species actually release their catabolic enzymes into the community, by pinching off vesicles from the outer membrane. These outer membrane vesicles contain enzymes that break down lipids, carbohydrates, and peptides into small molecules. Vesicle formation increases the surface area for catabolism, but it also makes the catabolites available to unrelated bacteria. This unexpected resource sharing is important for the gut microbiome.

Source: Wael Elhenawy et al. 2014. *mBio* **5**:e00909-14.

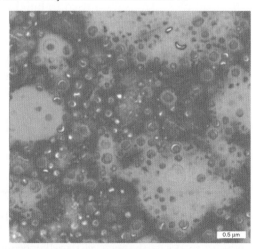

0.5 µm

AN INTERVIEW WITH

WAEL ELHENAWY, MICROBIOLOGY GRADUATE STUDENT, UNIVERSITY OF ALBERTA

MOHAMED ADEL, FROM UNIVERSITY OF ALBERTA

What is so exciting about intestinal bacteria exporting catabolic enzymes in vesicles?

Bacteroides species digest a broad spectrum of polysaccharides in the human gut, including host-derived glycans. We showed that the bacteria selectively secrete a large number of sugar hydrolases via outer membrane vesicles. The liposomal nature of vesicles stabilizes the secreted hydrolases compared to the soluble form. Moreover, the secretion of hydrolases into vesicles enables the bacteria to use out-of-reach nutrients.

How do you think our assumptions will change about patterns of catabolism in the microbiome?

Bacteroides fragilis can regulate the level of hydrolases packed into outer membrane vesicles in response to the kinds of glycans available. These results suggest that outer membrane vesicle secretion is a regulated process that helps *Bacteroides* adapt to its niche. Other researchers have shown that *Bacteroides* vesicle hydrolases also target key molecules in the host signaling pathways. In addition, vesicle sulfatases play a fundamental role in the inflammatory response triggered by *Bacteroides thetaiotaomicron* during colitis.

The intestines of humans and other animals support vast communities of microbes that digest much of the food we consume. Many of our gut microbes break down food molecules using enzymes that our bodies don't have—thus extending the repertoire of our genome. The microbes grow as mutualists, while our bodies provide a home in return for breakdown of complex foods into small molecules that we can digest. Microbes **catabolize** (break down for energy) many carbon sources that animals cannot. For example, soil bacteria such as *Pseudomonas* species may catabolize benzene and chlorinated pollutants.

But who would have thought that digestive bacteria are also mutualists of each other? We have long assumed that the vast number of intestinal microbes compete with each other for food. Yet recent research shows that bacteria such as *Bacteroides* species actually share their catabolic enzymes with the community. The enzymes are shared by vesicles pinched off from the bacterium's outer membrane (see the Current Research Highlight). The outer membrane (discussed in Chapter 3) contains enzyme complexes that break down polysaccharides to oligosaccharides (short sugar chains) and transport the oligosaccharides to the periplasm (**Fig. 13.1**). These enzymes and transporters comprise the "starch utilization system" (SUS). Within the periplasm, oligosaccharides are further degraded to monosaccharides, which then cross the inner membrane to the cytoplasm. At the same time, some of the outer membrane pinches off to form outer membrane vesicles. These vesicles include some of the SusD starch-binding proteins, as well as glycosidases (sugar breakdown), lipases (fat digestion), and peptidases (protein breakdown). Their breakdown products diffuse back to *Bacteroides* cells, as well as to other bacteria.

Why share catabolic enzymes? Amazingly, *Bacteroides* species possess hundreds of SUS homologs for polysaccharide catabolism and transport. **Catabolism** is reactions that break down complex organic substrates (discussed in Section 13.4). Different versions of SUS proteins catabolize different types of polysaccharides, such as those with sulfated sugars. Suppose two different species possess enzymes for different steps of a pathway. Individually, the two bacteria might not digest a certain substrate, but together they may complete its breakdown. The surprising nutritional cooperation of our microbiome is discussed further in Chapter 21.

Chapter 13 explains how catabolism and other kinds of energy-yielding reactions enable microbes to do the work of growing new cells. The energy released from a reaction is transferred by enzymes to other reactions that build simple molecules into a complex cell. Chapter 14 explores electron transfer through pathways of organic respiration (oxidation of organic nutrients), lithotrophy (oxidation of inorganic nutrients), and photosynthesis. The energy from all these pathways is used to build cells by **anabolism** (biosynthesis), which is the focus of Chapter 15. Finally, Chapter 16 explores commercial applications of microbial metabolism to produce food and beverages, as well as industrial products and pharmaceuticals.

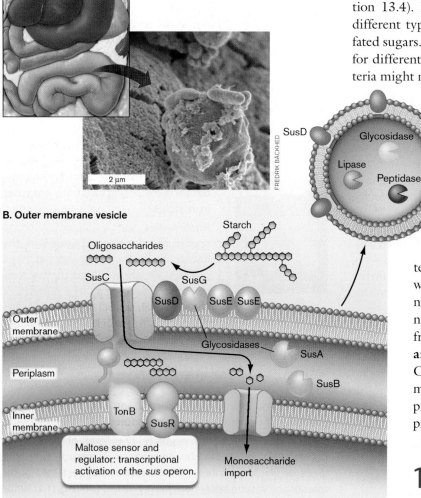

FIGURE 13.1 ■ Communal catabolism by intestinal bacteria.
A. *Bacteroides thetaiotaomicron* (colorized yellow) attached to a starch food particle within mouse intestine. **B.** Starch degradation enzymes and transporters within the *Bacteroides* envelope. **Blowup:** The outer membrane vesicle carries catabolic enzymes.

13.1 Energy and Entropy for Life

Every form of life, from a composting microbe to a human body, uses energy. **Energy** is the ability to do work, such as flagellar propulsion

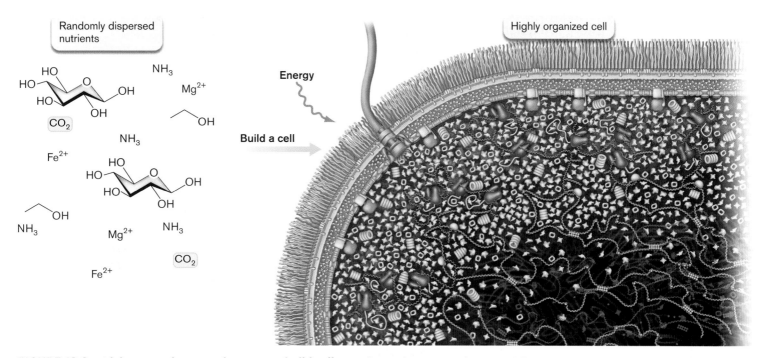

FIGURE 13.2 ■ Living organisms need energy to build cells. Cell growth must spend energy while synthesizing biomass and increasing order.

or cell growth. Energy is used to organize proteins, maintain ion gradients, and build biomass (**Fig. 13.2**). Yet the second law of thermodynamics tells us that systems tend to become less ordered and that **entropy**—the disorder, or randomness, of the universe—always increases. For life to build order out of disorder, cells need energy.

Many Sources of Energy

Collectively, microbes use more diverse energy sources than do multicellular organisms (**Table 13.1**). Recall from Chapter 4 that an organism may gain energy from chemical rearrangement of molecules (prefix "chemo-") or from light absorption (prefix "photo-"). Chemotrophy yields energy from electron transfer between chemicals, yielding products that are more stable. Chemotrophy in which organic compounds donate electrons to yield energy is known as organotrophy or chemoorganotrophy. Organic compounds include the foods we eat—which our intestinal bacteria help catabolize. Most organotrophs are also heterotrophs, organisms that use preformed organic compounds for biosynthesis (a process called **heterotrophy**). But chemotrophy also includes lithotrophy or chemolithotrophy (literally "rock eating"), obtaining energy from inorganic reactions. For example, the archaeon *Pyrodictium occultum* gains energy by oxidizing hydrogen gas with sulfur.

Phototrophy yields energy from light absorption. Phototrophs such as marine cyanobacteria are autotrophs. A photoautotroph gains energy solely from light and builds biomass solely from CO_2. Other bacteria, such as purple proteobacteria, practice photoheterotrophy, in which energy from light supplements chemotrophy. Phototrophy and lithotrophy are discussed in Chapter 14.

But all sources of energy pose the challenges of how to obtain the energy, how to avoid losing it, and how to convert it for cell growth and function.

Note: Distinguish the following prefixes for "-trophy" terms.

Carbon source for biomass

> **Auto-** CO_2 is fixed and assembled into organic molecules.
> **Hetero-** Preformed organic molecules are acquired from outside and assembled into new organic molecules.

Energy source

> **Photo-** Light absorption captures energy.
> **Chemo-** Chemical reactions yield energy without absorbing light.

Electron source

> **Litho-** Inorganic molecules donate electrons.
> **Organo-** Organic molecules donate electrons.

Microbes Use Energy to Build Order

Since the universe overall becomes more disordered, how can microbes use energy to assemble less-ordered molecules into a complex cell? As order increases, the cell is said to decrease in entropy, or disorder. But the decrease in entropy is local to the cell, and temporary. Ultimately, the cell's energy must be spent as heat, which radiates away,

TABLE 13.1	Energy Acquisition in Bacteria and Archaea			
Energy source	**Class of metabolism**	**Examples of energy-yielding reactions**	**Electron acceptor**	**Systems for energy acquisition**
CHEMICAL				
Chemoorganotrophy Organic compounds (at least one C–C bond) donate electrons	**Fermentation** Catabolism	$C_6H_{12}O_6 \rightarrow 2C_3H_6O_3$ (or other small molecules)	Organic	Glycolysis and other catabolism
	Organic respiration Catabolism with inorganic electron acceptor, or with small organic electron acceptor	$C_6H_{12}O_6 + 6H_2O + 6O_2 \rightarrow 6CO_2 + 12H_2O$ $C_6H_{12}O_6 + 6H_2O + 12NO_3^- \rightarrow 6CO_2 + 12H_2O + 12NO_2^-$	O_2 $NO_3^-, SO_4^{2-}, Fe^{3+},$ or other	Glycolysis and other catabolism, TCA cycle, and electron transport systems
Chemolithotrophy Inorganic compounds donate electrons	**Lithotrophy or chemolithoautotrophy**	Electron donor for respiration is H_2, Fe^{2+}, H_2S, NH_4^+	O_2, NO_3^-, or other	Electron transport system
	Methanogenesis	Electron donor is H_2, $CO_2 + 4H_2 \rightarrow CH_4 + 2H_2O$	CO_2	Methanogenesis
LIGHT				
Phototrophy Light absorption provides electrons	**Photoautotrophy** Light absorption drives CO_2 fixation	Photolysis of H_2O $6CO_2 + 12H_2O \rightarrow C_6H_{12}O_6 + 6H_2O + 6O_2$	CO_2	Photosystems I and II
		Photolysis of H_2S, HS^-, or Fe^{2+} $6CO_2 + 12H_2S \rightarrow C_6H_{12}O_6 + 6H_2O + 12S$	CO_2	Photosystem I or II
	Photoheterotrophy Light absorption without CO_2 fixation	Photolysis of H_2S, HS^-, or light-driven H^+ pump. Usually supplements organotrophy.	Organic	Photosystem I or II; bacteriorhodopsin or proteorhodopsin

causing entropy to increase. In other words, the local, temporary gain of energy enables a cell to grow. Continued growth requires continual gain of energy and continual radiation of heat. We see this release of heat, for example, in a compost pile, where heat is produced faster than it dissipates. The temperature of compost typically rises to 60°C.

Similarly, throughout Earth's biosphere, the total metabolism of all life must dissipate energy as heat. Biological heat production is not always obvious, because soil and water provide a tremendous heat sink. But overall, Earth's biosphere behaves as a giant thermal reactor (**Fig. 13.3**). As solar radiation reaches Earth, a small fraction is captured by photosynthetic microbes and plants. The fraction captured is largely in the range of visible light (**Fig. 13.4**), the wavelengths at which photon energies are appropriate for the controlled formation and dissociation of molecular bonds. Some bacteria conduct photosynthesis using ultraviolet and near-infrared radiation. At shorter wavelengths (X-rays), chemical bonds are broken indiscriminately; at longer wavelengths (microwaves and radio waves), the quantum energy is too low to drive chemical reactions.

Microbial and plant photosynthesis generates biomass, which is catabolized by consumers and decomposers. The consumers store a small fraction of their energy in biomass. At each successive level, the majority of energy is lost, radiated from Earth as heat. Thus, despite the growth of living organisms on Earth, the universe as a whole becomes more disordered. The complex roles of microbial metabolism in global ecosystems are discussed in Chapters 21 and 22.

Note: The principles of energy change, discussed next, apply to all the reactions of Table 13.1. These reactions are covered in detail in Chapters 13 and 14.

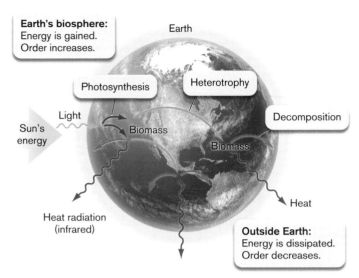

FIGURE 13.3 ■ **Solar energy.** Solar radiation reaches Earth, where a small fraction is captured by photosynthetic microbes and plants. The microbial and plant biomass enters heterotrophs and decomposers, which convert a small fraction to biomass at each successive level. At each level, the majority of energy is lost, radiated from Earth as heat.

Gibbs Free Energy Change

From **Table 13.1** we see that microbes can harness an enormous variety of chemical reactions for growth. But what determines whether a given reaction can support life?

Gibbs free energy change (ΔG). To provide energy to a cell, a biochemical reaction must go forward from reactants to products. The direction of a reaction can be predicted by a thermodynamic quantity known as the **Gibbs free energy**

change, **ΔG** (also known as free energy change or Gibbs energy change). The ΔG value of a reaction determines how much energy is potentially available to do work, such as to drive rotary flagella, to build a cell wall, or to store accurate information in DNA. The sign of the free energy change, ΔG, determines whether a process may go forward. If ΔG is negative, the process may go forward, whereas positive values mean that the reaction will go in reverse. The sign of ΔG determines which "foods" a microbe can eat or, more precisely, which reactions between available molecules can be harnessed for microbial growth.

Under defined laboratory conditions, calculating the ΔG value of a reaction can predict how much biomass microbes will build (**Fig. 13.5**). **Figure 13.5A** lists various reactants for catabolic reactions performed by our gut bacteria, and by bacteria in soil. When oxygen is available, oxidative catabolism of sugars or of short-chain fatty acids yields relatively large amounts of energy. Once oxygen is used up, some bacteria can oxidize food with an alternative electron acceptor, such as nitrate (NO_3^-)—a process called **anaerobic respiration**. Alternatively, bacteria can ferment the carbon source with no mineral oxidant. Whatever the reaction, a linear relationship appears between ΔG and the biomass produced (**Fig. 13.5B**). In natural environments, however, the changing concentrations of substrates and products make the ΔG calculation more complicated.

ΔG includes enthalpy and entropy. The free energy change ΔG has two components:

- ΔH = change in **enthalpy**, the heat energy absorbed or released as reactants become products at constant pressure. When reactants absorb heat from their surroundings as they convert to products, ΔH is positive. When, instead, heat energy is released, ΔH is negative. Release of heat (negative value of ΔH) can yield energy for the cell to use. An example of a reaction with negative ΔH is the oxidation of glucose by O_2.

- ΔS = change in **entropy**, or disorder. Entropy is based on the number of states of a system, such as the number of possible conformations of a molecule. If a cellular reaction splits one molecule into two, all else being equal, entropy increases; the system is more disordered, and ΔS is positive. Most catabolic reactions have a positive value of entropy change. A positive value of ΔS makes ΔG more negative and increases the potential energy yield of a reaction.

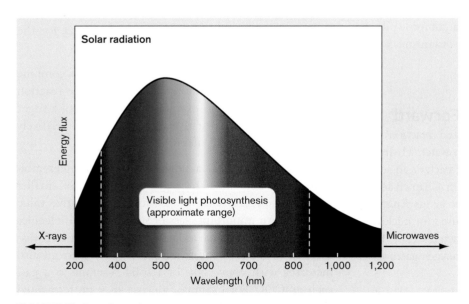

FIGURE 13.4 ■ **The solar spectrum.** The Sun radiates across the spectrum, but the intensity of solar radiation reaching Earth peaks in the range of visible light.

A.

Carbon source	Oxidant	ΔG (kJ/mol)	Biomass (g/mol)
Glucose	O_2	−2,883.0	70.54
Propionate	O_2	−1,487.0	32.54
Lactate	O_2	−1,333.0	32.83
Ethanol	O_2	−1,308.0	26.32
Acetate	O_2	−847.0	18.72
Formate	O_2	−234.0	4.99
Glucose	NO_3^-	−2,774.0	60.00
Lactate	NO_3^-	−1,282.0	20.20
Ethanol	NO_3^-	−1,257.0	24.00
Acetate	NO_3^-	−813.0	16.95
Ethanol	None	−14.6	3.50

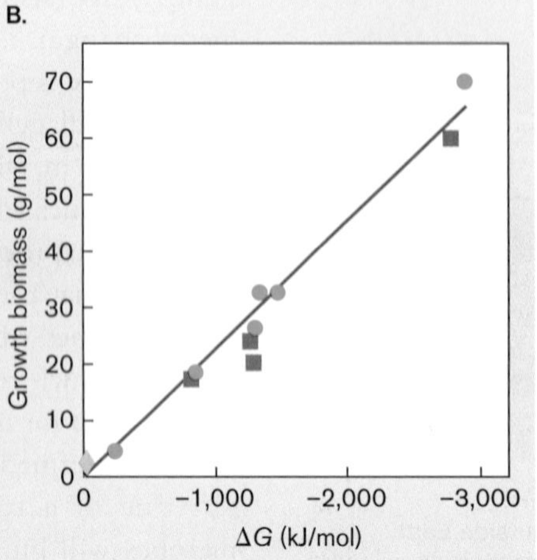

B.

FIGURE 13.5 ■ **Bacterial growth biomass depends on free energy change of metabolic reactions.** **A.** Energy data used for panel B. Oxidant types (O_2, NO_3^-, none) are color-coded. **B.** Observed biomass of cells grown on given pairs of carbon source and oxidant is plotted as a function of free energy change (ΔG). Product is CO_2, except for ethanol without oxidant (green diamond), whose products are acetate and hydrogen gas. *Source:* Data from Eric E. Roden and Qusheng Jin. 2011. *Appl. Environ. Microbiol.* **77**:1907.

The relationship of the free energy change ΔG with ΔH and ΔS is given by:

$$\Delta G = \Delta H - T\Delta S$$

The overall sign of ΔG depends on its two components: ΔH (the absorption or release of heat energy) and $-T\Delta S$ [the negative product of entropy change (ΔS) and temperature (T)]. In living organisms, a sufficiently negative ΔH (energy lost as heat) often overrides $-T\Delta S$, the term for increase in order (negative value of ΔS, positive value of $-T\Delta S$). Thus, a living organism, whose development entails increasing order and decreasing ΔS, can grow as long as the sum of its metabolism has a sufficiently negative value of ΔH. The heat loss associated with ΔH is obvious in a compost pile, where temperature rises. Similarly, all living organisms and communities lose heat.

Negative ΔG Drives a Reaction Forward

An example of a thermodynamically favored reaction is the oxidation of hydrogen gas (H_2) to form water. Hydrogen is oxidized for energy by many kinds of bacteria in soil and water (a form of lithotrophy discussed in Chapter 14). For example, hydrogenotrophic bacteria of the genus *Ralstonia* have been isolated from ultrapure water used for nuclear fuel storage, where radioactive ionization generates H_2. The chemical reaction of dissolved hydrogen and oxygen gases is:

$$2H_2 + O_2 \rightarrow 2H_2O$$

In this reaction, two molecules of hydrogen gas donate four electrons to oxygen, forming water. Under conditions

of standard temperature (298 kelvins, or K) and pressure (sea level), $\Delta H =$ −572 kilojoules per mole (kJ/mol). The ΔH is strongly negative (much heat is released) because the bonds of the product H_2O are much more stable than those of the substrates H_2 and O_2.

However, entropy decreases because the three molecules are replaced by two—a more ordered state. Thus, ΔS is negative [$\Delta S =$ −0.327 kJ/(mol · K)]. In the Gibbs equation the negative sign on the entropy term $-T\Delta S$ makes its contribution to ΔG positive—unfavorable for reaction. So which term wins: ΔH or $-T\Delta S$?

$$\begin{aligned} \Delta G &= \Delta H - T\Delta S \\ &= -572 \text{ kJ/mol} - (298 \text{ K}) \\ &\quad [-0.327 \text{ kJ/(mol · K)}] \\ &= -572 \text{ kJ/mol} + 97 \text{ kJ/mol} \\ &= -475 \text{ kJ/mol} \end{aligned}$$

Overall, ΔG is negative, so bacteria with the appropriate enzyme pathways can use the reaction of hydrogen gas with oxygen to provide energy.

Note: The **joule (J)** is the standard SI unit to denote energy. 1 kilojoule (kJ) = 1,000 joules. Another unit commonly used is the kilocalorie (kcal). The conversion factor is: 1 kJ = 0.239 kcal.

Enthalpy and Entropy in Metabolism

How do ΔH and $-T\Delta S$ affect biochemical reactions? Each reaction in a living cell is associated with these two types of energy change, but to differing degrees depending on the reaction. In general, a reaction yields energy for the cell if:

■ **Molecular stability increases.** When reactants combine to form products with more stable bonds, the reaction has a negative ΔH. For example, the reaction of a sugar with oxygen has a negative ΔH because the relatively unstable oxygen molecules are reduced to H_2O.

■ **Entropy increases.** Reactions in which a complex molecule is broken down to a greater number of smaller molecules increase entropy (positive ΔS, negative value of $-T\Delta S$). An important class of such product molecules is carboxylic acids, such as lactic acid, in which a H^+ dissociates from a carboxylate ion (R–COO^-). Entropy also increases with conversion of a solid reactant to a gas, such as CO_2. For example, glucose may be fermented to ethanol and CO_2, as in the production of alcoholic beverages.

The relative contributions of ΔH and $-T\Delta S$ are important because they determine the effect of temperature on microbial growth. ΔH-dependent reactions such as glucose oxidation release a lot of heat, causing, for example, the rise in temperature of an aerated compost pile. By contrast, the entropy component of Gibbs energy, $-T\Delta S$, does not reflect heat loss. The magnitude of $-T\Delta S$ grows larger as temperature increases. An extreme case of entropy-driven metabolism is that of acetate conversion to methane and CO_2, conducted by soil archaea such as *Methanosarcina barkeri*. This methanogen's growth is driven solely by the increase in entropy ($-T\Delta S$). But the sign of ΔH is positive; thus the organism actually absorbs heat and cools its environment. Despite the positive ΔH, the larger magnitude of the entropy change $-T\Delta S$ allows the reaction to proceed, yielding net free energy.

How do we measure ΔG and its components, ΔH and $-T\Delta S$? The amount of energy released by a reaction is measured in an instrument called a **calorimeter**. A calorimeter may be designed to measure thermal energy (heat) released by a reaction of pure chemicals in a test tube, or of metabolizing bacteria—or even a large farm animal (**Fig. 13.6**). The cow's thermal energy includes microbial digestion coupled to its own digestive processes, and to its own cellular biosynthesis; what is measured is residual energy lost as heat. In a calorimeter, the release of heat during a reaction (or collection of reactions) maintained at constant temperature gives a measure of ΔH. Measurement conducted at various temperatures (T) yields the temperature dependence of heat release, from which we calculate $-T\Delta S$. The sum of the two components yields ΔG.

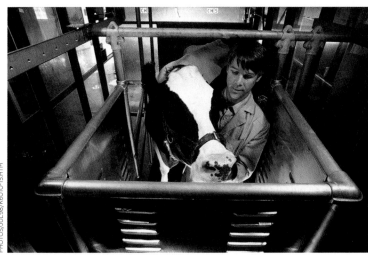

FIGURE 13.6 ■ Calorimeter for animal feed. A cow is prepared for a feeding test within a large-animal calorimeter at the U.S. Department of Agriculture (USDA) in Beltsville, Maryland. The test will show how much energy the cow and its digestive microbes obtain from feeding on ground corn.

To Summarize

■ **Energy** enables cells to build ordered structures out of simple molecules from the environment.

■ **Cells release heat continually** because transfer of energy is never perfectly efficient.

■ **The free energy change (ΔG)** includes enthalpy (ΔH), the heat energy absorbed or released; and $-T\Delta S$, the negative product of temperature and entropy change (ΔS). Most reactions include both ΔH changes (such as oxidation-reduction) and ΔS changes (such as breakdown to a larger number of products).

■ **Negative values of ΔG** show that a reaction can drive the cell's metabolism. The sign of ΔG depends on the relative magnitude of ΔH and $-T\Delta S$.

■ **ΔH-driven reactions release heat.**

■ **$T\Delta S$-driven reactions do not release heat,** but they cause a greater change in free energy (ΔG) at higher temperature.

Thought Questions

13.1 Consider glucose catabolism in your blood, where the sugar is completely oxidized by O_2 and converted to CO_2:

$$C_6H_{12}O_6 + 6O_2 \rightarrow 6CO_2 + 6H_2O$$

Do you think this reaction releases greater energy as heat, or by change in entropy? Explain.

13.2 The bacterium *Lactococcus lactis* was voted the official state microbe of Wisconsin because of its importance for cheese. During cheese production, *L. lactis* ferments milk sugars to lactic acid:

$$C_6H_{12}O_6 \rightarrow 2C_3H_6O_3 \rightleftharpoons 2C_3H_5O_3^- + 2H^+$$

Large quantities of lactic acid are formed, with relatively small growth of bacterial biomass. Why do you think biomass is limited? Cheese making usually runs more efficiently at high temperature; why?

13.2 Energy in Biochemical Reactions

For a given reaction in a given environment, many factors determine ΔG. These factors fall into two classes—those intrinsic to the reaction, and those dependent on the environment:

■ **Intrinsic properties of a reaction.** The intrinsic properties of a reaction are the changes of ΔH and ΔS contributing to ΔG. We can define standard values of these properties relative to arbitrary standard conditions such as concentration and temperature.

■ **Concentrations and environmental factors.** The direction of the reaction depends on the concentrations of reactants and products. An excess of reactants over products makes ΔG more negative (favoring the forward reaction), whereas an excess of product makes ΔG more positive (favoring the reverse reaction). Reaction direction also depends on environmental factors such as temperature, pressure, and ionic strength (salt concentrations).

Standard Reaction Conditions

Scientists commonly present thermodynamic values under standard conditions for temperature, pressure, and concentration. The standard conditions enable scientists to compare the intrinsic properties of reactions. The standard Gibbs free energy change is designated $\Delta G°$. The standard conditions for $\Delta G°$ are as follows:

■ The temperature is 298 K (25°C).

■ The pressure is 1 atm (standard atmospheric pressure).

■ All concentrations of substrates and products are 1 molar (M).

For every chemical compound, a standard $\Delta G°$ of formation can be determined by chemical measurement. We obtain the $\Delta G°$ for a molecular reaction under standard conditions by summing the $\Delta G°$ values of the products, and subtracting the sum of $\Delta G°$ values of the reactants. A table of standard $\Delta G°$ values and a sample calculation are provided in Appendix 1.

But the $\Delta G°$ values reported in data tables hold only for isolated reactions under "standard reaction conditions." Standard conditions differ greatly from the actual conditions of living cells. These conditions include temperature, ionic strength, and gas pressure (in the case of gaseous components, such as CO_2), as well as the concentrations of reactants and products. To account for some differences, biochemists add special standard conditions: the hydrogen ion concentration at pH 7, because living cells commonly maintain their cytoplasm within a unit of neutral pH; and the water concentration of 55.5 M for dilute solutions. The free energy change in biochemistry is thus designated $\Delta G°'$.

The additivity of energy change is central to all living metabolism. Additivity makes it possible to do work by coupling an energy-yielding reaction to an energy-spending reaction (see Section 13.3). **Figure 13.7** shows an example, the initial reaction of glycolysis (see Section 13.5), in which ATP phosphorylates glucose to glucose 6-phosphate, catalyzed by the enzyme hexokinase. In this reaction, the loss of a phosphoryl group from ATP yields energy, some of

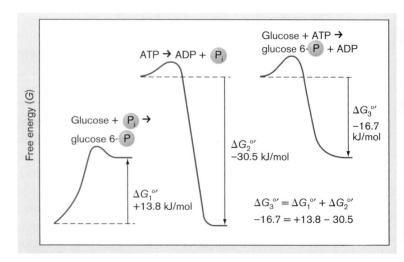

FIGURE 13.7 ■ **Calculating the standard free energy of a reaction.** The $\Delta G°'$ value for the phosphorylation of glucose by ATP equals the sum of the values for the two coupled reactions: phosphorylation of glucose ($\Delta G_1°'$) and ATP hydrolysis ($\Delta G_2°'$).

which is captured by the enzyme to transfer the phosphoryl group onto glucose. The reactions and their $\Delta G°'$ values are summed as follows:

Phosphorylation of glucose

$$C_6H_{12}O_6 + H_2PO_4^- \rightarrow C_6H_{12}O_6 - PO_3^- + H_2O$$

$$\Delta G_1°' = +13.8 \text{ kJ/mol}$$

ATP hydrolysis

$$ATP + H_2O \rightarrow ADP + H_2PO_4^- + H^+$$

$$\Delta G_2°' = -30.5 \text{ kJ/mol}$$

ATP phosphorylation of glucose

$$C_6H_{12}O_6 + ATP \rightarrow C_6H_{12}O_6 - PO_3^- + ADP + H^+$$

$$\Delta G_3°' = -16.7 \text{ kJ/mol}$$

Note: Distinguish these forms of the Gibbs free energy term:

ΔG = change in Gibbs free energy for a reaction under defined conditions.

$\Delta G°$ = ΔG at standard conditions of temperature (298 K) and pressure (1 atm, sea level), with all reactants and products at a concentration of 1 M. (Table of values in Appendix 1.)

$\Delta G°'$ = ΔG at standard temperature, pressure, and concentrations for $\Delta G°$, plus the biochemically relevant conditions of pH 7 (H^+ concentration of 10^{-7} M) and water (H_2O concentration of 55.5 M; activity = 1). Some biochemists standardize additional factors, such as magnesium ion concentration ($[Mg^{2+}] = 1$ mM).

Concentrations of Reactants and Products

In living cells, the concentrations of reactants and products usually differ from 1 M; for example, in the *E. coli* cytoplasm, the concentrations of ATP and P_i are about 8 mM. Thus, the actual ΔG of reactions within a cell differs from ΔG° or $\Delta G^{\circ\prime}$.

Consider a reaction in which reactants A and B are reversibly converted to products C and D:

$$A + B \rightleftharpoons C + D$$

Higher concentration of reactants (A or B) drives the reaction forward, whereas higher concentration of products drives it in reverse. So, ΔG includes the ratio of products to reactants:

$$\Delta G = \Delta G^\circ + RT \ln \frac{[C][D]}{[A][B]}$$

$$= \Delta G^\circ + 2.303 RT \log \frac{[C][D]}{[A][B]}$$

where R is the gas constant [$R = 8.315 \times 10^{-3}$ kJ/(mol · K)] and T is the absolute temperature in kelvins (298 K, or 25°C). The factor 2.303 converts the logarithm of the ratio of products to reactants from base e (ln) to base 10 (log). Note that in $\Delta G^{\circ\prime}$ calculations, the water "activity" (concentration modified by a constant) is set at 1. The logarithm of 1 equals zero, so the water activity falls out of the equation.

Table 13.2 shows how the concentration ratio affects ΔG. In reactions at medium temperature (25°C–40°C), a 100-fold increase in the ratio of products to reactants adds about 11 kJ/mol to ΔG. ΔG is then less negative and the reaction less favorable. On the other hand, a 100-fold <u>decrease</u> in the concentration ratio makes ΔG more <u>negative</u> by −11 kJ/mol.

In some environments, a highly negative concentration term can override a positive ΔG°, resulting in a reaction with negative ΔG that microbes can use for energy. For example, in an iron mine the high concentration of reduced iron favors iron-oxidizing microbes. Alternatively, a high temperature may increase the magnitude of the term $2.303RT \log$ [products]/[reactants] when it is negative,

until it overrides a positive ΔG°. This temperature dependence of ΔG is observed in thermophiles such as *Sulfolobus*, which metabolizes sulfur at 90°C; these sulfur reactions could not go forward at lower temperatures.

Another way the direction of reaction may change is for a second metabolic process to remove one of the reaction's products (C or D) as fast as it is produced. The decrease in product concentration could then change the ΔG of the first reaction to a negative value. For example, when our intestinal bacteria break down glucose to pyruvate (see Section 13.5), many of the individual reactions have $\Delta G^{\circ\prime}$ values smaller than 5 kJ/mol. Their actual direction of reaction in the cell depends on concentrations of products and reactants. Glycolytic reactions with near-zero $\Delta G^{\circ\prime}$ are reversible and can, in fact, participate in the biosynthetic pathway of gluconeogenesis (glucose biosynthesis, discussed in Chapter 15).

A surprising discovery has been that many bacteria and archaea in natural environments grow extremely slowly, using energy-yielding metabolism with values of ΔG approaching zero (that is, near thermodynamic equilibrium). When the actual ΔG (under actual reaction conditions) equals zero, a reaction proceeds equally forward and in reverse, and there is no net change in energy. At equilibrium, the ratio of product and reactant concentrations exactly cancels ΔG° or $\Delta G^{\circ\prime}$:

$$\Delta G = 0 = \Delta G^\circ + 2.303 RT \log \frac{[C][D]}{[A][B]}$$

$$\Delta G^\circ = -2.303 RT \log \frac{[C][D]}{[A][B]}$$

Living cells can never grow exactly at equilibrium ($\Delta G = 0$). But most soil bacteria and archaea gain energy from anaerobic metabolism with near-zero values of ΔG. Such organisms must form biomass very slowly, but if no other metabolism is available, they may outgrow competitors. The discovery of low-ΔG energetics has opened new possibilities for environmental remediation previously thought impossible, such as the anaerobic digestion of complex organic pollutants in contaminated soil (see Section 13.6).

Some of the near-zero anaerobic pathways involve **syntrophy**, an intimate metabolic relationship between two species. Jessica Sieber at the University of Illinois and Michael McInerney at the University of Oklahoma, Norman, show how syntrophy works for microbes that break down pollutants (**Special Topic 13.1**). For syntrophy, a bacterium catabolizes reactants by a reaction that has a positive $\Delta G^{\circ\prime}$ value, while H_2-oxidizing organisms such as methanogens complete the reaction with a net negative $\Delta G^{\circ\prime}$. The bacterial catabolism provides H_2 gas, which the partner species consumes, thus keeping H_2 concentration low enough to drive the syntrophic reaction. Analogous partnerships may link many catabolic reactions of gut bacteria (see Chapter 21).

TABLE 13.2	Effect of the Concentration Ratio on ΔG	
Initial ratio of products to reactants: $\frac{[C][D]}{[A][B]}$	Change in ΔG (kJ/mol) at standard temperature (298 K) and atmospheric pressure	Result of change from standard concentrations
10^{-4}	−23	Products increase
10^{-2}	−11	Products increase
1	0	$\Delta G = \Delta G^\circ$
10^2	+11	Reactants increase
10^4	+23	Reactants increase

SPECIAL TOPIC 13.1 Microbial Syntrophy Cleans Up Oil

Have you ever wondered how to clean up polluted soil and water? Michael McInerney at the University of Oklahoma, Norman, thinks about cleanup from the point of view of microbial energetics (**Fig. 1**). As the son of a wastewater treatment engineer in Chicago, he learned early on about pollution in the Great Lakes. Later, in Oklahoma, he realized that there was a growing problem of water supplies contaminated by sewage and waste petroleum. He began to study how microbes catabolize sewage components, such as the short-chain acids

propionate and butyrate, and aromatic oil compounds such as benzoate.

A fundamental challenge to degrading these pollutants is the lack of oxygen. The molecules seep into the depths of soil or water, where any oxygen is quickly used up by microbes capable of respiration. Jessica Sieber, as a postdoctoral fellow, took up the challenge of elucidating anaerobic pathways to degrade aromatics. Since her move to the University of Illinois, Sieber has kept her focus on anaerobic life, tweeting as @SyntrophySieber, "I tune out when O_2 is mentioned!"

As microbes use up the long-chain components of sewage or oil, the remaining short-chain molecules and aromatic rings can be broken down only by transferring leftover electrons onto hydrogen ions, forming H_2. For example, *Syntrophus aciditrophicus* and *Syntrophomonas wolfei* catabolize benzoate into acetate and bicarbonate ions, releasing H_2 (**Table 1**).

But the H_2-releasing reactions of *Syntrophus* and *Syntrophomonas* have a positive value of $\Delta G°$ (see **Table 1**), and thus they cannot go forward to yield energy. So how can these bacteria obtain energy? The ΔG can be shifted by rapidly removing H_2 gas, thus lowering the concentration of this key product. Once the equations are balanced, negative values of ΔG are obtained. H_2 can be removed by other microbes that oxidize hydrogen using alternative electron acceptors such as bicarbonate (the archaeal methanogen *Methanospirillum*), sulfate or nitrate (the bacterium *Desulfovibrio*) (**Fig. 2A**).

Hydrogen-producing bacteria have evolved to grow in close proximity to the hydrogen-oxidizing bacteria or archaea. **Figure 2B** shows a section through a syntrophic community

FIGURE 1 ■ **Jessica Sieber at the University of Illinois collaborates on syntrophy with Michael McInerney at the University of Oklahoma, Norman.**

NEIL WOFFORD, UNIVERSITY OF OKLAHOMA

TABLE 1	Syntrophy: Reactions Producing and Consuming H_2			

Genus	Reaction	$\Delta G°'$ (kJ/mol)[a]	$\Delta G'$ (kJ/mol)[b]	
H_2 reduces an oxidant (HCO$_3^-$, SO$_4^{2-}$, or NO$_3^-$)				
Methanospirillum	$4H_2 + HCO_3^- + H^+ \rightarrow CH_4 + 3H_2O$	–136		
Desulfovibrio	$4H_2 + SO_4^{2-} + H^+ \rightarrow HS^- + 4H_2O$	–152		
Desulfovibrio	$4H_2 + NO_3^- + 2H^+ \rightarrow NH_4^+ + 3H_2O$	–600	**Oxidant for syntrophy:**	

			HCO$_3^-$	SO$_4^{2-}$	NO$_3^-$
Syntrophic catabolism producing H_2					
Syntrophus	Benzoate + $7H_2O \rightarrow$ 3 acetate + HCO$_3^-$ + $3H^+$ + $3H_2$	+70	–25	–33	–45
Syntrophomonas	Butyrate + $2H_2O \rightarrow$ 2 acetate + H^+ + $2H_2$	+49	–5	–13	–18

[a]$\Delta G°'$ is the standard free energy change at pH 7 when the concentrations of reactants and products are the same.

[b]$\Delta G'$ was the measured free energy change at pH 7 when syntrophic metabolism stopped.

Sources: Michael McInerney et al. 2007. *PNAS* **104**:7600; Ralf Cord-Ruwisch et al. 1988. *Arch. Microbiol.* **149**:350; Bradley E. Jackson and Michael J. McInerney. 2002. *Nature* **415**:454.

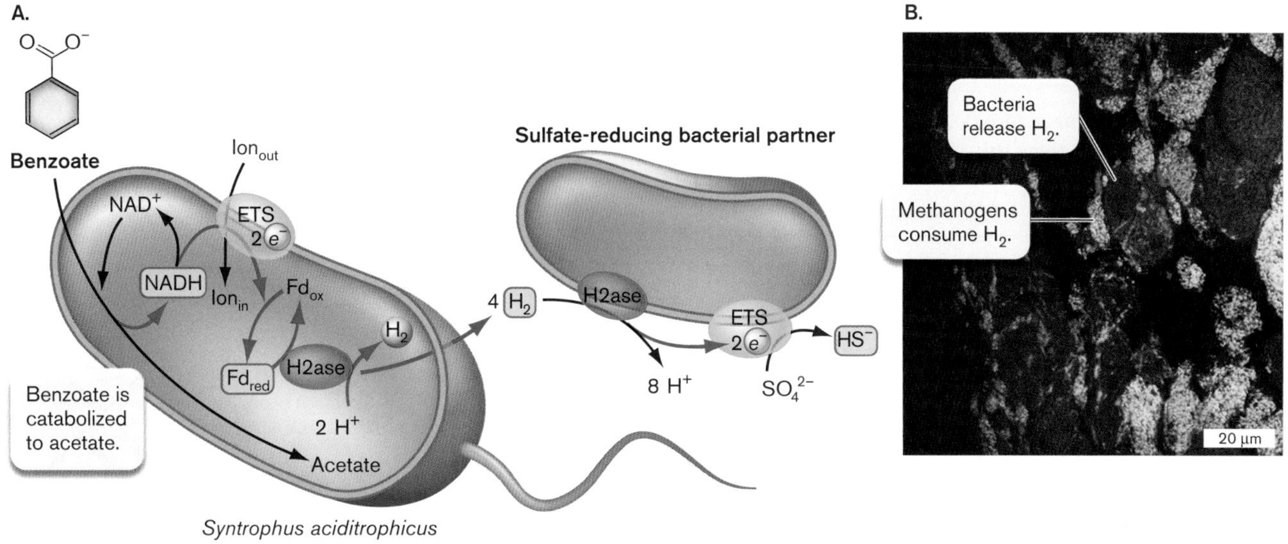

FIGURE 2 ■ Syntrophy between *Syntrophus aciditrophicus* and a sulfate-reducing bacterium, *Desulfovibrio*.
A. *S. aciditrophicus* catabolizes benzoate to acetate, transferring electrons back onto the hydrogen ions removed from the substrate. The H_2 formed must donate electrons to an electron transport system (ETS), despite a positive ΔG value (reverse electron transfer). Rapid removal of H_2 by the sulfate-reducing partner (*Desulfovibrio* sp.) provides energy, making the net ΔG value negative.
B. Syntrophic community of a wastewater sludge granule, labeled by fluorescence in situ hybridization (FISH).

within a wastewater sludge granule, where anaerobic catabolism must degrade sewage components to CO_2. The granule is stained with fluorescent markers for the H_2-producing bacteria (red) and their partner methanogen (yellow). In order for the syntrophy to proceed, the H_2 from the catabolizing bacteria must be transferred quickly to the methanogenic partner, by processes that are not yet understood. Genomic analysis of *Syntrophus aciditrophicus* reveals genes for flagellar motility, suggesting that flagellar motility helps the cell find its partner in order to form an association. Microscopy reveals intercellular nanotubes that may transmit electrons from the H_2 producer to the partner.

Under laboratory conditions, Sieber and McInerney demonstrated syntrophic growth in the presence of defined concentrations of reactants and products (**Table 1**). Their calculations showed that cells grew with negative values of free energy change. Yet even in the presence of the syntrophic partner, these anaerobic pathways have free energy changes that are remarkably low (about –20 kJ/mol), often below the theoretical minimum required to generate ATP (about –60 kJ/mol). Such extremely low energy levels require special arrangements called "reverse electron transport," in which electrons are transferred from one carrier to the next, despite a positive ΔG value (requiring energy input). The elec-

tron transfer must be coupled with a compensating reaction that spends enough energy to yield a net negative ΔG.

Syntrophic partnerships are considered "thermodynamic extremophiles" because they grow at such low ΔG values. Nonetheless, these organisms are of global importance because they play an essential role in recycling the small-molecule products of anaerobic catabolism. Furthermore, they present exciting industrial applications. For instance, the methanogenic partnership between H_2-producing bacteria and *Methanococcus maripaludis* might be used to attack unextractable oil residues and convert the carbon into methane for recovery as natural gas. Other syntrophic interactions may be used to increase the efficiency of sewage mineralization during water treatment.

RESEARCH QUESTION

How do hydrogen and formate transfer enable environmental bacteria to digest fatty acids? What would happen to an environment where none of the fatty acids got catabolized?

Jessica R. Sieber, Huynh M. Le, and Michael J. McInerney. 2014. The importance of hydrogen and formate transfer for syntrophic fatty, aromatic and alicyclic metabolism. *Environmental Microbiology* **16**:177–188.

Thought Question

13.3 The thermophilic bacteria *Thermus* species grow in deep-sea hydrothermal vents at 80°C. It was proposed that they metabolize formate to bicarbonate ion and hydrogen gas:

$$HCOO^- + H_2O \rightarrow HCO_3^- + H_2$$

But the standard $\Delta G^{\circ\prime}$ is near zero (−2.6 kJ/mol). Under actual conditions, do you think the reaction yields energy? Assume concentrations of 150 mM formate, 20 mM bicarbonate ion, and 10 mM hydrogen gas.

A. Diffusion: positive ΔS

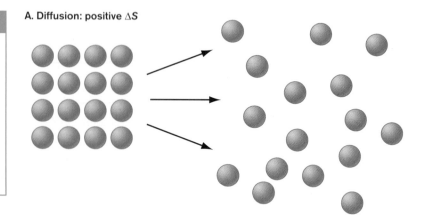

Concentration Gradients

So far, our discussion of energy has assumed an isotropic system, in which concentrations are the same everywhere. But a living cell needs to obtain its molecules from outside, such as sugars, amino acids, and inorganic ions (discussed in Chapter 4). Suppose a blood pathogen needs to obtain a scarce substance such as iron (Fe^{2+}). How does the microbe move the iron "uphill" against its concentration gradient?

A substance dissolved in water diffuses by random movements until its distribution has the same concentration throughout (**Fig. 13.8A**). The random distribution of molecules at uniform concentration represents the state of greatest entropy. Diffusion in the environment ultimately brings nutrients into contact with microbial cells, even cells that lack chemotactic motility to hunt for food. For example, sugars in the food we eat diffuse through our saliva to reach the bacteria growing in biofilms on our teeth.

The cell membrane contains transporters for useful molecules, allowing nutrient molecules to cross. Entropy favors their movement from higher to lower concentration (**Fig. 13.8B**). In most environments, however, the nutrients are at lower concentrations than inside the cell. To obtain these molecules from outside, the bacterial cell must transport them against their gradient—that is, from lower to higher concentration—increasing the concentration difference and thus decreasing entropy. Uptake against a concentration gradient requires an energy source to power transport proteins embedded in the membrane (as shown in Chapter 4). Alternatively, a transmembrane gradient of ions can store energy for the cell. The most important ion gradient is the H^+ gradient, a component of the proton motive force (to be explained in Chapter 14).

To Summarize

■ **Intrinsic properties of the reaction determine the standard value of ΔG°.** Properties include the molecular

B. Transmembrane gradient

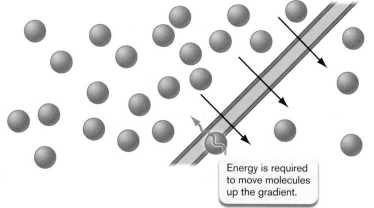

Energy is required to move molecules up the gradient.

FIGURE 13.8 ■ **Diffusion and transport.** **A.** Water-soluble molecules diffuse to uniform concentration throughout the solution: ΔS is positive, $-T\Delta S$ negative; the entropy term favors the process. **B.** If a membrane separating two compartments is permeable, molecules move from a compartment with high concentration to one with low concentration. Energy is required to move molecules up their concentration gradient.

stability of reactants and products, and the entropy change associated with product conversion to reactants. By convention, standard conditions are set to define ΔG° (for biochemists, $\Delta G^{\circ\prime}$).

■ **Concentrations of reactants and products affect the actual value of ΔG.** The lower the concentration ratio of products to reactants, the more negative the value of ΔG. Environmental factors such as temperature and pressure also affect ΔG.

■ **Within living cells, energy-yielding reactions are coupled with energy-spending reactions.** The measurement of energy flow under changing conditions requires calculations more complex than those shown here.

■ **A concentration gradient stores energy.** Solutes run down their concentration gradient unless energy is applied to reverse the flow. Gradient energy can be interconverted with energy from chemical reactions.

13.3 Energy Carriers and Electron Transfer

Our ΔG equations show only the total energy of a reaction such as glucose oxidation. If all the energy were released at once, however, it would dissipate as heat without building biomass. In living cells, glucose is never oxidized in one step. Instead, the energy yield is divided among a large number of stepwise reactions with smaller energy changes. In this way, the cell can be thought of as "making change" by converting a large energy source to numerous smaller sources that can be "spent" conveniently for cell function and biosynthesis. The "spending" of energy is controlled by enzymes that couple all of the energy-providing reactions to specific energy-spending reactions.

Energy Carriers Gain and Release Energy

Many of the cell's energy transfer reactions involve **energy carriers**. Examples of energy carriers are ATP (adenosine triphosphate) and NADH (the reduced form of nicotinamide adenine dinucleotide). Energy carriers are molecules that gain and release small amounts of energy in reversible reactions. Energy carriers are used to transfer energy in a wide range of biochemical reactions.

Some energy carriers, such as NADH, transfer energy associated with electrons received from a food molecule. A molecule that transfers, or "donates," electrons to another molecule is called an **electron donor** or a reducing agent; a molecule that receives, or "accepts," electrons is called an **electron acceptor**. For example, during glucose catabolism a molecule of glyceraldehyde 3-phosphate transfers a pair of electrons ($2e^-$) with a hydrogen ion (H^+) to NAD^+, forming NADH. NAD^+ is an electron acceptor that receives the electrons; it then becomes the electron donor NADH. Electron donors such as NADH transfer electrons from reduced food molecules to a terminal electron acceptor such as oxygen (see Chapter 14). Energy carriers that transfer electrons are needed for all energy-yielding pathways and for biosynthesis of cell components such as amino acids and lipids (see Chapter 15).

Note that in living cells, all energy transfer reactions are coupled by enzymes to specific biochemical processes. Without enzyme coupling, energy dissipates and would be lost from the living system.

ATP Carries Energy

Adenosine triphosphate, or **ATP (Fig. 13.9A)**, is composed of a base (adenine), a sugar (ribose), and three phosphoryl groups. Note that adenine-ribose-phosphate (adenosine nucleotide) is equivalent to a nucleotide of

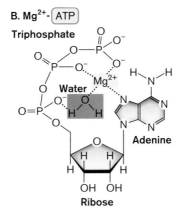

A. ADP phosphorylation to ATP

B. Mg²⁺-ATP

FIGURE 13.9 ■ ADP plus inorganic phosphate makes ATP. The reaction requires energy input (positive ΔG) because the negatively charged oxygens of the phosphoryl groups are forced to interact. **A.** The chemical reaction phosphorylating ADP (adenosine diphosphate) to ATP (adenosine triphosphate). **B.** Model of Mg²⁺-ATP. The multiple negative charges of ATP are stabilized by binding a magnesium ion plus a water molecule.

RNA. The base adenine is a fundamental molecule of life, one that forms spontaneously from methane and ammonia in experiments simulating the origin of life on early Earth (discussed in Chapters 1 and 17). Like the sugar ribose, ATP is an ancient component of cells, found in all living organisms.

Under physiological conditions, ATP always forms a complex with Mg^{2+} (**Fig. 13.9B**). The magnesium cation partly neutralizes the negative charges of the ATP phosphates, stabilizing the structure in solution. Most enzyme-binding sites for ATP actually bind Mg^{2+}-ATP. This is one reason why magnesium is an essential nutrient for all living cells.

ADP phosphorylation to ATP. During cell metabolism, ATP is generated by **phosphorylation**, the condensation of inorganic phosphate with adenosine diphosphate (ADP):

$\Delta G^{\circ\prime} = 31$ kJ/mol

The phosphorylation of ADP to form ATP requires energy input (positive ΔG).

Why does ATP formation require energy? The inorganic phosphate molecule has four oxygen atoms that share a negative charge. When phosphate reacts with another phosphate to form a bond, the charged oxygens of adjacent phosphates are forced to interact despite charge repulsion. The charge repulsion in ATP limits rotation of oxygens, and thus decreases entropy, which results in a negative ΔG of hydrolysis. Hydrolysis of each phosphoryl group yields energy. The formation and hydrolysis of ATP can be shown as:

$$A–P{\sim}P + H^+ + P_i \rightleftharpoons A–P{\sim}P{\sim}P + H_2O$$

where $\sim$ designates each energy-storing phosphoryl bond and P_i designates inorganic phosphate; or, in more concise shorthand:

$$ADP + P_i \rightleftharpoons ATP + H_2O$$

ATP is a "medium-sized" energy carrier, since the cell contains many phosphorylated molecules that yield greater energy upon hydrolysis. **Table 13.3** summarizes examples of free energy change associated with hydrolysis of various phosphoryl groups. As we will see, phosphoryl group hydrolysis with $\Delta G^{\circ\prime}$ values larger than –31 kJ/mol can yield energy that is stored by ATP formation. ATP

TABLE 13.3	Hydrolysis of Phosphoryl Groups: Values of $\Delta G^{\circ\prime}$ (pH 7.0, 25°C)
Reaction of hydrolysis	**$\Delta G^{\circ\prime}$ (kJ/mol)**
Glucose 6-P + $H_2O \rightarrow$ glucose + P_i	–14
Fructose 1,6-bis-P + $H_2O \rightarrow$ fructose 6-P + P_i	–16
ATP + $H_2O \rightarrow$ ADP + P_i + H^+	–31
ADP + $H_2O \rightarrow$ AMP + P_i + H^+	–33
ATP + $H_2O \rightarrow$ AMP + PP_i (pyrophosphate) + H^+	–46
PP_i (pyrophosphate) + $H_2O \rightarrow$ 2 P_i + H^+	–19
1,3-Bis-P-glycerate + $H_2O \rightarrow$ 3-P-glycerate + P_i	–49
Phosphoenolpyruvate + $H_2O \rightarrow$ pyruvate + P_i	–62

hydrolysis can yield energy to phosphorylate molecules at $\Delta G^{\circ\prime}$ values smaller than –31 kJ/mol.

ATP transfers energy. ATP can transfer energy to cell processes in three different ways: hydrolysis releasing phosphate, hydrolysis releasing pyrophosphate (diphosphate), and phosphorylation of an organic molecule. Each process serves different functions in the cell.

■ **Hydrolysis releasing phosphate.** The **hydrolysis** of ATP at the terminal phosphate consumes H_2O to produce ADP and P_i, releasing energy. The energy released by ATP hydrolysis can be transferred to a coupled reaction of biosynthesis, such as building an amino acid. The two reactions are coupled by an enzyme with binding sites specific for ATP and the substrate.

■ **Hydrolysis releasing pyrophosphate.** ATP can hydrolyze at the middle phosphate, releasing pyrophosphate (PP_i). The pyrophosphate usually hydrolyzes shortly afterward to make 2 P_i. Overall, the release of 2 P_i from ATP yields approximately twice as much energy as the release of 1 P_i. Pyrophosphate release and subsequent hydrolysis drives a reaction strongly forward, because twice as much energy would be required to reverse the reaction. Pyrophosphate is released in reactions that must avoid reversal—for example, the incorporation of nucleotides into growing chains of RNA.

■ **Phosphorylation of an organic molecule.** ATP can transfer its phosphate to the hydroxyl group of a molecule such as glucose to activate the substrate for a subsequent rapid reaction. No inorganic phosphate appears, and no water molecule is consumed:

$$ATP + glucose \rightarrow ADP + glucose\ 6\text{-}P$$

Some enzymes catalyze ATP transfer of phosphate to activate sugar molecules for catabolism. Other enzymes couple the phosphorylation of a sugar to its transport across the cell membrane; consequently, these enzymes make up the phosphotransferase system (PTS). The PTS enzymes play a critical role in determining which nutrients from the environment some microbes can acquire and catabolize (discussed in Chapter 4).

Thought Questions

13.4 When ATP phosphorylates glucose to glucose 6-phosphate, what is the net value of $\Delta G°$? What if ATP phosphorylates pyruvate? Can this latter reaction go forward without additional input of energy? (See **Table 13.3**.)

13.5 Linking an amino acid to its cognate tRNA is driven by ATP hydrolysis to AMP (adenosine monophosphate) plus pyrophosphate. Why release PP_i instead of P_i?

ATP produced by glucose catabolism. A large number of ATPs can be formed by coupling ATP synthesis to the step-by-step breakdown and oxidation of a food molecule such as glucose. In theory, complete oxidation of glucose through respiration can produce as many as 38 ATP molecules. The overall $\Delta G°'$ of the coupled reactions is:

$$C_6H_{12}O_6 + 6H_2O + 6O_2 \rightarrow$$
$$12H_2O + 6CO_2 \quad \Delta G°' = -2,878 \text{ kJ/mol}$$

$$38[ADP + P_i \rightarrow ATP + H_2O] \quad \Delta G°' = 38 \times (+31 \text{ kJ/mol})$$

$$\text{Net } \Delta G°' = -1,700 \text{ kJ/mol}$$

The difference in $\Delta G°'$ for the coupled reactions is the energy lost as heat and entropy—in this case, $-1,700$ kJ/mol ($-2,878$ kJ/mol $+ 1,178$ kJ/mol). Thus, the maximal efficiency of energy capture by ATP is about 40%, a level that may be approached by highly efficient systems such as mitochondria. When ΔG values are corrected for cellular concentrations of reactants and products, the actual efficiency may be greater than 50%. Under conditions such as low oxygen concentration, much smaller amounts of ATP are made per molecule of glucose. For comparison, the efficiency of a typical machine, such as an internal combustion engine, is about 20%.

Thought Question

13.6 In the microbial community of the bovine rumen, the actual ΔG value has been calculated for glucose fermentation to acetate:

$$C_6H_{12}O_6 + 2H_2O \rightarrow 2C_2H_3O_2^- + 2H^+ + 4H_2 + 2CO_2$$
$$\Delta G = -318 \text{ kJ/mol}$$

If the actual ΔG for ATP formation is +44 kJ/mol and each glucose fermentation yields four molecules of ATP, what is the thermodynamic efficiency of energy gain? Where does the lost energy go?

Note that, besides ATP, other nucleotides carry energy. Guanosine triphosphate (GTP) provides energy for ribosome elongation of proteins. And the phosphodiester bonds of all four nucleotide triphosphates, as well as their corresponding deoxyribonucleoside triphosphates, carry energy for their own incorporation into RNA and DNA, respectively.

NADH Carries Energy and Electrons

Another major energy carrier is **nicotinamide adenine dinucleotide**, or **NAD** (NADH, reduced; NAD$^+$, oxidized). Unlike ATP, NADH carries energy associated with two electrons that reduce a substrate (**Fig. 13.10**). NADH carries two or three times as much energy as ATP, depending on cell conditions. During sugar catabolism, NADH carries electrons from breakdown products of glucose. Its oxidized form, NAD$^+$, receives two electrons ($2e^-$) plus a hydrogen ion (H$^+$) from a food molecule; a second H$^+$ from the food molecule enters the solution. Overall, reduction of NAD$^+$ consumes two hydrogen atoms to make NADH:

$$NAD^+ + 2H^+ + 2e^- \rightarrow NADH + H^+ \quad \Delta G°' = +62 \text{ kJ/mol}$$

For this reaction, $\Delta G°'$ is positive; therefore, it requires input of energy from the catabolism of the food molecule. The reduced energy carrier NADH can then reverse this reaction by donating two electrons ($2e^-$) to another molecule, regenerating NAD$^+$.

Note: A hydrogen ion, or "proton" (H$^+$), does not exist free in solution. In water, a H$^+$ combines with H$_2$O to form a hydronium ion (H$_3$O$^+$), but for clarity we use H$^+$. An atom of hydrogen removed from a C–H bond consists of a proton (H$^+$) plus an electron (e^-). In a reaction, the proton and electron may be transferred to one molecule or to separate molecules.

NADH structure and function. NAD$^+$ consists of an ADP molecule attached to nicotinamide instead of a third phosphate. The nicotinamide mononucleotide, like a ribonucleotide, contains a nitrogenous base attached to a sugar phosphate. In NAD$^+$, the nicotinamide has a ring structure (shaded orange in **Fig. 13.10A**) that forms a stable cation.

NAD$^+$ is a relatively stable structure because the ring electrons are **aromatic**; that is, the bonding electrons delocalize around the ring, as in benzene. Aromatic rings that contain noncarbon atoms are said to be heteroaromatic. Many biologically active molecules are heteroaromatic, including adenine and other nucleotide bases. A heteroaromatic ring is stable, but its disruption requires less energy than the disruption of benzene. Thus, it is possible to disrupt the ring by adding two electrons with H$^+$ eliminating one double bond. The donation of electrons eliminates the

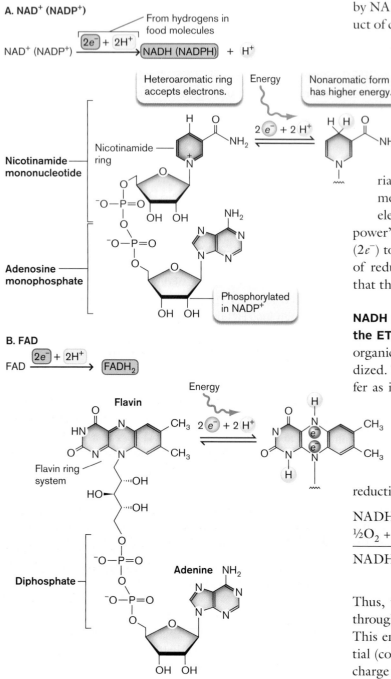

FIGURE 13.10 ■ Reduction of NAD⁺ and FAD. A. NAD⁺ reduction: The nicotinamide ring (shaded pink) loses a double bond as two electrons are gained from an electron donor. Two hydrogen atoms are consumed; one bonds to NADH, while the other ionizes. B. FAD reduction: The flavin ring system gains two electrons associated with two hydrogens.

by NADH can be spent is to transfer $2H^+ + 2e^-$ onto a product of catabolism. For example, in ethanolic fermentation to make wine or beer, NADH reduces pyruvate to ethanol. In this case, however, the energy is lost to the cell. Alternatively, NADH can transfer its electrons to one of a series of electron carrier molecules known as the **electron transport system** (**ETS**), also called the electron transport chain (ETC). Electron transport within bacteria can actually be used to generate electricity for commercial power (discussed in Chapter 14). Examples of electron transfer reactions are shown in the "tower of power" in **Table 13.4**. For example, adding two electrons ($2e^-$) to NAD⁺ to make NADH has a highly negative value of reduction potential $E^{\circ\prime}$ (positive $\Delta G^{\circ\prime}$), which means that the reaction requires energy input.

NADH oxidation and reduction participate in reactions of the ETS. The ETS includes a series of proteins and small organic molecules that can be reduced and cyclically reoxidized. The redox reactions store energy from electron transfer as ion gradients across the membrane of the cell or an organelle. At the end of the redox series, the electrons are transferred to a **terminal electron acceptor** whose product leaves the cell. For example, as a terminal electron acceptor, molecular oxygen (O_2) is reduced to H_2O. The reaction of O_2 reduction to H_2O may be coupled to oxidation of NADH:

$$NADH + H^+ \rightarrow NAD^+ + 2H^+ + 2e^- \qquad \Delta G^{\circ\prime} = -62 \text{ kJ/mol}$$
$$\tfrac{1}{2}O_2 + 2H^+ + 2e^- \rightarrow H_2O \qquad \Delta G^{\circ\prime} = -158 \text{ kJ/mol}$$

$$NADH + H^+ + \tfrac{1}{2}O_2 \rightarrow NAD^+ + H_2O$$
$$\Delta G^{\circ\prime} = -220 \text{ kJ/mol}$$

Thus, the total energy released during NADH oxidation through the ETS is –220 kJ/mol (–62 kJ/mol – 158 kJ/mol). This energy is converted to transmembrane proton potential (composed of the H^+ concentration difference plus the charge difference across the membrane). The proton potential (see Chapter 14) drives nutrient transport, motility, and synthesis of ATP.

Note: Reduction potentials, electron transport, and proton motive force are discussed further in Chapter 14.

Other energy carriers that transfer electrons. Different steps of metabolism utilize different but related energy carriers. For example, NADPH differs from NADH only in its extra phosphate attached to the 2′ carbon of adenine nucleotide; the amount of energy carried is the same. Some enzymes can utilize both NADPH and NADH, whereas other enzymes use only one or the other.

ring's aromatic nature, generating NADH, which carries energy in an amount useful for cell reactions.

The electrons transferred to NADH eventually must be put somewhere else, onto the next electron acceptor. If NADH builds up in a cell, no NAD⁺ remains to continue oxidizing food molecules. One way that the energy stored

TABLE 13.4	Standard Reduction Potentials*				
Electron acceptor	**→**		**Electron donor**	$E^{\circ\prime}$ (mV)	$\Delta G^{\circ\prime}$ (kJ)
$2H^+ + 2e^-$	→		H_2	−420	+81
$NAD^+ + 2H^+ + 2e^-$	→		$NADH + H^+$	−320	+62
$FAD + 2H^+ + 2e^-$	→		$FADH_2$	−220	+42
$FMN + 2H^+ + 2e^-$	→		$FMNH_2$	−190	+37
Menaquinone $+ 2H^+ + 2e^-$	→		Menaquinol	−74	+14
Fumarate $+ 2H^+ + 2e^-$	→		Succinate	+33	−6
Ubiquinone $+ 2H^+ + 2e^-$	→		Ubiquinol	+110	−21
$NO_3^- + 2H^+ + 2e^-$	→		$NO_2^- + H_2O$	+420	−81
$\frac{1}{2}O_2 + 2H^+ + 2e^-$	→		H_2O	+820	−158

*For a more extensive list, see Table 14.1.

Another related energy carrier is **flavin adenine dinucleotide**, or **FAD** ($FADH_2$, reduced; FAD, oxidized), in which flavin substitutes for nicotinamide. The flavin nucleotide includes a ring structure whose aromaticity is eliminated by its receiving two electrons (**Fig. 13.10B**). The redox function of the flavin isoalloxazine ring system is similar to that of NADH:

$$FAD + 2H^+ + 2e^- \rightarrow FADH_2$$

Like NADH, $FADH_2$ donates $2e^-$ to an electron acceptor. $FADH_2$ is a weaker electron donor than NADH, but when $FADH_2$ is combined with a strong electron acceptor such as O_2, electrons are transferred and significant energy is released:

$$\Delta G^{\circ\prime} = -42 - 158 = -200 \text{ kJ/mol}$$

Why do different kinds of reactions use different energy carriers?

■ **Different redox levels.** Food molecules may have more or fewer electrons (level of reduction/oxidation) than those associated with the cell structure. For example, lipids are more highly reduced than glucose. Thus, lipid catabolism requires a greater proportion of electron-accepting energy carriers (such as NAD^+ or $NADP^+$) than does glucose catabolism, and it makes relatively few ATP molecules directly. A combination of energy carriers with different redox states enables cells to balance their overall redox potential while transferring energy.

■ **Different amounts of energy.** Biochemical reactions yield different amounts of energy—that is, different values of ΔG. Suppose a reaction can provide more than enough energy to generate ATP from ADP (31 kJ), but not quite enough to generate NADH from NAD^+ (62 kJ). An example is the conversion of succinate to fumarate in the TCA cycle (see Section 13.6). Succinate

conversion provides the energy to reduce FAD to $FADH_2$, whose oxidation by O_2 can yield two molecules of ATP. Thus, the use of $FADH_2$ enables the cell to make more efficient use of its food than if generation of ATP or NADH were the only choices.

■ **Regulation and specificity.** Specific energy carriers can direct metabolites into different pathways serving different functions. For example, in many bacteria NADH is directed into the ETS, whereas NADPH, the 2′-phosphorylated form of NADH, is directed into biosynthesis of cell components such as amino acids and lipids.

The concentrations and reduction level of energy carriers provide much information on the state of a cell, such as the effects of environmental stress on cell metabolism. But energy carriers such as NADH undergo rapid turnover (interconversion with NAD^+). How can we observe NADH and ATP concentrations within living cells? One method uses nuclear magnetic resonance (NMR) spectroscopy (discussed in **eTopic 13.1**). The use of NMR to observe living cells led to the development of magnetic resonance imaging (MRI) to observe the entire human body.

Enzymes Catalyze Metabolic Reactions

In living cells, each reaction must occur only as needed, in the right amount at the right time. The rate of a reaction is determined by the **activation energy** (E_a), the input energy needed to generate the high-energy transition state on the way to products (**Fig. 13.11**). Most biochemical reactions require an activation energy that exceeds the average kinetic energy of the reactant molecules colliding. Thus, no matter how negative the ΔG, the reaction will proceed at a significant rate only when the activation energy is lowered by interaction with a catalyst, an agent that participates in a reaction without being consumed.

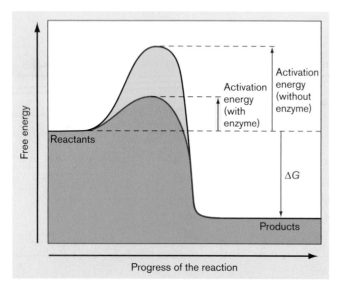

FIGURE 13.11 ▪ Enzymes lower the activation energy of the transition state. In the presence of an enzyme, the activation energy of the reaction is decreased, allowing rapid conversion of reactants to products.

Biological reactions are catalyzed by **enzymes**, structures composed of protein (or, in some cases, RNA) that bind substrates of a specific reaction. The enzyme lowers the activation energy by bringing the substrates in proximity to one another and by correctly orienting them to react. In some cases, enzymes provide a reactive amino acid residue to participate in a transition state between reactants and products. Microbial enzymes have growing importance in industry; they are used for food production, fabric treatment, and drug therapies (discussed in Chapter 16).

Enzymes couple specific energy-yielding reactions (such as those of glucose breakdown) with the cell's reactions requiring energy, such as making ATP. An example of coupled reactions is shown in **Figure 13.12**. The substrate phosphoenolpyruvate (PEP), a phosphorylated breakdown product of glucose, is converted to pyruvate while the phosphate is added to ADP to generate ATP:

The $\Delta G°'$ of phosphate cleavage from PEP is −62 kJ/mol, whereas the $\Delta G°'$ of ATP formation is only +31 kJ/mol. Thus, the net $\Delta G°'$ is negative (−31 kJ/mol), and the reaction goes forward to pyruvate. But a high activation energy makes the reaction extremely slow; without help, it rarely goes forward on the timescale of life. The reaction proceeds only when the two reacting substrates (PEP and ADP) are coupled by the enzyme pyruvate kinase (**Fig. 13.12A**). Pyruvate kinase has specific binding sites for each of its substrates: PEP and ADP. ADP and PEP are brought together by the enzyme and positioned so as to lower the activation energy of phosphate transfer.

In addition to the catalytic site, the enzyme has an **allosteric site** (a regulatory site distinct from the substrate-binding site) for its activator, fructose 1,6-bisphosphate. The difference between a substrate-binding site and an allosteric site can be seen in the molecular model based on X-ray crystallography (**Fig. 13.12B**). The allosteric site is found at a distance from the substrate-binding site, but its interaction with the regulator, fructose 1,6-bisphosphate, alters the conformation of the entire enzyme, increasing the rate of reaction. As we will see, fructose 1,6-bisphosphate is a central intermediate of glycolysis; thus, it makes sense that as this molecule builds up, it activates pyruvate kinase to remove products farther down the chain of catabolic reactions, maintaining the steady flow of metabolites.

> **Thought Question**
>
> **13.7** What would happen to the cell if pyruvate kinase catalyzed PEP conversion to pyruvate but failed to couple this reaction to ATP production?

Note that pyruvate kinase is actually named for its reverse reaction: pyruvate phosphorylation to PEP. The enzyme may have been named for activity that was originally observed in the reverse direction, before the cell function was understood. Enzymes can catalyze both forward and reverse reactions. The predominant direction of catalysis depends on the concentrations of substrates and products (which determine ΔG) and on allosteric regulators.

To Summarize

- **Catabolic pathways organize the breakdown of large molecules** in a series of sequential steps coupled to reactions that store energy in small carriers such as ATP and NADH.

- **ATP and other nucleotide triphosphates store energy** in the form of phosphodiester bonds.

- **NADH, NADPH, and FADH$_2$** each store energy associated with an electron pair that carries reducing power.

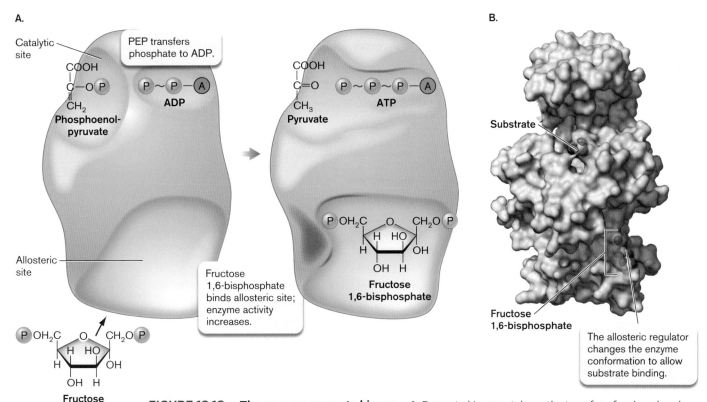

FIGURE 13.12 ■ **The enzyme pyruvate kinase.** **A.** Pyruvate kinase catalyzes the transfer of a phosphoryl group from PEP to ADP, generating pyruvate and ATP. The enzyme possesses separate binding sites for substrates and allosteric effectors. **B.** The molecular model of pyruvate kinase, based on a crystal structure of the enzyme bound to a substrate analog and an allosteric effector, fructose 1,6-bisphosphate. (PDB code: 1A3W)

■ **Enzymes catalyze reactions by lowering the ΔG** required to reach the transition state. They couple energy transfer reactions to specific reactions of biosynthesis and cell function.

13.4 Catabolism: The Microbial Buffet

In the early twentieth century, it was thought that microbes could catabolize a limited subset of naturally occurring organic molecules, such as sugars. Molecules not known to be catabolized were termed "xenobiotics," especially if they were "synthetic" products of human industry. We have since found that virtually any organic molecule can be catabolized by a microbe that has evolved the appropriate enzymes. Catabolism by microbes in our own digestive system helps us digest complex food molecules. Catabolism also plays key roles in microbial disease; for example, the causative agent of acne, *Propionibacterium acnes*, degrades skin cell components such as lipids.

Note: Besides releasing energy, the breakdown of organic food molecules provides substrates for biosynthesis. Biosynthesis is covered in Chapter 15.

Substrates for Catabolism

Here we present several important classes of catabolic substrates (**Fig. 13.13**). While in principle virtually any organic constituent may be catabolized, certain kinds of substrates are used more rapidly because they require less activation energy or fewer types of enzymes to break them down. Many of these substrates form products important to our nutrition and technology. Others, such as aromatic components of petroleum, are environmental pollutants that only bacteria and fungi can degrade (see Section 13.6).

Carbohydrates. Carbohydrates (sugars and sugar polymers, or **polysaccharides**) (**Fig. 13.13A**), are important as structural components of cells, and for their central role in human digestion. Microbial catabolism is central to our production of food and drink (presented in Chapter 16). Glucose as such is rarely available to microbes, except to pathogens growing within a host. But the pathways of glucose catabolism complete the digestion of a diverse range of food molecules found in the environment. The main carbohydrates digested by human enzymes are **starches**, glucose polymers in which an acetal (O–COH) condenses with a hydroxyl group, releasing H_2O.

But intestinal bacteria such as *Bacteroides* species can digest a far greater range of polysaccharides, such as the cellulose of plant cell walls and the pectin of fruit. Such

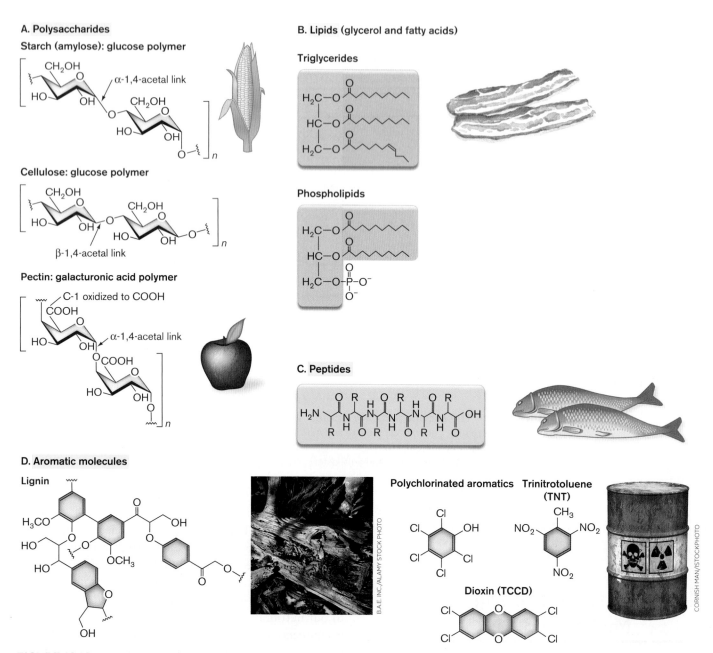

FIGURE 13.13 ■ Complex carbon sources for catabolism. **A.** Polysaccharides such as starch, cellulose, and pectin are hydrolyzed to glucose. **B.** Lipids are broken down to acetate. **C.** Peptides are hydrolyzed to amino acids and then broken down to acetate, amines, and other molecules. **D.** Complex aromatic molecules such as lignins and halogenated aromatic pollutants are broken down to acetate and other molecules.

polysaccharides were originally defined as "fiber"—that is, polymers indigestible by humans. We now know that partial digestion of fiber contributes significant caloric content to the human diet. Members of the gut microbiome are so important that human breast milk contains specific types of carbohydrates to feed them.

Plant-derived polysaccharides are long molecules that cannot be taken up by cells. Instead, microbes secrete enzymes that initiate extracellular digestion (see **Fig. 13.1**). Sugar chains are broken down by microbial enzymes, first to short chains (oligosaccharides), then to two-sugar units (disaccharides), and then to monosaccharides such

as glucose or fructose. In some sugars, the aldehyde is replaced by a hydroxyl (sorbitol, mannitol) or a carboxylate (gluconate, glucuronate). Polysaccharides are hydrolyzed to products that enter central catabolic pathways such as glycolysis (**Fig. 13.14**).

Lipids. Many bacteria catabolize lipids (**Fig. 13.13B**) from sources such as milk, animal fats, and nuts. The oxidation of lipid catabolites causes the rancid odor of spoiled meat or butter (issues of food microbiology, discussed in Chapter 16). Microbes catabolize lipids by hydrolysis to glycerol and fatty acids (**Fig. 13.14**).

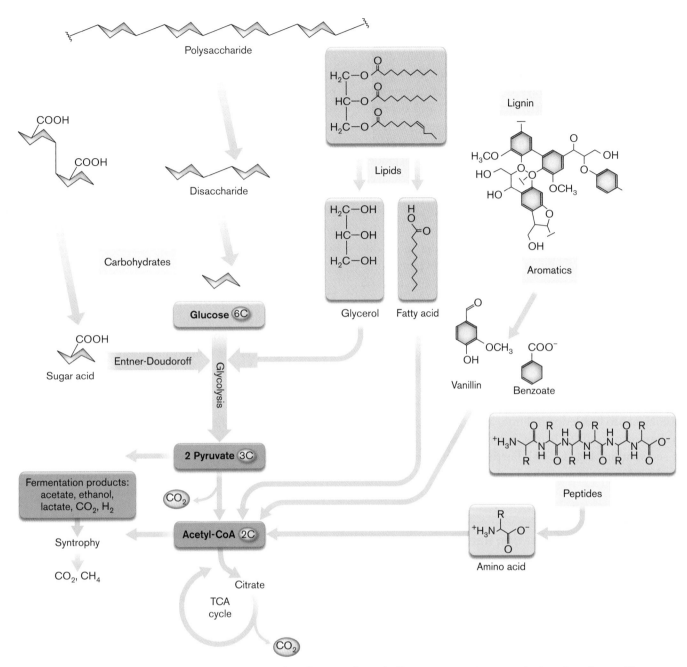

FIGURE 13.14 ■ **Many carbon sources enter central pathways of catabolism.** Carbohydrates are broken down by specific enzymes to disaccharides and then to monosaccharides such as glucose. Glucose and sugar acids are converted to pyruvate, which releases acetyl groups. Acetyl groups or acetate are also the breakdown products of fatty acids, amino acids, and complex aromatic plant materials such as lignin.

Glycerol, a three-carbon triol (a compound with three–OH groups), can be considered a three-carbon sugar; it commonly enters catabolism as an intermediate of glycolysis. Alternatively, other pathways break down glycerol to acetate. Fatty acids, more highly reduced than glycerol, undergo oxidative breakdown by the fatty acid degradation (FAD) pathway, forming acetyl groups. These acetyl groups enter the TCA cycle when a terminal electron acceptor is available; alternatively, they enter fermentation or anaerobic syntrophy.

Peptides. We commonly think of proteins as essential parts of a cell. However, when present in excess, proteins can be catabolized to provide energy. Initially, proteins are broken into peptides (**Fig. 13.13C**) by sequence-specific proteases. Peptides are then hydrolyzed to individual amino acids (**Fig. 13.14**). Some pathogens catabolize specific amino acids as part of the disease process. For example, *Legionella pneumophila*, the cause of legionellosis, catabolizes threonine. Threonine catabolism is required for the pathogen to grow within macrophages, a type of white blood cells.

Thus, threonine catabolism may be a target for new drugs against *L. pneumophila*.

Specific enzymes catalyze the early steps in the degradation of each amino acid until products are formed that can enter common pathways of carbohydrate catabolism or the TCA cycle. The initial step of amino acid degradation is one of two kinds: decarboxylation (removal of CO_2) to produce an amine, or deamination (removal of NH_3) to produce a carboxylic acid. Carboxylic acids are degraded through the TCA cycle. Amine products, such as cadaverine and putrescine, are often excreted, causing the noxious odor of decomposing flesh.

Aromatic molecules. Aromatic compounds are more difficult to digest than sugars because of the exceptional stability of aromatic ring structures. Yet many bacteria metabolize benzene derivatives, and even polycyclic aromatic molecules, either partly or all the way to CO_2. A particularly important aromatic substance found in nature is **lignin** (**Fig. 13.13D**), which forms the key structural support of trees and woody stems. A discouragingly complex molecule, lignin is made from six-carbon sugars converted to benzene rings, with ether connections that are difficult for enzymes to break down. Fungi and soil bacteria catabolize lignin to oxidized benzene derivatives such as benzoate and vanillin (**Fig. 13.14**). Further breakdown produces acetyl-CoA, which enters the TCA cycle.

Today, the environment contains increasing amounts of human-made aromatic compounds, produced for herbicides and other industrial uses, that are highly toxic pollutants. These include halogenated aromatics, such as polychlorinated biphenyls, the source of highly toxic dioxins. Halogenated aromatics turn out to be catabolized by a number of soil microbes, which are promising candidates for bioremediation of polluted environments (described in Section 13.6). Other types of aromatic molecules catabolized by microbes are polycyclic aromatic hydrocarbons (PAHs). Found in petroleum, the multiple fused rings of PAH compounds build up in natural environments—but with time, microbes can catabolize them and remediate the soil (see Section 13.6).

Products of Catabolism

What is the ultimate fate of all the bits of organic carbon chewed up by microbial catabolism? The answer has myriad consequences for human nutrition and environmental cycling, as well as food and industrial production. From **Table 13.1**, recall two major forms of catabolism: **fermentation**, in which all the electrons from organic substrates are put back onto the organic products; and **respiration**, in which the electrons removed are ultimately transferred to an inorganic electron acceptor such as oxygen or nitrate (anaerobic respiration). In a third case, photoheterotrophy,

TABLE 13.5	Fermentation Reactions in Bacteria (Examples)		
Reaction		**ATP produced**	**Species of bacteria**
$C_6H_{12}O_6 \rightarrow 2CH_3CH_2OH + 2CO_2$ [ethanolic]		2	*Zymomonas* sp.
$C_6H_{12}O_6 \rightarrow 2CH_3CH_2OCOO^- + 2H^+$ [lactate]		2	*Lactobacillus acidophilus*
$C_6H_{12}O_6 \rightarrow CH_3CH_2OH + CO_2 + CH_3CH_2OCOO^- + H^+$ [heterolactic]		2	*Leuconostoc mesenteroides*
$C_6H_{12}O_6 \rightarrow$ succinate, 2-oxoglutarate, acetate, ethanol, formate, lactate, CO_2, H_2 [mixed-acid; equation unbalanced]		Varies	*Escherichia coli*
$C_6H_{12}O_6 \rightarrow$ butanol, acetone, butyric acid, isopropanol, CO_2 [equation unbalanced]		Varies	*Clostridium acetobutylicum*
$2CH_3CH_2OCOO^-$ (lactate) $\rightarrow$ $CH_3(CH_2)COO^-$ (propionate) + $CH_3COO^- + CO_2 + 2H^+$		3	*Propionibacterium freudenreichii*
$2C_2H_2$ (acetylene) + $3H_2O \rightarrow CH_3CH_2OH + CH_3COO^- + H^+$		1	*Pelobacter acetylenicus*
2 Citrate^{3-} + $H^+ \rightarrow$ 2 succinate^{2-} + $CH_3COO^- + 2CO_2$		1	*Providencia rettgeri*
CH_3CHNH_2COOH (alanine) + $2NH_2CH_2COOH$ (glycine) + $2H_2O \rightarrow$ $3CH_3COO^- + 3NH_4^+ + CO_2$ [Stickland reaction]		3	*Clostridium sporogenes*
$COOH(CH_2)_2COO^-$ (succinate) $\rightarrow CH_3(CH_2)COO^-$ (propionate) + CO_2 [$Na^+_{in} \rightarrow Na^+_{out}$]		0	*Propionigenium modestum*

bacteria gain energy from light while using organic carbon substrates for catabolism and/or biosynthesis.

Fermentation. First identified by Louis Pasteur as *la vie sans air* ("life without air"), fermentation is the partial breakdown of organic food without net transfer of electrons to an inorganic terminal electron acceptor. Thus, food ferments without oxygen. Fermentation has a negative ΔG, owing to breakdown of a large molecule to several smaller products, which are usually more stable as well. Examples of fermentation pathways include ethanolic fermentation producing alcoholic beverages, and lactate fermentation by lactic acid bacteria producing cheese and yogurt. (We will discuss more examples; also see **Table 13.5**).

Respiration. Respiration combines catabolic breakdown of organic molecules with electron transfer to a terminal electron acceptor such as oxygen. Respiration yields far more energy from catabolism than does fermentation. For humans, respiration is synonymous with breathing; but in the absence of O_2, many microbes use alternative electron acceptors such as nitrate (see Chapter 14).

Catabolism: Molecular and Cell Biology

In natural ecosystems, how do cells manage to break down complex carbon sources? Free-living bacteria and fungi may catabolize thousands of different carbon sources, each requiring specific transporters and enzymes for initial breakdown. Most of these, particularly complex plant constituents, are not digestible by animals, whose genomes fail to encode the necessary enzymes. The only polysaccharides that human enzymes can digest are starch, lactose, and sucrose. Yet human breast milk includes a mixture of complex milk oligosaccharides, including branched chains of *N*-acetylglucosamine, sialic acid, and fucose. Amazingly, our own breast milk has evolved to support bacteria with specific catabolic abilities to colonize the infant gut.

It is well known that animals such as cattle require microbes living within their digestive tracts to ferment cellulose from grasses or wood. We now know that humans, too, require related bacteria to digest a variety of plant-derived fibers. For example, both cattle and humans contain anaerobes of the Gram-negative genus *Bacteroides* and the Gram-positive genus *Ruminococcus* (originally named for the bovine rumen). By evolving a symbiosis with microbes, animals avoid the need to acquire new catabolic genes in their own genome. The microbial genomes are functionally part of the human metagenome, the total sequence of genomes of a community of organisms.

The genome sequence of our microbiome determines our ability to digest common foods such as lettuce and tomatoes (**Fig. 13.15**). The major polysaccharide fibers of these

FIGURE 13.15 ■ Lettuce and tomatoes are made of xyloglucans. Theresa Rogers, now at California Lutheran University **(A)**, and Johan Larsbrink **(B)** study xyloglucan catabolism. **(C)**. Lettuce xyloglucans are beta-linked polymers of D-glucose (Glc) with side chains of xylose (Xyl), galactose (Gal), and fucose (Fuc). In tomatoes, xyloglucan side chains also have arabinose (Ara).

vegetables are xyloglucans, beta-linked glucose polymers with side chains containing xylose, galactose, and fucose or arabinose. Theresa Rogers at the University of Michigan, Ann Arbor, and Johan Larsbrink at the KTH Royal Institute of Technology, Stockholm, investigated how lettuce and tomato fibers are digested in our microbiome.

Each type of xyloglucan requires a slightly different set of genes, called a polysaccharide utilization locus (PUL). The PULs evolved from a common ancestral starch utilization system (SUS). Most gut bacteria possess a number of PULs distributed around their genomes, showing evidence of horizontal gene transfer and evolution of genomic islands (discussed in Chapter 9). **Figure 13.16A** shows the xyloglucan PUL for each of four different species of *Bacteroides* found in the human gut. The DNA sequences of these species show synteny—that is, sufficient similarity

A.

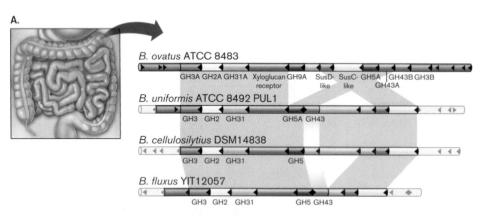

FIGURE 13.16 ■ Xyloglucan catabolism by gut bacteria. **A.** Some *Bacteroides* species have PUL sets of genes encoding xyloglucan degradation. The PULs show synteny, evidence of descent from a common ancestor. **B.** Xyloglucan degradation enzymes and transporters within a *Bacteroides* envelope. Xyloglucan cleavage by enzyme GH9A releases short sugar chains that cross the outer membrane via a SusC-like complex. Periplasmic enzymes further break down the chains. Protein colors are matched to their genes in (A).

B.

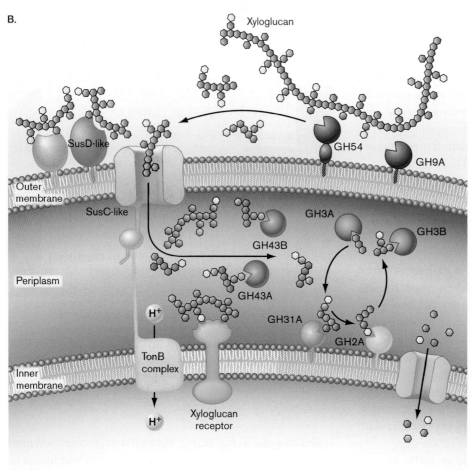

of sequence and map position to predict common ancestry (discussed in Chapter 17). Surprisingly, only a small minority of our gut microbiome possesses the ability to digest xyloglucans. Thus, the caloric content we receive from lettuce and tomatoes depends on just a few key members of our gut community.

How do our gut bacteria actually handle complex food sources such as xyloglucans? Since bacteria have solid cell walls and are incapable of phagocytosis, a bacterium cannot take up a large food particle, but it can secrete enzymes or place enzymes on the outer membrane to break down the

fibers into short oligosaccharides (**Fig. 13.16B**). The PUL genes encode outer membrane–inserted enzymes (GH5A and GH9A in **Fig. 13.16B**) that specifically cleave xyloglucans. Other genes encode oligosaccharide outer membrane transporters (SusC and SusC-related proteins). Within the periplasm, various amylases break down the oligosaccharides to monosaccharides. The monosaccharide products are then transported across the inner membrane into the cytoplasm, where they undergo glycolysis (discussed in the next section). Metagenome analysis reveals hundreds of Sus homologs, called Sus-like systems, that target different polysaccharides; these comprise, for example, 18% of the genome of *Bacteroides thetaiotaomicron*.

In a complex environment, with numerous potential substrates for catabolism, how does a given microbe decide what to eat first? Organisms select substrates on the basis of their availability and energy efficiency. Substrate selection involves gene regulation, as discussed in Chapter 10. For example, in *E. coli* the sugar lactose induces transcription of genes that encode beta-galactosidase (*lacZ*) and lactose permease (*lacY*). But in the presence of glucose, a preferred carbon source, *lac* transcription is halted. Halting *lac* transcription enables preferential catabolism of glucose. The process of prioritized consumption of substrates is known as **catabolite repression** (presented in Chapter 10).

As we saw in **Figure 13.14**, the products of catabolism of many diverse substrates ultimately funnel into a few common pathways of metabolism. The remainder of this chapter presents key pathways in detail: glycolysis and other pathways of glucose catabolism, the TCA cycle, and the catechol pathway of benzoate catabolism. These pathways play key roles in medical microbiology and in industrial fields such as bioremediation.

To Summarize

- **Carbohydrates, or polysaccharides**, are broken down to disaccharides, and then to monosaccharides. Sugars and sugar derivatives, such as amines and acids, are catabolized to pyruvate.

- **Pyruvate and other intermediary products of sugar catabolism** are fermented, or they are further catabolized to CO_2 and H_2O through the TCA cycle (in the presence of a terminal electron acceptor) or to CO_2, H_2, and CH_4 through fermentation and syntrophy.

- **Lipids and amino acids** are catabolized to glycerol and acetate, as well as other metabolic intermediates.

- **Aromatic compounds such as lignin and benzoate derivatives** are catabolized to acetate through different pathways, such as the catechol pathway.

- **Fermentation and respiration** complete the process of catabolism. In fermentation, the catabolite is broken down to smaller molecules without an inorganic electron acceptor. Respiration requires an inorganic terminal electron acceptor such as O_2.

- **Catabolism of complex substrates requires specialized enzymes and transporters.** Different bacteria contain homologous SUS or PUL complexes that fit different carbon sources.

13.5 Glucose Breakdown and Fermentation

For microbes feasting on complex carbohydrates, the main breakdown product is glucose. Glucose catabolism is important as a widespread source of energy, and also as a source of key substrates for biosynthesis, such as five-carbon sugars to build nucleic acids (discussed in Chapter 15).

Glucose and related sugars are catabolized through a series of phosphorylated sugar derivatives. A common theme in sugar catabolism is the splitting of a six-carbon substrate into two three-carbon products. The three-carbon products may form two molecules of pyruvate:

$$C_6H_{12}O_6 \rightarrow 2C_3H_4O_3 + 4H\ (2\ NADH + 2H^+)$$

Under anaerobiosis—the prevailing condition of many microbial habitats—the pyruvate obtained from sugar breakdown must be converted to compounds that receive electrons from NADH, in order to restore the electron-accepting form NAD^+. Different microbes reduce pyruvate to different end products of fermentation. Alternatively, through respiration, NADH may reduce an electron acceptor such as oxygen or nitrate, allowing pyruvate to feed into the TCA cycle (discussed in Section 13.6).

Note: The carboxylic acid intermediates of metabolism exist in equilibrium with their dissociated, or ionized, forms, identified by the suffix "-ate." For example, lactic acid dissociates to lactate; acetic acid dissociates to acetate. We use the "-ate" terms for acids whose ionized form predominates under typical cell conditions (around pH 7).

To catabolize glucose, bacteria and archaea use three main routes to pyruvate (**Fig. 13.17**):

- **Glycolysis**, or the **Embden-Meyerhof-Parnas (EMP) pathway**, in which glucose 6-phosphate isomerizes to fructose 6-phosphate, ultimately forming two molecules of pyruvate. EMP is used by many bacteria, eukaryotes, and archaea. From each glucose, the pathway generates net 2 ATP and 2 NADH.

- **Entner-Doudoroff (ED) pathway**, in which glucose 6-phosphate is oxidized to 6-phosphogluconate, a phosphorylated sugar acid. Alternatively, sugar acids may be converted directly to 6-phosphogluconate. Sugar acids often derive from intestinal mucus, and the ED pathway is essential for enteric bacteria to colonize the intestinal epithelium. The ED pathway generates only 1 ATP, 1 NADH, and 1 NADPH.

- **Pentose phosphate pathway (PPP)**, also known as the pentose phosphate shunt, in which glucose 6-phosphate is oxidized to 6-phosphogluconate, and then

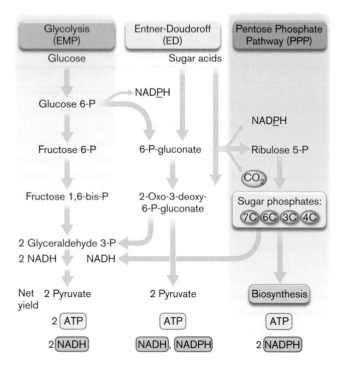

FIGURE 13.17 ■ **From glucose to pyruvate: three pathways.** The Embden-Meyerhof-Parnas (EMP) pathway of glycolysis, the Entner-Doudoroff (ED) pathway, and the pentose phosphate pathway (PPP) catabolize carbohydrates by related but different routes.

decarboxylated to a five-carbon sugar (pentose), ribulose 5-phosphate. The PPP produces sugars of three to seven carbons, which serve as precursors for biosynthesis or are converted to pyruvate as needed. The PPP generates 1 ATP plus 2 NADPH, the reducing cofactor most commonly associated with biosynthesis.

Glycolysis: The EMP Pathway

The **Embden-Meyerhof-Parnas** (**EMP**) **pathway**, or **glycolysis**, is the form of glucose catabolism most commonly studied in introductory biology; it is central for animals and plants, as well as many bacteria. In the EMP pathway, one molecule of D-glucose undergoes stepwise breakdown to two molecules of pyruvic acid (or its anion, pyruvate) (**Fig. 13.18**). Glucose is broken down in two stages. In the first stage, the glucose molecule is primed for breakdown by two steps of sugar phosphorylation by ATP. Each ATP phosphotransfer step spends Gibbs free energy, as we saw for glucose 6-phosphate formation in **Table 13.3**. The phosphoryl groups tag two sides of the glucose molecule for subsequent splitting into two three-carbon molecules of glyceraldehyde 3-phosphate (G3P).

In the second stage, each glyceraldehyde 3-phosphate is oxidized by NAD+ through steps leading to pyruvate. Each conversion of glyceraldehyde 3-phosphate to pyruvate forms 2 ATP—one ATP molecule from dephosphorylation of the substrate, and one from addition of inorganic phosphate. Since 2 ATP were spent originally to phosphorylate glucose, the net gain of energy carriers from each glucose is 2 NADH plus 2 ATP.

Most of the conversion steps are associated with a small change in energy—so small that the sign of ΔG depends on the concentrations of substrates or products; thus, some individual steps are reversible. In the cytoplasm, however, as intermediate products form they are quickly consumed by the next step, so the pathway flows in one direction. The direction of flow is determined by the key irreversible reactions that consume ATP. These steps drive the pathway by spending energy.

Phosphorylation and splitting of glucose. In the first stage of the EMP pathway, the six-carbon sugar is activated by two phosphorylation steps (**Fig. 13.19**, left). The first phosphoryl group (phosphate) is added at carbon 6 of glucose. (In some species of bacteria, the first phosphoryl group is added by phosphoenolpyruvate instead of ATP, but the net effect is the same.) The next enzyme-catalyzed step—rearrangement of glucose 6-phosphate to fructose 6-phosphate—involves no significant change in energy but prepares the sugar to receive the second phosphoryl group, forming fructose 1,6-bisphosphate.

The sugar then splits into two three-carbon sugars (trioses), each tagged with one of the two phosphates. The

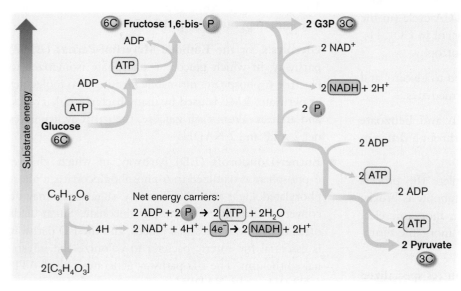

FIGURE 13.18 ■ **Substrate energy changes during the Embden-Meyerhof-Parnas pathway (glycolysis).** Glucose is activated through two substrate phosphorylations by ATP. The breakdown of glucose to two molecules of pyruvate is coupled to net production of 2 ATP and 2 NADH. Phosphoryl groups are shown as (P).

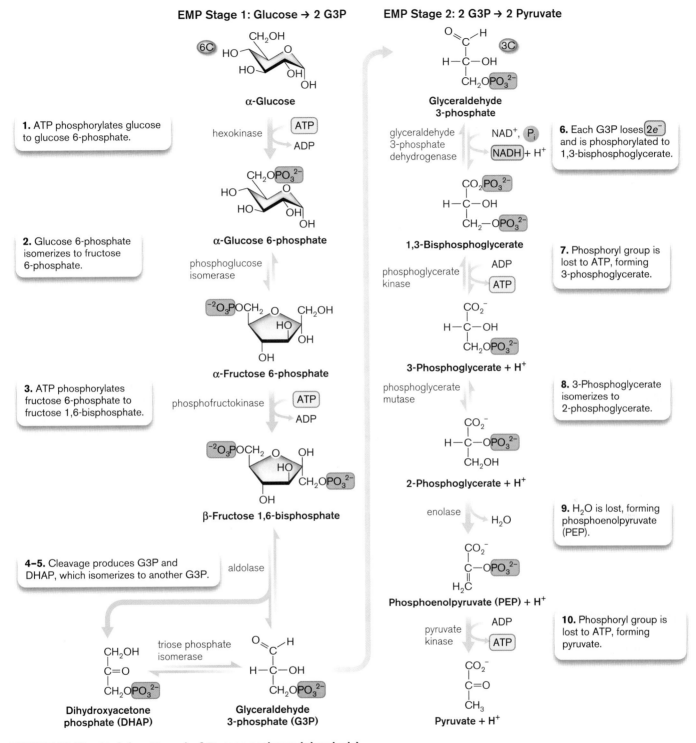

EMP Stage 1: Glucose → 2 G3P

EMP Stage 2: 2 G3P → 2 Pyruvate

α-Glucose

1. ATP phosphorylates glucose to glucose 6-phosphate.

hexokinase

α-Glucose 6-phosphate

2. Glucose 6-phosphate isomerizes to fructose 6-phosphate.

phosphoglucose isomerase

α-Fructose 6-phosphate

3. ATP phosphorylates fructose 6-phosphate to fructose 1,6-bisphosphate.

phosphofructokinase

β-Fructose 1,6-bisphosphate

4–5. Cleavage produces G3P and DHAP, which isomerizes to another G3P.

aldolase

triose phosphate isomerase

Dihydroxyacetone phosphate (DHAP)

Glyceraldehyde 3-phosphate (G3P)

Glyceraldehyde 3-phosphate

glyceraldehyde 3-phosphate dehydrogenase

6. Each G3P loses $2e^-$ and is phosphorylated to 1,3-bisphosphoglycerate.

1,3-Bisphosphoglycerate

phosphoglycerate kinase

7. Phosphoryl group is lost to ATP, forming 3-phosphoglycerate.

3-Phosphoglycerate + H^+

phosphoglycerate mutase

8. 3-Phosphoglycerate isomerizes to 2-phosphoglycerate.

2-Phosphoglycerate + H^+

enolase

9. H_2O is lost, forming phosphoenolpyruvate (PEP).

Phosphoenolpyruvate (PEP) + H^+

pyruvate kinase

10. Phosphoryl group is lost to ATP, forming pyruvate.

Pyruvate + H^+

FIGURE 13.19 ■ Embden-Meyerhof-Parnas pathway (glycolysis).

splitting of this molecule has a favorable entropy change (ΔS), but its enthalpy change (ΔH) is unfavorable, largely canceling out the energy yield. The two triose phosphates—glyceraldehyde 3-phosphate (G3P) and dihydroxyacetone phosphate (DHAP)—have nearly the same energy, so an enzyme interconverts them reversibly. Interconversion is necessary because only G3P proceeds further in the pathway.

Note: In Chapters 13–16, every substrate conversion shown requires catalysis by an enzyme. For the EMP pathway, the enzymes are shown, but for other pathways, the enzyme names are omitted.

ATP generation. In the second stage of the EMP pathway, each three-carbon glyceraldehyde 3-phosphate is directed into an energy-yielding pathway to pyruvate (**Fig. 13.19**, right).

First, the aldehyde (R–CHO) is converted to the carboxylate ion (R–COO⁻) plus H⁺. Conversion to a carboxylate releases a substantial amount of energy (negative ΔG). This oxidation of the aldehyde represents the major source of energy in glycolysis—the step at which the energy obtained is used to transfer a pair of electrons onto NAD^+, forming NADH (with an ionized H^+). In addition, sufficient energy is released to add a phosphoryl group (from inorganic phosphate, P_i) to the carboxylate, generating 1,3-bisphosphoglycerate.

In subsequent steps, the added phosphoryl group will be transferred to ADP, yielding net 1 ATP per pyruvate (2 ATP per glucose). The transfer of a phosphoryl group from an organic substrate to make ATP is called **substrate-level phosphorylation**. With subtraction of the initial two ATP molecules invested, the net energy carriers gained from each glucose molecule are as follows:

$$2\ NAD^+ + 4e^- + 4H^+ \rightarrow 2\ NADH + 2H^+$$
$$2\ ADP + 2\ P_i \rightarrow 2\ ATP + 2H_2O$$

Regulation of glycolysis. Enzymes of catabolism are regulated at the level of transcription of the enzyme. In addition, the activities of certain enzymes in long pathways require allosteric regulation. Allosteric regulation by enzyme substrates and products ensures that excess intermediates do not build up, and it avoids releasing more energy than the cell can use at a given time. Glycolysis is regulated so that its reactions go forward only when the cell needs energy, not when the cell is trying to synthesize glucose. The regulation occurs at steps where the products are consumed so rapidly that their forward reaction is effectively irreversible. Irreversible steps are shown as unidirectional arrows in **Figure 13.19**.

Thought Question

13.8 Some bacteria make an enzyme, dihydroxyacetone kinase, that phosphorylates dihydroxyacetone to dihydroxyacetone phosphate (DHAP). How could this enzyme help the cell yield energy?

Enzymes that catalyze irreversible steps are regulated so as to maintain consistent levels of intermediates in the pathway. The most important irreversible reaction in glycolysis is the phosphorylation of fructose 6-phosphate to fructose 1,6-bisphosphate, mediated by the enzyme phosphofructokinase. This enzyme is activated allosterically by ADP and inhibited by ATP or by the alternative phosphoryl donor, phosphoenolpyruvate.

What happens when the cell needs to reverse glycolysis in order to make glucose? Most of the intermediate reactions are reversible, so the same enzymes can be used for biosynthesis. Pathways that participate in both catabolism

(breakdown) and anabolism (biosynthesis) are called **amphibolic**.

The amphibolic pathway of glycolysis includes enzymes that can run in either direction, such as phosphoglucose isomerase. The direction of the pathway at a given time is determined by key enzymes that operate only in the catabolic direction or in the anabolic direction. For example, in glycolysis the ATP phosphorylation of fructose 6-phosphate is catalyzed by the enzyme phosphofructokinase, whereas in biosynthesis this step is reversed by a different enzyme, fructose bisphosphatase. Instead of regenerating ATP, fructose bisphosphatase removes the second phosphoryl group as inorganic phosphate—a step yielding energy and thus driving the whole pathway in reverse (toward biosynthesis of sugar). The two enzymes are regulated differently; the catabolic enzyme phosphofructokinase is activated by ADP, a signal of energy need, whereas the biosynthetic enzyme is inhibited by such signals.

The Entner-Doudoroff Pathway

The **Entner-Doudoroff (ED) pathway** offers a slightly different route to catabolize sugars, as well as sugar acids (sugars with acidic side chains). The ED pathway was originally studied for its role in production of the Mexican beverage *pulque*, or "cactus beer," by *Zymomonas* fermentation of the blue agave plant. Later, Tyrrell Conway and colleagues discovered genes encoding the Entner-Doudoroff enzymes in the genomes of many bacteria and archaea. In the human colon, the ED pathway enables *E. coli* and other enteric bacteria to feed on mucus secreted by the intestinal epithelium (**Fig. 13.20**). Some gut flora, such

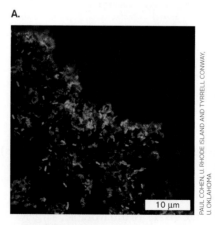

FIGURE 13.20 ■ Intestinal bacteria use the Entner-Doudoroff pathway. **A.** Intestinal *E. coli* (orange) feed primarily on gluconate from mucous secretions (fluorescence micrograph). **B.** Tyrrell Conway, at the University of Oklahoma, used genomics and genetic analysis to dissect the role of the Entner-Doudoroff pathway in the enteric bacterial catabolism of sugar acids from intestinal mucus.

Entner-Doudoroff Pathway

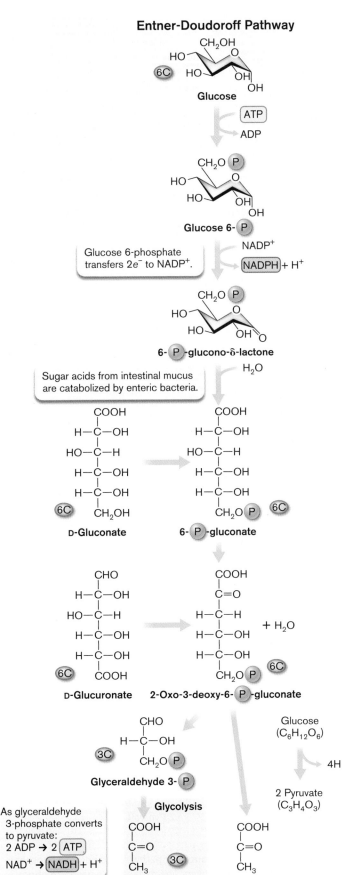

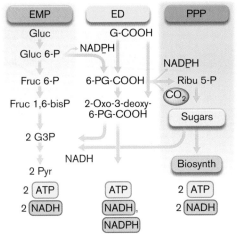

FIGURE 13.21 ■ Entner-Doudoroff pathway. Glucose 6-phosphate is oxidized to 6-phosphogluconate, with one pair of electrons transferred to NADPH. The 6-phosphogluconate is dehydrated and cleaved to form one pyruvate plus one glyceraldehyde 3-phosphate (G3P) that enters the EMP pathway to pyruvate.

as *Bacteroides thetaiotaomicron*, actually induce colonic production of the mucus that they consume. These bacteria that "farm" intestinal mucus enhance human health by preventing pathogen colonization, and by stimulating the immune system.

The ED pathway appears to have evolved earlier than the EMP pathway, because it involves fewer substrate phosphorylation steps and produces less ATP, and it is found in a wider range of prokaryotes. As in the EMP pathway, glucose is phosphorylated to glucose 6-phosphate (**Fig. 13.21**). The next step, however, involves oxidation by NADP+ at carbon 1, with loss of two hydrogens to form 6-phosphogluconate, a sugar acid. Gluconate is also found in intestinal mucus; the sugar acid can be phosphorylated to enter the ED pathway.

During ED, the hydrogens and electrons from glucose 6-phosphate are transferred to NADP+ instead of NAD+ as in the EMP pathway. This step differs from the EMP pathway in two respects: The carrier used is NADP+ instead of NAD+, and the electrons are transferred earlier, without a second ATP-consuming phosphorylation step. When the six-carbon substrate is eventually split into two three-carbon products, one of the three-carbon products is glyceraldehyde 3-phosphate, which enters the second stage of glycolysis (see **Fig. 13.19**). NADH is made, and 2 ATP are made by substrate-level phosphorylation. The remaining three-carbon product, however, is pyruvate. This one-step production of pyruvate short-circuits the

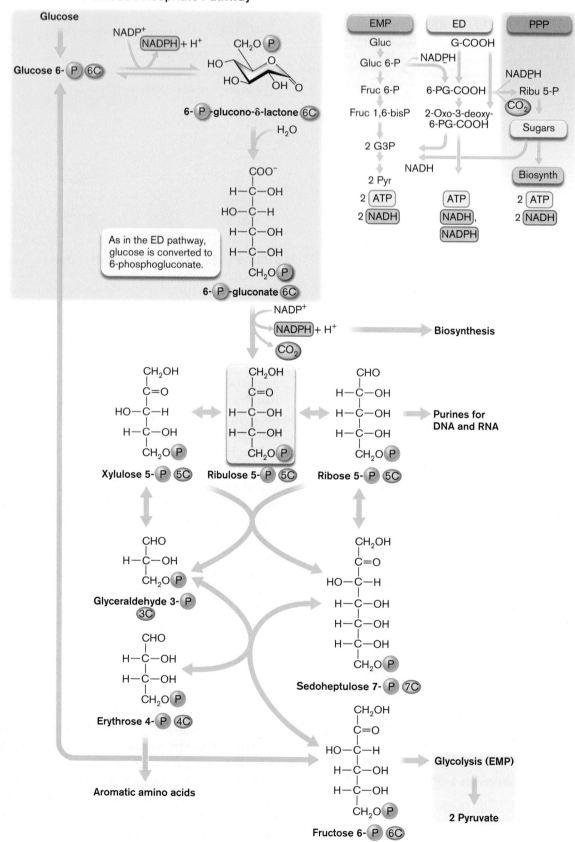

FIGURE 13.22 ▪ Pentose phosphate pathway of glucose catabolism. Like the Entner-Doudoroff pathway, the pentose phosphate pathway forms 6-phosphogluconate. One CO_2 is released, and 2 NADPH are produced for biosynthesis. The pathway can generate ribose 5-phosphate for purine synthesis or erythrose 4-phosphate to synthesize aromatic amino acids.

catabolic pathway, missing the formation of an ATP. The unused potential energy is released as waste heat.

The net ATP gain from the Entner-Doudoroff pathway is only 1 ATP per molecule of glucose—half that of the EMP pathway (see **Fig. 13.21**, inset). The electrons transferred, however, are equivalent: Instead of 2 NADH, the Entner-Doudoroff pathway generates 1 NADH and 1 NADPH.

What is the significance of NADPH, compared to NADH? In most cases, NADH transfers electrons to the ETS to store energy, whereas NADPH is used for biosynthesis (see Chapter 15). Thus, enzymes for amino acid biosynthesis will use NADPH but not NADH. The ratio between NADH and NADPH enables cells to balance their need for energy with their need to build biomass.

Thought Question

13.9 Explain why the ED pathway generates only 1 ATP, whereas the EMP pathway generates 2 ATP. What is the consequence for cell metabolism?

The Pentose Phosphate Pathway

A third pathway of glucose catabolism is the **pentose phosphate pathway** (**PPP**), which forms the key intermediate **ribulose 5-phosphate**, a five-carbon sugar. The pentose phosphate pathway generates 1 ATP with no NADH, but 2 NADPH for biosynthesis (**Fig. 13.22**). In addition, the PPP generates a complex series of intermediates that can be redirected as substrates for biosynthesis of diverse cell components such as amino acids and vitamins.

The pentose phosphate pathway starts like the Entner-Doudoroff pathway: Glucose 6-phosphate gives up two electrons to form NADPH and is oxidized to 6-phosphogluconate. The next step involves a second oxidation by $NADP^+$, with loss of a carbon as CO_2. The loss of CO_2 generates the five-carbon sugar ribulose 5-phosphate (hence the pathway name "pentose phosphate"). In succeeding steps, pairs of sugars, such as sedoheptulose 7-phosphate and glyceraldehyde 3-phosphate, exchange short carbon chains, giving rise to sugar phosphates of various lengths—for example, ribose 5-phosphate and erythrose 4-phosphate, which are precursors of purines and aromatic amino acids, respectively. Alternatively, if these routes to biosynthesis are not taken, the intermediates convert to fructose 6-phosphate and reenter the EMP pathway, where ATP and NADH are produced.

Fermentation Completes Catabolism

None of the pathways from glucose to pyruvate constitute completed pathways of catabolism, because NADH (and for some pathways NADPH) remains to be recycled. In the absence of oxygen or other electron acceptors, heterotrophic cells must transfer the hydrogens from NADH + H^+ back onto the products of pyruvate, forming partly oxidized fermentation products with the same redox level (balance of O and H) as the original glucose had (see **Table 13.5**). For example, the pyruvate may be converted to lactate by adding two electrons (with two hydrogen atoms) at the ketone, generating an alcohol group. The product, lactic acid, has a number of atoms ($C_3H_6O_3$) equal to half of the original glucose ($C_6H_{12}O_6$). Overall, one glucose is converted to two molecules of lactic acid—a process called **lactic acid fermentation**:

Glucose
$C_6H_{12}O_6$

2 Lactic acid
$C_3H_6O_3$

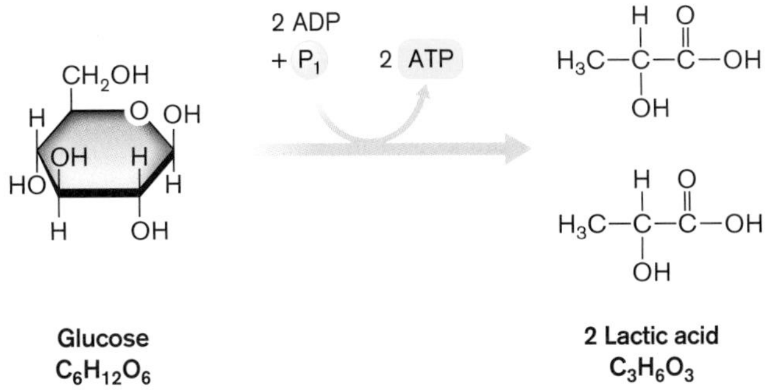

How does lactic acid fermentation yield energy to be stored as ATP? While two molecules of lactic acid contain the same number of atoms and electrons as those found in glucose, the molecular bonds are rearranged to form two carboxylic acids. The carboxylate/carboxylic acid has multiple states and increased entropy compared to the hydroxyl groups of glucose. Overall, glucose conversion releases free energy, and the reaction has a net negative value of ΔG. The ATP formation in such a reaction is called **substrate-level phosphorylation** because it involves only substrate reactions, no proton pumping across membranes (discussed in Chapter 14).

Alternative pathways of fermentation produce two molecules of ethanol plus two CO_2 molecules (**ethanolic fermentation**) or one lactic acid, one ethanol, and one CO_2 (heterolactic fermentation). Other kinds of fermentation are shown in **Table 13.5**. In all cases, fermentation products must be excreted from the cell. The large quantities of substrate consumed in fermentation generate large amounts of waste products to be excreted. These bacterial wastes are actually useful to human "fermentation industries" such as the production of alcoholic beverages (ethanolic fermentation) or cheese (lactic acid fermentation).

Why do fermenting bacteria give up such large quantities of waste products that retain usable energy? The reason is that in the absence of oxygen (or another electron acceptor), the fermentation products cannot yield energy. Most fermentation pathways do not generate ATP beyond that produced by substrate-level phosphorylation, the direct transfer of a phosphate group from an organic phosphate

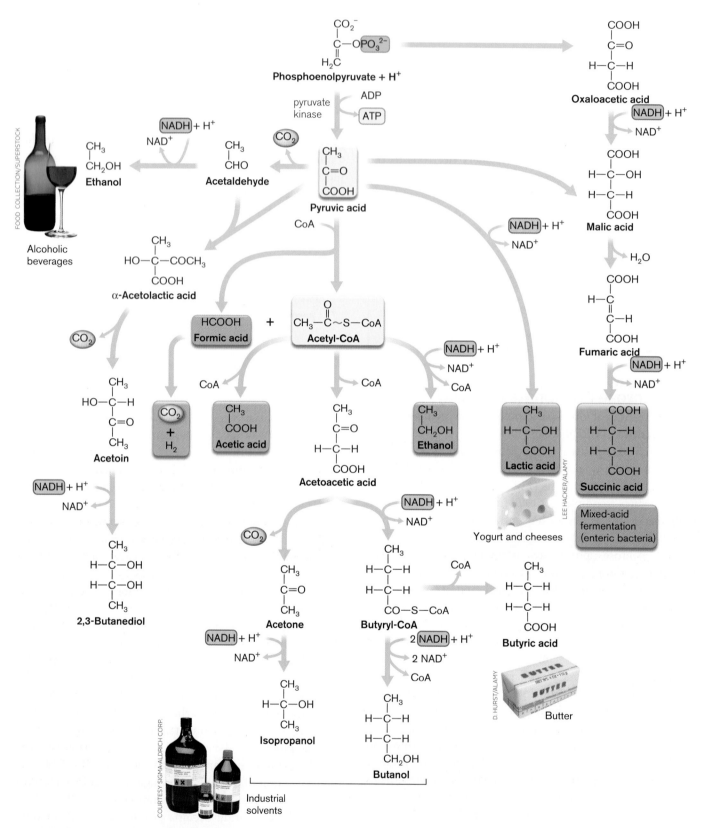

FIGURE 13.23 ▪ Fermentation pathways. Alternative pathways from pyruvate and phosphoenolpyruvate to end products, many of which we use for food or industry. Different species conduct different portions of the pathways shown.

to ADP. An example is the final step of glycolysis, catalyzed by pyruvate kinase (see **Fig. 13.19**). Much of the energy available from glucose remains unspent or is lost as heat. Nevertheless, fermentation is essential for microbes in environments such as anaerobic soil or animal digestive tracts. Even aerated cultures of bacteria start fermenting once their demand for oxygen exceeds the rate of oxygen dissolving in water. Microbes compensate for the low efficiency of fermentation by consuming large quantities of substrate and excreting large quantities of fermentation products. An advantage of fermentation is that the rapid accumulation of acids or ethanol can inhibit growth of competitors.

Mixed-acid fermentation. Numerous pathways have evolved to dispose of the waste in different forms (**Fig. 13.23**). *E. coli* ferments by a combination of routes known collectively as **mixed-acid fermentation**, forming acetate, formate, lactate, and succinate, as well as ethanol, H_2, and CO_2. Hydrogen (H_2) and carbon dioxide (CO_2) are the main gases passed by the human colon.

Colonic H_2 plus CO_2 can yield further energy for methanogenic microbes, through conversion to methane (discussed in Chapter 14). Excess hydrogen and methane gases can cause problems for medical procedures such as colonoscopy. For colonoscopy, the colon may need to be flushed with a carbohydrate solution, which bacteria may ferment, releasing hydrogen. When a polyp is removed from the colon by electrocautery (high-frequency electric current), the gas may ignite, causing an explosion. Colonic explosion can be avoided by flushing out the gases before electrocautery.

During mixed-acid fermentation, the proportions of products vary with pH. Low pH favors ethanol and lactate (which minimize acidification) over formate and acetate. *Clostridium* species produce alcohols (butanol, isopropanol); *Porphyromonas gingivalis*, a cause of periodontal disease, produces short-chain acids (propionate, butyrate).

Many fermentation products share key intermediates, such as acetyl-CoA. Acetyl-CoA is a versatile two-carbon intermediate that serves as the "Lego block" of metabolism. The molecule consists of an acetyl group esterified to **coenzyme A** (**CoA**) (**Fig. 13.24**), an important coenzyme whose discovery won Fritz Lipmann the 1953 Nobel Prize in Physiology or Medicine. CoA has a thiol (SH) that exchanges its hydrogen for an acyl group, thus activating the molecule for transfer in various metabolic pathways.

For example, the enzyme pyruvate formate lyase splits pyruvate to form acetyl-CoA plus formate:

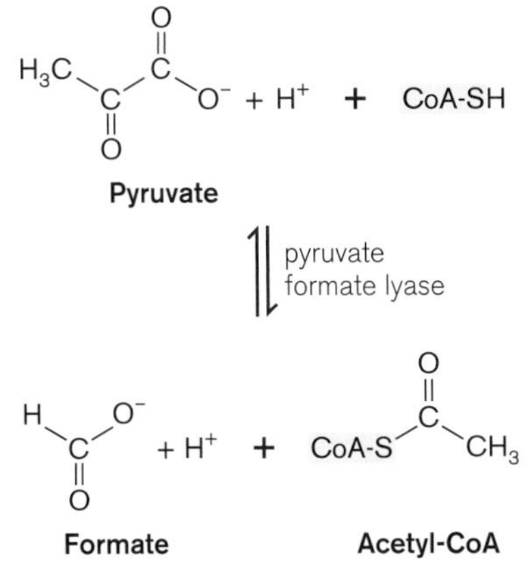

The SH group of CoA accepts the acetyl group to form acetyl-CoA. The hydrogen ($H^+ + e^-$) of the SH group is transferred onto the carboxyl carbon of pyruvate, generating formate.

Acetyl-CoA can be converted to various fermentation products by several different pathways. The simplest is an exchange of water with CoA, forming acetate:

$$CH_3CO—S\text{-}CoA + H_2O \rightleftharpoons CH_3COO^- + H^+ + HS\text{-}CoA$$

Incorporation of water restores the hydrogen to the thiol of CoA, and the OH to acetate. The acetate thus formed may then be excreted by the cell.

Note that the excreted acids and alcohols are readily recovered by cells when an electron acceptor becomes available for oxidation or when these fermentation products are needed as building blocks for biosynthesis. Alternatively, the excreted "wastes" may be utilized by other species capable of metabolizing them further. For example, the H_2 released by gut bacteria through mixed-acid fermentation can be oxidized to water by the gastric pathogen *Helicobacter pylori*. Alternatively, H_2 plus CO_2 can be used by gut methanogens to yield energy with release of methane.

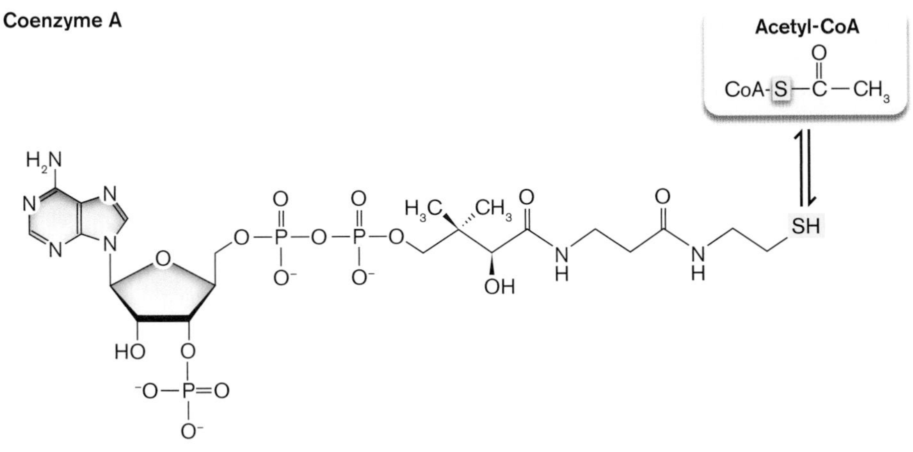

Coenzyme A

FIGURE 13.24 ■ **Structure of coenzyme A.** The thiol (SH) forms an ester link with the COOH of acetic acid, generating acetyl-CoA.

In soil ecosystems, various other fermentations have evolved to utilize available substrates (see **Table 13.5**). *Clostridium* species generate butanol and other solvents of great industrial value. Other species ferment pairs of amino acids, in a type of mechanism called the **Stickland reaction**. Fermenting a pair of amino acids avoids generating H_2 (a loss that would waste reduction potential) and produces as much as 3 ATP. By contrast, the succinate fermentation of *Propionigenium modestum* is an anaerobic reaction that proceeds with too small a ΔG value to generate a single ATP. Instead, the decarboxylation step (catalyzed by methylmalonyl-CoA decarboxylase) is coupled to the pumping of sodium ions across the cell membrane. The sodium pump stores energy (discussed in Chapter 14).

Food and Industrial Applications

The "waste products" of fermentation retain much of their organic structure and food value. Their carbon-hydrogen bonds retain the ability to reduce oxygen, and their organic structures can be used for biosynthesis. Thus, fermentation products have proved extremely useful in human culture and technology. For thousands of years, ethanolic fermentation by yeast has been used to produce wine and beer, while lactic acid fermentation has been used to produce yogurt and cheese (see Chapter 16). The minor product butyric acid (butyrate) lends taste to butter.

Fermentations involving amino acids generate small amounts of products with intense odors and flavors. Products of amino acid catabolism confer some of the distinctive flavors of fermented foods and beverages, such as cheese. Swiss cheese is produced in two fermentation stages. In the first stage, at high temperature, thermophilic lactic acid bacteria such as *Lactobacillus helveticus* and *Streptococcus salivarius* ferment the milk sugar lactose, a disaccharide of glucose and galactose, generating mainly lactic acid. In the second stage, *Propionibacterium freudenreichii* converts lactate to propionate, acetate, and CO_2, which bubbles to form the "eyes" (**Fig. 13.25**). The propionate contributes to the distinctive flavor of Swiss cheese, along with other minor side products that have intense flavors. For more on Swiss-cheese fermentation, see **eTopic 13.2**. Microbial food production is discussed in Chapter 16.

In the chemical industry, microbial fermentation produces industrial solvents such as butanol and acetone. Acetone production had historic impact during World War I, when Britain needed a source of acetone to manufacture gunpowder. Acetone and butanol were produced as fermentation products by *Clostridium acetobutylicum*, a bacterium identified by the biochemist Chaim Weizmann. Weizmann was a Russian-born Jew who sought a Jewish homeland in Palestine. At the end of the war, Weizmann's process for acetone production helped earn him the British government's support for the national homeland of Israel, where the biochemist later became the country's first president.

Today, microbial fermentation produces many "commodity chemicals," such as ethanol, butanol, and glycerol. Compared to industrial alternatives, such as production from petroleum, microbial culture is environmentally desirable and energy efficient. Furthermore, microbes are "smart" producers of small but complex pharmaceuticals such as vitamins, amino acids, and antibiotics, molecules that would require many steps of organic synthesis. Applications of microbial biosynthesis are discussed further in Chapters 15 and 16.

Another important application of fermentation lies in diagnostic microbiology. To quickly identify the microbe causing a disease and prescribe an effective antibiotic,

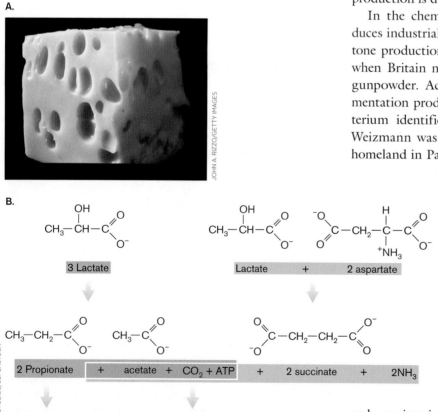

A. Swiss cheese (Emmentaler).

JOHN A. RIZZO/GETTY IMAGES

©PHOTODISC/SUPERSTOCK

3 Lactate

Lactate + 2 aspartate

2 Propionate + acetate + CO_2 + ATP + 2 succinate + 2NH_3

Swiss-type flavor

"Eyes" in cheese

FIGURE 13.25 ■ Swiss-cheese production involves fermentation of lactate to propionate. **A.** Swiss cheese (Emmentaler). **B.** Lactate is fermented by *Propionibacterium freudenreichii* to propionate, acetate, and CO_2. Concurrent fermentation of lactate and aspartate generates additional CO_2, increasing the size and number of eyes.

A. Phenol red broth test

Gas

B. Sorbitol MacConkey agar

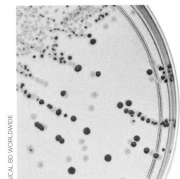

FIGURE 13.26 ■ **Clinical tests based on fermentation.**
A. Phenol red broth test with Durham tube. From left to right: *Escherichia coli* gives acidic fermentation products (yellow) and gas (CO_2 and H_2) in a Durham tube; *Alcaligenes faecalis* does not ferment, and the tube turns deep red without gas; uninoculated control is orange-red. **B.** Sorbitol fermentation test for pathogen *E. coli* O157:H7. White colonies (strain O157:H7) fail to ferment sorbitol, unlike red colonies (nonpathogenic *E. coli*).

hospitals use rapid and inexpensive biochemical tests (Chapter 28). A pH indicator added to growth media can detect the acid produced when a substrate is fermented, resulting in a color change (**Fig. 13.26A**). Phenol red is a pH indicator that is orange-red at neutral pH. It turns yellow in media acidified by fermentation acids (below pH 6.8) and red at higher pH (above pH 7.4). A culture of *Escherichia coli*, which ferments quickly, turns phenol red to a medium yellow after 24 hours. *Alcaligenes faecalis*, meanwhile, ferments poorly on sugar but converts peptides in the broth to alkaline amines; this culture turns deep red.

More specific tests depend on the microbe's ability to ferment specific sugars. Different species possess different enzymes able to convert different sugars and sugar derivatives to glucose, which then enters the common fermentation pathway. A well-known example is the use of sorbitol MacConkey agar to test for *E. coli* O157:H7, a lethal pathogen contaminating beef and vegetables. The O157:H7 strain of *E. coli* has a set of virulence genes absent from nonpathogenic *E. coli* strains that are present among our normal colon biota. But the pathogen also happens to lack genes present in normal *E. coli*, such as a gene encoding an enzyme to ferment sorbitol. Thus, failure to ferment sorbitol indicates a high probability that the strain is *E. coli* O157:H7. On sorbitol MacConkey agar, bacteria that ferment sorbitol during growth produce acids. The acidity causes a dye to turn red (the opposite of the phenol red test). Failure to ferment sorbitol (observed as pale colonies) indicates high probability of *E. coli* O157:H7 (**Fig. 13.26B**).

To Summarize

- **Embden-Meyerhof-Parnas (EMP) pathway:** Glucose is activated by two substrate phosphorylations, and then cleaved to two three-carbon sugars. Both sugars eventually are converted to pyruvate. The pathway produces 2 ATP and 2 NADH.

- **Entner-Doudoroff (ED) pathway:** Glucose is activated by one phosphorylation, and then dehydrogenated to 6-phosphogluconate. 6-Phosphogluconate is cleaved to pyruvate and a three-carbon sugar, which enters the EMP pathway to form pyruvate. The ED pathway produces 1 ATP, 1 NADH, and 1 NADPH.

- **Pentose phosphate pathway (PPP):** Glucose is dehydrogenated to 6-phosphogluconate, and decarboxylated to ribulose 5-P. A series of intermediate sugars may serve as substrates for biosynthesis. The PPP may produce 1 ATP and 2 NADPH.

- **Glucose catabolism is reversible**, enabling cells to build glucose from small molecules.

- **Fermentation is the completion of catabolism** <u>without</u> the electron transport system and a terminal electron acceptor. The electrons from NADH are restored to pyruvate or its products in reactions that generate fermentation products, including alcohols and carboxylates, as well as H_2 and CO_2. Energy is stored in the form of ATP.

- **Fermentation** has applications in food, industrial, and diagnostic microbiology.

13.6 The TCA Cycle and Aromatic Catabolism

The products of sugar breakdown can be catabolized to CO_2 and H_2O through the **tricarboxylic acid (TCA) cycle**. The TCA cycle generates molecules of NADH and $FADH_2$, which donate electrons to an electron transport system (ETS) with a terminal electron acceptor such as O_2 (see Chapter 14). But sugars are not the cycle's only substrates. Lignins, phenolics, and even polycyclic aromatic toxins are degraded to products that enter the TCA cycle.

The TCA cycle is also known as the Krebs cycle, named for Hans Krebs (1900–1981), who shared the 1953 Nobel Prize in Physiology or Medicine with Fritz Lipmann. Krebs and his colleagues at Sheffield University, England, studied catabolism by observing the oxidizing activities of crude enzyme preparations from sources such as pigeon breast muscle, beef liver, and cucumber seeds. In all of these animal and plant tissues, the TCA cycle is conducted by mitochondria, using virtually the same process as their bacterial ancestors used.

Pyruvate Is Converted to Acetyl-CoA

Glucose catabolism connects with the TCA cycle through the breakdown of pyruvate to acetyl-CoA and CO_2. Recall from **Figure 13.14** that acetyl-CoA is also generated from many other catabolic pathways, including those for breakdown of fatty acids, amino acids, and lignin fragments in soil. Regardless of its source, acetyl-CoA enters the TCA cycle by condensing with the four-carbon intermediate oxaloacetate to form citrate (**Fig. 13.27**). Citrate undergoes two steps of oxidative decarboxylation, in which CO_2 is released and two hydrogens with electrons are transferred to make $NADH + H^+$ or $FADH_2$. The TCA cycle, in whole or in part, is found in all microbial species except for degenerately evolved pathogens that are dependent on host metabolism, such as *Treponema pallidum*, the cause of syphilis (**eTopic 13.3**).

Thought Question

13.10 If a cell respiring on glucose runs out of oxygen and other electron acceptors, what happens to the electrons transferred from the catabolic substrates?

We present first the connecting step between pyruvate and acetyl-CoA, followed by the details of the TCA cycle. Bacteria and archaea use at least ten known variations on the TCA cycle, conducted by diverse species under various environmental conditions. You will no doubt be relieved to hear that we present only one: the Krebs pathway, which is found in most Gram-negative and Gram-positive bacteria, and in most eukaryotes. Other pathways are detailed in online resources such as the KEGG Pathway Database.

Pyruvate is converted to acetyl-CoA through removal of CO_2 and transfer of $2e^-$ onto NAD^+. The removal of CO_2 and transfer of two electrons is known as oxidative decarboxylation. The oxidative decarboxylation of pyruvate, coupled to CoA incorporation, is performed by an unusually large multisubunit enzyme called the **pyruvate dehydrogenase complex**, or **PDC**. PDC is a key component of metabolism in bacteria and mitochondria, the first molecular player to direct sugar catabolism into respiration. In human mitochondria, defects in PDC affect organs that have a high metabolic rate, such as heart and brain, causing myocardial malfunction and heart failure, and neurodegeneration. A structural model for PDC, and the details of its reaction mechanism, are shown in **eTopic 13.4**.

The overall reactions catalyzed by PDC are:

$$CH_3COCOO^- + H^+ + HS\text{-}CoA \rightarrow$$
$$CH_3CO\text{-}S\text{-}CoA + CO_2 + 2H^+ + 2e^-$$
$$NAD^+ + 2H^+ + 2e^- \rightarrow NADH + H^+$$

The removal of stable CO_2 yields energy for the electron transfer to NADH. The protons dissociated from pyruvate and from the thiol (SH) of CoA ($2H^+$ in total) are balanced by the net gain of protons by $NADH + H^+$.

The activity of PDC is increased by high concentrations of its substrates (CoA and NAD^+) and inhibited by its products (acetyl-CoA and NADH). The product acetyl-CoA may enter one of several pathways. In *E. coli*, when glucose is plentiful, acetyl-CoA is mostly converted to acetate via the intermediate acetyl phosphate (**Fig. 13.27**). Acetyl phosphate is a global signaling molecule that indicates to the cell the quantity and quality of carbon source available. As glucose decreases, the cell starts to reclaim acetate, converting it back to acetyl-CoA for entry into the TCA cycle. At the level of gene expression, PDC responds to environmental conditions. As would be expected, PDC gene expression is repressed by carbon starvation and by low levels of oxygen.

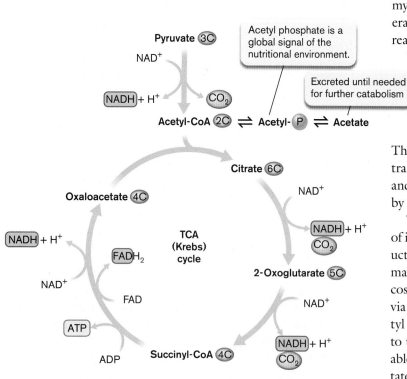

FIGURE 13.27 ■ Acetyl-CoA feeds into the TCA cycle. Pyruvate undergoes oxidative decarboxylation and incorporates CoA to form acetyl-CoA. Depending on the state of the cell, either acetyl-CoA is converted to acetate for excretion, or it enters the TCA cycle.

Acetyl-CoA Enters the TCA Cycle

Recall that acetyl-CoA can participate in various fermentations (see **Fig. 13.23**). But when a strong terminal electron acceptor is available (such as O_2), acetyl-CoA can enter the TCA cycle to transfer its electrons to electron carriers (**Figs. 13.27** and **13.28**). First, the acetyl group condenses with oxaloacetate, a four-carbon dicarboxylate (double acid). The condensation forms citrate, a six-carbon tricarboxylate. An advantage of intermediates with two or more acidic groups is that the concentration of the fully protonated form is extremely low; thus, the molecule is unlikely to be lost from the cell by diffusion across the membrane, as are monocarboxylic acids, such as acetate.

Through the rest of the cycle, citrate loses two carbons as CO_2 by a series of reactions that transfer increments of energy to 3 NADH, 1 $FADH_2$, and 1 ATP. Each reaction step couples energy-yielding to energy-storing events (**Fig. 13.28**).

Step 1. As the acetyl group condenses with oxaloacetate, the hydrolysis of acetyl-CoA consumes a molecule of H_2O to restore HS-CoA. The removal of HS-CoA yields energy to condense acetate with oxaloacetate, forming citrate.

Step 2. Citrate undergoes two rearrangements (with little energy change) to form isocitrate. Isocitrate then undergoes oxidative decarboxylation. As we saw for pyruvate, removal of CO_2 yields energy to transfer $2H^+ + 2e^-$ to form NADH + H^+, producing 2-oxoglutarate (alpha-ketoglutarate).

Step 3. 2-Oxoglutarate undergoes decarboxylation to release CO_2 and make another NADH + H^+. In this case, CoA is incorporated, making succinyl-CoA.

Step 4. Succinyl-CoA releases CoA, yielding energy to phosphorylate ADP to ATP. To form fumarate, $2H^+ + 2e^-$ are transferred to FAD to form $FADH_2$—a reaction involving negligible free energy change. FAD is reduced instead of NAD^+ because electron donation from succinyl-CoA does not yield enough energy to reduce NAD^+.

Step 5. Fumarate incorporates water across its double bond, forming the hydroxy acid malate. The increasing stability from fumarate to malate and from malate to oxaloacetate yields enough energy to form the final NADH + H^+. Oxaloacetate is the original intermediate of the cycle, which again accepts the next acetyl-CoA.

Note: Some textbooks state that succinyl-CoA synthetase, the enzyme catalyzing step 4 in the TCA cycle (Fig. 13.28), phosphorylates GDP to GTP. According to the primary literature, ADP phosphorylation predominates in *E. coli* (Margaret Birney et al. 1996. *J. Bacteriol.* **178**:2883), whereas in *Pseudomonas* species, various nucleoside diphosphates are phosphorylated (Vinayak Kapatral et al. 2000. *J. Bacteriol.* **182**:1333). Human mitochondria have two enzymes, which form ATP and GTP, respectively (David Lambeth et al. 2004. *J. Biol. Chem.* **279**:36621).

Observing the TCA cycle intermediates. How were all the TCA intermediates identified? The main experimental approach available to Krebs and his contemporaries was to guess at dozens of short-chain acids known to exist in cells, and then to add each individually to an enzyme preparation and test for TCA cycle activity. In aerobic organisms, the TCA cycle is tightly tied to respiration, so an increase in uptake of oxygen signaled a TCA intermediate. A major experimental advance was the use of tracer isotopes such as ^{14}C, discovered by Martin Kamen in 1940 (see **eTopic 15.1**).

Radioisotope tracers answered an important question about the TCA cycle. As the two acetyl carbons cycle through, which two carbons of each intermediate are removed as CO_2? Which carbons are lost first: the acetyl carbons or those of the original oxaloacetate? This question was answered by use of substrates radiolabeled with ^{14}C (highlighted green in **Fig. 13.28**). In the experiment, bacteria are fed ^{14}C-radiolabeled acetate, which enters the TCA cycle. The radiolabeled carbons are captured by the TCA intermediates and retained into the next cycle, whereas two COOH carbons from the original oxaloacetate are lost. Thus, the four-carbon intermediate does not recycle intact, but breaks down and re-forms each time it passes through the cycle.

Loss of the second CO_2 forms succinyl-CoA, which is converted to succinate. Succinate is a symmetrical molecule; thus, the former identities of the acetyl and oxaloacetate moieties are now erased, and the carbons are equally likely to disappear from either half of the molecule. Further conversion steps regenerate oxaloacetate—half its carbon from the acetyl group and half from the original oxaloacetate.

The TCA cycle and oxidative phosphorylation. In all, each acetate generates 3 NADH molecules, 1 $FADH_2$, and 1 ATP; and all the carbons from pyruvate (ultimately from glucose)

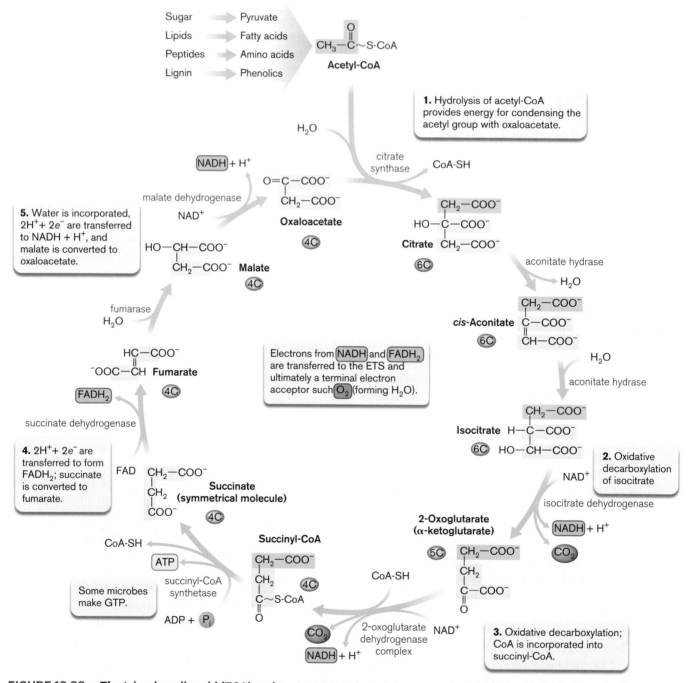

FIGURE 13.28 ■ The tricarboxylic acid (TCA) cycle. Acetyl-CoA derived from pyruvate and other catabolic pathways enters the TCA cycle. Green highlighting shows the fate of labeled acetate incorporated into the TCA cycle. Some forms of isocitrate dehydrogenase reduce $NADP^+$ instead of NAD^+.

have been released as waste CO_2. From the standpoint of the carbon skeleton, the glucose breakdown is now complete. But do we have a completed metabolic pathway? No, because all of the NADH and $FADH_2$ need to be recycled by donating their electrons onto a terminal electron acceptor.

The process of electron transfer from NADH and $FADH_2$ is mediated by the electron transport system (ETS, discussed in Chapter 14). In the ETS, electrons are transferred from reduced proteins and cofactors to more oxidized proteins and cofactors, as in the examples in **Table 13.4**.

Some of the membrane proteins use the energy of electron transfer to pump protons, generating a gradient of hydrogen ions across the membrane (**Fig. 13.29**)—a process discussed in detail in Chapter 14.

Assuming a theoretical maximum yield of 3 ATP generated per NADH and 2 ATP per $FADH_2$, the hydrogen ion gradient then drives the membrane ATP synthase to synthesize as many as 34 ATP. Another 4 ATP come from glucose breakdown and the TCA cycle (38 total per glucose). Under actual conditions, however, bacteria make less ATP;

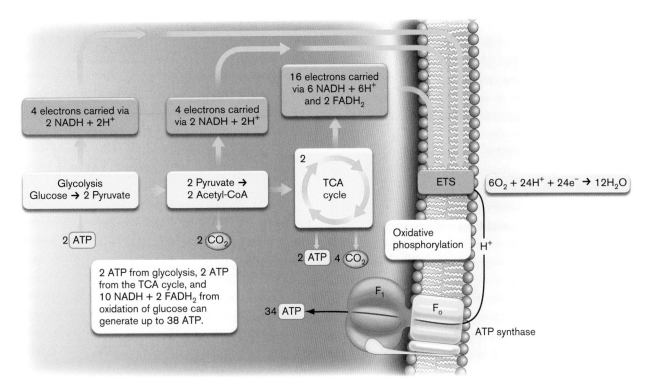

FIGURE 13.29 ■ **Complete oxidation of glucose.** Glucose catabolism generates ATP through substrate-level phosphorylation and through the electron transport system's pumping of H⁺ ions to drive the ATP synthase. The complete oxidative breakdown of glucose to CO_2 and H_2O could theoretically generate up to 38 ATP. Under actual conditions, the number is smaller.

about 20 ATP per glucose are made by a well-aerated culture of *E. coli*. Bacterial cells make trade-offs for flexibility, spending energy to maintain a stable proton potential during extreme changes in external pH and redox levels.

The overall process of electron transport and ATP generation is termed **oxidative phosphorylation**. The overall process of oxidative catabolism from substrate breakdown to oxidative phosphorylation is a form of **respiration**. The overall equation for the aerobic respiration of glucose is:

$$C_6H_{12}O_6 + 6H_2O + 6O_2 \rightarrow 12H_2O + 6CO_2$$

Glucose respiration can generate a relatively large number of ATPs per glucose—far more than fermentation can. In bacteria, however, the actual number of ATP molecules generated varies widely with availability of carbon source and oxygen. For example, as oxygen decreases in the environment, the ability to oxidize NADH decreases, so the cell may make only 1 or 2 ATP per NADH (discussed in Chapter 14). In addition, the enzymes of the TCA cycle are regulated extensively by substrate activation and product inhibition, and their expression is induced by high levels of oxygen and glucose.

Glyoxylate bypass. What happens when glucose is scarce and cells need carbon both for energy and for biosynthesis? Bacteria may catabolize lipids instead of glucose, breaking down the fatty acids into acetyl-CoA for the TCA cycle. But oxygen may be limited, and carbons would be lost as CO_2. Some bacteria can switch to using a modified TCA cycle called the **glyoxylate bypass** (**Fig. 13.30A**). The glyoxylate bypass allows lung pathogens such as *Pseudomonas aeruginosa* and *Mycobacterium tuberculosis* to catabolize fatty acids, via their breakdown to acetyl-CoA.

The glyoxylate bypass consists of two enzymes that divert isocitrate to glyoxylate, and then incorporate a second acetyl-CoA to form malate. The malate can then regenerate oxaloacetate to complete the bypass cycle, donating $2e^-$ to form NADH. The net reaction is:

2 Acetyl-CoA + oxaloacetate + NAD⁺ →
succinate + malate + 2 CoA + NADH + H⁺

Alternatively, malate or oxaloacetate can be diverted into biosynthesis of glucose (gluconeogenesis)—a pathway that reverses much of glucose catabolism (see Chapter 15). Most bacteria need sugar biosynthesis to build their cell walls.

The glyoxylate bypass cuts out all loss of CO_2 and electron transfer to energy carriers, with the exception of 1 $FADH_2$ and possibly 1 NADH from malate to oxaloacetate. Thus, limited energy is released, but two carbons can be diverted to biosynthesis. **Figure 13.30B** shows how the pathogen *Mycobacterium tuberculosis* uses the glyoxylate bypass. A mystery of the disease tuberculosis is the ability of *M. tuberculosis* to persist for long periods, growing slowly within macrophages. What carbon source and metabolism does the pathogen use? The question was addressed by labeling infected cells with carbon sources enriched for the isotope ¹³C. The intracellular bacteria were then isolated

A.

B. *M. tuberculosis* in macrophages

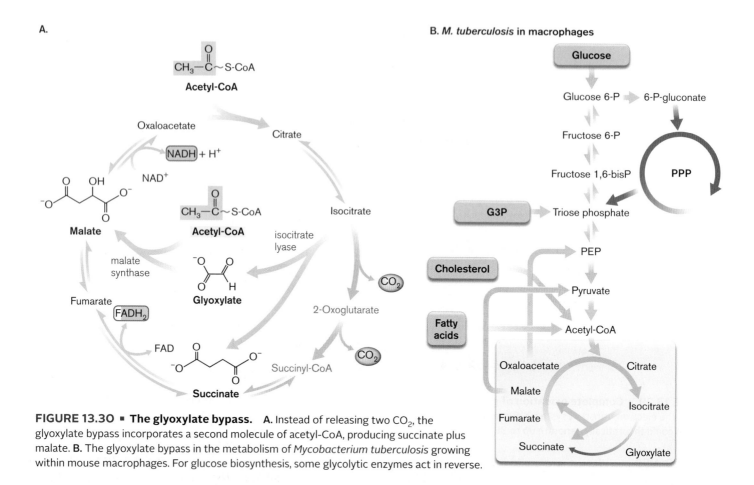

FIGURE 13.30 ■ **The glyoxylate bypass.** **A.** Instead of releasing two CO_2, the glyoxylate bypass incorporates a second molecule of acetyl-CoA, producing succinate plus malate. **B.** The glyoxylate bypass in the metabolism of *Mycobacterium tuberculosis* growing within mouse macrophages. For glucose biosynthesis, some glycolytic enzymes act in reverse.

by cell fractionation (discussed in Chapter 3) and analyzed for ^{13}C isotope enrichment of various components. It turns out that persistent intracellular *M. tuberculosis* catabolizes host lipids via the glyoxylate bypass, diverting much of the carbon to build sugars and amino acids for bacterial cell growth. Thus, key enzymes of the glyoxylate bypass, and their regulators, offer targets for new antibiotics.

The TCA cycle for amino acid biosynthesis. Analysis of pathway evolution indicates that the TCA cycle originally evolved to provide substrates for building amino acids (see Chapter 15). For example, the TCA cycle intermediate 2-oxoglutarate (alpha-ketoglutarate) is aminated to form glutamate, which leads to glutamine. The amine group comes from nitrogen gas fixed to ammonium ion (discussed with biosynthesis in Chapter 15). Oxaloacetate is aminated to form aspartate, entering pathways to purines and pyrimidines as well. The TCA cycle, like glycolysis, is an amphibolic pathway that provides substrates for biosynthesis. Many bacteria use the TCA cycle and glycolytic enzymes to build their sugars and amino acids, as discussed in Chapter 15. Others, such as *Treponema pallidum*, the cause of syphilis, have lost the TCA cycle by reductive evolution. They must obtain amino acids synthesized by their host organism (see **eTopic 13.3**).

Catabolism of Benzene Derivatives

The TCA cycle plays a key role in the breakdown of aromatic carbon sources such as the benzene ring. Natural sources of aromatic carbon include lignin from wood, and polycyclic aromatic hydrocarbons (PAHs) from petroleum. Aromatic components of petroleum cause much of the environmental damage during oil spills such as the *Deepwater Horizon* spill in 2010. Soil is often polluted by industrial aromatic compounds, such as nitrate explosives, aniline dyes, and the solvent toluene. Even the most remote places on Earth, such as Antarctica, show traces of such pollutants.

To remove aromatic pollutants from water and soil, we depend on microbial catabolism. Aromatic molecules are notoriously difficult to catabolize because of the stability of the benzene ring. The benzene ring breaks down slowly, but over time, bacteria and fungi catabolize a wide range of aromatic molecules. **Figure 13.31** shows an example of research to study microbial digestion of phenanthrene (a tricyclic PAH) in the soil of Antarctica. Because the Antarctic Treaty forbids introduction of exogenous organisms, any bioremediation of pollutants must be accomplished by native microbes.

Uchechukwu Okere and colleagues at Lancaster University sought evidence for phenanthrene biodegradation by soil microbial communities at Livingston Island, off the tip of the Antarctic Peninsula. The temperatures at this

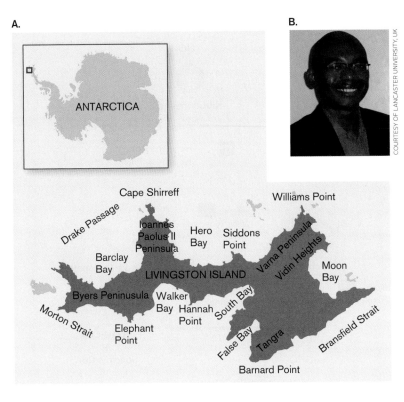

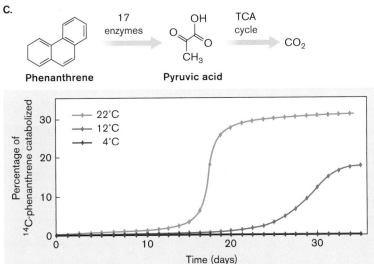

FIGURE 13.31 ■ Phenanthrene bioremediation by Antarctic psychrotrophs. A. Livingston Island, Antarctic Peninsula. **B.** Uchechukwu Okere, environmental microbiologist. **C.** Phenanthrene broken down to CO_2 by indigenous psychrotrophs. *Source: Uchechukwu Okere et al. 2012. FEMS Microbiol. Lett.* **329**:69.

northernmost end of the continent are moderately cold, between 3°C in southern summer and –11°C in winter. Okere incubated soil samples with phenanthrene substrates containing ^{14}C-radiolabeled carbons. The incubation was performed in a respirometer with a CO_2 trap, and the ^{14}C radioactivity was measured in the trapped CO_2. Okere found that over 10–30 days, microbes degraded as much as 30% of the phenanthrene to CO_2. The degradation rate was higher at higher temperatures; thus, the catabolic microbes

appear to be psychrotrophs (cold tolerant) rather than psychrophiles, which require cold for growth. Psychrotrophs may take advantage of a temporary rise in temperature to outgrow their competitors.

How do microbes catabolize such tough carbon sources? As we saw earlier (**Fig. 13.14**), fungi and bacteria break down lignin and PAHs to form single-ring aromatic compounds such as benzoate and phenols. Some of these products have commercial uses—benzoate is a food preservative; vanillin, a flavor additive; phenol, an industrial solvent. In nature, the single-ring compounds are further catabolized to acetyl-CoA, which enters the TCA cycle. Benzoate, activated to benzoyl-CoA, has a central role in the catabolism of aromatic molecules, comparable to that of glucose in polysaccharide catabolism. Bacteria such as *Pseudomonas* and *Rhodococcus* species degrade benzoate and related molecules aerobically or anaerobically. Anaerobic degradation takes much longer, but it is critical because the volume of anaerobic habitat (such as soil) greatly exceeds that of oxygenated habitat.

Aerobic benzene catabolism. Benzene and related aromatic compounds, such as toluene (methylbenzene), chlorobenzoate, and nitrobenzene, can be catabolized via sequential oxidation steps, requiring the presence of an ETS that terminates with O_2 (**Fig. 13.32**). Early in the pathways, enzymes must remove substituents such as chlorides or nitrates. The methyl group of toluene is oxidized to carboxylate ($R–COO^-$), whose removal then drives a key breakdown step. Aerobic degradation commonly proceeds through the intermediate catechol, a benzene ring bearing two adjacent hydroxyl groups. Each benzene derivative is converted to catechol by a specific dioxygenase, an enzyme that coordinately oxygenates two adjacent ring carbons. Next follows ring cleavage and breakdown to pyruvate (forming NADH) and to acetyl-CoA, which may enter the TCA cycle and respiration.

The intermediate catechol then undergoes another key oxidation by a catechol dioxygenase, which adds two more oxygens while cleaving the ring. In different bacterial species, the enzyme may oxidize catechol at either the 1,2 positions or the 2,3 positions, as shown in **Figure 13.32**. Typical products are succinyl-CoA and acetyl-CoA, which enter the TCA cycle, completing breakdown to CO_2. The details of benzene catabolism vary among bacterial species, and many diverse mechanisms continue to be discovered.

Anaerobic benzene catabolism. A challenge for biodegradation is that many pollution sources reach deep underground, where the soil is anoxic. Thus, oxygen is unavailable to conduct the conversions shown in **Figure 13.32**, particularly the formation of carboxylate ($R–COO^-$) and the introduction of hydroxyl groups. How can microbes catabolize benzene and benzoate derivatives without oxygen?

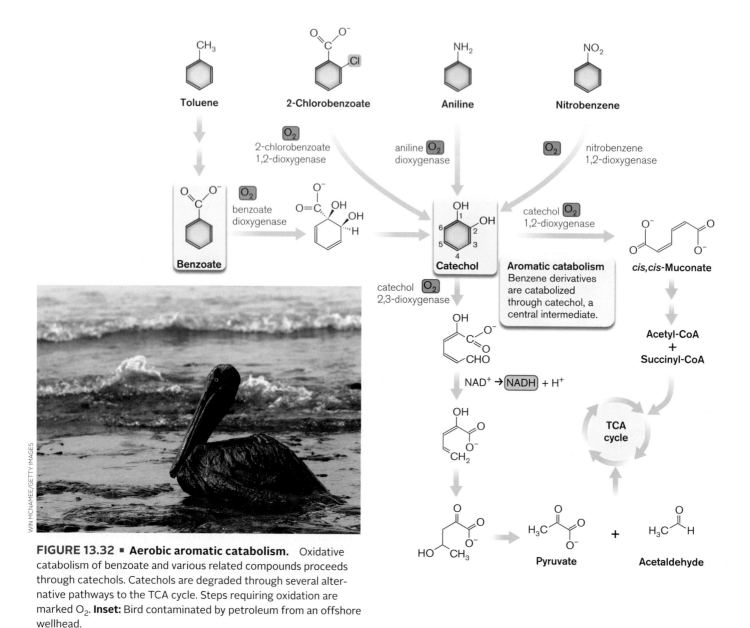

FIGURE 13.32 ■ Aerobic aromatic catabolism. Oxidative catabolism of benzoate and various related compounds proceeds through catechols. Catechols are degraded through several alternative pathways to the TCA cycle. Steps requiring oxidation are marked O_2. **Inset:** Bird contaminated by petroleum from an offshore wellhead.

In early stages of anaerobic catabolism (**Fig. 13.33**), benzene and naphthalene incorporate CO_2, forming the carboxylate group of benzoate. Because CO_2 is a very weak oxidant, these reactions require input of energy by hydrolysis of ATP. The carboxylate is then activated by HS-CoA, forming benzoyl-CoA, the same key intermediate as for aerobic benzene catabolism.

Anaerobically, the benzoyl-CoA must use reducing energy from NADH to hydrogenate its ring carbons, thus breaking the aromaticity. Some bacteria spend ATP as well, whereas others, such as the iron-reducing bacterium *Geobacter metallireducens*, can break the ring without spending ATP. The hydroxyl groups that enable shifting of the double-bond positions are introduced by incorporation of H_2O. Catabolism continues, forming three acetyl groups

activated by HS-CoA, plus one CO_2. Aromatic catabolism requires high initial investment of reducing energy—one reason the process operates slowly. For soil bacteria, the energy invested must come from anaerobic phototrophy or from anaerobic respiration.

The discovery of effective benzene degraders is of great interest for bioremediation of sites such as the South Platte River, north of Denver, where a spill from the Suncor oil refinery in 2011 released benzene (**Fig. 13.34A**). To identify novel benzene-degrading microbes, Nidal Abu Laban and colleagues at the Helmholtz Centre, Munich, performed benzene enrichment culture of soil from a coal gasification site. From the enrichment culture, the predominant organism was identified as *Pelotomaculum* species of clostridia, Gram-positive endospore-forming bacteria. The

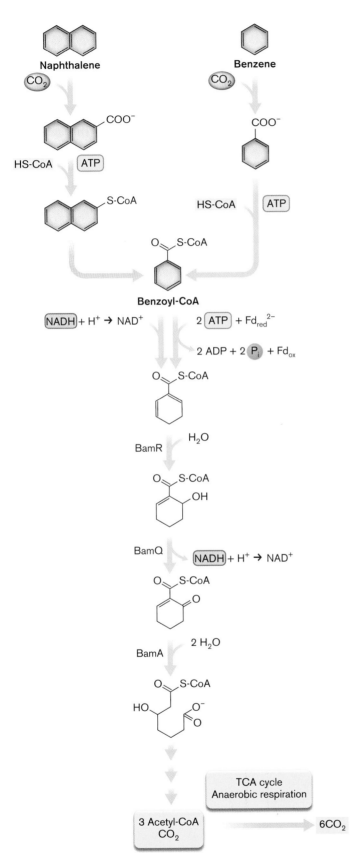

FIGURE 13.33 ■ Anaerobic benzoate catabolism. A. In the absence of oxygen, benzene and naphthalene incorporate CO_2 and are activated by HS-CoA to form benzoyl-CoA. Benzoyl-CoA breakdown requires hydrolysis (incorporating H_2O) and reduction by ferredoxin (Fd) and NADH. *Source:* Modified from Georg Fuchs et al. 2011. *Nat. Rev. Microbiol.* **9**:803.

genus *Pelotomaculum* was identified based on 16S rRNA gene sequence (for methods, see Chapter 17).

Pelotomaculum bacteria are related to other clostridia that use the anaerobic electron acceptor sulfate (SO_4^{2-}). So Abu Laban hypothesized that sulfate could be used during benzene catabolism. In the experiment whose results are plotted in **Figure 13.34B**, $^{13}CO_2$ was measured by gas chromatography–mass spectrometry (GC-MS), a technique that measures molecular masses and can thus distinguish "heavy isotopes" such as ^{13}C from the more common isotope ^{12}C. When benzene enrichment was performed in the presence of sulfate, the ^{13}C-labeled benzene showed conversion to $^{13}CO_2$, while at the same time the sulfate concentration decreased. The autoclaved controls showed no change in $^{13}CO_2$ or sulfate concentration. These data are consistent with anaerobic breakdown of benzene via the TCA cycle and ETS, with sulfate oxidation.

To Summarize

- **The pyruvate dehydrogenase complex (PDC)** removes CO_2 from pyruvate, generating acetyl-CoA. PDC activity is a key control point of metabolism, induced when carbon sources are plentiful, and repressed under carbon starvation and low oxygen.

- **The tricarboxylic acid cycle (TCA cycle)** converts the acetyl group to $2CO_2$ and $2H_2O$ in the presence of a terminal electron acceptor such as O_2 to receive the electrons associated with the hydrogen atoms.

- **Acetyl-CoA enters the TCA cycle by condensing with oxaloacetate to form citrate.** A series of enzymes sequentially removes carbon dioxide and water molecules and generates 3 NADH, 1 $FADH_2$, and 1 ATP. Each reaction step couples energy-yielding to energy-storing events.

- **The glyoxylate bypass** provides a way to gain limited energy from the TCA cycle while avoiding CO_2 loss; thus the bypass diverts intermediates to sugar biosynthesis.

- **Catabolism of aromatic molecules** by bacteria and fungi recycles lignin and PAHs within ecosystems. Aromatic metabolism is used for bioremediation.

- **Benzoate undergoes aerobic catabolism to catechol.** The catechol ring is cleaved, generating acetyl-CoA, which enters the TCA cycle.

- **Anaerobic catabolism of benzoate** involves activation by HS-CoA and reduction by NADH. The energy for reduction comes from anaerobic respiration or phototrophy.

A.

B.

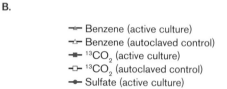

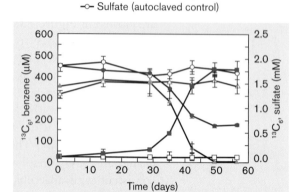

FIGURE 13.34 ▪ **Benzene catabolism for bioremediation.** **A.** South Platte River, Denver, Colorado, contaminated in 2011 by benzene from the Suncor oil refinery. **B.** Anaerobic oxidation of benzene by sulfate. *Source:* Nidal Abu Laban et al. 2009. *FEMS Microbiol. Ecol.* **68**:300.

Concluding Thoughts

In this chapter we have seen how energy-yielding reactions enable living cells to generate order from disorder. In fact, the result of cell metabolism is to increase entropy (chaos) in the universe as a whole, even while complexity (order) increases in the living cell. Living organisms acquire energy by transferring it from reactions with negative ΔG to cell-building reactions with positive ΔG. The fundamental energy-yielding pathways of Earth's biosphere are those of photosynthesis, and the biomass generated through photosynthesis provides materials for catabolism. Much of human civilization has been built on harnessing microbial catabolism for waste treatment, food production, and biotechnology.

All these forms of energy-yielding metabolism involve chemical exchange of electrons through oxidation and reduction. In this chapter we frequently mentioned oxidation-reduction reactions leading to an electron transport system. In Chapter 14 we focus on the mechanism of electron transport between donor and acceptor molecules, through membrane-embedded complexes that generate ion gradients across the membrane. This transport of electrons is fundamental to respiration, the oxidative completion of catabolism to CO_2 and water. It also underlies the inorganic sources of energy: the gain of energy from light (phototrophy) and from mineral oxidation (lithotrophy).

CHAPTER REVIEW

Review Questions

1. Why must the biosphere continually take up energy from outside? Why can't all the energy be recycled among organisms, like the fundamental elements of matter?

2. Explain how a biochemical reaction can be driven by a change in enthalpy, ΔH. Explain how a different reaction can be driven by a change in entropy, ΔS. In each case, explain the role of the free energy change, ΔG.

3. Why do some biochemical reactions release energy only above a threshold temperature?

4. How do organisms determine which of their catabolic pathways to use? How does catabolism depend on environmental factors?

5. Beer is produced by yeast fermentation of grain to ethanol. Why must the process of beer production be anaerobic? Why are such large quantities of ethanol produced with a relatively small production of yeast biomass?

6. Explain the three different routes to catabolize glucose to pyruvate. Why is it necessary to start by spending one or two molecules of ATP?

7. Explain how the TCA cycle incorporates an acetyl group. How are the two CO_2 molecules removed?

8. Compare and contrast aerobic and anaerobic processes of benzoate catabolism.

Thought Questions

1. Why are glucose catabolism pathways ubiquitous, even in bacterial habitats where glucose is scarce? Give several reasons.

2. In glycolysis, explain why bacteria have to return the hydrogens from NADH back onto pyruvate to make fermentation products. Why can't NAD^+ serve as a terminal electron acceptor, like O_2?

3. Why does catabolism of benzene derivatives yield less energy than sugar catabolism? Why is benzene-derivative catabolism nevertheless widespread among soil bacteria?

4. Why do environmental factors regulate catabolism? For example, why are amino acids decarboxylated at low pH? Cite other examples.

Key Terms

activation energy (E_a) (493)
adenosine triphosphate (ATP) (489)
allosteric site (494)
amphibolic (504)
anabolism (478)
anaerobic respiration (481)
aromatic (491)
calorimeter (483)
catabolism (478)
catabolite repression (501)
catabolize (478)
coenzyme A (CoA) (509)
electron acceptor (489)
electron donor (489)
electron transport system (ETS) (492)
Embden-Meyerhof-Parnas (EMP) pathway (501, 502)
energy (478)
energy carrier (489)

enthalpy (481)
Entner-Doudoroff (ED) pathway (501, 504)
entropy (479, 481)
enzyme (494)
ethanolic fermentation (507)
fermentation (498)
flavin adenine dinucleotide (FAD) (493)
Gibbs free energy change (ΔG) (481)
glycolysis (501, 502)
glyoxylate bypass (515)
heterotrophy (479)
hydrolysis (490)
joule (J) (482)
lactic acid fermentation (507)
lignin (498)
mixed-acid fermentation (509)
nicotinamide adenine dinucleotide (NAD) (491)

oxidative phosphorylation (515)
pentose phosphate pathway (PPP) (501, 507)
phosphorylation (490)
phototrophy (479)
polysaccharides (495)
pyruvate dehydrogenase complex (PDC) (512)
respiration (498, 515)
ribulose 5-phosphate (507)
starch (495)
Stickland reaction (510)
substrate-level phosphorylation (504, 507)
syntrophy (485)
terminal electron acceptor (492)
tricarboxylic acid (TCA) cycle (511)

Recommended Reading

Bunge, Michael, Lorenz Adrian, Angelika Kraus, Matthias Opel, Wilhelm G. Lorenz, et al. 2003. Reductive dehalogenation of chlorinated dioxins by an anaerobic bacterium. *Nature* **421**:357–360.

Eisenreich, Wolfgang, Thomas Dandekar, Jürgen Heesemann, and Werner Goebel. 2010. Carbon metabolism of intracellular bacterial pathogens and possible links to virulence. *Nature Reviews. Microbiology* **8**:401–412.

Fuchs, Georg, Matthias Boll, and Johann Heider. 2011. Microbial degradation of aromatic compounds—From one strategy to four. *Nature Reviews. Microbiology* **9**:803–816.

Head, Ian M., D. Martin Jones, and Wilfred F. M. Röling. 2006. Marine microorganisms make a meal of oil. *Nature Reviews. Microbiology* **4**:173–182.

Hickey, William J., Shicheng Chen, and Jiangchao Zhao. 2012. The phn island: A new genomic island encoding catabolism of polynuclear aromatic hydrocarbons. *Frontiers in Microbiology* **3**:125.

Jackson, Bradley E., and Michael J. McInerney. 2002. Anaerobic microbial metabolism can proceed close to thermodynamic limits. *Nature* **415**:454–456.

Koropatkin, Nicole M., Elizabeth A. Cameron, and Eric C. Martens. 2012. How glycan metabolism shapes the human gut microbiota. *Nature Reviews. Microbiology* **10**:323–335.

Ladas, Spiros D., George Karamanolis, and Emmanuel Ben-Soussan. 2007. Colonic gas explosion during therapeutic colonoscopy with electrocautery. *World Journal of Gastroenterology* **13**:5295–5298.

Larsbrink, Johan, Theresa E. Rogers, Glyn R. Hemsworth, Lauren S. McKee, Alexandra S. Tauzin, et al. 2014. A discrete genetic locus confers xyloglucan metabolism in select human gut Bacteroidetes. *Nature* **506**:498–502.

Martens, Eric C., Elisabeth C. Lowe, Herbert Chiang, Nicholas A. Pudlo, Meng Wu, et al. 2011. Recognition and degradation of plant cell wall polysaccharides by two human gut symbionts. *PLoS Biology* **9**:e1001221.

McInerney, Michael J., Jessica R. Sieber, and Robert P. Gunsalus. 2011. Microbial syntrophy: Ecosystem-level biochemical cooperation. *Microbe* **6**:479–485.

Okere, Uchechukwu V., Ana Cabrerizo, Jordi Dachs, Kevin C. Jones, and Kirk T. Semple. 2012. Biodegradation of phenanthrene by indigenous microorganisms in soils from Livingstone Island, Antarctica. *FEMS Microbiological Letters* **329**:69–77.

Peekhaus, Norbert, and Tyrrell Conway. 1998. What's for dinner? Entner-Doudoroff metabolism in *Escherichia coli*. *Journal of Bacteriology* **180**:3495–3502.

Sieber, Jessica R., Huynh M. Le, and Michael J. McInerney. 2014. The importance of hydrogen and formate transfer for syntrophic fatty, aromatic and alicyclic metabolism. *Environmental Microbiology* **16**:177–188.

Tang, Hongzhi, Yuxiang Yao, Lijuan Wang, Hao Yu, Yiling Ren, et al. 2012. Genomic analysis of *Pseudomonas putida*: Genes in a genome island are crucial for nicotine degradation. *Scientific Reports* **2**:377.

Yoshida, Shosuke, Kazumi Hiraga, Toshihiko Takehana, Ikuo Taniguchi, Hironao Yamaji, et al. 2016. A bacterium that degrades and assimilates poly(ethylene terephthalate). *Science* **351**:1196–1199.

CHAPTER 14
Electron Flow in Organotrophy, Lithotrophy, and Phototrophy

Microbes transfer energy by moving electrons—the equivalent of an electric current. Electrons move from reduced food molecules onto energy carriers, from energy carriers onto membrane proteins called cytochromes, and from cytochromes onto oxygen or oxidized minerals. Bacteria use electron flux to drive protons across the membrane and generate a proton motive force. The proton motive force stores energy to make ATP. Similar electron transport systems power our own mitochondria and plant chloroplasts. In Earth's crust, lithotrophic bacteria use electron flow to deposit iron and gold.

CURRENT RESEARCH highlight

Bacterial electricity. In this colorized SEM, *Geobacter* bacteria cling to iron oxide. *Geobacter* can donate electrons outside the cell to iron (Fe^{3+}) or to a metal electrode. Jessica Smith and Derek Lovley showed that *Geobacter metallireducens* can transfer electrons from ethanol to *Geobacter sulfurreducens*. *G. sulfurreducens* then donates the electrons to fumarate. Together, the two kinds of bacteria obtain energy to grow—a form of syntrophy based on electricity. Interspecies electron transfer may help bioremediation of organic wastes and the generation of bioelectricity in fuel cells.

Source: Geobacter.org; Jessica Smith et al. 2015. Front. Microbiol. 6:121. Photo: Eye of Science/Science Source.

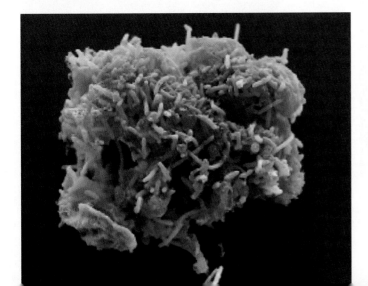

AN INTERVIEW WITH

JESSICA SMITH, POSTDOCTORAL RESEARCHER, UNIVERSITY OF MASSACHUSETTS, AMHERST

UNIVERSITY OF MASSACHUSETTS AMHERST

What is so exciting about microbial electricity?

Geobacter species are ubiquitous in the environment. They have the ability to transfer electrons located on the outer cell surface to extracellular electron acceptors such as insoluble metal oxides and oxidized metals. These microorganisms can produce electricity by transferring electrons to anodes in microbial fuel cells, thereby earning them the title of "electric microbes."

How do you think our assumptions will change about redox metabolism in environmental microbial communities?

Geobacter can transfer electrons to partner species through direct electrical connections—a process known as direct interspecies electron transfer. Interspecies electron transfer in anaerobic environments has paved the way for a thrilling area of research in microbiology. A fuller understanding of the process and its role in anaerobic environments will help us devise better strategies for wastewater digestion or methane emission control in environments such as landfills, rice paddies, and animal intestinal systems. The potential for discovery promises an "electrifying" future!

Electric current in living organisms has fascinated scientists ever since the eighteenth century, when Luigi Galvani (1737–1798) showed that a voltage caused a dead frog limb to flex. The idea of biological electricity inspired Mary Shelley's famous novel *Frankenstein: or, the Modern Prometheus*, in which a physician is imagined to "create life" by jolting body parts with an electric shock. Today we know that electric current flows in all cells. Bacteria such as *Geobacter* species (see the Current Research Highlight) can transfer electrons from an electrode onto hydrogen ions, forming hydrogen gas. First isolated by Derek Lovley at the University of Massachusetts, Amherst, *Geobacter* can extract electrons from organic molecules, iron, and even uranium. In a soil community, *Geobacter* and other microbes transfer electrons from cell to cell, in a syntrophy that enhances growth of all the cells involved. For example, Lovley's students showed how *Geobacter metallireducens* can form an "electrical" coculture with *Methanosaeta*, a methanogen (**Fig. 14.1**). The *Geobacter* cells oxidize ethanol and then transfer the electrons to the methanogen, which reduces CO_2 to methane. This electrical syntrophy supports digestion in water treatment plants, and it can be a source of usable electric current.

How do cells make electricity? The energy-yielding reactions we saw in Table 13.1 involve transfer of electrons from one molecule to another. Much of the energy yield comes from successive redox steps within an **electron transport system** (ETS) of carriers in a membrane, whose energy-yielding redox reactions are coupled to transport of ions such as protons (H^+). Types of metabolism that use an ETS include **organotrophy** (organic electron donors), **lithotrophy** (inorganic electron donors), and **phototrophy** (light absorption excites electrons). Some bacteria and archaea are specialists, whereas others use more than one of these types of metabolism. Even organisms with very different forms of metabolism show common themes in their mechanisms of electron transfer. In fact, the protein complexes for processes as different as organotrophy and phototrophy show homology, indicating that they evolved from a common ancestral ETS.

Chapter 14 presents electron flow through the ETS in organotrophy, lithotrophy, and phototrophy. In each step of the ETS, a molecule becomes "reduced" (gains an electron) while the molecule donating the electron becomes "oxidized" (loses an electron). In various kinds of metabolism, some of the energy from electron transfer is stored in the form of an electrochemical potential (voltage) across a membrane. The potential is composed of the chemical concentration gradient of H^+ ions plus the charge difference across the membrane. The proton potential drives ATP synthesis. Ion potentials also drive nutrient uptake, mediate pathogenesis of infective microbes, and shape the chemistry surrounding

FIGURE 14.1 ■ ***Geobacter metallireducens* transfer electrons to methanogens.** **A.** *G. metallireducens* cells (green probe, Cy5) cocultured with *Methanosaeta* (red probe, Cy3) and analyzed by fluorescence in situ hybridization microscopy (FISH). Image obtained in the laboratory of Derek Lovley by Fanghua Liu **(B)**, now at Yantai Institute of Coastal Zone Research, Chinese Academy of Sciences; and Amelia-Elena Rotaru **(C)**, now at the University of Southern Denmark, Odense.

a microbe. Microbial electron transfer reactions, and the ion potentials they generate, have profound consequences for ecosystems; they acidify or alkalinize soil and water, and they can contribute or remove nitrogen and minerals.

14.1 Electron Transport Systems

As we learned in Chapter 13, energy to support life is obtained through reactions that convert substrates to products of lower energy—that is, reactions in which ΔG (free energy change) is negative. Many reactions involve transfer of electrons from a reduced **electron donor** to an oxidized **electron acceptor**. The simplest path of electron flow is found in fermentation, where electrons from the fermented substrate, such as glucose, are transferred onto NAD^+ to make NADH (reduction), and then returned to the glucose breakdown products. More complex kinds of metabolism, such as aerobic respiration, transfer electrons through a series of membrane-soluble carriers called an **electron transport system** (**ETS**), also known as an **electron transport chain** (**ETC**).

Unlike cytoplasmic redox reactions, a membrane-embedded ETS can convert its energy into an ion potential or electrochemical potential between two compartments separated by the membrane. The ion potential is most commonly a proton (H^+) potential, also known as proton motive force (PMF). The proton motive force drives essential cell processes such as synthesis of ATP (discussed in Section 14.2).

Note: Recall the following prefixes.

Energy source:

Photo-: Light absorption captures energy and excites electron.

Chemo-: Chemical electron donors are oxidized.

Electron source:

Litho-: Inorganic molecules donate electrons.

Organo-: Organic molecules donate electrons.

Electron Donors and Acceptors

Different types of metabolism use electron transport systems with different components embedded in the bacterial membrane. For chemotrophy (oxidizing an electron donor to yield energy), each ETS must accept electrons from an initial electron donor. The electron donor may be an organic "food" molecule, such as glucose (organotrophy), or a reduced mineral, such as Fe^{2+} (lithotrophy). Next, the ETS proteins and cofactors act sequentially as electron donors and acceptors, whose oxidation-reduction reactions are coupled to H^+ transport across the membrane.

Each ETS finally transfers its electrons to a terminal electron acceptor molecule, whose product may exit the cell.

Metabolism using an ETS is classified based on the nature of the initial electron donors and terminal electron acceptors. In this chapter we cover three major classes of prokaryotic energy acquisition involving an ETS (their relationships are summarized in Table 13.1):

■ **Organotrophy** (or **chemoorganotrophy**) is a form of metabolism in which organic molecules donate electrons. If the electrons are donated to a membrane-embedded electron transport system, and ultimately reduce a terminal electron acceptor, the overall pathway is called "respiration." The terminal electron acceptor may be O_2 for aerobic respiration, or an alternative electron acceptor such as nitrate (NO_3^-) for anaerobic respiration. Some terminal electron acceptors are organic (such as fumarate, $^-O_2C–CH=CH–CO_2^-$).

■ **Lithotrophy** (or **chemoautotrophy**) is the oxidation of <u>inorganic</u> electron donors, such as Fe^{2+} or H_2. The electron acceptor may be O_2 or an anaerobic alternative, such as NO_3^-. A strong electron donor, such as H_2, can use organic electron acceptors such as fumarate. Some chemoorganotrophs that catabolize many substrates are "facultative lithotrophs"—for example, oxidizing sulfur compounds. But many lithotrophs are obligate autotrophs, fixing CO_2 for biosynthesis (discussed in Chapter 15). These specialist organisms are called chemolithoautotrophs. For example, the bacterium *Nitrosomonas europaea* obtains nearly all its energy by oxidizing ammonia.

■ **Phototrophy** involves light capture by chlorophyll, usually coupled to splitting of H_2S or H_2O (photolysis). Photoautotrophs couple photolysis to CO_2 fixation for forming biomass (discussed in Chapter 15). Photoheterotrophs absorb light to generate ATP, while incorporating preformed carbon compounds for biosynthesis. Photoheterotrophs may also capture light energy to supplement catabolism, which provides electron donors (photoorganotrophy).

Note: The term **respiration** refers to metabolism in which chemical electron donors yield energy through an ETS. In this book we reserve the term "respiration" for an ETS with organic electron donors (organotrophy). The use of inorganic electron donors for an ETS is called lithotrophy. Other fields use the term "respiration" differently. In medical physiology, "respiration" refers to catabolism with an ETS, or to the process of carbon dioxide exchange with oxygen through breathing. In ecology, "respiration" refers to carbon flux in an ecosystem, or throughout the atmosphere.

Energy Storage

In Chapter 13 we showed how the free energy change ΔG determines whether energy can be obtained from a given reaction of substrates to products. Similar considerations of ΔG govern the reactions of electron flow through an ETS and the storage of energy in transmembrane ion gradients.

The reduction potential. To obtain energy, biochemical reactions require a negative ΔG. In oxidation-reduction reactions (redox reactions), the ΔG values are proportional to the reduction potential (E) between the oxidized form of a molecule (electron acceptor) and its reduced form (electron donor). The reduction potential represents the tendency of a compound to accept electrons, measured in volts (V) or millivolts (mV). **A positive value of E has a negative ΔG**, so the gain of electrons yields energy. A negative value of E means that the reverse reaction (loss of electrons) yields energy.

The oxidized and reduced states of a compound are called a **redox couple**. An example of a redox couple is the electron acceptor O_2 and the electron donor H_2O. Because O_2 is a strong electron acceptor (readily gains $2e^-$), the redox couple $\frac{1}{2}O_2/H_2O$ has a high positive value of E. So when O_2 plus $2H^+ + 2e^-$ forms H_2O, the reaction yields a large amount of energy.

Standard values of E (E°) are known as the **standard reduction potential**. The standard reduction potential $E^{\circ\prime}$ assumes a concentration of 1 M for all components at pH 7. For example, the value of $E^{\circ\prime}$ for H_2 formation from $2H^+ + 2e^-$ is -420 mV. This large negative reduction potential means that it takes

a lot of energy to add the two electrons to $2H^+$. Thus, the reverse reaction donating $2e^-$ from H_2 to an appropriate electron acceptor yields a lot of energy. An example of microbes that use this reaction for energy is hydrogen-oxidizing bacteria growing in oral biofilms, causing gum disease.

Reduction potentials represent a form of the standard free energy change, $\Delta G^{\circ\prime}$. The value of $\Delta G^{\circ\prime}$ (in kilojoules per mole, or kJ/mol) for an electron transfer reaction with a given reduction potential $E^{\circ\prime}$ is given by

$$\Delta G^{\circ\prime} = -nFE^{\circ\prime}$$

where n is the number of electrons transferred, F is Faraday's constant ($F = 96.5$ kJ/V · mol), and $E^{\circ\prime}$ is the reduction potential (V) at 1 M concentration, 25°C, and pH 7. Thus, $E^{\circ\prime}$ represents the standard reduction potential per electron, whereas $\Delta G^{\circ\prime}$ gives the overall free energy change. Note that the reaction is favored for positive values of E, which correspond to negative values of ΔG.

The standard reduction potential $E^{\circ\prime}$ gives the potential difference between the oxidized and reduced states under standard conditions, when both oxidized and reduced forms are at equal concentrations—namely, 1 M. Assuming equal concentrations at 1 M, we may compare $E^{\circ\prime}$ of various molecules available in the environment for microbial respiration. Such a comparison generates an "electron tower" representing the reduction potential $E^{\circ\prime}$ of molecules that may act as electron acceptors or donors (**Table 14.1**). The

TABLE 14.1	"Electron Tower" of Standard Reduction Potentials			
Electron acceptor	$\rightarrow$	**Electron donor**	**$E^{\circ\prime}$ (mV)[a]**	**$\Delta G^{\circ\prime}$ (kJ)**
$CO_2 + 4H^+ + 4e^-$	$\rightarrow$	$[CH_2O]$ glucose $+ H_2O$	-430	$+166$
$2H^+ + 2e^-$	$\rightarrow$	H_2	-420	$+81$
$NAD^+ + 2H^+ + 2e^-$	$\rightarrow$	$NADH + H^+$	-320	$+62$
$S^0 + H^+ + 2e^-$	$\rightarrow$	HS^-	-280	$+27$
$CO_2 + 2H^+ + 3H_2 + 2e^-$	$\rightarrow$	$CH_4 + 2H_2O$	-240	$+46$
$SO_4^{2-} + 10H^+ + 8e^-$	$\rightarrow$	$H_2S + 4H_2O$	-220	$+170$
$FAD + 2H^+ + 2e^-$	$\rightarrow$	$FADH_2$	-220[b]	$+42$
$FMN + 2H^+ + 2e^-$	$\rightarrow$	$FMNH_2$	-190	$+37$
Menaquinone $+ 2H^+ + 2e^-$	$\rightarrow$	Menaquinol	-74	$+14$
Fumarate $+ 2H^+ + 2e^-$	$\rightarrow$	Succinate	$+33$	-6
$Fe^{3+} + e^-$	$\rightarrow$	Fe^{2+} (at pH 7)	$+200$	-19
Ubiquinone $+ 2H^+ + 2e^-$	$\rightarrow$	Ubiquinol	$+110$	-21
$NO_3^- + 2H^+ + 2e^-$	$\rightarrow$	$NO_2^- + H_2O$	$+420$	-81
$NO_2^- + 8H^+ + 6e^-$	$\rightarrow$	$NH_4^+ + 2H_2O$	$+440$	-255
$MnO_2 + 4H^+ + 2e^-$	$\rightarrow$	$Mn^{2+} + 2H_2O$	$+460$	-89
$NO_3^- + 6H^+ + 5e^-$	$\rightarrow$	$\frac{1}{2}N_2 + 3H_2O$	$+740$	-357
$Fe^{3+} + e^-$	$\rightarrow$	Fe^{2+} (at pH 2)	$+770$	-74
$\frac{1}{2}O_2 + 2H^+ + 2e^-$	$\rightarrow$	H_2O	$+820$	-158

[a]Reduction potentials for redox couples when all concentrations are 1 M at 25°C and pH 7.

[b]This value is for free FAD; FAD bound to a specific flavoprotein has a different $E^{\circ\prime}$ that depends on its protein environment.

oxidized state of each redox couple in the table is shown in red, the reduced state in blue.

Note: A more positive $E^{\circ\prime}$ means that <u>reducing</u> the electron <u>acceptor</u> yields more energy. A more negative value of $E^{\circ\prime}$ means that <u>oxidizing</u> the electron <u>donor</u> yields more energy. Either form of half reaction, in the favored direction, is associated with a negative value of $\Delta G^{\circ\prime}$.

In the electron tower, the more negative values of $E^{\circ\prime}$ represent couples (half reactions) with a stronger electron donor (H_2, NADH), whereas the more positive values represent stronger electron acceptors (O_2, NO_3^-). For example, in the redox couple $NAD^+/NADH + H^+$, the oxidized form is NAD^+ and the reduced form is NADH. The reduction potential is strongly negative (–320 mV), so NADH is a strong electron donor; that is, the reverse reaction, $NADH + H^+ \rightarrow NAD^+$, releases a lot of energy (+320 mV).

A complete redox reaction combines two redox couples: one accepting electrons (red column), the other donating electrons (blue column, arrow reversed). Redox couples with more negative values of $E^{\circ\prime}$ can provide electron donors (blue column) for electron acceptors with more positive values of $E^{\circ\prime}$ (red column). The $E^{\circ\prime}$ for the overall reaction is then given by adding the reversed reaction of the electron donor $E^{\circ\prime}$ to the electron acceptor $E^{\circ\prime}$. A positive value of $E^{\circ\prime}$ means that the reaction may proceed to provide energy. For example, the oxidation of NADH by O_2 results from combining two couples: O_2/H_2O (forward) and $NADH/NAD^+$ (reversed from **Table 14.1**). For the reversed couple, the sign of $E^{\circ\prime}$ changes:

	$E^{\circ\prime}$	$\Delta G^{\circ\prime}$
Electron acceptor is reduced:		
$\frac{1}{2}O_2 + 2e^- + 2H^+ \rightarrow H_2O$	+820 mV	–158 kJ/mol
Electron donor is oxidized:		
$NADH + H^+ \rightarrow NAD^+ + 2e^- + 2H^+$	–(–320 mV)	–(+62 kJ/mol)
$NADH + H^+ + \frac{1}{2}O_2 \rightarrow H_2O + NAD^+$	+1,140 mV	–220 kJ/mol

The aerobic oxidation of NADH pairs a strong electron donor (NADH) with a strong electron acceptor (O_2). Thus, NADH oxidation via the ETS provides the cell with a huge amount of potential energy (comparable to a 1-V battery cell) to make ATP and generate ion gradients. NADH is oxidized by many electron transport systems of bacteria and archaea, as well as mitochondria. Whenever we digest food, we make NADH for our mitochondria to oxidize.

Bacteria and archaea have evolved many alternative donor-acceptor systems; the one they use depends on what their environment provides. For example in anaerobic marine sediment, the proteobacterium *Shewanella* can donate electrons to many alternative oxidized minerals

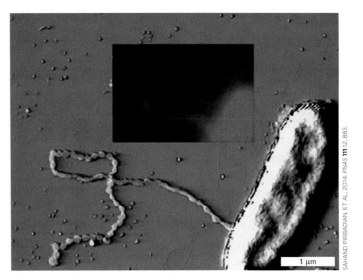

SAHAND PIRBADIAN, ET AL, 2014. *PNAS* **111**:12, 883

FIGURE 14.2 ■ ***Shewanella oneidensis* grows a nanowire.** The bacterium extends vesicles of outer membrane, which lengthen into nanowires. **Inset:** Fluorescence microscopy, with a membrane-specific fluorophore.

as terminal electron acceptors, including iron, lead, and nitrate. How does *Shewanella* gain access to immobilized minerals? The bacterium grows "nanowires" to transport electrons some distance from the cell (**Fig. 14.2**). The nanowires develop as a chain of outer membrane vesicles that lengthen into narrow tubes. A membrane-specific fluorophore shows that the nanowire is composed of membrane.

Electron donors can also be oxidized by organic molecules. In the human colon, which lacks molecular oxygen, *Escherichia coli* can oxidize NADH by transferring $2e^-$ onto fumarate, which is reduced to succinate. From **Table 14.1**, we find that:

$$NADH + H^+ + fumarate \rightarrow NAD^+ + succinate$$
$$E^{\circ\prime} = 320 \text{ mV} + 33 \text{ mV} = 353 \text{ mV} \qquad \Delta G^{\circ\prime} = -68 \text{ kJ/mol}$$

The fumarate reduction potential is small but yields some energy for growth. When oxygen reappears in the environment, the succinate can donate electrons to O_2, forming fumarate and water:

$$\frac{1}{2}O_2 + succinate \rightarrow H_2O + fumarate$$
$$E^{\circ\prime} = 820 \text{ mV} - 33 \text{ mV} = 787 \text{ mV} \qquad \Delta G^{\circ\prime} = -152 \text{ kJ/mol}$$

Thought Questions

14.1 *Pseudomonas aeruginosa*, a cause of pneumonia in cystic fibrosis patients, oxidizes NADH with nitrate (NO_3^-) to nitrite (NO_2^-) at neutral pH. What is the value of $E^{\circ\prime}$?

14.2 Could a bacterium obtain energy from succinate as an electron donor with nitrate (NO_3^-) as an electron acceptor?

Concentrations of electron donors and acceptors. A high concentration of an electron donor can enhance metabolism using reactions with small values of $E^{\circ\prime}$. This is particularly common for lithotrophic reactions such as iron oxidation. The standard reduction potential $E^{\circ\prime}$ assumes 1 M concentrations of reactants and products at 25°C and pH 7, but actual values of E within cells depend on actual reactant concentrations.

In Chapter 13 we saw that ΔG includes a term incorporating the ratio of product concentrations (C and D) to reactant concentrations (A and B):

$$\Delta G = \Delta G^{\circ\prime} + 2.303RT \log \frac{[C][D]}{[A][B]}$$

where R is the gas constant [8.315 joules per kelvin-mole, or J/(mol · K)], and T is the temperature in kelvins. The reduction potential E of a redox reaction depends on the ratio of products to reactants:

$$E = \frac{-\Delta G}{nF}$$

$$= E^{\circ\prime} - \frac{2.303RT}{nF} \times \log \frac{[C][D]}{[A][B]}$$

$$= E^{\circ\prime} - 60 \ (mV/n) \log \frac{[C][D]}{[A][B]}$$

where n is the number of electrons transferred and F is the Faraday constant. As the ratio of products to reactants increases, E decreases; in other words, there is

less potential to do work. At moderate temperatures (25°C–40°C), the constants in the concentration term $(2.303RT/F)$ combine to yield approximately 60 mV per electron when there is a tenfold ratio of products to reactants. Thus, a tenfold ratio of products to reactants subtracts about 60 mV from E (decreases energy yield), whereas a tenfold excess of reactants adds 60 mV to E (increases energy yield).

An ETS Functions within a Membrane

An electron transport system transfers electrons onto membrane-embedded carriers in a series of increasing reduction potential, ending at a terminal electron acceptor such as O_2, Fe^{3+}, or NO_3^-. Some electron transfer steps yield energy to pump ions across the membrane. To store this energy, the ETS must maintain an ion gradient across a membrane that fully separates two aqueous compartments. In bacteria, the cytoplasmic membrane separates the cytoplasm and external medium. Gram-negative bacteria such as *E. coli* have their ETS in the inner (cytoplasmic) membrane, which separates the cytoplasm from the periplasm (see Chapter 3). The Gram-negative outer membrane surrounding the periplasm is permeable to protons and other small molecules; thus, the outer membrane does not store energy. **Figure 14.3A** shows the inner membrane of *Helicobacter pylori*, a Gram-negative gastric pathogen. The inner membrane contains the ETS

A.

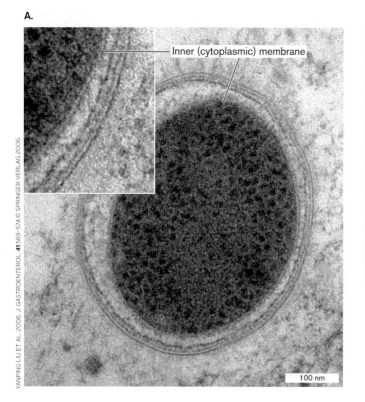

B.

FIGURE 14.3 ■ **Respiratory membranes. A.** Electron transport occurs in the inner (cytoplasmic) membrane of *Helicobacter pylori*. The cytoplasmic membrane and cell wall are surrounded by periplasm and outer membrane. **B.** Mitochondrion within the brain stem neuron of a cat, modeled by cryo-EM tomography, shown with one EM section through the cell. The mitochondrial outer membrane (blue) encloses inner membrane organized in pockets called cristae (other colors).

YANPING LIU ET AL. 2006. *J. GASTROENTEROL.* **41**:569–574.© SPRINGER-VERLAG 2006.

GUY A. PERKINS ET AL. 2010. *J. NEUROSCI.* **30**:1015

complexes; the cell wall and outer membrane do not participate in respiration.

How do bacteria obtain their electron acceptors? Soluble electron acceptors such as O_2 or NO_3^- can reach the cell by diffusion, but metals such as Fe^{3+} are stuck within insoluble particles. Metal-reducing bacteria may grow adherent to particles of iron oxide (recall *Geobacter* in the Current Research Highlight). Others secrete organic carrier molecules to dissolve the metal. Still others, such as *Shewanella*, form outer membrane nanowire extensions full of extracellular cytochromes. These cytochromes help metal-reducing bacteria gain access to insoluble minerals—or to the surface of a fuel-cell electrode (see **Special Topic 14.1**).

Some respiratory bacteria, such as the nitrite oxidizer *Nitrospira*, pack their ETS within intracytoplasmic pockets called "lamellae." Respiratory membranes similar to bacterial lamellae are found within our own mitochondria. Mitochondrial respiration uses only O_2 as a terminal electron acceptor. The ETS proteins are embedded in folds of the mitochondrial inner membrane called "cristae" (singular, crista) (**Fig. 14.3B**). The inner membrane separates the inner mitochondrial space from the intermembrane space (between the inner and outer membranes). The mitochondrial inner membrane, including its electron transport proteins, evolved from the cell membrane of an endosymbiotic bacterial ancestor. Mitochondrial evolution is discussed further in Chapter 17.

Electron carrier molecules include proteins and small organic cofactors bound to the proteins. The protein components of an ETS were first discovered in the 1930s at Cambridge University by Russian entomologist David Keilin (1887–1963), who studied insect mitochondria. The mitochondrial inner membranes contain proteins called **cytochromes**, which were named for their deep colors, typically red to brown. The colors derive from absorption of visible light due to the relatively small energy transitions of electrons in a cytochrome.

In prokaryotes, cytochromes are often found in the cell membrane. **Figure 14.4** shows the absorbance spectrum of a cytochrome from the cell membrane of *Haloferax volcanii*, a halophilic archaeon isolated from the Dead Sea. In the reduced cytochrome, light absorption peaks in the blue range (440 nm) and in the orange (607 nm). Upon oxidation, however, the cytochrome loses its absorption peak at 607 nm, and the 440-nm peak shifts to a shorter wavelength. Different species of bacteria make many different cytochromes, with different absorption peaks, but all cytochromes show spectral changes with a change in reduction state.

A membrane ETS typically includes several different cytochromes with different reduction potentials. Keilin proposed that the cytochromes pass electrons sequentially from each protein complex to the next-stronger electron acceptor, with

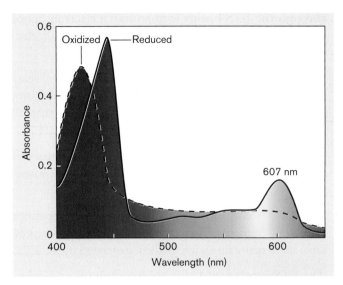

FIGURE 14.4 ■ Light absorbance spectrum of a cytochrome. Absorption peaks shift between the oxidized and reduced forms of a cytochrome from the electron transport system of *Haloferax volcanii*, a halophilic archaeon.

each step providing a small amount of energy to the organism (**Fig. 14.5**). The electron transport proteins are called **oxidoreductases** because they oxidize one substrate (removing electrons) and reduce another (donating electrons). Thus, they couple different half reactions in the electron tower (see **Table 14.1**). Oxidoreductases consist of multiprotein complexes that include cytochromes as well as noncytochrome proteins. The structure and function of ETS oxidoreductase complexes are discussed in Section 14.3.

To Summarize

■ **An electron transport chain (ETS)** consists of a series of electron carriers that sequentially transfer electrons to the carrier of next-higher reduction potential E (that is, the next-stronger electron acceptor). Electron flow

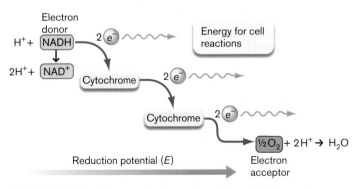

FIGURE 14.5 ■ Electron transport system. Keilin's model for electron transfer through cytochromes. Each cytochrome in sequence receives electrons from a stronger electron donor and transfers them to a stronger electron acceptor.

through the ETS begins with an initial electron donor from outside the cell and ultimately transfers all electrons to a terminal electron acceptor.

■ **The reduction potential E for a complete redox reaction must be positive to yield energy for metabolism.** The standard reduction potential $E^{\circ\prime}$ assumes that all reactant concentrations equal 1 M, at 25°C and pH 7.

■ **Concentrations of electron donors and acceptors** in the environment influence the actual reduction potential E experienced by the cell.

■ **The ETS is embedded in a membrane that separates two compartments.** Two aqueous compartments must be separate to maintain an ion gradient generated by the ETS.

■ **The ETS is composed of protein complexes and cofactors.** Protein complexes called oxidoreductases include cytochromes and noncytochrome proteins. Cytochromes are colored proteins whose absorbance spectrum shifts with a change in redox state.

14.2 The Proton Motive Force

The sequential transfer of electrons from one ETS protein to the next yields energy to pump ions (in most cases H^+) across the membrane. Proton pumping generates a **proton motive force** (or **proton potential**) composed of the H^+ concentration difference plus the charge difference across the membrane. The proton motive force (PMF, or Δp) drives many different cellular processes, such as ATP synthesis and flagellar rotation. In pathogens, the PMF drives drug efflux pumps that confer resistance to antibiotics such as tetracycline.

Note that in water, H^+ never occurs as such, because it combines with a water molecule to form hydronium ion, H_3O^+. Current data, however, are consistent with the passage of hydrogen nuclei (protons) through the proton pumps of the electron transport system. Within the pump complex, the proton associates with one chemical group after another; for example, H^+ may combine with the amine of an amino acid (RNH_2) to form an ammonium ion (RNH_3^+), and then transfer to a different proton-accepting group within the protein.

Note: In this book we use "hydrogen ions" and "H^+" interchangeably with "protons."

The ETS Pumps Protons

The discovery of the proton potential radically changed the field of biochemistry. Early in the twentieth century, David Keilin and other scientists knew that the energy acquired

B.

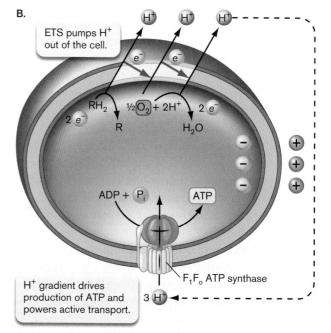

FIGURE 14.6 ■ Peter Mitchell and Jennifer Moyle discovered the proton motive force (Δp). A. Mitchell and Moyle proposed and tested the chemiosmotic theory. **B.** An electron transport system pumps protons out of the cell. The resulting electrochemical gradient of protons (proton motive force) drives conversion of ADP to ATP through ATP synthase.

by electron transport proteins was used to make ATP, but they did not know how. Most were convinced that electron transport was somehow directly coupled to ATP synthesis. The actual means of coupling was proposed and demonstrated by one of Keilin's students—Peter Mitchell (1920–1992)—and Mitchell's colleague Jennifer Moyle (**Fig. 14.6A**). In 1961, Mitchell proposed an astonishing explanation for the coupling of electron transport to ATP synthesis. The chemiosmotic theory states that the energy from electron transfer between membrane proteins is used to pump protons across the membrane, accumulating a higher H^+ concentration in the compartment outside. The pump generates the proton motive force Δp, which stores energy that can be used to make ATP (**Fig. 14.6B**).

Other biologists questioned how a proton potential could be coupled to ATP synthesis at an enzyme complex separate from the ETS, at a distant location on the membrane. They were skeptical that the H^+ concentration gradient and charge difference could exist everywhere on the membrane separating the two compartments. Because the membrane is impermeable to hydrogen ions, the H^+ current can flow back only through a proton-driven complex such as the membrane ATP synthase (discussed further in Section 14.3). In effect, the proton potential is a "proton battery," analogous to the electron potential of an electrical battery.

The chemiosmotic theory proved so controversial that Mitchell left Cambridge University to found his own laboratory, Glynn House, where Moyle joined him to test it. A key requirement was to show that the ETS generates a proton potential. Moyle's experiment showed that respiration of mitochondria is associated with proton efflux: Mitochondria isolated from rat livers were exposed to oxygen, causing efflux of hydrogen ions. The H^+ efflux occurred as electrons were transferred across the ETS and hydrogen ions were expelled by the proton pumps. In Moyle's experiments, the number of protons extruded per electron transferred down the ETS was consistent with the chemiosmotic model. Similar results were obtained with vesicles made first from mitochondrial membranes and later from the membranes of chloroplasts and bacteria.

A greater challenge was to demonstrate that a proton motive force could, in fact, drive the ATP synthase and other ion transporters without any undiscovered intermediate. A friend of Mitchell and Moyle, André Jagendorf, at Johns Hopkins University, tested the effect of a proton gradient imposed on spinach chloroplasts (**Fig. 14.7A**). Chloroplasts contain ATP synthase directed outward (that is, driven by proton flow from inside to outside). The chloroplasts were partly opened by osmotic shock, and then suspended in medium containing concentrated hydrogen ions (pH 4). With the osmotic balance restored, the chloroplast membranes closed again, with increased $[H^+]$ trapped inside. When the pH outside was raised to pH 8.3—that is, lower $[H^+]$—protons from the acidic interior of the vesicles flowed out through the ATP synthase, and ADP and inorganic phosphate were converted to ATP. Jagendorf interpreted this result as consistent with the chemiosmotic theory for the mechanism of phosphorylation.

Other researchers showed that ATP synthesis could be driven by a charge difference across a vesicle membrane (**Fig. 14.7B**). An artificial electrical potential ($\Delta\psi$) was applied by loading vesicles with potassium ions (K^+). The potassium ions were conducted across the membrane by the ionophore valinomycin, a small cyclic peptide that binds to an ion and solubilizes it in the membrane. Valinomycin specifically binds K^+, while its hydrophobic side

A. A pH difference, ΔpH, drives ATP synthesis.

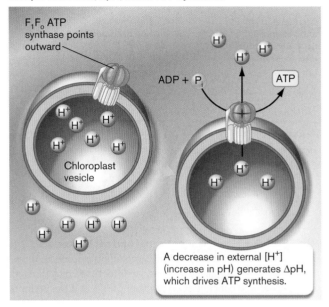

B. A charge difference, Δψ, drives ATP synthesis.

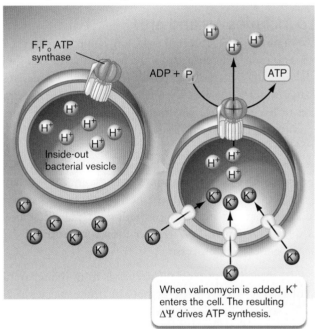

FIGURE 14.7 ■ Either ΔpH or Δψ drives ATP synthesis.
A. A pH difference imposed across the chloroplast inner membrane drives a proton current through an outwardly directed ATP synthase. **B.** A charge difference generated by K^+ influx drives a proton current through ATP synthase.

chains solubilize it in the membrane; thus the molecule conducts K^+ across the membrane down its concentration gradient. In the experiment shown, the vesicles are "inside-out" bacterial membrane vesicles, in which the ATP synthase points outward instead of inward. The K^+ influx adds positive charge, which drives H^+ out through ATP synthase, catalyzing formation of ATP. Thus, $\Delta\psi$ drives formation of ATP.

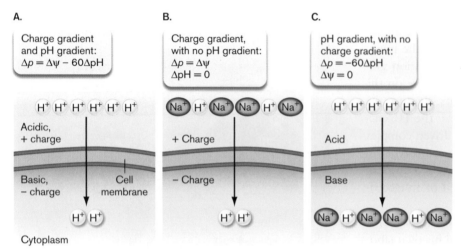

A.

Charge gradient
and pH gradient:
$\Delta p = \Delta\psi - 60\Delta pH$

H⁺ H⁺ H⁺ H⁺ H⁺ H⁺

Acidic,
+ charge

Basic,   Cell
– charge   membrane

H⁺ H⁺

Cytoplasm

B.

Charge gradient,
with no pH gradient:
$\Delta p = \Delta\psi$
$\Delta pH = 0$

Na⁺ H⁺ Na⁺ Na⁺ H⁺ Na⁺

+ Charge

– Charge

H⁺ H⁺

C.

pH gradient, with no
charge gradient:
$\Delta p = -60\Delta pH$
$\Delta\psi = 0$

H⁺ H⁺ H⁺ H⁺ H⁺ H⁺

Acid

Base

Na⁺ H⁺ Na⁺ Na⁺ H⁺ Na⁺

FIGURE 14.8 ■ Electrical potential and pH difference. Proton motive force drives protons into the cell (arrow). **A.** The proton motive force Δp is composed of the transmembrane electrical potential $\Delta\psi$ (the charge difference), plus the transmembrane pH difference ΔpH (the chemical concentration gradient of H⁺). **B.** If the pH inside and outside the cell is equal, then $\Delta p = \Delta\psi$. **C.** If the electrical charge inside and out is equal, then $\Delta p = -60\Delta pH$.

Δp Includes $\Delta\psi$ and ΔpH

The transfer of H⁺ through a proton pump generates a H⁺ concentration difference across the membrane. Since H⁺ carries a positive charge, the proton transfer also generates a charge difference across the membrane. The H⁺ concentration difference (ΔpH) plus the charge difference ($\Delta\psi$) make a proton potential (Δp), also called a proton motive force (PMF).

Thus, when protons are pumped across the membrane, the proton motive force stores energy in two different forms—the separation of charge, or electrical potential, and the gradient of H⁺ concentration, or pH difference— as shown in **Figure 14.7**. Either form (or both) can drive Δp-dependent cell processes:

- **The electrical potential** ($\Delta\psi$) arises from the separation of charge between the cytoplasm (more negative) and the solution outside the cell membrane (more positive). For many bacteria, this "battery" potential is about –50 to –150 mV.

- **The pH difference** (ΔpH) is the difference between internal and external pH [pH(in) – pH(out)]. For example, if the bacterial internal pH is 7.5 and the external pH is 6.5, the ΔpH is 1.0, and the ratio of [H⁺]$_{out}$ to [H⁺]$_{in}$ is 10. A ΔpH of 1.0 corresponds to a proton potential of approximately –60 mV when the temperature is 25°C.

The relationship between electrical and chemical components of the proton potential Δp (in mV) is given by:

$$\Delta p = \Delta\psi - (2.3 RT/F)\Delta pH$$

or approximately:

$$\Delta p = \Delta\psi - 60\Delta pH$$

For cells grown at neutral pH, all three terms (Δp, $\Delta\psi$, and $-60\Delta pH$) usually have a negative value, meaning that their force drives protons inward from outside.

In living cells, the relative contributions of $\Delta\psi$ and $-60\Delta pH$ vary depending on other sources of charge difference and protons, as shown in **Figure 14.8**. Both the $\Delta\psi$ and ΔpH components of Δp are influenced by other factors besides ETS proton transport. For example:

- **The charge difference** $\Delta\psi$ includes charges on other ions, such as K⁺ and Na⁺. These ions are pumped or exchanged by ion-specific membrane transport proteins.

- **The pH difference** ΔpH is affected by metabolic generation of acids, such as fermentation acids, or by pH changes outside the cell. Permeant acids can run down the gradient through the membrane, collapsing the ΔpH to zero.

Overall, the cell uses its various membrane pumps and metabolic pathways to adjust and maintain its proton motive force at a size sufficient to drive ATP synthesis but not so great as to disrupt the membrane.

Thought Questions

14.3 What do you think happens to $\Delta\psi$ as the cell's external pH increases or decreases? What could happen to the Δp of bacteria that are swallowed and enter the extremely acidic stomach?

14.4 How could a simple experiment provide evidence that a proton pump in the bacterial cell membrane drives efflux of antibiotics such as tetracycline?

What about other ions, such as the sodium ions in **Figure 14.8**? Can a sodium gradient drive cell processes? For some bacteria, a gradient of Na⁺ concentration can drive transport of a nutrient through a transporter protein embedded in the cell membrane. Bacteria such as *Vibrio cholerae* (the cause of cholera) can use a Na⁺ concentration gradient to power ATP synthesis by a Na⁺-dependent ATP synthase.

Dissipation of proton motive force. What happens if something disrupts the membrane so that protons leak through? Even a small leak can quickly dissipate all the energy stored as ΔpH.

Many fermentation products are weak acids, such as acetic acid, whose protonated form can dissolve in the membrane and

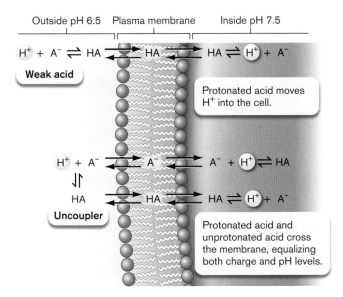

FIGURE 14.9 ■ **Membrane-permeant weak acids and uncouplers. Top:** A weak acid crosses the membrane in the protonated form and then dissociates, acidifying the cell. **Bottom:** An uncoupler crosses the membrane in both protonated and unprotonated forms, cyclically acidifying the cell and dissipating the charge difference (electrical potential, $\Delta\psi$).

then dissociate on the other side (**Fig. 14.9**, top) (discussed also in Chapter 3). A membrane-permeant weak acid conducts protons through the membrane until the ΔpH is dissipated. Other weak acids cross the membrane in both charged and uncharged forms. The weak acid 2,4-dinitrophenol (DNP) dissociates to an anion whose charge is relatively evenly distributed around the molecule; thus, it remains hydrophobic enough to penetrate the membrane (**Fig. 14.9**, bottom). Because both protonated (uncharged) and unprotonated (negatively charged) forms cross the membrane, DNP can cyclically bring protons into the cell and collapse both $\Delta\psi$ and ΔpH. Such molecules are called **uncouplers** because they uncouple electron transport from ATP synthesis through the membrane ATP synthase. Uncouplers are highly toxic to bacteria, as well as to human cells. All cells need to maintain ion gradients and potentials.

Δp Drives Many Cell Functions

Besides ATP synthesis, Δp drives many cell processes directly (**Fig. 14.10**). An important proton-driven function is the rotation of flagellar motors (discussed in Chapter 3). Proton flux is also coupled to transport of ions such as K^+ or Na^+ through parallel transport (symport) or oppositely directed transport (antiport), as discussed in Chapter 4. Ion flux then drives uptake of nutrients such

as amino acids or efflux of molecules such as antibacterial drugs. Drug efflux pumps are a major problem in hospitals, where the pump proteins are encoded by genes carried on plasmids that spread among virulent strains (Chapter 9).

Thought Question

14.5 Suppose that de-energized cells of *E. coli* ($\Delta p = 0$) with an internal pH of 7.6 are placed in a solution at pH 6. What do you predict will happen to the cell's flagella? What does this demonstrate about the function of Δp?

To Summarize

■ **The ETS complexes generate a proton motive force** (Δp). The force is usually directed inward, driving protons into the cell.

■ **The proton potential drives ATP synthase** and other functions, such as ion transport and flagellar rotation.

■ **The proton potential (Δp, measured in millivolts)** is composed of the electrical potential ($\Delta\psi$) and the hydrogen ion chemical gradient (ΔpH): $\Delta p = \Delta\psi - 60\Delta$pH.

■ **Uncouplers are molecules taken up by cells in both protonated and unprotonated forms.** Uncouplers can collapse the entire proton potential, thus uncoupling respiration from ATP synthesis.

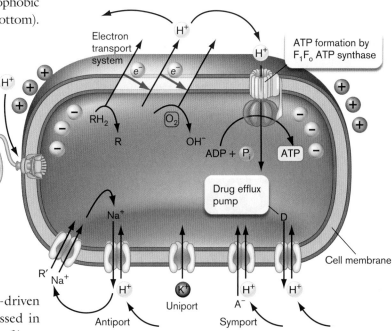

FIGURE 14.10 ■ **Processes driven by the proton motive force.** Processes powered by proton potential include ATP synthesis through the F_1F_o ATP synthase, flagellar rotation, uptake of nutrients, and efflux of toxic drugs. "R" refers to an organic nutrient.

14.3 The Respiratory ETS and ATP Synthase

We now discuss the carrier molecules and proteins that comprise the electron transport system (ETS) of oxidative respiration. For aerobic bacteria, and for the mitochondrial inner membrane, the respiratory ETS receives electrons from NADH and $FADH_2$ and transfers electrons ultimately to O_2, producing H_2O. In between, the series of membrane-embedded carriers harvests the reducing potential of electrons in small steps.

Cofactors Allow Small Energy Transitions

ETS proteins such as cytochromes associate electron transfer with small, reversible energy transitions. The energy transitions are mediated by **cofactors**, small molecules that associate with the protein. **Figure 14.11** shows the protein structure of a cytochrome from the pathogen *Pseudomonas aeruginosa*. The protein, cytochrome *c*, transfers electrons for nitrate respiration. Its reddish color derives from the buried cofactor called **heme**, a ring of conjugated double bonds surrounding an iron ion (Fe^{2+} or Fe^{3+}). The heme plays a key role in acquiring and transferring electrons, with an Fe^{2+}/Fe^{3+} transition. A metal-reducing bacterium such as *Geobacter* may make over a hundred different types of cytochromes in its envelope, where electrons accumulate until the bacterium finds an oxidant to accept them.

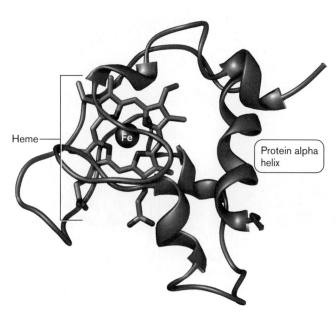

FIGURE 14.11 ■ Cytochrome c. Portion of the protein structure of cytochrome *c* containing one heme cofactor, from the pathogen *Pseudomonas aeruginosa*. (PDB code: 2PAC)

A.

FIGURE 14.12 ■ Reaction centers for electron transport. A. Flavin mononucleotide (FMN). **B.** Iron-sulfur clusters: [2Fe-2S] and [4Fe-4S]. **C.** Heme *b*. The side chains of the ring vary among hemes, yielding different levels of redox potential. **D.** Ubiquinone, which is reduced to ubiquinol.

Flavin mononucleotide (FMN)

B.

Iron-sulfur cluster [2Fe-2S]

Iron-sulfur cluster [4Fe-4S]

C.

Heme *b*

D.

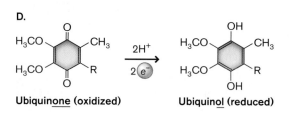

Ubiquinone (oxidized)   Ubiquinol (reduced)

Cofactors such as heme allow small, reversible redox changes. Other examples include flavin mononucleotide (FMN) and ubiquinone (**Fig. 14.12**). The structure of each cofactor must allow transition of an electron between closely spaced energy levels to avoid "spending it all in one place." If all of the energy were spent in one transition, most of it would be lost as heat instead of being converted to several small processes, such as pumping H^+ across the membrane.

Small energy transitions typically involve these kinds of molecular structures:

■ **Metal ions such as iron or copper,** coordinated (and hence held in place) with amino acid residues. Iron is often coordinated by sulfur atoms of cysteine residues in the protein; examples shown in **Figure 14.12B** are [2Fe-2S] and [4Fe-4S]. Transition metals make useful electron carriers because their outer electron shell has several closely spaced energy levels, facilitating small energy transitions.

■ **Conjugated double bonds and heteroaromatic rings**, such as the nicotinamide ring of $NAD^+/NADH$, also provide narrowly spaced energy transitions. Membrane-soluble carriers such as **quinones** (reduced to **quinols**) allow even smaller energy transitions than does $NAD^+/NADH$.

The major protein complexes of electron transport each have one or more redox centers containing either metal ions or conjugated double bonds or both. In FMN, the conjugated double bonds allow a small energy transition (**Fig. 14.12A**). In iron-sulfur clusters—[2Fe-2S] and [4Fe-4S]—the metal atoms provide the site for a small energy transition, its size dependent on the cluster's connections within the associated protein (**Fig. 14.12B**). The heme group found in cytochromes and other oxidoreductases contains extensive conjugated double bonds, coordinated around the metal Fe^{3+} (**Fig. 14.12C**). Reduction by a transferred electron converts the iron to Fe^{2+}. The branching of side chains on the ring varies among different hemes, altering the magnitude of $E^{\circ\prime}$ for the redox couple Fe^{3+}/Fe^{2+}.

Some electron carriers are small molecules that associate loosely with a protein complex and then come off to diffuse freely within the membrane. Mobile electron carriers include quinones such as ubiquinone (**Fig. 14.12D**), which can be reduced to quinols:

$$Q + 2H^+ + 2e^- \rightarrow QH_2$$

Quinols carry electrons and protons laterally within the membrane between the proton-pumping protein complexes of the ETS. The hydrophobic quinols never leave the membrane; thus their electrons are kept in the membrane until transfer out of the ETS. After transferring their electrons to the next protein complex, quinols revert to quinones, capable of accepting electrons again.

Note: *Dehydrogenases, reductases,* and *oxidases* are all **oxidoreductases**, which oxidize one substrate (remove electrons) and reduce another (donate electrons). Oxidoreductases that accept electrons from NADH or $FADH_2$ are also called dehydrogenases, because their reaction releases hydrogen ions.

Oxidoreductase Protein Complexes

A respiratory electron transport system includes at least three functional components: an initial substrate oxidoreductase (or dehydrogenase), a mobile electron carrier, and a terminal oxidase. Microbes make alternative versions of each component using alternative electron-donating substrates and terminal electron acceptors, depending on what is available in the environment. Here we present a typical bacterial ETS receiving electrons from NADH and transferring them to oxygen.

Initial substrate oxidoreductase. A respiratory ETS begins with an initial oxidoreductase that receives a pair of electrons from an organic substrate such as NADH. Note that NADH forms by receiving two electrons plus $2H^+$ from an organic product of catabolism (designated RH_2) (**Fig. 14.13A**). The $2H^+$ are ultimately balanced by $2H^+$ from the cytoplasm combining with O_2 (or another terminal electron acceptor) at the end of the ETS. The two electrons ($2e^-$) from NADH enter an ETS protein complex embedded in the membrane.

NADH donates electrons to NADH dehydrogenase (NADH:quinone oxidoreductase, NDH-1) (**Fig. 14.13A**). In bacteria, the NDH-1 complex includes 14 different subunits. The NDH-1 complex includes the cofactor FMN, as well as several iron-sulfur clusters—typically 7[4Fe-4S] and 2[2Fe-2S]. The cofactors "hand off" electrons to each other through adjacent connections; see, for example, the placement of FMN and the first [4Fe-4S] within the peptide coils of NDH-1 (**Fig. 14.13B**). Each electron from NADH travels through FMN and the iron-sulfur series. At the end of the chain, the electrons and $2H^+$ from solution are transferred to a quinone, which is thus reduced to a quinol. Quinones are designated Q; and quinols, QH_2.

Within the NDH-1 complex, the oxidation of NADH and reduction of Q to QH_2 yields energy to pump up to $4H^+$ across the membrane. A crystallographic model shows four apparent proton channels through transmembrane alpha helices of the protein subunits (**Fig. 14.13A**). Within each channel, a hydrogen ion hops along a series of amino acid residues. The H^+ translocation is driven by a conformational change in the alpha helices throughout the protein, arising from the initial two-electron reduction by NADH. The hydrogen ions pumped across the membrane contribute to the proton potential Δp. In human mitochondria, the NADH dehydrogenase (aka "complex I") is critical for health; genetic defects in complex I are associated with diseases such as Parkinson's and some forms of diabetes.

Note that the $4H^+$ pumped across the membrane are distinct from the $2H^+$ acquired by the quinone ($Q \rightarrow QH_2$). In some halophilic bacteria, $4Na^+$ are pumped instead of $4H^+$.

Note: In our figures, protons that cross the membrane by the end of the ETS (and thus contribute to Δp) are highlighted yellow.

A. NDH-1 complex

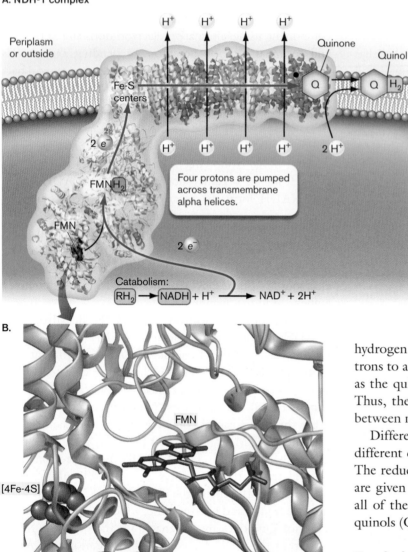

B.

FIGURE 14.13 ▪ NADH:quinone oxidoreductase complex (NDH-1). A. NDH-1 transfers two electrons from NADH onto NDH-1. The energy from oxidizing NADH is coupled to pumping 4H$^+$ across the cell membrane. **B.** Within the NDH-1 complex, FMN lies adjacent to the first iron-sulfur center, [4Fe-4S]. (PDB code: 1NOX) *Source:* Part A modified from R. Efremov and L. Sazanov. 2010. *Nature* **476**:414; part B modified from L. A. Sazanov and P. Hinchliffe. 2006. *Science* **311**:1430–1436.

Not all substrate oxidoreductases pump protons. For example, *E. coli* has an alternative NADH dehydrogenase (NDH-2) that transfers two electrons to Q without pumping additional protons across the membrane. (The unused energy is lost as heat.) NDH-2 functions during rapid growth, when the cell must limit its proton potential to avoid membrane breakdown. Other complexes, such as succinate dehydrogenase, transfer electrons from substrates in a reaction that lacks sufficient energy to pump extra protons.

Quinone pool. A quinone can receive $2e^-$ from the substrate oxidoreductase, along with $2H^+$ from solution, to balance the negative charges, yielding a quinol (**Fig. 14.13A**). The quinols diffuse within the membrane and carry reduction energy to other ETS components. After transferring $2e^-$ to the next protein complex, $2H^+$ are released. Usually the $2H^+$ released are on the opposite side of the membrane from where $2H^+$ were originally picked up (**Fig. 14.14**). Thus, besides electron transfer, a quinol may contribute two protons to the transmembrane proton potential. The reoxidized carriers then recycle back as quinones.

Each quinone can bind to a substrate dehydrogenase, pick up a pair of electrons and hydrogen ions, and then diffuse away and carry the electrons to a reductase. The quinones and quinols, referred to as the quinone pool, diffuse freely within the membrane. Thus, the quinones/quinols are able to transfer electrons between many different redox enzymes.

Different oxidoreductase complexes interact with slightly different quinones, such as ubiquinone and menaquinone. The reduction potentials of ubiquinone and menaquinone are given in **Table 14.1**. For clarity, this chapter refers to all of them as quinones (Q), and their reduced forms as quinols (QH$_2$).

Terminal oxidase. A terminal oxidase complex receives electrons from a quinol (QH$_2$) and transfers them to a terminal electron acceptor, such as O$_2$ (**Fig. 14.14**). The complex usually includes a cytochrome that accepts electrons from quinols. Cytochromes of comparable function are designated by letters—for example, cytochrome *b* (*E. coli* and mitochondria) and cytochrome *c* (mitochondria). The cytochrome is bound to an oxidase complex containing a series of electron-transferring carriers: two iron-centered hemes and three copper atoms. This unique center couples electron transfer and proton pumping.

The *E. coli* cytochrome *bo* quinol oxidase consists of cytochrome *b* plus oxidase complex *o*. The cytochrome *b* subunit receives two electrons from a quinol (QH$_2$ → Q) and releases the 2H$^+$ out to the periplasm. Each electron from quinol travels through the two hemes of the oxidase complex. Because the two quinol hydrogens originated from 2H$^+$ in the cytoplasm (see **Fig. 14.13**) and 2H$^+$ were released outside by a quinol, there is net efflux of 2H$^+$. In addition, the transfer of $2e^-$ between the two oxidase hemes is coupled to pumping 2H$^+$ from the cytoplasm across to the periplasm (**Fig. 14.14**).

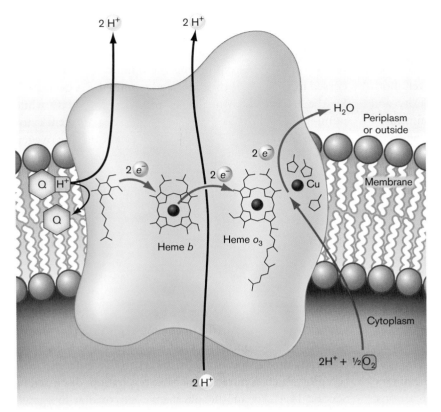

FIGURE 14.14 ■ Cytochrome *bo* quinol oxidase complex. Each quinol (QH₂) transfers two electrons to heme *b*. The two H⁺ from each quinol are expelled to the periplasm (or outside the bacterial cell). The 2*e*⁻ from the quinols are transferred through the complex onto 2H⁺ plus an oxygen atom from O₂, forming water. *Source: Based on the structure determined by Jeff Abramson et al. 2000. Nat. Struct. Biol. **7**:910–917.*

The hydrogen ions leak through the outer membrane, effectively outside the cell.

The second heme of the oxidase (heme o_3) acts to reduce an atom of oxygen from O_2. Each oxygen atom receives two electrons and combines with two protons ($2H^+$) from the cytoplasm to form H_2O. The $2H^+$ consumed balances the $2H^+$ released by catabolism to make NADH + H^+.

Note that the oxidase is conventionally shown as obtaining two electrons from the cytoplasm and donating them to an atom of an oxygen molecule($\frac{1}{2}O_2$). A full reaction cycle of cytochrome *bo* quinol oxidase actually puts four electrons from two quinols (originally 2 NADH) onto O_2, taking up $4H^+$ from the cytoplasm to make two molecules of H_2O.

Besides cytochrome *bo* quinol oxidase, bacteria express different terminal oxidases that differ with respect to the ratio of cytoplasmic protons pumped to electrons transferred. For example, the alternative cytochrome *bd* quinol oxidase reduces O_2 to water but pumps no extra protons. Although it pumps no protons, cytochrome *bd* quinol oxidase can bind O_2 at much lower concentrations and donate electrons, completing the respiratory circuit. Thus, the *bd* oxidase enables *E. coli* to respire within low-oxygen habitats such as the mammalian intestine.

Figure 14.15 ▶ summarizes a complete ETS for oxidation of NADH by $\frac{1}{2}O_2$ in the inner membrane of *E. coli*. Overall, an ETS for respiration on organic substrates includes

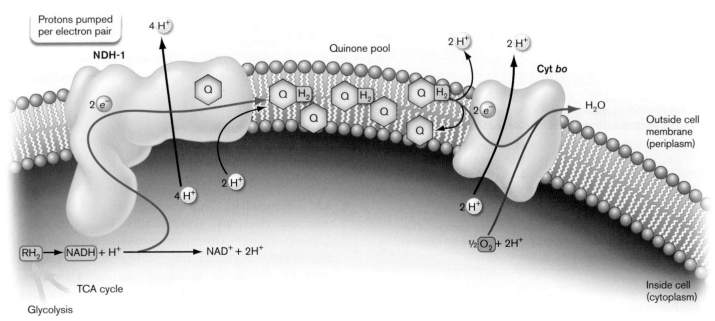

FIGURE 14.15 ■ A bacterial ETS for aerobic NADH oxidation. In *E. coli*, electrons from NDH-1 are transferred to quinones, generating quinols, which transfer electrons onto cytochrome *bo* (Cyt *bo*). For each NADH oxidized, up to 8H⁺ may be pumped across the membrane. ▶

at least three phases of electron transfer, such as: (1) NADH from an organic substrate donates electrons to an initial oxidoreductase; (2) the electrons are transferred to a quinone, which is reduced to a quinol; (3) the quinol transfers electrons to a terminal oxidase and releases $2H^+$ outside the cell. Both enzymes and quinones show considerable complexity and diversity among different species in different environments.

The entering carrier NADH carries two electrons with protons obtained from catabolized food molecules. The two electrons $(2e^-)$ from NDH-1 and two protons $(2H^+)$ from solution are transferred onto Q (quinone), converting it to QH_2 (quinol). The transfer of $2e^-$ from NADH yields sufficient energy to pump as many as $4H^+$ across the membrane. The exact number depends on cell conditions, such as the concentrations of NADH and the terminal electron acceptor.

The QH_2 diffuses within the membrane until it reaches a terminal oxidase complex, such as the cytochrome *bo* quinol oxidase. The $2H^+$ from QH_2 are released outside the cell, while the $2e^-$ enter the oxidase, reducing the two hemes. The two electrons join $2H^+$ from the cytoplasm, combining with an oxygen atom to make H_2O. The reaction is coupled to pumping of $2H^+$ across the membrane plus a net increase of $2H^+$ outside through redox reactions.

Overall, the oxidation of NADH exports about $8H^+$ per $2e^-$ transferred through the ETS to make H_2O. The export of protons generates a proton potential Δp.

Note: Protons (hydrogen ions, H^+) can have three different fates in the ETS:

1. Protons are <u>pumped</u> across the membrane (H^+) by an oxidoreductase complex. Contributes to Δp.

2. Protons are <u>consumed</u> from the cytoplasm by quinone/quinol, while other protons are <u>released</u> outside the membrane (H^+). Contributes to Δp.

3. Protons are consumed by <u>combining with the terminal electron acceptor</u> (O_2). If the loss balances protons released by catabolism, it does <u>not</u> affect Δp.

Thought Questions

14.6 In **Figure 14.15**, what is the advantage of the oxidoreductase transferring electrons to a pool of mobile quinones, which then reduce the terminal reductase (cytochrome complex)? Why does each oxidoreductase not interact directly with a cytochrome complex?

14.7 In **Figure 14.15**, why are most electron transport proteins fixed within the cell membrane? What would happen if they "got loose" in aqueous solution?

Environmental modulation of the ETS. The ETS just described represents optimal conditions, when food and oxygen are unlimited. What happens when food (that is,

electron donors) or oxygen is scarce? When the environment changes, bacteria adjust the efficiency of their ETS by expressing alternative oxidoreductases. For example, at low concentrations of oxygen (microaerophilic conditions, discussed in Chapter 5) the reduction potential E is decreased. So the ETS may be unable to reduce O_2 to H_2O while pumping four protons. Instead, *E. coli* uses a different cytochrome quinol oxidase that has higher affinity for oxygen but pumps fewer protons. Thus, the bacteria will gain less energy, but they will still be able to grow. Environmental regulation of ETS is discussed further in **eTopic 14.1**.

Bacteria also have alternative oxidoreductases to serve different electron donors and acceptors. Some enzymes take electrons from donors lacking the potential of NADH; for example, succinate dehydrogenase catalyzes the one step of the TCA cycle that yields $FADH_2$—a step that provides not quite enough energy to produce NADH (discussed in Chapter 13). Succinate dehydrogenase is the only TCA enzyme embedded in the membrane as an ETS component. When O_2 concentration is so low as to be thermodynamically unavailable, some bacteria can use other electron acceptors, such as nitrate (anaerobic respiration, discussed in Section 14.4). Yet another alternative is to oxidize inorganic electron donors (lithotrophy, discussed in Section 14.5).

Mitochondrial respiration. In contrast to *E. coli*, mitochondria have only a single ETS, optimized for a relatively uniform intracellular environment (**Fig. 14.16**). Protected by the eukaryotic cytoplasm, mitochondria do not need to use alternative versions of their ETS to respire under different conditions. Instead, a set of just four electron-carrying complexes has evolved so as to maximize the energy obtained from NADH, minimizing energy lost as heat. This mitochondrial ETS is nearly universal throughout the cells of animals, plants, and most eukaryotic microbes. In addition, some bacteria, such as *Paracoccus denitrificans*, have an ETS whose organization resembles that of mitochondria.

The mitochondrial ETS has homologs (proteins encoded by genes with a common ancestor) of bacterial ETS components, including NADH dehydrogenase, succinate dehydrogenase, and cytochrome *c* oxidase. However, the mitochondrial ETS differs from that of *E. coli* in these respects:

■ **An intermediate cytochrome oxidase complex transfers electrons.** Besides NADH dehydrogenase (complex I) and cytochrome *c* oxidase (complex IV), mitochondria show an intermediate step of electron transfer to ubiquinol:cytochrome *c* oxidoreductase (complex III, shown in **Fig. 14.16**). The intermediate electron transfer step pumps an additional $2H^+$. Another $2e^-$ and $2H^+$ come from succinate dehydrogenase (complex II), which forms $FADH_2$ through the TCA cycle.

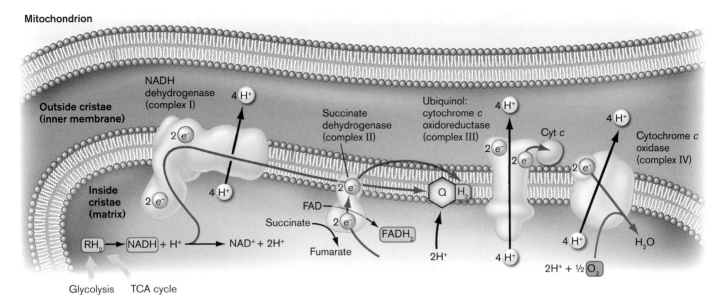

FIGURE 14.16 ■ **Mitochondrial electron transport.** In addition to NADH dehydrogenase and a terminal cytochrome oxidase, mitochondria possess ubiquinol:cytochrome c oxidoreductase, which provides an intermediate electron transfer step. As a result, mitochondrial membranes export 10–12 H^+ per NADH.

- **The mitochondrial ETS pumps more protons per NADH.** As many as 10–12 protons may be pumped per NADH, in contrast to 2–8 protons in *E. coli*.

- **Homologous complexes have numerous extra subunits.** Mitochondria have evolved additional nonhomologous subunits specific to eukaryotes. For example, the homologous subunits of cytochrome c oxidoreductase are enveloped by a series of eukaryotic proteins.

The Proton Potential Drives ATP Synthase

The proton potential drives synthesis of ATP by the membrane ATP synthase, also known as the F_1F_o ATP synthase. The proton-driven synthesis of ATP completes the cycle of **oxidative phosphorylation**, in which hydrogen ions pumped by the ETS drive phosphorylation of ADP to ATP by the ATP synthase. The same ATP synthase can use the proton potential generated by lithotrophy (see Section 14.5) or by phototrophy (Section 14.6). Surprisingly, despite the homology of ATP synthase across all forms of life, the complex is a target for antibiotics. For example, the ATP synthase of *Mycobacterium tuberculosis*, the cause of tuberculosis, is inhibited specifically by bedaquiline, an antibiotic approved in 2012 for patients whose disease resists all other drugs.

The F_1F_o ATP synthase is a protein complex highly conserved in the bacterial cell membrane, the mitochondrial inner membrane, and the chloroplast thylakoid membrane. An elegant molecular machine, the ATP synthase is composed of two complexes—F_o and F_1—that rotate relative to each other (**Fig. 14.17**). The F_o complex translocates protons across the membrane. Twelve c subunits form a cylinder embedded in the membrane, stabilized by subunits a and b.

The F_1 complex consists of six alternating subunits of types alpha and beta surrounding a gamma subunit that acts

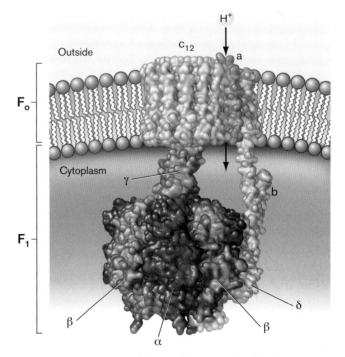

FIGURE 14.17 ■ **Bacterial membrane-embedded ATP synthase (F_1F_o ATP synthase).** The F_o complex is embedded in the bacterial plasma membrane, whereas the F_1 complex catalyzes ATP synthesis. (PDB codes: 1B9U, 1C17, 1E79, 2CLY)

A.

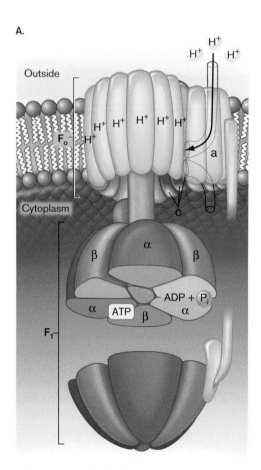

B.

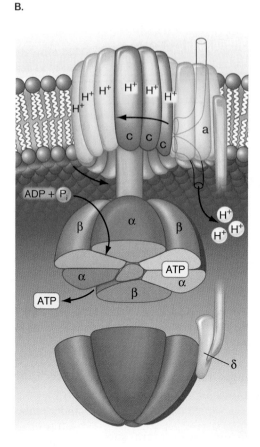

FIGURE 14.18 ▪ H⁺ flux drives ATP synthesis. A. Three protons enter c subunits of the F_o complex. The number of c subunits varies from 8 to 15 among bacterial species; 12 are shown here. **B.** The ring of c subunits rotates one-third turn relative to F_1. Flux of three protons through F_o is coupled to F_1 converting ADP + P_i to ATP.

as a drive shaft. Each "third" of the F_1 complex (an alpha plus a beta subunit) interconverts ADP + P_i with ATP + H_2O. The gamma subunit connects the tripartite "knob" of F_1 to the membrane-embedded F_o. Proton transport through F_o drives ATP synthesis by F_1.

One proton at a time enters the subunit-a channel and moves into a c subunit of F_o (**Fig. 14.18** ▶). The proton potential directed inward ensures that protons more often enter from the outside than from the cytoplasm. Each entering proton causes a c subunit to rotate around the axle, with release of a bound proton to the cytoplasm. The flux of three protons through F_o is coupled to forming one molecule of ATP by one alpha-beta-gamma unit of F_1. During each cycle generating ATP, the ring of c subunits of the F_o rotor rotates in the membrane one-third of one turn relative to the F_1 in the cytoplasm.

Note that the generation of ATP is completely reversible, so ATP hydrolysis by F_1 can pump protons back through F_o across the membrane. In the absence of a proton potential, a high ATP concentration can drive the F_o in reverse, actually pumping protons to generate Δp. This reversal of ATP synthase is used, for example, by *Enterococcus faecalis*, intestinal Gram-positive cocci that generate ATP mainly by fermentation. *E. faecalis* can operate the membrane-embedded F_1F_o ATP synthase in reverse, consuming ATP and thus generating a proton potential for nutrient uptake and ion transport.

Na⁺ Pumps: An Alternative to H⁺ Pumps

While a proton potential provides primary energy storage for most species, some bacteria generate an additional potential of sodium ions. A sodium motive force (ΔNa^+) is analogous to the proton motive force in that it includes the electrical potential $\Delta\psi$ plus the sodium ion concentration gradient (log ratio of the Na^+ concentration difference across the membrane). For extreme halophilic archaea, which grow in concentrated NaCl, the sodium potential entirely substitutes for the proton potential to drive ATP synthesis. These "haloarchaea" make use of the high external Na^+ concentration to store energy in the form of a sodium potential.

In some bacteria, an ETS oxidoreductase pumps Na^+ instead of H^+. For example, the proton-pumping NADH dehydrogenase can be supplemented by an NADH dehydrogenase that pumps Na^+ out of the cell. This primary sodium pump is found in many pathogens, including *Vibrio cholerae* (the cause of cholera) and *Yersinia pestis* (the cause of bubonic plague). These pathogens use the sodium-rich blood plasma to store energy in a sodium potential.

To Summarize

- **Electron carriers** containing metal ions and/or conjugated double-bonded ring structures are used for electron transfer. For example, cytochrome *bo* quinol oxidase has two hemes and three copper ions.

- **A substrate dehydrogenase** receives a pair of electrons from a particular reduced substrate, such as NADH. NADH dehydrogenase (NADH:quinone oxidoreductase) typically has an FMN carrier and nine iron-sulfur clusters.

- **Quinones receive electrons** from the substrate dehydrogenase and become reduced to quinols. Typically, a quinol receives $2H^+$ from the cytoplasm, and then releases $2H^+$ across the membrane upon transfer of $2e^-$ to an electron acceptor complex.

- **Protons are pumped** by substrate dehydrogenases (oxidoreductases) and terminal oxidases. The number of protons pumped by a bacterial ETS is determined by environmental conditions, such as the concentrations of substrate and terminal electron acceptor.

- **Protons are consumed** by combining with the terminal electron acceptor, such as combining with oxygen to make H_2O.

- **The proton potential drives ATP synthesis** through the membrane-embedded F_1F_o ATP synthase. Three protons drive each F_1F_o cycle, synthesizing one molecule of ATP. Some bacteria use a similar ATP synthase driven by Na^+.

14.4 Anaerobic Respiration

Nearly all multicellular animals and plants require electron transport to oxygen. Likewise, some bacteria are called obligate aerobes because they grow only using O_2 as a terminal electron acceptor; examples include important nitrogen-fixing bacteria, such as *Sinorhizobium meliloti* and *Azotobacter vinelandii*. However, other organotrophic bacteria and archaea use a wide range of terminal electron acceptors, including metals, oxidized ions of nitrogen and sulfur, and chlorinated organic molecules. This **anaerobic respiration** generally takes place in environments where oxygen is scarce, such as wetland soil and water, and the human digestive tract.

E. coli possesses several different terminal oxidoreductases to reduce alternative electron acceptors (**Fig. 14.19**). These enzymes, although comparable to cytochrome quinol oxidases, are conventionally termed "reductases" to emphasize reduction of the alternative electron acceptor. Some of the electron acceptors are inorganic, such as nitrate (NO_3^-) reduced to nitrite (NO_2^-), or NO_2^- reduced to NO (nitric oxide). Others are organic products of catabolism; for example, the TCA cycle intermediate fumarate can be reduced to succinate. Organic electron acceptors play important roles in food decomposition. For example, the substance trimethylamine oxide, used by fish as an osmoprotectant against sea salt, is reduced by bacteria to trimethylamine—the main cause of the "fishy" smell.

At the top of the ETS, alternative dehydrogenases (or substrate oxidoreductases) receive electrons from different organic electron donors, as well as from molecular hydrogen (H_2). The enzymes "connect" to various terminal oxidoreductases through the pool of quinones. Note that the various electron donors differ greatly in their reduction potential and hence in their capacity to generate proton potential

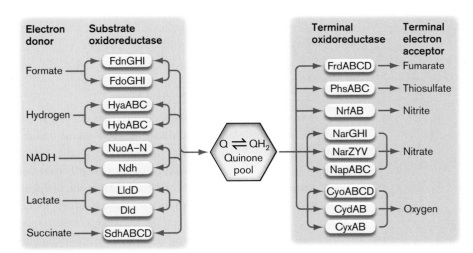

FIGURE 14.19 ■ Alternative electron donors and electron acceptors. *Escherichia coli* can oxidize various foods (electron donors) while reducing various electron acceptors. Each electron donor may use alternative substrate oxidoreductases, depending on the environmental conditions.

(see **Table 14.1**). In a given environment, bacteria use the strongest electron donor and the strongest electron acceptor available. The best donor and best acceptor usually induce the expression of genes encoding their respective redox enzymes. For example, in the presence of nitrate, genes encoding nitrate reductase are expressed. At the same time, nitrate represses the expression of reductases for poorer electron acceptors, such as fumarate. (The mechanisms of gene induction and repression are discussed in Chapter 10.)

Oxidized Forms of Nitrogen

Respiration using oxidized forms of nitrogen is widespread among bacteria and archaea. Most eukaryotes must breathe oxygen, but some eukaryotic microbes respire on nitrate and nitrite. In anaerobic soil, many yeasts and filamentous fungi can reduce nitrate to nitrite, and nitrite to nitrous oxide.

The nitrogen series offers an abundant source of strong electron acceptors. Reduction of oxidized states of nitrogen for energy yield is called **dissimilatory denitrification**. Dissimilatory nitrate reduction to ammonium (DNRA) contributes to respiration, whereas <u>assimilatory</u> reduction of nitrate generates ammonium ion for fixation into biomass (discussed in Chapters 15 and 22). Some pathogens are dissimilatory denitrifiers, such as *Neisseria meningitidis* (a cause of meningitis) and *Brucella* species that cause brucellosis in cattle, sheep, and dogs. In the dissimilatory nitrogen redox series, a given oxidation state can serve as an acceptor in one redox couple but as a donor in the next. The redox couples are summarized here:

$$NO_3^- \xrightarrow{2e^-} NO_2^- \xrightarrow{e^-} NO \xrightarrow{e^-} \tfrac{1}{2}N_2O \xrightarrow{e^-} \tfrac{1}{2}N_2$$
Nitrate   Nitrite   Nitric   Nitrous   Nitrogen
                        oxide   oxide   gas

Each nitrogen state requires a specialized reductase, such as nitrate reductase or nitrite reductase, to receive electrons from the ETS. During the reactions, oxygen atoms are removed and combined with protons to form water. The full series of nitrate reduction to N_2 plays a crucial role in producing the nitrogen gas of Earth's atmosphere (discussed in Chapter 22).

An alternative option for many soil bacteria, such as *Bacillus* species, is to reduce nitrite to ammonium ion (NH_4^+):

$$NO_2^- + 8H^+ + 6e^- \rightarrow NH_4^+ + 2H_2O$$

Dissimilatory nitrate and nitrite reduction to ammonium ion can increase the soil pH. The high pH will precipitate metals such as iron.

The presence of a particular reductase can be used as a diagnostic indicator for clinical isolates of bacteria. For example, the chemical test for nitrate reduction is a key step in the diagnosis of *Neisseria gonorrhoeae*, the causative agent of gonorrhea. *N. gonorrhoeae* happens to lack the terminal reductase for nitrate; hence, it tests negative, whereas several closely related species test positive.

Oxidized Forms of Sulfur

The redox potentials for sulfur oxyanions are generally lower than those for oxidized nitrogen. Nevertheless, with the appropriate oxidoreductases, sulfate and sulfite receive electrons from many kinds of electron donors, including acetate, hydrocarbons, and H_2. Sulfate-reducing bacteria and archaea are widespread in the ocean, from the Arctic waters to submarine thermal vents. The ubiquity of sulfate reduction may be related to the high sulfate content of seawater, in which SO_4^{2-} is the most common anion after chloride.

The major oxidized forms of sulfur that serve as bacterial electron acceptors include:

$$SO_4^{2-} \xrightarrow{2e^-} SO_3^{2-} \xrightarrow{2e^-} \tfrac{1}{2}S_2O_3^{2-} \xrightarrow{2e^-} S^0 \xrightarrow{2e^-} H_2S$$
Sulfate   Sulfite   Thiosulfate Elemental Hydrogen
                          sulfur   sulfide

A surprising use of a sulfur oxyanion by a pathogen is tetrathionate respiration by *Salmonella enterica* (**Fig. 14.20**). When *S. enterica* infects the gut epithelium, it induces host inflammation and activates white blood cells (discussed in Chapter 25). *Salmonella* catabolizes amino acids, releasing hydrogen sulfide (H_2S), which becomes oxidized to thiosulfate $(S_2O_3^{2-})$. White blood cells called neutrophils attempt to kill the bacteria by secreting toxic reactive oxygen species (ROS), or oxidants. The ROS convert thiosulfate to tetrathionate $(S_4O_6^{2-})$. But, surprisingly, *S. enterica* can use tetrathionate as a terminal electron acceptor for its ETS to yield energy. With this extra energy source unavailable to competitors, the pathogenic *S. enterica* can outgrow the host's native microbiome and cause disease.

Dissimilatory Metal Reduction

An important class of anaerobic respiration involves the reduction of metal cations, or **dissimilatory metal reduction**. The term "dissimilatory" indicates that the metal reduced as a terminal electron acceptor is excluded from the cell. This is in contrast to minerals reduced for the purpose of incorporation into cell components (assimilatory metal reduction). Metal-reducing bacteria such as *Geobacter* or *Shewanella* offer the intriguing prospect of making electricity in a fuel cell (**Special Topic 14.1**). The metals most commonly reduced through anaerobic respiration are iron $(Fe^{3+} \rightarrow Fe^{2+})$ and manganese $(Mn^{4+} \rightarrow Mn^{2+})$, but virtually any metal with multiple redox states can be reduced

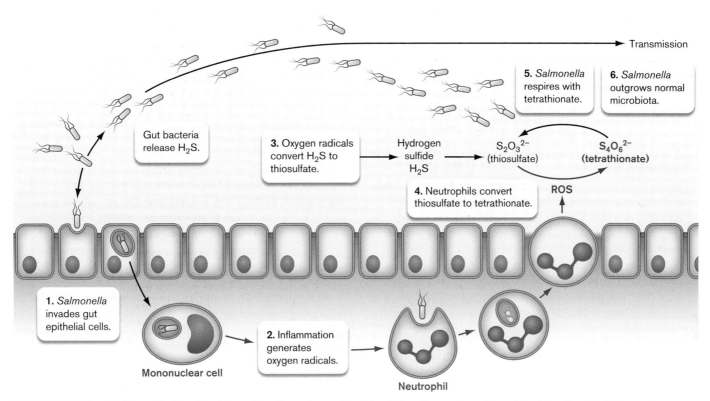

FIGURE 14.20 ▪ *Salmonella* invades the gut and respires with tetrathionate produced by white blood cells. *Salmonella* releases H_2S, oxidized to thiosulfate, which neutrophils (white blood cells) then oxidize to tetrathionate. The tetrathionate serves as a terminal electron acceptor for *Salmonella*.

or oxidized by some bacteria. For example, *Geobacter* can reduce uranium ($U^{6+} \rightarrow U^{4+}$) to respire on acetate. Reduced uranium is insoluble in water, precipitating as uranite (UO_2). Thus, *Geobacter* respiration with uranium can remediate uranium-contaminated water. The U.S. Department of Energy used uranium-reducing bacteria for remediation at Rifle, Colorado, where uranium contamination threatened the Colorado River. In this process, acetate is pumped into the water table, where *G. metallireducens* oxidizes the acetate to CO_2 by reducing U^{6+} to U^{4+}. The reduced uranium then precipitates out of the water into the soil, where it can be collected and removed, while the cleansed water flows through.

Metal electron acceptors require a specific oxidoreductase (or reductase). Metal ions in solution interact directly with reductases, but oxidized metals such as Fe^{3+} are often barely soluble in water. How does a bacterial reductase interact with an insoluble source of metal outside the cell? Soil bacteria may "shuttle" electrons to extracellular metals using quinone-like degradation products of lignin, a complex aromatic substance that forms the bulk of wood and woody stems. Lignin degradation products are also called "humics" because of their presence in humus, the organic components of soil (discussed in Chapter 21). Humics accumulate in anaerobic environments, where decomposition is slow. Alternatively, bacteria may contact insoluble metals using surface-bound

cytochromes—and even electron-conducting "nanowires" such as those of *Shewanella* (seen earlier, in **Fig. 14.2**).

Anoxic or low-oxygen environments, such as the sediment of a lake or wetland, offer a series of different electron acceptors (**Fig. 14.21**). The stronger electron acceptors are consumed, in turn, by species that have the terminal oxidases

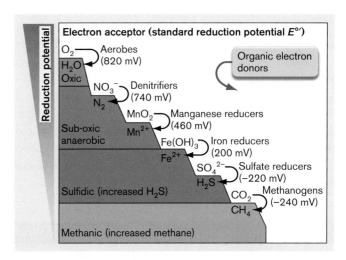

FIGURE 14.21 ▪ Anaerobic respiration in a lake water column. As each successive terminal electron acceptor is used up, its reduced form appears. The next-best electron acceptor is then used, generally by a different species of microbe.

SPECIAL TOPIC 14.1   Bacterial Electric Power

In natural biofilms, bacteria transfer electrons to metals and even between neighboring bacteria. What if we could harness bacterial electron transfer to power our own electrical devices? Kenneth Nealson (**Fig. 1A**) and students at the University of Southern California are trying to do just that.

The concept of bacterial electric power is surprisingly recent. Up to 20 years ago, it was thought that bacteria could reduce only soluble ions such as nitrate. But Nealson, like the famous nineteenth-century microbial ecologist Sergei Winogradsky, was fascinated by the metal transformations he observed in wetlands and lake sediments. Nealson was particularly intrigued by the high levels of reduced manganese (Mn^{2+}) found in Lake Oneida, New York. He reasoned that only microbial activity could account for so much reduced manganese, compared to other lakes where manganese is in the oxidized form (Mn^{4+}, as the insoluble mineral MnO_2). In 1988, Nealson and his colleague Charles Myers discovered a bacterium that could respire anaerobically by donating electrons to MnO_2, releasing soluble Mn^{2+}. The bacterium, named *Shewanella oneidensis* MR-1 (for "manganese reducer") contains a large number of different cytochromes. The cytochromes help the bacterium donate electrons to different kinds of metals, even insoluble forms such as iron and cobalt embedded within clay.

How can bacteria reduce a substance outside their cell envelopes? Nealson's colleague Mohamed El-Naggar found that *Shewanella* bacteria transfer electrons via nanowires (see **Fig. 14.2**). The nanowire extensions of periplasm help bacteria connect with each other in a biofilm, and transfer a current to the metal electron acceptor. In an electrolytic fuel cell, the bacteria form a biofilm on the anode (electron-attracting electrode) (**Fig. 1B**). As the bacteria oxidize organic fuel, they cause charge separation between electrons and hydrogen ions. The electrons then pass within a current to the cathode.

The "fuel" for the cell can be a mixture of organic substances derived from any kind of food waste or sewage. Organic waste includes small organic molecules such as lactate, acetate, and even formaldehyde. The biofilm bacteria on the anode can remove hydrogens from these organic molecules, separating the electrons and hydrogen ions (**Fig. 1C**). The hydrogen ions migrate through a polymer membrane, whereas the electrons enter the anode leading to an electrical wire. The remaining carbon and oxygen atoms of the fuel are released as CO_2. The process is similar to natural respiration, except that instead of molecular oxygen, the electron acceptor is an electrode made of graphite. To complete the circuit, the electrons from the wire current ultimately react with oxygen and hydrogen ions to form water, as in aerobic respiration.

So far, microbial fuel cells have been able to generate milliamps of current, enough to drive small devices, such as clocks and marine data sensors. Microbial fuel cells offer a clean way to make electricity, with inexpensive fuel. The main challenge is generating larger currents. Many researchers are working to ramp up the currents and make the dream of bacterial electricity a reality.

RESEARCH QUESTION

How might you test the question of whether bacteria require nanowires to generate electricity in a fuel cell?

Pirbadian, Sahand, Sarah E. Barchinger, Kar Man Leung, Hye Suk Byon, et al., and Mohamed Y. El-Naggar. 2014. *Shewanella oneidensis* MR-1 nanowires are outer membrane and periplasmic extensions of the extracellular electron transport components. *Proceedings of the National Academy of Sciences USA* **111**:12883–12888.

to use them. For example, as oxygen grows scarce, denitrifying bacteria reduce nitrate to nitrogen gas ($NO_3^- \rightarrow N_2$). As nitrate is used up, other species reduce manganese ($Mn^{4+} \rightarrow Mn^{2+}$), iron ($Fe^{3+} \rightarrow Fe^{2+}$), and sulfate ($SO_4^{2-} \rightarrow H_2S$). At the bottom of the lake or sediment, methanogenic archaea reduce carbon dioxide to methane (see Section 14.5). Microbial reduction of minerals plays a critical role in every ecosystem and participates in the geochemical cycling of elements throughout Earth's biosphere (discussed in Chapter 22).

Thought Question

14.9 The proposed scheme for uranium removal requires injection of acetate under highly anoxic conditions, with less than 1 part per million (ppm) dissolved oxygen. Why must the acetate be anoxic?

To Summarize

■ **Anaerobic terminal electron acceptors**, such as nitrogen and sulfur oxyanions, oxidized metal cations, and oxidized organic substrates, accept electrons from a specific reductase complex of an ETS.

■ **Nitrate** is successively reduced by bacteria to nitrite, nitric oxide, nitrous oxide, and ultimately nitrogen gas. Alternatively, nitrate and nitrite may be reduced to ammonium ion, a product that alkalinizes the environment.

■ **Sulfate** is successively reduced by bacteria to sulfite, thiosulfate, elemental sulfur, and hydrogen sulfide. Sulfate reducers are especially prevalent in seawater.

■ **Oxidized metal ions** such as Fe^{3+} and Mn^{4+} are reduced by bacteria in soil and aquatic habitats—a form of

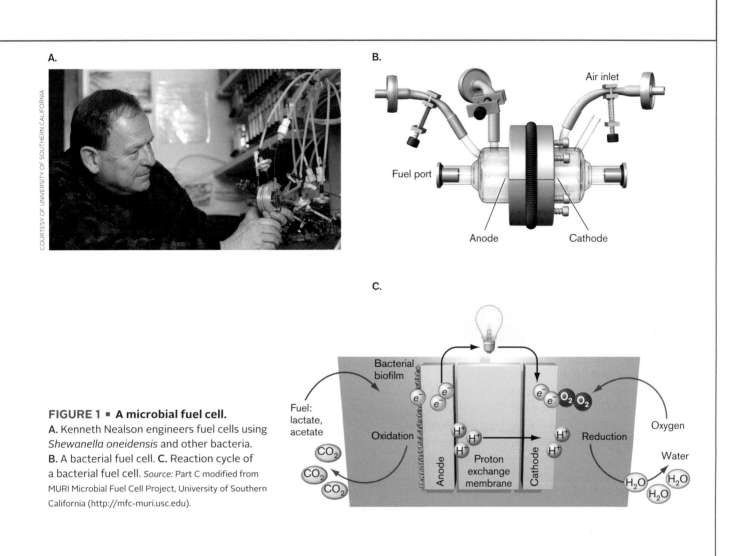

FIGURE 1 ■ A microbial fuel cell.
A. Kenneth Nealson engineers fuel cells using *Shewanella oneidensis* and other bacteria. **B.** A bacterial fuel cell. **C.** Reaction cycle of a bacterial fuel cell. *Source:* Part C modified from MURI Microbial Fuel Cell Project, University of Southern California (http://mfc-muri.usc.edu).

anaerobic respiration also known as dissimilatory metal reduction.

■ **The geochemistry of natural environments** is shaped largely by anaerobic bacteria and archaea.

14.5 Lithotrophy and Methanogenesis

Many reduced minerals and single-carbon compounds can serve as electron donors for an ETS, in the energy-yielding form of metabolism known as **lithotrophy** (or **chemolithotrophy**). Some organotrophs have alternative oxidoreductases that conduct lithotrophy by oxidizing H_2 or Fe^{2+}; for example, the gastric pathogen *Helicobacter pylori* oxidizes hydrogen gas released by mixed-acid-fermenting bacteria in the colon. Other bacteria are "obligate lithotrophs"; that is, they oxidize only inorganic molecules. All lithotrophs are bacteria or archaea. Thus, bacteria and archaea fill many key niches in ecosystems that eukaryotes cannot.

Lithotrophy includes many kinds of electron donors, from metals and anions to single-carbon groups (**Table 14.2**). Each type of electron donor, such as H_2, NH_4^+, or Fe^{2+}, requires a specialized electron-accepting oxidoreductase. Most inorganic substrates other than H_2 are relatively poor electron donors compared to organic donors such as glucose. Therefore, the terminal electron acceptor is usually a strong oxidant, such as O_2, NO_3^-, or Fe^{3+}. Obligate lithotrophs (organisms that conduct only lithotrophy) consume no organic carbon source; they build biomass by fixing CO_2 (a process discussed in Chapter 15).

TABLE 14.2 Lithotrophy: Electron Donors and Acceptors

Type of lithotrophy	Species example	Electron donor	Electron acceptor
Hydrogenotrophy	*Aquifex aeolicus*	$H_2 \rightarrow 2H^+ + 2e^-$	$O_2 \rightarrow H_2O$
Sulfate reduction (hydrogenotrophy)	*Desulfovibrio vulgaris*	$H_2 \rightarrow 2H^+ + 2e^-$	$SO_4^{2-} \rightarrow S^O, HS^-$
Methanogenesis (hydrogenotrophy)	*Methanocaldococcus jannaschii*	$H_2 \rightarrow 2H^+ + 2e^-$	$CO_2 \rightarrow CH_4$
Iron oxidation	*Acidithiobacillus ferrooxidans*	$Fe^{2+} \rightarrow Fe^{3+} + e^-$	$O_2 \rightarrow H_2O$
Ammonia oxidation (nitrosification)	*Nitrosomonas europaea*	$NH_3 \rightarrow NO_2^-$	$O_2 \rightarrow H_2O$
Nitrification	*Nitrobacter winogradskyi*	$NO_2^- \rightarrow NO_3^-$	$O_2 \rightarrow H_2O$
Anammox	"*Candidatus* Kuenenia stuttgartiensis"	$NH_3 + 4H^+ + 4e^- \rightarrow N_2$	$NO_2 \rightarrow N_2$
Methylotrophy	*Methylococcus capsulatus*	$CH_4 \rightarrow CO_2 + 4H^+ + 4e^-$	$O_2 \rightarrow H_2O$
Carboxidotrophy	*Carboxydothermus hydrogenoformans*	$CO \rightarrow CO_2 + e^-$	$H_2O \rightarrow H_2$
Sulfide oxidation	*Sulfolobus solfataricus*	$HS^- \rightarrow S^O + H^+ + 2e^-$	$O_2 \rightarrow H_2O$
Sulfur oxidation	*Acidithiobacillus thiooxidans*	$S^O + 2H^+ + 4e^- \rightarrow H_2SO_4$	$O_2 \rightarrow H_2O$

Iron Oxidation

Reduced metal ions such as Fe^{2+} and Mn^{2+} provide energy through oxidation by O_2 or NO_3^-. Bacteria perform these lithotrophic reactions in soil where weathering exposes reduced minerals. They generate metal ions with higher oxidation states (such as Fe^{3+} or Mn^{4+}), which other bacteria use for anaerobic respiration. Environments such as ponds and wetlands that experience frequent shifts between oxygen availability and oxygen depletion are likely to host a variety of metal-oxidizing lithotrophs, as well as metal-reducing anaerobic heterotrophs. The roles of lithotrophy in ecology are discussed further in Chapters 21 and 22.

An example of lithotrophy is iron oxidation by the bacterium *Acidithiobacillus ferrooxidans* (**Fig. 14.22**). These bacteria oxidize Fe^{2+} (ferrous ion) using the reduction potential between Fe^{3+}/Fe^{2+} and O_2/H_2O:

	$E^{o'}$ (pH 2)	$G^{o'}$ (pH 2)
$2Fe^{2+} \rightarrow 2Fe^{3+} + 2e^-$ (pH 2)	$-(+770)$ mV	$+149$ kJ/mol
$\frac{1}{2}O_2 + 2H^+ + 2e^- \rightarrow H_2O$ (pH 2)	$+1,100$ mV	-212 kJ/mol
$2Fe^{2+} + \frac{1}{2}O_2 + 2H^+ \rightarrow 2Fe^{3+} + H_2O$	$+330$ mV	-63 kJ/mol

The removal of electrons from Fe^{2+} requires an input of energy, which is compensated for by the larger yield of energy from reducing oxygen to water. The net reduction potential is small; thus, *A. ferrooxidans* must cycle large quantities of iron in order to grow. Because the reaction consumes H^+, it goes forward only at low pH, such as that of acid mine drainage where other microbes oxidize sulfur to sulfuric acid. The bacterial cell must keep its cytoplasmic pH considerably higher (pH 6.5) by transmembrane exchange of H^+ with cations, and inverting the electrical potential $\Delta\psi$ (positive inside). Thus, the proton potential Δp exists entirely in the form of ΔpH. The ΔpH drives ATP synthesis.

Unlike organic electron donors, metals such as iron must be accessed from insoluble particles outside the cell. In *A. ferrooxidans*, the electrons are collected from iron outside by an outer membrane cytochrome c_2. The cytochrome is associated with a periplasmic protein called rusticyanin. Rusticyanin collects electrons while excluding the potentially toxic metal from the cell. The electrons from rusticyanin can be transferred to an inner membrane cytochrome complex that reduces oxygen to water.

For iron oxidation, the ΔG of reaction is too small to pump protons. So how do cells obtain enough energy to form NADPH for biosynthesis? An alternative pathway directs electrons to an enzyme complex that uses proton influx (spends Δp) to convert $NADP^+$ to NADPH. This pathway is called **reverse electron flow**, because it reverses the flux of electrons seen in an ETS that spends NADH (or NADPH) to form NAD^+ (or $NADP^+$). In reverse electron flow, a relatively poor electron donor (such as $2Fe^{2+}$) reduces an ETS with an unfavorable reduction potential, requiring input of energy to generate NADH or NADPH. For iron oxidation, the energy input comes from the large ΔpH across the inner membrane (pH 6.5 inside, and pH 2 outside). Reverse electron flow is seen in some kinds of lithotrophy, in syntrophy (discussed in Chapter 13), and in phototrophy (discussed in Section 14.6).

Thought Question

14.10 Use **Figure 14.15** to propose a pathway for reverse electron flow in an organism that spends ATP from fermentation to form NADH. Draw a diagram of the pathway.

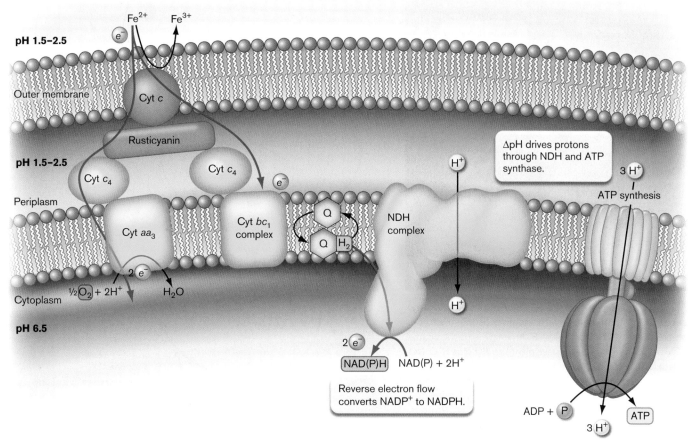

FIGURE 14.22 ■ **Iron oxidation ETS.** Fe^{2+} is oxidized to Fe^{3+} outside the cell, at about pH 2. The periplasm also is at about pH 2. Electrons from exogenous iron are collected by rusticyanin, which feeds electrons to the cytochrome aa_3 complex to reduce O_2 to H_2O. Alternatively, some electrons from rusticyanin reduce the cytochrome bc_1 complex. ΔpH drives reverse electron flow through NDH to form NADPH. *Source:* Modified from Raquel Quatrini et al. 2009. *BMC Genomics* **10**:394.

Nitrogen Oxidation

One kind of lithotrophy essential for the environment is oxidation of nitrogen compounds:

$$NH_4^+ \xrightarrow{\frac{1}{2}O_2} NH_2OH \xrightarrow{O_2} HNO_2 \xrightarrow{\frac{1}{2}O_2} HNO_3$$

Ammonium   Hydroxylamine   Nitrous acid   Nitric acid
                                    (nitrite)   (nitrate)

Reduced forms of nitrogen, such as ammonium ion derived from fertilizers, support growth of **nitrifiers**, bacteria that generate nitrites or nitrates (forming nitrous acid or nitric acid, respectively). Acid production can degrade environmental quality. In soil treated with artificial fertilizers, nitrifiers decrease the ammonium ions obtained from such fertilizer and produce toxic concentrations of nitrites, which leach into groundwater. Still, nitrifiers can be useful in sewage treatment, where they eliminate ammonia that would harm aquatic life (discussed in Chapter 22).

Ammonia/ammonium and nitrite are relatively poor electron donors compared to organic molecules. Thus, their oxidation through the ETS pumps fewer protons, and the bacteria must cycle relatively large quantities of substrates to grow. As we saw for iron oxidation, reduced cofactors such as NADPH must be obtained through reverse electron flow involving redox carriers with different amounts of energy.

What happens to ammonium ion from detritus that accumulates in anaerobic regions, such as the bottom of a lake? Surprisingly, NH_4^+ can yield energy through oxidation by nitrite (a product of nitrate respiration):

$$NH_4^+ + NO_2^- \rightarrow N_2 + 2H_2O \qquad \Delta G^{\circ\prime} = -357 \text{ kJ/mol}$$

Under conditions of high ammonium and extremely low oxygen, nitrite oxidation of ammonium ion supports growth of bacteria. Known as the **anammox reaction**, anaerobic ammonium oxidation plays a major role in wastewater treatment, where it eliminates much of the ammonium ion from sewage breakdown. In the oceans, anammox bacteria cycle as much as half of all the nitrogen gas returned to the atmosphere.

A.

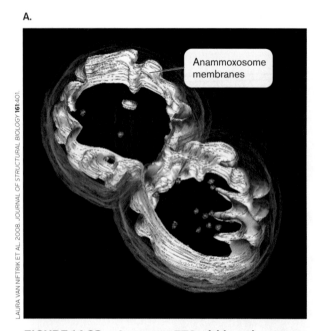

FIGURE 14.23 ▪ Anammox ETS within a planctomycete. A. A planctomycete with anammoxosome membranes (cryo-EM tomography). **B.** An enzyme of the anammoxosomal membrane reduces nitrite to NO plus H_2O. NO is reduced by NH_4^+ to form hydrazine (N_2H_4). Hydrazine is a toxic molecule that gets trapped in the anammoxosome by the membrane composed of ladderanes. Hydrazine donates electrons to reduce nitrite, and gives off N_2. Protons may be pumped by the hydrazine synthase complex.

Anammox is conducted by planctomycetes, irregularly shaped bacteria with unusual membranous organelles that fill much of the cell (**Fig. 14.23A**). The central compartment is called the anammoxosome. The anammoxosome membrane is composed of unusual ladder-shaped lipids called ladderanes. An enzyme of the anammoxosomal membrane reduces nitrite to NO plus H_2O (**Fig. 14.23B**, step 1). Another membrane-embedded enzyme, hydrazine synthase, catalyzes NO reduction by NH_4^+ to form hydrazine (N_2H_4) (step 2). Hydrazine is a high-energy compound that engineers use for rocket fuel; the substance is highly toxic, so the planctomycete keeps it sequestered within the specialized anammoxosome membrane.

The hydrazine is further oxidized to N_2 (**Fig. 14.23B**, step 3), and its protons are released within the anammoxosome compartment. The protons then drive an ATP synthase embedded in the anammoxosome membrane. The hydrazine oxidation enzyme obtains additional electrons from catabolism of organic substrates. Thus, anammox bacteria are examples of microbes whose metabolism combines lithotrophy and organotrophy.

Note: Distinguish between lithotrophy (<u>oxidation</u> of reduced minerals, usually by O_2, nitrate, or nitrite) and anaerobic respiration (<u>reduction</u> of oxidized minerals, usually by organic food molecules).

Sulfur and Metal Oxidation

Major sources of lithotrophic electron donors are minerals containing reduced sulfur, such as hydrogen sulfide and sulfides of iron and copper. As we saw for nitrogen compounds, each sulfur compound that undergoes partial oxidation may serve as an electron donor and be further oxidized.

$$\overset{\tfrac{1}{2}O_2}{H_2S} \rightarrow \overset{\tfrac{1}{2}O_2}{S^0} \rightarrow \overset{O_2 + H_2O}{\tfrac{1}{2}S_2O_3^{2-}} \rightarrow H_2SO_4$$

Hydrogen sulfide   Elemental sulfur   Thiosulfate   Sulfuric acid

Sulfur oxidation produces the strong acid sulfuric acid (H_2SO_4), which dissociates to produce an extremely high H^+ concentration. In the early twentieth century, no one would have believed that living organisms could grow in concentrated sulfuric acid, much less produce it. The *Star Trek* science fiction episode "The Devil in the Dark"

A. *Star Trek*: an imaginary creature produces H_2SO_4

B. A hot spring supports sulfur-oxidizing archaea

70–80°C

FIGURE 14.24 ■ **Organisms produce sulfuric acid: science and science fiction. A.** In the *Star Trek* episode "The Devil in the Dark," starship officers encounter an imaginary creature, called the Horta, that tunnels through rock by producing corrosive acid. **B.** Volcanic rocks and hot springs support growth of sulfur-oxidizing thermophilic archaea such as *Sulfolobus* species, whose growth at pH 2 colors the rocks. **C.** *Sulfolobus acidocaldarius* grows as irregular spheres (TEM).

C. *Sulfolobus acidocaldarius*

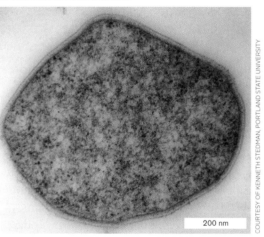

200 nm

portrayed an imaginary alien creature, the Horta, that produced corrosive acid in order to eat its way through solid rock (**Fig. 14.24A**). In actuality, of course, no creatures beyond the size of a microbe are yet known to grow in sulfuric acid. But archaea such as *Sulfolobus* species oxidize hydrogen sulfide to sulfuric acid and grow at pH 2, often in hot springs at near-boiling temperatures (**Fig. 14.24B** and **C**). *Sulfolobus* makes irregular Horta-shaped cells without even a cell wall to maintain shape; how it protects its cytoplasm from disintegration is not yet understood.

Microbial sulfur oxidation can cause severe environmental acidification, eroding concrete structures and stone monuments. The problem is compounded by sulfur oxidation in the presence of iron, such as in iron mine drainage. For example, the sulfur-oxidizing archaeon *Ferroplasma acidarmanus* was discovered in the abandoned mine Iron Mountain, in northern California, by Katrina Edwards (1968–2014; University of Southern California) and colleagues from the University of Wisconsin–Madison. *Ferroplasma* oxidizes ferrous disulfide (pyrite) with ferric iron (Fe^{3+}) and water:

$$FeS_2 + 14Fe^{3+} + 8H_2O \rightarrow 15Fe^{2+} + 2SO_4^{2-} + 16H^+$$

This reaction generates large quantities of sulfuric acid. The acidity of the mine water is near pH 0—one of the most acidic environments found on Earth. As *Ferroplasma* grows, it forms biofilms of thick streamers in the mine drainage, which poison aquatic streams.

Anaerobic reactions between sulfur and iron cause hidden hazards for human technology, such as the corrosion of steel in underwater bridge supports. Anaerobic corrosion was long considered a mystery, since iron was known to rust by means of spontaneous oxidation by O_2. In anaerobic conditions, however, sulfur-reducing bacteria can corrode iron (**Fig. 14.25**). In one pathway, the bacteria reduce elemental sulfur (S^0) with H_2 to hydrogen sulfide (H_2S). H_2S then combines with iron metal (Fe^0), which gives up two electrons to form Fe^{2+}, precipitating as iron sulfide (FeS). The displaced $2H^+$ combine with the $2e^-$ from iron, regenerating H_2—now available to reduce sulfur once again. In an alternative mechanism, bacteria use Fe^0 to reduce sulfate directly to FeS. These damaging processes may resemble the iron-based metabolism of Earth's most ancient life-forms.

Nevertheless, like the imaginary Horta that ended up helping miners with their excavations, acid-producing microbes are now used to supplement commercial mining, a process called "biomining." Lithotrophs such as *Acidithiobacillus ferrooxidans* oxidize sulfides of iron and copper

A.

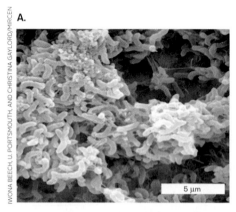

B.

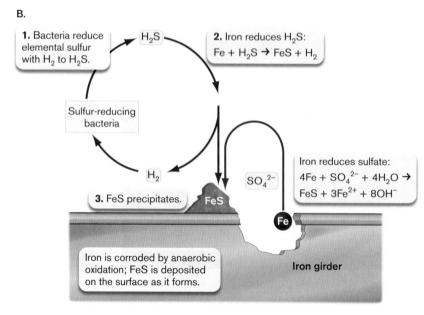

1. Bacteria reduce elemental sulfur with H_2 to H_2S.

H_2S

2. Iron reduces H_2S:
$Fe + H_2S \rightarrow FeS + H_2$

Sulfur-reducing bacteria

H_2

Iron reduces sulfate:
$4Fe + SO_4^{2-} + 4H_2O \rightarrow FeS + 3Fe^{2+} + 8OH^-$

SO_4^{2-}

3. FeS precipitates.

FeS

Fe

Iron is corroded by anaerobic oxidation; FeS is deposited on the surface as it forms.

Iron girder

FIGURE 14.25 ■ **Anaerobic iron corrosion. A.** Sulfate-reducing bacteria corrode iron. **B.** Anaerobic corrosion of iron is accelerated by sulfur-reducing bacteria. Alternatively, in other bacteria, iron reduces sulfate. *Source: Part A from Derek Lovley. 2000. Environmental Microbe-Metal Interactions, p. 163.*

found in minerals such as chalcopyrite ($CuFeS_2$), chalcocite (Cu_2S), and covellite (CuS). The oxidation of Cu^+ to Cu^{2+}, as well as the acidification resulting from production of sulfate, dissolves the metal from the rock. Cu^+ can be oxidized aerobically with O_2 or anaerobically with NO_3^- from the soil. Other metals oxidized by *A. ferrooxidans* include selenium, antimony, molybdenum, and uranium. The process of metal dissolution from ores is called leaching. Leaching of minerals has been a part of mining since ancient times, long before the existence of microbes was known. Today, over 10% of the copper supply in the United States is provided by biomining ores too low in copper to smelt directly (**Fig. 14.26**). A similar process is being developed to biomine gold.

Hydrogenotrophy Uses H_2 as an Electron Donor

The use of molecular hydrogen (H_2) as an electron donor is called **hydrogenotrophy** (**Table 14.3**). An example is oxidation of H_2 by sulfur to form H_2S, performed by *Pyrodictium brockii*, an archaeon that grows at thermal vents above 100°C. Hydrogen gas is a stronger electron donor than most organic foods, so it can be oxidized with the full range of electron acceptors. It is available in anaerobic communities as a fermentation product.

Hydrogenotrophy is hard to categorize. Because H_2 is inorganic, oxidation by O_2 is considered lithotrophy; yet many species that oxidize H_2 with molecular oxygen are

A. A copper mine

B. *Acidithiobacillus ferrooxidans*

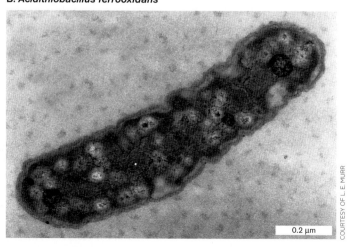

FIGURE 14.26 ■ **Copper mining. A.** Bingham Canyon copper mine near Salt Lake City, Utah, where copper is leached from low-grade ores by *Acidithiobacillus ferrooxidans* (**B**), a Gram-negative rod that oxidizes copper and iron sulfides (TEM).

TABLE 14.3	Hydrogenotrophy: Examples
Specific reaction	**General description**
$O_2 + 2H_2 \rightarrow 2H_2O$	Aerobic oxidation of H_2
Fumarate + $H_2 \rightarrow$ succinate	Organic + $H_2 \rightarrow$ organic
$2CO_2 + 4H_2 \rightarrow CH_3COOH + 2H_2O$	Mineral + $H_2 \rightarrow$ organic
$2H^+ + SO_4^{2-} + 4H_2 \rightarrow H_2S + 4H_2O$	Mineral + $H_2 \rightarrow$ mineral
$CO_2 + 4H_2 \rightarrow CH_4 + 2H_2O$	Methanogenesis

organotrophs that also catabolize organic foods. When hydrogen reduces an organic electron acceptor, such as fumarate, the process may be considered either fermentation or anaerobic respiration. When hydrogen reduces a mineral such as sulfur or sulfate, the process is anaerobic lithotrophy.

A form of hydrogenotrophy with enormous potential for bioremediation is "dehalorespiration," in which halogenated organic molecules serve as electron acceptors for H_2 (**Fig. 14.27A**). Chlorinated molecules such as chlorobenzenes, perchloroethene, and polyvinyl chloride (known as PVC) are highly toxic environmental pollutants. In dehalorespiration, the chlorine is removed as chloride anion and replaced by hydrogen—a reaction requiring input of two electrons from H_2. Many soil bacteria conduct dehalorespiration; a particularly unusual cell wall–less bacterium, *Dehalococcoides* (**Fig. 14.27B**), was discovered that dechlorinates chlorobenzene, a highly stable aromatic molecule.

Methanogenesis

Hydrogen is such a strong electron donor that it can even reduce the highly stable carbon dioxide to methane. Reduction of CO_2 and other single-carbon compounds, such as formate, to methane is called **methanogenesis**. Methanogenesis supports a major group of archaea known as methanogens, many of which grow solely by autotrophy, generating methane.

The simplest form of methanogenesis involves hydrogen reduction of CO_2:

$$CO_2 + 4H_2 \rightarrow CH_4 + 2H_2O \qquad E^{\circ\prime} = 180 \text{ mV}$$

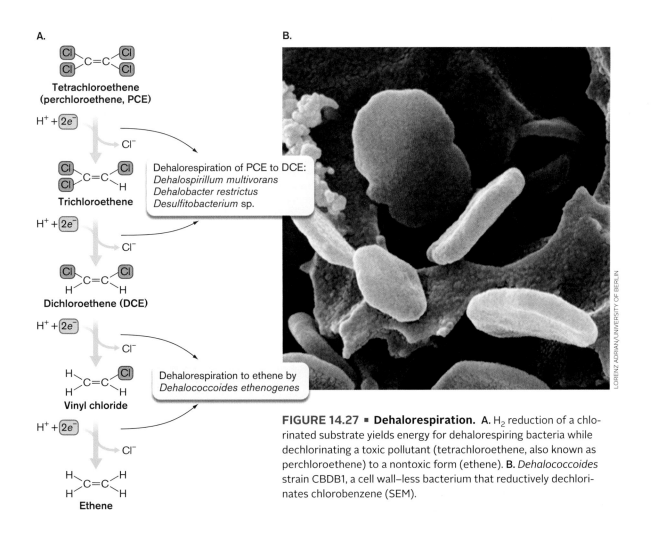

A.

Tetrachloroethene
(perchloroethene, PCE)

$H^+ +$ 2e$^-$

Cl$^-$

Trichloroethene

Dehalorespiration of PCE to DCE:
Dehalospirillum multivorans
Dehalobacter restrictus
Desulfitobacterium sp.

$H^+ +$ 2e$^-$

Cl$^-$

Dichloroethene (DCE)

$H^+ +$ 2e$^-$

Cl$^-$

Vinyl chloride

Dehalorespiration to ethene by
Dehalococcoides ethenogenes

$H^+ +$ 2e$^-$

Cl$^-$

Ethene

LORENZ ADRIAN/UNIVERSITY OF BERLIN

FIGURE 14.27 ■ **Dehalorespiration. A.** H_2 reduction of a chlorinated substrate yields energy for dehalorespiring bacteria while dechlorinating a toxic pollutant (tetrachloroethene, also known as perchloroethene) to a nontoxic form (ethene). **B.** *Dehalococcoides* strain CBDB1, a cell wall–less bacterium that reductively dechlorinates chlorobenzene (SEM).

Because both sides of the equation contain a weak electron acceptor (CO_2, H_2O) and a strong electron donor (H_2, CH_4), it was surprising that such a reaction could yield energy for growth. But the presence of sufficient carbon dioxide and hydrogen supports enormous communities of methanogens. Such conditions prevail wherever bacteria grow by fermentation and their gaseous products are trapped, such as in landfills, where methane can be harvested as natural gas, as well as in the digestive systems of cattle and humans. Variants of methanogenesis also include pathways by which H_2 reduces various one-carbon and two-carbon molecules, such as methanol, methylamine, and acetic acid. Diverse methanogens are found in all kinds of environments (discussed in Chapter 19).

A simplified pathway of methanogenesis from CO_2 is shown in **Figure 14.28**. The CO_2 undergoes stepwise hydrogenation, and each oxygen is reduced to water. The increasingly reduced carbon is transferred through a series of unique cofactors (methanofuran, tetrahydromethanopterin, coenzyme M-SH). Three of the hydrogenation steps involve a membrane ETS complex that includes the carrier coenzyme F_{420}, whose reaction generates a proton potential. Note, however, that the final step of methane production generates a transmembrane sodium potential (ΔNa^+), which drives ATP synthesis by a sodium-powered ATP synthase embedded in the cell membrane. Further details of methanogenesis are presented in Chapter 19.

> **Thought Question**
>
> **14.11** Hydrogen gas is so light that it rapidly escapes from Earth. Where does all the hydrogen come from to be used for hydrogenotrophy and methanogenesis?

Methylotrophy

Many forms of catabolism release reduced single-carbon molecules such as methanol (CH_3OH) or methylamine (CH_3NH_2). Oxidation of single-carbon molecules via an ETS is called **methylotrophy**, a form of metabolism conducted by many soil bacteria. In addition, the methane released by methanogenesis provides a niche for **methanotrophy**, a form of methylotrophy in which bacteria and archaea oxidize methane. In deep-ocean sediment, the activity of methanogens is so high that enormous quantities of methane become trapped on the seafloor in the form of water-based crystals known as methane hydrates. If all the methane were to be released at once, it would greatly accelerate global warming.

How is this methane recycled? Until recently, it was thought that most methanotrophy required O_2; indeed, many bacteria are aerobic methanotrophs. However, the deep-ocean sediments contain other species of bacteria and

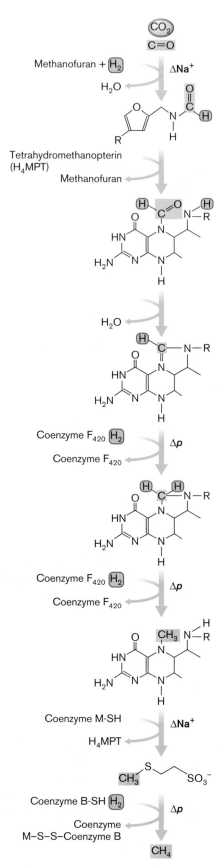

FIGURE 14.28 ■ Methanogenesis. A methanogen reduces carbon dioxide with hydrogen, generating methane (CH_4). The incorporation of hydrogen contributes to both a proton potential and a sodium potential (discussed in Chapter 19).

archaea that oxidize methane using nitrite or sulfate. This anaerobic methane oxidation may be critical for the global carbon cycle, since it suggests a mechanism for removal of deep-sea methane (discussed in Chapters 21 and 22).

To Summarize

- **Lithotrophy (chemolithotrophy)** is the acquisition of energy by oxidation of inorganic electron donors.

- **Reverse electron flow** powered by Δp can generate NADH or NADPH.

- **Sulfur oxidation** includes oxidation of H_2S to sulfur or to sulfuric acid by sulfur-oxidizing bacteria, often accompanied by iron oxidation. Sulfuric acid production leads to extreme acidification.

- **Nitrogen oxidation** includes successive oxidation of ammonia to hydroxylamine, nitrous acid, and nitric acid. Anammox, the oxidation of ammonium ion by nitrite, returns half the ocean's N_2 to the atmosphere.

- **Hydrogenotrophy** uses hydrogen gas as an electron donor. Hydrogen (H_2) has sufficient reducing potential to donate electrons to nearly all biological electron acceptors, including chlorinated organic molecules (through dehalorespiration).

- **Methanogenesis** is the oxidation of H_2 by CO_2, releasing methane. Methanogenesis is performed only by the methanogen group of archaea.

- **Methylotrophs** use O_2, nitrite, or sulfate to oxidize single-carbon compounds such as methane, methanol, or methylamine. A class of methylotrophs called **methanotrophs** specifically oxidize methane.

14.6 Phototrophy

On Earth today, the ultimate source of electrons driving metabolism is phototrophy, the use of photoexcited electrons to power cell growth. Every year, photosynthesis converts more than 10% of atmospheric carbon dioxide to biomass, most of which then feeds microbial and animal heterotrophs. Most of Earth's photosynthetic production, especially in the oceans, comes from microbes (discussed in Chapter 21). Even the frigid seas of Antarctica support vast communities of phototrophic algae and bacteria, at –2°C beneath pack ice (**Fig. 14.29**).

In phototrophy, the energy of a photoexcited electron is transferred through an ETS to pump protons. Different kinds of phototrophy include the bacteriorhodopsin proton pump; the single-cycle chlorophyll-based photosystems I

FIGURE 14.29 ■ Phototrophy beneath Antarctic pack ice. Underwater view of the underside of ice showing yellow-green algae illuminated by sunlight.

and II; and the double-cycle Z pathway of oxygenic photosynthesis in cyanobacteria and chloroplasts. Diverse kinds of phototrophic bacteria are presented in Chapter 18.

Retinal-Based Proton Pumps

In most ecosystems, the dominant source of carbon and energy is photosynthesis based on chlorophyll. At the same time, many halophilic archaea and marine bacteria supplement their metabolism with a simpler, more ancient form of phototrophy based on a single-protein light-driven proton pump containing the pigment retinal. Many varieties of the retinal-based proton pump have evolved, known as **bacteriorhodopsin** in haloarchaea and as **proteorhodopsin** in bacteria. We present bacteriorhodopsin as a relatively simple form of phototrophy, followed by the more complex ETS-based photolysis in bacteria and chloroplasts.

Bacteriorhodopsin and proteorhodopsin. Bacteriorhodopsin is a small membrane protein commonly found in halophilic archaea (or haloarchaea) such as *Halobacterium salinarum*, a single-celled archaeon that grows in evaporating salt flats containing concentrated NaCl (discussed in Chapter 19). For many years, bacteriorhodopsin-like proton pumps were thought to be limited to extreme halophilic archaea. As bacterial genomes were sequenced, however, homologs of the protein appeared in several species of proteobacteria; the homologs were termed proteorhodopsin. The proteorhodopsin genes appear to have entered bacteria by horizontal transfer from halophilic archaea. In 2005, Oded Béjà and colleagues from Israel, Austria, Korea, and the United States surveyed the genomes of unculturable bacteria from the upper waters of the Mediterranean and Red seas. They found that 13% of the marine bacteria contain proteorhodopsins, accounting for a substantial—and previously unrecognized—fraction of marine phototrophy.

Bacteriorhodopsin absorbs light with a broad peak in the green range; thus, the organisms containing large amounts of bacteriorhodopsin reflect blue and red, appearing purple. The protein consists of seven hydrophobic alpha helices surrounding a molecule of **retinal**, the same cofactor bound to light-absorbing opsins in the vertebrate retina (**Fig. 14.30A**). In bacteriorhodopsin, the retinal is attached to the nitrogen end of a lysine residue (**Fig. 14.30B**).

Retinal has a series of conjugated double bonds that absorb visible light. Upon absorbing a photon, an electron in one of the double bonds is excited to a higher energy level. This process is called photoexcitation. As the electron falls back to the ground state, the double bond shifts position from *trans* (substituents pointing opposite) to *cis* (substituents pointing in the same direction). This change in shape of the retinal alters the conformation of the entire protein, causing it to pick up a proton from the cytoplasm. Eventually, the retinal switches back to its original *trans* configuration. The reversion to *trans* is coupled to the release of a proton from the opposite end of the protein facing outside of the cell. Thus, photoexcitation of bacteriorhodopsin is coupled to the pumping of one H^+ across the membrane.

The proton gradient generated by bacteriorhodopsin drives ATP synthesis by a typical F_1F_o ATP synthase. Light capture by bacteriorhodopsin supplements catabolism for energy and heterotrophy for carbon source. This combination of light absorption and heterotrophy is a form of **photoheterotrophy**.

Purple membrane captures light rays. One problem every phototroph needs to solve is how to "capture" light rays. For chemotrophy, food molecules diffuse in solution and can be picked up by receptors for transport into a cell. Phototrophy, however, requires a photon to impinge on one point of the cell, where the photon either is absorbed or passes through. Thus, the only way to absorb a high percentage of photons is to spread light-absorbing pigments over a wide surface area. To maximize light absorption, *Halobacterium salinarum* archaea pack their entire cell membranes with bacteriorhodopsin. The protein assembles in trimers that pack in hexagonal arrays, forming the "purple membrane" (**Fig. 14.31**).

Although the bacteriorhodopsin cycle is much simpler than chlorophyll-based photosynthesis (discussed next), it

A.

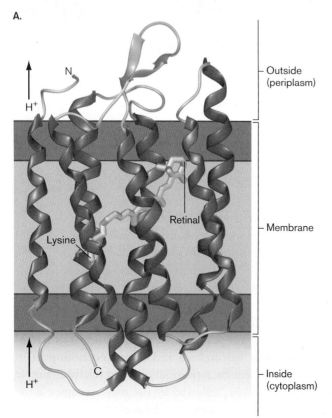

B.

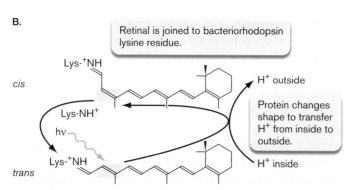

FIGURE 14.30 ▪ **The light-driven cycle of bacteriorhodopsin.** **A.** Bacteriorhodopsin contains seven alpha helices that span the membrane in alternating directions and surround a molecule of retinal, which is linked to a lysine residue. (PDB code: 1FBB) **B.** A photon (hv) is absorbed by retinal, which shifts the configuration from *trans* to *cis*. The relaxation back to the *trans* form is coupled to pumping 1H^+ across the membrane. (PDB code: 1MOL) *Source: Purple Membrane: Theoretical Biophysics Group, VMD Image Gallery, NIH Resource for Macromolecular Modeling and Bioinformatics.*

FIGURE 14.31 ▪ **Bacteriorhodopsin purple membrane.** Trimers of bacteriorhodopsin (monomers shown red, blue, and green) are packed in hexagonal arrays, forming the "purple membrane." (PDB code: 1MOL) *Source: Purple Membrane: Theoretical Biophysics Group, VMD Image Gallery, NIH Resource for Macromolecular Modeling and Bioinformatics.*

nevertheless illustrates several principles that apply to more complex forms of phototrophy:

■ A photoreceptor absorbs light, causing excitation of an electron to a higher energy level, followed by return to the ground state.

■ To maximize light collection, large numbers of photoreceptors are packed throughout a membrane.

■ The photocycle (absorption and relaxation of the light-absorbing molecule) is coupled to energy storage in the form of a proton gradient.

Note: Distinguish among these terms of phototrophy:

■ **Photoexcitation** means light absorption that raises an electron to a higher energy state, as in bacteriorhodopsin.

■ **Photoionization** means light absorption that causes electron separation.

■ **Photolysis** means light absorption coupled to splitting a molecule.

■ **Photosynthesis** means photolysis with CO_2 fixation and biosynthesis.

Chlorophyll Photoexcitation and Photolysis

Cyanobacteria and chloroplasts, as well as other kinds of bacteria, obtain energy by photoexcitation of chlorophylls. Figuring out their "light reactions"—the fundamental source of energy for Earth's biosphere—was one of the most exciting projects of the twentieth century. Among hundreds of important contributors, we note two major figures: the married couple Roger Stanier (1916–1982) and Germaine Cohen-Bazire (1920–2001) (**Fig. 14.32A** and **B**). Stanier, a Canadian microbial physiologist at UC Berkeley, clarified the nature of cyanobacteria as phototrophic prokaryotes distinct from eukaryotic algae, and he helped distinguish the water-based photosynthesis of cyanobacteria from the sulfide metabolism of purple bacteria. Cohen-Bazire was a French bacterial geneticist who had studied *lac* operon regulation with Nobel laureate Jacques Monod at the Pasteur Institute in Paris. Cohen-Bazire applied her genetics skills to phototrophs, and she conducted the first genetic analysis of photosynthesis in purple bacteria and cyanobacteria.

Cyanobacteria, the only oxygen-producing bacteria, appear green, like algae or plants (**Fig. 14.32C**). Their green color arises from their chlorophyll, which absorbs blue and red but reflects green. Cyanobacteria include a wide range of species, such as the ocean's major producers, the submicroscopic *Prochlorococcus marinus*, barely visible under a light microscope. Other cyanobacteria have cells as large as eukaryotic algae and form complex developmental structures with important symbiotic associations (see Chapters 18 and 21). Cyanobacteria are among the most successful and diverse groups of life on

A. Roger Stanier

B. Germaine Cohen-Bazire

C. *Merismopedia sp.*

200 μm

FIGURE 14.32 ■ Roger Stanier and Germaine Cohen-Bazire studied cyanobacterial photosynthesis. A. Stanier pioneered the study of cyanobacterial physiology. **B.** Cohen-Bazire performed the first studies of genetic regulation of bacterial photosynthesis. **C.** Green colonies of the cyanobacterium *Merismopedia* sp.

Earth. Along with the chloroplasts of algae and plants, cyanobacteria produce all the oxygen available for aerobic life.

Overview of photolysis. The energy for photosynthesis derives from the photoexcitation of a light-absorbing pigment. Photoexcitation leads to photolysis, the light-driven separation of an electron from a molecule coupled to an ETS. The components of the ETS are often homologous to those of respiratory electron transport, and they share common electron carriers such as iron-sulfur clusters.

In plant chloroplasts and in cyanobacteria, photolysis is known as the "light reactions," coupled to the "light-independent reactions" of carbon dioxide fixation. Note, however, that many of the sulfur- or organic-based bacterial phototrophs, such as *Rhodospirillum rubrum*, combine photolysis with heterotrophy instead of with CO_2 fixation. At the same time, lithotrophic bacteria such as *Nitrospira* and *Acidithiobacillus* fix CO_2 using energy from mineral oxidation instead of photolysis. In this chapter we focus on photolysis in cyanobacteria. We discuss CO_2 fixation along with other biosynthetic pathways later, in Chapter 15.

Chromophore

Key

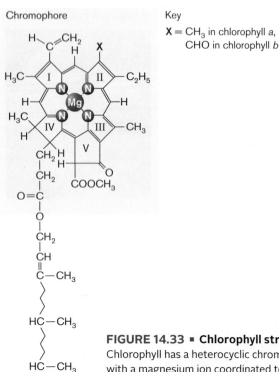

X = CH_3 in chlorophyll *a*,
CHO in chlorophyll *b*

Chlorophyll

FIGURE 14.33 ■ **Chlorophyll structure.**
Chlorophyll has a heterocyclic chromophore
with a magnesium ion coordinated to four
nitrogens. Different chlorophyll types differ
mainly in small substituents of the chromophore; for example, an aldehyde replaces the
ring II methyl group of chlorophyll *a* to make
chlorophyll *b*.

In ETS-based photosynthesis, photoexcitation leads to
separation of an electron from a donor molecule such as
H_2O or H_2S. Each electron is then transferred to an ETS,
whose components show common ancestry with respiratory ETS proteins. The ETS generates a proton potential
and the reduced cofactor NADPH. The proton potential drives ATP synthesis through an F_1F_o ATP synthase,
similar to the one for respiration.

Chlorophylls absorb light. The main light-absorbing pigments are **chlorophylls**. Each type of chlorophyll contains
a characteristic **chromophore**, a light-absorbing electron
carrier. The chlorophyll chromophore consists of a heteroaromatic ring complexed to a magnesium ion (Mg^{2+})
(**Fig. 14.33**). As we saw for ETS electron carriers, metal
ions and aromatic molecules offer electrons with relatively
narrow energy transitions. The chromophore absorbs a photon through a reversible energy transition, such that the
chlorophyll can alternate between excited and ground states.

Chlorophyll molecules differ slightly in their substituent
groups around the ring; for example, chlorophyll *a* of chloroplasts has a methyl group in ring II, whereas chlorophyll *b*
has an aldehyde. Both chlorophylls *a* and *b* are made by chloroplasts and by cyanobacteria, their nearest bacterial relatives.
Because they absorb red and blue, they reflect the middle
range of the spectrum and so appear green (**Fig. 14.34A**).

A.

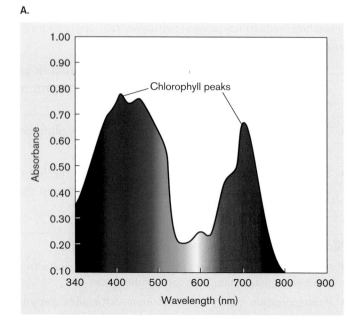

B.

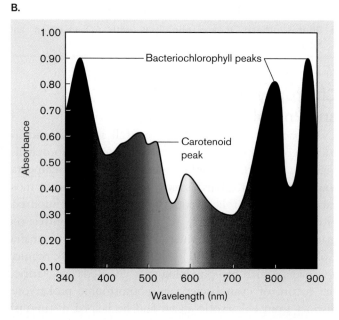

FIGURE 14.34 ■ **Absorbance spectra of photosynthetic
pigments. A.** Absorption by chloroplasts, including chlorophyll *a*,
chlorophyll *b*, and carotenoid accessory pigments. The middle range
(green) is reflected. **B.** Absorption by purple photosynthetic bacteria
includes bacteriochlorophyll and carotenoids.

By contrast, the chlorophylls of anaerobic phototrophs, or
"purple bacteria," such as *Rhodobacter* and *Rhodospirillum*,
absorb most strongly in the far-red (infrared) and, in some
cases, ultraviolet ranges (**Fig. 14.34B**). Their chlorophylls
are specifically named **bacteriochlorophylls**. The purple
bacteria grow in pond water or sediment. Bacteriochlorophyll absorption over an extended range of wavelengths
helps capture light missed by the cyanobacteria and algae at
the water's surface. In purple bacteria, bacteriochlorophylls
are supplemented by accessory pigments called **carotenoids**,
which absorb light of green wavelengths and transfer the

A. Antennas receive electromagnetic energy

FIGURE 14.35 ▪ Antenna complexes surround the reaction center. **A.** Antennas, like these VLA (Very Large Array) dishes in New Mexico, receive electromagnetic energy. **B.** The antenna complex of *Rhodopseudomonas viridis* contains 9 chlorophylls with chromophores facing parallel to the membrane (gold) and 18 with chromophores facing out from the ring (red). **C.** Multiple rings of chlorophyll-protein antenna complexes surround the reaction center (RC), like a funnel collecting photons. (PDB codes: 1PYH, 2FKW) *Source: Quantum Biology of the PSU, NIH Resource for Macromolecular Modeling and Bioinformatics—BChl antenna.*

B. Light-harvesting antenna complex (LH-II)

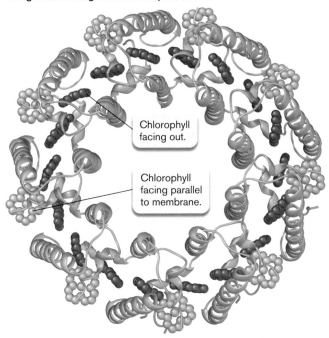

Chlorophyll facing out.

Chlorophyll facing parallel to membrane.

C. Antenna complexes surround the reaction center (RC)

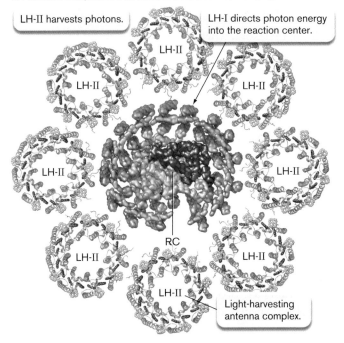

LH-II harvests photons.

LH-I directs photon energy into the reaction center.

Light-harvesting antenna complex.

energy to bacteriochlorophyll. The combination of green-absorbing carotenoids and infrared-absorbing bacteriochlorophylls makes cultures appear deep purple or brown.

The infrared radiation absorbed by bacteriochlorophylls is too weak to permit splitting H_2O to produce oxygen. Thus, purple bacteria are limited to photolysis of H_2S and small organic molecules; many are photoheterotrophs. On the other hand, infrared rays are available in water below the oxygenic phototrophs absorbing red and blue. Thus, anaerobic phototrophs grow at depths where their more high-powered oxygenic relatives do not.

***

Note: Distinguish among these classes of photopigments:

- **Bacteriorhodopsin** is a retinal-containing proton pump.
- **Chlorophyll** is a charge-separating photopigment (usually referring to chloroplasts and cyanobacteria).
- **Bacteriochlorophyll** is a charge-separating chlorophyll of anaerobic purple and green bacteria (also known generically as chlorophyll).
- **Carotenoid** is an accessory pigment that absorbs the midrange of light wavelengths but does not directly conduct photolysis.

Antenna complex and reaction center. Light photons cannot be transported or concentrated in a compartment. Instead, they must be captured by absorption. The larger the array of absorptive molecules, the more photons will be captured.

To maximize light collection, many molecules of chlorophyll are grouped in an **antenna complex**. These antenna complexes are arranged like a satellite dish within the plane of the membrane in an elaborate cluster around accessory proteins (**Fig. 14.35A** and **B**). The clusters then associate in a ring around the **reaction center** (**RC**), the protein complex in which chlorophyll photoexcitation connects to the ETS (**Fig. 14.35C**). Throughout the complex of bacteriochlorophylls and accessory pigments, whichever pigment molecule happens to be in the right place at the right time captures the photon. The energy from the photon then transfers at random from one chromophore to the next, until it arrives at the reaction center for electron transfer to the ETS. Other kinds of bacteria have other forms of antenna complexes, such as the "phycobilisome" of cyanobacteria (discussed shortly).

Purple bacteria and cyanobacteria increase their efficiency of photon uptake by extensive backfolding of the photosynthetic membranes in oval pockets stacked like pita breads (**Fig. 14.36**). These oval pockets are called **thylakoids**. The extensive packing of thylakoids gives an incident photon hundreds of chances to meet a chlorophyll at just the right angle for absorption.

The thylakoids are connected by tubular extensions, so that there exists one interior space, the lumen, separated topologically from the regular cytoplasm, or stroma. Protons are pumped from the stroma across the thylakoid membrane into the lumen. The F_1F_o complex is embedded in the thylakoid, where it makes ATP using the proton current running through it into the cytoplasm. The F_1 knob of ATP synthase appears to face "outward" in photosynthetic organelles (as opposed to "inward" in respiratory chains). In each case, however, the proton current and ATP motor face in the same direction with respect to the cytoplasm (stroma). The proton potential is more negative in the cytoplasm (stroma), thus drawing protons through the ATP synthase to generate ATP.

The photolytic electron transport system. In photolysis, the absorption of light by chlorophyll or bacteriochlorophyll drives the separation of an electron. The chlorophyll may then gain an electron from the ETS, as in *Rhodospirillum* or *Rhodobacter*, or it may remove an electron from H_2S or H_2O, depending on the photosystem of a given bacterial species. In either case, the excited electron carries energy gained from the light absorbed. The excited electron enters a membrane-embedded ETS of oxidoreductases and quinones/quinols, as we saw for the ETS of respiration and lithotrophy.

Diverse kinds of photosynthesis in different environmental niches include oxygenic, sulfur-based, iron-dependent, and even heterotrophic photolysis. Nevertheless, all forms of photolysis share a common design:

1. **Antenna system.** The antenna system (**Fig. 14.35**) maximizes photon capture. A phototrophic antenna system is a large complex of chlorophylls that captures photons and transfers their energy among the

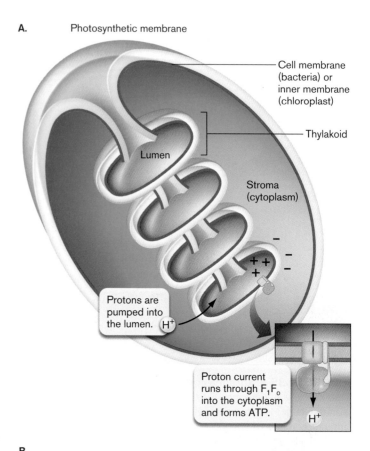

A. Photosynthetic membrane

Cell membrane (bacteria) or inner membrane (chloroplast)

Thylakoid

Lumen

Stroma (cytoplasm)

Protons are pumped into the lumen. H^+

Proton current runs through F_1F_o into the cytoplasm and forms ATP.

H^+

B.

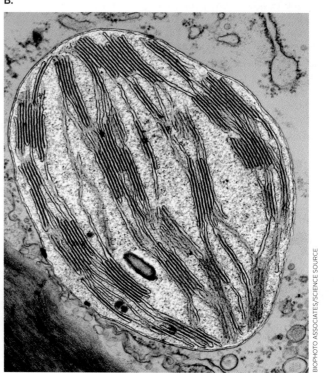

BIOPHOTO ASSOCIATES/SCIENCE SOURCE

FIGURE 14.36 ■ Photosynthetic membranes. **A.** The photosynthetic membranes of bacteria and chloroplasts appear as hollow disks with tubular interconnections. The disks are called thylakoids. Topologically, the membrane separates the cytoplasm (stroma) from the interior space (lumen). **B.** TEM section of a chloroplast, showing the stacked thylakoids.

photopigments until it reaches a reaction center. A complex of chlorophylls and accessory pigments in the photosynthetic membrane collects photons. Energy from each photoexcited electron is transferred among antenna pigments and eventually to the reaction center.

2. **Reaction center complex.** In the reaction center, the photon energy is used to separate an electron from chlorophyll. The electron is replaced by one from a small molecule such as H_2S (**photosystem I**, or **PS I**) or from the ETS (**photosystem II**, or **PS II**).

3. **Electron transport system.** Each photoexcited electron enters an ETS. In PS I, electrons separated from H_2O or H_2S are transferred to $NADP^+$ to form NADPH. In PS II, the electron separated from bacteriochlorophyll is replaced by an electron returned from the ETS. In the **oxygenic Z pathway** (H_2O photolysis), electrons flow from PS II into PS I, ultimately releasing O_2 from H_2O.

4. **Energy carriers.** In PS I, electrons are used to make NADPH. In PS II, electron transfer provides energy to pump protons and drive the synthesis of ATP. The Z pathway makes both NADPH and ATP, which are used to fix CO_2.

Photosystems I and II

The steps of photolysis and electron transport occur in three different kinds of systems, in different classes of bacteria:

- **Anaerobic photosystem I** receives electrons associated with hydrogens from H_2S, HS^-, or H_2, or even from reduced iron (Fe^{2+}). Anaerobic PS I is found in chlorobia ("green sulfur" bacteria) and in chloroflexi (filamentous green bacteria).

- **Anaerobic photosystem II** returns an electron from the ETS to bacteriochlorophyll. Anaerobic PS II is found in

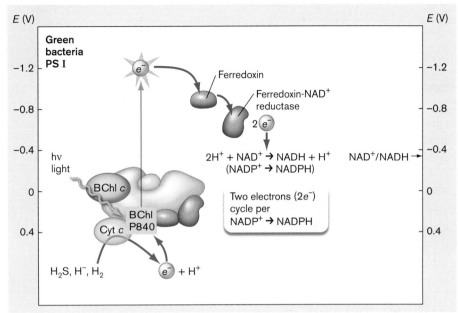

A. Photosystem I

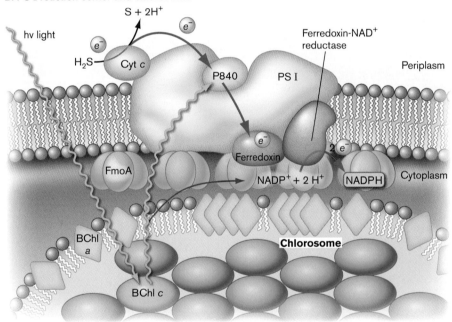

B. PS I reaction center and chlorosome

FIGURE 14.37 ■ **Photosystem I separates electrons from sulfides and organic molecules.** **A.** In green sulfur bacteria, photoexcitation of P840 transfers e^- to a quinone (phylloquinone, PQ), at high reduction potential E. From PQ the electron is transferred to ferredoxin (Fd). Ferredoxin is oxidized by ferredoxin-NAD^+ reductase (FNR), donating the $2e^-$ to NAD^+ (or the energetically equivalent $NADP^+$) to form NADH (or NADPH). **B.** The chlorosome antenna complex transfers photon energy to the PS I reaction center.

alphaproteobacteria, "purple nonsulfur" bacteria, and other proteobacteria.

- **Oxygenic Z pathway** includes homologs of photosystems I and II. Two pairs of electrons are received from

A. Photosystem II

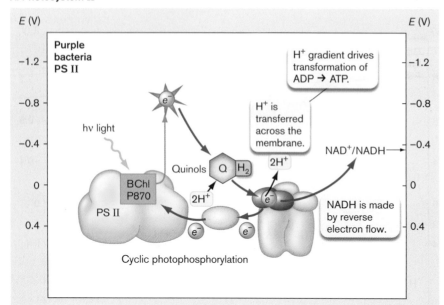

FIGURE 14.38 ▪ Photosystem II separates an electron from bacteriochlorophyll. A. In purple bacteria, BChl P870 donates an energized electron to a quinone (Q). Two of these donated electrons complete the conversion of quinone to quinol (QH_2). Electrons flow through cytochrome *bc*, coupled to pumping of protons. The proton potential drives synthesis of ATP. The cytochrome *bc* complex transfers the electrons back to P870. **B.** PS II reaction center.

B. PS II reaction center and F_1F_o ATP synthase

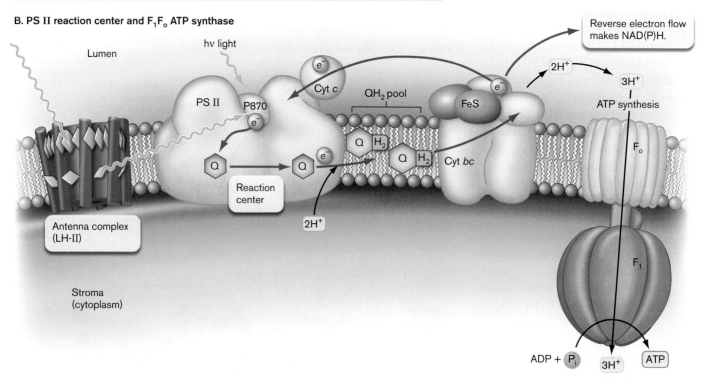

two water molecules to generate O_2. The Z pathway is found in cyanobacteria and in the chloroplasts of green plants.

The components of photosystems I and II (PS I and PS II) share common ancestry. Each system runs anaerobically, producing sulfur or oxidized organic by-products, but not O_2. Each photosystem shows more recent homology with the respective PS I and PS II components of the oxygenic Z pathway (so called because the electron path through a diagram of the two photosystems traces a Z). The Z pathway ultimately generates O_2—the source of nearly all the oxygen we breathe.

Note: In photolysis, each quantum of light excites a single electron. Some ETS components, such as the quinones/quinols (Q → QH_2) actually process two of these electrons in completing their redox cycle. The single-electron intermediate states of quinones are not shown in Figures 14.37, 14.38, and 14.40.

Photosystem I in chlorobia. Bacteria such as *Chlorobium* species use PSI (**Fig. 14.37**). The reaction center (RC) contains bacteriochlorophyll P840, named for its peak absorption at 840 nm—that is, the near-infrared, actually beyond the range humans can see. But P840 and the chlorophylls of the antenna complex also absorb light over shorter wavelengths, in the range of 400–550 nm.

The *Chlorobium* antenna complex consists of a membrane compartment called a chlorosome (**Fig. 14.37B**). A single chlorosome may contain 200,000 molecules of bacteriochlorophyll, harvesting photons with nearly 100% efficiency. The chlorosome is so sensitive that some chlorobia actually harvest thermal radiation from deep-sea thermal vents.

When any one bacteriochlorophyll absorbs a photon, the energy transfers among the photopigments until it reaches the PS I reaction center (**Fig. 14.37**). The photon yields sufficient energy to donate the high-potential electron to a high-potential quinone: phylloquinone/phylloquinol (PQ). Phylloquinol donates the electron to **ferredoxin**, an FeS protein. Ferredoxin transfers the electron to the enzyme ferredoxin-NAD⁺ reductase; two of these electrons then reduce NAD^+ or the energetically equivalent $NADP^+$. The reduced carrier (NADH or NADPH) provides reductive energy for CO_2 fixation and biosynthesis (discussed in Chapter 15).

Chlorobium is a true autotroph, fixing CO_2 for biosynthesis (see Chapter 15). However, the phototrophic ETS is supplemented by lithotrophy, the donation of electrons from sulfur or H_2. The PS I electron flow generates a net proton gradient by consuming H^+ inside and generating H^+ outside the cell, thus providing proton motive force to drive ATP synthesis.

Photosystem II in alphaproteobacteria. Phototrophic alphaproteobacteria such as *Rhodospirillum rubrum* and *Rhodopseudomonas palustris* are typically found in wetlands and streams, where they capture light not used by other phototrophs. Their antenna complex was shown earlier, in **Figure 14.35**. The peak wavelength absorbed by bacteriochlorophyll P870 lies so far into the infrared

(800–1,100 nm) that the photon energy is insufficient to reduce NAD(P) to NAD(P)H. The electrons from P870 are transferred by low-potential quinols to a terminal cytochrome oxidoreductase (**Fig. 14.38**). The quinols pass their electrons to cytochromes, while moving $2H^+$ across the membrane. When the electrons reach cytochrome *c*, they flow back to bacteriochlorophyll, where they can be reexcited by photon energy from the antenna complex. Because the electron path traces back to its source, the ETS of photosystem II leading to ATP synthesis is called **cyclic photophosphorylation**.

Since the reduction potential is too small to reduce $NADP^+$ to NADPH, photosystem II requires reverse electron flow. As we saw for iron oxidation (see Section 14.5), in reverse electron flow a low-potential electron donor reduces an ETS, requiring input of energy. Purple bacteria obtain this energy by spending ATP to increase the proton potential, or from a pathway outside photolysis, such as catabolism of organic compounds. The organic compounds also provide substrates for biosynthesis. Thus, most purple bacteria are photoheterotrophs.

Purple bacteria such as *Rhodopseudomonas palustris* are the ultimate generalists, often combining several major classes of metabolism (**Fig. 14.39**). Caroline Harwood and colleagues at the University of Washington showed that *R. palustris* is a "photolithoheterotroph," capable of photosynthesis, catabolism on many substrates, and lithotrophy (**eTopic 14.2**). Anaerobically, *R. palustris* is "the bacterium that eats everything," catabolizing sugars, fatty acids, and lignin to obtain reduced electron carriers. Meanwhile, the bacteria make ATP through anoxygenic photosynthesis (cyclic photophosphorylation). In the presence of oxygen, *R. palustris* can fix CO_2 by lithotrophy. These bacteria also fix atmospheric nitrogen (discussed in Chapter 15). During nitrogen fixation (powered by

FIGURE 14.39 ■ *Rhodopseudomonas palustris*: a photolithoheterotroph. **A.** Caroline Harwood, professor of microbiology. **B.** *R. palustris* conducts photoheterotrophy and fixes nitrogen, while producing hydrogen gas. The bacteria make ATP through photosynthesis but obtain electrons from organic molecules, which also provide carbon to build biomass. Excess electrons are used by nitrogenase to generate hydrogen gas (H_2). *Source:* Part B modified from Larimer et al. 2003. *Nat. Biotechnol.* **22**:55–61; and from James B. McKinley and Caroline S. Harwood. 2011. *Microbe* **6**:345.

A.

COURTESY OF CAROLINE HARWOOD

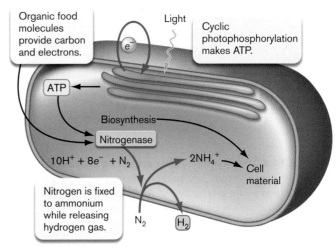

B.

Organic food molecules provide carbon and electrons.

Light

Cyclic photophosphorylation makes ATP.

ATP

Biosynthesis

Nitrogenase

$10H^+ + 8e^- + N_2$

$2NH_4^+$

Cell material

Nitrogen is fixed to ammonium while releasing hydrogen gas.

N_2

H_2

A. Z pathway of oxygenic photosynthesis

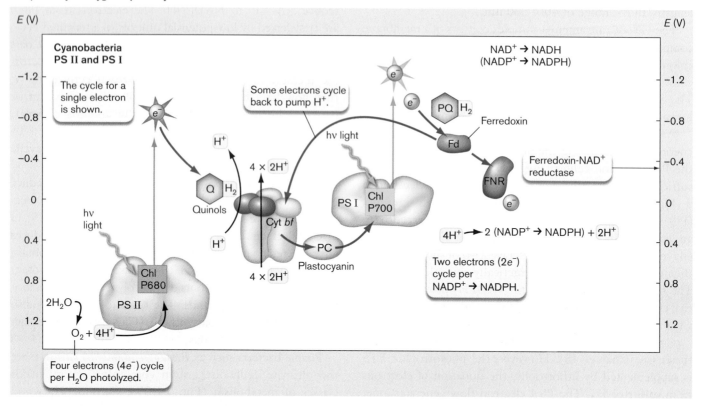

B. Z pathway reaction complexes and ATP synthase

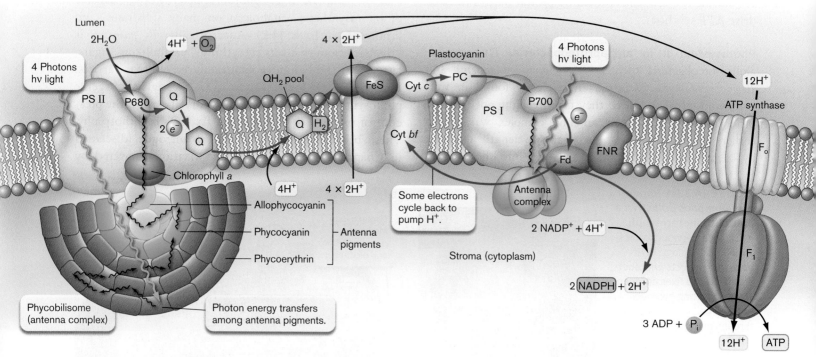

FIGURE 14.40 ■ Oxygenic photosynthesis in cyanobacteria and chloroplasts. A. Each H_2O is photolyzed via the Z pathway of PS II and PS I. The $2e^-$ from each water molecule ($4e^-$ in all) are transferred to the quinone pool, which transfers $4e^-$ to cytochrome bf. Cytochrome bf pumps $4 \times 2H^+$ across the membrane and transfers $4e^-$ via plastocyanin to a PS I containing chlorophyll P700. A second photon excites P700, enabling transfer of e^- to ferredoxin ($4e^-$ per O_2 formed), and from there to $NADP^+ \rightarrow NADPH$. **B.** The Z pathway within a photosynthetic membrane. *Source:* Part B based on crystallographic data from thermophilic cyanobacteria, from Genji Kurisu et al. 2003. *Science* **302**:1009. ▶

photosystem II) the nitrogenase enzyme transfers electrons onto hydrogen ions, forming hydrogen gas. Harwood's laboratory is engineering strains of *R. palustris* to increase the light-driven hydrogen output. Hydrogen production offers an exciting potential source of "clean" biofuel, using photosynthesis and catabolism of sewage waste.

Thought Question

14.12 Suppose you discover bacteria that require a high concentration of Fe^{2+} for photosynthesis. Can you hypothesize what the role of Fe^{2+} may be? How would you test your hypothesis?

Oxygenic photolysis. The "Z" pathway of photolysis found in cyanobacteria and chloroplasts combines key features of both PS II and PS I (**Fig. 14.40** ▶). Both reaction centers, however, contain chlorophylls that absorb at shorter wavelengths (higher energy) than those of the respective purple or green homologs: P680 instead of P870 (PS II), and P700 instead of P840 (PS I). Thus, the cyanobacterial reaction centers can split water—a highly stable molecule that cannot be photolyzed by anaerobic phototrophs. Photolysis of water requires greater energy input, but ultimately yields greater energy overall, and produces molecular oxygen. The energy potentials are high enough to generate NADPH and fix CO_2 into biomass (presented in Chapter 15). Oxygenic phototrophs dominate the shallow water depths, whereas anaerobes grow at lower depths, using light at wavelengths unused by cyanobacteria and algae near the surface.

Cyanobacteria harvest light via an exceptionally efficient antenna complex called the phycobilisome (**Fig. 14.40B**). In the PS II reaction center, the photoexcitation of chlorophyll P680 yields enough energy to split H_2O. The entire cycle of water splitting involves $4H^+$ removed from $2H_2O$, $4e^-$ transferred to carriers, and the formation of O_2. The net reaction forming oxygen is:

$$2H_2O \rightarrow 4H^+ + 4e^- + O_2$$

Through the ETS, the electrons are transferred to quinones. As in respiration, each $2e^-$ reduction of quinone to quinol requires pickup of $2H^+$ from the stroma (equivalent to cytoplasm). Thus, the four electrons transferred generate a net change in the proton gradient of $4H^+$. Furthermore, the energy of electron transfer to cytochrome *bf* enables pumping of an additional $2H^+$ across the membrane. Thus, in all, for each conversion of $2H_2O$ to O_2 the net protons transferred across the thylakoid membrane include $4H^+$ (water photolysis) plus $4 \times 2H^+$ (quinones to quinols) through cytochrome *bf*, to yield a total of $12H^+$ for the proton gradient. The proton gradient drives the ATP synthase to make approximately 3 ATP per O_2 formed.

The electrons from PS II do not cycle back to the PS II reaction center, as they do in purple bacteria. Instead they are transferred to PS I by a protein called plastocyanin. The energy of the electron transferred by plastocyanin is augmented through absorption of a second photon by the chlorophyll of PS I. Subsequent electron flow through ferredoxin can now generate NADH or NADPH. Some of the electron flow instead cycles back to cytochrome *bf*, where it contributes to pumping protons.

The overall equation for energy yield of oxygenic photolysis can be represented as:

$$2H_2O + 2\,NADP^+ + 3\,[ADP + P_i] \rightarrow$$
$$O_2 + 2\,[NADPH + H^+] + 3\,ATP + 3H_2O$$

To release one O_2 and fix one CO_2 requires absorption of between 8 and 12 photons. The efficiency—that is, the proportion of photon energy converted to CO_2 fixation—is estimated to be 20%–30%. To generate one molecule of glucose by CO_2 fixation, we need six rounds of the photolysis equation—one per CO_2 molecule "fixed" into sugar ($C_6H_{12}O_6$) (discussed in Chapter 15). The CO_2 fixation equation works out to:

$$12H_2O + 12\,NADP^+ + 18\,[ADP + P_i] \rightarrow$$
$$6O_2 + 12\,[NADPH + H^+] + 18\,ATP + 18H_2O$$

$$\mathbf{6CO_2} + 12\,[NADPH + H^+] + 18\,ATP + 18H_2O \rightarrow$$
$$\mathbf{C_6H_{12}O_6} + 6H_2O + 12\,NADP^+ + 18\,[ADP + P_i]$$

$$12H_2O + \mathbf{6CO_2} \rightarrow \mathbf{C_6H_{12}O_6} + 6H_2O + 6O_2$$

To Summarize

- **Bacteriorhodopsin** and **proteorhodopsin** are forms of a light-driven proton pump that contains retinal, found in haloarchaea and in bacteria, respectively. The energy gained from light absorption supplements heterotrophy.

- **The antenna complex** of chlorophylls and other photopigments captures light for transfer to the reaction center in chlorophyll-based photosynthesis.

- **Thylakoids** are folded membranes within phototrophic bacteria or chloroplasts. The membranes extend the area for chlorophyll light absorption, and they separate two compartments to form a proton gradient.

- **Photosystem I** obtains electrons from H_2S or HS^-. The electrons are transferred through an ETS to form NADH or NADPH.

- **Photosystem II** transfers an electron through an ETS and pumps H^+ to generate ATP. An electron ultimately returns to bacteriochlorophyll through cyclic photophosphorylation. Reverse electron flow generates NADPH or NADH.

- **The oxygenic Z pathway in cyanobacteria and chloroplasts** includes homologs of photosystems I and II. Eight photons are absorbed, and two electron pairs are removed from $2H_2O$, ultimately producing O_2.

- **Oxygenic photosynthesis generates 3 ATP + 2 NADPH** per $2H_2O$ photolyzed and O_2 produced. The ATP and NADPH are used to fix CO_2 into biomass.

Concluding Thoughts

Seemingly disparate biochemical means of nutrition, including organotrophy, lithotrophy, and photosynthesis, share common mechanisms of electron flow and proton transfer through an ETS. A recurring theme is that all forms of metabolism involve electron transfer reactions that yield energy for cell function. As molecules are rearranged by transfer of electrons from one substrate to another, energy is provided to form ion gradients and energy carriers. The energy carriers always need to be balanced between redox-neutral carriers, such as ATP, and reducing carriers, such as NADH or NADPH, for biosynthesis (as discussed in Chapter 15).

Whatever the means of obtaining energy, ultimately cells must spend energy for biosynthesis. Chapter 15 presents how microbes construct the fundamental "nuts and bolts" of their cells—including products surprisingly useful for biotechnology.

CHAPTER REVIEW

Review Questions

1. Explain the source of electrons and the sink for electrons (terminal electron acceptor) in respiration, lithotrophy, and photolysis.
2. How do bacteria combine redox couples for a metabolic reaction that yields energy? Cite examples, calculating the reduction potential.
3. How do environmental conditions affect the reduction potential of a metabolic reaction?
4. Explain the role of cytochromes and redox cofactors in electron transport systems. What features of a molecule make it useful for redox biochemistry?
5. Explain how a proton potential is composed of a chemical concentration difference plus a charge difference. Explain how each component of Δp can drive a cellular reaction.
6. Explain the role of the substrate dehydrogenase (oxidoreductase), the quinones, and the cytochrome oxidase (oxidoreductase) in the respiratory ETS.
7. Compare the ETS function in lithotrophy with that in respiration.
8. Summarize the inorganic redox couples that can be used in anaerobic respiration and those that can be used in lithotrophy. What constraints determine whether a given molecule can serve as electron acceptor or as electron donor?
9. How do diverse forms of anaerobic respiration and lithotrophy contribute to ecosystems?
10. Explain the differences and common features of bacteriorhodopsin phototrophy and chlorophyll phototrophy.
11. Explain the differences and common features of photosystems I and II. Explain how the two photosystems combine in the Z pathway. Why can the Z pathway generate oxygen, whereas PS I and PS II cannot?

Thought Questions

1. The lung pathogen *Pseudomonas aeruginosa*, which also grows in soil, can respire aerobically or else anaerobically using nitrate. Under what conditions would *P. aeruginosa* use each form of metabolism? What part of its ETS would need to change to accommodate the different forms?
2. What environments favor oxygenic photosynthesis, versus sulfur phototrophy? Explain.
3. In pathogens, which components of the ETS do you think would make good targets for new antibiotics, and why?
4. Devise a form of energy-yielding metabolism in which fumarate is converted to succinate; and a different form, in which succinate is converted to fumarate. Explain why the two reactions are reasonable and, on the Internet, try to find actual organisms that obtain energy through these reactions.

Key Terms

anaerobic respiration (541)
anammox reaction (547)
antenna complex (557)
bacteriochlorophyll (556)
bacteriorhodopsin (553)
carotenoid (556)
chlorophyll (556)
chromophore (556)
cofactor (534)
cyclic photophosphorylation (560)
cytochrome (529)
dissimilatory denitrification (542)
dissimilatory metal reduction (542)
electron acceptor (525)
electron donor (525)
electron transport system (ETS)
    (electron transport chain)
    (524, 525)

ferredoxin (561)
heme (534)
hydrogenotrophy (550)
lithotrophy (chemolithotrophy)
    (524, 525, 545)
methanogenesis (551)
methanotrophy (552)
methylotrophy (552)
nitrifier (547)
organotrophy (chemoorganotrophy)
    (524, 525)
oxidative phosphorylation (539)
oxidoreductase (529, 535)
oxygenic Z pathway (559)
photoexcitation (555)
photoheterotrophy (554)
photoionization (555)
photolysis (555)

photosynthesis (555)
photosystem I (PS I) (559)
photosystem II (PS II) (559)
phototrophy (524, 525)
proteorhodopsin (553)
proton motive force
    (proton potential) (530)
quinol (535)
quinone (535)
reaction center (RC) (558)
redox couple (526)
respiration (525)
retinal (554)
reverse electron flow (546)
standard reduction potential
    ($E^{\circ\prime}$) (526)
thylakoid (558)
uncoupler (533)

Recommended Reading

Andries, Koen, Peter Verhasselt, Jerome Guillemont, Hinrich W. H. Göhlmann, Jean-Marc Neefs, et al. 2005. A diarylquinoline drug active on the ATP synthase of *Mycobacterium tuberculosis*. *Science* **307**:223–227.

Beatty, J. Thomas, Jörg Overmann, Michael T. Lince, Ann K. Manske, Andrew S. Lang, et al. 2005. An obligately photosynthetic bacterial anaerobe from a deep-sea hydrothermal vent. *Proceedings of the National Academy of Sciences USA* **102**:9306–9310.

Farha, Maya A., Chris P. Verschoor, Dawn Bowdish, and Eric D. Brown. 2013. Collapsing the proton motive force to identify synergistic combinations against *Staphylococcus aureus*. *Chemistry & Biology* **20**:1168–1178.

Ferreira, Kristina N., Tina M. Iverson, Karim Maghlaoui, James Barber, and So Iwata. 2004. Architecture of the photosynthetic oxygen-evolving center. *Science* **303**:1831–1838.

Ishii, Shun'ichi, Takefumi Shimoyama, Yasuaki Hotta, and Kazuya Watanabe. 2008. Characterization of a filamentous biofilm community established in a cellulose-fed microbial fuel cell. *BMC Microbiology* **8**:6.

Jones, Shari A., Fatema Z. Chowdhury, Andrew J. Fabich, April Anderson, Darrel M. Schreiner, et al. 2009. Respiration of *Escherichia coli* in the mouse intestine. *Infection and Immunity* **75**:4891–4899.

Kracke, Frauke, Igor Vassilev, and Jens O. Krömer. 2015. Microbial electron transport and energy conservation—the foundation for optimizing bioelectrochemical systems. *Frontiers in Microbiology* **6**:575.

Kuenen, J. Gijs. 2008. Anammox bacteria: From discovery to application. *Nature Reviews. Microbiology* **6**:320–326.

Lower, Brian H., Liang Shi, Ruchirej Yongsunthon, Timothy C. Droubay, David E. McCready, et al. 2007. Specific bonds between an iron oxide surface and outer membrane cytochromes MtrC and OmcA from *Shewanella oneidensis* MR-1. *Journal of Bacteriology* **189**:4944–4952.

Pfeffer, Christian, Steffen Larsen, Jie Song, Ming-dong Dong, Flemming Besenbacher, et al. 2012. Filamentous bacteria transport electrons over centimetre distances. *Nature* **491**:218–221.

Raghoebarsing, Ashna A., Arjan Pol, Katinka T. van de Pas-Schoonen, Alfons J. P. Smolders, Katharina F. Ettwig, et al. 2006. A microbial consortium couples anaerobic methane oxidation to denitrification. *Nature* **440**:918–921.

Seeger, Marcus A., André Schiefner, Thomas Eicher, François Verrey, Kay Diedrichs, et al. 2006. Structural asymmetry of AcrB trimer suggests a peristaltic pump mechanism. *Science* **313**:1295.

Shrestha, Pravin M., and Amelia-Elena Rotaru. 2014. Plugging in or going wireless strategies for interspecies electron transfer. *Frontiers in Microbiology* **5**:237.

Winter, Sebastian E., Parameth Thiennimitr, Maria G. Winter, Brian P. Butler, Douglas L. Huseby, et al. 2010. Gut inflammation provides a respiratory electron acceptor for *Salmonella*. *Nature* **467**:426–429.

Yun, Jiae, Nikhil S. Malvankar, Toshiyuki Ueki, and Derek R. Lovley. 2016. Functional environmental proteomics: Elucidating the role of a *c*-type cytochrome abundant during uranium bioremediation. *ISME Journal* **10**:310–320. doi:10.1038/ismej.2015.113.

CHAPTER 15

Biosynthesis

How do microbes build their cells? Some bacteria and archaea build themselves entirely from carbon dioxide and nitrogen gas, plus a few salts. The cell assimilates carbon and nitrogen into small molecules and then assembles more complex structures using many enzymes. How do cells organize their biosynthesis to build precisely the forms they need? How do bacteria avoid wasting energy on excess production? New genomes reveal new microbial capacities for biosynthesis, ranging from antibiotics and pesticides to surgical materials.

CURRENT RESEARCH highlight

Actinomycetes make antibiotics. Hee-Jeon Hong tests actinomycete extracts for novel activity. In the extracts, a cell wall–attacking antibiotic stimulates expression of a cell wall stress gene. Extracts from various actinomycetes (white disks) are placed on an agar medium containing kanamycin, which prevents growth of a *Streptomyces* tester strain. The tester strain expresses the stress gene fused to a kanamycin resistance reporter gene. The extract-induced kanamycin resistance enables *Streptomyces* to grow, forming a ring of red colonies. The known antibiotic teicoplanin generates a ring of colonies. The halo for the extract from the species *Amycolatopsis* MJM2582 reveals a new antibiotic, ristocetin.

Source: Andrew Truman et al. 2014. Antimicrob. Agents Chemother. **58**:5687.

AN INTERVIEW WITH

HEE-JEON HONG, MICROBIOLOGIST, CAMBRIDGE UNIVERSITY

CHRIS GREEN, UNIVERSITY OF CAMBRIDGE, DEPT. OF BIOCHEMISTRY

What is the most interesting thing about a new actinomycete producing the antibiotic ristocetin?

The new actinomycete we discovered provides the first known gene sequences for enzymes responsible for ristocetin production. Discovery of the cluster is a proof of principle for our two-step bioassay system, verifying that it is actually working in a real-world screening for glycopeptide natural products.

What new discoveries about antibiotics do you expect to see?

We would expect that if we extend the screening to cover many more strains from different collections, we would be able to discover entirely new glycopeptide antibiotics and the genes that encode their synthesis. We also anticipate that some of the enzymes in the ristocetin cluster will find use in helping to make novel combinatorial glycopeptide structures, which may have novel activity against pathogens.

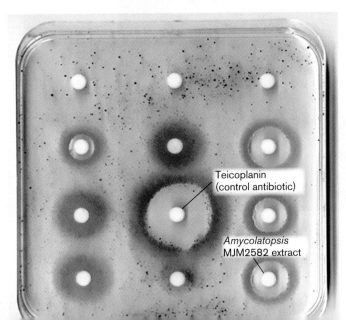

Teicoplanin (control antibiotic)

Amycolatopsis MJM2582 extract

In Chapters 13 and 14 we learned how microbes use chemical reactions to gain energy, storing it in ion gradients and in small molecules such as ATP and NADPH. Chapter 15 shows how the microbes spend this energy for biosynthesis. An example is *Amycolatopsis* MJM2582, the actinomycete bacterium isolated by South Korean biochemist Hee-Jeon Hong and colleagues at the University of Cambridge, including her student Min Jun Kwun (**Fig. 15.1**). The white fuzz on the actinomycete colonies is full of arthrospores (see Chapters 5 and 18), which can blow off to another location. When the cell filaments enter stationary phase (discussed in Chapter 4), they begin to produce dozens of different secondary products, such as the antibiotic ristocetin (**Fig. 15.1**, inset). Such antibiotics show extraordinary diversity and complexity, often including multiple aromatic rings—a surprising biosynthetic feat for an organism that's running out of resources. But the antibiotics enable the producer to outcompete other bacteria, killing them to scavenge their substrates for biosynthesis. Antibiotic synthesis requires elaborate enzyme complexes and multistep reaction pathways. Hong's lab identified such a pathway in the *Amycolatopsis* genome. The genome reveals a span of 79,000 base pairs comprising thirty-nine genes for enzymes, regulators, and transporters—all to produce and export the antibiotic ristocetin.

Chapter 15 presents the fundamental ways that microbes perform biosynthesis. Autotrophs assimilate elements such as carbon and nitrogen to build useful carbon skeletons, key functions in the food web. Microbial enzyme factories build remarkably complex biomolecules, including vitamins and antibiotics important for human health. Industrial applications of microbial biosynthesis are described in Chapter 16.

15.1 Overview of Biosynthesis

Biosynthesis is the building of complex biomolecules, also known as **anabolism**, the reverse of catabolism. **Figure 15.2** presents an overview of anabolism, or biosynthesis, and how it relates to catabolic pathways we saw in Chapter 13. Enzyme pathways such as the TCA cycle are reversed to build up carbon skeletons, incorporating nitrogen to form amino acids. Ultimately, sugar monomers and amino acids are assembled to form polysaccharides, polypeptides, and the cell wall material, peptidoglycan. Besides the biomass of the cell proper, cells synthesize and secrete products such as antibiotics, toxins for pathogenesis, quorum signals for cooperation, and matrix for biofilm (**eTopic 15.1**). All this biosynthesis requires substrates and energy.

Biosynthesis Requires Substrates

For their biosynthesis, where do microbes obtain their organic substrates, such as acetyl-CoA? Some microbes synthesize all their organic parts from minerals such as

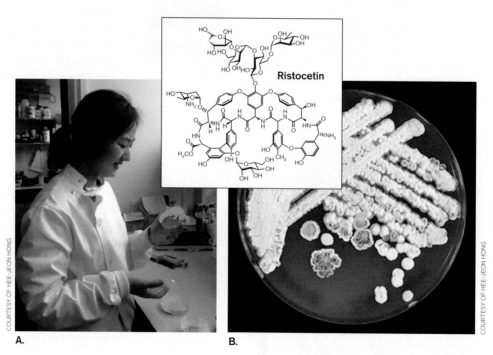

FIGURE 15.1 ■ **Discovering a new antibiotic-producing actinomycete.** **A.** Min Jung Kwun discovered the actinomycete *Amycolatopsis* MJM2582. **B.** *Amycolatopsis* MJM2582 cultured for 7 days at 30°C on mannitol soya agar, a medium that encourages sporulation (white fuzz). The actinomycete produces ristocetin (inset), a glycopeptide antibiotic.

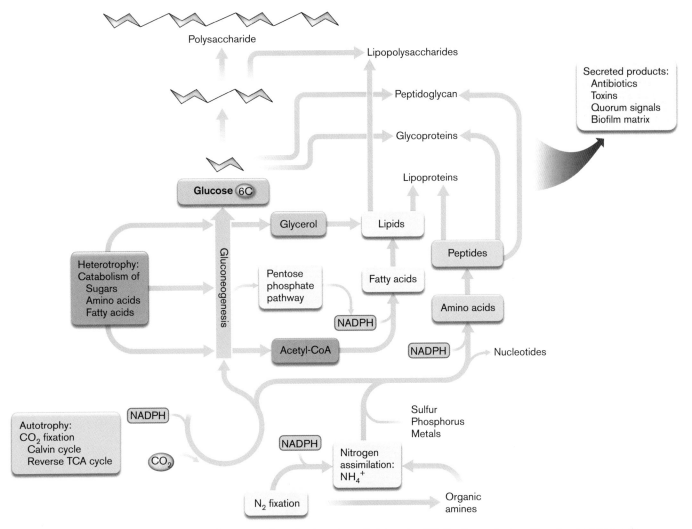

FIGURE 15.2 ■ Biosynthesis: an overview. Microbes obtain carbon skeletons by CO_2 fixation or by breaking down compounds formed by other organisms. For nitrogen, microbes fix N_2 or take up nitrates or organic amines. All biosynthesis requires energy from ATP and from reducing cofactors such as NADPH.

carbonate and nitrate (autotrophy). Two fundamental classes of autotrophy are photosynthesis and chemosynthesis. In **photosynthesis**, or **photoautotrophy**, single carbon molecules (usually CO_2) are fixed into organic biomass, using energy from light absorption (discussed in Chapter 14). **Chemosynthesis** fixes carbon dioxide via similar pathways, but <u>without</u> light. The energy for chemosynthesis usually comes from oxidation of minerals (lithotrophy; see Chapter 14). Other microorganisms must obtain organic substrates from their environment (heterotrophy). Many soil microbes are capable of both autotrophy and heterotrophy, depending on what their environment provides.

For all organisms, biosynthesis requires:

■ **Essential elements.** Biosynthesis requires carbon, oxygen, hydrogen, nitrogen, and other essential elements. Carbon is obtained either through CO_2 fixation (autotrophy) or through acquisition of organic molecules made by other organisms (heterotrophy). Autotrophy

assembles carbon and water into small molecules such as acetyl-CoA, which then serve as substrates or building blocks for the cell. By contrast, heterotrophy breaks down larger molecules such as carbohydrates and peptides to release acetyl-CoA and other small substrates. In addition to carbon, biosynthesis must assimilate nitrogen and sulfur for proteins, phosphorus for DNA, and metals for metal-containing enzymes.

■ **Reduction.** Biosynthesis usually reduces the substrate, by hydrogenation and by removing oxygen. Cell components such as lipids and amino acids are more reduced than substrates such as CO_2 and acetate, so their biosynthesis requires a reducing agent such as NADPH.

■ **Energy.** Building complex structures, with or without reduction, requires energy. Biosynthetic enzymes spend energy by coupling their reactions to the hydrolysis of ATP, the oxidation of NADPH, or the flux of ions down a transmembrane ion gradient.

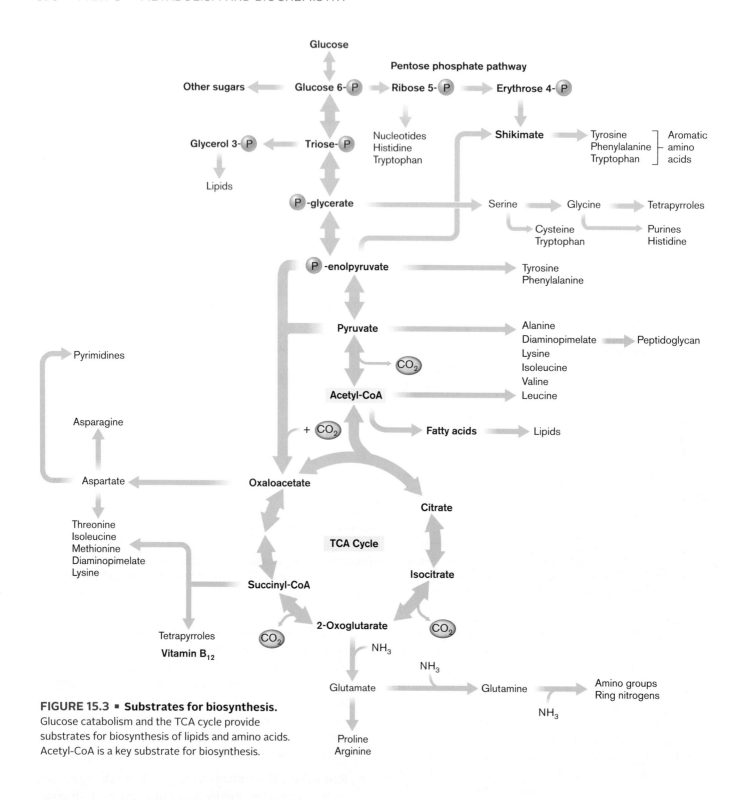

FIGURE 15.3 ■ Substrates for biosynthesis.
Glucose catabolism and the TCA cycle provide substrates for biosynthesis of lipids and amino acids. Acetyl-CoA is a key substrate for biosynthesis.

Heterotrophs such as *E. coli* and actinomycetes obtain many substrates from glucose catabolism and the tricarboxylic acid (TCA) cycle, central catabolic pathways discussed in Chapter 13 (**Fig. 15.3**). For example, the succinyl-CoA molecules from the TCA cycle serve as the foundation of several kinds of amino acids, as well as vitamin B_{12}. Glycerol 3-phosphate provides the glyceride backbone of lipids. Pyruvate provides the backbone of several amino acids with aliphatic side chains. From the pentose phosphate pathway erythrose 4-phosphate contributes to the ring structures of aromatic amino acids. Other amino acids derive their carbon skeleton from TCA cycle intermediates oxaloacetate and 2-oxoglutarate, which incorporate nitrogen in the form of ammonium ions (NH_4^+).

Note that glucose catabolism and the TCA cycle are both reversible. Some autotrophs can synthesize entire sugar molecules, starting with CO_2 and working up through the reverse TCA cycle and reverse glycolysis. (The reversal of glycolysis is called **gluconeogenesis**). Thus, many common metabolites are available both to autotrophs and to heterotrophs, as well as to microbes capable of mixed metabolism.

Biosynthesis Spends Energy

Biosynthesis costs energy in several ways. The organism's genome must maintain the DNA that encodes all the enzymes that catalyze all the steps of the pathway. The ribosomes spend energy to make enzymes. And the enzymes couple synthetic reactions to energy-releasing reactions. Collectively, these genomic and energetic costs generate enormous selective pressure for microbes to evolve mechanisms that control these costs. The diversity of these cost-cutting mechanisms generates surprising ecological relationships.

■ **Regulation.** Biosynthesis is regulated at multiple levels, such as gene transcription, protein synthesis, and allosteric enzyme control. In general, energy-expensive products block the expression or function of their biosynthetic enzymes. For example, as we saw in Chapter 10, if the cell's tryptophan concentration increases, the molecule binds a corepressor to block transcription of the genes for tryptophan biosynthesis; and the aminoacyl-tRNA stalls the ribosome midtranslation. In this way, microbes avoid making more than they need under given conditions. Microbes regulate their biosynthesis at the levels of transcription and translation of enzymes, as well as through feedback inhibition of enzyme activity. An important example of molecular regulation is nitrogen fixation, discussed in Section 15.5.

■ **Competition and predation.** Free-living bacteria secrete antimicrobial agents that inhibit or kill competitors. The victors scavenge the spoils for raw materials. Actinomycetes, such as those discussed at the start of the chapter, possess large genomes that commonly encode enzymes for multiple antibiotics. High nutrient levels inhibit antibiotic production, but as the bacterial filaments grow more slowly in stationary phase, they start producing antibiotics.

■ **Genome loss and cooperation.** A drastic way to avoid the expense of biosynthesis is to lose the enzymes through reductive evolution (discussed in Chapter 17). But the species then depends on intricate relations with other community members. Parasites and mutualists that grow only within a host cell show the most extensive loss of biosynthetic genes. Each species takes an evolutionary path of maintaining the expense of a particular biosynthetic pathway, such as N2 fixation, or of losing the pathway and becoming dependent on partner organisms.

Surprisingly, many marine phototrophs no longer encode certain biosynthetic pathways and depend on synthesis by partner heterotrophs. For example, on the sea ice of the Southern Ocean (temperatures around 0°C), Andrew Allen at the J. Craig Venter Institute discovered a fundamental mutualism of Antarctic diatoms and bacteria (**Fig. 15.4**). The Southern Ocean is highly productive for biomass and oxygen, because of the high solar irradiance and iron

A.

B.

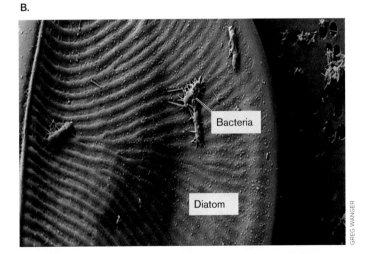

FIGURE 15.4 ■ Bacteria provide vitamin B$_{12}$ for Antarctic diatom. A. Explorers collect microbes on the ice edge, McMurdo Sound. Typical water temperature in December is –1.9°C. **B.** The diatom *Amphiprora* sp. hosts vitamin-producing bacteria, in sea ice of the Southern Ocean (SEM).

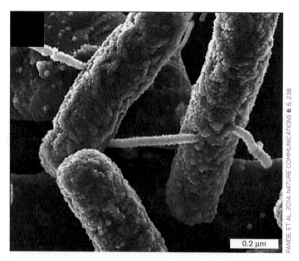

FIGURE 15.5 ■ Sharing amino acids via nanotubes. *E. coli* strains connect with nanotubes to obtain the amino acid each lacks but the other strain overproduces.

in which each overproduced an amino acid (histidine or tryptophan). Pande found that when the growth medium lacked the two complementary biosynthesis pathways (for His or Trp), the two bacterial strains formed nanotubes of phospholipids derived from the cell envelope (**Fig. 15.5**). But when the medium provided the amino acids needed, the nanotubes failed to form. Thus, even within the competitive gut microbial community, bacteria find ways to share excess carbon substrates for biosynthesis—and regulate their sharing to avoid helping a competitor when a ready source is available.

In this chapter we present the pathways by which autotrophs assimilate carbon and nitrogen into simple building blocks, such as acetyl-CoA. Then we present pathways of construction of the key parts of the cell, such as fatty acids and amino acids. Finally, we consider exciting recent discoveries in the biosynthesis of secondary products such as antibiotics.

availability. Many of the sea ice phytoplankton (phototrophic algae) fail to synthesize vitamin B_{12}, a fundamental cofactor for enzymes but one of the most complicated biomolecules to synthesize. Allen's group found that the Antarctic diatoms obtain vitamin B_{12} from adherent bacteria. The major bacteria providing the vitamin are Oceanospirillales, a family of proteobacteria known for cleaning up oil spills (see Chapter 21).

Closer to home, our gut bacteria also evolve ingenious ways to share substrates and minimize biosynthesis. Samay Pande and coworkers at the Max Planck Institute for Chemical Biology, Jena, Germany, investigated resource sharing in *E. coli* by constructing a complementary pair of mutants

15.2 CO_2 Fixation: The Calvin Cycle

The fundamental significance of **carbon dioxide fixation** by green plants and algae was recognized in the early twentieth century. "Fixation" refers to the covalent incorporation of a small molecule into larger biochemical material. Fixing carbon requires tremendous energy input, as well as a large degree of reduction to incorporate hydrogen atoms.

The majority of the biomass on Earth consists of carbon fixed by chloroplasts and bacteria through the

A.

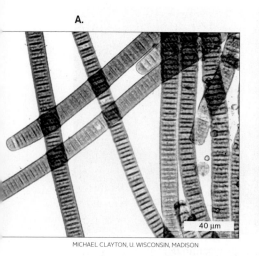

MICHAEL CLAYTON, U. WISCONSIN, MADISON

B.

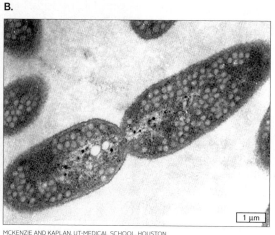

MCKENZIE AND KAPLAN, UT-MEDICAL SCHOOL, HOUSTON

C.

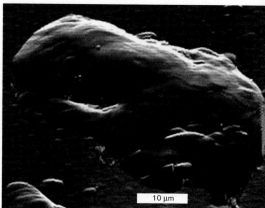

S. N. TAN AND M. CHEN. 2012. *HYDROMETALLURGY* **119–120**: 87–94

FIGURE 15.6 ■ Phototrophy and lithotrophy fix carbon via the Calvin cycle. A. *Oscillatoria*, a filamentous cyanobacterium, fixes CO_2 by oxygenic photosynthesis (light micrograph). **B.** *Rhodobacter sphaeroides*, purple photoheterotrophs with spherical photosynthetic membranes (TEM). **C.** *Acidithiobacillus ferrooxidans* corrodes concrete while oxidizing sulfur and iron (AFM). This form of lithotrophy uses the Calvin cycle to fix CO_2.

TABLE 15.1	Carbon Dioxide Fixation Pathways		
	Organisms in which pathways occur		
Pathway	**Bacteria**	**Archaea**	**Eukaryotes**
Photoautotrophs, photoheterotrophs, and lithoautotrophs (Section 15.2)			
Calvin cycle	Cyanobacteria; purple phototrophs; lithotrophs	Rubisco homologs appear, but their function is unclear	Chloroplasts
Lithoautotrophic bacteria and archaea (Section 15.3)			
Reductive (reverse) TCA cycle	Green sulfur phototrophs (*Chlorobium*); thermophilic epsilonproteobacteria	Hyperthermophilic sulfur oxidizers (*Thermoproteus* and *Pyrobaculum*)	Anaplerotic reactions fix CO_2 to regenerate TCA intermediates
Reductive acetyl-CoA pathway	Anaerobes: acetogenic bacteria and sulfate reducers	Methanogens; other anaerobes	None known
3-Hydroxypropionate cycle	Green phototrophs (*Chloroflexus*)	Aerobic sulfur oxidizers (*Sulfolobus*)	None known

reductive pentose phosphate cycle, which recycles a pentose phosphate intermediate. This cycle is also known as the **Calvin cycle**, for which Melvin Calvin (1911–1997) was awarded the 1961 Nobel Prize in Chemistry. The full Calvin cycle is found only in bacteria and in chloroplasts, which evolved from bacteria. It has not been found in archaea or in the cytoplasm of eukaryotes. (Archaea and some lithoautotrophic bacteria fix carbon by other pathways, as described in Section 15.3.) Thus, the Calvin cycle appears to have evolved after the divergence of the three domains of life.

The mechanism of the pentose phosphate cycle was solved by Calvin with colleagues Andrew Benson and James Bassham, at UC Berkeley. The importance of the Calvin cycle to the global ecosystem can scarcely be overestimated; it plays a major role in removing atmospheric CO_2. The Calvin cycle's responsiveness to CO_2, temperature, and other factors must be considered in all models of global warming.

Note: The Calvin cycle is also known as the **Calvin-Benson cycle**, the **Calvin-Benson-Bassham cycle**, or the **CBB cycle**. This book uses the terms "Calvin cycle" and "CBB cycle."

Several categories of bacteria conduct the Calvin cycle (**Fig. 15.6** and **Table 15.1**). Photoautotrophs such as cyanobacteria—and all plant chloroplasts—use the Calvin cycle to fix CO_2. Some photoheterotrophs, such as *Rhodobacter sphaeroides*, use the Calvin cycle only when light is available; in the dark, they use preformed substrates (heterotrophy). Still other bacteria fix CO_2 using energy from metal oxidation (lithoautotrophy) (see Chapter 14). For example,

Acidithiobacillus ferrooxidans (**Fig. 15.6C**) oxidizes sulfur and iron, generating acid that corrodes concrete.

■ **Oxygenic phototrophic bacteria (cyanobacteria) and the chloroplasts of algae and multicellular plants fix CO_2** by the Calvin cycle coupled to oxygenic photosynthesis. The Calvin cycle is also called the "dark reactions," or "light-independent reactions," of oxygenic photosynthesis. Cyanobacteria are believed to generate the majority of the oxygen gas in Earth's atmosphere.

■ **Facultatively anaerobic purple bacteria**, including sulfur oxidizers and photoheterotrophs such as *Rhodospirillum* and *Rhodobacter*, use the Calvin cycle to fix CO_2. These bacteria also obtain carbon through catabolism.

■ **Lithoautotrophic bacteria** fix CO_2 through the Calvin cycle, using NADPH and ATP provided by the oxidation of minerals. Mineral oxidation actually consumes oxygen.

Overview of the Calvin Cycle

Decades of biochemical and genetic experiments established the details of CO_2 fixation in the Calvin cycle. Early in the twentieth century, biochemists tried to figure out the mechanism of CO_2 fixation, believing that agricultural photosynthesis could be made more efficient. With the tools then available, however, researchers had no hope of sorting out the intermediate products through which CO_2 was fixed. A fundamental breakthrough was the use of tracer radioisotopes, which are specific compounds labeled with radioactivity. The discovery of the carbon isotope ^{14}C by Martin

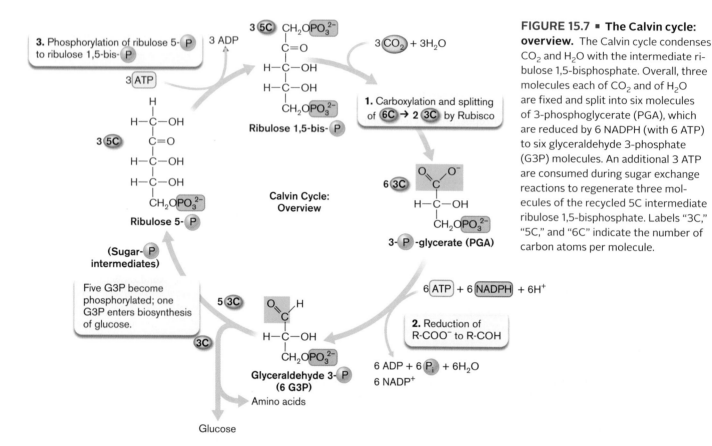

FIGURE 15.7 ■ **The Calvin cycle: overview.** The Calvin cycle condenses CO_2 and H_2O with the intermediate ribulose 1,5-bisphosphate. Overall, three molecules each of CO_2 and of H_2O are fixed and split into six molecules of 3-phosphoglycerate (PGA), which are reduced by 6 NADPH (with 6 ATP) to six glyceraldehyde 3-phosphate (G3P) molecules. An additional 3 ATP are consumed during sugar exchange reactions to regenerate three molecules of the recycled 5C intermediate ribulose 1,5-bisphosphate. Labels "3C," "5C," and "6C" indicate the number of carbon atoms per molecule.

Kamen (1913–2002) in 1940 revolutionized biochemistry, enabling the discovery of all kinds of cellular metabolism (**eTopic 15.2**). Another key technique was paper chromatography, a means of separating labeled compounds based on differential migration in a solvent. This technique reveals short-lived intermediate compounds of a cycle. Calvin used paper chromatography for his Nobel-winning experiments to figure out the cycle that bears his name (**eTopic 15.3**).

Figure 15.7 shows an overview of the key steps of the Calvin cycle. The cycle begins with CO_2 fixation by ribulose 1,5-bisphosphate, a key intermediate that can derive from the pentose phosphate cycle. This reaction is catalyzed by the enzyme **Rubisco** (ribulose 1,5-bisphosphate carboxylase/oxygenase), believed to be the most abundant protein on Earth. (Further details of the Calvin cycle biochemistry, including sugar regeneration, appear in **Figure 15.9**.)

In each "turn" of the cycle, one molecule of CO_2 is condensed (combined, forming a new C–C bond) with the five-carbon sugar ribulose 1,5-bisphosphate. The resulting six-carbon intermediate splits into two molecules of 3-phosphoglycerate (PGA). The fixed CO_2 ultimately ends up as a carbon of glyceraldehyde 3-phosphate (G3P), reduced by $2H^+ + 2e^-$ from NADPH + H$^+$. Recall from Chapter 13 that NADPH is a phosphorylated derivative of NADH commonly associated with biosynthesis. The phosphoryl group of NADPH however does <u>not</u> participate in energy transfer.

How does the fixed CO_2 become one "corner" of a glucose molecule? For every three turns of the cycle, fixing three molecules of CO_2, the cycle feeds one molecule of G3P ($C_3H_5O_3$–PO_3^{2-}) into biosynthesis:

$$3CO_2 + 6\ NADPH + 6H^+ + 9\ ATP + 9H_2O \rightarrow$$
$$C_3H_5O_3\text{–}PO_3^{2-} + 6\ NADP^+ + 9\ ADP + 8\ P_i$$

The H_2O and the phosphoryl group of G3P are ultimately recycled during biosynthetic assimilation of G3P. Two molecules of G3P may condense (that is, form a new C–C bond) in a pathway to synthesize glucose. The overall condensation of $6CO_2 \rightarrow 2$ G3P $\rightarrow$ glucose yields:

$$6CO_2 + 12\ NADPH + 12H^+ + 18\ ATP + 18H_2O \rightarrow$$
$$C_6H_{12}O_6 + 12\ NADP^+ + 18\ ADP + 18\ P_i$$

Alternatively, instead of conversion to glucose, G3P can enter the biosynthetic pathways of amino acids, vitamins, and other essential components of cells. Glyceraldehyde 3-phosphate is the fundamental unit of carbon assimilation into biomass.

Ribulose 1,5-Bisphosphate Incorporates CO_2

Each turn of the Calvin cycle has three main phases: carboxylation and splitting into two three-carbon (3C) intermediates, reduction of two molecules of PGA to two molecules of G3P, and regeneration of ribulose 1,5-bisphosphate (see **Fig. 15.7**):

1. **Carboxylation and splitting: 6C → 2[3C].** Ribulose 1,5-bisphosphate condenses with CO_2 and H_2O, mediated by Rubisco. Rubisco generates a six-carbon intermediate, which immediately hydrolyzes (splits into two parts by incorporating H_2O). The split produces two molecules of PGA, one of which contains the CO_2 fixed by this cycle.

2. **Reduction of PGA to G3P.** The carboxyl group of each PGA molecule is phosphorylated by ATP. The phosphorylated carboxyl group is then hydrolyzed and reduced by NADPH, forming G3P.

3. **Regeneration of ribulose 1,5-bisphosphate.** Of every six G3P, resulting from three cycles of fixing CO_2, five G3P enter a complex series of reactions (including hydrolysis of 3 ATP) to regenerate three molecules of ribulose 1,5-bisphosphate. The net conversion of five G3P molecules to three molecules of ribulose 1,5-bisphosphate releases $2H_2O$, restoring two of the

$3H_2O$ fixed with $3CO_2$. The remaining sixth G3P exits the cycle, available to be used in the biosynthesis of sugars and amino acids. Thus, three fixed carbons lead to one three-carbon product.

Overall, each CO_2 fixed sends one carbon into biosynthesis and regenerates one ribulose 1,5-bisphosphate.

Rubisco catalyzes CO_2 incorporation. Rubisco is the key enzyme of the Calvin cycle. The CO_2-fixing enzyme is unique to bacteria and chloroplasts. Some archaeal genomes do show a homolog of Rubisco, but the archaeal enzyme does not fix CO_2. Instead, the archaeal Rubisco forms PGA out of adenine, through a "scavenging pathway" of catabolism.

The structure of Rubisco is highly conserved across bacterial and chloroplast domains. It consists of two types of subunits, designated small and large (**Fig. 15.8A**). The large (L) subunit contains the active (catalytic) site. The function of the small (S) subunit remains unclear; some bacteria, such as *Rhodospirillum rubrum*, have a Rubisco with no small subunits. Different species contain

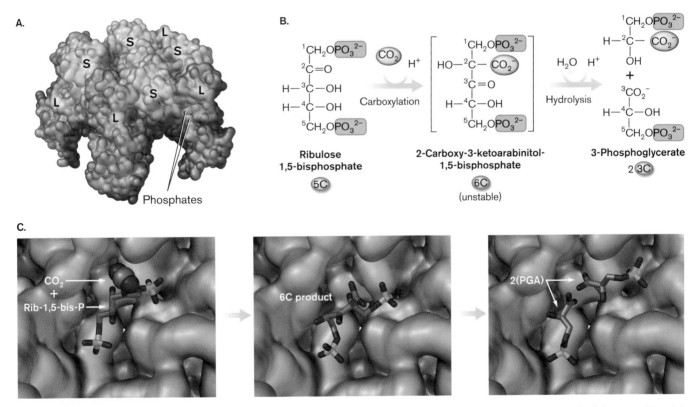

FIGURE 15.8 ■ The mechanism of Rubisco. A. Rubisco from *Alcaligenes eutrophus* consists of eight large subunits (L) crowned by eight small subunits (S). Only the upper four L and S subunits are shown. On each L subunit, a catalytic site contains two phosphates (orange), which compete with the substrates for binding. (PDB code: 1BXN) **B.** Rubisco adds CO_2 to ribulose 1,5-bisphosphate to give an unstable six-carbon intermediate bound to the enzyme. The bound intermediate hydrolyzes to form two molecules of 3-phosphoglycerate (PGA). **C.** The mechanism of CO_2 fixation at the catalytic site. CO_2 adds to the ketone, generating the 6C intermediate, which splits into two molecules of PGA. (PDB code: 1RUS)

different multiples of the small and large subunits: eight (in lithotrophs, green phototrophs, and some algae), six (in plant chloroplasts and some algae), or two (in purple phototrophs). Nevertheless, the fundamental mechanism of CO_2 fixation in all these organisms appears to be similar.

Initially, Rubisco adds CO_2 to ribulose 1,5-bisphosphate to give an unstable six-carbon intermediate that remains bound to the enzyme (**Fig. 15.8B and C**). The intermediate hydrolyzes into two molecules of PGA, each held at the active site by its phosphate. Only one PGA molecule contains a carbon from the newly fixed CO_2; nevertheless, as both molecules dissociate from the enzyme, they enter the cellular pool of PGA and behave equivalently in the rest of the cycle.

Rubisco has a high affinity for CO_2, and the typical concentration of Rubisco active sites (one per large subunit, six per complex) in plant chloroplasts is 4 mM—about 500 times greater than the concentration of CO_2. Thus, considerable energy is invested in CO_2 absorption. Yet the efficiency of carbon fixation by Rubisco is lowered by the existence of a competing reaction with O_2 that leads to 2-phosphoglycolate instead of 3-phosphoglycerate. This oxygenation reaction is called photorespiration. The function of photorespiration has been studied intensively, but remains unclear.

> **Thought Question**
>
> **15.2** Speculate on why Rubisco catalyzes a competing reaction with oxygen. Why might researchers be unsuccessful in attempting to engineer Rubisco without this reaction?

Regeneration of ribulose 1,5-bisphosphate. Overall, each glyceraldehyde 3-phosphate (G3P) arises from three rounds of CO_2 fixation by ribulose 1,5-bisphosphate. Thus, to maintain the cycle, for each G3P provided to biosynthesis, five other molecules of G3P must be recycled into three five-carbon molecules of ribulose 1,5-bisphosphate. Details of regenerating the intermediate are shown in the lower half of **Figure 15.9**.

The regeneration pathway is summarized here:

■ Two molecules of G3P condense to form the six-carbon sugar fructose 6-phosphate.

■ Fructose 6-phosphate condenses with a third G3P. The nine-carbon molecule splits to form a five-carbon sugar (xylulose 5-phosphate) and a four-carbon sugar (erythrose 4-phosphate). Xylulose 5-phosphate rearranges to ribulose 5-phosphate.

■ Erythrose 4-phosphate condenses with a fourth G3P (rearranged to dihydroxyacetone 3-phosphate) to make a seven-carbon sugar (sedoheptulose 7-phosphate).

■ The seven-carbon sugar condenses with the fifth G3P and splits into two molecules of the ribulose 5-phosphate (via xylulose 5-phosphate and ribose 5-phosphate).

■ Each of the three five-carbon sugars receives a second phosphoryl group from ATP, generating ribulose 1,5-bisphosphate. Each ribulose 1,5-bisphosphate is now ready for Rubisco to fix another CO_2.

Why would a cycle have evolved requiring so many enzymatic steps to so many different intermediates? First, a cycle with many steps breaks down the energy flow into numerous reversible conversions with near-zero values of ΔG. The nearer to equilibrium, the more energy is conserved in the conversion. Second, the multiple different intermediates provide substrates for biosynthesis. For example, some molecules of erythrose 4-phosphate and ribose 5-phosphate are withdrawn from the cycle to build aromatic amino acids and nucleotides.

> **Thought Questions**
>
> **15.3** Why does ribulose 1,5-bisphosphate have to contain two phosphoryl groups, whereas the other intermediates of the Calvin cycle contain only one?
>
> **15.4** Which catabolic pathway (see Chapter 13) includes some of the same sugar-phosphate intermediates that the Calvin cycle has? What might these intermediates in common suggest about the evolution of the two pathways?

Regulation of the Calvin Cycle

How is the Calvin cycle organized and regulated? Expression of the CO_2-fixing enzymes varies with CO_2 concentration, light levels (for phototrophs), and temperature. The concentration of CO_2 is a special problem because CO_2 diffuses readily through phospholipid membranes. Thus, cells cannot concentrate this substrate across the cell membrane to reach the level needed to drive Rubisco. The gas concentration problem is solved by some species through enzymatic conversion of CO_2 to bicarbonate (HCO_3^-), which is trapped in the cytoplasm, unable to leak out of the cell membrane. This enzyme system is called the **carbon-concentrating mechanism** (**CCM**). Other bacteria use alternative CO_2-fixing systems adapted to different CO_2 concentrations.

Carboxysomes contain Rubisco. Many organisms that fix CO_2 contain the Rubisco complex within subcellular

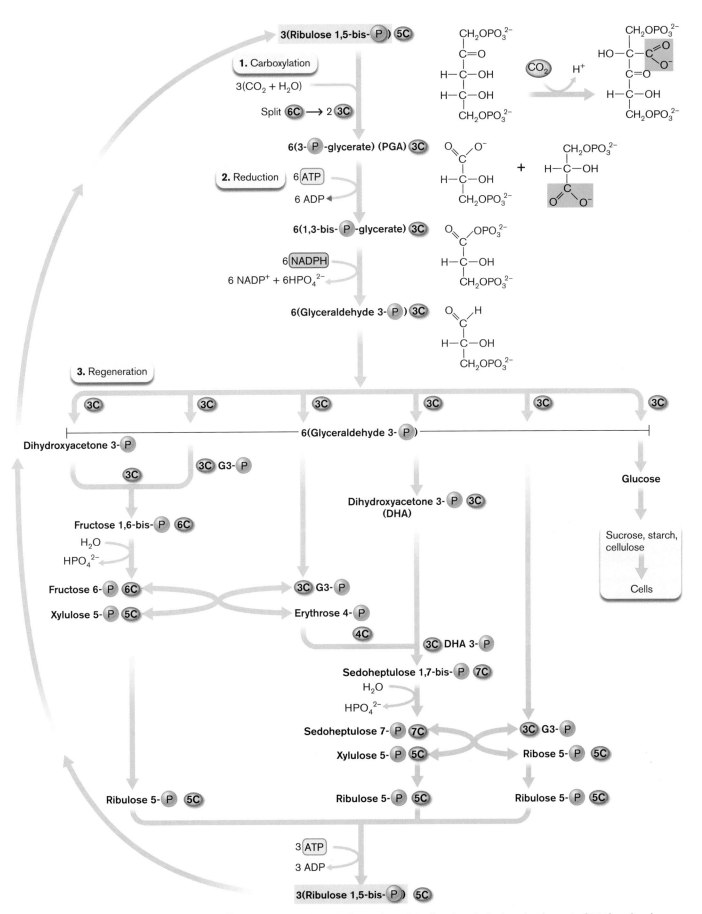

FIGURE 15.9 ■ The Calvin cycle in detail. The Calvin cycle assimilates <u>three</u> CO_2, forming <u>six</u> 3-phosphoglycerate (PGA) molecules reduced to glyceraldehyde 3-phosphate (G3P), and converts <u>five</u> G3P into <u>three</u> molecules of ribulose 1,5-bisphosphate. Labels "3C," "5C," "6C," and so on indicate the number of carbon atoms per molecule.

A.

B.

ADAPTED FROM TSAI, ET AL. 2007. PLOS BIOLOGY 5. E144. MICROGRAPHS COURTESY OF SABINE HEINHORST.

ADAPTED FROM TSAI, ET AL. 2007. PLOS BIOLOGY 5. E144. MICROGRAPHS COURTESY OF SABINE HEINHORST.

200 nm

100 nm

FIGURE 15.10 ■ Carboxysomes. A. *Halothiobacillus neapolitanus*, a sulfur-oxidizing lithoautotroph. Thin section (TEM) showing polyhedral carboxysomes (arrows). **B.** Isolated carboxysomes, packed with Rubisco complexes (TEM).

structures called **carboxysomes** (**Fig. 15.10**). Carboxysomes are found within CO_2-fixing lithotrophs, as well as within cyanobacteria and chloroplasts. A carboxysome consists of a polyhedral shell of protein subunits surrounding tightly packed molecules of Rubisco. The carboxysome takes up bicarbonate (converted from CO_2) by an unknown process. Once inside the carboxysome, the bicarbonate is immediately converted to CO_2 by the enzyme carbonic anhydrase. The CO_2 is then fixed by Rubisco to PGA—the first step of CO_2 fixation (shown in **Fig. 15.9**). The PGA exits the carboxysome to complete the Calvin cycle in the cytoplasm. Mutant strains of bacteria lacking carboxysomes can fix CO_2 only at high concentration (5%), much higher than the atmospheric CO_2 concentration (0.037%).

When CO_2 levels decline, genes are induced to express transmembrane transporters of CO_2 and bicarbonate (HCO_3^-) (**Fig. 15.11**). The inducible genes encode transport systems that participate in the carbon-concentrating mechanism. The CCM enables rapid concentration of HCO_3^- into the carboxysome for dehydration to CO_2. The CCM genes encode a low-affinity CO_2 transporter (NdhF4), a high-affinity CO_2 transporter (NdhF3), and two high-affinity HCO_3^- transporters (CmpABCD and SbtA). The low-affinity CO_2 transporter is effective only when CO_2 concentration is high. By contrast, the high-affinity transporters are induced by CO_2 starvation.

Induction of transporters by low CO_2 levels is shown by reverse transcription quantitative PCR (RT-qPCR), a technique for detection of

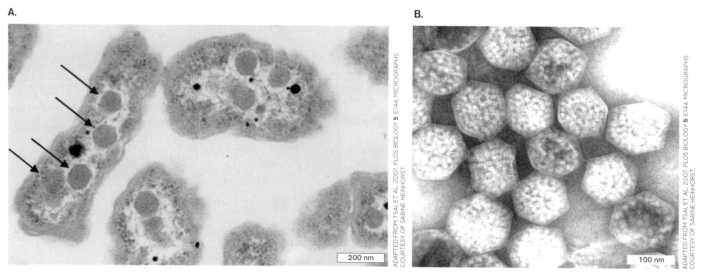

FIGURE 15.11 ■ The carbon-concentrating mechanism (CCM) in cyanobacteria. CO_2 and HCO_3^- are brought into the cell by transport complexes in the inner membrane or thylakoid membranes of *Synechocystis* species. The HCO_3^- is concentrated in the carboxysome and converted to CO_2 by carbonic anhydrase. *Source:* Model from Dean Price and colleagues, Australian National University at Canberra; Patrick McGinn et al. 2003. *Plant Physiol.* **132**:218.

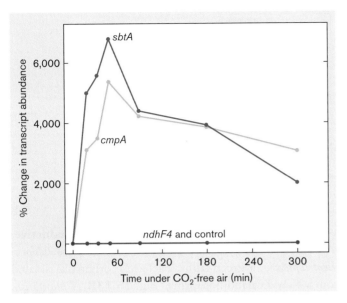

FIGURE 15.12 ■ **Time course of CCM gene transcription.** When CO_2 is limiting, *Synechocystis* genes encoding the carbon uptake complexes show induced transcription. Induction is measured by reverse transcription quantitative PCR (RT-qPCR). The *sbtA* and *cmpA* genes show especially high induction, whereas the constitutive system (*ndhF4*) shows no induction. *Source:* Patrick McGinn et al. 2003. Plant Physiol. 132:218.

minute quantities of mRNA using the enzyme reverse transcriptase followed by the polymerase chain reaction (discussed in Section 12.2). In RT PCR, the RNA transcripts are reverse-transcribed to DNA; the DNA is then amplified by PCR to generate a detectable product. RNA transcripts are obtained for genes expressed before and after CO_2 limitation. The *sbtA* and *cmpA* transcripts, encoding HCO_3^- transporters, show particularly strong induction by CO_2 (**Fig. 15.12**). By contrast, the transcript for the constitutive low-affinity CO_2 transporter, *ndhF4*, shows no induction.

Instead of carboxysomes, some phototrophs possess alternative systems of CO_2 uptake and concentration adapted to different levels of CO_2. For example, the purple phototroph *Rhodobacter sphaeroides* has two unlinked operons, each encoding a different set of CO_2 fixation genes, called form I and form II. Form I and form II each encodes all the key enzymes, such as Rubisco. When CO_2 is limiting, form I enzymes work best; when CO_2 levels are saturating, form II enzymes work best.

To Summarize

■ **The Calvin cycle** fixes CO_2 by reductive condensation with ribulose 1,5-bisphosphate. The Calvin cycle is used by cyanobacteria and chloroplasts and by some lithotrophs and photoheterotrophs.

■ **^{14}C radiolabeling and paper chromatography** were first used to identify intermediates of the Calvin cycle.

■ **Rubisco** catalyzes the condensation of CO_2 with ribulose 1,5-bisphosphate. The six-carbon intermediate immediately splits into two molecules of 3-phosphoglycerate (PGA), which are activated by ATP and reduced by NADPH to glyceraldehyde 3-phosphate (G3P).

■ **One of every six G3P is converted to glucose** or amino acids. The other five molecules of G3P undergo reactions to regenerate ribulose 1,5-bisphosphate.

■ **Carboxysomes sequester and concentrate CO_2** for fixation by the Calvin cycle. CO_2 levels regulate gene expression for carboxysome transporters and the Calvin cycle.

15.3 CO_2 Fixation: Diverse Pathways

We have seen how oxygenic phototrophs and some lithoautotrophs fix CO_2 by the Calvin cycle. Other bacteria, however, fix CO_2 by alternative means (see **Table 15.1**). From the standpoint of evolution and phylogenetic distribution, the most ancient pathways are the reductive, or reverse, TCA cycle of anaerobic phototrophs and the reductive acetyl-CoA pathway of methanogens. Diverse kinds of CO_2 fixation play essential roles in soil, aquatic, and wetland ecosystems, as well as in the digestive systems of animals such as cattle.

The Reductive, or Reverse, TCA Cycle

All major groups of organisms run the TCA cycle, or some part of it (discussed in Section 13.6). Most of the individual reaction steps of the TCA cycle are reversible because each has a relatively small ΔG of reaction. Thus, the steps that release CO_2 in the forward direction may assimilate small amounts of CO_2 in the reverse direction. Furthermore, all organisms, including humans, fix small amounts of CO_2 through special reactions that regenerate TCA cycle intermediates. These regeneration steps are called **anaplerotic reactions**. Common anaplerotic reactions are the formation of oxaloacetate from phosphoenolpyruvate (PEP), catalyzed by PEP carboxylase; and the formation of malate from pyruvate, catalyzed by malic enzyme.

In some anaerobic bacteria and archaea, the entire TCA cycle functions in "reverse," reducing CO_2 to generate

A.

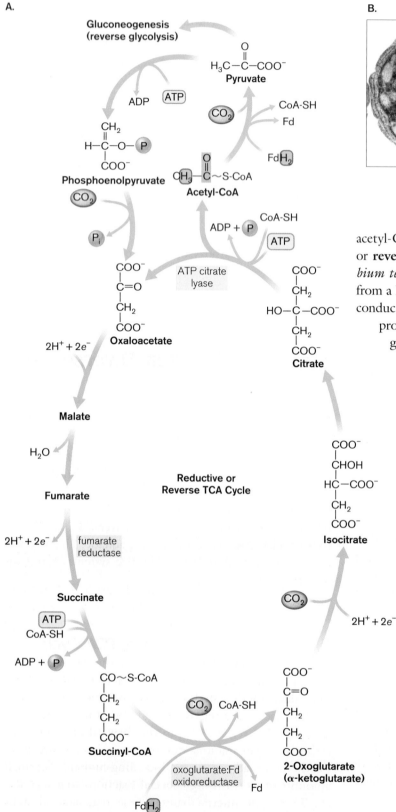

B.

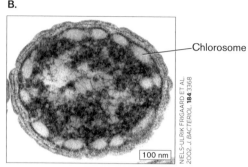

Chlorosome

NIELS-ULRIK FRIGAARD ET AL. 2002. J. BACTERIOL. 184:3368

100 nm

FIGURE 15.13 ■ The reductive, or reverse, TCA cycle for CO₂ fixation. A. Anaerobic phototrophs and archaea use the reverse TCA cycle to fix carbon into biomass. CO₂ is fixed by several intermediates, including succinyl-CoA, 2-oxoglutarate, and acetyl-CoA. Reduction (addition of $2H^+ + 2e^-$) is performed by NADPH or NADH and by reduced ferredoxin (FdH₂). **B.** *Chlorobium tepidum*, a green sulfur bacterium that fixes CO₂ by the reductive TCA cycle (TEM). Chlorosome membranes contain the reaction centers of photolysis.

acetyl-CoA and build sugars (**Fig. 15.13A**). The **reductive**, or **reverse**, **TCA cycle** is used by bacteria such as *Chlorobium tepidum*, a green sulfur bacterium originally isolated from a New Zealand hot spring (**Fig. 15.13B**). *Chlorobium* conducts anoxygenic photosynthesis by photolyzing H₂S to produce elemental sulfur, which collects in extracellular granules. Sulfur granule formation is of interest for the development of a process to remove H₂S from sulfide-generating industries such as refining of coal and oil. The reductive TCA cycle is also used by epsilonproteobacteria of hydrothermal vent communities and by sulfur-reducing archaea, such as *Thermoproteus* and *Pyrobaculum*.

Reversal of the TCA cycle uses four or five ATP to fix four molecules of CO₂ and generates one molecule of oxaloacetate. The enzymes used are the same as for the "forward" cycle, except for three key enzymes that spend ATP to drive the reaction in reverse: ATP citrate lyase, 2-oxoglutarate:ferredoxin oxidoreductase, and fumarate reductase. For example, ATP citrate lyase catalyzes the cleavage of citrate into acetyl-CoA and oxaloacetate. CO₂ assimilation requires reduction by NADPH. The reductive TCA cycle is believed to be the most ancient means of CO₂ fixation and amino acid biosynthesis, the original cycle of biomass generation in the ancestors of all three living domains.

From CO₂ to acetyl-CoA, the reverse TCA cycle essentially reverses the overall cycle of catabolism:

$$2CO_2 + 2\ ATP + 8H^+ + 8e^- + HS\text{-}CoA \rightarrow$$
$$CH_3CO\text{-}S\text{-}CoA + 3H_2O + 2\ ADP + 2\ P_i$$

The coenzyme A is recycled as the acetyl group enters biosynthesis. For example, acetyl-CoA can assimilate another CO₂ with reduction to pyruvate. Pyruvate then builds up to glucose and related sugars by **gluconeogenesis** (reverse glycolysis), with expenditure of ATP and NADPH. Alternatively, acetyl-CoA can enter biosynthetic pathways to produce fatty acids or amino acids.

Unlike the Calvin cycle, the reverse TCA cycle provides several different molecular intermediates that can assimilate CO_2. Here are the key steps:

- Succinyl-CoA assimilates CO_2 to form 2-oxoglutarate (alpha-ketoglutarate).

- 2-Oxoglutarate assimilates CO_2 to form isocitrate.

- Acetyl-CoA (produced by the reverse TCA cycle) assimilates CO_2 to form pyruvate.

Each CO_2 assimilation requires one or more reduction steps. 2-Oxoglutarate is reduced by the protein ferredoxin (FdH_2). FdH_2 also mediates acetyl-CoA reduction to pyruvate. Other reduction steps may be accomplished by NADPH or NADH.

The Acetyl-CoA Pathway

Yet another route for CO_2 assimilation into acetyl-CoA and pyruvate is the **reductive acetyl-CoA pathway** (**Fig. 15.14**). In the acetyl-CoA pathway, two CO_2 molecules are condensed through converging pathways to form the acetyl group of acetyl-CoA. The acetyl-CoA pathway is used by anaerobic soil bacteria, such as *Clostridium thermoaceticum*, and by most autotrophic sulfate reducers, such as *Desulfobacterium autotrophicum*. It also provides the main source of biomass for methanogens. Most of the CO_2 absorbed by methanogens is used to yield energy by reducing CO_2 to methane (discussed in Chapter 14). However, 5% of the CO_2 absorbed by a methanogen enters the acetyl-CoA pathway for biosynthesis, generating the entire biomass of the organism.

The substrates and ultimate products of the acetyl-CoA pathway are the same as those of the reverse TCA cycle, except that the reducing agent is H_2 instead of NADPH. The acetyl-CoA pathway is slightly more efficient than the TCA cycle, requiring one less ATP:

$$2CO_2 + ATP + 4H_2 + HS\text{-}CoA \rightarrow$$
$$CH_3CO\text{-}S\text{-}CoA + 3H_2O + ADP + P_i$$

Nevertheless, the order of the pathway and its intermediate compounds are completely different from those of the TCA cycle. The reductive acetyl-CoA pathway is linear, with no recycled intermediates.

Reduction of the first CO_2. The first CO_2 enters the linear pathway by reduction to formate. The formate is transferred onto a complex carrier cofactor. In bacteria, the cofactor is tetrahydrofolate (THF), a reduced form of **folate**. Folate (folic acid) is a heteroaromatic cofactor; it is an essential vitamin required by many organisms, including humans. A different cofactor, methanopterin (MPT), carries the formate in methanogenic archaea. In either case, the formate carbon is reduced in successive steps to a methyl group ($-CH_3$).

Reductive CO_2 incorporation into acetyl-CoA. The second CO_2 is reduced to carbon monoxide (CO) by the enzyme carbon monoxide dehydrogenase. This same enzyme is used in reverse reaction by some lithotrophs to gain energy from carbon monoxide (CO) as an electron donor. For fixation, the CO is condensed with the methyl group carried by the vitamin B_{12}–like cofactor TB to form acetyl-CoA. The acetyl-CoA then enters pathways of biosynthesis, as it does when formed by the reverse TCA cycle.

FIGURE 15.14 ■ The reductive acetyl-CoA pathway of CO_2 fixation. A. The first CO_2 is reduced to formate and is transferred onto the cofactor tetrahydrofolate (THF). After three reduction steps, the methyl group is transferred to a vitamin B_{12}–like cofactor (TB). The second CO_2 is reduced to carbon monoxide (CO) by the enzyme carbon monoxide dehydrogenase, and then incorporated into acetyl-CoA. **B.** *Methanocaldococcus jannaschii*, a thermophilic marine methanogen that fixes CO_2 by the reductive acetyl-CoA pathway (SEM).

A.

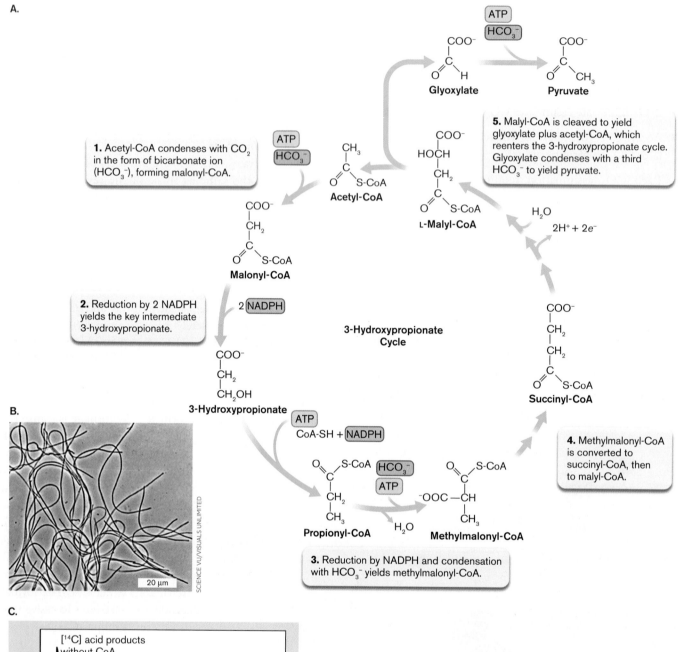

1. Acetyl-CoA condenses with CO_2 in the form of bicarbonate ion (HCO_3^-), forming malonyl-CoA.

2. Reduction by 2 NADPH yields the key intermediate 3-hydroxypropionate.

5. Malyl-CoA is cleaved to yield glyoxylate plus acetyl-CoA, which reenters the 3-hydroxypropionate cycle. Glyoxylate condenses with a third HCO_3^- to yield pyruvate.

3-Hydroxypropionate Cycle

4. Methylmalonyl-CoA is converted to succinyl-CoA, then to malyl-CoA.

3. Reduction by NADPH and condensation with HCO_3^- yields methylmalonyl-CoA.

B.

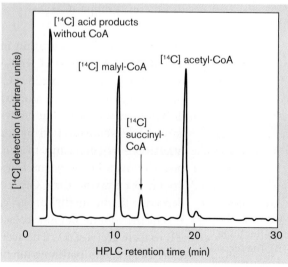

SCIENCE VU/VISUALS UNLIMITED

20 μm

C.

FIGURE 15.15 ■ The 3-hydroxypropionate cycle of CO_2 fixation. A. In the 3-hydroxypropionate cycle, acetyl-CoA condenses with CO_2 in the form of bicarbonate ion (HCO_3^-), forming malonyl-CoA (step 1). Reduction by 3 NADPH and condensation with a second HCO_3^- ultimately lead to malyl-CoA (steps 2–4). In step 5, malyl-CoA is cleaved to yield glyoxylate plus acetyl-CoA, which reenters the cycle. Glyoxylate condenses with a third HCO_3^- to form pyruvate. **B.** *Chloroflexus*, a green phototroph that performs the 3-hydroxypropionate cycle. **C.** Products of [^{14}C] acetate in the 3-hydroxypropionate cycle were separated by high-performance liquid chromatography (HPLC). *Source:* Part A modified from Silke Friedmann et al. 2006. *J. Bacteriol.* **188**:2646; part C modified from Sylvia Herter et al. 2002. *J. Biol. Chem.* **277**:20277.

One place where methanogens fix CO_2 and release methane is the bovine rumen, or digestive fermentation chamber (see Chapter 21). Within the rumen, bacteria convert cattle feed into CO_2 and H_2, and methanogens divert a significant portion of these gases into methane. The methane then escapes through burping and flatulence. If methanogen growth could be prevented, the CO_2 and H_2 could be fixed by other microbes into fermentation acids assimilated by the rumen. Beef production would then release less of the greenhouse gas methane.

The 3-Hydroxypropionate Cycle

The green photoheterotroph *Chloroflexus aurantiacus* fixes CO_2 yet another way: by the **3-hydroxypropionate cycle** (**Fig. 15.15A**). *Chloroflexus* bacteria are filamentous thermophiles that absorb light at longer wavelengths than cyanobacteria and often grow below cyanobacteria in thick mats of biofilm (**Fig. 15.15B**). The 3-hydroxypropionate cycle is also used by archaea, such as the thermoacidophile *Acidianus brierleyi* and the aerobic sulfur oxidizer *Sulfolobus metallicus*.

In the 3-hydroxypropionate cycle, CO_2 is fixed by acetyl-CoA into 3-hydroxypropionate (**Fig. 15.15A**). Acetyl-CoA condenses with hydrated CO_2 (bicarbonate ion, HCO_3^-) and is reduced by 2 NADPH to 3-hydroxypropionate. Condensation with a second molecule of HCO_3^- forms methylmalonyl-CoA, followed by several intermediates to malyl-CoA. These intermediates were detected by $[^{14}C]$ acetate incorporation into the metabolism of *Chloroflexus* (**Fig. 15.15C**). The intermediates were separated by high performance liquid chromatography (HPLC).

Malyl-CoA releases a molecule of acetyl-CoA to renew the cycle, plus glyoxylate. Glyoxylate condenses with a third HCO_3^- to form pyruvate, which enters standard routes for biosynthesis of sugars and amino acids. Thus, in all, the cycle fixes three molecules of CO_2 into one molecule of pyruvate. Pyruvate then serves as a substrate to build carbohydrates and amino acids.

To Summarize

- **The reductive, or reverse, TCA cycle** fixes CO_2 in some anaerobes and archaea.

- **Anaplerotic reactions** in all organisms regenerate TCA cycle intermediates by fixing CO_2.

- **The acetyl-CoA pathway** in anaerobic bacteria and methanogens fixes CO_2 by condensation to form acetyl-CoA.

- **The 3-hydroxypropionate cycle** in *Chloroflexus* and in some hyperthermophilic archaea fixes CO_2 in a cycle generating the intermediate 3-hydroxypropionate.

15.4 Biosynthesis of Fatty Acids and Polyketides

The biosynthesis of cell parts begins with small, versatile substrates such as acetyl-CoA. Acetyl-CoA is a product of catabolic pathways such as glycolysis and benzoate catabolism (see Sections 13.5 and 13.6). The acetyl group also forms anabolically from the reverse TCA cycle (**Fig. 15.13**). A key pathway of biosynthesis assembles acetyl-CoA into fatty acids. Fatty acids then condense with glycerol to form the phospholipids of cell membranes and the lipid components of envelope proteins. Fatty acid biosynthesis provides targets for antimicrobial drugs such as triclosan, an inhibitor of enoyl-ACP reductase. Related condensation pathways yield materials such as polyesters and polyketide antibiotics, such as tetracycline.

The repeating form of fatty acids, polyketides, and polyesters enables long-chain molecules to be synthesized with a relatively small number of different enzymes. By contrast, other cell components, such as amino acids, have more complex nonrepeating structures whose biosynthetic enzymes require long operons (presented in Section 15.6).

Fatty Acids Are Built from Repeating Units

Fatty acids exemplify the construction of cell components from repeating units (**Fig. 15.16**). The construction of molecules based on repeating units requires a cyclic pathway that feeds its products back repeatedly as substrates for further synthesis. The advantage of a cyclic process is that large polymers can be made with a limited number of enzymes; for example, the mycolic acids of *Mycobacterium tuberculosis* are extended to lengths of over 100 carbons.

While fatty acid structures used by cells include many different forms (discussed in Chapter 3), their synthesis always begins with the successive addition of two-carbon acetyl groups. Thus, most fatty acids contain an even number of carbon atoms.

The fatty acid synthase complex. The cyclic process of fatty acid synthesis is managed by the fatty acid synthase complex. The complex contains all the enzymes and component binding proteins bound together in proximity, so that all steps proceed in one place without "losing" the unfinished molecule. The main bacterial version of this complex is designated FASII. All of the components are essential for viability and thus are potential drug targets. Analogous multienzyme complexes are used by actinomycete bacteria to synthesize other long-chain products, such as polyketide antibiotics.

A.

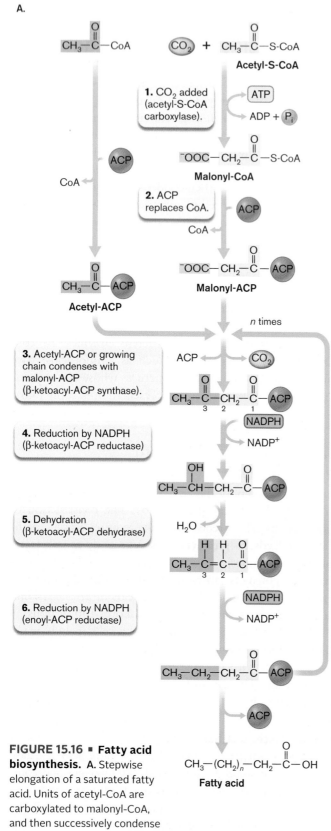

B.

FIGURE 15.16 ■ Fatty acid biosynthesis. A. Stepwise elongation of a saturated fatty acid. Units of acetyl-CoA are carboxylated to malonyl-CoA, and then successively condense to form the long chain of a fatty acid. Each two-carbon unit requires reduction by 2 NADPH. **B.** An alkene "kink" can be left unsaturated during chain extension. The double bond then forms between the third and fourth carbons, rather than the second and third. In this position, the double bond escapes reduction as the chain lengthens through subsequent cycles of malonyl addition.

Activation of acetyl-CoA. Most acetyl groups within the cell are "tagged" with coenzyme A (CoA), a cofactor that directs acetate into catabolic pathways such as the TCA cycle. To redirect acetyl groups into fatty acid biosynthesis, acetyl-CoA molecules are first tagged at the "back end" by condensation with CO_2 (**Fig. 15.16A**, step 1), catalyzed by acetyl-CoA carboxylase. The addition of CO_2 in this step does not count as carbon fixation, because the CO_2 added does not permanently add to the carbon skeleton. Instead, like phosphate or CoA, the CO_2 exists to be displaced when its function is no longer required. The CO_2-tagged acetyl-CoA is called malonyl-CoA. The coenzyme A is replaced by **acyl carrier protein** (**ACP**), making malonyl-ACP (step 2).

Malonyl-ACP condenses with the growing chain. In step 3 (**Fig. 15.16A**), malonyl-ACP hooks onto the head of an acetyl-ACP or a longer-growing chain. The process of hooking onto the new unit involves displacement of CO_2 from the back end of malonyl-ACP, which replaces ACP from the front end of the growing chain. The growing chain now contains a ketone (at carbon 3, from the former acetyl group). The ketone is reduced by NADPH and dehydrated, forming an alkene (C=C, unsaturated double bond between adjacent carbons 2 and 3). The alkene is further reduced by a second NADPH (steps 4–6). In all, the chain gains $4H^+ + 4e^-$ and is now fully hydrogenated (saturated).

Once hydrogenated, the chain is ready to take on the next malonyl-ACP, which, as before, loses CO_2 and replaces the ACP from the front of the growing chain (return to step 3). Successive addition can continue many times to build a saturated fatty acid.

Unsaturation. Certain fatty acids require unsaturated "kinks" in the chain (alkenes) to serve a structural purpose, such as to increase the fluidity of the membrane. Unsaturation can be generated during a cycle of elongation (**Fig. 15.16A**, step 5). During the first four cycles of acetyl addition, an alkene bond forms between the second and third carbons of the fatty acid. After the fifth two-carbon addition, however, a special dehydratase enzyme generates the alkene double bond between the third and fourth carbons (**Fig. 15.16B**). This alkene fails to be hydrogenated further. Instead, several additional malonyl units are added, generating a long-chain fatty acid with a kink of a *cis* double bond.

Regulation of fatty acid synthesis. The synthesis of fatty acids consumes enormous quantities of reducing energy; thus, cells must regulate it to avoid waste. From a structural standpoint, production of fatty acids incorporated into membranes must be balanced with growth of the cytoplasm. Furthermore, the many different variants of fatty acids are regulated in response to particular environmental needs. Some mechanisms of regulation include:

■ **Acetyl-CoA carboxylase represses its own transcription.** Transcription of an operon encoding two subunits of acetyl-CoA carboxylase (AccB, AccC) is repressed by one of its subunits, protein AccB. As AccB increases in concentration, it binds the promoter of the *accBC* operon, repressing further transcription. Thus, initiation of fatty acid biosynthesis (**Fig. 15.16A**, step 1) is always limited by the amount of AccB and AccC enzyme subunits that are present.

■ **Starvation blocks fatty acid biosynthesis.** Starvation for carbon sources blocks fatty acid biosynthesis through the "stringent response." Blockage is mediated by the polyphosphorylated nucleotide ppGpp (guanosine tetraphosphate), the global regulator of the stringent response (discussed in Section 10.2).

■ **Temperature regulates fatty acid composition.** Bacterial fatty acid composition is regulated by environmental factors such as temperature. In *E. coli*, low temperature favors unsaturated fatty acids because they are less rigid and maintain membrane flexibility. Low temperature induces expression of the gene *fabA* encoding the dehydratase enzyme that desaturates the fatty acid bond. As the dehydratase activity is increased, more unsaturated fatty acids are made.

Thought Question

15.5 For a given species, uniform thickness of a cell membrane requires uniform chain length of its fatty acids. How do you think chain length may be regulated?

Cyclic pathways of elongation comparable to fatty acid biosynthesis are used to build other kinds of polymers for energy storage. For example, many bacteria synthesize polyesters such as polyhydroxybutyrate. The term "polyester" indicates the multiple ester groups that are formed by repeated esterification of the carboxylic acid group of the chain with the hydroxyl group of a new alkanoate unit. Polyesters are insoluble in water, so they collect as storage granules within the bacterial cell, ready to release stored energy when needed. Polyester granules are synthesized by human pathogens such as *Legionella pneumophila*, the cause of legionellosis. Polyester storage helps *L. pneumophila* survive in water sources such as air-conditioning units. Commercially, polyesters produced by soil bacteria such as *Ralstonia eutropha* are used to manufacture biodegradable surgical sutures.

Polyketide Antibiotics

The cyclic biosynthesis of fatty acids provides clues to the biosynthesis of a famous family of antibiotics, the polyketides. A well-known polyketide is erythromycin, a broad-spectrum antibiotic prescribed for bacterial pneumonia and chlamydia infections (**Fig. 15.17**). Erythromycin blocks bacterial translation at the step of peptide elongation (see Chapter 8). Polyketides are secondary products produced by actinomycetes, as discussed earlier.

The synthesis of polyketide antibiotics involves cyclic elongation by an enormous enzyme complex called a **modular enzyme**, which consists of multiple modules that add similar but nonidentical units to a growing chain. Like fatty acids, polyketides are built by the successive condensation of malonyl-ACP units, but each malonyl group carries a unique extension, or R group, instead of a CO_2. In erythromycin, the R groups are all methyl groups. In other polyketides, the R groups range from hydroxyl groups and amides to aromatic rings, some of which undergo secondary reactions and interconnections.

How is the order of different R groups determined? The modular enzyme contains a series of active sites, or modules, each of which catalyzes the addition of one of a series of units. Each module contains all the enzyme activities needed to add the unit and recognizes a unit with one specific R group. The modular polyketide synthase may consist of a single protein with a series of domains, or it may consist of a complex of several protein subunits. Either the multidomain enzyme or the complex acts as an assembly line to generate the specific polyketide.

Figure 15.17A and **B** show the chain extension synthesis of a polyketide. The initial two domains of enzyme 1 are acyltransferase (AT) and an acyl carrier protein (ACP) domain, analogous to the ACP of fatty acid biosynthesis but specialized to accept one component of the polyketide.

A. Modular polyketide synthase

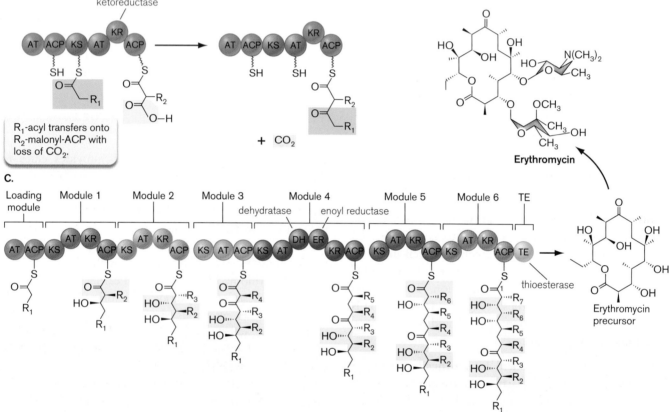

FIGURE 15.17 ■ **Synthesis of the erythromycin ring.** **A.** An acyltransferase (AT) transfers R_1-acetyl-CoA onto an ACP, with release of coenzyme A. **B.** The R_1-acetyl group is then transferred onto a ketosynthase (KS). The R_1-acetyl group condenses with R_2-malonyl-ACP, with release of CO_2. **C.** Modular subunits of the polyketide synthase complex elongate the polyketide to form the ring precursor of erythromycin. In this example, all R groups are methyl (CH_3). Some modules include reducing enzymes such as ketoreductase (KR), dehydratase (DH), and enoyl reductase (ER). Elongation is terminated by thioesterase (TE). *Source:* Based on David E. Cane et al. 1998. *Science* **282**:63.

The acyltransferase transfers the acyl group (R_1-acetyl) from R_1-acetyl-CoA onto the ACP, with release of coenzyme A (HS-CoA). The R_1-acetyl is then transferred to a ketosynthase domain (KS).

Meanwhile, a malonyl group with its own R group (R_2-malonyl) has been transferred by a second AT onto the second ACP domain (**Fig. 15.17B**). The R_1-acetyl group then leaves KS and condenses with R_2-malonyl-ACP, releasing CO_2. The R_1R_2 acyl chain subsequently undergoes a series of extensions by R-malonyl groups (**Fig. 15.17C**). For erythromycin, all R groups are methyl; other polyketides have R groups that are more complex. The initial AT and ACP domains constitute a "loading module" for the first acyl group, whereas subsequent sets of domains constitute "extender modules" that each add a different R-malonyl group. Extender modules can include secondary activities such as dehydratase (DH; removes OH) and ketoreductase

(KR; hydrogenates a ketone to OH). In principle, the modular approach can construct a limitless range of products with diverse antibiotic properties.

Elongation of the polyketide is terminated by thioesterase (TE), which hydrolyzes the thioester bond to the final ACP. This polyketide chain is an erythromycin precursor. The precursor requires additional enzymes (not shown) to add extra components, including two sugars (glycosylation). Sugar addition completes the erythromycin.

Another remarkable class of antibiotics formed on modular enzymes consists of polypeptides made without a ribosome. An example of a nonribosomal peptide antibiotic is ristocetin (see the Current Research Highlight). The biosynthesis of an important nonribosomal peptide antibiotic, vancomycin, is presented in **eTopic 15.4**. Modular enzyme biosynthesis of peptide antibiotics offers exciting prospects for the design of new drugs (see **Special Topic 15.1**).

To Summarize

- **Fatty acid biosynthesis** involves successive condensation of malonyl-ACP groups formed from acetyl groups tagged with acyl carrier protein (ACP) and a carboxylate. Each successive malonyl group is transferred onto the growing acyl chain, with release of CO_2.

- **The growing acyl chain is hydrogenated.** Each added unit is hydrogenated by two molecules of NADPH, unless an unsaturated kink is required.

- **Some fatty acids are partly unsaturated.** An unsaturated kink may be generated by a special dehydratase, which generates the alkene double bond between the third and fourth carbons.

- **Fatty acid biosynthesis is regulated** by the levels of acetyl-CoA carboxylase and by the stringent response to carbon starvation. Bond saturation is regulated by temperature and other environmental factors.

- **Polyesters** for energy storage are synthesized by cyclic elongation of alkanoates.

- **Polyketide antibiotics are synthesized by modular enzymes.** Nonribosomal peptide antibiotics are also made by modular enzymes.

15.5 Nitrogen Fixation and Regulation

The synthesis of amino acids, cell walls, and other cofactors requires an additional element we have not yet considered—namely, nitrogen. In principle, nitrogen should be more accessible to cells than carbon is, since nitrogen gas (N_2) comprises more than three-quarters of our atmosphere. In fact, however, N_2, with its triple bond, is one of the most stable molecules in nature. An enormous input of energy is required to reduce nitrogen to ammonia for assimilation into carbon skeletons. Early in evolution, all cells may have fixed their own N_2, but today only certain species of bacteria and archaea retain the ability. All other organisms depend on reduced or oxidized forms of nitrogen, which ultimately derive from N_2. Thus, all living organisms, directly or indirectly, depend on N_2-fixing prokaryotes within the biosphere (discussed in Chapters 21 and 22). Nitrogen limits microbes in many ecosystems, such as aquatic algae in lakes.

Note: Bacteria and archaea play essential roles in global cycling of nitrogen, sulfur, and phosphorus. The geochemical cycling of these elements is discussed in Chapter 22.

Nitrogen Assimilation

To fix nitrogen into biomass, bacteria or archaea must reduce the nitrogen completely to ammonia (NH_3). At pH 7, ammonia is mostly protonated to ammonium ion (NH_4^+). Unlike carbon, nitrogen rarely appears in oxidized form in complex biomolecules. Inorganic forms, such as nitric oxide (NO), are used as defense mechanisms against invading pathogens (see Chapter 23) or as signaling molecules. In macromolecules, however, virtually all the nitrogen is reduced; organic compounds containing oxidized nitrogen are generally toxic.

While N_2 is the ultimate source and sink of biospheric nitrogen, several oxidized or reduced forms are found in the environment, produced by living organisms (**Fig. 15.18**). Most free-living bacteria can acquire nitrate (NO_3^-) or nitrite (NO_2^-) for reduction to ammonium ion. Even nitrogen-fixing legume symbionts, such as *Rhizobium*, can use nitrate.

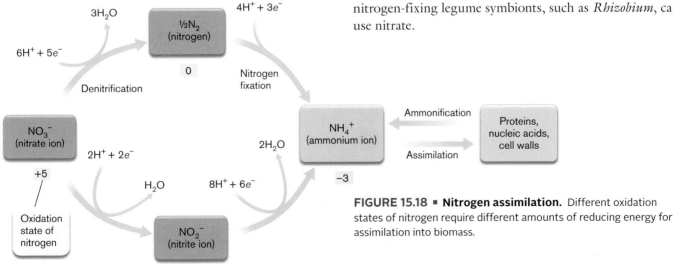

FIGURE 15.18 ■ Nitrogen assimilation. Different oxidation states of nitrogen require different amounts of reducing energy for assimilation into biomass.

SPECIAL TOPIC 15.1 Mining a Bacterial Genome for Peptide Antibiotics

Bacteria synthesize a vast array of peptide products that serve the functions of development, communication, and combat. These peptides have unique modifications that depart from the standard 20 amino acid residues. Some peptides are translated by ribosomes and then modified, whereas others, the nonribosomal peptides, are synthesized entirely by modular enzymes comparable to those that make polyketides. Nonribosomal peptides include powerful antibiotics such as vancomycin (**eTopic 15.4**) and immunosuppressant drugs such as cyclosporine.

As resistance evolves, in our arms race with pathogens, we always need more new antibiotics. The stendomycins are an emerging class of lipopeptides (peptides attached to lipid) that have antifungal activity (**Fig. 1**). Where do we find them? We can mine the genomes of the ever-evolving strains of bacteria in our environment, which undergo their own arms races with each other. The challenge is (1) to detect previously unknown antibiotics within cells—a chemical "needle in the haystack"—and (2) to reveal the genes encoding their biosynthetic enzymes. The genes can then be cloned in an engineered fermenter strain (discussed in Chapter 16).

So which clues come first—the peptide or the biosynthetic genes? Either the peptide or the gene sequence may offer a clue, which then serves to interrogate the other side.

A multistep interactive approach between peptide and gene was devised by Pieter Dorrestein and colleagues at UC San Diego (**Fig. 2**). The approach is called peptidogenomics. **Figure 3** outlines the peptidogenomic approach that identified the structure and biosynthetic genes for stendomycin I–VI, an antifungal agent produced by the actinomycete *Streptomyces hygroscopicus*.

Stendomycin was the first antibiotic identified in *S. hygroscopicus* by a version of mass spectrometry called MALDI-TOF (matrix-associated laser desorption/ionization–time-of-flight). In the MALDI technique, a UV laser beam is trained on a sample embedded in a chemical matrix. A thin portion of sample plus matrix is "desorbed"—that is, ionized—and the ions are taken up into the mass spectrometer. The "time of flight" into the detector offers a measure of the size of the ionized molecules.

For Dorrestein's work, the MALDI laser was aimed at a matrix-embedded colony of *S. hygroscopicus*. Fragments of thousands of molecules were generated, among them components with sizes characteristic of certain uniquely modified amino acids that appear only in nonribosomal peptides, such as dehydroalanine. In addition, the fragments predicted the existence of a peptide containing eight amino acids, whose sequence did not match any translated sequence within the *S. hygroscopicus* genome. The data were consistent

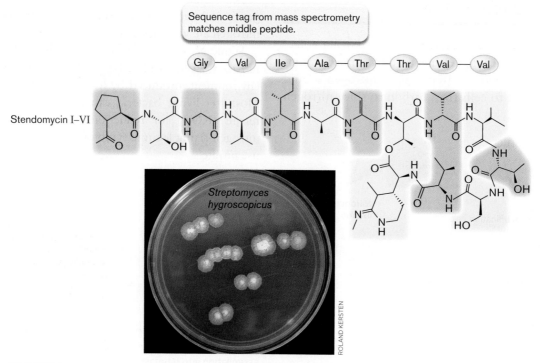

FIGURE 1 ■ **A nonribosomal peptide antibiotic, stendomycin I–VI, is synthesized by *Streptomyces hygroscopicus*.**

FIGURE 2 ■ **Pieter Dorrestein discovers new antibiotic peptides by peptidogenomics.**

with eight slightly different versions of the peptide, called a sequence tag (**Fig. 3**).

The peptide sequence tag was then used to query the genome for genes encoding modular enzymes that might catalyze the synthesis of the amino acids in the peptide. Previous bioinformatic studies characterized many families of modular enzymes that evolved from common ancestors. In seeking signature patterns for such genes, a cluster of genes

for modular enzymes was identified that appeared likely to synthesize one version of the peptide: Gly-Val-Ile-Ala-Thr-Thr-Val-Val. This peptide proved to be the backbone of the lipopeptide stendomycin I–VI (**Fig. 1**). Note that in the mature lipopeptide, other enzymes have catalyzed various modifications, such as an ester linkage between the carboxyl terminus and the hydroxyl group of a threonine residue.

The full structure of stendomycin I–VI was identified and confirmed by back-and-forth testing between models predicted by the mass spectrometry fragments and by gene cluster analysis. The method is now being used to identify many other peptide products. We can only hope that our rate of drug discovery outpaces the rate of resistance arising in pathogens.

RESEARCH QUESTION

The genomic cluster shown in **Figure 3** encodes three modular enzymes for stendomycin biosynthesis, one of which synthesizes the tag peptide. What do the other two enzymes synthesize? How could you test this question using peptidogenomics?

Kersten, Roland D., Yu-Liang Yang, Yuquan Xu, Peter Cimermancic, Sang-Jip Nam, et al. 2011. A mass spectrometry–guided genome mining approach for natural product peptidogenomics. *Nature Chemical Biology* **7**:794–802.

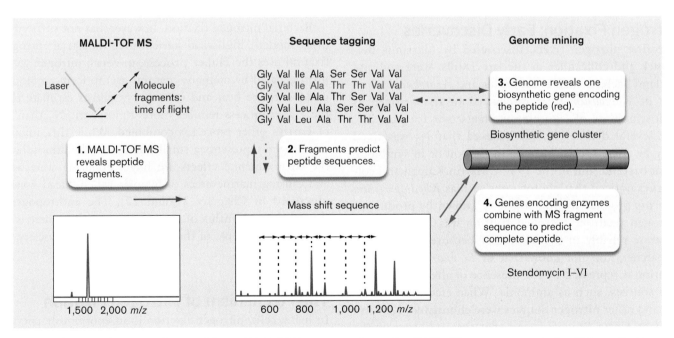

FIGURE 3 ■ **Peptidogenomics combines MALDI-TOF mass spectrometry (MS) with sequence analysis of biosynthetic gene clusters to reveal the structure of novel antibiotics.** m/z = mass-to-charge ratio (mass per charge).

FIGURE 15.19 ■ **Nitrogen fixation requires specialized structures.** *Anabaena spiroides*, a filamentous cyanobacterium, segregates N_2 fixation in heterocysts, cells that do not photosynthesize and thus maintain anoxic conditions.

In natural environments, most potential sources of nitrogen are subject to competition from dissimilatory metabolism in which the molecule is oxidized or reduced for energy, as discussed in Chapter 14. For example, anaerobic respirers convert nitrate and nitrite to N_2 (denitrification), whereas lithotrophs oxidize NH_4^+ to nitrite and nitrate (nitrification). An important consequence of nitrification is that most of the commercial ammonia fertilizer spread on agricultural fields is soon oxidized by lithotrophs to nitrates and nitrites. High concentrations of nitrates in water are harmful because they combine with hemoglobin, generating a form that cannot take up oxygen. When infants drink water with high nitrite, they may become ill with "blue baby syndrome."

Nitrogen Fixation: Early Discoveries

The first nitrogen fixers, discovered by Martinus Beijerinck and colleagues in the late 1800s, were soil and wetland bacteria such as *Beggiatoa* and *Azotobacter*. Species of *Rhizobium* were discovered to fix nitrogen as endosymbionts of leguminous plants (see Chapter 21). For several decades, it was believed that N_2 was fixed only by a few special bacteria in the soil or in symbiotic plant bacteria. But in the 1940s, Martin Kamen and colleagues noticed that phototrophs such as *Rhodospirillum rubrum* produce hydrogen gas—a known by-product of nitrogen fixation. Nitrogen fixation was hard to demonstrate reliably in the laboratory because at that time researchers did not know that in *R. rubrum*, nitrogen fixation is repressed by the presence of alternative nitrogen sources, such as ammonia. When traces of ammonia and other nitrogen sources were eliminated, Kamen's student Herta Bregoff found that these photosynthetic bacteria actually fix nitrogen.

Nitrogen is now known to be fixed by most phototrophic bacteria (green and purple bacteria, as well as cyanobacteria) and by many archaea. Marine cyanobacteria fix a large proportion of both the nitrogen and carbon dioxide assimilated by our biosphere. To fix N_2, aquatic cyanobacteria such as *Anabaena* develop special cells called **heterocysts**, in which photosynthesis is turned off to maintain anaerobic conditions (**Fig. 15.19**). Land ecosystems require nitrogen-fixing bacteria and archaea in the soil, often in mutualistic relationships with plants. For example, the bacterium *Bradyrhizobium japonicum* associates with the roots of soybean plants, causing them to grow nodules. The nodules contain "bacteroids," forms of the bacteria that grow within the root cells. The bacteroids release extra nitrogen into the soil; for this reason, farmers alternate soy crops with nitrogen-intensive crops such as corn. Nitrogen-fixing mutualism is discussed in Chapter 21.

Bacterial nitrogen fixation, however, has not sufficed to drive modern high-yield agriculture. Industrial nitrogen fixation uses the **Haber process**, in which nitrogen gas is hydrogenated by methane (natural gas) to form ammonia, under extreme heat and pressure. Scientists estimate that the Haber process reduces more atmospheric N_2 than all of Earth's other processes combined. While this amount of nitrogen represents a small fraction of the atmosphere, the environmental effects are huge, polluting waterways and causing marine areas of hypoxia called "dead zones" (discussed in Chapters 21 and 22). The anthropogenic (human-caused) influx of nitrogen into the biosphere is an astonishing example of the influence of human society on our planet.

The Mechanism of Nitrogen Fixation

In living cells, nitrogen fixation is an enormously energy-intensive process. The mechanism is largely conserved across species:

$$N_2 + 8H^+ + 8e^- + 16\,ATP \rightarrow 2NH_3 + H_2 + 16\,ADP + 16\,P_i$$

The electrons are donated by NADH, H_2, or pyruvate, or obtained through photosynthesis. The total energy investment includes approximately 3 ATP-equivalents per $2e^-$, plus 16 ATP molecules as shown. That makes 12 + 16 = 28 ATP in all—a large part of the energy gained from oxidation of glucose. The production of H_2 is surprising, as it consumes extra ATP. Hydrogen loss results from initiating the cycle of nitrogen reduction (discussed shortly). Some bacteria have secondary reactions to reclaim the lost hydrogen with part of its lost energy.

Note: Although nitrogen fixation is commonly represented as producing ammonia (NH_3), under most conditions of living cells the predominant form is the protonated ammonium ion (NH_4^+).

Nitrogenase reaction mechanism. The overall conversion of nitrogen gas to two molecules of ammonia is catalyzed in four cycles by **nitrogenase**, an enzyme highly conserved in nearly all nitrogen-fixing species. Nitrogenase probably evolved once in a common ancestor of all nitrogen-fixing organisms.

The mechanism of nitrogenase is of intense interest to agricultural scientists because of the potential benefits of improving efficiency of plant growth and of extending nitrogen-fixing symbionts to nonleguminous plants such as corn. In 1960, scientists at the DuPont laboratory first isolated the nitrogenase complex from a bacterium, *Clostridium pasteurianum*. They measured its activity by incorporating heavy-isotope nitrogen gas, $^{15}N_2$, into $^{15}NH_3$. Since then, the detailed structure of nitrogenase has been solved (**Fig. 15.20**).

The active nitrogenase complex includes two kinds of subunits encoded by different genes: Fe protein (shaded green in **Fig. 15.20**), containing a [4Fe-4S] center; and FeMo protein (shaded cyan), containing iron and molybdenum. The Fe protein contains a typical [4Fe-4S] structure to facilitate electron transfer. As we learned in Chapter 14, metal atoms help to transfer electrons because their orbitals are closely spaced and transitions between these orbitals involve relatively small amounts of energy. The electrons are funneled through a second Fe-S cluster (the P cluster) to the FeMo (iron-molybdenum) cluster. The FeMo cluster is an unusual structural characteristic of nitrogenase, in which trios of sulfur atoms alternate with trios of iron atoms. One end of the cluster is capped with another iron, and the other end is capped with an atom of molybdenum (Mo). A consequence of the nitrogenase structure is that most nitrogen-fixing

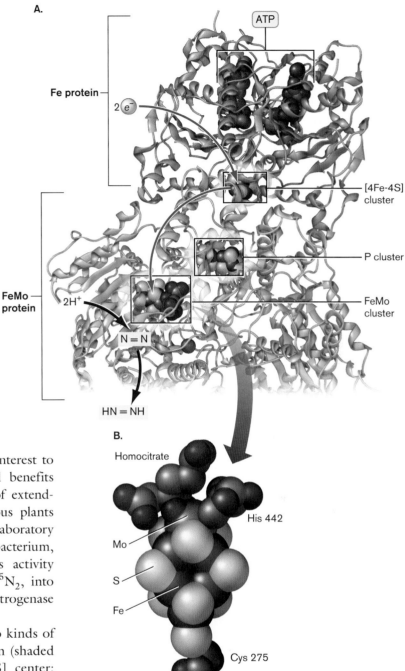

FIGURE 15.20 ■ Structure of the nitrogenase complex.
A. In each active site of the Fe protein–FeMo protein complex, the Fe protein binds ATP and receives electrons from the electron donors. The electrons are subsequently channeled down through the [4Fe-4S] cluster and the P cluster (Fe-S cluster) to the FeMo cluster, where they reduce the N_2. (PDB code: 1N2C) **B.** The metal cluster (Fe_7S_9Mo) is held in place by coordination with a molecule of homocitrate plus two amino acid residues of nitrogenase: His 442 and Cys 275.

organisms (and their plant hosts, for leguminous symbionts) require the element molybdenum for growth. Some bacteria make an alternative nitrogenase that substitutes vanadium for molybdenum.

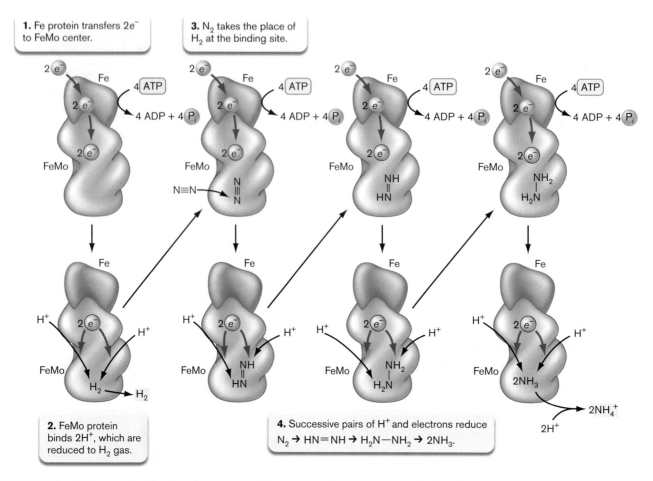

1. Fe protein transfers $2e^-$ to FeMo center.

3. N_2 takes the place of H_2 at the binding site.

2. FeMo protein binds $2H^+$, which are reduced to H_2 gas.

4. Successive pairs of H^+ and electrons reduce $N_2 \rightarrow HN=NH \rightarrow H_2N{-}NH_2 \rightarrow 2NH_3$.

FIGURE 15.21 ■ Nitrogen fixation by nitrogenase. The enzyme nitrogenase successively reduces nitrogen by electron transfer, ATP hydrolysis, and H^+ incorporation.

Four cycles of reduction. Nitrogen fixation requires four reduction cycles through nitrogenase (**Fig. 15.21**).

To initiate the first cycle of N_2 reduction, $2e^-$ from an electron donor such as NADH or H_2 are transferred by a ferredoxin to Fe protein. The reduced Fe protein transfers each electron to the FeMo center, with energy supplied by four molecules of ATP (**Fig. 15.21**, step 1). The FeMo protein binds $2H^+$, which are reduced to H_2 (step 2). Only then does N_2 bind to the active site, by displacing the H_2 (step 3).

In the next reduction cycle, two electrons are transferred to Fe protein, where they reduce the iron. The reduced Fe protein transfers each electron to the FeMo center, near the binding site for N_2. The electron transfer requires expenditure of 2 ATP per electron, or 4 ATP per $2e^-$. Two hydrogen ions join the N_2 and receive the two electrons, forming $HN=NH$.

A third pair of electrons enters the Fe protein, which then joins another $2H^+$, reducing N_2 to $H_2N{-}NH_2$. The cycle again requires hydrolysis of 4 ATP. A final cycle of electron transfer and H^+ uptake reduces $H_2N{-}NH_2$ to $2NH_3$. At typical cytoplasmic pH values (near pH 7), NH_3 is protonated to NH_4^+.

The loss of H_2 during nitrogen fixation is puzzling because it represents lost energy in a highly energy-intensive process. The actual fate of the H_2 varies among species. *Klebsiella pneumoniae*, a Gram-negative organism in which nitrogen regulation has been much studied, gives off the H_2 without further reaction. On the other hand, *Azotobacter* uses an irreversible hydrogenase enzyme to convert H_2 back to $2H^+$ and recover the transferred electrons. In leguminous rhizobial symbionts such as *Sinorhizobium* species, H_2 recovery varies surprisingly among strains and can affect the efficiency of plant growth.

Nitrogen fixation is expensive, both in terms of protein synthesis and in terms of reducing energy and ATP. Even bacteria capable of nitrogen fixation repress the process in the presence of other nitrogen sources, such as nitrate (NO_3^-), nitrite (NO_2^-), ammonia (NH_3), and nitrogenous organic molecules acquired from dead cells.

Anaerobiosis and N_2 Fixation

The reductive action of nitrogenase is extremely sensitive to oxygen because of the large reducing power needed to make NH_4^+. Thus, cells can fix nitrogen only in an anaerobic environment. This is no problem for anaerobes. But oxygenic phototrophs such as cyanobacteria face an obvious problem, as do bacterial symbionts of oxygenic plants. Aerobic and oxygenic organisms have developed several solutions to the problem of fixing nitrogen in aerobic environments:

- **Protective proteins.** Aerobic *Azotobacter* species synthesize protective proteins that stabilize nitrogenase and prevent attack by oxygen. For rhizobia, on the other hand, the plant hosts produce leghemoglobin (named for "legume" plants), a form of hemoglobin that sequesters oxygen away from the bacteria.

- **Temporal separation of photosynthesis and N_2 fixation.** Some species of cyanobacteria fix nitrogen only at night, when they do not conduct photosynthesis and thus release no oxygen.

- **Specialized cells for N_2 fixation.** Filamentous cyanobacteria develop specialized nitrogen-fixing cells called heterocysts (see **Fig. 15.19**). The heterocysts lose their photosynthetic capacity entirely and specialize in nitrogen fixation. Heterocyst development is directed by a complex genetic program induced by nitrogen starvation. In natural environments, the heterocysts leak organic acids that attract heterotrophic bacteria, which use up all the oxygen around the heterocyst, generating ideal anaerobic conditions for nitrogen fixation.

A. Sydney Kustu

SYDNEY KUSTU, UC BERKELEY

FIGURE 15.22 ■ **Regulation of nitrogen fixation.** **A.** Sydney Kustu has made important discoveries about NtrC and other forms of nitrogen regulation in the Gram-negative soil bacterium *Klebsiella pneumoniae*. **B.** When cellular levels of NH_4^+ are low, NtrC is phosphorylated and activates expression of *nifLA*. Expression is coactivated by the nitrogen starvation sigma factor, sigma-54. The NifA protein (and sigma-54) activates expression of *nif* genes encoding nitrogenase. When oxygen levels are high, however, nitrogen fixation cannot occur, so NifA is blocked by binding NifL. (Numbers refer to positions upstream of the translation site.)

Molecular Regulation of N_2 Fixation

Nitrogen fixation costs substantial energy, and therefore the process is regulated with extraordinary fine tuning. Multiple factors determine the expression of genes encoding nitrogenase (*nifHDKTY*) and other nitrogen-fixation proteins. Oxygen represses the expression of *nif* genes because nitrogenase is inactivated by oxygen; and high NH_4^+ depresses expression because enough fixed nitrogen is already available to the cell. Responses are mediated by several molecular regulators, including a nitrogen starvation sigma factor (sigma-54) and the NtrB-NtrC two-component signal transduction system. (Sigma factors and two-component regulators are discussed in Chapter 10.)

The NtrB-NtrC system was dissected in *Klebsiella pneumoniae* by Sydney Kustu and colleagues at UC Berkeley (**Fig. 15.22A**). When NH_4^+ is low (**Fig. 15.22B**), the cell needs to fix nitrogen. The nitrogen sensor kinase NtrB autophosphorylates (obtains a phosphoryl group from ATP). NtrB-P then phosphorylates NtrC, forming NtrC-P, the nitrogenase activator. The NtrC-P phosphoprotein binds an upstream enhancer sequence to activate expression of *nifLA*, making NifL and NifA proteins. Expression of *nifLA* is coactivated by sigma-54. Sigma-54 and the NifA protein together activate expression of *nifHDKTY* to make nitrogenase. Thus, low NH_4^+ concentration turns on nitrogen fixation.

When cellular levels of NH_4^+ are high, NtrC remains unphosphorylated. The *nifLA* operon is not transcribed, no activators bind the *nifHDKTY* promoter, and the nitrogen fixation genes are not expressed.

Nitrogenase activity is stopped by oxygen, so high oxygen levels block expression of the gene. When oxygen is high, NifA binds NifL and is prevented from binding the nitrogenase promoter. Low oxygen allows NifA to bind the promoter and express nitrogenase to fix nitrogen. Overall, NtrC governs response to NH_4^+, whereas NifA governs response to oxygen. Being responsive to multiple environmental signals is typical of energy-expensive biosynthetic pathways.

NtrB and NtrC also regulate nitrogen usage at the level of biosynthesis of amino acids, glutamate and glutamine (discussed in next section).

To Summarize

- **Oxidized or reduced forms of nitrogen**, such as nitrate, nitrite, and ammonium ions, can be assimilated by bacteria and plants. Assimilation competes with dissimilatory reactions that obtain energy.

- **Nitrogen gas (N_2) is fixed into ammonium ion (NH_4^+)** only by some species of bacteria and archaea—never by eukaryotes.

- **Nitrogenase enzyme** includes a protein containing an iron-sulfur core (Fe protein) and a protein containing a complex of molybdenum, iron, and sulfur (FeMo protein). Electrons acquired by Fe protein (with energy from ATP) are transferred to FeMo protein to reduce nitrogen.

- **Four cycles of reduction by NADPH** or an equivalent reductant reduce one molecule of N_2 to two molecules of NH_3. At neutral pH, NH_3 is protonated to NH_4^+.

- **Oxygen inhibits nitrogen fixation.** Bacteria and plants have various means of separating nitrogen fixation from aerobic respiration, such as heterocyst development or temporal separation.

- **Nitrogen and oxygen regulate transcription of nitrogenase.**

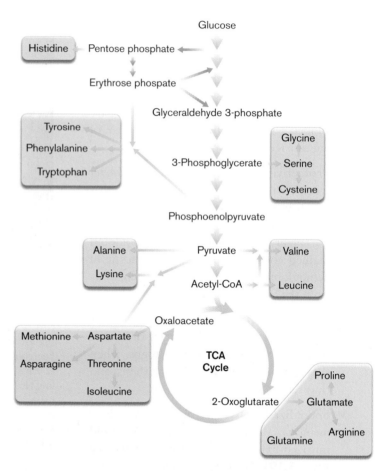

FIGURE 15.23 ■ **Major pathways of amino acid biosynthesis.** Some amino acids arise from key metabolic intermediates, whereas others must be synthesized out of other amino acids.

15.6 Biosynthesis of Amino Acids and Nitrogenous Bases

Where do microbes obtain amino acids to make their proteins and cell walls, as well as nitrogenous bases to synthesize DNA and RNA? When possible, microbes obtain these molecules from their environment through membrane-embedded transporters. But competition for such valuable nutrients is high, especially for free-living microbes in soil or water. Most free-living microbes and plants have the ability to make all the standard amino acids and bases of the genetic code, as well as nonstandard variants used for cell walls and transfer RNAs. We can use microbial biosynthesis of amino acids in the industrial production of food supplements, for ourselves and for farm animals.

Amino Acid Synthesis

Like fatty acid biosynthesis, synthesis of amino acids and nitrogenous bases requires the input of large amounts of reducing energy. These compounds pose additional challenges because of their unique and diversified forms, which cannot be made by the cyclic processes that generate molecules from repeating units. Synthesis of complex, asymmetrical molecules such as amino acids requires many different conversions, each mediated by a different enzyme. Nevertheless, some economy is gained by an arrangement of branched pathways in which early intermediates are utilized to form several products (**Fig. 15.23**). For example, oxaloacetate is converted to aspartate, which can be converted to four other amino acids.

The carbon skeletons of amino acids arise from diverse intermediates of metabolism (see **Fig. 15.23**). As in fatty acid biosynthesis, precursor molecules are channeled into amino acid biosynthesis by specialized cofactors and reducing energy carriers such as NADPH. Note that certain amino acids arise directly from key metabolic intermediates (for example, glutamate from 2-oxoglutarate), whereas others must be synthesized from preformed amino acids (for example, glutamine, proline, and arginine from glutamate).

FIGURE 15.24 ■ **The Murchison meteorite.** Fragments of a meteorite that fell in Murchison, Australia, in 1969 were shown to contain the five fundamental amino acids of biosynthetic pathways (glutamate, aspartate, valine, alanine, and glycine).

Some amino acids can arise from more than one source; for example, leucine and isoleucine can be made from succinate as well as from pyruvate.

It has been hypothesized that the amino acids arising in just one or two steps from central intermediates are more ancient in cell evolution than those requiring more complex pathways. Five of these "ancient" amino acids—glutamate, aspartate, valine, alanine, and glycine—are the same as those detected in meteorites, whose composition resembles that of prebiotic Earth (**Fig. 15.24**). These same five amino acids also appear in early-Earth simulation experiments in which methane, ammonia, and water are heated under reducing conditions and subjected to electrical discharge. Thus, we speculate that the first amino acids that early cells evolved to make were the same as those that arose spontaneously in the prebiotic chemistry of our planet.

A more subtle effect of evolution has been to adjust the amino acid composition of proteins based on the energetic cost of biosynthesis. Proteins that are secreted or that project outside the cell cannot be recycled; their amino acids are ultimately lost to the cell. Thus, secreted and externally projecting proteins have evolved to contain the "cheaper" amino acids—that is, the amino acids whose synthesis requires spending fewer molecules of ATP and NADPH. This effect can be seen in the diagram of a bacterial flagellum and its attached motor, which contains extracellular as well as cytoplasmic components, colored on a scale based on biosynthetic "expense" (**Fig. 15.25**). The external and secreted components favor less expensive amino acids, compared to the cytoplasmic components.

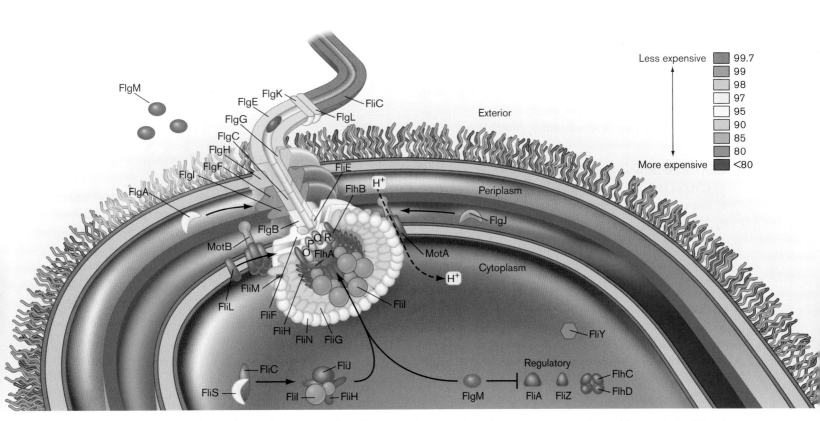

FIGURE 15.25 ■ **External proteins and proteins extending outside the cell use less expensive amino acids.** Diagram of flagellar proteins (FlgE, FlgK, FlgL, FlgG, FliC) attached to the motor complex. The external and secreted structures (FlgE, FliC, FlgM) are composed of amino acids that are less energy-expensive. *Source:* Modified from Daniel Smith and Matthew Chapman. 2010. *MBio* **1**:e00131.

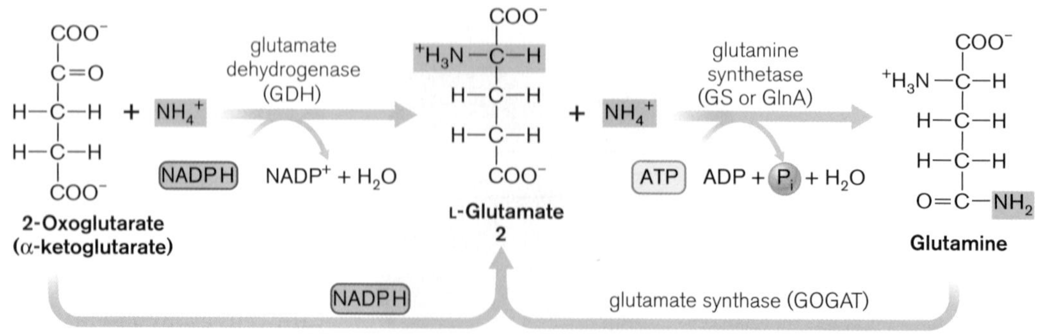

FIGURE 15.26 ■
Assimilation of NH$_4$$^+$ into glutamate and glutamine. The key TCA cycle intermediate 2-oxoglutarate (alpha-ketoglutarate) incorporates one molecule of NH$_4$$^+$ at the ketone to form glutamate. Glutamine can combine with oxoglutarate to produce two molecules of glutamate. These reactions provide sources of nitrogen to feed other pathways of amino acid synthesis.

Assimilation of NH$_4$$^+$

Unlike sugars and fatty acids, amino acids must assimilate another key ingredient: nitrogen. The NH$_4$$^+$ produced by N$_2$ fixation or by nitrate reduction is the key source of nitrogen for biosynthesis. But NH$_4$$^+$ always exists in equilibrium with NH$_3$, which is toxic to cells. Moreover, NH$_3$ travels freely through membranes, making it difficult to store NH$_4$$^+$ within a cell. Deprotonation to NH$_3$ increases as pH rises; by pH 9.2, deprotonation reaches 50%. Even at neutral pH, a very small equilibrium concentration of NH$_3$ can drain NH$_4$$^+$ out of the cell. Thus, cells avoid storing high levels of NH$_4$$^+$; instead, the fixed nitrogen is incorporated immediately into organic products.

2-Oxoglutarate and glutamate condense with NH$_4$$^+$. In most bacteria, the main route for NH$_4$$^+$ assimilation is the condensation of NH$_4$$^+$ with 2-oxoglutarate to form glutamate, or with glutamate to form glutamine (**Fig. 15.26**). Three key enzymes interconvert these substrates:

- **Glutamate dehydrogenase (GDH)**, actually named for its reverse activity, condenses NH$_4$$^+$ with 2-oxoglutarate to form glutamate. The condensation requires reduction by NADPH.

- **Glutamine synthetase (GS or GlnA)** condenses a second NH$_4$$^+$ with glutamate to form glutamine—a process driven by spending one ATP.

- **Glutamate synthase (GOGAT, glutamine:2-oxoglutarate aminotransferase)** converts 2-oxoglutarate plus glutamine into two molecules of glutamate. Different variants of this enzyme use different reducing agents: NADPH, NADH, or ferredoxin. The NADPH variant is shown in **Figure 15.26**.

Through these reactions, all three substrates exchange amino groups readily. High nitrogen levels induce GDH to take up NH$_4$$^+$, but repress GS and GOGAT. GS has a higher affinity than GDH for NH$_4$$^+$. Low nitrogen levels induce GS (to make glutamine) and GOGAT (to convert some glutamine to glutamate as needed).

Both glutamate and glutamine contribute an amine, as well as their carbon skeletons, to the synthesis of other amino acids in the biosynthetic "tree." The transfer of ammonia between two metabolites such as glutamate and glutamine is called **transamination**. Many other pairs of amino acids and metabolic intermediates undergo transamination. For example, glutamate transfers NH$_3$ to oxaloacetate, making aspartate and 2-oxoglutarate; the reaction can also be reversed. In another example, valine transfers an amine group to pyruvate, generating alanine and 2-ketoisovalerate.

> **Thought Question**
>
> **15.6** Suggest two reasons why transamination is advantageous to cells.

Glutamate and Glutamine Signal Nitrogen Availability

The cellular levels of glutamate and glutamine act as indicators of nitrogen availability. When nitrogen is scarce, NtrC is phosphorylated to NtrC-P as we saw earlier in **Figure 15.22**. Along with nitrogenase, NtrC-P up-regulates expression of glutamine synthetase (**Fig. 15.27**), as well as a high-affinity ammonia transporter, and transporters for organic sources of nitrogen such as amino acids, oligopeptides, and cell wall fragments containing amino sugars. All these molecules can be "scavenged" to obtain nitrogen for biosynthesis.

When ammonium ion (NH$_4$$^+$) is abundant, cells store the nitrogen safely as glutamine, via glutamine synthetase (GlnA) (**Fig. 15.27**). GlnA uses the energy released from ATP hydrolysis to assimilate NH$_4$$^+$ into glutamic acid and produce glutamine. But what if the cell makes too much active GlnA? Then all of the cell's glutamate

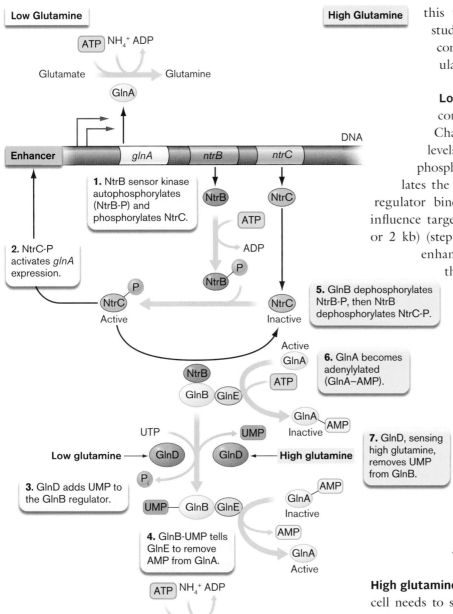

FIGURE 15.27 ■ **Glutamine regulation.** The NtrB-NtrC two-component system regulates glutamine synthetase (GlnA) expression. GlnB, D, E regulate GlnA activity.

this very complex regulatory system. As you study the system, imagine how such molecular controls interconnect all the enzymes and regulators throughout a bacterium.

Low glutamine. NtrB and NtrC form a two-component signal transduction system (see Chapter 10). In **Figure 15.27**, when glutamine levels are low, the sensor kinase NtrB autophosphorylates (NtrB-P) and then phosphorylates the response regulator NtrC (step 1). NtrC-P regulator binds to an enhancer sequence, which can influence target gene expression from great distances (1 or 2 kb) (step 2). As a result of NtrC-P binding to its enhancer, the bacterium makes glutamine synthetase, assimilates nitrogen (in the form of ammonium ion), and incorporates the ammonium ion as glutamine. Maintaining GlnA requires active NtrC-P, and active NtrB-P. When nitrogen level increases, NtrB-P gets dephosphorylated by GlnB (next section).

Glutamine synthetase (GlnA) control is so important that after the enzyme is synthesized, its activity is also regulated. To regulate GlnA activity, GlnB gets modified by another protein, GlnD, which adds uridine monophosphate (UMP). GlnB-UMP then activates GlnA by removing AMP from the deactivated form, GlnA-AMP (discussed next).

High glutamine. When glutamine levels are in excess, the cell needs to stop making GlnA and inactivate whatever still remains. To halt *glnA* transcription, GlnB dephosphorylates NtrB-P, and then NtrB dephosphorylates NtrC-P (step 5). NtrC cannot bind the enhancer, and thus *glnA* transcription stops.

Meanwhile, GlnB also compels GlnE to add an AMP to GlnA (adenylylation, step 6). This modification inactivates whatever GlnA remains in the cell. The reaction is amplified by GlnD, which, under high glutamine, reverses its reaction, removing UMP from GlnB (step 7). The result is that ammonium ion no longer condenses with glutamate, and the glutamate/glutamine ratio is maintained.

Building Complex Amino Acids

Seven "fundamental" amino acids have relatively simple biosynthetic pathways. Others require longer pathways involving numerous enzymes. One amino acid produced

will be converted to gluta<u>mine</u>, and not enough glutamate will remain to build proteins. To prevent a glutamate shortage, excess glutamine signals the cell to stop making GlnA and inactivate whatever GlnA is already present. To maintain the balance requires an intricate series of regulators responding to <u>low glutamine</u> or to <u>high glutamine</u> (**Fig. 15.27**).

Glutamine acts at the levels of transcription and post-translational control. Here we describe only a portion of

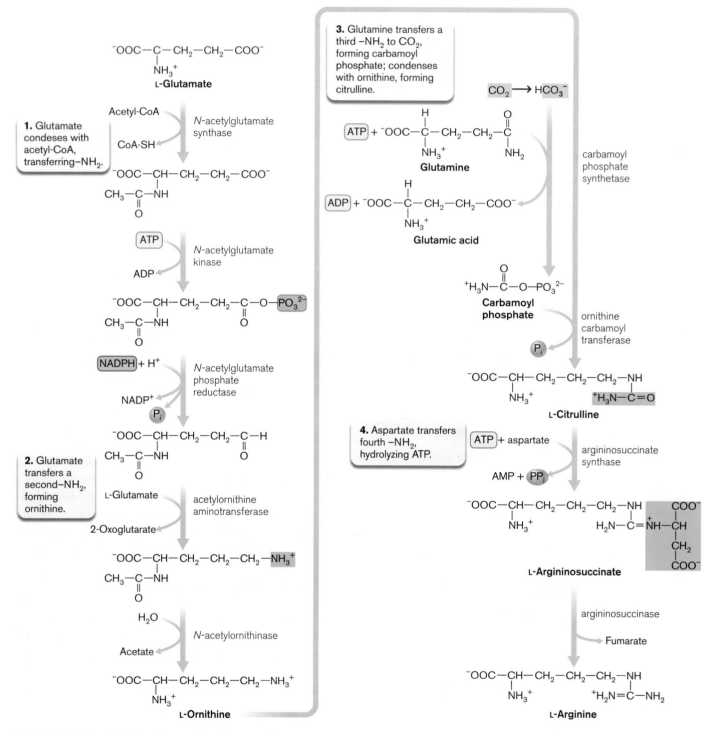

FIGURE 15.28 ■ **Arginine biosynthesis in *E. coli.*** In arginine biosynthesis, two glutamate molecules condense with acetyl-CoA, and nitrogen is donated by glutamate and glutamine.

by a complex pathway is arginine (**Fig. 15.28**). Bacterial arginine biosynthesis generally involves about a dozen different enzymes distributed among four to eight operons. Operon expression is regulated by an arginine repressor that binds the promoter of each operon to prevent transcription in the presence of sufficient arginine for cell needs. In some species, the arginine repressor also activates

enzymes of arginine catabolism, making the excess amino acid available as a carbon source.

Arginine biosynthesis. Arginine synthesis begins with the condensation of glutamate with acetyl-CoA (**Fig. 15.28,** step 1), transferring an amino group to the arginine precursor. Three more amino groups (–NH₂) are transferred subsequently by glutamate, glutamine, and aspartate. A second glutamate transfers an amino group to the arginine precursor (step 2), while its own carbon skeleton cycles back as 2-oxoglutarate (alpha-ketoglutarate). This transfer of NH_4^+ between two organic intermediates is an example of transamination. The original acetyl group from acetyl-CoA is then hydrolyzed, producing ornithine. Ornithine is a central intermediate, used by cells to synthesize proline and various polyamines, as well as arginine.

Thought Question

15.7 Which energy carriers (and how many) are needed to make arginine from 2-oxoglutarate?

Ornithine receives a third amino group (**Fig. 15.28**, step 3) from glutamine via conversion of CO_2 to carbamoyl phosphate [$H_2N–CO(PO_4)^{2-}$]:

$$CO_2 + H_2O \rightarrow H^+ + HCO_3^-$$

$$HCO_3^- + ATP + glutamine \rightarrow$$
$$H_2N–CO(PO_4)^{2-} + ADP + glutamate$$

Carbamoyl phosphate provides a way to assimilate one nitrogen plus one carbon into a structure.

The fourth nitrogen is acquired from aspartate in a two-step process requiring ATP release of pyrophosphate (PP_i) (**Fig. 15.28**, step 4). The release of fumarate, a TCA cycle intermediate, yields arginine.

Aromatic amino acids. The complexity of aromatic amino acids requires particularly energy-expensive biosynthesis. The number of different enzymes required, however, is minimized by the presence of a common core pathway branching to several different amino acids (**Fig. 15.29**).

The three aromatic amino acids—phenylalanine, tyrosine, and tryptophan—each require a common precursor, chorismate. The pathway to chorismate starts with simple glycolytic intermediates (phosphoenolpyruvate and erythrose 4-phosphate), but its synthesis requires 10 enzymes encoded in a single operon, *aroABCDEFGHKL*. From chorismate to phenylalanine or tyrosine requires three additional enzymatic steps. From chorismate to tryptophan, the most complex amino acid specified by the genetic code, requires another five enzymes, expressed by *trpABCDE*. Thus, at least 15 enzymes encoded by different genes are required to make tryptophan out of simple substrates.

Not surprisingly, the presence of tryptophan in the cell represses expression of its own biosynthetic enzymes. Repression of the *trp* operon includes two mechanisms: the Trp repressor, effective over moderate to high levels of tryptophan; and an RNA loop mechanism called attenuation, sensitive to lower levels of tryptophan (discussed in Chapter 10). Attenuation involves the destabilization of the transcription complex by formation of a specific stem loop on the nascent *trp* mRNA. The mechanism of attenuation in *E. coli* was discovered in 1981 by Charles Yanofsky, winner of a National Medal of Science in 2003. Several other amino acids commonly show attenuation in bacteria—particularly histidine, threonine, phenylalanine, valine, and leucine.

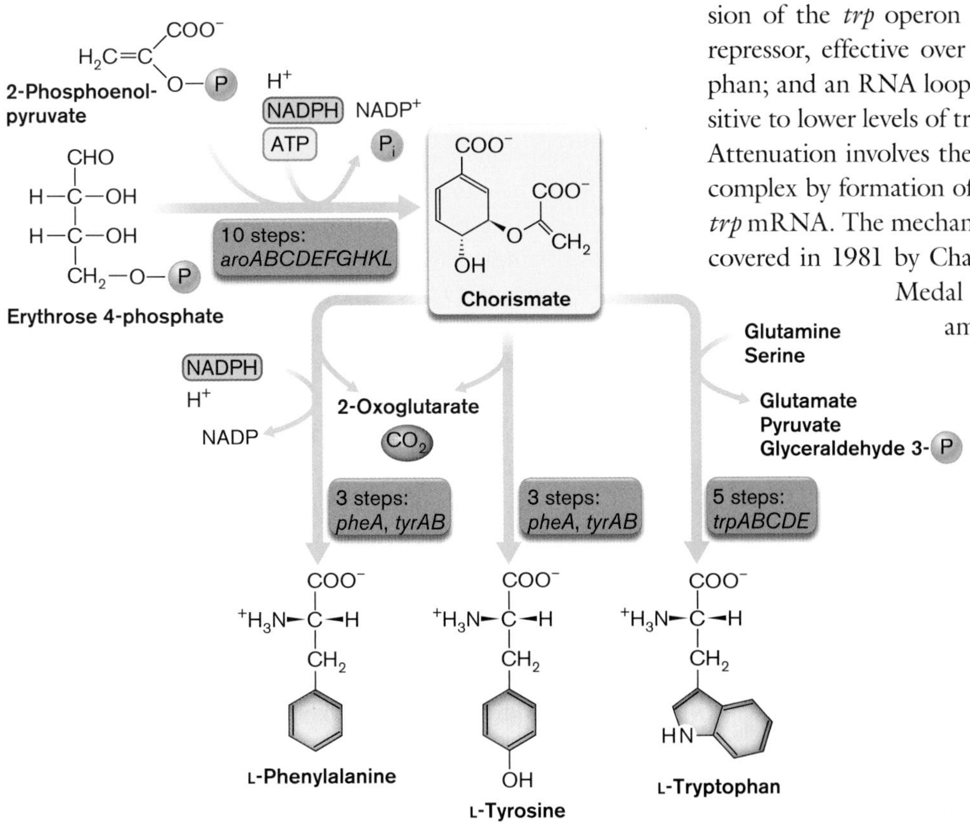

Purine and Pyrimidine Synthesis

Nitrogenous bases include purines and pyrimidines, the essential coding components of DNA and RNA. The accuracy of the entire genetic code requires accurate synthesis of these bases. Besides their role in nucleic acids, purine and pyrimidine nucleotides such as ATP serve as energy carriers.

FIGURE 15.29 ■ Biosynthesis of aromatic amino acids. Aromatic amino acids are assembled out of various carbon skeletons. In *E. coli*, the pathways to chorismate and tryptophan are encoded by operons of contiguous genes, *aro* and *trp*.

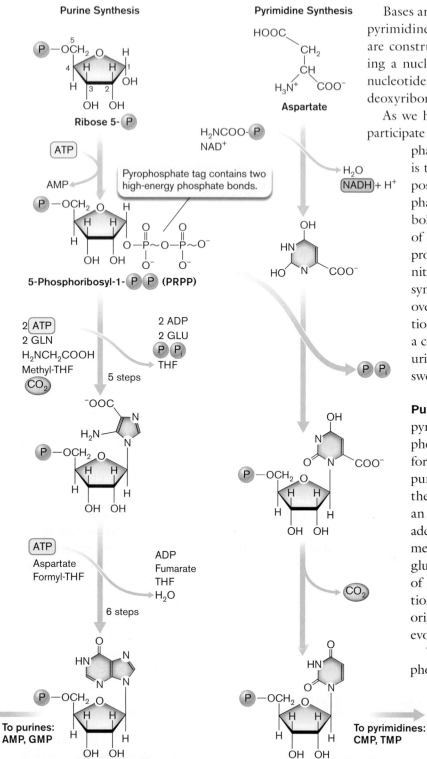

Purine Synthesis

Pyrimidine Synthesis

Bases are built onto ribose 5-phosphate. The purine and pyrimidine bases are not built as isolated units; rather, they are constructed on a ribose 5-phosphate substrate, forming a nucleotide (sugar-base-phosphate). Note that ribonucleotides are synthesized first and then converted to deoxyribonucleotides by enzymatic removal of the 2′ OH.

As we have seen, ribose 5-phosphate and related sugars participate in numerous metabolic pathways. Ribose 5-phosphate is directed into nucleotide synthesis when it is tagged with pyrophosphate at the carbon 1 (C-1) position, forming 5-phosphoribosyl-1-pyrophosphate (PRPP) (**Fig. 15.30**). PRPP is a major metabolic intermediate, the starting point for the synthesis of purine and pyrimidine nucleotides. PRPP is also produced by "scavenger" pathways, in which excess nitrogenous bases are broken down and recycled. Its synthesis is regulated by feedback inhibition to avoid overproduction of purines. In humans, overproduction of PRPP is one mechanism that leads to gout, a condition in which the purine breakdown product uric acid precipitates in the joints, causing painful swelling of wrists and feet.

Purine synthesis. The hydrolysis of PRPP releases pyrophosphate, thus spending two high-energy phosphate bonds. This irreversible reaction drives forward the subsequent reactions to make the purine (**Fig. 15.30**, left). To construct the purine, the pyrophosphate at carbon 1 (C-1) is replaced by an amine from glutamine. The purine is built up by addition of a series of single-carbon groups (formyl, methyl, and CO_2) alternating with nitrogens from glutamine and carbamoyl phosphate. The number of single-carbon assimilations, including assimilation of CO_2, is striking. It may reflect the ancient origin of the purine synthetic pathway, which evolved in CO_2-fixing autotrophs.

The first purine constructed is inosine monophosphate. Inosine is the "wobble" purine found at the third position of tRNA anticodons. Inosine monophosphate is subsequently converted to adenosine monophosphate (AMP) or guanosine monophosphate (GMP).

Pyrimidine synthesis. The pyrimidine is built by a slightly different route (**Fig. 15.30**, right). First, the six-membered pyrimidine ring forms from aspartate plus carbamoyl phosphate. The pyrimidine ring then displaces the pyrophosphate of PRPP, attaching by a nitrogen to the ribosyl carbon 1. The first pyrimidine built is uracil (as UMP), which can be converted to cytosine or thymine—as cytidine monophosphate (CMP) or thymidine monophosphate (TMP), respectively.

FIGURE 15.30 ■ Biosynthesis of purines and pyrimidines. Both purines and pyrimidines are built onto ribose 5-phosphate. Ribose 5-phosphate is directed into purine and pyrimidine biosynthesis by the activating step of ATP → AMP, converting the sugar to 5-phosphoribosyl-1-pyrophosphate (PRPP). The purine ring is built up out of successive additions of amines, one-carbon units, plus a formyl group from formyltetrahydrofolate (formyl-THF). The purine ring is modified to yield AMP and GMP. Likewise, the pyrimidine ring is modified to yield CMP or TMP.

Nonribosomal Peptide Antibiotics

In addition to the standard amino acids used by ribosomes, there exist hundreds of different amino acids used by cells for functions such as the cross-bridges of peptidoglycan (discussed in Chapter 3). Some of these are made by one-step modification of standard amino acids, such as epimerization (altering the chirality from L to D) or halogenation with chlorine or fluorine. Others have complex carbon skeletons, including unsaturated bonds and three-membered rings. Both standard and nonstandard amino acids are used by actinomycetes to build secondary metabolites with antimicrobial activity. These secondary metabolites are called **nonribosomal peptide antibiotics** because their peptide backbone is constructed by an enzyme without ribosomes. Nonribosomal peptide antibiotics are synthesized by modular "assembly-line" enzymes, analogous to those that build polyketides. An example is vancomycin, the main antibiotic used against the lethal hospital-borne pathogen *Clostridium difficile*. Vancomycin biosynthesis is presented in **eTopic 15.4**.

Amino acids also serve as precursors to form larger structures such as vitamins. For example, bacteria may assemble eight molecules of glutamic acid into a giant ring structure called a tetrapyrrole. Tetrapyrrole derivatives include essential cofactors of energy transduction and biosynthesis, such as chlorophylls, hemes of cytochromes and hemoglobin, and vitamin B_{12}. Tetrapyrrole biosynthesis and its control by riboswitches are described in **eTopics 15.5 and 15.6**.

To Summarize

- **Amino acid biosynthesis** requires numerous different enzymes to catalyze many unique conversions. Structurally related amino acids branch from a common early pathway.

- **Metabolic intermediates** from glycolysis and the TCA cycle initiate amino acid biosynthetic pathways.

- **Ammonium ion is assimilated by TCA intermediates**, such as 2-oxoglutarate into glutamate. Glutamate assimilates ammonium ion to form glutamine. Transamination is the donation of NH_4^+ from one amino acid

to another, such as glutamine transferring ammonia to 2-oxoglutarate to make aspartate.

- **Arginine biosynthesis** requires multiple steps of NH_3 transfer and carbon skeleton condensation.

- **Aromatic amino acids** are built from a common pathway that branches out. Their biosynthesis is regulated tightly at both transcriptional and translational levels.

- **Purines are built as nucleotides** attached to a ribose phosphate. Several single-carbon groups are assimilated, including CO_2—a phenomenon suggesting an ancient pathway. Pyrimidines are made from aspartate, and then added onto PRPP.

- **Nonribosomal peptide antibiotics** are built by modular enzymes analogous to those that build polyketides.

Concluding Thoughts

In living cells, no enzymatic pathway occurs in isolation. All pathways of energy acquisition and biosynthesis occur together, sharing common substrates and products. For example, acetyl-CoA is produced by glycolysis or by fatty acid degradation; it is then utilized by pathways that synthesize fatty acids, amino acids, and nucleobases. Other precursors to these molecules arise from the TCA cycle. At the same time, cells need the TCA cycle to yield energy through catabolism. The microbial cell regulates these competing needs on an extremely rapid timescale: Within seconds of the disappearance or appearance of a nutrient, such as an amino acid, its biosynthesis is turned on or shut off.

A given microbe needs to build as much biomass as it can by the most efficient means available. For a pathogen growing within a host, this means using preformed host compounds for rapid growth and replication. Free-living microbes, however, face intense competition for organic nutrients; so, in addition, they need to fix essential elements into readily accessible forms. Marine microbes build complex biological molecules out of single-carbon and single-nitrogen sources, but they also obtain molecules such as vitamins from their community. In the soil, actinomycetes build antibiotics to eliminate competitors and extract their nutrients. Those antibiotics offer powerful tools for human medicine.

Microbial metabolism has many applications in food preparation and preservation, and in industrial production of pharmaceuticals. Food and industrial microbiology are discussed in Chapter 16. Later, in Chapters 21 and 22, we will see how microbial metabolism contributes to communities of organisms, and ultimately to Earth's global cycles.

CHAPTER REVIEW

Review Questions

1. What are the sources of substrates for biosynthesis? From what kinds of pathways do they arise?
2. How do microbial species economize by synthesizing only the products they need? Cite long-term as well as short-term mechanisms.
3. Compare and contrast the different cycles of carbon dioxide fixation. What classes of organisms conduct each type?
4. How is ribulose 1,5-bisphosphate consumed and re-formed through the Calvin cycle? What key products emerge to form sugars and amino acids?
5. How do oxygenic phototrophs maintain CO_2 at sufficient levels to conduct the Calvin cycle?
6. Explain the cyclic process of chain extension in fatty acid biosynthesis. Explain the generation of occasional unsaturated "kinks" in the chain.
7. Explain the different kinds of regulation of fatty acid biosynthesis.
8. What are the different environmental sources of nitrogen? Explain how and why microbes use these different sources.
9. Explain the process by which nitrogenase converts N_2 to $2NH_4^+$. Why is H_2 formed?
10. Explain the different ways that microbes maintain anaerobic conditions for nitrogenase.
11. Explain the molecular basis for regulation of nitrogen fixation and nitrogen scavenging.
12. Compare and contrast the general scheme of the biosynthesis of amino acids with that of fatty acids.
13. Outline the interconversions of 2-oxoglutarate, glutamate, and glutamine that provide nitrogen for amino acid biosynthesis.
14. Compare and contrast the processes of purine and pyrimidine biosynthesis. What is the role of the sugar ribose in each case?

Thought Questions

1. Why do some soil microbes fix N_2, whereas others depend on available nitrate, ammonium ion, or organic nitrogen? What environmental conditions would favor each strategy?
2. Disease-causing bacteria vary widely in their ability to synthesize amino acids. What kinds of pathogens would be likely to make their own amino acids, and what kinds would not?
3. *Mycoplasma genitalium*, an organism growing in human skin, lacks the ability to synthesize fatty acids. How do you think it makes its cell membrane? How could you test this?

Key Terms

acyl carrier protein (ACP) (584)
anabolism (568)
anaplerotic reaction (579)
biosynthesis (568)
Calvin cycle (Calvin-Benson cycle, Calvin-Benson-Bassham cycle, CBB cycle) (573)
carbon-concentrating mechanism (CCM) (576)
carbon dioxide fixation (572)
carboxysome (578)
chemosynthesis (569)

folate (581)
gluconeogenesis (571, 580)
glutamate dehydrogenase (GDH) (596)
glutamate synthase (glutamine: 2-oxoglutarate aminotransferase) (GOGAT) (596)
glutamine synthetase (GS or GlnA) (596)
Haber process (590)
heterocyst (590)
3-hydroxypropionate cycle (583)

modular enzyme (585)
nitrogenase (591)
nonribosomal peptide antibiotic (601)
photoautotrophy (569)
photosynthesis (569)
reductive acetyl-CoA pathway (581)
reductive pentose phosphate cycle (573)
reductive (reverse) TCA cycle (580)
Rubisco (574)
transamination (596)

Recommended Reading

Bertranda, Erin M., John P. McCrow, Ahmed Moustafa, Hong Zheng, Jeffrey B. McQuaida, et al. 2015. Phytoplankton–bacterial interactions mediate micronutrient colimitation at the coastal Antarctic sea ice edge. *Proceedings of the National Academy of Sciences USA* **112**:9938–9943.

Gollnick, Paul, Paul Babitzke, Alfred Antson, and Charles Yanofsky. 2005. Complexity in regulation of tryptophan biosynthesis in *Bacillus subtilis*. *Annual Review of Genetics* **39**:47–68.

Herter, Sylvia, Jan Farfsing, Nasser Gad'On, Christoph Rieder, Wolfgang Eisenreich, et al. 2001. Autotrophic CO_2 fixation by *Chloroflexus aurantiacus*: Study of glyoxylate formation and assimilation via the 3-hydroxypropionate cycle. *Journal of Bacteriology* **183**:4305–4316.

Iancu, Cristina V. H., Jane Ding, Dylan M. Morris, D. Prabha Dias, Arlene D. Gonzales, et al. 2007. The structure of isolated *Synechococcus* strain WH8102 carboxysomes as revealed by electron cryotomography. *Journal of Molecular Biology* **372**:764–773.

Kerfeld, Cheryl A., William B. Greenleaf, and James N. Kinney. 2010. The carboxysome and other bacterial microcompartments. *Microbe* **5**:257–263.

Pande, Samay, Shraddha Shitut, Lisa Freund, Martin Westermann, Felix Bertels, et al. 2015. Metabolic cross-feeding via intercellular nanotubes among bacteria. *Nature Communications* **6**:6238.

Rehm, Bernd H. A. 2010. Bacterial polymers: Biosynthesis, modifications and applications. *Nature Reviews. Microbiology* **8**:578–592.

Schirmer, Andreas, Rishali Gadkari, Christopher D. Reeves, Fadia Ibrahim, Edward F. DeLong, et al. 2005. Metagenomic analysis reveals diverse polyketide synthase gene clusters in microorganisms associated with the marine sponge *Discodermia dissoluta*. *Applied and Environmental Microbiology* **71**:4840–4849.

Xu, Ying, Roland D. Kersten, Sang-Jip Nam, Liang Lu, Abdulaziz M. Al-Suwailem, et al. 2012. Bacterial biosynthesis and maturation of the didemnin anticancer agents. *Journal of the American Chemical Society* **134**:8625–8632.

Yang, Yu-Liang, Yuquan Xu, Paul Straight, and Pieter C. Dorrestein. 2009. Translating metabolic exchange with imaging mass spectrometry. *Nature Chemical Biology* **5**:885–887.

Zimmer, Daniel P., Eric Soupene, Haidy L. Lee, Volker F. Wendisch, Arkady B. Khodursky, et al. 2000. Nitrogen regulatory protein C–controlled genes of *Escherichia coli*: Scavenging as a defense against nitrogen limitation. *Proceedings of the National Academy of Sciences USA* **97**:14674–14679.

CHAPTER 16
Food and Industrial Microbiology

Microbes have nourished humans for centuries, generating cheese, bread, wine and beer, tempeh, and soy sauce. Yet from the moment of harvest, microbes on the food's surface or from the air colonize the food. Their growth can render food rancid or putrid. The need for food preservation has led to drying, salting, smoking, and adding spices, all of which retard microbial growth. The principles of food microbiology extend to a growing field of industrial microbiology. Industrial microbiology includes the development of microbial antibiotics and enzymes. Giant fermentors grow transgenic microbes that make human proteins such as insulin and anti-inflammatory factors.

CURRENT RESEARCH highlight

Immobilized lipase from an Antarctic psychrophile. The enzyme lipase is used in industrial microbiology to interconvert alcohols and esters. Lipase reactions can separate the right and left mirror forms of a drug precursor. A particularly valuable lipase was identified in *Candida antarctica*, a fungus isolated from Lake Vanda in Antarctica. The Antarctic enzyme functions over a broad range of temperature for industrial processes (10°C–110°C). The enzyme can be immobilized on large carrier molecules composed of acrylic resin. The product of the reaction can then be purified easily by simply removing the carrier.

Source: Novozymes Product Sheet "Enzymes for Biocatalysis, Immobilized Lipases."

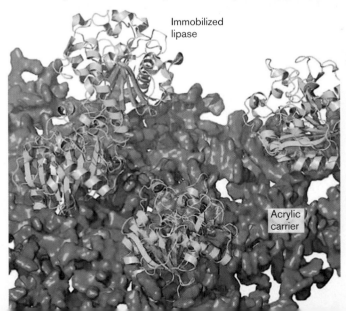

Immobilized lipase

Acrylic carrier

AN INTERVIEW WITH

MICHAEL BARNHART, INDUSTRIAL SCIENTIST, NOVOZYMES

COURTESY OF MICHAEL BARNHART

What is the most exciting thing about working at a biotechnology company?
Working in the biotechnology private sector gives me an opportunity to apply cutting-edge advances in science directly to issues that impact our environment and our health. Watching these discoveries become solutions that society can immediately apply to impact the world is very meaningful and fulfilling.

What advances do you foresee for microbial products?
The development of "omics" technologies and the rapid lowering of sequencing costs provide scientists, both public and private, the tools to explore the intricacies between microbes and their environment. The results will lead to increased understanding of the microbes and their effects on humans, plants, and the environment, and provide targeted treatments or amendments for diseases, improved health, sustainable crop production, and a healthier planet.

The industrial enzyme lipase B (see the Current Research Highlight) is just one of many lucrative commercial products of the microbial **fermentation industry**, a term for the culturing of microbes to make products for market. Today, bioprospecting reveals microbial sources for products in remote locations around the world, particularly those of extreme and unusual environments. Lipase B, with its wide temperature range, was isolated from Lake Vanda, a hypersaline Antarctic lake covered by 3 meters of ice. Yet the deep water of the lake maintains temperature as high as 25°C, sustained in part by solar heating. Thus, microbes from this lake experience a wide range of temperatures; and their enzymes may show activity over a wide temperature range, which is useful for industrial processes.

While today's industry involves sophisticated biotechnology, the roots of the fermentation industry reach back to 5000 BCE, the date of pottery jars excavated from a Neolithic mud-brick kitchen in the Zagros Mountains of modern Iran (**Fig. 16.1A**). The jars contain residue showing chemical traces of grapes fermented to wine. Besides making wine, Neolithic people leavened bread, another staple food requiring microbial growth. For thousands of years, microbial fermentation has been a daily part of human life and commerce.

Today, microbial products are made by bacteria grown in giant fermentation vessels (**Fig. 16.1B**) (discussed in Section 16.6). Industrial "fermentation" generally uses respiratory metabolism to maximize microbial growth. Microbial growth drives companies such as DuPont Industrial Biosciences, now earning hundreds of millions of dollars in annual revenues from engineering microbial enzymes for industrial use. DuPont's microbial products range from contact lens cleansing agents to fashion finishes for denim at manufacturing centers around the world. The fermentation industry originated from long-standing microbial associations with human food. Until the relatively recent invention of steam-pressure sterilization, all foods contained live microbes. Microbial metabolism could spoil food—or it could improve the food by adding flavor, preserving valuable nutrients, and preventing growth of pathogens. As we learned in Chapter 1, one of the first great microbiologists, Louis Pasteur, began his career as a chemist investigating fermentation in winemaking. In Chapter 16 we see how the many kinds of microbial biochemistry presented in Chapters 13–15 give rise to the characteristics of food so familiar to us—from the taste of cheese and chocolate to the rising of bread dough and the physiological effects of alcoholic beverages. Industrial research continues to improve food through <u>food biochemistry</u>, discovering the molecular basis for the flavor and texture of microbial foods; <u>food preservation</u>, eliminating undesirable microbial decay by preservative methods; and <u>food engineering</u>, improving the quality, shelf life, and taste of microbial foods.

16.1 Microbes as Food

Certain kinds of microbial bodies have long been eaten as food, especially the fruiting bodies of fungi and the fronds of marine algae. Single-celled algae and cyanobacteria are also used as food supplements. These microbes can provide important sources of protein, vitamins, and minerals.

FIGURE 16.1 ■ The fermentation industry: then and now.
A. A Neolithic jar that contained wine, from 5000 BCE. Excavated at Hajji Firuz Tepe, Iran. **B.** An industrial fermentation apparatus for growing microbes to produce enzyme products, at GeneFerm Biotechnology Co., Ltd., in Taipei, Taiwan.

Edible Fungi

In the children's classic *Homer Price*, by Robert McCloskey, settlers on the Ohio frontier save themselves from starvation when they discover "forty-two pounds of edible fungus, in the wilderness a-growin'." Fungal fruiting bodies, multicellular reproductive structures that generate spores, are commonly known as mushrooms (see Chapter 20). Mushrooms offer a flavorful source of protein and minerals. The protein content of edible mushrooms can be as high as 25% dry weight, comparable to that of whole milk (percent dry weight), and includes all essential dietary amino acids.

Mushrooms contributed to the survival of preindustrial humans,

FIGURE 16.2 ■ Commercial mushroom production. Farming of *Agaricus bisporus* mushrooms on horse manure compost, by Penn State graduate student Kelly Ivors.

while killing those unlucky enough to consume varieties that were poisonous. Less than 1% of mushrooms are poisonous, but those few are deadly, such as the amanita, or "destroying angel," which produces toxins including the RNA polymerase inhibitor alpha-amanitin. People who ingest the amanita typically die of liver failure. Other kinds of mushrooms actually invite consumption in order to disperse their spores. For example, the underground truffles prized in European cooking produce odorant molecules that attract animals to dig them up. Truffle odorants include sex pheromones such as androstenol, found in human perspiration.

Many varieties of edible mushrooms are farmed and marketed for human fare. **Figure 16.2** shows the culturing of *Agaricus bisporus*, the mushroom variety most commonly sold in the United States as button mushrooms and portobellos. Button mushrooms are harvested at an early stage, whereas portobellos are harvested later, when the gills are fully exposed and some of the moisture has evaporated. The decrease in moisture in portobellos concentrates the flavor and gives them a dense, meaty texture; the mushrooms are often served in gourmet sandwiches as a vegetarian alternative to hamburger. *Agaricus* culture was first developed around 1700 in France, where the mushrooms were grown in underground caves. In modern mushroom farms, *Agaricus* mushrooms are cultured on composted horse manure or chicken manure in chambers controlled for temperature and humidity.

Other mushroom varieties from China and Japan are grown on logs or wooden blocks. Wood-grown mushrooms include the strong-flavored *Lentinula edodes* (black forest mushroom, or shiitake), *Pleurotus* (oyster mushrooms), and *Flammulina velutipes* (enoki mushrooms), with long, thin, white stalks and delicate flavor.

Edible Algae

Several kinds of seaweed (marine algae) are cultivated, most notably in Japan. The red alga *Porphyra*, a eukaryotic "true alga" (discussed in Chapter 20), forms large multicellular fronds cultured for **nori** (**Fig. 16.3**). Nori is best known for its use in wrapping rice, fish, and vegetables to form sushi. Another edible alga, kelp, is the source of alginate, an important food additive (see Section 16.6).

For nori production, the red algae are grown as "seeds," or starter cultures, in enclosed tanks. The starter cultures are then distributed on nets in a protected coastal area, usually an estuary. The cultures grow until they hang heavy from the nets, when they are harvested for processing into sheets. The sheets are toasted, turning dark green.

Edible Bacteria and Yeasts

Since prehistoric times, consumption of single-celled fungi or yeasts has provided a supplemental source of protein and vitamins. Yeasts such as *Saccharomyces* species were grown to high concentration in fermented milks and grain beverages. Traditional beers contained only a low percentage of alcohol with a thick suspension of nutrient-rich yeasts. Especially important was the content of vitamin B_{12}, an essential substance for the human diet that cannot be obtained from plant sources.

Few bacteria are edible as isolated organisms, mainly because their small cells contain a relatively high proportion of DNA and RNA. The high nucleic acid content is a problem because nucleic acids contain purines, which the human digestive system converts to uric acid. Because we humans lack the enzyme urate oxidase, the uric acid cannot be metabolized. Consumption of more than 2 grams per day of nucleic acids causes uric acid to precipitate, resulting

FIGURE 16.3 ■ Nori production for sushi.
A. Nori grows from nets in seawater. **B.** Toasted sheets of nori wrap rice, vegetables, and fish to make sushi.

A.

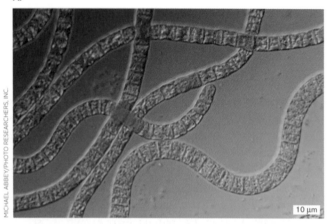

B.

FIGURE 16.4 ▪ **Single-celled protein.** **A.** *Spirulina* filaments are processed into protein-rich food supplements (LM, 250x). **B.** Scene from *Soylent Green* (1973), a science fiction film about a future Earth whose overgrown population is forced to eat a form of food called Soylent, supposedly based on soybeans and single-celled protein.

in painful conditions such as gout and kidney stones. For this reason, most bacteria can be consumed only as a minor component of food mass, as in fermented foods.

An exception is the cyanobacterium *Spirulina*, whose purine content is low enough to include as a modest part of the human diet (**Fig. 16.4A**). *Spirulina* consists of spiral-shaped cells that grow photosynthetically in freshwater. *Spirulina* is sold as a food additive rich in protein, vitamin B_{12}, and minerals. It also contains antioxidant substances that may prevent cancer. *Spirulina* is grown with illumination in special ponds lined for food production. The final product is collected and vacuum-dried to form a dark green powder of flour-like consistency.

In the food industry, *Spirulina* is classified as a form of **single-celled protein**, a term for edible microbes of high food value. Other kinds of single-celled proteins include eukaryotes such as yeast and algae. The twentieth century saw the development of single-celled protein as a food source for impoverished populations. The yeast *Saccharomyces*

cerevisiae was grown for protein by Germany during World War I, using inexpensive molasses for culture; and during World War II, *Candida albicans* was grown on paper mill wastes. Single-celled protein was later promoted by Western countries as a food for rapidly expanding populations of developing countries. This idea inspired the science fiction film *Soylent Green* (1973), in which people of a future overpopulated Earth are forced to eat "Soylent," a food supposedly based on soybeans and single-celled protein, though its true source is recycled humans (**Fig. 16.4B**).

To Summarize

- **Fungal fruiting bodies** such as mushrooms and truffles are consumed as a protein-rich food.

- **Edible algae** include nori (toasted red algae, used to wrap sushi) and kelp.

- *Spirulina* is an edible cyanobacterium, a source of single-celled protein. Most bacteria, however, are inedible in isolation because of their high concentration of nucleic acids.

- **Yeasts** have been grown as an economical protein and vitamin supplement.

16.2 Fermented Foods: An Overview

Virtually all human cultures have developed varieties of **fermented foods**, food products that are modified biochemically by microbial growth. The purposes of food fermentation include the following:

- **To preserve food.** Certain microbes, particularly the lactobacilli, metabolize only a narrow range of nutrients before their waste products build up and inhibit further growth. Typically, the waste fermentation products that limit growth are carboxylic acids, ammonia (alkaline), or alcohol. Buildup of these substances renders the product stable for much longer than the original food substrate.

- **To improve digestibility.** Microbial action breaks down fibrous macromolecules and tenderizes the product, making it easier for humans to digest. Meat and vegetable products are tenderized by fermentation.

- **To add nutrients and flavors.** Microbial metabolism generates vitamins, particularly vitamin B_{12}, as well as flavor molecules such as esters and sulfur compounds.

Different societies have devised thousands of different kinds of fermented foods. Examples are given in **Table 16.1**.

TABLE 16.1 Fermented Foods and Beverages

Product (origin)	Description	Microbial genera
Acid fermentation of dairy products, meat, and fish		
Buttermilk (Asia, Europe)	Bovine milk; lactic fermented	*Lactococcus*
Yogurt (Asia, Europe)	Bovine milk; lactic fermented and coagulated	*Lactobacillus, Streptococcus*
Kefir (Russia)	Bovine or sheep's milk; mixed fermentation, acid with some alcoholic	*Lactobacillus, Streptococcus*, yeasts, others
Sour cream (Asia, Europe)	Bovine cream; lactic fermented	*Lactococcus*
Cheese (Asia, Europe)	Milk (bovine, sheep, or goat); lactic fermented, coagulated, and pressed; in some cases cooked; mold ripened (spiked or coated)	Acid fermentation: *Lactobacillus, Streptococcus, Propionibacterium*; Mold ripening: *Penicillium*
Sausage (Asia, Europe)	Ground beef and/or pork encased with starter culture; lactic fermented, dried, or smoked	*Lactobacillus, Pediococcus, Staphylococcus*, others
Fermented fish (Africa, Asia)	Many kinds of fish; mixed fermentation, acid and amines produced	Halotolerant bacteria and haloarchaea
Acid fermentation of vegetables		
Tempeh (Indonesia)	Soybean cakes; fungal fermentation	*Rhizopus oligosporus*
Miso (Japan)	Soy and rice paste; fungal fermentation	*Aspergillus*
Soy sauce (China)	Extract of soy and wheat; fungal fermentation, brined, bacterial fermentation	*Aspergillus*, followed by halotolerant bacteria and haloarchaea
Kimchi (Korea)	Cabbage, peppers, and other vegetables, with fish paste, brined; container is buried	*Leuconostoc*, other bacteria
Sauerkraut (Europe)	Cabbage; fermented, making lactic and acetic acids, ethanol, and CO_2	*Leuconostoc, Pediococcus, Lactobacillus*
Pickled foods (Asia)	Cucumbers, carrots, fish; brined, then fermented	*Leuconostoc, Pediococcus, Lactobacillus*
Kenkey (western Africa)	Maize; fermented, wrapped in banana leaves and cooked	Unknown
Chocolate (South America)	Cocoa beans; soaked and fermented before processing to chocolate	*Lactobacillus, Bacillus, Saccharomyces*
Alkaline fermentation		
Pidan (China, Japan)	Duck eggs; coated in lime (CaO), aged, producing ammonia and sulfur odorants	*Bacillus*
Natto (China, Japan)	Whole soybeans; fermented	*Bacillus natto*
Dawadawa (Africa)	Locust beans; fermented	*Bacillus*
Ogiri (Africa)	Melon seed paste; fermented	*Bacillus*
Leavened bread dough		
Yeast breads (Asia, Europe)	Ground grain; dough leavened by yeast	*Saccharomyces*
Sourdough (Egypt)	Ground grain; dough leavened by starter culture from previous dough	*Saccharomyces, Torulopsis, Candida*
Injera (Ethiopia)	Ground teff grain; dough leavened and fermented 3 days by organisms from the grain	*Candida*
Alcoholic fermentation		
Wine (Asia, Europe)	Grape juice; yeast fermented, followed by malolactic fermentation	*Saccharomyces, Oenococcus*
Beer (Asia, Europe, Africa)	Barley and hops; yeast fermented	*Saccharomyces*
Sake (Japan)	Rice extract; yeast fermented	*Saccharomyces*
Tequila (Mexico)	Blue agave; yeast fermented and distilled	*Saccharomyces*
Whiskey (United Kingdom)	Barley or other grains or potatoes; fermented and distilled	*Saccharomyces*

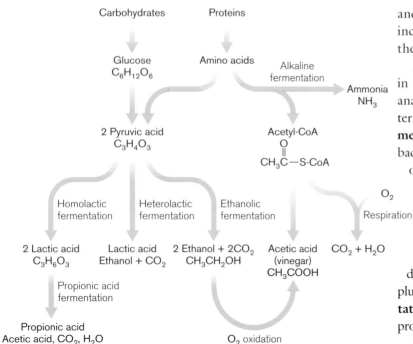

FIGURE 16.5 ▪ **Major chemical conversions in fermented foods.**

Fermented foods that are produced commercially include dairy products such as cheese and yogurt, soy products such as miso (from Japan) and tempeh (from Indonesia), vegetable products such as sauerkraut and kimchi, and various forms of cured meats and sausages. Alcoholic beverages are made from grapes and other fruits (wine), grains (beer and liquor), and cacti (tequila). Other kinds of foods require microbial treatment for special purposes, such as leavening by yeast (for bread) or cocoa bean fermentation (for chocolate). Besides commercial production, numerous fermented products are homemade by traditional methods thousands of years old. Such products are known as "traditional fermented foods." Occasionally, a traditional fermented food enters commercial production and becomes widespread. For example, soy sauce, a traditional Japanese product, was marketed by the Kikkoman company and achieved global distribution in the twentieth century.

The nature of fermented foods depends on the quality of the fermented substrate, as well as on the microbial species and the type of biochemistry performed. Traditional fermented foods usually depend on **indigenous microbiota**—that is, microbes found naturally in association with the food substrate; or on starter cultures derived from a previous fermentation, as in yogurt or sourdough fermentation. Commercial food-fermenting operations use highly engineered microbial strains to inoculate their cultures, although in some cases indigenous microbiota still participate. For example, wines

and cheeses aged in the same caves for centuries often include fermenting organisms that persist in the air and the containers used.

Major classes of fermentation reactions are summarized in **Figure 16.5**. The most common conversions involve anaerobic fermentation of glucose, as discussed in Chapter 13. Glucose is fermented to lactic acid (**lactic acid fermentation**) in cheeses and sausages, primarily by lactic acid bacteria such as *Lactobacillus*. A second-stage fermentation of lactic acid to propionic acid (**propionic acid fermentation**) by *Propionibacterium* generates the special flavor of Swiss and related cheeses. Some kinds of vegetable fermentation, as in sauerkraut, involve production of lactic acid and CO_2, with small amounts of acetic acid and ethanol. This **heterolactic fermentation** is conducted primarily by *Leuconostoc*. Fermentation to ethanol plus carbon dioxide without lactic acid (**ethanolic fermentation**) is conducted by yeast during bread leavening and production of alcoholic beverages.

In some food products, particularly those fermented by *Bacillus* species, proteolysis and amino acid catabolism generate ammonia in amounts that raise pH (**alkaline fermentation**). For example, alkaline fermentation forms the soybean product natto. Other products require the growth of mold, such as the mold-spiked Roquefort cheese and the soy product tempeh. Mold growth requires some oxygen for aerobic respiration. Respiration must be limited, however, to avoid excessive decomposition of food substrate and loss of food value.

Note that the conversions cited here include only the major reactions in achieving the food product. In addition, thousands of minor or secondary reactions occur, some of which produce tiny amounts of potent odorants and flavors. While these flavor molecules have less nutritional consequence than the main fermentation products, they provide the complex, "sophisticated" taste for which fine cheeses, wines, and soy products are known.

Thought Questions

16.1 Why do the lipid components of food experience relatively little breakdown during anaerobic fermentation?

16.2 Why does oxygen allow excessive breakdown of food, compared with anaerobic processes?

To Summarize

- **Anaerobic fermentation of food** enhances preservation, digestibility, nutrient content, and flavor.

- **Acid fermentations** lead to organic acid fermentation products, such as lactate and propionate.

- **Alkaline fermentations** produce ammonia and break down proteins to peptides.

- **Ethanolic fermentation** produces ethanol and carbon dioxide.

- **Lipids are relatively stable** under anaerobic conditions of fermentation.

16.3 Acid- and Alkali-Fermented Foods

Many food fermentations produce acids or bases. An acid or base serves as an effective preservative because the pH change is unlikely to be reversed, and because animal or plant bodies grown at near-neutral pH are unlikely to support growth of acidophiles or alkaliphiles, which grow at extreme pH conditions.

Acid Fermentation of Dairy Products

The conversion of milk to solid or semisolid fermented products dates far back in human civilization. The practice of milk fermentation probably arose among herders who collected the milk of their pack animals but had no way to prevent the rapid growth of bacteria. The milk had to be stored in a portable container such as the stomach of a slaughtered animal. After hours of travel, the combined action of lactic acid–producing bacteria and stomach enzymes caused the coagulation of milk proteins into **curd**. The curd naturally separated from the liquid portion, called **whey**. Both curds and whey can be eaten, as in the nursery rhyme "Little Miss Muffet." The curds, however, are particularly valuable for their concentrated protein content.

Curd formation. A **cheese** is any milk product from a mammal (usually cow, sheep, or goat) in which the milk protein coagulates to form a semisolid curd. Curd formation results from two kinds of processes: acidification, usually as a result of the microbial production of lactic acid; and treatment by proteolytic enzymes such as rennet. The curd may then be separated and processed to varying degrees, depending on the type of cheese.

How does milk coagulate? The major organic components of cow's milk are butterfat (about 4% unless skimmed), protein (3.3%), and the sugar lactose (4.7%). Milk starts out at about pH 6.6, very slightly acidic. At this pH, the milk proteins are completely soluble in water; otherwise they would clog the animal's udder as the milk came out. Fermentation generally begins with bacteria

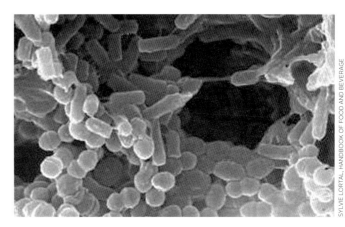

FIGURE 16.6 ■ Bacterial community within Emmentaler cheese. Bacterial species include *Lactobacillus helveticus* (rods, 2.0–4.0 µm in length) and *Streptococcus thermophilus* (cocci).

such as *Lactobacillus* and *Streptococcus* (**Fig. 16.6**). As bacteria ferment lactose to lactic acid, the pH starts to decline. The dissociation constant of lactic acid (pK_a = 3.9) allows greater deprotonation than with other fermentation products, such as acetate (pK_a = 4.8). Thus, lactic acid rapidly acidifies the milk product to levels that halt further growth of bacteria. Halting bacterial growth minimizes the oxidation of amino acids, thereby maintaining food quality.

Milk contains micelles (suspended droplets) of hydrophobic proteins called caseins. As the pH of milk declines below pH 5, the acidic amino acid residues of caseins become protonated, eventually destabilizing the tertiary structure. As the casein molecules unfold (or "denature"), they expose hydrophobic residues that regain stability by interacting with other hydrophobic molecules. The intermolecular interaction of caseins generates a gel-like network throughout the milk, trapping other substances, such as droplets of butterfat. This protein network generates the semisolid texture of **yogurt**, a simple product of milk acidified by lactic acid bacteria.

In most kinds of cheese formation, an additional step of casein coagulation is accomplished by proteases such as rennet. Rennet derives from the fourth stomach of a calf, although modern versions are made by genetically engineered bacteria. Calf rennet includes two proteolytic enzymes: chymosin and pepsin. Chymosin specifically cleaves casein into two parts, one of which is charged and water-soluble, the other hydrophobic. The hydrophobic portion forms a firmer curd than intact casein and results in the harder texture of solid cheeses. The water-soluble portion, about one-third of the total casein, enters the whey and is lost from the curd. Processing of some cheese varieties includes exposure to high temperature, which denatures even the whey protein, so it is retained in the curd.

FIGURE 16.7 ▪ Cheese varieties. A. Cottage cheese, an unripened perishable cheese. **B.** Emmentaler Swiss cheese, with eyes produced by carbon dioxide fermentation. **C.** Feta cheese, a soft cheese from goat's milk, preserved in brine. **D.** Roquefort, a medium-hard cheese ripened by spiking with *Penicillium roqueforti*.

Varieties of cheese. An extraordinary number of cheese varieties have been devised (**Fig. 16.7**). These fall into several categories based on particular steps in their production.

■ **Soft, unripened cheeses**, such as cottage cheese and ricotta, are coagulated by bacterial action, without rennet. The curd is cooked slightly, and the whey is partly drained, but their water content is 55% or greater. These cheeses spoil easily; there are no steps of aging, or **ripening**.

■ **Semihard, ripened cheeses**, such as Muenster and Roquefort, include rennet for firmer coagulation, and the curd is cooked down to a water content of 45%–55%. The cheese is aged for several months.

■ **Hard cheeses**, such as Swiss cheese and cheddar, are concentrated to even lower water content. Extra-hard varieties, such as Parmesan and Romano, have a water content as low as 20%. These cheeses are aged for many months, even several years.

■ **Brined cheeses**, such as feta, are permeated with brine (concentrated salt), which limits further bacterial growth and develops flavor. Harder cheeses, such as Gouda, may be brined at the surface.

■ **Mold-ripened cheeses** are inoculated with mold spores that germinate and grow during the ripening, or aging, process to contribute texture and flavor. The mold may be inoculated on the surface, to form a crust (as in Brie and Camembert), or it may be spiked deep into the cheese (as in blue cheese or Roquefort).

Cheese production. Commercial production of cheese involves a standard series of steps (**Fig. 16.8**). At each of these steps, choices of treatment lead to very different varieties. Key steps are illustrated for the example of Gouda cheese in **Figure 16.9**.

In the first step, the milk is filtered to remove particulate objects, such as straw, and microfiltered or centrifuged to remove potentially pathogenic bacteria and spores. Most

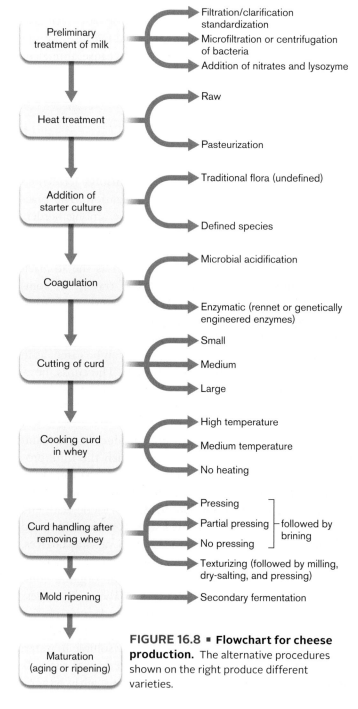

FIGURE 16.8 ▪ Flowchart for cheese production. The alternative procedures shown on the right produce different varieties.

A. Milk fermentation

B. Cheddaring the curd

C. Shaping the curd

D. Brining

FIGURE 16.9 ■ **Cheese production.** Gouda cheese is produced at the Henri Willig factory in the Netherlands. **A.** Milk is poured into a fermentation tub with a bacterial starter culture and rennet. **B.** The milk curd is cut, or "cheddared." **C.** The curds are shaped into round molds and then pressed to remove whey. **D.** The solidified curds are floated in brine to form the characteristic rind of Gouda cheese. The cheese then dries and ripens on the shelf.

modern production includes flash pasteurization (brief heating to 72°C) (discussed in Chapter 5), although some traditional cheeses continue to be made from unpasteurized milk. Unpasteurized milk in cheese has been linked to illness, particularly from *Listeria*, bacteria that grow at typical refrigeration temperatures.

The fermenting microbes are added as a **starter culture** (**Fig. 16.9A**). The starter was traditionally derived from a sample of previous fermentation, in which case the flora were undefined. Commercial cheese production now uses defined species. In all but the soft cheeses, bacterial coagulation and curd formation are supplemented by rennet or by genetically engineered proteases.

The solid curd is then cut, or **cheddared** (hence the name "cheddar" cheese) (**Fig. 16.9B**). The finer the pieces, the more whey that can be pressed out and the harder the cheese produced. Curd is then heat treated, with or without the whey; if whey is included, more protein is retained. Brining at this stage leads to a salty cheese, such as feta.

The pressed curd is then shaped into a mold (**Fig. 16.9C**), which determines the ultimate shape of the cheese. Before ripening (or aging), the cheese may be floated in brine to generate a rind (**Fig. 16.9D**); or it may be coated or spiked with a *Penicillium* mold. The ripening period then allows flavor to develop. Texture also changes; for example, where fermentation has produced CO_2, the trapped gas forms "eyes," or holes.

> **Thought Questions**
>
> **16.3** In an outbreak of listeriosis from unpasteurized cheese, only the refrigerated cheeses were found to cause disease. Why would this be the case?
>
> **16.4** Cow's milk contains 4% lipid (butterfat). What happens to the lipid during cheese production?

Flavor generation in cheese. In all fermented foods, microbial metabolism generates by-products that confer a characteristic aroma and flavor. In some cases, particular species confer distinctive flavors; for example, *Propionibacterium* ferments lactate or pyruvate to propionate, a flavor component of Swiss cheese. All bacteria generate a surprising range of side reactions, forming trace products that confer distinctive flavors. For example, while *Lactobacillus* converts most of the lactose to lactic acid, a small fraction of the pyruvate is converted to acetoin, acetaldehyde, or acetic acid, which contributes flavor. Most amino acids are retained intact, but traces are converted to flavorful alcohols, esters, and sulfur compounds (**Fig. 16.10**). For example, methanethiol (CH_3SH) contributes to the desirable flavor of cheddar cheese. Lipids are not significantly metabolized by *Lactobacillus*, but in mold-ripened cheeses such as Camembert, *Penicillium* oxidizes a small amount of the lipids to flavorful methylketones, alcohols, and lactones.

Acid Fermentation of Vegetables

Many kinds of vegetable products are based on microbial fermentation. Commercial products marketed globally include pickles, soy sauce, and sauerkraut. Other products provide staple foods for particular nations or regions, such as Indonesian tempeh and Korean kimchi.

Soy fermentation. Soybeans offer one of the best sources of vegetable protein and are indispensable for the diet of millions of people, particularly in Southeast Asia. In North America, soy products are important for vegetarian consumption and as a milk substitute, as well as for animal feed. But soybeans also contain substances that decrease their nutritive value. Phytate, or inositol hexaphosphate, chelates

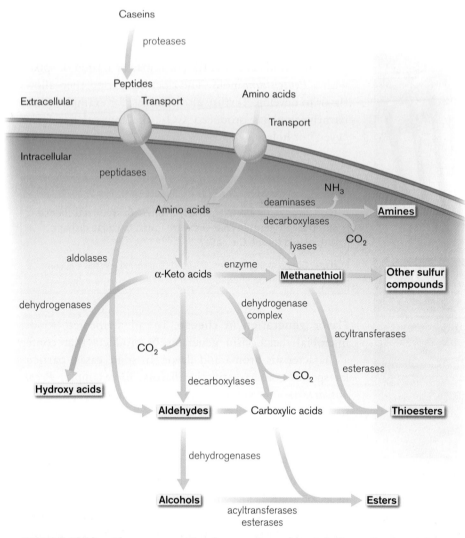

FIGURE 16.10 ■ **Flavor generation from amino acid catabolism.** Casein catabolism generates flavor molecules (highlighted). Extracellular enzymes break down casein into peptides and amino acids, which are taken into the bacterial cell by membrane transporters. The amino acids are fermented to volatile alcohols and esters. In some cases, they combine with sulfur to form methanethiol and other sulfur-containing odorants characteristic of cheese.

minerals such as iron, inhibiting their absorption by the intestine. Lectins are proteins that bind to cell-surface glycoproteins within the human body. At high concentration, soybean lectins may upset digestion and induce autoimmune diseases. Soy protease inhibitors interfere with digestive enzymes chymotrypsin and trypsin, thus decreasing the amount of protein that can be obtained from soy-based food.

All of these drawbacks of soybeans are diminished by microbial fermentation, while the protein content remains comparable to that of the unfermented bean (40%). A variety of fermented soy foods have been developed. Most soy fermentation involves mold growth, supplemented by bacteria that contribute vitamins, including vitamin B_{12}.

A major fermented soy product is **tempeh**, a staple food of Indonesia, the world's fourth-most-populous country, as well as of other countries in Southeast Asia (**Fig. 16.11**). Tempeh consists of soybeans fermented by *Rhizopus oligosporus*, a common bread mold. Besides decreasing the negative factors of soy, the mold growth breaks down proteins into more digestible peptides and amino acids. During World War II, tempeh was fed to American prisoners of war held by the Japanese. The tempeh was later

A.

B.

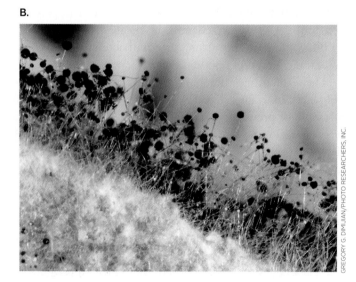

FIGURE 16.11 ■ **Tempeh, a mold-fermented soy product.** A. Fried tempeh. B. *Rhizopus oligosporus* mold, used to make tempeh.

credited with saving the lives of prisoners whose dysentery and malnutrition had impaired their ability to absorb intact proteins.

Tempeh is commonly produced in home-based factories in Indonesia. The soybeans are soaked in water overnight, allowing initial fermentation by naturally present lactic acid bacteria; in some cases, a crude "starter" may be introduced from the water of soybeans soaked previously. This pre-fermentation allows bacterial generation of vitamins and produces mild acid that promotes growth of mold. The soaked beans are then hulled, cooked, and cooled to room temperature for inoculation with *R. oligosporus* spores or with a previous tempeh culture. The inoculated beans are wrapped in banana leaves or in perforated plastic bags and then allowed to incubate for 2 days. The mold grows as a white mycelium that permeates the beans, joining them into a solid cake. The final product has a mushroom-like taste and is often served fried or grilled like a hamburger.

Other soy products undergo acid fermentation by the mold *Aspergillus oryzae*. The Japanese condiment **miso** is made from ground soy and rice, salted and fermented for 2 months by *A. oryzae*. Soy sauce is made from *jiang*, a Chinese condiment similar to miso in which the rice starter culture is replaced by wheat. The fermentation generates glutamic acid, a flavor-enhancing compound known popularly in the form of its salt—monosodium glutamate, or MSG.

Fermentation of cabbage and other vegetables. Various leaf vegetables are fermented by traditional societies, originally as a means of storage over the winter months. In Europe and North America, the best-known fermented products include sauerkraut and pickles. Sauerkraut production involves heterolactic fermentation by *Leuconostoc mesenteroides*. In heterolactic fermentation, each fermented sugar molecule yields lactic acid, as well as ethanol and carbon dioxide. The culture is used to inoculate shredded cabbage, which is layered in alternation with salt. The salt helps limit the number of species and the extent of microbial growth. A similar brine-enhanced fermentation process is used to pickle cucumbers, olives, and other vegetables.

An important food based on brine-fermented cabbage is Korean **kimchi** (**Fig. 16.12**). Kimchi is prepared from Chinese cabbage, salted and layered with radishes, peppers, onions, and other vegetables. The vegetables are covered with a paste of fish, rice, and chili peppers. Pickled seafoods such as shrimp or oysters may be included. The entire mixture is stored in a pot, traditionally buried underground for several months. The main fermentation organism is *Leuconostoc mesenteroides*, although *Streptococcus* and *Lactobacillus* species participate.

FIGURE 16.12 ■ Kimchi. A. To make kimchi, cabbage leaves are layered alternating with chili paste containing salted fish and vegetables. **B.** The salted layers are packed, to be buried and aged for 2 months.

Chocolate from Cocoa Bean Fermentation

Cocoa and coffee beans both require fermentation within the juice of the fruit before the beans are dried and processed. Chocolate, the product of the cocoa bean, *Theobroma cacao* (or "food of the gods"), requires one of the most complex fermentations of any food. For all the commercialization of chocolate production, totaling 2.5 billion kilos per annum worldwide, no "starter culture" has yet been standardized to ferment the cocoa bean. The beans cannot be exported and fermented later; the fermentation must occur immediately where the beans are harvested (**Fig. 16.13**).

The cocoa beans harvested in Africa or South America are heaped in mounds upon plantain leaves for fermentation by indigenous microorganisms, essentially the same way cocoa has been processed for thousands of years (**Fig. 16.14**). The microbial fermentation actually takes place outside the cocoa bean, within the pulp that clings to the beans after they are removed from the cocoa fruit. The pulp contains approximately 15% sugars and pectin (a branched polysaccharide), 2% citric acid, plus a rich supply of amino acids and minerals. These nutrients support growth of many kinds of microbes. Brazilian microbiologist Rosane Schwan (**Fig. 16.13A**), of the University of Lavras, Brazil, analyzed the microbial community. She defined three stages of succession: yeasts, lactic acid bacteria, and acetic acid bacteria (**Fig. 16.14B**).

Yeasts (anaerobic). Citric acid acidifies the pulp (pH 3.6). The acidity favors growth of yeasts, including *Candida*, *Kloeckera*, and *Saccharomyces*. The yeasts consume sugars and citric acid, increasing pH. They also degrade pectin into glucose and fructose, allowing the pulp to liquefy. As the liquefied pulp drains, yeast ferments the remaining sugars to ethanol, CO_2, and mixed acids such as acetate. The reactions release heat, increasing the temperature and accelerating metabolism. But the acetate, a permeant acid (discussed in Chapter 13), crosses the yeast cell membrane, where it lowers cell pH and inhibits growth. The ethanol and acetate also penetrate the bean embryo, killing the cells and releasing enzymes that generate key flavor molecules of chocolate.

FIGURE 16.13 ▪ Cocoa beans.
A. Rosane Schwan harvests cocoa beans for her research on cocoa fermentation. **B.** Cocoa fruit, showing beans encased in mucilage.

Lactic acid bacteria (anaerobic). The consumption of citric acid by yeast increases the bean pH to pH 4.2, encouraging growth of lactic acid bacteria, such as *Lactobacillus plantarum*. *Lactobacillus* species convert sugars to lactic acid, as well as acetate and CO_2. As fermentable substrates disappear, however, lactic acid bacteria are inhibited.

Acetic acid bacteria (aerobic). After 2 days, the beans are turned over and mixed periodically to permit access to oxygen. Aerobic acetic acid bacteria such as *Acetobacter* now oxidize the ethanol and acids to CO_2. The consumption of acids neutralizes undesirable acidity. Heat is released,

increasing the temperature to as high as 50°C. Oxygen penetrates the bean, oxidizing key components such as polyphenols. Polyphenol oxidation generates the brown color of cocoa and contributes flavor.

After fermentation and pulp drainage, the beans are dried and roasted. The roasting process completes the transformation of cocoa substances that contribute flavor. Cocoa liquor and cocoa butter are extracted from the beans and then recombined with sugar and other components to make "cocoa mass" (**Fig. 16.15**). The cocoa mass is stirred for several days to achieve a smooth texture; then it is molded into the decorative forms known

FIGURE 16.14 ▪ Microbial succession during cocoa pulp fermentation. A. A heap of cocoa beans covered by plantain leaves. Fruit pulp ferments, liquefies, and drains away, while the beans acidify and turn brown. **B.** Yeasts ferment citrate to alcohol; then, lactic acid bacteria convert sugars to lactate. With aeration, acetic acid bacteria oxidize ethanol to CO_2. CFUs = colony-forming units. *Source:* Data in part B from R. Schwan.

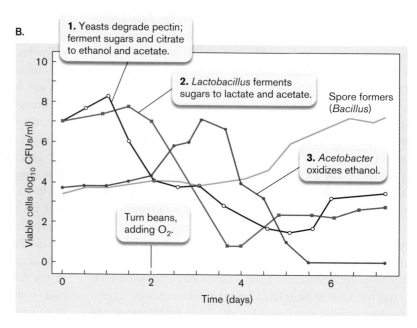

FIGURE 16.15 ■ **Chocolate manufacture.** Cocoa mass contains cocoa butter and liquor extracted from the cocoa beans, mixed with sugar and other ingredients.

as chocolate. But without the preceding fermentation process, no flavor would develop. Schwan and other researchers are working to develop a defined starter culture, a community of microbes to produce a predictable high quality of flavor. At present, however, one of the world's most refined and highly prized commercial food products still depends on the indigenous microbial fermentation of cocoa pulp.

Alkaline Fermentation: Natto and Pidan

In Western countries, food-associated fermentation is almost synonymous with acidification. In Southeast Asia and in Africa, however, many food products involve increased pH. Such fermentations typically release small amounts of ammonia, which raises the pH to about pH 8, retarding growth of all but alkali-tolerant bacteria. The fermenting bacteria are usually *Bacillus*, aerobic species tolerant of moderate alkali and capable of extensive proteolysis and amino acid decomposition. The fermentation needs to be controlled to limit the loss of protein content, but the end result is a highly stable food product.

Natto. The Japanese soybean product **natto** is prepared by a process similar to that for tempeh. Soybeans are washed, pre-fermented, and cooked briefly before incubation with the starter organism *Bacillus natto*. The fermenting beans are incubated in a shallow, ventilated container at a slightly raised temperature (40°C).

B. natto secretes numerous extracellular enzymes, including proteases, amylases, and phytases. These enzymes decrease the undesirable components of soy, such as phytates and lectins, while liberating more easily digestible peptides and amino acids. Some of the amino acids are deaminated, generating ammonia; a well-ventilated natto chamber allows most of this gas to escape. In addition, *B. natto* synthesizes extracellular polymers such as polyglutamate (a peptide chain consisting exclusively of glutamic acid residues). Polyglutamate generates long elastic strings that bind the beans together. The stretching of these strings from chopsticks is considered a sign of a good natto (**Fig. 16.16A**).

Alkali-fermented vegetables. In Africa, numerous vegetable products are based on alkaline fermentation predominantly by *Bacillus* species. An example is dawadawa, a paste of fermented locust beans common in western Africa. The locust beans are washed and supplemented with potash (potassium hydroxide) originally obtained from wood ashes. This addition of alkali retards growth of bacteria other than *Bacillus* species, which predominate at higher pH. The beans are fermented by indigenous bacteria (bacteria already present in the beans). The fermented beans are sun-dried, releasing most of the ammonia, and pounded into cakes for storage. Similar alkali-fermented vegetables include ogiri, from melon seeds, and ugba, from oil beans.

Pidan. An ancient means of preserving eggs led to the famous Chinese delicacy pidan, or "century egg," now a favorite at dim sum restaurants (**Fig. 16.16B**). To make

A.

B.

FIGURE 16.16 ■ **Alkali-fermented foods.** **A.** Natto consists of soybeans fermented by *Bacillus natto*. The fermentation generates long strings of polyglutamate. **B.** Pidan, or "century egg," consists of duck eggs coagulated by sodium hydroxide and fermented by *Bacillus* species. One egg here is cut open, revealing the transformed yolk, which develops a greenish color.

pidan, duck eggs are covered with a mixture of brewed tea, lime (CaO), and sodium carbonate (Na_2CO_3). The lime and sodium carbonate react to form NaOH, which penetrates the eggshells, raising pH and coagulating the egg white proteins. The eggs are buried in mud for several months, during which time the combined action of alkali and *Bacillus* fermentation generates dark colors and interesting flavors.

> **Thought Question**
>
> **16.5** In traditional fermented foods, without pure starter cultures, what determines the kind of fermentation that occurs?

To Summarize

- **Milk curd** forms by lactic acid fermentation and rennet proteolysis, rendering casein insoluble. The cleaved peptides coagulate to form a semisolid curd. The main fermentative organisms are lactic acid bacteria.

- **Cheese varieties** include unripened cheeses, semihard and hard cheeses that are cooked down and ripened, brined cheeses, and mold-ripened cheeses.

- **Cheese flavors** are generated by minor side products of fermentation, such as alcohols, esters, and sulfur compounds.

- **Soy fermentation** to tempeh and other products improves digestibility and decreases undesirable soy components such as phytates and lectins. The fermentative agent of tempeh is the bread mold *Rhizopus oligosporus*.

- **Vegetables** are fermented and brined to make sauerkraut and pickles. Cabbage and supplementary foods are fermented and brined to make kimchi.

- **Cocoa fermentation** for chocolate requires complex fermentation of cocoa beans within the fruit pulp, including anaerobic fermentation by yeast and lactic acid bacteria, and aerobic respiration by *Acetobacter* species.

- **Alkali-fermented vegetables** include the soy product natto, the egg product pidan, and the locust bean product dawadawa. The main fermenting organisms are *Bacillus* species.

16.4 Ethanolic Fermentation: Bread and Wine

Some of our most nutritionally significant and culturally important foods, including most bread and alcoholic beverages, require ethanolic fermentation by yeast fungi. Ethanolic fermentation converts pyruvic acid to ethanol and carbon dioxide:

$$C_3H_4O_3 \rightarrow CH_3CH_2OH + CO_2$$

The most prominent yeast used is *Saccharomyces cerevisiae*, known as baker's yeast or brewer's yeast (**Fig. 16.17**). A hardy organism, *S. cerevisiae* easily survives on a grocery shelf for home use and is genetically tractable for fundamental research. The yeast has been studied since the time of Pasteur, who used it to prove the biological basis of fermentation (presented in Chapter 1). *S. cerevisiae* today is a major model system of cell biology, yielding the molecular secrets of human cancer and other diseases.

Bread making depends on carbon dioxide to generate air spaces that **leaven** the dough, making its substance easier to chew and digest. The small amount of ethanol produced is eliminated during baking. For alcoholic beverages, however, ethanol is the key product, accompanied by carbon dioxide bubbles for "fizz," known as carbonation.

Bread Dough Is Leavened by Microbial CO₂

Bread is made in many different forms (**Fig. 16.18**) and from diverse kinds of flour, or ground grain. The earliest breads probably arose from grain mush naturally contaminated by

A. **B.**

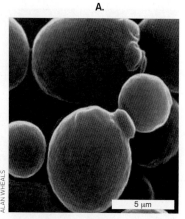 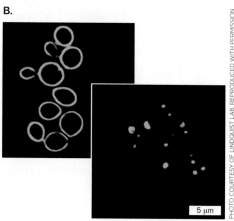

5 μm   5 μm

FIGURE 16.17 ▪ **Baker's yeast, the "champion" fermenter.** **A.** *Saccharomyces cerevisiae* cells budding; some show bud scars (SEM). **B.** *S. cerevisiae* is used to study the function of human proteins such as alpha-synuclein (green fluorescence), which plays a role in Parkinson's disease. Yeast cells engineered to express one copy of the gene (left panel) show the protein normally within their cell membrane. Two gene copies (right panel) cause the protein to clump and kill the cells.

FIGURE 16.18 ■ **Yeast bread.** Many varieties of bread are made.

yeast. Later, bread makers learned to include yeast left over from wine or beer production as a starter culture.

Yeast bread production. The preparation of all forms of yeast bread requires the same fundamental steps. A starter culture of yeast is included in the dough. The yeast can be commercial baker's yeast, or it can be **sourdough** starter, an undefined microbial population derived from a previous batch of dough. Analysis of sourdough shows mainly yeasts and lactobacilli, which release acids that favor the growth of the yeasts. The dough is kneaded to develop a fine network of air pockets and allowed to rise, expanding with production of the carbon dioxide gas (**Fig. 16.19**). The finest-textured breads are made from wheat flour, which contains gluten, a protein complex that forms a fine molecular network supporting the rising dough.

FIGURE 16.19 ■ **Making bread.** As yeast fermentation generates carbon dioxide gas, the dough rises.

Modern bread sold commercially is produced on an industrial scale. The dough is made in huge vats and then cut into regular chunks that are placed into the bread pans. The loaves are baked under hot-air convection, a method that greatly decreases the baking time. In the United States, most commercial bread is sliced mechanically—an invention that dates back to 1912. Sliced bread, however, has a shorter shelf life unless preservative chemicals prevent growth of mold on the exposed interior.

> **Thought Question**
>
> **16.6** Compare and contrast the role of fermenting organisms in the production of cheese and bread.

Injera: extended fermentation. Most kinds of bread involve only a short fermentation period, just long enough to produce enough gas for leavening. A prolonged fermentation, with more extensive microbial activity, unfolds in the dough for an Ethiopian bread called **injera** (**Fig. 16.20**).

A.

B.

FIGURE 16.20 ■ **Injera.** **A.** After 3 days of fermentation, injera dough is baked in a ceramic pan upon a "Mirte" charcoal stove. **B.** Injera forms an edible tablecloth for a variety of Ethiopian foods.

A.

B.

FIGURE 16.21 ▪ **Beer production: ancient and modern. A.** Making beer in ancient Egypt, circa 3000 BCE. The mash was stirred in earthen jars. **B.** Fermentors in a modern brewery.

The high microbial content provides a substantial source of vitamins not found in quick-rising breads.

Injera is made from teff (*Eragrostis tef*), a grain with small, round kernels that have high protein and lack gluten. Teff grows in arid regions and is now being cultivated in the United States for gluten-free products. Because teff lacks gluten, it cannot rise as much as wheat flour, but it makes a kind of flatbread. The dough is spread into a wide pancake, and the organisms present in the grain and air are allowed to ferment it for 3 days. The fermentation includes a succession of species, usually dominated by the yeast *Candida*. The extended fermentation generates a complex range of by-products that confer exceptional flavors in the baked product. In Ethiopia, injera forms the basis of an entire meal, served with other food items placed upon it as an edible tablecloth. Diners wrap samples of each food in a fold of injera and consume them together.

Alcoholic Beverages: Beer and Wine

Ethanolic fermentation of grain or fruit was important to early civilizations because it provided a drink free of waterborne pathogens. Traditional forms of beer also provided essential vitamins in the unfiltered yeasts.

Ethanol is unique among fermentation products in that it provides a significant source of caloric intake, but it is also a toxin that impairs mental function. A modest level of ethanol enters the human circulation naturally from intestinal flora, equivalent to a fraction of a drink per day. The human liver produces the enzyme alcohol dehydrogenase, which detoxifies ethanol. This enzyme in a healthy liver can metabolize small amounts of alcohol without harm.

However, excess alcohol consumption can overload the liver's capacity for detoxification and permanently damage the liver and brain.

Beer: alcoholic fermentation of grain. Beer production is one of the most ancient fermentation practices and is depicted in the statuary of ancient Egyptian tombs dated to 5,000 years ago (**Fig. 16.21A**). The earliest Sumerian beers were made from bread soaked in water and fermented. Today, most beer is produced commercially by fermenting barley using giant vats (**Fig. 16.21B**). Production of high-quality beer involves complex processing with many steps, including germination of barley grains, mashing in water and cooking, and introduction of hops for flavor (**eTopic 16.1**).

First the barley grains must germinate; that is, the seed embryos must start to grow. The germinating embryo makes enzymes needed to break down the barley starch to maltose (disaccharide) and glucose. Most of the sugars are fermented by yeast to ethanol and carbon dioxide. However, minor side products contribute flavors—or unpleasant off-flavors if present in too great an amount (**Fig. 16.22**). For example, off-flavors may result from the presence of small amounts of oxygen that oxidize some ethanol to acetaldehyde.

The yeast ferment most sugars to ethanol, but a small fraction is drawn off to make amino acids via TCA cycle intermediates, as discussed in Chapter 15. The 2-oxo acids of the TCA cycle are analogous to pyruvate, with the methyl group replaced by extended carbon chains (R group). A tiny amount of the 2-oxo acids is converted to long-chain alcohols, which add desirable flavor to beer.

Thought Question

16.7 Compare and contrast the role of low-concentration by-products in the production of cheese and beer.

Wine: alcoholic fermentation of fruit. The fermentation of fruit gives rise to wine, another class of alcoholic products of enormous historical and cultural significance. Grapes

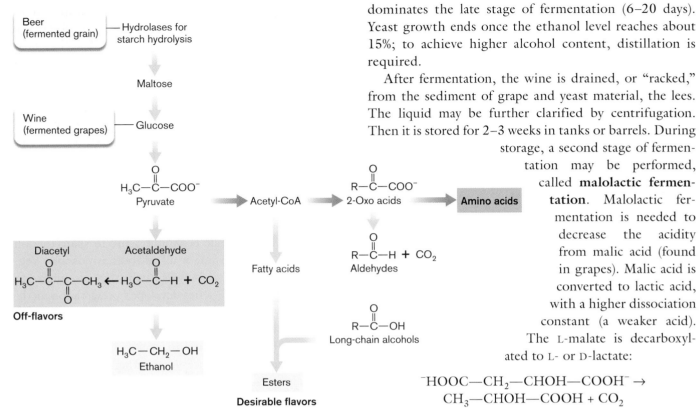

FIGURE 16.22 ■ Alcoholic fermentation in beer and wine. Yeast fermentation generates ethanol in substantial quantities. The biosynthesis of amino acids generates by-products that contribute both desirable flavors (long-chain alcohols) and off-flavors (acetaldehyde and diacetyl).

dominates the late stage of fermentation (6–20 days). Yeast growth ends once the ethanol level reaches about 15%; to achieve higher alcohol content, distillation is required.

After fermentation, the wine is drained, or "racked," from the sediment of grape and yeast material, the lees. The liquid may be further clarified by centrifugation. Then it is stored for 2–3 weeks in tanks or barrels. During storage, a second stage of fermentation may be performed, called **malolactic fermentation**. Malolactic fermentation is needed to decrease the acidity from malic acid (found in grapes). Malic acid is converted to lactic acid, with a higher dissociation constant (a weaker acid). The L-malate is decarboxylated to L- or D-lactate:

$$^-HOOC{-}CH_2{-}CHOH{-}COOH^- \rightarrow$$
$$CH_3{-}CHOH{-}COOH + CO_2$$

The wine is seeded with *Oenococcus oeni* bacteria, which ferment L-malate (deprotonated L-malic acid).

As in beer production, yeast fermentation of wine produces numerous minor products contributing flavor, such as long-chain alcohols and esters. At the same time, overgrowth of yeast or the growth of undesired species can produce excess amounts of these compounds, such as sulfides and phenolics, giving rise to off-flavors. Some undesired species require oxygen exposure, whereas others can grow during storage and bottling. The balance of microbial populations is challenging to control and has a major role in determining the quality of a given wine vintage.

produce the best-known wines, but wines and distilled liquors are also made from apples, plums, and other fruits. The key difference between fermentation of fruits and fermentation of grains is the exceptionally high monosaccharide content in fruits. Grape juice, for example, can contain concentrations of glucose and fructose as high as 15%. The availability of simple sugars allows yeast to begin fermenting immediately, with no need for preliminary breakdown of long-chain carbohydrates, as in the malting and mashing of beer.

Most modern wine production uses strains of the grape *Vitis vinifera*. The grapes are crushed to release juices, usually in the presence of antioxidants such as sulfur dioxide (**Fig. 16.23**). For white wine, the skins are removed before juice is fermented. For red wine, the skins are included in early fermentation to extract the red and purple anthocyanin pigments, as well as phenolic flavor compounds. The first few days of fermentation are dominated by indigenous species of yeast naturally present on the grapes, such as *Kloeckera* and *Hanseniaspora* species. Commercial producers usually inoculate with standard *Saccharomyces cerevisiae*, whose population

To Summarize

- **Bread is leavened** by yeasts conducting limited ethanolic fermentation, producing enough carbon dioxide gas to expand the dough.

- **Injera** bread dough undergoes more extensive fermentation by indigenous organisms and, as a result, generates multiple flavors.

- **Beer** requires alcoholic fermentation of grain. Barley grains are germinated, allowing enzymes to break down the starch to maltose for yeast fermentation.

- **Secondary products of grain fermentation**, such as long-chain alcohols and esters, generate the special flavors of beer.

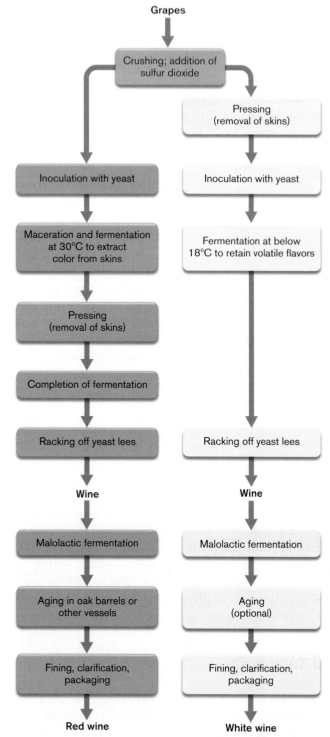

FIGURE 16.23 ■ **Production of red and white wines. Left:** For red wine, the grapes are fermented with the skins at a temperature that increases extraction of color and tannins. **Right:** For white wine, the skins are removed before addition of yeast starter, and the temperature is kept low to retain volatile flavors. Both kinds of wine usually undergo malolactic fermentation by *Oenococcus oeni*, special lactic acid bacteria that consume malic acid.

■ **Wine** derives from alcoholic fermentation of fruit, most commonly grapes. The grape sugar (glucose) is fermented by yeast to alcohol. A secondary product, malate, undergoes malolactic fermentation by *Oenococcus oeni* bacteria.

16.5 Food Spoilage and Preservation

We humans have always competed with microbes for our food. When early humans killed an animal, microbes commenced immediately to consume its flesh. Because meat perished so fast, it made economic sense to share the kill immediately and consume it all as soon as possible. Vegetables might last longer, but eventually they succumbed to mold and rot. Later societies developed preservation methods, such as drying, smoking, and canning, that enabled humans to survive winters and dry seasons on stored food.

Modern food preservation depends on **antimicrobial agents** (chemical substances that either kill microbes or slow their growth), as well as physical preservative measures, such as treatment with heat and pressure. These general principles of microbial control are described in Chapter 5. Here we focus on microbial contamination and food preservation from the perspective of the food industry.

Food Spoilage and Food Contamination

After food is harvested, several kinds of chemical changes occur. Some begin instantly, whereas others take several days to develop. Some changes, such as meat tenderizing, may be considered desirable; others, such as putrefaction, render food unfit for consumption. The major classes of food change include:

■ **Enzymatic processes.** Following the death of an animal, its flesh undergoes proteolysis by its own enzymes. Limited proteolysis tenderizes meat. Plants after harvest undergo other changes; for example, in harvested corn the sugar rapidly converts to starch. That is why vegetables taste sweetest immediately after harvest.

■ **Chemical reactions with the environment.** The most common abiotic chemical reactions involve oxidation by air—for example, lipid autooxidation, which generates rancid odors. Much research addresses oxidation, and the development of technologies to prevent it (**Fig. 16.24**). For example, produce may be packaged under an anaerobic atmosphere, in a film that prevents oxygen transmission.

■ **Microbiological processes.** Microbes from the surface of the food begin to consume it—some immediately,

© PEGGY GREB

FIGURE 16.24 ■ Research to prevent oxidation of produce. At the U.S. Department of Agriculture, microbiologist Arvind Bhagwat withdraws a gas sample from bagged lettuce, to test an anaerobic packaging method.

others later in succession—generating a wide range of chemical products. In meat, internal organs of the digestive tract are an important source of microbial decay.

Microbial activity can aid food production, but it can also have various undesirable effects. Two different classes of microbial effects are distinguished: food spoilage and food contamination with pathogens.

Food spoilage refers to microbial changes that render a product obviously unfit or unpalatable for consumption. For example, rancid milk or putrefied meats are unpalatable and contain metabolic products that may be deleterious to human health, such as oxidized fatty acids or organic amines. Even in these cases, however, the definition of spoilage depends partly on cultural practice. What is sour milk to one person may be buttermilk to another; what one society considers spoiled meat, another may consider merely aged.

Different pathways of microbial metabolism lead to different kinds of spoilage. Sour flavors result from acid fermentation products, as in sour milk. Alkaline products generate bitter flavor. Oxidation, particularly of fats, causes **rancidity**, whereas general decomposition of proteins and amino acids leads to **putrefaction**. The particularly noxious odors of putrefaction derive from amino acid breakdown products that often have apt names, such as the amines cadaverine and putrescine and the aromatic product skatole.

"Food contamination," or **food poisoning**, refers to the presence of microbial pathogens that cause human disease—for example, rotaviruses that cause gastrointestinal illness. Other pathogens, such as *Clostridium botulinum*, produce toxins with deadly effects (discussed in Chapter 25). Pathogens usually go unnoticed as food is consumed, because their numbers are very low, and they may not even grow in the food. Even the freshest-appearing food may cause serious illness if it has been contaminated with a small number of pathogens.

Since our intestines are full of beneficial bacteria, what makes a pathogen? Enteric pathogens are often closely related to members of our gut microbiome, which normally outcompete invaders (discussed in Chapters 21 and 23). Nevertheless, a pathogen may overcome our defenses, either by use of its virulence factors (see Chapter 25) or by taking advantage of host weakness due to immunosuppression or to antibiotics depleting the normal microbiome.

How Food Spoils

Different foods spoil in different ways, depending on their nutrient content, the microbial species, and environmental factors such as temperature. **Table 16.2** summarizes common forms of spoilage.

Dairy products. Milk and other dairy products contain carbon sources, such as lactose, protein, and fat. In fresh milk, the nutrient most available for microbial catabolism is lactose, which commonly supports anaerobic fermentation to sour milk. Fermentation by the right mix of microbes, however, leads to yogurt and cheese production, as previously described.

Under certain conditions, bitter off-flavors may be produced by bacterial degradation of proteins. The release of amines causes a rise in pH. Protein degradation is most commonly caused by psychrophiles, species that grow well at cold temperatures, such as those of refrigeration.

Cheeses are less susceptible than milk to general spoilage, because of their solid structure and lowered water activity. However, cheeses can grow mold on their surface. Historically, the surface growth of *Penicillium* strains led to the invention of new kinds of cheeses. But other kinds of mold, such as *Aspergillus*, produce toxins and undesirable flavors.

Meat and poultry. Meat in the slaughterhouse is easily contaminated with bacteria from hide, hooves, and intestinal contents. Muscle tissue offers high water content, which supports microbial growth, as well as rich nutrients, including glycogen, peptides, and amino acids. The breakdown of peptides and amino acids produces the undesirable

TABLE 16.2 Food Spoilage (Examples)

Food product	Signs of spoilage	Microbial cause
Dairy products		
Milk	Sour flavor	Lactic acid bacteria produce lactic and acetic acids.
Milk	Coagulation	Lactic acid bacteria produce proteases that destabilize casein and lower pH, causing coagulation.
Milk	Bitter flavor	Psychrophilic bacteria degrade proteins and amino acids.
Cheese	Open texture, fissures	Lactic acid bacteria produce carbon dioxide.
Cheese	Discoloration and colonies	Molds such as *Penicillium* and *Aspergillus* grow on the cheese.
Meat and poultry		
Meat and poultry	Rancid flavor	Psychrotrophic bacteria produce fatty acids that become oxidized.
Meat and poultry	Putrefaction	*Pseudomonas* and other aerobes degrade amino acids, producing amines and sulfides.
Meat and poultry	Discolored patches	Molds such as *Mucor* and *Penicillium* grow on the surface.
Eggs	Pink or greenish egg white	*Pseudomonas* and related bacteria grow on albumin, producing water-soluble pigments.
Eggs	Sulfurous odor	Bacterial growth on albumin releases hydrogen sulfide.
Seafood		
Fish	Fishy smell	Anaerobic psychrophiles such as *Photobacterium* convert trimethylamine oxide to trimethylamine.
Fish	Odor of putrefaction	*Pseudomonas* and other Gram-negative species degrade amino acids, producing amines and sulfides.
Shellfish	Odor of putrefaction	*Vibrio* and other marine bacteria decompose the protein.
Fruits, vegetables, and grains		
Plants before harvest	Rotting or wilting	Plant pathogens, most commonly fungi such as *Alternaria*, *Aspergillus*, and *Penicillium*.
Stored plant foods	Rotting or wilting	Molds or bacteria produce degradative enzymes, such as pectinases and cellulases.
Apples, pears, cherries	Geosmin off-flavor	*Penicillium* mold.
Peeled oranges	Discoloration and off-flavor	*Enterobacter* and *Pseudomonas* spp.
Pasteurized fruit juices	Medicine-like phenolic off-flavor	Acid- and heat-tolerant spore former, *Alicyclobacillus* sp., produces 2-methoxyphenol (guaiacol).
Bread	Ropiness	*Bacillus* spp. grow, forming long filaments.
Bread	Red discoloration	*Serratia marcescens*.

odorants that define spoilage (for example, cadaverine and putrescine).

Meat also contains fat, or adipose tissue, but the lipids are largely unavailable to microbial action because they consist of insoluble fat (triacylglycerides). Instead, meat lipids commonly spoil abiotically by autooxidation (reaction with oxygen) of unsaturated fatty acids, independent of microbial activity. Thus, when meats are exposed to air during storage, they turn rancid—particularly meats such as pork, which contains highly unsaturated lipids. Autooxidation can be prevented by anaerobic storage, such as vacuum packing, which also prevents growth of aerobic microorganisms. The absence of the suppressed organisms,

however, favors growth of lactic acid bacteria and facultative anaerobes such as *Brochothrix thermosphacta*. These organisms generate short-chain fatty acids, which taste sour.

In industrialized societies, the most significant factor determining microbial populations in meat spoilage is the practice of refrigeration. Refrigeration prolongs the shelf life of meat because contaminating microbes are predominantly mesophilic (grow at moderate temperatures, as discussed in Chapter 5). But ultimately, the few psychrotrophs initially present do grow; typically, these are *Pseudomonas* species. The pseudomonads are also favored by the low pH of meat (pH 5.5–7.0), which results from the accumulation of lactic acid in the muscle.

Seafood. Fish and other seafood contain substantial amounts of protein and lipids, as well as amines such as trimethylamine oxide. Fish spoils more rapidly than meat and poultry for several reasons. First, fish do not thermoregulate, and they inhabit relatively low-temperature environments. Because fish grow in low-temperature environments, their surface microorganisms tend to be psychrotrophic and thus grow well under refrigeration. In addition, marine fish contain high levels of the osmoprotectant trimethylamine oxide, which bacteria reduce to trimethylamine, a volatile amine that gives seafood its "fishy" smell. Finally, the rapid microbial breakdown of proteins and amino acids leads to foul-smelling amines and sulfur compounds, such as hydrogen sulfide and dimethyl sulfide.

> **Thought Question**
>
> **16.8** Why would bacteria convert trimethylamine oxide (TMAO) to trimethylamine? Would this kind of spoilage be prevented by exclusion of oxygen?

Plant foods. Fruits, vegetables, and grains spoil differently from animal foods because of their high carbohydrate content and their relatively low water content. The low water content of plant foods usually translates into considerably longer shelf life than for animal-based foods. Carbohydrates favor microbial fermentation to acids or alcohols that limit further decomposition, and this microbial action can be managed to produce fermented foods, as described in Section 16.2.

Plant pathogens rarely infect humans but may destroy the plant before harvest. Most plant pathogens are fungi, although some are bacteria, such as *Erwinia* species. Historically, plant pathogens have caused major agricultural catastrophes, such as the Irish potato famine, caused by a fungus-like pathogen. Plant pathogens continue to devastate local economies and cause shortages worldwide; for example, the witches'-broom fungus *Crinipellis perniciosa* causes a fungal disease of cocoa trees that has drastically cut Latin American cocoa production.

After harvest, various molds and bacteria can soften and wilt plant foods by producing enzymes that degrade the pectins and celluloses that give plants their structure. In general, the more processed the food, the greater the opportunities for spoilage. For example, citrus fruits generally last for several weeks, but peeled oranges are susceptible to spoilage by Gram-negative bacteria.

Baked bread usually resists spoilage, except for surface molds. In rare cases, however, improperly baked bread can show contamination. The appearance of red bread, caused by the red bacterium *Serratia marcescens*, is believed to have been the source of the "blood" observed in communion bread during a Catholic mass in the Italian town of Bolsena in 1263—an event that became known as the Miracle of Bolsena.

Pathogens Contaminate Food

Intestinal pathogens spread readily because microbes can be transmitted through food without any outward sign that the food is spoiled. The U.S. Centers for Disease Control and Prevention (CDC) estimates that there are 76 million cases of gastrointestinal illness a year in this country, usually spread through water or food. Thus, one in four Americans experiences gastrointestinal illness in a given year. In 2013, under President Obama, the Food and Drug Administration implemented new controls and rules as part of the 2011 Food Safety Modernization Act, the largest reform of the U.S. food safety laws in 70 years. The new laws require, for example, that crop irrigation water be free of pathogens and that agricultural workers have access to bathroom facilities.

An example of food contamination is the 2008 outbreak of *Salmonella enterica* from peanut products. Peanuts contaminated at one processing plant led to an epidemic that sickened 700 people across the United States (**Fig. 16.25A**). The first cases of *Salmonella* infection were reported to the CDC on September 1, 2008. Most infected individuals developed diarrhea, fever, and abdominal cramps 12–72 hours after infection, and symptoms lasted 4–7 days. Over the next 6 months, cases were reported from nearly all U.S. states. The curve of the outbreak (cases rising and then falling) followed the profile of a single-source epidemic, in which all infections are ultimately traced back to one source. (Epidemics are discussed in Chapter 28.)

What was the original source of the widespread outbreak? The CDC researchers compared the food intake histories of ill persons against matched controls. They found a statistical association between illness and intake of peanut butter, eventually narrowed to a specific brand of peanut butter sold to institutions. As the epidemic grew, cases emerged in which the contaminated food product was crackers filled with peanut butter cream. Ultimately, the peanut butter and cream were traced back to peanuts from a single factory in Georgia. At the food plant, the source of *Salmonella* contamination could not be identified, but the plant records showed that product samples had tested positive for *Salmonella*. Instead of discarding the product, the plant had retested the samples until they "tested negative." Numerous health violations were cited, including gaps in the walls and dirt buildup throughout the plant.

A.

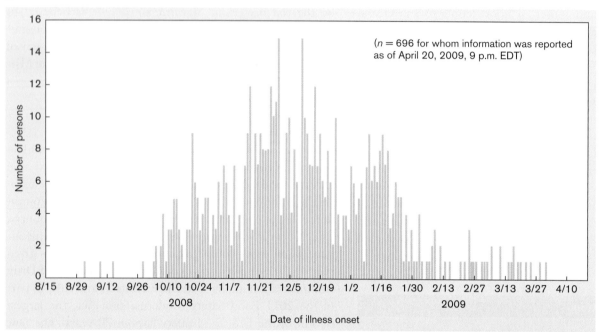

B.

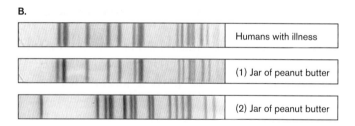

Humans with illness

(1) Jar of peanut butter

(2) Jar of peanut butter

FIGURE 16.25 ■ *Salmonella enterica* **outbreak from contaminated peanut butter.** **A.** Infected individuals reported to the CDC from September 1, 2008, through April 20, 2009. **B.** Electrophoretic separation of restriction-digest DNA fragments from bacteria strains isolated from humans with illness. (1) Peanut butter containing the same strain; (2) peanut butter containing a different strain of *S. enterica*. *Source: www.cdc.gov.*

Salmonella bacteria are very common pathogens. Did all of the cases of illness result from a common strain? The CDC used DNA analysis to show that all patients carried a common strain of *S. enterica* serovar Typhimurium. (A serovar is a strain whose surface proteins elicit a distinctive immune response.) The strain was identified by analysis of its genomic DNA cleaved by restriction endonucleases (see Section 7.6). Each restriction endonuclease cleaves DNA at sequence-specific positions. Strains that differ at key restriction sites generate cleavage fragments of differing length, which are separated by pulsed-field electrophoresis (**Fig. 16.25B**). In electrophoresis, applied voltage causes DNA fragments to migrate different distances according to size; the pulsed field optimizes separation of the largest sizes. The distance each fragment moves is visualized as a band in the gel. The band pattern, or "fingerprint," of *Salmonella* DNA from infected patients showed the same fragment lengths as *Salmonella* DNA from the peanut butter sample, labeled "(1)" in **Figure 16.25B**. The band pattern from this peanut butter sample (1) was different from that of another infected peanut butter sample (2); the bacterium in (2) proved unrelated to the *Salmonella* outbreak.

This case illustrates several troubling features of food contamination in modern society. It shows the consequence of a food production plant's failure to follow regulations, and the failure of health inspection to enforce them. The contaminated product shipped out to a diverse array of institutions such as schools and to secondary producers such as cookie manufacturers, who incorporated the peanut butter cream ingredient. The bacteria then remained viable in contaminated food products for many months, sickening people long after the contamination event occurred.

Food-Borne Pathogens Emerge from Environment and Agriculture

Our environment and agriculture present constant sources of potential pathogens. Agricultural microbiologists continually assess the potential contamination of produce. For example, student Julia DeNiro worked with Douglas Doohan at the Ohio State University to study contamination of lettuce and tomatoes fertilized by manure (**Fig. 16.26**). DeNiro assessed the microbial contamination of produce

A.

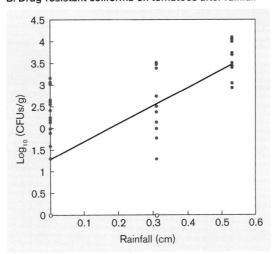

COURTESY OF JULIA DENIRO

B. Drug-resistant coliforms on tomatoes after rainfall

FIGURE 16.26 ■ **Produce acquires bacterial contamination.** **A.** Julia DeNiro studies bacterial contamination of lettuce and tomatoes grown with manure fertilizer. **B.** Colony-forming units (CFUs) from the surface of tomatoes increase with increasing rainfall.

dehydration. By contrast, the spore-forming pathogen *Clostridium botulinum* causes only about 100 cases of botulism per year. The incidence is relatively low, but if untreated, the fatality rate is 50%. In this case, the low incidence of botulism actually enhances its danger because the condition is likely to go undiagnosed.

What distinguishes a pathogen from a spoilage organism? Pathogens possess highly specific mechanisms for host colonization, as discussed in Chapter 25. **Figure 16.28** shows intestinal crypt cells covered with *Escherichia coli* O157:H7 bacteria, an emergent pathogen first recognized in 1982 in fast-food hamburgers. Since then, *E. coli* O157:H7 has also been found to contaminate spinach and other vegetables. The bacteria can actually grow as **endophytes** (plant endosymbionts) within the plant transport vessels. By 2010, six lesser-known strains of *E. coli* had sickened people through contaminated lettuce or beef.

Bacterial factors that contribute to disease are often encoded together in the genome in a region known as a

as a function of many factors, such as distance from the manure, temperature, and rainfall. An example of her results is shown (**Fig. 16.26B**). The surface of tomatoes was tested for colony-forming units of coliforms (intestinal bacteria) resistant to the antibiotics ampicillin, chloramphenicol, and streptomycin. With increasing rainfall, the tomatoes showed increasing numbers of drug-resistant colonies. Thus, rainfall is identified as a factor to consider when assessing produce contamination. The growing presence of drug resistance is concerning for the future biosafety of our food.

Food-borne pathogens can arise from a surprising variety of sources. Consider for example the transmission routes of *Listeria monocytogenes*, a psychrotrophic pathogen that invades the cells of the intestinal epithelium, causing listeriosis (**Fig. 16.27**). Psychrotrophic organisms grow optimally at moderate temperatures but also grow slowly at lower temperatures, typically 0°C–30°C. *Listeria* can be transferred from soil and feed to cattle, whose manure then cycles it back to soil. From cattle, the pathogen contaminates milk and meat, where it can eventually infect human consumers. Because *Listeria* is a psychrotroph, it outcompetes other food-borne bacteria under refrigeration.

The U.S. Public Health Service judges the importance of food-borne illnesses by their incidence and/or severity (**Table 16.3**). For example, the Norwalk-type viruses, or noroviruses, infect 180,000 Americans per year; their spread is very difficult to control, especially in close quarters, such as a cruise ship, where an outbreak can easily infect a large proportion of passengers. The course of illness is usually short, but it can lead to complications from

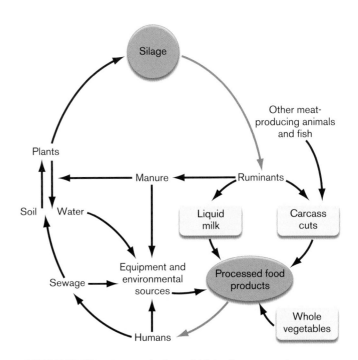

FIGURE 16.27 ■ **Transmission of *Listeria monocytogenes*.** Transmission occurs through various routes, including passage through food products. Colored arrows indicate transmission of disease.

TABLE 16.3	Food-Borne Pathogens in the United States[a]	

Pathogen	Incidence and transmission	Course of illness
Norovirus (Norwalk and Norwalk-like viruses)	Most common cause of diarrhea; also called "stomach flu" (no connection with influenza). 180,000 cases per year are estimated. Transmitted mainly by virus-contaminated food and water. Infection rates are highest under conditions of crowding in close quarters, such as inside a ship or a nursing home.	Disease lasts 1 or 2 days. Includes vomiting, diarrhea, and abdominal pain; headache and low-grade fever may occur.
Salmonella	Most common food-borne cause of death; more than 1 million cases per year; estimated 600 deaths per year. Transmission nearly always through food—raw, undercooked, or recontaminated after cooking, especially eggs, poultry, and meat; also contaminates dairy products, seafood, fruits, and vegetables.	Gastrointestinal disease that includes diarrhea, fever, and abdominal cramps lasting 4–7 days. Fatal cases are most common in immunocompromised patients.
Campylobacter	More than 1 million cases of campylobacteriosis per year; estimated 100 deaths per year. Grows in poultry without causing symptoms. Transmission is mainly through raw and undercooked poultry; contaminates half of poultry sold. Occurs less often in dairy products or in foods contaminated after cooking.	In humans, usually severe bloody diarrhea, fever, and abdominal cramps lasting 7 days. Fatal cases are most common in immunocompromised patients.
Escherichia coli O157:H7	An emerging pathogen, first recognized in a hamburger outbreak in 1982; now known to infect 73,000 people yearly, including 60 deaths per year. Grows in cattle without causing symptoms. Transmitted through ground beef; also through unpasteurized cider and from plant produce, where it grows as an endophyte.	In humans, usually severe bloody diarrhea and abdominal cramps lasting 5–10 days. About 5% of patients, especially children and elderly, develop hemolytic uremic syndrome, in which the red blood cells are destroyed and the kidneys fail.
Clostridium botulinum	Causes about 100 cases per year of botulism, with a 50% fatality rate if untreated. Grows in improperly home-canned foods, more rarely in commercially canned low-acid foods and improperly stored leftovers such as baked potatoes. Spores occur in honey, endangering infants under 2 years of age.	Botulinum toxin from growing bacteria causes progressive paralysis, with blurred vision, drooping eyelids, slurred speech, difficulty swallowing, and muscle weakness. Infant botulism causes lethargy and impaired muscle tone, leading to paralysis.
Listeria monocytogenes	Listeria bacteria grow in animals without causing symptoms. Animal feces may contaminate water, which is then used to wash vegetables. Transmission occurs mainly through vegetables washed in contaminated water and through soft cheeses. Listeria is psychrotrophic, growing at refrigeration temperatures.	Listeriosis involves fever, muscle aches, and sometimes gastrointestinal symptoms. In pregnant women, symptoms may be mild but lead to serious complications for the unborn child.
Shigella	Infects about 18,000 people a year in the United States; in developing countries, Shigella infections are endemic in most communities. Transmission occurs through fecal-oral contact or from foods washed in contaminated water.	Shigellosis involves gastrointestinal symptoms such as diarrhea, fever, and stomach cramps, usually lasting 7–10 days. Complications are rare.
Staphylococcus aureus	Best known as the cause of skin infections transmitted through open wounds. However, can also be transmitted through high-protein foods such as ham, dairy products, and cream pastries.	S. aureus causes toxic shock syndrome. Can also cause food poisoning via preformed toxins.
Toxoplasma gondii	A parasite believed to infect 60,000 people annually, most with no symptoms. In a few cases, serious disease results. Transmitted through contact with feces of infected animals, particularly cats, or through contaminated foods such as pork.	Toxoplasmosis causes mild flu-like symptoms; but in pregnant women, its transmission to the unborn child can lead to severe neurological defects, including death. Neurological complications also occur in immunocompromised patients.
Vibrio vulnificus	A free-living marine organism that contaminates seafood or open wounds. About 40 cases per year are reported.	V. vulnificus can infect the bloodstream, causing septic shock. Mainly threatens people with preexisting conditions such as liver disease.

[a]Ten major food-borne pathogens highlighted by the U.S. Public Health Service (USPHS).

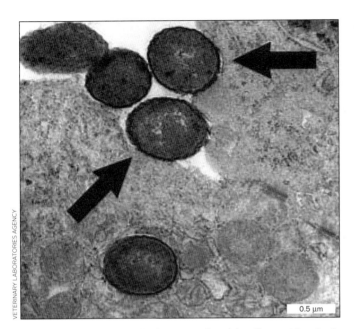

FIGURE 16.28 ■ **Intestinal crypt cells with adherent bacteria,** *Escherichia coli* **strain O157:H7.** A gnotobiotic (germ-free) piglet was infected with the bacteria (arrows) (TEM).

pathogenicity island. A pathogenicity island consists of a set of genes and operons that function coordinately (as discussed in Section 9.6). The colocalization of the genes enables transfer of virulence capability to other species as pathogens evolve. **Figure 16.29** shows an example of a pathogenicity island in *Salmonella*. Four of its operons contribute to the type III secretion complex, which secretes toxins and host colonization factors (discussed in Chapter 25). Other genes encode outer membrane proteins that counteract host defenses, as well as regulators of virulence gene expression.

The most dangerous consequence of infection by foodborne pathogens is the production of a potentially fatal toxin. For example, *E. coli* O157:H7 infection of the intestines can be overcome, but the bacteria produce Shiga toxin, which can destroy the kidneys. In cases of adult botulism, *Clostridium botulinum* does not usually grow within the patient; the botulinum toxin comes from bacteria that grew previously in improperly sterilized food. The botulinum toxin has a highly specific effect, inhibiting synaptic vesicle fusion in the terminals of peripheral motor neurons (**Fig. 16.30**). Synaptic inhibition prevents activation of muscle cells, causing flaccid paralysis. Microbial toxins are discussed further in Chapter 25.

Food Preservation

Cultural practices and cuisines have long evolved to limit food spoilage. Such practices include cooking (heat treatment), addition of spices (chemical preservation), and fermentation (partial microbial digestion). In modern commercial food production, spoilage and contamination are prevented by numerous methods based on fundamental principles of physics and biochemistry that limit microbial growth (discussed in Chapter 5).

Physical means of preservation. Specific processes that preserve food based on temperature, pressure, or other physical factors include:

- **Dehydration and freeze-drying.** Removal of water prevents microbial growth. Water is removed either by application of heat or by freezing under vacuum (known as **freeze-drying**, or **lyophilization**). Drying is especially effective for vegetables and pasta. The disadvantage of drying is that some nutrients are broken down.

- **Refrigeration and freezing.** Refrigeration temperature (typically –2°C to 16°C) slows microbial growth, as shown in an experiment comparing bacterial growth in ground beef at different temperatures (**Fig. 16.31**). Nevertheless, refrigeration also selects for psychrotrophs, such as *Listeria*. Freezing halts the growth of most microbes, but preexisting contaminant strains often survive to grow again when the food is thawed. This is why deep-frozen turkeys, for example, can still cause *Salmonella* poisoning, especially if the interior is not fully thawed before roasting.

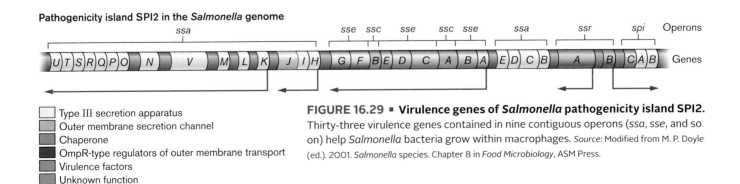

Pathogenicity island SPI2 in the *Salmonella* genome

FIGURE 16.29 ■ **Virulence genes of *Salmonella* pathogenicity island SPI2.** Thirty-three virulence genes contained in nine contiguous operons (*ssa*, *sse*, and so on) help *Salmonella* bacteria grow within macrophages. *Source:* Modified from M. P. Doyle (ed.). 2001. *Salmonella* species. Chapter 8 in *Food Microbiology*, ASM Press.

A.

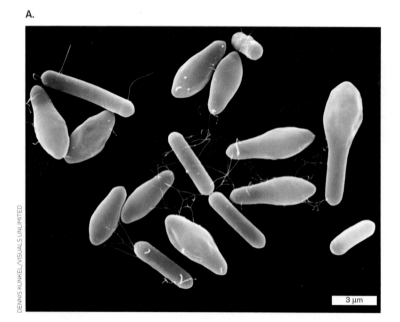

B.

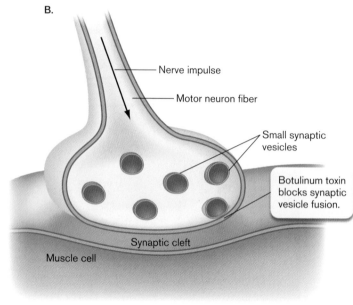

FIGURE 16.30 ■ *Clostridium botulinum* **produces botulinum toxin.** **A.** Club-shaped morphology of *C. botulinum* cells containing endospores. **B.** Botulinum toxin inhibits synaptic vesicle fusion in the terminal of a peripheral motor neuron, preventing activation of the muscle cell.

- **Controlled or modified atmosphere.** Food can be packed under vacuum or stored under atmospheres with decreased oxygen or increased CO_2. Controlled atmospheres limit abiotic oxidation, as well as microbial growth. For example, CO_2 storage is particularly effective for extending the shelf life of apples.

- **Pasteurization.** Invented by Louis Pasteur, pasteurization is a short-term heat treatment designed to decrease microbial contamination with minimal effect on food value and texture (discussed in Chapter 5). For example, milk is commonly pasteurized at 63°C for 30 minutes, followed by quick cooling to 4°C. Pasteurization is most effective for extending the shelf life of liquid foods with consistent, well-understood microbial flora, such as milk and fruit juices.

- **Canning.** In canning, the most widespread and effective means of long-term food storage, food is cooked under pressure to attain a temperature high enough to destroy endospores (typically 121°C). Commercial canning effectively eliminates microbial contaminants, except in very rare cases. The main drawback of canning is that it incurs some loss of food value, particularly that of labile biochemicals such as vitamins, as well as loss of desirable food texture and taste.

- **Ionizing radiation.** Exposure to ionizing radiation, known as food irradiation, effectively sterilizes many kinds of food for long-term storage. The main concerns about food irradiation are its potential for unknown effects on food chemistry and the hazards of the irradiation process itself for personnel involved in food processing. Nevertheless, irradiation has proved highly effective at eliminating pathogens that would otherwise cause serious illness.

Often, two or more means of preservation are used in combination, such as acid treatment and refrigeration. For example, **Figure 16.32** shows results of a typical experiment measuring the effect of pH on microbial survival in refrigerated food—in this case, *E. coli* O157:H7 in Greek eggplant salad. Note the critical threshold pH required to decrease bacterial counts. At pH 4.0, about the pH of lemon juice, the bacteria show a steep exponential death curve, whereas at pH 4.5 the bacteria remain viable for many days.

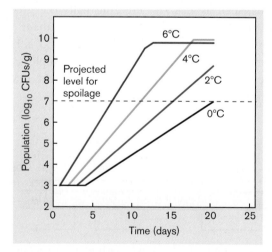

FIGURE 16.31 ■ **Bacterial growth in ground beef.** The growth rate of total aerobic bacteria in ground beef declines at lower storage temperatures. CFUs = colony-forming units. *Source:* Modified from M. P. Doyle (ed.). 2001. *Meat, poultry, and seafood.* Chapter 5 in *Food Microbiology,* ASM Press.

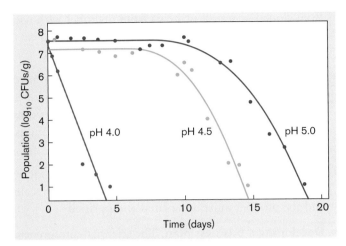

FIGURE 16.32 ■ **Bacteria die sooner at lower pH.** Survival curves of *E. coli* O157:H7 in eggplant salad stored at 5°C at pH 4.0, pH 4.5, and pH 5.0. CFUs = colony-forming units. *Source:* Modified from Panagiotis N. Skandamis and George-John E. Nychas. 2000. *Appl. Environ. Microbiol.* **66**:1646.

How does the food industry know how long to treat food for sterilization? Several measurements indicate the efficiency of heat killing. The D-value (decimal reduction time) was described in Chapter 5. An additional measure is 12D, the amount of time required to kill 10^{12} spores (or to decrease a population by 12 $\log_{10}$ units). A third measure is the z-value, the increase in degrees Celsius needed to lower the D-value to 1/10 of the time. If, for example, D_{100} (the D-value at 100°C) and D_{110} (the D-value at 110°C) for a given organism are 20 minutes and 2 minutes, respectively, then $12D_{100}$ equals 240 minutes (that is, 20 minutes × 12), and the z-value is 10°C (because a 10°C increase in temperature reduced the D-value to 1/10, from 20 minutes to 2 minutes). These measurements are determined empirically for each organism. The values are extremely important to the canning industry, which must ensure that canned goods do not contain spores of *Clostridium botulinum*, the anaerobic soil microbe that causes the paralyzing foodborne disease botulism (see **Fig. 16.30**).

Because the tastes of certain foods suffer if they are overheated, z-values and 12D-values are used to adjust heating times and temperatures to achieve the same sterilizing result. Take the following example, where D_{121} is 10 minutes and $12D_{121}$ is 120 minutes. Sterilizing at 121°C for 120 minutes might result in food with a repulsive taste, whereas decreasing the temperature and extending the heating time might yield a more palatable product. The D-values and z-values are used to adjust conditions for sterilization at a lower temperature. If D_{121} is 15 minutes (the time needed to kill 90% of cells) and the z-value is known to be 10°C (the temperature change needed to change D-value tenfold), then decreasing temperature by 10°C, to 111°C, will mean D_{111} is 150 minutes (10 × D_{121}).

Therefore, the value of 12D (required to decrease a population by 12 logs) is $12D_{111}$ = 1,800 minutes. Sterilization may take longer, but food quality is likely to remain high, because the sterilizing temperature is lower.

Chemical means of preservation. Many kinds of chemicals are used to preserve foods. Major classes of chemical preservatives include:

■ **Acids.** While microbial fermentation can preserve foods by acidification, an alternative approach is to add acids directly. Organic acids commonly used to preserve food include benzoic acid, sorbic acid, and propionic acid. The acids are generally added as salts: sodium benzoate, potassium sorbate, sodium propionate. These acids act by crossing the cell membrane in the protonated form and then releasing their protons at the higher intracellular pH. For this reason, they work best in foods that already have moderate acidity (pH 5–6), such as dried fruits and processed cheeses.

■ **Esters.** The esters of organic acids often show antimicrobial activity whose basis is poorly understood. Examples include fatty acid esters and parabens (benzoic acid esters). They are used to preserve processed cheeses and vegetables.

■ **Other organic compounds.** Numerous organic compounds, both traditional and synthetic, have antimicrobial properties. For example, cinnamon and cloves contain the benzene derivative eugenol, a potent antimicrobial agent.

■ **Inorganic compounds.** Inorganic food preservatives include salts, such as phosphates, nitrites, and sulfites. Nitrites and sulfites inhibit aerobic respiration of bacteria, and their effectiveness is enhanced at low pH. These substances, however, may have harmful effects on humans; nitrites can be converted to toxic nitrosamines, and sulfites cause allergic reactions in some people.

Thought Question

16.9 Is it possible for physical or chemical preservation methods to completely eliminate microbes from food? Explain.

To Summarize

■ **Food spoilage** refers to chemical changes that render food unfit for consumption. Food spoils through degradation by enzymes within the food, through spontaneous chemical reactions, and through microbial metabolism.

- **Food contamination, or food poisoning,** refers to the presence of microbial pathogens that cause human disease, or toxins produced by microbial growth. Food harvesting, food processing, and shared consumption are all activities that spread pathogens.

- **Dairy products** can be soured by excessive fermentation or made bitter by bacterial proteolysis.

- **Meat and poultry** are putrefied by decarboxylating bacteria, which produce amines with noxious odors.

- **Fish and other seafood** spoil rapidly because their unsaturated fatty acids rapidly oxidize; they harbor psychrotrophic bacteria that grow under refrigeration; and their trimethylamine oxide is reduced by bacteria to the fishy-smelling trimethylamine.

- **Vegetables** spoil by excess growth of bacteria and molds. Plant pathogens destroy food crops before harvest.

- **Food preservation** includes physical treatments, such as freezing and canning, as well as the addition of chemical preservatives, such as benzoates and nitrites.

16.6 Industrial Microbiology

The production and preservation of food is only one field of **industrial microbiology**, the commercial exploitation of microbes. Industrial microbiology is commonly understood to include a broad range of commercial products derived from microbes, including vaccines and clinical devices; industrial solvents and catalytic enzymes; bioinformatic analysis of genomes; and genetically modified plants and animals using microbial vectors.

Other important fields of industrial microbiology are wastewater treatment, bioremediation, and environmental management, covered in Chapters 21 and 22.

Industrial Microbiology Aims for Commercial Success

The practical application of a microbial product or device may arise out of an industrial laboratory, or it may be conceived by a research scientist with the aim of meeting a compelling need in society. In all cases, however, the key goal is to succeed in the marketplace—that is, to generate a product that customers adopt over alternative technologies. The product's sales must cover the costs of raw materials and production and (in a for-profit company) generate a profit for the shareholders. Success requires:

- **Identifying a useful product.** Possible products include small molecules, such as antibiotics; human proteins from cloned genes; or proteins from microbes with useful properties, such as thermostability or increased catalytic activity for a chemical process.

- **Isolating a microbe to produce the product.** A novel product, such as an antibiotic, is generally developed from a naturally occurring microbe. The genes encoding product biosynthesis may then be cloned into an industrial vector.

- **Scaling up production in quantity.** The producer microbial strain must be grown on an industrial scale, and the product must be isolated and purified.

- **Developing a business plan.** The scientist-entrepreneur must obtain partners skilled at industrial management, finance, and marketing. Patents must be filed to protect intellectual property rights.

- **Safety and efficacy testing.** Human consumption, environmental introduction, or consumer use requires many levels of testing prior to commercialization, including, in some cases, approval by government agencies.

- **Effective marketing.** The benefits of the new product must be communicated effectively to convince customers of its superiority to current products or processes.

Failure at any of the preceding tasks spells doom for the product. Thus, a prudent business plan includes having multiple alternative products in development. Although the failure rate of new products is high, all the products we use had to overcome these risks. How does a bench scientist convert a microbial concept into a profitable product? Some scientists acquire partners to found their own company. Others join a mature company that offers many kinds of expertise in business and production.

Case Example: A Microbiologist Founds a Company

Carol Nacy (**Fig. 16.33A**) studied tropical infectious diseases for 17 years at the Walter Reed Army Institute of Research. She saw the need to fight tuberculosis, a disease that is the world's number one killer of women aged 15–44 and the leading killer of men, after traffic accidents. The United States spends $1 billion yearly to treat 13,000 incident cases; yet the standard antibiotics for tuberculosis were developed before 1970, and the main diagnostic test available (the tuberculin skin test) dates to 1880, the time of Robert Koch. The best available vaccine—the Bacille Calmette-Guérin (BCG) live, attenuated vaccine—is only 50% effective.

Nacy obtained business experience by working for several years as a chief scientific officer for the pharmaceutical

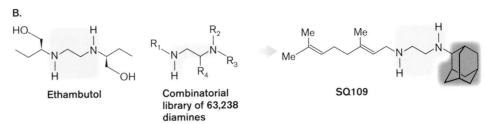

Ethambutol

Combinatorial library of 63,238 diamines

SQ109

FIGURE 16.33 ■ **Sequella develops antibiotics to fight tuberculosis.** **A.** Carol Nacy founded the Sequella company. **B.** A new antibiotic for TB is obtained by screening ethambutol analogs. A combinatorial library of compounds containing the ethambutol diamine core (yellow) was screened for antibacterial effect against *Mycobacterium tuberculosis*. The most promising agent screened was SQ109, with an unusual carbon-cage side group (pink).

greatest efficacy and fewest side effects was SQ109, a molecule with an unusual cage-like side group of three fused rings. SQ109 is now in human clinical trials.

company EntreMed, Inc. She then founded her own company, Sequella, to develop innovative drugs and treatment devices. Sequella targets innovative ideas with high risk but also high potential to improve performance, rapidity, and safety of diagnosing and treating TB infections. Nacy and her cofounders scanned the academic community for novel ideas that had succeeded against TB in "proof of principle" animal models. The most promising ideas were developed for improved antibiotics, rapid and less invasive tests for TB exposure, and devices to measure the extent of pulmonary infection. Because any one idea had a high risk of failure, researchers pursued multiple prospects in each category. The company also took on multiple target diseases, including methicillin-resistant *Staphylococcus aureus* (MRSA) and vancomycin-resistant *Enterococcus* (VRE).

Sequella's most promising antibiotic, SQ109, was discovered by high-throughput screening of a chemical library, in collaboration with Clifton Barry at the National Institutes of Health. The chemical library consisted of over 60,000 analogs of a known TB antibiotic, ethambutol (**Fig. 16.33B**). Ethambutol is part of the current standard treatment for TB, whose 6-month time course has a poor compliance rate. It is hoped that improved drugs will shorten the time course and improve compliance, thereby decreasing the appearance of drug-resistant strains. The analog molecules were selected for their common diamine core, with different combinations of side chains. The 60,000 compounds in the library were subjected to combinatorial screening, a mathematically intensive analysis based on numerous tests. Of the compounds tested, 2,796 showed activity against *Mycobacterium tuberculosis* in the test tube. The 69 best compounds were tested for cytotoxicity in tissue culture, activity in TB-infected macrophages, and activity in infected animals. The compound with the

Microbial Products

Companies such as Sequella seek agents to combat microbial pathogens, but increasingly, industry seeks beneficial microbes from natural sources to produce useful products. The microbe is cultured to synthesize a commercially valuable chemical substance such as the industrial enzyme lipase B (see the Current Research Highlight), or a material for manufacturing. An example of a microbial material used for manufacture is alginate (**Fig. 16.34A**). Alginate is a carboxylated polysaccharide, which has long been harvested commercially from giant kelp, *Macrocystis pyrifera* (**Fig. 16.34B**). The kelps harvested for alginate are "brown algae," eukaryotic microbes (discussed in Chapter 20). The product alginate has many commercial uses, as a thickener for food products from ice cream to faux caviar, a textile printing agent, and a modeling material for dental materials and life casting (**Fig. 16.34C**). Alginate is just one of a mushrooming number of products in what we call the **fermentation industry**.

Examples of commercial microbial products are listed in **Table 16.4**. Each product requires a gene or operon of genes encoding either the product itself or the enzymes for the product's biosynthesis. For example, the Danish company Novozymes markets over 700 microbial enzymes for products ranging from laundry detergents to agricultural inoculants (discussed in **Special Topic 16.1**). Novozymes also produces numerous enzymes as biocatalysts for green chemistry. "Green chemistry" refers to environmentally friendly procedures for reactions in organic chemistry, typically using water solution in place of petroleum-derived organic solvents.

We may distinguish between two fundamentally different sources of products: cloned genes from human, animal, or plant sources; and native microbial products

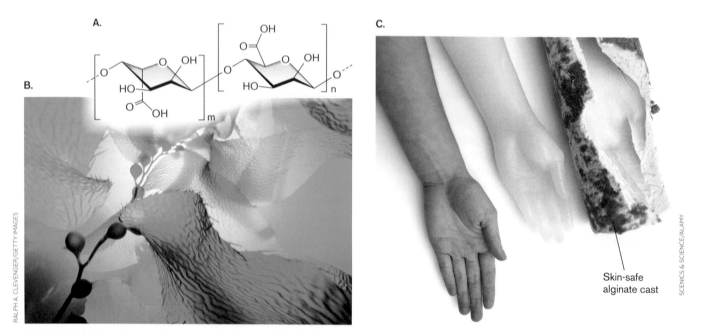

FIGURE 16.34 ■ **Giant kelp produce industrial alginate.** **A.** Alginate polysaccharide. **B.** Giant kelp, *Macrocystis pyrifera*. **C.** Alginate is used for skin-safe life casting.

from newly discovered species in the environment, often from extreme conditions. Cloned human genes typically encode a protein of valuable function in the human body. For example, Therabron Therapeutics produces the recombinant human protein CC10, a lung development protein that is often deficient in the lungs of premature infants. The recombinant protein, produced and purified from a recombinant bacterium, can be used to reduce lung inflammation in premature infants, as well as patients with chronic obstructive pulmonary disease (COPD). Cloning gene products in recombinant organisms was discussed in Chapter 12.

To identify new microbial products, companies screen thousands of microbial strains from diverse ecosystems. The search for organisms with potential commercial applications is called **bioprospecting**. Bioprospecting can be done anywhere, from one's backyard to Yellowstone National Park. Unique ecosystems are the most promising sources of previously unknown microbial strains with valuable properties. Extreme environments such as the hot springs of Yellowstone are particularly promising because their products may tolerate higher temperatures that are required for industrial use. Psychrophiles from extremely cold environments such as Antarctica are a useful source for enzymes that are active in a wide range of temperatures, such as the lipase from *Candida antarctica* (see the Current Research Highlight). Unfortunately, many such unique ecosystems are endangered by pollution, human-introduced invasive species, and global climate change.

An important aspect of bioprospecting is "mining the genome." Once a promising source strain is obtained, its genome is sequenced to investigate the genetic sequences that encode the useful product and regulate its expression. The operons encoding the product (or enzymes for its production) can then be cloned for optimal production (see Chapter 12). The cloned genes are transferred into an **industrial strain**, a strain whose growth characteristics are well studied and optimized for industrial production. An industrial strain must possess the following attributes:

■ **Genetic stability and manipulation.** The industrial strain must reproduce reliably, without major DNA rearrangements. It must also have an efficient gene transfer system by which vectors can introduce genes of interest into its genome.

■ **Inexpensive growth requirements.** Industrial strains must grow on low-cost carbon sources with minimal special needs, such as vitamins, and at easily maintained conditions of temperature and gases.

■ **Safety.** Industrial strains must be nonpathogenic and must not produce toxic by-products.

■ **High level of product expression.** The strain or recombinant vector must possess an efficient gene expression system to generate the desired product as a high proportion of its cell mass.

■ **Ready harvesting of product.** Either the product must be secreted by the cell or, if the product is intracellular, the cells must be easily breakable to liberate the product.

Common species for industrial strains include the bacterium *Bacillus subtilis*, the yeast *Candida utilis*, and the

TABLE 16.4	Commercial Products from Microbes	
Class of product	**Product**	**Producer microbe[a]**
Agricultural products	Gibberellin (plant hormone)	*Fusarium moniliforme* (fungus)
	Fungicides	*Coniothyrium minitans* (fungus)
	Insecticides (live pathogens)	*Bacillus thuringiensis*
Enzymes	Alpha-amylase	*Bacillus subtilis*
	Amyloglucosidase	*Aspergillus niger* (fungus)
	Lactase (beta-galactosidase)	*Kluyveromyces lactis* (fungus)
	Lipases	*Candida cylindraceae* (yeast)
	Alkaline protease	*Aspergillus oryzae* (fungus)
Food supplements	L-Lysine	*Brevibacterium lactofermentum*
	L-Tryptophan	*Klebsiella aerogenes*
	Monosodium glutamate (MSG)	*Corynebacterium ammoniagenes*
	Vitamin B_{12}	*Pseudomonas denitrificans*
	Vitamin C (ascorbic acid)	*Acetobacter suboxidans*
Fuels and solvents	Acetone	*Clostridium* spp.
	Butanol	*Clostridium acetobutylicum*
	Glycerol	*Zygosaccharomyces rouxii* (yeast)
	Methane (natural gas)	Methanogens (archaea)
Organic acids	Acetic acid	*Acetobacter xylinum*
	Citric acid	*Aspergillus niger* (fungus)
	Lactic acid	*Lactobacillus delbruckii*
Pharmaceuticals: antibiotics	Streptomycin	*Streptomyces griseus*
	Penicillin	*Penicillium chrysogenum* (fungus)
	Erythromycin	*Saccharopolyspora erythraea*
	Tetracycline	*Streptomyces aureofaciens*
Pharmaceuticals: other drugs	Lysergic acid (LSD, hallucinogen)	*Claviceps paspali* (fungus)
	Cyclosporine (immunosuppressant)	*Trichoderma polysporum* (fungus)
	Steroids	*Arthrobacter* spp.
Pharmaceuticals: cloned human proteins	Insulin	Recombinant *Escherichia coli*
	Interferon	Recombinant *E. coli* or *Saccharomyces cerevisiae* (yeast)
	Human growth hormone	Recombinant *E. coli*
	CC10 lung development protein	Recombinant *E. coli*
	Antibodies	Baculovirus-infected caterpillars
	Cancer regulators	Baculovirus-infected caterpillars

[a]Bacteria, unless stated otherwise.

Source: M. J. Waites. 2001. *Industrial Microbiology*, Blackwell Science, table 4.1, pp. 76–77.

filamentous fungus *Aspergillus niger*. Each of these species is safe; grows to high density on inexpensive carbon sources, such as molasses; and expresses desired products at high concentration.

Thought Question

16.10 Why would different industrial strains or species be used to express different kinds of cloned products?

Fermentation Systems

Commercial success requires optimizing every detail of the fermentation system. "Fermentation" in industrial terms refers not just to anaerobic metabolism, but to all means of growth of microbes on an industrial scale. In an **industrial fermentor**, the growth vessel and all its environmental supports, such as temperature control and oxygenation, must be scaled up to thousands of liters (**Fig. 16.35**). This increase in scale generates many problems of quality control, such as maintaining uniform temperature, pH, and

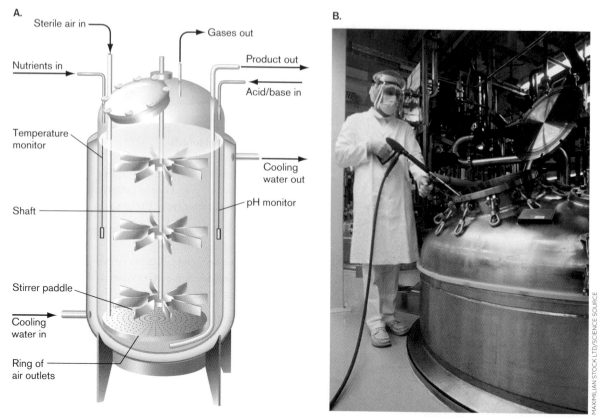

FIGURE 16.35 ▪ An industrial fermentor. **A.** Industrial production of microbial products requires scaled-up culture of the production microorganism. **B.** An industrial fermentor.

oxygenation throughout the vessel and minimizing foaming of the culture liquid. A small change in any of the growth factors can impact production costs and profit margin. Another major concern is to avoid contamination by other organisms.

The fermentor is the core of the first half of industrial production, known as **upstream processing** (**Fig. 16.36**, top). Upstream processing refers to the culturing of the industrial microbe to produce large quantities of product or cell mass. All aspects of the process must be controlled to maximize the final concentration of product, which in most cases peaks at a specific time in the microbial growth cycle. Following microbial growth, the culture must be harvested and the product purified. These processes constitute **downstream processing** (**Fig. 16.36**, bottom). The first step of downstream processing is to separate the microbial cells from the culture fluid, by centrifugation or by filtration. Next, the **primary recovery** of product follows one of two different pathways, depending on whether the product is maintained within the cells or secreted into the culture fluid. Many kinds of subsequent purification and finishing steps are necessary before the product has acceptable quality for its desired use. Again, failure of any detail can render the entire product unusable.

Products designed for human consumption or use in the environment face formidable hurdles in clinical testing, toxicological studies, and, finally, approval by the appropriate regulatory agency, such as the U.S. Food and Drug Administration (FDA). After millions of dollars are invested in process development, the product may still fail one of these late-stage hurdles and never come to market. Not surprisingly, a company must research thousands of potential products before achieving one that makes a profit. The consumer cost inevitably includes the development costs not only of the one successful product, such as recombinant insulin, but also of all the products that failed.

Despite all the hurdles to overcome, companies such as Novozymes sell hundreds of products and make billion-dollar profits. Examples of commercially successful products are described in **Special Topic 16.1**.

Production in Plant or Animal Host Systems

Some products require subtle processing that occurs correctly only within eukaryotic cells. For example, protein products may require posttranslational modifications such as glycosylation (attachment of polysaccharide chains).

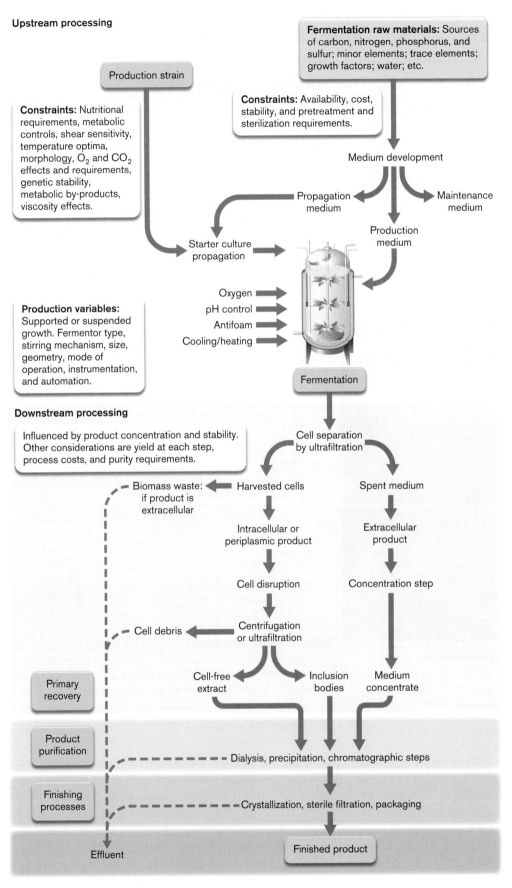

Upstream processing

Production strain

Constraints: Nutritional requirements, metabolic controls, shear sensitivity, temperature optima, morphology, O_2 and CO_2 effects and requirements, genetic stability, metabolic by-products, viscosity effects.

Fermentation raw materials: Sources of carbon, nitrogen, phosphorus, and sulfur; minor elements; trace elements; growth factors; water; etc.

Constraints: Availability, cost, stability, and pretreatment and sterilization requirements.

Medium development

Propagation medium

Maintenance medium

Production medium

Starter culture propagation

Oxygen
pH control
Antifoam
Cooling/heating

Production variables: Supported or suspended growth. Fermentor type, stirring mechanism, size, geometry, mode of operation, instrumentation, and automation.

Fermentation

Downstream processing

Influenced by product concentration and stability. Other considerations are yield at each step, process costs, and purity requirements.

Cell separation by ultrafiltration

Biomass waste: if product is extracellular

Harvested cells

Spent medium

Intracellular or periplasmic product

Extracellular product

Cell disruption

Concentration step

Cell debris

Centrifugation or ultrafiltration

Primary recovery

Cell-free extract

Inclusion bodies

Medium concentrate

Product purification

Dialysis, precipitation, chromatographic steps

Finishing processes

Crystallization, sterile filtration, packaging

Effluent

Finished product

FIGURE 16.36 ■ Details of upstream and downstream processing. Top: Upstream processing consists of engineering the microbial strain and large-scale growth to generate the product. **Bottom:** Downstream consists of product concentration and purification.

SPECIAL TOPIC 16.1 Microbial Enzymes Make Money

How can we clean our clothes with less pollution and keep their colors bright? Can we remove carcinogens from our food and increase crop yields with less environmental impact? The answers to these questions lie in the microbial genomes. Microbial genes encode enzymes with properties worth hundreds of billions of dollars. But it takes a sophisticated research company to discover enzymes, develop them, and make a profit.

One of the foremost microbial companies today is Novozymes. Founded in 1925, the company produced insulin and trypsin by extraction from animal tissues—a laborious, expensive process. In 1963, Novozymes introduced its first product of microbial fermentation: the detergent enzyme Alcalase. In the 1980s the company took up genetic modification (recombi-nant DNA) to develop dozens of profitable microbial enzymes for cleaning, including Lipolase, a lipase (fat cleavage) from the fungus *Thermomyces lanuginosus*; and Protamex, a protease (protein cleavage) from *Bacillus* bacteria. These enzymes break down and solubilize various components of food stains. Active at room temperature, these enzymes precisely remove the most common food stains while avoiding the energy-expensive heating of water. They also avoid the need for phosphate detergents, which pollute waterways.

Some cleaning enzymes actually improve the appearance of clothing. The cellulase Carezyme, for example, hydrolyzes short, frayed ends of cellulose that project from a cotton garment after many cleanings (**Fig. 1**). Cotton consists of

Washed 25 times

Washed again with Carezyme

FIGURE 1 ■ **Carezyme Premium restores color.** During cleaning, the cellulase cleaves frayed ends of material, removing the fuzzy appearance and improving the garment's color.

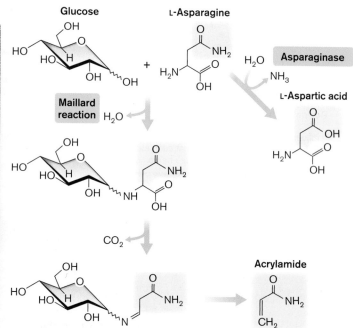

FIGURE 2 ■ **Asparaginase prevents formation of acrylamide.** Asparaginase converts asparagine into aspartic acid. Without the action of asparaginase, asparagine combines with sugar at high temperature (the Maillard reaction) and then releases acrylamide.

In such cases, the expressed genes must be transferred from a bacterial or viral vector into an animal, fungal, or plant tissue culture or to a transgenic organism. The microbially transformed organism may itself be the industrial product. Here we present an example of microbial production involving the plant engineering bacterium *Agrobacterium tumefaciens*. A second example, that of an insect caterpillar virus, is presented in **eTopic 16.2**.

Bacteria and viruses provide the vectors for transferring genes into multicellular organisms, as discussed in Chapter 12. A bacterium of major industrial importance is *A. tumefaciens*, a tumor-inducing plant pathogen that conducts natural genetic engineering on dicot plants (**Fig. 16.37** ▶). *Agrobacterium* strains have now been adapted by scientists to engineer even monocots. Long before scientists invented "recombinant DNA," *A. tumefaciens* had evolved a gene transfer system by which it induces infected plant cells to generate food molecules to feed the pathogen. This highly efficient gene transfer system is readily modified to insert genes conferring

crystalline fibers of cellulose (a glucose polysaccharide, discussed in Chapter 13). Cutting the frayed ends removes the fuzzy appearance, restoring the solid color of the garment. But why does this powerful enzyme not destroy the entire garment? The enzyme has been engineered for high specificity: It cleaves only the highly disordered stray fibrils, not the ordered cellulose of the intact fabric. The enzyme was originally isolated from a species of *Bacillus* that uses it to catabolize plant material.

Novozymes develops a number of enzymes with exciting applications for food production. In 2002, food scientists reported in *Nature* that high-temperature cooking of many foods results in formation of acrylamide, a carcinogen. The acrylamide forms in foods such as bread crusts and crackers, French fries, and coffee. It was shown to arise from a reaction between sugar and amino acids, particularly the amino acid asparagine (**Fig. 2**). First the asparagine condenses with the sugar by a mechanism known as the Maillard reaction. The molecule then rearranges, releasing acrylamide.

How can the acrylamide production be prevented? Novozymes scientists reasoned that most of the acrylamide production could be avoided by pretreatment of the food with the enzyme asparaginase, which deaminates asparagine to aspartic acid. They obtained the asparaginase gene from a strain of the fungus *Aspergillus*. An industrial organism was engineered to produce the enzyme in quantity, and marketed as Acrylaway. **Figure 3** outlines the use of Acrylaway to avoid acrylamide production in French fries. The enzyme is added to potato strips during the dipping step, before the potatoes are dried and fried. Because the enzyme is completely "natural," it introduces no new toxin. Industrial trials completed in 2012 confirmed that Acrylaway effectively protects many kinds of food from carcinogenic acrylamide.

RESEARCH QUESTION

The Novozymes product Acrylaway solves a particular problem for a specific industry. Search on the Internet for another industrial problem that might be solved by a microbial enzyme. Outline the steps you might take to develop such an enzyme product.

Stadler, Richard H., Imre Blank, Natalia Varga, Fabien Robert, Jörg Hau, et al. 2002. Food chemistry: Acrylamide from Maillard reaction products. *Nature* **419**:449–450.

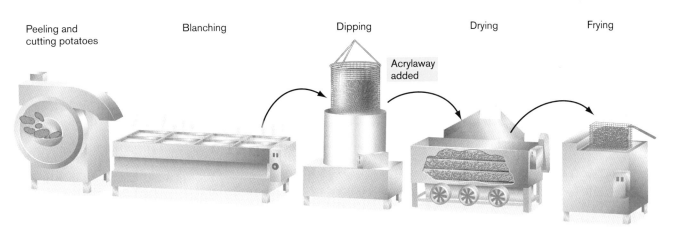

FIGURE 3 ■ Asparaginase in French fries. Asparaginase added at the dipping step prevents acrylamide formation in French fries.

traits of interest, such as herbicide resistance, into plant genomes.

Tumorigenic strains of *A. tumefaciens* possess a special plasmid for engineering plant cells, called the **Ti plasmid** (tumor-inducing plasmid). The Ti plasmid is the source of the genetic material that gets transferred into a plant cell through a process mediated by bacterial proteins, similar to conjugation (see Section 9.1). The Ti plasmid includes the *vir* operons, which encode a virulence system, as well as the T-DNA (transferable DNA), a set of genes that will be transferred to the host plant and recombined into its genome. T-DNA encodes tumor induction genes, such as auxin synthesis genes, as well as enzymes for biosynthesis of a carbon and nitrogen source called an opine. Opines are specialized amino acids made by a one-step synthesis from arginine, typically by amination of a central metabolite such as pyruvate or 2-oxoglutarate. A given strain of *A. tumefaciens* typically provides one type of opine synthesis enzyme and has the ability to metabolize the corresponding opine.

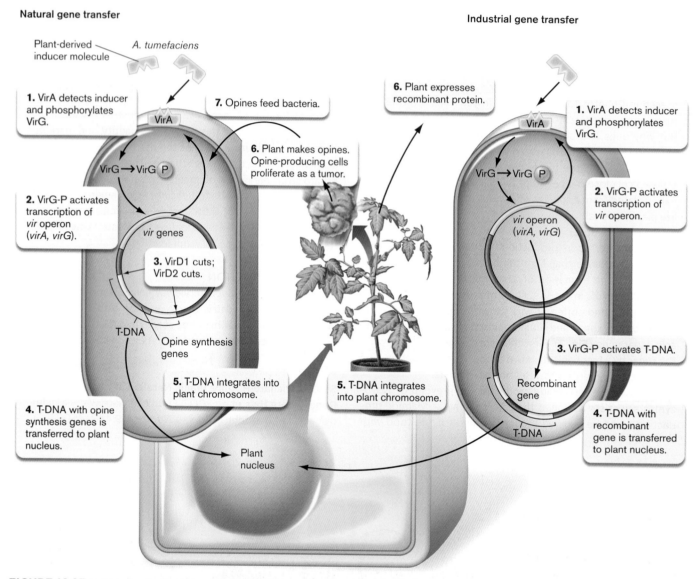

Natural gene transfer

Plant-derived inducer molecule — *A. tumefaciens*

1. VirA detects inducer and phosphorylates VirG.

VirA

7. Opines feed bacteria.

VirG → VirG P

6. Plant makes opines. Opine-producing cells proliferate as a tumor.

2. VirG-P activates transcription of *vir* operon (*virA, virG*).

vir genes

3. VirD1 cuts; VirD2 cuts.

T-DNA

Opine synthesis genes

5. T-DNA integrates into plant chromosome.

4. T-DNA with opine synthesis genes is transferred to plant nucleus.

Plant nucleus

Industrial gene transfer

6. Plant expresses recombinant protein.

1. VirA detects inducer and phosphorylates VirG.

VirA

VirG → VirG P

2. VirG-P activates transcription of *vir* operon.

vir operon (*virA, virG*)

3. VirG-P activates T-DNA.

Recombinant gene

4. T-DNA with recombinant gene is transferred to plant nucleus.

T-DNA

5. T-DNA integrates into plant chromosome.

FIGURE 16.37 ■ *Agrobacterium tumefaciens:* **a natural gene transfer vector for plants. Left:** *A. tumefaciens* transfers T-DNA containing opine synthesis genes into a plant, which then produces opines to feed the bacteria. **Right:** *A. tumefaciens* can be engineered to transfer T-DNA containing a recombinant gene of interest into the plant genome. A recombinant strain of *A. tumefaciens* has the Ti plasmid divided into two separate plasmids: one containing the *vir* operon conducting DNA transfer, the other containing T-DNA with most of its genes substituted by a desired recombinant gene. ▶

The *Agrobacterium* vector transfers its DNA to the plant by means of the *vir* gene products. The *vir* gene products from the Ti plasmid detect the presence of a plant host and stimulate plasmid transfer (see **Fig. 16.37,** left). In the cell envelope, VirA protein detects a chemical signal from a wounded plant, which is capable of being infected. VirA then activates VirG to induce expression of other *vir* genes, encoding endonucleases VirD1 and VirD2, and the VirB secretion complex. The VirD endonucleases cleave the left and right ends of the T-DNA and direct its transfer into the plant cell through the VirB system. Within the plant cell nucleus, the T-DNA becomes integrated into the plant genome, where it induces opine production. The opine-producing cells proliferate, forming a tumor.

For industrial use (**Fig. 16.37,** right), a recombinant strain of *A. tumefaciens* has the Ti plasmid divided into two separate plasmids. One contains the *vir* operons conducting DNA transfer, while the other contains T-DNA with its left and right ends intact but its genes substituted by the desired recombinant genes. Upon infection, the virulence system induces transfer of the T-DNA to the plant cell without any tumor-inducing genes, allowing genomic integration of the recombinant genes with their desired traits, without tumor induction or opine production.

Microbe as Product

Microbes as biological agents possess enormous potential for environmental use. The most well known example is the bacterial insecticide *Bacillus thuringiensis* (discussed in Chapter 12). Such microbes, derived from natural sources, may improve crop production and enhance agricultural sustainability. Most methods used by farmers to improve crop yields require large amounts of fertilizer and pest management, including herbicides and fungicides. To alleviate the financial and environmental impacts of large-scale farming, scientists are working to develop microbial solutions that can supplement chemical additives and pesticides, resulting in higher yields to farmers.

Scientists at Novozymes partner with Monsanto to conduct large-scale bioprospecting. The scientists screen thousands of microbes for positive impacts on agriculture. Some of the microbial products result in higher corn yields across North America. One example is JumpStart, which consists of the mold *Penicillium bilaii* (**Fig. 16.38A**). The mold solubilizes phosphate from the soil and makes it available to the roots of plants, whose exudates feed the mold. This phosphate-solubilizing inoculant improves growth for a variety of crops.

Another microbial product is Actinovate, a biological fungicide used for the suppression of root rot and damping-off fungi, as well as the suppression or control of leaf fungal pathogens. The active ingredient of Actinovate is the actinomycete *Streptomyces lydicus* (**Fig. 16.38B**). *S. lydicus* grows as a mutualist associated with the roots of crop plants. The bacteria feed off plant exudates while secreting antimicrobial substances that suppress pathogens. These examples are the start of a growing field of microbial applications for improving agriculture in an environmentally sustainable manner.

To Summarize

- **Industrial microbiology** includes the production of vaccines and clinical devices, industrial solvents and pharmaceuticals, and genetically modified organisms.

- **Bioprospecting** is the mass screening of new microbial strains for promising traits and potential use in a product. Thermophiles and psychrophiles are particularly important sources of new strains with potentially interesting new properties.

- **Microbial products must be competitive** with alternative technologies. Developing a competitive new molecular product requires identifying a useful molecule, isolating and developing a fermentation technique to produce it, scaling up for production in quantity, developing a business plan, and testing for safety.

  - **Industrial strains**, commonly *Escherichia coli* or *Bacillus subtilis*, are often used to incorporate the newly discovered genes into an industrially useful microbe.

  - **Upstream processing** refers to the culturing of the industrial microbe to produce large quantities of product. **Downstream processing** involves product recovery and purification. **Posttranscriptional processing** may be required for human or plant gene products.

  - **Microbes may be developed as products** such as a fungicide or a microbial mutualist.

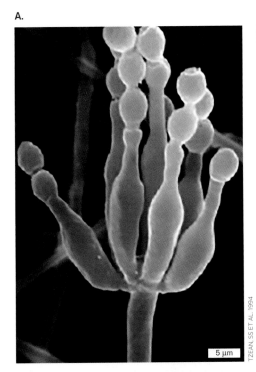

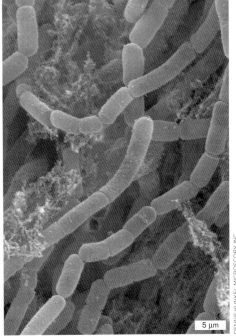

FIGURE 16.38 ■ Bacterial supplements for agriculture. A. *Penicillium bilaii*, active ingredient of JumpStart, solubilizes phosphate for roots of crop plants (SEM). B. *Streptomyces lydicus*, active ingredient of Actinovate, suppresses plant pathogens (SEM).

Concluding Thoughts

In this chapter we have seen how microbes from our environment contribute products for human use, from foods and vitamin supplements to industrial enzymes and highly specific protein therapeutics. These microbial applications incorporate much of the natural metabolism and biochemistry introduced in Chapters 13–15. Human inventiveness and industrial and financial management must meet many challenges to develop new microbial products.

What does the future hold for industrial microbiology? Entirely new microbes—viruses, bacteria, algae—are being engineered or "synthesized" to perform functions in an industrial system, or even within the human body. The field of synthetic biology is just beginning to develop such microbial partners for industry and for sustainable agriculture.

CHAPTER REVIEW

Review Questions

1. What kinds of microbes are consumed as food? Why are most bacteria inedible for humans?
2. What are the advantages of fermentation for a food product?
3. What are the differences between traditional fermented foods and commercial fermented foods?
4. How do acid fermentations contribute to the formation of different kinds of cheeses? What is the role of different kinds of metabolism performed by different microbial species?
5. Compare and contrast acid and alkaline fermentation processes. What different kinds of foods are produced?
6. Compare and contrast the role of ethanolic fermentation in bread making and winemaking.
7. How do relatively minor fermentation reactions contribute to the flavor of food?
8. Explain the differences between food spoilage and food poisoning.
9. What are the most important food-borne pathogens, based on infection rates? Based on mortality rates?
10. What are the major means of preserving food? Compare and contrast their strengths and limitations.
11. What tasks must be accomplished to develop a microbial product for commercial marketing?
12. What are the sources of potential new microbial products? What is the difference between a source strain and an industrial strain?
13. Explain upstream processing and downstream processing.
14. Explain why genes encoding industrially useful products may be transferred into plant or animal systems for production. What are the roles of microbes in these systems?

Thought Questions

1. In cheese production, how do different kinds of fermenting microbes generate different flavors?
2. If you were a food safety regulator, which pathogen on the list in **Table 16.3** would you consider your top priority? Defend your answer by citing factors such as numerical incidence of infection, severity of disease, and economic losses due to illness.
3. Suppose you undertake industrial production of a recombinant glycoprotein for human therapy. Would you synthesize your product in bacteria, in yeast, or in caterpillars? Explain the advantages and limitations of your choice.

Key Terms

alkaline fermentation (610)
antimicrobial agent (622)
bioprospecting (634)
cheddared (613)
cheese (611)

curd (611)
downstream processing (636)
endophyte (627)
ethanolic fermentation (610)
fermentation industry (606, 633)

fermented food (608)
food poisoning (623)
food spoilage (623)
freeze-drying (629)
heterolactic fermentation (610)

Recommended Reading

Centers for Disease Control and Prevention. 2008. Outbreak of *Listeria monocytogenes* infections associated with pasteurized milk from a local dairy—Massachusetts, 2007. *Morbidity and Mortality Weekly Report* **57**:1097–1100.

Centers for Disease Control and Prevention. 2009. Multistate outbreak of *Salmonella* infections associated with peanut butter and peanut butter–containing products—United States, 2008–2009. *Morbidity and Mortality Weekly Report* **58**:1–6.

Chang, Shu-Ting, and Philip G. Miles. 2004. *Mushrooms: Cultivation, Nutritional Value, Medicinal Effect, and Environmental Impact.* 2nd ed. CRC Press, New York.

Doyle, Michael P., and Robert L. Buchanan (eds.). 2013. *Food Microbiology: Fundamentals and Frontiers.* 4th ed. ASM Press, Washington, DC.

Giraffa, Giorgio. 2004. Studying the dynamics of microbial populations during food fermentation. *FEMS Microbiological Reviews* **28**:251–260.

Hui, Y. H., Lisbeth Meunier-Goddick, Åse S. Hansen, Jytte Josephsen, Wai-Kit Nip, et al. (eds.). 2004. *Handbook of Food and Beverage Fermentation Technology.* Marcel Dekker, New York.

Leggett, Mary, Nathaniel K. Newlands, David Greenshields, Lee West, S. Inman, et al. 2014. Maize yield response to a phosphorus-solubilizing microbial inoculant in field trials. *Journal of Agricultural Science* **153**:1464–1478. doi:10.1017/S0021859614001166.

Ma, Zhenkun, Christian Lienhardt, Helen McIlleron, Andrew J. Nunn, and Xiexiu Wang. 2010. Global tuberculosis drug development pipeline: The need and the reality. *Lancet* **375**:2100–2109.

Marilley, L., and M. G. Casey. 2004. Flavours of cheese products: Metabolic pathways, analytical tools and identification of producing strains. *International Journal of Food Microbiology* **90**:139–159.

Mills, David A., Helen Rawsthorne, C. Parker, D. Tamir, and K. Makarova. 2005. Genomic analysis of *Oenococcus oeni* PSU-1 and its relevance to winemaking. *FEMS Microbiological Reviews* **29**:465–475.

Schwan, Rosane F. 1998. Cocoa fermentations conducted with a defined microbial cocktail inoculum. *Applied Environmental Microbiology* **64**:1477–1483.

Schwan, Rosane F., and Alan E. Wheals. 2004. The microbiology of cocoa fermentation and its role in chocolate quality. *Critical Reviews in Food Science and Nutrition* **44**:205–221.

CHAPTER 17
Origins and Evolution

Microbial life appeared as early as 3.8 billion years ago, soon after our planet, Earth, formed out of dust surrounding the young Sun. Since then, microbes have evolved into forms adapted to diverse ways of life, from psychrophiles beneath the ice of Antarctica to anaerobes in the human colon. All living plants and animals, including ourselves, are descendants of those early microbes. How did microbes originate and evolve? How did their evolving metabolism shape the chemistry of Earth's crust and atmosphere?

CURRENT RESEARCH highlight

Bacteria evolve before our eyes. Two *Escherichia coli* clones compete, marked by genetic traits that make colonies appear red or white with an indicator dye. The proportion of colonies with each color shows which clone outcompetes the other. For example, if red colonies can metabolize citrate in the growth medium, whereas white colonies do not, then the red colonies will outnumber the white colonies, indicating greater fitness. This competition assay measures the relative fitness of bacteria from Richard Lenski's Long-Term Evolution Experiment. In this experiment, *E. coli* evolved a surprising way to metabolize the medium's citrate buffer. Caroline Turner discovered that some clones evolved a cooperative relationship in which one clone obtains nutrients produced by a different clone that catabolizes citrate.

Source: Caroline B. Turner, et al. 2015 *PLoS One* **10(11)**: e0142050

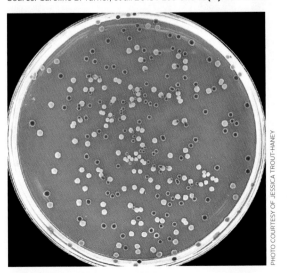

PHOTO COURTESY OF JESSICA TROUT-HANEY

AN INTERVIEW WITH

CAROLINE TURNER, MICROBIAL ECOLOGIST, MICHIGAN STATE UNIVERSITY

AMAR BHAGWAT

How did citrate-minus bacteria persist in the presence of bacteria that evolved to use citrate?

The persistence of the citrate-minus bacteria is remarkable because the citrate-plus bacteria had such a big advantage. They had sole access to a large pool of resources. To me, the survival of the citrate-minus bacteria is a beautiful example of the power of ecology—how having a distinct ecological niche (metabolizing waste products of citrate catabolism) can allow organisms to coexist.

In the future, what other novel phenotypes do you think might emerge from the Long-Term Evolution Experiment?

What other new traits might the bacteria evolve? Could they evolve some means of genetic exchange? Might the within-population competitive interactions ever take a turn toward predation? Who knows? Only time will tell—and only if we allow time, the bacteria, and future generations of scientists to do the work of evolution and science.

Living organisms generate offspring of their own kind—but slightly different. Despite the high accuracy of DNA replication, a few genes mutate, yielding diverse progeny. Some of the mutants reproduce more than others, either by chance or because certain genetic traits work better in the given environment. The more successful variants show greater fitness, for the environment tested. This process of evolution generates all the different kinds of life on Earth. Amazingly, evolution continues today, in every living organism. Bacteria rapidly generate large populations, which allow us to watch evolution as it happens. The Current Research Highlight shows a snapshot of competition between two clones of *E. coli* with slightly different traits. Even a small fitness difference, over many generations, can reshape the traits of a population.

If fitness advantage determines traits, how do organisms ever evolve to cooperate? Caroline Turner's discovery in Richard Lenski's lab shows one way in which cooperation may evolve. In the Long-Term Evolution Experiment, some clones had unexpectedly evolved the ability to metabolize the medium's citrate buffer (see Section 17.4). Turner found that a different clone evolved to use organic acids released by the citrate metabolizers. Within a population, the environment for any given individual includes other members of the population. If other members possess a trait useful to the first individual—such as excretion of a food molecule—the individual may sustain reproductive success, as long as the partner members are present. Such relationships explain, for example, the reproductive success of *E. coli* in the intestine, where the enteric bacteria feed on sugars released by *Bacteroides* species that break down complex polysaccharides (discussed in Chapter 21). Often, such cross-feeding relationships turn out to be reciprocal. For example, marine cyanobacteria release organic molecules that attract aerobic respirers to consume the oxygen produced by photosynthesis. Their aerobic respiration actually helps the cyanobacteria get rid of toxic oxygen.

Chapter 17 explores evidence for the origin of the earliest cells, as well as the challenges in interpreting data from so long ago. We show how molecular techniques reveal deep similarities among all life-forms, such as the core macromolecular apparatus of DNA, RNA, and proteins. We watch the mechanisms of ongoing microbial evolution emerge from laboratory experiments (discussed in Section 17.4). We present:

■ The origin of life on Earth and the nature of the earliest cells.

■ The divergence of microbes from common ancestors, modified by gene transfer and symbiosis.

■ The mechanisms of microbial evolution, as it unfolds in nature and in the laboratory.

17.1 Origins of Life

For centuries, observers of the natural world have wondered where life came from. Medieval alchemists in Europe argued that life arose spontaneously from inert matter—

A. Glacier meltwater bubbling with cyanobacteria

10°C

COURTESY OF RUTH C. HEINDEL

COURTESY OF JESSICA TROUT-HANEY

1 cm

B. Microbial mat

COURTESY OF JESSICA TROUT-HANEY

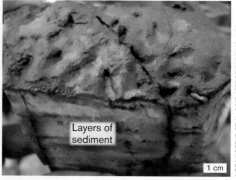

C. Fossil mat, Australia, 3.48 Gyr old

Layers of sediment

1 cm

NORA NOFFKE, ET AL. 2013 ASTROBIOLOGY 13: 1103.

FIGURE 17.1 ■ Microbial mat from Antarctica, and fossil mat from Australia. A. Cyanobacteria bubble oxygen in glacier meltwater at height of summer, McMurdo Dry Valleys, Antarctica. **Inset:** Jessica Trout-Haney studies environmental change in Antarctica. **B.** Cyanobacterial mat. **C.** This fossil microbial mat shows sedimentary layers, from the Dresser Formation, Australia, 3.48 billion years (Gyr) ago. *Source:* Nora Noffke.

a concept called spontaneous generation (discussed in Chapter 1). Others argued against spontaneous generation and devised experiments to show that even microbes have "parents." Today, genetic evidence overwhelmingly confirms that all life on Earth, including microbes, arises from preexisting life. But these experiments do not address the origin of the very first living cells—or how early life gave rise to multicellular plants and animals.

As early as 1802, the naturalist Erasmus Darwin, grandfather of Charles Darwin, wrote:

> Organic life beneath the shoreless waves
> Was born and nurs'd in ocean's pearly caves;
> First forms minute, unseen by spheric glass,
> Move on the mud, or pierce the watery mass;
> These, as successive generations bloom,
> New powers acquire and larger limbs assume.

Thus, nineteenth-century biologists developed the idea that "minute" life-forms arose in the ocean—and that all organisms evolved from microbes, perhaps even from cells too small to be seen with the "spheric glass" of a microscope. Even without the tools of genetics, thoughtful observers recognized the commonalities among all living cells, such as the membrane-enclosed compartment of cytoplasm and common metabolic pathways like sugar metabolism. Today, lines of evidence from geology, biochemistry, and genetics overwhelmingly support the microbial origin of life.

What might the first life-forms have looked like?

Early Life

Microbiologists find clues to the nature of early life in a place where today only the most rudimentary life-forms exist: the Dry Valleys of Antarctica (**Fig. 17.1**). The valleys' cold, dry winters (−40°C) preclude most multicellular life. But in the austral summer (January), temperatures rise above zero and glacier melt trickles into streams. Long-dormant cyanobacteria come to life, painting the streams orange (**Fig. 17.1A**). They photosynthesize throughout the 24-hour daylight, fixing CO_2 and bubbling oxygen. The cyanobacteria can grow into thick masses of biofilm, called **microbial mats** (**Fig. 17.1B**). The mats support communities of protists and nematodes. Microbial ecologists such as Jessica Trout-Haney, Dartmouth College, study what happens to microbial mats as the climate warms.

Antarctic microbial mats bear a striking resemblance to Australian fossils that date to 3.48 billion years (gigayears, or Gyr) before today (**Fig. 17.1C**). These fossils, from the Dresser Formation, Pilbara Craton, were identified by geologist Nora Noffke, from Old Dominion University. The fossils formed as silicate grains sedimented in the mat and gradually replaced its organic structure. The sedimentary layers and wrinkled surface remain visible after billions of years. Remarkably, similar rock formations appear on Mars, offering evidence that our neighbor planet, too, supported ancient microbial life (shown in Chapter 1).

Note: In geological description, a billion years (10^9) is a giga-year, or Gyr. A million years (10^6) is a megayear, or Myr.

Ancient microbes formed even more complex microbial communities, called **stromatolites** (**Fig. 17.2A**). A stromatolite is a bulbous mass of layered limestone (calcium carbonate, $CaCO_3$) accreted by microbial mats. The mat layers can build over centuries, even more than a thousand years, reaching heights over 2 meters. The outermost layers of the mat contain oxygenic phototrophs, such as diatoms and filamentous cyanobacteria, that exude bubbles of oxygen. A few millimeters below the surface, red light supports bacteria photolyzing H_2S to sulfate, which is then reduced

A.

B.

FIGURE 17.2 ■ Stromatolites: ancient life-forms in modern seas. A. Cyanobacteria form colonial stromatolites, present-day structures that resemble the earliest forms of life on Earth. Shark Bay, Western Australia. **B.** Section through a 3.4-billion-year-old fossil stromatolite from the Strelley Pool Chert, Pilbara Craton, Australia.

by still lower layers of bacteria. This shared metabolism is another kind of cross-feeding, like that shown by Caroline Turner in the Long-Term Evolution Experiment.

Stromatolites today grow mainly in tidal pools whose high salt concentration excludes predators, as in Hamlin Pool, Shark Bay, Australia. But 3 billion years ago, stromatolites covered shallow seas all over Earth. Fossil stromatolites appear in Pilbara Craton's 3.4-Gyr granite (**Fig. 17.2B**). The rock layers preserve the wavy form of the microbial mats, similar to living stromatolites.

Thought Question

17.1 Evolution by natural selection is based on competition, yet the earliest fossil life shows organized structures such as a stromatolite built by cooperating cells. How could this be explained?

Conditions for life. How did Earth's very first life-forms arise out of inert molecules? This process remains one of the great mysteries of science. But we know certain things that life required:

■ **Essential elements.** Because all life on Earth is composed of molecules, the origin of life required the fundamental elements that compose organic molecules.

■ **Continual source of energy.** The generation of life requires continual input of energy, which ultimately is dissipated as heat. The main source of energy for life is nuclear fusion reactions within the Sun.

■ **Temperature range permitting liquid water.** Above 150°C, life's macromolecules fall apart; below the freezing point of water, metabolic reactions cease. Maintaining the relatively narrow temperature range conducive to life depends on the nature of our Sun, our planet's distance from the Sun, and the heat-trapping capacity of our atmosphere.

Elements of Life

For life to arise and multiply, elements such as carbon and oxygen needed to be available on Earth. The planet Earth coalesced during formation of the solar system 4.5 Gyr ago. Central to the solar system is our Sun, a "yellow" star of medium size and surface temperature (5,770 K). The Sun's surface temperature generates electromagnetic radiation across the spectrum, peaking in the range of visible light. As we learned in Chapter 13, the photon energies of visible light are sufficient to drive photosynthesis but not so energetic that they destroy biomolecules. Thus, the stellar class of our Sun makes organic life possible.

The Sun's surface temperature and luminosity are generated by nuclear fusion reactions in which hydrogen nuclei fuse to form helium nuclei. (Be careful to distinguish nuclear reactions, involving nuclei, from chemical reactions, involving electrons.) Besides hydrogen and helium, 2% of the solar mass consists of heavier elements, such as carbon, nitrogen, and oxygen, as well as traces of iron and other metals—elements that compose Earth, including its living organisms. Where did these heavier elements come from? To answer this question, we must look to other stars in the universe at different stages of their development (**Fig. 17.3**).

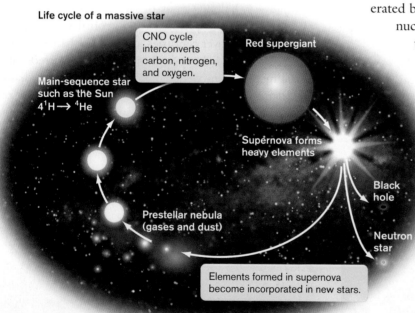

Life cycle of a massive star

CNO cycle interconverts carbon, nitrogen, and oxygen.

Red supergiant

Main-sequence star such as the Sun
$4\,^1H \longrightarrow \,^4He$

Supernova forms heavy elements

Black hole

Neutron star

Prestellar nebula (gases and dust)

Elements formed in supernova become incorporated in new stars.

FIGURE 17.3 ■ Stellar origin of atomic nuclei that form living organisms. In young stars, hydrogen nuclei fuse to form helium. In older stars, fusion of helium forms carbon, nitrogen, oxygen, and all the heavier elements up through iron. Massive stars explode as supernovas, spreading all the elements of the periodic table across space. These elements are picked up by newly forming stars, such as our own Sun.

Elements of life formed within stars. Throughout the universe, young stars such as our Sun fuse hydrogen to form helium. As stars age, they use up all their hydrogen. With hydrogen gone, the aging star contracts and its temperature rises, enabling helium nuclei to fuse, forming carbon (see **Fig. 17.3**). Carbon drives a cyclic nuclear reaction, the CNO (carbon-nitrogen-oxygen) cycle, to form isotopes of nitrogen and oxygen. Subsequent nuclear reactions generate heavier elements through iron (Fe). In this way, the major elements of biomolecules were formed within stars that aged before our solar system was born.

The later nuclear reactions of aging stars generate heavier nuclei, as large as that of iron. The aging star expands, forming a red giant (see **Fig. 17.3**). When a star of sufficient mass expands (becoming a supergiant), it explodes as a supernova. The explosion of a supernova generates in a brief time all the heaviest elements and ejects the entire contents of the star at near light speed. Billions of years before our Sun was born, the first stars aged and died, spreading all the elements of the periodic table across the universe. Some of these elements coalesced with our Sun and formed the planets of our solar system. In effect, all life on Earth is made of stardust, the remains of stars long gone.

Thought Question

17.2 What would have happened to life on Earth if the Sun were a different stellar class, substantially hotter or colder than it is?

Elemental composition of Earth. When our solar system formed, individual planets coalesced out of matter attracted by the force of gravity. Because of Earth's small size, most of the hydrogen gas escaped Earth's gravity very early. The most abundant dense component of Earth was iron (**Fig. 17.4**). Much of Earth's iron sank to the center to form the core. The core is surrounded by a mantle, composed primarily of iron combined with less dense

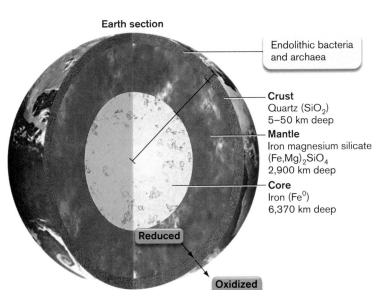

Earth section

Endolithic bacteria and archaea

Crust
Quartz (SiO_2)
5–50 km deep

Mantle
Iron magnesium silicate
$(Fe,Mg)_2SiO_4$
2,900 km deep

Core
Iron (Fe^0)
6,370 km deep

Reduced

Oxidized

FIGURE 17.4 ■ **Geological composition of Earth.** This cross section of Earth shows the core, the mantle, and the thin outer crust. The core and mantle are rich in iron; oxygen content increases toward the crust. The crust is composed primarily of silicates such as quartz (SiO_2). Crustal rock supports endolithic bacteria and archaea.

crystalline minerals such as silicates of iron and magnesium: $(Fe,Mg)_2SiO_4$. The mantle is coated by Earth's thin outer crust. The crust is composed primarily of silicon dioxide, SiO_2, also known as quartz or chert. Crustal rock contains smaller amounts of numerous minerals, including the carbonates and nitrates that provided the essential elements for life. Overall, the crust shows a redox gradient, reducing in the interior and oxidizing at the surface.

The crust provides a habitat for microbes to surprising depths, such as within gold mines excavated down to 3 kilometers (km). Some endolithic microbes (microbes living within rock) metabolize by oxidizing electron donors generated through decay of radioactive metals. The discovery of endolithic organisms deep in the crust was of great interest to NASA scientists seeking life on Mars. Of the planets, Mars most closely resembles Earth in geology and distance from the Sun, and its crust might provide a habitat similar to Earth's.

The outer surface of the crust and the atmosphere above it support the remainder of the **biosphere**, the sum total of all life on Earth. The biosphere generates oxidants (electron acceptors), most notably O_2. Oxygen-breathing organisms can live only on the outer surface, where O_2 is produced by photosynthesis.

Earth's atmosphere. From the crust and the mantle of early Earth, volcanic activity released gases such as carbon dioxide and nitrogen, which formed Earth's first atmosphere, while volcanic water vapor formed the ocean. The composition of this first atmosphere, before life evolved, looked much like that of Mars: thin, about 1% as dense as that of Earth today, consisting primarily of CO_2. But unlike Mars, Earth developed living organisms that filled the atmosphere with gaseous N_2 and O_2 and that continue to produce these gases today. Organisms also produce CO_2, as well as fixing it into biomass. Some CO_2 and N_2 arise from geological sources such as volcanoes, but their contribution is small compared to that of biological cycles (discussed in Chapter 22). The overall composition of Earth's atmosphere is determined by living organisms, primarily microbes.

Temperature. Another important aspect of Earth's habitat, determined by the atmospheric density and composition, is temperature. Atmospheric gases absorb light and convert the energy to heat, raising the temperature of the surface and atmosphere. This rise in temperature is known as the **greenhouse effect**. Because carbon dioxide is an especially potent greenhouse gas, the CO_2-rich atmosphere of early Earth could have heated the planet to temperatures approaching those of Venus, eliminating the possibility of life. Instead, microbial consumption of CO_2 and generation of nitrogen and oxygen gases limited Earth's surface temperature to an average of 13°C. The cooling effect may have led to an ice age, possibly reversed by rising methane from methanogens. One way or another, the history of Earth's atmosphere is intimately related to the history of microbial evolution.

Geological Evidence for Early Life

Evidence for life in the geological record is called a **biosignature**, or **biological signature**. Biosignatures have been found that are even earlier than the oldest fossils. Their significance is limited, however, because it is hard to rule out nonbiogenic explanations, so researchers seek additional, corroborating evidence based on independent principles.

When in Earth's geological record do the first biosignatures appear? Little evidence appears in the very earliest period of Earth's existence, ranging from 4.6 to 4.0 Gyr ago; it is called the **Hadean eon**, named for Hades, the ancient Greek world of the dead. During the Hadean eon, repeated bombardment by meteorites vaporized the oceans, which then cooled and recondensed. Meteor bombardment may have killed off incipient life more than once before living microbes finally became established. Still, scientists speculate on whether some forms of life might have survived Hadean conditions, perhaps growing 3 km below Earth's surface. Like the dead spirits imagined by the Greeks to have populated Hades, Earth's earliest cells may have reached deep enough within the crust that they were protected from the heat and vaporization at the surface.

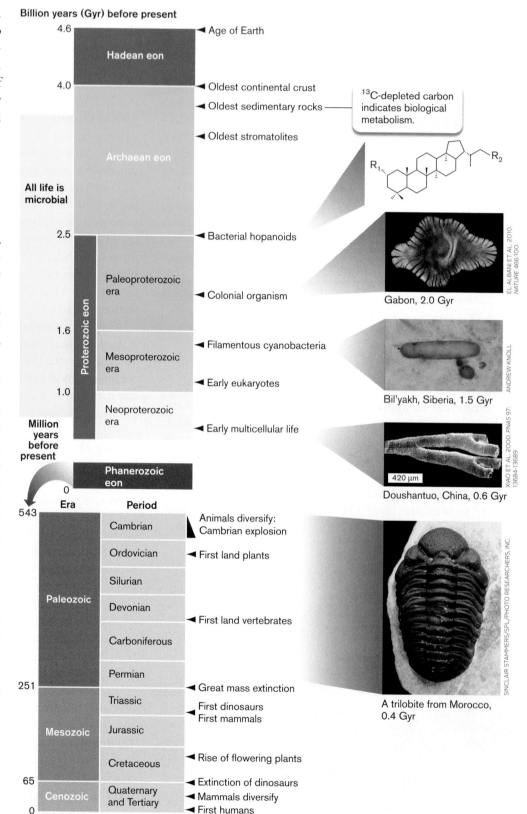

FIGURE 17.5 ■ **Geological evidence for early life.** The geological record shows biosignatures of microbial life early in Earth's history, 3 Gyr before the first multicellular forms.

The Archaean eon. The earliest geological evidence for life that is generally accepted dates to 4.0–2.5 Gyr ago, in the **Archaean eon** (**Fig. 17.5**). In the Archaean, meteor

TABLE 17.1	Geological Evidence of Early Life	
Type of evidence	**Advantages**	**Limitations**
Stromatolites		
Layers of phototrophic microbial communities grew and died, and their form was filled in by calcium carbonate or silica.	Fossil stromatolites appear in the oldest rock of the Archaean eon. Their distinctive shapes resemble those of modern living stromatolites.	Some layered formations attributed to stromatolites have been shown to be formed by abiotic processes.
Microfossils		
Early microbial cells decayed, and their form was filled in by calcium carbonate or silica. The size and shape of microfossils resemble those of modern cells.	Microfossils are visible and measurable under a microscope, offering direct evidence of cellular form.	Microscopic rock formations require subjective interpretation. Some formations may result from abiotic processes.
Isotope ratios		
Microbes fix $^{12}CO_2$ more readily than $^{13}CO_2$. Thus, limestone depleted of ^{13}C must have come from living cells. Similarly, sulfate-respiring bacteria cause depletion of ^{34}S compared with ^{32}S.	Isotope ratios are a highly reproducible physical measurement. Isotope ratios generated by key biochemical reactions can calibrate the time lines of phylogenetic trees.	We cannot prove absolutely that no abiotic process could generate a given isotope ratio. Isotope ratios tell us nothing about the shape of early cells or how they evolved.
Biosignatures		
Certain organic molecules found in sedimentary rock are known to be formed only by certain microbes. These molecules are used as biosignatures.	Biosignatures such as hopanoids are complex molecules specific to bacteria.	A biosignature thought specific to one kind of organism may be discovered in others. In the oldest rocks, organic biosignatures are eliminated by metamorphic processes.
Oxidation state		
The oxidation state of metals such as iron and uranium indicates the level of O_2 available when the rock formed. Banded iron formations suggest intermittent oxidation by microbial phototrophs.	Oxidized metals offer evidence of microbial processes even in highly deformed rock.	It is hard to rule out abiotic causes of oxidation. Even if the oxidation was biogenic, it does not reveal the kind of metabolism.

bombardment was less frequent, and Earth's crust had become solid. The Archaean marked the first period with stable oceans containing the key ingredient of life: liquid water. Water is a key medium for life because it remains liquid over a wide range of temperatures and because it dissolves a wide range of inorganic and organic chemicals. Rock strata dating to the Archaean eon reveal the first evidence of living organisms and their metabolic processes.

Note: The term "Archaean" refers to the earliest geological eon when life existed, whereas "archaeal" is the adjective referring to the taxonomic domain Archaea. The domain Archaea (originally "Archaebacteria") was named by Carl Woese based on his theory that members of this domain most closely resembled the earliest life-forms of the Archaean eon. In fact, early life may have encompassed diverse traits later associated with archaea, bacteria, and eukaryotes.

How and when did living cells arise out of inert materials? Without a time machine to take us back 4 Gyr, we must rely on evidence from Earth's geology. Interpreting geology is a challenge because most forms of evidence for early life are indirect and subject to multiple interpretations. The further back in time, the more the rock has changed, and the greater the difficulties are. One way to meet this challenge, however, is to compare the results from different kinds of biosignatures (**Table 17.1**). If two or more kinds of evidence (such as microfossils and isotope ratios) point to life in the same location, the conclusion is strengthened.

Stromatolites. Fossil stromatolites date as early as 3.4 Gyr ago (see **Fig. 17.2**). These rock formations appear similar to the layered forms of stromatolites today, but the ancient rock is too deformed to reveal the detailed structure of microscopic cells.

Microfossils. The most convincing evidence for early microbial life is the visual appearance of **microfossils**, microscopic fossils in which minerals have precipitated and

Microfossils

A. Filamentous prokaryotes

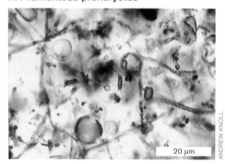

C. Colonial cyanobacteria

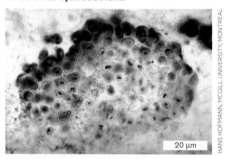

E. Algae (eukaryote)

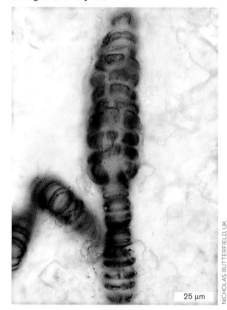

Modern species

B. *Leptothrix* sp.

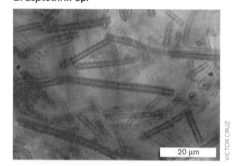

D. *Entophysalis* sp.

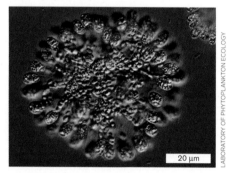

F. *Bangia* sp., red algae

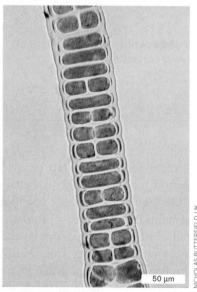

FIGURE 17.6 ■ **Microfossils compared with modern bacteria.** A. Filamentous prokaryotes, 2.0 Gyr old, from the Gunflint Formation, Ontario, Canada. B. Modern *Leptothrix* filamentous bacteria. C. Colonial cyanobacteria, about 2.0 Gyr old, from Belcher Islands, Canada. D. Modern *Entophysalis* cyanobacteria. E. Filamentous algae, 1.2 Gyr old, from Arctic Canada. F. Modern red algae, a eukaryote, *Bangia* sp.

to show regular 3D patterns of cells that resemble those of modern living cells, and that cannot be ascribed to abiotic (nonbiological) causes. The fossil needs to appear in nonmetamorphic rock—that is, rock whose fundamental form has not been reshaped by heat and pressure. The geochemistry must be consistent with rock deposition into organic materials.

The earliest microfossils accepted as biogenic are dated at 2.0 Gyr ago. Microfossils dated to 2.0 Gyr include filamentous prokaryotes in the Gunflint Formation, Ontario, Canada (**Fig. 17.6A**). The Gunflint outcrops consist of chert, a kind of silicate formed by precipitation from an ancient sea. The sea was rich in carbonates and reduced iron, a good combination for redox metabolism. Some of the fossils resemble the form of filamentous iron-metabolizing bacteria today, such as *Leptothrix* species (**Fig. 17.6B**). Other microfossils dated at 2.0 Gyr ago resemble colonial cyanobacteria (compare **Fig. 17.6C** with **17.6D**). More recent strata, dated at 1.2 Gyr ago, contain larger fossil cells comparable to those of modern eukaryotes such as algae (**Fig. 17.6E** and **F**).

If life existed even earlier than 3.4 Gyr, where are the microfossils? Archaean rock is metamorphic, greatly modified by temperature and pressure. We can identify the macroscopic contours of a stromatolite, but identifying microfossils is more problematic. In the early 1990s, microfossils of cyanobacteria dated to 3.85 Gyr ago by William Schopf, a paleobiologist at UCLA, were accepted and described in many textbooks (**Fig. 17.7**). But these fossils were later reinterpreted by English paleobiologist Martin Brasier (1947–2014) as nonbiogenic artifacts (caused by abiotic processes). The form of the proposed Archaean microfossils is less regular and convincing than the form of later specimens, particularly when observed at different angles not shown in the original publication. Identifying microfossils remains a challenge because of the lack of quantitative criteria.

filled in the form of ancient microbial cells (**Fig. 17.6**). Microfossils are dated by the age of the rock formation in which they are found, which in turn is based on evidence such as radioisotope decay. To be accepted as **biogenic** (formed from living organisms), a microfossil needs

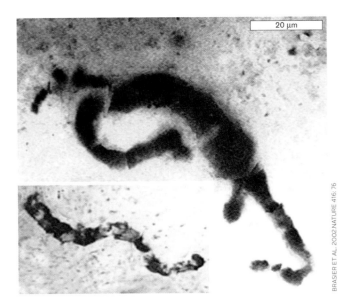

FIGURE 17.7 ■ **Microfossils or artifacts?** Structures originally identified as cyanobacterial microfossils from Western Australia, dated to 3.85 Gyr ago. Further testing indicated that the structures are nonbiological artifacts.

Isotope ratios. An **isotope ratio** provides a biosignature with quantitative evidence (**Fig. 17.8**). The source is biogenic if the ratio between certain isotopes of a given element is altered by biological activity. Enzymatic reactions, unlike abiotic processes, are so selective for their substrates that their rates may differ for molecules containing different isotopes. For example, the carbon-fixing enzyme Rubisco, found in chloroplasts, preferentially fixes CO_2

containing ^{12}C rather than ^{13}C. The carbon dioxide fixed into microbial cells eventually is converted to calcium carbonate in sedimentary rock. Thus, the calcium carbonate deposited by CO_2-fixing autotrophs (such as cyanobacteria) shows lower ^{13}C content than does calcium carbonate deposited by abiotic processes (**Fig. 17.8A**). The difference, $\delta^{13}C$, is defined by the fractional difference (in parts per thousand) between the $^{13}C/^{12}C$ ratios in a sample versus a standard inorganic rock:

$$\delta^{13}C = \frac{^{13}C/^{12}C \text{ (experimental)} - ^{13}C/^{12}C \text{ (standard)}}{^{13}C/^{12}C \text{ (standard)}} \times 1,000$$

Typical $\delta^{13}C$ values are shown in **Figure 17.8A**. Organisms on land and sea show $\delta^{13}C$ values of −10 to −30 parts per thousand, representing a significant ^{13}C depletion through carbon fixation. A comparable $\delta^{13}C$ is observed in fossil fuels, which formed from plant and animal bodies decomposed by bacteria.

The most ancient mineral samples showing a substantial $\delta^{13}C$ (about −18 parts per thousand) are graphite granules in the Isua rock bed of West Greenland, dated at 3.7 Gyr old. The graphite granules were analyzed by Minik Rosing, a native Greenlander at the Danish Lithosphere Centre (**Fig. 17.8B**). The graphite grains derive from microbial remains buried within sediment that subsequently metamorphosed, driving out the water content but leaving behind the telltale carbon. By contrast, nonbiogenic carbonate rock from the same formation shows a $\delta^{13}C$ near zero.

A. ^{13}C **isotope depletion**

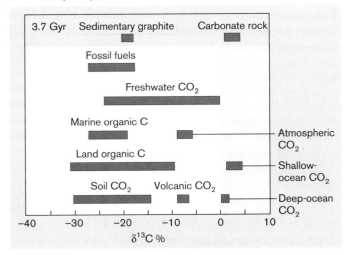

B. Minik Rosing

FIGURE 17.8 ■ **Carbon isotope depletion. A.** ^{13}C isotope depletion (negative $\delta^{13}C$) occurs in biomass as a result of the Calvin cycle. Negative $\delta^{13}C$ is observed at 3.7 Gyr in sedimentary graphite, which may derive from sedimented phototrophs. Little or no isotope depletion is seen in carbonate rock, which has no biological origin. **B.** Minik Rosing (left), in Greenland, shows the Isua rocks whose carbon isotope ratios indicate photosynthesis at 3.8 Gyr ago.

A.

B.

FIGURE 17.9 ■ **Banded iron formations. A.** Jim Crowley, U.S. Geological Survey, studies a banded iron formation in Dales Gorge, Australia. **B.** The BHP Billiton Iron Ore mine at Newman, Western Australia.

Chemical biosignatures. A different kind of biosignature is given by organic molecules specific to a particular life-form. Certain organic molecules may last within rock for hundreds of millions of years. A particularly durable class of molecules consists of membrane lipids. Recall from Chapter 3 that some bacterial cell membranes contain steroid-like molecules called hopanoids. A hopanoid consists of four or five fused rings of hydrocarbon with side groups that vary depending on the kind of bacteria. The hopanoid derivative 2-methylhopane is found in sedimentary rock of the Hamersley Basin of Western Australia, dated to 2.5 Gyr ago. This biosignature offers evidence that some kind of bacteria existed by the end of the Archaean eon.

Banded Iron Formations Reveal Oxidation by O_2

An extraordinary event in the planet's history was the evolution of the first oxygenic phototrophs: cyanobacteria that split water to form O_2. The entry of O_2 into Earth's biosphere is often portrayed as a sudden event that would have been disastrous to microbial populations lacking defenses against its toxicity. In fact, geological evidence shows that oxygen arose gradually in the oceans, starting about 2 Gyr ago, and may have arisen and disappeared numerous times before reaching a high, steady-state level in our atmosphere. The mechanism of the oxygen fluctuation is unknown, but it caused cycles of aerobic and anaerobic microbial metabolism.

Evidence for oxygen in the biosphere comes from the oxidation state of minerals, particularly those containing iron. The bulk of crustal iron is in the reduced form (Fe^{2+}), which is soluble in water and reached high concentrations in the anoxic early oceans. Sedimentary rock, however, contains many fine layers of oxidized iron (Fe^{3+}), which is insoluble and forms a precipitate, such as iron oxide (Fe_2O_3). The layers of iron oxide suggest periods of alternating oxygen-rich and anoxic conditions. The layered rock is called a **banded iron formation** (**BIF**) (**Fig. 17.9A**). A common form of banded iron consists of gray layers of silicon dioxide (SiO_2) alternating with layers colored red by iron oxides and iron oxyhydroxides [$FeO_x(OH)_y$]. Banded iron formations are widespread around the world and provide our major sources of iron ore (**Fig. 17.9B**).

Banded iron formations are often found in rock strata containing signs of past life such as ^{13}C depletion and other biomarkers. For example, the Isua formations (Greenland) and Hamersley formations (Western Australia), which both show ^{13}C depletion, also contain extensive banded iron. Calculations indicate that the layers of oxidized iron could result from biological metabolism involving iron oxidation. One possibility is that chemolithotrophs oxidized the iron, using molecular oxygen produced by cyanobacteria. The Archaean and early Proterozoic eons experienced fluctuating levels of molecular oxygen in the atmosphere. These fluctuations could have led to oscillating levels of iron oxide, thus producing bands in the sediment, as microbes used up all the oxygen.

Alternatively, Dianne Newman, at the California Institute of Technology, proposes that the iron oxides arose directly from anaerobic photosynthesis, in which the reduced iron served as the electron donor (**Fig. 17.10**). In iron phototrophy, light excites an electron from Fe^{2+}, oxidizing the ion to Fe^{3+}, while the excited electron cycles through an electron transport system to yield energy (discussed in Chapter 14). Newman discovered iron phototrophy in modern purple bacteria such as *Rhodopseudomonas palustris*. In the ancient Earth, photosynthetic oxidation of Fe^{2+} to Fe^{3+}, or "photoferrotrophy," could have occurred in cycles until the marine iron was all oxidized, generating sedimentary layers of iron oxides and iron oxyhydroxides. Today, photoferrotrophs thrive in deep anoxic lakes that resemble the anoxic Archaean ocean.

By 2.3 Gyr ago, the prevalence of oxidized iron and other minerals indicates the steady rise of oxygen from photosynthesis in Earth's atmosphere. Most of the dissolved Fe^{2+} from the ocean floor was oxidized, leaving the oceans in the iron-poor state that persists today. As the most efficient electron acceptor, molecular oxygen enabled the evolution of aerobic respiratory bacteria. Aerobic bacteria gave rise to mitochondria, which enabled the evolution of eukaryotes and, ultimately, multicellular organisms (**Fig. 17.11**).

Remarkably, modern cells still consist mainly of reduced molecules, highly reactive with oxygen—a relic of the time when our ancestral cells evolved in the absence of oxygen. The conditions under which such cells may have evolved can be simulated in the laboratory. Conditions resembling our model of early Earth spontaneously generate some of life's most common molecules, such as adenine and simple amino acids (presented earlier, in Special Topic 1.1). These simulation experiments can never prove the actual conditions under which life began, but they can suggest testable models with intriguing implications.

To Summarize

■ **Elements of life** were formed through nuclear reactions within stars that exploded into supernovas before the birth of our own Sun.

■ **Reduced molecules** compose Earth's interior. Oxidized minerals are found only near the surface. Early Earth had no molecular oxygen (O_2).

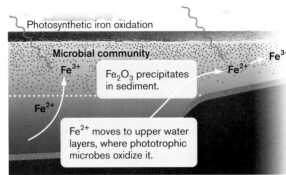

FIGURE 17.10 ■ **Iron phototrophy. A.** Dianne Newman proposes that early iron phototrophs caused the iron oxide deposition generating banded iron formations. **B.** Photosynthetic oxidation of Fe^{2+} to Fe^{3+} may have generated sedimentary layers of Fe_2O_3 and $FeO_x(OH)_y$.

Origin and evolution of life on Earth

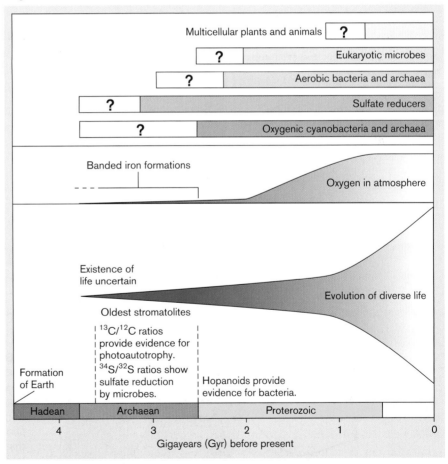

FIGURE 17.11 ■ **Proposed time line for the origin and evolution of life.** The planet Earth formed during the Hadean eon (about 4.5 Gyr ago). The environment was largely reducing until cyanobacteria pumped O_2 into the atmosphere. When the O_2 level reached sufficient levels (about 0.6 Gyr ago), multicellular animals and plants evolved. Question marks designate periods when evidence for given life-forms is uncertain.

- **Archaean rocks show evidence for life** based on fossil stromatolites, isotope ratios, and chemical biosignatures. Fossil stromatolites appear in chert formations formed 3.4 Gyr ago. Isotope ratios for carbon indicate photosynthesis at 3.7 Gyr ago. Bacterial hopanoids appear at 2.5 Gyr ago.

- **Microfossils of filamentous and colonial prokaryotes** date to 2.0 Gyr ago. At 1.2 Gyr ago, larger fossil cells resemble those of modern eukaryotes.

- **Banded iron formations** reflect the cyclic increase and decrease of oxygen produced by cyanobacteria and consumed through reaction with reduced iron. After all the ocean's iron was oxidized, oxygen increased gradually in the atmosphere.

17.2 Early Metabolism

How did the earliest life-forms gain energy without oxygen gas to respire and without the complex machinery of photosynthesis? The nature of the first metabolism is unknown, but geochemistry and modern metabolism indicate several features of the earliest living systems (**Fig. 17.12A**).

Oxidation-reduction reactions. The early oceans contained oxidized forms of nitrogen, sulfur, and iron that could interact with reduced minerals from the crust. For example, nitrate (NO_3^-) or sulfate (SO_4^{2-}) could be reduced by hydrogen gas to yield energy (hydrogenotrophy; discussed in Chapter 14). The oxidized molecules were generated by reactions driven by ultraviolet radiation, which penetrated

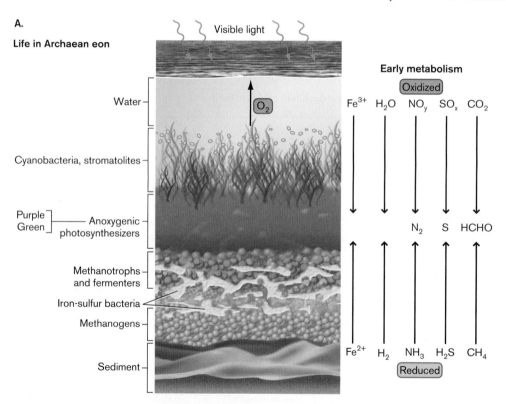

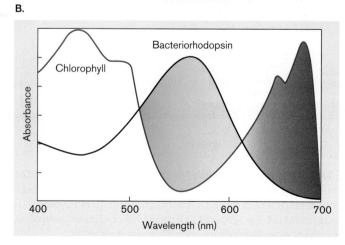

FIGURE 17.12 ■ Early metabolism. **A.** Oxidized minerals diffused down from the air and water and reduced minerals in the sediment, upwelling from hydrothermal vents. **B.** Early photosynthesis may have resembled that of haloarchaea, whose bacteriorhodopsin absorbs light in the central range of sunlight (green-yellow, 500–600 nm) between the blue and red chlorophylls of cyanobacteria and plants. *Source:* Part B adapted from Shil DasSarma, University of Maryland.

the atmosphere in the absence of the ozone layer. Sulfur isotope ratios ($^{34}S/^{32}S$) suggest the growth of sulfate-reducing bacteria as early as 3.47 Gyr ago.

Light-driven ion pumps. A simple light-driven pump, such as the bacteriorhodopsin of haloarchaea (halophilic archaea), could have conducted the first kind of phototrophy. The absorption spectrum of bacteriorhodopsin matches the peak of solar radiation reaching the upper layers of ocean (**Fig. 17.12B**), in contrast to the chlorophylls of cyanobacteria and anoxygenic bacteria, which absorb the outer ranges of blue and red. Perhaps these chlorophyll-using bacteria evolved in the presence of haloarchaea, filling the unexploited photochemical niches.

Methanogenesis. Climate models of early Earth suggest a methane atmosphere, produced by methanogenic archaea. Methanogenesis involves reaction of H_2 and CO_2 producing CH_4 and H_2O. Methanogens show highly divergent genomes—a finding that suggests early evolution of their common ancestor. Their biochemistry and evolution (discussed in Chapter 19) are consistent with proposed models of ancient life.

Models for the First Cells

Various models have been proposed to explain how the first living cells originated from nonliving materials, and how they replicated and evolved. Models for early life attempt to address the following questions: In what kind of environment did the first cells form? What kind of metabolism did the first cells use to generate energy? What was their hereditary material?

Three kinds of models attempt to explain how nonliving materials gave rise to a cell. The **prebiotic soup** model proposes that **abiotic** (nonliving) reactions involving simple reduced chemicals such as ammonia and methane could have assembled complex organic molecules of life. These complex macromolecules eventually acquired the apparatus needed for self-replication and membrane compartmentalization. Metabolist models propose that central components of intermediary metabolism, including using the TCA cycle to generate amino acids, arose from self-sustaining chemical reactions based on inorganic chemicals. These abiotic reactions then acquired self-replicating macromolecules and membranes. Finally, the **RNA world** is a model of early life in which RNA performed all the informational and catalytic roles of today's DNA and proteins. The concept of an RNA world draws upon genome sequences, which reveal thousands of catalytic and structural RNAs.

The prebiotic soup. In the mid-twentieth century, biochemists Aleksandr Oparin at Moscow University, and Stanley Miller and Harold Urey at the University of Chicago, showed that organic building blocks of life such as amino acids could arise abiotically out of a mixture of water and reduced chemicals, including CH_4, NH_3, and H_2 (**Fig. 17.13**). The mixture was subjected to an electrical discharge, similar to the lightning discharges that arise from volcanic eruptions, which would have been common in the late Hadean or early Archaean eon. The chemical reaction produced fundamental amino acids such as glycine and alanine. Similar experiments by Juan Oró, at the University of Houston, showed the formation of adenine by condensation of ammonia and hydrogen cyanide.

The same amino acids and nucleobases arising from early-Earth simulation are also found in meteorites, which are believed to retain the chemistry of the early solar system "frozen" in time. But unlike the chemical experiments, the meteorites show a remarkable predominance of left-handed

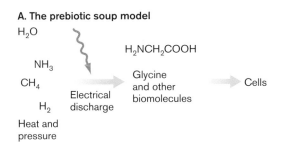

A. The prebiotic soup model

H_2O

NH_3

CH_4 H_2NCH_2COOH

H_2 Glycine
             and other   Cells
Electrical   biomolecules
discharge

Heat and
pressure

FIGURE 17.13 ■ The prebiotic soup model for the origin of life. A. In the prebiotic soup, inorganic molecules could have reacted to form complex macromolecules that eventually acquired the apparatus for self-replication and membrane compartmentalization. **B.** Lightning accompanies eruption of the Galunggung Volcano in West Java, Indonesia, in 1982. The first biomolecules may have formed as a result of lightning triggered by volcanic eruption.

mirror forms (L-enantiomers) of the amino acids—the same L form utilized in the protein synthesis of all life on Earth. Some researchers hypothesize that organic compounds formed in outer space were restricted to the L form by an unknown process and then "seeded" the propagation of L-form compounds in Earth's prebiotic soup.

The original model conditions for the prebiotic soup assumed that molecules in the early Archaean ocean were largely reduced, with little or no oxygen present in the atmosphere. More recent geochemical evidence suggests that the early ocean actually included oxidized forms of nitrogen, sulfur, and iron that arose through reactions driven by ultraviolet radiation, which penetrated the atmosphere in the absence of the ozone layer. These oxidized minerals could have reacted with the reduced crustal minerals, releasing energy to drive production of more complex biomolecules.

Forming the first cell, or "proto-cell," must have required enclosing the first biochemical reactants within a membrane-like compartment. Jack Szostak and colleagues at Harvard Medical School investigate how such compartments can arise spontaneously from fatty acid glycerol esters. The fatty acid derivatives are "amphipathic"; that is, they possess both hydrophobic portions that associate together and hydrophilic portions that associate with water. In water, the fatty acid derivatives collect in "micelles"—small, round aggregates in which the hydrophobic portions associate in the interior and the hydrophilic portions associate with water. Under certain conditions, micelles can aggregate to form hollow vesicles of membrane. Szostak showed how the vesicles can take up molecules such as

RNA, suggesting a primitive cell-like form. These spontaneous processes of membrane formation offer models for how the first living cells may have arisen.

Metabolist models. Other models attempt to explain the origin of biosynthesis on the basis of CO_2 fixation, the fundamental metabolism of all life today. Proponents of a metabolist model, including Harold Morowitz at George Mason University, and Günter Wächterhäuser at the University of Regensburg, propose that CO_2-based metabolism originated through self-sustaining reactions (**Fig. 17.14A**). Simulation experiments suggest that such premetabolic reactions could have been catalyzed by metal sulfides prevalent in the early ocean. For example, abiotic polymerization of CO_2 into TCA cycle intermediates may be catalyzed by FeS.

How did simple inorganic reactions lead to biochemical cycles involving more complex biopolymers, such as nucleic acids and proteins? A major challenge is to explain the genetic code by which nucleic acid codons are assigned to unique amino acids. The genetic code may have arisen from an earlier pretranslational mechanism of amino acid biosynthesis (**Fig. 17.14B**). In this mechanism, proposed by Morowitz and colleagues, each amino acid was originally synthesized from a TCA cycle acid complexed to a dinucleotide. The dinucleotide later evolved into the first two nucleotides of the codon specifying the amino acid. The proposed link between a specific dinucleotide and each amino acid would explain certain features of the genetic code, such as the fact that most amino acids specified by a codon starting with the same nucleotide are synthesized from the same TCA cycle acid (discussed in Chapter 15).

FIGURE 17.14 ■ Metabolist models for the origin of life. A. Self-sustaining abiotic chemical reactions, such as polymerization of CO_2 and H_2, could have formed the basis of cellular metabolism. The reductive TCA cycle arose out of intermediates with small thermodynamic transitions to CO_2. **B.** The genetic code may have originated from synthesis of an amino acid complexed to a dinucleotide. This dinucleotide could later have evolved into the first two nucleotides of the codon specifying the amino acid.

The RNA world. Neither the prebiotic soup model nor the metabolist models account for the evolution of macromolecules that encode complex information, such as nucleic acids and proteins. A candidate for life's first "informational molecule" is RNA. RNA is a relatively simple biomolecule, with only four different "letters," compared to the 20 standard amino acids of proteins. Its purine base adenine arises spontaneously from ammonia and carbon dioxide under conditions believed to resemble those of the Archaean eon. Its ribose sugar is a fundamental building block of living cells, with key roles in numerous biochemical pathways, such as the Calvin cycle. Most exciting, in 2009, John Sutherland and co-workers discovered an organic reaction pathway that generates a sugar-base ribonucleotide in the absence of any enzyme. This discovery suggests a way that the first RNA molecules could have formed before there were living cells.

For several reasons, RNA is a better candidate than DNA for the earliest information molecule. Compared to DNA, RNA requires less energy to form and degrade. RNA's pyrimidine base uracil is formed early by biochemical pathways; only later is it converted to the thymine used by DNA.

Most important, RNA molecules have been shown to possess catalytic properties analogous to those of proteins. Catalytic RNA molecules are called **ribozymes**. The first ribozyme, discovered by Nobel laureate Thomas Cech in the protist *Tetrahymena*, can splice introns in mRNA. Other ribozymes actually catalyze synthesis of complementary strands of RNA, suggesting a model for early replication of RNA chromosomes. The most elaborate example of catalytic RNA is found in the ribosome. X-ray crystallography of the ribosome reveals that the key steps of protein synthesis, such as peptide bond formation, are actually catalyzed by the RNA components, not proteins (discussed in Chapter 8). The ribosomal proteins possess relatively little catalytic function; their main role seems to be protection and structural support of the RNA.

Could RNA molecules have composed the earliest cells? In 2009, Tracey Lincoln and Gerald Joyce at the Scripps Research Institute devised a system in which two RNA ribozymes catalyze each other's synthesis. The ribozyme model system suggests how, in the earliest cells, RNA might have fulfilled the key functions that are today filled by DNA and proteins, including information storage, replication, and catalysis (**Fig. 17.15**). This model is known as the "RNA world."

The prominent function of RNA in the ribosome, one of life's most ancient and conserved molecular machines, suggests a model for the transition from an RNA world to the modern cell. In the ribosome, the actual steps of catalysis, such as forming the peptide bond, are conducted by RNA subunits acting as ribozymes. The ribozymes are stabilized

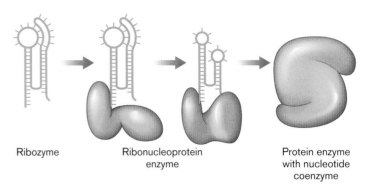

Ribozyme   Ribonucleoprotein enzyme   Protein enzyme with nucleotide coenzyme

FIGURE 17.15 ■ From the RNA world to proteins. Earliest cells may have been composed of RNA enzymes (ribozymes). As the RNA cells evolved, ribozymes acquired protein subunits that eventually assumed most of their catalytic functions. Remnants of the original RNA may persist as nucleotide cofactors such as NADH.

by protein subunits. Thomas Cech, Sidney Altman, and colleagues propose that the earliest RNA components of cells evolved by acquiring proteins to enhance stability. The proteins helped prevent the tendency of RNA to hydrolyze (come apart in reaction with water). As cells evolved, their peptide components increased through natural selection, and the RNA subunits may have shrunk by reductive evolution (the evolutionary loss of unneeded parts). A few complexes, such as the ribosome, still maintain their ribozymes; for others, perhaps all that remains are one or two nucleotides. Dinucleotide cofactors such as NADH persist in enzymes today, perhaps representing vestigial remnants of the RNA world. For more on RNA evolution, and surprising applications for medicine, see **eTopic 17.1**.

The RNA-world model explains the central role of RNA throughout the history of living cells. Yet the role of RNA offers little clue as to the origins of cell compartmentalization and metabolism. No single model of life's origin yet addresses all the requirements for a living cell—metabolism, membrane compartmentalization, and hereditary material. Each model does, however, offer important insights into the evolutionary history and mechanisms of life today.

Thought Question

17.3 Outline the strengths and limitations of each model of the origin of living cells. Which aspects of living cells does each model explain?

Unresolved Questions about Early Life

Overall, geology and biochemistry provide compelling evidence that organisms resembling today's cyanobacteria lived on Earth at least 2.5 Gyr ago, possibly 3.7 Gyr ago, and that bacteria with anaerobic metabolism evolved as early or earlier. Many intriguing questions remain. We outline here three unresolved questions regarding the temperature of early Earth, the role of methane in the early atmosphere, and the actual source of Earth's first cells.

Thermophile or psychrophile? The apparent existence of life so soon after Earth cooled suggests a thermophilic origin. Thermophily is supported by the fact that in the domains Bacteria and Archaea, the deepest-branching clades (that is, kinds of organisms that diverged the earliest from others in the domain) are thermophiles. Such organisms could have thrived at hydrothermal vents, which offer a continuous supply of H_2S and carbonates.

On the other hand, after meteoric bombardment abated, early Earth should have become glacially cold. In the Archaean, solar radiation was 20%–30% less intense than it is today, and the thin CO_2 atmosphere was insufficient to increase the temperature by a greenhouse effect. A colder habitat would support psychrophiles. Psychrophiles might have had an advantage in an RNA world, given the thermal instability of RNA compared to DNA and proteins.

A world of methane? Among Earth's earliest life-forms were methanogens (methane producers). Methanogens are one of the most widely divergent groups of organisms and persist today in environments ranging from anaerobic sediment to the human intestine. Methanogenesis requires only carbon dioxide and hydrogen gas, which would have been plentiful in early anoxic sediment. The production of methane, an extremely potent greenhouse gas, could have greatly increased Earth's temperature during the Archaean eon. Thus, methanogenesis could explain how Earth escaped the permanent freeze of Mars. Overheating would have been halted when the oxygen gas produced by cyanobacteria enabled growth of methanotrophs, bacteria that oxidize methane. The decline of methane and the rise of CO_2 then would have brought about relative thermal stability.

The debate over the temperature and climate of early Earth has interesting implications for Earth today, when we again face the prospect of massive global climate change. Human agriculture favors explosive growth of methanogens, which threaten to accelerate global warming faster than the biosphere can moderate it. Understanding the climate of early Earth may help us better understand and manage our own climate, discussed in Chapter 22.

Origin on Earth or elsewhere? Emerging evidence from fossils and geochemistry has inexorably pushed back the earliest known dates for several kinds of metabolism closer to 3.7 Gyr ago (see **Fig. 17.11**). This implies that as soon

as Earth cooled to a temperature suitable for life, all the fundamental components of cells evolved almost immediately. How could life, with all its diverse kinds of metabolism, have arisen so quickly?

The idea that life-forms originated elsewhere and "seeded" life on Earth is called **panspermia**. Theories of panspermia remain highly speculative. One hypothesis is that microbial cells originated on Mars and were then carried to Earth on meteorites. As the solar system formed, Mars would have cooled sooner than Earth, and because it is also smaller than Earth, Mars's weaker gravity would have generated less bombardment by meteorites. Martian rocks ejected into space by meteor impact have reached Earth, and calculations based on simulated space habitats show that microbes could survive such a journey. A Martian origin, however, gains us only about half a billion years; it does not really explain the origin of life's complexity and diversity. Did life-forms come from still farther away, perhaps borne on interstellar dust from some other solar system?

A more likely explanation is that life evolved on early Earth faster than it does today. Today, RNA viruses such as influenza and HIV mutate and evolve much faster than modern cells (discussed in Chapter 11). If early cells with RNA genomes mutated as fast as viruses, they would have evolved and diverged faster than cellular organisms that we know today.

> **Thought Question**
>
> **17.4** Suppose a NASA rover discovered living organisms on Mars. How might such a find shed light on the origin and evolution of life on Earth?

To Summarize

- **Early metabolism** involved anaerobic oxidation-reduction reactions. Likely forms of early metabolism include sulfate respiration, light-driven ion pumps, iron phototrophy, and methanogenesis.

- **Prebiotic soup models** propose that the fundamental biochemicals of life arose spontaneously through condensation of reduced inorganic molecules.

- **Metabolist models** propose that components of intermediary metabolism arose from self-sustaining chemical reactions that connected nucleotides with amino acids, forming the basis of the genetic code.

- **The RNA-world model** proposes that in the first cells, RNA performed all the informational and catalytic roles of today's DNA and proteins.

■ **Thermophile or psychrophile?** Classic models of early life assume thermophily, but Earth may actually have been cold when the first cells originated.

■ **A world of methane?** If the first cells were methanogens, methane production could have led to the first greenhouse effect, warming Earth and enabling evolution of other kinds of life.

■ **Origin on Earth or elsewhere?** Isotope ratios suggest the presence of complex metabolism by 3.7 Gyr ago, shortly after Earth cooled (4.0 Gyr ago). Simpler cells existing before 4.0 Gyr ago may have evolved much faster than life today. A more speculative possibility is that life first evolved on another planet.

17.3 Microbial Phylogeny and Gene Transfer

The unifying assumption of modern biology is genetic relatedness, or molecular phylogeny. Phylogeny generates a series of branching groups of related organisms called **clades**. Each clade is a **monophyletic group**—that is, a group of organisms that share a common ancestor not shared by any kind of organism outside the clade. Each monophyletic group then branches into smaller monophyletic groups, and ultimately **species**, the fundamental kind of organism (discussed in Section 17.5). The full description of branching divergence of a species is called its **phylogeny**. Phylogeny, however, consists of more than branching divergence; the ancestry of all life-forms includes **horizontal gene transfer** (discussed in Chapter 9). Both vertical gene flow (parent to offspring) and horizontal gene transfer shape all life-forms, including humans.

Divergence through Mutation and Natural Selection

Populations of organisms diverge from each other through several fundamental mechanisms of evolution. These include:

■ **Random mutation.** DNA sequences change through rare mistakes (in bacteria and archaea, typically one out of a million base pairs) as the chromosome replicates (discussed in Chapter 7). Replication errors result in mutation. Most mutations are neutral; that is, they have no effect on gene function.

■ **Natural selection and adaptation.** In a given environment, natural selection favors organisms that produce greater numbers of offspring in that environment

(discussed in Section 17.4). Genes encoding traits under selection pressure may show mutation frequencies much higher or lower than those generated by the random mutation rate. Natural selection enables a population to adapt to a changing environment.

■ **Reductive evolution (degenerative evolution).** In the absence of selection for a trait, the genes encoding the trait accumulate mutations without affecting the organism's reproductive success. Because mutations that decrease function are more common than mutations that improve function, accumulating mutations without selection pressure leads to decline and ultimately loss of the trait. The loss or mutation of DNA encoding unselected traits is called **reductive evolution** (or **degenerative evolution**).

Random mutations with neutral effects that are not subject to selection tend to accumulate at a steady rate over generations because the error rate of the DNA replication machinery stays about the same. The resulting "genetic drift" causes sequences in separate populations to diverge over time. The constancy of the mutation rate (within limits) provides a tool for us to measure the time of divergence of organisms based on their DNA sequences.

Thought Question

17.5 What kinds of DNA sequence changes have no effect on gene function? (*Hint:* Refer to the table of the genetic code, Figure 8.10.)

Molecular Clocks Are Based on Mutation Rate

An important conceptual advance of the twentieth century was that information contained in a macromolecule such as protein or DNA could measure the history of a species. The temporal information contained in a macromolecular sequence is called a **molecular clock**. Molecular clocks have revolutionized our understanding of the emergence of all living organisms, including human beings. The first molecular clocks, based on protein sequencing and DNA hybridization, were developed in the 1960s. Subsequently, the sequences of ribosomal RNA (rRNA) were used by Carl Woese to reveal the divergence of three domains of life. The rRNA sequence is particularly useful because ribosome structure and function are highly similar across all organisms. Genome sequences now offer many other genes to measure divergence at different levels of classification.

A molecular clock is based on the acquisition of new random mutations in each round of DNA replication. In

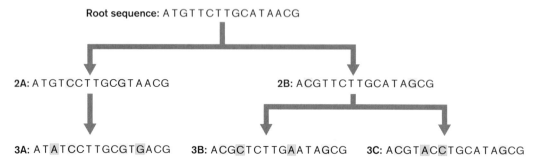

Root sequence: A T G T T C T T G C A T A A C G

2A: A T G T C C T T G C G T A A C G

2B: A C G T T C T T G C A T A G C G

3A: A T A T C C T T G C G T G A C G

3B: A C G C T C T T G A A T A G C G

3C: A C G T A C C T G C A T A G C G

FIGURE 17.16 ■ **The molecular clock.** As genetic molecules reproduce, the number of mutations (shaded in yellow and green) accumulated at random is proportional to the number of generations and thus the time since divergence. In each sequence designation (for example, "2A"), the number indicates the generation and the letter identifies a specific strain.

Figure 17.16, each offspring in generation 2 acquires two new mutations; their sequences now differ by 25%. In the next generation, each individual propagates the earlier mutant sequences while acquiring two more random mutations. Strain 3A now differs by 50% from strains 3B and 3C, which differ by 25% from each other. (Actual mutation frequencies, of course, are much lower—about one base per million per generation.) We assume that the chromosomes of offspring acquire a consistent number of random mutations from their parents, and therefore the number of sequence differences between two species should be proportional to the time of divergence between them.

Ideally, the molecular clock works best for a particular gene sequence with the following features:

■ **The gene has the same function across all types of organisms compared.** That is, all versions are orthologous; they have not evolved to serve different functions. Functional difference may lead to different rates of change.

■ **The generation time is the same for all organisms compared.** Shorter generation times (more frequent reproductive cycles) lead to overestimates of the overall time of divergence because of the increased opportunity for DNA mutation.

■ **The average mutation rate remains constant among organisms and across generations.** If different kinds of life mutate at different rates, then organisms with more rapid rates of mutation will appear to have diverged over a longer time than is actually the case.

In practice, these requirements are never fulfilled exactly, but we do the best we can. Genes that show the most consistent measures of evolutionary time encode components of the transcription and translation apparatus, such as ribosomal RNA and proteins, tRNA, and RNA polymerase. The most widely used molecular clock is the gene encoding the **small-subunit rRNA** (**SSU rRNA**). The SSU rRNA is also known by its sedimentation coefficient: 16S rRNA (bacteria)

or 18S rRNA (eukaryotes). (Sedimentation coefficients are discussed in Chapter 3.) The SSU rRNA is particularly useful because certain portions of its sequence are remarkably conserved across all forms of life. These portions can be used to define primers to amplify DNA of the gene encoding the rRNA, using the polymerase chain reaction (PCR; discussed in Section 7.6). The gene sequence lying between the pair of highly conserved rDNA sequences will show greater variation, allowing distinction between different clades. PCR can be used to amplify genes even from a mixture of uncultured organisms.

Use of a molecular clock requires the alignment of homologous sequences in divergent species or strains (**Fig. 17.17** ▶). Alignment is the correlation of portions of two gene sequences that diverged from a common ancestral sequence (homologous sequence). The process requires assumptions and decisions about base substitutions, as well as insertions and deletions. For example, in **Figure 17.17** the best alignment requires us to assume that three bases were lost from the fourth sequence (or else inserted into

A. SSU rDNA sequences from uncultured soil bacteria

A A A T G T T G G G C T T C C G G C A G T A G T G A G T G
A A A T G T T G G G A T T C C G G A A G T A G T G A G T G
A A A T G C T G G G C T T C C G G A A G T A G C G A G T G
A A A T G A T G G C T T T C C G G G G G C G A G T G C C C

B. Alignment

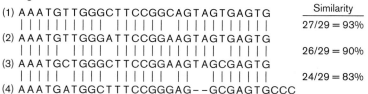

	Similarity
(1) A A A T G T T G G G C T T C C G G C A G T A G T G A G T G	
(2) A A A T G T T G G G A T T C C G G A A G T A G T G A G T G	27/29 = 93%
(3) A A A T G C T G G G C T T C C G G A A G T A G C G A G T G	26/29 = 90%
(4) A A A T G A T G G C T T T C C G G G A G – – G C G A G T G C C C	24/29 = 83%

C. Phylogenetic tree

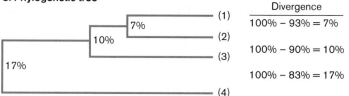

	Divergence
(1)	100% – 93% = 7%
(2)	
(3)	100% – 90% = 10%
(4)	100% – 83% = 17%

FIGURE 17.17 ■ **DNA sequence alignment. A.** SSU rRNA sequences from different organisms can be aligned at homologous regions. **B.** The best alignment is that which minimizes mismatches. **C.** A possible phylogenetic tree of divergence of the four sequences. ▶

the ancestor of the other sequences). The relative differences among the sequences can be used to propose a tree of divergence (**Fig. 17.17C**). In the tree, the length of each branch is proportional to the number of differences between two sequences. The number of differences, or divergence, is given by 100% minus the percent similarity between the aligned sequences.

In practice, a much larger amount of sequence with multiple differences is needed to calculate a phylogenetic tree. All trees are based on probability, with ambiguities depending on the assumptions of how sequences change. The data are calculated by computer programs, which may yield different results, depending on their assumptions. Standard assumptions include the following:

- **Minimum number of changes.** The best alignment between two sequences is that which assumes the smallest number of mutational changes.

- **Functional sequences change more slowly.** Sequences that encode essential catalytic portions of the gene product are maintained by selection pressure and change more slowly than portions of the molecule without essential function.

- **Third-base codon positions show more random change.** In a protein-coding gene, many codons have multiple anticodons that differ at the third base, so third-base changes are least likely to change the amino acid. Thus, third-base nucleotides offer the best molecular clock information.

Phylogenetic Trees

Once homologous sequences are aligned, the frequency of differences between them can be used to generate a **phylogenetic tree** that estimates the relative amounts of

evolutionary divergence between the sequences. If divergence rate over time, or mutation rate, is the same for all sequences compared, then divergence data can be used to infer the length of time since two organisms shared a common ancestor.

A phylogenetic tree is a model. All phylogenetic trees depend on complex mathematical analysis to measure degrees of divergence and propose a tree that most probably connects present sequences with their common ancestor. Because we can never know the exact tree with certainty, different types of calculations may lead to slightly different results. Two common approaches are:

- **Maximum parsimony.** Evolutionary distances are computed for all pairs of taxa based on the numbers of nucleotide or amino acid substitutions between them. A proposed common ancestor, or "ancestral state," is reconstructed. All possible trees comparing relative time of divergence from the common ancestor are computed. The "best fit" tree is defined as the one requiring the fewest mutations to fit the data (that is, the one that is most "parsimonious"). A limitation of parsimonious reconstruction is that more than one tree may produce results consistent with the data.

- **Maximum likelihood.** For each possible tree, one calculates the likelihood (probability) that such a tree would have produced the observed DNA sequences. The probability of given mutations is based on complex statistical calculations. Maximum-likelihood methods require large amounts of computation but obtain the most information from the data, usually generating one tree or a small set of probable trees.

A portion of a computed phylogenetic tree is shown in **Figure 17.18 ▶**. The sequence data were obtained from PCR-amplified sequences of SSU rRNA from isolates in

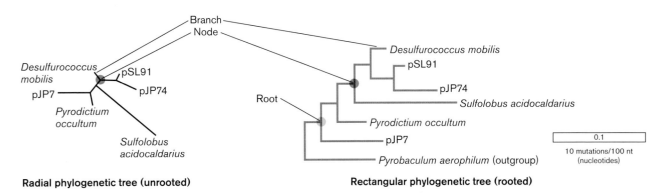

Radial phylogenetic tree (unrooted)

Rectangular phylogenetic tree (rooted)

FIGURE 17.18 ■ Phylogenetic trees: rooted and unrooted. Comparison of rRNA sequences was used to generate a phylogenetic tree for thermophiles isolated from Obsidian Pool in Yellowstone National Park. **Left:** The radial phylogenetic tree shows the degrees of relatedness among taxa based on the number of molecular substitutions on each branch (see scale bar). **Right:** The rectangular tree shows the same divergence distances lined up in parallel. This rectangular tree includes a root, the distance to an outgroup organism. A rooted tree indicates the position of a common ancestor. *Source:* Susan M. Barnes et al. 1996. *PNAS* **93**: 9188. ▶

Obsidian Pool, a thermal spring at Yellowstone National Park. The sequences were analyzed by Susan Barns and colleagues at Indiana University in the laboratory of Norman Pace, a pioneering investigator of extreme habitats. Note that some of the Obsidian Pool samples are known species, such as the thermophile *Pyrodictium occultum*, whereas others are uncharacterized organisms given alphanumeric designations, their existence known only by the rRNA sequences reported here. In fact, any microbial habitat surveyed by PCR, including a human body, will yield previously unknown species (as in the shower curtain biofilm described in **eTopic 17.2**).

The Obsidian Pool rRNA sequences were aligned by the method illustrated in **Figure 17.17**, and divergence distances were analyzed by maximum likelihood. In each tree, the total distance (in terms of base-substitution frequency) between any two sequences is approximated by the distance from each sequence and its branch point, or **node** (**Fig. 17.18**). A node for a group of branches is called a **root**—that is, a common ancestor. Moving from the root to the tips of the branches means moving forward in time.

Phylogenetic trees can be drawn in different ways. The tree at left in **Figure 17.18** is a radial tree, in which branches indicate distance outward from their nodes. The tree at right shows the same divergence distances from nodes, drawn as a rectangular tree. In a rectangular tree, all branches run in parallel. A rectangular tree helps compare divergent branches, although it takes up more space than a radial tree, especially when large numbers of organisms are compared.

If the length of each branch corresponds to a given length of time, we expect all the branches to add up to the same total length. In some trees this condition is approximated, but in microbial trees the lengths differ greatly. In **Figure 17.18**, for example, *Sulfolobus acidocaldarius* appears to have evolved much more than *Desulfurococcus mobilis*. The reason is that *S. acidocaldarius* and its recent ancestors have accumulated mutations faster. In every tree, some lineages accumulate mutations faster than others. The difference in rate arises from differences in mutation rate and from differences in generation time between organisms whose sequences are compared. Thus, our molecular clocks are inevitably distorted, especially for distantly diverged organisms with disparate mutation rates.

A phylogenetic tree can compare any set of organisms, even from multiple habitats. **Figure 17.19** ▶ shows the phylogeny of selected bacteria from the human intestine, within a larger clade (the gammaproteobacteria, presented in Chapter 18). The phylogenetic tree is based on a set of protein sequences of "housekeeping" genes—that is, genes encoding functions essential for all cells, and usually transmitted vertically (parent to offspring). *Escherichia*, *Shigella*, and *Salmonella* are closely related genera, including pathogenic strains, as well as normal members of the gut microbiome. The genus *Klebsiella* includes species that grow outside the gut, causing pneumonia. *Photorhabdus luminescens* is a nematode bacterium that helps its host parasitize insects. *Erwinia carotovora* infects carrots and other plants; it is closely related to *Yersinia pestis*, the cause of bubonic plague. *Photobacterium profundum* is a marine barophile (high pressure) growing in a deep-sea trench. Overall, the tree shown is rooted by *Shewanella*, a genus of metal reducers used to construct fuel cells. The tree shows how bacteria of bewildering diversity, from human pathogens to deep-ocean dwellers, diverged relatively recently from a common ancestor.

One group, the genus *Buchnera*, appears to have diverged much faster than the other bacteria. *Buchnera* species are intracellular endosymbionts of aphids (**Fig. 17.19** inset). The intracellular symbionts have undergone reductive evolution, losing many functions supplied by their host. Intracellular endosymbionts typically mutate much faster than free-living bacteria. In addition, they undergo intense selection pressure for adaptation to their obligate host. And yet, other bacteria (*Regiella*) within the same host evolved the opposite way— they gained genes encoding parasitic factors. Endosymbiosis is discussed further in Section 17.6.

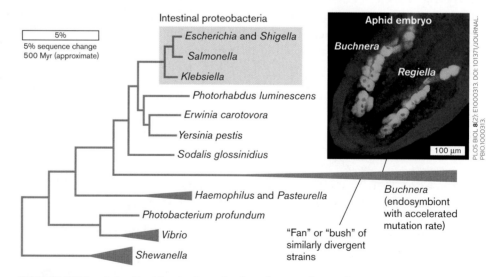

FIGURE 17.19 ■ Intestinal bacteria and related proteobacteria. The phylogenetic tree was derived from concatenated sequences of highly conserved "housekeeping" proteins. The scale bar corresponds to 5% amino acid sequence divergence. **Inset:** Aphid embryo contains bacterial symbionts, with DNA labeled by fluorescence in situ hybridization (FISH). Green = *Buchnera*; pink = *Regiella*; blue = aphid nuclei. *Source:* Phylogeny modified from Morgan Price et al. 2008. *Genome Biol.* **9**:R4 and from Fabia Battistuzzi et al. 2004. *BMC Evol. Biol.* **4**:44. ▶

Buchnera and several other lineages show a widening branch. The widening branch indicates a "fan" or "bush" of strains equally distant from a common node (branch point). The cause of "bushy" branches is debated; some researchers argue that all branches become bushy once enough strains have been sequenced.

How do we calibrate a phylogenetic tree—that is, relate the number of mutations to the time since divergence? We need an external measure of time. A tree can be calibrated if some kind of fossil evidence or geological record exists to confirm at least one branch point of the tree. But for microbes, such fossil calibrations remain speculative. One convincing method of calibration is to correlate the divergence of microbial species growing only inside of particular host species with the divergences of their hosts, based on the host fossil record. For example, the exceedingly rapid divergence of *Buchnera* species (**Fig. 17.19**) can be calibrated on the basis of their host insects. Such calculations, however, reveal vastly divergent rates of evolutionary change in different bacterial taxa.

Could a life-form mutate even faster than *Buchnera*? Some RNA viruses, such as HIV, mutate so fast that they form multiple strains within one host (discussed in Chapter 11).

> **Thought Question**
>
> **17.6** What are the major sources of error and uncertainty in constructing phylogenetic trees?

Divergence of Three Domains of Life

Carl Woese first used SSU rRNA phylogeny to reveal the existence of a third kind of life, Archaea, roughly as distant from bacteria as from eukaryotes (**Fig. 17.20**). The three fundamental groups of life-forms—Archaea, Bacteria, and Eukarya—are termed **domains**. How was an entire domain of life missed in the past? Many archaea grow in habitats previously thought inhospitable for life; for example, *Thermoplasma* species grow at 60°C and pH 2 (**Fig. 17.21**). Others, such as methanogens and halophiles, were long known to microbial ecologists, but without tools for genetic analysis they were simply classified among bacteria.

Rooting the tree of life. Where is the root of the tree, the position of the last universal common ancestor of all life-forms? Which of the three domains (Bacteria, Archaea, Eukarya) was the first to diverge from the other two? The question has profound importance for biology because the research community bases its "model systems" for study on their commonalities with organisms of importance to humans. For example, investigators of intron splicing in

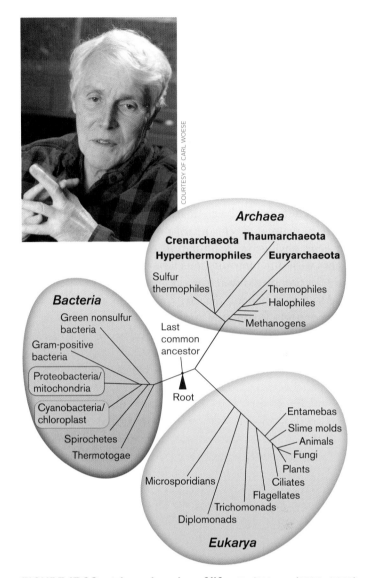

FIGURE 17.20 ▪ Three domains of life. Carl Woese (1928–2012) (inset) used SSU rRNA sequencing to reveal three equally distinct domains of life: Bacteria, Eukarya (eukaryotes), and Archaea.

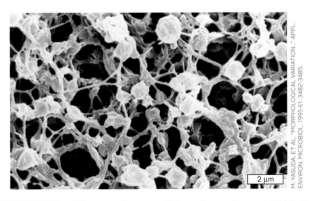

FIGURE 17.21 ▪ *Thermoplasma*. This archaeon lives at 60°C at pH 2—with no cell wall, only a cell membrane.

archaea argue that archaea represent a model system for related processes in complex eukaryotes such as humans.

The root, however, can be found only by measuring divergence relative to an outside group of organisms. Since no "outgroup" exists for the tree of all life, how is the tree rooted? One approach is to compare a pair of homologous genes within one organism—homologs that diverged from a common ancestral gene and acquired distinct functions (paralogs). The pair of paralogs chosen for analysis must have diverged within the common ancestral cell before the divergence of the three domains of life. The early divergence is determined by comparing the pair's sequence similarity in many different organisms and constructing a tree of gene phylogeny. A limitation of this approach is that genes with different functions have been subjected to natural selection in addition to random variation.

Most phylogenetic data so far indicate a root between Bacteria and the common ancestor of Archaea and Eukarya; that is, Bacteria diverged from Archaea and Eukarya before Archaea and Eukarya diverged from each other. While the position of the root remains controversial, the three major divisions of life based on rRNA phylogeny have been confirmed by sequence data from the numerous genomes published to date. Furthermore, structural and physiological comparison of the three domains largely confirms the rRNA tree (**Table 17.2**). Note first that all living cells on Earth share profound similarities. All cells consist of membrane-enclosed compartments that shelter the same fundamental apparatus of cell production: the DNA-RNA-protein machine. The fundamental components of this machine appear to have evolved before the three domains diverged from their last universal common ancestor. From a molecular standpoint, all cells on Earth appear more similar than they are different.

Nonetheless, important differences emerge between each pair of domains; indeed, each domain shows distinctive traits absent or scarce in the other two. We summarize here the major features common to most members of each domain. Various classes and species within each domain are explored in Chapters 18–20.

Archaea, bacteria, and eukaryotes. Of the three domains, the eukaryotes stand out as having a nucleus and other complex membranous organelles (see **Table 17.2**). Eukaryotic organelles include mitochondria and chloroplasts, which evolved from internalized bacteria. Bacteria and archaea possess no nucleus and have relatively simple intracellular membranes. Their size is limited by diffusion across the cell membrane, with occasional exceptions, such as the "giant bacterium" *Epulopiscium fishelsoni*. The larger size and complexity of eukaryotic cells mean that they generally require the most highly powered sources of energy, such as aerobic respiration and oxygenic photosynthesis, although

some protists conduct fermentation. By contrast, the prokaryotes (bacteria and archaea) employ a wider range of metabolic alternatives, including lithotrophy and anaerobic respiration. Finally, eukaryotic plants and animals have attained a degree of multicellular complexity unknown in the prokaryotic domains.

On the other hand, eukaryotes share key traits with archaea that distinguish both from bacteria. The core information machinery of eukaryotes more closely resembles that of archaea. The two domains share closely related components of the central DNA-RNA-protein machine: RNA polymerase, ribosomes, and transcription factors. Even such hallmarks of eukaryotes as intragenic introns, the splicing machinery, and the "RNA interference" regulatory complexes are found in archaea. All this explains why rRNA trees place archaea closer to eukaryotes than to bacteria.

Nevertheless, eukaryotes share fundamental structures with bacteria that differ from those in archaea. Archaea possess unique cell membrane components, such as their ether-linked lipids (see Chapter 19). Outside the archaea, ether-linked membrane lipids are found only in deep-branching bacterial species that share habitat (and exchange genes) with hyperthermophilic archaea. Only the domain Archaea includes species capable of growth in the most "extreme" environments of temperature (above 110°C or below −20°C) and pH (below pH 1). At the same time, many other archaeal species grow well at mesophilic temperatures in soil or water. Perhaps the most striking distinction of archaea is the complete absence of archaeal pathogens of animals or plants. Even the many methanogens that live within animal digestive tracts have never been shown to cause disease.

Overall, the three-domain phylogeny divides life usefully into three distinctive groups. Yet the tree also shows signs of gene flow unaccounted for by monophyletic descent. The eukaryotes contain mitochondria derived from assimilated bacteria whose genomes persist within the organelle. And pathogenic bacteria such as *Agrobacterium* species transfer DNA into the genomes of plants. Moreover, sequenced genomes reveal evidence of gene transfer between bacteria and archaea sharing high-temperature habitats. What if the tree of life is not strictly monophyletic?

Horizontal Gene Transfer

In retrospect, the very first demonstration of the molecular basis of heredity—the transformation of avirulent *Pneumococcus* to virulence in 1928—involved **horizontal gene transfer** (or lateral gene transfer). Horizontal gene transfer is the acquisition of a piece of DNA from another cell, as distinguished from **vertical gene transfer**, the transmission of an entire genome from parent to offspring.

TABLE 17.2	Three Domains of Life		
Characteristic	**Traits of living organisms**		
	All cells on Earth resemble each other in these traits:		
Chromosomal material	Double-stranded DNA		
RNA transcription	Common ancestral RNA polymerase		
Translation	Universal genetic code; common ancestral rRNAs and elongation factors		
Protein	Common ancestral functional domains		
Cell structure	Aqueous cell compartment enclosed by a membrane		
	COMPARISON OF DOMAINS		
	Bacteria	**Archaea**	**Eukarya**
	Archaea resemble bacteria in these traits:		
Cell volume	1–100 μm^3 (usually)		1–10^6 μm^3
DNA chromosome	Circular (usually)		Linear
DNA organization	Nucleoid		Nucleus with membrane
Gene organization	Multigene operons		Single genes
Metabolism	Denitrification, N$_2$ fixation, lithotrophy, respiration, and fermentation		Respiration and fermentation
Multicellularity	Simple		Simple or complex
		Archaea resemble eukaryotes in these traits:	
Intron splicing	Introns are rare	Introns are common	
RNA polymerase	Bacterial homologs	Eukaryotic homologs	
Transcription factors	Bacterial homologs	Eukaryotic homologs	
Ribosome sensitivity to chloramphenicol, kanamycin, and streptomycin	Sensitive	Resistant	
Translation initiator	Formylmethionine	Methionine (except mitochondria use formylmethionine)	
	Bacteria resemble eukaryotes and differ from archaea in these traits:		
Methanogenesis	No	Yes	No
Thermophilic growth	Up to 90°C	Up to 120°C	Up to 80°C
Photosynthesis	Many species; bacteriochlorophyll (proteorhodopsin)	Haloarchaea only; bacteriorhodopsin (shares homology with proteorhodopsin)	Many species; chlorophyll (bacterial origin)
Light absorption	Red and blue (chlorophyll absorption)	Green (central range of solar spectrum; no chlorophyll)	Red and blue (chloroplasts of bacterial origin)
Membrane lipids (major)	Ester-linked fatty acids	Ether-linked isoprenoids	Ester-linked fatty acids
Pathogens infecting animals or plants	Many pathogens	No pathogens	Many pathogens

Among bacteria and archaea, DNA is transferred horizontally by plasmids, transposable elements, and bacteriophages, as well as through the process of transformation, as discussed in Chapter 9. For example, drug resistance genes are transferred from harmless human-associated bacteria to pathogens. Another example is the transfer of genes encoding light-driven proton pumps (bacteriorhodopsin) from halophilic archaea into marine bacteria. Such transfer events are relatively rare, occurring perhaps once in a million generations. But over time, the number of such "rare" events can accumulate.

How can we tell when a genome contains DNA sequence "transferred" from a distant relative? One sign of horizontal transfer is a DNA sequence whose GC/AT ratio (proportion of GC and AT base pairs) differs from the rest of the genome. A surprising proportion of genomic DNA can show "spikes" of GC content that differ from the GC/AT ratio of neighboring sequences. These regions of anomalous GC/AT ratio indicate origin elsewhere, even from species of a different domain. In some archaea, particularly the hyperthermophiles *Pyrococcus* and *Aeropyrum*, 10%–20% of the genes appear to come from bacteria that share their high-temperature environment. On the other hand, the thermophilic bacterium *Thermotoga maritima* shows a number of genes transferred from archaea.

Between more closely related taxa, genes migrate even faster. For example, the *Escherichia coli* genome acquired about 18% of its genes from closely related species after its relatively recent divergence from the close relative *Salmonella enterica*. Some medically important genera, such as *Neisseria* (which causes gonorrhea and meningitis), are particularly "recombinogenic." Rapid gene exchange enables pathogens to avoid the host immune system by expressing novel proteins not recognized by host antibodies. Furthermore, the genomes of eukaryotes include many genes acquired from bacteria, most of them from the endosymbiotic ancestors of their mitochondria and chloroplasts (discussed in Section 17.6).

How does horizontal gene transfer affect microbial phylogeny? In 1999, Ford Doolittle, at Dalhousie University, redrew the standard tree of life with a bewildering array of cross-cutting lineages to show how actual phylogeny combines horizontal and vertical transfer (**Fig. 17.22**). Perhaps the "last common ancestor" was actually a last common <u>community</u> of diverse life-forms that contributed different parts of our genetic legacy.

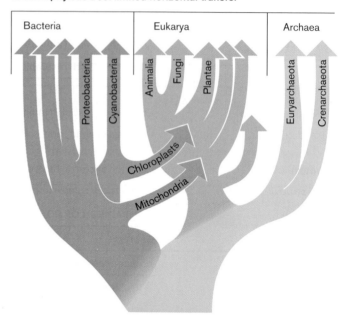

A. Monophyletic tree: limited horizontal transfer

Bacteria — Proteobacteria, Cyanobacteria

Eukarya — Animalia, Fungi, Plantae

Archaea — Euryarchaeota, Crenarchaeota

Chloroplasts

Mitochondria

Last common ancestor

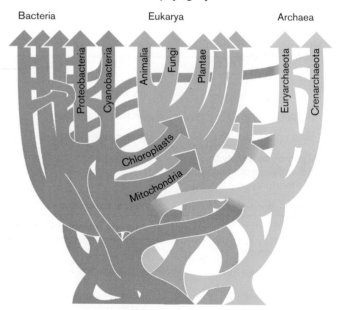

B. Horizontal transfer obscures phylogeny

Bacteria — Proteobacteria, Cyanobacteria

Eukarya — Animalia, Fungi, Plantae

Archaea — Euryarchaeota, Crenarchaeota

Chloroplasts

Mitochondria

Last common community

FIGURE 17.22 ▪ **Vertical and horizontal gene transfer.**
A. The traditional view of phylogeny: Most gene transfer is vertical, and horizontal transfer is limited to rare cases such as transfer of mitochondria and chloroplasts from bacteria into eukaryotes.
B. An alternative view: Horizontal gene transfer occurs so often that it obscures monophyletic distinctions between taxa. *Source:* Modified from W. Ford Doolittle. 1999. *Science* **284**:2124.

Reconciling Vertical and Horizontal Gene Transfer

One approach to sorting out vertical and horizontal gene transfer was proposed by James Lake and colleagues at UCLA and developed further by Doolittle. Lake's approach

---

Thought Question

17.7 What are the limits of evidence for horizontal gene transfer in ancestral genomes? What alternative interpretation might be offered?

assumes that some classes of genes nearly always transmit vertically—particularly "informational genes," which specify products essential for transcription and translation. Informational genes include RNA polymerase, as well as ribosomal RNAs such as SSU rRNA, and elongation factors. Informational genes need to interact directly in complex ways with large numbers of cellular components; thus, their capacity for horizontal transfer is limited. On the other hand, "operational genes" are those whose products govern metabolism, stress response, and pathogenicity. Operational genes function with relative independence from other cell components and consequently move more easily among distantly related organisms, particularly organisms that share the same habitat. An important category of such movable genes is virulence factors. Pathogens show extensive horizontal transfer of virulence genes and genes encoding resistance to host defenses and antibiotics.

Groups of such genes are often transferred together on plasmids or on **genomic islands**, regions of DNA of foreign origin that confer special properties, such as virulence or nitrogen fixation (discussed in Section 9.6). Examples of a virulence plasmid and a genomic island are found in the deadly pathogen *E. coli* O157:H7 (**eTopic 17.3**). But in some cases, the group of genes transferred is so large that it makes up a substantial chunk of an organism's genome. For example, the Aquificales group of thermophilic bacteria includes a large section derived from Proteobacteria similar to *E. coli*—so large that the Aquificales phylogeny remains unclear.

A balanced view of vertical and horizontal gene transfer is shown in **Figure 17.23**. The row of arrows designates vertical transfer of the bulk of the genomes of the two diverging species. Black lines represent vertically transferred genes, such as ribosomal RNA; gray lines indicate genes with more flexible function. At various levels, colored lines indicate horizontal transfer, where genes enter each lineage by processes such as conjugation or transformation. Gray lines peel out, indicating loss of a gene by mutation. Overall, the vertical lineage persists for most of the core informational genes (black lines), while horizontally acquired genes enter the genome. The persistence of genes showing vertical inheritance in nearly all life (such as SSU rRNA) may reflect the phylogeny of the organism as a whole, despite the other genes that are transferred horizontally. But to fully assess phylogeny, we must sequence entire genomes.

To Summarize

■ **Phylogeny is the divergence of related organisms.** Organisms diverge through random mutation, natural selection, and reductive evolution.

■ **Molecular clocks are based on mutation rate.** Given a constant mutation rate and generation time, the degree

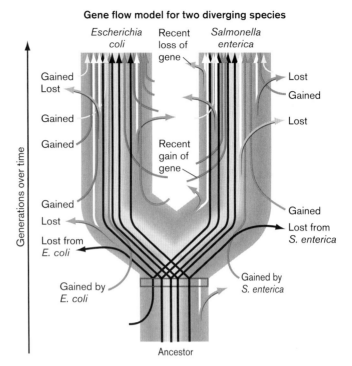

Gene flow model for two diverging species

FIGURE 17.23 ■ Genes enter and leave genomes. Both rRNA trees and whole-genome trees consistently reflect monophyletic descent (black lineages) down to the genus level. Below the genus level, monophyletic descent may be obscured by high rates of horizontal transfer (colored lines entering, gray lines departing). White spaces indicate genes lost by reductive evolution.

of difference between two DNA sequences correlates with the time since the two sequences diverged from a common ancestor.

■ **Different sequences diverge at different rates.** Under selection pressure, the actual divergence rates of sequences depend on structure and function of the RNA or protein products.

■ **Phylogenetic trees are based on sequence analysis.** The more different the two sequences are, the longer is the branch representing time since divergence from the common ancestor. Rooting a tree requires comparison with an outgroup.

■ **The tree of life diverges to three domains: Bacteria, Archaea, and Eukarya.** Eukaryotes are distinguished by the nucleus, which is lacking in archaea and bacteria. Archaea possess ether-linked isoprenoid lipids that are rare or absent in bacteria and eukaryotes, and they are never pathogens. The machinery of archaeal gene expression resembles that of eukaryotes.

■ **Genes transfer between different species.** Horizontal transfer is most frequent between closely related species or between distantly related species that share a common habitat. Informational genes with many molecular interactions transfer vertically, whereas operational genes that function independently of other components are more likely to transfer horizontally.

■ **Horizontal gene transfer is important for adaptation to new environments and for pathogenesis.** Gene transfer among pathogenic and nonpathogenic strains leads to emergence of new pathogens.

17.4 Adaptive Evolution

In discussing phylogeny, we focused on the accumulation of random mutations that cause genome sequences to diverge at a steady rate. The rate of nonselected sequence changes enables us to measure evolutionary time. But how does evolution help life survive in a new environment? Adaptive evolution requires **natural selection**. In natural selection, the genetic variants that arise by random mutation differ in their chance of survival. Those variants that survive to leave more offspring undergo "selection"; that is, their traits are overrepresented in the next generation. Over many generations, the descendants develop traits very different from those of their ancestors.

How can we study adaptive evolution—especially the evolution of different species, which takes place over millions of years? We have several kinds of evidence:

■ **Genomic analysis.** Comparing gene sequences, both within and between genomes, enables us to track how organisms adapted in the past.

■ **Strongly selective environments.** Environments under intensive selective pressure, such as high-temperature habitats or exposure to antibiotics, lead to rapid evolution.

■ **Experimental evolution.** Experimental strategies reveal evolution in the laboratory, enabling us to test predictive models.

Genomic Analysis

As discussed in Section 17.3, the sequence of genomes reveals descent over time based on steady accumulation of mutations. But other mutations undergo selection that provides a cell with new functions. A common way this happens is by **gene duplication**.

As cells replicate their DNA over many generations (discussed in Chapter 9), occasionally the DNA polymerase will make a duplicate copy of a gene. With duplicate copies of a gene, one copy may acquire mutations that change its function without detriment to the organism because the "backup" copy still functions. Now, suppose the mutated copy gains additional mutations that further alter its function, providing a new function to the cell. For example, a gene encoding a transporter for one sugar may now encode a transporter that

better "fits" a different sugar. The two **paralogous genes**, or **paralogs**, have evolved to serve different functions. Paralogs are a major source of raw material that contributes to new functions arising through evolution. We can detect paralogs in a genome, through sequence relatedness; for example, the genome of the hyperthermophilic bacterium *Thermotoga maritima* shows paralogous ABC transporters for lactose, cellobiose, mannose, and xylose, among others. Evolution of paralogs provides an important way for organisms to enhance fitness in a complex environment.

Does a population ever lose some of its paralogs? Certain environments select for loss of many genes—a phenomenon called **degenerative**, or **reductive**, **evolution**. The selection for loss is "positive" because organisms save energy by avoiding replication and expression of unneeded genes. Intracellular pathogens and obligate symbionts often lose large chunks of their genome, by processes such as intracellular recombination of the chromosome. The lost genes typically encoded functions supplied by the host, such as capture of nutrients and generation of energy. Lost genes can be identified by comparison with free-living species. For example, *Treponema pallidum*, the cause of syphilis, has a genome of 1.1 million base pairs, which is about a quarter the size of the *E. coli* genome. *T. pallidum* has lost all of its enzymes used in the TCA cycle, respiration, and amino acid biosynthesis. These genes remain in the genomes of free-living treponemes.

In other species, dilute, nutrient-poor natural environments select for gene reduction. Marine cyanobacteria, a major source of global photosynthesis, have lost numerous genes that confer little advantage in the open ocean. Missing genes encode transporters for sugars and amino acids (which are scarce in the ocean) as well as proteins for flagellar motility and pili. More surprising, *Prochlorococcus* species have lost catalase, an enzyme that destroys the hydrogen peroxide they produce. Their hydrogen peroxide is detoxified by catalase-positive bacteria that share their marine habitat. Erik Zinser at the University of Tennessee–Knoxville showed that *Prochlorococcus* could be cultured on agar only in the presence of partner bacteria that supply catalase. Given that catalase is an expensive enzyme to produce, *Prochlorococcus* may have gained an advantage when this enzyme was deleted during its evolution in the nutrient-poor open ocean.

Strongly Selective Environments

Evolution requires many generations, but certain microbes may produce 40 generations in a day. When such microbes are under strong selection pressure, such as the presence of an antibiotic, evolution may occur surprisingly fast. In the case shown in **Figure 17.24**, a hospitalized patient was infected by MRSA, a deadly strain of *Staphylococcus aureus* that had already evolved resistance to drugs such as methicillin. The

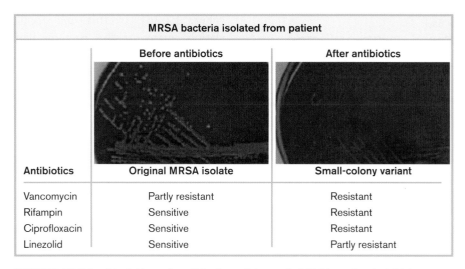

MRSA bacteria isolated from patient		
	Before antibiotics	**After antibiotics**
Antibiotics	**Original MRSA isolate**	**Small-colony variant**
Vancomycin	Partly resistant	Resistant
Rifampin	Sensitive	Resistant
Ciprofloxacin	Sensitive	Resistant
Linezolid	Sensitive	Partly resistant

FIGURE 17.24 ■ **Evolution of antibiotic resistance in MRSA. Left:** The initial MRSA isolate infecting the patient showed partial resistance to vancomycin but was sensitive to rifampin, ciprofloxacin, and linezolid. **Right:** Following exposure to these antibiotics, a new strain was isolated that grew more slowly (the small-colony variant) but was at least partly resistant to all the antibiotics. *Source:* Wei Gao et al. 2010. *PLoS Pathog.* **6**:e1000944.

patient was an elderly man with a weakened immune system, unable to eliminate even small populations of an opportunistic pathogen such as MRSA. The original culture from the patient's blood showed bacteria sensitive to four antibiotics. But prolonged exposure resulted in natural selection for resistance. Ultimately, following a total of 12 weeks' exposure to linezolid (one of the last antibiotics effective against MRSA), the bacteria isolated showed resistance to three other antibiotics plus partial resistance to linezolid.

In **Figure 17.24**, note that the latest drug-resistant strain actually made smaller colonies with or without the drug—the "small-colony variant." In other words, natural selection yielded a population that was the "fittest" under a particular environmental condition—the presence of linezolid in an immunocompromised host—where even a slow-growing strain could persist. In the absence of the drug, the original strain would outcompete the small-colony variant.

In the case just described, what was the molecular basis of the multidrug resistance? Researchers obtained DNA from the patient's original MRSA, as well as the later small-colony variant. They sequenced the two genomes and compared them. Just three point mutations in the small-colony variant accounted for the drug resistance, as well as the retarded growth. One of these mutations derepressed a stress regulon (group of genes under one regulator), causing accumulation of the stress signal ppGpp (guanosine tetraphosphate). The ppGpp stress regulon includes expression of many protective genes that enable a cell to survive in the presence of antibiotics. But the cost to the cell of expressing this regulon is a slower growth rate; like a community under "terror alert," the cell's normal everyday processes are slowed by the demands of the stress response.

The MRSA example illustrates two key points:

- **The "fittest" trait depends on the environment in which selection occurs.** The presence of an antibiotic selects individuals that are resistant, despite slower growth (small-colony size). Without the antibiotic, the faster-growing individuals prevail.

- **Rapid adaptive evolution may be enhanced by disabling regulation.** In this case, the ppGpp stress regulon was derepressed, enabling stress responses that normally would be turned off because they inhibit cell growth.

Experimental Evolution in the Laboratory

The process of evolution can be studied in the laboratory—an approach known as **experimental evolution** or **laboratory evolution**. A landmark experiment on evolution in the laboratory was performed by Richard Lenski at Michigan State University (**Fig. 17.25** and **eTopic 17.4**).

FIGURE 17.25 ■ **Rich Lenski and Zack Blount conduct the Long-Term Evolution Experiment (LTEE).** At Michigan State, Lenski (inset) opens a box of evolved clones stored in the –80°C freezer. Blount meditates before a tower of petri plates that represent 20 years of competition assays for relative fitness of evolved clones.

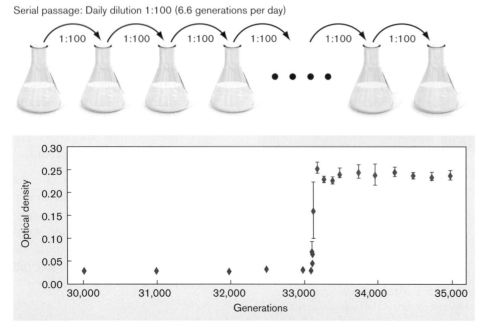

Serial passage: Daily dilution 1:100 (6.6 generations per day)

FIGURE 17.26 ▪ **Evolution of an aerobic, citrate-utilizing strain of *E. coli*.** The bacteria were diluted daily in a glucose-limited growth medium. The medium also contained citrate to allow the bacteria to take up iron. After 33,000 generations (estimated cell doublings), the bacterial population suddenly grew to a much greater density because of the rise to high frequency of a strain that had evolved the ability to catabolize the citrate at about 31,000 generations. *Source: Adapted from Zachary Blount et al. 2008. PNAS **105**:7899.*

In 1988, Lenski began the Long-Term Evolution Experiment (LTEE) by founding 12 populations of *E. coli* from a single clone. The 12 populations were cultured in a medium in which growth was limited by a small amount of glucose (**Fig. 17.26**). Every 24 hours, 1% of each population was transferred to fresh medium and then grew for about 6.6 generations per day. Every 500 generations, samples of each population were frozen for later study. The populations have now evolved for more than 60,000 generations, and students continue the daily dilutions, led by postdoctoral fellow Zachary Blount.

The bacteria in Lenski's frozen populations remain alive and can be revived at any time for analysis and use in experiments. Experiments include the sequencing of genomic DNA of clones isolated from these frozen time points. With laboratory evolution experiments, we can test predictions of how new traits appear and use genome sequences to examine the mutations that led to the new traits. Evolution experiments can also include many replicate populations, making it possible to examine the repeatability of evolution.

During the LTEE, Blount and Lenski have seen many changes, such as the evolution of bacteria that can grow faster than the ancestor and form larger cells. After generation 33,000, Blount found the most striking change: One of the 12 populations suddenly became much denser

(**Fig. 17.26**). In this one population, many of the bacteria had evolved a new phenotype: They were able to grow aerobically on citrate—a trait found in other species but rare in *E. coli*. The bacteria are called citrate plus, or Cit⁺.

Since citrate has been available in the medium from the beginning, why did the Cit⁺ mutation take so long to evolve? By testing *E. coli* from earlier generations that he had stored in a freezer, Blount was able to show that over the first 31,000 generations, no bacteria could metabolize citrate, but then a rare mutation occurred that enabled citrate uptake and catabolism. Over a few generations, the Cit⁺ cells had increased their proportion of the population, until they became the predominant strain.

But was the overall phenotype really so sudden? Could any *E. coli* cell mutate to catabolize citrate? Lenski proposed that the earlier generations had acquired some kind of "potentiating" mutations that somehow enabled *E. coli* to gain the Cit⁺ mutation. He called this the "historical contingency" hypothesis—that evolution of a new trait may require some other unknown trait to appear first. In support of the hypothesis, Lenski noted that the Cit⁺ strains were extremely rare, arising in only one of the 12 populations. Therefore, mathematically it was likely that Cit⁺ required the presence of some other rare mutation.

To test the hypothesis, Blount delved into the fossil record of the evolving populations and isolated clones from various time points. He and Lenski then tested the ability of these clones, as well as of the ancestral strain, to mutate to Cit⁺. They found that only clones isolated from time points after 20,000 generations reevolved the new Cit⁺ phenotype; the original strain did not. This result supported the hypothesis that some other mutation was needed first. To gain additional information about the early and late mutations, Blount sequenced the genomes of 29 clones isolated from the population's frozen fossil record at numerous time points through 40,000 generations (**Fig. 17.27**). The strains were measured for citrate utilization, and for ability to evolve Cit⁺. The data fit the following model of stages of evolving a new trait: (1) potentiation to achieve useful mutations; (2) actualization of a novel mutant phenotype; and (3) refinement, or increasing the degree of the phenotype.

A. Evolution of Cit⁺ (citrate catabolism)

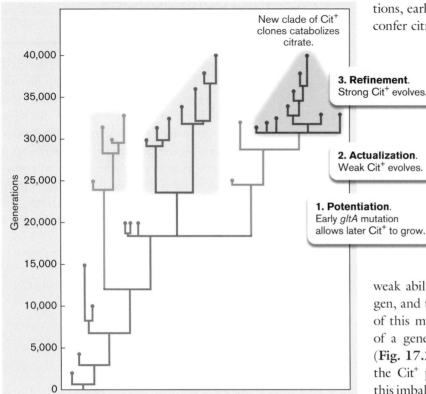

B. Tandem duplication fuses *rnk* promoter upstream of *citT*

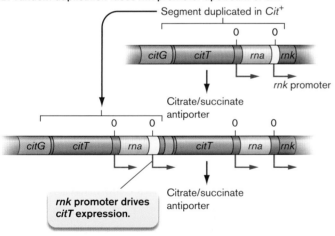

FIGURE 17.27 ■ **Evolution of citrate utilization in *E. coli*.**
A. One or more of the mutations accumulated over 30,000 generations potentiated evolution of the Cit⁺ trait. The Cit⁺ trait first appeared in about generation 31,000. Subsequent mutations refined the phenotype, increasing the rate of citrate utilization. Ball symbols indicate genomes from the population's history that were sequenced for analysis. **B.** The Cit⁺ mutation. After 31,000 generations, a tandem duplication event placed a copy of the *rnk* promoter, which directs expression when oxygen is present, upstream of the *citT* gene encoding a citrate/succinate antiporter that is normally repressed by oxygen. This new *rnk-citT* module now expresses CitT, which takes up citrate, allowing bacteria to respire on citrate from the medium. *Source:* Adapted from Zachary Blount et al. 2012. *Nature* **489**:513.

Potentiating mutations. During the first 31,000 generations, early mutations occurred in some cells that did not confer citrate catabolism but somehow "potentiated" the ability of cells to gain a Cit⁺ mutation later. These mutations led to the appearance of Cit⁺ cells (**Fig. 17.27A**) and to the appearance of similar mutations in "replayed" evolution from late-generation strains. One potentiating mutation was shown to be *gltA*, a gene encoding citrate synthase, which synthesizes citrate in the TCA cycle. This mutation allowed growth on acetate, which is excreted during growth on glucose (discussed in Chapter 13).

Actualization of the novel phenotype.

The Cit⁺ mutation arose initially as a weak ability to transport citrate in the presence of oxygen, and therefore to respire on citrate. The genetic basis of this mutation was found to be a tandem duplication of a gene encoding a citrate/succinate antiporter, *citT* (**Fig. 17.27B**). Because the antiporter excretes succinate, the Cit⁺ phenotype causes a metabolic imbalance—but this imbalance gets corrected by the potentiating mutation for acetate utilization, *gltA*. The duplicated *citT* (upstream of the first copy) happens to include a promoter for an adjacent gene, *rnk*, which encodes a regulator of nucleic acid metabolism, expressed in the presence of oxygen. Thus, the duplication places *citT* expression under control of the copied *rnk* promoter. The *rnk* promoter now expresses *citT* in the presence of oxygen, enabling the cell to respire on citrate.

The history of the initial Cit⁺ clone thus required two rare events: the potentiating mutation, *gltA*; and the tandem duplication of *citT-rnk*, which placed *citT* expression under control of a promoter active in the presence of oxygen.

Refinement of the phenotype. The earliest Cit⁺ clones were very poor at growing on the citrate. But later clones evolved as a result of mutations that increased the rate of citrate utilization—that is, by evolutionary refinement of the Cit⁺ trait. The researchers showed that this improvement was due to an increase in the copy number of the duplicated segment that produces the *rnk-citT* module, which increased the number of expressed copies of *citT*. This improved growth on citrate eventually led the Cit⁺ clones to dominate the population. Interestingly, though, even the "refined" Cit⁺ cells never fully took over the cultures. Some cells always remained that had adapted to limiting glucose by other means. We saw the example of Caroline Turner's mutant (Current Research Highlight), which evolved to metabolize the succinate and other four-carbon acids excreted by Cit⁺ clones.

The results of Lenski's Long-Term Evolution Experiment raise this question: If *E. coli* is evolving new traits, is it in the process of evolving a new species? We address the species concept next (Section 17.5). Meanwhile, beyond the philosophical debates, industrial laboratories now use advanced forms of experimental evolution. Evolution can improve commercial enzymes—and even develop new products. For example, Bernhard Palsson and colleagues at UC San Diego used laboratory evolution to evolve a strain of *Streptomyces* to produce high concentrations of a novel antibiotic against MRSA. Others take evolution to an extreme: Could we generate in vitro all the possible mutants of an organism, and set them against each other all at once? Such an approach is called **directed evolution**. An example of directed evolution is the development of hyperthermophilic forms of a xylanase enzyme used for paper production (discussed in **Special Topic 17.1**).

To Summarize

- **Adaptive evolution** requires natural selection.

- **Genomic analysis** tracks how organisms adapted in the past. Gene duplications provide the opportunity to evolve paralogous genes with different functions.

- **Degenerative (reductive) evolution** happens when unneeded genes are lost from the genome. The organism saves energy by avoiding their replication and expression.

- **Strongly selective environments**, such as antibiotic exposure, reveal rapid evolution.

- **Selective pressure depends on the particular environment.** A trait favored in one environment may be disadvantageous in another.

- **Experimental evolution in the laboratory enables** us to test hypotheses about natural selection. Richard Lenski's Long-Term Evolution Experiment shows potentiation to achieve useful mutations; actualization of a novel mutant phenotype; and refinement, or increasing the degree of the phenotype. Laboratory evolution is used widely in industrial development.

17.5 Microbial Species and Taxonomy

What is a species? Among eukaryotes, a species is defined by the principle that members of different species do not normally interbreed. Thus, the failure to interbreed is the traditional property that distinguishes species. Bacteria and archaea, however, transfer genes horizontally between distantly related clades. So the definition of bacterial species is complex, subject to heated debate among microbiologists. Also much debated is the classifying of life-forms into different kinds (**classification**) and the naming of species (**taxonomy**).

Note: Distinguish these terms:
Classification is the sorting of life-forms into bins (categories) based on genetic relatedness and traits of the organisms.
Nomenclature is the naming of categories, including species.
Taxonomy is the classification and naming of organisms.

Defining a Species

As genome sequence data became available for microbes, scientists hoped that quantitative measures of divergence could provide a consistent basis for defining microbial species. But for some organisms, such as *Helicobacter pylori*, the genomes of different strains that cause the same disease (gastritis) differ by as much as 7%. On the other hand, strains of *Bacillus* with nearly identical genomes cause completely different diseases, such as anthrax (*B. anthracis*) and caterpillar infection (the biological pesticide *B. thuringiensis*). Even more puzzling, hyperthermophilic bacteria such as *Aquifex aeolicus* show a high proportion of genes from the domain Archaea! Some researchers argue that the species concept lacks meaning for microbes.

Amid the debates, microbiologists generally agree on the importance of two perspectives: phylogeny (based on DNA relatedness) and ecology (based on shared traits and ecological niche).

Phylogenetic relatedness. A species is a group of individuals that share relatedness of a key set of "housekeeping genes," typically informational genes such as ribosomal and transcriptional components. Ideally, these genes should all be orthologs (genes with a common origin and function), not paralogs (which diverged from a common ancestor but now differ in function). Within a genus, species that cannot be distinguished by SSU rRNA alone may be defined by analysis of multiple genes. For example, analysis of multiple gene loci effectively distinguishes *Neisseria meningitidis* (the cause of meningitis) from *Neisseria lactamica*, a harmless resident of the nasopharynx. The multigene approach has proved successful even in the case of highly "recombinogenic" organisms known to acquire and rearrange genes readily.

Ecological niche (ecotype). Besides a high degree of genomic relatedness, a species should include individuals that share common traits and an ecological niche, or "ecotype." Shared traits should include cell shape and nutritional requirements, and there should be a common habitat and life history (for example, causing the same disease). By these criteria, highly divergent strains of *Helicobacter pylori* causing gastritis make up one species, whereas *Bacillus anthracis* (anthrax) and *Bacillus thuringiensis* (caterpillar infection) are different species despite their highly similar genomes.

A working definition of species. While the debate goes on, many microbiologists accept the following criteria for a working definition of a microbial species:

- **SSU rRNA identity ≥95%.** Two organisms with 95% or greater similarity in SSU rRNA sequence generally are considered to share the same genus. Beyond the genus level, rRNA sequence lacks resolution.

- **Whole-genome similarity: average nucleotide identity (ANI) of orthologs ≥95%.** Within whole genomes, we can define all the orthologous genes (orthologs, genes of the same function) that two strains share. If the strains share 95% or greater ANI for their orthologs, they may be considered the same species.

- **Shared ecotype.** If two organisms with 95% or greater identity share a common habitat and metabolism, or cause the same disease, they are considered the same species.

This working definition classifies most known bacteria in a way that reconciles phylogeny with ecotype. For example, *Helicobacter pylori* genomes with a common gastric pathology show exceptional variation due to horizontal gene transfer, even including their SSU rRNA genes. Nevertheless, a set of orthologs can be defined that share 95% ANI.

At the same time, new questions arise from the availability of multiple sequenced genomes for a single species. Suppose that every time we sequence a new isolate from nature, or from a clinical specimen, we find that every new genome sequenced has a few new genes absent from previously sequenced isolates. How, then, do we define the gene map? This situation would be unheard of for animals and plants, in which the gene map is fixed by Mendelian recombination. It is common, however, for bacteria such

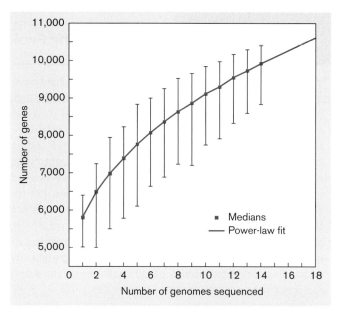

FIGURE 17.28 ■ Pan-genome of *Bacillus cereus*. The median number of genes found is shown as a function of the number of *B. cereus* genomes sequenced. Error bars indicate the range of gene number found upon sequencing different combinations of genomes. *Source:* Modified from Hervé Tettelin et al. 2008. *Curr. Opin. Microbiol.* **12**:472.

as *Bacillus cereus*, a Gram-positive soil bacterium and food pathogen (**Fig. 17.28**). For each new genome sequenced, several hundred new genes are identified that are absent from all the other genomes so far. In a pathogen, some of these new genes could be involved in pathogenesis, with profound implications for medical therapy and vaccine development.

Should we now attempt to define bacterial and archaeal genomes not just by a single sequenced genome, but by the sum total of all expected genes in all possible isolates? This theoretical total is called the **pan-genome**. The pan-genome includes genes present in all sequenced genomes of a species (known as its **core genome**) plus "accessory genes" present in one or more sequenced isolates. But how do we estimate the size of a pan-genome for a given species? A statistical model can predict the chance of finding new genes every time we sequence another genome. A few species, such as the anthrax agent *Bacillus anthracis*, have a "closed" pan-genome that appears to be defined by a relatively small number of natural isolates. But for *Bacillus cereus* (**Fig. 17.28**), after the first ten genomes, each new genome still reveals another 250 genes. Amazingly, the *B. cereus* pan-genome appears to be infinite in size! This is called an "open" pan-genome. Other species with open pan-genomes include the pneumonia pathogen *Streptococcus pneumoniae*, the marine cyanobacterium *Prochlorococcus marinus*, and the halophilic archaeon *Haloquadratum walsbyi*. Open pan-genomes may well predominate in nature;

SPECIAL TOPIC 17.1   Jump-Starting Evolution of a Hyperthermophilic Enzyme

What commercial product is used to make chicken feed more digestible, to improve the texture of cotton fabric, to bake better bread, and to make paper using less bleach? The answer is xylanase, an enzyme that cleaves plant polysaccharides containing a variety of sugar monomers, such as the five-carbon sugar xylose. Xylanases are made by many different types of bacteria and fungi, and different xylanases have evolved naturally to function best under different conditions. But for commercial use, we need to fine-tune evolution to optimize the enzyme for particular uses: a xylanase for animal feed, another xylanase for baking, and yet another for treatment of paper pulp. The paper pulp application imposes particularly demanding conditions of high pH and high temperature, and requires an enzyme that is stable above 100°C.

To design types of xylanases and other enzymes for industrial conditions, scientists at Verenium Corporation (now BASF) in San Diego developed a technology of directed evolution. As introduced in Chapter 12, directed evolution "jump-starts" the evolution process by generating a large number of possible mutations—in some cases, the entire collection of single mutations possible in a given gene. Large-scale screening then "selects" for the mutant strains that have optimal activity. While directed evolution may involve more than one round of mutation and screening, it avoids the need for hundreds or thousands of rounds of growth and selection.

To generate a hyperthermophilic xylanase, Verenium scientists first undertook "discovery," a process of collecting and testing genes encoding xylanase from many naturally occurring bacteria and fungi from all over the world. The most promising xylanase producers were collected from an alkaline hot spring of the Russian Kamchatka Peninsula, including bacteria such as *Bacillus* species and fungi such as *Trichoderma*. The scientists identified 200 different xylanases having promising levels of activity and stability. Both bacterial and fungal xylanases showed homology—members of a group called "family 11," which shares a common ancestor and related peptide structure.

From the hot-spring collection, several genes were selected that encoded the best thermophilic xylanases active at high pH. The top genes were used for "gene assembly," in which parental genes are recombined to form a highly diverse set of daughter genes (**Fig. 1**). In effect, the gene assembly process jump-starts evolution by forming in one generation a collection of recombinations that in nature would require many generations to occur by chance.

In gene assembly, the parental genes were cleaved by restriction endonucleases into several fragments (using methods described in Chapters 7 and 12). The gene fragments were then shuffled like a deck of cards by recombination in the test tube, generating a library of many different daughter genes (**Fig. 1A**). For example, six parental genes cleaved into five fragments could recombine (assuming one fragment of each type) into 5^6 = 15,625 different recombinants. The library was then screened in *E. coli* for activity of all the expressed enzymes, using a robotic screening system. Verenium scientists selected the gene encoding the xylanase that was most active at the highest temperatures, up to 76°C.

The product of gene assembly was then subjected to a secondary method of directed evolution, called gene site saturation mutagenesis (GSSM). GSSM was performed in collaboration with another lab—that of Claire Dumon and Harry Gilbert at Newcastle University, England.

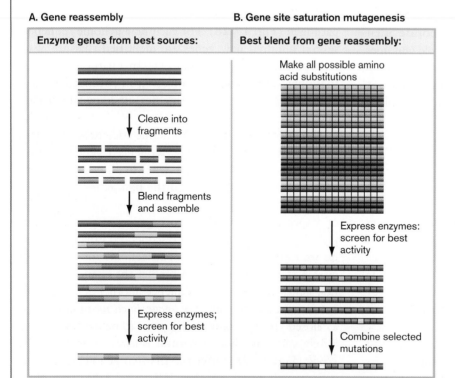

A. Gene reassembly

Enzyme genes from best sources:

Cleave into fragments

Blend fragments and assemble

Express enzymes; screen for best activity

B. Gene site saturation mutagenesis

Best blend from gene reassembly:

Make all possible amino acid substitutions

Express enzymes: screen for best activity

Combine selected mutations

FIGURE 1 ■ Directed evolution by gene reassembly and in vitro mutagenesis. A. Candidate genes are cleaved, and then the fragments are blended and assembled (gene reassembly). **B.** The best product of gene reassembly is subjected to gene site saturation mutagenesis (GSSM), in which all possible amino acid substitutions are made. The mutants are screened and subjected to a combinatorial process that generates all possible mutant combinations.

The GSSM method also imitates nature, but it yields in one "generation" a pool of all possible single-codon mutations that, in nature, might occur, but only over many generations. For GSSM, the evolved thermophilic xylanase gene was copied by PCR using primers with 64-fold degeneracy—that is, short DNA sequences that replace each codon in the gene sequence with all 64 possible codons (**Fig. 1B**). For example, in a gene of 300 codons, making all the codon substitutions with 20 different amino acids at each position results in 300 × 20 = 6,000 different single-substitution mutations. The mutant genes were cloned in an expression vector and transformed into bacteria. To ensure that greater than 95% of the library was sampled, 70,000 different colonies were screened for xylanase activity and thermostability. Clones were selected that increased enzyme stability at 76°C from 10% to 50%.

The thermostable mutant genes were then assembled and combined by a multisite combinatorial assembly process that generates all possible combinations of the single-codon substitutions. **Figure 2** shows a distribution of the number of combined mutations in combinatorial library isolates with enzyme activity at temperatures as high as 86°C. The combinatorial process ultimately yielded an enzyme (**Fig. 3**) that is active at even higher temperatures, up to 101°C. A compelling result is that none of the mutations found would have been predicted from crystal structure to increase thermostability; they arose only from the screen of all possible mutants and, in combination, yield dramatic enhancement.

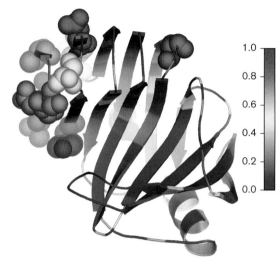

FIGURE 3 ■ **Hyperthermophilic xylanase.** The crystallographic model shows the position of mutations conferring enhanced thermostability. Amino acid color represents the proportion of the GSSM combinatorial library strains that include a substitution at that position. The amino acid substitutions that were combined in the best hyperthermophilic xylanase are shown as space-filled atoms. *Source:* Modified from Claire Dumon et al. 2008. *J. Biol. Chem.* **283**:22557–22564, fig. 5.

The hyperthermophilic xylanase is now used for paper pulp manufacture, enabling the pulp to be processed with much lower levels of bleach, which is toxic to the environment. Thus, directed evolution in the test tube jump-started the development of a product to enhance industrial sustainability.

RESEARCH QUESTION

Compare Claire Dumon's approach of DNA shuffling and gene site saturation mutagenesis with the new approaches described by Packer and Liu (2015, *Nat. Rev. Genet.* **16**:379–394). How could directed evolution make an enzyme with an activity not known to exist in nature?

Dumon, Claire, Alexander Varvak, Mark A. Wall, James E. Flint, Richard J. Lewis, et al. 2008. Engineering hyperthermostability into a GH11 xylanase is mediated by subtle changes to protein structure. *Journal of Biological Chemistry* **283**:22557–22564.

Packer, Michael S., and David R. Liu. 2015. Methods for the directed evolution of proteins. *Nature Reviews. Genetics* **16**:379–394.

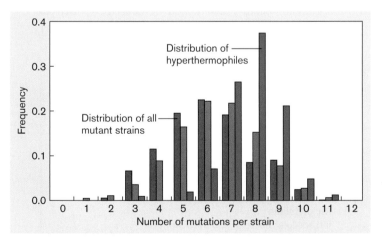

FIGURE 2 ■ **Distribution of mutants in the combinatorial library.** The frequencies of combination strains are plotted as a function of number of mutations per strain. Blue bars represent the frequencies of hyperthermophiles; red bars represent the distribution of all combination strains. *Source:* Modified from Claire Dumon et al. 2008. *J. Biol. Chem.* **283**:22557–22564, fig. 2.

thus, most bacterial and archaeal species have access to their core genome plus an uncountable number of possible accessory genes.

Classification and Nomenclature

Defining a species is part of the task of **taxonomy**, the classifying of life-forms into different categories with shared traits. Taxonomy is critical for every microbiological pursuit, from diagnosing a patient's illness to understanding microbial ecosystems.

Classification generates a hierarchy of **taxa** (groups of related organisms; singular, **taxon**) based on successively narrow criteria. The fundamental basis of modern taxonomy is DNA sequence similarity, but the use of DNA arises within a long historical tradition of phenotypic description that shapes the views and practice of microbial taxonomy. Historically, taxa have been defined and named on the basis of a combination of genetic and phenotypic traits. The nomenclature of microbes remains surprisingly fluid: As new traits are identified and the genetic sequence of a microbe is established, species are all too frequently renamed. Current information on microbial names is available through online databases such as the List of Prokaryotic Names with Standing in Nomenclature (LPSN).

Traditional classification designates levels of taxonomic hierarchy, or rank, such as phylum, class, order, and family (**Table 17.3**). Some levels of rank are designated by certain suffixes—for example, "-ales" (order) and "-aceae" (family). The ultimate designation of a type of organism is that of **species**, which includes the capitalized name of the **genus** (group of closely related species) followed by the uncapitalized species name—for example, the well-known species *Streptomyces coelicolor*. *S. coelicolor* is a member of the Actinomycetales (informally called actinomycetes), filamentous Gram-positive bacteria that produce many kinds of antibiotics. Actinomycetes are subdivided into families and genera, such as the genus *Streptomyces*.

Microbiologists continually reveal previously unknown taxa. In 2002 a group of uncultured marine bacteria, the SAR11 cluster, was found to comprise a quarter of all marine microbial cells. SAR11 was originally defined by its SSU rRNA sequence. Surprisingly, the rRNA sequence phylogeny places these marine oligotrophs in the order Rickettsiales, most of whose known members are obligate intracellular parasites, such as *Rickettsia*, the cause of Rocky Mountain spotted fever—and also the endosymbiont ancestor of our mitochondria. The SAR11 cluster has since been classified as part of the order Pelagibacterales, and a few species have been cultured, such as *Pelagibacter ubique* (see **Table 17.3**).

Note: Taxonomic categories generally have two forms: formal and informal. The formal term is capitalized, with a latinized suffix: Actinomycetales, *Pseudomonas*, *Micrococcus*. The informal term is lowercase, in some cases with an anglicized ending; and informal references to genera are not italicized: actinomycetes, pseudomonads, micrococci.

Emerging Clades: Unclassified and Uncultured Bacteria

As we discover new microorganisms, how do we decide what to call them? The term **emerging** is used to refer to an organism recently discovered or described, such as

TABLE 17.3	Taxonomic Hierarchy of Classification	
Taxon rank	**A long-studied taxon**	**An emerging taxon**
Domain	Bacteria	Bacteria
Division (phylum)	Actinobacteria High-GC Gram-positive	Proteobacteria Gram-negative purple bacteria and heterotrophs
Class	Actinobacteria	Alphaproteobacteria
Subclass	Actinobacteridae	Rickettsidae Intracellular bacteria and mitochondria
Order	Actinomycetales Filamentous; acid-fast stain	Pelagibacterales Planktonic marine bacteria
Family	Streptomycetaceae Hyphae produce spores	Pelagibacteraceae
Genus	*Streptomyces*	*Pelagibacter*
Species (date first described)	*S. coelicolor* (1908)	*P. ubique* (2002)

Pelagibacter (**Table 17.3**). If the new organism causes a disease, it is called an "emerging pathogen." An emerging organism that cannot be cultured may be known only by its habitat and its small-subunit rRNA sequence. Such organisms require a culture method and a description of essential cell structure and metabolism before designation as a new species. "Incompletely described" emerging organisms are designated as follows:

- **Unclassified/uncultured organism.** An unclassified organism, or uncultured organism, is assigned to a taxonomic rank based on SSU rRNA sequence but has not yet been grown in pure culture—for example, an "uncultured actinomycete."

- **Environmental sample.** An environmental sample is designated by its habitat and assigned a rank based on SSU rRNA. Examples of environmentally defined actinomycetes in the NCBI database include "oil-degrading bacterium AOB1" and "glacier bacterium FJS11."

- **Candidate species.** A cultured organism with some physiological characterization beyond DNA sequence may be published with a provisional status of **candidate species**, designated by the prefatory term "*Candidatus.*" For example, "*Candidatus* Nostocoida limicola" was published as "a filamentous bacterium from activated sludge."

Nongenetic Categories for Medicine and Ecology

Genetic relatedness is the standard for classifying and naming organisms in all fields of biology. At the same time, several nongenetic systems of categorization serve a practical purpose in certain fields. These systems include:

- **Phenotypic categories for identification.** Categories such as pigmentation or cell shape (rod or coccus) may have minor genetic significance but are useful for practical identification of organisms isolated from field or clinical sources.

- **Ecological categories.** In ecology, the niche filled by an organism may be more important than its phylogeny. For example, cyanobacteria and sequoia trees both fill the role of photosynthetic producers of biomass. Both are primary producers—a trophic category of organisms that feed other organisms in the food web. Ecological categories are discussed further in Chapter 21.

- **Disease categories.** In medical microbiology, microorganisms are categorized according to the type of disease they cause or the host organ system they inhabit. For example, *Mycoplasma pneumoniae* (a cell wall-less bacte-

rium) and influenza virus are both pulmonary pathogens, whereas *Escherichia coli* and *Bacteroides thetaiotaomicron* are both normal members of the gut microbiome. Disease categories are discussed further in Chapters 25 and 26.

Naming a Species

A commonly accepted set of names for species and higher taxa is essential for research and communication. The accepted rules for naming species and taxa have been determined by the International Committee on Systematics of Prokaryotes (ICSP). The ICSP establishes minimal criteria for designating species, genera, classes, and other taxa of bacteria and archaea. To establish a new species, a previously unknown form of microbe must be isolated and grown in pure culture; this cultured organism is known as an **isolate**. The isolate's unique genetic and phenotypic traits are published, with a proposed species name designated "*Candidatus*" for **candidate species**. An example of a candidate species is "*Candidatus* Nitrosopumilus maritimus,*" a marine archaeon that oxidizes ammonia, first isolated in 2005. The candidate species becomes accepted as an official species upon publication in the *International Journal of Systematic and Evolutionary Microbiology*, the official journal of record for novel prokaryotic taxa.

Historically, the taxonomic categories and species descriptions are compiled in *Bergey's Manual of Determinative Bacteriology*, a multivolume reference work of prokaryotic taxonomy. Up-to-date taxonomy of bacteria and archaea, as well as eukaryotes, is maintained in the "Taxonomy" browser of the interactive National Center for Biotechnology Information (NCBI) online database, supported by the U.S. government.

In recent years, the standard practice for accepting and naming species has been overwhelmed by the number of new isolates, as well as uncultured microbes, reported in the literature. Many uncultured microbes are known only by their SSU rRNA sequences. Increasingly, however, uncultured organisms have large portions of their genomes sequenced and assembled using **metagenomics** (presented in Chapters 7 and 21). Metagenomics is the sequencing of multiple genomes in an environmental community. The community can be sampled from any kind of habitat, such as the human digestive tract or the Sargasso Sea.

Identification

Once a species has been described and classified, we need a way to identify future members of the species isolated from natural environments. **Identification** requires the recognition of the class of a given microbe isolated in pure culture. Identification poses special difficulties with microbes,

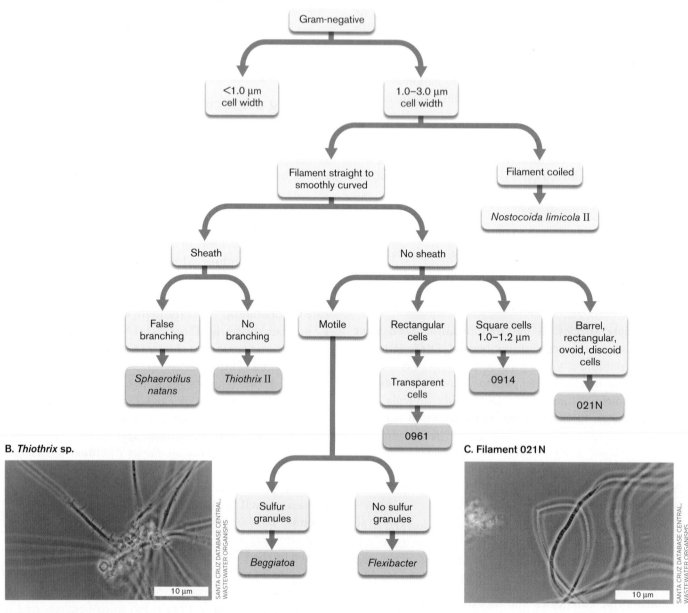

A. Dichotomous key for wastewater bacteria

FIGURE 17.29 ■ Dichotomous key for wastewater bacteria. **A.** A section of a dichotomous key for identifying Gram-negative bacteria from wastewater. The designations O961, O914, and O21N refer to unidentified species. **B.** *Thiothrix* is a filamentous wastewater bacterium. **C.** Filament O21N is an unclassified organism that forms starburst filaments. *Source:* Part A modified from Santa Cruz Database Central, Wastewater.

which, by definition, are invisible to the unaided eye. Even under the electron microscope, thousands of divergent species may possess similar shape and form.

Because the definition of a species is based on its genetic sequence, the most consistent and straightforward way to identify an isolate is to sequence part or all of its genome or to attempt hybridization of its DNA with a labeled sequence probe (discussed in Section 7.6). In clinical practice, such DNA-based methods are used increasingly. Nevertheless, even in the clinical lab—as well as in field and environmental microbiology—it is convenient to narrow down the possibilities using various easily determined traits, such as cell shape, staining properties, and metabolic reactions. Thus,

practical identification is based on a combination of phylogeny (relatedness based on DNA sequence divergence) and phenetic, or phenotypic, traits.

A common strategy of practical identification is the **dichotomous key**, in which a series of yes/no decisions successively narrows down the possible categories of species. **Figure 17.29** shows an example of a dichotomous key used to identify filamentous bacteria isolated from a wastewater treatment facility. Filamentous bacteria are important in wastewater because their entangled filaments interfere with the settling of bacteria out of the treated water. Note that in the key shown, most of the traits are phenotypic, such as cell size and motility. Most of the branch decisions have two choices, although one juncture (cell motility and shape) has four. The key identifies some organisms down to the species level (*Sphaerotilus natans*), others to the genus level (*Flexibacter*), and others only to numbered samples whose characterization remains incomplete (0914, 021N).

> **Thought Question**
>
> **17.9** Using **Figure 17.29**, how would you identify a straight, nonsheathed, Gram-negative bacterium that has sulfur granules, is motile, and is 1.0 μm wide? What would happen if you assigned the bacterium a width of 0.9 μm?

A disadvantage of the dichotomous key is that it requires a series of steps, each of which takes time. In the clinical setting, time is critical in identifying a potential pathogen and prescribing appropriate treatment. An alternative means of identification is the probabilistic indicator. A probabilistic indicator is a battery of biochemical tests performed simultaneously on an isolated strain. The indicator requires a predefined database of known bacteria from a well-studied habitat, such as Gram-negative bacteria from the human intestinal tract. To build such a database, numerous strains of each species must be isolated in pure culture and tested for all the traits in the database. The fraction of isolates that test positive for each trait, for each species, is noted in the database as the probability of obtaining a positive result. The database can be used to identify a specimen isolated from a patient (discussed in Chapter 28). But, like the dichotomous key, the probabilistic indicator works only if the isolate coincides with a member of the predefined database.

To Summarize

- **Microbial species** are defined by sequence similarity of vertically transmitted genes such as SSU rRNA sequences and multiple orthologous genes. The species definition should be consistent with the ecological niche or pathogenicity.

- A **pan-genome** includes core genes possessed by all isolates of a species plus accessory genes found in some isolates but not others. A pan-genome may be open (infinite number of genes) or closed (finite set of available genes).

- **Taxonomy** is the description and organization of life-forms into classes (taxa). Taxonomy includes classification, nomenclature, and identification.

- **Classification** is traditionally based on a hierarchy of ranks. Groups of organisms long studied tend to have many ranks, whereas recent isolates have few.

- **DNA sequence relatedness** defines microbial taxa. Below genus level, however, the definition of bacterial species can be problematic.

- **Emerging taxa** are types of organisms recently discovered or described. They may be uncultured, and their phylogeny may be uncertain.

- **Practical identification** is based on phenotypic and genetic traits. Methods of identification include the dichotomous key and the probabilistic test battery. Both methods assume a predefined set of organisms.

17.6 Symbiosis and the Origin of Mitochondria and Chloroplasts

So far, we have largely considered single species in isolation. In fact, however, all organisms evolve in the presence of other kinds of species, with whom they share interactions, both positive and negative. A major engine of evolution is **symbiosis**, the intimate association of two unrelated species. The ecology and behavioral adaptations of microbial symbiosis are discussed in Chapter 21; here we focus on the role of symbiosis in the evolution of cells.

Evolution of Endosymbiosis

The word "symbiosis" is popularly understood to mean **mutualism**, a relationship in which both partners benefit and may absolutely require each other. Biologists, however, recognize **parasitism**, in which one partner is harmed, as a relationship equally as intimate as mutualism; both mutualism and parasitism are forms of symbiosis. Intimate relationships between species, either negative or positive, lead to **coevolution**, the evolution of two species in response to one another, showing parallel phylogeny. An important bacterial mutualism is that of nitrogen fixation, in which rhizobia form intracellular "bacteroids" within legume plants that cannot fix nitrogen on their own. Both rhizobia and their plant hosts are highly evolved to respond to each other chemically and develop the nitrogen-fixing system. Another remarkable example of coevolution is that of leaf-cutter ants, which cultivate both fungal and bacterial partners (discussed in **eTopic 17.5**).

The most intimate kind of symbiosis is **endosymbiosis**, in which one partner population grows within the body of another organism. Endosymbiosis includes communities of microbes within the digestive tracts of animals, such as the human intestinal microbiome (discussed in Chapters 21 and 23). The internalized endosymbiont can also

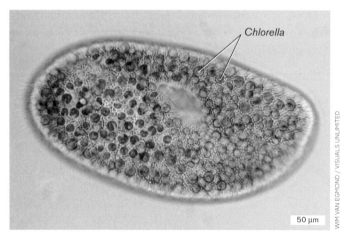

FIGURE 17.30 ■ Endosymbiosis. *Paramecium bursaria*, a ciliate protist with endosymbiotic *Chlorella* algae.

FIGURE 17.31 ■ Filariasis. Patient suffering from a form of filariasis known as elephantiasis.

be intracellular, as in the case of rhizobial bacteroids within legume tissues. Rhizobia retain the genetic capacity for independence, growing readily in soil. Other intracellular endosymbionts, however, become wholly dependent on their host cells. Pathogenic endosymbionts, such as chlamydias, evolve specialized traits enabling their growth at the expense of the host and evasion of the host immune system. But endosymbionts also undergo drastic reductive evolution, evolving ever-deeper interdependence with their host cells.

A simple example of intracellular endosymbiosis is that of the alga *Chlorella* growing within *Paramecium bursaria* (**Fig. 17.30**). The algae conduct photosynthesis and provide nutrients to the paramecium, which in turn shelters the algae from predators and viruses. The relationship is highly specific—only certain species of algae and paramecia participate—and the algal growth is limited to a population that avoids harming the host. This relationship may give clues to how the bacterial ancestor of chloroplasts began its intracellular existence.

Despite this intimate relationship, *Chlorella* retains its ability to multiply outside the paramecium; thus this symbiosis is reversible. Moreover, under conditions of starvation in the absence of light, the paramecium may start to digest its endosymbionts as prey. Thus, the nature of the symbiosis (mutualistic or predatory) depends on the environment.

Many invertebrate animals, often themselves parasites of animals or plants, possess obligate bacterial endosymbionts. For example, the fruit fly, *Drosophila*, carries a parasitic endosymbiont, the bacterium *Wolbachia pipientis*. The *Wolbachia* strains that infect *Drosophila* cells are transmitted only through egg cells; they cannot exist outside the insect. Other invertebrate endosymbionts, however, are mutualists. In fact, 15% of insect species depend on intracellular bacteria to produce essential nutrients, such

as certain amino acids or vitamins. In these mutualisms, both partners have lost essential traits by reductive evolution, and each now requires the partner species to provide the lost function.

A surprising number of human invertebrate parasites, such as filarial nematodes and *Anopheles* mosquitoes, carry bacterial endosymbionts required for host growth. This discovery has exciting implications for treatment of parasitic diseases. Invertebrate parasites such as filarial nematode worms invade human lymph nodes, causing forms of disease (filariasis) that are notoriously difficult to treat. Few anti-nematode compounds are sufficiently selective for worm metabolism versus human metabolism, since both are eukaryotic animals. A form of filariasis is elephantiasis, in which a limb expands with edema because the lymph ducts are blocked by worms (**Fig. 17.31**). Filariasis afflicts more than 120 million people worldwide, largely in the Indian subcontinent and in Africa.

A filarial nematode, *Brugia malayi*, harbors endosymbiotic *Wolbachia* (a different strain from the one that infects *Drosophila*) (**Fig. 17.32**). *Wolbachia* may have entered the nematode originally as a pathogen or parasite, and then persisted because of its metabolic contributions to the host. The nematode's *Wolbachia* are mutualists; the nematode needs them for embryonic development. The bacteria are found within tissue layers beneath the nematodes' cuticle and within the uterine tubes of females, where they enter the developing offspring (**Fig. 17.32**). When human patients infected by the nematodes are treated with antibiotics such as tetracycline, the bacteria disappear from worm tissues. The worm burden gradually decreases, and no offspring are produced. Antibacterial antibiotics eliminate the worms sooner and more completely than does treatment with anti-nematode agents.

The genome of a *Wolbachia* strain from a filarial nematode reveals extensive reductive evolution. With barely a

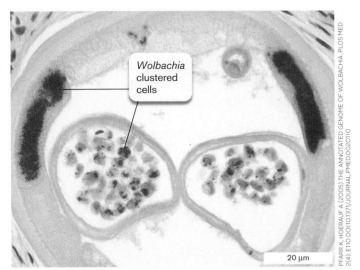

FIGURE 17.32 ■ **The filarial endosymbiont *Wolbachia*.** Cross section of the nematode *Brugia malayi*, showing *Wolbachia* bacteria (stained pink) within the worm's dermis and uterine tubes.

million base pairs, the *Wolbachia* genome has lost many metabolic pathways. It retains glycolysis and the TCA cycle but has lost the pathways for biosynthesis of all amino acids and most vitamins. It nonetheless retains pathways to make purines, pyrimidines, and the coenzymes riboflavin and FAD—essential pathways lost by its host nematode. Overall, *Wolbachia* appears to be evolving into an organelle of its host, like the ancestors of mitochondria and chloroplasts.

Mitochondria and Chloroplasts

As Lynn Margulis and colleagues showed, the assimilation of endosymbionts as mitochondria and chloroplasts played a central role in the evolution of eukaryotes (**Fig. 17.33**). Many similar endosymbioses are known today, such as the free-living algae *Chlorella* acquired by the protist predator *Paramecium bursaria* (see **Fig. 17.30**). Like *Wolbachia*, mitochondria evolved from a bacterium related to the rickettsias (intracellular pathogens). The mitochondrial ancestor must have entered the eukaryotic lineage as, or shortly after, the eukaryotes diverged from archaea, since all known eukaryotes retain mitochondria or vestigial remnants of mitochondrial genomes (discussed in Chapter 20). Mitochondria provide the cell with the essential functions of electron transport and respiration. The electron transport system (ETS) is found in the mitochondrial inner

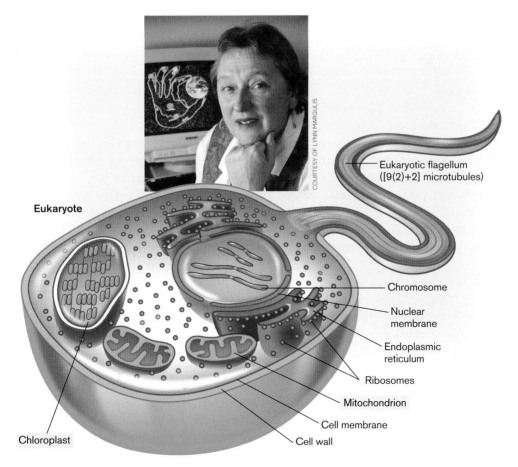

FIGURE 17.33 ■ **Endosymbiotic cells evolved into mitochondria and chloroplasts.** Eukaryotic cells contain mitochondria and chloroplasts, organellar remnants of ancient endosymbioses. **Inset:** Lynn Margulis (1938–2011).

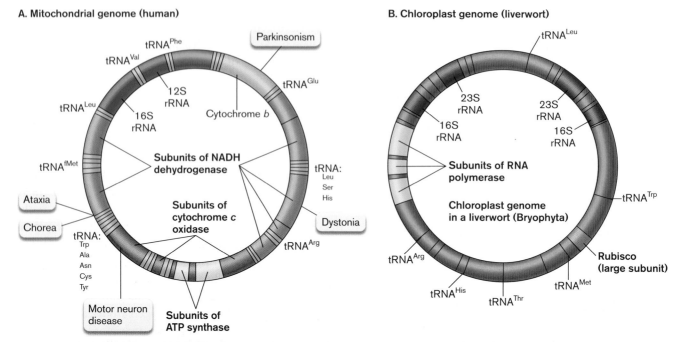

A. Mitochondrial genome (human)

B. Chloroplast genome (liverwort)

FIGURE 17.34 ■ **Genomes of mitochondria and chloroplasts.** **A.** The mitochondrial genome retains large- and small-subunit rRNAs, five tRNA genes, plus subunits of the respiratory electron transport chain. Text bubbles indicate human diseases associated with mitochondrial defects. **B.** The chloroplast genome retains large- and small-subunit rRNAs (23S and 16S), several tRNA and RNA polymerase genes, plus Rubisco (large subunit) and components of photosystems I and II (dispersed around the circle, not labeled in the figure).

membrane, believed to derive from the cell membrane of the ancestral bacterium. The outer membrane may derive from the invaginating membrane of the host cell that originally engulfed the endosymbiont.

Chloroplasts arose from cyanobacteria at some point before the divergence of red and green algae (discussed in Chapter 20). A model for cyanobacterial uptake can be seen in the protist *Glaucocystophyta*, whose cyanobacterial endosymbionts retain cell walls and some metabolism. Like mitochondria, chloroplasts possess inner and outer membranes, believed to derive from the ancestral endosymbiont and host, respectively. Photosynthetic complexes are located in the thylakoid membranes, similar to those of modern cyanobacteria.

The genomes of mitochondria and chloroplasts both show extreme reduction (**Fig. 17.34**)—even more extreme than that of any known endosymbiotic bacteria. The few genes that remain in the organellar genome include remnants of the central transcription-translation apparatus, such as rRNA and tRNAs, as well as a handful of genes whose products are essential for survival of the host cell: for respiration (mitochondria) or photosynthesis (chloroplasts).

Mitochondrial genome. In human mitochondria, the mitochondrial genome encodes key parts of the respiratory

chain, including subunits of NADH dehydrogenase, cytochrome oxidase, and ATP synthase (**Fig. 17.34A**). Mutations in these key genes lead to serious diseases; for example, damage to mitochondrial genes of respiration is associated with motor neuron disease, parkinsonism, and forms of ataxia.

But thousands of genes encoding ETS subunits, as well as other essential parts of mitochondria, have migrated from the mitochondrion to the nucleus. The nuclear acquisition probably occurred through accidental copying of mitochondrial genes into the nuclear genome. Reductive evolution then occurred, faster in the mitochondrial copy because of the faster mutation rate. Some of these nuclear-acquired mitochondrial genes show tissue-specific expression, resulting in different mitochondrial types associated with different tissues. Thus, some mitochondrial defects are actually inherited through the nuclear genome. The mitochondria have evolved as integral parts of the host cell.

Chloroplast genome. In chloroplasts, the organellar genome encodes essential products for photosynthesis, including photosystems I and II and the ATP synthase. The chloroplast genome shown in **Figure 17.34B** retains the large subunit of Rubisco, whereas the gene encoding the small subunit has migrated to the nucleus.

Remarkably, the process of symbiogenesis, the generation of new symbiotic associations, continues in many protists. Protist-algae, also known as "secondary symbiont" algae, result from symbiogenesis in which an alga (containing a chloroplast) was engulfed by an ancestral protist. The protist cell contains the degenerate remains of the algal endosymbiont (**Fig. 17.35**). The algal mitochondrion was lost through reductive evolution, and the nucleus shrank to a nucleomorph, the vestigial remains of a nucleus containing a small amount of the chromosomal DNA of the original algal nucleus. But the algal chloroplast was maintained, "enslaved" by its new host. The result is a new protist-alga species, *Guillardia theta*, capable of both phototrophy and heterotrophy. Its chloroplast has a double membrane derived from the original cyanobacterial ancestor of the chloroplast and from the algal ancestor, surrounded by another double membrane derived from the cell membranes of the alga and a secondary host.

Other secondary endosymbiont algae, such as kelps, diatoms, and dinoflagellates, are discussed in Chapter 20. In some species, tertiary symbiosis has been documented, in which a secondary-endosymbiont alga has been swallowed in turn by another protist.

FIGURE 17.35 ■ A protist-alga results from secondary endosymbiosis. The protist *Guillardia theta* contains the primary algal chromosome surrounded by the remains of the primary algal cytoplasm and nucleus, which has shrunk to a nucleomorph. *Source: Modified from Paul R. Gilson. 2001. Genome Biol.* **2**:1022.

Thought Question

17.10 Besides mitochondria and chloroplasts, what other kinds of entities within cells might have evolved from endosymbionts?

To Summarize

■ **Symbiosis is the intimate association of two unrelated species.** A symbiosis in which both partners benefit is called **mutualism**. If one partner benefits while harming the other, the symbiosis is called **parasitism**.

■ **Symbiotic partners undergo coevolution**, the evolution of two species in response to one another. Coevolution involves reductive (degenerative) evolution, in which each partner species loses some functions that the other partner provides.

■ **An endosymbiont lives inside a much larger host species.** Many microbial cells harbor endosymbiotic bacteria whose metabolism yields energy for their hosts.

■ **Many invertebrates harbor endosymbiotic bacteria.** The bacteria are required for host survival and, in some cases, for pathology caused by a parasitic invertebrate.

■ **Mitochondria evolved from endosymbionts.** The ancestor of mitochondria was an alphaproteobacterium related to rickettsias.

■ **Chloroplasts evolved from endosymbionts.** The chloroplast ancestor was a cyanobacterium.

Concluding Thoughts

The Antarctic microbial landscape provides a glimpse of what Earth might have looked like 3.4 billion years ago, as the first microbial communities evolved, their mats building rock layers that persist today. From those early microbes, all subsequent life evolved. It is hard to say which is more astonishing: the overall commonalities of all living cells, including membrane-enclosed support systems for genomes of shared ancestry, or the subsequent evolution of organisms with vastly different adaptations to exploit every possible niche of our planet. Our next three chapters explore these diverse adaptations: Chapter 18, bacterial diversity; Chapter 19, archaeal diversity; and Chapter 20, diversity among microbial eukaryotes, including fungi, algae, and protozoa.

CHAPTER REVIEW

Review Questions

1. What was the composition of Earth's early crust and atmosphere? What processes changed their composition to that found today?
2. What kinds of evidence support the presence of life in the Archaean eon? What are the advantages and limitations of each kind of evidence?
3. What kinds of metabolism are believed to have existed in Archaean life? What kinds of evidence support their existence?
4. Compare and contrast three models for the origin of the first cells. Which features of life does each model explain, and which features are unexplained?
5. Explain the roles of classification, nomenclature, and identification for microbial taxonomy.
6. Why is the definition of species in bacteria and archaea more problematic than in eukaryotes? What is generally considered the present basis for defining prokaryotic species?

7. Discuss the roles of mutation, natural selection, and reductive evolution in the divergence of microbial species. Cite specific examples.
8. Explain the basis of a "molecular clock" for measuring microbial evolution. What fundamental properties must be met by a gene to function as a molecular clock? What are the limitations of a molecular clock?
9. Explain the basis of a phylogenetic tree. Why is the fundamental tree at the divergence of bacteria, archaea, and eukaryotes unrooted?
10. How does horizontal gene transfer determine genomic content? What kinds of genes are likely to undergo horizontal transfer?
11. Explain three different ways that we can test questions about adaptive evolution.
12. Explain how endosymbiosis can lead to obligate association. Explain how reductive evolution and gene transfer lead to the evolution of organelles that are inseparable from host cells.

Thought Questions

1. How convincing are the microfossils in **Figure 17.6**? What criteria do you think would define a microfossil?
2. In the phylogeny shown here, where are the root and the outgroup? How does the outgroup organism differ from the others? Which two organisms are the most closely related? Which node represents the last common ancestor of *Neisseria* and *Haemophilus*? Which genome has evolved much faster than the others, and why?

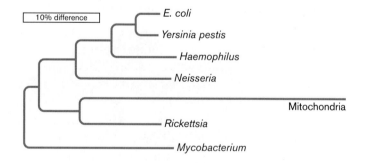

3. Design an evolution experiment in the laboratory to evolve a bacterium that breaks down a dangerous pollutant such as dioxin. How would you select the starting organism, and what experimental steps would you perform?

Key Terms

abiotic (657)
Archaean eon (650)
banded iron formation (BIF) (654)
biogenic (652)
biosignature (biological
    signature) (650)
biosphere (649)
candidate species (679)
clade (661)
classification (674)
coevolution (681)
core genome (675)
degenerative evolution (661, 670)
dichotomous key (680)
directed evolution (674)
domain (665)
emerging (678)
endosymbiosis (681)
experimental evolution (671)

gene duplication (670)
genomic island (669)
genus (678)
greenhouse effect (649)
Hadean eon (650)
horizontal gene transfer (661, 666)
identification (679)
isolate (679)
isotope ratio (653)
laboratory evolution (671)
metagenomics (679)
microbial mat (647)
microfossil (651)
molecular clock (661)
monophyletic group (661)
mutualism (681)
natural selection (670)
node (664)
nomenclature (674)

pan-genome (675)
panspermia (660)
paralogous gene (paralog) (670)
parasitism (681)
phylogenetic tree (663)
phylogeny (661)
prebiotic soup (657)
reductive evolution (661, 670)
ribozyme (659)
RNA world (657)
root (664)
small-subunit (SSU) rRNA (662)
species (661, 678)
stromatolite (647)
symbiosis (681)
taxa (678)
taxon (678)
taxonomy (674, 678)
vertical gene transfer (666)

Recommended Reading

Achtman, Mark, and Michael Wagner. 2008. Microbial diversity and the genetic nature of microbial species. *Nature Reviews. Microbiology* **6**:431–440.

Blount, Zachary D., Jeffrey E. Barrick, Carla J. Davidson, and Richard E. Lenski. 2012. Genomic analysis of a key innovation in an experimental *Escherichia coli* population. *Nature* **489**:513–518.

Blount, Zachary D., Christina Z. Borland, and Richard E. Lenski. 2008. Historical contingency and the evolution of a key innovation in an experimental population of *Escherichia coli*. *Proceedings of the National Academy of Sciences USA* **105**:7899–7906.

Brasier, Martin D., Owen R. Green, Andrew P. Jephcoat, Annette K. Kleppe, Martin J. Van Kranendonk, et al. 2002. Questioning the evidence for Earth's oldest fossils. *Nature* **416**:76–81.

Charusanti, Pep, Nicole L. Fong, Harish Nagarajan, Alban R. Pereira, Howard J. Li, et al. 2012. Exploiting adaptive laboratory evolution of *Streptomyces clavuligerus* for antibiotic discovery and overproduction. *PLoS One* 7:e33722.

Currie, Cameron R., Michael Poulsen, John Mendenhall, Jacobus J. Boomsma, and Johan Billen. 2006. Coevolved crypts and exocrine glands support mutualistic bacteria in fungus-growing ants. *Science* **311**:81–83.

Didelot, Xavier, A. Sarah Walker, Tim E. Peto, Derrick W. Crook, and Daniel J. Wilson. 2016. Within-host evolution of bacterial pathogens. *Nature Reviews. Microbiology* **14**:150–162.

Fraser, Christophe, Eric J. Alm, Martin F. Polz, Brian G. Spratt, and William P. Hanage. 2009. The bacterial species challenge: Making sense of genetic and ecological diversity. *Science* **322**:741–746.

Gevers, Dirk, Frederick M. Cohan, Jeffrey G. Lawrence, Brian G. Spratt, Tom Coenye, et al. 2005. Reevaluating prokaryotic species. *Nature Reviews. Microbiology* **3**:733–739.

Hanage, William P., Christophe Fraser, and Brian G. Spratt. 2005. Fuzzy species among recombinogenic bacteria. *BMC Biology* **3**:6.

Jiao, Yongqin, Andreas Kappler, Laura R. Croal, and Dianne K. Newman. 2005. Isolation and characterization of a genetically-tractable photoautotrophic Fe(II)-oxidizing bacterium, *Rhodopseudomonas palustris* strain TIE-1. *Applied and Environmental Microbiology* **71**:4487–4496.

Lapierre, Pascal, and J. Peter Gogarten. 2009. Estimating the size of the bacterial pan-genome. *Trends in Genetics* **25**:107–110.

Lincoln, Tracey A., and Gerald F. Joyce. 2009. Self-sustained replication of an RNA enzyme. *Science* **323**:1229–1232.

Moran, Nancy A., John P. McCutcheon, and Atsushi Nakabachi. 2008. Genomics and evolution of heritable bacterial symbionts. *Annual Reviews in Genetics* **42**:165–190.

Nora Noffke, Daniel Christian, David Wacey, and Robert M. Hazen. 2013. Microbially induced sedimentary structures recording an ancient ecosystem in the ca. 3.48 billion-year-old Dresser Formation, Pilbara, Western Australia. *Astrobiology* **13**:1103–1124.

Ochman, Howard. 2005. Genomes on the shrink. *Proceedings of the National Academy of Sciences USA* **102**:11959–11960.

Salzberg, Steven L., Julie C. Dunning Hotopp, Arthur L. Delcher, Mihai Pop, Douglas R. Smith, et al. 2005. Serendipitous discovery of *Wolbachia* genomes in multiple *Drosophila* species. *Genome Biology* **6**:R23.

Tettelin, Hervé, David Riley, Ciro Cattuto, and Duccio Medini. 2008. Comparative genomics: The bacterial pan-genome. *Current Opinion in Microbiology* **12**:472–477.

CHAPTER 18
Bacterial Diversity

Bacteria vary tremendously in their cell structure and metabolism. They include heterotrophs, phototrophs, and lithotrophs—and some species can be classified as all three. There are obligate aerobes, anaerobes, and microaerophiles. Their cell shapes include rods, cocci, spirals, and budding forms. Ecologically, bacteria include mutualists, pathogens, and organisms that cannot be cultured in our laboratories. Every day we discover new species, ranging from metal metabolizers buried beneath Antarctic mountains to filamentous mutualists clinging to our intestinal epithelium.

CURRENT RESEARCH highlight

Multibacterial biofilm invades colorectal tumor. Fluorescence in situ hybridization (FISH) of specific DNA probes shows surprisingly diverse bacteria growing upon a tumor of the right ascending colon. Three different DNA probes hybridize with 16S rRNA of distinct bacterial taxa: Bacteroidetes (green), Lachnospiraceae (magenta), Fusobacteria (cyan). By contrast, the lining of a noncancerous colon shows a thicker layer of protective mucus, colonized by thinner biofilms with a healthy distribution of commensal bacteria.

Source: Christine M. Dejea et al. 2014. *PNAS* **111**:18321.

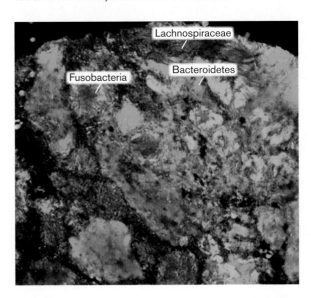

AN INTERVIEW WITH

CYNTHIA SEARS, MEDICAL MICROBIOLOGIST, JOHNS HOPKINS UNIVERSITY

What different kinds of bacteria do you find associated with colorectal tumors?

We find that the biofilms that are associated with sporadic colon cancer are composed of a diverse array of bacteria. One set of biofilm tumors is associated predominantly with Bacteroidetes and Lachnospiraceae; a second subset has at least 25% of the sequences mapping to *Fusobacterium* species; and a third subset, so far seen in only one tumor, was dominated by Proteobacteria.

What surprised you most about the bacterial biofilms found with tumors?

The big surprise was the geographic localization of biofilm-positive tumors to the right colon. As the work was being done, we noted some tumors were biofilm-positive, some not so. The day we decided to plot them on a cartoon of the colon based on the location of the tumor is the day the pattern revealed itself. As yet, we have not fully unraveled why this dramatic, nearly categorical difference exists for biofilm formation between right-colon and left-colon cancers.

Bacteria evolve a bewildering array of life-forms that colonize every habitat on Earth, from the ocean depths to our own skin and digestive tract. You might think that, by now, we would know at least all the major phyla of bacteria, just as we know the major kinds of vertebrate animals and vascular plants. Yet we are constantly discovering new clades of bacteria never seen before. At the Johns Hopkins School of Medicine, Cynthia Sears discovers new bacteria that colonize our colonic epithelium (see the Current Research Highlight). In Chapter 13 we learned that many colonic bacteria are normal residents that aid our digestion. But other bacteria, including members of the Lachnospiraceae, Bacteroidetes, and Fusobacteria, act as pathogens. Some take advantage of cancer pathology to invade the tissue of colon tumors.

Elsewhere, microbe hunters discover bacteria in remote, pristine sites such as the frozen brine of Antarctic Lake Vida (**Fig. 18.1**). The deepest hypersaline layer of Lake Vida remains trapped without light or oxygen, at –13°C. From this deep anoxic layer, Alison Murray and colleagues from the Desert Research Institute recovered growing bacteria that are ultrasmall—some less than 200 nm in diameter. We now predict that all water and soil habitats, as well as parts of the human body, might also host ultrasmall microbes, perhaps showing yet other unique traits to be discovered.

How do we begin to describe bacterial diversity, when even in familiar habitats the vast majority of species remain unknown? Chapter 18 surveys the bacteria that we do know something about. For instance, many kinds of Gram-negative Proteobacteria inhabit our digestive tract, or grow in ponds by photosynthesis, or colonize legumes as nitrogen-fixing mutualists. Well-known bacteria are generally those cultured in the laboratory, subject to experiments under controlled conditions. At the same time, comparing genome sequences clarifies the relatedness of major groups of bacteria.

Chapter 18 emphasizes major taxa that are of physiological, ecological, and medical importance. We organize our discussion by phylogeny (evolutionary relatedness of taxa), as well as key traits of a taxon, such as the Gram-positive cell wall of Firmicutes and the oxygenic photosynthesis of cyanobacteria. We show how diverse bacteria contribute to communities. Microbial communities and ecology are explored further in Chapter 21, and their roles in global biogeochemical cycles are discussed in Chapter 22.

For each major taxonomic group, we describe a few key species to represent the spectrum of diversity. Additional genera and species are referenced in an expanded table found in Appendix 2 and eAppendix 4.

18.1 Bacterial Diversity at a Glance

To survey bacterial diversity in one chapter is like touring all the countries of a continent in a single day. Like countries, bacterial taxa have complex traits and histories, and often contested borders. But overall, bacteria share major traits in common. We review these common traits of bacteria and then go on to explore their differences.

A.

B.

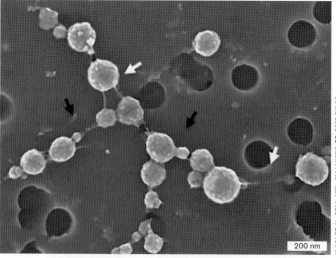

FIGURE 18.1 ■ Small bacterial cells in an Antarctic brine lake. A. Lake Vida, Antarctica. Below the ice is concentrated salt. B. Bacterial cells connected by filaments, dehydrated on 0.2-µm-pore-size filter (SEM). These cells were obtained by Alison Murray and colleagues from the Desert Research Institute, Reno. *Source:* Emanuele Kuhn et al. 2014. *Appl. Environ. Microbiol.* **80**:3687.

Bacteria: Common Traits and Diverging Phylogeny

Chapter 17 summarized the differences and similarities of the three major domains—Bacteria, Archaea, and Eukarya (see Table 17.2). A common feature of bacteria is their central apparatus for gene expression, particularly their RNA polymerases, ribosomal RNAs, and translation factors. Bacterial gene expression complexes differ more from those of Archaea or Eukarya than the complexes of Archaea or Eukarya differ from each other. This subtle point of molecular biology has a profound consequence for human medicine and agriculture: It underlies the selective activity of many antibiotics, such as streptomycin, that attack only bacteria, without affecting animals or plants.

Another trait distinguishing bacteria from archaea and eukaryotes is that most bacterial cells possess a cell wall of peptidoglycan (discussed in Chapter 3). Peptidoglycan is composed of disaccharide-peptide chains that can cross-link in three dimensions; key enzymes that build the peptide links are blocked by antibiotics such as penicillin and vancomycin. Some archaea possess analogous sugar-peptide structures called "pseudopeptidoglycan" (discussed in Chapter 19), but their structural details and antibiotic sensitivity differ fundamentally from those of bacterial peptidoglycan. Eukaryotes such as fungi and plants have cell walls of polysaccharides such as cellulose and chitin (discussed in Chapter 20).

In bacteria, variant forms of peptidoglycan distinguish different species. For instance, the Gram-positive pathogen *Staphylococcus aureus* has cell wall peptides cross-linked by pentaglycine (a chain of five glycine residues). Some species, such as mycoplasmas, lack peptidoglycan altogether, although they arose by reductive evolution from bacteria that possess it.

The phylogeny of known bacteria is presented in **Figure 18.2**. The names of well-studied phyla are lettered in blue. A **phylum** (plural, **phyla**) is defined as a group of organisms sharing a common ancestor that diverged <u>early</u> from other bacteria, based on small-subunit rRNA (SSU rRNA) sequence (discussed in Chapter 17). Increasingly,

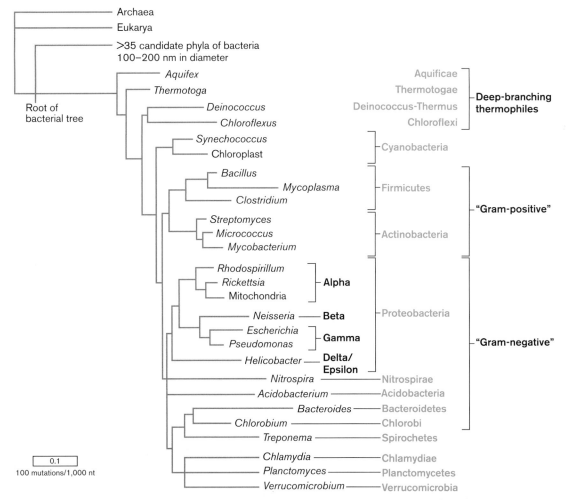

FIGURE 18.2 ■ Bacterial phylogeny. A phylogenetic tree of representative Bacteria based on 16S rRNA sequence comparison. The tree is rooted with respect to Archaea and Eukarya. Bold labels correspond to major groups in Table 18.1. Phylum names are lettered blue.

whole-genome data reveal new kinds of bacteria faster than we can figure out how to culture or characterize them. We continually discover very different new kinds of bacteria, known as emerging clades (see Chapter 17). Some microbiologists estimate that there exist 1,500 phyla, many as yet undiscovered. Apart from their genome sequences, we know little about those newly discovered bacteria. Lineages whose genome sequences diverged early, at or before the well-known phyla, are called **deep-branching taxa**. In addition, other environmental DNA sequences of uncultivated bacteria (not shown) branch from all parts of this tree.

Phyla and other major divisions are also defined on the basis of historical convention and consensus of the research community. Generally, a well-studied bacterial phylum comprises species that share key traits as well as ancestry. While sharing key traits, the member species often show remarkable diversity in other ways. Some phyla, such as Cyanobacteria, share a unique form of metabolism (oxygenic photosynthesis), yet have evolved many diverse cell shapes and grow in diverse habitats. Other phyla, such as Spirochetes, share a unique cell structure while diverging in habitat and metabolism.

Table 18.1 introduces seven major groups of bacteria that appear most frequently in the literature. **An expanded version of this table, including many important species, appears in Appendix 2.2 and eAppendix 4.2.**

Well-Studied Bacterial Phyla

Certain groups of bacteria have been studied for a century or more, and their traits are known in detail. A major group that serves all ecosystems is the **Cyanobacteria** (see Section 18.2). Cyanobacteria conduct photosynthesis by splitting water and releasing oxygen (O_2). Some species adapt to use H_2S when available. Oxygenic photosystems evolved from ancestral sulfide photosystems (discussed in Chapter 14).

Cyanobacteria and chloroplasts (eukaryotic organelles that evolved from Cyanobacteria) are the only life-forms that produce oxygen gas. These bacteria have a unique two-photosystem apparatus for oxygenic photosynthesis arranged in lamellar arrays of membranes called thylakoids (discussed in Chapter 14). Unique to Cyanobacteria and chloroplasts are the chlorophylls, most notably chlorophylls *a* and *b*; these molecules are distinct from the bacteriochlorophylls used by non-oxygenic bacterial phototrophs. Cyanobacteria share a common metabolism, yet the cell shape and organization of cyanobacterial species show an immense range of different forms, including chains of cells, square arrays, and globular colonies.

For soil and sediment, major phyla include the **Firmicutes** and **Actinobacteria**, which stain Gram-positive

(see Section 18.3). A few Firmicutes (the mycoplasmas) fail to stain Gram-positive because they lack cell walls, and some Actinobacteria possess a thick waxy coat that excludes the Gram stain. Most Firmicutes and Actinobacteria have an exceptionally thick cell wall, with several layers of peptidoglycan threaded by supporting molecules such as teichoic acids or mycolic acids. The thick, reinforced cell wall is what retains the Gram stain (discussed in Chapter 2). In addition, most Gram-positive species possess a well-developed S-layer of protein with glycan strands. By contrast, in other bacterial phyla the S-layer is either absent or present in diminished form.

Firmicutes and Actinobacteria differ genetically in their **GC content**—that is, the proportion of their genomes consisting of guanine-cytosine base pairs (as opposed to adenine-thymine). Firmicutes (low GC) generally grow as well-defined rods or cocci, isolated or in simple filaments consisting of cells that divide but remain attached end to end. Many Firmicutes form **endospores**, inert heat-resistant spores that can remain viable for thousands of years. Endospores are the most durable type of spore formed by bacteria.

Actinobacteria have a relatively high GC content and are known as the "high-GC Firmicutes." These bacteria include the **actinomycetes** (order Actinomycetales), which undergo complex life cycles, forming filamentous hyphae and arthrospores. Other groups closely related to actinomycetes grow as isolated rods and cocci, often with variable shape, such as the corynebacteria. Actinomycete relatives include the well-known causative agents of tuberculosis (*Mycobacterium tuberculosis*) and leprosy (*M. leprae*).

Proteobacteria are called "Gram-negative" because their single layer of peptidoglycan fails to retain the Gram stain (see Section 18.4). All Proteobacteria have an outer membrane consisting of lipopolysaccharides (LPS) and porins. All species of Proteobacteria have an LPS outer membrane, but their cell shape may be rods, cocci, or spiral rods. In contrast to Cyanobacteria, the Proteobacteria show an immense range of diverse metabolism (heterotrophy, lithotrophy, and anaerobic phototrophy). Most species are aerobic, facultative, or microaerophilic (requiring a low level of O_2; inhibited at higher levels).

Proteobacteria include five major classes, named Alphaproteobacteria, Betaproteobacteria, Gammaproteobacteria, Deltaproteobacteria, and Epsilonproteobacteria. (Some sources separate the Greek letter from the name "Proteobacteria"—for example, "Alpha Proteobacteria," "Beta Proteobacteria," and so on.) Proteobacteria include famous model organisms and pathogens, such as *Escherichia coli*, *Salmonella enterica*, and *Yersinia pestis*. Others are human or animal symbionts, either as mutualists or as pathogens—including *Rickettsia*, the genus most closely related to mitochondria.

The **Spirochetes** (see Section 18.6) have evolved a unique and complex cellular form of a flexible, extended spiral, resembling an old-style telephone cord. The cytoplasm

TABLE 18.1 Bacterial Diversity* (Expanded Table Appears in Appendix 2.2 and eAppendix 4.2)

Deep-branching thermophiles. Thermophilic bacteria that diverged early from archaea and eukaryotes. Many genes transferred laterally from archaea.

Aquificae. Hyperthermophiles (70°C–95°C). Oxidize H_2. *Aquifex.*

Chloroflexi. Filamentous phototrophs, often with chlorosomes. *Chloroflexus aurantiacus.*

Deinococcus-Thermus. Radiation-resistant species and thermophiles. *Deinococcus radiodurans.*

Thermotogae. Thermophiles (55°C–100°C). Anaerobic heterotrophs. *Thermotoga* sp.

Cyanobacteria. Oxygenic photoautotrophs with thylakoid membranes. Share ancestry with chloroplasts.

- **Chroococcales.** Square colonies based on two division planes.
- **Gloeobacterales.** Lack thylakoids; conduct photosynthesis in cell membrane.
- **Nostocales.** Filamentous chains with N_2-fixing heterocysts. Some grow symbiotically with corals or plants. *Nostoc* sp.
- **Oscillatoriales.** Filamentous chains with motile hormogonia (short chains). *Oscillatoria* sp.
- **Pleurocapsales.** Globular colonies; reproduce through baeocytes.
- **Prochlorales.** Tiny single cells, elliptical or spherical (0.5 μm). *Prochlorococcus* sp.

Firmicutes and Actinobacteria (Gram-positive). Peptidoglycan multiple layers, cross-linked by teichoic acids.

Firmicutes. Low-GC, Gram-positive rods and cocci.

- **Bacillales.** Aerobic or facultative anaerobes. *Bacillus subtilis.*
- **Clostridiales.** Anaerobic rods. *Clostridium botulinum* and *C. tetani.*
- **Lactobacillales.** Non–spore formers. Facultative anaerobes. Ferment, producing lactic acid. *Lactobacillus lactis.*
- **Tenericutes (Mollicutes).** Lack cell wall; require animal host. *Mycoplasma pneumoniae.*

Actinobacteria. High-GC, Gram-positive bacteria with moderate salt tolerance.

- **Actinomycetales.**

   Actinomycetaceae. Filamentous, producing aerial hyphae and spores. Actinomycetes such as *Streptomyces coelicolor.*

   Corynebacteriaceae. Irregularly shaped rods. *Corynebacterium diphtheriae.*

   Micrococcaceae. Small airborne cocci. *Micrococcus luteus.*

   Mycobacteriaceae. Exceptionally complex cell walls. *Mycobacterium tuberculosis.*

   Propionibacteriaceae. Propionic acid fermentation. *Propionibacterium shermanii.*

- **Bifidobacteriaceae.** Ferment without gas. *Bifidobacterium* sp.

Proteobacteria. Gram-negative bacteria with diverse cell forms and metabolism.

Alphaproteobacteria (class)

- **Caulobacterales.** Aquatic oligotrophs; alternate stalk and flagellum. *Caulobacter crescentus.*
- **Rhizobiales.** Plant mutualists and pathogens. *Sinorhizobium meliloti.*
- **Rhodobacterales, Rhodospirillales.** Flagellated photoheterotrophs.
- **Rickettsiales.** Includes intracellular parasites; related to mitochondria. *Rickettsia rickettsii* causes Rocky Mountain spotted fever.
- **Sphingomonadales.** Heterotrophs and photoheterotrophs; some opportunistic pathogens. *Sphingomonas* sp.

Betaproteobacteria (class)

- **Burkholderiales.** *Burkholderia pseudomallei* causes melioidosis.
- **Neisseriales.** Diplococci. *Neisseria gonorrhoeae* causes gonorrhea; *N. meningitidis* causes meningitis.

Gammaproteobacteria (class)

- **Acidithiobacillales.** Lithotrophs. *Acidithiobacillus ferrooxidans* oxidizes iron and sulfur.
- **Aeromonadales.** Aquatic heterotrophs such as *Aeromonas hydrophila.*
- **Enterobacteriales.** Facultative anaerobes; colonize the human colon.
- **Legionellales.** *Legionella pneumophila* causes legionellosis pneumonia.
- **Pseudomonadales.** Rods; aerobic or respire on nitrate; catabolize aromatics. *Pseudomonas aeruginosa* infects lungs in cystic fibrosis patients.
- **Thiotrichales.** Lithotrophs and heterotrophs.
- **Vibrionales.** Marine heterotrophs. *Vibrio cholerae* causes cholera.

Deltaproteobacteria (class)

- **Bdellovibrionales.** Periplasmic predators.
- **Desulfobacterales.** Reduce sulfate.
- **Myxococcales.** Gliding bacteria that form fruiting bodies.

Epsilonproteobacteria (class)

- **Campylobacterales.** Spirillar pathogens. *Campylobacter jejuni* causes gastroenteritis. *Helicobacter pylori* causes gastritis.

Deep-branching Gram-negative phyla.

Acidobacteria. Acidophiles and thermophiles. "*Candidatus* Chloracidobacterium" species.

Bacteroidetes. Obligate anaerobes. Heterotrophs that feed on diverse carbon sources. *Bacteroides thetaiotaomicron.*

Chlorobi. Green sulfur-oxidizing phototrophs. *Chlorobium tepidum.*

Fusobacteria. Gram-negative anaerobic bacteria found in septicemia and in skin ulcers. *Fusobacterium nucleatum.*

Nitrospirae. Oxidize nitrite; aerobic or facultative. *Nitrospira marina.*

Spirochetes (Spirochaeta). Narrow, coiled cells with axial filaments, encased by sheath. Polar flagella beneath sheath double back around cell.

- *Spirochaetales.* Aquatic free-living; endosymbionts and pathogens (width <0.5 μm; length 10–20 μm).

   Borrelia. *B. burgdorferi* causes Lyme disease, transmitted by ticks.

   Hollandina. Termite gut endosymbionts.

   Leptospira. L-shaped animal pathogens; cause leptospirosis.

   Spirochaeta. Aquatic, free-living heterotrophs.

   Treponema. *T. pallidum* causes syphilis.

Chlamydiae, Planctomycetes, and Verrucomicrobia. Irregular cells lacking peptidoglycan, with subcellular structures analogous to those of eukaryotes.

Chlamydiae. Intracellular cell wall–less pathogens of animals or protists. *Chlamydia trachomatis* causes sexually transmitted disease and trachoma (eye infection).

Planctomycetes. Nucleoid has double membrane analogous to eukaryotic nuclear membrane and other intracytoplasmic subcellular membrane compartments. *Brocadia* species anaerobically oxidize ammonium and release N_2 (anammox reaction). *Pirellula* species inhabit marine sediment.

Verrucomicrobia. Stalk-like appendages contain actin filaments. Aquatic oligotrophs. *Prosthecobacter* sp.

*Bulleted terms are representative orders within the phylum (unless stated otherwise).

and cell membrane are contained within an outer membrane called the sheath. Between the sheath and the cell membrane extend flagella doubled back from each pole. The rotation of these flagella is coordinated so as to twist and flex the helical body, generating motility and chemotaxis. Spirochetes include many free-living forms in aquatic systems, as well as digestive endosymbionts and pathogens.

Three phyla of bacteria have unusual cell shapes: **Chlamydiae**, **Planctomycetes**, and **Verrucomicrobia** (see Section 18.7). Most members of these groups have lost their peptidoglycan cell walls, but they show complex structural adaptations and developmental forms. The Chlamydiae (best-known genus *Chlamydia*) are intracellular parasites that lose most of their cell envelope during intracellular growth. The replicating parasites generate multiple spore-like "elementary bodies" that escape to infect the next host.

Planctomycetes, by contrast, are free-living aquatic bacteria, with stalked cells that reproduce by budding. Each planctomycete cell contains an extra double membrane surrounding its nucleoid, analogous to a eukaryotic nuclear membrane, though it evolved independently. Verrucomicrobia, also free-living, are bacteria with wart-like protruding structures containing actin.

Deep-Branching Thermophiles

Deep-branching thermophiles show unusually large genetic divergence from other bacterial clades. This large divergence makes the thermophiles appear to have separated earlier from other kinds of bacteria. Other evidence, however, implicates high mutation rates and gene transfer between distant relatives as the source of genetic divergence of thermophiles.

The deep-branching taxa include extremophiles such as *Aquifex aeolicus* (growing at up to 95°C at marine thermal vents). These organisms also show rapid growth and high mutation rates, which may have accelerated their molecular clock. These hyperthermophilic bacteria also share their high-temperature habitats with Archaea—and show surprising archaeal traits. For example, *Aquifex* species have archaeal ether-linked membrane lipids (discussed in Chapters 3 and 19). The genes encoding these "archaeal" lipids appear to have entered bacteria by horizontal transfer from archaea to bacteria sharing the high-temperature habitat.

Aquificae and Thermotogae. Most members of the thermophilic phyla Aquificae and Thermotogae are hydrogenotrophs, oxidizing hydrogen gas with molecular oxygen to make water. *Aquifex pyrophilus*, a flagellated rod, was first discovered by extremophile microbiologist Karl Stetter at the University of Regensburg in a submarine hydrothermal vent north of Iceland (for an interview, see **eTopic 18.1**). Stetter named the bacterium *Aquifex*, Latin

for "water maker," because it oxidizes hydrogen to form water. *Aquifex* species are obligate autotrophs, fixing CO_2 into biomass using the reverse TCA cycle. Other genera of the phylum Aquificae oxidize small organic molecules or sulfides, and form filamentous mats in hydrothermal springs. *Thermocrinis ruber* grows at 82°C–88°C as a mat of pink filamentous streamers in the outflow channel of Octopus Spring, in Yellowstone National Park. A related phylum of thermophiles (growing at 50°C–80°C) is Thermotogae. *Thermotoga maritima* is a sulfur-reducing respirer, originally isolated from a geothermal vent in Vulcano, Italy. The cells have a loosely bound sheath, or "toga," for which the genus is named.

Both Aquificae and Thermotogae show remarkable mosaic genomes. Nearly a quarter of the *T. maritima* genome derives from archaea. The Aquificae, too, show large proportions of archaeal DNA sequence and traits, which are ascribed to the sharing of the high-temperature habitat. But Aquificae also possess large portions of sequence related to clostridia and to Epsilonproteobacteria (discussed shortly). This degree of mosaicism leads some researchers to argue that Aquificae cannot be said to branch from one clade; or if they do, that we can never know, statistically, which clade that is.

Thought Question

18.1 What taxonomic questions are raised by the apparent high rate of gene transfer between archaea and thermophilic bacteria?

Chloroflexi. Chloroflexi bacteria are filamentous photoheterotrophs, supplementing heterotrophy with photosystem II (PS II) to generate ATP (discussed in Chapter 14). Together with other thermophiles, Chloroflexi species form massive microbial mats in the hot springs of Yellowstone (**Fig. 18.3**). Most species of *Chloroflexus* contain their photosynthetic apparatus within membranous organelles called **chlorosomes**. The process of photosynthesis by chlorosomes is presented in Chapter 14. Chloroflexi are informally called "green nonsulfur bacteria" to distinguish them from Chlorobi, a phylum of green phototrophs that are strict anaerobes (presented shortly, with the deep-branching Gram-negative phyla). However, *Chloroflexus* species may appear red or yellow, owing to accessory pigments.

Deinococcus-Thermus. The phylum Deinococcus-Thermus possesses a unique structural trait: the substitution of L-ornithine for diaminopimelic acid in the peptidoglycan cross-bridge. *Thermus* species (growing at 70°C–75°C) are heterotrophs commonly isolated from hot tap water. *Deinococcus* species, however, are not thermophilic. *D. radiodurans* bacteria resist extremely high doses of ionizing radiation (discussed in Chapters 5 and 9). A heterotroph,

A.

BOB LINDSTROM

B.

COURTESY OF L.L. JAHNKE

C.

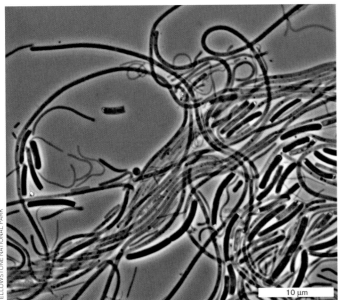

10 μm

YELLOWSTONE NATIONAL PARK

FIGURE 18.3 ■ **A deep-branching thermophile:** *Chloroflexus*.
A. This hot-spring bacterial mat in Yellowstone National Park contains *Chloroflexus* species and other thermophilic bacteria.
B. Bacterial mat section showing layers of *Chloroflexus*. **C.** *Chloroflexus* bacteria (LM).

D. radiodurans was originally isolated from cans of meat supposedly sterilized with a radiation dose of several megarads. *Deinococcus* bacteria are found suspended in air, both indoors and high in the atmosphere; they resist drought.

Deep-Branching Gram-Negatives

Several deep-branching phyla stain Gram-negative but diverge distantly from the Proteobacteria. They show diverse metabolism and morphology (see Section 18.5). Most members of the phyla **Bacteroidetes** and **Chlorobi** are obligate anaerobes. Within Bacteroidetes, *Bacteroides* species ferment complex carbohydrates, serving as the major mutualists of the human gut. By contrast, the closely related Chlorobi species are anaerobic "green sulfur" phototrophs that photolyze sulfides or hydrogen. The **Nitrospirae** largely resemble proteobacteria in form. Most oxidize nitrite (NO_2^-) to nitrate (NO_3^-). Nitrite oxidation is a lithotrophic conversion essential for ecosystems (see Chapter 22). Other Gram-negative clades include **Acidobacteria**, an important phylum of soil bacteria; and the phylum **Fusobacteria**, which includes human pathogens.

Emerging Clades

Clades of microbes that are recently defined or characterized are referred to as **emerging**. Emerging clades are discovered by field microbiologists who devise ever-more-creative screens, finding new bacteria with unexpected traits. For instance, in 2015 Jill Banfield's group at UC Berkeley discovered ultrasmall bacteria from the Rifle, Colorado, aquifer. These bacteria passed through a 0.2-μm filter—barely big enough for ribosomes. Cryo-TEM shows tiny cells that contain about fifty ribosomes, with a cell wall enclosed by an S-layer and long pili (**Fig. 18.4**). The cells' genomes are smaller than those of most known bacteria. Their DNA sequences show deeply branching phylogeny—that is, very distant relatedness to all the bacteria known. Some of their 16S rRNA sequences are so distant from those of other taxa that they cannot be amplified by so-called universal bacterial primers for PCR. We know little about these microbes—or others yet unknown, which future microbe hunters may discover.

Thought Questions

18.2 Which taxonomic groups in **Table 18.1** stain Gram-positive, and which stain Gram-negative? Which group contains both Gram-positive and Gram-negative species? For which groups is the Gram stain undefined, and why?

18.3 Which groups of bacterial species share common structure and physiology within the group? Which groups show extreme structural and physiological diversity?

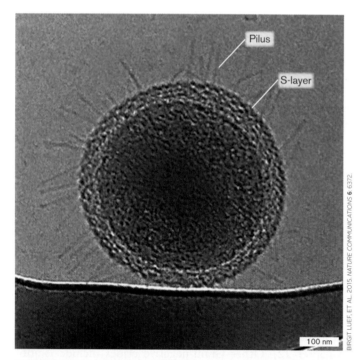

FIGURE 18.4 ■ **Ultrasmall bacteria.** Bacteria capable of passing through a 0.2-µm filter were discovered in 2015 by Jill Banfield and her colleagues in an aquifer in Rifle, Colorado. The tiny cells possess approximately fifty ribosomes, a cell wall surrounded by an S-layer, and long pili (cryo-TEM).

Note: Formal names for taxa such as phyla, orders, and genera are capitalized (phylum Cyanobacteria, genus *Streptococcus*). Informal forms are lowercase roman (cyanobacteria, streptococci), as are adjectival forms (cyanobacterial, streptococcal).

Following our brief tour of the bacterial domain, we now explore some of the diverse species within well-studied phyla of **Table 18.1**. For reference, additional species of interest are compiled in **Appendix 2 and eAppendix 4**.

To Summarize

- **Cyanobacteria** conduct oxygenic photosynthesis and interact closely with heterotrophs. Species vary widely in their cell shape and ecological niche.

- **Firmicutes and Actinobacteria** have a thick cell wall and generally stain Gram-positive.

- **Proteobacteria and deep-branching Gram-negative phyla** have a thin cell wall and an LPS-containing outer membrane. They show diverse metabolism and ecological adaptations.

- **Spirochetes** have flexible, spiral-shaped cells with complex intracellular architecture.

- **Chlamydiae, Planctomycetes, and Verrucomicrobia** have irregularly shaped cells with complex intracellular form and development.

- **Deep-branching thermophiles** such as *Aquifex* and *Thermotoga* species share traits and habitats with thermophilic archaea. Their genomes are highly mosaic, including many archaeal genes. *Deinococcus* species are highly resistant to ionizing radiation. *Chloroflexus* species are thermophilic photoheterotrophs.

- **Emerging organisms** continually reveal previously unknown clades of bacteria.

Note: The main organizing principle used in Chapters 18–20 is that of phylogeny based on DNA relatedness. Characteristic traits described for each branch apply to the majority of its known species, though many exceptions have evolved, such as members of Spirochetes that lack spiral form.

18.2 Cyanobacteria: Oxygenic Phototrophs

All of the oxygen gas in Earth's atmosphere comes from Cyanobacteria, and from plant chloroplasts that evolved from an ancient cyanobacterium. The phylum Cyanobacteria (also called Cyanophyta) is named for the blue phycocyanin accessory pigments possessed by some genera, giving them a bluish tint. Cyanobacteria commonly appear green because of the predominant blue and red absorption by chlorophylls *a* and *b* (see **Table 18.2**). Essentially the same chlorophylls are found in plant chloroplasts, which evolved from internalized cyanobacteria (discussed in Chapter 17). Some cyanobacteria, however, appear red because of the accessory pigment phycoerythrin, which absorbs blue-green light in a range missed by cyanobacterial chlorophylls. Cyanobacteria are the only prokaryotes that use both photosystems I and II, photolyzing water to produce oxygen, as explained in Chapter 14. Under anoxic conditions, most cyanobacteria can also photolyze hydrogen, reduced sulfur compounds, and organic compounds. Use of H_2S allows flexibility in habitats such as wetlands that alternate between aerobic and anoxic conditions.

The success of oxygenic phototrophy is shown by its fundamental similarity across all members of Cyanobacteria, despite their amazing variety of cell form, behavior, and genetics. Even two cyanobacterial genera such as *Synechococcus* and *Anabaena* share only 37% of their genes. Cyanobacteria are found in all habitats, from tropical soils to Antarctica. Some kinds are eaten by humans, such as in *Spirulina* salad or *Nostoc* soup.

How does the oxygenic photosynthesis of cyanobacteria compare with the non-oxygenic photosynthesis of thermophilic Chloroflexi, or of sulfur-based phototrophs among the Proteobacteria? **Table 18.2** summarizes key features

TABLE 18.2	Phototrophic Bacteria

Taxon	Energy generation	Cell structure and photopigments	Absorption spectrum (whole cell)	Reaction center
Chloroflexi "Green nonsulfur" *Chloroflexus*	$+O_2$ Heterotrophy $-O_2$ Photoheterotrophy or use reduced sulfur	Chlorosome, BChl *a*, BChl *c/d*	400 600 800 1,000	PS II
Cyanobacteria "Blue-greens" *Anabaena*	$+O_2$ Oxygenic phototrophy $-O_2$ Photolithoautotrophy on reduced sulfur	Thylakoid, Chl *a/b*	400 600 800 1,000	PS I and PS II
Firmicutes "Sun bacteria" *Heliobacterium*	$+O_2$ $-O_2$ Photoheterotrophy	Forms endospore, BChl *g*	400 600 800 1,000	PS I
Proteobacteria "Purple nonsulfur" (Alpha and Beta classes) BChl *a* *Rhodospirillum*	$+O_2$ Heterotrophy	BChl *a*	400 600 800 1,000	PS II
BChl *b* *Blastochloris*	$-O_2$ Photoheterotrophy	BChl *b*	400 600 800 1,000	
"Purple sulfur" (Gamma class) *Chromatium*	$+O_2$ $-O_2$ Photolithoautotrophy on reduced sulfur	BChl *a/b*	400 600 800 1,000	PS II
"Proteorhodopsin" (Alpha and Gamma classes) *Pelagibacter* (SAR11) SAR86	$+O_2$ Heterotrophy $-O_2$ Photoheterotrophy	Proteorhodopsin	400 600 800 1,000	PR
Chlorobi "Green sulfur" *Chlorobium*	$-O_2$ only Photolithoautotrophy on reduced sulfur	Chlorosome, BChl *a*, BChl *c/d/e*	400 600 800 1,000	PS I

Sources: J. Overman and F. Garda-Pichet. 2001. The phototrophic way of life. In *Prokaryotes.* 2002. Springer-Verlag; Oded Beja et al. 2001. *Nature* **411**:786–789.

of phototrophy in clades throughout the bacterial domain, including those presented in this chapter. All light-harvesting complexes descend from one of two ancestral sources: the chlorophyll/bacteriochlorophyll with electron transport (PS I, PS II) or the proteorhodopsin proton pump. These photosystems have diverged through vertical inheritance, as well as by horizontal transfer between different clades.

Cyanobacterial Cell Structure

The photosynthetic apparatus of cyanobacteria is organized within thylakoids, pockets of membrane resembling flattened spheres packed with reaction centers. The thylakoids may be distributed through the cell, as in filamentous genera such as *Nostoc* (**Fig. 18.5A**), or they may encircle the cell in concentric layers, as in the single-celled marine species *Prochlorococcus marinus* (**Fig. 18.5B**). *Prochlorococcus* is one of the smallest and most abundant oxygen producers in the biosphere, accounting for 40%–50% of all marine phototrophic biomass. In both large cells and small, the thylakoids are completely separate from the plasma membrane, unlike the attached chlorosomes of Chloroflexi and the plasma membrane extensions of "purple" Proteobacteria (see **Table 18.2**). Cyanobacterial thylakoids resemble the thylakoids of eukaryotic chloroplasts; they are the most complex and specialized form of photosynthetic apparatus.

Cyanobacteria have several other subcellular structures (**Fig. 18.5**). **Carboxysomes** (also known as polyhedral bodies) are rich in the enzyme Rubisco, and they fix CO_2 (discussed in Chapter 15). Cyanobacteria store energy-rich compounds in lipid bodies. To maintain height in the water column and thus access to sunlight, cyanobacteria have **gas vesicles**, whose buoyancy enables cells to float. Their external structures include a thick peptidoglycan cell wall, similar to that of Gram-positive cells, plus several external layers that vary with different species. Many species move by "gliding," a form of motility whose mechanism is poorly understood.

Besides fixing CO_2, most cyanobacteria fix N_2. Because nitrogen fixation requires the absence of oxygen, cyanobacteria have to solve the problem of maintaining anaerobic biochemistry while producing huge quantities of highly toxic O_2. Different species solve this problem in different ways:

- Formation of specialized nitrogen-fixing cells called **heterocysts**. Heterocysts exclude oxygen, which associated heterotrophic bacteria consume.

- Temporal separation, alternating between photosynthesis during daylight and nitrogen fixation at night.

- Accumulation of large aggregates of cells in which the interior becomes sufficiently anaerobic for nitrogenase to function, while the exterior continues oxygenic photosynthesis.

- Symbiosis with fungi (as in lichens) or plants that consume oxygen or otherwise maintain anoxic conditions.

Single-Celled, Filamentous, and Colonial Cyanobacteria

Single-celled species include *Synechococcus* and *Prochlorococcus*, the most abundant phototrophs in the oceans. Another single-celled cyanobacterium is *Microcystis*, a colonial microbe found in freshwater that produces dangerous toxins (microcystins). Blooms of *Microcystis* form in the spring and summer when nitrogen and phosphorus runoff from agricultural fertilizer and animal waste enters lakes and rivers. A *Microcystis* bloom in Lake Erie in 2014 resulted in microcystin levels great enough that the water supply of Toledo was contaminated. Since cyanobacterial species can have positive or negative impacts on their habitat, environmental scientists survey the cyanobacterial communities of lakes and ponds in order to assess their health. **Figure 18.6C** shows undergraduate Alex Schaal sampling a

A.

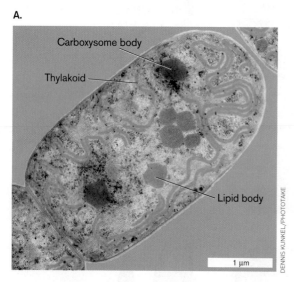

Carboxysome body

Thylakoid

Lipid body

1 μm

DENNIS KUNKEL/PHOTOTAKE

B.

Cell envelope   Carboxysomes   Thylakoids

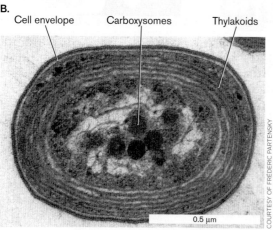

0.5 μm

COURTESY OF FREDERIC PARTENSKY

FIGURE 18.5 ▪ Cyanobacterial cell structure. A. Intracellular organelles of *Nostoc*, a typical filamentous cyanobacterium (colorized TEM). B. Intracellular organelles of *Prochlorococcus*, a prochlorophyte cyanobacterium, the smallest known phototroph. *Prochlorococcus* accounts for 40%–50% of marine phototrophic biomass.

A. *Nostoc*

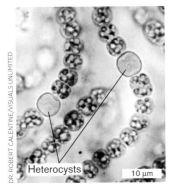

Heterocysts

10 µm

DR. ROBERT CALENTINE/VISUALS UNLIMITED

B. *Oscillatoria*

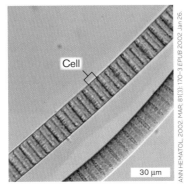

Cell

30 µm

ANN HEMATOL. 2002 MAR; 81(3):170-3 EPUB 2002 Jan 26.

C.

JOAN SLONCZEWSKI, KENYON COLLEGE

FIGURE 18.6 ■ **Pond cyanobacteria. A.** *Nostoc* filamentous cyanobacteria form heterocysts that fix nitrogen. **B.** *Oscillatoria* filaments consist of platelike cells. **C.** Alex Schaal, undergraduate at Kenyon College, samples water from a pond in Knox County, Ohio. Cyanobacteria and algae are filtered to obtain DNA for sequencing and identification.

pond in Ohio to obtain cyanobacteria and eukaryotic algae. From these samples, she will sequence DNA to identify the species, and measure the concentration of microcystins to determine whether the water is safe for human use.

Other cyanobacterial genera, such as *Oscillatoria* and *Nostoc*, form multicellular filaments. Such filaments may contain hundreds or even thousands of cells (**Fig. 18.6A** and **B**). *Oscillatoria* cells are stacked like plates, wider than they are long. To disseminate their cells beyond the biofilm, the filaments produce **hormogonia** (singular, **hormogonium**),

short motile chains of three to five cells. Many filamentous species, such as *Nostoc*, develop heterocysts to fix nitrogen (**Fig. 18.6A**) (discussed in Chapter 15).

Under environmental stress, such as light limitation or phosphate starvation, filamentous cyanobacteria such as *Anabaena* form specialized spore cells called akinetes. An akinete forms as a long, oval cell adjacent to a heterocyst, where it stores nitrogen and develops a thickened envelope. Like other types of spores, akinetes resist desiccation and remain viable for long periods. **Table 18.3** compares

TABLE 18.3	Spore Types in Bacteria			
Spore type	**Bacteria that produce the spore**	**Initiation of spore formation**	**Formation of the spore**	**Properties of the spore**
Akinete	Filamentous cyanobacteria	Light limitation Cold temperature Phosphate starvation	An akinete develops next to a heterocyst, as a large oval cell with a multilayered envelope.	Desiccation and cold resistant Viable for decades
Arthrospore	Actinomycetes	Carbon starvation Phosphate starvation	At the tip of an aerial mycelium, cells undergo vegetative division and pinch off as arthrospores.	Desiccation resistant Heat resistant
Elementary body	Chlamydias	Completion of intracellular life cycle	Intracellular chlamydia reticular bodies replicate and then develop into elementary bodies with cross-linked outer membrane proteins. Elementary bodies survive outside the host cell.	Survives outside host Desiccation resistant
Endospore	Firmicutes	Carbon starvation Nitrogen starvation Phosphate starvation Low pH Peptide antibiotics	Individual cell develops mother cell and forespore. The forespore develops into an endospore with a spore coat reinforced by keratin and calcium dipicolinate.	Highly heat and desiccation resistant Viable for centuries
Myxospore	Myxobacteria	Nutrient starvation Heat shock Glycerol Dimethyl sulfoxide	Myxobacteria aggregate to form a fruiting body in which vegetative cell division forms a mass of myxospores.	Desiccation resistant UV resistant

A. *Gloeocapsa*

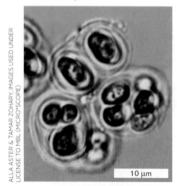

B. *Merismopedia*

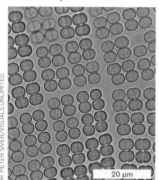

C. *Pleurocapsa*

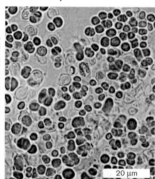

FIGURE 18.7 ▪ **Colonial cyanobacteria.** **A.** *Gloeocapsa* is surrounded by mucus. Cells grow as single cells, doublets, or quartets. **B.** *Merismopedia* forms extended quartets, octets, and so on. **C.** *Pleurocapsa* forms enormous aggregates that release baeocytes.

the properties of akinetes with those of other spore types (which are discussed along with Firmicutes and other taxa). Akinetes lie dormant but viable until improved conditions permit germination and growth of new vegetative filaments. In lake water, akinete germination may cause toxic blooms of *Anabaena*.

Some filamentous cyanobacteria, such as *Lyngbya* and *Trichodesmium*, form algal blooms in the ocean. *Trichodesmium* can form giant blooms visible from outer space, covering many square kilometers of ocean surface. Such a bloom can be triggered by an influx of iron carried by wind from a dust storm blowing off the Sahara desert (discussed in Chapter 22).

Yet other species, called colonial cyanobacteria, divide to form small groups or larger colonies (**Fig. 18.7**). *Gloeocapsa* and *Chroococcus* form doublets or quartets, encased in a thick protective mucous slime (**Fig. 18.7A**). Others, such as *Merismopedia*, continue cell division in two planes, extending to form long, square sheets of attached cells (**Fig. 18.7B**). Colonial genera such as *Myxosarcina* and *Pleurocapsa* reproduce by multiple fission, forming large cell aggregates (**Fig. 18.7C**). As the aggregate matures, some cells continue to divide and release single cells called baeocytes. Each baeocyte reproduces and develops into a new cell aggregate. The aggregate group maintains anoxic conditions at the center for nitrogen fixation.

Multicellularity

Cyanobacterial filaments and colonies, in effect, represent multicellular organisms. Different bacteria have evolved diverse means of generating multicellularity (**Fig. 18.8**). As we have seen, filamentous cyanobacteria such as *Anabaena* or *Nostoc* form chains by serial cell division. At intervals, a cell differentiates into a heterocyst, which performs nitrogen fixation—a specific function for the filament as a whole. In other clades, we will see very different mechanisms. Firmicute *Bacillus* cells (discussed in Section 18.3) assemble by attachment to a substrate. The cells grow forming a biofilm that releases endospores. Actinomycetes (see Section 18.3) form branching filaments, aerial and into the soil, where they produce antibiotics. The older cells undergo programmed cell death, while aerial filaments generate exospores. Gram-negative myxobacteria (Section 18.4) assemble by chemotaxis and differentiate into a fruiting body to release myxospores.

Cyanobacterial Communities

Cyanobacteria share many kinds of mutualistic associations with animals, plants, fungi, and protists. Sponges growing on coral reefs may harbor communities of cyanobacteria that provide the sponges with nutrients from photosynthesis. The products of photosynthesis supplement the nutrients obtained by the sponges from their filter feeding, and these extra nutrients greatly augment the sponge growth rate within the competitive coral reef environment. Sponge symbionts produce many pharmaceutically active compounds.

In salt marshes and sand flats, cyanobacteria participate in multilayered microbial mats, such as the section shown in **Figure 18.9**, cut from the sand flats of Great Sippewissett Salt Marsh, on Cape Cod, Massachusetts. The high concentration of sulfides in the sediment supports growth of high populations of sulfur phototrophs. Typically, cyanobacteria and eukaryotic algae such as diatoms form the upper green layer. Below, the purple layer consists of "purple sulfur" proteobacteria, whose photopigments (primarily BChl *a*) absorb at longer wavelengths (discussed in Section 18.5). The pale-colored layer below the purple layer consists of proteobacteria with BChl *b*, which absorbs farther into the infrared.

Thought Questions

18.4 What are the relative advantages and disadvantages of propagation by hormogonia, as compared with akinetes?

18.5 What are the relative advantages and disadvantages of the different strategies for maintaining separation of nitrogen fixation and photosynthesis?

FIGURE 18.8 ▪ Bacterial multicellularity. Multicellularity arises by various means. **A.** Filamentous cyanobacteria form chains by serial cell division. At intervals, a cell differentiates into a heterocyst. **B.** *Bacillus* species assemble by attachment to a substrate. They grow as a biofilm that produces antibiotics, then releases endospores. **C.** Actinomycetes form branching filaments that produce antibiotics, then undergo programmed cell death. Aerial filaments generate exospores. **D.** Myxobacteria assemble by chemotaxis of swarmer cells. They differentiate into a fruiting body and release myxospores.

A. *Anabaena* spp.

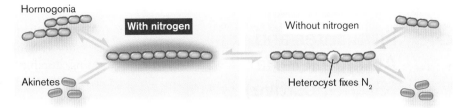

Hormogonia — With nitrogen — Without nitrogen

Akinetes

Heterocyst fixes N₂

B. *Bacillus subtilis*

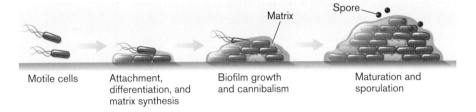

Motile cells — Attachment, differentiation, and matrix synthesis — Biofilm growth and cannibalism — Matrix — Spore — Maturation and sporulation

C. *Streptomyces* spp.

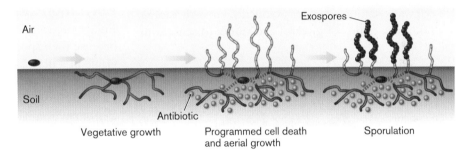

Air — Soil — Exospores — Antibiotic

Vegetative growth — Programmed cell death and aerial growth — Sporulation

D. *Myxococcus xanthus*

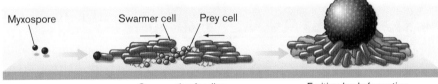

Myxospore — Swarmer cell — Prey cell

Cooperative feeding — Fruiting-body formation

Cyanobacteria and diatoms
Purple sulfur proteobacteria
Long-wavelength purple sulfur bacteria

3 cm

FIGURE 18.9 ▪ Cyanobacteria in microbial mats. Cutaway through a multilayered microbial mat from the sand flats of Great Sippewissett Salt Marsh (Cape Cod, Massachusetts). Cyanobacteria and diatoms form the upper green layer, above layers of purple sulfur proteobacteria.

To Summarize

- **Cyanobacteria** are the only oxygenic prokaryotes. They conduct photosynthesis in thylakoids, fix CO_2 in carboxysomes, and maintain buoyancy using gas vesicles. They exhibit gliding motility.

- **Single-celled cyanobacteria** such as *Prochlorococcus* are among the smallest and most abundant phototrophic producers in the oceans. *Microcystis* species produce microcystins that poison lakes polluted by agricultural runoff.

- **Filamentous cyanobacteria** such as *Nostoc* and *Oscillatoria* are common in freshwater lakes. They form heterocysts to fix nitrogen, and reproduce by hormogonia or by akinetes.

- **Colonial cyanobacteria** such as *Myxosarcina* produce large cell aggregates with an anaerobic core for nitrogen fixation. The colonies reproduce through baeocytes.

- **Symbiotic associations** of cyanobacteria occur with animals, fungi, and plants.

18.3 Firmicutes and Actinobacteria (Gram-Positive)

Which bacteria have the toughest, thickest cell walls? Most bacteria of the phylum Firmicutes, meaning "tough skin" bacteria, have thick peptidoglycan cell walls that retain the Gram stain (discussed in Chapter 3). The thick cell wall helps exclude antibiotics and antibacterial agents from competitors in the environment. Members of a related phylum, Actinobacteria, also stain Gram-positive. As shown in **Table 18.1**, we define Firmicutes as the phylum of Gram-positive bacteria that contains "low-GC" species (having a low ratio of GC/AT base pairs), whereas Actinobacteria are the "high-GC" species. Species in both phyla have thick cell walls reinforced by teichoic acids, cross-threading phosphodiester chains of glycerol and ribitol (discussed in Chapter 3).

Many firmicutes, such as *Bacillus* and *Clostridium* species, survive unfavorable environmental conditions by forming durable endospores. Non–spore formers such as *Lactobacillus* and *Streptococcus* may have evolved from a common Firmicutes ancestor that formed endospores. In many cases, the machinery to form endospores is discovered in the genomes of firmicutes previously thought to be non–spore formers, such as *Carboxydothermus*, soil bacteria that oxidize CO (carbon monoxide) to CO_2. On the other hand, actinobacteria of the order Actinomycetales (actinomycetes), such as *Streptomyces*, do not form endospores, but they develop filaments that disperse arthrospores (see **Table 18.3**).

Firmicutes Include Endospore-Forming Rods

Endospore-forming bacteria are common in soil and air because their spore forms resist desiccation and can remain viable in a dormant state for thousands of years. The best-known orders are Bacillales (mainly aerobic respirers) and Clostridiales (obligate anaerobes). Both groups include species of environmental and economic importance, as well as causative agents of well-known diseases.

Bacillales. The genus *Bacillus* was one of the first bacterial genera to be classified, in the nineteenth century (**Fig. 18.10**). Colonies soon appear on a nutrient agar plate exposed to air. *Bacillus* species can be isolated from soil or food by suspending a sample in water and heating at 80°C for half an hour. Vegetative cells (that is, cells undergoing binary fission) and non–spore formers are killed at that temperature. The remaining endospores will germinate and grow on a beef broth agar plate at 25°C–30°C. *Bacillus* species isolated in this way include over a thousand characterized strains, all but a few of them harmless to humans.

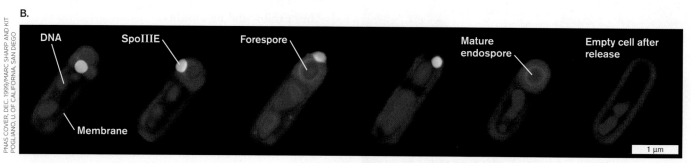

FIGURE 18.10 ▪ ***Bacillus* species: Gram-positive endospore formers.** **A.** Gram-stained *Bacillus* species, sporulating culture. Endospores stain green with malachite green. **B.** Correlated fluorescence imaging of membrane migration, protein translocation, and chromosome localization during *B. subtilis* sporulation. Membranes were stained with red fluorescent FM4-64. Chromosomes were localized with the blue fluorescent nuclear counterstain DAPI. The small, green fluorescent patches indicate the localization of a green fluorescent protein (GFP) gene fusion to Spo III E, a protein essential for both initial membrane fusion and forespore engulfment. Progression of the engulfment is shown from left to right.

The large, rod-shaped **vegetative cells** (growing and replicating form) of *Bacillus* species are easily stained and visualized. A species of particular scientific importance is *Bacillus subtilis*, the best-studied Gram-positive organism, a "model system" for Firmicutes. The *B. subtilis* genome sequence reveals a large number of transporters for carbon sources and many secretory complexes for industrially important enzymes and drug resistance proteins. It also reveals several integrated prophages (integrated phage genomes). Phage genomes contribute to bacterial evolution by transferring genes between different strains and species, as discussed in Chapter 17.

B. subtilis is used as a model system to study stress response. Enormous changes in protein expression accompany starvation and general stress conditions. As the vegetative cells run short of nutrients on an agar plate, they begin a program to "sporulate"—that is, develop inert endospores (**Fig. 18.10A**). The life cycle of endospore production (see Chapter 4) involves a coordinated developmental plan, in which the cell divides near the pole instead of at the cell equator (**Fig. 18.10B**). The polar compartment develops as the **forespore**, directed by unique regulatory proteins such as SpoIIIE. The larger compartment, called the **mother cell**, provides DNA and nutrients to the growing forespore and disintegrates after release of the mature endospore. When the released endospore encounters favorable conditions of moisture and nutrients, it germinates and restarts vegetative growth. The sporulation cycle can proceed out of a biofilm (see **Fig. 18.8B**).

A *Bacillus* species of great economic importance is *B. thuringiensis*, the most successful biological control agent yet produced (**Fig. 18.11**). *B. thuringiensis* was originally discovered in 1901 by Japanese bacteriologist Shigetane Ishiwata as a cause of disease in silkworms. The organism proved easy to culture on agar-based medium, and its spores are now applied as an insecticide against the gypsy moth caterpillar. During sporulation, *B. thuringiensis* generates a diamond-shaped crystal adjacent to its endospore (**Fig. 18.11A**). The crystal contains an insecticidal protein known as delta endotoxin. The crystals are ultimately released from the cells (**Fig. 18.11B**). The toxin is activated only at high pH in

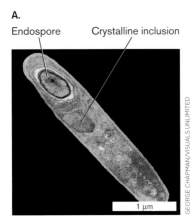

A. Endospore Crystalline inclusion

B.

FIGURE 18.11 ■ *Bacillus thuringiensis:* **the biological insecticide. A.** Sporulating cell of *B. thuringiensis*, showing crystalline inclusion of insecticidal toxin (colorized TEM). **B.** Toxin crystals (colorized blue) embedded in *B. thuringiensis* colony.

the digestive tracts of insect larvae; thus, it is safe for animals with acidic digestive tracts.

Many *Bacillus* species are extremophiles, growing at high pH (*B. alkalophilus*), at high temperature (*B. thermophilus*), or in high salt (*B. halodurans*). Some species combine alkaliphily with thermophily, as in the case of the "alkalithermophile" *B. alkalophilus*. Genomic research has focused on the surprisingly subtle differences that distinguish an extremophile from a closely related mesophile. For example, thermostability of proteins can be determined by comparing the protein sequences of a thermophile with those of a related mesophile. The thermophilic sequence shows specific patterns of amino acid residues that confer thermostability. Such amino acid substitutions provide useful information for industrial engineering of enzymes.

Clostridiales. This anaerobic family of spore formers includes the genus *Clostridium*, species of which cause botulism (*C. botulinum*) and tetanus (*C. tetani*) (**Fig. 18.12A**). The botulism toxin (botulinum, or "Botox") is famous for its therapeutic use to relax muscle spasms and smooth

A. *Clostridium* sp.

B. Botox treatment

Before   After

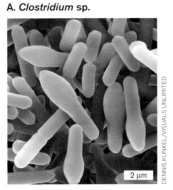
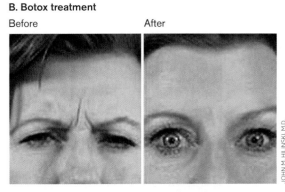

FIGURE 18.12 ■ *Clostridium* **species: spore-forming anaerobes. A.** As *Clostridium* cells sporulate, the endospore swells, forming a characteristic "drumstick" appearance. **B.** The deadly botulinum toxin (Botox) from *C. botulinum* is used to relax muscle spasms. This woman was unable to open her eyelids fully (left frame)—a condition that was cured by injection of Botox (right frame).

wrinkles in skin (**Fig. 18.12B**) (discussed in Chapter 26). Other species, such as *C. acetobutylicum*, have economic importance as producers of industrial solvents such as butanol and acetone (for industrial microbiology, see Chapter 16). The butanol pathway is an example of the diverse fermentative strategies found among clostridia. Unlike *Bacillus* (a monophyletic clade with a common ancestor), the *Clostridium* group is polyphyletic, representing many clades that branch among different genera.

The best-known *Clostridium* species, such as *C. botulinum* and *C. tetani*, sporulate in a form distinct from that of *Bacillus* (see **Fig. 18.12A**). The growing *Clostridium* endospore swells the end of the cell, forming a "drumstick" appearance. Mature clostridial spores are generally less heat resistant than the spores of *Bacillus* and thus more difficult to isolate in pure culture from natural environments. Nevertheless, *Clostridium* spores are found in soil and water, ready to germinate and grow when the environment becomes anoxic. Many harmless species are found in the human colon, particularly in infants. Pathogenic *C. botulinum* can grow within the colon of very young infants, and infant botulism has been implicated in some cases of sudden infant death syndrome.

Another species, *Clostridium difficile*, is a life-threatening intestinal pathogen, resistant to most antibiotics. *C. difficile* grows in patients treated with antibiotics that eliminate normal enteric bacteria, allowing growth of the pathogen. The pathogen can be so hard to eradicate that patients require fecal bacteriotherapy, or "fecal transplant" from a healthy person's colon. The healthy colon residents recolonize the gut and outcompete the pathogen.

Related to *Clostridium* are the Heliobacteriaceae, or "Sun bacteria," the only known phototrophs in Firmicutes (see **Table 18.2**). Their name derives from their yellowish color, caused by the shift of their red-peak absorbance into the infrared, which allows transmission of red light plus green light (perceived as yellow). The heliobacteria are photoheterotrophs, with a PS I reaction center providing a modest boost of energy to supplement heterotrophic metabolism of simple organic compounds. Most heliobacterial cells are elongated rods with peritrichous flagella, and they form endospores—the only known endospore-forming phototrophs.

Note: Distinguish *Heliobacterium* from *Helicobacter*, a helical species of the Gammaproteobacteria.

Variant sporulation and "live birth." The order Clostridiales includes species of exceptionally large bacteria that grow only within the digestive tract of specific animal hosts. These species have evolved intriguing variations of the endospore former life cycle, as revealed by Esther Angert and colleagues at Cornell University (**Fig. 18.13A**). *Metabacterium polyspora* (size 15–20 μm) grows throughout the

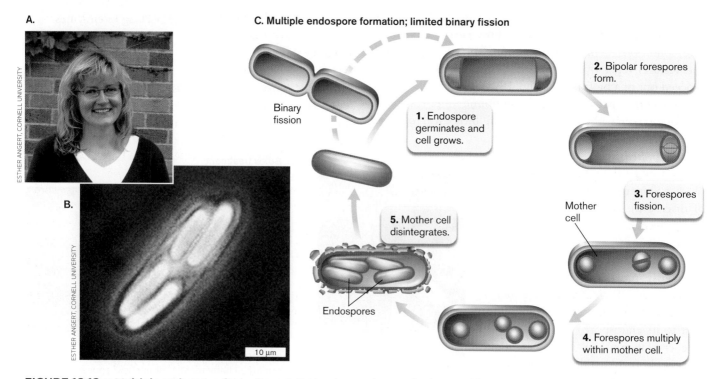

FIGURE 18.13 ■ Multiple endospore formation. A. Esther Angert characterized unusual forms of sporulation and reproduction in exceptionally large firmicute bacteria. **B.** *Metabacterium polyspora* forms multiple endospores (phase-contrast LM). **C.** A forespore forms at each pole. Forespores fission and multiply within the mother cell and then are released. Germinated cells undergo limited or no binary fission.

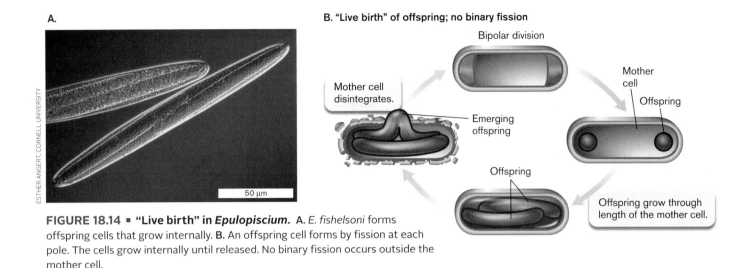

FIGURE 18.14 ■ **"Live birth" in *Epulopiscium*.** **A.** *E. fishelsoni* forms offspring cells that grow internally. **B.** An offspring cell forms by fission at each pole. The cells grow internally until released. No binary fission occurs outside the mother cell.

digestive tract of guinea pigs (**Fig. 18.13B**). Endospores ingested from feces germinate in the upper intestine but rarely undergo binary fission. Instead, the growing cell forms forespores at both poles (**Fig. 18.13C**). The forespores actually multiply within the mother cell to form several endospores, which are released in the colon before defecation.

An even larger enteric endosymbiont is *Epulopiscium fishelsoni*, found in the digestive tract of surgeonfish (**Fig. 18.14**). These bacteria are large enough to be seen by eye—about the size of the period at the end of this sentence. As in *M. polyspora*, Angert showed that *E. fishelsoni* reproduction is synchronized with the digestive cycle of its host, but it has gone even further in transformation of the sporulation cycle. Binary fission is eliminated; the cell must fission at both poles. Each polar fission generates an intracellular daughter cell that grows to nearly the full length of the mother cell. From two to seven intracellular offspring ultimately emerge in "live birth" from the mother cell, which then disintegrates.

An even more bizarre "live birth" species related to clostridia was discovered in the mammalian intestine. These bacteria, provisionally named "*Candidatus* Savagella," grow as filaments attached to the intestinal epithelial cells (**Special Topic 18.1**).

Non-Spore-Forming Firmicutes

Many Firmicute species do not form spores. These non–spore formers include species of Bacillales and Clostridiales, as well as other Gram-positive orders such as Lactobacillales. In these taxa, the endospore-forming system was probably lost by reductive evolution. Non-spore-forming firmicutes include human pathogens such as *Listeria* and *Streptococcus* species, as well as lactic acid bacteria that are important for food production.

Listeria species are facultative anaerobic bacilli, named for the British surgeon Joseph Lister (1827–1912), who was the first to promote antisepsis during surgery. They include enteric pathogens, such as *L. monocytogenes*, that contaminate cheese and sauerkraut (discussed in Chapter 16). Unlike other food-associated organisms, *Listeria* grows at temperatures as low as 4°C. Under preindustrial conditions of food preparation, *L. monocytogenes* was generally outcompeted by other flora. The era of refrigeration led to the emergence of *Listeria* as the cause of listeriosis, a severe gastrointestinal illness that can progress to the nervous system. *L. monocytogenes* cells are taken up by macrophages into phagocytic vesicles, but they avoid digestion and escape the vesicles. The bacteria then multiply as they travel through the host cytoplasm, generating "tails" of actin (**Fig. 18.15** ▶). The actin tails eventually project the cells of *Listeria* out of the original host cell and enable it to penetrate a neighboring host cell.

Lactic acid bacteria. The lactic acid bacteria (order Lactobacillales) are aerotolerant (capable of growth in the presence of oxygen), though they do not use oxygen to respire. Most lactic acid bacteria are obligate fermenters; that is, they generate ATP by substrate-level phosphorylation (discussed in Chapter 13). They ferment primarily by converting sugars to lactic acid (a fermentation pathway discussed in Chapter 13). As the acid builds up, the pH decreases until it halts bacterial growth; thus, the carbon source retains much of its food value for human consumption. This is the basis of yogurt and cheese production. *Lactococcus* and *Lactobacillus* species are extremely important for the dairy industry (discussed in Chapter 16). Other common genera of lactic acid bacteria include *Leuconostoc*, which often spoils meat, and *Pediococcus*, found in sauerkraut and fermented bean products, as well as meat products such as sausage.

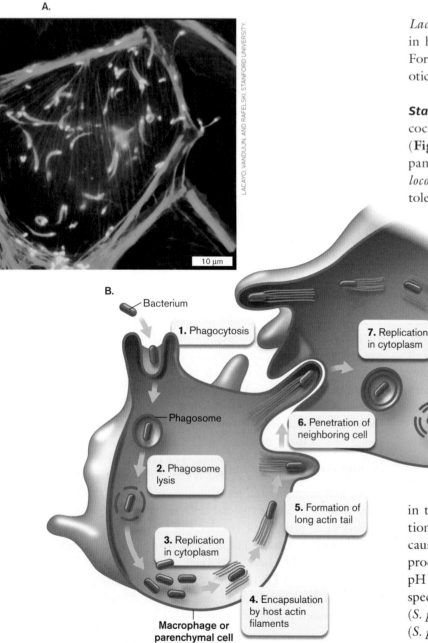

A.

10 μm

LACAYO, VANDIJN, AND RAFELSKI, STANFORD UNIVERSITY.

B.

- Bacterium
- 1. Phagocytosis
- 7. Replication in cytoplasm
- Phagosome
- 6. Penetration of neighboring cell
- 2. Phagosome lysis
- 5. Formation of long actin tail
- 3. Replication in cytoplasm
- 4. Encapsulation by host actin filaments
- Macrophage or parenchymal cell

FIGURE 18.15 ■ *Listeria monocytogenes:* **intracellular pathogen travels on tails of actin.** **A.** Fluorescent antibodies mark the tails of polymerized actin (green) behind the *Listeria* cell bodies (red) traveling within an infected macrophage. **B.** Invading bacteria encapsulate themselves in actin. Actin tails propel the bacteria through the host cytoplasm and out through the cell membrane to invade a neighboring cell. *Source:* Part B from U. South Carolina, Microbiology and Immunology On Line. ▶

Lactobacillus acidophilus, are believed to play a positive role in human health by inhibiting the growth of pathogens. For this reason *L. acidophilus* may be ingested as a probiotic therapy.

Staphylococcus and Streptococcus. The staphylococci are facultative aerobic cocci that grow in clusters (**Fig. 18.16A**), often packed in hexagonal arrays (see panel A). They include common skin flora such as *Staphylococcus epidermidis.* The staphylococci are generally salt tolerant, and their fermentation generates short-chain fatty acids that inhibit growth of skin pathogens. Certain species, however, are themselves serious pathogens. *Staphylococcus aureus* causes impetigo and toxic shock syndrome, as well as pneumonia, mastitis, osteomyelitis, and other diseases (discussed in Chapter 26). It is a major cause of nosocomial (hospital-acquired) infections, especially contamination of surgical wounds. The most dangerous strains, now resistant to most known antibiotics, are termed MRSA (methicillin-resistant *S. aureus*).

Streptococcus species generally form chains instead of clusters, because their cells divide in a single plane (**Fig. 18.16B**). They are aerotolerant (grow in the presence of oxygen) but metabolize by fermentation. Many live on oral or dental surfaces, where they cause caries (tooth decay). Their fermentation of sugars produces such high concentrations of lactic acid that the pH at the tooth surface can fall to pH 4. *Streptococcus* species cause many serious diseases, including pneumonia (*S. pneumoniae*), strep throat, erysipelas, and scarlet fever (*S. pyogenes* or group A streptococci).

The streptococci are less salt tolerant than *Staphylococcus* species, which tolerate as much as 5%–15% NaCl. Another genus whose size and fermentative metabolism resembles that of streptococci is *Enterococcus. E. faecalis* is a common member of the intestinal flora, and related strains are enteric pathogens.

Anaerobic dechlorinators. Some firmicutes from the soil show promising abilities to degrade chlorinated pollutants, such as dry-cleaning solvents that are biodegraded very slowly in the environment. The chlorinated molecules are reduced as alternative electron acceptors. *Dehalobacter restrictus*, a flagellated rod, was isolated as an anaerobe capable of respiring by donating electrons to chlorine atoms in tetrachloroethene (**Fig. 18.17**). Thus, the discovery of *D. restrictus* has promising potential for bioremediation of chlorinated pollutants.

The shape of lactate-producing bacteria varies among species, from long, thin rods to curved rods and cocci. Most lactic acid bacteria have fastidious growth requirements and need many amino acids and vitamins. They can be isolated from pasture grasses incubated anaerobically in moderate acid (pH 5). The human intestinal flora include species of lactic acid bacteria. Certain species, particularly

A. *Staphylococcus*

B. *Streptococcus*

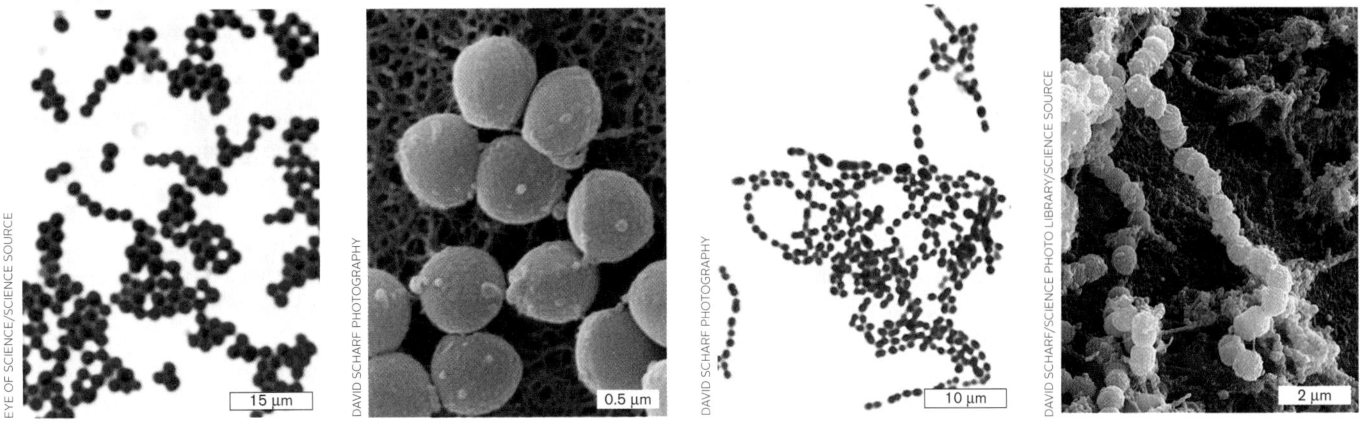

FIGURE 18.16 ■ **Staphylococci and streptococci.** **A.** *Staphylococcus* species: Gram stain (left); colorized SEM (right). **B.** *Streptococcus* species: Gram stain (left); colorized SEM (right).

While genetic analysis places *D. restrictus* among the clostridia, the bacterium actually stains Gram-negative, perhaps because its peptidoglycan layer is relatively thin. Nevertheless, the species shows no Gram-negative outer membrane. It does possess a thick S-layer of hexagonally tiled proteins, typical of Gram-positive bacteria.

Mycoplasmas lack a cell wall. The mycoplasmas are cell wall–less bacteria, classified as **Tenericutes** or **Mollicutes** (Latin for "soft skin"). Mycoplasmas have completely lost their cell wall and S-layer through reductive evolution, retaining only their cell membrane. Presumably, the loss of these energy-expensive structures enhanced the reproductive rate of cells in a protected host environment. The Mollicutes comprise many genera of flexible wall-less cells that maintain a shape through some kind of cytoskeleton (**Fig. 18.18A** and **B**). On agar they form colonies of a characteristic "fried-egg" appearance (**Fig. 18.18C**).

The best-known genus of Mollicutes is *Mycoplasma*, although its species now appear to fall in several distantly related branches. Mycoplasmas are found as parasites of every known class of multicellular organism, including vertebrates, insects, and vascular plants; in humans, they cause pneumonia and meningitis. Another medically important mycoplasma is *Ureaplasma urealyticum*, an opportunistic pathogen inhabiting the genital tracts of men and women. Most individuals are unaware they harbor the organism, but *U. urealyticum* has been associated with urethritis, amniotic infections, and pulmonary infections.

Some *Mycoplasma* species remain adherent to the host cell, whereas others penetrate and grow intracellularly. Most mycoplasma cells have a rounded cell shape with

A.

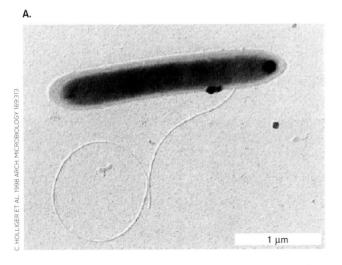

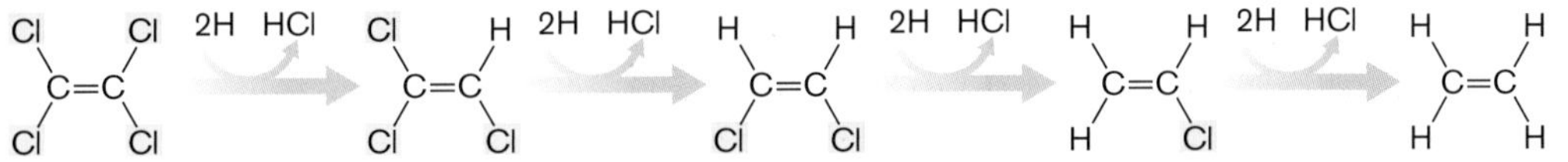

FIGURE 18.17 ■ *Dehalobacter restrictus* **conducts anaerobic respiration by dechlorination.** **A.** *Dehalobacter restrictus*, a flagellated rod related to clostridia. **B.** *D. restrictus* donates electrons to remove chlorine from tetrachloroethene, a major industrial pollutant.

B. Dechlorination of tetrachloroethene by *Dehalobacter restrictus*

SPECIAL TOPIC 18.1 | Gut Bacterial Hair Balls

A bizarre kind of bacteria sit attached to cells of mammalian gut epithelium (**Fig. 1A**). Related to clostridia, these bacteria form long, segmented filaments that interact symbiotically with their host cell. These "segmented filamentous bacteria" were first described by American microbiologist Dwayne Savage. They were mistakenly identified as *Arthromitus* (a different kind of bacterium) but have now been provisionally named for their discoverer, as "*Candidatus* Savagella." Because the name is provisional, microbiologists have a rule that "*Candidatus*" (Latin for "candidate") is italicized, whereas the provisional name ("Savagella") is not.

For over 50 years, microscopists have seen these segmented filamentous bacteria in the gut of mice and humans. The bacteria require their host for growth, and in turn, they provide essential services by inducing development of the host immune system. Attached to the epithelium above gut lymph tissues, the bacteria actually help protect their host from infection.

Recently, Pamela Schnupf and Philippe Sansonetti (**Fig. 1B** and **C**) at the Pasteur Institute cultured Savagella filaments attached to mouse gut cells in tissue culture. To do this, Schnupf isolated the multicell filaments by filtration at 5 μm, a pore size large enough to exclude most single-celled gut bacteria. The filaments were inoculated into germ-free mice. They were also used to inoculate cultured TC7 cells, a cell line of gut epithelial tumor cells. The bacteria attached to the cultured cells and grew massive "hair balls" of filaments.

FIGURE 1 ▪ Segmented filamentous bacteria (Savagella) adhere to mouse gut epithelium. **A.** Savagella filaments adhere to mouse gut epithelium (SEM). **B, C.** Pamela Schnupf and Philippe Sansonetti, at the Pasteur Institute, Paris. *Source:* Part A from Ivaylo Ivanov (Columbia University Medical Center), Dan Littman (NYU Langone Medical Center), and Doug Wei (Carl Zeiss Smt, Inc.).

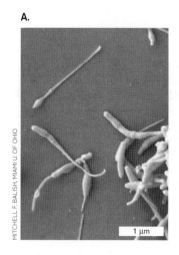

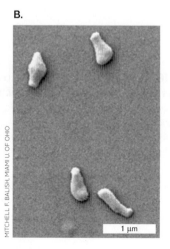

FIGURE 18.18 ▪ Mycoplasmas: parasites without cell walls. **A.** *Mycoplasma penetrans* cells have an elongated tip used for attachment to the host (SEM). **B.** *M. mobile* (SEM). **C.** Mycoplasmas cultured on agar show a "fried-egg" colony shape.

The in vitro system enabled Schnupf to observe the bacteria's unique process of differentiation and production of intracellular offspring (**Fig. 2**). Electron microscopy showed that each filament begins with a single "newborn" cell whose holdfast attaches to the surface of the host epithelial cell. The cell then grows and divides, forming a multicellular filament (primary segment cells). When the filament length exceeds 50 µm, the distal-segment cells start to differentiate into mother and daughter cells. Each daughter cell becomes engulfed by a mother cell. The engulfed daughter cell then divides and differentiates into two intracellular offspring. Finally, the mother cell breaks open, releasing the two offspring. Each offspring cell possesses a holdfast to attach a new site on the gut epithelium.

How do the attached bacterial filaments influence our gut epithelium? Schnupf has shown that the filament-attached epithelial cells have elevated transcription of genes that govern innate immune responses (discussed in Chapter 23). Thus, this bacterium's unusual life cycle plays a key role in the function of its mammalian host.

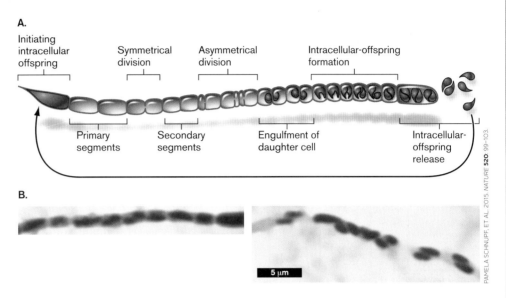

FIGURE 2 ■ **Savagella growth and reproduction. A.** Primary segments divide and differentiate until offspring bacteria develop within each cell. Offspring are released, to attach to an epithelial cell at a new position and grow a new filament. **B.** Intracellular offspring develop within cells of the filament (Gram stain).

PAMELA SCHNUPF, ET AL. 2015. NATURE **520** 99–103.

RESEARCH QUESTION

How do you think Savagella filaments communicate with the host immune system? Can you design an experiment to test your hypothesis?

Schnupf, Pamela, Valérie Gaboriau-Routhiau, Marine Gros, Robin Friedman, Maryse Moya-Nilges, et al. 2015. Growth and host interaction of mouse segmented filamentous bacteria in vitro. *Nature* **520**:99–103.

one or two extended tips. In *M. penetrans*, an opportunistic pathogen infecting AIDS patients, the cell's attachment tip is coated with adhesion molecules that enable attachment to a host cell surface (**Fig. 18.18A**). The attachment tip penetrates deep into epithelial tissues. By contrast, the fish pathogen *M. mobile* has no attachment tip, but glides along a surface at a rate of seven cell lengths per second (**Fig. 18.18B**). The basis of mycoplasma motility is not understood, although it may involve gliding or cytoskeletal contraction. In some species, motility involves a specialized cell tip called the "terminal organelle," but species that lack this organelle are equally motile. Species of *Spiroplasma* have a spiral shape and undergo corkscrew motion; the basis for this motion is also unknown.

Mycoplasma genomes are among the smallest in known cellular organisms. They also lack biosynthetic pathways for amino acids and phospholipids, which instead must be acquired from the host. Mycoplasmas have unique nutritional requirements, such as cholesterol, a membrane component typical of eukaryotes but rare for prokaryotes. For these reasons, mycoplasmas are difficult to grow in pure culture, although they readily infect tissue cultures. In fact, mycoplasma contamination of tissue culture is so prevalent that it has compromised major studies of cancer and AIDS.

A.

B.

C.

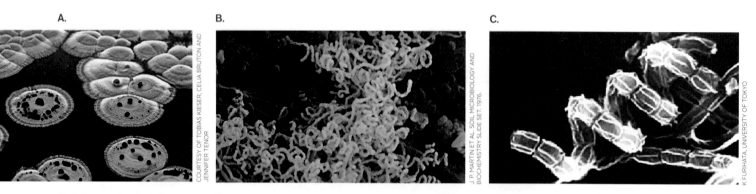

COURTESY OF TOBIAS KIESER, CELIA BRUTON AND JENNIFER TENOR

J. P. MARTIN ET AL, SOIL MICROBIOLOGY AND BIOCHEMISTRY SLIDE SET, 1976

K. FURIHATA, UNIVERSITY OF TOKYO

D. Telomere (end of chromosome)

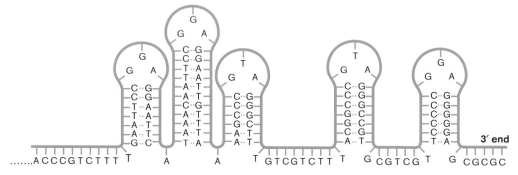

FIGURE 18.19 ■ *Streptomyces* bacteria. **A.** Colonies of *S. coelicolor* show sky-blue mycelia. **B.** *Streptomyces* cells form coiled filaments (filament width approx. 0.5 μm) (SEM). **C.** Close-up of a coiled filament, showing individual cells (SEM). **D.** Hairpin-looped telomere end of the linear chromosome of *S. griseus*. *Source:* Part D from Prokaryotes, Springer.

Actinomycetes Form Multicellular Filaments

The phylum Actinobacteria, the "high-GC Gram-positives," comprises several orders, including Actinomycetales and Bifidobacteriales. Members of the order Actinomycetales are "actinomycetes." Actinomycetes include filamentous spore formers such as *Streptomyces* and *Frankia*, as well as short-chain organisms such as *Mycobacterium*, and irregularly shaped cells such as *Corynebacterium*. Members of an emerging group of marine actinomycetes, such as *Salinispora*, are isolated from sediment and from sponges. *Salinispora* species form numerous exotic secondary products that show promise as pharmaceutical agents.

The actinomycetes form complex multicellular filaments superficially resembling the branched "fuzzy" form of fungi (fungi are discussed in Chapter 20). Profoundly important for medicine, actinomycetes and fungi produce most of our antibiotics. Researchers mine the soil of remote environments to find actinomycetes producing antibiotics for which resistance genes are not yet widespread.

***Streptomyces*: filamentous spore formers.** The best-studied actinomycetes are *Streptomyces* bacteria, which form multicellular filaments that generate dispersible spores. Streptomycetes play a major role in the ecosystems of soil (discussed in Chapter 21). Decaying

Streptomyces cells produce the compound geosmin, which causes the characteristic odor of soil and can affect the taste of drinking water. In culture, the best-known species, *S. coelicolor* (Latin, "sky color"), forms strikingly blue colonies (**Fig. 18.19A**). The blue color derives from several pigments, including actinorhodin, a polyketide antibiotic. Other species produce filaments that are red, orange, green, or gray, depending on their distinctive products, many of which are antibiotics.

> **Thought Question**
>
> **18.6** Why would *Streptomyces* produce antibiotics targeting other bacteria?

Streptomyces species are obligate aerobes, requiring access to air to complete their multicellular life cycle (see earlier, **Fig. 18.8C**). When a *Streptomyces* spore germinates, it extends **vegetative mycelia**, branched filaments that grow into the substrate. Some of the filaments then grow upward into the air, where they develop into **aerial mycelia**. The aerial mycelia in some species grow in tightly coiled spirals (**Fig. 18.19B** and **C**). As the mycelial colony runs out of nutrients, older cells of the filament age and lyse, releasing nutrients that are absorbed by the younger cells. The nutrients also attract other scavengers, which may be killed by antibiotics produced by the aging streptomycete cells. The dead scavengers, too, release nutrients that feed the growing tip of the mycelium.

As mycelia mature, they fragment into smaller cells called exospores (also called arthrospores). Exospores are vegetative cells, not dormant like *Bacillus* endospores (see **Table 18.3**). The exospores separate and are dispersed by the wind, enabling them to colonize a new location. Streptomycete mycelia can be obtained from natural habitats by burying a glass slide in soil and then waiting several days for spores to germinate, covering the slide with mycelia. They are challenging to isolate in pure culture, however, because their coiled filaments trap cells of other bacteria.

The *S. coelicolor* genome contains over 8 million base pairs—one of the largest prokaryotic genomes. Streptomycete chromosomes are linear with special "telomeres," single-stranded end sequences that double back to form hairpin loops (**Fig. 18.19D**). Much of the lengthy genome of a streptomycete encodes catabolism of a rich array of diverse organic components of decaying plant and animal matter, including even lignin. Other genes encode extensive operons for production of diverse secondary products (see Chapter 15), including antibiotics. More than half of the antibiotics currently used in medicine derive from *Streptomyces* species.

Note: Distinguish *Streptomyces*, filamentous rod-shaped actinomycetes, from nonactinomycete *Streptococcus*, Gram-positive cocci that form short, unbranched chains.

Actinomycetes associated with animals and plants. Many marine actinomycetes associate with sponges, where they produce antibiotics that may help the sponge resist pathogens. Pharmaceutical companies now investigate these sponge symbionts for new drugs. Some *Streptomyces* species maintain a mutualism with leaf-cutter ants. The ants culture the bacteria on special organs to produce antibiotics against parasites of their fungal gardens (discussed in **eTopic 17.5**). A few actinomycetes are animal pathogens; for example, *Actinomyces* species cause actinomycosis, a form of skin abscesses in humans and cattle.

Actinomycetes have many mutually beneficial associations with plants. For example, *Frankia* associates with the alder tree, which develops orange-yellow-colored nodules on its roots to fix nitrogen (**Fig. 18.20**). *Frankia* species live as **endophytes**, endosymbionts of vascular plants. This nitrogen fixation mutualism is analogous to the better-known symbiosis between legumes and rhizobia (discussed along with Alphaproteobacteria, in Section 18.4). Other *Frankia* species benefit wheat by conferring

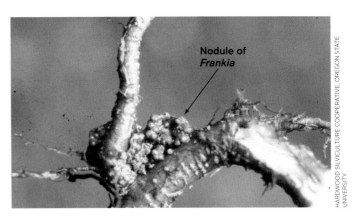

FIGURE 18.20 ■ ***Frankia* species associate endophytically with plants.** Root of alder tree (*Alnus glutinosa*) bears the orange-yellow-colored nodules (arrowhead) containing *Frankia*.

resistance to major fungal pathogens, such as the fungus that causes "take-all" disease, as well as to certain parasites and insects. Still other *Frankia* species grow as saprophytes, consuming dead leaf litter. A few actinomycetes cause plant diseases. For example, potato scab disease is caused by *Streptomyces scabies*.

Nonmycelial Actinobacteria: *Mycobacterium* and *Corynebacterium*. These actinobacteria share the thick acid-fast cell wall of *Streptomyces* but lack the mycelial lifestyle. Two genera that cause dreaded diseases are *Mycobacterium* (**Fig. 18.21**) and *Corynebacterium*. Both genera have thick cell envelopes containing **mycolic acids** and phenolic glycolipids. The mycolic acids of *M. tuberculosis* are extremely diverse and include some of the longest-chain acids known, up to 90 carbons (**Fig. 18.21C**). The mycolic acids are linked to arabinogalactan, a polymer of arabinose and galactose built on the peptidoglycan (discussed in Chapter 3). The mycolyl-arabinogalactan-peptidoglycan complex forms a waxy coat that impedes the entry of nutrients through porins and thus limits growth rate, but it also protects the bacterium from host defenses and antibiotics. For this reason, to cure tuberculosis requires an exceptionally long course of antibiotic therapy.

Mycobacterium includes the species *M. tuberculosis* and *M. leprae*, as well as lesser-known pathogens such as *M. ulcerans*. Cells of *M. tuberculosis* can be detected by the **acid-fast stain** as tiny rods associated with sloughed cells in sputum (**Fig. 18.21A**). In the acid-fast stain, cells are penetrated with a dye that is retained under treatment with acid alcohol (discussed in Chapter 2). The acid-fast property is associated with unusual cell wall lipids, such as mycolic acids. *M. tuberculosis* bacteria are challenging to culture; they form crinkled colonies after 2 weeks of growth on agar-based media (**Fig. 18.21B**).

A.

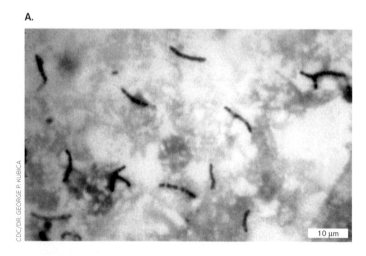

B.

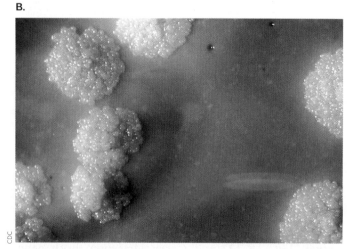

FIGURE 18.21 ■ *Mycobacterium tuberculosis* **causes tuberculosis.** **A.** Acid-fast stain of tissue sample containing *M. tuberculosis* (chains of pink rods). **B.** Crinkled appearance of *M. tuberculosis* colonies. **C.** Mycolic acids coat the cell wall of *M. tuberculosis*.

C.

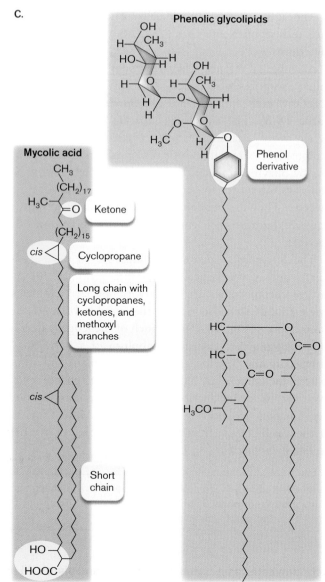

The closely related species *M. leprae* causes the disfiguring disease leprosy (**Fig. 18.22A**). The species has one of the longest known doubling times of any pathogen (about 14 days), and it can take a year to grow enough cells in the laboratory for observation. Growth of *M. leprae* requires lower temperature; for this reason, leprosy attacks the extremities (hands and feet) whose temperature is lower than that of the body core. Culture on artificial media is impossible; the bacteria can be grown only within low-temperature animals, such as armadillos, or within genetically immunodeficient mice.

The genomes have been sequenced for both *M. tuberculosis* and *M. leprae*. *M. tuberculosis* has surprisingly few recognizable pathogenicity genes but a large number of environmental stress components, including 16 environmental sigma factors (discussed in Chapter 8), as well as 250 genes for its complex lipid metabolism. Over half of the *M. leprae* genome consists of pseudogenes, homologs of *M. tuberculosis* genes undergoing reductive evolution (**Fig. 18.22B**). Thus, *M. leprae* appears to be an evolving pathogen "caught in the act" of losing many genes no longer needed in its sheltered host environment. How it lost the need for so many genes preserved in *M. tuberculosis* remains a mystery. *M. leprae* causes disease worldwide, including 250 cases of leprosy annually in the United States. In 2013, a new test was approved that reveals leprosy infection a year before symptoms appear—enabling it to be cured with antibiotics before nerve damage is irreversible.

Mycobacteria also include a much larger number of harmless commensals, such as *M. smegmatis*, isolated from human skin. Species of mycobacteria can be isolated from soil and water, as well as from various animal sources. Their culture is difficult because of their slow growth rates, but isolation can be enhanced by treatment with a base (NaOH or KOH) at concentrations that kill most other bacteria.

Note: Distinguish *Mycobacterium*, rods whose cell walls contain mycolic acids, from *Mycoplasma*, firmicutes lacking cell walls, related to *Bacillus* and *Clostridium*.

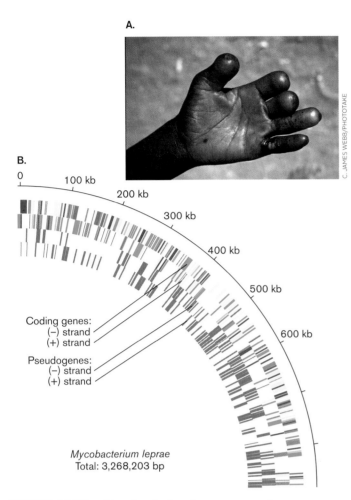

FIGURE 18.22 ■ *Mycobacterium leprae* **causes leprosy.**
A. Hand disfigured by leprosy. **B.** The genome of *M. leprae* shows a high content of decaying pseudogenes (gray bars), most of which correspond to functional genes in the genome of *M. tuberculosis*.

Irregularly shaped actinomycetes. Several nonmycelial actinomycetes show unusual cell shapes. Members of the genus *Corynebacterium* include soil bacteria, as well as pathogens such as *C. diphtheriae*, the cause of the lung disease diphtheria. *Corynebacterium* species grow as irregularly shaped rods, which may divide by a "half-snapping" mechanism in which one side of the cell remains attached like a hinge (**Fig. 18.23A**). Related soil bacteria include the genera *Nocardia* and *Rhodococcus*.

Soil bacteria of the genus *Arthrobacter* exhibit an unusual cell cycle in which coccoid stationary-phase cells sprout into rods, which eventually run out of nutrients and revert to the coccoid form. The growing rods form irregular branched filaments (**Fig. 18.23B**). An *Arthrobacter* species was discovered conducting anaerobic respiration by reduction of hexavalent chromium (Cr^{6+}), a toxic metal pollutant, to a less toxic oxidation state. *Arthrobacter* now shows potential as an agent of bioremediation of hexavalent chromium.

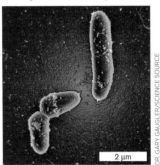

A. *Corynebacterium diphtheriae*

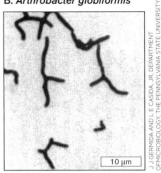

B. *Arthrobacter globiformis*

FIGURE 18.23 ■ **Irregularly shaped actinomycetes:** ***Corynebacterium*** **and** ***Arthrobacter.*** **A.** *Corynebacterium diphtheriae* divides by snapping off one side while remaining attached at the other; the result is a typical V shape or "Chinese letter" arrangement (colorized SEM). **B.** *Arthrobacter globiformis* cultures form coccoid cells in stationary phase. With added nutrients, the coccoid cells grow out as rods.

Micrococcaceae. Relatives of *Arthrobacter* are nonfilamentous cocci such as *Micrococcus. M. luteus* is one of the most widespread of soil bacteria, appearing readily as yellow colonies on agar plates exposed to air. Historically, the genus *Micrococcus* in the family Micrococcaceae was classified with *Staphylococcus*, but its DNA sequence data now place *Micrococcus* and most of the Micrococcaceae within the actinomycetes.

Micrococci are aerobic heterotrophs, commonly isolated from air and dust, although their habitat of choice is human skin. Micrococci grow in square or cuboid formations, dividing in two or three planes (**Fig. 18.24**); cuboid clusters are known as **sarcinae** (singular, **sarcina**). *M. luteus* is harmless to humans and grows well at room temperature, so it makes an excellent laboratory organism for observation by students.

To Summarize

- **Firmicutes**, the low-GC Gram-positive bacteria, include endospore-forming genera such as *Bacillus* and *Clostridium*. The cycle of endospore formation was probably present in the common ancestor of this phylum.

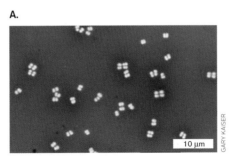

A.

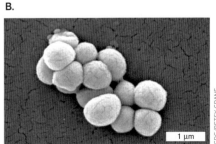

B.

FIGURE 18.24 ■ ***Micrococcus*** **species grow in tetrads, or sarcinae.** **A.** *M. luteus* growing in tetrads (negative or indirect stain with nigrosin). **B.** *M. luteus* bacteria grow in clumps when cultured on a solid substrate (SEM).

- **Nonsporulating firmicutes** include pathogenic rods such as *Listeria*, as well as food-producing bacteria such as *Lactobacillus* and *Lactococcus*.

- *Staphylococcus* and *Streptococcus* are Gram-positive cocci that include normal human flora, as well as serious pathogens causing toxic shock syndrome, pneumonia, and scarlet fever.

- **Mycoplasmas** belong phylogenetically to Firmicutes but lack the cell wall and S-layer. They have flexible cytoskeletons and show ameboid motility. Mycoplasma species cause diseases such as meningitis and pneumonia.

- **Actinobacteria** (order Actinomycetales) include mycelial spore-forming soil bacteria, such as the actinomycete *Streptomyces*. Other actinobacteria have irregularly shaped cells, such as *Corynebacterium* species that cause diphtheria.

- **Mycobacteria** are actinobacterial rods whose cell envelope contains a diverse assemblage of complex mycolic acids. Mycobacterial species cause tuberculosis and leprosy. They stain acid-fast.

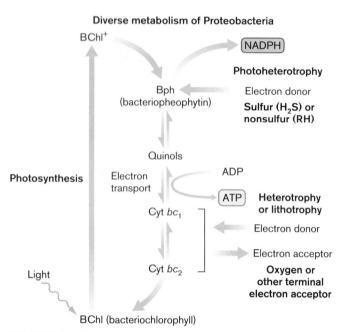

FIGURE 18.25 ■ The protean metabolism of Proteobacteria. In many Gram-negative species, metabolic diversity arises through minor "add-ons" of biochemical modules such as light absorption by bacteriochlorophyll, use of sulfide or organic electron donors, and use of oxygen or alternative (anaerobic) electron acceptors.

18.4 Proteobacteria (Gram-Negative)

Proteobacteria—the "protean," or "many-formed," bacteria—show more diverse metabolism than any other phylum. Their lifestyles range from plant and animal pathogens to soil lithotrophs that reduce nearly any oxidized metal. Nevertheless, all proteobacteria share a common structure: their Gram-negative cell envelope, which consists of an outer membrane, peptidoglycan cell wall permeated by the periplasm, and inner membrane (plasma membrane) (discussed in Chapter 3). The outer membrane is packed with receptor proteins and porins, comprising two-thirds of the mass of the membrane. Porins evolve so as to admit nutrients while excluding antibiotics. The outer membrane lipids contain long sugar polymer extensions (lipopolysaccharide, or LPS). In pathogens, LPS repels phagocytosis and has toxic effects when released by dying cells.

The Protean Metabolism of Proteobacteria

The phylum Proteobacteria is defined to include five monophyletic classes, labeled Alpha through Epsilon (see **Table 18.1**). Even within each closely related group, we see nearly as wide a range of cell shape and metabolic strategies as we see in Proteobacteria as a whole.

A closer look at proteobacterial metabolism shows that processes that at first seem very different actually connect through linked biochemical modules (**Fig. 18.25**) (metabolism is discussed in detail in Chapter 14). Many proteobacteria can oxidize H_2 (hydrogenotrophy) and small organic acids, as well as complex organic molecules with aromatic rings. Some "photoheterolithotrophs" carry out nearly all the fundamental classes of metabolism, depending on environmental conditions such as availability of light, oxygen, and nutrients. An example is *Rhodopseudomonas palustris*, the species of Alphaproteobacteria characterized and sequenced by Caroline Harwood at the University of Washington (see Fig. 14.39 and **eTopic 14.2**). In such a "protean" organism, the core of all proteobacterial energy acquisition is a respiratory chain of electron donors and acceptors. Electrons may enter the chain from a photoexcited chlorophyll, from organic electron donors such as sugars or benzoates, or from a mineral electron donor such as reduced sulfur or iron (lithotrophy). In facultative anaerobes, the electron acceptor may be a mineral such as nitrate or sulfate (anaerobic respiration). The capabilities of a given organism depend on which oxidoreductases its genome encodes.

Photoheterotrophy: light-supplemented heterotrophy. In the Alpha-, Beta-, and Gammaproteobacteria, diverse forms of light absorption have evolved from a common ancestor of photosystems I and II (see **Table 18.2**). The various bacteriochlorophylls of Proteobacteria peak in two ranges—in the blue and in the red or infrared. Bacteriochlorophyll *b*, in *Blastochloris viridis*, peaks well beyond 1,000 nm. Species with different photopigments often grow together in stratified layers of sediment or wetland—the infrared absorbers below, where they capture the longer wavelengths "left over" from the shorter-wavelength absorbers above.

The prevalence of homologous photosystems suggests that the common ancestor of the Gammaproteobacteria was a photoheterotroph. At the same time, other evidence supports horizontal transfer of photosystems among proteobacterial branches. The evidence is particularly strong in the case of **proteorhodopsin**, a homolog of the retinal protein light-driven proton pump first characterized in halophilic archaea (discussed in Chapters 14 and 19). The proteorhodopsin light pump in *Pelagibacter*, a marine relative of *Rickettsia*, absorbs green and yellow (wavelength 500–600 nm), the midrange of solar radiation reaching Earth's surface. Different species have proteorhodopsins with very different absorption ranges, apparently adapted to different niches in the marine ecosystem.

In the literature, proteobacterial phototrophs are historically called "purple bacteria." The actual colors of all the phototrophs range from purple-red through yellow and brown. "Purple sulfur bacteria" are species that photolyze reduced forms of sulfur, such as H_2S, HS^-, S^0, or $S_2O_3^{2-}$ (thiosulfate),

whereas "purple nonsulfur bacteria" photolyze H_2 or use photosystem II for cyclic photophosphorylation (discussed in Chapter 14). However, it turns out that many purple bacteria also use photosystem II. Most proteobacterial phototrophs conduct photosynthesis only in the absence of oxygen, reverting to aerobic heterotrophy when oxygen is present. Some, however, are obligate anaerobes. Still others, found in marine environments, surprisingly photolyze sulfide or organic compounds only in the presence of oxygen. Making sense of this extraordinary diversity is the job of marine and aquatic microbiologists, particularly those calculating global cycles of carbon and oxygen (discussed in Chapters 21 and 22).

> **Thought Question**
>
> **18.7** Why might genes for the proteorhodopsin light-powered proton pump be more likely to transfer horizontally than the bacteriochlorophyll-based photosystems PS I and PS II?

Lithotrophy: inorganic electron donors. Many different Proteobacteria oxidize inorganic electron donors (**Table 18.4**). The ability to oxidize or reduce minerals

---

TABLE 18.4 Lithotrophy and Methylotrophy (Examples)

Class/phylum	Example species	Reaction	$\Delta G°'/2e^-$ (kJ)	Class of reaction
Alphaproteobacteria	*Nitrobacter winogradskyi*	$NO_2^- + \frac{1}{2}O_2 \rightarrow NO_3^-$	−54	Nitrite oxidation (nitrification to nitrate)
	Paracoccus pantotrophus	$H_2S + 2O_2 \rightarrow 2H^+ + SO_4^{2-}$		Sulfide oxidation to sulfate
	Roseobacter litoralis	$CO + H_2O \rightarrow CO_2 + 2H^+ + 2e^-$		CO oxidation
	Hyphomicrobium	$CH_3OH + O_2 \rightarrow CO_2 + H_2O + 2H^+ + 2e^-$		Methylotrophy
Betaproteobacteria	*Nitrosomonas europaea*	$NH_3 + O_2 \rightarrow HNO_2^- + 3H^+ + 2e^-$		Ammonia oxidation (nitrification to nitrite)
	Ralstonia eutropha	$2H_2 + \frac{1}{2}O_2 \rightarrow 2H_2O$	−237	Hydrogen oxidation
	Thiobacillus denitrificans	$3S^0 + 4NO_3^- \rightarrow 3SO_4^{2-} + 2N_2$		Anaerobic sulfur oxidation
Gammaproteobacteria	*Acidithiobacillus ferrooxidans*	$4FeS_2 + 15O_2 + 14H_2O \rightarrow$ $4Fe(OH)_3 + 16H^+ + 8SO_4^{2-}$	−164	Iron-sulfur oxidation
	Acidithiobacillus thiooxidans	$2S^0 + 3O_2 + 2H_2O \rightarrow 2SO_4^{2-} + 4H^+$	−196	Sulfur oxidation
	Methylococcus capsulatus	$CH_4 + 2O_2 \rightarrow HCO_3^- + H^+ + H_2O$	−203	Methanotrophy
	Beggiatoa alba	$2H_2S + O_2 \rightarrow 2S^0 + 2H_2O$	−210	Sulfide oxidation
	Chromatium	$4Fe^{2+} + CO_2 + 11H_2O + h\nu \rightarrow$ $4Fe(OH)_3 + [CH_2O] + 8H^+$		Iron phototrophy (photoferrotrophy)
Deltaproteobacteria	*Desulfovibrio*	$4S^0 + 4H_2O \rightarrow SO_4^{2-} + 3HS^- + 5H^+$	−11.3 (at pH 8)	Sulfur oxidation
Nitrospirae	*Nitrospira*	$NO_2^- + \frac{1}{2}O_2 \rightarrow NO_3^-$	−54	Nitrite oxidation (nitrification to nitrate)
Planctomycetes	*Brocadia anammoxidans*	$NH_4^+ + NO_2^- \rightarrow N_2 + 2H_2O$	−238	Anaerobic ammonium oxidation (anammox)

evolved multiple times in different clades by modification of the electron transport chain. Many lithotrophic reactions generate acid, which conveniently breaks down rock containing additional reduced minerals. Bacterial and archaeal lithotrophy has tremendous significance for global cycling of elements in the biosphere (discussed in Chapter 22). Some lithotrophs are autotrophs, fixing carbon dioxide into biomolecules by the Calvin cycle (in carboxysomes) or the reverse TCA cycle. Others have alternative pathways of heterotrophy when organic foods are available. An interesting kind of lithotrophy performed by many Proteobacteria is carbon monoxide (CO) oxidation (**eTopic 18.2**). Oxidizing CO is important because this toxic substance is a byproduct of animal and plant metabolism.

We now survey species of the major Greek-lettered classes of Proteobacteria.

Alphaproteobacteria: Photoheterotrophs, Methylotrophs, and Endosymbionts

The class Alphaproteobacteria includes photoheterotrophs and heterotrophs, as well as bacteria metabolizing single-carbon compounds—along with intracellular mutualists and pathogens.

Photoheterotrophs. Most alphaproteobacterial photoheterotrophs are unicellular. Their cell shapes range from flagellated spirilla to rounded rods (*Rhodobacter sphaeroides*) (**Fig. 18.26A**), wide spirals (*Rhodospirillum rubrum*), and stalked cells (*Rhodomicrobium*) (**Fig. 18.26B**). The TEM of *R. sphaeroides* in **Figure 18.26A** shows that the cell is packed with photomembranes (membranes containing the photosynthetic complex) arranged in folds or tubes invaginated from the cytoplasmic membrane. These intracellular photomembranes expand the surface area for photon capture, but during growth with oxygen, the photomembranes

disappear. The outer surface of the cell shows LPS filaments extending from the outer membrane.

The metabolism of *Rhodospirillum rubrum* shifts drastically with oxygen. Anaerobically, *R. rubrum* bacteria grow dark red because they synthesize membrane containing BChl *a*, which absorbs green and infrared and reflects primarily red. With oxygen, however, *R. rubrum* fails to synthesize BChl *a*, and the cells grow white as their metabolism switches to straight heterotrophy. Heterotrophy is supplemented by oxidizing small molecules such as carbon monoxide.

Alphaproteobacterial photoheterotrophs that require oxygen (but do <u>not</u> photolyze water) were found unexpectedly in genomic surveys of marine bacteria. Originally thought to be straight heterotrophs, these organisms revealed genes encoding bacteriochlorophyll. Their aerobic heterotrophy was shown to be driven by light absorption. The reason for the oxygen requirement is unclear, since their photosystems do not involve oxygen. Examples of O_2-requiring phototrophs have since been found in most of the proteobacterial clades, and species have been isolated from all major habitats, including freshwater, marine, and soil ecosystems. Some show unusual cell morphology, such as the Y shape of *Citromicrobium* (**Fig. 18.26C**).

The aerobic photoheterotroph *Erythromicrobium ramosum* can reduce toxic metal compounds such as tellurite ion (TeO_3^{2-}). The cells generate tellurium crystals that take up 30% of their cell weight—a trait that we might use to remove tellurite from liquid waste. Other toxic metals reduced by aerobic photoheterotrophs include selenium and arsenic. Thus, these obscure bacteria now have a promising future in remediation of metal-contaminated industrial wastes.

Aquatic and soil oligotrophs. Alphaproteobacteria also include many nonphototrophic heterotrophs of soil and water. Many aquatic and soil heterotrophs are oligotrophs adapted to extremely low nutrient concentrations; an example is *Caulobacter crescentus*, the stalk-to-flagellum organism discussed in Chapters 3 and 4. Oligotrophic bacteria

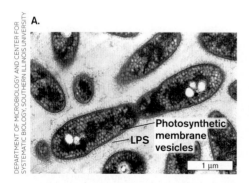
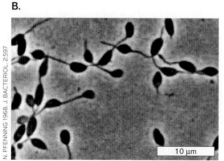
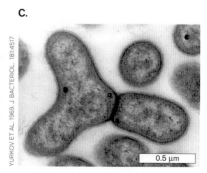

FIGURE 18.26 ■ Alphaproteobacterial photoheterotrophs. A. *Rhodobacter sphaeroides*, showing intracellular photosynthetic membranes and outer membrane LPS filaments (TEM). **B.** *Rhodomicrobium vannielii* with stalked cells (phase contrast). **C.** *Citromicrobium* species, an aerobic photoheterotroph, forms highly pleomorphic shapes, including this Y shape (TEM).

often have unusual extended shapes enhancing nutrient uptake, such as the starlike cell aggregates of *Seliberia stellata*. In *Seliberia* species, the individual tightly coiled rods generate oval or spherical reproductive cells by a budding process. The budding reproductive cells germinate into rods, which then form new aggregates.

Some heterotrophs common in soil are pathogens. *Brucella* species are intracellular pathogens of animals that can also infect humans. The soil pathogen *Granulibacter bethesdensis* is associated with chronic granulomatous disease, an inherited disorder of the phagocyte oxidase system that leaves patients susceptible to infection. Other pathogens are carried by insects or animal hosts; for example, *Bartonella henselae* causes cat scratch disease.

Methylotrophy and methanotrophy. **Methylotrophy** is the ability of an organism to oxidize reduced single-carbon compounds such as methanol, methylamine, or methane. The Alphaproteobacteria include several genera of methylotrophs, which are found in all environments, including soil, freshwater, and the ocean. Most methylotrophs can grow on both single-carbon and organic compounds. One such versatile genus, *Methylobacterium*, is equally at home in soil and water, on plant surfaces, and as a contaminant of facial creams and purified water for silicon chip manufacture. Other species are restricted to single-carbon compounds, incapable of metabolizing organic compounds with carbon-carbon bonds. Methylotrophs that grow solely on methane (CH_4) are called **methanotrophs**.

Methane-oxidizing bacteria of both Alpha and Gamma classes contribute to aquatic ecosystems, serving as major food sources for zooplankton. They eliminate much of the methane produced by methanogens before it reaches the atmosphere, where it has a potent greenhouse effect (see Chapter 22).

An interesting methylotroph is *Hyphomicrobium*, a bacterium with an unusual stalk-to-flagellum transition similar to that of the alphaproteobacterium *Caulobacter crescentus*, whose life cycle is described in Chapter 3. *Hyphomicrobium* species are found in environments as diverse as wastewater sludge and Antarctic island soil. For example, the Antarctic soil species *H. sulfonivorans* metabolizes sulfur compounds such as dimethyl sulfone (**Fig. 18.27**).

Like *Caulobacter*, a flagellated cell of *Hyphomicrobium* species may lose its flagellum to form a stalk. But unlike *Caulobacter*, *Hyphomicrobium* then forms a daughter cell from the opposite end of the stalk! As the DNA replicates, one daughter nucleoid must migrate all the way through the stalk to reach the daughter cell. The daughter cell forms a flagellum and septates, separating from the parent (**Fig. 18.27B**).

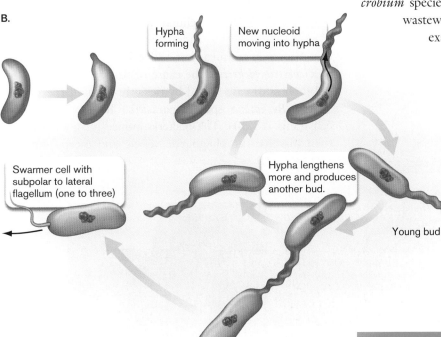

S. AZRA MOOSVI ET AL. 2005. SYSTEMATIC AND APPLIED MICROBIOLOGY 28:541.

1 μm

A.

B.

Hypha forming

New nucleoid moving into hypha

Swarmer cell with subpolar to lateral flagellum (one to three)

Hypha lengthens more and produces another bud.

Young bud

FIGURE 18.27 ■ *Hyphomicrobium sulfonivorans* **from Signy Island, Antarctica. A.** These bacteria with coiled stalks catabolize dimethyl sulfone. **B.** Life cycle of *Hyphomicrobium* species.

Thought Question

18.8 Can you hypothesize a mechanism for migration of the daughter nucleoid of *Hyphomicrobium* through the stalk to the daughter cell? For possibilities, see Chapter 3 and consider the various molecular mechanisms of cell division and shape formation.

Endosymbionts: mutualists and parasites. The Alphaproteobacteria include many highly evolved intracellular symbionts—some mutualists, others parasites. As isolated bacteria, they are generally rod-shaped with aerobic metabolism, but their shape is transformed within the host cell. Intracellularly, they need to solve special problems, such as the exclusion of oxygen from nitrogen fixation (a process requiring anaerobiosis) or, in the case of pathogens, the need to resist host defenses.

Nitrogen-fixing endosymbionts of plants include genera such as *Rhizobium*, *Bradyrhizobium*, and *Sinorhizobium*. The nomenclature has undergone many changes, but the species are generally referred to as rhizobia. Though they can live freely in the soil, rhizobia prefer to colonize plants, usually legumes such as peas or alfalfa, where they form distinctive nodule structures. Each bacterial species colonizes and infects a particular host range. Complex chemosensory processes attract the bacteria to the surface of the host root, where their presence induces root hairs to form curls around the bacteria (**Fig. 18.28**). The curling of the root hairs enables the bacteria to form infection threads, chains of rod-shaped bacteria that invade the root cells. Within the host cells, the bacteria lose their cell wall and become rounded **bacteroids**, specialized for nitrogen fixation. The host plant cells provide the bacteroids with nutrients, as well as protective components such as **leghemoglobin**, an oxygen-binding protein that maintains anaerobiosis within infected cells. Leghemoglobin turns the nodule interior pink (**Fig. 18.28A**).

The genus *Agrobacterium* includes plant pathogens closely related to the rhizobia. *A. tumefaciens* is known for its ability to convert host cells to a form that produces tumors called galls (**Fig. 18.29**). The ability of *A. tumefaciens* to insert its DNA into plant genomes has made it a major tool for plant biotechnology (see Chapter 16).

Rickettsias are famous intracellular pathogens. Short coccoid rods, rickettsias lack flagella and can grow only within a host cell. The best-known rickettsia is *Rickettsia rickettsii*, the cause of Rocky Mountain spotted fever, a disease spread by ticks throughout the United States. *Rickettsia* species parasitize human endothelial cells (**Fig. 18.30**). The bacteria induce phagocytosis and then dissolve the phagocytic vesicle and escape into the

A.

B.

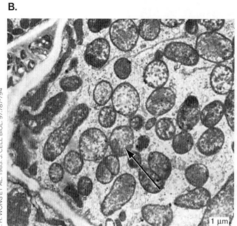

C. *Rhizobium*-legume symbiosis early steps

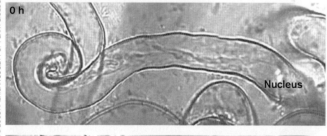

0 h

Nucleus

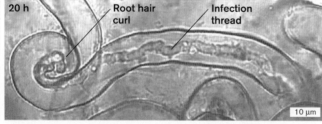

20 h

Root hair curl

Infection thread

10 μm

FIGURE 18.28 ■ Rhizobia: legume endosymbionts.
A. Legume nodules cut open to show pink regions where the plant cells produce leghemoglobin to maintain anaerobic conditions for bacteroid nitrogen fixation. **B.** Within the host cells, *Sinorhizobium meliloti* cells form bacteroids (arrow) that fix nitrogen while receiving nutrients from the plant. **C.** Clover root hair curls around infecting *S. meliloti*. The bacteria enter the curl and grow down the root hair as an infection thread that penetrates the legume cells, enabling the bacteria to colonize in the form of bacteroids.

FIGURE 18.29 ■ Plant gall induced by *Agrobacterium tumefaciens*. Crown gall on chrysanthemum plant.

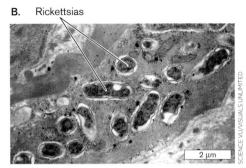
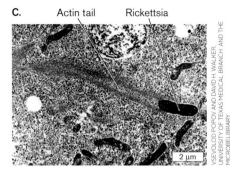

A.

B. Rickettsias

C. Actin tail   Rickettsia

2 µm

2 µm

FIGURE 18.30 ■ **Rickettsias are obligate intracellular parasites. A.** The Rocky Mountain tick carries *Rickettsia rickettsii*, the cause of Rocky Mountain spotted fever. **B.** *Rickettsia* species parasitize human endothelial cells (TEM). **C.** The rickettsias propel themselves through the host cell by polymerizing cytoplasmic actin behind them (TEM).

cytoplasm. Some rickettsias propel themselves through the host cell by polymerizing cytoplasmic actin behind them—a process similar to that of *Listeria* (described earlier, along with Firmicutes). The actin tails eventually project outward as filopodia, extensions of host cytoplasm and membrane that protect the bacteria from host defenses while enabling them to invade adjacent cells.

Genetic analysis shows that the rickettsia clade includes mitochondria. All eukaryotic mitochondria appear to be descendants of an ancient rickettsial parasite whose respiratory apparatus ultimately became essential to power eukaryotic cells.

Betaproteobacteria: Photoheterotrophs, Lithotrophs, and Pathogens

The Betaproteobacteria include photoheterotrophs such as *Rhodocyclus*, as well as a diverse range of lithotrophs (see **Table 18.4**). Several are important pathogens that infect humans and other animals.

Lithotrophs: nitrifiers and sulfur oxidizers. An important group of nitrogen lithotrophs consists of the **nitrifiers**. Nitrifiers oxidize ammonia (NH_3) to nitrite (NO_2^-), or nitrite to nitrate (NO_3^-). Typically, different species conduct the two reactions separately while coexisting in soil and water. Nitrifiers are of enormous economic and practical importance for wastewater treatment because they decrease the reduced nitrogen content of sewage. Special systems have been developed to retain nitrifier bacteria behind filters as one stage of water treatment. In the system shown in **Figure 18.31A**, nitrifying bacteria are encapsulated in pellets to retain them within the bioreactor while the treated water flows through a filter.

The most commonly isolated ammonia oxidizers are *Nitrosomonas* species of the Beta class. *Nitrosomonas* cells conduct electron transport through extensive internal membranes

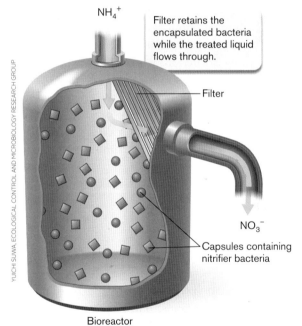

A. Encapsulated nitrifiers

NH_4^+

Filter retains the encapsulated bacteria while the treated liquid flows through.

Filter

NO_3^-

Capsules containing nitrifier bacteria

Bioreactor

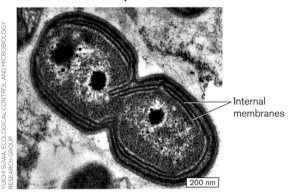

B. *Nitrosomonas europaea*

Internal membranes

200 nm

FIGURE 18.31 ■ **Nitrifiers. A.** Wastewater treatment uses nitrifier bacteria to remove ammonia. **B.** *Nitrosomonas europaea*, of the Betaproteobacteria, oxidizes ammonia to nitrite (TEM). Internal membranes contain the electron transport complexes.

that are either stacked or invaginated (**Fig. 18.31B**). Related genera of ammonia oxidizers include *Nitrosolobus* and *Nitrosovibrio*, some with peritrichous or polar flagella. One ammonia-oxidizing genus of the Gamma group has been found: *Nitrosococcus.* Major nitrite oxidizers include the nonproteobacterial phylum Nitrospirae. Although genetically outside the Proteobacteria, the unusual spiral-shaped cells of Nitrospirae have a Gram-negative cell envelope.

Pathogens. The Betaproteobacteria include aerobic heterotrophic cocci such as *Neisseria.* The cocci of *Neisseria* species form distinctive pairs known as **diplococci**. Most *Neisseria* species are harmless commensals of the nasal or oral mucosa, but *N. gonorrhoeae* causes the sexually transmitted disease gonorrhea. *N. gonorrhoeae* is actually a microaerophile, requiring a narrow range of oxygen concentration; it has fastidious growth requirements, necessitating cultivation on a special blood-based medium. A related organism, *N. meningitidis*, may be carried asymptomatically by as much as a quarter of the human population, but occasionally it causes meningitis, which can be fatal. Other neisserias, such as *N. sicca*, are easily isolated from skin and often presented to undergraduates as an unknown for identification.

Other members of the Beta class are important animal and plant pathogens, such as *Burkholderia. B. cepacia* was originally isolated from onions as a cause of bulb rot; it is now known to be a major opportunistic invader of the lungs of cystic fibrosis patients.

Gammaproteobacteria: Photolithotrophs, Enteric Flora, and Pathogens

The Gammaproteobacteria include vast numbers of marine organisms that catabolize complex organic pollutants such as petroleum hydrocarbons. However, the best-known Gammaproteobacteria are the Enterobacteriaceae family of facultative anaerobes found in the human colon—the family that includes *Escherichia coli.* Gammaproteobacteria also include unusual phototrophs that oxidize iron and nitrite.

Sulfur lithotrophs. The sulfur-oxidizing genus *Beggiatoa* was one of the first kinds of lithotrophs described by pioneer microbial ecologist Sergei Winogradsky (discussed in Chapter 1). *Beggiatoa* species oxidize H_2S to elemental sulfur, which collects as sulfur granules within the periplasm. *Beggiatoa* also stores carbon in cytoplasmic granules of polyhydroxybutyrate—a common strategy among proteobacteria. The cells of *Beggiatoa* grow as extended filaments with visible sulfur granules, forming biofilms on sulfide-rich sediment. Another sulfur oxidizer, the genus *Thioploca*, forms thick biomats in aquatic sediment. The bacteria grow as long filaments that glide through hollow tubes full

20 µm

FIGURE 18.32 ▪ *Thioploca* **filaments emerging from a sheath.** *Thioploca* species oxidize sulfur in marine sediment, using oxygen or nitrate; this sample is from the northeastern Arabian Sea off Pakistan.

of mucous secretions. **Figure 18.32** shows filaments of *Thioploca* species emerging from a sheath.

Bacteria of the genus *Acidithiobacillus* oxidize iron or sulfur (**Fig. 18.33A**). Iron-oxidizing bacteria commonly form the brown stains found inside plumbing. Most species are short rods or vibrios (comma-shaped). *Acidithiobacillus* and other sulfur-oxidizing genera can undergo a number of different reactions oxidizing H_2S to S^0 and S^0 to SO_4^{2-} (see **Table 18.4**). Sulfate production makes an environment acidic enough to erode stone monuments and the interior surface of concrete sewer pipes. Sulfur oxidation is often coupled to oxidation of iron, $Fe^{2+} \rightarrow Fe^{3+}$. The bacterium *A. ferrooxidans* is known for its role in acidification of mine water, which contributes to leaching of iron, copper, and other minerals (**Fig. 18.33B**).

Sulfur and iron phototrophs. The Gamma group phototrophs, such as *Chromatium* species (**Fig. 18.34A**), mainly utilize sulfide and produce sulfur, which is deposited as intracellular granules visible within the cytoplasm. Their phototrophy is entirely anaerobic. Some of the Gamma group are true autotrophs and do not use organic substrates. Some of these actually conduct phototrophy using iron (Fe^{2+}) to donate electrons. Iron phototrophy, or "photoferrotrophy" (**Fig. 18.34B**), is considered an intriguing possibility for the metabolism of early life (discussed in Chapter 17). *Thiocapsa* uses NO_2^- and was the first nitrogen-based phototroph discovered.

Enterobacteriaceae: intestinal fermenters and respirers. The family Enterobacteriaceae, facultative anaerobes of the

all bacterial species—is the model organism *Escherichia coli*. Some strains of *E. coli* grow normally in the human intestine, feeding on our mucous secretions and producing vitamins, such as vitamin K. They may grow in symbiosis with anaerobic fermenters such as *Bacteroides* species, which release short-chain sugars that *E. coli* digests (discussed in Chapter 21). But other strains, such as *E. coli* O157:H7, cause serious illness. A large proportion of the world's children die of *E. coli*–related intestinal illness before the age of 5. Related pathogens include *Salmonella enterica* and *Shigella flexneri*.

The Enterobacteriaceae are Gram-negative rods, although as nutrients diminish, their size dwindles almost to a coccoid form. They grow singly, in chains, or in biofilms. Many species are motile, with numerous flagella. Most strains grow well with or without oxygen, by either respiration (aerobic or anaerobic) or fermentation. They ferment rapidly on carbohydrates, generating fermentation acids, ethanol, and gases (CO_2 plus H_2) in varying proportions, depending on the species. Their presence in the intestine supports the growth of organisms utilizing these gases, including methanogens (discussed in Chapter 19). Many strains form biofilms. Biofilm formation explains the persistence of drug-resistant infections, such as those associated with urinary catheters in long-term hospital patients.

FIGURE 18.33 ■ *Acidithiobacillus:* **iron oxidizers. A.** *A. ferrooxidans* leaches copper and iron from molybdenite ore (SEM). The hexagonal object is a molybdenite crystal. **B.** In a copper mine, *Acidithiobacillus* species can oxidize copper ores, leaching the copper into solution for retrieval.

Thought Question

18.9 Why do you think it took many years of study to realize that *Escherichia coli* and other Proteobacteria can grow as a biofilm?

Gamma subdivision, include some of the most intensively studied species of all bacteria. Species are readily isolated from the contents of the human digestive tract and easily grown on laboratory media based on human food. The best-known species of Enterobacteriaceae—indeed, the most studied of

Because they are easy to cultivate and have been studied extensively in clinical laboratories, many genera of Enterobacteriaceae are familiar to students in introductory microbiology laboratory courses. A common laboratory exercise is to distinguish *Enterobacter* and *Klebsiella* species from *E. coli* by their fermentation to the pH-neutral product butanediol, which tests positive in the Voges-Proskauer test. (Fermentation is discussed in Chapter 13.) *Enterobacter* species occur more frequently in freshwater streams than in the human body, although a few species colonize the intestine and cause illness.

Proteus mirabilis and *P. vulgaris* cause bladder and kidney infections, particularly as a complication of surgical catheterization. *Proteus* species are heavily flagellated and display a remarkable **swarming** behavior

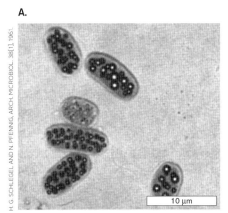

FIGURE 18.34 ■ **Chromatium: sulfur and iron phototrophs. A.** *Chromatium* forms single-flagellated rods full of sulfur granules. **B.** Photoferrotrophy. Under illumination, color develops over time (1–5) as a *Chromatium* isolate oxidizes Fe^{2+} to Fe^{3+}.

A.

ROBERT BELLAS/BELLA LAB

2 μm

FIGURE 18.35 ■ **The enteric rod *Proteus mirabilis*: isolated swimmer or cooperative swarmer.**
A. A thickly flagellated swarmer cell (TEM). Cell length can reach 20 μm.
B. Swarmer rafts of *P. mirabilis* migrate through blood agar.

B.

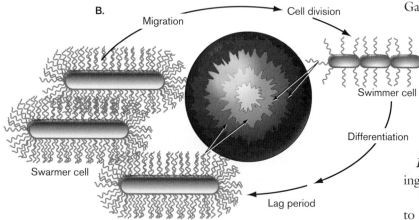

Migration • Cell division
Swimmer cell
Differentiation
Swarmer cell • Lag period

(**Fig. 18.35**). In response to an environmental signal, the flagellated rods grow into long-chain swarmer cells. The swarmers gather together, forming "rafts" that swim together and grow into a complex biofilm.

Besides symbionts of animals, Enterobacteriaceae include plant pathogens, such as *Erwinia carotovora* and related species. *Erwinia* species cause wilts, galls, and necrosis of a wide variety of plants, including bananas, tomatoes, and orchids.

Related facultative rods in soil and water conduct anaerobic respiration by donating electrons from organic substrates to a variety of metals, such as iron and magnesium. *Shewanella oneidensis* and other metal reducers are used to make electricity in fuel cells (discussed in Chapter 14).

Aerobic rods. Closely related to Enterobacteriaceae are several genera of rod-shaped bacteria that are obligate respirers. Many are obligate aerobic respirers (requiring O_2 for growth), although some can use alternative electron acceptors such as nitrate. These genera metabolize an extraordinary range of natural compounds, including aromatic derivatives of lignin; thus, they have important roles in natural recycling and soil turnover.

The Pseudomonadaceae are a large and amorphous group. As the "pseudo-" prefix suggests, their taxonomic unit is poorly defined and includes species whose DNA sequence has necessitated their reassignment to other groups. The Gamma class pseudomonads, such as *Pseudomonas aeruginosa* and *P. fluorescens*, respire on oxygen or nitrate and are vigorous swimmers with single or multiple polar flagella. *P. aeruginosa* can swim throughout a standard agar plate (much to the chagrin of students attempting to isolate colonies). Nevertheless, under appropriate environmental conditions, pseudomonad cells give up their motility and develop biofilms (**Fig. 18.36**) (discussed in Chapter 4). Biofilms of *P. aeruginosa* cause lethal infections of the pulmonary lining of cystic fibrosis patients.

Some pseudomonad pathogens have the unusual ability to infect both plants and animals. For example, *P. aeruginosa* commonly infects plants as well as humans. *P. fluorescens* infects seedlings and causes rotting of citrus fruit, while it also appears as an opportunistic pathogen of immunocompromised cancer patients. Other pseudomonad species, however, are harmless residents of soil or sewage.

Legionella pneumophila is a well-publicized pathogen related to the pseudomonads. Incapable of growth on sugars, *L. pneumophila* requires oxygen to respire on amino acids. The organism exhibits an unusual dual lifestyle, alternating between intracellular growth within human macrophages and intracellular growth within freshwater amebas (**Fig. 18.37**). Growth within amebas facilitates transmission through

A.

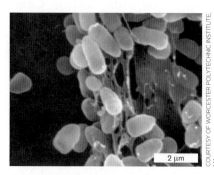

COURTESY OF WORCESTER POLYTECHNIC INSTITUTE, MA

2 μm

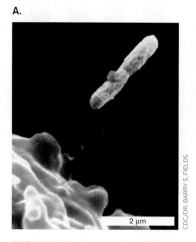

CDC/DR. BARRY S. FIELDS

2 μm

B.

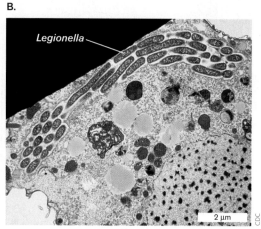

Legionella

CDC

2 μm

FIGURE 18.36 ■ ***Pseudomonas* species form biofilms.** *P. fluorescens* biofilm on a plant surface.

FIGURE 18.37 ■ ***Legionella pneumophila* colonizes an ameba.** **A.** *L. pneumophila* cell caught by an ameba's pseudopod (colorized SEM). **B.** *L. pneumophila* cells have colonized the ameba (TEM).

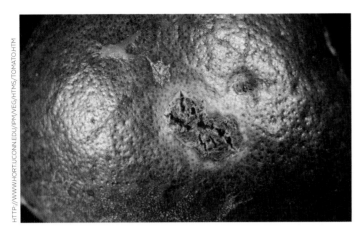

HTTP://WWW.HORT.UCONN.EDU/IPM/VEG/HTMS/TOMATO.HTM

FIGURE 18.38 ■ **Bacterial spot disease of orange fruit, caused by *Xanthomonas* species.**

aerosols into the human lung. *L. pneumophila* is an environmental pathogen that takes advantage of our lifestyle (the prevalence of large-scale air-conditioning units that contain unfiltered water).

A related bacterium, *Coxiella burnetii*, causes Q fever, a respiratory illness of livestock and humans. *C. burnetii* converts to a spore-like form that persists in soil. *C. burnetii*

resembles a rickettsia in some of its strategy of pathogenesis, but it is genetically closer to *Legionella*.

Plant pathogens. Gammaproteobacteria include important plant pathogens, such as *Xanthomonas* species. *Xanthomonas* is a flagellate rod that colonizes a wide range of agricultural plants, such as tomatoes, potatoes, onions, broccoli, and citrus fruits (**Fig. 18.38**). The disease may mottle the leaves and fruit, and it may spread throughout the plant.

Deltaproteobacteria: Lithotrophs and Multicellular Communities

The Deltaproteobacteria include important sulfur and iron reducers, such as the fuel-cell bacterium *Geobacter metallireducens* (discussed in Chapter 14). Other species have complex life cycles that include multicellular developmental forms. The best-studied example is the myxobacteria. Myxobacteria have exceptionally large genomes, such as that of *Sorangium cellulosum* (12.6 Mb).

Myxobacteria. The myxobacteria, such as *Myxococcus xanthus*, are free-living soil bacteria that can grow as isolated cells but come together to form a multicellular structure for the purpose of spore dispersal (**Fig. 18.39**; also see **Fig. 18.8D**). When nutrients are plentiful, the myxobacteria grow and divide as individual cells. As nutrients run out, the cells begin to attract each other, moving into parallel formations. Myxobacteria have no flagella, but they move along a surface by using a form of motility called gliding.

The formations develop into a bulging mass, and the aggregating cells coalesce to develop a **fruiting body**. The fruiting body may be branched with bulbous ends, as in *Stigmatella* species (**Fig. 18.40**). Analogous fruiting bodies

A.

WOLGEMUTH ET AL. 2002. CURR. MICROBIOL. 12:369

10 µm

B.

KUNER, ET AL. 1982. BACTERIOLOGY 151:458.

50 µm

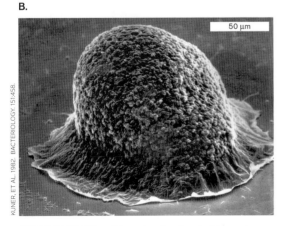

FIGURE 18.39 ■ **Myxobacteria. A.** *Myxococcus xanthus* cells glide while producing trails of slime (dark-field LM). **B.** Upon starvation, cells come together to generate a fruiting body packed with spherical myxospores (SEM).

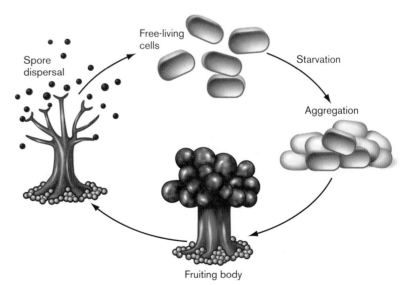

Free-living cells

Spore dispersal

Starvation

Aggregation

Fruiting body

FIGURE 18.40 ■ **Life cycle of a myxobacterium (*Stigmatella*).** Starving *Stigmatella* cells glide toward each other to aggregate. The aggregation generates a fruiting body with bulges packed with small, spherical myxospores.

also form in slime molds, a class of eukaryotic microbe (discussed in Chapter 20). The myxobacteria, however, have evolved their process independently. Within the myxobacterial fruiting body, durable spherical cells called **myxospores** develop (**Fig. 18.40**). Ultimately, the fruiting body releases the myxospores to be carried on the wind or by insects to a more favorable location.

Thought Question

18.10 Compare and contrast the formation of cyanobacterial akinetes, firmicute endospores, actinomycete arthrospores, and myxococcal myxospores.

Note: Distinguish my<u>x</u>obacteria from *My<u>c</u>obacterium*, the Gram-positive genus that includes the causative agents of tuberculosis and leprosy.

Bdellovibrios parasitize bacteria. Bacteria can be parasitized or preyed on by smaller bacteria. The Delta class includes *Bdellovibrio* species, which attack proteobacterial host cells. The structure of the "attack cell" is a small, comma-shaped rod with a single flagellum. The attack cell attaches to the envelope of its host and then penetrates into the periplasm, where it uses host resources to grow (**Fig. 18.41**). The growing cell produces enzymes that cross the inner membrane to degrade host macromolecules and make their components available to the bdellovibrio. The entire host cell loses its shape and becomes a protective incubator for the predator. This stage is called the bdelloplast.

Within the periplasm, the invading bdellovibrio elongates as a spiral filament while replicating several copies of its DNA. When most nutrients have been exhausted, the filament septates into multiple short cells. The cells develop flagella, and the bdelloplast bursts, releasing the newly formed attack cells (**Fig. 18.41B**).

Bdellovibrios can be isolated from sewage, soil, or marine water—any environmental source of Gram-negative prey bacteria. Different species infect *E. coli* and *Pseudomonas*, as well as *Agrobacterium* and *Rhizobium* in their free-living state. They can be cultured as plaques on a top-agar plate containing host bacteria—the same procedure used to isolate bacteriophages (discussed in Chapter 6).

Epsilonproteobacteria: Microaerophilic Helical Pathogens

Epsilonproteobacteria include the genera *Campylobacter* and *Helicobacter*. *Helicobacter pylori* is well known today as the causative agent of gastritis and stomach ulcers (**Fig. 18.42**). Yet, until recently, most microbiologists believed that bacteria could not live in the acidic stomach. The discovery of *H. pylori* as the cause of gastritis earned Barry Marshall and J. Robin Warren, of the University of Western Australia,

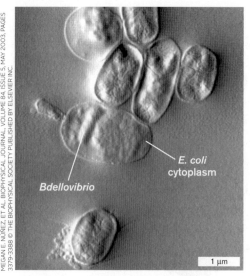

A. *Bdellovibrio bacteriovorus* attacking *Escherichia coli*

E. coli cytoplasm

Bdellovibrio

1 μm

MEGAN E. NÚÑEZ, ET AL. BIOPHYSICAL JOURNAL, VOLUME 84, ISSUE 5, MAY 2003, PAGES 3379-3388 © THE BIOPHYSICAL SOCIETY PUBLISHED BY ELSEVIER INC.

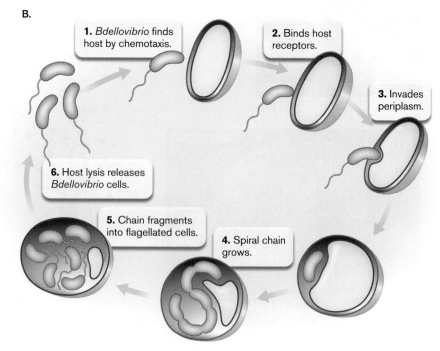

B.

1. *Bdellovibrio* finds host by chemotaxis.

2. Binds host receptors.

3. Invades periplasm.

4. Spiral chain grows.

5. Chain fragments into flagellated cells.

6. Host lysis releases *Bdellovibrio* cells.

FIGURE 18.41 ■ Parasites of bacteria: Bdellovibrio. A. *Escherichia coli* under attack by *B. bacteriovorus* (note the *Bdellovibrio* cell within the *E. coli* periplasm) (AFM). **B.** Life cycle of a bdellovibrio.
Source: Part A from Megan E. Núñez et al. 2003. *Biophys. J.* **84**:3379.

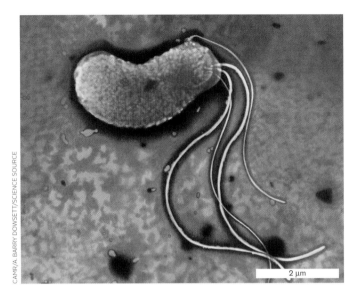

FIGURE 18.42 ■ *Helicobacter,* **a neutralophile growing within the acidic stomach.** *H. pylori* is a short spirillum with unusual knobbed flagella projecting from one end.

the 2005 Nobel Prize in Physiology or Medicine. Marshall famously swallowed a *Helicobacter* culture to prove these bacteria cause gastritis.

H. pylori and related species grow primarily on the stomach epithelium, at about pH 6, which is less acidic than the gastric contents (pH 2–4). The bacteria bury themselves in the epithelial layer and neutralize their acidic surroundings by making urease enzyme, which converts urea to ammonia and carbon dioxide. *Helicobacter* species form wide spiral cells (spirilla) with an unusual grouping of flagella at one end. The metabolism of *H. pylori* is microaerophilic, requiring a low level of oxygen. The bacteria can be isolated through biopsy of the gastric mucosa.

Several groups of related Epsilonproteobacteria are sulfur oxidizers and sulfur reducers, found in marine and aquatic habitats. *Thiovulum* species oxidize sulfides aerobically, using oxygen available in marine water. In deep-ocean sediment, near hydrothermal vents, hydrogenotrophs *Nautilia* and *Hydrogenimonas* oxidize H_2 using sulfur or nitrate. These bacteria enrich the marine habitat by cycling carbon, nitrogen, and sulfur. Surprisingly, genes from Epsilonproteobacteria are found in the sequenced genomes of deep-branching taxa, the *Aquifex* species (see Section 18.1). Like other hyperthermophiles, the Epsilons "get around," with their genes migrating into distantly related organisms.

To Summarize

■ **Proteobacteria** stain Gram-negative, with a thin cell wall and an outer membrane containing LPS. They show wide diversity of form and metabolism, including phototrophy, lithotrophy, and heterotrophy on diverse organic substrates.

■ **Alphaproteobacteria** include photoheterotrophs (such as *Rhodospirillum*) and heterotrophs, as well as methylotrophs. They include intracellular mutualists such as rhizobia, and pathogens such as the rickettsias. Rickettsias share ancestry with mitochondria.

■ **Betaproteobacteria** include photoheterotrophs (*Rhodocyclus*), as well as nitrifiers (*Nitrosomonas*) and iron-sulfur oxidizers (*Thiobacillus*). Pathogenic diplococci include *Neisseria gonorrhoeae*, the cause of gonorrhea.

■ **Gammaproteobacteria** include sulfur and iron bacteria (*Acidithiobacillus, Chromatium*). The most famous species are the Enterobacteriaceae found in the human colon. Intracellular pathogens include *Salmonella* and *Legionella*. The pseudomonads, aerobic rods, can respire on a wide range of complex organic substrates.

■ **Deltaproteobacteria** include sulfur and iron reducers (*Geobacter*), fruiting-body bacteria (*Myxococcus*), and bacterial predators (*Bdellovibrio*).

■ **Epsilonproteobacteria** are spirillar pathogens such as *Helicobacter pylori*, the cause of gastritis.

18.5 Deep-Branching Gram-Negative Phyla

Do other phyla besides Proteobacteria possess an outer membrane and stain Gram-negative? Some Gram-negative bacteria branch more deeply from the Proteobacteria than the five Greek-lettered classes do from each other, and they are thus classified as separate phyla. Most members of these phyla (though not all) are obligate anaerobes. They have diverse lifestyles and habitats, from aquatic phototrophs to human pathogens. Major phyla include Acidobacteria, Bacteroidetes, Chlorobi, Fusobacteria, and Nitrospirae.

Acidobacteria

Acidobacteria are abundant in soils, where they metabolize a wide range of organic substrates using diverse electron acceptors. Many species grow in extreme conditions, such as in the presence of acid and metals—for example, in soils contaminated by uranium mining. Others grow at high temperature, such as isolate K22 (**Fig. 18.43**) obtained from the Taupo Volcanic Zone, New Zealand, by GNS Science, a geoscience prospecting company. The thin section of isolate K22 in **Figure 18.43B** shows its extensive outer membrane.

The candidate species *Chloracidobacterium thermophilum* is a thermophilic aerobic phototroph, isolated from Octopus Spring, at Yellowstone National Park. This acidobacterium is a photoheterotroph that supplements

A.

B.

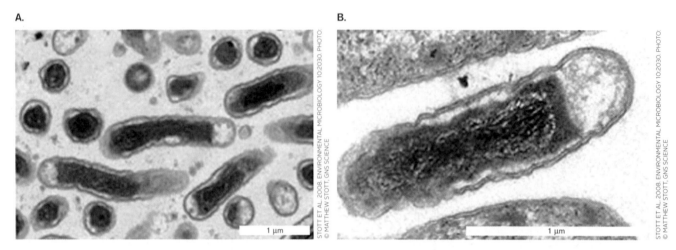

FIGURE 18.43 ■ Acidobacteria isolate K22. **A.** A thermophilic acidophile, showing a Gram-negative outer membrane. **B.** Thin section of isolate K22 (TEM).

catabolism with light absorption by photosystem I (PS I). It grows chlorosomes similar to those of Chloroflexi (see Section 18.1). These distantly related organisms may have shared photosynthesis genes via horizontal gene transfer.

Bacteroidetes

The phylum Bacteroidetes includes genera such as *Bacteroides* and *Flavobacterium*. *Bacteroides* species, such as *B. fragilis* and *B. thetaiotaomicron*, are the major inhabitants of the human colon (**Fig. 18.44**). Their envelope polysaccharides help the bacteria evade the immune system. *Bacteroides* species grow anaerobically under extremely low oxygen concentrations such as those found in the human intestine (1 ppm), yet they can actually use the oxygen they find; thus they have been called "nanoaerobes."

Their main source of energy is fermentation of a wide range of sugar derivatives from plant material, compounds that are indigestible by humans and potentially toxic (discussed in Chapter 13). *Bacteroides* bacteria convert these substances into simple sugars and fermentation acids, some of which are absorbed by the intestinal epithelium; others feed associated gut bacteria such as *E. coli* (discussed in Chapter 21). Thus, *Bacteroides* species serve important functions for their host: They break down potential toxins in plant foods, and their fermentation products make up as much as 15% of the caloric value we obtain from food. Yet another benefit of *Bacteroides* is their ability to remove side chains from bile acids, enabling the return of bile acids to the hepatic circulation. In effect, our gut bacteria, including *Bacteroides*, constitute a functional organ of the human body.

Bacteroides species cause trouble, however, when they reach parts of the body not designed to host them. During abdominal surgery, bacteria can escape the colon and invade the surrounding tissues. The displaced bacteria can form an abscess, a localized mass of bacteria and pus contained in a cavity of dead tissue. The interior of the abscess is anaerobic and often impenetrable to antibiotics.

Chlorobi

The Chlorobi (mainly *Chlorobium* species) are known informally as green sulfur bacteria. While they are genetically close to *Bacteroides*, their metabolism is surprisingly different. Chlorobi are strict photolithotrophs, using PS I to split electrons from H_2, H_2S, or other reduced sulfur compounds (**Fig. 18.45**; see also **Table 18.2**). During the oxidation of sulfide, *Chlorobium* species deposit elemental sulfur extracellularly, forming attached sulfur globules (**Fig. 18.45A**). In contrast, the sulfur granules of Gammaproteobacteria such as *Chromatium* are deposited internally.

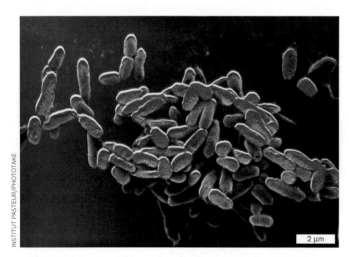

FIGURE 18.44 ■ *Bacteroides fragilis* cells colonize the human colon. *B. fragilis* causes intestinal infections (colorized SEM).

A.

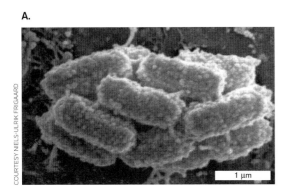

COURTESY NIELS-ULRIK FRIGAARD

1 μm

B.

JORG OVERMANN, U. OF MUNICH

Chlorosomes

100 μm

FIGURE 18.45 ■ *Chlorobium:* **Gram-negative green sulfur bacteria. A.** *Chlorobium* sp. cells covered with sulfur globules; cells form a cluster surrounding a nonphototrophic symbiotic bacterium. **B.** *C. tepidum* cell containing chlorosomes.

Chlorobium species require extensive membrane systems full of photopigments for light absorption. The chlorophyll reaction centers of *Chlorobium* are contained within chlorosomes associated with the cytoplasmic membrane (**Fig. 18.45B**), similar to chlorosomes of the deep-branching phylum Chloroflexi. The photopigment of *Chlorobium* is predominantly BChl *c*, which absorbs in the blue (460 nm) and near-infrared (750 nm), thus reflecting the middle range, brownish green.

Note: Distinguish phylum Chlorobi (sulfur photoautotrophs) from phylum Chloroflexi (thermophilic filamentous photoheterotrophs). Both taxa are phototrophs containing green photopigment complexes arranged in chlorosomes, but they are deeply divergent genetically.

Fusobacteria

Fusobacteria is an emerging taxon of surprisingly virulent pathogens. *Fusobacterium nucleatum* was identified from dental plaque; this bacterium is the second most frequent cause of human abscesses (after *Bacteroides*). Within biofilms such as dental plaque, *Fusobacterium* species form carbohydrate bridges with numerous other kinds of bacteria, such as spirochetes, proteobacteria, and firmicutes, as well as eukaryotic pathogens such as fungi. These distantly related biofilm partners have transferred exceptional numbers of genes to the *F. nucleatum* genome.

Nitrospirae

The phylum Nitrospirae consists of Gram-negative spiral bacteria that oxidize nitrite ion to nitrate (NO_2^- to NO_3^-). Their cell structure resembles that of the Proteobacteria, although their phylogenetic branch is deep enough for assignment to a separate phylum. Most species, such as *Nitrospira* species (see **Tables 18.1** and **18.4**), are true lithotrophs or autotrophs, fixing carbon in the form of carbon dioxide or carbonate using carboxysomes. *Nitrospira* species are generally found in freshwater or salt water. Their removal of excess nitrite makes a key contribution to aquatic ecosystems.

Another important genus is *Leptospirillum*, which includes acidophilic iron oxidizers. *Leptospirillum* species are also strict autotrophs, fixing carbon by using Fe^{2+} as their electron donor and O_2 as the electron acceptor. Their metabolism generates acid, contributing to acid mine drainage in iron mines at Iron Mountain, California, where they grow in massive pink biofilms.

To Summarize

- **Acidobacteria** are Gram-negative soil bacteria, including many acidophiles.

- **Bacteroidetes** are anaerobes that ferment complex plant materials in the human colon. They may enter body tissues through wounds and cause abscesses.

- **Chlorobi** are green sulfur phototrophs, obligate anaerobes incapable of heterotrophy.

- **Fusobacteria** are pathogens that cause septicemia and skin ulcers.

- **Nitrospirae** are Gram-negative spiral bacteria that oxidize nitrite to nitrate (*Nitrospira*).

18.6 Spirochetes: Sheathed Spiral Cells with Internalized Flagella

Spirochetes (or Spirochaeta) is a unique phylum of heterotrophic bacteria that form tightly coiled spirals. While different species of spirochetes conduct a broad range of heterotrophy, from aerobic to anaerobic, all spirochetes share a distinctive cell structure consisting of a long, tight spiral that is flexible like an old-style telephone cord

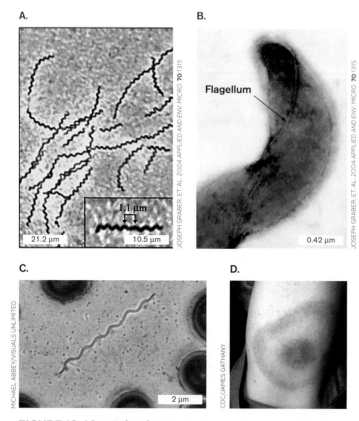

FIGURE 18.46 ■ Spirochetes. **A.** *Treponema azotonutricium* fixes nitrogen in the termite gut (TEM). **B.** Periplasmic flagellum of *T. azotonutricium*. **C.** *Borrelia recurrentis*, shown here in a blood film, is the cause of relapsing fever (stained LM). **D.** Lyme disease rash, caused by tick-borne *Borrelia burgdorferi*.

(**Fig. 18.46A**). For many species, the spiral is so thin that its width cannot be resolved by bright-field microscopy, and the organisms can pass through a filter of pore size 2 μm.

too narrow (0.2 μm) to visualize by bright-field microscopy; instead, dark-field, fluorescence, or electron microscopy must be used. *T. pallidum* is not culturable in the laboratory, but its genome sequence reveals much about its physiology. The genome of 1.1 million base pairs (presented in **eTopic 13.3**) is highly degenerate, lacking nearly all components of biosynthesis and of the TCA cycle; its only system for ATP production is glycolysis.

The spirochete genus *Borrelia* includes two pathogens that cause serious tick-borne diseases in the United States. *B. recurrentis* causes relapsing fever (**Fig. 18.46C**). The related species *B. burgdorferi* causes Lyme disease known for the distinctive bull's-eye rash (**Fig. 18.46D**). A unique feature of *Borrelia* species is their multipartite genome. Each species possesses a linear main chromosome of less than a million base pairs, plus a number of linear and circular plasmids. *B. burgdorferi*, for example, has a linear chromosome of 910,725 base pairs, with at least 17 linear and circular plasmids that total an additional 533,000 base pairs. The reason for this unusual fragmentation of *Borrelia* genomes is unknown.

Spirochetes include important pathogens of animals, such as *Leptospira*, the cause of leptospirosis, a form of nephritis (kidney inflammation) with complications in the liver and other organs. *Leptospira* cells are known for the peculiar L shape of the cell, as each end of the spirochete turns out at an angle. Members of the genus *Treponema* also cause digital dermatitis in cattle and sheep. As in other bacterial phyla, however, the pathogenic species are far outnumbered by harmless organisms.

Note: Distinguish the spirochete *Leptospira* from the nitrite-oxidizing proteobacterium *Leptospirillum*.

Spirochete Diversity

Spirochetes grow in a wide range of habitats, from ponds and streams to the digestive tracts of animals. Aquatic systems carry free-living sugar fermenters of the genus *Spirochaeta*. A particularly interesting spirochete community is found in the termite gut, where the organisms form elaborate symbiotic associations with protists and assist in the digestion of cellulose. *Treponema azotonutricium* actually fixes nitrogen for its host termite (**Fig. 18.46A** and **B**). Other members of the genus *Treponema* are normal residents of the human and animal oral, intestinal, and genital regions.

The best-known spirochetes are those that cause human and animal diseases. The causative agent of the sexually transmitted disease syphilis is the spirochete *Treponema pallidum*. The cell of *T. pallidum* is

Cell Structure of Spirochetes

The spirochete cell is surrounded by a thick outer sheath of lipopolysaccharides and proteins. The spirochete sheath is similar to a proteobacterial outer membrane, except that the periplasmic space completely separates the sheath from the plasma membrane. At each end of the cell, one or more polar flagella extend and double back around the cell body within the periplasmic space (**Fig. 18.47**). The periplasmic

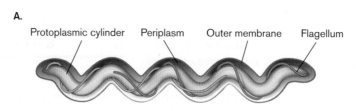

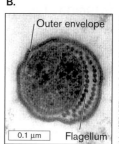

FIGURE 18.47 ■ Spirochete structure. **A.** Spirochete cell structure, showing the arrangement of periplasmic flagella. **B.** Cross section through a human gingival spirochete, showing outer envelope and flagella (axial fibrils) (TEM).

flagella (axial fibrils) rotate on proton-driven motors, as do regular flagella; but because they twine back around the cell body, their rotation forces the entire cell to twist around, corkscrewing through the medium.

This corkscrew motion of the cell body turns out to have a physical advantage in highly viscous environments, such as human mucous secretions or agar culture medium. Few nonspirochetes can swim through agar at standard culture concentrations (1.5% agar); thus, growth within agar provides a way to isolate anaerobic spirochetes from environmental sources. When an environmental sample is inoculated into agar, other bacteria grow and concentrate at the injection point, whereas spirochetes migrate outward in a "veil" through the agar.

To Summarize

- **The spirochete cell** is a tight coil, enclosed by a sheath and periplasmic space containing periplasmic flagella.

- **Spirochete motility** is driven by a flexing motion caused by rotation of the periplasmic flagella, propagated the length of the coil.

- **Spirochetes grow in diverse habitats.** Some are free-living fermenters in water or soil. Others are pathogens, such as *Treponema pallidum*, the cause of syphilis. Still others are endosymbionts of an animal digestive tract, such as the termite gut.

18.7 Chlamydiae, Planctomycetes, and Verrucomicrobia: Irregular Cells

Several related phyla of bacteria have either no cell walls or diminished cell walls. These organisms evolved independently of the mycoplasmas, and their alternative cell forms show diverse environmental adaptations.

Chlamydiae: Intracellular Parasites

In the phylum Chlamydiae, the genera *Chlamydia* and *Chlamydophila* evolved complex developmental life cycles of parasitizing host cells. *Chlamydia trachomatis* causes the most prevalent sexually transmitted infection among young people in the United States. The same pathogen also causes trachoma, an eye disease dating back to records in ancient Egypt. The related species *Chlamydophila pneumoniae* causes pneumonia and has been implicated in cardiovascular disease.

A.

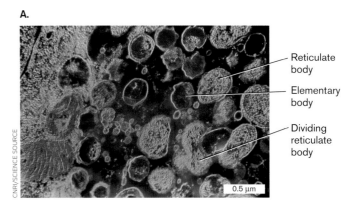

Reticulate body

Elementary body

Dividing reticulate body

0.5 μm

CNRI/SCIENCE SOURCE

B. *Chlamydia* developmental cycle

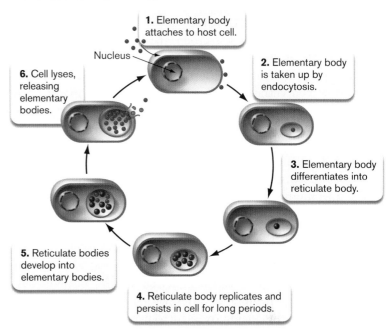

1. Elementary body attaches to host cell.

Nucleus

6. Cell lyses, releasing elementary bodies.

2. Elementary body is taken up by endocytosis.

3. Elementary body differentiates into reticulate body.

5. Reticulate bodies develop into elementary bodies.

4. Reticulate body replicates and persists in cell for long periods.

FIGURE 18.48 ■ ***Chlamydia* life cycle. A.** *C. trachomatis*, multiplying within a human cell (colorized TEM). Infected cell, equivalent to step 5 in part (B), contains reticulate bodies (yellow) growing and dividing, as well as newly formed elementary bodies (red). **B.** *Chlamydia* species persist outside the host as a spore-like "elementary body." Upon endocytosis, the elementary body avoids lysosomal fusion and develops into a "reticulate body" in which the DNA, now uncondensed, has a reticular (netlike) appearance. The reticulate body replicates within the host cytoplasm and then develops into new elementary bodies that are released when the cell lyses.

Chlamydiae alternate between two developmental stages with different functions: elementary bodies and reticulate bodies (**Fig. 18.48A**). The form of chlamydia transmitted outside host cells is called an **elementary body**. Like endospores, elementary bodies are metabolically inert, with a compacted chromosome. While lacking a cell wall, they possess an outer membrane whose proteins are cross-linked by disulfide bonds, making a tough coat that provides osmotic stability. The elementary body adheres to a host cell surface and is endocytosed (**Fig. 18.48B**).

To reproduce, the elementary body must transform itself into a **reticulate body**, named for the netlike appearance of its uncondensed DNA. The reticulate body has an active metabolism and divides rapidly, but outside the cell it is incapable of infection and vulnerable to osmotic shock. To complete the infection cycle, therefore, reticulate bodies must develop into new elementary bodies before exiting the host. When the host cell lyses, the elementary bodies are released to infect new cells. Chlamydiae infect a wide range of host cell types, from respiratory epithelium to macrophages.

Planctomycetes: A Nucleus-Like Compartment

Another cell wall–less group, the Planctomycetes, evolved largely as free-living organisms. Planctomycetes are oligotrophs, heterotrophs requiring nutrients at extremely low concentrations. They grow in aquatic, marine, and saline environments. Their mechanism of osmoregulation remains poorly understood.

Planctomycete cells possess multiple internal membrane compartments of unknown function (**Fig. 18.49A**). An internal membrane just inside the cell membrane divides the cytoplasm into concentric portions. A double membrane surrounds the entire nucleoid, analogous to the double membrane surrounding the eukaryotic nucleus—a remarkable example of independent analogous evolution between bacteria and the eukaryotes. The actual function of the nucleoid-surrounding membrane varies with different species; for example, in anammox planctomycetes, a single membrane enclosing the nucleoid contains the complexes conducting the anammox reaction (anaerobic ammonium oxidation; discussed in Chapter 14).

Like eukaryotic protists (discussed in Chapter 20), the planctomycetes have flexible cell bodies that can assume diverse forms. For example, some *Planctomyces* species have rotary flagella (**Fig. 18.49B**), whereas *P. bekefii* cells have stalks that attach to each other to generate a starlike aggregate (**Fig. 18.49C**). Most planctomycetes reproduce by budding, a strategy typical of eukaryotic yeasts.

Verrucomicrobia: Wrinkled Microbes

Verrucomicrobia, or "wrinkled microbes," are irregularly shaped bacteria found in a wide variety of aquatic and terrestrial environments, as well as in the mammalian gastrointestinal tract (see **Table 18.1**). Most Verrucomicrobia are oligotrophs, growing heterotrophically in low-salt habitats. Some are ectosymbionts of protists, attached to the cell surface of the eukaryote, where they eject harpoon-like objects. They are rarely cultured, and until recently they

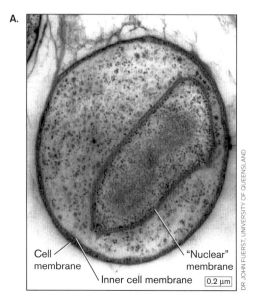

A.

Cell membrane — Inner cell membrane — "Nuclear" membrane

0.2 μm

DR JOHN FUERST, UNIVERSITY OF QUEENSLAND

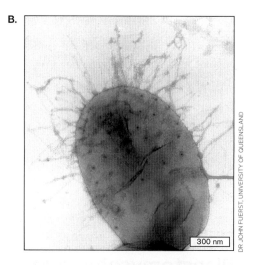

B.

300 nm

DR JOHN FUERST, UNIVERSITY OF QUEENSLAND

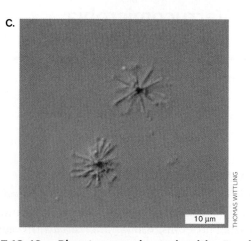

C.

10 μm

THOMAS WITTLING

FIGURE 18.49 ▪ *Planctomyces:* **bacteria with a "nuclear membrane."** **A.** Section through the planctomycete *Gemmata obscuriglobus*, showing membrane compartmentalization (TEM). The DNA is contained within a double membrane analogous to a eukaryotic nuclear membrane. **B.** A swarmer cell of *Planctomyces* sp. with multiple flagella (TEM). **C.** *Planctomyces bekefii* cells form stalks that attach to each other at the center to yield a starlike form (LM).

failed to show up in PCR amplification of natural isolates because their rDNA sequences are poorly amplified by the standard primer sequences. Using appropriate primer pairs, Verrucomicrobia comprise 5% of all surveyed microbial sequences in some natural environments.

Verrucomicrobia have peptidoglycan cell walls, but their shape is dominated by wart-like projections. The wart-like cytoskeleton appears to contain tubulin, a cytoskeletal protein previously believed to exist only in eukaryotes. In 2002, genes encoding tubulin were found in the partly sequenced genome of *Prosthecobacter dejongeii*, a free-living member of Verrucomicrobia. The genes appear so similar to those of eukaryotes that they must have undergone horizontal transfer from a eukaryotic genome. This horizontal transfer of a eukaryotic trait may be contrasted with the independent development of a nucleus-like structure in planctomycetes.

To Summarize

■ **Chlamydiae** are obligate intracellular parasites that undergo a complex developmental progression, culminating in a spore-like form called an elementary body that can be transmitted outside the host cell. Chlamydiae lack cell walls.

■ **Planctomycetes** lack cell walls, and they have evolved a membrane enclosing the nucleoid, analogous to the eukaryotic nuclear membrane.

■ **Verrucomicrobia** have cell projections containing tubulin. Their tubulin genes are thought to have arisen by horizontal transfer from a eukaryote.

Concluding Thoughts

Several themes emerge in the diversity of the domain Bacteria. Some phyla, such as Proteobacteria and Actinobacteria, have evolved highly diverse metabolism and cell form. Others show remarkable uniformity in cell structure (Spirochetes) or in metabolism (Cyanobacteria). At the same time, our attempts to generalize about any given clade inevitably run up against the unexpected appearance of exceptions, such as the heliobacteria, photosynthetic endospore formers that unaccountably branch from clostridia. Given the range of bacterial phenotypes, it is hard to imagine that yet more diverse forms of microbial life exist, but they do, as we will find among the archaea (Chapter 19) and the microbial eukaryotes (Chapter 20).

CHAPTER REVIEW

Review Questions

1. Which deep-branching phyla include hyperthermophiles? Why is the actual branch position of these groups controversial?

2. Compare and contrast Chloroflexi and Cyanobacteria with respect to habitat, cell structure, and means of photosynthesis.

3. Compare and contrast the colonial and filamentous cyanobacteria with respect to life cycle and means of nitrogen fixation.

4. Name and describe three genera of Firmicutes that form endospores. Discuss the difference between single and multiple spore formers. Explain the existence of non-spore-forming species of Firmicutes.

5. Compare and contrast firmicute endospores and actinomycete arthrospores with respect to their means of production, their resistance properties, and their dispersal mechanisms.

6. Describe the diverse kinds of metabolism available to different species of Proteobacteria. Explain how it is possible for some species to perform many different kinds of energy-gaining metabolism.

7. What do species of Bacteroidetes and Chlorobi have in common, and how do they differ?

8. Explain the structure and mechanism of motility typical in species of Spirochetes.

9. Discuss the properties of as many intracellular parasites as you recall from various phyla. What do they have in common, and how do they differ?

10. Describe the unique cell structure of *Planctomyces*. Explain how the traits of planctomycete cells appear analogous to aspects of eukaryotic cells.

11. How do microbiologists seek to find previously unknown kinds of bacteria? What techniques reveal unknown microorganisms?

Thought Questions

1. Why are the Proteobacteria metabolically diverse? How is it possible that so many closely related organisms use such different molecules to yield energy?

2. How do you think *Mycobacterium tuberculosis* manages to grow, despite its thick envelope screening out most nutrients?

3. For motility, what are the relative advantages of external flagella versus the flexible spiral cells of spirochetes?

4. Why do different species of microbes grow together in layered biofilms (a) on a sand flat and (b) on the surface of human teeth?

Key Terms

acid-fast stain (711)
Acidobacteria (695)
Actinobacteria (692)
actinomycete (692)
aerial mycelia (710)
bacteroid (718)
Bacteroidetes (695)
carboxysome (698)
Chlamydiae (694)
Chlorobi (695)
chlorosome (694)
Cyanobacteria (692)
deep-branching taxa (692)
diplococcus (720)
elementary body (729)
emerging (695)

endophyte (711)
endospore (692)
Firmicutes (692)
forespore (703)
fruiting body (723)
Fusobacteria (695)
gas vesicle (698)
GC content (692)
heterocyst (698)
hormogonium (699)
leghemoglobin (718)
methanotroph (717)
methylotrophy (717)
Mollicutes (707)
mother cell (703)
mycolic acid (711)

myxospore (724)
nitrifier (719)
Nitrospirae (695)
phylum (691)
Planctomycetes (694)
Proteobacteria (692)
proteorhodopsin (715)
reticulate body (729)
sarcina (713)
Spirochetes (692)
swarming (721)
Tenericutes (707)
vegetative cell (703)
vegetative mycelia (710)
Verrucomicrobia (694)

Recommended Reading

Andersson, Siv G. E., Alireza Zomorodipour, Jan O. Andersson, Thomas Sicheritz-Pontén, U. Cecilia M. Alsmark, et al. 1998. The genome sequence of *Rickettsia prowazekii* and the origin of mitochondria. *Nature* 396:133–140.

Angert, Esther. 2006. Beyond binary fission: Some bacteria reproduce by alternative means. *Microbe* 1:127–131.

Beatty, J. Thomas, Jörg Overmann, Michael T. Lince, Ann K. Manske, Andrew S. Lang, et al. 2005. An obligately photosynthetic bacterial anaerobe from a deep-sea hydrothermal vent. *Proceedings of the National Academy of Sciences USA* 102:9306–9310.

Brown, Christopher T., Laura A. Hug, Brian C. Thomas, Itai Sharon, Cindy J. Castelle, et al. 2015. Unusual biology across a group comprising more than 15% of domain Bacteria. *Nature* 523:208–211.

Bryant, Donald A., and Niels-Ulrik Frigaard. 2006. Prokaryotic photosynthesis and phototrophy illuminated. *Trends in Microbiology* 14:488–496.

Campbell, Barbara J., Annette Summers Engel, Megan L. Porter, and Ken Takai. 2006. The versatile epsilon-proteobacteria: Key players in sulphidic habitats. *Nature Reviews. Microbiology* 4:458–468.

Chouari, Rakia, Denis Le Paslier, Patrick Daegelen, Philippe Ginestet, Jean Weissenbach, et al. 2003. Molecular evidence for novel planctomycete diversity in a municipal wastewater treatment plant. *Applied and Environmental Microbiology* 69:7354–7363.

Claessen, Dennis, Daniel E. Rozen, Oscar P. Kuipers, Lotte Søgaard-Andersen, and Gilles P. van Wezel. 2014. Bacterial solutions to multicellularity: A tale of biofilms, filaments and fruiting bodies. *Nature Reviews. Microbiology* 12:115–124.

Frigaard, Niels-Ulrik, Asuncion Martinez, Tracy J. Mincer, and Edward F. DeLong. 2006. Proteorhodopsin lateral gene transfer between marine planktonic Bacteria and Archaea. *Nature* 439:847–850.

Geissinger, Oliver, Daniel P. R. Herlemann, Erhard Mörschel, Uwe G. Maier, and Andreas Brune. 2009. The ultramicrobacterium "*Elusimicrobium minutum*" gen. nov., sp. nov., the first cultivated representative of the termite group 1 phylum. *Applied and Environmental Microbiology* **75**:2831–2840.

Human Microbiome Project Consortium. 2012. Structure, function and diversity of the healthy human microbiome. *Nature* **486**:207–214.

Jenkins, Cheryl, Ram Samudrala, Iain Anderson, Brian P. Hedlund, Giulio Petroni, et al. 2002. Genes for the cytoskeletal protein tubulin in the bacterial genus *Prosthecobacter*. *Proceedings of the National Academy of Sciences USA* **99**:17049–17054.

Jensen, Paul R., Philip G. Williams, Dong-Chan Oh, Lisa Zeigler, and William Fenical. 2007. Species-specific secondary metabolite production in marine actinomycetes of the genus *Salinispora*. *Applied and Environmental Microbiology* **73**:1146–1152.

Kindaichi, Tomonori, Tsukasa Ito, and Satoshi Okabe. 2004. Ecophysiological interaction between nitrifying bacteria and heterotrophic bacteria in autotrophic nitrifying biofilms as determined by microautoradiography-fluorescence in situ hybridization. *Applied and Environmental Microbiology* **70**:1641–1650.

King, Gary M., and Carolyn F. Weber. 2007. Distribution, diversity and ecology of aerobic CO-oxidizing bacteria. *Nature Reviews. Microbiology* **5**:107–118.

Kuhn, Emanuele, Andrew S. Ichimura, Vivian Peng, Christian H. Fritsen, Gareth Trubl, et al. 2014. Brine assemblages of ultrasmall microbial cells within the ice cover of Lake Vida, Antarctica. *Applied and Environmental Microbiology* **80**:3687–3698.

Luef, Birgit, Kyle R. Frischkorn, Kelly C. Wrighton, Hoi-Ying N. Holman, Giovanni Birarda, et al. 2015. Diverse uncultivated ultra-small bacterial cells in groundwater. *Nature Communications* **6**:6372.

Masaki, Toshihiro, Jinrong Qu, Justyna Cholewa-Waclaw, Karen Burr, Ryan Raaum, et al. 2013. Reprogramming adult Schwann cells to stem cell-like cells by leprosy bacilli promotes dissemination of infection. *Cell* **152**:51–67.

Moosvi, S. Azra, Ian R. McDonald, David A. Pearce, Donovan P. Kelly, and Ann P. Wood. 2005. Molecular detection and isolation from Antarctica of methylotrophic bacteria able to grow with methylated sulfur compounds. *Systematic and Applied Microbiology* **28**:541–554.

Schlieper, Daniel, María A. Oliva, José M. Andreu, and Jan Löwe. 2005. Structure of bacterial tubulin BtubA/B: Evidence for horizontal gene transfer. *Proceedings of the National Academy of Sciences USA* **102**:9170–9175.

Schnupf, Pamela, Valérie Gaboriau-Routhiau, Marine Gros, Robin Friedman, Maryse Moya-Nilges, et al. 2015. Growth and host interaction of mouse segmented filamentous bacteria in vitro. *Nature* **520**:99–103.

Tolli, John D., S. M. Sievert, and C. D. Taylor. 2006. Unexpected diversity of bacteria capable of carbon monoxide oxidation in a coastal marine environment, and contribution of the *Roseobacter*-associated clade to total CO oxidation. *Applied and Environmental Microbiology* **72**:1966–1973.

Venter, J. Craig, Karin Remington, John F. Heidelberg, Aaron L. Halpern, Doug Rusch, et al. 2004. Environmental genome shotgun sequencing of the Sargasso Sea. *Science* **304**:66–74.

Wang, Jenny, Cheryl Jenkins, Richard I. Webb, and John A. Fuerst. 2002. Isolation of *Gemmata*-like and *Isosphaera*-like planctomycete bacteria from soil and freshwater. *Applied and Environmental Microbiology* **68**:417–422.

Wu, Martin, Qinghu Ren, A. Scott Durkin, Sean C. Daugherty, Lauren M. Brinkac, et al. 2005. Life in hot carbon monoxide: The complete genome sequence of *Carboxydothermus hydrogenoformans* Z-2901. *PLoS Genetics* **1**:e65.

CHAPTER 19
Archaeal Diversity

Archaea are found in all soil and water habitats, in symbiosis with animals and plants, and in extreme environments that exclude bacteria and eukaryotes. Archaea include hyperthermophiles inhabiting Earth's hottest habitats, as well as Arctic and Antarctic psychrophiles. Methanogens inhabit the digestive tracts of humans and other animals, as well as anaerobic soil and water, with consequences for global climate. Many archaea collaborate with bacteria in multispecies biofilms. No archaeon is yet known to cause disease, but archaea do interact with our innate immune system.

CURRENT RESEARCH highlight

Archaea with grappling hooks. A newly discovered archaeon, *Altiarchaeum hamiconexum*, was discovered in marsh water rich in sulfides, from the Sippenauer Moor, Germany. The archaea fix CO_2 by an unusual pathway related to methanogenesis. Their cells use a unique appendage, pilus-like grappling hooks that connect neighboring cells in a biofilm matrix. The matrix of grappling hooks also coats filaments of sulfide-oxidizing bacteria that may cooperate with *Altiarchaeum*.

Source: Alexandra K. Perras et al. 2014. *Front. Microbiol.* **5**:397.

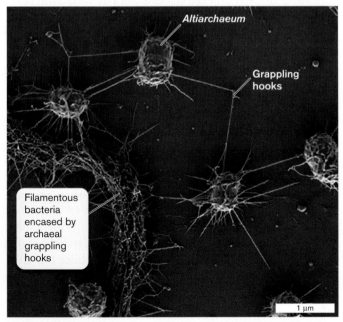

Altiarchaeum

Grappling hooks

Filamentous bacteria encased by archaeal grappling hooks

1 μm

AN INTERVIEW WITH

CHRISTINE MOISSL-EICHINGER, EXTREMOPHILE MICROBIOLOGIST, MEDICAL UNIVERSITY OF GRAZ

©ANNA AUERBACH

What is most surprising about the new kind of archaeon you discovered?

The ultrastructural analyses of *Altiarchaeum* were breathtaking, especially the discovery of the "hami." The hami are filamentous cell-surface appendages that carry a tiny grappling hook at their end. Such features have never been found for any cellular appendage of eukaryotic or prokaryotic cells.

How did your research on archaea prepare you to investigate microbes on board the International Space Station?

Our methods found that Archaea are a substantial part of the human skin microbiome. I am driven by the idea that Archaea are everywhere, and that we just need to use the right methods to understand their biology. I obtained an award from the European Space Agency for a comprehensive microbiome analysis of the International Space Station in order to find interesting Archaea and Bacteria there and to identify the potential adaptations to this very interesting habitat. Such analysis will help us understand the mechanisms of an encapsulated ecosystem, the impact of microbes on long-term spaceflight, and the adaptation capacity of a microbial community.

In 1977, Carl Woese revealed the existence of a third kind of life—the Archaea, a domain of life-forms very different from the plants, animals, and bacteria that had long been known. Today, the Archaea continue to yield surprises. An example is the candidate species *Altiarchaeum hamiconexum*, first known as the SM1 euryarchaeon discovered in 2004 by Rudolph Huber and his students at the University of Regensburg, Germany. The marsh-dwelling microbe is an archaeon related to methanogens, and it appears to fix CO_2, but its precise metabolism remains unknown. Christine Moissl-Eichinger (see the Current Research Highlight) showed how this organism forms netlike biofilms by use of grappling-hook appendages, called **hami** (singular, **hamus**) (**Fig. 19.1**). Each grappling hook contains paired barbs along an extended protein filament many times the length of the cell. The filament ends with a triple fishhook that enables filaments to clasp each other. No other kind of cell is known to make this type of appendage.

Archaea are known for extremophiles, such as hyperthermophiles growing above 100°C and hyperacidophiles at below pH 1. Extremophiles are of interest to biotechnology because their enzymes may catalyze reactions under industrially useful conditions, such as high salt or acidity. But we also find archaea throughout mesophilic soil and water, and within the digestive tracts of humans and other animals. Moissl-Eichinger even discovered archaea on human skin, comprising more than 4% of the skin's prokaryotic microbiome. She argues that archaea exist everywhere, in every livable habitat; we just need the right methods to find them.

How do archaea in "moderate" habitats interact with bacteria and eukaryotes? Many archaea join bacteria within biofilms in soil and on marine particles. For example, *Altiarchaeum* envelops sulfide-oxidizing bacteria in its grappling appendages to form thick biofilms like pearls on a string. Other archaea cooperate with eukaryotes, for example, by colonizing the surface of plant roots. Even in these shared habitats, archaea show unique traits, such as energy-yielding methanogenesis and cyclic diether membranes. Perhaps the most compelling unique feature of archaea, from the human point of view, is that no archaea are known to cause disease. Recent reports find gut methanogens associated with the inflammatory response, but no causal relationship has yet been demonstrated.

In Chapter 19 we explore the diversity of archaea. We emphasize the unique structures and metabolic pathways of archaeal cells. While surveying their major taxonomic categories, we introduce research techniques used to study organisms in extreme environments and those with unique forms of metabolism, such as methanogenesis.

19.1 Archaeal Traits and Phylogeny

Key traits such as ether-linked membrane lipids are found only in archaea, and in a few bacteria that received them by horizontal gene flow from archaea. Other traits, such as diverse pathways of redox metabolism, archaea share with bacteria. But archaea share core traits of DNA-RNA machinery and transcription factors with eukaryotes. Genomic analysis suggests that eukaryotes possess an archaeal ancestor.

Archaeal Cell Structure and Metabolism

Distinctive features of archaea, sometimes called "archaeal signatures," include components of the cell membrane and envelope and certain metabolic pathways to gain energy (**Table 19.1**).

Isoprenoid membranes. The most distinctive structure of archaea is their membrane (**Fig. 19.2**). The membrane lipids of archaea differ profoundly from those of bacteria and eukaryotes, with the exception of a few thermophilic bacteria that obtained ether lipids horizontally from archaea. Most features of archaeal lipids increase lipid stability in extreme environments such as high temperature or extreme acidity. Nevertheless, ether lipids are also widespread in mesophiles (grow at moderate temperatures). Some researchers argue that the original archaea were hyperthermophiles but later evolved as mesophiles by acquiring genes from bacteria.

Archaeal lipids contain distinctive structures that differ from those of bacteria and eukaryotes:

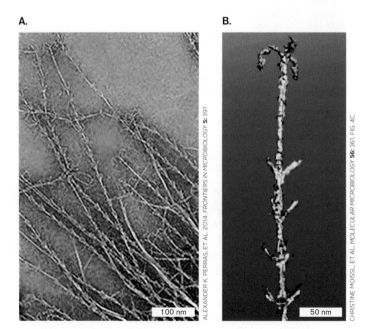

A.

B.

ALEXANDER K. PERRAS, ET AL. 2014. FRONTIERS IN MICROBIOLOGY **5**: 397.

CHRISTINE MOISSL, ET AL. MOLECULAR MICROBIOLOGY **56**: 361, FIG. 4C.

100 nm

50 nm

FIGURE 19.1 ■ Grappling hooks (hami) of *Altiarchaeum* species. A. Hooked appendages extending from a cell (TEM). **B.** Cryo-EM tomography model of a grappling hook (hamus).

TABLE 19.1 Archaeal Traits Distinct from Bacteria and Eukaryotes

Traits of archaea	Archaea showing the trait	Alternative traits of bacteria and/or eukaryotes
Cell envelope		
Membrane lipids: isoprenoid L-glycerol ethers or diethers	All archaea	Membrane lipids: D-glycerol hydrocarbon diesters
Membrane lipid chains stiffened by covalent cross-links or by pentacyclic rings	Most archaea	Chains stiffened by saturation
S-layer of glycoprotein	Crenarchaeota and Thaumarchaeota	Peptidoglycan (bacteria); cellulose and other polysaccharides (eukaryotes)
S-layer of protein, methanochondroitin, or sulfated polysaccharide	Methanogens and Haloarchaea	Peptidoglycan (bacteria); cellulose and other polysaccharides (eukaryotes)
Pseudomurein sacculus contains talosaminuronic acid; peptide bridges contain only L-amino acids	Methanobacteriales, Methanopyrales	Peptidoglycan (bacteria); cellulose and other polysaccharides (eukaryotes)
Metabolism		
Nonphosphorylated intermediates of sugar catabolism and synthesis (EMP glycolysis in some cases)	All archaea	EMP pathway of glycolysis, with phosphorylation mediated by NAD or NADP (or Entner-Doudoroff)
Methanogenesis from H_2 and CO_2; or from CO, methanol, methyl sulfides, formate, or acetate	Methanogens	Anaerobic metabolism such as fermentation; no methane production
Archaeal coenzymes: cofactors F_{420} and F_{430}, and coenzyme M	Most methanogens	Flavin mononucleotide, coenzyme A, others
Retinal-associated light-driven membrane pumps for H^+ or Na^+	Haloarchaea	Chlorophyll-based photosynthesis (bacteria and eukaryotic chloroplasts)
Nucleic acid structure and function		
Positive superturns generated by reverse gyrase protect DNA from extreme acid.	Hyperthermophilic archaea	Negative superturns generated by gyrase
Unique base structures in tRNA, such as the guanine analog archaeosine	Most archaea	tRNA bases, such as queuosine, found only in bacteria and eukaryotes

Sources: O. Kandler and H. König. 1998. *Cell. Mol. Life Sci.* **54**:305–308; C. Bullock. 2000. *Biochem. Mol. Biol. Ed.* **28**:186–191.

■ **Glycerol 1-phosphate.** Archaeal membrane lipids incorporate glycerol 1-phosphate, rather than the mirror-symmetrical form glycerol 3-phosphate, which is used by bacteria and eukaryotes. The two chiral forms show similar thermal stability, but their biochemistry requires different enzymes, and thus they represent a deep divergence in ancestry. In some archaea the glycerol is extended by six carbons, forming nonitol (nine OH groups).

■ **Ether linkage.** The glycerol units are linked to side chains by ether links (R–O–R) instead of the ester links (R–COO–R) found in bacteria and eukaryotes (**Fig. 19.2A**). Ether links are much more stable than ester links; in other words, breaking them requires more energy.

■ **Isoprenoid chains.** The side chains of archaeal lipids are branched at every fourth carbon. The methyl branches arise by condensation (C–C bond formation) of units of isoprene (see **Fig. 19.2A**). Condensed isoprene chains

are called **isoprenoid** or diphytanyl chains; thus, the overall lipid is diphytanylglycerol diether. Isoprenoid branched chains increase membrane stability by hooking each other in place.

■ **Cross-linked lipids.** In some hyperthermophiles, the ends of side chains are linked covalently, either to each other (**Fig. 19.2B**) or to a lipid on the opposite side of the membrane (**Fig. 19.2C**). Two pairs of lipid chains cross-linked across the membrane form a **tetraether**, so called because the complex contains four ether links in all. In some cases, an additional covalent bond links the two linked pairs of side chains across the middle (**Fig. 19.2D**).

■ **Cyclopentane rings.** In some archaea, the lipid's methyl branches cyclize, forming cyclopentane rings (**Fig. 19.2E**). Cyclopentane rings strengthen membranes at high temperature.

A.

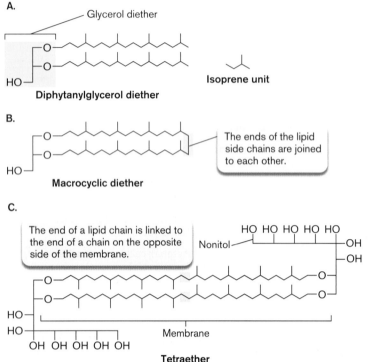

Glycerol diether

Isoprene unit

Diphytanylglycerol diether

B.

The ends of the lipid side chains are joined to each other.

Macrocyclic diether

C.

The end of a lipid chain is linked to the end of a chain on the opposite side of the membrane.

Nonitol

Membrane

Tetraether

D.

Cross-link is formed by an additional covalent bond between the lipid chains.

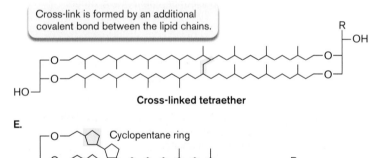

Cross-linked tetraether

E.

Cyclopentane ring

FIGURE 19.2 ■ **Branched-chain ether lipids are characteristic of archaeal membranes.** *Source:* Modified from R. M. Daniel and D. A. Cowan. 2000. *Cell. Mol. Life Sci.* **57**:250–264.

S-layer and cell wall. Most archaea possess a single membrane, without any outer membrane such as that of Gram-negative bacteria (shown in Chapter 3). Many, such as the thermophilic crenarchaeotes, possess no cell wall at all, but only an S-layer of proteins plugged into the tetraether membrane (**Fig. 19.3**). The S-layer proteins of *Sulfolobus* species lock together to form a sturdy array (**Fig. 19.3A**). While helping to keep the cell intact, the S-layer protein array nonetheless allows a flexible cell structure, in contrast to the more rigid structures of most bacteria.

Some archaea do have a cell wall, but the structure differs fundamentally from the bacterial peptidoglycan. The methanogen *Methanothermus* (**Fig. 19.3B**) has a peptide-sugar sacculus that is composed not of murein (see Chapter 3), but of a related macromolecule called **pseudomurein** or **pseudopeptidoglycan**. Pseudopeptidoglycan contains chains of alternating sugar derivatives including *N*-acetylglucosamine, as in bacterial peptidoglycan. The bacterial *N*-acetylmuramic acid, however, is replaced by a related sugar, *N*-acetyltalosaminuronic acid; and the sugar linkage is β(1,3) instead of β(1,4) as in bacteria. As a result, these archaea are resistant to lysozyme, which degrades bacterial cell walls at the β(1,4) sugar linkage. The peptide cross-bridges of pseudopeptidoglycan differ as well, causing resistance to penicillin (discussed in Chapter 3).

Similarly, many archaea possess filamentous protein structures for attachment and motility that are analogous to bacterial pili and flagella. Despite functional similarity, the archaeal protein structures are very different from those of bacteria. Archaeal flagella have rotary motors, but their distinctive form earns them the name "archaella" (singular, archaellum).

Unique metabolic pathways. Archaea utilize many distinctive metabolic pathways. Glucose is catabolized by several variants of the Entner-Doudoroff (ED) and

A. *Sulfolobus*

S-layer

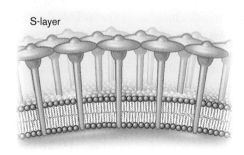

B. *Methanothermus*

Pseudopeptidoglycan

S-layer

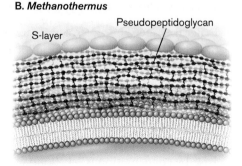

FIGURE 19.3 ■ **Envelopes of archaea.** **A.** *Sulfolobus* species have S-layer proteins plugged into a tetraether membrane. **B.** *Methanothermus* species have a cell wall of pseudopeptidoglycan between their membrane and S-layer.

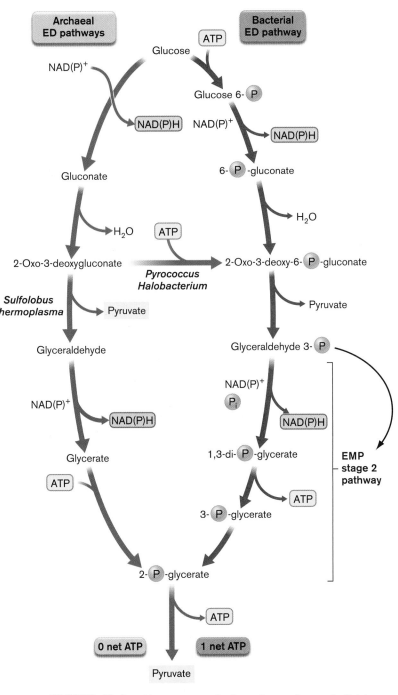

FIGURE 19.4 ■ Glucose catabolism in archaea. *Sulfolobus* and *Thermoplasma* species catabolize glucose to pyruvate by a modified ED pathway without phosphorylating glucose and produce no net ATP. *Halobacterium* species phosphorylate 2-oxo-3-deoxygluconate and produce one net ATP by the EMP stage 2 pathway. *Pyrococcus furiosus* oxidizes glyceraldehyde 3-phosphate using ferredoxin instead of NAD$^+$ and avoids phosphorylation.

Embden-Meyerhof-Parnas (EMP) pathways that rarely occur in bacteria (**Fig. 19.4**). For example, the sulfur thermophiles *Sulfolobus* and *Thermoplasma* convert glucose to gluconate without phosphorylation, ultimately generating pyruvate with no net production of ATP. (ATP is still produced by further breakdown of pyruvate.) On the other hand, halophilic archaea such as *Halobacterium* phosphorylate the dehydrated product of gluconate (2-oxo-3-deoxygluconate), which enters the "standard" ED pathway. The ED pathway generates one molecule of pyruvate and one molecule of 3-phosphoglycerate, which produces one net ATP through the second stage of the EMP pathway. A unique variant of the EMP pathway is seen in the vent thermophile *Pyrococcus furiosus*, which oxidizes glyceraldehyde 3-phosphate using ferredoxin instead of NAD$^+$ and avoids phosphorylation. The reduced ferredoxin is then used to reduce 2H$^+$ to H$_2$ in an energy-yielding reaction.

The energy-yielding process of **methanogenesis** occurs only in archaea. Methane production by methanogenic archaea (methanogens) makes a growing contribution to global warming, as discussed in Chapter 22. Methanogenesis includes a unique pathway of carbon fixation, called the carbon monoxide reductase pathway because the key enzyme can fix CO as well as CO$_2$. A CO group is fixed into acetyl-CoA by condensation with a methyl group generated by methanogenesis. Acetyl-CoA then enters the TCA cycle. Methanogenesis also uses unique cofactors that largely replace NAD, FAD, and FMN (discussed in Section 19.4).

The only known form of phototrophy in archaea is that of retinal-based ion pumps. Membrane protein complexes known as **bacteriorhodopsin** and halorhodopsin contain retinal and pumps for H$^+$ or Na$^+$ (discussed in Chapter 14). Bacteriorhodopsins are found in haloarchaea (formerly called halobacteria) such as *Halobacterium halobium*; the species was named before it was known to be an archaeon (discussed in Section 19.5). It turns out that haloarchaea have transferred genes encoding light-absorbing proton pumps into many marine bacteria. By contrast, the chlorophyll-based phototrophy found in bacteria and plants is completely unknown in archaea.

Archaeal Genome Structure and Regulation

How do archaeal genomes compare with those of bacteria and eukaryotes? The genomes of archaea generally resemble those of bacteria in size and gene density, and genes of related function are often arranged in operons, like those of bacteria. Nevertheless, certain features of genome structure are unique to archaea, and some traits more closely resemble those of eukaryotes than those of bacteria.

A distinctive feature of archaeal chromosome function is the "reverse gyrase" enzyme found in hyperthermophiles. For comparison, all bacteria and eukaryotes, as well as mesophilic archaea, have gyrase to maintain their DNA in an "underwound" negatively supercoiled state (presented in Chapter 7). But hyperthermophilic archaea with reverse gyrase maintain positive supercoiling to stabilize their DNA. The positive superturns "overwind" the DNA, preventing the helix from melting into separate strands at high temperature.

A unique feature of archaeal RNA is the distinctive modified bases in their tRNA molecules. In particular, the guanosine analog archaeosine (7-formamidino-7-deaza-guanosine) is used by nearly all archaea, but not by any bacteria or eukaryotes. Other unusual tRNA bases, such as queuosine, are found only in bacteria and eukaryotes, not in archaea.

The genes encoding most archaeal proteins contain uninterrupted coding sequences, as in bacteria, but certain tRNA gene sequences are interrupted by introns (nontranslated sequence), similar to the tRNA introns found in eukaryotes. Furthermore, the archaeal apparatus for DNA and RNA polymerases, transcription factors, and protein synthesis show remarkable similarity to those of

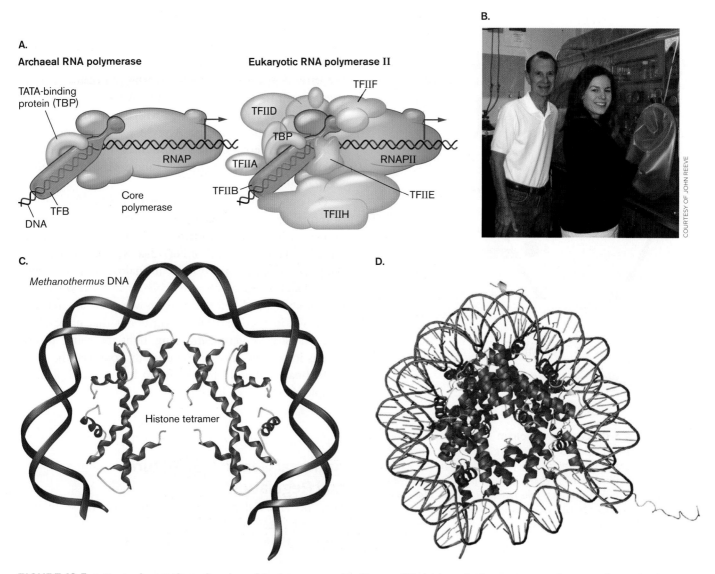

FIGURE 19.5 ■ **Central genetic molecules of Archaea resemble those of Eukarya.** **A.** The core RNA polymerase (RNAP) subunits of the archaeon *Sulfolobus* show homology to those of the eukaryotic RNA polymerase II (RNAPII). The TATA-binding protein (TBP, green) and transcription factor B (TFB, pink) also show homology to eukaryotic counterparts. **B.** John Reeve, pictured here with a graduate student, studies the molecular biology of methanogens using an anaerobic glove box. **C.** *Methanothermus fervidus* DNA binds to histone tetramers that show homology to the histones of eukaryotes. (PDB code: 1A7W) **D.** Eukaryotic histone octamer. (PDB code: 1AOI) *Source:* Part C from Kathryn A. Bailey et al. 2002. *J. Biol. Chem.* **277**:9293; part D from K. Luger et al. 1997. *Nature* **389**:251.

eukaryotes. **Figure 19.5A** compares the components of an archaeal RNA polymerase (from *Sulfolobus*) with those of eukaryotic RNA polymerase II, the enzyme that catalyzes synthesis of messenger RNA. The archaeal polymerase possesses two transcription factors (regulatory protein components) that are found in eukaryotes: TATA-binding protein (TBP, a subunit of transcription factor II D, TFIID) and transcription factor II B (TFIIB, which in archaea is designated TFB). By contrast, bacteria have no homologs of these factors. A consequence of the eukaryotic-like transcription and translation in archaea is that archaea are resistant to antibacterial antibiotics that target transcription and translation.

Another eukaryotic structure for which archaeal homologs were discovered is that of **histones**, the fundamental packaging proteins of DNA. The histone complex found in eukaryotic chromosomes contains a histone $(H3 + H4)_2$ tetramer flanked by two histone (H2A + H2B) dimers. The archaeal homologs form an $(H3 + H4)_2$ tetramer, with no (H2A + H2B) homologs. John Reeve and colleagues at the Ohio State University (**Fig. 19.5B**) showed that isolated DNA of specific sequences could be bound and curved around a histone tetramer from the methanogen *Methanothermus fervidus* (**Fig. 19.5C**). The DNA has AT-rich sequences that specifically fit the histone complex. The key sequences of the histones (the ones that bind to DNA) show homology to eukaryotic histones. Histones have since been found in many species of archaea.

Phylogeny of Archaea

What are the major kinds of Archaea? The domain includes three phyla, also called divisions, that have been extensively characterized (**Fig. 19.6**). These phyla are **Crenarchaeota**, **Thaumarchaeota**, and **Euryarchaeota**. In addition, a growing number of phyla emerge through exploring the sequences of environmental samples, such as the Ancient Archaeal Group (AAG) of vent hyperthermophiles and benthic subsurface communities. Major taxa are outlined in **Table 19.2**. **An expanded version of this table appears in Appendix 2.**

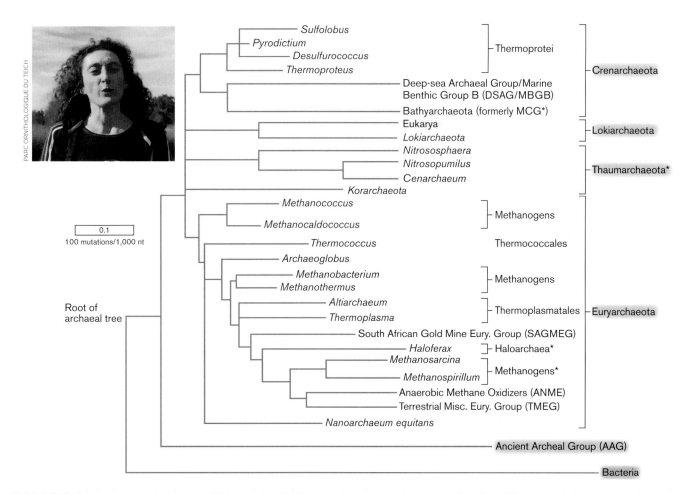

FIGURE 19.6 ■ **Archaeal phylogeny.** Major clades, divisions, and metagenomic groups of Archaea. Divergence is based approximately on SSU rRNA and genomic sequences. **Inset:** Purificación López-García argues that hyperthermophiles evolved into mesophiles by acquiring genes from bacteria. Starred taxa contain >20% bacterial genes.

TABLE 19.2 Archaeal Diversity* (Expanded Table Appears in Appendix 2.2 and eAppendix 4.2)

Crenarchaeota. Tetraether membranes surrounded by S-layer. Wide range of growth temperatures.

Bathyarchaeota (formerly MCG). Widely divergent crenarchaeotes in diverse soil and water.

Deep-Sea Archaeal Group/Marine Benthic Group B (DSAG/MBGB). Anoxic marine sediments: deep-sea methane hydrates, hydrothermal vents, coastal regions.

Thermoprotei. Many thermophiles in hot springs and marine vents. Also includes marine mesophiles and psychrophiles. *Thermosphaera.*

- **Caldisphaerales.** Thermoacidophilic heterotrophs grow in hot springs. *Caldisphaera* spp.
- **Desulfurococcales.** Anaerobic sulfur reduction with organic electron donors. Irregularly shaped cells with glycoprotein S-layer; no cell wall (*Aeropyrum pernix, Desulfurococcus fermentans*). *Ignicoccus islandicus. Pyrodictium abyssi, Pyrodictium occultum, Pyrolobus fumarii* grow at marine thermal vents, up to 110°C.
- **Psychrophilic marine crenarchaeotes (uncharacterized).** Marine water, deep sea, and Antarctica.
- **Sulfolobales.** Aerobic acidophiles; moderate thermophiles. Oxidize H_2S to H_2SO_4. *Sulfolobus, Sulfurisphaera, Acidianus.*
- **Thermoproteales.** *Pyrobaculum, Thermoproteus, Vulcanisaeta.*

Thaumarchaeota. Tetraether membranes include crenarchaeol. Marine and soil archaea that oxidize NH_3 with O_2. Important marine sources of nitrates for phytoplankton, and of methylphosphonate ($CH_3-PO_3^{2-}$), which bacteria convert to methane. Includes members of Marine Group 1.1a.

- **Cenarchaeales.** Sponge symbionts; grow at 10°C. *Cenarchaeum symbiosum.*
- **Nitrosopumilales.** Ammonia-oxidizing archaea (AOA), marine and soil. *Nitrosopumilus maritimus, Nitrososphaera gargensis.*
- **Psychrophilic marine thaumarchaeotes (uncharacterized).** Marine water, deep sea, and Antarctica. Anaerobic heterotrophs, sulfate reducers, nitrite-reducing methanotrophs.

Euryarchaeota. Metabolism includes methanogenesis, halophilic photoheterotrophy, and sulfur and hydrogen oxidation; acidophiles and alkaliphiles. Methanogens and halophiles have rigid cell walls.

Altiarchaeota. Includes *Altiarchaeum.* Form grappling-hook biofilms in cold sulfidic water.

Anaerobic Methane Oxidizers (ANME). Anoxic marine sediments. Oxidize methane from methanogens, in syntrophy with sulfate-reducing bacteria.

Archaeoglobi. Order **Archaeoglobales**. Hyperthermophiles; sulfate oxidation of H_2 or organic hydrogen donors; reverse methanogenesis. *Archaeoglobus fulgidus.*

Haloarchaea. Order **Halobacteriales**. Halophiles; grow in brine (concentrated NaCl). Conduct photoheterotrophy, with light-driven H^+ pump and Cl^- pump. *Haloarcula, Halobacterium, Haloferax* grow in salterns and salt lakes. *Haloquadra (Haloquadratum)* are square-shaped. *Halorubrum lacusprofundi* is an Antarctic psychrophile. *Natronococcus* spp. are alkaliphiles in soda lakes.

Methanogens (many classes). Generate methane from CO_2 and H_2, formate, acetate, other small molecules; strict anaerobes. Pseudopeptidoglycan or sulfated chondroitin cell walls. Grow in anaerobic soil, water, or animal digestive tracts. Deep-sea psychrophiles generate methane hydrates.

- **Methanobacteriales.** Lack cytochromes; reduce CO_2, formate, or methanol with H_2 by electron bifurcation. *Methanobrevibacter smithii* and *Methanosphaera stadtmanae* inhabit human digestive tract.
- **Methanomicrobiales, Methanococcales, and Methanopyrales.** Lack cytochromes; reduce CO_2, formate, or methanol with H_2 by electron bifurcation. *Methanocaldococcus jannaschii*, vent thermophile.
- **Methanosarcinales.** Possess cytochromes; reduce methylamines and acetate (as well as CO_2 and formate) with H_2.

Nanoarchaeota. Vent hyperthermophiles; obligate symbionts attached to *Ignicoccus. Nanoarchaeum equitans.*

South African Gold Mine Euryarchaeotal Group (SAGMEG). Deep subsurface. Anaerobic heterotrophs.

Terrestrial Miscellaneous Euryarchaeotal Group (TMEG). Deep subsurface and surface soils, marine and freshwater sediments.

Thermococci. Hyperthermophiles (grow above 100°C) and barophiles (up to 200 atm pressure). Anaerobes; reduce sulfur. Order **Thermococcales**: *Thermococcus, Pyrococcus abyssi, P. furiosus.*

Thermoplasmata. Extreme acidophiles; oxidize sulfur from pyrite (FeS_2), generating sulfuric acid. Mesophiles or moderate thermophiles. Order **Thermoplasmatales**: *Ferroplasma acidiphilum* and *F. acidarmanus* grow at 37°C–50°C. Oxidize sulfur from FeS_2, generating ambient pH as low as pH 0. No cell wall. *Thermoplasma acidophilum* grows at 59°C and pH 2.

Deeply branching groups of uncultured Archaea.

Ancient Archaeal Group (AAG). Vent hyperthermophiles and benthic subsurface communities.

Korarchaeota. Hyperthermophiles in Yellowstone hot springs and in deep-sea thermal vents.

Lokiarchaeota. Found in marine sediment near thermal vent; shares 3% eukaryotic genes. May represent ancient ancestor of Eukarya.

Marine Hydrothermal Vent Group (MHVG). Vent hyperthermophiles.

*Bulleted terms are representative orders (unless stated otherwise).

For several reasons, the phylogeny of Archaea is a challenge to define. Most archaea are uncultured and are known solely through small-subunit (SSU) rRNA sequence or through metagenomics (the genome sequences of a microbial community). Many of their genomes are highly "recombinogenic." For example, two different samples of *Ferroplasma acidarmanus* show 99% identical SSU rRNA, yet their overall genomes differ by 22%, implying extensive horizontal gene exchange. And many archaeal genomes, particularly those of mesophiles, include large portions of DNA transferred from bacteria. (In **Figure 19.6**, the starred taxa contain more than 20% bacterial genes.) Microbial ecologist Purificación López-García, Université Paris-Sud, proposes that hyperthermophilic archaeal species evolved into mesophiles by acquiring genes from bacteria. According to her model, the bacterial genes provided key metabolic pathways and cell structures optimized for growth at moderate or cold temperatures.

Finally, deeply branching clades such as the Ancient Archaeal Group (AAG) have such divergent SSU rRNA that their sequence fails to amplify with "standard" PCR primers based on the most highly conserved regions of the gene. Much "dark matter" remains unknown, as yet undetected by our current molecular tools.

Note: The phylogeny of Archaea undergoes frequent revision, as new taxa are discovered and novel genomes are sequenced. For example, the taxa now designated Thaumarchaeota were originally classified with the Crenarchaeota, until multilocus genome comparison revealed the deeper division between the two clades.

Crenarchaeota. The Crenarchaeota include a substantial proportion of soil, marine, and benthic (marine sediment) microbial communities and are found in the microbiomes of plants and animals. Understanding these organisms is important for assessing their contribution to the global carbon cycle.

A fascinating group of crenarchaeotes is the thermophiles. Many thermophilic crenarchaeotes metabolize sulfur, either by anaerobic reduction (such as by H_2 to form H_2S) or by aerobic oxidation (by O_2 to form sulfuric acid). Anaerobic sulfur metabolizers include moderate thermophiles (growth range about 60°C–80°C), as well as hyperthermophiles (90°C–120°C). Many of the hyperthermophiles are also barophiles, growing under high pressure at hydrothermal vents on the ocean floor; an example is *Pyrodictium abyssi*. Yet other crenarchaeotes are mesophiles (growing at moderate temperatures) and even psychrophiles (growing at temperatures below 20°C), found in Antarctic lakes. Overall, the Crenarchaeota encompass the widest range of growth temperature of any division of life.

Thaumarchaeota. The phylum Thaumarchaeota includes mesophiles originally classified with Marine Group I Crenarchaeota but now recognized as a distinct clade very important for marine environments. Thaumarchaeota include ammonia-oxidizing archaea (AOA), which oxidize ammonia to nitrite. These ammonia oxidizers play a major role in the nitrogen cycle (discussed in Chapters 21 and 22). Other Thaumarchaeota include mesophilic heterotrophs and sulfur oxidizers in soil and water. Some are key symbionts of invertebrate animals such as deep-sea sponges.

Lokiarchaeota. A recently characterized clade of deep-sea archaea is the Lokiarchaeota, branching from the Deep-Sea Archaeal Group/Marine Benthic Group B (DSAG/MBGB). First identified at a thermal vent named Loki's Castle, the Lokiarchaeota genomes share surprising traits with eukaryotes. Eukaryotic-like genes found in Lokiarchaeota include those associated with an actin cytoskeleton and phagocytosis. These features suggest that Lokiarchaeota may descend from a common ancestor of a cell that contributed to the ancestral eukaryote.

Euryarchaeota. The phylum Euryarchaeota also includes members throughout soil and water, and associated with plants and animals. They show a greater range of metabolism than the Crenarchaeota. The most highly divergent group of Euryarchaeota is the methanogens. Methanogens serve a key energetic role in ecosystems by offering an anaerobic mechanism for removing excess H_2 and other small-molecule reductants. Despite their common energetic pathway, methanogens branch deeply among other euryarchaeotic groups, and they show a wide range of cell shape and environmental adaptations. One branch of euryarchaeotes is the Haloarchaea, extreme halophiles that are the only form of life to grow in concentrated brine (NaCl). Most haloarchaea, such as *Halobacterium* species NRC-1, are photoheterotrophs that can supplement their metabolism with light-driven ion pumps.

Euryarchaeotes include extreme acidophiles, as well as extreme alkaliphiles. With respect to pH, the euryarchaeotes show the widest range of any clade, from *Ferroplasma* (order Thermococcales), growing at pH 0, to *Natronococcus*, growing at pH 10.

Note: In Table 19.2, certain archaeal names contain a "bacteria" component—for example, the genus *Halobacterium*. Such organisms were known and named as bacteria before 1977, when the category "archaea" was first defined.

Emerging phyla. Besides the major phyla Crenarchaeota, Thaumarchaeota, and Euryarchaeota, metagenomic analysis continues to reveal new clades that branch near the root of the tree. The Miscellaneous Crenarchaeota Group (MCG) was reclassified in 2015 as the deep-branching group Bathyarchaeota, capable of methanogenesis. The Bathyarchaeota are the only known methanogens outside of the Euryarchaeota. An even deeper branch is the Ancient Archaeal Group (AAG), discovered in hydrothermal vents and deep benthic sediment. The AAG organisms fail to be detected by PCR primers based on known SSU rRNA gene sequences. Instead, their detection requires reverse transcription of rRNA from uncultured cells. The reverse transcripts have shown that the rRNA sequence of AAG organisms differs by 20% from the most conserved regions of other known archaeal rRNA.

Thought Question

19.1 Suppose two deeply diverging clades each show a wide range of growth temperature. What does this suggest about the evolution of thermophily or psychrophily?

To Summarize

■ **Archaeal membranes are composed of L-glycerol diether or tetraether lipids,** with isoprenoid side chains that may include cross-links or pentacyclic rings.

■ **Many archaea have no peptidoglycan cell wall, but an S-layer of protein.** Some methanogens have cell walls of pseudomurein.

FIGURE 19.7 ■ Thermophiles colonize a hot spring. Morning Glory Pool, a hot spring in Yellowstone National Park, supports thermophilic archaea and bacteria.

■ **Glucose is catabolized by variants of the ED pathway.** Other metabolic pathways found in archaea include methanogenesis and retinal-associated light-driven ion pumps such as bacteriorhodopsin.

■ **Central genetic functions of archaea resemble those of eukaryotes,** as seen in the structure of DNA and RNA polymerases and of histone-like DNA-binding proteins.

■ **The most studied phyla or divisions of archaea are Crenarchaeota, Thaumarchaeota, and Euryarchaeota.** Crenarchaeota include sulfur hyperthermophiles, mesophiles, and psychrophiles. Thaumarchaeota include ammonia oxidizers, marine invertebrate symbionts, and others. Euryarchaeota include methanogens, halophiles, and acidophiles.

■ **The deep-sea clade Lokiarchaeota** includes genes shared by an ancestor of the eukaryotes.

■ **Metagenomics reveals uncultured organisms.** We continue to discover deeply branching clades of archaea by designing new PCR primers, reverse-transcribing community rRNA, and sequencing metagenomes.

19.2 Crenarchaeota across the Temperature Range

How were the first crenarchaeotes discovered and described? The name Crenarchaeota means "scalloped archaea," derived from the amorphous scalloped cell shapes of the first hyperthermophiles discovered. The first members described were hyperthermophiles such as *Thermosphaera* species (see **Table 19.2**). Since their discovery, however, numerous mesophiles and psychrophiles have been identified in soil and marine habitats.

Thought Question

19.2 What are the advantages of naming a microbial lineage after a morphological feature? What problems might arise later, as more members of the lineage are characterized?

Habitats for Thermophiles

Thermophiles and hyperthermophiles commonly grow in hot springs and geysers, such as those of Yellowstone National Park (**Fig. 19.7**) or the Solfatara volcanic area near Naples, Italy. A hot spring occurs where water seeps underground above a magma chamber, which heats the water to near boiling. The heated water expands and is

TABLE 19.3	Hyperthermophilic Crenarchaeota			
Representative species	**Growth temperature (°C)**	**Growth pH**	**Cell shape**	**Metabolism**
Aeropyrum pernix	70–100°	pH 5–9	Cocci	O_2 respiration
Desulfurococcus fermentans	78–87°	pH 6	Flagellated cocci	Anaerobic S^0 respiration or fermentation
Ignicoccus islandicus	70–98°	pH 5–7	Flagellated cocci with periplasmic space	Anaerobic lithotrophy, S^0 oxidation of H_2
Pyrodictium abyssi	80–110°	pH 5–7	Disks linked by cannulae	Anaerobic oxidation of H_2 by S^0 or $S_2O_3^{2-}$ or fermentation
Sulfolobus solfataricus	50–87°	pH 2–4	Irregular cocci	O_2 respiration on S^0, producing H_2SO_4
Thermosphaera aggregans	65–90°	pH 5–7	Flagellated cocci in aggregates	Anaerobic fermentation

forced upward through fissures, coming out in a heated spring. In a geyser, the water is heated under pressure. As the water escapes upward, it turns into steam, which expands and jets upward, falling into a heated pool. These heated pools and their surrounding edges generate extreme ranges of temperature, mineral content, and acidity. They support a diverse range of microbial life, including thermophilic cyanobacteria and firmicutes, as well as archaea.

Several features of hot springs and geysers are important for thermophiles:

■ **Reduced minerals.** The heated water dissolves high concentrations of sulfides and other reduced minerals. When the water emerges and cools, the minerals precipitate. These reduced minerals serve as rich energy sources for autotrophs.

■ **Low oxygen content.** At higher temperatures, the oxygen concentration of water declines. Therefore, hyperthermophiles tend to be anaerobic, although there are important exceptions, such as the aerobic sulfur oxidizer *Sulfolobus.*

■ **Steep temperature gradients.** The temperature of the water falls dramatically within a short distance from the source, forming a steep gradient. Different species of thermophiles are adapted to different temperatures and grow in separate patches at the different temperatures, causing a variegated pattern.

■ **Acidity.** Some hot-spring environments show extreme acidity. The acidity results from oxidation of sulfur or iron in reactions that generate strong inorganic acids such as sulfuric acid (H_2SO_4).

A special subcategory of volcanic hot-spring habitats is that of submarine hydrothermal vents on the ocean floor. The vent thermophiles must evolve to grow at high pressure

under several kilometers of ocean. Pressure increases by approximately 100 atm per kilometer of ocean depth. Organisms that grow only at high pressure are called **barophiles** (discussed in Chapter 5).

Desulfurococcales: Reducing Sulfur from Hot Springs

Species of the order **Desulfurococcales** show distinctive cell structures and forms of metabolism (**Table 19.3**). All possess elaborate S-layers but lack a cell wall; their diphytanylglycerol membranes contain a combination of diethers and tetraethers. Many take advantage of the high temperatures that increase the thermodynamic favorability of sulfur redox reactions. An example is *Desulfurococcus fermentans* (**Fig. 19.8A**), a flagellated coccoid cell isolated from hot springs; it grows optimally at 85°C. *D. fermentans* respires anaerobically by reducing elemental sulfur (S^0) to sulfide (HS^-). Sulfur reduction is coupled to oxidation of small organic molecules such as sugars.

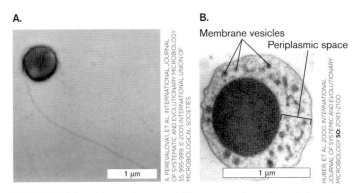

A.
1 μm

A. PEREVALOVA1, ET AL. INTERNATIONAL JOURNAL OF SYSTEMATIC AND EVOLUTIONARY MICROBIOLOGY 55, 995-999, © 2005 INTERNATIONAL UNION OF MICROBIOLOGICAL SOCIETIES.

B.
Membrane vesicles
Periplasmic space
1 μm

HUBER, ET AL. 2000, INTERNATIONAL JOURNAL OF SYSTEMIC AND EVOLUTIONARY MICROBIOLOGY **50**: 2093-2100.

FIGURE 19.8 ■ Hyperthermophilic crenarchaeotes. A. *Desulfurococcus fermentans* (shadow EM). **B.** *Ignicoccus islandicus* (TEM). Its unique periplasmic space contains membrane vesicles.

Another flagellated coccus, *Ignicoccus islandicus*, has an unusual cell architecture (**Fig. 19.8B**). *Ignicoccus* has an outer membrane surrounding its cytoplasmic membrane, with a large aqueous compartment between them. This outer compartment contains membrane-enclosed vesicles of unknown function. The evolution of this compartment, similar to the extra membranes of the bacterial genus *Planctomyces* (see Chapter 18), suggests a model for an intermediate stage of evolution of the eukaryotic nucleus. *I. islandicus*, unlike *Desulfurococcus*, is a marine organism, growing at temperatures as high as 98°C. It is a lithotroph, oxidizing hydrogen with sulfur:

$$H_2 + S^0 \longrightarrow H_2S$$

Most of the cultured species of Desulfurococcales are obligate anaerobes. An exception is *Aeropyrum pernix*, one of the first archaea to have its genome sequenced. *A. pernix* is an aerobic heterotroph, respiring with O_2 on complex compounds during growth at 70°C–100°C.

Barophilic Vent Hyperthermophiles

The most extreme hyperthermophiles are barophiles adapted to grow near hydrothermal vents at the ocean floor. The high pressure beneath several kilometers of ocean allows water to remain liquid at temperatures above 100°C; the highest known temperature for growth of an organism is 121°C.

A common feature of thermal vents is the **black smoker** (**Fig. 19.9**). A black smoker is a chimney-like structure resulting from the upwelling of seawater superheated by an undersea magma chamber. As in a geyser above ground, the heated water is forced upward through a small opening. Because the thermal vent is under steam pressure, the water can reach temperatures of over 400°C, enabling it to dissolve high concentrations of minerals such as iron II sulfide (FeS). When the rising water escapes, however, it immediately cools, depositing iron sulfide around the edge of the vent chimney and precipitating iron sulfide particles that cloud the water—hence the term "black smoker." While no organism can grow at 400°C, various species of archaea are adapted to grow in the range of 100°C–120°C, where the vent stream meets the seawater and minerals precipitate (**Fig. 19.9B**).

How do we study organisms under such extreme conditions? To study hyperthermophiles from black smoker vents requires specialized equipment. The isolation of such organisms is a challenge because their habitats endanger our own survival. Undersea vent systems must be approached by a special submersible device with a robotic arm. An example is the Environmental Sample Processor from the Monterey Bay Aquarium Research Institute (**Fig. 19.10A**). The robotic system samples temperature and other properties of fluid emerging from a black smoker hydrothermal

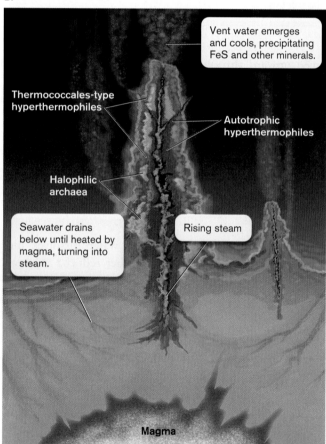

FIGURE 19.9 ■ Extreme temperature and pressure: black smoker vents. A. Black smoker vents with steam escaping from "chimneys" of sulfide minerals at the Juan de Fuca Ridge on the ocean floor. **B.** Different parts of the smoker vent system support different classes of archaea.

vent. It can then sample organisms for study. An advanced version of this device can actually process the organism's DNA. Thus, the DNA can be obtained from vent-adapted microbes that could not survive transfer to a laboratory at sea level. The robotic sample processor is supported by NASA as a model for a future space probe to explore one of

A.

COURTESY OF MONTEREY BAY AQUARIUM RESEARCH INSTITUTE. MARK GREISE © 2001 MBARI

FIGURE 19.10 ■ Robotic sampling from a black smoker vent. **A.** Engineer Gene Massion, from the Monterey Bay Aquarium Research Institute, deploys the submersible Environmental Sample Processor with robotic collection arm at an ocean site off the coast of Maine. **B.** The Deep Aquarium, a pressurized device for cultivation of vent organisms.

B.

IMAGES COURTESY OF JAPAN AGENCY FOR MARINE-EARTH SCIENCE AND TECHNOLOGY.

Jupiter's moons, Europa, considered a possible source of extraterrestrial life.

Organisms that do survive transport to sea level must nonetheless be maintained at high pressure and temperature to ensure viability. In the laboratory, all devices for microscopy and cultivation must be kept under pressure and at high temperature. Organisms are cultured in a pressurized cell such as the "Deep Aquarium" (**Fig. 19.10B**). The culture must be provided with reduced minerals and gases needed for the growth of vent microbes.

Vent-adapted crenarchaeotes include *Pyrodictium abyssi* (**Fig. 19.11**), *P. occultum*, and *P. brockii*; the latter is named for Thomas Brock of the University of Wisconsin–Madison, a pioneering researcher of hyperthermophiles. For energy, *Pyrodictium* species reduce sulfur to H_2S, either with molecular hydrogen or with organic compounds. A membrane-embedded sulfur-reducing complex and a proton-translocating ATP synthase have been isolated from *P. abyssi*. The complexes are extremely heat stable, exhibiting a temperature optimum of 100°C.

Pyrodictium species grow as flat, disk-shaped cells that can be as thin as 0.1 μm. The cells contain a periplasm and outer membrane with an S-layer that is coated with zinc sulfide, a mineral that precipitates from the vent. The cell disks are interconnected by glycoprotein tubules called **cannulae** (singular, **cannula**). The cannulae can extend to more than 0.1 mm, forming complex networks of connections (**Fig. 19.11**). In liquid culture, the networks grow into white balls up to 10 mm in diameter. Cryo-electron tomography of a *Pyrodictium* cell shows that the cannulae bridge the periplasm between cells, but not the cytoplasm. The cannulae may enable *Pyrodictium* cells to share nutrients and maintain a biofilm, while keeping their cellular identity distinct with separated cytoplasm.

What happens when a *Pyrodictium* cell divides? Cells of *P. abyssi* generate new cannulae as they undergo fission (**Fig. 19.12**). Some of the new cannulae form as loops

Cannulae

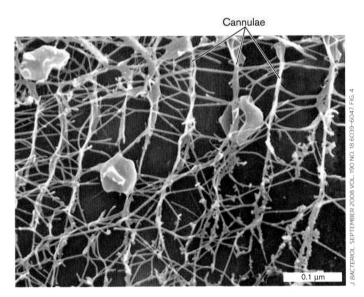

J. BACTERIOL. SEPTEMBER 2008 VOL. 190 NO. 18 6039–6047. FIG. 4

0.1 μm

FIGURE 19.11 ■ *Pyrodictium abyssi* **growing as networks of cells linked by cannulae.** SEM.

Cannula

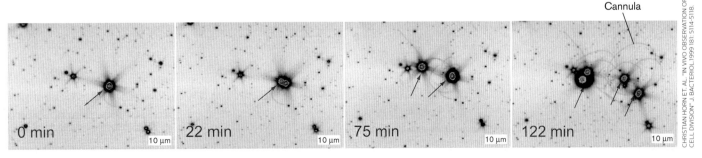

0 min   22 min   75 min   122 min
10 μm   10 μm   10 μm   10 μm

CHRISTIAN HORN ET. AL. "IN VIVO OBSERVATION OF CELL DIVISION" J. BACTERIOL. 1999 181: 5114–5118.

FIGURE 19.12 ■ *Pyrodictium abyssi* **undergoing cell division.** Cells of *P. abyssi* generate new interconnecting cannulae as they divide.

connecting the two daughter cells while simultaneously pushing the two cells apart. In this fashion, the cell division process expands the cell network.

Note: Two genera of vent thermophiles have similar names but only distant genetic relatedness: *Pyrodictium abyssi*, a crenarchaeote; and *Pyrococcus abyssi*, a euryarchaeote.

The intricate cell network of *P. abyssi* is an example of a single-species biofilm. Other forms of biofilms are found at hydrothermal vents. *Thermosphaera aggregans* forms colonies so tightly bound that they cannot be dissociated by protease treatment or sonication (**Fig. 19.13**). Multispecies biofilms of hyperthermophiles line the chimneys of black smoker vents.

> ### Thought Question
>
> **19.3** What might be the advantages of flagellar motility for a hyperthermophile living in a thermal spring or in a black smoker vent? What would be the advantages of growth in a biofilm?

Sulfolobales: High Temperature and Extreme Acid

Can some archaea grow in extreme acid, as well as high temperature? The crenarchaeote order **Sulfolobales** includes species that respire by oxidizing sulfur (instead of reducing it as *Desulfurococcus* does). These organisms, such as *Sulfolobus* species, grow at 80°C–90°C within hot springs and solfataras (volcanic vents that emit only gases). Ken Stedman and colleagues at Portland State University

FIGURE 19.14 ■ Isolating *Sulfolobus* at Yellowstone.
A. A researcher holds a collecting tube at length to obtain samples from a steaming spring, Rabbit Creek, at Yellowstone National Park. **B.** *Sulfolobus* species grow at 80°C at pH 2–3.

study *Sulfolobus solfataricus*, a species that grows at 80°C and pH 2 (**Fig. 19.14**). *Sulfolobus* species oxidize organic compounds with oxygen, or they oxidize S^0 or H_2S to sulfuric acid:

$$2S^0 + 3O_2 + 2H_2O \longrightarrow 2H_2SO_4 \longrightarrow 4H^+ + 2SO_4^{2-}$$

As a result of sulfuric acid production, the pH of the organism's surroundings falls to pH 2–3, effectively excluding all but acidophiles. While other archaea grow at higher temperatures (up to 120°C) or lower pH (below pH 0), *Sulfolobus* is of interest as a "double extremophile," requiring both high temperature and extreme acidity simultaneously.

Sulfolobus cells have a membrane composed mainly of tetraethers with cyclopentane rings (see **Fig. 19.2E**). Tetraether membranes are commonly seen in acidophilic thermophiles, probably because they are exceptionally impermeable to protons. Remarkably, *Sulfolobus* has no cell wall, but only an S-layer of glycoprotein (**Fig. 19.15A**; also see **Fig. 19.3A**). Like all archaea, these organisms are nonpathogenic to animals, yet they secrete toxins deadly to competitor strains of *Sulfolobus*.

Sulfolobus species are not obligate autotrophs; they can also grow heterotrophically on sugars or amino acids. In fact, many species are easily cultured in tryptone broth at 80°C, pH 3. Their internal pH is typically pH 6.5; thus, they maintain more than three units of pH difference across their membrane. The full metabolic potential of this organism is revealed by annotation of its genome, which contains homologs of sugar and amino acid transporters, as well as the non-ATP-forming Entner-Doudoroff pathway of glucose catabolism. Genes are present for enzymes

FIGURE 19.13 ■ Hyperthermal biofilm. *Thermosphaera aggregans* forms a dense aggregate of cells (SEM).

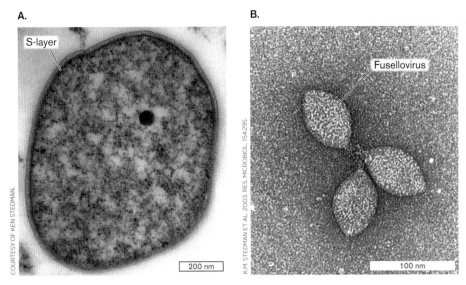

FIGURE 19.15 ■ ***Sulfolobus.*** A. *Sulfolobus* cell with thick S-layer. B. Fusellovirus, a double cone–shaped virus that infects *S. solfataricus.*

to oxidize HS^- and $S_2O_3^{2-}$, as well as S^0. The main redox carrier for respiration appears to be ferredoxin (instead of NADH, which is relatively unstable at high temperature).

Thought Question

19.4 What problem with cell biochemistry is faced by acidophiles that conduct heterotrophic metabolism?

Because *Sulfolobus* species are readily cultured, their metabolism and ecology have been studied extensively, revealing unexpected features. Some species show multiple origins of replication of their DNA; for example, *S. solfataricus* and *S. acidocaldarius* each use three active replication origins. The presence of multiple origins is yet another feature of archaeal DNA management that resembles that of eukaryotes.

Another interesting discovery in *Sulfolobus* was that of archaeal viruses. *Sulfolobus* species are attacked by a number of viruses, including fuselloviruses (**Fig. 19.15B**). Fuselloviruses generally resemble bacteriophages in size and function, but their capsid is spindle-shaped, forming a cone at each end. Spindle-shaped capsids are common in archaeal viruses, but never seen in bacteriophages.

How does a virus infect an archaeon? The process of a viral infection has been observed in *S. solfataricus* (**Fig. 19.16**). Cells were infected with *Sulfolobus* turreted icosahedral virus (STIV), a virus isolated from a boiling acid hot spring in Yellowstone National Park. Each icosahedral particle of STIV has 12 turret-like projections, as shown in the cryo-EM image reconstruction in **Figure 19.16A**. Mature virions contain a double-stranded DNA genome coated with lipid within the capsid. The thin section of an infected cell (**Fig. 19.16B**) shows progeny virions packed into a hexagonal array, portions of which poke through the host cell's S-layer in pyramidal forms. After lysis (**Fig. 19.16C**), the S-layer complex is all that remains of the empty cell. The S-layer appears surprisingly intact, showing that it provides a sturdy cell covering perhaps comparable in strength to a cell wall.

This lytic cycle resembles lytic and fast-release cycles of bacterial and eukaryotic viruses, although the capsid "turrets" and the pyramidal bulges of the lysing cell are unique to archaea. The structure of a capsid protein component shows surprising homology to both bacterial and eukaryotic viral proteins. This finding supports the hypothesis that modern viruses derive from an ancient reproductive form predating the divergence of the three domains of life (discussed in Chapter 11).

Numerous other archaeal viruses have been discovered in high-temperature environments, including icosahedral, tailed, and filamentous forms. All known archaeal viruses

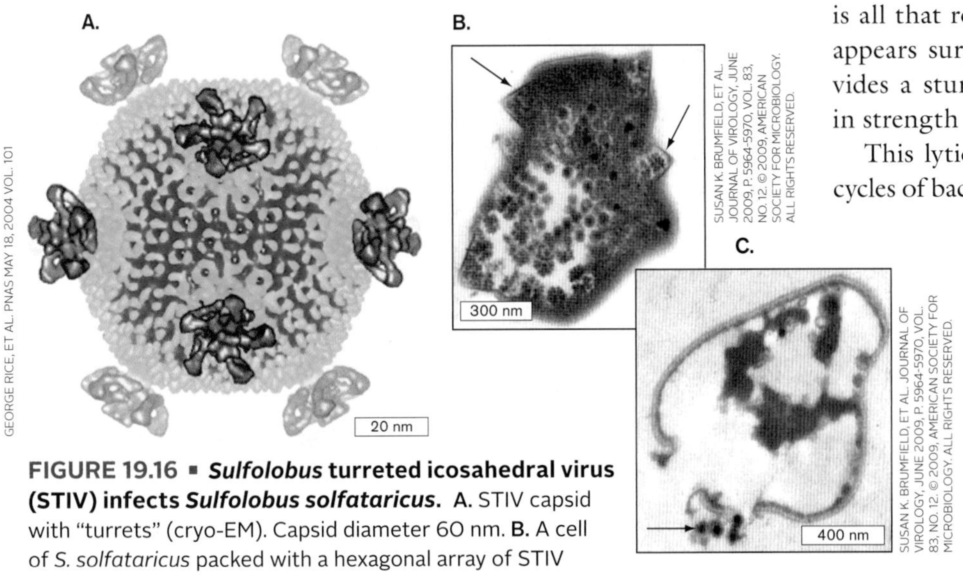

FIGURE 19.16 ■ ***Sulfolobus* turreted icosahedral virus (STIV) infects *Sulfolobus solfataricus.*** A. STIV capsid with "turrets" (cryo-EM). Capsid diameter 60 nm. B. A cell of *S. solfataricus* packed with a hexagonal array of STIV particles. Arrows point to pyramidal bulges where virus arrays poke through a breach in the S-layer. C. Empty cell membrane and S-layer following lysis and viral release. Arrow points to released virus particles.

FIGURE 19.17 ■ Crenarchaeotes growing on the surface of a tomato root. Crenarchaeotes were detected by fluorescence in situ hybridization (FISH, explained in eAppendix 3). Red fluorescence arises from a fluorophore-DNA probe that hybridizes to an rRNA sequence specific to Crenarchaeota.

10 µm

H. SIMON, OREGON HEALTH AND SCIENCE U. BEAVERTON, OR

have genomes consisting of double-stranded DNA, suggesting that only double-stranded DNA is stable enough for virus particles to persist at high temperature. In 2012, however, a possible archaeal RNA-dependent RNA transcriptase (discussed in Chapter 11) was identified from a Yellowstone hot spring. This finding suggests that there may exist positive-strand RNA viruses that infect hyperthermophilic archaea.

Thought Question

19.5 What hypotheses might be proposed about archaeal evolution if viruses of mesophilic archaea are found to have RNA genomes? What if, instead, all archaeal viruses have DNA genomes only?

Besides Sulfolobales, the class Thermoprotei includes other orders of thermoacidophiles. Caldisphaerales is an order of thermoacidophiles first isolated from a Philippine hot spring at Mount Makiling. Members of Caldisphaerales, such as *Caldisphaera*, typically grow at pH 3 up to 80°C. Unlike *Sulfolobus* species, *Caldisphaera* species are

anaerobes or microaerophiles, tolerating only low concentrations of oxygen. They grow by fermentation or anaerobic respiration. Another order, Thermoproteales, includes hyperthermoacidophiles isolated from marine vents; thus, they survive extreme pressure, as well as heat and acid! *Thermoproteus* species grow at temperatures up to 97°C and at pH values lower than pH 3. Their rod-shaped cells are less than 0.3 µm in length—one of the smallest cell types known. Their metabolism is autotrophic, gaining energy by reducing sulfur with H_2 to H_2S.

Mesophilic Crenarchaeota

Because archaea were first isolated from extreme habitats, it came as a surprise when SSU rRNA probes revealed crenarchaeotes in moderate habitats throughout the biosphere. The mesophiles may have evolved from hyperthermophiles by gene acquisition from mesophilic bacteria (see Section 19.1). Today, many crenarchaeotes—perhaps the majority, in nature—grow in water or soil, or in association with plants; for example, **Figure 19.17** shows crenarchaeotes growing on the surface of a tomato root. Others are marine mesophiles. The first marine mesophiles were found in 1992 by Edward DeLong and colleagues in samples from the Pacific Ocean (**Fig. 19.18A**). A survey in 2001 of marine archaea at the Hawaii Ocean Time-series Station found high numbers of crenarchaeotes (**Fig. 19.18B**). The abundance of crenarchaeotes varied according to season and increased with depth, typically comprising 40% of the total microbial population at depths of 1,000 meters, where temperatures are cold.

A.

COURTESY OF ED DELONG.

B.

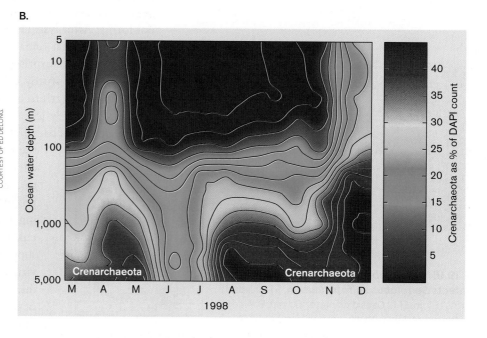

FIGURE 19.18 ■ Crenarchaeota in the Pacific Ocean. A. Ed DeLong, now at the Massachusetts Institute of Technology, discovered marine crenarchaeotes (some of which have since been reclassified as thaumarchaeotes). **B.** The proportion of Crenarchaeota (color profile) is shown as a function of depth and season, measured at the Hawaii Ocean Time-series Station. Crenarchaeotes were identified by FISH (explained in eAppendix 3). A fluorescein-labeled DNA probe specific to Crenarchaeota was hybridized to a cellular rRNA sequence; DNA was detected using the DNA-binding fluorophore DAPI.
Source: Part B from Markus Karner et al. 2001. *Nature* **409**:507.

Some of the organisms identified as crenarchaeotes were later reassigned as thaumarchaeotes (discussed in the next section).

Psychrophilic Crenarchaeota

Genetic surveys also reveal crenarchaeotes that are psychrophiles—species adapted to low temperatures. Alison Murray and colleagues from the Desert Research Institute found crenarchaeotes growing in sea ice off Antarctica (**Fig. 19.19**). Organisms were collected from ice and seawater at a temperature of –1.8°C; the water remains liquid below zero because of the high salt concentration. Other crenarchaeotes are adapted to high pressure as well as cold. At the ocean floor, at temperatures of about 2°C, benthic strata show crenarchaeotes that include anaerobic heterotrophs, sulfate reducers, and anaerobic methanotrophs (methane oxidizers). Methanotrophs are particularly crucial in recycling the methane produced by seafloor methanogens (euryarchaeotes, discussed in Section 19.4).

Another source of psychrophilic archaea is Ace Lake, Antarctica, a well-studied cold, high-salt habitat with temperatures in the range of 14°C–24°C and bottom anoxic layers never warmer than 2°C. Ace Lake supports crenarchaeotes, as well as euryarchaeotes such as methanogens (discussed in Section 19.4).

To Summarize

- **Crenarchaeote species are found in a wide range of extreme and moderate habitats.**

- **Habitats for hyperthermophiles** include hot springs and submarine hydrothermal vents. Vent organisms are barophiles as well as thermophiles. Anaerobic hyperthermophilic acidophiles include Caldisphaerales and Thermoproteales.

- **Desulfurococcales includes diverse thermophiles.** Most are anaerobes that use sulfur to oxidize hydrogen or organic molecules.

- *Pyrodictium* species are disk-shaped cells interconnected by cytoplasmic bridges called cannulae.

- *Sulfolobus* species oxidize sulfur or H_2S to sulfuric acid, and they catabolize organic compounds.

- **Oceans, soil, plant roots, and animals** provide habitats for mesophilic and psychrophilic crenarchaeotes. Psychrophilic crenarchaeotes in marine sediment play a key role in recycling methane hydrates produced by methanogens.

19.3 Thaumarchaeota: Symbionts and Ammonia Oxidizers

The first mesophilic crenarchaeotes were identified by the sequence of their SSU rRNA. But more recent analysis combined the sequences of multiple genes, such as those encoding ribosomal proteins. This more finely tuned analysis showed that some of the organisms, the Marine Group I Crenarchaeota, branched deeply from the rest. These deep-branching archaea were reclassified as a new clade, the Thaumarchaeota. Thaumarchaeotes synthesize a distinctive tetraether lipid called crenarchaeol, containing

FIGURE 19.19 ■ Collecting psychrophiles in Antarctica. **A.** Alison Murray studies psychrophilic microbes in Antarctica. **B.** Microbes are collected off Bonaparte Point, Antarctica, from seawater samples down to 45 meters. **C.** Antarctic seawater samples are filtered to concentrate psychrophilic archaea and bacteria.

FIGURE 19.20 ■ Crenarchaeol: a biosignature for thaumarchaeotes. Crenarchaeol is a diphytanylglycerol diether containing a six-membered cyclic ring.

a six-membered cyclic ring (**Fig. 19.20**). Crenarchaeol was originally named for "Crenarchaeota" but now appears to be found mainly in those organisms reclassified as Thaumarchaeota. Thaumarchaeota include intriguing symbionts of marine sponges, as well as ammonia oxidizers that perform a key role in cycling nitrogen.

Ammonia-Oxidizing Thaumarchaeotes Cycle Global Nitrogen

From a global standpoint, the most significant thaumarchaeotes discovered are the ammonia-oxidizing archaea (AOA), such as those of the order Nitrosopumilales. Ammonia-oxidizing archaea gain energy by aerobically oxidizing ammonia to nitrite:

$$2NH_3 + 3O_2 \longrightarrow 2NO_2^- + 2H_2O + 2H^+$$

This lithotrophic reaction yields redox energy, enabling the microbe to fix CO_2 for biomass (discussed in Chapter 14). In marine environments, excreted ammonia can build up to high levels, until it is oxidized by archaea and bacteria. In some habitats, the Thaumarchaeota perform most of the recycling of ammonia, such as the ammonia excreted by fish in your aquarium. The thaumarchaeotes thus help balance the ecology of intertidal pools and beaches. The reaction plays a key role in the global nitrogen cycle, as the first step of returning organic nitrogen to atmospheric N_2 (discussed in Chapters 21 and 22). It also provides a major source of nitrite for marine phytoplankton.

Habitats for free-living ammonia-oxidizing archaea (AOA). Ammonia-oxidizing archaea are found in marine environments, such as the seafloor of Tanoura Bay, Japan (**Fig. 19.21**). Tatsunori Nakagawa and colleagues at Nihon University, Japan, used enrichment culture (described in Chapter 4) to hunt for new kinds of AOA. Samples of sand from the seafloor were serially diluted in medium containing ammonium sulfate and adjusted to pH 8, a pH high enough for significant deprotonation of ammonium ion (NH_4^+) to ammonia (NH_3). During the enrichment culture, ammonia was progressively consumed and converted to nitrite (NO_2^-). In the culture, the gene encoding ammonia oxidase (*amoA*) was identified by PCR amplification. The ammonia-oxidizing archaea (AOA) were identified by fluorescence in

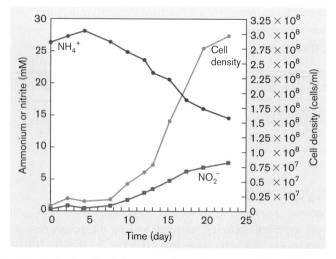

FIGURE 19.21 ■ Enrichment culture for ammonia oxidizers. Ammonia-oxidizing microbes from eelgrass seafloor samples convert ammonia to nitrite. *Source:* Naoki Matsutani et al. 2011. *Microbes and Environments* 26:23–29.

situ hybridization (FISH) using a fluorophore attached to DNA that hybridizes to rRNA of *Nitrosopumilus* species.

The best-studied ammonia-oxidizing thaumarchaeote, *Nitrosopumilus maritimus*, was isolated from a marine aquarium in 2005 by David Stahl and colleagues at the University of Washington, Seattle. *Nitrosopumilus* and related AOA oxidize ammonia at extremely low concentrations. Thus, AOA quickly remove a substance toxic to fish. AOA have since been found throughout marine and freshwater communities, as well as in soil, where they perform the important function of ammonia removal. They appear in industrial waste sludge, revealed by fluorescent DNA probes (**Fig. 19.22**). While bacteria also oxidize ammonia, in many habitats the AOA are the dominant oxidizers. Ammonia-oxidizing thaumarchaeotes continue to be discovered in other habitats, including Antarctic lakes down to temperatures of –20°C.

Symbiotic ammonia-oxidizing archaea. Are AOA found on humans? Christine Moissl-Eichinger discovered AOA in the microbiome of human skin, where they are the major archaeal component of the skin microbiome. The human body continually emits ammonia, so AOA may help metabolize this toxic substance while controlling skin pH for the microbial community. The finding of ammonia-oxidizing thaumarchaeotes on the skin, as well as methanogens in the gut, may help explain another discovery made by Moissl-Eichinger—namely, the DNA signatures of thaumarchaeotes and methanogens in hospital intensive care units and industrial clean-room facilities.

In the deep ocean, some of the psychrophilic thaumarchaeotes found by DeLong and colleagues live as endosymbionts of marine animals such as the sponge. The thaumarchaeote *Cenarchaeum symbiosum* inhabits the sponge *Axinella mexicana* (**Fig. 19.23A**). Like other

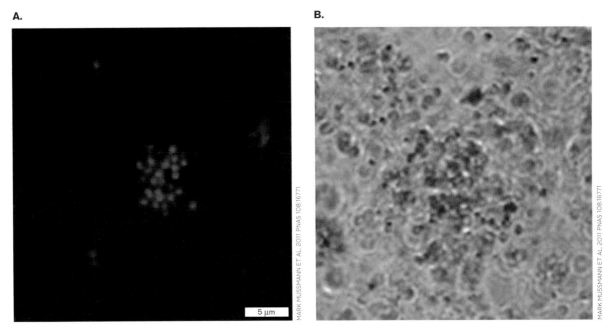

FIGURE 19.22 ■ **Ammonia-oxidizing Thaumarchaeota in nitrogen-rich industrial sludge.** **A.** FISH probe hybridizes with rRNA sequence specific to thaumarchaeotes. **B.** Light micrograph.

marine thaumarchaeotes, *C. symbiosum* has yet to be grown in pure culture. But the presence of the microbes is shown by the fluorescence of a DNA probe that hybridizes to sequences specific to *C. symbiosum* (**Fig. 19.23B**). How the microbe benefits its sponge host is unknown, but the sponge and its endosymbionts can be cocultured in an aquarium for many years. A hypothesis investigated by researchers is that *C. symbiosum* produces antimicrobial agents that protect the sponge from pathogens. Some of the products of *C. symbiosum* are being tested for their pharmacological properties.

Further metagenomic studies of soil and water continue to reveal additional novel kinds of Thaumarchaeota, including thermophiles. Overall, the clade may turn out to be as diverse as the Crenarchaeota.

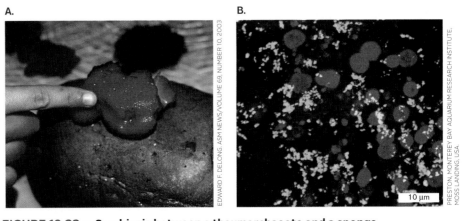

FIGURE 19.23 ■ **Symbiosis between a thaumarchaeote and a sponge.** **A.** *Cenarchaeum symbiosum* inhabits the sponge *Axinella mexicana.* **B.** Differential fluorescent staining of *C. symbiosum* present in sponge tissue (green fluorescence) visualized by FISH (fluorescent probes that hybridize rRNA; discussed in Chapter 21). Host cell nuclei fluoresce red (DAPI stain).

To Summarize

- **Ammonia-oxidizing archaea (AOA)** contribute to global nitrogen cycling in water and soil.

- **Ammonia-oxidizing archaea comprise a significant part of the human skin microbiome.**

- **Thaumarchaeotes include symbionts of marine sponges.** Emerging thaumarchaeotes include mesophiles, psychrophiles, and thermophiles in many environments.

19.4 Methanogenic Euryarchaeota

A third major branch of known archaea is Euryarchaeota, the "broad-ranging archaea" (see **Table 19.2**). The euryarchaeotes include several independent branches of **methanogens**, species that derive energy through reactions producing methane. Methanogens release methane from soil and animal digestive habitats throughout Earth. For the methanogen, methane is just a by-product of its energy-yielding metabolism, but this by-product is a greenhouse gas with profound consequences for our biosphere (discussed in Chapter 22).

The phylogeny of **Figure 19.6** shows how clades of methanogens branch deeply, as diverse as all the euryarchaeotes together. Two possible explanations of this phylogeny would be (1) the ancestral euryarchaeote was a methanogen, but many descendant lineages lost the metabolism; or (2) divergent euryarchaeotes acquired the genes for methanogenesis later, independently by gene transfer. As yet, we lack sufficient genomic data to determine which model is correct. Furthermore, methanogenesis now appears in one deep-branching clade outside the Euryarchaeota, now called the Bathyarchaeota (formerly the Miscellaneous Crenarchaeota Group).

Biogenic Paths to Methane

Biogenic methane formation has actually been observed since 1776, when the Italian priest Carlo Campi and physicist Alessandro Volta (1745–1827) investigated bubbles of "combustible air" from a wetland lake. Volta wrote:

Being in a little boat on Lake Maggiore, and passing close to an area covered with reeds, I started to poke and stir the bottom with my cane. So much air emerged that I decided to collect a quantity in a large glass container. . . . This air burns with a beautiful blue flame.

Volta noted that methane arose from wetlands containing water-saturated decaying plant material. In 1882, the German medical researcher Hermann von Tappeiner (1847–1927) combined plant materials with ruminant stomach contents (a source of methanogens) and showed that both components were essential to produce methane. During the late nineteenth century, in England, methane was collected from manure and sewage and used as a fuel for street lamps.

Different species generate methane from different substrates, such as CO_2 and H_2 or small organic compounds, generally fermentation products of bacteria. Each pathway of methanogenesis generates a small free energy change ($\Delta G°'$) that is just enough to drive processes of carbon fixation and biosynthesis. All types of methanogenesis are poisoned by molecular oxygen and therefore require extreme anaerobiosis. Major substrates and reactions include:

Carbon dioxide: $CO_2 + 4H_2 \longrightarrow CH_4 + 2H_2O$
Formic acid: $4CHOOH \longrightarrow CH_4 + 3CO_2 + 2H_2O$
Acetic acid: $CH_3COOH \longrightarrow CH_4 + CO_2$
Methanol: $4CH_3OH \longrightarrow 3CH_4 + CO_2 + 2H_2O$
Methylamine: $4CH_3NH_2 + 2H_2O \longrightarrow$
$$3CH_4 + CO_2 + 4NH_3$$
Dimethyl sulfide: $2(CH_3)_2S + 2H_2O \longrightarrow$
$$3CH_4 + CO_2 + 2H_2S$$

In CO_2 reduction, the most common form of methanogenesis, carbon dioxide plus molecular hydrogen combine to form water and methane. CO_2-reducing methanogens are autotrophs, growing solely on CO_2 and H_2 with a source of nitrogen and other minerals. Methanogenesis from other carbon sources, such as formate, acetate, or methanol, is heterotrophic and generates CO_2 as a product in addition to methane. The nitrogen-containing substrate methylamine generates ammonia in addition to methane and CO_2, whereas the sulfur-containing substrate dimethyl sulfide generates hydrogen sulfide.

Note that only a narrow range of substrates supports methanogenesis. For unknown reasons, most methanogens lack the vast array of energy-yielding pathways found in soil bacteria such as *Rhodopseudomonas* or *Streptomyces*. Thus, methanogens generally require close association with bacterial partners to provide their substrates—a relationship called **syntrophy** (discussed in Chapter 14). Syntrophic methanogens usually grow in habitats with minimal resource flux, where hydrogen and carbon dioxide gases can be trapped for their use. Removal of these gases then enhances bacterial metabolism.

Many kinds of methanogens can be cultured in the laboratory. Their culture poses special challenges because the reactions of methanogenesis are halted by oxygen. Thus, most methanogens are strict anaerobes, although a few that tolerate oxygen have been found. To exclude oxygen, all growth and manipulation of cultures are performed in an anaerobic chamber. Furthermore, methanogens that use CO_2 and H_2 as substrates must receive a steady supply of these gases. Even more challenging is the culture of hyperthermophilic methanogens, such as *Methanopyrus*, a deep-ocean vent hyperthermophile that grows at scorching temperatures up to 122°C. For these organisms, we must maintain both high temperature and high pressure.

Methanogens include at least four classes, of which five major orders are shown in **Table 19.4**. The amount of divergence within an order can be nearly as great as that between orders. Particularly striking is the wide range of growth temperatures among closely related species. Thermophiles and even hyperthermophiles branch from closely related mesophiles; for example, the order Methanobacteriales includes *Methanobrevibacter ruminantium* (37°C–39°C), *Methanobacterium thermoautotrophicum* (50°C–75°C), and *Methanothermus fervidus* (60°C–97°C).

Methanogens Show Diverse Cell Forms

Despite their metabolic similarity, methanogens display an astonishing diversity of form, perhaps as diverse as the entire domain of bacteria (**Fig. 19.24**). For example, *Methanocaldococcus jannaschii* cells grow as cocci with

TABLE 19.4	Methanogens				
Representative species	**Growth temperature (°C)**	**Growth pH**	**Cell shape**	**Substrates for methanogenesis**	
Methanobacteriales					
Methanobrevibacter ruminantium	37–39°	pH 6–9	Chains of short rods	H_2 and CO_2	
Methanobacterium thermoautotrophicum	50–75°	pH 7–8	Filaments of long rods	H_2 and CO_2, formate	
Methanothermus fervidus	60–97°	pH 6–7	Rods	H_2 and CO_2	
Methanococcales					
Methanococcus vannielii	20–40°	pH 7–9	Cocci with flagella	H_2 and CO_2	
Methanocaldococcus jannaschii	48–94°	pH 6–7	Cocci with flagella	H_2 and CO_2	
Methanomicrobiales					
Methanoculleus olentangii	30–50°	pH 6–8	Cocci	Acetate, complex nutrients	
Methanospirillum hungatei	20–45°	pH 6–7	Spirilla	Acetate	
Methanopyrales					
Methanopyrus kandleri	84–122°	pH 6–8	Long rods (2–14 µm)	H_2 and CO_2	
Methanosarcinales					
Methanosarcina barkeri	20–50°	pH 5–7	Aggregates of cocci	H_2 and CO_2, methanol, methylamine, acetate	
Methanosaeta concilii	10–45°	pH 6–8	Filaments of rods	Acetate	
Methanohalophilus zhilinae	45°	pH 8–10	Cocci	Methanol, methylamine, dimethyl sulfide	

Source: Harald Huber and Karl O. Stetter. 2002. *The Prokaryotes.* Springer-Verlag.

numerous flagella attached to one side, while *Methanosarcina mazei* forms peach-shaped cocci lacking flagella. *Methanothermus fervidus* cells are short, fat rods without flagella, and *Methanobacterium thermoautotrophicum* grows as elongated rods reminiscent of the bacterial genus *Bacillus*. Still others, such as *Methanospirillum hungatei*, form wide spirals. The morphological diversity of methanogens may be explained in part by their rigid cell walls, which can maintain a distinctive shape.

The composition of methanogen cell walls is much more diverse than that of bacteria. *Methanobacterium* species have a cell wall composed of pseudopeptidoglycan, a structure in

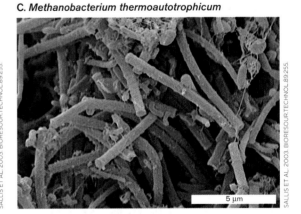

A. *Methanosarcina mazei*

RALPH ROBINSON/VISUALS UNLIMITED AND D. L. MAEDER ET AL. 2006. J. BACTERIOL. 188:7922

B. *Methanothermus fervidus*

SALLIS ET AL. 2003. BIORESOUR.TECHNOL. 89:255.

C. *Methanobacterium thermoautotrophicum*

SALLIS ET AL. 2003. BIORESOUR.TECHNOL. 89:255.

10 µm

5 µm

5 µm

FIGURE 19.24 ■ **Methanogens show a wide range of shapes.** A. *Methanosarcina mazei*, a lobed coccus form lacking flagella (SEM). B. *Methanothermus fervidus*, a short bacillus (SEM). C. *Methanobacterium thermoautotrophicum*, an elongated bacillus (SEM).

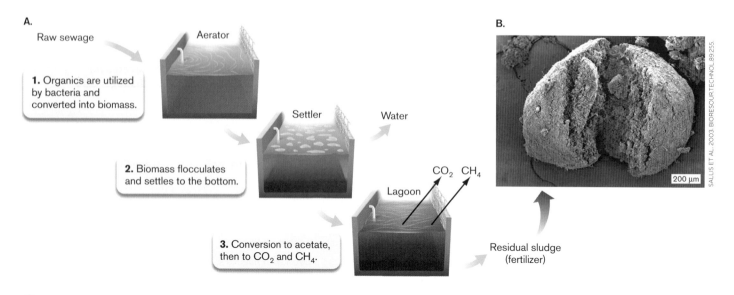

A.

Raw sewage → Aerator

1. Organics are utilized by bacteria and converted into biomass.

Settler → Water

2. Biomass flocculates and settles to the bottom.

Lagoon

CO_2 CH_4

3. Conversion to acetate, then to CO_2 and CH_4.

Residual sludge (fertilizer)

B.

200 μm

SALLIS ET AL. 2003. BIORESOUR.TECHNOL.89:255.

C.

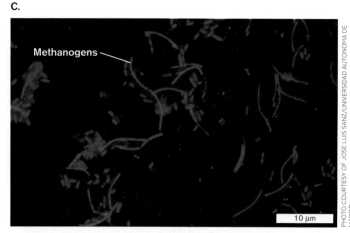

Methanogens

10 μm

PHOTO COURTESY OF JOSE LUIS SANZ/UNIVERSIDAD AUTONOMA DE MADRID.

FIGURE 19.25 ■ Filamentous methanogens bind bacterial communities in waste treatment. **A.** Raw sewage is aerated and decomposed by bacteria, followed by anaerobic incubation. Under anaerobiosis, the bacterial waste products are converted to methane and CO_2. The bacteria are packed together by filamentous methanogens to form sludge. **B.** Sludge (wastewater sediment) forms granules packed with bacteria and methanogens (discussed in Chapter 22). **C.** Filamentous methanogens entangle bacteria, forming "flocs" that settle in the wastewater tank. Methanogens fluoresce red with a DNA probe for Methanomicrobiales, while other wastewater microbes fluoresce blue (DAPI stain).

which chains of alternating amino sugars are linked by peptide cross-bridges analogous to those of peptidoglycan. By contrast, *Methanosarcina* species have a cell wall composed of sulfated polysaccharides. The genera *Methanomicrobium* and *Methanococcus* have protein-derived cell walls.

Filamentous methanogens form chains of large cells similar to those of filamentous cyanobacteria. Filamentous methanogens, such as *Methanosaeta*, perform a key function in the treatment of sewage waste (**Fig. 19.25**). In waste treatment, the raw sewage first undergoes aerobic respiration by bacteria, converting the organic materials to small molecules such as CO_2 and acetate, and then anaerobic digestion of the remainder. The bacteria performing anaerobic decomposition become trapped in filaments of *Methanosaeta*, which convert bacterial fermentation products such as acetate into methane and CO_2. The methanogenic filaments serve a key function by trapping bacteria into granules that settle out from the liquid.

Anaerobic Habitats for Methanogens

Methanogens grow in soil, in animal digestive tracts, and in marine floor sediment. A major methanogenic

environment is the anaerobic soil of wetlands (**Fig. 19.26**). The wetlands that generate the most methane are typically disturbed or artificial wetlands, particularly rice paddies, which contain high levels of added fertilizer that bacteria convert to the substrates used by methanogens. From the standpoint of the organism, methane is an incidental byproduct, but this product has great significance for our biosphere as a greenhouse gas (discussed in Chapter 22).

COURTESY OF CHANGSHENG LI

FIGURE 19.26 ■ Rice paddies release methane. Chinese farmers plant rice in waterlogged soil, an anaerobic habitat for methanogens.

Methanogenesis in soil and landfills. Chinese farmers have found that one way to decrease methane production is to drain the soils used for rice production. Changsheng Li and colleagues at the University of New Hampshire showed in experimental plots that drained soil produces less methane. The drained soil receives more oxygen, which blocks methanogenesis. Draining soil also stimulates rice root development and accelerates decomposition of organic matter in the soil to release nitrogen. Sufficiently dry rice paddy soil can actually become a sink for atmospheric methane, thanks to methane-oxidizing bacteria.

Another major source of methanogens is landfills. Landfills such as those outside New York City are among the largest human-made structures on Earth. They are rich in organic wastes, which bacteria ferment to CO_2, H_2, and short-chain organic molecules that methanogens convert to methane. As a result, large amounts of gas can build up and spontaneously combust, causing explosions. To avoid explosions, the methane needs to be piped out. In some cases, the gas can be collected and used to generate electricity.

As global temperatures increase, the largest release of methane may come from the polar regions, where permafrost thaws and glaciers retreat. In Canada and Alaska, many lakes and melting permafrost soils produce methane so fast that one can ignite a flame (**Fig. 19.27**). In Antarctica, scientists such as Ricardo Cavicchioli from the University of New South Wales, Australia, have discovered psychrophilic

A.

B.

FIGURE 19.28 ■ **Antarctic lakes support psychrophilic methanogens.** **A.** Ace Lake now opens to the sea, containing anoxic sulfidic layers. **B.** Ric Cavicchioli (left) and Torsten Thomas process microbial samples collected at Ace Lake.

FIGURE 19.27 ■ **Polar methane release.** Katey Anthony at the University of Alaska Fairbanks demonstrates how bubbles of methane from methanogens beneath the ice burst into the atmosphere, where the methane ignites into flame.

methanogens (**Fig. 19.28**). The near-freezing water of Ace Lake is saturated with methane from methanogens present at high numbers. Cavicchioli showed that the Ace Lake methanogens have membrane lipids with many double bonds; the unsaturated form allows more fluidity at low temperature (see Chapter 3).

Rising temperatures accelerate methanogenesis, and the methane bubbles up so fast that it melts holes in the ice sheets covering Arctic lakes. In 2012, Andrew McDougall and colleagues at the University of Victoria, British Columbia, modeled the permafrost carbon release and its feedback effect on global warming. They project that permafrost could release more than a quarter of its carbon stores by the year 2100, and that this amount could add another 1.5°C to the global temperature by the year 2300.

SPECIAL TOPIC 19.1   Methanogens for Dinner

If our gut is full of methanogens, what are they doing there? Do they just pass through, picking up hydrogen and CO_2 along the way; or do they interact specifically with their host, and with their neighboring microbes in the gut microbiome? Elizabeth Hansen and colleagues at Washington University in St. Louis set out to address these questions. They selected a well-studied gut methanogen, *Methanobrevibacter smithii* (**Fig. 1**). *M. smithii* can be cultured inside the cecum (entrance to the large intestine) of a germ-free mouse. When cultured in the mouse, the archaeal cell forms a thick capsule of polysaccharide. By

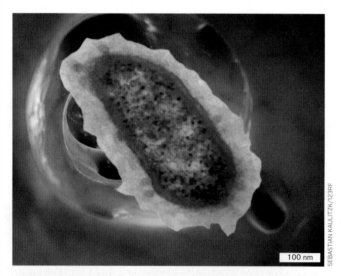

FIGURE 1 ■ *Methanobrevibacter smithii* **grown in gut shows a thick capsule.** TEM superimposed upon image of colon. *Source:* Buck Samuel et al. 2007. *PNAS* **104**:10643.

contrast, *M. smithii* cultured in a batch fermentor forms cells whose capsule is very thin. This difference suggests that capsule formation is a response to the animal host.

Hansen asked, does the existence of methanogens depend on the genetic identity of the specific host? This question is of interest because only some humans release methane, while others do not; thus, we infer that only some of us are colonized by methanogens. We do not know whether this difference arises from host genetics or from environmental factors. So, Hansen conducted a study comparing the methanogen prevalence between pairs of monozygotic twins (identical twins, from the same fertilized egg) and dizygotic twins (fraternal twins, from two different zygotes) (**Fig. 2**). Methanogen prevalence was measured by quantitative PCR amplification (qPCR) of the signature gene *mcrA*. The gene encodes methyl-coenzyme M reductase, the enzyme that converts methyl-coenzyme M (CH_3–S-CoM) and thiol-coenzyme B (HS-CoB) to methane and the CoM-CoB disulfide (CoM-S-S-CoB) (shown in **Fig. 19.31B**). A remarkably high correlation was found between the methanogen levels of the monozygotic twin pairs. The dizygotic twin pairs, however, showed no significant correlation. This finding suggests that methanogens are highly sensitive to the host molecular biology.

Hansen then asked whether the presence of methanogens depends on specific bacterial members of the gut microbiome. The researchers sequenced metagenomes (sets of partial genomes from a microbial community; presented in Chapter 21). From the metagenomes, 22 partial bacterial genomes were assembled that showed significant correlation with the presence of *M. smithii*. Remarkably, 20 of the 22 partial

Methane hydrates on the seafloor. Psychrophilic marine methanogens grow at or beneath the seafloor. These seabed methanogens generate large volumes of methane that seep up slowly from the sediment. Under the great pressure of the deep ocean, the methane becomes trapped as **methane gas hydrates**, which are crystalline materials in which methane molecules are surrounded by a cage of water molecules. The methane hydrates accumulate in vast quantities. Methane hydrates are of interest as a potential source of natural gas. But if a large part of this methane were to be released, it could greatly accelerate global warming (discussed in Chapter 22).

Fortunately, much of the methane produced by seafloor methanogens is oxidized to CO_2 by **anaerobic methane oxidizers** (**ANME**). The ANME euryarchaeotes oxidize methane in syntrophy with sulfur-reducing bacteria such as *Desulfurococcus* (discussed in Chapter 18). The bacteria reduce sulfate (abundant in seawater) to sulfide. Coupled together, the reactions of methane oxidation and sulfate reduction have the negative value of ΔG needed to drive metabolism for both kinds of organisms.

Digestive methanogenic symbionts. Numerous methanogens also grow within the digestive fermentation chambers of animals such as termites and cattle (**Fig. 19.29A**) (discussed in Chapter 21). Termites have to aerate their mounds continually in order to remove methane; when rainfall temporarily clogs the mound, the mound can be ignited by lightning and explode. Cattle support methanogenesis within their rumen and reticulum (a common veterinary trick is to insert a tube into the rumen and ignite the escaping methane gas). Bovine methanogenesis diverts carbon from meat production, and it makes a significant contribution to global methane.

Research on rumen microbiology focuses on attempts to suppress growth of methanogens. For example, Steven Ragsdale and his graduate student Bree DeMontigny at the University of Nebraska–Lincoln tested potential chemical inhibitors of methanogenesis on samples of rumen fluid (**Fig. 19.29B**). The rumen fluid was maintained in vials that serve as artificial rumens. The artificial rumens allow researchers to assess how well the proposed methane-blocking compounds might perform in an actual bovine rumen.

A. Monozygotic twins

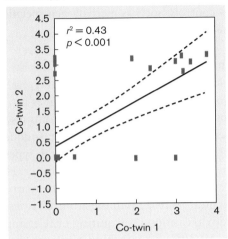

B. Dizygotic twins

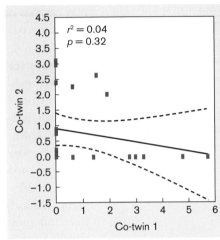

FIGURE 2 ■ **Correlation of intestinal methanogen levels between twins.** The amount of methanogens in fecal samples was measured by qPCR amplification of *mcrA*. A. Monozygotic twins show similar methanogen levels. B. Dizygotic twins show no correlation. *Source:* Elizabeth Hansen et al. 2011. *PNAS* **108**:4599.

genomes were assigned to the order Clostridiales. This is the first such specific methanogen-bacterial correlation to be shown across multiple human host metagenomes. The basis of the connection is not known, but Hansen speculates that specific methanogens couple with specific clostridia-like bacteria to provide the hydrogen gas needed for methanogenesis. In effect, methanogens may bring preferred "dinner guests" to the intestinal buffet.

RESEARCH QUESTION

Similar experiments have now been performed testing the colonization requirements of other gut methanogens, such as *Methanomassiliicoccus* species. How do the results differ, and why?

Elizabeth E. Hansen, Catherine A. Lozupone, Federico E. Rey, Meng Wu, Janaki L. Guruge, et al. 2011. Pan-genome of the dominant human gut-associated archaeon, *Methanobrevibacter smithii*, studied in twins. *Proceedings of the National Academy of Sciences USA* **108**:4599–4606.

Sonja Vanderhaeghen, Christophe Lacroix, and Clarissa Schwab. 2015. Methanogen communities in stools of humans of different age and health status and co-occurrence with bacteria. *FEMS Microbiology Letters* **362**. doi:10.1093/femsle/fnv092.

Methanogens also contribute to human digestion. Species such as *Methanobrevibacter smithii* and *Methanosphaera stadtmanae* may constitute about 10% of gut anaerobes. *Methanobrevibacter smithii* increases the efficiency of digestion by consuming excess reduced products (formate and H_2) from bacterial mixed-acid fermentation (discussed in Chapter 13). High levels of H_2 inhibit bacterial NADH dehydrogenases and thus decrease the proton potential for ATP production. *Methanosphaera stadtmanae* consumes methanol, a potentially toxic by-product of the bacterial degradation of pectin, a major fruit polysaccharide. Thus, methanogens may enhance the growth of human enteric bacteria (see **Special Topic 19.1**).

If methanogens influence the fermentation efficiency of gut flora, do

A.

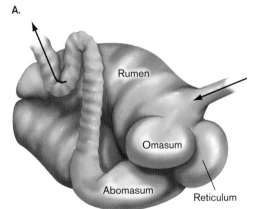

Rumen

Omasum

Abomasum

Reticulum

B.

FIGURE 19.29 ■ **Methane production in the bovine rumen.** A. Methanogens grow with the community of fermentative bacteria in the rumen and reticulum of ruminating animals. B. Bree DeMontigny, a graduate student in animal science who worked with Stephen Ragsdale at the University of Nebraska, is shown here pipetting rumen fluid samples into vials that serve as "artificial rumens." The vials were then incubated and assayed for the amount of methane formed. Ragsdale is now at the University of Michigan, Ann Arbor.

they affect the caloric content we obtain from food? Experiments with gnotobiotic mice test the interactions between methanogens and *Bacteroides thetaiotaomicron*, an anaerobic bacterium that ferments complex polysaccharides from plant foods. The mice were colonized in the presence or absence of *Methanobrevibacter smithii*, which consumes H_2 through methanogenesis. Mice colonized with both *B. thetaiotaomicron* and *M. smithii* were found to make more fat after consuming the same quantity of food than those colonized solely with *B. thetaiotaomicron*. Following up this result, a study of humans found a correlation between methane breath release and obesity. Thus, methanogenesis may somehow interact with human gut bacterial digestion to increase the efficiency of fat storage. How the composition of intestinal microbiota may influence obesity is discussed in Section 23.2.

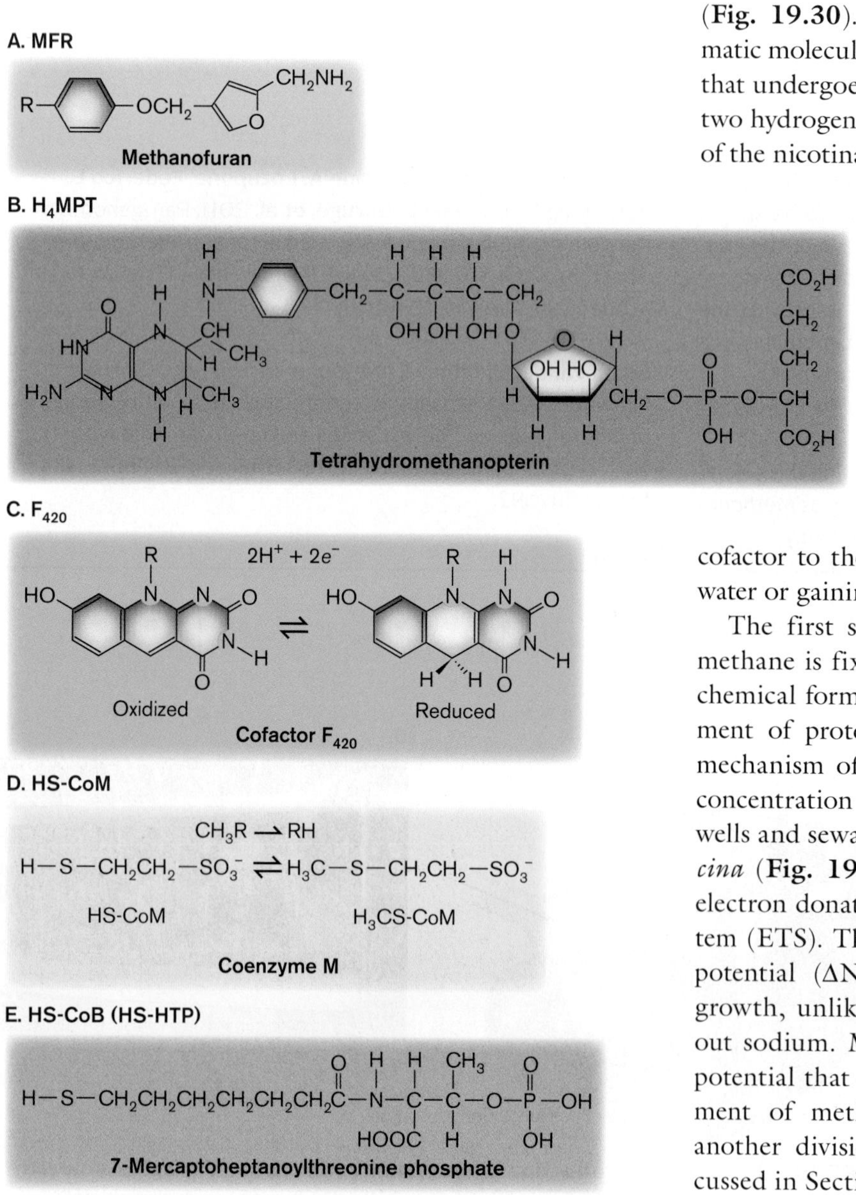

A. MFR

Methanofuran

B. H₄MPT

Tetrahydromethanopterin

C. F₄₂₀

$2H^+ + 2e^-$

Oxidized   Reduced

Cofactor F₄₂₀

D. HS-CoM

$CH_3R \longrightarrow RH$

$H{-}S{-}CH_2CH_2{-}SO_3^- \rightleftharpoons H_3C{-}S{-}CH_2CH_2{-}SO_3^-$

HS-CoM   H₃CS-CoM

Coenzyme M

E. HS-CoB (HS-HTP)

7-Mercaptoheptanoylthreonine phosphate

FIGURE 19.30 ■ **Cofactors for methanogenesis.** Cofactors specific for methanogenesis transfer the hydrogens and the increasingly reduced carbon to each enzyme in the pathway.

Biochemistry of Methanogenesis

Methanogenesis is extremely important for our environment because the process releases a potent greenhouse gas. Thus, much effort goes into better understanding how methanogenesis works. Knowledge of methanogenic biochemistry may help us develop controls, such as methanogen-specific antibiotics to minimize methane output from cattle.

Cofactors for methanogenesis. The process of methanogenesis uses a series of specific cofactors to carry each carbon from CO_2 (or other substrates) as it becomes progressively reduced by hydrogen. The hydrogen atoms also require redox carriers. Most of the cofactors are unique to methanogens, although the general structural types resemble those of redox cofactors we saw in Chapter 14 (**Fig. 19.30**). For example, cofactor F_{420} is a heteroaromatic molecule (containing nitrogens in the aromatic rings) that undergoes a redox transition by acquiring or releasing two hydrogens (**Fig. 19.30C**)—a transition similar to that of the nicotinamide ring of NADH.

Methanogenesis from CO_2. The formation of methane from CO_2 and H_2 (**Fig. 19.31**) is technically a form of anaerobic respiration in which H_2 is the electron donor and CO_2 is the terminal electron acceptor (discussed in Chapter 14). The process fixes CO_2 onto the cofactor methanofuran (MFR) and then passes the carbon stepwise from one cofactor to the next, each time losing an oxygen to form water or gaining a hydrogen carried by another cofactor.

The first step in the conversion of CO_2 and H_2 to methane is fixing the carbon onto methanofuran (for its chemical formula, see **Fig. 19.30A**). This requires placement of protons onto the oxygen to form water. The mechanism of methanogenesis depends on the available concentration of H_2. High-H_2 environments, such as oil wells and sewage sludge, favor genera such as *Methanosarcina* (**Fig. 19.31A**). The high H_2 concentration allows electron donation to CO_2 from an electron transport system (ETS). The ETS may provide energy from a sodium potential (ΔNa^+). Most methanogens require Na^+ for growth, unlike bacteria, many of which can grow without sodium. Methanogenesis ultimately generates a Na^+ potential that drives ATP synthesis. The sodium requirement of methanogens is something they share with another division of Euryarchaeota, the halophiles (discussed in Section 19.5).

Other methanogens, such as *Methanobacterium* and *Methanococcus*, use much lower concentrations of H_2 (<1/10,000 atm) common in soil or water. At lower

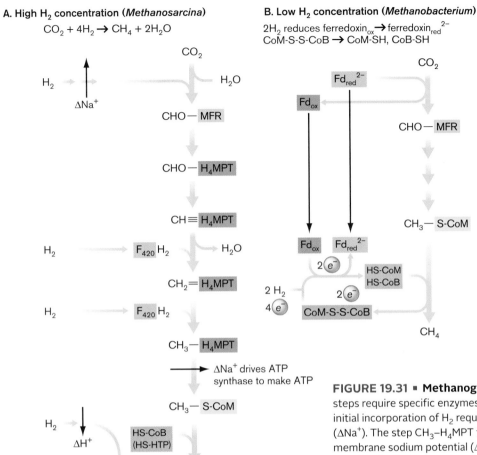

A. High H_2 concentration (*Methanosarcina*)

$$CO_2 + 4H_2 \rightarrow CH_4 + 2H_2O$$

B. Low H_2 concentration (*Methanobacterium*)

$$2H_2 \text{ reduces ferredoxin}_{ox} \rightarrow \text{ferredoxin}_{red}{}^{2-}$$
$$\text{CoM-S-S-CoB} \rightarrow \text{CoM-SH, CoB-SH}$$

FIGURE 19.31 ■ Methanogenesis from CO_2 and H_2. All steps require specific enzymes (not shown). **A.** At high $[H_2]$: The initial incorporation of H_2 requires a coupled sodium potential (ΔNa^+). The step CH_3–H_4MPT to CH_3–S-CoM generates a transmembrane sodium potential (ΔNa^+), which drives ATP synthesis. **B.** At low $[H_2]$: Electron donation to ferredoxin (Fd) requires energy input from coupled electron donation to CoM-S-S-CoB.

concentrations of substrate, the free energy (ΔG) of reaction becomes less favorable (discussed in Chapters 13 and 14). So the methanogen needs to couple CO_2 reduction to an energy-spending reaction, the reduction of CoM-S-S-CoB (**Fig. 19.31B**). This coupling of an energy-spending electron transfer (CO_2 reduction via ferredoxin) to an energy-yielding electron transfer (CoM-S-S-CoB reduction) is called **electron bifurcation**. The cost of electron bifurcation (compared to the ETS used with high H_2, see **Fig. 19.31A**) is that fewer ATPs can be made; but it allows many more types of methanogens to grow in a much wider range of natural habitats.

Later steps in methanogenesis reduce the carbon with H_2 to methane. In the final step, CoM-S-S-CoB serves as an anaerobic terminal electron acceptor for an ETS accepting electrons from H_2. Overall, H_2 reduces CoM-S-S-CoB back to the two cofactors HS-CoM and HS-CoB through an ETS, generating a proton motive force. The proton motive force drives ATP synthase.

Another important feature of methanogenesis is that the enzymes catalyzing each step require several transition metals. For example, the hydrogenase that reduces F_{420} requires both nickel and iron. The enzyme catalyzing the reaction

$$CO_2 + \text{methanofuran} + 3H \longrightarrow$$
$$\text{CHO-methanofuran} + H_2O$$

requires either molybdenum or tungsten, depending on the species; some species have two alternative enzymes, depending on which metal is available. Another metal, cobalt, is required for a B_{12}-related cofactor that participates in methane production from methanol and methylamines.

Thought Question

19.7 What do the multiple metal requirements suggest about how and where the early methanogens evolved?

Methanogenesis from acetate. Methane production from acetate is particularly important for wastewater treatment, where most of the substrate consists of short-chain bacterial fermentation products. The acetate methanogenesis pathway is not fully understood, but the initial incorporation of H_2 probably requires a coupled gradient of sodium ion, as it does for CO_2 methanogenesis. The two carbons from acetate enter different pathways, which eventually converge:

$$CH_3\text{–S-CoM} + HS\text{-CoB} \longrightarrow CH_4 + CoM\text{-S-S-CoB}$$

To Summarize

- **Methanogens gain energy through redox reactions that generate methane** from CO_2 and H_2, formate, and acetate. They require association with bacterial species that generate the needed substrates.

- **Methanogens have rigid cell walls of diverse composition in different species,** including pseudopeptidoglycan, protein, and sulfated polysaccharides.

- **Species of methanogens show a wide range of different shapes,** including rods (single or filamentous), cocci (single or clumped), and spirals.

- **Methanogens inhabit anaerobic environments** such as wetland soil, marine benthic sediment, and animal digestive organs.

- **Biochemical pathways of methanogenesis** involve transfer of the increasingly reduced carbon to cofactors that are unique to methanogens.

environments are extremely halophilic archaea, members of the class **Haloarchaea**. The halophilic archaea require at least 1.5-M NaCl or equivalent ionic strength, and most grow optimally at near saturation (about 4.3 M, seven times the concentration of seawater). Salted foods such as meat and fish and salt-cured hides can be spoiled by haloarchaea.

Most haloarchaea belong to the order **Halobacteriales**, which was named before the archaea were classified as distinct from bacteria. Halobacteriales includes species of haloarchaea that diverge among the methanogens (see **Fig. 19.6**). The halophiles, however, do not conduct methanogenesis, but grow as photoheterotrophs, using light energy to drive a retinal-based ion pump to establish a proton potential. Their photopigments color salterns, brine pools that are evaporated to mine salt (**Fig. 19.33A**). The red pigment bacterioruberin protects cells from damage by light (**Fig. 19.33B**).

Haloarchaeal Form and Physiology

Halophilic microbes need a way to maintain turgor pressure—that is, to avoid cell shrinkage as cytoplasmic water runs down the osmotic gradient toward external high salt. Most bacterial halophiles compensate for high external salt by uptake or synthesis of other kinds of osmolytes, such as small organic molecules. Haloarchaea, however,

19.5 Halophilic Euryarchaeota

In the saturated salt water of Utah's Great Salt Lake or Israel's Dead Sea, swimmers float with their heads well above the dense brine, but few living things can grow (**Fig. 19.32**). The main inhabitants of high-salt

FIGURE 19.32 ■ A high-salt environment. A swimmer effortlessly floats with her head up in the highly concentrated brine of the Dead Sea, in Israel.

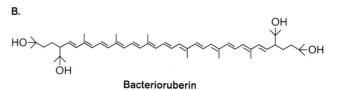

Bacterioruberin

FIGURE 19.33 ■ A saltern for salt production. A. Aerial view of a solar saltern facility in Grantsville, Utah. **B.** The red pigment bacterioruberin protects haloarchaea from damage by light.

TABLE 19.5	Halophilic Archaea					
Representative species	**NaCl range (M)**	**Growth temperature (°C)**	**Growth pH**	**Cell shape**	**Location of isolate**	
Haloarcula quadrata	2.7–4.3	53°	pH 7	Square flat; pleomorphic	Sabkha, Sinai, Egypt	
Haloarcula valismortis	3.5–4.3	40°	pH 7.5	Pleomorphic rods	Salt pools, Death Valley, CA	
Halobacterium salinarum	3.0–5.2	35–50°	pH 7	Rods	Salted cow hide	
Halococcus morrhuae	2.5–5.2	30–45°	pH 7	Cocci	Dead Sea, Israel	
Haloferax volcanii	1.5–5.2	40°	pH 7	Pleomorphic dish-shaped	Dead Sea, Israel	
Halorubrum lacusprofundi	1.5–5.2	1–44°	pH 7	Long rods (12 µm)	Deep Lake, Antarctica	
Natronococcus occultus	1.4–5.2	30–45°	pH 9.5	Cocci	Lake Magadi, Kenya	
Natronomonas pharaonis	2.0–5.2	45°	pH 9–10	Rods	Wadi El Natrun, Egypt	

Sources: Aharon Oren, The order Halobacteriales, *The Prokaryotes*, Springer-Verlag; the *Halorubrum lacusprofundi* growth temperature range comes from Ricardo Cavicchioli. 2006. *Nat. Rev. Microbiol.* **4**:331.

adapt to high external NaCl by maintaining a high intracellular concentration of potassium chloride (about 4-M KCl). Potassium ion concentrations are moderately high in most microbial cells (commonly 200 mM), but the exceptionally high KCl concentration within haloarchaea requires major physiological adaptations:

■ **High GC content of DNA.** High salt concentration decreases the fidelity in base pairing of DNA. Because the triple hydrogen bonds of GC pairs hold more strongly than those of AT pairs, the exceptionally high GC content of haloarchaea (above 60% for most species) may protect their DNA from denaturation in high salt.

■ **Acidic proteins.** Most haloarchaeal proteins are highly acidic, with an exceptional density of negative charges at their surface. The high negative charge maintains a layer of water in the form of hydrated potassium ions (K^+). This unusual hydration layer keeps the acidic proteins soluble in the high salt within the cytoplasm.

The high salt content of haloarchaea provides a convenient way to lyse cell contents for analysis: Simply transfer cells into low-salt buffer, and they fall apart owing to osmotic shock. This technique provides a quick way for beginning students to isolate DNA. This and other techniques have made the organism *Halobacterium* a model system for molecular biology education (**eTopic 19.1**).

The properties of diverse halophiles are presented in **Table 19.5**. The haloarchaea are generally uniform with respect to temperature range (mesophilic). With respect to pH, some halophiles grow at neutral pH, whereas others are alkaliphiles, growing in soda lakes above pH 9, such as Lonar Lake, Maharashtra State, India. The cell envelopes of most haloarchaea contain rigid cell walls of glycoprotein, as do some of the methanogens. Most species possess flagella for phototaxis and **gas vesicles** for maintaining their buoyancy in the upper layer of the water column. Unlike phospholipid vesicles, gas vesicles are made entirely of protein, and they are filled with air.

Haloarchaeal gas vesicles earned an unusual grant from the Bill & Melinda Gates Foundation. Shiladitya DasSarma at the University of Maryland School of Medicine proposes to use recombinant gas vesicles of *Halobacterium* NRC-1 as a delivery vehicle for recombinant vaccines against typhoid bacteria. The recombinant haloarchaeal vaccine can be embedded in salt crystals (**Fig. 19.34**)—a cheap

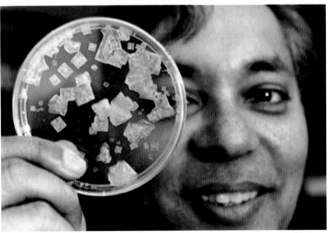

FIGURE 19.34 ■ Pink-pigmented salt crystals deliver haloarchaeal vaccine. *Halobacterium* NRC-1 embedded in salt crystals contains gas vesicles formed of a recombinant protein vaccine. Crystals are held here by Shiladitya DasSarma.

COURTESY OF PRIYA DASSARMA

and convenient delivery method for developing countries where typhoid is widespread.

With respect to cell shape, haloarchaea display considerable diversity. Some species form symmetrical rods, as in *Halobacterium* NRC-1. Other species form pleomorphic cells, flattened like pancakes; and still others form regular cocci (*Haloferax mediterranei*) (**Fig. 19.35A**). A few species grow as flattened squares—the only microbial cells known to be square (*Haloquadratum walsbyi*) (**Fig. 19.35B** and **C**). The mechanism that maintains the various shapes is unknown. In oligotrophic (low-nutrient) environments, it is likely that the greater surface-to-volume ratio gives flattened cells a competitive edge in obtaining nutrients.

Different Hypersaline Habitats Support Different Haloarchaea

Not all hypersaline (high-salt) habitats are alike. They differ with respect to pH, temperature, and the presence of other minerals, such as magnesium ion. Different kinds of hypersaline habitats support different species of haloarchaea (see **Table 19.5**), as well as halophilic bacteria. Major types of habitats include:

- **Thalassic lakes.** Thalassic lakes (from the Greek *thalassa*, meaning "ocean"), such as the Great Salt Lake in Utah, contain saturated salts with essentially the same ionic proportions as the ocean: Na^+ and Cl^- (NaCl) predominate, followed by Mg^{2+}, K^+, and SO_4^{2-}.

Thalassic lakes (also called brine lakes) support genera such as *Halobacterium*. Antarctic brine lakes support growth of cold-adapted halophiles, such as *Halorubrum lacusprofundi*, at temperatures as low as −18°C.

- **Athalassic lakes.** Athalassic lakes, such as the Dead Sea in Israel, contain higher proportions of magnesium ions. These habitats favor genera such as *Haloarcula*, which require 100 mM Mg^{2+} for growth and grow best at magnesium concentrations above 1 M.

- **Solar salterns.** These are artificial pools of brine, or saturated NaCl, that evaporate in sunlight, precipitating halite (salt crystals) for commercial production. Commercial evaporation pools for salt actually benefit from the red microbes, whose light absorption accelerates heating and evaporation.

- **Brine pools beneath the ocean.** Undersea brine pools collect near geothermal vents, as in the Gulf of Mexico, Mediterranean Sea, and Red Sea. These hypersaline regions contain salts brought up by vent water. They support hyperthermophilic halophiles, most of which have not yet been cultured.

- **Alkaline soda lakes**, such as Lake Magadi in Kenya, where carbonate salts drive the pH above pH 9. These lakes support alkaliphilic haloarchaea such as *Halobacterium*.

- **Underground salt deposits**, which contain micropockets of salt-saturated water where halophiles may survive for thousands of years.

A. *Haloferax mediterranei*

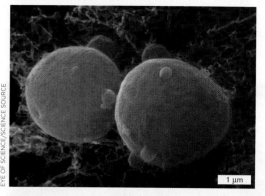

EYE OF SCIENCE/SCIENCE SOURCE

1 μm

B. Square haloarchaea

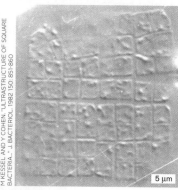

M. KESSEL AND Y. COHEN, "ULTRASTRUCTURE OF SQUARE BACTERIA." J. BACTERIOL. 1982 150: 851-860

5 μm

C. *Haloquadratum walsbyi*

W. STOECKENUS. J. BACTERIOL. 1981 148: 352-360

1 μm

FIGURE 19.35 ■ **Diverse haloarchaea.** **A.** *Haloferax mediterranei* (SEM). **B.** Square haloarchaea (*Haloquadratum walsbyi*) during their sixth round of cell division, which alternates in vertical and horizontal directions (photomicrograph, Nomarski optics). **C.** Cross-sectioned cells of square haloarchaea show their thinness (TEM).

Retinal-Based Photoheterotrophy

Most haloarchaea are photoheterotrophs, in which light-directed energy acquisition supplements respiration on complex carbon sources. Respiration occurs with oxygen or anaerobically with nitrate. A typical habitat for haloarchaea starts out as a pool with moderate salt content, containing a range of bacterial and archaeal species with varied tolerance to salt. As the pool evaporates, the salt concentrates, and the various bacteria die and lyse, releasing cell components that the haloarchaea can metabolize. The halophiles supplement their utilization of organic substrate energy by using light-driven ion pumps.

Light is captured by retinal-containing transmembrane proteins

A. Bacteriorhodopsin proton pump

B. Light-absorbing pumps and sensors

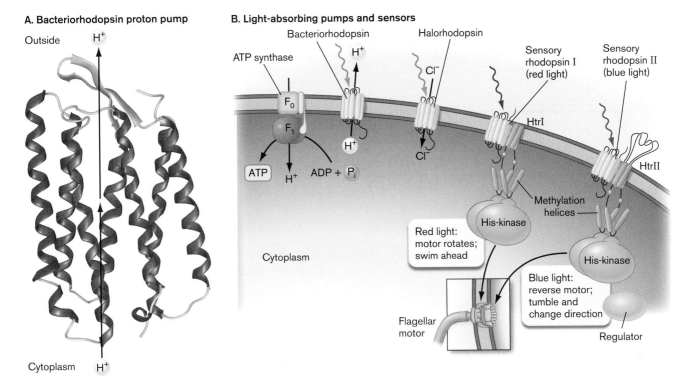

FIGURE 19.36 ■ Light-driven ion pumps and sensors. **A.** Bacteriorhodopsin absorbs light and pumps H⁺ out of the cell (PDB code: 1QKO), whereas light-activated halorhodopsin pumps chloride into the cell (not shown). **B.** Rhodopsin family molecules in the membrane of *Halobacterium*: bacteriorhodopsin, light-driven proton extrusion; halorhodopsin, light-driven chloride intake; sensory rhodopsins I and II, with their signal transduction proteins HtrI and HtrII. The sensory transduction proteins phosphorylate and dephosphorylate a protein that regulates the direction of rotation of the flagellar motor. ▶

called **bacteriorhodopsin** and the chloride pump **halorhodopsin** (**Fig. 19.36** ▶). (The role of light-driven ion pumps in phototrophy is discussed in Chapter 14.) The proton pump bacteriorhodopsin was named to indicate that it is a prokaryotic version of the better-known eukaryotic retinal rhodopsins; the name was chosen before the domain Archaea was recognized. The chloride pump halorhodopsin was discovered and named in reference to the chloride ion (a halide). The two proteins are homologs with similar structure and mechanism. In the cell membrane, they form complexes that aggregate in patches called purple membrane.

Each bacteriorhodopsin proton pump contains seven alpha helices that traverse the membrane (**Fig. 19.36A**), surrounding a buried molecule of retinal, the same light-absorbing molecule found in photoreceptors of the human retina. Light absorption triggers a conformational change that enables a proton from the cytoplasmic face to be picked up by an aspartate residue. The proton is then transferred stepwise through several other amino acid residues in the protein, leading to release of the proton outside the cell. The net result of proton transfer by bacteriorhodopsin is generation of a proton motive force that

can run a proton-driven ATP synthase, storing energy as ATP (**Fig. 19.36B**).

Halorhodopsin has a similar structure, in which chloride (instead of H⁺) is pumped <u>into</u> the cell instead of outward. Because chloride is negatively charged, this chloride transport contributes to the proton motive force. Light-driven chloride pumps are unique to haloarchaea.

Haloarchaea also possess homologs of bacteriorhodopsin that serve as sensory devices: the sensory rhodopsins I and II. These proteins, too, each have seven alpha helices containing retinal. When the sensory rhodopsins absorb light, they signal the cell to swim using its flagella (discussed in Chapter 3). The activated sensory rhodopsin I directs the cell to swim toward red light, the optimum range for photosynthesis. Sensory rhodopsin II is activated to reverse the flagellar motor and make the cell swim away from blue and ultraviolet light, which causes photooxidative damage to DNA. The rhodopsins signal through the chemotaxis machinery to the flagellar motor, using histidine kinase enzymes that phosphorylate regulatory proteins (see **Fig. 19.36B**). This response to light, or phototaxis, is an important model for archaeal molecular regulation.

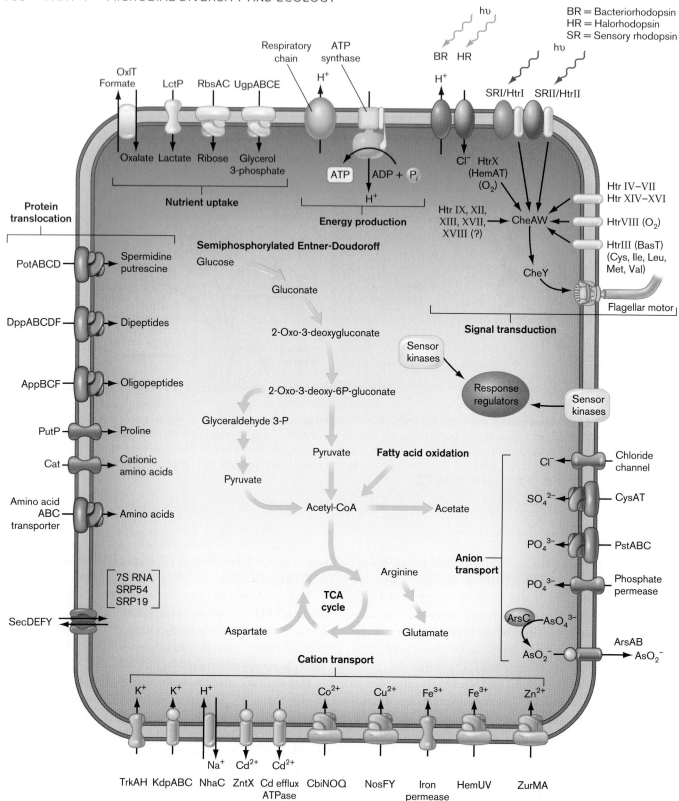

FIGURE 19.37 ■ Metabolic pathways from the genome of *Halobacterium* NRC-1. Components of metabolism predicted by the genome, including light-driven energy circuit and sensory apparatus. *Source:* Modified from Wailap V. Ng et al. 2000. *PNAS* **97**:12176.

The photosynthesis and phototaxis systems can be seen in the context of the overall metabolic map of *Halobacterium* NRC-1 (**Fig. 19.37**). In the metabolic map, the proton-translocating ATP synthase is driven by the proton potential generated through bacteriorhodopsin. Alternatively, a proton potential can be generated by the respiratory chain, through consumption of organic foods. Glucose is catabolized by the semiphosphorylated Entner-Doudoroff pathway, in which the intermediate 2-oxo-3-deoxygluconate is formed and then phosphorylated, producing one net molecule of ATP.

The *Halobacterium* genome reveals homologs for many genes that encode transporters of organic food molecules such as sugars and amino acids. In addition, the genome contains a number of cation and anion transporter genes. Some transporters move in needed molecules such as phosphate, whereas others pump out harmful substances such as arsenate. Efflux of inorganic toxins is particularly critical for growth in evaporating brine, where the concentrations of trace metals are increasing.

To Summarize

- **Haloarchaea are extreme halophiles**, growing in at least 1.5-M NaCl. They are isolated from salt lakes, solar salterns, underground salt micropockets, and salted foods.

- **Haloarchaea show diverse cell shapes**, including slender rods (*Halobacterium*), cocci (*Halococcus*), and flat squares (*Haloquadratum*). The cell envelopes of most haloarchaea contain rigid cell walls of glycoprotein.

- **Gas vesicles provide buoyancy**, enabling haloarchaea to remain near the top of the water column.

- **Molecular adaptations to high salt** include DNA of high GC content and acidic proteins (proteins with a high number of negatively charged residues).

- **Retinal-based photoheterotrophy** involves the proton pump bacteriorhodopsin or the chloride pump halorhodopsin. Both pumps contain retinal for light absorption.

19.6 Extremophilic Euryarchaeota and Deeply Branching Divisions

Besides the methanogens and halophiles, the phylum Euryarchaeota includes a number of hyperthermophiles that superficially resemble the crenarchaeotes, although their genetic divergence from these organisms is deep. Some of the euryarchaeotes are acidophiles that grow under the most extreme acidic conditions on Earth.

Thermococcales

A major group of hyperthermophilic euryarchaeotes is the order Thermococcales, including genera such as *Thermococcus* and *Pyrococcus* (**Fig. 19.38A**). These genera are most commonly isolated from submarine solfataras, volcanic vents that emit only gases. Most are anaerobes fermenting complex carbon sources, such as peptides or carbohydrates. Their growth is accelerated by the use of elemental sulfur as a terminal electron acceptor for anaerobic respiration:

$$2H^+ + 2e^- + S^0 \longrightarrow H_2S$$

Alternatively, these species can oxidize molecular hydrogen with sulfur:

$$H_2 + S^0 \longrightarrow H_2S$$

Most *Thermococcus* and *Pyrococcus* species grow at temperatures well above 90°C; *Pyrococcus woesei*, for example, grows at temperatures as high as 105°C.

Species of Thermococcales are the source of "vent polymerases" for PCR amplification. They now replace the enzyme Taq polymerase obtained from the Yellowstone hot-spring bacterium *Thermus aquaticus*, which grows only at temperatures up to 90°C. The Taq enzyme has only limited stability at or above 95°C, but DNA polymerases with greater stability at higher temperatures and increased accuracy in replication have been produced from vent-dwelling archaea such as *Pyrococcus furiosus* and *Thermococcus litoralis*. These vent polymerases now allow higher-temperature denaturation and synthesis of GC-rich sequences that were difficult to amplify with the original thermostable enzyme.

The first member of Thermococcales whose genome was sequenced was *Pyrococcus abyssi*. (Remember to distinguish *Pyrococcus abyssi* from the crenarchaeote *Pyrodictium abyssi*.) The sequence was completed in 1999 by Daniel Prieur,

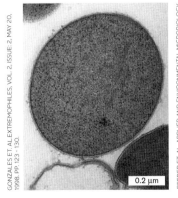

A. *Pyrococcus horikoshii*

0.2 µm

GONZALES ET. AL EXTREMOPHILES, VOL. 2, ISSUE. 2, MAY 20, 1998. PP. 123 - 130.

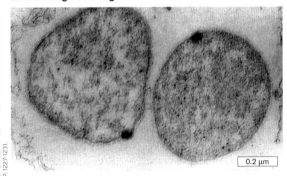

B. *Archaeoglobus fulgidus*

0.2 µm

BEEDER ET. AL. APPLIED AND ENVIRONMENTAL MICROBIOLOGY, 1994, P. 1227-1231.

FIGURE 19.38 ■ **Hyperthermophilic Euryarchaeota.** **A.** *Pyrococcus horikoshii*, isolated from a hydrothermal vent at the Okinawa Trough (TEM). **B.** *Archaeoglobus fulgidus* (TEM), isolated from hot water in the North Sea oil field.

Patrick Forterre, and colleagues at Genoscope, the French National Sequencing Center. The genome of *Pyrococcus abyssi* reveals an especially high number of eukaryotic homologs for DNA replication, transcription, and protein translation. Examples include the eukaryotic-like primase, helicase, and endonuclease for generation of Okazaki fragments in the lagging strand of DNA synthesis. At the same time, the genome also shows bacterial homologs for cell division and DNA repair. Thus, *P. abyssi* offers a striking view of mixed heritage from the common ancestor of the three domains.

Pyrococcus species possess enzymes conducting most of the classic conversions of the EMP pathway of glycolysis, but several steps use enzymes unrelated to those in bacteria. Examples include two ADP-dependent enzymes—glucokinase and phosphofructokinase—as well as the phosphate-independent enzyme glyceraldehyde 3-phosphate ferredoxin oxidoreductase. The glyceraldehyde 3-phosphate conversion is unique in that it sidesteps the use of inorganic phosphate, which is required at this step by bacterial glycolysis (shown earlier, in **Fig. 19.4**).

Pyrococcus and other members of Thermococcales have enzymes requiring tungsten, a metal rarely required outside the archaea. (Tungsten is present in elevated concentrations at hydrothermal vents.) Another unusual feature of

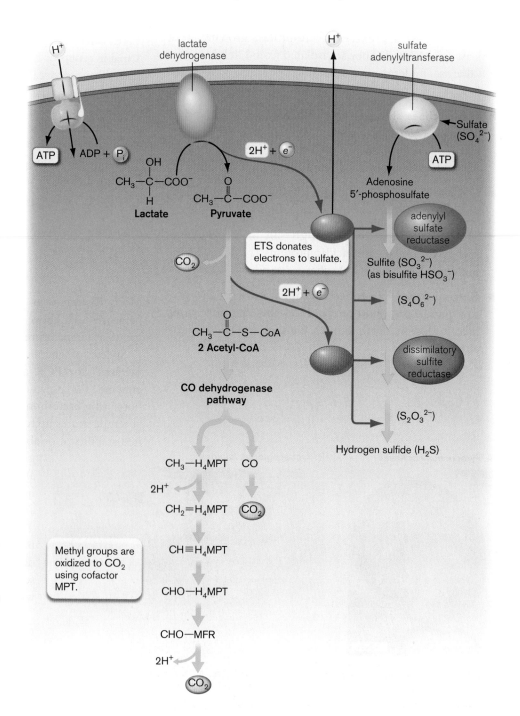

FIGURE 19.39 ■ **Metabolism of Archaeoglobus fulgidus.** *A. fulgidus* gains energy by sulfate respiration on small organic molecules such as lactate. This process drives a unique acetyl-CoA degradation pathway involving reversal of methanogenesis (compare with Fig. 19.31A). *Source:* H. P. Klenk et al. 1997. *Nature* **390**:364.

Pyrococcus is that both proton motive force generation and the ATP synthase appear to involve a Na^+/H^+ antiport system. This dependence on sodium is a trait shared with the methanogens and the halophiles.

Archaeoglobus Partly Reverses Methanogenesis

While many archaea reduce sulfur, only one group is known to reduce sulfate ion (SO_4^{2-}) without a bacterial partner. The order Archaeoglobales includes *Archaeoglobus fulgidus*, a hyperthermophile with unusual metabolic characteristics. **Figure 19.38B** shows *A. fulgidus* isolated from water at 75°C beneath an oil rig in the North Sea.

The genome of *A. fulgidus* reveals an exceptionally complex array of metabolic pathways, including an acetyl-CoA degradation pathway that reverses part of methanogenesis (**Fig. 19.39**). *A. fulgidus*, a deep-sea-vent hyperthermophile, oxidizes methyl groups back to CO_2 using the methanogenic cofactor methanopterin (MPT) (see **Fig. 19.31A**). The methyl groups are oxidized by reduction of sulfate. *Archaeoglobus* is the only archaeon known to reduce sulfate without a bacterial partner. This archaeon may play an important role in limiting ocean release of methane from methanogens.

A. fulgidus uses several enzymes and cofactors, such as MPT, present in methanogens. But instead of generating methane from CO_2, they run the reactions in reverse, converting the methyl group of acetate to CO_2. The energy required is gained through fermentation of carbohydrates and through anaerobic respiration with sulfate. The sulfate is successively reduced to sulfite and sulfide. When organic substrates are available, sulfate reduction yields more energy than methanogenesis.

Thermoplasmatales

The order Thermoplasmatales includes thermophiles and acidophiles with no cell wall and no S-layer, but only a plasma membrane. How they maintain their cells against extreme conditions is poorly understood. An example is the moderate thermophile isolated from self-heating coal refuse piles, *Thermoplasma acidophilum*, growing at 59°C and pH 2. The cells have flagella and are motile, but it is unclear how the flagella maintain a torque against the membrane without a rigid envelope for support. The metabolism of *T. acidophilum*, like that of *Pyrococcus*, is based on S^0 respiration of organic molecules.

The genome sequence of *T. acidophilum* contains just 1.5 million bp—one of the smallest known for a free-living organism. It shows substantial evidence of horizontal transfer from the crenarchaeote *Sulfolobus*, a hyperthermophile that shares the same range of habitats. Horizontal transfer between distant relatives turns out to be common among the hyperthermophiles, causing some controversies in their classification.

Another acidophilic genus of the order Thermoplasmatales is *Ferroplasma* (**Fig. 19.40**). *Ferroplasma* species are found in mines containing iron pyrite ore (FeS_2). Using dissolved Fe^{3+} as an oxidizing agent in the presence of H_2O, they oxidize the sulfur to sulfuric acid. The chemical equation is:

$$FeS_2 + 14Fe^{3+} + 8H_2O \longrightarrow 15Fe^{2+} + 2SO_4^{2-} + 16H^+$$

The reaction generates pH values below pH 0 (1-M H^+). This degree of acidity is enough to dissolve a metal shovel within a day.

The amorphous cells of *Ferroplasma acidarmanus* grow in biofilms that form long streamers into water draining from the mine (**Fig. 19.40B**). The oxidative disintegration of iron-bearing ores can be useful for leaching of minerals, but it also causes acid mine drainage into aquatic systems (discussed in Chapter 22).

A.

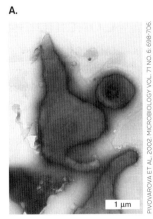

B.

FIGURE 19.40 ■ **The extreme acidophile *Ferroplasma*.** **A.** *F. acidiphilum* grows at pH 0 (TEM). **B.** Streamers of *F. acidarmanus* anchored to deposits of pyrite within the Iron Mountain mine in California. The stream is about a meter across, its water about pH 0. This level of acidity will dissolve a metal shovel in a day.

PIVOVAROVA ET AL. 2002. MICROBIOLOGY VOL. 71 NO. 6: 698-706.

K. J. EDWARDS ET AL. 2000. SCIENCE 287:1796.

Nanoarchaeota

The smallest known euryarchaeotes are the Nanoarchaeota, recognized so far for a single species, *Nanoarchaeum equitans* (**Fig. 19.41**). The organism consists of exceptionally small cells that are obligate symbionts of the crenarchaeote *Ignicoccus*. The *N. equitans* cell in **Figure 19.41** is attached to *I. hospitalis* by a membrane bridge (black arrow). The host cell may harbor up to four of the smaller cells. Both host and symbiont genomes have been sequenced, revealing extensive coevolution of the two.

The *N. equitans* genome is exceptionally small (less than 500 kb) and shows evidence of rapid degenerative evolution. The diminished genome is typical of a dependent organism that has lost numerous genes for functions now provided by a host. It remains unclear, however, whether *N. equitans* is parasitic on *Ignicoccus* or if it makes a metabolic contribution to its host (which is able to grow without the attached *N. equitans*).

Deeply Branching Divisions

We continue to identify emerging clades of archaea through PCR-amplified rDNA probes. Some of these clades, such as the Lokiarchaeota, emerge within existing phyla, such as Euryarchaeota. Others, however, diverge more deeply than the divergence between Crenarchaeota and Euryarchaeota.

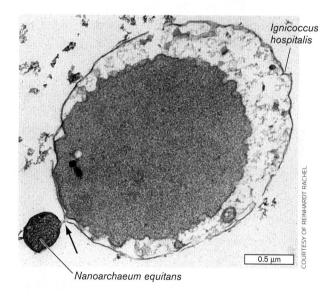

Nanoarchaeum equitans

FIGURE 19.41 ■ *Nanoarchaeum equitans*, a small obligate symbiont attached to *Ignicoccus hospitalis*. TEM.

Most deeply branching organisms are uncultivated, and little is known of them other than their rRNA sequence. But for some organisms, new techniques of genome sequencing enable us to read most of a genome from a single cell.

An example of a deeply branching division is the Ancient Archaeal Group (AAG) of hyperthermophiles. AAG includes the Korarchaeota vent hyperthermophiles isolated by Susan Barns and Norman Pace. The organisms have not been isolated in pure culture, but in 2008 the genome of one korarchaeote was sequenced from an enriched mixed culture at 80°C–90°C at Obsidian Pool, Yellowstone National Park. The organism, provisionally named "*Candidatus* Korarchaeum cryptofilum," grows in long, thin filaments less than 200 nm wide. Its genes suggest that it gains energy mainly from anaerobic peptide fermentation, yet we still do not know enough to culture this organism in the laboratory.

To Summarize

- **Thermococcales includes hyperthermophiles and acidophiles.** Most species use sulfur to oxidize complex organic substrates.

- ***Pyrococcus* and *Thermococcus* are the source of vent polymerases** for the polymerase chain reaction (PCR). Thermococcal enzymes notably use tungsten at their active site.

- ***Archaeoglobus* species reduce sulfate.** The methyl group of acetate is oxidized to CO_2 by a pathway using methanogenic cofactors.

- **Thermoplasmatales includes extreme acidophiles.** *Ferroplasma* oxidizes iron pyrite ore (FeS_2) in a process that generates concentrated sulfuric acid, causing acid mine drainage.

- ***Nanoarchaeum equitans*** is a tiny obligate symbiont of *Ignicoccus hospitalis*.

Concluding Thoughts

Overall, species of archaea inhabit a wider range of environments than either bacteria or eukaryotes do—from extreme heat to extreme cold, from high pH to extreme acid, as well as temperate environments such as the oceans. Crenarchaeotes are found in all soil and water habitats, and as endosymbionts of marine animals. Archaea show unique cell structures such as grappling-hook appendages, and unique forms of metabolism such as methanogenesis. Methanogens colonize all soil and water environments, as well as the human digestive tract. Yet we probably know less about the actual scope of Archaea than we do about the other two domains because so many members remain uncultured. As of this writing, we recognize three major divisions: the crenarchaeotes, including sulfur-metabolizing hyperthermophiles, as well as marine mesophiles and psychrophiles; the thaumarchaeotes, including ammonia oxidizers and marine endosymbionts; and the euryarchaeotes, including methanogens, halophiles, and sulfur-metabolizing extremophiles. New deeply branching isolates continue to be discovered, potentially representing entire new divisions. Much of what we call Archaea remains to be explored.

CHAPTER REVIEW

Review Questions

1. What distinctive structures are seen in the archaeal cell membrane and envelope?
2. Which aspects of archaeal genetics resemble the genetics of bacteria, and which aspects have more in common with eukaryotes?
3. Compare and contrast diverse members of the Crenarchaeota and Thaumarchaeota.
4. Outline the genetic phylogeny and key traits of these groups of archaea: Haloarchaea, methanogens, Thermococcales, Thermoplasmatales.
5. What are some specific physiological adaptations found in hyperthermophiles? Psychrophiles? Extreme acidophiles?
6. Outline three different specific types of mutualism involving an archaeal symbiont.
7. Explain how different archaea contribute to cycling of nitrogen and sulfur in ecosystems.
8. Explain what is known and what is unknown about the following groups of archaea: marine pelagic crenarchaeotes; marine benthic anaerobes; soil and plant root–associated crenarchaeotes. What kinds of experiments may reveal additional traits of these organisms?
9. How do methanogens interact metabolically with communities of bacteria, within a host animal or within a soil environment?

Thought Questions

1. What approaches can you use to discover previously unknown deeply branching groups of archaea? Explain the strengths and limitations of each method.
2. Why do you think we have found no archaea that are pathogens of animals or plants?
3. Why do you think methanogens appear in many branches among different groups, whereas haloarchaea branch as a single group?

Key Terms

anaerobic methane oxidizers (ANME) (758)
bacteriorhodopsin (739, 765)
barophile (745)
black smoker (746)
cannula (747)
Crenarchaeota (741)
Desulfurococcales (745)
electron bifurcation (761)

Euryarchaeota (741)
gas vesicle (763)
Haloarchaea (762)
Halobacteriales (762)
halorhodopsin (765)
hamus (736)
histone (741)
isoprenoid (737)
methane gas hydrate (758)

methanogen (753)
methanogenesis (739)
pseudopeptidoglycan (pseudomurein) (738)
Sulfolobales (748)
syntrophy (754)
tetraether (737)
Thaumarchaeota (741)

Recommended Reading

Bang, Corinna, Katrin Weidenbach, Thomas Gutsmann, Holger Heine, and Ruth A. Schmitz. 2014. The intestinal archaea *Methanosphaera stadtmanae* and *Methanobrevibacter smithii* activate human dendritic cells. *PLoS One* **9**:e99411.

Biddle, Jennifer F., Julius S. Lipp, Mark A. Lever, Karen G. Lloyd, Ketil B. Sørensen, et al. 2006. Heterotrophic Archaea dominate sedimentary subsurface ecosystems off Peru. *Proceedings of the National Academy of Sciences USA* **103**:3846–3851.

Bolduc, Benjamin, Daniel P. Shaughnessy, Yuri I. Wolf, Eugene Koonin, Francisco F. Roberto, et al. 2012. Identification of novel positive-strand RNA viruses by metagenomic analysis of archaea-dominated Yellowstone hot springs. *Journal of Virology* **86**:5562–5573.

Brumfield, Susan K., Alice C. Ortmann, Vincent Ruigrok, Peter Suci, Trevor Douglas, et al. 2009. Particle assembly and ultrastructural features associated with replication of the lytic archaeal virus *Sulfolobus* turreted icosahedral virus. *Journal of Virology* **83**:5964–5970.

DasSarma, Priya, Regie C. Zamora, Jochen A. Müller, and Shiladitya DasSarma. 2012. Genome-wide responses of the archaeon *Halobacterium* sp. strain NRC-1 to oxygen limitation. *Journal of Bacteriology* **194**:5530. doi:10.1128/JB.01153-12.

DeMaere, Matthew Z., Timothy J. Williams, Michelle A. Allen, Mark V. Brown, John A. E. Gibson, et al. 2013. High level of intergenera gene exchange shapes the evolution of haloarchaea in an isolated Antarctic lake. *Proceedings of the National Academy of Sciences USA* **110**:16939–16944.

Edwards, Katrina J., Philip L. Bond, Thomas M. Gihring, and Jillian F. Banfield. 2000. An archaeal extreme acidophile involved in acid mine drainage. *Science* **287**:1796–1799.

Golyshina, Olga V., and Kenneth N. Timmis. 2005. *Ferroplasma* and relatives, recently discovered cell wall-lacking archaea making a living in extremely acid heavy metal-rich environments. *Environmental Microbiology* **7**:1277–1288.

Karr, Elizabeth A., Joshua M. Ng, Sara M. Belchik, W. Matthew Sattley, Michael T. Madigan, et al. 2006. Biodiversity of methanogenic and other archaea in the permanently frozen Lake Fryxell, Antarctica. *Applied and Environmental Microbiology* **72**:1663–1666.

Könneke, Martin, Anne E. Bernhard, José R. de la Torre, Christopher B. Walker, John B. Waterbury, et al. 2005. Isolation of an autotrophic ammonia-oxidizing marine archaeon. *Nature* **437**:543–546.

López-García, Purificación, Yvan Zivanovic, Philippe Deschamps, and David Moreira. 2015. Bacterial gene import and mesophilic adaptation in archaea. *Nature Reviews. Microbiology* **13**:447–456.

Pernthaler, Annelie, Anne E. Dekas, C. Titus Brown, Shana K. Goffredi, Tsegereda Embaye, et al. 2008. Diverse syntrophic partnerships from deep-sea methane vents revealed by direct cell capture and metagenomics. *Proceedings of the National Academy of Sciences USA* **105**:7052–7057.

Perras, Alexandra K., Gerhard Wanner, Andreas Klingl, Maximilian Mora, Anna K. Auerbach, et al. 2014. Grappling archaea: Ultrastructural analyses of an uncultivated, cold-loving archaeon, and its biofilm. *Frontiers in Microbiology* **5**:397.

Pester, Michael, Christa Schleper, and Michael Wagner. 2011. The Thaumarchaeota: An emerging view of their phylogeny and ecophysiology. *Current Opinion in Microbiology* **14**:300–306.

Richter, Ingrid, Craig W. Herbold, Charles K. Lee, Ian R. McDonald, John E. Barrett, et al. 2014. Influence of soil properties on archaeal diversity and distribution in the McMurdo Dry Valleys, Antarctica. *FEMS Microbiological Ecology* **89**:347–359.

Ruepp, Andreas, Werner Graml, Martha-Leticia Santos-Martinez, Kristin K. Koretke, Craig Volker, et al. 2000. The genome sequence of the thermoacidophilic scavenger *Thermoplasma acidophilum*. *Nature* **407**:508.

Samuel, Buck S., and Jeffrey I. Gordon. 2006. A humanized gnotobiotic mouse model of host–archaeal–bacterial mutualism. *Proceedings of the National Academy of Sciences USA* **103**:10011–10016.

Sauder, Laura A., Katja Engel, Jennifer C. Stearns, Andre P. Masella, Richard Pawliszyn, et al. 2011. Aquarium nitrification revisited: Thaumarchaeota are the dominant ammonia oxidizers in freshwater. *PLoS One* **6**:e23281.

Spang, Anja, Jimmy H. Saw, Steffen L. Jørgensen, Katarzyna Zaremba-Niedzwiedzka, Joran Martijn, et al. 2015. Complex archaea that bridge the gap between prokaryotes and eukaryotes. *Nature* **521**:173–179.

Stams, Alfons J. M., and Caroline M. Plugge. 2009. Electron transfer in syntrophic communities of anaerobic bacteria and archaea. *Nature Reviews. Microbiology* **7**:568–577.

CHAPTER 20
Eukaryotic Diversity

The domain Eukarya encompasses a breathtaking range of size and shape, from giant whales and sequoias to microbial fungi, algae, and protists. Fungi include multicellular forms such as mushrooms, as well as unicellular yeasts and filamentous *Penicillium*. Algae conduct photosynthesis using chloroplasts; they include broad sheets of kelp, as well as unicellular phytoplankton. Protozoa include amebas and stalked vorticellae. Most protists are free-living predators that use cilia to catch their prey, but some are parasites that cause diseases such as malaria and sleeping sickness.

CURRENT RESEARCH highlight

Antarctic choanoflagellates form proto-multicellular life-forms. Choanoflagellates are protists with a single flagellum surrounded by a collar of interconnected microvilli (tiny cell extensions). Choanoflagellates may alternate between unicellular free-living and multicellular colonial life stages. The ability to form complex colonies of interconnected cells suggests that these protists might be showing characteristics of early-stage transition to multicellular life. Their DNA sequence and cellular traits suggest that choanoflagellates are the most appropriate protozoan reference for studies of animal origins.

Source: Wei Li and Rachael Morgan-Kiss, Miami University of Ohio, 2015.

10 μm

AN INTERVIEW WITH

WEI LI, ANTARCTIC MICROBIOLOGIST, MIAMI UNIVERSITY OF OHIO

WEI LI, MIAMI UNIVERSITY OF OHIO

Why is it interesting to find choanoflagellates in a lake in Antarctica?

In the McMurdo Dry Valleys, perennially ice-covered lakes are the only source of year-round liquid water for life. Most of the lake food webs are microbially dominated and virtually lack metazoa. Microbial eukaryotes play important roles as primary producers and consumers. Photosynthetic protists represent the major producers of organic matter in McMurdo lake ecosystems, while heterotrophic flagellates and ciliates are the predators of bacteria and smaller protists. In Lake Bonney, heterotrophic flagellates (such as choanoflagellates) are major bacterivores and a key component in the food web.

How do you think choanoflagellates resemble the earliest form of animal life?

Over 600 million years ago, the multicellular ancestor of modern animals evolved from a unicellular flagellate. The transition to multicellularity that triggered the evolution of metazoa from protozoa is a critical question regarding animal origins. Recent phylogenetic studies indicate that metazoa are monophyletic and their common ancestor was shared by single-celled choanoflagellates.

When we think of eukaryotes, we think first of plants and animals consisting of complex multicellular bodies. But eukaryotes also include vast numbers of microbes, such as protists (algae and protozoa). How did complex animals such as ourselves evolve from a protist ancestor? Certain protists called choanoflagellates show remarkable genetic similarity to animals. Choanoflagellates can be induced to form a multicellular colony called a rosette (shown in the Current Research Highlight). The rosette is formed of interconnected cells and can grow and fission like a unified organism. Both single-celled and rosette choanoflagellates are ubiquitous in marine and freshwater habitats. Even in the subfreezing lakes of Antarctica, these protists grow with their hint of life's ancient transition to metazoa (multicellular animals). Experiments on algae suggest models for the evolution of multicellular life (discussed in Section 20.3).

In Chapter 20 we present the phylogeny of major groups of microbial eukaryotes, along with the branching of animals and plants from the microbial family tree. We explore the form and function of algae, such as the calcium carbonate–plated coccolithophores that bloom over vast stretches of ocean (**Fig. 20.1A**). Fungi range from yeast to mushrooms to common molds such as *Penicillium*, our source of penicillin (**Fig. 20.1B**). Protozoa include a dozen deep-branching clades, from free-living amebas to diarrhea-causing parasites such as *Giardia* (**Fig. 20.1C**). We get to know the eukaryotic microbes as essential partners in ecosystems and as infectious agents that cause some devastating diseases.

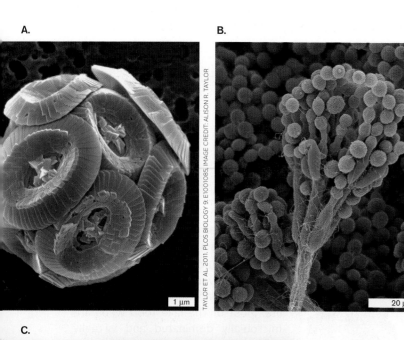

A. *(scale bar: 1 μm)* TAYLOR ET AL. 2011. PLOS BIOLOGY 9: E1001085; IMAGE CREDIT: ALISON R. TAYLOR

B. *(scale bar: 20 μm)* DENNIS KUNKEL MICROSCOPY, INC.

C. *(scale bar: 2 μm)* CDC/JANICE HANEY CARR

FIGURE 20.1 ■ Microbial eukaryotes. A. *Coccolithus pelagicus*, a marine coccolithophore, a single-celled alga that forms scales of calcium carbonate (colorized SEM). **B.** *Penicillium notatum* conidiophore, an airborne fungus (colorized SEM). **C.** *Giardia intestinalis*, a waterborne parasite (SEM).

20.1 Phylogeny of Eukaryotes

With all their diversity, eukaryotic cells share a common structure defined by the presence of the nucleus and other membrane-enclosed organelles (for review, see eAppendix 2). The extensive compartmentalization of the eukaryotic cell, including the nucleus and endomembrane system, enables eukaryotic cells to grow a thousandfold larger than prokaryotic cells. Yet these organelles are found within even the tiniest of eukaryotes, such as *Ostreococcus tauri*, a green alga less than 2 μm across (**Fig. 20.2**). This alga consists of a flat, disk-shaped cell containing one or two of each major type of organelle: a mitochondrion, a chloroplast, and a stack of Golgi, all packed within the cell's small volume. The genome of *O. tauri* is also downsized; at only 8 Mb, the genome is barely twice as long as that of the bacterium *Escherichia coli*. Discovered only recently, such tiny bacteria-sized eukaryotes are believed to be the most numerous and ubiquitous forms of eukaryotic life.

Note that, despite their extraordinary range of form, the metabolism of eukaryotes is less diverse than that of either bacteria or archaea. Most eukaryotes conduct either oxygenic photosynthesis or heterotrophy—or both. All have descended from an ancestral cell that engulfed a bacterial endosymbiont, giving rise to mitochondria, the source of aerobic respiration. And all phototrophs descended from a cell that engulfed the bacterial ancestor of chloroplasts.

A.

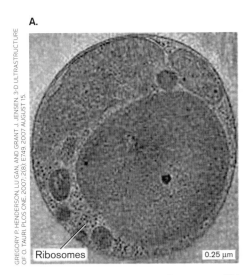

GREGORY P. HENDERSON, LU GAN, AND GRANT J. JENSEN. 3-D ULTRASTRUCTURE OF O. TAURI. PLOS ONE. 2007;2(8): E749. 2007 AUGUST 15.

Ribosomes
0.25 μm

B.

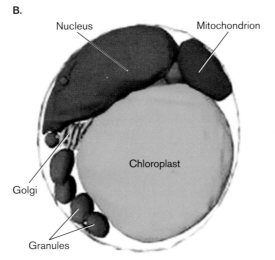

Nucleus
Mitochondrion
Chloroplast
Golgi
Granules

FIGURE 20.2 ▪ A tiny eukaryote still has organelles. *Ostreococcus tauri*, visualized by cryo-EM **(A)** and in 3D by tomography **(B)**.

Historical Overview of Eukaryotes

For most of human history, life was understood in terms of macroscopic multicellular eukaryotes: animals (creatures that move to obtain food) and plants (rooted organisms that grow in sunlight). **Fungi** (singular, **fungus**), which lack photosynthesis, were nonetheless considered a form of plant because they grow on the soil or other substrate. Thus, **mycology**, the study of fungi, was often included with botany, the study of plants. But the basis of fungal growth was poorly understood, and its mystery was often associated with magic. For example, people were mystified by the sudden growth of mushrooms in a ring, which they called a "fairy ring" (**Fig. 20.3A**). The mushrooms actually arise as fruiting bodies from the tips of fungal hyphae (filaments of cells) that propagate from a single spore and extend radially underground.

In the eighteenth and nineteenth centuries, microscopists came to recognize microscopic forms of fungi such as filamentous hyphae and unicellular yeasts. Other unicellular life-forms, such as amebas and paramecia, were motile and appeared more like microscopic animals. These animal-like organisms were called **protozoa** (singular, **protozoan**) (**Fig. 20.3B**). The cellular dimensions of protozoa were typically ten- to a hundredfold larger than those of bacteria, and their form and motility offered intriguing subjects for observation. So did single-celled phototrophs such as diatoms and dinoflagellates, which were called **algae** (singular, **alga**). Algae were thought of as unicellular plants, although simple multicellular algae were known. The unicellular and microscopic

forms of fungi, protozoa, and algae came to be included in the subject of microbiology.

Discoveries in physiology led us to redefine these organisms. For example, the motile organisms defined as protozoa often contain chloroplasts and fit the classification of algae. On the other hand, slime molds, originally classified with fungi, show form and motility more typical of protozoa. By the mid-twentieth century, naturalists classified protozoa, unicellular algae, and undifferentiated colonial forms as **protists**. Researchers including Herbert Copeland, Robert Whittaker, and Lynn Margulis attempted to refine the definition of "protist" to better distinguish microbial life-forms.

Today, molecular phylogeny shows that protists comprise several clades equally distant from each other as

A.

HTTP://WWW.UGAURBANAG.COM/CONTENT/FAIRY-RING

FIGURE 20.3 ▪
Traditional views of fungi and protozoa (protists).
A. Basidiomycete mushrooms growing in a "fairy ring." **B.** A nineteenth-century depiction of ciliated protozoa, by Rudolf Leuckart.

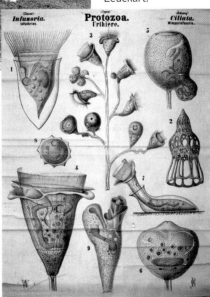

THE MBL/WHOI LIBRARY

B.

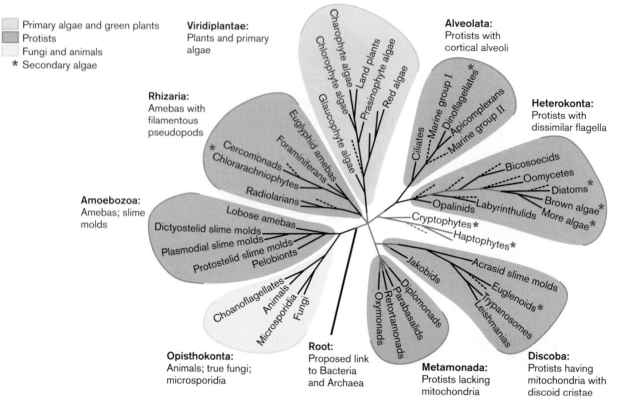

FIGURE 20.4 ■ Eukaryotic phylogeny. A phylogenetic tree of major clades based on DNA sequence data. Dotted lines denote emerging little-known taxa. Green asterisks denote secondary algae whose ancestor engulfed a primary alga with chloroplast. Gray branches indicate less certain position. *Source:* Modified from Sandra L. Baldauf. 2003. *Science* **300**:1703.

from animals and plants (**Fig. 20.4**). In the terminology used today:

- **Protist** refers to single-celled and colonial eukaryotes other than fungi. Protists include many diverse clades of algae and protozoa.

- **Protozoa** are protists that are single-celled heterotrophs. They include environmental consumers, as well as medically important parasites such as *Giardia*.

The algae include two major kinds. Those of the first kind are derived from a single endosymbiotic event and are closely related to green plants. These algae are called **primary algae** (Viridiplantae in **Fig. 20.4**). By contrast, various groups of heterotrophic protists later incorporated algae in a second event of symbiogenesis. These are **secondary algae** (green asterisks in **Fig. 20.4**).

Note: Medical textbooks also cover invertebrate animal parasites such as worms and mites as eukaryotic agents of disease, although they are not considered microbes.

Challenges for Classification

Classifying eukaryotes presents a challenge, for several reasons. Complex eukaryotic cells frequently lose structures through reductive (degenerative) evolution. Thus,

for example, a clade originally defined by possession of flagella often includes members that lack flagella. In addition, superficially similar forms of organisms have evolved independently in distantly related taxa; this is called convergent evolution. For example, the "water molds" that grow on aquarium fish superficially resemble fungi, but they actually evolved in a clade that includes brown algae and diatoms.

Another challenge for classification is the size and complexity of eukaryotic genomes, which has delayed the completion of genome sequences. Eukaryotic genomes are typically severalfold larger than those of bacteria and archaea, and 50%–90% of their DNA consists of noncoding sequences. Thus, eukaryotic genomes take longer to sequence and are more challenging to annotate than those of prokaryotes.

Furthermore, the evolution of eukaryotes includes multiple events of **endosymbiosis**, in which an engulfed cell evolved into an essential organelle. An endosymbiotic incorporation of a proteobacterium by the ancestor of all eukaryotes gave rise to mitochondria. Later, incorporation of a cyanobacterium by the ancestor of plants and algae gave rise to chloroplasts. Much later, several lineages of protists took up chloroplast-bearing algae, which now show varying stages of evolution as organelles.

A consensus view of eukaryotic phylogeny (**Fig. 20.4**) is based on DNA sequence comparison, protein trees, and the appearance of specific gene fusions and deletions. The distinct clades of protists include amebas and slime molds (Amoebozoa); amebas with needlelike pseudopods (Rhizaria); ciliates and dinoflagellates (Alveolata); oomycetes, brown algae, and diatoms (Heterokonta); and other groups, such as Metamonada.

We will now look at key traits of each major group. **Table 20.1** summarizes representative clades of

---

TABLE 20.1 Eukaryotic Microbial Diversity* (Expanded Table Appears in Appendix 2)

Opisthokonta (fungi and metazoan animals). Single flagellum on reproductive cells. Includes multicellular animals.

Metazoa (animals). Multicellular organisms with motile cells and body parts. Includes colonial animals, both invertebrates and vertebrates. *Homo sapiens.*

Choanoflagellata. Single flagellum and collar of microvilli. Resemble sponge choanocytes. Possible link to common ancestor of multicellular animals.

 Eumycota (fungi). Cells form hyphae with cell walls of chitin.

Ascomycota. Fruiting bodies form asci containing haploid ascospores. Includes opportunistic pathogens and bread-making yeasts. *Penicillium, Aspergillus, Saccharomyces cerevisiae, Stachybotrys.* Lichens are a mutualism between an ascomycete and green algae (*Trebouxia*) or cyanobacteria (*Nostoc*).

Basidiomycota. Basidiospores form primary and secondary mycelia; some generate mushrooms. May be edible (*Lycoperdon*) or toxic (*Amanita*). Plant pathogens (*Ustilago maydis* causes corn smut). Human pathogens (*Cryptococcus neoformans*).

Chytridiomycota. Motile zoospores with a single flagellum. Saprophytes or anaerobic rumen fungi. *Allomyces.* Frog pathogens (*Batrachochytrium dendrobatidis*), bovine rumen digestive endosymbionts (*Neocallomastix*).

Glomeromycota. Mutualists of plant roots, forming arbuscular mycorrhizae, filamentous networks that share nutrients with and among diverse plants.

Microsporidia. Single-celled parasites that inject a spore through a tube into a host cell, causing microsporidiosis. *Encephalitozoon* species. Commonly infect AIDS patients.

Zygomycota. Sexual hyphae (1N) grow toward each other and fuse to form the zygote (zygospore). Saprophytes or insect parasites. Some form mycorrhizae.

 Viridiplantae (primary endosymbiotic algae and plants). Includes green algae and multicellular land plants. Chloroplasts arose from a primary endosymbiont.

Charophyta. Multicellular algae with rhizoids that adhere to sediment. *Spirogyra.*

Chlorophyta (green algae). Chlorophyll *a* confers green color. Inhabit upper layer of water.

- **Unicellular with paired flagella.** *Chlamydomonas, Volvox* (colonial).
- **Multicellular.** *Ulva* grow in sheets; *Cymopolia* forms calcified stalks with filaments.

Glaucophyta. Unicellular algae whose chloroplasts have peptidoglycan.

- **Picoeukaryotes.** *Ostreococcus* and *Micromonas* are unicellular algae.
- **Siphonous algae.** *Caulerpa* species consist of a single cell with multiple nuclei, growing to indefinite size.

Rhodophyta (red algae). Phycoerythrin obscures chlorophyll, colors the algae red. Absorption of blue-green light enables colonization of deeper waters. *Porphyra* forms sheets edible by humans; *Mesophyllum* is a coralline alga, hardened by calcium carbonate crust; resembles coral.

 Amoebozoa (amebas and slime molds). Lobe-shaped (lobose) pseudopods driven by sol-gel transition of actin filaments.

Amebas. Unicellular. No microtubules to define shape. Life cycle is primarily asexual. Predators in soil or water. Giant free-living amebas (*Amoeba proteus*); parasites (*Entamoeba histolytica*).

Mycetozoa. Slime molds. Cellular slime molds (*Dictyostelium*). Upon starvation, amebas aggregate to form a fruiting body, which produces spores. Plasmodial slime molds (*Physarum polycephalum*) undergo meiosis, producing spores that germinate to ameboid gametes.

 Rhizaria (amebas with filament-shaped pseudopods). Filament-shaped (filose) pseudopods. Some species have a test (shell) of silica or other inorganic materials.

Possess flagella or pseudopods (Cercozoa); form spiral tests (Foraminifera); form thin pseudopods called filopodia (Radiolaria).

 Alveolata (having cortical alveoli). Cortex contains flattened vesicles called alveoli, reinforced below by lateral microtubules.

Ciliophora. Common aquatic predators. Undergo sexual exchange by conjugation, in which micronuclei are exchanged, then regenerate macronuclei. Covered with cilia (*Paramecium*); mouth ringed with cilia (*Vorticella*); suctorians (*Acineta*).

Dinoflagellata. Secondary or tertiary endosymbiont algae, from engulfment of primary or secondary algae. Cortical alveoli contain stiff plates. Pair of flagella, one wrapped around the cell. Free-living aquatic (*Peridinium*); zooxanthellae, endosymbionts of coral (*Symbiodinium*).

Apicomplexa (formerly Sporozoa). Parasites with complex life cycles. Lack flagella or cilia; possess apical complex for invasion of host cells. Vestigial chloroplasts. *Plasmodium falciparum* causes malaria; *Toxoplasma gondii* causes feline-transmitted toxoplasmosis; *Cryptosporidium parvum* is a waterborne opportunistic parasite.

 Heterokonta. Paired flagella of dissimilar form, one much shorter than the other.

Diatoms (Bacillariophyceae); kelps (Phaeophyceae); golden algae (Chrysophyceae); coccolithophores (Prymnesiophyceae); water molds (Oomycetes).

 Discoba (having disk-shaped cristae). Disk-shaped cristae of mitochondria.

Free-living flagellates (Euglenida); trypanosomes (*Trypanosoma brucei*, sleeping sickness; *T. cruzi*, Chagas' disease).

 Metamonada (vestigial mitochondria). Parasitic or symbiotic flagellates. Mitochondria and Golgi degenerated through evolution.

Human intestinal parasites (*Giardia intestinalis*, or *G. lamblia*); symbionts of termite gut (*Pyrsonympha*).

*Bulleted terms are representative groups within the phylum.

EF-1α peptide sequence alignment

FIGURE 20.5 ■ **Alignment of peptide sequences for EF-1α reveals insertion in opisthokont species.** A DNA insertion encoding 11–17 amino acid residues appears in the EF-1α sequence of animals, fungi, and microsporidians (known collectively as opisthokonts) but not in plants or most protists. This finding suggests that opisthokonts have a common ancestor that branched early from the eukaryotes and diverged later to animals and fungi. *Source:* Modified from Sandra L. Baldauf. 2003. *Science* **300**:1703.

microbial eukaryotes. **For a more extensive table, see Appendix 2.**

Opisthokonts: Animals and Fungi

Where do humans and other multicellular animals fit into this taxonomy? The position of animals ("metazoa") among the eukaryotic microbial clades is of interest because it suggests which contemporary microbes most closely resemble our own cells. The degree of relatedness can help define microbial model systems for probing key questions of human cell biology. For example, the baker's yeast *Saccharomyces cerevisiae* shares so much of its genetic machinery with humans that it provides a model for cancer, cell-trafficking defects, and even the basis of human aging.

As we saw in Chapter 17, the first measure of species relatedness is based on DNA sequences encoding ribosomal RNA, particularly small-subunit rRNA. For eukaryotes, however, rRNA sequences yield ambiguous results, in part because of differing rates of change among different clades. So researchers have searched the genomes for stronger clues to relative divergence—that is, which clade diverged earliest from the others. A particularly strong clue would be the appearance of a gene insertion or deletion unique to a particular clade. Such a sequence indicates that all species containing the gene insertion or deletion must have diverged after their

common ancestor branched from the outgroup (most distantly related taxon).

Surprisingly, genomic analysis relates animals more closely to fungi (such as yeasts) than to motile microbial eukaryotes (such as paramecia or amebas). The true **fungi**, or **Eumycota**, are heterotrophs, either single-celled or growing in nonmotile filaments of cells called hyphae. Nevertheless, animals and fungi share several key gene insertions and deletions, as discovered by Sandra Baldauf and colleagues. Baldauf focused on a short sequence insertion in a key gene encoding a protein translation factor, elongation factor 1α (EF-1α) (**Fig. 20.5**). The inserted DNA, encoding 12 amino acids, appears in all sequenced genomes of animals and fungi and is absent from all plants and most protists. It is present in organisms, such as microsporidians, that were previously considered protists but whose genomes now indicate classification under fungi. The EF-1α gene sequence, as well as several others, supports the grouping of animals and fungi (including microsporidians) together in one clade (see **Fig. 20.4**).

Animal and fungal cells also share a structural feature distinguishing them from protists: the presence of an unpaired flagellum. Both animals and fungi include species whose life cycle has a uniflagellar stage, in contrast to other microbial eukaryotes whose flagella are paired (for example, euglenas). In the case of humans, the uniflagellar stage is the spermatozoan. Similarly, some species of fungi

A. Choanoflagellate

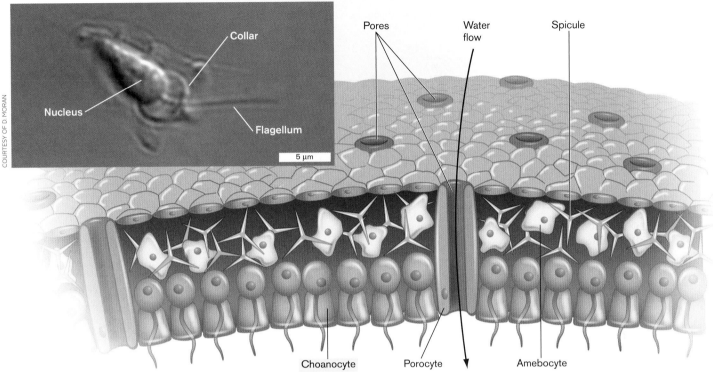

B. Sponge body wall

FIGURE 20.6 ■ **Choanoflagellates resemble sponge choanocytes. A.** A choanoflagellate, *Acanthocorbis unguiculata*. **B.** Sponge choanocytes resemble choanoflagellates. Within the sponge, choanocytes assist the circulation of water and the uptake of nutrients.

generate uniflagellar reproductive cells called **zoospores**. As a whole, the clade including single-flagellum members is termed "opisthokont," based on the Greek words meaning "backward-pointing pole," because the flagellum points backward like an oar.

Among opisthokonts, the microbes that diverged most recently from animals (600 million years ago) appear to be the choanoflagellates. Genetic studies of choanoflagellates reveal several genes found only in animals. The prefix "choano-" (meaning "funnel") refers to the collar of filaments surrounding the flagellum. The collared cells of choanoflagellates closely resemble the choanocyte cells of colonial sponges, an ancient form of animal (**Fig. 20.6**). Furthermore, choanoflagellates form colonies that crudely resemble the colonial structure of sponges. Thus, choanoflagellates may represent a "missing link" between animals and the microbial eukaryotes.

Note: Eukaryotic flagella are whiplike organelles composed of microtubules and surrounded by a membrane; their action is powered by ATP along the entire filament. Distinguish them from bacterial and archaeal flagella, which are rotary helical filaments composed entirely of protein subunits; their rotation is powered at the base by proton motive force.

Fungi (Eumycota) consist of cells with chitinous cell walls that grow in chains called **hyphae** (singular, **hypha**). Fungi range from single-celled organisms, such as yeasts, to complex multicellular forms such as mushrooms. The deepest-branching clade of fungi, Chytridiomycota, produces the uniflagellar zoospores that define opisthokonts. Other fungi, however, generate nonmotile reproductive cells, a result of reductive evolution.

Several taxa that historically were grouped with fungi (for example, slime molds) are now classified genetically as protists. Slime molds generate populations of cells that migrate into a unified structure called a **fruiting body** to make reproductive cells. Slime molds are now grouped with amebas (Amoebozoa, discussed in Section 20.4). Water molds, which are plant and animal pathogens of the class Oomycetes, are now recognized as heterokont protists (see eTopic 20.1).

Algae Evolved by Engulfing Phototrophs

Algae are commonly defined as single-celled plants and simple multicellular plants lacking true stems, roots, and leaves. Algal cells contain **chloroplasts**, membrane-enclosed organelles of photosynthesis that evolved from a

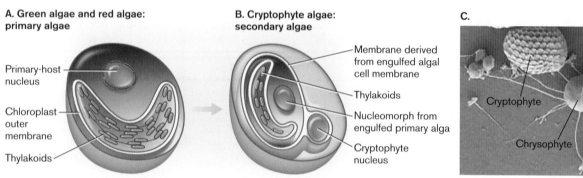

FIGURE 20.7 ■ Chloroplast evolution: primary and secondary endosymbiosis. A. Green algae (Chlorophyta) and red algae (Rhodophyta) contain chloroplasts (green) that evolved from engulfed cyanobacteria. **B.** Cryptophyte algae contain chloroplast (green), vestigial primary-host cytoplasm (yellow), and vestigial nucleus or nucleomorph (brown) from the engulfed primary endosymbiont. **C.** Cryptophyte and chrysophyte algae from Antarctic Lake Fryxell.

cyanobacterium (**Fig. 20.7**). In Earth's biosphere, algae plus bacterial phototrophs feed all marine and freshwater ecosystems, producing the majority of oxygen and biomass available for Earth's consumers.

How did ancestral algae get their chloroplasts? The "green plants" (Viridiplantae) include **primary algae** descended from a common ancestor containing a chloroplast. The chloroplast evolved by endosymbiosis between an ancient protist and a cyanobacterium (discussed in Chapter 17). In primary algae, the chloroplast is enclosed by two membranes (**Fig. 20.7A**): the inner membrane (from the ancestral phototroph's cell membrane) and the outer membrane (from the host cell membrane as it enclosed its prey). Both **green algae** (**chlorophytes**) and **red algae** (**rhodophytes**) are primary algae. Their chloroplasts diverged from their common ancestor to utilize pigments absorbing different ranges of the light spectrum.

Surprisingly, several other taxa traditionally considered to be algae evolved through a secondary endosymbiosis with a second protist host. The symbiotic history of these **secondary algae** is most evident in the cryptophytes (**Fig. 20.7B**), which still retain a vestigial nucleus, or **nucleomorph**, derived from the engulfed alga. In secondary algae, the chloroplast is surrounded by two extra membranes, one from the primary algal cell membrane and one from the secondary host. Other secondary algae include the **chrysophytes**, such as kelps and diatoms. The **dinoflagellates**, of the protist clade Alveolata, are secondary or tertiary algae, descended from a flagellate that consumed one or more types of algae. Dinoflagellates also engage in

"kleptoplasty," or "chloroplast stealing," in which the chloroplast of a digested prey is retained long enough to derive some photosynthetic energy, but ultimately consumed. The variety of endosymbiosis among protists provides clues as to how the original chloroplast evolved within the ancestral algae.

Note: "Algae" may refer to primary algae, or "true algae," as well as secondary algae that derive from protist clades. Fungi (the "true fungi," or Eumycota) are opisthokonts. Several fungus-like organisms have been reclassified within protist clades.

Protists Include Many Divergent Clades

The protists actually include several distantly related categories of eukaryotes (see **Table 20.1**). All protists are heterotrophs, commonly predators or parasites, although many also conduct photosynthesis as primary or secondary algae. Protists are important producers and consumers in marine, freshwater, and soil food webs. In ecology, phototrophic protists are termed "phytoplankton" and heterotrophs are termed "zooplankton," although many, in fact, are "mixotrophs" that act as both producers and consumers.

Amebas are unicellular organisms of highly variable shape that form **pseudopods**, locomotory extensions of cytoplasm enclosed by the cell membrane. Their size can reach several millimeters, and they can eat small invertebrates. There are two major groups of amebas: Amoebozoa and Rhizaria (see **Table 20.1**). The Amoebozoa, the most familiar kind of amebas, have lobed pseudopods, pseudopods that extend lobes of cytoplasm through cytoplasmic streaming. Most lobed amebas are free-living in aquatic habitats, but some cause human diseases such as dysentery. Still other kinds of lobed amebas are cellular slime molds, in which individual amebas converge to form a fruiting body. Amebas of the second group, Rhizaria, have thin, filamentous pseudopods, often radially arranged like a star, as in heliozoan amebas. Some Rhizaria, such as the foraminiferans, form inorganic shells called tests. Fossil foraminiferan tests are common in rock formations derived from ancient seas. Foraminiferan shells formed the White Cliffs of Dover in Britain and the stone used to build the Egyptian pyramids.

Note: DNA sequencing led to reclassification of the filamentous amebas as Rhizaria (formerly Cercozoa). "Cercozoa" now denotes a subgroup under Rhizaria.

Alveolates (Alveolata) include ciliated protists (ciliates), dinoflagellates, and apicomplexans. An example is the ciliated protist *Paramecium* (**Fig. 20.8A**). Alveolates are known for their complex outer covering, or cortex. The cortex contains networks of vesicles called **cortical alveoli** (**Fig. 20.8B**). Alveoli store calcium ions, and in some species protective plates form from them. Organisms equipped with paired flagella or cilia are known as **flagellates** and **ciliates**, respectively. **Flagella** (singular, **flagellum**) and **cilia** (singular, **cilium**) are essentially equivalent organelles composed of microtubules and enveloped by the cell membrane (**Fig. 20.8C**). Cilia are shorter than flagella and more numerous, and usually cover a broad surface. Alveolata also includes a major group of parasites whose major forms lack flagella, the **apicomplexans**. A well-known apicomplexan parasite is *Plasmodium falciparum*, which causes malaria.

Heterokonts (Heterokonta) are named for their pairs of differently shaped flagella (see **Table 20.1**). Heterokonts are thus distinguished from opisthokonts, which possess a single unpaired flagellum (if any). Flagellated heterokonts possess two flagella of unequal length and different structure. They include many voracious zooplankton. Other heterokonts include the oomycetes, or water molds (formerly classified with fungi), as well as diatoms and kelps (secondary endosymbiotic algae with chloroplasts). **Diatoms** are single cells with unique bipartite shells that fit together like a petri dish. The shells of diatoms form an infinite variety of different patterns for different species. **Kelps**, also known as brown algae, extend multicellular sheets floating at the water's surface.

The Discoba include free-living protists such as euglenas, as well as parasites showing extensive evolutionary reduction. Most Discoba have mitochondria with distinctive disk-shaped cristae (membrane pockets). Parasitic Discoba include the trypanosomes that cause sleeping sickness and Chagas' disease. A related clade, Metamonada, includes the waterborne parasite *Giardia intestinalis* (*G. lamblia*). In metamonads the mitochondria have lost their genomes and degenerated into **mitosomes** (discussed in Section 20.6).

Emerging Eukaryotes

Are there still eukaryotes that we haven't yet discovered? Genes from natural communities continually reveal new species of microbial eukaryotes in previously unknown divisions. Many of the new isolates are single cells as small as bacteria, designated nanoeukaryotes (3–10 μm) and picoeukaryotes (0.5–3 μm).

Genetic analysis shows that similar miniaturized eukaryotes branch deeply from all groups in the phylogenetic tree,

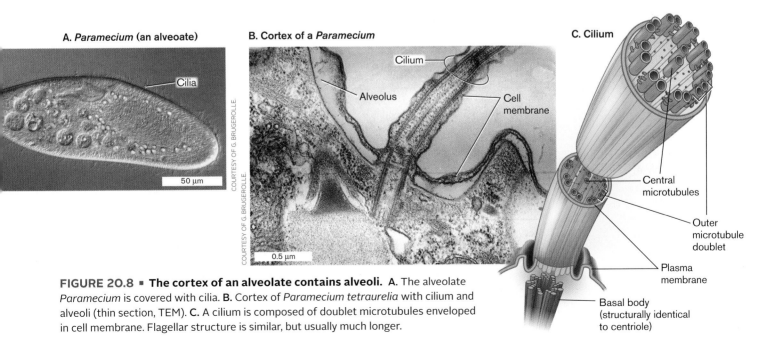

A. *Paramecium* (an alveoate) — Cilia — 50 μm — COURTESY OF G. BRUGEROLLE

B. Cortex of a *Paramecium* — Cilium — Alveolus — Cell membrane — 0.5 μm — COURTESY OF G. BRUGEROLLE

C. Cilium — Central microtubules — Outer microtubule doublet — Plasma membrane — Basal body (structurally identical to centriole)

FIGURE 20.8 ■ The cortex of an alveolate contains alveoli. A. The alveolate *Paramecium* is covered with cilia. **B.** Cortex of *Paramecium tetraurelia* with cilium and alveoli (thin section, TEM). **C.** A cilium is composed of doublet microtubules enveloped in cell membrane. Flagellar structure is similar, but usually much longer.

potentially doubling the known number of major eukaryotic taxa. Furthermore, our metagenomic analysis reveals eukaryotes in environments previously believed to be restricted to bacteria and archaea, such as anaerobic submarine sediments, and the hyperacidic Tinto River in Spain. In every new habitat tested, new deep-branching clades of eukaryotes emerge, with new implications for ecology and global cycling (discussed in Chapters 21 and 22). The wealth of new genomic data is reshaping our understanding of the domain Eukarya.

To Summarize

- **Opisthokonta** includes true fungi (Eumycota) and multicellular animals (Metazoa), as well as certain kinds of protists.

- **Viridiplantae** includes green plants and primary algae.

- **Protist groups** include Amoebozoa, the unshelled amebas with lobe-shaped pseudopods; Rhizaria, the filopodial amebas; Alveolata, ciliates and flagellates with complex cortical structure; Heterokonta, the kelps, diatoms, and flagellates with nonequivalent paired flagella; and Discoba and Metamonada, primarily parasites.

- **Many protists are phototrophs as well as heterotrophs**, based on secondary or tertiary endosymbiosis derived from engulfed algae.

- **Metagenomic analysis reveals new clades of microbial eukaryotes.** New kinds of microbial eukaryotes emerge from all habitats.

20.2 Fungi

Fungi provide essential support for all communities of multicellular organisms. Fungi recycle the biomass of wood and leaves, including substances such as lignin, which other organisms may be unable to digest. Underground fungal filaments called mycorrhizae extend the root systems of most plants, forming a nutritional "internet" that interconnects the plant community (discussed in Chapter 21). Mycorrhizae may have inspired the fictional underground tree network depicted in the film *Avatar* (2009).

Within the ruminant digestive tract, fungi ferment plant materials. On the other hand, pathogenic fungi infect plants and animals, and they contribute to the death of immunocompromised human patients. Still other fungi produce antibiotics such as penicillin, as well as food products such as wine and cheeses (discussed in Chapter 16).

Shared Traits of Fungi

Most fungi share these distinctive traits:

- **Absorptive nutrition.** Most fungi cannot ingest particulate food, as do protists, because their cell walls cannot part and re-form, as do the flexible pellicles of amebas and ciliates. Instead they secrete digestive enzymes and then absorb the broken-down molecules from their environment.

- **Hyphae.** Most fungi grow by extending multinucleate cell filaments called hyphae (**Fig. 20.9A**). As a hypha extends, its nuclei divide mitotically without cell division, generating a multinucleate cell. Later, septa may form to partition the hypha into cells. Hyphae grow by cytoplasmic extension and branching. A branched mass of extending hyphae is called a **mycelium** (plural, **mycelia**).

- **Cell walls contain chitin.** Chitin is an acetylated amino-polysaccharide of immense tensile strength, stronger than steel (**Fig. 20.9B**). Its strength derives from multiple hydrogen bonds between fibers. Chitinous cell walls enable fungi to penetrate plant or animal cells, including tough materials such as wood. Inhibitors of chitin synthesis, such as the polyoxins and nikkomycins, are used as antibiotics against fungal infections.

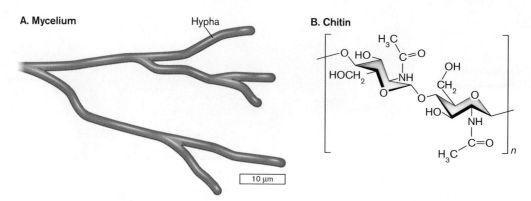

FIGURE 20.9 ■ **Fungi grow hyphae with cell walls of chitin.** **A.** Fungal hyphae extend and form branches, generating a mycelium. **B.** Chitin consists of beta-linked polymers of *N*-acetylglucosamine.

■ **Membranes contain ergosterol.** Ergosterol is an analog of cholesterol not found in animals or plants. Ergosterol is so distinctive to fungi that its presence can be used as a measure of fungal content in plant material such as grains. Inhibitors of ergosterol biosynthesis, such as the triazoles, are used to treat fungal infections. Another antifungal agent, nystatin, specifically binds ergosterol and forms membrane pores that leak K^+ ions.

Fungal Hyphae Absorb Nutrients

How do fungal hyphae grow and form colonies of "mold"? A fungal hypha expands at the tip. Cytoplasmic expansion is driven by turgor pressure against the chitin cell wall—the force that enables fungi to penetrate tough materials such as wood. Fungal hyphae can extend as fast as half a centimeter per hour.

The cytoplasmic turgor pressure is regulated by uptake of hydrogen ions in exchange for potassium ions (**Fig. 20.10**). Loss of turgor pressure—for example, by puncture of the hypha—leads to accelerated K^+ uptake and water influx, restoring turgor. Molecules that cause loss of K^+, such as nystatin, serve as antifungal agents.

At the hypha's growing tip, turgor pressure pushes the cell membrane forward, and the membrane expands by incorporating vesicles generated from the endoplasmic reticulum (as seen in **Fig. 20.10A**). The ER and mitochondria store Ca^{2+}, whose release triggers vesicle fusion with the plasma membrane. The fused vesicles provide phospholipids and proteins to extend the membrane surface area as the cytoplasm expands.

Just behind the hypha's growing tip lies its absorption zone (**Fig. 20.10B**). The absorption zone takes in nutrients from the surrounding medium, such as the cytoplasm of an invaded animal cell. Behind the absorption zone, the older part of the hypha collects and stores nutrients. As the storage zone expands, the nucleus divides multiple times. Septa form across the hypha, partly compartmentalizing the cytoplasm. As the older part of the hypha ages, its tubular form begins to lyse, releasing cell constituents. This

A.

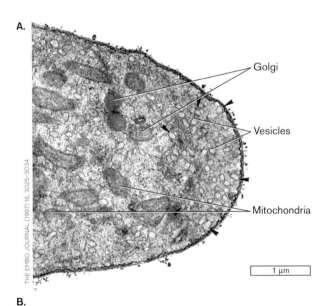

THE EMBO JOURNAL (1997) 16, 3025–3034

- Golgi
- Vesicles
- Mitochondria

1 μm

B.

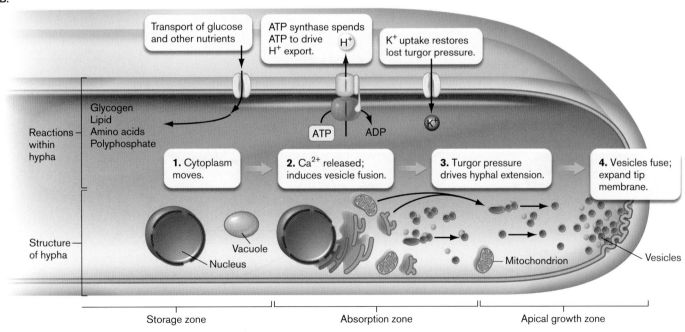

Transport of glucose and other nutrients

ATP synthase spends ATP to drive H^+ export.

K^+ uptake restores lost turgor pressure.

Reactions within hypha

Glycogen
Lipid
Amino acids
Polyphosphate

ATP ⟶ ADP

K^+

1. Cytoplasm moves. → **2.** Ca^{2+} released; induces vesicle fusion. → **3.** Turgor pressure drives hyphal extension. → **4.** Vesicles fuse; expand tip membrane.

Structure of hypha

Vacuole
Nucleus
Mitochondrion
Vesicles

Storage zone | Absorption zone | Apical growth zone

FIGURE 20.10 ■ **Cellular basis of hyphal extension. A.** Section through the growing tip of a hypha (TEM). Vesicles collect at the tip, where they fuse into the cell membrane, enabling extension. **B.** The absorption zone takes in nutrients. Cytoplasm moves toward the tip of the apical growth zone, driven by turgor pressure. Turgor pressure is regulated by H^+ export and K^+ uptake. Ca^{2+} released by the ER and mitochondria induces vesicles to fuse and to expand the plasma membrane at the growing tip.

aging part of the hypha is called the senescence zone (not shown in the figure).

As hyphae grow, branches extend from their sides. The hyphae branch and extend radially, forming the mycelium. The mycelium forms the characteristic round, fuzzy colony of a fungus or "mold." On a substrate such as wood or agar, mycelia grow in two forms: aerial mycelium, which extends out into the air; and surface mycelium, which grows into and along the surface of the substrate.

Unicellular Fungi

Despite the advantages of multinucleate hyphae, some fungi are unicellular, known as **yeasts**. Yeast forms evolved in many different fungal taxa. The yeast *Saccharomyces cerevisiae* (**Fig. 20.11A**) is used to leaven bread and to brew wine and beer (for more on food microbiology, see Chapter 16). *S. cerevisiae* reproduces by **budding**, in which mitosis of the mother cell generates daughter cells of smaller size. The mother cell acquires a bud scar where the smaller

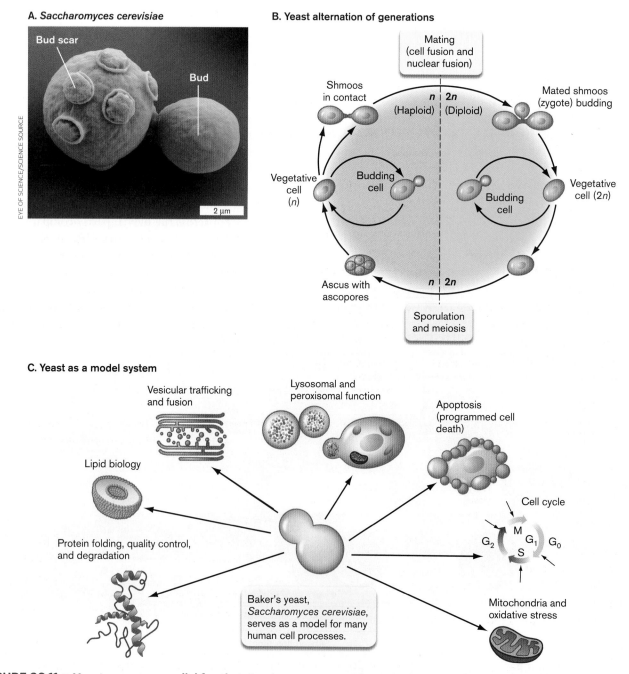

FIGURE 20.11 ■ **Yeasts are nonmycelial fungi.** **A.** *Saccharomyces cerevisiae*, or baker's yeast, reproduces by budding (SEM). Upper cell shows six bud scars. **B.** In the life cycle of *S. cerevisiae*, haploid cells reproduce many generations by budding. **C.** *S. cerevisiae* serves as a model for human cell processes. *Source:* Part C modified from Vikram Khurana and Susan Lindquist. 2010. *Nat. Rev. Neurosci.* **11**:436–449.

one pinched off (**Fig. 20.11A**). After generating a limited number of buds, the mother cell senesces and dies. Thus, yeasts provide a unicellular model system for the process of aging.

Other yeasts, such as *Candida albicans*, are important members of human vaginal flora but can cause opportunistic infections. Some are important opportunistic pathogens, occurring frequently in AIDS patients; for example, *Pneumocystis jirovecii* (formerly *P. carinii*) is a yeast-form ascomycete, whereas *Cryptococcus neoformans* is a yeast-form basidiomycete (discussed shortly). Pathogens such as *Candida albicans* can grow either as single cells (yeast form) or as mycelia; because they exist in two different forms, these are known as "dimorphic" fungi. The yeast form grows normally in the mucosa, but germination of mycelia leads to disease. A very different dimorphic fungal pathogen is *Blastomyces dermatitidis*, the cause of blastomycosis, a type of pneumonia. *B. dermatitidis* forms a mycelium in culture and in soil environments, but it grows as a yeast within the infected lung.

Yeast as a model organism for research. The yeast *Saccharomyces cerevisiae* is a model organism for research in eukaryotic biology (**Fig. 20.11C**). Its cells grow rapidly, in haploid and diploid forms, and are amenable to genetic recombination and transformation. The yeast genome of 6,000 genes includes many human homologs, such as the *ras* proto-oncogene (a gene involved in cancer). Often entire networks of proteins interacting in yeast have human homologs; thus, *S. cerevisiae* has been called a "single-celled human." Susan Lindquist at the Whitehead Institute for Biomedical Research uses yeast as a model for brain neuronal diseases such as Alzheimer's (**Special Topic 20.1**).

> **Thought Question**
>
> **20.1** Why would yeasts remain unicellular? What are the relative advantages and limitations of hyphae?

Yeast reproductive cycles. Some yeasts are asexual, whereas others can undergo sexual **alternation of generations** (**Fig. 20.11B**). This life cycle alternates between generation of a haploid population, with a single copy of each chromosome (*n*), and a diploid population, with a diploid chromosome number (*2n*). The haploid form develops gametes to fertilize each other, making a *2n* zygote. After vegetative (nonsexual) divisions, the *2n* form undergoes meiosis, regenerating the haploid form. (The process of meiosis is reviewed in eAppendix 2.) Alternation of generations allows an organism to respond genetically to environmental change by reassorting its genes through meiosis, and by recombining them through fertilization. Gene reassortment and recombination provide new genotypes, some of which may increase survival in the changed environment.

In baker's yeast (*Saccharomyces cerevisiae*), haploid spores divide and proliferate by mitosis. Under environmental signals such as starvation, mating factors induce the haploid cells to differentiate into gamete forms called "shmoos" (see **Fig. 20.11B**). Gametes of two different mating types fuse, and their nuclei combine to form a zygote. In the diploid generation, the zygote divides mitotically, generating a population of diploids that appear superficially similar to haploid cells. Under stress, particularly desiccation, the diploids undergo meiosis to reassort their genes for combinations that may better survive the changed environment. Meiosis generates an **ascus** (plural, **asci**) that contains four haploid spores.

Many fungi and protists undergo modified versions of alternation of generations, utilizing a wide variety of haploid and diploid structures to accomplish essentially the same genetic tasks. In many fungi, the haploid form predominates; for example, ascomycetes such as *Aspergillus* and *Neurospora* form mainly haploid mycelia. In some fungi, no sexual reproduction has been observed, probably because the inducing conditions are unknown. Species that lack a known sexual cycle are called **mitosporic fungi**, also known as "imperfect" fungi. Mitosporic species are found in many different clades. An example is the famous *Penicillium* mold, an ascomycete, from which we discovered the antibiotic penicillin.

> **Thought Question**
>
> **20.2** Why would some fungi conduct asexual reproduction under most conditions? What are the advantages and limitations of sexual reproduction?

Mycelia, Mushrooms, and Mycorrhizae

Different species of fungi show vastly different forms, from the familiar mushrooms (fruiting bodies that can weigh several pounds) to the mycelia of pathogens and the symbiotic partners of algae in lichens. Major clades of fungi include Chytridiomycota, Zygomycota, Ascomycota, and Basidiomycota.

Note: The major groups of fungi are also known by names with the alternative suffix "-etes": Chytridiomycetes, Zygomycetes, Ascomycetes, Basidiomycetes.

SPECIAL TOPIC 20.1   Yeast: A Single-Celled Human Brain?

The yeast *Saccharomyces cerevisiae* provides a model for human diseases such as cancer because it shares so many homologs of our genes. Most of these genes encode fundamental parts of cells, such as actin and cell growth regulators. But of course the single-celled fungus lacks the differentiated parts and connectors of human cells such as neurons. So yeast could not serve as a model for complex diseases of the brain. Or could it?

Neurodegenerative diseases such as Alzheimer's can result from disorders of cell function, such as lysosomal storage and degradation. A possible cause is that a protein called beta-amyloid gets cleaved to a peptide that is secreted; the secreted peptide returns to the cell by endocytosis. The endocytosed beta-amyloid peptide then interferes with intracellular trafficking by an unknown mechanism. To investigate the mechanism of beta-amyloid toxicity, Susan Lindquist at the Whitehead Institute for Biomedical Research (**Fig. 1**) developed a yeast model.

First, Lindquist and her students constructed a yeast strain that expresses beta-amyloid peptide from a plasmid. (We describe methods to construct such strains in Chapter 12.) In this yeast strain, the beta-amyloid gets secreted, but it

returns through endocytosis to the endoplasmic reticulum (ER) for trafficking—just as it does in a human cell. So what happens to the yeast? The yeast grows more slowly because beta-amyloid disrupts ER trafficking (**Fig. 2**). The controls—a yeast protein expressed on a plasmid, and a plasmid vector expressing no protein—grow normally.

The disruption of trafficking by beta-amyloid was consistent with its proposed role in Alzheimer's disease. But to strengthen the connection—and to reveal other parts of the process—Lindquist used her yeast model to screen for yeast proteins that could overcome beta-amyloid interference. So she transformed her yeast model strain with a library of yeast

FIGURE 1 ■ **Susan Lindquist develops yeast models of human disease.**

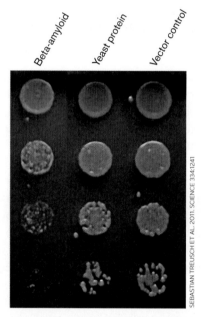

FIGURE 2 ■ **Yeast expresses human beta-amyloid peptide.** Yeast cultures were serially diluted (vertical direction) and spread on agar to grow patches. A plasmid expressing beta-amyloid (left) causes slower growth (thinner patches) than a plasmid expressing a normal yeast protein or no protein (vector control) (middle and right, respectively).

Chytridiomycota: motile zoospores. The deepest-branching clade of fungi is Chytridiomycota (the chytrids), which possess motile, flagellated reproductive forms called zoospores. The zoospore form has been lost by other fungi.

Chytrid species include bovine rumen inhabitants whose hyphae penetrate tough plant material, facilitating digestion. An example is *Neocallomastix* (**Fig. 20.12A**). Unlike most fungi, *Neocallomastix* is an obligate anaerobe whose

mitochondria have evolved into hydrogenosomes, organelles that ferment carbohydrates in a pathway generating H_2. Hydrogenosomes are a unique adaptation of certain anaerobic fungi and protists.

Other chytrids are aerobic animal pathogens. **Figure 20.12B** shows the skin of a frog infected by the chytridiomycete *Batrachochytrium dendrobatidis*. *B. dendrobatidis* has caused a widespread die-off of frogs in

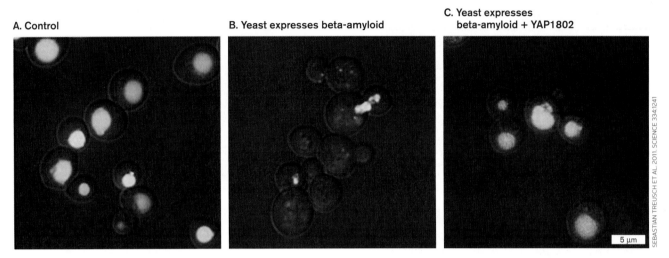

A. Control
B. Yeast expresses beta-amyloid
C. Yeast expresses beta-amyloid + YAP1802

5 μm

SEBASTIAN TREUSCH ET AL. 2011. SCIENCE 334:1241

FIGURE 3 ■ **Beta-amyloid disrupts traffic of yeast protein.** **A.** A yeast protein fused to yellow fluorescent protein (YFP) is trafficked normally to the vacuole (fluorescence microscopy). **B.** Beta-amyloid expression interferes with transport. **C.** YAP1802 expression restores transport. .

genes that were "overexpressed" (that is, expressed on a plasmid at levels severalfold greater than normal). She identified 23 suppressors, genes whose overexpression suppressed the effect of beta-amyloid and restored normal yeast growth. The suppressors also restored normal trafficking of a reporter protein, a fluorescent YFP fusion protein expressed by the yeast genome (**Fig. 3**). The control cells showed reporter protein fluorescence localized normally to the vacuole (**Fig. 3A**). Beta-amyloid expression prevented trafficking to the vacuole (**Fig. 3B**), but expression of the suppressor protein YAP1802 restored movement to the vacuole (**Fig. 3C**).

Are these suppressors of the yeast model relevant to humans? Six of the suppressor genes turned out to have homologs in the human genome already identified as possible risk factors for Alzheimer's disease. The YAP1802 human homolog is a protein called PICALM. Lindquist tested whether PICALM overexpression could protect rat brain neurons from beta-amyloid toxicity. She provided PICALM to the neurons by expression on a lentiviral vector—that is, a gene therapy vector derived from HIV virus (discussed in Chapter 11). Remarkably, PICALM restored normal growth to neurons that had

taken up beta-amyloid. For the future, Lindquist and her colleagues are using the yeast model to reveal possible therapeutic agents for neurodegeneration.

RESEARCH QUESTION

How would you go about using the yeast model to identify therapeutic agents for Alzheimer's disease?

Treusch, Sebastian, Shusei Hamamichi, Jessica L. Goodman, Kent E. S. Matlack, Chee Yeun Chung, et al. 2011. Functional links between Ab toxicity, endocytic trafficking, and Alzheimer's disease risk factors in yeast. *Science* **334**:1241–1245.

Matlack, Kent E. S., Daniel F. Tardiff, Priyanka Narayan, Shusei Hamamichi, Kim A. Caldwell, et al. 2014. Clioquinol promotes the degradation of metal-dependent amyloid-β (Aβ) oligomers to restore endocytosis and ameliorate Aβ toxicity. *Proceedings of the National Academy of Sciences USA* **111**:4013–4018.

Central and South America, in an epidemic associated with global warming. The mycelium of *B. dendrobatidis* grows within the frog skin, producing capsules full of diploid zoospores called zoosporangia. Each zoosporangium protrudes through the skin surface, ready to expel zoospores in search of a new host.

The life cycle of a chytrid includes mycelia that are haploid (gametophyte) or diploid (sporophyte) (**Fig. 20.12C**).

Haploid mycelia produce motile gametes that detect each other by sex-specific attractants. The gametes fuse to produce a motile zygote. The zygote forms a cyst, a cell with arrested metabolism that can persist for long periods. In a favorable environment, the cyst germinates to form a diploid mycelium, or sporophyte. The sporophyte generates zoosporangia full of zoospores. There are two alternative forms of zoosporangia: those that produce diploid

A. Chytridiomycete fermenter in sheep rumen

B. Chytridiomycete infection of frog skin

C. Chytridiomycete life cycle

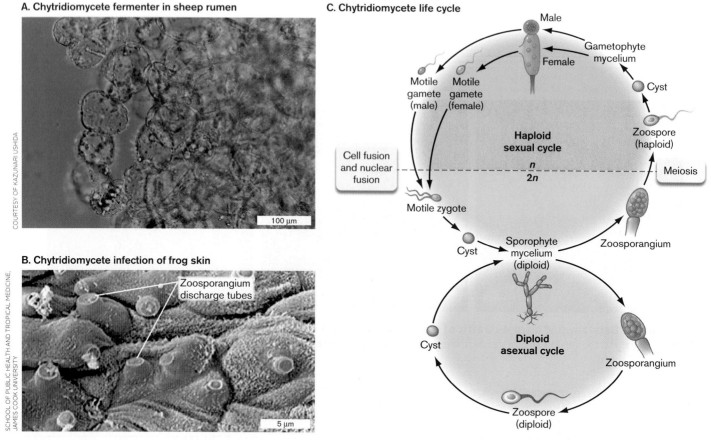

COURTESY OF KAZUNARI USHIDA

SCHOOL OF PUBLIC HEALTH AND TROPICAL MEDICINE, JAMES COOK UNIVERSITY

FIGURE 20.12 ■ **Chytridiomycete form and life cycle.** **A.** Chytrids such as *Neocallomastix* species ferment complex plant material for ruminant animals such as cattle and sheep. **B.** A pathogenic chytrid, *Batrachochytrium dendrobatidis*, infects the skin of a frog. Frog skin cells are penetrated by discharge tubes of diploid zoosporangia about to release zoospores. **C.** Life cycle of a chytrid. The diploid mycelium produces motile zoospores that form cysts in a poor environment. Alternatively, the diploid mycelium undergoes meiosis to form a haploid mycelium (gametophyte) that produces motile gametes.

zoospores, which form cysts and regenerate the diploid mycelium; and those that undergo meiosis to produce haploid zoospores. The haploid zoospores generate a haploid mycelium (gametophyte) capable of producing haploid gametes.

Zygomycota: nonmotile sporangia. The **zygomycetes** and other nonchytridiomycete fungi generate nonmotile spores. Nonmotile spores require transport by air or water, or ballistic expulsion (expulsion under pressure) from a spore-bearing organ, called the **sporangium** (plural, **sporangia**). A common zygomycete is the bread mold *Rhizopus* (**Fig. 20.13A**). Most zygomycetes, such as *Mucor* species, are soil molds that decompose plant material or other fungi or the droppings of animals (**Fig. 20.13B**). These modest molds fill important niches in all terrestrial ecosystems.

Like chytrids, the life cycle of a zygomycete alternates between haploid (*n*) and diploid (2*n*) forms (**Fig. 20.13C**). The mechanics differ, however, owing to the lack of motile gametes. The haploid spore (sporangiospore) is disseminated through air currents. A sporangiospore does not

directly undergo sexual reproduction; it grows into a haploid mycelium. The haploid mycelium then forms special hyphae whose tips differentiate into gamete cells. The gametes cannot separate from the filament; instead, two gamete-bearing hyphae must grow toward each other in order to fuse and form a **zygospore**. The zygospore undergoes meiosis and generates the sporangium, a haploid structure that releases **sporangiospores**.

> **Thought Question**
>
> **20.3** What are the advantages and limitations of motile gametes, as compared to nonmotile spores?

Ascomycota: mycelia with paired nuclei. The **ascomycete** fungi are famous in the history of science, as well as in the culinary arts (**Fig. 20.14**). The bread mold *Neurospora* was used by George Beadle and Edward Tatum in the 1940s to formulate the one gene–one protein theory.

A. Bread mold, *Rhizopus*

GREGORY G. DIMIJIAN / SCIENCE SOURCE

B. *Mucor* **diploid hyphae form zygospores**

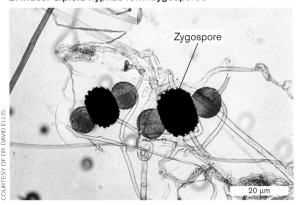

Zygospore

20 μm

COURTESY OF DR. DAVID ELLIS

C. Zygomycete life cycle

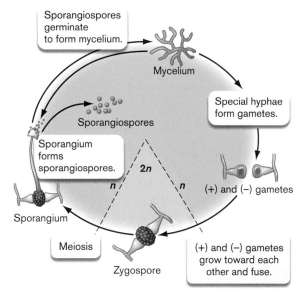

Sporangiospores germinate to form mycelium.

Mycelium

Special hyphae form gametes.

Sporangiospores

Sporangium forms sporangiospores.

2n

n   n

(+) and (−) gametes

Sporangium

Meiosis

(+) and (−) gametes grow toward each other and fuse.

Zygospore

FIGURE 20.13 ■ Zygomycete fungi form nonmotile sporangia.
A. *Rhizopus* (bread mold) haploid sporangia contain sporangiospores.
B. Diploid hyphae of *Mucor* species terminate in zygospores. **C.** The life cycle of zygomycetes involves primarily haploid mycelia. Special hyphae form gametes at their tips. Gametes of different mating types fuse to form the diploid zygospore. The zygospore undergoes meiosis, regenerating haploid cells that form sporangia. The sporangia release sporangiospores, which germinate to form new mycelia.

A. Asci containing ascospores

20 μm

ED RESCHKE/PETER ARNOLD/GETTY IMAGES

B. Morel (an ascomycete fruiting body)

ED RESCHKE/PETER ARNOLD/GETTY IMAGES

C. Ascomycete life cycle

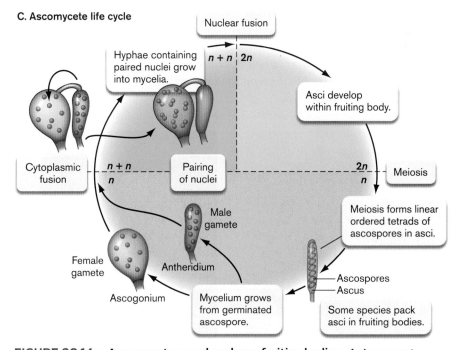

Nuclear fusion

Hyphae containing paired nuclei grow into mycelia.

n + n   2n

Asci develop within fruiting body.

Cytoplasmic fusion

n + n
n

Pairing of nuclei

2n
n

Meiosis

Meiosis forms linear ordered tetrads of ascospores in asci.

Male gamete

Female gamete

Antheridium

Ascogonium

Mycelium grows from germinated ascospore.

Ascospores
Ascus

Some species pack asci in fruiting bodies.

FIGURE 20.14 ■ Ascomycetes produce large fruiting bodies. A. Ascomycete asci containing ascospores (stained red). **B.** The culinary delicacies known as morels are fruiting bodies of the species *Morchella hortensis*. The dark pits of the morel are lined with asci. **C.** The life cycle of an ascomycete alternates between the diploid and haploid forms. The diploid mycelium produces asci, within which the haploid ascospores are formed.

In *Neurospora*, meiosis produces pods (asci) of **ascospores** aligned in rows that reflect the ordered tetrads of meiotic division (**Fig. 20.14A**). The tetrad patterns were used by geneticists to demonstrate the segregation and independent assortment of chromosomes. In other species, by contrast, the asci are packed in large mushroom-like fruiting bodies known as morels (*Morchella hortensis*) (**Fig. 20.14B**) and truffles (*Tuber aestivum*). The ascospores of such fruiting bodies are spread by animals attracted by their delicious flavor. Human collectors traditionally use muzzled pigs to detect and unearth the famous underground truffles.

The ascomycete life cycle (**Fig. 20.14C**) includes a phase in which each cell possesses a pair of separate nuclei, one from each parent (chromosome number is designated $n + n$). The "dikaryotic" (paired-nuclei) phase is generated by haploid mycelia in which male and female reproductive structures fuse, followed by migration of all the male nuclei into the female structure. The paired nuclei then undergo several rounds of mitotic division while migrating into the growing mycelium. In the mycelial tips, the paired nuclei finally fuse (becoming $2n$), and the mycelial tips develop into asci. Each ascus then undergoes meiosis in which the haploid products segregate in the same order that the meiotic chromosomes separated.

Some ascomycetes, such as *Aspergillus* and *Penicillium* species, form small asexual fruiting bodies called conidiophores for airborne spore dispersal (**Fig. 20.15A**). *Penicillium* is known for producing penicillin, the first antibiotic in widespread use; forms of penicillin are still used today (discussed in Chapter 1). *Aspergillus* (**Fig. 20.15A** and **B**) is a growing medical problem as an opportunistic pathogen of immunocompromised patients. *Aspergillus* can produce toxins (called mycotoxins) such as aflatoxin. Aflatoxin poisoning commonly affects livestock and, in some cases, agricultural workers; the toxin causes liver damage, immunosuppression, and cancer.

Conidiophore-forming ascomycetes such as *Aspergillus* and *Stachybotrys* are the major form of mold associated with dampness in human dwellings; for example, they caused massive damage to homes flooded in the wake of Hurricane Katrina in 2005 (**Fig. 20.15C**). The flooding of homes full of drywall made ideal conditions for the growth of mold, which commonly consists of airborne ascomycete mycelia. Mold grew not only on materials submerged, but also on the surface above, exposed to water-saturated air, up to 3 feet above the flood line (the highest level submerged). Unfortunately, most homeowner insurance policies covered damage only "up to the flood line."

Many ascomycetes are pathogens of animals or plants. For example, *Microsporum* and *Trichophyton* species cause ringworm skin infection, whereas *Magnaporthe oryzae* causes rice blast, the most serious disease of cultivated rice. Other species, however, are beneficial symbionts of plants, including crop plants such as beans, cucumbers, and cotton. *Trichoderma* species grow on the roots, or in some cases within the vascular tissue, of the plant. The fungi share nutrients with the plant, and they induce plant defenses against pathogens. *Trichoderma* even has a commercial use in cloth processing; the fungus is used to make "stonewashed jeans," as its cellulase enzymes partly digest the cotton.

Thought Question

20.4 Compare the life cycle of an ascomycete (**Fig. 20.14C**) with that of a chytridiomycete (**Fig. 20.12C**). How are they similar, and how do they differ?

A.

B.

C.

FIGURE 20.15 ■ **Ascomycete molds. A.** *Aspergillus* forms a microscopic asexual fruiting structure called a conidiophore, containing spores in its spherical tip. **B.** Colony of *Aspergillus nidulans* on an agar plate. **C.** Black mold (such as *Stachybotrys*) grows above the flood line on a kitchen wall. An undergraduate volunteer shows the presence of mold in New Orleans, 6 months after flooding caused by Hurricane Katrina.

A. *Amanita*

B. *Aleuria*

FIGURE 20.16 ■ Mushrooms and other basidiomycetes form large, complex fruiting bodies. A. *Amanita phalloides* makes one of the most dangerous toxins known: alpha-amanitin, an inhibitor of RNA polymerase II. **B.** *Aleuria* mushrooms releasing spores from their gills, to be carried on the wind.

Basidiomycota: cells with paired nuclei form mushrooms. The **basidiomycetes** form large, intricate fruiting bodies known as "true" mushrooms (**Fig. 20.16**). Mushrooms produce some of the world's deadliest poisons, such as alpha-amanitin, which inhibits RNA polymerase II. Alpha-amanitin is produced by the amanita, or "destroying angel" (**Fig. 20.16A**), a taste of which is usually fatal. (The amanita's own RNA polymerase is insensitive to the toxin.) Other mushroom species include some of the world's most prized

culinary delights, such as the portobello. Many grow in soil, while others, such as *Piptoporus*, grow on tree bark. Some mushrooms have evolved elaborate insect-attracting structures and odors, such as the "starfish stinkhorn," with its ring of bright red horns.

The mushroom itself is only the fruiting body of the basidiomycete, whose life cycle involves transitions among *n*, *n* + *n*, and *2n* (**Fig. 20.17A**) similar to those of ascomycetes (see **Fig. 20.14C**). In the basidiomycete, however,

A. Basidiomycete (mushroom) life cycle

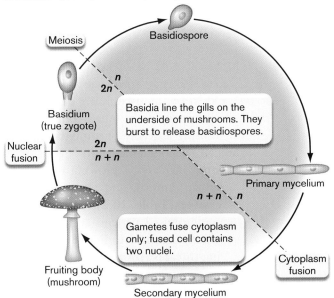

B.

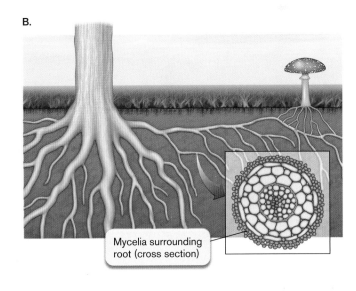

FIGURE 20.17 ■ Mushroom life cycle. A. Haploid basidiospores generate primary mycelium underground, where they form gametes. Gametes of opposite mating types fuse their cytoplasm only, forming secondary mycelium. The parental nuclei remain separate throughout many generations of mitosis during development of the fruiting body (mushroom). As the mushroom matures, the basidia undergo nuclear fusion and meiosis, forming progeny basidiospores. **B.** The underground secondary mycelia of some mushrooms form mycorrhizae with tree roots. Mycorrhizae enhance and extend the absorptive power of the tree roots, while obtaining plant sugars for the fungus.

the fruiting body consists largely of cells with paired nuclei ($n + n$). A few of the paired nuclei fuse to form diploid cells ($2n$) called basidia (singular, basidium), which line the gills of the mushroom. The basidia undergo meiosis to form haploid **basidiospores** (n). Some types of basidia can release basidiospores under pressure, whereas other basidiospores are dispersed by wind (see **Fig. 20.16B**).

The basidiospores germinate to form underground mycelium. This haploid "primary mycelium" generates gametes that ultimately fuse to form $n + n$ "secondary mycelium." The primary and secondary mycelia may radiate underground, unseen, until their tips generate mushrooms aboveground, at points approximately equidistant from the origin. The result is a mysterious "fairy ring" of mushrooms (see **Fig. 20.3A**). In a forest, these invisible underground hyphae of basidiomycetes and glomeromycota contribute

mycorrhizae (singular, **mycorrhiza**) that gain sugars from trees while extending the tree root systems (**Fig. 20.17B**).

Glomeromycota form arbuscular mycorrhizae. A remarkable phylum of fungi is the Glomeromycota, all of which are obligate mutualists of plants. These fungi, such as *Glomus* species, form extensive networks of filamentous connections with plant roots similar to the mycorrhizae formed by basidiomycetes; but unlike the basidiomycete mycorrhizae, the mycorrhizae formed by glomeromycota are **arbuscular mycorrhizae**, in which the fungal filaments actually penetrate plant cells in a most intimate symbiosis (**Fig. 20.18**). Mycorrhizae expand the roots' absorptive capacity, while obtaining plant sugars for the fungus. More than 90% of all land plants, including trees, depend on these fungal interconnections, which share nutrients among many unrelated plants, as well as the fungi (discussed in Chapter 21).

In arbuscular mycorrhizae, a fungal hypha first grows in between plant cells without breaching the plant cell wall (**Fig. 20.18B**). Next, the hypha branches into the plant cytoplasm; the plant cell membrane invaginates to accommodate the branch, while maintaining a "periarbuscular space" between the plant cell membrane and the fungal plasma membrane. This highly regulated invaginating branch is called an "arbuscule"—hence the term "arbuscular mycorrhiza." Arbuscular formation is regulated by plant hormones called strigolactones. The arbuscule expands the surface area to exchange sugars from the plant for ammonium and phosphate from the fungus.

Emerging Fungal Pathogens

The dominant role of fungi in our biosphere is positive—as decomposers, recyclers, and symbiotic partners within lichens and mycorrhizae (discussed further in Chapter 21). But some fungi are important pathogens, such as *Histoplasma capsulatum*, an ascomycete fungus that infects healthy people who inhale contaminated dust, causing a deadly pneumonia. And as human demographics shift and climate change increases, a growing number of human, animal, and plant pathogens emerge (**Table 20.2**). For immunocompromised patients, especially the elderly confined to hospitals, a growing threat is *Aspergillus* species, which can colonize the lungs and other tissues. In 2012, contaminated steroid injections led to an outbreak of infections by *Exserohilum rostratum* and other previously rare opportunists. Other fungal pathogens cause massive mortality of animals and plants.

A.

B.

USDA

Arbuscule

Fungus

Fungal cell wall

Plant cell membrane

Fungal cytoplasm

Fungal plasma membrane

Periarbuscular space

Periarbuscular membrane

Plant cytoplasm

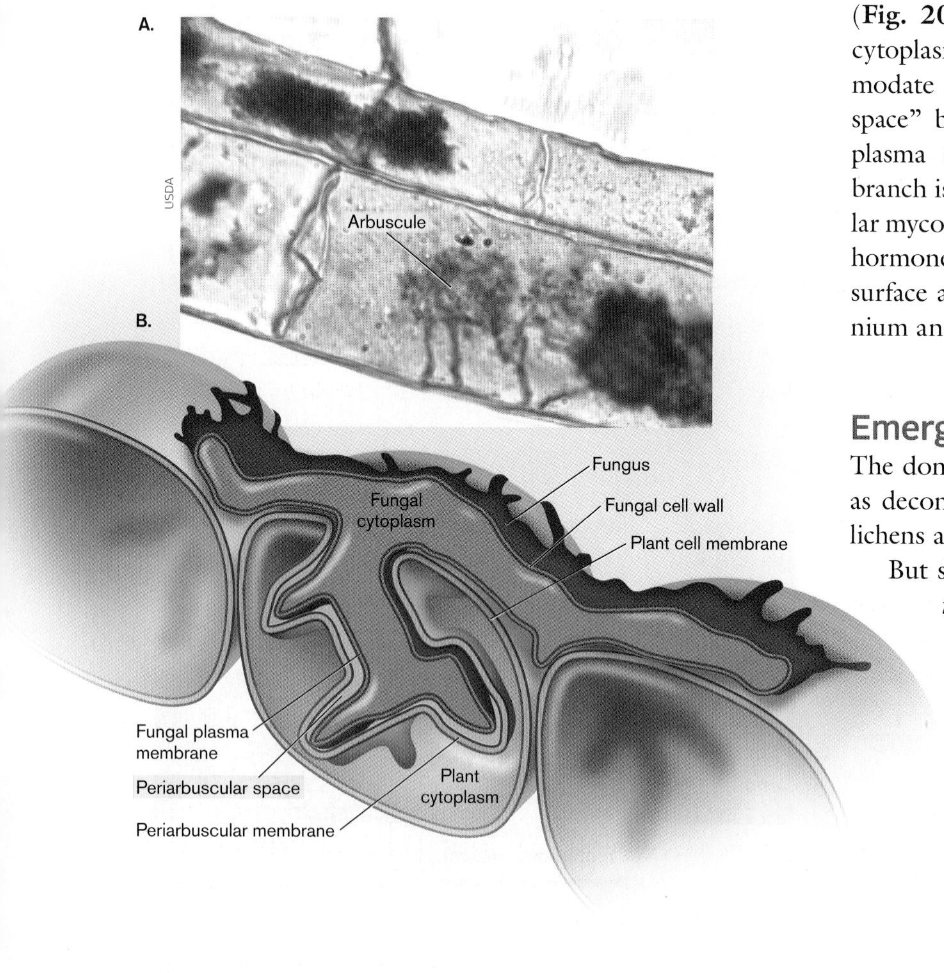

FIGURE 20.18 ■ **Arbuscular mycorrhiza. A.** Glomeromycota fungi invade corn root cells, as part of a mutualistic symbiosis to exchange nutrients. **B.** The fungal hypha grows between the plant cells, and then into a plant cell to form an arbuscule. The arbuscule expands surface area for exchange while maintaining the plant cell wall intact. *Source:* Part B modified from Martin Parniske. 2008. *Nat. Rev. Microbiol.* **6**:763.

TABLE 20.2	Emerging Fungal Pathogens		
Species	**Phylum**	**Host**	**Disease**
Aspergillus fumigatus	Ascomycota	Humans (immunocompromised)	Aspergillosis of lung, and elsewhere in the body
Cryptococcus neoformans	Basidiomycota	Humans (immunocompromised)	Meningitis and meningoencephalitis
Encephalitozoon intestinalis	Microsporidia	Humans with AIDS	Microsporidiosis (intestinal)
Exserohilum rostratum	Ascomycota	Humans	Wound and skin infections; contaminated injections
Histoplasma capsulatum	Ascomycota	Humans, dogs, cats	Histoplasmosis (lung)
Pneumocystis jirovecii	Ascomycota	Humans (immunocompromised)	Pneumocystis pneumonia
Stachybotrys chartarum	Ascomycota	Humans	Black mold disease; respiratory damage
Batrachochytrium dendrobatidis	Chytridiomycota	Amphibians (frogs and toads)	Chytridiomycosis
Fusarium solani	Ascomycota	Sea turtles (loggerheads)	Hatch failure
Geomyces destructans	Ascomycota	Brown bats	White nose disease
Puccinia graminis	Basidiomycota	Wheat	Wheat stem rust
Magnaporthe oryzae	Ascomycota	Rice	Rice blast disease
Nosema species	Microsporidia	Honeybees	Colony collapse disorder

Several parasitic organisms originally classified as protozoa because of their superficial appearance have now been shown to be fungi or fungus-related, based on their genome sequence and biochemistry. The reclassification has important consequences for research, taxonomy, and therapy. An example is the ascomycete *Pneumocystis jirovecii*. The organism was first described in 1909, when it was thought to be a life stage of a trypanosome causing Chagas' disease (discussed in Section 20.6). When the organism was recognized as a distinct species of protists infecting animals, it was named *Pneumocystis carinii*. In the 1970s, a strain of *Pneumocystis* was renamed *P. jirovecii* for causing pneumonia in immunocompromised humans. In the 1980s, the rise of AIDS led to a sudden increase in infections. Sequencing the organism's genome revealed it to be an ascomycete fungus. As a result, the organism's name was called into question by fungus researchers; its developmental forms and biochemistry were reevaluated, and medical research was redirected to culture and treat the organism based on fungal physiology.

A major clade of parasites now shown to be fungi is Microsporidia. Microsporidia have relatively small genomes; their mitochondria have lost their DNA and are nonfunctional. Microsporidia form spores as small as a few micrometers in size, which can infect animal cells. The microsporidian spore extrudes a specialized invasion complex called the polar tube that penetrates the host cell, typically a macrophage. In humans, microsporidians are opportunistic pathogens, such as the intestinal parasite *Encephalitozoon intestinalis*, emerging with the rise of AIDS and the growth of elderly and immunocompromised populations.

Other taxa originally assigned as fungi, based on superficial appearance, are now classified as protists, based on genetic similarity. The **oomycetes**, or "water molds," were originally classified as fungi because of their fungus-like filaments, which infect plants and animals. Oomycete *Phytophthora infestans* devastated Irish potato crops in the 1840s, causing the Great Irish Famine, in which a million people died. Today, a related species, *Phytophthora ramosum*, causes "sudden oak death," a disease killing tens of thousands of oaks and other trees in the western United States. Oomycetes are discussed further in **eTopic 20.1**.

To Summarize

■ **Fungi form hyphae with cell walls of chitin.** Hyphae absorb nutrients from decaying organisms or from infected hosts. Some fungi remain unicellular; these are called yeasts or mitosporic fungi.

■ **Chytridiomycete fungi have motile zoospores.** Motile reproductive forms are a trait shared with animals. Flagellar motility has been lost by other fungi through reductive evolution.

■ **Zygomycete fungi form haploid mycelia.** Hyphal tips differentiate into gametes and grow toward each other to undergo sexual reproduction. Some zygomycetes form mycorrhizae that connect the roots of plants.

■ **Ascomycete fungal mycelia form paired nuclei.** Within the hyphal cells, the paired nuclei fuse, followed by meiosis and development of ascospores. Some ascomycetes form fruiting bodies called conidiophores.

■ **Basidiomycete fungi form mushrooms.** Cells with paired nuclei (secondary mycelium) form large fruiting bodies called mushrooms. The paired nuclei fuse to form the diploid basidium, which generates haploid basidiospores. The basidiospores develop underground hyphae or mycorrhizae that interconnect plant roots.

■ **Glomeromycota form arbuscular mycorrhizae.** These fungi form intimate mutualistic networks of connections with plant roots, exchanging minerals for plant sugars.

■ **Emerging fungal pathogens threaten humans, plants, and animals.**

20.3 Algae

Algae are CO_2-fixing producers in all ecosystems, most crucially freshwater and marine habitats. In freshwater and marine ecology, the algae, together with photosynthetic bacteria, are known as **phytoplankton** (see Chapter 21). All algae possess chloroplasts.

The primary algae are products of a single ancestral endosymbiosis that also gave rise to land plants. The biochemistry and cell structures of true algae and land plants are similar; thus, the alga *Chlorella* was the model organism of choice for pioneers of photosynthesis research Martin Kamen, Melvin Calvin, and others (discussed in Chapter 15).

Other photosynthetic eukaryotes, or secondary endosymbiotic algae, arose from protists that engulfed a primary or secondary alga. Secondary algae often show "mixotrophic" nutrition, involving both phototrophy and heterotrophy. For example, dinoflagellate "algae" are voracious predators of smaller protists. The heterokont algae (diatoms, coccolithophores, and kelps) are covered in this section. Dinoflagellates are covered under alveolates (Section 20.5).

Primary Algae

The primary endosymbiotic algae include two major clades: Chlorophyta, or green algae; and Rhodophyta, or red algae—although not all members of each group appear green or red, respectively. Rhodophyta that appear red have a secondary pigment called phycoerythrin, in addition to green chlorophyll.

Green algae (Chlorophyta). Many green algae are unicellular. An important model system for genetics and phototaxis is *Chlamydomonas reinhardtii*, a unicellular chlorophyte common in freshwater systems as well as Antarctic pools. The genetics of cell cycle regulation in *C. reinhardtii* provides clues to formation of human tumors.

C. reinhardtii has a symmetrical pair of flagella—a common pattern for green algae and their gametes (**Fig. 20.19A**). The alga swims forward by bending its flagella back toward the cell, like a breaststroke. *Chlamydomonas* cells are mostly haploid, reproducing by asexual cell division (**Fig. 20.19B**). For sexual reproduction, opposite mating types fuse to form a zygote, which loses flagella and grows a spiny protective coat. The zygote undergoes meiosis to regenerate haploid cells.

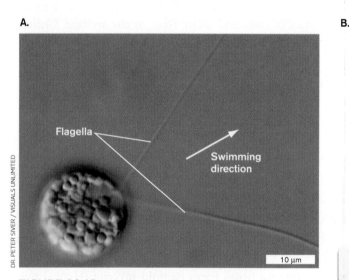

A.

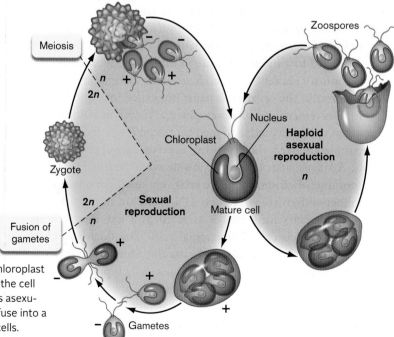

B.

FIGURE 20.19 ■ **A single-celled green alga:** ***Chlamydomonas reinhardtii.*** **A.** *Chlamydomonas* has a green chloroplast and grows as a photoautotroph. A symmetrical pair of flagella pulls the cell forward. **B.** Alternation of generations. *Chlamydomonas* reproduces asexually as haploid cells. Alternatively, the alga generates gametes that fuse into a zygote that immediately undergoes meiosis to regenerate haploid cells.

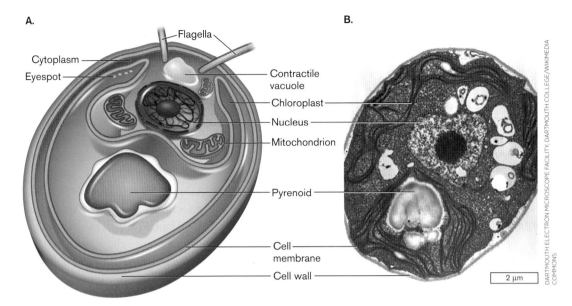

A.

- Cytoplasm
- Eyespot
- Flagella
- Contractile vacuole
- Chloroplast
- Nucleus
- Mitochondrion
- Pyrenoid
- Cell membrane
- Cell wall

B.

2 µm

DARTMOUTH ELECTRON MICROSCOPE FACILITY, DARTMOUTH COLLEGE/WIKIMEDIA COMMONS

FIGURE 20.20 ■ Cell structure of *Chlamydomonas reinhardtii*. A. The cell membrane is surrounded by a cell wall of cellulose and glyco-proteins. The single chloroplast fills much of the cell and wraps around the nucleus. Within the chloroplast lies a pyrenoid, a structure for concentrating bicarbonate ion for conversion to CO_2. The pyrenoid is surrounded by starch bodies that store high-energy compounds. A contractile vacuole maintains constant osmotic pressure. **B.** Electron micrograph of *C. reinhardtii* shows the nucleus, chloroplast, pyrenoid, and other organelles.

The cell ultrastructure of *Chlamydomonas* is typical of algal cells (**Fig. 20.20**). The nucleus is cupped by a single chloroplast, which is surrounded by a double membrane. The double membrane indicates primary algae; the inner membrane derives from the ancestral bacterium, and the outer membrane derives from the engulfing host. Within the chloroplast is a pyrenoid, an organelle that concentrates bicarbonate (HCO_3^-) and converts it to CO_2 for fixation. The pyrenoid is surrounded by one or more starch bodies, which are used for energy storage. The starch is broken down to sugars as needed, followed by glycolysis and respiration in the mitochondria. Osmolarity is maintained by the contractile vacuole. The *Chlamydomonas* cell is encased in a cell wall composed predominantly of glycoprotein. Other green algae have cellulose cell walls similar to those of plants.

While *C. reinhardtii* is unicellular in nature, an exciting experiment showed that this alga can quickly evolve into a multicellular form. In 2013, William Ratcliff at the Georgia Institute of Technology (**Fig. 20.21B**) and Michael Travisano at the University of Minnesota

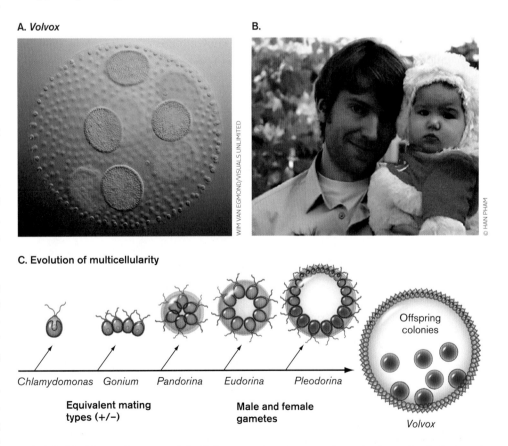

A. *Volvox*

WIM VAN EGMOND/VISUALS UNLIMITED

B.

© HAN PHAM

C. Evolution of multicellularity

Chlamydomonas — **Equivalent mating types (+/−)**

Gonium

Pandorina

Eudorina — **Male and female gametes**

Pleodorina

Offspring colonies

Volvox

FIGURE 20.21 ■ *Volvox* is multicellular. A. *Volvox* forms a spherical colony of cells (1–3 mm) that generates offspring colonies within. **B.** William Ratcliff, who studies origins of multicellularity, with his own multicellular offspring. **C.** Different types of algae suggest a model for the evolution of multicellularity from a unicellular ancestor (like *Chlamydomonas*) to a multicellular colony with female and male gametes (*Volvox*).

showed that *C. reinhardtii* could undergo selection by repeated subculturing in a standing tube. In each culture, a few cells would clump at the bottom, and the clumps were selectively cultured further. Eventually, a strain was isolated that grew regularly in connected cell clusters that alternated with dispersal.

Some marine and freshwater algae that resemble *Chlamydomonas* grow naturally in intricate multicellular colonies. Colonial algae such as *Volvox* (**Fig. 20.21A**) generate geodesic spheres of biflagellate cells. Their flagella point outward from the colony, propelling it forward and drawing nutrients across its surface. Each cell of *Volvox* connects by cytoplasmic bridges to five or six of its neighbors. The colony reproduces by generating daughter colonies within the sphere, which grow until the outer sphere falls apart, liberating the daughters.

A survey of algal diversity from *Chlamydomonas* through *Volvox* suggests a model for the progression of evolving multicellularity (**Fig. 20.21C**). First, cells might have evolved to grow in small ordered structures, as is seen for *Gonium* and *Pandorina* species. The structures then might have evolved into hollow spheres of cells with some differentiation, as seen in *Eudorina*. Male and female gametes then evolved (*Pleodorina*), and ultimately colonies that generate colonial offspring (such as *Volvox*).

Other species of algae grow as multicellular filaments. An example is *Spirogyra*, a common pond dweller known for its spiral chloroplasts (**Fig. 20.22A**). As with *Chlamydomonas*, the haploid form predominates; but unlike the unicellular alga, *Spirogyra* forms neither flagellated gametes nor zoospores. Instead, its sexual reproduction requires alignment of two filaments of opposite mating type (**Fig. 20.22B**). Cells conjugate (form cytoplasmic bridges) between the two filaments. The "male" gametes are those whose cytoplasm inserts through the conjugation bridge to join that of the "female" gamete. As the gametes fuse, a row of empty cell walls is left behind (**Fig. 20.22C**). The zygote eventually hatches and germinates a new chain of cells.

Some marine algae grow in undulating sheets. An example familiar to beach bathers is the "sea lettuce" *Ulva* (**Fig. 20.23A**). Sheets of *Ulva* can extend over many square meters, although they are only two cells thick (see **Fig. 20.23A** inset). The *Ulva* life cycle shows classic alternation of generations between haploid and diploid forms (**Fig. 20.23B**). Haploid sheets of cells (the gametophyte) produce symmetrically biflagellate gametes, similar to unicellular *Chlamydomonas*. But when *Ulva* gametes fuse, the zygote grows into an immense diploid sheet of cells. This sporophyte (diploid multicellular body) appears similar in form to the gametophyte. The sporophyte eventually undergoes meiosis, releasing haploid zoospores with two pairs of flagella. The zoospores undergo mitosis and regenerate the gametophyte.

Algae serve as symbiotic partners for many important living systems. For example, certain algae grow in intimate association with fungi to form unified structures called **lichens**, important colonizers of dry and cold habitats (discussed in Chapter 21). A form of algal-fungal ground cover similar to lichens is **cryptogamic crust**, common on desert soil. Still other algae grow within the cells of paramecia and hydras, providing photosynthetic nutrition in exchange for protection.

Thought Question

20.5 What are the relative advantages of being unicellular or multicellular?

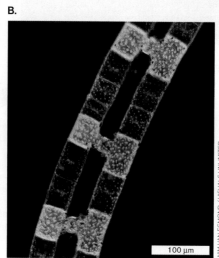
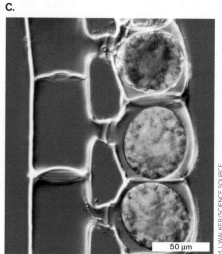

FIGURE 20.22 ■ Filaments with spiral chloroplasts: *Spirogyra*. A. *Spirogyra* species grow in long multicellular filaments, typically 25 μm wide and several centimeters long. Each cell contains one or more chloroplasts that spiral around the cytoplasm. **B.** Sexual reproduction involves conjugation between cells of two mating types. The cytoplasm from each male cell exits from its cell wall and enters the female cell. **C.** Gamete fusion is complete, generating zygotic spores.

Sheet of cells, two cells thick

FIGURE 20.23 ■ **Multicellular algae: *Ulva*.** **A.** *Ulva* species generate large, undulating sheets of cells in double layers (inset). **B.** *Ulva* undergoes symmetrical alternation of generations. The haploid and diploid forms appear very similar. *Source:* Part B modified from Christine Bobin-Dubigeon et al. 1997. *J. Sci. Food Agric.* **75**:341–351.

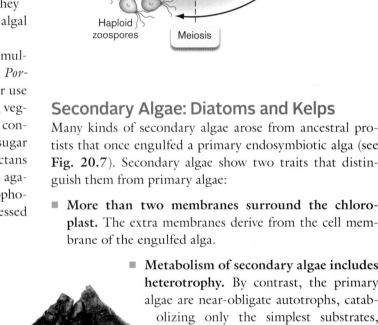

Red algae (Rhodophyta). Red algae, or rhodophytes, are colored red by the photopigment phycoerythrin. Phycoerythrin absorbs efficiently in the blue and green range, which green algae fail to absorb. Because the red algae absorb wavelengths missed by the green algae, they can colonize deeper marine habitats below the green algal populations.

Rhodophytes include unicellular, filamentous, and multicellular forms. Several kinds are human food sources. *Porphyra* forms large sheets that are harvested in Japan for use as nori for wrapping sushi, a delicacy that includes rice, vegetables, and uncooked fish (**Fig. 20.24**). Red algae contain valuable polymers called sulfated polygalactans (sugar polymers with sulfate side chains). Sulfated polygalactans include agar, used to solidify microbial growth media; agarose, a processed sugar derivative used to form electrophoretic gels; and carrageenan, an additive used in processed foods.

Secondary Algae: Diatoms and Kelps

Many kinds of secondary algae arose from ancestral protists that once engulfed a primary endosymbiotic alga (see **Fig. 20.7**). Secondary algae show two traits that distinguish them from primary algae:

■ **More than two membranes surround the chloroplast.** The extra membranes derive from the cell membrane of the engulfed alga.

■ **Metabolism of secondary algae includes heterotrophy.** By contrast, the primary algae are near-obligate autotrophs, catabolizing only the simplest substrates, such as acetate.

Several major groups of secondary algae are heterokonts (stramenopiles). These include the diatoms (Bacillariophyceae); the brown algae (Phaeophyceae), such as kelps; the golden algae (Chrysophyceae), mainly flagellates; the yellow-green algae (Xanthophyceae); and

FIGURE 20.24 ■ **Red algae: *Porphyra*.** **A.** *Porphyra* forms large, red, multicellular sheets. **B.** When the sheets are harvested and toasted, they are known as "nori." Nori are used to wrap sushi, a Japanese delicacy.

A.

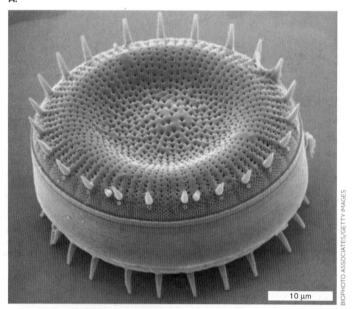

10 μm

BIOPHOTO ASSOCIATES/GETTY IMAGES

B.

JAN HINSCH/SCIENCE SOURCE

FIGURE 20.25 ■ Diatoms of various species. A. *Stephanodiscus astraea*, a centric diatom. **B.** Diverse centric and pennate diatoms.

other less-studied forms of phytoplankton. Flagellated species of heterokont algae generally show a pair of differently shaped flagella. Typically, the flagellum that drives the cell through the environment is brush-like with lateral hairs, while the other is usually shorter and sometimes missing altogether. In some cases it functions like a rudder.

Diatoms (Bacillariophyceae). Diatoms are unicellular algae found ubiquitously in both fresh and marine waters (**Fig. 20.25**). They conduct a fifth of all photosynthesis on Earth, and they fix as much biomass as all the terrestrial rain forests. A diatom grows a unique kind of bipartite shell called a **frustule**. The frustule is composed of silica (cross-linked silicon dioxide, SiO_2). The silicate frustules protect diatoms from many kinds of predators. Diatoms

are nonetheless consumed by flagellates and amphipods (shrimplike invertebrates) and are infected by viruses.

Frustules of different species form an extraordinary range of shapes with intricate pore formations (**Fig. 20.25A**). The shapes fall into two classes: centric, with radial symmetry; or pennate, with bilateral symmetry (**Fig. 20.25B**). Frustules of decomposed diatoms eventually sediment on the ocean floor, where they build sedimentary rock strata more than a kilometer thick. This "diatomaceous earth" is used in insulation material and in toothpaste. Diverse species of diatoms are highly sensitive to environmental factors such as pH, and the frequency of their shells in sediment can be used to track a lake's environmental history.

Diatoms pose a unique challenge to cell division (**Fig. 20.26**). As the diatom grows and fissions, each daughter cell receives one parental half of the frustule while forming a new half fitting within the parental half, like the bottom dish of a petri plate. Thus, each generation results in an inexorable decline in size of the organism. As its cell size reaches a critical point, the diatom must undergo meiosis to generate gametes. Maria Vernet at the Scripps Institution of Oceanography studies diatoms of the polar oceans, such as the Antarctic diatom *Corethron criophilum*. Centric diatoms such as *Corethron* form egg cells and flagellated sperm. When the gametes fuse, they form a special kind of zygote called an auxospore. The auxospore generates a frustule of the same size as the original diatom.

Coccolithophores (Prymnesiophyceae). Coccolithophores are secondary algae of the phylum Haptophyta (phylogeny shown in **Fig. 20.4**). Coccoliths superficially resemble diatoms in that their cells have a solid mineral exoskeleton (**Fig. 20.27**). Instead of silicate, however, their exoskeleton is composed of calcium carbonate ($CaCO_3$). The calcium carbonate grows in multiple scales (unlike the unitary exoskeleton of a diatom). The plates may extend radially in all directions, such as the "trumpet" shapes of *Discosphaera tubifera* (**Fig. 20.27**), or form multiple layers of flat, oval shapes encasing the cell, as in *Emiliana huxleyi*. These scales are the "coccoliths" that give coccolithophores their name. The calcium carbonate scales protect the tiny cell from some predators. Coccolithophores such as *E. huxleyi* generate huge, milky blooms that appear in NASA satellite images. The blooms are dissipated by predation or by virus infection.

Coccolithophores are increasingly recognized as major players in the ocean's carbon cycle (discussed in Chapter 22). They sequester large amounts of carbon from CO_2 into their carbonate shells. Unfortunately, their calcium carbonate is sensitive to acidification caused by the global rise in CO_2. For this reason, intensive research is focused

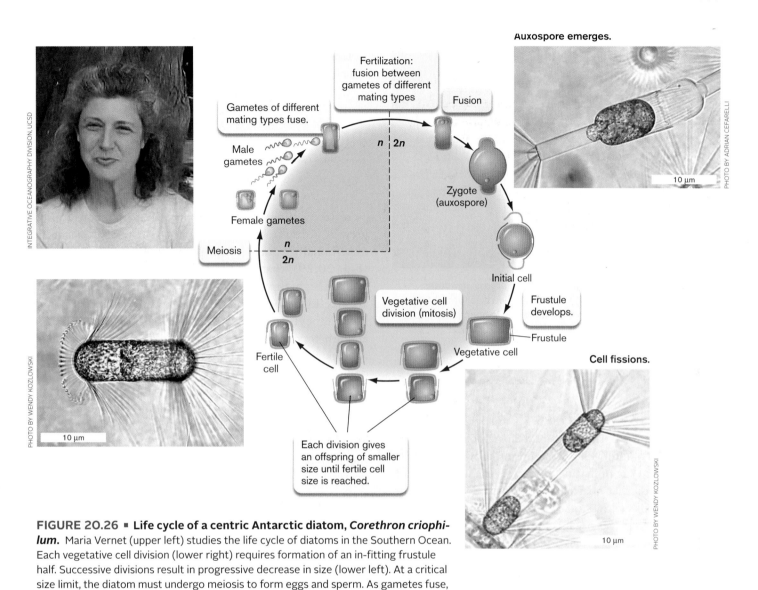

FIGURE 20.26 ■ Life cycle of a centric Antarctic diatom, *Corethron criophilum*. Maria Vernet (upper left) studies the life cycle of diatoms in the Southern Ocean. Each vegetative cell division (lower right) requires formation of an in-fitting frustule half. Successive divisions result in progressive decrease in size (lower left). At a critical size limit, the diatom must undergo meiosis to form eggs and sperm. As gametes fuse, they form an auxospore (upper right), which regenerates a frustule of the original size.

on understanding the response of coccolithophores such as *E. huxleyi* to pH change. In 2011, a large-scale study conducted by several European universities concluded that increasing CO_2 levels correlate with a decline in the overall mass of marine coccoliths. At the same time, however, certain strains of *Emiliana* were shown to grow despite low pH, so many uncertainties remain.

Brown algae (Phaeophyceae). Brown algae, such as kelps, possess vacuoles of leucosin, a polysaccharide for energy storage whose color gives the organism a brown or yellow

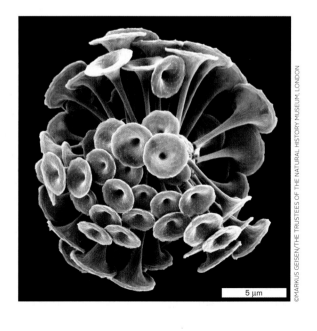

FIGURE 20.27 ■ A coccolithophore, *Discosphaera tubifera*. The tiny cell is surrounded by a much larger volume of trumpet-shaped coccoliths. From the Alboran Sea, western Mediterranean.

Sargassum weed

FIGURE 20.28 ■ **Sargassum forests.** *Sargassum natans* forms the basis of the Sargasso Sea. The brown alga forms stalks with leaflike blades and round gas bladders to keep the alga afloat. Sargassum forests support animals such as the sea turtle.

tint. Kelps are familiar to ocean bathers as the long, dark brown blades that root near the beach until the surf rips the blades off and tosses them ashore. Kelps support important communities of multicellular organisms known as kelp forests. Another type of marine "forest" is the unrooted **sargassum weeds** that float on the Sargasso Sea. Sargassum consists of stalks with photosynthetic blades and round gas bladders to keep the organism afloat (**Fig. 20.28**). Sargassum supports a complex food web of invertebrate and vertebrate animals, including worms, crabs, fish, and sea turtles.

To Summarize

- **Chlorophyta (green algae)** grow near the top of the water column. Green algae include unicellular, filamentous, and sheet forms. They offer models for research on the evolution of multicellularity.

- **Rhodophyta (red algae)** have the accessory photopigment phycoerythrin, which absorbs green and longer-wavelength blue light, enabling growth at greater depths. Red algae include species of diverse forms, many of which are edible for humans.

- **Secondary algae** are derived from protists that had engulfed primary algae. They are mixotrophs, combining phototrophy and heterotrophy.

- **Diatoms are heterokonts with silicate shells called frustules.** Diatoms replicate by an unusual division cycle generating successively smaller frustules.

- **Kelps are heterokonts that grow in long, sheetlike fronds.** Kelps play an important role in the ecology of the coastal ocean, as well as the ecology of marine beaches.

20.4 Amebas and Slime Molds

The ameba (alternative spelling "amoeba") is familiar to most of us as an apparently amorphous form of microscopic life, capable of engulfing and consuming prey in a dramatic fashion. While the ameba's shape is exceptionally variable, the pseudopods, or "false feet" (**Fig. 20.29**), that it extends are complex structures that undertake highly controlled movements. Most amebas are free-living predators in soil or water, engulfing prey by **phagocytosis**. They range in size up to 5 mm, large enough to phagocytose bacteria, algae, ciliates, smaller amebas, and even invertebrates such as rotifers. A few are dangerous parasites of humans or animals. Furthermore, free-living amebas can harbor bacterial pathogens such as *Legionella pneumophila*, which contaminates water supplies and air ducts. The bacteria cause legionellosis, an often fatal form of pneumonia. The host ameba enables the pathogen's persistence and transmission to human hosts.

Free-living amebas such as *Acanthamoeba* species are common predators in the soil microbial community. They cause problems when they contaminate contact lens cleaning solutions, causing keratitis (infection of the cornea). Wearers of contact lenses have an increased risk of *Acanthamoeba* keratitis.

Thought Question

20.6 What can happen when an ameba phagocytoses algae?

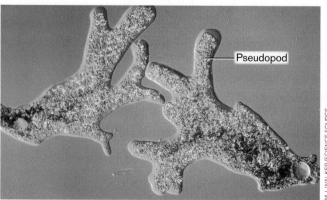

Pseudopod

FIGURE 20.29 ■ ***Amoeba proteus* moves by extending its pseudopods.**

Note: The taxonomy of amebas and slime molds remains problematic, with diverse views as to the number of clades, their relatedness, and their degree of divergence.

Pseudopod Motility

Many different species form ameba-like cells. Some species persist as an ameba throughout all or most of the life cycle. Others can convert to flagellated forms, particularly when the habitat fills with water—a low-viscosity condition favoring flagellar motility. Still other protist species, such as dinoflagellates, never become fully ameboid, but they can extend a pseudopod to engulf prey.

Different kinds of amebas have different kinds of pseudopods. "Classic" amebas (Amoebozoa) have lobe-shaped pseudopods (**Fig. 20.29**). Lobe-shaped pseudopods have the most variable shape. A different form is the sheetlike pseudopod, or lamellar pseudopod, used by dinoflagellates. Lamellar pseudopods are extended by dinoflagellates. Similar lamellar pseudopods are generated by human white blood cells such as leukocytes. Finally, needlelike pseudopods, or filopodia, are thin extensions reinforced by parallel actin filaments. These are typical of Rhizaria amebas, although some Rhizaria instead make long, thin pseudopods supported by microtubules. The most famous protists with microtubule-supported pseudopods also have mineralized supporting structures, such as Foraminifera (spiral shells) and Radiolaria (radial-form "skeleton").

The extension of lobe-shaped and lamellar pseudopods has been studied closely for its relevance to human white blood cells (for more on white blood cells as host defenses, see Chapter 23). The mechanism of pseudopod motility remains poorly understood, but it is known to involve a sol-gel transition between cortical cytoplasm (just beneath the cell surface) and the cytoplasm of the deeper interior (**Fig. 20.30**). The tip of a pseudopod contains a gel of polymerized actin beneath its cell membrane. From the center of the ameba, liquid cytoplasm (sol) containing actin subunits streams forward along microtubular "tracks," powered by ATP hydrolysis. The actin subunits stream into the pseudopod, where they polymerize, forming a gel. The gel region grows,

pushing the membrane forward and extending the pseudopod. As the gel is pushed backward, it resolubilizes to continue the cycle.

Amebas can have one nucleus or multiple nuclei. They are usually haploid and reproduce asexually by nuclear mitosis, without dissolution of the nuclear membrane, followed by fission of the cytoplasm. Some species do have developmental alternatives, such as cyst formation, gamete fusion and meiosis, and even growth of flagella in a favorable habitat.

Thought Question
20.7 What kind of habitat would favor a flagellated ameba?

Ameba genetics is poorly understood, but at least one ameba genome has been sequenced—that of the intestinal parasite *Entamoeba histolytica*. The sequence contains 20 Mb of DNA in 14 chromosomes; some of these are linear, whereas others are circular. Closely related strains show considerable variation in organization, suggesting that ameba genomes undergo extensive rearrangement.

Slime Molds

Some amebas conduct a life cycle in which thousands of individuals (all members of one species) aggregate into a complex differentiated fruiting body (**Fig. 20.31A**). Such an organism is called a **cellular slime mold**. A cellular slime mold forms from ameboid cells that aggregate into a multicellular "slug." Slime molds, as the name implies, were originally classified with fungi because their fruiting bodies superficially resemble fungal reproductive forms.

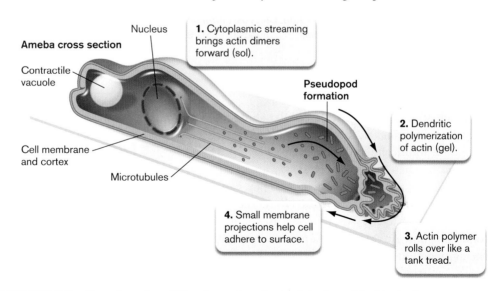

FIGURE 20.30 ■ Pseudopod motility. A pseudopod extends by flow of liquid cytoplasm (sol state) followed by actin polymerization (gel state). As actin polymerizes, the cell rotates down toward the substrate like a tank tread.

A.

500 μm

OWEN GILBERT

B.

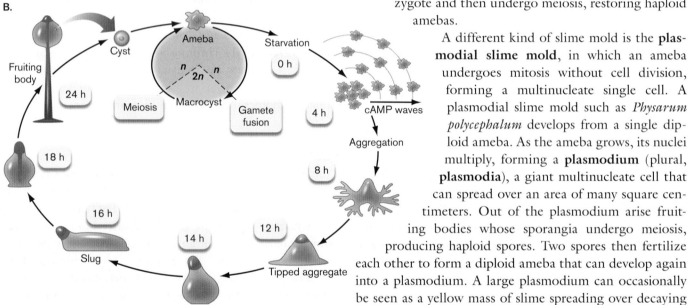

FIGURE 20.31 ■ **A cellular slime mold: *Dictyostelium discoideum*.** **A.** Fruiting bodies of *D. discoideum* (composite SEM). **B.** Life cycle of *D. discoideum*.

A well-studied example of a cellular slime mold is *Dictyostelium discoideum*, historically an important model system for multicellular development (**Fig. 20.31**). *D. discoideum* amebas are relatively small, about 10 μm, but large enough to consume bacteria. They can be cocultured on a plate with *Escherichia coli*. As the haploid amebas consume bacteria, they divide asexually until their food runs out. At this point, a few amebas begin to emit the aggregation signal molecule cyclic AMP (cAMP). An ameba emitting cyclic AMP attracts other amebas nearby, which move toward the center and begin emitting cyclic AMP as well. Successive waves of cyclic AMP continue to attract thousands of amebas to the center, where they pile

on top of each other to form a slug as long as 1 mm. The slug then migrates, attracted by light and warmth, to find an appropriate place to form a fruiting body and disperse its spores.

The slug at last differentiates into a fruiting body, a spherical sporangium supported on a stalk of largely empty cells that emerges from a basal disk. The sporangium then releases spores (also called cysts), which are dispersed on air currents and can remain viable for several years. When a spore detects chemical signals from bacteria, it germinates an ameba to feed on them.

Note that the entire reproductive cycle just described is asexual; the amebas and their differentiated structures remain haploid throughout. *D. discoideum* amebas do have a sexual alternative (illustrated in **Fig. 20.31B**), in which cells of opposite mating type can fuse to form a diploid zygote and then undergo meiosis, restoring haploid amebas.

A different kind of slime mold is the **plasmodial slime mold**, in which an ameba undergoes mitosis without cell division, forming a multinucleate single cell. A plasmodial slime mold such as *Physarum polycephalum* develops from a single diploid ameba. As the ameba grows, its nuclei multiply, forming a **plasmodium** (plural, **plasmodia**), a giant multinucleate cell that can spread over an area of many square centimeters. Out of the plasmodium arise fruiting bodies whose sporangia undergo meiosis, producing haploid spores. Two spores then fertilize each other to form a diploid ameba that can develop again into a plasmodium. A large plasmodium can occasionally be seen as a yellow mass of slime spreading over decaying wood.

Note: Distinguish the term "plasmodium" (a large multinucleate cell) from the genus *Plasmodium* (an apicomplexan parasite, such as *Plasmodium falciparum*, which causes malaria).

Filamentous and Shelled Amebas

Amebas of a distinct clade (Rhizaria) form needlelike pseudopods. Most of these amebas are encased by mineral shells called tests. A major group of shelled amebas is the **radiolarians**, whose skeleton-like tests are made of silica perforated with numerous holes through which pseudopods appear to radiate in all directions (**Fig. 20.32A**). Many different kinds of radiolarians exist today, and many can be recognized in fossil rock. A second group of shelled amebas is the **foraminiferans** (**Fig. 20.32B**). The foraminiferans, or forams, generate shells of calcium

A. Radiolarian tests

EYE OF SCIENCE/SCIENCE SOURCE

B. A foraminiferan

CUSHMAN FOUNDATION FOR FORAMINIFERAL RESEARCH, INC., 1987.

500 µm

100 µm

FIGURE 20.32 ■ **Filamentous and shelled amebas.**
A. Shells (tests) of radiolarians (stereoscope). **B.** A live foraminiferan, *Globigerinella aequilateralis* (dark-field).

carbonate as chambers laid down in helical succession. Their pseudopods all extend from one opening in the most recent chamber.

Radiolarians and shelled forams grow in marine and freshwater habitats. Their shells make up a large part of reef formations, sedimentary rock, and beach sand. Forams, in particular, are used in geological surveys as indicators of petroleum deposits.

To Summarize

- **Amebas** move using pseudopods. In different species, pseudopods are lobe-shaped, lamellar, or filamentous (filopodia).

- **Cytoplasmic streaming** through cycles of actin polymerization and depolymerization drives the extension and retraction of pseudopods.

- **Slime molds** show an asexual reproductive cycle in which a fruiting body produces spores. In cellular slime molds, amebas aggregate to form a slug. In plasmodial slime molds, a single ameba develops into a multinucleate cell.

- **Radiolarians** have silicate shells penetrated by filamentous pseudopods.

- **Foraminiferans** have calcium carbonate shells with helical arrangement of chambers. The most recent chamber opens to extend filamentous pseudopods.

20.5 Alveolates: Ciliates, Dinoflagellates, and Apicomplexans

The Alveolata include voracious predators such as the ciliated protists (**Fig. 20.33A**), as well as a major group of algae, and perhaps the most successful group of parasitic microbial eukaryotes. Alveolates are named for the flattened vacuoles called alveoli (singular, alveolus) within their outer cortex (**Fig. 20.33B**; see also **Fig. 20.8A**). Some alveoli contain plates of stiff material, such as protein, polysaccharide, or minerals. Besides alveoli, most alveolate protists possess other kinds of cortical organelles, such as extrusomes for delivery of enzymes or toxins, bands of microtubules for reinforcement, and whiplike cilia or flagella. The alveolate cell form is highly structured, in contrast to the amorphous shape of amebas. Major groups of alveolates include ciliates, dinoflagellates, and apicomplexans.

Ciliates

A diverse group of alveolates known as Ciliophora, or ciliates, possess large numbers of cilia, short projections containing [9(2)+2] microtubules. Their whiplike action is driven by ATP (for review, see eAppendix 2). The cilia beat in coordinated waves that maximize the efficiency of motility. Cilia serve two functions:

- **Cell propulsion.** Coordinated waves of beating cilia, usually covering the cell surface, propel the cell forward.

- **Food acquisition.** By generating water currents into the mouth of the cell, a ring of cilia around the mouth brings food into the cell.

A.

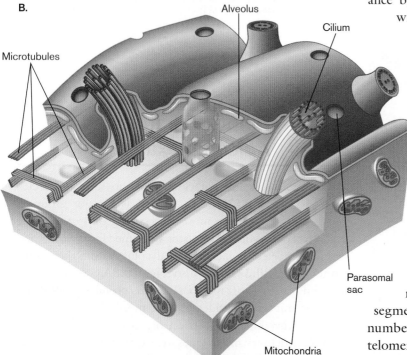

B.

C.

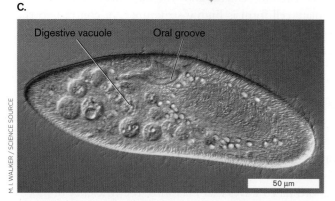

FIGURE 20.33 ■ **Ciliated protists. A.** *Didinium*, consuming *Paramecium* (SEM). **B.** Cortical structure of a ciliate. Beneath the outer cell membrane lie flattened sacs called alveoli. The cilia, composed of [9(2)+2] microtubules, are rooted in a complex network of microtubules. Parasomal sacs take up nutrients and form endocytic vesicles. **C.** A *Paramecium* has digestive vacuoles and an oral groove for ingestion (differential interference contrast, colorized).

Ciliate cell structure. *Paramecium* is one of the most studied ciliates. Paramecia feed on bacteria and in turn are consumed by other ciliates, such as *Didinium* (**Fig. 20.33A**). They can also take up smaller particles through endocytosis by specialized pores in their cortex, called parasomal sacs (**Fig. 20.33B**).

The cell structure of *Paramecium* includes an **oral groove** for uptake of food driven by the beating cilia (**Fig. 20.33C**). Once ingested through the oral groove, a food particle travels within the digestive vacuole in a circuit around the cell. The digestive vacuole ultimately empties into the cytoproct, a specialized vacuole for the discharge of waste outside the cell. Paramecia maintain osmotic balance by means of a **contractile vacuole**, a vacuole that withdraws water from the cytoplasm to shrink it or contracts to expand it. Contractile vacuoles are widespread among protists and algae, but their mode of action has been most studied in paramecia.

Genetics and reproduction. Most ciliates have a complex genetic system involving one or more **micronuclei** and **macronuclei**. The micronucleus contains a diploid set of chromosomes that undergoes meiosis for sexual exchange (a process reviewed in eAppendix 2). For gene expression, however, one of the micronuclei develops into a macronucleus that forms hundreds of copies of its DNA. The DNA copies in the macronucleus are rearranged and fragmented to small segments. The small DNA segments generate a large number of "telomeres," chromosome ends. In humans, telomere shortening is associated with aging; thus, ciliate macronucleus formation provides a model system for study of human aging (**eTopic 20.2**).

In ciliates, only macronuclear genes are transcribed to RNA and translated to protein. When a ciliate reproduces asexually, the micronucleus undergoes mitosis, whereas the macronucleus divides by a different mechanism that is poorly understood. Cell division occurs across the long axis, necessitating generation of a new oral groove for the posterior daughter cell and a new cytoproct for the anterior daughter cell—again, a process poorly understood.

Most ciliates are diploid and never produce haploid gamete cells. Instead, their sexual reproduction involves exchange of micronuclei. The two ciliates of a mating pair exchange haploid micronuclei by **conjugation**. In conjugation, two cells of opposite mating type form a cytoplasmic bridge and exchange their nuclear products of meiosis (**Fig. 20.34**). While the cells connect, the micronucleus of each cell undergoes meiosis to form four haploid nuclei. Three out of four of the haploid nuclei disintegrate, as

FIGURE 20.34 ■ Conjugation. **A.** Two paramecia conjugating (LM). **B.** In conjugation, two paramecia of opposite mating type form a cytoplasmic bridge. The 2n micronucleus of each cell undergoes meiosis. Each macro-nucleus, as well as three out of four meiotic products, disintegrates. The haploid micronu-clei undergo mitosis, forming two daughter micronuclei. Daughter nuclei from each cell are exchanged across the cytoplasmic bridge and then fuse with their respective counterparts, restoring 2n micronuclei. The micronuclei fission several times, and one transforms into a new macronucleus.

A. Conjugating paramecia

50 μm

MICHAEL ABBEY/VISUALS UNLIMITED

B. Conjugation between ciliates

Micronucleus (2n)

Macronucleus

Conjugation

Meiosis

Macronuclei disintegrate.

Three of four meiotic products disintegrate, leaving one micronucleus (n).

Haploid micronuclei replicate and divide (mitosis).

Haploid micronuclei exchange.

Haploid micronuclei fuse, becoming diploid (2n).

Micronucleus fissions; one transforms into macronucleus.

does the entire macronucleus. The haploid micronuclei then undergo mitosis, and one of each daughter nuclei is exchanged across the cytoplasmic bridge. Each transferred nucleus then fuses with its haploid counterpart, restoring diploidy. The two cells come apart, and each recombined micronucleus fissions several times. One of the daughter micronuclei then transforms into the new macronucleus.

> **Thought Questions**
>
> **20.8** Compare and contrast the process of conjugation in ciliates and bacteria (see Chapter 9).
>
> **20.9** For ciliates, what are the advantages and limitations of conjugation, as compared with gamete production?

Stalked ciliates. Some ciliates adhere to a substrate and use their cilia primarily to obtain prey. **Stalked ciliates** such as *Stentor* and *Vorticella* have a ring of cilia surrounding a large mouth (**Fig. 20.35A**). The ciliary beat is specialized to draw large currents of water and whatever prey it carries. Stalked ciliates are commonly found in pond sediment and in waste-water during biological treatment by microbial digestion, where they are attached to flocs of filamentous bacteria.

Another group of stalked ciliates, the suctorians (**Fig. 20.35B**), possess cilia for only a short period after a daughter cell is released by the stalked cell. The daughter cell swims by ciliary motion until it finds a good habitat in which to settle, whereupon its cilia are replaced by knobbed tentacles similar to the filopodia of shelled amebas. Suctori-ans prey on swimming ciliates such as paramecia.

Dinoflagellates Are Phototrophs and Predators

The dinoflagellates (Dinoflagellata) are a major group of marine phytoplankton, essential to marine food webs. Like ciliates, they are highly motile, but instead of numerous short cilia, dinoflagellates possess just two long flagella,

A. *Stentor* (stalked ciliate)

B. *Acineta* (suctorian)

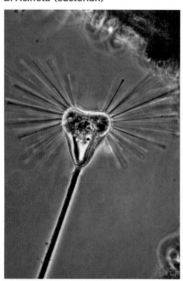

FIGURE 20.35 ■ Stalked ciliates. A. *Stentor*, a ciliate with a flexible stalk, 1.5–2.0 mm in length (phase contrast). The oral ring of cilia generates currents drawing food into the mouth. **B.** The suctorian *Acineta* replaces cilia with knobbed tentacles (LM).

one of which wraps along a crevice encircling the cell (**Fig. 20.36A**). Some dinoflagellates possess elaborate hornlike extensions (**Fig. 20.36B**). The cell extensions increase the range of nutrient uptake, and they may deter predation.

Dinoflagellates are secondary or tertiary algae. They have a chloroplast derived from a red alga, which in some species was later replaced by a heterokont alga, itself a secondary alga (**Fig. 20.36C**). Some dinoflagellates possess carotenoid pigments that confer a red color. Blooms of red dinoflagellates cause the famous red tide, which may have inspired the biblical story of the plague in which water turns to blood (**Fig. 20.36D**). Dinoflagellates release toxins that can be absorbed by shellfish, poisoning consumers.

The armor-plated appearance of a dinoflagellate results from its stiff alveolar plates, composed of cellulose (**Fig. 20.36C**). The complex outer cortex includes various extrusomes (organelles that extrude a defensive substance) and endocytic pores, as well as a species-specific pattern of alveolar plates. Dinoflagellates supplement their photosynthesis by

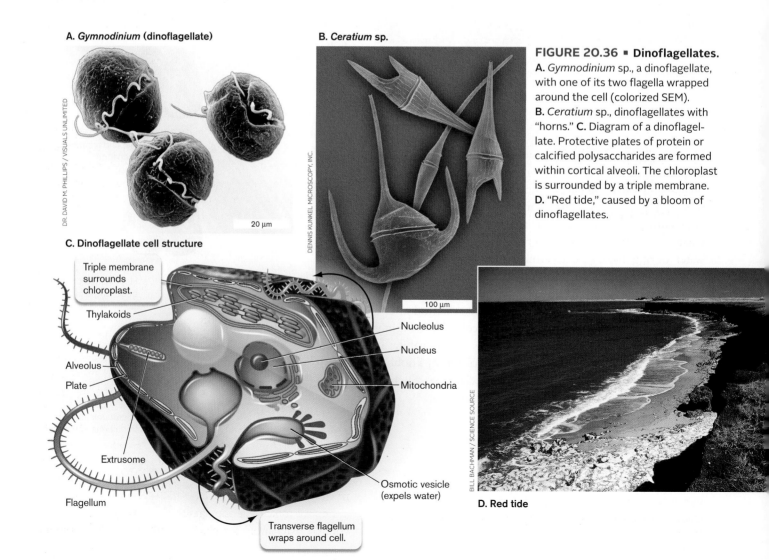

A. *Gymnodinium* (dinoflagellate)

B. *Ceratium* sp.

20 µm

C. Dinoflagellate cell structure

Triple membrane surrounds chloroplast.

Thylakoids

Alveolus

Plate

Extrusome

Flagellum

Transverse flagellum wraps around cell.

100 µm

Nucleolus

Nucleus

Mitochondria

Osmotic vesicle (expels water)

FIGURE 20.36 ■ Dinoflagellates.
A. *Gymnodinium* sp., a dinoflagellate, with one of its two flagella wrapped around the cell (colorized SEM).
B. *Ceratium* sp., dinoflagellates with "horns." **C.** Diagram of a dinoflagellate. Protective plates of protein or calcified polysaccharides are formed within cortical alveoli. The chloroplast is surrounded by a triple membrane.
D. "Red tide," caused by a bloom of dinoflagellates.

D. Red tide

predation, extending a special type of pseudopod to engulf prey. Many dinoflagellates have evolved to lose their chloroplasts altogether, becoming obligate predators or parasites.

Some dinoflagellates inhabit other organisms as endosymbionts, providing sugars from photosynthesis in exchange for a protected habitat. Their hosts include shelled amebas, sponges, sea anemones, and, most important, reef-building corals. Coral endosymbionts, known as **zooxanthellae** (singular, **zooxanthella**), are vital to reef growth. The zooxanthellae are temperature sensitive, and their health is endangered by global warming. Rising temperatures in the ocean lead to coral bleaching (the expulsion of zooxanthellae), after which the coral dies.

Apicomplexans Are Specialized Parasites

Apicomplexans include many human parasites, such as the intestinal parasite *Cryptosporidium*, which infected hundreds of people in the United States in 2013. Apicomplexan cells have an apical complex, a highly specialized structure that facilitates entry of the parasite into a host cell. Another important apicomplexan is *Toxoplasma gondii*, a parasite commonly carried by cats and transmissible to humans, where it can harm a developing fetus. Like the ciliates and dinoflagellates, apicomplexans possess an elaborate cortex composed of alveoli, pores, and microtubules. But as parasites, apicomplexans have undergone extensive reductive evolution, losing their flagella or cilia. They possess a unique organelle called the apicoplast, derived by genetic reduction from an endosymbiotic chloroplast. No capacity for photosynthesis remains, but the apicoplast provides one essential function in fatty acid metabolism.

The best-known apicomplexan is *Plasmodium falciparum*, the main causative agent of **malaria**, the most important parasitic disease of humans worldwide. **Figure 20.37** shows

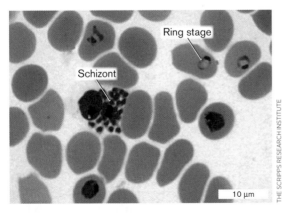

FIGURE 20.37 ■ *Plasmodium falciparum*, a cause of malaria. Red blood cells infected with *P. falciparum*, which is stained purple with a dye that interacts with DNA (LM). One late-stage infected blood cell (schizont) can be seen bursting, unleashing parasites on surrounding cells.

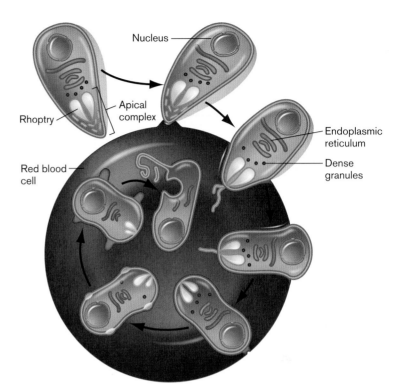

FIGURE 20.38 ■ Merozoite form of *Plasmodium falciparum* invades a red blood cell. The apical complex facilitates invasion and then dissolves as the merozoite transforms into an intracellular form.

red blood cells in the early "ring stage" of infection, and in the "schizont" stage, which bursts, releasing progeny parasites. *P. falciparum* is carried by mosquitoes, which transmit the parasite to humans when the insect's proboscis penetrates the skin. The disease is endemic in areas inhabited by 40% of the world's population; it infects hundreds of millions of people and kills half a million African children each year.

The transmitted parasites invade the liver and then develop into the **merozoite** form that invades red blood cells (**Fig. 20.38**). The merozoite first contacts a red blood cell with its apical complex. The apical complex contains secretory organelles called rhoptries that inject enzymes to aid entry by the parasite. The cone-like tip of the apical complex penetrates the host cell, enabling secretion of lipids and proteins that facilitate invasion. Eventually, the entire merozoite enters the host cell, leaving no traces of the parasite on the host cell surface. Thus, the internalized parasite becomes invisible to the immune system until its progeny burst out.

P. falciparum acquires resistance rapidly, and many strains no longer respond to drugs, such as quinine, that nearly eliminated the disease half a century ago. The life cycle and molecular properties of *P. falciparum* have been studied extensively for clues to aid in the development of new antimalarial drugs and vaccines. The elaborate life cycle of *P. falciparum* and other apicomplexan parasites involves several common features:

■ **Schizogony**, mitotic reproduction of a haploid form (in the mammalian host) to achieve a large population within a host tissue. Usually, the nuclei multiply first, followed by separation of individual nucleated cells.

■ **Gamogony**, the differentiation of haploid cells into male and female gametes capable of fertilization.

■ **Mitosis** and meiosis of the diploid zygote (within the insect) turns it into a haploid, spore-like form transmissible to the next host.

In the case of malaria (**Fig. 20.39** ▶), whip-shaped sporozoites injected by the mosquito invade the liver, where they undergo **schizogony** (nuclear multiplication followed by cell separation). The cell products of schizogony are called merozoites. The merozoites from the liver then invade red blood cells, where they feed on hemoglobin. An early infected blood cell appears as a "ring stage" (**Fig. 20.39**). The parasite multiplies, filling the host cell, now called a "schizont." The schizont bursts, liberating progeny merozoites that invade another round of red cells. The bursting of red blood cells also releases cell fragments that trigger the cyclic fevers characteristic of malaria.

Some of the merozoites in the bloodstream develop into pre-gamete cells, or gametocytes. The gametocytes are acquired by bloodsucking mosquitoes and then multiply and mature (gamogony) in the mosquito's midgut. The gametocytes develop into female eggs and thin male cells with flagella (this is the only stage in the apicomplexan life cycle that has flagella). The male cells fertilize the egg cells, and the resulting zygotes undergo meiosis and differentiate into sporozoites, which enter the mosquito's salivary gland for transmission to the next human host.

The nuclear genome of *P. falciparum* consists of 24 Mb contained in 14 chromosomes. Sequence annotation and expression studies predict 5,300 protein-encoding open reading frames (ORFs), comparable to the number in a yeast genome. The parasite has lost many genes encoding enzymes and transporters while expanding its repertoire of proteins involved in antigenic diversity. In addition, the parasite contains two smaller nonnuclear genomes: that of its mitochondria and that of the chloroplast-derived apicoplast.

How can we treat malaria and eradicate the disease? The malarial genome reveals promising targets for drug design. For example, the fatty acid biosynthesis within the apicoplast is targeted by triclosan and other antibiotics. Other promising targets for antimalarial drugs are the unique proteases required to digest hemoglobin within the *P. falciparum* food vacuole.

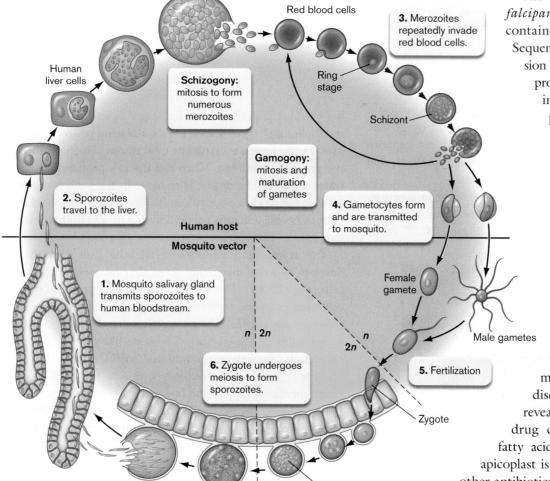

FIGURE 20.39 ■ Malaria: cycle of *Plasmodium falciparum* transmission between mosquito and human. ▶

A. *Leishmania* infection

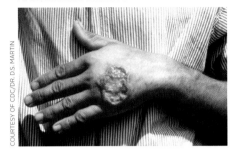

B. *Leishmania major*

10 µm

C. *Trypanosoma brucei*

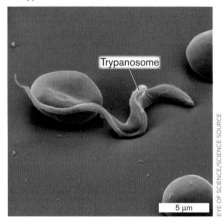

Trypanosome

5 µm

FIGURE 20.40 ■ Trypanosomes.
A. Patient suffering from *Leishmania* infection (leishmaniasis). **B.** Cluster of *L. major* undergoing schizogony within the sand fly (colorized SEM). **C.** *Trypanosoma brucei*, seen here among red blood cells, cause of African sleeping sickness (colorized SEM).

To Summarize

■ **Ciliates are covered with numerous cilia.** Cilia provide motility and help capture prey. Ciliates undergo complex reproductive cycles involving exchange of micronuclei through conjugation.

■ **Dinoflagellates are phototrophic predators.** Dinoflagellates are tertiary endosymbiotic algae. Their alveoli contain calcified plates; they have paired flagella, one of which is used for propulsion. Predation occurs by extension of a lamellar pseudopod.

■ **Apicomplexans are parasites that penetrate host cells.** The apicoplast is a specialized organ for cell penetration. Apicomplexans such as *Plasmodium falciparum* conduct complex life cycles within mammalian and arthropod hosts.

20.6 Parasitic Protozoa

Surprisingly, many protists harmlessly inhabit the human gut as commensals. Such commensals include members of diverse clades, such as *Entamoeba coli* (an ameba), *Blastocystis* (a heterokont), and *Retortamonas intestinalis* (a metamonad). Such protists might provide positive benefits to their host. Nevertheless, closely related organisms, such as *Entamoeba histolytica*, may cause deadly disease. The previous section introduced apicomplexan parasites of major importance, especially those that cause malaria. Other major clades of parasites (and related harmless protozoa) are the trypanosomes and metamonads. The human intestine harbors many kinds of protozoa, including both harmless symbionts and deadly parasites.

Trypanosomes

The group Euglenida includes flagellated protists such as *Euglena*, with chloroplasts arising from secondary endosymbiosis. Like other algal protists, euglenas combine photosynthesis and heterotrophic nutrition. The Euglenida, however, also include a group of obligate parasites called **trypanosomes**. Trypanosomes consist of an elongated cell with a single flagellum. The trypanosome has a unique organelle called the "kinetoplast," consisting of a mitochondrion containing a bundle of multiple copies of its circular genome, usually placed near the base of the flagellum.

Trypanosomes cause some of the most gruesome and debilitating conditions known to humanity, such as leishmaniasis (**Fig. 20.40A**). *Leishmania major* (**Fig. 20.40B**) causes skin infections that may enter the internal organs. If untreated, leishmaniasis can lead to swelling and decay of the extremities (**Fig. 20.40A**) and eventually death. Carried by sand flies, *Leishmania* infects 1.5 million people annually, in South America, Africa and the Middle East, and southern Europe. *Leishmania* often infects Americans serving in Iraq; for this reason, returning veterans from Iraq are permanently restricted from donating blood.

Another major disease caused by trypanosomes is trypanosomiasis, also known as African sleeping sickness. The parasite, *Trypanosoma brucei* (**Fig. 20.40C**), is carried by the tsetse fly. *T. brucei* multiplies in the bloodstream of the host animal, causing repeated cycles of proliferation and fever that ultimately lead to death, if untreated. This trypanosome is known for its extraordinary degree of antigenic variation. Its genome includes 200 different active versions of its variant surface glycoprotein (VSG), the antigen inducing the immune response, as well as 1,600 different "silent" versions that can recombine with "active" VSG to make further variations. In effect, the trypanosome overwhelms the host immune system by continually

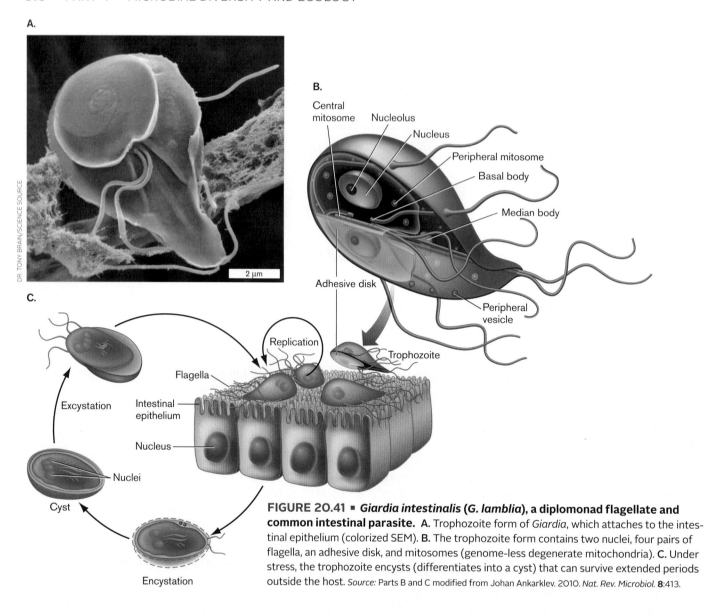

FIGURE 20.41 ▪ *Giardia intestinalis (G. lamblia)*, **a diplomonad flagellate and common intestinal parasite. A.** Trophozoite form of *Giardia*, which attaches to the intestinal epithelium (colorized SEM). **B.** The trophozoite form contains two nuclei, four pairs of flagella, an adhesive disk, and mitosomes (genome-less degenerate mitochondria). **C.** Under stress, the trophozoite encysts (differentiates into a cyst) that can survive extended periods outside the host. *Source:* Parts B and C modified from Johan Ankarklev. 2010. *Nat. Rev. Microbiol.* **8**:413.

generating new antigenic forms until the host repertoire of antibodies is exhausted. In order to infect its human host, the trypanosome needs to interconvert among several different forms of its life cycle. The molecular basis of conversion offers targets for drug therapy. An example of research on the trypanosome life cycle is described in **eTopic 20.3**.

A related trypanosome, *T. cruzi*, is carried by reduviid bugs, a kind of blood-feeding insect. *T. cruzi* causes Chagas' disease, a debilitating infection of the heart and other internal organs. Chagas' disease is prevalent in South and Central America, and global warming is expected to expand its range north.

Metamonads

The Metamonada are another major group of parasites and symbionts. The metamonad parasites include the diplomonads (named for their double nuclei), such as

Giardia intestinalis (*G. lamblia*), a common intestinal parasite (**Fig. 20.41**). *Giardia* is a frequent nemesis of day-care centers, but it also occurs in freshwater streams visited by bears and other wildlife. *Giardia* occasionally contaminates community water supplies in the United States and has become endemic in major Russian cities. *Giardia* and other metamonads are noted for their anaerobic metabolism and their degenerated organelles, reflecting their adaptation to the anaerobic intestinal environment.

The *Giardia* life cycle alternates between two major forms: the trophozoite and the dormant cyst. The trophozoite (**Fig. 20.41B**) has two nuclei (and, therefore, 4*n* chromosomes). There are four pairs of flagella, and an "adhesive disk" enabling the parasite to adhere to the intestinal epithelium. The cell body contains no Golgi, and its mitochondria have degenerated to "mitosomes." Mitosomes lack mitochondrial genomes. When the trophozoite experiences stress conditions, such as high levels of bile and a high pH, the organism encysts (**Fig. 20.41C**). The cyst detaches from the intestine and is expelled from the host. It remains dormant until ingestion by a new host, where stomach acid triggers differentiation into a trophozoite.

Intestinal Parasites

Giardia is just one of many unpleasant intestinal visitors acquired by humans and animals. The ameba *Entamoeba histolytica* grows in the human colon, causing amebiasis (**Fig. 20.42A**). The disease includes diarrhea and possible damage to the intestinal wall; in some cases, the parasite can invade the blood and internal organs. Worldwide, *E. histolytica* kills tens of thousands of people per year. The organism is challenging to diagnose, as it appears very similar to a harmless ameba, *Entamoeba dispar*, which grows normally in the intestine.

The apicomplexan *Cryptosporidium parvum* (**Fig. 20.42B**) commonly contaminates water supplies in the United States; it caused the nation's largest waterborne disease outbreak to date, sickening more than 400,000 people in Milwaukee, in 1993. *Cryptosporidium* is especially dangerous to immunocompromised patients.

An important ciliate parasite is *Balantidium coli* (**Fig. 20.42C**). *Balantidium* is transmitted by a fecal-oral route, most commonly in malnourished individuals whose stomach acid is low, thus failing to kill the pathogen. Infection may be without symptoms, or it can lead to diarrhea and damage the colon.

Microsporidians (**Fig. 20.42D**) were once thought to be protozoa, but genetically and physiologically they are fungi (see Section 20.2). *Encephalitozoon intestinalis* is an obligate fungal parasite of the intestine, causing problems especially for immunocompromised patients. More discussion of intestinal pathogens is found in Chapters 23 and 26.

To Summarize

- **Trypanosomes** include free-living euglenas, as well as the important parasites *Leishmania* (cause of leishmaniasis), *Trypanosoma brucei* (cause of African sleeping sickness), and *Trypanosoma cruzi* (cause of Chagas' disease).

- **Metamonads** include parasites such as *Giardia intestinalis*, a frequent contaminant of natural freshwater environments, and frequently transmitted among children.

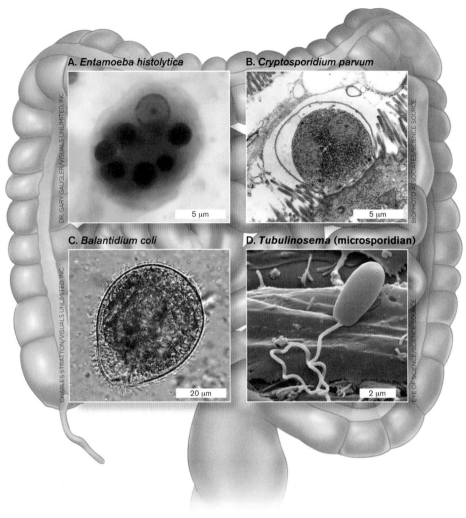

FIGURE 20.42 ■ Human intestinal protozoa.

- **Other intestinal parasites** include the ameba *Entamoeba histolytica*, the apicomplexan *Cryptosporidium parvum*, the ciliate *Balantidium coli*, and the microsporidian (fungal parasite) *Encephalitozoon intestinalis*.

Concluding Thoughts

The microbial eukaryotes, including fungi, algae, and many kinds of protozoa, serve many diverse roles in our ecosystems. A survey of protozoa may leave an impression that most exist primarily to parasitize humans. However, recent genetic surveys based on DNA sequence analysis of environmental communities suggest the existence of twice as many protozoan clades in nature as we have studied to date. Most of these unknown clades have no direct connection with humans, yet they may fill crucial niches in the ecosystems on which human existence depends. Chapters 21 and 22 emphasize the interconnections among many kinds of microbes that form the foundations of Earth's biosphere.

CHAPTER REVIEW

Review Questions

1. Discuss the evidence for the branching of fungi and animals within one clade, the Opisthokonta, which is distinct from algae and protists.
2. How do primary symbiont algae differ from secondary and tertiary symbiont algae? Compare with respect to cell structure and nutritional options.
3. Compare and contrast the molecular basis of motility in amebas and ciliates. Cite particular species.
4. Compare and contrast kelps, diatoms, and dinoflagellates in terms of cell structure, colony organization, and nutritional options.
5. Summarize the key traits of fungi. What do fungi have in common with protists, and how do they differ?
6. Outline the life cycles of the major phyla of fungi: Chytridiomycota, Zygomycota, Ascomycota, and Basidiomycota. Explain their ecological significance.
7. Compare and contrast the traits of green and red algae.
8. Outline the life cycle of the slime mold *Dictyostelium discoideum*. Compare and contrast its features with that of fungi that produce fruiting bodies, such as basidiomycetes.
9. Outline the complex parasitic life cycles of an apicomplexan parasite and a trypanosome. Cite evidence of reductive evolution, as well as of evolution of elaborate specialized structures to facilitate the parasite's life cycle.

Thought Questions

1. Compare and contrast eukaryotic microbes that have inorganic shells or plates. What is their composition, and how do they grow?
2. Explain mixotrophy. Why are so many marine eukaryotes mixotrophs?
3. Why do eukaryotes show such a wide range of cell size? What selective forces favor large cell size, and what favors small cell size?
4. Do eukaryotic parasites have genomes that are larger or smaller than those of free-living organisms? Explain.

Key Terms

alga (775, 779)
alternation of generations (785)
alveolate (781)
ameba (780)
apicomplexan (781)
arbuscular mycorrhizae (792)
ascomycete (788)
ascospore (790)
ascus (785)
basidiomycete (791)
basidiospore (792)
budding (784)
cellular slime mold (801)
chlorophyte (780)
chloroplast (779)
chrysophyte (780)
ciliate (781)
cilium (781)
conjugation (804)
contractile vacuole (804)
cortical alveolus (781)

cryptogamic crust (796)
diatom (781)
dinoflagellate (780)
endosymbiosis (776)
Eumycota (778)
flagellate (781)
flagellum (781)
foraminiferan (802)
fruiting body (779)
frustule (798)
fungus (775, 778)
gamogony (808)
green alga (780)
heterokont (781)
hypha (779)
kelp (781)
lichen (796)
macronucleus (804)
malaria (807)
merozoite (807)
micronucleus (804)

mitosis (808)
mitosome (781)
mitosporic fungus (785)
mycelium (782)
mycology (775)
mycorrhizae (792)
nucleomorph (780)
oomycete (793)
oral groove (804)
phagocytosis (800)
phytoplankton (794)
plasmodial slime mold (802)
plasmodium (802)
primary algae (776, 780)
protist (775)
protozoan (775)
pseudopod (780)
radiolarian (802)
red alga (780)
rhodophyte (780)
sargassum weed (800)

schizogony (808)

secondary algae (776, 780)

sporangiospore (788)

sporangium (788)

stalked ciliate (805)

trypanosome (809)

yeast (784)

zoospore (779)

zooxanthella (807)

zygomycete (788)

zygospore (788)

Recommended Reading

Ankarklev, Johan, Jon Jerlström-Hultqvist, Emma Ringqvist, Karin Troell, and Staffan G. Svärd. 2010. Behind the smile: Cell biology and disease mechanisms of *Giardia* species. *Nature Reviews. Microbiology* 8:413–422.

Armbrust, E. Virginia. 2009. The life of diatoms in the world's oceans. *Nature* 459:185–192.

Armbrust, E. Virginia, John A. Berges, Chris Bowler, Beverley R. Green, Diego Martinez, et al. 2004. The genome of the diatom *Thalassiosira pseudonana*: Ecology, evolution, and metabolism science. *Science* 306:79–86.

Baldauf, Sandra L. 2003. The deep roots of eukaryotes. *Science* 300:1703–1706.

Baum, Jake, Anthony T. Papenfuss, Buzz Baum, Terence P. Speed, and Alan F. Cowman. 2006. Regulation of apicomplexan actin-based motility. *Nature Reviews. Microbiology* 4:621–628.

Beaufort, L., I. Probert, T. de Garidel-Thoron, E. M. Bendif, D. Ruiz-Pino, et al. 2011. Sensitivity of coccolithophores to carbonate chemistry and ocean acidification. *Nature* 476:80–83.

Cefarelli, Adrián O., Martha E. Ferrario, and Maria Vernet. 2015. Diatoms (Bacillariophyceae) associated with free-drifting Antarctic icebergs: Taxonomy and distribution. *Polar Biology* 39:443–459. doi:10.1007/s00300-015-1791-z.

Fisher, Matthew C., Daniel A. Henk, Cheryl J. Briggs, John S. Brownstein, Lawrence C. Madoff, et al. 2012. Emerging fungal threats to animal, plant and ecosystem health. *Nature* 484:186–194.

Gardner, Malcolm J., Shamira J. Shallom, Jane M. Carlton, Steven L. Salzberg, Vishvanath Nene, et al. 2002. The genome sequence of the human malaria parasite *Plasmodium falciparum*. *Nature* 419:498–511.

Haldar, Kasturi, Sophien Kamoun, N. Luisa Hiller, Souvik Bhattacharje, and Christiaan van Ooij. 2006. Common infection strategies of pathogenic eukaryotes. *Nature Reviews. Microbiology* 4:922–931.

Harman, Gary E., Charles R. Howell, Ada Viterbo, Ilan Chet, and Matteo Lorito. 2004. *Trichoderma* species—Opportunistic, avirulent plant symbionts. *Nature Reviews. Microbiology* 2:43–56.

Henderson, Gregory P., Lu Gan, and Grant J. Jensen. 2007. 3-D ultrastructure of *O. tauri*: Electron cryotomography of an entire eukaryotic cell. *PLoS One* 8:e749.

Keeling, Patrick J. 2010. The endosymbiotic origin, diversification and fate of plastids. *Philosophical Transactions of the Royal Society B* 365:1541.

Kronstad, James W., Rodgoun Attarian, Brigitte Cadieux, Jaehyuk Choi, Cletus A. D'Souza, et al. 2011. Expanding fungal pathogenesis: *Cryptococcus* breaks out of the opportunistic box. *Nature Reviews. Microbiology* 9:193–203.

Lew, Roger R. 2011. How does a hypha grow? The biophysics of pressurized growth in fungi. *Nature Reviews. Microbiology* 9:509–518.

Lukeš, Jules, Christen R. Stensvold, Kateřina Jirků-Pomajbíková, and Laura W. Parfrey. 2015. Are human intestinal eukaryotes beneficial or commensals? *PLoS Pathogens* 11:e1005039. doi:10.1371/journal.ppat.1005039.

Parfrey, Laura W., and Laura A. Katz. 2010. Dynamic genomes of eukaryotes and the maintenance of genomic integrity. *Microbe* 5:156–163.

Parniske, Martin. 2008. Arbuscular mycorrhiza: The mother of plant root endosymbioses. *Nature Reviews. Microbiology* 8:763–775.

Pounds, J. Alan, Martin R. Bustamante, Luis A. Coloma, Jamie A. Consuegra, Michael P. L. Fogden, et al. 2007. Widespread amphibian extinctions from epidemic disease driven by global warming. *Nature* 439:161–167.

Ratcliff, William C., Matthew D. Herron, Kathryn Howell, Jennifer T. Pentz, Frank Rosenzweig, et al. 2013. Experimental evolution of an alternating uni- and multicellular life cycle in *Chlamydomonas reinhardtii*. *Nature Communications* 4:2742.

Wilson, Richard A., and Nicholas J. Talbot. 2009. Under pressure: Investigating the biology of plant infection by *Magnaporthe oryzae*. *Nature Reviews. Microbiology* 7:185–196.

CHAPTER 21
Microbial Ecology

Microbes dominate all habitats on Earth, from the Antarctic Southern Ocean to the human intestine. Microbial communities form functional components of all plants and animals, including human beings. While most microbes cannot be cultured, we reveal their secrets by sequencing the DNA of their metagenomes, and by fluorescence microscopy of their biofilms. For their plant and animal hosts, microbes provide nutrition, enhance development, and even control behavior. Chapter 21 explores how microbes interact with each other and with their many diverse habitats on Earth.

CURRENT RESEARCH highlight

Crab farms chemolithoautotrophic bacteria. A male specimen of the yeti crab (*Kiwa tyleri*) was found at a depth of 2.5 kilometers at the Antarctic Southern Ocean hydrothermal vents of the East Scotia Ridge. Bacteria (white tufts) grow on the crab's legs and pincers. The bacteria oxidize sulfides from the vent and fix CO_2 into biomass. The crab waves its claws through the vent water but keeps enough distance to avoid scalding by the superheated fluid. As its bacteria grow, the crab grazes on them.

Source: Sven Thatje et al. 2015. *PLoS One* **10**:e0127621.

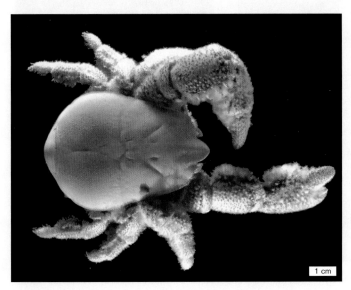

1 cm

AN INTERVIEW WITH

SVEN THATJE, MARINE ECOLOGIST, NATIONAL OCEANOGRAPHY CENTRE, SOUTHAMPTON

SVEN THATJE, UNIVERSITY OF SOUTHAMPTON, UK

Why do chemosynthetic bacteria associate with the yeti crab? How does each partner benefit?

The yeti crab *Kiwa tyleri* feeds entirely on chemosynthetic bacteria. The association has not yet been proved to benefit the bacteria, although the crab keeps the bacteria within a region of survivable temperature.

What aspect of the yeti crab ecology do you find most surprising?

Yeti crabs need the warm water surrounding the hydrothermal vents to survive. The surrounding deep sea is freezing cold (around 0°C) and crabs cannot survive under such conditions. Their chemosynthetic way of life and the need for warm water constrains the crabs to a very limited habitat of a few cubic meters of seawater at each chimney system.

Microbes have evolved to colonize every habitat of our biosphere, including soil, water, air, and the bodies of plants and animals. The yeti crab characterized by Sven Thatje, Katrin Linse, and colleagues shows a dramatic example of microbial interaction with an animal partner (see the Current Research Highlight). The crab exhibits a distinctive behavior of "dancing" into vent waters to obtain the reduced nutrients for its bristle bacteria to grow (**Fig. 21.1**). Katrin Linse, senior biodiversity biologist at the British Antarctic Survey (**Fig. 21.1A**), leads an interdisciplinary team investigating the ecology and biogeochemistry of invertebrate-associated vent microbes (discussed in Section 21.5). To obtain evidence that the crab actually consumes its bacteria as food, Linse analyzed the crab tissues for their isotope ratios (a method introduced in Chapter 17). She found that the crab's biomass isotope ratios were consistent with those of the bacteria. The bacteria's biomass isotope ratios resulted from their mechanisms of CO_2 fixation with chemolithoautotrophy (discussed in Chapter 15). The crab's partner chemolithoautotrophs include various species of bacteria, like the Proteobacteria *Sulfurovum* species of sulfur oxidizers (a form of metabolism described in Chapter 14). To identify the bacterial species, Linse studied the crab's **metagenome**, the collective DNA sequences of the crab's associated bacteria. Metagenomic analysis is our focus in Section 21.1.

Chapter 21 explores the unique roles of microbes in their **ecosystem**. "No man is an island," nor is any microbe. All organisms evolve within an ecosystem, which consists of populations of species plus their habitat or environment. A **population** is a group of individuals of one species living in a common location. The sum of all the populations of different species constitutes a **community**. Microbial communities critically impact other organisms in all habitats, from oceans and forests to the interstices of rock. Microbes recycle organic material in aquatic and terrestrial ecosystems, providing resources for plants and animals; and deep below Earth's surface, microbes shape the rock of Earth's crust.

The first problem of microbial ecology is: How do we find and identify a habitat's microorganisms? The great eighteenth-century taxonomist Carl von Linné (known also as Carolus Linnaeus) called microbes "chaos" because he thought we would never be able to distinguish one from another (see Chapter 1). In the nineteenth century, Robert Koch developed techniques of pure culture, and Sergei Winogradsky developed enrichment culture methods that isolate microbial species and reveal their metabolic traits (discussed in Chapters 1 and 4). Yet the vast majority of microbes remain uncultured, and perhaps undiscovered. In 2015, a single publication by Jill Banfield's group revealed 35 new phyla, expanding the domain of all known bacteria by 15% (see Chapter 18).

To find the microbes in a particular habitat, we need specialized tools. Section 21.1 begins with the tools of metagenomic analysis—and beyond, the new emerging tools of microbial ecology. In subsequent sections, we evaluate the functions of those microbes in their habitats and microbiomes, including symbioses with partner animals and plants, as well as the larger communities found in the oceans and Earth's crust.

Note: This chapter presents the interactions of microbes and partner organisms within communities. Chapter 22 presents the role of microbes in global fluxes of nutrients and climate change.

A.

B.

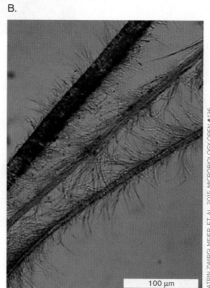

C.

FIGURE 21.1 ■ Bacteria farmed by a crab. **A.** Katrin Linse studies the hydrothermal vent ecosystem. **B.** Bristles (setae) of *Kiwa* sp. crab from East Scotia Ridge, Southern Ocean. **C.** Epibiotic (surface-colonizing) bacteria revealed by SYBR green fluorescent stain for DNA.

21.1 Metagenomes—and Beyond

A major breakthrough in microbial ecology was the discovery that we can identify uncultured microbes by their DNA. In 1991, one of the first to sequence genes from environmental samples was Norman Pace, then at Indiana University, who cloned ribosomal RNA genes from plankton (floating microbes) filtered from the Pacific Ocean. The small-subunit (SSU) rRNA sequence became the standard for identifying environmental taxa (discussed in Chapter 17). But SSU rRNA is just one highly conserved gene, which tells us little about the rest of the organism. In 1998, Jo Handelsman and colleagues, then at the University of Wisconsin–Madison, used shotgun sequencing to analyze large portions of genomes from a soil microbial community—a mixture of species that was previously thought impossible to interpret. Handelsman coined the term "metagenome" to refer to the DNA sequence obtained directly from a mixture of genomes.

Today, with advanced sequencing technologies, we are addressing exciting questions:

- **Who is there?** What microbes inhabit a given environment or human body part?

- **What are they doing?** What function do the microbes contribute? Do they fix carbon, break down toxic waste, or produce antibiotics?

- **How do the microbes vary under different conditions?** As climate changes, what happens to the microbial community structure of forests and soil?

Examples of metagenomic studies are listed in **Table 21.1**. For instance, in 2013 an international consortium sequenced the metagenome of human intestinal microbiomes from 207 individuals in three countries. They found that individual gut microbiomes show distinctive species profiles that are stable over time. This finding suggests that understanding individual microbiomes could help fine-tune drug therapies for different patients.

TABLE 21.1	Examples of Metagenomes		
Microbial target community	**Discovery**	**Sequencing analysis**	**Reference**
Rumen microbiome of fistulated cows fed switchgrass	Assembled 15 bacterial genomes with 27,700 carbohydrate-digesting enzymes.	Genome libraries; Illumina sequencing by synthesis.	Matthias Hess et al. 2011. *Science* **331**:463.
Human gut microbiomes of 22 individuals from four countries	Microbial genomes vary with human age; taxa clusters are shared by all human subjects.	Genome libraries; ABI capillary sequencer.	Manimozhiyan Arumugam et al. 2011. *Nature* **473**:174.
Antarctic and Arctic microbial mats	Antarctic mats show more osmotic stress genes, while Arctic mats show more copper response (pollution related).	Roche 454 pyrosequencer; MG-RAST functional annotation.	Thibault Varin et al. 2012. *Appl. Environ. Microbiol.* **78**:549.
Soils sampled from Antarctic desert, hot desert, and temperate regions	Desert soil genomes show more osmotic stress genes but fewer antibiotic resistance genes than nondesert genomes.	10 Mb per sample, Illumina sequenced; MG-RAST functional annotation.	Noah Fierer et al. 2012. *PNAS* **209**:21390.
Human gut microbiomes of 207 individuals from three countries	High coverage reveals within-species variation; human individuals possess distinct microbial strains, stable over time. May lead to personalized chemotherapies.	1.5 terabases (10^{12} base pairs); Illumina sequenced.	Siegfried Schloissnig et al. 2013. *Nature* **493**:45.
Global ocean microbiome sampled at 68 locations	Ocean microbial catalog compiled; core genes vary with depth and temperature.	7.2 terabases, Illumina sequenced; MOCAT assembly.	Shinichi Sunagawa et al. 2015. *Science* **348**:1261359.
River sediments, Rifle, Colorado	Rare organisms (<0.1%) in new candidate phyla detected from diverse communities.	Illumina: multi-kb reads compared to short reads. Long-read sequencing improves genome assembly.	Itai Sharon et al. 2015. *Genome Res.* **25**:534.
Coal-based methane wells, Queensland, Australia	Archaeal phylum Bathyarchaeota contains previously unknown methanogens.	Illumina sequence; MetaBAT multisample assembly reveals reference-independent genomes.	Paul Evans et al. 2015. *Science* **350**:435.

Sampling the Environment

Metagenomic sequencing poses challenges far beyond those of sequencing a single intact genome. To sequence a metagenome requires a series of steps, each of which presents important choices. Our first choice is to define a **target community** from which to obtain DNA. Extreme environments offer attractive targets because their communities often show low diversity (thus a less complex metagenome) and because the native microbes are likely to show novel traits. Such environments may pose safety threats for researchers; for example, the acid mine drainage site sampled by Banfield (**Fig. 21.2A**) is acidic enough to dissolve an iron shovel. Nonetheless, Banfield sequenced community DNA including the genome of the hyperacidophile *Leptospirillum* (**Fig. 21.2B**). Other target communities require interaction with a host animal, such as the rumen (fermentation organ) of a cow (**Fig. 21.2C and D**). The microbes associated with a host animal or plant are called the **microbiome**. Every multicellular organism possesses a microbiome.

Sampling the target community. Sequencing a metagenome requires several steps (**Fig. 21.3**). The cells of the target community must be separated from their surroundings without loss of DNA (**Fig. 21.3**, step 1). For example, sampling a soil community requires removal of humic acids (wood breakdown products), which inhibit DNA polymerases. Suppose the target community inhabits a host plant or animal; what additional separation is required? Before DNA extraction, we need to dislodge host-associated microbes from their host. Otherwise, the host DNA could contaminate the microbial DNA pool.

Filtering the sample. A different problem arises when sampling a large, dilute target community, such as an ocean. Which community interests you—the protists and small invertebrates? the bacteria and archaea? the viruses? If you sample the entire community, one fraction may yield most of the DNA but overwhelm other community members. To manage the complexity, you may decide to filter or fractionate the target community (**Fig. 21.3**, step 2). Filtering can be physical—for example, by pore size or by cell sorting through flow cytometry. But bacteria, protists, and even viruses may overlap considerably in size. Alternatively, later, after we've sequenced the DNA, we can use "computational filtration" to eliminate sequences. For example, when we are focusing on bacteria, we can filter out the sequences known to come from protists. However, computational filtering may also exclude bacterial sequences of interest because they are new and match nothing reported before in the databases. Every procedure has trade-offs.

Isolating DNA. Once separated from the physical habitat, the cells of the target community must be opened in such a way that all of the DNA is released with minimal breakage of the strands (**Fig. 21.3**, step 3). We can lyse the cells by "bead beating" or by sonication (methods discussed in Chapter 3). The DNA can then be purified by precipitation with phenol and ethanol, or by binding to special filters. But—unlike the analysis of a single-species genome—for a metagenome we must account for different species that possess different kinds of enzyme inhibitors, as well as envelope, sheath, and S-layers of diverse composition.

A. PAUL WILMES, UC BERKELEY
B. CHRISTOPHER BELNAP/UC BERKELEY
C. PROFESSOR MIGUEL LICONA, NEW MEXICO STATE UNIVERSITY
D. DENNIS KUNKEL MICROSCOPY, INC./VISUALS UNLIMITED, INC.

1 μm
2 μm

FIGURE 21.2 ■ Sampling a target community of microbes. A. Jillian Banfield samples "pink slime" of archaea growing in acid mine drainage at Iron Mountain, California, one of the largest Superfund cleanup sites. **B.** *Leptospirillum* sp., bacteria whose genome emerged out of Banfield's metagenomic analysis. **C.** An undergraduate student samples rumen contents from a fistulated (cannulated) cow. The closable opening (see the plug sitting on the cow's back) does not harm the animal. **D.** *Ruminococcus albus* bacteria digest plant fibers within the rumen of a cow.

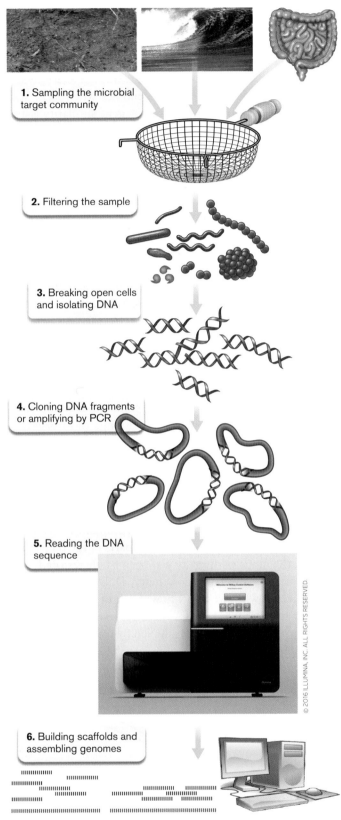

1. Sampling the microbial target community

2. Filtering the sample

3. Breaking open cells and isolating DNA

4. Cloning DNA fragments or amplifying by PCR

5. Reading the DNA sequence

6. Building scaffolds and assembling genomes

FIGURE 21.3 ■ Sequencing a metagenome. Select a target community to sample, from a habitat such as soil, water, or host plant or animal. Remove the sampled microbes from their environment (step 1) and filter the desired class of organisms (step 2). Lyse the cells and isolate pure, intact DNA (step 3). Copy the DNA, either by cloning or by PCR amplification (step 4). Read the DNA sequence using a high-throughput sequencer (step 5). Use computational software to build scaffolds and assemble genomes (step 6).
Source: John Wooley et al. 2010. *PLoS Computational Biol.* **6**:e1000667.

How can we ensure that our protocol will be optimal for all the thousands of species in the community? In fact, no single best way exists to extract metagenomic DNA. Different researchers argue for one of two main approaches:

■ **Use a single, universally applied method of DNA extraction for all target communities.** If all research groups sample metagenomes using a common DNA extraction method, then results may be compared across all projects.

■ **Use multiple DNA extraction methods for each target.** If a research group uses multiple DNA extraction methods to sample one target community, then they have the best chance to maximize coverage of all the microbial genomes in the sample.

> **Thought Question**
>
> **21.1** Suppose you are conducting a metagenomic analysis of soil sampled from different parts of a wetland. Your budget for soil analysis covers only ten sample analyses. Would you use one DNA extraction method or multiple methods?

Metagenome Sequencing and Assembly

Once the DNA samples are obtained, how do we sequence them? In most cases, the sample DNA must be amplified (**Fig. 21.3**, step 4) followed by high-throughput sequencing (step 5). The choice of sequencing method depends on several factors.

■ **DNA concentration.** How concentrated are the DNA samples relative to other cell components? Samples such as a human tissue biopsy, or groundwater, will contain DNA in such small quantities that it must be amplified. Amplification is achieved by cloning or PCR, either of which can introduce sequence errors and sample loss.

■ **Sample diversity.** Species in the community may differ in their fractional representation by several orders of magnitude. How important are the community's rarest members? This question is hard to answer without first taking a trial run at the genome to establish a rarefaction curve (discussed shortly).

■ **Functional interest.** Is our aim a complete description of the target community, such as the microbiota inhabiting the human stomach? Or do we focus on a narrower goal, such as identifying soil actinomycete genes that produce novel antibiotics? A narrower focus allows quicker analysis of the DNA—unless those organisms are among the community's rarer members.

Assessing diversity. Before investing in large-scale DNA sequencing, we may perform a preliminary screen for sequence diversity of the sample by sequencing amplified SSU rRNA genes (discussed in Chapter 17). For bacteria or archaea, this process is commonly called 16S rRNA amplicon analysis or iTAG analysis. (For eukaryotic microbes,

18S rRNA primers would be used.) An advantage of 16S rRNA amplicon sequencing is that we may quickly identify many taxa present in a community, including relatively rare members that do not yield full genomes when large-scale sequencing is performed. SSU rRNA gene similarity is used to define **operational taxonomic units** (**OTUs**), a working metagenomic definition of "species." SSU rRNA gene sequencing reveals the general categories of microbes found in a community, their relative abundance in a community, and the overall diversity of the sample.

An example of diversity analysis was an assessment of seawater following the *Deepwater Horizon* oil well blowout, which released 4 million barrels of oil into the Gulf of Mexico in 2010 (**Fig. 21.4**; **eTopic 21.1**). During several months that followed the spill, Molly Redmond and David Valentine from UC Santa Barbara sampled microbes from Gulf seawater. They used PCR to amplify 16S rRNA, as well as genes from water contaminated by the plume of oil rising from the leak and from uncontaminated Gulf seawater. The collection of amplified sequences shows the relative abundance of various marine taxa at different times after the spill (**Fig. 21.4B**). A sample of nonplume (uncontaminated) water, taken in May, showed a broad range of diverse taxa, including Proteobacteria, Cyanobacteria, and Bacteroidetes. By contrast, samples of oil-contaminated water, taken in May and again in June, showed a marked shift to particular taxa: a novel clade of Oceanospirillales (bright blue bars), and the genus *Colwellia* (blue-gray bars), a benthic gammaproteobacterium that catabolizes propane and benzene. By September, when the most digestible parts of the oil had dissipated, the range of diversity had been partly restored.

In assessing diversity, how much DNA do we need to sequence? How do we know how much of the actual diversity our sampled SSU rRNA reads represent? We can estimate the degree of diversity sampled by using a **rarefaction curve** (**Fig. 21.5**). A rarefaction curve plots the number of

A.

B.

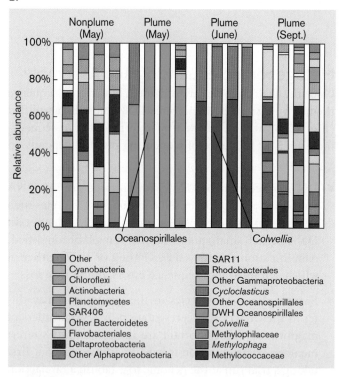

FIGURE 21.4 ■ **Bacterial diversity shifts following the *Deepwater Horizon* oil spill. A.** Petroleum from the *Deepwater Horizon* well blowout in 2010 contaminated the Louisiana coast. **B.** Bacterial relative abundance in 16S rRNA genome libraries. Samples were obtained from the plume of oil spreading through seawater, and from nonplume seawater. During May and June, major taxa shifted to oil-consuming Oceanospirillales and *Colwellia*. By September, after bacteria had consumed much of the oil, the taxon distribution appeared more similar to that before the oil spill. *Source:* Modified from Molly C. Redmond and David L. Valentine. 2012. *PNAS* **109**:20292–20297.

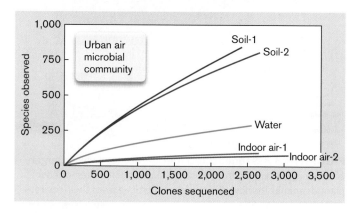

FIGURE 21.5 ■ **Rarefaction curves of indoor air, outdoor soil, and river water were sampled from Singapore.** Rarefaction curves show the number of species, or OTUs, found as a function of the number of clones sequenced for the metagenome. *Source:* Modified from Susannah Tringe et al. 2008. *PLoS One* **3**:e1862.

different OTUs (species) found, as a function of increasing sample size. If resampling continues to reveal new OTUs, then the community diversity has not yet been fully sampled. On the other hand, if the number of OTUs reaches a plateau, then the community diversity may be well represented. **Figure 21.5** shows rarefaction curves for metagenomes from samples of urban office interiors (indoor air-1, indoor air-2) compared to those from nearby samples of soil and river water (soil-1, soil-2, water). Indoor air microbiology is of interest because indoor microbes may colonize human occupants—and because human occupants may "mark" the room with their own telltale microbes. In **Figure 21.5**, the curves for outdoor soil and water show a continual increase in OTUs with resampling, implying high diversity beyond the samples obtained. However, the indoor air samples approach a plateau after about 2,500 clones. Thus, the indoor-air microbial diversity is well represented by the air samples.

A limitation is that SSU rRNA sequences show taxa only down to the level of genus. Furthermore, the so-called "universal" primers for bacteria and archaea miss many uncharacterized taxa, such as the ultrasmall bacteria discovered by Banfield in subsurface water (discussed in Chapter 18). An improvement over single-gene analysis is to select a collection of marker genes that are widespread in numerous taxa, and then combine the data to achieve a consensus diversity survey. Such tools, however, require large-scale DNA sequence information, comparable to the sequence volumes required for full metagenomes.

Alignment of short reads

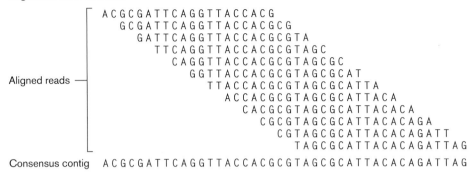

Reference genome sequence

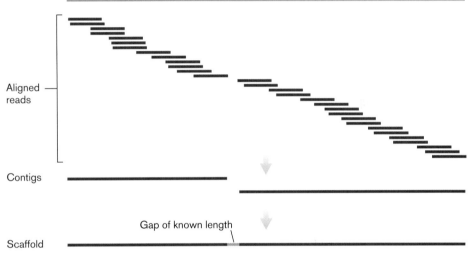

FIGURE 21.6 ■ Assembly of reads into contigs and scaffolds. Overlapping reads generate a contig. Contigs matched to a reference genome generate a scaffold. Scaffolds may still contain gaps of unknown sequence.

Thought Question

21.2 Could you use DNA sequences to identify human occupants of a room, based on the room's airborne microbiome? For background, read James Meadow, 2015, *PeerJ* **3**:e1258.

Sequencing technologies. Different methods of DNA sequencing (**Fig. 21.3**, step 5) differ with respect to the amount of PCR amplification, the length of the sequence read, and the accuracy of the sequence (discussed in Chapter 7). The classic sequencing method, ABI capillary technology, is based on Sanger dye termination. Sanger sequencing provides the longest sequence reads (600–900 bp) and the most accurate base calls. But Sanger sequencing is impractical for the high volumes required for metagenomes. The method of choice for metagenome sequencing is Illumina sequencing by synthesis (described in Chapter 7). Illumina generates short sequence reads (150–300 bp) but offers a 1,000-fold greater volume of data for comparable cost. The shorter reads are more challenging to assemble, typically requiring large-scale computation on a supercomputer (discussed next).

Assembling genomes. Once we obtain a large volume of Illumina DNA sequence (typically from 10^9 to 10^{12} bp), how do we assemble the short reads into genomes (**Fig. 21.3**, step 6)? Recall from Chapter 7 that reading the genome of a single organism requires **assembly** of overlapping fragments into **contigs**, regions of contiguous DNA sequence without gaps (**Fig. 21.6**). If the contigs show high relatedness to the genome of a known organism, this

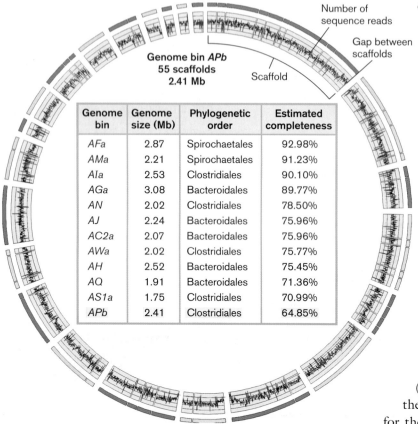

Genome bin	Genome size (Mb)	Phylogenetic order	Estimated completeness
AFa	2.87	Spirochaetales	92.98%
AMa	2.21	Spirochaetales	91.23%
AIa	2.53	Clostridiales	90.10%
AGa	3.08	Bacteroidales	89.77%
AN	2.02	Clostridiales	78.50%
AJ	2.24	Bacteroidales	75.96%
AC2a	2.07	Bacteroidales	75.96%
AWa	2.02	Clostridiales	75.77%
AH	2.52	Bacteroidales	75.45%
AQ	1.91	Bacteroidales	71.36%
AS1a	1.75	Clostridiales	70.99%
APb	2.41	Clostridiales	64.85%

FIGURE 21.7 ■ Partial genomes assembled from a cow rumen microbiome. Map of genome bin *APb* (order Clostridiales) showing assembled scaffolds. Inner rings indicate fold coverage (number of sequence reads) backward and forward; red lines indicate 25-fold coverage. **Center:** Genome bins were matched to orders by comparison with reference sequences. *Source:* Modified from Matthias Hess et al. 2011. *Science* **331**:463.

reference genome may be used to match contigs together in **scaffolds** containing gaps of presumed length. In effect, the assembly of contigs resembles a jigsaw puzzle with thousands of pieces. But a metagenome requires assembly of thousands of jigsaw puzzles, whose pieces are all jumbled together. In practice, metagenomes rarely yield complete genomes of individual species, but they can yield partial genomes of previously unknown organisms with interesting properties. These partial genomes are called operational taxonomic units (OTUs).

Genome assembly requires a **computational pipeline**, a linear series of programs that combine mathematical tools with biological assumptions to propose assembled genomes. A mathematical tool commonly used is the de Bruijn graph, which compiles overlaps between sequences of symbols. One of the first pipelines to use the de Bruijn graph approach for metagenomes is MetaVelvet assembler,

developed by a team at Keio University. One common technique is to align the contigs and scaffolds with known reference genomes. Reference genomes can be effective at "pulling out" genomes of organisms present in low abundance.

An example of a metagenome study is offered by the cow rumen microbiome, a highly diverse community sequenced by Matthias Hess and colleagues at the Joint Genome Institute (see **Table 21.1**). Hess's study yielded 15 partial genomes (OTUs) ranging from 60% to 93% estimated completeness; 12 are listed in **Figure 21.7**. The completeness of each genome was estimated from the number of "core genes" identified. Core genes are defined as those genes found in nearly all members of a given taxonomic order, such as the order Clostridiales for the group of genomes (genome bin) *APb* (highlighted row in **Fig. 21.7**). The fraction of the core genes found, divided by those expected for the order, gives an estimate of completeness of the genome sequence.

The OTUs that Hess found, however, account for only a tiny fraction of the species present, out of 268 gigabases (billions of base pairs) sequenced. Each partial genome assembled from the rumen community represents a **bin**—that is, a set of sequence reads from closely related members of one taxonomic unit, showing a given level of similarity. Since no two individual organisms possess exactly the same sequence, the investigator must decide how much difference to allow in defining a taxon. **Binning**, the sorting of sequences into taxonomic bins, requires further computational analysis. Factors for computation may include, for example, similarity of base composition and similarity to reference database sequences.

Figure 21.7 shows an example of a binned partial genome, *APb*, as a ring composed of scaffolds matched to a Clostridiales reference genome. Between the scaffolds, there remain unsequenced gaps. In the wheel, the jagged trace represents the degree of coverage—that is, the number of sequence reads that cover a given region. Multiple copies of overlapped sequence represent a reliable assembly; typically, 30-fold coverage is considered good. But the remainder of the rumen sequences are unattached scaffolds and single-copy reads out of thousands of unknown genomes. These unattached sequences encode tens of thousands of novel enzymes for carbohydrate digestion, and are of great interest to the biofuel industry.

Limitations of metagenomic analysis. Metagenomic analysis is limited by the challenge of assigning DNA sequence reads correctly to their shared genomes. All computational pipelines must choose assumptions about binning, sequence gaps, and other sequence characteristics. For this reason, different pipelines often predict different partial genomes. An improvement is the method of long-read assembly, in which portions of sample DNA are isolated and Illumina-sequenced separately and then the sequences are assembled into long reads for further assembly into OTUs. Long-read Illumina assembly improves accuracy and reveals rare community members that would otherwise be missed because of low sequence coverage. For example, long-read assembly enabled Jill Banfield's group to sequence genomes of rare Chloroflexi and Deltaproteobacteria with unusual metabolic properties (see **Table 21.1**).

Another limitation is that pipelines based on reference genomes restrict the genomes discovered to close relatives of known organisms. New pipelines aim to discover new kinds of taxa by reference-independent approaches; an example is the MetaBAT pipeline developed by Zhong Wang and colleagues at the Joint Genome Institute. MetaBAT bins taxa based on the relative abundance of reads from multiple samples of a target community. The assumption is that different samples contain slightly different proportions of each species in the community; thus, a given genome will have scaffolds present in the same abundance within one sample but a different abundance in a different sample. For Paul Evans and colleagues, the MetaBAT pipeline revealed a new clade of methanogens growing in Australian gas wells (see **Table 21.1**).

A modification of metagenomic analysis is that of single-cell genome sequencing. In this approach, single cells from a microbial community are isolated by fluorescent cell sorting or by a microfluidic chamber. The DNA from a single cell is then amplified by a special kind of PCR called multiple displacement amplification (MDA). The MDA-amplified genome is then sequenced by Illumina and assembled as a single genome. Thus, in single-cell sequencing (described in Chapter 7), multiple genomes from a community are isolated and assembled separately, without needing to sort out a polymicrobial mixture of DNA.

> **Thought Question**
>
> **21.3** Suppose you plan to sequence a marine metagenome for the purpose of understanding carbon dioxide fixation and release, to improve our model for global climate change. Do you focus your resources on assembling as many complete genomes as possible, or do you focus on identifying all the community's enzymes of carbon metabolism?

Functional Annotation

Once we have our metagenomic DNA sequence reads organized into scaffolds and partial genomes, how do we "call" the genes? No single method works best to ensure that we recognize all the actual genes encoding functional products—or that we don't mistakenly define some noncoding sequences as genes (false positives). Many bioinformatic tools are used to call genes. Some approaches include:

■ **Gene structure.** The six reading frames are searched for start and stop codons that bracket sequences of appropriate length to encode proteins (open reading frames, or ORFs). To be transcribed and translated to a functional product, each ORF must have an upstream promoter sequence, as well as a ribosomal start site (discussed in Chapters 8 and 10).

■ **Homologs and motifs.** The metagenomic reads may match homologs of related genes from databases of previously sequenced organisms. In addition, the databases include short, recurring peptide sequences called "motifs" that are common to functional classes of proteins.

■ **Comparison with previously processed metagenomes.** Metagenome comparison is particularly useful in the clinical setting, where diagnostic tests aim to compare the microbiomes of diseased patients with a reference collection of "normal" human microbiomes.

Note that all of these gene-calling (annotation) approaches are incomplete because they miss truly novel genes for which no homologs or motifs exist in the databases. Nonetheless, functional genes offer a basis for comparing two microbial communities. An example of a comparative overview of functional genes is shown in **Figure 21.8**. This figure compares the abundance of functional gene categories between microbes from desert soils and nondesert soils. Functional categories were determined from Illumina-sequenced metagenomes using the MG-RAST pipeline (Argonne National Laboratory), which assigns functions to genes based on sequence comparison with a functional gene reference database. The most striking difference between the two is that the desert soil genomes contain far fewer genes for virulence and defense, such as antibiotic production and resistance. This finding confirms that microbes in deserts are more isolated than those in moist soil communities; thus, the survival of desert microbes depends less on competition with fellow microbes than on the ability to tolerate extreme conditions of dryness. On the other hand, the desert genomes encode a greater number of products for protein metabolism. This finding suggests that growth under deprivation may

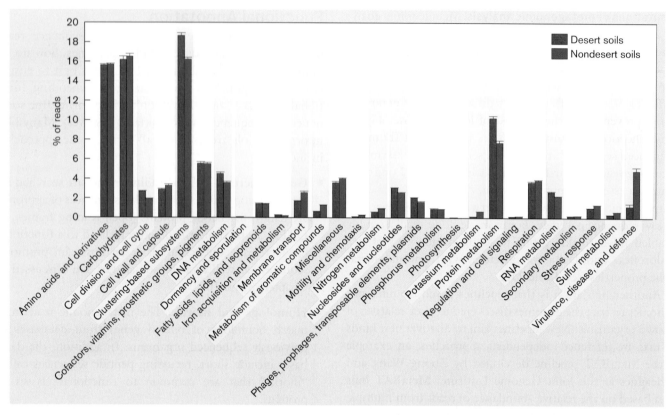

FIGURE 21.8 ■ Functional gene categories: relative abundance compared between desert and nondesert soils. Highlighted categories denote significant differences in abundance of genes in functional categories. Sixteen desert soil samples from hot deserts and Antarctic cold deserts were compared with seven nondesert soils from forests and grasslands. (Error bars represent standard error of the mean.) *Source:* Modified from Noah Fierer. 2012. *PNAS* **109**:21390, fig. 3.

require a greater range of metabolic options for growth of cell biomass.

Beyond the Metagenome

Metagenomics is our most powerful tool to identify the microbes that inhabit an ecosystem. Yet serious limitations remain. In a study of soil in an apple orchard, Jo Handelsman and colleagues compared the sampling rate of a metagenomic screen versus plate culture. For isolation of the soil community, they used a sophisticated culture medium tailored to support growth of bacteria from the rhizosphere (associated with plant roots) and to exclude fast-growing fungi. From the same locations, the researchers sequenced the metagenome. To their surprise, the metagenomes failed to detect more than 60% of the cultured bacteria.

How could these cultured organisms be missed in the metagenome? The researchers hypothesize that the cultured organisms represent organisms of the "rare biosphere"—that is, species of such low abundance in the soil that the metagenomic screen does not pick them up. But upon culturing, these rare organisms are favored by

the sudden provision of concentrated nutrients—a condition that inhibits the more abundant soil organisms, many of which are oligotrophs (discussed in Chapter 4). These "copiotrophs" or "weed organisms" may normally be rare but prevail when nutrients suddenly appear. Even our latest advances, like long-read assembly and primer-independent sequencing, surely miss unknown numbers of species.

Another question unanswered by metagenomes is the spatial organization and interaction of microbes within a habitat. However, the sequence data can be used to construct probes to answer such questions. A key technique combines probes derived from sequencing with fluorescence microscopy, called **fluorescence in situ hybridization,** or **FISH** (**Fig. 21.9A**). The technique makes use of a fluorophore-labeled oligonucleotide probe (usually a short DNA sequence) that hybridizes to microbial DNA or rRNA. Hybridizing to rRNA increases sensitivity because rRNA is present in approximately 100-fold to 10,000-fold excess over DNA. In a typical procedure, the cells of a sample are fixed to a slide (**Fig. 21.9A**, step 1) by a chemical treatment that maintains cell integrity while permeabilizing the cell so that the fluorophore-labeled DNA probe can enter (step 2). Next, the fixed cells are incubated in a

A.

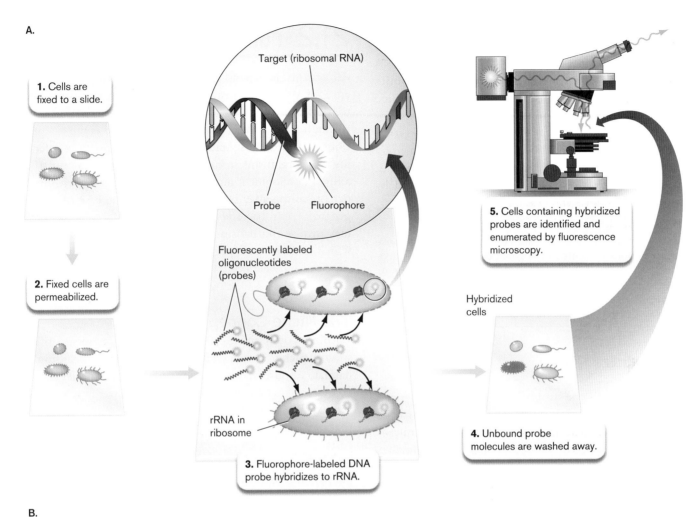

1. Cells are fixed to a slide.

2. Fixed cells are permeabilized.

Target (ribosomal RNA)

Probe   Fluorophore

Fluorescently labeled oligonucleotides (probes)

5. Cells containing hybridized probes are identified and enumerated by fluorescence microscopy.

Hybridized cells

rRNA in ribosome

4. Unbound probe molecules are washed away.

3. Fluorophore-labeled DNA probe hybridizes to rRNA.

B.

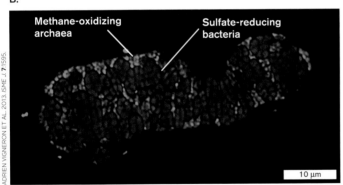

Methane-oxidizing archaea

Sulfate-reducing bacteria

10 µm

ADRIEN VIGNERON ET AL. 2013. *ISME J.* **7**:1595.

FIGURE 21.9 ■ Fluorescence in situ hybridization (FISH).
A. Fluorophore-labeled DNA oligonucleotide hybridizes to a taxon-specific sequence of rRNA molecules within bacteria that are fixed and permeabilized on a microscope slide. **B.** Syntrophy between anaerobic methane-oxidizing archaea (green FISH) and sulfate-reducing bacteria (red FISH). From deep-sea cold seep, Guaymas Basin. *Source:* Part A modified from Rudolf Amann and Bernhard M. Fuchs. 2008. *Nat. Rev. Microbiol.* **6**:339, fig. 1.

hybridization buffer containing the probe, at a temperature designed to maximize specificity of binding to the sequence of the desired taxa (step 3). A probe with broad specificity might hybridize to all bacterial rRNA, but not to archaeal or eukaryotic rRNAs. For greater specificity, a probe may have a sequence complementary to a sequence found only in rRNA of a given bacterial taxon. Following hybridization and a wash (step 4), the cells containing hybridized probes are observed by fluorescence microscopy (step 5).

Figure 21.9B shows an example of FISH used to reveal the deep-sea methanotrophic consortium that oxidizes 90% of the methane emitted by methanogens—a major

contribution to global climate (discussed in Chapter 22). The target consists of a mixed biofilm of anaerobic methane-oxidizing archaea (ANME) and sulfate-reducing bacteria (SRB), obtained from a methane seep at the Guaymas Basin in the Gulf of California, located 2 km below sea level. The biofilm was labeled with oligonucleotide fluorescent probes specific for ANME (red) and for bacteria (green). The FISH image shows the two kinds of cells grouped together in a remarkably regular pattern. The organization of these cells facilitates their syntrophy, enabling ANME to transfer electrons from methane to SRB, which reduce sulfate to sulfide.

Beyond metagenomes, how do we know which genes are actually expressed by the target community? Emerging technologies sample not just the DNA, but also the RNA and protein pools of a community. **Metatranscriptomics** is the study of the RNA transcripts (or RNAseq) obtained from an environmental community. The "metatranscriptome" gives a snapshot of gene expression activity of a community at a given point in time. A metatranscriptome of a marine water community will be presented in Section 21.2. **Metaproteomics** is the study of proteins synthesized by environmental samples and cultures.

Note: Microbial ecology is an interdisciplinary enterprise that uses methods presented throughout this book, such as fluorescence microscopy (Chapter 2), enriched medium (Chapter 4), metagenome sequencing (Chapter 7), isotope labeling (Chapter 13), and molecular phylogeny (Chapter 17). Additional methods, such as gel electrophoresis, are described in eAppendix 3.

To Summarize

- **A metagenome is the sum total of all DNA sequenced from a microbial community.** The community may be any natural or human-made environment, or the microbiome (host-associated microbes) of a plant or animal.

- **Samples are obtained from a target community of a defined environment.** Sampling requires separating the microbes from their physical environment, filtering the organisms of interest, breaking open the cells, and purifying the DNA. Different sampling procedures may lead to different views of a metagenome.

- **Species diversity of a community is assessed by SSU rRNA amplification.** Rarefaction curves estimate the completeness of sampling the diversity.

- **Metagenomic DNA is sequenced.** High-throughput sequencing methods, such as Illumina sequencing by synthesis, generate large amounts of data, which require computational analysis.

- **The sequence reads are assembled into scaffolds and binned into partial genomes.** Assembly requires a computational pipeline that incorporates mathematical tools and biological assumptions. Different pipelines make different assumptions and may predict different taxonomic bins.

- **Functional annotation offers clues as to the ecological functions contributed by the microbes.**

- **Emerging technologies go beyond metagenomics.** FISH reveals spatial organization of community partners. Gene expression and function are investigated by metatranscriptomics and metaproteomics.

21.2 Functional Ecology

All organisms depend, directly or indirectly, on the presence of other organisms. How do microbes contribute to these interactions? Microbes cycle essential nutrients through a food web. They also serve more complex functions that we are just beginning to discover, such as defending host organisms from pathogens, and even modulating animal development and behavior. Cooperation with partner organisms may be incidental, as in the case of hydrogen-oxidizing bacteria using H_2 from fermenters; or it may involve mutualism, a highly developed partnership in which two or more species coevolve to support each other (discussed in Sections 21.3 and 21.4).

The Niche Concept

Within a community, each population of organisms fills a specific **niche**. The niche is a set of conditions, including its habitat, resources, and relations with other species of the ecosystem, that enable an organism to grow and reproduce. For example, the niche of *Anabaena*, a cyanobacterium, is that of a filamentous or mat-forming marine organism that fixes CO_2 into biomass while fixing nitrogen via specialized cells called heterocysts (see Chapter 18). *Anabaena*'s photosynthesis releases molecular oxygen that is used by swarms of respiring bacteria. The habitat of *Anabaena* is fresh or brackish water; its biomass provides food for invertebrates and fish. Despite autotrophy, *Anabaena* needs the other organisms too. The cyanobacteria grow best in the presence of heterotrophic Proteobacteria, whose respiration removes oxygen gas from *Anabaena*'s heterocysts, which need anaerobiosis to fix nitrogen. Thus, organisms do more than fill a niche; they construct niches for other kinds of organisms. Organisms perform **niche construction** by shaping the biochemical dimensions of their habitat.

All Ecosystems Require Microbes

The role of microorganisms in all ecosystems was originally formulated by the Dutch microbiologist Cornelius B. van Niel (1897–1985). Van Niel was the first to show that bacteria in the soil and water can conduct photosynthesis without producing oxygen, using electron donors such as H_2S instead of H_2O. This surprising discovery revealed one of many kinds of metabolism unique to microbes, and unknown in plants or animals. Other unique forms of microbial metabolism (discussed in Chapters 13–15) include nitrogen fixation by bacteria and archaea, and the degradation of lignin by bacteria and fungi. Moreover, microbial metabolism provides ecosystems with their sole source of key elements such as sulfur, phosphorus, and iron.

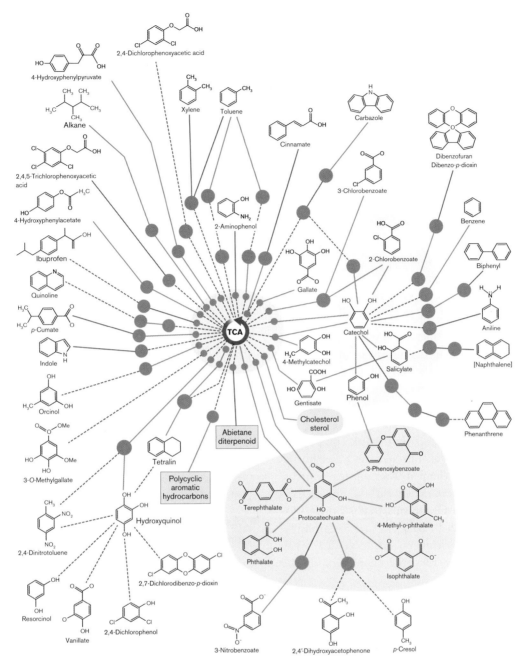

FIGURE 21.10 ■ **Bacteria consume oil.** Petroleum-polluted marine waters contain bacteria whose enzymes (orange balls) catabolize components of crude oil, revealed by bacterial metagenomes and metatranscriptomes. *Source:* Modified from Rafiela Bargela et al. 2015. *Sci. Rep.* **5**:11651.

These examples of microbial metabolism provide unique roles for microbes in ecosystems. Van Niel expressed this principle in two hypotheses of microbial ecology:

■ **Every molecule existing in nature can be used as a source of carbon or energy by a microorganism somewhere.** Any molecule found in the environment can participate in some kind of energy-yielding reaction. If an energy-yielding reaction exists, some microbe will evolve to use it.

■ **Microbes are found in every environment on Earth.** Every possible habitat for life supports microbes. In fact, the largest part of the biosphere (below Earth's surface) is inhabited solely by microbes.

The van Niel hypotheses imply that a limitless variety of species carry out different energy-yielding reactions, depending on what their environment has to offer. For example, oil-contaminated water from the *Deepwater Horizon* spill region shows bacteria with enzymes that catabolize numerous components of crude petroleum (**Fig. 21.10**). In effect, a pipeline spill acts as a kind of **enrichment culture**, in which the addition of a particular class of nutrient favors growth of microbes that can use that nutrient (discussed in

Chapter 4). The catabolic enzymes were revealed by functional annotation of metagenomes and metatranscriptomes. Many of the substrate molecules, such as naphthalene, were once considered nondegradable "xenobiotic" molecules. But all these substrates can ultimately be degraded to CO_2 via the TCA cycle (discussed in Chapter 13), although the rate of breakdown may be slow. The energy gained by microbial metabolism, as well as the elements that microbes assimilate into biomass, eventually circulate throughout the ecosystem. Even the microbe's own genes encoding useful pathways may circulate by horizontal gene transfer (see Section 9.6).

Carbon Assimilation and Dissimilation: The Food Web

The interactions between microbes and their ecosystems include two common roles of metabolic input and output, often called assimilation and dissimilation, respectively. Most of these metabolic processes were discussed in Chapters 13, 14, and 15, but ecology offers a different perspective.

Assimilation refers to processes by which organisms acquire an element, such as carbon from CO_2, to build into cells. When the environment lacks organic compounds containing an element such as nitrogen or phosphorus, microbes may assimilate the element from mineral sources. Common kinds of assimilation include carbon dioxide fixation and nitrogen fixation. Organisms that produce biomass from inorganic carbon (usually CO_2 or bicarbonate ions) are called **primary producers**. Producers are a key determinant of productivity for other members of the ecosystem.

Dissimilation is the process of breaking down organic nutrients to inorganic minerals such as CO_2 and NO_2^-, usually through oxidation. Microbial dissimilation releases minerals for uptake by plants and other microbes, and it provides the basis of wastewater treatment (discussed in Chapter 22). But microbial dissimilation can decrease habitat quality by removing organic nitrogen. When soil bacteria break down amines (RNH_2) to ammonium ion (NH_4^+), nitrifying bacteria such as *Nitrosomonas* oxidize the ammonium to nitrite and nitrate. These highly soluble anions are then washed out of soil into the groundwater.

This chapter covers microbial assimilation and dissimilation of carbon and nitrogen in association with the plants and animals of an ecosystem. The cycles of other key elements, and their effects on the global biosphere, are explored in Chapter 22.

The major interactions among organisms in the biosphere are dominated by the production and transformation of **biomass**, the bodies of living organisms. To obtain energy and materials for biomass, all organisms participate in **food webs** (**Fig. 21.11**). A food web describes the ways in which various organisms produce and consume biomass. Levels of consumption are called **trophic levels**. Organisms at each trophic level consume biomass of organisms from another level. At each trophic level, the fraction of biomass retained by the consumer is small; most is released as CO_2, through respiration to provide energy.

Every food web depends on primary producers for two things:

■ **Absorbing energy from outside the ecosystem.** A key source of energy is sunlight, which drives production by photoautotrophy.

■ **Assimilating minerals into biomass.** The biomass of producers is then passed on to subsequent trophic levels.

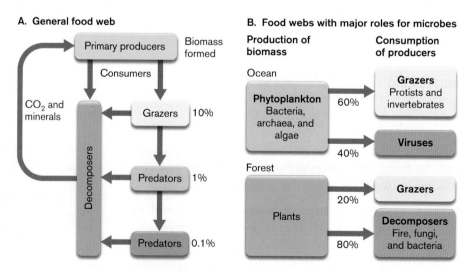

FIGURE 21.11 ■ Microbes within food webs. A. Biomass production and carbon recycling. Percentages indicate the fraction of original CO_2 converted to biomass at each trophic level. **B.** In marine ecosystems, the primary producers are bacteria, archaea, and algae. Viruses break down both producer and consumer microbes. In a forest, the major producers are trees, while the main decomposers are fungi and bacteria.

The majority of carbon in Earth's biosphere is assimilated by oxygen-producing phototrophs such as cyanobacteria, algae, and plants. Certain important ecosystems are founded on lithoautotrophs—for example, the hydrothermal vent communities, in which bacteria oxidize hydrogen sulfide to fix CO_2, capturing both gases as they well up from Earth's crust. The vent communities use oxygen generated by phototrophs living in the euphotic zone (see Section 21.5).

In addition to producers, all ecosystems include **consumers**, which acquire nutrients from producers and ultimately dissimilate biomass by respiration, returning carbon back to the atmosphere (see **Fig. 21.11A**). Consumers constitute several trophic levels based on their distance from the primary producers. The first level of consumers, generally called **grazers**, directly feed on producers. Grazers usually convert 90% of the producer carbon back to CO_2 through respiratory metabolism, yielding energy. The next level of consumers, often called **predators**, feed on the grazers, again converting 90% back to atmospheric CO_2. In microbial ecosystems, the trophic relationships are often highly complex, because a given species may act as both producer and consumer.

At each trophic level, some of the organisms die, and their bodies are consumed by **decomposers**, returning carbon and minerals back to the environment for use by producers. All decomposers are microbes (fungi or bacteria). Decomposers have particularly versatile digestive enzymes capable of breaking down complex molecules such as lignin. Without decomposers, carbon and minerals needed by phototrophs would be locked away by ever-increasing mounds of dead biomass. Instead, all biomass is recycled somewhere in the biosphere. As we learned in Chapter 13, the energy gained by ecosystems can be cycled in part, but all is eventually lost as heat.

In the function of ecosystems, phylogenetic distinctions among the domains Bacteria, Archaea, and Eukarya are of less significance than the biological and biochemical consequences of an organism's presence in the community—that is, what the organism produces or consumes. Thus, in this chapter we place greater emphasis on trophic roles than on phylogenetic distinctions.

The relative significance of microbial and multicellular producers and consumers varies considerably among different habitats. A major difference appears between marine and terrestrial ecosystems (**Fig. 21.11B**). In the oceans, the smallest inhabitants, phototrophic bacteria, perform most of the CO_2 fixation and biomass production. The main consumers are protists and viruses. Viruses are the most numerous replicating forms in the ocean—and they lyse most marine cells before any multicellular predators get a chance to consume them. In terrestrial ecosystems, by contrast, the major primary producers and fixers of CO_2

are multicellular plants. Plants generate **detritus**, discarded biomass such as leaves and stems, that requires decomposition by fungi and bacteria. While viruses are important, multicellular consumers such as worms and insects play a greater role in decomposition.

The differences between the food webs of ocean and dry land explain why most of the food we harvest from the ocean consists of predators at the higher trophic levels (fish), whereas most food harvested on land consists of producers and first-level consumers (plants and herbivores). Fish depend on a large number of trophic levels that dissipate the carbon assimilated by a vast array of microbial producers. Thus, the numbers of fish remain limited, despite the seemingly huge volume of ocean.

Oxygen and Other Electron Acceptors

The availability of oxygen and other electron acceptors is the most important factor that determines how nutrients containing carbon, nitrogen, and sulfur are assimilated and dissimilated. Examples of aerobic and anaerobic metabolism are presented in **Table 21.2**. In aerobic environments, microbes use molecular oxygen as an electron acceptor to respire on organic compounds (abbreviated CHO) produced by other organisms (discussed in Chapters 13 and 14). Aerobic respiration on organic compounds is highly dissimilatory in that it tends to break compounds down to CO_2. Microbes also use oxygen to respire on reduced minerals such as NH_3, H_2S, and Fe^{2+} (lithotrophy). In most cases, lithotrophy is coupled to CO_2 fixation and is therefore assimilatory metabolism.

In anaerobic environments, such as deep soil, microbes use minerals such as Fe^{3+} and NO_2^- to oxidize organic compounds supplied by other organisms (anaerobic respiration) or reduced minerals (anaerobic lithotrophy). Some metals may be reduced to counteract their toxicity; for example, *Cupriavidus metallidurans* reduces gold ion (Au^{3+}) to form

TABLE 21.2	**Aerobic and Anaerobic Metabolism**	
Element	Oxidized by O_2 (lithotrophy)	Reduced by CHO* or H_2 (anaerobic respiration)
Nitrogen	$NH_3 + O_2 \rightarrow NO_3^-$	$NO_3^- + CHO \rightarrow N_2$
Manganese	$Mn^{2+} + O_2 \rightarrow Mn^{4+}$	$Mn^{4+} + CHO \rightarrow Mn^{2+}$
Iron	$Fe^{2+} + O_2 \rightarrow Fe^{3+}$	$Fe^{3+} + CHO \rightarrow Fe^{2+}$
Sulfur	$H_2S + O_2 \rightarrow SO_4^{2-}$	$SO_4^{2-} + CHO \rightarrow H_2S$
Carbon	$CH_4 + O_2 \rightarrow CO_2$	$CO_2 + H_2 \rightarrow CH_4$

*CHO = organic material.

A.

B.

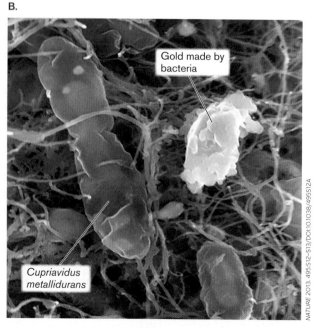

Gold made by bacteria

Cupriavidus metallidurans

FRANK REITH HTTP://WWW.ADELAIDE.EDU.AU/DIRECTORY/FRANK.REITH

NATURE 2013. 495:S12–S13/DOI:10.1038/495S12A

FIGURE 21.12 ■ **Bacteria make gold.** **A.** Frank Reith first showed that *Cupriavidus metallidurans* transfers electrons to gold and may form natural gold deposits. **B.** *C. metallidurans* reduces Au^{3+} to gold metal (Au^0) (SEM).

particles of gold metal (**Fig. 21.12**). Frank Reith, at the University of Adelaide, proposes that most of Earth's gold ore was originally generated by bacterial reduction. Anaerobic environments require much slower rates of assimilation and dissimilation than occur in the presence of oxygen. The total biomass of anaerobic microbial communities, however, far exceeds that of our oxygenated biosphere.

Temperature, Salinity, and pH

Other abiotic factors can profoundly affect the environment for microbes and other members of the food chain, either directly or by impacting other factors. The effects are most obvious for **extremophiles**, species that grow in environments considered extreme by human standards (**Table 21.3**), discussed in Chapter 5. Temperature limits the rate of metabolism. Higher temperatures found in hot springs (80°C–100°C) enable some of the fastest growth rates measured (doubling times as short as 10 minutes for some hyperthermophiles). On the other hand, temperature limits the oxygen concentration in water, so hyperthermophiles often reduce sulfur instead. Extreme cold such as that of polar regions may limit diversity and exclude all but microorganisms and microscopic invertebrates. An example of a cold-adapted ecosystem is the Antarctic lake cyanobacterial mats described in **Special Topic 21.1**.

High salt concentration limits the growth of microbes adapted to freshwater conditions. By contrast, many microbial species have adapted to high salinity (they are called

TABLE 21.3	Extremophiles
Class of extremophile	**Typical environmental conditions for growth**
Acidophile	Acidic environments at or below pH 3
Alkaliphile	Alkaline environments at a range of pH 9–14
Barophile	High pressure, usually at ocean floor, from 200 to 1,000 atm
Endolith	Within rock crystals down to a depth of 3 km
Halophile	High salt, typically above 2-M NaCl
Hyperthermophile	Extreme high temperature, above 80°C
Oligotroph	Low carbon concentration, below 1 ppm
Psychrophile or cryophile	Low temperature, below 15°C
Thermophile	Moderately high temperature, 50°C–80°C
Xerophile	Desiccation, water activity below 0.8

halophiles). The haloarchaea, for instance, bloom in population as a body of water shrinks and becomes hypersaline.

Acidity is important geologically because a high concentration of hydrogen ions accelerates the release of reduced minerals from exposed rock. Extreme acidity is often produced by lithotrophs (discussed in Section 14.5). The increasing acidity releases minerals to be oxidized, while excluding acid-sensitive competitors. Other kinds of habitats, such as soda lakes, show extreme alkalinity resulting from high sodium carbonate.

As our understanding of ecosystems deepens, we discover other remarkable ways that organisms benefit each other. For example, plant-associated bacteria and fungi enhance the uptake of nutrients and protect the host from pathogens. Animal-associated microbes protect the host from pathogens, enhance digestion, and modulate the immune system. These processes are detailed in the next section, and in the sections on animal and plant microbial communities (Sections 21.4 and 21.6).

To Summarize

- **Microbial populations fill unique niches in ecosystems.** Microbes are found in every environment. Every chemical reaction that may yield free energy can be utilized by some kind of microbe.

- **Microbes fix or assimilate essential elements into biomass**, which recycles within ecosystems. Important elements, such as nitrogen, are fixed solely by bacteria and archaea.

- **Consumers and viruses break down the bodies of producers**, generating CO_2 and releasing heat energy. Dissimilation is the process of breaking down nutrients to inorganic minerals such as CO_2 and NO_3^-, usually through oxidation.

- **Primary producers fix single-carbon units, usually** CO_2. Microbial primary producers include algae, cyanobacteria, and lithotrophs.

- **Decomposers such as fungi and bacteria release nutrients from dead organisms.**

- **Microbial activity depends on levels of oxygen, carbon, nitrogen, and other essential elements**, as well as environmental factors such as temperature, salinity, and pH. The largest volume of our biosphere contains anaerobic bacteria and archaea. Beneath Earth's surface, most metabolism is anaerobic.

21.3 Symbiosis

One of the most fascinating features of evolution is how organisms adapt to the presence of others. Some relationships of microbes occur at a distance; for example, the oxygen gas released by marine cyanobacteria is breathed by organisms around the globe. Other relationships require intimate association between two or more partners. An intimate association between organisms of different species is called **symbiosis** (plural, **symbioses**). Symbiotic associations include a full range of both positive and negative relationships (**Table 21.4**). Whether positive or negative, both partners evolve in response to each other. Symbiosis may involve two or more partner species, even thousands of partners, as in animal digestive communities (see Section 21.4).

TABLE 21.4	**Types of Symbiotic Associations Involving Microbial Species**	
Type of interaction	**Effects of interaction**	**Example**
Mutualism	Two organisms grow in an intimate species-specific relationship in which both partner species benefit and may fail to grow independently.	Lichens consist of fungi and algae (in some cases, cyanobacteria) growing together in a complex layered structure. Each species requires the presence of the other.
Synergism	Both species benefit through growth, but the partners are easily separated and either partner can grow independently of the other.	Human colonic bacteria ferment, releasing H_2 and CO_2, which methanogens convert to methane. The methanogens gain energy, and the bacteria benefit energetically from the removal of their fermentation products.
Commensalism	One species benefits, while the partner species neither benefits nor is harmed.	In wetlands, *Beggiatoa* bacteria oxidize H_2S for energy. Removal of H_2S enables growth of other microbes for whom H_2S is toxic. The other microbes are not known to benefit *Beggiatoa*.
Amensalism	One species benefits by harming another. The relationship is nonspecific.	In the soil, *Streptomyces* bacteria secrete antibiotics that lyse other species, releasing their cell contents for *Streptomyces* to consume.
Parasitism	One species (the parasite) benefits at the expense of the other (the host). The relationship is usually obligatory for the parasite.	*Legionella pneumophila*, the cause of legionellosis, parasitizes amebas in natural aquatic habitats. Within the human lung, *L. pneumophila* parasitizes macrophages.

SPECIAL TOPIC 21.1   Antarctic Cyano Mats: Have Ecosystem, Will Travel

What kind of life grows in an ice-covered Antarctic lake—and how does it escape? To reach Antarctica from New Zealand, scientists must take an 8-hour flight in an air force transport plane. The plane typically carries all kinds of workers and equipment, prefabricated buildings, and all-terrain vehicles—in effect, a sample of the human "ecosystem" needed to sustain science at McMurdo Station and other research locations. It turns out that lake microbes have invented a similar system. The cyanobacterial mats of many Antarctic lakes support entire ecosystems limited to microbes and meiofauna (microscopic invertebrates). But within these lakes, the mats have evolved a surprising cycle that flies them out to colonize new locations—along with multiple components of their lake ecosystem.

The lakes of the McMurdo Dry Valleys are isolated, without outlet, and their main source of water is glacier melt in the summer. Thus, each lake has little circulation or turnover, and the water is covered by about 5 meters of ice. The ice surface is sculpted by katabatic winds, from cold air masses that fall down the mountain slope reaching speeds of 200 km/h. These winds carve stone and ice into fantastic forms (**Fig. 1A**).

But through the meters of ice, enough light penetrates to power photosynthesis. The deepest water, where light barely appears, supports mats of cyanobacteria such as *Microcoleus*. The cyanobacteria absorb every photon, growing slowly over the years, in some cases building flame-like forms a few centimeters tall (**Fig. 2**). To photograph these spectacular bacterial gardens, divers like Tyler Mackey from UC Davis (**Fig. 1B**) must drill through the ice and scuba dive (in water much colder than off California). The cyano mat photosynthesis releases a steady supply of oxygen that aerates the upper water, in a way that suggests how oxygenation could have begun on early Earth 2 billion years ago. As producers, the cyanobacteria support a community of consumers including bacteria, algae, and flagellated protists (discussed in Chapter 20). These protists are collected by Rachael Morgan-Kiss and her students at Miami University of Ohio. The eukaryotes include meiofauna such as rotifers, nematodes, and tardigrades (water bears)—top predators of the lake, where no fish can live.

All Antarctic lake organisms must survive 6 months of frozen dark, when water is below zero and surface temperatures reach –40°C. In the summer, with 24-hour sunlight, the phototrophs and consumers come back to life. The ice cover persists, except around the edge where a "moat" melts through. Meanwhile, 10 meters below the ice, cyanobacteria are bubbling oxygen. The oxygen bubbles lift the mat, until a scrap breaks off and floats toward the surface (**Fig. 2**). The scrap of cyanobacterial filaments carries with it all kinds of associated bacteria and protists, nematodes and water bears—in effect, a sample of the whole lake ecosystem. The scrap halts beneath the ice, where it freezes; and come winter, new ice freezes

A.

B.

FIGURE 1 ■ Exploring an Antarctic lake. A. Rachael Morgan-Kiss on the wind-sculpted surface of Lake Hoare, McMurdo Dry Valleys. **B.** Tyler Mackey prepares to dive below 5 meters of ice to observe the benthic cyanobacterial mats.

beneath. But above, wind ablates the ice, wearing it down, as annual winter freezes more below. Thus, over the years the frozen scrap of life travels up the "ice elevator" to the surface. There, as the ice breaks open, the wind carries the scrap away. The scrap may fall in the moat, ready to colonize anew. Amazingly, even the nematodes and tardigrades come to life again with all their phototrophs ready for photosynthesis. And

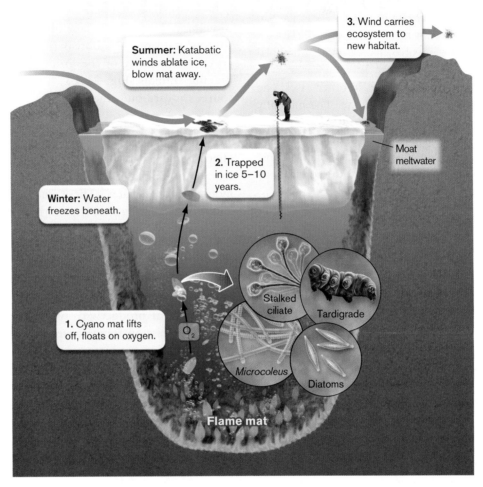

FIGURE 2 ■ Cyanobacterial mats carry an ecosystem. The cyanobacterial mats lift off upon bubbles of oxygen, carrying protists, heterotrophic bacteria, and meiofauna. In winter, water freezes the mat into the ice. Over several years, the mat works its way through the ice until katabatic winds scrape it out and carry the freeze-dried ecosystem to a new habitat.

some lucky scraps are borne away to a moat in some other lake, ready to colonize a new home.

Morgan-Kiss and colleagues investigate these flyaway ecosystems to discover new life-forms. They sampled Lake Fryxell cyano mats emerging from the ice, and isolated community DNA. Metagenomes were assembled by Don Kang and Zhong Wang at the Joint Genome Institute using the MetaBAT computational pipeline. Analysis showed cyanobacteria (*Microcoleus* and *Nostoc*), oil-eating betaproteobacteria (such as *Rubrivivax*), and actinobacteria (such as *Pseudonocardia*). Actinobacteria could produce new antibiotics—perhaps antibiotics that never encountered human pathogens. Thus, this remote ecosystem could provide unexpected benefits for medicine.

RESEARCH QUESTION

How do Antarctic lake microbes and meiofauna adapt to the seasonal extremes of oxygen and light, followed by freezing anaerobiosis and dark?

Sumner, Dawn Y., Ian Hawes, Tyler J. Mackey, Anne D. Jungblut, and Peter T. Doran. 2015. Antarctic microbial mats: A modern analog for Archean lacustrine oxygen oases. *Geology* **43**:887–890.

Slonczewski, Joan L., and Rachael Morgan-Kiss. 2015. Cyanobacterial communities of Antarctic Lake Fryxell liftoff mats and glacier meltwater. Award #1936, Genome Portal of the Department of Energy Joint Genome Institute.

Mutualism Involves Partner Species That Require Each Other

In the most highly evolved forms of symbiosis, partner species <u>require</u> each other for survival—a phenomenon termed **mutualism**. A striking case of microbial mutualism is the interdependence of luminescent bacteria *Aliivibrio fischeri* and their host squid, *Euprymna scolopes* (described in Section 10.5). Mutualism can involve two or more microbial partners. It can also involve one or more microbial partners with a plant or animal host. In some cases, both partners absolutely require each other; in other cases, one or the other is incapable of growing alone. The mutually beneficial relationship is maintained by numerous genetic responses that regulate each partner, avoiding damage to the other. Mutualisms such as nitrogen-fixing rhizobia within legumes can have enormous practical applications (discussed in Section 21.6).

Lichens. A highly evolved form of mutualism is the **lichen** (**Fig. 21.13**). Lichens consist of an intimate symbiosis between a fungus and an alga or cyanobacterium—sometimes both. The symbiosis requires compatible partner species. The alga or bacterium provides photosynthetic nutrition, while the fungus provides minerals and protection. Lichens grow very slowly, but they tolerate extreme desiccation.

Lichens show a surprising variety of form. Different species may form a flat crust, branched filaments, or leaflike lobes. A cross section of the leaflike lichen *Lobaria pulmonaria* reveals a layer of algae (green) covered by fungal mycelium (white), which protects the algae from ultraviolet light damage (**Fig. 21.13C**). In addition to the algae, this lichen includes patches of cyanobacteria, which fix nitrogen. Thus, the fungus, algae, and cyanobacteria form a three-way mutualism. For dispersal, the lichen forms asexual clumps of algae wrapped in fungal mycelium. The clumps flake off and are carried by wind to new locations. In boreal (northern) forests, lichens cover the majority of the ground and provide food for grazing animals (**Fig. 21.14**). Lichens are a winter food source for caribou, which dig beneath the snow to obtain them.

Microbiome of corals and sea anemones. Many invertebrate animals acquire endosymbiotic algae, most commonly dinoflagellates. These endosymbionts are called zooxanthellae (singular, zooxanthella). The algae receive protection from predators, while the animal receives photosynthetic products. Examples include anemones, clams, and corals (**Fig. 21.15** and **eTopic 21.2**). Coral zooxanthellae are extremely important to the biosphere because healthy coral is required for reef formation and much of the biological productivity of coastal shelf ecosystems. The slight rise in temperature that has occurred from global warming has already caused severe problems with **coral bleaching**, in which the algal symbionts die or are expelled. The coral turns white and soon dies, unless its symbionts return.

A.

B.

C. Lichen cross section

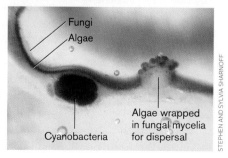

Fungi
Algae
Cyanobacteria
Algae wrapped in fungal mycelia for dispersal

FIGURE 21.13 ■ **Lichens. A.** A tombstone at Kenyon College Cemetery in Ohio, encrusted with lichens (pale green) and mosses (dark green), a nonvascular plant. **B.** Close-up of the lichens in part (A). **C.** Section through a lichen (*Lobaria pulmonaria*) shows fungal, algal, and bacterial symbionts.

FIGURE 21.14 ■ **Lichens cover the ground in boreal forests.**

FIGURE 21.15 ■ **Coral polyps carry mutualistic algae (zooxanthellae).**

The most common algal partners are members of the dinoflagellate genus *Symbiodinium*. A given coral or anemone may harbor several different species of *Symbiodinium*, which show different preferences for light or shade and different tolerances for temperature change. Studies of coral bleaching due to temperature increase suggest that corals containing diverse species of symbionts are more likely

to survive, because one of their species may happen to be resistant to a rise in temperature.

Symbiosis Involves Varying Degrees of Cooperation and Parasitism

Symbiosis between organisms involves a range of interdependence, from obligate mutualism (cooperation) to obligate parasitism (see **Table 21.4**). In a gut community, some of the microbial members may enhance each other's growth, but they can also grow independently. Their optional cooperation is called **synergism**, in which both species benefit but can grow independently and show less specific cellular communication. For example, human colonic bacteria produce fermentation products that colonic methanogens metabolize to methane. The methanogens gain energy, and the fermenting bacteria benefit energetically through the steady-state removal of their end products. The bacteria found within marine sponges may offer an example of synergism. These bacteria fix carbon or secrete defense chemicals that protect the host sponge (see **eTopic 21.3**).

In other cases, one species derives benefit from another without return; for example, some wetland bacteria derive benefit from *Beggiatoa* because *Beggiatoa* bacteria oxidize H_2S, which inhibits growth of other species. An interaction that benefits one partner only is called **commensalism**. Commensalism is difficult to define in practice, since "commensal" microbes often provide a hidden benefit to their host. For example, gut bacteria such as *Bacteroides* species were considered commensals until it was discovered that their metabolism aids human digestion.

An interaction that harms one partner nonspecifically, without an intimate symbiosis, is called **amensalism**. An example of amensalism is actinomycete production of antimicrobial peptides that kill surrounding bacteria. The dead bacterial components are then catabolized by the actinomycete.

Finally, **parasitism** is an intimate relationship in which one member (the parasite) benefits while harming a specific host. Many microbes have evolved specialized relationships as parasites, including intracellular parasitic bacteria such as the rickettsias, which cause diseases such as Rocky Mountain spotted fever.

The distinction between mutualist and parasite is often subtle. Lichens consist of a mutualistic association between fungus and algae, but environmental change can convert the fungus to a parasite. On the other hand, parasitic microbes may coevolve with a host to the point that each depends on the other for optimal health. For example, the high incidence of human allergies is proposed to correlate with lack of exposure to parasites that stimulate development of the immune system.

A.

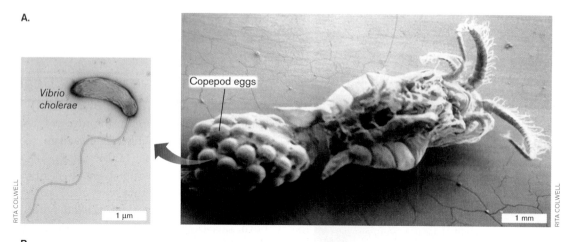

B.

FIGURE 21.16 ▪ *Vibrio cholerae* colonizes copepods. A. Copepod (SEM) with case of eggs to be "hatched" by *V. cholerae* bacteria (TEM). **B.** Anwar Huq compares water with and without filtration by sari cloth, which prevents passage of contaminated copepods and thus prevents transmission of cholera.

A parasite of one host may be a mutualist of another. For example, the human cholera pathogen *Vibrio cholerae* is a mutualist of copepods (**Fig. 21.16**). This relationship was discovered by Anwar Huq, from Bangladesh, who worked with Rita Colwell, then director of the U.S. National Science Foundation (interviewed in **eTopic 1.1**). Huq and Colwell showed that in natural water systems, most *V. cholerae* cells do not swim freely but colonize the surfaces of copepods. The copepods actually depend on these bacteria to eat through the chitin of their egg cases, releasing their young. Thus, bacterial mutualists of copepods are virulent pathogens of humans, whose cholera diarrhea returns the organism to the water full of copepods. To decrease the incidence of cholera, Huq found that drinking water contaminated by *V. cholerae* could be partly decontaminated by filtering out the copepods through sari cloth. The elucidation of this complex microbial partnership reaped benefits for people threatened by cholera.

Multiple partner species can form a complex web of positive and negative dependence. An example is the leaf-cutter-ant symbiosis with fungi, which includes a fungal mutualist, a fungal parasite, and a bacterial mutualist that counteracts the parasite (presented in **eTopic 17.5**).

To Summarize

- **Symbiosis** is an intimate association between organisms of different species.

- **Mutualism** is a form of symbiosis in which each partner species benefits from the other. The relationship is usually obligatory for growth of one or both partners.

- **Lichens** are a mutualistic community of algae or cyanobacteria with fungi. Lichens are essential producers for dry soil habitats.

- **Corals and sea anemones harbor zooxanthellae,** algae that provide products of photosynthesis in exchange for a protected habitat.

- **Parasitism** is a form of symbiosis in which one species grows at the expense of another, usually much larger, host organism.

- **Interactions of multiple species can include both mutualism and parasitism.**

21.4 Animal Digestive Microbiomes

All animals have microbial communities on their surfaces or within certain internal organs. Many microbes have beneficial effects, such as enhancing digestion or generating protective substances in the skin. The beneficial properties are so essential that an animal is now considered a **holobiont**, an entity composed of multiple types of organisms, including microbes. By contrast, a relatively small proportion of animal-associated microbes cause disease. Human microbial interactions with the immune system are discussed in Chapter 23. Pathogenesis and disease in humans are discussed in Chapters 25 and 26.

The digestive tracts of animals harbor particularly complex and ecologically interesting microbial communities. Digestive chambers, such as those of the termite hindgut, the bovine rumen, and the human colon, support thousands of species of bacteria, protists, and archaea.

Termite Wood-Digesting Microbiome

A particularly complex metabolic mutualism is that of termites, whose digestive tract contains bacteria that catabolize wood polysaccharides such as cellulose. The termite feeds on wood and is completely dependent on its symbiotic bacteria.

Wood particles ingested by the termite consist of cellulose and hemicellulose sugar chains entwined with complex aromatic polymers called lignin. Within the termite digestive organ (called the hindgut), its bacteria form highly complex associations with protists such as *Mixotricha paradoxa* (**Fig. 21.17**), which can be as long as half a millimeter.

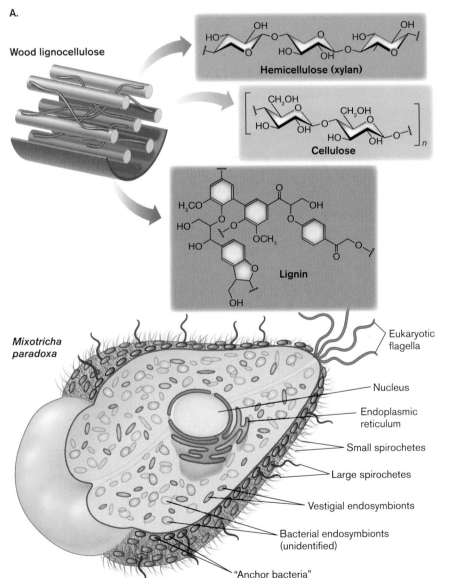

A.

Wood lignocellulose

Hemicellulose (xylan)

Cellulose

Lignin

Mixotricha paradoxa

Eukaryotic flagella

Nucleus

Endoplasmic reticulum

Small spirochetes

Large spirochetes

Vestigial endosymbionts

Bacterial endosymbionts (unidentified)

"Anchor bacteria"

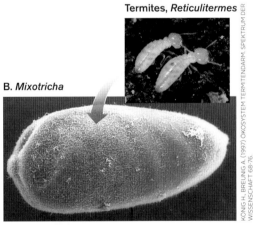

Termites, *Reticulitermes*

B. *Mixotricha*

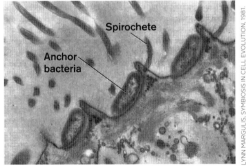

C. Surface symbionts of *Mixotricha*

Spirochete

Anchor bacteria

FIGURE 21.17 ■ *Mixotricha paradoxa*: a multiple symbiont. **A.** Wood lignocellulose is degraded by *Mixotricha paradoxa*. The flagellated protist possesses attached spirochetes (large and small species), "anchor bacteria," and two kinds of bacterial endosymbionts. **B.** Soldier termites, *Reticulitermes flavipes*, contain *Mixotricha* (SEM) and other gut endosymbionts that digest wood cellulose. **C.** TEM section through *Mixotricha*'s pellicle, including anchor bacteria and attached spirochetes.

Mixotricha and other metamonad protists (see Chapter 20) partly break down the lignin component of wood fibers. It is not clear whether *Mixotricha* gains energy from lignin, but the breakdown of fibers makes cellulose available for bacterial catabolism.

M. paradoxa is covered with cilia and flagella, and possesses several kinds of bacterial symbionts. The protist takes up termite-ingested wood particles by phagocytosis, and then the protist's intracellular bacteria digest the wood polysaccharides. *Mixotricha* also has organelles that appear to be vestigial remnants of another endosymbiont, diminished by reductive evolution. On the protist's surface, four kinds of bacteria are attached. Two of the attached species are spirochetes, one significantly larger than the other. The spirochetes extend from the protist's cell membrane; they are flagellated, and their flagellar motility propels the protist cell.

Two other species of "anchor bacteria" are Gram-negative rods attached to knobs of the protist surface. All members of the partnership have evolved an obligate relationship, although the basis for some relationships remains unclear.

The relationship among microbial symbionts within the termite gut community generates a complex series of metabolic fluxes (**Fig. 21.18**). This kind of metabolic cooperation is called **syntrophy**, which means, "feeding together." For syntrophy, the fluxes within the community must balance energetically with a negative value of ΔG, as they would for a single free-living organism (discussed in Chapter 13). Most commonly, one species produces a product that is consumed by the second species, which, if left to build up in high concentration, would result in an unfavorable ΔG for its continued production.

In the simplified model shown in **Figure 21.18**, the wood polysaccharides are hydrolyzed by bacteria and

FIGURE 21.18 ■ Endosymbiont metabolic fluxes within termite hindgut. Bacteria and their protist symbionts ferment wood polysaccharides to lactic acid, formic acid, acetic acid, H_2, and CO_2 within the hindgut of a termite, *Reticulitermes santonensis*. Some hydrogen is lost from the gut, some is converted to methane, and some is converted to acetic acid. Acetic acid is absorbed through the outer lining and feeds the termite.

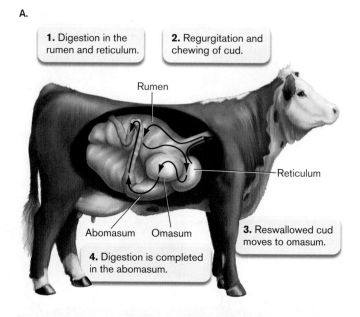

A.

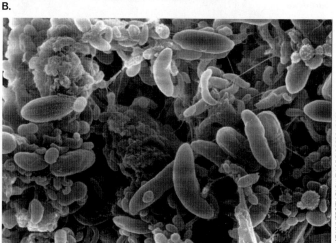

B.

FIGURE 21.19 ■ The bovine rumen. A. The rumen is the largest of four chambers in the bovine stomach. **B.** Bacteria and fungi growing within the rumen.

fermented to short-chain fatty acids, such as acetic acid, which is absorbed by the termite. (The termite's metabolism then oxidizes the acetate to CO_2.) Other products of the termite bacterial fermentation include CO_2 and H_2, which can be converted to methane by methanogenic archaea. In some termites the H_2 builds up to levels as high as 30%. The termite microbial mutualism is being studied as a model system for production of hydrogen biofuel.

The Bovine Rumen Fermenter

All vertebrate animals possess digestive microbiomes. The best-known microbiomes are those of ruminants, such as cattle, sheep, and caribou. Throughout most of human civilization, ruminants have provided protein-rich food, textile fibers, and mechanical work. A historical reference is the biblical injunction against consuming an animal that "is cleft-footed and chews the cud"—that is, "ruminates," or redigests its food in the fermentation chamber known as the **rumen** (**Fig. 21.19**).

The microbial community of the rumen enables herbivores to acquire nutrition from complex plant fibers. From a genomic standpoint, such an arrangement makes evolutionary sense. If the animal had to digest all the diverse polysaccharide chains encountered in nature, its own genome would have to encode a wide array of different enzyme systems. Instead, the ruminant relies on diverse microbial species to conduct various kinds of digestion. As we saw for termites, microbes that partly digest a substrate provide short-chain fatty acids (SCFAs) that the host animal can absorb and digest to completion by aerobic respiration.

> **Thought Question**
>
> **21.5** How does ruminant fermentation provide food molecules that the animal host can use? How is the animal able to obtain nourishment from waste products that the microbes could not use?

The bovine gut system has four chambers (**Fig. 21.19A**). The rumen initially digests the feed and then passes it to the reticulum. The reticulum breaks the feed into smaller pieces and traps indigestible objects, such as stones or nails. After initial digestion, feed is regurgitated for rechewing and then returned to the rumen, by far the largest of the chambers. In the rumen, feed is broken down to small

particles and fermented slowly by thousands of species of microbes. The partially digested feed passes to the omasum, which absorbs water and short-chain acids produced by fermentation. The abomasum then decreases pH and secretes enzymes to digest proteins before sending its contents to the colon for further nutrient absorption and waste excretion.

In the twentieth century, Robert Hungate (1906–2004) at UC Davis pioneered techniques of anaerobic microbiology. One of Hungate's methods still in use today is that of obtaining anaerobic cultures from a **fistulated**, or **cannulated**, **cow**—that is, a cow in which an artificial connection is made between the rumen and the animal's exterior (see **Fig. 21.2C**). The cow is unharmed by the fistula and rumen sampling.

Metagenomics coupled with bioenergetics studies shows how different microbes fill different niches in ruminal metabolism (**Fig. 21.20**). Cattle grown on relatively poor forage (that is, forage high in complex plant content) show a high proportion of ruminal fungi, the chytridiomycetes

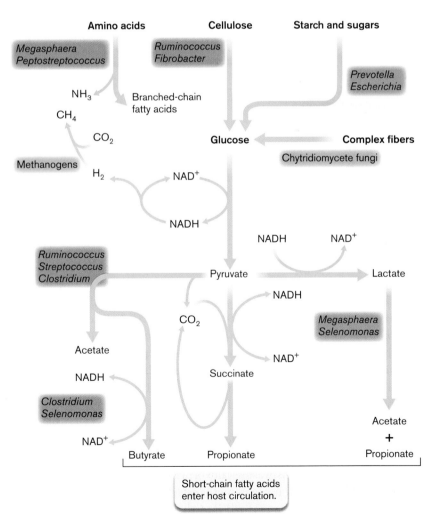

FIGURE 21.20 ■ **Ruminal metabolism.** Various microbes participate in digesting food, ultimately producing short-chain fatty acids that are absorbed by the bovine gut epithelium. *Source:* Based on J. B. Russell and J. L. Rychlik. 2001. *Science* **292**:1119–1122.

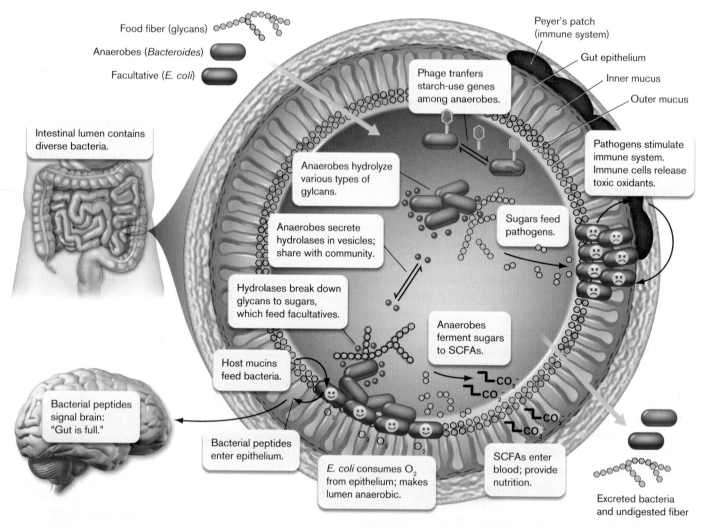

Food fiber (glycans)

Anaerobes (*Bacteroides*)

Facultative (*E. coli*)

Intestinal lumen contains diverse bacteria.

Phage tranfers starch-use genes among anaerobes.

Peyer's patch (immune system)

Gut epithelium

Inner mucus

Outer mucus

Anaerobes hydrolyze various types of gylcans.

Pathogens stimulate immune system. Immune cells release toxic oxidants.

Anaerobes secrete hydrolases in vesicles; share with community.

Sugars feed pathogens.

Hydrolases break down glycans to sugars, which feed facultatives.

Anaerobes ferment sugars to SCFAs.

Host mucins feed bacteria.

Bacterial peptides signal brain: "Gut is full."

Bacterial peptides enter epithelium.

E. coli consumes O_2 from epithelium; makes lumen anaerobic.

SCFAs enter blood; provide nutrition.

Excreted bacteria and undigested fiber

FIGURE 21.21 ■ **Human gut microbiome digests our food.** In a "restaurant" mixed-species biofilm, *Bacteroides* bacteria break down glycans into sugars that *E. coli* catabolizes, consuming oxygen from the host blood supply. The sugars may also feed pathogens, which stimulate an immune response.

(discussed in Chapter 20). Chytridiomycete mycelia appear on ruminal food particles, and their motile zoospores—formerly mistaken for protists—swim through rumen fluid. By contrast, cattle fed a high-cellulose diet, such as hay, grow faster and show cellulolytic bacteria such as *Ruminococcus albus* and *Fibrobacter flavefaciens*. The cellulolytic bacterial metabolism also requires the presence of *Megasphaera* and *Peptostreptococcus* species, which release small amounts of branched-chain fatty acids. Thus, while bacteria compete for food, they also share in a complex web of syntrophy.

Ruminal fermentation includes production of H_2 and CO_2, which support methanogens. From a farmer's point of view, methanogenesis wastes valuable carbon from feed and emits the greenhouse gas methane. So much methane forms that a cannula inserted into the rumen liberates enough of the gas to light a flame. Other vertebrate digestive tracts produce less methane than cattle do. For example, kangaroos have a shorter gut retention time, so there is less time for action of slow-growing microbes such as

methanogens. Switching from cattle to kangaroo meat has been proposed in Australia as a way to decrease greenhouse gas emissions.

Thought Question

21.6 How do you think cattle feed might be altered or supplemented to decrease methane production?

The Human Colon

In contrast to ruminants, humans (and other vertebrates) conduct much of their microbial fermentation at a much later stage of digestion, in the colon, which (unlike the rumen) resides near the end of the digestive tract (**Fig. 21.21**). The genomic coding capacity of human gut microbes may exceed that of the human genome by a factor of 100. Colonic fermentation favors bacteria capable of digesting complex plant materials that pass undigested through the small intestine. For example, Bacteroidetes genera such as

Bacteroides and *Prevotella* ferment mucopolysaccharides, pectin, and arabinogalactan, among many others. Because the colonic oxygen pressure is low, most bacterial digestion is fermentative, releasing acetate and other SCFAs that are absorbed by the intestinal epithelium, thus providing up to 15% of our caloric intake. Amazingly, our bodies also use amino acids synthesized by gut bacteria.

Our gut community varies among individuals, though certain major taxa predominate. Besides *Bacteroides*, other major colonic bacteria include Firmicutes such as *Clostridium* and *Lactococcus*, Actinobacteria such as *Bifidobacterium*, and Verrucomicrobia such as *Akkermansia* (bacterial diversity is surveyed in Chapter 18). The ratio of Firmicutes to Bacteroidetes evolves over the human lifetime, with Firmicutes predominating in adults while Bacteroidetes dominate in infants and the elderly. As in the bovine rumen, diverse colonic bacteria both compete with each other and collaborate through syntrophy. *Bacteroides* species release outer membrane vesicles of hydrolases that break down various types of glycans (polysaccharides) to release oligosaccharides (short sugar polymers). These oligosaccharides may be picked up by other species of *Bacteroides* and *Bifidobacterium* that break them down further to SCFAs. Some fermentation products support methanogens such as *Methanobrevibacter smithii*.

A problem for the colon is the relatively short retention time of the human digestive tract. To remain in the colon with continual resupply of nutrients, *Bacteroides* and other anaerobes may form mixed biofilms with *E. coli* that adhere to the epithelium in the outer mucus layer (**Fig. 21.21**). The mucus layer turns over every 2 hours, but *E. coli* growth can outpace the rate of shedding. The mixed biofilm benefits both bacterial partners, as *E. coli* gains access to sugars while reducing oxygen that leaks in from the epithelium, sustaining the anaerobic environment for fermenters. These patches of mixed biofilm are termed "restaurants" by Tyrrell Conway at Oklahoma State University, who developed the restaurant hypothesis to explain colonic community structure. The mixed biofilm has the added benefit of outcompeting pathogens. When competition fails, however, pathogens can get through the mucus layer and invade the epithelium, stimulating the immune system. The positive and negative interactions of gut bacteria with the immune system are discussed in Chapter 23.

An exciting research question is the interaction of gut bacteria with the brain. For example, as food reaches *E. coli* in the colon, the bacteria start to grow exponentially and release certain peptides into the epithelium. These peptides circulate through the blood and signal the brain that the gut is full. Another intriguing finding is that bacterial catabolism of amino acids produces neurotransmitters such as GABA and agmatine. These bacterial neurotransmitters have been proposed to stimulate the vagus nerve, which modulates anxiety and may influence psychiatric disorders.

Gut microbiomes even help establish circadian rhythms in mammals, and may influence our experience of "jet lag." A study of mice showed that mouse microbial populations vary over the daily cycle, and that restoring normal gut populations can overcome a genetic defect in the molecular clock that governs animal activity. In humans, jet lag induces abnormal fluctuations in the gut microbiome. Perhaps someday we will develop probiotic therapies for jet lag.

The mechanisms by which gut microbes influence our gut epithelium and immune system are discussed in Chapter 23.

Thought Question

21.7 How could you design an experiment to test the hypothesis that an animal makes use of amino acids synthesized by its gut bacteria?

To Summarize

- **Animals harbor digestive microbial communities.** The microbes possess numerous digestive enzymes absent in the host genome.

- **Termite gut mutualists** include protists that break down lignin, as well as bacteria that catabolize cellulose. A major by-product is hydrogen gas.

- **Syntrophy** is a metabolic association between (at least) two species, requiring both partners in order to complete the metabolism with a negative value of ΔG.

- **The rumen of ruminant animals is a complex microbial digestive chamber.** Rumen microbes, including bacteria, protists, and fungi, digest complex plant materials. The microbial digestion generates short-chain fatty acids that are absorbed by the intestinal epithelium.

- **The human gut microbiome contributes to our digestion.** Anaerobes such as *Bacteroides* and facultative respirers such as *E. coli* form mixed biofilms. The anaerobes break down large glycans to sugars utilized by facultatives, and to short-chain fatty acids that enter the intestinal epithelium.

- **Commensal members of the gut microbiome normally outcompete pathogens.** Pathogenic bacteria can invade the inner mucus layer and harm the gut epithelium.

- **Human gut microbes communicate with their host.** Commensal bacteria send chemical signals to our immune tissues and to our brain.

21.5 Marine and Freshwater Microbes

What microbial communities inhabit the oceans? Oceans cover more than two-thirds of Earth's surface, reaching depths of several kilometers and forming an immense habitat. Both oceans and freshwater support huge quantities of bacteria and algae, which drive vast ecosystems. Marine microbes contribute half the planet's productivity (production of biomass). Marine and freshwater microbes form the base of the food chain for seafood, with enormous impact on humans.

Marine Habitats

Marine water has a salt concentration averaging 3.5%. The major ions are Na^+ and Cl^-, with significant levels of sulfate and iodide. The salt concentration is high enough to prevent growth of many aquatic and terrestrial bacteria, such as *E. coli*. Nevertheless, salt-tolerant organisms such as *Vibrio cholerae* grow well over a broad range of salt concentration.

The ocean varies considerably with respect to temperature, pressure, light penetration, and concentration of organic matter. In the open ocean, the water column (known as the **pelagic zone**) is subdivided into distinct regions (see **Fig. 21.22**):

■ **Neuston (about 10 μm).** The neuston is the air-water interface. Although extremely thin, the neuston layer contains the highest concentration of microbes. Many algae and protists have evolved so as to "hang" from the layer of surface tension that forms at the air-water interface.

■ **Euphotic zone (100–200 m).** The **euphotic zone**, or **photic zone**, is the upper part of the water column, which receives light for phototrophs. In the open ocean, the euphotic zone extends down a couple hundred meters, whereas at the **coastal shelf** (<200 meters to the ocean floor), a higher concentration of silt and organisms decreases the photic zone to as little as 1 meter.

■ **Aphotic zone.** Below the reach of sunlight, in the aphotic zone, only heterotrophs and lithotrophs can grow.

■ **Benthos.** The benthos includes the region where the water column meets the ocean floor, as well as sediment below the surface. Organisms that live in the benthos, such as those of thermal vent communities, are called **benthic organisms**.

Another important determinant of marine habitat is the **thermocline**, a depth at which temperature decreases steeply and water density increases. A thermocline typically exists in an unmixed region. At the thermocline, a population of heterotrophs will peak, feeding on organic matter that settles from above. An experiment showing the thermocline, conducted by undergraduates, is described in **eTopic 21.4**.

Coastal regions show the highest concentration of nutrients and living organisms, and the least light penetration. By contrast, the open ocean is largely oligotrophic (having an extremely low concentration of nutrients and organisms). The concentration of heterotrophic microorganisms determines the **biochemical oxygen demand** (**BOD**; also called **biological oxygen demand**), the amount of oxygen removed from the water by aerobic respiration. Normally, the open ocean has such a low concentration of organisms that the BOD is extremely low; therefore, the dissolved oxygen content is high. This explains why enough oxygen reaches the ocean floor to serve chemolithoautotrophs such as sulfide-oxidizing bacteria. The BOD rises, however,

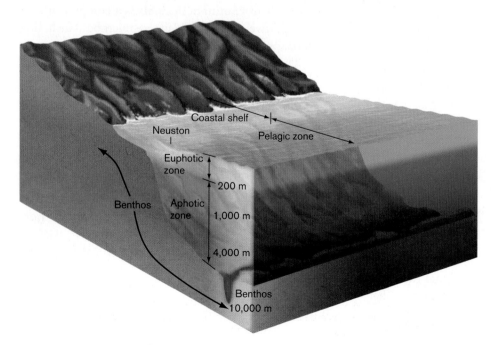

FIGURE 21.22 ■ **Regions of marine habitat.** The marine habitat subdivides into several categories. The coastal shelf region is defined as the water extending from the shoreline out to a depth of 200 meters. The pelagic zone (open ocean) includes several depth regions: the neuston, the microscopic interface between water and air; the euphotic zone of light penetration, where phototrophs can grow, down to 100–200 meters; the aphotic zone, which supports only heterotrophs and lithotrophs; and the benthos, at the ocean floor.

A. Holger Jannasch

B. Submersible *Alvin*

FIGURE 21.23 ■ **Holger Jannasch discovered unculturable marine bacteria.** **A.** Holger Jannasch pioneered the study of marine microbes in the open sea and on the seafloor. **B.** A recent model of the submersible *Alvin* used by researchers from WHOI for sampling deep marine organisms.

when excess sewage or petroleum is present, such as that spilled by the *Deepwater Horizon* oil rig in 2010. Microbial consumption of these wastes at first deprives fish of oxygen, but the process is the only way the ocean recovers (see **Figs. 21.4** and **21.10**).

Culturing the Uncultured

To nineteenth-century microbiologists, the oceans appeared virtually free of bacteria. Trained in the tradition of Robert Koch (discussed in Chapter 1), microbiologists attempted to isolate marine bacteria by plate culture, considered the definitive way to study a microbial species. But the colonies that grew on traditional plate media were extremely few. Then, in 1959, the German microbial ecologist Holger Jannasch (1927–1998) (**Fig. 21.23A**) showed that many more bacteria could be seen by light microscopy than could be grown on plates. Later, with colleagues at the Woods Hole Oceanographic Institution (WHOI), Jannasch went on to study life at the ocean floor, using the famous submersible vessel *Alvin* (**Fig. 21.23B**).

Jannasch was one of the first to show that most marine microorganisms cannot be cultured in the laboratory. His observations sparked decades of controversy. How can the organisms be alive if they cannot be cultured? In 1985, Rita Colwell (**Fig. 21.24A**), at the University of

A.

B.

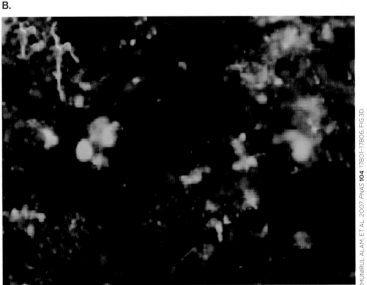

FIGURE 21.24 ■ **Uncultured bacteria.** **A.** Rita Colwell, former director of the National Science Foundation. **B.** Uncultured cells of *Vibrio cholerae* in a biofilm (green fluorescence). Fluorescence microscopy.

Maryland, introduced the concept of **viable but noncul-turable** (**VBNC**), referring to environmental organisms that metabolize but cannot be cultured. In some cases, previously cultured bacteria, such as the cholera pathogen *Vibrio cholerae*, were observed to have entered such a state (**Fig. 21.24B**)—a phenomenon of great importance for medicine. In other cases, environmental organisms were found in an unculturable state, with no known means of culture. Today we term such environmental organisms **uncultured**—with the understanding that their culture requirements may yet be discovered.

In fact, emerging culture methods now enable many uncultured organisms to grow in the laboratory. Often,

microbial growth requires hidden synergy or mutual-ism with other organisms in the natural environment. For example, the tiny cyanobacterium *Prochlorococcus* is the ocean's most abundant oxygenic phototroph, accounting for half the ocean's photosynthesis. The abundance map in **Figure 21.25**, obtained from marine samples worldwide, shows that in most regions, *Prochlorococcus* outnumbers its relative *Synechococcus* by tenfold; the dominance of the tiny phototroph is due in part to its unusual chlorophylls, which absorb more of the blue light from the solar spectrum. Nev-ertheless, *Prochlorococcus* species are very difficult to grow in pure culture. Erik Zinser and colleagues at the University of Tennessee–Knoxville showed that culturing *Prochlorococ-cus* in the laboratory requires the presence of a "helper bacterium" that catalyzes hydrogen peroxide, a by-product of oxygenic photo-synthesis. The helpers produce catalase, which *Prochlorococcus* has lost through degenerative evolution, enabling more efficient growth when heterotrophs are present. Thus, in the ocean—as in gut microbiomes—microbes require surprising synergisms and symbioses, in the midst of com-petition. Synergism, rather than single-species growth, may be the more common condition for microbes in nature.

A. *Prochlorococcus*

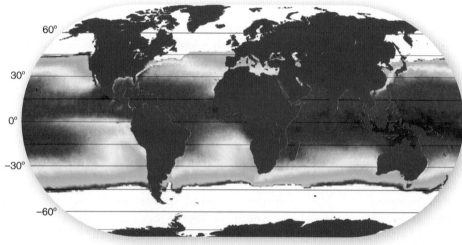

B. *Synechococcus*

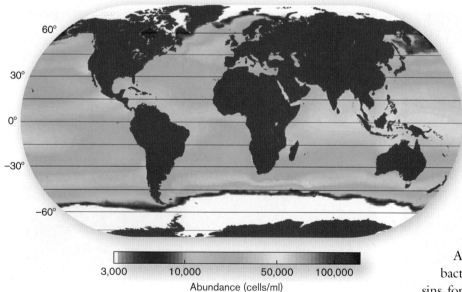

Marine Metagenomes

Cultured or not, the marine microbial communities are now emerging through metagenomes. In 2015, an international study led by Shi-nichi Sunagawa at the European Molecular Biology Laboratory (EMBL) sequenced seven tera-bases (7×10^{12} bp), including sam-ples of all the world's oceans. The study shows that, as abundant as *Prochlorococcus* and *Synechococcus* are, these phototrophs constitute only a small fraction of the microbes present. Abundant heterotrophs include Proteo-bacteria, many of which have proteorhodop-sins for photoheterotrophy. Also abundant are the Thaumarchaeota, which oxidize ammonia released by fish and other organisms.

As discussed in Section 21.1, even metagenomes miss community members that fill important niches. To find rare

FIGURE 21.25 ■ Global distribution of marine phototrophs at the sea surface. A. *Prochlorococcus* abundance peaks in warmer water. B. *Synechococcus* is less abundant overall but reaches colder regions. *Source:* Pedro Flombaum et al. 2013. *PNAS* **110**:9824.

organisms with functional importance, researchers set out to explore expressed genes—a "metatranscriptome." Sallie Chisholm and Edward DeLong, then at the Massachusetts Institute of Technology, sequenced the mRNAs expressed by marine bacteria obtained from a Hawaiian research station. The RNA molecules were polyadenylated (a string of adenines was added) using an enzyme that favors mRNA and excludes ribosomal RNAs because of their high degree of secondary structure (intramolecular folding). The resulting RNA pool, enriched for expressed mRNA, was reverse-transcribed to DNA (called cDNA for "complementary DNA"). The cDNA was then amplified and sequenced.

The sequences of the cDNA (representing expressed genes) were compared to those amplified from metagenomic DNA of the same microbial community. The relative abundance of expressed and metagenomic sequences gives a measure of expression levels of all genes, which are plotted in declining rank order (**Fig. 21.26**). The most highly expressed genes included those encoding functional elements of photosynthesis, such as light-harvesting proteins and Rubisco. Genes for DNA repair were also highly expressed, presumably to correct UV damage in the open ocean. But the most highly expressed genes were some of the rarest in the metagenomes (colored red in **Fig. 21.26**). And 40% of all the cDNA (expressed genes) had no counterpart in the genomic DNA (not shown in the figure). Thus, expression analysis does indeed reveal organisms undetected in the metagenome, presumably too rare for detection by our current sequencing methods. It also suggests that vast ranges of organisms may remain unreached by either genome or transcriptome sequencing.

Beyond DNA, microbial ecology addresses organisms in their relationships with their habitat and with each other. We will now introduce approaches to measuring microbial populations and their interactions within marine communities. These approaches have enormous practical applications, from the management of fisheries to the assessment of greenhouse gas fluxes (a topic pursued in Chapter 22).

Measuring Planktonic Communities

In marine science, the term **plankton** refers to organisms that float passively in water. Microbiologists use the term loosely, as microbial "plankton" include motile bacteria and protists. Marine phytoplankton (microbial phototrophs) produce a substantial part of the world's oxygen and consume much of the atmospheric CO_2.

Microbial plankton, or **microplankton**, include numerous members of the three domains discussed in Chapters 18, 19, and 20: bacteria (cyanobacteria and proteobacteria), eukaryotes (diatoms and dinoflagellates), and archaea (crenarchaeotes and euryarchaeotes). Nanoplankton (about 2–20 μm) arbitrarily include smaller algae and flagellated protists, as well as filamentous cyanobacteria. Picoplankton (about 0.2–2 μm) consist of bacteria and the smaller eukaryotes, including the smallest living cells known. Besides these size classes of cells, the term "femtoplankton" may refer to marine viruses, the smallest detectable particles capable of reproduction. Note, however, that the use of these prefixes ("nano-," "pico-," "femto-") is approximate and does <u>not</u> refer to size units such as nanometers.

Despite the definition of "plankton," not all marine microbes float independently. Many form biofilms on colonial algae such as kelps, or on suspended inorganic particles known as **marine snow**. As many as half of all marine bacteria are associated with particulate substrates from broken-down organisms. As observed by Farooq Azam at the Scripps Institution of Oceanography, "Seawater is an organic matter continuum, a gel of tangled polymers with embedded strings, sheets, and bundles of fibrils and particles, including living organisms, as 'hot spots.'" Thus, while the ocean as a whole has a low average nutrient concentration, marine waters include a suspension of concentrated tangles of nutrient-rich substrates. Collectively, their global volume is so large that their sedimentation rate influences calculations of Earth's carbon cycle.

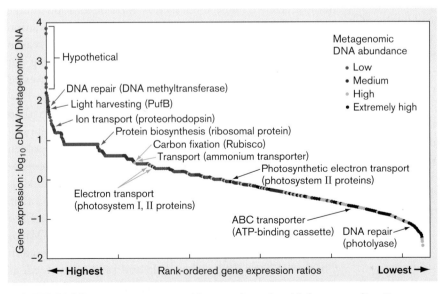

FIGURE 21.26 ■ Genes expressed in a marine microbial community. Expression of each gene is measured as RNA (copied to cDNA) divided by the gene's relative abundance in the metagenome. The y-axis represents the ratio of cDNA to metagenomic DNA for all expressed genes, plotted in rank order (from highest to lowest). The relative abundance of the DNA sequence within the metagenome is color-coded. The genes most highly expressed tend to be rarest in the genome (colored red).

Population size can be estimated in different ways: the number of individual reproductive units, the total organic biomass available to consumers, or the rate of productivity or assimilation of key nutrients, such as carbon dioxide. All ways of measuring these quantities present challenges; there is no single right method, but different approaches answer different questions.

Fluorescence microscopy. Marine microbes of all sizes, even viruses, can be detected and counted under the microscope using a DNA-intercalating fluorescent dye. Recall that fluorescence enables detection even of particles whose size is below the resolution limit defined by the wavelength of light (discussed in Chapter 2). Fluorescence microscopy with a DNA-binding fluorophore such as DAPI is used to detect planktonic microbes. Specific taxa can be detected by FISH (see **Fig. 21.9**). FISH uses a fluorophore attached to a DNA probe, a short sequence that can hybridize only to DNA or RNA of a particular taxon.

Biomass. The amount of biomass can be determined by standard chemical assays of protein and other forms of organic matter. Net biomass of a population, however, does not indicate productivity within an ecosystem, because it misses the amount of carbon cycled through respiration. Marine microbes have an extremely rapid rate of turnover, but what appears to be a small population may nonetheless conduct tremendous rates of carbon fixation into biomass, which is rapidly consumed by the next trophic level.

Incorporation of radiolabeled substrates. The measured amount of biomass does not indicate the rate of biomass production. The rate of production can be estimated by the cells' incorporation of a radiolabeled substrate. Uptake of $^{14}CO_2$ indicates the rate of carbon fixation—a highly significant property for the study of global warming. Uptake of ^{14}C-thymidine measures the rate of DNA synthesis, an indicator of the rate of cell division (example shown in **eTopic 21.4**). These procedures can be conducted in seawater under native conditions, without requiring laboratory cultivation.

A limitation of these methods is that addition of the radiolabeled substrate may raise a nutrient concentration to artificially high levels that distort the naturally occurring rates of activity. In the case of labeled thymidine, another limitation is that not all growing cells incorporate exogenous thymidine into their DNA.

Planktonic Food Webs

Marine plankton interact with each other as producers and consumers in a food web. A simplified outline of the food web in the open ocean is shown in **Figure 21.27A**. The diagram of a food web is often called a "spaghetti diagram" because so many trophic interactions cross each other in various directions.

The phototrophic producers are known as **phytoplankton**. A phytoplankton of great importance is *Prochlorococcus*, but this trophic category also includes myriad unrelated organisms, such as eukaryotic algae, diatoms, and dinoflagellates. In the food web, bacteria and protists are consumed by larger protists, which feed small invertebrates, which feed larger invertebrates and, ultimately, vertebrates such as fish.

All levels of microbial plankton undergo intense predation by viruses. The degree of viral predation is difficult to measure, but recent studies indicate that cell lysis by viruses breaks down about half of microbial biomass (discussed in Chapter 6). Virus particles represent a significant sink for carbon and nitrogen. They accelerate the return of minerals to producers and necessitate a larger base of producers to sustain the ecosystem. Some marine viruses are highly host specific, infecting only certain species of dinoflagellates or cyanobacteria. Their presence selects for diverse communities containing numerous scattered species. Other viruses attack many hosts—and they transfer genes from one host to another, such as the genes encoding photosystems. Thus, marine viruses are a dominant force determining community species distribution and genome content.

> **Thought Question**
>
> **21.8** How do viruses select for increased diversity of microbial plankton?

Ocean ecosystems contain a large number of trophic levels, of which the top consumers include fish and humans. Because each trophic level spends 90% of its food intake for energy, ecosystems require a huge lower foundation to sustain the highest-level consumers. This is one reason why fisheries worldwide are now in danger of running out of fish for human consumption: Fish are being harvested faster than the ecosystem can replace them.

Note that actual food webs are far more complex than the one shown in **Figure 21.27A**. One source of complexity is that many algal protists, such as chrysophytes (golden algae) and dinoflagellates, are actually **mixotrophs**, organisms that both fix CO_2 through photosynthesis and catabolize organic compounds from outside. Mixotrophy offers the opportunity to grow at night, without light, and to acquire scarce minerals, such as iron, from prey. Mixotrophs include secondary endosymbiont algae (discussed in Chapters 17 and 20), such as kelp and sargassum weed, as well as protists containing cyanobacterial or algal endosymbionts. A scheme for a food web that includes mixotrophs is shown in **Figure 21.27B**. The scheme is further complicated by the existence of elaborate mutualistic associations between different kinds of protists, algae, and bacteria.

FIGURE 21.27 ■ The pelagic marine food web. A. In the marine water column, the vast majority of carbon transfer occurs among microbes. Arrows lead from organisms consumed and point to their consumers. The primary producers are phototrophic bacteria and algae, with a smaller contribution from lithotrophic bacteria and archaea. Grazers include heterotrophic bacteria and protists. Predators include protist flagellates and ciliates. About 50% of bacterial and protist biomass is degraded by viral lysis. Multicellular organisms cycle a relatively small fraction of biomass. **B.** A more complex view of the marine food web includes mixotrophs (organisms that combine phototrophic and heterotrophic nutrition) and symbiotic associations between protists and phototrophic bacteria or algae.

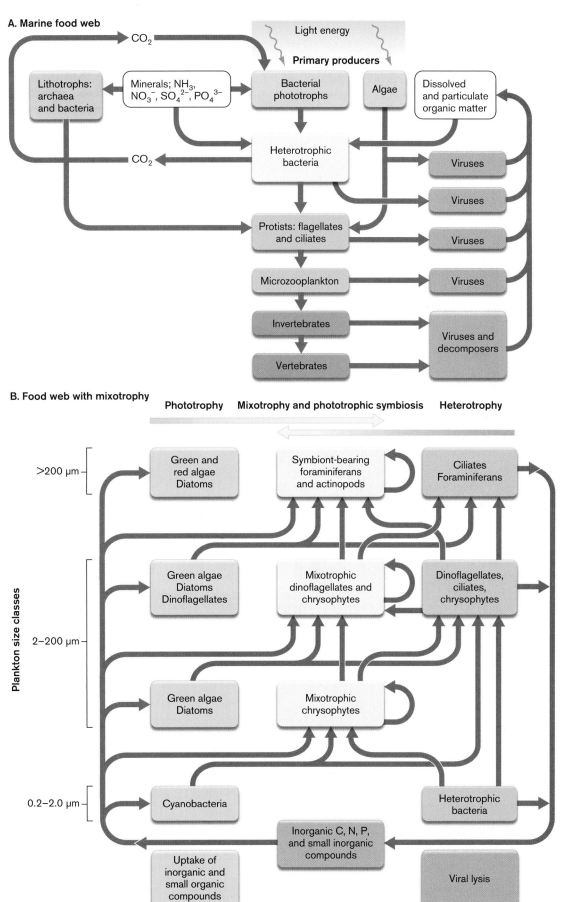

The Ocean Floor

The ocean floor (benthos) experiences extreme pressure beneath several kilometers of water. Most organisms that live there are pressure-dependent species, known as **barophiles** (discussed in Chapter 5). Barophiles require high pressure for growth (200–1,000 atm), failing to grow when cultured at sea level. Cold temperatures (about 2°C) select for **psychrophiles** (cold-adapted species), whose rate of growth is relatively slow. In the absence of light, heterotrophic bacteria depend on detritus from above as their carbon source, but by the time any material reaches the benthos, much of the organic carbon has been depleted by prior consumers. A metagenome of the benthos (deep ocean) reported in 2011 by Emiley Eloe, Douglass Bartlett, and colleagues at UC San Diego revealed primarily Proteobacteria, Bacteroidetes, and Planctomycetes. Their genomes encode a large number of heavy-metal transport and resistance proteins, which help cells survive toxic metals upwelling from the ocean floor. These metal resistance genes may have practical applications for bioremediation.

Among the first to investigate life on the ocean floor, in the 1960s, were Holger Jannasch (see **Fig. 21.23**) and colleagues at WHOI. Using the submersible research vessel *Alvin*, they discovered entirely new forms of benthic bacteria and archaea. Some of these microbes grow in thick biofilms at black smoker **thermal vents**, or

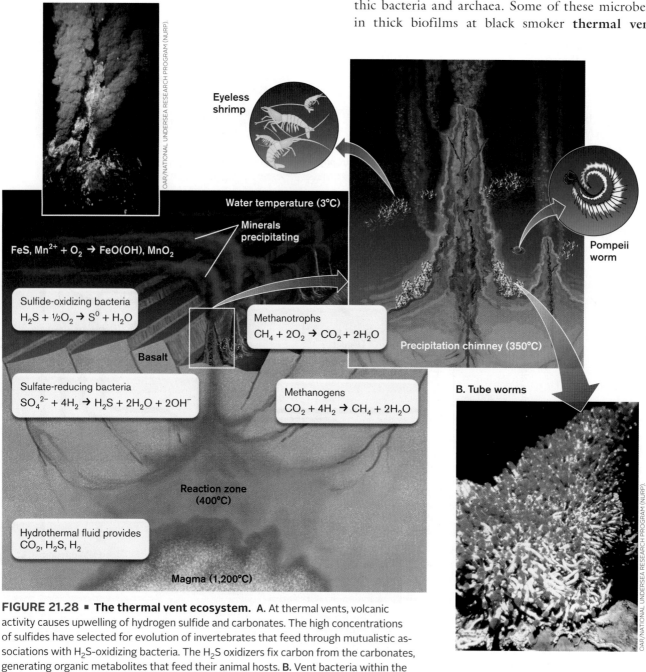

A. Thermal vent

Eyeless shrimp

Water temperature (3°C)

Minerals precipitating

$FeS, Mn^{2+} + O_2 \rightarrow FeO(OH), MnO_2$

Sulfide-oxidizing bacteria
$H_2S + \frac{1}{2}O_2 \rightarrow S^0 + H_2O$

Basalt

Methanotrophs
$CH_4 + 2O_2 \rightarrow CO_2 + 2H_2O$

Sulfate-reducing bacteria
$SO_4^{2-} + 4H_2 \rightarrow H_2S + 2H_2O + 2OH^-$

Methanogens
$CO_2 + 4H_2 \rightarrow CH_4 + 2H_2O$

Precipitation chimney (350°C)

Pompeii worm

Reaction zone (400°C)

Hydrothermal fluid provides CO_2, H_2S, H_2

Magma (1,200°C)

B. Tube worms

FIGURE 21.28 ■ The thermal vent ecosystem. A. At thermal vents, volcanic activity causes upwelling of hydrogen sulfide and carbonates. The high concentrations of sulfides have selected for evolution of invertebrates that feed through mutualistic associations with H_2S-oxidizing bacteria. The H_2S oxidizers fix carbon from the carbonates, generating organic metabolites that feed their animal hosts. **B.** Vent bacteria within the gills of tube worms provide energy from oxidation of hydrogen sulfide. *Source:* Part A from Raina Maier et al. 2000. *Environmental Microbiology.* Academic Press. Reprinted by permission of Elsevier, Ltd.

hydrothermal vents, such as those of Guaymas Basin, in the Gulf of California (**Fig. 21.28**) (discussed in Chapter 19). The clouds rising from the thermal vent are minerals precipitating as the superheated solution meets cold seawater. The reduced minerals from the vents support entire ecosystems of lithotrophic bacteria, including mutualists of uniquely evolved giant clams and worms. Many vent-dependent microbes are **thermophiles**, adapted to high temperatures (discussed in Chapter 5). Many are hyperthermophilic archaea such as *Pyrodictium occultum*, growing at temperatures above 100°C, which are reached only at high pressure (1,000 atm; discussed in Chapter 19).

The ocean floor provides reduced inorganic minerals such as iron sulfide (FeS) and manganese (Mn^{2+}) that can combine with dissolved oxygen to drive chemolithoautotrophy (discussed in Chapter 14). Chemolithoautotrophy by microbes in the sediment generates a permanent voltage potential between the reduced sediment and the oxidizing water. As a result, the entire benthic interface between floor and water acts as a charged battery that can actually generate electricity.

The benthic redox gradient is enhanced dramatically at hydrothermal vents, where volcanic activity causes upwelling of hydrogen sulfide (H_2S), H_2, and carbonates (**Fig. 21.28A**). At a hydrothermal vent, these reduced minerals are brought up by seawater that seeps through the sediment until it reaches a magma pool, where it becomes superheated and rises to the surface as steam. Sulfate-reducing bacteria reduce sulfate from seawater with H_2 upwelling from the vent fluids, to form H_2S. As the H_2S rises, it is oxidized by sulfur-oxidizing bacteria such as *Thiomicrospira*. Nearby anaerobic sediment supports methanogens, and the methane they produce seeps up into the oxygenated water, where it is oxidized by methanotrophs.

The high concentrations of sulfides near thermal vents have selected for a remarkable evolution of invertebrate species that feed through mutualistic associations with H_2S-oxidizing bacteria, such as the yeti crab (see the Current Research Highlight). The H_2S oxidizers fix carbon from the carbonates, generating organic metabolites that feed their animal hosts—species of worms, anemones, and giant clams, all closely related to surface-dwelling species that would be poisoned by H_2S. The tube worm *Riftia* is colored bright red by a pigment carrying H_2S and O_2 in its circulatory fluid (**Fig. 21.28B**). The worm has evolved such a complete dependence on its symbionts that it has lost its own digestive tract.

Surprisingly, microbial communities related to those at thermal vents are found in unheated benthic habitats known as cold seeps, where hydrogen and methane seep slowly through cracks in the rock. Cold-seep endosymbiosis within animals is presented in **eTopic 21.2**.

Freshwater Microbial Communities

Many features of marine habitats also apply to freshwater systems, such as lakes and rivers. Freshwater habitats, of course, contain much lower salt, usually less than 0.1%. The study of freshwater systems is called limnology. Like the marine water column, the profile of a lake water column is defined by depth, aeration, and temperature (**Fig. 21.29**).

Large, undisturbed lakes are usually **oligotrophic**, with dilute concentrations of nutrients and microbes (**Fig. 21.29A**). The warm upper water layer above the thermocline is called the epilimnion. The epilimnion water is warm, well mixed, and oxygenated relative to the lower layers of water, which are isolated from the atmosphere. It supports oxygenic phototrophs such as algae and cyanobacteria. Below the epilimnion lies a thermocline, a steep transition zone to colder, denser water below.

The epilimnion reaches only about 10 meters in depth, much shallower than the marine euphotic zone. In a lake, the depth of light penetration varies greatly, depending on the concentration of microbes and particulate matter. At the edge of the lake, where the upper layer becomes shallow enough for rooted plants, is the **littoral zone**. In the deeper water, below the epilimnion, lies the hypolimnion, a region that becomes anoxic.

A lake that receives large concentrations of nutrients, such as runoff from agricultural fertilizer or septic systems, becomes **eutrophic** (**Fig. 21.29B**). In a eutrophic lake, the nutrients support growth of algae to high densities, causing an **algal bloom** (**Fig. 21.29C**). The algal bloom is rapidly consumed by heterotrophic bacteria, whose respiration removes all the oxygen. This oxygen loss causes the anoxic hypolimnion to reach nearly up to the surface of the lake.

But all fish and invertebrates need access to the epilimnion for oxygen. In a eutrophic lake, fish die off owing to lack of oxygen, which heterotrophic microbes have consumed. Such a lake is said to have a high level of biochemical oxygen demand (BOD). For example, in 2010, fertilizer runoff caused eutrophication of Grand Lake St. Marys in Ohio. The lake was covered with thick green algae that released toxins, depleted oxygen, ruined boat hulls, and prevented human use of the lake.

Common causes of eutrophication include:

■ **Phosphates.** Because phosphorus is commonly a **limiting nutrient** (nutrient in shortest supply) for algae, addition of phosphates from detergents and fertilizers can lead to an algal bloom.

■ **Nitrogen** from sewage effluents and agricultural fertilizer runoff can lead to algal blooms by relieving nitrogen limitation.

■ **Organic pollutants** from sewage effluents overfeed heterotrophic bacteria, depleting the epilimnion of oxygen.

A. Oligotrophic lake (dilute nutrients)

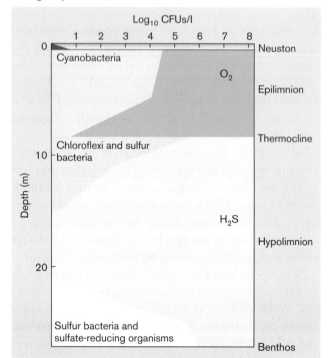

B. Eutrophic lake (rich nutrients)

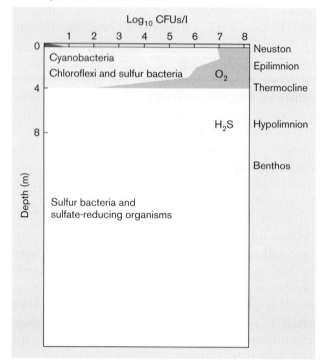

C. Algal bloom

FIGURE 21.29 ▪ **Microbial metabolism in freshwater lakes. A.** Depth profile of an oligotrophic lake. The oxygenated zone (epilimnion) reaches to the thermocline, below which only anaerobes grow. CFU = colony-forming unit. **B.** Depth profile of a eutrophic lake. The oxygenated epilimnion is much shallower, and microbial concentrations are tenfold higher than in an oligotrophic lake. **C.** Algal bloom from fertilizer runoff fills a channel into the reservoir at Grand Lake St. Marys State Park, Ohio, in 2010.

In a eutrophic lake, the lower layers have become depleted of oxygen as a result of overgrowth of microbial producers and consumers. Thermal stratification may break up, and the lake may mix one to several times per year, thus reoxygenating the entire lake. A permanently eutrophic lake typically supports ten times the microbial concentrations of an oligotrophic lake (see **Fig. 21.29A** and **B**), but shows greatly decreased animal life.

Anoxic water supports only anaerobic microbes. These include anaerobic phototrophs that do not produce O_2. Enough light may penetrate to support anaerobic

H_2S-oxidizing phototrophs such as *Chlorobium* and *Rhodopseudomonas*. Although H_2S photolysis provides less energy than oxygenic H_2O photolysis, these bacteria have evolved to use chlorophylls whose spectrum extends into the infrared (**Fig. 21.30**). Light in the infrared portions of the spectrum cannot be used by oxygenic phototrophs, because the photon energy is insufficient to split water. The less efficient H_2S photolyzers, however, can harness the energy of red and infrared radiation (discussed in Chapter 14). H_2S-oxidizing phototrophs overlap metabolically with anaerobic heterotrophs and lithotrophs that reduce oxidized minerals. Some

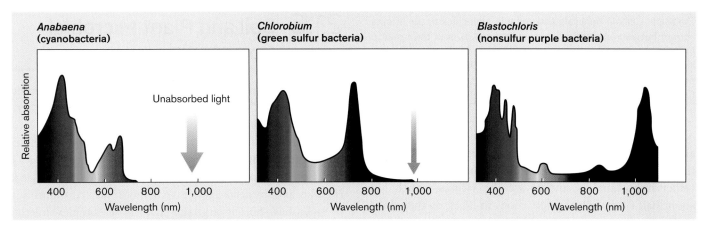

FIGURE 21.30 ■ **Absorbance spectra of chlorophyll from lake phototrophs.** In the upper waters, algae and cyanobacteria absorb primarily blue and red. Below, where red has been absorbed by microorganisms above, anaerobic phototrophs such as *Blastochloris* absorb infrared (wavelengths beyond 750 nm).

bacteria, such as *Rhodospirillum rubrum*, can grow anaerobically with or without light; others grow only by anaerobic metabolism, unassisted by light. These low-oxygen bacteria form a flourishing community, but they cannot support oxygen-breathing consumers such as fish.

At the bottom of the water column, the water meets the sediment (benthos). In the benthic sediment, gradients develop in which successive electron acceptors become reduced by anaerobic respirers and lithotrophs (**Fig. 21.31**). Electron acceptors that yield the most energy are consumed first; as each in turn is depleted, the electron acceptor with the most energy is consumed next. First, molecular oxygen is used to oxidize organic material

and reduced minerals such as NH_4^+. Below, as molecular oxygen falls off, bacteria use nitrate (NO_3^-) from oxidized ammonium ion as an electron acceptor to respire on remaining organic material. As the nitrate is used up, still other bacteria use manganese (Mn^{4+}) as an electron acceptor, followed by iron (Fe^{3+}) and sulfate (SO_4^{2-}). Reduction of sulfate leads to H_2S, which eventually returns to the upper layers supporting anaerobic photolysis. Reduction of CO_2 by H_2 produces methane (CH_4). Methane collects below and sometimes ignites when it escapes to the surface. Methane from freshwater lakes and streams is emerging as a major contributor to global warming (discussed in Chapter 22).

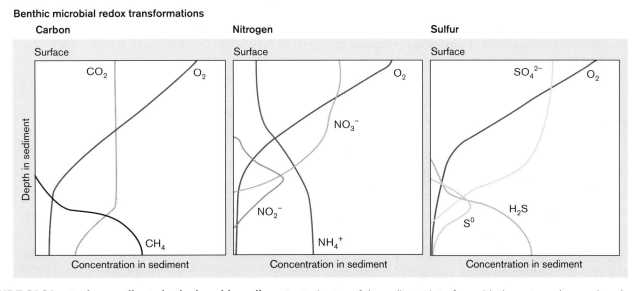

FIGURE 21.31 ■ **Redox gradients in the benthic sediment.** At the top of the sediment interface with the water column, minerals are first oxidized lithotrophically by O_2; then, as O_2 declines, the oxidized minerals are used as alternative electron acceptors for anaerobic respiration.

Note: Certain minerals and organic molecules occur in equilibrium between ionized and un-ionized states over the range of pH typical of most common habitats (pH 4–9). Examples include ammonia (NH_3), which protonates to ammonium ion (NH_4^+), and organic acids such as acetic acid (CH_3COOH), which deprotonates to acetate (CH_3COO^-). In this chapter we refer to the form most prevalent at pH 7 unless stated otherwise.

Throughout the water column, protists consume algae and bacteria, while fungi decompose detritus. Protists and fungi also interact with invertebrates and fish as parasites. Other important consumers are the viruses, which lyse about half of microbial populations in lakes, as they do in the ocean. Viruses limit the number of microbes and keep the water clear enough for light to penetrate.

To Summarize

- **The euphotic zone of the ocean is the upper part of the water column that receives light for phototrophs.** Below, in the aphotic zone, only heterotrophs and lithotrophs can grow. The benthos includes the region where the water column meets the ocean floor, as well as sediment below the surface.

- **Most marine microbes either cannot be cultured or require special culturing methods.** Some can be cultured using conditions that mimic their natural habitat, and allow intimate association with other species that provide growth factors such as siderophores.

- **Plankton are small floating organisms, including swimming microbes.** Phytoplankton are phototrophs such as cyanobacteria and algae. Microbial consumers include protists and viruses. Many marine protists are mixotrophs (producers and consumers in one).

- **Picoplankton include bacteria and small microbial eukaryotes.** They are measured by fluorescence microscopy. Their biochemical rate of production is estimated by uptake of radiolabeled nutrients.

- **Benthic microbes are barophiles.** The seafloor supports psychrophiles, whereas hydrothermal vents support thermophiles. Vents and cold seeps support sulfur- and metal-oxidizing bacteria, sulfur-reducing bacteria, methanogens, and methanotrophs. Bacteria that oxidize H_2S and methane feed symbiotic animals such as tube worms.

- **Freshwater lakes have stratified water levels.** As depth increases, minerals become increasingly reduced. Anaerobic forms of metabolism predominate, with the more favorable alternative electron acceptors used in turn.

- **Lakes may be oligotrophic or eutrophic.** Eutrophic lakes may show such high biochemical oxygen demand (BOD) that the oxygen concentration falls to levels too low to support vertebrate life.

21.6 Soil and Plant Microbial Communities

In contrast to the ocean, where the base of producers is almost entirely microbial, the major producers of terrestrial ecosystems are macroscopic plants. Plants vary greatly in size and form, from mosses and bryophytes to prairie grasses and forest trees, but most terrestrial plants are rooted in **soil**. Soil is a complex mixture of decaying organic and mineral matter that feeds vast communities of microbes. Soil is arguably the most complex microbial ecosystem on our planet. And soil-based agriculture is the major source of food for our planet's human inhabitants. The qualities of a given soil—oxygenated or water saturated, acidic or alkaline, salty or fresh, nutrient-rich or -poor—define what food can be grown and whether the human community will eat or starve.

Soil Microbiology

The general structure of soil (**Fig. 21.32**) includes a series of layers called "horizons" that arise as a result of soil-forming factors such as rainfall, temperature variation, wind, and biological activity. Note that the soils of different habitats, such as prairie, forest, and desert, vary greatly as to the depth and quality of each layer.

The surface layer of soil we see is the organic horizon (O horizon). The organic horizon consists of dark, organic detritus, such as shreds of leaves fallen from plants. The detritus of the organic horizon is in the earliest stages of decomposition by microbes, primarily fungi and bacteria such as actinomycetes. Early-stage decomposition is defined loosely as a state in which the origin of the detritus may be still recognizable.

Beneath the organic horizon lies the lighter-colored aerated horizon (A horizon), in which organic particles in more advanced stages of decomposition combine with minerals from rock at lower levels. In the aerated horizon, the source of the organic particles is no longer recognizable, and decomposers have broken down some of the more difficult-to-digest plant structural components, such as lignin (a complex aromatic polymer found in wood; discussed in the next section). This partly decomposed material is often sold by garden stores as peat or topsoil.

In well-drained soil, both the organic and aerated horizons are full of oxygen, as well as nutrients liberated by the decomposers and used by plants. Soil consists of a complex assemblage of organic and inorganic particles (**Fig. 21.32**). Between the soil particles are air spaces that provide access to oxygen, allowing aerobic respiration. Each particle of soil supports miniature colonies, biofilms, and filaments of bacteria and fungi that interact with each other and with the roots of plants (**Fig. 21.33**).

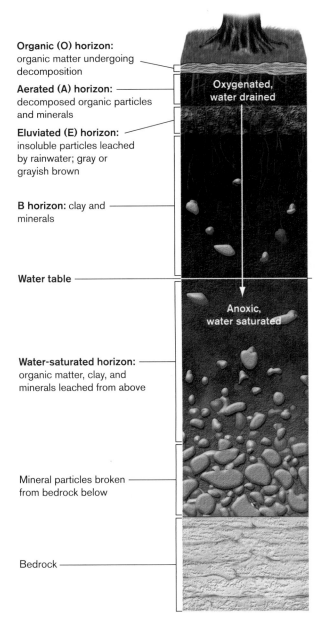

FIGURE 21.32 ■ The soil profile. Soil forms layers in which decomposing organic material predominates at the top, and minerals toward the bottom, at bedrock. The top layers are aerated, providing heterotrophs with access to O_2, whereas the bottom layers are water saturated and anaerobic.

Below the aerated horizon, the eluviated horizon (E horizon) experiences periods of water saturation from rain (see **Fig. 21.32**). Rainwater leaches (dissolves and removes) some of the organic and mineral nutrients from the upper layers. Below the eluviated horizon lie increasing proportions of minerals and rock fragments broken off from bedrock below. These lower water-saturated layers form the **water table**. This anoxic, water-saturated region contains mainly lithotrophs and anaerobic heterotrophs.

The soil layers finally end at bedrock, a source of mineral nutrients such as carbonates and iron. Interestingly, bedrock is permeated with microbes. Core samples show that

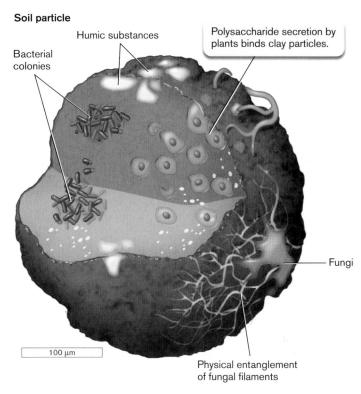

FIGURE 21.33 ■ Microbes in soil and rock. Soil particles support growth of complex assemblages of microbes. A soil particle contains bacterial colonies, biofilm associations, and microbes associated with fungi and plant roots.

crustal rock as deep as 3 km down contains **endoliths**, bacteria growing between crystals of solid rock. What energy source feeds microbes trapped within rock? For some endoliths, a surprising answer may be the radioactive decay of uranium. Uranium-238 decay generates hydrogen radicals that combine to form hydrogen gas. The hydrogen gas combines with CO_2 from carbonate rock, providing an electron donor and a carbon source for methanogens and other endolithic lithotrophs.

The Soil Food Web

The top horizons of soil feature a food web of extraordinary complexity (**Fig. 21.34**). The major producers are green plants, whose leaves generate detritus and whose root systems feed predators, scavengers, and mutualists. Some carbon is also fixed by lithotrophs oxidizing reduced nitrogen (NH_3), hydrogen sulfide (H_2S), and iron (Fe^{2+}). In well-aerated soil, however, the proportion of carbon fixed by lithotrophs is small compared to that fixed by plants.

Plant matter is decomposed by many species of fungi, such as *Mycena* species, and by bacteria such as the actinomycetes. The actinomycetes include *Streptomyces*, a genus famous for the production of antibiotics and for generating chemicals whose odors give soil its characteristic smell

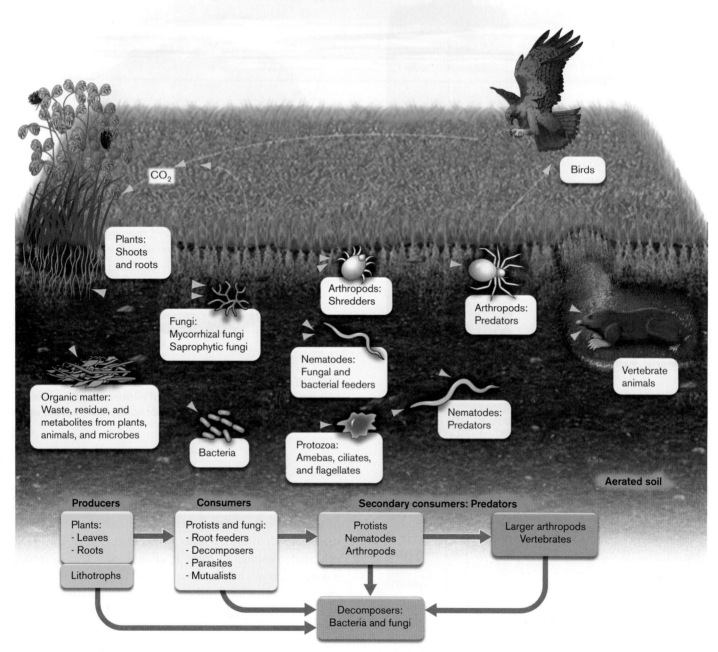

FIGURE 21.34 ■ The soil food web (aerated zone). Plants are the major producers, although some production also occurs from lithotrophs such as ammonia oxidizers. Detritus from plants is decomposed by fungi and bacteria, which feed protists and small invertebrates such as nematodes. Protists and small invertebrates are consumed by larger invertebrates and vertebrate animals.

(**Fig. 21.35A**). Besides fallen leaves, another source of organic matter from plants is the **rhizosphere**, the region of soil surrounding plant roots. The rhizosphere contains proteins and sugars released by roots, as well as sloughed-off plant cells. These materials feed large numbers of bacteria, which then cycle minerals back to the plant. Bacteria in the rhizosphere may also discourage growth of plant pathogens.

In the aerated zone, bacteria feeding on leaf detritus and root exudates are then preyed on by protists and

nematodes. Many complex interactions occur. The nematode *Heterorhabditis bacteriophora* (**Fig. 21.35B**) carries symbiotic *Photorhabdus luminescens* bacteria, which it alternately consumes and transmits as a mutualist when it infects an insect. *Vampirella* protists drill holes in fungal hyphae to suck out their nutrients. Parasitic fungi prey on plants or invertebrates; some actually capture and strangle nematodes. Mycorrhizal fungi extend the absorptive surface area of plant roots. Microbes ultimately feed invertebrates, which then feed larger invertebrates and vertebrate

A. *Streptomyces griseus*

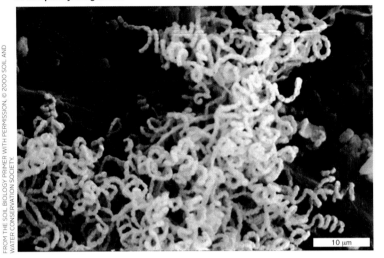

B. Nematode with symbiotic bacteria

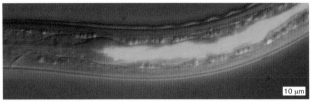

FIGURE 21.35 ▪ Soil microbes. A. Hyphae of actinomycete bacteria, *Streptomyces griseus*. *Streptomyces* bacteria give the soil its characteristic odor. **B.** The nematode *Heterorhabditis bacteriophora* carries luminescent *Photorhabdus luminescens* bacteria, which it alternately consumes and transmits to a host insect.

A. Lignin

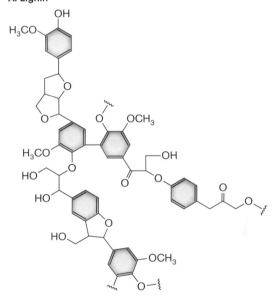

B. White rot fungi

FIGURE 21.36 ▪ Microbial degradation of lignin. A. Lignin is a complex organic polymer that is a component of wood and bark. **B.** *Xylobolus frustulatus*, a white rot fungus, growing on a willow log. The fungus degrades lignin.

predators. Some predators, such as earthworms and burrowing animals, enhance the soil quality by turning over the matter, thus aerating the soil particles and helping to mix the organic matter from above with the mineral particles from below.

A critical role of fungal decomposers (also known as saprophytes) is the breakdown of extremely complex structural components of vascular plants such as grass and trees (**Fig. 21.36**). Trees, in particular, accumulate vast stores of biomass in forms that are difficult to digest, such as **lignin**. Lignin is a highly complex and diverse covalent polymer composed of interlinked phenolic groups (benzene rings with OH or related oxygen-bearing side groups) (**Fig. 21.36A**). We saw earlier that termites host protozoa that help degrade the lignin portion of wood particles. In soil, fungi and bacterial decomposers possess enzyme systems to degrade lignin and other complex components of plants. Examples of decomposers include white rot fungi (**Fig. 21.36B**) and actinomycete soil bacteria. The prevalence of lignin is one reason that decomposition by fungi plays a much larger role in terrestrial ecosystems than in marine ecosystems.

The first phase of microbial degradation (about 50% of the carbon) is complete within a year of deposition in the soil. The remaining phenolics, however, may be degraded at less than 5% per year, and some samples dated by ^{14}C isotope ratios have been shown to last 2,000 years. These phenolic molecules are called **humic material** or **humus**. Because of its slow degradation, humic material provides a steady slow-release supply of nutrients for plant growth. But forests whose rate of microbial decomposition is particularly low—for example, the New Jersey Pine Barrens—depend on fire to clear the mounting layers of humus and return its minerals to the ecosystem.

Microbes Associated with Roots

The presence of plant roots provides yet another level of complexity to soil communities (**Fig. 21.37**). Plant roots influence the surrounding soil by taking up nutrients and by secreting organic substances and molecules that modulate their surroundings. The environment surrounding a plant root can be further subdivided into two categories: the rhizoplane, the root surface; and the rhizosphere, the region of soil outside the root surface but still influenced by plant exudates (materials secreted by the plant). Particular bacterial species are adapted to these environments. For example, in anaerobic wetland soil, the rhizoplane and rhizosphere of plant roots provide oxygen for methanotrophs that oxidize methane produced by methanogens.

The rhizoplane and rhizosphere also provide the environment for symbiotic fungi that generate mycorrhizae. At least 80% of plants in nature, including 90% of forest trees, require mycorrhizae for optimal growth.

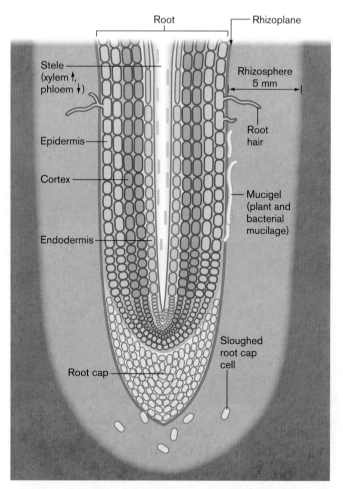

FIGURE 21.37 ■ Plant roots offer special habitats for microbes. The rhizoplane is the region of soil directly contacting the plant root surface. The rhizosphere is the soil outside the rhizoplane that receives substances from the root, such as mucilage, sloughed cells, and exudates.

Mycorrhizae: The Fungal Internet

The function and significance of mycorrhizae for plant growth is just beginning to be understood. **Mycorrhizae** (singular, **mycorrhiza**; from *myco*, "fungal," and *rhiza*, "root") consist of fungal mycelia that associate intimately with the roots of plants, extending access to minerals while obtaining in return the energy-rich products of plant photosynthesis. Mycorrhizae were first discovered in the 1880s by German truffle hunters who sought to cultivate the prized delicacy, the fruiting body of an ascomycete (for review of fungi, see Chapter 20). The propagation of truffles was investigated by mycologist Albert Frank at the Agricultural University of Berlin. To his surprise, Frank found that the truffles extended their mycelia far beyond the site of the fruiting body, and that the mycelia formed an impenetrable tangle with plant roots. Frank called the tangled mycelia "fungus-roots" or mycorrhizae. A century later, we are beginning to appreciate that these mysterious fungus-root tangles offer a vast interconnected network for exchange of nutrients among fungi and many different plants, like an internet connecting countless sites.

Two different kinds of mycorrhizae are observed: ectomycorrhizae and endomycorrhizae. **Ectomycorrhizae** colonize the rhizoplane, the surface of plant rootlets—the most distal part of plant roots (**Fig. 21.38**). The fungal mycelia never penetrate the root cells. They form a thick mantle surrounding the root and growing between the root cells, and then extend long mycelia away from the root to absorb nutrients. Numerous kinds of fungi form ectomycorrhizae, including ascomycetes (such as truffles) and basidiomycetes, known by their mushrooms (such as stinkhorns). Plants grown with ectomycorrhizae invest less of their body mass in roots and more in the aboveground stems and leaves—an important consideration for agriculture, in which the aboveground plant is usually the part harvested.

Endomycorrhizae form a more intimate association, in which the fungal hyphae penetrate plant cells deep within the cortex (**Fig. 21.39**). The penetrating hyphae form knobbed branches that resemble microscopic "trees," or arbuscules, within the root cells. Some of the hyphae form specialized vesicles within the plant that store nutrients. Another name for this kind of mycorrhizae is **vesicular-arbuscular mycorrhizae** (**VAM**).

Endomycorrhizae are more specialized than ectomycorrhizae. They comprise a relatively small number of fungal species, such as members of the Glomeromycota genus *Glomus*, and they show obligate dependence on their host plants. Their presence in nature and their importance in the ecosystem, however, are actually greater. Endomycorrhizal species exist entirely underground (they do not form mushrooms), and they completely lack sexual cycles. They

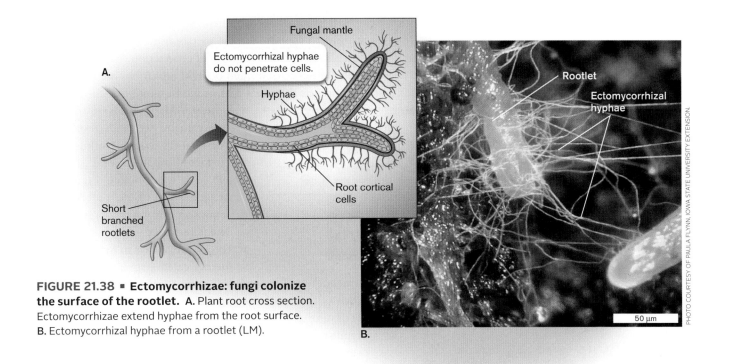

FIGURE 21.38 ■ **Ectomycorrhizae: fungi colonize the surface of the rootlet.** **A.** Plant root cross section. Ectomycorrhizae extend hyphae from the root surface. **B.** Ectomycorrhizal hyphae from a rootlet (LM).

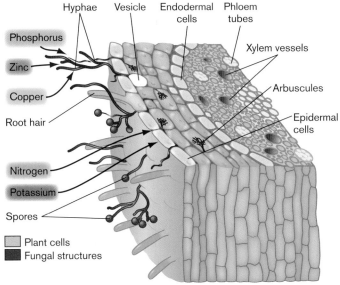

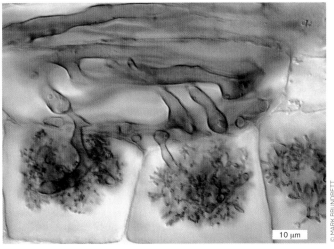

FIGURE 21.39 ■ **Endomycorrhizae: fungi invade root cells, forming arbuscules.** **A.** Plant root cross section. Endomycorrhizal hyphae penetrate cells deep within the root cortex. **B.** The penetrating hypha forms an arbuscule within a root cell.

may acquire 25% of the photosynthetic product of their hosts in exchange for tremendously expanding access to soil resources.

Mycorrhizae greatly enhance the plant's uptake of water, as well as minerals such as nitrogen and phosphorus. In addition, the hyphae sequester toxins, and they actually distribute organic substances from one plant to another.

Mycorrhizae may join many different plants, even different species, in a vast, nutrient-sharing network.

Thought Question

21.9 Design an experiment to test the hypothesis that the presence of mycorrhizae enhances plant growth in nature.

A.

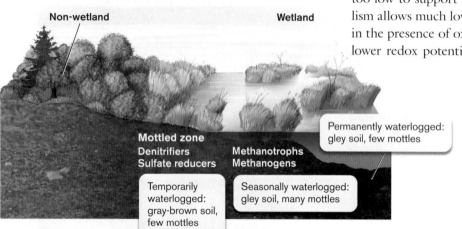

Non-wetland Wetland

Mottled zone
Denitrifiers
Sulfate reducers

Methanotrophs
Methanogens

Permanently waterlogged:
gley soil, few mottles

Temporarily
waterlogged:
gray-brown soil,
few mottles

Seasonally waterlogged:
gley soil, many mottles

B.

COURTESY OF CHIEN-LU PING, UNIV. OF ALASKA, ANCHORAGE.

1 cm

FIGURE 21.40 ■ Wetland soil. A. A true wetland experiences periods of water saturation alternating with dry soil. Soil that is water saturated becomes anaerobic because O_2 diffuses slowly through water. **B.** Alternating periods of water saturation and dryness give the soil a mottled color. The red-orange-colored portions around root holes result from oxidized iron (Fe^{3+}), whereas the gray portions ("gley") indicate that water has washed iron away after reduction (Fe^{2+}).

Wetland Soils

So far, we have considered the interface between ground and water (the benthic sediment beneath oceans and lakes) and the interface between ground and air (aerated soil). An interesting case that combines the two is wetlands (**Fig. 21.40A**). A **wetland** is defined as a region of land that undergoes seasonal fluctuations in water level, so that sometimes the land is dry and oxygenated, and at other times water saturated and anaerobic. Wetlands provide many crucial functions; for example, the Everglades filter much of the water supply for Florida communities.

Wetland soil that undergoes such periods of anoxic water saturation is known as **hydric soil**. Hydric soil is characterized by "mottles," patterns of color and paleness (**Fig. 21.40B**). The reddish brown portions (for example, surrounding a plant root) result from oxidized iron (Fe^{3+}). The gray portions indicate loss of iron in its water-soluble reduced form (Fe^{2+}) generated by anaerobic respiration.

Soil becomes anoxic when the rate of oxygen diffusion is too low to support aerobic metabolism. Anaerobic metabolism allows much lower rates of production than metabolism in the presence of oxygen because anaerobes use oxidants of lower redox potential and limited quantity, such as sulfate and nitrate. Many kinds of anaerobic bacteria inhabit wetlands. For example, denitrifiers (bacteria using nitrate to oxidize organic food) remove nitrate from water before it enters the water table—one of the ways that wetlands protect our water supply.

The alternative oxidants (electron acceptors) are always in limited supply, so further catabolism is carried out by fermentation. Fermentation allows only incomplete breakdown of food molecules, generating a rich and diverse supply of nutrients for a variety of consumers, including aerobic organisms when the water recedes. Thus, despite its lower overall productivity, anaerobic soil contributes to the nutritional diversity of wetland ecosystems. The relatively slow rate of decomposition can lead to accumulation of high levels of organic carbon, particularly rich for plant growth.

The anoxic conditions of wetlands also favor methanogenesis. Methanogenesis is performed solely by archaea (discussed in Chapters 14 and 19). Methanogenesis occurs when fermenting bacteria generate H_2, CO_2, and other one- or two-carbon substrates that methanogens convert to methane. Methane is a more potent greenhouse gas than CO_2, and although the current methane concentration in our atmosphere is low, it is rising exponentially.

In water-filled anaerobic soils, particularly those of rice fields, significant quantities of methane escape to the air through air-conducting channels in the roots of the rice plant. The roots of rice plants, like those of other wetland vascular plants, contain channels to carry oxygen. These same channels, however, allow methane to escape from anaerobic soil. Fortunately, the rhizosphere of the roots supports methanotrophs that oxidize methane, so research is being done to maximize methanotroph activity and minimize the release of methane. There is some evidence that natural wetlands actually take in more carbon than they put out, whereas disturbed wetlands (wetlands altered by human activity) generate net efflux of CO_2 and CH_4. The role of wetlands in global cycling is discussed further in Chapter 22.

Plant Endophytic Communities

The plant interior provides a special home for microbes termed **endophytes**. The plant's vascular system of phloem tubes conducts photosynthesized sugars down from the leaves, while xylem tubes bring water and minerals up from the roots. Unlike the sterile blood vessels of animals, plant transport vessels are normally colonized by endophytic fungi

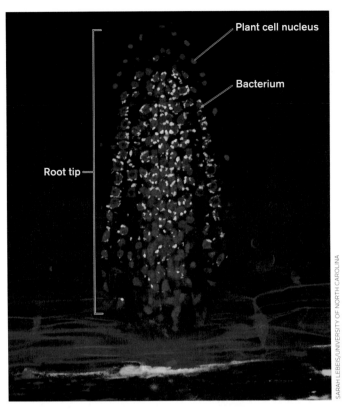

FIGURE 21.41 ■ **Endophytic bacteria within plants.** Communities of bacteria colonize a new root tip of the plant *Arabidopsis thaliana*. The bacteria (green) and the root cell nuclei (blue) are visualized by FISH with probes hybridized to rRNA, using confocal laser scanning microscopy. Endophytes include species of Actinobacteria and Proteobacteria that may help the plant resist pathogens and survive environmental stresses such as high salt or temperature. Distinct clades of bacteria grow on the root surface or within the plant tissues. *Source: Derek Lundberg et al. 2012. Nature* **488**:86.

and bacteria. Thus, when we eat plants, we also consume all their endophytes, or endophytic microbes (**Fig. 21.41**).

Some endophytes are obligate (can live only within the plant), whereas others, such as pseudomonads, have alternative lifestyles in the soil. The microbial partners may grow as mutualists, commensals, or parasites. Plants tolerate endophytes because they confer substantial benefits. For example, prairie grasses called fescue, grazed by cattle in the southeastern United States, host fungal epiphytes. The fungi, *Neotyphodium coenophialum*, produce alkaloids that deter insect predators, pathogens, and root-feeding nematodes. Unfortunately, some of the alkaloids poison cattle, but agricultural scientists have engineered fungal strains that still protect the plant from pathogens while allowing cattle to graze.

Human pathogens such as *E. coli* O157:H7 and *Salmonella enterica* can grow endophytically in crop plants such as spinach and alfalfa. Endophytic pathogens pose a problem for the food industry because the bacteria cannot be "washed off" from raw produce. On the other hand, endophytes such as *Stenotrophomonas* species have potentially valuable uses. *Stenotrophomonas* bacteria produce enzymes that deter many plant pathogens. The bacteria also absorb and concentrate toxic metals such as arsenic, suggesting a possible use for bioremediation of metal-contaminated soil. They produce promising antibiotics and proteases for cleansing agents. Other kinds of endophytes protect plants from heat, salt, and drought—contributions of growing importance as our global climate changes.

Rhizobia Fix Nitrogen for Legumes

A form of bacteria-plant mutualism critical for agriculture is that of nitrogen fixation by **rhizobia** (singular, **rhizobium**), a group of soil-dwelling Alphaproteobacteria (discussed in Chapter 18). Major rhizobial genera include *Rhizobium*, *Bradyrhizobium*, and *Sinorhizobium*. Rhizobia, associated with legumes such as peas and beans, fix more nitrogen than the plants absorb from soil, actually increasing the soil's nitrogen content. For this reason, farmers often alternate crops such as corn with soybeans to restore nitrogen to the soil. The rhizobial bacteria develop specialized forms within plant cells, called **bacteroids**. Bacteroids lack cell walls and are unable to reproduce; their function is specialized for nitrogen fixation. The rhizobial infection of root hairs induces the formation of nodules within which the nitrogen-fixing bacteroids are sequestered (**Fig. 21.42**).

Thought Question
21.10 How do you think symbiotic rhizobia reproduce? Why do bacteroids develop if they cannot proliferate?

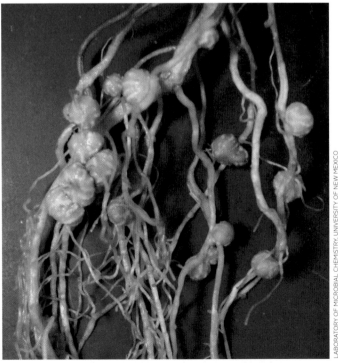

FIGURE 21.42 ■ **Rhizobium nodules on pea plant roots.** Rhizobia induce legume roots to form nitrogen-fixing nodules.

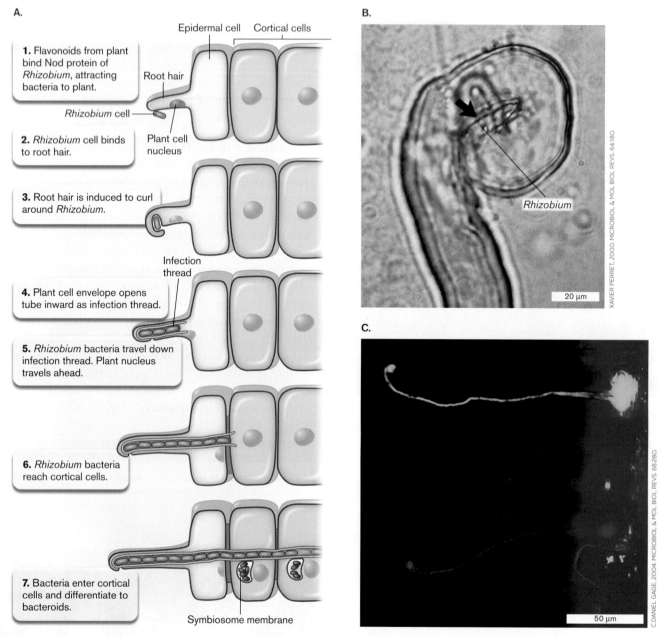

A.

1. Flavonoids from plant bind Nod protein of *Rhizobium*, attracting bacteria to plant.

Epidermal cell   Cortical cells

Root hair

Rhizobium cell

2. *Rhizobium* cell binds to root hair.

Plant cell nucleus

3. Root hair is induced to curl around *Rhizobium*.

Infection thread

4. Plant cell envelope opens tube inward as infection thread.

5. *Rhizobium* bacteria travel down infection thread. Plant nucleus travels ahead.

6. *Rhizobium* bacteria reach cortical cells.

7. Bacteria enter cortical cells and differentiate to bacteroids.

Symbiosome membrane

B.

Rhizobium

20 µm

XAVIER PERRET, 2000. MICROBIOL. & MOL BIOL. REVS. 64:180.

C.

50 µm

C. DANIEL GAGE, 2004. MICROBIOL. & MOL. BIOL. REVS. 68:280.

FIGURE 21.43 ▪ The infection thread. A. Rhizobia are attracted to the legume by chemotaxis toward exuded flavonoids. A bacterium induces an epidermal root hair to curl around it and take it up into the infection thread, a tube of plant cell wall material. The thread eventually penetrates cortical cells, where the bacteria lose their cell walls and become nitrogen-fixing bacteroids. **B.** Root hair curling around *Rhizobium*, and the formation of an infection thread (LM). **C.** Infection threads invading the cortex (fluorescence microscopy). Bacteria express either DsRed (pink) or green fluorescent protein (green).

The initiation, development, and maintenance of the rhizobia-legume symbiosis poses intriguing questions of genetic regulation. How does the association begin? How do host plant and bacterium recognize each other as suitable partners? The legume exudes signaling molecules called flavonoids into its rhizosphere. Flavonoids resemble steroid hormones such as estrogen and have similar effects on animals; they are also called phytoestrogens. The flavonoids are detected by rhizobial bacteria, which respond by chemotaxis, swimming toward the root surface. Flavonoids then induce bacterial expression of Nod factors, molecules

composed of chitin with lipid attachments. Nod factors communicate with the host plant and help establish species specificity between bacterium and host.

The entry of the bacteria into the host involves a fascinating interplay between bacterial and plant cells (**Fig. 21.43A**). First, a bacterium is attracted by flavonoids to the surface of a root hair extended by a root epidermal cell (**Fig. 21.43A**, step 1). The bacterial Nod factor induces the root hair to curl around it and ultimately surround the bacterium with plant cell envelope (steps 2 and 3). The bacterium then induces growth of a tube poking into the plant cell

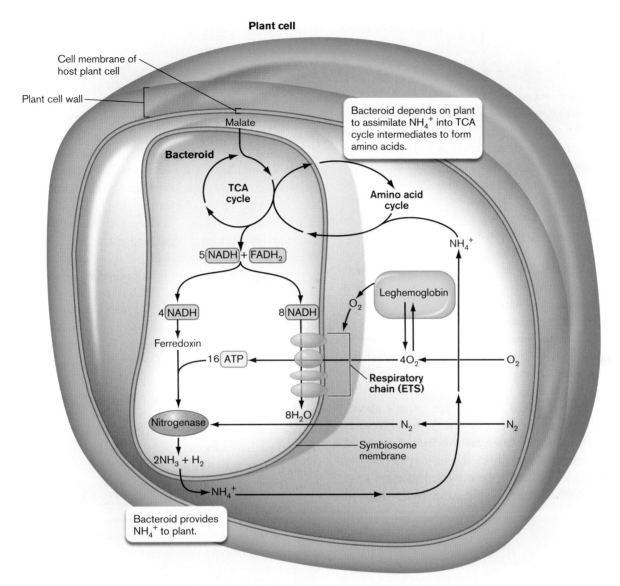

FIGURE 21.44 ■ Energy and oxygen regulation during nitrogen fixation. The bacteroid receives photosynthetic products from the plant, such as malate and oxygen, to generate ATP for nitrogen fixation. The amount of oxygen is regulated closely by leghemoglobin. The bacteroid provides nitrogen (fixed as ammonium ion) to the plant cell.

(step 4). The tube growth is directed by the plant nucleus, which migrates toward the plant cortex (step 5). As the tube grows, bacteria proliferate, forming a column of cells that projects down the tube (steps 6 and 7). This column of cells is known as the infection thread. The infection thread can be visualized by light microscopy (**Fig. 21.43B**).

As the infection thread develops, signals from the bacteria induce the cortical cells (below the epidermis) to prepare to receive the bacteria. The bacteria induce further tube formation into the cortical cells and continue penetration as they grow. The penetration of the infection thread into the cortex is shown in fluorescence micrographs (**Fig. 21.43C**) in which the bacteria are engineered to express a fluorescent protein.

The cortical cells invaded by bacteria are induced to proliferate in an organized manner, forming nodules. Within the nodules, most of the infecting bacteria differentiate into wall-less bacteroids that will fix nitrogen. A few bacteria fail to differentiate; their fate is unclear. The bacteroids remain sequestered within a sac of plant-derived membrane known as the symbiosome. The symbiosome membrane contains special transporters that mediate the exchange of nutrients between the bacteroid and its host cell, sustaining bacteroid metabolism while preventing harm to the host.

From the plant cytoplasm, the bacteroid receives catabolites such as malate, which enter the TCA cycle and donate electrons for respiration. The oxygen for respiration comes from the plant's photosynthesis, regulated by leghemoglobin to maintain levels low enough to allow nitrogen fixation (**Fig. 21.44**). Nitrogen fixation consumes about a fifth of the plant's photosynthetic products. As discussed in Chapter 15, nitrogen is fixed by the nitrogenase enzyme into ammonium ions:

$$N_2 + 10H^+ + 8e^- \longrightarrow 2NH_4^+ + H_2$$

The reaction requires expenditure of eight NADPH or NADH, plus 16 ATP, which are generated by respiration. But respiration requires oxygen, which poisons nitrogenase. Thus, oxygen needs to be delivered to the bacteroid only as needed, and in an amount just enough to run respiration. The oxygen is sequestered and brought to the bacteroid by leghemoglobin, an iron-bearing plant protein related to blood hemoglobin.

Overall, the nitrogen fixation symbiosis is kept in balance by several regulatory mechanisms. The presence of ammonium or nitrate ions inhibits symbiosis and nitrogen fixation. The bacteroids cannot synthesize their own amino acids; instead, they must provide ammonium to the plant cytoplasm for assimilation into amino acids, some of which cycle back to the bacteroid. But what is known of regulation is dwarfed by the unanswered questions: How is the infection thread formed? How does the plant allow infection while preventing uncontrolled growth of bacteria? What determines how much of the plant's photosynthetic products are harvested by the bacteria? Why is hydrogen gas released, and how can this loss of potential energy be prevented? What limits the host specificity of rhizobia to legumes, and can it be extended to other crop plants, such as corn? Research on these questions is critical for agriculture.

Thought Question

21.11 High levels of nitrate or ammonium ion corepress the expression of Nod factors (see **Fig. 21.43**). What is the biological advantage of Nod regulation?

Plant Pathogens

We have seen many ways in which bacteria and fungi interact positively with plants, but other species act as pathogens. In any environment, pathogens are always outnumbered by the vast community of neutral or helpful microbes. Nevertheless, when a pathogen does colonize a plant, its growth can have effects ranging from minimal to devastating (**Fig. 21.45**). A relatively harmless plant virus was associated with a famous historical phenomenon: the sixteenth-century tulip craze in the Netherlands. The virus caused streaking of tulip petals (**Fig. 21.45A**), a pattern much admired by tulip fanciers. Other viruses, however, can cause devastating blights and epidemics. (For more on viruses, see Chapters 6 and 11.)

A pathogenic relative of rhizobia is *Agrobacterium tumefaciens*, a bacterium whose DNA transforms plant cells to form crown gall tumors (**Fig. 21.45B**). *A. tumefaciens* has an unusually broad host range, and its natural genetic transformation system has been applied widely for commercial plant engineering (discussed in Chapter 16). The tumors remain largely confined and have relatively little effect on plant growth. Other bacterial pathogens, particularly species of *Erwinia* and *Xanthomonas*, severely damage plants.

The most common plant pathogens are fungi. Fungal diseases such as anthracnose (**Fig. 21.45C**) cause substantial losses in agriculture, affecting cucumbers, tomatoes, and other vegetables. Dutch elm disease, which has wiped out nearly all the native elms of the United States, is caused by the fungus *Ophiostoma novo-ulmi*. The fungus is carried by bark beetles, which bore into the xylem, damaging the plant's transport vessels and allowing access for fungal spores.

Some fungal pathogens generate specialized structures to acquire nutrients from plants. As a hypha grows across the plant epidermis, its tip can penetrate the plant cell

---

A. Virus infection of tulips

COURTESY KATHY NICHOLS

B. Crown gall tumor on rose stem

ALAMY

C. Anthracnose fungus on white oak leaf

PAULA FLYNN, IOWA STATE EXTENSION SERVICE

FIGURE 21.45 ■ Plant diseases range from innocuous to devastating. A. Striped tulips result from a virus. **B.** Crown gall tumor on rose stem, caused by *Agrobacterium tumefaciens*. **C.** White oak leaf spotted by anthracnose fungus.

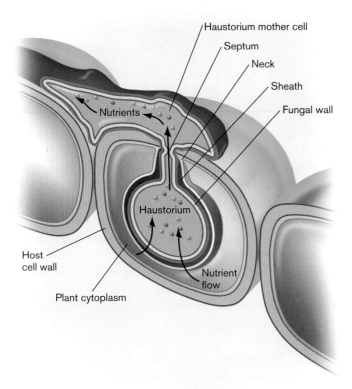

FIGURE 21.46 ■ **A fungal pathogen inserts haustoria into plant cells.** The haustorium is surrounded by an invagination of the plant's cell membrane. Plant nutrients such as sucrose flow into the haustorium and are transferred out to the fungal mycelia.

wall, followed by ingrowth of a bulbous extension called a **haustorium** (plural, **haustoria**) (**Fig. 21.46**). The haustorium never penetrates the plant cell membrane, thus avoiding leakage and loss of plant cytoplasm. Instead, it causes the membrane to invaginate, while expanding into the volume of the plant cell. The haustorium takes up nutrients such as sucrose, generated by adjacent chloroplasts. Depending on the species of fungus, haustorial parasitism can lead to mild growth retardation, or it can rapidly kill the plant.

Thought Question

21.12 Compare and contrast the processes of plant infection by rhizobia and by fungal haustoria.

To Summarize

- **The uppermost horizons of soil** consist of detritus and are largely aerated. Below the aerated layers, the eluviated horizon experiences water saturation. Lower layers are anoxic.

- **The soil food web** includes a complex range of microbial producers, consumers, predators, decomposers, and mutualists. Microbial mats and biofilms are complex multispecies communities of bacteria, fungi, and other microbes.

- **Fungi decompose lignin**, a complex aromatic tree component that is challenging to digest. Lignin decomposition forms humus.

- **Certain fungi form symbiotic associations** with plant roots called mycorrhizae. Mycorrhizae transport soil nutrients among many different kinds of plants.

- **Wetland soils** alternate aerated (dry) with anoxic (water-saturated) conditions. Anoxic wetland soil favors methanogenesis. Wetland soils are among the most productive ecosystems.

- **Endophytes** are bacteria or fungi that grow within plant transport vessels, conferring benefits such as resistance to pathogens.

- **Rhizobia induce legume roots** to nodulate for nitrogen fixation. The bacteria enter the root as infection threads. Some of the bacteria enter root cells and develop into nitrogen-fixing bacteroids, which gain energy from plant cell respiration but must remain anaerobic.

- **Plant pathogens** include bacteria, fungi, and viruses. Some pathogens only mildly affect the plant, whereas others cause devastation. Some fungi invade plants using haustoria, which grow into the plant cell by penetrating the cell wall and invaginating the cell membrane to avoid leakage of cytoplasm.

Concluding Thoughts

This chapter has introduced the challenge of characterizing microbial ecosystems, with the promise and pitfalls of the metagenomes. We have seen how microbes colonize a vast array of habitats, from the deserts and oceans to the roots of plants and the digestive tracts of animals. Microbes are found from the upper layers of our planet's atmosphere down to the deepest rock strata we can reach. Wherever found, microbes both respond to and modify the environment that surrounds them. Thus, microbes both fill available niches and construct their own. While this chapter has focused on food webs in local habitats, Chapter 22 takes a global perspective of microbial ecology and its roles in the cycling of Earth's essential elements, such as nitrogen and iron. We will see how microbes interact with global climate change and examine the evidence for existence of microbial life beyond Earth.

CHAPTER REVIEW

Review Questions

1. How do we analyze a metagenome of a microbial community? What are the advantages and pitfalls of metagenomics, compared to culturing microbes?
2. What unique functions do microbes perform in ecosystems?
3. Explain the difference between carbon assimilation and dissimilation.
4. What kinds of microbial metabolism are favored in aerated environments? In anoxic environments?
5. What are examples of microbial producers in ecosystems? Include phototrophs as well as lithotrophs. Can a microbe be both a producer and a consumer? Explain.
6. Explain the microbial relationships in various forms of symbiosis, including mutualism, commensalism, and parasitism. Outline an example of each, detailing the contributions of each partner.
7. Compare and contrast the digestive communities of the bovine rumen and the human colon with respect to taxa, metabolic pathways, and contributions to the host.
8. Compare and contrast the marine food web with the soil food web. What kinds of organisms are the producers and consumers? How many trophic levels are typically found?
9. Explain how microbes interact with each other in marine and soil habitats. Are these habitats typically uniform or patchy? How does "patchiness" affect microbial growth?
10. Compare and contrast the roles of microbes in photic and aphotic marine communities.
11. Compare and contrast the microbial activities in aerated and waterlogged soils.
12. Explain how anaerobic microbial metabolism can enrich soil for plant cultivation.
13. What are mycorrhizae, and how are they important for plant growth?
14. Explain how bacterial mutualists fix nitrogen for plants.

Thought Questions

1. Explain what you can learn about a marine microbial community from a metagenome, as compared with a metatranscriptome. How and why might the two approaches yield different results?
2. Explain, with specific examples, what mutualism and parasitism have in common, and how they differ.

Describe an example of a relationship that combines aspects of both.
3. The photic zone and the benthic zone pose challenges and opportunities for marine microbes. What challenges do they have in common, and how do they differ?

Key Terms

algal bloom (849)
amensalism (835)
assembly (821)
assimilation (828)
bacteroid (859)
barophile (848)
benthic organism (842)
bin, binning (822)
biochemical oxygen demand (biological oxygen demand) (BOD) (842)
biomass (828)

cannulated cow (839)
coastal shelf (842)
commensalism (835)
community (816)
computational pipeline (822)
consumer (829)
contig (821)
coral bleaching (834)
decomposer (829)
detritus (829)
dissimilation (828)
ecosystem (816)

ectomycorrhizae (856)
endolith (853)
endomycorrhizae (856)
endophyte (858)
enrichment culture (827)
euphotic zone (842)
eutrophic (849)
extremophile (830)
fistulated cow (839)
fluorescence in situ hybridization (FISH) (824)
food web (828)

grazer (829)
haustorium (863)
holobiont (837)
humic material (humus) (855)
hydric soil (858)
hydrothermal vent (849)
lichen (834)
lignin (855)
limiting nutrient (849)
littoral zone (849)
marine snow (845)
metagenome (816)
metaproteomics (826)
metatranscriptomics (826)
microbiome (818)
microplankton (845)
mixotroph (846)
mutualism (834)

mycorrhizae (856)
niche (826)
niche construction (826)
oligotrophic (849)
operational taxonomic unit
   (OTU) (820)
parasitism (835)
pelagic zone (842)
photic zone (842)
phytoplankton (846)
plankton (845)
population (816)
predator (829)
primary producer (828)
psychrophile (848)
rarefaction curve (820)
rhizobium (859)
rhizosphere (854)

rumen (839)
scaffold (822)
soil (852)
symbiosis (831)
synergism (835)
syntrophy (838)
target community (818)
thermal vent (848)
thermocline (842)
thermophile (849)
trophic level (828)
uncultured (844)
vesicular-arbuscular mycorrhizae
   (VAM) (856)
viable but nonculturable
   (VBNC) (844)
water table (853)
wetland (858)

Recommended Reading

Azam, Farooq, and Francesca Malfatti. 2007. Microbial structuring of marine ecosystems. *Nature Reviews. Microbiology* **5**:782–792.

Brune, Andreas. 2014. Symbiotic digestion of lignocellulose in termite guts. *Nature Reviews. Microbiology* **12**:168–180.

Conway, Tyrrell, and Paul S. Cohen. 2014. Commensal and pathogenic *Escherichia coli* metabolism in the gut. *Microbiology Spectrum* **3**:1–15.

DeLong, Edward F. 2009. The microbial ocean from genomes to biomes. *Nature* **459**:200–212.

Donaldson, Gregory P., S. Melanie Lee, and Sarkis K. Mazmanian. 2016. Gut biogeography of the bacterial microbiota. *Nature Reviews. Microbiology* **14**:20–32.

Dubilier, Nicole, Claudia Bergin, and Christian Lott. 2008. Symbiotic diversity in marine animals: The art of harnessing chemosynthesis. *Nature Reviews. Microbiology* **6**:725–740.

Evans, Paul N., Donovan H. Parks, Grayson L. Chadwick, Steven J. Robbins, Victoria J. Orphan, et al. 2015. Methane metabolism in the archaeal phylum Bathyarchaeota revealed by genome-centric metagenomics. *Science* **350**:434–438.

Fierer, Noah, and Robert B. Jackson. 2006. The diversity and biogeography of soil bacterial communities. *Proceedings of the National Academy of Sciences USA* **103**:626–631.

Gage, Daniel J. 2004. Infection and invasion of roots by symbiotic, nitrogen-fixing rhizobia during nodulation of temperate legumes. *Microbiology and Molecular Biology Reviews* **68**:280–300.

Huq, Anwar, Mohammed Yunus, Syed Salahuddin Sohel, Abbas Bhuiya, Michael Emch, et al. 2010. Simple sari filtration is sustainable and continues to protect villagers from cholera in Matlab, Bangladesh. *mBio* **1**:e00034-10.

Hurwitz, Bonnie L., and Matthew B. Sullivan. 2013. The Pacific Ocean Virome (POV): A marine viral metagenomic dataset and associated protein clusters for quantitative viral ecology. *PLoS One* **8**:e57355.

Kang, Dongwan D., Jeff Froula, Rob Egan, and Zhong Wang. 2015. MetaBAT, an efficient tool for accurately reconstructing single genomes from complex microbial communities. *PeerJ* **3**:e1165.

Kiers, E. Toby, Marie Duhamel, Yugandhar Beesetty, Jerry A. Mensah, Oscar Franken, et al. 2011. Reciprocal rewards stabilize cooperation in the mycorrhizal symbiosis. *Science* **333**:880–882.

Mariat, Denis, Olivier Firmesse, Florence Levenez, Valeria D. Guimarães, Harry Sokol, et al. 2009. The Firmicutes/Bacteroidetes ratio of the human microbiota changes with age. *BMC Microbiology* **9**:123.

Meadow, James, Adam E. Altrichter, Ashley C. Bateman, Jason Stenson, G. Z. Brown, et al. 2015.

Humans differ in their personal microbial cloud. *PeerJ* 3:e1258.

Parniske, Martin. 2008. Arbuscular mycorrhiza: The mother of plant root endosymbioses. *Nature Reviews. Microbiology* 6:763–775.

Redmond, Molly C., and David L. Valentine. 2012. Natural gas and temperature structured a microbial community response to the *Deepwater Horizon* oil spill. *Proceedings of the National Academy of Sciences USA* 109:20292–20297.

Rodrigues, Jorge L. M., Vivian H. Pellizari, Rebecca Mueller, Kyunghwa Baek, Ederson da C. Jesus, et al. 2013. Conversion of the Amazon rainforest to agriculture results in biotic homogenization of soil bacterial communities. *Proceedings of the National Academy of Sciences USA* 110:988–993.

Sato, Tomoyuki, Yuichi Hongoh, Satoko Noda, Satoshi Hattori, Sadaharu Ui, et al. 2009. *Candidatus* Desulfovibrio trichonymphae, a novel intracellular symbiont of the flagellate *Trichonympha agilis* in termite gut. *Environmental Microbiology* 11:1007–1015.

Sharon, Itai, Michael Kertesz, Laura A. Hug, Dmitry Pushkarev, Timothy A. Blauwkamp, et al. 2015. Accurate, multi-kb reads resolve complex populations and detect rare microorganisms. *Genome Research* 25:534–543.

Sunagawa, Shinichi, Luis Pedro Coelho, Samuel Chaffron, Jens Roat Kultima, Karine Labadie, et al. 2015. Structure and function of the global ocean microbiome. *Science* 348:1261359.

Thaiss, Christoph A., David Zeevi, Maayan Levy, Gili Zilberman-Schapira, Jotham Suez, et al. 2014. Transkingdom control of microbiota diurnal oscillations promotes metabolic homeostasis. *Cell* 159:514–529.

Vigneron, Adrien, Perrine Cruaud, Patricia Pignet, Jean-Claude Caprais, Marie-Anne Cambon-Bonavita, et al. 2013. Archaeal and anaerobic methane oxidizer communities in the Sonora Margin cold seeps, Guaymas Basin (Gulf of California). *ISME Journal* 7:1595–1608.

CHAPTER 22
Microbes in Global Elemental Cycles

Microbes throughout the biosphere recycle carbon, nitrogen, and other elements essential for all life. Through their biochemical transformations, diverse microbial activities largely determine the quality of soil, air, and water. Today, all of these geochemical cycles are altered profoundly by human activity. But microbes can help us manage environmental change. From local wastewater treatment to the control of greenhouse gases, microbes are our hidden partners on Earth. And Earth's microbial cycles lead us to wonder whether biospheres exist on other worlds.

CURRENT RESEARCH highlight

Nitrifying bacteria reshape the nitrogen cycle. Microcolonies of *Nitrospira* (yellow) and other bacteria (green) within nitrifying activated sludge. The microorganisms were detected by fluorescence in situ hybridization with 16S rRNA–targeted oligonucleotide probes specific for *Nitrospira* or all bacteria, respectively. *Nitrospira* in activated sludge are canonical nitrite oxidizers or complete nitrifiers.

Source: Image by Holger Daims.

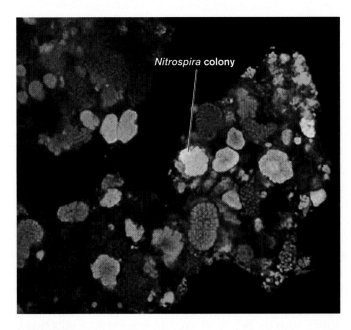

Nitrospira **colony**

AN INTERVIEW WITH

HOLGER DAIMS, MICROBIAL ECOLOGIST, UNIVERSITY OF VIENNA

© ALLEN TSAO

What is surprising about finding the complete nitrification pathway in *Nitrospira*?

Since the first description of nitrifying bacteria in 1892, a central hypothesis of nitrification research has been that different microorganisms catalyze ammonia or nitrite oxidation. Microbiologists speculated about the existence of completely nitrifying microbes, but no such organism was found for more than a century. The discovery of *Nitrospira* species that complete nitrification was a great surprise, especially because *Nitrospira* are widely distributed in nature but were thought to be merely nitrite oxidizers incapable of ammonia oxidation.

How will this discovery revise our view of the global nitrogen cycle?

Nitrification is a key process of the nitrogen cycle in all oxic ecosystems. The identification of complete nitrifiers, which were overlooked for decades, means that current concepts of nitrifier ecology and physiology need a thorough reevaluation. We should take this into account in future studies of microbial ecology, agriculture, wastewater treatment, and global climate change research.

The global cycle of nitrogen impacts every form of life on Earth, and we invest vast resources in nitrogen fertilizers for agriculture. Much of the nitrogen cycle depends on microbes. We continue to discover unexpected microbial contributors, such as the *Nitrospira* species that oxidize ammonia all the way to nitrate (see the Current Research Highlight). Nitrate is a key form of nitrogen that living organisms may assimilate. Surprisingly, we find nitrate on the planet Mars—suggesting a possible source of nitrogen for life, if it should (or did) exist there.

The challenge of global cycles is quantification, adding up all the inputs and outputs of key molecules. Ammonia oxidation is important to quantify because one of the pathway's intermediate forms is nitrous oxide (N_2O), one of three major **greenhouse gases** whose increase correlates with the industrial age (**Fig. 22.1**). Greenhouse gases trap solar radiation as heat—an effect known as the greenhouse effect. Levels of these gases from thousands of years ago are measured from air bubbles trapped in ice in Antarctica. The gases are generated by bacteria, methanogens, and other life-forms, but human technology accelerates their release.

Burning petroleum releases CO_2 from a product that bacteria and plants took millions of years to form. At the same time, phototrophic bacteria and plants fix much of the global carbon dioxide into biomass. But the added human output of CO_2 has outpaced the rate of plant and microbial CO_2 fixation. Global warming accelerates methane release by methanogens. And bacteria oxidize sewage nitrogen to N_2O. Will the human-induced global climate change cause mass extinctions of a majority of Earth's species—as did

the rise of ancient cyanobacteria? Or can we use our knowledge of microbial ecology to channel microbial activities into recovering the balance—for example, by increasing microbial CO_2 fixation?

Throughout most of this book, we present microbial biochemistry in the context of growth of individual organisms. Here in Chapter 22, we show how the collective metabolic activities of microbial populations contribute to global cycles of elements throughout Earth's biosphere. We consider the ways that we humans enlist microbes for **bioremediation**, the use of microbial metabolism to reverse pollution and restore chemical cycles, such as the water cycle in the Florida Everglades (see Section 22.3). Finally, we explore how our awareness of global microbiology has renewed our quest for life beyond Earth. For example, the discovery of nitrate on Mars shows the surprising possibility of nitrogen "fertilizer" available for life on that planet (see Section 22.6).

22.1 Biogeochemical Cycles

Two billion years ago, ancient cyanobacteria began to photolyze water and produce molecular oxygen, as discussed in Chapter 17. Oxygen is a powerful oxidant, lethal to most life in that anoxic era, when all living organisms were anaerobic microbes. To survive, anaerobes had to evolve mechanisms of protection against oxygen, such as antioxidant molecules and enzymes, or else retreat to anoxic habitats. Since then, microbes have shaped our biosphere by releasing oxygen, by fixing nitrogen and returning it to the atmosphere, and by fixing and producing carbon dioxide. These and countless other microbial processes have made Earth's atmosphere what it is, and they continue their roles in its homeostasis.

Microbes Cycle Essential Elements

Microbes have exceptional abilities to interconvert molecules (discussed in Chapter 21). But even microbes face the limits of the actual elements themselves: No metabolism can convert one element to another. Because organisms cannot form their own elements, they need to get them from their environment. While small amounts of matter enter the biosphere from outer space, and gases continually escape Earth, the rates of these changes are tiny compared to the rates of living processes. So, organisms acquire their elements either from nonliving components of their environment, such as by fixing atmospheric CO_2, or from other organisms, by grazing, predation, or decomposition. Furthermore, all organisms recycle their components back to the biosphere. The partners in this recycling include **abiotic** entities such as air, water, and minerals, as well as **biotic** entities such as predators and decomposers. Collectively,

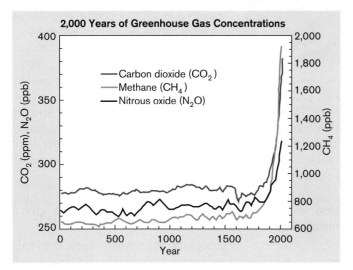

FIGURE 22.1 ■ Global greenhouse gases. Atmospheric levels of CO_2, methane, and nitrous oxide were measured in Antarctic ice cores. The increase of these greenhouse gases since 1750 accompanies the rise in fossil fuel burning and fertilizer-intensive agriculture. *Source:* Modified from U.S. Global Change Research Program. 2009. Global Climate Change Impacts in the United States. Cambridge University Press.

TABLE 22.1	Typical Elemental Composition of Gram-Negative Bacteria
Element	**Dry weight (%)**
Carbon	50
Oxygen	20
Nitrogen	14
Hydrogen	8
Phosphorus	3
Sulfur	1
Potassium	1
Sodium	1
Calcium	0.5
Magnesium	0.5
Chlorine	0.5
Iron	0.2
All others	0.3

the metabolic interactions of microbial communities with the biotic and abiotic components of their ecosystems are known as **biogeochemistry** or **geomicrobiology**.

Which elements need to be recycled and made available for life? In most organisms, six elements predominate: carbon, oxygen, nitrogen, hydrogen, phosphorus, and sulfur (discussed in Chapter 4). **Table 22.1** summarizes the elemental composition typical of Gram-negative bacteria. In other microbes, elemental proportions vary with species and growth conditions. Some species include inorganic components such as the silicate shells of diatoms or the sulfur granules of phototrophic bacteria. All of these elements flow in **biogeochemical cycles** of nutrients throughout the biotic and abiotic components of the biosphere. The environmental levels of key elements can limit biological productivity. For example, the concentration of iron can limit marine populations of phytoplankton, which depend on iron supplied by wind blowing dust off a land mass. Other elements, such as zinc and copper, are **micronutrients**, nutrients required in still smaller amounts. Micronutrients, too, must be cycled, although their flux is challenging to measure.

Sources and Sinks of Essential Elements

Biogeochemical cycles include both biological components (such as phototrophs that consume CO_2 and heterotrophs that release CO_2) and geological components (such as volcanoes that release CO_2 and oceans that absorb CO_2). The major parts of the biosphere containing significant amounts of an element needed for life are called **reservoirs** of that element (**Table 22.2**). Each reservoir acts both as a **source** of that element for living organisms and as a **sink** to which the element returns. For example, the ocean is an important reservoir for carbon (CO_2 in equilibrium with HCO_3^-, bicarbonate ion). The ocean's carbon cycles rapidly; thus, the ocean serves as both a source and a sink for carbon.

As elements cycle from sources to sinks, microbial metabolism generates a series of redox changes (discussed in Chapter 14). The major oxidation states of carbon, nitrogen, and sulfur are summarized in **Table 22.3**. Biospheric

TABLE 22.2	Global Reservoirs of Carbon, Nitrogen, and Sulfur[a]					
Reservoir	**Carbon**	**Rate of cycling**	**Nitrogen**	**Rate of cycling**	**Sulfur**	**Rate of cycling**
Atmosphere	700 (CO_2)	Fast	3,900,000 (N_2)	Slow	0.0014 (SO_2, H_2S)	Fast
Ocean						
Biomass	4	Fast	0.5	Fast	0.15	Fast
Organic molecules	2,100	Fast	300	Fast	—	—
Inorganic molecules	38,000 (HCO_3^-, CO_3^{2-})	Fast	20,000 (N_2) 690 (NO_3^-, NO_2^-, NH_4^+)	Slow Fast	1,200,000 (SO_4^{2-})	Slow
Land						
Biomass	500	Fast	25	Fast	8.5	Fast
Organic matter (soil)	1,200	Fast	110	Slow	16	Fast
Crust (below land and ocean)	120,000,000	Slow	770,000	Slow	18,000,000	Slow
Fossil fuel (oil, coal, natural gas)	13,000	Fast				

Source: R. Maier et al. 2000. *Environmental Microbiology.* Academic Press, San Diego, CA.

[a]Units are 10^9 metric tons, or 10^{12} kg.

TABLE 22.3	Oxidation States of Cycled Compounds					
Oxidation state	**Carbon**		**Nitrogen**		**Sulfur**	
−4	CH_4	Methane				
−3			NH_3, NH_4^+	Ammonia, ammonium ion		
−2	CH_3OH, $(CH_2)_n$	Methanol, hydrocarbon	H_2N-NH_2	Hydrazine	H_2S, HS^-	Sulfides
−1			NH_2OH	Hydroxylamine		
0	$(CH_2O)_n$	Carbohydrate	N_2	Nitrogen	S^0	Elemental sulfur
+1			N_2O	Nitrous oxide		
+2	$CHOOH$	Formic acid	NO	Nitric oxide	$S_2O_3^{2-}$	Thiosulfate
+3			HNO_2, NO_2^-	Nitrous acid, nitrite ion		
+4	CO_2, HCO_3^-	Carbon dioxide, bicarbonate ion	NO_2	Nitrogen dioxide	SO_3^{2-}	Sulfite
+5			HNO_3, NO_3^-	Nitric acid, nitrate ion		
+6					H_2SO_4, SO_4^{2-}	Sulfuric acid, sulfate

carbon can be found as CH_4 generated by methanogens (completely reduced, −4), as CO_2 produced by respiration and fermentation (completely oxidized, +4), or as one of various intermediate states of oxidation. Nitrogen is excreted by many organisms in its most reduced form, ammonia (NH_3), which is protonated to ammonium ion (NH_4^+). Ammonium is oxidized by lithotrophic bacteria through several stages to nitrite (NO_2^-) and nitrate (NO_3^-), which serve as terminal electron acceptors for anaerobic respiration. Sulfides (H_2S, HS^-) serve as electron donors for respiration and sulfur phototrophy, whereas oxidized forms, such as sulfite (SO_3^{2-}) and sulfate (SO_4^{2-}), serve in anaerobic respiration.

Thought Question

22.1 Why is oxidation state critical for the acquisition, usability, and potential toxicity of cycled compounds? Cite examples based on your study of microbial metabolism (see Chapter 14).

How do we study microbial cycling on a global scale? How do we figure out whether ecosystems are net sources or sinks of CO_2? Does microbial activity enhance or limit availability of nitrogen? Rates of flux of elements in the biosphere are very difficult to measure, yet the questions have enormous political and economic implications.

To measure environmental carbon, nitrogen, and other elements, various methods are used. These methods fall under the following categories:

- **Chemical and spectroscopic analysis.** Bulk quantities of CO_2, nitrates, and other chemicals can be determined by sophisticated chemical instrumentation. Atmospheric CO_2 is measured by infrared absorption spectroscopy, applied to samples from towers such as those of the NASA FLUXNET study (**Fig. 22.2**). Gas chromatography is used to separate and quantify various gases, including oxygen, nitrogen, sulfur dioxide, and carbon monoxide. Mass spectrometry detects extremely small quantities of different molecules, even distinguishing between elemental isotopes.

- **Radioisotope incorporation.** The influx and efflux of CO_2 can be measured by the uptake of ^{14}C-labeled substrates in a small, controlled model ecosystem called a **mesocosm**. Alternatively, CO_2 flux can be measured with radioisotope tracers in the field using a field chamber.

- **Stable isotope ratios.** Some enzyme reactions show a preference for one isotope over another, such as ^{14}N versus [15]N. For example, denitrifiers (bacteria that metabolize nitrate) strongly prefer the ^{14}N isotope, leaving behind nitrate enriched in [15]N. The ^{14}N/^{15}N ratio is measured using mass spectrometry. Measuring nitrogen isotope ratios can indicate whether denitrifiers could have conducted metabolism in the sample.

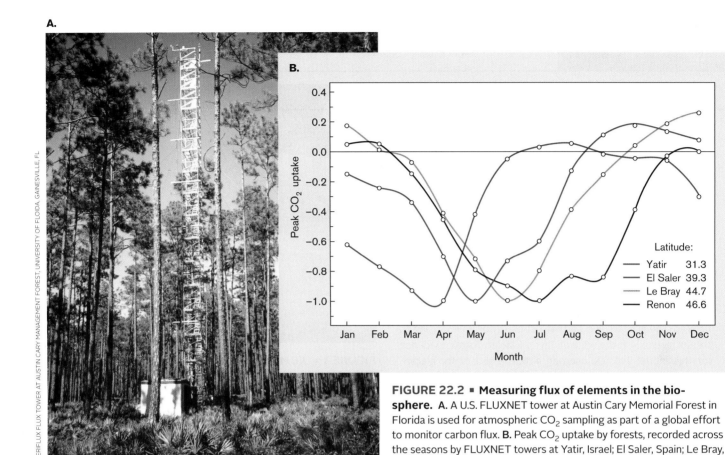

A.

B.

AMERIFLUX FLUX TOWER AT AUSTIN CARY MANAGEMENT FOREST, UNIVERSITY OF FLOIDA, GAINESVILLE, FL

FIGURE 22.2 ▪ Measuring flux of elements in the biosphere. **A.** A U.S. FLUXNET tower at Austin Cary Memorial Forest in Florida is used for atmospheric CO_2 sampling as part of a global effort to monitor carbon flux. **B.** Peak CO_2 uptake by forests, recorded across the seasons by FLUXNET towers at Yatir, Israel; El Saler, Spain; Le Bray, France; and Renon, Italy. *Source:* Kadmiel Maseyk. 2013. *FluxLetter* **5**:15.

Most flux measurements apply to gases, which are the easiest forms to sample, but the data leave unanswered many questions about the deep ocean and subsurface. Subsurface studies require drilling for samples, or more exotic kinds of remote sensing such as airborne imaging of magnetic resistivity to reveal underground hydrology in Antarctica (described in **Special Topic 22.1**).

To Summarize

▪ **Microbes cycle essential elements in the biosphere.** Many key cycling reactions are performed only by bacteria and archaea.

▪ **Elements cycle between organisms and abiotic sources and sinks.** The most accessible source of carbon and nitrogen is the atmosphere. Earth's crust stores large amounts of key elements, but their availability to organisms is limited.

▪ **Environmental flux of elements is measured through chemistry.** Methods include infrared spectroscopy and mass spectrometry, gas chromatography, radioisotope incorporation, and the measurement of isotope ratios.

22.2 The Carbon Cycle and Bioremediation

The foundation of all food webs involves influx and efflux of carbon. The major reservoirs of carbon are shown in **Table 22.2**. In theory, carbonate rock forms the largest reservoir of carbon. But Earth's crust is the source least accessible to the biosphere as a whole. Crustal rock provides carbon only to organisms at the surface and to subsurface microbes that grow extremely slowly. Thus, subsurface carbon turnover is very slow. The carbon reservoir that cycles most rapidly is that of the atmosphere, a source of CO_2 for photosynthesis and chemolithoautotrophy. The atmosphere also acts as a sink for CO_2 produced by heterotrophy and by geological outgassing from volcanoes.

The atmospheric reservoir is much smaller than other sources, such as the oceans, crustal rock, and fossil fuels. For this reason, the industrial burning of fossil fuels has

SPECIAL TOPIC 22.1   An Underground River in Antarctica

Microbial communities shape our planet's geology and geo-chemistry in surprising ways. From a glacier in the Antarctic Taylor Valley spurts a bizarre red stream known as Blood Falls (**Fig. 1**). For several years, Jill Mikucki and her students from the University of Tennessee–Knoxville have investigated Blood Falls and its mysterious microbial inhabitants. The microbes turn out to be lithotrophs, reducing iron (*Shewanella frigidi-marina*) or oxidizing sulfur (*Thiomicrospira arctica*). These iron and sulfur metabolizers are also halophiles, growing in iron-rich brine. The brine concentrated over 5 million years from ancient marine water that once flowed there. The halophilic iron and sulfur metabolizers fix CO_2 into biomass by chemosynthesis, a process that plays a role in our planet's carbon budget.

But where does the Blood Falls brine come from? Mikucki hypothesized that a subsurface river of brine flows deep beneath the glaciers and permafrost of Taylor Valley, rising to the surface only at the edge of Taylor Glacier. To gain evidence for this model, Mikucki used an airborne sensor for electri-cal resistivity, a property that increases as water freezes but decreases with salt concentration. A giant airborne electro-magnetic sensor (**Fig. 2A**) was carried by helicopter above the length and breadth of Taylor Valley (**Fig. 2B**). Traveling back and forth, the sensor mapped electrical resistivity of the geological strata below the lakes (**Fig. 2C**). The mapping showed a dramatic drop in electrical resistivity in sediments 500 meters below the glacier surface, indi-cating salt-saturated groundwater. But this layer of brine rises dramatically at the glacier's forward edge, emerging to the surface at Blood Falls. Thus, the concentrated remains of a million-year-old river flow underground, bearing with them a vast community of iron and sulfur bacteria. The existence of such a community suggests a possible mode of existence for subsurface microbes on the planet Mars.

RESEARCH QUESTION

What is the productivity of subsurface microbes, and how much carbon do they fix in our biosphere?

Mikucki, Jill A., E. Auken, S. Tulaczyk, R. A. Virginia, C. Schamper, et al. 2015. Deep ground-water and potential subsurface habitats beneath an Antarctic dry valley. *Nature Communications* **6**: 6831. doi:10.1038/ncomms7831.

FIGURE 1 ■ Jill Mikucki (left) and student at Blood Falls. The red-colored flow from Taylor Glacier at Lake Bonney is where iron-rich brine wells up from groundwater. Surface tem-perature is commonly 2°C in summer and –40°C in winter.

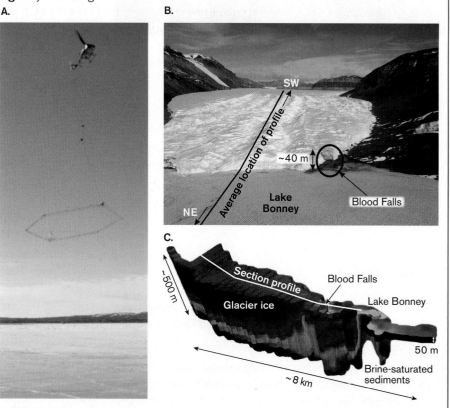

FIGURE 2 ■ Electrical resistivity reveals an underground river. A. A helicopter carries a giant electromagnetic sensor to map subsurface electrical resistivity. **B.** Blood Falls at Taylor Glacier shows iron-rich brine leaking up from an underground source. **C.** A map of electrical resistivity reveals the underground flow of brine. *Source:* (A-C) Jill A. Mikucki, et al. 2015. *Nature Communications.* DOI: 10.1038/ncomms 7831.

perturbed the balance between atmospheric CO_2 and larger reservoirs, such as the ocean. The ocean actually absorbs a good part of the extra CO_2, in equilibrium with carbonates; and microbial photosynthesis traps carbon in biomass. But despite the ocean's carbon capture, atmospheric CO_2 continues to increase at an annual rate of 2 parts per million (ppm). This is twice the rate observed in the past century, during the decade of the 1960s. Thus, atmospheric CO_2 is rising at an ever-faster rate. In 2013, the Mauna Loa Observatory in Hawaii measured a level of CO_2 at 400 ppm—believed to be the highest level on Earth for the past 3 million years.

On land, terrestrial plants, particularly forest trees, sequester significant amounts of carbon—perhaps 10%–20% of the CO_2 released by burning fossil fuels. Forest carbon sequestration is shown in the international FLUXNET data (**Fig. 22.2B**). This experiment compares the seasonal patterns of net CO_2 uptake (the negative values on the *y*-axis) for forests at four different latitudes. It shows that the higher the latitude of the forest, the later in summer its CO_2 uptake peaks. Such measurements provide the basis for modeling global climate change, and for negotiating agreements to "trade" pollution for forest growth.

Note, however, that being a greenhouse gas does not make CO_2 inherently "bad" for the environment. In fact, if heterotrophic production of CO_2 were to cease altogether, phototrophs would run out of CO_2 in roughly 300 years, despite the vast quantities of carbon present in the ocean and crust. Thus, both CO_2 fixers and heterotrophs need each other for a continuous cycle.

Carbon Cycles Depend on Oxygen

The global cycle of carbon in the biosphere is closely linked to the cycles of oxygen and hydrogen, elements to which most carbon is bonded. Overall, carbon cycles between carbon dioxide (CO_2) and various reduced forms of carbon, including biomass (living material). Note that the results of carbon cycling differ greatly, depending on the presence of molecular oxygen.

Marine carbon cycling. The largest aerated ecosystem is the photic zone of oceans (**Fig. 22.3A**). In the aerated (or oxic) habitat of the marine photic zone, the ecosystem absorbs enough light for the rate of photosynthesis to exceed the rate of heterotrophy. Photosynthesis drives what is called the **biological carbon pump**. Microbial and plant photosynthesis fixes CO_2 into biomass, designated by the shorthand $[CH_2O]$. Phototrophs include bacteria and protists. Marine phototrophs such as diatoms and coccolithophores trap a substantial amount of carbon in biomass (**Fig. 22.3B**). A portion of their biomass sinks to the ocean floor through the weight of their silicate or carbonate exoskeletons. Photosynthetic CO_2 fixation is accompanied by release of O_2.

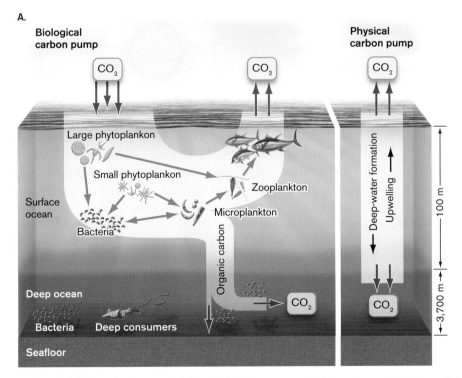

FIGURE 22.3 ■ The ocean's biological carbon pump. A. CO_2 enters equilibrium with carbonates, releasing H^+ ions; as a result, the ocean acidifies. In the biological carbon pump, CO_2 is fixed by phytoplankton. Aerobic respiration by zooplankton and mixotrophs converts some biomass back to CO_2, while other biomass sinks to the deep ocean. **B.** Diatoms in lake sediment store carbon (colorized SEM).

A.

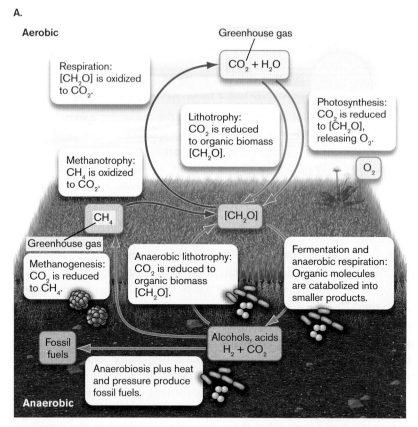

B.

FIGURE 22.4 ■ **The terrestrial carbon cycle: aerobic and anaerobic.** **A.** Aerobic and anaerobic conversions of carbon. Blue = reduction of carbon; red = oxidation of carbon; orange = fermentation; [CH$_2$O] = organic biomass. In an aerobic environment (top), photosynthesis generates molecular oxygen (O$_2$), which enables the most efficient metabolism by heterotrophs, methanotrophs, and lithotrophs. **B.** *Cortinarius armillatus* mushrooms are fruiting bodies of mycorrhizal fungi that share mutualism with birch trees. **Inset:** Karina Clemmensen studies mycorrhizal sequestration of carbon.

The O$_2$ is then used by heterotrophs (such as bacteria, protists, and animals) to convert biomass [CH$_2$O] back to CO$_2$. In the presence of light, a net excess of O$_2$ is released. Biomass is also produced through lithotrophy or chemolithoautotrophy—the oxidation of hydrogen, hydrogen sulfide, ferrous iron (Fe^{2+}) and other reduced minerals, and even carbon monoxide.

Terrestrial carbon cycling: aerobic. Another large aerobic habitat is the oxygenated layer of soil (discussed in Chapter 21) (**Fig. 22.4A**, top). While plants perform most of the terrestrial photosynthesis, microbial phototrophs also fix CO$_2$ into biomass. Microbial lithotrophy fixes carbon in soil and in weathered areas of crustal rock. Lithotrophy is performed solely by bacteria and archaea, essential microbial partners in these ecosystems. Unlike photosynthesis, lithotrophy usually consumes O$_2$ instead of producing it. Much carbon is stored as carbon polymers such as cellulose and lignin. But microbial and animal consumers release a lot of CO$_2$ through respiration.

In 2013, another important microbial carbon sink was discovered by Karina Clemmensen, Björn Lindahl, and colleagues at Uppsala BioCenter, Sweden: the mycorrhizal fungi (presented in Chapter 21). Mycorrhizal fungi (**Fig. 22.4B**) form mutualistic associations with forest plant roots and are especially important for the roots of trees. Clemmensen showed that in some forests, the roots and fungi can sequester as much as 22 kilograms of carbon per square meter of forest soil, which may be 70% of the total carbon sequestered. Thus, mycorrhizal fungi play an important role in minimizing release of the greenhouse gas CO$_2$.

Anaerobic carbon cycling. Anoxic environments support lower rates of biomass production than do oxygen-rich environments because they depend on oxidants of lower redox potential and limited quantity, such as Fe^{3+}. Anaerobic conversion of CO$_2$ to biomass is done mainly by bacteria and archaea. Vast, permanently anaerobic habitats extend several kilometers below Earth's surface, encompassing a greater volume than the rest of the biosphere put together (**Fig. 22.4A**, bottom). In these habitats, endolithic bacteria inhabit the interstices of rock crystals (discussed in Chapter 21).

FIGURE 22.5 ■ **Researchers ignite a bubble of methane on Alaska's Seward Peninsula.**

In soil and water, anaerobic metabolism includes fermentation of organic carbon sources, as well as respiration and lithotrophy with alternative electron acceptors such as nitrate, ferric iron (Fe^{3+}), and sulfate. Anaerobic decomposition by microbes is one stage in the formation of fossil fuels such as oil and natural gas (primarily methane). In soil, anoxic conditions (extremely low in O_2) favor incomplete breakdown of organic material. This characteristic of anaerobic soil actually enriches ecosystems, particularly those of wetlands, which undergo periodic cycles of aeration and hydration. The partly decomposed matter becomes available for further decomposition. Anoxic environments favor production of methane from the H_2, CO_2, and other fermentation products of anaerobes.

A major source of concern for global greenhouse gases is the methane hydrates accumulating in deep marine sediments, generated by huge benthic communities of methanogens (discussed in Chapters 19 and 21). Warming of methane hydrate releases gaseous methane. Some methane is oxidized by methanotrophs, but an unknown amount rises to the atmosphere (**Fig. 22.5**). Geological evidence suggests that rapid methane release accompanied the retreat of the glaciers during ice age transitions. A rapid methane release today could accelerate global warming.

At the ocean floor, many of the methane hydrates are oxidized by microbial mats of sulfate-reducing bacteria and anaerobic methane-oxidizing archaea (ANME), discussed in Section 19.4. The sulfate reducers plus the methane oxidizers conduct a syntrophic reaction, for which the overall ΔG value is negative:

$$CH_4 + SO_4^{2-} \longrightarrow HCO_3^- + HS^- + H_2O$$

In this metabolism, methane is the initial electron donor oxidized by the ANME partner, and sulfate is the terminal electron acceptor reduced by the bacteria. The high sulfate concentration of marine water drives this reaction in anoxic sediment, where all of the O_2 has been consumed by microbes that oxidize upwelling reduced minerals (such as sulfide oxidizers at thermal vents and cold seeps, discussed in Chapter 21).

In 2012, researchers reported a different source of marine methane, from aerated water: the aerobic ammonia-oxidizing archaea, such as *Nitrosopumilus* (discussed in Chapter 19). William W. Metcalf and colleagues at the University of Illinois at Urbana-Champaign showed that *Nitrosopumilus* species conduct reactions that degrade toxic phosphonates, organic compounds containing a direct carbon-phosphorus bond. The reactions release methylphosphonate, a compound that many kinds of bacteria convert to methane in order to acquire the scarce phosphate for phospholipids and nucleic acids. In addition, the sequence of the *Nitrosopumilus* enzyme for methylphosphonate production revealed homologs in the Global Ocean Sampling Expedition (GOS) metagenome—for example, in the bacterium *Pelagibacter*. Thus, both archaeal and bacterial methylphosphonate may be sources of methane from aerated habitats.

The Global Carbon Balance

What determines atmospheric levels of CO_2? The global balance of biological CO_2 fixation and release largely determines the level of CO_2. Since the beginning of the industrial age, however, the release of CO_2 has accelerated significantly. The major part of this increase comes from the combustion of **fossil fuels**, which adds about 6×10^{15} grams of carbon annually to the atmosphere. Fossil fuels are the product of microbial anaerobic digestion of plant and animal remains, reduced to hydrocarbons by the pressure and heat of Earth's crust. When burned as fuel, carbon that had accumulated over millions of years is rapidly returned to the atmosphere as CO_2. Increased CO_2 fixation and ocean absorption compensate for some of the CO_2 flux, but about a tenth of the carbon remains in the atmosphere.

Another factor in the increase of atmospheric CO_2 is microbial decomposition in the soil (discussed in Chapter 21). Microbes release carbon through respiration (aerobic and anaerobic), as well as fermentation in anoxic strata, and methanogenesis. A major concern today is the increased rate of microbial activity in Arctic permafrost as temperatures increase. The Arctic permafrost is estimated to store more than twice the amount of carbon that is in the atmosphere. As permafrost melts, so do methane hydrates from methane produced earlier by methanogens. In 2012, Andrew McDougall and colleagues at the University of Victoria (British Columbia) modeled both the response of permafrost carbon release and its feedback effect on global warming. They project that permafrost could release more than a quarter of its carbon stores by the year 2100, and that this amount could add another 1.5°C to the global temperature by the year 2300. The microbial release of methane and CO_2 could increase the rate of global warming, which in turn thaws the permafrost faster.

The contribution of ocean ecosystems to carbon flux is even more difficult to measure. Studies of carbon and oxygen flux based on isotope ratios suggest a much greater level of biological production than earlier estimates suggested. The oceans conduct more than half of the global biological uptake of carbon from the atmosphere. Metagenomic surveys are revealing previously unrecognized communities of marine phototrophs, such as submicroscopic cyanobacteria, the prochlorophytes (*Prochlorococcus* species). In the deep ocean, similar studies reveal new methane-oxidizing archaea (ANME) that play a critical role in removing benthic methane hydrates.

From a political standpoint, ecosystems such as forests that act as carbon sinks by fixing carbon into stable biomass are considered desirable because they lessen the rate of CO_2 input into the atmosphere. Ecosystems that act as sources of carbon dioxide may be viewed with disfavor because they contribute to global warming. But what if CO_2-generating ecosystems provide other environmental benefits? For example, wetlands are among Earth's most productive ecosystems, supporting vast amounts of plant and animal life, although they also release significant amounts of CO_2 and methane. An example of research on wetland carbon cycles is presented in **eTopic 22.1**.

To Summarize

- **The most accessible carbon reservoir is the atmosphere (CO_2).** Atmospheric carbon is severely perturbed by the burning of fossil fuels.

- **Cyanobacteria and other phytoplankton cycle much of biospheric CO_2 and release O_2.** Oxygen released by phototrophs is used by aerobic heterotrophs and lithotrophs. Marine plankton, as well as terrestrial trees and mycorrhizal fungi, serve as major carbon sinks.

- **Anaerobic environments cycle carbon through bacteria and archaea.** Bacteria conduct fermentation and anaerobic respiration, while methanogens convert CO_2 and H_2 to methane. Methane is oxidized to CO_2 by methane-oxidizing bacteria and archaea.

- **In oxygenated water, bacteria may release methane via methylphosphonate.**

- **Microbial decomposition returns CO_2 to the atmosphere.** Microbial decomposition is the main source of accelerated CO_2 flux from the soil, especially from melting permafrost.

22.3 The Hydrologic Cycle and Wastewater Treatment

The fate and distribution of complex carbon compounds are largely functions of the **hydrologic cycle**, or **water cycle**, the cyclic exchange of water between atmospheric water vapor and Earth's ecosystems (**Fig. 22.6A**). A vast reservoir of water is supplied by the ocean. In the hydrologic cycle, water precipitates as rain, which is drawn by gravity into groundwater, rivers, lakes, and ultimately the ocean. All along this route, of course, evaporation returns water to the air. Human communities interact with the hydrologic cycle by drawing water for drinking and other purposes, and by returning wastewater. Before wastewater can be safely returned to the hydrologic cycle, organic contaminants must be removed. Key parts of that treatment are performed by microbes.

Biochemical Oxygen Demand

What determines the health of an aquatic ecosystem? As discussed in Chapter 21, a major factor in the health of ecosystems is the balance between the level of oxygen and the levels of reduced organic nutrients. Organic contamination destabilizes marine, freshwater, and terrestrial ecosystems. Water passing through the ground and aquatic ecosystems carries organic carbon material from humus, sewage, and fertilizer runoff. A sudden influx of rich carbon substrates accelerates respiration by aquatic microbes. Microbial respiration then competes with respiration by fish, invertebrates, and amphibians for the limited supply of oxygen dissolved in water, raising the **biochemical oxygen demand** (**BOD**), also called the biological oxygen demand (BOD). The higher the concentration of organic substances, the higher the BOD arising from microbial oxygen consumption.

High BOD can cause a massive die-off of fish and other aquatic animals. Thus, a routine part of monitoring the health of lakes and streams is the measurement of BOD. A standard value of BOD is defined by measuring the rate of oxygen uptake in a water sample by a defined set of heterotrophic bacteria (**Fig. 22.6B**). Oxygen uptake is observed in a BOD analyzer (**Fig. 22.6C**), which detects dissolved oxygen in water. The rate of decrease of dissolved oxygen measured by the BOD analyzer is approximately proportional to the amount of dissolved organic matter available for respiration. Note, however, that the BOD in a natural environment will depend on the microbes actually present, as well as the plants and animals competing for oxygen and other resources.

Until recently, BOD was considered a local issue, affecting the health of lakes and rivers in a community. But today

A. Hydrologic cycle

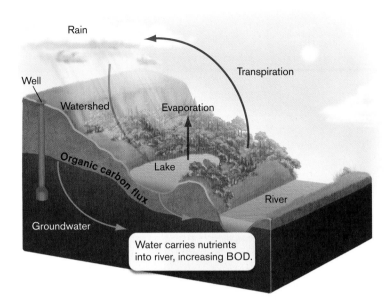

B. BOD measurement

C. BOD analyzer

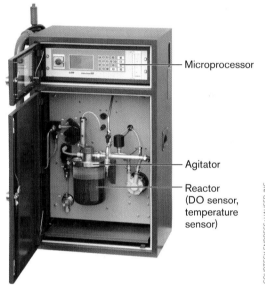

FIGURE 22.6 ■ **The hydrologic cycle interacts with the carbon cycle.** **A.** The hydrologic cycle carries bacteria and organic carbon into groundwater and aquatic systems. **B.** Bottled water samples are measured for dissolved oxygen over time; the rate of decrease of dissolved oxygen indicates biochemical oxygen demand (BOD). The rate of decrease of dissolved oxygen in water samples is approximately proportional to the concentration of organic matter available for respiration. **C.** A microprocessor-controlled BIOX-1010 BOD analyzer measures rate of respiration. Water samples are mixed with a concentrated microbial biomass, and a dissolved-oxygen (DO) sensor measures small rates of oxygen decrease over time.

we recognize huge impacts of rising BOD in the oceans (**Fig. 22.7**). Ocean oxygen levels are high near the surface, where phototrophs release oxygen, but organic nutrients are so scarce that respiration is limited. Oxygen is also high in the deep benthos, because most organic nutrients have been consumed, and because the sheer volume of water can hold dissolved oxygen in large amounts. But near the coastal shelf, currents may carry sediment up to a middle region where organic nutrients meet the oxygen. This combination supports rapid bacterial respiration. The result is an **oxygen minimum zone** (**OMZ**), a region of low or near-zero oxygen sandwiched between the upper and lower oxygenated layers. Above and below the anoxic water, there is a steep oxycline (gradient of oxygen concentration). A well-known OMZ is located off the coast of Oregon, where crabs and other animals are found dying as they try to escape asphyxiation.

Oxygen minimum zones are worsened by increasing temperatures, which accelerate respiration, and by influx of sewage and agricultural waste. The zone expands upward

toward the surface and downward to the sediment, trapping crabs and fish. Today, large regions of ocean have become **dead zones**, or **zones of hypoxia**, devoid of most fish and invertebrates. A major dead zone is a region in the Gulf of Mexico off the coast of Louisiana where the Mississippi River releases about 40% of the U.S. drainage to the sea (**Fig. 22.8A**). Over its long, meandering course, which includes inputs from the Ohio and Missouri rivers, the Mississippi builds up high levels of organic pollutants, as well as nitrates from agricultural fertilizer. When these nitrogen-rich substances flow rapidly out to the Gulf in the spring, they lift the nitrogen limitation on algal growth and feed massive algal blooms. The algal population then crashes, and their sedimenting cells are consumed by heterotrophic bacteria. The heterotrophs use up the available oxygen, causing **hypoxia**. Hypoxia kills off the fish, shellfish, and crustaceans over a region equivalent in size to the state of New Jersey.

In 2010 the Gulf of Mexico dead zone was expanded by the unprecedented spill of oil from the *Deepwater Horizon*

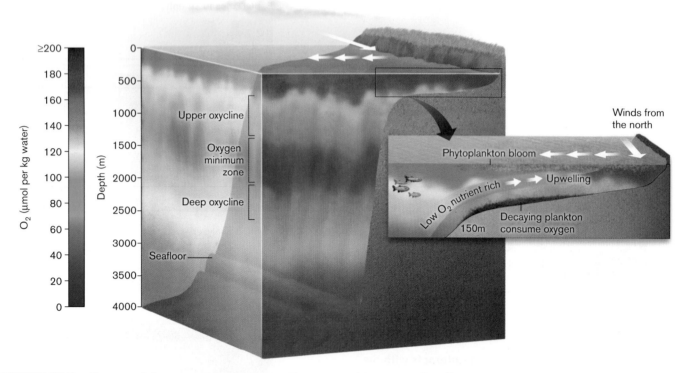

FIGURE 22.7 ■ **Oxygen minimum zone in the ocean.** The oxygen minimum zone (OMZ) occurs in the region of the coastal shelf where nutrients from below meet the oxygen produced by phototrophs above. In this region, microbial respiration consumes oxygen faster than the organic nutrients. The OMZ can expand upward when phytoplankton blooms decay, consuming the surface oxygen.

oil well blowout (**Fig. 22.8B**). The offshore oil platform exploded, releasing millions of barrels of oil into the Gulf over 3 months. The leaked oil killed wildlife throughout the Gulf, causing unprecedented environmental damage. Workers tried to contain the spill, but ultimately most of the oil was consumed by marine bacteria. Only bacteria and archaea possess the enzymes needed to degrade the complex organic mixture of petroleum, which includes waxy aromatic compounds such as paraffins (discussed in Chapter 21). Many of the pollutants were metabolized to CO_2 by oil-eating bacteria that feed naturally on oil seeping slowly from the Gulf sediment. But the sudden large input of oil raised microbial respiration and BOD levels throughout the Gulf, thus lowering oxygen available to wildlife. Furthermore, certain oil components linger in the environment for decades, including toxins such as polycyclic aromatic hydrocarbons (PAHs).

Dead zones now occur along the coasts of industrial and developing countries throughout the world, including India and Australia. Dead zones deplete habitat for much marine life—for example, forcing sharks to swim out of hypoxic regions and nearer the shore. Dead zones contribute to the crash of fisheries worldwide, removing a critical food resource for human populations. To avoid such dead zones, release of industrial pollutants and fertilizers must be prevented. In addition, all the communities throughout the river drainage areas must treat their wastes to eliminate nitrogenous wastes before disposal.

There are two common approaches to community wastewater treatment, both of which involve microbial partners: wastewater treatment plants and wetland filtration. Both approaches depend on microbes to remove organic carbon and nitrogen from water before it returns to aquatic systems and ultimately the ocean.

Wastewater Treatment

In industrialized nations, all municipal communities use some form of **wastewater treatment** (**Fig. 22.9**). The purpose of wastewater treatment is to decrease the BOD and the level of human pathogens before water is returned to local rivers. The treatment process includes microbial metabolism. A modern wastewater treatment plant can convert sewage into water that exceeds all government standards for humans to drink.

The wastewater treatment plant is the final destination for all household and industrial liquid wastes passing through the municipal sewage system. A typical plant includes the following stages of treatment, illustrated in **Fig. 22.9A**:

- **Preliminary treatment** consists of screens that remove solid debris, such as sticks, dead animals, and feminine hygiene items.

- **Primary treatment** includes fine screens and sedimentation tanks that remove insoluble particles. The particles

A. Mississippi River basin with Gulf of Mexico hypoxia

B. *Deepwater Horizon* **oil spill**

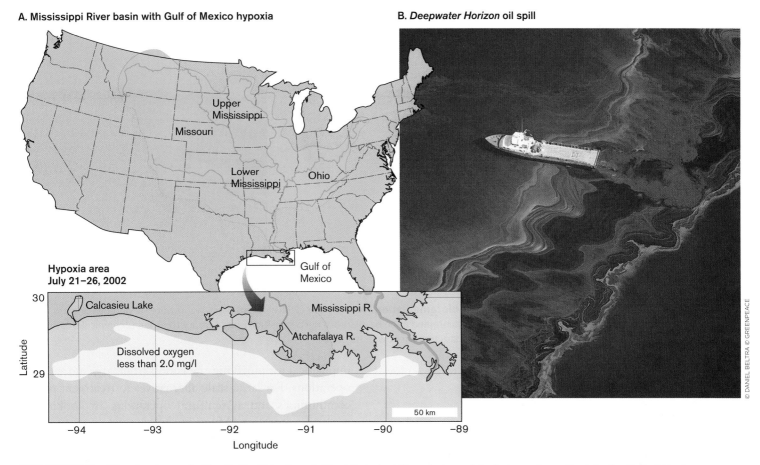

FIGURE 22.8 ■ The dead zone in the Gulf of Mexico. A. The Mississippi River drainage (shaded green) empties into the Gulf of Mexico. Every summer, drainage high in organic carbon and nitrogen causes algal blooms, leading to hypoxia and death of fish. **B.** Aerial view of the oil leaked from the *Deepwater Horizon* oil wellhead in the Gulf of Mexico, 2010. *Source:* A: National Oceanographic and Atmospheric Administration.

eventually are recombined with the solid products of wastewater treatment to form what is known as **sludge**. The sludge ultimately is used for fertilizer or landfill.

■ **Secondary treatment** consists primarily of microbial ecosystems that decompose the soluble organic content of wastewater, by aerobic and anaerobic respiration. The nutrient removal process may include biological removal of nitrogen and phosphorus. If included at this point, nitrogen is removed by nitrifying bacteria that oxidize ammonium, and phosphorus is removed by polyphosphate-accumulating bacteria. The microbes form particulate **flocs** of biofilm. The flocs are sedimented as sludge, also known as activated sludge, owing to their microbial activity.

■ **Tertiary (advanced) treatment** includes filtration of particulates from the microbial flocs of secondary treatment, and may include chemical processes to decrease nitrogen and phosphorus. The final step involves disinfection to eliminate pathogens, usually by a chemical process such as chlorination. The treated water is then returned to local freshwater sources.

The microbial ecosystems of secondary treatment require continual aeration to maximize breakdown of molecules to carbon dioxide and nitrates. The floc size and composition must be monitored for optimal performance. Floc microbes typically include bacilli such as *Zoogloea*, *Flavobacterium*, and *Pseudomonas*, as well as filamentous species such as *Nocardia* (**Fig. 22.10A**). Optimal treatment depends on the ratio of filamentous to single-celled bacteria: enough filaments to hold together the flocs for sedimentation, but not so many as to trap air and cause flocs to float and foam, preventing sedimentation.

Besides bacteria, the ecosystem of activated sludge includes filamentous methanogens (discussed in Chapter 19), which metabolize short-chain molecules such as acetate within the anaerobic interior of flocs. Thus, wastewater treatment generates methane, often in quantities that can be recovered as fuel.

In addition, the bacteria are preyed upon by protists such as stalked ciliates (**Fig. 22.10B**), swimming ciliates, and amebas, as well as invertebrates such as rotifers and nematodes. The predators serve the valuable function of limiting the numbers of planktonic single-celled

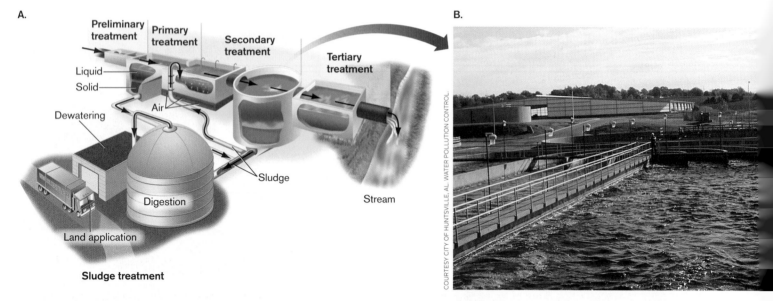

FIGURE 22.9 ■ Wastewater treatment with bioremediation. A. In a municipal treatment plant, wastewater undergoes primary treatment (filtering and settling), secondary treatment (bioremediation by microbial decomposition), and tertiary treatment (chemical treatments including chlorination). **B.** Aeration basin for secondary treatment with microbes.

bacteria, enabling the bulk of the biomass to be removed by sedimentation.

> **Thought Question**
>
> **22.2** What would happen if wastewater treatment lacked microbial predators? Why would the result be harmful?

Wastewater treatment plants are remarkably effective at converting human wastes to ecologically safe water and ultimately human drinking water. The plants are, however, impractical for purifying the runoff from large agricultural operations. For large-scale alternatives to treatment plants, communities and agricultural operations are looking to wetland restoration. Much of our current water supply is already filtered and purified by natural wetlands, such as the Florida Everglades (**Fig. 22.11A**). Wetlands remove nitrogen through the action of denitrifying bacteria. In wetlands, rainwater and river water trickle slowly through vast stretches of soil, where microbial conversion acts as the foundation of a macroscopic ecosystem including trees and vertebrates. Thus, much of the carbon and nitrogen is fixed into valuable biomass. In the wetlands of the Everglades, the remaining water filters slowly through limestone into

A. *Nocardia* sp.

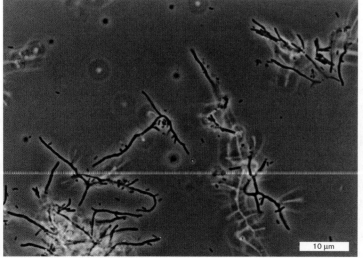

B. Flocs

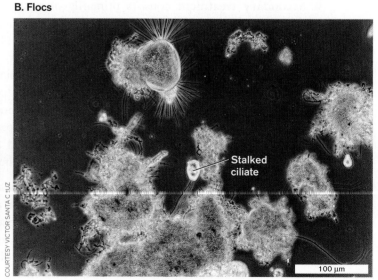

Stalked ciliate

FIGURE 22.10 ■ Microbes in wastewater bioremediation. A. Filamentous *Nocardia* sp. bacteria from flocs formed during secondary treatment (LM). **B.** Flocs with a stalked ciliate, which preys on bacteria (LM).

A. Everglades

COURTESY OF EVERGLADES NATIONAL PARK

B. Artificial wetlands

TIM MCCABE/USDA/NRCS

FIGURE 22.11 ■ **Water filtration by wetlands. A.** The marshes of the Everglades act as natural filters that bioremediate water entering the aquifers of southern Florida. **B.** The filtering system that Steve Kerns installed on his hog farm in Taylor County, Iowa, consists of a series of hillside terraces that form wetlands containing bacteria that metabolize hog manure and wastewater.

underground aquifers, which ultimately provide water through wells to human communities.

Some agricultural operations are building artificial wetlands to replace treatment plants. **Figure 22.11B** shows a hog farm where a series of terraced wetlands was built to drain liquefied manure. The wetlands were found to produce fewer odors and to remove organics more efficiently and at a lower cost than do traditional filtration plants.

A much greater challenge is that of industrial runoff, in which toxic wastes from factory chemistry enter an aquifer, at levels endangering human health. These chemicals may poison the treatment plants designed for human effluents. Large regions of land surrounding such a plant may require remediation that is expensive or impractical. And traces of pollutants such as chlorinated aromatics reach every spot on the globe; for example, dioxins are found in the tissues of Antarctic penguins.

How can we remediate industrial pollutants before they poison the local countryside and spread throughout the globe? One way is to harness naturally occurring bacteria for bioremediation. The bacterial community's reaction rate can be enhanced through selection by enrichment culture (described in Chapter 4). In the culture, organisms from the polluted site are cultivated repeatedly in the presence of added pollutant, and then tested for activity using radiolabeled substrates. An example of successful bioremediation is the case of Aberdeen Proving Ground, where a microbial mat was used to catabolize chlorinated hydrocarbons left by weapons manufacture (**eTopic 22.2**).

To Summarize

- **The hydrologic cycle is the cyclic exchange of water between the atmosphere and the biosphere.** Water precipitates as rain, which enters the ground and ultimately flows to the oceans. Along the way, some of the water evaporates, returning to the atmosphere.

- **Water carries organic carbon that generates biochemical oxygen demand (BOD).** High BOD accelerates heterotrophic respiration and depletes oxygen needed by fish.

- **Oxygen minimum zones are hypoxic regions of ocean sandwiched between upper and lower oxygenated layers.** Low oxygen levels are found at middle depth, where the oxygen from phototrophs meets upwelling organic nutrients.

- **Dead zones occur where sewage and agricultural runoff expand the oxygen minimum zones.** Dead zones exclude all aerobic life.

- **Wastewater treatment cuts down BOD.** Secondary treatment involves formation of flocs, microbial communities that decompose the soluble organic content.

- **Wetlands filter water naturally.** Wetland filtration helps purify groundwater entering aquifers.

- **Industrial effluents are highly toxic and can reach all parts of the globe.** Bioremediation with microbes may eliminate such toxins.

22.4 The Nitrogen Cycle

Besides carbon, oxygen, and hydrogen, another major element that cycles largely by microbial conversion is nitrogen (**Fig. 22.12**). The nitrogen cycle is notable in three respects:

- **Many oxidation states.** A larger number of oxidation states exist for biological nitrogen than for any other major biological element (see **Table 22.3**). Conversion between these oxidation states requires metabolic processes such as N_2 fixation, lithotrophy, and anaerobic respiration.

- **Dependence on prokaryotes.** Many steps of the nitrogen cycle require bacteria and archaea. Without bacteria and archaea, the nitrogen cycle would not exist.

- **Extreme perturbation by human technology.** Half the nitrogen in the biosphere now comes from anthropogenic (human-generated) sources.

Thought Question

22.3 How many kinds of biomolecules can you recall that contain nitrogen? What are the usual oxidation states for nitrogen in biological molecules?

Sources of Nitrogen

Where is nitrogen found on Earth? As we saw for carbon, a significant amount of nitrogen is found in Earth's crust, in the form of ammonium salts in rock (see **Table 22.2**). Unlike crustal carbon, however, the nitrogen found in rock is largely inaccessible to microbes. The major accessible source of nitrogen is the atmosphere, of which dinitrogen gas (N_2) constitutes 79%. However, N_2 is a highly stable molecule that requires an enormous input of reducing energy before assimilation is possible (see Chapter 15). Thus, for many natural ecosystems, and most forms of agriculture, nitrogen is the limiting nutrient for primary productivity.

Until recently in Earth's history, nitrogen was fixed entirely by nitrogen-fixing bacteria and archaea. In the twentieth century, however, the Haber process was invented for artificial nitrogen fixation to generate fertilizers for agriculture. The process was devised by German chemist Fritz Haber, who won the 1918 Nobel Prize in Chemistry. In the Haber process, N_2 is hydrogenated by methane under extreme heat and pressure. Today, the Haber process for producing fertilizers accounts for approximately 30%–50% of all nitrogen fixed on Earth. Other human activities, such as fuel burning and use of nitrogenous fertilizers, contribute to oxidized nitrogen pollutants such as nitrous oxide (N_2O), a potent greenhouse gas.

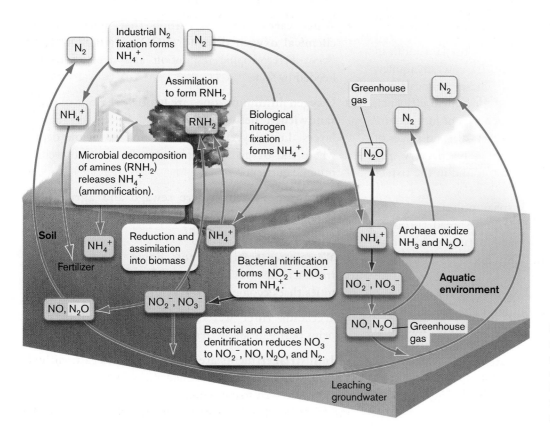

FIGURE 22.12 ■ The global nitrogen cycle. Bacteria and archaea interconvert forms of nitrogen throughout the biosphere. Blue = reduction; red = oxidation; orange = redox-neutral.

A. The nitrogen triangle

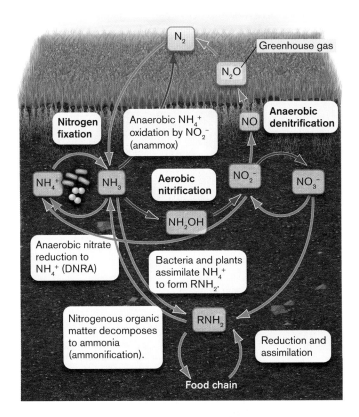

B. *Nitrobacter winogradskyi*

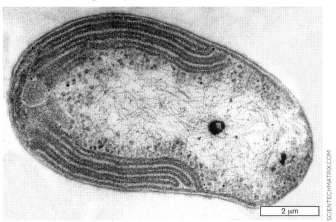

FIGURE 22.13 ■ **The nitrogen cycle: fixation, nitrification, denitrification. A.** The "nitrogen triangle" consists of nitrogen fixation and assimilation, reductive dissimilation of nitrate (<u>denitrif</u>ication; blue), and oxidation (<u>nitr</u>ification; red). Denitrification includes production of the potent greenhouse gas nitrous oxide (N_2O). Assimilation into biomass is often reductive; virtually all nitrogen in biomolecules is highly reduced. Oxidation of ammonia generates nitrites and nitrates, whose runoff can pollute water supplies. **B.** *Nitrobacter winogradskyi* oxidizes nitrite to nitrate. Folded layers of membrane contain the electron transport complexes.

Because of extensive fertilizer use, the nitrogen cycle today is the most perturbed of the major biogeochemical cycles. The ultimate effects of perturbation are not yet clear. Some models show that increased nitrogen fixation could fertilize CO_2 fixation by marine and terrestrial ecosystems, partly decreasing net CO_2 emissions. The amount, however, will be too small to prevent global warming. Other models show that accelerated nitrogen use could limit nitrogen availability in the global biosphere, with unknown consequences.

The "Nitrogen Triangle"

The numerous oxidation states available for microbial metabolism of nitrogen generate a complex cycle of conversions. One way to organize the complexity is to envision a "nitrogen triangle" whose corners include three key forms: atmospheric N_2; the reduced forms NH_3 and NH_4^+; and the oxidized forms NO_2^- and NO_3^- (**Fig. 22.13A**). At the base of the triangle, both reduced and oxidized forms of nitrogen are assimilated into biomass.

Thought Question

22.4 The nitrogen cycle has to be linked with the carbon cycle, since both contribute to biomass. How might the carbon cycle of an ecosystem be affected by increased input of nitrogen?

We will consider in turn the three sides of the nitrogen triangle:

■ **N_2 fixation to ammonium (NH_4^+)**, a form assimilated into biomass by microbes and plants.

■ **Ammonia oxidation, in which ammonia (NH_3) is oxidized aerobically to nitrite (NO_2^-) and nitrate (NO_3^-)**, which are also assimilated by microbes and plants. Aerobic oxidation to nitrite or nitrate is called nitrification. In the absence of oxygen gas, anaerobic ammonium oxidation by nitrite generates N_2—a reaction called anammox.

■ **Denitrification of NO_2^- and NO_3^- back to N_2** (or, in carbon-rich habitats, reduction to NH_3).

Nitrogen fixation. The main avenue for entry of nitrogen into the biosphere is bacterial and archaeal **nitrogen fixation**—specifically, fixation of dinitrogen, or nitrogen gas (N_2), into NH_3, protonated to NH_4^+. Ammonium ion is rapidly assimilated by bacteria and plants, typically by combination with TCA cycle intermediates to form key amino acids, such as glutamate (discussed in Chapter 15). The ability to assimilate NH_4^+ into nitrogenous organic molecules is found in virtually all primary producers.

Fixation of nitrogen requires enormous energy because the triple bond of N_2 is exceptionally stable. Breaking the triple bond to generate ammonia requires a series of reduction steps involving high input of energy:

Dinitrogen

$$N{\equiv}N \longrightarrow HN{=}NH \longrightarrow H_2N{-}NH_2 \longrightarrow 2NH_3$$
$$\phantom{N{\equiv}N \longrightarrow} 2H^+ + 2e^- 2H^+ + 2e^- 2H^+ + 2e^-$$

All the intermediate reactions of nitrogen fixation are tightly coupled, so the intermediate compounds rarely become available for other uses. The final product, ammonia, exists in ionic equilibrium with ammonium ion:

$$NH_3 + H_2O \leftrightarrow NH_4^+ + OH^-$$

Half the ammonia is protonated at about pH 9.3, so at near-neutral pH, most exists in the protonated form, NH_4^+.

Nitrogen fixation is catalyzed by the enzyme nitrogenase (discussed in Chapter 15). The highly reductive reactions of nitrogenase require exclusion of oxygen, yet they also require a tremendous input of energy, usually provided by aerobic metabolism. Therefore, nitrogen-fixing bacteria generally isolate anaerobic nitrogen fixation from aerobic metabolism by one of several mechanisms, such as the heterocysts of cyanobacteria (also discussed in Chapter 15).

Given the energy expense and the need to exclude oxygen, many species of bacteria, as well as all eukaryotes, lack the nitrogen fixation pathway. But all ecosystems, both marine and terrestrial, include some species of bacteria and archaea that fix N_2 into ammonia. Nitrogen-fixing bacteria in the soil include obligate anaerobes, such as *Clostridium* species, and facultative Gram-negative enteric species of *Klebsiella* and *Salmonella*, as well as obligate respirers such as *Pseudomonas*. The rhizobia form nitrogen-fixing endosymbionts of legume plants. In oceans and in freshwater systems, cyanobacteria are the major nitrogen fixers. After fixation, all these organisms assimilate the reduced nitrogen into essential components of their cells. Within an ecosystem, nitrogen fixers ultimately make the reduced nitrogen available for assimilation by nonfixing microbes and plants, either directly through symbiotic association (such as that of rhizobia and legumes, discussed in Chapter 21) or indirectly through predation (marine cyanobacteria) and decomposition (soil bacteria).

If nitrogen fixation is ubiquitous in soil and water, then why is nitrogen limiting? Nitrogen fixation is extremely energy intensive; thus, the rate of fixation usually fails to meet the potential demand of other members of the ecosystem. The one exception is legume symbiosis with rhizobia, which provide ample nitrogen for their hosts.

In agriculture, symbiotic nitrogen fixation by rhizobial bacteria increases the yield of crops such as soybeans (**Fig. 22.14A**). To enhance colonization by nitrogen fixers, farmers apply molecules called isoflavonoids, which mimic the natural plant-derived attractants for rhizobia.

A. Soybeans grow with symbiotic rhizobia

THOMAS HOVLAND/ALAMY

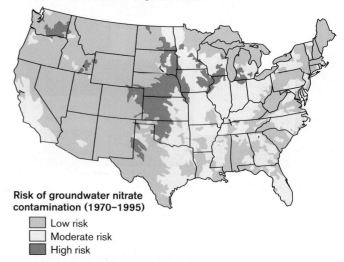

B. Nitrate contamination of groundwater

Risk of groundwater nitrate contamination (1970–1995)

- Low risk
- Moderate risk
- High risk

FIGURE 22.14 ■ Agricultural benefits and consequences of microbial nitrogen metabolism. A. Soybean field in Maine. Rhizobial nitrogen fixation enhances growth of soybeans and other major crops. **B.** Nitrate in drinking water is especially prevalent in agricultural regions of the United States. Ammonification of fertilizer, followed by nitrification, generates nitrate and nitrite. Nitrate and nitrite runoff from oxidized nitrogenous fertilizers pollutes streams and groundwater.

Nitrification. Free ammonia in soil or water is quickly oxidized for energy by **nitrifiers**, bacterial species that possess enzymes for oxidation of ammonia to nitrite (NO_2^-), or of nitrite to nitrate (NO_3^-). This process is called **nitrification**. Nitrification of ammonia is a form of lithotrophy, an energy generation pathway involving oxidation of minerals. The nitrification pathway generates the red base of the triangle in **Figure 22.13A**. The pathway of nitrification includes:

Ammonia Nitrite Nitrate

$$NH_3 \longrightarrow NH_2OH \longrightarrow NO_2^- \longrightarrow NO_3^-$$
$$+\tfrac{1}{2}O_2 +O_2 + +\tfrac{1}{2}O_2$$
$$ H_2O + H^+$$

The pathway includes two separate energy generation mechanisms: (1) ammonia through NH_2OH to nitrite, and (2) nitrite to nitrate. Typically, soil contains both kinds of microbes, collaborating to complete the conversion to nitrate, while some species of *Nitrospira* perform the entire pathway oxidizing ammonia to nitrate. Nitrifying genera include *Nitrosomonas*, which oxidizes ammonia to nitrite, as well as *Nitrobacter* (**Fig. 22.13B**) and *Nitrospira*, which oxidize nitrite to nitrate. Nitrite can also serve as an electron donor for photosynthesis by *Thiocapsa* species. Note that the production of both nitrite and nitrate generates acid, which can acidify the soil.

Thought Question

22.5 In the laboratory, which bacterial genus would likely grow on artificial medium including NH_2OH as the energy source: *Nitrosomonas* or *Nitrobacter*?

Nitrate produced in the soil is assimilated by plants and bacteria nearly as quickly as ammonium ion, although extra energy is needed to reduce nitrate to NH_4^+ for incorporation into biomass. Nitrate assimilation to biomass is called **assimilatory nitrate reduction**. Assimilatory nitrate reduction differs from the nitrate reduction involved in anaerobic respiration, which releases the reduced nitrogen and yields energy. In agriculture, intensive fertilization generates a large excess of ammonia resulting from **ammonification**, the breakdown of organic nitrogen, releasing ammonia. Ammonia is oxidized rapidly by lithotrophic bacteria (nitrifiers). The excess ammonia leads to a buildup of nitrites and nitrates, which are highly soluble in water, and they readily diffuse into aquatic systems. Aquatic nitrate reacts with organic compounds to form toxic nitrosamines. Nitrate influx also relieves the nitrogen limit on algae, causing algal blooms and raising BOD. Chronic nitrate influx leads to eutrophication and die-off of fish.

Consumption of nitrate in drinking water can lead to methemoglobinemia, a blood disorder in which hemoglobin is inactivated. Methemoglobinemia is a problem for infants because their stomachs are not yet acidic enough to inhibit growth of bacteria that convert nitrate to nitrite. Nitrite oxidizes the iron in hemoglobin, eliminating its capacity to carry oxygen. The failure to carry oxygen leads to a bluish appearance—one cause of "blue baby syndrome." Nitrite-induced blue baby syndrome is a problem in intensively cultivated agricultural regions, such as Kansas and Nebraska (**Fig. 22.14B**).

Agricultural runoff also contributes to oxygen minimum zones and dead zones, such as those off the Oregon coast and in the Gulf of Mexico. The amines in sewage are removed as ammonia (ammonification), and the ammonia is oxidized by lithotrophs. The major marine ammonia oxidizers include the Thaumarchaeota, such as *Nitrosopumilus* (discussed in Chapter 19).

Denitrification. N_2 is regenerated by anaerobic respiration (see **Fig. 22.13A**), in which an oxidized form of nitrogen, such as nitrate or nitrite, receives electrons from organic electron donors. Bacteria and archaea reduce nitrate through a series of decreased oxidation states back to atmospheric nitrogen:

$$\begin{array}{ccccccc}
& & & & & \text{Nitric} & \\
\text{Nitrate} & & \text{Nitrite} & & \text{oxide} & \\
2NO_3^- & \longrightarrow & 2NO_2^- & \longrightarrow & 2NO & \longrightarrow \\
& 4H^+ + 4e^- & + & 4H^+ + 2e^- & + & 2H^+ + 2e^- \\
& 2H_2O & & 2H_2O & &
\end{array}$$

$$\begin{array}{ccc}
\text{Nitrous} & & \\
\text{oxide} & & \text{Dinitrogen} \\
N_2O & \longrightarrow & N_2 \\
+ \quad 2H^+ + 2e^- \quad + \\
H_2O & & H_2O
\end{array}$$

In terms of elemental flux through ecosystems, nitrate and nitrite reduction are known as **denitrification** or **dissimilatory nitrate reduction**. (In contrast, assimilatory nitrate reduction incorporates nitrogen into biomass.) Anaerobic respirers in soil or water use an oxidized form of nitrogen as an alternative electron acceptor in the absence of O_2 (discussed in Chapter 14). All types of nitrogen-based anaerobic respiration are repressed in the presence of oxygen, a more favorable electron acceptor; therefore, denitrification is limited to anoxic habitats.

In undisturbed environments, the products of nitrate respiration rarely build up to levels that harm the ecosystem. Heavy fertilization, however, causes buildup of excess nitrate and so increases the environmental rate of denitrification. During this process, some of the nitrogen escapes as nitrous oxide gas (N_2O). A highly potent greenhouse gas, N_2O generates 200 times the warming effect of CO_2; thus, relatively small amounts of N_2O can make a disproportionate contribution to global warming. Furthermore, N_2O in the upper atmosphere reacts catalytically with ozone, depleting the ozone layer. Thus, the atmospheric effects of bacterial denitrification are a serious concern in agricultural and waste treatment processes.

Nitrous oxide also builds up in marine dead zones, where the nitrates from ammonium oxidation are subsequently used as electron acceptors. Syed Wajih Naqvi and colleagues at the National Institute of Oceanography in Goa, India, investigated denitrification in the zone of hypoxia off the coast of India (**Fig. 22.15A**). The study revealed unexpectedly high levels of N_2O production at a series of stations within the zone. In one experiment, introduction of nitrate into water samples from the dead zone led quickly to production of nitrite and N_2O (**Fig. 22.15B**), a process conducted by denitrifying bacteria. Studies in 2012 suggested additional N_2O production by ammonia-oxidizing archaea such as *Nitrosopumilus*. Thus, bacterial denitrification and archaeal ammonia oxidation in polluted ocean waters may contribute significantly to global warming.

Dissimilatory nitrate reduction to ammonia. Most environmental nitrate and nitrite are reduced via N_2O as described previously. Certain conditions, however, favor the alternate route of dissimilatory nitrate reduction to ammonia (DNRA) (see **Fig. 22.13A**). Nitrate reduction to ammonia is a form of anaerobic lithotrophy or hydrogenotrophy in which nitrate serves as an electron acceptor and hydrogen gas (H_2) is the electron donor:

$$NO_3^- + 4H_2 + H^+ \longrightarrow NH_3 + 3H_2O$$

Bacteria reduce nitrate to ammonia mainly in anaerobic environments rich in organic carbon and H_2 generated by fermentation, but low in reduced nitrogen, such as sewage sludge and stagnant water. Another carbon-rich habitat favoring this pathway is the rumen, the digestive tract of cattle, goats, and other ruminant animals (see Chapter 21). Ruminants consume grasses whose cellulose requires digestion by mutualistic microbes. Ruminants also depend on their digestive microbes to assimilate nitrate into ammonia and synthesize amino acids.

Anaerobic ammonium oxidation (anammox). For many years, denitrification was considered the main way that nitrogen compounds return N_2 to the atmosphere. In 2003, two research groups from Europe and Costa Rica, led by Tage Dalsgaard and Marcel Kuypers, showed that in anoxic deep-sea water the major source of N_2 is bacteria conducting anaerobic ammonium oxidation by nitrite (the anammox reaction, discussed in Chapter 14):

$$NH_4^+ + NO_2^- \longrightarrow N_2 + 2H_2O$$

In **anammox**, ammonium ion serves as electron donor, and nitrite serves as anaerobic electron acceptor—an unusual combination of two different oxidation states of the same key element, nitrogen. The reaction is a kind of anaerobic lithotrophy.

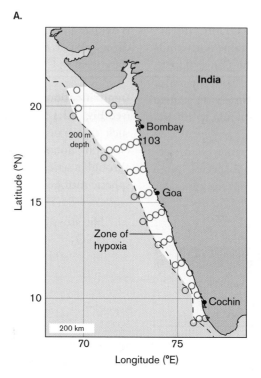

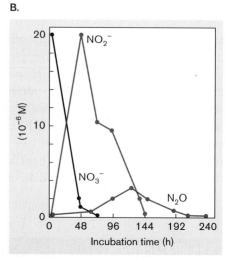

FIGURE 22.15 ■ **N_2O production from a coastal dead zone. A.** Zone of hypoxia off the coast of India. Circles represent sample-collection stations. **B.** Levels of NO_3^-, NO_2^-, and N_2O following addition of NO_3^- to a water sample from the zone of hypoxia. The sequential rise of NO_2^- and N_2O indicates metabolism of denitrifying bacteria.

Anammox by bacteria has now been observed in all kinds of anaerobic habitats, including terrestrial soil and aquatic sediment. The reaction accounts for a majority of all N_2 returned to the atmosphere. Among bacteria, the main anammox contributors are planctomycete genera such as *Kuenenia* and *Scalindua* (**Fig. 22.16**). Anammox planctomycetes (discussed in Chapter 18) are an unusual kind of cell wall–less bacteria with a special interior membrane that segregates the toxic intermediates of the anammox reaction. Promising anammox microbes have been identified from wastewater sludge, in the hope of using them for more effective removal of excess nitrogen from wastewater.

To Summarize

■ **Nitrogen in ecosystems is found in a wide range of oxidation states.** Interconversion of most of these states requires bacteria or archaea.

■ **The main source and sink of nitrogen is the atmosphere.** N_2 is fixed into NH_4^+ by some bacteria and archaea. Denitrifying bacteria reduce NO_3^- successively back to N_2 and return it to the atmosphere.

■ **Nitrogen fixation is conducted by symbiotic bacteria in association with specific plants.** Legume-associated nitrogen fixation is critical for agriculture.

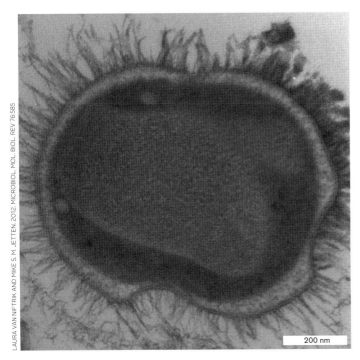

200 nm

FIGURE 22.16 ■ **Anammox bacterium: *Scalindua*.** The planctomycete *Scalindua* has an interior compartment to undergo anammox lithotrophy (discussed in Chapter 14). Pili surround the envelope.

- **Nitrification ($NH_4^+ \longrightarrow NO_2^- \longrightarrow NO_3^-$)** is aerobic oxidation of ammonia to nitrite and nitrate. Nitrification yields energy for lithotrophic bacteria in soil and water, and consumes oxygen, thus expanding marine oxygen minimum zones.

- **Bacteria conduct anaerobic respiration**, reducing nitrate to N_2 and nitrous oxide (N_2O), a greenhouse gas. In the deep ocean, NO_3^- may be reduced by hydrogen gas to NH_3.

- **Under anaerobic conditions, bacteria oxidize NH_4^+ with NO_2^-, generating nitrogen gas (the anammox reaction).**

22.5 Sulfur, Phosphorus, and Metals

Besides carbon and nitrogen, many other elements participate in biochemical cycles that have important consequences for the biosphere, as well as for human environments. Sulfur undergoes a "triangle" of redox conversions analogous to that of nitrogen. Phosphorus, unlike other biological elements, is generally assimilated in the oxidized state, and phosphate is often a limiting nutrient for plants. Iron cycles in complex interactions with sulfur and phosphate. Beyond

these macronutrients, many toxic metals in the environment, such as mercury and arsenic, are either bioactivated or detoxified by microbes.

The Sulfur Cycle

Sulfur is a major component of biomass, including proteins and cofactors. Like carbon and nitrogen, sulfur can be assimilated in either mineral or organic form. At the same time, assimilation competes with dissimilation. Reduced and oxidized forms of sulfur (comparable to those of nitrogen; see **Table 22.3**) offer electron donors and acceptors for dissimilatory reactions that generate energy. Reduced forms of sulfur, such as H_2S, can also participate in photolysis through reactions analogous to those of H_2O. Oxidized forms, such as sulfate, serve as anaerobic electron acceptors. As with nitrogen, most of these redox reactions are performed solely by bacteria and archaea. The biochemistry of sulfur cycling has important environmental consequences, such as the corrosion of concrete and iron.

In the ocean, sulfate is the second most common anion after chloride. Marine sulfate turns over slowly and constitutes an essentially limitless supply. Thus, sulfur is rarely a limiting nutrient. Marine algae release various reduced forms of sulfur, such as dimethylsulfoniopropionate (DMSP). Bacteria metabolize DMSP to dimethyl sulfide [$(CH_3)_2S$], which causes part of the "salty sea smell." Dimethyl sulfide enters the atmosphere, where it becomes oxidized to aerosols that may help water condense and form clouds. In the atmosphere, the overall amount of sulfur (mainly sulfur dioxide) is small. Nevertheless, some sulfur compounds generate toxic effects, as well as acidic pollution. Thus, both biochemical and industrial sources of atmospheric sulfur are of concern.

Competing assimilatory and dissimilatory sulfur reactions in the biosphere form a "sulfur triangle" including H_2S, S^0, and SO_4^{2-} (**Fig. 22.17A**). The oxidation and reduction pathways are analogous to those of nitrogen (compare **Fig. 22.13A**). Bacterial sulfur metabolism includes additional options of anaerobic phototrophy, such as H_2S phototrophy. In an aquatic system, H_2S arises from spring waters welling up from the sediment and from decomposition of detritus. In decomposition, anaerobic respirers such as *Desulfovibrio* species convert sulfate to sulfur, then to H_2S. As H_2S rises to the oxygenated surface water, it is readily oxidized by sulfur-oxidizing bacteria such as *Acidithiobacillus* and *Beggiatoa*. *Beggiatoa* species are known as "white sulfur bacteria" because they form white mats (**Fig. 22.17B**). Their appearance is due to the sulfur granules generated by sulfide oxidation. Microbial sulfide oxidation is helpful for environments because it removes H_2S, which is highly toxic to most nonsulfur bacteria and plants. Because of the toxicity of H_2S, autotrophs more readily assimilate

A. The sulfur triangle

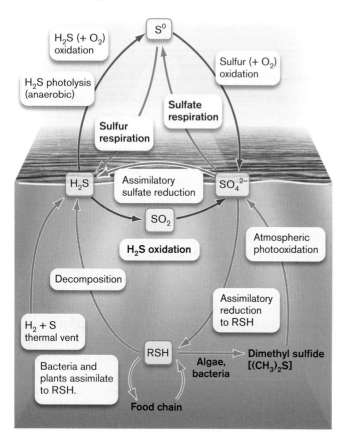

B. White sulfur bacteria

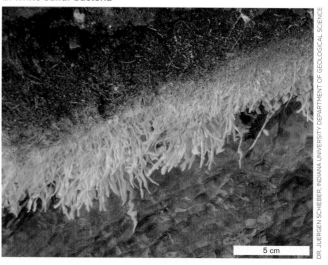

DR. JUERGEN SCHIEBER, INDIANA UNIVERSITY DEPARTMENT OF GEOLOGICAL SCIENCE

FIGURE 22.17 ■ The sulfur cycle. A. The "sulfur triangle." With oxygen, bacteria and archaea oxidize H_2S to sulfur dioxide, and then to sulfate. Anaerobically, H_2S may be photolyzed to sulfur. Sulfate serves as an electron acceptor for anaerobic respiration, or it may be assimilated into biomass, with reduced sulfur groups (RSH). Algae form DMSP, which bacteria convert to dimethyl sulfide. **B.** Microbial mat of white, sulfur-oxidizing bacteria (probably *Beggiatoa*) growing at a sulfide spring.

SO_4^{2-} into biomass, despite the extra energy needed to reduce SO_4^{2-} to the thiol form found in proteins.

If light is available, anaerobic phototrophs such as *Rhodopseudomonas* species will use light energy to oxidize H_2S. Some phototrophic bacteria further oxidize the S^0 generated to sulfate. In some sulfur-rich lakes, such as the Russian Lake Sernoye, underground springs pump in so much H_2S that most of the sulfur is photolytically converted to S^0, which forms up to 5% of the sediment. This elemental sulfur can be mined commercially.

In thermal vent ecosystems, several sulfur-based reactions drive metabolism. For example, thermal vent archaea such as *Pyrodictium* species use sulfur to oxidize hydrogen gas to H_2S:

$$H_2 + S^0 \longrightarrow H_2S$$

This type of sulfur-based hydrogenotrophy is enhanced by the vent conditions of extreme pressure and temperature (100°C)—conditions under which elemental sulfur exists in a molten state more accessible to microbes than solid sulfur.

Decomposition of biomass generates various organic sulfur compounds, many of which are volatile. Odors of certain microbial sulfur products contribute to the smell of rotting eggs, but others enhance the taste of cheeses (discussed in Chapter 16).

Sulfur is reduced by sulfur-reducing bacteria in many subsurface environments, such as beneath deposits of oil and coal (**Fig. 22.18A**). Oil and coal contain sulfur in the form of SO_4^{2-} and provide a carbon source for sulfate respirers (sulfate-reducing bacteria that respire using sulfate as a terminal electron acceptor). The products of sulfate respiration include S^0 and organic sulfur. When the fuel is burned, these forms of sulfur cause severe pollution. Before burning, however, the S^0 and organic sulfur can be oxidized by sulfur-oxidizing bacteria. Fuel processors are now turning to sulfur-oxidizing microbes for experimental use in "desulfuration," the removal of sulfur from coal.

One habitat that exhibits the entire range of sulfur oxidation states is a sewer pipe. In a concrete sewer pipe, the alternation between anaerobic and oxygenated sulfur biochemistry causes severe corrosion (**Fig. 22.18B**). The microbial decomposition of sewage yields large quantities of toxic H_2S, which then volatilizes to high levels that endanger sewer workers. The H_2S is then oxidized to sulfuric acid by *Acidithiobacillus ferrooxidans*, a bacterium that colonizes the surface of the concrete. The sulfuric acid (H_2SO_4) decreases the pH at the concrete surface to pH 2. In the concrete surface, sulfuric acid converts calcium hydroxide to soluble calcium sulfate. Over several years, this corrosion can eat away half the thickness of a sewer pipe.

A. Sulfate-reducing bacteria underground

B. Sewer pipe corrosion by sulfur bacteria

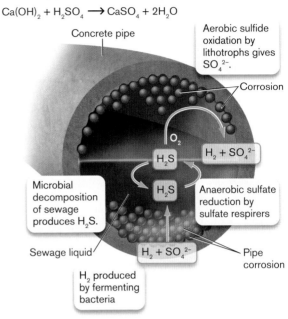

$$Ca(OH)_2 + H_2SO_4 \longrightarrow CaSO_4 + 2H_2O$$

FIGURE 22.18 ■ **Consequences of the sulfur cycle. A.** Oil- or coal-bearing geological strata provide rich electron donors for sulfate-reducing bacteria (sulfate respirers). Eventually, various forms of sulfur contaminate the oil or coal, which, when burned as fuel, generates first SO_2 and ultimately sulfuric acid (acid rain). **B.** In a sewer pipe, sulfate-reducing bacteria (sulfate respirers) generate H_2S. Sulfur-oxidizing bacteria then oxidize H_2S to sulfate in the form of sulfuric acid. The sulfuric acid reacts with calcium hydroxide in the concrete, thus corroding the interior surface of the pipe.

Sulfur metabolism shows important connections with metabolism of metals. For example, sulfur-oxidizing bacteria such as *A. ferrooxidans* oxidize iron as well (discussed shortly).

> **Thought Question**
>
> **22.6** Compare and contrast the cycling of nitrogen and sulfur. How are the cycles similar? How are they different?

The Phosphate Cycle

Phosphorus is a fundamental element of nucleic acids, phospholipids, and phosphorylated proteins. Unlike sulfur and nitrogen, which are found in several different oxidation states, phosphorus is cycled mainly in the fully oxidized state of phosphate (**Fig. 22.19**). The absence of fully reduced phosphorus in ecosystems may be due to the fact that reduced phosphorus (phosphine, PH_3) undergoes spontaneous combustion in the presence of oxygen. Nevertheless, some anaerobic decomposers use phosphate as a terminal electron acceptor, reducing it to phosphine. In marshes and graveyards, where extensive decomposition occurs, phosphine emanates from the ground, where it ignites with a green glow. Microbial respiration of phosphate might be the cause of such "ghostly" apparitions.

Where is phosphate available? Although phosphate is abundant in Earth's crust, its availability in ecosystems is limited by its tendency to precipitate with calcium, magnesium, and iron ions. Thus, dissolved phosphate in water and soil is often a limiting nutrient for productivity. In natural ecosystems, the available phosphate is taken up rapidly by bacteria and phytoplankton, and then consumed by grazers and predators and dispersed by decomposers.

Marine water is extremely limited for phosphate, because of the distance from sediment minerals. Thus, the genomes of marine phototrophs encode many systems for acquiring phosphate from organic sources. Some actually acquire phosphorus in a partly reduced form, such as phosphite (PO_3^{3-}) or phosphonate (HPO_3^{2-}), both of which are available in marine water. In addition, marine phototrophs show an unusual ability to substitute nonphosphorus lipids for phospholipids, thus cutting their phosphorus needs by half. For example, the cyanobacteria *Prochlorococcus*, *Synechococcus*, and *Trichodesmium* can replace membrane phospholipids with sulfonated lipids. Similarly, algae such as *Thalassiosira* species can replace phosphatidylcholine with betaine, which contains no phosphate but includes a carboxylate group.

Another microbial response to phosphate scarcity is to replace membrane phosphates with organic phosphonates, in which the phosphorus atom bonds directly to carbon ($R–PO_3^{2-}$), instead of the easily hydrolyzed phosphoester

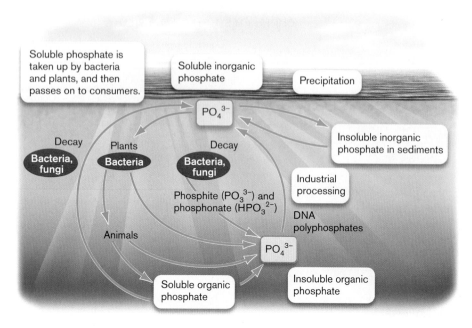

FIGURE 22.19 ▪ The phosphate cycle. Phosphorus in the biosphere occurs entirely in the form of inorganic or organic phosphate. Most phosphate precipitates as insoluble salts in sediment. The small amount of soluble phosphate is taken up by plants and bacteria, which may then be taken up by consumers. Decomposers return phosphate to the environment.

bond. These organic phosphonates, however, are cleaved by marine archaea, the ammonia oxidizers such as *Nitrosopumilus*. The archaea release methylphosphonate, which many bacteria can cleave to obtain phosphate. Phosphate scavenging from methylphosphonate releases methane, thus interacting with the carbon cycle.

In agriculture, phosphate is often added as a fertilizer. Phosphate fertilizer is obtained by treating calcium phosphate rock with sulfuric acid, producing calcium sulfate (gypsum) and phosphoric acid. Excess phosphate from fertilizer or industry may drain into streams and lakes where phosphorus is the limiting element. The sudden influx of phosphate causes an algal bloom. The overgrowth of algae leads to overgrowth of heterotrophs, depletion of oxygen, and destruction of the food chain.

Thought Question

22.7 Compare and contrast the cycling of nitrogen and phosphorus. How are the cycles similar? How are they different?

The Iron Cycle

Why do organisms need iron? As a micronutrient, iron forms a negligible part of biomass but is essential for growth (discussed in Chapter 4). Organisms require iron as a cofactor for enzymes and an essential component of oxygen carrier molecules such as hemoglobin. Other common micronutrients required for enzymes and cofactors are zinc, copper, and selenium.

Iron in soil and sediment. Iron is a major component of Earth's crust, and substantial quantities are present in most soil and aquatic sediment. Yet the availability of iron to organisms is limited by its extremely low solubility in the oxidized form. In microbial biochemistry, the oxidized form, ferric iron (Fe^{3+}), interconverts with the reduced form, ferrous iron (Fe^{2+}). In the presence of oxygen, iron metal (Fe^0) "rusts" and is therefore available to organisms mainly as Fe^{3+}. Ferric iron is especially insoluble at high pH, precipitating with hydroxide ions as ferric hydroxide [$Fe(OH)_3$] or with phosphate ions as ferric phosphate ($FePO_4$) (**Fig. 22.20A**). At neutral pH, iron is oxidized by bacteria such as *Gallionella*, *Leptothrix*, and *Mariprofundus*.

In iron mines, where pyrite (FeS_2) is exposed to air, spontaneous oxidation releases sulfuric acid:

$$2FeS_2 + 7O_2 + 2H_2O + 4H^+ \longrightarrow 2Fe^{2+} + 4H_2SO_4$$

Lithotrophs such as *Acidithiobacillus ferrooxidans* can catalyze iron oxidation to yield energy, thus increasing the rate of production of sulfuric acid. Rapid sulfuric acid production leads to acid mine drainage. Where acid mine drainage enters an aquatic system, the reduced iron is oxidized by lithotrophs to ferric hydroxide, forming an orange precipitate, seen in home plumbing (**Fig. 22.20B**). Iron mine drainage causes severe pollution of streams (**Fig. 22.21**).

In anoxic habitats, such as benthic sediment, Fe^{3+} is largely reduced to Fe^{2+} by anaerobic respiration. Reduction leads to loss of the reddish color, generating gray-colored sediment, as in wetland soil (discussed in Chapter 21). The reduction of Fe^{3+} to Fe^{2+} is one of the enriching contributions of anaerobic sediment to freshwater wetlands and coastal estuaries, because reduced iron is more available to plants and bacteria. Oxidized iron, in aerobic soils, can be taken up only by bacteria. Bacteria synthesize special iron uptake systems, including molecules called **siderophores** that bind ferric ion outside and then are taken up by the cell (see Section 4.2). Bacterial iron then becomes available to consumers in the ecosystem.

Iron cycling often connects with the sulfur cycle in ways that can prove unfortunate for human engineering.

A. The iron cycle

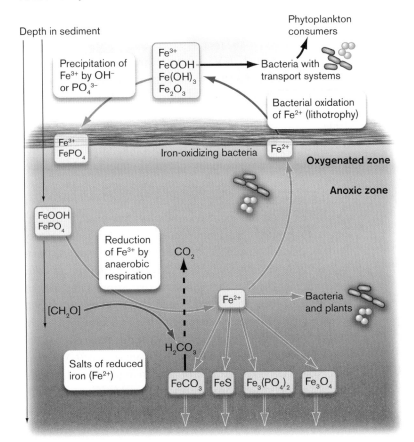

B. Iron precipitation due to lithotrophy

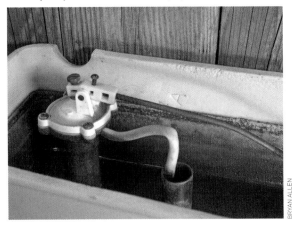

BRYAN ALLEN

FIGURE 22.20 ■ **The iron cycle. A.** Ferric iron (oxidized iron, Fe^{3+}) precipitates with hydroxide or phosphate. Only bacteria can assimilate Fe^{3+}. In anoxic sediment, bacterial respiration reduces Fe^{3+} to Fe^{2+}, a more soluble form available to plants. **B.** Home plumbing shows signs of lithotrophic oxidation of Fe^{2+} to Fe^{3+} (orange precipitate).

The most serious problem is anaerobic corrosion of iron. Iron corrodes spontaneously in the presence of oxygen. In anoxic environments, however, little spontaneous corrosion takes place. Instead, iron is corroded by sulfate-reducing bacteria (**Fig. 22.22**). Bacteria such as *Desulfovibrio* and *Desulfobacter* can grow on the iron surface, forming an anaerobic biofilm. Within the biofilm, iron is oxidized to Fe^{2+}, and sulfate dissolved in the water is reduced to ferrous sulfide:

$$4Fe^0 + SO_4^{2-} + 4H_2O \longrightarrow FeS + 3Fe^{2+} + 8OH^-$$

The FeS flakes away, exposing more iron to react with water, resulting in cyclic corrosion.

Marine iron. In the upper layers of most oceans, iron is extremely scarce. This is because the benthic sediment containing iron is so distant that its iron is largely inaccessible. The main source of marine iron is eolian (wind-borne) dust from dry land, such as from windstorms in the Sahara desert. Most of the wind-borne iron is particulate, oxidized, and unavailable to eukaryotic phytoplankton (algae). Thus, bacteria that acquire and reduce ferric iron provide the main entry of iron into the ecosystem. This may explain why so many marine algae are mixotrophs: Although their photosynthesis can fix plenty of

carbon for biomass, they consume bacteria as a source of iron. Another source of marine iron, discovered in 2013, is hydrothermal vents.

Iron is thus a limiting nutrient for marine phytoplankton. As discussed in Chapter 21, when higher quantities

THOMAS R. FLETCHER/ALAMY

FIGURE 22.21 ■ **Mine drainage enters a stream.** Lithotrophic oxidation of reduced iron yields the orange material polluting this stream in Preston County, West Virginia.

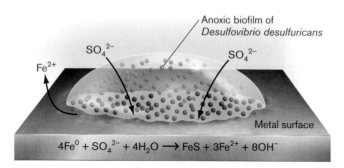

Anoxic biofilm of *Desulfovibrio desulfuricans*

$$4Fe^0 + SO_4^{2-} + 4H_2O \longrightarrow FeS + 3Fe^{2+} + 8OH^-$$

FIGURE 22.22 ■ **Anaerobic corrosion.** Iron-sulfur bacteria convert Fe^0 to FeS. FeS flakes off, exposing iron to further oxidation and rapid corrosion. *Source:* Modified from Hang T. Dinh et al. 2004. *Nature* **427**:829–832.

of a limiting nutrient enter an ecosystem, the limited populations rapidly increase. The rise of algal populations following iron addition was demonstrated in an "iron fertilization" experiment conducted in 2007 by Phillip Boyd and colleagues at the University of Otago, New Zealand. Several thousand kilograms of ferrous sulfate ($FeSO_4$) was released in the Southern Ocean, near Antarctica. Within a week after the iron release, there was a bloom of phytoplankton. The dominant phytoplankton in the bloom were diatoms of the species *Fragilariopsis kerguelensis.* The diatom bloom was so large that it was detected by a NASA satellite as a region of increased color reflected by chlorophyll.

The dramatic effects of iron fertilization led some researchers to propose that increasing iron throughout the oceans could cause blooms of diatoms, removing CO_2 from the atmosphere. If a sufficiently large proportion of the diatoms were to escape predation and fall to the ocean floor, they would effectively remove their carbon from circulation. An alternative outcome, however, would likely be eutrophication, causing an ecological collapse comparable to a dead zone.

While iron fertilization is unlikely to meet our need for CO_2 removal, microbiology offers other approaches to offset the perturbation of Earth's carbon cycle by human technology. For example, bacteria and algae may be engineered to provide biofuels, such as hydrogen gas, whose use yields energy without releasing CO_2.

Our view of Earth's biosphere increasingly is moving toward environmental management—the concept that wilderness as such no longer exists, but only a biosphere to be managed for better or worse. Microbes will have many roles as partners in environmental management.

Other Metals in the Environment

Another concern for environmental management is that of toxic metals and metalloids such as mercury and arsenic.

Besides iron, numerous trace metals interact with bacterial species as either electron donors or acceptors (**Table 22.4**). Bacterial conversion of metals can either produce toxic species or remove toxic metals from ecosystems. For example, chromium-6 [Cr(VI)] is an extremely toxic pollutant that can be reduced by soil bacteria to the much less toxic Cr(III). Microbial metal cycling is discussed further in **eTopic 22.3**.

How can bioremediation remove toxic minerals from groundwater? An example is the bioremediation of arsenic, a critical water contaminant in densely populated countries such as India, Bangladesh, and Cambodia. Arsenic causes cancers and crippling skin disorders. Arsenic and other metals can be removed from groundwater via bioremediation. Bhaskar Sengupta and colleagues at Queen's University, Belfast, developed a bioreactor technology for subterranean arsenic removal, or SAR (**Fig. 22.23**). This process takes advantage of the fact that different oxidation states of arsenic differ in their solubility. For the bioreactor, groundwater is cycled through an oxygen pump and returned to the source. Naturally occurring lithotrophic bacteria oxidize arsenite (AsO_3^{3-}, water-soluble) to arsenate (AsO_4^{3-}, insoluble). Bacteria also oxidize iron and manganese to their less soluble forms, as discussed in Chapter 14. The negatively and positively charged ions precipitate together, effectively removed from the water.

Multiple Limiting Factors Modulate Complex Ecosystems

A common assumption of terrestrial and aquatic ecology is that a single nutrient, such as phosphorus or iron, is limiting for a given ecosystem. The requirement for more than one limiting factor is known as **resource colimitation**. In highly oligotrophic marine water, populations often depend on multiple limiting factors. For example, in the Baltic Sea, both nitrogen and phosphorus are present in such low concentrations that they must be added together in order to stimulate a phytoplankton bloom. In another example, the North Atlantic is limited for both phosphorus and iron. Addition of both P and Fe stimulates growth of nitrogen-fixing cyanobacteria.

To Summarize

- **Sulfate is abundant in marine water.**

- **Oxidized and reduced forms of sulfur are cycled in ecosystems.** Sulfate and sulfite serve as electron acceptors for respiration. Hydrogen sulfide serves as an electron donor. Sulfur oxidation to sulfuric acid causes acid mine drainage and pipe erosion.

TABLE 22.4	Microbial Metabolism of Metals		
Metal	**Major conversions**	**Microbial genera (examples)**	**Effects within environment**
Arsenic (As)	$AsO_3^{3-} \longrightarrow AsO_4^{3-}$	Alcaligenes, Pseudomonas	Removal of poisonous AsO_3^{3-} via use as terminal e^- acceptor, converting to insoluble AsO_4^{3-}.
	$AsO_4^{3-} \longrightarrow AsO_2^-$	Bacillus, Chrysiogenes, Pyrobaculum	AsO_4^{3-} (arsenate) used as terminal e^- acceptor.
	$AsO_4^{3-} \longrightarrow (CH_3)_3As$	Candida (a fungus), Scopulariopsis (a fungus)	Methylarsines. Poisonous; inhalation of moldy wallpaper with arsenic pigment.
Chromium (Cr)	$CrO_4^{2-} \longrightarrow Cr^{3+}$	Aeromonas, Arthrobacter, Desulfovibrio	CrO_4^{2-} [Cr(VI)] is mutagenic and carcinogenic; as e^- acceptor, reduced to Cr^{3+} [Cr(III)], less toxic.
Manganese (Mn)	$Mn^{2+} \longrightarrow Mn^{4+}$	Hyphomicrobium, Arthrobacter	Mn is a trace element required for enzymes.
	$Mn^{4+} \longrightarrow Mn^{2+}$	Geobacter, Pseudomonas	Anaerobic respiration in sediment.
Mercury (Hg)	$Hg^{2+} \longrightarrow Hg^0$	Acidithiobacillus	Hg^0 volatilizes; little harm.
	$Hg^{2+} \longrightarrow (CH_3)Hg^+$	Desulfovibrio	$(CH_3)Hg^+$ is a severe neurotoxin; accumulates at higher trophic levels, such as fish.
Selenium (Se)	$Se^0 \longrightarrow SeO_3^{2-}$	Bacillus, Micrococcus	Se is a trace element; small amounts in food help remove mercury. Larger amounts are toxic.
Uranium (U)	$UO_2^{2+} \longrightarrow UO_2$	Veillonella, Shewanella, Geobacter	UO_2^{2+} [U(VI)], soluble, is respired to uranium dioxide, UO_2 [U(IV)], insoluble; used for cleanup of radioactive uranium.
Vanadium (V)	$VO_3^- \longrightarrow VO(OH)$	Veillonella, Desulfovibrio, Clostridium	V is a trace element for some nitrogenases and invertebrate blood pigments.
			VO_3^- (vanadate) is oxidized as an e^- donor.

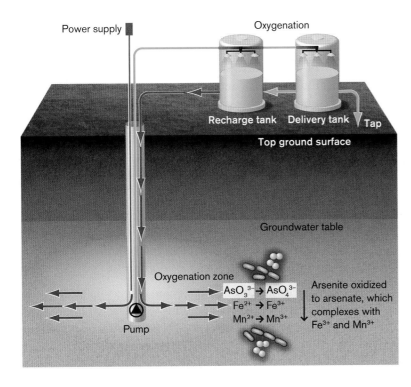

FIGURE 22.23 ▪ Subterranean arsenic removal. Bioremediation of arsenic-contaminated groundwater by lithotrophic bacteria. Water cycles through oxygenation tanks back to the source, where bacteria oxidize arsenite, iron, and manganese. The oxidized form of arsenic, which is arsenate, precipitates with iron and manganese for convenient removal.

■ **Phosphorus cycles primarily in the fully oxidized form (phosphate).** Phosphate limits growth of phototrophic bacteria and algae in some aquatic and marine systems.

■ **Iron cycles in oxidized and reduced forms.** Oxidized iron (Fe^{3+}) serves as a terminal electron acceptor in anaerobic soil and water. Reduced iron (Fe^{2+}) from rock is oxidized through weathering or mining. Bacterial lithotrophy accelerates iron oxidation, leading to acidification.

■ **Metal toxins can be metabolized by bacteria.** Bacterial metabolism may either increase or decrease toxicity. Metabolic conversion to an insoluble form, such as arsenite to arsenate, offers an effective means of bioremediation.

■ **Marine habitats show resource colimitation.** Multiple resources may be limiting for the phytoplankton community or for different populations.

22.6 Astrobiology

As we come to appreciate the ubiquitous contributions of microbes to shaping our planet, in all its diverse habitats, increasingly we wonder whether microbes exist on worlds beyond Earth. Is Earth unique in supporting life, or have living cells evolved as well on Mars or Venus or on Jupiter's planet-sized moons? Astrobiology is the study of life in the universe, including its origin and possible existence beyond Earth. The discovery of life beyond Earth would arguably be the most significant advance in science since a human set foot on the moon.

If the same physical and chemical laws govern the universe everywhere, then it is hard to suppose that only one of the billions of stars would have a planet supporting life. On the other hand, we have no idea how many planets have been capable of developing and sustaining a biosphere. As Isaac Asimov said, "There are two possibilities. Maybe we're alone. Maybe we're not. Both are equally frightening."

If life exists elsewhere, is it built on the same fundamental elements as ours? Many lines of evidence suggest that the biochemistry of life elsewhere would resemble that of Earth. Terrestrial life is founded on macroelements in the first two rows of the periodic table, including carbon, nitrogen, oxygen, phosphorus, and sulfur (see **Table 22.1**; for the periodic table, see Appendix 1). The valence numbers (that is, the number of electrons in the outer electron shell) of these elements from the middle of the periodic table enable them to form complex molecular structures

with strong covalent bonds. The same fundamental molecules that appear in early-Earth simulation experiments, such as adenine and glycine, also appear in meteorites. Thus, we suspect that the fundamental building blocks for biochemistry are universal. On the other hand, if life could indeed be founded on some other basis, how would we recognize it?

If life is found on other planets, it will almost certainly include microbes. In fact, a case can be made that the majority of biospheres in our galaxy would consist entirely of microbial life, since microbes inhabit a wider range of conditions than do multicellular plants and animals. Even on Earth itself, the largest bulk of our biosphere—including deep sediments and rock strata—consists of microbial ecosystems.

Did Mars Once Support a Biosphere?

For several reasons, the most studied candidate for extraterrestrial life has been the planet Mars:

■ **Geology.** Of all the solar planets, Mars seems the most similar to Earth in its topography; indeed, some areas of Mars remarkably resemble desertscapes on Earth. Martian rock contains the fundamental elements needed for life on Earth.

■ **Day and year length.** Mars has a day length similar to that of Earth, and a year only twice as long as ours.

■ **Temperature.** The average temperature on Mars is 220 K ($-53°C$), too cold for most biochemistry on Earth, but its temperature rises above freezing at the equator. By contrast, the torrid heat of Venus (460°C) would exclude stable macromolecules.

■ **Atmosphere.** Overall, Mars has an atmospheric pressure of 6 millibars (mbar), barely a hundredth that of Earth (1,013 mbar), and it lacks molecular oxygen. Thus, aerobes could not grow. But the Martian atmosphere does include carbon dioxide, actually at 20 times the CO_2 content on Earth, so there would be plenty for photoautotrophic production of biomass.

■ **Water.** Surface water freezes out of the Martian atmosphere, without existing as a liquid. But mineral formations suggest that liquid water flowed in the past. Liquid water may yet exist deep underground, supporting life forms similar to the endoliths of Earth's crustal rock.

The existence of liquid water is a key question because on Earth, wherever liquid water exists—even brine (concentrated salt) at $-20°C$—there we find microbial life. In 2015, NASA's *Mars Reconnaissance Orbiter* (*MRO*) found evidence of flowing water on Mars (see Chapter 1). Water condenses

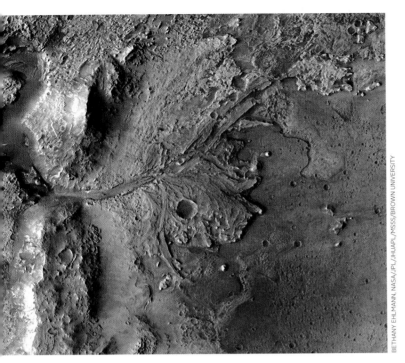

FIGURE 22.24 ■ Evidence of past water on Mars. Sedimentary deposits in the Mars Jezero crater, mapped by the *Mars Reconnaissance Orbiter* in 2007. The flow patterns resemble a river delta, best explained by a model involving the flow of liquid water. Green indicates claylike mineral deposits.

and freezes out of the Martian atmosphere, generating ice clouds and snow. The Martian soil composition showed high levels of perchlorate, a chemical that attracts water to form liquid solution and can support the metabolism of some Earth microbes.

Other evidence for liquid water in the past on Mars comes from geological formations surveyed by the *Mars Reconnaissance Orbiter* in 2007. The orbiter mapped sedimentary deposits in Mars's Jezero crater (**Fig. 22.24**). The crater formations reveal flow patterns typical of a river delta flowing into a lake. Claylike mineral deposits (false-colored green in **Fig. 22.24**) may have trapped organic compounds needed for life. The layered patterns are best explained by a model involving fluid flow, such as the flow of water.

If life once existed on Mars, what became of it? Two main possibilities are considered:

- **Life developed and existed until the planet froze.** Under this scenario, life originated much as it did on Earth, during a time of heavy bombardment from space and outgassing of nitrogen and carbon dioxide. Unfortunately, however, life failed to generate sufficient atmosphere for a greenhouse effect to sustain temperate conditions. No oxygenic organisms produced molecular oxygen and an ozone layer.

- **Life developed and still exists underground.** In the absence of an ozone layer, the Martian surface is sterilized by cosmic radiation. Nevertheless, microbes may yet exist deep underground, similar to the endolithic prokaryotes on Earth, or the subsurface brine communities of Antarctica (see **Special Topic 22.1**). Subsurface microbes are protected from cosmic radiation.

Do Biosignatures Indicate Life?

If microbes do exist on Mars, they should be detectable. But the detection of unknown life, even on Earth, remains a challenge. The advent of PCR sequence detection of species based on ribosomal genes has revealed thousands of unknown species, most of which cannot be grown or recognized by other methods. Other species may well be missed if their rRNA sequences fail to amplify with our probes. On Mars, assuming life evolved independently of Earth life-forms, we would have no way to define sequence probes for detection.

Instead, researchers try to define **biosignatures**, chemical and physical signs that could only have been formed by life. Most of the proposed biosignatures are based on types of evidence for life on Earth, either as fossils of ancient life (discussed in Chapter 17) or as signs of current life in extreme habitats. Proposed biosignatures include the following:

- **Microfossils.** The mineralization of microbial cells leads to the formation of structures that can be visualized under a microscope. The cell fossils must be sufficiently distinct to establish that no abiotic process could have formed them.

- **Isotope ratios.** Certain biochemical reactions preferentially use one isotope of an atom over another; for example, Rubisco, the key enzyme of carbon dioxide fixation, uses ^{12}C in preference to ^{13}C. Thus, both photosynthetic and chemolithoautotrophic use of ^{12}C can decrease the $^{12}C/^{13}C$ ratio in subsequent carbonate deposits. Isotope ratios of nitrogen, oxygen, and sulfur are also used as biosignatures.

- **Mineral deposits.** Certain mineral formations are observed to be caused only by microbial activity. For example, insoluble manganese oxides such as Mn_2O_3 are almost always the result of microbial oxidation of reduced manganese. Reduced manganese ions are very stable, and their abiotic oxidation rate is extremely slow.

- **Metabolic activity.** Samples of soil can be incubated with radioactive tracer substances such as $^{14}CO_2$ and tested for metabolic conversion or incorporation into biomass. It must be established that the conversion could not have occurred abiotically and that no organisms from Earth were present.

Various kinds of evidence for life on Mars have been reported, such as possible microfossils within a Martian meteorite that landed in Antarctica. Metabolic activity was tested in samples obtained by the NASA *Viking* lander in 1975. As of this writing, however, no evidence has proved conclusive for active or past living microorganisms on Mars.

Could Mars Be Terraformed to Support Earth Life?

If no life exists—or if it exists only in the form of microbes deep underground—should we consider human intervention, or **terraforming**, to make Mars habitable for life from Earth? Scenarios for terraforming, long explored in science fiction, are now receiving serious thought among space scientists. Terraforming Mars would require increasing the temperature and air pressure. The temperature of the atmosphere might be increased by release of greenhouse gases. In addition, sufficient carbon dioxide and nitrogen gases must be made available from the Mars surface rock. In principle, microbes from Earth could be seeded to grow and generate an atmosphere containing nitrogen, oxygen, and CO_2.

The dilemmas and consequences of terraforming are depicted in the novel *Red Mars* (1993), by Kim Stanley Robinson. In favor of terraforming, it is argued that Mars offers enormous natural resources of potential benefit for humanity, especially as terrestrial resources are used up. Human settlements on Mars would be a major step forward for space exploration. On the other hand, it is argued that the planet Mars is a natural monument, a place with its own right to exist as such. It should be allowed to remain in its natural state for future generations to appreciate. As a practical matter, terraforming remains unfeasible for the near future. For example, the amount of chlorofluorocarbons required to raise Martian temperature is calculated to be 100 times greater than our global capacity to produce such substances.

Does Europa Have an Ocean?

Farther out in the solar system, surprising candidates for microbial life are the moons of Jupiter. In 2000, the *Galileo* space probe passed several of Jupiter's moons, including Ganymede, Callisto, and Europa (**Fig. 22.25**). While their distance from the Sun results in extreme cold, these bodies receive extra heat from friction generated by tidal forces from the giant planet Jupiter. In the case of Europa, the tidal forces have been calculated to provide enough heat to liquefy water without boiling it off. Furthermore, measurement of Europa's magnetic field suggests that its composition includes a dense iron core surrounded by 15% water.

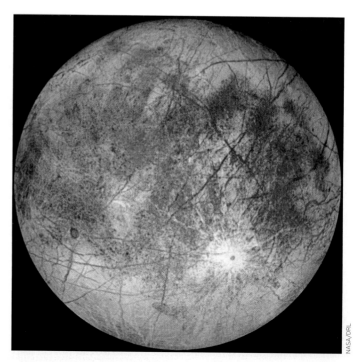

NASA/JPL

FIGURE 22.25 ■ Jupiter's moon Europa. Does life exist beneath Europa's ice, in a salty sea?

Most of the water must be locked in ice, but tidal heating could melt enough water for an underlying ocean of brine. On Earth, similar brine lakes beneath the ice of Antarctica harbor halophilic archaea that grow at –20°C.

If such oceans do exist, how could they support life without photosynthesis? Photosynthesis is impossible at Jupiter's distance from the Sun. However, an alternative source of chemical energy might be the influx of charged particles accelerated in Jupiter's magnetic field. Charged particles entering Europa's ice can react with water to form hydrogen peroxide (H_2O_2). The H_2O_2 then breaks down, releasing molecular oxygen. Oxygen reaching the brine layer below could combine with electron donors from crustal vents and power metabolism. Alternatively, molecular hydrogen (H_2) could be generated from water ionized by decay of radioisotopes. A similar source of H_2 for life has been proposed to occur on Earth in rock strata several kilometers below the surface, where it may support lithotrophs. On Europa, the hydrogen could then combine with oxygen gas or other oxidants to power life. These schemes are highly speculative—but tantalizing enough to encourage future NASA missions to take a closer look at Europa and its sister moons.

Does Life Exist on Planets of Distant Stars?

The past decade of astronomy has seen an extraordinary growth in our knowledge of solar systems beyond our own. Now that we know that so many other stars possess planets, we can only wonder whether they also possess biospheres of life-forms we would recognize.

In recent years, astronomers have found several hundred "extrasolar" planets orbiting distant stars, including some the size of Earth. Could we ever hope to detect signs of life on an extrasolar planet? A possible means of detection is suggested by William Sparks and colleagues at the Space Telescope Science Institute in Baltimore, and collaborating institutions. They show that light scattered by marine phytoplankton exhibits a property called "circular polarization." Circular polarization arises from substances that are homochiral—that is, present in only one of two mirror forms, such as the L and D forms of an amino acid. Homochirality is a strong biosignature, typical of proteins and metabolites. If someday we have telescopes capable of detecting light from distant planets, circular polarization might provide evidence of life.

Figure 22.26 shows an infrared photograph through the Spitzer Space Telescope of nebula RCW-49, a gaseous cloud full of newborn stars. Recall from Chapter 17 that as stars form, they take up dust of supernovas that includes all the elements needed to form biomolecules. Spectroscopic observation of RCW-49 reveals stars surrounded by disks coalescing into planets. The planetary disks contain icy particles full of organic molecules such as methanol, glycine, and ethylene glycol, a reduced form of sugar. Could there be biospheres in the making?

FIGURE 22.26 ■ A stellar nursery. Nebula RCW-49 contains more than 300 newborn stars, from which NASA's Spitzer Space Telescope detected spectroscopic signals of common organic constituents of life (infrared photograph).

■ **Jupiter's moon Europa is proposed as another possible site for life.** Europa is bathed in a sea of brine similar to terrestrial habitats for halophiles.

Concluding Thoughts

Our observations of distant molecules, as well as our analysis of meteorites (discussed in Chapter 17), suggest that the biomolecules of Earth fit into a universal pattern of interstellar chemistry. Whether life exists elsewhere or we are alone, we must remember that our Earth is the only place we know of at this time that can support humans and the forms of life we require for our own survival. The survival of our entire biosphere depends on our microbial partners cycling key elements and acquiring energy to drive the food web. For the first time in history, human technology now rivals the ability of microbes to alter fundamental cycles of biogeochemistry. But to manage and moderate our alterations—for our own survival and that of the biosphere—our fate still depends on the microbes.

To Summarize

■ **Astrobiology is the study of life in the universe, including possible habitats outside Earth.**

■ **The search for extraterrestrial life is based on methods similar to those used to seek early life on Earth.** Evidence includes chemical and physical biosignatures, isotope ratios, microfossils, and metabolic activity.

■ **Mars is the planet whose geology most closely resembles that of Earth.** Geological features strongly support the past existence of flowing water, a prerequisite for microbial life.

CHAPTER REVIEW

Review Questions

1. Identify the major sources and sinks of carbon, nitrogen, and sulfur. Which sources recycle rapidly, and why?
2. Explain how the carbon cycle differs in oxygenated and anoxic environments.
3. Explain two different chemical methods of measuring the environmental levels of carbon and nitrogen.
4. Outline the hydrologic cycle. Explain the role of biochemical oxygen demand (BOD) in water quality and how it may be perturbed by human pollution.
5. Outline the functions of a wastewater treatment plant. Include the phases of primary, secondary, and tertiary treatment. Explain the roles of microbes in these phases of water treatment.
6. Outline the main transformations of the nitrogen cycle. Which reactions are carried out only by microbes?
7. Outline the main transformations of the sulfur cycle. Which features are comparable to the nitrogen cycle, and which are unique to the sulfur cycle?
8. For the iron cycle, explain aerobic and anaerobic processes of microbial transformation. Explain how microbial iron transformation may be linked to sulfur transformation.
9. Explain how bacteria may convert metals to toxic forms. Alternatively, explain how bacteria may detoxify metal-polluted sediment.
10. Offer several arguments for and against the existence of life on planets beyond Earth.

Thought Questions

1. How does influx of nitrogen-rich fertilizers to soil ecosystems increase the rate of CO_2 efflux from wetlands?
2. In the past 50 years, most of the wetlands off the coast of Louisiana have been destroyed. How is wetland destruction related to formation of the dead zone in the Gulf of Mexico? How could the dead zone be revived, and why would the cost of restoration be projected to be billions of dollars?
3. What nutrients are limiting in the marine benthos, and why?

Key Terms

abiotic (868)
ammonification (885)
anammox (886)
assimilatory nitrate reduction (885)
biochemical oxygen demand (BOD) (876)
biogeochemical cycle (869)
biogeochemistry (869)
biological carbon pump (873)
bioremediation (868)
biosignature (895)
biotic (868)
dead zone (877)

denitrification (885)
dissimilatory nitrate reduction (885)
floc (879)
fossil fuel (875)
geomicrobiology (869)
greenhouse gas (868)
hydrologic cycle (876)
hypoxia (877)
mesocosm (870)
micronutrient (869)
nitrification (884)
nitrifier (884)
nitrogen fixation (883)

oxygen minimum zone (OMZ) (877)
reservoir (869)
resource colimitation (892)
siderophore (890)
sink (869)
sludge (879)
source (869)
terraforming (896)
wastewater treatment (878)
water cycle (876)
zone of hypoxia (877)

Recommended Reading

Arrigo, Kevin R. 2009. Marine microorganisms and global nutrient cycles. *Nature* **437**:349–355.

Canfield, Donald E., Alexander N. Glazer, and Paul G. Falkowski. 2010. The evolution and future of Earth's nitrogen cycle. *Science* **330**:192–193.

Clemmensen, Karina E., Adam Bahr, Otso Ovaskainen, Anders Dahlberg, Alf Ekblad, et al. 2013. Roots and associated fungi drive long-term carbon sequestration in boreal forest. *Science* **339**:1615–1618.

Gruber, Nicolas, and James N. Galloway. 2008. An Earth-system perspective of the global nitrogen cycle. *Nature* **451**:293–296.

Johnston, Andrew W. B., Jonathan D. Todd, and Andrew R. J. Curson. 2012. Microbial origins and consequences of dimethyl sulfide. *Microbe* **4**:181–185.

King, Gary M. 2015. Carbon monoxide as a metabolic energy source for extremely halophilic microbes: Implications for microbial activity in Mars regolith. *Proceedings of the National Academy of Sciences USA* **112**:4465–4470.

MacDougall, Andrew H., Christopher A. Avis, and Andrew J. Weaver. 2012. Significant contribution to climate warming from the permafrost carbon feedback. *Nature Geoscience* **5**:719–721.

Melton, Emily D., Elizabeth D. Swanner, Sebastian Behrens, Caroline Schmidt, and Andreas Kappler. 2014. The interplay of microbially mediated and abiotic reactions in the biogeochemical Fe cycle. *Nature Reviews. Microbiology* **12**:797–808.

Montzka, S. A., E. J. Dlugokencky, and J. H. Butler. 2011. Non-CO_2 greenhouse gases and climate change. *Nature* **476**:43–50.

Naqvi, Syed Wajih A., D. A. Jayakumar, P. V. Narvekar, H. Naik, V. V. S. S. Sarma, et al. 2000. Increased marine production of N_2O due to intensifying anoxia on the Indian continental shelf. *Nature* **408**:346–349.

Pollard, Raymond T., Ian Salter, Richard J. Sanders, Mike I. Lucas, C. Mark Moore, et al. 2009. Southern Ocean deep-water carbon export enhanced by natural iron fertilization. *Nature* **457**:577–581.

Sohm, Jill A., Eric A. Webb, and Douglas G. Capone. 2011. Emerging patterns of marine nitrogen fixation. *Nature Reviews. Microbiology* **9**:499–508.

Sparks, William B., James Hough, Thomas A. Germer, Feng Chen, Shiladitya DasSarma, et al. 2009. Detection of circular polarization in light scattered from photosynthetic microbes. *Proceedings of the National Academy of Sciences USA* **106**:7816–7821.

Stern, Jennifer C., Brad Sutter, Caroline Freissinet, Rafael Navarro-González, Christopher P. McKaye, et al. 2015. Evidence for indigenous nitrogen in sedimentary and aeolian deposits from the Curiosity rover investigations at Gale crater, Mars. *Proceedings of the National Academy of Sciences USA* **112**:4245–4250.

Van Mooy, Benjamin A. S., Helen F. Fredricks, Byron E. Pedler, Sonya T. Dyhrman, David M. Karl, et al. 2009. Phytoplankton in the ocean use non-phosphorus lipids in response to phosphorus scarcity. *Nature* **458**:69–72.

Wright, Jody J., Kishori M. Konwar, and Steven J. Hallam. 2012. Microbial ecology of expanding oxygen minimum zones. *Nature* **10**:381–394.

CHAPTER 23
Human Microbiota and Innate Immunity

The human body teems with microbes (the microbiome) vital to our existence and is under constant attack from pathogens. How do we survive? Obstacles such as skin and stomach acid, which are nonspecific defenses, will repel most microorganisms. But for microbes able to breach these physical barriers, there await innate and adaptive immune defenses. Chapter 23 explores the microbiome and describes the series of barriers and elaborate innate immune defenses that keep microbes at bay.

CURRENT RESEARCH highlight

Inflammasomes. An inflammasome is a multimeric disk that forms in the eukaryotic cytoplasm to initiate inflammation. Hao Wu studies inflammasome assembly using cryo-EM. She found that a single NOD-like receptor (NLR) protein (NAIP2) can initiate disk assembly after binding to a *Salmonella* secretion protein. Activated NAIP2 then nucleates the circular assembly of 10–11 monomeric blades of another NLR (NLRC4) to form the disk. At the base of the disk, the caspase-1 protease oligomerizes. Caspase-1 proteolytically activates precursors of cytokines IL-1β and IL-18. Cytokine secretion begins the process of inflammation.

Source: Zhang et al. 2015. Science **350**:404–409.

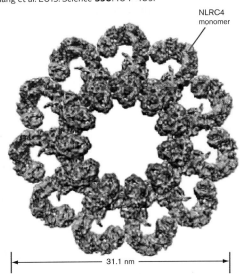

NLRC4 monomer

31.1 nm

AN INTERVIEW WITH

HAO WU, STRUCTURAL BIOLOGIST, HARVARD MEDICAL SCHOOL

HARVARD MEDICAL SCHOOL

How has your work changed the way we think about inflammasome assembly?

The field presumed that inflammasome assembly resembles that of the multisubunit apoptosome, a molecular machine that initiates programmed cell death in damaged cells. Every subunit of an apoptosome must bind to a ligand before assembly. Here we, as well as Jijie Chai's group, showed that only one subunit of the inflammasome (NAIP2) needs to bind ligand. The conformationally activated sensor NAIP then triggers stepwise assembly of NLRC4 subunits through ATPase-mediated polymerization. NLRC4 does not need to bind microbial ligands.

Would interfering with the inflammasome have an effect on infection or cancer?

An acute infection would likely worsen if we inhibited inflammasome assembly, because reduced inflammation means the microbe could grow more. Fortunately, other inflammasome-independent mechanisms can still initiate inflammation. In the context of long-term, chronic inflammation, inhibiting inflammasome assembly may have hugely beneficial effects, such as in inflammatory diseases and autoimmunity. In contrast, activating inflammasomes could, in some cases, enhance antitumor immunity, as suggested by the protective role of caspase-1 protease in a model of colorectal cancer.

Two vacationing tourists, unknown to each other, were swimming in the Gulf of Mexico along the Alabama coast. One, a 12-year-old girl, had a cut on her leg caused by the sharp edge of a scooter. The other, a man 66 years of age, had recently cut his arm on a nail protruding from a wooden handrail. Within moments of entering the water, several small Gram-negative microbes invaded both of their bodies through those cuts. Four days later, the young girl was heading back to Birmingham, sitting in the back of the family car, oblivious to the battle recently waged in her bloodstream. At the same time, the man lay dead of an aggressive blood infection.

Both swimmers were attacked by the same pathogen, *Vibrio vulnificus*. Why did one live and the other die? As is the case for most healthy people, a variety of nonspecific, innate immune factors present in the girl's body killed the invading pathogen before it could multiply. The man, on the other hand, was an alcoholic with liver disease. He was deficient in several of those defense mechanisms—mechanisms that made the difference between life and death. What are these powerful innate factors? How can they be nonspecific, able to attack many different microbes without ever previously encountering any of them? And why do they kill the invaders and not our own cells? The inflammasome described in the Current Research Highlight is one of those important innate factors.

Before exploring questions of immunity and self-defense, we must understand that each of us is a self-contained ecosystem, home to numerous microbes inhabiting all sorts of body niches. The day-to-day presence of these microbial guests both shapes and sharpens our immune system, but it is a benefit that comes with considerable risk.

23.1 Human Microbiome

From the moment of our birth to the time of our death, we are constantly populated by microbes. These include bacteria, archaea, fungi, viruses, and even some protozoa. The consortium of colonizing microbes has been dubbed the human **microbiota** or **microbiome**.

Note: Usage of the terms "microbiota" and "microbiome" has blurred. "Microbiota" originally referred to actual microbes inhabiting a body site, whereas "microbiome" referred to the gene pool within those microbes.

To illustrate the significance of the microbiome, consider that our bodies carry about 10 times as many bacterial cells as nucleated human cells and about 100 times more nonredundant bacterial genes than human genes. (A recent estimate that includes non-nucleated red blood cells places the bacterial/human cell ratio closer to between 1:1 and 2:1.) Microbes colonize wherever our body meets the external environment

(for example, skin, mouth, gastrointestinal tract, and parts of the genitourinary tract). Most internal organs, blood, and cerebrospinal fluid are sterile. The presence of any microbes at these sites is considered an infection. The majority of species that make up our microbiota are unknown and have never been grown in the laboratory. Consequently, metagenomic strategies (described in Chapters 7 and 21) are being employed to identify our resident microbes. In 2016, the U.S. government announced the National Microbiome Initiative to advance understanding of how microbiomes contribute to our health and the environment.

By and large, the body's barriers of defense work well to prevent incursions by microbiota or, in the event of an incursion, to kill the invader. Unfortunately, these defenses break down when an immune system is compromised by medical treatments (such as anticancer drugs) or by diseases (for instance, a deficiency in complement factors, discussed in Section 23.6). Such a person is described as a **compromised host** and can be repeatedly infected by certain normal biota. Organisms causing disease in this situation are called **opportunistic pathogens**.

We now understand that bacteria colonizing our bodies may be as important collectively as a kidney or liver. For instance, Gary Siuzdak and his colleagues at the Scripps Research Institute showed that our body cells are continually bathed in metabolites produced by gut microbes. It is now thought that these metabolites, which circulate in our blood, can have positive as well as negative effects on human health and development.

Microbiome Acquisition and the Hygiene Hypothesis

When and how does a person develop a microbiome? Recent evidence suggests that the microbiome begins to develop before birth. The first stool (meconium) passed by a newborn, <u>before</u> its first meal, contains some bacteria. The source of those microbes may be the placenta, which has its own microbiome.

Once the baby breaks out of the embryonic membrane, however, it is exposed to a dizzying array of microbes residing in the birth canal and the outside world. Contributions to the baby's skin, oral, gut, and genitourinary microbiomes come from the food they eat, the air they breathe, and the people, places, and things they touch. Young babies actually have microbiomes that are more diverse than those of adults. But by 3 years old, the diversity decreases in complexity to assume adult composition.

Figure 23.1 presents the major human body sites (skin, respiratory, digestive, and genitourinary tracts) that are colonized by microbes, and illustrates the relative makeup of bacteria, fungi, and viruses that comprise each microbiome. As you view the figure, notice the dramatic difference

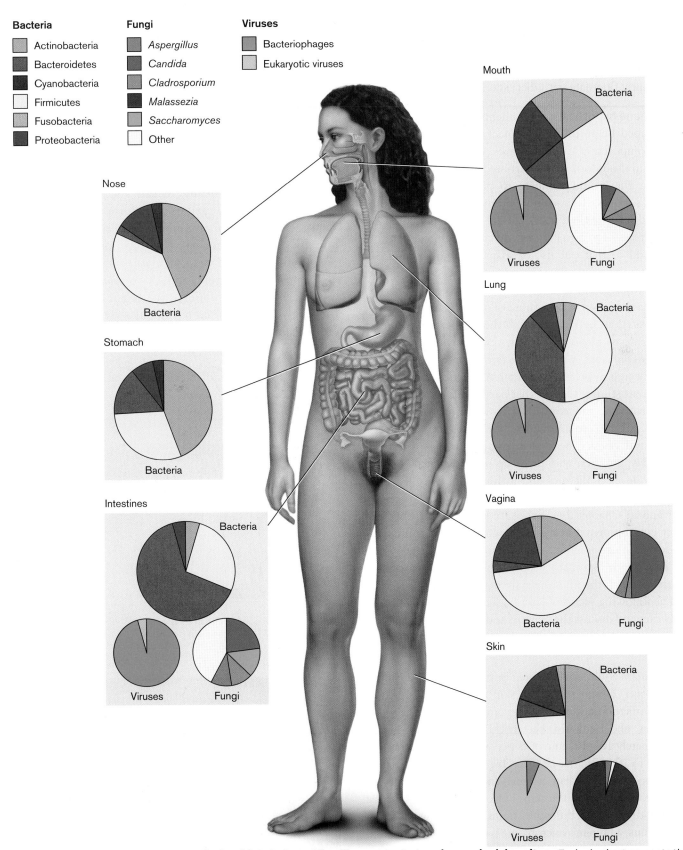

FIGURE 23.1 ■ **Relative amounts of microbial phyla and families present at various colonizing sites.** Each pie chart presents the relative compositions of bacterial phyla, genera of fungi, or viruses present in each microbiome as determined by the sequencing of genes encoding 16S ribosomal RNA (see Sections 7.6 and 15.5). Classes of microbes are identified by color in the legend. *Source:* Data from Marsland and Gollwitzer. 2014. *Nat. Rev. Immunol.* **14**: 827.

in levels of Bacteroidetes when comparing nasal and intestinal microbiomes, or of *Candida* in the microbiomes of the mouth and vagina. Composition within an individual's microbiome remains relatively constant over time but can significantly fluctuate with diet, age, geography, or drugs.

The hygiene hypothesis. For millennia, human microbiota were shaped by an intimate contact with natural environments composed of animals, caves, dirt, poop, and bugs. This natural outdoor world harbored a vast array of microbial taxa that could compete to populate our skin and mucosal surfaces. Today we are an indoor species spending almost all of our time inside closed buildings, segregated from nature—an arguably less diverse microbial environment. Add to that our use of soaps, antibiotics, and disinfectants and you can appreciate how severely we have restricted our access to microbes. As a result, our microbiota appear less diverse than those of our long-ago ancestors.

Studies that have compared the modern-day human microbiome with those of closely related wild African apes and of uncontacted Amerindians (see Chapter 27) bear this out. Although improved hygiene limits exposure to pathogens, recent studies suggest that narrowing the diversity of our microbiome can contribute to inflammatory diseases such as asthma, inflammatory bowel disease, colorectal cancer, and obesity. Several intriguing studies find that, even today, children who grow up on dairy farms are less likely to develop allergies and asthma. The reason is that our microbiome helps train our immune system (discussed later and in Chapter 24). Exposure to more microbes and other environmental antigens, especially early in life, may produce a more tolerant, well-controlled immune system less prone to inflammatory and autoimmune diseases (disorders in which the immune system reacts against the self).

The following sections describe the microbiota of various body sites and discuss the benefits and risks of a microbiome.

Skin

The human adult, on average, is covered with 2 square meters (over 21 square feet) of skin (**epidermis**) populated by 10^{12} microorganisms. Skin microbiota include aerobes, anaerobes, and facultative bacteria. There are, for instance, approximately 10^4–10^5 microbes per sweat gland at a ratio of 1 aerobe to 10 facultative or anaerobic species. As with all colonized body sites, resident (normal) and transient members of the microbiota inhabit the skin. But even a resident microbe exhibits diversity as different strains colonize at different times. Thus, a human's microbiome is not a static population, but represents a vibrantly changing mix of strains and species.

Several features of epidermis make it difficult to colonize. Large expanses of skin are subject to drying, although some areas harbor enough moisture to support microbial growth; moist areas include scalp, ear, armpit,

and genital and anal regions. The skin has an acidic pH (pH 4–6) owing to the secretion of organic acids by oil and sweat glands. As noted in Section 5.4, organic acids inhibit microbial growth by lowering bacterial cytoplasmic pH. Epidermal secretions are high in salt and low in water activity (see Section 5.3), and they contain enzymes, such as lysozyme, that degrade bacterial peptidoglycan.

Despite these hurdles, many species of bacteria manage to colonize the epidermal habitat. Most of these organisms are Gram-positive, because Gram-positive bacteria tend to be more resistant to salt and dryness. *Staphylococcus epidermidis*, various *Bacillus* species, and yeasts such as *Candida* are common examples of normal skin microbes. Though normal, they are not always innocuous. One member of the skin's resident microbiota—the Gram-positive, anaerobic rod *Propionibacterium acnes*—causes acne, a very visible plague of adolescence. During the teen years, increased hormonal activity stimulates oil production by the sebaceous glands (**Fig. 23.2**). *P. acnes*

A.

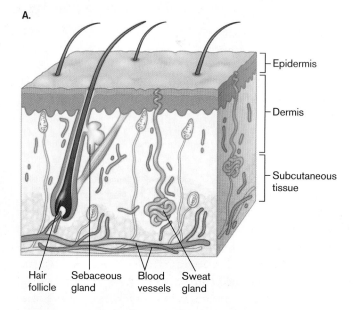

Hair follicle — Sebaceous gland — Blood vessels — Sweat gland

Epidermis — Dermis — Subcutaneous tissue

B.

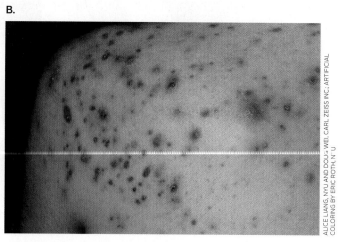

FIGURE 23.2 ■ Microbiology of skin and the development of acne. A. The location of sebaceous glands in skin. **B.** Acne.

readily degrades the triglycerides in this oil, turning them into free fatty acids that then promote inflammation of the gland. One consequence of the inflammatory response is the formation of a blackhead, a plug of fluid and keratin that forms in the gland duct. The result is the typical skin eruptions of acne. Because of its microbial basis, treatments for acne include tetracycline (oral) or clindamycin (topically applied) to kill the bacteria.

Eye

The eye is exposed to the outside environment, so is it heavily colonized? Actually, it's not, because colonization is inhibited by the presence of antimicrobial factors, such as lysozyme, in the tears that continually rinse the eye surface (conjunctiva). Despite this protection, a few transient commensal bacteria can be found on the conjunctiva. Skin microbiota such as *Staphylococcus epidermidis* and diphtheroids (Gram-positive rods that look like clubs), as well as some Gram-negative rods, such as *Escherichia coli*, *Klebsiella*, and *Proteus*, manage, at least temporarily, to make the eye their home without causing damage. (The traits of these genera were discussed in Chapter 18.) Occasionally, bacteria such as *Streptococcus pneumoniae* and *Haemophilus influenzae*, as well as some viruses, can cause an ocular disease known as pinkeye, in which the eye becomes reddened with a watery discharge.

Oral and Nasal Cavities

Within hours after birth, a human infant's mouth becomes colonized with nonpathogenic *Neisseria* species (Gram-negative cocci), *Streptococcus*, *Actinomyces*, *Lactobacillus* (all Gram-positive), and some yeasts. These organisms come from the environment surrounding the newborn, such as the mother's skin and garments. As teeth emerge in the newborn, the anaerobic space between teeth and gums supports the growth of anaerobes, such as *Prevotella* and *Fusobacterium* (**Fig. 23.3B** and **C**). Whatever the organism, colonizers of the oral cavity must be able to adhere to surfaces, like teeth and gums, to avoid mechanical removal and flushing into the acidic stomach. The teeth and gingival crevices are colonized by over 500 species of bacteria.

Organisms such as *Streptococcus mutans* (which attaches to tooth enamel) and *Streptococcus salivarius* (which binds gingival surfaces) form a glycocalyx that enables them to firmly adhere to oral surfaces and to each other. They are but two of the microbes that lead to dental plaque formation. The acidic fermentation products of these organisms demineralize teeth and cause dental caries (tooth decay).

Important microbial habitats of the throat include the **nasopharynx**, which is the area leading from the nose to the oral cavity, and the **oropharynx**, which lies between the

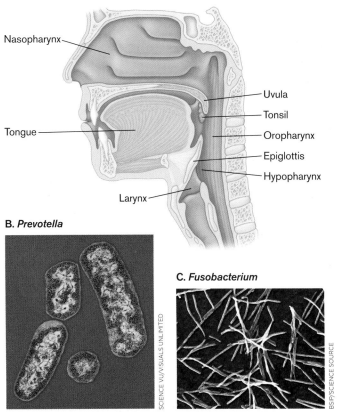

A. Oral and nasal cavities

Nasopharynx

Uvula

Tonsil

Oropharynx

Epiglottis

Hypopharynx

Tongue

Larynx

B. *Prevotella*

SCIENCE VU/VISUALS UNLIMITED

C. *Fusobacterium*

BSIP/SCIENCE SOURCE

D. Periodontal disease

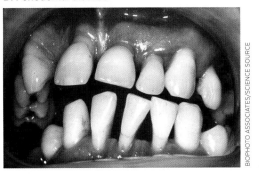

BIOPHOTO ASSOCIATES/SCIENCE SOURCE

FIGURE 23.3 ■ Examples of anaerobic oral microbiota and periodontal disease. **A.** Structures of the oral and nasal cavities. **B.** *Prevotella* (colorized TEM; each cell approx. 2 µm long). **C.** *Fusobacterium* (SEM; each cell approx. 1–5 µm long). **D.** Symptoms of periodontal disease include red, swollen gums; bleeding gums; gum shrinkage; and teeth drifting apart.

soft palate and the upper edge of the epiglottis (**Fig. 23.3A**). Organisms like *Staphylococcus aureus* and *Staphylococcus epidermidis* populate these sites. The nasopharynx and oropharynx can also harbor relatively harmless streptococci such as *Streptococcus salivarius* and *Streptococcus oralis*, as well as *Streptococcus mutans*, a major cause of tooth decay. Other oropharyngeal organisms include a large number of diphtheroids and the small Gram-negative rod *Moraxella catarrhalis*. Within the tonsillar crypts (small pits along the tonsil surface) lie anaerobic species such as *Prevotella*, *Porphyromonas*, and *Fusobacterium*.

A.

B.

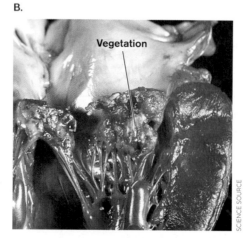

Vegetation

FIGURE 23.4 ▪ **Gross pathology of subacute bacterial endocarditis involving the mitral valve.** **A.** An open, normal mitral valve shows cords that tether it to the heart wall. **B.** The left ventricle of the heart has been opened to show mitral valve fibrin vegetations due to infection. These growths are not present in a normal heart.

The oral microbiota are normally harmless, but they can cause disease. Dental procedures, for instance, will often cause these organisms to enter the bloodstream, producing what is called **bacteremia** (infection of the bloodstream). Normal immune mechanisms typically clear these transient bacteremias quite easily; but in patients who have a mitral valve prolapse (heart murmur), microbes can become trapped in the defective valve and form bacterial vegetations. Vegetations are biofilms that contain a large number of bacterial cells encased within glycocalyx (a polysaccharide or peptide polymer secreted by the organism) and fibrin (produced by clotting blood). Because the onset of disease is often insidious (slow), it is called subacute bacterial endocarditis (**Fig. 23.4**). Once ensconced within a vegetation, the microbes are extremely difficult to kill with antibiotics. Immunocompetent individuals are at very low risk of valve infection. Immunocompromised patients, however, are at much higher risk and receive prophylactic treatment with antibiotics before any dental procedure.

Thought Question

23.1 How can an anaerobic microorganism grow on skin or in the mouth, both of which are exposed to air?

Respiratory Tract

The respiratory tract has a surface area of approximately 246 square feet (75 square meters) directly exposed to the environment. Originally thought to be sterile, the lungs and trachea are now recognized to harbor normal microbiota. Estimates in the lower respiratory tract are 10–100 bacteria per 1,000 human cells, the most prominent members being from the phyla Bacteroidetes and Firmicutes.

Many organisms entering the nasopharynx become trapped in the nose by cilia that beat toward the pharynx. The microbes are propelled toward the acidic stomach and death. Microorganisms that slip into the trachea are trapped by mucus produced by ciliated epithelial cells lining airways. Cilia usher the microbes up and away from the lungs. The ciliated mucous lining of the trachea, bronchi, and bronchioles makes up the **mucociliary escalator** (**Fig. 23.5**). The mucociliary escalator constantly sweeps foreign particles up and out of the lungs. This action is extremely important for preventing respiratory infections. When the mucociliary escalator fails, as when it is covered with tar from years of heavy smoking, or when it is overwhelmed by the inhalation of too many infectious microbes, infections such as the common cold (for example, rhinovirus) or pneumonia (for example, *Streptococcus pneumoniae*) can result.

Genitourinary Tract

Much of the genitourinary tract is normally free from microbes. These areas include the kidneys (which remove waste products from the blood) and the ureters (which remove urine from the kidneys). The urinary bladder, which holds urine until it is excreted, was thought to be sterile but is now known to harbor microbes, mainly anaerobes. The distal urethra, however—because of its

FIGURE 23.5 ▪ **Mucociliary escalator.** Movement of these hairlike cilia ushers particles up and out of the trachea and lungs (colorized SEM; diameter between 0.5 and 1 μm).

proximity to the outside world—normally contains *Staphylococcus epidermidis, Enterococcus* species, and some members of Enterobacteriaceae. Some of these organisms can cause bladder disease (known as urinary tract infection, or UTI) if they make their way into the bladder (for example, via catheterization).

The large surface area and associated secretions of the female genital tract make it a rich environment for microbes, including bacteria and yeasts. Firmicutes are the largest bacterial contributor to the vaginal microbiome; however, composition changes with the menstrual cycle, owing to changing nutrients and pH. The mildly acidic nature of vaginal secretions (approximately pH 4.5) discourages the growth of many bacteria. As a result, the acid-tolerant *Lactobacillus crispatus* is among the most populous vaginal species. Healthy women appear to fall into two broad categories: 70% have lactobacillus as the primary member of their vaginal microbiota, while 30% have mixed species with few lactobacilli. Women in the latter group appear to be more susceptible to sexually transmitted infections.

The balance between different species that comprise vaginal microbiota is crucial to preventing disease. Antibiotic therapy to treat an infection anywhere in the body can also affect the vaginal microbiota. The resulting imbalance can allow overgrowth of *Candida albicans*, otherwise known as a yeast infection. *C. albicans*, as a fungus, is not susceptible to antibiotics designed to kill bacteria.

Stomach

We have known for a hundred years that the stomach contents are acidic and that gastric acidity can kill bacteria. Just how important that acidity is for protection against microbes is illustrated by the infection caused by *Vibrio cholerae*, the causative agent of cholera. Cholera is a severe diarrheal disease endemic to many of the poorer countries of the world. Although cholera actually affects the intestines, not the stomach, the bacteria must survive passage through the stomach to reach the intestines. The organism, however, is extremely acid sensitive. Healthy volunteers must ingest a trillion organisms before contracting disease. Despite the high infectious dose, cholera epidemics in developing countries kill tens of thousands of people every year. Part of the reason is that the poor, malnourished populations in these countries suffer from hypochlorhydria (decreased stomach acid). The less-acidic stomach gives ingested microbes more time to enter the intestine, where they can thrive and cause devastating disease. Compare the astronomical number of acid-sensitive *V. cholerae* needed to cause disease to the mere ten organisms needed to develop disease from *Shigella*, a very acid-resistant pathogen.

Although the stomach contents are very acidic, the mucous lining of the stomach is much less so. It is there that some bacteria can take refuge, primarily Actinobacteria and Firmicutes

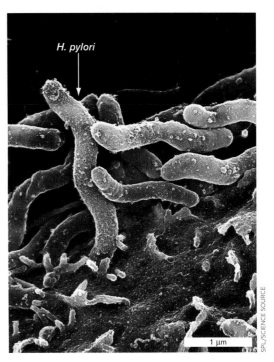

H. pylori

FIGURE 23.6 ■ *Helicobacter pylori* defies stomach acidity. *H. pylori* growing in the mucus on stomach epithelium (colorized SEM). *H. pylori* bacteria attach to gastric epithelial cells and induce specific changes in cellular function, such as an increase in the expression of laminin receptor 1, a protein associated with malignancy. Peptic ulcers can result.

(see **Fig. 23.1**). In fact, the stomach harbors a diverse microbiota, as detected using cultural and molecular techniques. Estimates are between 10 and 1,000 organisms per gram.

A classic stomach pathogen is *Helicobacter pylori*. This organism has a remarkable ability to resist acidic pH. (It survives at pH 1, using the enzyme urease to generate ammonia, which neutralizes acid.) However, *H. pylori* will not grow in strong acid pH conditions, but it can grow in the mucous lining of the stomach, where the pH is closer to 5 or 6 (**Fig. 23.6**). The U.S. Centers for Disease Control and Prevention (CDC) estimates that *Helicobacter* colonizes the stomachs of half the world's population. Most of the time, *Helicobacter* does not cause any apparent problem, but on occasion, the organism can produce gastric ulcers and even cancer. On the plus side, recent data suggests that *H. pylori* colonization in children is associated with a reduced risk for allergic disease.

Intestine

The intestine is an extremely long tube consisting of several sections, each of which supports the growth of different combinations of bacterial species. Pancreatic secretions (at pH 10) enter the intestine at a point just past the stomach and raise the intestinal pH to about pH 8, which immediately relieves the acid stress placed on microbes. The relatively high pH and bile content of

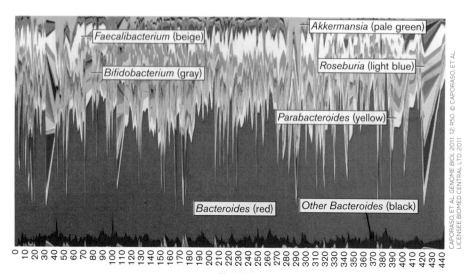

Faecalibacterium (beige)

Bifidobacterium (gray)

Akkermansia (pale green)

Roseburia (light blue)

Parabacteroides (yellow)

Bacteroides (red)

Other Bacteroides (black)

CAPORASO, ET AL. GENOME BIOL 2011 12:R50. © CAPORASO, ET AL. LICENSEE BIOMED CENTRAL LTD. 2011

Days

FIGURE 23.7 ▪ **Variability of gut microbiota over time.** Daily fecal samples taken from one individual were analyzed to identify and quantify resident bacterial families by sequencing 16S rRNA. Selected genera are identified. Colored segments at each time point represent the proportions of specific genera relative to the entire population.

the duodenum and jejunum allow colonization by only a few resident, mostly Gram-positive, bacteria (enterococci, lactobacilli, and diphtheroids). These particular Gram-positive organisms (Firmicutes phylum) possess a bile salt hydrolase that helps them grow in the presence of intestinal bile salts. The distal parts of the human intestine (ileum and colon) have a slightly acidic pH (pH 5–7) and a lower concentration of bile salts—conditions that support a more vibrant ecosystem. The intestine actually contains roughly 10^{11}–10^{13} bacteria per gram of feces. This generally anaerobic environment is populated by both anaerobic and facultative microbes in a ratio of 1,000 anaerobes to one facultative organism. Why is the intestinal lumen anaerobic? The small amount of oxygen that diffuses from the intestinal wall into the lumen is immediately consumed by the facultative bacteria, such as *E. coli*, which renders the environment anaerobic (discussed in Chapter 21).

Acquisition of gut microbiota. Infant intestines are initially colonized by large numbers of *E. coli* and streptococci that quickly generate a reducing environment able to support growth of strict anaerobic species—mainly *Bifidobacterium*, *Bacteroides*, *Clostridium*, and *Ruminococcus*. The intestines of breast-fed babies are dominated by bifidobacteria, possibly because of growth factors present in breast milk. In contrast, the microbiota of formula-fed infants is more diverse, with high numbers of Enterobacteriaceae, enterococci, bifidobacteria, *Bacteroides*, and clostridia. As seen in **Figure 23.1**, by adulthood over 90% of the bacteria in the intestine are composed of just two

phyla: Bacteroidetes (for example, *Bacteroides* species) and Firmicutes (for example, Clostridiales clusters XIV and IV) followed by Proteobacteria (for example, *Escherichia coli*) and Actinobacteria (for example, *Bifidobacterium* species). Besides bacteria, other inhabitants are the yeast species *Candida albicans*, and protozoa such as *Trichomonas hominis* and *Entamoeba hartmanni*.

Some 200 different bacterial species and a few methanogenic archaea comprise the intestinal microbiome. One reason the intestine can support such a large and eclectic mix of species is that different bacteria attach to different host cell receptors. Another reason is that many different food sources are available to support diverse groups of microbes (discussed previously, in Chapter 21). Even human breast milk evolved to encourage the growth of specific subpopulations of microbiota. Breast milk contains carbohydrates that nourish babies (lactose), but also includes complex carbohydrates that only our microbiota can digest.

Makeup of the gut microbiome within a single individual is relatively constant but still varies over time. **Figure 23.7** follows the fecal biota from a single individual over a period of 14 months. The relative proportions of bacterial families fluctuate significantly but tend to return to typical adult composition. Across the spectrum of the human population, however, the composition of intestinal microbiota is a continuum of many different combinations of microbes (enterotypes). At one end of the spectrum are microbiomes dominated by *Bacteroides*, while at the other end are communities dominated by *Prevotella* (another Gram-negative anaerobic species). One study showed, for example, that people with high protein and animal-fat diets have fecal communities enriched with the *Bacteroides*-dominant enterotype, while those with carbohydrate-rich diets have fecal communities dominated by *Prevotella*. The significance of this difference is not yet clear.

Keeping our microbiome at bay. Clearly, our intestines are packed with microbes. Why don't they cause chronic intestinal inflammation? Lora Hooper (**Fig. 23.8A**) and colleagues at the University of Texas Southwestern Medical Center discovered part of the answer: an antimicrobial "force field" made of antimicrobial lectins (carbohydrate-binding proteins). Whenever bacteria contact epithelial cells, these lectins are secreted into the mucus layer lining the intestine and kill bacteria that get too close. **Figure 23.8B** shows that microbes (green) are kept away from

A.

B.

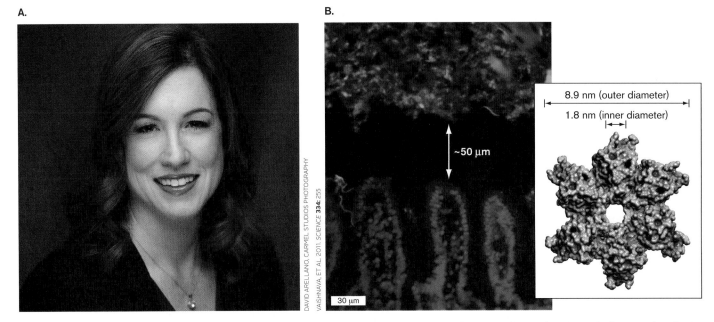

FIGURE 23.8 ■ **Keeping the gut microbiome at bay.** **A.** Lora Hooper. **B.** Separation of the microbiome from intestinal mucosal surface. Bacteria are green (FISH using a DNA probe that hybridizes to bacterial 16S rRNA genes), and the nuclei of intestinal mucosal cells are blue (DAPI stained). **Inset:** Model of the RegIIIα pore that forms in Gram-positive cell membranes.

the intestinal villi (blue) of the mouse intestine, in part because of these lectins. The separation between microbiome and host mucosal cells minimizes microbiome activation of an immune response that could severely damage the intestinal lining. Two of the antimicrobial lectins are RegIIIα and RegIIIγ. Hooper's group discovered that antimicrobial lectin RegIIIα (**Fig. 23.8B,** inset; cryo-EM map) forms a membrane-penetrating pore in Gram-positive microbes. RegIIIα and -γ, therefore, promote peaceful coexistence between host and microbiome. Other immune mechanisms that control microbiota in the intestinal mucosa, such as secretory IgA antibodies, are described in Chapter 24.

Thought Questions

23.2 Why do many Gram-positive microbes that grow on the skin, such as *Staphylococcus epidermidis*, grow poorly or not at all in the gut?

23.3 Why can the colon be considered a fermenter?

To Summarize

■ The **normal microbiota** present on skin and mucosal surfaces is acquired at birth but changes over a lifetime. **Skin microbiota** consists primarily of Gram-positive microbes, including *Propionibacterium acnes*, which can cause acne. **Oral and nasal surfaces** are colonized by aerobic and anaerobic microbes. **Vaginal micro-**

biota influences susceptibility to sexually transmitted infections.

■ **Microbe-free areas of the body** include blood, cerebrospinal fluid, and internal organs.

■ **Normal microbiota can cause disease** if organisms gain access to the circulation or deeper tissues.

■ **Opportunistic pathogens** infect only compromised hosts.

■ **The intestine** is populated by 10^{11}–10^{13} microbes per gram of feces. Principal phyla are Firmicutes, Bacteroidetes, Proteobacteria, and Actinobacteria. The ratio of anaerobes to facultative bacteria is 1,000:1. The gut microbiome includes bacteria, archaea, fungi, and viruses.

23.2 Benefits and Risks of Microbiota

Our colonizing microbes are not hostile armies camped at our body's gates (also called portals of entry) waiting to invade. These microorganisms are of great benefit to us, and we, as good hosts, reciprocate. Microbes in our gut, for instance, have enzymes to catabolize foods that we cannot digest. Some bacteria synthesize vitamins that we cannot make (*E. coli*, for instance, makes vitamin B_{12}). In addition, chemical signals released by members of the microbiome promote host tissue development. Germ-free animals raised in laboratories, for instance, have very thin intestinal walls. The short-chain fatty acid products of fermentation affect host gene expression and can influence host immune system function. Acetate, for instance, tends to promote inflammation, whereas butyrate tends to inhibit inflammation.

We, for our part, maintain our gut microbiome by secreting complex carbohydrates that members of the microbiome can use, and chemical factors, such as hormones, that can alter microbiome gene expression.

Examples of Beneficial Gut Microbiota

An important gut microbe that promotes host tissue rejuvenation is *Akkermansia muciniphila*, a Gram-negative, anaerobic bacterium (Verrucomicrobia phylum) that degrades mucin, the glycosylated protein layer that covers gut epithelium. Mucin degradation by *A. muciniphila* fosters continuous rejuvenation of the protective mucin layer. The short-chain fatty acids produced by this organism feed epithelial metabolism and modulate inflammatory responses of immune cells.

A well-studied model anaerobe, *Bacteroides thetaiotaomicron*, benefits us by breaking down the complex carbohydrates we eat. Much of this organism's genome is dedicated to taking many of the complex carbohydrates we eat and breaking them down into products that can be absorbed by the body or used by other members of the microbiome. We humans actually absorb 15%–20% of our daily caloric intake in this way. A detailed example is discussed in Chapter 21.

Other benefits our microbiota provide include limiting infections by competing with pathogens for food sources and attachment receptors on host cells. Colonizing bacteria can also make antimicrobial compounds that limit growth of potential pathogens. For example, nonpathogenic gut *E. coli* produces bacteriocin (a secreted protein toxin), which directly inhibits growth of the related pathogen enterohemorrhagic *E. coli* (EHEC; see Chapter 25).

Gut microbes also influence the development and efficiency of our immune system. A polysaccharide produced by *Bacteroides fragilis* stimulates production of an anti-inflammatory cytokine that will prevent gastrointestinal colitis caused by *Helicobacter hepaticus*. **Cytokines** are small, secreted host proteins that bind to cells of the immune system and regulate their function. (Cytokines are discussed in Sections 23.5 and 24.3.)

The intestinal microbe *Faecalibacterium prausnitzii*, a member of the order Clostridiales (Firmicutes phylum), elicits powerful anti-inflammatory effects on human immune cells through secreted bacterial factors and interactions with Toll-like receptors (see Section 23.5). For example, it inhibits induction of the inflammatory cytokine IL-8 and increases the population of an anti-inflammatory T-cell lymphocyte population (T regulatory cells; see Chapter 24). The absence of this organism in human intestines has been linked to painful inflammatory intestinal diseases such as Crohn's disease.

Another important group of intestinal anaerobes consists of the segmented filamentous bacilli (SFBs) in mice and probably humans (see Special Topic 18.1). These spore formers are important for developing a healthy immune system. We will discuss the connections between gut microbiota and the immune system more fully in Chapter 24. The anti-inflammatory effects of these and other microbes protect the intestinal mucosa from an overzealous immune system.

Containment Breaches and Disrupted Balance

Everything works well, as long as the composition of our microbiota is balanced and stays where it belongs. However, the accidental penetration of certain organisms beyond a site of colonization or an imbalance in microbiome composition, called **dysbiosis**, can cause infections and inflammatory diseases, respectively. A cancerous lesion in the colon, for example, can provide a passageway for microbes to enter deeper tissues and cause infection. *Bacteroides fragilis*, a harmless anaerobe in the gut, can invade tissues through surgical wounds, causing intra-abdominal abscesses and even gangrene after abdominal surgery (**Fig. 23.9**). Other infections caused by specific gut microbes escaping the intestine include urinary tract infections (cystitis), septicemia, and meningitis.

Emotional stress, a change in diet, or antibiotic therapy can alter gut microbiome balance. The resultant dysbiosis can lead to poor digestion, pseudomembranous enterocolitis [an infectious disease caused by *Clostridioides difficile* (formerly *Clostridium difficile*)] (**Fig. 23.10**), or an extremely painful

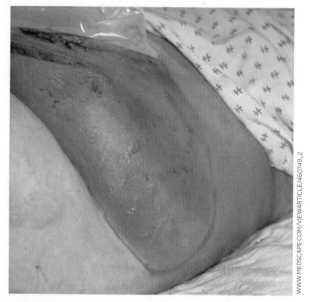

FIGURE 23.9 ■ **Anaerobic gas gangrene of the abdominal wall caused by *Bacteroides*.** This infection was an unfortunate result of bowel surgery. Organisms escaped from the intestine and initiated infection in the abdominal wall.

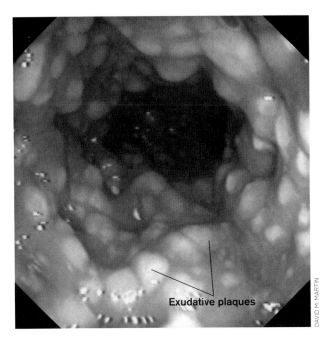

FIGURE 23.10 ▪ **Pseudomembranous enterocolitis caused by** *Clostridioides difficile* **(formerly** *Clostridium difficile***).** The organism is often a part of the normal intestinal microbiota, but antibiotic treatment can kill off competing microbes, leaving *C. difficile* to grow unabated. *C. difficile* toxins kill host cells, causing exudative plaques to form on the intestinal wall. The small plaques coalesce to form a large pseudomembrane that can slough off into the intestinal contents.

inflammatory disease such as inflammatory bowel disease (IBD) or Crohn's disease.

Some people favor restoring natural microbial balance by orally ingesting living microbes such as the lactobacilli present in yogurt. Taking these supplements, called **probiotics** (from the Greek meaning "for life"), is thought to restore balance to the microbial community and return the host to good health. The most commonly used probiotic genera are *Lactobacillus* and *Bifidobacterium* (**Fig. 23.11**). The potential mechanisms by which they improve intestinal health include competitive bacterial interactions (normal biota prevent growth of pathogenic bacteria), production of antimicrobial compounds, and immunomodulation (an ability to change the activity of cells of the immune system).

Claire Fraser and colleagues at the University of Maryland School of Medicine recently showed that ingestion of a single probiotic microbe, *Lactobacillus rhamnosus* GG, does not alter microbiome composition but will orchestrate broad transcriptional changes in other members of the microbiome—functions that could promote anti-inflammatory pathways in the resident microbes.

Probiotics are now being tested for treating several gastrointestinal disorders, including inflammatory bowel disease (IBD). Realize, though, that using probiotics such as yogurt to treat complex diseases such as IBD is still very controversial.

An extreme, but clinically proven, form of delivering probiotic microbes to the intestine is the so-called fecal transplant (fecal bacteriotherapy). In its original forms fecal transplant transferred the intestinal microbiome of a healthy person to a relative with a severe intestinal disease such as *pseudomembranous enterocolitis*. Restoring a "normal" microbiome in this way has successfully "cured" patients with repeated *C. difficile* infections that did not respond to antibiotics. Today the procedure uses a "super donor" rather than relatives and may someday be a personalized bacterial cocktail delivered in pill form. Probiotics in the form of vaginal suppositories are also used to treat patients prone to vaginal infections.

The Link between Obesity and Dysbiosis

Obesity is epidemic in industrialized and developing countries. Over 500 million people worldwide are obese and predisposed to develop related diseases such as diabetes, cardiovascular disease, nonalcoholic fatty liver disease, cancer, and some immune disorders. Obesity involves complex, puzzle-like interactions among genes, diet, and a long-term imbalance

A.

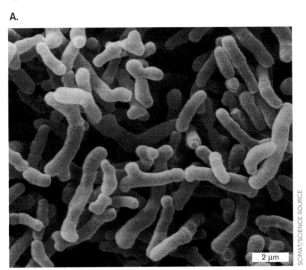

B.

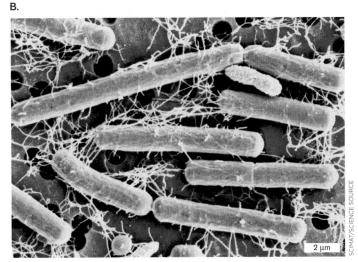

FIGURE 23.11 ▪ **Commonly used probiotic microorganisms.** A. *Bifidobacterium* (note the Y shape of some cells). B. *Lactobacillus acidophilus.* Colorized SEMs.

between energy intake and expenditure, which together yield an excessive increase in body fat. Compelling evidence now points to gut microbiota playing a prominent role in obesity by influencing nutrient acquisition, regulating energy metabolism, and affecting fat storage.

Many studies showing connections between gut microbiota and obesity use gnotobiotic animals. A **gnotobiotic animal** is an animal that is germ-free or one in which <u>all</u> the microbial species present are <u>known</u>. To develop a gnotobiotic colony of animals, offspring must be delivered by cesarean section under aseptic conditions in an isolator. The newborn is moved to a separate isolator where all entering air, water, and food are sterilized. Once gnotobiotic animals are established, the colony is maintained by normal mating between the members. Fecal organisms and fecal content can be transplanted orally to these animals by syringe and stomach tube (gavage). After the transplant, the effects on animal physiology and disease can be observed. Fecal transplants include mouse-to-mouse transfers or even human-to-mouse transfers (discussed later).

Early evidence suggesting a link between the gut microbiome and obesity included finding that germ-free (GF) mice have less fat and are more resistant to diet-induced obesity, compared to conventionally raised (microbiome-containing) mice. Furthermore, treating obese mice with antibiotics reduces their body weight and improves glucose metabolism. Livestock farmers and cattle ranchers have known for decades that adding antibiotics in low doses to animal feed actually <u>increases</u> animal growth. These studies suggest that altering an animal's microbiome with antibiotics can increase or decrease weight depending on circumstances.

Newer metagenomic studies find that obese humans and mice carry a less diverse gut microbiome and have

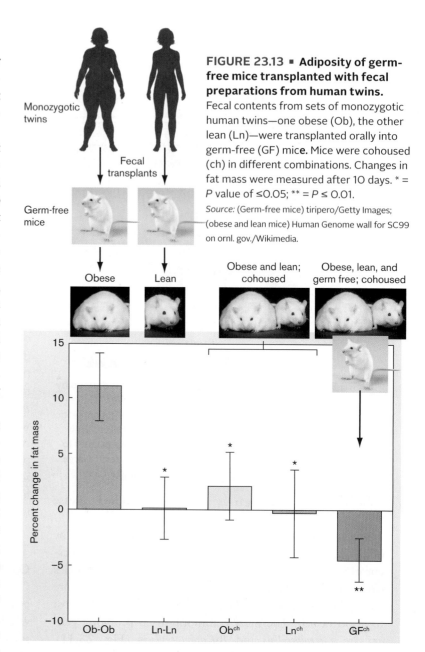

FIGURE 23.13 ▪ **Adiposity of germ-free mice transplanted with fecal preparations from human twins.** Fecal contents from sets of monozygotic human twins—one obese (Ob), the other lean (Ln)—were transplanted orally into germ-free (GF) mice. Mice were cohoused (ch) in different combinations. Changes in fat mass were measured after 10 days. * = P value of ≤ 0.05; ** = $P \leq 0.01$. *Source:* (Germ-free mice) tiripero/Getty Images; (obese and lean mice) Human Genome wall for SC99 on ornl. gov./Wikimedia.

FIGURE 23.12 ▪ **Jeffrey I. Gordon, MD.** Gordon (second from left) is a recipient of the 2015 Keio Medical Science Prize. He stands with some of the researchers who work in his lab at Washington University School of Medicine (from the left: Tarek Salih, Lihui Feng, Mark Charbonneau, Sid Venkatesh, and Laura Blanton).

an increased capacity to absorb energy compared to their lean counterparts. The ratios of major bacterial phyla in the colon also seem to differ between obese and lean individuals. For instance, the gut microbiomes of obese children have Firmicutes-to-Bacteroidetes phylum ratios that are higher than those of lean children. Studies looking at this ratio among adults are less clear. Some find that obese adults have higher Firmicutes-to-Bacteroidetes ratios compared to lean adults. Other studies indicate the opposite, and some show no difference. The real key to microbiome effects on obesity may be the difference in species makeup and metabolic output rather than broad phylum differences.

Can your microbiome keep you thin? A landmark study from Jeffrey Gordon (**Fig. 23.12**; Washington University School of Medicine in St. Louis) and colleagues demonstrated that obesity in humans is clearly linked to their microbiomes. The scientists collected fecal contents from sets of monozygotic human twins, one of whom was obese and the other lean, and then orally transplanted the fecal preps into germ-free mice (**Fig. 23.13**). Mice receiving fecal microbiota from the lean human co-twin remained lean, as measured by change in fat mass ("Ln" in **Fig. 23.13**). However, mice receiving fecal microbiota from the obese co-twin gained adiposity over a mere 10-day period ("Ob" in **Fig. 23.13**). Even more remarkable, when Ob and Ln co-twin fecal-transplanted mice were cohoused ("ch") in a single cage, the Ob^{ch} mice gained <u>less</u> weight than the control Ob mice. Metagenomic analysis of fecal contents from the mice revealed that the microbiome of Ob^{ch} mice came to resemble that of the Ln^{ch} mice. Essentially, the microbiota of the Ln^{ch} mouse transferred to the Ob^{ch} mouse and was maintained. Transfer took place because mice are naturally coprophagous (eat feces). Even though Ln^{ch} mice also consumed Ob^{ch} mouse feces, the Ln^{ch} fecal microbiome did not change. The most successful Ln^{ch} invaders of the Ob^{ch} microbiota were *Bacteroides* species.

Another aspect of this study asked what would happen to <u>germ-free mice</u> if they were cohoused with Ob and Ln fecal-transplanted mice. Up to 5 days after cohousing, the microbiome acquired by GF^{ch} mice resembled that of noncohoused Ob mice, but surprisingly, after 5 days the profile dramatically shifted to that of their Ln^{ch} cage mates. GF^{ch} mice then lost weight (see **Fig. 23.13**). This result confirmed that the Ln-derived microbial taxa had greater fitness over the Ob-derived taxa. From these studies, Gordon concluded that a higher Firmicutes-to-Bacteroidetes ratio favors obesity, while the opposite favors leanness. His studies also found that as fat people lose weight, the proportion of Bacteroidetes to Firmicutes in the gut microbiome increases.

Bacteria may not be the only microbial link to obesity. The fungal mycobiome was also found to differ between obese and thin human subjects. *Mucor racemosus* and *M. fuscus* were the species most represented in non-obese subjects compared to obese counterparts. The abundance of *Mucor* species increased when obese subjects lost weight, much as was found for Bacteroidetes.

How might the microbiome influence obesity? Microbiota can influence obesity in two major ways. One is the harvesting of energy from ingested foods; the other is the triggering of intestinal inflammation.

As noted in Chapter 21, the metabolic dexterity of intestinal microbes allows them to digest many foods that we cannot. In the process, these microbes produce short-chain fatty acids (SCFAs) that our cells use in a variety of ways. The major SCFAs are butyrate, propionate, and acetate. Numerous investigations suggest that the amounts and ratios of these SCFAs influence obesity (**Fig. 23.14**).

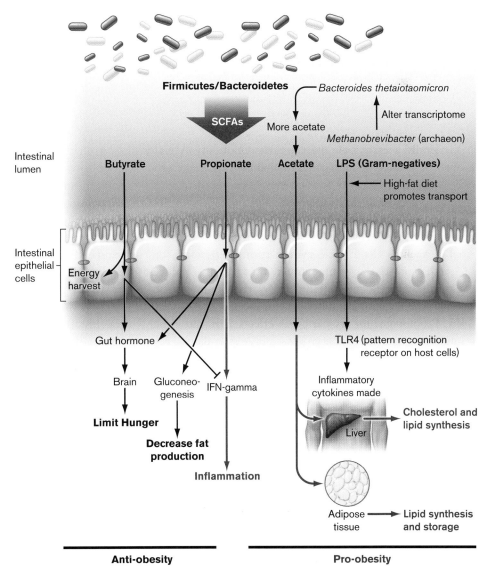

FIGURE 23.14 ■ Effects of microbiome products on weight gain. Members of the gut microbiome ferment ingested complex carbohydrates and produce short-chain fatty acids (SCFAs). SCFAs have varied effects on hunger, fat production, and inflammation. LPS from Gram-negative bacteria can be transported across the mucosal barrier and interact with pattern recognition receptor TLR4. The inflammatory cytokines made can affect liver metabolism. Arrows and terms in red promote weight gain.

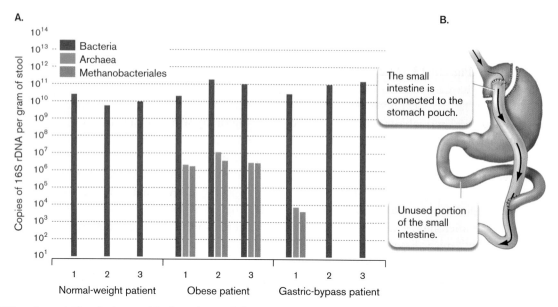

FIGURE 23.15 ▪ **Association between Methanobacteriales, obesity, and gastric bypass.** **A.** Relative numbers of distal colon Bacteria, Archaea in general, and Archaea of the order Methanobacteriales (primarily *Methanobrevibacter smithii*) in normal-weight, obese, and gastric-bypass patients. Organisms were quantified using real-time PCR to detect 16S ribosomal RNA genes. **B.** A gastric bypass.

Generally, acetate is thought to promote obesity. Acetate is absorbed well by intestinal epithelial cells, travels to the liver where it is used for lipid synthesis, and is distributed throughout the body to be used as a substrate for cholesterol synthesis (particularly in adipose tissues). Butyrate, however, seems to prevent obesity. It is used primarily by intestinal epithelial cells for energy and stimulates production of gut hormones that limit hunger. Butyrate also reduces the synthesis of the pro-inflammatory cytokine interferon-gamma (IFN-gamma), described in Sections 23.5 and 24.3. Propionate, for its part, has effects that can either stimulate or prevent obesity. This SCFA stimulates IFN-gamma synthesis, which promotes inflammation associated with obesity. But propionate is also involved in gluconeogenesis, which diverts fatty acids away from cholesterol synthesis and toward glucose synthesis. It also stimulates the production of gut hormones that limit hunger, both of which are anti-obesity effects. So, is there a correlation between SCFA levels in the colon and obesity?

A study from Elena Comelli's lab at the University of Toronto found that a cohort of obese or overweight patients generally had higher levels of SCFAs (propionate, butyrate, and acetate) than did healthy lean patients on similar diets, and that SCFA levels did correlate to Firmicutes/Bacteroidetes ratios. The more Firmicutes there were, the more SCFAs were produced. However, obese and lean subjects actually had similar Firmicutes/Bacteroidetes ratios, suggesting that the most important factor is which species of Firmicutes and Bacteroidetes are present (and what they are doing), rather than the overall phylum ratios.

Archaea in the gut might also tilt metabolic balance toward obesity. About 50% of humans have significant numbers of methane-producing archaea, primarily *Methanobrevibacter smithii* (discussed in Chapter 21). This archaeon can comprise up to 10% of the anaerobes in the colons of healthy adults. Bruce Rittmann and colleagues from Arizona State University compared the numbers of Bacteria and Archaea in normal-weight, obese, and post-gastric-bypass patients (**Fig. 23.15**). They discovered that all three groups had highly diverse species of bacteria, but the obese group had significantly more archaea than either the lean or gastric-bypass patients. This finding raised the possibility that Archaea may play a role in obesity. But how?

One possible explanation comes from another study by Jeffrey Gordon's laboratory. His team colonized germ-free mice with either *Bacteroides thetaiotaomicron* or *Methanobrevibacter smithii*, or with both species together, and found that when co-colonized, the methanogen dramatically altered the transcriptome of *B. thetaiotaomicron*. The methanogen shifted the carbohydrate priority of *Bacteroides* from glucans to fructans and enticed more acetate production as a result. Higher levels of acetate could enhance adiposity as outlined already. How the archaeon modulated the anaerobe's metabolism is not known, but it may involve some form of quorum sensing.

In addition to energy harvesting, inflammation induced by gut microbes appears to influence the development of obesity and diabetes. Obese individuals are in a chronic state of low-grade inflammation, characterized by elevated blood and tissue levels of pro-inflammatory cytokines and markers of inflammation such as C-reactive protein (discussed in Section 23.6). The source of inflammation is thought to be outer membrane lipopolysaccharides (LPS), also called endotoxin, produced by Gram-negative members of the gut microbiome. A high-fat diet promotes absorption of endotoxin across the intestinal epithelium and its distribution to tissues via blood. LPS interacts with a cell receptor called TLR4, present in a variety of host cells (see Section 23.5). LPS-TLR4 interaction induces production of pro-inflammatory cytokines, such as IL-6, and reactive oxygen species that interfere with the normal metabolism of liver, adipose, and skeletal muscle; all of which are important to the regulation of glucose and lipid homeostasis. While some gut bacterial groups correlate with energy intake, others, such as *Faecalibacterium prausnitzii*, influence obesity by altering inflammation. The presence of *F. prausnitzii*, for example, reduces the low-grade inflammation associated with obesity and diabetes.

Can individual members of a lean-associated microbiome treat obesity? We have gained tremendous insight into host-microbe relationships from the various microbiome projects under way. The data obtained thus far suggest that different blends of microbiota influence the efficiency of calorie harvesting from the diet, alter how derived energy is stored, and modulate chronic inflammatory processes. But can we pick out individual microbes or blends of microbes to treat obesity or other gastrointestinal diseases?

Research is under way to identify individual members of the microbiome that might individually, or in combination, treat diseases such as IBS (irritable bowel syndrome), IBD, or perhaps obesity. For example, Section 7.6 describes the identification of the intestinal Firmicute *Christensenella minuta* from human twins, which prevents weight gain when transferred to germ-free mice. There is great hope that we may one day cure or prevent these disorders using this novel form of probiotic treatment.

To Summarize

■ **Benefits of the microbiome** include interfering with pathogen colonization, producing immunomodulatory proteins, metabolizing foods that the host cannot process (energy harvesting), producing vitamins that the host cannot make, and honing our immune system.

■ **Infections such as sepsis, abscesses, cystitis, and meningitis** can be caused by some members of the microbiota if they breach the body's containment mechanisms.

■ **Dysbiosis of the microbiome** can contribute to infection, obesity, and inflammatory and autoimmune diseases.

■ **Probiotics** are foods, or solutions, containing helpful members of a microbiome. Different probiotics can restore balance to the intestinal or vaginal microbiomes.

■ **Obesity has been linked to intestinal dysbiosis.** An increase in species able to harvest energy from ingested foods and a decrease in species that temper inflammation are involved.

23.3 Overview of the Immune System

Given that we are surrounded by and host numerous bacteria, how is it we are not constantly infected? Here we begin our discussion of the many layers of protection designed to prevent infection and disease. As you will see, numerous physical and chemical barriers provide an effective first line of defense against infection by invading microbes. As such, they play a critical role in managing our resident ecosystems. But these barriers are not unbreachable. Organisms can still slip through. Consequently, humans, as well as other mammals, have a more aggressive defense called the immune system. The **immune system** is an integrated system of organs, tissues, cells, and cell products that differentiates self from nonself and neutralizes potentially pathogenic organisms or substances. This complex collection of cells and soluble proteins is capable of responding to nearly any foreign molecular structure.

Innate and Adaptive Immunity

There are two broad types of immunity: **innate immunity** (often called **nonadaptive immunity**) and **adaptive immunity** (discussed in detail in Chapter 24). Innate and adaptive immunity have several key differences. Innate immunity includes physical barriers such as skin, chemical barriers such as stomach acid, and relatively nonspecific cellular responses to infection that engage if the physical and chemical barriers are breached. The cellular innate responses are triggered by microbial structures such as peptidoglycan and lipopolysaccharides. Innate immunity is essentially "hardwired" into the body. It is present at birth, so that mechanisms of innate immunity exist before the body ever encounters a microbe. The protection afforded by innate immunity is nonspecific, capable of blocking or attacking many different types of foreign substances and organisms.

In contrast to innate immunity, adaptive immunity is designed to react to very specific structures called **antigens**. An antigen is any chemical, compound, or structure foreign to the body that elicits an immune response. The adaptive immune response to a specific antigen is not launched until the body "sees" that antigen. Adaptive immune mechanisms can recognize at least 10^{10} different antigenic structures and specifically launch a directed attack against each one. Once such an attack has been activated, the organism keeps a "memory" of the exposure in the form of specific memory cells, and an encounter with the microbe years later will reactivate the memory cells specific to the antigen.

The two types of immunity are illustrated by the response of the immune system to infection by the microorganism *Neisseria gonorrhoeae*, which causes the sexually transmitted disease gonorrhea. A component of the innate immune response is **complement**, composed of several soluble protein factors constantly present in the blood. Within moments of an initial infection, especially upon entering the bloodstream, complement proteins form holes in the bacterial membrane, thereby killing the microbe (see Section 23.6). Over time, the adaptive immune response will generate specific antibodies made to the cells of *N. gonorrhoeae* that escaped the innate mechanisms. These antibodies, however, are not made until well after the organism infects a person. Together, innate and adaptive immunity can help ward off disease caused by this organism. Unfortunately, *N. gonorrhoeae* infection does not generate good "memory," so reinfection with this organism is possible.

It is important to know that the innate and adaptive immune systems are interconnected. Antibodies made by the adaptive immune system will trigger parts of the innate immune system, such as the complement cascade that we will discuss later. Likewise, activation of innate resistance mechanisms will cause the release of small immunomodulatory peptides (cytokines) that influence the type and strength of adaptive immunity brought to bear. In military terms, it is similar to the army coordinating its actions with those of the air and naval forces.

Infection versus Disease

Contact with an infectious agent does not guarantee that a person will actually contract the disease. If the number of infecting organisms is small and the immune system (innate and adaptive) is effective, the individual may not develop disease, although a person known to have been exposed to certain microorganisms will be treated with antibiotics as a preventive measure. Chapter 26 more fully discusses the difference between being infected and having a disease.

Any microbe that launches a successful attack on a human or other animal and causes disease must first breach the host's physical and chemical barriers to gain entrance to the body. It must then survive the innate defense mechanisms and begin to multiply. Finally, the microbe must surmount the last line of defense, adaptive immunity, which begins to respond as the microbe struggles to overcome

innate immune defenses. The rest of this chapter will discuss the various innate defense mechanisms. Adaptive immunity is discussed in Chapter 24.

Although innate and adaptive immunity are often treated as separate entities, certain kinds of cells and organs play a role in both types of immunity. We thus introduce various cells and organs of the immune system as a whole before focusing on innate immunity. Our coverage of innate immunity will include physical and chemical barriers to infection, inflammation and nonspecific killing through phagocytosis, neutrophil extracellular traps, interferon, natural killer cells, and complement.

Cells of the Immune System

Blood is composed of red blood cells, white blood cells (also generally known as leukocytes), and platelets (**Fig. 23.16**). The many types of white blood cells are formed by the differentiation of stem cells produced in bone marrow (**Fig. 23.17**). Among these white blood cells are various components of innate immunity that differentiated from myeloid stem cells. These differentiated cells include:

- Polymorphonuclear leukocytes (PMNs)
- Monocytes
- Macrophages
- Dendritic cells
- Mast cells

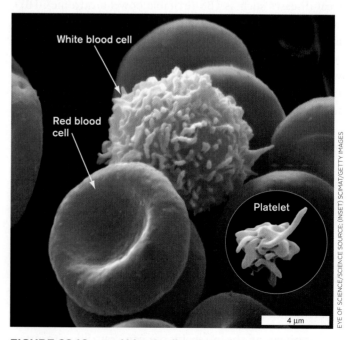

FIGURE 23.16 ▪ Red blood cells, white blood cell, and a platelet. This colorized scanning electron micrograph illustrates the relative sizes and 3D morphologies of these components of blood.

EYE OF SCIENCE/SCIENCE SOURCE; (INSET) SCIMAT/GETTY IMAGES

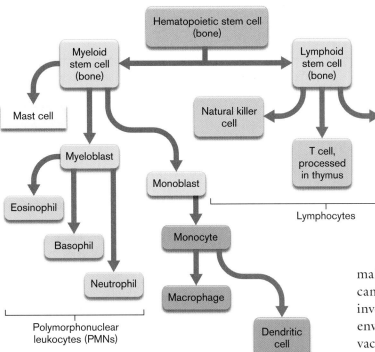

FIGURE 23.17 ■ **Development of white blood cell components of the immune system.** Pluripotent stem cells in bone marrow divide to form two lineages. One lineage consists of the myeloid stem cells, which develop into polymorphonuclear leukocytes (PMNs) and monocytes. These cells function primarily as part of innate immunity. The other branch consists of the lymphoid stem cells, which ultimately form natural killer cells, B cells, and T cells. Final maturation into B cells and T cells (the principal cells involved in adaptive immunity) occurs in the bone marrow and thymus, respectively. Colors indicate a group of differentiated cells that arise from the same progenitor.

making up the vast majority of white cells in the blood, can engulf microbes by phagocytosis. Phagocytosis involves the extrusion of pseudopods that attach to and envelop the pathogen, which ends up in a phagosome vacuole. The phagocyte then kills the organism by fusing enzyme-gorged lysosomes with the **phagosomes** (**Fig. 23.19** ▶). Enzymes spilling from the lysosome

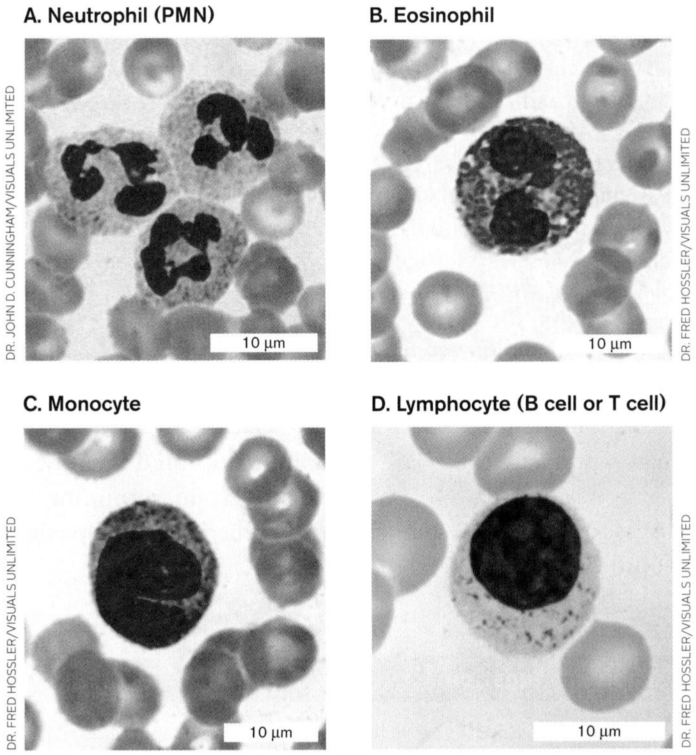

PMNs, also called granulocytes, are known for their conspicuous multilobed nuclei and enzyme-rich lysosome organelles. PMNs differentiate from an intermediate cell called the myeloblast. There are several types of PMNs, named for their different staining characteristics. Each cell type has a different function. **Neutrophils** (**Fig. 23.18A**),

A. Neutrophil (PMN)

B. Eosinophil

C. Monocyte

D. Lymphocyte (B cell or T cell)

FIGURE 23.18 ■ **Types of white blood cells.** **A.** Neutrophil (PMN). **B.** Eosinophil. **C.** Monocyte. **D.** Lymphocyte (B cell or T cell).

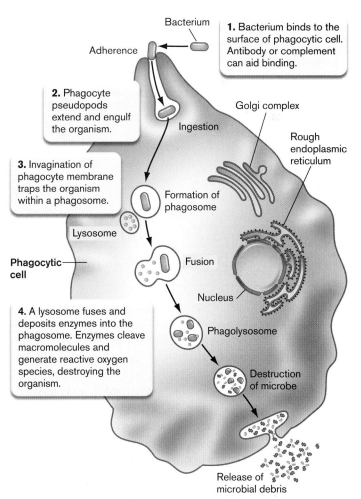

1. Bacterium binds to the surface of phagocytic cell. Antibody or complement can aid binding.

2. Phagocyte pseudopods extend and engulf the organism.

3. Invagination of phagocyte membrane traps the organism within a phagosome.

4. A lysosome fuses and deposits enzymes into the phagosome. Enzymes cleave macromolecules and generate reactive oxygen species, destroying the organism.

FIGURE 23.19 ■ **Phagosome-lysosome fusion.** ▶

SPECIAL TOPIC 23.1   Are NETs a Cause of Lupus?

It started with slight joint pain in her hands. Then, within weeks, Annette developed searing pain in all her joints. Blood tests revealed that Annette had systemic lupus erythematosus (SLE), an autoimmune disease in which the immune system can mistakenly attack any part of the body. The name "lupus" (Greek for "wolf") comes from a butterfly-shaped facial rash that resembles the pattern of fur on a wolf's face (**Fig. 1**). In addition to joint pain and skin rashes, SLE patients may have heart arrhythmias, abdominal pain, difficulty breathing, fatigue, fever, and swollen lymph nodes—depending on which parts of the body are under attack.

An important test used to diagnose lupus looks for antinuclear antibodies in the patient's blood. Antibodies are small proteins that usually bind substances foreign to the body (see Chapter 24). Lupus patients, however, are autoreactive, so they make antibodies to their own DNA and other "self" proteins.

The mechanisms underlying SLE are complex and not fully defined. One component that seems important is the production of type I interferons. In addition to their role in inhibiting viral replication, type I interferons increase the lytic potential of natural killer cells (see Section 23.5) and stimulate the development of cytotoxic lymphocytes (via T_H1 cells, as discussed in Chapter 24). Both cell types can cause tissue damage. Type I interferons can also lower the "threshold" for

making autoreactive antibodies. But what leads to increased interferon production in lupus patients?

Recent evidence suggests that a unique subset of neutrophils present in all lupus patients can initiate a cycle of pronounced interferon production via NET formation. **Figure 2** illustrates that these special neutrophils, called low-density granulocytes (LDGs), more easily undergo NETosis, as compared to typical healthy control neutrophils or to normal-density neutrophils from lupus patients. NETosis generates an ample source of autoantigen, such as chromatin, to which autoantibodies can be made, and helps propagate tissue damage, especially to blood vessel endothelial cells, by NET-entangled peroxidases and proteases.

But what stimulates interferon production? Autoantibodies that bind to antimicrobial peptides like LL-37 embedded in NETs promote transport of DNA into circulating plasmacytoid dendritic cells, a primary source of interferon-alpha. The subsequent interaction between DNA and Toll-like receptor 9 (TLR9 binds to CpG sequences in DNA) on these cells stimulates interferon-alpha production. Interferon-alpha not only enhances natural killer (NK) cell and cytotoxic lymphocyte activity, but also amplifies NET release by other LDGs (**Fig. 3**). The end result is a cycle of inflammation and tissue damage that contributes to the symptoms of SLE. Stopping NET release by LDGs could be a way to lessen those symptoms.

RESEARCH QUESTION

How would you determine whether NET formation by LDGs really stimulates interferon-alpha production by plasmacytoid dendritic cells? Design an experiment using components of the system and wells of a microtiter plate.

Villanueva, E., S. Yalavarthi, C. C. Berthier, J. B. Hodgin, R. Khandpur, et al. 2011. Netting neutrophils induce endothelial damage, infiltrate tissues, and expose immunostimulatory molecules in systemic lupus erythematosus. *Journal of Immunology* **187**:538–552.

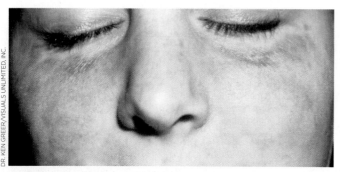

FIGURE 1 ▪ **Characteristic butterfly rash of lupus.**

into the phagosome will destroy various components of the microbe and, ultimately, the microbe itself (see Section 23.5).

Neutrophils also "throw" **neutrophil extracellular traps** (**NETs**) around nearby pathogens. After interacting with bacteria, neutrophils can undergo an unusual form of cell death, called NETosis, that spews a latticework of DNA (chromatin) impregnated with antimicrobial compounds into the immediate area (**Fig. 23.20**). Much as a fisherman's net traps fish, NETs trap pathogens and prevent them from spreading. Antimicrobial compounds

that impregnate the NET then kill the captured microbes. NETs have also been implicated in certain autoimmune diseases, such as systemic lupus erythematosus (**Special Topic 23.1**).

Thought Question

23.4 **Figure 23.20** shows how a neutrophil extracellular trap can ensnare a nearby pathogen. What bacterial structure might blunt the microbicidal effect of NETs?

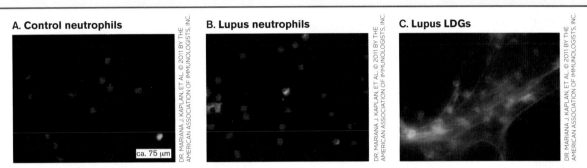

FIGURE 2 ■ **Circulating lupus low-density granulocytes undergo increased NETosis.** Representative images of control neutrophils **(A)**, lupus normal-density neutrophils **(B)**, and lupus low-density granulocytes (LDGs) **(C)** isolated from peripheral blood. Cells were fixed to coverslips and stained for nuclei and NETs. Blue structures are fluorescently stained nuclei (Hoechst stain); NETs are stained green by a fluorochrome-tagged antibody that binds to a protein (elastase) embedded in NET strands. Even without provocation, lupus LDGs release NETs.

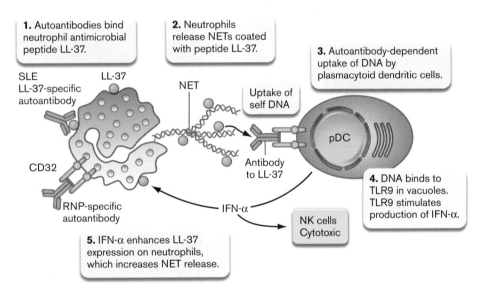

FIGURE 3 ■ **NET stimulation of IFN-alpha.** In systemic lupus erythematosus (SLE), autoantibodies specific for ribonucleoproteins (RNPs) and the antimicrobial peptide LL-37 accelerate the release of NETs by neutrophils (low-density granulocytes). The progressive increase in proteases and peroxidases released by NETs and the activation of natural killer (NK) cells and cytotoxic lymphocytes enhance tissue damage. (CD32 is a membrane receptor for antibodies.) *Source:* Modified from Alberto Mantovanti et al. 2011. *Nat. Rev. Immunol.* **11**:519–531.

Basophils, which stain with basic dyes, and **eosinophils (Fig. 23.18B)**, which stain with the acidic dye eosin, do not phagocytose microbes or throw NETs, but release products, such as major basic protein, that are toxic to the microbe. These two types of blood cells also release chemical mediators (called vasoactive agents) that affect the diameter and permeability of blood vessels (the significance of which is discussed in Section 23.5). Eosinophils play a major role in killing multicellular parasites such as helminth worms. **Mast cells** are similar to basophils in structure but differentiate in a lineage separate from

PMNs. Unlike PMNs, mast cells are residents of connective tissues and mucosa and do not circulate in the bloodstream. Basophils and mast cells also contain high-affinity receptors for a class of antibody called immunoglobulin E (IgE) associated with allergic responses (detailed in Section 24.7).

Note: Because neutrophils constitute the vast majority of PMNs, the terms "neutrophil" and "PMN" are often used interchangeably.

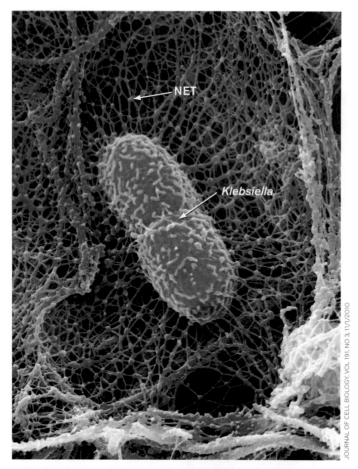

FIGURE 23.20 ▪ **Neutrophil extracellular trap (NET).** An infected mouse lung shows a *Klebsiella pneumoniae* bacterium (colorized; size approx. 1 μm) snared in a neutrophil extracellular trap (green), a web of decondensed chromatin released by neutrophils to catch and kill pathogens.

A. Macrophage engulfing bacteria

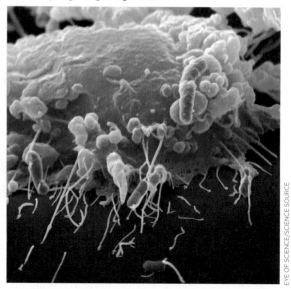

B. Dendritic cell

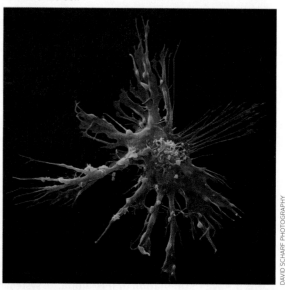

FIGURE 23.21 ▪ **The innate immune system depends on white blood cells called macrophages and dendritic cells.** **A.** Membrane protrusions from a macrophage (20 μm long) detecting and engulfing bacteria (pink; *E. coli*, 1.5 μm long). **B.** Dendritic cell. Colorized SEMs.

Monocytes (**Fig. 23.18C**) are white blood cells with a single nucleus (not multilobed like a PMN); they engulf (phagocytose) foreign material. Monocytes circulating in the blood can migrate out of blood vessels into various tissues and differentiate into **macrophages** and **dendritic cells** (**Fig. 23.21**). Macrophages are phagocytic and form a major part of the amorphous **reticuloendothelial system**, which is widely distributed throughout the body. The reticuloendothelial system is a group of cells with the ability to take up and sequester particles. In addition to macrophages, the system is composed of specialized endothelial cells lining the sinusoids (special capillaries) of the liver (Kupffer cells), spleen, and bone marrow, as well as reticular cells of lymphatic tissue (macrophages) and of bone marrow (fibroblasts). The function of macrophages and the reticuloendothelial system is to phagocytose microorganisms.

Macrophages are present in most tissues of the body and are the cells most likely to make first contact with invading pathogens. They have two functions. As part of innate immunity, they kill invaders directly. Protrusions from the macrophage surface extend and clasp nearby bacteria, pulling them into the cell (**Fig. 23.21A**). Once ingested by the macrophage, the bacteria are destroyed. Subsequently, as the first step in adaptive immunity, the remnants are processed (degraded) into smaller peptides (called antigens), which are then presented on the macrophage cell surface. Specific white blood cells called T cells (a type of lymphocyte; see Chapter 24) can bind the displayed antigens and become activated as part of the adaptive immune response. Thus, macrophages are considered **antigen-presenting cells** (**APCs**; see Chapter 24).

Macrophages are not the only APCs; dendritic cells (**Fig 23.21B**) are also APCs. Present in the spleen and lymph nodes, they, like macrophages, can take up, process, and present small antigens on their cell surface. Dendritic cells are different from macrophages in structure and in the fact that dendritic cells primarily take up small soluble antigens from their surroundings in addition to phagocytosing whole bacteria.

Platelets are small, puzzle piece–shaped cell fragments (they lack a nucleus) that derive from megakaryocytes, cells that differentiate from the precursors of red blood cells. Platelets, which circulate in the bloodstream, are required for efficient blood clotting. When activated by damaged endothelial cells (cells lining blood vessels), platelets clump, become trapped by fibrin, and form plugs that stop bleeding. Several factors produced by platelets also help wound repair. In addition to mediating blood clots and wound repair, recent evidence shows that platelets are part of the innate immune system. Invading bacteria can bind platelets and trigger the release of antimicrobial peptides (see Section 23.4). Bacterially activated platelets can also induce NET formation by neutrophils.

Natural killer (NK) cells derive from the lymphoid stem cells shown in **Figure 23.17** and are also part of innate immunity. Instead of killing microbes, however, the mission of NK cells is to kill host cells that harbor microorganisms or that have been transformed into cancer cells (**Fig. 23.22**). Natural killer cells recognize changes in cell-surface proteins of infected or cancer cells (MHC class I molecules; see Section 24.3) and then degranulate to release chemicals that kill those cells.

Natural killer cells recognize their targets in two basic ways: by the absence of MHC class I molecules on host cells, or by the presence of antibodies on host cells. A normal host cell displays two classes of MHC molecules on the outside of the cell membrane. MHC I is an indicator of "self." (MHC II molecules are discussed in Chapter 24.) NK cells have specific receptors that bind to self MHC I molecules on the surfaces of other cells in the body. An NK cell that "touches" a self MHC I–containing cell from the same person will not attack that cell. However, if a host cell lacks MHC class I molecules, NK cells perceive the target as foreign and a potential threat. Host cells can lose their MHC molecules during infection or as a result of malignant transformation. When an NK cell encounters a host cell lacking these markers, the NK cell inserts a pore-forming protein (**perforin**) into the membrane of the target cell, through which cytotoxic enzymes are delivered.

Natural killer cells also contain antibody Fc receptors on their cell surface. The second killing mechanism, called **antibody-dependent cell-mediated cytotoxicity (ADCC)**, is activated when the Fc receptor on the NK cell links to an antibody-coated host cell. Why would host cells be coated with antibodies? During their replication, many viruses place viral proteins in the membrane of the infected cell. Antibodies to those viral proteins will coat the compromised cell, tagging it for ADCC. Once the compromised cell has been targeted, it is killed by the NK cell in the same way as described for cells lacking MHC—namely, by insertion of a perforin molecule. This killing mechanism is another example of cooperation between innate immunity (NK cells) and adaptive immunity (antibody-producing lymphocytes).

Diagnostic value of white blood cell ratios. Many diseases, including infections, can alter the ratios of white blood cells (WBCs). A WBC differential is a test that physicians often order to help them diagnose infections and other syndromes. **Table 23.1** presents general guidelines for interpreting a WBC differential, indicating which cell types increase or decrease in response to infections with bacteria, viruses, or parasites (protozoa or worms). Notice that total WBC counts increase in each case, but the type of WBC that increases in number differs with the infectious agent.

Lymphoid Organs

The lymphoid stem cells shown in **Figure 23.17** produce lymphocytes as well as natural killer cells, described earlier. **Lymphocytes** (**Fig. 23.18D**), which are the main participants in adaptive immunity, are present in blood at about 2,500 cells per microliter, accounting for over one-third of all peripheral white blood cells. However, an individual lymphocyte spends most of its life within specialized solid tissues (lymphoid organs) and enters the bloodstream only periodically, where it migrates

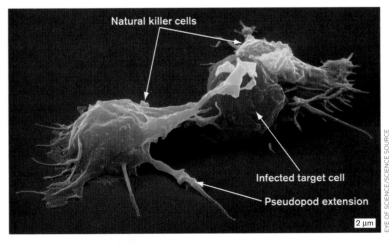

Natural killer cells

Infected target cell

Pseudopod extension

2 μm

EYE OF SCIENCE/SCIENCE SOURCE

FIGURE 23.22 ▪ Natural killer cells. Natural killer (NK) cells attack eukaryotic cells infected by microbes, not the microbes themselves. Perforin produced by the NK cell punctures the membranes of target cells, causing them to burst (colorized SEM).

TABLE 23.1	Guidelines for Interpreting White Blood Cell (WBC) Counts*	
	Normal	**Elevated during:**
Total WBC count (total leukocyte count)	4,500–11,000 per mm^3	Bacterial infection (12,000–30,000 per mm^3); also viral and parasitic infections, and allergy
Differential:		
Neutrophils	54%–62%	Bacterial infection (also high immature forms, band cells)
Eosinophils	1%–3%	Parasitic infection
Basophils	0%–0.75%	Allergy
Lymphocytes	25%–33%	Viral infection
Monocytes	3%–7%	Chronic infections

*These are general guidelines. Individual organisms or certain noninfectious medical conditions or immunological defects can alter the findings.

from one place to another, surveying tissues for possible infection or foreign antigens. Consequently, most lymphocytes are found in the lymph nodes or spleen. No more than 1% of the total lymphocyte population circulates in the blood at any one time.

The tissues of the immune system, where the great majority of lymphocytes are found, are classified as primary or secondary lymphoid organs or tissues, depending on their function (**Fig. 23.23**). The primary lymphoid organs and tissues are where immature lymphocytes mature into antigen-sensitive B cells and T cells. **B cells**, which ultimately produce antibodies (see Chapter 24), develop in bone marrow tissue. whereas **T cells**, which modulate various facets of adaptive immunity, develop in the thymus, an organ located above the heart.

The secondary lymphoid organs serve as stations where lymphocytes can encounter antigens. These encounters lead to the differentiation of B cells into antibody-secreting plasma cells, and of T cells into antigen-specific helper cells, as will be discussed in Chapter 24. The spleen is an example of a secondary lymphoid organ. It is designed to filter blood directly and detect microorganisms. Macrophages in the spleen engulf these organisms and destroy them, and then migrate to other secondary lymphoid organs and present pieces of the microbe (called antigens) to the B and T cells, which then become activated.

The **lymph nodes** are another kind of secondary lymphoid organ. Lymph nodes are arranged to trap organisms from local tissues, not from blood. Lymph nodes are situated at various sites in the body where lymphatic vessels converge (for example, under the armpits). Lymphoid tissues are also present in the mucosal regions of the gut and respiratory tracts (for example, Peyer's patches and gut-associated lymphoid tissue, or GALT, discussed later). Other secondary lymphoid organs are the tonsils, adenoids, and appendix.

To Summarize

- **The immune system** consists of both innate and adaptive mechanisms that recognize and eliminate pathogens.

- **Innate immunity** includes physical and chemical barriers and some cellular responses to various microbial structures.

- **Adaptive immunity** is a cellular response to specific structures (antigens) in which a memory of exposure is produced.

- **Myeloid bone marrow stem cells** differentiate to form cells of the innate immune system—namely, phagocytic PMNs, monocytes, macrophages, antigen-presenting dendritic cells, and mast cells. **Platelets** are derived from a different cell line.

- **Lymphoid stem cells** differentiate into natural killer cells (part of the innate immune system) and lymphocytes (cells of the adaptive immune system). Lymphocytes are classified as B cells, which ultimately produce antibodies, and T cells, which regulate adaptive immunity.

- **Natural killer (NK) cells** are a class of white blood cells that continually patrol tissues for cancer cells, cells infected with microbes, or host cells coated with antibody (antibody-dependent cell-mediated cytotoxicity, ADCC). **NK cells kill by inserting pores of perforin** into target cell membranes.

- **Primary lymphoid organs** include bone marrow (where B cells develop) and the thymus (where T cells develop). **Secondary lymphoid organs** (spleen, lymph nodes, Peyer's patches, tonsils, appendix) are where lymphocytes encounter antigens.

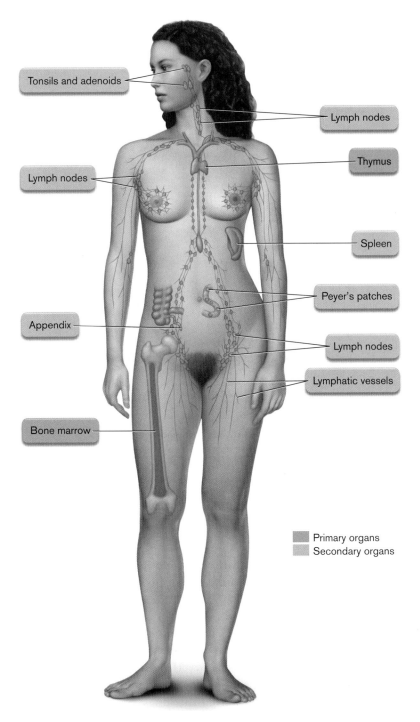

Tonsils and adenoids

Lymph nodes

Thymus

Lymph nodes

Spleen

Peyer's patches

Appendix

Lymph nodes

Lymphatic vessels

Bone marrow

Primary organs
Secondary organs

FIGURE 23.23 ■ Lymphoid organs.

23.4 Physical and Chemical Defenses against Infection

When defending a castle in medieval times, the first lines of defense included physical barriers (the castle wall), chemical barriers (boiling oil tossed onto invaders trying to scale the wall), and finally, hand-to-hand combat once the wall was breached. Similarly, the body's initial defenses against

infectious disease are composed of physical, chemical, and cellular barriers designed to prevent a pathogen's access to host tissues. Although generally described as nonspecific, some innate defense systems are more specific than others.

Physical Barriers to Infection

The first line of defense against any potential microbial invader (either commensal or pathogenic) is found where parts of the body interface with the environment. These interfaces (skin, lung, gastrointestinal tract, genitourinary tract, and oral cavities) have similar defense strategies, although each has unique characteristics. A defense common to all host surfaces involves **tight junctions**—watertight adhesions that link adjacent epithelial cells at mucosal surfaces and endothelial cells lining blood vessels. Tight junctions prevent bacteria, and even host cells, from moving between internal and external host compartments. The "glue" that holds tight junctions together is a series of interconnecting glycoprotein molecules (**Fig. 23.24**).

Skin. Few microorganisms can penetrate skin, because of the thick keratin armor produced by closely packed cells called keratinocytes. Keratin protein is a hard substance (hair and fingernails are made of it) that is not degraded by known microbial enzymes. An oily substance (sebum) produced by the sebaceous glands also covers and protects the skin. Its slightly acidic pH inhibits bacterial growth. Competition between species also limits colonization by pathogens, and microorganisms that manage to adhere are continually removed by the constant shedding of outer epithelial skin layers.

Other, more specialized cells just under the skin can recognize microbes managing to slip through the physical barrier. They are part of a consortium of cells called **skin-associated lymphoid tissue** (**SALT**). **Langerhans cells** make up a significant portion of SALT. They are specialized dendritic cells that can phagocytose microbes. Once a lymphoid Langerhans cell has ingested a microbe, the cell migrates by ameboid movement to nearby lymph nodes and presents parts of the microbe to the immune system to activate antimicrobial immunity.

Note: Do not confuse phagocytic Langerhans cells with the pancreatic "islets of Langerhans," which secrete insulin.

Mucous membranes. Mucosal surfaces form the largest interface (200–300 square meters) between the human host and the environment (the intestine alone is 7–8 meters, or

23–26 feet, long). Mucous membranes in general present a containment problem. They must be selectively permeable in order to exchange nutrients, as well as to export products and waste components. At the same time, they must constitute a barrier against invading pathogens. Mucosal membranes are covered with special tightly knit epithelial layers that support this barrier function. The mucus secreted from stratified squamous epithelial cells coats mucosal surfaces and traps microbes. Compounds within the mucus can serve as a food source for some commensal organisms, but other secreted compounds—like the enzymes lysozyme (which cleaves cell wall peptidoglycan) and lactoperoxidase (which produces superoxide radicals)—can kill an organism trapped in the mucus.

Semispecific innate immune mechanisms are also associated with mucosal surfaces. Host cells, even epithelial cells, in mucosa have evolved mechanisms to distinguish harmless compounds from microorganisms. Patterns of conserved structures on microbes, called **microbe-associated molecular patterns** (**MAMPs**), are recognized by host cell-surface receptors such as various Toll-like receptors and CD14 (discussed in Section 23.5). Once a MAMP has been recognized, the host cell sends out chemicals that can activate immune system cells involved with innate and adaptive immune mechanisms.

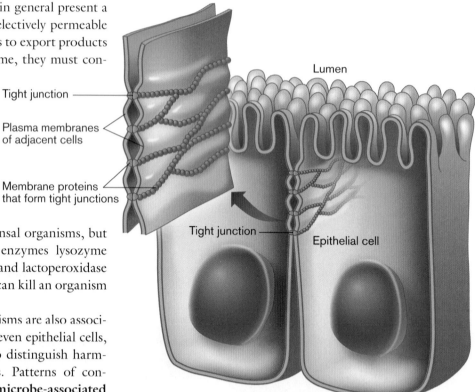

FIGURE 23.24 ■ Tight junctions hold adjacent cells together. Tight junctions are formed by an interconnected series of glycoproteins (blowup). By tightly linking adjacent cell membranes together, tight junctions produce a barrier through which bacteria, viruses, and other host cells cannot easily pass.

Note: MAMPs were previously called PAMPs, for "pathogen-associated molecular patterns." Because the structures recognized are also present on nonpathogenic bacteria and viruses, the term was changed to MAMP.

Like skin, the gastrointestinal system possesses an innate mucosal immune system, in this case called **gut-associated lymphoid tissue** (**GALT**). GALT includes tonsils, adenoids, and Peyer's patches (**Fig. 23.25A**). These tissues include specialized **M cells** that dot the intestinal surface and are wedged between epithelial cells. "M" stands for "microfold," which describes their appearance (**Fig. 23.25B**). These are fixed cells that take up microbes (microbiota or pathogens) from the intestine and release them, or pieces of them, into a pocket formed on the opposite, or basolateral, side of the cell. Other cells of the innate immune system, such as macrophages (which migrate through tissues), gather here and collect the organisms that emerge. Macrophages engulf and try to kill the organism and then place small, degraded components of the organism on the macrophage cell surface, where other immune system cells can recognize them. As a result, M cells are extremely important for the development

of mucosal immunity to pathogens. However, M cells can also serve as a portal for some pathogens to gain entry to the body.

The lungs. The lungs also have a formidable defense. In addition to the mucociliary escalator discussed in Section 23.1, microorganisms larger than 100 μm become trapped by hairs and cilia lining the nasal cavity and trigger a forceful expulsion of air from the lungs (a sneeze). The sneeze is designed to clear the organism from the respiratory tract. Organisms that make it to the alveoli are met by phagocytic cells called **alveolar macrophages**. These cells can ingest and kill most bacteria.

Another important factor that prevents lung infections is an epithelial membrane protein called **cystic fibrosis transmembrane conductance regulator** (**CFTR**). CFTR is a membrane chloride channel that regulates chloride movement across the membrane—an essential part of hydrating mucus in healthy individuals. Cystic fibrosis patients have a defective CFTR and are much more susceptible to lung infections, especially those caused by *Pseudomonas aeruginosa*. The *P. aeruginosa* strains posing the most serious

A. Peyer's patch

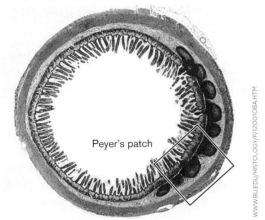

Peyer's patch

WWW.BU.EDU/HISTOLOGY/P/12001OBA.HTM

B. M cell

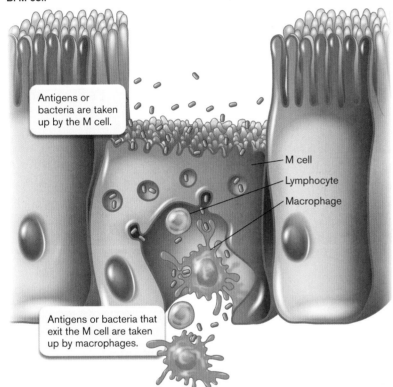

Antigens or bacteria are taken up by the M cell.

M cell

Lymphocyte

Macrophage

Antigens or bacteria that exit the M cell are taken up by macrophages.

FIGURE 23.25 ■ Gut-associated lymphoid tissue (GALT).
A. A Peyer's patch located on the small intestine. **B.** Diagram of an M cell (microfold cell).

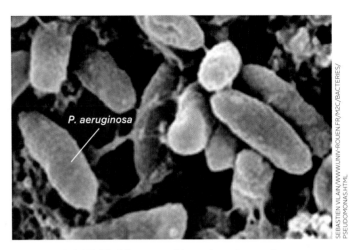

P. aeruginosa

SEBASTIEN VILAIN/WWW.UNIV-ROUEN.FR/M2C/BACTERIES/ PSEUDOMONAS.HTML

FIGURE 23.26 ■ Mucoid strains of *Pseudomonas aeruginosa* that infect cystic fibrosis patients. The organism is shown growing in a biofilm with strings of dehydrated exopolysaccharide (EPS) connecting cells. This EPS material gives colonies of these strains a very mucoid appearance. (Cells 2–3 μm long; SEM.)

Chemical Barriers to Infection

Some examples of chemical barriers were mentioned previously, such as the acidic pH of the stomach, lysozyme in tears, and generators of superoxide. In addition, a variety of human cells generate small antimicrobial, cationic (positively charged) peptides called defensins (**Table 23.2**). These antimicrobial peptides are important components of innate immunity against microbial infections.

Defensins range in length from 29 to 47 amino acids and are found in mammals, birds, amphibians, and plants. Defensins (and other antimicrobial peptides) destroy an invading microbe's cytoplasmic membrane and are effective against Gram-positive and Gram-negative bacteria, fungi, and even some viruses (those with membranes, like HIV). To kill, the peptides must bind the negatively charged outer membrane lipopolysaccharides (LPS) of Gram-negative bacteria and move into the periplasm. Defensins are then pulled into the cytoplasmic membrane of either Gram-positive or Gram-negative bacteria by the transmembrane electrical potential, about –150 mV in bacteria (that is, the cell interior is more negative than the exterior). The peptides assemble into channels that destroy the cytoplasmic membrane barrier, killing the bacterial cell. Defensins generally do not affect eukaryotic cells, which have a much lower membrane potential (–15 mV). Antimicrobial peptides are produced by many human cells, including cells of the skin, lungs, genitourinary tract, and gastrointestinal tract (**Fig. 23.27**). Specific defensins produced by different animals may partially explain pathogen-host specificity (see **eTopic 23.1**).

threat produce a thick slimy material (alginate) that retards other lung clearance mechanisms (**Fig. 23.26**). How CFTR helps protect lungs from infection is not clear. One hypothesis is that normal CFTR is a receptor for *P. aeruginosa* lipopolysaccharide (LPS), which enables endocytosis and killing of the microbe by phagolysosome mechanisms. Another hypothesis is that normal CFTR helps maintain an airway pH optimal for the activity of secreted antimicrobial peptides. Either way, CFTR is part of the innate barrier guarding against lung infection.

TABLE 23.2	Categories of Natural Antimicrobial Peptides
Class (Examples)	**Major presence in humans (Source)**
Alpha defensins (-1, -2, -3, -4)[a]	Neutrophils (Stored in granules)
Alpha defensins (-5, -6)	Paneth cells, small intestine (Stored in granules)
Cathelicidins (LL-37, hCAP18)	Neutrophils (Secreted)
Histatins	Saliva (Secreted)
Beta defensins (HBD-1, -2)	Epithelia (Secreted)
Kinocidins (tPMP,[b] PF-4)	Platelets (Secreted)
Other species	
Maganins	Frogs
Protegrins	Pigs
Indolicidin	Cattle

[a]Alpha defensins are named alpha defensin-1, alpha defensin-2, etc.

[b]tPMP = thrombin-induced platelet microbicidal protein.

One form of vertebrate defensins (alpha) are stored in membrane-enclosed granules within neutrophils and in Paneth cells in the small intestine (**Fig. 23.27A**). When stimulated, these cells **degranulate** (release their granule contents) by fusing their granule membranes to cytoplasmic or vacuolar membranes, dumping their contents into the surroundings or into phagocytic vacuoles, where the alpha defensins can destroy engulfed microbes (**Fig. 23.27B**). In contrast, the beta defensins are not stored in cytoplasmic granules. The synthesis of beta defensins is activated only after contact with bacteria or their products. Other cationic antimicrobial peptides are listed in **Table 23.2**. Neutrophil cathelicidins are discussed in **eTopic 23.2**.

Thought Question

23.5 Why do defensins have to be so small? Do defensins kill normal microbiota?

To Summarize

- **Skin defenses** against invading microbes include closely packed keratinocytes and a SALT lymphoid system made up largely of phagocytic Langerhans cells.

- **Mucous-membrane defenses** involve secreted enzymes, cytokines, and GALT tissues, such as Peyer's patches, that contain phagocytic M cells.

- **M cells** in gut-associated lymphoid tissues sample bacterial cells at their surface and release pieces of them to immune system cells.

- **Phagocytic alveolar macrophages** inhabit lung tissues, contributing to nonspecific defense.

- **Chemical barriers against disease** include cationic defensins, acid pH in the stomach, and superoxide produced by certain cells.

- **Microbe-associated molecular patterns (MAMPs)** are recognized by Toll-like receptors (TLRs) found on many host cells, such as macrophages. Binding triggers release of chemical signaling molecules that activate innate and adaptive immune mechanisms.

A.

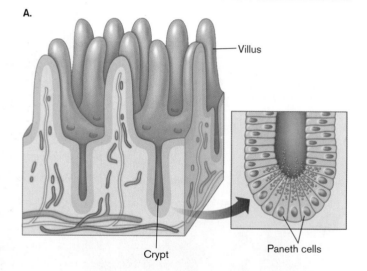

Villus

Crypt

Paneth cells

B.

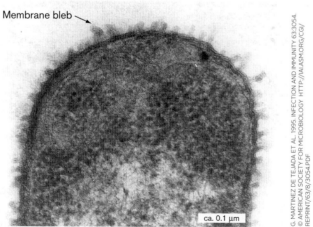

Membrane bleb

ca. 0.1 µm

G. MARTINEZ DE TEJADA ET AL. 1995, INFECTION AND IMMUNITY 63:3054. © AMERICAN SOCIETY FOR MICROBIOLOGY HTTP://IAI.ASM.ORG/CGI/ REPRINT/63/8/3054.PDF

FIGURE 23.27 ■ Defensins. A. Certain defensins are produced in the crypts of the intestine. The crypts contain granule-rich Paneth cells (blowup) that discharge their granules into the crypt lumen in response to the entry of bacteria or as a result of food-related stimulation by acetylcholine. **B.** Effect of cationic peptides on *E. coli* O111. Polymyxin B is a small cationic peptide antibiotic that mimics the action of defensins. In this micrograph, polymyxin B causes blebs of membrane to ooze from the surface of the cell.

23.5 Innate Immunity: Surveillance, Cytokines, and Inflammation

The boil shown in **Figure 23.28** is an inflammatory response triggered by infection with the organism *Staphylococcus aureus*. Inflammation is a critical innate defense in the war between microbial invaders and their hosts. It provides a way for phagocytic cells (such as neutrophils) normally confined to the bloodstream to gain access to

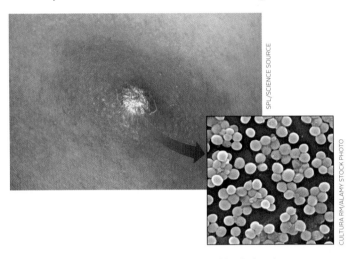

FIGURE 23.28 ▪ **Inflammation caused by infection.** Boil resulting from infection of a hair follicle by *Staphylococcus aureus* (size 0.5–1.0 µm). **Blowup:** Colorized SEM.

infected sites within tissues. Movement of these cells out of blood vessels is called **extravasation** or diapedesis, which we discuss shortly. Once at the infection site, the neutrophils begin engulfing microbes. The white pus associated with an infection is teeming with these white blood cells. The five cardinal signs of inflammation, first described over 2,000 years ago, are: redness, warmth, pain, swelling, and altered function at the affected site. In this section we describe the pathophysiology of these signs and other aspects of innate immunity.

Although many things can trigger inflammation, we focus here on how microbes cause the response. The process begins with the infection itself. Microorganisms introduced into the body—for example, on a wood splinter—will begin to grow and produce compounds that host cells sense or that damage host cells (**Fig. 23.29** ▶). Resident macrophages that wander into the infected area engulf these organisms and then release inflammatory mediators (chemoattractants) that "call" for more help. These mediators include **vasoactive factors** such as leukotrienes, platelet-activating factor, and prostaglandins, which act on blood vessels of the microcirculation, increasing blood volume and capillary permeability to help deliver white blood cells to the area. In addition, small protein molecules called cytokines are secreted, diffusing to the vasculature and stimulating the expression of specific receptors (selectins) on the endothelial cells of capillaries and venules. But how are cytokines made in response to an infection?

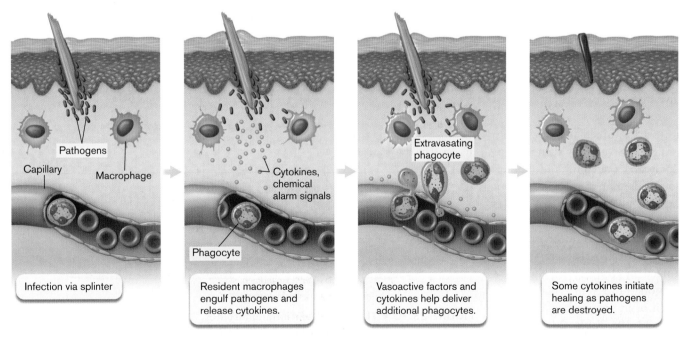

FIGURE 23.29 ▪ **Basic inflammatory response.** Neutrophils (a type of phagocyte) circulate freely through blood vessels and can squeeze between cells in the walls of a capillary (extravasation) to the site of infection. They then engulf and destroy any pathogens they encounter. ▶

TABLE 23.3 **Examples of Toll-Like Receptors and NOD-Like Receptors**

Receptor	MAMPs recognized	Source	Host cells	Location
TLR1	Lipopeptides	Bacteria	Monocytes/macrophages, dendritic cells, B cells	Cell surface
TLR2	Glycolipids, lipoteichoic acids	Bacteria	Monocytes/macrophages, dendritic cells, mast cells	Cell surface
TLR3	Double-stranded RNA	Viruses	Dendritic cells, B cells	Cell compartment
TLR4	Lipopolysaccharide, heat-shock proteins	Bacteria	Monocytes/macrophages, dendritic cells, mast cells, intestinal epithelium	Cell surface
TLR5	Flagellin	Bacteria	Monocytes/macrophages, dendritic cells, intestinal epithelium	Cell surface
TLR6	Diacyl lipopeptides	*Mycoplasma*	Monocytes/macrophages, mast cells, B cells	Cell surface
TLR9	Unmethylated CpG residues in DNA	Bacteria	Monocytes/macrophages	Cell compartment
NOD 1	Component of Gram-negative peptidoglycan	Gram-negative bacteria	Many cell types	Inflammasome
NOD 2	Peptidoglycan component	Bacteria	Macrophages, dendritic cells, epithelia of lung and GI tract	Inflammasome
NLRP-3	Peptidoglycan	Bacteria	Many cell types	Inflammasome
NLRP-4	Flagellin, CpG, ATP, dsRNA	Bacteria	Many cell types	Inflammasome

Sensing the Invader: Pattern Recognition Receptors and Cytokines

The faster the body can detect the presence of pathogens, the more quickly it can begin to deal with them. The more quickly it deals with them, the better the outcome of an infection. It takes time, however, for the adaptive immune system to make antibodies specific for a microbe (see Chapter 24). All the while, the pathogen can grow and cause disease. Fortunately, bacteria and viruses possess unique structures that immediately tag them as foreign. Structures such as peptidoglycan, flagellin, and lipoteichoic acids are not normally present in tissues unless bacteria or viruses are present. These structures have microbe-associated molecular patterns (MAMPs) that can be recognized by Toll-like or NOD-like receptors (TLRs or NLRs) present on or in various host cell types (**Fig. 23.30; Table 23.3**). TLRs and NLRs are tantamount to burglar alarm systems that activate

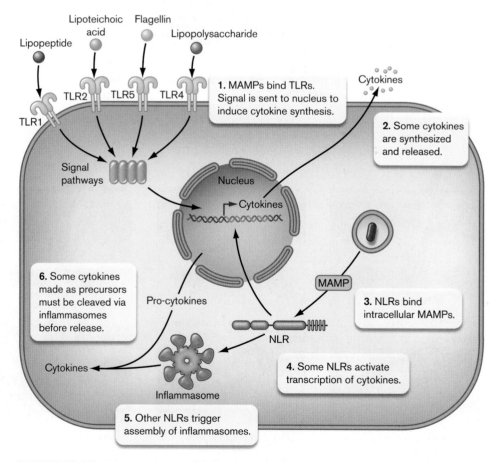

FIGURE 23.30 ■ TLRs, NLRs, and inflammasomes. Stimulation of TLRs and some NLRs activates transcription factors to induce production of cytokines and other factors. Some cytokines are directly released (for instance, IL-1, IL-8). Others are made as inactive precursors (procytokines) that must be cleaved by inflammasomes. Inflammasome assembly is orchestrated by certain NLRs.

upon encountering an intruder. Collectively, TLRs, NLRs, and similar proteins are called **pattern recognition receptors** (**PRRs**) because they recognize MAMPs.

Toll-like receptors. First discovered in insects and named Toll receptors, **Toll-like receptors** (**TLRs**) are evolutionarily conserved cell-surface glycoproteins present on the cells of many eukaryotic genera (**Fig. 23.30**). The term "Toll" came from Christiane Nüsslein-Volhard's 1985 exclamation, "That's crazy!" (in German, "Das ist ja toll!"), when shown the underdeveloped posterior of a mutant fruit fly. Thus, the gene was dubbed "Toll." TLRs in mammals display sequence similarities to the Toll genes involved with insect embryogenesis (Nüsslein-Volhard was co-winner of the 1995 Nobel Prize in Physiology or Medicine for her research on the control of embryonic development). TLRs are transmembrane proteins with an extracellular Toll/interleukin 1 receptor domain (TIR domain). Humans have numerous Toll-like receptors, each of which recognizes different MAMPs present on pathogenic microorganisms, making them an innate defense mechanism with some degree of specificity. For example, TLR2 binds to lipoarabinomannan from mycobacteria, zymosan from yeasts, lipopolysaccharide (LPS) from spirochetes, and peptidoglycan. TLR4, on the other hand, binds LPS, as well as host proteins released at sites of infection (for example, heat-shock protein 60). CD14, another host cell-surface protein, serves as a coreceptor for LPS. Note that these receptors bind fragments of structures after they are released from the microbe. They don't interact with the whole organism.

Once bound to their ligands, the TLRs trigger an intracellular transcription regulatory cascade via their TIR domain, causing the host cell to make and release cytokines that diffuse away from the site, bind to receptors on various cells of the immune system (see **Table 23.3**), and direct them to engage the invader. Cytokines are discussed further here and in Chapter 24. The cells that respond to cytokines can be part of innate immunity, adaptive immunity, or both. TLR recognition of MAMPs can also trigger autophagy in infected cells. Bruce Beutler and Jules Hoffman shared the 2011 Nobel Prize in Physiology or Medicine for their work on the role of TLRs in immunity.

NOD-like receptors. While TLRs are important sensors of external MAMPs (or of MAMPs in endosomes), **NOD-like receptor** (**NLR**) proteins are important cytoplasmic sensors of MAMPs (see **Table 23.3**). NLRs are structurally similar to a family of plant proteins called NODs (nucleotide-binding oligomerization domains) that provide resistance to pathogens. In mammals, some NLRs bound to a MAMP send a signal to the nucleus to activate cytokine production. Other NLRs become part of large, intracellular complexes of proteins called **inflammasomes** on which the protease caspase-1 oligomerizes (see the Current Research Highlight). Some cytokines (IL-1β and IL-18) are made as

large, inactive precursor proteins (procytokines) that must be processed by caspase-1 cleavage to make an active cytokine. The array of cytokines released stimulates inflammation and activates adaptive immune mechanisms.

Type I Interferons Are "Intruder-Alert" Cytokines

When a community is threatened by a thief, a neighbor who has been robbed alerts others to take precautions. Something similar happens during viral infections. In 1957, it was discovered that cells exposed to inactivated viruses produce at least one soluble factor that can "interfere" with viral replication when applied to newly infected cells. The term **interferon** was coined to represent these molecules. Interferons are low-molecular-weight cytokines (14–20 kDa) produced by many eukaryotic cells in response to intracellular infection. The action of interferons is usually species specific (interferon from mice will not work on human cells) but virus nonspecific (human interferon will help protect against both poliovirus and influenza virus, for example).

There are two general types of interferons, which differ in the receptors they bind and the responses they generate. Type I interferons have high antiviral potency; they consist of IFN-alpha (IFN-α), IFN-beta (IFN-β), and IFN-omega (IFN-ω). Type II interferon, IFN-gamma (IFN-γ), has more of an immunomodulatory function, discussed in Chapter 24.

Type I interferons can bind to specific receptors on uninfected host cells and render those cells resistant to viral infection. The host cell becomes resistant because interferon induces the intracellular production of two classes of proteins. One class encompasses double-stranded RNA-activated endoribonucleases that can cleave viral RNA. Proteins in the second class are protein kinases that phosphorylate and inactivate eukaryotic initiation factor 2 (eIF2), which is required to translate viral RNA. These mechanisms affect both RNA and DNA viruses because protein synthesis is required for the propagation of all viruses. However, RNA viruses are better than DNA viruses at inducing interferon. Type I interferons are used to treat certain viral infections (for example, hepatitis C).

Type II interferon functions by activating various white blood cells—for example, macrophages, natural killer cells, and T cells—to, among other things, increase the number of **major histocompatibility complex** (**MHC**) antigens on their surfaces. MHC proteins are important for recognizing self and for presenting foreign antigens to the adaptive immune system. They are discussed more fully in Chapter 24.

Acute Inflammation

In medicine, the term "acute" means "rapid onset." Acute inflammation, for instance, is inflammation that develops rapidly, usually within a day, after a foreign object such as

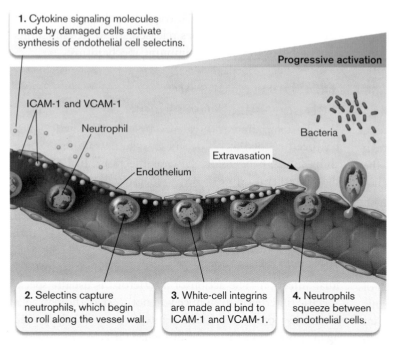

1. Cytokine signaling molecules made by damaged cells activate synthesis of endothelial cell selectins.

Progressive activation

ICAM-1 and VCAM-1

Neutrophil

Bacteria

Extravasation

Endothelium

2. Selectins capture neutrophils, which begin to roll along the vessel wall.

3. White-cell integrins are made and bind to ICAM-1 and VCAM-1.

4. Neutrophils squeeze between endothelial cells.

FIGURE 23.31 ■ Mechanism of extravasation. Extravasation is the process by which neutrophils move from the bloodstream into surrounding tissues.

selectins snag neutrophils zooming by in the bloodstream, slow them down, and cause them to roll along the endothelium (**Fig. 23.31**). Rolling neutrophils that encounter inflammatory mediators are activated to produce and display integrin adhesion molecules on their surface. Integrins on neutrophils lock onto the endothelial adhesion molecules ICAM-1 (intercellular adhesion molecule 1) and VCAM-1 (vascular cell adhesion molecule 1). Binding to these endothelial adhesion molecules stops the neutrophils from rolling and initiates extravasation in which the white blood cells squeeze through the endothelial wall and into the tissues.

The passage of neutrophils through vascular walls also requires relaxation of adhesion between endothelial cells. Vasoactive factors such as **bradykinin,** a nine-amino-acid polypeptide released by damaged tissue cells and macrophages, increase vascular permeability so that neutrophils can pass through vessel walls (**Fig. 23.32**). Increased vascular permeability also allows blood plasma to escape into tissues, causing swelling (edema). Even though tight junctions

a splinter or infectious agent is introduced into the body. <u>Chronic</u> inflammation develops over long periods of time. The function of acute inflammation is to wall off, kill, or digest the intruder, thereby quickly resolving an infection and promoting healing of the affected area. Neutrophils play a central role in the inflammatory process.

Neutrophils, however, normally reside in the circulation. How do they find the infection site? Certain cytokines synthesized after MAMPs bind to PRRs are called **chemokines** (examples are IL-8 and MCP-1). Chemokines diffuse from a site of infection toward nearby capillaries and form a concentration gradient. Then, like a fox following a prey's scent, neutrophils leave the bloodstream and follow the gradient back to the infection. But how do neutrophils pass through blood vessel walls?

The events leading to extravasation begin when the cytokines **interleukin 1 (IL-1)** and **tumor necrosis factor alpha (TNF-alpha)**, released by macrophages, stimulate the production of adhesion molecules (**selectins**) on the inner lining of the capillaries. P-selectin is produced first, followed by E-selectin. The

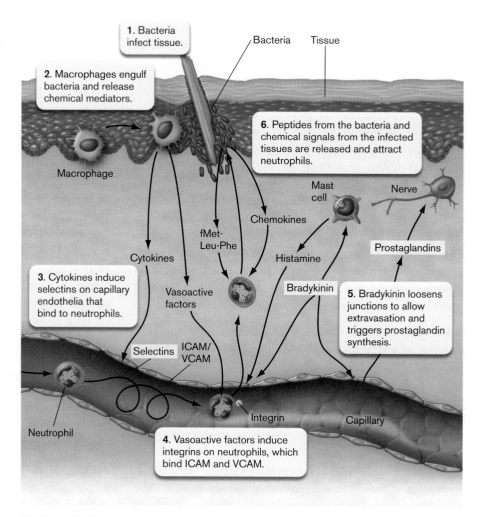

1. Bacteria infect tissue.

Bacteria   Tissue

2. Macrophages engulf bacteria and release chemical mediators.

6. Peptides from the bacteria and chemical signals from the infected tissues are released and attract neutrophils.

Macrophage

Mast cell

Nerve

Chemokines

fMet-Leu-Phe

Cytokines

Histamine

Prostaglandins

3. Cytokines induce selectins on capillary endothelia that bind to neutrophils.

Vasoactive factors

Bradykinin

5. Bradykinin loosens junctions to allow extravasation and triggers prostaglandin synthesis.

Selectins   ICAM/ VCAM

Integrin   Capillary

Neutrophil

4. Vasoactive factors induce integrins on neutrophils, which bind ICAM and VCAM.

FIGURE 23.32 ■ Summary of the inflammatory response.

are loosened, to actually pass between endothelial cells, activated neutrophils need the enzyme sialidase to temporarily break carbohydrate linkages that hold tight junctions together. Bradykinin also triggers degranulation of mast cells. The histamine released from mast cells further loosens the endothelial cell junctions. More fluid enters tissues and accumulates.

In addition to increasing vascular permeability, bradykinin and histamine relax smooth muscles within blood vessel walls so that vessel diameter increases (vasodilation). Vasodilation slows blood flow and, as a result, increases blood volume in the affected area. Increased vascular permeability and vasodilation cause the localized swelling, redness, and heat associated with inflammation.

Thought Question

23.6 As illustrated in **Figure 23.31**, integrin is important for neutrophil extravasation. Some individuals, however, produce neutrophils that lack integrin. What is the likely consequence of this genetic disorder?

What causes the pain of inflammation? Bradykinin induces capillary cells to make prostaglandins that cause pain by stimulating nerve endings in the area. A key enzyme involved in prostaglandin synthesis is cyclooxygenase (COX). Aspirin, ibuprofen, and the anti-inflammatory agent naproxen are COX inhibitors that prevent the synthesis of prostaglandins and thus reduce inflammatory pain.

Once neutrophils have passed through the vascular wall, chemokines lure them to the proper location. In addition, neutrophils can sense certain microbial chemoattractants—fMet-Leu-Phe peptide, for instance. Many bacterial proteins have fMet (*N*-formylmethionine) as their N-terminal amino acid, but mammalian proteins, in general, do not. Bacteria will often cleave off the fMet peptide, which can then diffuse away from the bacterium. The fMet peptide binds to neutrophil receptors and stimulates pseudopod projections aimed toward the microbe. As a result, the white blood cell migrates in the direction of the infection. Once phagocytes arrive at the site of infection, they begin devouring microbes. Realize, however, that much of the damage caused by an infection is not due directly to the microbe but is the result of the body's reaction to its presence.

Thought Question

23.7 What happens to all the neutrophils that enter a site of infection once the infection has resolved?

Phagocytes Recognize Alien Cells and Particles

For phagocytosis to proceed, macrophages and neutrophils must first recognize the surface of a particle as foreign (**Fig. 23.33**). When a phagocyte surface interacts with

A. White blood cell attacking bacteria

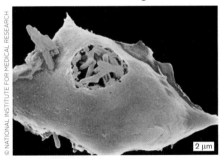

B. Contacts between phagocyte and target microbe

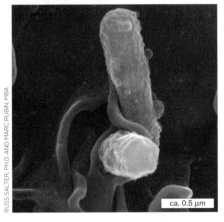

C. Macrophage engulfing bacteria

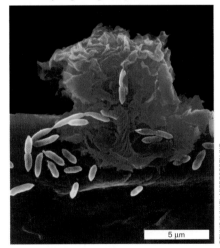

D. *Streptococcus pneumoniae* and capsule

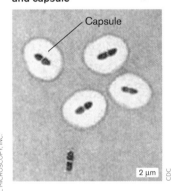

FIGURE 23.33 ■ **Images of phagocytosis.** **A.** A white blood cell phagocytosing *Mycobacterium* cells (green, 2 μm long; colorized SEM). **B.** Contacts between phagocyte and target microbe, illustrating how phagocytes "grab" the target bacterium. **C.** A macrophage engulfing bacteria on the outer surface of a blood vessel (SEM, magnification 1,315×). **D.** *Streptococcus pneumoniae* and capsule (India ink preparation). The slippery nature of the polysaccharide capsule makes phagocytosis more difficult.

the surface of another body cell, the phagocyte becomes temporarily paralyzed (unable to form pseudopods). Paralysis allows the phagocyte to evaluate whether the other cell is friend or foe, self or nonself. Self recognition involves glycoproteins located on the white blood cell membrane binding inhibitory glycoproteins present on all host cell membranes. The inhibitory glycoprotein on human cells is called CD47. Because invading bacteria lack these inhibitory surface molecules, they can readily be engulfed.

Although many bacteria, such as *Mycobacterium* or *Listeria* species, are easily recognized and engulfed by phagocytosis (**Fig. 23.33A–C**), others, such as *Streptococcus pneumoniae*, possess polysaccharide capsules that are too slippery for pseudopods to grab (**Fig. 23.33D**). This is where innate immunity and adaptive immunity join forces. Adaptive immunity produces anticapsular antibodies that aid the innate immune mechanism of phagocytosis through a process known as **opsonization** (**Fig. 23.34**). The anticapsular antibodies coat the surface of the bacterium, leaving the tail end of the antibodies, called the Fc region (see Section 24.2), pointing outward. These bacteria are said to be "opsonized." The Fc regions of these antibodies are recognized and bound by specific receptors on phagocyte cell surfaces. As a result, the antibodies stick the opsonized bacteria to phagocytic Fc receptors, which allow the phagocyte to more easily engulf the invader.

Thought Question

23.8 If NK cells can attack infected host cells coated with antibody, why won't neutrophils?

Oxygen-Independent and -Dependent Killing Pathways

During phagocytosis, the cytoplasmic membrane of the phagocyte flows around, and then engulfs, the bacterium, producing an intracellular phagosome, as described earlier. Subsequent phagosome-lysosome fusion (producing a phagolysosome; see **Fig. 23.19**) results in both oxygen-independent and oxygen-dependent killing pathways. Mechanisms independent of oxygen include enzymes like lysozyme to destroy the cell wall, compounds such as lactoferrin to sequester iron away from the microbe, and defensins, small cationic antimicrobial peptides (described in Section 23.4).

Oxygen-dependent mechanisms are activated through Toll-like receptors (discussed later in this section) and kill through the production of various oxygen radicals. NADPH oxidase, myeloperoxidase, and nitric oxide synthetase in the phagosome membrane are extremely important. NADPH oxidase yields superoxide ion ($^\bullet O_2-$), hydrogen peroxide, and, ultimately, hydroxyl radicals ($^\bullet OH$) and hydroxide ions (OH^-).

$$NADPH + 2O_2 \xrightarrow{\text{NADPH oxidase}} NADP^+ + H^+ + 2^\bullet O_2^- \xrightarrow{\text{superoxide dismutase}}$$

$$H_2O_2 + Cl^- \xrightarrow{\text{myeloperoxidase}} OH^- + HOCl$$

Myeloperoxidase, present only in neutrophils, converts hydrogen peroxide and chloride ions to hypochlorous acid (HOCl).

Macrophages, mast cells, and neutrophils also generate reactive nitrogen intermediates that serve as potent cytotoxic agents. Nitric oxide (NO) is synthesized from arginine by NO synthetase. Further oxidation of NO by oxygen yields nitrite (NO_2^-) and nitrate (NO_3^-) ions. All of these reactive oxygen

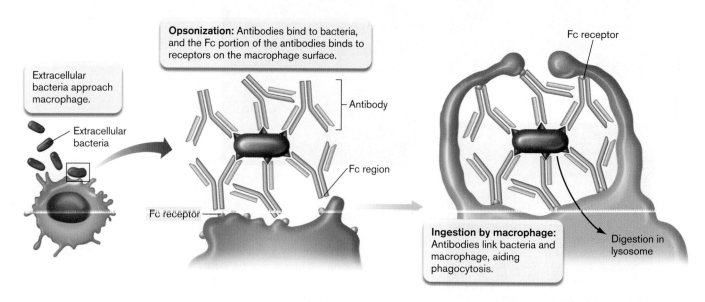

FIGURE 23.34 ■ **Opsonization.** Opsonization is a process that facilitates phagocytosis. Here, macrophage Fc receptors bind to the Fc region of antibodies binding to bacteria.

species attack bacterial membranes and proteins. These mechanisms greatly increase oxygen consumption during phagocytosis, called the **oxidative burst**. The reactive chemical species formed during the oxidative burst do little to harm the phagocyte because the burst is limited to the phagosome and because the various reactive oxygen species, such as superoxide, are very short-lived. Although phagocytes are very good at clearing infectious agents, many bacteria have developed ways to outsmart this aspect of innate immunity.

Autophagy and Intracellular Pathogens

Intracellular pathogens that grow in eukaryotic cytoplasm can be a serious problem for the host. Many pathogens, such as *Mycobacterium tuberculosis*, the cause of tuberculosis, enter the host cell in ways that bypass endosome formation or, if they do enter via an endosome, can escape from that compartment. These pathogens block normal host cell clearance pathways. To circumvent this problem, eukaryotic host cells (not just phagocytes) use a process that normally degrades damaged organelles (called **autophagy**) to

clear themselves of intracellular pathogens. During autophagy, the cell constructs a double membrane around the organism (or damaged organelle). This structure, called the autophagosome, sequesters the microbe from the nutrient-rich cytosol. Lysosomes then fuse with the autophagosome, depositing degradative enzymes that digest the organism. Ever-adapting intracellular microbes, however, have found ways to suppress autophagy and survive.

Chronic Inflammation Causes Permanent Damage

Inflammation that persists over months or years is called **chronic inflammation** and is provoked by the long-term presence of a causative stimulus. Chronic inflammation inevitably causes permanent tissue damage, even though the body attempts repair. The causes of chronic inflammation are many. For example, infectious organisms such as *Mycobacterium tuberculosis*, *Actinomyces bovis*, and various protozoan parasites can avoid or resist host defenses (**Fig. 23.35**).

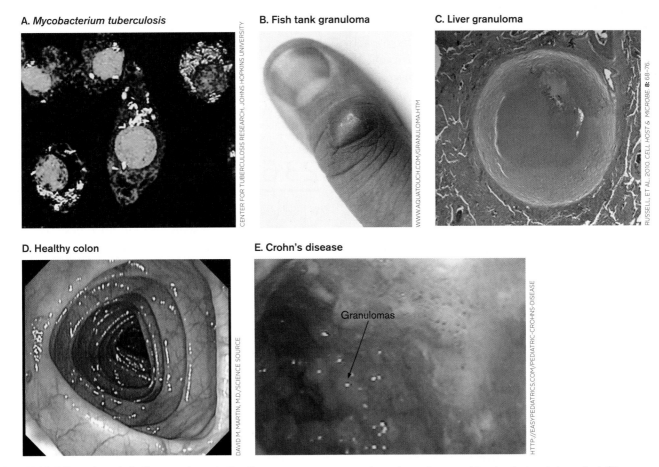

FIGURE 23.35 ■ **Chronic inflammation. A.** The fluorescent orange organisms shown here are *Mycobacterium tuberculosis* (2 μm long; fluorescence microscopy) within macrophages in a tuberculosis abscess. **B.** Fish tank granuloma. *Mycobacterium marinum* can cause tuberculosis-like infection in fish. The organism accidentally enters the human body through an open wound or abrasion during cleaning of an aquarium containing infected fish. The infection is first noted as a lesion that heals very slowly on the hand or forearm. Once it does heal, it often forms a granuloma at the site that contains live organisms. **C.** Cross section of liver showing a necrotizing granuloma caused by *M. tuberculosis* that spread through the bloodstream to this site after escaping the lung. **D.** Healthy colon. **E.** Intestinal granulomas of Crohn's disease.

As a result, they persist at the site and continually stimulate the basic inflammatory response. The continual stimulation of an inflammatory response leads to chronic inflammation.

Nonliving, irritant material like wood splinters, inhaled asbestos particles, or surgical implants can also cause chronic inflammation. Autoimmune diseases are another important cause. Autoimmunity (reaction against self) occurs when there is a failure to regulate some aspect of adaptive immunity (see Chapter 24). As a result, the immune system recognizes a part of the body as foreign (not self) and begins to react against it. Rheumatoid arthritis is one example of the body attacking itself.

Whatever causes chronic inflammation, macrophages and lymphocytes are continually recruited from the circulation. The body may attempt to "wall off" the site of inflammation by forming a **granuloma**. A granuloma begins as an aggregation of the mononuclear inflammatory cells surrounded by a rim of lymphocytes. The body then deposits fibroconnective tissue around the lesion, causing tissue hardening known as fibrosis.

Several forms of granulomas are shown in **Figure 23.35**. *M. tuberculosis*, for example, has a thick, waxy cell wall that protects mycobacteria against the mechanisms used by macrophages to destroy microorganisms. As a result, the organisms live for prolonged periods within macrophages (**Fig. 23.35A**). Resistance to host defenses can produce long-term chronic infections and granulomas. For example, skin infections caused by *Mycobacterium marinum* will produce skin granulomas (**Fig. 23.35B**), while *M. tuberculosis* can cause liver granulomas (**Fig. 23.35C**). Another disease thought to involve granuloma formation is Crohn's disease, which commonly manifests as abdominal pain, frequent bowel movements, and rectal bleeding. Crohn's disease has been attributed to an autoimmune reaction possibly activated by intestinal microbiota. In this case, the intestinal bacteria are thought to cause a chronic inflammation resulting in characteristic granulomas (compare **Fig. 23.35D** and **E**).

To Summarize

- **Acute inflammation begins** when host cells are damaged or infected. The process walls off and digests invading microbes, and initiates the healing of affected areas. Damaged cells or tissue macrophages release vasoactive factors that produce vasodilation and increased vascular permeability, cytokines that stimulate the production of blood vessel selectin receptors, and chemoattractant molecules that cause neutrophil movement (extravasation) from the bloodstream into infected tissues. **Bradykinin** causes the release of prostaglandins, which

produce pain in the affected area. The cardinal signs of inflammation include redness, warmth, swelling, pain, and loss of function.

- **Surface Toll-like receptors (TLRs)** and cytoplasmic **NOD-like receptors (NLRs)** in most host cells recognize microbe-associated molecular patterns (MAMPs) and synthesize cytokine proteins that diffuse and activate other cells of the immune system.

- **Interferons are one group of cytokines** that can nonspecifically interfere with viral replication (type I) or modulate the immune system (type II).

- Phagocytic cells have **oxygen-independent and oxygen-dependent mechanisms of killing** initiated by the fusion of lysosomes and bacteria-containing phagosomes.

- **The oxidative burst**, a large increase in oxygen consumption during phagocytosis, results in the production of superoxide ions, nitric oxide, and other reactive oxygen species.

- **Autophagy** is a process by which intracellular bacteria can be sequestered from the cytoplasm (via an autophagosome) and killed following fusion with a lysosome.

- **Chronic inflammation** results from the persistent presence of a foreign object.

23.6 Complement and Fever

White blood cells (WBCs) engulf and kill pathogens, but can simple serum proteins also kill microbes? Yes, a series of 20 serum proteins (complement factors) that make up the complement cascade can also attack bacterial invaders. Complement was first discovered as a heat-labile component of blood that enhances (or complements) the killing effect of antibodies on bacteria. Several complement factors are proteases that sequentially form and cleave other complement factors. (The liver is the main source of complement proteins.) Once a complement cascade is triggered, several things happen. Pores are inserted into bacterial membranes, causing cytoplasmic leaks, while pieces of some complement proteins attract WBCs and facilitate phagocytosis (opsonization).

Complement Activation Pathways

The three routes to complement activation are officially known as the classical pathway, the alternative pathway, and the lectin pathway. The classical complement pathway depends on antibody, so it is part of adaptive rather than innate immunity and is discussed in the next chapter (Section 24.4). The lectin pathway requires the synthesis of mannose-binding lectin by the liver in response to macrophage cytokines. Lectin coats the surface of invading microbes and activates complement without antibody. The lectin pathway connects to the classical complement pathway. We focus here in Chapter 23 on the alternative pathway because it is a well-characterized part of innate resistance. Like the lectin pathway, the alternative pathway is a nonspecific defense mechanism that does not require antibody for activation. The alternative complement pathway can attack invading microbes long before a specific immune response can be launched.

One goal of the complement cascade is to insert pores into target microbial membranes. The pores destroy membrane integrity, and thus kill the cell. The alternative complement pathway begins with the complement factor C3. In blood, C3 slowly cleaves into C3a and C3b (**Fig. 23.36**, step 1). C3b, under normal circumstances,

is rapidly degraded—a process that thwarts inadvertent complement activation. However, if C3b meets LPS on an invading Gram-negative microbe, the bound C3b becomes stable and binds another factor, designated factor B (step 2), and makes factor B susceptible to cleavage by yet another protein, factor D (step 3). The resulting complex, called C3bBb, has two roles. It is a C3 convertase that can quickly cleave more C3 to amplify the cascade and is changed by another serum protein (properdin) into what is called C5 convertase (step 4). From this point on, all complement pathways are identical. C5 convertase cleaves C5 in serum to C5a and C5b (step 5), and C5b then forms a prepore complex by binding to C6 and C7 (step 6). The resulting C5bC6C7 complex binds to membranes. Finally, C8 and C9 factors join in to form the **membrane attack complex** (**MAC**), becoming a destructive pore (step 7).

In Gram-negative bacteria, MAC pores first form in the outer membrane (**Fig. 23.36**, step 7). Lysozyme (present in serum) enters through the MAC outer

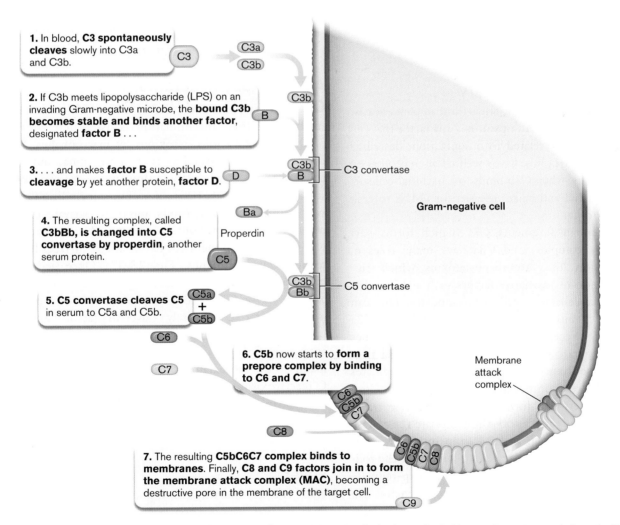

1. In blood, **C3 spontaneously cleaves** slowly into C3a and C3b.

2. If C3b meets lipopolysaccharide (LPS) on an invading Gram-negative microbe, the **bound C3b becomes stable and binds another factor**, designated **factor B** . . .

3. . . . and makes **factor B** susceptible to **cleavage** by yet another protein, **factor D**.

4. The resulting complex, called **C3bBb, is changed into C5 convertase by properdin**, another serum protein.

5. C5 convertase cleaves C5 in serum to C5a and C5b.

6. C5b now starts to **form a prepore complex by binding to C6 and C7**.

7. The resulting **C5bC6C7 complex binds to membranes**. Finally, **C8 and C9 factors join in to form the membrane attack complex (MAC)**, becoming a destructive pore in the membrane of the target cell.

C3 convertase

C5 convertase

Gram-negative cell

Membrane attack complex

FIGURE 23.36 ■ **The alternative complement pathway.** Although called "alternative," this complement cascade is part of the first-line innate defense.

membrane pores and cleaves peptidoglycan, making the cell membrane more susceptible to the membrane attack complex. The inner membrane MAC pores destroy membrane integrity and, with it, proton motive force. Gram-positive bacteria are resistant to complement because they lack an outer membrane (and therefore have no LPS to efficiently start the cascade) and have a thick peptidoglycan layer that hinders access of complement components. Even in the absence of LPS, however, there are ways to activate complement that involve antigen-antibody complexes (see Section 24.4). Eukaryotic cells infected with viruses and coated with antibody can also activate complement via the antibody-dependent classical complement pathway.

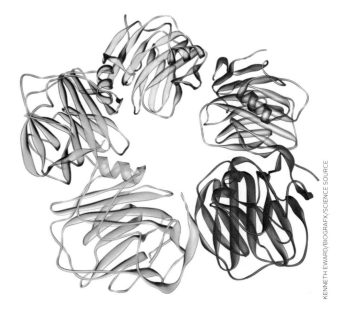

FIGURE 23.37 ▪ **C-reactive protein.** The 3D structure of pentameric human C-reactive protein (CRP) is unique. CRP is synthesized as a 206-amino-acid polypeptide that folds to form a flattened jelly roll structure, which then assembles into a radially symmetrical pentamer that circulates in serum. This structure can bind bacterial cell surfaces and C3 complement factor.

Thought Question

23.9 Figure 23.36 shows how the complement cascade can destroy a bacterial cell. Factor H (not shown) is a blood protein that regulates complement activity. Factor H binds host cells and inhibits complement from attacking our cells by accelerating degradation of C3b and C3bBb. How could bacteria take advantage of factor H?

Other Roles for Complement Peptides in Innate Immunity

Factor C3b, in addition to initiating the complement cascade, is a potent opsonin. An **opsonin** is any factor that can promote phagocytosis (related to opsonization, described earlier). PMNs (neutrophils) have specific C3b receptors on their surface. Thus, when C3b binds to a bacterial cell surface, it tags that cell and makes it easier for PMNs to grab and engulf the organism.

The complement fragments C5a (which forms part of the prepore complex) and C3a have many roles in immune function. They are anaphylatoxins, which trigger degranulation of vasoactive factors such as histamine from endothelial cells, mast cells, or phagocytes. They can also stimulate chemotaxis of immune cells. C5a and C3a bind to separate receptors on WBC plasma membranes and activate separate signal cascade pathways. C5a triggers Ca^{2+} release from intracellular stores and stimulates the actin polymerization needed for cell migration. Conversely, C3a triggers Ca^{2+} influx, which facilitates extravasation. These peptide mediators also stimulate the release of certain cytokines, such as IL-4 from monocytes and mast cells, that prepare capillary endothelium for the rolling and adhesion of neutrophils. C5a will also up-regulate P-selectin and ICAM-1. Thus, some complement factors not only help directly destroy target bacterial cells, but also facilitate phagocytosis and contribute directly to inflammation.

Acute-Phase Reactants, Complement, and Heart Disease

As noted earlier, inflammation is associated with the production of various cytokines by macrophages. Some of these cytokines (such as IL-1, TNF-alpha, and IL-6) travel to the liver, where they stimulate synthesis of several so-called acute-phase reactant proteins, including **C-reactive protein.** Named for its ability to activate complement, C-reactive protein is an acute-phase reactant that will bind to components of bacterial cell surfaces but not to host cell membranes (**Fig. 23.37**). Once tied to the bacterial cell surface, C-reactive protein will bind complement factor C1q of the classical complement pathway (see Chapter 24), ultimately converting factor C3 to C3b, propagating the complement cascade. C-reactive protein accelerates C3b production at the bacterial surface, where it can do the most damage. Note that an elevated level of C-reactive protein in serum is considered a general indicator of inflammation, not just inflammation caused by infection.

Fever

What is fever, and why is it a good thing? To understand fever, we first need to know how the body controls its temperature (themoregulation). Heat in a human body is produced as a consequence of metabolic reactions. The liver and muscles are the major generators of heat and will warm the blood that passes through them. In a healthy person, body temperature is kept between 36°C and 38°C (97°F and 100°F), despite large differences in surrounding temperature and physical activity. Fever is defined as an oral temperature above 100.4°F.

How do we normally maintain a body temperature of 37°C? Heat sensors located throughout the skin and large organs and along the spinal cord send information about the body's temperature to the thermoregulatory center in the hypothalamus, a small organ located in the brain near the brain stem. The hypothalamus acts as a thermostat by controlling blood flow through the skin and subcutaneous areas. Vasoconstriction (tightening of blood vessel diameter) prevents the release of body heat when we are cold, whereas vasodilation secures its quick release when we are hot. If skin temperature is too high, the hypothalamus directs vasodilation to accelerate heat release. If body temperature is too low, blood flow will decrease to conserve heat, and shivering begins as a way to generate more heat.

Fever is a natural reaction to infection and is usually accompanied by general symptoms such as sweating, chills, and the sensation of being cold. Substances that cause fever are known as **pyrogens**. Exogenous pyrogens (for example, certain bacterial toxins) originate outside the body, whereas internal or endogenous pyrogens (such as tumor necrosis factor and the interferon IL-6) are made by the body itself. External pyrogens generally cause fever by inducing the release of internal cytokine pyrogens.

Pyrogenic cytokines cross the blood-brain barrier and bind to neurons in the thermoregulatory center of the anterior hypothalamus. Cytokine-receptor interaction stimulates the production of phospholipase A2, an enzyme required to make prostaglandins. Prostaglandin E2 is made and changes the responsiveness of thermosensitive neurons. In other words, prostaglandin E2 turns up the thermostat. The body "thinks" it is cold, so the hypothalamus sends signals via the autonomic nervous system (which acts below the level of consciousness) to constrict peripheral blood vessels (vasoconstriction). Heat is not released and builds up to cause fever. Involuntary muscle contractions (shivering and chills) also generate heat.

What are the advantages of fever? Because the ideal growth temperature for many microbes is 37°C, elevated temperature can place the pathogenic organism outside its "comfort zone" of growth. There is also evidence that fever reduces iron availability to bacteria (cytokine release causes an increase in iron storage protein). Slower growth of the pathogen "buys" the immune system time to subdue the infection before it is too late. Consequently, interventions that reduce a <u>moderate</u> fever caused by infection may actually slow recovery.

Thought Question

23.10 If increased fever limits bacterial growth, why do bacteria make pyrogenic toxins?

To Summarize

- **Complement** is a series of 20 proteins naturally present in serum.

- **Activation of the complement cascade** results in a pore being introduced into target membranes.

- **The three pathways for activation** are the classical, alternative, and lectin pathways.

- **The alternative activation pathway** begins when complement factor C3b is stabilized by interaction with the LPS of an invading microbe.

- **The cascade of protein factors**, C3b $\longrightarrow$ factor B $\longrightarrow$ factor D $\longrightarrow$ properdin $\longrightarrow$ C5 $\longrightarrow$ C6 $\longrightarrow$ C7 $\longrightarrow$ C8 and C9, results in the formation of a membrane attack complex in target membranes.

- **C-reactive protein** in serum is activated when bound to microbial structures and will convert C3 to C3b, which can start the complement cascade.

- **The hypothalamus** acts as the body's thermostat.

- **Exogenous and endogenous pyrogens** elevate body temperature by stimulating prostaglandin production.

- **Prostaglandins** change the responsiveness of thermosensitive neurons in the hypothalamus.

Concluding Thoughts

The human body plays host to numerous species of microbes (our microbiota), many of which contribute to our health and well-being. The various "hardwired" innate immune mechanisms described here in Chapter 23 keep our microbiota at bay and provide an effective first line of defense against potential pathogens. The next chapter addresses what happens when microbes breach these innate defenses. Unlike the general protective mechanisms just discussed, the system of adaptive immunity generates a molecular defense specifically tailored to a given pathogen. Keep in mind that innate and adaptive immune systems "talk" to each other and collaborate to present the most effective response to a given pathogen, whether bacterial, viral, or parasitic.

CHAPTER REVIEW

Review Questions

1. Name some sterile body sites.
2. What body sites are colonized by normal microbiota?
3. Under what circumstances can commensal organisms cause disease?
4. Why are commensal organisms beneficial to the host?
5. Name and describe various types of innate immunity.
6. What are probiotics? How do they help maintain health?
7. How do the lungs avoid being colonized?
8. What is a gnotobiotic animal?
9. Describe GALT and SALT.
10. Describe some chemical barriers to infection.
11. Discuss the different types of white blood cells.
12. What is a lymphoid organ?
13. Outline the process of inflammation.
14. Explain why phagocytes do not indiscriminately phagocytose body cells.
15. What is interferon?
16. Describe antibody-dependent cell-mediated cytotoxicity.
17. How does complement kill bacteria?
18. Why might fever be helpful in fighting infection?

Thought Questions

1. Is it common for microbial pathogens to pass the placental barrier (called transplacental transmission) and infect the fetus? Consider papillomavirus, *Listeria monocytogenes*, *Escherichia coli*, HIV, *Treponema pallidum*, *Neisseria gonorrhoeae*, and *Staphylococcus aureus*.
2. The vagina contains competing, commensal microbes that contribute to the health of the organ. So, is "cleaning" the vagina by douching actually unhealthy and likely to lead to an increased chance of infection (vaginosis)?
3. Why have microbes not altered their structures to avoid being recognized by Toll-like receptors?

Key Terms

adaptive immunity (915)
alveolar macrophage (924)
antibody-dependent cell-mediated cytotoxicity (ADCC) (921)
antigen (915)
antigen-presenting cell (APC) (920)
autophagy (933)
B cell (922)
bacteremia (906)
basophil (919)
bradykinin (930)
C-reactive protein (936)
chemokine (930)
chronic inflammation (933)
complement (916)
compromised host (902)
cystic fibrosis transmembrane conductance regulator (CFTR) (924)
cytokine (910)
defensin (925)
degranulate (926)

dendritic cell (920)
dysbiosis (910)
eosinophil (919)
epidermis (904)
extravasation (927)
gnotobiotic animal (912)
granuloma (934)
gut-associated lymphoid tissue (GALT) (924)
immune system (915)
inflammasome (929)
innate immunity (915)
interferon (929)
interleukin 1 (IL-1) (930)
Langerhans cell (923)
lymph node (922)
lymphocyte (921)
M cell (924)
macrophage (920)
major histocompatibility complex (MHC) (929)

mast cell (919)
membrane attack complex (MAC) (935)
microbe-associated molecular pattern (MAMP) (924)
microbiome (microbiota) (902)
monocyte (920)
mucociliary escalator (906)
nasopharynx (905)
natural killer (NK) cell (921)
neutrophil (917)
neutrophil extracellular trap (NET) (918)
NOD-like receptor (NLR) (929)
nonadaptive immunity (915)
opportunistic pathogen (902)
opsonin (936)
opsonization (932)
oropharynx (905)
oxidative burst (933)
pattern recognition receptor (PRR) (929)

perforin (921)
phagosome (917)
platelet (921)
probiotic (911)
pyrogen (937)

reticuloendothelial system (920)
selectin (930)
skin-associated lymphoid tissue
  (SALT) (923)
T cell (922)

tight junction (923)
Toll-like receptor (TLR) (929)
tumor necrosis factor alpha (TNF-
  alpha) (930)
vasoactive factor (927)

Recommended Reading

Bloes, Dominik Alexander, Dorothee Kretschmer, and Andreas Peschel. 2015. Enemy attraction: Bacterial agonists for leukocyte chemotaxis receptors. *Nature Reviews. Microbiology* **13**:95–104.

Browne, Hilary P., Samuel C. Forster, Blessing O. Anonye, Nitin Kumar, B. Anne Neville, et al. 2016. Culturing of "unculturable" human microbiota reveals novel taxa and extensive sporulation. *Nature* **533**:543–546. doi:10.1038/nature17645.

Brubaker, Sky W., Kevin S. Bonham, Ivan Zanoni, and Jonathan C Kagan. 2015. Innate immune pattern recognition: A cell biological perspective. *Annual Reviews in Immunology* **33**:257–290.

Chakraborti, Chandra Kanti. 2015. New-found link between microbiota and obesity. *World Journal of Gastrointestinal Pathophysiology* **6**:110–119.

Chow, Jonathan, Kate M. Franz, and Jonathan C. Kagan. 2015. PRRs are watching you: Localization of innate sensing and signaling regulators. *Virology* **479–480**:104–109.

Chung, Hachung, Sunje J. Pamp, Jonathan A. Hill, Neeraj K. Surana, Sanna M. Edelman, et al. 2012. Gut immune maturation depends on colonization with a host-specific microbiota. *Cell* **149**:1578–1593.

Dogra, Shaillay, Olga Sakwinska, Shu-E Soh, Catherine Ngom-Bru, Wolfram M. Brück, et al. 2015. Dynamics of infant gut microbiota are influenced by delivery mode and gestational duration and are associated with subsequent adiposity. *MBio* **6**:e02419–14.

Eldridge, Matthew J., and Avinash R. Shenoy. 2015. Antimicrobial inflammasomes: Unified signalling against diverse bacterial pathogens. *Current Opinions in Microbiology* **23**:32–41.

Hathaway, Lucy J., and Jean-Pierre Kraehenbuhl. 2000. The role of M cells in mucosal immunity. *Cellular and Molecular Life Science* **57**:323–332.

Hewetson, J. T. 1904. The bacteriology of certain parts of the human alimentary canal and of the inflammatory processes arising therefrom. *British Medical Journal* **2**:1457–1460.

Kaplan, Mariana J., and Marko Radic. 2012. Neutrophil extracellular traps: Double-edged swords of innate immunity. *Journal of Immunology* **189**:2689–2695.

Le Chatelier, Emmanuelle, Trine Nielsen, Junjie Qin, Edi Prifti, Falk Hildebrand, et al. 2016. Richness of human gut microbiome correlates with metabolic markers. *Nature* **500**:541–546.

Lecuit, Marc, and Marc Eloit. 2013. The human virome: New tools and concepts. *Trends in Microbiology* **21**:510–515.

Ogawa, Michinaga, and Chihiro Sasokawa. 2006. Bacterial evasion of the autophagic defense system. *Current Opinion in Microbiology* **9**:62–68.

Orvedahl, Anthony, and Beth Levine. 2009. Eating the enemy within: Autophagy in infectious diseases. *Cell Death and Differentiation* **16**:57–69.

Sender, Ron, Shai Fuchs, and Ron Milo. 2016. Are we really vastly outnumbered? Revisiting the ratio of bacterial to host cells in humans. *Cell* **164**:337–340.

Stetson, Daniel B., and Ruslan Medzhitov. 2006. Type I interferons in host defense. *Immunity* **25**:373–381.

Stuart, Lynda M., and R. Alan B. Ezekowitz. 2005. Phagocytosis: Elegant complexity. *Immunity* **22**:539–550.

van Sorge, Nina M., and Kelly S. Doran. 2012. Defense at the border: The blood-brain barrier versus bacterial foreigners. *Future Microbiology* **7**:383–394.

Wang, Miao, Zequin Gao, Zhongwang Zhang, Li Pan, and Yonguang Zhang. 2014. Roles of M cells in infection and mucosal vaccines. *Human Vaccine Immunotherapy* **10**:3544–3551.

Wikoff, William R., Andrew T. Anfora, Jun Liu, Peter G. Schultz, Scott A. Lesley, et al. 2009. Metabolomics analysis reveals large effects of gut microflora on mammalian blood metabolites. *Proceedings of the National Academy of Sciences USA* **106**:3698–3703.

Zhou, Wuding. 2012. The new face of anaphylatoxins in immune regulation. *Immunobiology* **217**:225–234.

CHAPTER 24
The Adaptive Immune Response

Once an infecting pathogen eludes our innate immune defenses, it faces an even more daunting foe: an adaptive immune system that stirs only when challenged. Adaptive immunity is composed of special lymphocytes called B cells and T cells that multiply during an infection and "remember" the specific invader for years. These specially trained lymphocytes are poised to respond quickly to a second infection by the pathogen. Left unregulated, however, adaptive immunity can turn on us, producing autoimmune disease.

CURRENT RESEARCH highlight

Lymphocyte traffic. Lymph node follicles contain separate T-cell and B-cell zones. For a B cell to become a plasma cell, antigen-activated helper T cells must move across the T-B border and directly interact with antigen-activated B cells (see figure below). Chemokines emanating from the B-cell zone start recruitment, but Hai Qi discovered that unactivated "bystander" B cells were also needed. ICOS ligand (ICOSL) protein on bystander B cells must interact with ICOS (inducible T-cell <u>costimulator</u>) on T cells to initiate coordinated movement of T-cell pseudopods. Migration of T cells fails if the B cells lack ICOSL.

Source: Zu et al. 2013. *Nature* **496**:523–527.

ICOSL$^{+/+}$ B cells   ICOSL$^{-/-}$ B cells

B-cell zone
T-cell zone
Migrating T-cell
No migration
100 µm

■ B cells
■ T cells
■ Antigen-activated T cells

AN INTERVIEW WITH

HAI QI, IMMUNOLOGIST, TSINGHUA-PEKING CENTER FOR LIFE SCIENCES, BEIJING, CHINA

SCHOOL OF MEDICINE, TSINGHUA UNIVERSITY

Why are chemokines and B-cell ICOSL needed for T-cell migration into the B-cell zone?

Available data indicate that a factor made in the follicle <u>antagonizes</u> chemokine-dependent phosphatidylinositol 3-kinase (PI3K) activation signals in T cells needed for optimal cytoskeleton reorganization and motility. Consequently, PI3K activated only by the follicle-homing chemokine CXCL13 is insufficient for T-cell movement into the B-cell zone. This is why ICOSL-mediated activation of ICOS, which strongly activates the PI3K pathway, becomes necessary for T-cell migration at the T-B border.

The B-cell ICOSL$^{-/-}$ chimeric mice you constructed failed to move T cells into the B-cell zone. Were these mice able to make antibody?

These mice do not efficiently make antibody to T-dependent antigens. However, they are still able to respond to T-independent antigens.

Do you foresee a potential clinical application for your findings?

Yes. ICOSL can be massively induced in inflamed tissues where <u>effector</u> T cells are recruited. This is beneficial to fighting infections but detrimental to tissue homeostasis. Understanding how bystander B-cell ICOS activation regulates T-cell migration may provide new ways to therapeutically augment or reduce inflammatory infiltration.

Our immune system evolved to render harmless any microbe that gains entry into our body. The system is complex but controlled by many checks and balances to prevent overreaction. The Current Research Highlight illustrates one small but very important part of immunity: the segregation of T-cell and B-cell lymphocytes in the lymph node and the recruitment of T cells into the B-cell zone. Once recruited, the T cell helps a cognate, antigen-stimulated B cell to become an antibody-secreting plasma cell. This and other parts of the immune system are needed to prevent and cure infections, but its complexity makes us susceptible to potentially catastrophic genetic defects. A dramatic example is David Philip Vetter, better known as "the bubble boy."

David Vetter was born in 1971 without an immune system. He was the only human known to live in a plastic, germ-free bubble for his entire 12 years of life (1971–1984) (**Fig. 24.1A**). His predicament stemmed from a genetic disease known as severe combined immunodeficiency (SCID). In the most serious form of SCID (caused by a defective cytokine receptor), the patient forms no T cells and has dysfunctional (or sometimes no) B cells. Severe SCID patients cannot launch a meaningful immune defense against any invading microbe—whether pathogen or normal microbiota. Because David's older brother had died of SCID, physicians knew David might have the disorder too. Within seconds of his birth, David was transferred to a sterile environment to await a bone marrow transplant. Water, air, food, diapers, clothes—all were disinfected with special cleaning agents before entering his sterile plastic bubble. He lived for 12 years physically isolated in this plastic, sterile environment, venturing out only in a NASA-designed sterile space suit. A bone marrow transplant was attempted in 1984, but his defective immune system failed to protect him from Epstein-Barr virus, which went undetected in the transplanted cells. He died just before his thirteenth birthday.

Today, bone marrow transplants from antigenically compatible donors are the only approved treatment of SCID. However, an experimental gene therapy pioneered by Dr. Adrian Thrasher of University College London (**Fig. 24.1B**) is a promising new approach to curing the disease. In this strategy, the gene encoding the normal cytokine receptor is inserted into a defective virus that can infect cells without causing disease. The debilitated virus is used to deliver the normal cytokine receptor gene into the patient's stem cells. Stem cells are undifferentiated, and capable of changing into many different cell types, including B cells and T cells (see Section 23.3). Once the genetic

FIGURE 24.1 ■ Living without an immune system. A. David Vetter, the bubble boy, inside his environmental bubble. The tube behind him is a port that was used to introduce sterile food, clothes, and other items. **B.** Adrian J. Thrasher conducted a successful gene therapy trial on children with severe combined immunodeficiency (SCID).

material enters the nucleus, the healthy copy of the gene functions and the corrected stem cells are reintroduced into the patients. If the gene replacement is successful, the child makes his own T cells with the correct receptor and develops a functional immune system. SCID dramatically illustrates the importance of our immune system and how fragile its development is. All it takes is a small defect in a single gene to subvert the entire process.

In Chapter 24 we describe the two types of adaptive immunity: antibody-dependent immunity in which B cells differentiate into plasma cells that make antibodies, and cell-mediated immunity whereby specific T-cell lymphocytes develop to directly kill infected host cells. You will also learn how certain types of T cells balance antibody and cell-mediated responses to a given infection. Along the way we will explore how gut mucosal immunity mediates coexistence with our microbiome. Ultimately, you will appreciate that adaptive immunity is a major reason humans still exist.

24.1 Overview of Adaptive Immunity

Recall from Chapter 23 that the immune system has both nonadaptive and adaptive mechanisms. Nonadaptive (innate) immune mechanisms are present from birth, whereas adaptive immunity develops as the need arises. For instance, adaptive immunity against malaria does not develop until the individual has encountered the plasmodial parasite that causes the disease. The adaptive immune response—**adaptive immunity**—is a complex, interconnected, and cross-regulated defense network.

> **Note:** The terms "adaptive immune response" and "immune response" are often used interchangeably.

Antibody-Dependent versus Cell-Mediated Immunity

Two types of adaptive immunity are recognized: humoral immunity and cell-mediated immunity. In **humoral immunity**, **antibodies** are produced that directly target microbial invaders. The term "humoral" means "related to body fluids." Thus, antibodies are proteins that circulate in the bloodstream and recognize foreign structures called antigens. An **antigen** (also called an **immunogen**) is any molecule that, when introduced into a person, will elicit the synthesis of antibodies that specifically bind the antigen. Antigens stimulate B cells (B lymphocytes) to differentiate into antibody-producing cells. **Cell-mediated immunity**, the second type of adaptive immunity, employs teams of T cells (T lymphocytes) that recognize antigens and then destroy host cells infected by the microbe possessing the antigen. The humoral and cellular immune responses are intertwined, each relying on some facet of the other to work efficiently. T cells serve a central role in adaptive immunity by determining whether humoral or cell-mediated mechanisms predominate in response to a specific antigen.

What triggers an immune response, and how long does it take? Adaptive immunity develops over a 3- to 4-day period after exposure to an invading microbe. The immune system does not recognize the <u>whole</u> microbe, but innumerable tiny <u>pieces</u> of it. Each small segment of an antigen that is capable of eliciting an immune response is called an **antigenic determinant** or **epitope**. Many single-protein antigens are recognized when the larger antigen is broken into smaller segments upon being phagocytosed.

Even distinct tertiary (3D) shapes within a protein may be counted as antigenic determinants if they produce a specific response. Immune responses to 3D shapes are possible when stretches of amino acids far removed from each other in a protein's primary sequence align side by side in 3D space after folding (**Fig. 24.2**). Such a 3D structure may be recognized by the immune system as a single entity or antigen. Besides proteins, other structures in the cell, such as complex polysaccharides, can have linear and 3D epitopes. So the immune response to a microbe is really a composite of thousands of B-cell responses to different epitopes. The response to each individual epitope is **clonal**; that is, it gives rise to a population of cells that originate from a single B cell. This means that each clone of B cells will target a unique epitope.

The humoral immune response requires several cell types and cell-to-cell interactions. What are those interactions, and where do they take place? As illustrated in **Figure 24.3**, the process begins with an infection somewhere in the body. Dendritic cells and macrophages patrolling the area gather up the foreign antigens and present them on their cell surface. Any phagocytic cell that degrades large antigens into smaller antigenic determinants and places those determinants on their cell surface is called an **antigen-presenting cell (APC)**. Many types of cells can be antigen-presenting. They include "professional" APCs, such as macrophages (monocytes), mast cells, or dendritic

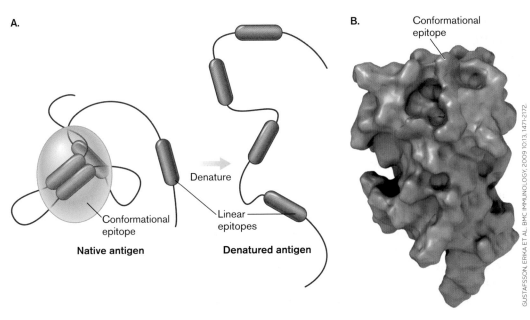

A.

Denature →

Conformational epitope

Linear epitopes

Native antigen

Denatured antigen

B.

Conformational epitope

GUSTAFSSON, ERIKA ET AL. BMC IMMUNOLOGY, 2009 10:13, 1471-2172.

FIGURE 24.2 ■ Antigens and epitopes. A. Native proteins fold into a 3D shape, where several regions separated in the linear sequence can reside next to each other to form a conformational epitope. Denaturing the protein with the detergent sodium dodecyl sulfate (SDS) and reducing agents like dithiothreitol (to remove disulfide bonds) will unfold the protein and separate the various amino acid stretches that formed the conformational epitope. **B.** A 3D protein structure showing, in red, four amino acids that form a conformational epitope.

A.

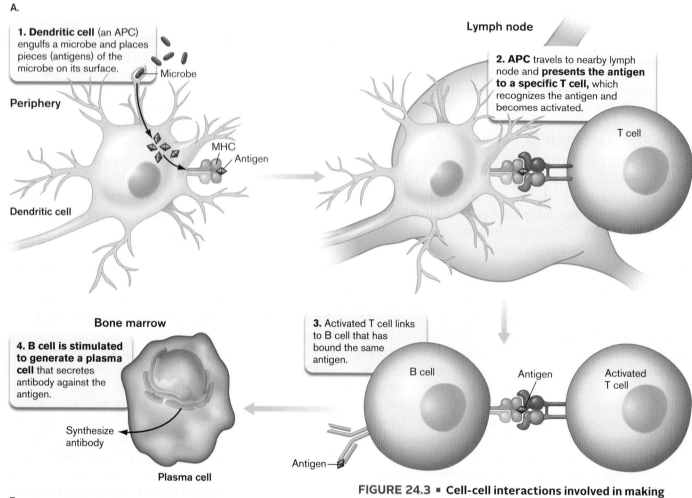

1. Dendritic cell (an APC) engulfs a microbe and places pieces (antigens) of the microbe on its surface.

Microbe

Periphery

MHC
Antigen

Dendritic cell

Lymph node

2. APC travels to nearby lymph node and **presents the antigen to a specific T cell,** which recognizes the antigen and becomes activated.

T cell

Bone marrow

4. B cell is stimulated to generate a plasma cell that secretes antibody against the antigen.

Synthesize antibody

Plasma cell

3. Activated T cell links to B cell that has bound the same antigen.

B cell

Antigen

Activated T cell

Antigen

FIGURE 24.3 ■ Cell-cell interactions involved in making antibody. **A.** Two basic cell-cell interactions are required to make an antibody: (1) An APC (dendritic cell) presents antigen to a helper T cell. (2) The activated T cell links to and activates a B cell bound to the same antigen. The two interactions can be broken down into four steps, as illustrated. **B.** Ralph Steinman discovered dendritic cells.

B.

cells; and, under the right circumstances, "nonprofessional" APCs (for example, endothelial cells or fibroblasts). Because dendritic cells are so important to the immune response, their discoverer, Ralph Steinman (1943–2011), was awarded the Nobel Prize in Physiology or Medicine in 2011 (**Fig. 24.3B**).

Once decorated with antigen, the professional antigen-presenting cell travels to secondary lymphoid organs (lymph nodes) where B cells and T cells await. Specific T cells in the node then link to the antigen presented on the APC and become activated T cells. One type of

activated T cell then binds to and activates a lymph node B cell that encountered the same antigen. That interaction authorizes B cells to generate plasma cells (usually short-lived) able to pump out large amounts of specific antibody. (Note that plasma cells are <u>not</u> B cells.) The activated B cells also produce long-lived memory B cells that remember the exposure and stand ready to quickly generate plasma cells, should the antigen be encountered months or years later. Once formed, the memory B cells and most plasma cells leave the lymph node and migrate to the bone marrow. Other plasma cells remain in the lymph node. In the bone marrow, memory B cells patiently wait to be called to future sites of infection.

The cell-mediated immune response shares some aspects of the humoral immune response. A type of T cells, called cytotoxic T cells, also bind to microbial antigens presented on an APC and become activated. Activated cytotoxic

T cells seek out and directly kill any host cell infected with the microbe. In addition to directly killing infected cells, cytotoxic T cells synthesize and secrete growth factors, called cytokines (see Sections 23.2 and 23.5), that incite nearby macrophages to indiscriminately attack cells in the local area. Thus, cellular immunity, in general, is critical for dealing with intracellular pathogens such as viruses, whereas humoral immunity is most effective against extracellular bacterial pathogens like *Streptococcus pyogenes*, one cause of wound infections and sore throat.

As we proceed through this chapter, we will reveal in layers how the immune system functions, with each layer building on the previous one. We will periodically return to a particular aspect of the immune response—for example, B-cell differentiation into plasma cells—to integrate seemingly distinct parts of the immune system into a unified concept of immunity.

Thought Question

24.1 Two different stretches of amino acids in a single protein form a 3D antigenic determinant. Will the specific immune response to that 3D antigen also respond to one of the two amino acid stretches alone?

Immunogenicity

Immunogenicity measures the effectiveness by which an antigen elicits an immune response. One antigen can be more immunogenic than another. For example, proteins are the strongest antigens, but carbohydrates can also elicit immune reactions. Nucleic acids and lipids are usually weaker antigens, in part because these molecules are both made of relatively uniform repeating units that are very flexible. The flexible units present a variable 3D structure that does not easily interact with antibodies. Proteins are more effective antigens for three reasons: Different proteins have different shapes, they maintain their tertiary structure, and they are made of many different amino acids that can be assembled in many different combinations. These features provide stronger interactions with antibodies in the bloodstream and enable better recognition by lymphocytes, the cellular workhorses of the immune system.

Several other factors contribute to the immunogenicity of proteins (see **eTopic 24.1**). For example, the larger the antigen, the more likely it is that phagocytic cells will "see" and engulf it. This is important because, as noted earlier, an immune response cannot occur until phagocytic antigen-presenting cells (such as macrophages and dendritic cells) first engulf large antigens and degrade them, presenting the epitopes on their cell surface. The immune response begins when a B cell binds to a foreign peptide and a T lymphocyte binds to the same foreign peptide displayed on the surface of an antigen-presenting cell. The B cell–APC and T cell–APC binding events are independent of each other.

The presentation of antigens on APCs forces the antigen to be placed on a membrane surface protein structure called the **major histocompatibility complex**, or **MHC** (discussed later). The more tightly an antigen can bind to these MHC surface proteins, the more immunogenic it is. The stronger the binding, the easier it is for T cells to recognize the complex.

Each specific antigen shows a different **threshold dose** needed to generate an optimal response. A dose higher or lower than that threshold will generate a weaker immune response. Lower doses activate only a few B cells, whereas exceedingly high doses of antigen can cause **B-cell tolerance**, a state in which B cells have been overstimulated to the point that they do not respond to subsequent antigen exposures and make antibody. Tolerance is part of the reason your immune system does not react against your own protein antigens.

As you might expect, the body must regulate the immune system carefully so that a response is not leveled against itself. In effect, the immune system must become "blind" to its own antigens; as a result, the host will often be blind to foreign antigens that resemble epitopes of its own cells. Therefore, the more complex the foreign protein is, the more likely it will possess antigenic determinants that a lymphocyte can recognize as nonself. The farther an antigen is from "self," the greater its immunogenicity will be.

Immunological Specificity

Our earliest clues into the nature of immunity came from smallpox. Smallpox is a devastating disease that caused enormous suffering and killed millions of people in the seventeenth and eighteenth centuries (see Chapter 1 and **Fig. 24.4**). There was no cure, and the only available preventive treatment was to take dried material from the lesions of a previous smallpox sufferer, place it on a healthy person, and hope the person survived. Those who survived were protected from subsequent bouts of smallpox but were still susceptible to other diseases. This early observation gave rise to the idea of **immunological specificity**, which means that an immune response to one antigen is not effective against a different antigen. In other words, the immune response to smallpox will not protect someone against the plague bacillus (*Yersinia pestis*), which is antigenically different from the smallpox virus.

While immunological specificity is important, it is not absolute. As described in Chapter 1, an English country physician named Edward Jenner (see Fig. 1.18B) in the late eighteenth century (long before viruses were discovered) learned to protect townsfolk from deadly smallpox disease

A. Smallpox patient

B. Variola major

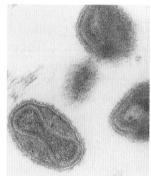

FIGURE 24.4 ■ Immunological specificity is the basis of vaccination. A. Smallpox patient covered with white pox pustules. **B.** The smallpox virus, variola major (300 nm long; TEM). The photo shows the dumbbell-shaped, membrane-enclosed nucleic acid core. **C.** The vaccinia virus that causes cowpox (360 nm long; EM). Edward Jenner recognized the similarity between the deadly smallpox and less severe cowpox diseases and used cowpox scrapings to vaccinate humans against smallpox.

C. Vaccinia

(caused by variola virus) by inoculating them with scrapings from lesions produced by cowpox (a tamer disease caused by vaccinia virus; see **Fig. 24.4C**). This story illustrates that an immune reaction against one organism or virus may be sufficient to protect against an antigenically related, if not identical, organism. The technique of exposing individuals to "tame" microbes to protect them against pathogens, now generally called **vaccination**, has been used to protect humans against many bacterial and viral pathogens (**Table 24.1**). Most vaccinations today involve administering crippled (live, attenuated) strains of the pathogenic microbe or inactivated microbial toxins (for example, diphtheria toxin).

Cross-protection, in which immunization against one microbe protects against a second, will work only if two proteins critical to the pathogenesis of the two different microorganisms share key antigenic determinants. Cross-protection will not take place if these determinants differ significantly. A good example is the common cold, which is caused by hundreds of closely related rhinovirus strains (rhinitis, a runny nose, is one of the symptoms of this viral disease). Infection with one strain will not immunize the victim against a second strain. The reason is that the structures of rhinovirus proteins that attach to the ICAM-1 surface protein on host cells differ dramatically between different strains of rhinovirus (**Fig. 24.5A**). Antibodies called neutralizing antibodies, which bind to the attachment protein on one strain of rhinovirus, will prevent infection by that strain (**Fig. 24.5B**) but will not bind an antigenically distinct ICAM-1 receptor protein from a different strain. A key to one lock will not work in a different lock.

Thought Question

24.2 How does a neutralizing antibody that recognizes a viral coat protein prevent infection by the associated virus?

A. Rhinovirus

B. Antibody-coated rhinovirus

Cell receptor (ICAM-1)

Cell

Antibody to virus receptor protein

FIGURE 24.5 ■ Antibodies prevent rhinovirus attachment to cell receptors. A. The complex rhinovirus capsid is pictured here attaching to the cell-surface molecule ICAM-1 (intercellular adhesion molecule, shown in reddish brown). (PDB code: 1rhi) **B.** Rhinovirus coated with protective (neutralizing) antibodies (green) that block the ICAM-1 receptors on the virus. As a result, the virus fails to attach to and infect the host cell. (PDB code: 1RVF)

Antigens and Immunogens

Antigens that, by themselves, can elicit antibody production are called **immunogens**. Molecules of molecular weight less than 1,000, however, are generally not immunogenic, because they do not bind MHC molecules. Nevertheless, these small molecules, called **haptens** (from the Greek word meaning "to fasten" and the German word for "stuff"), will elicit the production of specific antibodies if they are covalently attached to a larger carrier protein or other molecule (**Fig. 24.6**). Haptens can be thought of as small, incomplete antigens. An example of a hapten is the antibiotic penicillin, a serious

TABLE 24.1	Vaccines Against Viral and Bacterial Pathogens	
Disease	**Vaccine**	**Vaccination recommended for:**
Viral diseases		
Chickenpox	Attenuated strain (will still replicate)	Children 12–18 months
Hepatitis A	Inactivated virus (will not replicate)	Children 12 months
Hepatitis B	Viral antigen	Newborns
Influenza	Inactivated virus or antigen	Everyone, after 6 months old, yearly
Measles, mumps, rubella (MMR)	Attenuated viruses; MMR combined vaccine	Children 12 months
Polio	Inactivated (injection, Salk)	Children 2–3 months
Rabies	Inactivated virus	Persons in contact with wild animals
Yellow fever	Attenuated virus	Military personnel
Bacterial diseases		
Anthrax	*Bacillus anthracis*, toxin components; unencapsulated strain	Agricultural and veterinary personnel; key health care workers
Cholera	Killed *Vibrio cholerae*, toxin components	Travelers to endemic areas
Diphtheria	Toxoid (inactivated toxin)	Children 2–3 months
Lyme disease	*Borrelia burgdorferi*, lipoproteins OspA and OspC surface antigens	Canines; human vaccine discontinued
Meningitis caused by *Haemophilus influenzae* type b (Hib)	Bacterial capsular polysaccharide	Children under 5 years
Meningococcal disease	*Neisseria meningitidis*, bacterial capsular polysaccharides	Children >2 years; adults >50 years
Pertussis	Acellular *Bordetella pertussis*	Children 2–3 months
Pneumococcal pneumonia	*Streptococcus pneumoniae*, bacterial capsular polysaccharides	Children; adults >50 years
Tetanus	Toxoid	Children 2–3 months
Tuberculosis (*Mycobacterium tuberculosis*)	Attenuated *Mycobacterium bovis* [BCG (Bacille Calmette-Guérin) vaccine]	Exposed individuals
Typhoid fever	Killed *Salmonella* Typhi	Individuals in endemic areas
Typhus	Killed *Rickettsia prowazekii*	Medical personnel in endemic areas; scientists; discontinued

cause of immune hypersensitivity reactions in some individuals (see Section 24.7).

Consider the following: A protein carrier such as bovine serum albumin (BSA) injected into a mouse elicits antibodies that react against BSA (**Fig. 24.6**). BSA antigen is an immunogen because it elicits an immune response to itself. In contrast, a mouse injected with the hapten benzene sulfate fails to produce antibodies to benzene sulfate. However, when the hapten chemical is attached to BSA protein and the hapten-BSA complex is injected, the mouse produces not only antibodies that react to the carrier (BSA), but also other antibodies that react against the benzene hapten. The reason for this carrier effect will become evident later in the chapter, when we describe how antigens are processed by cells of the immune system. Thus, antigens include immunogens that elicit an immune response by themselves, and haptens that must be attached to an immunogen in order to generate an immune response. Karl Landsteiner uncovered the antigen-immunogen-hapten relationship in the early 1900s, along with discovering the ABO blood group system that first defined the concept of immunological specificity (**eTopic 24.2**).

Thought Question

24.3 The attachment proteins of different rhinovirus strains all bind to ICAM-1. How can all these proteins be immunologically different if they find the same target (ICAM-1)? Why won't antibodies directed against one rhinovirus strain block the attachment of other rhinovirus strains?

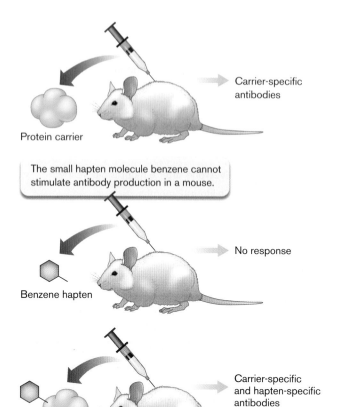

The small hapten molecule benzene cannot stimulate antibody production in a mouse.

Attaching the hapten to a larger carrier molecule (e.g., BSA protein) will result in production of antibodies to both the carrier and the hapten molecules.

FIGURE 24.6 ■ This basic schematic shows how haptens can elicit antibody production.

To Summarize

- **An antigen** can elicit an antibody response. An antigen usually consists of many different epitopes (antigenic determinants), each of which binds to a different, specific antibody.

- **Humoral immunity** against infection is the result of antibody production originated by B cells. A subgroup of T cell lymphocytes stimulate B cells to become antibody-secreting plasma cells.

- **Cell-mediated (cellular) immunity** involves another subgroup of T cells called cytoxic T cells that can directly kill host cells.

- **Proteins are better immunogens** than nucleic acids and lipids because proteins have more diverse chemical forms.

- **Antigen-presenting cells (APCs),** such as phagocytes, degrade microbial pathogens and present distinct pieces on their cell-surface MHC proteins.

- **Immunological specificity** means that antibody made to one epitope will not bind to different epitopes (antibodies can cross-bind weakly to similar epitopes; for example, antibody to cowpox virus will bind to a similar epitope on smallpox virus).

- **A hapten** is a small compound that must be conjugated to a larger carrier antigen to elicit the production of an antibody.

24.2 Antibody Structure, Diversity, and Synthesis

What are antibodies, and why are they important? Antibodies, also called **immunoglobulins**, are members of the larger immunoglobulin superfamily of proteins. Made by the body in response to an antigen, antibodies are the keys to immunological specificity. Members of the immunoglobulin superfamily of proteins have in common a 110-amino-acid domain with an internal disulfide bond. The immunoglobulin superfamily includes antibodies and other important binding proteins, such as the major histocompatibility proteins and parts of the B-cell receptor described later.

Note: Antibodies are called immunoglobulins and belong to the immunoglobulin superfamily of proteins. However, this superfamily includes many other cell-surface-binding proteins, such as the major histocompatibility complex (MHC) proteins and various cytokine receptor proteins. These latter proteins are not antibodies and are not called immunoglobulins. The term "immunoglobulins" is reserved for antibodies.

Like miniature "smart bombs," antibody immunoglobulins individually circulate through blood, ignoring all antigens except those for which they were designed. When an antibody finds its antigenic match, it binds to the antigen and initiates several events designed to destroy the target. Antibodies, in addition to being free-floating, are strategically situated on the surface of B cells, where they enable the B lymphocytes to recognize specific antigens (B-cell receptors are discussed later in this section).

A typical antibody consists of four polypeptide chains. There are two large **heavy chains** and two smaller **light chains** (**Fig. 24.7**). The four polypeptides combine to form a Y-shaped tetrameric structure held together by disulfide bonds. Two bonds connect the two identical heavy chains to each other. One light chain is then attached near its carboxyl end to the middle of each heavy chain by a single disulfide bond. The antigen-binding sites are formed at the amino-terminal ends of the light and heavy chains. One antibody molecule possesses two identical antigen-binding sites, one on each "arm" of the molecule.

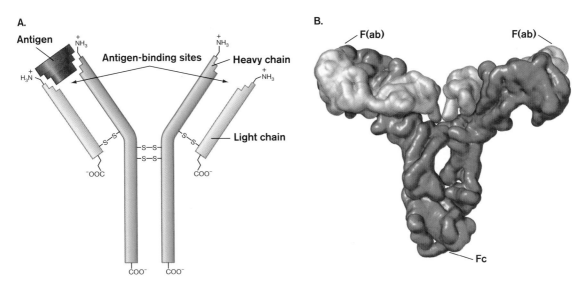

FIGURE 24.7 ■ Basic antibody structure. **A.** Each antibody contains two heavy chains and two smaller, light chains held together by disulfide bonds. The Y-shaped structure contains two antigen-binding sites, one at each arm of the molecule. These two sites are formed by the amino-terminal regions of the heavy- and light-chain pairs. **B.** The 3D structure of an antibody. The heavy chains are shown in green, the light chains in yellow. The F(ab) regions represent the antigen-binding sites. The Fc portion points downward and is used to attach the antibody to different cell-surface molecules. (PDB code: 1R7O)

Antibody Structure Enables Immunoprecipitation

A typical antibody molecule has two antigen-binding sites, each of which binds identical antigens. Because a single antibody can bind two identical epitopes, antibodies can cross-link antigens in solution, ultimately forming complexes that are too large to remain soluble, so they fall out of solution (**Fig. 24.8**). The phenomenon, called **immunoprecipitation**, is normally observed only in vitro, where the concentrations of antigen and antibody can be manipulated experimentally.

Immunoprecipitation occurs only with appropriate ratios of antigen and antibody molecules. Too many antigen molecules (antigen excess; **Fig. 24.8A**) or too few

antigen molecules (antibody excess; **Fig. 24.8B**) result in complexes too small to immunoprecipitate. Large complexes are formed only at an appropriate antigen/antibody ratio, called **equivalence** (**Fig. 24.8C**). Equivalence is the point where the number of antigenic sites is roughly equal to the number of antigen-binding sites.

Immunoprecipitation is the basis for many experimental immunological techniques. For example, the concentration of an antigen can be determined in vitro, or antibodies can be used to identify and remove specific antigens from a complex mixture because the specific antigen-antibody complex precipitates. eAppendix 3 describes some experimental uses for immunoprecipitation, such as radial immunodiffusion and western blotting.

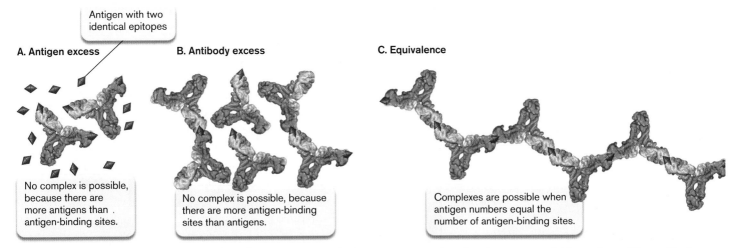

FIGURE 24.8 ■ Basis of immunoprecipitation. Immunoprecipitation cannot occur when there is an excess of either antigens (**A**) or antibodies (**B**). Only when the numbers of epitopes and antigen-binding sites are roughly equivalent (**C**) will a large complex form and fall out of solution.

Antibodies Have Constant and Variable Regions

There are five classes of antibodies, defined by five different types of heavy chains, called alpha (α), mu (μ), gamma (γ), delta (δ), and epsilon (ε). The heavy-chain classes are distinguished one from another by regions of highly conserved amino acid sequences, known as **constant regions** (denoted C_H for the heavy chain) (**Fig. 24.9**). Antibodies containing gamma heavy chains are called IgG; those with alpha, mu, delta, and epsilon heavy chains are called IgA, IgM, IgD, and IgE, respectively. Each antibody class serves a specific purpose in the immune system.

In contrast to heavy chains, there are only two classes of light chains—kappa (κ) and lambda (λ)—which are defined by their own constant regions (C_L). A single antibody of any heavy-chain class (for example, IgG) may contain two kappa light chains or two lambda light chains, but never one of each. Two-thirds of all antibody molecules carry kappa chains; the rest have lambda chains.

The antigen-binding part of an antibody is formed by highly variable amino acid sequences situated at the amino-terminal ends of the light and heavy chains. These **variable regions** are referred to as the V_L and V_H regions, respectively (**Fig. 24.9**). The rest of each immunoglobulin chain is composed of the highly conserved constant regions. C_H1, C_H2, and C_H3 denote the three different constant regions in each heavy chain. Shortly, we will discuss the genes that code for these regions and the combinatorial possibilities underlying the formation of different antibody molecules.

In addition to the two "arms" that bind antigens, every antibody contains a "tail" called the **Fc region** (**Fig. 24.9**). The Fc region is not involved in antigen recognition but is important for anchoring antibodies to the surface of certain host cells and for binding components of the complement system.

Thought Question

24.4 Can an F(ab′)$_2$ antibody fragment prevent the binding of rhinovirus to the ICAM-1 receptor on host cells? And can an F(ab′)$_2$ antibody fragment facilitate phagocytosis of a microbe?

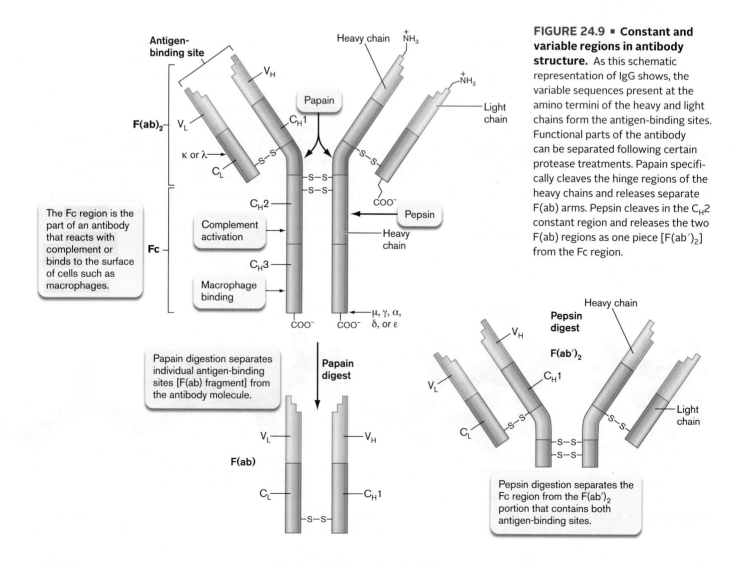

FIGURE 24.9 ■ Constant and variable regions in antibody structure. As this schematic representation of IgG shows, the variable sequences present at the amino termini of the heavy and light chains form the antigen-binding sites. Functional parts of the antibody can be separated following certain protease treatments. Papain specifically cleaves the hinge regions of the heavy chains and releases separate F(ab) arms. Pepsin cleaves in the C_H2 constant region and releases the two F(ab) regions as one piece [F(ab′)$_2$] from the Fc region.

The Fc region is the part of an antibody that reacts with complement or binds to the surface of cells such as macrophages.

Papain digestion separates individual antigen-binding sites [F(ab) fragment] from the antibody molecule.

Pepsin digestion separates the Fc region from the F(ab′)$_2$ portion that contains both antigen-binding sites.

Isotypes, Allotypes, and Idiotypes

Antibody diversity involves a hierarchy of protein sequence differences found among antibodies.

- **Isotype** sequences define the various heavy chains of a species.

- **Allotype** sequences define individuals within a species.

- **Idiotype** sequences define antibodies within an individual.

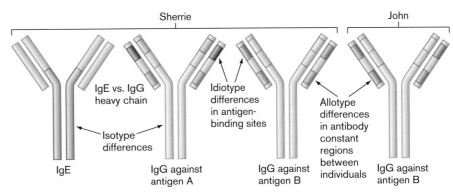

FIGURE 24.10 ■ Isotype, idiotype, and allotype differences on antibodies. Colors indicate different amino acid sequences (or epitopes) within an individual (Sherrie) or between individuals (Sherrie and John).

Understanding how an antibody's isotype, allotype, and idiotype differ is important because these terms reflect how antibodies carry out different functions in a host and bind different antigens.

"Isotype" refers to the amino acid differences in the constant regions of heavy-chain classes (what makes IgG different from IgE). These differences will be the same in all members of a species (for example, *Homo sapiens*). "Allotype" refers to any amino acid sequence differences in the constant regions of either heavy or light chains that make IgG from one person different from IgG in another person. "Idiotype" refers to amino acid sequence differences between two molecules of the same isotype (IgG, for example) in a single person. Idiotype differences usually occur in the antigen-binding region of an antibody.

In real terms, consider two different people, John and Sherrie. Blood of both John and Sherrie carries the human IgG isotype. However, all of the IgG molecules circulating in John have small, allotypic amino acid differences from the IgG antibodies circulating in Sherrie (John and Sherrie have different IgG allotypes) (**Fig. 24.10**). Within Sherrie herself, the antigen-binding site of an IgG molecule that binds a herpes virus epitope possesses idiotypic amino acid differences from an IgG molecule in her that binds to a rhinovirus epitope. The synthesis and functional purpose of these different levels of antibody diversity will be described in the next several sections.

Note that antibodies are proteins too, and are themselves antigens. Thus, the amino acid differences found within a single class of antibody—IgG, for example—also represent different antigenic epitopes within that class. Isotypic, allotypic, and idiotypic epitopes in antibody molecules can be detected using the immunological techniques described in eAppendix 3.

> **Thought Question**
>
> **24.5** (refer to **Fig. 24.10**) What types of antibodies will IgG taken from Sherrie raise when injected into John (for example, anti-isotype, anti-allotype, or anti-idiotype)?

Antibody Isotype Functions and "Super" Structures

All antibody isotypes have the same basic structure. However, each isotype has a unique superstructure (for example, monomer or dimer), and each is designed to carry out a different task. Some key properties of the five different immunoglobulin classes are listed in **Table 24.2**.

TABLE 24.2	**Properties of Human Immunoglobulins**							
	IgG							
Property	**IgG1**	**IgG2**	**IgG3**	**IgG4**	**IgA**	**IgM**	**IgD**	**IgE**
Serum half-life (days)	21	20	7	21	6	10	3	2
% Total serum Ig	70%				15%–20%	5%–10%	0.2%	0.002%
Antigen-binding sites	2				2–4	2–10	2	2
Produced by fetus	Poor, if at all				Poor, if at all	Yes	?	Poor, if at all
Transmitted across placenta	Yes				No	No	No	No
Binds complement	Yes				No	Yes	No	No
Opsonizing	Yes				No	No	No	No
Binds mast cells	No				No	No	No	Yes

A. IgM

IgM forms a pentamer held together by disulfide bonds and the J chain. IgM can bind ten antigens.

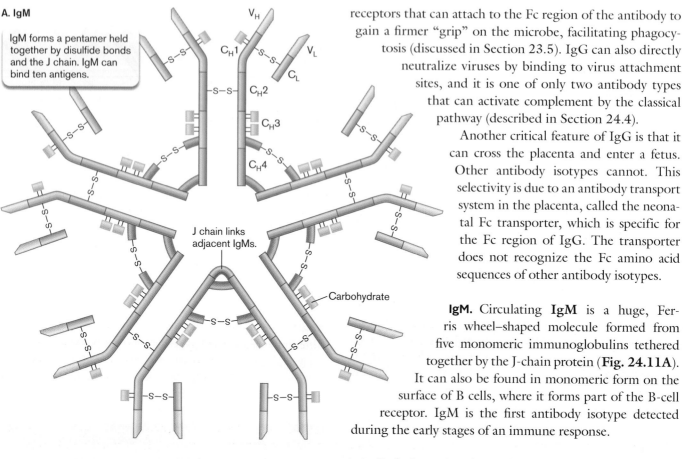

J chain links adjacent IgMs.

Carbohydrate

B. IgA

IgA is secreted as a dimer held together by the secretory piece and J chain and can bind four identical antigens.

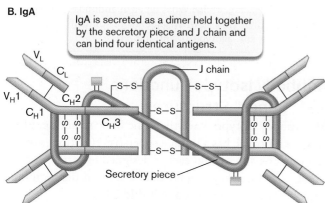

J chain

Secretory piece

FIGURE 24.11 ■ Structures of IgM and IgA. The antibodies are made as multimers of five immunoglobulin molecules, in the case of IgM (**A**); or two, in the case of IgA (**B**).

IgG. The simplest and most abundant antibody in blood and tissue fluids is **IgG** (its superstructure is shown in **Fig. 24.9**). It is made as a monomer but has four subclasses. Each subclass varies in its amino acid composition and number of interchain cross-links. IgG molecules carry out several missions for the immune system. First, they bind and **opsonize** microbes; that is, they make microbes more susceptible to phagocytes. Opsonizing IgG antibodies use their antigen-binding sites to stick to microbes, thereby causing their Fc regions to protrude outward. Phagocytes possess surface Fc receptors that can attach to the Fc region of the antibody to gain a firmer "grip" on the microbe, facilitating phagocytosis (discussed in Section 23.5). IgG can also directly neutralize viruses by binding to virus attachment sites, and it is one of only two antibody types that can activate complement by the classical pathway (described in Section 24.4).

Another critical feature of IgG is that it can cross the placenta and enter a fetus. Other antibody isotypes cannot. This selectivity is due to an antibody transport system in the placenta, called the neonatal Fc transporter, which is specific for the Fc region of IgG. The transporter does not recognize the Fc amino acid sequences of other antibody isotypes.

IgM. Circulating **IgM** is a huge, Ferris wheel–shaped molecule formed from five monomeric immunoglobulins tethered together by the J-chain protein (**Fig. 24.11A**). It can also be found in monomeric form on the surface of B cells, where it forms part of the B-cell receptor. IgM is the first antibody isotype detected during the early stages of an immune response.

IgA. **IgA** is secreted across mucosal surfaces (linings of the respiratory, gastrointestinal, urinary, and reproductive tracts) and is most commonly found as a dimer (**Fig. 24.11B**). This explains why IgA can bind four molecules of antigen (each monomer can bind two). The components of the IgA dimer are linked by disulfide bonds to a protein called the J chain, which joins two IgAs by their Fc regions. A sixth protein, the secretory piece, is wrapped around the IgA dimer during the secretion process. The secreted molecule, now called sIgA (secretory IgA), is found in tears, breast milk, and saliva, and on other mucosal surfaces. The molecule sIgA is important for mucosal immunity against pathogens that infect mucosal linings.

IgD. The last two antibody isotypes, IgD and IgE, are present at very low levels in the blood. **IgD** is a monomer that can neither bind complement nor cross the placenta. IgD molecules, however, are abundant on the surface of B cells. Attached to B-cell surfaces by Fc regions, IgD, along with monomeric IgM, can bind antigen and signal B cells to differentiate and make antibody.

IgE. **IgE**, also present in only trace amounts in the blood, is found more prominently bound to the surfaces of mast cells and basophils, where it has potent biological activity. Mast cells and basophils contain granules loaded with inflammatory mediators. The primary role of IgE is to

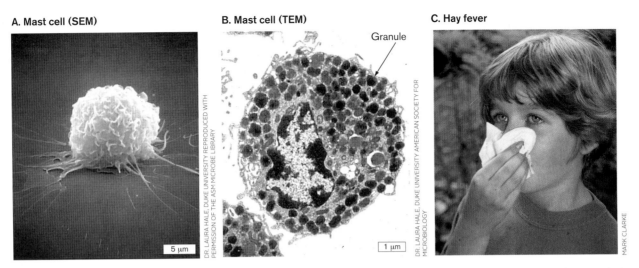

A. Mast cell (SEM)

B. Mast cell (TEM)

Granule

5 μm

DR. LAURA HALE, DUKE UNIVERSITY REPRODUCED WITH PERMISSION OF THE ASM MICROBE LIBRARY

1 μm

DR. LAURA HALE, DUKE UNIVERSITY AMERICAN SOCIETY FOR MICROBIOLOGY

C. Hay fever

MARK CLARKE

FIGURE 24.12 ■ Mast cells. A. Mast cell (SEM). **B.** Granules (arrow) inside a mast cell (TEM). **C.** Hay fever is the result of degranulation of IgE-coated mast cells, which release histamine and other pharmacological mediators.

amplify the body's response to invaders. Once secreted into serum, IgE attaches to mast cells (**Fig. 24.12A and B**), again by way of its Fc region, and, like a Venus flytrap, waits until its matched antigen binds to its antigen-binding site. When two surface IgE molecules on a mast cell are cross-linked by antigen, a signal is sent internally that triggers degranulation (see Section 24.7). The subsequent release of histamine and other pharmacological mediators from these granules helps orchestrate the acute inflammation that takes place during early host responses to microbial infection (that is, while the antibody response is gearing up). The system also causes severe allergic hypersensitivities (such as anaphylaxis) and milder forms like hay fever (**Fig. 24.12C**).

differentiate into plasma cells and secrete antibody. Plasma cells are much larger than B cells because of an enormous increase in protein synthesis and secretion machinery.

The net result of the primary antibody response is the early synthesis and secretion of pentameric IgM molecules specifically directed against the antigen (also called the immunogen). Later during the primary response, a process known as **isotype switching** (**class switching**) occurs, and the predominant antibody type produced becomes IgG rather than IgM (discussed shortly). Antibodies made during the primary response, while specific for the immunogen, are actually <u>not</u> of the highest affinity. Later responses by memory B cells increase antibody affinity (see the next section).

Primary and Secondary Antibody Responses

Once you have been infected with a microorganism or have been given a vaccine, what happens? After a lag period of several days, antibodies begin to appear in the **serum** (the fluid that remains after blood clots). During the lag period, a series of molecular and cellular events causes a distinct subset of B cells (located in lymph nodes and spleen) to proliferate and differentiate into antibody-secreting **plasma cells** and **memory B cells**. Each B cell is genetically programmed to make antibodies to one antigen or epitope—a process called the **primary antibody response**.

A subsequent exposure to the antigen, which can take place months or years after the initial encounter, will trigger a rapid, almost instantaneous increase in the production of antibodies and is called the **secondary antibody response** (**Fig. 24.13**). This quick response occurs thanks to the memory B cells formed during the primary response. Once stimulated, memory B cells rapidly

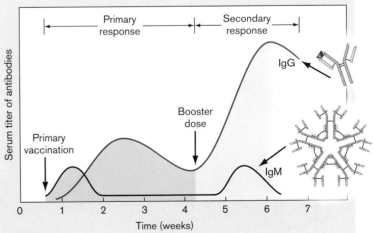

FIGURE 24.13 ■ Primary versus secondary antibody response. Primary vaccination or infection leads to the early synthesis of IgM, followed by IgG. Reinfection or a second, booster dose of a vaccine results in a more rapid antibody response consisting mainly of IgG due to memory B cells formed during the primary response. Note that the time course and level of antibody made vary with the immunogen and the host.

As the immunogen is cleared from the body, the levels of both IgG and IgM decline because the plasma cells that produced them die. Plasma cells have an average life span of only 100 days. However, the immune system has been primed to respond more aggressively to the immunogen, should it encounter it again. A rapid response to a second encounter is possible because memory B cells produced during the primary response are maintained in lymph nodes or bone marrow and continue to divide. If a person later encounters the same antigen, these memory B cells quickly proliferate and differentiate into plasma cells with no lag phase. Thus, memory B cells rapidly initiate the secondary antibody response (or anamnestic response, from the Greek *anamnesis*, meaning "remembrance"). Memory B cells comprise approximately 40% of the B-cell population.

During the secondary response, memory B cells that have undergone isotype switching from IgM to IgG become plasma cells that secrete copious amounts of IgG antibody. These antibodies have a higher specificity for the antigen than do the antibodies produced during the primary response (see "Making Memory B Cells" at the end of this section). The higher specificity—a result of hypermutation—causes some plasma cells to produce antibodies that bind their antigens more strongly than those produced

by the clonal ancestor. Small amounts of IgM are also produced from the few memory cells that did not undergo isotype switching during the primary response.

The secondary antibody response is why vaccinations work. Before vaccinations, the only way to be protected from an infection by a given pathogen was to have already been infected by that pathogen. Memory cells made during the primary response will protect you against a second infection, but pathogens can do considerable harm during the lag phase of the primary response. To avoid this harm, an innocuous version of a pathogen, or a harmless piece of it, can be injected into a person to trigger a primary response without producing disease (or, at worst, producing only a mild form of the illness). Immunization thus primes the immune system to respond efficiently and without delay upon encountering the real pathogen. **Table 24.1** lists a variety of viral and bacterial diseases for which immunizations are available.

Thought Question

24.6 The mother of a newborn was found to be infected with rubella, a viral disease. Infection of the fetus could lead to serious consequences for the newborn. How could you determine whether the newborn was infected in utero?

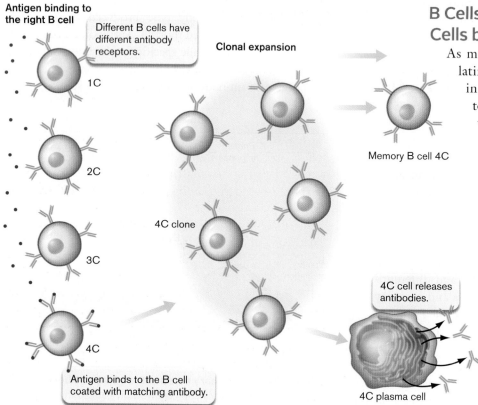

Antigen binding to the right B cell

Different B cells have different antibody receptors.

Clonal expansion

1C

2C

4C clone

3C

4C

Antigen binds to the B cell coated with matching antibody.

Memory B cell 4C

4C cell releases antibodies.

4C plasma cell

FIGURE 24.14 ■ Clonal selection. The B-cell population is composed of individuals that have specificity for different antigens. When a B cell contacts its cognate antigen, an intracellular signal is generated, leading to proliferation and differentiation of that clone (clonal expansion). Plasma cells and memory B cells result.

B Cells Differentiate into Plasma Cells by Clonal Selection

As mentioned earlier, each B cell circulating throughout the body or nestled in a lymphoid organ is programmed to synthesize antibody that reacts with a single epitope (for example, a small portion of a protein). In a process called **clonal selection**, an invading antigen will inadvertently select which B-cell clone will proliferate to large numbers and differentiate into antibody-producing plasma cells or memory B cells. In this way, large amounts of antibody specific for the antigen are made. The mechanism of clonal selection begins with the antigen binding to a matching B cell (a B cell preprogrammed to bind to that antigen; see the discussion of B-cell receptors that follows) (**Fig. 24.14**).

Mature but naive B cells (those that have not previously encountered antigen) can produce only

IgM and IgD, which have identical antigen specificities. These two antibody classes are displayed like tiny satellite dishes on the B-cell surface, anchored by their Fc regions through hydrophobic transmembrane segments. These surface antibodies are the keys to stimulating the proliferation and differentiation of B cells into antibody-producing plasma cells or memory B cells. Upon binding to its corresponding antigen via these surface antibodies, the B cell is said to become <u>activated</u>, whereby it multiplies and differentiates into a plasma cell that ultimately synthesizes only one antibody isotype (for example, IgG1 or IgA2). Clonal selection has begun. In addition to antigen binding, most B cells require help from T cells to become plasma cells and memory B cells (discussed later).

B-cell receptor. Each antibody bound to a B-cell membrane is associated with two other membrane proteins, called Igα and Igβ (these are <u>not</u> immunoglobulins, but they are designated "Ig" because they <u>associate</u> with the surface antibody). The complex is called the **B-cell receptor**, or **BCR** (**Fig. 24.15**). Each B cell may have upward of 50,000 B-cell receptors. A microbe generally has multiple copies of the same epitope on its surface (think about a virus capsid). Each epitope can bind adjacent B-cell receptors on one B cell. Once bound, surface B-cell receptors begin to cluster in a process called capping. **Capping** activates Igα and Igβ to initiate a phosphorylation signal cascade directed into the nucleus (**Fig. 24.16**). In a phosphorylation cascade, a phosphate group donated by ATP is passed from one protein to another, usually ending up on, and activating, a transcriptional regulator. The transcription factors at

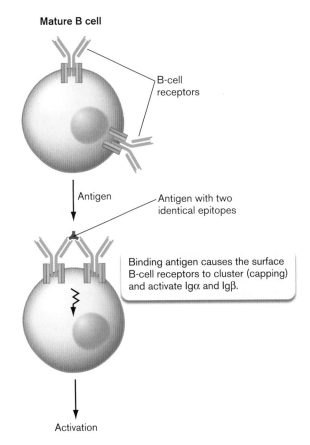

Mature B cell

B-cell receptors

Antigen

Antigen with two identical epitopes

Binding antigen causes the surface B-cell receptors to cluster (capping) and activate Igα and Igβ.

Activation

FIGURE 24.16 ■ **Capping and activation of the B cell.** Two B-cell receptors can bind two identical epitopes on a pathogen. The resulting capping process initiates a signal cascade that activates differentiation and proliferation of the B cell independent of T-cell help. Antigens with repeating epitopes, such as polysaccharides, can directly cross-link B-cell receptors.

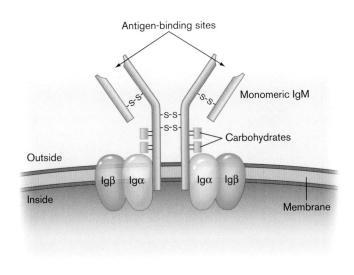

Antigen-binding sites

Monomeric IgM

Carbohydrates

Outside

Igβ Igα   Igα Igβ

Inside

Membrane

FIGURE 24.15 ■ **B-cell receptor.** The B-cell receptor is formed as a complex consisting of a monomeric IgM plus Igα and Igβ in the membrane.

the end of the BCR cascade stimulate transcription of genes that contribute to cellular proliferation. Some of these activated B cells can differentiate into plasma cells and secrete IgM antibody as part of the primary immune response.

T cell–dependent and –independent antibody production. There are two routes by which antigens can stimulate B cells to differentiate into plasma cells. In one, called the T cell–independent route, antigens that possess multiple repeating epitopes (for example, polysaccharide antigens) can directly cross-link B-cell receptors (the capping process)—a step necessary for triggering differentiation (see **Fig. 24.16** and **Special Topic 24.1**). Proteins, however, which are the largest group of antigens, do not contain multiple repeating units. Proteins possess many small, discrete, single epitopes, making the cross-linking of B-cell receptors difficult. B-cell responses to these types of antigens require help from specific T cells, and this constitutes the second route to B-cell activation. Thus, B cells usually require multiple signals to initiate a primary response. How

SPECIAL TOPIC 24.1   Can Retroviruses Help B Cells?

The differentiation of B cells into antibody-producing plasma cells first requires that antigen bind to B-cell receptors (BCRs). However, a second signal to the B cell is required to complete B-cell activation. For most antigens, the second signal involves B-cell interaction with helper T cells (T cell–dependent antigens; see Section 24.3). However, some antigens successfully stimulate B-cell differentiation <u>without</u> T-cell help. These antigens are called T cell–independent (TI) antigens and come in two forms: TI-1 antigens bind BCR and engage Toll-like receptors (TLRs) as the second signal. TI-2 antigens are large molecules with highly repetitive structures; for example, bacterial capsular polysaccharides and viral capsids. These highly repetitive structures cause extensive cross-linking of surface BCR molecules. Until recently, the second signal for TI-2 antigens was unknown.

In a surprising finding, Bruce Beutler (who received the 2011 Nobel Prize for Physiology or Medicine for the discovery of Toll-like receptors) (**Fig. 1A**) and his colleagues at the University of Texas Southwestern Medical Center discovered that endogenous retroviruses provide the help B cells need to respond to TI-2 antigens. Endogenous retroviral DNA sequences are the "footprints" of previous retroviral infections (see Chapter 11). Mammalian genomes, including human, are rife with latent retroviruses, some of which can still replicate to form virus-like particles. It is startling that these remnants of past infection may also play a role in immunity. So, what led the scientists to this conclusion?

Suspecting that an innate immune signaling pathway in the B cell was involved in TI-2 responses, Beutler's team screened a series of genetically engineered mice missing different pattern recognition receptors (such as TLRs and NLRs) or signaling pathways for an ability to make IgM when challenged with a TI-2 antigen called NP (**Fig. 1B**). Of all the pathways tested, only mice missing the cGAS-STING or Mavs pathways were deficient in producing NP-specific IgM. So, too, were *Ikbkg* mutant mice that are defective in the synthesis of the transcription factor NF-κB. The STING and cGAS proteins are components of a cytoplasmic DNA sensor, whereas the Mavs protein is an adaptor for the cytoplasmic RNA sensor RIG-I. The results indicated that cytoplasmic RNA or DNA is required for TI-2 antigen stimulation of B cells.

Where did these cytoplasmic nucleic acids come from? An external source was highly unlikely, but could the cytoplasmic nucleic acids come from endogenous retroviruses (ERVs) lying latent in the cell nucleus? When tested, NP-specific B cells isolated from mice immunized with NP had higher levels of various ERV RNAs bound to the RNA sensor RIG-I than did nonspecific B cells (**Fig. 2A**). The scientists then showed that cross-linking BCR with anti-IgM antibodies also induced ERV replication (**Fig. 2B**) and that inhibiting reverse transcriptase before NP immunization limited anti-NP antibody production. Retroviral reverse transcriptase synthesizes ERV RNA from retroviral DNA templates.

A.

B.

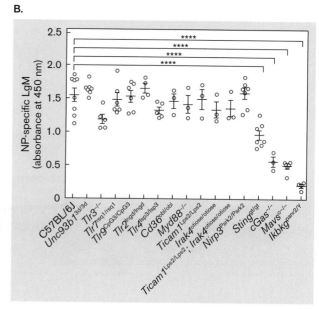

FIGURE 1 ■ NP-specific IgM production in mice (C57BL/6J) defective in various pattern recognition receptors or pathways.
A. Bruce Beutler (right) and postdoc Ming Zeng. **B.** Mutant strains of mice are identified on the *x*-axis. Superscript designations such as "3d/3d" or "−/−" indicate that the mouse is homozygous for the defective gene. The *y*-axis shows NP-specific IgM measured in blood 4.5 days after NP injection. Each dot represents a single mouse. **** = statistical *P* value ≤0.0001.

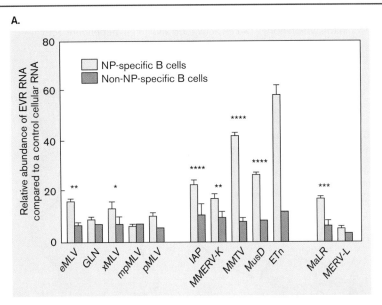

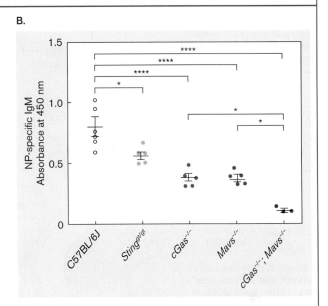

FIGURE 2 ■ Endogenous retroviruses stimulate B-cell response to TI-1 antigens. **A.** Immunoprecipitation (see eAppendix 3) of ERV RNAs bound to RIG-I protein. B cells were collected from mouse spleens and separated into NP-specific and nonspecific B cells. PCR was used to identify the specific ERV RNA transcripts identified on the x-axis. **B.** The Mavs and cGAS-STING pathways collaborate in stimulating NP-specific IgM production. Statistical P values: * = ≤0.05; ** = ≤0.01; *** = ≤0.001; **** = ≤0.0001.

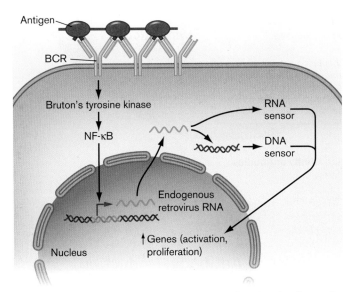

FIGURE 3 ■ Proposed retrovirus participation in the TI-2 antibody response. TI-2 antigens are presented by APCs to B cells. Extensive cross-linking activates a signal cascade leading to transcription of retroviral DNA. Resulting viral RNA and/or cDNA activate the cytoplasmic RNA-sensing (e.g. Rig-1) and/or DNA-sensing (e.g. cGAS-STING) pathways. Resulting signals induce the activation and expansion of B cells and differentiation into plasma cells.

From these and many other experiments, the authors propose the following events after TI-2 immunization (**Fig. 3**): (1) TI-2 antigens cross-link BCRs on B cells, setting off a signal cascade leading to the production of the NF-κB transcription factor. (2) NF-κB stimulates transcription of ERV proviruses, and the resulting RNAs are transported to the cytoplasm. (3) ERV mRNAs are translated to produce reverse transcriptase along with other viral proteins. (4) Reverse transcriptase produces cDNA from ERV mRNA. (5) ERV mRNAs and cDNAs present in the cytoplasm activate RIG-I and cGAS, respectively. The result is B-cell activation and differentiation to IgM-producing plasma cells.

One possible ramification of this finding is that the continuous stimulation of mucosal B cells by microbiota might increase the expression of endogenous retroviruses. The effect that retroviral expression might then have on antibody production at mucosal surfaces is unclear.

RESEARCH QUESTION

This report provides compelling, but circumstantial, evidence of retroviral involvement in TI-2 antibody responses. What "heroic" experiment would definitively prove that ERVs produce the second signal needed for this response?

Zeng, Ming, Zeping Hu, Xiaolei Shi, Xiaohong Li, Xiaoming Zhan, et al. 2014. MAVS, cGAS, and endogenous retroviruses in T-independent B cell responses. *Science* **346**:1486–1492.

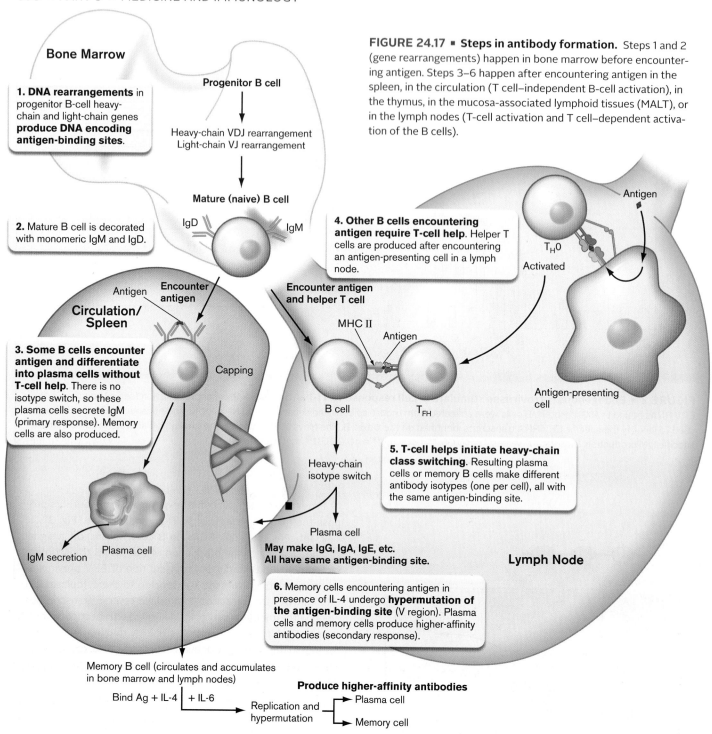

Bone Marrow

1. DNA rearrangements in progenitor B-cell heavy-chain and light-chain genes **produce DNA encoding antigen-binding sites**.

Progenitor B cell

Heavy-chain VDJ rearrangement
Light-chain VJ rearrangement

Mature (naive) B cell

2. Mature B cell is decorated with monomeric IgM and IgD.

IgD   IgM

Circulation/ Spleen

Antigen

Encounter antigen

3. Some B cells encounter antigen and differentiate into plasma cells without T-cell help. There is no isotype switch, so these plasma cells secrete IgM (primary response). Memory cells are also produced.

Capping

IgM secretion   Plasma cell

Memory B cell (circulates and accumulates in bone marrow and lymph nodes)

Bind Ag + IL-4 | + IL-6

Replication and hypermutation

Produce higher-affinity antibodies
→ Plasma cell
→ Memory cell

FIGURE 24.17 ■ Steps in antibody formation. Steps 1 and 2 (gene rearrangements) happen in bone marrow before encountering antigen. Steps 3–6 happen after encountering antigen in the spleen, in the circulation (T cell–independent B-cell activation), in the thymus, in the mucosa-associated lymphoid tissues (MALT), or in the lymph nodes (T-cell activation and T cell–dependent activation of the B cells).

Antigen

4. Other B cells encountering antigen require T-cell help. Helper T cells are produced after encountering an antigen-presenting cell in a lymph node.

T_H0

Activated

Encounter antigen and helper T cell

MHC II   Antigen

B cell T_{FH}

Antigen-presenting cell

Heavy-chain isotype switch

5. T-cell helps initiate heavy-chain class switching. Resulting plasma cells or memory B cells make different antibody isotypes (one per cell), all with the same antigen-binding site.

Plasma cell

May make IgG, IgA, IgE, etc. All have same antigen-binding site.

Lymph Node

6. Memory cells encountering antigen in presence of IL-4 undergo **hypermutation of the antigen-binding site** (V region). Plasma cells and memory cells produce higher-affinity antibodies (secondary response).

T cells help foster B-cell activation will be discussed in Section 24.3. During B-cell activation, antibody heavy-chain isotype switching (or class switching) can occur at the gene level. Class, or isotype, switching changes the class of antibodies produced. For example, some B cells will switch from making the IgM isotype to making the IgA isotype.

Figure 24.17 summarizes the basic steps of antibody formation leading to the production of plasma cells and memory cells. More detailed descriptions of these events follow in the next section.

Genetics of Antibody Production

Before we discuss how T cells influence antibody production, let's jump ahead and look at the process leading to antibody diversity. It is estimated that each human can synthesize 10^{11} different antibodies. Given that each B cell displays antibodies to only one antigenic determinant, it follows that there are 10^{11} different B cells in the body. We have learned, however, that each person possesses only about a thousand genes or gene segments involved in antibody formation. How are 10^{11} different antibodies made from only 10^3 genes? Susumu Tonegawa was awarded the 1987 Nobel Prize in Physiology or Medicine for discovering that antibody genes can move and rearrange themselves within the genome of a differentiating cell. Three steps are involved: (1) the rearrangement of antibody gene segments (or cassettes), (2) the random introduction of somatic mutations, and (3) the generation of different codons during antibody gene splicing. In humans, antibody diversity is generated continually over a lifetime.

Making the antigen-binding site. The first step in making a specific antibody occurs during the formation of a B cell from a progenitor stem cell (progenitor B cell) in bone marrow, before a foreign antigen is encountered. This process happens throughout a person's life. Immunoglobulin genes in a bone marrow progenitor B cell have many gene segments that can rearrange in many possible combinations. During differentiation into a mature B cell, DNA segments are deleted in a process called **gene switching**, which decreases the number of gene segments in the mature B-cell DNA. The process starts at the 5′ end of an immunoglobulin gene cluster corresponding to the ultimate antigen-binding site (**Fig. 24.18**).

In both the heavy- and light-chain genes are a number of tandem gene cassettes encoding potential variable regions (antigen-binding sites) separated by **recombination signal sequences** (**RSSs**). RSSs allow recombination to bring two widely separated gene segments together. There are approximately 170 variable region (V-region) gene segments for the heavy and light chains. The light-chain V-region gene cluster lies upstream of a cluster of J-region ("J" for "joint" or "joining") genes that will eventually join a variable region to a light-chain constant region (C). Genes for the constant region reside farther downstream in the DNA. The arrangement of the heavy-chain genes is slightly more complex. In this case, the heavy-chain V cluster is followed by a D (diversity) cluster and then the J region.

Note: Do not confuse the J-region gene segments used to make heavy- and light-chain proteins with the J-chain protein that holds together IgM and IgA multimers. They are completely different and unrelated. The J-region gene segments do not encode the J chain.

A summary of the genetic processes leading to antibody formation is shown in **Figure 24.18**. Antibody formation begins with a recombination event between RSS sites at one D and one J segment, which deletes all the intervening D and J segments (**Fig. 24.18**, step 1). Next, the new DJ region joins to one of the V segments, deleting all of the intervening V and D segments (step 2). The result is a joined VDJ DNA sequence, which can then be transcribed (step 3). Each V segment has its own promoter. However, if extra V segments remain upstream of the rearranged VDJ sequence on the DNA, then the only promoter that will fire is the one immediately upstream of the rearranged VDJ segment. The primary RNA transcript will then undergo RNA splicing to remove any J-segment RNA sequences that remain downstream of the VDJ RNA sequence (step 3). The result is a mature B cell that can synthesize a specific antibody (steps 4–6). Remember, all of the DNA recombination and RNA splicing happens before the B cell ever "sees" the antigen. Consequently, the mature B cell is called naive because it has not yet been stimulated by antigen.

A similar sequence of events occurs for the light chains, except that the product is VJ. **Table 24.3** illustrates the amount of antibody diversity that can be achieved in humans simply by this combinatorial re-joining (a total of about 5×10^6 antigens can be recognized).

Where does the rest of antigen-binding-site diversity arise? After recombination, the V regions of the germ lines are susceptible to high levels of somatic mutation (called hypermutation), resulting in the hypervariable regions. Hypermutation happens every time a memory B cell is exposed to the antigen; the memory B cells divide and the hypervariable regions mutate. Additional diversity comes from the junctions of VJ and VDJ, where recombinational joining can occur between different nucleotides. Each gene recombination event can generate additional codons, so the resulting peptides will differ by one or more amino acids. The interactions between the light- and heavy-chain hypervariable regions in an antibody form the antigen-binding sites.

In sum, a combination of gene recombination and random mutations gives us the remarkable level of antigen-binding-site diversity that we each possess. As noted earlier, the human body is capable of responding to 10^{11} antigens, yet there are only 10^8 antigens in nature. This apparent overkill suggests that the immune system is well prepared to cope with any possible antigen it could encounter. Unfortunately for humans, enterprising microbes, such as the trypanosomes that cause sleeping sickness, can stay one step ahead of the immune system by changing the structure of key surface antigens. Changing the antigenic structure of a protein renders useless those antibodies made to the previous structure.

It isn't hard to understand why multicellular organisms, like humans, need the capacity to make any one of billions of different antibodies quickly. Pathogens can undergo many generations of growth within the single life span of a human host. So, humans have to generate recombinant clones of cells quickly to overcome rapidly dividing pathogens.

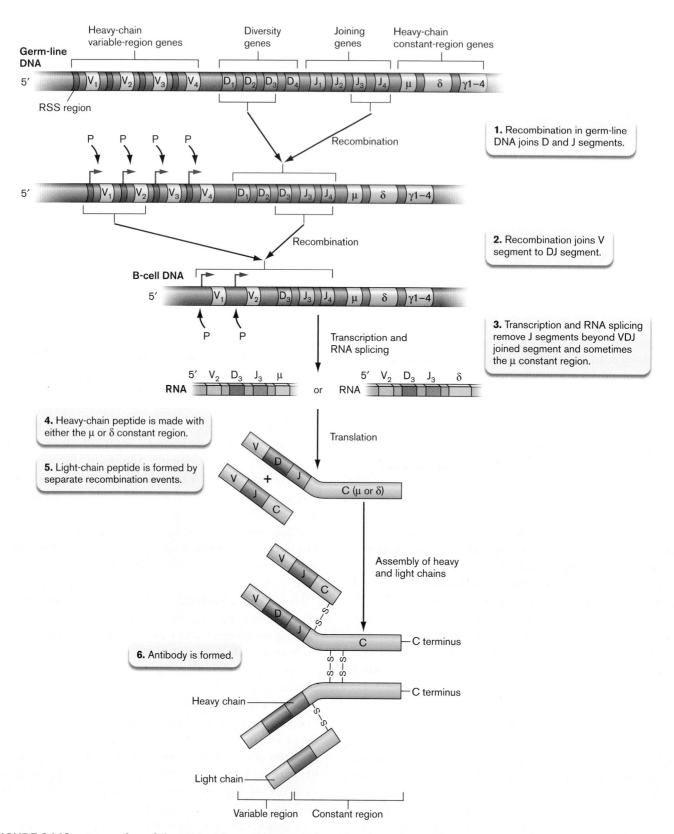

FIGURE 24.18 ■ Formation of the VDJ regions of heavy chains. Note that only a small subset of the V, D, and J genes listed in Table 24.3 is shown in this model. RSS = recombination signal sequence.

TABLE 24.3	Antibody Diversity Attributed to Combinatorial Joining in the Human Germ Line			
	Number of:			
Chain type	**V regions**	**D regions**	**J regions**	**Number of combinations**
λ Light chains	30	0	4	$30 \times 4 = 120$
κ Light chains	40	0	5	$40 \times 5 = 200$
Heavy chains	100	27	6	$100 \times 27 \times 6 = 16,200$
Number of possible antibodies	16,200 heavy-chain combinations × 120 λ-chain combinations = 1.94×10^6 16,200 heavy-chain combinations × 200 κ -chain combinations = 3.24×10^6 $(1.94 \times 10^6) + (3.24 \times 10^6) = 5.18 \times 10^6$ combinations			

The isotype class switch. We just described how a progenitor B cell becomes a mature (naive) B cell (see **Fig. 24.17**, bone marrow). The mature B cell (a B cell that has not yet "seen" the antigen but has already assembled its immunoglobulin VDJ binding site) produces both IgM and IgD B-cell receptors (so the naive B cell is referred to as IgM⁺ IgD⁺). The primary immune response begins when a mature (naive) B-cell receptor finds its matched antigen (see **Fig. 24.17**, spleen). In the early stages of the response, plasma cells produced from these B cells secrete only IgM, but they eventually secrete IgM and IgD. If the activated B cell reaches a lymph node and receives appropriate signals from a certain type of T cell known as a **helper T cell (T_H cell)**, then further immunoglobulin isotype switching will occur (**Fig. 24.17**, lymph node); the switched B cell may then make IgG, IgA, or IgE, each with the same antigen recognition domain (variable region) but different constant regions. The type of switch is influenced by small peptides called **cytokines** that are secreted by helper T cells.

What "flips" the switch? Notice in **Figures 24.18** and **24.19** that the constant-region gene segments defining different antibody heavy-chain classes are arranged in tandem <u>after</u> a VDJ region. The mechanism by which a B cell switches to make IgG (C regions gamma 1–4), IgE (C-epsilon), or IgA (C-alpha 1 and 2) is very similar to VDJ formation. Each constant segment, except delta, contains a repeating DNA base sequence called a **switch region**. Recombination between these switch regions will delete the intervening

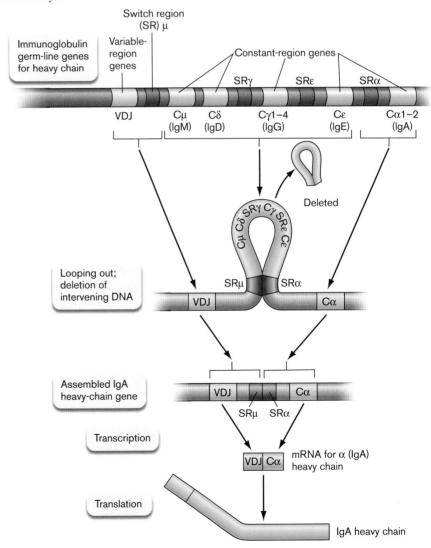

FIGURE 24.19 ■ Heavy-chain class switching. As a B cell becomes activated, a switch in antibody isotype will occur. The switch involves recombination between isotype cassettes that brings one heavy-chain constant region (Cα in the example, for IgA) in tandem with a VDJ sequence.

DNA between the VDJ region and one of the constant regions. The primary RNA transcript produced after this recombination event has all of its introns spliced out before translation. Because the VDJ region is the same regardless of which C_H gene is selected, the antibody produced will have the same antigenic specificity as the original IgM. However, the heavy-chain switch selection process is <u>not</u> random. The type of cytokine present at the time of the switch will influence which C_H gene is selected.

Note that any heavy-chain peptide (alpha, gamma, delta, and so on) can combine with any light chain (lambda or kappa). However, a single, mature B cell can make only <u>one</u> type of heavy chain and <u>one</u> type of light chain.

Thought Questions

24.7 B cells in early stages have both IgM and IgD surface antibodies, but the delta region has no switch region. Why does the delta region have no switch region?

24.8 Why do individuals with type A blood have anti-B and not anti-A antibodies?

Making memory B cells. Having encountered an antigen, the antigen-activated B cell will divide to make memory B cells, as well as plasma cells (see **Figs. 24.14** and **24.17**). Memory B cells are B cells that have already undergone class switching (committed to making IgG, for instance) but are very long-lived. Their long lives provide immunological memory. But why do they survive longer than regular B cells? Part of the reason is the production of what one might call an "anti-suicide" protein called Mcl-1. B cells that do not encounter their matched antigen eventually undergo a process of programmed cell death (apoptosis). However, if the B-cell receptor binds to its matched (cognate) antigen, then a signal is sent to the nucleus to make the Mcl-1 anti-apoptosis protein. The memory-B-cell line can now propagate for many years, guarding the person against microbial assault.

During their lifetimes, memory B cells, like other B cells, hypermutate the antigen-binding regions of antibody genes (the VDJ and VJ regions). These mutations increase the affinity of the antibody produced during the secondary response; that is, the hypermutation helps sharpen the antigen-binding site, making it even more specific toward a given microbe or antigen. Note, too, that as an immune response clears an infection, the antigen becomes scarce. This means that only B cells with the highest-affinity antibodies as part of their B-cell receptors will remain activated. This process is known as **affinity maturation**.

To Summarize

■ **Antibodies, or immunoglobulins, are Y-shaped molecules** that contain two heavy chains and two light chains. **There are five classes (isotypes) of antibodies.** Each antibody isotype is defined by the structure of the heavy chain. Antibodies are members of the immunoglobulin superfamily.

■ **Each antibody molecule contains two antigen-binding sites.** Each binding site is formed by the hypervariable ends of a heavy- and light-chain pair. **The Fc portion ("tail") of an antibody can bind to specific receptors on host cells.** This binding is antigen independent.

■ **Antibody diversity (idiotype)** occurs via a complex series of splicing events between adjacent DNA cassettes, as well as mutational events in DNA sequences encoding the hypervariable regions of heavy and light chains.

■ **Mature (naive) B cells** have completed VDJ rearrangements in their DNA to make the specific antigen-binding site for antibodies. The naive B cell places IgM and IgD antibodies on its surface (part of B-cell receptors).

■ **A B-cell receptor** consists of a membrane-embedded antibody in association with the Igα and Igβ proteins. Binding of antigen to the B-cell receptor triggers B-cell proliferation and differentiation.

■ **Clonal selection** is the rapid proliferation of a subset of B cells during the primary or secondary antibody response.

■ **The primary antibody response** to an antigen begins when B cells differentiate into antibody-producing plasma cells and memory B cells. IgM antibodies are generally the first class of antibodies made during the primary response.

■ **The secondary antibody response** occurs during subsequent exposures to an antigen. Memory B cells are activated, hypermutate the VDJ region, and then rapidly proliferate and differentiate into antibody-secreting plasma cells. IgG is the predominant antibody made.

■ **Isotype switching** (or class switching) from IgM production to other antibody isotypes occurs after naive B cells bind their target antigen. The B-cell receptors on a single B cell are made of a single antibody isotype that specifically binds one epitope.

24.3 T cells Link Antibody and Cellular Immune Systems

Do the two arms of the adaptive immune system (antibody-based and cell-mediated) know each other exist? Because different types of infections tilt the immune response one way or the other, the systems must communicate with each other to provide balance. T-cell lymphocytes, with integral roles in both antibody production and cell-mediated immunity, manage the balance.

Although derived from the same progenitor stem cell as B cells (see Fig. 23.17), T cells develop in the thymus (rather than in the bone marrow, where B cells develop), and they contain surface protein antigens different from those of B cells. T cells come in several varieties marked by the presence of different cell differentiation (CD), or cluster of differentiation, surface proteins and by the types of cytokines they produce. Two critically important groups are helper T cells (already mentioned) and **cytotoxic T cells** (**T$_C$ cells**). Helper T cells display the surface antigen CD4, while cytotoxic T cells display CD8. Cytotoxic T cells are the "enforcers" of the cell-mediated immune response. They destroy the membranes of host cells infected with viruses or bacteria.

Our understanding of helper T cells has evolved over the last few years. Helper T cells come in several models, the first of which is the T$_H$0 cell. T$_H$0 cells are precursors to the other types, including:

■ T$_{FH}$ (follicular helper T cells), a heterogeneous set of CD4 T cells that drive B-cell differentiation into antibody-secreting plasma cells. Different T$_{FH}$ cell subsets secrete different combinations of cytokines that trigger antibody isotype class switching.

■ T$_H$1 cells assist in the activation of cytotoxic T cells.

■ T$_H$2 cells recruit eosinophils to combat parasitic infections and can inhibit T$_H$1 proliferation.

■ T$_H$17 cells stimulate inflammation by secreting IL-17, a pro-inflammatory cytokine. IL-17 helps trigger recruitment of neutrophils and macrophages to an infection.

■ Treg (regulatory T cells) dampen the inflammatory response and secrete the anti-inflammatory cytokine IL-10.

The ratios of these cells and the cytokines they produce tilt the immune system toward either more humoral or more cell-mediated responses. To be of any use, however, T cells first must be activated by antigen.

T-Cell Activation Requires Antigen Presentation

Unlike B cells, T cells never bind free-floating antigen. T cells are activated only by antigen bound to another cell's surface. These other cells are collectively called **antigen-presenting cells** (**APCs**) because they present antigens to T cells. The APC surface proteins that hold and present the antigen are known as **major histocompatibility complex** (**MHC**) proteins (**Fig. 24.20A**). MHC proteins differ between species and between individuals within a species. They help determine whether a given antigen is recognized

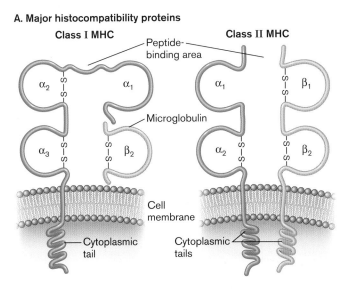

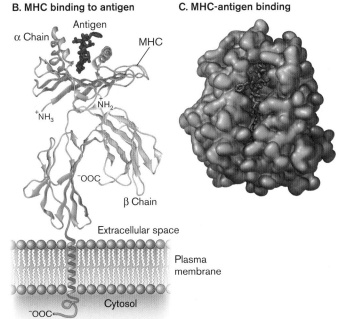

FIGURE 24.20 ■ Major histocompatibility proteins.
A. MHC class **I** molecules are composed of a 45-kDa chain and a small peptide called β$_2$-microglobulin (12 kDa). MHC class **II** molecules contain an alpha chain (30–34 kDa) and a beta chain (26–29 kDa). The peptide-binding regions of both classes show variability in amino acid sequence that yields different shapes and grooves. Peptide antigens nestle in the grooves and are held there awaiting interaction with T-cell receptors. CD8 T cells recognize antigen peptides associated with class **I** molecules, while CD4 T cells recognize peptides bound to class **II** molecules. **B.** Antigen binding to an MHC molecule. **C.** Top view of antigen (red) nestled in the MHC peptide-binding site. (PDB code: 1BII)

as coming from the host (a self antigen) or from another source (a foreign antigen), in a phenomenon called histocompatibility (hence the name "major histocompatibility complex"). The salient feature of all MHC molecules is that they bind antigen that has entered the host cell, and then present the antigen back on the cell surface. MHC molecules are critical to the immune system because the

T cell, to be activated, must first recognize a foreign antigen attached to an MHC molecule on an APC.

Two classes of MHC molecules are found on cell surfaces. Both classes belong to the immunoglobulin family of proteins, but they are not immunoglobulins (antibodies). **Class I MHC molecules** are found on all nucleated cells, whereas **class II MHC molecules** have a more limited

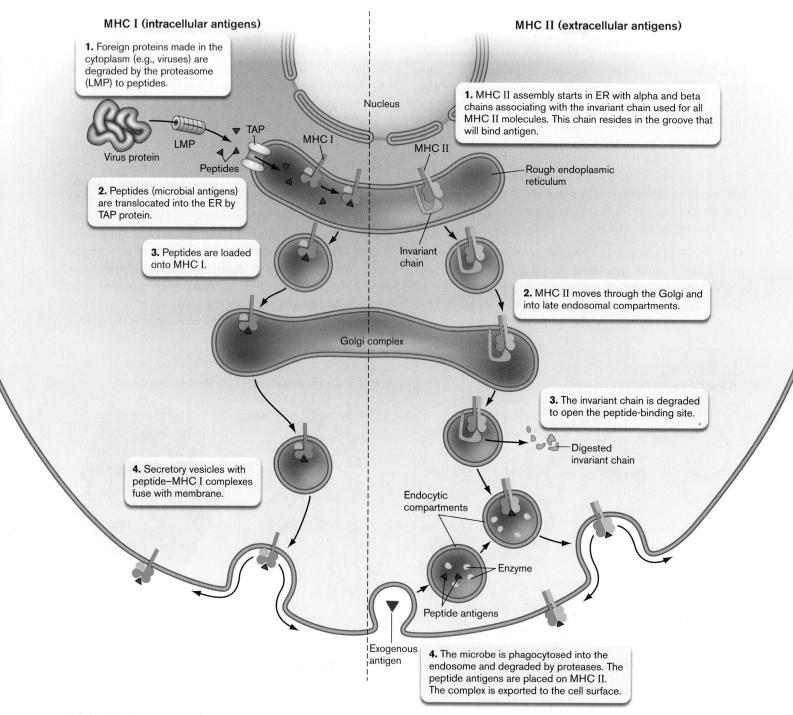

FIGURE 24.21 ■ Processing and presentation of antigens by antigen-presenting cells on class I and class II MHC proteins.
Left: Microbial proteins made in the host cytoplasm are degraded and peptides are placed on MHC class I molecules in the endoplasmic reticulum (ER). **Right:** Microbial proteins made outside the cell are endocytosed (bottom), degraded in the endosome, and placed on MHC class II molecules. LMP = low-molecular-mass polypeptide component of the proteasome; TAP = transporter of antigen peptides.

distribution on professional APCs such as dendritic cells, B cells, and macrophages.

The surface CD proteins CD8 and CD4 mentioned earlier help T cells distinguish between MHC class I and class II molecules on antigen-presenting cells. The CD8 molecules on T_C cells selectively bind MHC class I, while CD4 molecules on T_H cells selectively bind MHC class II. Antigens presented on class I MHC molecules generally arise from intracellular pathogens, such as viruses and some bacteria that require cell-mediated immunity for resolution. In contrast, antigen peptides presented on class II MHC molecules originate from extracellular infections, which are resolved by antibody and are recognized only by CD4 T_H cells.

APCs Receive, Process, and Present Antigens by Two Paths

How are foreign antigens placed, or presented, on host cell surfaces? In the initial stages of an immune response, antigen-presenting cells internalize the pathogen, such as a virus or a bacterium. Inside the APCs, pathogen proteins are degraded into smaller peptides (epitopes). These epitopes are placed within the MHC-binding clefts of MHC I or II molecules and transported back to the cell surface.

Whether an antigen peptide binds to class I or class II MHC molecules generally depends on how the antigen initially entered the cell (**Fig. 24.21**). Endogenous antigens, which are synthesized by viruses and intracellular bacteria as they grow within the cytoplasm of an APC, will attach to class I MHC molecules on the endoplasmic reticulum and are moved to the cell surface (**Fig. 24.21** left). In contrast, exogenous antigens, which are produced <u>outside</u> of the APC (as are most bacterial antigens), will enter the cell via phagocytosis and attach to class II MHC molecules moved to the acidic phagosome or lysosome (**Fig. 24.21** right). The MHC class II–peptide complex is then carried to the cell surface. Once the antigen is presented on the APC cell surface, T cells can interact with the antigen-MHC complex via the T-cell receptors (discussed next). Interactions between antigen-presenting cells and naive T cells occur within the lymph nodes, the spleen, or Peyer's patches in the gut. So APCs must make their way to those locations to generate an immune response.

Note: MHC molecules present only peptide antigens. Carbohydrate antigens cannot bind MHC binding clefts but will trigger antibody production independent of T cells (Special Topic 24.1).

The antigen-binding clefts of different MHC molecules differ markedly and have distinct binding affinities for different antigenic peptides (**Fig. 24.20B and C**).

T-Cell Receptors

T-cell receptors (**TCRs**) are the antigen-binding molecules present on the surfaces of T cells (**Fig. 24.22**). Unlike BCRs, TCRs are not antibodies. A TCR on helper T cells will bind only to antigens attached to MHC surface proteins on antigen-presenting cells, as noted earlier. However, the TCRs of cytotoxic T cells can bind viral antigens present on any virus-infected cell, whether or not that cell is normally considered an APC.

The T-cell receptor is composed of several transmembrane proteins. Two molecules—alpha and beta—make up the part that recognizes antigen. Much like antibodies, the alpha and beta proteins of the TCR are formed from gene clusters that undergo gene rearrangements analogous to, but different from, the immunoglobulin genes.

The TCR alpha and beta proteins are found in a complex with four other peptides, which together form the CD3 complex (see **Fig. 24.22**). When stimulated by binding antigen, these ancillary CD3 complex proteins recruit and activate intracellular protein kinases and launch a phosphorylation cascade that triggers proliferation of the T cell and key immunological events that we describe later in the chapter.

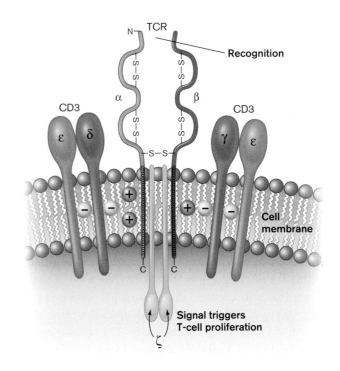

FIGURE 24.22 ■ The T-cell receptor (TCR) and CD3 complex. T-cell receptor proteins are associated with CD3 proteins at the cell surface. Antigen binds to the alpha and beta subunits. The positive and negative charges holding the complex together come from amino acids in the peptide sequences. Once bound to antigen, the complex transduces a signal into the cell. This signal triggers T-cell proliferation.

Why don't your T cells <u>see</u> your body's antigens? Are they tolerant somehow? Actually, T cells that strongly bind self antigens are not made tolerant; they are killed in a process called T-cell education.

T-Cell Education and Deletion

Greek mythology tells of Narcissus, a young man punished by the gods for scorning the women who fell in love with him. One day Narcissus saw a beautiful face in a pool of water. Not knowing it was his own reflection, he fell in love, tumbled into the pool, and drowned. His inability to recognize himself is analogous to the danger posed by an immune system. Immune systems must be able to distinguish what is self (meaning antigens present in our own tissues) from what is not self. Otherwise our immune system would constantly attack our own cells. Innate immunity accomplishes this important task, in part, through the use of pattern recognition proteins such as the Toll-like and NOD-like receptors (see Section 23.5). How adaptive immune mechanisms avoid recognizing self is equally important. At the outset of development, the immune system is fully capable of reacting against self.

One mechanism the body uses to avoid attacking itself is to delete any T cells that strongly react against self antigens. T cells undergo a two-stage selection or "education" process in the thymus to recognize self versus nonself. T cells bearing T-cell receptors (TCRs) that <u>weakly</u> recognize self MHC proteins displayed on thymus epithelial cells are allowed to survive (**positive selection**) and leave the thymus to seed secondary lymphoid organs such as the spleen. T cells in the thymus that recognize self MHC peptides <u>too strongly</u>, however, are killed, or deleted from the population (**negative selection**). Almost 95% of T cells entering the thymus die during these positive and negative selection processes.

This weeding out is important because if our T-cell repertoire included cells that bound self MHC too tightly, then our T cells would constantly react to our own MHC molecules, regardless of which antigen peptides were attached. Note, however, that some self-reactive T cells are allowed to survive and are converted into **regulatory T cells** (**Tregs**). Regulatory T cells can block the activation of harmful, self-reactive (autoimmune) lymphocytes that escape deletion and enter the circulation. Note that this is the first way that Tregs can be made. The second involves APC activation of T_H0 cells, discussed later.

Why is positive selection that promotes <u>weak</u> binding to MHC molecules important? The positive selection process is needed because T cells must be able to recognize <u>self</u> MHC proteins before the T-cell receptor can bind the MHC-associated antigen. Consequently, T cells from one individual normally do not recognize peptides placed on antigen-presenting cells from another individual—a property known as **MHC restriction**. CD4 T_H cells are class II MHC restricted, while CD8 T_C cells are class I restricted.

Rolf Zinkernagel (University of Zurich) and Peter Doherty (St. Jude Children's Research Hospital, Memphis) discovered the MHC restriction phenomenon in 1976 with a simple experiment. They demonstrated that mouse cytotoxic T cells would efficiently kill virus-infected target cells having an MHC I surface protein identical to that of the T cell, but failed to kill infected cells containing a different MHC I protein from a different strain of mouse. Thus, T-cell recognition has two components: self (the MHC molecule) and nonself (the antigen). This discovery won Zinkernagel and Doherty the 1996 Nobel Prize in Physiology or Medicine.

Why transplanted organs with one MHC type are prone to rejection by a recipient of a different MHC type is explained in **eTopic 24.3**.

Where are T cells "educated"? In contrast to negative selection for B cells, which occurs in bone marrow, T-cell education is limited to the thymus. But if the thymus expresses only thymus antigens, how can T cells that respond to antigens expressed on other host cells (for example, heart cells) be removed? The answer is a special gene activator called AIRE in thymus cells that stimulates the synthesis of <u>all</u> human proteins in small amounts. This expression is necessary to complete T-cell education within the thymus.

You might ask how someone whose thymus has been removed (done as a treatment for myasthenia gravis) can live if the organ is critical for T-cell maturation and for deleting self-reactive T cells. Actually, the thymus begins losing function shortly after birth, such that very little function remains in adults. Fortunately, a large amount of T-cell education occurs during fetal development. Once a T cell is educated, reserve pools of these T cells are maintained throughout life outside of the thymus. Some T cells, apparently, can also mature in secondary lymphoid tissues. Thus, adults without a thymus can live relatively normal lives. Babies born without a thymus, however, have a severe, life-threatening T-cell deficit.

APCs Activate T_H0 Helper T cells

Figure 24.23 ▶ summarizes the steps that lead to T-cell activation in the lymph node and highlights how activated T cells influence the two types of adaptive immunity: humoral (antibody) and cellular. The different classes of T cells, T_H and T_C, require different but interrelated activation programs. First we consider T_H cell activation. T_H cells require two molecular signals to become active. T-cell receptors on precursor T_H0 cells first recognize and link to

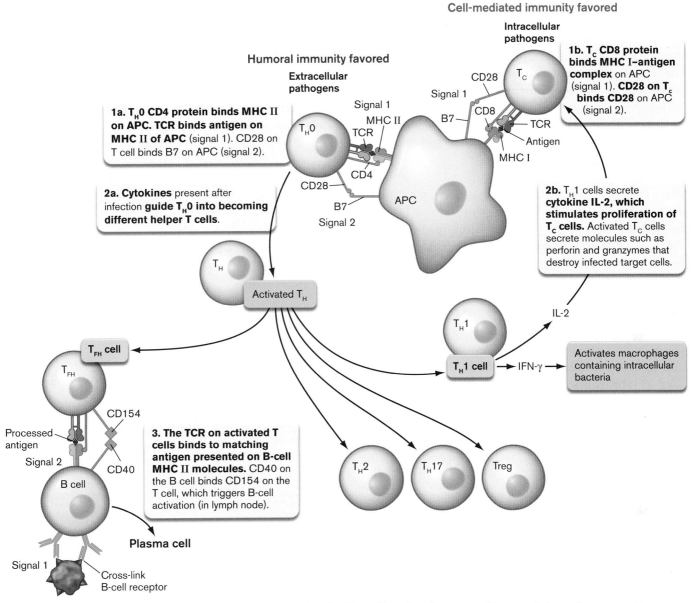

Cell-mediated immunity favored

Humoral immunity favored

1a. T$_H$0 CD4 protein binds MHC II on APC. TCR binds antigen on MHC II of APC (signal 1). CD28 on T cell binds B7 on APC (signal 2).

1b. T$_C$ CD8 protein binds MHC I–antigen complex on APC (signal 1). **CD28 on T$_c$ binds CD28 on APC** (signal 2).

2a. Cytokines present after infection **guide T$_H$0 into becoming different helper T cells.**

2b. T$_H$1 cells secrete cytokine IL-2, which stimulates proliferation of T$_C$ cells. Activated T$_C$ cells secrete molecules such as perforin and granzymes that destroy infected target cells.

Intracellular pathogens

Extracellular pathogens

Signal 1

Signal 1

CD28

T$_C$

MHC II

B7   CD8

TCR

T$_H$0

TCR

Antigen

CD28

MHC I

CD4

B7

APC

Signal 2

Activated T$_H$

T$_H$

IL-2

T$_FH$ cell

T$_H$1

T$_FH$

T$_H$1 cell ─ IFN-γ ─

Activates macrophages containing intracellular bacteria

CD154

Processed antigen

Signal 2

CD40

3. The TCR on activated T cells binds to matching antigen presented on B-cell MHC II molecules. CD40 on the B cell binds CD154 on the T cell, which triggers B-cell activation (in lymph node).

B cell

T$_H$2   T$_H$17   Treg

Plasma cell

Signal 1 ─ Cross-link B-cell receptor

FIGURE 24.23 ■ Summary of the activation of humoral and cell-mediated pathways. Left: Extracellular pathogens tend to activate humoral immunity (B cells). **Right:** Intracellular pathogens generally activate cell-mediated immunity by stimulating cytotoxic T cells (CD8 marker). The balance between cell-mediated and humoral immune responses to a given infection is regulated by the balance between the production of T$_H$1 (cell-mediated) versus other T$_H$ (antibody) helper T cells. This balance is influenced by whether the foreign antigen was made by intracellular pathogens (T$_H$1-favored) or extracellular pathogens. T$_H$1 cells will encourage activation of cytotoxic T cells (cell-mediated immunity; best for killing intracellular pathogens), while T$_FH$ cells will promote antibody production (humoral immunity; best for attacking extracellular pathogens). See text for more details. ▶

antigen-MHC proteins on antigen-presenting cells (APCs) (**Fig. 24.23**, step 1a). In the case of T$_H$0 cells, the MHC molecules are class II. The second signal needed to activate a T$_H$0 cell is the binding of a CD28 molecule present on the T$_H$0 cell surface to a B7 protein on the APC cell surface (also step 1a). Different cytokines produced by mast cells, macrophages, epithelial cells, and so on, during infection then convert activated T$_H$0 cells to T$_FH$, T$_H$1, T$_H$2, T$_H$17, or Treg cells (step 2a). The activation of cytotoxic T cells (T$_C$ cells) will be examined later.

T$_FH$ Cells Activate B Cells

During the <u>early phase</u> of a primary response, one type of B cell in the lymph node can become activated after binding antigen if the B cell also binds to complement C3 on a bacterium (see Section 23.6), or if a microbe-associated molecular pattern (MAMP) binds to a Toll-like or NOD-like receptor on or in the B cell (see Section 23.5). This activation does not result in heavy-chain class switching; only IgM is secreted by the resulting plasma cell. This is partly why the primary response starts with IgM (see **Fig. 24.13**).

During the later stages of a primary response, however, a helper T cell is required to activate most B cells after the B-cell receptor binds antigen (**Fig. 24.23**, step 3). This assistance is called T-cell help. Interactions between B cells and T cells primarily take place in areas of lymph nodes called follicles (see the Current Research Highlight). The follicular helper T cell (T_{FH}) is the primary T cell involved in driving B-cell activation. (Actually, any T_H cell other than T_H0 and Treg can activate a B cell to make antibody, but to a lesser extent.)

T-cell help is specific to the antigen and triggers heavy-chain class switching. But how does a B cell gain specific T-cell help? The specific T_{FH} cell must have a TCR able to bind the same antigen that the B cell binds. This way the T_{FH} cell "knows" which B cell to help. As part of the B-cell receptor capping mechanism (antigen cross-linking of BCR), some antigen bound by B-cell receptors becomes internalized and processed to be presented back on the B cell's surface MHC receptor (**Fig. 24.23**, step 3). For instance, a T_{FH} cell that was activated by dendritic cells presenting antigen A (step 1a) can also use its T-cell receptor to bind antigen A presented on a B-cell MHC II molecule (step 3). This contact allows CD40 on the B cell to bind CD154 on the T cell—an interaction that completes the activation of the B cell. The B cell will now differentiate into a plasma cell (**Fig. 24.23**, step 3).

The isotype class switch from IgM to another antibody type is directed by cytokines secreted by the T_{FH} cell. For instance, IL-4 and IFN-gamma trigger the switch from IgM to IgE and IgG, respectively.

During the secondary immune response, memory B cells also need T-cell help to become plasma cells, but direct contact with helper T cells is not required. Memory B cells that have antigen bound to their B-cell receptors can respond to the soluble IL-4 and IL-6 cytokines secreted by activated helper T cells without having direct contact with the T_H cell. IL-4 stimulates B-cell proliferation, while IL-6 directs differentiation into antibody-secreting plasma cells.

Thought Question

24.9 What would happen to someone lacking CD154 on T_H2 cells because of a genetic mutation?

T_H1 Cells Activate Cytotoxic T Cells

Because they directly attack host cells, CD8 T_C cells are the major "enforcers" of the cellular immune system, along with macrophages and natural killer (NK) cells. T_C cells, however, are not actually cytotoxic until they are activated. As with helper T cells, activation of cytotoxic T cells takes place in the lymph node and requires two signals. The first

signal is the binding of T-cell receptors (TCRs) to antigen–MHC <u>class I</u> complexes on an antigen-presenting cell (**Fig. 24.23**, step 1b). CD8 on the T_C cell recognizes an MHC I–antigen complex on the APC. Because class I MHC molecules are found on all nucleated cells, any infected cell can potentially activate T_C cells, converting them to cells that ultimately kill the infected cell. **Figure 24.24A** takes a closer look at signal 1 interaction.

A.

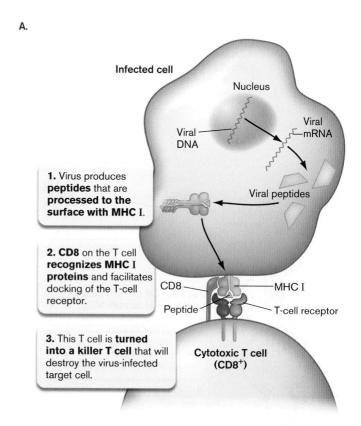

B.

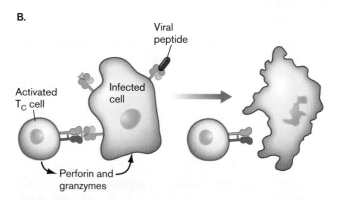

FIGURE 24.24 ■ **Presentation of a viral antigen to a T cell and cytotoxic T-cell action.** **A.** CD8 protein on a T cell directs the interaction between a T-cell receptor and a viral antigen bound to a class I MHC protein on an antigen-presenting cell. Signal 2 for activation of cytotoxic T cells is an interaction between B7 (on an APC) and CD28 on a T_C cell, described in the text (not shown). **B.** The activated T_C cell then leaves the lymph node and migrates to the site of infection, where it can recognize the viral peptide presented on the MHC class I receptor on an infected cell. This interaction authorizes the cytotoxic T cell to kill the infected cell.

The second signal needed to activate T_C cells is a B7-to-CD28 interaction that further links the APC to the T_C cell. This is the same second signal described earlier for helper T cells. Once activated, the T_C cell gains cytotoxic activity and places <u>receptors</u> for the cytokine IL-2 on its surface. IL-2 secreted by the T_H1 class of helper T cells (activated earlier) binds to the IL-2 receptors and stimulates proliferation of the cytotoxic T cell (**Fig. 24.23**, step 2b). The resulting "platoon" of activated cytotoxic T cells moves from the lymph node to the site of infection, killing any cell bearing the same peptide–MHC class I complex (for example, cells infected with the same virus that triggered T_C production) (**Fig. 24.24B**).

Activated cytotoxic T cells kill infected target cells by releasing the contents of their granules, which contain the proteins **granzyme** and **perforin**. Perforin produces a pore in the target cell membrane through which granzymes can enter. In the cytoplasm, granzymes cleave and activate caspase proteins in the infected cell. The activated caspases then trigger apoptosis and cell death. The infected host cell is sacrificed for the good of the whole animal, human or otherwise.

Why are cytotoxic T cells more effective than B cells and antibodies at clearing viral infections? A major reason is that intracellular pathogens (for example, viruses) hide inside host cells, where they are protected from antibody. Consequently, these pathogens are best killed when the harboring host cell is also sacrificed (via cellular immunity). Because all nucleated cells have class I MHC molecules, any infected cell can be recognized and killed by an appropriate T_C cell for the good of the host.

T_H17 Cells Promote, and Treg Cells Limit, Inflammation

The immune system is full of checks and balances. Two sets of T cells critical to regulating the immune response are **T_H17 cells** that amplify inflammation, and regulatory T cells (Tregs) that dampen inflammation. T_H17 cells are derived from T_H0 cells and are characterized by their secretion of pro-inflammatory cytokine IL-17. T_H17 cells are the first subset of T cells generated during an infection. Receptors for IL-17 are expressed on fibroblasts, epithelial cells, and keratinocytes. Contact with IL-17 causes the production of IL-6 to initiate fever, chemokines like CXCL8 (IL-8) to recruit neutrophils and macrophages, and granulocyte-macrophage colony-stimulating factors (GM-CSFs) to enhance synthesis of these phagocytes in bone marrow. IL-22, also produced by T_H17 cells, cooperates with IL-17 to induce synthesis of antimicrobial peptides, such as beta-defensins, in epidermal keratinocytes. Altogether, T_H17 cells enhance the innate acute inflammatory response to infection.

However, the problem with inflammation is that it also causes damage to local tissues. This is where Treg cells become crucial.

Treg cells either form during T-cell selection in the thymus (mentioned earlier), or differentiate from T_H0 cells in a lymph node. Treg cells are involved in shutting down immune responses after an invading organism has successfully been eliminated, and can prevent autoimmunity. Once activated by an antigen presented to its TCR receptor, Treg cells secrete the anti-inflammatory cytokine IL-10. IL-10 inhibits inflammatory cytokine secretion by macrophages (TNF, IL-1, and IL-12), reduces MHC II expression in macrophages, blocks macrophage activation by IFN-gamma, and blocks cytokine IL-2 production from T_H1 cells, thereby limiting production of cytotoxic T cells (see **Fig. 24.23**).

Ratios of T_H17 to Treg cells vary depending on the severity of an infection, either exacerbating or quelling inflammation. The cytokines produced by APCs and other cells during an infection can tilt the balance one way or the other. In addition, Treg cells themselves can inhibit T_H17 cell functions.

CD4 Helper T Cells and Cytokines Balance the Immune Response

Antigens presented on class I MHC molecules elicit only cell-mediated immunity (via CD8 T cells). Antigens presented on class II molecules can stimulate both arms of the immune system by activating CD4 helper T cells. CD4 T_H cells do not kill host cells directly, but instead release various cytokines that incite other cells to do the killing. IL-8, for instance, attracts additional white blood cells to the area. Activated T_H1 cells secrete cytokines that stimulate cytotoxic T_C cells (see **Fig. 24.23**, step 2b), as well as macrophages and natural killer cells. However, sets of activated T_{FH} cells secrete different combinations of cytokines to trigger B cells to switch immunoglobulin classes and differentiate into plasma cells that secrete antibody (see **Fig. 24.23**, step 3). **Table 24.4** lists a fraction of the many different cytokines produced by various cell types and describes their influence over the immune system.

But what determines which class of T_H cell predominates during an infection? Part of the answer is found in the blend of cytokines generated by macrophages or mast cells during different types of infection. Infections by viruses and intracellular bacteria, for instance, generate cytokines, such as IFN-gamma, that favor the production of T_H1 cells. T_H1 cells promote cell-mediated immunity. Infections by extracellular bacteria and parasites, however, generate IL-4, which <u>inhibits</u> T_H0 differentiation into T_H1 cells, so antibody-mediated immunity is favored (see **Fig. 24.23**, step 2a).

TABLE 24.4	Select Cytokines That Modulate the Immune Response*	
Cytokine	**Sample sources**	**General functions**
IL-1	Many cell types, including endothelial cells, fibroblasts, neuronal cells, epithelial cells, macrophages	Pro-inflammatory; affects differentiation and activity of cells in inflammatory response; acts as endogenous pyrogen in the central nervous system
IL-2	T_H1 cells	Stimulates T-cell and B-cell proliferation
IL-3	T cells, mast cells, keratinocytes	Stimulates production of macrophages, neutrophils, mast cells, others
IL-4	T_H2 cells, mast cells	Promotes differentiation of CD4 and T cells into T_H2 helper T cells; promotes proliferation of B cells, class switch to IgE
IL-5	T_H2 cells	Chemoattracts eosinophils; activates B cells and eosinophils
IL-6	T_H2 cells, macrophages, fibroblasts, endothelial cells, hepatocytes, neuronal cells	Stimulates T-cell and B-cell growth; stimulates production of acute-phase proteins
IL-8 (CXCL8)	Monocytes, endothelial cells, T cells, keratinocytes, neutrophils	Chemoattracts PMNs; promotes migration of PMNs through endothelium
IL-10	T_H2 cells, B cells, macrophages, keratinocytes	Inhibits production of IFN-γ, IL-1, TNF-α, and IL-6 by macrophages
IL-17	T_H17 cells (helper T cells unique from T_H1 and T_H2)	Recruits macrophages and neutrophils to sites of inflammation
IFN-α/β	T cells, B cells, macrophages, fibroblasts	Promotes antiviral activity
IFN-γ	T_H1 cells, cytotoxic T cells, NK cells	Activates T cells, NK cells, macrophages, B-cell class switch to IgG
TNF-α	T cells, macrophages, NK cells	Exerts wide variety of immunomodulatory effects
TNF-β	T cells, B cells	Exerts wide variety of immunomodulatory effects

*IFN = interferon; IL = interleukin; NK = natural killer; PMN = polymorphonuclear leukocyte; TNF = tumor necrosis factor.

As a result, once the immune system starts to go down one pathway, the other pathway is held back. But this is not an all-or-none response. For instance, a viral infection will also result in T_{FH} cells that stimulate antibody production. The tilt toward T_H1 or T_{FH} predominance occurs because the body realizes which pathway—cell-mediated (T_H1) or humoral (T_{FH})—will more effectively clear a particular infection.

The importance of CD4 T cells in immunity is tragically illustrated by acquired immunodeficiency syndrome (AIDS; Chapter 11). The human immunodeficiency virus (HIV) binds to CD4 molecules to infect CD4 T cells. As the disease progresses, the number of CD4 T cells declines below its normal level of about 1,000 per microliter. When the number of CD4 T cells drops below 400 per microliter, the ability of the patient to mount an immune response is jeopardized. The patient not only becomes hypersusceptible to infections by pathogens, but also becomes susceptible to infections by commensal organisms. Most AIDS patients actually die from these opportunistic infections.

Superantigens Do Not Require Processing to Activate T Cells

To stimulate T-cell responses, normal antigens require processing by antigen-presenting cells. Each peptide produced by antigen processing must be placed on an MHC molecule and presented to a cognate T cell. However, when an antigen is introduced into a host, only a few T cells have the proper TCR needed to recognize that antigen—in the range of 1–100 cells per million. Proliferation of the T cells through antigen-dependent activation increases their number and increases the immune response to that antigen.

Some proteins, called **superantigens**, indiscriminately stimulate T cells by bypassing the normal route of antigen processing. In fact, recognition as an antigen is not even involved. Certain microbial toxins, such as staphylococcal toxic shock syndrome toxin (TSST), are actually superantigens. As illustrated in **Figure 24.25**, these proteins simultaneously bind to the outside of T-cell receptors on T cells and to the MHC molecules on APCs (for example, macrophages). This joining of the T cell and macrophage activates more T cells than a typical immune reaction does, stimulating the release of massive amounts of inflammatory cytokines from both cell types.

The effect of superantigens can be devastating because cytokines like **tumor necrosis factor** (**TNF**) will overwhelm the host immune system's regulatory network and cause severe damage to tissues and organs. The result is disease and sometimes death. Jim Henson, the creator of Kermit the Frog and other Muppet characters, died in

A.

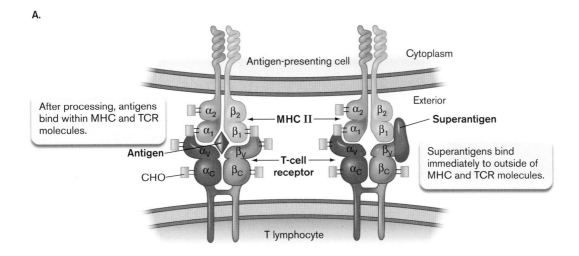

B.

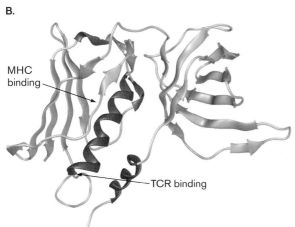

FIGURE 24.25 ■ **Superantigens. A.** The difference between the presentation of antigen and superantigen. Antigen presentation requires the antigen to bind within the binding pockets of MHC and TCR molecules. Superantigens do not require processing. They can bind directly to the outer aspects of the TCR and MHC proteins, linking and activating the two cell types. **B.** Staphylococcal toxic shock syndrome toxin is one example of a potent superantigen. Alpha helices are shown as red ribbons, while sections of the beta sheet are shown as blue ribbons. (PDB code: 2QIL)

1990 from complications of pneumonia caused by a potent superantigen produced by *Streptococcus pyogenes*. This example is but one of many ways that overreaction by the immune system causes morbidity (disease) and mortality (death)—effects more pronounced than the direct effects of the pathogen involved.

Microbial Evasion of Adaptive Immunity

As efficient as our immune system is, numerous viral and bacterial pathogens have developed effective means for avoiding the adaptive immune response. Many viruses produce proteins that down-regulate production of class I MHC molecules on infected cell surfaces. Down-regulation of MHC I will limit antigen presentation, since MHC I is needed for that process. On the other hand, losing MHC I should expose the infected host cell to natural killer cells because NK cells attack peers that lack MHC I (discussed in Section 23.5). To surmount this obstacle, for example, human cytomegalovirus places a decoy MHC I–like molecule on the surface of infected cells. These decoys are thought to bind inhibitory receptors on NK surfaces that block NK cell cytotoxicity.

Like viruses, bacteria are masters of illusion when it comes to the immune system. A major cause of gastric ulcers, *Helicobacter pylori*, expresses proteins from a cluster of pathogenicity genes that trigger apoptosis (programmed cell death) of T cells. Other bacteria have evolved mechanisms that interfere with signal transduction pathways controlling the expression of cytokines. For example, YopP from *Yersinia enterocolitica*, one cause of gastroenteritis, inhibits a specific signal transduction pathway needed to produce TNF, IL-1, and IL-8. Thus, *Yersinia* avoids the detrimental effects of those pro-inflammatory cytokines.

Another method of immune avoidance is employed by various mycobacteria, some of which cause tuberculosis and leprosy (**Fig. 24.26**). These bacteria induce the production of anti-inflammatory cytokines, which dampen the immune response. *Mycobacterium*-infected macrophages produce IL-6, which inhibits T-cell activation, and IL-10, which down-regulates the production of MHC II molecules needed for specific activation of T cells. Fortunately for humans, in most cases the immune system catches on to these tricks and, through redundant humoral and cellular mechanisms, manages to resolve these infections.

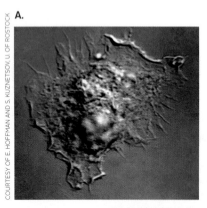
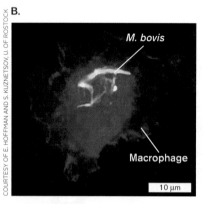

FIGURE 24.26 ■ *Mycobacterium bovis* **growing within cultured macrophages (cell line J774).** **A.** Differential interference contrast image of an infected macrophage. **B.** The same cell viewed by fluorescence microscopy. *M. bovis* is carrying green fluorescent protein. The macrophage was stained for F-actin (red) and the nucleus (blue).

Activated T$_H$1 Cells Also Activate Macrophages

Macrophages, too, must be activated to become highly effective killers of microbes. Activation of macrophages also requires two signals. One activation pathway involves interferon-gamma (IFN-γ) produced by nearby infected or damaged cells, followed by binding of the macrophage to lipopolysaccharide (LPS) or other microbial components via Toll-like or NOD-like receptors (TLRs or NLRs). Once activated, the macrophage becomes aggressive in terms of phagocytosis and increases production of numerous antimicrobial reactive oxygen intermediates.

As noted earlier, some bacterial pathogens, such as mycobacteria, the causative agents of tuberculosis and leprosy, grow primarily in the phagolysosomes of macrophages. There they are shielded from antibodies and cytotoxic T cells. Intracellular pathogens live in the usually hostile environment of the phagocyte either by inhibiting the fusion of lysosomes to the phagosomes in which they grow or by preventing the acidification of the vesicles needed to activate lysosomal proteases. However, a macrophage activated by a T$_H$1 cell can rid itself of such pathogens. Even unactivated macrophages are able to process some of the intracellular bacteria and place antigen from them on their class II MHC molecules. T$_H$1 cells then activate these macrophages through binding of their TCR molecules to the macrophage MHC II–antigen complex and through secretion of IFN-gamma. The activated macrophages can then kill any bacteria that may be growing within them.

Given that activated macrophages are such effective assassins of microbial pathogens, why are macrophages not always kept in an active state? A major reason is that once activated, macrophages also damage nearby host tissue through the release of reactive oxygen radicals and proteases. Thus, effective killing of microbial pathogens comes at the expense of host tissue damage.

To Summarize

■ The major histocompatibility complex (MHC) consists of membrane proteins with variable regions that can bind antigens. Class I MHC molecules are on all nucleated cells, while antigen-presenting cells contain both class I and class II MHC molecules.

■ Antigen-presenting cells (APCs) such as dendritic cells present antigens synthesized during an intracellular infection on their surface class I MHC molecules, but place antigens from engulfed microbes or allergens on their class II MHC molecules.

■ Activation of a T$_H$0 cell requires two signals: TCR-CD4 binding to an MHC II–antigen complex on an antigen-presenting cell, and B7-CD28 interaction.

■ T$_H$0 cell differentiation to T$_{FH}$, T$_H$1, T$_H$2, T$_H$17, or Treg cells is influenced by different cytokine "cocktails" secreted by macrophages and NK cells during an infection.

■ Activation of a B cell into an antibody-producing plasma cell usually requires two signals: antigen binding to a B-cell receptor and binding to a T$_{FH}$ cell activated by the same antigen.

■ Activation of cytotoxic T cells requires two signals: TCR-CD8 molecules that recognize MHC I–antigen complexes on APCs, and B7-CD28 interactions. IL-2 secreted from activated T$_H$1 cells stimulates cytotoxic-T-cell proliferation. The activated cytotoxic T cells, in turn, destroy infected host cells.

■ Superantigens abnormally stimulate T cells by directly linking TCR on a T cell with MHC on an APC without undergoing APC processing and surface presentation.

Thought Questions

24.10 Transplant rejection is a major consideration when transplanting most tissues, because host T$_C$ cells can recognize allotypic MHC on donor cells. Why, then, are corneas easily transplanted from a donor to just about any other person?

24.11 Why does attaching a hapten to a carrier protein allow antihapten antibodies to be produced?

24.4 Complement as Part of Adaptive Immunity

Chapter 23 describes how complement, in what is called the alternative pathway, can attack invading microbes before an adaptive immune response is launched (see Section 23.6). Factor C3b binds to LPS and sets off a reaction cascade ending with a membrane attack complex (MAC) pore composed of C5b, C6, C7, C8, and C9 proteins (see Fig. 23.36). However, antibody made during the adaptive response to a pathogen offers another route to activate complement called the **classical complement pathway**. Dubbed "classical" because it was the first complement pathway to be discovered, it requires a few additional proteins before reaching C3, the linchpin factor connecting the two pathways.

The classical cascade begins when a complement C1 protein complex binds to the Fc region of an antibody bound to a bacterial or viral pathogen (**Fig. 24.27**). The bound C1 complex then cleaves two other complement factors, C2 and C4, not used in the alternative pathway. Two fragments, one each from C2 and C4, combine to form another protease, called C3 convertase, that cleaves C3 into C3a and C3b. C3b in the alternative pathway is stabilized by interacting with LPS. In the classical pathway, however, C3b combines with C3 convertase to make a C5 convertase (note that this C5 convertase is different from the C5 convertase formed in the alternative pathway; see Fig. 23.36). The subsequent steps leading to formation of a membrane attack complex (MAC) are the same as in the alternative pathway. As before, C5b binds to a target membrane and is joined by C6, C7, and C8. Multiple C9 proteins then assemble around the MAC and form a pore that compromises the integrity of the target cell.

In Section 23.6 we noted the existence of another complement activation pathway that does not require antigen-antibody complexes. The lectin activation pathway is initiated by lectins (produced by the liver), which recognize and coat the sugar structures that decorate the surfaces of infectious organisms. The lectin-coated polysaccharides trigger cleavage of factor C4 to C4a and C4b. C4b can cleave C2 and combine with one of the fragments to form C3 convertase. C3 convertase then continues through the classical activation pathway just described.

With all the other immune responses at our disposal, why do we need complement? The need becomes evident in individuals with complement deficiencies. These patients are extremely susceptible to recurrent septicemic (blood) infections by organisms such as *Neisseria gonorrhoeae* (the cause of gonorrhea) that normally do not survive forays into the bloodstream, because they are killed by complement. Why does reinfecting *Neisseria* not quickly succumb to

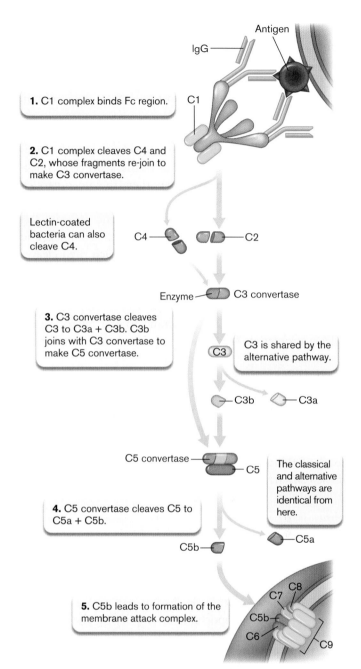

1. C1 complex binds Fc region.

2. C1 complex cleaves C4 and C2, whose fragments re-join to make C3 convertase.

Lectin-coated bacteria can also cleave C4.

C4 — C2

Enzyme — C3 convertase

3. C3 convertase cleaves C3 to C3a + C3b. C3b joins with C3 convertase to make C5 convertase.

C3 — C3 is shared by the alternative pathway.

C3b — C3a

C5 convertase — C5 — The classical and alternative pathways are identical from here.

4. C5 convertase cleaves C5 to C5a + C5b.

C5b — C5a

5. C5b leads to formation of the membrane attack complex.

C7 — C8

C5b — C6 — C9

FIGURE 24.27 ■ Classical complement cascade.

all the other defenses, such as antibodies and cytotoxic T cells? It turns out that key antigens on the bacterial surface change shape over generations in a process called phase variation (see Section 10.1). The new antigens are "invisible" to antibodies made against earlier versions.

Regulating Complement Activation

How do normal body cells prevent self-destruction following complement activation? The sequential assembly of the membrane attack complex provides several places where regulatory factors can intervene. One such factor is the host cell-surface protein CD59. CD59 will bind any C5b-C8 complex trying to form in the membrane and prevent C9 from polymerizing. Thus, no pore is formed and the host cell is spared. (Complement will not normally attack uninfected or infected host cells, since both contain CD59.)

Another regulatory mechanism hinges on a normal serum protein called **factor H**. Factor H prevents the inadvertent activation of complement in the absence of infection. If factor H is unavailable, then the uncontrolled activation of complement can damage host cells despite the presence of CD59. To short-circuit the cascade, factor H binds to C3bBb, displaces Bb from the complex, and acts as a cofactor for factor I protease, which then cleaves C3b. Without C3b, the complement cascade stops. Some bacteria, such as certain strains of *Neisseria gonorrhoeae*, have learned to protect themselves from complement by binding factor H. Factor H bound to the microbe provides a protective "force field" against local complement activation.

Microbiota can also assist host cells in resisting complement. The intestinal microbe *Bacteroides thetaiotaomicron* protects host cells from complement-mediated cytotoxicity by up-regulating a **decay-accelerating factor** present in host cell membranes. Decay-accelerating factor stimulates decay of complement factors and prevents their deposition at the cell surface.

Other Roles for Complement Fragments

The C3a, C4a, and C5a cleavage fragments generated by the complement cascade do not participate in MAC formation but are important for amplifying the immune reaction. These peptides act as chemoattractants to lure more inflammatory cells into the area. C3b, which does participate in MAC formation, also can act as an opsonin when it is bound to a bacterial cell. An opsonin is a protein factor that can facilitate phagocytosis. Because phagocytes contain C3b receptors on their surfaces, it is easier for them to grab and engulf cells coated with C3b. Additional roles for complement fragments were discussed in Section 23.6.

Thought Questions

24.12 Do bone marrow transplants in a patient with severe combined immunodeficiency (SCID) require immunosuppressive chemotherapy?

24.13 How can a stem cell be differentiated from a B cell at the level of DNA?

To Summarize

- **The classical pathway for complement activation** begins with an interaction between the Fc portion of an antibody bound to an antigen and C1 factor in blood. (The alternative pathway does not need antibody but begins when C3b binds to LPS on a bacterium.)

- **The Fc-C1 complex** reacts with C2 and C4, leading to production of C3b and a novel C5 convertase specific to the classical pathway.

- **After C5 convertase cleaves C5**, the classical pathway is the same as the alternative pathway, resulting in the formation of a membrane attack complex (MAC).

- **CD59 and/or factor H** prevents inappropriate MAC formation in host cells.

24.5 Gut Mucosal Immunity and the Microbiome

The mucosal surface of the gastrointestinal tract is exposed to a vast number of food, chemical, and other ingested antigens. The gut also harbors an expansive microbiome whose members must be controlled and nurtured. Monitoring this large antigen load requires localized and highly integrated innate and adaptive immune systems, and a system of gut-associated lymphoid tissues called GALT (see Section 23.4). The innate immune mechanisms that keep our gut microbes at bay were described in Chapter 23, but the intestine also has a dedicated adaptive immune system that monitors and shapes our microbiota, usually without causing excessive inflammation. In this section we describe how the innate and adaptive immune systems in the gut shape the microbiome while protecting the host.

The Gut Immune System

We discussed in Chapter 21 how the gut microbial community helps digest our food and outcompetes pathogens (see Fig. 21.21). Chapter 23 described how we acquire gut microbiota and how dysbiosis can lead to intestinal inflammatory diseases and obesity. Now in Chapter 24 we see how the microbiome trains the immune system and, in turn, how the immune system shapes our microbiome. But first we must discuss the gut immune system, where the training and shaping take place.

As described in Chapter 23, the intestinal epithelial barrier is one cell thick and composed mainly of columnar epithelial cells [**Fig. 24.28**, step 1, intestinal epithelial cells (IECs)] and goblet cells (step 2). Goblet cells secrete

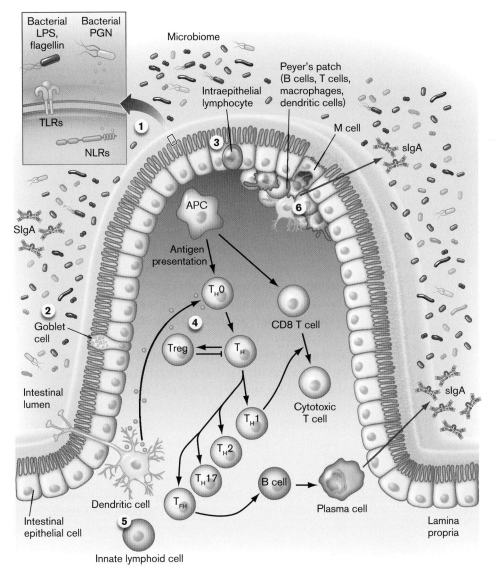

FIGURE 24.28 ■ Mucosal immune system. Depicted is the area of an intestinal villus. Mucosal immunity functions independently of regional lymph nodes to control inflammation induced by microbiota or to stimulate inflammation when pathogens are present. (1) Intestinal epithelial cells (IECs) contain membrane-associated TLRs and cytoplasmic NLRs that engage microbiome MAMPs. Cytokines released from IECs affect immune cell function (2). Goblet cells secrete the mucin layer and antimicrobial peptides. (3) Epithelial cells, macrophages, and dendritic cells present antigens to T_H0 cells located in the lamina propria or in Peyer's patches. (4) The local cytokine mixture influences whether helper T cells become Treg, T_{FH}, T_H1, T_H2, or T_H17. IgA-secreting plasma cells are formed following T_{FH}–B cell interaction. (5) Innate lymphoid cells (ILCs) lack BCRs and TCRs. ILCs include natural killer cells and cells that can secrete IL-17 and IL-22. (6) In Peyer's patches, M cells sample antigens from gut lumen and pass them to APCs (macrophages and dendritic cells) in the lymphoid tissue. APCs activate T cells to become T_{FH}, T_H1, T_H2, Treg, or T_H17. PGN = peptidoglycan.

mucin and antimicrobial peptides. Epithelial cells express numerous pattern recognition receptors (PRRs), including TLRs and NLRs that can be activated by MAMPs shed from our resident microbes. Once activated, epithelial cells will synthesize and secrete a variety of cytokines that stimulate cells of the immune system—for instance, chemokines that attract neutrophils and lymphoid cells. The epithelial barrier is also punctuated by cells called **intraepithelial lymphocytes** (**IELs**), mostly T cells, capable of secreting cytokines, including IFN-gamma. IELs exert protective and inflammatory effects in the intestine (step 3).

Beneath the epithelial layer is the lamina propria, rich in B cells and T cells. Epithelial cells can process antigens they encounter and place them onto MHC I and MHC II receptors for presentation to T cells residing in the lamina propria. (Although we normally don't think of epithelial cells as having MHC II, they will express the molecule if IFN-gamma is made during an infection.) After being

presented with gut antigens, T cells can rapidly stimulate inflammatory responses via differentiation to T_H1, T_H2, and T_H17 cells. Alternatively, the T cells can inhibit inflammatory responses by converting to Treg cells. In addition, dendritic cells in the lamina propria use their dendrite extensions to "reach" between epithelial cells and capture antigens in the intestinal lumen (**Fig. 24.28**, step 4). One class of dendritic cells stimulates inflammation by producing cytokines TNF-alpha and IL-6. A separate set of dendritic cells, however, promotes the generation of regulatory T cells whose cytokine repertoire can dampen an inflammatory response. This is important for limiting the host inflammatory response to microbiota.

Other important cells in the lamina propria include macrophages and **innate lymphoid cells** (**ILCs**). Macrophages were discussed earlier. ILCs look like lymphocytes, but because they lack antigen-specific B-cell or T-cell receptors, they are really part of the innate immune system (**Fig. 24.28**, step 5). A diverse range of stimuli can activate ILCs, including neuropeptides, hormones, and cytokines produced by epithelial cells responding to microbiome MAMPs. ILCs include natural killer cells (described in Chapter 23). Other ILCs secrete the pro-inflammatory cytokine IL-17 to summon neutrophils, and cytokine IL-22 to stimulate epithelial cells to secrete antimicrobial peptides. IL-22 also promotes tissue repair and regeneration in an inflamed intestine. Consequently, ILCs contribute to immunity, inflammation, tissue homeostasis, and microbiome control.

Peyer's patches, first described in Section 23.4, are lymphoid tissues made up of IgA$^+$ B-cell and T-cell centers. Peyer's patches are ideal sites for adaptive immune responses along the intestinal tract (**Fig. 24.28**, step 6). M cells in the epithelium above a Peyer's patch will sample microbes and other antigens floating in the gut and offer them to underlying macrophages for engulfment and presentation to Peyer's patch T cells. T$_{FH}$–B cell interactions influenced by local cytokines will trigger a B-cell class switch to IgA production and development of IgA-secreting plasma cells. Secretion of dimeric sIgA into the intestinal lumen from Peyer's patches and from plasma cells present in lamina propria inhibits penetration of microbes through the epithelium and modulates the composition of the microbiome.

Determining Friend or Foe

How might the gut immune system determine whether a microbe is indigenous microbiota or dangerous pathogen? The distinction, in large part, involves the strategic distribution of pattern recognition receptors (TLRs and NLRs) on and within epithelial cells. TLRs are primarily on the cell surface, whereas NLRs are cytoplasmic. Pathogenic bacteria possess virulence factors that allow them to attach to or invade host cells, thereby introducing MAMPs into the cytosol, where they are recognized by NLRs. However, bacteria of the indigenous microbiota are noninvasive and therefore less potent activators of NLRs. When the epithelium is intact, these "friendlies" interact only with the apical surface TLRs of epithelial cells, which are less responsive than TLRs on the host side (basolateral side) of epithelial cells.

However, an interaction with TLRs on the basolateral side of epithelial cells signals a barrier breach, which could be caused by either a pathogen or an indigenous microbe. The resultant cytokine cocktail will trigger inflammation. Nevertheless, members of the microbiome evolved to lessen inflammation by dampening epithelial cell TLR signaling (**eTopic 24.4**). The repertoire of cytokines secreted by epithelium also supports microbiome tolerance by conditioning a subset of dendritic cells to become tolerogenic. Tolerogenic dendritic cells will present antigen to T cells but convert those T cells into regulatory

T cells that suppress inflammatory responses. Treg cells are essential for maintaining tolerance to the microbiota. Without Treg cells, effector T-cell responses (for instance, those of cytotoxic T cells) go unopposed and produce inflammatory bowel diseases.

Compartmentalization of the Gut Immune System

It is important to note that adaptive immunity generated during breaches in the gut is typically limited to mucosal tissues. Systemic immunity is not generated. Compartmentalization is important because it allows the intestinal immune system to become tolerant of the microbiome without causing the systemic immune system to become tolerant too. We must remain capable of mounting a vigorous systemic immune response to a gut microbe in case the organism escapes to the bloodstream.

Separation of the intestinal and systemic immune systems is possible, in part, because secreted IgA antibodies along the intestine will trap bacterial antigens on mucosal surfaces. The trapping decreases antigen movement into lymph nodes, which minimizes systemic antibody responses. In addition, adaptive immune cells produced in Peyer's patches or in mesenteric lymph nodes lining the intestine are programmed to travel back to the mucosa. Programmed immune cells can circulate throughout the body, but they express homing receptors that bind molecules (**addressins**) selectively present on vascular endothelial cells in different tissues (for example, in the mucosa). Interactions between homing receptors and addressins cause the immune cells to reside longer in a certain tissue.

Can intestinal microbiota influence immune responses at distal mucosal sites—in the lung, for instance? The mucociliary escalator of the upper airways naturally sweeps microparticles that enter the lung upward into the oral cavity, where the particles are ingested. Once in the intestine, the gut immune system can respond to those antigens and the activated immune cells generated will home to mucosal areas throughout the body, including lung. Gut microbiota, therefore, can influence immune responses in the lung. For instance, gut microbiota have been shown to influence the generation of influenza virus–specific CD4 and CD8 T cells and affect antibody responses to respiratory influenza virus infections.

The Microbiome Shapes Gut Immunity and Human Health

We have known for decades that germ-free animals devoid of intestinal microbiota exhibit major defects in the organization and activity of immune structures in the gut.

Germ-free mice display an "underdeveloped" innate and adaptive immune system that includes reduced expression of antimicrobial peptides, reduced IgA production, fewer T-cell types, and increased susceptibility to microbial infections. Proper gut immunity can be restored, however, by simple microbial stimulation brought about by fecal transplant.

Several gut microbiota with beneficial effects for human health were discussed in Chapters 21 and 23. Organisms such as *Akkermansia muciniphila* rejuvenate the mucin layer. *Bacteroides thetaiotaomicron* contributes greatly to carbohydrate utilization and to the production of short-chain fatty acids (SCFAs) that induce Treg cell production. Also contributing to Treg cell production are organisms such as *Bacteroides fragilis*, via its capsular polysaccharide antigen (PSA). *Faecalibacterium prausnitzii* and a group of Clostridiales called segmented filamentous bacilli (SFBs) induce IgA-secreting cells in Peyer's patches and stimulate the production of T_H17 cells. *Bifidobacterium* and *Lactobacillus* species also contribute to the shaping of the gut immune system.

As described earlier, the gut microbiome influences **mucosal immunity** through encounters with TLRs, NLRs, and other pattern recognition receptors on epithelial cells and a variety of immune cells. Influence is also provided through antigen interactions with dendritic cells and subsequent antigen presentation to T cells. The various blends of cytokines released can tilt immunity toward tolerance or inflammation and can even affect how strongly immune cells react to different antigens. For example, antibiotic-induced alterations in the microbiota will change not only the numbers of Treg cells, but also the T-cell receptor (TCR) repertoire on the remaining Treg cells, suggesting that gut microbiome composition can influence the response of Treg cells to various antigens. Evidence that microbiota can dampen inflammation is described in eTopic 24.4.

It has also become clear that the influence of the gut microbiome extends beyond the intestinal tract. Immune-related disorders such as inflammatory bowel disease (IBD), cancer, diabetes, allergies, and even obesity (see Chapter 23) may result from dysbiosis of the commensal microbial communities. Bacteria can produce neurotoxic metabolites such as D-lactic acid and ammonia. Even beneficial metabolites such as SCFAs may exert neurotoxicity. Gut microbes can also produce hormones and neurotransmitters that are identical to those produced by humans. Consequently, gut bacteria can directly stimulate afferent neurons of the enteric nervous system to send signals to the brain via the vagus nerve. Through these varied mechanisms, gut microbes can shape the architecture of sleep and stress reactivity of the hypothalamic-pituitary-adrenal axis. Believe it or not, your gut microbes can influence your memory, mood, and cognition.

To Summarize

- **The gut immune system** includes intestinal epithelial cells (IECs); goblet cells; innate lymphoid cells (ILCs); lamina propria, which contains B cells, T cells, and dendritic cells; and Peyer's patches, which contain M cells, IgA⁺ B cells, and T cells.

- **Secretion of sIgA** helps control the composition and balance of the gut microbiome.

- **Intestinal epithelial cell** TLRs and NLRs can distinguish normal microbiota from pathogens.

- **Microbiota composition** influences the ratios of T_{FH}, T_H1, T_H2, T_H17, and Treg cells to tilt the gut immune response between inflammatory and anti-inflammatory.

- **Mucosal immunity** is an immune system compartment mostly separate from the rest of the immune system.

- **Development of the mucosal immune system** is guided by bacterial members of the gut microbiota.

24.6 Immunization

Most of you reading this have been vaccinated with many of the vaccines listed in **Table 24.1**. The adaptive immune response is why vaccines work. In this section we will describe the various types of vaccines, note when and how they are given, and discuss why some children are not vaccinated.

Vaccines

Vaccines come in three basic types. As noted in **Table 24.1**, some vaccines utilize killed organisms (examples are the hepatitis A vaccine and the Salk inactivated polio vaccine), while others contain live, but attenuated, microbes (BCG for tuberculosis, or the Sabin live polio vaccine). The third type of vaccine consists of purified components (subunits) of an infectious agent (such as capsular antigens from *Streptococcus pneumoniae* and *Haemophilus influenzae* type b). Some vaccines are injected in combination as polyvalent vaccines to control multiple diseases (for example, MMR for measles, mumps, and rubella), while others are given individually but may change from year to year (influenza viral envelope proteins).

Multiple vaccines can be given simultaneously and safely. Most vaccines are administered during childhood, when the diseases can be the most devastating. **Table 24.5** provides the current immunization schedule recommended for children and adolescents by the Centers for Disease Control

TABLE 24.5 Recommended Immunization Schedule for Children and Adolescents[a]

Vaccine	Birth	1 month	2 months	4 months	6 months	9 months	12 months	15–18 months	24 months	4–6 years	11–12 years
Hepatitis B[b]	HepB	HepB			HepB					HepB series	
Rotavirus			Rota	Rota	Rota						
Diphtheria, tetanus, acellular pertussis[c]			DTaP	DTaP	DTaP			DTaP		DTaP	Tdap
Haemophilus influenzae type b[d]			Hib	Hib	Hib		Hib				
Inactivated poliovirus			IPV	IPV	IPV					IPV	
Measles, mumps, rubella							MMR			MMR	MMR
Varicella							Varicella			Varicella	
Meningococcal[e]							Administer to high-risk children				MCV4
Pneumococcal[f]			PCV13	PCV13	PCV13		PCV13		PPSV23		
Influenza[g]					Influenza (yearly)						
Hepatitis A[h]							HepA series			HepA	
Human papillomavirus[i]											HPV

[a]This schedule indicates the recommended ages for routine administration of currently licensed childhood vaccines, as of January 1, 2013. ■ Range of recommended ages; ■ Catch-up immunization; ■ Assessment at age 11–12 years.

[b]Hepatitis B vaccine (HepB). *At birth:* All newborns should receive monovalent HepB, administered soon after birth and before hospital discharge.

[c]Diphtheria and tetanus toxoids and acellular pertussis vaccine (DTaP). Tdap is a modified vaccine with lower doses of diphtheria and tetanus toxoids.

[d]*Haemophilus influenzae* type b (Hib) conjugate vaccine.

[e]Meningococcal conjugate vaccine (MCV4). MCV4 should be administered to all children at age 11–12 years, as well as to unvaccinated adolescents at high school entry (age 15 years). Vaccine contains four types of capsules.

[f]Pneumococcal vaccine. The 13-valent pneumococcal conjugate vaccine (PCV) is recommended for all children aged 2–23 months and for certain children aged 24–59 months. The final dose in the series should be administered at age ≥12 months. Pneumococcal polysaccharide vaccine (PPSV) is a 23-valent vaccine recommended in addition to PCV for certain high-risk groups. PPSV is also recommended for people over 65.

[g]Inactivated influenza vaccine should be administered annually starting at age 6 months. Live, attenuated influenza vaccine should not be given until 2 years, and not to immunocompromised individuals.

[h]Hepatitis A vaccine (HepA). HepA is recommended for all children at age 1 year (12–23 months).

[i]Human papillomavirus (HPV) vaccine. [HPV4 (Gardasil) and HPV2 (Cervarix).] A 3-dose series of HPV vaccine should be administered on a schedule of 0, 1–2, and 6 months to all adolescents aged 11–12 years. Either HPV4 or HPV2 may be used for females, and only HPV4 may be used for males.

Source: Adapted from the Centers for Disease Control and Prevention website (http://www.cdc.gov).

and Prevention. Notice that all of these vaccines are given in multiple doses, called booster doses. The exception to this rule is the influenza vaccine, which is given in a single dose but changes every year. The reason for multiple doses, as noted earlier, is that secondary exposure to an antigen provides a more robust and long-lasting immunity. However, most vaccines are not administered until 2 months of age, because maternal antibody crossing through the placenta to the fetus (or to a newborn through breast milk) will persist for a short time in the newborn, temporarily protecting the baby from disease and possibly dampening the response to a vaccine antigen administered during that time.

Herd Immunity

You might wonder whether all members of a community must be vaccinated against a given microbe to lower the risk of disease for every individual in that community. It depends on the disease. For infections spread by person-to-person contact, the risk of disease to an unvaccinated person can be lowered dramatically even when only about two-thirds of the community is vaccinated. Vaccinating a large percentage of a community effectively conveys community (or herd) immunity by interrupting transmission of the disease. If one individual contracts the disease, the chance that he or she will come into contact with another unvaccinated person and transmit the disease is much reduced. Thus, the risk of disease to any single unvaccinated person is lessened as a result of community immunity. Gardasil, the vaccine against human papillomavirus (the cause of genital warts), is a good example of a vaccine that can provide herd immunity (see **eTopic 26.1**).

Herd immunity works well for diseases such as diphtheria, whooping cough (pertussis), measles, and mumps. However, herd immunity will not lower the risk that an unvaccinated person will contract tetanus, which is not spread by person-to-person contact. *Clostridium tetani*, the agent whose toxin causes tetanus, is a ubiquitous soil organism transmitted through punctured skin. The risk of tetanus for an unvaccinated person doesn't change even if every other person in the community is vaccinated against tetanus.

Vaccines and Immune System Compartments

What difference does it make which type of vaccine you take? Attenuated, killed, subunit—don't they all generate immunity? Yes, but generally, a live, attenuated vaccine is better because when a crippled but live microbe replicates at its normal body target site, an immune response most appropriate to that site develops. Take the case of the Salk (killed) and Sabin (attenuated) polio vaccines. Poliovirus typically enters the body through ingestion, replicates in the mucosa, moves to the regional lymph nodes, and produces a viremia. Eventually, the virus can attack the central nervous system and cause paralysis.

The killed Salk vaccine (inactivated polio vaccine, or IPV) is injected and generates an antibody response in the bloodstream. The antibody response is capable of preventing the viremia and paralytic consequences of a natural infection, but it does not generate a mucosal immunity capable of preventing mucosal replication of the natural virus. Consequently, mild disease can result if wild poliovirus infects someone vaccinated with the Salk vaccine. These immunized people will shed poliovirus in their feces, enabling the virus to spread to others who might not be vaccinated. In contrast, the live virus in the Sabin vaccine, which is administered orally (oral polio vaccine, or OPV), will replicate in the mucosal lymphatic system of the gut, where secretory IgA antibodies are best generated (discussed earlier). Anti-polio IgA antibodies secreted in these mucosal areas will prevent wild-type poliovirus from replicating in the intestinal mucosa of a vaccinated person, so no virus is shed and there are no symptoms. However, OPV has not been used in the United State since 2000 because polio in this country is rare and fecal spread unlikely.

Can vaccines stimulate cell-mediated immunity? Live, attenuated vaccines can, but killed-organism vaccines generally do not. Classical inactivated virus or subunit vaccine components, because they are large, must be endocytosed to be processed by APCs. Endocytosed and processed antigens are presented only on MHC II receptors and recognized by CD4⁺ T cells. Consequently, killed or subunit vaccines will activate antibody responses but not cytotoxic-T-cell responses that require MHC I presentation to CD8⁺ cells (see **Fig. 24.21**). Since cytotoxic T cells are important to protect against, or resolve, many viral infections, there is a need for CD8 T cell–inducing vaccines. As mentioned already, attenuated vaccines can activate CD8 T cells, but live vaccines are often too toxic to use.

How can a subunit vaccine be made to generate cytotoxic T cells? One way is to include small peptides (<20 amino acids) in a vaccine. Small peptides do not need processing and can be presented on MHC I or MHC II molecules. Consequently, specific cytotoxic-T-cell responses can be induced. Currently, the only peptide vaccine approved for human use is one for human papillomavirus (HPV). The vaccine is designed to target HPV strains that can cause genital warts, as well as cervical or penile cancer. Peptide vaccines for other diseases are under study.

Thought Question

24.14 Why do immunizations lose their effectiveness over time?

Are Vaccines Dangerous?

The vast majority of people who receive vaccines suffer no, or only mild, reactions, such as fever or soreness at the injection site. Very rarely do more serious side effects, such as allergic reactions, occur. Vaccines are extremely safe. Unsettling, erroneous information circulates on the Internet purporting a link between vaccinations and other diseases, such as diabetes or autism, but no well-controlled scientific study supports these claims. The risks, including death, associated with these preventable infectious diseases are far greater than the minimal risk associated with being vaccinated against them. As proof of this point, the undervaccination of children in the United States in the 1980s and '90s has led to an increase in cases of whooping cough, caused by *Bordetella pertussis*. The number of cases rose from 6,586 in 1993 to 28,639 in 2013. Successful vaccination programs carried out in the United States have come close to eradicating many diseases once feared, such as polio, rubella, and diphtheria, and have dramatically lowered the morbidity and mortality of many others.

The current risk of infection for some diseases (because they are so rare) is lower than the risk of an adverse reaction to immunization. However, failing to vaccinate, as just mentioned, will result in a population susceptible to the microbe and a resurgence of serious disease.

To Summarize

■ **Vaccines** can be made from live, attenuated organisms; killed organisms; or purified microbe components.

■ **Herd immunity** can help protect unimmunized persons from diseases transmitted person-to-person.

■ **Vaccine makeup and delivery** influence whether the immune response is primarily humoral or cellular in nature.

■ **Serious side effects** from immunizations are very rare.

24.7 Hypersensitivity and Autoimmunity

Sometimes the immune system overreacts to certain foreign antigens and causes more damage than the antigen (microbe) alone might cause. Furthermore, some foreign antigens possess structures similar in shape to host structures and can trick the immune system into reacting against self. These immune miscues are called allergic hypersensitivity reactions, and the antigen causing the reaction is called an **allergen**. There are four types of hypersensitivity reactions (**Table 24.6**). The first three are antibody mediated; the fourth is cell mediated.

Case History: Type I Hypersensitivity

A bee stings a 9-year-old boy walking with his mother at the zoo. Within minutes, the boy begins sweating and itching. His chest then starts to tighten, and he has tremendous difficulty breathing. Terrified, he looks to his equally frightened mother for help.

This is a classic and severe example of **type I** (**immediate**) **hypersensitivity**, called **anaphylaxis**, in which smooth muscle contracts and capillaries dilate in response to the release of pharmacologically active substances. Type I hypersensitivity occurs within minutes of a <u>second</u> exposure to an allergen when the allergen reacts with IgE-coated mast cells. Recall that the primary role of IgE is to amplify the body's response to invaders by causing mast cells to release inflammatory mediators (see Section 24.2). Allergic individuals with type I hypersensitivity produce an excessive amount of IgE to the allergen. On initial exposure, an allergen elicits the production of IgE antibodies specific to the allergen (bee venom in this example). The Fc portions of these antibodies bind to Fc receptors on the surfaces of mast cells. IgE-coated mast cells are then said to be <u>sensitized</u>.

During a second exposure to the allergen, identical antigenic sites on the allergen bind to adjacent surface IgE molecules pointing out from the mast cell. The result is a bridge between adjoining binding sites (**Fig. 24.29**). This cross-linking inhibits **adenylate cyclase** in the mast cell. The resulting decrease in cellular cyclic adenosine monophosphate (cAMP) level causes mast cell granules to move to the cell surface, where they release their contents in a process called degranulation. Understanding this process is clinically important because anaphylaxis inhibitors such as epinephrine (also known as adrenaline) <u>stimulate</u>

Type	Description	Time of onset	Mechanism[a]	Manifestations
I	IgE-mediated hypersensitivity	2–30 min	Ag induces cross-linking of IgE bound to mast cells with release of vasoactive mediators.	Systemic anaphylaxis, local anaphylaxis, hay fever, asthma, eczema
II	Antibody-mediated cytotoxic hypersensitivity	5–8 h	Ab directed against cell-surface antigens mediates cell destruction via ADCC or complement.	Blood transfusion reactions, hemolytic disease of the newborn, autoimmune hemolytic anemia
III	Immune complex–mediated hypersensitivity	2–8 h	Ag-Ab complexes deposited at various sites induce mast cell degranulation via Fc receptor; PMN degranulation damages tissue (localized reaction; **eTopic 24.4**).	Systemic reactions, disseminated rash, arthritis, glomerulonephritis
IV	Cell-mediated hypersensitivity	24–72 h	Memory T_H1 cells release cytokines that recruit and activate macrophages.	Contact dermatitis, tubercular lesions

TABLE 24.6 **Summary of Hypersensitivity Reactions**

[a]Ag = antigen; Ab = antibody; ADCC = antibody-dependent cell-mediated cytotoxicity; PMN = polymorphonuclear leukocyte.

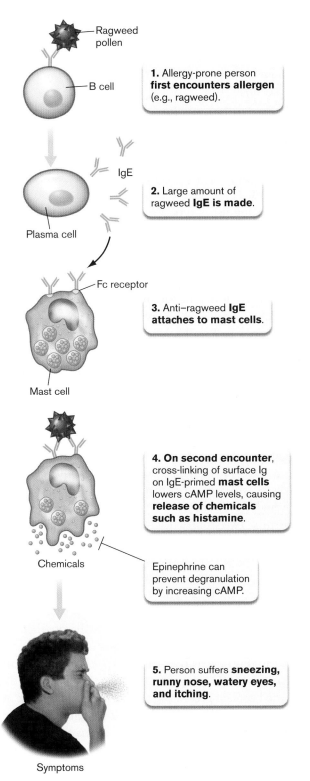

1. Allergy-prone person **first encounters allergen** (e.g., ragweed).

IgE

2. Large amount of ragweed **IgE is made**.

3. Anti–ragweed **IgE attaches to mast cells**.

4. On second encounter, cross-linking of surface Ig on IgE-primed **mast cells** lowers cAMP levels, causing **release of chemicals such as histamine**.

Epinephrine can prevent degranulation by increasing cAMP.

5. Person suffers **sneezing, runny nose, watery eyes, and itching**.

FIGURE 24.29 ■ Events leading to type I hypersensitivity reactions.

adenylate cyclase activity, stopping the anaphylaxis cascade in its tracks.

Mast cell degranulation releases chemicals with potent pharmacological activities. The most important of these is histamine, which binds to histamine receptors (H1 receptors) present on most body cells. Antihistamines have

a structure similar to histamine and work by preventing histamine from binding (antagonist) the H1 receptor or by dampening the basal activity of H1 receptors. Upon binding to H1 receptors on smooth muscle, histamine triggers production of a signaling molecule (inositol trisphosphate) that initiates smooth-muscle contraction to constrict small blood vessels. Histamine also weakens contacts between adhesion proteins (VE-cadherin) on vascular endothelial cells, causing gaps between cells through which blood fluids can seep. As a result, histamine-induced constriction of small blood vessels causes fluid to be forced from the circulation into the tissues. The immediate consequence is swelling (**edema**) in the joints and around the eyes and a rash (similar to hives), with burning and itching of the skin due to nerve involvement. In the clinical example involving the bee sting, the contraction of lung smooth muscles also led to breathing difficulties.

The bee sting in our example caused a severe type I allergic reaction, but type I hypersensitivity reactions do not usually involve the whole body. Most type I reactions are more localized and cause what is called atopic ("out of place") disease. Hay fever, or allergic rhinitis, is a common manifestation of atopic disease that can be caused by the inhalation of dust mite feces (**Fig. 24.30**), animal

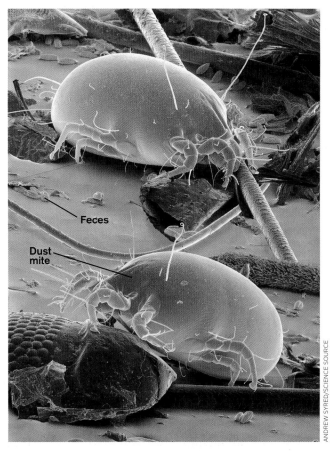

FIGURE 24.30 ■ Inhaled allergen. Dust mite (200–600 μm) and dust mite feces (colorized SEM).

skin or hair (dander), and certain types of grass or weed pollens. This disease affects the eyes, nose, and upper respiratory tract. Atopic asthma, another form of type I hypersensitivity resulting from inhaled allergens, affects the lower respiratory tract and is characterized by wheezing and difficulty breathing.

The administration of antihistamines is an effective treatment of allergic rhinitis, but the more important chemical mediators of asthma are the leukotrienes produced during what is called **late-phase anaphylaxis**. In late-phase anaphylaxis, mast cells release chemotactic factors that call in eosinophils. Eosinophils entering the affected area produce large amounts of leukotrienes that, like histamine, cause vasoconstriction and inflammation. At this point, antihistamines have little effect. Effective treatment includes inhaled steroids to minimize inflammation and bronchodilators (for example, albuterol) to widen bronchioles and facilitate breathing. One medication, Singulair, works by blocking leukotriene receptors in the lung. During severe asthma attacks, injected epinephrine is critical. Epinephrine will open airways to ease breathing through direct hormonal action.

People sensitized to allergens are not necessarily doomed to suffer with the allergy their entire life. A clinical treatment called **desensitization** can sometimes be used to prevent anaphylaxis. Desensitization involves injecting small doses of allergen over a period of months. This process is thought to produce IgG molecules that circulate, bind, and neutralize allergens before they contact sensitized mast cells. Desensitization has been useful in cases of asthma, food allergies, and bee stings, although results are variable and by no means guaranteed.

Another treatment for allergic asthma, omalizumab (Xolair), can actually prevent sensitization. Omalizumab is a monoclonal antibody (an antibody preparation of one antibody type that binds only one epitope). Administered by injection, omalizumab selectively binds IgE. Given to allergic patients once a month, the monoclonal antibody prevents IgE from binding to Fc receptors on mast cells, and thus prevents sensitization. A subsequent encounter with an allergen will not trigger an asthmatic attack.

Why don't all antigens generate type I hypersensitivity? It turns out that most antigens do not elicit high levels of IgE antibodies. It is unclear what gives certain antigens this capability and what makes different individuals prone to different allergies.

Case History: Type IV Hypersensitivity

The patient is an 8-year-old girl from Argentina. She recently received the BCG vaccine for tuberculosis before coming to the United States. Upon entering school, she is required to take a skin test for tuberculosis. She tries to refuse but is told it is a requirement and allows the nurse to apply the test to her arm. Three days later, the test site has a large, red lesion and the skin is starting to slough (peel off).

Type IV hypersensitivity, also known as **delayed-type hypersensitivity** (DTH), is the only class of hypersensitivity that is triggered by antigen-specific T cells. Because T cells have to react and proliferate to cause a response, it generally takes 24–48 hours before a reaction is noticed. This delay distinguishes type IV hypersensitivity from the more rapid antibody-mediated allergic reactions (types I–III). The girl in the case study was initially sensitized with BCG (an attenuated strain of *Mycobacterium bovis*, a close relative of *M. tuberculosis*) as a result of vaccination. Because the organism is intracellular, the vaccination produced a cell-mediated immunity, complete with preactivated memory T cells (similar to memory B cells). When she was reinoculated by the skin test, the memory T cells activated and elicited a localized reaction at the site of injection. (Note: About 6 months after vaccination, the hypersensitivity usually diminishes and the tuberculin skin test becomes useful once again.)

Type IV hypersensitivity develops in two stages. In the first stage (sensitization), antigen is processed and presented on <u>cutaneous</u> dendritic cells (called Langerhans cells). These APCs travel to the lymph nodes, where T_H0 cells can react to them as described earlier (see **Fig. 24.23**), generating activated T cells and a subset of memory T cells. On second exposure, two routes leading to DTH are possible. In the first pathway (**Fig. 24.31A**), memory T_H1 cells bind antigen that is complexed to class II MHC receptors (this happens at the site of infection) and release IFN-gamma, TNF-beta, and IL-2. These cytokines recruit macrophages and PMNs to the site and activate macrophages and natural killer cells to release inflammatory mediators that damage innocent, uninfected bystander host cells. In the second pathway (**Fig. 24.31B**), memory T_C cells recognize antigen on class I MHC receptors, become activated, and directly kill the host cell presenting the antigen. A hallmark of type IV hypersensitivity is that white blood cells, not serum, can transfer the sensitivity to a naive animal. This is because T cells, not serum antibody, cause the reaction.

Forms of DTH reactions include contact dermatitis (such as poison ivy) and allograft rejection. The antigens involved with contact dermatitis are usually small haptens that have to bind and modify normal host proteins to become antigenic (**Fig. 24.32**). In this case, the hapten-modified protein is processed, and fragments containing the bound hapten are presented on the surfaces of antigen-presenting cells.

There are two other types of hypersensitivity, both of which involve antigen-antibody complexes. **Type II**

A.

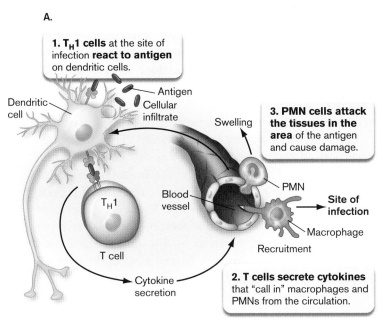

1. **T$_H$1 cells** at the site of infection **react to antigen** on dendritic cells.

Dendritic cell

Antigen
Cellular infiltrate

3. **PMN cells attack the tissues in the area** of the antigen and cause damage.

Swelling

T$_H$1

Blood vessel

PMN

Site of infection

T cell

Macrophage

Recruitment

Cytokine secretion

2. **T cells secrete cytokines** that "call in" macrophages and PMNs from the circulation.

B.

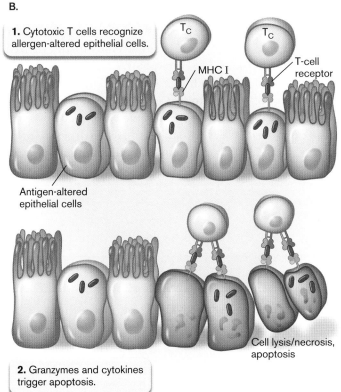

1. Cytotoxic T cells recognize allergen-altered epithelial cells.

T$_C$

T$_C$

MHC I

T-cell receptor

Antigen-altered epithelial cells

Cell lysis/necrosis, apoptosis

2. Granzymes and cytokines trigger apoptosis.

FIGURE 24.31 ■ **Mechanisms of damage in delayed-type hypersensitivity (type IV).** **A.** T$_H$1 cells react to antigen on dendritic cells. **B.** Antigen-sensitized cytotoxic T cells recognize allergen haptens attached to proteins of other cells. This recognition triggers the release of granzymes and production of cytokines, which cause apoptosis of the target cell.

hypersensitivity occurs when antibody binds to host cell-surface antigens. ABO blood group incompatibility, discussed in **eTopic 24.2**, is an example of type II hypersensitivity. **Type III hypersensitivity** happens when large complexes are formed between antibody and small

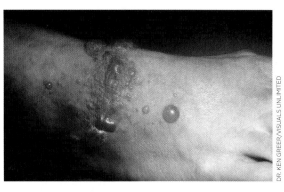

An unsaturated pentadecylcatechol

FIGURE 24.32 ■ **Contact dermatitis.** Pentadecylcatechols are chemicals present on the surface of poison ivy leaves. They are haptens that bind to proteins in dermal cells, where they can activate T cells. The result of the ensuing delayed-type hypersensitivity is contact dermatitis.

soluble antigens. The complexes can trigger complement activation that leads to rash formation. Specific case histories and discussion of these types of hypersensitivity can be found in **eTopic 24.5**.

Autoimmunity: An Inability to Recognize Self

The ability to distinguish between self antigens and foreign antigens is crucial to our survival; without this ability, our immune system would constantly attack us from within (see **eTopic 24.3**). Normally, the body develops tolerance to self; occasionally, however, an individual loses immune tolerance against some self antigens and the body mounts an abnormal immune attack against its own tissues. The attack can involve antibodies or T cells and is called an **autoimmune response**. Autoimmune responses may or may not be associated with pathological changes (autoimmune disease). Almost 30% of the population will have an autoimmune antibody by age 65, but many will not exhibit disease. The mechanisms that lead to autoimmune disease are essentially hypersensitivity reactions.

An autoimmune response results in autoimmune disease when an autoantibody or autoimmune lymphocyte (cytotoxic T cell) damages tissue components. Tissue damage can ensue when antibody to self activates antibody-dependent cell cytotoxicity (see Section 23.5). In this scenario, autoantibodies affixed to host cells can bind to NK cell Fc receptors and cause the NK cell to kill the tissue cell. In contrast, autoreactive cytotoxic T cells that escaped negative selection in the thymus will directly kill host cells that express the cognate self antigen.

How might autoimmune antibodies be formed? One proposed mechanism starts with the occasional escape of self-reacting B cells from the negative selection process. Fortunately, self-reacting B cells that escape negative selection are usually not a problem, because the specific helper T cells needed to activate them were deleted from the T-cell population. How, then, can a self-reacting B cell be activated? The B-cell receptor on autoreactive B cells can, of course, take up and process a self antigen or a foreign antigen that mimics a self antigen. A foreign antigen that resembles self, however, will also contain one or more nonself epitopes that "piggyback" with the foreign epitope into the self-reacting B cell. As a result, the nonself epitope will also be presented on a B-cell MHC molecule. Helper T cells specific to the nonself epitope (these T cells were not deleted in the thymus) will then bind the nonself epitope presented by the self-reacting B cell. Helper-T-cell binding activates the B cell to make antibody, but the antibody, and that of its descendant plasma cells, is directed to the self antigen, not the nonself antigen. Once made, the self-reacting antibodies begin attacking host antigen on host tissues.

A good example of autoimmunity caused by molecular mimicry is the M protein (pili) of *Streptococcus pyogenes*, a microbe that causes "strep throat" and the autoimmune disease rheumatic fever. M pili contain an epitope that resembles cardiac antigen, but the cardiac-like epitope is flanked by epitopes that are not related to the human host. The cardiac-like epitope can bind surface antibody on an escaped self-reacting B cell and be taken up. Because the nonself epitope and the heart-like epitope are contained within the same protein, the nonself epitope will "piggyback" its way into the B cell. The self epitope is processed and placed on the B-cell surface MHC. Once this happens, T cells specific to the nonself epitope (M protein in this example) will recognize the nonself peptide and activate the B cell. Because the B cell is programmed to make antibody to the cardiac antigen, those self-reacting antibodies are made, begin to bind to cardiac tissue, and trigger autoimmune disease. In the case of *S. pyogenes*, cardiac tissue is damaged and rheumatic fever results. This similarity between the epitope in M protein and cardiac tissue is an example of antigenic mimicry.

Other autoimmune diseases involve the production of autoantibodies that bind and block certain cell receptors. Graves' disease (hyperthyroidism), for example, occurs when an autoantibody is made that binds to the thyroid-stimulating hormone (TSH) receptor. This receptor binding mimics actual TSH binding and continually stimulates the production of two hormones—triiodothyronine (T3) and thyroxine (T4)—which generally increases metabolic rate. The thyroid is not destroyed, but in fact enlarges to form a goiter. Examples of other important autoimmune diseases are listed in **Table 24.7**.

TABLE 24.7	Examples of Autoimmune Diseases	
Disease	**Autoantigen**	**Pathology**
Type II hypersensitivity—mediated by antibody to cell-surface antigens		
Acute rheumatic fever	Streptococcal M protein, cardiomyocytes	Myocarditis, scarring of heart valves
Autoimmune hemolytic anemia	Rh blood group	Destruction of red blood cells by complement, phagocytosis
Goodpasture's syndrome	Basement membrane collagen	Pulmonary hemorrhage, glomerulonephritis
Graves' disease	Thyroid-stimulating hormone (TSH) receptor	Antibody stimulates T3, T4 production, hyperthyroidism
Myasthenia gravis	Acetylcholine receptor	Interrupts electrical transmission, progressive muscular weakness
Type III hypersensitivity—mediated by antibody complexes with small soluble antigens (immune complex)		
Systemic lupus erythematosus	DNA, histones, ribosomes	Arthritis, vasculitis, glomerulonephritis
Type IV hypersensitivity—mediated by antigen-specific T cells		
Type 1 diabetes	Pancreatic beta cell antigen	Beta cell destruction
Multiple sclerosis	Myelin protein	Demyelination of axons

To Summarize

- **Allergens** cause the host immune system to overrespond or react against self.

- **Type I hypersensitivity** involves IgE antibodies bound to mast cells by the antibody Fc region. Binding of antigen to the mast cell–attached IgE causes mast cell degranulation. This type of hypersensitivity can occur within minutes of exposure.

- **Type IV hypersensitivity** (delayed-type hypersensitivity, or DTH) involves antigen-specific T cells. T_H1 cells release cytokines that activate macrophages and NK cells. T_C cells can directly kill cells that present the antigen. Reaction is seen within a few days of exposure.

- **Autoimmune disease** is caused by the presence of lymphocytes that can react to self. These autoreactive lymphocytes escaped negative selection in the thymus.

- **Autoreactive B cells** (B cells that make antibody directed against self epitopes) can be activated if their BCR takes up a self-mimic epitope linked to a nonself epitope and presents the nonself epitope on the surface MHC. T cells that recognize the nonself epitope can then activate the B cell, which then secretes antibody against the self epitope. NK cells can target host cells decorated with those autoantibodies.

- **Cytotoxic T cells can produce autoimmune disease** by killing host cells expressing the cognate self antigen.

Concluding Thoughts

Immunity is a remarkable feat of nature. The idea that any one person's immune system can recognize and respond to virtually any molecular structure and yet remain selectively "blind" to his or her own antigens is hard to comprehend. But even more remarkable, pathogenic microbes have managed ways to outmaneuver the interconnected redundancies and safeguards of the immune system. In some cases, the pathogen evolves to become less harmful and coexists with the host; indeed, pathogens in nature are far outnumbered by closely related strains that are harmless or even beneficial to their hosts. In other cases, however, evolution generates a never-ending arms race between pathogen and host. The next chapter describes some of the microbial strategies that contribute to the success of a pathogen.

CHAPTER REVIEW

Review Questions

1. Define "antigen," "epitope," "hapten," and "antigenic determinant."
2. What is the basic difference between humoral immunity and cellular immunity?
3. Why are proteins better immunogens than nucleic acids are?
4. What makes IgA antibody different from IgG?
5. What is immunoprecipitation?
6. Explain isotypic, allotypic, and idiotypic differences in antibodies.
7. How is IgE involved in allergic hypersensitivity?
8. Discuss differences in the primary and secondary antibody responses.
9. Outline the basic steps that turn a B cell into a plasma cell.
10. What is isotype switching, and how is antibody diversity achieved?

11. What signals are needed to activate helper T cells?
12. What signals activate cytotoxic T cells?
13. Discuss the differences between transplant rejections of nucleated cells and the rejection of blood cells.
14. How do superantigens activate T cells?
15. Discuss the differences between the alternative and classical pathways of complement activation.
16. How does the host prevent membrane attack complexes from being formed in host cells?
17. Describe the development of type I hypersensitivity.
18. How does a B cell programmed to make an antibody against self become activated in the absence of specific T-cell help?
19. Explain key features of the gut immune system.
20. Discuss the types of immune responses generated by different types of vaccines.

Thought Questions

1. Cytotoxic T cells lyse the membranes of host cells carrying viruses or bacteria. Why doesn't this lysis facilitate the spread of those organisms rather than help clear the infection?

2. Immunity raised by the poliovirus vaccine will prevent subsequent infection by poliovirus, but someone with a *Staphylococcus aureus* infection (and the attendant immune response) can be reinfected many times by *S. aureus*. Why does the immune system work so well against infection by some pathogens (for example, poliovirus) but not others (for example, *S. aureus*)?

3. (refer to **eTopic 24.2**) Why are blood group type O people "universal <u>donors</u>" (meaning they can donate their red blood cells to type O, type A, type B, or type AB individuals), but not universal acceptors of red blood cells from type O, A, B, or AB individuals? Which blood type person might be considered a universal <u>recipient</u>?

4. IgM is the first antibody produced during a primary immune response. Of the five different types of antibodies produced, why would the body want IgM to be the first?

5. Some pathogens, such as rubella virus, can cross the placental barrier from mother to the fetus and cause a dangerous infection. Which antibody isotype against rubella virus would you look for in a newborn to diagnose congenital rubella syndrome—IgG or IgM?

Key Terms

adaptive immunity (943)
addressin (976)
adenylate cyclase (980)
affinity maturation (962)
allergen (980)
allotype (951)
anaphylaxis (980)
antibody (943)
antigen (943)
antigen-presenting cell
    (APC) (943, 963)
antigenic determinant (943)
autoimmune response (983)
B-cell receptor (BCR) (955)
B-cell tolerance (945)
capping (955)
cell-mediated immunity (943)
class I MHC molecule (964)
class II MHC molecule (964)
class switching (953)
classical complement pathway (973)
clonal (943)
clonal selection (954)
constant region (950)
cytokine (961)
cytotoxic T cell (T$_C$ cell) (963)
decay-accelerating factor (974)
delayed-type hypersensitivity
    (DTH) (982)

desensitization (982)
edema (981)
epitope (943)
equivalence (949)
factor H (974)
Fc region (950)
gene switching (959)
granzyme (969)
hapten (946)
heavy chain (948)
helper T cell (T$_H$ cell) (961)
humoral immunity (943)
idiotype (951)
immediate hypersensitivity (980)
immunogen (943, 946)
immunogenicity (945)
immunoglobulin (948)
    IgA (952)
    IgD (952)
    IgE (952)
    IgG (952)
    IgM (952)
immunological specificity (945)
immunoprecipitation (949)
innate lymphoid cell (ILC) (975)
intraepithelial lymphocyte (IEL) (975)
isotype (951)
isotype switching (953)
late-phase anaphylaxis (982)

light chain (948)
major histocompatibility complex
    (MHC) (945, 963)
memory B cell (953)
MHC restriction (966)
mucosal immunity (977)
negative selection (966)
opsonize (952)
perforin (969)
plasma cell (953)
positive selection (966)
primary antibody response (953)
recombination signal sequence
    (RSS) (959)
regulatory T cell (Treg) (966)
secondary antibody response (953)
serum (953)
superantigen (970)
switch region (961)
T-cell receptor (TCR) (965)
T$_H$17 cell (969)
threshold dose (945)
tumor necrosis factor (TNF) (970)
type I hypersensitivity (980)
type II hypersensitivity (982)
type III hypersensitivity (983)
type IV hypersensitivity (982)
vaccination (946)
variable region (950)

Recommended Reading

Artis, David, and Hergen Spits. 2015. The biology of innate lymphoid cells. *Nature* **517**:293–301.

Attaf, Meriem, Mateusz Legut, David K. Cole, and Andrew K. Sewell. 2015. The T cell antigen receptor: The Swiss army knife of the immune system. *Clinical and Experimental Immunology* **181**:1–18.

Bonneville, Marc, Rebecca L. O'Brien, and Willi Born. 2010. Gammadelta T cell effector functions: A blend of innate programming and acquired plasticity. *Nature Reviews. Immunology* **10**:467–478.

Cyster, Jason G. 2010. B cell follicles and antigen encounters of the third kind. *Nature Immunology* **11**:989–996.

Dunston, Christopher R., Rebecca Herbert, and Helen R. Griffiths. 2015. Improving T cell-induced response to subunit vaccines: Opportunities for a proteomic systems approach. *Journal of Pharmacy and Pharmacology* **67**:290–299.

Hale, J. Scott, and Rafi Ahmed. 2015. Memory T follicular helper CD4 T cells. *Frontiers of Immunology* **6**:16.

Hevia, Arancha, Susana Delgado, Borja Sanchez, and Abelardo Margolles. 2015. Molecular players involved in the interaction between beneficial bacteria and the immune system. *Frontiers in Microbiology* **6**:1285.

Iwasaki, Aikiko, and Rusian Medzhitov. 2015. Control of adaptive immunity by the innate immune system. *Nature Immunology* **16**:343–353.

Joffre, Olivier P., Elodie Segura, Ariel Savina, and Sebastian Amigorena. 2012. Cross-presentation by dendritic cells. *Nature Reviews. Immunology* **12**:557–569.

McDermott, Andrew J., and Gary B. Huffnagle. 2014. The microbiome and regulation of mucosal immunity. *Immunology* **142**:24–31.

Noriega, Vanessa, Veronika Redmann, Thomas Gardner, and Domenico Tortorella. 2012. Diverse immune evasion strategies by human cytomegalovirus. *Immunologic Research* **54**:140–151.

O'Keeffe, Meredith, Wai Hong Mok, and Kristen J. Radford. 2015. Human dendritic cell subsets and function in health and disease. *Cellular and Molecular Life Sciences* **72**:4309–4325.

Spencer, Jo, Linda S. Klavinskis, and Louise D Fraser. 2012. The human intestinal IgA response; burning questions. *Frontiers in Immunology* **3**:108. doi:10.3389/fimmu.2012.00108.

Weill, Jean-Claude, Sandra Weller, and Claude-Agnès Reynaud. 2009. Human marginal zone B cells. *Annual Review of Immunology* **27**:267–285.

CHAPTER 25
Microbial Pathogenesis

How does a pathogen differ from natural microbiota? The answer depends on the pathogen. Some, for instance, avoid phagocytosis, whereas others actively encourage it. Other pathogens develop a mysterious undetectable latent stage in the host, only to emerge later to cause disease. Some infectious agents, such as Ebola virus, efficiently slay their victims. Others persist for many years, causing only minor symptoms. In Chapter 25 we explore the diverse strategies that pathogenic bacteria and viruses use to infect hosts, subvert immune responses, and cause disease.

CURRENT RESEARCH highlight

Hostile takeover. *Yersinia pestis*, the pathogen that causes plague, survives inside host macrophages by altering phagosome maturation. The result is a *Yersinia*-containing vacuole (YCV). Matthew Lawrenz examined how *Y. pestis* might manipulate host pathways to form the YCV. He focused on the host protein Rab1b, a target needed by other pathogens for intracellular survival. Lawrenz discovered that *Yersinia* recruits Rab1b to the YCV. Rab1b then inhibits phagosome acidification and prevents fusion to lysosomes. As shown here, a YCV fuses with a macrophage lysosome only if the cellular Rab1b level is <u>reduced</u> by antisense RNA.

Source: M. G. Connor et al. 2015. PLoS Pathog. **11**:e1005241.

Macrophages infected with *Yersinia*

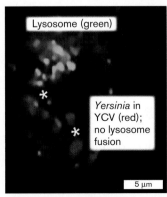

Lysosome (green)

Yersinia in YCV (red); no lysosome fusion

5 μm

Normal Rab1b level

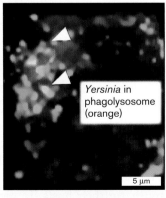

Yersinia in phagolysosome (orange)

5 μm

Decreased Rab1b

AN INTERVIEW WITH

MATTHEW LAWRENZ, MICROBIOLOGIST, UNIVERSITY OF LOUISVILLE SCHOOL OF MEDICINE

JESSICA INDRIGO

What additional questions do you have about how *Yersinia* manipulates the Rab1b system?

Moving forward, we have two important questions we would like to answer to better understand the role of Rab1b in the generation of the YCV. First, what bacterial factors are produced by *Yersinia pestis* to recruit Rab1b to the YCV? Second, how does recruitment of Rab1b inhibit phagosome acidification?

What special precautions are needed to work with this dangerous pathogen?

Y. pestis can be transmitted via aerosols and requires special biocontainment laboratories for safe handling. All of our work with virulent *Y. pestis* is performed in a biosafety level 3 (BSL-3) laboratory. BSL-3 laboratories are engineered to control the airflow in the lab, and workers wear specific personal protective equipment (PPE) that includes respiratory protection.

A 15-year-old girl from Georgia was rushed to the emergency room by her father. The girl was lethargic and confused, had muscle aches, and was registering a fever of 104°F. She also had a very painful, enlarged inguinal lymph node that the attending physician called a bubo. The physician quickly ordered a blood culture to look for bacterial pathogens and an antibiotic susceptibility profile of the pathogen once it was found. When asked about travel, the father said they had just returned from camping at Yosemite National Park, where they had encountered numerous wild animals close to their campsite. The child's symptoms, combined with recent travel to the western United States, raised the suspicion of plague. Plague is a life-threatening disease caused by *Yersinia pestis*, which is found in some wild animals in the Southwest. Fleas feeding on the infected animals can transmit the pathogen to humans (described further in Section 26.5).

Because the child was severely ill, the doctor immediately admitted her to the hospital and initiated antibiotic treatment with gentamicin and ciprofloxacin (see Chapter 27). Within two days, the lab confirmed the presence of gentamicin-susceptible *Y. pestis* in the girl's blood. After 14 days on gentamicin, the girl recovered.

Yersinia pestis is a Gram-negative rod similar in appearance to many other microbes, such as nonpathogenic strains of *Escherichia coli*, a member of the gut microbiome. So, why does one Gram-negative rod kill and another does not when they are visually indistinguishable? In the case of *Y. pestis*, the difference between friend and foe includes plasmid gene products (called effector proteins) that are injected into host epithelial cells to hijack host cell functions (see the Current Research Highlight).

Pathogens use a vast array of molecular tools to gain access to nutrients within hosts. In this chapter we discuss various relationships between pathogens and their victims, and the factors that contribute to **pathogenesis**, the process by which microbes cause disease. The degree of harm that results depends on the virulence mechanisms that the pathogen wields and the immune response to its presence.

25.1 Host-Pathogen Interactions

How long have we humans suffered with infections? Millions of years, it turns out. Paleopathologists found evidence of brucellosis in a skeleton from an *Australopithecus africanus* male, a predecessor of *Homo sapiens* that lived over 2 million years ago. The vertebrae of this "person" exhibited damage characteristic of disease caused by a pathogenic species of *Brucella*. Even though we have long been plagued by infectious diseases, the idea that these diseases can be caused by tiny living organisms invading our bodies became apparent only 150 years ago, when Robert Koch discovered the microbial cause of anthrax.

To Catch a Pathogen

How are the microscopic agents of infectious disease discovered? The revelation that *Bacillus anthracis* causes anthrax led Koch to propose a set of steps, or postulates, (discussed in Section 1.3) needed to prove that a specific microbe causes a specific disease. Koch's postulates state that the organism must be present in every case of a disease, must be propagated in pure culture, must cause the same disease when inoculated into a naive host, and must be recovered from the newly diseased host.

Viruses, however, cannot grow in pure culture. Thus, viruses causing disease cannot satisfy Koch's postulates. To accommodate viral diseases, Koch's criteria were modified by Thomas Rivers in 1937 to include cultivating the agent in host cells (rather than in pure culture), proving that the agent passes through a 0.2-μm filter (known bacteria did not), and demonstrating an immune response to the virus in patients. These steps were used in 2003 to rapidly discover the virus causing severe acute respiratory syndrome (SARS) and in 2013 to discover a related virus causing Middle East respiratory syndrome (MERS). Both viruses were grown in a macaque monkey model, fulfilling River's postulates.

Fulfilling Koch's and Rivers's postulates remains the most persuasive evidence of causation, but there are some problems with this standard. Many agents cannot be cultured, and there may be no suitable animal model in which disease can be reproduced. In these situations we must resort to a statistical association between organism and disease based on the presence of the agent or its footprints (nucleic acid, antigen, and preferably, an immune response; see Chapter 28). Statistical association was used to link Zika virus and newborn microcephaly, first observed in South America in 2015. This link was strengthened by finding Zika virus in the tissues of stillborn infants.

The Language of Pathogenesis

Before discussing the microbial mechanisms of disease, we should first establish the vocabulary of infectious disease. The term **parasites**, in the broadest sense, includes bacteria, viruses, fungi, protozoa, and worms that colonize and harm their hosts. In practice, however, only disease-causing protozoa and worms are called parasites, whereas bacterial, viral, and fungal agents of disease are referred

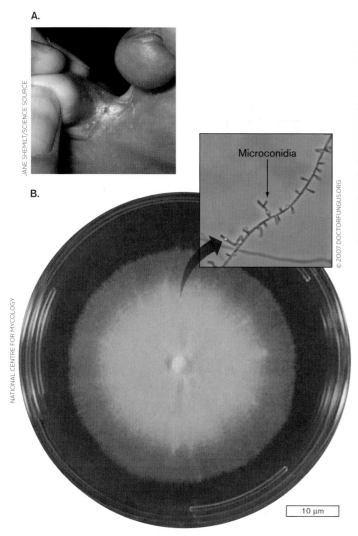

FIGURE 25.1 ■ An ectoparasite. A. Athlete's foot can be caused by the fungus *Trichophyton rubrum*. **B.** Colony morphology and microscopic, branching conidia (blowup) of *T. rubrum*. Conidia are asexual spores that grow on stalks called conidiophores (see Chapter 20).

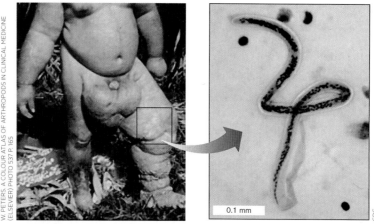

FIGURE 25.2 ■ An endoparasite. The disease filariasis, commonly known as "elephantiasis" for obvious reasons, is caused by the worm *Wuchereria bancrofti* (blowup), which enters the lymphatics and blocks lymphatic circulation. Adult worms are threadlike and measure 4–10 cm in length. The young microfilariae (inset) are approximately 0.5 mm in length. Though not a problem in the United States, *W. bancrofti* and elephantiasis are found throughout middle Africa, Asia, and New Zealand.

to as **pathogens**. Pathogens and parasites infect their animal and plant hosts in a variety of ways and enter into a variety of host-pathogen relationships, depending on the site of colonization. For example, organisms that live on the surface of a host are called **ectoparasites**. The fungus *Trichophyton rubrum*, one cause of athlete's foot, is an ectoparasite (**Fig. 25.1**). *Wuchereria bancrofti*, the worm parasite that causes elephantiasis, is an **endoparasite** because it lives inside the body (**Fig. 25.2**).

An infection occurs when a pathogen or parasite enters or begins to grow on a host. But the term **infection** does not necessarily imply overt disease. Any potential pathogen growing in or on a host is said to cause an infection, but that infection may be only transient, if immune defenses kill the pathogen before noticeable disease results. Indeed, most infections go unnoticed. For example, every

time you have your teeth cleaned by a dentist, your gums bleed and your resident oral microbes transiently enter the bloodstream, but you rarely suffer any consequences.

Primary pathogens are disease-causing microbes with the means to breach the defenses of a healthy host. *Shigella flexneri*, the cause of bacillary dysentery, is a primary pathogen. When ingested, it can survive the natural barrier of an acidic (pH 2) stomach, enter the intestine, and begin to replicate. **Opportunistic pathogens**, on the other hand, cause disease only in a compromised host. *Pneumocystis jirovecii* (previously *P. carinii*) is an opportunistic pathogen that causes life-threatening infections in AIDS patients, whose immune systems have been eroded (**Fig. 25.3A**).

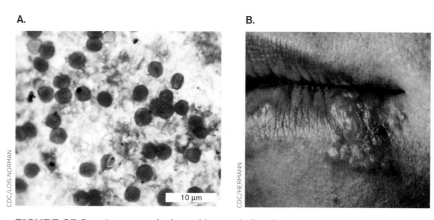

FIGURE 25.3 ■ Opportunistic and latent infections. A. *Pneumocystis jirovecii* cysts in bronchoalveolar material. Notice that the fungi look like crushed Ping-Pong balls. **B.** Cold sore produced by a reactivated herpes virus hiding latent in nerve cells.

A.

B.

FIGURE 25.4 ■ **Highly virulent viruses.** **A.** Ebola virus (approx. 1 μm long; TEM). **B.** The body of a victim of Marburg virus is placed in a coffin for safe burial in Angola. Marburg and Ebola cause hemorrhagic infections in which patients bleed from the mouth, nose, eyes, and other orifices. They have a 70%–80% mortality rate.

infectious dose 50% (ID_{50})—can be measured. ID_{50} is the dose required to cause disease symptoms in half of an experimental group of hosts.

Although it might be possible to measure the infectious dose rather than lethal dose for a lethal pathogen, it is not typically done. LD_{50} gives a clear end point, so it is much easier to use when trying to determine the effectiveness of a given treatment (an antibiotic, for example) or to quantify the role of a given gene in pathogenesis.

Some microbes even enter into a **latent state** during infection, in which the organism cannot be found by culture. Herpes virus, for instance, can enter the peripheral nerves and remain dormant for years, and then suddenly emerge to cause cold sores (**Fig. 25.3B**). The bacterium *Rickettsia prowazekii* causes epidemic typhus, but it can also enter a latent phase and then, months or years later, cause a disease relapse called recrudescent typhus.

Pathogenicity is an organism's ability to cause disease. It is defined in terms of how easily an organism causes disease (infectivity) and how severe that disease is (virulence). Pathogenicity, overall, is shaped by the genetic makeup of the pathogen. In other words, an organism is more—or less—pathogenic, depending on the tools at its disposal (such as toxins) and their effectiveness.

Virulence is a measure of the degree, or severity, of disease. For instance, Ebola virus and the closely related Marburg virus have case fatality rates near 70%, so they are highly virulent (**Fig. 25.4**). By contrast, rhinovirus, the cause of the common cold, is very effective at causing disease but almost never kills its victims, so it is highly infective but has a low virulence. Both organisms are pathogenic, but with one you live and the other you die.

One way to measure virulence is to determine how many bacteria or virions are required to kill 50% of an experimental group of animal hosts. This is called the **lethal dose 50% (LD_{50})**. A pathogen with a low LD_{50}, in which very few organisms (or viruses) are required to kill 50% of the hosts, is more virulent than one with a high LD_{50} (**Fig. 25.5**). For organisms that colonize but do not kill the host, the infectious dose needed to colonize 50% of the experimental hosts—that is, the

> **Thought Question**
>
> **25.1 Figure 25.5** presents the association between LD_{50} and virulence. Is a microbe with an LD_{50} of 5×10^4 more or less virulent than a microbe with an LD_{50} of 5×10^7?

Infection Cycles Can Be Direct or Indirect

Pathogens must pass somehow from one person or animal to another, if a disease is to spread. The route of transmission an organism takes is called the **infection cycle**. A cycle of infection can be simple or complex (**Fig. 25.6**). Organisms that spread directly from person to person, such as rhinovirus,

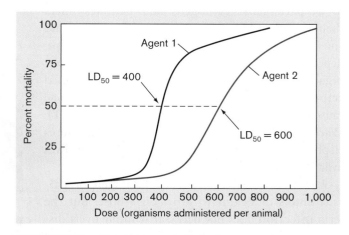

FIGURE 25.5 ■ **Measurement of virulence.** Each LD_{50} measurement requires infecting small groups of animals with increasing numbers of the infectious agent and observing how many animals die. The number of microbes that kill half the animals is called the LD_{50} dose. In this example, agent 1 is more virulent than agent 2.

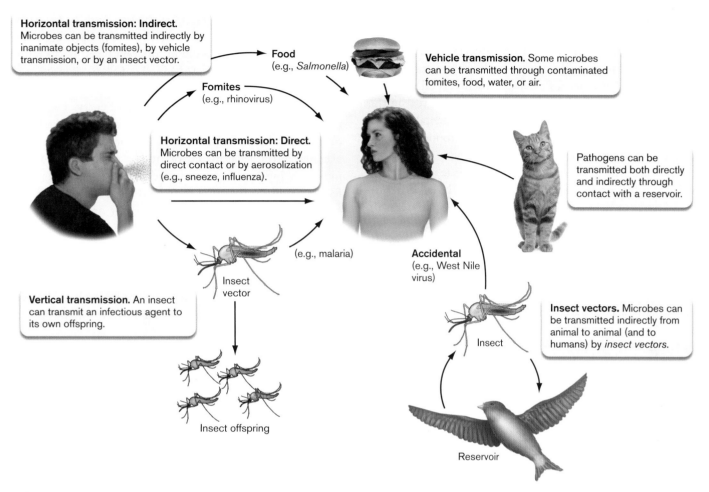

Horizontal transmission: Indirect. Microbes can be transmitted indirectly by inanimate objects (fomites), by vehicle transmission, or by an insect vector.

Food (e.g., *Salmonella*)

Fomites (e.g., rhinovirus)

Vehicle transmission. Some microbes can be transmitted through contaminated fomites, food, water, or air.

Horizontal transmission: Direct. Microbes can be transmitted by direct contact or by aerosolization (e.g., sneeze, influenza).

Pathogens can be transmitted both directly and indirectly through contact with a reservoir.

(e.g., malaria)

Accidental (e.g., West Nile virus)

Insect vector

Vertical transmission. An insect can transmit an infectious agent to its own offspring.

Insect vectors. Microbes can be transmitted indirectly from animal to animal (and to humans) by *insect vectors*.

Insect

Insect offspring

Reservoir

FIGURE 25.6 ■ Infection cycles. Infectious agents can be transmitted horizontally from one member of a species to another by a variety of means: direct contact between people; or indirect transmission by inanimate objects (fomites) or arthropod (insect or tick) vectors, or by a vehicle such as air, food, or water. Transmission can also occur through contact with a reservoir, such as cats harboring *Bartonella hensalae* (cat scratch disease). Vertical transmission is the passage of a pathogen from parent to offspring. In insects, vertical transmission occurs because the egg itself is infected during formation. In humans, vertical transmission occurs when the organism passes through the placenta to the fetus or to the newborn during birth. Accidental transmission happens when a host that is not part of the normal infection cycle unintentionally encounters that cycle.

have a simple infection cycle. Rhinovirus spreads person-to-person through virus-laden aerosols, or droplet nuclei, that are produced by sneezing and then inhaled by someone else (**airborne transmission**). A person can also transmit some infectious agents indirectly to another person by contaminating food, water, or an inanimate object. Inanimate objects that transmit disease are called **fomites** (for example, the tissue used to stifle a sneeze). Indirect transmission via fomites, food, or water is also called **vehicle transmission**.

More complex cycles often involve **vectors**, usually insects or ticks (arthropods), as intermediaries. Vectors carry infectious agents from one animal to another. A mosquito vector, for example, transfers the virus causing yellow fever from infected to uninfected individuals (**Fig. 25.7**) in what is called

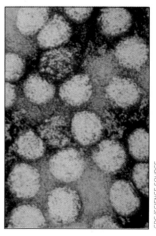

FIGURE 25.7 ■ Insect vector and yellow fever. The mosquito *Aedes aegypti* carries the yellow fever virus (inset; colorized TEM). The virus varies in size from 50 to 90 nm. Yellow fever remains endemic in the northern part of South America and in central Africa.

horizontal transmission. The mosquito can also bequeath this particular virus to its offspring via infected eggs in a form of **vertical transmission** called **transovarial transmission**. Although yellow fever is not a problem in the United States today, West Nile virus, a flavivirus closely related to yellow fever virus, currently claims several victims each year in this country. Because insects and ticks are instrumental in transmitting many pathogens, killing these arthropod vectors is an important way to halt the spread of disease.

Another critical factor in an infection cycle is the "reservoir" of infection. A **reservoir** is an animal, bird, or arthropod (insect or tick) that normally harbors the pathogen. In the case of yellow fever, the mosquito is not only the vector, but the reservoir as well, because the insect can pass the virus to future generations of mosquitoes through vertical transmission. The virus causing eastern equine encephalitis (EEE), however, uses birds as a reservoir. The EEE virus is normally a bird pathogen and is transmitted from bird to bird via a mosquito vector. The virus does not persist in the insect, but transmission by the insect vector keeps the virus alive by passing it to new avian hosts. Humans or horses entering geographic areas that harbor the disease (called endemic areas) can also be bitten by the mosquito. When this happens, they become accidental hosts and contract disease. The virus does not replicate to high titers in mammals, which means that horses and humans are poor reservoirs for it. EEE virus, however, does replicate to high numbers in the avian host.

Reservoirs are critically important for the survival of a pathogen and as a source of infection. If the eastern equine encephalitis virus had to rely on humans to survive, the virus would cease to exist because of limited replication potential and limited access to mosquitoes. Note that the reservoir of a given pathogen might not exhibit disease.

Even a simple infection cycle can become more complex. For example, rhinovirus can be spread from person to person by a sneeze (airborne) or through the sharing of inanimate objects (fomites), such as contaminated utensils (fork, pen), towels, cloth handkerchiefs, and doorknobs. Handshaking is also an efficient means of transferring some pathogens. Imagine that one person in a city of 100,000 people has a cold and sneezes on her hands. Then, without washing them, she goes through the day shaking hands with 10 people, and each of those people shakes hands with another 10 people per day, and so on. If there were no repeat handshakes, and if none of the contacts washed their hands, it would take only 4 days to spread the virus throughout the population.

In this example, the entire populace of the city would eventually come in contact with the virus, but not everyone would actually contract disease. Additional factors influence whether the virus successfully replicates in a given individual.

Portals of Entry

How do infectious agents enter the body? Each organism is adapted to enter the body in different ways. Food-borne pathogens (for example, *Salmonella*, *E. coli*, *Shigella*, and rotavirus) are ingested by mouth and ultimately colonize the intestine. They have an oral portal of entry. These gastrointestinal microbes are also described as having a **fecal-oral route of transmission** in which pathogens or parasites excreted in the fecal matter of an infected person are then indirectly ingested by an uninfected person. For example, fecal organisms are often found on hands or surfaces and can then be transferred through touch or contaminated foods to someone else. That person can unknowingly ingest the virus, bacterium, or parasite, completing the cycle.

Airborne organisms, in contrast, infect through the respiratory tract (rhinovirus, *Mycobacterium tuberculosis*). Some microbes enter through the conjunctiva of the eye; others enter through the mucosal surfaces of the genital and urinary tracts.

Agents that are transmitted only by mosquitoes or other insects enter their human hosts via the **parenteral route**, meaning injection into the bloodstream. Wounds and needle punctures can also serve as portals of entry for many microbes. Thus, shared needle use between drug addicts has been an important factor in the spread of HIV.

Immunopathogenesis of Infectious Diseases

Although we focus in this chapter on the mechanisms microbes use to cause disease (toxins, for example), it is often "friendly fire" by our immune system reacting to a pathogen that causes major tissue and organ damage. The immune response to any infection involves activating a complex network of cell types and soluble factors (discussed in Chapters 23 and 24) that may inadvertently damage the host to such a degree that it causes illness and even death. This collateral damage, or "immunopathology," is a calculated risk taken by the host in its haste to eradicate the pathogen. The term **immunopathogenesis** applies when the immune response to a pathogen is a contributing cause of pathology and disease.

The disease dengue hemorrhagic fever is a case in point. Caused by the dengue virus and transmitted by the *Aedes* mosquito, dengue fever manifests as a severe headache, muscle and joint pain, fever, and rash. The symptoms can also include abdominal pain, nausea, and vomiting. However, these symptoms are more a consequence of immunopathogenesis than they are a direct result of viral replication. Replication of the virus in host cells will produce a massive activation of T cells (CD4$^+$ and CD8$^+$). Activation triggers a cytokine cascade (some call it a "storm") that targets vascular endothelial cells and produces an endothelial "sieve" effect leading to fluid and protein leakage. Cytokines such as TNF-alpha, IL-1, and IFN-gamma (discussed in Chapters 23 and 24), along with many others that are released, contribute to inflammation by attracting and activating neutrophils and macrophages. The disease symptoms mentioned earlier can result. So, to fully understand any infectious disease, you must be aware of the pathogenic mechanisms wielded by the pathogen, but also realize that some disease symptoms may be due to immunopathogenesis.

To Summarize

■ **Fulfilling Koch's and Rivers's postulates** is important for identifying the microbial cause of a new disease, but they must be supplemented with molecular or disease-tracking tools if one of the postulates cannot be satisfied.

■ **Infection does not equal disease.** The immunocompetence of the host and the virulence potential of the pathogen influence whether an infection causes disease.

■ **Primary pathogens** have mechanisms that help the organism circumvent host defenses in a healthy host, whereas **opportunistic pathogens** cause disease only in a compromised host.

■ **Pathogenicity** refers to the mechanisms a pathogen uses to produce disease and how efficient the organism is at causing disease, whereas **virulence** is a measure of disease severity.

■ **Diseases can be spread** by direct or indirect contact between infected and uninfected persons/animals or by insect vectors.

■ **Pathogens use portals of entry** best suited to their mechanisms of pathogenesis.

■ **Immunopathogenesis** is damage to host tissues caused by the immune system's response to an infection.

25.2 Virulence Factors and Pathogenicity Islands

To cause disease, all pathogens must enter a host; find their unique niche; avoid, circumvent, or subvert normal host defenses; multiply; and eventually be transmitted to a new susceptible host. Pathogens can be distinguished from their avirulent counterparts by the presence of **virulence factors** that help accomplish these goals. Virulence factors, which are encoded by virulence genes, include toxins, attachment proteins, capsules, and other devices used to avoid host innate and adaptive immune systems.

As you learn more about pathogens, their virulence mechanisms, and the inflammatory responses they trigger,

realize that a pathogen's growth will also affect the host's microbiota. For example, diarrhea can reduce overall numbers of gut microbiota by causing more forceful and frequent evacuations. Intestinal pathogens can change microbiome diversity by occupying limited host binding sites or altering available nutrients. Microbiota may also be killed more effectively than the pathogen by infection-induced inflammation, depending on the pathogen's virulence mechanisms. Aspects of inflammation can even generate nutrients that feed the pathogen but not normal microbiota. All of these mechanisms affect competition between microbial species and, ultimately, species diversity. Finally, as victims recover from a disease, their gut microbiota may not always achieve preinfection balance. The resulting dysbiosis, as discussed in Chapters 23 and 24, can have negative health effects beyond those of the infection itself.

Note: Because intestinal infections can drastically alter the gut microbiome, this chapter uses the gut microbiome icon to highlight the virulence mechanisms of intestinal pathogens.

So, virulence factors are important to pathogens trying to outcompete established microbiota for a host niche. But how do we measure virulence or identify a virulence gene?

Molecular Koch's Postulates

Virulence is a measurable phenotype (see Section 25.1), but how do we identify genes that encode virulence factors? Identifying genes for a metabolic or biosynthetic pathway is relatively easy because it requires merely observing a clear phenotype on an agar plate. If a gene involved with an amino acid biosynthetic pathway is defective, then the amino acid is not made and must be added to the medium, or else the mutant will not grow. Finding a virulence gene is more difficult because the screen involves growth in a host. There is no in vitro way to identify a virulence phenotype using an agar plate. Virulence mutants can be recognized only if they fail to survive or cause disease in animals.

Numerous clever techniques have been developed over past decades to identify potential virulence genes. Examples are discussed in **eTopics 25.1** and **25.2**. Regardless of how the suspected virulence gene was found, it can be confirmed as having a role in virulence or pathogenicity only if it fulfills a set of "molecular Koch's postulates" originally formulated by Stanley Falkow, a preeminent infectious disease scientist. The molecular postulates are as follows:

1. The phenotype under study should be associated with pathogenic strains of a species.
2. Specific inactivation of the suspected virulence gene(s) should lead to a measurable loss in virulence or pathogenicity. The gene(s) should be isolated by molecular methods.

3. Reversion or replacement of the mutated gene should restore pathogenicity.

A variety of other molecular questions can suggest the role of a virulence protein in pathogenesis. For example:

■ Does moving the virulence gene into an avirulent strain impart a pathogenicity trait on the avirulent strain? For instance, does moving a gene for attachment from a pathogen to a nonpathogen allow the nonpathogen to attach to host cells?

■ Does the suspected virulence protein bind to important host proteins? Binding to a host protein could indicate the target of the microbial protein.

■ Does the microbial protein resemble the sequence or structure of an important host protein? Such resemblance might indicate that the microbial protein mimics the function of the orthologous host protein.

■ Does introducing the suspected virulence protein (or its encoding gene) into a host cell alter the host cell's physiology, or does a GFP-tagged version of the protein localize to a specific host cell compartment or organelle? A positive finding in either case could reveal a target for the suspected virulence protein.

In addition to the experimental approaches just described, there are bioinformatic ways to identify potential pathogenicity genes.

Pathogenicity Islands

Extensive sequencing efforts have allowed us to compare genomes of many pathogens and expose some "footprints" of their evolution. For example, in bacterial pathogens, most chromosomes are dotted with clusters of pathogenicity genes that encode virulence functions. These gene clusters, called **pathogenicity islands**, can be considered the toolboxes of pathogens (originally discussed in Section 9.6). Many virulence genes reside in pathogenicity islands, although many others do not. Some virulence genes reside on plasmids (for example, the genes for the diarrhea-producing labile toxin of certain *E. coli* strains) or in phage genomes (such as the genes encoding the diphtheria toxin of *Corynebacterium diphtheriae*).

Most pathogenicity islands appear to have been horizontally transmitted (via conjugation or transduction; discussed in Section 9.1) from organisms long ago extinct into the ancestors of today's pathogens. Horizontal gene transfers move whole blocks of DNA (more than 10 kb) from one organism to another, placing the blocks directly in the chromosome in what is called a **genomic island** (see Chapter 9 and **eTopic 25.3**). If the island increases the "fitness" (virulence) of a microorganism (pathogen) that interacts with a host, it is called a pathogenicity island. Genomic

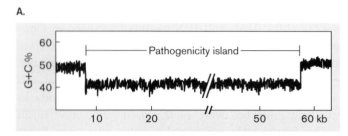

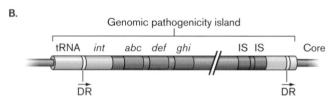

FIGURE 25.8 ■ Model pathogenicity island. A. The guanine + cytosine (GC) content of the island is different from that of the core genome. **B.** Schematic model of a pathogenicity island. The DNA block is linked to a tRNA gene and flanked by direct repeats (DRs) that may be "footprints" of a transposon or viral-mediated transfer. The integrase gene (*int*) and insertion sequences (ISs) may also be remnants of transposition.

islands generally reveal themselves by several anomalies that they possess with respect to the rest of the host genome:

■ Genomic islands are often linked to a tRNA gene and generally have a GC/AT ratio very different from that of the rest of the chromosome (**Fig. 25.8A**). For example, a plot of GC content along the length of a chromosome may reveal that most of the genome has a 50% GC content. But somewhere in the middle, a 50-kb region sticks out on the graph, showing a content of 40%. This deviation probably reflects the GC content of the microbe that donated the island. The reason tRNA genes are often the targets for insertion of pathogenicity islands is not known. One hypothesis is that the conserved secondary structure of tRNA genes facilitates integration by an integrase.

■ Genomic islands are typically flanked by genes with homology to phage or plasmid genes (**Fig. 25.8B**). This arrangement is thought to reflect the transfer vector used to move the island from one organism to another.

But what do the pathogenicity genes do? Some genes encode molecular "grappling hooks," such as pili that attach to host cells. Once attached, microbes can secrete toxins that injure the host cell. Other bacteria wall themselves off to prevent damage by host inflammatory responses. Some bacterial pathogens are even capable of what could be called "host cell reprogramming." These organisms inject proteins directly into the host cell to disrupt normal signaling pathways. This reprogrammed target cell can be made to do one of several things: engulf the bacterium; "commit suicide" (undergo apoptosis); engineer a tighter, more intimate attachment platform at the cell surface; or alter the amount or types of cytokines that the affected cell produces.

TABLE 25.1	Examples of Pathogenicity Islands		
Pathogenicity island	**Function**	**Organism**	**Disease**
HPI (high-pathogenicity island)	Iron uptake	*Yersinia* spp.	Plague, enterocolitis
VPI (*Vibrio* pathogenicity island)	Toxin production	*Vibrio cholerae*	Cholera
PAI III (pathogenicity island III)	Encoding adhesins	Uropathogenic *E. coli*	Urinary tract infection
SPI-1 and SPI-2 (*Salmonella* pathogenicity islands)	Type III secretion	*Salmonella enterica*	Gastroenteritis
SHI-1 and SHI-2 (*Shigella* islands)	Type III secretion	*Shigella flexneri*	Bloody diarrhea
YSA (*Yersinia* secretion apparatus)	Type III secretion	*Yersinia* spp.	Plague, enterocolitis
cag PI (cytotoxin-associated gene pathogenicity island)	Type IV secretion	*Helicobacter pylori*	Gastric ulcers, gastric cancer
icm/dot (intracellular multiplication)	Type IV secretion	*Legionella pneumophila*	Legionnaires' disease

Table 25.1 lists several pathogenicity islands and their functions. It provides examples of pathogenicity islands present in different bacteria, names a key function of the products of the island, and identifies the disease caused. Various pathogenicity island functions will be described as the chapter proceeds.

Caught in the Act: Examples of Pathogen Evolution by Horizontal Gene Transfer

Escherichia coli as a species includes strains that are part of normal gut microbiota but has also evolved many pathovars through horizontal gene transfers. These pathovars cause diseases ranging from urinary tract infections, diarrhea, sepsis, and meningitis. A recent example of *E. coli* evolving via horizontal gene transfer involves the enteroaggregative hemorrhagic strain O104:H4, which caused the frightening 2011 outbreak of diarrhea and hemolytic uremic syndrome that began in Germany and spread through much of Europe. The lethality associated with the production of Shiga toxin by this strain and its resistance against many antibiotics was incredible. The presence of several new virulence traits, including antibiotic resistance, in this strain that are absent in its closest relative suggested the involvement of horizontal gene transfer in its evolution.

Streptococcus pyogenes. Another dangerous pathogen caught in the act of evolving is *S. pyogenes*, otherwise known as group A *Streptococcus* (GAS), a strict human pathogen that can cause sore throat, scarlet fever, and necrotizing fasciitis (the so-called "flesh-eating disease"; see Section 26.2). For a century, GAS strains have been sorted into serological types by amino acid differences present in a cell-surface molecule called M protein. James Musser from the Houston Methodist Research Institute and his colleagues recently determined through large-scale genome sequencing that epidemics of streptococcal disease are caused by clonal replacement events (horizontal gene transfers) rather than by reemergence of preexisting clones. For instance, the origin of the most recent GAS M1 global pandemic strain was traced to around 1983, when a horizontal gene transfer event exchanged a 36-kb chromosomal region encoding several proteins. Two of those proteins, NAD glycohydrolase (SPN; the gene is *nga*) and streptolysin O (SLO; gene *slo*), are potent toxins (**Fig. 25.9A**).

Compared to preepidemic strains, all of the epidemic strains contained the same three single nucleotide polymorphisms (SNPs) within the 36-kb region. Two SNPs were located in the upstream promoter region of the SPN operon. These two promoter mutations increased expression of the SPN and SLO proteins (**Fig. 25.9B**) and dramatically increased virulence. The third SNP was a missense mutation in SPN that changed a glycine to aspartate and restored NAD glycohydrolase enzymatic activity. The restored activity of NAD glycohydrolase was also required to maximize virulence of the epidemic GAS strain. Up-regulating SPN and SLO in the pandemic strain enhanced tissue destruction, heightened resistance to killing by polymorphonuclear leukocytes, and helped propagate GAS infections around the world.

Evolutionary features of two other pathogens—*Streptococcus agalactiae* and *Staphylococcus aureus*—are described in **eTopic 25.4**.

Sections 25.3–25.6 describe some of the specific tools that pathogens use to undermine the integrity of the body. From attachment to toxins to intracellular invasion, the infection process is like a chess match, with each side, human and microbe, trying to outmaneuver the other.

To Summarize

■ **Fulfilling the molecular Koch's postulates** validates the identity of a virulence gene.

■ **Pathogenicity islands** are DNA sequences within a species that are acquired by horizontal gene transfer from a different species.

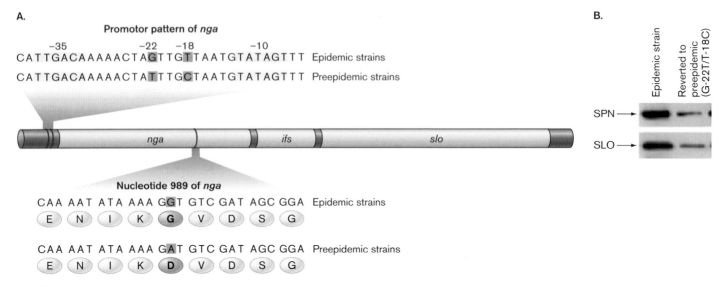

FIGURE 25.9 ■ Emergence of a pandemic clone of group A streptococci. **A.** Sequence comparison in the *nga-slo* operons of pre-epidemic and epidemic strains of *Streptococcus pyogenes* revealed three consistent single nucleotide polymorphisms (SNPs). The *nga* and *slo* genes encode NAD glycohydrolase (SPN) and streptolysin O, respectively. The *ifs* gene encodes a regulator of SPN. **B.** Western blot of secreted proteins, showing that changing the SNPs in the promoter of an epidemic strain back to those found in a preepidemic strain reduced secretion of SPN and SLO. *Source for B:* Luchang Zhu, et al. **125(9)**: 3545–3559. ©2015, American Society for Clinical Investigation.

- **Virulence genes** encode products that enhance the disease-causing ability of the organism. Many virulence genes can be found within pathogenicity islands, but some are located outside of an obvious genomic island or reside in plasmids.

- **Pathogenicity islands** contain distinct features, such as GC content and the remnants of phages or plasmids, that mark them as being different from the rest of the genome.

- **Horizontal gene transfer mechanisms** move virulence factor genes and pathogenicity islands among bacterial strains and species.

25.3 Microbial Attachment: First Contact

Regardless of the disease, pathogens must reach a colonization site either through their own motility or by hitchhiking with a vector. Once at the site, the pathogen needs attachment mechanisms to stay there.

The human body has many ways to exclude pathogens. The lungs use a mucociliary escalator (see Fig. 23.5) to rid themselves of foreign bodies, the intestine uses peristaltic action to ensure that its contents are constantly flowing, and the bladder uses contraction to propel urine through the urethra with tremendous force. How do bacteria ever manage to stick around long enough to cause disease? Like a person grasping a telephone pole during a hurricane,

successful pathogens moving through the body manage to grab on to host cells and tenaciously hold on. Thus, the first step toward infection is attachment, also called adhesion. Any microbial factor that promotes attachment is called an **adhesin**.

Viruses attach to the host through their capsid or envelope proteins, which bind to the specific host cell receptors discussed in Chapters 6 and 11. Bacteria have a variety of similar strategies. They can use hairlike appendages called **pili** (also called **fimbriae**), whose tips contain receptors for mammalian cell-surface structures. Or they can use a variety of adherence proteins or other molecules that are not part of a pilus. Sometimes they use both. **Table 25.2** summarizes bacterial attachment strategies. Note that different pili can impart tissue specificity for attachment (for example, uropathogenic versus diarrheagenic *E. coli*).

Pili and Pilus Assembly

Pili on different bacterial species are now classified by protein sequence homology rather than by their binding site—an old classification method. In this chapter, we consider two groups, called type I and type IV pili. Note that pili and other adhesins can be virulence factors on some organisms but nonpathogenic structures on others.

How bacteria assemble pili on their cell surfaces is an engineering marvel. The shafts of pili are cylindrical structures composed of identical pilin protein subunits. Several different

TABLE 25.2	Specific Attachments of Bacteria to Cell or Tissue Surfaces			
Bacterium	**Adhesin**	**Host receptor**	**Attachment site**	**Disease**
Streptococcus pyogenes	Protein F	Amino terminus of fibronectin	Pharyngeal epithelium	Sore throat
Streptococcus mutans	Glucan	Salivary glycoprotein	Pellicle of tooth	Dental caries
Streptococcus salivarius	Lipoteichoic acid (?)	Unknown	Buccal epithelium of tongue	None
Streptococcus pneumoniae	Cell-bound protein	N-acetylhexosamine galactose disaccharide	Mucosal epithelium	Pneumonia
Staphylococcus aureus	Clumping factors A and B	Fibronectin	Mucosal epithelium	Various
Neisseria gonorrhoeae	N-methylphenylalanine pili	Glucosamine galactose carbohydrate	Urethral/cervical epithelium	Gonorrhea
Enterotoxigenic *E. coli*	Type I fimbriae (pili)	Species-specific carbohydrate(s)	Intestinal epithelium	Diarrhea
Uropathogenic *E. coli*	Type I fimbriae (pili)	Complex carbohydrate	Urethral epithelium	Urethritis
	P pili (pyelonephritis-associated pili)	P blood group	Upper urinary tract	Pyelonephritis
Bordetella pertussis	Pili ("filamentous hemagglutinin")	Galactose on sulfated glycolipids	Respiratory epithelium	Whooping cough
Vibrio cholerae	N-methylphenylalanine pili	Fucose and mannose carbohydrate	Intestinal epithelium	Cholera
Treponema pallidum	Peptide in outer membrane	Surface protein (fibronectin)	Mucosal epithelium	Syphilis
Mycoplasma	Membrane protein	Sialic acid	Respiratory epithelium	Pneumonia
Chlamydia	Lipooligosaccharide, OmcB surface protein	Sulfonated glycosaminoglycans	Conjunctival or urethral epithelium	Conjunctivitis or urethritis
Corynebacterium diphtheriae	Pili	Unknown	Pharyngeal epithelium	Diphtheria

proteins adorn the tip, including one at the very apex, FimH, that binds to host receptors (**Fig. 25.10A**). How the tip protein manages to "hold on" in the face of tremendous shear forces (such as those exerted during urination) is described in **eTopic 25.5**. In addition to the structural components of pili, numerous other proteins collaborate to assemble the structure. Genes encoding a given pilin protein and the associated assembly apparatus are typically arranged on the chromosome as an operon (**Fig. 25.10B**).

A. FimH adhesin tip

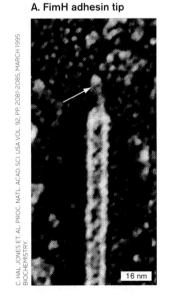

C. HAL JONES ET. AL. PROC. NATL. ACAD. SCI USA VOL. 92, PP. 2081-2085, MARCH 1995 BIOCHEMISTRY.

16 nm

B. Type I pilus gene cluster

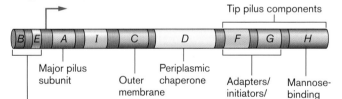

Regulation

Major pilus subunit

Outer membrane usher

Periplasmic chaperone

Tip pilus components

Adapters/initiators/terminators

Mannose-binding adhesin

FIGURE 25.10 ■ Attachment pili and encoding operon. **A.** High-resolution micrograph showing a type I pilus (TEM). The FimH adhesin at the tip (arrow) is the protein that binds to the cell receptor. **B.** Genetic organization of the type I gene cluster, which includes genes involved in pilus assembly. The genes are designated *fimA–I*.

Assembly of type I pili. Pyelonephritis-associated pili (Pap) of the uropathogenic *E. coli* are type I pili that bind to a digalactoside present on host urinary tract surfaces called the P-blood-group antigen. "Pyelonephritis" is the term for kidney infection. Pap pili are essential for uropathogenic *E. coli* to cause this disease. **Figure 25.11** illustrates

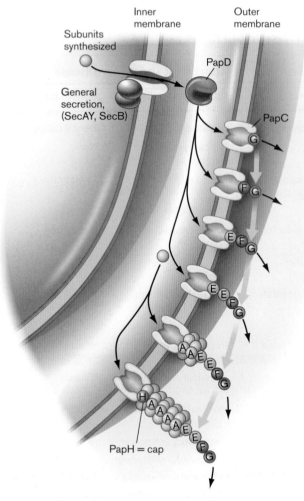

1. Proteins from the cytoplasm are secreted by the Sec system to the periplasm, where they are chaperoned by PapD to the site of assembly.

2. PapC, also called the usher, assembles the individual proteins in the proper order. Assembly starts with the tip protein, PapG, marked at the far right, which ultimately binds to carbohydrates on host membranes.

FIGURE 25.11 ▪ Assembly of type I pili. The pyelonephritis-associated pilus (Pap) assembly illustrated is representative of other type I pili, such as in Fim in Figure 25.10. Only the names of the proteins will differ. The subunits fit together like pieces of a jigsaw puzzle. The arrow at the head of the elongating pilus indicates the direction of pilus growth.

how the type I pilus from uropathogenic *E. coli* (Pap) is assembled. The mechanism is representative of other type I pili; only the names of the proteins will differ for each system. Protein components synthesized in the cytoplasm are secreted into the periplasm by the SecA-dependent general secretory system (discussed in Section 8.5). Once in the periplasm, the subunits are chaperoned by PapD to the membrane site of assembly, which is marked by the presence of the usher protein PapC. PapC proteins form channels in the outer membrane large enough to accommodate individual pilus subunits and, like an usher in a theater, direct the subunits to their proper places.

Chaperoning of the pilin building blocks by PapD is necessary to prevent pilin subunits from inadvertently assembling in the periplasm. As illustrated in **Figure 25.11**, appropriate assembly of pili at the usher site starts with the tip protein, PapG, which will ultimately bind to

carbohydrates on host membranes after the pilus is complete. After PapG, the ushers add PapF and PapE, forcing PapG farther away from the surface. Then, identical PapA pilin subunits are strung together in a series to form the shaft. PapA subunits assemble by sequentially sharing a domain from each other, linking together like pieces of a jigsaw puzzle. Once assembled, type I pili are static. They simply stick to the host receptor. Type IV pili are more dynamic, continually extending and contracting.

Assembly of type IV pili. Other pili with an important role in pathogenesis are the type IV pili. Type IV pili are found in a broad spectrum of Gram-negative bacteria and share amino acid homology in their major pilin structure (**Fig. 25.12**). All type IV pili use similar secretion and assembly machinery involving least a dozen proteins. One major difference between the assembly of type IV and type I pili is that type IV pilus proteins are never free in the periplasm; instead, they are transported directly from the cytoplasm through a channel in the outer membrane (**Fig. 25.12A**). Thus, type IV pilus assembly is SecA independent. Species with type IV pili include *Vibrio cholerae*, *Pseudomonas aeruginosa* (**Fig. 25.12B**), certain pathogenic strains of *E. coli*, *Neisseria meningitidis* (**Fig. 25.12C**), and *Neisseria gonorrhoeae*. Over the course of evolution, the genes for type IV pili were duplicated and modified into a protein secretion mechanism designated type II secretion. Despite having evolved from genes encoding type IV pili, type II secretion systems export virulence proteins unrelated to pili (discussed in Section 25.5).

Type IV pilus assembly can actually make cells move by a process called twitching motility. The assembly process involves the reiterative elongation and retraction of the pili. The pilus elongates, attaches to a surface, and then depolymerizes from the base, which shortens the pilus and pulls the cell forward. This mechanism is akin to using a grappling hook to scale a building. Similarly, the slime mold *Myxococcus xanthus* uses type IV pili to mediate gliding motility. The type IV pili of *Neisseria meningitidis*, shown in **Figure 25.12C**, are essential for crossing the blood-brain barrier and causing bacterial meningitis.

The retraction of type IV pili, at least in diarrhea-causing strains of *E. coli*, helps disrupt the tight junctions connecting adjacent cells that line the intestine. Tight junctions normally form a permeability barrier between the intestinal lumen and the intestine, through which nutrients, ions, and water are absorbed. Disrupting tight junctions prevents the absorption of water and electrolytes, which accumulate in the intestine and contribute to the diarrhea.

Nonpilus Adhesins

Bacteria also carry adhesin proteins that bind host tissues but do not form pili (**Fig. 25.13**). Some examples include

Bordetella pertactin (which binds to host cell integrin), *Streptococcus* protein F (which binds to fibronectin), *Streptococcus* M protein (which binds to fibronectin and complement regulatory factor H), and intimin of enteropathogenic *E. coli* (which binds to Tir; see Section 25.5). Many Grampositive bacteria have surface-exposed proteins resembling pili that contain serine-rich repeats able to bind sialic acid or keratin.

Fimbriae (pili) often mediate the initial binding between bacterium and host, after which a more intimate attachment is formed by an afimbriate attachment protein. In the case of *Neisseria gonorrhoeae*, once the type IV pilus has attached to the surface of the mucosal epithelial cell, the filamentous pilus contracts, pulling the bacterium down onto the host cell membrane. Tight secondary interactions are then mediated by the neisserial Opa membrane proteins, another example of an afimbriate adhesin (Opa gets its name from the opacity it adds to colony appearance).

MAM7, a promiscuous adhesin. Although many different bacterial adhesins are known, most are species specific, and many are not produced until after an infection starts, making these molecules less suitable for initial binding to host cells. Could there be a single microbial adhesion molecule used as a nonspecific "first binder" by multiple pathogens? Kim Orth and colleagues at the University of Texas Southwestern Medical Center have described one such adherence factor, called multivalent adhesion

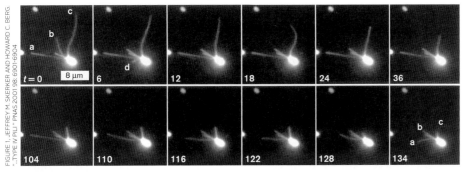

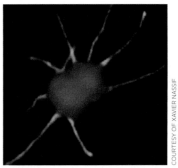

FIGURE 25.12 ■ **Type IV pili.** **A.** Model of pilus assembly and disassembly. In this example, PilA is the pilin protein, and PilC1 and Y1 form the attachment tip. Filament approx. 6 nm in diameter. Assembly and disassembly require the hydrolysis of nucleoside triphosphate (NTP) and take place at the inner membrane, not in the periplasm. **B.** Photographic evidence of type IV pilus extension and retraction in cells of *Pseudomonas aeruginosa*. Filament b retracts; then filament d extends at 6 seconds and retracts. Filament c attaches briefly at its distal tip (note straightening at 24 seconds) and then begins to retract. Fluorescent microscopy. *t* = time in seconds. **C.** Type IV pili are essential for the interaction of *Neisseria meningitidis* with brain endothelial cells. Type IV pili are green in this SEM. Diplococcal cells approx. 1.6 μm in diameter.

Source: Part A modified from Bardy et al. 2003. *Microbiology* **149**:295–304.

A. M protein

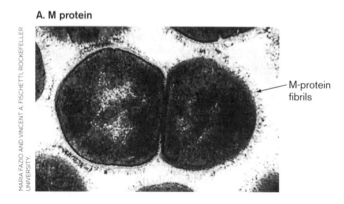

B. *B. pertussis* colonizing the trachea

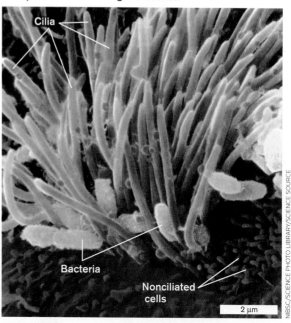

FIGURE 25.13 ▪ Nonpilus adhesins. A. M-protein surface fibrils on *Streptococcus pyogenes* (TEM). Cells 0.5–1 µm in diameter. **B.** Colonization of tracheal epithelial cells by *Bordetella pertussis* (colorized SEM). This organism uses a surface protein called pertactin, as well as a pilus called filamentous hemagglutinin (FHA), to bind bronchial cells.

molecule 7 (MAM7). It is used by a variety of Gram-negative bacterial pathogens but not found in Gram-positives. MAM7 is an outer membrane protein containing seven so-called mammalian cell entry domains (initially described for *Mycobacterium tuberculosis*).

The mammalian cell entry domains of MAM7 interact with host membrane phosphatidic acids (PAs) and host cell fibronectin. MAM7 binding to phosphatidic acid causes PA clustering, a process that activates a host GTPase called RhoA. Activation of RhoA disrupts cell-cell tight junctions and compromises the intestinal membrane barrier. As a result, the microbe has easier access to deeper tissues. Because of its role in pathogenesis, MAM7 could be an attractive general target for antimicrobials.

Host susceptibility. Why are some people susceptible to certain infections while others are not? Part of the reason is immunocompetence, but another is receptor availability. Pathogens rely on key host surface structures such as gangliosides to recognize and attach to the correct host cell by the mechanisms just described. But a host species can evolve to become resistant to infection when the gene encoding the receptor (or receptor synthesis) mutates. The mutation could lead to complete loss of the protein or a change in its shape to prevent recognition or alter its function. An example is the T-cell surface protein CCR5, which acts as a receptor for HIV (see Section 11.3). Individuals with a genetic defect that eliminates CCR5 are resistant to HIV infection, so even without methods for preventing and curing HIV infection, humans would eventually evolve a level of resistance to HIV. Differences in attachment receptors also explain, at least in part, why some

pathogens have broad host specificity while others more narrowly target their host. It can also explain aspects of tissue specificity.

Biofilms and Infections

As first discussed in Section 4.5, bacteria in most environments form organized, high-density communities of cells called biofilms that are embedded in self-produced exopolymer matrices. Biofilm development is an ancient prokaryotic adaptation that allows microorganisms to adhere to any surface, living or nonliving, and facilitates survival in hostile environments. Within a single biofilm you can find localized differences in the expression of surface molecules, antibiotic resistance, nutrient utilization, and virulence factors. Bacteria in biofilms also coordinate their behavior through cell-cell communication using secreted chemical signals.

Biofilms are important features in chronic infections found on oral, lung, and urogenital (bladder) tissues. *Pseudomonas aeruginosa* causes a life-threatening, chronic lung infection in individuals with cystic fibrosis (CF). This microbe has been found growing as aggregates enclosed in a matrix within mucus from CF patients. It is thought that insufficient mucociliary clearance contributes to *P. aeruginosa* biofilm formation. Biofilms are also important in periodontitis (gum disease), indwelling catheter infections, infections of artificial heart valves, chronic urinary tract infections, recurrent tonsillitis, rhinosinusitis, chronic otitis media (middle ear infection), chronic wound infections, and osteomyelitis (bone infection). **Figure 25.14A** shows a scanning EM of a chronic infection of tonsil tissue from a pediatric patient. The confocal fluorescent image in

A.

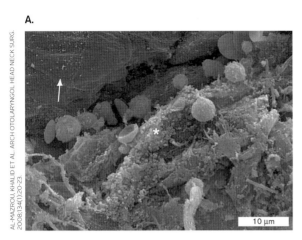

B.

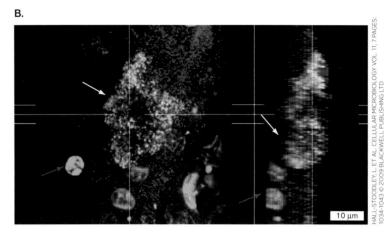

FIGURE 25.14 ■ **A bacterial biofilm infection.** **A.** Scanning EM showing mixed bacteria (asterisk) and adherent biofilm on the surface epithelium (arrow) of an infected tonsil. **B.** Confocal micrograph (described in Chapter 2) showing biofilm clusters (white arrows) consisting of rods and cocci on the mucosa of a pediatric adenoid. Removal of the adenoid is a routine treatment for recurrent otitis media (middle ear infection). Specimens were treated with nucleic acid stains using the LIVE/DEAD *Bac*Light Bacterial Viability Kit, in which live bacteria stain green and dead bacteria stain red. Host inflammatory cells (red arrows) were also stained green, but their nuclei appear much larger than the bacteria. The mucosal surface (blue) was imaged using reflected light.

Figure 25.14B shows a 3D view of a biofilm on adenoid tissue with live cells stained green and dead cells stained red.

The chronic presence of these organisms in biofilms will continually stimulate innate immune mechanisms through interactions with Toll-like receptors and cause chronic inflammation as a result. A biofilm infection may linger for months, years, or even a lifetime. If the infection is associated with an implanted medical device (such as an artificial knee), the device may have to be replaced. Today, vascular catheter-related bloodstream infections are the most serious and costly health care–associated infections.

Biofilm infections are very important clinically because bacteria in biofilms exhibit tolerance to antimicrobial compounds and persistence in spite of sustained host defenses. Thus, biofilm infections are hard to cure. Tolerance to antibiotics may be caused by poor nutrient penetration through the exopolymer matrix into the deeper regions of the biofilm, leading to a stationary phase–like dormancy. Bacterial factors important to biofilm formation include type IV pili, structural genes and regulators controlling cell-cell signaling (quorum sensing), and extracellular matrix synthesis. Interfering with cell-cell signaling is effective in preventing or limiting biofilm formation and may provide a target for new antimicrobial therapies.

To Summarize

- **Bacteria use pili and nonpilus adhesins** to attach to host cells.

- **Type I pili** produce a static attachment to the host cell, whereas **type IV pili** continually assemble and disassemble. Pili assemble starting at the tip.

- **Nonpilus adhesins** are bacterial surface proteins, or other molecules, that can tighten interactions between bacteria and target cells.

- **Biofilms** play an important role in chronic infections by enabling persistent adherence and resistance to bacterial host defenses and antimicrobial agents.

25.4 Toxins Subvert Host Function

Following attachment, many microbes secrete protein toxins (called **exotoxins**) that kill host cells and unlock their nutrients (because dead host cells ultimately lyse). Bacterial pathogens have developed an impressive array of toxins that take advantage of different key host proteins or structures. All Gram-negative bacteria also possess a toxic compound called **endotoxin** (a part of lipopolysaccharide, LPS) that can hyperactivate host immune systems to harmful levels. Do not confuse <u>en</u>dotoxin with protein <u>exo</u>toxins.

Categories of Microbial Exotoxins

Microbial exotoxins fall into nine categories based on their mechanisms of action (**Table 25.3**). These classes are summarized here, and several are illustrated in **Figure 25.15**.

- **Plasma membrane disruption.** Members of the first class, exemplified by the alpha (α) toxin of *Staphylococcus*

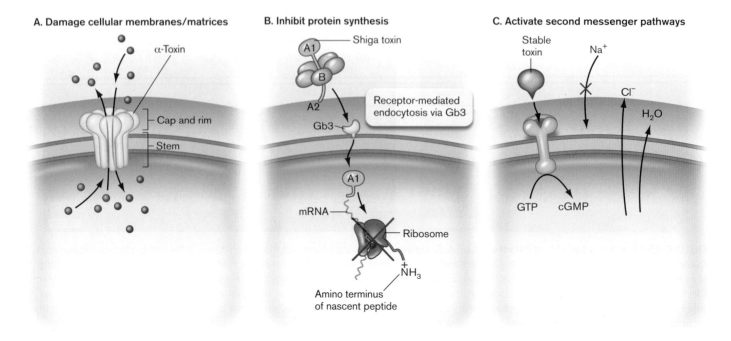

FIGURE 25.15 ▪ **Three classes of microbial exotoxins.** These classes are defined by mode of action. **A.** Pore-forming toxins assemble in target membranes and cause leakage of compounds into and out of cells. **B.** Shiga toxin attaches to ganglioside Gb3, enters the cell, and cleaves 28S rRNA in eukaryotic ribosomes to stop translation. **C.** Enterotoxigenic *E. coli* heat-stable toxin affects cGMP production. The result is altered electrolyte transport—inhibition of Na⁺ uptake and stimulation of Cl⁻ transport. In response to the resulting electrolyte imbalance, water leaves the cell.

aureus, form pores in host cell membranes and cause leakage of cell constituents (**Fig. 25.15A**).

▪ **Cytoskeleton alterations.** A second class of toxins can stimulate either actin polymerization or actin depolymerization. For example, *Vibrio cholerae* RTX toxin depolymerizes actin by cross-linking actin fibers. Modifying actin polymerization can alter cell shape or cause cell membranes to wrap around the pathogen, inviting the organism into the cytoplasm.

▪ **Protein synthesis disruption.** A third class, exemplified by diphtheria and Shiga toxins, targets eukaryotic ribosomes and destroys protein synthesis (**Fig. 25.15B**).

▪ **Cell cycle disruption.** These toxins either stop (*E. coli* cytolethal distending toxin, or CLDT) or stimulate (*Pasteurella multocida* toxin) host cell division.

▪ **Signal transduction disruption.** The fifth type of toxin subverts host cell second messenger pathways. Cholera toxin and *E. coli* ST (stable toxin), for instance, cause runaway synthesis of cyclic adenosine monophosphate (cAMP; described later) and cyclic guanosine monophosphate (cGMP) (**Fig. 25.15C**), respectively, in target cells. Elevated cAMP or cGMP levels, in turn, trigger critical changes in ion transport and fluid movement.

▪ **Cell-cell adherence.** These exotoxins are proteases that cleave proteins binding host cells to one another. One

such toxin, exfoliative toxin of *Staphylococcus aureus*, breaks the connection between dermis and epidermis, giving victims the gruesome appearance of scalded skin.

▪ **Vesicle traffic.** The major toxin in this class (VacA of *Helicobacter pylori*) actually has several modes of action, depending on the host cell. The most visually striking effect is its ability to cause vacuolization, which is the fusion of numerous intracellular vesicles.

▪ **Exocytosis.** The eighth class of exotoxin includes two protease toxins that alter the movement of nerve cell cytoplasmic vesicles to membranes where they release neurotransmitters. One example is tetanus toxin, which cleaves host proteins required for exocytosis of the inhibitory neurotransmitter gamma-aminobutyric acid (GABA).

▪ **Superantigens.** Members of the ninth and last class of toxins activate the immune system without being processed by antigen-presenting cells (discussed in Section 24.3). The pyrogenic (fever-producing) toxins of *Staphylococcus aureus* (such as toxic shock syndrome toxin, TSST) and *Streptococcus pyogenes* are examples of superantigen toxins.

Mechanisms of selected exotoxins are described in the following sections. We focus on toxins that disrupt membranes and a set of exotoxins collectively called AB subunit exotoxins that target either protein synthesis or signal transduction. Superantigens were described in Section 24.3, and exotoxins that affect exocytosis (tetanus and botulism toxins) will be discussed in Section 26.7.

Membrane Disruption

Toxins that disrupt membranes include pore-forming proteins that bind cholesterol and insert themselves into target membranes, and phospholipases that hydrolyze membrane phospholipids into fatty acids. General descriptive terms for these toxins are **hemolysins**, which lyse red blood cells (but other cells as well), and **leukocidins**, which more specifically lyse white blood cells. One example is streptolysin O produced by *Streptococcus pyogenes*. This pathogen can cause pharyngitis (sore throat) and necrotizing fasciitis, also known as "flesh-eating" disease.

Pore-forming toxins. A classic example of a pore-forming exotoxin is the hemolytic alpha toxin produced by *Staphylococcus aureus*, an organism that causes boils and blood infections. Alpha toxin forms a transmembrane, oligomeric (seven-member) beta barrel pore in target cell plasma membranes. It is easy to see how the resulting leakage of cell constituents and influx of fluid cause the target cell to burst. To form the pore, hydrophobic areas of each monomer face the lipids of the membrane, and hydrophilic residues face the channel interior. A completed pore and a cutaway view exposing the channel are illustrated in **Figure 25.16A** and **B**. Diagnostic microbiology laboratories visualize hemolysins such as alpha toxin by inoculating bacteria onto agar plates containing sheep red blood cells (**Fig. 25.16C**). The clear, yellow zones around the *S. aureus* colonies growing on blood agar indicate that the microbe secretes a hemolysin. A particularly potent pore-forming leukocidin is Panton-Valentine toxin, produced by most strains of methicillin-resistant *Staphylococcus aureus*.

Phospholipase toxins. Phospholipase C of *Clostridium perfringens*, a cause of gas gangrene, cleaves phosphatidylcholine in host plasma membranes. At high concentrations the toxin causes membrane disruption, but at sublethal concentrations the exotoxin will generate signaling molecules from membrane lipids that can activate cytokine production. Some phospholipases, also called lecithinases, can increase the permeability of capillaries to cause edema (fluid accumulation in tissues).

Escaping vacuoles. Other membrane-disrupting exotoxins specifically target vacuolar membranes. Many bacterial pathogens enter eukaryotic host cells by inducing phagocytosis and then end up in a phagosome vacuole. Some of these intracellular pathogens need to break out of the phagosome to grow in the cytoplasm. One example is *Listeria monocytogenes*, a gastrointestinal pathogen that can also cause meningitis (described later). This Gram-positive rod uses listeriolysin (a pore-forming exotoxin) and two phospholipases to escape the phagosome to grow in the cytoplasm.

Two-Subunit AB Exotoxins

As noted earlier, exotoxins target a variety of different host mechanisms. But despite the diversity of their targets, many exotoxins share a common structural theme; they have two subunits, usually called A and B. These two-subunit complexes are called AB exotoxins. The actual toxic activity in AB exotoxins resides within the A subunit. The role of the B subunit is to bind host cell receptors. Thus, the B subunit for

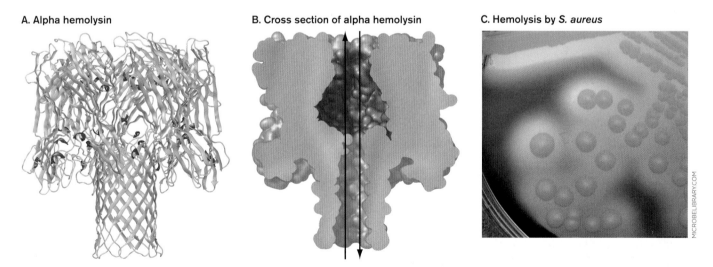

A. Alpha hemolysin

B. Cross section of alpha hemolysin

C. Hemolysis by *S. aureus*

MICROBELIBRARY.COM

FIGURE 25.16 ■ Alpha hemolysin of *Staphylococcus aureus*. **A.** 3D image of the pore complex, comprising seven monomeric proteins. (PDB code: 7AHL) **B.** Cross section showing the channel. Arrows indicate movement of fluids through the pore. **C.** A blood agar plate inoculated with *S. aureus*. The alpha toxin is secreted by the organism and diffuses away from the producing colony. It forms pores in the red blood cells embedded in the agar, causing the cells to lyse—a process that is visible as a clear area surrounding each colony.

TABLE 25.3	Characteristics of Bacterial Exotoxins[a]			
Toxin	**Organism**	**Mode of action**	**Host target**	**Disease**	**Toxin implicated in disease[b]**
Damage membranes					
Aerolysin	*Aeromonas hydrophila*	Pore former	Glycophorin	Diarrhea	(Yes)
Perfringolysin O	*Clostridium perfringens*	Pore former	Cholesterol	Gas gangrene[c]	(Yes)
Hemolysin[d]	*Escherichia coli*	Pore former	Plasma membrane	UTIs	(Yes)
Listeriolysin O	*Listeria monocytogenes*	Pore former	Cholesterol	Food-borne systemic illness, meningitis	Yes
Alpha toxin	*Staphylococcus aureus*	Pore former	Plasma membrane	Abscesses[c]	(Yes)
Panton-Valentine leukocidin	*Staphylococcus aureus*	Pore former	Plasma membrane	Abscesses, necrotizing pneumonia	(Yes)
Pneumolysin	*Streptococcus pneumoniae*	Pore former	Cholesterol	Pneumonia[c]	(Yes)
Streptolysin O	*Streptococcus pyogenes*	Pore former	Cholesterol	Strep throat, scarlet fever	Unknown
Disrupt cytoskeletons					
Vibrio cholerae RTX	*Vibrio cholerae*	Actin depolymerization	Actin and Rho GTPase	Cholera	Unknown
C2 toxin	*Clostridium botulinum*	ADP-ribosyltransferase	Monomeric G-actin	Botulism	Unknown
Iota toxin	*Clostridium perfringens*	ADP-ribosyltransferase	Actin	Gas gangrene[c]	(Yes)
Inhibit protein synthesis					
Diphtheria toxin	*Corynebacterium diphtheriae*	ADP-ribosyltransferase	Elongation factor 2	Diphtheria	Yes
Shiga toxins	*E. coli/Shigella dysenteriae*	N-glycosidase	28S rRNA	HC and HUS	Yes
Exotoxin A	*Pseudomonas aeruginosa*	ADP-ribosyltransferase	Elongation factor 2	Pneumonia[c]	(Yes)
Disrupt cell cycle					
CLDT	*E. coli, Campylobacter, Haemophilus ducreyi, others*	DNase	DNA damage (triggers G_2 cell cycle arrest)	Diarrhea, chancroid, others	Unknown
Pasteurella multocida toxin	*Pasteurella multocida*	Mitogen (also activates Rho GTPases)	Nucleus (encourages cell division)	Wound infection	(Yes)
Activate second messenger pathways					
CNF	*E. coli*	Deamidase	Rho G proteins	UTIs	Unknown
LT	*E. coli*	ADP-ribosyltransferase	G proteins	Diarrhea	Yes
ST[d]	*E. coli*	Stimulates guanylate cyclase	Guanylate cyclase receptor	Diarrhea	Yes
EAST	*E. coli*	ST-like?	Unknown	Diarrhea	Unknown
Edema factor	*Bacillus anthracis*	Adenylate cyclase	ATP	Anthrax	Yes

Toxin	Organism	Enzymatic activity	Target	Disease	Role[b]
Dermonecrotic toxin	*Bordetella pertussis*	Deamidase	Rho G proteins	Rhinitis	(Yes)
Pertussis toxin	*Bordetella pertussis*	ADP-ribosyltransferase	G protein(s)	Pertussis (whooping cough)	Yes
C3 toxin	*Clostridioides botulinum* (formerly *Clostridium*)	ADP-ribosyltransferase	Rho G protein	Botulism	Unknown
Toxin A	*Clostridioides difficile*	Glucosyltransferase	Rho G protein(s)	Diarrhea/PC	(Yes)
Toxin B	*Clostridium difficile*	Glucosyltransferase	Rho G protein(s)	Diarrhea/PC	Unknown
Cholera toxin	*Vibrio cholerae*	ADP-ribosyltransferase	G protein(s)	Cholera	Yes
Lethal factor	*Bacillus anthracis*	Metalloprotease	MAPKK1/MAPKK2	Anthrax	Yes
Disrupt cell-cell adherence					
Exfoliative toxins	*Staphylococcus aureus*	Serine protease, superantigen	Desmoglein, TCR, and MHC II	Scalded skin syndrome[c]	Yes
Bacteroides fragilis toxin	Enterotoxigenic *Bacteroides fragilis*	Metalloprotease	E-cadherin (indirect?)	Diarrhea, inflammatory bowel disease	Yes
Alter vesicle traffic					
VacA	*Helicobacter pylori*	Large vacuole formation, apoptosis	Receptor-like protein tyrosine phosphatase, sphingomyelin	Gastric ulcers, gastric cancer	(Yes)
Block exocytosis					
Neurotoxins A–G	*Clostridium botulinum*	Zinc metalloprotease	VAMP/synaptobrevin, SNAP-25 syntaxin	Botulism	Yes
Tetanus toxin	*Clostridium tetani*	Zinc metalloprotease	VAMP/synaptobrevin	Tetanus	Yes
Superantigens (activate immune response)					
Enterotoxins	*Staphylococcus aureus*	Superantigen	TCR and MHC II, medullary emetic center (vomit center)	Food poisoning[c]	Yes
Exfoliative toxins	*Staphylococcus aureus*	See "Disrupt cell-cell adherence" above			
Toxic shock syndrome toxin	*Staphylococcus aureus*	Superantigen	TCR and MHC II	Toxic shock syndrome[c]	Yes
Pyrogenic exotoxins	*Streptococcus pyogenes*	Superantigens	TCR and MHC II	Toxic shock syndrome, scarlet fever	Yes
Lethal factor	*Bacillus anthracis*	Metalloprotease	MAPKK1/MAPKK2	Anthrax	Yes

[a]**Abbreviations:** CLDT = cytolethal distending toxin; CNF = cytotoxic necrotizing factor; EAST = enteroaggregative *E. coli* heat-stable toxin; HC = hemorrhagic colitis; HUS = hemolytic uremic syndrome; LT = heat-labile toxin; MAPKK = mitogen-activated protein kinase; MHC II = major histocompatibility complex class II; PC = antibiotic-associated pseudomembranous colitis; SNAP-25 = synaptosomal-associated protein; ST = heat-stable toxin; TCR = T-cell receptor; UTI = urinary tract infection; VAMP = vesicle-associated membrane protein.

[b]Yes = strong causal relationship between toxin and disease; (Yes) = role in pathogenesis has been shown in animal model or appropriate cell culture.

[c]Other diseases are also associated with the organism.

[d]Toxin is also produced by other genera of bacteria.

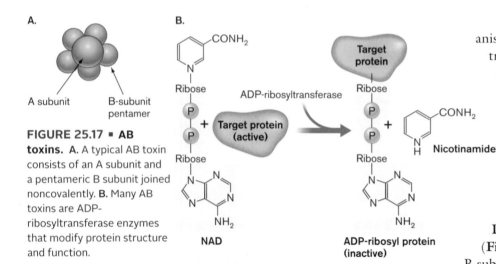

FIGURE 25.17 ▪ AB toxins. A. A typical AB toxin consists of an A subunit and a pentameric B subunit joined noncovalently. **B.** Many AB toxins are ADP-ribosyltransferase enzymes that modify protein structure and function.

each toxin delivers the A subunit to the host cell. Many AB toxins have five identical B subunits arranged as a ring, in the center of which is nestled a single A subunit (**Figs. 25.15B** and **25.17A**).

AB exotoxins of one major subclass have **ADP-ribosyltransferase** enzymatic activity as part of their A subunits. These enzyme toxins transfer the ADP-ribose group from an NAD molecule to a target protein (**Fig. 25.17B**). Cholera toxin modifies an arginine residue in a GTPase. Pertussis toxin alters a cysteine in a different GTPase. The ADP-ribosylated protein has an altered function. Sometimes the function is destroyed (for example, protein synthesis is destroyed by diphtheria toxin); other times the protein is locked into an active form insensitive to regulatory feedback control (for example, cAMP synthesis continues unchecked in the presence of cholera toxin).

AB Exotoxins That Target Signal Transduction

Cholera and *E. coli* labile toxins. *Vibrio cholerae* (**Fig. 25.18A**) produces a severe diarrheal disease called cholera that generally afflicts malnourished people populating poor countries like Bangladesh, or countries where access to clean water has been disrupted by war or natural disasters, such as the aftermath of the 2010 earthquake in Haiti. *V. cholerae* produces a gastrointestinal AB enterotoxin nearly identical to one produced by some strains of *E. coli* associated with what is known as "traveler's diarrhea." Enterotoxins specifically affect the intestine. The *E. coli* enterotoxin is called **labile toxin** (**LT**) because it is easily destroyed by heat. Cholera toxin (CT) and labile toxin are both AB toxins with identical modes of action, which is to increase the level of cAMP made inside the host cell.

Normally, intestinal transport mechanisms absorb NaCl and other ions (electrolytes), as well as water from food material moving through the intestine. These actions produce well-formed feces with very little water and salt content. CT and LT, however, reverse this process by secreting water and electrolytes into the intestinal lumen. After the bacteria attach to the cells lining the intestinal villi (**Fig. 25.18C** and **D**), they secrete their AB exotoxins (**Fig. 25.18B**). Both exotoxins have five B subunits arranged as a ring around a single A subunit. The B subunits bind to ganglioside GM1 on eukaryotic cell membranes and deliver the A subunit to the target cell (**Fig. 25.19** ▶, steps 1–4). The A subunit possesses the toxic part of the molecule, an ADP-ribosyltransferase, which must be activated by the host.

The binding of CT or LT to GM1 triggers endocytosis and the formation of a toxin-containing vacuole, which is transported to the endoplasmic reticulum. During this time, the A subunit is cleaved by a host protease into two fragments, called A1 and A2, which are still held together by a disulfide bond. The reducing environment in the vacuole reduces that bond and frees the A1 peptide containing active ADP-ribosyltransferase into the endoplasmic reticulum, which exports the toxin into the cytoplasm.

The mission of the A1 peptide is to modify (that is, ADP-ribosylate) a membrane-associated GTPase (called a G protein or G factor) that binds to adenylate cyclase and controls its activity (**Fig. 25.19**, step 5). Human cells have two types of G-factor complexes that stimulate (G_s) or inhibit (G_i) adenylate cyclase, respectively, when bound to GTP (see **eTopic 25.6** for details). An intrinsic GTPase in each G factor hydrolyzes GTP and prevents continual stimulation of adenylate cyclase by G_s, or its inhibition by G_i. Cholera toxin (and *E. coli* labile toxin) ADP-ribosylate G_s and thereby inhibit the intrinsic GTPase activity (step 5). As a result, adenylate cyclase is constantly stimulated to produce cAMP.

The high amount of cAMP stimulates a host protein kinase that activates various ion transport channels, including the cystic fibrosis transmembrane conductance regulator (CFTR), so named because a defect in this protein manifests as the lung disease cystic fibrosis. CFTR controls chloride transport in several cell types, including intestinal epithelia (discussed in Section 23.4). As a result of CFTR activation, chloride, sodium, and other ions leave the cell, and in an attempt to equilibrate osmolarity, water leaves as well. Because the affected cells line the intestine,

A. *Vibrio cholerae*

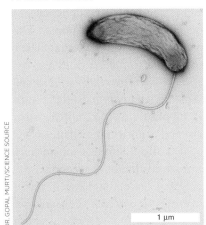

B. Cholera toxin

A subunit

5 B subunits

GM1   Intestinal cell surface

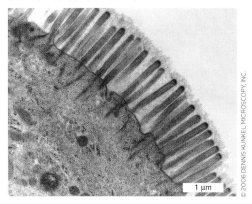

C. Brush border of intestine

D. *V. cholerae* **attachment**

V. cholerae

FIGURE 25.18 ■ **Pathogenesis of cholera. A.** *Vibrio cholerae* (SEM). Note the slight curve of the cell and the presence of a single polar flagellum. **B.** 3D structure of cholera toxin, binding ganglioside GM1 on the intestinal cell surface. (PDB code: 1S5F) **C.** Brush border of intestine (TEM). *V. cholerae* binds to the fingerlike villi on the apical surface. **D.** *V. cholerae*, binding to the surface of a host cell (SEM). Note that *V. cholerae* does not invade the host cell.

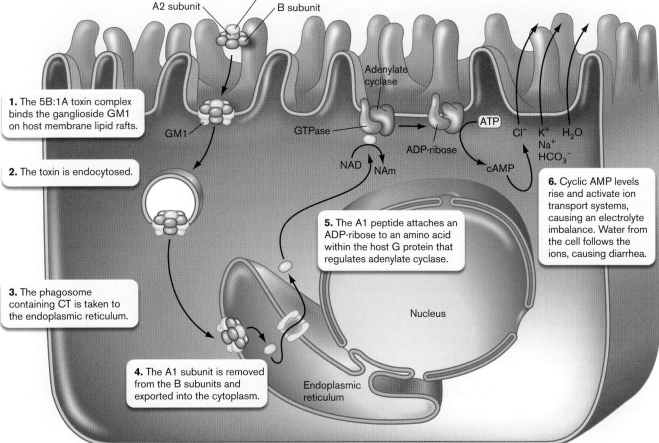

1. The 5B:1A toxin complex binds the ganglioside GM1 on host membrane lipid rafts.

2. The toxin is endocytosed.

3. The phagosome containing CT is taken to the endoplasmic reticulum.

4. The A1 subunit is removed from the B subunits and exported into the cytoplasm.

5. The A1 peptide attaches an ADP-ribose to an amino acid within the host G protein that regulates adenylate cyclase.

6. Cyclic AMP levels rise and activate ion transport systems, causing an electrolyte imbalance. Water from the cell follows the ions, causing diarrhea.

A2 subunit   A1 subunit   B subunit

GM1

Adenylate cyclase

GTPase

ADP-ribose

NAD   NAm

ATP

Cl^- K^+ H_2O
   Na^+
   HCO_3^-

cAMP

Nucleus

Endoplasmic reticulum

FIGURE 25.19 ■ **Cholera toxin mode of action.** Delivery of cholera toxin into target cells and deregulation of adenylate cyclase activity. NAm = nicotinamide. ▶

the escaping water enters the intestinal lumen, leading to watery stools, or diarrhea (**Fig. 25.19**, step 6).

Thought Question

25.2 **Figure 25.19** illustrates how cholera toxin works to cause diarrhea. To develop a vaccine that generates protective antibodies, which subunit of cholera toxin should be used to best protect a person from the toxin's effects?

How is diarrhea a benefit to the microorganism? The more diarrhea is produced and expelled, the more organisms are made and disseminated throughout the environment. In this way, diarrhea increases the chance that another host will ingest the organism, ensuring survival of the species. Diarrhea can also benefit a pathogen by decreasing competition with other organisms as they are swept away. In fact, the vast majority of organisms found in the diarrheal fluids of cholera patients are *V. cholerae* bacteria.

Different pathogens have discovered alternative ways to alter host cAMP levels. The Gram-negative pathogen *Bordetella pertussis* causes a childhood respiratory infection called whooping cough, so named for the whooping sound made as a child tries to take a breath after a fit of coughing. This microbe also secretes an ADP-ribosylating toxin, but this toxin modifies G_i (the inhibitory factor that normally downregulates adenylate cyclase activity). ADP-ribosylation in this instance prevents that inhibition, with the result (again) of runaway cAMP synthesis. Pertussis toxin, however, is structurally different from cholera toxin. In addition, the organism makes a "stealth" adenylate cyclase that is secreted from the bacterium but remains inactive until it enters host cells, where it becomes active by binding the calcium-binding protein calmodulin. The resulting increase in cAMP levels causes inappropriate triggering of certain host cell signaling pathways and damages lung cells.

Anthrax. A century ago, anthrax (caused by *Bacillus anthracis*; **Fig. 25.20A**) was mainly a disease of cattle and sheep. Humans acquired the disease only accidentally. Today we fear the deliberate shipment of *B. anthracis* through the mail (as happened in 2001) or its dispersion from the air ducts of heavily populated buildings. What makes this Gram-positive, spore-forming microbe so dangerous? In large part, its lethality is due to the secretion of a plasmid-encoded tripartite toxin (a variant of the AB exotoxin theme). The core subunit of the toxin is called **protective antigen** (**PA**) because immunity to this protein protects hosts from disease. Protective antigen is made as a single peptide but then binds to the host cell surface (there are multiple receptors), where a human protease cleaves off a fragment (**Fig. 25.20B**). The remaining part of PA can

self-assemble in the membrane to form heptameric (seven-membered) and octomeric (eight-membered) pores. The other two components of anthrax toxin—**edema factor** (**EF**) and **lethal factor** (**LF**)—bind to the PA rings and are carried into the cell (**Fig. 25.20C**). The complex is endocytosed, and the two proteins carried in are passed through the pore into the host cytoplasm. Proton motive force helps unfold the exotoxins and translocates them across the endocytic membrane. PA is the B subunit for anthrax toxin, while EF and LF represent different A subunits.

Edema factor and lethal factor are the toxic parts of anthrax toxin. Both are enzymes that attack the signaling functions of the cell. Edema factor is an adenylate cyclase that remains inactive until entering the cytoplasm, where it binds calmodulin. This binding activates adenylate cyclase, resulting in a huge production of cAMP, and inactivates calmodulin from its normal function in the cell.

Lethal factor is actually a protease that cleaves several host protein kinase kinases, each of which is part of a critical regulatory cascade affecting cell growth and proliferation. A protein kinase kinase is an enzyme that phosphorylates, and thereby activates, another protein kinase that can then phosphorylate one or more subsequent target proteins. One consequence of subverting these phosphorylation cascades is a failure to produce signals that recruit immune cells to fight the infection.

Thought Question

25.3 How might you experimentally determine whether a pathogen secretes an exotoxin? (*Hint:* You can learn more about how new toxins are found in **eTopic 25.8**.)

AB Exotoxins That Target Protein Synthesis

Shiga toxin: *Shigella* and *E. coli* O157:H7. *Shigella dysenteriae* and *E. coli* O157:H7 (also known as enterohemorrhagic *E. coli*) cause food-borne diseases whose symptoms include bloody diarrhea. These organisms produce an important toxin known as Shiga toxin (or Shiga-like toxin). The gene (*stx*) encoding Shiga toxin is part of a phage genome integrated into the bacterial chromosome. The toxin has five B subunits for binding and one A subunit imbued with toxic activity. The A subunit, upon entry, destroys protein synthesis by cleaving 28S rRNA in eukaryotic ribosomes. Strains that produce high levels of this toxin are associated with acute kidney failure, known as hemolytic uremic syndrome.

Note: *Shigella dysenteriae* is the only *Shigella* species that produces Shiga toxin.

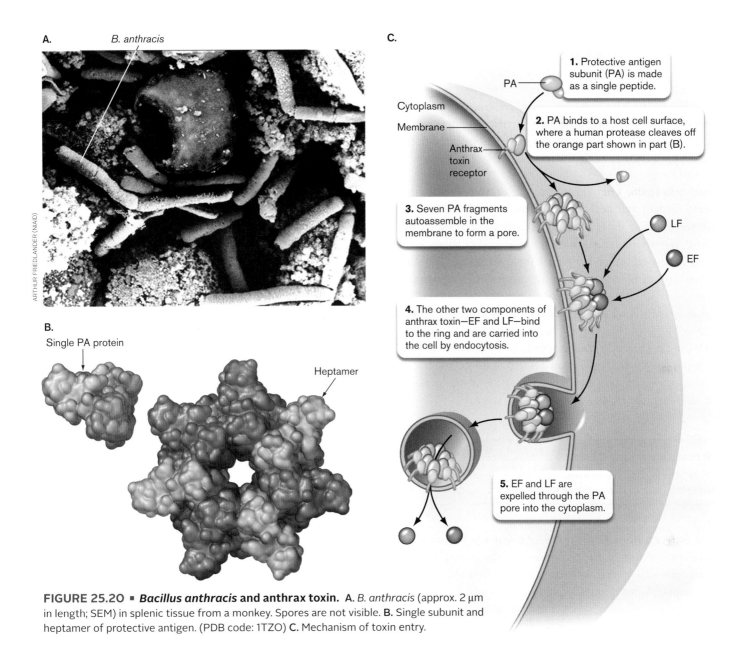

FIGURE 25.20 ■ *Bacillus anthracis* **and anthrax toxin.** **A.** *B. anthracis* (approx. 2 μm in length; SEM) in splenic tissue from a monkey. Spores are not visible. **B.** Single subunit and heptamer of protective antigen. (PDB code: 1TZO) **C.** Mechanism of toxin entry.

E. coli O157:H7 is a recently emerged pathogen. The organism can colonize cattle intestines without causing bovine disease; as a result, undetected bacteria can easily contaminate meat products following slaughter, as well as irrigation water from farms (see Chapter 28). The first large U.S. outbreak of O157:H7 disease was associated with fast-food hamburgers at a Washington State Jack in the Box restaurant in 1993. Both *E. coli* and *Shigella* have remarkable acid resistance mechanisms that rival that of the gastric pathogen *Helicobacter pylori*. This acid resistance makes these pathogens infectious at a very low infectious dose.

What activates *stx* expression? Iron availability is a key factor for inducing the expression of *stx* and many other virulence genes in pathogens. The body holds its iron tightly in proteins such as lactoferrin and transferrin. To an invading organism, the body is a very iron-poor environment. In the presence of low iron, expression of Shiga toxin increases, the toxin kills host cells, and dead cells release their iron. Shiga toxin, then, offers a way to rob the host of its iron stores.

An intriguing question often pondered by scientists is, What roles do virulence factors play in the natural ecology of these bacteria? Surely, factors such as Shiga toxin did not evolve after *Shigella* started infecting humans. William Lainhart, Gino Stolfa, and Gerald Koudelka from the State University of New York at Buffalo discovered that Shiga toxin is actually a natural defense against *Tetrahymena thermophila*, a ciliated protist that grazes on bacteria. Lainhart and his colleagues found that *Tetrahymena* was killed when cocultured with Shiga toxin–producing bacteria.

They proposed that reactive oxygen species produced by *Tetrahymena* induce the SOS response in *Shigella*, which then activates phage replication and the expression of *stx* (SOS induction is discussed in Section 9.4). This is another example of altruistic behavior by members of a bacterial species in which some individuals sacrifice themselves (as a result of phage lysis) to save the population. For further discussion of the beneficial roles of viruses, see Chapter 6.

Diphtheria toxin. The classic example of an exotoxin that targets protein synthesis is diphtheria toxin, produced by *Corynebacterium diphtheriae*, the cause of the respiratory disease diphtheria. The two-component diphtheria exotoxin, discussed more fully in **eTopic 25.7**, kills cells by ADP-ribosylating eukaryotic protein synthesis elongation factor 2 (EF2). The vaccine used to prevent diphtheria (the "D" in DTaP) is an inactivated form of this exotoxin (see Section 24.7). *C. diphtheriae* is another example of a pathogen whose toxin gene (*dtx*) is regulated by iron availability and is part of a prophage genome integrated into the bacterial chromosome. Recall that a prophage is phage genome that has integrated into a bacterial chromosome.

> **Thought Question**
>
> **25.4** Would patients with iron overload (excess free iron in the blood) be more susceptible to infection?

We have examined only a few of the many protein exotoxins employed by pathogens. Some of the others, including tetanus and botulism toxins, will be described in the next chapter. What should be apparent from our brief sampling is the evolutionary ingenuity that pathogens have used to try to tame the human host. A discussion of how new microbial toxins are discovered is found in **eTopic 25.8**.

> **Thought Question**
>
> **25.5** Use an Internet search engine to determine which other toxins are related to the cholera toxin A subunit. (*Hint:* Start by searching on "protein" at PubMed to find the protein sequence; then use a BLAST program.)

Endotoxin (LPS) Is Made Only by Gram-Negative Bacteria

Another important virulence factor common to all Gram-negative microorganisms is endotoxin present in the outer membrane (discussed in Chapter 3). Not to be confused with secreted exotoxins, endotoxin is an embedded part of the bacterial cell surface and an important contributor to disease. Endotoxin, otherwise called lipopolysaccharide (LPS), is composed of lipid A, core glycolipid, and a repeating polysaccharide chain known as the O antigen (**Fig. 25.21**). LPS molecules form the outer leaflet of the Gram-negative outer membrane (discussed in Chapter 3). As bacteria die, they release endotoxin. Endotoxin is a microbe-associated molecular pattern (MAMP) molecule that can bind to certain Toll-like receptors on macrophages or B cells and trigger the release of TNF-alpha, interferon, IL-1, and other cytokines (MAMPs, Toll-like receptors, and cytokines are discussed in Chapters 23 and 24). The release of these active agents causes a variety of symptoms, such as:

- Fever

- Activation of clotting factors, leading to disseminated intravascular coagulation

A. LPS membrane

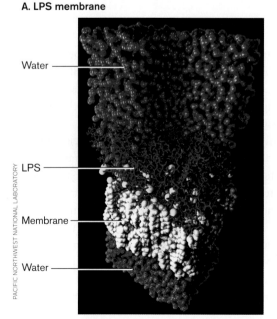

Water

LPS

Membrane

Water

B. Gram-negative bacterial endotoxin (lipopolysaccharide, LPS)

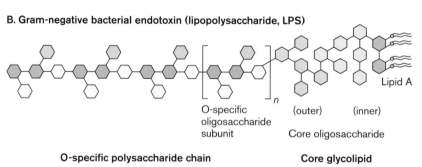

Lipid A

O-specific oligosaccharide subunit

(outer) (inner)

Core oligosaccharide

O-specific polysaccharide chain

Core glycolipid

FIGURE 25.21 ■ **Endotoxin. A.** Model of a lipopolysaccharide (LPS) membrane of *Pseudomonas aeruginosa*, consisting of 16 lipopolysaccharide molecules (red) and 48 ethylamine phospholipid molecules (white). **B.** Basic structure of endotoxin, showing the repeating O-antigen side chain that faces out from the microbe and the membrane proximal core glycolipid and lipid A (contains endotoxic activity).

TABLE 25.4	Major Distinctions between Bacterial Exotoxins and Endotoxins	
Property	**Exotoxin**	**Endotoxin**
Producing organism	Gram-positive or -negative	Gram-negative only
Chemical	Protein (size 50–1,000 kDa)	Lipopolysaccharide (lipid A moiety; size 10 kDa)
Denatured by boiling	Yes, if boiled long enough	No
Mode of action	Some exotoxins target specific features of eukaryotic cells (membrane, protein synthesis, signal transduction, etc.); others are superantigens	Bind Toll-like receptor 4; activate cytokine production
Enzyme activity	Often	No
Toxicity	High (1-μg quantities)	Low (>100-μg quantities), but the primary cause of Gram-negative sepsis (serious blood infections)
Immunogenicity	Highly antigenic	Poorly antigenic
Vaccine	Toxoids can be made for some	Toxoids cannot be made
Fever production (pyrogenicity)	Occasionally	Yes

- Activation of the alternative complement pathway

- Vasodilation, leading to hypotension (low blood pressure)

- Shock due to hypotension

- Death, when other symptoms are severe

The major differences between exotoxins and endotoxins are summarized in **Table 25.4**.

Note: Lipopolysaccharide O antigens forming part of the outer membrane are used to classify different strains of *E. coli* (for example, *E. coli* O157 versus *E. coli* O111), as well as other Gram-negative organisms.

The role of endotoxin can be seen in infections with the Gram-negative diplococcus *Neisseria meningitidis* (**Fig. 25.22A**), a major cause of bacterial meningitis. *N. meningitidis* has, as part of its pathogenesis, a septicemic phase in which the organism can replicate to high numbers in the bloodstream. The large amount of endotoxin present causes a massive depletion of clotting factors, which leads to internal bleeding, most prominently displayed to a physician as small pinpoint hemorrhages called **petechiae** on the patient's hands and feet (**Fig. 25.22B**). Capillary bleeding near the surface of the skin causes petechiae. One danger of treating massive Gram-negative sepsis with antibiotics is that the enormous release of endotoxin from dead bacteria could well kill the

A.

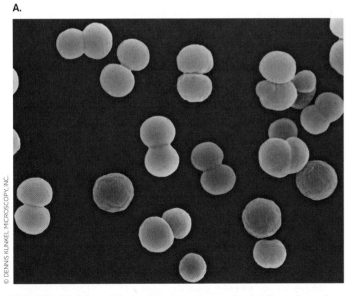

B.

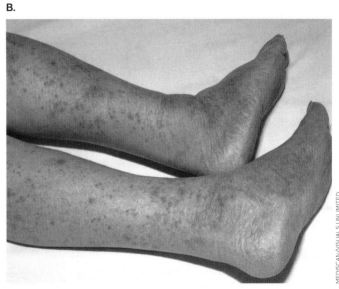

FIGURE 25.22 ▪ **Effect of *Neisseria meningitidis* endotoxin.** **A.** *N. meningitidis* (cell 0.8–1 μm in diameter; SEM). **B.** Petechial rash caused by *N. meningitidis*.

patient. Untreated Gram-negative sepsis is, however, almost always fatal, so its treatment, albeit risky, is imperative.

An approach to prevent endotoxic shock currently under study is based on the knowledge that LPS must bind to Toll-like receptor TLR4 to cause endotoxic shock. What if we could neutralize TLR4 by antibody and prevent it from binding LPS during an infection? Several investigators have shown that injecting antibody raised to TLR4 (anti-TLR4) into infected mice will successfully block TLR4 receptors and protect the mice from *E. coli*–induced septic shock. Unfortunately, this approach has not yet proved successful in human trials.

To Summarize

■ **There are nine categories of protein exotoxins** based on mode of action. These include toxins that disrupt membranes, inhibit protein synthesis, or alter the synthesis of host cell signaling-molecules, as well as toxins that are superantigens or target-specific proteases.

■ **Many bacterial toxins are two-component, AB subunit toxins.** The B subunit promotes penetration through host cell membranes, whereas the A subunit has toxic activity.

■ ***Staphylococcus aureus* alpha toxin** forms pores in host cell membranes.

■ **Cholera toxin**, *E. coli* labile toxin, and pertussis toxin are AB toxins that alter host cAMP production by adding ADP-ribose groups to different G-factor proteins.

■ **Shiga toxin** is an AB toxin that cleaves host cell 28S rRNA in host cell ribosomes.

■ **Anthrax toxin** is a three-part AB toxin with one B subunit (protective antigen) and two different A subunits that affect cAMP levels (edema factor) and cleave host protein kinases (lethal factor).

■ **Lipopolysaccharide (LPS)**, known as endotoxin, is an integral component of Gram-negative outer membranes and an important virulence factor that triggers massive release of cytokines from host cells. The indiscriminate release of cytokines can trigger fever, shock, and death.

25.5 Deploying Toxins and Effectors

A recurring theme among bacterial pathogens is the secretion of proteins that destroy, cripple, or subvert host target cells. The bacterial toxins described in Section 25.4 are secreted into the surrounding environment, where they float randomly until chance intervenes and they hit a membrane-binding site. However, many pathogens attach to tissue cells and inject bacterial proteins (called effectors) directly into the host cell cytoplasm. The proteins may not kill the cell, but they redirect host signaling pathways in ways that benefit the microbe.

Protein secretion pathways were introduced in Section 8.5, focusing on ATP-binding cassette (ABC) proteins as a model. Additional secretion models are described here in their critical role of delivering pathogenicity proteins such as toxins. A particularly interesting aspect of these secretory systems is that many of them evolved from, and bear structural resemblance to, other cell structures that serve fundamental cell functions. The molecular processes that are evolutionarily related to secretion include:

■ Type IV pilus biogenesis (homologous to type II protein secretion)

■ Flagellar synthesis (homologous to type III protein secretion)

■ Conjugation (homologous to type IV protein secretion)

TABLE 25.5	Secretion Systems for Bacterial Toxins	

Secretion type	Features	Examples
I	SecA dependent, one effector per system	*E. coli* alpha hemolysin, *Bordetella pertussis* adenylate cyclase
II	SecA dependent, similar to type IV pili	*Pseudomonas aeruginosa* exotoxin A, elastase, cholera toxin
III	SecA independent, syringe, related to flagella, multiple effectors secreted	*Yersinia* Yop proteins, *Salmonella* Sip proteins, enteropathogenic *E. coli* (EPEC) EspA proteins, TirA
IV	Related to conjugational DNA transfers, multiple effectors secreted	*B. pertussis* toxin, *Helicobacter* CagA
V	Autotransporter, SecA dependent to periplasm, self-transport through outer membrane, one effector per system	Gonococcal and *Haemophilus influenzae* IgA proteases
VI	Related to phage tails, single effector, harpoon mechanism	*Burkholderia* and *Vibrio cholerae* VgrG
VII	Unrelated to other systems	*Mycobacterium tuberculosis* Esx and Esp

Table 25.5 lists features of the seven export systems of Gram-negative bacteria and examples of associated virulence effector proteins. Secretion system types II, III, and IV will be described in more detail.

Type II Secretion Resembles Type IV Pilus Assembly

Cholera toxin, discussed in Section 25.4, is a well-known example of a toxin secreted by a **type II secretion system**. Type II secretion offers a clear example of how nature has modified the blueprints of one system to do a very different task. DNA sequence analysis has revealed that the genes used for type IV pilus biogenesis (see Section 25.3) were duplicated at some point during evolution and redesigned to serve as a protein secretion mechanism. Type IV pili have the unusual ability to extend and retract from the outer membrane—a property that produces the gliding motility of *Myxococcus* (see Section 4.6) and the twitching motility of *Neisseria* and *Pseudomonas*. As you might guess, assembly/disassembly of these appendages is quite complex.

Type II protein secretion mechanisms mirror this complexity. Proteins to be secreted first make their way, via the SecA-dependent general secretion pathway, to the periplasm, where they are folded and then encounter the appropriate type II secretion system. Type II secretion systems cyclically assemble and disassemble the pilus-like structure, using it as a piston to ram folded toxins or effector proteins through an outer membrane pore structure and into the surrounding void (**Fig. 25.23**).

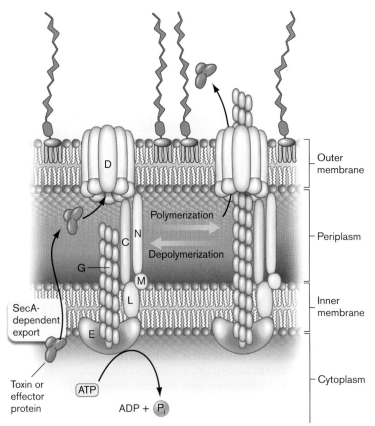

FIGURE 25.23 ■ Type II secretion. C, D, E, G, L, M, and N are protein components of the secretion system.

Type III Secretion Injects Effector Proteins into Host Cells

Yersinia, *Salmonella*, and *Shigella* are the etiological agents of Black Death and various forms of diarrhea. In the 1990s it was discovered that these organisms could somehow take the bacterial virulence proteins made in their cytoplasm and drive them directly into the eukaryotic cell cytoplasm without the protein ever getting into the extracellular environment. Direct delivery is a good idea because it eliminates the dilution that happens when a toxin is secreted into media. Another advantage of this strategy is that it avoids the need to tailor the toxin to fit a preexisting host receptor.

What kind of molecular machine can directly deliver cytoplasmic bacterial proteins into target cells? Some microbes use tiny molecular syringes embedded in their membranes to inject proteins directly into the host cytoplasm; this mechanism of delivery is called a **type III secretion system (T3SS)**. **Figure 25.24** ▶ shows electron micrographs of type III secretion needles and a model of the complex. Genes encoding type III systems are actually

related to flagellar genes, whose products export the flagellin proteins through the center of a growing flagellum (discussed in Section 3.6). It appears that flagellar genes were evolutionarily reengineered to encode proteins that act more like molecular syringes (**Fig. 25.24C**).

The bacterial virulence proteins secreted by type III systems subvert normal host cell signaling pathways, some of which cause dramatic rearrangements of host cytoskeleton at the cell membrane that lead to engulfment of the microbe (**Fig. 25.25**). The genes encoding type III systems are usually located within pathogenicity islands inherited from other microbial sources. Many bacterial pathogens use this type of secretion system, including plant pathogens such as *Pseudomonas syringae* (the cause of blight, a disease of many plants in which leaves or stems develop brown spots). Secretion is normally triggered by cell-cell contact between host and bacterium.

E. coli **pathogens use a T3SS to "inject" their own receptor into host cells.** Enteropathogenic *E. coli* (EPEC) and enterohemorrhagic *E. coli* (EHEC) are two diarrhea-producing forms of *E. coli*. These pathogens use pili to initially bind to the host's intestinal epithelial cells; however, the bacterium must establish a more intimate attachment to these cells to cause disease. The bacterial outer membrane

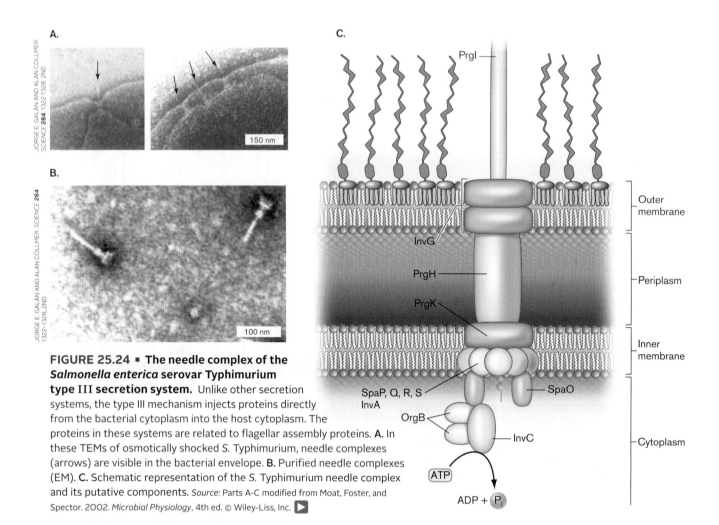

JORGE E. GALÁN AND ALAN COLLMER. *SCIENCE* **284** 1322–1328. 2ND

JORGE E. GALÁN AND ALAN COLLMER. *SCIENCE* **284** 1322–1328. 2ND

FIGURE 25.24 ■ **The needle complex of the *Salmonella enterica* serovar Typhimurium type III secretion system.** Unlike other secretion systems, the type III mechanism injects proteins directly from the bacterial cytoplasm into the host cytoplasm. The proteins in these systems are related to flagellar assembly proteins. **A.** In these TEMs of osmotically shocked *S.* Typhimurium, needle complexes (arrows) are visible in the bacterial envelope. **B.** Purified needle complexes (EM). **C.** Schematic representation of the *S.* Typhimurium needle complex and its putative components. *Source:* Parts A-C modified from Moat, Foster, and Spector. 2002. *Microbial Physiology*, 4th ed. © Wiley-Liss, Inc. ▶

protein called **intimin** mediates this intimate attachment. The only problem is that the host lacks a receptor for intimin. No problem. EPEC and EHEC use a type III secretion system to insert their own receptor, Tir (for translocated intimin receptor), into the target cells. (**Fig. 25.26A**). Brett Finlay and his colleagues in Vancouver, British Columbia, discovered this system and found that the genes encoding intimin, Tir, and the secretion apparatus are all part of an EPEC pathogenicity island.

Once injected and placed in the host membrane, Tir binds intimin on the bacterial surface (**Fig. 25.26B**). Think of Tir as a wall anchor you poke into a board in order to attach something to it. The result is a tighter, more intimate adherence between the bacterium and host cell surface, which is required for infection to proceed. In addition, host protein kinases phosphorylate Tir at tyrosine residue 474. Phosphorylated Tir directly triggers a remarkable reorganization of host cellular cytoskeletal components (actin, alpha-actinin, ezrin, talin, and myosin light chain) such that a membrane "pedestal" is formed, raising the microbe up (**Fig. 25.26C**). The result of this attachment is the characteristic attaching and effacing (A/E) lesion,

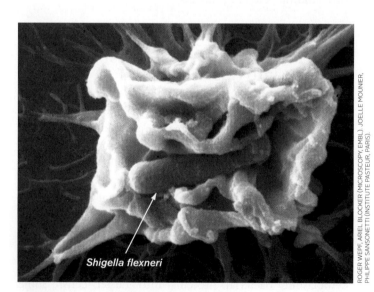

ROGER WEPF, ARIEL BLOCKER (MICROSCOPY, EMBL), JOELLE MOUNIER, PHILIPPE SANSONETTI (INSTITUTE PASTEUR, PARIS).

FIGURE 25.25 ■ ***Shigella* invades a host cell ruffle produced as a result of type III secretion.** *Shigella flexneri* entering a HeLa cell (SEM). The bacterium (small diameter, approx. 1 μm) interacts with the host cell surface and injects (via its type III secretion apparatus) its invasin proteins, which choreograph a local actin-rich membrane ruffle at the host cell. The ruffle engulfs the bacterium and eventually disassembles, internalizing the bacterium.

A.

B.

C.

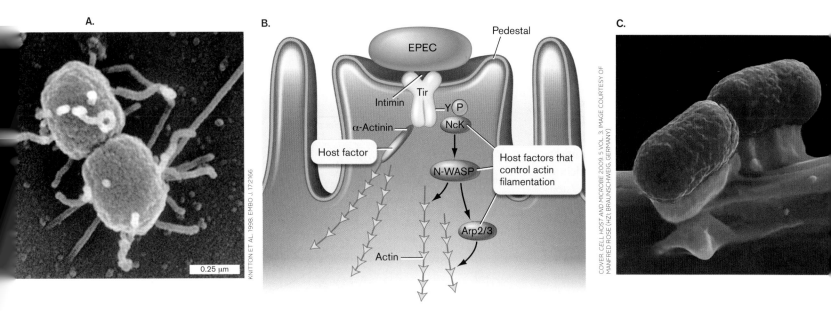

FIGURE 25.26 ■ *E. coli* type III secretion and cell-cell interaction. A. EspA filaments form a bridge between enteropathogenic *E. coli* (EPEC) and an epithelial cell during the early stages of attaching and effacing (A/E) lesion formation (SEM). These filaments are part of the type III secretion apparatus, functioning as a molecular syringe to inject proteins from pathogenic *E. coli* into the host cell. B. Model of cytoskeletal components within the EPEC pedestal. EPEC injects Tir protein into the host cell, where it moves to the membrane and acts as a receptor for intimin. Tir also communicates through phosphorylation with other host proteins to cause a change in actin cytoskeleton, which leads to pedestal formation. C. Pedestal formation (colorized SEM).

characterized by destruction of the microvilli and pedestal formation. By placing itself on a "pedestal," EPEC avoids engulfment and the perils of the phagolysosome.

Salmonella pathogenesis. The Gram-negative bacterium *Salmonella enterica* is currently the most common bacterial food-borne pathogen in the United States. Its transmission has been linked to everything from cantaloupe and peanut butter (in 2012) to poultry (pick any year). As part of its pathogenesis, *Salmonella* uses type III secretion systems to invade the eukaryotic host cell and replicate intracellularly. Following ingestion of contaminated food or water, this bacterium attaches to and invades M cells that are interspersed along the wall of the small intestine. M cells are specialized intestinal epithelial cells (see Fig. 23.25B) that sample normal intestinal microbes and transfer pathogens across the epithelial barrier for recognition by the immune system. *Salmonella* subverts the normal function of M cells and causes an inflammatory response that leads to diarrhea.

During its evolutionary journey toward becoming a pathogen, *Salmonella* has acquired as many as 14 pathogenicity islands. Five of those islands can be found in all *S. enterica* serovars, but only two will be discussed here. *Salmonella* pathogenicity island 1 (SPI-1) encodes a type III protein secretion system that delivers a cocktail of at least 13 different protein toxins (called effector proteins) directly into the cytosol of host epithelial cells in the gut (**Fig. 25.27A,** ▶ step 1). Inside epithelial cells, these effector proteins interfere with signal transduction cascades and

modulate the host response. One mission of these effectors is to induce cytoskeletal actin rearrangements that cause ruffling of the eukaryotic membrane around the microbe (step 2; also see **Fig. 25.25**). The membrane ruffling starts the process of engulfment. *Salmonella* induces this response as a way to avoid the normal endocytic process.

While engulfment is under way, some SPI-1 effectors act to loosen tight junctions that hold together adjacent epithelial cells (**Fig. 25.27A,** step 3). Uncontrolled chloride secretion (triggered by another effector) produces diarrhea as water leaves infected cells to compensate for an electrolyte imbalance. Other SPI-1 effectors activate transcription pathways that alter cytokine expression (step 4). One of the cytokines made, IL-8, helps lure phagocytic neutrophils (PMNs) to the area of infection, which initiates inflammation (step 5).

Initiating inflammation may seem counterintuitive, but *Salmonella*'s strategy is to employ the neutrophils as "mercenaries" to kill competitors among the microbiota. The neutrophils squeeze between epithelial cells to reach the gut lumen, where they begin to engulf microbiota and produce reactive oxygen species (ROS) as part of their oxidative burst (see Chapter 23). Besides killing competitors, the oxidative burst converts thiosulfate, a compound made by gut microbiota, into tetrathionate which *Salmonella*, but not its competitors, can use as an alternative electron acceptor (**Fig. 25.27A,** step 6). Tetrathionate provides a competitive advantage to *Salmonella* growing in the lumen.

Once *Salmonella* enters an epithelial cell or a macrophage, it finds itself in a vacuole called the *Salmonella*-containing vacuole, or SCV (**Fig. 25.27B,** step 1). In the normal course of events, an enzyme-packed lysosome would fuse with the phagosome and release its contents in an effort to kill the invader. *Salmonella*, however, possesses a second pathogenicity island, called SPI-2, which subverts this host response. SPI-2 uses another type III secretion system to inject proteins that alter vesicle remodeling (step 2) and vesicle trafficking (step 3), thereby reducing phagosome-lysosome fusion so that the intracellular bacteria are spared.

Some SPI-2 effectors down-regulate production of IL-8 and TNF-alpha to limit inflammation; other effectors inhibit migration of dendritic cells, which will subvert antigen presentation (**Fig. 25.27B**, step 4). *Salmonella* eventually escapes from the initial host cell when the SCV traffics to the cell periphery and fuses with the host membrane. In addition, some SPI-2 effectors will trigger host cell apoptosis (step 5), a programmed cell death that kills the host cell in a way that prevents further inflammation. Freed from the initial cell, *Salmonella* can now infect other cells, including macrophages.

Because of their importance to virulence, type III secretion systems are the subject of intensive research designed to exploit them as potential drug targets.

Type IV Secretion Resembles Conjugation Systems

Many bacteria can transfer DNA from donor to recipient cells via a cell-cell contact system known as conjugation (see Section 9.1). The conjugation systems of some pathogens have been modified, through evolution, into new systems that transport proteins, or proteins plus DNA, directly into target cells. *Agrobacterium tumefaciens*, for example, uses its Vir system to transfer the tumor-producing Ti plasmid and some effector proteins into plant

A.

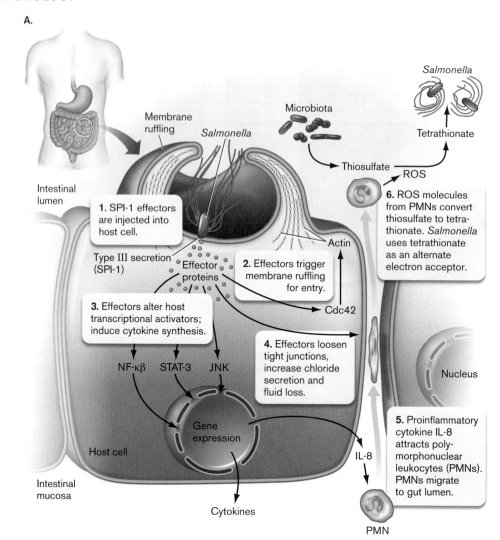

B.

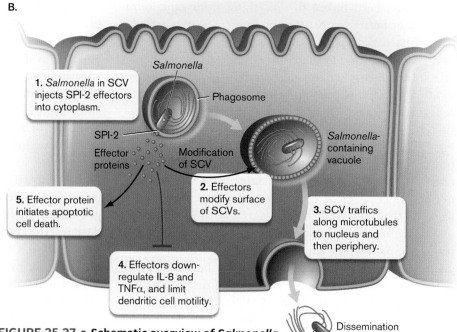

FIGURE 25.27 ■ **Schematic overview of *Salmonella* pathogenesis.** Effector proteins injected by *Salmonella* into a host M cell affect many aspects of host physiology. **A.** SPI-1 effector functions. **B.** SPI-2 effector functions. PMN = polymorphonuclear leukocyte; ROS = reactive oxygen species; SCV = *Salmonella*-containing vacuole; SPI = *Salmonella* pathogenicity island. ▶

cells. The result is a plant cancer called crown gall disease. The bacterium that causes whooping cough in humans, *Bordetella pertussis*, also uses a type IV secretion system, to export pertussis toxin, but it simply exports the toxin, without injecting it into the host (**Fig. 25.28**). Type IV systems also differ with respect to whether the protein is taken directly from the cytoplasm, like CagA from *Helicobacter*, or from the periplasm, as with pertussis toxin. In the latter case, the SecA-dependent general secretory system first delivers the toxin to the periplasmic space.

Another unique toxin delivery system, type VI secretion, resembles a harpoon or blowgun (see **Special Topic 25.1**).

> **Thought Question**
>
> **25.6** Protein and DNA have very different structures. Why would a protein secretion system be derived from a DNA-pumping system?

To Summarize

- **Many pathogens** use specific protein secretion pathways to deliver toxins.

- **Type II secretion** systems use a pilus-like extraction/retraction mechanism to push proteins out of the cell.

- **Type III secretion** uses a molecular syringe to inject proteins from the bacterial cytoplasm into the host cytoplasm.

- **Type IV secretion** utilizes a group of proteins homologous to conjugation machinery to secrete proteins from either the cytoplasm or the periplasm.

25.6 Surviving within the Host

Once inside a host, how does a successful pathogen avoid detection and destruction? Many of the virulence factors in the pathogen's arsenal help the microbe escape or resist innate immune mechanisms. Others are dedicated to stealth—that is, hiding from the immune system. But before discussing how these organisms survive in a host, we must ask how the pathogen knows it is in a host.

Where Am I?

A pathogen that can grow either outside or inside a host must adjust its physiology to match its whereabouts. Why make a type III secretion system if there are no host cells around? But how do microbial pathogens know whether they are in a host or in a pond? And which bacterial genes are expressed exclusively while in a host?

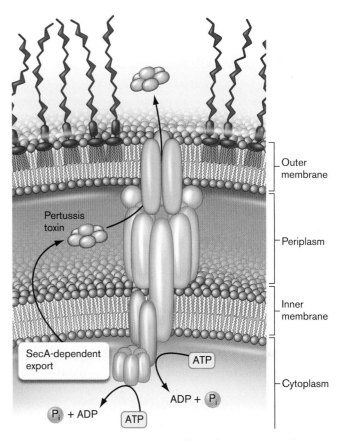

FIGURE 25.28 ■ Type IV secretion of pertussis toxin. Evolutionarily related to conjugation systems, this type IV system in *Bordetella pertussis* takes pertussis toxin from the periplasm (moved there by SecA-dependent transport) across the outer membrane.

The same types of regulatory mechanisms that sense environmental conditions in a pond are used by the microbe to intuit its whereabouts in a host. That is, various sensing systems act in concert to recognize a specific environmental niche. Two-component signal transduction systems, discussed in Section 10.1, are used to monitor magnesium concentrations, which are characteristically low in a host cell vacuole. Other regulators measure pH, which will be low (acidic) in the same vacuole. The point is that there is no single in vivo sensor system. The various regulators collaborate to trigger the expression of virulence genes.

Enterohemorrhagic *E. coli* uses positive and negative signals to determine where in a host it should express virulence genes. As mentioned earlier, EHEC uses a T3SS to inject effector proteins into gastrointestinal cells, so synthesis of the T3SS apparatus is best induced only in the intestine. One signal that induces T3SS synthesis is fucose. Fucose ends up in the intestine because fucosidases from the nonpathogenic gut microbe *Bacteroides thetaiotaomicron* cleave fucose from the host glycans present on intestinal epithelial cells. EHEC senses the fucose via a two-component system and responds by synthesizing the T3SS. However, other metabolites in the host can <u>inhibit</u> expression of the T3SS apparatus

SPECIAL TOPIC 25.1   Type VI Secretion: Poison Darts

Pygmies living in Central Africa and the Yagua from Peru both use an ancient weapon for hunting—namely, blowguns. Blowguns are carefully crafted, hollowed-out tubes designed to shoot curare-tipped poison darts that can penetrate and paralyze animal prey. What do blowguns and microbes have in common? Many Gram-negative microbes, such as *E coli*, *Vibrio cholerae*, *Pseudomonas*, and *Burkholderia*, make incredible blowgun-like devices that deliver either virulence proteins to eukaryotic prey, or antimicrobials into bacterial competitors to kill them.

These molecular weapons, called type VI secretion systems (T6SSs), comprise at least 13 proteins and are present in at least one copy in about 25% of all sequenced Gram-negative bacteria. The genomes of some bacteria, such as *Burkholderia*, carry five or six distinct T6SS gene clusters.

Structure and function. The components of T6SSs appear to be repurposed phage tail parts normally used to deliver phage DNA into bacterial host cells (see Fig. 6.11A). VipA/B proteins of *V. cholerae* T6SS, and their orthologs in other species, are homologous to the phage T4 tail sheath, a structure that contracts and drives the T4 needlelike core through the host cell wall. T4 DNA is delivered through the core and into the cytoplasm. Hcp proteins of T6SSs are homologs of the phage T4 needle proteins, while VgrG proteins are structural homologs of the T4 spike complex that sits atop the T4 core. The T4 spike complex initially punctures the target bacterial cell surface. T6SSs actually resemble an inverted phage tail inside a bacterial cell. T6SSs translocate substrates from one cell directly into a target cell only after physical contact with the target.

Bacterial T6SSs kill or maim by translocating (firing) toxic effector molecules into bacterial competitors or into eukaryotic target cells. But do they actually fire like phage tails? Grant Jensen (California Institute of Technology; **Fig. 1A**) and John Mekalanos (Harvard University; **Fig. 1B**) wanted to find out.

The team and their colleagues used time-lapse fluorescence light microscopy to visually track what happens to a *Vibrio cholerae* T6SS sheath tagged with green fluorescent protein (VipA-GFP). As shown in **Figure 1C**, the sheaths of the type VI secretion system cycled from assembly to quick contraction to disassembly and then reassembly. Real-time video of this cycling can be seen at the links provided in the supplementary data of the paper cited at the end of this Special Topic.

When examined using cryo-EM imaging, T6SS appeared as straight, intracellular tubular structures aligned perpendicular to the cytoplasmic membrane. One structure was long, thin, and stretched almost the entire width of the cell. The second

A.

B.

FIGURE 1 ■ Extension and contraction of type VI secretion system. Grant Jensen (standing, with Martin Pilhofer at the electron microscope) **(A)** and John Mekalanos **(B)** headed the study to visualize the T6SS mechanism of *Vibrio cholerae*. **C.** Time-lapse images showing the extension of the VipA-GFP structure from one side of the cell to another (arrows), followed by a contraction event and apparent disassembly of the contracted VipA-GFP structure.

C.

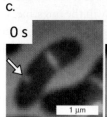
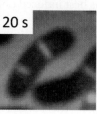

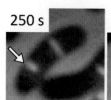
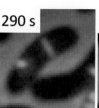

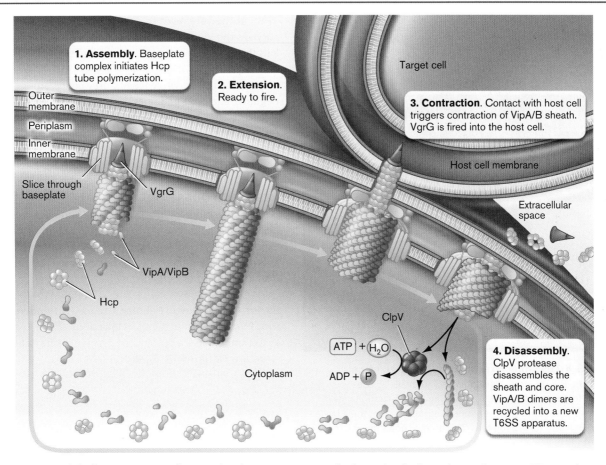

FIGURE 2 ■ Model of type VI secretion mechanism. A Gram-negative bacterium is shown here using a type VI secretion system to attack another Gram-negative bacterium. A similar series of events takes place when a pathogen attacks a eukaryotic host cell.

form was shorter, wider, and appeared contracted. The researchers concluded that these structures corresponded to what was seen expanding and contracting in the fluorescent microscopic study. They further suggested that the rapid contraction (firing) of the type VI secretion system sheath provides the energy needed to shoot proteins out of the bacterium and into an adjacent target cell.

After being fired, the VipA/B sheath and Hcp core proteins are depolymerized by another T6SS gene product, called ClpV, a type of chaperone. The authors of the study propose that the disassembled Hcp and VipA/B monomers and dimers are reused to assemble a new microbial "blowgun." The model for T6SS assembly (loading), extension (ready to fire), contraction (firing), and disassembly is illustrated in **Figure 2**.

Secretion. What do type VI secretion systems secrete? When engaged with eukaryotic cells, VgrG proteins appear to be the main effector, although some effectors can bind to VgrG and hitchhike into the target cell. VgrG proteins carry extended C-terminal regions that function in the eukaryotic host cell

cytoplasm. For example, the C-terminal extension of *Vibrio cholerae* VgrG1 carries an RtxA toxin domain that, after being fired into target cells, cross-links actin. VgrG1 from *Aeromonas hydrophila* induces host cell toxicity by ADP-ribosylating host cell actin, which ultimately triggers apoptotic cell death. Depending on the pathogen, T6SS-mediated phenotypes include aspects of adhesion, internalization, cytoskeletal rearrangements, cytotoxicity, phagocytosis, and giant-cell formation (**Fig. 3**).

In addition to delivering VgrG effector "darts" into eukaryotic cells, the T6SSs have been characterized as mediators of competitive interbacterial interactions. The only antimicrobial T6SS effector characterized to date is the *Pseudomonas aeruginosa* protein Tse2, which appears to be the toxin component of a toxin immunity system (see **eTopic 10.4**). Bacterial cells lacking Tsi2, the cognate immunity protein to Tse2, suffer arrested growth when Tse2 is expressed intracellularly. Some organisms can also sense when they are under attack by another T6SS. *Pseudomonas*, for instance, can detect a T6SS assault from *Vibrio cholerae* and assemble a retaliatory

SPECIAL TOPIC 25.1 (continued)

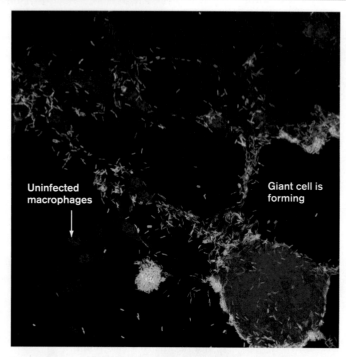

FIGURE 3 ■ T6SS-mediated multinucleate giant-cell formation. *Burkholderia thailandensis* (green, GFP) uses the T6SS-5 effector to induce the fusion of infected macrophages. Nuclei are stained blue; actin is stained red. Notice that the intracellular *Burkholderia* form actin tails. This image shows many infected macrophages fusing into a giant cell. *Source:* Paul Brett, University of South Alabama.

T6SS at the very point of attack. *Pseudomonas* somehow survives the *Vibrio* attack and launches a counterstrike that kills the *Vibrio*. How *Pseudomonas* senses the attack is still unclear.

With its role in virulence, its prevalence among Gram-negative bacterial genomes, and its predicted phage-like structure, the type VI secretion system has become an exciting area of research. Identification of pathogen-specific T6SS components and substrates may reveal novel targets for treating bacterial disease.

RESEARCH QUESTION

Before searching for, or designing, a potential antimicrobial that would target one of the T6SS components, what experiment(s) could you do to determine whether a single antimicrobial compound could potentially target T6SS from many different organisms?

Basler, Marek, Martin Pilhofer, Gregory P. Henderson, Grant J. Jensen, and John J. Mekalanos. 2012. Type VI secretion requires a dynamic contractile phage tail-like structure. *Nature* **483**:182–186.

---

in <u>unfavorable</u> body sites. Andrew Roe and his team at the University of Glasgow found that the host metabolite D-serine can inhibit EHEC T3SS expression. D-Serine is found at high concentrations at extraintestinal sites such as the urinary tract and brain, two sites that EHEC does not infect. How D-serine inhibits T3SS gene expression is not clear.

Many bacterial pathogens will sense the concentration of free iron, which is typically very low in the host, to regulate virulence genes. For instance, *Corynebacterium diphtheriae* sensing a low iron environment will induce synthesis of diphtheria toxin (see **eTopic 25.7**; Figure 7.2). The toxin kills host cells to release their iron, which the bacteria can now use. For other pathogens, iron concentration alone is usually not enough to provoke virulence. Regulators sensitive to other in vivo signals must be activated to achieve a successful infection.

Cell-cell communication is also important during infections. *Pseudomonas aeruginosa*, for example, has at least two quorum-sensing systems that detect secreted autoinducers. As the number of bacteria in a given space increases, so, too, does the concentration of the chemical autoinducer. When the autoinducer reaches a critical concentration, it diffuses back into the bacterium or binds to a surface receptor and triggers the expression of bacterial target genes. Genes included in the *P. aeruginosa* quorum-sensing regulons encode *Pseudomonas* exotoxin A and other secreted proteins, such as elastase, phospholipase, and alkaline protease.

Why would a pathogen employ quorum sensing to regulate virulence factors? One reason may be to prevent alerting the host that it is under attack before enough microbes can accumulate through replication. Tripping the host's alarms too early would make eliminating infection easy. Waiting until a large number of bacteria have amassed before releasing toxins and proteases will increase the chance that the host can be overwhelmed.

Extracellular Immune Avoidance

This topic was first covered in the discussions of host defense (Chapter 23) and immunology (Chapter 24). Many bacteria, such as *Streptococcus pneumoniae* and *Neisseria meningitidis*, produce a thick polysaccharide capsule that envelops the cell. Capsules help organisms resist phagocytosis in several ways.

Recall that phagocytes must recognize bacterial cell-surface structures or surface-bound C3b complement factor to begin phagocytosis (see Section 23.6). A capsule will cover bacterial cell wall components and mannose-containing carbohydrates that phagocytes normally use for attachment. The uniformity and slippery nature of capsule composition make it difficult for phagocytes to lock on to the bacterial cell.

But what about complement factor C3b, which can bind to the bacterial cell? Phagocytes have surface C3b receptors that latch on to C3b molecules. Capsules will envelop any C3b complement factor that binds to the bacterial surface, thereby hiding it from the phagocyte. Fortunately, immune defense mechanisms can eventually circumvent this avoidance strategy by producing opsonizing antibodies (IgG) against the capsule itself (see Section 24.2). The Fc regions of antibodies that bind to the capsule point away from the bacterium, so they are free to bind Fc receptors on phagocyte membranes. Binding of the Fc region to the phagocyte's Fc receptor triggers phagocytosis.

Pathogens can also use proteins on the cell surface to avoid phagocytosis. *Staphylococcus aureus* has a cell wall protein called **protein A** that binds to the Fc region of antibodies, hiding the bacteria from phagocytes. This works in two ways. Protein A can bind to the Fc region of any antibody before the antibody-binding site ever finds its bacterial target. Thus, the business end of the antibody—that is, the part with antigen-binding sites—is pointing away from the microbe. However, even if the antibody finds its antigen target on one bacterium, protein A from a second bacterium can bind to that Fc region, once again blocking phagocyte recognition.

Some microbes can trigger apoptosis in target host cells. Proteins made by the pathogen enter the host cell and trigger this programmed cell death. How does this help the pathogen? If a macrophage is targeted for self-destruction, it cannot destroy the microbe.

Another immune avoidance strategy used by microorganisms, both the extracellular and intracellular types, is to change their antigenic structure. Genes encoding flagella, pili, and other surface proteins often use site-specific gene inversions to express alternative proteins (for example, *Salmonella* phase variation) or slipped-strand mispairing (see eTopic 10.1) to add or remove amino acids from a sequence (see Section 10.1). You can think of these processes as shape-shifting to avoid recognition.

Intracellular Pathogens

In an effort to escape both innate and humoral immune mechanisms (see Chapters 23 and 24), many bacterial pathogens, called **intracellular pathogens**, seek refuge by invading host cells. (Viruses, of course, are, by definition, intracellular pathogens.) Hiding within a host cell temporarily provides the pathogen safe harbor from antibodies and phagocytic cells. Some bacteria dedicate their entire lifestyle to intracellular parasitism and are called **obligate intracellular pathogens**. *Rickettsia*, for example, for reasons unknown, will not grow outside a living eukaryotic cell. Other microbes, such as *Salmonella* and *Shigella*, are considered **facultative intracellular pathogens** because they can live either inside host cells or free. We have already discussed how intracellular pathogens get into cells, but how do they withstand intracellular attempts to kill them?

Once inside the phagosome, intracellular pathogens have three options to avoid being killed by a phagolysosome (**Fig. 25.29**). They can escape the phagosome, prevent phagosome-lysosome fusion, or prefer growth inside the phagolysosome.

Escaping the phagosome. The Gram-negative bacillus *Shigella dysenteriae* and the Gram-positive bacillus *Listeria monocytogenes*, both of which cause food-borne gastrointestinal disease, use hemolysins to break out of the phagosome vacuole before fusion (described earlier). By escaping the phagosome, the bacteria completely avoid lysosomal enzymes. Once free in the cytoplasm, they enjoy unrestricted growth. Yet even in the cytoplasm, these microbes have found a way to redirect host cell function to their own ends.

A fascinating aspect of escaping the phagosome involves motility. *Shigella* and *Listeria* are both nonmotile at 37°C in vitro; however, they both move around inside the host cell, even though they have no flagella. How do they move? These species are equipped with a special device at one end of the cell that mediates host cell actin polymerization. The polymerizing actin, called a "rocket tail," propels the organism forward through the cell (**Fig. 25.30A**) until it reaches a membrane. The membrane is then pushed into an adjacent cell, where the organism once again ends up in a vacuole (**Fig. 25.29**, fate 3). This strategy allows the microbe to spread from cell to cell without ever encountering the extracellular environment, where it would be vulnerable to attack. Actin motility is also a feature of some species of *Rickettsia* (**Fig. 25.31**), *Mycobacterium*, and *Burkholderia* (see Fig. 3 in **Special Topic 25.1**).

Inhibiting phagosome-lysosome fusion. Some intracellular pathogens avoid lysosomal enzymes by preventing lysosomal fusion with the phagosome. *Salmonella*, *Mycobacterium*, *Legionella*, and *Chlamydia* are good examples. For example, *Legionella pneumophila* grows inside alveolar macrophage phagosomes and produces the potentially fatal Legionnaires' disease, so named for the veterans group that suffered the first recognized outbreak, in 1976. The organism, once inside a phagosome (called a *Legionella*-containing vacuole, or LCV), uses a type IV secretion system to secrete proteins through the vesicle membrane and into the cytoplasm. These bacterial proteins interfere with the cell signaling pathways that cause phagosome-lysosome fusion. A bacterial protein called LegC7, for instance, inhibits endosome trafficking that would lead to lysosome fusion. The result is that *L. pneumophila* can grow in a friendlier vesicle. Interestingly, *L. pneumophila* is actually a soil and water microbe. Its ability to survive inside macrophages evolved from its ability to survive in amebas, which serve as a natural reservoir for this pathogen.

Salmonella Typhi is another pathogen that prevents phagosome-lysosome fusion, but this organism eventually leaves the host cell (exocytosis) and enters extracellular tissue spaces where macrophages await (**Fig. 25.29**, fate 2). The bacterium is once again engulfed, by a macrophage this time, but again survives inside the macrophage phagosome. The infected macrophage homes to a regional lymph node, enters the bloodstream, and, like a Trojan horse, disseminates *Salmonella* throughout the body.

Thought Question

25.7 Figure 25.31A shows *Shigella* forming an actin tail at one pole. Why do organisms such as *Shigella* and *Listeria* assemble actin-polymerizing proteins at only one pole?

Thriving under stress. In what could be called the "bring it on" strategy, some intracellular pathogens prefer the harsh environment of the phagolysosome. *Coxiella burnetii*, for example, grows well in the very acidic phagolysosome environment (**Fig. 25.30B**). This obligate intracellular organism (an organism that grows only inside another living cell) causes a flu-like illness called Q fever (query fever). The symptoms of Q fever include sore throat, muscle aches, headache, and high fever. The illness has a mortality rate of about 1%, so most people recover to good health. The organism allows phagosome-lysosome fusion because the acidic environment that results is needed for it to survive and grow.

Thought Question

25.8 How can you determine whether a bacterium is an intracellular pathogen?

Why some bacteria are obligate intracellular pathogens is unclear. One intracellular bacterium, *Rickettsia prowazekii*, a cause of epidemic typhus, appears to be an "energy

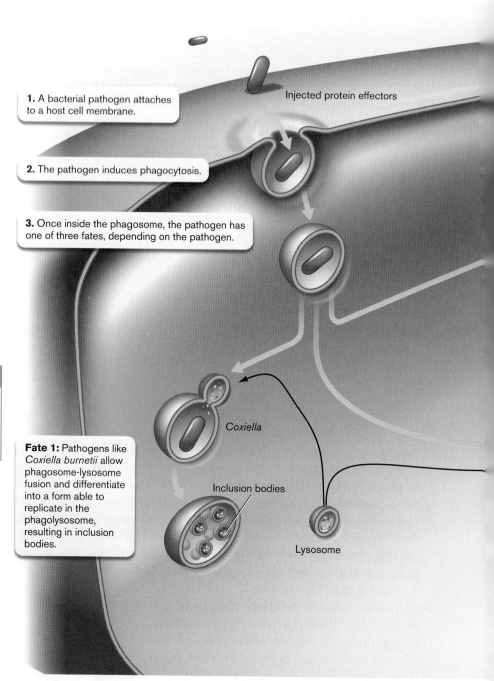

1. A bacterial pathogen attaches to a host cell membrane.

Injected protein effectors

2. The pathogen induces phagocytosis.

3. Once inside the phagosome, the pathogen has one of three fates, depending on the pathogen.

Coxiella

Fate 1: Pathogens like *Coxiella burnetii* allow phagosome-lysosome fusion and differentiate into a form able to replicate in the phagolysosome, resulting in inclusion bodies.

Inclusion bodies

Lysosome

FIGURE 25.29 ■ Alternative fates of intracellular pathogens. Different pathogens have different strategies for surviving in a host cell. Some tolerate phagolysosome fusion (*Coxiella*), others prevent phagolysosome fusion (*Salmonella*), and still others escape the phagosome to replicate in the cytoplasm (*Shigella* and *Listeria*).

parasite" that can transport ATP from the host cytoplasm and exchange it for spent ADP in the bacterium's cytoplasm. But this does not explain its obligate intracellular status, since giving *Rickettsia* ATP outside a host does not allow the bacterium to grow. Other factors remain to be discovered.

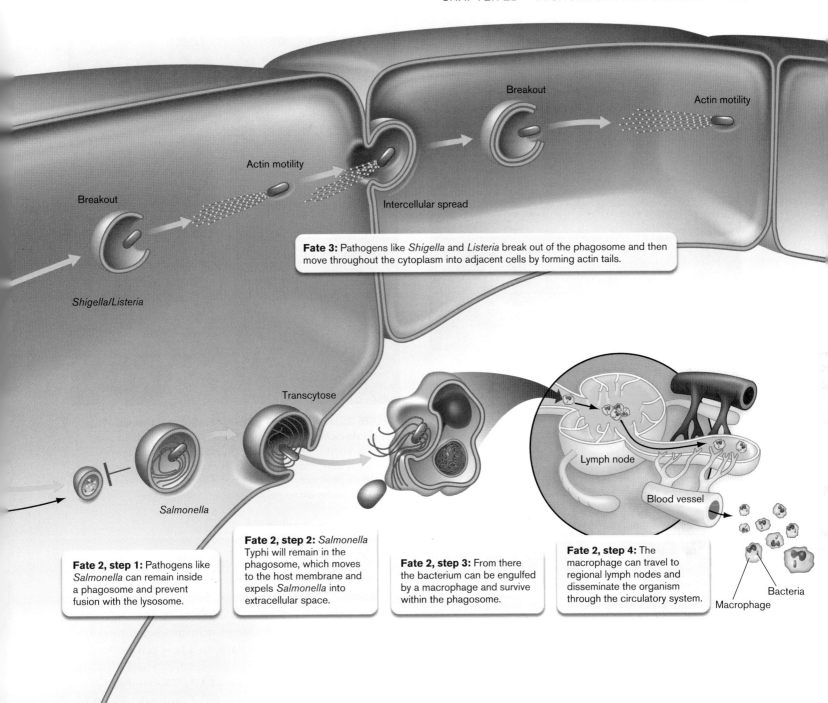

Breakout

Actin motility

Breakout

Actin motility

Actin motility

Intercellular spread

Fate 3: Pathogens like *Shigella* and *Listeria* break out of the phagosome and then move throughout the cytoplasm into adjacent cells by forming actin tails.

Shigella/Listeria

Transcytose

Salmonella

Lymph node

Blood vessel

Fate 2, step 1: Pathogens like *Salmonella* can remain inside a phagosome and prevent fusion with the lysosome.

Fate 2, step 2: *Salmonella* Typhi will remain in the phagosome, which moves to the host membrane and expels *Salmonella* into extracellular space.

Fate 2, step 3: From there the bacterium can be engulfed by a macrophage and survive within the phagosome.

Fate 2, step 4: The macrophage can travel to regional lymph nodes and disseminate the organism through the circulatory system.

Bacteria

Macrophage

Thought Question

25.9 Why might killing a host be a bad strategy for a pathogen?

Sleeping with the Enemy

As just described, many bacterial and, of course, viral pathogens find safe haven by growing inside host cells. However, from our knowledge of innate and adaptive immune responses, it is not intuitively obvious why this is so. After all, infected cells present microbial antigens on their class I or class II MHC receptors to alert the innate and adaptive immune systems that the infected host cell must be killed to resolve the infection. In addition, pieces of intracellular microbes (flagella, LPS, peptidoglycan) will bind pattern recognition receptors (TLRs and NLRs) that activate intracellular inflammasomes. Inflammasomes trigger production of pro-inflammatory cytokines that mediate inflammation.

So how do intracellular pathogens avoid destruction? It turns out that these invaders employ a variety of molecular tricks that misdirect the immune system much as a magician misdirects an audience. All of these strategies, summarized

A. *Shigella flexneri* (red) and actin tails (green)

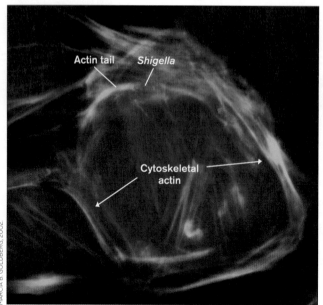

Actin tail *Shigella*

Cytoskeletal actin

B. *Coxiella burnetii*

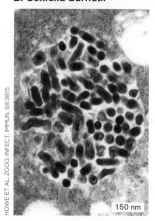
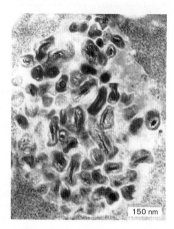

150 nm

150 nm

FIGURE 25.30 ▪ **Intracellular pathogens: *Shigella* motility and *Coxiella* development.** **A.** Intracellular *Shigella flexneri* (fluorescence microscopy), fluorescently stained red (1 μm in length), moves through the host cytoplasm propelled by actin tails, stained green. **B.** A typical vacuole in J774A.1 mouse macrophage cells infected with *Coxiella burnetii* at 2 hours (left) and 6 hours (right) postinfection (TEM). The organism lives in an acidified vacuole and undergoes a form of differentiation that changes its shape and alters its interactions with the host cell.

in **Figure 25.32**, are designed to buy the microbe more time to overwhelm the host.

Molecular mimicry and subverting antigen presentation.

A variety of bacteria and viruses use mimicry to confuse the immune system (discussed in Section 24.7). In some cases, microbial proteins are made that look like cytokines or that bind to host cells and hitchhike via normal host trafficking to the nucleus, where the bacterial protein interferes with cytokine gene expression. One example involving *Shigella* is discussed in **eTopic 25.9**. These factors can manipulate

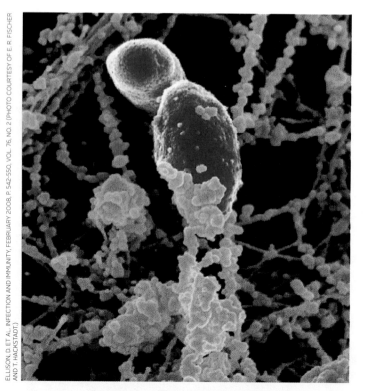

FIGURE 25.31 ▪ **The obligate intracellular pathogen *Rickettsia rickettsii.*** This SEM shows *R. rickettsii* (blue, approx. 0.7 μm in length), the cause of Rocky Mountain spotted fever, in association with host actin (gold). Several pathogens propel themselves through host cytoplasm by polymerizing host actin at one pole of the bacterial cell.

the balance of helper T cells, for example, and send immunity down the wrong path for combating the microbe.

Microbes, especially viruses, can also interfere with antigen presentation on the surfaces of infected cells. Recall from Chapter 24 that proteins made by infectious agents growing in the cytoplasm of a host cell are broken down in the cytosol and transported into the ER by the transporter of antigen peptides (TAP). The viral or bacterial antigens are loaded onto MHC I molecules found on the ER membrane, and the complexes are sent to the cell surface. Surface MHC I presents those peptide antigens to roaming CD8 cytotoxic T cells that will then kill the infected cell. Any scheme that interrupts MHC I presentation will spare the infected cell and its infectious cargo from destruction. So how do pathogens derail MHC presentation?

Collectively, pathogens have four basic ways to subvert antigen presentation: (1) Make proteins that resist digestion by host proteasomes so that antigens are not processed. (2) Make a protein that blocks the TAP protein so that processed antigens cannot be loaded onto MHC I (herpes viruses such as HSV, CMV, and VZV are famous for this approach). (3) Make a protein that induces TAP

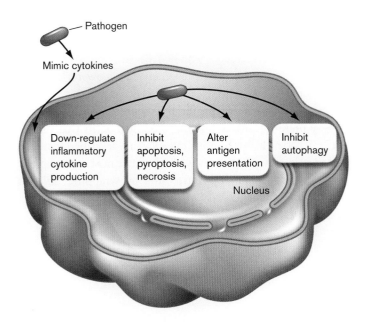

FIGURE 25.32 ■ **Summary of microbial strategies that misdirect the immune system.** Bacteria and viruses produce different molecules that can mimic cytokines or transcriptional regulators, alter cytokine production, prevent programmed cell death, alter antigen presentation, or inhibit autophagy.

degradation via ubiquitylation (*Pseudomonas aeruginosa*). (4) Induce degradation of MHC molecules via ubiquitylation. Ubiquitylation strategies that coerce the host to degrade MHC proteins are described later.

Flipping cytokine profiles. Many pathogens defy death by growing inside macrophages that are "armed" with numerous antimicrobial weapons. Pathogens that can grow inside macrophages include bacteria such as *Salmonella, Yersinia, Listeria, Mycobacterium tuberculosis, Francisella tularensis*, and *Chlamydia*; as well as protozoans such as *Leishmania* and *Trypanosoma cruzi*. How do they survive? These, and many other pathogens, interrupt host cell signaling pathways designed to activate macrophage antimicrobial mechanisms. To perform this trick, most intracellular pathogens inhibit production of pro-inflammatory cytokines, such as TNF-alpha, IL-8, IL-12, and IFN-gamma, but encourage production of anti-inflammatory cytokines like IL-10, TGF-beta (transforming growth factor beta), and IL-4.

Mycobacterium, T. cruzi, and *Leishmania* also down-regulate the expression of host membrane receptors for IFN-gamma (thereby inhibiting inflammation), and interfere with downstream regulators that activate the production of MHC class I proteins, which dulls host immune mechanisms. Reduced MHC I production means less antigen presentation, fewer activated T_H1 helper cells, and, thus, fewer cytotoxic T cells to attack the infected cell.

Stopping programmed cell death. A macrophage that fails to eliminate infecting pathogens will eventually give up and try to kill itself and the infecting pathogens through one of three programmed-cell-death pathways (apoptosis, necrosis, or pyroptosis). This is a last resort to clear the infection. Apoptotic cells maintain membrane integrity during apoptotic death and are engulfed by nearby phagocytes. Inflammation is not provoked, because cytokines are not released. In contrast, necrosis and pyroptosis initiate rapid inflammatory responses by secreting inflammatory cytokines. All of these mechanisms can kill the intracellular pathogen.

To "keep hope alive" and retain their intracellular niche, intracellular microbes can prevent host cell suicide by interfering with the molecular signals that initiate host death programs, or they can activate pro-survival mechanisms. Some pathogens even synthesize microbial mimics of host anti-apoptotic proteins. *Yersinia enterocolitica* and *Mycobacterium tuberculosis*, for instance, prevent pyroptotic cell death by inhibiting inflammasome formation and caspase-1 activation. Caspase-1 is a protease that initiates pyroptosis.

Herpes viruses (discussed in Chapter 11), which cause cold sores, genital herpes, mononucleosis, and some cancers, can enter into a latent state. Latency requires shutting down viral replication and integrating viral DNA into a host genome. But why doesn't the virus trigger apoptosis in the host cell before becoming latent? Herpes viruses produce a number of small RNA molecules called microRNAs (miRNAs) that can interfere with the apoptosis program. One such miRNA prevents the translation of two host cell proteins required for cell death. The miRNAs bind to regions in the apoptosis mRNAs and promote their degradation. Once latent, the virus no longer produces any viral proteins, so the immune system of the infected individual cannot detect the infected cell.

Some intracellular pathogens (*Trypanosoma cruzi, Leishmania*, and *Yersinia pseudotuberculosis*) actually tilt macrophage suicide pathways toward apoptosis to promote pathogen dissemination (recall that apoptotic cells are engulfed by other phagocytes). Some intracellular pathogens both inhibit and activate host cell death—just not at the same time. Early in infection, *M. tuberculosis* (the cause of tuberculosis) orchestrates the inhibition of host cell suicide pathways to enable the organism to grow, but then later it promotes suicide as a way to disseminate.

Autophagy is a highly regulated mechanism by which eukaryotic cells form intracellular vesicles around damaged organelles to scavenge them for nutrients. Autophagy is also used as a universal innate defense mechanism to fight intracellular pathogens (**Fig. 25.33**). Autophagic vacuoles (autophagosomes) can encase these pathogens and deliver them to degradative lysosomes for destruction. Pathogen components are then sent to endosomes,

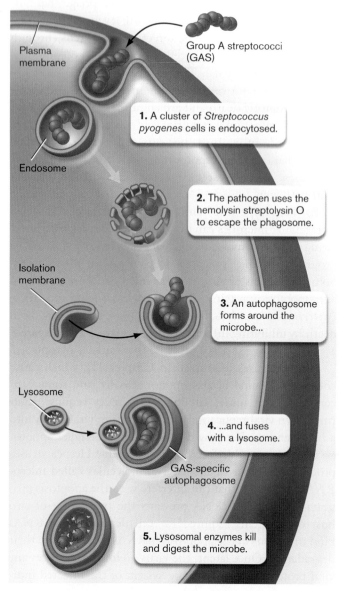

1. A cluster of *Streptococcus pyogenes* cells is endocytosed.

2. The pathogen uses the hemolysin streptolysin O to escape the phagosome.

3. An autophagosome forms around the microbe...

4. ...and fuses with a lysosome.

5. Lysosomal enzymes kill and digest the microbe.

Group A streptococci (GAS)

Plasma membrane

Endosome

Isolation membrane

Lysosome

GAS-specific autophagosome

FIGURE 25.33 ■ **Autophagy as an innate immune mechanism.** When a pathogen escapes the phagosome, the host cell will try to form an intracellular vacuole (autophagosome) around the organism in a second attempt to kill it.

where microbial structures are recognized by endosomal Toll-like receptors that trigger the innate immune system (see Chapter 23). The microbial components (antigens) are also sent to cell compartments rich in major histocompatibility complex II (MHC II) molecules. MHC II molecules then rise to the cell surface and present the microbial antigens to the adaptive immune system as described in Chapter 24.

Intracellular pathogens, however, have evolved mechanisms that can prevent autophagy and prolong their survival. For example, the Nef protein of human immunodeficiency virus (HIV) and protein M2 of influenza prevent autophagosome formation by targeting beclin 1, a

protein central to autophagosome production. Interactome studies indicate that RNA viruses, as a group, encode proteins that can interact with 35% of autophagy-associated proteins, suggesting that autophagy is widely targeted by pathogens.

Shigella, a cause of bacterial dysentery, avoids autophagy by making tails of polymerized host actin that propel the microbe through host cytoplasm (described earlier). This tactic may sometimes work, but the host can stop the bacterium from making actin tails by wrapping *Shigella* in septin filaments, as shown in the 3D rendering in **Figure 25.34**. Septin cages initially require actin to form, but then they inhibit actin polymerization. The septin-caged microbe is now trapped and marked for autophagy. One of the discoverers of autophagy, Yoshinori Ohsumi, won the 2016 Nobel Prize for Physiology or Medicine.

Redirecting host ubiquitylation signals. How do bacteria and viruses misdirect the immune system? One way is to target a regulatory mechanism common to many host systems—namely, ubiquitylation (ubiquitination). Ubiquitin is a highly conserved, 76-amino-acid polypeptide in eukaryotes that can be covalently attached to other proteins. Attachment involves an enzymatic cascade of three enzymes: E1, E2, and E3 (reviewed in **eTopic 8.5**). Depending on where ubiquitin is placed on a protein, the protein can be activated or, alternatively, tagged for destruction by the host proteasome. While there are only a few E1 and E2 enzymes, there are hundreds of E3 ubiquitin ligase enzymes that recognize and tag target proteins with polyubiquitin. There are also deubiquitylation enzymes that can reverse the process.

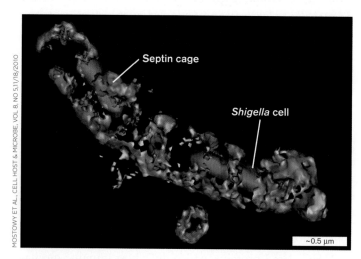

Septin cage

Shigella cell

~0.5 μm

FIGURE 25.34 ■ **Entrapment of intracytoplasmic bacteria by septin cage-like structures.** *Shigella* may move in the host cytoplasm by actin tails or may become trapped by a septin cage and marked for autophagy.

In contrast to their eukaryotic hosts, viral and bacterial pathogens lack ubiquitylation systems but have evolved E3 ligases and deubiquitylases that effectively subvert normal host ubiquitylation pathways (**Fig. 25.35**). Like someone changing signs on a highway, the viral enzymes cause host signaling systems, and the immune system, to veer off course.

There are several ways ubiquitylation normally directs the immune system. In the innate immune system, TLR pathway proteins are ubiquitylated as a way to induce production of inflammatory cytokines. When a MAMP binds to a TLR, ubiquitylation of signal induction proteins will activate the signal. In contrast, proteins that inhibit the pathway are marked by ubiquitin for destruction. The result is an activated signal pathway that leads to the formation of inflammatory cytokines (**Fig. 25.35**, step 1).

To subvert this system, some pathogens produce their own E3 ligases that divert the normal signal induction pathways. For example, rotavirus, a major cause of infant diarrhea, produces an E3 ligase that adds ubiquitin to activators of NF-kappaB, the transcription regulator that induces cytokine production (**Fig. 25.35**, step 2). The ubiquitylated proteins are destroyed, and NF-kappaB is not activated. Using a different strategy, the bacterial pathogen *Salmonella enterica* produces a deubiquitylase that

removes ubiquitin from a normal inhibitor of NF-kappaB. The inhibitor is not degraded, which means that NF-kappaB is not activated and inflammatory cytokines are not made (step 3). In both instances, inflammation is minimized, allowing the pathogen to survive.

Ubiquitylation is an important mechanism for adaptive immunity too. For example, the process regulates when cell-surface MHC class I and II molecules are expressed in dendritic cells (**Fig. 25.36**). Before dendritic cells mature, MHC molecules are polyubiquitylated and degraded, limiting their placement on cell surfaces. After maturation, however, the MHC molecules are no longer ubiquitylated and, as a result, accumulate on the cell surface, making dendritic cells better equipped for antigen presentation. A number of viruses exploit this process by producing viral E3 ligases that polyubiquitylate MHC proteins (**Fig. 25.36**). The MHC molecules that would present viral proteins to the immune system are degraded.

Some pathogens even make E3 ligases that ubiquitylate host proteins that are not normally tagged. One example

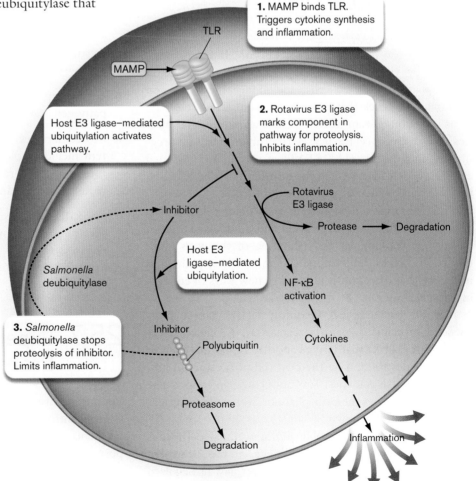

FIGURE 25.35 ■ Microbial E3 ligases and deubiquitylases alter innate immune systems. When a MAMP binds to a Toll-like receptor (TLR), a series of events, including a host E3 ligase–mediated ubiquitylation, activates the transcriptional regulator NF-kappaB, which activates transcription of inflammatory cytokine genes. The cytokines are secreted and initiate inflammatory processes. Microbial E3 ligases (for example, rotavirus E3 ligase) can ubiquitylate other components of the NF-kappaB activation pathway and mark them for destruction. As a result, cytokine synthesis is inhibited and inflammation is limited. *Salmonella* makes a deubiquitylase that removes polyubiquitin from an inhibitor of NF-kappaB activation. Deubiquitylation saves the inhibitor from destruction, allowing continued inhibition of NF-kappaB activation.

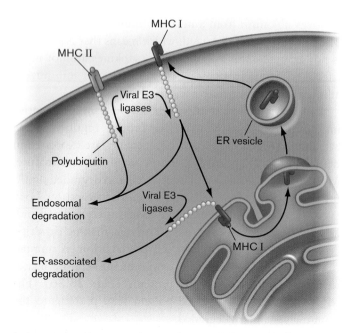

FIGURE 25.36 ■ **Microbial E3 ligases and deubiquitylases alter adaptive immune systems.** MHC class I and class II molecules present antigens on host cell surfaces to helper T cells. Some viral E3 ligases can ubiquitylate the MHC molecules, marking them for degradation via endosomal pathways. Other E3 ligases can ubiquitylate MHC I molecules while in the endoplasmic reticulum (ER), leading to degradation of the MHC I receptors before they can be placed on the cell surface.

is found in the interferon signal cascade. Interferons are secreted by virus-infected host cells, bind to interferon receptors on uninfected cells, and induce the expression of numerous genes that protect the uninfected cell from virus infection. Paramyxoviruses such as mumps and measles viruses produce E3 ligases that ubiquitylate key regulatory components (JAK-STAT) of the interferon signal cascade, which marks them for destruction. The host cell then becomes quite vulnerable to virus attack.

These examples demonstrate that ubiquitylation is a crucial part of immune regulation and a popular target of pathogens. Gaining an understanding of the ways ubiquitylation directs the immune system and how pathogens influence that direction could lead to new ways of enhancing host defense.

Not all bacterial E3 ubiquitin ligases target host-encoded proteins. *Legionella* secretes an E3 ligase effector protein (LubX) that targets another *Legionella* effector (SidH) for destruction by the host cell proteasome. LubX triggers SidH degradation late during the infection. The role of SidH during the early phase of *Legionella* infection, however, is not yet known.

Clearly, facultative and obligate intracellular pathogens have developed many ingenious strategies to take control of host cell defenses, but the host, for its part, has also evolved effective countermeasures.

Thought Question

25.10 You have discovered an E3 ligase that specifically modifies surface proteins of *Listeria monocytogenes*. Using what has been presented about toxin entry and protein secretion in Chapter 25, suggest some strategies for delivering this enzyme into host cells infected with this pathogen.

To Summarize

■ **Two-component signal transduction systems** can regulate virulence gene expression in response to the host environment.

■ **Quorum sensing** may prevent pathogens from releasing toxic compounds too early during infection.

■ **Extracellular pathogens** evade the immune system by hiding in capsules, by changing their surface proteins, or by triggering apoptosis.

■ **Intracellular bacterial pathogens** attempt to avoid the immune system by growing inside host cells. They use different mechanisms to avoid intracellular death.

■ **Hemolysins** are used by certain pathogens to escape from the phagosome and grow in the host cytoplasm.

■ **Actin tails** are used by some microbes to move within and between host cells.

■ **Inhibiting phagosome-lysosome fusion** is one way pathogens can survive in phagosomes.

■ **Molecular mechanisms for avoiding the immune system** include molecular mimicry, altering cytokine profiles, stopping programmed host cell death, interfering with autophagy, and redirecting ubiquitylation signals.

■ **Specialized physiologies** enable some organisms to survive in the normally hostile environment of fused phagolysosomes.

25.7 Experimental Tools That Probe Pathogenesis

For any scientist studying pathogenic microbes, it is essential to identify virulence genes and determine what they do. As mentioned in Section 25.2, Falkow's molecular Koch's postulates argue that once a gene has been singled out as a possible virulence factor, proof requires knocking out the gene and observing a decrease in virulence, followed by restoring virulence by replacing the mutant gene. But how do you identify which genes to test? Earlier, in Section 25.2 (and see also **eTopics 25.1** and **25.2**), we discussed how virulence genes in microbes can be identified using some clever selection techniques. Today there exist amazing molecular techniques that can broadly, and quickly, reveal a microbe's overall pathogenic strategy and help scientists focus on likely virulence genes. In addition, powerful techniques can now deeply probe how pathogens and hosts respond to each other during an infection.

Genomics

The sequencing of a pathogen's genome (discussed in Chapter 7), followed by bioinformatic analysis of the sequence (described in Chapter 8), can yield valuable information about a pathogenic organism's metabolism and can identify potential pathogenicity islands and virulence genes.

Clues about potential virulence genes can also be gained by comparing the sequences of virulent and attenuated strains of a pathogen. For example, *Leptospira* species are spirochetes that cause a zoonotic renal disease in humans called leptospirosis. Dereck Fouts (J. Craig Venter Institute), Joseph Vinetz (UC San Diego), and colleagues recently compared the genome sequences of numerous *Leptospira* species differentially classified as pathogenic, intermediate, and nonpathogenic. They discovered a subset of genes whose distribution was restricted to the pathogenic strains, meaning that these genes potentially contribute to *Leptospira* pathogenesis or host adaptation. The pathogen-unique genes included, among others, a catalase, a protease able to degrade complement, a large family of virulence-modifying proteins of unknown function, and a CRISPR-Cas system. The results provided many new directions for research that probes leptospiral pathogenesis.

DNA sequencing can also reveal how to grow so-called obligate intracellular pathogens in the laboratory. The Gram-positive microorganism *Tropheryma whipplei* is an intracellular pathogen that causes the gastrointestinal illness called Whipple's disease. The symptoms of Whipple's disease include diarrhea, intestinal bleeding, abdominal pain, loss of appetite, weight loss, fatigue, and weakness. Though identified, the causative agent had never been grown outside of fibroblast cells, making it nearly impossible to study its physiology. Once the complete genome sequence of *T. whipplei* became available, however, Didier Raoult and colleagues at the Marseille School of Medicine discovered that the organism lacks the machinery to make several amino acids. The investigators were then able to design a cell-free culture medium that supported the growth of *T. whipplei*. Similar approaches were used to coax cell-free growth of the so-called obligate intracellular

pathogen *Coxiella burnetii*. *C. burnetii* causes the respiratory infection Q fever. Genomic strategies like these should lead to other successes in growing previously unculturable intracellular pathogens.

Transcriptomics

Other major goals of infectious disease research include learning how a pathogen causes an infection, how the host responds to the infection, and how the pathogen responds to the host's response. The situation is much like watching two armies waging war with numerous weapons at their disposal. Is there a way to view the molecular drama of host-pathogen interactions? There is now.

Dual RNA sequencing. RNAseq analysis has been used to separately monitor transcript modulation in pathogens during the course of an infection, or to view the host's transcriptional response to a pathogen. Jörg Vogel (**Fig. 25.37A**) at the University of Würzburg and colleagues recently used RNAseq to <u>simultaneously</u> profile the changing host and pathogen transcriptomes during *Salmonella* infection of human host cells. The scientists started by infecting human cells with GFP-labeled *Salmonella* and using fluorescence-activated cell sorting (FACS; see Section 4.3) to select host cells at two different stages of infection: 4 hours (10 bacteria per cell) and 24 hours (75 bacteria per cell) (**Fig. 25.37B**, see also **Fig. 25.27**). The researchers then used Illumina technology (see Sections 7.6 and 10.6) to simultaneously sequence pathogen and host cell transcripts. Their analysis showed that 4 hours postinfection, bacterial transcripts of SPI-1 genes <u>decreased</u> relative to levels just prior to infection, while the transcript levels of SPI-2 genes <u>increased</u> (**Fig. 25.37C**). Their analysis also found that host genes activated by NF-kappaB were elevated at 4 hours postinfection (**Fig. 25.37D**). NF-kappaB is a transcriptional activator of cytokine genes. The NF-kappaB result illustrates the host response to *Salmonella* invasion.

The scientists then refined the strategy to identify *Salmonella* small RNA transcripts that became induced during infection. The most highly induced was the 80-nt sRNA called PinT (<u>P</u>hoP-induced sRNA in <u>int</u>racellular *Salmonella*). This sRNA was found to control the transition from SPI-1 to SPI-2 expression. By altering the timing of this transition, PinT also influenced many other aspects of host cell physiology, including production of cytokine IL-8 and mitochondrial gene expression. **Figure 25.38** shows how the loss of PinT led to relocalization of mitochondria (red) in infected host cells.

The simultaneous examination of host and pathogen transcriptional profiles by dual RNAseq techniques will reveal other hidden gene functions in pathogens with profound effects on host cells.

A.

B.

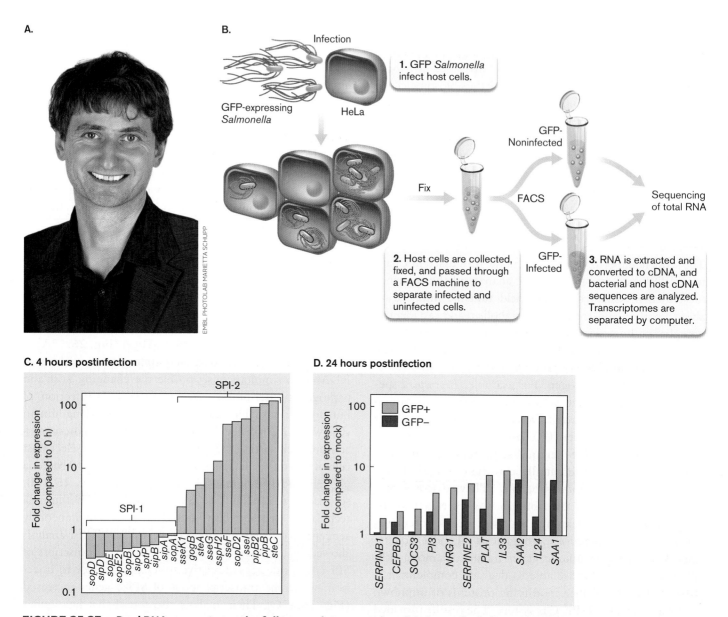

EMBL PHOTOLAB MARIETTA SCHUPP

C. 4 hours postinfection

D. 24 hours postinfection

FIGURE 25.37 ▪ **Dual RNAseq captures the full transcript repertoire of *Salmonella*-infected human cells.** **A.** Jörg Vogel uses dual RNAseq to study the interactions between pathogens and their hosts. **B.** Dual RNAseq work flow. **C.** Compared to extracellular *Salmonella* (0 h), intracellular bacteria at 4 hours postinfection repress SPI-1 and induce SPI-2 effector genes. **D.** Invaded (GFP⁺) host cells at 24 hours postinfection activate NF-kappaB-associated immunity genes. *Source:* Alexander J. Westermann et al. 2016. *Nature* **529**:496–501, fig. 1A, C, E.

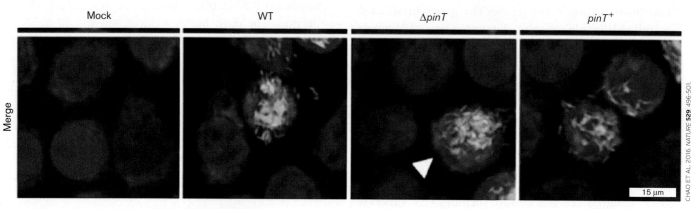

CHAO ET AL. 2016. *NATURE* **529**:496–501.

FIGURE 25.38 ▪ **Impact of PinT on host mitochondria.** Uninfected (mock) host cells were compared to host cells infected with wild-type (WT) *Salmonella*, Δ*pinT Salmonella*, or complemented *pinT⁺ Salmonella* (a Δ*pinT* strain containing a separate copy of wild-type *pinT⁺*). The white arrowhead points to a prominent cluster of relocalized mitochondria. Green = *Salmonella*; red = mitochondria; blue = nuclei.

Cell Biology

You have seen numerous examples throughout this and previous chapters in which fluorescent stains, proteins (GFP), and antibodies combined with fluorescent microscopy were used to identify dramatic alterations in host cell structures, as well as the movement/location of bacterial or host cell proteins during the course of an infection. A relatively new fluorescent technology has been developed that can identify which host cells during an infection have been targeted by translocated microbial effector proteins. The affected host cells fluoresce and can be sorted by a FACS machine for closer examination.

The technique relies on **fluorescence resonance energy transfer**, **FRET** (**Fig. 25.39**). FRET employs a fluorescent probe molecule composed of a donor fluorophore linked to an acceptor fluorophore. The emission spectrum of the donor overlaps with the absorption spectrum of the acceptor, essentially quenching donor fluorescence. Breaking the link between the two fluorophores, however, separates them. The donor fluorophore, now free from the quenching molecule, will produce strong, detectible emission upon excitation. This FRET strategy has been used to identify subsets of host cells targeted by any of the type III, IV, or VI secretion systems that pathogens employ to translocate effector proteins into host cells.

The technique involves constructing a translational fusion between a gene encoding a bacterial effector protein and a beta-lactamase gene (**Fig 25.39**, step 1; see Chapter 12). Beta-lactamase is an enzyme that cleaves the antibiotic penicillin and will also cleave the fluorescent probe (see Chapter 27). Once expressed, the fusion protein can be translocated from the pathogen directly into host cells (step 2), but not all types of host cells are targeted. Host cells that receive the protein can be identified by adding the fluorescent reporter CCF4-AM. The quenched fluorescent reporter enters all host cells, but fluorophore and quencher can be separated only by host cells that received the beta-lactamase-effector fusion protein (step 3). Fluorescence microscopy or FACS analysis can then identify those cells.

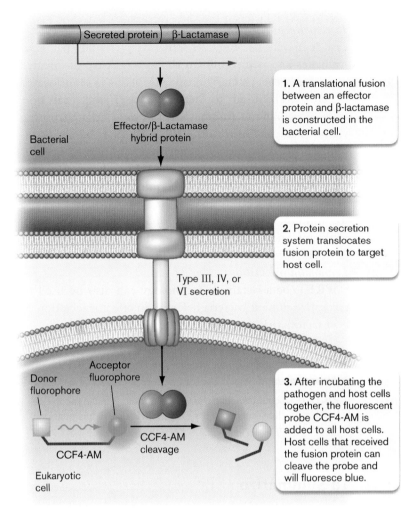

1. A translational fusion between an effector protein and β-lactamase is constructed in the bacterial cell.

2. Protein secretion system translocates fusion protein to target host cell.

3. After incubating the pathogen and host cells together, the fluorescent probe CCF4-AM is added to all host cells. Host cells that received the fusion protein can cleave the probe and will fluoresce blue.

FIGURE 25.39 ■ **FRET detection of bacterial effector protein translocation to host cells.** The emission wavelength of the released donor fluorophore is detected (blue). The emission wavelength of the acceptor fluorophore (green) is different from that of the donor.

Sunny Shin (**Fig. 25.40A**) and colleagues at the University of Pennsylvania used this technology to identify which in vivo host cells are targeted by the T4SS of

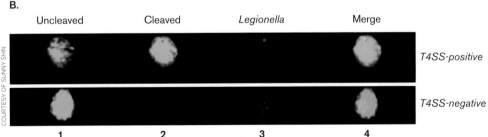

FIGURE 25.40 ■ *Legionella pneumophila* **translocates BlaM-RalF into mouse macrophages. A.** Sunny Shin studies how bacterial pathogens manipulate host defenses. **B.** Macrophage monolayers were infected with *dot*⁺ (T4SS-positive) and Δ*dot* (T4SS-negative) *Legionella* and treated with CCF4-AM fluorescent probe. **Column 1:** Uninfected cells were scanned for uncleaved probe (green). **Column 2:** Infected host cells were scanned for cleaved probe (blue). **Column 3:** Infected cells were immunostained for *Legionella* (red). The immunostain does not differentiate live from dead bacteria. **Column 4:** Merged images from columns 1, 2, and 3.

Legionella pneumophila, the cause of the respiratory disease legionellosis (**Fig. 25.40B**). They found that macrophages and neutrophils in the airway space of infected mice were the primary recipients of T4SS-translocated effector RalF(fused to beta-lactamase) and that host cells receiving the fusion were the only cells harboring viable bacteria. **Figure 25.40B** shows that the T4SS system (encoded by the *dot* genes) was required to translocate the effector protein into the host cell. Shin's group also found that host cells containing viable *L. pneumophila* secreted TNF-α and IL-1α, two cytokines needed for immune system clearance of infection. Thus, macrophages and neutrophils play a dual role as intracellular niche for these bacteria and immune mediator during *Legionella* infection.

FRET/beta-lactamase technology can also be used to recognize mutant bacteria defective in effector secretion or host factors that contribute to effector translocation.

Thought Question

25.11 How could you use the FRET/beta-lactamase technique to identify chemical inhibitors of type II, IV, or VI translocation?

To Summarize

- **Genome sequence analysis** of pathogenic and non-pathogenic strains of a species can identify potential virulence genes.

- **DNA sequencing** revealed growth requirements for the intestinal pathogen *Tropheryma whipplei*.

- **Dual RNA sequencing** of pathogen and host transcripts from infected and uninfected host cells can expose a complex, yet orchestrated, response to infection.

- **Fluorescence resonance energy transfer (FRET)** approaches can identify host cells targeted by pathogens for type III, type IV, and type VI effector protein delivery.

Concluding Thoughts

This chapter has only scratched the surface of all that is known about microbial pathogenesis. It should be clear at this point that transmission, attachment, immune avoidance, and subversion of host signaling pathways are common goals of most successful pathogens, whether bacterial, fungal, viral, or parasitic. But even with all we know, there is much left to learn.

The remaining chapters deal with the basic principles used to diagnose disease and eradicate offending pathogens, but many pathogens are still hard to detect and difficult, if not impossible, to kill. So, the more we know about the mechanisms that microbes use to cause disease, the better we will be at developing effective countermeasures.

As you will learn in the coming chapters, new pathogens are constantly emerging. Over the last few decades, we have seen the development of Zika virus, HIV, SARS, avian flu, hantavirus, West Nile virus, *E. coli* O157:H7, and the reemergence of flesh-eating streptococci, to name but a few. Can we ever stop pathogens from emerging? Probably not. For every countermeasure we develop, nature designs a counter-countermeasure. Our hope is that continuing research into the molecular basis of pathogenesis and antimicrobial pharmacology will keep us one step ahead.

CHAPTER REVIEW

Review Questions

1. Describe the differences between infection and disease; pathogenicity and virulence; LD_{50} and ID_{50}.
2. What is meant by direct versus indirect routes of infection?
3. What are the characteristics of a good reservoir for an infectious agent?
4. Name the various portals of entry for infectious agents, and a disease associated with each.
5. Describe the basic features of a pathogenicity island.
6. Explain various ways in which bacteria can attach to host cell surfaces.
7. Describe the basic steps by which pili are assembled on the bacterial cell surface. How do type I and type IV pili differ?
8. Explain the nine broad categories of toxin mode of action.
9. What is ADP-ribosylation, and how does it contribute to pathogenesis?
10. Explain the differences between exotoxins and endotoxins.

11. Explain the mechanisms of secretion carried out by type II and type III protein secretion systems. What are the paralogous origins of these systems?

12. Describe the key features of *Salmonella* pathogenesis.

13. How can genomic approaches help identify pathogens in an infection?

14. What different mechanisms do intracellular pathogens use to survive within the infected host cell?

15. Describe different molecular strategies that microbes use to avoid the immune system.

16. How do bacteria determine whether they are in a host environment?

17. Discuss the relationships between ubiquitylation and intracellular pathogens.

Thought Questions

1. Why do new versions of swine and avian flu often originate in Asia?

2. How can you modify Koch's postulates to prove that a bacterial gene is a virulence factor?

3. How would you determine whether a particular pilus on group A streptococci (GAS) is required for the organism's pathogenesis? Use a tissue culture model.

4. Why have humans not developed resistance to microbial toxins?

5. You want to make a live oral vaccine for cholera. But, because *Vibrio cholerae* is an acid-sensitive organism, the person to be immunized would have to ingest large numbers of organisms. *E. coli*, on the other hand, is very acid resistant and able to survive stomach acidity for long periods of time. You think that moving the acid resistance system from *E. coli* to *V. cholerae* will solve this problem. Is there an ethical issue to consider when trying to move the acid resistance system from *E. coli* into *V. cholerae*?

Key Terms

adhesin (998)
ADP-ribosyltransferase (1008)
airborne transmission (993)
autophagy (1027)
ectoparasite (991)
edema factor (EF) (1010)
endoparasite (991)
endotoxin (1003)
exotoxin (1003)
facultative intracellular pathogen (1023)
fecal-oral route of transmission (994)
fimbria (998)
fluorescence resonance energy transfer (FRET) (1033)
fomite (993)
genomic island (996)
hemolysin (1005)

horizontal transmission (994)
immunopathogenesis (994)
infection (991)
infection cycle (992)
infectious dose 50% (ID_{50}) (992)
intimin (1016)
intracellular pathogen (1023)
labile toxin (LT) (1008)
latent state (992)
lethal dose 50% (LD_{50}) (992)
lethal factor (LF) (1010)
leukocidin (1005)
obligate intracellular pathogen (1023)
opportunistic pathogen (991)
parasite (990)
parenteral route of transmission (994)
pathogen (991)
pathogenesis (990)

pathogenicity (992)
pathogenicity island (996)
petechia (1013)
pilus (998)
primary pathogen (991)
protective antigen (PA) (1010)
protein A (1023)
reservoir (994)
transovarial transmission (994)
type II secretion system (1015)
type III secretion system (T3SS) (1015)
vector (993)
vehicle transmission (993)
vertical transmission (994)
virulence (992)
virulence factor (995)

Recommended Reading

Basler, Marek, Martin Pilhofer, Gregory P. Henderson, Grant J. Jensen, and John J. Mekalanos. 2012. Type VI secretion requires a dynamic contractile phage tail-like structure. *Nature* **483**:182–186.

Bierne, Hélène, and Pascale Cossart. 2012. When bacteria target the nucleus: The emerging family of nucleomodulins. *Cellular Microbiology* **14**:622–633.

Casadevall, Arturo. 2008. Evolution of intracellular pathogens. *Annual Review of Microbiology* **62**:19–33.

Che, Dongsheng, Mohammad Shabbir Hasan, and Bernard Chen. 2014. Identifying pathogenicity islands in bacterial pathogenomics using computational approaches. *Pathogens* **3**:36–56.

Chen, John, and Richard P. Novick. 2009. Phage-mediated intergeneric transfer of toxin genes. *Science* **323**:139–141.

Costa, Tiagro R., Caterina Felisberto-Rodrigues, Amit Meir, Marie S. Prevost, Adam Redzej, et al. 2015. Secretion systems in Gram-negative bacteria: Structural and mechanistic insights. *Nature Reviews. Microbiology* **13**:343–359.

Didelot, Xavier, A. Sarah Walker, Tim E. Peto, Derrick W. Crook, and Daniel J. Wilson. 2016. Within-host evolution of bacterial pathogens. *Nature Reviews. Microbiology* **14**:150–162.

Fouts, Derrick E., Michael A. Matthias, Haritha Adhikarla, Ben Adler, Luciane Amorim-Santos, et al. 2016. What makes a bacterial species pathogenic?: Comparative genomic analysis of the genus *Leptospira*. *PLoS Neglected Tropical Diseases* **10**:e0004403.

Fredlund, Jennifer, and Jost Enninga. 2014. Cytoplasmic access by intracellular bacterial pathogens. *Trends in Microbiology* **22**:128–137.

Huibregtse, Jon, and John R. Rohde. 2014. Hell's BELs: Bacterial E3 ligases that exploit the eukaryotic ubiquitin machinery. *PLoS Pathogens* **10**:e1004255.

Keilberg, Daniella, and Karen M. Ottermann. 2016. How *Helicobacter pylori* senses, targets and interacts with the gastric epithelium. *Environmental Microbiology* **18**:791–806. doi:10.1111/1462-2920.13222.

Krachler, Anne Marie, Hyeilin Ham, and Kim Orth. 2011. Outer membrane adhesion factor multivalent adhesion molecule 7 initiates host cell binding during infection by Gram-negative pathogens. *Proceedings of the National Academy of Sciences USA* **108**:11614–11619.

Liu, Qianhong, Wenyu Han, Changjiang Sun, Liang Zhou, Limin Ma, et al. 2012. Deep sequencing-based expression transcriptional profiling changes during *Brucella* infection. *Microbial Pathogenesis* **52**:267–277.

Malik-Kale, Preeti, Carrie E. Jolly, Stephanie Lathrop, Seth Winfree, Courtney Luterbach, et al. 2011. *Salmonella*—At home in the host cell. *Frontiers in Microbiology* **2**:125.

Pechous, Roger D., and William E. Goldman. 2015. Illuminating targets of bacterial secretion. *PLoS Pathogens* **11**:e1004981. doi:10.1371/journal.ppat.1004981.

Persat, Alexandre, Yuki F. Inclan, Joanne Engel, Howard A. Stone, and Zemer Gitai. 2015. Type IV pili mechanochemically regulate virulence factors in *Pseudomonas aeruginosa*. *Proceedings of the National Academy of Sciences USA* **112**:7563–7568.

Renesto, Patricia, Nicolas Crapoulet, Hiroyuki Ogata, Bernard La Scola, Guy Vestris, et al. 2003. Genome-based design of a cell-free culture medium for *Tropheryma whipplei*. *Lancet* **362**:447–449.

Rutherford, Steven T., and Bonnie L. Bassler. 2012. Bacterial quorum sensing: Its role in virulence and possibilities for its control. *Cold Spring Harbor Perspectives in Medicine* **2**:a012427.

Santos, José Carlos, and Jost Enninga. 2016. At the crossroads: Communication of bacteria-containing vacuoles with host organelles. *Cellular Microbiology* **18**:330–339. doi:10.1111/cmi.12567.

Thi, P. Emily, Ulrike Lambertz, and Neil E. Reiner. 2012. Sleeping with the enemy: How intracellular pathogens cope with a macrophage lifestyle. *PLoS Pathogens* **8**:e1002551.

van de Weije, Michael L., Rutger D. Luteijn, and Emmanuel J. H. J. Wiertz. 2015. Viral immune evasion: Lessons in MHC class I antigen presentation. *Seminars in Immunology* **27**:125–137.

Westermann, Alexander J., Konrad U. Förstner, Fabian Amman, Lars Barquist, Yanjie Chao, et al. 2016. Dual RNA-seq unveils noncoding RNA functions in host–pathogen interactions. *Nature* **529**:496–501.

CHAPTER 26

Microbial Diseases

The impact of infectious disease on the world's population is staggering. The number of deaths caused directly by infections or as a consequence of underlying conditions such as cancer, smoking, or drug abuse, makes infectious disease the leading contributor to death in the world. There are over 1,400 different species of viruses, bacteria, fungi, and protozoa that infect humans. In Chapter 26 we explore the major types and etiologies of infections and introduce the art of diagnosis.

CURRENT RESEARCH highlight

Zika virus. Initially discovered in 1947 from the Zika Forest of Uganda, this *Aedes* mosquito–borne virus was thought to cause only mild rashes and fever. In May 2015, however, Zika virus spread from central Africa to South America, where it infected thousands. Many pregnant women became infected and delivered babies with very small heads, a consequence of incomplete brain development. Could the virus have caused neonatal microcephaly? Recently, Tatjana Avšič Županc and colleagues discovered direct evidence of the link when they found Zika virus in the brain of a microcephalic fetus (arrows in the figure).

Source: Mlakar et al. *N. Engl. J. Med.* **374**:951–958.

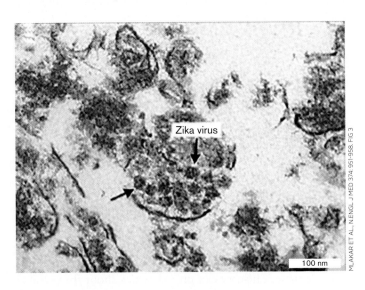

MLAKAR ET AL. N.ENGL. J MED 374:951-958. FIG 3

AN INTERVIEW WITH

TATJANA AVŠIČ ŽUPANC, MICROBIOLOGIST, UNIVERSITY OF LJUBLJANA, SLOVENIA

UNIVERSITY OF LJUBLJANA

Zika virus is not common in Slovenia, where you work. What compelled you to study this virus?

We began studying the link between Zika and microcephaly after receiving autopsy material from a fetus suspected of having an intrauterine infection. The mother was a volunteer in Brazil when she became pregnant. At 28 weeks of pregnancy she returned to Europe, where ultrasound uncovered fetal anomalies. By week 32, there were clear signs of intrauterine growth retardation, microcephaly, and brain calcifications. The fetus was given a poor prognosis, and the mother decided to terminate the pregnancy.

What is the significance of recovering the entire genome of Zika virus in fetal brain tissue?

Our results present strong evidence of the teratogenic potential of Zika virus. While not definitive proof, demonstrating viral particles and a high load of Zika virus RNA in the brain samples, noting the absence of other pathogens, and recovering a complete Zika genome sequence represent compelling evidence that congenital brain malformations associated with Zika virus infection during pregnancy are a consequence of viral replication in the fetal brain.

Infectious diseases have challenged human existence for millions of years. Our tenacity as a species is due in no small way to our growing ability to diagnose, treat and prevent known infections, and to identify new emerging ones. The Current Research Highlight is an example of the latter. By discovering the viral genome in the brain of a microcephalic fetus, Tatjana Avšič Županc and neuropathologist Mara Popović (**Fig. 26.1**) provided a direct link between Zika virus and microcephaly. The story of Zika virus is still unfolding; we don't yet know how discovering Zika virus in fetal brain tissue will impact diagnosis or treatment. But ultimately it will.

In Chapter 26 we discuss a wide variety of infectious microbes by describing their infectious routes, the pathologies they cause, and the symptoms that result. We also examine prevention strategies such as vaccinations when available. Antimicrobial treatments are surveyed later in Chapter 27, while laboratory diagnostics and epidemiological considerations are examined in Chapter 28.

What we present in this chapter is not an exhaustive compendium of microbial illnesses, but a representative sampling of infections that illustrate key aspects of microbial disease. The material is arranged and presented so as to integrate your knowledge of microbiology and immunology within the framework of the practice of medicine and the study of infectious disease.

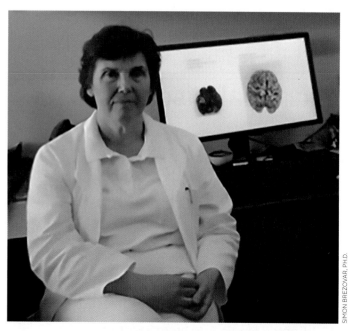

FIGURE 26.1 ■ **Linking Zika and microcephaly.** Neuropathologist Mara Popović (pictured) worked with Tatjana Avšič Županc to provide an important link between the Zika virus and neonatal microcephaly. The background image shows a fetal brain affected by Zika (left), which is threefold smaller than the age-matched control (right).

26.1 Diagnosing Microbial Diseases

What is the best way to classify microbial diseases? Microbial diseases are often presented from the microbe's point of view; for example, where in the body does *E. coli* cause disease? Classifying diseases by organism is useful when examining the ways a given species can cause disease in different organ systems. For instance, *E. coli* can cause genitourinary tract infections, gastrointestinal infections, and meningitis. How do the various strains causing these diseases differ from one another?

Pathogens can also be classified by their route of infection (or portal of entry; see Chapter 25). In this approach, pathogens are grouped as food-borne, airborne, bloodborne, or sexually transmitted; or grouped by transmission such as fecal-oral, respiratory, or insect-borne. However, once having entered the body, different pathogens can again infect widely diverse organ systems. In Chapter 26, we look at infectious agents from the vantage point of the infected organ. We ask, for example, what pathogens cause lung infections, and how are they differentially diagnosed?

The organ approach for classifying microbial diseases is more attuned to how health care workers interact with a patient. An individual who goes to a clinician complaining of a fever, cough, and chest pain does not typically report having encountered a particular bacterium. The clinician must determine from the patient that the disease is localized to the chest and then theorize that it may be a respiratory tract infection. Appropriate samples are then collected and sent to a clinical microbiology laboratory, where data are generated to confirm or dispute the presence of a specific etiological agent, such as *Streptococcus pneumoniae* or other lung pathogens. So how <u>is</u> the infectious cause of a disease diagnosed?

The Art of Diagnosis

Diagnosing an infectious disease is not just a matter of taking a specimen, ordering a test, and prescribing a drug. The clinician first needs to figure out which microbes, out of hundreds, are possible causes. When you go to the clinic, what does a clinician first ask? "What brings you here today?" followed by "How long have you had these symptoms?" and perhaps "Where have you traveled recently?" This is not idle conversation; the clinician is taking a patient history.

Because many infectious diseases display similar symptoms, a patient history can provide clues about the possible

A.

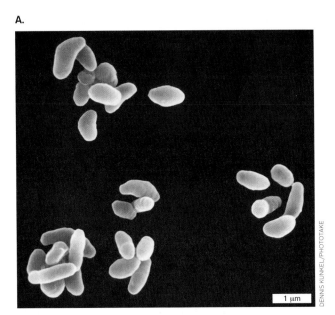

B.

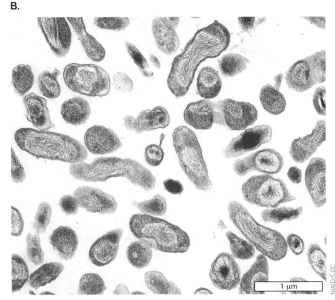

FIGURE 26.2 ■ **Examples of bacteria that cause zoonotic diseases.** **A.** *Francisella tularensis*, the cause of tularemia, a highly infectious disease spread usually via a tick vector but also through cuts (colorized SEM). **B.** *Coxiella burnetii*, the cause of Q fever (colorized TEM). This irregularly shaped organism undergoes developmental stages that range in size from 0.2 to 1 μm.

culprit. For example, *Vibrio cholerae* and enterotoxigenic *E. coli* both produce diarrheal diseases characterized by cramps, lethargy, and liters of watery stool each day. So when a patient presents these symptoms, how does the clinician know what microbes to look for? Here is where a good patient history can make all the difference. Although cholera is not commonly seen in the United States, a clinician might suspect cholera if the patient recently traveled to or emigrated from an endemic area where the disease is regularly observed. This travel information is not gained by examining the patient but comes from talking with the patient and taking a patient history.

Many questions asked during the course of taking a patient's social history can seem irrelevant or intrusive to the patient, who only wants relief from the symptoms. For instance: "Do you have any hobbies? What is your occupation? What foods have you eaten recently? Does your child attend day care? Has anyone in your family had similar symptoms?" All of those questions address possible sources of the infectious agent. Other questions can reveal high-risk behavior that can lead to certain infections. "Do you smoke, drink, or take recreational drugs? Have you had multiple sex partners? Do you use contraception?"

Answers to these questions can provide important diagnostic clues to the astute clinician. For example, learning that a man suffering from enlarged glands, fever, and headaches also likes to hunt and recently killed rabbits can be important. The patient may have been exposed to the Gram-negative, rod-shaped bacterium *Francisella tularensis* (**Fig. 26.2A**), a facultative intracellular pathogen that infects various wild animals and is the cause of tularemia,

an illness also known as "rabbit fever." The patient could have accidentally infected a cut while cleaning his kill. Tularemia is an example of a **zoonotic disease**, an infection that normally affects animals but can be transmitted to humans.

Or consider the case of a woman with acute pneumonia. Upon learning that the woman is a sheep farmer, the clinician considers *Coxiella burnetii* as one of the possible causes of the infection (the list of possible causes is called the **differential**). *C. burnetii* is a pathogenic bacterial species that infects sheep and is shed in large quantities in the animal's amniotic fluid or placenta (**Fig. 26.2B**). Dried soil contaminated with *C. burnetii* can become aerosolized whenever the dirt is disturbed, and any human (such as a farmer) who inhales the dried particles can develop the lung infection Q fever. If the woman were an accountant, *C. burnetii* would be lower on the differential.

Because patient histories are so helpful in diagnosing infectious diseases, we will present several case histories in the following sections and segue into discussions of various microbes that can infect each organ system. Key aspects of infectious diseases will emerge as we proceed.

To Summarize

- **Pathogens can be classified** as food-borne, airborne, blood-borne, or sexually transmitted.

- **Understanding infectious disease** requires knowledge of the organ system, the portal of entry, and the infectious organism.

■ **Patient histories** are vital in diagnosing microbial diseases.

■ **Zoonotic diseases** are animal diseases accidentally transmitted to humans.

26.2 Skin, Soft-Tissue, and Bone Infections

The basic shape and appearance of the human body is dictated by bones, skin, and soft tissues such as muscle and underlying connective tissues. Because these organ systems and tissues collectively shape and maintain body architecture, infections that threaten their integrity will be discussed together. Infections that can affect one of these tissues can also spread directly to another tissue.

Skin and soft-tissue infections range from simple boils to severe, complicated, so-called flesh-eating diseases that can be caused by a variety of bacteria, fungi, and viruses (see **Table 26.1**). Recall that the integrity of the skin, as well as the presence of normal skin microbiota, prevents infection. However, even minor insults to the skin (such as a paper cut) can result in infections, most of which are caused by the Gram-positive pathogen *Staphylococcus aureus*. Healthy individuals develop infections of the skin only rarely, whereas people with underlying immunosuppressive diseases such as diabetes are at much higher risk.

Boils

Staphylococcus aureus (**Fig. 26.3A**) is a common cause of painful skin infections called boils, or carbuncles. This Gram-positive organism, often a normal inhabitant of the nares (nostrils), can infect a cut or gain access to the dermis via a hair follicle. It possesses a number of enzymes that contribute to disease, including coagulase, which helps coat the organism with fibrin, thereby walling off the infection from the immune system and antibiotics. As a result, boils generally require surgical drainage in addition to antibiotic therapy. As noted earlier, some strains of *S. aureus* also produce toxic shock syndrome toxin, a superantigen that can lead to serious systemic symptoms. Recall that a superantigen links and activates antigen-presenting cells and T cells by binding to the <u>outside</u> of MHC class II receptors and T-cell receptors. Antigen recognition is not required. As a result, many different T cells become activated to release a flood of cytokines (see Chapters 24 and 25).

A particularly dangerous strain of *Staphylococcus aureus* is called methicillin-resistant *S. aureus* (MRSA). *S. aureus* infections are commonly treated with penicillin-like drugs, such as oxacillin (a form of methicillin). MRSA has developed resistance to methicillin (oxacillin) and many other penicillin-like drugs through a mutation that alters one of the proteins (called a penicillin-binding protein, PBP) involved in cell wall synthesis (see Chapter 27). Because methicillin is normally used as a first line of defense against staphylococcal infections, treatment failure of MRSA can be life-threatening, and we need to use alternative drugs, such as vancomycin.

MRSA, originally seen mainly as hospital-acquired (**nosocomial**) infections, are no longer contained only in hospitals. Individuals who have not been in a hospital are being infected with MRSA in what are called community-acquired MRSA (CA-MRSA) infections. CA-MRSA infections are occurring at an epidemic rate in the United States (an incidence of approximately 20 per 100,000 population).

A.

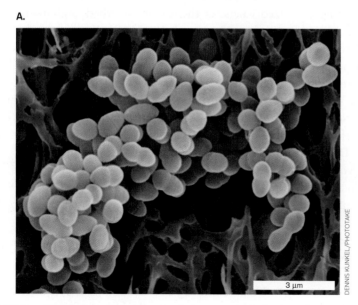

B.

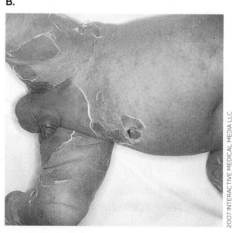

FIGURE 26.3 ■ *Staphylococcus aureus.* A. *S. aureus* (colorized SEM). **B.** Exfoliative toxin from some strains of *S. aureus* causes scalded skin syndrome.

TABLE 26.1	Common Infectious Diseases of the Skin		

Disease	Symptoms	Etiological agent[a]	Virulence factors
Bacterial			
Folliculitis	Boils	*Staphylococcus aureus* (G+ cocci); fibrin wall around abscess renders it poorly accessible to antibiotics	Coagulase, protein A, TSST, leukocidin, exfoliative toxin
Scalded skin syndrome	Peeling skin on infants, systemic toxin	*S. aureus* (G+ cocci)	Disseminated exfoliation exotoxin
Impetigo	Skin lesions on the face, mostly children	*S. aureus* or *Streptococcus pyogenes* (G+ cocci)	Various, as for boils
Scarlet fever	Sore throat, fever, rash	*Streptococcus pyogenes* (G+ cocci)	M-protein pili, C5a peptidase, hemolysin, pyrogenic toxins, others
Erysipelas	Skin lesions, usually facial, that spread to cause systemic infection	*S. pyogenes*	As for scarlet fever (see above)
Cellulitis	Uncomplicated infection of the dermis	*S. aureus*, *S. pyogenes*	As for scarlet fever (see above)
Necrotizing fasciitis	Rapidly progressive cellulitis	*S. aureus*, *S. pyogenes*, *Clostridium perfringens*	As for scarlet fever (see above); a variety of toxins
Viral			
Rubella[b]	Discolored, pimply rash; mild disease unless congenital	Rubella virus [ssRNA(+)]	Envelope proteins E1 and E2
Measles[b]	Severe disease, fever, conjunctivitis, cough, rash	Rubeola virus [ssRNA(−)]	V protein (interferes with interferon signaling)
Chickenpox[b]	Generalized discolored lesions	Varicella-zoster (dsDNA)	Glycoprotein B (fusion of viral and cellular membranes)
Shingles	Pain and skin lesions, usually on trunk in adults		Glycoprotein E (required for cell-cell fusion)
Smallpox[c]	Raised, crusted skin rash, highly contagious	Variola major (dsDNA)	SPICE (smallpox inhibitor of complement enzymes)
Warts[b]	Rapid growth of skin cells	Papillomaviruses (dsDNA)	E6 and E7 oncoproteins (Chapter 25)
Fungal			
Dermatophytosis	Dry, scaly lesions like athlete's foot (tinea pedis)	Dermatophytes (*Epidermophyton*, *Tricophyton*, *Microsporum*)	Unclear
Sporotrichosis	Granulomatous, pus-filled lesions; can disseminate to lungs or other organs	*Sporothrix schenckii*	Melanin
Blastomycosis	Granulomatous, pus-filled lesions; can disseminate to lungs or other organs	*Blastomyces dermatitidis*	BAD1 adherence
Candidiasis	Patchy inflammation of mouth (thrush) or vagina; can disseminate in immunocompromised patients	*Candida albicans*; *Candida glabrata*	Proteinase, phospholipase, Ssn6/Tup1 regulators, others
Aspergillosis	Infected wounds, burns, cornea, external ear	*Aspergillus* spp.	PacC/FOS1 regulators, gliotoxin
Zygomycosis	Oropharyngeal infections; affects mainly diabetic patients; can rapidly disseminate	*Mucor* and *Rhizopus* spp.	Iron acquisition (rhizoferrin); rhizoxin

[a]G+ = Gram-positive; G− = Gram-negative.

[b]Vaccine is available against this agent.

[c]Vaccine is no longer in use, because disease has been eradicated. The exceptions are highly restricted laboratories.

Seventy percent of staphylococcal skin infections are now caused by MRSA. Thus, a physician can no longer assume that a patient walking into the office with a staphylococcal infection will respond to methicillin-like drugs such as oxacillin. The doctor must assume that it may be MRSA. As a result, treatment regimens around the country, and the world, are being forced to change. Today, vancomycin and linezolid are typically used to treat staph infections. Antibiotics are discussed in Chapter 27.

Other staphylococcal diseases are caused by toxin-producing strains in which the organism remains localized but the toxin disseminates. We have already mentioned TSST, but there are other toxins. For example, some strains of *S. aureus* produce a toxin called exfoliative toxin, which causes a blistering disease in children called staphylococcal scalded skin syndrome (**Fig. 26.3B**). Exfoliative toxin, like TSST, is a superantigen (described in Section 24.3). In addition, exfoliative toxin cleaves a skin cell adhesion molecule that, when inactivated, results in blisters.

Thought Question

26.1 Does *Staphylococcus aureus* have to disseminate through the circulation to produce the symptoms of scalded skin syndrome (SSS)? Explain why or why not.

Table 26.1 presents this and other common infections of the skin and soft tissues.

Case History: Necrotizing Fasciitis by "Flesh-Eating" Bacteria

*One weekend in June, Cassi was camping with her three children. She suffered a minor cut on her finger, which she bandaged properly. She also injured the left side of her body while playing sports with her kids. Not thinking much of either of her minor injuries, she went to bed. Two days later, Cassi was extremely ill. Her symptoms included vomiting, diarrhea, and a fever. She was also in severe pain where she had injured her side, and the area had begun to bruise (the skin was not broken). By the next day she could barely get out of bed, and by the end of that night she was having difficulty breathing and could not see. Her side began to leak fluid and blood. Cassi was admitted to the hospital in septic shock, with no detectable blood pressure. An infectious disease specialist diagnosed the problem as necrotizing fasciitis, and she was rushed into surgery. In an effort to save her life, about 7% of her body surface was removed. Because the large wound infection in her side would need to resolve before a skin graft could be performed to repair it, the hole in Cassi's body was left wide open (**Fig. 26.4B**). Long-term efforts to maintain her blood pressure included vasopressors that led to gangrene in her fingers (**Fig. 26.4C**). After nearly 3 months and several operations, Cassi recovered.*

What kind of organism can cause this type of devastating disease? The disease **necrotizing fasciitis**, also known incorrectly as "flesh-eating disease," is rare and is often caused by the Gram-positive coccus *Streptococcus pyogenes* (**Fig. 26.4A**), a microbe normally associated with throat infections (pharyngitis). Although sometimes described as a recently emerging infectious disease, necrotizing fasciitis was first discovered in 1783, in France. Its incidence may have risen recently owing to the increased use of nonsteroidal, anti-inflammatory drugs (such as ibuprofen), which increase a person's susceptibility to infection (see Section 25.2).

In this case history, Cassi probably had this organism on her skin when the injury to her side occurred. The injured area probably suffered an invisible microabrasion, providing

A.
B.
C.

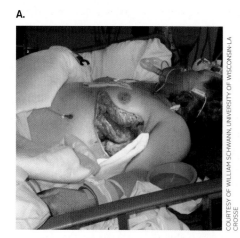
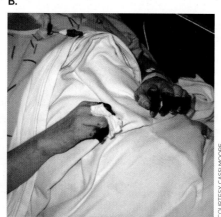
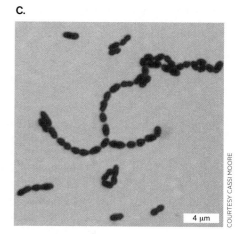

4 µm

COURTESY OF WILLIAM SCHWANN, UNIVERSITY OF WISCONSIN-LA CROSSE
COURTESY CASSI MOORE
COURTESY CASSI MOORE

FIGURE 26.4 ■ **Flesh-eating *Streptococcus pyogenes*.** **A.** Flesh was removed from patient Cassi in an effort to stop the spread of necrotizing fasciitis. **B.** Sepsis resulting from the infection caused Cassi's blood pressure to fall to critical levels. The gangrene of her fingers was an unavoidable consequence of the vasopressor agents given to maintain her blood pressure. Vasopressors raise blood pressure by constricting blood vessels. Unfortunately, long-term use of the vasopressors limited blood flow to Cassi's fingers, causing death of her tissues (necrosis). **C.** Gram stain of *S. pyogenes*. Each cell is approx. 1 µm in diameter.

a good growth environment for the organism, leading to the secretion of potent toxins and death of surrounding tissues. The bacteria will spread in subcutaneous tissue, destroying fat and fascia without initially harming the skin itself. Fascia is the sheath of thin, fibrous tissue that covers muscles and organs.

Rapid, aggressive surgical removal of affected tissue and antibiotic treatment is required in these extreme cases, even before the clinical microbiology lab has had time to identify the organism. In this approach to antibiotic treatment, called **empiric therapy**, one or more antibiotics are given to "cover" (kill) the most likely causative agents. Therapy can include clindamycin and metronidazole (which act against anaerobes and Gram-positive cocci) and gentamicin (a drug particularly effective against Gram-negative microbes). (Chapter 27 further discusses these and other antibiotics.) Often, however, antibiotic treatment of patients with necrotizing fasciitis is ineffective because of insufficient blood supply to the affected tissues.

A less aggressive but similar skin infection is **cellulitis**. Cellulitis is a non-necrotizing inflammation of the dermis that does not involve the fascia or muscles, but is characterized by localized pain, swelling, tenderness, erythema, and warmth. *S. pyogenes* is the most frequent cause of cellulitis in immunocompetent adults, but a number of other bacteria, including *Staphylococcus aureus*, Gram-negative bacilli, and anaerobes, can also cause this skin infection.

Streptococcus pyogenes wields many different virulence factors, including M protein (pilus-like), superantigen exotoxins, and secreted enzymes such as hyaluronidase and DNase (some of these are described in Chapter 25). The sources of many established virulence factor genes in *S. pyogenes* are the numerous prophages (phage genomes) integrated into the bacterial genome. Prophages constitute approximately 10% of the organism's genome. One study found that a soluble factor produced by human pharyngeal cells can facilitate activation of at least some of these phages and cause horizontal transfer of the associated virulence factors between strains of this pathogen. Prophage activation and phage production are discussed in Chapter 6.

> **Thought Question**
>
> **26.2** Why would treatment of some infections require multiple antibiotics?

Osteomyelitis

Osteomyelitis is a bone infection caused by bacteria with accompanying inflammation and bone destruction. All bones can be infected, but the lower extremities are most commonly involved. Bone can be infected in several ways. Acute trauma or surgery can directly introduce organisms into affected bone. Organisms from an adjacent soft-tissue infection can spread to nearby bone. Organisms can also spread from peripheral sites of infection to bone through the circulation (hematogenous seeding), most commonly the vertebrae because they are well vascularized.

Organisms that commonly cause osteomyelitis include staphylococci (*Staphylococcus aureus*), streptococci (*Streptococcus pyogenes*), Gram-negative bacilli (*Pseudomonas, Escherichia*), and some anaerobes (*Bacteroides*). Bone biopsy is usually needed to identify the organism. Host factors that contribute to osteomyelitis include diabetes (when a diabetic foot wound goes untreated), children with sickle-cell disease, joint prosthetics, and intravenous drug use. Antibiotic treatment of osteomyelitis is challenging because most bones are not well vascularized. So debridement is also necessary.

Viral Diseases Causing Skin Rashes

Several viruses can produce skin rashes, although their route of infection is usually through the respiratory tract. Measles, for example (see Section 6.1), is a highly contagious viral infection caused by a paramyxovirus, whose hallmark symptom is skin rash (see Fig. 6.2D). The first signs of measles, also known as rubeola, are fever, cough, runny nose, and red eyes occurring 9–12 days after exposure. A few days later, spots (Koplik's spots) appear in the mouth, along with a sore throat. Then a skin rash develops that typically starts on the face and spreads down the body. The virus replicates in the lymph nodes and spreads to the bloodstream (viremia), where it can infect endothelial cells of the blood vessels. The rash occurs when T cells begin to interact with these infected cells.

Although skin rash is the main symptom of measles, infection can also cause respiratory symptoms and complications, including pneumonia, bronchitis, croup, and even a fatal encephalitis in immunocompromised patients. In the United States, measles has been almost completely eliminated by the measles, mumps, and rubella (MMR) multivalent vaccine. Worldwide, however, measles is still a serious problem in places where vaccinations are not routine.

Rubella virus, a togavirus, causes a maculopapular rash known as German measles, or three-day measles. The rash is similar to but less red than that of measles (**Fig. 26.5**). German measles is an infection that affects primarily the skin and lymph nodes and is usually transmitted from person to person by aerosolization of respiratory secretions. It is not dangerous in adults or children; the virus can, however, cross the placenta in a pregnant woman and infect her fetus. If the virus crosses the placenta within the first trimester, the result is congenital rubella syndrome, which can cause death or serious congenital defects in the developing fetus.

Other viruses affecting the skin, such as chickenpox, the related disease shingles, smallpox, and human papillomavirus, are included in **Table 26.1** and discussed in **eTopic 26.1**. The vaccination schedule for preventing these diseases is provided in Table 24.1.

A.

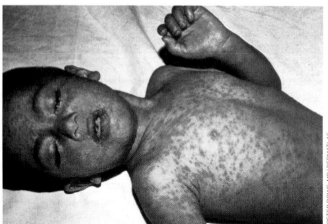

B.

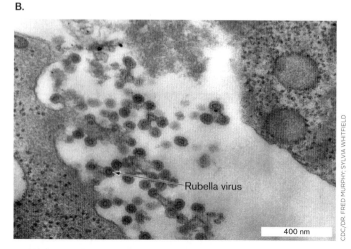

Rubella virus

400 nm

FIGURE 26.5 ▪ **German measles. A.** Skin rash caused by rubella virus. **B.** Rubella virus budding from the cell surface to form an enveloped virus particle (approx. 50–70 nm; TEM).

To Summarize

- *Staphylococcus aureus* and *Streptococcus pyogenes* are common bacterial causes of skin infections. The organisms usually infect through broken skin.

- **Methicillin-resistant** *Staphylococcus aureus* (**MRSA**) has become an important cause of community-acquired staphylococcal infections.

- **Necrotizing fasciitis** is usually caused by *Streptococcus pyogenes*, but it can be the result of other infections.

- **Infections of the skin** can disseminate via the bloodstream to other sites in the body.

- **Osteomyelitis**, usually caused by *Staphylococcus aureus*, begins with direct bone trauma, hematogenous seeding, or contact with a nearby soft-tissue infection.

- **Rubeola and rubella viruses** infect through the respiratory tract, but their main manifestation is the production of similar maculopapular skin rashes.

26.3 Respiratory Tract Infections

Lung and upper respiratory tract infections are among the most common diseases of humans. Many different bacteria, viruses, and fungi are well adapted to grow in the lung. Successful lung pathogens come equipped with appropriate attachment mechanisms and countermeasures to avoid various lung defenses (such as alveolar macrophages). One reemerging bacterial pathogen, *Bordetella pertussis*, the cause of whooping cough, inhibits the mucociliary escalator by binding to lung cilia (**eTopic 26.2**). Although many microbes can infect the lung, most respiratory diseases are of viral origin, and most viral infections (such as the common cold) do not spread beyond the lung. Fortunately, viral diseases by and large are self-limiting and typically resolve within 2 weeks; however, the damage caused by a primary viral infection can lead to secondary infections by bacteria.

Bacterial infections of the lung, whether of primary or secondary etiology, require intervention. Today this means antibiotic therapy. Before the advent of antibiotics, the only recourse was to insert a tube into the patient's back to drain fluid accumulating in the pleural cavity around the lung (a pathological process known as pleural effusion). Unless released, the pressure on the lung will collapse the alveoli and make breathing difficult.

Viral infections predispose patients to secondary bacterial infections in several ways. Viral lung infections cause the patient to dehydrate, which increases mucus viscosity in the airways. Increased mucus viscosity limits motility of the mucociliary escalator (described in Section 23.1), making it harder to eliminate bacterial pathogens. To keep the escalator moving, cold sufferers are advised to drink plenty of fluids to decrease mucus viscosity. Another factor leading to secondary bacterial infections is that viruses can inhibit key aspects of lung innate-immune mechanisms that prevent bacterial growth. Note that many deaths resulting from viral influenza are caused by secondary bacterial infections.

Case History: Bacterial Pneumonia

In March, James, an 80-year-old resident of a New Jersey nursing home, had a fever accompanied by a productive cough with brown sputum (mucous secretions of the lung that

A. Lobar pneumonia

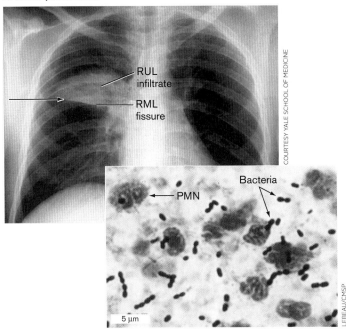

B. *Streptococcus pneumoniae*

C. Relative incidence of pneumonia

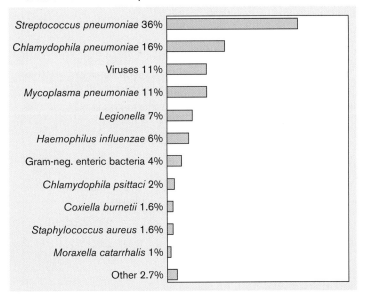

FIGURE 26.6 ■ **Pneumonia caused by *Streptococcus pneumoniae*.** **A.** X-ray view of a patient with lobar pneumonia. Infiltrate in the right upper lobe (RUL) is caused by *S. pneumoniae*. The sharp lower border represents the upper boundary of the right middle lobe (RML) fissure (arrow). **B.** Micrograph of *S. pneumoniae*. Sputum sample showing numerous PMNs and extracellular diplococci in pairs and short chains. Bacteria range from 0.5 to 1.2 μm in diameter. **C.** Relative incidence of pneumonia caused by various microorganisms.

can be coughed up). He reported to the attending physician that he had pain on the right side of his chest and suffered from night sweats. Blood tests revealed that his white blood cell (WBC) count was 14,000/μl (normal is 5,000–10,000/μl) with a makeup of 77% segmented forms (polymorphonuclear

*leukocytes, PMNs; normal range 40%–60%) and 20% bands (immature PMNs; normal range 0%–5%). The chest radiograph revealed a right-upper-lobe infiltrate with cavity formation (**Fig. 26.6A**). From this information, the clinician diagnosed pneumonia. Microscopic examination of the patient's sputum revealed Gram-positive cocci in pairs and short chains surrounded by a capsule (**Fig. 26.6B**). Bacteriological culture of his sputum and blood yielded* Streptococcus pneumoniae.

Pneumonia is a disease that can be caused by many different microbes (**Table 26.2**). The pneumococcus *Streptococcus pneumoniae* accounts for about 25% of community-acquired cases of pneumonia, but pneumococcal pneumonia occurs mostly among the elderly and immunocompromised, including smokers, diabetics, and alcoholics. A breakdown of pneumonia cases by causative organism is shown in **Figure 26.6C**.

The noses and throats of 30%–70% of a given population can contain *S. pneumoniae*. The microbe can be spread from person to person by sneezing, coughing, or other close, personal contact. Pneumococcal pneumonia may begin suddenly, with a severe shaking chill usually followed by high fever, cough, shortness of breath, rapid breathing, and chest pains.

Pneumococcal lung infection begins when the pneumococcus is aspirated into the lung. Once in the lung, the microbe grows in the nutrient-rich edema fluid of the alveolar spaces. Neutrophils and alveolar macrophages then arrive to try to stop the infection. They are called into the area from the circulation by chemoattractant chemokines released by damaged alveolar cells. The thick polysaccharide capsule of the pneumococcus, however, makes phagocytosis very difficult (see Section 25.6). In an otherwise healthy adult, pneumococcal pneumonia usually involves one lobe of the lungs; thus, it is sometimes called lobar pneumonia. The infiltration of PMNs and fluid leads to the typical radiological findings of diffuse, cloudy areas. In contrast, infants, young children, and elderly people more commonly develop an infection in other parts of the lungs, such as around the air vessels (bronchi), causing bronchopneumonia.

The white blood cell count in the case history is telling. The patient had an elevated WBC count (normal is 5,000–10,000/μl) and an elevated proportion of band cells (normal is 0%–5%). These increases are indicative of a bacterial, not viral, infection. Neutrophils (PMNs), the front-line combatants against infection, rise in response to bacterial infections and are first released from bone marrow as immature band cells, whose presence is a sure sign of bacterial infection.

Several outbreaks of pneumococcal pneumonia have occurred over recent years in nursing homes, where numerous residents have been affected. These incidents underscore

TABLE 26.2 Selected Respiratory Tract Infectious Diseases

Agent/disease	Key symptoms	Virulence properties	Source	Treatment
Bacterial				
Bacillus anthracis [Anthrax]	Hypotension, respiratory failure	Peptide capsule; PA, LF, and EF toxins	Soil/airborne	Ciprofloxacin [Vaccine (military)]
Corynebacterium diphtheriae [Diphtheria]	Tracheal pseudomembrane	Diphtheria toxin	Humans	Penicillin [Vaccine]
Streptococcus pneumoniae [Pneumonia]	Fever, chills, cough, chest pain	Capsule, pneumolysin	Humans	Macrolides, quinolones, ceftriaxone [Vaccine]
Bordetella pertussis [Whooping cough]	Violent cough, inhalation "whoop"	Adenylate cyclase toxin, filamentous hemagglutinin (adhesin)	Humans	Erythromycin [Vaccine]
Pseudomonas aeruginosa [Pneumonia]	Infects cystic fibrosis patients	Exotoxin A, phospholipase C, exopolysaccharide, others	Water, soil	Quinolones, aminoglycosides
Legionella pneumophila [Legionnaire's disease]	Chest pain, cough, muscle pain, vomiting	Intracellular growth, hemolysin, cytotoxin, protease	Water towers/ inhalation	Macrolides
Chlamydophila pneumoniae [Pneumonia]	Sore throat, chest pain	Obligate intracellular growth; prevents phagolysosome fusion	Humans	Tetracycline, erythromycin
Chlamydophila psittaci [Psittacosis]	Sore throat, chest pain	Obligate intracellular growth; prevents phagolysosome fusion	Bird droppings/ dust	Quinolones
Mycobacterium tuberculosis [Tuberculosis]	Cough, bloody sputum, fatigue, weight loss	Cord factor, wax D, intracellular growth	Humans	Rifampin, isoniazid, ethambutol, pyrazinamide) [Vaccine (BCG[a])]
Mycoplasma pneumoniae	Sore throat, nonproductive cough	Adhesin tip	Humans	Erythromycin
Viral				
Cytomegalovirus (CMV)	Cough, chest pain	Reduces MHC I presentation	Humans	Ganciclovir
Respiratory Syncytial Virus (RSV)	Cough, chest pain	Prevent T cell activation	Humans	Treat symptoms/ ribavirin
Influenza	Cough, chest pain	Neuraminidase; hemaglutinin	Humans	Zanamivir [Vaccine]
Severe Acute Respiratory Virus (SARS)	Cough, chest pain	IRF-3 (antagonist of interferon)	Humans	Treat symptoms
Fungi				
Aspergillus spp. [Aspergillosis]	Lungs, sinuses; breathing difficulty	Dimorphism; gliotoxin	Environment	Amphotericin B, voriconazole
Histoplasma capsulatum [Histoplasmosis]	Flu-like	Dimorphism; calcium-binding protein	Bird, chicken, bat droppings	Amphotericin B, itraconazole
Coccidioides immitis [Coccidiomycosis]	Flu-like	Dimorphism; arginase 1	Environment	Amphotericin B, fluconazole
Blastomyces dermatitidis [Blastomycosis]	Flu-like	Dimorphism; BAD1	Environment	Amphotericin B, itraconazole
Pneumocystis jerovicii [Pneumocystosis]	Chest pain, cough, skin lesion	Unknown	Environment	Bactrim

[a]BCG = Bacille Calmette-Guerin (a weakened strain of the bovine tuberculosis strain).

the importance of elderly people receiving the pneumococcal polysaccharide vaccine (PPSV) as a hedge against infection. While there are over 80 antigenic types of pneumococcal capsular polysaccharides, the injected vaccine contains only the 23 types that are most often associated with disease (see Section 24.6). A vaccine that is formulated to respond to multiple antigens is called a "multivalent vaccine." The resulting immune response will protect vaccinated individuals against infection by those antigenic types. The vaccine is recommended for individuals over 65, as well as for those who are immunocompromised. The patient in this case history failed to receive the vaccine.

In addition to causing serious infections of the lungs (pneumonia), *S. pneumoniae* can invade the bloodstream (bacteremia) and the covering of the brain (meningitis). The death rates for these infections are about one out of every 20 who get pneumococcal pneumonia, about four out of 20 who get bacteremia, and six out of 20 who get meningitis. Individuals with special health problems, such as liver disease, AIDS (caused by HIV), or organ transplants, are even more likely to die from the disease, because of their compromised immune systems.

An emerging infectious disease problem throughout the United States and the world is the increasing resistance of *S. pneumoniae* to antibiotics. At least 30% of the strains isolated are already resistant to penicillin, the former drug of choice for treating the disease. Chapter 27 discusses why antibiotic resistance is on the rise for this and other microbes.

Case History: Disseminated Disease from a Fungal Lung Infection

A 35-year-old male boxer named Tyrrell, who installs home insulation for a living, was admitted to a Maryland hospital when he presented with difficulty walking, fever, chills, night sweats, and a recent 10-kg (22-lb) weight loss. He denied having prior pneumonia, sinus infection, arthritis, hematuria (blood in the urine), numbness, or muscle weakness. Tyrrell also denied any history of intravenous drug use and had been in a monogamous relationship for 4 years. There was no history of travel outside of the area for the past 4 years. Tyrrell's past medical history showed that he had visited the emergency department 6 months earlier with flu-like symptoms, a chronic cough that produced blood-tinged white sputum, shortness of breath, loss of appetite, and weight loss. One month prior to the current admission, he had developed some painless subcutaneous nodules and had become so short of breath that he could no longer continue boxing. At that time, an X-ray taken in the emergency room showed right-upper-lobe infiltrate, indicating pneumonia (Fig. 26.7A). A tuberculosis skin test was negative. He was given a prescription for the antibiotic azithromycin (a macrolide antibiotic

A. Pneumonia infiltrate

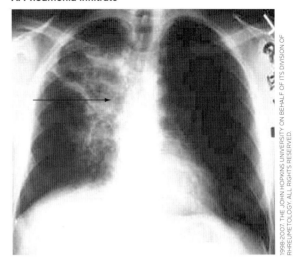

B. Leg lesion

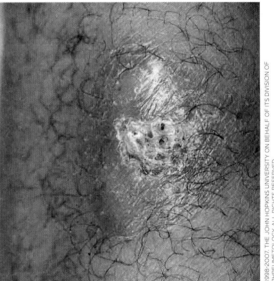

C. Colony of *Blastomyces dermatitidis*

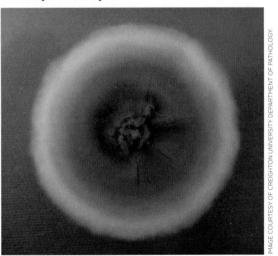

FIGURE 26.7 ■ Pneumonia and metastatic disease caused by *Blastomyces dermatitidis*. A. Diffuse infiltrate in the right lung (arrow). **B.** Metastatic leg lesion at the tibia. **C.** Fungal colony of *B. dermatitidis*.

commonly used to treat bacterial infections of the respiratory tract) and discharged. Despite antibiotic treatment, the cough and weight loss continued. Physical exam at the time of admission revealed several painful subcutaneous nodules (one filled with pus) and a tender tibia (indicating bone involvement) that prevented him from walking (Fig. 26.7B). A CBC (complete blood count) obtained at the time of current admission showed normal counts and differential.

In this case, the expression of frankly purulent material (pus) from the left-leg nodule suggested an infectious process. The infection in this patient probably started in the lung (clued by the cough), after which the organism spread throughout the body via the bloodstream. Fungus is a probable cause, given the chronic nature of the patient's symptoms. An alternative possibility would be tuberculosis, caused by the bacterium *Mycobacterium tuberculosis. M. tuberculosis* also causes a chronic lung infection and might have been suspected, except for the negative TB skin test. In the tuberculin skin test, a small amount of mycobacterial antigen called PPD (purified protein derivative) is injected under the skin of the lower arm. A person who has been infected with *M. tuberculosis* will exhibit a localized delayed-type hypersensitivity reaction at the site of injection, although this does not equate to currently active disease.

Cryptococcus, an encapsulated yeast, is not the likely cause, since it typically requires an immunocompromised host to cause disease—not the case in this instance. (*Cryptococcus*, which causes cryptococcosis, is an opportunistic pathogen that commonly infects AIDS patients.) The most prevalent clinical form of cryptococcosis is meningoencephalitis, although disease can also involve the skin, lungs, prostate gland, urinary tract, eyes, myocardium, bones, and joints.

The most likely fungal causes of infection in this case history are the endemic mycoses, such as histoplasmosis, blastomycosis, and coccidioidomycosis. This patient had never traveled to the western United States, where coccidioidomycosis is endemic, so exposure to *Coccidioides* was ruled out. Histoplasmosis most commonly presents as a flu-like pulmonary illness, with erythema nodosum (tender bumps on skin) and arthritis (swollen joints) or arthralgia (joint pain), none of which the patient had. Blastomycosis can disseminate to the lung, skin, bone, and genitourinary tract, consistent with the pattern of organ involvement seen in this patient.

Amphotericin B, an antifungal agent (discussed in Section 27.5), was given to this patient. His fever lowered almost immediately, and the skin nodules diminished. After 2 weeks, a fungus was found in the cultures of the nodule biopsy, bronchoalveolar lavage (washes), and urine. This fungus was identified by a DNA probe as *Blastomyces dermatitidis* (**Fig. 26.7C**), confirming the diagnosis of blastomycosis.

B. dermatitidis is a dimorphic fungus that resides in the soil of the Ohio and Mississippi river valleys and the southeastern United States. The portal of entry is the respiratory tract, and infection is usually associated with occupational and recreational activities in wooded areas along waterways, where there is moist soil with a high content of organic matter and spores. The incubation period ranges from 21 to 106 days. This patient most likely inhaled conidia (fungal spores) from the soil while crawling underneath houses installing insulation. The physician learned that the patient used only a T-shirt to cover his mouth and nose—not an effective method of keeping spores from entering the respiratory tract. He should have worn a respirator.

Several critical features of this case help differentiate it from the preceding case of pneumococcal pneumonia. First, the initial macrolide antibiotic, azithromycin, should have killed most bacterial sources of infection. Second, the X-ray finding of diffuse infiltrate is more indicative of fungal lung infection than bacterial infection, which in a patient of this age would likely be confined to one lobe. The patient was young and in good health prior to the infection, making it unlikely to be pneumococcal pneumonia. The blood count was also a clue. Fungal infections do not usually cause an increase in WBCs or an increase in band cells. Finally, the **metastatic lesions** (infectious lesions that develop at a secondary site away from the initial site of infection) on the leg were in no way consistent with *Streptococcus pneumoniae*. They arise when the organism moves through the bloodstream from the primary site of infection to another body site, where it can begin to grow. Many infectious diseases start out as a localized infection but end up disseminating throughout the body to cause metastatic lesions.

Note: The term "metastasis" means the spread of disease from one organ to another, noncontiguous organ. Only microbial infections and malignant cancer cells can metastasize.

Tuberculosis as a Reemerging Disease

Tuberculosis, caused by the acid-fast bacillus *Mycobacterium tuberculosis* (**Fig. 26.8** inset), was once considered of passing historical significance to physicians practicing in the developed world. In 1985, however, owing primarily to the newly recognized HIV epidemic and a growing indigent population, TB resurfaced, especially in inner-city hospitals. In 1991, highly virulent multidrug-resistant (MDR) strains of *M. tuberculosis* were reported. These strains not only produced fulminant (rapid-onset) and fatal disease among patients infected with HIV (the time from TB exposure to death is 2–7 months), but also proved highly infectious. Tuberculin skin test conversion rates of up to 50% were reported in exposed health care workers (the conversion rate today is less than 1%).

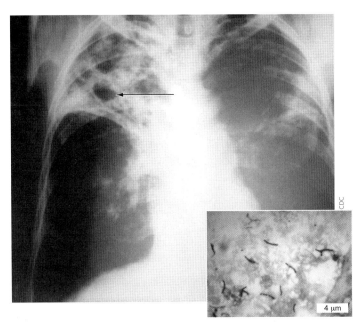

FIGURE 26.8 ▪ Calcified Ghon complex of tuberculosis. The arrow points to a Ghon complex in a patient's right upper lobe. Note the difference in appearance compared to Figure 26.7A. **Inset:** Acid-fast stain of *Mycobacterium tuberculosis*. (Cells approx. 2–3 μm in length.)

A positive tuberculin skin test (also known as the Mantoux test) is seen as a delayed-type hypersensitivity reaction to *M. tuberculosis* proteins (called purified protein derivative, or PPD) injected under the skin. Note, however, that a positive tuberculin skin test does not signify <u>active</u> disease, but only that the person was infected at one time. The bacterium may have been killed by the immune system without having caused disease or may lie dormant, waiting to reactivate. MDR-TB cases in 2014 represented approximately 1.3% of new TB cases (9,421). More information on mycobacterial structure can be found in Section 18.3, and on its pathogenesis, in Section 25.6.

M. tuberculosis causes primarily a respiratory infection (pulmonary TB), but it can disseminate through the bloodstream to produce abscesses in many different organ systems (extrapulmonary TB). Extrapulmonary TB is sometimes called miliary TB because the size of the infected nodules (1–5 mm), called tubercles, approximates the size of millet seeds. Tubercles in the lung are filled with *M. tuberculosis* cells that drain into bronchial tubes and the upper respiratory tract. Aerosolization of respiratory secretions (coughing) directly spreads the tubercle bacilli from someone with active disease to an uninfected person (person-to-person transmission, no animal reservoir).

Once inhaled into the lung, the bacilli are subject to three possible outcomes: They can die, they can produce primary disease, or they can become latent. In all three cases the bacteria are initially phagocytosed by alveolar macrophages, and if not killed, they survive ensconced within modified phagolysosomes. The bacilli can multiply in these vacuoles, kill the macrophage (via induced apoptosis), and then be released to infect other macrophages. Four to six weeks postinfection, an infected person can develop active primary disease marked by a productive cough that generates sputum. The patient also experiences fever, night sweats, and weight loss. Patients with active disease are contagious and produce a delayed-type hypersensitivity that makes them tuberculin-positive.

An alternative to primary disease is an asymptomatic latent tuberculosis infection (LTBI). Here the tubercle bacilli remain hidden in macrophages and may even become dormant. The delayed-type hypersensitivity response to the organism builds small, hard tubercles around these cells. Over time, the tubercles develop into caseous lesions that have a cheese-like consistency and can calcify into the hardened Ghon complexes seen on typical X-rays (**Fig. 26.8**). The bacteria can remain dormant in these tubercles for many years. Even primary tuberculosis can resolve and become latent tuberculosis, unless the organisms were killed with antibiotics.

In a patient with LTBI, *M. tuberculosis* can sometimes, years after initial infection, overcome the imposed confinement of the immune system. The bacteria begin to multiply and cause **secondary**, or **reactivation**, **tuberculosis**. Secondary TB commonly occurs in immunocompromised people (such as HIV patients). The symptoms of secondary TB are more serious than those of primary TB and include severe coughing, greenish or bloody sputum, low-grade fever, night sweats, and weight loss. The gradual wasting of the body led to the older name for tuberculosis: "consumption."

Ten drugs are currently approved by the Food and Drug Administration (FDA) for treatment of tuberculosis. Initial treatment of active disease is aggressive and involves a four-drug regimen of what are called first-line drugs (or drugs of choice)—isoniazid, rifampin, pyrazinamide, and ethambutol—given over a course of several months. MDR strains are defined as being resistant to two or more first-line drugs. An MDR strain is treated with a regimen of four to five drugs that do not include the first-line drugs to which it is resistant. Extensively drug-resistant tuberculosis (XDR-TB) strains are resistant to three or more second-line drugs (drugs used when drugs of choice fail), as well as two or more first-line drugs. These strains are almost untreatable.

Thought Question

26.3 Explain why patient noncompliance (failure to take drugs as directed) is thought to have led to XDR-TB.

Viral Diseases of the Lung

Numerous viruses can cause lung infections (see **Table 26.2**). Influenza virus and rhinovirus are described in Chapters 6 and 11, and compared in **eTopic 26.3**. The recently emerged infection SARS (severe acute respiratory syndrome) is discussed in Chapter 28. However, an important viral lung infection not discussed elsewhere is respiratory syncytial disease, caused by respiratory syncytial virus (RSV). A negative-sense, single-stranded RNA, enveloped virus, RSV is the most common cause of bronchiolitis and pneumonia among infants and children under 1 year of age.

Illness begins most frequently with fever, runny nose, cough, and sometimes wheezing. RSV is spread from respiratory secretions through close contact with infected persons or by contact with contaminated surfaces or objects. Infection can occur when the virus contacts mucous membranes of the eyes, mouth, or nose and possibly through the inhalation of droplets generated by a sneeze or cough. Unlike rubella or rubeola, which infect the respiratory tract and disseminate though the body, RSV remains localized in the lung.

The majority of children hospitalized for RSV infection are under 6 months of age. RSV can cause repeated infections throughout life, usually associated with moderate to severe cold-like symptoms. Severe lower respiratory tract disease may occur at any age, especially among the elderly or people with compromised cardiac, pulmonary, or immune systems. As yet, a vaccine to control this disease is not available. However, a monoclonal antibody called palivizumab, which binds to an RSV epitope, can prevent RSV in high-risk infants that were born premature or with medical problems such as congenital heart failure.

Table 26.2 presents many other bacterial, fungal, and viral microbes that can cause respiratory tract infection. Be aware that very different diseases can produce similar symptoms. For instance, people constantly confuse influenza (the flu) with the common cold. Symptomatically, they may start out similarly, but there are telling differences. Influenza is characterized by fever, myalgia (muscle aches), pharyngitis (sore throat), and headache (viral infection is discussed in Chapter 11). A runny nose is <u>not</u> one of the symptoms. The common cold, however, manifests as a runny nose, nasal congestion, sneezing, and throat irritation. No myalgia. The observant clinician will note the difference.

To Summarize

- **Lung infections are mostly viral**, but most <u>deadly</u> lung infections are caused by bacteria.

- **The mucociliary escalator** is a primary defense mechanism used by the lung to avoid infection.

- **An elevated white cell count** in blood is an indicator of bacterial infection.

- **Pneumococcal vaccine** should be administered to the elderly because they are often immunocompromised.

- **Fungal agents** commonly cause long-term, chronic infections.

- **Localized bacterial infections in the lung can disseminate** via the bloodstream to form metastatic lesions at other body sites.

- **Tuberculosis** is an ancient bacterial disease with an increasing mortality rate resulting from multidrug-resistant strains, the susceptibility of HIV patients, and an increasing indigent population.

- **Respiratory syncytial virus** is one of several viruses that can cause lung disease, but it rarely disseminates.

26.4 Gastrointestinal Tract Infections

Nearly everyone has experienced diarrhea, a condition characterized by frequent loose bowel movements accompanied by abdominal cramps. Hundreds of millions of cases occur each year in the United States and are a major cause of death in developing countries. As with respiratory tract infections, most diarrheal disease is viral in origin, with rotavirus being the primary culprit. Among the bacteria, the Gram-negative rod *Salmonella enterica* serovar Typhimurium and the spiral-shaped or curved bacillus *Campylobacter* are the most frequent causes of self-limiting diarrheal disease.

You might wonder why some microbes prefer the chaos of diarrhea to the relative stability of a nice commensal relationship. After all, isn't finding a niche and sticking to it the goal of every microbe? The simple answer is that diarrhea enables the dissemination of microbes that might otherwise kill their host or be killed themselves after too aggressively provoking the host's immune system. Dissemination allows the microbe to reach new hosts and proliferate. A diarrhea-causing microbe is sort of like a serial bank robber fleeing from city to city to avoid capture and find more banks to rob.

Types of Diarrhea

What causes diarrhea? Normally, intestinal mucosal cells absorb water (8–10 liters per day) from the intestinal contents, essentially drying out stool. Diarrhea occurs when water is not absorbed or when it actually leaves intestinal cells and enters the intestinal lumen. The excess water loosens stool, and diarrhea results. There are several types of diarrhea.

- **Osmotic diarrhea** follows the intake of nonabsorbable substrates, such as lactulose, a synthetic sugar, used to treat constipation. Lactulose increases osmolarity in the intestine, which causes water to leave mucosal cells. The result is osmotic diarrhea. Infectious organisms (such as rotavirus) that prevent nutrient absorption can cause osmotic diarrhea. Osmosis is discussed in eAppendix 2.

- **Secretory diarrhea** develops when microbes cause mucosal cells to increase ion secretion, as seen with cholera toxin. Mucosal cells expel water to try to equilibrate the resulting electrolyte imbalance.

- **Inflammatory diarrhea** forms when an infectious agent triggers the production of inflammatory cytokines that attract PMNs. Subsequent damage to the intestinal wall will limit water and nutrient absorption, and cause red and white blood cells to enter the stool (bloody diarrhea or dysentery). *Shigella, Salmonella* species, and some strains of *E. coli* can cause infections leading to inflammatory diarrhea.

- **Motility-related diarrhea** can develop when enterotoxins made by pathogens such as rotavirus cause intestinal hypermotility. Food moves so quickly through the intestine (hypermotility) that there isn't sufficient time to absorb water or nutrients.

Terms used to describe the inflammation of different parts of the GI tract include:

- **Gastritis** (inflammation of the stomach lining, for instance, ulcers);

- **Gastroenteritis** (nonspecific term for any inflammation along the gastrointestinal tract);

- **Enteritis** (inflammation mainly of the small intestine);

- **Enterocolitis** (inflammation of the colon and small intestine); and

- **Colitis** (inflammation of the large intestine—colon).

Diarrhea due to viral growth, bacterial growth, or toxin production can cause large amounts of water and electrolytes to leave the intestinal cells and enter the intestinal lumen. As a result, the patient not only suffers diarrhea but can become dangerously dehydrated. Most deaths resulting from infectious diarrhea are the result of dehydration.

Staphylococcal Food Poisoning

We have all heard of the local church picnic where scores of people become violently ill within hours of eating unrefrigerated potato salad. *Staphylococcus aureus* is the usual cause

of these disasters, but it is <u>not</u> an infection. The culprit is an enterotoxin (an exotoxin that affects the gastrointestinal tract) secreted by some strains of *S. aureus* into tainted foods such as pies, turkey dressing, or potato salad. After ingestion, the toxin travels to the intestine, where it enters the bloodstream and stimulates the vagus nerve leading to the vomit center in the brain.

Because the toxin is preformed, symptoms occur quickly after ingestion. Within 2–6 hours, the poisoned patient will begin vomiting and may also experience diarrhea. The disease, though violent, is not life-threatening and usually resolves spontaneously within 24–48 hours. In contrast, diarrhea caused by infectious agents, such as *Salmonella enterica*, which must first grow in the victim, do not occur until 12–24 hours after ingestion, sometimes longer. A clinician noting quick onset of symptoms in a patient will immediately suspect staphylococcal food poisoning. Obviously, antibiotic treatment is not needed for staph food poisoning, but it may be indicated for other gastrointestinal infections. Staphylococcal enterotoxins are also heat resistant, so simply heating a food already containing enterotoxin will not destroy the toxic activity.

It is important to recognize that the most important treatment for diarrhea is **rehydration therapy**. Vomiting and diarrhea can cause huge losses in water volume. Because the exit of water from tissues causes an osmotic imbalance, salts also leave the cells, to try to reestablish that balance. The problem is that the resulting electrolyte imbalance can severely affect cardiovascular, respiratory, and renal systems. Consequently, replacing body fluids orally or by IV is imperative to prevent death. Rehydration solutions, therefore, must contain glucose as well as sodium and potassium salts in proper balance (for example, Pedialyte).

Antibiotics are often inappropriate when treating diarrhea. Although antibiotic treatment of infectious gastroenteritis seems intuitive, it is rarely used and actually contraindicated. Most gastrointestinal infections are viral (norovirus or rotavirus), so antibiotics are ineffective. Likewise, gastroenteritis caused by bacteria usually resolves spontaneously, without antibiotic treatment. However, severe systemic disease stemming from gastroenteritis can develop, often in the young or elderly. Diseases such as typhoid fever (*Salmonella* Typhi) or bacillary dysentery (*Shigella dysenteriae*) respond well to antibiotics.

In some cases, antibiotic treatment can actually <u>trigger</u> gastrointestinal disease. For example, many antibiotics used to treat infectious diseases (especially clindamycin) can kill most normal intestinal bacteria, except the naturally resistant Gram-positive anaerobe *Clostridioides difficile* (formerly *Clostridium difficile*), the causative agent of pseudomembranous enterocolitis. Unrestrained by microbial competition, *C. difficile* growing at the epithelial surface of the intestine will produce specific toxins that damage and kill intestinal cells. The organism's growth leads to inflammation and the formation of exudative plaques along the intestinal wall (refer to Fig. 23.10). The plaques eventually coalesce into larger pseudomembrane structures that block the intestinal mucosa. The blockage causes the malabsorption of nutrients and water, which results in diarrhea. As the pseudomembrane enlarges, it begins to slough off and pass into the stool. Diagnosis of this disease involves PCR identification of the organism or immunological identification of the toxin in fecal samples.

Some patients suffer recurrent *C. difficile* infections. It is not clear why, but one suggestion is that spores lodged in colon folds may escape clearance by peristalsis. Another hypothesis is that patients with recurrent disease have an impaired response to the *C. difficile* toxins. How do you prevent recurrences? In some instances, after vegetative *C. difficile* has been killed by antibiotic treatment, a procedure known as a fecal transplant (see Section 23.2) can be used to restore a healthy gastrointestinal microbiota that prevents the recurrence of *C. difficile* disease.

Case History: Enterohemorrhagic *E. coli*

Tammy, a 6-year-old girl from Montgomery County, Pennsylvania, arrived at the ER with bloody diarrhea, a temperature of 39°C (102.2°F), abdominal cramping, and vomiting. She was admitted to the hospital 5 days after a kindergarten field trip to the local dairy farm. When the parents were questioned about Tammy's activities during the trip, they said she had purchased a snack while at the farm. Upon laboratory analysis, a fecal smear was positive for leukocytes, and isolation of organisms confirmed the presence of Gram-negative rods that produced Shiga toxins 1 and 2. In subsequent testing by pulsed-field gel electrophoresis, the isolate was indistinguishable from E. coli *O157:H7. By this time, Tammy had developed additional problems. Her face and hands became puffy, she had decreased urine output despite being given IV fluids (suggesting kidney damage), and she was beginning to develop some neurological abnormalities. Laboratory analyses of blood samples revealed thrombocytopenia (reduced blood platelet count) and confirmed hemolytic uremic syndrome (HUS; renal failure). The child was treated by IV fluid and electrolyte replacement. Antibiotics were not administered. Tammy eventually recovered.*

In this case history, the presence of leukocytes in a fecal smear is a sign that the intestinal pathogen may have invaded the epithelial mucosa of the intestine (or severely damaged it). Breaching this barrier sends out a chemical call to neutrophils, which then enter the area and, in an effort to kill the pathogen, also damage the intestinal cells.

Shigella dysenteriae, Salmonella enterica, and enteroinvasive *E. coli* (EIEC) actually invade enterocytes and are considered intracellular pathogens. Enterohemorrhagic *E. coli* (EHEC), which also produces leukocytes and blood in stools, is not an intracellular parasite (it is not invasive), but causes damaging attachment and effacing lesions, described in Section 25.3, which destroy the mucosal epithelium. The resulting inflammation, in conjunction with damage caused by the Shiga toxin it produces, leads to blood and white cells in the stool. *E. coli* O157:H7, the etiological agent in the case history, is a common serotype of EHEC.

There are at least six different classes of pathogenic *E. coli* that differ in their repertoire of pathogenicity islands, plasmids, and virulence factors. They include EIEC and EHEC, already mentioned; enterotoxigenic *E. coli* (ETEC) and uropathogenic *E. coli* (UPEC), both described in Chapter 25; as well as enteropathogenic *E. coli* (EPEC) and enteroaggregative *E. coli* (EAEC). All but UPEC causes gastrointestinal disease. (A new diarrheagenic strain is described in **Special Topic 26.1**) To tell these strains apart, each group has telltale O and H antigens that can be identified using serology.

Note: "O antigen" is part of the bacterium's LPS, while "H antigen" is a flagellar protein. Thus, "O157:H7" denotes the specific version of LPS (O157) and flagellar protein (H7) found on *E. coli* O157:H7. Other pathogenic strains of *E. coli* have different O and H antigens.

Shiga toxin. *Shigella* and EHEC, the agent in the preceding case history, both produce toxins, called Shiga toxins 1 and 2, that are encoded by genes of bacteriophage genomes embedded in the bacterial chromosome. These toxins inhibit host protein synthesis and, in the process, damage endothelial cells in the intestine, kidney, and brain. Shiga toxin–induced death of vascular endothelial cells in the intestine causes the breakdown of blood vessel linings, followed by hemorrhage that manifests as bloody diarrhea. Shiga toxin 2 also triggers the release of pro-inflammatory cytokines.

Endothelial damage initiates the formation of platelet-fibrin microthrombi (clots) that occlude blood vessels in the various organs, leading to two major syndromes: hemolytic uremic syndrome (HUS) and thrombotic thrombocytopenic purpura (TTP). HUS occurs when the microthrombi are limited to the kidney. The microclots clog the tiny blood vessels in this organ and cause decreased urine output, ultimately leading to kidney failure and death. In TTP, the clots occur throughout the circulation, causing reddish skin hemorrhages called petechiae and purpuras. Neurological symptoms (for example, confusion, severe headaches, and possibly coma) then arise from microhemorrhages in the brain. The hemorrhaging occurs because platelets needed for normal clotting have been removed from the circulation as they form the microthrombi. The decreased number of platelets is called thrombocytopenia.

The toxins, which are absorbed through the intestine and disseminated via the bloodstream, have five B subunits used to bind to target cell membranes, and one A subunit imbued with toxic activity (see Section 25.4). The A subunit, upon entry, destroys protein synthesis by cleaving an adenine from 28S rRNA in eukaryotic ribosomes.

Enterohemorrhagic *E. coli* (EHEC). *E. coli* O157:H7 is a recently emerged pathogen that can colonize cattle intestines at the recto-anal junction without affecting the animal and, as a result, can contaminate meat products following slaughter. Initially identified in 1982, the organism came to national prominence during a large-scale U.S. outbreak, in 1993, linked to a Washington State Jack in the Box restaurant in which 732 people were sickened.

E. coli O157:H7 rarely affects the health of the reservoir animal. But when an infected steer is slaughtered, the carcass can become contaminated with EHEC-containing feces despite manufacturers' considerable efforts to prevent it. Grinding the tainted meat into hamburger distributes the microbe throughout. Cooking burgers to 160°C is essential to kill any existing EHEC. Cross-contamination between foods is possible too. Using the same cutting board to prepare meat and salad is a great way to contaminate the salad, which will not be cooked.

Despite EHEC's common association with hamburger, vegetarians are not safe from this organism. During heavy rains, waste from a cattle farm can easily wash into nearby vegetable fields unless precautions are taken. If the cattle waste contains *E. coli* O157:H7, the crops become contaminated. One such outbreak occurred in 2006, when spinach from certain areas of California were contaminated with this pathogen, prompting a nationwide recall of bagged spinach and a month without spinach salad.

Early on, the remarkably low infectious dose of *E. coli* O157:H7 mystified researchers. However, we have since learned that *E. coli* has an impressive level of acid resistance, rivaling that of the gastric pathogen *Helicobacter pylori*. Acid resistance mechanisms permit *E. coli* to survive in the acidic stomach and enable a mere 10–100 individual organisms to cause disease.

As already noted, EHEC strains produce two toxins that are identical to the Shiga toxins produced by *Shigella* species. The toxins cleave host ribosomal RNA, thereby halting translation. As discussed earlier, one consequence of Shiga toxin is HUS. The development of HUS, as in the case history described, is a common consequence of *E. coli* O157:H7 infection. Unfortunately, HUS can be treated only with supportive care, such as blood transfusions and dialysis throughout the critical period until kidney function resumes. Antibiotic treatment can increase the release of Shiga toxins from the organisms and actually trigger HUS. Thus, antimicrobial therapy is not recommended.

In contrast to the case just described, many gastrointestinal infections do not produce fecal leukocytes or blood in the stool. Diarrheal diseases caused by *Vibrio cholerae* (cholera) or enterotoxigenic *E. coli* (ETEC), which produces a cholera-like disease, do not involve invasion of the intestinal lining by the microbe, and they yield copious amounts of watery diarrhea. In these two toxin-driven diseases, the bacteria attach to cells lining the intestine and secrete toxins that are imported into the target cells (see Section 25.4).

Epidemiology of EHEC. Epidemiology is the study of factors and mechanisms involved in the spread of disease. Researchers at the U.S. Centers for Disease Control and Prevention (CDC) used epidemiological principles (discussed later, in Chapter 28) to identify the risk factors associated with the case study presented earlier. They interviewed 51 infected patients and 92 controls (children who visited the farm but did not become ill). Infected patients were more likely than controls to have had contact with cattle, an important reservoir for *E. coli* O157:H7. All 216 cattle on the farm were sampled by rectal swab, and 13% yielded *E. coli* O157:H7 with a DNA restriction pattern indistinguishable from that isolated from the patients. This finding indicated that the cattle were the source of infection.

Furthermore, separate areas were not established for eating and interactions with farm animals. Visitors could touch cattle, calves, sheep, goats, llamas, chickens, and a pig while eating and drinking. Hand-washing facilities were unsupervised and lacked soap, and disposable hand towels were out of the children's reach. All of these circumstances provided opportunity for infection.

How can we prevent disease caused by enterohemorrhagic *E. coli*? Industry approaches include thoroughly washing carcasses before processing, maintaining cold temperatures, and testing for possible contamination. In addition, the use of gamma irradiation to sterilize beef, spinach, and lettuce has been approved. Recent outbreaks of EHEC disease caused by contaminated hamburger have declined dramatically because of industry practices and USDA inspections. Irradiation could eliminate the problem (described in Chapter 5), but less than 1% of hamburger meat is currently irradiated in the United States.

Type III secretion and diarrhea. Type III secretion systems were first described in Section 25.5, where we discussed the pathogenesis of *Salmonella enterica*, but these systems are present in numerous Gram-negative pathogens, such

SPECIAL TOPIC 26.1   Sprouts and an Emerging *Escherichia coli*

In May 2011, people in Germany began dying from hemolytic uremic syndrome (HUS) (**Fig. 1**). As of June, 3,228 cases and 35 deaths had been reported. By then, a massive epidemiological hunt was under way for the cause of the disease and its source. The organism isolated in each case was a form of *E. coli* rarely seen before. Its LPS and flagellar serotype was O104:H4. The organism secreted type 2 Shiga toxin, which contributed to HUS but, strangely, did not produce the attaching and effacing proteins often associated with enterohemorrhagic *E. coli* (for instance, strain O157:H7). Unfortunately, this unique pathovar also harbored a plasmid that conveyed resistance to all penicillins, cephalosporins, and co-trimoxazole. Another feature was the presence of enteroaggregative adherence fimbriae (*aaf*), which allow *E. coli* to adhere like stacked bricks to host cells. Again, AAF factor is not usually seen in Shiga toxin–producing *E. coli* (called STEC), but it is a defining feature of enteroaggregative *E. coli* (EAEC).

One more odd thing about this HUS outbreak was that a majority of the patients (90%) were adults rather than children, the usual victims of EHEC or other STEC strains. All of this evidence, plus extensive DNA sequence analysis, indicates that the deadly O104:H4 strain is a genetic hybrid between enterohemorrhagic and enteroaggregative strains of *E. coli*. As a result, it is officially referred to as EAHEC O104:H4. What series of unfortunate events produced this "chimeric" patho-gen? The *stx-2* gene is part of a prophage that can horizontally transfer between strains of *E. coli*, but the trafficking of the *aaf* and drug resistance characteristics is less apparent and under investigation.

A particularly hideous aspect of disease produced by O104:H4 is that the prevalence of HUS symptoms in diarrheagenic patients far exceeds the usual rate for typical EHEC victims. But why? One hypothesis is that this variant produces high amounts of Shiga toxin (Stx-2) as compared to other STEC strains. Victor Gannon and his colleagues at the University of Lethbridge and the Public Health Agency of Canada demonstrated this in vitro using mitomycin C. Mitomycin C is a DNA intercalating agent that can trigger reactivation of prophages via the SOS response (see Chapter 9), and in so doing, it stimulates the expression of phage-encoded toxin genes. **Figure 2** shows that after mitomycin C treatment, the O104:H4 strain produced five to six times more Shiga toxin than did other EHEC strains. It is thought that conditions in the gastrointestinal tract could induce a similar response.

The HUS outbreak started in May 2011. By early June, the finger-pointing had begun. Germany initially blamed Spanish cucumbers, which led to a huge economic loss in Spain as foreign countries refrained from importing these vegetables. Later it was shown that Spain was not the source of this organism. Finally, the German agricultural minister of Lower Saxony announced that an organic farm near Uelzen, which produces a variety of sprouted foods, was the likely source of the *E. coli* outbreak (**Fig. 3**). Before the source was found, however, several tourists visiting Germany ingested the organism and returned to their home countries, including the United States and Canada, where they then developed HUS. Fortunately, as of this writing, no further major outbreaks caused by this pathogen have been reported.

RESEARCH QUESTION

Which features of the O104:H4 strain might be suitable targets for antimicrobial drug design? Explain why they might be suitable and what pitfalls there may be to using these drugs in a patient.

Laing, C. R., Y. Zhang, M. W. Gilmour, V. Allen, R. Johnson, et al. 2012. A comparison of Shiga-toxin 2 bacteriophage from classical enterohemorrhagic *Escherichia coli* serotypes and the German *E. coli* O104:H4 outbreak strain. *PLoS One* **7**:e37362. [Online.] http://www.plosone.org/article/info%3Adoi%2F10.1371%2Fjournal.pone.0037362.

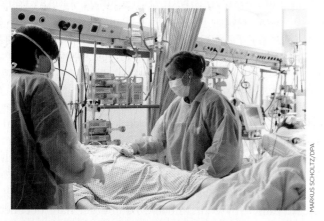

FIGURE 1 ■ Hemolytic uremic syndrome patient in Germany. This patient was being treated in the medical intensive care unit of the University Hospital Schleswig-Holstein in Lübeck. Note the biosafety precautions taken by the medical personnel (gloves, gown, and mask).

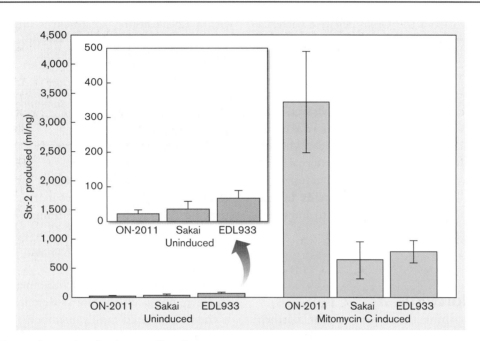

FIGURE 2 ■ **Shiga toxin production by *E. coli* pathovars.** Stx-2 production by *E. coli* O104:H4 outbreak-related strain ON-2011 and *E. coli* O157:H7 strains EDL933 and Sakai in uninduced and mitomycin C–induced states as measured by an Stx-2-specific ELISA. Error bars represent standard deviations from three independent replicates. Mitomycin C is a DNA intercalating agent that can trigger reactivation of prophages. Lytic replication can stimulate expression of phage-encoded toxin genes. *Source:* Adapted from Laing, C. R. et al. 2012. *PLoS One* **7(5)**:e37362. © dpa.

FIGURE 3 ■ **What's the source?** **A.** Early on, the source of the new strain of *E. coli* was unclear, prompting this cartoon by Martin Sutovec, depicting a police lineup composed of the suspects: cucumbers, sprouts, and avocados. **B.** Ultimately, fenugreek sprouts grown at an organic farm in Germany were identified as the source. These sprouts are commonly used in salads.

as EHEC in our case history. Recall that type III protein secretion systems directly inject proteins from the cytoplasm of a bacterial pathogen into the cytoplasm of a target eukaryotic host cell. The system delivers proteins across three membranes—two from the Gram-negative bacterial pathogen and one surrounding the target cell.

In addition to stimulating bacterial entry into host cells, bacterial proteins injected by type III transport systems cause host cells to secrete pro-inflammatory cytokines. The cytokines then "call in" inflammatory cells and alter ion transport through the epithelial membrane. Excessive export of ions such as chloride causes water to leave the cell in an attempt to equilibrate the internal and external ionic concentrations. The water entering the intestine results in diarrhea. Type III–translocated effector proteins induce many host cellular responses that contribute to intestinal inflammation during an infection.

Rotavirus and Norovirus

Many people wrongly think that most cases of diarrhea are caused by a bacterial agent. Actually, two viruses— **rotavirus** and **norovirus**—cause more intestinal disease than any bacterial species. See Table 6.1: Rotavirus is classified in Group III (dsRNA viruses); norovirus is in Group IV (positive-sense single-stranded RNA viruses). Rotavirus is highly infectious, spreading by the fecal-oral route; it is endemic around the globe, and it affects all age groups, although children between 6 and 24 months are the most severely affected. It is estimated that by age 3, all children have had a rotavirus infection.

The incubation period is approximately 2 days, after which the victim commonly suffers frequent watery, dark green, explosive diarrhea. All of this may be accompanied by nausea, vomiting, and abdominal cramping. Severe dehydration and electrolyte loss due to the diarrhea will cause death unless supportive measures, such as fluid replacement, are undertaken. There is no cure, but most patients recover if rehydrated properly. Few deaths from rotavirus occur in the United States, but each year more than 400,000 children worldwide die from this viral diarrhea (2014 estimate). The mortality and incidence of this disease have decreased because of the introduction in 2006 of a safe and effective vaccine (see Section 24.6). In Mexico alone, the vaccine resulted in a 50% decline in diarrheal deaths.

With the success of the rotavirus vaccine, norovirus is set to become the most common worldwide cause of nonbacterial gastroenteritis. Norovirus infections have already surpassed rotavirus in the United States. The virus is perceived as the scourge of cruise ships and assisted-living facilities, but norovirus can also spread quickly in hotels or anywhere there are many people in a small area. The virus spreads by the fecal-oral route among children or adults, via contaminated food or person-to-person contact. Within 24 hours of infection, the victim experiences sudden vomiting, stomach cramps, and watery diarrhea that mercifully resolves within 12–24 hours. Treatment is similar to that for rotavirus, and there is no vaccine. Death is rare, but most common among infants and the elderly.

Note: Norovirus infection is sometimes called the "stomach flu," but that is a misnomer. It is not the flu and has nothing to do with influenza virus.

Diarrhea and the gut microbiome. An obvious question to ask about diarrhea is how does it affect the microbiome? A study headed by Shannon Manning (Michigan State University) found that the composition of intestinal microbiota of patients with diarrhea differs significantly from that of their healthy family members.

For one thing, the gut microbiomes of diarrhea patients who were not given antibiotics were less diverse than those of their uninfected family members. Abundance of Bacteroidetes and Firmicutes was higher in the healthy individuals, whereas Proteobacteria dominated the patient microbiomes. *Escherichia coli*, for instance, predominated in all patients regardless of the pathogen causing the infection. The composition of diarrheal microbiomes also varied with the bacterial cause of infection. As one example, the microbiome of *Campylobacter*-infected patients differed from that of patients infected with *Salmonella* or *Shigella*.

How does the intestine restore its bacterial population after being decimated by diarrhea or antibiotic treatment? Lawrence David (Duke University), Peter Turnbaugh (UC San Francisco) (**Fig 26.9A**), and their colleagues used metagenomic procedures to characterize the stools of 41 people in Bangladesh (children and adults) who had diarrhea caused by *E. coli* or *V. cholerae*. Stools were monitored before, during, and after diarrhea episodes.

The researchers identified a consistent succession of repopulation events in nearly every case, regardless of the cause of diarrhea (**Fig 26.9B**). After diarrhea (or antibiotic treatment for the diarrhea) clears out much of the microbiome, carbohydrates and oxygen accumulate in the gut. Carbohydrates and oxygen would normally be metabolized by gut microbiota. During the early stage of repopulation, facultative, oxygen-respiring and carbohydrate-utilizing bacteria (especially those using simple carbohydrates, such as *Escherichia*, *Enterococcus*, and *Streptococcus*) colonize the gut and consume these nutrients. Midstage recovery begins when the lack of simple sugars and oxygen (as well as increased phage predation) leads to a decline in the early-stage species. This decline allows succession to anaerobic, complex carbohydrate-fermenting bacteria (*Bacteroides*).

A.

PHOTO BY CHRISTINA CRUZ

STEPHANIE MITCHELL/HARVARD STAFF PHOTOGRAPHER

FIGURE 26.9 ■ Successive repopulation of the gut microbiome following *Vibrio cholerae* infection. A. Lawrence David (left) and Peter Turnbaugh unravel the complexities of the human microbiome. **B.** Fecal samples were taken at 1, 7, and 30 days past presentation of diarrhea (dpp). Patients were given a single dose of azithromycin on the day of presentation, which nearly eliminated *V. cholerae* by 1 dpp. Healthy contacts of patients are shown on the left. Letters indicate patient identifications; numbers indicate separate healthy contacts for each patient. Colored boxes reflect groups of different genera that are prominent during the infection and afterward during early, mid, and late stages of repopulation. The *x*-axis values reflect relative abundances of genera. "All swab" microbes were present only in rectal swab samples.

B.

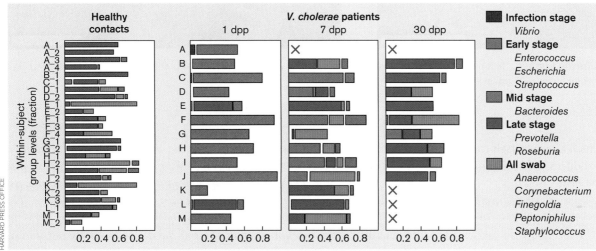

HARVARD PRESS OFFICE

Finally, in late-stage recovery, the gut microbiome once again resembles the complex community that existed prior to infection—the same composition seen in healthy contacts (**Fig. 26.9B**). The entire process takes about 30 days to complete but depends on a variety of factors, such as diet, antibiotic use, and duration of diarrhea.

Repopulation occurs, in part, by reingesting microbes from food. But exciting research suggests that the much maligned and trivialized appendix is, among other things, an important reservoir of gut microbes that can seed the intestine and reestablish the microbiota. Once properly reestablished, gut microbiota are capable of fending off pathogens such as *C. difficile*. As evidence, researchers found that patients with appendectomies were more than twice as likely to develop repetitive infections with *C. difficile*.

Protozoan Causes of Diarrheal Disease

As we learned in Chapter 20, some protozoa (also called protists) cause serious human diseases. For instance,

Entamoeba histolytica and *Cryptosporidium parvum* cause the diarrheal diseases amebic dysentery and cryptosporidiosis, respectively. In 2013, the CDC tallied 9,056 cases (up from 2,640 in 2005) of cryptosporidiosis, a reportable disease in the United States. Two other amebas, *Naegleria* and *Acanthamoeba*, cause amebic meningoencephalitis. Because some species of *Acanthamoeba* can infect the eye, soft-contact wearers should take precautions to prevent contamination of their lenses.

The flagellated protozoan *Giardia lamblia* is a major cause of diarrhea throughout the world. In the United States alone, *G. lamblia* caused 15,106 reported cases of giardiasis diarrhea in 2013 (over 19,000 in 2010) and likely caused thousands more that were not reported. *G. lamblia* enters a human or other host as a cyst present in drinking water contaminated by feces (**Fig. 26.10A**). Aside from humans, *G. lamblia* can be found in various rodents, deer, cattle, and even household pets. It is very infectious. Ingestion of as few as 25 cysts can lead to disease. Following ingestion, the hard, outer coating of the cyst is dissolved by the action of digestive juices to produce a trophozoite

A.

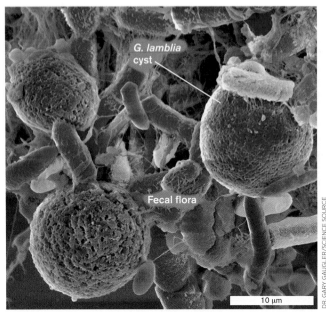

B.

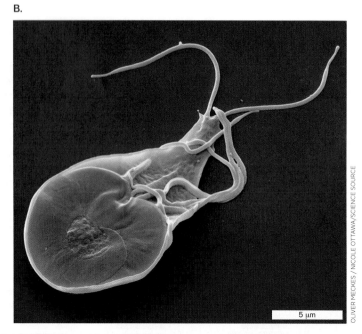

FIGURE 26.10 ■ ***Giardia lamblia.*** This protist is a major cause of diarrhea in the world. **A.** Cysts (7–14 μm) present in fecal matter (colorized SEM). **B.** Trophozoite form (5–15 μm in length; colorized SEM).

(**Fig. 26.10B**), which attaches itself to the wall of the small intestines and reproduces. Offspring quickly encyst and are excreted out of the host's body.

Asymptomatic carriers of *G. lamblia* are common; it has been estimated that anywhere from 1% to 30% of children in U.S. day-care centers are carriers. Disease usually manifests as greasy stools alternating between a watery diarrhea, loose stools, and constipation. However, some patients will experience explosive diarrhea. Diagnosis usually comes from observing the cysts or trophozoite forms of the protozoan in feces. Metronidazole is a drug often used to cure the disease. To prevent it in the first place, proper treatment of community water supplies is essential.

Although most gastrointestinal disease is intestinal in locale, specialized microbes can also target the stomach, with its harsh acidic environment.

Case History: Ulcers—It's Not What You Eat

Gary was a 34-year-old accountant who had immigrated to Nebraska from Poland 7 years earlier. Since his teenage years, he had been bothered periodically by episodes of epigastric pain (pain around the stomach), nausea, and heartburn. Antacids usually alleviated the symptoms. Over the years, he had received several courses of treatment with Tagamet or Pepcid to reduce acid secretion and provide relief. Recently, an upper-GI endoscopy had been performed, in which a long, thin tube tipped with a camera and light source was inserted into Gary's mouth and threaded down into his stomach. The view through the endoscope showed some reddened areas in the antrum (bottom part) of the stomach. The endoscope was also equipped with a small clawlike structure that obtained a small tissue sample from the lining of Gary's stomach. A urease test performed on the antral biopsy turned positive in 20 minutes. Histological examination of the biopsy confirmed moderate chronic active gastritis (inflammation of the stomach lining) and revealed the presence of numerous spiral-shaped organisms. Cultures of the antral biopsy were positive for Helicobacter pylori.

Painful and sometimes life-threatening gastric ulcers were for many years blamed on spicy foods and stress. These factors were believed to cause increased acid production that ate away at the stomach lining, even though the gastric mucosa is normally well protected from stomach acid, which can fall as low as pH 1.5. This protection argued against the model but was ignored. In the 1980s, after discovering odd, helical bacteria present in the biopsies of gastric ulcers, Australians J. Robin Warren and Barry Marshall (a medical intern at the Royal Perth Hospital at the time) (**Fig. 26.11A**) proposed that bacteria, not pepperoni, cause ulcers (**Fig. 26.11B** and **C**).

Their hypothesis was viewed with skepticism and declared as heresy by the established medical community.

A.

COURTESY OF BARRY MARSHALL.

B.

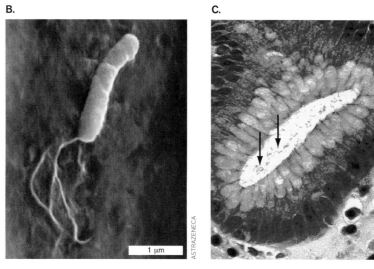

1 μm

ASTRAZENECA

C.

EMDICINE.COMN/A

FIGURE 26.11 ■ **A bacterial cause of gastric ulcers.** **A.** Physician Barry Marshall was so sure he was right about the cause of stomach ulcers that he swallowed bacteria to prove his point. **B.** *H. pylori* (SEM). Note the tuft of flagella at one pole. Cell length approx. 2 μm. **C.** *Helicobacter* (arrows) attached to gastric mucosa.

Faced with disbelief bordering on ridicule, the young intern drank a vial of the helical organisms and waited. A week later he began vomiting and suffered other painful symptoms of gastritis. Barry Marshall could not have been happier. He had proved his point. We now know that this curly microbe causes the vast majority of stomach ulcers.

The discovery of *Helicobacter pylori* and its association with gastric ulcer disease led to an upheaval in gastroenterology. Prior to this discovery, treatment had focused on suppressing acid production via proton pump inhibitors, which did not provide long-term relief. Within 1 year after acid suppression therapy, up to 80% of patients suffer a relapse of their ulcer. Therapy now includes antimicrobial treatment to kill the bacteria and acid suppression therapy to prevent further inflammation while the ulcer heals. Warren and Marshall, who recovered from his gastritis, received the 2005 Nobel Prize in Physiology or Medicine for their groundbreaking work.

The exact mechanism by which *H. pylori* causes gastric ulcers is not known, although a variety of virulence factors have been identified by gene array (see eAppendix 3) and signature-tagged mutagenesis (see **eTopic 25.1**). The basic scheme of *Helicobacter* pathogenesis is shown in **Fig. 26.12**. After the pathogen is ingested, *H. pylori* flagella propel the organism toward

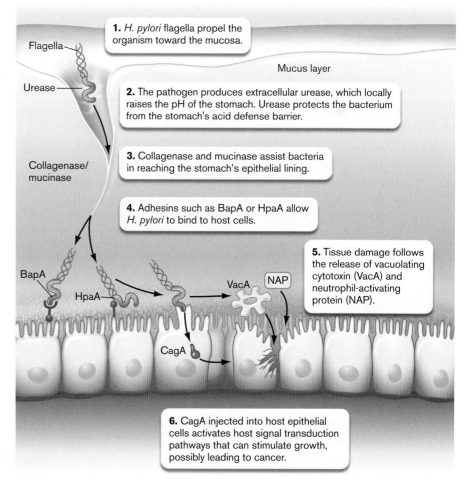

1. *H. pylori* flagella propel the organism toward the mucosa.

Flagella

Mucus layer

Urease

2. The pathogen produces extracellular urease, which locally raises the pH of the stomach. Urease protects the bacterium from the stomach's acid defense barrier.

Collagenase/mucinase

3. Collagenase and mucinase assist bacteria in reaching the stomach's epithelial lining.

4. Adhesins such as BapA or HpaA allow *H. pylori* to bind to host cells.

BapA

HpaA

VacA

NAP

5. Tissue damage follows the release of vacuolating cytotoxin (VacA) and neutrophil-activating protein (NAP).

CagA

6. CagA injected into host epithelial cells activates host signal transduction pathways that can stimulate growth, possibly leading to cancer.

FIGURE 26.12 ■ **Steps in *Helicobacter* pathogenesis.**

the mucosa (**Fig. 26.12**, step 1). As it approaches the mucosa, the pathogen produces intracellular and extracellular urease, an important virulence factor that converts urea to CO_2 and ammonia. The ammonia neutralizes acid around *Helicobacter* and allows the organism to survive the extreme acidity of the stomach (step 2).

Two other enzymes—collagenase and mucinase—then soften the mucous lining, which allows the bacteria to reach the stomach's epithelial lining (**Figs. 26.11C** and **26.12**, step 3). The epithelial lining is much less acidic than the lumen, so the organism can grow and divide. Once at the epithelium, *Helicobacter* produces various adhesins, such as BapA or HpaA, to bind host cells (step 4). After the organism has adhered, tissue damage develops with the release of vacuolating cytotoxin (VacA) and neutrophil-activating protein (NAP). NAP activates neutrophils and mast cells to damage local tissue (step 5). VacA forms a hexameric pore in the host membrane and induces apoptosis (programmed cell death) by damaging mitochondria. Apoptotic cells decrease the immune response, which will stabilize a chronic infection. Another protein, CagA, is injected into host epithelial cells, where it is phosphorylated; CagA then interacts with host signaling proteins and activates host signal transduction pathways that can stimulate growth, possibly leading to cancer (step 6).

Tools useful for diagnosing *H. pylori* include a fecal antigen test, rapid urease testing, and serology [for example, enzyme-linked immunosorbent assay (ELISA) to detect antibody to the CagA antigen]. ELISA is a common immunological tool to detect the presence, in serum, of antibodies to a specific organism—an indication of infection. The assay, described more fully in Section 28.2, is performed by coating wells in a plastic dish with an antigen (for example, *Helicobacter* CagA). Serum from the patient is then added to the well. If antibodies to CagA are present, they will bind to the CagA antigen. Unbound antibody is removed by washing, and a secondary antibody that binds human IgG is added to the well.

The anti–human IgG antibodies have an enzyme linked to them. A sandwich is formed as follows: [plastic dish]–[CagA protein]–[anti-CagA antibody]–[anti-human IgG antibody]–[enzyme]. When substrate is added to the well, the enzyme acts on it to produce light or a chromogenic (colored) product. The more anti-CagA antibody present in the serum, the more light or product that is produced by the linked enzyme. Applications of ELISA and further details of the methodology involved are described in Section 28.2.

Helicobacter pylori and cancer. Problems caused by *H. pylori* are not limited to gastric ulcers. The microbe has also been associated with gastric cancer. Here is some of the evidence establishing the connection:

- Many individuals with gastric cancer are also colonized by *H. pylori*.

- Gerbils infected with *H. pylori* develop gastric cancer.

- Human gastric cells infected with *H. pylori* downregulate genes required for DNA repair.

- When injected into gastric epithelial cells, *H. pylori* CagA is phosphorylated on a tyrosine residue and activates a regulatory cascade that causes the gastric cell to proliferate.

The link between *H. pylori* and gastric cancer is sobering when you consider that *H. pylori* can be detected in about two-thirds of the world's population, especially in impoverished countries.

We have examined in this section only a handful of the bacterial, viral, and protozoan microbes that cause gastrointestinal infection. Others are listed in **Table 26.3** and are described in Chapter 20.

To Summarize

- **Diarrhea leads to dehydration,** for which fluid replacement is a critical treatment. Antibiotic treatment is usually not recommended.

- **Staphylococcal food poisoning** is not an infection. It is a toxigenic disease.

- **Antibiotic treatments can sometimes cause gastrointestinal disease** (for example, pseudomembranous enterocolitis by *Clostridium difficile*).

- **The presence of red and white blood cells in fecal contents** is an indication of invasive bacterial infection by intracellular pathogens such as *Shigella*, *Salmonella*, and EIEC. **Bacteria that do not invade intestinal cells** <u>usually</u> produce watery diarrhea. EHEC is an exception: the attachment and effacing lesions it produces result in bloody stools.

- **Bacterial toxins** produced by bacterial enteric pathogens can cause systemic symptoms.

- **Rotavirus** is still the single greatest cause of diarrhea worldwide. Increasing use of rotavirus vaccine may eventually leave norovirus as the world's leading cause of diarrhea.

- *Giardia lamblia* is a major protozoan cause of diarrhea worldwide.

- **The bacterium *Helicobacter pylori*,** a common cause of gastric ulcers, lives in the stomach and is highly acid resistant.

TABLE 26.3	Selected Microbes That Cause Diseases of the Gastrointestinal Tract				
Etiological agent[a]	**Disease**	**Symptoms**	**Virulence factors**	**Source**	**Treatment**
Bacterial					
Campylobacter jejuni (G–)	Gastroenteritis	Fever, muscle pain, watery diarrhea, blood in stool, headache	Cytotoxin, enterotoxin, adhesin	Poultry, unpasteurized milk	Erythromycin
Clostridium botulinum (G+, anaerobe)	Botulism	Symptoms begin quickly; flaccid paralysis	Neurotoxin	Preformed toxin in foods	Antiserum
Clostridium difficile (G+)	Pseudomembranous enterocolitis	Fever, abdominal pain, diarrhea, pseudomembrane in colon	Cytotoxin, antibiotic resistance	Animals, normal microbiota	Vancomycin
Clostridium perfringens (G+)	Gastroenteritis	Watery diarrhea, nausea	Alpha toxin	Soil, food	Self-limiting
Enterohemorrhagic *E. coli*	Gastroenteritis	Bloody diarrhea, HUS	Intimin, Tir, type III secretion, Shiga toxin	Contaminated foods (hamburger) and crops	Oral rehydration; antibiotics if severe
Enterotoxigenic *E. coli*	Traveler's diarrhea	Watery diarrhea	Labile and stable toxins	Humans; food, water	Oral rehydration
Helicobacter pylori (G–)	Gastric ulcers	Abdominal pain, bleeding, heartburn	Adhesin, urease CagA, vacuolating toxin	?	Triple drug (omeprazole, clarithromycin, metronidazole)
Salmonella enterica (G–)	Salmonellosis	Symptoms after 18 h; abdominal pain, diarrhea; invade intestinal M cells	Type III secretion, intracellular growth	Chickens, other animals; fecal-oral route	Oral rehydration; antibiotics if severe
Salmonella enterica serovar Typhi (G–)	Typhoid fever	Headache, fever, chills, abdominal pain, rash (rose spots), hypotension, diarrhea in late stages	Type III secretion, intracellular growth, PhoPQ regulators, Vi antigen capsule	Human carriers (gallbladder reservoir); food, water	Quinolones
Shigella spp. (G–)	Shigellosis	Bloody diarrhea, HUS	Shiga toxin, type III secretion, intracellular growth, actin-based motility, escape phagosome	Humans; fecal-oral route	Oral rehydration; antibiotics if severe
Staphylococcus aureus (G+)	Staphylococcal food poisoning	Symptoms within 4 h of ingestion; nausea, vomiting, diarrhea	Enterotoxin	Preformed toxin in foods	Supportive
Vibrio cholerae (G–)	Cholera	Watery diarrhea	Cholera toxin, toxin-coregulated pili (TCPs), ToxR regulator	Human waste–contaminated water	Oral rehydration, antibiotics
Vibrio parahaemolyticus (G–)	Gastroenteritis	Diarrhea, blood in stool	Enterotoxin	Raw seafood	Self-limiting
Viral					
Norovirus (Norwalk virus)	Stomach "flu"	Nausea, vomiting, diarrhea	VP1	Fecal-oral route	Oral rehydration
Rotavirus (most common cause)	Stomach "flu"	Nausea, vomiting, diarrhea	NSP4	Fecal-oral route	Oral rehydration

[a]G+ = Gram-positive; G– = Gram-negative.

26.5 Genitourinary Tract Infections

Although the genital and urinary tracts are different organ systems with different purposes (procreation versus the filtering and excretion of waste products from blood), the close association of these organ systems in the body has often led them to be grouped together when discussing infections. Note that viruses and bacteria are capable of infecting the genital tract, but viruses rarely cause urinary tract infections. Also, very few pathogens can infect both organ systems.

Urinary Tract Infections

The urinary tract includes the kidneys, ureters, urinary bladder, and urethra. Infections anywhere along this route are called urinary tract infections (UTIs). UTIs are the second most common type of bacterial infection in humans, ranking in frequency just behind respiratory infections such as bronchitis or pneumonia. In the United States, bladder infections and other UTIs result in over 6 million patient visits annually, mostly by women. Estimates are that at least 25% of women between 20 and 40 years of age have experienced a UTI, and 20%–40% of those infected develop recurrent infections. UTIs result in 100,000 hospital admissions in the United States and $1.6 billion in medical expenses each year.

Urine, as produced in the kidneys, is normally sterile and was thought to be sterile when stored in the bladder. Studies now show that the bladder has a normal microbiome, although its role in human health is unclear. Why didn't we notice this earlier? All known bacterial causes of urinary tract infections are facultative anaerobes that grow under aerobic conditions. The microbiota of our urinary bladder are anaerobic and do not grow when urine is plated aerobically on blood agar—the normal procedure used to identify urinary tract pathogens. The current thought that bladder microbiota do not cause disease and do not grow under the conditions used to identify uropathogens is convenient for the clinical laboratory seeking pathogens.

Bacteria that cause UTIs are introduced into the bladder or kidney in one of four ways:

■ **Infection from the urethra to the bladder.** This is the most common route for bladder infections (called **cystitis**). Bacteria residing along the superficial urogenital membranes of the urethra can ascend to the bladder. This is a more common occurrence in women than in men. Uropathogenic bacteria colonizing the urethra can also be introduced into the bladder by means of mechanical devices such as catheters or cystoscopes that are passed through the urethra into the bladder.

■ **Deposition of bacteria from the bloodstream to the kidney.** Kidney infections (called **pyelonephritis**) can arise when microorganisms from infections elsewhere in the body disseminate via the bloodstream.

■ **Descending infection from the kidney to the bladder.** Descending infection occurs when bacteria from an infected kidney are shed into the ureters. The microbes are then carried by urine into the bladder.

■ **Ascending infection to the kidney.** In ascending infection, bacteria from an established infection in the bladder ascend along the ureter to infect the kidney.

Thought Question

26.4 Why do you think most urinary tract infections occur in women?

Urine is bacteriostatic to most of the commensal organisms inhabiting the perineum and vagina, such as *Lactobacillus, Corynebacterium*, diphtheroids, and *Staphylococcus epidermidis*. In contrast, many Gram-negative organisms thrive in urine. As a result, most urinary tract infections are caused by facultative Gram-negative rods from the GI tract. The most common etiological agents of UTIs are:

■ Certain serotypes of *E. coli* that comprise the uropathogenic *E. coli* (75% of all UTIs)

■ *Klebsiella, Proteus, Pseudomonas, Enterobacter* (20%)

■ *Staphylococcus aureus, Enterococcus, Chlamydia*, fungi, *Staphylococcus saprophiticus*, other (5%)

Case History: Classic Urinary Tract Infection

Lashandra was 24 years old and had been experiencing back pain, increased frequency of urination, and dysuria (painful or burning urination) over the previous 3 days. This was the first time Lashandra had ever suffered from persisting dysuria. She consulted her general practitioner, who requested a midstream specimen of urine. Upon microscopic examination, the urine was found to contain more than 50 leukocytes per microliter (normal is fewer than 5) and 35 red blood cells per microliter (normal is 3–20). No epithelial squamous cells (skin cells) were seen, indicating a well-collected midstream catch. The urine culture plated on agar medium yielded more than 10^5 colonies per milliliter (meaning more than 10^5 organisms per milliliter in the urine) of a facultative anaerobic Gram-negative bacillus capable of fermenting lactose.

The first question to ask in this case is whether the patient had a significant UTI. The purpose of the midstream urine collection is to provide laboratory data to make this determination. Even though urine in the bladder is normally considered sterile (see above), urine becomes contaminated with normal skin or GI microbiota that may adhere to the urethral wall. In a midstream collection, the patient urinates briefly, stops to position a collection jar, and resumes urinating to collect a midstream sample. This procedure minimizes the number of organisms in the sample by washing away organisms clinging to the urethra before actually collecting the sample. Nevertheless, the collected sample will still contain low numbers of organisms representing normal microbiota of the urethra.

A diagnosis of cystitis can be made in a <u>symptomatic</u> patient when the number of bacteria (single colony type) in a sample is at least 1,000/ml. The number of bacteria per milliliter of urine does <u>not</u> have to reach 10^5 to diagnose cystitis. In fact, some people (especially the elderly) are asymptomatic yet have bacterial counts of 10^5/ml. These patients have asymptomatic bacteriuria and are not usually treated with antibiotics, unless they are pregnant women. The patient in the case history is symptomatic and has sufficient numbers of bacteria in her urine to indicate a UTI.

The laboratory found the organism to be a Gram-negative bacillus that ferments lactose, suggesting *E. coli* as the likely culprit. Given that the normal habitat of *E. coli* is the gastrointestinal tract and this is the first UTI suffered by the patient, the infection is likely the result of an inadvertent introduction of the microbe into the urethra. The organism makes its way up the urethra and into the bladder. Another way the bladder can become infected is via a descending route from the kidney. Organisms from a kidney that was infected as a result of sepsis can descend along a ureter to cause a bladder infection.

Gram-negative rods not only thrive in urine but may be adapted to cause urinary tract infections through specialized pili. These pili have terminal receptors for glycolipids and glycoproteins present on urinary tract epithelial cells (**Fig. 26.13A**). Uropathogenic strains of *E. coli*, for example, typically have P-type pili, with a terminal receptor for the P antigen (a rather appropriate name for a bladder-specific virulence factor). The P antigen is a blood group marker found on the surface of cells lining the perineum and urinary tract; it is expressed by approximately 75% of the population. These individuals are particularly susceptible to UTIs.

Some patients, usually women, are also susceptible to recurrent bladder infections. These are thought to be caused by uropathogenic *E. coli* that invade urinary tract epithelial cells and form compact intracellular biofilms (discussed in **eTopic 26.4**).

Urinary tract infections are among those most frequently acquired during a hospital stay (so-called nosocomial, or

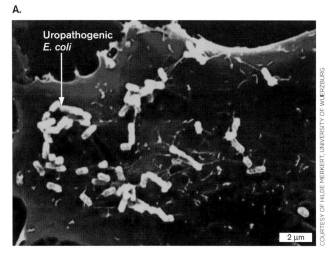

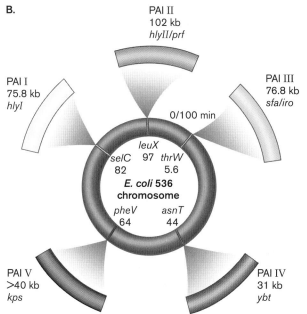

FIGURE 26.13 ■ **Uropathogenic** *E. coli*. **A.** Bladder cell with adherent uropathogenic *E. coli* (SEM). **B.** Distribution of pathogenicity islands (PAIs) in uropathogenic *E. coli*. The location of each insert is given in map units within the circle representing the genome. The 0–100 map units are called centisomes. Each centisome is approx. 44 kb of DNA. Zero is arbitrarily placed at the *thr* (threonine) gene. The origin of replication on this map is near 82 centisomes. A chromosomal gene flanking the insert is also provided. The size of each island is shown above the insert. A key virulence gene for each island is listed.

hospital-acquired, infections). In these cases, the causal organism is less likely to be *E. coli* and more likely to be another Gram-negative bacterium or *Staphylococcus*. Many UTIs resolve spontaneously, but others can progress to destroy the kidney or, via Gram-negative septicemia, the host. As a result, antibiotic therapy is recommended. In older patients, UTIs frequently show atypical symptoms, including delirium, which disappears when the UTI is treated.

TABLE 26.4	Common Sexually Transmitted Diseases				
Disease	Symptoms	Etiological agent[a]	Virulence factors	Treatment	Reported cases
Gonorrhea	Purulent discharge, burning urination; can lead to sterility	*Neisseria gonorrhoeae* (G–)	Type IV pili, phase variation	Ceftriaxone	333,004[b]
Syphilis	1°: chancre; 2°: joint pain, rash; 3°: gummata, aneurism, central nervous system damage	*Treponema pallidum* (spirochete)	Motility	Penicillin	56,471[b]
Nongonococcal urethritis	Watery or mucoid urethral discharge, burning urination	*Chlamydia trachomatis*	Intracellular growth; prevents phagolysosome fusion	Azithromycin	1.4 million[b]
Trichomoniasis	Vaginal itching, painful urination, strawberry cervix	*Trichomonas vaginalis* (protozoan)	Cytotoxin	Metronidazole	225,000[b]
Chancroid	Painful genital lesion	*Haemophilus ducreyi* (G–)	?	Erythromycin	10[b]
HIV diagnoses ranging from asymptomatic to AIDS	For AIDS, fever, diarrhea, cough, night sweats, fatigue, opportunistic infections	HIV	gp120, Rev, Nef, and Tat proteins	Azidothymidine (AZT), protease inhibitors, zidovudine	34,969[b]
Genital herpes	Painful ulcer on external genitals, painful urination	Herpes simplex 2	Cell fusion protein, complement-binding protein, latency	Acyclovir, iododeoxyuridine	306,000b (50 million[c])
Genital warts	Warts on external genitals	Human papillomavirus	E6, E7 proteins	Vaccine now available	404,000b (50 million[c])

[a]G+ = Gram-positive; G– = Gram-negative.

[b]CDC reported cases for 2013.

[c]Total estimated current cases.

Thought Question

26.5 Urine samples collected from six hospital patients were placed on a table at the nurses' station awaiting pickup from the microbiology lab. Several hours later, a courier retrieved the samples and transported them to the lab. The next day, the lab reported that four of the six patients had UTIs. Would you consider these results reliable? Would you start treatment based on these results?

What makes uropathogenic *E. coli* different from other strains of *E. coli*? This is a question still under investigation, but genomic analysis has exposed five pathogenicity islands unique to these strains (**Fig. 26.13B**). The functions of these pathogenicity islands are still under investigation.

Sexually Transmitted Diseases

Sexually transmitted diseases (STDs) are defined as infections transmitted primarily through sexual contact, which may include genital, oral-genital, or anal-genital contact. The organisms or viruses involved are generally very susceptible to drying and require direct physical contact with mucous membranes for transmission. Because sex can take many forms in addition to intercourse, these microbes can initiate disease in the urogenital tract, rectum, or oral cavities. Condoms can prevent transmission but do so, of course, only when used. Examples of common sexually transmitted diseases are listed in **Table 26.4**.

Case History: Secondary Syphilis

A pregnant 18-year-old woman came to the county urgent-care clinic with a low-grade fever, malaise, and headache. She was sent home with a diagnosis of influenza. She again sought treatment 7 days later, after she discovered a macular rash (flat, red) developing on her trunk, arms, palms of her hands, and soles of her feet. Further questioning of the patient revealed that 1 year earlier, she had had a painless ulcer on her vagina that healed spontaneously. She was diagnosed with secondary syphilis—a diagnosis confirmed by a serological test. She was given a single intramuscular injection of penicillin and told that her sexual partners had to be treated as well.

The vaginal ulcer, the long latent period, and secondary development of rash on the hands and feet described in the

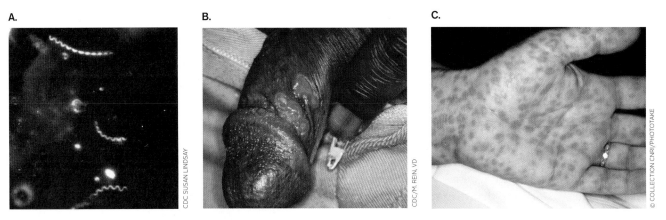

FIGURE 26.14 ■ **Syphilis. A.** *Treponema pallidum* (dark-field microscopy). Organisms are 10–25 μm long. **B.** Chancre of primary syphilis. **C.** Rash of secondary syphilis.

case history are classic symptoms of syphilis. Syphilis was recognized as a disease as early as the sixteenth century, but the organism responsible, the spirochete *Treponema pallidum*, was not discovered until 1905 (**Fig. 26.14A**). (Chapter 18 describes spirochete structure.) The illness has several stages. The disease has often been called the great imitator because its symptoms in the second stage, as exhibited in the case history, can mimic many other diseases. The incubation stage can last from 2 to 6 weeks after transmission, during which time the organism multiplies and spreads throughout the body.

Primary syphilis is an inflammatory reaction at the site of infection called a **chancre** (**Fig. 26.14B**). About a centimeter in diameter, the chancre is painless and hard, and it contains spirochetes. Patients are usually too embarrassed to seek medical attention and, because it is painless, hope it will just go away. It does go away after several weeks, and without scarring. The disease has now entered the primary latent stage. Over the next 5 years, symptoms may be absent, but at any time, as described in the case history, the infected person can develop the rash typical of **secondary syphilis** (**Fig. 26.14C**). The rash can be similar to rashes produced by many different diseases, which contributes to the "great imitator" label. The patient remains contagious in this stage. The symptoms eventually resolve, and the patient reenters a latent phase of syphilis. Some patients eventually progress over years to **tertiary syphilis** and develop many of the cardiovascular and nervous system–associated symptoms. Neurological symptoms resulting from syphilis at any stage of the disease are referred to as neurosyphilis. The patient can develop dementia and eventually dies from the disease.

The presence of *T. pallidum* in tissues can be detected with fluorescent antibody, but the initial screen is usually serological (that is, patient serum is tested for antibodies). Antibiotics are useful for eradicating the organism, but there is no vaccine, and cure does not confer immunity.

The disease is particularly dangerous in pregnant women. The treponeme can cross the placental barrier and infect the fetus to cause **congenital syphilis**. At birth, infected newborns will have notched teeth (visible on X-rays), perforated palates, and other congenital defects. Women should be screened for syphilis as part of their prenatal testing to prevent these congenital infections.

Columbus and the New World theory of syphilis. An outbreak of syphilis that spread throughout Europe soon after Christopher Columbus and his crew returned from the Americas (1493) led to the theory that Columbus brought the treponeme to Europe from the New World. However, the theory that Columbus brought syphilis to Europe has been difficult to prove.

Kristin Harper and colleagues used a phylogenetic approach to address the 500-year-old question. The group sequenced 26 geographically disparate strains of pathogenic *Treponema*. Of all the strains examined, the sexually transmitted syphilis-causing strains originated most recently and were more closely related to the nonsexually transmitted yaws-causing strains from South America, supporting the New World theory of syphilis. Thus, it seems the crew of the Columbus voyages brought smallpox and measles to America and did, in fact, return to Europe with syphilis.

The Tuskegee experiment. Unfortunately, much of what we know about syphilis is the result of the infamous Tuskegee experiment conducted in the 1930s in Alabama. The study was entitled "Untreated Syphilis in the Negro Male." Through dubious means and deception, a group of African-American males was enlisted in a study that promised treatment but whose real purpose was to observe how the disease progressed without treatment. Today, such experiments are barred, thanks to strict oversight by institutional review boards (IRBs) that require human subjects to sign informed consent forms. An interesting treatise on the Tuskegee experiment can be found on the Internet at the National Center for Case Study Teaching in Science (search the site for "Bad Blood").

Chlamydial Infections Are Often Silent

Chlamydia is the most frequently reported sexually transmitted infectious disease in the United States, according to the Centers for Disease Control and Prevention, but many people are unaware that they are infected. Three-fourths of infected women, for instance, have no symptoms.

The chlamydias are unusual Gram-negative organisms with a unique developmental cycle. (Chapter 18 describes chlamydial morphology.) They are obligate intracellular pathogens that start as a small, nonreplicating, infectious elementary body that enters target eukaryotic cells. Once inside vacuoles, they begin to enlarge into replicating reticulate bodies (**Fig. 26.15**). As the vacuole fills, the reticulate bodies divide to become new nonreplicating elementary bodies. *Chlamydia trachomatis* and *Chlamydophila pneumoniae* can both cause STDs, as well as other diseases, such as trachoma of the eye or pneumonia.

People most at risk of developing genitourinary tract infections with chlamydias are young, sexually active men and women; anybody who has recently changed sexual partners; and anybody who has recently had another sexually transmitted disease. The astute clinician knows that when one STD is discovered, others may also be present.

Left untreated, chlamydia can cause serious health problems. In women, the organism can produce pelvic inflammatory disease, a damaging infection of the uterus and fallopian tubes that can be caused by several different microbial species. The damage produced can lead to infertility, tubal pregnancies, and chronic pelvic pain. Men left untreated can suffer urethral and testicular infections and a serious form of arthritis.

Case History: Gonorrhea

A 22-year-old mechanic saw his family doctor for treatment of painful urination and urethral discharge. The patient was sexually active, with three regular and several "one time– good time" partners. Physical examination was unremarkable except for prevalent urethral discharge. The discharge was Gram-stained and sent for culture. The Gram stain revealed many pus cells, some of which contained numerous

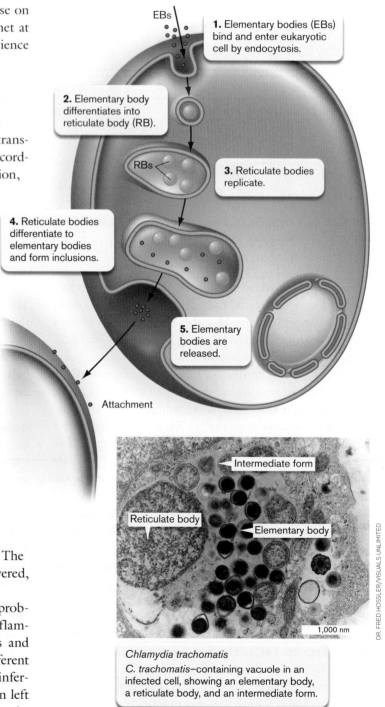

1. Elementary bodies (EBs) bind and enter eukaryotic cell by endocytosis.

2. Elementary body differentiates into reticulate body (RB).

3. Reticulate bodies replicate.

4. Reticulate bodies differentiate to elementary bodies and form inclusions.

5. Elementary bodies are released.

Attachment

Intermediate form

Reticulate body

Elementary body

1,000 nm

DR. FRED HOSSLER/VISUALS UNLIMITED

Chlamydia trachomatis
C. trachomatis–containing vacuole in an infected cell, showing an elementary body, a reticulate body, and an intermediate form.

FIGURE 26.15 ■ **Replication cycle of *Chlamydia*. Inset:** EM of a *C. trachomatis*–containing vacuole in an infected cell, showing a reticulate body, infectious elementary bodies, and intermediate forms.

*phagocytosed Gram-negative diplococci (**Fig. 26.16A**). Blood was drawn for syphilis serology, which proved negative. The patient was given an intramuscular injection of ceftriaxone (250 mg), and oral doxycycline (100 mg, twice a day) was prescribed for 7 days. The bacteriology lab was able to recover the bacteria seen in the Gram-stained smear of the urethral*

FIGURE 26.16 ▪ *Neisseria gonorrhoeae.* **A.** Within pus-filled exudates, the Gram-negative diplococci are found intracellularly inside PMNs. The intracellular bacteria in this case are no longer viable, having been killed by the antimicrobial mechanisms of the white cell. **B.** Colonies of *N. gonorrhoeae* growing on chocolate agar (agar plates containing heat-lysed red blood cells that turn the medium chocolate brown). **C.** *N. gonorrhoeae* binding to CD4+ T cells, inhibiting T-cell activation and proliferation, which may explain the ease of reinfection (colorized SEM).

*discharge. The organism produced characteristic colonies on chocolate agar (agar plates containing heat-lysed red blood cells that turn the medium chocolate brown) (**Fig. 26.16B**). The case was subsequently reported to the state public health department. When the patient came back for his return visit, his symptoms had resolved, and a repeat culture was negative.*

The disease here is classic gonorrhea caused by *Neisseria gonorrhoeae.* A characteristic that distinguishes *Neisseria* infections from *Chlamydia* infections is that bacterial cells are seen in gonorrheal discharges but not in chlamydial discharges. Gonorrhea has been a problem for centuries and remains epidemic in this country today. Symptoms generally occur 2–7 days after infection, but they can take as long as 30 days to develop. Most infected men exhibit symptoms; only about 10%–15% do not. Symptoms include painful urination, yellowish white discharge from the penis, and in some cases, swelling of the testicles and penis. The Greek physician Galen (AD 129–ca. 199) originally mistook the discharge for semen. This mistake led to the name *gonorrhea*, which means "flow of seed."

In contrast to men, most infected women (80%) do not exhibit symptoms and constitute the major reservoir of the organism. If they are asymptomatic, they have no reason to seek treatment and thus can spread the disease. When symptoms are present, they are usually mild. A symptomatic woman will experience a painful burning sensation when urinating and will notice vaginal discharge that is yellow or occasionally bloody. She may also complain of cramps or pain in her lower abdomen, sometimes with fever or nausea. As the infection spreads throughout the reproductive organs (uterus and fallopian tubes), pelvic inflammatory disease occurs (see earlier discussion of chlamydia). There is no serological test or vaccine for gonorrhea, because the organism frequently changes the structure of its surface

antigens (phase variation is discussed in Section 10.1 and **eTopic 10.1**).

Although *N. gonorrhoeae* is generally serum sensitive, owing to its sensitivity to complement, certain serum-resistant strains can make their way to the bloodstream and carry infection throughout the body. As a result, both sexes can develop purulent arthritis (joint fluid containing pus), endocarditis, or meningitis. An infected mother can also infect her newborn during parturition (birth), leading to a serious eye infection called ophthalmia neonatorum. Because of this risk and because most infected women are asymptomatic, all newborns receive antimicrobial eyedrops at birth.

Because adults engage in a variety of sexual practices, *N. gonorrhoeae* can also infect the anus or the pharynx, where it can develop into a mild sore throat. These infections generally remain unrecognized until a sex partner presents with a more typical form of genitourinary gonorrhea. Because no lasting immunity is built up, reinfection with *N. gonorrhoeae* is possible. Reinfection occurs in part because there is phase variation in various surface antigens and because the organism can apparently bind to CD4+ T cells, inhibiting their activation and proliferation to become memory T cells (**Fig. 26.16C**).

Over the decades, *N. gonorrhoeae* has incrementally developed resistance to many antibiotics used in its treatment, but there has always been a new, effective drug ready to take the place of the old drug. Soon, this may no longer be the case. To prevent treatment failures, the CDC recommends dual antibiotic therapy that includes an intramuscular injection of ceftriaxone and oral azithromycin or tetracycline. However, the incidence of ceftriaxone-resistant strains of *N. gonorrhoeae* overseas is increasing. The alarm has been raised that new antibiotics must be developed if we are to prevent an uncontrollable explosion of cases of this already epidemic disease.

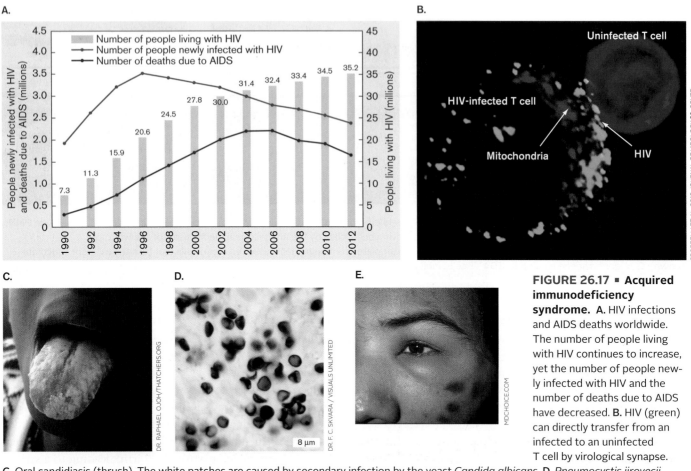

FIGURE 26.17 ■ **Acquired immunodeficiency syndrome.** **A.** HIV infections and AIDS deaths worldwide. The number of people living with HIV continues to increase, yet the number of people newly infected with HIV and the number of deaths due to AIDS have decreased. **B.** HIV (green) can directly transfer from an infected to an uninfected T cell by virological synapse.

C. Oral candidiasis (thrush). The white patches are caused by secondary infection by the yeast *Candida albicans*. **D.** *Pneumocystis jirovecii* infection of the lung. Note the cuplike appearance of the fungus, almost like crushed Ping-Pong balls. Organisms range from 2 to 6 μm in diameter. **E.** Kaposi's sarcoma (oval spots) and periorbital cellulitis infection.

> **Thought Question**
>
> **26.6** Aside from the CDC guidelines for treating gonorrhea, why else do you suppose the patient in this case was treated with doxycycline (a derivative of tetracycline)? And why can one person be infected repeatedly with *N. gonorrhoeae*?

HIV Causes AIDS, a Sexually Transmitted and Blood-Borne Disease

Though HIV (human immunodeficiency virus) is believed to have originated around the year 1900, it was not discovered until 1981, when the virus caused the greatest pandemic of the late twentieth century. HIV remains a serious problem today, especially in southern Africa, where it is estimated that 10%–30% of the population is infected (prevalence in the United States is less than 1%). HIV has claimed the lives of almost 2 million people per year worldwide (about 14,000 per year in the United States). You may be surprised to learn that in the United States, women make up about 19% of new AIDS cases per year, the majority resulting from heterosexual sex. The molecular biology and virulence of HIV are discussed in Chapters 11 and 25, and pathogenesis is covered in **eTopic 26.5**. This section focuses on the disease that HIV causes—namely, acquired immunodeficiency syndrome (AIDS).

HIV, a lentivirus in the retroviral family, is a prominent example of viruses that can be transmitted either sexually (vaginally, orally, anally—homosexually or heterosexually) or through direct contact with body fluids, such as occurs with blood transfusion or the sharing of hypodermic needles by intravenous drug users. HIV is <u>not</u> transmitted by kissing, tears, or mosquito bites. It can, however, be transferred from mother to fetus through the placenta (transplacental transfer). We discuss HIV infections in this section because sexual contact remains the major route of transmission. **Figure 26.17A** shows the worldwide decrease in the number of newly infected HIV patients and deaths due to AIDS. The number of people living with HIV has increased, however, because of the development of more effective antiviral treatments. The pathogenesis of HIV and the proportion of AIDS cases by sex and ethnicity are described in **eTopic 26.5**.

Once it has entered the bloodstream, HIV infects CD4$^+$ T cells and macrophages (which also have CD4 on their surface). The virus replicates very rapidly, producing a billion particles per day. The virus can also spread directly from cell to cell via virological synapses (**Fig. 26.17B**). When infected and uninfected T cells make contact, HIV virions and mitochondria (to supply energy) line up at the contact interface where transmission occurs. Viral replication starts to kill the CD4$^+$ T cells, which progressively decrease in number.

HIV infection has four stages.

1. A primary stage beginning with seroconversion (finding antibodies to the virus)
2. Clinical latency (slow steady loss of T cells without symptoms, or sometimes swollen lymph nodes)
3. Early symptomatic disease (formerly called AIDS-related complex)
4. Acquired immunodeficiency syndrome (AIDS)

Symptoms in the early symptomatic stage can include fever, headache, macular rash, and weight loss. The symptoms can manifest within a few months of infection, resolve within a few weeks, and then recur. As T-cell numbers decline, the debilitated immune system leaves the victim susceptible to secondary infections such as thrush, caused by the yeast *Candida albicans* (candidiasis) and related species (**Fig. 26.17C**).

The CDC defines AIDS as a CD4$^+$ cell count of less than 200 cells/µl or the presence of an AIDS indicator disease. For example, once the CD4$^+$ T-cell population falls below 500 cells/µl, opportunistic infections start to arise and disease processes begin. Opportunistic infections include pneumonia by *Mycobacterium avium-intracellulare* or *Pneumocystis jirovecii* (**Fig. 26.17D**), cryptococcal meningitis, *Histoplasma capsulatum* infection, and tuberculosis. The presence of these diseases indicates that the patient has AIDS, regardless of the CD4$^+$ count.

Various cancers are also on the list of AIDS indicator diseases. AIDS patients are more susceptible to cancers because their depressed immune systems cannot detect and destroy cancer cells generated by secondary agents. Kaposi's sarcoma (**Fig. 26.17E**), for instance, is a common cancer seen in AIDS patients that is caused by human herpes virus type 8 (HHV8). Kaposi's sarcoma originates in endothelial or lymphatic cells, but the resulting tumors can develop anywhere—gastrointestinal tract, mouth, lungs, skin, or brain.

Diagnosis of AIDS involves detecting anti-HIV antibodies and determining the CD4$^+$ T-cell count in a patient. An assay for HIV to determine viral load is done via quantitative PCR to detect HIV-specific genes such as *gag*, *nef*, or *pol*. Remember, a person who is HIV-positive does not necessarily have AIDS; the disease may take years to develop. A vaccine is not yet available to prevent AIDS, in part

because the envelope proteins of the virus (see Fig. 11.24) typically change their antigenic shape. However, progression of the disease can be controlled by antiretroviral drugs that inhibit two HIV enzymes critical to the replication of the virus: reverse transcriptase and protease. Antiretroviral drugs are discussed further in Chapter 27.

New treatment regimens (HAART; see Chapter 27) have made HIV an increasingly survivable infection. Consider this: In the past, nearly all HIV-infected individuals eventually became ill and died from AIDS-related diseases, but in the United States at least half of, if not most, HIV-positive people now die from diseases unrelated to HIV (for example, heart attack). Because HIV infection is now considered a treatable disease, the CDC recommends routine screening for HIV.

Can HIV infection be cure d? Because HIV DNA integrates into the genome of an infected cell, it has been impossible to cure anyone of the infection. However, a new study by Kamel Khalili has proved that the CRISPR-cas9 system described in Chapters 9 and 12 can be engineered to "surgically" remove proviral HIV DNA from infected T cells. Whether this strategy can be used in a patient remains to be seen.

Thought Question

26.7 Like the cause of plague, HIV is a blood-borne disease. Why, then, do fleas and mosquitoes fail to transmit HIV?

The Protozoan *Trichomonas vaginalis* Produces a Common Vaginal Infection

Trichomonas vaginalis is a flagellated protozoan (**Fig. 26.18**) that causes an unpleasant sexually transmitted vaginal disease called trichomoniasis. Approximately 2–3 million infections occur each year in the United States. Both men and women can be infected; however, men are usually asymptomatic. Even among infected women, 25%–50% are considered asymptomatic carriers.

There is no cyst in the life cycle of *T. vaginalis*, so transmission is via the trophozoite stage only (the form of a protozoan in the feeding stage). The female patient with trichomoniasis may complain of vaginal itching and/or burning and a musty vaginal odor. An abnormal vaginal discharge also may be present. Males will complain of painful urination (dysuria), urethral or testicular pain, and lower abdominal pain.

Owing to colonization by lactobacilli (which produce large amounts of acidic lactic acid), the normal, healthy vagina has a pH of less than 4.5. However, since *T. vaginalis* feeds on bacteria, the pH of the vagina rises as the numbers of lactobacilli decrease. Definitive diagnosis requires demonstrating by microscopy that the flagellated protozoan is in secretions. PMNs, which are the primary host

DAVID M. PHILLIPS/VISUALS UNLIMITED, INC.

1 μm

FIGURE 26.18 ▪ *Trichomonas vaginalis.* *T. vaginalis* is a protozoan that causes a common sexually transmitted disease (SEM).

defense against the organism, are also usually present. Like giardiasis, this disease is treated with metronidazole.

To Summarize

▪ **Urinary tract infections (UTIs)** include cystitis (bladder) and pyelonephritis (kidney). *E. coli* is the most common cause of UTIs.

▪ **Cystitis can develop** from bacteria ascending up the urethra (most common route) or descending down a ureter from an infected kidney.

▪ **Pyelonephritis can develop** from bacteria ascending along a ureter from an infected bladder or from bacteria in the bloodstream disseminating from an infection elsewhere in the body.

▪ **Syphilis, chlamydia, and gonorrhea** are the most common sexually transmitted diseases.

▪ **A patient with one sexually transmitted disease** often has another sexually transmitted disease too.

▪ **Complement prevents bloodstream dissemination** of *Neisseria gonorrhoeae*, which lacks a carbohydrate capsule. Because the organism frequently changes the structure of its surface antigens, no vaccine is available for *N. gonorrhoeae*.

▪ **HIV depletion of CD4$^+$ T cells** results in lethal secondary infections and cancers.

▪ *Trichomonas vaginalis* is a flagellated protozoan that causes a sexually transmitted vaginal disease. The reservoirs for this organism are the male urethra and female vagina.

26.6 Cardiovascular and Systemic Infections

The cardiovascular system, which is composed of the heart, arteries, veins, and capillaries, delivers oxygen, nutrients, and immune system components to all tissues of the body. There are pathogens that can directly infect the heart or the endothelial cells lining blood vessels. However, the blood itself can also become infected or serve as a mass transit system to spread pathogens throughout the body. Pathogens disseminated in this way can cause infections in many different organ systems. We will examine both types of infections in this section.

Infections of the cardiovascular system include septicemia, **endocarditis** (inflammation of the heart's inner lining), pericarditis (inflammation of the heart's outer lining), myocarditis (inflammation of heart muscle), and, possibly, atherosclerosis (deposition of fatty substances along the inner lining of arteries; **eTopic 26.6**). These are all life-threatening diseases.

Septicemia is, by strict definition, the presence of bacteria or viruses in the blood. The presence of viruses is a condition more specifically called **viremia**, while bacteria in the circulation is more specifically called **bacteremia**. In practice, however, the terms "septicemia" and "bacteremia" are often used interchangeably. Septicemia can develop from a local tissue infection situated anywhere in the body, although blood factors such as complement can nonspecifically kill many types of bacteria that enter the blood. Nevertheless, Gram-positives, Gram-negatives, aerobes, and anaerobes can all produce septicemia under the right conditions.

Endocarditis can be either viral or bacterial in origin. It can be a consequence of many bacterial diseases, such as brucellosis, gonorrhea, psittacosis, staphylococcal and streptococcal infections, candidiasis, and Q fever (*Coxiella*). Among the many viral causes are coxsackievirus, echovirus, Epstein-Barr virus, and HIV. Bacterial infections of the heart are always serious. Viral infections, although common, are rarely life-threatening in healthy individuals and are usually asymptomatic.

Case History: Bacterial Endocarditis

Elizabeth was 38 years old and had a history of mitral valve prolapse (a common congenital condition in which a heart valve does not close properly). She was also on immunosuppressive therapy following a kidney transplant. Recently, Elizabeth was admitted to the hospital complaining of fatigue, intermittent fevers for 5 weeks, and headaches for 3 weeks—symptoms the physician recognized as possible indications of endocarditis. Elizabeth reported having a dental procedure a few weeks prior to the onset of symptoms. A sample of her blood placed in a liquid bacteriological medium grew Gram-positive cocci, which turned out to be Streptococcus mutans, *a member of the viridans streptococci. With the finding of bacteria in the bloodstream, the diagnosis of bacterial endocarditis was confirmed. Elizabeth began a 1-month course of intravenous penicillin G and gentamicin therapy and eventually recovered to normal health.*

Endocarditis (inflammation of the heart) is traditionally classified as acute or subacute, depending on the pathogenic organism involved and the speed of clinical presentation. Subacute bacterial endocarditis (SBE) has a slow onset with vague symptoms. It is usually caused by bacterial infection of a heart valve (**Fig. 26.19**). SBE infections are usually (but not always) caused by a viridans streptococcus from the oral microbiota (for example, *Streptococcus mutans*, a common cause of dental caries). "Viridans streptococci" is a general term used for commensal streptococci whose colonies produce green alpha hemolysis on blood agar ("viridans" is from the Greek *viridis*, "to be green"). Most patients who develop infective endocarditis have mitral valve prolapse (90%), although this is frequently not the case when patients are intravenous drug abusers or have hospital-acquired (nosocomial) infections.

Subacute bacterial endocarditis can begin at the dentist's office, as it did in the case history presented here, although it very rarely does. Following a dental procedure (such as tooth restoration) or even while brushing your teeth, oral bacteria can transiently enter the bloodstream and circulate. *S. mutans*, which is not normally a serious health problem, can become lodged onto damaged heart valves, grow as a biofilm, and secrete a thick glycocalyx coating that encases the microbes and forms a vegetation on the valve, damaging it further. If untreated, the condition can be fatal within 6 weeks to a year.

A rapidly progressive (acute) and highly destructive infection can develop when more virulent organisms, such as *Staphylococcus aureus*, gain access to cardiac tissue. Symptoms of acute endocarditis include fever, pronounced valvular regurgitation (backflow of blood through the valve), and abscess formation.

Most patients with subacute bacterial endocarditis present with a fever that lasts several weeks. They also complain of nonspecific symptoms, such as cough, shortness of breath, joint pain, diarrhea, and abdominal or flank pain. Endocarditis is suspected in any patient who has a heart murmur and an unexplained fever for at least a week. It should also be considered in an intravenous drug abuser with a fever, even in the absence of a murmur. In either case, definitive diagnosis requires blood cultures that grow bacteria. Blood cultures involve taking samples of a

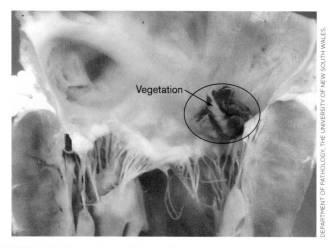

FIGURE 26.19 ■ **View of bacterial endocarditis.** Close-up of mitral valve endocarditis, showing vegetation.

patient's blood from two different locations (such as two different arms). Once collected, the blood is added to liquid culture medium and incubated at 37°C. Incubation should be done aerobically and anaerobically. Growth of the same organism in cultures taken from two body sites rules out inadvertent contamination with skin microbes, which would likely yield growth in only one culture.

Curing endocarditis is difficult because the microbes are usually ensconced in a nearly impenetrable glycocalyx. Consequently, eradicating microorganisms from the vegetations almost always requires hospitalization, where high doses of intravenous antibiotic therapy can be administered and monitored. Antibiotic therapy usually continues for at least a month, and in extreme cases, surgery may be necessary to repair or replace the damaged heart valve.

Patients with heart valve prolapse should not shy away from the dentist, however. As long as a healthy immune system is in place, the risk of infection is low. In fact, maintaining good oral hygiene can keep their risk of developing infectious endocarditis low. Immunocompromised patients, on the other hand, are at increased risk and should be treated prophylactically with oral penicillin or erythromycin 1 hour before a procedure, to kill oral bacteria that enter the bloodstream. This prophylactic treatment can prevent the development of endocarditis in these patients.

Viruses such as adenovirus and some enteroviruses can also cause endocarditis, as well as a condition known as myocarditis, an inflammation of the heart muscle.

Thought Question

26.8 A patient presenting with high fever and in an extremely weakened state is suspected of having a septicemia. Two sets of blood cultures are taken from different arms. One bottle from each set grows *Staphylococcus aureus*, yet the laboratory report states that the results are inconclusive. New blood cultures are ordered. What would make these results inconclusive?

Malarial Parasites Feed on Blood

Malaria is the most devastating infectious disease known. Each year, 300–500 million people develop malaria worldwide, and 1–3 million of these people, mostly children, die. In the United States the disease is relatively rare; only about 1,000 cases occur annually, and almost all are acquired as a result of international travel to endemic areas (**Fig. 26.20A**). Once again, this correlation illustrates the diagnostic importance of knowing a patient's travel history.

The disease is caused by four species of *Plasmodium*: *P. falciparum* (the most deadly), *P. malariae*, *P. vivax*, and *P. ovale*. The life cycle of *Plasmodium*, discussed in detail in Chapter 20, is complex and involves two cycles: an asexual erythrocytic cycle in the human, and a sexual cycle in the mosquito (see Fig. 20.39).

In the erythrocytic cycle, the organisms enter the bloodstream through the bite of an infected female *Anopheles* mosquito (the mosquito injects a small amount of saliva containing *Plasmodium*). The haploid sporozoites travel immediately to the liver, where they undergo asexual fission to produce merozoites. Released from the liver, the merozoites attach to and penetrate red blood cells, where *Plasmodium* consumes hemoglobin and enlarges into a trophozoite. The protist nucleus divides, so that the cell, now called a schizont, contains up to 20 or so nuclei. The schizont then divides to make the smaller, haploid merozoites (**Fig. 26.20B**). The glutted red blood cell eventually lyses, releasing merozoites that can infect new red blood cells (see Fig. 20.39).

Sudden, synchronized release of the merozoites and red blood cell debris triggers the telltale symptoms of malaria: violent, shaking chills followed by high fever and sweating.

The erythrocytic cycle, and thus the symptoms, repeats every 48–72 hours. After several cycles, the patient goes into remission lasting several weeks to months, after which there is a relapse.

Much of today's research focuses on why malarial relapse happens. Why does the immune system fail to eliminate the parasite after the first episode? When *Plasmodium* invades the red blood cells, it lines the blood cells with a protein, PfEMP1, which causes the parasite to stick to the sides of blood vessels. The parasite is thereby removed from circulation, but the protein cannot protect the parasite from patrolling macrophages, which eventually detect the invader and recruit other immune cells to fight it. So, during a malarial infection, a small percentage of each generation of parasites switches to a different version of PfEMP1 that the body has never seen before. In its new disguise, *Plasmodium* can invade more red blood cells and cause another wave of fever, headaches, nausea, and chills.

The antigenic shift happens when the parasite alters the PfEMP1 gene that is expressed. The body now has to repeat the recognition and attack responses all over again. The parasite has 60 cloaking genes, called *var*, that can be turned on and off individually, changing the organisms' antigenic structure, like a criminal repeatedly changing his disguise to elude police.

The *var* genes are regulated by chromosome packaging, which unwraps one gene for expression and packs away the inactive genes. DNA can be encased so securely by some proteins that other proteins cannot access the nucleic acid for transcription—a process known as epigenetic silencing. Becoming immune to all the types of malaria can take upwards of 5 years and requires constant exposure;

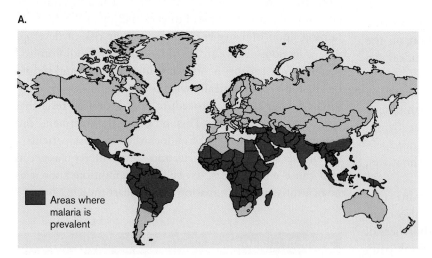

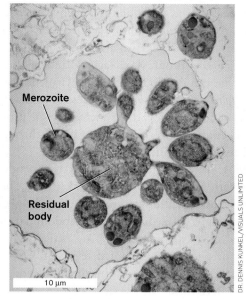

Merozoite

Residual body

10 µm

DR. DENNIS KUNKEL/VISUALS UNLIMITED

FIGURE 26.20 ▪ **Malaria is a major disease worldwide.** **A.** Endemic areas of the world where malaria is prevalent. **B.** *Plasmodium falciparum* (colorized TEM). Schizont after completion of division. A residual body of the organism (yellow-green) is left over after division. The erythrocyte has lysed and only a ghost cell remains; no cytoplasm is seen surrounding the merozoites just being released. Free merozoites are seen outside the membrane.

Areas where malaria is prevalent

otherwise the immunity is lost. Many children do not live long enough to gain immunity to malaria in all its forms.

Diagnosis of malaria involves microscopic demonstration of the protist within erythrocytes (Wright stain) or serology to identify antimalarial antibodies. Treatment regimens include artemisinin, whose mechanism of action is unknown; chloroquine or mefloquine, which kill the organisms in their erythrocytic asexual stages; and primaquine, effective in the exoerythrocytic stages by disrupting *Plasmodium* mitochondria.

The chloroquine family of drugs acts by interfering with the detoxification of heme generated from hemoglobin digestion. Malaria parasites accumulate the hemoglobin released from red blood cells in plasmodial lysosomes, where digestion occurs. The parasites use the amino acids from hemoglobin to grow but find free heme toxic. To prevent eating itself to death (from accumulating too much heme), the organism detoxifies heme via polymerization, which produces a black pigment. Many antimalarial drugs, such as chloroquine, prevent polymerization by binding to the heme. As a result, the increased iron level (from heme) kills the parasite. Unfortunately, *Plasmodium* has been developing resistance to these drugs, forcing the development of new ones.

An effective vaccine has so far proved elusive because of the antigenic shape-shifting carried out by this parasite. To prevent infection, antimalarial drugs are given prophylactically to persons traveling to endemic areas. Which drug is given depends on the destination.

Bear in mind that we have described only selected organisms that cause cardiovascular infections; there are many more (for example, *Rickettsia typhi*). There is even evidence that *Chlamydophila pneumoniae* may have a role in coronary artery disease (see **eTopic 26.6**).

Note: Babesiosis is an emerging disease caused by a protozoan (*Babesia microti*) that, like *Plasmodium*, infects red blood cells. *B. microti* is transmitted by the deer tick, the same insect that transmits the agent for Lyme disease. Typically, babesiosis is a mild, flu-like disease, but in its severe form it can present with symptoms similar to those of malaria.

Systemic Infections

Many pathogenic bacteria can produce septicemia as a way to disseminate throughout the body and infect other organs. These organisms cause what are considered systemic infections.

Case History: The Plague

*A 25-year-old New Mexico rancher was admitted to an El Paso hospital because of a 2-day history of headache, chills, and fever (40°C; 104°F). The day before admission, he began vomiting. The day of admission, an orange-sized, painful swelling in the right groin area was noted (**Fig. 26.21A**). A lymph node aspirate and a smear of peripheral blood were reported to contain Gram-negative rods that exhibited bipolar*

A.

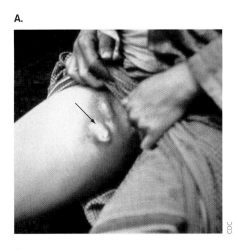

B.

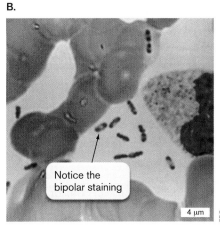

Notice the bipolar staining

4 μm

C.

D.

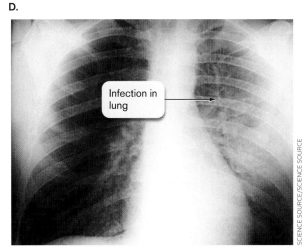

Infection in lung

FIGURE 26.21 ■ The plague. A. Classic bubo (swollen lymph node) of bubonic plague. **B.** *Yersinia pestis*, bipolar staining (length 1–3 μm). **C.** Prairie dogs are often hosts to fleas that carry plague bacilli. **D.** X-ray of pneumonic plague, showing bilateral pulmonary infection.

*staining (**Fig. 26.21B**). The patient's white blood cell count was 24,700/μl (normal is 5,000–10,000/μl), and his platelet count was 72,000/μl (normal is 130,000–400,000/μl). In the 2 weeks prior to becoming ill, the patient had trapped, killed, and skinned two prairie dogs, four coyotes, and one bobcat. The patient had cut his left hand shortly before skinning a prairie dog. PCR and typical biochemical testing of a Gram-negative rod isolated from blood cultures identified the organism as* Yersinia pestis, *the organism that causes plague. The patient received an antibiotic cocktail of gentamicin and tetracycline. He eventually recovered, after 6 weeks in intensive care.*

Plague is caused by the bacterium *Yersinia pestis*, which can infect both humans and animals. During the Middle Ages, the disease, known as the Black Death, decimated over a third of the population of Europe. Such was the horror it evoked that invading armies would actually catapult dead plague victims into embattled fortresses. This was probably the first reported case of biowarfare.

Y. pestis is present in the United States and is endemic in 17 western states. The microbe is normally transmitted from animal to animal, typically rodents like rats and even prairie dogs (**Fig. 26.21C**), by the bite of infected fleas. **Figure 26.22** illustrates the various infective cycles of the plague bacillus. Humans are not typically part of the natural infection cycle. However, in the absence of an animal host, the flea can take a blood meal from humans and thereby transmit the disease to them. During the Middle Ages, urban rats venturing back and forth to the countryside became infected by the fleas of wild rodents that served as a reservoir. Upon returning to the city, the rat flea passed the organism on to other rats, which then died in droves. The rat fleas, deprived of their normal meal, were forced to feed on city dwellers, passing the disease on to them.

Individuals bitten by an infected flea or accidentally infected through a cut while skinning an infected animal first exhibit the symptoms of **bubonic plague**. Bubonic plague emerges as the organism moves from the site of infection to the lymph nodes, producing characteristically enlarged nodes called buboes (see **Fig. 26.21A**). From the lymph nodes, the pathogen can enter the bloodstream, causing **septicemic plague**. In this phase the patient can go into shock from the massive amount of endotoxin in the bloodstream. Neither bubonic nor septicemic plague

FIGURE 26.22 ▪ The cycles of plague. The sylvatic cycle occurs in the wild, where fleas transmit the organism between rodents. An accidental interaction with urban rats can trigger a similar urban cycle. Humans can be infected through contact with infected fleas coming from either cycle. Flea bite transmission initiates bubonic plague symptoms that can progress to pneumonic plague. Pneumonic plague is highly infectious, which can cause epidemic spread of the disease.

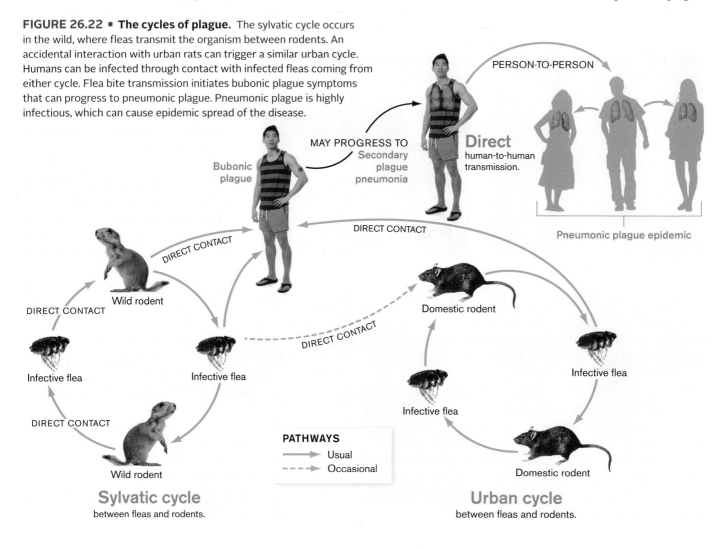

is passed from person to person. As the organism courses through the bloodstream, however, it will invade the lungs and produce **pneumonic plague** (**Fig. 26.21D**), which can be easily transmitted from person to person through aerosol droplets generated by coughing.

Pneumonic plague is the most dangerous form of the disease because it can kill quickly and spread rapidly through a population. Pneumonic plague is so virulent that an untreated patient can die within 24–48 hours. The organism is usually identified postmortem. It is now thought that the rapid spread of plague during the Middle Ages was the result of person-to-person transmission through respiratory aerosols.

Y. pestis has numerous virulence factors. For instance, YadA is a surface adhesin that binds collagen. Another factor, the F1 protein capsule surface antigen, plays a part in blocking phagocytosis in mammalian hosts. Certain biofilms formed by *Y. pestis* are also important. An extracellular matrix synthesized by *Y. pestis* produces an adherent biofilm in the flea midgut that contributes to flea-to-mammal transmission. The biofilm blocks flea digestion, making the flea feel "starved" even after a blood meal. Therefore, the flea jumps from host to host in a futile effort to feel full. As the flea tries to take a blood meal, the blockage causes the insect to regurgitate bacteria into the wound. This curious effect of *Y. pestis* on the insect vector is another unique aspect of how plague spreads so quickly.

Y. pestis also uses type III secretion systems to inject virulence proteins (YopB and YopD) into host cell membranes. Unlike *Salmonella*, which uses type III–secreted proteins to gain entrance into host cells, *Y. pestis* is not primarily an intracellular pathogen, although it can survive in macrophages. Injection of the Yop proteins disrupts the actin cytoskeleton and so helps the organism evade phagocytosis. By evading phagocytosis, the organism avoids triggering an inflammatory response and produces massive tissue colonization.

Plague has disappeared from Europe; the last major outbreak occurred in 1772. The reason for its disappearance is not known, but it was probably the result of multiple factors, not the least of which was human intervention. Although it wasn't until the nineteenth century that doctors understood how germs could cause disease, Europeans recognized by the sixteenth century that plague was contagious and could be carried from one area to another. Beginning in the late seventeenth century, governments created a medical boundary, or *cordon sanitaire*, between Europe and the areas to the east from which epidemics came. Ships traveling west from the Ottoman Empire were forced to wait in quarantine before passengers and cargo could be unloaded. Those who attempted to evade medical quarantine were shot.

Sepsis and Toxic Shock

Many bacterial pathogens can, under the right circumstances, infect the bloodstream to cause septicemia and ultimately sepsis; a life-threatening condition involving high fever, high white blood cell counts, rapid heart rate, and/or rapid breathing. The offending bacteria are typically found in the blood (see Chapter 28). Organisms that can cause sepsis include *Neisseria meningitidis*, *Escherichia coli*, *Enterococcus* species, and many others including *Yersinia pestis* from the preceding case. In severe cases sepsis can progress to septic shock in which the patient's blood pressure drops to dangerous levels.

Case History: Toxic Shock

In May 2013, a visibly ill 16-year-old girl was taken to her pediatrician, where she complained of diarrhea, high fever (39.9°C; 103.8°F), and vomiting. She told the physician she had been healthy until 2 days prior. Upon physical exam, the doctor noticed that the girl had a much lower than normal blood pressure (76/48 mm Hg; normal range 90–130/60–90), a rapid heart rate (120 beats per minute; normal range 60–100), and an erythematous (red) rash on her trunk. These are all signs of septic shock. Because of her deteriorating condition, the patient was admitted to the pediatric intensive care unit at the local hospital. However, blood cultures revealed no bacteria. She was immediately given intravenous fluids and IV antibiotics. The girl rallied and left the hospital one week later. Patient history taken upon admission revealed that the girl had started her menstrual period 4 days before becoming ill, providing an important clue as to the cause of the disease.

This potential tragedy reflects a larger story that emerged in the late 1970s and early 1980s, when women started dying from this dangerous, new (emerging) disease, now known as toxic shock syndrome. Scientists and physicians learned that it is caused by certain strains of *Staphylococcus aureus* that produce a superantigen type of toxin (toxic shock syndrome toxin, or TSST; see Section 24.3). Why had the disease not been recognized in previous decades? The answer turned out to be the use of one brand of superabsorbent tampons (since removed from the market). The tampons produced a rich growth environment for *S. aureus*. So, if the patient was colonized by a TSST-producing strain, huge amounts of toxin were released to circulate in the bloodstream.

Even though the superabsorbent tampons are no longer available, toxic shock syndrome is still sometimes associated with menstruation, as in the case history. Today, however, we recognize that toxic shock syndrome is a possible consequence of any *S. aureus* infection and can occur in both men and women. TSST-producing *S. aureus* can cause

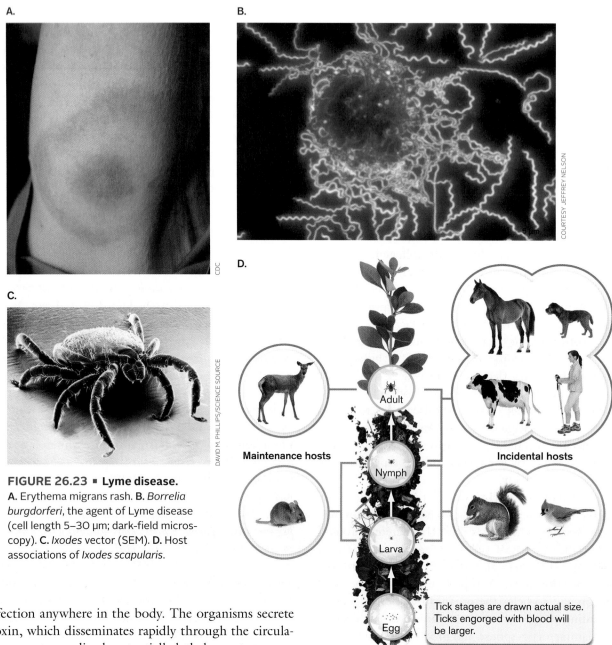

FIGURE 26.23 ■ Lyme disease.
A. Erythema migrans rash. **B.** *Borrelia burgdorferi*, the agent of Lyme disease (cell length 5–30 μm; dark-field microscopy). **C.** *Ixodes* vector (SEM). **D.** Host associations of *Ixodes scapularis*.

an infection anywhere in the body. The organisms secrete the toxin, which disseminates rapidly through the circulation to cause generalized, potentially lethal symptoms.

Knowledge that a protein toxin was responsible for the disease led to a recommendation that treatment of toxic shock syndrome include the antibiotic clindamycin. As you will learn in Chapter 27, clindamycin is an inhibitor of bacterial protein synthesis. By blocking protein synthesis, clindamycin decreased the amount of TSST made, while other antibiotics (vancomycin, for instance) killed the pathogen. The combination saved the girl's life. In Chapter 25 we discuss the weapons that microbes use to cause disease.

Case History: Lyme Disease

*Brad, a 9-year-old from Connecticut, developed a fever and a large (8-cm) reddish rash with a clear center (**erythema migrans**) on his trunk (**Fig. 26.23A**). He also had some left-facial-nerve palsy. Brad had returned a week previously from a Boy Scout camping trip to the local woods, where he had done a lot of hiking. When asked by his physician, Brad admitted finding a tick on his stomach while in the woods but thinking little of it. The doctor ordered serological tests for* Borrelia burgdorferi *(the organism that causes Lyme disease),* Rickettsia rickettsii *(which produces Rocky Mountain spotted fever), and* Ehrlichia equi *(which causes ehrlichiosis). The ELISA test for B. burgdorferi came back positive, confirming a diagnosis of Lyme disease. The boy was given a 3-week regimen of doxycycline (a tetracycline derivative), which resolved the rash and palsy.*

Lyme arthritis was first reported in Lyme, Connecticut, in the 1970s, but the causative organism, *Borrelia burgdorferi*, was not identified until 1982. Since then, **Lyme disease** (a form of **borreliosis**) has become the most common vector-borne illness in the United States and is considered an emerging infectious disease. The main endemic areas are the northeastern coastal area from Massachusetts to Maryland, Wisconsin and Minnesota, and northern California and Oregon. Lyme disease is also common in parts of Europe.

B. burgdorferi is a spirochete (**Fig. 26.23B**) transmitted to humans by ixodid ticks (hard ticks) (**Fig. 26.23C and D**). In the northeastern and central United States, where most cases occur, the deer tick *Ixodes scapularis* transmits the spirochete, usually during the summer months. In the western United States, *I. pacificus* is the tick vector.

During its nymphal stage (the stage after taking its first blood meal), *I. scapularis* is the size of a poppy seed. Its bite is painless, so it is easily overlooked. Infection takes place when the tick feeds, because the spirochete is regurgitated into the host. However, the organism grows in the tick's digestive tract and takes about 2 days to make its way to the tick's salivary gland, so if the tick is removed before that time, the patient will not be infected. Once it is transferred to the human, the microbe can travel rapidly via the bloodstream to any area in the body, but it prefers to grow in skin, nerve tissue, synovium (joint lining), and the conduction system of the heart.

Lyme disease has three stages. Similar to syphilis caused by the spirochete *Treponema palidum*, there are three general stages of Lyme disease. Stage 1 involves the localized spread of *Borrelia burgdorferi* between 3 and 30 days after the initial exposure. Approximately 75% of patients experience an erythema migrans rash, usually at the site of the tick bite. The appearance of the rash varies but is classically erythematous with central clearing ("bull's-eye" rash). This stage is often associated with constitutional symptoms such as fever, myalgia (muscle pain), arthralgia (joint pain), and headache.

Stage 2 occurs weeks to months after the initial infection as *B. burgdorferi* spreads from blood to other organs. In this stage, the patient can be quite ill with malaise, myalgia, and arthralgia, as well as neurological or cardiac involvement. Common neurological manifestations include Bell's palsy (facial paralysis), inflammation of spinal nerve roots, and chronic meningitis. The most common cardiac manifestation is an irregular heart rhythm.

Stage 3 borreliosis occurs months to years later and can involve the synovium, nervous system, and skin, though skin involvement is more common in Europe than in North America. Arthritis occurs in the majority of previously untreated patients; it is usually intermittent, involves the large joints, particularly the knee, and lasts from weeks to months in any given joint. Joint-fluid analysis typically shows a WBC count of 10,000–30,000/μl. Late neurological involvement may include peripheral neuropathy and encephalopathy, manifested as memory, mood, and sleep disturbances.

Treatment with antibiotics is recommended for all stages of Lyme disease but is most effective in the early stages. Treatment for early Lyme disease (stage 1) is a course of doxycycline for 14–28 days. Lyme arthritis is typically slow to respond to antibiotic therapy. Despite antimicrobial drug treatment, patients with persistent active arthritis and persistently positive PCR tests may have incomplete microbial eradication and may be the most likely to benefit from repeated treatment with injected antibiotics. However, the disease in some people does not resolve at all; why is not known.

Curiously, 25% of people infected with *B. burgdorferi* never experience erythema migrans, and many infected individuals are also unsure of tick bites. Hence, patients with Lyme disease may present with arthritis as their first complaint. Although this makes diagnosis extremely difficult, knowing that the patient lives in or recently traveled to an endemic area can provide a critical clue.

A startling example of how sinister Lyme disease can be is the case of Kris Kristofferson, a prolific singer, songwriter, and actor, now in his eighties. Kristofferson suffered for years with severe memory loss presumed to be a manifestation of Alzheimer's disease—until someone decided to test him for Lyme disease. To his doctor's surprise, and relief, the singer tested positive. Kristofferson immediately began antibiotic treatments and has since recovered much of his memory.

Many other bacteria can cause septicemia and systemic illness (**Table 26.5**). Gram-negative organisms like *E. coli*, *Salmonella* Typhi, and *Francisella*, and Gram-positive microbes like *Staphylococcus aureus*, *Enterococcus*, and *Bacillus anthracis*, can grow in the bloodstream if they can gain entrance. Even anaerobes that are normal inhabitants of the intestine (for example, *Bacteroides fragilis*) can be lethal if they escape the intestine and enter the blood, as might happen following surgery. This is one reason surgical patients are given massive doses of antibiotics immediately before and after surgery.

Hepatitis Viruses Target the Liver

Hepatitis is a general term meaning "inflammation of the liver." Hepatitis is caused by several viruses, including hepatitis A, B, C, and E viruses. Although these viruses are members of very different families, they all target the liver. We include them in this section on systemic infections because their infectious routes take them to the

TABLE 26.5	Selected Systemic Infectious Diseases					
Disease	Symptoms	Etiological agent[a]	Virulence properties	Source	Treatment	Vaccine
Lyme disease	Stage 1: rash; stage 2: chills, headache, malaise, systemic involvement; stage 3: neurological changes	*Borrelia burgdorferi* (spirochete)	Antigenic variation, OspE (binds complement)	Deer tick	Penicillin, tetracycline	No longer available
Brucellosis	Fever, weakness, sweats, splenomegaly, osteomyelitis, endocarditis, others	*Brucella abortis* (G– rod)	Intracellular, growth in monocytes	Animal products, unpasteurized milk	Doxycycline	Yes, for animals
Leptospirosis	Fever, photophobia, headache, abdominal pain, skin rash, liver involvement, jaundice	*Leptospira interrogans* (spirochete)	Burrowing motility	Urine of infected animals	Erythromycin, penicillin	Yes, for animals
Epidemic typhus	Chills, fever, headache, muscle pain, splenomegaly, coma	*Rickettsia prowazekii* (G– rod)	Obligate intracellular growth, escapes phagosome	Human louse, flying-squirrel flea	Tetracycline, chloramphenicol	
Tularemia	Fever, chills, headache, muscle pain, rash, bacteremia	*Francisella tularensis* (G– rod)	Intracellular	Rabbits, rodents, insect vectors	Gentamicin, streptomycin	Yes, but not currently in the U.S.
Typhoid fever	Septicemia, chills, fever, hypotension, rash (rose spots)	*Salmonella* Typhi (G– rod)	Type III secretion, intracellular growth, PhoPQ regulators, Vi antigen capsule	Gallbladder of human carrier	Ciprofloxacin, ceftriaxone	Vi antigen
	Septicemia, chills, fever, hypotension	*Salmonella choleraesuis* (G– rod)	Intracellular growth, invasin	Animals, poultry	Ceftriaxone	
Vibriosis	Serious with immunocompromised patients; fever, chills, multi-organ damage, death	*Vibrio vulnificus* (G– curved rod)	Cytolysin, capsule	Seawater, raw oysters	Tetracycline plus aminoglycoside	
Bubonic plague	Buboes (swollen lymph glands), high fever, chills, headache, cough, pneumonia, septicemia	*Yersinia pestis* (G– rod)	Intracellular growth, type III secretion of YOPs (*Yersinia* outer proteins), phospholipase D, toxin	Rodents, rodent fleas, human respiratory aerosol, potential bioterrorism agent	Streptomycin or tetracycline	Yes, but not available in the U.S.

[a]G+ = Gram-positive; G– = Gram-negative.

bloodstream before delivering them to the liver. The disease symptoms of fever, vomiting, dark urine, clay-colored stools, and often jaundice can be seen in all patients suffering from hepatitis, regardless of viral cause.

Hepatitis A virus (HAV) is a single-stranded RNA picornavirus (**Fig. 26.24A**) that causes an acute infection spread person-to-person by the fecal-oral route, but hepatitis A can also result from eating undercooked shellfish collected from contaminated waters. The virus replicates in the intestinal endothelium and is disseminated via the bloodstream to the liver. After replicating in hepatocytes, the progeny enter the bile and are released into the small intestine, explaining why stools are so infectious. Though the virus has an early viremic stage after leaving the intestine, it is rarely transmitted by transfusion, because the viremic stage is transient, ending after liver symptoms develop. In contrast, hepatitis B and C viruses produce persistent viremia and are readily transmitted by transfusion.

Many people who are infected with HAV are asymptomatic or exhibit very mild symptoms that include nausea, vomiting, diarrhea, low-grade fever, and fatigue. As the virus attacks the liver, patients may become jaundiced (from the accumulation of bilirubin in the skin), and their urine will turn dark brown. There is no specific treatment,

A.   **B.**

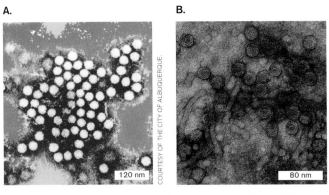

COURTESY OF THE CITY OF ALBUQUERQUE.

EYE OF SCIENCE/SCIENCE SOURCE

120 nm   80 nm

FIGURE 26.24 ■ Structures of hepatitis A and hepatitis B viruses. Both viruses are icosahedral in shape. **A.** Hepatitis A, spread by the fecal-oral route, is a single-stranded RNA virus in the *Picornaviridae* family (TEM). **B.** Hepatitis B is an enveloped, double-stranded DNA virus in the *Hepadnaviridae* family (TEM). The image also shows the tubular structures made from HepB surface antigen (HBsAg).

but the disease usually lasts for only a few months and then resolves without establishing a carrier state. Disease can be prevented, however, if immunoglobulin is given to someone who has had contact with an infected individual. A hepatitis A vaccine containing inactivated virus (called HepA vaccine) is administered after 1 year of age. For those not vaccinated, frequent hand washing is important for preventing the spread of the disease because it interrupts the fecal-oral cycle.

In contrast to HAV, **hepatitis B virus** (**HBV**) is a partially double-stranded circular DNA virus (family *Hepadnaviridae*) that causes diseases of varying severity. These include acute and chronic hepatitis, cirrhosis, and hepatocarcinoma. The virus also wears a membrane envelope donned when progeny viruses are released from infected cells. The virion coat protein, a surface antigen, is called HBsAg. The virus makes an excess amount of HBsAg, so it is sometimes extended as a tubular tail on one side of the virus particle and is often found in the blood of infected individuals in the form of noninfectious filamentous and spherical particles (**Fig. 26.24B**). The presence of HBsAg in blood is an indicator of HBV infection.

HBV is transferred primarily via blood transfusions, contaminated needles shared by IV drug users, and any human body fluid (saliva, semen, sweat, breast milk, tears, urine, or feces). It can even be transferred transplacentally to a fetus and can be sexually transmitted. Infection by HBV has two stages: a short-term acute phase and a long-term chronic phase that, if it extends beyond 6 months, may never resolve (chronic infection). Symptoms resemble those of the flu, but with jaundice and brown urine. Liver damage caused by HBV infection is due in large part to an efficient cell-mediated immune response. Cytotoxic T cells and natural killer cells cause immune lysis of infected liver cells. Over the long term, chronic hepatitis will lead to

a scarred and hardened liver (cirrhosis), the only recourse being a liver transplant. Fortunately, about 90% of those infected are able to fight off infection and never proceed to the chronic stage. A HepB vaccine (made from recombinant HBsAg) is available. Its administration is recommended after birth, followed by booster shots administered by 2 months and 18 months of age.

Hepatitis C virus (**HCV**) causes another form of hepatitis. HCV is a single-stranded positive-sense, linear RNA virus with a lipid coat; it is a member of the *Flaviviridae* family. It is transmitted by blood transfusions and causes 90% of transfusion-related cases of hepatitis. It can also be transmitted by needle sticks, razor blades, tattooing, and, less frequently, by sex. Over 100 million people worldwide are infected with HCV. Screening for HCV (serology or PCR) is recommended for anyone who exhibits signs of hepatitis or practices the risky behaviors noted here. In addition, the CDC recommends that anyone born between 1945 and 1965 (baby boomers) be tested because, for unknown reasons, baby boomers are five times more likely to be infected than other adults.

Most HCV-infected individuals (80%) do not exhibit symptoms, and in those who do, symptoms may not appear for 10–20 years. At least 75% of patients who exhibit symptoms ultimately progress to chronic hepatitis requiring a liver transplant or possibly to liver cancer. Fortunately, infection can be detected using ELISA. Liver biopsies of HCV patients are used to determine the extent of liver damage, which in turn helps establish the stage of disease.

Prevention of HBV or HCV infection for health care personnel includes avoiding inadvertent needle sticks. If such a stick should occur with HBV, anti-HBV immunoglobulin should be administered within 7 days. Currently no effective post-exposure prophylaxis is recognized for HCV. Chronic hepatitis can be treated with interferon, but more effective treatment now includes antiviral protease inhibitors that prevent the processing (by proteolysis) of an important HCV polyprotein. Though vaccines have been developed for HAV and HBV (see Section 24.6), no vaccine is yet available for HCV.

Note: Since hepatitis viruses can be spread via contaminated blood products, all blood donations collected by the Red Cross and other agencies are tested for the presence of these viruses, as well as for HIV. Thus, the blood supply is safe.

Ebola—The Perfect Pathogen or Too Deadly for Its Own Good?

How would you define the perfect pathogen? Would it be an organism that can kill its host with terrifying ease and quickness? If so, Ebola virus would fit the description. Ebola virus, a lipid-enveloped, threadlike RNA virus

(*Filoviridae*) (see Fig. 25.4), was first associated with an outbreak of 318 cases of a hemorrhagic disease in Zaire in 1976. Of the 318 people who contracted the disease, 280 died within days. The disease was characterized by acute (rapid) onset of fever, severe muscle pains, horrible bleeding from multiple orifices (nose, mouth, anus, and vagina), and ultimately death. Also in 1976, 284 people in Sudan were infected with the virus, and 156 of them died.

The Ebola virus has a frightening reputation. It spreads like wildfire through the body after infection, causing severe hemorrhagic fever, and typically kills 90% of its victims. Internal bleeding results in shock and acute respiratory distress, leading to death. The recent 2014–15 epidemic of Ebola in West Africa involved 28,683 cases, 40% of whom died (see Chapter 28).

The symptoms of Ebola (and of a related disease caused by the Marburg virus) reflect subversion of the innate immune system, coupled with uncontrolled viral replication, particularly in macrophages and dendritic cells. Ebola virus infection of these cells enhances production of proinflammatory cytokines, such as TNF-alpha, and inhibits stimulation of T-cell maturation by dendritic cells. Thus, Ebola infections stimulate inflammatory processes leading to tissue damage, but they shut down early immune responses and prevent activation of adaptive immune responses, thereby allowing unfettered viral replication.

Ebola viral proteins and their locations in the virion are shown in **Figure 26.25A**. Ebola VP35 protein is a component of the viral RNA polymerase complex, but it is also a potent inhibitor of host interferon (IFN) production. The cellular response to whichever IFN is made is inhibited by VP24, which blocks the nuclear accumulation of a regulatory protein called STAT1. STAT1 is critical to IFN-stimulated gene expression. These and other strategies allow rapid replication of the virus.

After replicating, Ebola offspring sprout from the cell surface in a mass of tangled threads (**Fig. 26.25B**). These new virions go on to attack new cells, riddling blood vessels and organs with damage as they go. Rapid release of new virions involves subverting another host mechanism, tetherin, designed to slow viral spread. Paul Bates and his colleagues at the University of Pennsylvania discovered that the cellular protein tetherin essentially <u>tethers</u> mature virus particles inside a cell so that they are unable to spread. Tetherin is IFN induced and can restrict the spread of structurally diverse enveloped viruses, including HIV (thus, it is part of the innate response). Ebola glycoprotein, however, counteracts tetherin, so that nothing slows down viral spread. The result is rapid release of massive numbers of virus particles that can then quickly spread infection to other organs and tissues.

The outlook for a patient infected with Ebola is dire. The incubation period is 4–16 days, and death occurs within 7–16 days. With good supportive care,

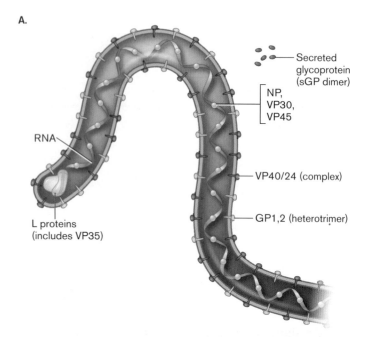

A.

Secreted glycoprotein (sGP dimer)

NP, VP30, VP45

RNA

VP40/24 (complex)

GP1,2 (heterotrimer)

L proteins (includes VP35)

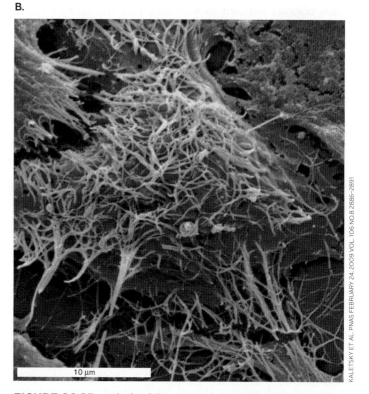

B.

10 µm

FIGURE 26.25 ■ **Ebola virion. A.** Composition of the virus. The ribonucleoprotein complex consists of the nucleoprotein (NP), the structural proteins VP30 and VP35, and the virion-associated RNA-dependent RNA polymerase (L proteins). The glycoprotein (GP-sGP) is an integral membrane protein that can be secreted. **B.** Threadlike Ebola virions budding from a cell (center).

including the replacement of coagulation factors, the mortality rate appears near 20%. Lacking such care, mortality can reach over 80%. There are no effective drugs or vaccines, although a recombinant monoclonal antibody directed against the envelope glycoprotein has neutralizing activity. Administering blood plasma from people who have recovered, anticoagulation agents to reduce hemorrhaging, and interferon has had limited success.

Chemical inhibitors of the host cathepsin proteases, which are required for viral replication, have been suggested as one possible antiviral therapy. Ebola enters a cell when the virus membrane glycoprotein attaches to host membranes. The virus is then taken up in an endosome. Cathepsins in the endosome cleave the viral glycoprotein and allow the virus membrane to fuse with the endosome membrane, releasing the uncoated virus into the cytoplasm. Cathepsin inhibitors prevent release and, thus, replication. There is also some hope that an attenuated, replication-deficient Ebola virus may serve as a vaccine.

Ebola epidemics result from person-to-person contact or from inadvertent laboratory exposures. Fortunately, Ebola outbreaks are self-contained because the viruses kill their victims quickly. Death comes before the virus can be transmitted to a new host. Some scientists therefore argue that efficient and quick killing is not the mark of a perfect pathogen. The better pathogen lets its host linger to ensure a home for itself and more opportunity to disseminate. So where does Ebola go when it is not infecting humans? The natural ecology of these viruses is largely unknown, although an association with monkeys and/or bats as possible reservoirs is suggested.

To Summarize

■ **Blood cultures** are useful in diagnosing septicemia and endocarditis.

■ **Septicemia** is caused by many Gram-positive and Gram-negative bacterial pathogens. It can start with the bite of an infected insect, introduction via a wound, escape from an abscess, or penetration of the mucosal epithelium by the pathogen (as through the intestine or vagina); it can lead to disseminated, systemic disease.

■ **Endocarditis** can have acute or subacute onsets. **Subacute bacterial endocarditis** is usually an endogenous infection of a heart valve caused by *Streptococcus mutans*.

■ **Malaria, caused by *Plasmodium* species**, manifests as repeated episodes of chills, fever, and sweating, owing to the organism's ability to alter the antigenic appearance of its surface proteins and evade the immune response.

■ **Plague, caused by *Yersinia pestis***, has sylvatic and urban infection cycles involving transmission between fleas. The bite of an infected flea leads to bubonic plague. Bubonic plague can progress to septicemic and pneumonic stages. Pneumonic plague can be spread directly from person to person (no insect vector) by **aerosolized respiratory secretions**.

■ **Lyme disease** is caused by the spirochete *Borrelia burgdorferi*, which is transmitted from animal reservoirs to humans by the bite of *Ixodes* ticks. The three stages of Lyme disease are characterized by a bull's-eye rash, called erythema migrans (stage 1); joint, muscle, and nerve pain (stage 2); and arthritis with WBCs in the joint fluid (stage 3).

■ **Hepatitis** is caused by several unrelated viruses; among them, HAV, HBV, and HCV account for most disease. **HAV** is transmitted by the fecal-oral route, does not establish chronic infection, and can be prevented by a vaccine. **HBV and HCV** can be transmitted by blood products (such as transfusions) and shared hypodermic needles and can lead to chronic hepatitis. Vaccine for HBV, but not HCV, is available.

■ **Ebola virus** spreads from human to human and kills its victims quickly. Its viral proteins alter cytokine production and facilitate virus release from infected cells.

26.7 Central Nervous System Infections

The brain and spinal cord are especially well protected against infection. Microbes cannot gain easy access to the brain in large measure because of the **blood-brain barrier**, a filter mechanism that allows only selected substances into the brain. The blood-brain barrier works to our advantage when harmful substances, such as bacteria, are prohibited from entering. However, it works to our disadvantage when substances that we want to enter the brain, such as antibiotics, are kept out. The barrier is not a single structure but a function of the way blood vessels, especially capillaries, are organized in the brain. Furthermore, the endothelial cells in those vessels have tight junctions that do not allow most compounds or microbes to cross. And yet, brain infections do occur.

Case History: Meningitis

*In April 2001, Laila, a 4-month-old infant from Saudi Arabia, was hospitalized with fever, tender neck, and purplish spots (purpuric spots) on her trunk (**Fig. 26.26A**). Suspecting meningitis, the clinician took a cerebrospinal fluid (CSF) sample and examined it by Gram stain. The smear revealed Gram-negative diplococci inside PMNs. The CSF was turbid with 900 leukocytes per microliter, and* Neisseria meningitidis *was confirmed by culture. The child was treated with cefotaxime (a cephalosporin antibiotic; see Chapter 27) and made a full recovery. Her father, Abdul, the person who brought her in, was clinically well. However, the meningococcus was isolated from his oropharynx, as well as from the throat of the patient's 2-year-old brother. Isolates from the*

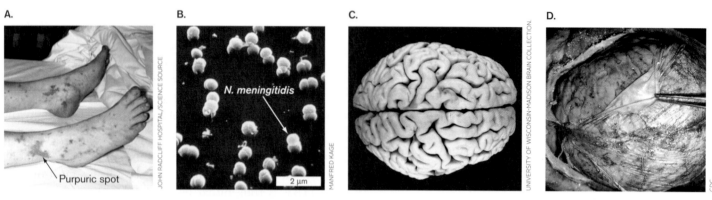

FIGURE 26.26 ■ Bacterial meningitis. A. Purpuric spots produced by local intravascular coagulation due to *Neisseria meningitidis* endotoxin. The rash in meningitis typically has petechial (small) and purpuric (large) components. **B.** *N. meningitidis* (diameter approx. 1 μm; SEM). **C.** Normal brain. **D.** Autopsy specimen of meningitis due to *Streptococcus pneumoniae*. Note the greening of the brain, compared with the pink normal brain.

patient, her father, and her brother were positive by aggluti-nation with meningococcal A/C/Y/W135 polyvalent reagent. Records showed that the father previously received a quad-rivalent meningococcal vaccine. All three isolates were con-firmed to be meningococcus serogroup W135. DNA analysis found the three isolates to be indistinguishable, meaning that the father and his children were infected with the same strain of N. meningitidis. *Why did Laila's brother not have menin-gitis? And why was Laila's vaccinated father colonized?*

Meningitis is an inflammation of the meninges, the mem-branes that surround the brain and spinal cord. Meningi-tis can be either bacterial or viral in origin. Sinus and ear infections can extend directly to the meninges, whereas septicemic spread requires passage through the blood-brain barrier. Viral meningitis is serious but rarely fatal in peo-ple with a normal immune system. The symptoms generally persist for 7–10 days and then completely resolve. Bacterial meningitis is usually caused by *Streptococcus pneumoniae*, *Neisseria meningitidis*, or *Haemophilus influenzae*. Symp-toms of bacterial meningitis can include sudden onset of fever, headache, neck pain or stiffness, painful sensitivity to strong light (photophobia), vomiting (often without abdominal complaints), and irritability. Prompt medical attention is extremely important because the disease can quickly progress to convulsions and death.

The meningococcus *N. meningitidis* (**Fig. 26.26B**) can colonize the human oropharynx, where it causes mild, if any, disease. At any given time, 10%–20% of the healthy population can be colonized and asymptomatic. The organism spreads directly by person-to-person contact or indirectly via droplet nuclei from sneezing or fomites. The problem arises when this organism enters the bloodstream. Unlike *N. gonorrhoeae*, the cause of gonorrhea, *N. men-ingitidis* is very resistant to complement, owing to its pro-duction of a polysaccharide capsule. The capsule allows the microbe to produce a transient blood infection (bactere-mia) and reach the blood-brain barrier.

How does *N. meningitidis* enter the bloodstream from the nasopharynx and then cross the blood-brain barrier?

The organism initially uses type IV pili to adhere to and enter nasopharyngeal endothelial cells (type IV pili are discussed in Section 25.3). The bacteria then cross the endothelial cell layer by **transcytosis**, a process by which an internalized pathogen passes through a host cell along microtubules to the opposite side. Ultimately, the patho-gen passes into the capillary lumen and is swept away to the brain. Once in the brain, type IV pili again adhere to endothelial cells, but this time they trigger the recruitment of host proteins that destabilize intercellular junctions. The bacteria then slip between the loosened junctions and enter the cerebrospinal fluid. Once in the CSF, microbes can multiply almost at will. **Figure 26.26C** and **D** show the remarkable damage (greening) that *N. meningitidis* and other microbes, such as *S. pneumoniae*, cause in the brain.

Several antigenic types of capsules, called type-specific capsules, are produced by different strains of pathogenic *N. meningitidis*: types A, B, C, W135, and Y. Types A, C, Y, and W135 are usually associated with epidemic infections seen among people kept in close proximity, such as col-lege students or military personnel. Type B meningococ-cus is typically involved in sporadic infections. However, large outbreaks of type B in the United States occurred at Princeton University and UC Santa Barbara in 2013. Antibodies to these capsular antigens are used to classify the capsular types of the organisms causing an outbreak. Knowing the capsular type of organism involved in each case helps determine whether the disease cases are related and where the infection may have started. In Abdul's case, a polyvalent reagent containing antibodies to the four main capsular types was used to confirm *N. meningitidis*.

Meningococcal meningitis is highly communicable. As a result, close contacts, such as the parents or siblings of any patient with meningococcal disease, should receive antimi-crobial prophylaxis within 24 hours of diagnosis; a single dose of ciprofloxacin (a quinolone antibiotic, discussed in Section 27.4) can be given to adults, and 2 days of rifam-picin given to children. Highly susceptible populations can be immunized with a vaccine containing four of the five capsular structures (type B capsule is not immunogenic; see

Section 24.6). However, a vaccine containing outer membrane vesicles and proteins from *N. meningitidis* type B (but not its capsule) was recently approved by the FDA.

Case History: Botulism— It Is What You Eat

In June, a 47-year-old resident of Oklahoma was admitted to the hospital with rapid onset of progressive dizziness, blurred vision, slurred speech, difficulty swallowing, and nausea. Findings on examination included drooping eyelids, facial paralysis, and impaired gag reflex. He developed breathing difficulties and required mechanical ventilation. The patient reported that during the 24 hours before onset of symptoms, he had eaten home-canned green beans and a stew containing roast beef and potatoes. Analysis of the patient's stool detected botulinum type A toxin, but no Clostridium botulinum *organisms were found. The patient was hospitalized for 49 days, including 42 days on mechanical ventilation, before being discharged.*

Imagine a disease that causes complete loss of muscle function. Using secreted exotoxins (neurotoxins), two microbes cause lethal paralytic diseases. In one instance, the victim suffers a flaccid paralysis in which the muscles go limp, as in the case history just given, causing paralysis and respiratory difficulty. Voluntary muscles fail to respond to the mind's will because botulinum toxin interferes with neural transmission. The disease (called **botulism**) is typically food-borne and is caused by an anaerobic, Gram-positive, spore-forming bacillus named *Clostridium botulinum* (see Chapter 18).

In striking contrast to botulism is tetanus, a very painful disease in which muscles continually and involuntarily contract (called tetany or spastic paralysis). Tetanus is caused by **tetanospasmin**, a potent exotoxin made by another anaerobic, Gram-positive spore-forming bacillus, called *Clostridium tetani* (**Fig. 26.27A**). Tetanospasmin interferes with neural transmission, but in contrast to botulism toxin, it causes <u>excessive</u> nerve signaling to muscles, forcing the victim's back to arch grotesquely while the arms flex and legs extend. The patient remains locked this way until death. Spasms can be strong enough to fracture the patient's vertebrae. **Figure 26.27B** shows the result of injecting a mouse's hind leg with just a tiny amount of tetanus toxin. In both botulism and tetanus, death can result from asphyxiation.

Botulism is typically caused by ingesting preformed toxin, although infected wounds or germination of

A. *Clostridium tetani*

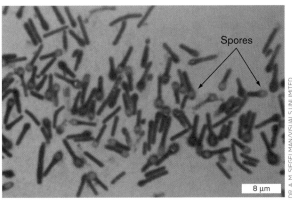

Spores

8 μm

DR. A. M. SIEGELMAN/VISUALS UNLIMITED

B. Spastic paralysis due to tetanus toxin

Tetany

COURTESY AMERICAN SOCIETY FOR MICROBIOLOGY

FIGURE 26.27 ■ Tetanus and botulism toxins. A. Photomicrograph of *Clostridium tetani* (cell length 4–8 μm). **B.** Mouse injected with tetanus toxin in left hind leg. **C.** Schematic diagram of tetanus and botulism toxins. **D.** 3D representation of the tetanus neurotoxin with the domains marked. (PDB code: 3BTA)

C. Basic structure of tetanus and botulism toxins

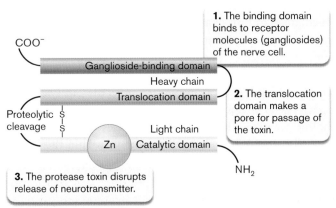

COO⁻

Ganglioside-binding domain

Heavy chain

Translocation domain

Proteolytic cleavage

Light chain

Zn

Catalytic domain

NH₂

1. The binding domain binds to receptor molecules (gangliosides) of the nerve cell.

2. The translocation domain makes a pore for passage of the toxin.

3. The protease toxin disrupts release of neurotransmitter.

D. Tetanus toxin structure

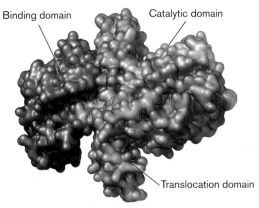

Binding domain

Catalytic domain

Translocation domain

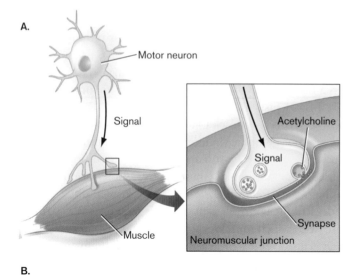

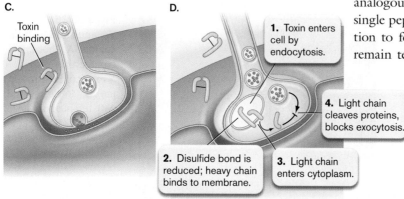

FIGURE 26.28 ■ Mechanism of action of botulism toxin.
A. The neuromuscular junction. The blowup shows vesicles filled with neurotransmitters. **B.** A series of proteins within the nerve is needed to allow synaptic vesicles to bind to the nerve endings. Fusion of the membranes releases acetylcholine into the neuromuscular junction. Botulism toxin types A and E cleave SNAP-25. Botulism toxins B, D, F, and G cleave VAMP. Botulism toxin C1 cleaves syntaxin and SNAP-25. **C, D.** Toxin binds via the heavy chain and is endocytosed into the nerve terminal. Once the toxin cleaves its target, the nerve terminal is no longer able to release acetylcholine.

ingested spores can also occasionally produce disease. The microbe germinates in the food and produces toxin. Because the organism is an anaerobe, an anaerobic

environment must be present. Home-canning processes are designed to remove oxygen in the canned food, as well as sterilize the food, but if the sterilization process is incomplete, then surviving spores germinate and produce toxin. The toxin is susceptible to heat but will remain active in improperly cooked food. After ingestion, the toxin is absorbed from the intestine. As in the case history given here, the organism is often absent from stool samples. This is why serological identification of the toxin is important.

A rare form of botulism, called infant botulism or "floppy head syndrome," can occur when infants (not older children) are fed honey. Honey can harbor *C. botulinum* spores that can germinate in the gastrointestinal tract, after which the growing vegetative cells will secrete toxin.

> **Thought Question**
>
> **26.10** Knowing the symptoms of tetanus, what kind of therapy would you use to treat the disease?

Toxin structure. Botulism and tetanus toxins share 30%–40% identity and have similar structures and nearly identical modes of action. Botulism and tetanus toxins are each composed of two peptides—a large, or heavy, fragment analogous to a B subunit of AB toxins and a small, or light, fragment analogous to an A subunit. Both toxins are initially made as single peptides (about 150 kDa) that are cleaved after secretion to form two fragments (heavy and light chains) that remain tethered by a disulfide bond (**Fig. 26.27C**). Each heavy chain includes the binding domain for receptor molecules (gangliosides) on the nerve cell membrane, and a translocation domain that makes a pore in the nerve cell through which the toxin passes. The light chains (catalytic domains) are proteases that disrupt the movement of exocytic vesicles containing neurotransmitters needed for contraction or relaxation. **Figure 26.27D** shows a 3D rendition of tetanus toxin with the three domains marked.

Mechanism of botulism neurotoxin. Figure 26.28A illustrates normal neurotransmission at a neuromuscular junction, the target of botulism toxin. A signal sent from a motor neuron to the neuromuscular junction causes the release of the neurotransmitter molecule acetylcholine from vesicles in the axon terminal. Once released, acetylcholine traverses the synapse and causes muscle contraction. **Figure 26.28B** shows the proteins within the vesicle and plasma membranes that allow the synaptic vesicle to fuse with the plasma membrane.

Once botulism toxin enters a peripheral nerve by endocytosis (**Fig. 26.28C**), the low pH that forms in the endosome reduces the disulfide bonds holding the two halves of the toxin together. The heavy chain assembles as a channel in the membrane through which the proteolytic light chain moves into the cytoplasm. The light chain then cleaves key host proteins, such as synaptobrevin (a vesicle-associated membrane protein, or VAMP), syntaxin, or synaptosomal-associated protein (SNAP-25) involved in the exocytosis of vesicles containing acetylcholine (**Fig. 26.28D**). Without acetylcholine to activate nerve transmission, muscles will not contract and the patient is paralyzed.

Although botulism disease is now rare, the toxin is considered a select biological agent of potential use to bioterrorists. As such, its use in laboratories is under strict governmental control. But botulism toxin also has important medical uses. Because the toxin can safely relax muscles in a localized area if injected in small doses, botulism toxin, or Botox, is used cosmetically by plastic surgeons to reduce facial wrinkles in some patients, and therapeutically by neurologists to treat migraine headaches.

Mechanism of tetanus neurotoxin. Earlier we mentioned that botulism and tetanus toxins are nearly identical in structure and mechanism. With such a high degree of mechanistic similarity, why do tetanus and botulism toxins have such drastically different effects? The answer is based on where each toxin acts in the nervous system.

Tetanus, in contrast to botulism, is not a food-borne disease. *Clostridium tetani* spores are introduced into the body by trauma (such as by stepping on the wrong end of a nail). Necrotic tissue then provides the anaerobic environment required for germination. Growing, vegetative cells release tetanospasmin, which enters the peripheral nerve cells at the site of injury. But rather than cleaving targets here, the toxin travels up axons in the direction opposite to nerve signal transmission until it reaches the spinal column, where it becomes fixed at the presynaptic inhibitory motor neuron. There the toxin also cleaves proteins like VAMP, but the function of vesicles in these nerves is to release <u>inhibitory</u> neurotransmitters that dampen nerve impulses. Tetanus toxin blocks release of the inhibitory neurotransmitters GABA and glycine into the synaptic cleft, leaving nerve impulses unchecked. As a result, impulses come too frequently and produce the generalized muscle spasms characteristic of tetanus. **Figure 26.29** ▶ illustrates the mechanism of tetanus toxin action. A tetanus toxoid vaccine is available and is administered to children as part of the DTaP (diphtheria, tetanus, and acellular pertussis) vaccine (see Section 24.6).

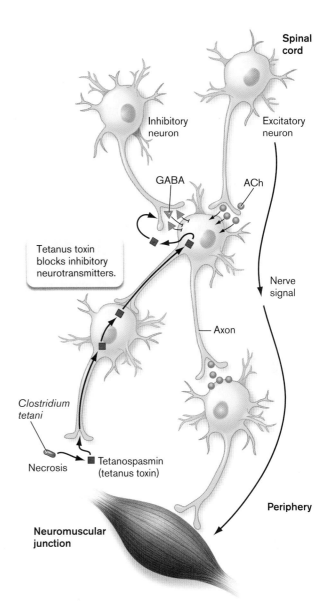

FIGURE 26.29 ■ **Retrograde movement of tetanus toxin to an inhibitory neuron.** Tetanus toxin enters the nervous system at the neuromuscular junction and travels retrogradely up the axons until reaching an inhibitory neuron located in the central nervous system. There it cleaves VAMP protein (Fig. 26.28B) associated with the exocytosis of vesicles containing inhibitory neurotransmitters. ACh = acetylcholine; GABA = gamma-aminobutyric acid. ▶

Thought Questions

26.11 **Figure 26.29** demonstrates that tetanus toxin has a mode of action (spastic paralysis) opposite to that of botulism toxin (flaccid paralysis). Since they have opposing modes of action, can botulism toxin be used to save a patient with tetanus?

26.12 How do the actions of tetanus toxin and botulinum toxin actually help the bacteria colonize or obtain nutrients?

26.13 If *Clostridium botulinum* is an anaerobe, how might botulism toxin get into foods?

TABLE 26.6 Selected Microbes That Cause Meningitis or Encephalitis

Type of disease	Etiological agent[a]	Virulence factors	Typical initial infection or source	Treatment	Vaccine
Bacterial (septic) meningitis	*Streptococcus pneumoniae* (G+)	Capsule, pneumolysin	Lung	Ampicillin	Multivalent, capsule
	Neisseria meningitidis (G–)	Capsule, IgA protease, endotoxin	Throat	Cephalosporin (3rd generation), ceftriaxone	Multivalent, capsule
	Haemophilus influenzae type b (G–)	Polyribitol capsule, IgA protease, endotoxin	Ear infection	Cephalosporin (3rd generation), ceftriaxone	Type b polysaccharide
	Other G– bacilli	Endotoxin	Septicemia		
	Group B streptococci (G+)	Sialic acid capsule, streptolysin, inhibition of alternate complement path	Neonate infected during parturition	Ampicillin	Capsular
	Listeria monocytogenes (G+)	Intracellular growth, PrfA regulator, actin-based motility	Mild GI disease of mother	Ampicillin plus gentamicin	
	Mycobacterium tuberculosis	Cord factor, wax D, intracellular growth	Lung	Combination therapy (rifampin, isoniazid, ethambutol, pyrazinamide)	BCG[b]
	Staphylococcus aureus (G+)	Coagulase, protein A, TSST, leukocidin	Septicemia	Methicillin, vancomycin	
	Staphylococcus epidermidis (G+)	Biofilm, slime	Complication of surgical procedure	Vancomycin, methicillin	
Aseptic meningitis[c]	Fungi (e.g., *Coccidioides*, *Cryptococcus*)		Sinusitis, direct spread to meninges	Ketoconazole, fluconazole	
	Amebas (e.g., *Naegleria*)		Swimming in contaminated waters	Amphotericin B	
	Treponema pallidum		Syphilis		
	Mycoplasmas	Adhesin tip	Respiratory	Erythromycin	
	Leptospira	Burrowing motility	Septicemia, water contaminated with animal urine	Erythromycin	Killed whole cell, animals only
Viral meningitis or encephalitis	Viruses (90% caused by enteroviruses)			Self-limiting	
Eastern equine encephalitis	EEE virus	?	Mosquito bite	None; often fatal	
West Nile disease	West Nile virus	?	Mosquito bite	Supportive therapy[d]	

[a]G+ = Gram-positive; G– = Gram-negative.

[b]BCG = Bacille Calmette-Guérin (a weakened strain of the bovine tuberculosis strain).

[c]For aseptic meningitis, bacterial sources cannot be isolated by ordinary means.

[d]Supportive therapy = hospitalization, intravenous fluids, airway management, respiratory support, and prevention of secondary infections.

Case History: Eastern Equine Encephalitis

In August, Mr. C brought his 21-year-old son, Rich, to a New Jersey emergency department. Rich appeared dazed and had trouble responding to simple commands. When questioned about his son's activities over the previous few months, Mr. C told the physician that Rich spent the month of July relaxing and sunning himself on New Jersey beaches and visiting a pond in a wooded area near a horse farm. On the afternoon prior to admission, Rich had become lethargic and tired. He returned home and went to bed. That evening, his father woke him for supper, but Rich was confused and had no appetite. By 11 p.m., Rich had a fever of 40.7°C (103.5°F) and could not respond to questions. A few hours later, when his father had trouble rousing him, he brought Rich to the ED. Over the next week, Rich's condition worsened to the point where his limbs were paralyzed. Two weeks later he died. Serum samples taken upon entering the hospital and a few days before he died showed a sixfold rise in antibody titer to eastern equine encephalitis (EEE) virus. Brain autopsy showed many small foci of necrosis in both the gray and white matter.

The EEE virus, a member of the family *Togaviridae*, is transmitted from bird to bird by mosquitoes. Horses contract the disease in the same way. Rich contracted it inadvertently while visiting the pond. The disease, an encephalitis, is often fatal (35% mortality) but fortunately rare. One reason human disease is rare is that the species of mosquito that usually transmits the virus between marsh birds does not prey on humans. But sometimes a human-specific mosquito bites an infected bird and then transmits EEE virus to humans. Another reason human disease is rare is that the virus generally does not fare well in the body. Many persons infected with EEE virus have no apparent illness, because the immune system thwarts viral replication. However, as already noted, mortality is high in individuals that do develop disease.

An interesting point in the case history is that diagnosis relied on detecting an increase in antibody titer to the virus. Typically, at the point that disease symptoms first appear, the body has not had time to generate large amounts of specific antibodies. After a week or so, often when the patient is nearing recovery (convalescence), antibody titers have risen manyfold. The rule of thumb is that a greater-than-fourfold rise in antibody titer between acute disease and convalescence (or in this case death) indicates that the patient has had the disease. While this knowledge could not help the patient in the case described here, it was valuable in terms of public health and prevention strategies.

Several bacterial, fungal, and viral causes of encephalitis and meningitis are listed in **Table 26.6**.

Prions Are Infectious Proteins

Imagine slowly losing your mind and knowing there is nothing you can do about it. An unusual infectious agent called the prion has been implicated as the cause of a series of relatively rare but invariably fatal brain diseases (**Table 26.7**). Prions are infectious agents that do not have a nucleic acid genome. The protein alone can mediate an infection. The **prion** is now recognized as an infectious, misfolded protein that resists inactivation by procedures that destroy proteins. The discovery that proteins alone can transmit an infectious disease came as a surprise to the scientific community. Diseases caused by prions are especially worrying, since prions resist destruction by many chemical agents and remain active after heating at extremely high temperatures. There have even been documented cases in which sterilized neurosurgical instruments, originally used on a prion-infected person, still held infectious agent and transmitted the disease to a subsequent surgery patient. Extremely rigorous decontamination procedures, such as autoclaving surgical instruments immersed in 1-M sodium hydroxide, will destroy prions.

How can a nonliving entity without nucleic acid be an infectious agent? Prions associated with human brain disease are thought to be aberrantly folded forms of a normal brain protein (PrP^C). The theory is that when a prion, designated PrP^{Sc}, is introduced into the body and manages to enter the brain, it will cause normally folded forms of the protein to refold incorrectly (**Fig. 26.30A**). The improperly folded proteins fit together like Lego blocks to produce damaging aggregated structures within brain cells.

TABLE 26.7	Prion Diseases			
Disease	**Susceptible animal**	**Incubation period**	**Disease characteristics**	
Creutzfeldt-Jakob (sporadic, familial, new variant [vCJD])	Human	Months (vCJD) to years	Spongiform encephalopathy (degenerative brain disease)	
Kuru	Human	Months to years	Spongiform encephalopathy	
Gerstmann-Straussler-Scheinker syndrome	Human	Months to years	Genetic neurodegenerative disease	
Fatal familial insomnia	Human	Months to years	Genetic neurodegenerative disease with untreatable insomnia	
Mad cow disease	Cattle	5 years	Spongiform encephalopathy	
Wasting disease	Deer	Months to years	Spongiform encephalopathy	

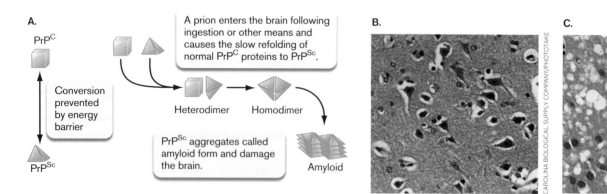

FIGURE 26.30 ■ Spongiform encephalopathies. A. Refolding model of prion diseases. PrP is a brain protein that can take two forms: the normal form (PrP^C), which is a natural brain protein; and the prion form (PrP^Sc). **B.** Normal brain section. **C.** Section of brain taken from a CJD victim. Note the Swiss-cheese appearance, indicating brain damage.

Prion diseases are often called **spongiform encephalopathies** because the postmortem appearance of the brain includes large, spongy vacuoles in the cortex and cerebellum. These are visible in a brain sample from a victim of one of these diseases, called Creutzfeldt-Jakob disease (CJD) (**Fig. 26.30B** and **C**). Most mammalian species appear to develop these diseases.

Since 1996, mounting evidence has pointed to a causal relationship between outbreaks in Europe of a disease in cattle called bovine spongiform encephalopathy (BSE, or "mad cow disease") caused by prions and a disease in humans called variant Creutzfeldt-Jakob disease (vCJD). Both disorders are invariably fatal. For humans to contract disease, prions from contaminated meat are ingested and penetrate the intestinal mucosa via the antigen-sampling M cells in Peyer's patches. The circulation then traffics the agent to the brain.

As of this writing, only four cases of BSE in cattle have been detected in the United States since 1993. Because of aggressive surveillance efforts in the United States and Canada (where 20 cases have been found), it is unlikely, but not impossible, that BSE will be a food-borne hazard to humans in this country. The CDC monitors the trends and current incidence of typical CJD (approximately 300 cases per year) and variant CJD (3 cases total) in the United States (see **eTopic 26.7**). The worldwide incidence of CJD is one per 1 million population.

There is no cure for CJD or vCJD. A test to detect PrP^Sc in asymptomatic primates has been developed but is not yet ready for human screening. The test is called protein-mediated cyclic amplification (PMCA). PMCA has successfully identified the presence of very tiny amounts of PrP^Sc in blood or urine. The process involves adding large amounts of normal PrP (PrP^C) to samples containing tiny amounts of PrP^Sc prions. The prions, if present in the sample, will seed the misfolding of PrP^C to PrP^Sc molecules that aggregate. Sonication is then used to partially break up the newly formed PrP^Sc aggregates into pieces that can further seed the misfolding and aggregation of normal PrP. Repeated cycles of sonication-incubation will exponentially amplify the amount of PrP^Sc present in a sample. The PrP^Sc is then identified by western blot (see Section 12.2).

To Summarize

- *Neisseria meningitidis* is resistant to serum complement because it produces a type-specific capsule. This allows the organism to reach and then cross the blood-brain barrier.

- In contrast to *Neisseria gonorrhoeae*, a **vaccine for N. meningitidis is available**.

- **Botulism toxin** causes flaccid paralysis.

- **Tetanospasmin** causes spastic paralysis.

- **Serological diagnosis of an infectious disease** is possible if specific pathogen antibody titer rises fourfold between the acute and convalescent stages.

- **Spongiform encephalopathies** are believed to be caused by nonliving proteins called prions.

Concluding Thoughts

In this chapter we have described key concepts of infectious disease using just a small sample of disease-causing pathogens. Other diseases are considered elsewhere in the book or on the accompanying website. These include anthrax, cholera, the common cold (caused by rhinovirus), dengue fever, diphtheria, gastric ulcers (*Helicobacter pylori*), herpes, AIDS, warts (human papillomavirus), influenza, salmonellosis, *Streptococcus agalactiae* infections, West Nile disease, and Whipple's disease (*Tropheryma whipplei*). The principal goal in this and the previous chapter was to illustrate how microbial metabolism can undermine the physiology of a human host. The next chapter will describe how humans fight back using pharmacology to sabotage the physiology of the infecting microbes.

CHAPTER REVIEW

Review Questions

1. How are pathogens classified by portal of entry?
2. Discuss some common skin infections.
3. What causes boils?
4. What are the symptoms of necrotizing fasciitis?
5. What is the difference between primary and secondary infections?
6. How do pneumococci avoid engulfment by phagocytes?
7. What causes the clouding seen in X-rays of infected lungs?
8. What are the key features of the pneumococcal vaccine?
9. Name the common fungal causes of lung disease.
10. What is a metastatic lesion?
11. Why is diarrhea watery?
12. What is the most common microbial cause of diarrhea?
13. List common bacterial agents that cause diarrhea.
14. Would you suspect *Salmonella* infection in a cluster of nauseated patients rushed to the hospital directly from a church picnic? Why or why not?
15. What is the significance of finding leukocytes in stool?
16. What is one reservoir of *E. coli* O157:H7?
17. How is a UTI diagnosed?
18. What is an important virulence determinant of uropathogenic *E. coli*?
19. What is the most common sexually transmitted disease?
20. How is gonorrhea different in men and women?
21. Why will *Neisseria gonorrhoeae* not usually disseminate in the bloodstream, while *N. meningitidis* will?
22. Name the major causes of bacterial meningitis. What are the two routes of infection?
23. If tetanus and botulism toxins have the same mode of action, why do they cause opposite effects on muscles?
24. What are some virulence factors of *Yersinia pestis*?

Thought Questions

1. Chickenpox is a disease of children and young adults caused by herpes virus 3 (varicella). It is characterized by a rash of fluid-filled vesicles that eventually become crusty. The rash starts on the trunk and spreads to the extremities. Illness usually resolves in 7–10 days, and the patient becomes immune to the disease. However, some individuals later in life (over 60 years old) develop a painful disease called shingles caused by the same virus—even if they have never been reexposed to the virus. The lesions are tender, persistent vesicles that form on the skin. Propose a plausible explanation, considering the age of the shingles victims and the occurrence of severe pain in this illness, for how they contracted shingles and why the lesions are painful.

2. A 5-year-old male was brought to the emergency room by his grandmother, who had found the boy on the floor of her apartment covered in bloody, loose feces. Patient history revealed that the boy attended day care regularly. The diagnostic laboratory determined the etiological agent to be a Gram-negative rod. This organism is a facultative intracellular pathogen that escapes the host cell vacuole and moves in and between cells by actin polymerization. From your reading of Chapters 25 and 26, what do you think are the most likely genus and species? Why is the fact that the boy attended day care significant?

3. Why are urinary tract infections among the most commonly acquired nosocomial (hospital-acquired) infections?

4. Visit the MMWR (Morbidity and Mortality Weekly Report) page of the CDC website that provides the "Summary of Notifiable Diseases" (the URL as of this writing is www.cdc.gov/mmwr/mmwr_nd). Study the table that summarizes the monthly incidence of infections in the United States and compare the data for gonorrhea and Lyme disease. Explain the differences in monthly incidence of these two diseases, and then view the incidence of those diseases in your state.

Key Terms

bacteremia (1070)
blood-brain barrier (1081)
borreliosis (1077)

botulism (1083)
bubonic plague (1074)
cellulitis (1043)

chancre (1065)
chlamydia (1066)
congenital syphilis (1065)

Recommended Reading

Boyett, Deborah, and Michael H. Hsieh. 2014. Wormholes in host defense: How helminths manipulate host tissues to survive and reproduce. *PLoS Pathogens* **10**:e1004014.

Chua, Caroline L., Graham Brown, John A. Hamilton, Stephen Rogerson, and Phillipe Boeuf. 2012. Monocytes and macrophages in malaria: Protection or pathology? *Trends in Parasitology* **29**:26–34.

Chun, Tae-Wook, and Antony S. Fauci. 2012. HIV reservoirs: Pathogenesis and obstacles to viral eradication and cure. *AIDS* **26**:1261–1268.

Coureuil, Mathieu, Olivier Join-Lambert, Herve Lecuyer, Sandrine Bourdoulous, Stefano Marullo, et al. 2012. Mechanism of meningeal invasion by *Neisseria meningitidis*. *Virulence* **3**:164–172.

David, Lawrence A., Ana Weil, Edward T. Ryan, Stephen B. Calderwood, Jason B. Harris, et al. 2015. Gut microbial succession follows acute secretory diarrhea in humans. *MBio* **6**:e00381-15.

Groppelli, Elisabetta, Shimona Starling, and Clare Jolly. 2015. Contact-induced mitochondrial polarization supports HIV-1 virological synapse formation. *Journal of Virology* **89**:14–24.

Hadjifrangiskou, Maria, and Scott J. Hultgren. 2012. What does it take to stick around? Molecular insights into biofilm formation by uropathogenic *Escherichia coli*. *Virulence* **3**:231–233.

Harper, Kristin N., Molly K. Zuckerman, and George J. Armelagos. 2014. Syphilis: Then and now. *Scientist*, February 1, 2014.

Kaminski, R., Y. Chen, T. Fischer, E. Tedaldi, A. Napoli, et al. 2016. Elimination of HIV-1 genomes from human T-lymphoid cells by CRISPR/Cas9 gene editing. *Scientific Reports* **6**:22555.

Misasi, John, Morgan S. A. Gilman, Masaru Kanekiyo, Miao Gui, Alberto Cagigi, et al. 2016. Structural and molecular basis for Ebola virus neutralization by protective human antibodies. *Science* **351**:1343–1346.

Neemann, Kari, D. D. Eichele, P. W. Smith, R. Bociek, M. Akhtari, et al. 2012. Fecal microbiota transplantation for fulminant *Clostridium difficile* infection in an allogeneic stem cell transplant patient. *Transplant Infectious Disease* **14**:E161–E165.

Olson, Patrick D., and David A Hunstad. 2016. Subversion of host innate immunity by uropathogenic *Escherichia coli*. *Pathogens* **5**:2. doi:10.3390/pathogens5010002.

Ruggiero, Paolo. 2012. *Helicobacter pylori* infection: What's new. *Current Opinion in Infectious Diseases* **25**:337–344.

Soon, J. M., P. Seaman, and R. N. Baines. 2012. *Escherichia coli* O104:H4 outbreak from sprouted seeds. *International Journal of Hygiene and Environmental Health* **216**:346–354.

CHAPTER 27
Antimicrobial Therapy

Antibiotics have played a major role in increasing life expectancy throughout the world. But antibiotics may soon become useless. Indiscriminant use of antibiotics to treat patients and boost animal growth has led to widespread antibiotic resistance in many pathogens. As a result, we are becoming vulnerable to infectious diseases once thought conquered. In Chapter 27 we discuss how antimicrobial agents work, how clinicians determine which antibiotics should be used, how microbes resist antibiotics, and how new antibiotics are being discovered.

CURRENT RESEARCH highlight

Antarctic antibiotic. Lindsey Shaw at the University of South Florida (USF) has been searching for novel antimicrobials that can kill antibiotic-resistant bacterial pathogens. A recent find is a diterpene chemical called darwinolide, which was extracted from the Antarctic sponge *Dendrilla membranosa* by USF colleague Bill Baker. Shaw's students showed that darwinolide has the novel ability to kill biofilms of methicillin-resistant *Staphylococcus aureus* (MRSA) fourfold better than it can kill planktonically growing MRSA. The scientists predict that the darwinolide structure could provide a unique scaffold upon which more potent anti-MRSA therapeutics can be built.

Source: J. L. von Salm et al. 2016. *Org. Lett.* **18**:2596–2599. doi:10.1021/acs.orglett.6b00979.

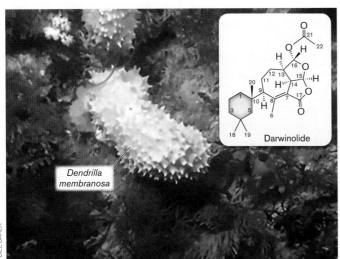
Dendrilla membranosa

© BILL BAKER

Darwinolide

AN INTERVIEW WITH

LINDSEY SHAW, MOLECULAR MICROBIOLOGIST, UNIVERSITY OF SOUTH FLORIDA

AIMEE BLODGETT/UNIVERSITY OF SOUTH FLORIDA

Why look for novel antibiotics in Antarctica?

Nature provided many antibiotics currently used in the clinic. Products from nature have excellent therapeutic properties and still represent viable sources of chemistry. Cures for many human diseases likely exist in nature, if only we can find them. But the potential of most easy-to-explore environments, like soil, has already been exhausted. We must now travel to exotic, hard-to-reach environments looking for new compounds. Antarctica is a perfect example. It possesses a rich biology we can explore for the next blockbuster antibiotic.

Darwinolide preferentially targets established MRSA biofilms. How?

We don't know, although we are working on it! Biofilms are resistant to antibiotics for many reasons (chemical microenvironments, stress responses, persister cells, and so on). Clearly, something about darwinolide is different from most drugs. Perhaps it targets something in biofilm persister cells missing in planktonic cells. It may be that darwinolide has a novel chemical structure that facilitates penetration of the biofilm. Or maybe biofilm bacteria undergoing stress response–type behavior engage a darwinolide-sensitive pathway, or target, that is lacking in active, free-growing cells.

Antibiotics undeniably have been a benefit to modern society. We live longer than in the past in large measure because of antibiotics. Infections we consider minor today often killed their victims just 50 or 60 years ago. But there has been a downside too: the development of genetically adapted, antibiotic-resistant pathogens that threaten to make current antibiotics obsolete. The Current Research Highlight illustrates the enormous efforts taking place in laboratories and at extreme locales throughout the world to discover replacement antibiotics.

How serious is the antibiotic resistance problem? Consider the case of a 56-year-old man with diabetes who entered the hospital for a heart transplant. The operation went well, but 1 week after the surgery, he developed a severe chest wound infection notable for exuded pus. Treatment with oxacillin, a penicillin derivative commonly used to combat infections, failed, and the patient became comatose. The diagnostic microbiology laboratory ultimately identified the agent as a methicillin (oxacillin)-resistant *Staphylococcus aureus* (MRSA), prompting the surgeon to immediately place the patient on intravenous vancomycin for 6 weeks. Vancomycin is an antibiotic, structurally different from methicillin and oxacillin, that can usually kill methicillin-resistant bacteria. Fortunately, this patient recovered; too often, they do not.

Where did the infection come from? Surprising as it may seem, this drug-resistant pathogen was a resident of the hospital itself. Nasal swabs taken of all hospital personnel revealed that several members of the surgical team harbored *S. aureus* as part of their resident microbiota. But which individual was the actual source of the patient's infection? In what could be described as "forensic microbiology," each strain was subjected to pulsed-field gel electrophoresis, and the resulting genomic restriction patterns were compared with that of the isolate from the infected patient. A match pointed to the source. The unwitting culprit turned out to be the perfusionist who manipulated the tubing used for cardiopulmonary bypass.

Hospital-acquired, or **nosocomial**, infections are not unusual. As many as 5%–10% of all patients admitted to acute-care hospitals acquire nosocomial infections, resulting in over 80,000 deaths each year. The deaths are due, in part, to the poor health of the patient and in part to the antibiotic-resistant nature of bacteria lurking in hospitals. In fact, as many as 60%–70% of staph infections that develop in a hospital setting are the result of methicillin-resistant *S. aureus* (MRSA). A foreboding report from the United Kingdom finds that one in four nursing-home residents is colonized by MRSA. A recent study in the United States found that the noses of nearly one in five nonhospitalized people are also colonized by MRSA. Antibiotic resistance is a problem that will only get worse.

We begin Chapter 27 with a discussion of the golden age of antibiotic discovery (1940–60) and move on to describe the basic concepts of antibiotics and their use and misuse. We will delve into the ways genomic and proteomic approaches broaden our ability to search for new antibiotics and targets—all part of our attempt to stay one step ahead of evolving, antibiotic-resistant pathogens. The urgency of this task cannot be overstated. In May 2016, an *E. coli* uropathogen resistant to a drug of last resort, colistin, arrived in the United States.

27.1 Fundamentals of Antimicrobial Therapy

Antibiotics (from the Greek meaning "against life") are compounds produced by one species of microbe that can kill or inhibit the growth of other microbes. However, the term "antibiotics" is also used for synthetic chemotherapeutic agents, such as sulfonamides, that are clinically useful but chemically synthesized.

We think of antibiotics as being a recent biotechnological development, but they have actually been used for centuries. Ancient remedies called for cloths soaked with organic material to be placed on wounds to allow them to heal faster. This organic material likely contained "natural antibiotics" that killed bacteria and prevented further infection. The medicinal properties of molds were also recognized for centuries. The ancient Chinese successfully treated boils with warm soil and molds scraped from cheeses, and in England a paste of moldy bread was a home remedy for wound infections up until the beginning of the twentieth century.

The Golden Age of Antibiotic Discovery

The modern antibiotic revolution began with the discovery of **penicillin** in 1928 by Sir Alexander Fleming (1881–1955). This discovery was actually a rediscovery, and was arguably one of the greatest examples of serendipity in science. Although Fleming generally receives the credit for discovering penicillin, a French medical student, Ernest Duchesne (1874–1912), originally discovered the antibiotic properties of *Penicillium* in 1896.

Duchesne observed that Arab stable boys at the nearby army hospital kept their saddles in a dark and damp room to encourage mold to grow on them. When asked why, they told him the mold helped heal saddle sores on the horses. Intrigued, Duchesne prepared a solution from the mold and injected it into diseased guinea pigs. All recovered.

Penicillium was forgotten in the scientific community until Fleming rediscovered it one day in the late 1920s. Petri dishes were glass in those days and could be rewashed

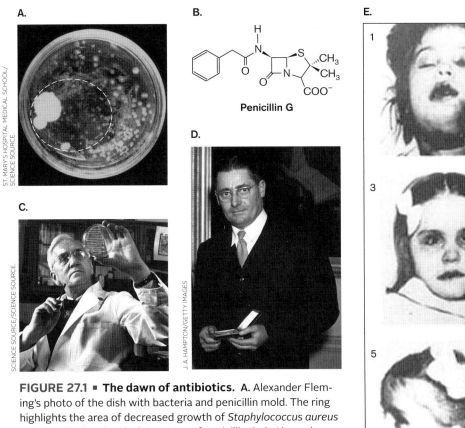

FIGURE 27.1 ■ The dawn of antibiotics. A. Alexander Fleming's photo of the dish with bacteria and penicillin mold. The ring highlights the area of decreased growth of *Staphylococcus aureus* colonies. **B.** The chemical structure of penicillin G. **C.** Alexander Fleming at work in his laboratory. **D.** Howard Florey. **E.** Pictures taken in 1942, shortly after the introduction of penicillin, show the improvement in a child suffering from an infection 4 days (panel 2) and 9 days (panel 4) after treatment. Panels 5 and 6 show her fully recovered.

and sterilized. Fleming was preparing to wash a pile of old petri dishes he had used to grow the pathogen *Staphylococcus aureus*. He opened and examined each dish before tossing it into a cleaning solution. He noticed that one dish had grown contaminating mold, which in and of itself was not unusual in old plates, but all around the mold the staph bacteria had failed to grow (**Fig. 27.1A**).

Fleming took a sample of the mold and found that it was from the penicillium family, later identified as *Penicillium notatum*. The mold appeared to have synthesized a chemical, now known as penicillin (**Fig. 27.1B**), which diffused through the agar, killing cells of *S. aureus* before they could form colonies. Fleming (**Fig. 27.1C**) presented his findings in 1929, but they raised little interest, since penicillin appeared to be unstable and would not remain active in the body long enough to kill pathogens.

As World War II began, Oxford professor Howard Florey (**Fig. 27.1D**) and his colleague Ernst Chain, rediscovered Fleming's work, thought it held promise, and set about purifying penicillin. To their amazement, the purified penicillin cured the mice infected with staphylococci or streptococci. Subsequent human trials proved successful (**Fig. 27.1E**), and penicillin gained wide use, saving countless lives during the war. Fleming, Florey, and Chain received the 1945 Nobel Prize in Physiology or Medicine for their work.

The next landmark discovery in antibiotics was made by Gerhard Domagk (1895–1964), a German physician at the Bayer Institute of Experimental Pathology and Bacteriology, who investigated antimicrobial compounds in the 1930s (**Fig. 27.2A**). In 1935, Domagk's 6-year-old child developed a serious streptococcal infection induced by an innocent pinprick to the finger. The infection spread to her axillary (armpit) lymph nodes and became so severe that lancing and draining the pus 14 times did little to help. The only remaining alternative was to amputate the arm. Unfortunately, even this option would probably not save her life.

Frustrated, Domagk took what would appear to be drastic measures. He administered a dose of a red dye (Prontosil) he was investigating. On agar plates (the usual medium

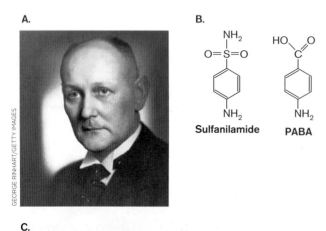

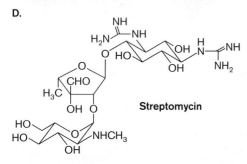

**FIGURE 27.2 ■ The discoverers of sulfanilamide and strep-
tomycin. A.** Gerhard Domagk discovered sulfanilamide. **B.** Chemical structure of sulfanilamide, an analog of *para*-aminobenzoic acid
(PABA), a precursor of the vitamin folic acid, which is necessary for
growth. Sulfanilamide inhibits one of the enzymes that converts
PABA into folic acid. **C.** Selman Waksman discovered streptomycin
in 1944. **D.** Chemical structure of streptomycin.

for testing antibiotics), Prontosil had shown absolutely no
ability to inhibit the growth of streptococcus. Domagk,
however, had tested the drug in animals, not agar plates,
where he found it could cure animals of infection. Remarkably, Domagk's daughter recovered completely.

Domagk discovered that Prontosil was metabolized by
the body into another compound, sulfanilamide, clearly
lethal to the streptococcus. This finding led to an entire
class of drugs, the **sulfa drugs**, that saved hundreds of
thousands of lives. The take-home message of this story is
that an antibiotic's activity on a plate, or lack thereof, does
not necessarily correlate with the drug's activity in a patient.

Sulfanilamide is an analog of *para*-aminobenzoic acid
(PABA), a precursor of folic acid, a vitamin necessary for
nucleic acid synthesis (**Fig. 27.2B**). Sulfanilamide and
other sulfa drugs bind to and inhibit the enzyme that converts PABA to folic acid. Without folic acid to make nucleic
acid precursors, the pathogen stops growing. Sulfa drugs
inhibit bacterial growth without affecting human cells
because folic acid is not synthesized by humans (it is a
dietary supplement instead), and because bacteria do not
transport folic acid (they must make it themselves).

In a dark turn, Domagk's possible participation in
human experimentation with concentration camp prisoners during World War II (ordered by his German employer)
was a source of controversy that haunted him long after the
war ended. Yet his contributions to medicine continued; he
also developed the first effective chemotherapy for tuberculosis, the thiosemicarbazones and isoniazid, which are still
used today.

During the same period of history, Selman Waksman
(1888–1973) at Rutgers University began screening 10,000
strains of soil bacteria and fungi for their ability to inhibit
growth or kill bacteria (**Fig. 27.2C**). In 1944, this hercu-
lean effort paid off with the discovery of streptomycin, an
antibiotic produced by the actinomycete *Streptomyces griseus* (**Fig. 27.2D**). Waksman's discovery of streptomycin
triggered the antibiotic gold rush that is still under way and
earned him the 1952 Nobel Prize in Physiology or Medicine.

Antibiotics Exhibit Selective Toxicity

As early as 1904, the German physician Paul Ehrlich
(1854–1915) realized that a successful antimicrobial compound would be a "magic bullet" that selectively kills or
inhibits the pathogen but not the host. This seemingly
obvious premise was innovative at the time. Ehrlich made
several discoveries based on this concept, the most celebrated of which was the arsenical compound known as Salvarsan. Salvarsan proved to be quite effective in killing the
syphilis agent *Treponema pallidum* (this was long before
penicillin was discovered). Syphilis, a sexually transmitted
disease, had been untreatable and the source of considerable long-term suffering. Ehrlich's "magic bullet" concept
is now known as **selective toxicity**. Although Ehrlich's
selective toxicity concept was right, Salvarsan itself was
not as selectively toxic as he thought. This arsenical compound harmed the host, but fortunately usually killed off
the treponemes before killing the patient.

Selective toxicity is possible because key aspects of a
microbe's physiology are different from those of eukaryotes. For example, suitable bacterial antibiotic targets
include peptidoglycan, which eukaryotic cells lack, and
ribosomes, which are structurally distinct between Bacteria and Eukarya. Thus, chemicals like penicillin, which

prevents peptidoglycan synthesis, and tetracycline, which binds to bacterial 30S ribosomal subunits, inhibit bacterial growth but are essentially invisible to host cells, since they do not interact with them at low doses.

Although their intended targets are bacterial cells, some antibiotics, particularly at high doses, can interact with elements of eukaryotic cells and cause side effects that harm the patient. For example, chloramphenicol, a drug that targets bacterial 50S ribosomal subunits, can interfere with the development of blood cells in bone marrow—a phenomenon that may result in aplastic anemia (failure to produce red blood cells). The toxicity of an antibiotic can also depend on the age of the patient. Ciprofloxacin, for instance, can cause defects in human bone growth plates and should not be administered to children. Problems can even arise if the drug does not directly impact mammalian physiology. For example, many people develop an extreme allergic sensitivity to penicillin, in which case the treatment of an infection may end up being worse than the infection itself. Physicians must be aware of these allergies and use alternative antibiotics to avoid harming their patients.

As a student of microbiology, you must be able to properly distinguish between the terms "drug susceptibility" and "drug sensitivity." A microbe is <u>susceptible</u> to the drug's action, but a human can develop an allergic <u>sensitivity</u> to the drug.

Spectrum of Activity

No single antimicrobial drug affects all microbes. As a result, antimicrobial drugs are classified by the types of organisms they affect. Thus, we have antifungal, antibacterial, antiprotozoan, and antiviral agents. The term "antibiotic" is usually reserved for compounds that affect bacteria. Even within a group, one agent might have a very narrow **spectrum of activity**, meaning it affects only a few species, while another antibiotic inhibits many species. For instance, penicillin has a relatively narrow spectrum of activity, killing primarily Gram-positive bacteria. However, ampicillin is penicillin with an added amino group that allows the drug to more easily penetrate the Gram-negative outer membrane. As a result of this chemically engineered modification, ampicillin kills Gram-positive and Gram-negative organisms, giving it a broader spectrum of activity than penicillin has. There are antimicrobials as well that exhibit extremely narrow activities. One example is isoniazid, which is clinically useful only against *Mycobacterium tuberculosis*, the agent of tuberculosis. The spectrum of activity of select antibiotics is explored in **eTopic 27.1**.

Patients typically believe that all antibiotics kill their intended targets. This is a misconception. Many drugs simply prevent growth of the organism and let the body's immune system dispatch the intruding microbe. Thus,

antimicrobials are also classified on the basis of whether or not they kill the microbe. An antibiotic is **bactericidal** if it kills the target microbe, whereas it is **bacteriostatic** if it merely prevents bacterial growth.

Antibiotics are good for treating infectious diseases, but do our microbiomes pay the price? As discussed in several prior chapters, we are increasingly aware that our natural microbiota contribute in important ways to human health and development. Many studies are now exploring the impact of antibiotic use on host-microbiota interactions (discussed in Chapters 23 and 24). We have known for decades that antibiotics—especially broad-spectrum antibiotics—can destroy the ecological balance of bacterial species in the gut (as well as at other body sites) and lead to gastrointestinal disease. Pathology can result when one species resistant to the antibiotic gains a growth advantage over various drug-susceptible species that ordinarily keep the pathogen in check (see the discussion of *Clostridioides difficile*, formerly *Clostridium difficile*, in Section 26.4). But what other effects might there be when the microbial balance of power in the intestine is disturbed? We may ultimately find that directing the toxicity of an antibiotic to a single bacterial species (while leaving all others alone) has benefits to human health that we do not currently understand.

Measuring Drug Susceptibility

One critical decision a clinician must make when treating an infection is which antibiotic to prescribe for the patient. There are several factors to consider, including:

■ **The relative effectiveness of different antibiotics on the organism causing the infection.** This determination will reveal whether the organism developed resistance to a drug.

■ **The average attainable tissue levels of each drug.** An antibiotic may work on an agar plate, but the concentration at which it affects bacterial growth may be too high to be safe in the patient. In fact, an important aspect of designing new antibiotics is to enhance the pharmacological activity of an existing drug—for example, modifying it so that the body does not break it down or quickly excrete it in urine.

Minimal inhibitory concentration. The in vitro effectiveness of an antimicrobial agent is determined by measuring how little of it is needed to stop growth. This amount is classically measured in terms of an antibiotic's **minimal inhibitory concentration (MIC)**, defined as the lowest concentration of the drug that will prevent the growth of an organism. But the MIC for any one drug will differ among different bacterial species. For example, the MIC of ampicillin needed to stop the growth of *Staphylococcus*

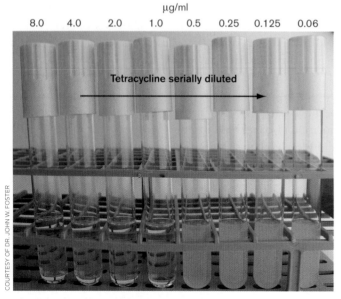

µg/ml

8.0   4.0   2.0   1.0   0.5   0.25   0.125   0.06

Tetracycline serially diluted

COURTESY OF DR. JOHN W. FOSTER

FIGURE 27.3 ▪ **Determining minimal inhibitory concentration (MIC).** In this series of tubes, tetracycline was diluted serially starting at 8 µg/ml. Each tube was then inoculated with an equal number of bacteria. Turbidity indicates that the antibiotic concentration was insufficient to inhibit growth. The MIC in this example is 1.0 µg/ml.

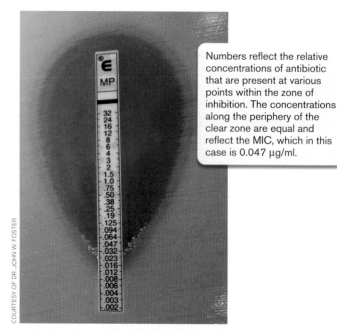

Numbers reflect the relative concentrations of antibiotic that are present at various points within the zone of inhibition. The concentrations along the periphery of the clear zone are equal and reflect the MIC, which in this case is 0.047 µg/ml.

COURTESY OF DR. JOHN W. FOSTER

FIGURE 27.4 ▪ **An MIC strip test.** The Etest (AB Biodisk) is a commercially prepared strip that produces a gradient of antibiotic concentration (in µg/ml) when placed on an agar plate. The MIC corresponds to the point where bacterial growth crosses the numbered strip.

aureus will be different from that needed to inhibit *Shigella dysenteriae*. The reasons that a drug may be more effective against one organism than another include the ease with which the drug penetrates the cell and the affinity of the drug for its molecular target.

So how do we measure MIC? As shown in **Figure 27.3**, an antibiotic is serially diluted along a row of test tubes containing nutrient broth. After dilution, the organism to be tested is inoculated at low, constant density into each tube, and the tubes are usually incubated overnight. Growth of the organism is seen as turbidity. In **Figure 27.3**, the tubes with the highest concentration of drug are clear, indicating no growth. The tube containing the MIC is the tube with the <u>lowest</u> concentration of drug that shows no growth. Note, however, that the MIC does <u>not</u> indicate whether a drug is bacteriostatic or bactericidal.

> **Thought Questions**
>
> **27.1 Figure 27.3** illustrates how MICs are determined. Test your understanding of how MICs are measured in the following example. The drug tobramycin is added to a concentration of 1,000 µg/ml in a tube of broth from which serial twofold dilutions are made. Including the initial tube (tube 1), there are a total of ten tubes. Twenty-four hours after all the tubes are inoculated with *Listeria monocytogenes*, turbidity is observed in tubes 6–10. What is the MIC?
>
> **27.2** What additional test performed on an MIC series of tubes will tell you whether a drug is bacteriostatic or bactericidal?

MIC determinations are very useful for estimating a single drug's effectiveness against a single bacterial pathogen isolated from a patient, but they are not very practical when trying to screen 20 or more different drugs. Dilutions take time—time that the technician, not to mention the patient, may not have. The time required to evaluate antibiotic effectiveness can be reduced by using a strip test (like the Etest shown in **Fig. 27.4**) that avoids the need for dilutions. The strip, containing a gradient of antibiotic, is placed on an agar plate freshly seeded with a dilute lawn of bacteria. While the bacteria are trying to grow, the drug diffuses out of the strip and into the media. Drug emanating from the more concentrated areas of the strip will travel faster and farther through the agar than will drug from the less concentrated areas of the strip. Thus, the drug's effect (killing or inhibiting the growth of cells) will extend farther away from the strip at locations of high concentration than at locations of lower concentration. The result is a **zone of inhibition** where the antibiotic has stopped bacterial growth. The MIC is the point at which the elliptical zone of inhibition intersects with the strip.

Kirby-Bauer Disk Susceptibility Test. Although the strip test eliminates the time and effort needed to make dilutions, it would take 20 or more plates to test an equal number of antibiotics for just one bacterial isolate. Clinical labs can receive up to 100 or more isolates in one day, so

A.

FIGURE 27.5 ■ The Kirby-Bauer disk susceptibility test.
A. Device used to deliver up to 12 disks to the surface of a Mueller-Hinton plate. **B.** Disks impregnated with different antibiotics are placed on a freshly laid lawn of methicillin-sensitive *Staphylococcus aureus* (MSSA) and incubated overnight. The clear zones around certain disks indicate growth inhibition. **C.** Results for methicillin-resistant *S. aureus* (MRSA). **D.** Results for *Streptococcus pneumoniae*. The brownish tint of the blood agar plates outside the zones of bacterial inhibition is caused by a hemolysin secreted by the lawn of pneumococci. C, chloramphenicol; CC, clindamycin; CZ, cefazolin; E, erythromycin; NOR, norfloxacin; OX, oxacillin; P, penicillin; RA, rifampin; SAM, sulbactam-ampicillin; SXT, sulfa-trimethoprim; TE, tetracycline; VA, vancomycin.

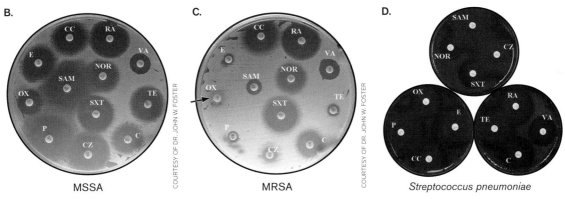

individual MIC determinations are impractical. A simplified agar diffusion test, however, which can test 12 antibiotics on one plate, makes evaluating antibiotic susceptibility a manageable task.

Named for its inventors, the **Kirby-Bauer assay** uses a series of round filter paper disks impregnated with different antibiotics. A dispenser (**Fig. 27.5A**) delivers up to 12 disks simultaneously to the surface of an agar plate covered by a bacterial lawn. Each disk is marked to indicate the drug used. During incubation, the drugs diffuse away from the disks into the surrounding agar and inhibit growth of the lawn to different distances (**Fig. 27.5B–D**). The zones of inhibition vary in width, depending on the antibiotic used, the concentration of drug in the disk, and the susceptibility of the organism to the drug. The diameter of the zone correlates to the MIC of the antibiotic against the organism tested. **Figure 27.5B** and **C** show the results for methicillin-sensitive and methicillin-resistant *Staphylococcus aureus* (MSSA and MRSA, respectively). Note the lack of inhibition by the oxacillin disk (arrow in **Fig. 27.5C**). This strain is resistant to both methicillin and oxacillin because they are structurally similar.

Correlations between MIC values and Kirby-Bauer zone sizes are made empirically. Every disk containing a given antibiotic is impregnated with a standard concentration of drug, and every antibiotic has a quantifiable MIC that will differ when tested against different bacterial species and strains. The outermost ring of the no-growth zone in a Kirby-Bauer disk test must, by definition, contain the minimal concentration of drug needed to prevent growth on agar. Thus, if species A and B have MIC values for penicillin of 4 µg/ml and 40 µg/ml, respectively, then species A will exhibit a proportionally larger zone of inhibition than species B in the disk test. A graph plotting MIC on one axis and zone diameter on the other provides the correlation.

After incubating agar plates in the Kirby-Bauer test, the diameters of the zones of inhibition around each disk are measured, and the results are compared with a table listing whether a zone is wide enough (meaning the MIC is low enough) to be clinically useful. **Table 27.1** shows susceptibility data for *S. aureus*. The concentration of antibiotic used and the zone size that is considered clinically significant are correlated with the average attainable tissue level for each antibiotic (discussed later). For the antibiotic to remain effective in vivo, it is important that the tissue concentration of the drug remain above the MIC; otherwise, invading bacteria will not be affected.

TABLE 27.1	Susceptibility Results for *Staphylococcus Aureus*			
		Zone of inhibition diameter (mm)		
Antibiotic	Quantity in disk (µg)	Resistant	Intermediate	Susceptible
Ampicillin	10	<12	12–13	>13
Chloramphenicol	30	<13	13–17	>17
Erythromycin	15	<14	14–17	>17
Gentamicin	10	≤12.5		>12.5
Streptomycin	10	<12	12–14	>14
Tetracycline	30	<15	15–18	>18

To ensure reproducibility, the Kirby-Bauer test was standardized over a half century ago. Reproducibility means that results from a laboratory in California will match those in Alabama, Ohio, or any other state. The following are standardizations used to make the test reproducible and easier.

- **Size of the agar plate.** The plates used (150 mm) are larger than standard agar plates (100 mm) to accommodate more disks and to maintain sufficient distance between disks so that zones of inhibition do not overlap.

- **Depth of the media.** Antibiotics diffuse out of impregnated disks in three dimensions. Because diffusion cannot occur very far downward in a thinly poured agar, the drug is forced to move more laterally. Thus, the zone of inhibition measured from a thinly poured agar plate will be larger than the zone from a thick agar plate.

- **Media composition.** Media used for testing antibiotic susceptibility should lack *para*-aminobenzoic acid (PABA), a compound used by cells to make the vitamin folic acid. Sulfonamide antibiotics are analogs of PABA that inhibit folic acid synthesis. However, standard laboratory media, such as nutrient agar, contain PABA. Grown on such a medium, the bacterial cell will be flooded with PABA, which will limit the ability of the sulfa drugs to inhibit growth. Thus, even though the drug might be effective in vivo, on nutrient agar there would be no zone of inhibition. The standardized medium used for the Kirby-Bauer test, called **Mueller-Hinton agar**, contains no PABA.

- **The number of organisms spread on the agar plate.** The more organisms that are placed on a plate, the faster visible growth will form. The faster visible growth forms, the less time there is for an antibiotic to diffuse from a disk to prevent growth. The result is that the diameter of the zone of inhibition for an antibiotic is inversely proportional to the number of organisms used to seed the agar plate. Thus, a standard optical density solution of each organism is prepared, and a cotton swab is used to spread organisms evenly over the entire agar surface.

- **Size of the disks.** A standard diameter of 6 mm means that all antibiotics start diffusing into the agar at the same point.

- **Concentrations of antibiotics in the disks.** The zone of inhibition for an antibiotic is proportional to the concentration of antibiotic in the disk. The higher the concentration of antibiotic in a disk, the faster that drug can diffuse through the plate and inhibit growth. To avoid differences between labs, the concentration of each drug impregnating a disk has been standardized.

- **Incubation temperature.** Incubation temperature will not affect growth and diffusion equally. To avoid differences, a temperature of 37°C is standard.

Automated MIC determinations. The methods just described to determine MIC are effective but take at least 24 hours to complete (the amount of time needed to see visible turbidity in broth or growth on agar). Modern clinical laboratories are equipped with automated machines (described in Chapter 28) that can determine an MIC within 6 hours. This speed is critical to a clinician wanting to quickly treat a serious infection with the most effective drug. The instruments work by monitoring growth of the infectious agent in microtiter plate wells and extrapolating MIC from growth curves recorded at different antibiotic concentrations.

Correlating antibiotic MIC with tissue level. The average attainable tissue level for a drug depends on how quickly the antibiotic is cleared from the body via secretion by the kidney or destruction in the liver. It also depends on

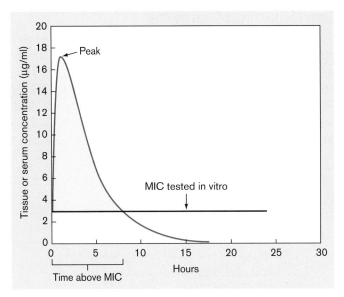

FIGURE 27.6 ■ Correlation between MIC and serum or tissue level of an antibiotic. This graph illustrates the serum level of ampicillin over time. The important consideration here is how long the serum level of the antibiotic remains higher than the MIC. Once the concentration falls below the MIC, owing to destruction in the liver or clearance through the kidneys and secretion, the infectious agent fails to be controlled by the drug—in this case, 7–8 hours after the initial dose. To maintain a serum level higher than the MIC, a second dose would be taken. The shaded area of the curve represents time above MIC.

when side effects of the drug start to appear. The graph in **Figure 27.6** shows that as long as the concentration of the drug in tissue or blood remains higher than the MIC, the drug will be effective. The concentration can be kept at sufficient levels either by administering a higher dose, which runs the risk of side effects, or by giving a second dose at a time when the levels from the first dose have declined. This is why patients are told to take doses of some antibiotics four times a day and other antibiotics only once a day.

Thought Questions

27.3 A patient with a bacterial lung infection was given the antibiotic represented in **Figure 27.6** and was told to take one pill twice a day. The pathogen is susceptible to this drug. Will the prescribed treatment be effective? Explain your answer.

27.4 You are testing whether a new antibiotic will be a good treatment choice for a patient with a staph infection. The Kirby-Bauer test using the organism from the patient shows a zone of inhibition of 15 mm around the disk containing this drug. Clearly, the organism is susceptible. But you conclude from other studies that the drug would be ineffective in the patient. What would make you draw this conclusion?

To Summarize

- **The importance of antibiotics** in treating disease was recognized in the early 1940s.

- **Some antimicrobial agents are initially inactive**, until converted by the body to an active agent.

- **Florey and Chain purified penicillin** and capitalized on Fleming's discovery of penicillin.

- **Antimicrobial agents** may be produced naturally or artificially.

- **Selective toxicity** refers to the ability of an antibiotic to attack a unique component of microbial physiology that is missing or distinctly different from eukaryotic physiology.

- **Antibiotic side effects** on mammalian physiology can limit the clinical usefulness of an antimicrobial agent.

- **Antibiotic spectrum of activity** is the range of microbes that a given drug affects.

- **Bactericidal antibiotics** kill microbes; **bacteriostatic antibiotics** inhibit microbial growth.

- **The spectrum of an antibiotic and the susceptibility of the infectious agent** are critical points of information required before prescribing antibiotic therapy.

- **Minimal inhibitory concentration (MIC)** of a drug, when correlated with average attainable tissue levels of the antibiotic, can predict the effectiveness of an antibiotic in treating disease.

- **MIC is measured** using tube dilution techniques, but it can be approximated using the Kirby-Bauer disk diffusion technique.

27.2 Antibiotic Mechanisms of Action

As noted in Section 27.1, selective toxicity of an antibiotic depends on enzymes or structures unique to the bacterial target cell. The following aspects of a microbe's physiology are classic targets:

- Cell wall synthesis
- Cell membrane integrity
- DNA synthesis
- RNA synthesis
- Protein synthesis
- Metabolism

Table 27.2 summarizes the general targets of common antibiotics. Chapters 3, 7, and 8 describe these cellular components and provide the basis for understanding how antibiotics work. The mechanisms of action for antibiotics affecting DNA, RNA, and protein synthesis are described in Chapters 7 and 8; they receive only brief mention here.

Cell Wall Antibiotics

Bacterial cell walls are the basis of selective toxicity for some antibiotics because peptidoglycan does not exist in mammalian cells; thus, antibiotics that target the synthesis of these structures should selectively kill bacteria. The following case history illustrates the use of two cell wall–targeting antibiotics and also reveals how bacteria can evolve to escape destruction.

Case History: Meningitis

A 3-year-old child was brought to the emergency room crying, with a stiff neck and high fever. Gram stain of cerebrospinal fluid revealed Gram-positive cocci, generally in pairs. The diagnosis was meningitis. The physician immediately prescribed intravenous ampicillin. Unfortunately, the child's condition worsened, so antibiotic treatment was changed to a third-generation cephalosporin (which more easily crosses the

TABLE 27.2	Targets of Antimicrobial Agents
Target	**Antibiotic examples**
Cell wall synthesis	Penicillins, cephalosporins, bacitracin, vancomycin
Protein synthesis	Chloramphenicol, tetracyclines, aminoglycosides, macrolides, lincosamides
Cell membrane integrity	Polymyxin, daptomycin, amphotericin, imidazoles (vs. fungi)
Nucleic acid function	Nitroimidazoles, nitrofurans, quinolones, rifampin; some antiviral compounds, especially antimetabolites
Intermediary metabolism	Sulfonamides, trimethoprim

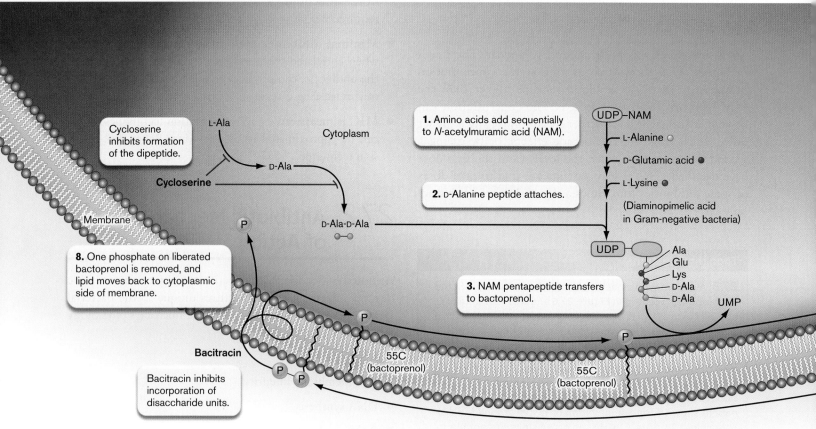

FIGURE 27.7 ■ Peptidoglycan synthesis in a Gram-positive bacterium (*Staphylococcus aureus*), and targets of antibiotics. Several small-molecular-weight compounds are sequentially joined to form a disaccharide unit that will be added to preexisting extracellular chains of this unit. Red lines indicate inhibition. Cycloserine inhibits ligation of the two D-alanines (step 2); bacitracin inhibits linking of the disaccharide units; vancomycin and the beta-lactams, such as penicillin, inhibit the peptide cross-linking of peptidoglycan side chains.

blood-brain barrier). The patient began to improve within hours and was released after 2 days. A report from the clinical microbiology laboratory identified the organism as Streptococcus pneumoniae.

Both of the antibiotics used in this case kill bacteria by targeting cell wall synthesis (introduced in Chapter 3). To synthesize peptidoglycan, sugar molecules called *N*-acetylglucosamine (NAG) and *N*-acetylmuramic acid (NAM) are made by the cell and linked together by a **transglycosylase** enzyme into long chains assembled at the cell wall (**Fig. 27.7**). *N*-acetylmuramic acid contains a short side chain of amino acids that is assembled enzymatically, not by a ribosome. The side chains from adjacent strands are cross-linked to make the structure rigid. The enzyme **transpeptidase** (D-alanyl-D-alanine carboxypeptidase/transpeptidase) catalyzes the cross-link. Several antibiotics target various stages of this assembly process.

Peptidoglycan synthesis. Before we can explain how cell wall antibiotics work, you need to know how the cell wall is made. Synthesis of peptidoglycan starts in the cytoplasm with a uridine diphosphate (UDP)–NAM molecule. The amino acids L-alanine, D-glutamic acid, and L-lysine [or diaminopimelic acid (DAP) in Gram-negative organisms] are individually and sequentially added to NAM (**Fig. 27.7**, step 1); and then a dipeptide of D-alanine is attached (step 2). Next, the NAM-pentapeptide is transferred to a membrane-situated, 55-carbon lipid molecule called bactoprenol (step 3), releasing uridine monophosphate (UMP). This structure is also called lipid I. Another sugar molecule, NAG, is then linked to NAM—once again through a UDP intermediate (step 4)—to make the bactoprenol structure called lipid II. All of this takes place on the cytoplasmic side of the membrane.

Bactoprenol then "flips," moving NAM-NAG to the outer side of the cytoplasmic membrane (**Fig. 27.7**, step 5), where transpeptidases and transglycosylases (two **penicillin-binding proteins**, or **PBPs**) bind to the D-Ala-D-Ala part of the pentapeptide. Transglycosylase attaches the new disaccharide unit to an existing peptidoglycan chain (step 6). Transpeptidase then links two peptide

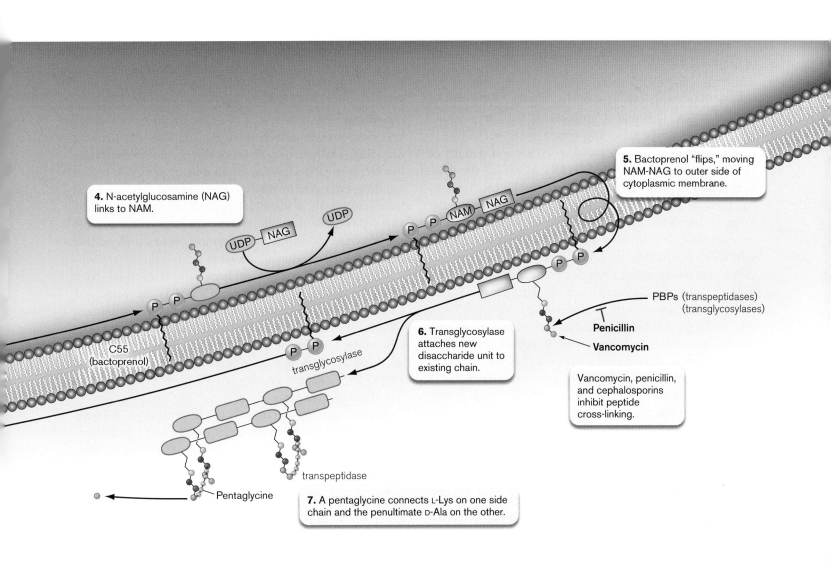

4. N-acetylglucosamine (NAG) links to NAM.

5. Bactoprenol "flips," moving NAM-NAG to outer side of cytoplasmic membrane.

C55 (bactoprenol)

transglycosylase

PBPs (transpeptidases) (transglycosylases)

Penicillin

Vancomycin

6. Transglycosylase attaches new disaccharide unit to existing chain.

Vancomycin, penicillin, and cephalosporins inhibit peptide cross-linking.

transpeptidase

Pentaglycine

7. A pentaglycine connects L-Lys on one side chain and the penultimate D-Ala on the other.

side chains of adjacent peptidoglycan molecules with a pentaglycine cross-link (in *Staphylococcus
aureus*). The pentaglycine connects L-Lys on one side chain and
the penultimate D-Ala on the other
side chain (step 7). The terminal
D-Ala is removed in the process.
Other bacteria do not use a pentaglycine cross-link, but directly
form a peptide bond between
L-Lys (DAP) and the penultimate
D-Ala. Cross-linking strengthens the cell wall. In the last part
of the cycle, one of the phosphates
on the now liberated bactoprenol
is removed and the lipid moves
back to the cytoplasmic side of the
membrane, ready to pick up and
taxi another unit of peptidoglycan
to the growing chain (step 8).

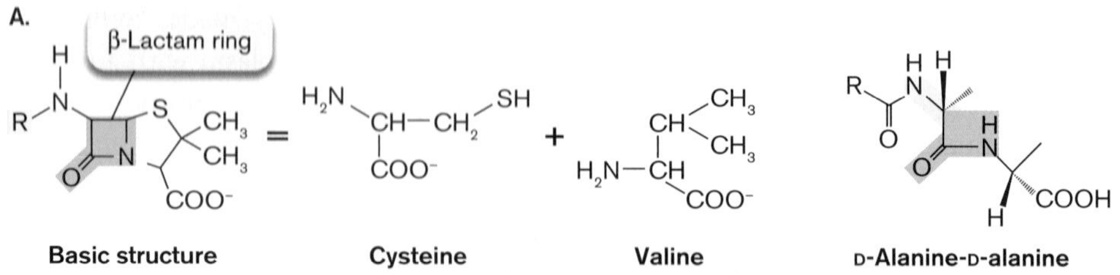

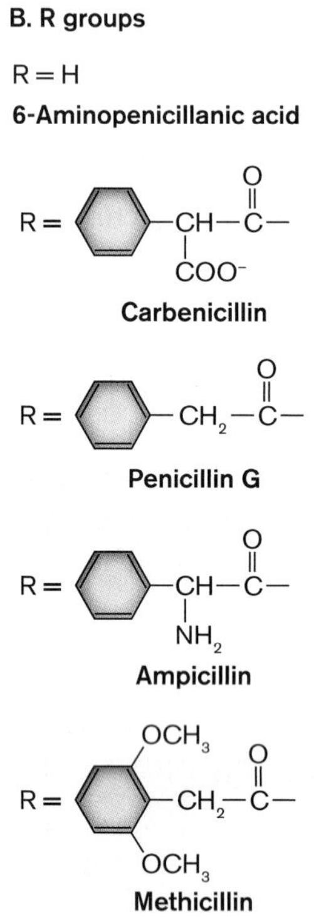

Beta-lactam antibiotics. Penicillin and other beta-lactam antibiotics target penicillin-binding
proteins. Penicillin is an antibiotic
derived from cysteine and valine,
which combine to form the beta-
lactam ring structure shown in
Figure 27.8A. Different R groups
can be added to the basic ring struc-
ture to change the antimicrobial spectrum and stability of
the derivative penicillin (**Fig. 27.8B**). Note that the beta-
lactam ring of penicillin chemically resembles the D-Ala-D-
Ala piece of peptidoglycan (**Fig. 27.8A**). This molecular
mimicry allows the drug to bind transpeptidase, transglycosylase, and several other proteins involved in cell wall
synthesis or remodeling (which is why the proteins are
called penicillin-binding proteins). Penicillins act by inhibiting transpeptidase-mediated cross linking between adjacent peptidoglycan chains. This activity makes the cell wall
very weak. In addition, penicillins can somehow activate
proteins in the cell wall that hydrolyze peptidoglycan. The
consequence is a disaster for bacteria that are trying to
grow larger and larger. Eventually, the growing cell bursts
for lack of cell wall restraint. Penicillin, then, is a bactericidal drug (unless the treated organism is suspended in an
isotonic solution). Note that in addition to cell lysis, there
is another explanation for why penicillin and other bactericidal drugs kill bacteria, which we will discuss at the end
of this section.

Penicillin is more effective against Gram-positive than
Gram-negative organisms because the drug has difficulty

FIGURE 27.8 ▪ The structure of penicillins. A. Penicillanic
acid (R = H) is derived from cysteine and valine. Also shown is the
D-alanine-D-alanine structure of peptidoglycan, which is structurally similar to the beta-lactam ring of penicillins (green shading).
B. The R group highlighted in (A) can be any one of a number of
different groups, some of which are shown here. Modifying this
group changes the pharmacological properties and antimicrobial
spectrum of the drug.

passing through the Gram-negative outer membrane.
Ampicillin, which was used in the case history, is a modified version of penicillin that more easily penetrates this
membrane and is more effective than penicillin against
Gram-negative microbes. Thus, ampicillin has a broader
spectrum of activity than penicillin.

Bacterial resistance to beta-lactam antibiotics. As noted
earlier, antibiotic resistance is a growing problem throughout the world. Bacteria develop resistance to penicillin in
two basic ways. The first is through inheritance of a gene
encoding one of the beta-lactamase enzymes, which cleave
the critical ring structure of this class of antibiotics. Beta-
lactamase is transported out of the cell and into the surrounding medium (for Gram-positives) or the periplasm
(for Gram-negatives), where it can destroy penicillin before
the drug even gets to the cell. Bacteria that produce beta-
lactamase are still susceptible to certain modified penicillins
and cephalosporins engineered to be poor substrates for
the enzyme. Methicillin, for example, works well against
beta-lactamase-producing microbes.

Unfortunately, a type of beta-lactamase called New
Delhi metallo-beta-lactamase-1 (NDM-1) has emerged
that confers resistance to almost all beta-lactam antibiotics.
Originating in India, NDM-1-containing plasmids are pro-
miscuously transferred, being found in various enterobac-
terial species, such as *Klebsiella pneumoniae* and *E. coli*, as
well as the nonenteric pathogens *Pseudomonas aeruginosa*
and *Acinetobacter baumannii*.

Aside from beta-lactamases, the second way a microbe
can become resistant to beta-lactam antibiotics is by acquiring a gene encoding an altered penicillin binding protein that no longer binds penicillin. Methicillin-resistant
Staphylococcus aureus (MRSA) uses this strategy. Resistance to methicillin in *S. aureus* is mediated by the *mecA*
gene, which is part of a mobile genetic element called

staphylococcal cassette chromosome *mec* (SCC*mec*). The *mecA* gene encodes an altered penicillin-binding protein (PBP2a or PBP2′) with low affinity for beta-lactam antibiotics. The low affinity provides resistance to all beta-lactam antibiotics, rendering them useless. Hospitals take special interest in MRSA because very few drugs can kill it.

One of the few remaining antibiotics effective against MRSA is vancomycin. Unfortunately, resistance is developing to this drug too. Vancomycin-resistant *S. aureus* strains are called VRSA. The penicillin-resistant *Streptococcus pneumoniae* in the preceding case history actually had an altered penicillin-binding protein. No beta-lactamase-producing *S. pneumoniae* has yet been found.

Beta-lactamase-resistant antibiotics. Cephalosporins are another type of beta-lactam antibiotic originally discovered in nature but modified in the laboratory to fight microbes that are naturally resistant to penicillins (especially *Pseudomonas aeruginosa*). Over the years, the structure of cephalosporin has undergone a series of modifications to improve its effectiveness against penicillin-resistant pathogens. Each modification adds complexity and produces what is called a new "generation" of cephalosporins. There are currently five generations of this semisynthetic antibiotic (**Fig. 27.9**). Unfortunately, the microbial world continually adapts and eventually becomes resistant to new antibiotics. In the case of the cephalosporins, new beta-lactamases evolve that can attack the sterically buried beta-lactam rings in these molecules. It is also important to note that because the core feature of these drugs is the beta-lactam ring, persons who are sensitive to penicillins may also suffer a hypersensitivity reaction to cephalosporins.

Treatment note: In the preceding case history, the infecting strain of *Streptococcus pneumoniae* turned out to be resistant to ampicillin. Had the patient been an adult, a fluoroquinolone (see Section 7.2) might have been the best secondary drug of choice because its target, a type II topoisomerase, is unrelated to cell wall synthesis. Quinolones are not recommended for children, as in this case, because of potential side effects. Other beta-lactam antibiotics, such as the third-generation cephalosporins, may still work on penicillin-resistant *S. pneumoniae*, because the modified antibiotic often can still bind the altered PBP. Nevertheless, cephalosporin-resistant strains of *S. pneumoniae* are now appearing, leaving vancomycin or an oxazolidinone as the best last choice.

Note: Archaeal pseudopeptidoglycan contains talosaminuronic acid instead of muramic acid and lacks the D-amino acids found in bacterial peptidoglycan. Archaea are thus insensitive to penicillins, which interfere with bacterial transpeptidases. This natural resistance is not a problem, because no archaea are known to be pathogens.

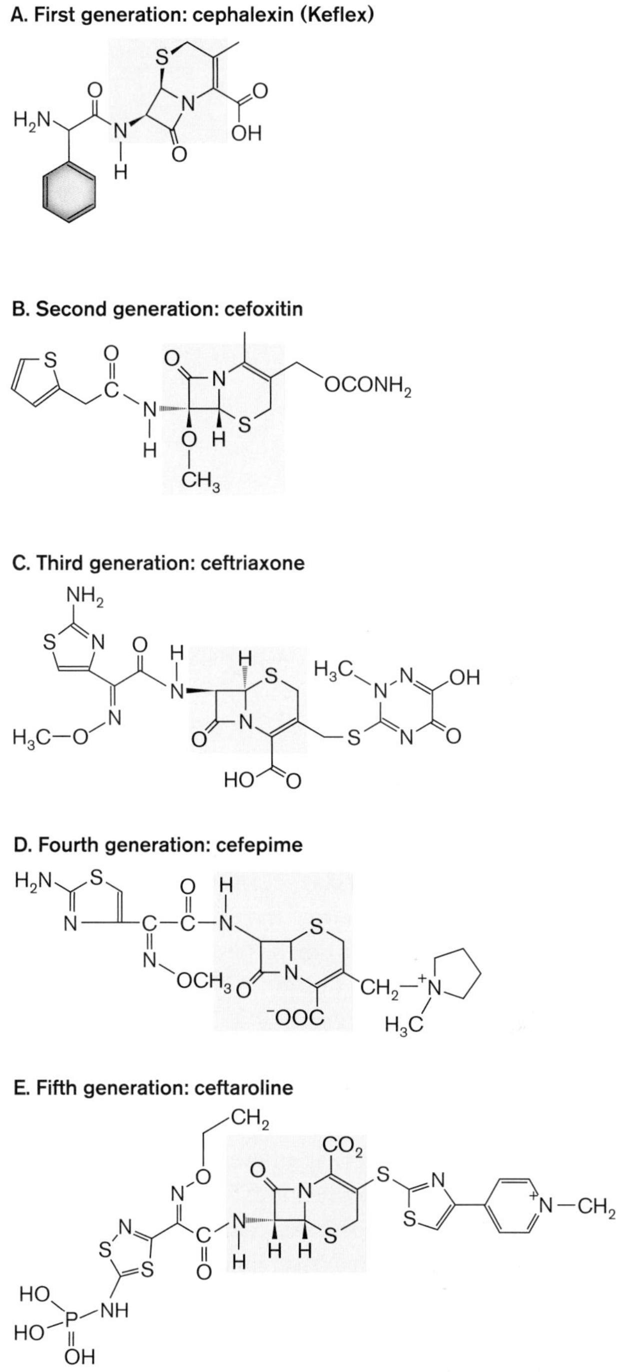

FIGURE 27.9 ■ **Cephalosporin generations.** Representative examples. With each successive generation, the side groups become more complex. Highlighted areas indicate the core structure of each of the cephalosporins, with beta-lactam rings.

Antibiotics that target other steps in peptidoglycan synthesis. Another antibiotic that affects cell wall synthesis is **bacitracin**, a large polypeptide molecule produced by *Bacillus subtilis* and *Bacillus licheniformis* (**Fig. 27.10A**). The antibiotic inhibits cell wall synthesis by binding to the bactoprenol lipid carrier molecule that normally transports monomeric units of peptidoglycan across the cell membrane and to the

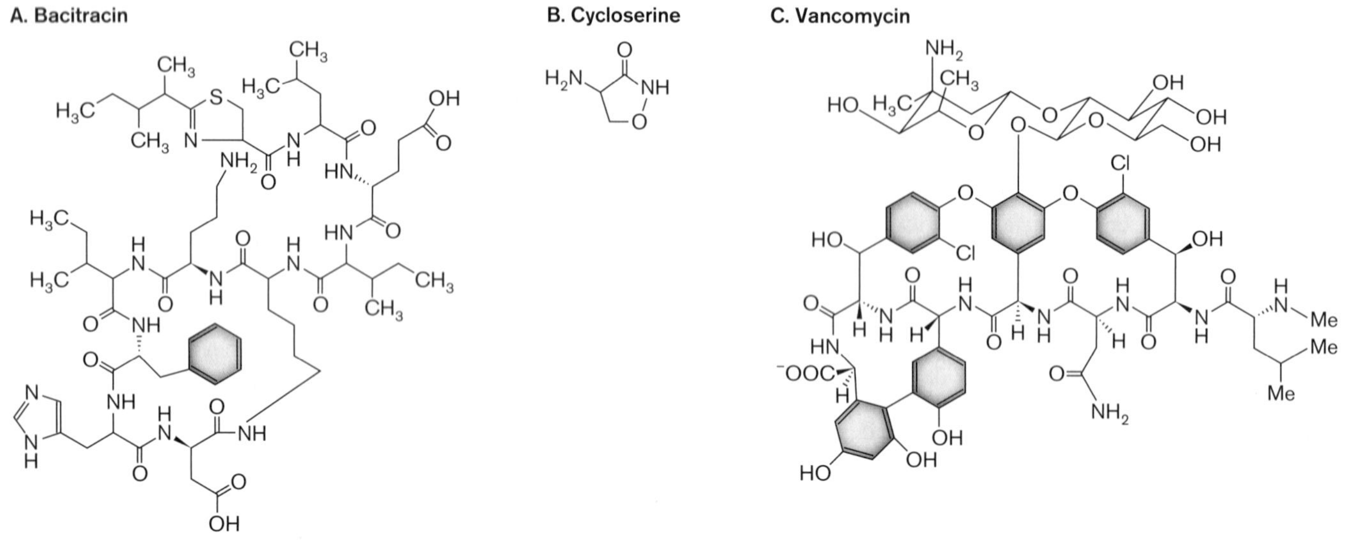

A. Bacitracin

B. Cycloserine

C. Vancomycin

FIGURE 27.10 ▪ **Other antibiotics that affect peptidoglycan synthesis. A.** Bacitracin is produced by *Bacillus subtilis*. It is generally used only topically to prevent infection. **B.** Cycloserine, an analog of D-alanine, is one of several drugs used to treat tuberculosis. **C.** Vancomycin is a cyclic polypeptide made by *Amycolatopsis orientalis*, previously classified as a streptomycete. These antibiotics, especially bacitracin and vancomycin, are synthesized by exceedingly complex biochemical pathways in the producing organisms. Me = methyl.

growing chain (see **Fig. 27.7**). Bacitracin binds to and inhibits dephosphorylation of the carrier, thereby preventing the carrier from accepting a new unit of UDP-NAM. Resistance to bacitracin can develop if the organism can rapidly recycle the phosphorylated lipid carrier molecule through dephosphorylation or if the organism possesses an efficient drug export system (discussed in Section 27.3). Normally, bacitracin is used only topically because of serious side effects, such as kidney damage, which can occur if bacitracin is ingested.

Cycloserine (made by *Streptomyces garyphalus*) is one of several antimicrobials used to treat tuberculosis (**Fig. 27.10B**). Relative to bacitracin, it acts at an even earlier step in peptidoglycan synthesis. Cycloserine inhibits the two enzymes that make the D-Ala-D-Ala dipeptide. As a result, the complete pentapeptide side chain on *N*-acetylmuramic acid cannot be made (see **Fig. 27.7**). Without these alanines, cross-linking cannot occur and peptidoglycan integrity is compromised.

Vancomycin, a very large and complex glycopeptide produced by the streptomycete *Amycolatopsis orientalis* (**Fig. 27.10C**), binds to the D-Ala-D-Ala terminal end of the disaccharide unit and prevents the action of transglycosylases and transpeptidases (see **Fig. 27.7**). The mechanism of resistance is very different for vancomycin and penicillin, which makes vancomycin particularly useful against penicillin-resistant bacteria. To prevent the development and spread of vancomycin-resistant bacteria, this antibiotic is typically used only as a drug of last resort.

Resistance to vancomycin can develop when products from a cluster of *van* genes collaborate to make D-lactate and incorporate it into the ester D-Ala-D-lactate, to which vancomycin cannot bind. Another enzyme in the

van gene cluster prevents the accumulation of D-Ala-D-Ala; as a result, D-Ala-D-lactate replaces D-Ala-D-Ala in peptidoglycan. Peptidoglycan containing D-Ala-D-lactate functions just fine, but the organism is resistant to the antibiotic because vancomycin cannot bind the D-lactate form. However, the antibiotic teixobactin, discussed in Special Topic 4.1, will bind to the D-Ala-D-lactate form and stop growth of vancomycin-resistant Gram-positive bacteria.

Note that antibiotics targeting cell wall biosynthesis generally kill only growing cells. These drugs do not affect static or stationary-phase cells, because in this state the cell has no need for new peptidoglycan.

> **Thought Question**
>
> **27.5** When treating a patient for an infection, why would combining a drug such as erythromycin with a penicillin be counterproductive? (Erythromycin is described in Section 8.3.)

Drugs That Affect Bacterial Membrane Integrity

Poking holes in a bacterial cytoplasmic membrane is an effective way to kill bacteria. There are a few compounds useful in this regard, among them a group called the peptide antibiotics, of which **gramicidin** is an example. Produced by *Bacillus brevis*, gramicidin is a cyclic peptide composed of 15 alternating D- and L-amino acids. It inserts into the membrane as a dimer, forming a cation

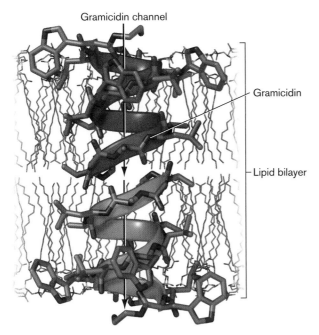

FIGURE 27.11 ■ **Gramicidin is a peptide antibiotic that affects membrane integrity.** As a dimer, gramicidin forms a cation channel across cell membranes through which H^+, Na^+, or K^+ can freely pass. (PDB code: 1GRM)

channel that disrupts membrane polarity (**Fig. 27.11**). Polymyxin (from *Bacillus polymyxa*), another polypeptide antibiotic, has a positively charged polypeptide ring that binds to the outer (lipid A) and inner membranes of bacteria, both of which are negatively charged. Its major lethal effect seems to be to destroy the inner membrane, much like a detergent. These antibiotics are used only topically to treat or prevent infection. Because they can also form channels across human cell membranes, they should never be ingested. Polymyxin has been fused to some bandage materials used to treat burn patients, who are particularly susceptible to Gram-negative infections (for example, *Pseudomonas aeruginosa*). Despite the drug's toxicity, polymyxin or colistin can be injected as a drug of last resort to treat certain multidrug-resistant bacterial infections (for instance, *Klebsiella pneumoniae*).

Daptomycin is a lipopeptide made by *Streptomyces roseosporus* using nonribosomal peptide synthetases. The drug aggregates in the membranes of Gram-positive bacteria to form an ion channel that leaks potassium ions. The resulting membrane depolarization leads to cell death. This drug is very effective against MRSA.

Drugs That Affect DNA Synthesis and Integrity

Bacteria generally make and maintain their DNA using enzymes that closely resemble those of mammals. Thus, you might think it impossible to selectively target bacterial DNA synthesis, but it is possible, as you will see.

Case History: Pneumonia Due to a Gram-Negative Anaerobe

A 23-year-old woman arrived at the emergency room by ambulance with fever, chills, and severe muscle aches. She developed a nonproductive cough, had difficulty breathing, had pleuritic chest pain, and became hypotensive (had low blood pressure). An X-ray showed lower-lobe infiltrate in the lungs, and the clinical laboratory reported the presence of the Gram-negative anaerobe Fusobacterium necrophorum *in blood cultures. The patient was diagnosed with pneumonia and treated with metronidazole, a DNA-damaging agent specific for anaerobes. She fully recovered.*

There are several classes of drugs, including sulfa drugs, quinolones, and metronidazole, that selectively affect the synthesis or integrity of DNA in microorganisms.

Sulfa drugs. The sulfa drugs, originally discovered by Gerhard Domagk, belong to a group of drugs known as antimetabolites because they interfere with the synthesis of metabolic intermediates. Ultimately, sulfa drugs inhibit the synthesis of nucleic acids. Drugs such as sulfamethoxazole or sulfanilamide work at the metabolic level to prevent the synthesis of tetrahydrofolic acid (THF), an important cofactor in the synthesis of nucleic acid precursors (**Fig. 27.12**).

All organisms use THF to synthesize nucleic acids, so why are the sulfa drugs selectively toxic to bacteria? The selectivity occurs because mammals do not synthesize folic acid, a precursor of THF. Higher mammals generally rely on bacteria and green leafy vegetables as sources of folic acid. Bacteria make folic acid from the combination of PABA, glutamic acid, and pteridine. Sulfanilamide (SFA), a structural analog of PABA, competes for one of the enzymes in the bacterial folic acid pathway and inhibits both folic acid and THF production (**Fig. 27.12C**). Because humans lack that pathway, sulfa drugs are selectively toxic toward bacteria.

Quinolones. Another group of drugs inhibits DNA synthesis by targeting microbial topoisomerases such as DNA gyrase and topoisomerase IV (the mechanism is discussed in Section 7.2). Because these enzymes are structurally distinct from their mammalian counterparts, drugs can be designed to selectively interact with them while not interfering with mammalian DNA metabolism. One such drug, nalidixic acid, was discovered in 1963. The drug targets bacterial DNA gyrase but has a very narrow antimicrobial spectrum, covering only a few Gram-negative organisms. However, adding various chemical modifications to nalidixic acid, such as fluorine and amine groups, has increased its antimicrobial spectrum and its half-life in the bloodstream. The result is the class of drugs known as the **quinolones**. (The mode of action of quinolones and fluoroquinolones was discussed in Section 7.2.)

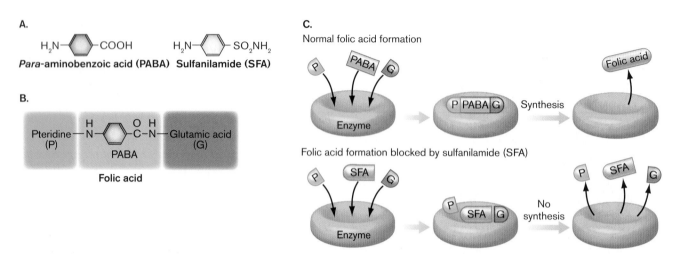

FIGURE 27.12 ■ **Mode of action of sulfanilamides.** **A.** The structures of PABA and sulfanilamide are very similar. **B.** PABA, pteridine, and glutamic acid combine to make the vitamin folic acid. **C.** Normal synthesis of folic acid requires that all three components engage the active site of the biosynthetic enzyme. The sulfa drugs replace PABA at the active site. The sulfur group, however, will not form a peptide bond with glutamic acid, and the size of sulfanilamide sterically hinders the binding of pteridine, so folic acid cannot be made.

Metronidazole. Also known as Flagyl, metronidazole is an example of a prodrug—a drug that is harmless until activated. Metronidazole is activated after it receives an electron (is reduced) from the microbial protein cofactors flavodoxin and ferredoxin, found in microaerophilic and anaerobic bacteria such as *Bacteroides* (**Fig. 27.13**). Once activated, the compound begins nicking DNA at random, thus killing the cell. Because the etiological agent in our case history was an anaerobe, metronidazole was an effective therapy. Metronidazole is also effective against protozoa such as *Giardia*, *Trichomonas*, and *Entamoeba*. Aerobic microbes, although in possession of ferredoxin, are incapable of reducing metronidazole, presumably because oxygen is reduced in preference to metronidazole.

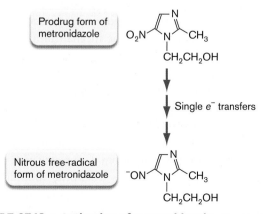

FIGURE 27.13 ■ **Activation of metronidazole.** Single-electron transfers are made by ferredoxin and flavodoxin from anaerobes. Ferredoxin and flavodoxin are reducing agents capable of reducing oxidized molecules in cells such as thioredoxin. They can also reduce the prodrug form of metronidazole.

> **Thought Question**
>
> **27.6** The enzyme DNA gyrase, a target of the quinolone antibiotics, is an essential protein in DNA replication. The quinolones bind to and inactivate this protein. Research has proved that quinolone-resistant mutants contain mutations in the gene encoding DNA gyrase. If the resistant mutants contain a mutant DNA gyrase and DNA gyrase is essential for growth, then why are these mutations not lethal?

RNA Synthesis Inhibitors

The mode of action of antibiotics that inhibit transcription, such as rifampin and actinomycin D (**Fig. 27.14**), was described in Chapter 8. These drugs are bactericidal and are most active against growing bacteria. The tricyclic ring of actinomycin D binds DNA from any source. As a result, it is not selectively toxic and not used to treat infections. Rifampin (also called rifampicin), on the other hand, does exhibit selective toxicity for bacterial RNA polymerase and is often prescribed for the treatment of tuberculosis or meningococcal meningitis. Curiously, because rifampin is reddish orange, it turns bodily secretions, including breast milk, orange. The astute physician will warn the patient of this highly visible but harmless side effect to avoid unnecessary anxiety when the patient's urine changes color.

Until recently, rifampin was the only clinically useful antibiotic known to directly target bacterial RNA polymerase (RNAP). Rifampin binding to RNAP prevents transcription only after the enzyme has started polymerization (it obstructs the exit tunnel for nascent RNA, as described in Chapter 8). Unfortunately, rifampin resistance has developed. However, two new classes of antibiotics have been discovered. Pyronins, represented by myxopyronin (produced

A. Rifampin

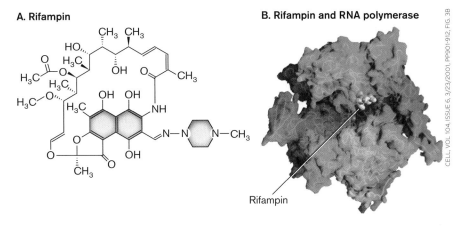

B. Rifampin and RNA polymerase

Rifampin

CELL, VOL. 104, ISSUE 6, 3/23/2001, PP.901-912, FIG. 3B

C. Actinomycin D

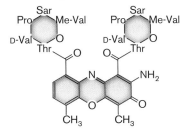

D. Actinomycin D and DNA

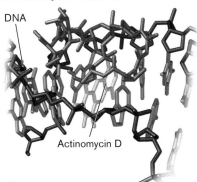

DNA

Actinomycin D

by *Myxococcus fulvis*), prevent RNAP from ever starting polymerization. The pyronins bind to RNA polymerase at a site called the hinge region, which is needed to separate (melt) DNA strands—a requirement to begin transcription.

The second drug, lipiarmycin, is an unusual macrolide antibiotic made by *Actinoplanes deccanensis*. Lipiarmycin binds to the same region as myxopyronin but completely stops closed RNA polymerase–DNA complexes from transitioning to the open forms. Because the binding site for rifampin is different from those of the other two antibiotics, rifampin-resistant RNAP molecules are still sensitive to the new antibiotics. This is exciting because pathogens such as rifampin-resistant strains of *Mycobacterium tuberculosis* or *Clostridioides difficile* can now be treated with a new drug that targets the same enzyme. Lipiarmycin has been approved by the FDA to treat *C. difficile*–associated diarrhea.

FIGURE 27.14 ▪ Antibiotics that inhibit transcription.
A. Rifampin. **B.** Rifampin-binding site on the RNA polymerase beta subunit. cyan = beta subunit; pink = beta-prime subunit; alpha subunits are behind the complex and not shown; the Mg²⁺ ion chelated at the active site is shown as a magenta sphere. **C.** Actinomycin D. **D.** Actinomycin D (yellow and red) interacting with DNA. Covalent intercalation of actinomycin interferes with DNA synthesis and transcription. (PDB code: 1DSC)

Protein Synthesis Inhibitors

The differences between prokaryotic and eukaryotic ribosomes account for the selective toxicity of antibiotics that specifically inhibit bacterial protein synthesis. How various antibiotics inhibit protein synthesis was discussed in Section 8.3. Recall that protein synthesis inhibitors can be classified into several groups based on structure and function (**Fig. 27.15**).

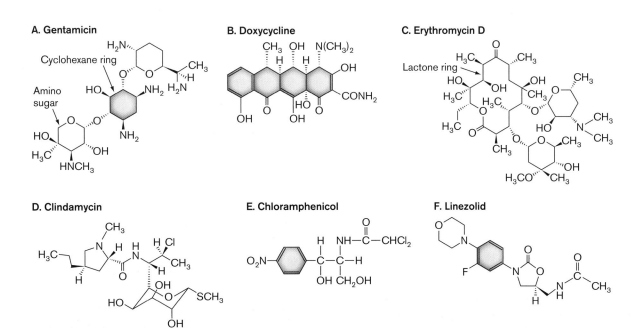

FIGURE 27.15 ▪ Protein synthesis inhibitors. **A.** The aminoglycoside gentamicin. **B.** The tetracycline doxycycline. **C.** The macrolide erythromycin. **D.** The lincosamide clindamycin. **E.** Chloramphenicol. **F.** The oxazolidinone linezolid.

Most of these antibiotics work by binding and interfering with the function of bacterial rRNA, which differs from eukaryotic rRNA. Recall, too, that protein synthesis inhibitors are, by and large, bacteriostatic (not bactericidal).

Case History: Erysipelas in a Penicillin-Sensitive Patient

Sixteen-year-old Jamal arrived at the emergency room after 2 days of fever, malaise, chills, and neck stiffness. His most notable symptom was a painful, red, rapidly spreading rash covering the right side of his face. The rash covered his entire cheek, which was swollen, and extended into his scalp. About 7 days earlier, Jamal had had a severe sore throat. Because it had subsided in 2 days, however, he was not clinically evaluated. Throat cultures taken on admission revealed group A Streptococcus pyogenes, suggesting that the rash was a case of erysipelas caused by this organism. Although penicillin would be the drug of choice, Jamal was known to be allergic to this antibiotic.

When a patient is known to be immunologically sensitive to the usual drug of choice, a structurally distinct drug is best. Often, that drug will be one that inhibits protein synthesis; in this case, the drug chosen was the macrolide azithromycin. Drugs that inhibit protein synthesis can be subdivided into different groups based on their structures and on which part of the translation machine is targeted.

Drugs That Affect the 30S Subunit

The classification of antibiotics affecting protein synthesis is initially based on the bacterial ribosomal subunit targeted. Thus, one class of antibiotics interferes with 30S subunit function, and the other scrambles 50S subunit activities.

Aminoglycosides. There is considerable variation in structure among different aminoglycosides, but all contain a cyclohexane ring and amino sugars (**Fig. 27.15A**). The aminoglycosides are unusual among protein synthesis inhibitors in that they are bactericidal rather than bacteriostatic. Most of them bind 16S rRNA and cause translational misreading of mRNA, which is why these drugs are bactericidal (see another explanation below). The resulting synthesis of jumbled peptides wreaks havoc with physiology and kills the cell.

Streptomycin and gentamicin (**Fig. 27.15A**) are two widely used drugs in this class. Ototoxicity (hearing damage) is a major, but uncommon, side effect of these antibiotics (approximately 0.5%–3% of patients treated with gentamicin suffer from this toxicity). Hearing is generally affected at frequencies above 4,000 Hz. The toxicity of aminoglycosides appears related to their ability to inhibit the function of <u>mitochondrial</u> ribosomes, which are evolutionarily related to bacterial ribosomes. Individuals with specific mutations in mitochondrial rRNA are more susceptible to aminoglycoside toxicity.

Tetracyclines. Tetracycline antibiotics are characterized by a structure with four fused cyclic rings—hence the name. **Figure 27.15B** shows one frequently used example, called doxycycline. Tetracyclines are bacteriostatic and work by binding to and distorting the ribosomal A site that accepts incoming charged tRNA molecules. Doxycycline is used to treat early stages of Lyme disease (caused by *Borrelia burgdorferi*), acne (*Propionibacterium acnes*), and other infections. An important adverse side effect of tetracyclines is that they can interfere with bone development in a fetus or young child. Tetracycline use by pregnant mothers will also cause yellow discoloration of the infant's teeth. As a result, this drug is not recommended for pregnant women or nursing mothers.

Drugs That Affect the 50S Subunit

Five classes of drugs subvert translation by binding to the 50S ribosomal subunit. Most of these drugs were discussed in Chapter 8 and are recapped here only briefly.

- **Macrolides**, all of which contain a 14- to 16-member lactone ring (**Fig. 27.15C**), inhibit translocation of the growing peptide (bacteriostatic action). Commonly prescribed examples are erythromycin and azithromycin. Azithromycin was the antibiotic used to treat the *Streptococcus pyogenes* infection in our case history, although other drugs could have been used. Because it is structurally dissimilar to any of the beta-lactam antibiotics, such as penicillin, it can be used safely in patients who are penicillin sensitive.

- **Lincosamides** (**Fig. 27.15D**), such as clindamycin, are similar to macrolides in function but have a different structure.

- **Chloramphenicol** (**Fig. 27.15E**) inhibits peptidyltransferase activity (bacteriostatic). Bone marrow depression leading to aplastic anemia is the most common serious side effect and limits its clinical use.

- **Oxazolidinones** (**Fig. 27.15F**) are a recently discovered class of synthetic antibiotics effective against many antibiotic-resistant microbes. In fact, this was the first new class of antibiotics discovered since the "golden age" of antibiotic discovery over 35 years ago. Oxazolidinones such as linezolid bind to the 23S rRNA in the 50S subunit of the prokaryotic ribosome and prevent formation of the protein synthesis 70S initiation complex. This is a novel mode of action; other protein synthesis inhibitors either block polypeptide extension or cause misreading of mRNA. Linezolid binds to the 50S subunit near where chloramphenicol binds, but it does not inhibit peptidyltransferase. Resistance is limited because most bacterial

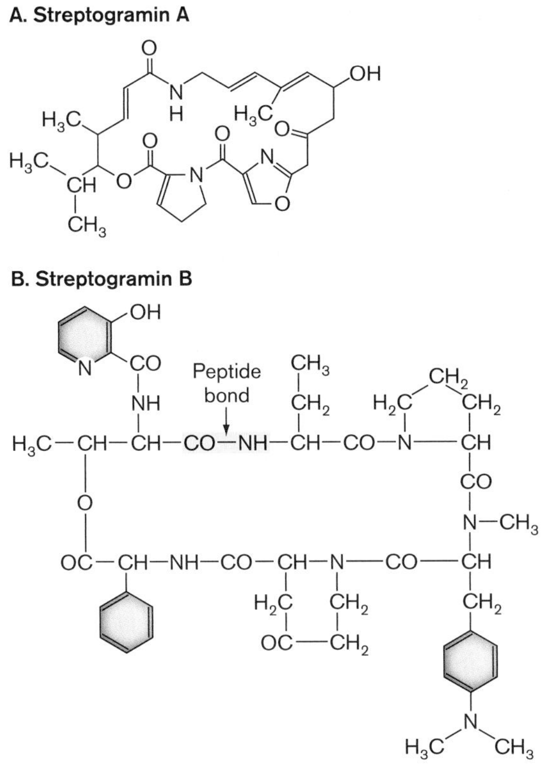

A. Streptogramin A

B. Streptogramin B

FIGURE 27.16 ■ **The streptogramins. A.** Streptogramin A is a large nonpeptide ring structure. **B.** Streptogramin B is a cyclic peptide.

genomes have multiple operons encoding 23S rRNA. Usually more than one of these genes must mutate to confer high-level resistance. The more mutant 23S rRNA genes there are relative to native 23S genes, the more oxazolidinone-resistant ribosomes will be present. Oxazolidinones are useful primarily against Gram-positive bacteria. Gram-negative bacteria are intrinsically resistant because of multidrug efflux pumps (see Section 27.3) and decreased permeability due to the outer membrane.

■ **Streptogramins** (**Fig. 27.16**), produced by some *Streptomyces* species, fall into two groups, designated A and B. Streptogramins belonging to group A have a large nonpeptide ring (**Fig. 27.16A**), whereas streptogramin B members are cyclic peptides (**Fig. 27.16B**). The two groups differ in their modes of action, although both inhibit bacterial protein synthesis by binding to the peptidyltransferase site. Group A streptogramins bind to the peptidyltransferase site and prevent binding of tRNA to the ribosome A site. In contrast, group B streptogramins are thought to narrow the peptide exit channel, preventing exit of the peptide and thereby blocking translocation. Natural streptogramins are produced as a mixture of A and B, the combination of which is more potent than either individual compound alone (an example of synergy). In tribute to this synergistic action, the drug combination is marketed under the name Synercid. Synergy between the two drugs occurs

because the A-type streptogramin alters the binding site for the B-type drug, increasing its affinity. Bacteria can develop resistance through ribosomal modification (the modification in 23S rRNA is the same one that provides resistance to macrolides), via the production of inactivating enzymes, or by active efflux of the antibiotic.

To Kill or Not to Kill: What Makes a Bactericidal Drug?

Bactericidal antibiotics include quinolones that bind to DNA topoisomerases; aminoglycosides that bind to the 30S ribosome subunit; rifampin, which binds to RNA polymerase; and penicillins that inhibit peptidoglycan synthesis. While these antimicrobials do halt critical cell processes, other antibiotics that inhibit some of these same processes are not bactericidal. For example, aminoglycosides and macrolides inhibit protein synthesis, but macrolides do not typically kill bacteria.

A new theory attempts to explain why particular antibiotics are bactericidal. Put simply, a common consequence of the modes of action for bactericidal antibiotics is the generation of highly reactive hydroxyl radicals, which damage DNA, protein, and lipids, leading to cell death. Partial evidence for this model came when James J. Collins and colleagues found that they could make cells more sensitive to bactericidal antibiotics by preventing induction of the SOS response that limits and repairs oxidative damage to DNA (see Section 9.4). The antibiotics would still inhibit growth but not kill the cells. A practical consequence of this finding is that determining how to inhibit the SOS response by pathogens could make them more susceptible to antibiotic therapy.

Thought Question

27.7 Why might a combination therapy of an aminoglycoside antibiotic and cephalosporin be synergistic?

To Summarize

■ **Antibiotic specificity** for bacteria can be achieved by targeting a process that occurs only in bacteria, not host cells; by targeting small structural differences between components of a process shared by bacteria and hosts; or by exploiting a physiological condition such as anaerobiosis present only in certain bacteria.

■ **Antibiotic targets** include cell wall synthesis, cell membrane integrity, DNA synthesis, RNA synthesis, protein synthesis, and metabolism.

■ **Antibiotics targeting the cell wall** bind to the transglycosylases, transpeptidases, and lipid carrier proteins involved with peptidoglycan synthesis and cross-linking.

- **Antibiotics interfering with DNA** include the anti-metabolite sulfa drugs that inhibit nucleotide synthesis; quinolones that inhibit DNA topoisomerases; and a drug, metronidazole, that, when activated, randomly nicks the phosphodiester backbone.

- **Inhibitors of RNA synthesis** target RNA polymerase (rifampin and pyronins) or bind DNA and inhibit polymerase movement (actinomycin D).

- **Aminoglycosides and tetracyclines** prevent protein synthesis by binding the 30S subunit of the prokaryotic ribosome.

- **A variety of antibiotics** bind the 50S ribosomal subunit and inhibit translocation (macrolides, lincosamides), peptidyltransferase (chloramphenicol), formation of the 70S complex (oxazolidinones), or peptide exit through the ribosome exit channel (streptogramins).

27.3 Challenges of Drug Resistance and Discovery

Why do microbes make antibiotics, and how do they avoid killing themselves in the process? The answers provide insight into the origin of antibiotic-resistant pathogens and our fight to halt their spread throughout the world.

Risky Business: Why Do Microbes Make Antibiotics?

Antibiotics are considered **secondary metabolites** because they often have no apparent primary use in the producing organism. This most likely means that a purpose, either current or ancestral, has yet to be identified. Certainly, antibiotic production today can help one microbe compete favorably with another in nature. Antibiotic production can also forge a mutualistic relationship between a microbe and a colonized host by protecting the host from deadly pathogens (as *Streptomyces* does for leaf-cutter ants; see Section 18.3 and **eTopic 17.4**). Whatever their use today, growth inhibition by antibiotics may not have been the original purpose of secondary metabolite production. The complexity of the biosynthetic pathways involved in making antibiotics suggests a more immediate purpose—for example, cell-cell signaling—that evolved into cross-species inhibition. Features of antibiotic biosynthetic pathways are discussed in **eTopic 27.2** (penicillin synthesis) and **eTopic 15.4** (vancomycin biosynthesis).

Given that microbes continue to make antibiotics for a reason, how does the producing microorganism avoid

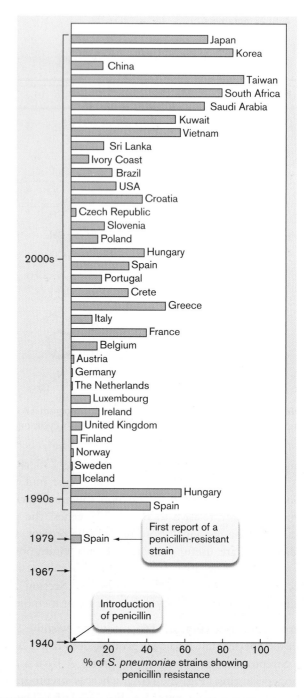

FIGURE 27.17 ▪ The rise of penicillin-resistant *Streptococcus pneumoniae* throughout the world. Numbers reflect the number of penicillin-resistant strains among clinical isolates (strains of disease-causing bacteria isolated from patients from different countries). No resistance among clinical isolates was noted until after 1967.

committing suicide? Fungi that make penicillin face no consequence for having done so because the organism does not contain peptidoglycan. Actinomycetes that produce compounds such as streptomycin or chloramphenicol, however, could be susceptible to their own secondary metabolite. Ribosomes isolated from *Streptomyces griseus*, for example, are fully sensitive to the streptomycin produced by this organism.

S. griseus avoids killing itself in two ways. First, the organism synthesizes an inactive precursor of streptomycin, 6-phosphorylstreptomycin, which is secreted from the cell and, once outside the mycelium, becomes activated by a specific phosphatase. In addition, this streptomycete has an enzyme that inactivates any streptomycin that may leak back into the mycelium. Other organisms protect themselves by methylating key residues on their rRNA to prevent drug binding or by setting up permeability barriers that thwart reentry of the antibiotic.

These and other strategies of self-preservation employed by antibiotic-producing microbes are clever. Unfortunately, these mechanisms have been shared via horizontal gene transfers, making many pathogens antibiotic resistant. Horizontal gene transfers are discussed in Sections 9.6 and 25.2.

Case History: Multidrug-Resistant Pneumonia

A 14-year-old boy with fever (39°C; 102.2°F), chills, and left-sided pleuritic chest pain was referred to a hospital emergency department by his general practitioner. A chest X-ray showed left-lower-lobe pneumonia. The boy reported that he was allergic to amoxicillin and cephalosporins (as a child he had developed a rash in response to these agents) and had been taking daily doxycycline (tetracycline) for the previous 3 months to treat mild acne. He was admitted to the hospital and treated with intravenous azithromycin (a macrolide antibiotic) because of his reported beta-lactam allergies, but he continued to feel sick. The day after admission, both sputum and blood cultures grew Streptococcus pneumoniae. *After 48 hours, antibiotic susceptibility results indicated that the microbe was resistant to penicillin, azithromycin, and tetracycline. Armed with this information, the clinician immediately changed antibiotic treatment to vancomycin. The boy's fever resolved over the next 12 hours, and he made a slow but full recovery over the next week.*

Unfortunately, the scenario presented in this case is far too common and has become an extremely serious concern. **Figure 27.17** shows the rapid rise of penicillin resistance among *Streptococcus pneumoniae* strains in the world. Another instance of emerging antibiotic resistance is unfolding in Europe and the Far East. The non-Enterobacteriaceae Gram-negative rod *Acinetobacter baumannii* is increasingly seen as a dangerous cause of nosocomial infections. It commonly colonizes hospitalized patients, particularly those in intensive care units. Before 1998, there were almost no cases of multidrug-resistant *A. baumannii*. The rate is now

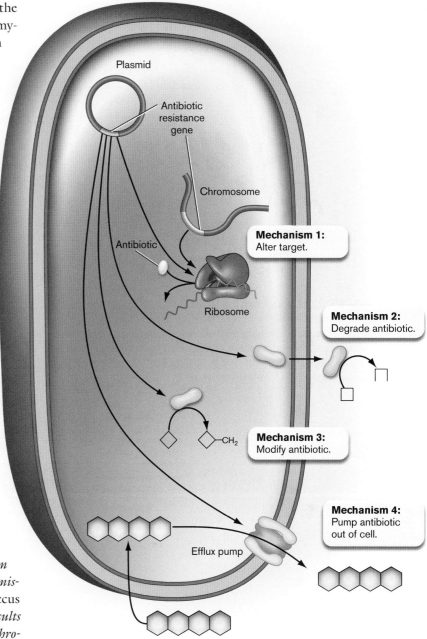

FIGURE 27.18 ■ Alternative mechanisms of antibiotic resistance. Antibiotic resistance genes can be plasmid-borne, or they can be part of the chromosome. Each specific antibiotic resistance gene product will use only one of the four mechanisms shown.

as high as 8%. The organism is resistant to drugs as diverse as ciprofloxacin (a quinolone), amikacin (an aminoglycoside), penicillins, third-generation cephalosporins, tetracycline, and chloramphenicol. Imipenem, one of a relatively new class of beta-lactam drugs, is currently useful, but resistance to it is also likely to develop.

There are four basic forms of antibiotic resistance (**Fig. 27.18**). The resistant organism can:

■ **Modify the target so that it no longer binds the antibiotic.** Mutations in key penicillin-binding proteins and

A.

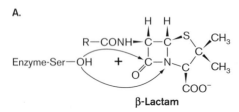

β-Lactam

B.

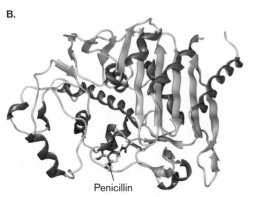

Penicillin

FIGURE 27.19 ■ Destroying penicillin. **A.** Beta-lactamase (or penicillinase) cleaves the beta-lactam ring of penicillins and cephalosporins. There are two types of penicillinases, based on where the enzyme attacks the ring. In either type, a serine hydroxyl group launches a nucleophilic attack on the ring. **B.** Structure of a beta-lactamase and location of the penicillin-binding site. (PDB code: 1XX2)

ribosomal proteins, for instance, can confer resistance to methicillin and streptomycin, respectively. These mutations occur spontaneously and are not typically transferred between organisms.

- **Destroy the antibiotic before it enters the cell.** For example, the enzyme beta-lactamase (or penicillinase) is made exclusively to destroy penicillins (see Section 27.3). The sites of ring cleavage and the structure of the enzyme are illustrated in **Figure 27.19**.

- **Add modifying groups that inactivate the antibiotic.** For instance, there are three classes of enzymes that modify and inactivate aminoglycoside antibiotics. The results of these types of enzyme modifications are illustrated for kanamycin in **Figure 27.20**.

- **Pump the antibiotic out of the cell using specific transporters (for example, tetracycline export) and transport complexes.** This strategy works because the pumps bail drugs out of the cell faster than the drugs can enter. Some are single-component pumps present in the cytoplasmic membrane of Gram-negative and Gram-positive bacteria (for example, NorA in *Staphylococcus*

A.

> Aminoglycoside acetyltransferase (AAC) catalyzes acetyl-CoA-dependent acetylation of an amino group.

AAC (6′)-Ii

CoA-SH
Ac-CoA

FIGURE 27.20 ■ Aminoglycoside-inactivating enzymes. Different enzymes can inactivate aminoglycoside antibiotics.

B.

> Aminoglycoside phosphotransferase (APH) catalyzes ATP-dependent phosphorylation (yellow) of a hydroxyl group.

Kanamycin

APH

ATP   ADP

Kanamycin 3′-phosphate

C.

> Aminoglycoside adenylyltransferase (ANT) catalyzes ATP-dependent adenylylation (yellow) of a hydroxyl group.

Kanamycin

ANT

ATP   P Pi

4′-Adenylylkanamycin

aureus, PmrA in *Streptococcus pneumoniae*, and the TetA and B proteins in Gram-negatives). Other drug efflux pumps are multicomponent systems present in Gram-negative bacteria only (discussed shortly). Efflux in either case is usually energized by proton motive force.

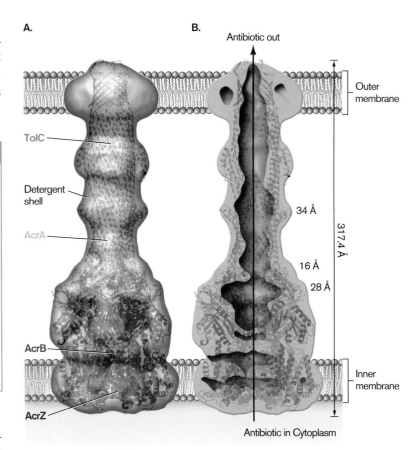

FIGURE 27.22 ■ **Structure of the *E. coli* AcrAB multidrug resistance efflux pump.** **A.** TolC (blue ribbon) and AcrB and Z (orange and green ribbons) are homotrimers linked by 6 protomers of AcrA (yellow ribbons). Transport of antibiotics is driven by proton motive force. **B.** A slice through the model shows the continuous conduit that runs from AcrB through the TolC porin domain, spanning the inner and outer membranes. AcrZ is a small peptide that affects substrate preference.

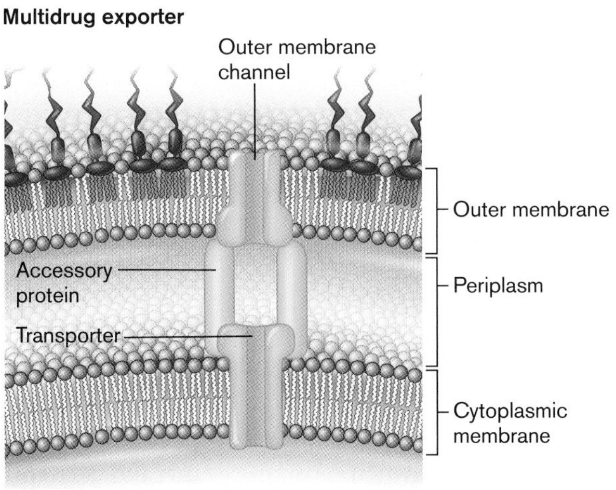

> ## Thought Questions
>
> **27.8** Fusaric acid is a cation chelator that normally does not penetrate the *E. coli* membrane, which means *E. coli* is typically resistant to this compound. Curiously, cells that develop resistance to tetracycline become <u>sensitive</u> to fusaric acid. Resistance to tetracycline is usually the result of an integral membrane efflux pump that pumps tetracycline out of the cell. What might explain the development of fusaric acid sensitivity?
>
> **27.9** Mutations in the ribosomal protein S12 (encoded by *rpsL*) confer resistance to streptomycin. Would a cell containing both *rpsL*$^+$ and *rpsL*^{R} genes be streptomycin resistant or sensitive? (Recall that genes encoding ribosome proteins for the <u>s</u>mall subunit are designated *rps*. "+" indicates the wild-type allele, while "R" indicates a gene whose product is resistant to a certain drug.)

A particularly dangerous type of drug resistance is mediated by what are called **multidrug resistance (MDR) efflux pumps** (**Fig. 27.21**). A single pump in this class can export many different kinds of antibiotics with little regard to structure. MDR pumps of Gram-negative microbes are similar to the ABC export systems described in Section 4.2. They include three proteins: an inner membrane pump protein (fueled by proton motive force—a distinction from true ABC exporters), an outer membrane

channel connected to the pump protein, and an accessory protein that may link the other two proteins. For instance, the AcrAB transporter (**Fig. 27.22**) almost indiscriminately binds antibiotics in a large central cavity within AcrB (a promiscuous binding site) and uses proton motive force to move those compounds through the AcrB pore and out a funnel (AcrA) that connects to an outer membrane channel, TolC.

Antibiotic efflux pumps contribute significantly to bacterial antibiotic resistance because of the very broad variety of substrates they recognize and because of their expression in important pathogens. Strains of the pathogen *Mycobacterium tuberculosis*, for instance, have developed multidrug-resistant phenotypes in part because of MDR pumps. Approximately 2 million people die from tuberculosis annually, mostly in developing nations. What is even more alarming is that an increasing number of *M. tuberculosis* strains isolated from patients exhibit multidrug resistance. Although most of the antibiotic resistance in the majority of *M. tuberculosis* multidrug-resistant strains is

FIGURE 27.21 ■ **Basic structure of a multidrug resistance efflux pump in Gram-negative bacteria.** These efflux systems have promiscuous binding sites that can bind and pump a wide range of drugs out of the bacterial cell.

due to the accumulation of independent mutations in several genes, MDR pumps are thought to increase the level of resistance. Chemists typically try to tweak the structure of an antibiotic to overcome a specific type of resistance mechanism, but the MDR pumps act on an exceptionally wide range of antibiotics, almost without regard to structure.

How Does Drug Resistance Develop?

As discussed in previous chapters, nature has engineered a certain degree of flexibility in the way genomes are replicated and passed from one generation to the next. DNA repair pathways involving lesion bypass polymerases (for example, UmuDC; see Section 9.4) are thought to play a large role in randomized adaptive and evolutionary processes. For instance, at some point during evolution, gene duplication and mutational reshaping generated a gene product able to cleave the beta-lactam ring, producing an organism resistant to penicillin.

However, de novo antibiotic resistance through gene duplication and/or mutation does not occur in all species. Why reinvent the wheel—or, in this case, drug resistance? Gene transfer mechanisms such as conjugation, described in Chapter 9, can move antibiotic resistance genes from one organism to another and from one species to another. In fact, several drug resistance genes found in pathogenic bacteria actually had their start in the chromosomes of the drug-producing organisms and were passed on through gene swapping. For instance, *Streptomyces clavuligerus* produces a beta-lactamase (encoded by the *bla* gene) that protects this organism from the penicillin it produces. Transfer of a drug resistance gene is particularly evident if the gene has been incorporated into a plasmid and that plasmid is found in a new species.

Antibiotic resistance within uncontacted communities. Insight about the origins of antibiotic resistance in our human microbiome was provided in 2015 by Maria Dominguez-Bello (New York University and the University of Puerto Rico) and her collaborators, who reported on the fecal, oral, and skin microbiomes of uncontacted Amerindians in the Amazon. The Amerindians had no previous contact with Westerners who could be a source of antibiotic-resistant microbes, or to antibiotics that could provide selective pressure to develop antibiotic resistance. Without such contact, you might expect the microbiomes of this culture to lack antibiotic resistance genes. You would be wrong.

What the scientists found by DNA sequencing was that bacteria in the Amerindian microbiome did contain antibiotic resistance genes, possibly obtained by exchange with antibiotic-producing soil microbes. To test their functionality, the genes were cloned, placed under the control of a constitutive plasmid promoter to ensure expression, and transferred to *E. coli*. Many of the Amerindian microbiome genes conferred resistance to natural as well as synthetic antibiotics. The authors, however, suspect that many of these genes are silenced in the native microbes of the Amerindian microbiota. Yet introduction of antibiotic selective pressure could readily select for regulatory mutations that activate those genes.

It is also worth noting that the microbiomes of the Amerindians were unprecedented in their high bacterial and functional diversity. The diversity is thought to have developed, at least in part, in response to the more intimate contact that Amerindians have with the natural environment compared to Westerners.

Multidrug-resistant pathogens. An interesting case study of antibiotic resistance is provided by the Gram-positive bacterium *Enterococcus faecalis*, a natural inhabitant of the mammalian gastrointestinal tract that can cause life-threatening disease if granted access to other body sites (as in subacute bacterial endocarditis; see Section 26.6). *E. faecalis* is naturally resistant to numerous antibiotics, making disease treatment particularly challenging.

Vancomycin is one of the last lines of defense for treating serious *E. faecalis* infections. Unfortunately, increasing numbers of vancomycin-resistant strains (called VREs) have arisen in recent years. The completed genome sequence of one vancomycin-resistant strain illustrates the reason. The organism has an incredible propensity for incorporating mobile genetic elements that encode drug resistance. About a quarter of the genome consists of mobile or exogenously acquired DNA, including 7 probable phages, 38 insertion elements, numerous transposons, and integrated plasmid genes. One such mobile element encodes vancomycin resistance in the sequenced strain.

Multidrug resistance in various microbes (for example, *Salmonella enterica*) has also been attributed to the presence of integrons. Integrons are gene expression elements that account for rapid transmission of drug resistance because of their mobility and ability to collect resistance gene cassettes (see **eTopic 9.5**).

We should note that the development of antibiotic resistance is not without consequence to the bacterium. Altering one aspect of an organism's physiology for the better may weaken another area. Bacteria may become resistant to a certain antibiotic, but that resistance comes at a price. For example, the altered DNA gyrase that affords resistance to quinolones may not function as well as the "normal" gyrase does. Thus, when both resistant and susceptible organisms cohabit the same environment, the wild-type (sensitive) strain may grow faster and eventually overwhelm the mutant strain—unless fluoroquinolone is present, of course.

Methods to Identify Drug-Resistant Pathogens

The faster a clinical laboratory can identify a pathogen's antibiotic susceptibility pattern, the more quickly a clinician can prescribe a narrow-spectrum antibiotic. The traditional MIC method described in Section 27.1 can take 3 days—one day to grow the organism on an agar plate from a clinical sample, one day to prep the MIC tubes, and another day for the organisms to grow and for the technician to read the results. The automated MIC can cut one day off of that timeline. Meanwhile, the seriously ill patient is subjected to empiric therapy in which very broad-spectrum antibiotics—sometimes two or three drugs—are used to "cover" as many pathogens as possible until the organism's identity and antibiotic susceptibility pattern are known.

Fortunately, more rapid tests that can provide answers in less than a day are slowly being introduced into the clinical laboratory. For instance, multiplex PCR platforms are available that can detect pathogen-specific or drug resistance genes within an hour directly from respiratory or stool samples. A miniature magnetic resonance machine has also been developed that can detect pathogens at concentrations as low as one organism per milliliter of blood within a few hours. Combining these technologies could mean that a sample brought to the lab at 8 a.m. could be ready by lunch. Such speed could quickly lead to more focused, pathogen-directed therapy. Unfortunately, we have a long way to go before realizing this ideal.

How Did We Get into This Mess?

Consider the following case: A grandmother brings her 4-year-old grandchild to the physician. The child is screaming because he has an extremely painful sore throat. Simply looking at the throat is not diagnostic. The raw tissue could mean the child is suffering from a bacterial infection, in which case antibiotics are needed. Alternatively, a virus could be the cause—a situation in which antibiotics do nothing but pacify the grandparent, or parent. More often than not, the clinician will prescribe an antibiotic without ever knowing the cause of disease. The problem is this: The more an antibiotic is used, the more opportunities there are to select for an antibiotic-resistant organism.

The presence of drug does not <u>cause</u> resistance, but it will kill off or inhibit the growth of competing bacteria that are sensitive, thereby allowing the resistant organism to grow to high numbers. Resistance can form not only in the pathogen itself, but also in a member of the normal microbiota. The danger, then, is that the gene imparting resistance might be horizontally transferred to other bacteria—some of them pathogens. A study in 2016 from the Centers for Disease Control and Prevention found that nearly one in three antibiotic prescriptions in the United States is inappropriate. For acute respiratory infections, only half of the prescriptions for antibiotics were deemed appropriate.

When <u>should</u> antimicrobials be administered? Certainly, in life-threatening situations where time is of the essence, antibiotics should be administered even before the cause of the infection is known. On the other hand, the most prudent course to take when a patient has a simple infection is to confirm a bacterial etiology, and then prescribe. An exception may be in an elderly or otherwise immunocompromised individual, who may be more susceptible to secondary bacterial infections that can occur subsequent to viral disease.

Another proposed source of antibiotic resistance is the widespread practice of adding antibiotics to animal feed. Giving animals subtherapeutic doses of antibiotics in their food makes for larger, and therefore more profitable, animals. The reason antibiotics promote livestock growth is unclear but may be the result of altering the diversity of gut microbiota. Some estimates suggest that 80% of all antibiotics used in the United States are fed to healthy livestock. The consequence is that the animals may serve as incubators for the development of antibiotic resistance. Even if the resistance develops in nonpathogens, the antibiotic genes produced can be transferred to pathogens.

Feeding growth-promoting antibiotics to cattle can also stimulate the spread of pathogenicity genes between bacteria. The antibiotics are added to feed at subinhibitory concentrations that do not kill the bacteria, but stress them instead. Stress can trigger SOS responses (see Chapter 9) that reactivate prophages embedded into bacterial chromosomes. If those phages carry toxin genes, the new phage can transfer toxin production to new strains or species cohabiting the intestine. For example, the Shiga toxin genes (*stx*) carried by certain pathovars of *E. coli* are associated with prophage genomes. Growth-promoting antibiotics such as carbadox [methyl-3-(2-quinoxalinylmethylene)-carbazate-N^1, N^4-dioxide] or monensin (not used in humans) have been shown to reactivate these *stx*-carrying prophages and foster distribution of *stx* to other strains. The unusual O104:H4 Shiga toxin–producing *E. coli* that in 2011 caused a European outbreak of hemolytic uremic syndrome may have been generated by this mechanism. Bradley Bearson of the National Laboratory for Agriculture and the Environment (Ames, Iowa) has also shown that carbadox will induce prophage replication in pathogenic *Salmonella* and *Shigella* strains. The phage can subsequently transfer virulence or antibiotic

A.

© KELLIE WINTER

B.

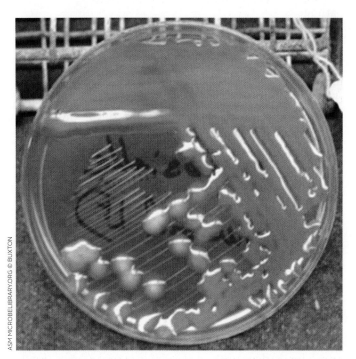

FIGURE 27.23 ■ **Carbadox induces prophage replication in enteric pathogens. A.** Bradley Bearson showed that carbadox, an antibacterial compound formally used to promote growth in swine, activates prophage replication in several enteric bacteria. A phage example is shown on the screen behind him. **B.** Carbadox (0.5 µg/ml) was added to growing cultures of lysogens for 3 hours (closed symbols). At the indicated times, cells were killed with chloroform and supernatants were titrated for phage on susceptible hosts. The data illustrate that carbadox-treated lysogens generated 100- to 1,000-fold higher titers of phage.

resistance genes to other bacteria (**Fig. 27.23**). Efforts to curb the addition of antibiotics to animal feed have been under way since the 1960s.

Many, or all, of these events conspired to produce an incredibly dangerous bacterium that is resistant to almost every antibiotic known. The organism, a *Klebsiella pneumoniae*, was first isolated in 2008 in Örebro, Sweden, from a Swedish citizen returning from New Delhi, India (**Fig. 27.24**). This organism is resistant to most commonly used antibiotics, such as aminopenicillins, beta-lactam/beta-lactamase inhibitors, aminoglycosides, fluoroquinolones, cephalosporins, tigecycline (structurally similar to tetracycline), and carbapenems. It carries the NDM-1 plasmid noted earlier, and numerous other antibiotic resistance genes. Despite attempted treatment with linezolid, the patient died.

The Infectious Diseases Society of America (IDSA) coined the term ESKAPE pathogens almost a decade ago, referring to the six bacterial species that collectively cause about two-thirds of all U.S. nosocomial infections and have effectively escaped the ability to be treated by existing drugs. These bacteria are *Enterococcus faecium*, *Staphylococcus aureus*, *Klebsiella pneumoniae*, *Acinetobacter baumannii*, *Pseudomonas aeruginosa*, and *Enterobacter* species. Considerable effort has been made to identify new antibiotics capable of killing or stopping the growth of these pathogens.

Thought Questions

27.10 Figure 27.24 shows a distinctive-looking colony morphology. Why are the colonies on this agar plate red and mucoid?

27.11 Could genomics ever predict the drug resistance phenotype of a microbe? If so, how?

ASM MICROBELIBRARY.ORG © BUXTON

FIGURE 27.24 ■ **MacConkey agar plate containing a sputum culture of *Klebsiella pneumoniae*.** *K. pneumoniae* carrying the antibiotic resistance gene *bla*$_{NDM-1}$ is emerging as a dangerous, drug-resistant pathogen.

Fighting Drug Resistance

Several strategies are being used to stay one step ahead of drug-resistant pathogens. In some instances, dummy target compounds that inactivate resistance enzymes have been developed. Clavulanic acid, for example, is a compound sometimes used in combination with penicillins such as amoxicillin. Clavulanic acid is a beta-lactam compound with no antimicrobial effect. It is, however, a chemical decoy that competitively binds to beta-lactamases secreted from penicillin-resistant bacteria. Because the enzyme releases bound clavulanic acid very slowly, the amoxicillin remains free to enter and kill the bacterium. Another strategy is to alter the structure of the antibiotic in a way that sterically hinders the access of modifying enzymes. **Figure 27.25** illustrates how adding a side chain to gentamicin, converting it to amikacin, blocks the activity of various aminoglycoside-modifying enzymes. Of course, we are now seeing resistance to amikacin developing (this resistance is ribosome based, involving mutational alteration of the S12 protein or 16S rRNA).

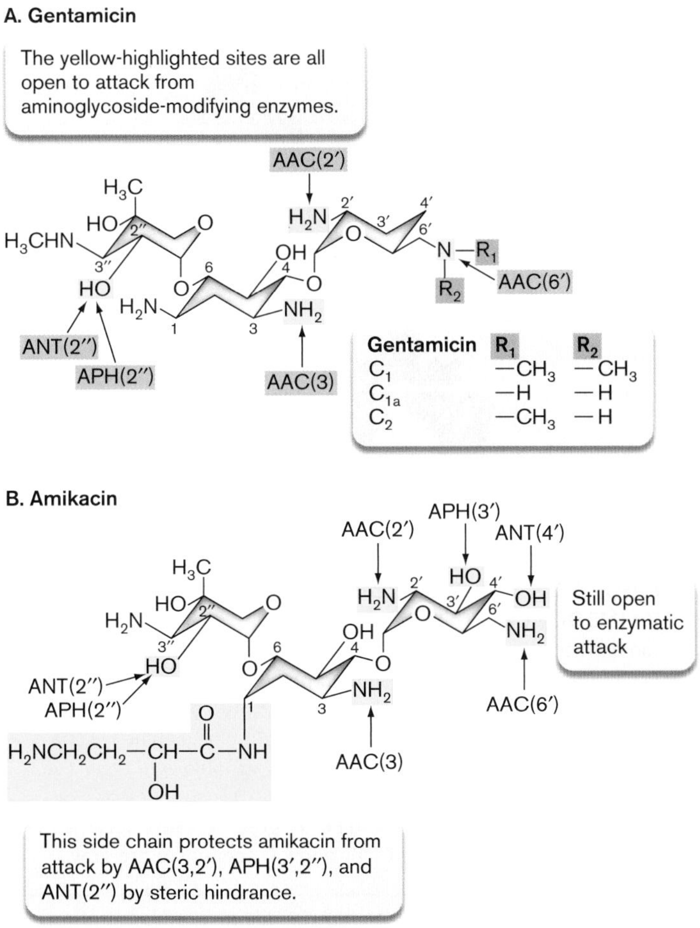

Oxazolidinone   **Quinolone**

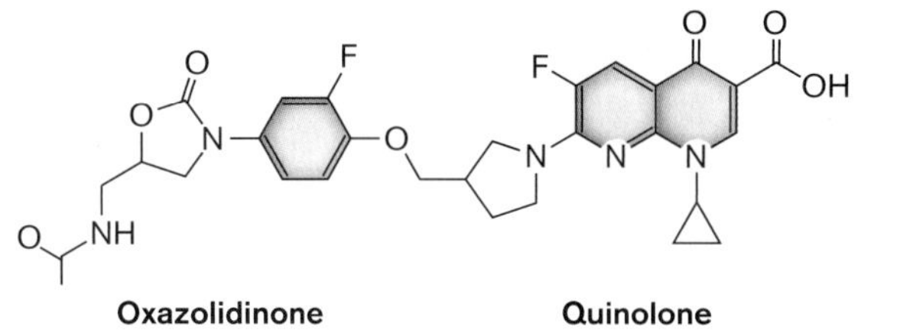

FIGURE 27.26 ■ **Quinolone-oxazolidinone hybrid.** Physically combining two different antibiotics may help reduce the emergence of drug-resistant bacteria.

Linking antibiotics is another strategy currently used to limit resistance. Recent advances have been made in linking a quinolone to an oxazolidinone to form a hybrid antibiotic with dual modes of action (**Fig. 27.26**). Because it has two modes of action, this hybrid antibiotic may limit the development of antibiotic resistance. Here's why: The rate of spontaneous resistance to a given antibiotic is roughly one out of 10^7 cells. For spontaneous resistance to develop to two antibiotics, that probability rises to one out of 10^{14}, making it very unlikely that an organism can become doubly drug resistant. However, multidrug resistance efflux pumps, integron cassettes, and plasmids carrying multiple antibiotic resistance genes can overwhelm that approach.

Scientists are also thinking "outside the box" by developing new, more pathogen-specific, antimicrobials. One such avenue, described in **Special Topic 27.1**, involves synthetically engineering bispecific antibodies that contain binding sites for two different virulence factors.

A. Gentamicin

The yellow-highlighted sites are all open to attack from aminoglycoside-modifying enzymes.

B. Amikacin

This side chain protects amikacin from attack by AAC(3,2′), APH(3′,2″), and ANT(2″) by steric hindrance.

FIGURE 27.25 ■ **Fighting drug resistance. A.** Sites where gentamicin is vulnerable to enzymatic inactivation. AAC = aminoglycoside acetyltransferase; ANT = aminoglycoside adenylyltransferase; APH = aminoglycoside phosphotransferase. Inset shows the R groups for different gentamicin compounds: C_1, C_{1a}, and C_2. **B.** Gentamicin can be chemically modified at the highlighted sites to prevent loss of activity due to enzyme action. The side groups block access to enzyme active sites by steric hindrance (that is, the added groups prevent the active site from interacting with its target structure) but do not inactivate the antibiotic.

Biofilms, Persisters, and the Mystery of Antibiotic Tolerance

Why do some infections return after bactericidal antibiotic treatment is discontinued? The reason is a subpopulation of dormant organisms, called **persister cells**, that arise within a population of antibiotic-susceptible bacteria. The stalled metabolism of persisters renders them tolerant to bactericidal antibiotics during treatment. Removing the drug allows persisters to grow and reestablish infection. The strategy is analogous to a hiker's tactic, on accidentally encountering a bear in the woods, of "playing dead" to avoid being attacked.

Persistence is a long-recognized mystery of microbiology. Joseph Bigger in 1944 noticed that penicillin would lyse a growing culture of *Staphylococcus aureus*, but a small number of persister cells always survived. These persisters were not mutants made permanently resistant through mutation. They acted as though dormant. Persister cells that tolerate antibiotic treatment can be found in any biofilm or population of late-exponential-phase cells. In addition to causing antibiotic treatment failures, persistence may be the reason for latent bacterial infections such as recrudescent typhus or latent tuberculosis.

SPECIAL TOPIC 27.1   Are Designer Antibodies the Next Antibiotics?

The use of broad-spectrum antibiotics to treat an infectious disease could be called the "nuclear option" of antimicrobial therapy. Not only do these antibiotics kill the intended pathogen, but they also wipe out much of our microbiota. Nearly indiscriminate clinical use of broad-spectrum antibiotics has spawned widespread resistance to all classes of antimicrobials, sparking fear that we are fast approaching a postantibiotic world in which traditional antibiotics no longer work. The situation, though dire, has fostered interest in developing new, more focused, pathogen-specific antimicrobial strategies to treat infections—something that would minimize the use of broad-spectrum antibiotics. One such option is monoclonal antibody technology.

A monoclonal antibody (mAb) originates from a single antibody-producing B-cell clone and is specific for a single epitope. Once identified in the laboratory, a B cell that produces a desired antibody is fused to an immortal line of cancer cells. The result, called a hybridoma, can live indefinitely and secrete antigen-specific monoclonal antibodies on demand. Monoclonal antibodies directed against virulence factors can disrupt a target microbe's pathogenesis and facilitate clearance of the pathogen by the immune system (for example, via opsonization).

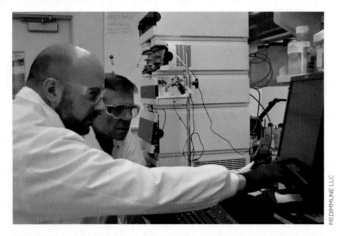

FIGURE 1 ■ **Developing bispecific antibodies.** MedImmune's senior director Kendall Stover (right) and senior scientist Antonio DiGiandomenico developed the bispecific antibody directed against *Pseudomonas aeruginosa*.`

A recent advance in this technology is the ability to construct <u>bispecific</u> antibodies that can bind two <u>different</u> epitopes. Bispecific antibodies are usually made by engineering single-chain variable fragment (scFv) domains that are specific for different epitopes and fusing them to the free termini of mAb heavy-chain or light-chain sequences. Because two different epitopes are targeted on a pathogen, bispecific antibodies should be more effective antimicrobials than are monoclonal antibodies.

C. Kendal Stover and colleagues at MedImmune, LLC, in Gaithersburg, Maryland (**Fig. 1**), have constructed novel bispecific antibodies that target the dangerous, multiantibiotic-resistant pathogen *Pseudomonas aeruginosa*, a major cause of lung and burn infections. The chimeric antibody, called BiS4αPa (or MEDI3902), is shown in **Figure 2**.

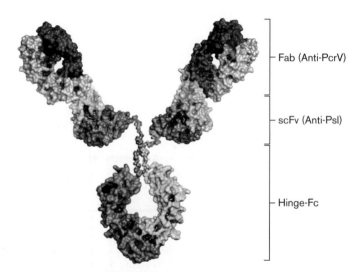

FIGURE 2 ■ **Anti-*Pseudomonas* bispecific antibody BiS4α Pa (MEDI3902).** This bispecific antibody has binding sites for Psl (an extracellular sugar polymer implicated in biofilm formation) and PcrV (a component of the *Pseudomonas* type III secretion system). It also includes an Fc region that can promote opsonization and complement binding. Fab light chain is light blue; Fab heavy chain is dark blue; PcrV binding site is red; scFv variable light chain (VL) is light orange and variable heavy chain (VH) is dark orange; Psl-binding site is salmon; hinge and Fc are green. Linkers between VH and VL in scFv and between scFv and IgG sequences are gray.

The mechanisms that cause persistence in the face of bactericidal antibiotics are varied. The laboratory of Kim Lewis recently published evidence that stochastic depletion of ATP in *Staphylococcus aureus* cells underlies persister formation. In Gram negative bacteria so-called toxin-antitoxin modules play a role. One example involves the *hipA* and *hipB* genes in uropathogenic *E. coli*. HipA is the toxin part of a toxin-antitoxin module (see **eTopic 10.4**)

in which the antitoxin, HipB, binds to and neutralizes HipA. A delicate counterbalance between HipA and HipB levels allows cells to grow normally. However, because HipB antitoxin is less stable than HipA, a portion of HipA can become active if antitoxin synthesis lags. As it is freed of antitoxin, HipA will phosphorylate and inactivate some EF-Tu molecules, thereby slowing translation (Chapter 8 explains the role of EF-Tu in translation). Active HipA

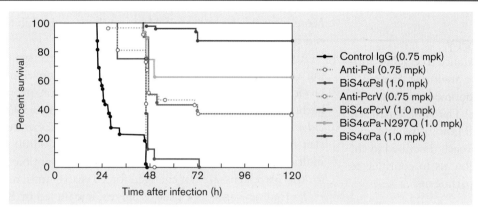

FIGURE 3 ▪ Effect of bispecific antibody BiS4αPa on survival in an acute-pneumonia model. Mice were treated with molar-equivalent doses of the indicated antibodies (mpk = milligrams per kilogram of body weight) followed by intranasal infection with wild-type *Pseudomonas* strain 6206 (1.1×10^6 CFUs). Mouse survival results were compiled from six independent experiments. BiS4αPa-N297Q contains an amino acid substitution of glutamine (Q) for asparagine (N) at residue 297 in the Fc region. The mutation diminishes Fc binding to phagocyte Fc receptors and complement.

BiS4αPa includes antibody-binding sites for PcrV (a component of the *Pseudomonas* type III secretion system) and Psl (an extracellular sugar polymer that coats *Pseudomonas* and is implicated in biofilm formation). The bispecific antibody was constructed by taking the monoclonal antibody against PcrV and genetically inserting a Psl-binding scFv domain within the hinge region connecting the anti-PcrV Fab segment to the Fc region.

BiS4αPa was then shown to inhibit *Pseudomonas* biofilm formation by blocking Psl, as well as by interfering with the secretion of *Pseudomonas* effector proteins by blocking the type III secretion system component, PcrV. Furthermore, because of its Fc region, the bispecific antibody promotes opsonization of *Pseudomonas* by phagocytes. All of these properties should translate into effective antipseudomonad activity for the bispecific antibody. **Figure 3** shows that this antibody readily protected mice from *Pseudomonas* in a mouse model of acute pneumonia. Notice, too, how much better than anti-PcrV alone BiS4αPa was at protecting mice.

Why was the bispecific antibody better than the monospecific anti-PcrV at protecting mice? The bispecific antibody initially binds to the high-abundance Psl polysaccharide coating the organism and produces a "cloud" of antibody. The antibody cloud increases the likelihood that the anti-PcrV binding site of the bispecific antibody will find and inactivate the less abundant surface PcrV molecules. This idea was tested by comparing how well BiS4αPa and anti-PcrV prevented the *Pseudomonas* type III secretion–dependent killing (cytotoxicity) of host cells. *P. aeruginosa* injects a phospholipase (ExoU) through its type III secretion system into host cells. ExoU then kills host cells by lysing host cell membranes. Consistent with the "cloud" hypothesis, the bispecific antibody prevented the cytotoxic effect at far lower concentrations than did monospecific anti-PcrV antibody.

These studies suggest that pathogen-targeted, bispecific antibodies could be used in at least two ways: prophylactically in severely ill patients at high risk for contracting lethal infections, or as a direct treatment for acute drug-resistant *Pseudomonas* infections for which few antibiotic options remain. Who knows? Designer antimicrobial antibodies may, indeed, be our next-generation antibiotics.

RESEARCH QUESTION

How would you test whether *Pseudomonas aeruginosa* can develop resistance to BiS4αPa? Discuss why, or why not, resistance could develop.

DiGiandomenico, Antonio, Ashley E. Keller, Cuihua Gao, Godfrey J. Rainey, Paul Warrener, et al. 2014. A multifunctional bispecific antibody protects against *Pseudomonas aeruginosa*. *Science Translation Medicine* **6**:262ra155.

will also phosphorylate and inactivate glutamyl-tRNA synthetase. The lack of charged glutamyl-tRNA also stalls translation, but in a way that activates synthesis of (p)ppGpp/ppGpp by RelA (part of the stringent response discussed in Chapter 10). Elevated ppGpp stimulates a protease called Lon protease to degrade a series of ten or more different antitoxin proteins, including HipB. The toxins that are unleashed encode mRNA nucleases that

further inhibit translation and cell growth. The result is a dormant cell.

How might dormancy explain antibiotic tolerance? If persisters are dormant and have little or no cell wall synthesis, translation, or topoisomerase activity, then even if bactericidal antibiotics can bind to their targets, target function cannot be corrupted. Tolerance, then, provides antibiotic resistance at the price of not growing. An antibiotic can kill all susceptible bacteria in an infection, but the remaining persister cells serve as a source of population regrowth (and reinfection) once the antibiotic is removed.

The Future of Drug Discovery

The pervasive nature of bacterial resistance to current antibiotics has led many to declare that we are in the "postantibiotic era." But is humankind really doomed to a future in which antibiotics will no longer work? The hope is that the prudent use of current antibiotics and innovative strategies for finding new ones, such as the approach described in the Current Research Highlight, will allow us to continue to effectively control evolving bacterial pathogens.

How do you find new antibiotics? Certainly, the classic approach—in which microbes, plants, and even animals collected from around the world are screened for their abilities to make new antibiotics—is still valid and remains a fruitful source of new potential drugs. Even previously unculturable bacteria have been screened for new antibiotics, as described in Special Topic 4.1.

The antibiotic platensimycin, made by *Streptomyces platensis* is a recent example in which old brute-force screening of natural products led to a new antibiotic. The screening method, although laborious, was novel. Merck scientists screened 250,000 natural product extracts for an ability to specifically inhibit bacterial fatty acid biosynthesis. Fatty acid biosynthesis is an attractive target for antibiotic development because the process and proteins involved are different from those of eukaryotes. The strategy specifically targeted the protein FabF because it is an essential component of fatty acid synthesis and is conserved among key pathogens such as *Staphylococcus aureus.*

Scientists engineered a strain of *S. aureus* to contain a gene expressing an antisense RNA to *fabF* mRNA. When the antisense RNA was induced, it bound to *fabF* mRNA and prevented efficient translation. As a result, the level of FabF protein in the cell decreased. The strain could still grow but would be exquisitely sensitive to any compound that targeted the remaining FabF protein. Fewer molecules of FabF present per cell means that fewer molecules of an active ingredient in a natural product are needed to stop fatty acid synthesis. Thus, zones of inhibition on an agar plate will be wider than they would be if cells produced normal amounts of FabF. This novel screening method led to the discovery of platensimycin.

Platensimycin binds FabF and exhibits bacteriostatic, broad-spectrum activity, acting on Gram-positive and Gram-negative bacteria. It is only the third entirely new antibiotic developed in the last four decades (daptomycin and linezolid, an oxazolidinone, are the other two). The novel chemical structure of platensimycin and its unique mode of action provide a great opportunity to develop a new class of critically needed antibiotics—a class that selectively targets fatty acid biosynthesis. However, as of this writing, the antibiotic has not won FDA approval for use. More information on fatty acid biosynthesis and inhibitors can be found in Chapter 15.

Another recent success story paired brute-force screening with combinatorial chemistry to identify a novel antibiotic. Almost 30% of Earth's humans (2 billion people) harbor latent tuberculosis infections (*Mycobacterium tuberculosis*). What makes the situation even more desperate is that of the 9 million new TB cases each year, 500,000 are multidrug-resistant (MDR). Unfortunately, the antibiotics for this pathogen have not changed for 40 years—until now.

To find new anti-tuberculosis agents, a team led by Belgian scientist Koen Adries screened 70,000 compounds for antimicrobial effects on *Mycobacterium smegmatis*, a fast-growing relative of *M. tuberculosis*. One compound that significantly affected growth was chemically modified to increase its efficacy. After exhaustive clinical trials, the new drug, now called bedaquiline, was approved by the FDA in 2013 to treat MDR-TB. The antibiotic selectively targets the organism's energy-generating ATP synthase—a novel mode of action—and starves the pathogen of energy. The hope is that this new antibiotic can finally stem the rising tide of TB. Tempering that hope, however, is the knowledge that bedaquiline-resistant mutants of *M. tuberculosis* have already been isolated in the laboratory.

Newer strategies of drug discovery have centered on genome sequence analysis and associated genetic techniques to identify potential bacterial molecular targets. Once a target is identified, clever screening techniques are used to find natural antibiotics, and molecular modeling is used to synthetically design potential inhibitor molecules. High-throughput biochemical screens of large collections of synthetic chemicals have also been attempted. Although many promising lead drugs have been identified, unfortunately only a rare few have proved therapeutically useful. Examples of metagenomic and peptidogenomic approaches to drug discovery are described in eTopic 15.4 and Special Topic 15.1, respectively.

So now what? The functions of genes that constitute potential new drug targets fall into three broad categories: those required only for growth of bacteria in laboratory media (in vitro–expressed genes), those required only for bacterial infection (virulence genes or in vivo–induced genes; the latter are discussed in eTopic 25.2), and those required for bacterial growth both in vitro and in vivo (essential housekeeping genes). The second and third categories clearly constitute new potential drug targets, but there is also a subcategory of in vitro–expressed genes that could be drug targets. These are conditionally essential genes needed only under some in vitro conditions (for instance, minimal defined medium) and some in vivo conditions.

A recent example showing that conditionally essential genes might be useful as drug targets was reported by Marvin Whiteley at the University of Texas at Austin. His group studies *Pseudomonas aeruginosa*, a potentially lethal lung pathogen that plagues cystic fibrosis patients. This bacterium readily grows in the sputum of these patients, so the scientists wondered which bacterial genes might be essential for growth in this substance. Using a massive pool of Tn5 transposon insertions, Whiteley's group discovered that insertion mutants defective in vitamin production grew poorly, if at all, in CF sputum. CF sputum must have a vitamin deficit that forces the microbe to synthesize those cofactors. The inference is that *Pseudomonas* enzymes required for vitamin synthesis are rational targets for small inhibitors of molecules. Such inhibitors would limit growth of *Pseudomonas* in vivo but have no effect on growth in the nutrient media typically used to identify new antibiotics and determine MIC values.

Virulence genes and proteins of pathogens hold great promise as targets for new drugs because these genes, too, are required for in vivo growth. However, these proteins are also conditionally essential, so inhibitors fail to inhibit growth in vitro. Virulence proteins that can be potential drug targets are easy to find by measuring loss of virulence when the gene is missing (for instance, the gene encoding the tip of a type III secretion system), but how do you screen for specific inhibitors of these proteins? New high-throughput screening platforms that measure <u>in vivo</u> growth need to be developed to find chemical inhibitors of virulence. Currently, it is easier to make monoclonal antibodies that target virulence proteins than it is to design and find chemical inhibitors. Monoclonal antibodies are now being tested as pathogen-specific antimicrobials (see **Special Topic 27.1**).

Other intriguing ideas may lead to novel antimicrobial therapies. One involves using photosensitive chemicals that can penetrate the microorganism and generate toxic reactive oxygen species (such as superoxide) when exposed to specific wavelengths of visible light (obviously good only for topical use). Another clever approach is to interfere with quorum-sensing mechanisms of pathogens (see **eTopic 27.3**). Finally, the promise and flexibility of synthetic-biology approaches discussed in Chapter 12 could revolutionize the process of antibiotic discovery and production.

To Summarize

- **Antibiotic resistance** is a growing problem worldwide.
- **Certain microbes make antibiotics** to eliminate competitors in the environment and prevent self-destruction

by means of various antibiotic resistance mechanisms. Genes encoding some of these drug resistance mechanisms have been transferred to pathogens.

- **Mechanisms of antibiotic resistance** include modifying the antibiotic, destroying the antibiotic, altering the target to reduce affinity, and pumping the antibiotic out of the cell. **Multidrug resistance efflux pumps** use promiscuous binding sites to bind antibiotics of diverse structure.

- **Antibiotic resistance can arise spontaneously** through mutation, can be inherited by gene exchange mechanisms, or can arise de novo through gene duplication and mutational reengineering.

- **Indiscriminate use of antibiotics** has significantly contributed to the rise in antibiotic resistance.

- **Measures to counter antibiotic resistance** include synthetically altering the antibiotic, using combination antibiotic therapy, and adding a chemical decoy.

- **Persister cells** in a population have stopped actively growing, making them tolerant to bactericidal antibiotics.

- **The quest to discover novel antibiotics** includes **designing candidate antimicrobial compounds** to interact with and inhibit the active site of a known microbial enzyme, and **screening previously unculturable microbes** for new antibiotics.

- **Potential targets for rational antimicrobial drug design** include proteins expressed only in vivo or proteins expressed both in vivo and in vitro.

27.4 Antiviral Agents

A father pleading with a physician to give his child antibiotics when the infant is suffering with a cold is an all-too-common dilemma faced by the general practitioner, but there is nothing of substance the physician can do. The common cold is caused by the rhinovirus, and no antibiotic designed for bacteria can touch it. Why are there so few antiviral agents in the clinician's arsenal? The reason is that applying the principle of selective toxicity is much harder to achieve for viruses than it is for bacteria. Viruses routinely usurp host cell functions to make copies of themselves. Thus, a drug that hurts the virus is likely to also harm the patient. Nevertheless, there are several useful antiviral agents for which selective viral targets have been found and exploited. Some of these agents are listed in **Table 27.3**. Select examples are discussed in this section. All of the molecular mechanisms of viruses presented in Chapter 11 are studied as potential drug targets.

TABLE 27.3	Examples of Antiviral Agents		
Virus	**Agent**	**Mechanism of action**	**Result**
Influenza virus	Amantadine	Inhibits viral M2 protein	Prevents viral uncoating
	Zanamivir	Neuraminidase inhibitor (nasal spray)	Prevents viral release
	Oseltamivir (Tamiflu)	Neuraminidase inhibitor (oral prodrug)	Prevents viral release
Herpes simplex virus and varicella-zoster virus (shingles)	Acyclovir	Guanosine analog	Halts DNA synthesis
	Famciclovir	Prodrug of penciclovir, a guanosine analog	Halts DNA synthesis
Cytomegalovirus	Ganciclovir	Similar to acyclovir	Halts DNA synthesis
	Foscarnet	Analog of inorganic phosphate	Binds and inhibits virus-specific DNA polymerase
Respiratory syncytial virus and chronic hepatitis C virus	Ribavirin	RNA virus mutagen	Causes catastrophic replication errors
Hepatitis C virus	Sofosbuvir (Sovaldi)	Analog inhibitor of HCV polymerase	Halts RNA synthesis
	Simeprevir (Olysio)	Protease inhibitor	Prevents viral maturation
HIV	Zidovudine (AZT)	Nucleoside analog; resembles thymine	Inhibits reverse transcriptase
	Nevirapine	Binds to allosteric site	Inhibits reverse transcriptase
	Nelfinavir	Protease inhibitor	Prevents viral maturation
	Raltegravir	Integrase inhibitor	Prevents integration into host genome
	Maraviroc	CCR5 entry inhibitor	Prevents virus entry into host cells

Antiviral Agents That Prevent Virus Uncoating or Release

Membrane-coated viruses are vulnerable at two stages: when the virus is invading the host cell and after viral propagation, when the progeny viruses release from the host cell. The flu virus presents a good example of both.

Case History: Antiviral Treatment of Infant Influenza

A 9-month-old infant arrived at the Johns Hopkins Hospital with an acute onset of fever, cough, regurgitation from his gastrostomy feeding tube, and dehydration. This illness followed a series of chronic problems, including bronchiolitis (infection and inflammation of the bronchioles) caused by respiratory syncytial virus, and neonatal group B streptococcal sepsis. (Neonatal sepsis is often caused by Lancefield group B Streptococcus agalactiae; Lancefield group classification is described in Chapter 28.) Physical exam revealed fever, a severe cough resulting in respiratory distress, a rapid heart rate, and moderate dehydration. Nasopharyngeal aspirate was positive for influenza A antigen. The patient was treated with oseltamivir when influenza was diagnosed. He gradually improved and was discharged home 4 days after admission.

In 2003, an unusually severe form of influenza (H3N2) spread across the United States. Most states reported a higher-than-normal number of influenza-related deaths of children and young adults. Several factors, however, helped keep the outbreak from becoming an epidemic of larger proportions. First, the administration of flu vaccine afforded the population what is called herd immunity (discussed in Section 24.6). Herd immunity occurs when only a portion of the population is immunized. The vaccinated individuals will not become infected and so cannot spread the disease to others. At the very least, this slows the progression of infection throughout the population.

Note: It is impossible to immunize all humans against any given disease. However, it is estimated that immunizing about 80% of a population for influenza can halt an epidemic by cutting off transmission. Unfortunately, this level of immunization is rarely achieved.

The second factor that prevented an influenza pandemic was the availability of antiviral agents that can limit the disease course. As a result of studying the molecular biology of influenza, scientists discovered two selective targets. Influenza virus (200 nm) is encased in a membrane envelope donned when the virion buds from an infected cell. As described in Section 11.2, the envelope contains the viral proteins neuraminidase (NA) and hemagglutinin (HA). Spikes of hemagglutinin bind to sialic acid receptors on the host cell and trigger receptor-mediated endocytosis. The virus ends up inside the resulting endosome. After

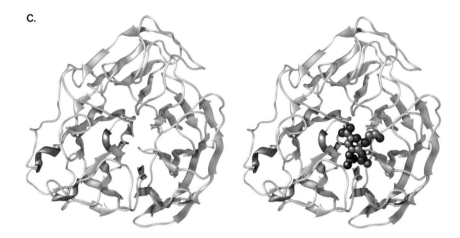

FIGURE 27.27 ■ Inhibitors of influenza proteins. A. Amantadine inhibits the M2 protein. **B.** Zanamivir inhibits neuraminidase. **C.** Neuraminidase without (left) and with (right) bound inhibitor. (PDB: 2HTQ)

the endosome is formed, proton pumps in the membrane acidify endosomal contents. The drop in pH changes the structure of hemagglutinin on the viral membrane so that it can now bind to receptors on the endocytic membrane. The result is fusion of the two membranes and release of the virion into the cytoplasm.

For hemagglutinin structure to change, the <u>interior</u> of the enveloped virion must also be acidified. Acidification is mediated by a membrane channel formed by the virus-encoded M2 protein. The drug amantadine (**Fig. 27.27A**) is a specific inhibitor of the influenza M2 protein that prevents M2 channel formation, which, in turn, prevents viral uncoating. Unfortunately, amantadine-resistant strains of influenza have developed, in part because of the wide-spread use of amantadine by Chinese poultry farmers. As a result, the drug is no longer recommended as a treatment for influenza.

The second target of the influenza virus is the envelope protein neuraminidase. The newer antiflu drugs, such as zanamivir (Relenza) and oseltamivir (Tamiflu), are **neuraminidase inhibitors** that act against types A and B influenza strains (**Fig. 27.27B** and **C**). Neuraminidase on the viral envelope allows virus particles to leave the cell in which they were made. The enzyme cleaves any sialic acid on the cell surface that may have bound to the hemagglutinin on the virus. Unencumbered, the virus can leave the cell surface and infect another cell. Neuraminidase inhibitors prevent sialic acid cleavage, causing the virus particles to aggregate at the cell surface. Surface aggregation reduces the number of virus particles released. The contributions of different NA and HA genes to the severity of influenza are discussed in **eTopic 27.4**.

The neuraminidase inhibitors, when used within 48 hours of disease onset, decrease shedding and reduce the duration of influenza symptoms by approximately one day. However, flu symptoms generally last only 3–10 days. While this does not sound like a substantial benefit, short-ening the course of the flu in the elderly can minimize damage to the lungs, which in turn reduces the chance of

developing life-threatening secondary bacterial infections such as pneumonia and bronchitis.

Antiviral DNA Synthesis Inhibitors

Most antiviral agents work by inhibiting viral DNA synthesis. These drugs chemically resemble normal DNA nucleosides, molecules containing deoxyribose and analogs of adenine, guanine, cytosine, or thymine. Viral enzymes then add phosphate groups to these deoxynucleoside analogs to form deoxynucleotide analogs. The deoxynucleotide analogs are then inserted into the growing viral DNA strand in place of a normal nucleotide. Once inserted, however, new nucleotides cannot attach to the nucleotide analogs, and DNA synthesis stops. These DNA chain–terminating analogs (**Fig. 27.28**) are selectively toxic because viral

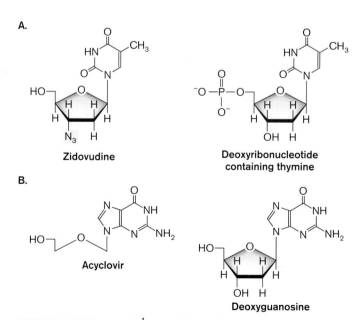

FIGURE 27.28 ■ Antiviral inhibitors that prevent DNA synthesis. Zidovudine (AZT) **(A)** and acyclovir **(B)** are analogs of thymine and guanine nucleotides, respectively. Because the analogs have no 3′ OH to which another nucleotide can add, chain elongation ceases.

polymerases are more prone to incorporate nucleotide analogs into their nucleic acid than are the more selective host cell polymerases. Antiviral DNA synthesis inhibitors work on DNA viruses or retroviruses, but not on viruses such as influenza, with its RNA genome.

Antiretroviral Therapy

Retroviruses are RNA viruses that use viral reverse transcriptase to make DNA and then use viral integrase to insert that DNA into the eukaryotic host cell genome (see Chapter 11). The integrated provirus can then be activated to make retroviral RNA. The retroviral RNA travels to the cytoplasm and directs synthesis of more virus particles. One of the most devastating retroviruses is human immunodeficiency virus (HIV), the cause of acquired immunodeficiency syndrome (AIDS) (see Section 11.3).

Case History: Treatment of HIV

A married couple came to the community clinic for prenatal care. He was 20 years old. She was 19 and reportedly 2 months pregnant with her first child. She denied intravenous (IV) drug use or a history of other sexual partners and had no history of sexually transmitted disease; however, a routine prenatal HIV antibody screen was reported as positive for HIV-1. Careful questioning of the patient and her husband elicited from him a history of IV drug use 5 years earlier. An HIV antibody screen for him was also positive. The laboratory results indicated that the wife might not yet require therapy (she had a low viral load—that is, less than 1,000 copies per milliliter of blood—and a high CD4 T-cell count), but since she was pregnant, a short course of antiretroviral therapy (AZT) would be helpful in preventing transmission of the virus to her child. The husband had an HIV viral load of 10,000 copies per milliliter and was started on combination antiretroviral therapy, including a protease inhibitor.

Being diagnosed with HIV is no longer a death sentence. Advances in antiretroviral therapy (ART) over the past 10 years have transformed HIV into a manageable chronic condition. At least half, if not most, of the HIV-positive people in the United States now live long enough to die from diseases of aging, such as heart attacks or strokes. Because it is such a treatable infection, the CDC even recommends that everyone be tested for HIV. The reason for this recommendation is that many people do not know they are infected, but treating an asymptomatic, HIV-positive person reduces the risk of sexually transmitting the virus. Antiretroviral therapy can also prevent transplacental transmission from an HIV-positive pregnant woman to her fetus, as in our case history. The drugs described next are important components of effective treatment.

Nucleoside, nucleotide, and nonnucleoside reverse transcriptase inhibitors. As an RNA retrovirus, HIV uses a reverse transcriptase to make DNA that then integrates into host nuclear DNA (discussed in Section 11.3). The antiretroviral drug zidovudine (abbreviated ZDV or AZT), which was used in the case history, is a nucleoside analog recognized by reverse transcriptase. Once incorporated into a replicating HIV DNA molecule, the DNA chain–terminating property (lack of a 3′ hydroxyl group) of AZT prevents further DNA synthesis. Nucleoside inhibitors must undergo three successive phosphorylation steps inside the cell to become an active trinucleotide form of the drug. Nucleotide inhibitors are essentially monophosphorylated analogs that require only two phosphorylation steps for activation.

In the case history described, AZT therapy was used to prevent transplacental transmission of the virus from mother to fetus. Because HIV transmission from the mother to the neonate can also occur at delivery or by breast-feeding, treatment of the mother and the child post delivery is important for preventing transmission.

In addition to nucleoside inhibitors, there are also nonnucleoside reverse transcriptase inhibitors. For example, the drug delavirdine binds directly to reverse transcriptase and allosterically inactivates the enzyme.

Protease inhibitors. To make optimal use of its limited provirus DNA sequence, HIV generates long, nonfunctional polypeptide chains that are proteolytically cleaved into the actual proteins and enzymes used to replicate and produce new virions. For example, the *gag* and *pol* genes reside next to each other in the HIV genome and are transcribed as a single mRNA molecule (see Section 11.3). The Gag and Pol open reading frames overlap but are offset by one base. This mRNA produces two polyproteins, called Gag and Gag-Pol, the latter being the result of a shift in reading frames that takes place during translation. Once made, both polyproteins are cleaved by HIV protease. The Gag protein is proteolytically cleaved to make different capsid components (p17, p24, and p15, which is further cleaved to make nucleocapsid protein p7) (**Fig. 27.29A**). Gag-Pol is cleaved to make reverse transcriptase and integrase.

Protease inhibitors such as Viracept and Lopinavir belong to a powerful class of drugs that block the HIV protease (**Fig. 27.29B**). When the protease is inactivated, the polyproteins remain uncleaved and the virus cannot mature, even though new virus particles are made. Because immature HIV particles cannot infect other cells, progress of the disease stalls. Note that protease inhibitors do not cure AIDS; they can only decrease the number of infectious copies of HIV.

A.

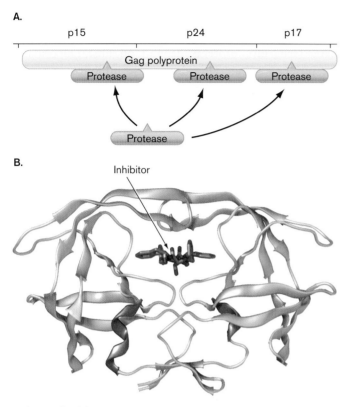

B.

FIGURE 27.29 ■ **HIV protease inhibitor. A.** Representation of HIV protease cleavage of a single Gag polyprotein into multiple, smaller proteins. **B.** The protease enzyme is shown here as a ribbon structure, while the protease inhibitor BEA 369 is shown as a stick model. (PDB code: 1EBY)

Antiviral therapy with protease inhibitors is recommended for patients with symptoms of AIDS and for asymptomatic patients with HIV viral loads above 30,000 copies per milliliter of blood. Treatment should be considered even for patients with viral loads as low as 5,000 copies per milliliter, as for the husband in the case history presented earlier.

Entry inhibitors. Another way to stop HIV is to prevent the virus from infecting cells in the first place. Drugs called entry inhibitors do just that: They stop entry. There are two types of entry inhibitors. CCR5 inhibitors block virus envelope protein gp120 (also known as SU) from binding to host surface protein CCR5, a coreceptor that, together with CD4, is needed for virus binding (see Chapter 11 and Figure 11.24 for details). As a result of CCR5 inhibition, the virus never attaches. Fusion inhibitors, in contrast, do not prevent initial binding, but prevent HIV membranes from fusing with T-cell membranes and thereby halt viral entry. Imagine trying to enter a room through a door. CCR5 inhibitors are like removing the doorknob from the door so there is nothing to grab. Fusion inhibitors, however, are like gluing the door shut. You can grab the knob but still cannot open the door.

Treatment regimens. Because HIV can mutate rapidly and become resistant to single-drug therapies (see Section 11.3),

treatment today involves administering combinations of three or more antiretroviral drugs. This therapeutic strategy is called **highly active antiretroviral therapy (HAART)**. Most current HAART regimens include three drugs—usually two nucleoside reverse transcriptase inhibitors, plus a protease inhibitor, a nonnucleoside reverse transcriptase inhibitor, or an integrase inhibitor. Integrase inhibitors, first approved in the United States in 2007, block the enzyme needed to insert viral DNA into the host genome.

You might wonder why HIV cannot be eliminated from an infected person if the available antiretroviral drugs are so effective. In 2011, Timothy Schaker at the University of Minnesota, Twin Cities, examined patients undergoing ART who had undetectable blood levels of HIV. His group found evidence that the virus in these individuals still remained trapped in lymphatic tissues that were poorly penetrated by the drugs. So, even though ART can lower HIV to undetectable levels in blood, tissue pockets of HIV remain, able to reestablish infection if ART is stopped. New strategies designed to more effectively force drugs into tissues might be the cure we have long awaited.

Remarkably, "cures" of HIV have recently been accomplished through early and aggressive therapy. In 2013, a newborn baby in rural Mississippi was treated aggressively with antiretroviral drugs starting about 30 hours after birth—something that was not usually done (unfortunately, the HIV virus has since been found in this baby). Then, in that same year, it was revealed that a group of 14 adults treated within 4–10 weeks of infection, but whose treatment had been inadvertently interrupted for 3 years, did not relapse while off the drugs. The "cured" patients still had traces of HIV in their blood, but these levels were easily kept in check by their immune systems.

HIV treatment as prevention. AIDS can be a devastating disease. Fortunately, we have effective antivirals that can prevent HIV replication and AIDS. Could treating at-risk populations with antivirals before exposure be effective at preventing infection? This strategy is known as preexposure prophylaxis. In fact, the FDA has approved the daily use of an HIV medicine, Truvada (tenofovir/emtricitabine), by healthy but high-risk people hoping to lower their risk of infection by a sexual partner. Both drugs are nucleoside analogs of adenosine and cytosine, respectively, and are reverse transcriptase inhibitors. Although this is an approved strategy, it is controversial even among physicians. Some fear the drug will encourage risky behavior.

Future Antivirals May Target Host Functions

As you can see from the preceding discussion, most antiviral drugs approved for clinical use target viral proteins

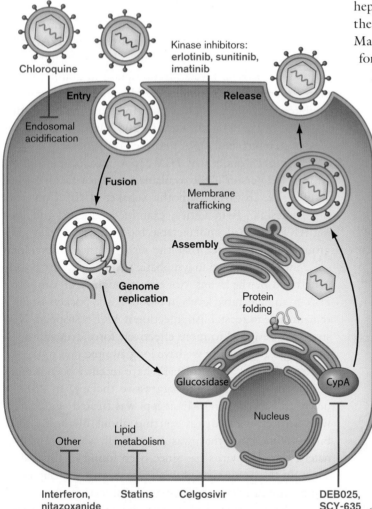

FIGURE 27.30 ■ Potential host targets for broad-spectrum antivirals. Different stages of viral development are shown (entry, fusion, genome replication, assembly). Examples of broad-spectrum compounds are connected to the corresponding targeted proteins or pathways by blunt arrows. CypA = cyclophilin A.

(proteases, polymerases, entry proteins) because they afford some measure of selective toxicity. However, a perceived limitation of these direct-acting antivirals is their narrow spectrum of virus coverage. Narrow-spectrum antivirals cannot provide adequate protection against newer, rapidly emerging viral threats. Examples include flavivirus dengue, coronaviruses SARS-CoV and MERS-CoV, and the filovirus Ebola. Finding new broad-spectrum antivirals could address this need. But what do we target?

A novel approach is to ignore the virus but target host cell pathways that are required by multiple viruses for replication (**Fig. 27.30**). For example, cyclophilin A is involved in host and viral protein folding. Cyclophilin A inhibitors such as alisporivir impair the folding of viral proteins and augment innate immune responses. These drugs affect a variety of DNA and RNA viruses (dengue,

hepatitis C, HIV, SARS-CoV). Another potential target is the enzyme alpha-glucosidase in endoplasmic reticulum. Many virus glycoproteins depend on host glucosidases for proper folding. The glucosidase inhibitor celgosivir has, indeed, proved effective against many unrelated viruses in vitro and in rodent models, but not yet in humans. Nevertheless, the enzyme warrants further investigation. Host kinases that regulate intracellular virus trafficking are also potential drug targets.

A more pathogen-specific antiviral agent (ABX464) affects HIV replication by binding to the host cap-binding complex (CBC). ("Cap" refers to the 5′ 7-methylguanosine cap added to eukaryotic and viral mRNAS.) The CBC controls export of host mRNA from the nucleus to the cytoplasm where mRNA is translated to protein. The HIV Rev protein brings HIV mRNA to the CBC. ABX464 binds to the CBC and specifically blocks Rev-mediated export of viral RNA while not interfering with host transcripts. Because the drug binds to a host protein rather than a viral protein, it is less likely that HIV can develop resistance. This antiviral agent is currently in clinical trials.

Figure 27.30 also shows some host-targeted drugs already approved for other purposes, such as chloroquine (antimalarial agent) and statins (anti-cholesterol metabolism). These drugs are now being evaluated for their usefulness against emerging viruses. Chloroquine inhibits endosome acidification, which is needed by some enveloped viruses to escape the vacuole and enter the host cytoplasm (dengue virus and Zika virus, for instance). Statins interfere with lipid metabolism needed for the life cycles of viruses such as hepatitis C virus. How therapeutically effective these drugs will be as antiviral agents is unclear.

There are many challenges to the host-as-target approach to antiviral therapy. Host proteins function in a complex network of interactions, so elucidating drug mechanisms of actions is difficult. Toxicity is another worry, although it may be possible to identify a therapeutic concentration window in which the drug inhibits virus replication but has minimal toxic effects on the patient.

Thought Question

27.12 The text states that cells and viruses would have difficulty developing resistance to antiviral drugs that target host proteins rather than viral proteins. Why would targeting a host protein decrease the likelihood of developing resistance? Propose a mechanism by which the drug could still become ineffective against a virus.

To Summarize

■ **Fewer antiviral agents** than antibacterial agents are available because it is harder to identify viral targets that provide selective toxicity.

■ **Preventing viral attachment to, or release from, host cells** is a mechanism of action for antiviral agents, such as amantadine and zanamivir, used to treat influenza virus.

■ **Inhibiting DNA synthesis** is the mode of action for most antiviral agents, although they work only for DNA viruses and retroviruses.

■ **HIV treatments** include reverse transcriptase inhibitors that prevent the synthesis of DNA, protease inhibitors that prevent the maturation of viral polyproteins into active forms, and entry inhibitors that either prevent HIV from binding to host membranes or inhibit fusion of the HIV envelop to the host cell membrane.

■ **Future antiviral agents** may target host proteins needed by the virus to replicate. To be useful, such drugs must exhibit concentration-dependent selective toxicity against viral replication rather than host function.

27.5 Antifungal Agents

Fungal infections are much more difficult to treat than bacterial infections, in part because fungal physiology is more similar to that of humans than bacterial physiology is. The other reason is that fungi have an efficient drug detoxification system that modifies and inactivates many antibiotics. Thus, to have a fungistatic effect, repeated applications of antifungal agents are necessary to keep the level of unmodified drug above MIC levels.

Case History: Blastomycosis

A 37-year-old male presented to the emergency department of a Florida hospital with persistent fever, malaise, and a painful right-arm mass. He denied trauma to the arm. White blood cell count was elevated at 27,000/μl, and chest X-ray revealed a left-lung infiltrate. Bronchoscopy revealed granulomatous inflammation containing a single yeastlike mass. Incision and drainage were performed on the arm mass, and cultures were obtained. Serum cryptococcal antigen tests were negative, as were tests for Bartonella henselae *and* Toxoplasma. *Cultures from the right arm grew out a fungal form similar to that identified from the bronchoscopy specimens.*

A tentative diagnosis of Blastomyces dermatitidis *was confirmed using PCR. The patient was placed on amphotericin B, and his fevers and leukocytosis subsequently subsided. His medication was changed to fluconazole for a recommended duration of 6 months.*

Superficial mycoses (fungal infections) such as athlete's foot, and systemic mycoses such as blastomycosis, require very different treatments. Imidazole-containing drugs (clotrimazole, miconazole) are often used topically in creams for superficial mycoses (**Fig. 27.31A**). Others, such as itraconazole, are administered orally. Superficial mycoses include infections of the skin, hair, and nails, as well as *Candida*

FIGURE 27.31 ■ Examples of antifungal agents. A. Clotrimazole belongs to the group of imidazole antifungals, so named because they all contain an imidazole ring. **B.** Griseofulvin is produced by *Penicillium griseofulvum*. **C.** Nystatin is a polyene macrolide produced by *Streptomyces noursei*. **D.** Amphotericin B is a polyene produced by *Streptomyces nodosus*.

TABLE 27.4 **Major Antifungal Agents and Their Common Uses**

| | Clinical applications[a] | | | |
| | Systemic mycoses | | | |
Drug/Mode of administration	Coccidioidomycosis	Histoplasmosis	Blastomycosis	Paracoccidioidomycosis
Polyenes (fungicidal)				
Amphotericin B/Intravenous	+	+	+	+
Nystatin/Oral; topical	−	−	−	−
Natamycin/Topical; eyedrops	−	−	−	−
Azoles (fungistatic): imidazoles				
Clotrimazole/Topical	−	−	−	−
Miconazole (Monistat)/Topical	−	−	−	−
Ketoconazole/Topical	−	−	+	−
Azoles (fungistatic): triazoles				
Itraconazole/Oral	+	+	+	+
Fuconazole/Oral	+	+	+	?[c]
Voriconizole	−	−	−	−
Allylamines (fungicidal)				
Terbinafine (Lamisil)/Topical or oral	−		−	−
Griseofulvin (fungistatic)				
Griseofulvin/Oral	−	−	−	−
Echinocandins				
Caspofungin/Intravenous	−	−	−	−
Antimetabolites (fungistatic or fungicidal)				
5-Fluoro-cytosine[d] (flucytosine)/Oral	−	−	−	−

[a]+ indicates that the drug inhibits growth of the disease-causing agent; − indicates that the drug is not useful for the disease.

[b]mc = mucocutaneous but not systemic candidiasis.

[c]Insufficient data.

[d]Used only in combination with amphotericin B.

infections of moist skin and mucous membranes (for example, vaginal yeast infections). The imidazole-containing drugs appear to disrupt the fungal membrane by inhibiting sterol synthesis. Lamisil (a terbinafine compound) is a different class of agent that selectively inhibits ergosterol synthesis by fungi. Humans do not make ergosterol nor use it in their cell membranes. Lamisil is currently a popular antifungal agent used to treat superficial mycoses.

More chronic dermatophytic infections typically require another antifungal agent, called **griseofulvin**, produced by a *Penicillium* species (**Fig. 27.31B**). Griseofulvin disrupts the mitotic spindle and derails cell division (called metaphase arrest). This action does not kill the fungus, but as the hair, skin, or nails grow and are replaced, the fungus is shed.

Vaginal yeast infections caused by *Candida* are often treated with nystatin, a polyene antifungal agent synthesized by *Streptomyces* that forms membrane pores

(**Fig. 27.31C**). The name "nystatin" came about because two of the people who discovered it worked for the laboratory of the New York State Public Health Department (now the Wadsworth Center).

The serious, sometimes fatal, consequences of systemic mycoses require more aggressive therapy. The drugs used in these instances include **amphotericin B** (produced by *Streptomyces*; see **Fig. 27.31D**) and fluconazole. Amphotericin B binds to the sterols in fungal membranes and destroys membrane integrity. It has a high affinity for ergosterol, which is prevalent in fungal but not mammalian membranes. Fluconazole, on the other hand, inhibits the synthesis of ergosterol. Thus, fungal cells grown in the presence of fluconazole make defective membranes. Typically, curing systemic fungal infections requires long-term treatment to prevent disease relapse. **Table 27.4** lists a number of other commonly used antifungal agents.

	Clinical applications[a]			
	Opportunistic mycoses			
Aspergillosis	Candidiasis[b]	Cryptococcosis	Dermatophytosis	Other
+	+	+	–	
–	+mc	–	–	
+ Superficial	+ Superficial	–	–	Mycotic keratins
–	+mc	–	+	
–	+mc	–	+	
–	+	–	+	Dandruff shampoo
+	+ mc	+	+	Sporotrichosis
–	+	+	+	Sporotrichosis
+	+	–	–	Fusarium
–	–	–	+	
–	–	–	+	
+ (Fungistatic)	+ (Fungicidal)	–	–	
+	+	+	–	Phaeohyphomycosis

To Summarize

- **Fungal infections** are difficult to treat because of similarities in human and fungal physiologies.

- **Imidazole-containing** antifungal agents inhibit sterol synthesis.

- **Griseofulvin** inhibits mitotic spindle formation.

- **Nystatin** produces membrane pores.

- **Amphotericin B** binds to membranes and destroys membrane integrity.

Concluding Thoughts

Antibiotics have greatly improved the health and well-being of humans and animals around the globe. Millions of people live today because antimicrobial drugs cured infectious diseases that we've dreaded for centuries. Unfortunately, by the late twentieth century, a sense of complacency about infections and antibiotics was pervasive among citizens of developed countries. This complacency led to the irresponsible use of antibiotics, and now a crisis of antibiotic resistance looms. We may even be entering a postantibiotic era. Despite government and industry throwing considerable amounts of money and human effort at the problem, very few new, clinically useful antibiotics have been discovered over the past four decades. And when one is found, pathogens keep developing resistance. Nevertheless, hope remains. New approaches to antimicrobial discovery are continually being conceived and tested. With any luck, one or more of the strategies will pay off, keeping us one step ahead of the microbes.

CHAPTER REVIEW

Review Questions

1. What is selective toxicity? Provide examples.
2. Explain the difference between antibiotic susceptibility and antibiotic sensitivity.
3. What does the term "spectrum of antibiotic activity" mean?
4. Provide examples of bacteriostatic and bactericidal antibiotics.
5. What is the Kirby-Bauer test? Does it indicate whether a drug is bacteriostatic or bactericidal?
6. Give examples of drugs that target cell wall synthesis; RNA synthesis; protein synthesis; DNA replication. What are their modes of action?
7. What mechanism do producing organisms use to synthesize peptide antibiotics?
8. How do antibiotic-producing microorganisms prevent "suicide"?
9. Why is antibiotic resistance a growing problem?
10. What are the four basic mechanisms of antibiotic resistance?
11. Explain the basic concept of an MDR efflux pump.
12. Discuss the current concepts of the origin of antibiotic resistance.
13. What are some mechanisms used to combat the development of drug resistance?
14. Why are there few antiviral agents available to treat disease?
15. What is herd immunity?
16. How does oseltamivir inhibit influenza?
17. Discuss the general modes of action of antifungal agents.

Thought Questions

1. A cephalosporin and clindamycin are often used to treat a patient with toxic shock syndrome caused by a TSST-producing *Staphylococcus aureus*. One agent is bactericidal; the other is bacteriostatic. Usually this combination is discouraged because the bacteriostatic agent dampens the effectiveness of the bactericidal agent, which kills only growing cells. Why would an exception to this rule be considered when treating toxic shock?
2. A patient presented to the emergency room complaining of a nonproductive cough (no sputum) that had persisted for 6 weeks. The clinician prescribed a 7-day course of a cephalosporin. After day 7 the patient returned, no better than before he started the treatment. Laboratory tests later showed that the infection was caused by *Mycoplasma pneumoniae*. Explain why the antibiotic did not work.
3. How would you determine the MIC of an obligate intracellular pathogen such as *Rickettsia prowazekii*, the cause of typhus?
4. You already know that some antimicrobial compounds must be converted by human metabolism into an active drug. Imagine how else an antibiotic might be more effective in vivo than in vitro, even if you do not know of any specific examples of your hypothesis.

Key Terms

amphotericin B (1128)
antibiotic (1092)
bacitracin (1103)
bactericidal (1095)
bacteriostatic (1095)
chloramphenicol (1108)
cycloserine (1104)
daptomycin (1105)
gramicidin (1104)
griseofulvin (1128)
highly active antiretroviral therapy (HAART) (1125)

Kirby-Bauer assay (1097)
lincosamide (1108)
macrolide (1108)
minimal inhibitory concentration (MIC) (1095)
Mueller-Hinton agar (1098)
multidrug resistance (MDR) efflux pump (1113)
neuraminidase inhibitor (1123)
nosocomial (1092)
oxazolidinone (1108)
penicillin (1092)

penicillin-binding protein (PBP) (1101)
persister cell (1117)
quinolone (1105)
secondary metabolite (1110)
selective toxicity (1094)
spectrum of activity (1095)
streptogramin (1109)
sulfa drug (1094)
transglycosylase (1101)
transpeptidase (1101)
vancomycin (1104)
zone of inhibition (1096)

Recommended Reading

Bearson, Bradley L., Heather K. Allen, Brian W. Brunelle, In Soo Lee, Sherwood R. Casjens, et al. 2014. The agricultural antibiotic carbadox induces phage-mediated gene transfer in *Salmonella*. *Frontiers in Microbiology* **5**:52.

Blair, Jessica M. A., Mark A. Webber, Alison J. Baylay, David O. Ogbolu, and Laura J. V. Piddock. 2015. Molecular mechanisms of antibiotic resistance. *Nature Reviews. Microbiology* **13**:42–49.

Brown, Eric D., and Gerard D. Wright. 2016. Antibacterial drug discovery in the resistance era. *Nature* **529**:336–343.

Clemente, Jose C., Erica C. Pehrsson, Martin J. Blaser, Kuldip Sandhu, Zhan Gao, et al. 2015. The microbiome of uncontacted Amerindians. *Science Advances* **1**:e500183.

Conlon, Brian P., Sarah E. Rowe, Autumn Brown Gandt, Austin S. Nuxoll, Niles P. Donegan, Eliza A. Zalis, et al. 2016. Persister formation in *Staphylococcus aureus* is associated with ATP depletion. *Nature Microbiology* **1**:Article number 16051.

Dawson, Clinton C., Chaidan Intapa, and Mary-Ann Jabra-Rizk. 2011. "Persisters": Survival at the cellular level. *PLoS Pathogens* **7**:e1002121.

DiGiandomenico, Antonio, and Brett R. Sellman. 2015. Antibacterial monoclonal antibodies: The next generation? *Current Opinion in Microbiology* **27**:78–85.

Fleming-Dutra, Katherine E., Adam L. Hersh, Daniel J. Shapiro, Monina Bartoces, Eva A. Enns, et al. 2016. Prevalence of inappropriate antibiotic prescriptions among US ambulatory care visits, 2010–2011. *Journal of the American Medical Association* **315**:1864–1873.

Hauser, Alan R., Joan Mecsas, and Donald T. Moir. 2016. Beyond antibiotics: New therapeutic approaches for bacterial infections. *Clinical Infectious Diseases* **63**:89–95. doi:10.1093/cid/ciw200

Holmes, Alison H., Luke S. P. Moore, Arnfinn Sundsfjord, Martin Steinbakk, Sadie Regmi, et al. 2015. Understanding the mechanisms and drivers of antimicrobial resistance. *Lancet* **387**:176–187.

Rogers, Geraint B., Mary P. Carroll, and Kenneth D. Bruce. 2012. Enhancing the utility of existing antibiotics by targeting bacterial behaviour? *British Journal of Pharmacology* **165**:845–857.

Sanyal, Gautam, and Peter Doig. 2012. Bacterial DNA replication enzymes as targets for antibacterial drug discovery. *Expert Opinion in Drug Discovery* **7**:327–339.

Sierra-Aragon, Salita, and Hauke Walter. 2012. Targets for inhibition of HIV replication: Entry, enzyme action, release and maturation. *Intervirology* **55**:84–97.

Tumpey, Terrence M., and Jessica A. Belser. 2009. Resurrected pandemic influenza viruses. *Annual Review of Microbiology* **63**:79–98.

Yount, Nannette Y., and Michael R. Yeaman. 2012. Emerging themes and therapeutic prospects for anti-infective peptides. *Annual Review of Pharmacology and Toxicology* **52**:337–360.

Zipperer, Alexander, Martin C. Konnerth, Claudia Laux, Anne Berscheid, Daniela J. Anek, et al. 2016. Human commensals producing a novel antibiotic impair pathogen colonization. *Nature* **535**:511–516.

CHAPTER 28
Clinical Microbiology and Epidemiology

nfections have killed more people than all of Earth's wars combined. Today the spread of disease is controlled in large part by worldwide surveillance agencies that can detect outbreaks quickly, before major epidemics develop. These agencies rely on smaller clinical microbiology laboratories scattered throughout the world that help clinicians diagnose infectious diseases. Here in Chapter 28, we discuss the principles of clinical microbiology and epidemiology that are used to identify, treat, and contain outbreaks, and explain how profiling our microbiota can predict diseases.

CURRENT RESEARCH highlight

Profiling gut microbiota. Dysbiosis of the gut microbiome is associated with Crohn's disease, ulcerative colitis, irritable bowel syndrome (IBS), inflammatory bowel disease (IBD), diabetes, and obesity. Christina Casén and colleagues at Genetic Analysis AS devised a diagnostic test for gut dysbiosis. The test uses 54 DNA probes targeting taxon-specific variable regions in 16S ribosomal DNA (V3–V9 in the figure). Probes that bind a region are fluorescently labeled and mixed with barcoded magnetic pull-down beads containing complementary probes. Fluorescence level reflects prevalence of a taxonomic group. "Normobiotic" is compared to IBS and IBD patients.

Source: Casén et al. 2015. *Aliment. Pharmacol. Ther.* **42**:71–83.

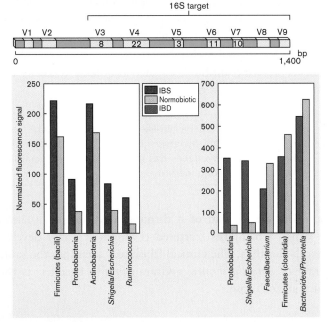

AN INTERVIEW WITH

CHRISTINA CASÉN, MOLECULAR BIOLOGIST, GENETIC ANALYSIS AS, OSLO, NORWAY

What fascinates you about intestinal microbiota?

I'm amazed that we can finally study the "black box" of intestinal microbiota in such detail. We've learned that intestinal microbiota affect so many facets of our lives, such as gastrointestinal disease, obesity, diabetes, mental illnesses, and even healthy brain function of infants. I'm also fascinated by how gut microbiota builds in an individual. Environmental surroundings, birth method, mother's microbiota (even during pregnancy)—all affect the infant's microbiota. I am most fascinated that severe *Clostridioides difficile* infections can be cured simply by fecal transplants from healthy donors.

How might profiling gut microbiota contribute to patient care?

IBS, for instance, is currently diagnosed in broad subgroups based on clinical symptoms (diarrhea, chronic constipation, or a mix). Patients within subgroups, however, harbor very different microbiota. So, profiling patient microbiota can lead to more specific diagnoses. Profiling microbiota can also indicate how a patient might respond to treatments such as anti–tumor necrosis factor. Fecal bacteriotherapy might then convert a nonresponder profile into responder. Finally, monitoring the profile of a recovered patient could signal a coming relapse. Preventive medication could shorten the period to remission.

1133

On January 12, 2010, the small country of Haiti on the Caribbean island of Hispaniola was violently shaken by a magnitude 7.0 earthquake. Over 100,000 people died, thousands of buildings collapsed, and the already fragile sanitation system was severely crippled. Almost immediately, volunteers from around the world poured in to help rescue the living, recover the dead, and rebuild what little infrastructure remained. Then, in October, a cholera epidemic struck. Over the next 5 years, 700,000 cases of severe diarrhea (see Chapter 25) ravaged the populace, and more than 9,000 died. Where did it come from? Haiti had never experienced cholera before.

Extensive epidemiological studies utilizing whole-genome sequencing strategies eventually traced the Haitian strain of *Vibrio cholerae* halfway around the world to a September 2010 cholera outbreak in Nepal. How did this faraway strain make it to Haiti so fast? Unwittingly, the United Nations sent troops infected with the organism from Nepal to Haiti to assist in the recovery. The Nepalese troops set up camp and used a nearby river for sanitation. The river became contaminated with *V. cholerae* and carried the pathogen through the town of Mirebalais, the site of the first reported cholera case. From there, the infection spread.

Such compelling stories underscore the integral roles that clinical microbiology and epidemiology play in health care. The Haitian cholera epidemic and the Current Research Highlight demonstrate how modern methods of DNA analysis have improved our ability to track disease around the world and even predict impending disease in an individual. In Chapter 28 we begin by explaining the basic strategies and methodologies used to diagnose infectious diseases, and we end with a discussion of how epidemiologists identify and track emerging pathogens. Our goal is not to catalog every infectious disease, but to demonstrate general principles and problem-solving approaches.

28.1 Clinical Microbiology: Specimen Collection and Handling

As in any good detective mystery, the first step in investigating an infectious disease is to identify the most likely suspects. We can accomplish this, in part, by observing the disease symptoms in a patient and knowing which organisms typically produce those symptoms. It also helps if the physician is aware of a similar disease outbreak under way in the community. Beyond these clues, the etiological agent must be identified through biochemical, molecular, serological, or antigen detection strategies.

Why Take the Time to Identify an Infectious Agent?

This is the first question many students ask when contemplating the effort and expense required to identify the genus and species of an organism causing an infection. Why not simply treat the patient with an antibiotic and be done with it? While this approach sounds appealing, there are several compelling reasons for identifying an infectious agent.

Antibiotic resistance. As discussed in Chapter 27, antibiotic resistance is an increasingly serious global problem. Characterizing a microbe's antibiotic resistance profile is part of any microbial identification process. Understanding which antibiotics are effective in treating an infectious agent will help the physician avoid prescribing an inappropriate drug. Furthermore, knowing which drugs are ineffective enables health organizations to track the spread of antibiotic-resistant strains. For example, before 1970, most strains of *Neisseria gonorrhoeae* were susceptible to penicillin. Today, most strains are resistant not only to penicillin, but also to many other antibiotics. Remember, too, that viral pathogens are not susceptible to antibiotics, so knowing that a patient has a viral disease can prevent the inappropriate use of antibiotics.

Pathogen-specific disease complications. Many diseases have serious complications that are common to a given organism or strain of organism. For example, children whose sore throats are caused by certain strains of *Streptococcus pyogenes* can develop serious complications affecting the heart and kidney long after the infection has resolved. These complications are the immunological consequence of bacterial and host antigen cross-reactivity; they are called **sequelae** (singular, **sequela**) because they occur <u>after</u> the infection itself is over. Life-threatening sequelae, such as rheumatic fever and acute glomerulonephritis caused by certain strains of *S. pyogenes*, produce severe damage to the heart and kidney, respectively. Knowing early on that *S. pyogenes* has caused a child's sore throat allows the physician to prescribe antibiotics that will quickly eradicate the infection and prevent the development of sequelae.

Note: Penicillin or a penicillin-like antibiotic is the usual treatment for *Streptococcus pyogenes* infection. Contrary to what you might expect, *S. pyogenes* has not developed any penicillin-resistant strains. It is not understood why.

Tracking the spread of a disease. Consider a situation in which ten infants scattered throughout a city develop bloody diarrhea. The clinical laboratory identifies the same strain of *Shigella sonnei*, a Gram-negative bacillus, as the cause in each case.

Immunological or nucleic acid amplification tests can be used to type a strain (discussed later). Finding the same strain in all cases suggests that they probably originated from the same source. Shigellosis is transferred from person to person by what is called the fecal-oral route. Carriers of *Shigella* will shed this organism in their feces. Inadequate hand washing after defecation will leave bacteria on the hands, which can then transfer the pathogen to foods or utensils, or to another person by touching.

Public health officials, armed with the knowledge that all the infants have the same strain of *Shigella*, then question the parents and learn that all of the children attend the same day-care center. By testing the other children and the workers in that center, officials can stop the infection from spreading. This investigative process is called **epidemiology** (covered in Section 28.3).

Specimen Collection and Processing

To diagnose an infectious disease, physicians must collect, and the laboratory must process, a wide variety of clinical specimens. The types of specimens range from simple cotton swabs of sore throats to urine and fecal samples. Here we describe how these samples should be collected. Note that, upon receiving and processing these specimens, the clinical microbiologist must wear latex gloves and use a laminar flow biosafety hood (see Fig. 5.28) as protection against potential infectious agents.

Some body sites should not contain any microorganisms when collected from a healthy individual. These include blood, cerebrospinal fluid (CSF), pleural fluid from the space that lines the outside of the lungs, synovial fluid from joints, and peritoneal fluid from the abdominal cavity. Because these sites are normally sterile, specimens can be plated onto nonselective agar media as well as selective media. Nonselective media, such as blood or chocolate agars, can be used because any organism found in these specimens is considered significant.

The following techniques are used to collect specimens from sterile body sites:

Cerebrospinal fluid. Lumbar punctures (spinal taps; **Fig. 28.1A**); testing for meningitis.

Blood samples. Generally taken by syringe from two body sites and placed in liquid media for aerobic and anaerobic culture. The same organism isolated from blood samples taken from the two sites is considered the likely etiological agent.

Pleural, synovial, and peritoneal fluid. Samples are aspirated by syringe. The fluid can be viewed microscopically for inflammatory cells and cultured on agar aerobically and anaerobically.

Identifying pathogens present at body sites containing normal microbiota is more challenging. A stool or fecal sample, for instance, is normally teeming with microbiota. These specimens are typically plated onto selective media—for example, MacConkey, Hektoen, or colistin–naladixic acid (CNA) agar (described in the next section)—to eliminate or decrease the number of normal microbiota that might contaminate the specimen. The following techniques are used to collect specimens from nonsterile body sites:

Swabs. Throat swabs (**Fig. 28.1B**), for example, should be placed in specialized liquid nutrient transport medium.

Sputum. Deep lung secretions expectorated for oral collection (**Fig. 28.1C**).

Stool samples. Cup or rectal swab; for identifying diarrhea-causing microbes.

Abscesses. Needle aspirations.

Urine samples. Midstream clean catch (will contain some microflora from urethra) or collected from catheters placed in the bladder (should be sterile); testing for urinary tract infections.

Note: The minimum standard for diagnosing a urinary tract infection used to be a finding of greater than 100,000 organisms per milliliter of urine. As noted in the text, this is no longer the case.

Urine is a special case. Urine in the bladder of a healthy individual was once considered sterile. Studies now show that the bladder does, indeed, have normal microbiota (at low numbers) although their role in human health is unclear. Why didn't we see them before? Most members are either anaerobic or difficult to grow on the standard media used to culture urine. Even though we now know that the bladder contains normal microbiota, for practical purposes clinical microbiologists still consider urine to be "sterile," or nearly so. The symptoms of urinary tract infections (which include increased frequency, painful urination, and suprapubic or flank pain) occur only when significant numbers of easily grown, aerobic or facultative microbes are present.

Urine is sampled in several ways. When collected from a catheterized patient, urine should contain few, if any, aerobically culturable organisms. Catheterization involves passing thin, sterile tubing through the urethra and directly into the bladder (**Fig. 28.1D**). (Catheterization is used primarily to assist urination by immobilized patients, but it also provides a convenient way to collect urine for bacteriological examination.) Unfortunately, the simple process of inserting the catheter through the nonsterile urethra can sometimes introduce organisms into the bladder and precipitate an infection. Also, when culturing urine from a catheterized patient, the urine should be collected from a port in the catheter, <u>never</u> from the collection bag. Urine may sit for hours in the collection

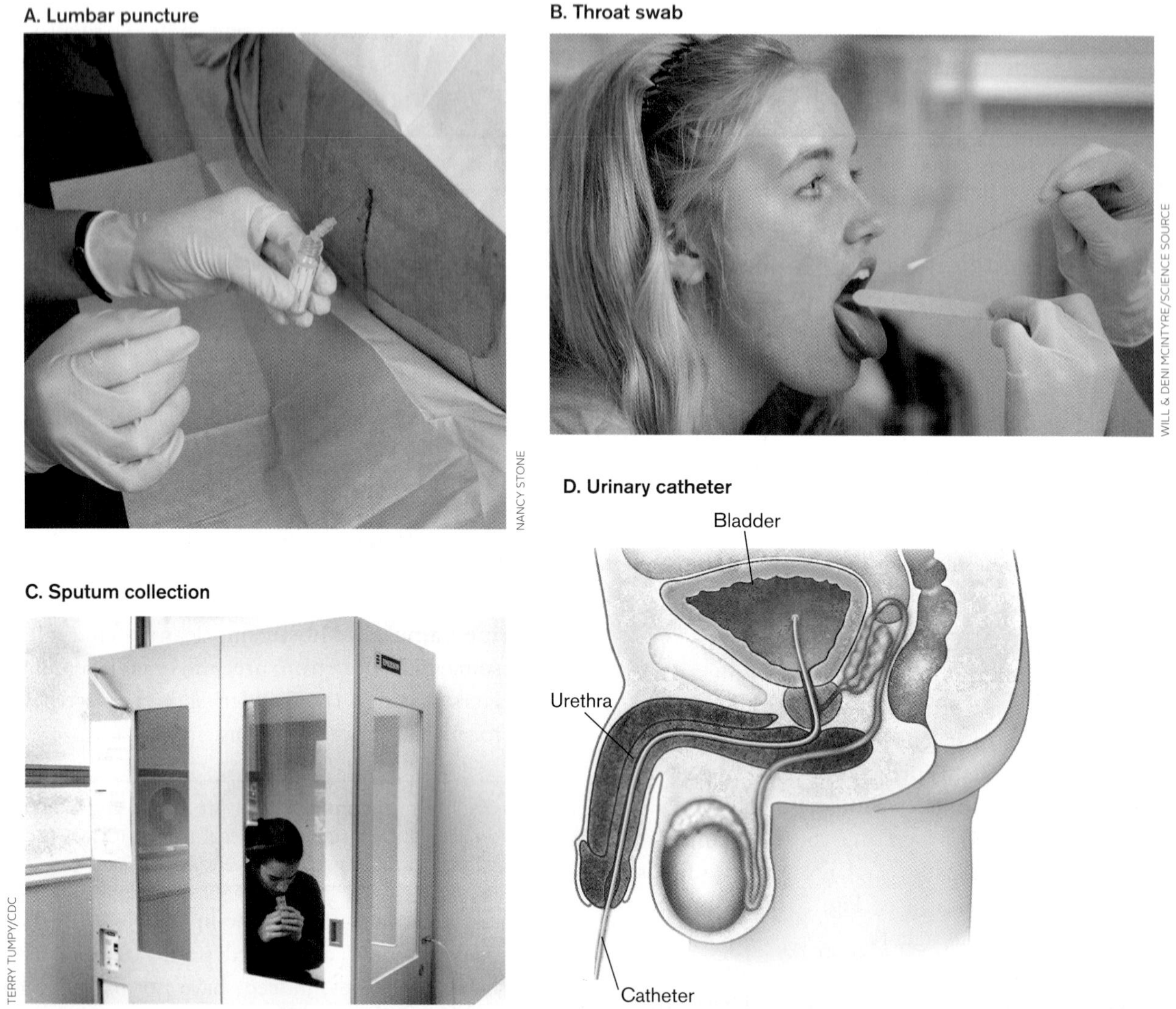

A. Lumbar puncture

B. Throat swab

C. Sputum collection

D. Urinary catheter

FIGURE 28.1 ■ **Specimen collection.** **A.** Urinary catheter, showing placement in the urethra. **B.** Throat swab. **C.** Sputum collection. A TB patient has coughed up sputum and is spitting it into a sterile container. The patient is sitting in a special sputum collection booth that prevents the spread of tubercle bacilli. The booth is decontaminated between uses. **D.** Lumbar puncture to obtain cerebrospinal fluid.

bag, so organisms initially present at insignificant numbers have time to replicate to high numbers even if the patient does not have a urinary tract infection (UTI).

When a catheter is not in place, urine is most commonly collected by what is called the midstream clean-catch technique, which is performed by the patient. In this procedure, the external genitals are first cleaned with a sterile wipe containing an antiseptic. The patient then partially urinates to wash as many organisms as possible out of the urethra and then collects 5–15 milliliters of the midstream urine in a sterile cup. This urine sample will usually not be sterile, because of urethral contamination, but the number of bacteria will be low. The clinical laboratory determines how many organisms per milliliter are present in the midstream catch and tells the physician whether an infection is present. Finding more than 1,000 organisms of a single species per milliliter of urine from a midstream clean catch is now considered indicative of an infection in a symptomatic patient.

Once a specimen from any body site has been properly collected, it must be transported to the clinical laboratory under conditions that will not undermine the viability of the pathogen. Then, after arriving in the lab, the sample must be processed quickly using protocols that ensure growth of likely pathogens.

Case History: Abdominal Abscess

A 4-year-old boy was admitted to the hospital for evaluation and treatment of persistent pain in the rectal area. His problem had begun about a week earlier with ill-defined pain in the same area. He had a white blood cell count of 24,900 (normal 4,000–11,000) with 87% neutrophils (normal 60%). An abdominal computed tomography (CT) scan revealed an abscess adjacent to his rectum. A needle aspiration drained 20 ml of yellowish, foul-smelling fluid from the abscess. Aerobic cultures of this specimen plated on blood and MacConkey agars were negative. Why didn't the infectious agent grow?

The problem in this instance is related to specimen collection and processing. Internal abscesses located near the gastrointestinal tract are often anaerobic infections, in this case caused by the Gram-negative rod *Bacteroides fragilis*, a strict anaerobe (**Fig. 28.2A**). Section 5.5 discusses anaerobes. Intestinal microbes, the majority of which are anaerobic, can sometimes escape the intestine if the organ is damaged in some way. The specimen in this instance should have been collected under anaerobic conditions by aspiration into a nitrogen-filled tube prior to transport to the clinical laboratory. Alternatively, a swab of the abscess material can be inserted into a special transport tube that has a built-in oxygen elimination system (**Fig. 28.2B**). Because the specimen was collected aerobically, many anaerobic microbes were probably killed by the oxygen.

Nevertheless, *B. fragilis* has a stress response system that permits survival of this anaerobe for 1 or 2 days in oxygen, so some of the bacteria may have survived transport. The laboratory still had a chance to find the organism, which raises the second problem in the case. After receiving the specimen, the lab cultured it only under aerobic conditions. The laboratory should have also incubated a series of plates anaerobically (see Section 5.5, Fig. 5.20). This case, therefore, illustrates the importance of both proper specimen collection and proper processing.

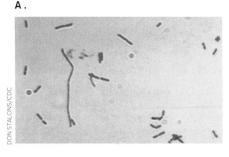

A.

B.

FIGURE 28.2 ■ **Anaerobic infection. A.** Gram stain of *Bacteroides fragilis* (1.5–4 μm in length). **B.** Vacutainer anaerobic specimen collector. Plunging the inner tube to the bottom will activate a built-in oxygen elimination system. The anaerobic indicator changes color when anaerobiosis has been achieved.

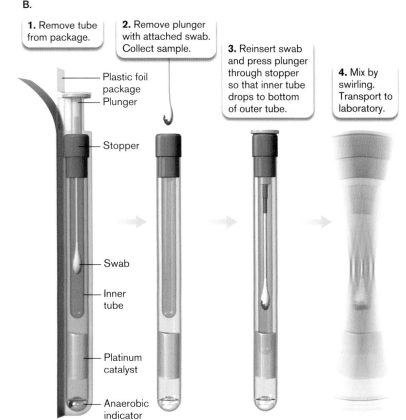

1. Remove tube from package.

2. Remove plunger with attached swab. Collect sample.

3. Reinsert swab and press plunger through stopper so that inner tube drops to bottom of outer tube.

4. Mix by swirling. Transport to laboratory.

Plastic foil package
Plunger

Stopper

Swab

Inner tube

Platinum catalyst

Anaerobic indicator

Biosafety Containment Procedures

Medical and laboratory personnel are exposed to extremely dangerous pathogens on a daily basis. When working with dangerous pathogens, clinical microbiologists must protect themselves from accidental infection and at the same time be certain the pathogen does not escape from the lab.

Case History: Fatal Meningitis

On July 15, an Alabama microbiologist was taken to the emergency room with acute onset of generalized malaise, fever, and diffuse myalgias. She was given a prescription for oral antibiotics and released. On July 16, she became tachycardic and hypotensive and returned to the hospital. She died 3 hours later. Blood cultures were positive for Neisseria meningitidis *serogroup C. Three days before the onset of symptoms, the microbiologist had prepared a Gram stain from the blood culture of a patient subsequently shown to have meningococcal disease; she had also handled agar plates containing cerebrospinal fluid (CSF) cultures from the same patient. Co-workers reported that fluids were aspirated from blood culture bottles at the open laboratory bench. No biosafety cabinets, eye protection, or masks were used for this procedure. Testing at CDC indicated that the isolates from both patients were indistinguishable. The laboratory at the hospital infrequently processed isolates of* N. meningitidis *and had not processed another meningococcal isolate during the previous 4 years.*

The microbiologist in this case did not take appropriate measures to protect herself and ended up with a laboratory-acquired infection leading to meningitis. The CDC has published a series of regulations designed to protect workers at risk of infection by human pathogens. Infectious agents are ranked by the severity of disease and ease of transmission. The more severe the disease or the more

TABLE 28.1	Biological Safety Levels and Select Agents[a]			
	Biosafety levels (BSLs)			
	BSL 1	**BSL 2**	**BSL 3**	**BSL 4**
Class of disease agent	Agents not known to cause disease.	Agents of moderate potential hazard; also required if personnel may have potential contact with human blood or tissues.	Agents may cause disease by inhalation route.	Dangerous and exotic pathogens with high risk of aerosol transmission; only 11 labs in the United States handle these.
Recommended safety measures	Basic sterile technique; no mouth pipetting.	Level 1 procedures plus limited access to lab; biohazard safety cabinets used; hepatitis vaccination recommended.	Level 2 procedures plus ventilation providing directional airflow into room, exhaust air directed outdoors; restricted access to lab (no unauthorized persons).	Level 3 procedures plus one-piece positive-pressure suits; lab is completely isolated from other areas present in the same building or is in a separate building.
Representative organisms in class	*Bacillus subtilis* *E. coli* K-12 *Saccharomyces* spp.	*Bordetella pertussis* *Campylobacter jejuni* *Chlamydia* spp. *Clostridium* spp. *Corynebacterium diphtheriae* *Cryptococcus neoformans* *Cryptosporidium parvum* Dengue virus Diarrheagenic *E. coli* *Entamoeba histolytica* *Giardia lamblia* *Haemophilus influenzae* *Helicobacter pylori* Hepatitis virus *Legionella pneumophila* *Listeria monocytogenes* *Mycoplasma pneumoniae* *Neisseria* spp. *Salmonella* spp. *Shigella* spp. *Staphylococcus aureus* *Toxoplasma* Pathogenic *Vibrio* spp. *Yersinia enterocolitica*	*Bacillus anthracis* (anthrax) *Brucella* spp. (brucellosis) *Burkholderia mallei* (glanders) California encephalitis virus *Coxiella burnetii* (Q fever) EEE (eastern equine encephalitis) virus *Francisella tularensis* (tularemia) Japanese encephalitis virus La Crosse encephalitis virus LCM (lymphocytic choriomeningitis) virus *Mycobacterium tuberculosis* Rabies virus *Rickettsia prowazekii* (typhus fever) Rift Valley fever virus SARS (severe acute respiratory syndrome) virus Variola major (smallpox) and other poxviruses VEE (Venezuelan equine encephalitis) virus West Nile virus Yellow fever virus *Yersinia pestis*	Ebola virus Guanarito virus Hantavirus Junin virus Kyasanur Forest disease virus Lassa fever virus Machupo virus Marburg virus Tick-borne encephalitis viruses

[a]Organisms in blue are on the list of CDC select agents that are considered possible agents of bioterrorism.

easily it is transmitted, the higher the risk category. On the basis of this ranking, four levels of biological containment are employed (**Table 28.1**).

Biosafety level 1 organisms have little to no pathogenic potential and require the lowest level of containment. Standard sterile techniques and laboratory practices are sufficient.

Biosafety level 2 agents have greater pathogenic potential, but vaccines and/or therapeutic treatments (for

example, antibiotics) are readily available. The pathogen in the case described here, *Neisseria meningitidis*, is in this risk group. These agents require more rigorous containment procedures, such as limiting laboratory access when experiments are in progress and using biological laminar flow cabinets if aerosolization is possible.

Biosafety level 3 pathogens produce a serious or lethal human disease. Vaccines or therapeutic agents may be available. To safely handle these organisms, level 2 procedures are

FIGURE 28.3 ■ Biosafety level 4 containment. Dr. Kevin Karem at the CDC performs viral plaque assays to determine the neutralization potential of serum from smallpox vaccination trials. He is protected by a positive-pressure suit working in a biosafety level 4 laboratory. The airflow into his suit is so loud that he must wear earplugs to protect his hearing. Note that this virus is used under extremely tight security at the CDC, one of only two places in the World allowed to do so.

supplemented with a lab design ensuring that ventilation air flows only <u>into</u> the room and that exhaust air vents directly to the outside, thus producing negative pressure. Negative pressure will keep any organism that may aerosolize from escaping into hallways. In addition, access to the lab is strictly regulated and includes double-door air locks at the entrance.

By law, extremely dangerous pathogens such as the Ebola virus, for which there is no treatment or vaccine, may be studied only at a <u>biosafety level 4</u> containment facility. Practices here dictate that lab personnel wear positive-pressure lab suits connected to a separate air supply (**Fig. 28.3**). The positive pressure ensures that if the suit is penetrated, organisms will be blown away from the breach and not sucked into the suit.

As reasonable as these regulations may seem, they were not always in effect. Prior to 1970, scientists had an almost cavalier approach toward handling pathogens. For instance, culture material was routinely transferred from one vessel to another by mouth pipetting (essentially using a glass or plastic pipette as a straw). This practice is now forbidden, for obvious reasons. As of this writing, there are 11 BSL4 laboratories operating in the United States.

It is important to note that most clinical microbiology laboratories are equipped to handle BSL2 organisms. Patient samples suspected to contain a level 3, or higher, agent are sent directly to regional reference laboratories or to the CDC for analysis.

Thought Questions

28.1 Two blood cultures, one from each arm, were taken from a patient with high fever. One culture grew *Staphylococcus epidermidis*, but the other blood culture was negative (no organisms grew out). Is the patient suffering from septicemia caused by *S. epidermidis*?

28.2 A 30-year-old woman with abdominal pain went to her physician. After the examination, the physician asked the patient to collect a midstream urine sample that would be sent to the lab across town for analysis. The woman complied and handed the collection cup to the nurse. The nurse placed the cup on a table at the nurses' station. Three hours later, a courier service picked up the specimen and transported it to the laboratory. The next day, the report came back: "greater than 200,000 CFUs/ml; multiple colony types; sample unsuitable for analysis." Why was this determination made?

To Summarize

- **Identifying a pathogen** enables clinicians to prescribe appropriate **antibiotics**, anticipate possible **sequelae**, and **track the spread** of the disease.

- **Specimen collection** is a critical first step in the process of identifying a pathogen.

- **Common specimens** include blood, pus, urine, sputum, throat swabs, stool, and cerebrospinal fluid.

- **Specimens** from sites containing normal microbiota must be handled differently from specimens taken from normally sterile body sites.

- **A specimen from an abscess or other infection that might contain anaerobes** must be collected under anaerobic conditions.

- **Various levels of protective measures** are used in handling potentially infectious biological materials.

- **Biosafety level 1** agents are generally not pathogenic and require the lowest level of containment.

- **Biosafety level 2** agents are pathogenic but not typically transmitted via the respiratory tract. Laminar flow hoods are required.

■ **Biosafety level 3** agents are virulent and transmitted by the respiratory route. They require laboratories with special ventilation and air-lock doors.

■ **Biosafety level 4** agents are highly virulent and require the use of positive-pressure suits.

28.2 Approaches to Pathogen Identification

Now that we have collected a sample, how do we identify the pathogen, and do it quickly? Pathogens can be identified on the basis of a variety of observations, including the patient's symptoms, the presence of organisms in stained clinical specimens, biochemical clues, and serology (the presence in a patient's serum of antibodies reactive against a specific microbe).

Conventional Approaches

Staining procedures. Some specimens, such as CSF and sputum, can be directly stained with the Gram stain procedure (described in Section 2.3) or the acid-fast stain. Knowing that an organism is Gram-positive, Gram-negative, or acid-fast guides which additional tests the clinical microbiologist must run. The case that follows illustrates how the acid-fast stain is critical for presumptively identifying mycobacteria.

Case History: Tuberculosis

A 31-year-old male presented to an emergency department (ED) in New York City after experiencing gross hemoptysis (blood in sputum). He had a 2-month history of productive cough, a 25-pound weight loss, night sweats, and fatigue. A chest X-ray revealed bilateral cavitary infiltrates. The initial sputum specimen was negative by Gram stain but positive for acid-fast bacilli (AFB). The specimen was submitted for a nucleic acid amplification assay (NAAT) to detect 16S rRNA, as well as for culture and sensitivity. The patient had a history of heavy alcohol and drug use.

The likely suspect in this case is *Mycobacterium tuberculosis*, although other mycobacterial species are possible causes. *M. tuberculosis* is presumptively diagnosed in the clinical laboratory

by the acid-fast stain, a technique first described in 1882 (called the Ziehl-Neelsen stain) that is used to find the tubercle bacillus in a patient's sputum. The acid-fast stain enables a technician to visualize bacteria such as *Mycobacterium* species that are <u>not</u> stained by the Gram stain. Mycobacteria have a very waxy outer coat composed of mycolic acid that resists penetration by most dyes—an obstacle the acid-fast stain was designed to overcome.

The original Ziehl-Neelsen acid-fast stain used phenol and heat to drive carbolfuchsin (a red dye) into mycobacterial cells on glass slides. Destaining with an acidic alcohol solution removes the stain from all cell types <u>except</u> mycobacteria. The slide is subsequently counterstained with methylene blue, after which the mycobacteria will be seen as curved, red rods (called acid-fast bacilli), while everything else will appear blue (**Fig. 28.4A**). A more modern version of the acid-fast stain uses the fluorochrome auramine O to stain the mycolic acid. This dye also resists removal by an acidic alcohol wash, so the mycobacteria will fluoresce bright yellow when observed under a fluorescence microscope.

Although the acid-fast stain is very useful, it detects organisms in the sputum of tuberculosis patients only about 60% of the time; and even if they are found, confirmatory tests are needed for a definitive diagnosis. Growth-dependent identification of *M. tuberculosis* typically starts with inoculating the sputum sample to a blue-green Löwenstein-Jensen medium (selective for mycobacteria) and waiting several weeks for the organism to grow before additional tests can be done (**Fig. 28.4B**). However, while the organism grows, a rapid, nucleic acid amplification test (NAAT) can be used to quickly detect even small amounts of the organism, even in AFB-negative sputum samples. The NAAT tests are described later in this section.

A. Acid-fast stain

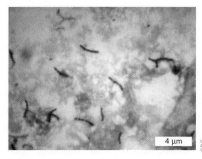

4 µm

CDC

B. *M. tuberculosis* colonies on Löwenstein-Jensen agar

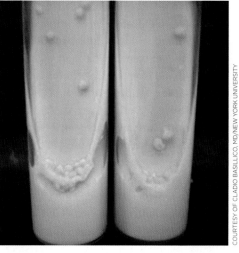

COURTESY OF CLADIO BASILLICO, MD/NEW YORK UNIVERSITY

FIGURE 28.4 ■ **Acid-fast stain and growth of *Mycobacterium tuberculosis*. A.** Acid-fast Ziehl-Neelsen stain of *M. tuberculosis*. **B.** Löwenstein-Jensen medium enables growth of mycobacterial species, some of which grow extremely slowly. The colonies have a "bread crumb–like" appearance.

Growth and biochemical testing. Once the Gram reaction of an organism is known, the conventional laboratory approach used to identify bacterial pathogens requires understanding basic microbial physiology and how diverse it is. Thousands of bacterial species are capable of causing disease, but no two species have the same biochemical "signature." The clinical microbiologist can look for reactions, or combinations of reactions, that are unique to a given species, as in the following case history.

Case History: Medical Detective Work

A 38-year-old woman with no significant previous medical history came to the emergency room complaining of a mild sore throat persisting for 3 days. Her symptoms included arthralgia (joint pain), myalgia (muscle pain), and a low-grade fever. The day before coming to the ER she had had a severe headache with neck stiffness, nausea, and vomiting. She was not taking any medications, had no known drug allergies, and did not smoke. She lived with her husband and two children, all of whom were well. Cerebrospinal fluid (CSF) was collected from a spinal tap. The CSF appeared cloudy (it should be clear) and contained 871 white blood cells per microliter (normal is 0–10/μl), the glucose level was 1 mg/dl (normal is 50–80), and the total protein level was 417 mg/dl (normal is less than 45). Gram stain of a CSF smear revealed Gram-negative rods. The CSF sample was sent to the diagnostic laboratory for microbial identification.

As discussed in Thought Question 26.9, low glucose and elevated protein levels in CSF are indicators of bacterial (not viral) infection. The increase in white blood cells revealed that the woman's immune system was trying to fight the disease. The presence of Gram-negative rods in the CSF smear confirmed a diagnosis of bacterial meningitis, since CSF should be sterile. Now it was up to the clinical laboratory to determine the etiological agent.

Algorithms to identify bacteria. Over the years, clinical microbiologists have developed algorithms (step-by-step problem-solving procedures) that expose the most likely cause of a given infectious disease. For instance, only a limited number of microbes are known to cause meningitis. The microbiologist poses a series of binary yes/no questions about the clinical specimen in the form of biochemical or serological tests. Typical questions in this case might include: Is an organism seen in the CSF of a patient with symptoms of meningitis? Is the organism Gram-positive or Gram-negative? Does it stain acid-fast? Answers to a first round of questions will then dictate the next series of tests to be used.

Because speed is of the essence in deciding how to treat the patient, a series of tests is not always carried out sequentially. To save time, a slew of tests are carried out simultaneously, but the results are interpreted sequentially using the algorithm. In our case history, for instance, consider the most common causes of bacterial meningitis: *Neisseria meningitidis*, *Streptococcus pneumoniae*, *Haemophilus influenzae*, and *Escherichia coli*.

The CSF sample was Gram-stained by a microbiologist and simultaneously plated onto three media: chocolate agar, blood agar, and Hektoen agar. Chocolate agar is an extremely rich medium that looks brown, owing to the presence of heat-lysed red blood cells (**Fig. 28.5A** and **B**). Because it is so nutrient-rich, all four organisms will grow on chocolate agar. However, nutritionally fastidious organisms such as *N. meningitidis* and *H. influenzae* will not grow well, if at all, on ordinary blood agar, because these bacteria cannot lyse red blood cells and release required nutrients. Less fastidious organisms, such as *S. pneumoniae* and *E. coli*, will grow on blood agar, but of these two, only *E. coli* can grow on Hektoen agar (**Fig. 28.5C–E**), which is a selective and differential medium for enteric Gram-negative rods. Hektoen is selective because bile salts and dyes inhibit the growth of Gram-positives. It is differential because the medium reveals organisms that ferment lactose or sucrose and produce hydrogen sulfide. Differential and selective media are described in Section 4.3.

In our case history, the Gram stain of the CSF revealed Gram-negative rods, which ruled out *N. meningitidis* (a Gram-negative diplococcus) and *S. pneumoniae* (a Gram-positive diplococcus). The organism in CSF did grow on blood agar, which eliminated *H. influenzae* (a Gram-negative, nonenteric rod) as a candidate. It also grew on Hektoen, where it produced orange, lactose-fermenting colonies. Thus, the organism was a Gram-negative, enteric rod, and likely *E. coli*. Additional biochemical tests confirming the identity of the organism had to be carried out, but this simple example shows how simultaneous tests can be interpreted.

Identifying Gram-negative bacteria. The Gram-negative bacterium in this case was subjected to a battery of 36 biochemical tests in an automated microbial identification instrument (**Fig. 28.6A**). Most clinical laboratories in the United States and Europe now use these automated identification systems. The organism is inoculated into the wells of a prepared microtiter dish in which each well contains materials that test an organism's ability to ferment different carbon sources or make different metabolic end products (**Fig. 28.6B** and **C**). By monitoring the growth or change in color produced in different wells, the instrument can quickly identify the pathogen's metabolic profile, sometimes within 5–6 hours. For each bacterial species, the system software "knows" the probability that a given reaction will be positive or negative, and from that information the program can identify the pathogen.

A. Chocolate agar

B. Colonies of *N. gonorrhoeae*

C. Hektoen agar

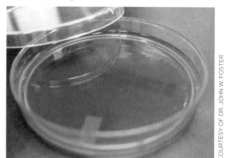

E. *S. enterica* colonies on Hektoen

D. *E. coli* colonies on Hektoen

FIGURE 28.5 ▪ Chocolate agar and Hektoen agar: two widely used clinical media. **A.** Uninoculated chocolate agar. Its color is due to gently lysed red blood cells that provide a rich source of nutrients for fastidious bacteria. **B.** Chocolate agar inoculated with *Neisseria gonorrhoeae*. This organism will not grow well on typical blood agar, because important nutrients remain locked within intact red blood cells. **C.** Uninoculated Hektoen agar, which contains lactose, peptone, bile salts, thiosulfate, an iron salt, and the pH indicators bromothymol blue and acid fuchsin; the bile salts prevent growth of Gram-positive microbes. **D.** Hektoen agar inoculated with *Escherichia coli*. This organism ferments lactose to produce acid-fermentation products that give the medium an orange color, owing to the pH indicators acid fuchsin and bromophenol blue. **E.** Hektoen agar inoculated with *Salmonella enterica*. This organism does not ferment lactose but grows instead on the peptone amino acids. The resulting amines are alkaline and produce a more intense blue color with bromothymol blue. *Salmonella* species also produce hydrogen sulfide gas from the thiosulfate. Hydrogen sulfide reacts with the medium's iron salt to produce an insoluble, black iron sulfide precipitate visible in the center of the colonies.

A.

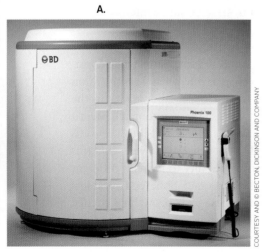

B.

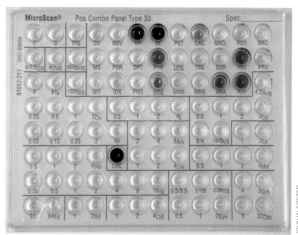

C.

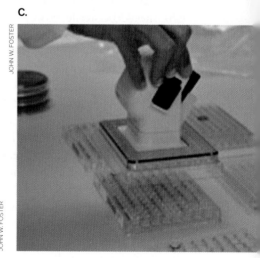

FIGURE 28.6 ▪ Automated microbiology system. **A.** The BD Phoenix system uses plates with numerous reaction wells and a computerized plate reader to automatically identify pathogenic bacteria. This type of instrument automatically generates and evaluates the numbers. **B.** Microbial identification plate. **C.** Loading a multiwall ID plate with a multichannel pipettor. All wells are simultaneously loaded with the same volume and number of bacteria.

A simplified example of the pathogen identification process is shown in **Figure 28.7**. The analytical profile index (API 20E) strip can test 20 metabolic processes. An uninoculated control and strips for two organisms are pictured in **Figure 28.7**. Overnight incubation is needed before the chamber reactions can be read, so the API strip does not deliver results as quickly as the automated systems do. Different-colored reactions in each chamber are scored as positive or negative, depending on the color (**Table 28.2** and **eTopic 28.1**). For example, in the indole chamber (well 9 in **Fig. 28.7B**), a red reaction at the top of the tube is positive and indicates that the organism can produce indole from tryptophan. A colorless chamber would be a negative result. Go to **eTopic 28.1** to see how API strips can also be used to generate a seven-digit number that identifies the bacterium.

The results of the API chambers can be used as a dichotomous key, by making a stepwise interpretation that begins with a key reaction. A simplified dichotomous key using a limited number of enteric Gram-negative species is shown in **Figure 28.8**. Often, the first reaction examined is lactose fermentation (tagged "ONPG" in **Table 28.2**). The lactose reaction is read as positive or negative, depending on the color. Then, following a printed flowchart, the technician goes to the next key reaction—say, indole production—and reads it as positive or negative. If the organism is a lactose fermenter and the indole test is positive, then the choices have been narrowed to *E. coli* or *Klebsiella* species (the other lactose-positive, indole-positive organisms in the figure do not cause meningitis). Another reaction is read to distinguish between the next two choices. The process continues until a single species is identified.

In reality, many more reactions than the 11 shown have to be used to make a definitive identification, because of species differences with respect to a single reaction. For example, **Figure 28.8** shows that *Klebsiella pneumoniae* and *Klebsiella oxytoca* (highlighted in yellow) exhibit opposite indole reactions, even though they are of the same genus. Even within a given species, only a certain percentage of strains might be positive for a given reaction. The inherent danger in using a dichotomous key is that one anomalous result can lead to an incorrect identification. Be aware, too, that the phenotypic dichotomous key does not mirror a phylogeny tree, in which the two *Klebsiella* species would branch from a common ancestor.

Identifying nonenteric Gram-negative bacteria. The procedures we have outlined will accurately identify members of Enterobacteriaceae, but there are pathogenic

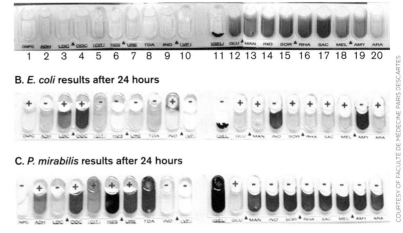

A. Uninoculated strip

B. *E. coli* **results after 24 hours**

C. *P. mirabilis* **results after 24 hours**

FIGURE 28.7 ■ API 20E strip technology for the biochemical identification of Enterobacteriaceae. A. Uninoculated API strip. Each well contains a different medium that tests for a specific biochemical capability. The well numbers correspond to Table 28.2. The color of each medium after 24-hour incubation indicates a positive or negative reaction (see Table 28.2). **B, C.** API results for *E. coli* (**B**) and *Proteus mirabilis* (**C**). Plus (+) and minus (−) indicate positive and negative reactions, respectively.

COURTESY OF FACULTE DE MEDECINE PARIS DESCARTES

TABLE 28.2	Reading the API 20E	
Well number	**Test**	**Reaction tested**
1	ONPG[a]	Beta-galactosidase
2	ADH	Arginine dihydrolase
3	LDC	Lysine decarboxylase
4	ODC	Ornithine decarboxylase
5	CIT	Citrate utilization
6	H₂S	H$_2$S production
7	URE	Urea hydrolysis
8	TDA	Tryptophan deaminase
9	IND	Indole production
10	VP	Acetoin production
11	GEL	Gelatinase
12	GLU	Glucose fermentation/oxidation
13	MAN	Mannitol fermentation/oxidation
14	INO	Inositol fermentation/oxidation
15	SOR	Sorbitol fermentation/oxidation
16	RHA	Rhamnose fermentation/oxidation
17	SAC	Sucrose fermentation/oxidation
18	MEL	Melibiose fermentation/oxidation
19	AMY	Amygdalin fermentation/oxidation
20	ARA	Arabinose fermentation/oxidation

[a]ONPG = *ortho*-nitrophenyl-beta-D-galactoside.

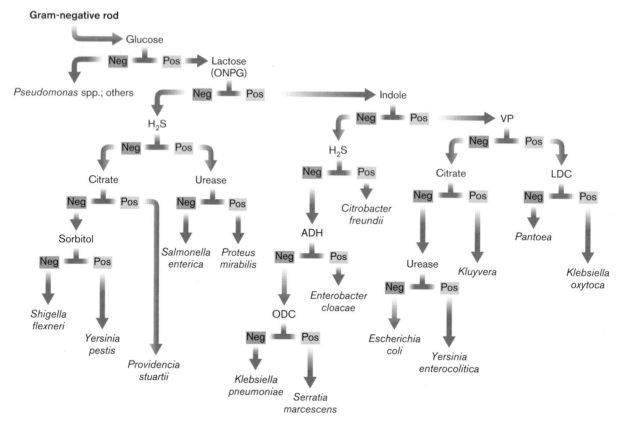

FIGURE 28.8 ▪ Simplified biochemical algorithm to identify Gram-negative rods. The diagram presents a dichotomous key using a limited number of biochemical reactions and selected organisms to illustrate how species identifications can be made using biochemistry. Abbreviations and reactions: **ADH** (arginine dihydrolase): stepwise degradation of arginine to citrulline and ornithine; **citrate** (CIT): citrate utilization as a carbon source; **glucose** (GLU): glucose fermentation to produce acid; **H_2S**: production of hydrogen sulfide gas; **indole** (IND): indole production from tryptophan; **lactose (ONPG)**: lactose fermentation to produce acid (*ortho*-nitrophenyl-beta-D-galactoside, cleaved by beta-galactosidase); **LDC** (lysine decarboxylase): cleavage of lysine to produce CO_2 and cadaverine; **ODC** (ornithine decarboxylase): cleavage of ornithine to make CO_2 and putrescine; **sorbitol** (SOR): sorbitol fermentation to produce acid; **urease** (URE): production of CO_2 and ammonia from urea; **VP** (Voges-Proskauer test): production of acetoin or 2,3-butanediol. (*Note:* "Neg" for *Pseudomonas* in terms of glucose indicates an inability to ferment glucose. *Pseudomonas* can still use glucose as a carbon source.)

Gram-negative bacilli found in other, nonenteric, phylogenetic families. For instance, one possibility in the preceding case history is that the Gram-negative bacillus seen in CSF smears would <u>not</u> grow on blood agar or on the other selective media, but would grow as small, glistening colonies on chocolate agar. This result would implicate the Gram-negative rod *Haemophilus influenzae*.

Meningitis caused by *H. influenzae* was a major problem prior to 1988, before the introduction of a vaccine containing *H. influenzae* type b capsular material. Growth of *H. influenzae* requires hemin (X factor) and NAD (V factor), so confirming the identity of *H. influenzae* involves growing the organism on agar medium containing hemin and NAD (nicotinamide adenine dinucleotide). This can be done by placing small filter paper disks containing these compounds on a nutrient agar surface (not blood agar) that has been covered with the organism. *H. influenzae* will grow only around a strip containing both X and V factors. Alternatively, X and V factors can be incorporated into Mueller-Hinton agar, as shown in **Figure 28.9A**. The

organism grows on chocolate medium because the lysed red blood cells release these factors. Although XV growth phenotype is still used for identification, fluorescent antibody staining is more specific (discussed shortly).

A completely different identification scheme would have been used in the preceding meningitis case if the laboratory had discovered the organism to be a Gram-negative diplococcus (rather than a Gram-negative rod). A Gram-negative diplococcus would suggest *Neisseria meningitidis*. *N. gonorrhoeae* is also possible, but it is less likely because the gonococcus lacks the protective capsule that meningococci use to survive in the bloodstream. Without this capsule, *N. gonorrhoeae* is not as resistant to serum complement and cannot disseminate to the meninges.

The first test to determine whether the organism is a species of *Neisseria* is the cytochrome oxidase test (**Fig. 28.9B**). Cytochrome oxidase is the terminal oxidoreductase for O_2 (discussed in Chapter 14). In this test, a few drops of the colorless reagent *N, N, N′, N′*-tetramethyl-*p*-phenylenediamine dihydrochloride are applied to the

A.

B.

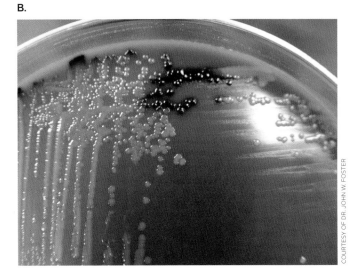

FIGURE 28.9 ■ ***Haemophilus influenzae* growth factors and *Neisseria meningitidis* oxidase reaction.** **A.** *H. influenzae* will grow on an agar plate (here, Mueller-Hinton agar) only when the medium has been fortified with both X factor (hemin) and V factor (NAD), but not either one alone. Nor will the organism grow on blood agar. **B.** Oxidase-positive reaction for *N. meningitidis.* Oxidase reagent (which is colorless) was dropped onto colonies of *N. meningitidis* grown on chocolate agar. The test is called the cytochrome oxidase test, but it really tests for cytochrome *c.*

suspect colonies. The reaction, which takes place only if the organism possesses both cytochrome oxidase and cytochrome *c*, turns the *p*-phenylenediamine reagent (and the colony) a deep purple/black. While many bacteria possess cytochrome oxidase, *Neisseria* is one of only a few genera that also contain cytochrome *c* in their membranes. Oxidase-positive organisms use cytochrome oxidase to oxidize cytochrome *c*, which then oxidizes *p*-phenylenediamine. Other oxidase-positive bacteria include *Pseudomonas*, *Haemophilus*, *Bordetella*, *Brucella*, and *Campylobacter* species—all of them Gram-negative rods. None of the Enterobacteriaceae, however, are oxidase-positive, because they lack cytochrome *c.*

An oxidase-positive, Gram-negative diplococcus is very likely a member of *Neisseria.* Differentiation between species of *Neisseria* is based on their ability to grow on certain carbohydrates, or it can be determined using immunofluorescent antibody staining tests that test for the presence of different capsule antigens in *N. meningitidis.*

Identifying Gram-positive pyogenic cocci. Recall from Section 26.2 the case of the woman with necrotizing fasciitis. How did the laboratory determine that the etiological agent was *Streptococcus pyogenes*? A sample algorithm, or flowchart (**Fig. 28.10**), shows how this is done. The physician sends a cotton swab containing a sample from a lesion to the clinical laboratory. The laboratory technician streaks the material onto several media: (1) blood agar (which will grow both Gram-positive and Gram-negative organisms); (2) blood agar containing the inhibitors colistin and nalidixic acid (called a CNA plate, this agar will grow <u>only</u>

Gram-positives); and (3) MacConkey agar (which will grow only Gram-negative organisms; see Section 4.3). The suspect organism in this case grows on the CNA and blood plates. Because it grows in the presence of the Gram-negative inhibitory compounds in CNA, one would immediately suspect the organism to be Gram-positive—an assumption borne out by the Gram stain.

Note: Though the skin is normally populated by many different microorganisms, samples from an infected lesion are overwhelmingly populated by the etiological agent. The pathogen predominates because it outgrows normal microbiota. Using selective media to isolate the infectious agent will further simplify diagnosis by decreasing the growth of any normal microbiota that may still be present.

The algorithm tells the laboratory technician that since the organism is a Gram-positive coccus, the next step is to test for catalase production. Catalase, which converts hydrogen peroxide (H_2O_2) to O_2 and H_2O, clearly distinguishes staphylococci from streptococci (remember, Gram stain morphology alone is insufficiently reliable to make that distinction). The catalase test is performed by mixing a colony with a drop of H_2O_2 on a glass slide. Effusive bubbling due to the release of oxygen indicates catalase activity (see **Fig. 28.10**). Staphylococci are catalase-positive; streptococci are catalase-negative. Note that many other organisms possess catalase activity, including the Gram-negative rod *E. coli.* According to the algorithm, however, *E. coli* would not be considered, because it does not grow on CNA agar and is not Gram-positive.

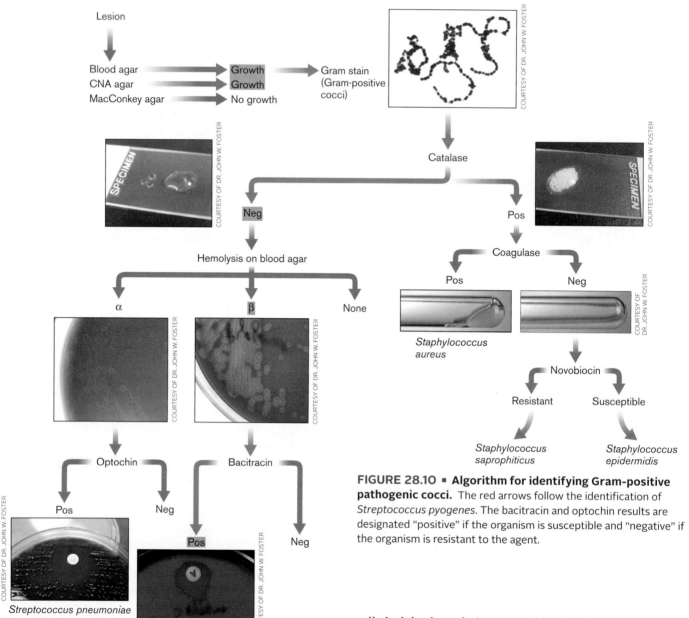

FIGURE 28.10 ▪ Algorithm for identifying Gram-positive pathogenic cocci. The red arrows follow the identification of *Streptococcus pyogenes*. The bacitracin and optochin results are designated "positive" if the organism is susceptible and "negative" if the organism is resistant to the agent.

Note: When performing a catalase test from colonies grown on blood agar, be sure not to transfer any of the agar, since red blood cells also contain catalase.

Having established that the organism is catalase-negative, the technician examines the blood plate for evidence of hemolysis. Three types of colonies are possible: nonhemolytic, alpha-hemolytic, and beta-hemolytic. Nonhemolytic streptococci do not produce any lytic zone. Alpha-hemolytic strains produce large amounts of hydrogen peroxide that oxidize the heme iron within intact red blood cells to generate a green product. As a result, alpha-hemolytic streptococci produce a green zone around their colonies called alpha hemolysis—even though the red blood cells remain intact. (For example, *Streptococcus mutans*, a cause of dental caries and subacute bacterial endocarditis, is alpha-hemolytic.) Still other streptococci produce a completely clear zone of true hemolysis surrounding their colony. This is called beta hemolysis. Red blood cells are hemolyzed completely by exported enzymes, called hemolysins, that lyse red-cell membranes. The flowchart in **Figure 28.10** indicates that the organism from the case history was beta-hemolytic.

The final relevant test in this flowchart involves susceptibility to the antibiotic bacitracin, which identifies the most pathogenic group of beta-hemolytic streptococci. The beta-hemolytic streptococci are subdivided into many different groups, known as Lancefield groups, based on differences in the composition of a carbohydrate antigen anchored to peptidoglycan. These cell wall differences are distinguished from each other immunologically and divide the

MARINE BIOLOGICAL LABORATORY

FIGURE 28.11 ■ Rebecca Lancefield. In 1918, Dr. Lancefield joined the Rockefeller Institute for Medical Research in New York City, where she studied the hemolytic streptococci, known then as *Streptococcus haemolyticus*. She was the first to use serum precipitation methods to classify *S. haemolyticus* into groups according to differences in cell wall carbohydrate antigens. The basic technique is still used today and in her honor is known as the Lancefield classification scheme.

streptococci into Lancefield groups A–U. Rebecca Lancefield (**Fig. 28.11**), for whom the classification scheme is named, was the first to use immunoprecipitation to group the streptococci (immunoprecipitation is described in Section 24.2 and eAppendix 3). The vast majority of streptococcal diseases are caused by group A̲ beta-hemolytic streptococci (also called GAS), defined as the species *Streptococcus pyogenes.*

Unfortunately, the Lancefield classification procedure is somewhat time-consuming and not readily amenable as a rapid identification method. However, the group A beta-hemolytic streptococci are uniformly susceptible to the antibiotic bacitracin. Thus, a simple antibiotic disk susceptibility test can be used to indicate GAS (that is, *S. pyogenes*). But beware—many bacteria are bacitracin sensitive, so, like the catalase test, to be useful for identification the bacitracin test must be used in conjunction with an algorithm. The technician must follow the appropriate algorithm before assigning importance to this or any other test result. It is irrelevant, for instance, if an alpha-hemolytic organism is bacitracin susceptible. Some of these may exist, but they are not associated with disease. The organism in our case of necrotizing fasciitis, however, was beta-hemolytic, so bacitracin susceptibility indicated that the organism was *S. pyogenes.*

The other tests named in **Figure 28.10** are equally important for identifying Gram-positive infectious agents.

For example, *Streptococcus pneumoniae* is an important cause of pneumonia. Like *S. pyogenes*, *S. pneumoniae* is a catalase-negative, Gram-positive coccus; but unlike *S. pyogenes*, it is alpha-hemolytic. Optochin susceptibility is a property closely associated with *S. pneumoniae*, while other alpha-hemolytic strains of streptococci are resistant to this compound. Thus, an optochin susceptibility disk test is a useful tool for identifying *S. pneumoniae.*

Coagulase catalyzes a key reaction used to distinguish the pathogen *Staphylococcus aureus*, which causes boils and bone infections, from other staphylococci, such as the normal skin species *Staphylococcus epidermidis*. To conduct a coagulase test, a tube of plasma is inoculated with the suspect organism. If the organism is *S. aureus*, it will secrete the enzyme coagulase. Coagulase will convert fibrinogen to fibrin and produce a clotted, or coagulated, tube of plasma. Coagulase-negative staphylococci can still be medically important, however. *Staphylococcus saprophiticus*, for instance, is an important cause of urinary tract infections. It can be distinguished from *S. epidermidis* by resistance to novobiocin.

> **Thought Question**
>
> **28.3** Use **Figure 28.10** to identify the organism from the following case: A sample was taken from a boil located on the arm of a 62-year-old man. Bacteriological examination revealed the presence of Gram-positive cocci that were also catalase-positive, coagulase-positive, and novobiocin resistant.

Rapid Techniques for Pathogen Identification

Conventional, petri plate–dependent methods used to identify pathogens take a minimum of 3 days to complete and may take several weeks, depending on the pathogen. *Brucella* species, for instance, can take 14–21 days to grow in blood culture bottles. The delay may be annoying for the physician, but agonizing for the patient awaiting a cure. Today, technologies that offer more rapid identifications, sometimes within minutes, have been introduced into the clinical laboratory.

Nucleic acid–based pathogen detection. Most diagnostic laboratories today use rapid DNA-based methods in addition to conventional petri dish microbiology. The DNA-based methods take mere hours to detect and identify bacteria and viruses. DNA/RNA detection methods are especially useful for viruses, which otherwise require elaborate electron microscopy to view morphology, or serology to detect an increased presence of antiviral antibodies. The problem with serology is that by the time these antibodies become detectable in blood, the patient is usually already recovering from the disease.

The polymerase chain reaction (PCR). PCR is the most widely used molecular method in the diagnostic toolbox (see Section 7.6, Fig. 7.28). DNA primers that bind to unique genes in a pathogen's genome can be used to specifically amplify DNA or RNA present in a clinical or environmental specimen thought to harbor a pathogen. Successful amplification is visualized as an appropriately sized DNA fragment (**amplicon**) in agarose gels following electrophoresis.

Why is PCR needed to detect the presence of these nucleic acids? Without the PCR amplification steps, clinical samples usually provide too little nucleic acid from infecting microorganisms to be detected. For example, sputum samples containing *Mycobacterium tuberculosis*, the cause of tuberculosis, yield minuscule amounts of bacterial DNA. Amplifying *M. tuberculosis* nucleic acid by PCR, however, will turn one copy of DNA into billions of copies.

PCR is very quick. Preparing a clinical specimen for PCR by extracting the DNA or RNA usually takes less than an hour. PCR itself is completed in 1–2 hours. Detection of the PCR-amplified DNA by DNA gel electrophoresis takes an additional 1–2 hours. So, what might take 2–3 days (or sometimes weeks) using biochemical algorithms may take less than a day using PCR.

The table in **eTopic 28.2** lists several instances in which DNA detection tests are useful. A specific example presented in **Figure 28.12** illustrates the use of PCR to

type different strains of the anaerobic pathogen *Clostridium botulinum*, the cause of food-borne botulism. These organisms are not typed serologically, as is the case for *Streptococcus pneumoniae*. *C. botulinum* is divided into different types based on the neurotoxin genes they possess. In this example, **multiplex PCR** was used to simultaneously search for these toxin genes. Multiplex PCR uses multiple sets of primers, one pair for each gene, combined in a single tube with a specimen. Care must be taken to be sure that the primers chosen make different-sized products, do not interfere with each other, and do not produce artifactual (biologically false) products that can confuse interpretation. Multiplex PCR can help identify sets of specific genes present in a single species or can screen for the presence of multiple pathogens in a clinical sample. In the latter, primer sets are designed to amplify genes unique to each pathogen.

PCR can quickly identify microbial DNA in clinical samples, but what about RNA? Many RNA viruses cause disease. Can molecular tools identify them?

Case History: West Nile Virus

A 55-year-old man was admitted to a local hospital complaining of headache, high fever, and neck stiffness. The man appeared confused and disoriented. He also complained of muscle weakness. History indicated he had received several mosquito bites approximately 2 weeks previously. A blood specimen was sent to the laboratory. The report the following day indicated that the patient was suffering from West Nile virus.

West Nile virus (WNV) is an RNA-containing virus that infects primarily birds and culicine mosquitoes (a group of mosquitoes that can transmit human diseases). Humans and horses serve only as incidental, dead-end hosts for the virus. Replication of virus in this bird-mosquito-bird cycle begins when adult mosquitoes emerge in early spring and continues until fall. Among humans, the incidence of disease peaks in late summer and early fall. Birds provide an efficient means of geographic spread of the virus. As a result, over the past two decades the virus has spread throughout much of the United States.

Isolating a disease-causing virus is extremely challenging. Most laboratories are not equipped for the special tissue culture techniques required to grow viruses. Consequently, many viral infections, including human West Nile virus infections, have been diagnosed by measuring the antibody response of the patient (described shortly). For instance, the presence of West Nile virus–specific IgM in cerebrospinal fluid is a good indicator of current WNV infection, but it is indirect and inconclusive. Real-time PCR is a molecular test that can quickly reveal the presence of the viral RNA.

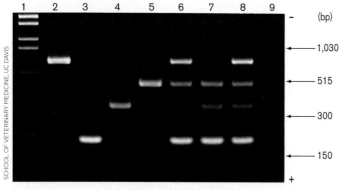

FIGURE 28.12 ■ **Multiplex PCR identification of *Clostridium botulinum.*** *C. botulinum* cells are typed by which toxin genes a strain possesses, not on the basis of surface antigens. Isolates can be typed in a single PCR reaction that includes primer pairs specific for each of the four major toxin genes: types A, B, E, and F. Because multiple products are sought in a single reaction, this is called multiplex PCR. Each lane in the agarose gel was loaded with multiplex products from different isolates of *C. botulinum* and subjected to electrophoresis. The slower-moving fragments (toward top of gel) are larger than those moving farther down the gel toward the positive pole. Lane 1, DNA size markers; lane 2, type A (*cntA*); lane 3, type B (*cntB*); lane 4, type E (*cntE*); lane 5, type F (*cntF*); lane 6, types A, B, and F; lane 7, types B, E, and F; lane 8, types A, B, E, and F.

Conversion of mRNA to cDNA by reverse transcription

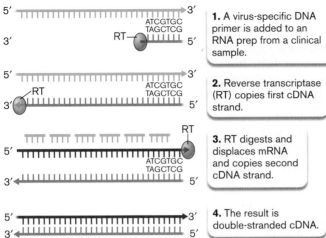

1. A virus-specific DNA primer is added to an RNA prep from a clinical sample.

2. Reverse transcriptase (RT) copies first cDNA strand.

3. RT digests and displaces mRNA and copies second cDNA strand.

4. The result is double-stranded cDNA.

FIGURE 28.13 ∎ **Identifying viruses by reverse transcription PCR.** Reverse transcriptase uses viral RNA as a template to make DNA. The DNA can then be amplified using standard PCR methods.

Reverse transcription quantitative PCR (RT-qPCR).

RT-qPCR is used routinely for the high-throughput diagnosis of many viral pathogens, including the West Nile virus. Because West Nile virus (*Flaviviridae* family) contains single-stranded RNA, its RNA must be converted to DNA using reverse transcriptase before PCR can be attempted. The quantitative advantage of RT-qPCR is that you can estimate the number of virus particles present in the sample by the number of viral RNA molecules there.

The basic technique is as follows: RNA is first extracted from the sample and a DNA primer specific to a viral RNA sequence is added. Once annealed, the primer allows reverse transcriptase to synthesize the first strand of a complementary DNA (cDNA) (**Fig. 28.13**). Reverse transcriptase then digests the initial RNA while synthesizing the opposite DNA strand. The more virus particles there are in the sample (viral load), the more viral RNA will be present and the more cDNA product will

A. Cycle number	Amount of DNA
0	1
1	2
2	4
3	8
4	16
5	32
6	64
7	128
8	256
9	512
10	1,024
11	2,048
12	4,096
13	8,192
14	16,384
15	32,768
16	65,536
17	131,072
18	262,144
19	524,288
20	1,048,576
21	2,097,152
22	4,194,304
23	8,388,608
24	16,777,216
25	33,554,432
26	67,108,864
27	134,217,728
28	268,435,456
29	536,870,912
30	1,073,741,824
31	1,400,000,000
32	1,500,000,000
33	1,550,000,000
34	1,580,000,000

be made. The cDNA is then amplified by PCR using two specific primers and a heat-stable DNA polymerase such as Taq polymerase (see Section 12.2, Fig. 12.13).

The trick is in the method used to quantify the result. In one method, a third, fluorescent oligonucleotide (called the probe) is added to the PCR reaction (see Fig. 12.13A). The probe contains a fluorescent dye at the 3′ end and a chemical dye at the 5′ end that quenches (absorbs) energy emitted from the fluorescent dye. As long as the two chemicals are kept in close proximity by the intact probe, no light is emitted. The probe is designed to anneal to a sequence between the binding sites of the two other primers (modifications on the ends of the probe prevent it from being used as a primer). So, in a successful amplification, Taq polymerase will synthesize DNA from the two outside primers and degrade the probe oligonucleotide as it passes through that area. This cleavage separates the dye from the quencher, and the dye begins to fluoresce. The greater the amount of cDNA there was to begin with, the fewer cycles it takes to register a fluorescence increase over background (**Fig. 28.14**).

> **Thought Question**
>
> **28.4** If the results in **Figure 28.14** came from testing for HIV RNA, then which patient would have the higher viral load in their blood?

B.

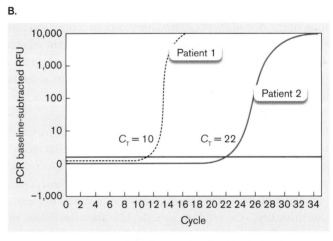

FIGURE 28.14 ∎ **Results of quantitative (real-time) PCR.** **A.** The exponential increase in PCR products after each cycle of hybridization and polymerization. The switch to yellow indicates the point where the increase in product plateaus because the primers have been exhausted. **B.** There is an increase in relative fluorescence units (RFUs) as the fluorescent dye is released from the dual-labeled probe during quantitative PCR. The red line indicates background fluorescence threshold. The cycle at which fluorescence rises above background is the cycle threshold (C_T). In patient 1, the dashed curve shows that fluorescent PCR product remained flat for the first ten cycles because the amount of DNA made, and therefore the level of fluorescent dye released remained below background. After cycle 10, fluorescence increased logarithmically: $C_T = 10$. For patient 2, the solid blue curve representing PCR product remained below detection for 22 cycles, and then increased: $C_T = 22$. The more starting DNA there is, the sooner RFU values will increase over background (that is, the fewer cycles will be needed to see the increase over background). The slope eventually decreases because the fluorescent probe has become limiting.

Rapid, next-generation DNA sequencing. In 2014, a 14-year-old boy with severe combined immunodeficiency (SCID) was in a medically induced coma at a Wisconsin hospital. His brain had swollen with fluid as a result of encephalitis, but its cause was unknown. The usual diagnostic tests and antibiotic treatments did not provide answers or relief. In a last attempt to find the infection's cause, the boy's physicians called Joseph DeRisi and Charles Chiu at UC San Francisco, who are renowned for their use of next-generation DNA sequencing (see Chapter 7) to identify pathogens. The scientists used this technology to perform an unbiased metagenomic analysis of all the DNA contained in the boy's cerebrospinal fluid. After sequencing 3 million DNA fragments, the answer was apparent. There, staring out from among the sequences of the boy's genome, were DNA sequences of the spirochete *Leptospira santarosai*, an easily treatable organism. After being given intravenous high-dose penicillin for 7 days, the boy eventually recovered.

Leptospirosis is a zoonotic disease. The organism is excreted in the urine of infected animals, usually into nearby bodies of freshwater. The child probably contracted the disease during a trip to Puerto Rico, where he swam in freshwater. Why didn't immunological tests initially identify antibodies to *Leptospira*? The immunocompromised state of the boy likely prevented him from making detectible amounts of anti-leptospiral antibodies. DNA, then, was the only way. This metagenomic strategy to detect pathogens is not widely used yet because of its complexity, but someday it could be used to diagnose any infection.

Rapid DNA sequencing by mass spectrometry. The future of nucleic acid–based diagnostic microbiology includes a new technology that combines PCR with mass spectrometry (MS). PCR can amplify DNA directly from clinical samples using primers that target conserved bacterial sequences (for instance, 16S rRNA genes). MS then analyzes the molecular weight of the "amplicon" and sequences it. As described for proteins in eAppendix 3, sequencing involves electron spray ionization to progressively shorten the PCR product one base at a time, and mass spectrometry to calculate the weight of each new, shortened fragment. The difference in molecular weights between the initial DNA fragment and the progressively shortened ones is used to determine the identity of each base removed and, thus, the sequence of the fragment. Because culturing would no longer be needed, this technology could eventually replace the time-honored petri dish.

Programmable RNA sensors. Zika virus is a fast-emerging, insect-borne pathogen that causes serious developmental defects, such as microcephaly, in infected fetal brains (see the Current Research Highlight in Chapter 26, and Section 28.4). The virus has been declared a public health emergency

that requires a quick and easy diagnostic test. Keith Pardee (University of Toronto) and Alexander Green (Arizona State University), shown in **Figure 28.15A**, with a team of scientists led by James Collins, have designed just such a test using synthetic-biology approaches (see Section 12.5). It is a paper-based test that takes only 3 hours to complete.

The basis of the test is illustrated in **Figure 28.15B**. The first step is to generate a "trigger" RNA that will ultimately flip a riboswitch sensor. To make sufficient amounts of the trigger RNA, a trigger primer that binds to a unique Zika virus RNA sequence is added to a serum sample extract that contains Zika virus RNA. Reverse polymerase uses the primer to generate a cDNA template. To amplify copies of trigger RNA, a second, hybrid primer is added whose 3′ end binds to the Zika virus cDNA and whose 5′ end includes a T7 phage promoter sequence. T7 RNA polymerase is then added to the new double-stranded cDNA to generate multiple copies of Zika trigger RNA. This amplification process is called isothermic DNA amplification because thermocycling, as used by PCR, is not needed. The trigger RNA copies derived from the original infecting virus will trip the riboswitch in the next phase of the test.

The synthetically constructed riboswitch sensor is shown in **Figure 28.15C**. The RNA switch contains a hairpin structure that blocks translation of downstream LacZ RNA by sequestering the LacZ ribosome-binding site (RBS) and start codon. This is called a "toehold" switch. When Zika trigger RNA (made previously) is added, the trigger RNA binds to the toehold switch and releases the LacZ RBS and start codon. Translation of the mRNA produces LacZ protein that will convert a yellow substrate to a purple product (a positive test). Once the trigger RNA is added to the switch, the mixture is placed on paper disks containing protein synthesis components, which, after an hour, are analyzed by an electronic reader. This rapid test product is not yet on the market, but its development provides a glimpse of what this technology can bring to the field of rapid diagnostics.

Serology-based pathogen detection. In addition to directly identifying a pathogen by culture isolation or by PCR detection, evidence for a pathogen can be found in a patient's serum, as seen in the following case history.

Case History: Ebola Outbreak

On December 26, 2013, an 18-month-old West African boy in the town of Meliandou, Guinea, fell ill with a mysterious disease of high fever, black stools, and vomiting. He died 2 days later. By January 2014, several members of the boy's family, as well as staff at the nearby Guéckédou hospital, had developed similar symptoms that included severe bloody diarrhea. All died. Then, in February, a member of the boy's family traveled to the capital city, Conakry, where he, too, fell ill and

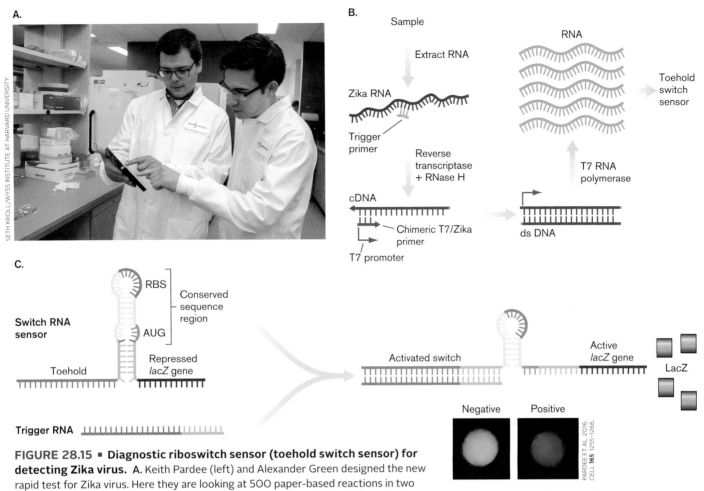

FIGURE 28.15 ■ Diagnostic riboswitch sensor (toehold switch sensor) for detecting Zika virus. A. Keith Pardee (left) and Alexander Green designed the new rapid test for Zika virus. Here they are looking at 500 paper-based reactions in two 384 well plates. **B.** Method to generate trigger RNA from Zika virus particles present in a clinical sample. **C.** Toehold switch sensor. Binding of trigger RNA to the toehold sensor releases the sequestered RBS and start codon and enables translation of the LacZ message. **Inset:** Negative (left) and positive (right) tests for Zika virus. *Source:* Pardee et al. 2016 Rapid, Low cost Detection of Zika virus using Programmable Biomolecular Components.

died. The cause remained unknown, and no precautions were taken. Consequently, over the next month the illness spread to four other prefectures in Guinea, killing at least half of those infected. In March, a World Health Organization (WHO) laboratory at the Pasteur Institute in France finally identified the agent as the Zaire strain of Ebola virus, the most deadly form of this filovirus. Laboratory confirmation tests included RT-qPCR, viral antigen detection, and antibody ELISA tests. Unfortunately, identification of the virus came too late to stop the horrific Ebola epidemic that swept through western Africa in 2014–15, killing thousands.

Ebola is another RNA virus, but one far more deadly than the West Nile virus identified in the earlier case history. Pathogenesis of Ebola was discussed in Section 26.6 (Fig. 26.25). This case includes many elements of epidemiology (discussed later), all of which rely on an ability to identify the presence of the virus. Diagnosis during this epidemic involved three techniques. We already discussed RT-qPCR, but what serological tests were used?

The ELISA test. Enzyme-linked immunosorbent assay (ELISA) is an immunological technique that can detect antigens or antibodies present in picogram quantities. (A picogram is one one-thousandth of a nanogram.) One form of ELISA detects serum antibodies. It is carried out in a 96-well microtiter plate, which allows multiple patient serum samples to be tested simultaneously. An antigen from the virus (Ebola in this case) is attached (adsorbed) to the plastic of the wells (**Fig. 28.16**). Albumin or powdered milk is used to block the remaining sites on the plastic that could result in false positives. Patient serum is then added.

Ebola-specific antibodies present in the serum will react with the antigen attached to the microtiter plate. The antigen-antibody complex is then reacted with rabbit antihuman IgG to which an enzyme has been attached, or conjugated (for example, horseradish peroxidase). The result is a chain of viral antigen connected to patient antibody connected to rabbit antibody-enzyme that links the enzyme to the well. The chromogenic substrate for the enzyme is added next (for example, tetramethylbenzidine).

Enzyme-linked immunosorbent assay (ELISA)

ELISA plate

Albumin

Ebola antigen

Albumin

Rinse off excess and add patient serum.

Human anti-Ebola antibody from patient serum binds to Ebola antigen.

Wash off unbound serum and add conjugated antibody.

Rabbit antihuman IgG antibody with attached (conjugated) enzyme

Wash off unbound conjugated antibody and add substrate.

Rate of conversion of substrate to colored product is proportional to the amount of anti-Ebola antibody that was present in the patient's serum.

FIGURE 28.16 ▪ Enzyme-linked immunosorbent assay (ELISA). ELISA to detect anti-Ebola antibodies circulating in patient serum. The 96-well plate can be used to make dilutions of a single patient's serum to more precisely determine the amount of anti-Ebola antibody, or it can be used to test samples from multiple patients.

Antigen capture

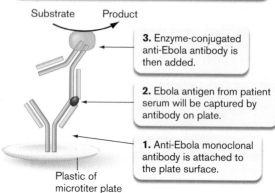

4. If Ebola antigen is present, the conjugated antibody will be captured by the complex. The addition of substrate will lead to production of a colored product.

Substrate   Product

3. Enzyme-conjugated anti-Ebola antibody is then added.

2. Ebola antigen from patient serum will be captured by antibody on plate.

1. Anti-Ebola monoclonal antibody is attached to the plate surface.

Plastic of microtiter plate

FIGURE 28.17 ▪ Antigen-capture ELISA. This ELISA technique captures Ebola antigens circulating in patient serum.

If enzyme-conjugated antibody has bound to human IgG captured by the antigen in the well, the enzyme will convert the substrate to a colored product (blue for tetramethylbenzidine). Enzyme activity can be measured with an ELISA plate reader. The amount of colored product formed, detected as absorbance with a spectrophotometer, will be an indication of the amount of anti-Ebola antibody present in the patient sample.

Thought Question

28.5 Why does adding albumin or powdered milk prevent false positives in ELISA?

Antigen capture is another ELISA technique, but in this instance, anti-Ebola antibody, not viral antigen, is adsorbed to the wells of a microtiter plate (**Fig. 28.17**). Patient serum is then added to the wells. If the serum contains Ebola antigen, the antigen will be captured by the antibody attached to the well. Then a second, enzyme-conjugated antibody against the Ebola antigen is added. The more antigen there is in the serum, the more enzyme-linked antibody will affix to the well. Addition of the appropriate chromogenic substrate will produce a colored product that can be measured.

Antibody against Ebola, or any other viral pathogen, may be easier to detect than viral antigen because antibodies will be present at higher levels than the virus itself. But since there is a delay between the time when the virus is first present in serum and when the body manages to make antibody, a speedier diagnosis can be made by directly detecting viral antigen.

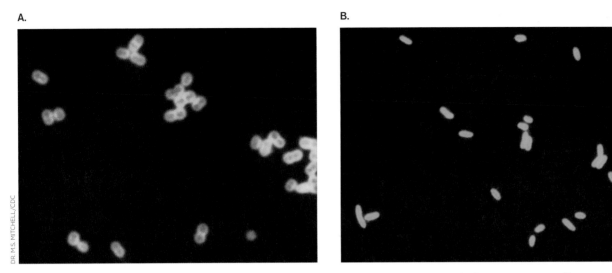

FIGURE 28.18 ■ **Fluorescent antibody stain. A.** *Streptococcus pneumoniae* capsules (cells approx. 0.8 μm; fluorescence microscopy). The capsule is the green halo. The center of the halo is the cell. **B.** *Legionella pneumophila* (cells approx. 1 μm in length) from a respiratory tract specimen.

Thought Questions

28.6 Why does finding IgM to West Nile virus indicate a current infection? Why wouldn't finding IgG do the same?

28.7 Specific IgG antibodies against an infectious agent can persist for years in the bloodstream, long after the infection resolves. How is it possible, then, that IgG antibody titers can be used to diagnose recently acquired diseases such as infectious mononucleosis? Couldn't the antibody be from an old infection?

Fluorescent antibody staining. Chapter 26 presents a case history involving an 80-year-old nursing-home resident who contracted pneumonia caused by *Streptococcus pneumoniae*. The laboratory diagnosis was probably made using the biochemical algorithm previously described. However, there are over 80 serological types of *S. pneumoniae*, each one containing a different capsular antigen. How can the lab identify which antigenic type has caused the infection? One way is to stain the organism with antibodies.

Figure 28.18A illustrates the result of staining a smear of the isolated streptococcus with fluorescently tagged antibodies directed against a specific antigenic type of capsule. Viewed under a fluorescence microscope, the organism is "painted" green when the right antibody binds to the capsule. In the case of pneumonia, this knowledge probably will not help in treating the patient, but its broader value is in determining whether a single type of organism is the cause of an outbreak of pneumonia, which in turn is of epidemiological value for identifying the source of the bacterium.

On the other hand, fluorescent antibody staining techniques are critically important for rapidly identifying organisms that are difficult to grow. Infected tissues can be subjected to direct fluorescent antibody staining. **Figure 28.18B**, for example, shows a direct fluorescent antibody stain of pleural fluid from a patient with Legionnaires' disease.

Identifying pathogens using mass spectrometry. Mass spectrometry (MS) was briefly described in Section 10.6 in the discussion of proteomics. This technology is now able to identify whole bacteria in clinical labs. One such technique is called matrix-associated laser desorption/ionization–time-of-flight mass spectrometry (MALDI-TOF-MS). Whole bacteria or an extract is mixed with a matrix solution and spotted on a metal plate (**Fig. 28.19**) that absorbs energy from a laser and converts it to heat. The heat ionizes proteins in the sample.

Matrix-associated laser desorption/ionization is a soft ionization technique that leaves proteins intact but ionized. An electrostatic field then accelerates the ionized proteins through a vacuum tube, and their time of flight before hitting a detector is measured. How long a protein takes to reach the detector depends on its mass-to-charge ratio. Large proteins with a high mass-to-charge ratio take longer to travel than small proteins with a small mass-to-charge ratio. The results are printed as a series of peaks. The pathogen can be identified by comparing the pattern of peaks with those in a database of organisms (**Fig. 28.20**).

MALDI-TOF-MS has been incorporated into the work flows of many diagnostic microbiology laboratories across the world. The procedure is very fast. Once a sample is loaded, an identification result can be delivered in approximately 15–30 seconds. In most cases of infection, a single

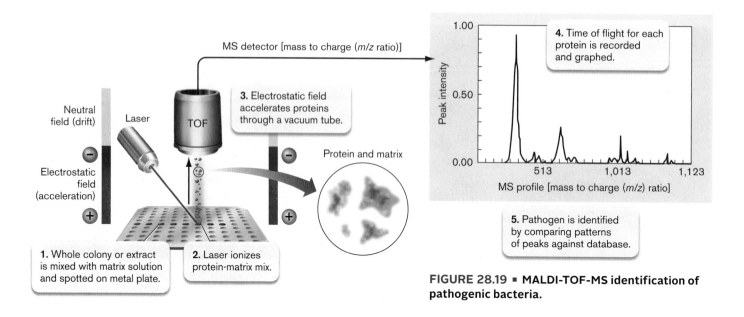

FIGURE 28.19 ▪ MALDI-TOF-MS identification of pathogenic bacteria.

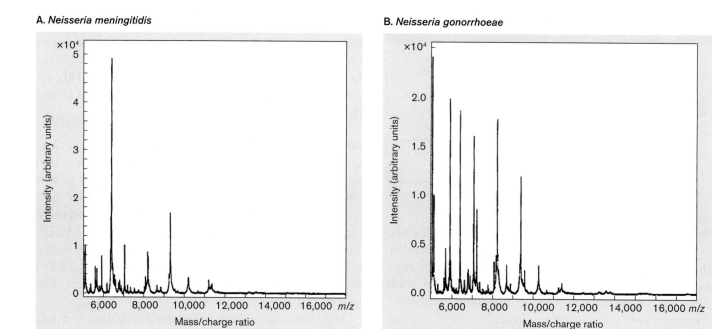

FIGURE 28.20 ▪ MALDI-TOF-MS comparison of pathogenic *Neisseria* species.

colony isolate must be used for identification, but the technology can now identify bacteria directly from blood (sepsis) or urine (cystitis).

Other Pathogens Identified by Conventional or Rapid Diagnostics

Space does not permit a complete listing of the methods used to identify microbial pathogens, but **Table 28.3** presents some additional examples. In general, bacteria that are easily cultured are grown in the laboratory, after which biochemical tests are performed. Bacterial, viral, and fungal species that are difficult to grow are typically identified using immunological techniques that either identify a microbial antigen present in infected tissues or measure a rise in antibody titer. Nucleic acid–based methodologies are also used to identify organisms that are difficult to grow. Eukaryotic microbial parasites such as *Plasmodium* species (the cause of malaria), *Giardia lamblia* (which causes giardiasis, a diarrheal disease), and *Entamoeba histolytica* (the cause of amebic dysentery) can be identified via their telltale morphologies under the microscope, making biochemical tests unnecessary.

TABLE 28.3	Identification Procedures for Selected Diseases	
Agent	**Disease**	**Means of identification**
Bacteria		
Corynebacterium diphtheriae	Diphtheria	Material from nose and throat cultured on a special medium; in vivo or in vitro tests for toxin
Bordetella pertussis	Pertussis (whooping cough)	Smears of nasopharyngeal secretions stained with fluorescent antibody; ELISA for toxin in respiratory secretions; culture on special media
Legionella pneumophila	Legionellosis, Legionnaires' disease, Pontiac fever	Culture on special medium is preferred method; antigen detection by ELISA; *Legionella* nucleic acid can be identified in clinical material using PCR and nucleic acid probes
Campylobacter spp.	Campylobacteriosis	Isolation of bacteria on selective media incubated at 42°C in atmosphere of nitrogen containing 5% oxygen and 10% carbon dioxide; then, biochemical testing performed
Leptospira interrogans	Leptospirosis	Serological tests early in the illness and after 2–3 weeks to detect rise in antibody titer
Listeria monocytogenes	Listeriosis	Culture of blood and spinal fluid; selective and enrichment cultures performed on food samples to grow potential pathogens; DNA probe for rapid identification of colonies
Chlamydia trachomatis	Chlamydial genital infections	Identification of *C. trachomatis* antigen in urine or pus using monoclonal antibody; nucleic acid probes
Treponema pallidum	Syphilis	Direct fluorescent antibody staining; serological tests
Francisella tularensis	Tularemia	Cultures using cysteine-containing media; fluorescent antibody stain of pus; detection of rise in antibody titer
Yersinia pestis	Plague	Identification of capsular antigen using fluorescent antibody or ELISA
Viruses		
Rhinovirus	Common cold	Strain identification requires use of specific antibodies
Influenza virus	Flu	Tests comparing influenza antibody levels in blood samples taken during acute and convalescent stages of illness
Hantavirus	Hantavirus pulmonary syndrome	Antigen detection in tissues using electron microscopy or monoclonal antibody; ELISA and western blot tests for IgG and IgM antibodies in victim's blood
Herpes simplex virus	Different strains cause cold sores, ocular lesions, and genital lesions	Identifying the viral antigen in clinical material using fluorescent antibody or DNA probes
Mumps virus	Mumps	Rise in antibody titer, or presence of IgM antibody to mumps virus in patient's blood
Rotavirus	Diarrhea	Electron microscopy or ELISA of diarrheal stool for virus
HIV	AIDS	Detection of antibody to HIV-1 in patient's blood
Fungi		
Coccidioides immitis	Coccidioidomycosis, infection of lung; can disseminate to almost any tissue	Observation of large, thick-walled, round spherules from clinical specimens; PCR identification
Histoplasma capsulatum	Histoplasmosis, intracellular infection of lung; sometimes disseminates	Stained material from pus, sputum, tissue, etc., examined for intracellular *H. capsulatum* yeast phase; blood tests for antibody to the organism

Point-of-Care Rapid Diagnostics

Conventional diagnosis of an infection often requires sending a clinical specimen to a faraway laboratory, followed by a considerable delay in obtaining results. Inevitably, some patients lose patience and fail to attend follow-up appointments. However, point-of-care (POC) laboratory tests are designed to be used directly at the site of patient care, such as physicians' offices, outpatient clinics, intensive care units, emergency departments, hospital laboratories, and even patients' homes. Most patients are happy to wait 40–50 minutes for a rapid POC test result in order to receive immediate treatment or reassurance.

A.

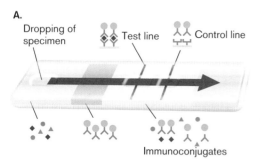

- ◆ C-ps derived from *S. pneumoniae*
- ▲ ● Other components
- ⊻ Antipneumococcal C-ps colloidal gold-labeled rabbit polyclonal antibody
- Y Y Antipneumococcal C-ps solid-phase rabbit polyclonal antibody
- ⊻⊻ Solid-phase goat antirabbit IgG antibody

B.

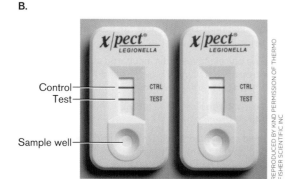

FIGURE 28.21 ■ **Principle of immunochromatographic rapid diagnostic tests. A.** The example is a test for the presence of *Streptococcus pneumoniae* in sputum. An extract of sputum is placed onto one end of the strip where *S. pneumoniae* capsular antigen (C-ps) will bind antipneumococcal C-ps polyclonal antibodies. The resulting immunoconjugates move by capillary action to the upper membrane and are captured by the antipneumococcal C-ps solid-phase polyclonal antibodies, thereby forming sandwich conjugates in the sample. A positive test result is indicated by the presence of both a test and a control line, whereas a negative result is indicated by the appearance of only a single control line. **B.** Immunochromatography test for anti–*Legionella* antibodies. Serum placed on strip 1 contained antibodies to *Legionella*.

The typical POC test approved for use involves an immunochromatographic assay. The test for *Streptococcus pneumoniae* capsular antigen shown in **Figure 28.21A** is one example. This particular immunochromatographic test, called a "red colloidal gold" test, involves extracting the relevant antigen (for example, capsule antigen) from a clinical specimen and placing a few drops of the extract on a test strip containing rabbit antibodies to the antigen. The antibodies have red colloidal gold particles attached to them. The antigen-antibody complexes, if present, move by capillary action to the upper level of the strip, where they are captured by a line of more anti-antigen antibodies embedded in the strip, forming an antibody-antigen-antibody sandwich. The colloidal gold particles accumulate and eventually produce a red test line, indicating the presence of the antigen. In contrast, rabbit antibodies that are not bound to the antigen pass through the test line but are captured by goat antirabbit IgG antibodies on a control line, once again forming a red control line that indicates

the strip components are working. **Figure 28.21B** shows results of a test for *Legionella* antibodies.

Two properties that are critical to any POC test are sensitivity and specificity. **Sensitivity** measures how often a test will be positive if a patient has the disease. Sensitivity reflects how small a concentration of antigen the test can detect. In contrast to sensitivity, **specificity** measures how often a test will be negative if a patient does not have the disease. Specificity reflects how well a test can distinguish between two closely related antigens (for example, *Streptococcus* strains carrying group A versus group B capsular antigens). Assays with high specificity and sensitivity are valuable diagnostic tools.

Commercial POC tests are widely available for the diagnosis of bacterial and viral infections and for parasitic diseases, including malaria (**Table 28.4**). However, as convenient as these tests are, sensitivity may be compromised in the quest for a speedy result. Some tests exhibit insufficient sensitivity and should therefore be coupled with confirmatory tests when the results are negative (one test that can produce false negatives is the *Streptococcus pyogenes* rapid antigen detection test). Other POC tests need to be confirmed when positive; for instance, a rapid malaria POC test can produce false positives.

There are several advantages and some disadvantages to POC rapid tests. The advantages are:

- Culturing is not required.

- The clinician can immediately initiate specific antibiotic therapy.

- The consumption of antibiotics is avoided in the case of a viral infection, and antiviral therapy, if possible, can be initiated.

- Infection chains among patients with similar symptoms are revealed.

- Compliance with patients who are difficult to reach is improved.

The disadvantages include:

- The tests provide no data on pathogen antibiotic sensitivity.

- There is a higher risk of the technician becoming infected.

- Double or multiple infections are more likely to be overlooked than in culture.

- The levels of false positive or false negative results can vary for different POC tests.

TABLE 28.4	Examples of Point-of-Care Rapid-Test Kits for Infectious Diseases[a]

Disease (pathogen)	Type of test	Sample	Indication	Performance[b]	Notes
Bacterial pathogens					
Chlamydia (*Chlamydia trachomatis*)	PCR	Vaginal swab, urine	Screening, suspicion of PID	Sensitivity: 83% Specificity: 99%	
Gonorrhea (*Neisseria gonorrhoeae*)	PCR	Urine	Screening, suspicion of PID	Sensitivity: 98% Specificity: 99%	
Syphilis (*Treponema pallidum*)	ICT	Blood	Screening	Sensitivity: 90%–95% Specificity: 90%–95%	
Strep throat (*Streptococcus pyogenes*)	EIA	Pharyngeal swab	Sore throat	Sensitivity: 53%–99% Specificity: 62%–100%	Confirmation of negative swabs
Legionellosis (*Legionella* spp.)	ICT	Urine	Severe pneumonia; risk factors for legionellosis	Sensitivity: 76% Specificity: 99%	Only serotype 1 reliably detected
Pneumococcal pneumonia (*Streptococcus pneumoniae*)	ICT	Urine; also pleural fluid or CSF	Severe pneumonia; also empyema, meningitis	Sensitivity: 66%–70% Specificity: 90%–100%	Detects capsule antigens
Pseudomembranous enterocolitis (*Clostridioides difficile*)	ICT	Stool	Antibiotic-associated diarrhea	Sensitivity: 49%–80% Specificity: 95%–96%	Notably less sensitive than cultures or PCR
Neonatal septicemia (*Streptococcus agalactiae*)	PCR	Vaginal swab	Peripartum detection of colonization	Sensitivity: 92% Specificity: 96%	
Protozoan pathogens					
Malaria (*Plasmodium falciparum*)	ICT	Blood	Fever in returning traveler	Sensitivity: 87%–100% Specificity: 52%–100%	Sensitivity better for *P. falciparum* (pan-malarial tests)
Trichomoniasis (*Trichomonas* spp.)	ICT	Vaginal swab	Symptoms of vaginitis	Sensitivity: 83% Specificity: 98%	
Viral pathogens					
Influenza (influenza virus)	ICT	Nasopharyngeal swab	Flu-like symptoms	Sensitivity: 20%–55% Specificity: 99%	Low sensitivity; probably not helpful during outbreaks; lower in adults
RSV disease (RSV)	ICT	Nasopharyngeal swab	Viral symptoms, especially during winter	Sensitivity: 59%–97% Specificity: 75%–100%	
HIV	ICT	Blood; also oral fluid	Screening, prevention of vertical transmission	Sensitivity: 99%–100% Specificity: 99%–100%	
Dengue fever (dengue virus)	ICT	Blood	Screening in endemic regions	Sensitivity: 90% Specificity: 100%	
Mononucleosis (Epstein-Barr virus)	ICT	Blood	Screening	Sensitivity: 90% Specificity: 100%	Detects IgM "heterophile antibodies"[c]
Diarrheal disease (rotavirus)	ICT	Stool	Diarrhea	Sensitivity: 88% Specificity: 99%	
Hepatitis B (HBV)	ICT	Blood	Prenatal or transfusion screening; or suspicion of acute or chronic carriage of HBV	Sensitivity: 99% Specificity: 100%	Detects HBs antigen
Rubella (rubella virus)	ICT	Blood	Pregnancy	Sensitivity: 99% Specificity: 99%	

[a]Abbreviations: AIDS, acquired immunodeficiency syndrome; CSF, cerebrospinal fluid; EIA, enzyme immunoassay; HBV, hepatitis B virus; HIV, human immunodeficiency virus; ICT, immunochromatographic test; PCR, polymerase chain reaction; PID, pelvic inflammatory disease; RSV, respiratory syncytial virus.

[b]Sensitivity = proportion of actual positives correctly identified. Specificity = proportion of actual negatives correctly identified.

[c]Epstein-Barr virus randomly infects B cells and causes them to secrete antibodies. Because many thousands of different antibodies are made, they are referred to collectively as "heterophile antibodies."

The newer nucleic acid–based tests described earlier exhibit better sensitivity and specificity than most immunochromatographic assays, but they require more expensive instrumentation and training. In the coming years, further evolution of POC tests may lead to new diagnostic approaches, such as panel testing that targets all possible pathogens suspected in a specific clinical setting. The development of next-generation serology-based and/or molecular-based multiplex tests will certainly facilitate quicker diagnosis and improved patient care. We also saw earlier how synthetic biology may revolutionize rapid diagnostics, as exemplified by the newly described Zika virus riboswitch test.

To Summarize

- **Direct Gram or acid-fast stains** are appropriate procedures to perform on some specimens. The results guide the direction of subsequent testing.

- **Knowing which patient specimens should be sterile and which should contain normal microbiota** informs the selection of media for processing the specimen. So, too, does understanding whether anaerobes may be expected.

- **Selective media** are used to prevent growth of normal microbiota while permitting growth of others (such as Gram-positive bacteria versus Gram-negative bacteria). **Differential media** exploit the unique biochemical properties of a pathogen to distinguish it from similar-looking nonpathogens.

- **Growth-dependent pathogen identification** uses numerous, simultaneously run biochemical tests. **An algorithm or a dichotomous key** is applied to the results to identify the species. Different algorithms (or decision trees) are applied to Gram-positive and Gram-negative bacteria.

- **Pathogens can be identified** with **biochemical analyses** (useful for bacteria), **molecular techniques** (for example, PCR; useful for bacteria or viruses), and/or **immunological methods** (for example, ELISA; useful for bacteria or viruses).

- **Fluorescent antibody staining** can rapidly identify organisms or antigens present in tissues.

- **Point-of-care (POC) diagnostic tests** can rapidly identify the cause of an infectious disease. Therapy can be initiated rapidly following a positive result. **Sensitivity** (the ability to detect small amounts of the pathogen) and **specificity** (the ability to distinguish one pathogen from another) for any point of care test are factors used to evaluate how accurate the test is at eliminating false negative and false positive reactions, respectively.

- **Immunochromatography** is the primary platform for POC testing.

- **A drawback to POC tests** is that simultaneous, multiple infections may be missed.

28.3 Principles of Epidemiology

In Section 25.1 we discussed how a new infectious disease can be identified using a version of Koch's postulates. But how do scientists track the spread of a new disease, or find and identify new variants of influenza virus that develop thousands of miles away, and then predict when that virus will arrive on our "doorsteps" many months in advance? We start our discussion of epidemiology with a case in which scientists had to trace a heinous criminal act back to its source.

Case History: Inhalation Anthrax

On October 16, 2001, a 56-year-old African-American U.S. Postal Service worker became ill with a low-grade fever, chills, sore throat, headache, and malaise. These symptoms were followed by minimal dry cough, chest heaviness, shortness of breath, night sweats, nausea, and vomiting. On October 19, the man arrived at a local hospital, where he presented with a normal body temperature and normal blood pressure. He was not in acute distress, but he had decreased breath sounds and rhonchi (dry sounds in lungs due to congestion). No skin lesions were observed, and he did not smoke. Total white blood cell count was normal, but there was a left shift in the differential—that is, more polymorphonuclear leukocytes (PMNs; see Section 26.3). A chest X-ray showed bilateral pleural effusions (accumulation of fluids in the lung) and a small, right-lower-lobe air space opacity. Within 11 hours, blood cultures taken upon admission grew Bacillus anthracis. Ciprofloxacin, rifampin, and clindamycin antibiotic treatments were initiated, and the patient recovered. His job at the post office was simply to sort mail.

From October 4 to November 2, 2001, the Centers for Disease Control and Prevention (CDC), and various state and local public health authorities reported 10 confirmed cases of inhalational anthrax and 12 confirmed or suspected cases of cutaneous anthrax in persons who worked in the District of Columbia, Florida, New Jersey, and New York. Many of them were postal workers. It was clear that a biological attack was in progress.

Painstaking detective work by federal agents and epidemiology scientists proved that the strain of *Bacillus anthracis* used in the 2001 anthrax attack had the same genetic signature as a strain used by Bruce Ivins, a scientist working at the army's Fort Detrick biodefense laboratory. Although his involvement was never proved, Dr. Ivins took his own life as agents were about to arrest him.

The word "epidemiology" is derived from the Greek meaning "that which befalls man." In scientific parlance, **epidemiology** examines the distribution and determinants of disease frequency in human populations. Put more simply, epidemiologists determine the source of a disease outbreak and the factors that influence how many individuals will succumb to the disease. Epidemiological principles are also used to determine the effectiveness of therapeutic measures and to identify new syndromes, such as SARS (severe acute respiratory syndrome), MERS (Middle East respiratory syndrome), and Lyme disease. Some of the basic concepts of epidemiology were already covered in Chapter 26 when we discussed infection cycles. Now we will explore how those principles are used to track disease.

Epidemiological early-warning systems require an extensive organization that coordinates information from many sources. In the United States, that duty falls to the Centers for Disease Control and Prevention (CDC). On the world stage, it is the World Health Organization (WHO). Any disease considered highly dangerous or infectious is first reported to local public health centers, usually within 48 hours of diagnosis. The local centers forward that information to their state agencies, which then report to the CDC in Atlanta. This is how authorities in 2001 quickly recognized that an outbreak of anthrax was under way.

Endemic, Epidemic, or Pandemic?

The terms "endemic" and "epidemic" are often used when referring to disease outbreaks. A disease is **endemic** if it is always present in a population at a low frequency. For example, Lyme disease, caused by the spirochete *Borrelia burgdorferi*, is endemic to the northeastern United States because the organism has found a reservoir in deer and ticks. Recall that a **reservoir** is an animal, bird, or insect that harbors the infectious agent and is indigenous to a geographic area. Humans become infected only when they come in contact with the reservoir. Thus, the disease incidence is low but relatively constant. A disease is **epidemic**, on the other hand, when larger-than-normal numbers of individuals in a population become infected over a short time. Epidemics arise, in part, because of rapid and direct human-to-human transmission. Food-source epidemics, for instance, are discussed in Chapter 16.

Figure 28.22A illustrates the difference in the frequency of cases observed between endemic and epidemic

A. Endemic versus epidemic

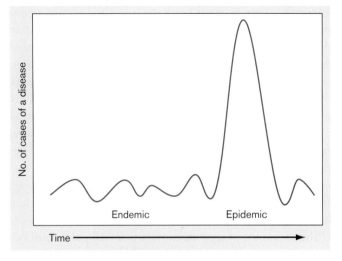

B.

FIGURE 28.22 ■ **The difference between endemic and epidemic disease. A.** An endemic disease is continually present at a low frequency in a population. A sudden rise in disease frequency constitutes an epidemic. **B.** A health care worker stands outside a quarantine area housing Ebola patients in the Ivory Coast. Epidemics can be minimized if infected persons are kept segregated from the general population (quarantined) to avoid spread of the infectious agent.

disease. An endemic disease can become epidemic if the population of the reservoir increases, allowing for more frequent human contact; or if the infectious agent evolves to spread directly from person to person, bypassing the need for a reservoir. This is the concern with the H5N1 avian flu virus, which is endemic in animals and birds in Asia (see Sections 11.2 and **eTopic 26.3**). A **pandemic** is an epidemic that occurs over a wide geographic area, usually the world. Pandemics may be long-lived, such as the bubonic plague pandemic in the fourteenth century and the AIDS pandemic in the late twentieth and early twenty-first centuries; or they may be short-lived, as with the 1918 flu pandemic.

When discussing a disease, epidemiologists distinguish between the prevalence and incidence of active cases. **Prevalence** describes the <u>total</u> number of active cases of a disease in a given location, regardless of when the case first developed. **Incidence**, however, refers to the number of <u>new</u> cases of a disease in that location over a specified time. Incidence reflects the risk a person has of acquiring the disease. Incidence rates can provide insight into whether efforts to limit a disease are working. Take a city of 100,000 people. If the incidence of <u>new</u> cases rises from 10 cases/100,000 population in one year to 30 cases/100,000 population the next year, then efforts to stem the disease are failing. Prevalence, reflecting the total burden on society, could also increase over those two years or might remain the same— say, 500 cases/100,000—if more people are being cured or perhaps die in that second year despite the increase in new cases. Endemic diseases maintain relatively constant prevalence and incidence rates. Outbreaks are typically marked by an increase in both parameters.

Finding Patient Zero

When trying to contain the spread of an epidemic, it is vital to track down the first case of the disease (known as the **index case** or patient zero) and then identify everyone who has had contact with that individual so that they can be treated or separated from the general population (**quarantined**) (**Fig. 28.22B**). When a new disease arises, the epidemiological search for the index case starts only after a number of patients have been diagnosed and a new disease syndrome declared. This is what happened with the Ebola epidemic in 2015.

Identifying an index case within a specific community is easier if the disease syndrome is already recognized, as was the case with the 2003 severe acute respiratory syndrome (SARS) outbreak in Singapore. According to the World Health Organization, a suspected case of SARS is defined as an individual who has a fever greater than 38°C (100.4°F), who exhibits lower respiratory tract symptoms, and who has traveled to an area of documented disease or has had contact with a person afflicted with SARS. The index case in Singapore was a 23-year-old woman who had stayed on the ninth floor of a hotel in Hong Kong while on vacation. A physician from southern China who stayed on the same floor of the hotel during this period is believed to have been the source of her infection, as well as that of the index patients who precipitated subsequent outbreaks in Vietnam and Canada.

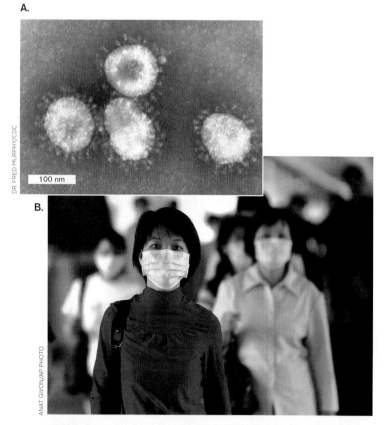

FIGURE 28.23 ■ **Severe acute respiratory syndrome (SARS).** **A.** The coronavirus that causes SARS (TEM). **B.** Citizens of China, including the military, donned surgical masks in 2003 to slow the spread of SARS.

During the last week of February, the woman, who had returned to Singapore, developed fever, headache, and a dry cough. She was admitted to Tan Tock Seng Hospital, Singapore, on March 1 with a low white blood cell count and patchy consolidation in the lobes of the right lung. Tests for the usual microbial suspects (*Legionella*, *Chlamydia*, *Mycoplasma*) were negative. Electron microscopy of nasopharyngeal aspirations showed virus particles with widely spaced club-like projections (**Fig. 28.23A**). At the time of her admission to the hospital, the clinical features and highly infectious nature of SARS were not known. Thus, for the first 6 days of hospitalization, the patient was in a general ward, without barrier infection control measures. During this period, the index patient infected at least 20 other individuals, including hospital staff, nearby patients, and visitors.

Within weeks, the WHO named the disease in China "SARS" and issued travel alerts (discussed further in Section 28.3). These alerts allowed Singapore health officials to rapidly identify the index patient and her contacts. As a result, they were able to limit spread of the illness. When all was said and done, SARS killed fewer than 1,000 victims

TABLE 28.5	Notifiable Infectious Diseases[a]

Bacterial

Anthrax	Hansen's disease (leprosy)	Salmonellosis (non–typhoid fever types)
Botulism	Legionellosis	Shigellosis
Brucellosis	Leptospirosis	Staphylococcal enterotoxin
Campylobacter infection	Listeriosis	Streptococcal invasive disease
Chlamydia infection	Lyme disease	*Streptococcus pneumoniae*, invasive disease
Cholera	Meningitis, infectious	Syphilis
Diphtheria	Pertussis	Tuberculosis
Ehrlichiosis	Plague	Tularemia
E. coli O157:H7 infection	Psittacosis	Typhoid fever
Gonorrhea	Q fever	Vancomycin-resistant *Staphylococcus aureus* (VRSA)
Haemophilus influenzae, invasive disease	Rocky Mountain spotted fever	

Viral

Dengue fever	Measles	Rubella and congenital rubella syndrome
Hantavirus infection	Mumps	Smallpox
Hepatitis, viral	Poliomyelitis	Varicella (chickenpox), fatal cases only (all types)
HIV infection	Rabies	Yellow fever

Fungal

Coccidioidomycosis

Parasitic

Amebiasis	Giardiasis	Trichinosis
Cryptosporidiosis	Malaria	
Cyclosporiasis	Microsporidiosis	

[a]For the latest list of notifiable agents, search on "CDC notifiable agents 2016" online.

worldwide, although thousands more became ill and recovered (**Fig. 28.23B**). Today, SARS remains a threat, but one of lesser concern. Methicillin-resistant *Staphylococcus aureus* (MRSA), H5N1 avian flu, and H1N1 swine flu are considered more pressing dangers.

Identifying Disease Trends

How do epidemiologists first recognize that an epidemic is under way, and then identify the agent and its source? Certain diseases, because of their severity and transmissibility, are called reportable, or notifiable, diseases (**Table 28.5**). Physicians are required to report instances of these diseases to a central health organization, such as the CDC in the United States and the WHO. This reporting allows the incidences of certain diseases within a population to be tracked and upsurges noted. An emerging disease not on the list of notifiable diseases can be detected as a cluster of patients with unusual symptoms or combinations of symptoms. Such detection is possible because diseases of

unknown etiology are also reported to health authorities. A new disease could manifest with common symptoms (for example, the cough and fever of SARS) that cannot be linked to a known disease agent by clinical tests. An upsurge in cases, of either a reportable disease or an emerging disease, will set off institutional "alarms" that initiate epidemiological efforts to determine the source and cause of the outbreak.

John Snow (1813–1858), the Father of Epidemiology

The first case in which the source of a disease outbreak was methodically investigated arose in the mid-nineteenth century. During a serious outbreak of cholera in the Soho district of London in 1854, John Snow (**Fig. 28.24A**) visited the addresses of all the diarrheal cases he learned about. The source of the infection was unknown, but Snow thought that if the cases clustered geographically, he might gain a clue as to its location. Water in that part of London

A.

B.

FIGURE 28.24 ■ **Early epidemiology. A.** John Snow. **B.** A map of London commissioned by Snow in 1855 shows the location of cholera victims of the 1854 outbreak. The map illustrates that each victim within the marked area lived closer to the Broad Street pump than to other nearby pumps. The Broad Street well is found within the red circle. Each black bar represents a death from cholera.

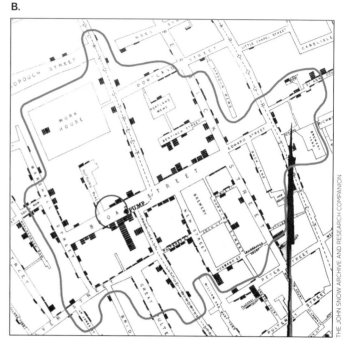

was pumped from separate wells located in the various neighborhoods. Snow realized there was a close association between the density of cholera cases and a single well located on Broad Street (**Fig. 28.24B**). Simply removing the pump handle of the Broad Street well put an end to the epidemic, proving that the well water was the source of infection (known as the "point source," since all infections originated from that point).

This approach succeeded brilliantly, even though the infectious agent that causes cholera, *Vibrio cholerae*, was not recognized until 1905, over 50 years later. Identifying clusters of patients afflicted with a given disease is still used to locate potential sources of infectious disease outbreaks. Outbreaks of disease that originate from a common source, as in the London cholera outbreak, are called common-source outbreaks. In contrast, a propagated outbreak involves person-to-person transmission (SARS) or transmission by insect vectors (such as West Nile virus).

Thought Questions

28.8 Methicillin, a beta-lactam antibiotic, is very useful in treating staphylococcal infections. The emergence of methicillin-resistant strains of *Staphylococcus aureus* (MRSA) is a very serious development because few antibiotics can kill these strains. Imagine a large metropolitan hospital in which there have been eight serious nosocomial infections with MRSA and you are responsible for determining the source of infection so that it can be eliminated. How would you accomplish this task using common bacteriological and molecular techniques?

28.9 What are some reasons why some diseases spread quickly through a population while others take a long time?

Genomic Strategies Help Identify Nonculturable Pathogens

Microbiologists have successfully developed many strategies to identify the causes of infectious diseases. As noted earlier, Robert Koch in the late nineteenth century devised a set of postulates that, when followed, can identify the agent of a new disease (discussed in Sections 1.3 and 25.1). One important tenet of Koch's postulates is that the suspected organism must be grown in pure culture. However, there are bacterial diseases for which the agent cannot be cultured. How might they be identified?

Case History: Whipple's Disease

In April, a 50-year-old man had an abrupt onset of watery diarrhea with a stool frequency of up to 10 times every 24 hours. This was the latest episode in a 6-year history of illness beginning with recurring fevers, flu-like symptoms, profuse night sweats, and painful joint swelling. The current bout of diarrhea was associated with gripping lower abdominal pain, especially after meals. No blood or mucus was found in the stools. Blood and stool cultures tested negative for known infectious agents. Serology was also negative for syphilis, brucellosis, toxoplasmosis, and leptospirosis. The man's weight fell rapidly from 83 to 73 kg within 4 weeks of onset. A flexible sigmoidoscopy showed only diffuse mild erythema in the bowel. However, the appearance of the small bowel was consistent with malabsorption, a disease in which nutrients are poorly absorbed by the intestine. A duodenal biopsy to look for the organisms of Whipple's disease showed large macrophages in the lamina propria (a layer of loose connective tissue beneath the epithelium of an organ). Bacterial rods characteristic of Whipple's disease were seen with electron micros-copy. PCR analysis of tissue samples confirmed the diagnosis.

First diagnosed by George Whipple in 1907, Whipple's disease is characterized by malabsorption, weight loss, arthralgia (joint pain), fevers, and abdominal pain. Any organ system can be affected, including the heart, lungs, skin, joints, and central nervous system. The cause of Whipple's disease went undiscovered for 85 years, but the disease was suspected to be of bacterial etiology, even though an organism was never successfully cultured. The agent was finally identified in 1992—not by culturing, but by blindly amplifying 16S rDNA sequences from biopsy tissues.

All bacterial 16S rRNA genes have some sequence regions that are highly homologous across species and other sequences that are unique to a species. Tissues from numerous patients diagnosed with Whipple's disease were subjected to PCR analysis using the common 16S rDNA primers. If a bacterial agent was present, it was predicted that PCR should success-fully amplify a DNA fragment corresponding to the agent's 16S rRNA gene. All tissues produced such a fragment, indicat-ing that there were bacteria in the tissues. DNA sequence anal-ysis of these fragments indicated that the organism was similar to actinomycetes but was unlike any of the known species.

The organism is actually a Gram-positive soil-dwelling actinomycete that was named *Tropheryma whipplei* in honor of the physician who first recognized the disease. Because of its bacterial etiology, Whipple's disease can be treated with antibiotics, usually trimethoprim sulfamethoxazole (Bactrim, Septra, Cotrim). **Figure 28.25** shows an in situ hybridization for *T. whipplei* RNA in a tissue biopsy.

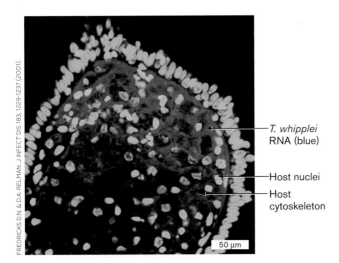

FREDRICKS D.N. & D.A. RELMAN. J INFECT DIS 183, 1229-1237 (2001).

T. whipplei
RNA (blue)

Host nuclei

Host
cytoskeleton

50 μm

FIGURE 28.25 ■ Whipple's disease. Fluorescence in situ hy-bridization of a small-intestine biopsy in a case of Whipple's disease (confocal laser scanning microscopy). In this test, a fluorescently tagged DNA probe that specifically hybridizes to *Tropheryma whip-plei* RNA is added to the tissue. Other fluorescent probes are used to visualize host nuclei and cytoskeleton. Blue = *T. whipplei* rRNA; green = nuclei of human cells; red = intracellular cytoskeletal protein vimentin. Magnification approx. 200×.

This story reveals that Koch's postulates (see Fig. 1.17) must sometimes be modified when identifying the cause of a new disease. In this instance, the organism could not be cultured in pure form in the laboratory, but it was found, via molecular techniques, in all instances of the disease. More recent studies have successfully cultured *T. whipplei* in vitro. The medium used was formulated on the basis of nutritional requirements deduced from knowing the DNA sequence of the *T. whipplei* genome (discussed in Section 25.7).

Molecular Approaches for Disease Surveillance

A worldwide pandemic of pulmonary tuberculosis currently affects over 2 billion people. Many of the *Mycobacterium tuberculosis* infections are caused by multidrug-resistant strains that are difficult, if not impossible, to kill with exist-ing antibiotics (see Section 26.3). This problem is especially serious among refugee populations attempting to flee war-torn countries. As a result, it is important to screen these refugees as they enter neighboring countries with low inci-dences of tuberculosis. Although chest X-rays are manda-tory in many cases, a positive image will be obtained only if the disease is at a relatively advanced stage. Actively infected individuals who have not developed the characteristic lung tubercles seen on X-ray will not be identified.

Unfortunately, the acid-fast staining of sputum sam-ples (discussed in Section 28.2) also fails to detect indi-viduals at an early stage of infection. Studies have shown, however, that PCR techniques are much more sensitive for detecting these individuals. As time progresses, PCR sur-veillance strategies will be used more often to track the worldwide ebb and flow of microbial diseases. PCR and restriction fragment length polymorphism (RFLP) strate-gies are already used for epidemiological purposes to type (that is, determine the relatedness of) different microbial isolates by generating a complex DNA profile that is spe-cific for a particular strain (see Section 28.2). For example, DNA profiles were used to link an outbreak of over 2,000 cases of salmonellosis in 2010 to a single strain of *Salmo-nella enterica* serovar Enteritidis. The results, conducted by a national network of public health agencies (PulseNet), led to the recall of over a half billion eggs.

Bioterrorism

"A wide-scale bioterrorism attack would create mass panic and overwhelm most existing state and local systems within a few days," said Michael T. Osterholm, director of the Center for Infectious Disease Research & Policy at the University of Minnesota, in October 2001. "We know this from simulation exercises."

Less than a month after the September 11 attack on the World Trade Center, a biodefense scientist working at the U.S. Army Medical Research Institute of Infectious Diseases (Fort Detrick, Maryland) allegedly sent weapons-grade anthrax spores through the U.S. mail (see the earlier inhalation anthrax case history). Thankfully, only 5 persons died, and a mere 25 became ill. But even though the efficiency of the attack was poor, the impact was enormous. Over 10,000 people took a 2-month course of antibiotics after possible exposure, and mail deliveries throughout the country were affected. The simple act of opening an envelope suddenly became a risky business.

As a result of this attack and other events, the CDC and the National Institutes of Health (NIH) assembled a list of select agents (marked in blue in **Table 28.1**) that could potentially be used as bioweapons. A bioweapon is considered to be any infectious agent or toxin that has high virulence and/or mortality rate. Microorganisms considered bioweapons can be used to conduct biowarfare, with the intent of inflicting massive casualties; or bioterrorism, which may result in only a few casualties but cause widespread psychological trauma.

Although the branding of select agents is recent, biowarfare is not new. In the Middle Ages, victims of the Black Death (plague caused by *Yersinia pestis*) were flung over castle walls using catapults. During the French and Indian War in the eighteenth century, British field marshal Jeffrey Amherst distributed smallpox-infected blankets to Native Americans. The Imperial Japanese Army during World War II experimented with infectious disease weapons using Chinese prisoners as guinea pigs. Even the United States participated by developing weapons-grade anthrax spores after World War II. That project was discontinued in the 1970s.

The 2001 anthrax attack described earlier was not the first act of bioterrorism in the United States. The first documented act occurred in 1984, when followers of the cult leader Bhagwan Shree Rajneesh tried to control a local election in The Dalles, Oregon, by infecting salad bars with *Salmonella*. Over 700 people became ill. Rajneesh was given a 10-year suspended sentence, fined $400,000, and deported.

How effective are bioweapons? The method by which a biological agent is dispersed plays a large role in its effectiveness as a weapon. Only a few people became ill during the 2001 anthrax attack—not only because of the epidemiological surveillance, but because anthrax is inherently difficult to disperse. Thousands of spores must be inhaled to contract disease, which means that effective dispersal of the spores is critical for the use of anthrax as a weapon. Once spores hit the ground, the threat of infection is limited. Weapons-grade spores are very finely ground so that they stay airborne longer. But, as we saw, the letter-borne dispersal system did not effectively generate large numbers of victims. Nevertheless, the potential threat of weapons-grade anthrax on the battlefield led the U.S. military in 2006 to resume vaccinating all soldiers serving in Iraq, Afghanistan, and South Korea.

An effective bioweapon would capitalize on person-to-person transmission. In an easily transmitted disease, one infected person could disseminate disease to scores of others within 1 or 2 days. So, in terms of generating massive numbers of deaths, anthrax was a poor choice. The goal of most terrorists, however, is not to kill large numbers of people, but to terrorize them. In that regard, the anthrax attack succeeded (**Fig. 28.26A**).

A.

B.

FIGURE 28.26 ■ **Dealing with bioterrorism.** **A.** Members of a hazardous-materials team near Capitol Hill during the anthrax attacks in 2001. **B.** An Illinois man suffering from smallpox in 1912.

The most effective bioweapon in terms of inflicting death (biowarfare) would have a low infectious dose, be easily transmitted between people, and be one to which a large percentage of the population was susceptible. Smallpox fits these criteria and would be the bioweapon of choice (**Fig. 28.26B**). Fortunately, however, smallpox has been eradicated (almost) from the face of the Earth and is not easily obtained. Two laboratories still harbor the virus—one in the United States and one in Russia. It is believed that the virus has been destroyed in all other laboratories. Since smallpox is the perfect biowarfare agent, it is imperative that the last two smallpox repositories remain secure.

While the good news is that smallpox disease has been eradicated, the bad news is that no individual born after 1970 has been vaccinated, with the exception of some military and laboratory personnel. As a result, anyone under 50 years of age is susceptible to smallpox. Even those of us who received the smallpox vaccination over 45 years ago are at risk, since our protective antibody titers have diminished. A terrorist attack with smallpox would cause terrible numbers of deaths. A new, safer vaccine does exist, however, and has been stockpiled. Were a smallpox attack to be launched, the vaccine would be rapidly administered to limit the spread of disease. Nevertheless, the economic and psychological impact of a smallpox epidemic would be devastating.

Research with organisms considered to be select agents is tightly regulated. Because *Yersinia pestis*, for instance, is a select agent, laboratory personnel working with it must now possess security clearance with the Department of Justice (even though the organism itself can be handled under biosafety level 2 conditions; see **Table 28.1**). The laboratory must also register with the CDC to legally possess this pathogen, and access to the lab and the organism must be tightly controlled.

Much has improved since Michael Osterholm offered his dire assessment of a wide-scale bioterrorism attack. Education and surveillance procedures have been bolstered, and new detection technologies are being developed (**Special Topic 28.1**; see also **eTopic 28.3**). We will probably never be fully protected from attack, biological or otherwise, but recent efforts have improved the situation.

To Summarize

- **John Snow** founded the discipline of epidemiology.

- **Epidemiology** examines factors that determine the distribution and source of disease.

- **Endemic, epidemic, and pandemic** are terms for different frequencies of disease in different geographic areas.

- **Finding patient zero (the index case)** is important for containing the spread of disease.

- **Molecular approaches using PCR and nucleic acid hybridization** are used to identify nonculturable pathogens and to track disease movements.

- **Bioweapons**, when they have been used, typically kill few people but incite great fear.

- **The CDC** has assembled a list of select agents with bioweapon potential.

28.4 Detecting Emerging Microbial Diseases

News, whether obtained from traditional sources or Twitter, almost always carries stories of new infectious diseases cropping up in the world, so-called emerging infectious diseases. This section discusses the concept of emerging diseases, how they develop, and how health organizations detect and track them.

Emerging and Reemerging Pathogens

A world map showing the general locations of emerging and reemerging diseases is shown in **Figure 28.27**. A reemerging disease is one thought to be under control but whose incidence has risen. For example, the incidence rate of tuberculosis dropped sharply in the 1950s, but the number of cases since 1980 has increased just as dramatically. Tuberculosis is a reemerging disease. The trigger for its reemergence was the AIDS pandemic. Immunocompromised AIDS patients are highly susceptible to infection by many organisms, including *Mycobacterium tuberculosis*.

M. tuberculosis is also reemerging among non-AIDS patients, owing to the development of drug-resistant strains. Drug resistance in *M. tuberculosis* developed largely because of noncompliance on the part of patients in completing their full courses of antibiotic treatment. Treatment usually involves three or more antibiotics to reduce the risk that resistance to any one drug will develop. Many patients failed to take all three drugs simultaneously, allowing the organism to develop resistance to one drug at a time, until it became resistant to all of them. These highly drug-resistant strains are almost impossible to kill. The link between noncompliance and the development of drug resistance is the primary reason for the current requirement that tuberculosis patients be in the presence of a medical staff member when taking the multiple antibiotics prescribed.

As described in Section 28.3, highly infectious diseases are identified using aggressive epidemiological surveillance. Communication among local, national, and world health organizations is critical and helped expose the rise in tuberculosis. To see for yourself how emerging diseases

SPECIAL TOPIC 28.1	What's Blowing in the Wind?

In April 1979, workers at a secret Soviet biological weapons facility neglected to replace a filter in a laboratory ventilation system, and a cloud of highly weaponized *Bacillus anthracis* spores quickly spewed into the air outside. The deadly plume drifted downwind, infecting humans and cattle across a wide area. The resulting anthrax outbreak ultimately killed more than 64 people. Had a rapid pathogen detection system been available and deployed at the facility, all those lives could have been saved. Unfortunately, 30 years ago such technology was only the stuff of science fiction. Today, that technology exists in small, portable forms.

Several multipathogen molecular detection platforms have been developed. Some involve machines that detect antigen-antibody interactions; others use automated PCR to amplify specific pathogen genes and detect the products through hybridization to immobilized oligonucleotides. The wet chemistries for all of these detection systems are carried out in a roughly 2-inch square called lab-on-a-chip.

An interesting addition to the field of quick pathogen detection is the BioFlash-E sensor. Initially called PANTHER (Pathogen Analyzer for Threatening Environmental Releases), the device was developed in 2008. BioFlash-E can detect and identify a pathogen in as little as 3 minutes—significantly faster than traditional methods requiring isolation and growth of the pathogen (which can take 48 hours or longer). The sensor (**Fig. 1A**) uses a cell-based technology called CANARY (Cellular Analysis and Notification of Antigen Risks and Yields) that can detect as few as a dozen particles of a pathogen per liter of air. Currently, it can detect 24 pathogens, including the potential bioterrorism agents of anthrax, plague, smallpox, and tularemia.

The device uses an array of B cells, each displaying antibodies specific to a particular bacterium or virus. The cells are engineered to emit photons of light when they detect their target pathogen. More specifically, the B cells have been bioengineered to express the gene that encodes aequorin, a calcium-sensitive bioluminescent protein. When a pathogen surface antigen cross-links the appropriate B-cell surface antibodies, a signal transduction cascade produces a rapid influx of calcium (**Fig. 1B**). Aequorin in the B cell will luminesce within seconds of calcium influx, and the light emitted is detected by a photon detector. The device then displays a list of any pathogens found.

Quick pathogen detection technologies are constantly being improved in the effort to defend against bioterrorism. Small, quick pathogen detection devices could be used in buildings, subways, and other public areas. Eventually, such technology is expected to supplant the classical clinical microbiology practices that require growth of the microbe to achieve identification. There is hope, for example, that such devices could be used on farms or in food-processing plants to test for contamination by *E. coli*, *Salmonella*, or other food-borne pathogens. Another potential application is in medical diagnostics, where the technology could be used to test patient samples, giving rapid results without having to send samples to a laboratory. Wherever it is used, successful quick pathogen detection should provide the clinician with better information to quickly prescribe the appropriate course of treatment.

RESEARCH QUESTION

The research article referenced below describes a method of biothreat detection that combines PCR with mass spectrometry (see Section 28.2). Why might this coupled technology be superior for detecting biowarfare agents as compared to

---

are monitored, search the Internet for a website called ProMED-mail. There you will find daily reports posted from around the world that describe new outbreaks of infectious diseases.

A posting on May 19, 2016, for example, reports 60 cases, including 10 deaths, of hemorrhagic fever syndrome (symptoms of unexpected bleeding, fever, fatigue, and vomiting) from South Sudan in Africa. Laboratory tests performed thus far have eliminated a variety of viruses, including Ebola and Marburg, that can cause these symptoms. Could this be the leading edge of a new, deadly epidemic? Hindsight, of course, is 20/20. A March 22, 2014, ProMED report from Guinea in Africa described an outbreak of a disease with similar symptoms that killed 60 people. A WHO laboratory had just identified it as Ebola. In the post, a physician working with the medical charity Doctors Without

Borders was quoted as saying: "The quicker we can contain this the fewer cases we'll have, then the smaller the scale of the epidemic. That's the idea of going in as strong as we can early on." We now know that this report represented the leading edge of the horrific Ebola epidemic that ravaged West Africa. As a fledgling microbiologist, tracking these ProMED reports allows you to see epidemics and new diseases emerge almost in real time.

Thought Question

28.10 On the ProMED-mail web page (www.promedmail.org), click on the interactive world map to view outbreaks recorded by WHO. What outbreaks happened throughout the world during the current year?

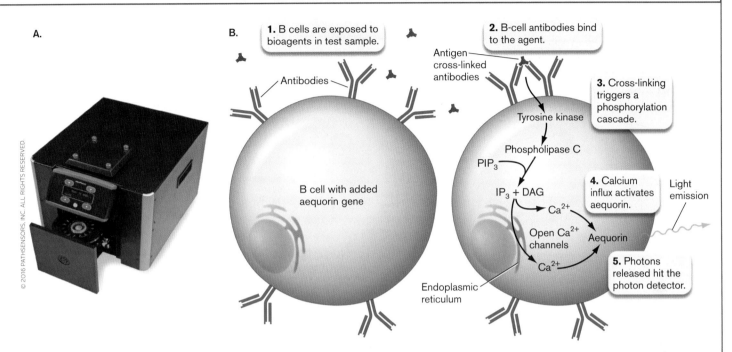

FIGURE 1 ■ A. The BioFlash-E detection system uses aerosol collection technology and CANARY detection technology. B. Principle of the bioelectronic sensor used in the BioFlash-E system. DAG = diacylglycerol; IP_3 = 1,4,5-inositol trisphosphate; PIP_3 = phosphatidylinositol 3,4,5-trisphosphate. *Source:* Modified from Martha S. Petrovick et al. 2008. *Lincoln Lab. J.* **17**(1).

the more conventional use of molecular probes to detect PCR amplicons? The authors also published the sequences of DNA primers that their system uses to amplify indicator genes of over 20 potential biowarfare agents. Discuss why this decision to publish was either good or bad.

Sampath, Rangarajan, Niveen Mulholland, Lawrence B. Blyn, Christian Massirel, Chris A. Whitehouse, et al. 2012. Comprehensive biothreat cluster identification by PCR/electrospray-ionization mass spectrometry. *PLoS One* **7**:e36528.

Tracking an Emerging Disease

Sometimes world and national health organizations quickly and efficiently identify and contain outbreaks of emerging infectious diseases, as was the case in 2003 with severe acute respiratory syndrome (SARS) (**eTopic 28.4**). In other instances, outbreaks take more time to contain, such as the 2014 Ebola epidemic in West Africa, described earlier. Agencies can also face difficulties in clearly identifying the causative agent of an outbreak, as in the current 2015–16 Zika virus outbreak in Brazil. The unfolding Zika virus story provides an instructive example of how new diseases are identified.

Case History: Zika Virus

Brazil, 2015. Around the fourth month of her pregnancy, a young woman from Recife, Brazil, awoke one morning with fever, a fiery red rash, and pain in her joints. Her symptoms went away within a few days, and she went about her life, oblivious to what was to come. Five months later, in September 2015, she delivered her baby and immediately knew from the look on the obstetrician's face that something was terribly wrong. Her son had an abnormally small head and incomplete brain development—something called microcephaly. The young mother was devastated.

In August 2015, Dr. Vanessa van der Linden, a neurologist in Recife, started noticing an increasing number of newborns with microcephaly (**Fig. 28.28**). Normally, months would pass without her seeing a single case. Now there were three or four in a single day. Using serological tests, she ruled out the usual suspects, such as rubella (German measles) and toxoplasmosis. Imaging studies, however, revealed

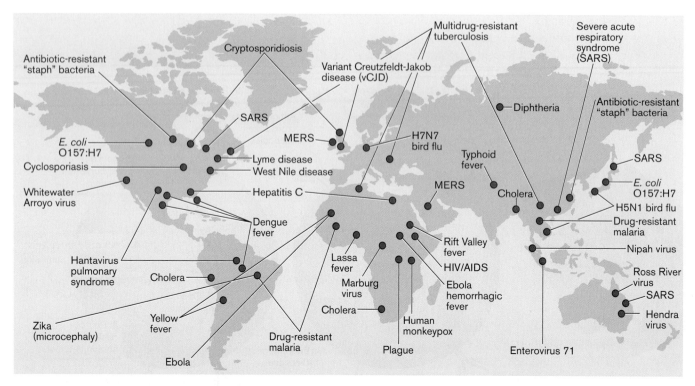

FIGURE 28.27 ■ **Locations of some emerging and reemerging infectious diseases and pathogens.** The examples given represent extreme increases in the reported cases over the last 20 years. Many of these diseases, such as HIV/AIDS and cholera, are widespread but show alarming increases in the areas indicated.

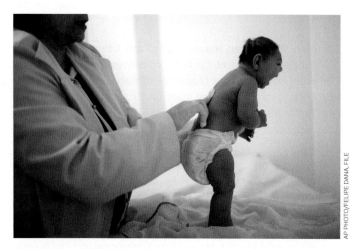

FIGURE 28.28 ■ **A child born with microcephaly in Brazil.**

unusual patterns of calcification in the affected brains (a possible sign of virus-induced necrosis). Vanessa knew something odd was going on. Then her mother called.

Vanessa's mother, Dr. Ana van der Linden, a neuropediatrician in another Recife hospital told her daughter that she had seen seven babies with microcephaly in just one day, and that some of the mothers remembered having a rash early during pregnancy. Could there be an infectious cause to these birth defects? In October, Vanessa alerted the state health secretary about the spike in microcephaly

cases. This was the first step in the epidemiological journey linking Zika virus, the cause of a normally benign, mosquito-transmitted disease in adults, to the development of microcephaly in Zika virus–infected fetal brains.

As the number of microcephaly cases grew, suspicion by public health officials initially fell on dengue virus, in part because the symptoms of Zika fever resemble those of dengue fever and because prior to 2014, Zika virus infections were rare in Brazil. Dengue virus was far more common and considered the most likely cause of the women's symptoms. Consequently, individuals who actually had Zika fever were not screened for Zika virus; they were screened by ELISA for anti–dengue virus antibodies. Unfortunately, patient antibodies to Zika virus can cross-react with dengue virus antigen, so a person infected with Zika can appear to test positive for dengue (see the discussion of test specificity in Section 28.2). This technical problem slowed efforts to identify Zika virus as the cause of the women's symptoms and obscured the virus's connection to the alarming rise in newborn microcephaly. Today, there are more specific RT-qPCR tests and virus isolations to confirm Zika virus diagnoses.

Both Zika and dengue viruses have positive-sense RNA genomes and are members of the *Flaviviridae* family of viruses. West Nile virus, mentioned earlier, is also a member of this family. Zika, dengue, and West Nile viruses are transmitted to humans by *Aedes* mosquito vectors in person-to-person or animal-to-human transmission cycles. West Nile virus uses birds as natural reservoirs, whereas Zika and dengue viruses infect primarily human and non-human primates. There are also some reports of sexual transmission of Zika virus in areas devoid of the *Aedes* mosquito vectors. As of July 2016, Zika virus was not considered endemic in the United States, even though there are people living in the United States infected with Zika, including 200–300 pregnant women initially infected elsewhere in the world.

How does a viral pathogen become endemic in a country? In order for Zika, and many other mosquito-borne viruses, to become endemic in a new location, four things must happen:

1. The mosquito vector must already be present in the area.
2. A person infected in an endemic area such as Brazil must travel to the nonendemic area.
3. A local mosquito must bite the infected traveler and transmit the virus to another victim.
4. Finally, the virus must also be transmitted to an animal reservoir that can maintain the virus within a geographical area. For example, West Nile virus, introduced into the United States in 1999, quickly infected the native bird population and is now transmitted from birds to humans by mosquito.

As of September 2016, Zika virus had fulfilled three of the four criteria in the United States. The *Aedes* mosquito vector inhabits areas of at least 30 states, stretching from California through the southern and northeastern states. There are hundreds of people infected with Zika virus now living in the United States. The virus has also been introduced into the native mosquito population around Miami, and the infected mosquitoes have transmitted the Zika virus to several humans in the area. Fortunately, the last requirement, finding a native animal host reservoir, is thought to be unlikely. Zika virus is not known to infect birds or animals other than nonhuman primates native to much of South America and the Caribbean but not to the United States.

How did Zika virus get to Brazil? Zika virus was previously recognized in Africa and Asia as causing mild disease but had never been reported in Brazil. Epidemiologists have used next-generation DNA sequencing of Brazilian Zika virus isolates to explore its origin. The data indicate that the Brazilian strain originated from an Asian strain that circulated in French Polynesia in 2013, but the event that introduced the Asian virus into Brazil is unclear. Regardless of how it got there, once the virus was introduced, native mosquitoes fed on the carrier(s), became infected and began transmitting the virus throughout the country. The disease

is now endemic to Brazil. There is no treatment for Zika fever, although furious efforts are under way to develop a vaccine. At the moment, the only protection against Zika is to prevent infection by avoiding or eradicating the *Aedes* mosquitoes.

So, have Koch's postulates been fulfilled? Does Zika virus infection of the mother lead to transplacental transmission of the virus to the fetus? And does virus replication interfere with fetal brain development? Evidence of the virus has been found in the blood of mothers who have delivered babies with microcephaly, and also in umbilical cord blood. The virus itself has been isolated from the brains of fetuses who died from microcephaly (see the Current Research Highlight in Chapter 26). Other studies have recapitulated the neurotropic nature of the virus and indicate that Zika virus can interfere with neurogenesis during human brain development. In sum, the evidence of a link is convincing but not yet ironclad. More epidemiological studies must be conducted to unequivocally prove the link.

As of this writing, at least 1,000 babies in Brazil have been born with microcephaly and other birth defects linked to Zika virus. But it is important to point out that despite the fear Zika virus brings to pregnant women, the vast majority of confirmed Zika virus–infected pregnant women deliver healthy babies. The World Health Organization (WHO) and the U.S. Centers for Disease Control and Prevention (CDC) are carefully monitoring the spread of Zika virus throughout the Americas by intensively screening mosquitoes.

Technology Helps New Infectious Agents Emerge and Spread

Despite all we know of microbes and despite the many ways we have to combat microbial diseases, our species, for all its cleverness, still lives at the mercy of the microbe. Lyme disease, MRSA, SARS, MERS-CoV, Ebola, *E. coli* O157:H7, HIV, "flesh-eating" streptococci, hantavirus, swine flu—all of these and many other new diseases and pathogens have emerged over the last 30 years. Worse yet, forgotten scourges, such as tuberculosis, have reappeared. Yet in the 1970s, medical science was claiming victory over infectious disease. What happened?

Part of the equation has been progress itself. Travel by jet, the use of blood banks, and suburban sprawl have all opened new avenues of infection. The spread of Zika virus from Africa to Asia and then the Americas is but one example. People unwittingly infected by a new disease in Asia or Africa can, traveling by jet, take the pathogen to any other country in the world within hours. A person may not even show symptoms until days or weeks after the trip. This means that diseases can spread faster and farther than ever before. In addition, newly emerged blood-borne pathogens can spread by transfusion. This was a major problem with HIV before an accurate blood test was developed to screen all donated blood.

Although human encroachments into the tropical rain forests have often been blamed for the emergence of new pathogens, one need go no farther than the Connecticut woodlands to find such developments. *Borrelia burgdorferi*, the spirochete that causes Lyme disease, lives on deer and white-footed mice and is passed between these hosts by the deer tick (see Fig. 26.23)—an infection cycle that has been going on for years. Humans crossed paths with these animals long before the disease erupted in our communities. Why have we suddenly become susceptible? The answer appears to be suburban development. In the wild, foxes and bobcats hunt the mice that carry the Lyme agent. These predators disappear as developers clear land and build roads and houses, leaving the infected mice and ticks to proliferate. Humans in these developed areas are more likely to be bitten by an infected tick and contract the disease than in prior decades. Luckily, many diseases that successfully leap from animals to humans find the new host to be a dead end, unable to spread the disease to others.

There are numerous examples in which technology and progress have had the unintended consequence of breeding disease. Here are just a few:

- **Mad cow disease.** Modern farming practices (North America and Europe) of feeding livestock the remains of other animals help spread transmissible spongiform encephalopathies, similar to Creutzfeldt-Jakob disease, associated with prions. Because the prion is infectious, the brain matter from one case of mad cow disease could end up infecting hundreds of other cattle, which in turn increases the chance that the disease could spread to humans.

- **Lyme disease.** Suburban development in the northeastern United States destroys predators of the mice that carry *Borrelia burgdorferi*.

- **Hepatitis C.** Transfusions and transplants spread this blood-borne disease.

- **Influenza.** Live poultry markets in Asia serve as breeding grounds for avian flu viruses that can jump to humans.

- **Enterohemorrhagic *E. coli* (for example, E. coli O157:H7).** Modern meat-processing plants can accidentally grind trace amounts of these acid-resistant, fecal organisms into beef while making hamburger.

Natural environmental events can also trigger upsurges in the incidence of unusual diseases. For example, an unprecedented outbreak of hantavirus pulmonary disease occurred in 1993 in the Four Corners area of the Southwest, where Arizona, New Mexico, Colorado, and Utah meet, when rain led to greater-than-normal increases in plant and animal numbers. The resulting tenfold increase in deer mice, which carries the virus, made it more likely that infected mice and humans would come in contact. A more recent outbreak of hantavirus occurred among campers at Yosemite National Park in 2012.

Climate Change Influences Emerging Diseases

The world's climate has been changing slowly but significantly over the last 50 years—largely as a result of humans pumping enormous amounts of greenhouse gases, such as carbon dioxide and methane, into the atmosphere. Greenhouse gases trapping heat in the atmosphere have caused the world's average yearly temperature to rise over 1.5°F since 1970. Effects of global warming on climate, depending on where you live, are evident as disappearing glaciers, increasing sea levels, mounting ferocity of hurricanes and tornadoes, extended droughts, increased rainfall (floods), dust storms, and heat waves. These climate changes are bad enough, but they can also affect the epidemiology of infectious disease.

Rising air temperatures, for example, can extend the habitat of mosquito and tick vectors to higher mountain elevations and wider latitudes. The result is that Zika virus, dengue virus, and *Plasmodium* (the protozoan cause of malaria), among other infectious agents, have increased their geographic range. Mathematical simulations predict that the area populated by *Anopheles* mosquito vectors that carry malaria will increase between 16% and 49% by 2030, depending on the vector. Extending a pathogen's range can bring it into contact with novel groups of hosts, possibly establishing new vector-pathogen transmission cycles or a situation that promotes pathogen evolution (such as influenza). Warmer temperatures can also accelerate host-parasite cycles that will further increase the incidence and prevalence of infections.

Increased rain and flooding promote the breeding of mosquito vectors that transmit disease—a situation that will worsen with climate change. Floods can also accelerate the spread of diarrheal diseases such as cholera when sanitation systems fail. As these events increase, so, too, will infectious disease.

The Arctic is particularly sensitive to climate change. It is home to diverse populations of plants, animals, and people. Approximately 10 million people live within the Arctic circle, a geographic area that encompasses parts of nine countries, including Canada and the United States. The Arctic has warmed twice as much as the global average of other parts of the world and is becoming increasingly fragile. The associated health risks for humans and animals include potential changes in pathogen and vector demographics that can affect disease patterns, degradation in the quality and availability of both drinking water and food, and changes in animal and plant species health.

Climate change will continue to create opportunities for some pathogens and may even limit the transmission of others. To stay ahead of these evolving infections, surveillance programs capable of detecting pathogen or disease emergence are essential. Make no mistake: Together, the effect of climate change on infectious diseases and the rise in antibiotic-resistant microbes pose a serious threat to the living world.

The One Health Initiative

We have provided numerous examples of how the natural environment can harbor pathogens and foster the emergence of new ones. Knowledge that pathogens have reservoirs in animals, arthropods, and plants has given rise to a new collaborative effort among clinicians, scientists, veterinarians, and ecologists called the One Health Initiative. The goal of the One Health Initiative is to control human health through animal health, and vice versa. For example, vaccinating wild rodents could decrease Lyme disease in humans. The initiative includes plant pathology because some bacteria are pathogens of both plants and humans (for example, *Pantoea agglomerans*, formerly *Enterobacter agglomerans*). In addition, some enteric bacteria can live within a plant vascular system (for example, *E. coli*, *Klebsiella*, and *Salmonella*).

Solving a mystery. An excellent example of how multidisciplinary collaboration resulted in better understanding of an infectious outbreak occurred in 2006 when approximately 200 people in 26 states were diagnosed with a particularly virulent case of *E. coli* O157:H7. Nearly half of the cases were hospitalized, and many suffered from hemolytic uremic syndrome (HUS, kidney failure described in Section 26.4). The source of the infection was contaminated spinach traced to the Salinas Valley of California. It turned out the organisms were contained within the vascular system of the spinach, so washing the spinach would not remove the pathogen. Had this outbreak been viewed through only the narrow lens of human health, efforts would have focused on morbidity, mortality, outbreak investigation, laboratory diagnosis, and clinical treatment. The origin of the disease would have remained a mystery.

Working together, epidemiologists and veterinarians found a genetically identical organism in cattle close to where the spinach was produced and in wild hogs that ran through the same fields. Ecologists and hydrologists understood that the groundwater and surface water in this region were being mixed because of a drought followed by heavy rains, and that irrigation systems were strained in the effort to keep up with intensified agricultural production. Eventually, the same *E. coli* strain was found in one of the water

ditches close to the spinach fields in the area. Scientists pondering these facts within the One Health framework deduced that cattle harboring the *E. coli* had defecated in a field, thereby contaminating wild hogs. The wild hogs, by running through the spinach fields, had contaminated those fields with their feces, and the irrigation water had then swept the pathogen into the plant vasculature.

Only by integrating our knowledge of the environment and ecology could this investigation be completely understood and appropriate intervention and prevention strategies implemented. This outbreak exemplifies the fact that human health and animal health are inextricably linked and that a holistic approach is needed to understand, protect, and promote the health of all species.

Stopping zoonotic diseases. A more recent example of the One Health approach is being carried out in the Ruaha region of Tanzania, Africa—a sprawling, wild area with scattered small villages and a rich wildlife. Thousands of children and adults in underdeveloped countries such as Tanzania die every day from diseases arising from the human-animal-environment interface. People in these areas often live in close contact with wild and domestic animals and are brought even closer as water becomes scarcer (**Fig. 28.29**). Sharing the same water sources for drinking, washing, and swimming facilitates zoonotic disease transmission to humans.

FIGURE 28.29 ■ **The One Health Initiative.** This photograph illustrates the close relationship that some residents of Tanzania have with their livestock, and the opportunities by which zoonotic infections can be transmitted to humans.

The One Health Initiative simultaneously addresses multiple and interacting causes of poor human health to try to stop or limit these zoonotic diseases. One underdiagnosed zoonotic disease that the collaborative is trying to address is bovine tuberculosis (BTB), a disease that normally affects animals but also infects humans. Tanzania suffers 40,000 new cases of tuberculosis per year caused by human, bovine, or atypical strains of mycobacteria. A majority of these TB patients are also infected with HIV. BTB in humans often progresses to extrapulmonary TB, making BTB a major focus of the initiative.

The approach in Tanzania, called the Health for Animals and Livelihood Improvement (HALI) Project, includes the testing of wildlife, livestock, and their water sources for zoonotic pathogens and disease. Water quality, availability, and use are monitored, as is the impact of livestock and human disease on farming households. New diagnostic techniques for disease detection are being introduced, and Tanzanians of all educational levels are trained about zoonotic diseases. Finally, new health and environmental policies have been developed to mitigate the impacts of zoonotic diseases.

The HALI Project has identified bovine tuberculosis and brucellosis in livestock and wildlife in the Ruaha ecosystem and has identified geographic areas where water availability increases the risk of transmission among wildlife, livestock, and people. In addition, *Salmonella*, *Escherichia coli*, *Cryptosporidium*, and *Giardia* species that cause disease in humans and animals have been isolated from multiple water sources used by people and frequented by livestock and wildlife. A major lesson of the HALI Project is that the causes and consequences of zoonotic diseases, as well as the interventions to mitigate them, are all cross-sectoral. Effective surveillance, assessments, and interventions are possible only by bridging organizational gaps among institutions that study and manage wildlife, livestock, water, and public health.

To Summarize

- **Emerging diseases** can spread quickly around the world as a result of air travel.

- **Modern technology** and urban growth have provided opportunities for new diseases to emerge.

- **Climate change** has significantly affected the epidemiology of infectious disease and the emergence of new pathogens.

- **Multidisciplinary collaboration** among ecologists, veterinarians, clinicians, and other scientists is necessary for devising appropriate strategies aimed at disease intervention and epidemic prevention.

Concluding Thoughts

In this chapter we have examined the basic principles used to collect, detect, and track pathogenic microorganisms. As you can see, the task of controlling the spread of disease is daunting and ever changing because microorganisms continue to evolve. Known pathogens change, eluding eradication efforts by the immune system and antibiotics, while new pathogens keep emerging. The world is depending on the next generation of microbiologists to face the growing threat of these evolving microbes.

CHAPTER REVIEW

Review Questions

1. Why is it important to identify the genus and species of a pathogen?
2. What is an API strip, and what is its use in clinical microbiology?
3. Describe three examples of selective media.
4. If a colony on a nutrient agar plate is catalase-positive, does this mean it is made up of Gram-positive microorganisms? Why or why not?
5. Describe the types of hemolysis visualized on blood agar.
6. What is the clinical significance of a group A, beta-hemolytic streptococcus?
7. How does one distinguish *Staphylococcus aureus* from *Staphylococcus epidermidis*?
8. Why are PCR identification tests preferable to biochemical approaches?
9. How is RT-qPCR performed?
10. Describe an ELISA.
11. Name some sterile and nonsterile body sites.
12. List seven common types of clinical specimens collected for bacteriological examination.
13. Describe key features of the four levels of biological containment.
14. How is a pandemic different from an epidemic?
15. How can genomics help identify nonculturable pathogens?
16. List and briefly describe four emerging diseases.
17. Name four select agents and the diseases they cause.

Thought Questions

The first four thought questions below are based on the following case history:

An infectious disease physician in Florida telephoned the CDC to report two possible cases of botulism. The two male patients presented with drooping eyelids, double vision, difficulty swallowing, and respiratory problems. The physician had drawn sera and collected stool specimens from the men to test for botulinum toxin, but no results were available.

1. What are the major concerns raised by these two possible cases of botulism?
2. How might you go about swiftly determining whether there is a link between the two cases and whether there are other cases of botulism?
3. The two patients ate only one food item in common: a cured and fermented fish called "moloha." How would you determine whether the men are indeed suffering from botulism and whether the fermented fish is the source of disease?
4. How could this anaerobic pathogen grow and make toxin in this fish product? Propose rational theories for what has hindered development of a vaccine against botulism.
5. Consider the following hypothetical case: *Five small outbreaks of Ebola occurred within days of one another in five different cities in the United States and Canada. Patient histories revealed that one person from each city had been in the Atlanta airport at the same time 2 days prior to becoming seriously ill. None had left the country recently; none flew on the same jet or even crossed paths while in the airport.* As an epidemiologist, conjure up a scenario, excluding bioterrorism, that could account for this scattered outbreak. You are aware that an outbreak recently occurred in the Congo.

Key Terms

amplicon (1148)
endemic (1159)
epidemic (1159)
epidemiology (1135, 1159)
incidence (1160)

index case (1160)
multiplex PCR (1148)
pandemic (1159)
prevalence (1160)
quarantine (1160)

reservoir (1159)
sensitivity (1156)
sequela (1134)
specificity (1156)

Recommended Reading

Bale, James F., Jr. 2012. Emerging viral infections. *Seminars in Pediatric Neurology* **19**:152–157.

Barczak, Amy K., James E. Gomez, Benjamin B. Kaufmann, Ella R. Hinson, Lisa Cosimi, et al. 2012. RNA signatures allow rapid identification of pathogens and antibiotic susceptibilities. *Proceedings of the National Academy of Sciences USA* **109**:6217–6222.

Bengis, R. G., F. A. Leighton, J. R. Fischer, M. Artois, T. Morner, et al. 2004. The role of wildlife in emerging and re-emerging zoonoses. *Reviews in Science and Technology* **23**:497–511.

Buchan, Blake W., and Nathan A. Ledeboer. 2014. Emerging technologies for the clinical microbiology laboratory. *Clinical Microbiology Reviews* **27**:783–822.

Faria, Nuno Rodrigues, Raimunda do Socorro da Silva Azevedo, Moritz U. G. Kraemer, Renato Souza, Mariana Sequetin Cunha, et al. 2016. Zika virus in the Americas: Early epidemiological and genetic findings. *Science* **352**:345–349.

Fraser, Christophe, Christl A. Donnelly, Simon Cauchemez, William P. Hanage, Maria D. Van Kerkhove, et al. 2009. Pandemic potential of a strain of influenza A (H1N1): Early findings. *Science* **324**:1557–1561.

Koser, Claudio U., Matthew J. Ellington, Edward J. Cartwright, Stephen H. Gillespie, Nicholas M. Brown, et al. 2012. Routine use of microbial whole genome sequencing in diagnostic and public health microbiology. *PLoS Pathogens* **8**:e1002824.

Kreft, Rachael, J. William Costerton, and Garth D. Ehrlich. 2013. PCR is changing clinical diagnostics. *Microbe* **8**:15–20.

Morse, Stephen S. 2012. Public health surveillance and infectious disease detection. *Biosecurity and Bioterrorism* **10**:6–16.

Pardee, Keith, Alexander A. Green, Mellisa K. Taka-hashi, Dana Braff, Guillaume Lambert, et al. 2016. Rapid, low cost detection of Zika virus using program-mable biomolecular components. *Cell* **165**:1255–1266.

Relman, David A., Thomas M. Schmidt, Richard P. MacDermott, and Stanley Falkow. 1992. Identifi-cation of the uncultured bacillus of Whipple's disease. *New England Journal of Medicine* **327**:283–301.

Semenza, Jan C., Elisabet Lindgren, Laszlo Balkanyi, Laura Espinosa, My S. Almqvist, et al. 2016. Deter-minants and drivers of infectious disease threat events in Europe. *Emerging Infectious Diseases* **22**:581–589.

Soto, S. M. 2009. Human migration and infec-tious diseases. *Clinical Microbiology and Infection* **15**(Suppl.1):26–28.

van Belkum, Alex, Geraldine Durand, Michel Peyret, Sonia Chatellier, Gilles Zambardi, et al. 2013. Rapid clinical bacteriology and its future impact. *Annals of Laboratory Medicine* **33**:14–27.

Wilson, Michael R., Samia N. Naccache, Erik Samayoa, Mark Biagtan, Hiba Bashir, et al. 2014. Action-able diagnosis of neuroleptospirosis by next-genera-tion sequencing. *New England Journal of Medicine* **370**:2408–2417.

Wu, Xiaoxu, Yongmei Lu, Sen Zhou, Lifan Chen, and Bing Xu. 2015. Impact of climate change on human infectious diseases: Empirical evidence and human adap-tation. *Environment International* **86**:14–23.

APPENDIX 1
Reference and Review

The cell is the basic unit of life. Cells are composed primarily of water and organic molecules: proteins, carbohydrates, nucleic acids, and lipids. The chemistry of life is organic chemistry, based on carbon (C) and several other elements: hydrogen (H), nitrogen (N), oxygen (O), phosphorus (P), and sulfur (S). Appendix 1 provides reference information on some essential aspects of chemistry, biochemistry, and cell biology likely covered in an introductory biology course. For further review of basic chemistry, biochemistry, and cell biology, see online eAppendices 1 and 2.

A1.1 A Periodic Table of the Elements

A periodic table organizes all of the known elements, grouping them by columns and rows according to their electronic and chemical properties (**Fig. A1.1**). Note that the six major elements of life—H, C, N, O, P, and S—are all nonmetals found in the top three periods (rows).

A1.2 Chemical Functional Groups

A functional group is a group of atoms that defines the structure and reactions of an organic compound (**Table A1.1**). Since organic chemistry underlies so much of biology, it is useful to be able recognize functional groups and their chemical properties.

Main-group elements

Atomic number
Chemical symbol
Atomic mass (average of all isotopes)

Transitional elements

Metals
Nonmetals
Metalloids
Noble gases

Main-group elements

1 1A																	18 8A
1 **H** 1.00794	2 2A											13 3A	14 4A	15 5A	16 6A	17 7A	2 **He** 4.00260
3 **Li** 6.941	4 **Be** 9.01218											5 **B** 10.811	6 **C** 12.011	7 **N** 14.0067	8 **O** 15.9994	9 **F** 18.9984	10 **Ne** 20.1797
11 **Na** 22.9898	12 **Mg** 24.3050	3 3B	4 4B	5 5B	6 6B	7 7B	8	9 8B	10	11 1B	12 2B	13 **Al** 26.9815	14 **Si** 28.0855	15 **P** 30.9738	16 **S** 32.066	17 **Cl** 35.4527	18 **Ar** 39.948
19 **K** 39.0983	20 **Ca** 40.078	21 **Sc** 44.9559	22 **Ti** 47.88	23 **V** 50.9415	24 **Cr** 51.9961	25 **Mn** 54.9381	26 **Fe** 55.847	27 **Co** 58.9332	28 **Ni** 58.693	29 **Cu** 63.546	30 **Zn** 65.39	31 **Ga** 69.723	32 **Ge** 72.61	33 **As** 74.9216	34 **Se** 78.96	35 **Br** 79.904	36 **Kr** 83.80
37 **Rb** 85.4678	38 **Sr** 87.62	39 **Y** 88.9059	40 **Zr** 91.224	41 **Nb** 92.9064	42 **Mo** 95.94	43 **Tc** (98)	44 **Ru** 101.07	45 **Rh** 102.906	46 **Pd** 106.42	47 **Ag** 107.868	48 **Cd** 112.411	49 **In** 114.818	50 **Sn** 118.710	51 **Sb** 121.76	52 **Te** 127.60	53 **I** 126.904	54 **Xe** 131.29
55 **Cs** 132.905	56 **Ba** 137.327	57 ***La** 138.906	72 **Hf** 178.49	73 **Ta** 180.948	74 **W** 183.84	75 **Re** 186.207	76 **Os** 190.23	77 **Ir** 192.22	78 **Pt** 195.08	79 **Au** 196.967	80 **Hg** 200.59	81 **Tl** 204.383	82 **Pb** 207.2	83 **Bi** 208.980	84 **Po** (209)	85 **At** (210)	86 **Rn** (222)
87 **Fr** (223)	88 **Ra** 226.025	89 **†Ac** 227.028	104 **Rf** (261)	105 **Db** (262)	106 **Sg** (263)	107 **Bh** (262)	108 **Hs** (265)	109 **Mt** (266)	110 **Ds** (271)	111 **Rg** (272)	112 **Cn** (277)	114 **Fl** (289)		116 **Lv** (293)			

*Lanthanide series

58 **Ce** 140.115	59 **Pr** 140.908	60 **Nd** 144.24	61 **Pm** (145)	62 **Sm** 150.36	63 **Eu** 151.965	64 **Gd** 157.25	65 **Tb** 158.925	66 **Dy** 162.50	67 **Ho** 164.930	68 **Er** 167.26	69 **Tm** 168.934	70 **Yb** 173.04	71 **Lu** 174.967
90 **Th** 232.038	91 **Pa** 231.036	92 **U** 238.029	93 **Np** 237.048	94 **Pu** (244)	95 **Am** (243)	96 **Cm** (247)	97 **Bk** (247)	98 **Cf** (251)	99 **Es** (252)	100 **Fm** (257)	101 **Md** (258)	102 **No** (259)	103 **Lr** (260)

†Actinide series

**Not yet named

Symbol	Name	Symbol	Name	Symbol	Name	Symbol	Name	Symbol	Name
Ac	Actinium	Cn	Copernicium	I	Iodine	Os	Osmium	Si	Silicon
Al	Aluminum	Cu	Copper	Ir	Iridium	O	Oxygen	Ag	Silver
Am	Americium	Cm	Curium	Fe	Iron	Pd	Palladium	Na	Sodium
Sb	Antimony	Ds	Darmstadtium	Kr	Krypton	P	Phosphorus	Sr	Strontium
Ar	Argon	Db	Dubnium	La	Lanthanum	Pt	Platinum	S	Sulfur
As	Arsenic	Dy	Dysprosium	Lr	Lawrencium	Pu	Plutonium	Ta	Tantalum
At	Astatine	Es	Einsteinium	Pb	Lead	Po	Polonium	Tc	Technetium
Ba	Barium	Er	Erbium	Li	Lithium	K	Potassium	Te	Tellurium
Bk	Berkelium	Eu	Europium	Lv	Livermorium	Pr	Praseodymium	Tb	Terbium
Be	Beryllium	Fm	Fermium	Lu	Lutetium	Pm	Promethium	Tl	Thallium
Bi	Bismuth	Fl	Flerovium	Mg	Magnesium	Pa	Protactinium	Th	Thorium
Bh	Bohrium	F	Fluorine	Mn	Manganese	Ra	Radium	Tm	Thulium
B	Boron	Fr	Francium	Mt	Meitnerium	Rn	Radon	Sn	Tin
Br	Bromine	Gd	Gadolinium	Md	Mendelevium	Re	Rhenium	Ti	Titanium
Cd	Cadmium	Ga	Gallium	Hg	Mercury	Rh	Rhodium	W	Tungsten
Ca	Calcium	Ge	Germanium	Mo	Molybdenum	Rg	Roentgenium	U	Uranium
Cf	Californium	Au	Gold	Nd	Neodymium	Rb	Rubidium	V	Vanadium
C	Carbon	Hf	Hafnium	Ne	Neon	Ru	Ruthenium	Xe	Xenon
Ce	Cerium	Hs	Hassium	Np	Neptunium	Rf	Rutherfordium	Yb	Ytterbium
Cs	Cesium	He	Helium	Ni	Nickel	Sm	Samarium	Y	Yttrium
Cl	Chlorine	Ho	Holmium	Nb	Niobium	Sc	Scandium	Zn	Zinc
Cr	Chromium	H	Hydrogen	N	Nitrogen	Sg	Seaborgium	Zr	Zirconium
Co	Cobalt	In	Indium	No	Nobelium	Se	Selenium		

FIGURE A1.1 ■ Periodic table of the elements. The atomic number (number of protons) and atomic mass are shown for each element.

TABLE A1.1 Common Functional Groups.

Functional group	General structure	Example	Comments
Aldehyde	$R-\overset{\overset{\displaystyle O}{\|\|}}{C}-H$	Acetaldehyde	Can react with alcohols
Alkane	$R-\overset{\overset{\displaystyle H}{\|}}{\underset{\underset{\displaystyle H}{\|}}{C}}-H$	Ethane	Nonpolar, tends to make molecules containing it hydrophobic; nonreactive
Amino	$R-N\overset{H}{\underset{H}{<}}$	Methylamine	Acts as a base by binding a proton; found in amino acids $\quad R-NH_2 + H^+ \rightarrow R-NH_3^+$
Carboxyl	$R-\overset{\overset{\displaystyle O}{\|\|}}{C}-O-H$	Acetic acid	Acts as an acid by releasing a proton; the ionized-form name ends in "-ate" (e.g., "acetate")
Ester	$R-\overset{\overset{\displaystyle O}{\|\|}}{C}-O-R$	Phosphodiester $R_1-O-\overset{\overset{\displaystyle O}{\|\|}}{\underset{\underset{\displaystyle O^-}{\|}}{P}}-O-R_2$	Common linkage found in lipids
Hydroxyl	$R-O-H$	Ethanol	Polar, makes compounds more soluble through hydrogen bonding; found in alcohols and sugars
Ketone	$R-\overset{\overset{\displaystyle O}{\|\|}}{C}-R$	Dihydroxyacetone	Found in many intermediates of metabolism
Phosphate	$R-O-\overset{\overset{\displaystyle O}{\|\|}}{\underset{\underset{\displaystyle O^-}{\|}}{P}}-O^-$	Glyceraldehyde 3-phosphate	When two or more phosphoryl groups are linked, a high-energy bond forms because the negative oxygens repel each other

Aldehyde comments reaction:

$$R-\overset{O}{\underset{H}{C}} + HO-R' \rightarrow R-\overset{OH}{\underset{H}{\overset{\|}{C}}}-OR'$$

Carboxyl comments reaction:

$$CH_3-\overset{O}{\underset{O^-}{C}} + H^+$$

Acetate   Proton

A1.3 Amino Acids

Proteins are made of amino acid monomers condensed to form polypeptide chains, which then fold into three-dimensional shapes. The conformation of a protein is largely determined by the side chains of the amino acids in that polypeptide. So, understanding protein structure and function requires knowing the structures and chemical properties of the twenty amino acids. **Figure A1.2** shows one way to organize the amino acids.

A1.4 The Genetic Code

The link between protein synthesis and the information encoded in DNA is the genetic code (**Fig. A1.3**). Three-letter codons in mRNA have counterparts in the tRNA anticodons. Transfer RNA molecules are the adapters connecting information in the mRNA transcript to the amino acid sequence in the growing polypeptide chain.

A. Positively charged R groups

Lysine (Lys or K) Arginine (Arg or R) Histidine (His or H)

B. Negatively charged R groups

Aspartate (Asp or D) Glutamate (Glu or E)

C. Polar, uncharged R groups

Asparagine (Asn or N) Glutamine (Gln or Q) Serine (Ser or S) Threonine (Thr or T)

D. Nonpolar, aliphatic R groups

Alanine (Ala or A) Valine (Val or V) Isoleucine (Ile or I) Leucine (Leu or L) Methionine (Met or M) Phenylalanine (Phe or F) Tyrosine (Tyr or Y) Tryptophan (Trp or W)

E. Aromatic R groups

Cysteine (Cys or C) Glycine (Gly or G) Proline (Pro or P)

FIGURE A1.2 ▪ Twenty common amino acids. The grouping of amino acids is based on their side chains, highlighted in yellow.

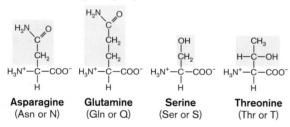
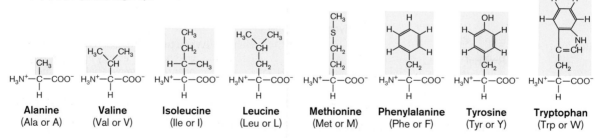

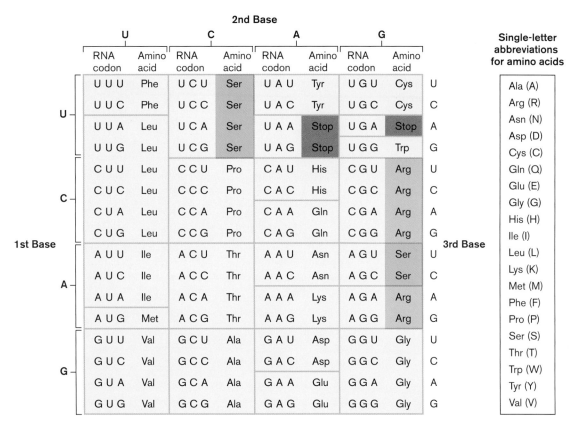

FIGURE A1.3 ▪ **The genetic code.** Codons within a single box encode the same amino acid. Color-highlighted amino acids are encoded by codons in two boxes. Stop codons are highlighted red.

A1.5 Calculating the Standard Free Energy Change, $\Delta G°$, of Chemical Reactions

For a chemical reaction, the net change in free energy ΔG determines whether the reaction will go forward. The value of ΔG at standard conditions of temperature (25°C), pressure (one atmosphere, P_a), and concentration (each reactant at 1 mole/liter), is designated $\Delta G°$. For a given reaction, how do we find the value of $\Delta G°$? The value is determined by summing the individual values for standard energy of formation ($\Delta G_f°$) for all the products, and then subtracting the sum of the values for all the reactants.

Consider the oxidation of glucose ($C_6H_{12}O_6$) to form water plus carbon dioxide:

$$C_6H_{12}O_6 + 6O_2 \rightarrow 6CO_2 + 6H_2O$$

We must assume that the glucose is present at 1-M concentration, that the O_2 and CO_2 are gases at 1 atmosphere, and that the water is liquid at 25°C (298 K). The value of $\Delta G°$ for this reaction is given by:

$$\Delta G° = [(6 \times \Delta G_f° \, CO_2) + (6 \times \Delta G_f° \, H_2O)]$$
$$- [(\Delta G_f° \, C_6H_{12}O_6) + 6 \times (\Delta G_f° \, O_2)]$$

TABLE A1.2　Free Energies of Formation ($\Delta G_f°$), in Units of kJ/mol.

Inorganic compounds	$\Delta G_f°$	Inorganic compounds (continued)	$\Delta G_f°$	Organic compounds	$\Delta G_f°$	Organic compounds (continued)	$\Delta G_f°$
CH_4	−50.8	Mn^{2+}	−228.1	Acetaldehyde	−127.6	Guanine	+47.4
CO	−137.2	$MnCl_2$	−490.8	Acetate	−369.7	Lactate	−517.8
CO_2	−394.4	$MnSO_4$	−972.8	Acetone	−152.7	Lactose	−1,567.0
H_2CO_3	−623.2	N_2	0	Arginine	−240.5	Malate	−845.1
HCO_3^-	−586.9	NO	+87.6	Aspartic acid	−730.7	Methanol	−166.6
CO_3^{2-}	−527.9	NO_2	+51.3	Benzene	+124.4	Methionine	−505.8
$HCOO^-$	−351.0	NO_2^-	−32.2	Benzoic acid	−245.3	Methylamine	+35.7
Cu^+	+50.0	NO_3^-	−111.3	1-Butanol	−162.5	Naphthalene	+201.6
Cu^{2+}	+65.5	NH_3	−26.6	Butyrate	+352.6	Oxalate	−697.9
CuS	−53.7	NH_4^+	−79.3	Citrate	−1,236.4	Oxaloacetate	−797.2
Fe^{2+}	−78.9	N_2O	+103.7	Cysteine	+339.8	2-Oxoglutarate	+797.1
Fe^{3+}	−4.7	N_2H_4	+149.3	Ethanol	−174.8	Phenol	−50.4
$FeCO_3$	−666.7	O_2	0	Ethylene	+68.4	Propionate	−361.1
FeS_2	−156.1	OH^-	−157.3	Fructose	−915.4	Pyruvate	+474.6
$FeSO_4$	−823.4	PO_4^{3-}	−1,018.8	Fumarate	−655.6	Ribose	−757.3
H_2	0	S^0	0	Gluconate	−1,128.3	Succinate	−690.2
H^+ (pH 0)	0	SO_3^{2-}	−486.5	Glucose	−910.4	Sucrose	−1,544.7
H^+ (pH 7)	−39.7	$S_2O_3^{2-}$	−522.5	Glutamic acid	−731.3	Toluene	+113.8
HCl	−131.3	H_2S	−27.9	Glutamine	−529.7	Trimethylamine	+93.0
H_2O	−237.1	HS^-	+12.1	Glyceraldehyde	+437.7	Tryptophan	−119.4
H_2O_2	−120.4	S^{2-}	+85.8	Glycerol	−477.0	Tyrosine	−385.7

Sources: James G. Speight. 2005. *Lange's Handbook of Chemistry,* 16th ed., McGraw-Hill, New York; WolframAlpha (http://www.wolframalpha.com); Rudolf K. Thauer. 1977. *Bacteriol. Rev.* **41**:100.

From **Table A1.2**, we can insert the values of $\Delta G_f°$ for each reactant and product. Note that one reactant, O_2, has a value of zero, because oxygen gas is the defined standard state for oxygen:

$$\Delta G° = [(6 \times −394.4 \text{ kJ/mol}) + (6 \times −237.1 \text{ kJ/mol})]$$
$$- [(−910.4 \text{ kJ/mol}) + 6 \times (0 \text{ kJ/mol})]$$
$$= −2,879 \text{ kJ/mol}$$

So the oxidation of glucose has a negative value of ΔG at standard conditions, and the reaction will go forward, releasing energy. The magnitude of the ΔG value suggests that enough energy could be released to form a number of energy carriers, such as ATP; but many additional factors need to be included before we know fully what a cell gains from this reaction.

A1.6　Generalized Cells

Prokaryotic cells and eukaryotic cells have several features in common. They each have a cell membrane that defines the boundary of the cell, ribosomes for protein synthesis,

and chromosomes made of DNA. Nevertheless, cells from different domains differ in a number of ways. **Figure A1.4** shows examples of generalized cells that highlight notable subcellular components.

A1.7　Semipermeable Membranes

The major functions of membranes (such as containing cytoplasmic components, regulating which substances enter and leave cells and organelles, and producing energy) depend on the semipermeable nature of membranes. Semipermeable (also called selectively permeable) membranes are permeable to some substances but not to others. In general, the cell membrane is permeable to hydrophobic molecules and impermeable to charged molecules (**Fig. A1.5**).

Cells are filled with and surrounded by water. Having water on both sides of the cell membrane presents a challenge for living cells because water moves across a

A. Prokaryotic cell (bacteria or archaea)

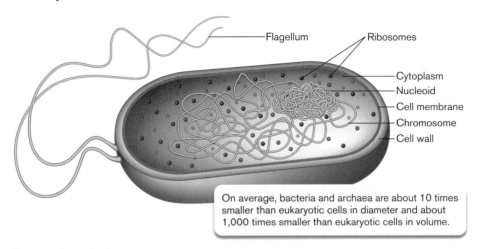

On average, bacteria and archaea are about 10 times smaller than eukaryotic cells in diameter and about 1,000 times smaller than eukaryotic cells in volume.

B. Generalized plant cell

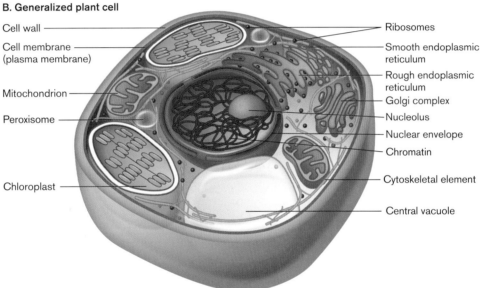

C. Generalized animal cell

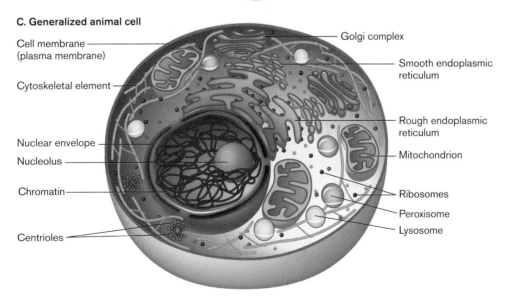

FIGURE A1.4 ▪ The prokaryotic cell and the eukaryotic cell. A. The prokaryotic cell typically contains a single compartment, and its DNA is organized in the nucleoid region. **B, C.** Eukaryotic cells are typically much larger than prokaryotic cells and contain organelles.

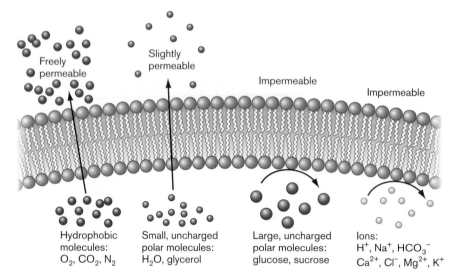

FIGURE A1.5 ■ **Selective permeability of cell membranes.**

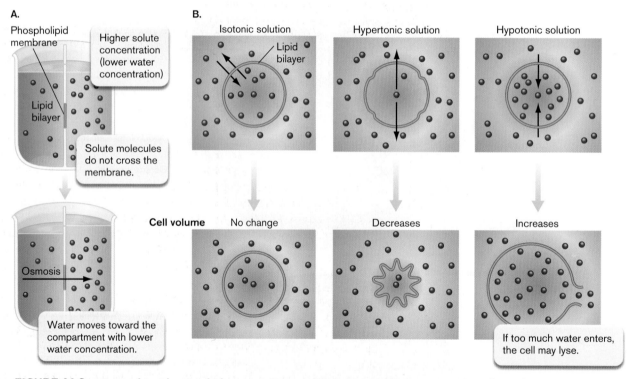

FIGURE A1.6 ■ **Osmosis and water balance.** A. Osmosis. B. Movement of water across the cell membrane, and shrinkage or expansion of the membrane in isotonic, hypertonic, and hypotonic environments. Black arrows indicate net water movement.

semipermeable membrane from regions of low solute concentration to regions of higher solute concentration (**Fig. A1.6**). Cells must maintain osmotic balance with their environment.

A1.8 The Eukaryotic Cell Cycle and Cell Division

Mitosis is the process that eukaryotic cells use to apportion their replicated DNA chromosomes evenly to their daughter cells. It is one phase in the cell cycle. The key steps of mitosis are summarized in **Figure A1.7**.

Meiosis is a special form of mitosis that generates haploid gametes. The key steps of meiosis are summarized in **Figure A1.8**.

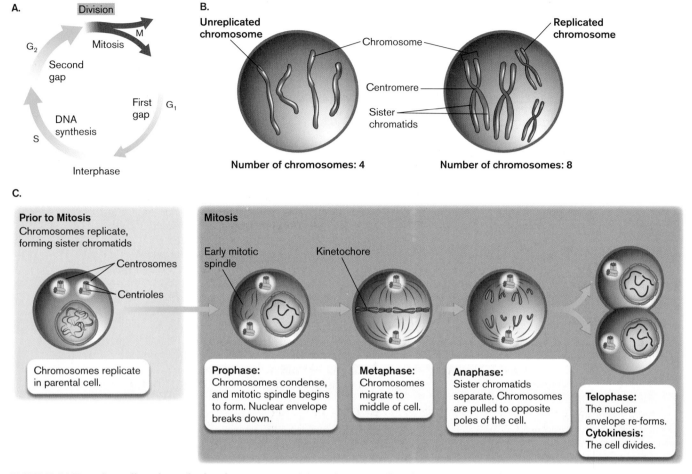

FIGURE A1.7 ▪ The cell cycle and mitosis. A. Stages of the eukaryotic cell cycle. G_1, S, and G_2 make up interphase (in blue). B. Duplication of the chromosomes during S phase. C. The phases of mitosis.

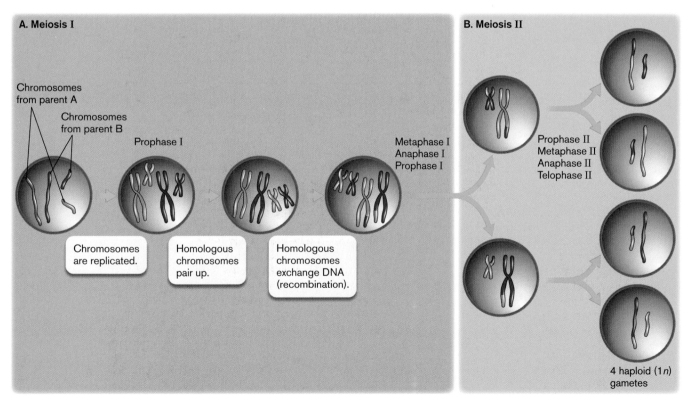

FIGURE A1.8 ▪ Meiosis. The chromosomes of a diploid (2n) cell undergo replication. A. In meiosis I, homologous chromosomes exchange DNA, and all the pairs are separated between two daughter cells. B. In meiosis II, homologous pairs are separated to produce haploid (1n) gametes.

APPENDIX 2
Taxonomy

Appendix 2 outlines the major known groups of microbial species. Most of these species are mentioned in Chapter 6 (Viruses) and in the diversity chapters: Chapter 18 (Bacteria), Chapter 19 (Archaea), and Chapter 20 (Eukaryotes). The species are grouped according to clades based on genome phylogeny. We indicate their major traits of metabolism, ecology, and relevance to medicine.

CURRENT RESEARCH highlight

Multispecies biofilm. A multispecies biofilm formed on an agar-coated slide immersed in a slurry of soil bacteria. The biofilm includes Gram-positive bacterial genera such as *Bacillus*, as well as Gram-negative genera such as *Brevundimonas*, *Flavobacterium*, and *Pseudomonas*.

Source: Mette Burmølle et al. 2007. *Microb. Ecol.* **54**:352.

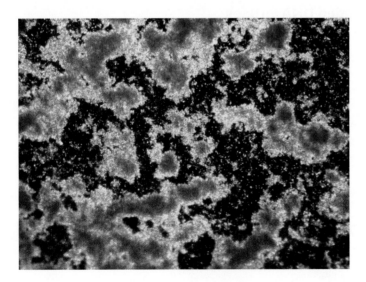

A2.1 Viruses

TABLE A2.1	Groups of viruses—Baltimore classification. (Expanded version of Table 6.1.)

Group I. Double-stranded DNA viruses
Replicate using host or viral DNA polymerase.

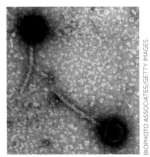

Bacteriophage lambda

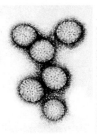

50 nm

Fusellovirus

BIOPHOTO ASSOCIATES/GETTY IMAGES

K.M. STEDMAN, ET AL. 2003, *RES. MICROBIOL* **154** 295.

Nonenveloped bacteriophages (structure includes head and tail)

Myoviridae. Bacteriophage T4 infects *Escherichia coli*.

Siphoviridae. Bacteriophages lambda and mu infect *E. coli*. Others infect Gram-positive hosts.

Tectiviridae. Infect enteric bacteria.

Nonenveloped viruses of animals and protists

Adenoviridae. Adenovirus generates tumors in humans.

Iridoviridae. Infect insects and amphibians.

Mimiviridae. Large viruses; infect *Acanthamoeba* and human macrophages.

Pandoraviridae. The largest known viruses; infect amebas.

Papillomaviridae. Papillomavirus causes genital warts.

Phycodnaviridae. Infect algae, controlling algal blooms. Some chloroviruses infect *Chlorella*, algal symbiont of paramecia and hydras.

Enveloped viruses of animals

Baculoviridae. Baculoviruses infect insects.

Herpesviridae. Herpes simplex causes oral and genital herpes; varicella-zoster virus causes chickenpox.

Poxviridae. Include smallpox and cowpox viruses.

Archaeal viruses

Ampullaviridae. Bottle-shaped viruses; infect *Acidianus*.

Fuselloviridae. Infect thermoacidophiles (*Sulfolobus*) and halophiles (*Haloarcula*).

Guttaviridae. Spindle-shaped viruses; infect *Sulfolobus*.

Haloviridae. Haloviruses infect haloarchea such as *Haloferax*, *Halobacterium*, and *Haloarcula*.

Rudiviridae, Lipothrixviridae. Infect *Sulfolobus* and *Thermoproteus*.

Group II. Single-stranded DNA viruses
Genome consists of (+) sense DNA; host DNA polymerase used; nonenveloped.

Geminivirus

ROBERT G. MILNE, CNR, ISTITUTO DE FITOVIROLOGICA APPLICATA, TORINO, ITALY

Bacteriophages

Inoviridae. Bacteriophage M13 infects *E. coli* and has a slow-release life cycle.

Microviridae. Bacteriophage φX174 infects *E. coli*.

Animal viruses

Anelloviridae. Found in human blood plasma; cause no known harm.

Circoviridae. Infect pigs and birds, causing immunosuppression.

Parvoviridae. Cause various diseases in cats, pigs, and other animals.

Plant viruses

Geminiviridae. Transmitted by aphids to tomato plants and other important crops. Their virions group in "twins," each member of the pair carrying one DNA circle with part of the genome.

Group III. Double-stranded RNA viruses
Require viral RNA-dependent RNA polymerase; usually package the polymerase before exiting host cell.

Rotavirus

GARY GAUGLER/VISUALS UNLIMITED

Nonsegmented, enveloped bacteriophages

Cystoviridae. Infect *Pseudomonas* species of bacteria.

Segmented, nonenveloped viruses of animals and plants

Birnaviridae. Infect marine and freshwater fish.

Reoviridae. Orthoreoviruses and rotaviruses infect humans and other vertebrates. Cypovirus infects insects. Rice dwarf virus (phytoreovirus) devastates rice crops worldwide.

Varicosaviridae. Infect plants.

Group IV. (+) sense single-stranded RNA viruses

Require viral RNA-dependent RNA polymerase to generate (−) template for progeny (+) genome; usually nonsegmented.

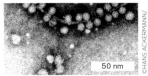

Bacteriophage MS2

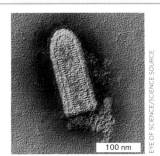

Rhinovirus 14

Nonenveloped bacteriophages

Leviviridae. Bacteriophages MS2 and Qβ infect *E. coli*.

Nonenveloped animal and plant viruses

Bromoviridae. Infect many kinds of plants and are often carried by beetles.

Picornaviridae. Poliovirus causes poliomyelitis. Rhinovirus causes the common cold. Aphthovirus causes foot-and-mouth disease in cattle and other stock.

Potyviridae. Viruses such as plum pox virus infect fruits, peanuts, potatoes, and other plants.

Tobamoviridae. Tobacco mosaic virus infects plants.

Enveloped animal and plant viruses

Coronaviridae. Include SARS (severe acute respiratory syndrome) and animal viruses.

Flaviviridae. Include West Nile virus, yellow fever virus, hepatitis C virus, and Zika virus.

Togaviridae. Include rubella virus and equine encephalitis virus.

Group V. (−) sense single-stranded RNA viruses

Require viral RNA-dependent RNA transcriptase. Genome often segmented. Some viruses include (+) and (−) strand regions.

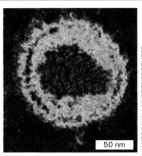

Rabies virus

Segmented, enveloped viruses

Orthomyxoviridae. Influenza virus causes major epidemics among humans and animals.

Nonsegmented, enveloped viruses

Filoviridae. Ebola virus causes outbreaks among humans and chimpanzees.

Paramyxoviridae. Infect humans and cause measles, mumps, and parainfluenza.

Rhabdoviridae. Rabies virus infects mammals (virion length 130–300 nm).

Segmented (+/−) strand, enveloped viruses

Arenaviridae. Spread by rodents; cause hemorrhagic fever and lymphocytic choriomeningitis.

Bunyaviridae. Hantaviruses are spread by rodents and infect humans. Tospoviruses are transmitted by thrips, infecting plants. Rift Valley fever virus infects livestock.

Group VI. Retroviruses (RNA reverse-transcribing viruses)

Viral reverse transcriptase copies RNA to DNA for integration into host chromosome.

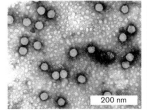

HIV-1

Retroviridae. Simple retroviruses include oncogenic retroviruses: feline leukemia virus, Rous sarcoma virus, avian leukosis virus. Lentiviruses (complex retroviruses) include human immunodeficiency virus (HIV, the cause of AIDS) and simian immunodeficiency virus (SIV). Lentiviruses are engineered to make lentiviral vectors (lentivectors) for gene therapy.

Group VII. Pararetroviruses (DNA reverse-transcribing viruses)

DNA transcribed to RNA; reverse-transcribed to DNA using host reverse transcriptase or packaged viral reverse transcriptase.

Cauliflower mosaic virus

Nonenveloped plant viruses

Badnaviridae. Infect bananas, cocoa plants, citrus, yams, and sugarcane.

Caulimoviridae. Transmitted by aphids; cauliflower mosaic virus and related viruses infect cauliflower, broccoli, groundnuts, soybeans, and cassava. The cauliflower mosaic virus promoter sequence is used to construct vectors to insert genes into transgenic plants.

Enveloped animal viruses

Hepadnaviridae. Hepatitis B virus causes widespread disease of the human liver.

A2.2 Bacteria

TABLE A2.2	Bacterial diversity. (Expanded version of Table 18.1.)

Note: Each trait applies to <u>most</u> members of the taxon described, but exceptions have evolved.

Deep-branching thermophiles

Thermophilic bacteria that diverged early from archaea and eukaryotes. Many genes transferred laterally from archaea.

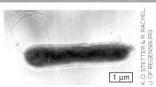

Aquifex

K. O. STETTER & R. RACHEL, U. OF REGENSBURG

Aquificae. Hyperthermophiles (70°C–95°C). Oxidize H_2, H_2S, or thiosulfate.

- **Aquificales.** Outer membrane; cells may be single with flagella or grow in filaments. *Aquifex, Thermocrinis.*

Chloroflexi. Filamentous green bacteria.

- **Chloroflexales.** Filamentous phototrophs; absorb light to oxidize organic compounds or H_2S. Most contain photosystem II in chlorosomes. *Chloroflexus aurantiacus* is thermophilic; forms mats in hot springs.

Chloroflexus aurantiacus

ANDREW SYRED/ SCIENCE SOURCE

Deferribacteres. Anaerobic vent thermophiles.

- **Deferribacterales.** Single rod or vibrio cells, flagellated. *Deferribacter autotrophicus* reduces Fe^{3+} with H_2; *D. abyssi* respires on small organic molecules.

Deinococcus-Thermus. Peptidoglycan contains ornithine. Diverse growth temperatures.

- **Deinococcales.** Thick envelope; stain Gram-positive. Not thermophilic, but extremely resistant to ionizing radiation and to desiccation. *Deinococcus.*
- **Thermales.** Species are filamentous clustered cells or unicellular. Grow at 70°C–75°C. *Thermus aquaticus* is the source of Taq polymerase for PCR.

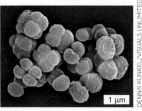

Deinococcus

DENNIS KUNKEL/VISUALS UNLIMITED

Thermotogae. Thermophiles or hyperthermophiles (55°C–100°C).

- **Thermotogales.** Outer membrane with large periplasm; anaerobic heterotrophs. *Petrotoga, Thermotoga.*

Cyanobacteria

Oxygenic photoautotrophs with thylakoid membranes. Fix CO_2 using Rubisco. Often mutualists. Share ancestry with chloroplasts.

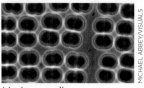

Merismopedia

MICHAEL ABBEY/VISUALS UNLIMITED

Cyanobacteria.
- **Chroococcales.** Square colonies based on two division planes.
    - ***Chroococcus***. Single, double, or quartet of cells (10–20 µm per cell). Grow in pond sediment.
    - ***Merismopedia***. Platelike colonies of elliptical cells (5 µm).
    - ***Synechococcus***. Single or double cells. Marine producer.
- **Gloeobacterales.** Lack thylakoids; conduct photosynthesis in cell membrane. *Gloeobacter violaceus.*
- **Nostocales.** Filamentous chains with N_2-fixing heterocysts. Some grow symbiotically with corals or plants.
    - ***Anabaena***. Aquatic. Grow in association with red water fern (*Azolla*).
    - ***Nostoc***. Aquatic. Grow independently; or mutualistically with fungi, as lichens; or as endosymbionts of *Gunnera* plant cells.

Anabaena

©CAROLINA BIOLOGICAL SUPPLY/VISUALS UNLIMITED

- **Oscillatoriales.** Filamentous chains with motile hormogonia (short chains) (filament width 50 µm).
    - ***Oscillatoria***. Freshwater or marine. Grow independently, or as sponge endosymbionts.
    - ***Spirulina***. Aquatic. Farmed as food supplement (filament width 5 µm).
    - ***Trichodesmium***. Major marine producers. Form "blooms" of overgrowth.
- **Pleurocapsales.** Globular colonies; reproduce through baeocytes. *Pleurocapsa, Myxosarcina.*
- **Prochlorales.** Tiny single cells, elliptical or spherical (0.5 µm).
    - ***Prochlorococcus***. Most abundant marine producers. One of the smallest cells of a free-living microbe, with a highly reduced genome.
    - ***Prochloron***. Tropical marine producers; endosymbionts of sea squirts.

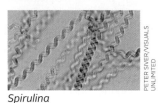

Spirulina

PETER SIVER/VISUALS UNLIMITED

Firmicutes and Actinobacteria (Gram-positive)

Peptidoglycan multiple layers, cross-linked by teichoic acids.

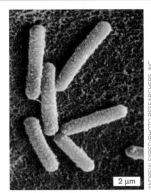

Bacillus subtilis

ANDREW SYRED/PHOTO RESEARCHERS, INC.

2 μm

Mycoplasma genitalium

DON W. FAWCETT/ SCIENCE SOURCE

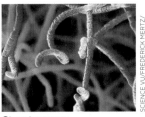

Streptomyces

SCIENCE VU/FREDERICK MERTZ/ VISUALS UNLIMITED

Micrococcus luteus

CDC/BETSY CRANE

1 μm

Firmicutes. **Low-GC, Gram-positive rods and cocci.**

- **Bacillales.** Aerobic or facultative anaerobes (2–10 μm).
  - *Bacillus*. Endospore-forming rods. Soil-growing: *B. subtilis, B. cereus, B. anthracis, B. thuringiensis*. Extremophiles: *B. alkalophilus, B. thermophilus, B. halodurans*.
  - *Listeria*. Non-spore-forming rods; intracellular pathogens. *L. monocytogenes* causes listeriosis.
  - *Staphylococcus*. Non-spore-forming cocci. Skin biota: *S. epidermidis. S. aureus* infects flesh.
- **Clostridiales.** Anaerobic rods.
  - *Carboxydothermus*. Oxidize carbon monoxide (CO) to CO_2. Form endospores.
  - *Clostridium*. Form endospores. *C. botulinum* and *C. difficile* are pathogens. *C. acetobutylicum* generates butanol.
  - *Dehalobacter*. Non-spore-forming rods. Dechlorinate chloroethenes.
  - *Epulopiscium*. Exceptionally large cells. Reproduce by "live birth."
  - *Heliobacterium, Heliophilum*. Endospore-forming photoheterotrophs.
  - *Metabacterium*. Exceptionally large cells; form multiple endospores. *M. polyspora*.
  - *Ruminococcus*. Digestive flora of ruminant animals.
- **Lactobacillales.** Non–spore formers. Facultative anaerobes. Ferment, producing lactic acid.
  - *Enterococcus*. Enteric cocci. *E. hirae*.
  - *Lactobacillus*. *L. acidophilus* is used for dairy culture.
  - *Lactococcus*. Used for dairy culture. *L. lactis*.
  - *Streptococcus*. Chains of cocci. Group A streptococci cause "strep throat."
- **Tenericutes (Mollicutes).** Lack cell wall; require animal host. *Mycoplasma genitalium* has one of smallest known genomes. *M. pneumoniae* causes pneumonia.

Actinobacteria. **High-GC, Gram-positive bacteria with moderate salt tolerance.**

- **Actinomycetales.**
  - **Actinomycetaceae.** Filamentous, producing aerial hyphae and spores (width 1 μm).
    - *Actinomyces*. *A. israelii* causes actinomycosis.
    - *Frankia*. Saprophytes; grow on leaf litter. Fix nitrogen for plants.
    - *Salinispora*. Sponge endosymbionts. Produce drug-like secondary products.
    - *Streptomyces*. Produce many antibiotics. *S. coelicolor, S. griseus*.
  - **Corynebacteriaceae.** Irregularly shaped rods. *Corynebacterium diphtheriae* causes diphtheria.
  - **Micrococcaceae.** Nonfilamentous soil bacteria; mostly obligate aerobes. Often airborne.
    - *Arthrobacter*. Rods. Respire on oxygen or on chlorinated aromatics.
    - *Micrococcus*. Aerobic cocci such as *M. luteus*. Grow in soil.
  - **Mycobacteriaceae.** Exceptionally thick cell envelope holds acid-fast stain.
    - *Mycobacterium*. *M. tuberculosis* causes tuberculosis; *M. leprae* causes leprosy.
  - **Propionibacteriaceae.** Propionic acid fermentation. *Propionibacterium acnes* causes acne; *P. freudenreichii* makes Swiss cheese.
- **Bifidobacteriales.** Ferment without gas. Enteric biota of breast-fed infants. *Bifidobacterium*.

Proteobacteria

Gram-negative. Outer membrane contains LPS. Diverse metabolism and cell forms (1–10 μm). Share ancestry with mitochondria.

Proteobacteria.

Alphaproteobacteria (class). Heterotrophic rods or spirilla.

- **Caulobacterales.** Aquatic oligotrophs; alternate stalk and flagellum. *Caulobacter*.
- **Rhizobiales.** Plant mutualists and pathogens, methyl oxidizers, and animal pathogens.
  - *Agrobacterium*. *A. tumefaciens* causes plant tumors; transgenic plant vector.
  - *Bartonella*. *B. quintana* causes trench fever; *B. henselae* causes cat scratch disease.
  - *Brucella*. Cause brucellosis in horses and sheep. *B. melitensis*.
  - *Methylobacterium*. Grow on single-carbon compounds.
  - *Nitrobacter*. Nitrite-oxidizing lithotrophs. *N. winogradskyi*.
  - *Rhizobium group*. *Bradyrhizobium* and *Sinorhizobium* fix nitrogen within legumes.
  - *Rhodopseudomonas, Rhodomicrobium*. Soil photoheterotrophs.

(continued)

TABLE A2.2 **Bacterial diversity. *(continued)***

Rhodospirillum rubrum

DAVID M. PHILLIPS/VISUALS UNLIMITED

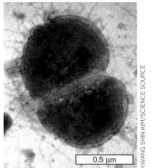

0.5 μm

Neisseria gonorrhoeae

KWANG SHIN KIM/SCIENCE SOURCE

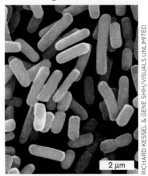

2 μm

Escherichia coli

RICHARD KESSEL & GENE SHIH/VISUALS UNLIMITED

Myxococcus

MICHIEL VOS, U. OF OXFORD

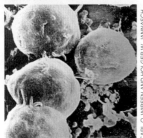

Thiovulum

CARL O. WIRSEN AND HOLGER W. JANNASCH. 1978. *J. BACTERIOL.* **136**:765

- **Rhodobacterales, Rhodospirillales.** Flagellated photoheterotrophs: *Rhodobacter sphaeroides*, *Rhodospirillum rubrum*, *Roseobacter* (oxidizes CO). Nonphototrophs: *Acetobacter* makes vinegar.
- **Rickettsiales.** Includes intracellular parasites; related to mitochondria.

  Rickettsia. Intracellular parasites. *R. rickettsii* causes Rocky Mountain spotted fever; *R. prowazekii* causes typhus.

  SAR11 cluster. Marine photoheterotrophs use proteorhodopsin. *Pelagibacter.*
- **Sphingomonadales.** Catabolize complex organics; used for bioremediation. Opportunistic pathogen: *Sphingomonas*. Aerobic photoheterotrophs: *Citromicrobium, Erythromicrobium*.

Betaproteobacteria (class). Phototrophs, lithotrophs, and pathogens.

- **Burkholderiales.** *Burkholderia pseudomallei* causes melioidosis in humans and farm animals.
- **Neisseriales.** Mucous-membrane normal flora and pathogens. Aerobic or microaerophilic diplococci. *Neisseria gonorrhoeae* causes gonorrhea; *N. meningitidis* causes meningitis.
- **Nitrosomonadales.** Lithotrophs. *Nitrosomonas europaea* oxidizes ammonia.
- **Rhodocyclales.** Soil heterotrophs and photoheterotrophs. *Azoarcus evansii* catabolizes complex aromatic molecules. *Rhodocyclus* species are purple photoheterotrophs.

Gammaproteobacteria (class). Facultative anaerobes and lithotrophs.

- **Acidithiobacillales.** Lithotrophs. *Acidithiobacillus ferrooxidans* oxidizes iron and sulfur.
- **Aeromonadales.** Aquatic heterotrophs. *Aeromonas hydrophila* infects fish and humans.
- **Alteromonadales.** Aquatic. *Shewanella oneidensis* reduces metals; used in fuel cells.
- **Chromatiales.** Lithotrophs: *Nitrococcus* species oxidize ammonia. Sulfur and iron phototrophs: *Chromatium*. Nitrite phototrophs: *Thiocapsa*.
- **Enterobacteriales.** Enterobacteriaceae. Facultative anaerobes; colonize the human colon.

  Escherichia coli. Strains include normal flora and enteric pathogens.

  Salmonella. *S. enterica* strains cause gastrointestinal infection and typhoid fever.

  Yersinia. *Y. pestis* causes bubonic plague.
- **Legionellales.** *Legionella pneumophila* causes legionellosis pneumonia. *Coxiella burnetii* causes Q fever.
- **Oceanospirillales.** Marine heterotrophs, including degraders of petroleum. *Alcanivorax*.
- **Pseudomonadales.** Rods; aerobic or respire on nitrate; catabolize aromatics. *Pseudomonas aeruginosa* infects lungs in cystic fibrosis patients. *P. fluorescens* suppresses plant diseases.
- **SAR86 cluster.** Marine photoheterotrophs use proteorhodopsin.
- **Thiotrichales.** Lithotrophs and heterotrophs. *Beggiatoa alba* and *Thiomargarita namibiensis* oxidize sulfur. *Cycloclasticus* species catabolize polycyclic aromatic hydrocarbons in petroleum.
- **Vibrionales.** Marine heterotrophs. *Vibrio cholerae* causes cholera.

Deltaproteobacteria (class). Lithotrophs and multicellular communities.

- **Bdellovibrionales.** Periplasmic predators. *Bdellovibrio bacteriovorus* consumes *E. coli*.
- **Desulfobacterales.** Reduce sulfate. *Desulfobacter, Desulfococcus*.
- **Desulfuromonadales.** Lithotrophs; reduce or oxidize sulfur or metals.

  Desulfuromonas. Reduce elemental sulfur.

  Geobacter metallireducens. Reduces iron oxide.
- **Myxococcales.** Gliding bacteria that form fruiting bodies. *Myxococcus*.

Epsilonproteobacteria (class).

- **Campylobacterales.** Spirillar pathogens.

  Campylobacter. *C. jejuni* causes food poisoning.

  Helicobacter. *H. pylori* causes gastritis.

  Nautilia, Hydrogenimonas, Sulfurimonas. Oxidize H_2 with sulfur or nitrate.

  Thiovulum. Oxidize sulfides, producing sulfur granules.

Deep-branching Gram-negative phyla

Gram-negative, with outer membrane. Most are anaerobes.

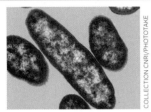

Bacteroides

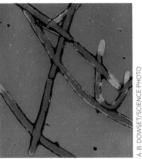

Fusobacterium nucleatum

Acidobacteria. **Gram-negative bacteria; include acidophiles, with diverse metabolism.**
    Acidobacterium capsulatum.
    Geothrix capsulatum. Oxidizes toluene with Fe III.
    Chloracidobacterium thermophilum. Phototrophic thermophile.

Bacteroidetes. **Obligate anaerobes. Heterotrophs that feed on diverse carbon sources.**
- **Bacteroidales.** Obligate anaerobes, soil or human flora.
    Bacteroides. Enteric species such as *B. thetaiotaomicron* digest complex plant carbohydrates. *B. fragilis* escaping the colon causes abscesses.
    Porphyromonas. Includes gingival pathogens such as *P. gingivalis*.
- **Flavobacteriales.** Aerobic or facultative heterotrophs in soil and water. *Flavobacterium psychrophilum* infects fish.

Chlorobi. **Green sulfur-oxidizing phototrophs.**
- **Chlorobiales.** Anaerobic H_2S photolysis, absorbing red and infrared. *Chlorobium tepidum*.

Fusobacteria. **Gram-negative anaerobic bacteria found in septicemia and in skin ulcers.**
    Fusobacterium nucleatum. Isolated from dental plaque.

Nitrospirae. **Nitrite oxidizers. Aerobic or facultative**.
- **Nitrospirales.** Tight spirilla. In soil and water, oxidize NO_2^- to NO_3^-. *Nitrospira, Leptospirillum*. Includes acidophilic Fe oxidizers.

Spirochetes (Spirochaeta)

Narrow, coiled cells with axial filaments, encased by sheath. Polar flagella beneath sheath double back, winding around cell.

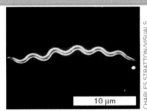

10 μm

Borrelia burgdorferi

Spirochetes.
- **Spirochaetales.** Aquatic free-living; endosymbionts and pathogens (width < 0.5 μm; length 10–20 μm).
    Borrelia. *B. burgdorferi* causes Lyme disease, transmitted by ticks.
    Hollandina. Termite gut endosymbionts.
    Leptospira. L-shaped animal pathogens; cause leptospirosis.
    Spirochaeta. Aquatic, free-living heterotrophs.
    Treponema. *T. pallidum* causes syphilis.

Chlamydiae, Planctomycetes, and Verrucomicrobia

Irregular cells lacking peptidoglycan, with subcellular structures analogous to those of eukaryotes.

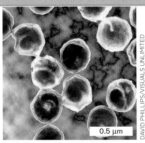

0.5 μm

Chlamydia trachomatis

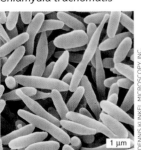

1 μm

Prosthecobacter fusiformis

Chlamydiae. **Intracellular cell wall–less pathogens of animals or protists.**
- **Chlamydiales.** Reticulate bodies form multiple spore-like transfer particles (elementary bodies) that infect the next host.
    Chlamydia. *C. trachomatis* causes sexually transmitted disease and trachoma (eye infection).
    Chlamydophila. *C. pneumoniae* causes pneumonia; *C. abortus* causes spontaneous abortion in animals; *C. psittaci* causes psittacosis in birds and humans.
    Parachlamydia. Chlamydia-like species that infect free-living amebas.

Planctomycetes. **Nucleoid has double membrane analogous to eukaryotic nuclear membrane and other intracytoplasmic subcellular membrane compartments.**
- **Planctomycetales.** Freshwater or marine. Flexible cell shape.
    Brocadia. Anaerobically oxidize ammonium and release N_2, using anammoxosomes.
    Gemmata. Nuclear double membrane, surrounded by additional intracellular membrane.
    Pirellula. Only one intracellular membrane. Has fimbriae and single flagellum. Marine habitat.
    Planctomyces. One intracellular membrane. Halophile.
    Scalindua. Conduct anammox in wastewater. Oxidize CO.

Verrucomicrobia. **Stalk-like appendages contain actin filaments. Aquatic oligotrophs.**
- **Verrucomicrobiales.**
    Prosthecobacter. Each cell has a polar cytoplasmic extension called a prostheca.
    Verrucomicrobium. Stellate cells have multiple prosthecae.

A2.3 Archaea

TABLE A2.3	Archaeal diversity. (Expanded version of Table 19.2.)

Note: Each trait applies to <u>most</u> members of the taxon described, but exceptions have evolved.

Crenarchaeota

Temperature: Hyperthermophiles and psychrophiles, as well as mesophiles.

Metabolism: Sulfur and hydrogen oxidizers, aerobic and anaerobic heterotrophs, ammonia oxidizers.

Envelope: Membrane lipids include tetraethers with crenarchaeol. Flexible S-layer surrounds membrane.

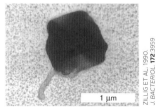

1 μm

ZILLIG ET AL. 1990. *J. BACTERIOL.* **172**:3959

Hyperthermus butylicus

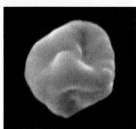

1 μm

OLIVER MECKES/NICOLE OTTAWA

Sulfolobus sp.

Bathyarchaeota (formerly MCG). Widely divergent clades. Terrestrial and marine, cold and hot, water and subsurface environments.

Deep-Sea Archaeal Group/Marine Benthic Group B (DSAG/MBGB). Anoxic marine sediments: deep-sea methane hydrates, hydrothermal vents, coastal regions. Metabolism uncertain.

Thermoprotei. Many thermophiles in hot springs and marine vents. Also includes marine mesophiles and psychrophiles.

- **Caldisphaerales.** Thermoacidophilic heterotrophs grow in hot springs. *Caldisphaera.*
- **Desulfurococcales.** Anaerobic sulfur reduction with organic electron donors. Irregularly shaped cells with glycoprotein S-layer; no cell wall.
  - ***Aeropyrum pernix, Desulfurococcus mobilis, Hyperthermus butylicus.***
  - ***Ignicoccus islandicus.*** Unique periplasmic space contains membrane vesicles.
  - ***Pyrodictium abyssi, Pyrodictium occultum, Pyrolobus fumarii.*** Grow at marine thermal vents, up to 110°C. Form 3D network of cells and extracellular cannulae.
  - ***Thermosphaera.*** Heterotrophic; inhibited by sulfur.
- **Psychrophilic marine crenarchaeotes (uncharacterized).** Marine water, deep sea, and Antarctica. Anaerobic heterotrophs, sulfate reducers, nitrite-reducing methanotrophs.
- **Sulfolobales.** Aerobic acidophiles; moderate thermophiles. Oxidize H_2S to H_2SO_4. *Sulfolobus, Sulfurisphaera, Acidianus.*
- **Thermoproteales.** *Pyrobaculum, Thermoproteus, Vulcanisaeta.*

Thaumarchaeota

Temperature: Mesophiles, thermophiles, and psychrophiles.

Metabolism: Marine and soil archaea that oxidize NH_3 with O_2. Important marine sources of nitrates for phytoplankton, and of methylphosphonate ($CH_3-PO_3^{2-}$), which bacteria convert to methane. Includes members of Marine Group 1.1a.

Envelope: Membrane lipids include tetraethers with crenarchaeol.

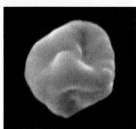

MICHAEL STIEGLMEIER AND NIKOLAUS LEISCH, DEPARTMENT OF ECOGENOMICS AND SYSTEMS BIOLOGY, UNIVERSITY OF VIENNA

Nitrososphaera viennensis

- **Cenarchaeales.** Sponge symbionts; grow at 10°C. *Cenarchaeum symbiosum.*
- **Nitrosopumilales.** Ammonia-oxidizing archaea (AOA), marine and soil. *Nitrosopumilus maritimus, Nitrososphaera gargensis.*
- **Psychrophilic marine thaumarchaeotes (uncharacterized).** Marine water, deep sea, and Antarctica. Anaerobic heterotrophs, sulfate reducers, nitrite-reducing methanotrophs.

Deeply branching groups of uncultured Archaea

Uncultured marine organisms, branching near root of archaeal tree.

Ancient Archaeal Group (AAG). Vent hyperthermophiles and benthic subsurface communities.

Korarchaeota. Hyperthermophiles in Yellowstone hot springs and in deep-sea thermal vents.

Lokiarchaeota. Found in marine sediment near thermal vent; shares 3% eukaryotic genes. May represent ancient ancestor of Eukarya.

Marine Hydrothermal Vent Group (MHVG). Vent hyperthermophiles.

Euryarchaeota

Temperature: Mesophiles and thermophiles; some psychrophiles.

Metabolism: Methanogens, halophilic photoheterotrophs, and sulfur and hydrogen oxidizers; acidophiles and alkaliphiles.

Envelope: Methanogens and halophiles have rigid cell walls of glycans, glycoproteins, or pseudopeptidoglycan.

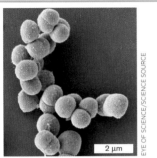

2 μm

Halobacteriales

EYE OF SCIENCE/SCIENCE SOURCE

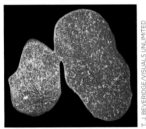

Methanoculleus nigri

T. J. BEVERIDGE/VISUALS UNLIMITED

Methanosarcina mazei

©RALPH ROBINSON/VISUALS UNLIMITED

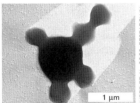

1 μm

Nanoarchaeum equitans
(attached to surface of
Ignicoccus sp.)

HUBER ET AL. 2003. *RESEARCH IN MICROBIOL.* **154**:165

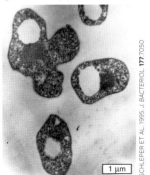

1 μm

Picrophilus oshimae

SCHLEPER ET AL. 1995. *J. BACTERIOL.* **177**:7050

Altiarchaeota. Includes *Altiarchaeum;* form grappling-hook biofilms in cold sulfidic water.

Anaerobic Methane Oxidizers (ANME). Anoxic marine sediments. Oxidize methane from methanogens, in syntrophy with sulfate-reducing bacteria. Important control of greenhouse gas.

Archaeoglobi. Hyperthermophiles; metabolize sulfur and use reverse methanogenesis.

- **Archaeoglobales.** Sulfate oxidation of H_2 or organic hydrogen donors. *Archaeoglobus fulgidus.*

Haloarchaea. Halophiles; grow in brine (concentrated NaCl). Conduct photoheterotrophy, with light-driven H^+ pump and Cl^- pump.

- **Halobacteriales.**

    Haloarcula, Halobacterium, Halococcus, Haloferax. Mesophiles at neutral pH. Grow in salterns and salt lakes. *Haloarcula marismortui, Halobacterium salinarum.*

    Haloquadra (Haloquadratum). Square-shaped mesophilic halophiles.

    Halorubrum lacusprofundi. Psychrophile; grows in subglacial Antarctic salt lakes.

    Natronococcus, Natronomonas. Alkaliphiles; grow above pH 9 in soda lakes.

Methanogens (four classes). Generate methane from CO_2 and H_2, formate, acetate, other small molecules. Strict anaerobes. Must associate with bacteria producing their substrates.

Cell walls of pseudopeptidoglycan or polysaccharides.

Wide variety of shapes: rods, filaments, cocci, and coccoid clusters.

Wide range of anaerobic habitats: wetlands, landfills, and intestinal tracts of animals. Psychrophiles in deep-ocean floor sediment generate methane hydrates.

- **Methanobacteriales.** Lack cytochromes; reduce CO_2, formate, or methanol with H_2.

    Methanobacterium. Have peptidoglycan cell walls.

    Methanobrevibacter smithii, Methanosphaera stadtmanae. Inhabit human digestive tract. Increase caloric output of human food by reducing methanol, a breakdown product of pectin.

    Methanothermus fervidus. Marine vent thermophile.

- **Methanomicrobiales, Methanococcales, and Methanopyrales.** Lack cytochromes; reduce CO_2, formate, or methanol with H_2.

    Methanocaldococcus jannaschii. Vent thermophile.

    Methanoculleus nigri. Budding cells, found in anaerobic wastewater digester.

    Methanogenium frigidum. Psychrophile.

    Methanopyrus kandleri. Marine vent thermophile.

- **Methanosarcinales.** Possess cytochromes; reduce methylamines and acetate (as well as CO_2 and formate) with H_2.

    Methanococcoides alaskense. Arctic marine habitat.

    Methanohalophilus. Halophile. Inhabits salt lakes, often at high pH.

    Methanosaeta concilii. Filamentous colonies.

    Methanosarcina acetivorans. Globular colonies; cell walls of sulfated polysaccharides.

Nanoarchaeota. Vent hyperthermophiles; obligate symbionts attached to *Ignicoccus*.

   Nanoarchaeum equitans.

South African Gold Mine Euryarchaeotal Group (SAGMEG). Terrestrial and marine deep subsurface. Anaerobic heterotrophs.

Terrestrial Miscellaneous Euryarchaeotal Group (TMEG). Terrestrial deep subsurface and surface soils, marine and freshwater sediments.

Thermococci. Hyperthermophiles (grow above 100°C) and barophiles (up to 200 atm pressure). Anaerobes; reduce sulfur.

- **Thermococcales.**

    Pyrococcus abyssi, P. furiosus. Use S^0 to oxidize H_2 or organic hydrocarbons. *Thermococcus* species are closely related.

Thermoplasmata. Extreme acidophiles; oxidize sulfur from pyrite (FeS_2), generating sulfuric acid. Mesophiles or moderate thermophiles.

- **Thermoplasmatales.**

    Ferroplasma acidiphilum, F. acidarmanus. Grow at 37°C–50°C. Oxidize sulfur from FeS_2, generating ambient pH as low as pH 0. No cell wall.

    Picrophilus torridus. Grows above 60°C. Oxidizes sulfur, generating acid. Possesses cell wall.

    Thermoplasma acidophilum. Grows at 59°C and pH 2.

A2.4 Eukarya

TABLE A2.4	Eukaryotic microbial diversity. (Expanded version of Table 20.1.)

Note: Each trait applies to <u>most</u> members of the taxon described, but exceptions have evolved.

Opisthokonta (fungi and metazoan animals)

Single flagellum on reproductive cells. Includes multicellular animals.

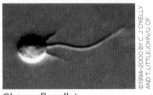

Choanoflagellate

©1994–2000 BY C. J. O'KELLY AND T. LITTLEJOHN/U OF MONTREAL

Metazoa (animals). Multicellular organisms with motile cells and body parts.
- **Colonial animals.** Sponges, jellyfish.
- **Invertebrates.** Hydras, mollusks, arthropods, worms.
- **Vertebrates.** Fish, amphibians, reptiles, birds, mammals, including *Homo sapiens*.

Choanoflagellata. Single flagellum with collar of microvilli. Resemble sponge choanocytes. Possible link to common ancestor of multicellular animals.

Eumycota (fungi)

Cells form hyphae with cell walls of chitin.

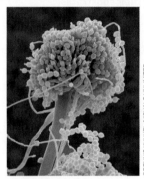

Aspergillus mold

©DENNIS KUNKEL/VISUALS UNLIMITED

Ascomycota. Fruiting bodies form asci containing haploid ascospores.
- **Filamentous species.** *Neurospora, Penicillium*.
    - *Aspergillus.* Opportunistic pathogen; produces aflatoxin.
    - *Geomyces.* Associated with bat mortality, white nose syndrome.
    - *Magnaporthe oryzae.* Causes rice blast, the most serious disease of cultivated rice.
    - *Microsporum, Trichophyton.* Cause ringworm skin infection.
    - *Trichoderma.* Endophyte within plants; provides nutrients and induces plant defenses.
- **Morels and truffles.** Large fruiting bodies are highly valued foods.
- *Stachybotrys.* Known as "black mold"; contaminates homes.
- **Yeasts.** Unicellular species have lost mycelial stages.
    - *Blastomyces dermatitidis, Candida albicans, Pneumocystis jirovecii.* Infect immunocompromised patients.
    - *Saccharomyces cerevisiae.* Ferments beer and bread dough.

Basidiomycota. Basidiospores form primary and secondary mycelia; generate mushrooms.
- *Amanita, Lycoperdon, Ustilago maydis. Amanita* is extremely poisonous; *Lycoperdon* is edible; *U. maydis* causes corn smut (a plant disease).
- *Cryptococcus neoformans.* Yeast-form basidiomycete; an opportunistic pathogen.

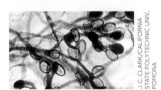

Allomyces zoospores

J. C. CLARK/CALIFORNIA STATE POLYTECHNIC UNIV, POMONA

Chytridiomycota. The deepest-branching fungal clade. Motile zoospores with a single flagellum resemble the gametes of animals. Saprophytes or anaerobic rumen fungi. *Allomyces.*
- *Batrachochytrium dendrobatidis.* Frog pathogen.
- *Neocallomastix.* Bovine rumen digestive endosymbiont.

Glomeromycota. Mutualists of plant roots, forming arbuscular mycorrhizae, filamentous networks that share nutrients with and among diverse plants.

Lichens. Mutualistic association between an ascomycete and green algae (*Trebouxia*) or cyanobacteria (*Nostoc*).

Microsporidia. Unicellular parasites that inject a spore through a tube into a host cell, causing microsporidiosis.
- *Encephalitozoon.* Commonly infects AIDS patients.

Zygomycota. Nonmotile gametes grow toward each other and fuse to form the zygote (zygospore). Saprophytes or insect parasites. Some form mycorrhizae.

Other opisthokonts: *Corallochytrium,* Ichthyosporea, Nucleariidae.

Viridiplantae (primary endosymbiotic algae and plants)

Includes algae and multicellular plants. Chloroplasts arose from a primary endosymbiont.

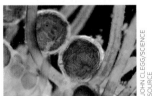
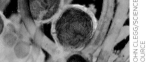

Charophyte

1 cm

Cymopolia barbata

Palmaria palmata

Plants. Adapted to growth on land.
- **Nonvascular plants.** Mosses, ferns.
- **Vascular plants.** Gymnosperms, angiosperms.

Charophyta. Multicellular algae with rhizoids that adhere to sediment. Form green mats with crust of calcium carbonate, giving the name "stoneworts."

Chlorophyta (green algae). Chlorophyll *a* confers green color. Inhabit upper layer of water.
- **Unicellular with paired flagella.** *Chlamydomonas* is a unicellular green alga; a model system for research on algae. *Volvox* forms colonies of flagellated cells.
- **Multicellular.** *Ulva* grows in large sheets. *Spirogyra* forms chains of cells. *Cymopolia* forms calcified stalks with filaments.
- **Picoeukaryotes.** *Ostreococcus* and *Micromonas* are unicellular algae <3 μm in diameter.
- **Siphonous algae.** *Caulerpa* species consist of a single cell with multiple nuclei, growing to indefinite size.

Rhodophyta (red algae). Phycoerythrin obscures chlorophyll, colors the algae red. Absorption of blue-green light enables colonization of deeper waters.

> **Porphyra.** Forms sheets edible by humans.
>
> **Sebdenia, Plocamium.** Form branched fronds. *Palmaria*.
>
> **Mesophyllum.** Forms coralline algae, hardened by calcium carbonate crust; resemble coral.

Amoebozoa (amebas and slime molds)

Lobe-shaped (lobose) pseudopods driven by sol-gel transition of actin filaments.

Chaos carolinense
(size 1–2 mm)

Amebas. Unicellular. No microtubules to define shape. Life cycle is primarily asexual. Predators in soil or water.

> **Amoeba proteus, Pelomyxa, Chaos chaos.** Giant free-living amebas in soil and water; they consume small invertebrates.
>
> **Entamoeba histolytica.** Intestinal parasite.
>
> **Acanthamoeba.** Soil predator, an opportunistic parasite causing keratitis and meningitis.

Mycetozoa. Slime molds. Upon starvation, amebas aggregate to form a fruiting body, which undergoes meiosis and produces spores.
- **Cellular slime molds.** Multicellular fruiting body alternates with unicellular amebas.
  > **Dictyostelium discoideum.** Provides an important model system for multicellular development.
- **Plasmodial slime molds.** Multinucleate plasmodium alternates with unicellular amebas.
  > **Physarum polycephalum.** In aqueous environment, amebas generate flagella.

Rhizaria (amebas with filament-shaped pseudopods)

Filament-shaped (filose) pseudopods. Some species have a test (shell) of silica or other inorganic materials.

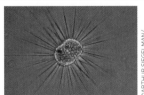

Actinophrys sol (cell size 30–90 μm)

Cercozoa. Amebas with flagella and/or filamentous pseudopods.
- **Euglyphida.** Amebas with a test.
- **Chlorarachniophyta.** Have algal endosymbionts.

Foraminifera. Form spiral tests. Fossil "forams" are an indicator of petroleum deposits.

Heliozoa. Form thin pseudopods. *Actinophrys sol*. Note: Phylogeny is disputed.

Radiolaria. Form thin pseudopods (filopodia) reinforced by microtubules, radiating starlike from the center. Some possess inorganic skeletons or spicules.

(continued)

Alveolata (having cortical alveoli)

Cortex contains flattened vesicles called alveoli, reinforced below by lateral microtubules.

Vorticella (length 1 mm)

Ciliophora. Common aquatic predators. Reproduce by conjugation, in which micronuclei are exchanged; then regenerate macronucleus for gene expression. Macronuclear DNA may be cut into thousands of segments.

> ***Didinium*.** Has two equatorial rings of flagella.
>
> ***Paramecium*, *Spirostomum*, *Oxytricha*.** Covered with cilia.
>
> ***Suctorians*.** Have knobbed tentacles. *Acineta*.
>
> ***Vorticella*, *Stentor*.** Stalked, with a mouth ringed by cilia.

Ceratium tripos and *Ceratium furca*

Dinoflagellata. Secondary or tertiary endosymbiont algae, from engulfment of red algae or diatoms. Cortical alveoli contain stiff plates. Pair of flagella, one wrapped around the cell.

- **Free-living marine and freshwater dinoflagellates** such as *Peridinium* and *Ceratium* supplement their photosynthesis with predation. Some produce bioluminescence. Blooms of marine dinoflagellates generate the "red tide." Cause paralytic shellfish poisoning.
- **Zooxanthellae** such as *Symbiodinium* are endosymbionts of corals, providing essential nutrition through photosynthesis. "Coral bleaching" is a serious condition in which the zooxanthellae are expelled under environmental stress. Zooxanthellae also inhabit clams, flatworms, mollusks, and jellyfish.

Apicomplexa (formerly Sporozoa). Parasites with complex life cycles. Lack flagella or cilia; possess apical complex for invasion of host cells. Vestigial chloroplasts.

> ***Cryptosporidium parvum*.** Waterborne opportunistic parasite.
>
> ***Plasmodium falciparum*.** Causes malaria.
>
> ***Toxoplasma gondii*.** Causes feline-transmitted toxoplasmosis.
>
> ***Babesia*.** Causes tick-transmitted babesiosis.

Heterokonta (having pair of dissimilar flagella)

Paired flagella of dissimilar form, one much shorter than the other.

Diatom

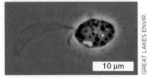

Paraphysomonas sp.

Secondary endosymbiotic algae. Free-living in marine and freshwater systems. Mixotrophic; combine phototrophy with heterotrophy.

- **Bacillariophyceae.** Diatoms. Possess intricate silicate shells with radial symmetry (centric) or bilateral symmetry (pennate).
- **Phaeophyceae.** "Brown algae" such as kelps and sargassum weed. Extend for many meters.
- **Chrysophyceae.** "Golden algae," pale-colored flagellates such as *Ochromonas* and *Paraphysomonas*.
- **Prymnesiophyceae.** Coccolithophores. Covered with intricate plates of $CaCO_3$.
- **Xanthophyceae.** "Yellow-green algae" and other less-studied forms of phytoplankton.

Oomycetes. **Water molds.** Superficially resemble fungi; often infect fish or plants. *Phytophthora infestans* destroyed potato crops and caused the Great Irish Famine.

Discoba (having disk-shaped cristae)

Usually possess a deep feeding groove. Disk-shaped cristae of mitochondria.

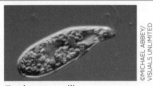

Euglena gracilis

Euglenida. Free-living flagellates. Some contain a secondary endosymbiotic chloroplast. *Euglena gracilis* is a common aquatic flagellate.

Jakobida. Free-living flagellates with lorica (stalk). *Reclinomonas americana*.

Trypanosomatidae. Parasites. *Trypanosoma brucei* causes sleeping sickness. *T. cruzi* causes Chagas' disease. *Leishmania* species cause leishmaniasis.

Metamonada (vestigial mitochondria)

Parasitic or symbiotic flagellates. Mitochondria and Golgi degenerated through evolution.

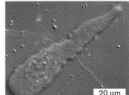

Pyrsonympha sp.

Diplomonadida and Retortamonadida. Human intestinal parasites. Highly degenerate cells, lacking mitochondria. *Giardia lamblia*. Cause of diarrhea from contaminated drinking water.

Oxymonadida and Parabasalia. Symbionts of termite gut. *Pyrsonympha*.

ANSWERS TO THOUGHT QUESTIONS

CHAPTER 1

1.1 The minimum size of known microbial cells is about 0.2 μm. Could even smaller cells be discovered? What factors may determine the minimum size of a cell?

ANSWER: The smallest cells known, about 0.2 μm in length, are cell wall–less bacteria called mycoplasmas—for example, *Mycoplasma pneumoniae*, a causative agent of pneumonia. Bacteria might be discovered that are smaller than 0.2 μm, but it is hard to see how their cell components, such as ribosomes (about a tenth this size), could fit inside such a small cell. The volume required for DNA and the apparatus of transcription and translation probably sets the lower limit on cell size.

1.2 If viruses are not functional cells, are they "alive"?

ANSWER: A traditional definition of a life-form includes the capability for metabolism and homeostasis (maintaining internal conditions of its cytoplasm), as well as reproduction and response to its environment. Viruses reproduce themselves and respond to the environment of the host cell, but they lack metabolism or homeostasis outside their host cell. Nevertheless, viruses such as herpes viruses contain numerous metabolic enzymes that participate in the metabolism of the host. Certain large viruses, such as the mimivirus, appear to have evolved from cells. Some microbiologists argue that viruses should be considered alive if reproduction is the main criterion and if the viral "environment" is considered the inside of the host cell.

1.3 Why do you think it took so long for humans to connect microbes with infectious disease?

ANSWER: For most of human history, we were unaware that microbes existed. Even after microscopy had revealed their existence, the incredible diversity of the microbial world and the difficulties in isolating and characterizing microbial organisms made it difficult to discern the specific effects of microbes. All healthy people contain microbes, and most disease-causing microbes are indistinguishable from normal microbiota by light microscopy. Not all microbial diseases can be transmitted directly from human to human; they may require complex cycles with intermediate hosts, such as the fleas and rats that carry bubonic plague.

1.4 How could you use Koch's postulates (**Fig. 1.17**) to demonstrate the causative agent of influenza? What problems not encountered with anthrax would you need to overcome?

ANSWER: Using Koch's postulates to demonstrate the causative agent of influenza would require an animal model host. Secretions from diseased patients could be applied to different animal species, such as monkeys and mice, in order to find an animal showing signs of the disease. To determine the causative agent of disease, the patient's secretions could be filtered in order to separate bacteria and viruses. Only the filtrate would cause disease, because it contains viruses (relevant to Koch's postulates 1 and 3). Viruses, however, are more difficult to isolate in pure culture than are bacteria (postulate 2)—a problem Koch did not address.

Furthermore, some viruses, such as HIV (human immunodeficiency virus) have no animal model; they grow only in human cells. Today, viruses are usually isolated in a tissue culture. Once isolated, the virus could be used to inoculate a new host animal (if an animal model exists) or a tissue culture and determine whether infection results (postulates 3 and 4). Another problem Koch did not address was the detection of infectious agents too small to be observed under a microscope. Today, antibody reactions are used to determine whether an individual has been exposed to a putative pathogen. An antibody test could be used to determine whether healthy and diseased individuals have been exposed to the isolated virus.

1.5 Why do you think some pathogens generate immunity readily, whereas others evade the immune system?

ANSWER: Some pathogens (microbes that cause disease) have external coat proteins that strongly stimulate the immune system and induce the production of antibodies. Other pathogens have evolved to avoid the immune system by changing the identity of their external proteins. Immunity also varies greatly with the host's status. The very young and very old generally have weaker immune systems than do people in the prime of life. Some pathogens, such as HIV, will directly attack the host's immune system, limiting the immune response to the pathogen.

1.6 How do you think microbes protect themselves from the antibiotics they produce?

ANSWER: Microbes protect themselves from their antibiotics by producing their own resistance factors. As discussed in later chapters, microbes may synthesize pumps to pump the antibiotics out; or they may make altered versions of the target macromolecule, such as the ribosome subunit; or they may produce enzymes to cleave the antimicrobial substance.

1.7 Why don't all living organisms fix their own nitrogen?

ANSWER: Nitrogen fixation requires a tremendous amount of energy, about 30 molecules of ATP per dinitrogen molecule converted to ammonia (discussed in Chapter 15). In a community containing adequate nitrogen sources, organisms that lose the nitrogen fixation pathway make more efficient use of their energy reserves than do those that spend energy to fix nitrogen from the atmosphere. Another consideration is that nitrogenase is an oxygen-sensitive enzyme, whereas plants, animals, and fungi are aerobes. In order to fix nitrogen, aerobic organisms need to develop complex mechanisms to exclude oxygen from nitrogenase.

1.8 What arguments support the classification of Archaea as a third domain of life? What arguments support the classification of archaea and bacteria together, as prokaryotes, distinct from eukaryotes?

ANSWER: The sequence of 16S rRNA (small-subunit rRNA) and other fundamental genes differs as much between archaea and bacteria as it does between archaea and eukaryotes. The composition of archaeal cell walls and phospholipids is completely distinct from that of bacteria and eukaryotes. Some aspects of gene expression, such as the RNA polymerase complex, are more similar between archaea and eukaryotes than between archaea and bacteria. On the other hand, archaeal and bacterial cells are prokaryotic; they both lack nuclei and complex membranous organelles. Archaeal metabolism and lifestyles are more similar to those of bacteria than to those of eukaryotes. Some archaea and bacteria sharing the same environment, such as high-temperature springs, have undergone horizontal transfer of genes that encode traits such as heat-stable membrane lipids.

1.9 Do you think engineered strains of bacteria should be patentable? What about sequenced genes or genomes?

ANSWER: Microbes—as well as multicellular organisms such as transgenic mice—have been patented, and the patents have held up in court. DNA sequences per se are not patentable, but specific plans for use of a DNA sequence can include the sequence as part of the patent. The reason for granting these patents is to encourage medical research by companies that need to earn a profit. The disadvantage of patents is that they restrict information flow and undercut competition. Furthermore, religious and philosophical arguments have been made that patenting live organisms cheapens life. Current laws aim for a balance among these concerns.

CHAPTER 2

2.1 As shown in Figure 2.2, the image passing through your cornea and lens is inverted on your retina. Why, therefore, does the world appear right side up?

ANSWER: The human brain interprets the image from the retina. Based on this interpretation, the brain knows that the image is upside down and inverts it to appear right side up. Researchers have tested what happens if an experimental subject wears special glasses that invert the image before the retina. After several days, the brain inverts the image perceived through the glasses, so it appears right side up. When the glasses are removed, the brain again takes time to restore its perception to right side up.

2.2 (refer to Fig. 2.7) You have discovered a new kind of microbe, never observed before. What kinds of questions about this microbe might be answered by light microscopy? What questions would be better addressed by electron microscopy?

ANSWER: Light microscopy could answer questions such as: What is the overall shape of this cell? Does it form individual cells or chains? Is the organism motile? Only light microscopy can visualize an organism alive. Electron microscopy can answer questions about internal and external subcellular structures. For example, does a bacterial cell possess external filamentous structures, such as flagella or pili? If the dimensions of the unknown microbe are smaller than the lower limits of a light microscope's resolution, EM may be the only way to observe the organism. Viruses are often characterized by shape, and this shape is observed by electron microscopy.

2.3 Explain what happens to the refracted light wave as it emerges from a piece of glass of even thickness. How do its new speed and direction compare with its original (incident) speed and direction?

ANSWER: The part of the wavefront that emerges first travels faster than the portion still in the glass, causing the wavefront to bend

toward the surface of the glass. Ultimately, the wave travels in the same direction and with the same speed as it did before entering the glass. The path of the emerging light ray is parallel to the path of the light ray entering the glass and is shifted over by an amount dependent on the thickness of the glass. This refraction will alter the path of the beam of light and decrease the amount of light reaching the lens of the microscope. Immersion oil has the same refractive index as glass and will limit the amount of light lost in this way.

2.4 (refer to Fig. 2.13) For a single lens, what angle θ might offer magnification even greater than 100X? What practical problem would you have in designing a lens to generate this light cone?

ANSWER: In theory, an angle theta (θ) of 90° would produce the highest resolution, even greater than 100°. However, a 90° angle of theta generates a cone of 180°, which would require the object to sit in the same position as the objective lens—in other words, to have a focal distance of zero. In practice, the cone of light needs to be somewhat less than 180°, to allow room for the object and to avoid substantial aberrations (light-distorting properties) in the lens material.

2.5 Under starvation, a bacterium such as *Bacillus thuringiensis*, the biological insecticide, packages its cytoplasm into a spore, leaving behind an empty cell wall. Suppose, under a microscope, you observe what appears to be a hollow cell. How can you tell whether the cell is indeed hollow or is simply out of focus?

ANSWER: You can tell whether the cell is out of focus or actually hollow by rotating the fine-focus knob to move the objective up and down while observing the specimen carefully. If the hollow shape appears to be the sharpest image possible, it is probably a hollow cell. If the hollow shape turns momentarily into a sharp, dark cell, it was probably out of focus before. Alternatively, you could use a confocal microscope to visualize the center of the hollow cell.

2.6 What experiment could you devise to determine the actual order of events in *Bacillus subtilis* DNA replication?

ANSWER: One way to track the movement of DNA during DNA replication would be to stain the DNA with a dye such as DAPI at various stages of cell division. Alternatively, green fluorescent protein (GFP) fused to a DNA-binding protein could be used to label a specific sequence of DNA and track its position. Another way to determine the order of events in sporulation could be to observe mutant strains of bacteria that contain defects in different proteins of the replication process. (DNA replication is discussed further in Chapter 7.)

2.7 Some early observers claimed that the rotary motions observed in bacterial flagella could not be distinguished from whiplike patterns, comparable to the motion of eukaryotic flagella. Can you imagine an experiment to distinguish the two and prove that the flagella rotate? *Hint:* Bacterial flagella can get "stuck" to the microscope slide or coverslip.

ANSWER: To prove that flagella rotate, you can "tether" a bacterium to the microscope slide by getting one of its flagella stuck to the slide. A simple way to tether bacteria is by using a slide coated with anti-flagellin antibody. When the flagellum is stuck to the slide, its motor continues to rotate; thus, the entire cell now rotates. The rotation of the cell body can easily be seen by video microscopy. If the flagella moved in a whiplike fashion, the tethered cell would move back and forth, not rotate.

2.8 Compare and contrast fluorescence microscopy with dark-field microscopy. What similar advantage do they provide, and how do they differ?

ANSWER: Both dark-field and fluorescence microscopy enable detection (but not resolution) of objects whose dimensions are smaller than the wavelength of light. Dark-field technique is based on light scattering, which detects all small objects without discrimination. Fluorescence, however, provides a means to label specific parts of cells, such as cell membrane or DNA, or particular species of microbes, using fluorescent antibody tags.

2.9 Like a light microscope, an electron microscope can be focused at successive powers of magnification. At each level, the image rotates at an angle of several degrees. Given the geometry of the electron beam, as shown in **Figure 2.37**, why do you think the image rotates?

ANSWER: The image rotates because the electron beam is not straight, as for photons, but travels in a spiral through the magnetic field lines. As magnification increases, the spiral expands, and it reaches the image plane at a slightly different angle than before.

2.10 What kinds of research questions could you investigate using SEM? What questions could you answer using TEM?

ANSWER: SEM could be used to examine the surface of cells: Do the cells possess a smooth surface, or does their surface contain protein complexes or bulges that serve special functions? How do pathogens attach to the surface of cells? TEM can be used to determine the intracellular structure of attachment sites, as well as of internal organelles. TEM can also visualize the shapes of macromolecular complexes such as flagellar motors or ribosomes.

CHAPTER 3

3.1 Which chemicals do we find in the greatest number in a bacterial cell? The smallest number? Why does a cell contain 100 times as many lipid molecules as strands of RNA?

ANSWER: The chemicals that occur in the greatest number in a prokaryotic cell are inorganic ions (250 million/cell). They are also the smallest in size. DNA molecules are found in the lowest number (one large molecule, branched during replication). A prokaryotic cell contains a hundred times as many lipid molecules as strands of RNA because lipids are small structural molecules, highly packed. RNA molecules are long macromolecules that either are packed into complexes (such as ribosomal RNA) or are temporary information carriers (messenger RNA), present only as needed to make proteins.

3.2 Suppose we wish to isolate flagellar motors, which are protein complexes that span the envelope from inner membrane to outer membrane (**Fig. 3.1**). How might we modify the cell fractionation procedure?

ANSWER: Several alternatives can be tried. Flagellar motors associate with both inner and outer membranes, complicating the cell fractionation. It is possible that the motors might associate more strongly with one membrane or the other, so the motor could be found primarily in one of the membrane fractions. A fluorescent antibody could identify fractions that contain the motors. Mild-detergent treatment of the fractions could strip away the membrane. Alternatively, the whole-membrane prep could be treated with mild detergent. The protein fraction could then be centrifuged through a sucrose gradient; because the motor complexes have a very specific size and density, they would be concentrated in one fraction. Electron microscopy could confirm the fraction containing the motors.

3.3 Amino acids have acidic and basic groups that can dissociate. Why are they <u>not</u> membrane-permeant weak acids or weak bases? Why do they fail to cross the phospholipid bilayer?

ANSWER: At neutral pH, amino acids each have both a positively charged amine and a negatively charged carboxylate; that is, they can act as either a weak acid or a weak base. Charged ions, no matter what their size, will not freely pass through a plasma membrane. In an amino acid, if either charged group becomes neutralized by acid or base, the other group remains charged, so the molecule as a whole will never cross the membrane.

3.4 What other ways can you imagine that bacteria might mutate to become resistant to vancomycin?

ANSWER: A common means of resistance to antibiotics is to pump them out of the cell. A protein pump that effluxes other molecules might mutate to capture vancomycin and export it from the cell. Another possibility is that an enzyme could modify the vancomycin by adding phosphoryl groups or acetyl groups, which would prevent the antibiotic from binding the alanine dipeptide. Still another possibility is that the bacteria might evolve a thicker cell wall that would exclude the vancomycin from the inner layers of peptidoglycan.

3.5 **Figure 3.16** highlights the similarities and differences between the cell envelopes of Gram-negative and Gram-positive bacteria. What do you think are the advantages and limitations of a cell's having one layer of peptidoglycan (Gram-negative) versus several layers (Gram-positive)?

ANSWER: Having multiple layers of peptidoglycan increases the cell's resistance to osmotic shock, to desiccation stress, and to enzymes that cleave the cell wall. On the other hand, it requires more energy and biomass to build the layers of peptidoglycan. In addition, a thick cell wall could slow the uptake of nutrients. The mycobacteria, which have exceptionally thick cell walls, grow relatively slowly.

3.6 Why would laboratory culture conditions select for evolution of cells lacking an S-layer?

ANSWER: Protective traits are often lost during culture of microbes that can produce 30 generations overnight. Their rapid reproductive rate gives ample opportunity for spontaneous mutations to accumulate over an experimental timescale. In the case of the S-layer, in a laboratory test tube free of predators or viruses, mutant bacteria that fail to produce the thick protein layer would save energy compared to S-layer synthesizers, and would therefore grow faster. Such mutants would quickly take over a rapidly growing population.

3.7 Why would proteins be confined to specific cell locations? Why would a protein not be able to function everywhere in the cell?

ANSWER: Proteins have evolved one or more specific functions often optimized for a specific part of the cell. For example, water-conducting porins are found solely in the inner membrane (cell membrane), which is otherwise impermeable to water. The outer membrane, which is water permeable, is the sole location for specific porins that transport small peptides and sugars. The sugars then need to be taken across the inner membrane by transport proteins that have evolved to function best in this location. Similarly, different chaperones (proteins that aid peptide folding) have evolved to function best in the environment of the cytoplasm or periplasm, membrane-enclosed regions that differ substantially in pH and ion concentrations. In a different chemical environment of the cell, a protein may denature and lose its functional structure. A protein may be active only as part of a complex of proteins. If the protein is placed in a different location within the cell, its protein partners may be absent, rendering the protein nonfunctional.

3.8 **Figure 3.33** presents data from an experiment that allows the function of the TipN protein of *Caulobacter* to be visualized by microscopy. Can you propose an experiment with mutant strains of *Caulobacter* to test the hypothesis that one of the proteins shown in **Figure 3.34** is required for one of the cell changes shown?

ANSWER: The diagram of **Figure 3.34** proposes that PodJ protein is required for a pole to develop a flagellum. Suppose we construct a mutant strain with a deletion of the gene *podJ*. This *podJ* mutant fails to express PodJ protein. When the *podJ* mutant is supplied with nutrients, the stalked cells should grow and fission, but the progeny from the plain pole should fail to grow a flagellum. The stalked progeny will continue to divide, producing a stalked cell and a cell with plain poles, lacking flagellum or stalk. Other results are possible, but the result described would be consistent with a requirement of PodJ for flagellar development.

3.9 How would a magnetotactic species have to behave if it was in the Southern Hemisphere instead of the Northern Hemisphere?

ANSWER: In the Northern Hemisphere, the field lines for magnetic north point downward; in the Southern Hemisphere, the opposite is true. Thus, if downward direction is the aim of magnetotaxis, bacteria existing in the two hemispheres would have to respond oppositely to the magnetic field; in the Southern Hemisphere, anaerobic magnetobacteria swim toward magnetic south. Near the equator, the proportions of north-seeking and south-seeking bacteria are roughly equal.

3.10 Most laboratory strains of *E. coli* and *Salmonella* commonly used for genetic research lack flagella. Why do you think this is the case? How can a researcher maintain a motile strain?

ANSWER: The motility apparatus requires 50 different genes generating different protein components. Cells that acquire mutations eliminating expression of the motility apparatus gain an energy advantage over cells that continue to invest energy in motors. In a natural environment, the nonmotile cells lose out in competition for nutrients, despite their energetic advantage; but in the laboratory, cells are cultured in isotropic environments such as a shaking test tube, where motility confers no advantage. These culture conditions lead to evolutionary degeneration of motility, as they do for the S-layer (see Thought Question 3.6). In order to maintain a motile strain, bacteria are cultured on a soft agar medium containing an attractant nutrient. As cells consume the attractant, they generate a gradient, and chemotaxis leads them to swim outward. By subculturing only bacteria from the leading edge of swimming cells, one can maintain a motile strain.

CHAPTER 4

4.1 In a mixed ecosystem of autotrophs and heterotrophs, what happens if the autotroph begins to outgrow the heterotroph and makes excess organic carbon?

ANSWER: At first, the growth of the heterotroph might outpace the growth of the autotroph, using the carbon sources faster than the autotroph can make them. As the organic carbon sources diminish through consumption, growth of the heterotroph decreases, but the CO_2 formed by the heterotroph will allow the autotroph to grow and make more organic carbon. Ultimately, the ecosystem comes into balance.

4.2 In what situation would antiport and symport be passive rather than active transport?

ANSWER: Antiport and symport are passive when both molecules are moving down their concentration gradients. A symport or an antiport system can do work to move a molecule from a low concentration to a high concentration (against a concentration gradient) as long as the cotransported molecule is moving from high to low concentration. When this happens, it is a form of active transport. If symport or antiport moves both molecules with the concentration gradient (from high concentration to low), it is passive transport, assuming the molecules are moving no faster than the rate of diffusion.

4.3 What kind of transporter, other than an antiporter, could produce electroneutral coupled transport?

ANSWER: Electroneutral coupled transport can occur by symport if molecules of opposite charge are symported—for example, Na^+ flux together with Cl^-.

4.4 What would be the phenotype (growth characteristic) of a cell that lacks the *trp* genes (genes required for the synthesis of tryptophan)? What would be the phenotype of a cell missing the *lac* genes (genes whose products catabolize the carbohydrate lactose)?

ANSWER: The difference lies in the function of the two pathways. The *trp* operon is a biosynthetic operon. Errors in the biosynthetic pathway will lead to a failure to produce tryptophan. Therefore, a *trp* auxotrophic mutant will grow on defined medium only if tryptophan is added. The lactose operon involves the catabolism of a carbon source, lactose. If any of these genes are damaged, the cells are no longer able to use lactose as a carbon source. A *lac* mutant will not grow on defined medium with lactose as the sole carbon source.

4.5 If lactose was left out of MacConkey medium (**Fig. 4.16**), would lactose-fermenting *E. coli* bacteria grow, and if so, what color would their colonies be?

ANSWER: Even without lactose in the medium, *E. coli* would grow nonfermentatively on the peptides present. The colonies would appear white because the cells do not make acidic products needed to bring neutral red into the colony.

4.6 The addition of sheep's blood to agar produces a very rich medium called blood agar. Do you think blood agar is a selective medium? A differential medium? *Hint:* Some bacteria can lyse red blood cells.

ANSWER: Blood agar can be considered differential, because different species growing on blood have different abilities to lyse the red blood cells in the agar. Some do not lyse, others completely lyse red blood cells (secreted hemolysin produces complete clearing around a colony), while still others only partially lyse the blood (the secreted hemolysin produces a greening around the colony). It will therefore differentiate between hemolytic and nonhemolytic bacteria. The medium is very rich and supports the growth of many species, so it is not considered selective.

4.7 Use the information in **Figure 4.19** to determine the concentration (in cells per milliliter) of bacteria shown.

ANSWER: 1.25×10^6 bacteria per ml.

SOLUTION:

- Each small square is 0.0025 mm^2 in size, and the depth from coverslip to surface is 0.2 mm.
- $0.0025 \text{ mm}^2 \times 0.2 \text{ mm} = 0.0005 \text{ mm}^3$ ($1 \text{ mm}^3 = 1 \text{ μl}$). Each square defines a volume of 0.0005 μl.
- 10 cells observed over 16 squares averages to 0.625 bacteria $/0.0005$ μl or 0.625 bacteria/square.
- $1,000$ μl/ml ÷ 0.0005 μl/square = 2×10^6 squares/ml.
- 2×10^6 squares/ml × 0.625 bacteria/square = 1.25×10^6 bacteria/ml.

4.8 A virus such as influenza virus might produce 800 progeny virus particles from one host cell infected by one virus. How would you mathematically represent the exponential growth of the virus? What practical factors might limit such growth?

ANSWER: In theory, the growth rate of the virus would be proportional to 800^n. In practice, however, it is unlikely that the 800 virus particles released from one host cell will find 800 different host cells to infect. Furthermore, it turns out that only a small proportion of the influenza virus progeny are viable (see Chapter 11).

4.9 Suppose you ingest 20 cells of *Salmonella enterica* in a peanut butter cookie. They all survive the stomach and enter the intestine. Suppose further that the sum total of subsequent bacterial replication and death (caused by the host) produces an average generation time of 2 hours and you will feel sick when there are 1,000,000 bacteria. How much time will elapse before you feel sick?

ANSWER: $N = \log_{10} (N_t/N_0)/0.301$; $\log_{10} (1,000,000/20)/0.301$; 15.61×2 hours = 31 hours.

4.10 Suppose one cell of the nitrogen fixer *Sinorhizobium meliloti* colonizes a plant root. After 5 days (120 hours), there are 10,000 bacteria fixing N_2 within the plant cells. What is the bacterial doubling time?

ANSWER: 9 hours. Note that this generation time is much longer than it would be if these same organisms were grown in a test tube containing suitable medium. In a suitable laboratory medium, the generation time is about 1.5 hours.

4.11 It takes 40 minutes for a typical *E. coli* cell to completely replicate its chromosome and about 20 minutes to prepare for another round of replication. Yet the organism enjoys a 20-minute generation time growing at 37°C in complex medium. How is this possible? *Hint:* How might the cell overlap the two processes?

ANSWER: After the DNA is replicated about halfway around the chromosome, each daughter half-chromosome initiates a second round of replication, so the time needed to divide from one cell to two is effectively halved. Most cells in a log-phase culture in rich medium actually have four copies of the DNA origin of replication, each with a separate attachment site on the cell envelope, the future midpoint of a cell two generations ahead (see Chapter 3).

4.12 The bacterium *Acidithiobacillus thiooxidans* is an extremophile that grows using sulfur as an energy source. (a) Draw the approximate growth curves that you would expect to see, extending from log phase to stationary phase, if four cultures with different starting numbers of bacteria were grown in the same concentration of sulfur. Use **Figure 4.23B** as the model, and 4×10^5, 4×10^6, 4×10^7, and 4×10^8 as the starting cell densities. Maximum growth yield is 10^9 cells per milliliter. (b) Draw a second graph showing how the curves would change if the initial population density was constant but the concentration of sulfur varied.

ANSWER: (a)

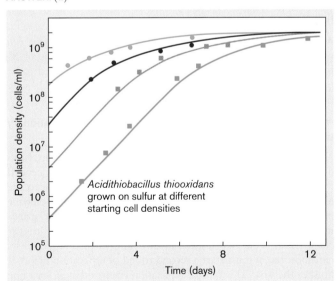

(b)

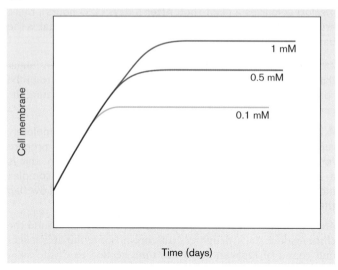

4.13 What can happen to the growth curve when a culture medium contains two carbon sources, if one is a preferred carbon source of growth-limiting concentration and the second is a nonpreferred source?

ANSWER: There are two possibilities. If the enzyme systems needed to utilize both carbon sources are always made, the growth curve will look normal because both will be used simultaneously. Usually the enzyme system for the nonpreferred carbon source is not produced until the preferred source is used up. In this case, a second lag phase will interrupt the exponential phase. This is called a diauxic growth curve and is commonly seen when cells are grown on both glucose and lactose. Lactose is the nonpreferred carbon source and is used second. The second lag phase marks the exhaustion of one nutrient and the gearing up by the cell to use the other (see Chapter 10).

4.14 How would you modify the equations describing microbial growth rate to describe the rate of death?

ANSWER: The death rate applies to a period of declining cell numbers. Therefore, the logarithm of the cell number ratio of N_1 to N_0 will be a negative number, and this factor will need to be preceded by a negative sign to convert it to a positive "halving time," or half-life of the culture.

4.15 Why are cells in log phase larger than cells in stationary phase?

ANSWER: Cells strive to maintain a certain DNA/mass ratio. In so doing, they balance the number of biochemical processes needed to sustain viability. If the cell mass becomes large relative to the number of copies of a given, critical gene, the amount of enzyme produced may not be sufficient to keep the cell alive and growing. In addition, the DNA/mass ratio serves as a signal to trigger cell division. Thus, when a cell divides faster than it replicates its chromosome, it must start a second round of replication before it finishes the first. This type of replication ensures that at least one chromosome duplication will be complete at the time of division. Because fast-growing cells contain more than one chromosome, they will increase in size to maintain the desired DNA/mass ratio. If the ratio were not maintained, cell division would not occur when needed.

4.16 How might *Streptomyces* and *Actinomyces* species avoid "committing suicide" when they make their antibiotics?

ANSWER: Bacteria that produce antibiotics need to make defenses against the antibiotic within their own cytoplasm. For example, their genes can express an altered form of the target molecule, such as a ribosomal subunit; or they can make pumps to pump the antibiotic out of the cell.

CHAPTER 5

5.1 Why haven't cells evolved so that all their enzymes have the same temperature optimum? If they did, wouldn't they grow even more rapidly?

ANSWER: An enzyme's function is not determined by temperature alone. There are other physicochemical constraints, based on the variety and complexity of functions that different enzymes must carry out. The thousands of different enzyme molecules must work in a coordinated fashion to support the basic functions of life. Having some enzymes work below or above their optimum temperatures will alter the rates of the reactions they catalyze. A population's evolution is based on an entire organism's ability to reproduce, not the speed at which each individual chemical reaction is carried out. The primary goal of a microbe is not just to grow fast but also to survive. Growing too fast could deplete food sources and produce toxic by-products too quickly.

5.2 If microbes lack a nervous system, how can they sense a temperature change?

ANSWER: Most bacteria respond to outside stimuli, such as heat, by altering their gene expression. They sense heat by monitoring the concentration of misfolded proteins, a consequence of excessively high temperature. The mechanism does not perceive heat per se, but recognizes the deleterious effects of moving outside the optimum growth temperature range so that the cell can launch an emergency response. The same mechanisms can sense other environmental stresses that misfold proteins, such as acid stress. See Chapter 10.

5.3 What could be a relatively simple way to grow barophiles in the laboratory?

ANSWER: High pressures can be maintained using a syringe. Scientists would place medium in a stainless steel syringe and maintain pressure with a calibrated vise.

5.4 How might the concept of water availability be used by the food industry to control spoilage?

ANSWER: Food preservation traditionally includes water exclusion by salt, as seen in hams, back bacon, and salted fish; or by high concentration of sugar, as in canned fruit or jellies. The lower a_w prevents microbial growth. Dehydrating foods will also prevent microbial growth.

5.5 Recall from Section 4.2 that an antiporter couples movement of one ion down its concentration gradient with movement of another molecule uphill, against its gradient. If this is true, how could a Na^+/H^+ antiporter work to bring protons into a haloalkaliphile growing in high salt at pH 10? Since the Na^+ concentration is lower inside the cell than outside and H^+ concentration is higher inside than outside (see **Fig. 5.15**), both ions are moving against their gradients.

ANSWER: In this situation the cell has to expend some energy to make the antiporter work. The energy involved is rooted in the charge difference between the inside of the cell (negative charge) and the outside of the cell (positive charge), called delta psi, $\Delta\psi$ (see Chapter 14). The antiporter in this case must also exchange a different number of Na^+ and H^+ ions, maintaining an electrical charge difference across the membrane; for example, export $2Na^+$ and import $1H^+$.

5.6   If anaerobes cannot live in oxygen, how do they incorporate oxygen into their cellular components?

ANSWER: Obligate anaerobes incorporate oxygen from their carbon sources (for example, CO_2 and carbohydrates such as glucose), all of which contain oxygen. This form of oxygen will not damage the cells.

5.7   How can anaerobes grow in the human mouth, where there is so much oxygen?

ANSWER: A synergistic relationship exists between facultatives and anaerobes within a tooth biofilm. The facultative anaerobes consume oxygen within the biofilm microenvironment, which allows the anaerobes to grow underneath them.

5.8   What evidence led people to think about looking for anaerobes? *Hint:* Look up "Spallanzani," "Pasteur," and "spontaneous generation" on the Internet.

ANSWER: The Italian priest Lazzaro Spallanzani (1729–1799), during his quest to disprove spontaneous generation, said, "Every beast on Earth needs air to live, and I am going to show just how animal these little animals are by putting them in a vacuum and watching them die." He dipped a glass tube into a culture, sealed one end, and attached the other to a vacuum. He was astonished to find that the microbes lived for weeks. He then wrote, "How wonderful this is. For we have always believed there is no living being that can live without the advantages air offers it." Fifty years later, Louis Pasteur observed that air could kill some organisms. After looking at a drop of liquid from a fermentation culture, he wrote, "There is something new here—in the middle of the drop they are lively, going every which way, but on the edge they were stiff as pokers."

5.9   Given a mixture of two microbes—one better at utilizing phosphate (microbe A) and the other better at utilizing nitrogen (microbe B)—what would happen if excess nitrogen was added to the mixed culture? What about both excess phosphate and excess nitrogen?

ANSWER: Excess nitrogen would give microbe A a nutrient advantage, and it might outgrow microbe B. An excess of both P and N would favor neither strain. In that case, other factors, such as relative growth rates, relative abilities to use alternative carbon sources, and so on, would play more dominant roles instead.

5.10   Bacteriostatic antibiotics do not kill bacteria; they only inhibit their growth. Why are they nevertheless effective at treating bacterial infections? *Hint:* Is the human body a quiet bystander during an infection?

ANSWER: The bacteriostatic agent stops growth of the bacteria and allows the host immune response to kill them.

5.11   If a disinfectant is added to a culture containing 1×10^6 CFUs per milliliter and the D-value of the disinfectant is 2 minutes, how many viable cells are left after 4 minutes of exposure?

ANSWER: 1×10^4 CFUs/ml (90% after 2 min = 1×10^5; 90% after another 2 min = 1×10^4).

5.12   How would you test the killing efficacy of an autoclave?

ANSWER: Construct a death curve by measuring survival of a known quantity of spores (for example, *Bacillus stearothermophilus*) after autoclaving for various lengths of time. Spores should be used because they are more resistant to heat than is any vegetative cell. Typically, autoclaves are regularly checked with spore strips that change color once the endospores are no longer viable.

CHAPTER 6

6.1   Suppose a certain virus depletes the population of an algal bloom. If some of the algae are genetically resistant, will they grow up again and dominate the producer community?

ANSWER: If some algae are resistant to the virus, they will reproduce and avoid infection. But their population growth will still face competition from other algal species that were never hosts of the virus. Furthermore, if the resistant algae form another bloom, eventually some other viral species will infect them and cut their population again.

6.2   Search for some specific viruses on the Internet. Which viruses do you think have a narrow host range, and which have a broad host range?

ANSWER: Examples of viruses with a narrow host range include poliovirus (poliomyelitis), which infects only humans and chimpanzees; smallpox virus, which infects only humans; and feline leukemia virus, which infects only cats. Examples of viruses with a broad host range include rabies virus, which infects numerous species of mammals; and influenza strains, which show preference for particular species but can jump between various mammals and birds.

6.3   What will happen if a virus particle remains intact within a host cell and fails to release its genome?

ANSWER: In most cases, a virus particle that fails to release its genome will be unable to reproduce, because DNA polymerases cannot reach its genome for reproduction and RNA polymerase cannot transcribe its genes to make gene products. An exception is double-stranded RNA viruses, which keep their genome partly enclosed in order to protect it from recognition by the host cell immune system.

6.4   For a viral capsid, what is the advantage of an icosahedron, as shown in Figure 6.7, instead of some other polyhedron?

ANSWER: The icosahedron is a polyhedron of 20 triangular faces, the largest number possible. Thus, the icosahedron turns out to be the largest and most economical form to enclose space with a small repeating unit. Natural selection probably favored viruses that could build the largest capsid from the smallest amount of genetic information.

6.5   How could viruses with different kinds of genomes (RNA versus DNA) combine and share genetic content in their progeny?

ANSWER: DNA viruses require messenger RNA intermediates to express their proteins. In a rare event, a DNA virus could mutate and evolve the ability to package its RNA transcript in a capsid, rather than its DNA. Alternatively, RNA retroviruses form DNA intermediates within their host cell; these DNA intermediates might recombine with the DNA genome of another virus.

6.6 What are the relative advantages and disadvantages (to a virus) of the slow-release strategy, compared with the strategy of a temperate phage, which alternates between lysis and lysogeny?

ANSWER: A disadvantage of slow release is that the phages can never reproduce progeny phages as rapidly as in a lytic burst. The drain on resources of the host cell infected by a slow-release virus causes it to grow more slowly compared with uninfected cells; in contrast, a lysogenized cell suffers little or no reproductive deficit compared with uninfected cells. An advantage of reproduction by slow release is the continuous release of phage, while avoiding the possibility of releasing all particles into an environment where no other host cell exists.

6.7 How might a phage evolve resistance to the CRISPR host defense (outlined in Fig. 6.25)?

ANSWER: The phage could have a mutation in its DNA sequence that was cleaved to form the spacer. Now when the Cas-crRNA complex forms, the crRNA will no longer base-pair correctly with the phage DNA and will not cleave the DNA of the infecting phage.

6.8 How could humans undergo natural selection for resistance to rhinovirus infection? Is such evolution likely? Why or why not?

ANSWER: Resistance to all rhinovirus infections might evolve through a mutation in the host gene encoding ICAM-1. The mutation would have to prevent rhinovirus binding without impairing the protein's ability to bind integrin. Such evolution is unlikely because of the importance of integrin binding and because rhinovirus infection is rarely fatal; thus, there is little selection pressure to evolve inherited resistance. Note, however, that the immune system rapidly generates immunity to particular strains of rhinovirus. Over a lifetime, most individuals acquire immunity to many rhinovirus strains but remain susceptible to others.

6.9 From the standpoint of a virus, what are the advantages and disadvantages of replication by the host polymerase, compared with using a polymerase encoded by its own genome?

ANSWER: An advantage of using the host polymerase is energetic: The virus avoids the energetic cost of manufacturing a polymerase to package with each virion. This is an advantage to the virus because its reproductive potential is limited by the energy resources of its host cell. Furthermore, because DNA and RNA polymerases are so central to cell function, the host species is unlikely to evolve a mutant form of the polymerase that resists the virus. On the other hand, the advantage of the virus making its own polymerase is that the viral polymerase can evolve traits that better meet the needs of its own replication, such as high speed and low accuracy to generate frequent variants. One disadvantage of a DNA virus using the host cell DNA polymerase is that the virus must gain access to the host cell nucleus where the polymerase is. In addition, if the host cell is fully differentiated, it has exited the cell cycle and is not replicating. If it is not replicating, a DNA polymerase may not be available unless the virus can force the host cell to start going through the cell cycle. These are two problems that the virus will not have to overcome if it brings in its own DNA polymerase.

6.10 Why does bacteriophage reproduction give a step curve, whereas cellular reproduction generates an exponential growth curve? Could you design an experiment in which viruses generate an exponential curve? Under what conditions does the growth of cellular microbes give rise to a step curve?

ANSWER: Lytic viruses appear to make a step curve because the number of progeny per infected cell is 100 or more, released simultaneously. After two or three generations the cell cycles would fall out of synchrony, and the curve would smooth out, but the later cell cycles are rarely observed in practice because by then, the supply of host cells is exhausted. If, however, an extremely low ratio of viruses to host cells is provided, the growth of virus particles will eventually generate an exponential curve. By contrast, the growth of cellular microbes is rarely observed during the first few doublings. By the time we measure the population, the cells are all undergoing different stages of division, and the population growth overall generates a smooth exponential curve. But if we observe the growth of a synchronized population of cells, we see a step curve of cell division too.

6.11 What kinds of questions about viruses can be addressed in tissue culture, and what questions require infection of an animal model?

ANSWER: Questions that can be answered using tissue culture would be how a virus binds to cell-surface receptors, and how it replicates progeny virions within a cell. Questions of viral transmission, however, would require the organ system of an animal, where the virions must undergo transport within and escape the immune system. Often, a virus passaged in tissue culture, such as influenza virus, will accumulate mutations that allow faster growth in the laboratory but poor infection of an animal host.

CHAPTER 7

7.1 What do you think happens to two single-stranded DNA molecules isolated from <u>different</u> genes when they are mixed together at very high concentrations of salt? *Hint:* High salt concentrations favor bonding between hydrophobic groups.

ANSWER: In high salt conditions, the stacking of hydrophobic bases is so strongly favored that two single strands of DNA will form a duplex no matter what the sequence of base pairs is.

7.2 How do the kinetics of denaturation and renaturation depend on DNA concentration?

ANSWER: The speed of denaturation does not depend on DNA concentration, but the speed of renaturation does. The higher the concentration of ssDNA, the more likely it is that complementary sequences will find each other and the faster the duplex can re-form.

7.3 DNA gyrase is essential to cell viability. Why, then, are nalidixic acid–resistant cells that contain mutations in *gyrA* still viable?

ANSWER: The *gyrA* mutations alter only the nalidixic acid–binding site on GyrA, not its gyrase activity. In other words, active DNA gyrase is still made, but the drug cannot bind to it.

7.4 Bacterial cells contain many enzymes that can degrade linear DNA. How, then, do linear chromosomes in organisms like *Borrelia burgdorferi* (the causative agent in Lyme disease) avoid degradation?

ANSWER: DNA-digesting exonucleases act on free 5′ or 3′ ends. The *Borrelia* linear chromosomes possess covalently closed hairpin ends called telomeres and do not possess free 5′ or 3′ groups.

7.5 Suppose you can label DNA in a bacterium by growing cells in medium containing either ^{14}N nitrogen or the heavier isotope, ^{15}N; you can isolate pure DNA from the organism; and you can subject DNA to centrifugation in a cesium chloride solution, a solution that forms a density gradient when subjected to centrifugal force, thereby separating the light (^{14}N) and heavy (^{15}N) forms of DNA to different locations in the test tube. Given these capabilities, how might you prove that DNA replication is semiconservative?

ANSWER: Grow bacteria in medium containing ^{15}N, so that all the DNA made by the cells will be in the heavy form. Transfer the cells to medium containing only ^{14}N and allow the cells to divide for one generation. Extract the DNA and centrifuge the preparation in cesium chloride. If replication is semiconservative, a hybrid DNA band will be seen at a position located between and equidistant from the point where heavy DNA and light DNA would be located. The hybrid band will be composed of one heavy strand and one light strand, making it of intermediate weight (density). This, in fact, was how Meselson and Stahl proved semiconservative replication in 1958, 5 years after the discovery that DNA is double-stranded (**eTopic 7.4**).

7.6 How fast does *E. coli* DNA polymerase synthesize DNA (in nucleotides per second) if the genome is 4,639,221 base pairs, replication is bidirectional, and the chromosome completes a round of replication in 40 minutes?

ANSWER: It takes *E. coli* 40 minutes (2,400 seconds) to complete the replication of its 4,639,221-bp chromosome. Since Pol III works as a dimer to synthesize the two strands simultaneously, we will consider only one strand in the calculation. The rate is 4,639,221 nucleotides (nt) per 2,400 seconds = 1,933 nt per second. But there are two replication forks, so each polymerase dimer synthesizes only one-half of the chromosome, which means that each polymerase still synthesizes a remarkable 800–1,000 nt per second.

7.7 Reexamine **Figure 7.15**. If you GFP-tagged a protein bound to the *ter* region, where would you expect fluorescence to appear in the cell with two origins?

ANSWER: The *ter* region is the last region of the chromosome to be replicated. It will appear at midcell, between the two origins.

7.8 Individual cells in a population of *E. coli* typically initiate replication at different times (asynchronous replication). However, depriving the population of a required amino acid can synchronize reproduction of the population. Ongoing rounds of DNA synthesis finish, but new rounds do not begin. Replication stops until the amino acid is once again added to the medium—an action that triggers simultaneous initiation in all cells. Why does this synchronization of reproduction happen?

ANSWER: Initiation requires synthesis of the initiator protein DnaA. Depriving the population of an amino acid prevents protein synthesis, which precludes synthesis of DnaA. Because DnaA is not required to complete already-initiated rounds of replication, all rounds already started are completed, but reinitiation cannot occur. Adding the amino acid once again will allow all cells to simultaneously make DnaA so that initiation is triggered in every cell at the same time.

7.9 The antibiotic rifampin inhibits transcription by RNA polymerase, but not by primase (DnaG). What happens to DNA synthesis if rifampin is added to a synchronous culture?

ANSWER: Initiation of DNA synthesis requires primer transcription at the origin by RNA polymerase, an enzyme sensitive to rifampin. Primase (DnaG), which synthesizes RNA primers in the lagging strand throughout DNA synthesis, is resistant to rifampin. So adding rifampin to a synchronized culture will prevent new rounds of DNA replication but will not affect already-initiated rounds.

7.10 Figure 7.25B describes how agarose gels separate DNA molecules. If you ran a highly supercoiled plasmid in one lane of the gel and the same plasmid with one phosphodiester bond cut in another lane, where would the two molecules end up relative to each other in this type of gel?

ANSWER: Removing one phosphodiester bond from a supercoiled plasmid will unravel (relax) all supercoils. Even though both plasmids will have the same DNA sequence, the highly supercoiled plasmid will be more compact than the relaxed plasmid and will move more quickly through the pores of the agarose gel. As a result, the supercoiled plasmid will be closer to the bottom of the gel than the relaxed form.

7.11 Given that *E. coli* possesses restriction endonucleases that cleave DNA from other species, how is it possible to clone a gene from one organism to another without the cloned gene sequence being degraded?

ANSWER: The best way to overcome the restriction barrier is to use a mutant strain of *E. coli* in which the restriction system has been inactivated. DNA entering the cell will not be degraded and, if the modification system is still in place, will be modified. Once modified, that DNA can be safely transferred to other *E. coli* strains that still have their restriction endonucleases.

7.12 PCR is a powerful technique, but a sample can be easily contaminated and the wrong DNA amplified—perhaps sending an innocent person to jail. What might you do to minimize this possibility?

ANSWER: Use sterile, filtered micropipette tips. The filters prevent contamination of the micropipette and subsequent reactions. In addition, perform a control PCR reaction without adding a DNA sample. If the water, buffers, or tubes are contaminated with a given DNA, a PCR product will appear even when DNA template has not been added deliberately to the reaction.

CHAPTER 8

8.1 If each sigma factor recognizes a different promoter, how does the cell manage to transcribe genes that respond to multiple stresses, each involving a different sigma factor?

ANSWER: In these situations, a given gene will have multiple promoters. Each promoter will be recognized by a different sigma factor and will begin transcription at different distances from the start codon of the gene.

8.2 Imagine two different sigma factors with different promoter recognition sequences. What would happen to the overall gene expression profile in the cell if one sigma factor were artificially overexpressed? Could there be a detrimental effect on growth?

ANSWER: Since sigma factors compete for the same site on core polymerase, overexpressing one sigma factor could displace the other sigma factors from the RNA polymerase population and compromise expression of those target genes. If those genes were important to survival, the cell could die.

8.3 **Why might some genes contain multiple promoters, each one specific for a different sigma factor?**

ANSWER: The gene might need to be expressed under multiple conditions at different levels. If a given condition increases expression of an alternate sigma factor, the target gene will need a promoter that the new sigma factor can recognize. As the need disappears and the sigma factor diminishes in concentration, a promoter that uses the housekeeping sigma factor will be needed. For example, the gene for DnaK heat-shock protein has promoters for RpoH (sigma-32) and RpoD (sigma-70), the housekeeping sigma factor. The level of protein needed during normal growth is supplied by sigma-70. Upon encountering heat stress, the RpoH sigma factor level increases and mediates an increase in DnaK production.

8.4 **How might the redundancy of the genetic code be used to establish evolutionary relationships between different species?** *Hints:* 1. Genomes of different species have different overall GC content. 2. Within a given genome one can find segments of DNA sequence with a GC content distinctly different from that found in the rest of the genome.

ANSWER: The codon preferences of different microorganisms are based in part on their GC content. Thus, an organism with an AT-rich genome will preferentially use codons for a given amino acid that have A's and T's over those with G's and C's. Evolutionarily, finding a long AT-rich region that encodes mRNA with AT codon bias within a chromosome that is otherwise GC-rich suggests that the AT-rich region was inherited by horizontal DNA transfer from another species. (See Chapter 9.)

8.5 **The synthesis of ribosomal RNAs and ribosomal proteins represents a major energy drain on the cell. How might the cell regulate the synthesis of these molecules when the growth rate slows because amino acids are limiting?** *Hints:* Predict what happens to any translating ribosome when an amino acid is limiting. Look up RelA protein on the Internet.

ANSWER: As energy levels in a cell fall, fewer amino acids are made. The decrease in amino acid production sets off a chain of events starting with fewer charged tRNA molecules being assembled because of a lack of amino acids. Reduced availability of charged tRNA molecules means ribosomes have to pause during translation until they find the right charged tRNA. This pause causes the synthesis of a signaling molecule called guanosine tetraphosphate (ppGpp). This signaling molecule is made from GTP and ATP by a ribosome-associated protein called RelA. Among other effects, ppGpp interacts with RNA polymerase and selectively stops transcription of the rRNA genes. The subsequent decrease in rRNA also stops new ribosome synthesis. How? Fewer rRNAs made means fewer available rRNA binding sites for ribosomal proteins. As a result, ribosomal proteins do not assemble in ribosomes and end up accumulating in the cytoplasm. These free ribosomal proteins bind to sequences in their own mRNA molecules that are similar to the target sequences they would bind to on rRNA. By binding to their own mRNA, these ribosomal proteins can prevent their own translation, so now fewer rRNA molecules, fewer

ribosomal proteins, and fewer new ribosomes are made in the cell (see Chapter 10). Finally, as cells continue to divide (even though at a slower rate because of amino acid deprivation), the remaining ribosomes will dilute to a lower steady-state level—that is, to a level suitable for sustaining the new growth rate.

8.6 **Figure 8.20 illustrates an operon and its relationship to transcripts and protein products. Imagine that a mutation that stops translation (TAA, for example) was substituted for a normal amino acid about midway through the DNA sequence encoding gene A. What happens to the expression of the gene A and gene B proteins?**

ANSWER: The part of the gene *A* protein only up until the stop codon will be made. The complete mRNA transcript, however, will be made and will include gene *B* (there are some rare exceptions to this). Gene *B* has its own translation start codon, so the complete gene *B* protein can be produced.

8.7 **While working as a member of a pharmaceutical company's drug discovery team, you find that a soil microbe snatched from the jungles of South America produces an antibiotic that will kill even the most deadly, drug-resistant form of *Enterococcus faecalis*, which is an important cause of heart valve vegetations in bacterial endocarditis. Your experiments indicate that the compound stops protein synthesis. How could you more precisely determine the antibiotic's mode of action?** *Hint:* Can you use mutants resistant to the antibiotic?

ANSWER: One way is to take a culture of bacteria susceptible to the antibiotic and isolate resistant mutants (bacteria that are not killed by the antibiotic), purify their ribosomes, and separate the 30S and 50S ribosomal subunits. Cross-mix subunits from sensitive and resistant cells (for example, mix 30S subunits from sensitive cells with 50S subunits from resistant cells). Then measure the ability of the hybrid ribosome to carry out protein synthesis with and without the drug. If resistance is due to an altered ribosomal protein or RNA, the subunit mix containing the altered component will make protein regardless of whether the drug is present. Once identified, the responsible ribosomal subunits from resistant and sensitive cells can be broken down further into their component parts, reconstituted in hybrid form, and again tested for an ability to make protein in the presence of drug. This reductive approach will likely, but not always, uncover the target ribosomal protein or rRNA.

8.8 **How might one gene code for two proteins with different amino acid sequences?**

ANSWER: One gene can code for two proteins with different amino acid sequences by having two different translation start sites in different reading frames. While this is not a common occurrence, it happens. Hepatitis B virus is one example.

8.9 **Why involve RNA in protein synthesis? Why not translate directly from DNA?**

ANSWER: Because transcription enables the cell to amplify the gene sequence information into multiple copies of RNA. Amplification means that more ribosomes can be engaged in translating the same protein, causing the concentration of the protein to rise more quickly than if only a single gene were used. The transcriptional process also makes it possible to more easily regulate the production of a protein.

8.10 Codon 45 of a 90-codon gene was changed into a translation stop codon, producing a shortened (truncated) protein. What kind of mutant cell could produce a full-length protein from the gene <u>without</u> removing the stop codon? *Hint:* What molecule recognizes a codon?

ANSWER: If a tRNA gene sequence corresponding to an anticodon is altered by mutation so that the anticodon of the tRNA "sees" the stop codon as an amino acid codon, then the mutant cell can produce a full-length protein from the gene. The mutated tRNA molecule will transfer its amino acid to the peptide chain. The stop codon is still there, but now it can direct the addition of an amino acid. The attached amino acid can be used to bridge the gap caused by the stop codon, and a full-length protein is made. These modified tRNAs are called suppressor tRNAs because they suppress the mutant phenotype.

8.11 What would happen if the tyrosine residue UAC in the mRNA-like domain of tmRNA was altered by mutation to UAA?

ANSWER: The peptide stuck on the ribosome would terminate early (UAA is a translation termination codon), but the shortened peptide tag would likely be insufficient to target the defective protein for degradation. The accumulation of this nonfunctional protein could have detrimental effects on cell growth.

8.12 An ORF 1,200 bp in length could encode a protein of what size and molecular weight? *Hint:* Find the molecular weight of an average amino acid.

ANSWER: There are three bases per codon, so the ORF can encode 400 amino acids. The average molecular weight of an amino acid is 110 Da. Therefore, the hypothetical protein will be approximately 44,000 Da (44 kDa).

CHAPTER 9

9.1 Figure 9.1 illustrates the process of transformation in *Streptococcus pneumoniae*. Will a mutant of *Streptococcus* lacking ComD be able to transform DNA?

ANSWER: No. The membrane sensor ComD detects the competence factor (CF) and initiates a cascade of events leading to transformasome construction. No ComD means no transformasome and no DNA transformation.

9.2 Transfer of an F factor from an F$^+$ cell to an F$^-$ cell converts the recipient to F$^+$. Why doesn't transfer of an Hfr do the same?

ANSWER: The last piece of an Hfr to transfer is the F factor and *oriT*. Only rarely will an entire chromosome transfer from one cell to another, so most Hfr transfers do not result in transfer of *oriT* and thus cannot initiate conjugation.

9.3 A prophage genome that is embedded in the chromosome of a lysogenic cell will prevent replication of any new phage DNA (from the same phage group) that has infected the cell. By preventing the replication of superinfecting phage DNA, the prophage genome protects itself from destruction by lysis. Is a lysogen also protected from generalized transduction carried out by a similar phage? For example, can a P22 lysogen of *Salmonella* be transduced by P22 grown on a <u>different</u> strain of *Salmonella*? Explain why or why not.

ANSWER: Yes, a P22 lysogen can be transduced by P22 phage grown on a different strain of *Salmonella*. Capsids of generalized transducing phages contain bacterial DNA instead of phage DNA, so the transducing particle has no phage DNA to replicate. Furthermore, transduction does not require replication of the transferred DNA before recombination.

9.4 How do you think phage DNA containing restriction sites evades the restriction-modification screening systems of its host?

ANSWER: Phage DNA will survive because sometimes the modification enzyme reaches the foreign DNA <u>before</u> the restriction endonuclease. Once methylated, the phage DNA will be shielded from restriction and the methylated molecule will replicate unchallenged. If, on the other hand, a foreign DNA fragment (not necessarily a phage) has been modified and the conditions are right, it might recombine into the host chromosome and convey a new character to the strain.

9.5 In a transductional cross between an $A^+B^+C^+$ genotype donor and an $A^-B^-C^-$ genotype recipient, 100 A^+ recombinants were selected. Of those 100, 15% were also B^+, while 75% were C^+. Is gene *B* or gene *C* closer to gene *A*?

ANSWER: Gene *C* is closer to gene *A*, because gene *C* was cotransduced with *A* at the higher frequency.

9.6 You have been asked to calculate the mutation rate for a gene in *Salmonella enterica*, so you dilute an 18-hour broth culture to 1×10^4 cells/ml and let it grow to 1×10^9 cells/ml. At that point you dilute and plate cells on the appropriate agar medium and estimate that there were 10,000,000 mutants in the culture. What you don't know is that the original dilution of 1×10^4 cells/ml already had 10 mutants defective in that gene. Does the presence of those 10 mutants affect your estimate of mutation rate? If so, then how?

ANSWER: Mutation rate is an estimate of how many <u>new</u> mutations arise per cell division as a culture grows. The division of preexisting mutants during growth could lead you to estimate a falsely high mutation rate. By knowing how many mutants existed in the beginning, you can compensate for how many mutants you would have as a result of mutant replication, and subtract that number from the total number of mutants you observe. This adjustment assumes that the growth rates of the mutant and wild-type bacteria are the same. If there were no mutations in the original dilute culture, then the mutation rate would be estimated as follows: (1×10^7 mutants) ÷ approximately 10^9 cell divisions = 10^{-2} mutation rate. This is a very high rate.

However, it takes about 17 generations (doublings) to go from 1×10^4 cells to 1×10^9 cells. If the diluted culture had 10 mutants to begin with, and if those mutants had the same generation time as the wild-type cells, then 10 cells after 17 generations would amount to about 10^6 cells. Subtracting this number from the actual numbers of mutants found gives the following: 1×10^7 (total mutants) $- 0.1 \times 10^7$ (sibling mutants) $= 0.9 \times 10^7$ newly derived mutants. So, (0.9×10^7 mutants) ÷ (10^9 cell divisions) $= 9 \times 10^{-3}$ mutation rate, a rate not much different from the original calculation. But if there were 100 mutants in the original 10,000-cell dilution, the situation would change. Now there would be 1×10^7 sibling mutants, which, subtracted from 1×10^7 total mutants, leaves no new mutants. The mutation rate would seem to be $<10^{-9}$, which is a huge difference from the previous calculation.

This exercise illustrates the importance of starting with as few cells as possible when calculating mutation rate, of doing controls to determine the number of preexisting mutants, and of performing replicates.

9.7 During transformation in *Streptococcus pneumoniae*, a single strand of donor DNA is transferred into the recipient. This single strand of DNA can be recombined into the recipient's genome at the homologous area. However, if the segment of recipient DNA had a mutation in it relative to the donor DNA strand, then the result would be a mismatch after recombination. Explain how, in the face of mismatch repair, the recipient cell ever retains the wild-type sequence from the donor.

ANSWER: Because the donor DNA was probably methylated when transferred, the methyl mismatch repair system should not distinguish between donor and recipient strands. Thus, there is equal chance that the system will repair the original recipient sequence (with a mutation) or the donor sequence. However, if the sequence was methylated differently (because there were different numbers of GATC sequences), then the mismatch repair system will preferentially remove the donor strand. The mismatch repair system in *S. pneumoniae* is called Hex and is orthologous to the *E. coli* Mut system.

9.8 It has been reported that hypermutable bacterial strains are overrepresented in clinical isolates. Out of 500 isolates of *Haemophilus influenzae*, for example, 2%–3% were mutator strains having mutation rates 100–1,000 times higher than lab reference strains. Why might mutator strains be beneficial to pathogens?

ANSWER: The mutator strains may speed microbial evolution, which could help the microbe outwit the immune system or escape the effects of administered antibiotics.

9.9 Diagram how a composite transposon like Tn*10* can generate inversions or deletions in target DNA during transposition. *Hint:* This happens when transposition within the chromosome occurs using the inverted repeats closest to the tetracycline resistance gene (see **Fig. 9.35**). If you draw it right, you will see, in the end, that the tetracycline resistance gene is lost.

ANSWER: The accompanying diagram shows that inversions or deletions occur when the inner rather than the outer ends of the inverted repeats are the targets of the transposase. The tetracycline resistance gene is lost, and the chromosome will contain a deletion or inversion, depending on the orientation of the chromosome (looped or unlooped) at the time of transposase action.

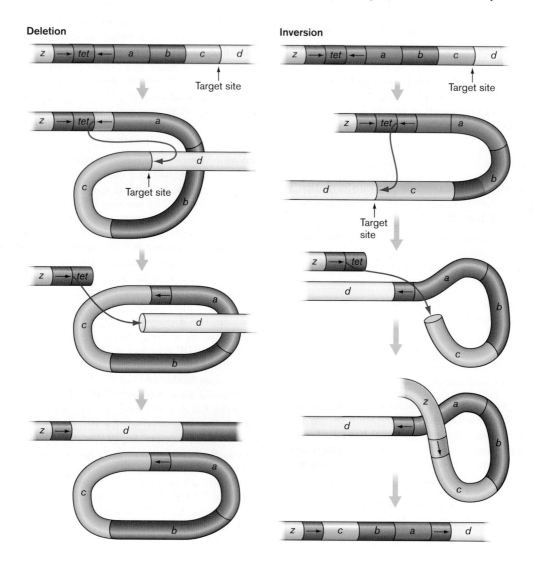

9.10 Gene homologs of *dnaK* encoding the heat-shock chaperone HSP70 exist in all three domains of life. All bacteria contain HSP70, but only some species of archaea encode a *dnaK* homolog. The archaeal homologs are closely related to those of bacteria. Knowing this information, how do you suppose *dnaK* genes arose in archaea?

ANSWER: The *dnaK* gene is thought to have been moved by some type of horizontal gene transfer mechanism from domain Bacteria to <u>some</u> members of domain Archaea.

CHAPTER 10

10.1 While viewing **Figure 10.1**, imagine the phenotype of a cell in which *fljB* has been deleted but *fljA* is still expressed. Would cells be motile? What type of flagella would be produced? Would the cells undergo phase variation? What would happen if *fliC* alone was deleted?

ANSWER: A *fljA* mutant lacks the repressor needed to turn off FliC. Thus, a cell will switch back and forth from making H2 flagellin to making both H1 and H2 flagellins. A *fliC* mutant, however, will switch from being motile to nonmotile. In one orientation, the invertible element will allow H2 flagellin to be made, but in the opposite orientation no flagellin will be made, at which point the cell will not be motile.

10.2 If the gene *lacZ* has a nonsense mutation in its open reading frame, will *lacY* still be translated?

ANSWER: *LacY* mRNA will still be translated, because each gene in this polycistronic mRNA (*lacZYA*) has its own ribosome-binding site and the *lacY* ribosome-binding site is functional. A translational stop codon in an upstream gene does not usually affect transcription of the downstream RNA.

10.3 Predict the effects of the following null mutations on the induction of beta-galactosidase by lactose, and predict whether the *lacZ* gene is expressed at high or low levels in each case. The inactivated, mutant genes to consider are *lacI*, *lacO*, *lacP*, *crp*, and *cya* (the gene encoding adenylyl cyclase). What effects will those mutations have on catabolite repression?

ANSWER: Loss of LacI repressor will lead to constitutive expression of *lacZ* and will partially affect catabolite repression. [Explanation: Because LacI is missing, allolactose inducer is not required, so the glucose effect on the LacY permease is irrelevant. What remains relevant is that the mechanism by which glucose transport reduces cAMP synthesis remains (see **eTopic 10.3**). Thus, decreased cAMP levels caused by growth on glucose will cause a decrease in cAMP-CRP-dependent activation of *lac* operon expression.] A *lacO* mutant will not bind LacI repressor, so the phenotype will mimic that of a *lacI* mutation. A *lacP* mutation will prevent expression of *lacZYA* because RNA polymerase will not bind. Mutations in *crp* or *cya* will partially prevent catabolite repression (see preceding explanation), but *lacZYA* induction by lactose will be normal. Without the cAMP-CRP complex, however, expression can never achieve maximal levels.

10.4 Predict what will happen to the expression of *lacZ* when a second copy of the *lac* operon region containing various mutations is present on a plasmid. The genotypes of these partial diploid strains are presented as chromosomal genes/ plasmid gene. (a) *lacI⁻ lacO⁺P⁺Z⁺Y⁺A⁺*/plasmid *lacI⁺*; (b) *lacO⁺ lacI⁺P⁺Z⁺Y⁺A⁺*/plasmid *lacO⁺*; (c) *crp⁻ lacI⁺O⁺P⁺Z⁺Y⁺A⁺*/ plasmid *crp⁺*.

ANSWER: (a) The *lacI⁺* gene on the plasmid will produce LacI repressor protein that can diffuse through the cytoplasm, bind chromosomal *lacO*, and repress the *lacZYA* operon. Because the complementing gene and the mutant gene are on different DNA molecules, the gene is said to work in *trans*. (b) Because the *lacO* gene does not produce a diffusible product (for example, protein or RNA), the plasmid *lacO⁺* cannot complement a *lacO* mutation in *trans*, and the strain will not make beta-galactosidase. Thus, the *lacO* gene functions only in *cis*—that is, when it resides next to the gene it regulates. (c) The *crp* gene produces a diffusible protein product, so it can function in *trans* and complement a *crp* mutation. The strain will make beta-galactosidase to the highest level, in the presence of inducer lactose.

10.5 Researchers often use isopropyl-β-D-thiogalactopyranoside (IPTG) rather than lactose to induce the *lacZYA* operon. IPTG resembles lactose, which is why it can interact with the LacI repressor, but it is not degraded by beta-galactosidase. Why do you think the use of IPTG is preferred in these studies?

ANSWER: There are at least two reasons IPTG is used. First, the level of IPTG inducer will not change, but the level of lactose inducer will continually decrease as it is consumed, affecting the kinetics of induction. Second, the act of degrading lactose produces glucose and galactose. Glucose, as the preferred carbon source, will catabolite-repress the *lacZYA* operon, once again affecting the kinetics of induction.

10.6 When the *lacI* gene of *E. coli* is missing because of mutation, the *lacZYA* operon is highly expressed regardless of whether lactose is present in the medium. Based on Figure 10.14 and its illustration of arabinose operon expression, what would happen to *araBAD* expression if AraC was missing? Why?

ANSWER: The operon would be poorly expressed because contact between AraC and RNA polymerase is needed to activate transcription.

10.7 Predict the phenotype of a *spoIIAA* mutant that completely lacks SpoIIAA anti-anti-sigma factor (see Fig. 10.18).

ANSWER: The *spoIIAA* mutant will have no way to efficiently remove anti-sigma F factor from sigma F. As a result, sigma F will almost never be active, even in the forespore. The cell can get as far as asymmetrical cell division to make what would be the forespore, but sigma F and anti-sigma F equally distribute in both compartments, so sigma F will not be activated and spores will not form.

10.8 The relationship between the small RNA RyhB, the iron regulatory protein Fur, and succinate dehydrogenase is shown in **Figure 10.22**. Based on this regulatory circuit, will a *fur* mutant grow on succinate?

ANSWER: No. Without Fur, RyhB is made whether or not iron is present, and RyhB sRNA causes the continuous degradation of the *sucCDAB* mRNA. Succinate dehydrogenase cannot be made, so the cell cannot grow on succinate.

10.9 Using antibody to flagella, we can tether a cell of *E. coli* to a glass slide via a single flagellum. Looking through a microscope, you will then see the <u>bacterium</u> rotate in opposite directions as the flagellar rotor switches from clockwise to counterclockwise rotation and back again (see Fig. 10.27). Which way will the bacillus rotate when an attractant is added? What would the interval between clockwise and counterclockwise switching be if you tethered the mutants *cheY*, *cheA*, *cheZ*, and *cheR* to the slide and then added attractant?

ANSWER: When an attractant is added to the slide, the rotor will turn counterclockwise for smooth swimming (CheA becomes less active, so there is less CheY-P and there are less frequent tumbles; thus, motor rotation is biased to counterclockwise). But because the flagellum is fixed to the slide, the bacillus will rotate in the opposite direction (clockwise). The following rotating phenotypes are expected from mutants fixed to slides: For mutant *cheY*, the rotor will turn mostly counterclockwise because there is no CheY-P, so there will be longer runs, but fixed cells will turn mostly clockwise. Mutant *cheA* will have the same phenotype as *cheY* (no CheY-P, longer runs). For mutant *cheZ*, the rotor will turn mostly clockwise, because there is more CheY-P (so there will be frequent tumbles and shorter runs), but the fixed cells will turn counterclockwise. For mutant *cheR*, no methylation of MCPs will reactivate CheA kinase after attractant is added, so there will be less CheY-P; therefore, the rotor will more frequently switch to clockwise, but fixed cells will turn counterclockwise.

10.10 Genes encoding luciferase can be used as "reporters." What would happen if the promoter for an SOS response gene was fused to the luciferase open reading frame?

ANSWER: It could serve as a real-time biosensor for an environmental stress response. You could fuse the luciferase gene to the *recA* promoter and examine a real-time increase in fluorescence after ultraviolet irradiation by inserting the whole culture into a spectrofluorometer.

10.11 What would happen if a culture was coinoculated with an *Aliivibrio fischeri luxI* mutant and a *luxA* mutant, neither of which produces light?

ANSWER: The *luxI* mutant would glow, and the *luxA* mutant would still make autoinducer as it grew. This autoinducer would accumulate in the culture medium and then diffuse and enter the *luxI* mutant cells, where it would trigger induction of the *lux* operon and production of luciferase.

10.12 Can tandem mass spectrometry (MS-MS) identify where modifications such as phosphorylation or acetylation happen on proteins?

ANSWER: Yes. Recall that the fragmentation of peptides by ionization will release single amino acids. Unambiguous identification of the amino acid lost is based on the mass difference between, for example, a five-amino-acid peptide and the four-amino-acid derivative peptide (the mass of alanine, for instance, is different from the mass of glycine). The presence of a side group like phosphate will change the weight of that amino acid and signal the presence of phosphate.

CHAPTER 11

11.1 Plaques from phage lambda quickly fill with resistant lysogens. Could there be a different way for the host cells to become resistant, without forming lysogens?

ANSWER: A gene encoding a host product essential for phage infection could mutate within the host bacteria. A common source of resistance is loss of the phage receptor, a host cell-surface protein. In the case of phage lambda, the receptor protein would be maltose porin, which could mutate to a form that no longer fits the phage protein. Other ways to become resistant could include specific cleavage of the phage DNA as it enters the cell, failure to interact with phage replication components, and the CRISPR-Cas memory defense system.

11.2 What advantages does rolling-circle replication offer a phage?

ANSWER: An advantage of rolling-circle replication is that many genome copies are made quickly from a single template. No proofreading occurs, and no methylation step distinguishes "old" from "new" DNA. Because viral genomes are relatively small, viruses tolerate a higher error rate per base pair than for cellular genomes.

11.3 A researcher adds phage lambda to an *E. coli* population whose cells fail to express maltose porin. After several days, the *E. coli* are now lysed by phage. What could be the explanation?

ANSWER: Several answers are possible. In practice, the most common explanation is that phage populations contain genetic variants, some of which can occasionally infect *E. coli* by binding a different porin for a different sugar. Such rare events lead to evolution of lambda mutants now adapted to bind a different receptor.

11.4 Suppose a mutant lambda phage lacks holin and endolysin. What will happen when this mutant infects *E. coli*?

ANSWER: The phage can undergo a lytic cycle, filling the *E. coli* cell with progeny phage particles. But the particles are trapped inside the cell wall. They cannot emerge to infect new host cells.

11.5 Could the influenza genome change by recombination of segments, rather than reassortment? What about the lambda phage genome?

ANSWER: The lambda phage genome consists of double-stranded DNA; thus it can recombine with the genome of a coinfecting phage by using the host cell's protein complex for homologous recombination (discussed in Chapter 9). The RNA segments of the influenza genome are less likely to recombine with segments of another genome, because the single-stranded RNA has no mechanism for maintaining homology with another strand. However, RNA viruses can undergo a different form of recombination, in which the RNA-dependent RNA polymerase starts transcribing one genome, then "hops off" onto the genome of a coinfecting virus. The polymerase thus completes a strand containing material from two different virions.

11.6 How could a swine "mixing bowl" enable avian influenza strain H7N9 to become a deadly pandemic strain?

ANSWER: Strain H7N9 is highly virulent in humans. Once the virions from an infected bird establish infection in a human, the virus rapidly spreads within the human body, causing disease with high mortality. However, the transmission rate between humans is low. Suppose H7N9 were to infect a swine that was simultaneously infected with strain H1N1. The H1N1 strain is transmitted between humans with high efficiency. During coinfection of swine, the H7N9 strain might acquire segments from

H1N1 that allow efficient transmission to humans. The reassortant strain might then escape into humans and spread, causing deadly disease.

11.7   How do attachment and entry of HIV resemble attachment and entry of influenza virus? How do attachment and entry differ between these two viruses?

ANSWER: Attachment of HIV requires the envelope spike proteins to bind receptors in the host plasma membrane, just as the influenza envelope protein hemagglutinin binds the sialic acid protein in the host membrane. However, the entry processes differ between the two viruses. Influenza virus induces formation of an endocytic vesicle, whose acidification triggers membrane fusion and release of the core contents into the host cytoplasm. HIV virions, however, do not induce endocytosis and do not require acidification to induce membrane fusion and release of the core into the cytoplasm. In both cases, receptor binding signals the major rearrangement of a viral envelope protein so that a fusion peptide is inserted into the host membrane: for HIV, it is the plasma membrane; for influenza virus, it is the endocytic membrane.

11.8   What do you think are the arguments for or against lentivectors editing the germ line?

ANSWER: Parents who carry two copies of a deleterious allele such as that for sickle-cell disease might wish to have their sperm or egg cells edited to replace the bad allele, thus allowing production of healthy children. It could be argued that the human genome is full of retroviral modifications that occurred during the history of our species, so why not intentionally add another one? As a practical matter, it is hard to know what would happen to the lentivector sequences in a germ line. Perhaps the process of sperm or egg maturation and embryogenesis would activate the viral sequences in unexpected ways. The future child would have no choice over this procedure. An alternative option that parents have is to produce children by selecting a sperm or egg donor.

11.9   Compare and contrast the fate of the HSV chromosome with that of the HIV chromosome.

ANSWER: HSV-1 contains a DNA chromosome, which is transported to the nuclear membrane within an intact capsid. HIV contains two RNA chromosomes, which are released in the cytoplasm upon dissolution of the capsid. The RNA chromosomes of HIV are copied to double-stranded DNA for transport into the nucleus. In HSV-1, the DNA chromosome circularizes and generates concatemeric duplicates by rolling-circle replication. In HIV, the replicated DNA circularizes but immediately integrates into the host chromosome. In both cases, the viral DNA can persist for decades as a latent infection.

CHAPTER 12

12.1   How would you modify the technique illustrated in Figure 12.2A to identify mutants defective in the synthesis of an amino acid, such as glutamic acid? *Hint:* You can use a minimal agar medium.

ANSWER: Cells subjected to transposon mutagenesis are plated onto rich agar medium. Colonies that grow are pressed onto sterile velveteen pads that cover a solid cylinder, as shown in the figure (a different pad must be used for each plate). The cylinder is used to transfer the colonies to minimal agar plates with and without glutamic acid. After growth, the plates are visually aligned, and any colonies able to grow on the glutamate plate but not on the unsupplemented plate are identified. These mutants require glutamate to grow and are called glutamate auxotrophs.

12.2   In this postgenomic era, how might you generally design a strategy to construct a strain of *E. coli* that cannot synthesize histidine?

ANSWER: The genome sequence of *E. coli* is known. We also know which genes encode the steps of histidine biosynthesis. The strategy today would be to engineer a plasmid containing DNA sequences (about 200 bp) that flank the histidine biosynthesis

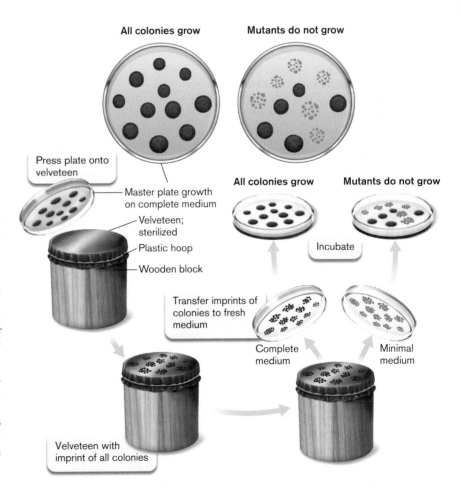

All colonies grow   Mutants do not grow

Press plate onto velveteen

Master plate growth on complete medium

Velveteen; sterilized

Plastic hoop

Wooden block

All colonies grow   Mutants do not grow

Incubate

Transfer imprints of colonies to fresh medium

Complete medium   Minimal medium

Velveteen with imprint of all colonies

gene you want to delete, but in place of the target gene you would instead insert an antibiotic resistance gene (for example, for kanamycin resistance). The plasmid used would be a "suicide" plasmid that cannot replicate in the strain you wish to mutate, but can replicate in other permissive strains (strains that produce a specific protein the plasmid needs to replicate). The permissive strain allows a lot of the plasmid to be made. The recombinant plasmid would then be moved into the nonpermissive target strain by transformation, and the transformation mixture would be plated on nutrient agar medium containing the antibiotic. Since the plasmid cannot replicate in this strain, the only way the target strain can become resistant to the antibiotic is if the chromosomal sequences flanking the drug marker in the plasmid undergo homologous recombination with the same DNA sequences on the actual chromosome. This double recombination event "surgically" removes the target gene from the chromosome and replaces it with the antibiotic resistance marker. In our example, the newly constructed mutant would be able to grow in the presence of antibiotic, but only if histidine was present.

12.3   You have monitored expression of a gene fusion in which *gfp* is fused to *gadA*, the glutamate decarboxylase gene. You find a 50-fold increase in expression of *gadA* (based on fluorescence level) when the cells carrying this fusion are grown in media at pH 5.5 as compared to pH 8. What additional experiments must you do to determine whether the control is transcriptional or translational?

ANSWER: To measure regulation of the *gadA* promoter only, you can fuse *gfp* containing its native ribosome-binding site (RBS) to the *gadA* promoter. This fusion will expose only transcriptional control of the *gadA* promoter. Translational control of the *gadA* message is absent because the *gadA* RBS (and all other *gadA* sequences) was replaced by the *gfp* RBS. To measure translational control, you must make a new fusion that starts with a constitutive promoter (a constantly expressed promoter), add a section of *gadA* containing the *gadA* ribosome-binding site, and follow that with *gfp* (minus its RBS) fused in-frame to the *gadA* sequence. Any condition that alters the expression of fusion 1 affects transcription of the *gadA* promoter. Any condition that alters the expression of fusion 2 must affect translation of the *gadA* message.

12.4   How could you quickly separate cells in the population that express sigma B–YFP from those that do not express this fusion? And what could you learn by separating these subpopulations? *Hint:* A technique presented in Chapter 4 will help.

ANSWER: The population of cells can be passed single file through a fluorescence-activated cell sorter (FACS; see Chapter 4). Cells that fluoresce can be collected in one tube, and nonfluorescent cells can be collected in another. The cells can then be analyzed using deep sequencing to determine the transcriptomes, and mass spectrometry can determine which proteins are produced in the different cell populations.

12.5   What would you conclude if the northern blot of the *gad* genes showed no difference in mRNA levels in cells grown at pH 5.5 and pH 7.7, but the western blot showed more protein at pH 5.5 than at pH 7.7?

ANSWER: Regulatory control would likely be at the posttranscriptional level, through increasing mRNA stability, translational efficiency, or protein stability.

12.6   Another regulator of glutamate decarboxylase (Gad) production does not affect the production of *gad* mRNA, but is required to accumulate Gad protein. What two regulatory mechanisms might account for this phenotype?

ANSWER: The two regulatory mechanisms are control of translation and control of protein degradation.

12.7   How would you have to modify standard quantitative PCR to quantify the level of a specific mRNA?

ANSWER: You would have to first convert the mRNA into cDNA by using the enzyme reverse transcriptase. (An oligonucleotide primer hybridizing to the 3′ end of the mRNA is acted on by reverse transcriptase to synthesize complementary DNA, or cDNA.) Normal qPCR techniques can then quantify the cDNA.

12.8   You suspect that proteins A and C can simultaneously bind to protein B, and that this interaction then allows A to interact with C in the complex. How could you use two-hybrid analysis to determine whether these three proteins can interact?

ANSWER: Place three plasmids in the cell. The first plasmid makes protein A fused to the GAL4 DNA-binding domain, the second plasmid makes protein C fused to the GAL4 activator domain, and the third plasmid makes simply protein B. If the yeast cell contains only the first two plasmids, then no interaction will occur and *lacZ* will not be activated. If all three plasmids are in the same yeast cell, then protein B will be the center of a sandwich, linking the A and C fusion proteins. The fusion proteins can then interact and activate *lacZ*.

12.9   Would insect resistance to an insecticidal protein be a concern when developing a transgenic plant? Why or why not? How would you design a transgenic plant to limit the possibility of insects developing resistance?

ANSWER: Insects have, in fact, developed resistance to single insecticidal proteins. The most common resistance mechanism involves a change in the membrane receptors in the midgut to which activated *Bt* toxins bind. Resistance can be due to a reduced number of *Bt* toxin receptors or to a reduced affinity of the receptor for the toxin. Some insects, such as the spruce budworm, can inactivate specific toxins by precipitating them with a protein complex present in the midgut.

While developing a transgenic plant, several steps could be taken to limit the development of resistance in an insect population. The insecticidal gene could be fused to a promoter that is expressed only when the plant is most susceptible to attack. Alternatively, the gene could be fused to a promoter that is expressed only in a tissue of the plant that is most vulnerable to attack. This would limit the time during which the insects can develop resistance. For instance, cotton plants attacked by bollworms could produce toxin only in young boll tissues, the most important part of the plant. In addition to specifically protecting the critical plant tissue, this strategy would affect only one generation of bollworms, avoiding the constant selection pressure that hastens evolution of resistance. Another technique would be to engineer two different insecticidal proteins into the plant genome that would not exhibit cross-resistance. In other words, even if an insect developed resistance to one toxin, it would still remain susceptible to the second.

12.10 You have just employed phage display technology to select for a protein that tightly binds and blocks the eukaryotic cell receptor targeted by anthrax toxin. When this receptor is blocked, the toxin cannot get into the target cell. You suspect that the protein may be a useful treatment for anthrax. How would you recover the gene following phage display and then express and purify the protein product?

ANSWER: The gene can be cut out of the phage genome using restriction endonucleases or amplified by PCR using known phage DNA sequences that flank the gene. The fragment can then be cloned into a His_6 tag expression vector. The sequence of the gene would have to be known so that the DNA encoding the His_6 tag could be placed in-frame with the open reading frame of the receptor-blocking protein. The plasmid containing the His_6-tagged protein gene is then induced to overexpress the protein in *E. coli*. The vector will contain an inducible promoter (for example, *lacP*) to drive expression of the gene. The His_6-tagged protein can then be purified by pouring cell extracts over a nickel column, as described in the text.

12.11 What would happen with a toggle switch if the genetic engineer could set repressor 1 and repressor 2 protein levels to be <u>exactly</u> equal? Imagine that this is done without either inducer present.

ANSWER: In a perfect system, both proteins would remain at the same level, but any small deviation in condition that tilts the balance between the two proteins by even a small amount will eventually lock the cell into either GFP on or GFP off. Because of intrinsic noise in the system, the population of cells could be half on and half off.

CHAPTER 13

13.1 Consider glucose catabolism in your blood, where the sugar is completely oxidized by O_2 and converted to CO_2:

$$C_6H_{12}O_6 + 6O_2 \rightarrow 6CO_2 + 6H_2O$$

Do you think this reaction releases greater energy as heat, or by change in entropy? Explain.

ANSWER: At first glance, the breakdown of glucose to six molecules of carbon dioxide seems to incur a large increase in entropy. But the reaction also consumes six molecules of oxygen, so the entropy gain is small. Furthermore, the oxidation reaction is associated with a large enthalpy change, approximately $\Delta H^{\circ\prime} = -2,540$ kJ/mol. (The degree symbol followed by prime connotes biochemical standard conditions.) At 37°C, the temperature-entropy term $-T\Delta S^{\circ\prime} = -(310 \text{ K})(0.973 \text{ kJ/K}) = -302$ kJ/mol. The overall value of free energy change $\Delta G^{\circ\prime} = -2,540$ kJ/mol $- 302$ kJ/mol $= -2,842$ kJ/mol. Thus, the entropy term yields some energy, but less than an eighth as much as the enthalpy ($\Delta H^{\circ\prime}$) term.

13.2 The bacterium *Lactococcus lactis* was voted the official state microbe of Wisconsin because of its importance for cheese. During cheese production, *L. lactis* ferments milk sugars to lactic acid:

$$C_6H_{12}O_6 \rightarrow 2C_3H_6O_3 \rightleftharpoons 2C_3H_5O_3^- + 2H^+$$

Large quantities of lactic acid are formed, with relatively small growth of bacterial biomass. Why do you think biomass is limited? Cheese making usually runs more efficiently at high temperature; why?

ANSWER: The lactic acid fermentation reaction does not involve a strong oxidant, but only breakdown of sugar to smaller molecules, so a larger proportion of the free energy yield is in the entropy term $-T\Delta S$. The overall energy yield is low, so a large amount of sugar must be cycled to lactic acid (lactate) for a relatively small amount of bacterial growth, as compared to bacterial growth with oxygen. Because fermentation depends on entropy change ($-T\Delta S$), the free energy yield can be increased by increasing the temperature.

13.3 The thermophilic bacteria *Thermus* species grow in deep-sea hydrothermal vents at 80°C. It was proposed that they metabolize formate to bicarbonate ion and hydrogen gas:

$$HCOO^- + H_2O \rightarrow HCO_3^- + H_2$$

But the standard $\Delta G^{\circ\prime}$ is near zero (−2.6 kJ/mol). Under actual conditions, do you think the reaction yields energy? Assume concentrations of 150 mM formate, 20 mM bicarbonate ion, and 10 mM hydrogen gas.

ANSWER: The ability to gain energy from formate will depend on the temperature and the concentrations of reactants and products. Remember that $[H_2O]$ equals 1. Consider the equation:

$$\Delta G = \Delta G^{\circ\prime} + 2.303RT \log[\text{products}]/[\text{reactants}]$$
$$= -2.6 \text{ kJ/mol} + 2.303\,[8.315 \times 10^{-3} \text{ kJ/(mol} \cdot \text{K)}]$$
$$(273 \text{ K} + 80 \text{ K}) \times \log[HCO_3^-][H_2]/[HCO_2^-][H_2O]$$
$$= -2.6 \text{ kJ/mol} + 6.760 \text{ kJ} \times \log[(0.02 \text{ M} \times 0.01 \text{ M})/0.15 \text{ M}]$$
$$= -22 \text{ kJ/mol}$$

The ΔG value is small, but it is enough to drive growth of some species of *Thermus*. Note that this simplified treatment omits the role of gas formation. (Data are based on Yun Jae Kim et al. 2010. *Nature* **467**:352.)

13.4 When ATP phosphorylates glucose to glucose 6-phosphate, what is the net value of $\Delta G^{\circ\prime}$? What if ATP phosphorylates pyruvate? Can this latter reaction go forward without additional input of energy? (See **Table 13.3**.)

ANSWER: Using **Table 13.3**, we see that the phosphorylation of glucose by ATP is composed of these two reactions:

Reaction	$\Delta G^{\circ\prime}$ (kJ/mol)
ATP + H_2O → ADP + P_i + H^+	−31
Glucose + P_i → glucose 6-P + H_2O	+14
ATP + glucose → ADP + glucose 6-P	−17

The net energy lost is −17 kJ/mol, so the phosphorylation can go forward. To phosphorylate pyruvate, however, the $\Delta G^{\circ\prime}$ of ATP hydrolysis (−31 kJ/mol) must be subtracted from the value of phosphoenolpyruvate formation (+62 kJ/mol), giving a net value of +31 kJ/mol. Since $\Delta G^{\circ\prime}$ is positive, this reaction cannot go forward without additional energy.

13.5 Linking an amino acid to its cognate tRNA is driven by ATP hydrolysis to AMP (adenosine monophosphate) plus pyrophosphate. Why release PP_i instead of P_i?

ANSWER: The formation of aminoacyl-tRNA must be irreversible until the ribosome is ready to release the tRNA. The pyrophosphate from ATP is immediately cleaved into 2 P_i, preventing the reversal of aminoacyl-tRNA formation.

13.6 In the microbial community of the bovine rumen, the actual ΔG value has been calculated for glucose fermentation to acetate:

$$C_6H_{12}O_6 + 2H_2O \rightarrow 2C_2H_3O_2^- + 2H^+ + 4H_2 + 2CO_2$$
$$\Delta G = -318 \text{ kJ/mol}$$

If the actual ΔG for ATP formation is +44 kJ/mol and each glucose fermentation yields four molecules of ATP, what is the thermodynamic efficiency of energy gain? Where does the lost energy go?

ANSWER: The energy efficiency is $(4 \times 44 \text{ kJ/mol})/(318 \text{ kJ/mol}) \times 100 = 55\%$. The remaining energy is dissipated as heat.

13.7 What would happen to the cell if pyruvate kinase catalyzed PEP conversion to pyruvate but failed to couple this reaction to ATP production?

ANSWER: If the bacterial cell were to convert PEP to pyruvate without coupling to ATP production, it would lose much of the energy available from glucose and other food substrates converted to glucose. The energy would be lost as heat.

13.8 Some bacteria make an enzyme, dihydroxyacetone kinase, that phosphorylates dihydroxyacetone to dihydroxyacetone phosphate (DHAP). How could this enzyme help the cell yield energy?

ANSWER: Bacteria can obtain dihydroxyacetone from their environment using a transporter protein. The bacterial kinase can then phosphorylate the substrate to DHAP and direct it into glycolysis. Some bacteria can grow on dihydroxyacetone as a sole carbon source.

13.9 Explain why the ED pathway generates only one ATP, whereas the EMP pathway generates 2 ATP. What is the consequence for cell metabolism?

ANSWER: The EMP pathway primes the six-carbon sugar with two phosphoryl groups. The sugar then splits into two three-carbon units (glyceraldehyde 3-phosphate), each of which generates 2 ATP for one of the original ATPs. By contrast, the ED pathway phosphorylates the sugar only once before it splits in two. The phosphorylated end yields glyceraldehyde 3-phosphate, which enters the EMP pathway to generate ATP, ending up as pyruvate. The unphosphorylated three-carbon unit yields pyruvate directly, with no ATP. The consequence for cell metabolism is that the ED pathway needs to cycle more substrate in order for a cell to grow the same amount of biomass as it would with the EMP pathway. Bacteria growing with the ED pathway may produce greater amounts of a valuable product such as ethanol.

13.10 If a cell respiring on glucose runs out of oxygen and other electron acceptors, what happens to the electrons transferred from the catabolic substrates?

ANSWER: The electrons from the catabolic substrates are transferred to NADH and FADH$_2$ during glycolysis and the TCA cycle. Without a terminal electron acceptor, the cytoplasmic electron carriers cannot use the electron transport system. Instead, they must transfer their electrons back onto pyruvate, acetate, and other products of catabolism in order to complete the reactions of fermentation.

13.11 Compare the reactions catalyzed by pyruvate dehydrogenase and pyruvate formate lyase (see Sections 13.5 and 13.6). What conditions favor each reaction, and why?

ANSWER: The pyruvate dehydrogenase complex (PDC) is favored in the presence of oxygen because the electrons transferred to NADH can enter the electron transport chain, eventually combining with oxygen to release energy. In the absence of oxygen, pyruvate formate lyase is favored to yield fermentation products that can be excreted from the cell without reducing more energy carriers. At high pH, formate and acetate production is especially favorable because the extra acid counteracts alkalinity.

13.12 Suppose a cell is pulse-labeled with ^{14}C-acetate (fed the ^{14}C label briefly, then "chased" with unlabeled acetate). Can you predict what will happen to the level of radioactivity observed in isolated TCA intermediates? Plot a curve showing your predicted level of radioactivity as a function of the number of rounds of the cycle.

ANSWER: The amount of radioactivity measured in TCA intermediates will rise steeply as labeled acetate is incorporated, and then will decrease by half with each succeeding cycle, as the order of the carbons is randomized by succinate. Succinate is a symmetrical molecule in which the two ends (labeled and unlabeled) are equivalent.

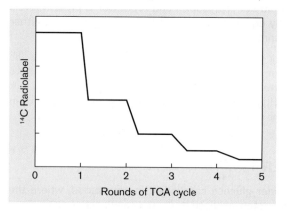

CHAPTER 14

14.1 *Pseudomonas aeruginosa*, a cause of pneumonia in cystic fibrosis patients, oxidizes NADH with nitrate (NO_3^-) to nitrite (NO_2^-) at neutral pH. What is the value of $E°'$?

ANSWER: To calculate the reduction potential $E°'$:

$$NADH + H^+ + NO_3^- \rightarrow NAD^+ + NO_2^- + H_2O$$
$$E°' = 320 \text{ mV} + 420 \text{ mV} = 740 \text{ mV}$$

14.2 Could a bacterium obtain energy from succinate as an electron donor with nitrate (NO_3^-) as an electron acceptor?

ANSWER: For nitrate reduction:

$$\text{Succinate} + NO_3^- \rightarrow \text{fumarate} + NO_2^- + H_2O$$
$$E°' = -33 \text{ mV} + 420 \text{ mV} = 387 \text{ mV}$$

Although succinate is a relatively poor electron donor, nitrate is a strong electron acceptor. This reaction should provide energy for bacterial metabolism.

14.3 What do you think happens to $\Delta\psi$ as the cell's external pH increases or decreases? What could happen to the Δp of bacteria that are swallowed and enter the extremely acidic stomach?

ANSWER: As external pH changes, ΔpH increases or decreases, affecting the magnitude of Δp. As enteric bacteria enter the stomach, they encounter pH values (pH 1.5–3.0) below their growth range (pH 5.0–9.0). At first, the cell's transmembrane ΔpH may be very large, as the cell tries to maintain its cytoplasmic pH above 5.0, the limit for viability. But since the cell no longer grows, it loses its energy supply and can no longer spend energy to maintain ΔpH. One way to maintain cytoplasmic pH homeostasis is to reverse the electrical potential $\Delta\psi$ (inside positive) so as to drive out some H^+ and maintain a small ΔpH. Oppositely directed ΔpH and $\Delta\psi$ enable the cell to keep cytoplasmic pH high enough to survive when external pH is extremely low. On the other hand, when the external pH is raised by pancreatic secretions (alkaline), the cell needs to compensate by inverting its ΔpH and maintaining a relatively large $\Delta\psi$.

14.4   How could a simple experiment provide evidence that a proton pump in the bacterial cell membrane drives efflux of antibiotics such as tetracycline?

ANSWER: Test the ability of the bacteria to form colonies on media buffered at a range of pH values. At lower pH, the ΔpH is increased, and thus a larger Δp is available to pump antibiotics out of the cell. At a higher pH, however, the ΔpH is inverted and actually subtracts from Δp. Thus, we expect the bacteria to grow with greater drug concentrations at low pH than at high pH.

14.5   Suppose that de-energized cells of *E. coli* (Δp = 0) with an internal pH of 7.6 are placed in a solution at pH 6. What do you predict will happen to the cell's flagella? What does this demonstrate about the function of Δp?

ANSWER: The flagella will rotate, driven solely by the ΔpH component of Δp. This result is consistent with the hypothesis that the transmembrane proton potential Δp drives flagellar rotation.

14.6   In Figure 14.15, what is the advantage of the oxidoreductase transferring electrons to a pool of mobile quinones, which then reduce the terminal reductase (cytochrome complex)? Why does each oxidoreductase not interact directly with a cytochrome complex?

ANSWER: The mobile quinone pool connects diverse electron donors with diverse electron acceptors. If each oxidoreductase had to interact specifically with a different terminal oxidase, the pathways of electron transport would be limited; for example, NADH might donate electrons only to O_2, whereas succinate might donate electrons only to nitrate. Instead, all potential electron donors can be coupled with all potential acceptors.

14.7   In Figure 14.15, why are most electron transport proteins fixed within the cell membrane? What would happen if they "got loose" in aqueous solution?

ANSWER: If the electron transport proteins came away from the membrane into aqueous solution, they could carry their energized electrons back into the cytoplasm or lose them outside the cell. In either case, they could no longer convert the flow of electrons into a proton gradient.

14.8   Would *E. coli* be able to grow in the presence of an uncoupler that eliminates the proton potential supporting ATP synthesis?

ANSWER: Yes. *E. coli* can grow with the proton gradient eliminated, but only with a rich supply of nutrients for substrate phosphorylation to generate ATP (for example, from glycolysis). In addition, the external pH and salt levels must be maintained close to those of the cytoplasm, to minimize the need for ion transport.

14.9   The proposed scheme for uranium removal requires injection of acetate under highly anoxic conditions, with less than 1 part per million (ppm) dissolved oxygen. Why must the acetate be anoxic?

ANSWER: Oxygen is the strongest terminal electron acceptor. If O_2 is present, bacteria will use it preferentially (instead of U^{6+}) to oxidize the acetate to CO_2.

14.10   Use Figure 14.15 to propose a pathway for reverse electron flow in an organism that spends ATP from fermentation to form NADH. Draw a diagram of the pathway.

ANSWER: In the ETS shown in **Figure 14.15**, a proton potential Δp would be generated by ATP synthase running in reverse as it spent ATP to pump protons out of the cell. The Δp would drive a cytochrome oxidase complex in reverse to transfer electrons from an electron donor onto quinones, forming quinols (**Fig. 14.12D**). The quinols would transfer electrons to an NDH complex, using energy from proton influx to reduce NAD^+ to NADH.

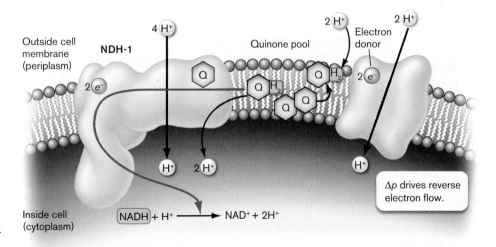

14.11   Hydrogen gas is so light that it rapidly escapes from Earth. Where does all the hydrogen come from to be used for hydrogenotrophy and methanogenesis?

ANSWER: Hydrogen is produced in substantial quantities as a by-product of fermentation. It may seem surprising that organisms would readily excrete quantities of energy-rich H_2, but in the absence of a good electron acceptor (or the enzymes to utilize electron acceptors), hydrogen may be just another waste product. Hydrogen gas that is trapped underground supports large communities of methanogens and hydrogenotrophs. The human colonic bacteria generate so much hydrogen that all parts of the body show traces of hydrogen gas.

14.12 Suppose you discover bacteria that require a high concentration of Fe^{2+} for photosynthesis. Can you hypothesize what the role of Fe^{2+} may be? How would you test your hypothesis?

ANSWER: The organism uses reduced iron as an electron donor for its photosystem ($Fe^{2+} \rightarrow Fe^{3+}$). To test this hypothesis, grow the organism on a defined concentration of Fe^{2+}. Measure the amount of iron oxidized and the amount of carbon fixed into biomass; if the Fe^{2+} is an electron donor for the photosystem, the two numbers should show a linear correlation.

CHAPTER 15

15.1 To run the TCA cycle and glycolysis in reverse, does a cell use the same enzymes or different ones? Explain why some enzymes might be used in both directions, whereas other steps require different enzymes for catabolic and anabolic directions.

ANSWER: Most enzymes are capable of catalyzing a reaction in either direction, depending on the relative amounts of substrates and products. In glycolysis and the TCA cycle, many individual steps of catalysis involve very small energy transitions, such as the interconversion of glucose 6-phosphate with fructose 6-phosphate. The direction of such reactions may be determined by the relative concentrations of substrates and products. (The contribution of reactant concentrations to free energy change ΔG is discussed in Chapter 13.) However, certain key steps of catabolism require a different enzyme for reversal, regulated by conditions that require catabolism or biosynthesis, respectively. For example, the catabolic enzyme phosphofructokinase phosphorylates fructose 6-phosphate to fructose 1,6-phosphate, spending ATP. The reversal of this key step requires fructose 1,6-bisphosphatase. These two enzymes are regulated by metabolites that signal whether the cell has a greater need for energy or for biosynthesis.

15.2 Speculate on why Rubisco catalyzes a competing reaction with oxygen. Why might researchers be unsuccessful in attempting to engineer Rubisco without this reaction?

ANSWER: The oxygenation reaction might have an essential function in regulation of metabolism. For example, it might help prevent excessive reduction of cell components or fixation of too much carbon to be used in biosynthesis. Given the universal existence of the oxygenation reaction in bacterial and chloroplast Rubiscos, it seems unlikely that oxygenation serves no purpose. For this reason, attempts to engineer Rubisco without oxygenation may not succeed. A possible function of photorespiration may be to serve to protect the cell from excess buildup of O_2 or of electron donors. Excess concentrations of these reactive molecules could damage the cell's DNA or proteins.

15.3 Why does ribulose 1,5-bisphosphate have to contain two phosphoryl groups, whereas the other intermediates of the Calvin cycle contain only one?

ANSWER: Only ribulose 1,5-bisphosphate needs to split into two molecules (3-phosphoglycerate). Each of the two products needs

to have its own phosphate as a tag for the enzymes to recognize it within the cycle.

15.4 Which catabolic pathway (see Chapter 13) includes some of the same sugar-phosphate intermediates that the Calvin cycle has? What might these intermediates in common suggest about the evolution of the two pathways?

ANSWER: The pentose phosphate pathway includes ribulose 5-phosphate, erythrose 4-phosphate, and sedoheptulose 7-phosphate in a similar series of carbon exchanges. Perhaps the pentose phosphate pathway and the Calvin cycle evolved from a common amphibolic pathway of sugar consumption and biosynthesis. Alternatively, the one pathway evolved earlier and then the sugar intermediates were available for evolution of the second pathway.

15.5 For a given species, uniform thickness of a cell membrane requires uniform chain length of its fatty acids. How do you think chain length may be regulated?

ANSWER: In *E. coli*, the chain length of a growing fatty acid appears to be limited by beta-ketoacyl-ACP synthase, which binds only precursor acyl-ACPs shorter than 18 carbons. Thus, only carbon chains of up to 18 carbons are synthesized.

15.6 Suggest two reasons why transamination is advantageous to cells.

ANSWER: Ammonia is toxic to cells. Transamination enables cells to store amine groups in nontoxic form, readily available for biosynthesis. The availability of multiple enzymes of transamination from different amino acids enables cells to quickly recycle existing resources into the amino acids most needed by the cell in a given environment. For example, if a sudden supply of glutamine appears, cells can immediately distribute its amines into all 20 amino acids.

15.7 Which energy carriers (and how many) are needed to make arginine from 2-oxoglutarate?

ANSWER: Arginine biosynthesis requires three ATP molecules and three NADPH molecules (including two for converting two molecules of 2-oxoglutarate to glutamate). An additional ATP is spent converting acetate to acetyl-CoA.

15.8 Why are purines synthesized onto a sugar-base-phosphate?

ANSWER: Purines are highly hydrophobic, insoluble in the cytoplasm. The ribose phosphate component solubilizes the molecule, enabling synthesis to occur in the cytoplasm, where purines are needed to make RNA and DNA.

15.9 Why are the ribosyl nucleotides synthesized first and then converted to deoxyribonucleotides as necessary? What does this order suggest about the evolution of nucleic acids?

ANSWER: Ribonucleic acid is believed to be the original chromosomal material of cells. Cells evolved to synthesize RNA first; then later, as DNA was used, pathways evolved to synthesize it by modification of RNA, which the cell already had the ability to make.

CHAPTER 16

16.1 Why do the lipid components of food experience relatively little breakdown during anaerobic fermentation?

ANSWER: Lipids are highly reduced molecules, largely hydrocarbon with relatively low oxidizing potential. Thus, lipids cannot undergo as many intramolecular redox reactions as do sugars, which readily generate energy through anaerobic fermentation.

16.2 Why does oxygen allow excessive breakdown of food, compared with anaerobic processes?

ANSWER: Oxygen functions as the terminal electron acceptor for the complete breakdown of all kinds of organic molecules to water and CO_2.

16.3 In an outbreak of listeriosis from unpasteurized cheese, only the refrigerated cheeses were found to cause disease. Why would this be the case?

ANSWER: In the cheeses kept at room temperature, other naturally occurring bacteria outgrew the pathogenic *Listeria*, whereas in the refrigerator only the *Listeria* could grow. (Note, however, that many other potential pathogens, such as *Salmonella*, are inhibited by refrigeration.)

16.4 Cow's milk contains 4% lipid (butterfat). What happens to the lipid during cheese production?

ANSWER: Lipids undergo little catabolism, because the fermentation conditions are anaerobic. During coagulation, lipid droplets become trapped in the network of denatured protein and are largely retained in the bulk of the cheese. "Low-fat" cheeses are made from skim milk, which eliminates the lipids before fermentation.

16.5 In traditional fermented foods, without pure starter cultures, what determines the kind of fermentation that occurs?

ANSWER: The fermentation can be controlled by introducing a crude starter culture obtained from a previous batch of the food product or from a natural source of a particular microbe; for example, rice straw is a source of *Bacillus natto* for natto production. The fermentation type can be manipulated by the addition of factors, such as brine, that retard growth of all but a few strains. In pidan, for example, the high concentration of sodium hydroxide limits bacterial growth to alkali-tolerant strains of *Bacillus*.

16.6 Compare and contrast the role of fermenting organisms in the production of cheese and bread.

ANSWER: In cheese production, fermentation causes major biochemical changes in the food, such as the buildup of acids and the breakdown of proteins to smaller peptides and amino acids. Minor by-products, such as methanethiols and esters, accumulate to levels that confer flavors. In yeast bread, by contrast, the only significant product of fermentation is the carbon dioxide that leavens the dough. The small amount of ethanol produced evaporates during cooking. A form of bread in which extended fermentation does generate flavor is injera, for which the dough ferments for 3 days.

16.7 Compare and contrast the role of low-concentration by-products in the production of cheese and beer.

ANSWER: In both cheese and beer, minor by-products such as esters contribute flavor. Oxidation of esters can lead to off-flavors. In cheese, however, the exclusion of oxygen usually prevents off-flavors. In beer, the yeast requires a low level of oxygen; thus, significant amounts of acetaldehyde and diacetyl are produced and must be eliminated by a secondary fermentation.

16.8 Why would bacteria convert trimethylamine oxide (TMAO) to trimethylamine? Would this kind of spoilage be prevented by exclusion of oxygen?

ANSWER: TMAO acts as a terminal electron acceptor—that is, an alternative to oxygen for anaerobic respiration, as discussed in Chapter 14. Exclusion of oxygen inhibits only aerobic bacteria; TMAO respirers continue to grow and can spoil the fish.

16.9 Is it possible for physical or chemical preservation methods to completely eliminate microbes from food? Explain.

ANSWER: Preservation methods either slow microbial growth or induce microbial death. Microbial death follows a negative exponential curve, as discussed in Chapter 5. In theory, the exponential curve never reaches zero, so total exclusion of microbes is impossible. In practice, there is a high probability of totally eliminating microbes if the treatment time extends several "half-lives" beyond the time at which microbial concentration declines to less than one per total volume.

16.10 Why would different industrial strains or species be used to express different kinds of cloned products?

ANSWER: Different industrial strains have biochemical systems that favor different products. Some fungi naturally possess the highly complex pathways to generate antibiotics, as well as regulatory timing to turn on these pathways after the culture has grown to high population density. On the other hand, bacteria such as *Bacillus subtilis* are the most genetically tractable and predictable in their growth cycles, and the easiest to manipulate to express recombinant products such as human genes.

16.11 Why would an herbicide resistance gene be desirable in an agricultural plant? What long-term problems might be caused by microbial transfer of herbicide resistance genes into plant genomes?

ANSWER: Introduction of an herbicide resistance gene allows application of higher amounts of herbicide to crops in order to control growth of weeds. But the higher concentrations of herbicide may also have greater side effects on animals and on human consumers of the crop. In the long run, the herbicide resistance gene is likely to escape into weed plants through natural genetic transfer mechanisms. Thus, eventually the weeds may require still higher concentrations of herbicide. While the costs versus benefits of new gene modifications remain poorly understood, it must be recognized that all modern crops today are the product of many generations of genetic manipulation.

CHAPTER 17

17.1 Evolution by natural selection is based on competition, yet the earliest fossil life shows organized structures such as a stromatolite built by cooperating cells. How could this be explained?

ANSWER: Individual microbes compete for resources in a given environment. Part of one's environment consists of other microbes—which may provide resources or favorable conditions. In the stromatolite, the adherence of cells to each other and to the substrate could help maintain their position in a favorable part of the sea, at an elevation with access to light. In addition, adherent microbes might resist predation. We cannot know the biochemistry of cells from ancient fossils, but some cells could have evolved specialized metabolism that led to cross-feeding. In modern stromatolites, phototrophic bacteria photolyze H_2S to sulfate, which is then reduced by lower layers of sulfate-reducing bacteria. Overall, the cells of the stromatolite might outcompete cells that grow individually.

17.2 **What would have happened to life on Earth if the Sun were a different stellar class, substantially hotter or colder than it is?**

ANSWER: If the Sun were hotter, too much ultraviolet and gamma radiation would reach Earth, breaking chemical bonds of living organisms so rapidly that life could not be sustained. If the Sun were colder, too little radiation with sufficient energy would be available to drive photosynthesis. In either case, life as we know it could not have evolved on Earth.

17.3 **Outline the strengths and limitations of each model of the origin of living cells. Which aspects of living cells does each model explain?**

ANSWER: The three models are complementary in that each explains aspects of modern cells not addressed by the others. The prebiotic soup model accounts for the major classes of compounds used by cells, such as nucleosides, TCA cycle intermediates, amino acids, and fatty acids. It also suggests the origin of membranes as soap bubble–like micelles. It does not, however, account for the evolution of metabolic pathways and replication of genetic information. The metabolist model accounts for the prevalence of major cellular reactions such as carbon fixation and the TCA cycle. It does not explain the evolution of membranes and genetic material. The RNA world accounts for the central role of RNA in living cells; of all molecular classes, RNA and ribonucleotides probably serve the widest range of functions as information carriers, agents of catalysis, and genetic regulators. Most RNA-world models do not address the origin of membranes.

17.4 **Suppose a NASA rover discovered living organisms on Mars. How might such a find shed light on the origin and evolution of life on Earth?**

ANSWER: A finding that life on Mars showed a completely different basis from that of Earth—for example, it was based on silicon polymers instead of carbon—would support the view that life originated independently on each planet, rather than traveling from one planet to the other; or that both planets were seeded from somewhere else. A finding that life on Mars was based on similar macromolecules, perhaps even showing the same genetic code, would support the view that life arose on Mars first, or that both planets were seeded from the same source.

17.5 **What kinds of DNA sequence changes have no effect on gene function? (*Hint:* Refer to the table of the genetic code, Figure 8.10.)**

ANSWER: Base substitutions that do not change the amino acid specified by the codon have no effect on gene function. For example, CUA → CUG still encodes leucine. In addition, a majority of the amino acids in any given protein can be replaced by an amino acid of similar form (for example, leucine → valine) without significantly affecting function of the gene.

17.6 **What are the major sources of error and uncertainty in constructing phylogenetic trees?**

ANSWER: Phylogenetic trees are affected by variability in the number of substitutions, or rate of mutation in different strains. The tree is distorted by errors in sequence alignment, and by systematic errors due to failure of the fundamental assumptions of the molecular clock. These assumptions include the constant rate of mutation for all branches, constant generation time, and true orthology of the gene chosen (that is, the encoded product has the same function and hence the same degree of selection pressure in all taxa under consideration).

17.7 **What are the limits of evidence for horizontal gene transfer in ancestral genomes? What alternative interpretation might be offered?**

ANSWER: Horizontal gene transfer is inferred from the appearance of genes in clade A that are absent from other members of the clade but present in clade B. The degree of similarity between genes in the two clades, however, must be high enough to exclude the possibility that the genes in question were retained from a common ancestor of the two clades but lost from other members of clade A. This possibility is difficult to exclude in the case of deep-branching clades, where all genes have had a long time to diverge. For example, the large number of archaeal genes present in deep-branching thermophilic bacteria such as *Thermotoga* may include some inherited from the last common ancestor, or they may represent archaeal genes that were horizontally transferred to bacteria sharing the high-temperature habitat.

17.8 **The evolution of clones requiring nutrients from other clones may be more common than expected. How might the mechanisms of evolution favor the emergence of dependent clones?**

ANSWER: The early mutations that arise under selection pressure confer imperfect phenotypes with deleterious side effects, such as excretion of valuable nutrients. Thus, an evolving population is likely to increase the excreted nutrients available in the medium, such as acetate and citrate. The increase of nutrients provides opportunities for new clones to achieve relative fitness within a population.

17.9 **Using Figure 17.29, how would you identify a straight, nonsheathed, Gram-negative bacterium that has sulfur granules, is motile, and is 1.0 μm wide? What would happen if you assigned the bacterium a width of 0.9 μm?**

ANSWER: The bacterium would key out as *Beggiatoa* sp. If the cell width were measured as 0.9 μm, however, the identification would proceed down a completely different, wrong track at the first step of the key. This is one disadvantage of the dichotomous key.

17.10 **Besides mitochondria and chloroplasts, what other kinds of entities within cells might have evolved from endosymbionts?**

ANSWER: Some of the large "megaplasmids" found in bacteria and protists are as large as genomic chromosomes and contain numerous housekeeping genes. These megaplasmids may have originated as endosymbiotic cells that lost all their membranes through reductive evolution. Similarly, some of the giant viruses, such as mimivirus and smallpox virus, as well as phages such as T4, possess a wide spectrum of housekeeping genes. These viruses may have originated as cellular parasites that underwent reductive evolution.

CHAPTER 18

18.1 What taxonomic questions are raised by the apparent high rate of gene transfer between archaea and thermophilic bacteria?

ANSWER: If gene transfer results in a species containing a quarter of its genes from organisms outside its domain, such a mosaic genome raises questions of how to define the species and the domain. How can a species be defined if its genome contains large portions from distantly related sources? Other interesting questions relate to the means of gene transfer. How do such distantly related organisms as bacteria and archaea maintain a compatible mechanism of gene transfer?

18.2 Which taxonomic groups in **Table 18.1** stain Gram-positive, and which stain Gram-negative? Which group contains both Gram-positive and Gram-negative species? For which groups is the Gram stain undefined, and why?

ANSWER: Most Firmicutes and Actinobacteria stain Gram-positive. These bacteria have relatively thick cell walls that retain the stain. The Proteobacteria, Nitrospirae, and Bacteroidetes/Chlorobi groups stain Gram-negative. The Cyanobacteria have an outer membrane and are considered Gram-negative, although their cell walls are very thick. The Deinococcus-Thermus group includes both Gram-positive- and Gram-negative-staining members. For Chlamydiae and Planctomycetes, the Gram stain is irrelevant because they lack the cell wall that retains the stain. For Spirochetes, many species are too narrow to observe the stain under light microscopy.

18.3 Which groups of bacterial species share common structure and physiology within the group? Which groups show extreme structural and physiological diversity?

ANSWER: All cyanobacteria carry out oxygenic photosynthesis within thylakoid membranes. Their overall cell structure and organization, however, take diverse forms. All spirochetes are a sheathed flexible spiral with internal flagella; most share anaerobic or facultative heterotrophy. Chlamydiae and Planctomycetes groups of species each share general structural features. Other groups, particularly Firmicutes and Actinobacteria, show considerable diversity of form and physiology. The Proteobacteria display more extreme diversity of metabolism than any other division.

18.4 What are the relative advantages and disadvantages of propagation by hormogonia, as compared with akinetes?

ANSWER: Hormogonia are motile, and thus capable of active chemotaxis toward a more favorable environment. On the other hand, hormogonia have active metabolism that requires nutrition; if the environment lacks nutrients, the hormogonia will die. Akinete cells can persist until environmental conditions improve, but they cannot actively seek out a new location.

18.5 What are the relative advantages and disadvantages of the different strategies for maintaining separation of nitrogen fixation and photosynthesis?

ANSWER: Temporal separation has the advantage that all cells possess the ability to perform both nitrogen fixation and photosynthesis. On the other hand, it eliminates the ability of a chain of cells to conduct both processes simultaneously—the benefit of heterocysts. Heterocysts face the problem of operating in close proximity to photosynthetic cells generating toxic oxygen. This problem may be solved by symbiosis with respiring bacteria. Globular clusters of cells can bury their nitrogen fixers within the cluster; this arrangement effectively excludes oxygen, but it may lack flexibility during environmental change. Endosymbiotic nitrogen fixation within a respiring eukaryote is probably the most effective strategy of all, because the host provides oxygen-removing proteins such as leghemoglobin. Endosymbiosis, however, requires the presence of an appropriate host organism.

18.6 Why would *Streptomyces* produce antibiotics targeting other bacteria?

ANSWER: *Streptomyces* species may produce antibiotics to curb the growth of bacterial competitors that have smaller genomes and faster rates of reproduction. The lysed cells release nutrients that feed growing mycelia of *Streptomyces*.

18.7 Why might genes for the proteorhodopsin light-powered proton pump be more likely to transfer horizontally than the bacteriochlorophyll-based photosystems PS I and PS II?

ANSWER: Proteorhodopsin requires only the one gene encoding the pump, plus one or two genes to produce retinal. A relatively small amount of sequence has to be transferred, and the encoded products generate proton potential on their own, without requiring interaction with recipient enzymes. By contrast, PS I and PS II each involve multiple electron carriers that must function together and interact with the recipient electron transport chain.

18.8 Can you hypothesize a mechanism for migration of the daughter nucleoid of *Hyphomicrobium* through the stalk to the daughter cell? For possibilities, see Chapter 3 and consider the various molecular mechanisms of cell division and shape formation.

ANSWER: One mechanism might involve polar localization similar to that seen in *Caulobacter*. A DNA-binding protein might pull the nucleoid through the stalk until it binds to a polar localization protein at the end of the daughter cell. Another mechanism might involve formation of a scaffold of cytoskeletal proteins similar to FtsZ or MreB. The cytoskeleton could act as a track for DNA movement through the stalk, perhaps powered by ATP hydrolysis.

18.9 Why do you think it took many years of study to realize that *Escherichia coli* and other Proteobacteria can grow as a biofilm?

ANSWER: *E. coli* and its relatives grow exceptionally well in liquid culture. Liquid culture is attractive because it enables quantitative measurement of defined aliquots of a microbial population. However, repeated subculturing in liquid medium selects for planktonic (nonbiofilm) cells. Eventually, the biofilm-forming property may be lost if nonbiofilm mutants evolve to grow faster than the original genotype in liquid medium.

18.10 Compare and contrast the formation of cyanobacterial akinetes, firmicute endospores, actinomycete arthrospores, and myxococcal myxospores.

ANSWER: An endospore forms as the daughter product (forespore) of a single cell. Within the same cell, endospore development is supported by the mother cell, which disintegrates after release of the endospore. Endospores have tough coatings of calcium dipicolinate; they are heat resistant. By contrast, arthrospores and myxospores are less durable and are not heat resistant, although they can persist in the environment for an extended period. Arthrospores form through binary fission of actinomycete filaments. Myxospores are formed by a multicellular fruiting body. In all three cases, spore formation can be induced by depletion of nutrients, and the spore-producing entity is left behind to die.

CHAPTER 19

19.1 Suppose two deeply diverging clades each show a wide range of growth temperature. What does this suggest about the evolution of thermophily or psychrophily?

ANSWER: The two clades diverged before temperature adaptation occurred. Adaptations to high or low temperature must have evolved independently in the two clades.

19.2 What are the advantages of naming a microbial lineage after a morphological feature? What problems might arise later, as more members of the lineage are characterized?

ANSWER: The advantage of naming a microbial taxon for a morphological feature such as cell shape is that it helps a researcher identify possible members of the taxon within an environmental sample or cell culture. Ideally, the cells identified can then be tested by single-cell DNA analysis. However, later researchers may discover that organisms sharing similar DNA sequence and metabolic properties differ greatly in shape. The crenarchaeotes turn out to include species with amorphous cells, cannulated cell networks, and rod shapes.

19.3 What might be the advantages of flagellar motility for a hyperthermophile living in a thermal spring or in a black smoker vent? What would be the advantages of growth in a biofilm?

ANSWER: Flagellar motility enables isolated cells to detect a new nutrient source, or an appropriate temperature range, and approach it through chemotaxis. Growth in a biofilm attached to a substrate prevents the microbes from floating away from the nutrient source, or from being carried away in the flow from the vent.

19.4 What problem with cell biochemistry is faced by acidophiles that conduct heterotrophic metabolism?

ANSWER: Heterotrophic metabolism generates fermentation products such as acetate and lactate, which act as permeant acids. Permeant acids become protonated outside the cell, at low pH; the protonated forms then permeate the membrane, returning into the cell. Given the high transmembrane pH difference maintained by *Sulfolobus*, one would expect even small traces of fermentation acids to cross the membrane in the protonated form, and then dissociate and accumulate to toxic levels of organic acids. It is unknown how *Sulfolobus* solves this problem.

19.5 What hypotheses might be proposed about archaeal evolution if viruses of mesophilic archaea are found to have RNA genomes? What if, instead, all archaeal viruses have DNA genomes only?

ANSWER: If mesophilic viruses show RNA genomes but thermophiles do not, then it is likely that only double-stranded DNA is sufficiently stable for viruses to persist in the environment of hyperthermophiles. Finding only double-stranded DNA viruses throughout the archaea would suggest that all archaeal viruses

evolved from viruses infecting a common ancestral cell that was a thermophile. The latter hypothesis would require supporting evidence from archaeal cell physiology and phylogeny.

19.6 What hypothesis might you propose to explain the reason for the time and depth distribution of marine Crenarchaeota seen in **Figure 19.18**? How might you test your hypothesis?

ANSWER: The crenarchaeotes appear to be most abundant in the deeper water (below 1,000 meters) and least abundant during the summer months of June and July. Perhaps the crenarchaeotes are adapted to low temperature, or to high pressure. Either hypothesis might be tested by obtaining a sample of seawater and incubating at different combinations of temperature and pressure. The crenarchaeotes and bacteria could then be sampled to see which temperature or pressure levels favor more rapid growth of crenarchaeotes, compared to bacteria.

19.7 What do the multiple metal requirements suggest about how and where the early methanogens evolved?

ANSWER: The requirement for so many different metals may suggest that methanogens evolved in habitats such as geothermal vents where superheated water carries up high concentrations of dissolved metal ions.

19.8 Compare and contrast the metabolic options available for *Pyrococcus* and for the crenarchaeote *Sulfolobus*.

ANSWER: *Sulfolobus* catabolizes sugars and amino acids aerobically, using O_2 as the terminal electron acceptor. *Pyrococcus abyssi* catabolizes sugars and amino acids anaerobically, using S^0 as the terminal electron acceptor. *P. abyssi* can also reduce sulfur lithotrophically with H_2 to form H_2S. By contrast, *Sulfolobus* uses molecular oxygen (O_2) to oxidize sulfur lithotrophically, from S^{2-} to S^0, SO_3^{2-}, and ultimately SO_4^{2-}.

19.9 Compare and contrast sulfur metabolism in *Pyrococcus* and *Ferroplasma*.

ANSWER: *Pyrococcus* species reduce S^0 with hydrogens from organic substrates, forming HS^- and H_2S. *Ferroplasma* species oxidize sulfur in the form of FeS_2 (using oxidant Fe^{3+}), forming sulfuric acid. The result is extreme acidification of their environment.

CHAPTER 20

20.1 Why would yeasts remain unicellular? What are the relative advantages and limitations of hyphae?

ANSWER: Yeasts grow in environments with sufficient dissolved nutrients to absorb from the medium. The advantage of forming hyphae is that they can penetrate solid substrates, such as soil or a host organism, and hence provide access to nutrients. On the other hand, hypha formation limits the rate of dispersal of progeny cells. Since yeasts grow in environments where dissolved nutrients can be absorbed from the medium, they do not need to produce hyphae and can proliferate more rapidly than mycelial fungi.

20.2 Why would some fungi conduct asexual reproduction under most conditions? What are the advantages and limitations of sexual reproduction?

ANSWER: The sexual life cycle involves significant genetic and metabolic costs to the organism. Asexual reproduction enables fungi to eliminate an energy drain and produce more offspring using fewer resources. Reductive evolution leading to loss of sexual reproduction might eliminate an energy drain and perhaps enable greater proliferation with fewer resources. On the other hand, the sexual life cycle provides a valuable means of generating diversity through genetic recombination, so that the population may respond to environmental change. Fungi reproducing asexually must rely on mutation and gene transfer by viruses and mobile sequence elements to generate genetic diversity.

20.3 **What are the advantages and limitations of motile gametes, as compared to nonmotile spores?**

ANSWER: Motile gametes have the advantage of rapid dispersal on their own and the potential for chemotaxis toward a food source or toward a gamete of the opposite mating type. On the other hand, motility uses up energy that could alternatively be invested in production of a greater number of nonmotile gametes. Motile gametes are especially useful in a watery habitat but are of little use in a terrestrial habitat, where air currents or animal hosts must be used for dispersal.

20.4 **Compare the life cycle of an ascomycete (Fig. 20.14C) with that of a chytridiomycete (Fig. 20.12C). How are they similar, and how do they differ?**

ANSWER: Both chytridiomycetes and ascomycetes undergo alternation of generations. Each has an alternative route of an asexual cycle of mitotic cell proliferation. In the chytridiomycete asexual cycle, the diploid form develops a mycelium, motile zoospores, and cysts. In the ascomycete, however, the haploid form undergoes mitotic divisions. The sexual cycles of the two fungal groups differ structurally. In the chytridiomycete, the zoosporangium forms motile zoospores that develop and release motile gametes, which fertilize each other to form motile zygotes. In the ascomycete, there are no motile forms. Instead of motile gametes, haploid hyphal antheridia undergo cytoplasmic fusion and form fruiting bodies, or asci, within which ascospores develop. The ascospores are not motile; they are carried by wind or water.

20.5 **What are the relative advantages of being unicellular or multicellular?**

ANSWER: Single-celled organisms require minimal nutrients to reproduce and can disperse rapidly, avoiding competition. These important advantages serve the vast majority of organisms on Earth, which are unicellular. On the other hand, a multicellular colony can work together to obtain a larger food source (by predation) and protect its cells from rapid changes in the habitat, such as change in salt concentration or pH. A multicellular organism can protect its offspring from predation, as in the case of *Volvox*. Cell differentiation (for example, in nitrogen-fixing heterocysts) can increase the efficiency of an organism exploiting its environment.

20.6 **What can happen when an ameba phagocytoses algae?**

ANSWER: If light is available, the algae may be retained as endosymbionts providing energy through photosynthesis. For example, *Chlorarachnion* possesses obligate chloroplast-bearing endosymbionts descended from green algae. Alternatively, the ameba may digest all but the algal chloroplast, which persists for some time, providing photosynthetic products.

20.7 **What kind of habitat would favor a flagellated ameba?**

ANSWER: A dilute watery habitat, because flagella allow more rapid propulsion than pseudopods do. Pseudopod motility requires a solid substrate, such as debris in the sediment of a pond.

20.8 **Compare and contrast the process of conjugation in ciliates and bacteria (see Chapter 9).**

ANSWER: Conjugation in ciliates is a completely different process from conjugation in bacteria, although the function (gene transfer) is similar. In ciliates, two cells form a bridge allowing cytoplasm to flow directly between them, along with micronuclei containing chromosomes. In bacteria, a donor cell attaches to another (by pili in some cases), and then a protein complex transfers DNA across both cell envelopes, without direct cytoplasmic contact. In bacterial conjugation, DNA is transferred unidirectionally from the donor cell to the recipient, whereas in ciliates there is reciprocal exchange of DNA. A donor bacterium generally transfers only part of its genome, whereas ciliates exchange entire copies of their respective genomes.

20.9 **For ciliates, what are the advantages and limitations of conjugation, as compared with gamete production?**

ANSWER: The process of conjugation avoids the necessity of dissolving the intricate cell structure of the ciliate in order to form gametes that fuse or fertilize each other. On the other hand, conjugation requires two diploid organisms to find each other and make contact for several hours, during which time feeding is suspended and the pair is vulnerable to predation.

CHAPTER 21

21.1 **Suppose you are conducting a metagenomic analysis of soil sampled from different parts of a wetland. Your budget for soil analysis covers only ten sample analyses. Would you use one DNA extraction method or multiple methods?**

ANSWER: The advantage of using one DNA extraction method is that it will maximize information about the differences between communities from different parts of the wetland. The advantage of multiple extraction methods applied to a given sample is that they will maximize the detection of diversity within a sample. Your choice of method (or a compromise design) may depend on how different you expect the microhabitats to be within the overall wetland.

21.2 **Could you use DNA sequences to identify human occupants of a room, based on the room's airborne microbiome? For background, read James Meadow, 2015, *PeerJ* 3:e1258.**

ANSWER: Meadow and colleagues sampled the airborne microbiome of a room occupied for 1.5–4 hours by various individual humans. The researchers filtered microbial particles from the air, extracted DNA, and obtained 16S rRNA amplicons (amplified the genes by PCR). Amplicons are not full metagenomes, but they offer a quick sample of diversity. The DNA was sequenced by Illumina and assembled by the QIIME pipeline. Samples were "rarified to 1,000 sequences per sample"; that is, the assembled DNA sequences were resampled at random until 1,000 rRNA sequences were obtained, in order to ensure the same depth of diversity for each sample. The researchers found that each of 11 occupants left a distinctive "signature" of rRNA amplicons from their microbes emitted in the room. In the future, it would be interesting to see how many human microbiomes could be distinguished, and how long the signature remains in the room. Obtaining full metagenomes (beyond rRNA amplicons) could increase the sensitivity of detection, though it would require greater resources.

21.3 Suppose you plan to sequence a marine metagenome for the purpose of understanding carbon dioxide fixation and release, to improve our model for global climate change. Do you focus your resources on assembling as many complete genomes as possible, or do you focus on identifying all the community's enzymes of carbon metabolism?

ANSWER: To maximize your yield of information about carbon flux, you are likely to learn more about the community as a whole by focusing on the enzymes involved in carbon metabolism. However, you will miss novel genes that were not previously known to participate in carbon flux. If your equipment and computational resources enable genome assembly, it may be informative to learn more about the most abundant species that participate in carbon flux.

21.4 From Chapters 13–15, give examples of microbial metabolism that fit patterns of assimilation and dissimilation.

ANSWER: Microbes can assimilate carbon either by reducing carbon dioxide or by oxidizing methane, and they can dissimilate carbon by fermentation and respiration. Nitrogen is assimilated by N_2 fixation, and by incorporation of NH_4^+ into glutamine and glutamate. Nitrogen is dissimilated by deamination of amino acids, and by lithotrophic oxidation.

21.5 How does ruminant fermentation provide food molecules that the animal host can use? How is the animal able to obtain nourishment from waste products that the microbes could not use?

ANSWER: The rumen interior is anaerobic. In the absence of oxygen as a terminal electron acceptor, microbes are forced to generate waste products in which the electrons are put back onto the electron donors (fermentation; see Chapter 13). When the short-chain fatty acid wastes enter the animal's bloodstream, the blood is full of oxygen, which enables complete digestion to CO_2 and water.

21.6 How do you think cattle feed might be altered or supplemented to decrease methane production?

ANSWER: Several methods have been proposed to limit methanogenesis. One is to feed cattle an inhibitor of a process that methanogens require, but not bacteria. An example would be inhibitors of sodium transport, which methanogens need to maintain a sodium potential. Another approach is to feed cattle an organic electron acceptor for H_2, such as fumarate, which bacteria use to generate short-chain acids instead of methane. These approaches have been used with only partial success—not surprisingly, given the complexity of the system.

21.7 How could you design an experiment to test the hypothesis that an animal makes use of amino acids synthesized by its gut bacteria?

ANSWER: Select a type of amino acid, such as lysine, that the animal cannot synthesize, and that is therefore essential to the diet of the test animal. The diet can then be altered to exclude the essential amino acid but include a nitrogen source labeled with a heavy isotope. (For a short period, the animal can grow without the added amino acid.) The animal's gut bacteria can be tested for incorporation of the heavy isotope into their amino acids, and the host animal can be tested for protein that incorporates the heavy isotope–labeled amino acid.

21.8 How do viruses select for increased diversity of microbial plankton?

ANSWER: Since viruses tend to infect only a narrow host range, their existence favors the evolution of a large number of different host species with highly dispersed populations. Broad dispersal of populations minimizes the chance of viral transmission from one host to another. Over generations, mutations that allow host microbes to "escape" viral infection will be selected for, while viral mutations that allow a virus to infect previously unsusceptible hosts will also be selected for.

21.9 Design an experiment to test the hypothesis that the presence of mycorrhizae enhances plant growth in nature.

ANSWER: Such an experiment requires a control based on the natural environment, where various unknown factors may be very different from those in the laboratory. One possibility is to compare the growth of seedlings in natural soil versus sterilized natural soil. However, this experiment would not prove that fungi are the cause of enhanced growth in unsterile soil. The sterilization procedure (usually involving heat and pressure) could break down key nutrients in the soil. A follow-up experiment might be to grow the plants in the presence of a fungus inhibitor in sterilized and unsterilized soil.

21.10 How do you think symbiotic rhizobia reproduce? Why do bacteroids develop if they cannot proliferate?

ANSWER: Various answers have been proposed. Not all of the invading bacteria become bacteroids; some continue to undergo cell division, particularly within senescing tissues of the plant. These bacteria benefit from plant growth, which is sustained by the bacteroids whose genes they share. Alternatively, the entire plant-bacteroid system may benefit rhizobia that grow just outside the plant, in the rhizosphere.

21.11 High levels of nitrate or ammonium ion corepress the expression of Nod factors (see Fig. 21.43). What is the biological advantage of Nod regulation?

ANSWER: Nitrate and ammonium ion are the main forms of nitrogen assimilated by plants. If they are abundant in the soil, the plant does not need rhizobial symbionts to fix N_2—a process that consumes much energy. The energy required to maintain the symbiosis comes from the plant, in the form of sugars and other nutrients; thus, it is more efficient not to have the symbiosis when fixed nitrogen is already available. Therefore, the presence of alternative nitrogen sources inhibits development of the rhizobia-legume symbiosis.

21.12 Compare and contrast the processes of plant infection by rhizobia and by fungal haustoria.

ANSWER: Both rhizobial bacteria and fungal haustoria penetrate the volume of a plant cell, but they keep the plant cell membrane intact, its invagination always surrounding the invading cell. Rhizobia establish a complex, highly regulated exchange of nutrients with the host, receiving catabolites and oxygen in exchange for ammonium, and cycling the components of amino acids. By contrast, haustoria establish one-way removal of nutrients such as sucrose, while providing no nutrients in return. Fungal pathogens weaken the structure of the host plant and decrease or halt its growth.

CHAPTER 22

22.1   Why is oxidation state critical for the acquisition, usability, and potential toxicity of cycled compounds? Cite examples based on your study of microbial metabolism (see Chapter 14).

ANSWER: Many examples can be cited. In the case of carbon, CO_2 can be fixed by many species with substantial input from photosynthesis or hydrogen donors. The reduced form methane, however, can be assimilated only by methanotrophic bacteria, usually with oxygen as electron acceptor. Nitrogen gas can be assimilated only by nitrogen-fixing bacteria and archaea, whereas NH_4^+ can be assimilated by many plants and microbes. But NH_4^+ can also be oxidized by lithotrophs to NO_3^-, a substance potentially toxic to humans.

22.2   What would happen if wastewater treatment lacked microbial predators? Why would the result be harmful?

ANSWER: Without predators, too many planktonic bacteria would remain in the wastewater after sedimentation of the sludge. The bacteria could be killed by chlorination, but the treated water would have significant BOD (biochemical oxygen demand) because the bacterial remains provide an organic carbon source for respirers.

22.3   How many kinds of biomolecules can you recall that contain nitrogen? What are the usual oxidation states for nitrogen in biological molecules?

ANSWER: Amino acids, nucleotide bases, polyamines for DNA stabilization, peptidoglycan (both amino sugar and peptide chains), and the heme derivatives of cytochromes, chlorophyll, and vitamin B_{12} all include nitrogen (as do many other biochemicals). The oxidation states of nitrogen in living organisms are nearly always reduced: R–NH_2, R=NH, or R–N=R. An exception is the neurotransmitter NO (nitric oxide).

22.4   The nitrogen cycle has to be linked with the carbon cycle, since both contribute to biomass. How might the carbon cycle of an ecosystem be affected by increased input of nitrogen?

ANSWER: One hypothesis is that the injection of nitrogen into an ecosystem accelerates growth of producers (phytoplankton in the ocean, or trees in a forest) and therefore facilitates net removal of CO_2 from the atmosphere. Overall, however, the additional fixed carbon ends up dissipated by consumers and decomposers.

22.5   In the laboratory, which bacterial genus would likely grow on artificial medium including NH_2OH as the energy source: *Nitrosomonas* or *Nitrobacter*?

ANSWER: *Nitrosomonas* is more likely to utilize NH_2OH, since it performs the intermediate oxidation of NH_2OH during nitrification of ammonia.

22.6   Compare and contrast the cycling of nitrogen and sulfur. How are the cycles similar? How are they different?

ANSWER: Cycling of both nitrogen and sulfur involves interconversion between different oxidation states. Most of these interconversion reactions are performed solely by microbes, many of them solely by bacteria. Examples include nitrification of ammonia and denitrification to N_2, as well as sulfide oxidation and photolysis. In both cases, oxidation produces strong acids (HNO_3, H_2SO_4). The major sources and sinks differ; nitrogen is obtained primarily from the atmosphere as N_2, whereas sulfur (in the form of sulfate) is at high levels in the ocean and soil. Sulfur is rarely limiting, whereas nitrogen frequently is. Sulfur participates extensively in phototrophy; nitrogen shows little involvement in phototrophy; phototrophy based on nitrate reduction has been observed.

22.7   Compare and contrast the cycling of nitrogen and phosphorus. How are the cycles similar? How are they different?

ANSWER: Nitrogen and phosphorus are both limiting nutrients in many ecosystems—marine, freshwater, and terrestrial. Addition of either element into an aquatic system may cause algal bloom and eutrophication. On the other hand, the two elements differ in their major sources: the atmosphere for nitrogen, and crustal rock for phosphate. Within biomass, nitrogen exists almost entirely in reduced form, whereas phosphorus is entirely oxidized. Phosphorus cycles through the biosphere mainly as inorganic or organic phosphates, whereas nitrogen cycles through a broad range of oxidation states, from NH_3 to NO_3^-.

CHAPTER 23

23.1   How can an anaerobic microorganism grow on skin or in the mouth, both of which are exposed to air?

ANSWER: Facultative organisms living in proximity to the anaerobes will deplete oxygen in the environment, especially around nooks and crannies (for example, between teeth and gums, in gingival pockets) that would ordinarily prevent anaerobes from growing. These small spaces have limited access to oxygen.

23.2   Why do many Gram-positive microbes that grow on the skin, such as *Staphylococcus epidermidis*, grow poorly or not at all in the gut?

ANSWER: Bile salts present in the intestine (not on the skin) easily gain access to and destroy cytoplasmic membranes of Gram-positive organisms (unless the organism possesses bile salt hydrolases). Gram-negative microbes have extra protection in the form of an outer membrane and so can survive better in the intestine.

23.3   Why can the colon be considered a fermenter?

ANSWER: The contents are continually flowing through the intestinal tube, with food containing substrates for fermentation ingested at one end and waste containing fermentation products removed from the other.

23.4   Figure 23.20 shows how a neutrophil extracellular trap can ensnare a nearby pathogen. What bacterial structure might blunt the microbicidal effect of NETs?

ANSWER: Bacterial capsules can prevent direct contact between the NET and the cell.

23.5   Why do defensins have to be so small? Do defensins kill normal microbiota?

ANSWER: Defensins need to be small so that they can get through the outer membrane of Gram-negative organisms and the thick peptidoglycan maze of Gram-positive organisms. Defensins do kill normal microbiota and, in fact, are part of what keeps levels of the normal intestinal microbiota in check. Research suggests that a decrease in intestinal defensin production can lead to an imbalance of gastrointestinal microbes in both number and species. This imbalance appears to contribute to conditions such as irritable bowel syndrome (IBS) or inflammatory bowel disease (IBD). Pathogens or normal microbiota that penetrate the intestinal mucosa probably encounter higher concentrations of defensins as they do so.

23.6   As illustrated in Figure 23.31, integrin is important for neutrophil extravasation. Some individuals, however, produce neutrophils that lack integrin. What is the likely consequence of this genetic disorder?

ANSWER: Individuals with neutrophils lacking integrin have leukocyte adhesion deficiency. Their neutrophils are defective in extravasation. These patients are more susceptible to infections because the neutrophils cannot easily get out of the bloodstream.

23.7   What happens to all the neutrophils that enter a site of infection once the infection has resolved?

ANSWER: Once bacteria at the site of infection have been killed, the tissue cells in the area stop making the cytokines and chemokines that attracted neutrophils in the first place, so neutrophils stop coming in, and some wander out. But the majority of neutrophils undergo a self-programmed cell death (called apoptosis) and are cleared by monocytes in the area through phagocytosis. The average life span of a neutrophil is only 5 days.

23.8   If NK cells can attack infected host cells coated with antibody, why won't neutrophils?

ANSWER: Neutrophils (PMNs) can attack infected cells coated with antibody, but the killing mechanism is different from ADCC. Human neutrophils do not make perforin or the other ADCC-related compounds, called granzymes, used by NK cells to kill target cells. In addition to that difference, NK cells possess a type of Fc receptor not found on neutrophils, which means the intracellular signaling pathways are different between NK cells and neutrophils. Neutrophils can be activated, however, when their Fc receptors bind antibody. Activated neutrophils make reactive oxygen products and can release a variety of peptides, including defensins, cathelicidins, and myeloperoxidase, which can all damage target cells.

23.9   Figure 23.36 shows how the complement cascade can destroy a bacterial cell. Factor H (not shown) is a blood protein that regulates complement activity. Factor H binds host cells and inhibits complement from attacking our cells by accelerating degradation of C3b and C3bBb. How could bacteria take advantage of factor H?

ANSWER: Some pathogens have structures on their cell surfaces that can bind factor H. This binding inhibits the alternative pathway from attacking the microbe.

23.10   If increased fever limits bacterial growth, why do bacteria make pyrogenic toxins?

ANSWER: The pyrogenic toxins have other effects that compromise and damage the host. The toxins can induce cytokines that damage local host cells (helping to provide the pathogen with nutrients) or confuse the immune system (allowing the pathogen to delay detection). Pyrogenic toxins include lipopolysaccharide and protein toxins such as toxic shock syndrome toxin (see Chapter 25).

CHAPTER 24

24.1   Two different stretches of amino acids in a single protein form a 3D antigenic determinant. Will the specific immune response to that 3D antigen also respond to one of the two amino acid stretches alone?

ANSWER: Most likely no. It is the 3D shape formed by the two stretches that is recognized as an antigen. A denatured protein that contains both amino acid stretches will not possess the 3D shape of the antigen. However, other specific immune responses involving different subsets of lymphocytes can recognize the separate amino acid stretches that together form the 3D antigenic determinant. As an analogy, take a computer image of a friend's face and shuffle the facial features. Turn the nose upside down, exchange the eyes with the mouth, and lower the ears. Since you are programmed to respond to the original facial configuration, you likely would not recognize the rearranged face as a whole. But you might find that the nose looks familiar.

24.2   How does a neutralizing antibody that recognizes a viral coat protein prevent infection by the associated virus?

ANSWER: Neutralizing antibodies usually bind attachment proteins on the virus and sterically prevent them from binding to host cell receptors (see Fig. 24.5). Some antibodies to enveloped viruses might trigger the complement cascade (see later in the chapter), thus destroying the membrane.

24.3   The attachment proteins of different rhinovirus strains all bind to ICAM-1. How can all these proteins be immunologically different if they find the same target (ICAM-1)? Why won't antibodies directed against one rhinovirus strain block the attachment of other rhinovirus strains?

ANSWER: The ICAM-1 binding sites on different rhinovirus strains have small, but immunologically significant, differences. ICAM-1 is "promiscuous" in its binding specificity toward the rhinovirus attachment proteins, whereas antibodies are more specific. Thus, ICAM-1 protein is like a master key that can fit dozens of different locks (rhinovirus-binding sites). Each lock (binding site) has a very specific key (antibody) that won't unlock any of the other locks. But the master key (ICAM-1) can turn all locks.

24.4   Can an F(ab′)$_2$ antibody fragment prevent the binding of rhinovirus to the ICAM-1 receptor on host cells? And can an F(ab′)$_2$ antibody fragment facilitate phagocytosis of a microbe?

ANSWER: An F(ab′)$_2$ antibody fragment can prevent binding of rhinovirus to the ICAM-1 receptor on host cell surfaces. The antigen-binding sites will block virus receptor access to ICAM-1. However, an F(ab′)$_2$ antibody fragment cannot facilitate phagocytosis of a microbe, because an opsonizing antibody needs its Fc region to bind Fc receptors on phagocytes. Thus, an $F(ab')_2$ antibody can bind to the microbe antigen but cannot link the microbe to a phagocyte's cell surface.

24.5 (refer to **Fig. 24.10**) What types of antibodies will IgG taken from Sherrie raise when injected into John (for example, anti-isotype, anti-allotype, or anti-idiotype)?

ANSWER: Since they are both human, Sherrie's IgG will not elicit anti-isotype antibodies in John. Sherrie's IgG will elicit anti-allotype antibodies and anti-idiotype antibodies in John. Thus, John will develop antibodies that react only to the amino acid differences between John's and Sherrie's IgG antibodies.

24.6 The mother of a newborn was found to be infected with rubella, a viral disease. Infection of the fetus could lead to serious consequences for the newborn. How could you determine whether the newborn was infected in utero?

ANSWER: Since maternal IgM antibodies cannot cross the placenta, finding IgM antibodies to rubella antigens in the newborn's circulation indicates that the fetus was infected and initiated its own immune response. If the newborn has only IgG antibodies to rubella (no IgM antibodies), then the child was not infected and maternal IgG crossed the placenta.

24.7 B cells in early stages have both IgM and IgD surface antibodies, but the delta region has no switch region. Why does the delta region have no switch region?

ANSWER: If the delta region had a switch region, then the B cell could make IgD only after DNA rearrangement. Then, no single B cell could have IgM and IgD at the same time. Recombination at the DNA level is not involved, because alternative RNA-splicing events after transcription determine whether an IgM or IgD molecule is made. B cells at the early stages have both IgM and IgD surface antibodies.

24.8 Why do individuals with type A blood have anti-B and not anti-A antibodies?

ANSWER: Because the B-cell population that would react to type A antigen was deleted during B-cell maturation.

24.9 What would happen to someone lacking CD154 on T_{FH} cells because of a genetic mutation?

ANSWER: The person's T-cell and B-cell numbers would remain normal, but the B cells would not undergo heavy-chain class switching. Thus, any plasma cells produced would make only IgM, causing serum levels of IgM to rise. No other antibody type would be secreted; the result is called hyper IgM syndrome.

24.10 Transplant rejection is a major consideration when transplanting most tissues, because host T_C cells can recognize allotypic MHC on donor cells. Why, then, are corneas easily transplanted from a donor to just about any other person?

ANSWER: The cornea is not normally vascularized. So even though corneal cells express MHC proteins, circulating host T cells do not have an opportunity to interact with them. The cornea will not be rejected. This is called an immune-privileged site.

24.11 Why does attaching a hapten to a carrier protein allow antihapten antibodies to be produced?

ANSWER: B cells with antihapten surface antibody (as part of the B-cell receptor) can take up hapten but cannot present the hapten to a helper T cell. The same B cell can also take up the hapten bound to a carrier molecule, and because the carrier molecule is larger than the hapten, the B cell will present the carrier epitope to the helper T cell. The helper T cell stimulates the B cell, which was already programmed to make antihapten antibody, to differentiate into plasma and memory B cells.

24.12 Do bone marrow transplants in a patient with severe combined immunodeficiency (SCID) require immunosuppressive chemotherapy?

ANSWER: Usually no. The SCID patient has no T cells to recognize foreign antigens, so the transplanted cells are not rejected. What we do worry about are the T cells present in the donor's bone marrow. Once transplanted, the donor's T cells can react against unrelated patient antigens and cause what is called graft-versus-host immune reactions. Consequently, donor bone marrow is rigorously depleted of T cells before transplanting, and immunosuppressive drugs are usually not needed. Using purified stem cells as the graft also avoids graft-versus-host reactions, because there are no T cells. The T cells that eventually develop from the transplanted stem cells become educated in the patient's own thymus and will be tolerized to the patient's antigens.

24.13 How can a stem cell be differentiated from a B cell at the level of DNA?

ANSWER: DNA recombination at switch regions will have taken place in the B cell but not the stem cell. Thus, the B cell will have fewer segments for each of the V, D, and J regions, while the stem cell will have all of them. PCR techniques can be used to view these differences.

24.14 Why do immunizations lose their effectiveness over time?

ANSWER: Because memory B cells eventually die. Without some exposure to antigen, those memory cells will not be replaced and the antibody already made will turn over within weeks.

24.15 How might you design/construct a more effective vaccine for an antigen (for instance, the *Yersinia pestis* F1 antigen, or hepatitis A) that would harness the power of a Toll-like receptor (TLR)? *Hint:* Look at TLR5 in Table 23.3.

ANSWER: By genetically fusing DNA encoding a hepatitis antigen to DNA encoding flagellin, a MAMP (microbe-associated molecular pattern) that is recognized by TLR5 on dendritic cells, you would end up with a chimeric protein that would activate innate immune mechanisms, via TLR5, and improve antigen presentation. The result is a better link between innate and adaptive immune mechanisms.

CHAPTER 25

25.1 **Figure 25.5** presents the association between LD_{50} and virulence. Is a microbe with an LD_{50} of 5×10^4 more or less virulent than a microbe with an LD_{50} of 5×10^7?

ANSWER: Because it takes fewer cells to cause disease, the microbe with the smaller LD_{50} (5×10^4) is more virulent.

25.2 **Figure 25.19** illustrates how cholera toxin works to cause diarrhea. To develop a vaccine that generates protective antibodies, which subunit of cholera toxin should be used to best protect a person from the toxin's effects?

ANSWER: Antibodies to the B subunit will be more protective. Inactivating the B subunit will prevent binding of the toxin to cell membranes. The A-subunit active site is typically sequestered in these toxins and inaccessible to antibody. Furthermore, once the A subunit has entered a host cell, antibodies cannot enter and neutralize it.

25.3 How might you experimentally determine whether a pathogen secretes an exotoxin? (*Hint:* You can learn more about how new toxins are found in eTopic 25.7.)

ANSWER: The microbe can be grown in liquid culture and the cells removed by either centrifugation or filtration. If the organism makes an exotoxin, it may well be present in the cell-free supernatant. The presence of a toxin can be determined by injecting the supernatant into an animal model (for example, mice) and examining the result (death or altered function). Alternatively, the supernatant can be administered to a layer of tissue culture cells and the health of the monolayer noted.

25.4 Would patients with iron overload (excess free iron in the blood) be more susceptible to infection?

ANSWER: With a few exceptions, withholding iron from potential pathogens is a host defense strategy, because when iron is plentiful, the microbe does not have to expend energy to get it and so can readily grow. On the other hand, low iron can also be a signal to express various virulence genes, so for some organisms, high iron might hinder infection.

25.5 Use an Internet search engine to determine which other toxins are related to the cholera toxin A subunit. (*Hint:* Start by searching on "protein" at PubMed to find the protein sequence; then use a BLAST program.)

ANSWER: Go to PubMed (http://www.ncbi.nlm.nih.gov/pubmed). At the top of the page, select **Protein** from the Search dropdown list. Then type "cholera enterotoxin A subunit" in the text box and click **Search**. Click on the appropriate result (any designated "258 aa protein"). Highlight and copy the entire amino acid sequence at the bottom of the page (the entire string of numbers and letters under "ORIGIN"). Then, back at the NCBI home page (http://www.ncbi.nlm.nih.gov), click on **BLAST** under "Popular Resources"; select **Protein BLAST**; paste the A-subunit sequence into the box; and click on **BLAST** at the bottom of the page. After some seconds, all the homology results will be displayed. The toxin of *Escherichia coli* possesses an A subunit that is about 80% identical to that of cholera.

25.6 Protein and DNA have very different structures. Why would a protein secretion system be derived from a DNA-pumping system?

ANSWER: Conjugation systems actually move DNA that is attached to a pilot protein at the 5′ end. A pilot protein is made by the conjugation system and then binds to the 5′ end of the DNA to be transferred and "pilots" the DNA through the conjugation pore. So, a modified conjugation system that moves only protein is not as much of a leap as might initially be thought.

25.7 **Figure 25.30A** shows *Shigella* forming an actin tail at one pole. Why do organisms such as *Shigella* and *Listeria* assemble actin-polymerizing proteins at only one pole?

ANSWER: If actin tails formed at both poles, the organism would spin aimlessly throughout the cell and would likely not be able to protrude into and infect an adjacent host cell. The organisms have mechanisms that localize the key bacterial actin polymerization protein at only one pole (the older pole) of the cell. The molecular basis for this unipolar localization remains unclear but is probably related in some way to polar aging (discussed in Chapter 3).

25.8 How can you determine whether a bacterium is an intracellular pathogen?

ANSWER: Microscopic examination to see whether bacteria are found within cultured mammalian cells is usually not satisfactory. The difficulty lies in determining whether the organism is inside the host cell or just bound to its surface—or, if it is inside, whether it is a live or dead bacterium. One commonly used approach is to add to infected cell monolayers an antibiotic that can kill the microbe but will not penetrate the mammalian cells. The protein synthesis inhibitor gentamicin is typically used. A bacterium that invades a host cell will gain sanctuary from gentamicin and grow intracellularly. Extracellular bacteria, as well as bacteria attached to the outside of the host cell, are killed. Counting viable colony-forming units of bacteria released from the mammalian cells by gentle detergent treatments at various times will reveal whether the organism grew intracellularly. However, this will work only if the microorganism is not an obligate intracellular parasite.

25.9 Why might killing a host be a bad strategy for a pathogen?

ANSWER: The goal of any microbe is to maintain its species. If a microbe did not have an opportunity to easily spread to a new host, killing its host would be tantamount to suicide.

25.10 You have discovered an E3 ligase that specifically modifies surface proteins of *Listeria monocytogenes*. Using what has been presented about toxin entry and protein secretion in Chapter 25, suggest some strategies for delivering this enzyme into host cells infected with this pathogen.

ANSWER: First, realize that *Listeria* is an intracellular pathogen that can escape the phagosome and avoid autophagy. Ubiquitylation of the organism's surface by the E3 ligase, however, may mark the organism for autophagy and clearance. There are several ways this new enzyme could be delivered into infected cells using components of microbial toxins. For instance, fuse the E3 ligase to a nontoxic form of the cholera toxin A subunit. If fused properly, this protein could interact with cholera toxin B subunits, which would bind to the infected host cell and deliver the E3 ligase. You might also be able to engineer the E3 ligase as part of a T3SS or T6SS effector protein. An avirulent form of another bacterium equipped with one of these secretion systems could then deliver the E3 ligase directly into the cytoplasm of the host cell. Finally, you might be able to engineer an attenuated virus with the E3 ligase gene. Once infected with the modified virus, the host would make the enzyme itself. Keep in mind that these scenarios are entirely hypothetical.

25.11 How could you use the FRET/beta-lactamase technique to identify chemical inhibitors of type II, IV, or VI translocation?

ANSWER: Microtiter plate wells containing human cell cultures can be used to screen a battery of potential inhibitor compounds collected from the environment or made by rational design in the laboratory. A pathogen capable of translocating a beta-lactamase-effector fusion protein into host cells would be added to the wells in the presence or absence of the potential inhibitor. After incubation you would add the fluorescent probe and scan for fluorescence. Wells in which host cells did <u>not</u> fluoresce would have <u>failed</u> to receive the beta-lactamase-effector fusion protein, suggesting that the compound added to that well may have inhibited the bacterial secretion system. You should add controls to test whether the test compound affects pathogen or host cell viability, or beta-lactamase activity.

CHAPTER 26

26.1 Does *Staphylococcus aureus* have to disseminate through the circulation to produce the symptoms of scalded skin syndrome (SSS)? Explain why or why not.

ANSWER: Because SSS symptoms are caused by an exotoxin, the infection can be in a single location [a focus of infection, such as the nares (nostrils)], but the toxin can disseminate through the circulatory system.

26.2 Why would treatment of some infections require multiple antibiotics?

ANSWER: Although not typically recommended for simple infections, with some organisms multiple antibiotic therapy is required to effect a cure. The organisms that require multidrug therapy are very hardy in vivo and able to grow in the presence of single antibiotics. Active infection with *Mycobacterium tuberculosis*, for example, is typically treated with isoniazid (thought to inhibit synthesis of mycolic acid) and rifampin (an RNA polymerase inhibitor), and sometimes a third drug, pyrazinamide (which inhibits fatty acid synthesis). Using multiple antibiotics will minimize the chance that the organism will develop resistance to any one antibiotic. Multiple antibiotic therapy is also recommended for *Helicobacter pylori* (ulcers, treated with lansoprazole, amoxicillin, and clarithromycin) and methicillin-resistant *Staphylococcus epidermidis* (MRSE; treated with vancomycin and gentamicin). MRSE can cause serious infections, such as meningitis.

26.3 Explain why patient noncompliance (failure to take drugs as directed) is thought to have led to XDR-TB.

ANSWER: In most cases, an organism causing an infection starts out being susceptible to a given antimicrobial agent. However, the more times an organism divides, the more likely it is that a spontaneous mutation producing drug resistance will arise. The amount of the antibiotic and the duration of its use are calibrated so that all organisms are killed—either by the drug itself or by the immune system. Often the drug is given to stop growth of the bacterium so that the immune system has enough time to actually do the killing. If a patient stops taking an antibiotic before the prescribed end of the treatment, the organism can start to replicate again, renewing the chance that a spontaneous drug-resistant variant will be produced. Even if drug therapy is resumed, the resistant organism will continue to grow and cause disease—and can be transmitted to other individuals.

26.4 Why do you think most urinary tract infections occur in women?

ANSWER: The major reason is anatomy. Most bladder UTIs come from access through the urethra. Since the urethra in men is longer than in women, it usually takes a catheter to introduce bacteria into a male bladder. The trip for the infecting organism in women is much shorter. In people older than 50, however, UTIs become more common in both men and women, with less difference between the sexes. The reason is as yet unclear.

26.5 Urine samples collected from six hospital patients were placed on a table at the nurses' station awaiting pickup from the microbiology lab. Several hours later, a courier retrieved the samples and transported them to the lab. The next day, the lab reported that four of the six patients had UTIs. Would you consider these results reliable? Would you start treatment based on these results?

ANSWER: Because urine is a good growth medium for many bacteria, the delay of several hours in picking up the samples gave the organisms time to replicate and increase their numbers. Consequently, the lab results should be viewed with suspicion.

26.6 Aside from the CDC guidelines for treating gonorrhea, why else do you suppose the patient in this case was also treated with doxycycline (a derivative of tetracycline)? And why can one person be infected repeatedly with *N. gonorrhoeae*?

ANSWER: Because STDs often travel in pairs and because the initial symptomology is similar, the clinician will want to "cover" the patient for possible chlamydia infection. Tetracycline (or azithromycin) will treat chlamydia infections. Chlamydias are not susceptible to ceftriaxone. The large, single dose of ceftriaxone (as opposed to smaller multiple injections given over days) was given because patients with gonorrhea are typically poorly compliant and fail to return for subsequent injections. A person who experiences gonorrhea is not protected from reinfection, because the organism's surface antigens undergo phase variations (see Section 10.1) and may down-regulate the immune system (see **Fig. 26.16C**).

26.7 Like the cause of plague, HIV is a blood-borne disease. Why, then, do fleas and mosquitoes fail to transmit HIV?

ANSWER: When insect vectors take a blood meal, they typically defecate or regurgitate simultaneously. So, theoretically, they could serve as a vector for HIV. Pathogens such as the West Nile virus that are transmitted by insect vectors actually grow in their insect hosts; however, HIV does not. Because it is unusually fragile, HIV will die quickly. There have been no cases of HIV transmitted by insect vectors.

26.8 A patient presenting with high fever and in an extremely weakened state is suspected of having a septicemia. Two sets of blood cultures are taken from different arms. One bottle from each set grows *Staphylococcus aureus*, yet the laboratory report states that the results are inconclusive. New blood cultures are ordered. What would make these results inconclusive?

ANSWER: There must be something different about the two strains of staphylococci grown in the separate bottles. For example, they may have different antibiotic susceptibility patterns when tested against a battery of antibiotics. One strain may be susceptible to penicillin, while the other strain is resistant. Since the expectation is that a single strain initiated the infection, both isolates should exhibit the same susceptibility pattern. The laboratory suspects contamination of the bottles from separate sources. These organisms are not the source of infection.

26.9 Normal cerebrospinal fluid is usually low in protein and high in glucose. The protein and glucose content does not change much during a viral meningitis, but bacterial infection leads to greatly elevated protein and lowered glucose levels. What could account for this?

ANSWER: There are several explanations. Bacteria and infiltrating PMNs will consume glucose, and alterations in the blood-brain barrier can lead to decreased transport of glucose into the spinal fluid. As a result, glucose levels plummet. Growth of the bacteria and infiltration of PMNs account for the increase in CSF protein levels. Viruses are very small and consist mostly of nucleic acids, so even at high numbers they will not significantly add to CSF protein content. Since viruses do not grow in CSF directly, they will not consume glucose. In addition, viral meningitis does not cause a great infiltration of PMNs into the CSF—another reason glucose levels remain high and protein levels remain low.

26.10 Knowing the symptoms of tetanus, what kind of therapy would you use to treat the disease?

ANSWER: At the first sign of muscle spasm, antitoxin should be given. If tetany is severe, then muscle relaxants can relieve spasms.

26.11 Figure 26.29 demonstrates that tetanus toxin has a mode of action (spastic paralysis) opposite to that of botulism toxin (flaccid paralysis). Since they have opposing modes of action, can botulism toxin be used to save a patient with tetanus?

ANSWER: When you first think about it, this sounds feasible. Tetanus toxin causes spastic paralysis, while botulinum toxin causes flaccid paralysis. However, the two toxins act at different places in the nervous system. Tetanus toxin produced by *Clostridium tetani* growing in a wound travels to the central nervous system to exert its effect, whereas botulism toxin works at the periphery. If you administer botulinum toxin intravenously to counteract the whole-body effect of tetanus toxin, then it will affect essentially all neuronal junctions, autonomic and voluntary alike. Thus, the dose of botulinum toxin required to counteract the effect of tetanus toxin on voluntary muscle contraction would likely kill the patient by stopping his or her breathing.

26.12 How do the actions of tetanus toxin and botulinum toxin actually help the bacteria colonize or obtain nutrients?

ANSWER: This is a difficult question to answer. Few scientists have speculated. Recall that the toxins are encoded by genes in resident bacteriophages that became part of the clostridial genome through horizontal transfer from some other source. Since the organisms (vegetative, as well as spores) normally reside in soil, the actual function of these toxins may have something to do with survival in that habitat. The toxin's effect on humans may simply be an unfortunate accident. With tetanus toxin, however, there may be benefit in that muscle spasms could limit oxygen delivery to infected tissues, enabling a more anaerobic environment for growth. Cell death may also release iron or other nutrients useful to *Clostridium tetani*.

26.13 If *Clostridium botulinum* is an anaerobe, how might botulism toxin get into foods?

ANSWER: The most common way botulism toxin gets into food today is via home canning. The canning process involves heating food in jars to very high temperatures. The heat destroys microorganisms and drives out oxygen; both processes help to preserve foods. If the jars are not heated to sterilization conditions, spores of *C. botulinum* will survive. When the jars are cooled for storage at room temperature, the spores germinate; the organism then grows in the anaerobic medium and releases the toxin. When the food is eaten, the toxin is eaten too.

CHAPTER 27

27.1 Figure 27.3 illustrates how MICs are determined. Test your understanding of how MICs are measured in the following example. The drug tobramycin is added to a concentration of 1,000 μg/ml in a tube of broth from which serial twofold dilutions are made. Including the initial tube (tube 1), there are a total of ten tubes. Twenty-four hours after all the tubes are inoculated with *Listeria monocytogenes*, turbidity is observed in tubes 6–10. What is the MIC?

ANSWER: The MIC is 62.5 μg/ml, the concentration in tube 5, the last tube with no growth. (Relative to tube 1, tube 5 has been diluted 2^4 times—a 16-fold dilution: $1,000 \to 500 \to 250 \to 125 \to 62.5$.)

27.2 What additional test performed on an MIC series of tubes will tell you whether a drug is bacteriostatic or bactericidal?

ANSWER: Streaking a portion of the broth from the dilution tubes showing no growth onto an agar plate. If the drug is bacteriostatic, colonies will form on the agar plate because during streaking, the bacteria are removed from the presence of the drug. If the antibiotic is bactericidal, no colonies will form, because the organisms are dead before plating. This method determines the minimum bactericidal concentration (MBC) of an antibiotic. MBC would be defined as the lowest dilution that did not yield viable cells.

27.3 A patient with a bacterial lung infection was given the antibiotic represented in Figure 27.6 and was told to take one pill twice a day. The pathogen is susceptible to this drug. Will the prescribed treatment be effective? Explain your answer.

ANSWER: No. Figure 27.6 shows that the antibiotic remains at effective serum levels for about 8 hours. If the patient takes two pills 12 hours apart, there will be two 4-hour periods (a total of 8 hours) during each 24-hour interval when serum levels fall below an effective MIC concentration. Each 4-hour window of low concentration will allow regrowth and a greater opportunity to select for antibiotic-resistant strains.

27.4 You are testing whether a new antibiotic will be a good treatment choice for a patient with a staph infection. The Kirby-Bauer test using the organism from the patient shows a zone of inhibition of 15 mm around the disk containing this drug. Clearly, the organism is susceptible. But you conclude from other studies that the drug would be ineffective in the patient. What would make you draw this conclusion?

ANSWER: If the average attainable tissue level of the drug is below the MIC, then the drug will be ineffective.

27.5 When treating a patient for an infection, why would combining a drug such as erythromycin with a penicillin be counterproductive? (Erythromycin is described in Section 8.3.)

ANSWER: Erythromycin, a bacteriostatic drug, will stop growth, which indirectly stops cell wall synthesis and renders the microbe insensitive to penicillin.

27.6 The enzyme DNA gyrase, a target of the quinolone antibiotics, is an essential protein in DNA replication. The quinolones bind to and inactivate this protein. Research has proved that quinolone-resistant mutants contain mutations in the gene encoding DNA gyrase. If the resistant mutants contain a mutant DNA gyrase and DNA gyrase is essential for growth, then why are these mutations not lethal?

ANSWER: The mutations cause changes in DNA gyrase that have little to no effect on its function but do prevent the drug from binding. Thus, the mutant organism will continue to twist its DNA and grow well with or without the drug.

27.7 Why might a combination therapy of an aminoglycoside antibiotic and cephalosporin be synergistic?

ANSWER: The two drugs given together could act synergistically because the cephalosporin can weaken the cell wall and allow the aminoglycoside easier access to the cell interior, where it can attack ribosomes. This synergism is especially useful in organisms that have some resistance to both drugs.

27.8 Fusaric acid is a cation chelator that normally does not penetrate the *E. coli* membrane, which means *E. coli* is typically resistant to this compound. Curiously, cells that develop resistance to tetracycline become <u>sensitive</u> to fusaric acid. Resistance to tetracycline is usually the result of an integral membrane efflux pump that pumps tetracycline out of the cell. What might explain the development of fusaric acid sensitivity?

ANSWER: Fusaric acid is imported by the tetracycline efflux pump. This phenomenon can be used to isolate mutants with deletions of transposons encoding tetracycline resistance. Transposons, such as Tn*10*, which carries tetracycline resistance, can spontaneously delete at a frequency of about 10^{-6}, but finding one out of a million tetracycline-susceptible cells is impossible without a positive selection. Fusaric acid provides that positive selection, since the cell with the Tn*10* deletion will be resistant to fusaric acid.

27.9 Mutations in the ribosomal protein S12 (encoded by *rpsL*) confer resistance to streptomycin. Would a cell containing both *rpsL*$^+$ and *rpsL*^{R} genes be streptomycin resistant or sensitive? (Recall that genes encoding <u>r</u>ibosome <u>p</u>roteins for the <u>s</u>mall subunit are designated *rps*. "+" indicates the wild-type allele, while "R" indicates a gene whose product is resistant to a certain drug.)

ANSWER: This merodiploid cell would contain two sets of ribosomes: One set, containing normal S12, would be sensitive to streptomycin; another set, containing the resistant S12, would be resistant. Because streptomycin causes mistranslation of mRNA on sensitive ribosomes, inappropriate proteins that can kill the cell would still be synthesized. Thus, the cell would remain sensitive to streptomycin. Note, however, that the recessive nature of antibiotic resistance seen in this case is not the norm. Resistance is a dominant trait in the majority of cases.

27.10 Figure 27.24 shows a distinctive-looking colony morphology. Why are the colonies on this agar plate red and mucoid?

ANSWER: MacConkey medium contains lactose. As described in Chapter 4, an organism that ferments lactose acidifies the medium and takes up neutral red, turning the colonies red. Organisms that produce a large capsule form slimy, mucoid colonies. *Klebsiella pneumoniae* is one such bacterium that also ferments lactose. A well-trained laboratory technician will immediately suspect *K. pneumoniae* upon seeing these colonies.

27.11 Could genomics ever predict the drug resistance phenotype of a microbe? If so, how?

ANSWER: Yes. If the organism's genome possesses genes whose deduced protein sequences harbor significant similarity to antibiotic resistance proteins from other organisms, then one can predict a similar drug resistance. Definitive proof of drug resistance requires actual in vitro testing.

27.12 The text states that cells and viruses would have difficulty developing resistance to antiviral drugs that target host proteins rather than viral proteins. Why would targeting a host protein decrease the likelihood of developing resistance? Propose a mechanism by which the drug could become ineffective against a virus.

ANSWER: The drug does not bind to a viral protein, so typical drug resistance mutations used by viruses to prevent drug binding are impossible. Likewise, incorporation of a gene in the virus genome that could inactivate the drug is unlikely because the virus would first have to replicate using the targeted host protein (which would be inactivated by the presence of the drug) in order to produce the drug-inactivating enzyme. A mutation in the host that could reduce host protein affinity for the drug is also unlikely because such a mutation would provide a selective advantage only to the virus, and not to the host. So, what kind of viral mutation could provide the virus with drug resistance against a drug that targets a host protein? The virus could conceivably mutate so that it no longer needed the original host target protein, perhaps using a different host protein. You might come up with other plausible scenarios.

CHAPTER 28

28.1 Two blood cultures, one from each arm, were taken from a patient with high fever. One culture grew *Staphylococcus epidermidis*, but the other blood culture was negative (no organisms grew out). Is the patient suffering from septicemia caused by *S. epidermidis*?

ANSWER: Probably not. *S. epidermidis* is a common inhabitant of the skin and could easily have contaminated the needle when blood was taken from the patient. The fact that only one of the two cultures grew this organism supports this conclusion. If the patient was really infected with *S. epidermidis*, then both blood cultures would have grown this organism.

28.2 A 30-year-old woman with abdominal pain went to her physician. After the examination, the physician asked the patient to collect a midstream urine sample that would be sent to the lab across town for analysis. The woman complied and handed the collection cup to the nurse. The nurse placed the cup on a table at the nurses' station. Three hours later, a courier service picked up the specimen and transported it to the laboratory. The next day, the report came back: "greater than 200,000 CFUs/ml; multiple colony types; sample unsuitable for analysis." Why was this determination made?

ANSWER: Although the CFU number is high enough to consider relevant, UTIs are typically caused by a single organism. The fact that the lab found many different colony types suggests a problem with specimen collection. In this case, not refrigerating the sample allowed the small number of urethral contaminants to overgrow the specimen.

28.3 Use **Figure** 28.10 to identify the organism from the following case: A sample was taken from a boil located on the arm of a 62-year-old man. Bacteriological examination revealed the presence of Gram-positive cocci that were also catalase-positive, coagulase-positive, and novobiocin resistant.

ANSWER: *Staphylococcus aureus*. The novobiocin test is irrelevant in this situation.

28.4 If the results in **Figure** 28.14 came from testing for HIV RNA, then which patient would have the higher viral load in their blood?

ANSWER: Patient 1. The detection threshold was crossed earlier for patient 1 (cycle 10) than for patient 2 (cycle 22), so patient 1's serum must have a higher number of HIV virus particles in it.

28.5 Why does adding albumin or powdered milk prevent false positives in ELISA?

ANSWER: Antibodies are proteins. They can stick to plastic just as easily as to the antigen being tested. If all the possible binding sites on plastic were not blocked with albumin (a major protein ingredient in milk), any antibody from the patient's serum could stick to the plastic instead of to the antigen and react with the secondary enzyme-conjugated antihuman antibody.

28.6 Why does finding IgM to West Nile virus indicate a current infection? Why wouldn't finding IgG do the same?

ANSWER: Upon infection with any organism, IgM antibodies are the first to rise. After a short time, the levels of IgM decline as IgG levels rise. IgG, however, can remain in serum for years, making it a poor indicator of current infection.

28.7 Specific IgG antibodies against an infectious agent can persist for years in the bloodstream, long after the infection resolves. How is it possible, then, that IgG antibody titers can be used to diagnose recently acquired diseases such as infectious mononucleosis? Couldn't the antibody be from an old infection?

ANSWER: During the course of a disease, the body's immune system increases the amount of antibody made specifically against the infectious agent. Thus, one compares the antibody titer in a blood sample taken from a patient in the active, or acute, phase of disease with the antibody titer several weeks later, when the patient is in the recovery, or convalescent, stage. Seeing a greater-than-fourfold rise in a specific antibody titer (for example, in mononucleosis) indicates that the patient's immune system was responding to the specific agent. Remember, simply finding IgG against an organism or virus in serum indicates only that the patient was exposed to that microbe at some time in the past.

28.8 Methicillin, a beta-lactam antibiotic, is very useful in treating staphylococcal infections. The emergence of methicillin-resistant strains of *Staphylococcus aureus* (MRSA) is a very serious development because few antibiotics can kill these strains. Imagine a large metropolitan hospital in which there have been eight serious nosocomial infections with MRSA and you are responsible for determining the source of infection so that it can be eliminated. How would you accomplish this task using common bacteriological and molecular techniques?

ANSWER: Samples from all the affected patients and from the hospital staff would be screened for the presence of *S. aureus* resistant to methicillin. Each strain would then be rapidly sequenced in part (multilocus sequencing) or in whole (whole-genome sequencing), using the techniques described in Chapter 7. Strains from all the patients would likely have identical restriction patterns or DNA sequences if they came from the same source. The source would then be identified by determining which staff member possesses MRSA with the same pattern. The source might also be inanimate, such as surgical equipment or ventilation apparatus. A connection between patients and specific staff members or instruments would also have to be demonstrated.

28.9 What are some reasons why some diseases spread quickly through a population while others take a long time?

ANSWER: There are several factors. One is mode of transmission; airborne diseases can spread more quickly than food-borne diseases, for instance. Sexually transmitted diseases spread more slowly still. Herd immunity is another factor. Herd immunity is based on the number of individuals within a population that are resistant to a disease. Someone immune to the disease cannot pass it on. The more immune people there are in a population (or herd), the slower the epidemic spreads to susceptible people.

28.10 On the ProMED-mail web page (www.promedmail. org), click on the interactive world map to view outbreaks recorded by WHO. What outbreaks happened throughout the world during the current year?

ANSWER: As of this writing (2016), there were outbreaks of yellow fever in Angola and the Democratic Republic of the Congo (over 1,100 cases); MERS-CoV virus in Saudi Arabia (over 1,400 cases and 600 deaths); meningococcal disease in 11 gay men from Los Angeles and Orange counties in California; a mysterious hemorrhagic fever involving 52 people in South Sudan; and, of course, Zika virus, among many others.

GLOSSARY

A

A site　See **acceptor site**.

ABC transporter　An ATP-powered transport system that contains an ATP-binding cassette.

aberration　An imperfection in a lens.

abiotic　Produced without living organisms; occurring in the absence of life.

absorption　In optics, the capacity of a material to absorb light.

acceptor site (A site)　The region of the ribosome that binds an incoming charged tRNA.

accessory protein　A protein found in the viral capsid or tegument that is needed early in the viral life cycle.

acid-fast stain　A diagnostic stain for mycobacteria, which retain the dye fuchsin because of mycolic acids in the cell wall.

Acidobacteria　A phylum of Gram-negative bacteria with an outer membrane, related to the Proteobacteria; often found in soil habitats.

acidophile　An organism that grows fastest in acid (generally defined as below pH 5).

ACP　See **acyl carrier protein**.

acquired immunodeficiency syndrome (AIDS)　A disease caused by HIV that leads to the destruction of T cells and the inability to fight off opportunistic infections.

actin filament　See **microfilament**.

Actinobacteria　A phylum of Gram-positive bacteria with high GC content.

actinomycete　A member of the group Actinomycetales, an order of Actinobacteria that includes branched spore formers such as *Streptomyces*, as well as irregularly shaped corynebacteria.

activation energy (E_a)　The energy needed for reactants to reach the transition state between reactants and products.

activator　A regulatory protein that can bind to a specific DNA sequence and increase transcription of genes.

active transport　An energy-requiring process that moves molecules across a membrane against their electrochemical gradient.

acyl carrier protein (ACP)　A protein that can carry an acetyl group for anabolic pathways such as fatty acid synthesis.

adaptive immunity　Immune responses activated by a specific antigen and mediated by B cells and T cells.

ADCC　See **antibody-dependent cell-mediated cytotoxicity**.

addressin　A tissue-specific protein that is selectively present on vascular endothelial cells in different tissues.

adenosine triphosphate (ATP)　A ribonucleotide with three phosphates and the base adenine. It has many functions in the cell, including precursor for RNA synthesis and energy carrier.

adenylate cyclase　An enzyme that converts ATP into cyclic adenosine monophosphate (cAMP).

adhesin　Any cell-surface factor that promotes attachment of an organism to a substrate.

ADP-ribosyltransferase　A bacterial toxin that enzymatically transfers the ADP-ribose group from NAD^+ to target proteins, altering the target protein's structure and function.

aerial mycelium　A mass of hyphae (branched filaments) that extend above the surface and produces spores at the tips.

aerobic respiration　The use of oxygen as the terminal electron acceptor in an electron transport chain. A proton gradient is generated and used to drive ATP synthesis.

aerotolerant anaerobe　An organism that does not use oxygen for metabolism but can grow in the presence of oxygen.

affinity chromatography　A chromatographic technique that uses the ability of biological molecules to bind to certain ligands specifically and reversibly.

affinity maturation　The process by which the antigen-binding site of an antibody gains increased affinity for its target antigen or epitope.

AFM　See **atomic force microscopy**.

agar　A polymer of galactose that is used as a gelling agent.

AIDS　See **acquired immunodeficiency syndrome**.

airborne transmission　In disease, the transfer of a pathogen via dust particles or on respiratory droplets produced when an infected person sneezes or coughs.

alga *pl.* **algae**　A microbial eukaryote that contains chloroplasts.

algal bloom　An overgrowth of algae on the water surface, caused by an increase in a limiting nutrient.

alkaline fermentation　Bacterial fermentation in conjunction with proteolysis and amino acid catabolism that generates ammonia in amounts that raise pH.

alkaliphile　An organism that grows fastest in alkali (generally defined as above pH 9).

allergen　An antigen that causes an allergic hypersensitivity reaction.

allograft　A tissue transplanted from one member of a species to a different member of the same species that possesses a different type of MHC molecule.

allosteric site　A regulatory site on a biological molecule distinct from the ligand/substrate-binding site.

allotype　An amino acid difference in the antibody constant region that distinguishes different individuals within a species.

alternation of generations　A life cycle that alternates between populations of haploid cells and diploid cells, which undergo meiosis and fertilization.

alveolar macrophage　A type of macrophage, located in the lung alveoli, that phagocytoses foreign material.

alveolate　A member of the eukaryotic group Alveolata—ciliated or flagellated protists with complex cortical structure.

ameba *or* **amoeba**　A protist that moves via pseudopods.

amensalism　An interaction between species that harms one partner but not the other.

amino acid　The monomer unit of proteins. Each amino acid contains a central carbon covalently bonded by a hydrogen, an amino group, a carboxyl group, and a side chain. An exception is proline, in which the side chain is cyclized with the central carbon.

aminoacyl-tRNA synthetase　An enzyme that attaches a specific amino acid to the correct tRNA, thereby charging the tRNA.

ammonification　The generation of ammonia from organic nitrogen.

amoeba　See **ameba**.

amphibolic   Describing a metabolic pathway that is reversible and can be used for both catabolism and anabolism.

amphipathic   Having both hydrophilic and hydrophobic portions.

amphotericin B   An antifungal drug that binds the fungus-specific sterol ergosterol and destroys membrane integrity.

amplicon   A specific PCR product in which a small DNA sequence is amplified (many copies are synthesized).

anabolism   Also called *biosynthesis*. The building up of complex biomolecules from smaller precursors.

anaerobic methane oxidizers (ANME)   A group of archaea that oxidize methane syntrophically with bacteria that reduce sulfate.

anaerobic respiration   The use of a molecule other than oxygen as the final electron acceptor of an electron transport chain.

anammox reaction   The anaerobic oxidation of ammonium to nitrogen gas, using nitrate as electron acceptor; yields energy.

anaphylaxis   A severe type I hypersensitivity reaction caused by chemically induced contraction of smooth muscles and dilation of capillaries.

anaplerotic reaction   A type of metabolic reaction, occurring in all organisms, that fixes small amounts of CO_2 to regenerate TCA cycle intermediates.

angle of aperture   The width of a light cone (theta, θ) that projects from the midline of a lens. Greater angles of aperture increase resolution.

anion   A negatively charged ion.

ANME   See **anaerobic methane oxidizers**.

annotation   The deciphering of genome sequences, including identification of genes and prediction of gene function.

antenna complex   A complex of chlorophylls and accessory pigments in the photosynthetic membrane that collects photons and funnels them to a reaction center.

anti-anti-sigma factor   A protein that inhibits an anti-sigma factor, allowing the target sigma factor to participate in initiating transcription.

anti-attenuator stem loop   An mRNA secondary structure whose formation prevents assembly of a downstream transcriptional termination (attenuator) stem loop. The anti-attenuator stem loop structure permits transcription of the downstream structural genes.

antibiotic   A molecule that can kill or inhibit the growth of selected microorganisms.

antibody   A host defense protein produced by B cells in response to a specific antigenic determinant. Antibodies, a type of immunoglobulin, bind to their corresponding antigenic determinants.

antibody stain   The attachment of a stain to an antibody to visualize cell components recognized by the antibody with high specificity.

antibody-dependent cell-mediated cytotoxicity (ADCC)   The process by which natural killer cells destroy viral protein expressing antibody-coated host cells.

anticodon   A set of three nucleotides in the middle loop of a tRNA that base-pairs with a codon in mRNA.

antigen   A compound, recognized as foreign by the cell, that elicits an adaptive immune response. See also **immunogen**.

antigen-presenting cell (APC)   An immune cell that can process antigens into antigenic determinants and display those determinants on the cell surface for recognition by other immune cells.

antigenic determinant   Also called *epitope*. A small segment of an antigen that is capable of eliciting an immune response. An antigen can have many different antigenic determinants.

antigenic shift   A major genetic alteration in a virus that results in a new pandemic strain.

antimicrobial agent   A chemical substance that can kill microbes or slow their growth.

antiparallel   Oriented such that the two strands are in opposite directions. Commonly refers to a nucleic acid double helix with one strand in the 5′-to-3′ orientation and the other strand in the 3′-to-5′ orientation.

antiport   Coupled transport in which the molecules being transported move in opposite directions across the membrane.

antisepsis   The removal of pathogens from living tissues.

antiseptic   Describing a chemical that kills microbes. Also, the chemical itself.

anti-sigma factor   A protein that inhibits a specific sigma factor, preventing transcription initiation.

AP site   A position in DNA where no base is attached to the sugar of the backbone.

APC   See **antigen-presenting cell**.

apicomplexan   A member of the eukaryotic group Apicomplexa—parasitic alveolates that possess an apical complex used for entry into a host cell.

apurinic site   A DNA site missing a purine base because the bond linking the base to the sugar has been hydrolyzed.

arbuscular mycorrhizae   Also called *vesicular-arbuscular mycorrhizae (VAM)*. Mutualistic associations between plant roots and certain fungi, involving hyphal penetration of plant root cells.

Archaea   One of the three domains of life, consisting of organisms with a last common ancestor not shared with members of Bacteria or Eukarya. Organisms are prokaryotic (lacking nuclei, unlike eukaryotes) and possess ether-linked phospholipid membranes (unlike bacteria).

Archaean eon   The second eon (major time period) of Earth's existence, from 4.0 or 3.8 to 2.5 gigayears (Gyr, 10^9 years) before the present. The earliest geological evidence for life dates to this eon.

archaeon *pl.* **archaea**   A prokaryotic organism that is a member of the domain Archaea, distinct from bacteria (domain Bacteria) and eukaryotes (domain Eukarya).

aromatic   A planar, unsaturated, ring-shaped organic molecule whose bonding electrons are delocalized equally around the ring.

artifact   A structure viewed through a microscope that is incorrectly interpreted.

ascomycete   A member of the eukaryotic group Ascomycota—fungi whose mycelia form paired nuclei. Haploid ascospores are produced in pods called asci.

ascospore   The spore produced by an ascomycete fungus.

ascus *pl.* **asci**   A spore-containing pod produced by ascomycete fungi.

aseptic   Free of microbes.

asexual reproduction   Reproduction of a cell by fission or by mitosis to form identical daughter cells.

asRNA   See *cis-antisense RNA*.

assembly   1. In a virus, the packaging of a viral genome into the capsid to form a complete virion. 2. In metagenomics, the piecing together of DNA sequence reads into contigs, and of contigs into scaffold.

assimilation An organism's acquisition of an element, such as carbon from CO_2, to build into body parts.

assimilatory nitrate reduction The uptake of nitrate and reduction to NH_4^+ by plants and bacteria for use in biosynthetic pathways.

atomic force microscopy (AFM) A technique that maps the 3D topography of a object using van der Waals forces between the object and a probe.

atomic mass The mass (in grams) of one mole of an element.

atomic number The number of protons in an atom; it is unique for each element.

ATP See **adenosine triphosphate**.

ATP synthase A protein complex that synthesizes ATP from ADP and inorganic phosphate using energy derived from the transmembrane proton potential. It is located in the prokaryotic cell membrane and in the mitochondrial inner membrane.

attenuator stem loop An intramolecular mRNA structure consisting of a base-paired stem connected by a single-stranded loop. The stem loop structure causes transcription to terminate. Its formation requires efficient translation of a leader peptide sequence.

autoclave A device that uses pressurized steam to sterilize materials by raising the temperature above the boiling point of water at standard pressure.

autoimmune response A pathology caused by lymphocytes that can react to self antigens.

autoinducer A secreted molecule that induces quorum-sensing behavior in bacteria.

autophagy Eukaryotic cell function normally used to degrade damaged organelles. Also used to kill intracellular pathogens.

autoradiography The visualization of a radioactive probe by exposing the probed material to X-ray film and then photographically developing the film.

autotrophy The metabolic reduction of carbon dioxide to produce organic carbon for biosynthesis.

auxotroph A mutant that has lost the ability to synthesize a substance required for growth. An auxotroph has a nutritional requirement not shared by the parent.

azidothymidine (AZT) A nucleotide analog that inhibits reverse transcriptase and was the first drug clinically used to fight HIV infections.

AZT See **azidothymidine**.

B

B cell An adaptive immune cell, developed in bone marrow tissue, that can give rise to antibody-producing cells.

B-cell receptor (BCR) A B-cell membrane protein complex containing an antibody in association with the Igα and Igβ immunoglobulins.

B-cell tolerance The exposure of B cells to a high antigen dose, preventing future antibody production against that antigen.

bacillus *pl.* **bacilli** A bacterium with a linear, or rod, shape.

bacitracin A topical antibiotic that affects cell wall synthesis.

bacteremia A bacterial infection of the blood.

Bacteria One of the three domains of life, consisting of organisms with a last common ancestor not shared with members of Archaea or Eukarya. Organisms are prokaryotic (lacking nuclei, unlike eukaryotes) and possess ester-linked phospholipid membranes (unlike archaea).

bactericidal Having the ability to kill bacterial cells.

bacteriochlorophyll The chlorophyll of anaerobic phototrophic bacteria; it absorbs photons most strongly in the far-red end of the light spectrum.

bacteriophage Also called *phage*. A virus that infects bacteria.

bacteriorhodopsin A haloarchaeal membrane-embedded protein that contains retinal and acts as a light-driven proton pump; it is homologous to the bacterial proteorhodopsin.

bacteriostatic Having the ability to inhibit the growth of bacterial cells.

bacterium *pl.* **bacteria** A prokaryotic organism that is a member of the domain Bacteria, distinct from archaea (domain Archaea) and eukaryotes (domain Eukarya).

bacteroid Cell wall–less, undividing, differentiated rhizobial cell within a plant cell. The bacteroid provides fixed nitrogen for the plant.

Bacteroidetes A phylum of Gram-negative bacteria; nearly all members are obligate anaerobes.

banded iron formation (BIF) A geological formation containing layers of oxidized iron (Fe^{3+}), which indicates formation under oxygen-rich conditions.

barophile Also called *piezophile*. An organism that requires high pressure to grow.

base excision repair (BER) A DNA repair mechanism that cleaves damaged bases off the sugar-phosphate backbone. After endonuclease activity at the AP site, a new, correct DNA strand is synthesized complementary to the undamaged strand.

basidiomycete A member of the eukaryotic group Basidiomycota—fungi that form mushrooms.

basidiospore A haploid spore formed by a basidiomycete through meiosis of a basidium, a reproductive cell of a mushroom.

basophil A white blood cell, stained by basic dyes, that secretes compounds that aid innate immunity.

batch culture The growth of bacteria in a closed system without additional input of nutrients.

BCR See **B-cell receptor**.

benthic organism An organism that lives on the ocean floor or within the sediment.

BER See **base excision repair**.

BIF See **banded iron formation**.

bin A set of sequences composed from metagenomic DNA reads showing a given level of similarity; defines an operational taxonomic unit.

binary fission The process of replication in which one cell divides to form two genetically equivalent daughter cells of equal size.

binning The sorting of metagenomic sequences into taxonomic bins.

biochemical oxygen demand (BOD) Also called *biological oxygen demand*. The amount of oxygen removed from an environment by aerobic respiration.

biofilm A community of microbes growing on a solid surface.

biogenic Formed by living organisms.

biogeochemical cycle The recycling of elements needed for life (such as carbon or nitrogen) through the biotic and abiotic components of the biosphere.

biogeochemistry Also called *geomicrobiology*. The metabolic interactions of microbial communities with the abiotic (mineral) components of their ecosystems.

bioinformatics A discipline at the intersection of biology and computing that analyzes gene and protein sequence data.

biological carbon pump The fixation of carbon dioxide by phototrophs and gravitational settling of biomass particles within the oceans.

biological oxygen demand See **biochemical oxygen demand**.

biological signature See **biosignature**.

biomass The mass found in the bodies of living organisms.

bioprospecting The search for organisms with potential commercial applications.

bioremediation The use of microbes to detoxify environmental contaminants.

biosignature Also called *biological signature*. A chemical indicator of life.

biosphere The region containing the sum total of all life on Earth.

biosynthesis See **anabolism**.

biotic Caused by living organisms.

black smoker An oceanic thermal vent containing high concentrations of minerals such as iron sulfide.

blood-brain barrier A selectively permeable membrane made up of tightly packed capillaries that supply blood to the brain and spinal cord. Large molecules and most pathogens cannot permeate the narrow spaces. Fat-soluble (lipophilic) molecules and oxygen can dissolve through the capillary cell membranes and are absorbed into the brain.

BOD See **biochemical oxygen demand**.

borreliosis Any of a variety of diseases caused by *Borrelia* species and transmitted by ticks or lice. Lyme disease is a form of borreliosis.

botulism A food-borne disease caused by a *Clostridium botulinum* toxin, involving muscle paralysis.

bradykinin A cell signaling molecule that promotes extravasation, activates mast cells, and stimulates pain perception.

Braun lipoprotein See **murein lipoprotein**.

bright-field microscopy A type of light microscopy in which the specimen absorbs light and appears dark against a light background.

brown alga See **kelp**.

bubonic plague A disease caused by the bacterium *Yersinia pestis*; characterized by swollen lymph nodes that often turn black.

budding A form of reproduction in which mitosis of the mother cell generates daughter cells of unequal size.

burst See *lysis*.

burst size The number of virus particles released from a lysed host cell.

C

C-reactive protein A peptide that stimulates the complement cascade, induced by cytokines in the liver. Elevated levels in the blood may be associated with heart disease.

calorimeter A device used to measure the amount of heat released or absorbed during a reaction.

Calvin cycle *or* **Calvin-Benson cycle** *or* **Calvin-Benson-Bassham (CBB) cycle** Also called *reductive pentose phosphate cycle*. The metabolic pathway of carbon fixation in which the CO_2-condensing step is catalyzed by Rubisco. Found in chloroplasts and in many bacteria.

candidate species A newly described microbial isolate that may become accepted as an official species.

cannula *pl.* **cannulae** A narrow tubule. For *Pyrodictium* species, glycoprotein cannulae interconnect cells at their periplasm.

cannulated cow See **fistulated cow**.

capping The clustering of B-cell receptor molecules on the surface of B cells after binding antigens or epitopes.

capsid The protein shell that surrounds a virion's nucleic acid. Within an enveloped virus, such as HIV, the capsid may be called a *core particle*.

capsule A slippery outer layer composed of polysaccharides that surrounds the cell envelope of some bacteria.

carbon-concentrating mechanism (CCM) The inducible expression of transporters of CO_2 and bicarbonate (HCO_3^-) into carboxysomes to enhance CO_2 levels near Rubisco.

carbon dioxide fixation The enzymatic covalent incorporation of inorganic carbon dioxide (CO_2) into an organic compound.

carboxysome A protein-enclosed compartment containing Rubisco to fix CO_2.

cardiolipin Diphosphatidylglycerol, a double phospholipid linked by glycerol.

carotenoid An accessory photosynthetic pigment that absorbs photons in the green end of the spectrum.

catabolism The cellular breakdown of large molecules into smaller molecules, releasing energy.

catabolite repression The inhibition of transcription of an operon encoding catabolic proteins in the presence of a more favorable catabolite, such as glucose.

catalytic domain The A subunit of a toxin, which carries the ADP-ribosyltransferase activity.

catalytic RNA Also called *ribozyme*. An RNA molecule that is capable of catalyzing reactions.

catenane Linked rings of DNA found immediately after replication of circular chromosomes.

cation A positively charged ion.

CBB cycle See **Calvin cycle**.

CCM See **carbon-concentrating mechanism**.

CCR See **chemokine receptor**.

cell fractionation A procedure to separate cell components that often includes ultracentrifugation.

cell-mediated immunity A type of adaptive immunity employing mainly T-cell lymphocytes.

cell membrane Also called *plasma membrane* or *cytoplasmic membrane*. The phospholipid bilayer that encloses the cytoplasm.

cell-surface receptor A transmembrane protein that senses a specific extracellular signal and may be the docking site for a specific virus.

cell wall A rigid structure external to the cell membrane. The molecular composition depends on the organism; it is composed of peptidoglycan in bacteria.

cellular slime mold A slime mold in which the individual cells retain their own cell membranes; not a fungus.

cellulitis A spreading infection (inflammation) of connective tissue just below the skin.

CFTR See **cystic fibrosis transmembrane conductance regulator**.

chain of infection The serial passage of a pathogenic organism from an infected individual to an uninfected individual, thus transmitting disease.

chancre A painless, hard lesion due to an inflammatory reaction at the site of infection with *Treponema pallidum*, the causative agent of syphilis.

chaperone *or* **chaperonin** A protein that helps other proteins fold into their correct tertiary structure.

cheddared curd Curd that has been cut and piled in order to remove the liquid whey.

cheese A solid or semisolid food product prepared by coagulating milk proteins. Its production commonly involves microbial fermentation.

chemical imaging microscopy A method of microscopy that maps the distribution of specific elements or chemicals within a sample.

chemiosmotic theory A theory stating that the products of oxidative metabolism store their energy in an electrochemical gradient that can drive cell processes such as ATP synthesis.

chemoautotrophy See **chemolithoautotrophy**.

chemoheterotrophy See **organotrophy**.

chemokine An attractant for white blood cells that is produced by damaged tissues.

chemokine receptor (CCR) A human T-cell membrane protein that binds chemokine hormones but is also used by HIV for attachment and infection.

chemolithoautotrophy Also called *chemoautotrophy*. Metabolism in which single-carbon compounds are fixed into organic biomass, using energy from chemical reactions without light absorption.

chemolithotrophy See **lithotrophy**.

chemoorganotrophy See **organotrophy**.

chemoreceptor See **methyl-accepting chemotaxis protein**.

chemostat A continuous culture system in which the introduced medium contains a limiting nutrient.

chemosynthesis The fixation of single-carbon molecules (usually carbon dioxide) into organic biomass, using energy from oxidation of inorganic electron donors.

chemotaxis The ability of organisms to move toward or away from specific chemicals.

chemotrophy Metabolism that yields energy from oxidation-reduction reactions without using light energy.

ChIP See **chromatin immunoprecipitation**.

ChIP sequencing (ChIP-seq) A procedure that scans a genome for all DNA sequences capable of binding to a specific DNA-binding protein.

chiral carbon A carbon bonded to four different types of functional groups; it can thus take two different forms that exhibit mirror symmetry.

chlamydia 1. An obligate, intracellular parasitic bacterium of the phylum Chlamydiae. 2. The most frequently reported sexually transmitted disease in the United States. Symptoms range from none, to a burning sensation upon urination, to sterility.

Chlamydiae A phylum of intracellular parasitic bacteria that grow only within a host cell and generate multiple spore-like structures that escape to infect the next host.

chloramphenicol A bacteriostatic antibiotic that acts by inhibiting peptidyltransferase activity of the bacterial ribosome.

Chlorobi A phylum of Gram-negative bacteria. They are obligate anaerobes, "green sulfur" phototrophs that photolyze sulfides or H_2.

chlorophyll A magnesium-containing porphyrin pigment that captures light energy at the start of photosynthesis.

chlorophyte See **green alga**.

chloroplast An organelle of endosymbiotic origin that conducts oxygenic photosynthesis; found in algae and plant cells.

chlorosome A membranous photosynthetic organelle found in some "green" bacteria of the phyla Chloroflexi and Chlorobi.

cholesterol A sterol lipid found in eukaryotic cell membranes.

chromatin Chromosomal DNA complexed with proteins. Usually refers to a eukaryotic chromosome.

chromatin immunoprecipitation (ChIP) An experimental method used to determine the DNA-binding sites on a chromosome to which a DNA protein binds.

chromophore A light-absorbing redox cofactor.

chronic inflammation Inflammation that has persisted over long periods of time, usually months or years.

chrysophyte A member of the eukaryotic group Chrysophyceae (also known as "golden algae")—flagellated heterokont protists possessing chloroplasts as secondary endosymbionts.

cilia *sing.* **cilium** Short, hairlike structures of eukaryotes that beat in waves to propel the cell; structurally similar to flagella.

ciliate An alveolate that has paired cilia.

cis-**antisense RNA (asRNA)** Noncoding RNA from transcription of a DNA sequence complementary to the template strand of a gene encoding a protein. A *cis*-antisense RNA binds and regulates the coding transcript made from the template strand of the gene. These regulatory RNAs can stop transcription, promote transcript degradation, or prevent translation.

clade Also called *monophyletic group*. A group of organisms that includes an ancestral species and all of its descendants.

class I MHC molecule Membrane surface protein on all nucleated cells of the human body (absent from red blood cells and platelets) that present intracellular foreign antigen epitopes to cytotoxic T cells.

class II MHC molecule Membrane surface protein on antigen-presenting cells (dendritic cells, macrophages) and lymphocytes that present phagocytosed extracellular foreign antigen epitopes to helper T cells.

class switching See **isotype switching**.

classical complement pathway An antibody-mediated pathway for complement activation.

classification The recognition of different forms of life and their placement into different categories.

clonal Giving rise to a population of genetically identical cells, all descendants of a single cell.

clonal selection The rapid proliferation of a subset of B cells during the primary or secondary antibody response.

cloning The insertion of DNA into a plasmid where it can be replicated.

CoA See **coenzyme A**.

coastal shelf Shallow regions of the ocean, less than 200 meters deep, that are adjacent to land.

coccus *pl.* **cocci** A spherically shaped bacterial or archaeal cell.

codon A set of three nucleotides that encodes a particular amino acid.

coenzyme A (CoA) A nonprotein cellular organic molecule that can carry acetyl groups and participates in metabolism.

coevolution The evolution of two species in response to one another.

cofactor A metallic ion or a coenzyme required by an enzyme to perform normal catalysis.

cointegrate A DNA molecule formed by a single-site recombination event joining two participating circular DNA molecules to form a larger single circle.

colony A visible cluster of microbes on a plate, all derived from a single founding microbe.

commensalism An interaction between two different species that benefits only one partner.

community The sum of all populations of organisms interacting within an ecosystem.

competent Able to take up DNA from the environment.

complement   Innate immunity proteins in the blood that form holes in bacterial membranes, killing the bacteria.

complex medium   Also called *rich medium*. A nutrient-rich growth solution including undefined chemical components such as beef broth.

compound microscope   A microscope with multiple lenses to compensate for lens aberration and increase magnification.

compromised host   An animal with a weakened immune system.

computational pipeline   A linear series of programs that combine mathematical tools with biological assumptions to propose assembled genomes.

concatemer   A long line of tandemly repeated genomes; commonly formed during rolling-circle replication.

condensation   In biochemistry, the joining of two molecules to release a water molecule.

condenser   In a microscope, a lens that focuses parallel light rays from the light source onto a small area of the specimen to improve the resolution of the objective lens.

confluent   Describing a mode of growth that results in a lawn of organisms completely covering a surface.

confocal laser scanning microscopy *or* **confocal microscopy**   A type of fluorescence microscopy in which the excitation and emitted light are laser beams focused together, producing high-resolution images.

congenital syphilis   Syphilis contracted in utero.

conjugation   Horizontal gene transfer involving cell-to-cell contact. In bacteria, pili draw together the donor and recipient cell envelopes, and a protein complex transmits DNA across. In ciliated eukaryotes, a conjugation bridge forms between two cells connecting their cytoplasm, through which micronuclei are exchanged.

conjugative transposon   A transposon that can be transferred from one cell to another via conjugation.

consensus sequence   A sequence of nucleotides or amino acids with a common function at many nucleic acid or protein positions. Consists of the base pair or amino acid most frequently found at each position in the sequence.

constant region   The region of an antibody that defines the class of a heavy chain or a light chain.

consumer   An organism that acquires nutrients from producers, either directly or indirectly.

contig   A sequence of overlapping fragments of cloned DNA that are contiguous along a chromosome.

continuous culture   A culture system in which new medium is continuously added to replace old medium.

contractile vacuole   An organelle in eukaryotic microbes that pumps water out of the cell.

contrast   Differential absorption or reflection of electromagnetic radiation between an object and a background that allows the object to be distinguished from the background.

coral bleaching   The death or expulsion of coral algal symbionts. One cause is an increase in temperature.

core genome   A set of genes shared by a group of related bacterial strains, showing stable inheritance.

core particle   A viral capsid that encloses its nucleic acid genome and is surrounded by an envelope.

coreceptor   A cell-surface receptor needed for viral entry along with a primary receptor.

corepressor   A small molecule that must bind to a repressor to allow the repressor to bind operator DNA.

cortical alveolus   One of the vesicles that forms a network in the outer covering of an alveolate protist.

counterstain   A secondary stain used to visualize cells that do not retain the first stain.

coupled transport   The movement of a substance against its electrochemical gradient (from lower to higher concentration, or from opposite charge to like charge) using the energy provided by the simultaneous movement of a different chemical down its electrochemical gradient.

covalent bond   A chemical bond in which two atoms share a pair of electrons.

Crenarchaeota   A major division of Archaea, containing sulfur hyperthermophiles, mesophiles, and psychrophiles.

CRISPR   Clustered regularly interspaced short palindromic repeats. It consists of short repeated DNA sequences in a bacterial or archaeal genome, derived from previous bacteriophage or viral infection and conferring protection from future infection; considered a prokaryotic "immune system."

cross-bridge   An attachment that links parallel molecules, such as the peptide link between glycan chains in peptidoglycan.

cryo-electron microscopy (cryo-EM)   Also called *electron cryomicroscopy*. Electron microscopy in which the sample is cooled rapidly in a cryoprotectant medium that prevents freezing. The sample does not need to be stained.

cryo-electron tomography   Also called *electron cryotomography*. A method of cryo-electron microscopy in which the electron beam generates multiple views in parallel planes through the specimen.

cryptogamic crust   A low-growing desert ground cover composed of cyanobacteria, lichens, and nonlichenous algae, fungi, and mosses.

curd   Coagulated milk proteins produced by the combined action of lactic acid–producing bacteria and stomach enzymes of certain mammals, such as cattle.

Cyanobacteria   A phylum of oxygen-producing photoautotrophic bacteria containing chlorophylls *a* and *b*. They share an ancient ancestor of chloroplasts.

cyclic photophosphorylation   A photosynthetic process in which chlorophyll serves as both the initial electron donor and the final electron acceptor. ATP is produced via the proton potential from an electron transport system, but no NADPH is generated.

cycloserine   A polypeptide antibiotic that inhibits peptidoglycan synthesis.

cystic fibrosis transmembrane conductance regulator (CFTR)   A chloride channel found in respiratory epithelia. Mutations in the CFTR gene lead to cystic fibrosis.

cystitis   Bladder infection.

cytochrome   A membrane protein that donates and receives electrons.

cytokine   A small, secreted host protein that binds to receptors on various endothelial and immune system cells, regulating the cells' responses.

cytoplasm   Also called *cytosol*. The aqueous solution contained by the cell membrane in all cells and outside the nucleus (in eukaryotes).

cytoplasmic membrane   See **cell membrane**.

cytoskeleton   A collection of filamentous proteins that impart structure to and aid movement of cells; in a eukaryote, these include microfilaments, intermediate filaments, and microtubules.

cytosol   See **cytoplasm**.

cytotoxic T cell (T_C cell) A T cell that expresses CD8 on its cell surface and can secrete toxic proteins such as perforin and granzymes.

D

D-value See **decimal reduction time**.

daptomycin A lipopeptide antibiotic that forms ion channels in Gram-positive bacteria.

dark-field microscopy The detection of microbes too small to be resolved by light rays, by observing the light they scatter.

dead zone Also called *zone of hypoxia*. An anoxic region of an ocean, devoid of most fish and invertebrates.

death phase A period of cell culture, following stationary phase, during which bacteria die faster than they replicate.

death rate The rate at which cells die; it is exponential during the death phase.

decay-accelerating factor A host cell membrane protein that stimulates the decay of complement factors and prevents their deposition at the cell surface.

decimal reduction time (D-value) The length of time it takes for a treatment to kill 90 percent of a microbial population, and hence a measure of the efficacy of the treatment.

decomposer An organism that consumes dead biomass.

deep-branching taxon A lineage whose genome sequences diverged early, at or before the well-known phyla.

defensin A type of small, positively charged peptide, produced by animal tissues, that destroys the cell membranes of invading microbes.

degenerative evolution See **reductive evolution**.

degranulate To release antimicrobial granule contents by fusing granule membranes to cytoplasmic or vacuolar membranes.

delayed-type hypersensitivity (DTH) See **type IV hypersensitivity**.

deletion The loss of nucleotides from a DNA sequence.

denature To lose secondary and tertiary structure in a protein or nucleic acid because of high temperature or chemical treatment.

dendritic cell An antigen-presenting white blood cell that primarily takes up small soluble antigens from its surroundings.

denitrification Also called *dissimilatory nitrate reduction*. Energy-yielding metabolism in which nitrate (NO_3^-) is reduced to nitrite (NO_2^-), diatomic nitrogen (N_2), and in some cases ammonia (NH_3).

depth of field In a microscope, a region of the optical column over which a specimen appears in reasonable focus.

derepression An increase in gene expression caused by the decrease in concentration of a corepressor.

desensitization A clinical treatment to decrease allergic reactions by exposing patients to small doses of an allergen.

Desulfurococcales An order of thermophilic archaea (Crenarchaeota) that metabolize sulfur and organic compounds.

detection The ability to determine the presence of an object.

detritus Discarded biomass that can be consumed by decomposers.

diapedesis See **extravasation**.

diatom A member of the eukaryotic group Bacillariophyceae—protists that possess intricate, silica-containing bipartite shells.

diauxic growth A biphasic cell growth curve caused by depletion of the favored carbon source and a metabolic switch to the second carbon source.

DIC See **differential interference contrast microscopy**.

dichotomous key A tool for identifying organisms, in which a series of yes/no decisions successively narrows down the possible categories of species.

differential In disease, the list of possible causes of an infection.

differential interference contrast microscopy (DIC) Also called *Nomarski microscopy*. A form of microscopy based on the interference pattern between two light beams split and polarized, one of which passes through a sample. The interference between the two beams generates contrast in a transparent sample.

differential medium A growth medium that can distinguish between various bacteria on the basis of metabolic differences.

differential RNA sequencing (dRNAseq) An RNA sequencing technique that differentiates the true start of an RNA transcript, which begins with a 5′ triphosphate, from an internal cleavage fragment that begins with a 5′ monophosphate.

differential stain A stain that differentiates among objects by staining only particular types of cells or specific subcellular structures.

diffusion The energy-independent net movement of a substance from a region of high concentration to a region of lower concentration.

dilution streaking A method of spreading bacteria on a plate in order to obtain colonies arising from an individual bacterium.

dinoflagellate A member of the eukaryotic group Dinoflagellata—secondary or tertiary endosymbiont algae, alveolates with two flagella, one of which is wrapped distinctively around the cell equator.

diplococcus The paired cocci of *Neisseria* species.

direct repeat Two identical sequences closely positioned in a DNA molecule, aligned in the same direction (e.g., 5′-ATCGATCGnnnnnnnATCGATCG-3′).

directed evolution The application of selective pressure in a laboratory to cause evolution of organisms with desirable properties, such as elevated production of an enzyme.

disinfection The removal of pathogenic organisms from inanimate surfaces.

dissimilation An organism's catabolism or oxidation of nutrients to inorganic minerals that are released into the environment.

dissimilatory denitrification Metabolic reduction of nitrate or nitrite to yield energy; anaerobic respiration of nitrate or nitrite.

dissimilatory metal reduction A type of anaerobic respiration that uses metal cations as terminal electron acceptors.

dissimilatory nitrate reduction See **denitrification**.

DNA-binding protein A protein that binds to DNA and modulates its function.

DNA control sequence A region of DNA, such as the promoter region, that controls the expression of structural genes but is not itself transcribed to RNA.

DNA-dependent RNA polymerase See **RNA polymerase**.

DNA ligase An enzyme that cells use to form a covalent bond at a nick in the phosphodiester backbone. It is also used in molecular biology laboratories to join pieces of DNA.

DNA microarray A technique, used for measuring the amount of specific mRNA molecules transcribed in cells, in which DNA fragments from every open reading frame in a genome are affixed to separate locations on a solid support surface (a DNA microchip), producing a grid, or array.

DNA replication The biological process of making an identical copy of double-stranded DNA using existing DNA as a template.

DNA reverse-transcribing virus   See **pararetrovirus**.

DNA sequencing   A technique to determine the order of bases in a DNA sample.

DNA shuffling   A method of artificially evolving new genes in which fragments of different examples (or orthologs) of a given gene are randomly mixed and PCR-amplified to make a new functional gene. It is an example of directed evolution.

domain   1. In taxonomy, one of three major subdivisions of life: Archaea, Bacteria, and Eukarya. 2. In protein structure, a portion of a protein that possesses a defined function, such as binding DNA. 3. In membranes, a region of membrane consisting of certain types of phospholipids that are distinct from surrounding lipids.

doubling time   The generation time of bacteria in culture. The amount of time it takes for the population to double.

downstream processing   The recovery and purification of a commercial product produced by industrial microbes.

dRNAseq   See **differential RNA sequencing**.

DTH   See **type IV hypersensitivity**.

dysbiosis   An imbalance in microbiome composition that can lead to disease.

E

E site   See **exit site**.

early gene   A viral gene expressed early in the infectious cycle.

eclipse period   The time in the viral life cycle after viral genome injection into a host cell but before complete virions are formed.

ecosystem   A community of species plus their environment (habitat).

ectomycorrhizae   Mycorrhizae that colonize the surface of plant roots. Their mycelia do not penetrate the root cells.

ectoparasite   A harmful organism that colonizes the surface of a host.

ED pathway   See **Entner-Doudoroff pathway**.

edema   Tissue swelling due to fluid accumulation.

edema factor (EF)   A component of anthrax toxin with adenylate cyclase activity.

EF   See **edema factor**.

electrochemical potential   A type of potential energy formed by the combined concentration gradient of a molecule and the electrical potential across a membrane.

electromagnetic radiation   Energy radiating in the form of alternating electrical and magnetic waves, quantized in photons.

electron acceptor   An oxidized molecule (e.g., NAD^+) that can accept electrons.

electron bifurcation   A biochemical reaction in which an electron transfer that yields energy is coupled to an electron transfer that consumes energy.

electron cryomicroscopy   See **cryo-electron microscopy**.

electron cryotomography   See **cryo-electron tomography**.

electron donor   Also called *reducing agent*. A reduced molecule (e.g., NADH) that can donate electrons.

electron microscope   A microscope that obtains high resolution and magnification by focusing electron beams on samples using magnetic lenses.

electron microscopy (EM)   A form of microscopy in which a beam of electrons accelerated through a voltage potential is focused by magnetic lenses onto a specimen.

electron transport system (ETS) *or* **electron transport chain (ETC)**   A series of membrane-embedded proteins that converts the energy of redox reactions into a proton potential.

electronegativity   The affinity of an atom for electrons. The greater the electronegativity, the stronger the attraction for electrons.

electrophoresis   A technique to separate charged proteins and nucleic acids based on how rapidly they migrate in an electrical field through a gel.

electrophoretic mobility shift assay (EMSA)   A technique to observe DNA-protein interactions that is based on the ability of a bound protein to slow the voltage-driven migration of DNA through a gel.

electroporation   A laboratory technique that temporarily makes the cell membrane more leaky to allow the uptake of DNA.

elementary body   The endospore-like form of chlamydias transmitted outside host cells.

EM   See **electron microscopy**.

Embden-Meyerhof-Parnas (EMP) pathway   See **glycolysis**.

emerging   Describing an organism that is newly isolated, defined, or recognized, as in "emerging clade" or "emerging pathogen."

emission wavelength   The wavelength of light emitted by a fluorescent molecule. It is of a lower energy and longer wavelength than the excitation wavelength.

EMP pathway   See **glycolysis**.

empiric therapy   An approach to treating infection before the infective organism is known in which multiple antibiotics are administered in an effort to kill the most likely causative agents.

empty magnification   Magnification without an increase in resolution.

EMSA   See **electrophoretic mobility shift assay**.

enantiomer   See **optical isomer**.

endemic   Describing a disease that is always present in a population, although the frequency of infection may be low.

endergonic reaction   A chemical reaction that requires an input of energy to proceed.

endocarditis   An inflammation of the heart's inner lining.

endocytosis   The invagination of the cell membrane to form a vesicle that contains extracellular material.

endogenous retrovirus   A retroelement (genome sequence descended from a retrovirus) that contains *gag*, *env*, and *pol* genes.

endogenous virus   A virus whose genome is encoded within the germ line of a host animal; may be considered a functional part of the host.

endolith   A bacterium that grows within the crystals of solid rock.

endomembrane system   A series of membranous organelles that organize uptake, transport, digestion, and expulsion of particles through a eukaryotic cell. It includes endosomes, lysosomes, endoplasmic reticulum, and the Golgi complex.

endomycorrhizae   Mycorrhizae whose fungal hyphae penetrate plant root cells.

endoparasite   A harmful organism that lives inside a host.

endophyte   An endosymbiont of vascular plants.

endosome   A vesicle formed from the pinching in of the cell membrane.

endospore   A durable, inert, heat-resistant spore that can remain viable for thousands of years.

endosymbiont   An organism that lives as a symbiont inside another organism.

endosymbiosis   An intimate association between different species in which one partner population grows within the body of another organism.

endotoxin A lipopolysaccharide in the outer membrane of Gram-negative bacteria that becomes toxic to the host after the bacterial cell has lysed.

energy The ability to do work.

energy carrier A molecule in the cell, such as ATP or NADH, that serves as energy currency. Energy carriers are produced during catabolic reactions and can be used to drive energy-requiring reactions.

enhancer A noncoding DNA region in eukaryotes that can lead to activation of transcription when bound by an appropriate transcription factor. Its location on the chromosome can be far removed from the regulated gene.

enriched medium A growth solution for fastidious bacteria, consisting of complex medium plus additional components.

enrichment culture The use of selective growth media to allow only certain microbes to grow.

enthalpy A measure of the heat energy in a system.

Entner-Doudoroff (ED) pathway A glycolytic pathway in which glucose 6-phosphate is initially oxidized to 6-phosphogluconate, and ultimately yields 1 pyruvate, 1 ATP, 1 NADH, and 1 NADPH.

entropy A measure of the disorder in a system.

envelope A structure external to the cell membrane, such as the cell wall or outer membrane of a bacterium. For a virus, the envelope is a membrane enclosing the capsid or core particle.

enzyme A biological catalyst; a protein or RNA that can speed up the progress of a reaction without itself being changed.

eosinophil A white blood cell that stains with the acidic dye eosin and secretes compounds that facilitate innate immunity.

epidemic A disease outbreak in which large numbers of individuals in a population become infected over a short time.

epidemiology The study of factors affecting the health and illness of populations.

epidermis The outer protective cell layer in most multicellular animals.

epifluorescence See *fluorescence*.

epitope See **antigenic determinant**.

EPS See **exopolysaccharides**.

equilibrium *pl.* **equilibria** A dynamic state in which there is no net change in a reaction.

equivalence The antigen/antibody ratio that leads to immuno-precipitation of large, insoluble complexes.

error-prone repair Low-accuracy DNA repair mechanisms that allow mutations.

error-proof repair DNA repair mechanisms that minimize the occurrence of mutations.

erythema migrans A bull's-eye rash characteristic of borreliosis (Lyme disease).

essential nutrient A compound that an organism cannot synthesize and must acquire from the environment in order to survive.

ETC See **electron transport system**.

ethanolic fermentation A fermentation reaction yielding 2 ethanol and $2CO_2$ as products.

ETS See **electron transport system**.

Eukarya One of the three domains of life, consisting of organisms with a last common ancestor not shared with members of Archaea or Bacteria. Cells possess nuclei, unlike cells of bacteria and archaea.

eukaryote An organism whose cells contain a nucleus. All eukaryotes are members of the domain Eukarya.

Eumycota True fungi, a taxonomic group of opisthokont eukaryotes with chitinous cell walls; the group most closely related to animals.

euphotic zone Also called *photic zone*. The region of the ocean that receives sunlight capable of supporting photosynthesis.

Euryarchaeota A major division of Archaea, containing methanogens, halophiles, and acidophiles.

eutrophic lake A lake in which overgrowth of heterotrophic microbes has eliminated oxygen, leading to a decrease in animal life.

eutrophication A sudden increase of a formerly limiting nutrient in an aquatic environment, leading to overgrowth of algae and grazing bacteria and subsequent oxygen depletion.

excitation wavelength The wavelength of light that must be absorbed by a molecule in order for the molecule to fluoresce. It is of a higher energy and shorter wavelength than the emission wavelength.

exergonic reaction A spontaneous chemical reaction that releases free energy.

exit site (E site) The region of the ribosome that holds the uncharged, exiting tRNA.

exocytosis The fusion of vesicles with the cell membrane to release vesicle contents extracellularly.

exon An expressed (protein-coding) portion of a eukaryotic gene.

exonuclease An enzyme that cleaves DNA from the end.

exopolysaccharides (EPS) Polysaccharides and entrapped materials that form a thick extracellular matrix around the microbes in a biofilm.

exotoxin A protein toxin, secreted by bacteria, that kills or damages host cells.

experimental evolution Also called *laboratory evolution*. The repeated culturing of a population in a laboratory under defined environmental conditions, leading to evolution of adapted genotypes.

exponential phase Also called *logarithmic (log) phase*. A period of cell culture during which bacteria grow exponentially at their maximal possible rate, given the conditions.

extravasation Also called *diapedesis*. The movement of cells of the immune system out of blood vessels and into surrounding infected tissue.

extremophile An organism that grows only in an extreme environment—that is, an environment including one or more conditions that are "extreme" relative to the conditions for human life.

eyepiece See **ocular lens**.

F

F⁻ cell The DNA recipient cell in bacterial conjugation.

F⁺ cell The DNA donor cell that transmits the fertility factor F⁺ to an F⁻ cell during bacterial conjugation.

F factor See **fertility factor**.

F-prime (F′) factor *or* **F-prime (F′) plasmid** A fertility factor plasmid that contains some host chromosomal DNA.

facilitated diffusion A process of passive transport across a membrane that is facilitated by transport proteins.

FACS See **fluorescence-activated cell sorter**.

factor H A normal serum protein that prevents the inadvertent activation of complement in the absence of infection.

facultative Able to grow in the presence or absence of a given environmental factor, such as oxygen.

facultative anaerobe   An organism that can grow in either the presence or absence of oxygen.

facultative intracellular pathogen   A pathogen that can replicate either inside host cells or outside host cells.

FAD   See **flavin adenine dinucleotide**.

Fc region   The region of an antibody that binds to specific receptors on host cells in an antigen-independent manner. It is found in the carboxy-terminal "tail" region of the antibody.

fecal-oral route of transmission   A method by which pathogens or parasites excreted in the fecal matter of an infected person are then indirectly ingested by an uninfected person.

fermentation   Also called *fermentative metabolism*. The production of ATP via substrate-level phosphorylation, using organic compounds as both electron donors and electron acceptors. Industrial fermentation is the production of microbial products that are made by microbes grown in fermentation vessels; it may include respiratory metabolism to maximize microbial growth.

fermentation industry   Commercial production of microbial products that are made by microbes grown in fermentation vessels; it may include respiratory metabolism to maximize microbial growth.

fermentative metabolism   See **fermentation**.

fermented food   Food products that are biochemically modified by microbial growth.

ferredoxin   An iron- and sulfur-containing protein that transfers electrons in electron transport systems.

fertility factor (F factor)   A specific plasmid (transferred by an F^+ donor cell) that contains the genes needed for pilus formation and DNA export.

filamentous virus   A viral structure type consisting of a helical capsid surrounding a single-stranded nucleic acid.

fimbria *pl.* **fimbriae**   See **pilus**.

Firmicutes   A phylum of low-GC-content Gram-positive bacteria.

FISH   See **fluorescence in situ hybridization**.

fistulated cow   Also called *cannulated cow*. A cow in which a hole in the skin has been connected surgically to a hole in the rumen and fitted with a cannula, allowing access for experimental analysis of the rumen.

fitness island   A type of genomic island in which the stretch of DNA expresses proteins that enable the organism's survival in new or harsh environments.

fixation   The adherence of cells to a slide by a chemical or heat treatment.

flagellate   A protist that has one or more flagella.

flagellum *pl.* **flagella**   A filamentous structure for motility. In prokaryotes, a helical protein filament attached to a rotary motor; in eukaryotes, an undulating membrane-enclosed complex of microtubules and ATP-driven motor proteins.

flavin adenine dinucleotide (FAD)   An energy carrier in the cell that can donate ($FADH_2$) or accept (FAD) electrons.

flesh-eating disease   See **necrotizing fasciitis**.

floc   Particulate matter formed by clumps of microbes during wastewater treatment.

fluid mosaic model   A model of the cell membrane in which proteins are free to diffuse laterally within the membrane.

fluorescence   Also called *epifluorescence*. The emission of light from a molecule that absorbed light of a shorter, higher-energy wavelength.

fluorescence-activated cell sorter (FACS)   A device that can count cells and sort them according to differences in fluorescence.

fluorescence in situ hybridization (FISH)   A technique to detect individual microbes in an ecological or clinical sample, using a fluorophore-labeled oligonucleotide probe (usually a short DNA sequence) that hybridizes to microbial DNA or rRNA.

fluorescence resonance energy transfer (FRET)   The detectable transfer of fluorescent energy from one molecule to another. Since the participating molecules must be near each other, FRET can be used to monitor protein-protein interactions in cells and is also used in real-time PCR.

fluorophore   A fluorescent molecule used to stain specimens for fluorescence microscopy.

focal point   The position at which light rays that pass through a lens intersect.

focus *pl.* **foci**   The point at which rays of energy converge; in light microscopy, the convergence of light rays maximizes the clarity of the optical image.

folate   Folic acid, a heteroaromatic cofactor that is required by some enzymes.

fomite   An inanimate object on which pathogens can be transmitted from one host to another.

food poisoning *or* **food contamination**   The presence of human disease-causing microbial pathogens or toxins in food.

food spoilage   Microbial changes that render a food unfit or unpalatable for consumption.

food web   A network of interactions in which organisms obtain or provide nutrients for each other—for example, by predation or by mutualism.

foraminiferan *or* **foram**   An ameba with a calcium carbonate shell and a helical arrangement of chambers.

forespore   In sporulation of Gram-positive bacteria, the smaller cell compartment formed through asymmetrical cell division; it develops into the endospore.

fossil fuel   Ancient organismal remains that have been converted to hydrocarbons (petroleum and natural gas) or sedimentary rock (coal) through microbial digestion followed by reduction under high pressure underground; fossil fuels are extracted and burned by humans for energy.

frameshift mutation   A gene mutation involving the insertion or deletion of nucleotides that cause a shift in the codon reading frame.

free energy change   See **Gibbs free energy change**.

freeze-drying   Also called *lyophilization*. The removal of water from food, by freezing under vacuum, to limit microbial growth.

FRET   See **fluorescence resonance energy transfer**.

fruiting body   A multicellular fungal or bacterial reproductive structure.

frustule   The silica bipartite shell produced by a diatom.

functional genomics   A field of molecular biology that utilizes data from genomic projects to describe gene and protein interactions.

functional group   A cluster of covalently bonded atoms that behaves with specific properties and functions as a unit.

fungus *pl.* **fungi**   A heterotrophic opisthokont eukaryote with chitinous cell walls. Includes Eumycota, but traditionally may refer to fungus-like protists such as the oomycetes.

fusion peptide   A portion of a viral envelope protein that changes shape to facilitate envelope fusion with the host cell membrane.

Fusobacteria   A phylum of anaerobic Gram-negative bacteria with an outer membrane, related to the Proteobacteria; includes human pathogens.

G

gain-of-function mutation A mutation that enhances the activity or allows new activity of a gene product.

GALT See **gut-associated lymphoid tissue**.

gamogony The differentiation of parasitic haploid cells into male and female gametes.

gas vesicle An organelle that traps gases to increase buoyancy of aquatic microbes.

GC content The proportion of an organism's genome consisting of guanine-cytosine base pairs.

GDH See **glutamate dehydrogenase**.

gene A distinct series of nucleotides within DNA that has a distinct function (regulatory) or whose product has a distinct function (protein). The functional unit of heredity.

gene duplication The formation of an extra copy of a gene within a genome.

gene fusion See **translational fusion**.

gene switching Switching between two (out of five) different classes of immunoglobulin genes (e.g., from IgM to IgG) during B-cell development.

gene transfer vector A mobile DNA engineered from a virus or plasmid, designed to insert a genetic sequence into the genome of an organism for experimental study or for medical therapy.

generalized recombination Recombination between two DNA molecules that share long regions of DNA homology.

generalized transduction A phage-mediated gene transfer process in which any donor gene can be transferred to a recipient cell.

generation time The species-specific time period for doubling of a population (e.g., by bacterial cell division) in a given environment, assuming no depletion of resources.

genome The complete genetic content of an organism. The sequence of all the nucleotides in a haploid set of chromosomes.

genomic island A region of DNA sequence whose properties indicate that it has been transferred from another genome. Usually comprises a set of genes with shared function, such as pathogenicity or symbiosis support.

genus A group of closely related species.

geochemical cycling The global interconversion of various inorganic and organic forms of elements.

geomicrobiology See **biogeochemistry**.

germ theory of disease The theory that many diseases are caused by microbes.

germicidal Able to kill cells but not spores.

germination The activation of a dormant spore to generate a vegetative cell.

Gibbs free energy change *or* **Gibbs energy change** *(ΔG)* Also called *free energy change*. In a chemical reaction, a measure of how much energy available to do work is released or required as the reaction proceeds.

gliding motility The movement of cells individually or as a collective over surfaces using pili.

gluconeogenesis The biosynthesis of glucose from single-carbon compounds.

glutamate dehydrogenase (GDH) An enzyme that condenses NH_4^+ with 2-oxoglutarate to form glutamate. The condensation requires reduction by NADPH.

glutamate synthase Glutamine:2-oxoglutarate aminotransferase (GOGAT), an enzyme that converts 2-oxoglutarate plus glutamine into two molecules of glutamate.

glutamine synthetase (GS) An enzyme that condenses NH_4^+ with glutamate to form glutamine.

glycan A polysaccharide chain composed of oxygen-linked (O-linked) monosaccharides.

glycolysis Also called *Embden-Meyerhof-Parnas (EMP) pathway*. The catabolic pathway of glucose oxidation to pyruvate, in which glucose 6-phosphate isomerizes to fructose 6-phosphate, ultimately yielding 2 pyruvate, 2 ATP, and 2 NADH.

glyoxylate bypass An alternative to the tricarboxylic acid cycle in which isocitrate is converted to glyoxylate and then malate; induced under low glucose conditions.

gnotobiotic animal An animal that is germ-free or colonized by a known set of microbes.

GOGAT See **glutamate synthase**.

Golgi complex *or* **Golgi apparatus** A series of membrane stacks that modifies proteins and helps sort them to the correct eukaryotic cell compartment.

Gram-negative Describing cells that do not retain the Gram stain.

Gram-positive Describing cells that retain the Gram stain and appear dark purple after staining.

Gram stain A differential stain that distinguishes cells that possess a thick cell wall and retain a positively charged stain (Gram-positive) from cells that have a thin cell wall and outer membrane and fail to retain the stain (Gram-negative).

gramicidin A peptide antibiotic that acts as a channel for monovalent cations to cross the cell membrane, thus collapsing the transmembrane ion gradients.

granuloma A thick lesion formed around a site of infection.

granzyme An enzyme, secreted by cytotoxic T cells, that damages target cells.

grazer A first-level consumer, feeding directly on producers.

green alga Also called *chlorophyte*. A member of the eukaryotic group Chlorophyta—microbes that have chloroplasts; closely related to plants (Viridiplantae).

greenhouse effect The trapping of solar radiation heat in the atmosphere by CO_2; a cause of global warming.

greenhouse gas A gas that transmits visible light but absorbs heat radiation—a property known as the greenhouse effect.

griseofulvin An antifungal antibiotic that inhibits cell division.

group translocation A form of active transport in which the transported molecule is modified after it enters the cell, thus keeping a favorable inward concentration gradient for the unmodified extracellular molecule.

growth factor A compound needed for the growth of only certain cells.

growth rate The rate of increase in population number or biomass.

growth rate constant The number of organismal generations per unit time; *k*.

GS See **glutamine synthetase**.

gut-associated lymphoid tissue (GALT) Lymphatic tissues such as tonsils and adenoids that are found in conjunction with the gastrointestinal tract and contain immune cells.

H

HAART See **highly active antiretroviral therapy**.

Haber process Industrial nitrogen fixation, in which dinitrogen is hydrogenated by methane (natural gas) under extreme heat and pressure to form ammonia.

Hadean eon The first eon (major time period) of Earth's existence, from 4.5 to approximately 4.0–3.8 gigayears (Gyr, 10^9 years) before the present.

half-life The amount of time it takes for one-half of a radioactive sample to decay.

Haloarchaea A class of extremely halophilic archaea, inhabiting high-salt environments.

Halobacteriales An order of Euryarchaeota that contains the class Haloarchaea.

halophile An organism that requires a high extracellular sodium chloride concentration for optimal growth.

halorhodopsin A haloarchaeal membrane-embedded protein that contains retinal and acts as a light-driven chloride pump; it is homologous to bacteriorhodopsin.

hamus *pl.* **hami** An archaeal cell appendage with grappling hooks that enable cells to connect with each other, adhere to a surface, and form biofilms.

hapten A small compound that must be conjugated to a larger carrier antigen in order to elicit production of an antibody that binds to it.

haustorium *pl.* **haustoria** A bulbous hyphal extension of a fungal plant pathogen into the host cell.

HAV See **hepatitis A virus**.

HBV See **hepatitis B virus**.

HCV See **hepatitis C virus**.

heat-shock protein (HSP) A chaperone protein whose synthesis is induced by high-temperature stress.

heat-shock response A coordinated response of cells to higher-than-normal temperatures. It includes changes in the membrane and expression of heat-shock genes.

heavy chain The larger of the two protein types that make up an antibody. Each antibody contains two heavy chains and two light chains.

helper T cell (T_H cell) A T cell that expresses CD4 on its cell surface and secretes cytokines that modulate B-cell isotype, or class, switching.

heme An organic molecule containing a ring of conjugated double bonds surrounding an iron atom. It is involved in redox reactions and oxygen binding.

hemolysin A toxin that lyses red blood cells.

hepatitis An inflammation of the liver, caused by infection or by exposure to a toxic substance.

hepatitis A virus (HAV) A single-stranded RNA picornavirus that causes an acute infection of the liver spread person-to-person by the fecal-oral route.

hepatitis B virus (HBV) A partially double-stranded circular DNA virus hepadnavirus that causes diseases of the liver of varying severity, including acute and chronic hepatitis, cirrhosis, and hepatocarcinoma.

hepatitis C virus (HCV) A single-stranded positive-sense, linear RNA flavivirus that is transmitted by blood transfusions and causes 90 percent of transfusion-related cases of hepatitis.

heterocyst In filamentous cyanobacteria, a specialized nitrogen-fixing cell that maintains a reducing environment and excludes O_2.

heteroduplex A double-stranded nucleic acid in which the two strands come from different sources.

heterokont A member of the eukaryotic group Heterokonta—protists possessing two flagella of unequal length.

heterolactic fermentation A fermentation reaction in which the products are lactic acid, ethanol, and CO_2.

heterotrophy The use of external sources of organic carbon compounds for biosynthesis.

Hfr strain A high-frequency recombination bacterial strain, caused by the presence of a chromosomally integrated F factor.

highly active antiretroviral therapy (HAART) A three-drug antiretroviral cocktail highly effective at inhibiting the replication of HIV in patients.

histone A protein that binds eukaryotic DNA and compacts chromosomes in nucleosomes.

HIV See **human immunodeficiency virus**.

Holliday junction A cross-like configuration of recombining DNA molecules that forms during generalized recombination.

holobiont An entity composed of multiple types of organisms, including microbes.

homolog 1. Also called *homologous gene*. A gene derived from a common ancestral gene. Homologs may be orthologs or paralogs. 2. A protein whose amino acid sequence is similar to that of another protein.

hopanoid *or* **hopane** A five-ringed hydrocarbon lipid found in bacterial cell membranes.

horizontal gene transfer *or* **horizontal transfer** Also called *lateral gene transfer*. The passage of genes from one genome into another, nonprogeny genome.

horizontal transmission In disease, the transfer of a pathogen from one organism into another, nonprogeny organism.

hormogonium *pl.* **hormogonia** A short, motile chain of three to five cells produced by filamentous cyanobacteria to disseminate their cells.

host factor A trait of an individual host that affects susceptibility to disease, in comparison with other individuals.

host range The species that can be infected by a given pathogen.

HSP See **heat-shock protein**.

human immunodeficiency virus (HIV) A human-specific retrovirus that causes AIDS.

humic material Also called *humus*. Phenolic molecules, derived from lignin, that are resistant to degradation and hence very stable in soil.

humoral immunity A type of adaptive immunity mediated by antibodies.

humus See **humic material**.

hybridization The annealing of a nucleic acid strand with another nucleic acid strand containing a complementary sequence of bases. The binding of one nucleic acid strand with a complementary strand.

hydric soil Soil that undergoes periods of anoxic water saturation.

hydrogen bond An electrostatic attraction between a hydrogen bonded to an oxygen or nitrogen and a second, nearby oxygen or nitrogen.

hydrogenotrophy The use of molecular hydrogen (H_2) as an electron donor for a variety of electron acceptors.

hydrologic cycle Also called *water cycle*. The cyclic exchange of water between atmospheric water vapor and Earth's bodies of liquid water.

hydrolysis The cleaving of a bond by the addition of a water molecule.

hydrophilic Soluble in water; either ionic or polar.

hydrophobic Insoluble in water; nonpolar.

hydrothermal vent Also called *thermal vent*. An opening in the seafloor through which superheated water arises, carrying high concentrations of reduced minerals such as sulfides.

3-hydroxypropionate cycle A carbon fixation process in which hydrated CO_2 condenses with acetyl-CoA to form 3-hydroxypropionate.

hyperthermophile An organism adapted for optimal growth at extremely high temperatures, generally above 80°C, and as high as 121°C.

hypertonic   Having more solutes than another environment separated by a semipermeable membrane. Water will tend to flow toward the hypertonic solution.

hypha *pl.* **hyphae**   The threadlike filament that forms the mycelium of a fungus.

hypotonic   Having fewer solutes than another environment separated by a semipermeable membrane. Water will tend to flow away from the hypotonic solution.

hypoxia   A state of lower-than-normal oxygen concentration.

I

icosahedral capsid   For a virus, a crystalline protein shell with 20 identical faces, enclosing the nucleic acid.

ID$_{50}$ See **infectious dose 50%**.

identification   The recognition of the species (or higher taxonomic category) of a microbe isolated in pure culture.

idiotype   An amino acid difference in the antigen-binding site (N terminus of heavy or light chains) that distinguishes different antibodies within an individual.

IEF   See **isoelectric focusing**.

IEL   See **intraepithelial lymphocyte**.

IgA   An antibody isotype that contains the alpha heavy chain. It can be secreted and found in tears, saliva, breast milk, and so on.

IgD   An antibody isotype that contains the delta heavy chain. It is found on B-cell membranes.

IgE   An antibody isotype that contains the epsilon heavy chain. It is involved in degranulation of mast cells.

IgG   An antibody isotype that contains the gamma heavy chain. It is found in serum.

IgM   The first antibody isotype detected during the early stages of an immune response. It contains the mu heavy chain and is found as a pentamer in serum.

IL-1   See **interleukin 1**.

ILC   See **innate lymphoid cell**.

immediate hypersensitivity   See **type I hypersensitivity**.

immersion oil   An oil with a refractive index similar to glass that minimizes light ray loss at wide angles, thereby minimizing wavefront interference and maximizing resolution.

immune system   An organism's cellular defense system against pathogens.

immunity   A body's resistance to a specific disease.

immunization   The stimulation of an immune response by deliberate inoculation with a weakened pathogen, in hopes of providing immunity to disease caused by the pathogen.

immunogen   An antigen that, by itself, can elicit antibody production.

immunogenicity   A measure of the effectiveness of an antigen in eliciting an immune response.

immunoglobulin   A member of a family of proteins that contain a 110-amino-acid domain with an internal disulfide bond. Members include antibodies and major histocompatibility proteins.

immunological specificity   The ability of antibodies produced in response to a particular epitope to bind that epitope almost exclusively. Antibodies made to one epitope bind only weakly, if at all, to other epitopes.

immunopathogenesis   The process by which an immune response or the products of an immune response cause disease.

immunoprecipitation   The antibody-mediated cross-linking of antigens to form large, insoluble complexes. Immunoprecipitation is used in research labs and is normally seen only in vitro.

in vivo expression technology (IVET)   Techniques that identify bacterial genes that are expressed only when the organism is growing inside a host.

incidence   The number of new cases of a disease in a given location over a specified time.

index case   Also called *patient zero*. The first case of an infectious disease, and an important piece of data for helping contain the spread of disease.

indigenous microbiota *or i***ndigenous microflora**   Microbes found naturally in a particular location, often in association with a food substrate.

inducer   A molecule that binds to a repressor and prevents it from binding to the operator sequence.

inducer exclusion   The ability of glucose to cause metabolic changes that prevent the cellular uptake of less favorable carbon sources that could cause unnecessary induction.

induction   Increased transcription of target genes because an inducer binds to a repressor and prevents repressor-operator binding.

industrial fermentor   The equipment used to grow microbes on an industrial scale.

industrial microbiology   The commercial exploitation of microbes.

industrial strain   A microbial strain whose characteristics are optimized for industrial use.

infection   The growth of a pathogen or parasite in or on a host.

infection cycle   The route a pathogen takes as it moves from one host into another.

infectious dose 50% (ID$_{50}$)   The number of bacteria or virions required to cause disease symptoms in 50 percent of an experimental group of hosts.

inflammasome   A cytoplasmic multiprotein complex that promotes the maturation of inflammatory cytokines IL-1β and IL-18. Inflammasome assembly is triggered by NLR interactions with microbe-associated molecular patterns (MAMPs).

information flow   The passing of genetic information from DNA to DNA (replication), or from DNA to RNA (transcription) to protein (translation).

injera   A highly fermented Ethiopian flat bread using the grain teff.

innate immunity   Also called *nonadaptive immunity*. Nonspecific mechanisms for protecting against pathogens.

innate lymphoid cell (ILC)   A lymphocyte-like cell in the intestinal lamina propria that lacks B- or T-cell receptors but secretes pro-inflammatory cytokines.

inner leaflet   The layer of the cell membrane phospholipid bilayer that faces the cytoplasm.

inner membrane   In Gram-negative bacteria, the membrane in contact with the cytoplasm, equivalent to the cell membrane.

insertion   The addition of nucleotides to the middle of a DNA sequence.

insertion sequence (IS)   A simple transposable element consisting of a transposase gene flanked by short, inverted-repeat sequences that are the target of transposase.

integral protein   See **transmembrane protein**.

integron   A large and complex mobile element that can accept, or capture, different antibiotic resistance gene cassettes to use later, when needed.

interactome   All of the proteins that interact with other proteins in a cell.

interference   The interaction of two wavefronts. Interference can be additive (amplitudes in phase, constructive) or subtractive (amplitudes out of phase, destructive).

interferon   A host-secreted immunomodulatory protein that inhibits viral replication.

interleukin 1 (IL-1)   A cytokine release by macrophages.

intermediate filament   A eukaryotic cytoskeletal protein that is composed of various proteins depending on the cell type.

internal ribosome entry site (IRES)   A site within an mRNA sequence where a ribosome can bind and initiate translation.

intimin   A pathogenic *E. coli* adhesion protein that binds tightly to an *E. coli*–produced receptor injected into host cells.

intracellular pathogen   A pathogen that lives within a host cell.

intraepithelial lymphocyte (IEL)   A lymphocyte embedded among epithelial cells that line the intestine.

intron   In eukaryotic genes, an intervening sequence that does not code for protein and is spliced out of the mRNA prior to translation.

inversion   A mutation in which a DNA fragment is flipped within a chromosome. It may allow or repress the transcription of a particular gene.

inverted repeat   A DNA sequence that is found in an identical but inverted form at two sites on the same double helix (e.g., 5′-ATCGATCGnnnnnnCGATCGAT-3′).

ion   An atom or molecule containing negative or positive charge—that is, a number of electrons, respectively, greater than or less than the number of protons.

ion gradient   A difference in concentration of an ion across a membrane.

ionic bond   A chemical bond between ions of positive and negative charge.

IRES   See **internal ribosome entry site**.

IS   See **insertion sequence**.

isoelectric focusing (IEF)   A technique that separates proteins according to their charge, via the migration of proteins to their isoelectric point in a pH gradient.

isoelectric point   The pH at which there is no net charge on an amino acid or a protein.

isolate   A microbe that has been obtained from a specific location and grown in pure culture.

isoprenoid   A condensed isoprene chain, found in archaeal membrane lipids.

isotonic   Being in osmotic balance, having equal concentrations of solutes on both sides of a semipermeable membrane. A cell in an isotonic environment will neither gain nor lose water.

isotope   An atom of an element with a specific number of neutrons. For example, carbon-12 (^{12}C), carbon-13 (^{13}C), and carbon-14 (^{14}C) are all isotopes of carbon.

isotope ratio   The ratio of amounts of two different isotopes of an element. May serve as a biosignature if the ratio between certain isotopes of a given element is altered by biological activity.

isotype   An antibody class within a species that is defined by the structure of the antibody's heavy chain. IgG, IgA, and IgE are examples of isotypes. An isotype (IgG, for example) from one species contains species-specific amino acid sequences present in the heavy chain of all members of that species.

isotype switching   Also called *class switching*. A change in the predominant antibody isotype produced by a cell.

IVET   See **in vivo expression technology**.

J

joule (J)   The standard SI unit for energy.

K

kelp   Also known as *brown alga*. A heterokont protist that grows as multicellular sheets at the water's surface.

kimchi   A popular Korean food based on brine-fermented cabbage.

Kirby-Bauer assay   A method for determining antibiotic susceptibility. Antibiotic-impregnated disks are placed on an agar plate whose surface has been confluently inoculated with a test organism. The antibiotic diffuses away from the disk and inhibits growth of susceptible bacteria. The width of the inhibitory zone is proportional to the susceptibility of the organism.

knockout mutation   A mutation that completely eliminates the activity of a gene product.

Koch's postulates   Four criteria, developed by Robert Koch, that should be met for a microbe to be designated the causative agent of an infectious disease.

Krebs cycle   See **tricarboxylic acid cycle**.

L

labile toxin (LT)   An *E. coli* enterotoxin, destroyed by heat, that increases cellular cAMP concentrations.

laboratory evolution   See **experimental evolution**.

lactic acid fermentation *or* **lactate fermentation**   A fermentation reaction that generates lactic acid from reduction of pyruvic acid.

lag phase   A period of cell culture, occurring right after bacteria are inoculated into new media, during which there is slow growth or no growth.

laminar flow biological safety cabinet   An air filtration appliance that removes pathogenic microbes from within the cabinet.

Langerhans cell   A specialized, phagocytic dendritic cell that is the predominant cell type in skin-associated lymphatic tissue.

late gene   A viral gene expressed late in the infectious cycle.

late-phase anaphylaxis   Anaphylaxis caused by leukotrienes released by eosinophils recruited by mast cells.

latent period   The time in the viral life cycle when progeny virions have formed but are still within the host cell.

latent state   A period of the infection process during which a pathogenic agent is dormant in the host and cannot be cultured.

lateral gene transfer   See **horizontal gene transfer**.

law of mass action   The tendency of a reaction at equilibrium to return to equilibrium after the concentrations of products or reactants are altered.

LD$_{50}$ See **lethal dose 50%**.

leaflet   One of the two lipid layers in a phospholipid bilayer. The inner leaflet of the cell membrane faces the cytoplasm.

leaven   For bread dough, to cause to rise by generating air spaces, usually through carbon dioxide production by microbial fermentation.

leghemoglobin   An iron-bearing plant protein that sequesters oxygen to maintain an anoxic environment for nitrogenase within cells containing bacteroids.

lens   An object composed of transparent, refractile material that bends light rays to converge at a focal point (or to diverge from an imaginary point). For electron microscopy, a series of magnets arranged as a magnetic lens bends electron beams to converge or diverge.

lentivector   Lentiviral vector, a gene transfer vector derived from a lentivirus such as HIV; designed to integrate genes into a host chromosome.

lentivirus A member of a family of retroviruses with a long incubation period. An example is HIV.

lethal dose 50% (LD$_{50}$) A measure of virulence; the number of bacteria or virions required to kill 50 percent of an experimental group of hosts.

lethal factor (LF) A component of anthrax toxin that cleaves host protein kinases.

leukocidin A toxin that lyses white blood cells.

LF See **lethal factor**.

lichen A simple multicellular organism formed by a mutualistic relationship between a fungus and an alga or cyanobacterium.

light chain The smaller of the two protein types that make up an antibody. Each antibody contains two heavy chains and two light chains.

light microscopy (LM) Observation of a microscopic object based on light absorption and transmission.

lignin A complex aromatic organic compound that forms the key structural support for trees and woody stems.

limiting nutrient A nutrient whose depletion generally restricts growth in a given ecosystem.

lincosamide Any of a class of bacteriostatic antibiotics (e.g., clindamycin).

lipid A The anchor lipid of lipopolysaccharide (LPS), composed of glucosamine plus six lipid chains.

lipopolysaccharides (LPS) Structurally unique phospholipids found in the outer leaflet of the outer membrane in Gram-negative bacteria. Many are endotoxins.

lithotrophy Also called *chemolithotrophy*. The metabolic oxidation of inorganic compounds to yield energy and fix single-carbon compounds into biomass.

littoral zone The upper water layer of aquatic habitats that light can penetrate.

LM See **light microscopy**.

logarithmic (log) phase See **exponential phase**.

long terminal repeat (LTR) A repeated nucleic acid sequence at the 5′ and 3′ ends of a provirus.

loss-of-function mutation A mutation that eliminates or decreases the function of the gene product.

LPS See **lipopolysaccharides**.

LT See **labile toxin**.

LTR See **long terminal repeat**.

Lyme disease One form of borreliosis. A tick-borne disease caused by *Borrelia burgdorferi*, which may involve skin lesions and arthritis.

lymph node A secondary lymphatic organ, formed by the convergence of lymphatic vessels, that traps foreign particles from local tissue and presents them to resident immune cells.

lymphocyte A mononuclear leukocyte (white blood cell) that is a product of lymphoid tissue and participates in immunity (e.g., B cell and T cell).

lyophilization See **freeze-drying**.

lysate The contents of broken cells; may include virus particles.

lysis Also called *burst*. The rupture of the cell by a break in the cell wall and membrane.

lysogeny A viral life cycle in which the viral genome integrates into and replicates with the host genome, but retains the ability to initiate host cell lysis.

lysosome An acidic eukaryotic organelle that aids digestion of molecules. Not found in plant cells.

M

M cell A phagocytic innate immune cell (microfold cell) found between intestinal epithelial cells.

MAC See **membrane attack complex**.

macrolide Any of a group of antibiotics containing a large lactone ring (e.g., erythromycin).

macronucleus A form of nucleus found in ciliates that is derived from gene amplification and rearrangement of micronuclear DNA; contains actively transcribed genes.

macronutrient A nutrient that an organism needs in large quantity.

macrophage A mononuclear, phagocytic, antigen-presenting cell of the immune system.

magnetosome An organelle containing the mineral magnetite that allows microbes to sense a magnetic field.

magnetotaxis The ability to direct motility along magnetic field lines.

magnification An increase in the apparent size of a viewed object as an optical image.

major histocompatibility complex (MHC) Transmembrane cell proteins important for recognizing self and for presenting foreign antigens to the adaptive immune system.

malaria A disease caused by the apicomplexan *Plasmodium falciparum*, transmitted by mosquitoes.

malolactic fermentation Fermentation of l-malate (a side product of glucose fermentation) by *Oenococcus oeni* bacteria; an important process in winemaking.

MAMP See **microbe-associated molecular pattern**.

marine snow Microbial biofilms on inorganic particles suspended in marine water.

mast cell A white blood cell that secretes proteins that aid innate immunity. Mast cells reside in connective tissues and mucosa and do not circulate in the bloodstream.

matrix protein A protein, found in some viruses, that is located between the capsid and the membrane envelope.

MCP See **methyl-accepting chemotaxis protein**.

MDR efflux pump See **multidrug resistance efflux pump**.

mechanical transmission A nonspecific mode of viral entry into damaged tissue.

meiosis A form of cell division by which a diploid eukaryotic cell generates haploid sex cells that contain recombinant chromosomes.

membrane attack complex (MAC) A cell-destroying pore produced in the membrane of invading bacteria by the host cell complement cascade.

membrane-permeant weak acid *or* **membrane-permeant organic acid** An acid that exists in charged and uncharged forms, such as acetic acid. The uncharged form can penetrate the membrane.

membrane-permeant weak base A base that exists in charged and uncharged forms, such as methylamine. The uncharged form can penetrate the membrane.

membrane potential Energy stored as an electrical voltage difference across a membrane.

memory B cell A long-lived type of lymphocyte preprogrammed to produce a specific antibody. After encountering their activating antigen, memory B cells differentiate into antibody-producing plasma cells.

merozoite The form of *Plasmodium falciparum*, the causative agent of malaria, that invades red blood cells.

mesocosm A small, controlled model ecosystem.

mesophile An organism with optimal growth between 20°C and 40°C.

messenger RNA (mRNA) An RNA molecule that encodes a protein.

metagenome The sum of genomes of all members of a community of organisms.

metagenomics The study of community genomes, or metagenomes.

metaproteomics The study of all the proteins synthesized by members of a community, known as a metaproteome.

metastatic lesion A lesion of infection, or of cancerous cells, that develops at secondary sites away from the initial site of infection.

metatranscriptomics The study of all the RNA transcripts expressed by members of a community, known as a metatranscriptome.

methane gas hydrate A crystalline material in which methane molecules are surrounded by a cage of water molecules. This molecular configuration is found in the deep ocean.

methanogen An organism that uses hydrogen to reduce CO_2 and other single-carbon compounds to methane, yielding energy.

methanogenesis An energy-yielding metabolic process that releases methane, commonly from hydrogen gas and oxidized one- or two-carbon compounds. It is unique to archaea.

methanotrophy The metabolic oxidation of methane to yield energy.

methyl-accepting chemotaxis protein (MCP) Also called *chemoreceptor*. A cell-membrane signal transduction protein that becomes methylated during adaptation to a chemotactic signal.

methyl mismatch repair A DNA repair system that fixes misincorporation of a nucleotide after DNA synthesis. The unmethylated daughter strand is corrected to complement the methylated parental strand.

methylotrophy The metabolic oxidation of single-carbon compounds such as methanol, methylamine, or methane to yield energy.

MHC See **major histocompatibility complex**.

MHC restriction The ability of T cells to recognize only those antigens complexed to self MHC molecules.

MIC See **minimal inhibitory concentration**.

microaerophilic Requiring oxygen at a concentration lower than that of the atmosphere, but unable to grow in high-oxygen environments.

microbe An organism or virus too small to be seen with the unaided human eye.

microbe-associated molecular pattern (MAMP) Formerly called *pathogen-associated molecular pattern (PAMP)*. Molecules associated with groups of microbes, both pathogenic and nonpathogenic, that are recognized by cells of the innate immune system.

microbial mat A complex biofilm of microbes, usually containing multiple layers.

microbiome *or* **microbiota** The total community of microbes found within a specified environment.

microfilament Also called *actin filament*. A eukaryotic cytoskeletal protein composed of polymerized actin.

microfossil A microscopic fossil in which calcium carbonate deposits have filled in the form of ancient microbial cells.

micronucleus A form of nucleus found in ciliates; contains a diploid set of chromosomes and undergoes meiosis for sexual exchange by conjugation.

micronutrient A nutrient that an organism needs in small quantity, typically a vitamin or a mineral.

microplankton Plankton consisting of microbes approximately 20–1,000 μm in diameter.

microscope A tool that increases the magnification of specimens to enable viewing at higher resolution.

microtubule A eukaryotic cytoskeletal protein composed of polymerized tubulin.

minimal defined medium A solution of chemically defined compounds for organismal growth that contains only the minimal components required for growth.

minimal inhibitory concentration (MIC) The lowest concentration of a drug that will prevent the growth of an organism.

miso A Japanese condiment, made from ground soy and rice, salted and fermented by the mold *Aspergillus oryzae*.

missense mutation A point mutation that alters the sequence of a single codon, leading to a single amino acid substitution in a protein.

mitochondrion *pl.* **mitochondria** An organelle of endosymbiotic origin that produces ATP through the use of an electron transport system to generate a proton potential. O_2 is the final electron acceptor to produce H_2O.

mitosis The orderly replication and segregation of eukaryotic chromosomes, usually prior to cell division.

mitosome An organelle derived from mitochondria, found in certain protists, such as *Giardia*, that lack mitochondria.

mitosporic fungus A species of fungus that generates spores by mitosis and lacks a known sexual cycle.

mixed-acid fermentation A bacterial fermentation process in which pyruvate is converted to several different organic acids, as well as ethanol, CO_2, and H_2O.

mixotrophy Metabolism that includes CO_2 fixation through photosynthesis and catabolism of organic compounds.

modular enzyme A multifunctional enzyme in which several domains or subunits conduct sequential steps to generate a product.

MOI See **multiplicity of infection**.

molarity A unit of concentration measured as the number of moles of solute per liter of solution.

molecular clock The use of DNA or RNA sequence information to measure the time of divergence among different species.

molecular formula A notation indicating the number and type of atoms in a molecule. For example, H_2O is the molecular formula for water.

Mollicutes See **Tenericutes**.

monocyte A white blood cell with a single nucleus that can differentiate into a macrophage or a dendritic cell.

monophyletic Having a single evolutionary origin—that is, diverging from a common ancestor.

monophyletic group See **clade**.

monosaccharide The monomer unit of sugars. Monosaccharides have a molecular formula of $(CH_2O)_n$.

mordant A chemical binding agent that causes specimens to retain stains better.

mother cell The larger cell that forms during the asymmetrical cell division leading to spore formation. The mother cell will engulf the forespore.

motility The ability of a microbe to generate self-directed movement.

mRNA See **messenger RNA**.

mucociliary escalator The ciliated mucous lining of the trachea, bronchi, and bronchioles that sweeps foreign particles up and away from the lungs.

mucosal immunity The portion of the innate and adaptive immune systems that protects the mucosa from microbial invasion.

Mueller-Hinton agar A specialized, standardized, *para*-aminobenzoic acid–free medium used for the Kirby-Bauer assay.

multidrug resistance (MDR) efflux pump A transmembrane protein pump that can export many different kinds of antibiotics with diverse structure.

multiplex PCR A polymerase chain reaction that uses multiple pairs of oligonucleotide primers to amplify several different DNA sequences simultaneously.

multiplicity of infection (MOI) The ratio of infecting virions to host cells.

murein See **peptidoglycan**.

murein lipoprotein Also called *Braun lipoprotein*. The major lipoprotein that connects the outer membrane of Gram-negative bacteria to the peptidoglycan cell wall.

mutagen A chemical that damages DNA and increases the rate of mutations.

mutation A heritable change in a DNA sequence.

mutation frequency The fraction of mutant cells (defective in a given gene) within the total cell population.

mutation rate The number of mutations introduced into DNA per generation (cell doubling).

mutator strain A strain of cells with a high mutation rate, usually due to a mutation in a DNA repair enzyme.

mutualism A symbiotic relationship in which both partners benefit.

mycelium *pl.* **mycelia** A single mass of fungal hyphae that projects into the air (aerial mycelium) or into the growth substrate (surface mycelium).

mycolic acid One of a diverse class of sugar-linked fatty acids found in the cell envelopes of mycobacteria such as *Mycobacterium tuberculosis*.

mycology The study of fungi.

mycorrhizae *sing.* **mycorrhiza** Fungi involved in an intimate mutualism with plant roots, in which nutrients are exchanged.

myxospore A durable spherical cell produced by the fruiting body of myxobacteria.

N

N-terminal rule The tendency of the N-terminal amino acid of a protein to influence protein stability.

NAD See **nicotinamide adenine dinucleotide**.

nanoscale secondary ion mass spectrometry (Nano-SIMS) A technique of chemical imaging in which an ionizing beam breaks off organic ions from a sample, which fly off the sample and are captured for analysis by a mass spectrometer.

nasopharynx The passage leading from the nose to the oral cavity.

native conformation The fully folded, functional form of a protein.

natto A soybean product, similar to tempeh, produced by alkaline fermentation.

natural killer (NK) cell A lymphocyte that does not need antigen stimulation to kill tumor or infected host cells by inserting granules containing perforin.

natural selection The change in frequency of genes in a population under environmental conditions that favor some genes over others.

necrotizing fasciitis Also known as *flesh-eating disease*. A severe skin infection usually caused by the Gram-positive coccus *Streptococcus pyogenes*.

negative selection In immunology, the destruction of T cells bearing T-cell receptors that bind strongly to self MHC proteins displayed on thymus epithelial cells.

negative stain A stain that colors the background and leaves the specimen unstained.

NER See **nucleotide excision repair**.

NET See **neutrophil extracellular trap**.

neuraminidase inhibitor Any of a class of anti-influenza drugs that target neuraminidase on the viral envelope and decrease the number of virus particles produced.

neutralophile An organism with an optimal growth range in environments between pH 5 and pH 8.

neutrophil A white blood cell of the innate immune system that can phagocytose and kill microbes.

neutrophil extracellular trap (NET) A net of chromatin and antimicrobial peptides expelled by dying neutrophils to trap and injure nearby pathogenic bacteria.

NHEJ See **nonhomologous end joining**.

niche An organism's environmental requirements for existence and its relations with other members of the ecosystem.

niche construction The actions of an organism that alter its environmental niche and change its chance of survival in that niche.

nicotinamide adenine dinucleotide (NAD) An energy carrier in the cell that can donate (NADH) or accept (NAD^+) electrons.

nitrification The oxidation of reduced nitrogen compounds to nitrite or nitrate.

nitrifier An organism that converts reduced nitrogen compounds to nitrite or nitrate.

nitrogen fixation The ability of some prokaryotes to reduce inorganic diatomic nitrogen gas (N_2) to two molecules of ammonium ion ($2NH_4^+$).

nitrogen-fixing bacterium A bacterium that can reduce diatomic nitrogen gas (N_2) to two molecules of ammonium ion (NH_4^+).

nitrogenase The enzyme that catalyzes nitrogen fixation.

nitrogenous base See **nucleobase**.

Nitrospirae A phylum of Gram-negative bacteria, many of which are lithotrophs, oxidizing nitrite to nitrate, or ammonia to nitrate.

NK cell See **natural killer cell**.

NLR See **NOD-like receptor**.

node In phylogenetic trees, the most recent common ancestor of branching descendents.

NOD-like receptor (NLR) A eukaryotic cytoplasmic protein that recognizes particular microbe-associated molecular patterns (MAMPs) present on microorganisms.

noise Variance from a mean in an assay. A system in which individual results vary greatly from a mean that was derived from many results is a noisy system.

Nomarski microscopy See **differential interference contrast microscopy**.

nomenclature The naming of different taxonomic groups of organisms.

nonadaptive immunity See **innate immunity**.

nonhomologous end joining (NHEJ) A pathway that repairs double-strand DNA breaks by direct ligation without the need for large regions of homology.

nonpolar covalent bond A covalent bond in which the electrons in the bond are shared equally by the two atoms.

nonribosomal peptide antibiotic A peptide with antimicrobial activity synthesized by modular enzymes and not by ribosomes.

nonsense mutation A mutation that changes an amino acid codon into a premature stop codon.

nori A Japanese food obtained from the red algae *Porphyra* species.

norovirus Also known as *Norwalk virus*. A nonenveloped ssRNA virus that causes severe diarrhea in children and adults.

northern blot A technique to detect specific RNA sequences. Sample RNA is subjected to gel electrophoresis, transferred to a blot, and probed with a labeled cDNA that will hybridize to target RNA sequences.

Norwalk virus See **norovirus**.

nosocomial Hospital-acquired; commonly refers to an infectious agent.

NP See **nucleocapsid protein**.

nucleobase Also called *nitrogenous base*. A planar, heteroaromatic nitrogen-containing base that forms a nucleotide of nucleic acids; it determines the information content of DNA and RNA.

nucleocapsid protein (NP) A protein that coats a viral genome.

nucleoid The looped coils of a bacterial chromosome.

nucleolus *pl.* **nucleoli** A region inside the nucleus where ribosome assembly begins.

nucleomorph A vestigial nucleus within a eukaryotic cell, evolved by genetic reduction from the nucleus of an endosymbiont.

nucleotide The monomer unit of nucleic acids, consisting of a five-carbon sugar, a phosphate, and a nitrogenous base.

nucleotide excision repair (NER) A DNA repair mechanism that cuts out damaged DNA. New, correctly base-paired DNA is synthesized by DNA polymerase I.

nucleus *pl.* **nuclei** A eukaryotic organelle that contains the DNA.

numerical aperture The product of the refractive index of the medium and sin θ. As numerical aperture increases, the magnification increases.

O

O antigen *or* **O polysaccharide** A sugar chain that connects to the core polysaccharide of lipopolysaccharides.

objective lens In a compound microscope, the lens that is closest to the specimen and generates the initial magnification.

obligate intracellular pathogen A pathogen that can replicate only inside host cells.

ocular lens Also called *eyepiece*. In a compound microscope, the lens situated closest to the observer's eye.

Okazaki fragments Short fragments of DNA that are synthesized on the lagging strand during DNA synthesis.

oligotrophic lake A lake having low concentration of organic nutrients; the opposite of a eutrophic lake.

OMZ See **oxygen minimum zone**.

oncogenic virus A virus that causes cancer.

oomycete A member of the eukaryotic group Oomycetes—heterokont protists whose life cycle resembles that of fungi; formerly classified as fungi (Oomycota).

open reading frame (ORF) A DNA sequence predicted to encode a protein.

operational taxonomic unit (OTU) A taxonomic group defined by a designated degree of similarity among members based on DNA sequence.

operon A collection of genes that are in tandem on a chromosome and are transcribed into a single RNA.

opportunistic pathogen A microbe that normally is not pathogenic but can cause infection or disease in an immunocompromised host organism.

opsonin An antibody that renders its target (e.g., bacteria) susceptible to phagocytosis.

opsonization The coating of pathogens with antibodies that aid pathogen phagocytosis by innate immune cells.

opsonize To bind IgG antibodies to microbes in order to enhance microbial phagocytosis by host immune cells.

optical density A measure of how many particles are suspended in a solution, based on light scattering by the suspended particles.

optical isomer Also called *enantiomer*. Either of two forms of a molecule that are mirror images of each other. Molecules that contain a chiral carbon can have optical isomers.

oral groove A ciliate structure for food uptake.

ORF See **open reading frame**.

organelle A membrane-enclosed compartment within eukaryotic cells that serves a specific function.

organic molecule A molecule that contains a carbon-carbon bond.

organism An individual form of life.

organotrophy Also called *chemoorganotrophy* or *chemoheterotrophy*. The metabolic oxidation of organic compounds to yield energy without absorption of light.

origin (oriC) The region of a bacterial chromosome where DNA replication initiates.

oropharynx The area between the soft palate and the upper edge of the epiglottis.

ortholog *or* **orthologous gene** A gene present in more than one species that derived from a common ancestral gene and encodes the same function.

osmolarity A measure of the number of solute molecules in solution.

osmosis The diffusion of water from regions of high water concentration (low solute) to regions of low water concentration (high solute) across a semipermeable membrane.

osmotic pressure Pressure exerted by the osmotic flow of water through a semipermeable membrane.

OTU See **operational taxonomic unit**.

outer leaflet The layer of the cell membrane phospholipid bilayer that faces away from the cytoplasm.

outer membrane In Gram-negative bacteria, a membrane external to the cell wall.

oxazolidinone One of a class of synthetic antibiotics that inhibit protein synthesis.

oxidative burst A large increase in the oxygen consumption of immune cells during phagocytosis of pathogens as the immune cells produce oxygen radicals to kill the pathogen.

oxidative phosphorylation An electron transport chain that uses diatomic oxygen as a final electron acceptor and generates a proton gradient across a membrane for the production of ATP via ATP synthase.

oxidoreductase An electron transport system protein that accepts electrons from one molecule (oxidizing that molecule), and donates electrons to a second molecule, thereby reducing the second molecule.

oxygen minimum zone (OMZ) The region of the marine water column in which oxygen is depleted by respiration; usually at a mid level, between the aerated upper water and the deeper oxygenated water where organic food is scarce.

oxygenic Z pathway An ATP-producing photosynthetic pathway consisting of photosystems I and II. Water serves as the initial electron donor (generating O_2), and $NADP^+$ is the final electron acceptor (generating NADPH).

P

P site See **peptidyl-tRNA site**.

PA See **protective antigen**.

palindrome A DNA sequence in which the top and bottom strands have the same sequence in the 5′-to-3′ direction.

PAMP See **microbe-associated molecular pattern**.

pandemic An epidemic that occurs over a wide geographic area.

pan-genome All the genes possessed by all individual members of a species.

panspermia The hypothesis that life-forms originated elsewhere in the universe and "seeded" life on Earth.

paralog *or* **paralogous gene** A gene that arises by gene duplication within a species and evolves to carry out a different function from that of the original gene.

pararetrovirus Also called *DNA reverse-transcribing virus.* A virus with a double-stranded DNA genome that generates an RNA intermediate and thus requires reverse transcriptase to generate progeny DNA genomes.

parasite Any bacterium, virus, fungus, or protozoan (protist) that colonizes and harms its host; the term commonly refers to protozoa and to invertebrates.

parasitism A symbiotic relationship in which one member benefits and the other is harmed.

parenteral route of transmission A method by which an infectious agent enters the body via injection into the bloodstream, often by a mosquito or other insect.

parfocal In a microscope with multiple objective lenses, having the objective lenses set at different heights that maintain focus when switching among lenses.

passive transport Net movement of molecules across a membrane without energy expenditure by the cell.

pasteurization The heating of food at a temperature and time combination that will kill spore-like structures of *Coxiella burnetii.*

pathogen A bacterial, viral, or fungal agent of disease.

pathogen-associated molecular pattern (PAMP) See **microbe-associated molecular pattern**.

pathogenesis The processes through which microbes cause disease in a host.

pathogenicity The ability of a microorganism to cause disease.

pathogenicity island A type of genomic island in which the stretch of DNA contains virulence factors and may have been transferred from another genome.

patient zero See **index case**.

pattern recognition receptor (PRR) A protein receptor that recognizes microbe-associated molecular patterns (MAMPs) and signals production of cytokines.

PBP See **penicillin-binding protein**.

PCM See **phase-contrast microscopy**.

PCR See **polymerase chain reaction**.

PDC See **pyruvate dehydrogenase complex**.

pelagic zone The water column of the open ocean, away from the shore and the ocean floor.

penicillin An antibiotic, produced by the *Penicillium* mold, that blocks cross-bridge formation during peptidoglycan synthesis.

penicillin-binding protein (PBP) A bacterial protein, involved in cell wall synthesis, that is the target of the penicillin antibiotic.

pentose phosphate cycle See **Calvin cycle**.

pentose phosphate pathway (PPP) *or* **pentose phosphate shunt (PPS)** An alternate glycolytic pathway in which glucose 6-phosphate is first oxidized and then decarboxylated to ribulose 5-phosphate, ultimately generating 1 ATP and 2 NADPH.

peptide bond The covalent bond that links two amino acid monomers.

peptidoglycan Also called *murein.* A polymer of peptide-linked chains of amino sugars; a major component of the bacterial cell wall.

peptidyltransferase The rRNA enzymatic ability to form peptide bonds.

peptidyl-tRNA site (P site) The ribosomal site that contains the growing protein attached to a tRNA.

perforin A cytotoxic protein, secreted by T cells, that forms pores in target cell membranes.

peripheral membrane protein A protein that is associated with a membrane but does not span the phospholipid bilayer.

periplasm In Gram-negative bacteria, the gel-like solution between the outer and inner membrane; contains the cell wall.

permease A substrate-specific carrier protein in the membrane.

peroxisome A eukaryotic organelle that converts hydrogen peroxide to water.

persister cell Any of a subpopulation of dormant organisms that arise within a population of antibiotic-susceptible bacteria and are tolerant to bactericidal antibiotics during treatment. On removal of the drug, persisters can grow and reestablish infection.

petechia *pl.* **petechiae** A pinpoint capillary hemorrhage due to the absence of clotting factors; may indicate the presence of endotoxin.

petri dish *or* **petri plate** A round dish with vertical walls covered by an inverted dish of slightly larger diameter. The smaller dish can be filled with a substrate for growing microbes.

PFU See **plaque-forming unit**.

phage See **bacteriophage**.

phage display A technique in which a phage particle contains recombinant coat proteins expressed by genes encoding a coat protein fused to the protein of interest, such as a vaccine antigen.

phagocytosis A form of endocytosis in which a large extracellular particle is brought into the cell.

phagosome A large intracellular vesicle that forms as a result of phagocytosis.

phase-contrast microscopy (PCM) Observation of a microscopic object based on the differences in the refractive index between cell components and the surrounding medium. Contrast is generated as the difference between refracted light and transmitted light shifted out of phase.

phase variation A gene regulatory mechanism that changes the amino acid sequence of a protein from one antigenic type to another. One mechanism involves site-specific recombination that flips a DNA sequence in a chromosome.

phenol coefficient test A test of the ability of a disinfectant to kill bacteria; the higher the coefficient, the more effective the disinfectant.

phosphodiester bond The bond that covalently attaches to adjacent nucleotides in a nucleic acid.

phosphorimaging The visualization of a radioactive probe by recording the energy emanated as light or radioactivity from gels or membranes.

phospholipid The major component of membranes. A typical phospholipid is composed of a core of glycerol to which two fatty acids and a modified phosphate group are attached.

phospholipid bilayer Two layers of phospholipids; the hydrocarbon fatty acid tails face the interior of the bilayer, and the charged phosphate groups face the cytoplasm and extracellular environment. The cell membrane is a phospholipid bilayer.

phosphorylation The enzyme-catalyzed addition of a phosphoryl group onto a molecule.

phosphotransferase system (PTS) A group translocation system that uses phosphoenolpyruvate to transfer phosphoryl groups onto the incoming molecule.

photic zone See **euphotic zone**.

photoautotrophy The fixation of single-carbon compounds into organic biomass, using light as an energy source.

photoexcitation Light absorption that raises an electron to a higher energy state, as in bacteriorhodopsin or in chlorophyll.

photoheterotrophy Metabolism that includes gain of energy from light absorption with biosynthesis from preformed organic compounds. Usually also includes organotrophy, gain of energy from reactions of organic compounds.

photoionization Light absorption that causes electron separation.

photolysis The first energy-yielding phase of photosynthesis; the light-driven separation of an electron from a molecule coupled to an electron transport system.

photoreactivation A light-induced, photolyase-catalyzed repair of pyrimidine dimers.

photosynthesis The metabolic ability to absorb and convert solar energy into chemical energy for biosynthesis. Autotrophic photosynthesis, or photoautotrophy, includes CO_2 fixation.

photosystem I (PS I) A protein complex that harvests light from a chlorophyll, donates an electron to an electron transport system, receives an electron from a small molecule such as H_2S or H_2O, and stores energy in the form of NADPH.

photosystem II (PS II) A protein complex that harvests light from a bacteriochlorophyll, donates an electron to an electron transport system, and stores energy in the form of a proton potential.

phototrophy The use of chemical reactions powered by the absorption of light to yield energy.

phylogenetic tree A diagram depicting estimates of the relative amounts of evolutionary divergence among different species.

phylogeny A measurement of genetic relatedness. The classification of organisms based on their genetic relatedness.

phylum *pl.* **phyla** The taxonomic rank one level below domain; a group of organisms sharing a common ancestor that diverged early from other groups.

phytoplankton Phototrophic marine bacteria, algae, and protists, the primary producers in pelagic food webs.

picornavirus A member of a medically important group of RNA viruses. An example is poliovirus.

piezophile See **barophile**.

pilus *pl.* **pili** Also called *fimbria*. A straight protein filament composed of a tube of protein monomers that extend from the bacterial cell envelope.

pinocytosis A form of endocytosis in which only extracellular fluid and small molecules are brought into the cell.

Planctomycetes A phylum of free-living bacteria that have stalked cells and reproduce by budding. Their nucleoid is surrounded by a membrane.

plankton Organisms that float in water.

planktonic cell An isolated cell, growing individually in a liquid without connections to other cells.

plaque A cell-free zone on a lawn of bacterial cells caused by viral lysis.

plaque-forming unit (PFU) A measure of the concentration of phage particles in liquid culture.

plasma cell A short-lived antibody-producing cell.

plasma membrane See **cell membrane**.

plasmid An extrachromosomal genetic element that may be present in some cells.

plasmodesma *pl.* **plasmodesmata** A membrane channel in plants that connects adjacent plant cells.

plasmodial slime mold A slime mold in which a fertilized zygote undergoes multiple nuclear divisions, generating a multinucleate single cell (plasmodium).

plasmodium *pl.* **plasmodia** The giant, multinucleate cell formed by a plasmodial slime mold.

platelet A small, cell fragment without a nucleus found in blood that is involved in clotting.

pneumonic plague A highly virulent and contagious *Yersinia pestis* lung infection.

point mutation A change in a single nucleotide within a nucleic acid sequence.

polar aging The process, during bacterial cell fission, in which septation forms two new poles while the original poles of the dividing cell age. With each new cell division, the preexisting poles age incrementally. The effects of aging vary among bacterial species.

polar covalent bond A covalent bond in which the electrons in the bond are distributed unequally between two atoms.

poliomyelitis *or* **polio** A paralytic disease caused by the poliovirus.

poliovirus A human-specific RNA virus that is the causative agent of poliomyelitis.

poliovirus receptor (PVR) A cell membrane glycoprotein, also known as CD155, found on cells of the intestinal epithelium and on the surface of motor neurons.

polymerase chain reaction (PCR) A method to amplify DNA in vitro using many cycles of DNA denaturation, primer annealing, and DNA polymerization with a heat-stable polymerase.

polyphyletic Having multiple evolutionary origins.

polyprotein A long peptide translated from one open reading frame but later cleaved into separate proteins with different functions.

polysaccharide A polymer of sugars. See also **glycan**.

polysome A cell structure consisting of multiple ribosomes performing translation on the same mRNA molecule.

population A group of individuals of one species living in a common location.

porin A transmembrane protein complex that allows movement of specific molecules across the cell membrane or the outer membrane.

portal of entry Site through which microorganisms enter the body. These can be natural orifices, cuts, or by injection.

positive selection In immunology, the survival of T cells bearing T-cell receptors that don't recognize self MHC proteins displayed on thymus epithelial cells.

pour plate technique A procedure in which organisms are added to melted agar cooled to 45°C–50°C and the mixture is poured into an empty petri plate. Colonies grow in the agar after the agar solidifies.

PPP See **pentose phosphate pathway**.

PPS See **pentose phosphate pathway**.

prebiotic soup A model for the origin of life based on the abiotic formation of fundamental biomolecules and cell structures such as membranes out of a "soup" of nutrients present on early Earth.

predator A consumer that feeds on grazers.

preliminary mRNA transcript (pre-mRNA) A eukaryotic messenger RNA prior to intron removal.

pre-mRNA See **preliminary mRNA transcript**.

prevalence The total number of active cases of a disease in a given location regardless of when the case first developed.

primary algae Algae that are derived from a single endosymbiotic event; closely related to green plants (Viridiplantae).

primary antibody The first antibody added in an immunological assay—enzyme-linked immunosorbent assay (ELISA) or western blot. This antibody binds to the antigen of interest.

primary antibody response The production of antibodies upon first exposure to a particular antigen. B cells become activated and differentiate into plasma cells and memory B cells.

primary pathogen A disease-causing microbe that can breach the defenses of a healthy host.

primary producer An organism that produces biomass (reduced carbon) from inorganic carbon sources such as CO_2.

primary recovery The initial isolation of commercial product from industrial microbes.

primary structure The first level of organization of polymers, consisting of the linear sequence of monomers—for example, the sequence of amino acids in a protein or nucleotides in a nucleic acid.

primary syphilis The initial inflammatory reaction (chancre) at the site of infection with *Treponema pallidum*.

primase An RNA polymerase that synthesizes short RNA primers complementary to a DNA template to launch DNA replication.

primer extension A technique to determine the 5′ end of an RNA transcript.

prion An infectious agent that causes propagation of misfolded host proteins; usually consists of a defective version of the host protein.

probiotic A food or nutritional supplement that contains live microorganisms and aims to improve health by promoting beneficial bacteria.

programmed cell death Cell death mediated by a regulated intracellular process.

prokaryote An organism whose cell or cells lack a nucleus; includes both bacteria and archaea.

promoter A noncoding DNA regulatory region immediately upstream of a structural gene that is needed for transcription initiation.

proofreading An enzymatic activity of some nucleic acid polymerases that attempts to correct mispaired bases.

prophage A phage genome integrated into a host genome.

propionic acid fermentation The fermentation of lactic acid to propionic acid by *Propionibacterium* species; used in the production of Swiss cheese.

protease inhibitor A molecule that inhibits a protease enzyme; some are used as anti-HIV drugs to block the virally encoded protease needed to complete HIV assembly.

protective antigen (PA) The core subunit of anthrax toxin, so called because immunity to this protein protects against disease.

protein A A *Staphylococcus aureus* cell wall protein that binds to the Fc region of antibodies, hiding the *S. aureus* cells from phagocytes.

protein fusion See **translational fusion**.

Proteobacteria A large, metabolically and morphologically diverse phylum of Gram-negative bacteria.

proteome All the proteins expressed in a cell at a given time. The "complete proteome" includes all the proteins the cell can express under any condition. The "expressed proteome" represents the set of proteins made under a given condition.

proteorhodopsin A bacterial membrane-embedded protein that contains retinal and acts as a light-driven proton pump; it is homologous to the archaeal protein bacteriorhodopsin.

protist A single-celled eukaryotic microbe, usually motile; not a fungus.

proton potential *or* **proton motive force** The potential energy of the concentration gradient of protons (hydrogen ions, H^+) plus the charge difference across a membrane.

protozoan *pl.* **protozoa** A heterotrophic eukaryotic microbe, usually motile, that is not a fungus.

provirus A viral genome that is integrated into the host cell genome.

PRR See **pattern recognition receptor**.

PS I See **photosystem I**.

PS II See **photosystem II**.

pseudogene A gene that is no longer functional.

pseudomurein See **pseudopeptidoglycan**.

pseudopeptidoglycan Also called *pseudomurein*. A peptidoglycan-like molecule composed of sugars and peptides that is found in some archaeal cell walls.

pseudopod A locomotory extension of cytoplasm bounded by the cell membrane.

psychrophile An organism with optimal growth at temperatures below 20°C.

PTS See **phosphotransferase system**.

pure culture A culture containing only a single strain or species of microorganism. A large number of microorganisms that all descended from a single individual cell.

purine A nitrogenous base with fused rings found in nucleotides; examples are adenine and guanine.

putrefaction Food spoilage due to the decomposition of proteins and amino acids.

PVR See **poliovirus receptor**.

pyelonephritis Kidney infection.

pyrimidine A single-ring nitrogenous base found in nucleotides; examples are cytosine, thymine, and uracil.

pyrogen Any substance that induces fever.

pyrosequencing A method of DNA sequencing that relies on the detection of pyrophosphate released upon nucleotide incorporation.

pyruvate dehydrogenase complex (PDC) The multisubunit enzyme that couples the oxidative decarboxylation of pyruvate to acetyl-CoA and NADH production.

Q

qPCR See **quantitative PCR**.

quantitative PCR (qPCR) Also called *real-time PCR*. A technique using fluorescence to detect the products of PCR amplification as the reaction progresses, in order to quantify the amount of DNA in a sample.

quarantine The separation of infectious individuals from the general population to limit the spread of infection.

quasispecies A collection of isolates (usually viruses) from a common source of infection that have evolved into many different types within one host.

quaternary structure The fourth (and highest) level of organization of proteins, in which multiple polypeptide chains interact and function together.

quinol A reduced electron carrier that can diffuse laterally within membranes.

quinolone A type of antibiotic drug that inhibits DNA synthesis by targeting bacterial topoisomerases such as DNA gyrase.

quinone An oxidized electron carrier that can diffuse laterally within membranes.

quorum sensing The ability of bacteria to sense the presence of other bacteria via secreted chemical signals called autoinducers.

R

radial immunodiffusion A technique in which a ring of precipitation is visualized in an agarose gel impregnated with antibody. Antigen placed within a well diffuses outward until reaching a zone of equivalence where antigen-antibody complexes precipitate and form a ring.

radiolarian A member of the eukaryotic group Radiolaria—amebas with a silicate shell penetrated by filamentous pseudopods.

rancidity Food spoilage due to the oxidation of fats; it may or may not involve microbial activity.

rarefaction curve A graph representing the number of different operational taxonomic units (species) found in a metagenome, as a function of increasing sample size.

RC See **reaction center**.

reaction center (RC) The complex containing a chlorophyll molecule that donates its excited electron to an electron transport system.

reactivation tuberculosis See **secondary tuberculosis**.

real-time PCR See **quantitative PCR**.

reassortment The packaging of viral chromosome segments from two different viruses into one progeny virion. Refers to separate segments from a segmented genome, without helix recombination.

receptor-binding domain The part of a secreted protein (e.g., exotoxin) that binds to a target cell membrane receptor.

recombination The process by which two DNA molecules exchange arms by cutting and splicing their helix backbones.

recombination signal sequence (RSS) A DNA region downstream of antibody heavy- and light-chain genes that allows recombination between widely separated gene segments.

recombinational repair A DNA repair mechanism that relies on recombination between an undamaged chromosome and a gap that occurred during replication of damaged DNA.

red alga Also called *rhodophyte*. An alga of the eukaryotic group Rhodophyta, which contain chloroplasts as primary algae, with red accessory photopigments.

redox couple The oxidized and reduced states of a compound. For example, NAD$^+$ and NADH form a redox couple.

redox reaction A chemical reaction in which one molecule or functional group becomes reduced and another becomes oxidized.

reducing agent See *electron donor*.

reductive acetyl-CoA pathway A carbon assimilation pathway in which two CO_2 molecules are condensed and reduced by 2 H_2 molecules to form an acetyl group.

reductive evolution Also called *degenerative evolution*. The loss or mutation of DNA encoding unselected traits.

reductive pentose phosphate cycle See **Calvin cycle**.

reductive TCA cycle Also called *reverse TCA cycle*. A CO_2 fixation pathway that generates acetyl-CoA through reversal of TCA cycle reactions. It requires ATP and NADPH.

reflection Deflection of an incident light ray by an object, at an angle equal to the incident angle.

refraction The bending and slowing of light as it passes through a substance.

refractive index The degree to which a substance causes the refraction of light; a ratio of the speed of light in a vacuum to its speed in another medium.

regulatory protein A protein that can bind DNA and modulate transcription in response to a metabolite.

regulatory T cell (Treg) A T cell that regulates the activity of another T cell, usually by suppressing its activity.

regulon A group of genes and operons located at different positions in a genome that are coordinately regulated and share a common biochemical function.

rehydration therapy A medical treatment for dehydration, in which a liquid solution of salts and glucose is delivered orally.

release factor A molecule that enters a ribosome A site containing an mRNA stop codon and initiates protein cleavage from the tRNA.

replication complex An intracellular complex of membranes and proteins that forms progeny virions.

replication fork During DNA synthesis, the region of the chromosome that is being unwound.

replisome A complex of DNA polymerase and other accessory molecules that performs DNA replication.

reporter gene A gene, such as *lacZ* (beta-galactosidase) or *gfp* (green fluorescent protein), whose protein product can be easily quantified; commonly used in a gene fusion.

repression The down-regulation of gene transcription.

repressor A regulatory protein that can bind to a specific DNA sequence and inhibit transcription of genes.

reservoir 1. The major part of the biosphere that contains a significant amount of an element needed for life. 2. An organism that maintains a virus or bacterial pathogen in an area by serving as a high-titer host.

resolution The smallest distance that two objects can be separated and still be distinguished as separate objects.

resource colimitation A situation in which a population size is limited by the lack of two different resources, such as nitrogen and phosphorus.

respiration The oxidation of reduced organic electron donors through a series of membrane-embedded electron carriers to a final electron acceptor. The energy derived from the redox reactions is stored as an electrochemical gradient across the membrane, which may be harnessed to produce ATP.

response regulator A cytoplasmic protein that is phosphorylated by a sensor kinase and modulates gene transcription depending on its phosphorylation state.

restricted transduction See **specialized transduction**.

restriction endonuclease *or* **restriction enzyme** A bacterial enzyme that cleaves double-stranded DNA within a specific short sequence, usually a palindrome.

restriction site A DNA sequence recognized and cleaved by a restriction endonuclease.

reticulate body The metabolically and reproductively active form of chlamydias.

reticuloendothelial system A collection of cells that can phagocytose and sequester extracellular material.

retinal A vitamin A–related cofactor in opsin proteins; it undergoes a conformational change upon absorbing a photon.

retrotransposon A retroelement that contains only partial retroviral sequences but may encode reverse transcriptase to allow further movement into the host genome.

retrovirus Also called *RNA reverse-transcribing virus*. A single-stranded RNA virus that uses reverse transcriptase to generate a double-stranded DNA.

reverse electron flow An enzyme-catalyzed redox reaction that couples an electron transfer with a positive ΔG (such as NAD^+ reduction to NADH) to a source of energy with a larger negative ΔG (such as a proton potential generated by an electron transport system).

reverse TCA cycle See **reductive TCA cycle**.

reverse transcriptase (RT) An enzyme that produces a double-stranded DNA molecule from a single-stranded RNA template.

reversion A mutation that changes a previous mutation back to its original state.

rhizobium *pl.* **rhizobia** A bacterial species of the order Rhizobiales that forms highly specific mutualistic associations with plants in which the bacteria develop into intracellular bacteroids that fix nitrogen for the plant.

rhizosphere The soil environment surrounding plant roots.

Rho factor A bacterial protein involved in terminating transcription.

rhodophyte See **red alga**.

ribosomal RNA (rRNA) An RNA molecule that includes the scaffolding and catalytic components of ribosomes.

ribosome-binding site Also called *Shine-Dalgarno sequence*. In bacteria, a stretch of nucleotides upstream of the start codon in an mRNA that hybridizes to the 16S rRNA of the ribosome, correctly positioning the mRNA for translation.

riboswitch A secondary structure (hairpin) within some mRNA transcripts that can interact with metabolites or antisense RNA molecules, change structure, and affect the production or translation of the mRNA.

ribozyme See **catalytic RNA**.

ribulose 5-phosphate The five-carbon sugar ribulose, phosphorylated at carbon 5.

rich medium See **complex medium**.

ripening The aging of cheese.

rise period During viral culture, the time when cells lyse and viral progeny enter the media.

RNA-dependent RNA polymerase An enzyme that produces an RNA complementary to a template RNA strand.

RNA polymerase Also called *DNA-dependent RNA polymerase*. An enzyme that produces an RNA complementary to a template DNA strand.

RNA reverse-transcribing virus See **retrovirus**.

RNA world A model of early life in which RNA performed all the informational and catalytic roles of today's DNA and proteins.

rolling-circle replication A form of DNA replication that proceeds in one direction around a circular template, making tandem copies in a linear array (concatemer). The copies are later cleaved and circularized.

root The earliest common ancestor of all members of a phylogenetic tree.

rotavirus One of a group of nonenveloped dsRNA viruses that cause severe diarrhea in children.

rRNA See **ribosomal RNA**.

RSS See **recombination signal sequence**.

RT See **reverse transcriptase**.

Rubisco Ribulose 1,5-bisphosphate carboxylase oxygenase, the enzyme that catalyzes the carbon fixation step in the Calvin cycle.

rumen The first chamber of the digestive tract of ruminant animals such as cattle; the main site for microbial digestion of feed.

S

S-layer A crystalline protein surface layer replacing or external to the cell wall in many species of archaea and bacteria.

sacculus *pl.* **sacculi** The bacterial cell wall, consisting of a single covalent molecule.

SALT See **skin-associated lymphoid tissue**.

Sanger sequencing A method of sequencing DNA based on the incorporation of chain-terminating dideoxynucleotides by a DNA polymerase.

sanitation The safe disposal of wastes hazardous to humans.

sarcina *pl.* **sarcinae** A cubical octad cluster of cells formed by septation at right angles to the previous cell division.

sargassum weed An unrooted kelp that floats in marine water and forms kelp forests.

scaffold The assembly of contigs (regions of contiguous sequence) into a large segment of a draft genome.

scanning electron microscopy (SEM) Electron microscopy in which the electron beams scan across the specimen's surface to reveal the 3D topology of the specimen.

scanning probe microscopy (SPM) A type of microscopy in which a physical probe scans the surface of a specimen and maps the topography by detecting a property such as electron tunneling current (scanning tunneling microscopy) or atomic force (atomic force microscopy).

scattering Interaction of light with an object that results in propagation of spherical light waves at relatively low intensity.

schizogony Mitotic reproduction of parasitic cells to achieve a large population within a host tissue.

second messenger A regulatory molecule such as cAMP that is produced in response to a primary signal. Second messengers typically affect the expression of numerous genes.

secondary algae Algae that evolved by engulfing primary algae in a second endosymbiotic event.

secondary antibody The second antibody added in an immunological assay—enzyme-linked immunosorbent assay (ELISA) or western blot. A secondary antibody carries a fluorescent or enzymatic tag and binds to the primary antibody in the assay.

secondary antibody response A memory B cell–mediated rapid increase in the production of antibodies in response to a repeat exposure to a particular antigen.

secondary metabolite A biosynthetic product that is not an essential nutrient but enhances nutrient uptake or inhibits competing species (e.g., an antibiotic).

secondary structure The second level of organization of polymers, consisting of regular patterns that repeat, such as the double helix in DNA or the beta sheet in proteins.

secondary syphilis A rash that may appear at some point after the primary latent stage of syphilis.

secondary tuberculosis Also called *reactivation tuberculosis*. A new round of serious disease that is caused by *Mycobacterium tuberculosis* in patients with latent tuberculosis who have become immunocompromised. Symptoms include severe cough, blood sputum, night sweats, and weight loss.

sedimentation rate The rate at which particles of a given size and shape travel to the bottom of a tube under centrifugal force. The rate depends on the particle's mass and cross-sectional area.

segmented genome A viral genome that consists of more than one nucleic acid molecule.

selectin One of a family of cell adhesion molecules.

selective medium A medium that allows the growth of certain species or strains of organisms but not others.

selective toxicity The ability of a drug, at a given dose, to harm the pathogen and not the host.

selectively permeable membrane See **semipermeable membrane**.

SEM See **scanning electron microscopy**.

semiconservative Describing the mode of DNA replication whereby each new double helix contains one old, parental strand and one newly synthesized daughter strand.

semipermeable membrane Also called *selectively permeable membrane*. A membrane that is permeable to some substances but impermeable to other substances.

sensitivity In diagnostic testing, a measure of how often a test will be positive if a patient has a particular disease, reflecting how small a concentration of antigen the test can detect.

sensor kinase A transmembrane protein that phosphorylates itself in response to an extracellular signal, and transfers the phosphoryl group to a receiver protein.

septation The formation of a septum, a new section of cell wall and envelope to separate two prokaryotic daughter cells.

septicemia An infection of the bloodstream.

septicemic plague Infection of the bloodstream by *Yersinia pestis*.

septum *pl.* **septa** A plate of cell wall and envelope that forms to separate two daughter cells.

sequela *pl.* **sequelae** A serious, harmful immunological consequence of bacterial and host antigen cross-reactivity that occurs after the infection itself is over. An example is rheumatic fever.

serial endosymbiosis theory The theory that mitochondria and chloroplasts were originally free-living prokaryotes that formed an internal symbiosis with early eukaryotes.

serum The noncellular, liquid component of the blood.

sexual reproduction Reproduction involving the joining of gametes generated by meiosis.

Shine-Dalgarno sequence See **ribosome-binding site**.

siderophore A high-affinity iron-binding protein used to scavenge iron from the environment and deliver it to a siderophore-producing organism.

sigma factor A protein needed to bind RNA polymerase for the initiation of transcription in bacteria.

signal recognition particle (SRP) A receptor that recognizes the signal sequence of peptides undergoing translation. The complex attaches to the cell membrane of prokaryotes (or the rough endoplasmic reticulum of eukaryotes), where it docks the protein-ribosome complex to the membrane for protein membrane insertion or secretion.

signal sequence A specific amino acid sequence on the amino terminus of proteins that directs them to the endoplasmic reticulum (of a eukaryote) or the cell membrane (of a prokaryote).

silent mutation A mutation that does not change the amino acid sequence encoded by an open reading frame. The changed codon encodes the same amino acid as the original codon.

simple stain A stain that makes an object more opaque, increasing its contrast with the external medium or surrounding tissue.

single-celled protein An edible microbe of high food value, such as *Spirulina* or some yeasts.

sink A part of the biosphere that can receive or assimilate significant quantities of an element; may be biotic (as in plants fixing carbon) or abiotic (as in the ocean absorbing carbon dioxide).

site-directed mutagenesis A technique of molecular biology that intentionally alters a specific sequence in a DNA molecule.

site-specific recombination Recombination between DNA molecules that do not share long regions of homology but do contain short regions of homology specifically recognized by the recombination enzyme.

skin-associated lymphoid tissue (SALT) Immune cells, such as dendritic cells, located under the skin that help eliminate bacteria that have breached the skin surface.

sliding clamp A protein that keeps DNA polymerase affixed to DNA during replication.

sludge The solid products of wastewater treatment.

small RNA (sRNA) A non-protein-coding regulatory RNA molecule that modulates translation or mRNA stability.

small-subunit (SSU) rRNA In bacteria and archaea, 16S rRNA; in eukaryotes, 18S rRNA. A ribosomal RNA found in the small subunit of the ribosome. Its gene is often sequenced for phylogenetic comparisons.

soil A complex mixture of decaying organic and mineral matter that covers the terrestrial portions of the planet.

SOS response A coordinated cellular response to extensive DNA damage. It includes error-prone repair.

source A part of the biosphere that stores a significant quantity of a given element; may be biotic (as in tree biomass, a source of carbon) or abiotic (as in carbonate rock).

sourdough An undefined yeast population, derived from a previous batch of dough, that is used in bread production.

Southern blot A technique (named for its inventor, Edward M. Southern) to detect specific DNA sequences. Sample DNA segments are separated by gel electrophoresis, transferred to a blot, and probed with a labeled DNA that will hybridize to complementary DNA sequences.

space-filling model A molecular model that represents the volume of the electron orbitals of the atoms, usually to the limit of the van der Waals radii.

specialized transduction Also called *restricted transduction*. Transduction in which the phage can transfer only a specific, limited number of donor genes to the recipient cell.

species A single, specific type of organism, designated by a genus and species name.

specificity In diagnostic testing, a measure of how often a test will be negative if a patient does not have a particular disease, reflecting how well a test can distinguish between two closely related antigens.

spectrum of activity The range of pathogens for which an antimicrobial agent is effective.

spheroplast A cell whose peptidoglycan is degraded by lysozyme; thus the cell loses its shape, forming a sphere.

spike protein A viral glycoprotein that connects the membrane to the capsid or the matrix and may be involved in viral binding to host cell receptors.

spirochete A bacterium with a tight, flexible spiral shape; a species of the phylum Spirochetes (Spirochaeta).

Spirochetes *or* **Spirochaeta** A phylum of bacteria with a unique morphology: a flexible, extended spiral that twists via intracellular flagella.

SPM See **scanning probe microscopy**.

spongiform encephalopathy   A brain-wasting disease caused by a prion.

spontaneous generation   The theory, much debated in the eighteenth century, that under current Earth conditions life can arise spontaneously from nonliving matter.

sporangiospore   A haploid spore that can germinate to form a haploid mycelium.

sporangium *pl.* **sporangia**   A fungal organ that releases nonmotile spores.

spore stain   A type of differential stain that is specific for the endospore coat of various bacteria, typically a firmicute species.

spread plate   A method to grow separate bacterial colonies by plating serial dilutions of a liquid culture.

sRNA   See **small RNA**.

SRP   See **signal recognition particle**.

SSU rRNA   See **small-subunit rRNA**.

staining   The process of treating microscopic specimens with a stain to enhance their detection or to visualize specific cell components.

stalk   An extension of the cytoplasm and envelope that attaches a microbe to a substrate.

stalked ciliate   A ciliate that adheres to a substrate and uses its cilia to obtain prey.

standard reduction potential ($E°$)   The reduction potential (tendency of a chemical to gain electrons and thereby become reduced) under standard conditions of 1-M concentration, 25°C temperature, and 1-atm pressure.

staphylococcus   A hexagonal arrangement of cells formed by septation in random orientations.

starch   A glucose polymer in which an acetal (O–COH) of each glucose has condensed with a hydroxyl group of the next glucose, releasing H_2O.

start codon   A codon (usually AUG) that signals the first amino acid of a protein.

starter culture   A mixture of fermenting microbes added to a food substrate to generate a fermented product.

stationary phase   A period of cell culture, following exponential phase, during which there is no net increase in replication.

sterilization   The destruction of all cells, spores, and viruses on an object.

stick model   A molecular model in which stick lengths represent the distances between bonded pairs of atomic nuclei.

Stickland reaction   An energy-yielding reaction between two amino acids in which one oxidizes the other. The reaction typically produces short organic acids plus $2NH_4^+$; it may also produce CO_2 and H_2.

stop codon   One of three codons (UAA, UAG, UGA) that do not encode an amino acid, and thus trigger the end of translation.

streptogramin   An antibiotic that binds 23S rRNA and blocks elongation of protein synthesis in bacteria.

strict aerobe   An organism that performs aerobic respiration and can grow only in the presence of oxygen.

strict anaerobe   An organism that cannot grow in the presence of oxygen.

stringent response   A cellular response to idle ribosomes (often indicating low carbon and energy stores) that includes a decrease in rRNA and tRNA production.

stromatolite   A mass of sedimentary layers of limestone produced by a marine microbial community over many years.

structural gene   A string of nucleotides that encodes a functional RNA molecule.

structural formula   A representation of molecular structure in which each covalent bond is shown as a line between atoms.

structural isomer   A molecule with the same molecular formula as a different molecule but a different arrangement of atoms.

substrate-level phosphorylation   The formation of ATP by the enzymatic transfer of phosphate from a substrate molecule onto ADP.

sulfa drug   An antibiotic that inhibits folic acid synthesis and, thus, nucleotide synthesis.

Sulfolobales   An order of thermophilic archaea (Crenarchaeota) that includes sulfur oxidizers.

superantigen   A molecule that directly stimulates T cells without undergoing antigen-presenting-cell processing and surface presentation.

super-resolution imaging   Techniques of microscopy that pinpoint the location of an object with a precision greater than the resolution of ordinary optical microscopy.

Svedberg coefficient   A measure of particle size based on the particle's sedimentation rate in a tube subjected to a high g force.

swarming   A behavior in which some microbial cells differentiate into large swarmer cells and swim together as a unit.

switch region   A repeating DNA sequence interspersed between antibody constant-region genes that serves as a recombination site during isotype, or class, switching.

symbiont   An organism that lives in a close association with another organism.

symbiosis *pl.* **symbioses**   The intimate association of two different species.

symbiosis island   A type of genomic island in which the stretch of DNA expresses proteins that enable a symbiotic relationship with another organism.

symport   Coupled transport in which the molecules being transported move in the same direction across the membrane.

synergism   Cooperation between species in which both species benefit but can grow independently. The cooperation is less intimate than symbiosis.

synthetic biology   The genetic construction of novel organisms with useful functions.

syntrophy   Metabolic cooperation between two different species; usually one member releases a product whose removal by the second species enables the pair to metabolize with a negative value of ΔG.

T

T cell   An adaptive immune cell, developed in the thymus, that can give rise to antigen-specific helper cells and cytotoxic T cells.

T-cell receptor (TCR)   A surface receptor on T cells that binds MHC-bound antigen on antigen-presenting cells.

T3SS   See **type III secretion system**.

tailed phage   A phage such as T4 that contains a genome delivery device called the tail.

tandem repeat   A stretch of directly repeating DNA sequence (direct repeats) without any intervening DNA.

TaqMan   A real-time PCR technique in which Taq polymerase, in the process of synthesizing DNA along a template, degrades a downstream fluorescent oligonucleotide probe. The increase in fluorescence indicates the production of an amplified DNA product.

target community   A community whose genomes are sequenced for metagenomic analysis.

taxon *pl.* **taxa**   A category of organisms with a shared genetic ancestor.

taxonomy The description of distinct life-forms and their organization into different categories.

T$_C$ cell See **cytotoxic T cell**.

TCA cycle See **tricarboxylic acid cycle**.

TCR See **T-cell receptor**.

tegument The contents of a virion between the capsid and the envelope.

teichoic acid A chain of phosphodiester-linked glycerol or ribitol that threads through and reinforces the cell wall in Gram-positive bacteria.

telomerase A reverse-transcriptase enzyme complex that reads RNA as a template to synthesize DNA.

telomere The DNA segment at either end of a eukaryotic chromosome.

TEM See **transmission electron microscopy**.

tempeh A mold-fermented soy product, popular as a food in parts of Asia.

temperate phage A phage capable of lysogeny.

template strand A DNA strand (or an RNA strand in some viruses) that is used as a template for the synthesis of mRNA.

Tenericutes Also called *Mollicutes*. A clade of bacteria lacking a cell wall; closely related to Gram-positive bacteria.

terminal electron acceptor The final electron acceptor at the end of an electron transport system.

termination (*ter*) site A sequence of DNA that halts replication of DNA by DNA polymerase.

terpenoid A branched lipid derived from isoprene that is found in hydrocarbon chains of archaeal membranes.

terraforming The idea of transforming the environment of another planet to make it suitable for life from Earth.

tertiary structure The third level of organization of polymers; the unique 3D shape of a polymer.

tertiary syphilis A final stage of syphilis, manifested by cardiovascular and nervous system symptoms.

tetanospasmin The tetanus-causing potent exotoxin produced by *Clostridium tetani*.

tetraether A molecule containing four ether links. An example is found in archaeal membranes, when two lipid side chains form ether linkages with a pair of side chains from the other side of the bilayer.

tetrapyrrole An essential primary product that contains four pyrroles (five-membered rings containing nitrogen), each with one or two double bonds. A precursor for many important cell cofactors, such as chlorophylls and vitamin B$_{12}$.

T$_H$ cell See **helper T cell**.

T$_H$17 cell A class of helper T cell that secretes the inflammatory cytokine IL-17.

Thaumarchaeota A major division of Archaea, containing ammonia oxidizers, marine invertebrate symbionts, and others.

thermal vent See **hydrothermal vent**.

thermocline A region of the ocean where temperature decreases steeply with depth, and water density increases.

thermophile An organism adapted for optimal growth at high temperatures, usually 55°C or higher.

threshold dose The concentration of antigen needed to elicit adequate antibody production.

thylakoid An intracellular chlorophyll-containing membrane folded within a phototrophic bacterium or a chloroplast.

Ti plasmid A plasmid found in tumorigenic strains of *Agrobacterium tumefaciens* that can be used as a vector to introduce DNA into plant cells.

tight junction A type of junction between the membranes of two adjacent vertebrate cells that form an impermeable barrier.

TLR See **Toll-like receptor**.

tmRNA A molecule resembling both tRNA and mRNA that rescues ribosomes stalled on damaged mRNAs lacking a stop codon.

TNF See **tumor necrosis factor**.

TNF-alpha See **tumor necrosis factor alpha**.

Toll-like receptor (TLR) A member of a eukaryotic transmembrane glycoprotein family that recognizes a particular microbe-associated molecular pattern (MAMP) present on pathogenic microorganisms.

tomography The acquisition of projected images of a transparent specimen from different angles that are digitally combined to visualize the entire specimen.

topoisomerase An enzyme that can change the supercoiling of DNA.

total magnification The magnification of the ocular lens multiplied by the magnification of the objective lens.

transamination The transfer of an ammonium ion between two metabolites.

transcript An RNA copy of a DNA template.

transcription The synthesis of RNA complementary to a DNA template.

transcriptional attenuation A transcriptional regulatory mechanism in which translation of a leader peptide affects transcription of downstream structural genes.

transcriptional fusion A technique to monitor transcriptional regulation by fusing a reporter gene containing its own ribosome-binding site to the 3′ end of an operon. Unlike the translational fusion technique, only the promoter of the target gene directs expression of the reporter.

transcriptome The set of transcribed genes in a cell at a given time. The "complete transcriptome" includes all the possible RNA transcription products from a given genome. The "expressed transcriptome" is the set of RNAs present during a given condition.

transcytosis The movement of a cell or substance from one side of a polarized cell to the other side, using an intracellular route.

transduction The transfer of host genes between bacterial cells via a phage head coat.

transertion The membrane insertion of a nascent polypeptide chain during translation of a messenger RNA undergoing transcription from DNA. Overall, the growing peptide-mRNA complex connects DNA to the membrane.

transfection Deliberate transfer of DNA (usually viral) into cells.

transfer RNA (tRNA) An RNA that carries an amino acid to the ribosome. The anticodon on the tRNA base-pairs with the codon on the mRNA.

transform To cause bacteria to take up exogenous DNA. In eukaryotes, to convert cultured cells into cancer cells.

transformasome A bacterial cell membrane protein complex that imports external DNA during transformation.

transformation The internalization of free DNA from the environment into bacterial cells.

transgene A gene that has been transferred by genetic engineering techniques from one organism to another.

transglycosylase An enzyme that links *N*-acetylglucosamine and *N*-acetylmuramic acid into chains during bacterial cell wall synthesis.

transition   A point mutation in which a purine is replaced by a different purine or a pyrimidine is replaced by a different pyrimidine.

translation   The ribosomal synthesis of proteins based on triplet codons present in mRNA.

translational control   A regulatory mechanism that modulates protein production by influencing the translation of mRNA.

translational fusion   Also called *gene fusion* or *protein fusion*. A technique to measure control of gene transcription or translation by inserting a reporter gene into a target gene. The reporter relies on both the promoter and the ribosome-binding site of the target gene.

translocation   The energy-dependent movement of the ribosome to the next triplet codon along an mRNA.

transmembrane domain   The membrane-spanning amino acid sequence within a membrane protein that connects intracellular and extracellular parts of the membrane protein.

transmembrane protein   Also called *integral protein*. A protein with a membrane-spanning region.

transmission electron microscopy (TEM)   Electron microscopy in which electron beams are transmitted through a thin specimen to reveal internal structure.

transovarial transmission   The transfer of a pathogen from parent to offspring by infection of the egg cell. Typically seen in insects.

transpeptidase   An enzyme that cross-links the side chains from adjacent peptidoglycan strands during bacterial cell wall synthesis.

transport protein *or* **transporter**   A membrane protein that moves specific molecules across a membrane.

transposable element   A segment of DNA that can move from one DNA region to another.

transposase   A transposable element–encoded enzyme that catalyzes the transfer of the transposable element from one DNA region to another.

transposition   The process of moving a transposable element from one DNA region to another.

transposon   A transposable DNA element that contains genes in addition to those required for transposition.

transversion   A point mutation in which a purine is replaced by a pyrimidine or vice versa.

Treg   See **regulatory T cell**.

tricarboxylic acid (TCA) cycle   Also called *Krebs cycle*. A metabolic cycle that catabolizes the acetyl group from acetyl-CoA to $2CO_2$ with the concomitant production of NADH, $FADH_2$, and ATP.

tRNA   See **transfer RNA**.

trophic level   A level of a food web representing the consumption of biomass of organisms from another level, usually closer to producers.

tropism   The ability of a virus to infect a particular tissue type.

true fungi   See **Eumycota**.

trypanosome   A parasitic excavate protist that has a cortical skeleton of microtubules culminating in a long flagellum.

tumor necrosis factor (TNF)   A cytokine released by several cell types (e.g., macrophages) in response to cell damage.

tumor necrosis factor alpha (TNF-alpha)   A cytokine involved in systemic inflammation.

twitching motility   A type of bacterial movement on solid surfaces in which a specific pilus extends and retracts.

two-component signal transduction system   A message relay system composed of a sensor kinase protein and a response regulator protein that regulates gene expression in response to a signal (usually an extracellular signal).

type I hypersensitivity   Also called *immediate hypersensitivity*. An IgE-mediated allergic reaction that causes degranulation of mast cells within minutes of exposure to the antigen. The severe reaction known as anaphylaxis is triggered by type I hypersensitivity.

type II hypersensitivity   An immune response in which antibodies bind to the patient's own cell-surface antigens or to foreign antigens adsorbed onto the patient's cells. Antibody binding triggers cell-mediated cytotoxicity or activation of the complement cascade.

type II secretion system   A bacterial protein secretion system that uses a type IV pilus-like extraction/retraction mechanism to push proteins out of the cell.

type III hypersensitivity   An immune reaction triggered when IgG antibody binds to an excess of soluble foreign antigen in the blood. The immune complexes deposit in small blood vessels, where they interact with complement to initiate an inflammatory response.

type III secretion system (T3SS)   A bacterial protein secretion system that uses a molecular syringe to inject bacterial proteins into the host cytoplasm.

type IV hypersensitivity   Also called *delayed-type hypersensitivity (DTH)*. An immune response that develops 24–72 hours after exposure to an antigen that the immune system recognizes as foreign. The response is triggered by antigen-specific T cells. It is delayed because the T cells need time to proliferate after being activated by the allergen.

U

ubiquitin (Ub)   A 76-amino-acid peptide clipped to eukaryotic proteins that mark the protein for degradation by the proteasome.

ultracentrifuge   A machine that subjects samples to high centrifugal forces and can be used to separate subcellular components.

uncoating   The release of a viral genome from its capsid, following entry of the virion into a host cell.

uncoupler   A molecule that makes a membrane permeable to protons, dissipating the proton motive force and uncoupling electron transport from ATP synthesis.

uncultured *or* **unculturable**   Describing an organism whose requirements for culture remain unknown.

upstream processing   The culturing of industrial microbes to produce large quantities of a desired product.

V

vaccination   Exposure of an individual to a weakened version of a microbe or a microbial antigen to provoke immunity and prevent development of disease upon reexposure.

VAM   See **arbuscular mycorrhizae**.

van der Waals force   A weak, temporary electrostatic attraction between molecules caused by shifting electron clouds.

vancomycin   A glycopeptide antibiotic that inhibits bacterial cell wall synthesis in a mechanism distinct from penicillin inhibition.

variable region   The amino-terminal portions of antibody light and heavy chains that confer specificity to antigen binding and define the antibody idiotype.

vasoactive factor   A cell signaling molecule that increases capillary permeability.

VBNC See **viable but nonculturable**.

vector 1. An organism (e.g., insect) that can carry infectious agents from one animal to another. 2. In molecular biology, a molecule of DNA into which exogenous DNA can be inserted to be cloned.

vegetative cell A metabolically active, replicating bacterial cell.

vegetative mycelium A mass of hyphae (branched filaments) produced by vegetative cells that expand into the substrate.

vehicle transmission In disease, the transfer of a pathogen when an infected person deposits it on a surface or in food or drink that another person touches or consumes.

Verrucomicrobia A phylum of free-living aquatic bacteria with wart-like, protruding structures containing actin.

vertical gene transfer The passage of genes from parent to off-spring through reproduction.

vertical transmission In disease, the transfer of a pathogen from parent to offspring. See also **transovarial transmission**.

vesicle A small, membrane-enclosed sphere found within a cell.

vesicular-arbuscular mycorrhizae (VAM) See **arbuscular mycorrhizae**.

viable Capable of replicating—for instance, by forming a colony on an agar plate.

viable but nonculturable (VBNC) Metabolically active but unable to replicate to form a colony on a plate by current means of culture.

viral shunt The release by viral lysis of cell contents as organic material available for microbial consumers in the upper region of the ocean.

viremia The presence of large numbers of virions in the bloodstream.

virion A virus particle.

viroid An infectious naked nucleic acid.

virome The genomes of all the viruses that inhabit a particular organism or environment.

virulence A measure of the severity of a disease caused by a pathogenic agent.

virulence factor A trait of a pathogen that enhances the pathogen's disease-producing capability.

virus A noncellular particle containing a genome that can replicate only inside a cell.

W

wastewater treatment A series of wastewater transformations designed to lower biological oxygen demand and eliminate human pathogens before water is returned to local rivers.

water activity A measure of the water that is not bound to solutes and is available for use by organisms.

water cycle See **hydrologic cycle**.

water table The layer of soil that is permanently saturated with water.

western blot A technique to detect specific proteins. Proteins are subjected to gel electrophoresis, transferred to a blot, and probed with enzyme-linked or fluorescently tagged antibodies that specifically bind the protein of interest.

wet mount A technique to view living microbes with a microscope by placing the microbes in water on a slide under a coverslip.

wetland A region of land that undergoes seasonal fluctuations in water level and aeration.

whey The liquid portion of milk after proteins have precipitated out of solution, usually during cheese production.

Winogradsky column A column containing a stratified environment that causes specific microbes to grow at particular levels; a type of enrichment culture for the growth of microbes from wetland environments.

X

X-ray crystallography *or* **X-ray diffraction analysis** A technique to determine the positions of atoms (atomic coordinates) within a molecule or molecular complex, based on the diffraction of X-rays by the molecule.

Y

yeast A unicellular fungus.

yeast two-hybrid system An in vivo technique to determine protein-protein interactions in which DNA sequences encoding proteins of interest are fused separately to the DNA-binding and activation domains of a yeast transcription factor. The recombinant yeast is then tested for expression of a reporter gene.

yogurt A semisolid food produced through acidification of milk by lactic acid–producing bacteria.

Z

zone of hypoxia See **dead zone**.

zone of inhibition A region of no bacterial growth on an agar plate due to the diffusion of a test antibiotic. Correlates to the minimal inhibitory concentration.

zoonotic disease An infection that normally affects animals but can be transmitted to humans.

zoospore A flagellated reproductive cell produced by chytridiomycete fungi.

zooxanthella *pl.* **zooxanthellae** A phototrophic coral endo-symbiont, most commonly a dinoflagellate of the genus *Symbiodinium*.

Z pathway See **oxygenic Z pathway**.

zygomycete A member of the eukaryotic group Zygomycota—fungi forming nonmotile haploid gametes that grow toward each other, fusing to form the zygospore.

zygospore In zygomycetes, the diploid structure formed by the fusion of two gamete-bearing hyphae.

FIGURE CREDITS

Figure 6.18: Didier Raoult, et al. Figure from "The 1.2-Megabase Genome Sequence of Mimivirus" *Science*, 19 Nov 2004: Vol. 306, Issue 5700, pp. 1344-1350. Copyright © 2004, The American Association for the Advancement of Science. Reprinted with permission from AAAS.

Figure 7.30B: Goodrich, et al. Figure 7B from "Human Genetics Shape the Gut Microbiome" *Cell* 159: 789-799, 2014. Copyright © 2014 Elsevier Inc. All rights reserved. Reprinted with permission.

Figure 8.37: Heinken, et al. Figure from "Functional Metabolic Map of Faecalibacterium prausnitzii, a Beneficial Human Gut Microbe." *Journal of Bacteriology* 196:3289-3302, 2014. Copyright @ 2014, American Society for Microbiology. Reprinted with permission.

Figure 8.38A: Jan-Hendrik Hehemann, et al. Figure 4 from "Transfer of carbohydrate-active enzymes from marine bacteria to Japanese gut microbiota." *Nature* 464, 908-912. April 2010. Copyright © 2010, Nature Publishing Group, a division of Macmillan Publisher Ltd. All rights reserved.

Figure 9.38: Kube, et al. Figure 2A from "Genome sequence and functional genomic analysis of the oil-degrading bacterium Oleispira Antarctica." *Nature Communications*, 4:2156. Copyright © 2013, Nature Publishing Group, a division of Macmillan Publisher Ltd. All rights reserved.

Figure 9.39: Crisp A, Boschetti C, Perry M, Tunnacliffe A, Micklem G. Figure 4 from "Expression of multiple horizontally acquired genes is a hallmark of both vertebrate and invertebrate genomes." *Genome Biology.* March 13, 2015; 16:50. © Crisp et al.; licensee BioMed Central. 2015.

Figure 10.26B: Jenny L. Baker, et al. Figure from "Widespread Genetic Switches and Toxicity Resistance Proteins for Fluoride," *Science* (13 Jan 2012), Vol. 335, Issue 6065, pp. 233-235. Copyright © 2012, The American Association for the Advancement of Science. Reprinted with permission from AAAS.

Figure 11.13A: Amie Eisfeld, et al. Figure from "At the centre: influenza A virus ribonucleoproteins," *Nature Reviews Microbiology* 13:28 November 2014. Copyright © 2014, Nature Publishing Group, a division of Macmillan Publisher Ltd. All rights reserved.

Figure 12.24A-C: Jarred M. Callura. Figure 1 from "Genetic switchboard for synthetic biology applications," *Proceedings of the National Academy of Sciences*, Vol. 109, No. 15. April 2012. Reprinted by permission of the National Academy of Sciences.

Figure 12.25A-E: Jarred M. Callura. Figure 3 from "Genetic switchboard for synthetic biology applications," *Proceedings of the National Academy of Sciences*, Vol. 109, No. 15. April 2012. Reprinted by permission of the National Academy of Sciences.

Figure 12.26: Jarred M. Callura. Figure 3A from "Tracking, tuning, and terminating microbial physiology using synthetic riboregulators," *Proceedings of the National Academy of Sciences* Vol. 107, No. 36. September 2010. Reprinted by permission of the National Academy of Sciences.

Figure 13.34B: Nidal Laban, et al. Figure from "Anaerobic benzene degradation by Gram-positive sulfate-reducing bacteria," *FEMS Microbiological Ecology* 68:300. 2009. Copyright @ 2009 John Wiley & Sons, Inc. Reprinted with permission.

Figure 15.25: Daniel Smith and Matthew Chapman. Figure 2 from "Economical evolution: microbes reduce the synthetic cost of extracellular proteins." *mBio* 1(3):e00131-10. (August 2010). © 2010, American Society for Microbiology. Reprinted with permission.

Figure 17.22: W. Ford Doolittle: Figure from "Phylogenetic Classification and the Universal Tree", *Science* 284:2124 (25 June 1999). Copyright © 1999, The American Association for the Advancement of Science. Reprinted with permission from AAAS.

Figure 17.26: Adapted from Zachary Blount, et al. Figure 1 from "Historical contingency and the evolution of a key innovation in an experimental population of *Escherichia coli*," *Proceedings of the National Academy of Sciences* 105:7899. June 2008. Reprinted by permission of the National Academy of Sciences.

Special Topic 17.2 Figure 2A: Scott T. Kelley. Figure 2 from "Molecular Analysis of Shower Curtain Biofilm Microbes," *Applied and Environmental Microbiology*, 2004 July; 70(7): 4187–4192. © 2004, American Society for Microbiology. Reprinted with permission.

Figure 18.8: Dennis Claessen, et al. Figure 3 from "Bacterial solutions to multicellularity: a tale of biofilms, filaments and fruiting bodies" *Nature Reviews Microbiology* 12:115. Copyright © 2014, Nature Publishing Group, a division of Macmillan Publisher Ltd. All rights reserved.

Figure 19.18B: Karner, DeLong, & Karl: Figure reprinted by permission from Macmillan Publishers Ltd. from "Archael Dominance in the Mesopelagic Zone of the Pacific Ocean," *Nature*, Vol. 409, No. 507. Copyright © 2001, Nature Publishing Group.

Figure 19.21: Matsutani, Naoki, Tatsunori Nakagawa, Kyoko Nakamura, Reiji Takahashi, Kiyoshi Yoshihara, et al. 2011. Figure from "Enrichment of a novel marine ammonia-oxidizing archaeon obtained from sand of an eelgrass zone." *Microbes and Environments* 26:23–29. © The Japanese Society of Microbial Ecology/ The Japanese Society of Soil Microbiology.

Special Topic 19.1.2A-B: Elizabeth Hansen, et al. Figure 1A and B from "Pan-genome of the dominant human gut-associated archaeon, Methanobrevibacter smithii, studied in twins," *PNAS* 108:4599. March 2011. Reprinted by permission of the National Academy of Sciences.

Figure 21.4B: Molly C. Redmond and David L. Valentine. Figure 1 from "Natural gas and temperature structured a microbial community response to the *Deepwater Horizon* oil spill," *Proceedings of the National Academy of Sciences* 109: 20292, December 2012. Reprinted by permission of the National Academy of Sciences.

Figure 21.5: Susannah Tringe, et al. 2008. Figure A from "The Airborne Metagenome in an Indoor Urban Environment," PLoS ONE 3:e1862. © 2008 Tringe et al.

Figure 21.8: Noah Fierer, Figure 3 from "Cross-biome metagenomic analyses of soil microbial communities and their functional attributes," *Proceedings of the National Academy of Sciences* 105:7899. December 2012. Reprinted by permission of the National Academy of Sciences.

Figure 21.17A: Lynn Margulis: Figure from *Symbiosis in Cell Evolution* by Lynn Margulis. Copyright © 1981, 1993 by W.H. Freeman and Company. Reprinted with permission from the Lynn Margulis Estate.

Figure 21.25A-B: Pedro Flombaum, et al. Figure from "Present and future global distributions of the marine Cyanobacteria Prochlorococcus and Synechococcus," *Proceedings of the National Academy of Sciences* 110:9824, June 2013. Reprinted by permission of the National Academy of Sciences.

Figure 21.28: Raina Maier, et al.: Figure from *Environmental Microbiology*, Copyright © 2000 by Academic Press. Reprinted by permission of Elsevier, Ltd.

eTopic 21.1: Molly C. Redmond and David L. Valentine. Figure 1 from "Natural gas and temperature structured a microbial community response to the *Deepwater Horizon* oil spill," *Proceedings of the National Academy of Sciences* 109: 20292, December 2012. Reprinted by permission of the National Academy of Sciences.

INDEX

Gut Microbiome
Shaping Human Health

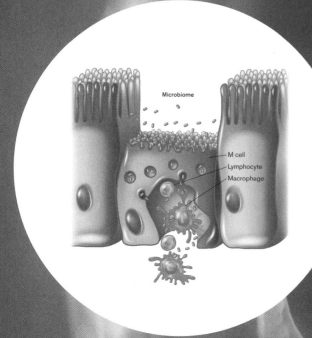

Intestinal microbiota interactions with gut epithelial cells shape our immune system.
Pages 925 and 974

Bacteriophage in the gut can kill members of the microbiota and cause a "dysbiosis" that deteriorates health.
Page 219

Intestinal anaerobes such as *Bacteroides* digest complex plant fibers and influence human obesity and leanness.
Pages 840 and 912

Ruth Ley found that the gut microbe *Christensenella minuta* was the most highly heritable species among lean human twins.
Page 269

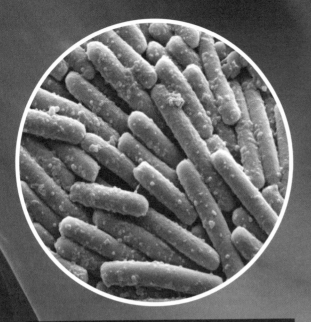

Diarrhea caused by *Clostridium difficile* is cured when a fecal transplant restores a normal gut microbiome.
Page 1052

Bradley Bearson explores how antibiotics activate prophage replication to initiate toxin and antibiotic resistance gene transfer between gut bacteria.
Page 1,116

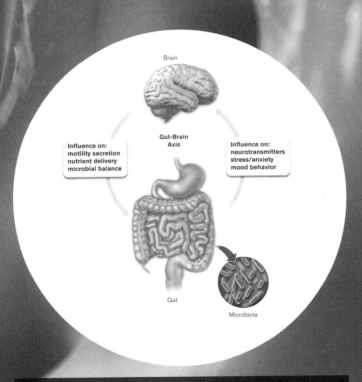

Gut microbiota produce molecules that affect the brain. The brain influences gut microbiota by adjusting gut motility and secretion.
Page 841

Throughout this book, examples of the gut microbiome are indicated with this icon to make them easier to find.